Organic Electronic Spectral Data
Volume XVIII   1976

Organic Electronic Spectral Data

III

Volume XVIII   1976

JOHN P. PHILLIPS, DALLAS BATES
HENRY FEUER & B. S. THYAGARAJAN

EDITORS

CONTRIBUTORS

Dallas Bates
H. Feuer
L. D. Freedman

C. M. Martini
F. C. Nachod
J. P. Phillips

AN INTERSCIENCE ® PUBLICATION
JOHN WILEY & SONS
New York • Chichester • Brisbane • Toronto • Singapore

An Interscience ® Publication

Copyright © 1982 by John Wiley & Sons, Inc.

All rights reserved. Published simultaneously in Canada.

Library of Congress Catalog Card Number: 60–16428

ISBN 0-471-87178-8

Printed in the United States of America

10 9 8 7 6 5 4 3 2 1

INTRODUCTION TO THE SERIES

In 1956 a cooperative effort to abstract and publish in formula order all the ultraviolet-visible spectra of organic compounds presented in the journal literature was organized through the enterprise and leadership of M.J. Kamlet and H.E. Ungnade. Organic Electronic Spectral Data was incorporated in 1957 to create a formal structure for the venture, and coverage of the literature from 1946 onward was then carried out by chemists with special interests in spectrophotometry through a page by page search of the major chemical journals. After the first two volumes (covering the literature from 1946 through 1955) were produced, a regular schedule of one volume for each subsequent period of two years was introduced. In 1966 an annual schedule was inaugurated.

Altogether, more than fifty chemists have searched a group of journals totalling more than a hundred titles during the course of this sustained project. Additions and subtractions from both the lists of contributors and of journals have occurred from time to time, and it is estimated that the effort to cover all the literature containing spectra may not be more than 95% successful. However, the total collection is by far the largest ever assembled, amounting to over 400,000 spectra in the eighteen volumes so far.

Volume XIX is in preparation.

v

PREFACE

Processing of the data provided by the contributors to Volume XVIII as to the last several volumes was performed at the University of Louisville.

John P. Phillips
Dallas Bates
Henry Feuer
B.S. Thyagarajan

ORGANIZATION AND USE OF THE DATA

The data in this volume were abstracted from the journals listed in the reference section at the end. Although a few exceptions were made, the data generally had to satisfy the following requirements: the compound had to be pure enough for satisfactory elemental analysis and for a definite empirical formula; solvent and phase had to be given; and sufficient data to calculate molar absorptivities had to be available. Later it was decided to include spectra even if solvent was not mentioned. Experience has shown that the most probable single solvent in such circumstances is ethanol.

All entries in the compilation are organized according to the molecular formula index system used by Chemical Abstracts. Most of the compound names have been made to conform with the Chemical Abstracts system of nomenclature.

Solvent or phase appears in the second column of the data lists, often abbreviated according to standard practice; there is a key to less obvious abbreviations on the next page. Anion and cation are used in this column if the spectra are run in relatively basic or acidic conditions respectively but exact specifications cannot be ascertained.

The numerical data in the third column present wavelength values in nanometers (millimicrons) for all maxima, shoulders and inflections, with the logarithms of the corresponding molar absorptivities in parentheses. Shoulders and inflections are marked with a letter s. In spectra with considerable fine structure in the bands a main maximum is listed and labelled with a letter f. Numerical values are given to the nearest nanometer for wavelength and nearest 0.01 unit for the logarithm of the molar absorptivity. Spectra that change with time or other common conditions are labelled "anom." or "changing", and temperatures are indicated if unusual.

The reference column contains the code number of the journal, the initial page number of the paper, and in the last two digits the year (1976). A letter is added for journals with more than one volume or section in a year. The complete list of all articles and authors thereof appears in the References at the end of the book.

Several journals that were abstracted for previous volumes in this series have been omitted, usually for lack of useful data, and several new ones have been added. Most Russian journals have been abstracted in the form of the English translation editions.

ix

ABBREVIATIONS

s	shoulder or inflection
f	fine structure
n.s.g.	no solvent given in original reference
$C_6H_{11}Me$	methylcyclohexane
C_6H_{12}	cyclohexane
DMF	dimethylformamide
DMSO	dimethylsulfoxide
THF	tetrahydrofuran

Other solvent abbreviations generally follow the practice of Chemical Abstracts.

Underlined data were estimated from graphs.

JOURNALS ABSTRACTED

Organic Electronic Spectral Data
Volume XVIII   1976

Compound	Solvent	$\lambda_{max}(\log \epsilon)$	Ref.
CBrClS			
Carbonothioic bromide chloride	isooctane	199(2.53),218(2.57), 250(2.53),270(2.31)	24-3432-76
CBr_2S			
Carbonothioic dibromide	isooctane	198(4.48),216(3.85), 236(3.71),270(3.69)	24-3432-76
CH_2O_2			
Formic acid	hexane	206.8(1.65)	44-3312-76
	H_2O	206(1.68),207s(--)	44-3312-76
	MeCN	214.5(1.68)	44-3312-76
CH_2S_3			
Carbonotrithioic acid, disodium salt	n.s.g.	226(3.93),320s(--), 332(4.26),400(--), 500(1.60)	139-0289-76B
barium salt	aq base	325(4.0),495(1.6)	139-0289-76B
CH_2S_4			
Carbono(dithioperoxo)dithioic acid,	H_2O	320(3.69),402(3.69)	139-0289-76B
disodium salt	M NaOH	321(3.43),401(3.65)	139-0289-76B
CH_3ClHg			
Mercury, chloromethyl-	H_2O	189(4.0)	3-0195-76
	6M HCl	206(3.73)	3-0195-76
CH_3Cl_2PSe			
Phosphonoselenoic dichloride, methyl-	hexane	195(4.03),234(3.77), 280(3.19)	70-0174-76
	EtOH	192(4.04),231(3.75), 273(3.12)	70-0174-76
CH_4N_2			
Diazene, methyl-	gas	360(0.7)	46-0559-76

Compound	Solvent	$\lambda_{max}(\log \epsilon)$	Ref.

$C_2BrF_3S_2$
   Carbonobromidodithioic acid, trifluoro-
     methyl ester
	isooctane	212(3.92),238(4.01),
		 269(3.72),288(3.75)
			24-3432-76

$C_2H_4N_2O_4$ Ethane, 1,1-dinitro-, potassium salt	pH 13	381(4.22)	104-2433-76
C_2H_4O Acetaldehyde	hexane H_2O EtOH	290(1.13) 277(0.90) 286(1.11)	140-0837-76 140-0837-76 140-0837-76
$C_2H_4OS_2$ 1,3-Dithietane, 1-oxide	EtOH	207(2.83),222(2.76), 266(2.00)	35-5715-76
$C_2H_4OS_3$ 1,2,4-Trithiolane, 1-oxide	n.s.g.	210(3.43),335(1.85)	44-2465-76
$C_2H_4O_2$ Acetic acid	hexane MeCN	202.6(1.70) 210.8(1.56)	44-3312-76 44-3312-76
$C_2H_4O_2S_2$ 1,3-Dithietane, 1,1-dioxide	EtOH	217(2.30),245s(1.48)	35-5715-76
$C_2H_4S_2$ 1,3-Dithietane	EtOH	216(2.93),293(1.58), 311s(1.30)	35-5715-76
$C_2H_5Cl_2PSe$ Phosphonoselenoic dichloride, ethyl-	hexane HCOOH CF_3COOH	192s(3.92),234(3.71), 276(3.16) 232(3.71),270(3.10) 227(4.08),270(3.48)	70-0174-76 70-0174-76 70-0174-76

$C_2H_6OS_2$
   Methanesulfinothioic acid, S-methyl
     ester
	hexane	260(3.29)	102-0187-76

$C_2H_6O_2S_2$
   Methanesulfonothioic acid, S-methyl
     ester
	hexane or
	 EtOH
	 H_2O
		238(1.81)
		
		 238(0.92)
			102-0187-76
			
			 102-0187-76

$C_2H_6S_2$ Disulfide, dimethyl	hexane	255(1.54)	102-0187-76

Compound	Solvent	λ_{max} (log ϵ)	Ref.
$C_3BrClFNS$ Thiazole, 5-bromo-4-chloro-2-fluoro-	EtOH	215(3.43),247(3.57)	4-1297-76
C_3BrCl_2NS Thiazole, 2-bromo-4,5-dichloro-	EtOH	220(3.46),268(3.76)	4-1297-76
C_3Br_2ClNS Thiazole, 2,5-dibromo-4-chloro-	EtOH	223(3.43),268(3.72)	4-1297-76
C_3Br_2FNS Thiazole, 4,5-dibromo-2-fluoro-	EtOH	215(3.51),247(3.59)	4-1297-76
C_3Br_3NS Thiazole, 2,4,5-tribromo-	EtOH	222(3.57),268(3.79)	4-1297-76
C_3Cl_2FNS Thiazole, 4,5-dichloro-2-fluoro-	EtOH	245(3.53)	4-1297-76

C_3Cl_3NS
  Isothiazole, 3,4,5-trichloro-
  Thiazole, 2,4,5-trichloro-
	EtOH
	EtOH
		235s(3.74),260(4.08)
		217(3.45),265(3.73)
			4-1297-76
			4-1297-76

$C_3Cl_4N_2$ 2H-Imidazole, 2,2,4,5-tetrachloro-	n.s.g.	212(3.94)	24-1625-76
C_3F_6O 2-Propanone, 1,1,1,3,3,3-hexafluoro-	gas	302(0.91)	60-1574-76

$C_3F_6S_5$
  Carbonobis(dithioperoxo)thioic acid,
    bis(trifluoromethyl) ester
	isooctane	203(4.15),215(4.11),
		259(4.09),317(3.59)
			24-3432-76

C_3HCl_2NOS 2(3H)-Thiazolone, 4,5-dichloro-	EtOH	220s(3.68),250(3.70)	4-1297-76
$C_3HCl_2NS_2$ 2-Thiazolethiol, 4,5-dichloro-	EtOH	252(3.42),328(4.19)	4-1297-76

$C_3HCl_3N_2$
  1H-Imidazole, 2,4,5-trichloro-
  Propanenitrile, 3,3,3-trichloro-2-imino-
	n.s.g.
	CHCl₃
		206(3.87),221(3.85)
		260.0(2.25)
			24-1625-76
			94-0912-76

C_3HF_5O 2-Propanone, 1,1,1,3,3-pentafluoro-	gas	303(1.03)	60-1574-76
$C_3H_2BrNS_2$ 2(3H)-Thiazolethione, 5-bromo-	dioxan	330(4.18)	104-2381-76
$C_3H_2Br_2N_2$ 2-Propenenitrile, 2-amino-3,3-dibromo-	CHCl₃	264.5(3.98)	94-0912-76
$C_3H_2Cl_2N_2$ 2-Propenenitrile, 2-amino-3,3-dichloro-	hexane EtOH CHCl₃	253.0(3.99) 263.5(3.95) 260.5(3.97)	94-0912-76 94-0912-76 94-0912-76
$C_3H_2F_4O$ 2-Propanone, 1,1,3,3-tetrafluoro-	gas	314(1.18)	60-1574-76

Compound	Solvent	$\lambda_{max}(\log \epsilon)$	Ref.
$C_3H_2N_2O_2S_2$ 2(3H)-Thiazolethione, 5-nitro-	dioxan	300(3.98)	104-2381-76
$C_3H_2N_6O_2$ 3H-Pyrazol-5-amine, 3-diazo-4-nitro- hydrochloride	H_2O 0.1N H_2SO_4 5% $NaHCO_3$	320(4.11) 305(4.02) 320(4.11)	103-1355-76 103-1355-76 103-1355-76
$C_3H_2O_3$ 2-Cyclopropen-1-one, 2,3-dihydroxy-, dilithium salt	H_2O	240s(2.61)	35-3641-76
$C_3H_2S_3$ 1,3-Dithiole-2-thione	benzene	360(4.22),420s(2.06)	101-0369-76Q
$C_3H_3BrO_3$ Propanoic acid, 2-bromo-3-oxo-	pH 13	267(3.72)	73-2059-76
$C_3H_3ClO_3$ Propanoic acid, 2-chloro-3-oxo-	pH 13	267(3.86)	73-2059-76
$C_3H_3F_3O$ 2-Propanone, 1,1,1-trifluoro-	gas	285(0.91)	60-1574-76
$C_3H_3IO_3$ Propanoic acid, 2-iodo-3-oxo-	pH 13	227(3.90),263(3.81)	73-2059-76
$C_3H_3NS_2$ 2(3H)-Thiazolethione	dioxan	320(4.35)	104-2381-76
$C_3H_3N_3O_2$ 1H-1,2,3-Triazole-4-carboxylic acid	H_2O	211(3.83)	44-1041-76
$C_3H_3N_5O_3S$ 3H-[1,2,5]Oxadiazolo[3,4-c][1,2,6]thia- diazin-7-amine, 5,5-dioxide	H_2O pH -2.0 pH 4.0 pH 14.0	204(3.97),262(3.57), 337(3.30) 265(3.72) 265(3.58),335(3.32) 285(3.48),335(3.42)	4-0793-76 4-0793-76 4-0793-76 4-0793-76
$C_3H_3N_7$ Cyanamide, (6-amino-1,2,3,5-tetrazin- 4-yl)-, monopotassium salt	MeOH	320(4.34)	18-1339-76
$C_3H_3N_7O_2$ Unknown compd. VII	H_2O	301(4.34)	18-1339-76
$C_3H_4F_2O$ 2-Propanone, 1,3-difluoro-	gas	285(1.01)	60-1574-76
$C_3H_4N_2$ Imidazole 2-Propenenitrile, 3-amino- Pyrazole 	pH 1 EtOH H_2O pH 1 EtOH	207(3.74) 210(3.70) 254(4.08) 214(3.73) 210(3.57)	103-0691-76 103-0691-76 149-0217-76B 103-0691-76 103-0691-76
$C_3H_4N_2O$ 4H-Pyrazol-4-one, 2,3-dihydro-	EtOH	234(3.857)	23-1752-76

Compound	Solvent	$\lambda_{max}(\log \epsilon)$	Ref.
$C_3H_4N_2O_2S$ 2H-1,2,4-Oxadiazin-3(4H)-one, dihydro-5-thioxo-	EtOH	274(4.20)	44-3128-76
$C_3H_4N_2O_3$ 2H-1,2,4-Oxadiazine-3,5(4H,6H)-dione	EtOH	220(3.10)	44-3128-76
$C_3H_4N_2S$ 2H-Imidazole-2-thione, 1,3-dihydro-	EtOH CH_2Cl_2	259(4.134) 270(4.458)	28-0143-76B 28-0143-76B
$C_3H_4N_4O$ 1H-1,2,3-Triazole-4-carboxamide	H_2O	197(3.97)	44-1041-76
C_3H_4O Cyclopropanone	gas	201(3.00),312(1.23)	35-2027-76
$C_3H_4O_3$ Propanoic acid, 3-oxo-	pH 13	258(4.12)	73-2059-76
$C_3H_4S_2Se$ 1,3-Dithiolane-2-selone	C_6H_{12}	217(3.65),298(3.96), 355(4.11),570(2.08)	44-0879-76
1,3-Thiaselenolane-2-thione	C_6H_{12}	208(3.97),303(4.10), 326(4.08),476(1.82)	44-0879-76
$C_3H_4S_3$ 1,3-Dithiolane-2-thione	benzene	325(4.17),450(1.92)	101-0369-76Q
$C_3H_5ClO_2$ Propanoic acid, 3-chloro-	hexane H_2O	208s(1.68) 207s(1.68)	44-3312-76 44-3312-76
C_3H_5FO 2-Propanone, 1-fluoro-	gas	278(1.08)	60-1574-76
$C_3H_5N_3O_2$ 2H-1,2,4-Oxadiazin-3(6H)-one, 5-amino- 6H-1,2,4-Oxadiazin-5(4H)-one, 3-amino-	EtOH EtOH	228(4.12) 234(3.89)	44-3128-76 44-3128-76
$C_3H_5N_5O_3S$ 4H-1,2,6-Thiadiazin-4-one, 3,5-diamino-, oxime, 1,1-dioxide	EtOH	247(3.84),292(3.86)	4-0793-76
$C_3H_6N_2O_4$ Propane, 1,1-dinitro-, potassium salt	pH 13	382(4.23)	104-2433-76
$C_3H_6N_2O_4S$ Sulfonium, dimethyl-, dinitromethylide	H_2O MeOH MeCN DMSO	254(3.94),333(4.04) 327(3.96) 329(3.98) 333(3.98)	70-1906-76 70-1906-76 70-1906-76 70-1906-76
$C_3H_6N_2S$ 2-Imidazolidinethione	EtOH CH_2Cl_2	240(4.215) 248.5(4.234)	28-0485-76A 28-0485-76A
$C_3H_6N_4O_2S$ 4H-1,2,6-Thiadiazine-3,5-diamine, 1,1-dioxide	pH 5.0 H_2O	204(4.34),225s(3.95) 209(4.07),226(3.85), 283(2.80)	4-0793-76 4-0793-76

Compound	Solvent	$\lambda_{max}(\log \epsilon)$	Ref.
4H-1,2,6-Thiadiazine-3,5-diamine, 1,1-dioxide (cont.)	pH 11.0	225s(3.83),282(4.21)	4-0793-76
$C_3H_6N_4S$ 1H-1,2,4-Triazolium, 4-amino-3-hydroxy-1-methyl-, hydroxide, inner salt	MeCN	234(4.52),248(4.56)	103-0361-76
C_3H_6O 2-Propanone	gas C_6H_{12}	277(1.11) 278(1.11)	60-1574-76 112-0777-76
C_3H_6OS Propanethial, S-oxide	hexane	254(3.71)	102-0187-76
$C_3H_6O_2$ Acetic acid, methyl ester	neat	276.5(-1.01)	44-3312-76
	hexane	210.2(1.74)	44-3312-76
	H_2O	202.7(1.74)	44-3312-76
	MeCN	206.5(1.80)	44-3312-76
Formic acid, ethyl ester	hexane	215.6(1.90)	44-3312-76
	H_2O	208.1(1.87)	44-3312-76
	MeCN	213.6(1.84)	44-3312-76
Propanoic acid	neat	273.6(-1.74)	44-3312-76
	hexane	204.6(1.76)	44-3312-76
	H_2O	202.8(1.72)	44-3312-76
	MeCN	211.4(1.58)	44-3312-76
$C_3H_7N_5O_2S$ 2H-1,2,6-Thiadiazine-3,4,5-triamine, 1,1-dioxide	H_2O	204(3.93),281(3.55)	4-0793-76
$C_3H_9O_2PSe$ Phosphonoselenoic acid, methyl-, O,O-dimethyl ester	hexane or EtOH	196s(3.96),217s(3.77)	70-0174-76
$C_3H_9S_3$ Sulfonium, methylbis(methylthio)-, hexachloroantimonate	CH_2Cl_2	272(3.75)	118-0323-76
C_3N_6O 4H-Pyrazol-4-one, 3,5-bis(diazo)-3,5-dihydro-	H_2O 90% H_2SO_4 50% H_2SO_4	250(4.75) 230(4.65) 250(4.75)	103-1355-76 103-1355-76 103-1355-76

Compound	Solvent	$\lambda_{max}(\log \epsilon)$	Ref.
$C_4Cl_2F_3NS$ Thiazole, 4,5-dichloro-2-(trifluoro-methyl)-	EtOH	222(3.45),254(3.65)	4-1297-76
$C_4HCl_2NO_2S$ 2-Thiazolecarboxylic acid, 4,5-dichloro-	EtOH	233(3.56),269(3.92), 350(--)	4-1297-76
$C_4HCl_7N_3OP$ 1,3,5,2-Triazaphosphorine, 2-chloro-1,2-dihydro-4,6-bis(trichloromethyl)-, 2-oxide	MeCN	288(3.23)	65-0239-76
C_4HN_5 1H-Imidazole-4-diazonium, 5-cyano-, hydroxide, inner salt	H_2O	200(4.15),268s(3.65), 311(4.00)	103-0465-76
$C_4H_2BrClO_2S_2$ 2-Thiophenesulfonyl chloride, 5-bromo-	C_6H_{12}	258(3.87),281(4.17)	32-0657-76
$C_4H_2BrIO_3$ 2(5H)-Furanone, 4-bromo-5-hydroxy-3-iodo-	MeOH	206(3.64),252(3.62)	39-2470-76C
$C_4H_2BrNO_3$ Furan, 2-bromo-5-nitro- 2H-1,3-Oxazine-2,6(3H)-dione, 5-bromo-	dioxan pH 2 borax	315(4.04) 206(3.86),283(3.90) 217(3.83),316(4.02)	103-1091-76 73-2059-76 73-2059-76
$C_4H_2ClIO_2S_2$ 2-Thiophenesulfonyl chloride, 5-iodo-	C_6H_{12}	262(3.98),285(4.20)	32-0657-76
$C_4H_2ClIO_3$ 2(5H)-Furanone, 4-chloro-5-hydroxy-3-iodo-	MeOH	205(3.71),248(3.81)	39-2470-76C
$C_4H_2ClNO_2$ 1H-Pyrrole-2,5-dione, 1-chloro-	MeOH	216(3.84),257(3.58)	117-0025-76
$C_4H_2ClNO_3$ Furan, 2-chloro-5-nitro- 2H-1,3-Oxazine-2,6(3H)-dione, 5-chloro-	dioxan pH 2 borax	313(5.01) 206(3.77),281(3.85) 214(3.74),315(3.96)	103-1091-76 73-2059-76 73-2059-76
$C_4H_2ClNO_4S_2$ 2-Thiophenesulfonyl chloride, 5-nitro-	C_6H_{12}	223(3.76),284(4.22)	32-0657-76
$C_4H_2ClN_5$ 1H-Imidazo[4,5-d]-1,2,3-triazine, 4-chloro-	H_2O	209(4.18),261(3.70)	103-0465-76
$C_4H_2Cl_2N_2O$ 3-Pyridazinol, 4,5-dichloro-	pH 6.0 pH 11.0	210s(4.20),214(4.22), 238s(3.15),289(3.47) 218(4.26),236s(3.48), 305(3.51)	39-1424-76C 39-1424-76C
$C_4H_2Cl_2O_2S_2$ 2-Thiophenesulfonyl chloride, 5-chloro-	C_6H_{12}	254(3.87),278(4.00)	32-0657-76

Compound	Solvent	$\lambda_{max}(\log \epsilon)$	Ref.
$C_4H_2INO_3$			
Furan, 2-iodo-5-nitro-	dioxan	328(4.09)	103-1091-76
2H-1,3-Oxazine-2,6(3H)-dione, 5-iodo-	pH 2	210(4.00),292(3.84)	73-2059-76
	borax	217(3.96),319(4.07)	73-2059-76
$C_4H_2N_4O_2$			
1H-Pyrazole-4-carbonitrile, 3-nitro-	MeOH	211(3.83),266(3.73)	104-1538-76
$C_4H_2O_2S_2$			
3-Cyclobutene-1,2-dione, 3,4-dimer-	H_2O	231(4.04),250(4.12),	44-3904-76
capto-, dipotassium salt		322s(4.30),347(4.42)	
$C_4H_3BrN_2O_2$			
2,4(1H,3H)-Pyrimidinedione, 5-bromo-	EtOH	276(4.14)	95-1094-76
$C_4H_3BrO_2$			
2(5H)-Furanone, 5-bromo-	EtOH	217(3.94)	33-0100-76
$C_4H_3Br_5O$			
3-Buten-2-ol, 1,1,1,4,4-pentabromo-	EtOH	215(4.08)	78-2843-76
$C_4H_3ClN_2O_2$			
3,6-Pyridazinediol, 4-chloro-	pH 2.0	211(4.31),232s(3.62),	39-1424-76C
		308(3.47)	
	pH 8.0	223(4.30),241s(3.87),	39-1424-76C
		337(3.39)	
	pH 14.7	241s(3.87),341(3.50)	39-1424-76C
$C_4H_3ClO_2S_2$			
2-Thiophenesulfonyl chloride	C_6H_{12}	247(3.88),267(3.84)	32-0657-76
$C_4H_3ClO_3S$			
2-Furansulfonyl chloride	C_6H_{12}	249(3.90)	32-0657-76
$C_4H_3Cl_2NOS$			
Thiazole, 4,5-dichloro-2-methoxy-	EtOH	250(3.77)	4-1297-76
$C_4H_3Cl_2N_3$			
4-Pyridazinamine, 3,6-dichloro-	pH -1.0	222s(3.96),281(4.10)	39-1424-76C
	pH 5.0	217(4.12),256(3.96),	39-1424-76C
		292(3.66)	
$C_4H_3FN_2O_2$			
2,4(1H,3H)-Pyrimidinedione, 5-fluoro-	pH 1	266(3.85)	35-7381-76
	pH 13	280(3.72)	35-7381-76
C_4H_3NOS			
2-Thiazolecarboxaldehyde	1% EtOH	237(2.48),295(2.60)	44-1952-76
$C_4H_3NO_3$			
2H-1,3-Oxazine-2,6(3H)-dione	H_2O	264(3.89)	87-0643-76
$C_4H_3N_5O$			
1,2,4-Triazolo[5,1-c][1,2,4]triazin-	EtOH	213(3.83),241(3.55),	39-1496-76C
7(4H)-one		246s(3.53),269s(3.53),	
		298(3.81)	
hydrate	EtOH	290(3.93)	39-1496-76C
Triazolo[3,2-c]triazin-7(4H)-one	EtOH	241(3.72),300(3.87)	22-i178-76
Triazolo[4,3-b]triazin-7(1H)-one	EtOH	235(3.94),295(3.75)	22-i178-76

Compound	Solvent	$\lambda_{max}(\log \epsilon)$	Ref.
$C_4H_4BrNO_4$ 2-Propenoic acid, 2-bromo-3-(carboxyamino)-	pH 13	276(4.13)	73-2059-76
$C_4H_4BrN_3O$ 4(1H)-Pyrimidinone, 2-amino-6-bromo- 4(1H)-Pyrimidinone, 6-amino-2-bromo-	MeOH MeOH	225(3.92),289(4.00) 216(4.36),259(3.73)	94-0507-76 94-0507-76
$C_4H_4Br_4O$ 3-Buten-2-ol, 1,1,4,4-tetrabromo-	EtOH	212.5(3.92)	78-2843-76
$C_4H_4ClNO_4$ 2-Propenoic acid, 3-(carboxyamino)-2-chloro-	pH 13	273(4.11)	73-2059-76
$C_4H_4ClN_3$ 3-Pyridazinamine, 6-chloro-	pH 1.0 pH 6.0	213(4.07),232(3.68), 301(2.99) 213s(3.67),239(3.76), 312(2.95)	39-1424-76C 39-1424-76C
$C_4H_4ClN_3O$ 4(1H)-Pyrimidinone, 6-amino-2-chloro-	MeOH	214(4.39),257(3.73)	94-0026-76
$C_4H_4ClN_3O_3S$ 1H-Imidazole-4-sulfonyl chloride, 5-(aminocarbonyl)-	H_2O	237(3.77),250(3.77)	103-0924-76
$C_4H_4FN_3O$ 2(1H)-Pyrimidinone, 4-amino-5-fluoro-	pH 1 pH 13	285(3.95) 292(3.88)	35-7381-76 35-7381-76
$C_4H_4INO_4$ 2-Propenoic acid, 3-(carboxyamino)-2-iodo-	pH 13	276(4.13),300s(3.76)	73-2059-76
$C_4H_4IN_3O$ 4(1H)-Pyrimidinone, 2-amino-6-iodo-	MeOH	227(4.17),294(3.94)	94-0507-76
$C_4H_4N_2O$ 1H-Imidazole-2-carboxaldehyde iminate anion 2(1H)-Pyrimidinone 4(3H)-Pyrimidinone	1% EtOH NaOH EtOH EtOH	215(3.60),287(3.70) 310(3.90) 298(3.67) 218(4.08),270(3.59)	44-1952-76 44-1952-76 35-0171-76 35-0171-76
$C_4H_4N_2OS$ 4(1H)-Pyrimidinone, 2,3-dihydro-2-thioxo-	EtOH	274(4.45)	95-1094-76
$C_4H_4N_2O_2$ 2,4(1H,3H)-Pyrimidinedione	pH 2 pH 5.5 pH 11 pH 12 EtOH	259.5(3.94) 259.5(3.93) 285(3.77) 284(3.81) 260(4.10)	88-2321-76 88-2321-76 46-0112-76 88-2321-76 95-1094-76
$C_4H_4N_2O_2S$ 5-Isothiazolecarboxamide, 2,3-dihydro-3-oxo-	MeOH	226(3.95),290(3.46)	1-0781-76

Compound	Solvent	$\lambda_{max}(\log \epsilon)$	Ref.
$C_4H_4N_4O_3S$			
Imidazo[4,5-e]-1,2,4-thiadiazin-3(2H)-one, 4,5-dihydro-, 1,1-dioxide	H_2O	227(4.17)	103-0924-76
$C_4H_4N_6O_3S$			
1H-Imidazole-4-carbonyl azide, 5-(aminosulfonyl)-	H_2O	274(4.16)	103-0924-76
$C_4H_4N_6S$			
[1,2,3]Thiadiazolo[5,4-d]pyrimidine-5,7-diamine	pH 1	235(4.23),274s(3.93), 283(3.94)	87-1186-76
	pH 7	243(4.09),310(4.00)	87-1186-76
	pH 13	243(4.10),310(4.00)	87-1186-76
7H-1,2,3-Triazolo[4,5-d]pyrimidine-7-thione, 5-amino-3,6-dihydro-	pH 1	224(4.15),258(3.87), 338(4.24)	87-1186-76
	pH 7	231(4.29),257(3.70), 282s(4.54),343(4.23)	87-1186-76
	pH 13	224(4.29),279(3.92), 328(4.17)	87-1186-76
$C_4H_4SSe_2$			
1,3-Thiaselenole-2-selone, 5-methyl-	hexane	255(3.79),300(3.00), 420(4.17),540(2.48)	88-0423-76
$C_4H_5BO_3$			
Boronic acid, 2-furanyl-	H_2O	233(4.00)	4-1265-76
Boronic acid, 3-furanyl-	H_2O	210(3.58)	4-1265-76
$C_4H_5BrN_4$			
4,6-Pyrimidinediamine, 2-bromo-	MeOH	224(4.43),263(3.76)	94-0026-76
$C_4H_5ClN_4$			
3,5-Pyridazinediamine, 6-chloro-	pH 2.0	204(4.29),232(4.00), 275s(3.77),302(3.98)	39-1424-76C
	pH 7.0	208(4.27),221(4.15), 259(3.75),297(3.76)	39-1424-76C
4,6-Pyrimidinediamine, 2-chloro-	MeOH	223(4.36),262(3.73)	94-0026-76
$C_4H_5NO_2S$			
Acetic acid, isothiocyanato-, methyl ester	pH 8.06	254(3.04)	73-2216-76
	dioxan	248(3.06)	73-2216-76
$C_4H_5NO_4$			
2-Propenoic acid, 3-(carboxyamino)-	pH 13	259(4.27)	73-2059-76
C_4H_5NS			
Isothiazole, 5-methyl-	EtOH	217(3.554),245(3.706)	48-0507-76
$C_4H_5N_3$			
3-Pyridazinamine	pH 2.0	227(3.76),289(3.42)	39-1424-76C
	pH 8.0	231(3.86),294(3.31)	39-1424-76C
4H-1,2,4-Triazole, 4-ethenyl-	pH 1	225(4.00)	103-0691-76
	EtOH	228(4.14)	103-0691-76
	dioxan	231(4.15)	103-0691-76
$C_4H_5N_3OS$			
4(1H)-Pyrimidinone, 6-amino-2,3-dihydro-2-thioxo-	EtOH	283(4.24)	95-1094-76

Compound	Solvent	$\lambda_{max}(\log \epsilon)$	Ref.
$C_4H_5N_3O_2$			
1H-Imidazole-1-carboxamide, 2,3-di- hydro-2-oxo-	pH 1 pH 7 pH 13	236(3.61) 236(3.61) 254(3.48)	35-8218-76 35-8218-76 35-8218-76
1H-1,2,3-Triazole-4-carboxylic acid, methyl ester	H_2O	219(3.91)	44-1041-76
1H-1,2,3-Triazole-4-carboxylic acid, 1-methyl-	H_2O	211(3.93)	44-1041-76
$C_4H_5N_3O_4$			
Butanenitrile, 4,4-dinitro-, potassium salt	pH 13	373(4.22)	104-2433-76
2,4(1H,3H)-Pyrimidinedione, 5-nitro-	pH 11	320(4.13)	87-1072-76
$C_4H_5N_5O_2S$			
1H,4H-Imidazo[2,3-c][1,2,6]thiadiazin- 7-amine, 5,5-dioxide	pH -2.0	223(3.95),245s(3.71), 279(3.72),305s(3.58)	4-0793-76
	pH 2.0	231(3.96),300(3.80)	4-0793-76
	pH 7.0	207(4.13),216s(4.09), 300(3.89)	4-0793-76
	H_2O	207(4.04),217(4.01), 301(3.86)	4-0793-76
$C_4H_5N_9$			

Guanidine, N-(6-amino-1,2,3,5-tetrazin-
    4-yl)-N'-cyano-
    potassium salt
	MeOH
	KOH
	H_2O
	KOH
		306(4.22),333(4.24)
		226(4.00),336(4.27)
		335(4.43)
		205(3.77),334(4.43)
			18-1339-76
			18-1339-76
			18-1339-76
			18-1339-76

$C_4H_6ClNO_2$			
Acetamide, N-acetyl-N-chloro-	MeOH MeCN	211(4.03) 206(3.99)	117-0025-76 117-0025-76
$C_4H_6ClN_5$			
Pyridazine, 5-amino-4-chloro-3-hydra- zino-	pH 4.0 pH 7.9 pH 11.0	232(4.36),285(3.75) 227(4.36) 259(3.91),295s(3.56)	39-1424-76C 39-1424-76C 39-1424-76C
$C_4H_6N_2$			
2-Butenenitrile, 3-amino-	H_2O	257(4.04)	149-0217-76B
1H-Imidazole, 1-methyl-, compound with iodine	CH_2Cl_2 CCl_4	370(3.49) 395(3.28)	65-0664-76 65-0664-76
$C_4H_6N_2OS$			
3-Isothiazolol, 5-(aminomethyl)-, zwitter ion	MeOH	264(3.68)	1-0781-76
hydrobromide	MeOH	262(3.65)	1-0781-76
2-Thiazolamine, 4-methyl-, 3-oxide	MeOH	264(3.69)	64-0251-76B
$C_4H_6N_2OSe$			
1,2,5-Selenadiazole, 3,4-dimethyl-, 2-oxide	EtOH	245(3.39),285(3.87)	1-0675-76
$C_4H_6N_2O_2$			
2-Butenedial, dioxime	MeOH	280(4.34),323(4.22), 347(4.24)	104-2039-76
$C_4H_6N_2O_3$			
2H-1,2,4-Oxadiazine-3,5(4H,6H)-dione, 2-methyl-	EtOH	225(4.21)	44-3128-76

Compound	Solvent	$\lambda_{max}(\log \epsilon)$	Ref.
$C_4H_6N_2S$			
1H-Imidazole, 2-(methylthio)-, mono-	EtOH	218-253(4.307-3.826)	28-0143-76B
hydriodide	CH_2Cl_2	226-255(3.802-3.384)	28-0143-76B
2H-Imidazole-2-thione, 1,3-dihydro-	EtOH	260(4.192)	28-0143-76B
1-methyl-	CH_2Cl_2	270(4.215)	28-0143-76B
$C_4H_6N_4$			
3,4-Pyridazinediamine	pH -2.1	226(3.97),299(3.93)	39-1424-76C
	pH 5.0	226(4.06),299(3.98)	39-1424-76C
	pH 10.0	214(4.13),254(3.72),	39-1424-76C
		291(3.77)	
3,5-Pyridazinediamine	pH 5.5	225(4.46),277(3.74)	39-1424-76C
	pH 10.0	218(4.49),257(3.54),	39-1424-76C
		278(3.54)	
3,6-Pyridazinediamine	pH -3.9	229(4.10),296(3.34)	39-1424-76C
	pH 2.0	233(4.26),332(3.27)	39-1424-76C
	pH 9.0	235(4.06),326(3.29)	39-1424-76C
4,5-Pyridazinediamine	pH 4.0	219(4.24),286s(3.79),	39-1424-76C
		314(3.97)	
	pH 9.0	214(4.38),281(3.94)	39-1424-76C
$C_4H_6N_4O$			
1H-1,2,3-Triazole-4-carboxamide,	H_2O	211(4.00)	44-1041-76
N-methyl-			
1H-1,2,3-Triazole-4-carboxamide,	H_2O	210(4.05)	44-1041-76
1-methyl-			
$C_4H_6N_4O_3S$			
1H-Imidazole-4-carboxamide, 5-(amino-	H_2O	206(3.87),244(3.96)	103-0924-76
sulfonyl)-			
C_4H_6O			
3-Buten-2-one	MeOH	211(3.85),321(1.41)	40-0166-76
Cyclobutanone	C_6H_{12}	273(1.20),280(1.00),	112-0777-76
		290(1.26)	
Cyclopropanecarboxaldehyde	C_6H_{12}	286(1.30)	39-0669-76B
	EtOH	279(1.30)	39-0669-76B
$C_4H_6O_2$			
2-Propenal, 3-hydroxy-2-methyl-	pH 1	250(4.01)	44-0294-76
	pH 6.8	275(4.17)	44-0294-76
	pH 13	275(4.17)	44-0294-76
	90% H_2SO_4	270(4.13)	44-0294-76
2-Propenoic acid, methyl ester	C_6H_{12}	192(4.16),246(1.97)	48-0745-76
	EtOH	195(4.15),239(2.04)	48-0745-76
	MeCN	193(4.20),242(1.90)	48-0745-76
$C_4H_6O_3$			
Propanoic acid, 2-formyl-	pH 13	266(3.89)	73-2059-76
$C_4H_6S_4$			
Ethanebis(dithioic) acid, dimethyl	C_6H_{12}	220(3.80),242(3.71),	97-0318-76
ester		266(3.50),312(3.53),	
		362(3.80)	
$C_4H_7ClO_2$			
Acetic acid, 2-chloroethyl ester	hexane	209.9(1.76)	44-3312-76
	H_2O	203.0(1.77)	44-3312-76
	MeCN	207.9(1.74)	44-3312-76

Compound	Solvent	λ_{max}(log ϵ)	Ref.
C_4H_7NO			
Acetamide, N-ethenyl-	EtOH	225(4.20)	35-5996-76
3-Buten-2-one, 4-amino-	C_6H_{12}	282(4.08)	35-2826-76
	MeOH	291(4.27)	35-2826-76
$C_4H_7NO_2$			
2,3-Butanedione, monooxime, anion	H_2O	360(1.76)	48-0001-76
	25% dioxan	365(1.73)	48-0001-76
	50% dioxan	370(1.75)	48-0001-76
	75% dioxan	370(1.62)	48-0001-76
$C_4H_7N_3O_2$			
2-Butenediamide, 2-amino-	MeOH	300(3.82)	1-0781-76
$C_4H_7N_5O_2$			
1H-1,2,4-Triazole-1-acetic acid, 3,5-	EtOH	215(1.4)	23-3620-76
diamino-			
$C_4H_7N_5O_2S_2$			
Methanethioamide, N-(3,5-diamino-2H-	H_2O	226s(3.88),280(4.10)	4-0793-76
1,2,6-thiadiazin-4-yl)-, S,S-dioxide			
$C_4H_8N_2O_2$			
2,3-Butanedione, dioxime	MeOH	228(4.25)	104-2039-76
$C_4H_8N_2S$			
1H-Imidazole, 2,5-dihydro-2-(methyl-	EtOH	221.5(4.432)	28-0485-76A
thio)-, monohydriodide	CH_2Cl_2	231.0(4.327)	28-0485-76A
2(1H)-Pyrimidinethione, tetrahydro-	EtOH	244.5(4.092)	28-0485-76A
	CH_2Cl_2	256(4.135)	28-0485-76A
$C_4H_8N_4O_2S$			
Methanesulfonamide, N-(1-methyl-1H-	EtOH	253(3.93)	32-0001-76
1,2,3-triazol-5-yl)-			
$C_4H_8N_4S$			
3H-1,2,4-Triazole-3-thione, 5-amino-	EtOH	233(3.99),271(4.13)	65-0174-76
1,2-dihydro-1,2-dimethyl-			
$C_4H_8N_6O_3S$			
Hydrazinecarboxamide, N-[5-(aminosul-	H_2O	236(4.18)	103-0924-76
fonyl)-1H-imidazol-4-yl]-			
C_4H_8O			
2-Butanone	C_6H_{12}	278(1.23)	35-0104-76
	C_6H_{12}	278(1.23)	112-0777-76
	CCl_4	275(1.34)	35-0104-76
C_4H_8OS			
2-Propanone, 1-(methylthio)-	hexane	241(2.57),302(2.34)	23-3026-76
	50% EtOH	243(2.52),295(2.46)	23-3026-76
$C_4H_8O_2$			
Acetic acid, ethyl ester	hexane	210.5(1.77)	44-3312-76
	hexane	210.2(1.77)	44-3312-76
	H_2O	202.7(1.76)	44-3312-76
	H_2O	202.9(1.80)	44-3312-76
	MeCN	206.5(1.80)	44-3312-76
	MeCN	209.0(1.77)	44-3312-76
Butanoic acid	neat	268.9(-1.28)	44-3312-76

Compound	Solvent	$\lambda_{max}(\log \epsilon)$	Ref.
Butanoic acid (cont.)	hexane	205.1(1.77)	44-3312-76
	H_2O	203.8(1.71)	44-3312-76
	MeCN	213.1(1.64)	44-3312-76
Propanoic acid, 2-methyl-	neat	270.2(-1.23)	44-3312-76
	hexane	206.1(1.87)	44-3312-76
	H_2O	206.2(1.84)	44-3312-76
	MeCN	212.9(1.73)	44-3312-76
$C_4H_8O_2S$			
3,4-Thiophenediol, tetrahydro-	EtOH	207(3.15)	39-2533-76C
$C_4H_8O_3S$			
3,4-Thiophenediol, tetrahydro-, 1-oxide, cis-α-	EtOH	209(3.00)	39-2533-76C
trans	EtOH	206(3.17),219s(2.96)	39-2533-76C
$C_4H_8O_4S$			
3,4-Thiophenediol, tetrahydro-, 1,1-dioxide, cis	EtOH	223(2.38)	39-2533-76C
$C_4H_9Cl_2PSe$			
Phosphonoselenoic dichloride, butyl-	C_6H_{12}	201s(3.90),239(3.71), 280(3.17)	70-0174-76
$C_4H_9NS_2$			
Carbamodithioic acid, dimethyl-, methyl ester	C_6H_{12}	223(4.13),247(3.95), 279(4.02),341(1.95)	97-0355-76
$C_4H_9N_3O_2$			
Acetamide, N-[(methylnitrosoamino)-methyl]-	H_2O	342(1.87)	94-0369-76
	EtOH	230(3.99),353(1.89), 356s(1.89),373s(1.71)	94-0369-76
	$CHCl_3$	360(2.03)	94-0369-76
$C_4H_{10}ClPSe$			
Phosphinoselenoic chloride, diethyl-	hexane	197(4.12),260(3.60)	70-0174-76
	EtOH	194(4.22),258(3.59)	70-0174-76
	HCOOH	251(3.40)	70-0174-76
	HOAc	256(3.59)	70-0174-76
	MeCN	255(3.55)	70-0174-76
	MeCN-HClO$_4$	254(3.58)	70-0174-76
	CF_3COOH	240(3.55)	70-0174-76
$C_4H_{22}B_{18}CoN$			
Cobalt, [[10,10'-μ-amidobis-[(7,8,9,10,11-η)decahydro-7,8-dicarbaundecaborato]](3-)]-	CH_2Cl_2	259(4.50),294s(--), 415(3.15),498(3.01)	73-3509-76

Compound	Solvent	$\lambda_{max}(\log \epsilon)$	Ref.
$C_5HCl_2F_2N$ Pyridine, 3,5-dichloro-2,6-difluoro-	MeOH	211(3.74),276(3.69)	33-0229-76
$C_5HCl_3N_2O_2$ Pyridine, 2,3,6-trichloro-5-nitro-	MeOH	252(3.68),294(3.61)	33-0190-76
$C_5H_2ClF_2N$ Pyridine, 3-chloro-2,6-difluoro-	MeOH	265.5(3.59)	33-0229-76
C_5H_2ClNO 2-Furancarbonitrile, 5-chloro-	EtOH	215(4.43),254(4.38)	73-1692-76
$C_5H_2Cl_3N$ Pyridine, 2,3,6-trichloro-	MeOH	224(3.65),279(4.01)	33-0190-76
$C_5H_2F_2N_2O_2$ Pyridine, 2,6-difluoro-3-nitro-	MeOH	235(3.65),271(2.78)	33-0229-76
$C_5H_2N_6O$ [1,2,4]Triazolo[5,1-c][1,2,4]triazine-6-carbonitrile, 4,7-dihydro-7-oxo-	EtOH	213(4.35),217s(4.35), 253(4.06),257s(4.05), 275(3.95),319(4.41)	39-1496-76C
pyridinium salt	EtOH	223(4.05),244s(3.58), 250(3.74),257(3.86), 263(3.89),273s(3.76), 332(4.11)	39-1496-76C
C_5H_3BrOS 2-Thiophenecarboxaldehyde, 5-bromo-	1% EtOH	270(3.70),302(4.08)	44-1952-76
$C_5H_3ClN_2$ 1H-Pyrrole-2-carbonitrile, 3-chloro-	H_2O	248(4.15)	33-2786-76
$C_5H_3ClN_4O$ 1H-Tetrazole, 5-(5-chloro-2-furanyl)-	EtOH	204(4.15),278(4.48)	73-1692-76
C_5H_3ClOS 2-Thiophenecarboxaldehyde, 4-chloro-	MeOH	256(3.96),296(3.63)	4-0393-76
$C_5H_3ClO_2S$ 2-Thiophenecarboxylic acid, 4-chloro-	MeOH	243(3.93),277(3.64)	4-0393-76
$C_5H_3NO_4$ 2-Furancarboxaldehyde, 5-nitro-	1% EtOH	227(3.83),310(4.04)	44-1952-76
$C_5H_3N_3O_2S$ 7H-1,3,4-Thiadiazolo[3,2-a]pyrimidin-7-one, 5-hydroxy-	isoPrOH	218(4.44),239(4.11), 275s(3.75)	4-0291-76
$C_5H_3N_5O_3$ [1,2,4]Triazolo[5,1-c][1,2,4]triazine-6-carboxylic acid, 4,7-dihydro-7-oxo-	EtOH	216(4.15),242s(3.62), 280s(3.71),312(3.97)	39-0421-76C
$C_5H_3N_7$ [1,2,4]Triazolo[5,1-c][1,2,4]triazine-6-carbonitrile, 7-amino-	EtOH	210(4.11),228(4.03), 279(3.78),284s(3.77), 332(3.90)	39-1496-76C
C_5H_4 1,2,3,4-Pentatetraene	hexane	201(5.00),242(4.23)	77-0235-76

Compound	Solvent	$\lambda_{max}(\log \epsilon)$	Ref.
$C_5H_4Br_2O_2$			
2(5H)-Furanone, 5-bromo-5-(bromomethyl)-	EtOH	212(4.21)	33-0100-76
2(5H)-Furanone, 5,5-dibromo-4-methyl-	EtOH	220(4.26)	33-2724-76
C_5H_4ClNO			
2(1H)-Pyridinone, 6-chloro-	H_2O	304(3.85)	35-0171-76
1H-Pyrrole-2-carbonyl chloride	$CHCl_3$	288(4.23)	44-3050-76
$C_5H_4ClNO_2$			
2(1H)-Pyridinone, 4-chloro-6-hydroxy-	MeOH	235(3.72),316(3.96)	39-2462-76C
$C_5H_4Cl_2N_2O$			
3(2H)-Pyridazinone, 5,6-dichloro-2-methyl-	pH 5.0	217(4.45),235s(3.67), 303(3.41)	39-1424-76C
$C_5H_4N_2O_3$			
3-Pyridazinecarboxylic acid, 1,6-di-hydro-6-oxo-	pH 0.5	247(3.87),281(3.38)	39-1424-76C
	pH 6.0	237(3.84),285(3.27)	39-1424-76C
	pH 12.5	253(4.06),295(3.55)	39-1424-76C
5-Pyrimidinecarboxaldehyde, 1,2,3,4-tetrahydro-2,4-dioxo-	H_2O	278(4.07)	44-0567-76
$C_5H_4N_2O_3S$			
4-Pyrimidinecarboxylic acid, 1,2,3,6-tetrahydro-6-oxo-2-thioxo- (2-thio-orotic acid)	pH 1	270(4.31),310(3.78)	104-1802-76
	pH 13	260(4.15),305(3.79)	104-1802-76
$C_5H_4N_2O_4$			
Orotic acid	pH 1	282(3.75)	104-1802-76
	pH 13	286(3.88)	104-1802-76
$C_5H_4N_2O_5$			
2-Furancarboxaldehyde, 4,5-dihydro-5-aci-nitro-4-oxo-, oxime, potassium salt	H_2O	405(4.26)	87-0729-76
$C_5H_4N_4$			
1H-1,2,3-Triazolo[4,5-c]pyridine	pH 1	261(3.66)	44-1449-76
	pH 11	264(3.64),279(3.66)	44-1449-76
	MeOH	258(3.63)	44-1449-76
	EtOH	258(3.66)	114-0405-76C
$C_5H_4N_4O$			
1H-Pyrazolo[3,4-b]pyrazine, 4-oxide	MeOH	231(4.2),296(3.9), 323(3.8)	104-0901-76
Pyrazolo[1,5-a][1,3,5]triazin-4(1H)-one	pH 1	258(3.84)	4-1305-76
	pH 7	263(3.87)	4-1305-76
	pH 13	266(3.89)	4-1305-76
1H-Tetrazole, 5-(2-furanyl)-	EtOH	259(4.41)	73-1692-76
3H-1,2,3-Triazolo[4,5-b]pyridine, 3-hydroxy-	EtOH	278(2.9)	103-1172-76
$C_5H_4N_4OS$			
Pyrazolo[1,5-a]-1,3,5-triazin-4(1H)-one, 2,3-dihydro-2-thioxo-	pH 0.1	225(3.82),246(4.00), 288(4.34)	4-0589-76
	pH 7.08	230(4.08),296(4.30)	4-0589-76
	pH 14	248(4.35),286(4.21)	4-0589-76
$C_5H_4N_4O_2S$			
1,2,3-Thiadiazolo[4,5-d]pyrimidine-	EtOH	212(4.23),250(3.87),	88-1129-76

Compound	Solvent	λ_{max}(log ϵ)	Ref.
5,7(4H,6H)-dione, 6-methyl- (cont.)		320(3.83)	88-1129-76
$C_5H_4N_4O_4$ [1,2,5]Oxadiazolo[3,4-d]pyrimidine- 5,7(4H,6H)-dione, 6-methyl-, 1-oxide	dioxan	268(4.06),312(3.23)	39-1327-76C
$C_5H_4N_6O_4$ Tetrazolo[1,5-c]pyrimidine-5,7(1H,6H)- dione, 6-methyl-8-nitro-	pH 1	265s(3.61),341(4.20)	39-1327-76C
C_5H_4OS 2-Thiophenecarboxaldehyde	1% EtOH	264(4.36),292(4.20)	44-1952-76
$C_5H_4O_2$ 2-Furancarboxaldehyde	gas 1% EtOH	330f(1.5) 278(4.13)	23-3089-76 44-1952-76
$C_5H_5BO_4$ Boronic acid, (2-formyl-3-furanyl)- Boronic acid, (3-formyl-2-furanyl)- Boronic acid, (4-formyl-2-furanyl)- Boronic acid, (5-formyl-2-furanyl)- Boronic acid, (5-formyl-3-furanyl)-	H_2O H_2O H_2O H_2O H_2O	237(3.54),288(4.11) 224(3.83),275(3.66) 260(3.67) 242(3.49),285(4.21) 238s(3.34),277(4.07)	4-1265-76 4-1265-76 4-1265-76 4-1265-76 4-1265-76
$C_5H_5BrN_2O_2$ 2,4(1H,3H)-Pyrimidinedione, 5-bromo- 6-methyl-	EtOH	275(3.89)	95-1094-76
$C_5H_5BrO_2$ 2-Cyclopenten-1-one, 3-bromo-2-hydroxy- 2(5H)-Furanone, 3-(bromomethyl)- 2(5H)-Furanone, 5-bromo-3-methyl- 2(5H)-Furanone, 5-bromo-4-methyl-	MeOH EtOH EtOH EtOH	262(4.21) 213(3.81) 212(4.10) 219(c.4.28)	104-0089-76 33-0100-76 35-3661-76 33-2724-76
$C_5H_5ClN_2O_2$ 3-Pyridazinone, 4-chloro-6-hydroxy- 2-methyl- 3-Pyridazinone, 5-chloro-6-hydroxy- 2-methyl-	pH 3.0 pH 8.0 pH 2.0 pH 7.0	211(4.29),230s(3.66), 314(3.56) 222(4.28),340(3.48) 214(4.22),312(3.36) 221(3.69),240s(3.21), 339(2.89)	39-1424-76C 39-1424-76C 39-1424-76C 39-1424-76C
$C_5H_5ClN_4O_2$ 2,6-Pyridinediamine, 3-chloro-5-nitro- 2,6-Pyridinediamine, 4-chloro-3-nitro-	MeOH pH 7	273(3.93),416(4.22) 224(4.03),273(3.84), 303(3.62),397(4.16)	33-0190-76 44-3784-76
$C_5H_5ClO_2$ 2,4-Pentadienoic acid, 5-chloro-	EtOH	256(4.41)	70-0652-76
$C_5H_5ClO_2S_2$ 2-Thiophenesulfonyl chloride, 5-methyl-	C_6H_{12}	250(3.82),277(3.94)	32-0657-76
$C_5H_5FN_2O_2$ 2,4(1H,3H)-Pyrimidinedione, 5-fluoro- 1-methyl-	acid base	273(3.92) 271(3.79)	35-7381-76 35-7381-76
C_5H_5N Pyridine	50% MeOH	255(3.27)	104-2433-76

Compound	Solvent	$\lambda_{max}(\log \epsilon)$	Ref.
C_5H_5NO			
2(1H)-Pyridinone	C_6H_{12}	295(3.69)	39-1428-76B
	MeOH	297(3.80)	39-1428-76B
	EtOH	305(3.78)	35-0171-76
	EtOH	299(3.87)	39-1428-76B
	$CHCl_3$	296(3.81)	39-1428-76B
	MeCN	303(3.86)	39-1428-76B
	C_6H_{12}-1% EtOH	300(3.65)	39-1428-76B
4(1H)-Pyridinone	MeOH	256(4.15)	39-1428-76B
	EtOH	257(4.17)	35-0171-76
	EtOH	257(4.26)	39-1428-76B
	$CHCl_3$	259(4.03)	39-1428-76B
	MeCN	256(4.18)	39-1428-76B
	C_6H_{12}-1% EtOH	258(2.77)	39-1428-76B
1H-Pyrrole-2-carboxaldehyde iminate anion	1% EtOH + NaOH	250(c.3.5),293(4.20) 265(c.3.0),315(4.34)	44-1952-76 44-1952-76
$C_5H_5NO_2S$			
4H-1,3-Thiazine-4,6(5H)-dione, 2-methyl-	EtOH	238(4.00),305(3.35)	103-0862-76
$C_5H_5NO_3$			
2H-1,3-Oxazine-2,6(3H)-dione, 4-methyl-	pH 2	206(3.76),266(3.97)	73-2059-76
	0.05M NaOH	216(3.76),295(4.05)	73-2059-76
2H-1,3-Oxazine-2,6(3H)-dione, 5-methyl-	pH 1	206(3.74),271(3.86)	73-2059-76
	pH 12	212(3.77),304(3.97)	73-2059-76
$C_5H_5NO_4S$			
1H-Pyrrole-3-sulfonic acid, 5-formyl-, monopotassium salt	H_2O	285(4.20)	33-1621-76
	M KOH	308(4.30)	33-1621-76
C_5H_5NS			
2(1H)-Pyridinethione	EtOH	287(4.01),362(3.69)	35-0171-76
4(1H)-Pyridinethione	EtOH	230(3.77),342(4.16)	35-0171-76
$C_5H_5N_5$			
Adenine	pH 1	263(4.11)	94-1331-76
	H_2O	261(4.12)	94-1331-76
	pH 13	269(4.09)	94-1331-76
Pyrazolo[1,5-a]-1,3,5-triazin-4-amine	pH 1	260(3.60),296(3.38)	4-1305-76
	pH 7	279(3.80)	4-1305-76
$C_5H_5N_5O$			
3H-Triazolo[4,5-b]pyridin-7(4H)-one, 5-amino-	pH 13	282(4.19),326(3.06)	44-3784-76
7H-1,2,3-Triazolo[4,5-b]pyridin-7-one, 5-amino-1,4-dihydro-	pH 1	264(3.85),297(4.17)	4-1365-76
	pH 11	284(4.22)	4-1365-76
Triazolo[3,2-c]triazin-7(4H)-one, 2-methyl-	EtOH	240(3.75),299(3.86)	22-1178-76
Triazolo[3,2-c]triazin-7(4H)-one, 4-methyl-	EtOH	240(3.58),296(3.80)	22-1178-76
Triazolo[3,2-c]triazin-7(4H)-one, 6-methyl-	EtOH	243(3.85),300(3.82)	22-1178-76
Triazolo[3,4-c]triazin-7(4H)-one, 4-methyl-	EtOH	244(3.92),331(3.79)	22-1178-76
Triazolo[3,4-c]triazin-7(4H)-one, 6-methyl-	EtOH	239(3.68),315(3.50)	22-1178-76
Triazolo[4,3-b]triazin-7(1H)-one, 1-methyl-	EtOH	233s(3.76),293(3.61)	22-1178-76

Compound	Solvent	$\lambda_{max}(\log \epsilon)$	Ref.
Triazolo[4,3-b]triazin-7(1H)-one, 3-methyl-	EtOH	230(3.82),300(3.62)	22-1178-76
Triazolo[4,3-b]triazin-7(1H)-one, 6-methyl-	EtOH	230(3.64),285(3.56)	22-1178-76
Triazolo[4,3-b]triazin-7(8H)-one, 8-methyl-	EtOH	249(3.54),295(3.09)	22-1178-76
1,2,4-Triazolo[5,1-c][1,2,4]triazin-7(4H)-one, 4-methyl-	EtOH	212(3.80),244(3.70), 250s(3.62),300(3.93)	39-0421-76C
$C_5H_5N_5OS$			
4H,6H-1,2,3-Triazolo[1,5-a][1,3,5]triazin-7(4H)-one, 5,6-dihydro-3-methyl-5-thioxo-	pH 0.1	227(3.85),248(3.93), 290(4.27)	4-0589-76
	pH 7.08	235(4.00),305(4.16)	4-0589-76
	pH 14	252(4.22),291(4.14)	4-0589-76
$C_5H_5N_5O_2$			
1,2,4-Oxadiazol-5(2H)-one, 3-(5-amino-1H-imidazol-4-yl)-	EtOH	276(3.99)	23-2804-76
$C_5H_5N_5S$			
1,2,3-Thiadiazolo[4,5-c]pyridine-4,6-diamine	pH 13	241s(4.01),256(4.09), 310(3.67),360(3.89)	44-3784-76
7H-1,2,3-Triazolo[4,5-b]pyridine-7-thione, 5-amino-1,4-dihydro-	pH 13	217(4.28),308(4.29)	44-3784-76
$C_5H_6BrN_3O$			
2-Pyrimidinamine, 4-bromo-6-methoxy-	MeOH	233(4.07),279(3.75)	94-0507-76
4-Pyrimidinamine, 2-bromo-6-methoxy-	MeOH	210(4.38),240(3.86)	94-0507-76
$C_5H_6ClN_3O$			
3-Pyridazinone, 5-amino-6-chloro-2-methyl-	pH -2.35	223(4.18),287(3.85)	39-1424-76C
	pH 4.0	282(3.73)	39-1424-76C
2-Pyrimidinamine, 4-chloro-6-methoxy-	MeOH	233(4.11),278(3.74)	94-0507-76
4-Pyrimidinamine, 2-chloro-6-methoxy-	MeOH	210(4.34),239(3.86)	94-0507-76
4(1H)-Pyrimidinone, 6-amino-2-chloro-5-methyl-	MeOH	213(4.31),267(3.85)	94-0026-76
$C_5H_6ClS_2$			
Cyclopropenylium, chlorobis(methylthio)-, hexachloroantimonate	MeCN	266(4.25)	35-4668-76
$C_5H_6N_2$			
2,4-Cyclopentadien-1-one, hydrazone	hexane	307(4.16)	39-0048-76C
1H-Imidazole, 1-ethenyl-	pH 1	218(3.95)	103-0691-76
	EtOH	202(3.90),230(4.03)	103-0691-76
	dioxan	234(4.09)	103-0691-76
1H-Pyrazole, 1-ethenyl-	pH 1	209(3.72),248(3.97)	103-0691-76
	EtOH	212(3.57),250(4.06)	103-0691-76
	dioxan	250(4.10)	103-0691-76
$C_5H_6N_2O$			
1H-Imidazole, 1-acetyl-	THF	244(3.62)	33-2738-76
Imidazole, 2-acetyl-	EtOH	277(4.10)	33-2738-76
Imidazole, 4-acetyl-	EtOH	255(4.12)	33-2738-76
1H-Imidazole-2-carboxaldehyde, N-methyl-	1% EtOH	223(2.60),289(3.11)	44-1952-76
Pyrimidine, 2-methoxy-	EtOH	264(3.68)	35-0171-76
Pyrimidine, 4-methoxy-	EtOH	248(3.50)	35-0171-76
2(1H)-Pyrimidinone, 1-methyl-	EtOH	302(3.73)	35-0171-76
4(1H)-Pyrimidinone, 1-methyl-	EtOH	242(4.20)	35-0171-76
4(3H)-Pyrimidinone, 3-methyl-	EtOH	218(3.88),275(3.57)	35-0171-76
at 190º	gas	275(--)	35-0171-76

Compound	Solvent	$\lambda_{max}(\log \epsilon)$	Ref.
$C_5H_6N_2OS$			
4(1H)-Pyrimidinone, 2,3-dihydro-	pH 1	278(4.02)	73-0311-76
5-methyl-2-thioxo-	pH 13	260(3.96),297s(3.71)	73-0311-76
4(1H)-Pyrimidinone, 2,3-dihydro-	pH 1	276.5(4.25)	73-0311-76
6-methyl-2-thioxo-	pH 13	260(4.25),313s(3.71)	73-0311-76
	EtOH	278(4.23)	95-1094-76
	isoPrOH	214(4.21),276(4.17), 280s(4.16)	4-0291-76
4(3H)-Pyrimidinone, 2-(methylthio)-	EtOH	231(3.95),289(3.88)	95-1094-76
$C_5H_6N_2O_2$			
1H-Imidazole-1-carboxylic acid, methyl ester	EtOH	209(4.08)	33-2738-76
1H-Imidazole-2-carboxylic acid, methyl ester	EtOH	258(4.11)	33-2738-76
1H-Imidazole-4-carboxylic acid, methyl ester	EtOH	236(4.05)	33-2738-76
2,4-Pentanedione, 3-diazo-	hexane	227(4.23),240s(4.11), 358s(1.59)	104-1245-76
	H_2O	229(4.21),238s(4.19), 343s(1.65)	104-1245-76
1H-Pyrazole-3-carboxylic acid, methyl ester	MeOH	217(4.08)	103-1026-76
4H-Pyrazol-4-one, 3-acetyl-1,5-dihydro-	EtOH	242(4.188),291(4.010)	23-1752-76
2,4(1H,3H)-Pyrimidinedione, 5-methyl-	pH 1	265(3.90)	73-0311-76
	pH 13	290(3.74)	73-0311-76
2,4(1H,3H)-Pyrimidinedione, 6-methyl-	pH 1	261(3.86)	73-0311-76
	pH 13	280(3.72)	73-0311-76
	EtOH	260(4.04)	95-1094-76
$C_5H_6N_2O_2S$			
5-Isothiazolecarboxamide, 2,3-dihydro-2-methyl-3-oxo-	MeOH	229(3.93),318(3.56)	1-0781-76
5-Isothiazolecarboxamide, 3-methoxy-	MeOH	224(3.98),286(3.50)	1-0781-76
$C_5H_6N_2O_3$			
Isoxazole, 3,5-dimethyl-4-nitro-	HOAc	264(3.74)	103-0851-76
	HOAc-17M H_2SO_4	223(4.00)	103-0851-76
5-Isoxazolecarboxamide, 3-methoxy-	MeOH	221(4.10)	1-0281-76
1H-Pyrazole-3-carboxylic acid, 5-hydroxy-1-methyl-	MeOH	223(4.00),254s(3.54)	24-0261-76
5-Pyrazolecarboxylic acid, 3-hydroxy-1-methyl-	MeOH	225(4.01),267(3.52)	24-0268-76
$C_5H_6N_2O_4$			
2,4(1H,3H)-Pyrimidinedione, 5-(hydroperoxymethyl)-	H_2O	261(3.88)	44-0567-76
$C_5H_6N_3O_2$			
Pyridine, 1-nitroimide	H_2O	224(3.85),260(3.64), 300(3.26)	39-1245-76C
	MeCN	232(4.11),265s(--), 320(3.51)	39-1245-76B
Pyridinium, 1-(nitroamino)-	80% H_2SO_4	213(3.59),256(3.77), 359(0.00)	39-1245-76B
$C_5H_6N_4O_2S$			
4(1H)-Pyridinethione, 2,6-diamino-3-nitro-	pH 7	234(3.93),267s(3.83), 338(3.69)	44-3784-76

Compound	Solvent	$\lambda_{max}(\log \epsilon)$	Ref.
$C_5H_6N_4O_3$			
Carbamic acid, (1H-1,2,3-triazol-4-yl-carbonyl)-, methyl ester	H_2O	228(4.03)	44-1041-76
4(1H)-Pyridinone, 2,6-diamino-3-nitro-	pH 7	253(3.94),334(3.94), 388(3.95)	44-3784-76
$C_5H_6N_6$			
1H-Imidazo[4,5-d]-1,2,3-triazin-4-amine, N-methyl-	H_2O	218(4.13),264(3.96), 303(3.75)	103-0465-76
4H-Imidazo[4,5-d]-1,2,3-triazin-4-imine, 3,5-dihydro-3-methyl-	H_2O	227(4.33),260(3.80), 302(3.62)	103-0465-76
$C_5H_6N_6O$			
8-Azaguanine, 1-methyl-	pH 1	252(3.75),264s(3.62)	87-1186-76
8-Azaguanine, 3-methyl-	pH 1	205(4.41),264(3.87)	87-1186-76
	pH 7	213(3.88),249(3.99), 267(4.05)	87-1186-76
	pH 13	249(3.99),267(4.05)	87-1186-76
8-Azaguanine, 7-methyl-	pH 1	210(4.37),270(3.73)	87-1186-76
	pH 7	210(4.41),240(3.85), 296(3.72)	87-1186-76
	pH 13	219(4.30),245s(3.74), 297(3.77)	87-1186-76
$C_5H_6N_{10}O$			
Urea, [6-[[(cyanoamino)iminomethyl]-amino]-1,2,3,5-tetrazin-4-yl]-, dipotassium salt	H_2O KOH	340(4.52) 214(4.00),340(4.53)	18-1339-76 18-1339-76
$C_5H_6OS_2$			
2-Cyclopropen-1-one, 2,3-bis(methyl-thio)-	CH_2Cl_2	264(4.26)	35-4668-76
$C_5H_6O_2$			
2-Cyclopenten-1-one, 4-hydroxy-	MeOH	210(4.04)	78-1713-76
1,3-Dioxepin	C_6H_{12}	264(3.85)	88-2113-76
2(5H)-Furanone, 4-methyl-	EtOH	219(4.05)	33-2724-76
$C_5H_6S_2$			
1,3-Dithiepin	EtOH	257(3.60),340(3.52)	88-1251-76
$C_5H_6S_3$			
2-Cyclopropene-1-thione, 2,3-bis(meth-ylthio)-	CH_2Cl_2	271(4.27)	35-4668-76
$C_5H_6Se_5$			
1,3-Diselenole-2-selone, 4,5-dimethyl-	C_6H_{12}	250(3.86),300(3.42), 430(3.68)	77-0148-76
$C_5H_7BO_3$			
Boronic acid, (5-methyl-2-furanyl)-	H_2O	242(4.04)	4-1265-76
$C_5H_7BrN_4$			
4,6-Pyrimidinediamine, 2-bromo-5-methyl-	MeOH	208(4.24),288(3.89)	94-0026-76
$C_5H_7ClN_4$			
4,6-Pyrimidinediamine, 2-chloro-5-meth-yl-	MeOH	218(4.43),268(3.91)	94-0026-76

Compound	Solvent	λ_{max}(log ϵ)	Ref.
$C_5H_7NOS_2$			
4H-1,3-Thiazin-4-one, tetrahydro-6-methyl-2-thioxo-	n.s.g.	259(4.07),310(4.09)	73-1388-76
$C_5H_7NO_2S$			
Acetic acid, isothiocyanato-, ethyl ester	C_6H_{12}	251(3.03)	5-1850-76
	pH 8.06	246(3.00)	73-2216-76
	MeOH	248(2.98)	73-2216-76
	dioxan	248(3.13)	73-2216-76
$C_5H_7NO_4$			
2-Propenoic acid, 3-(carboxyamino)-2-methyl-	pH 13	266(4.14)	73-2059-76
C_5H_7NS			
Isothiazole, 4,5-dimethyl-	EtOH	217(3.394),252(3.584)	48-0507-76
Thiazole, 2,5-dimethyl-	EtOH	246(3.76)	70-1353-76
Thiazole, 4,5-dimethyl-	EtOH	252(3.78)	70-1353-76
$C_5H_7N_3$			
3,4-Pyridinediamine	pH 1	285(3.87)	44-1449-76
	pH 11	246(3.76),283(3.54)	44-1449-76
	MeOH	253(3.61),295(3.76)	44-1449-76
$C_5H_7N_3O$			
3(2H)-Pyridazinone, 4-amino-2-methyl-	pH -2.35	222s(3.88),226(3.93), 235s(3.74),311(4.24)	39-1424-76C
	pH 4.0	287s(4.04),292(4.05), 309s(3.71)	39-1424-76C
3(2H)-Pyridazinone, 5-amino-2-methyl-	pH -2.35	220(4.14),259(3.82), 263(3.84),281(3.87)	39-1424-76C
	pH 4.0	276(3.78)	39-1424-76C
2(1H)-Pyrimidinone, 4-amino-1-methyl-	pH 1	213(4.00),283(4.09)	5-1395-76
	H_2O	230s(3.89),274(3.92)	5-1395-76
$C_5H_7N_3OS$			
4(1H)-Pyrimidinone, 6-amino-2-(methyl-thio)-	EtOH	230(4.24),276(3.86)	95-1094-76
$C_5H_7N_3O_2$			
4(1H)-Pyrimidinone, 6-amino-2-methoxy-	MeOH	211(4.35),262(4.23)	94-0026-76
1H-1,2,3-Triazole-4-carboxylic acid, 1-methyl-, methyl ester	H_2O	215(4.01)	44-1041-76
$C_5H_7N_3S$			
2(1H)-Pyrimidinethione, 4-amino-1-methyl-	H_2O	222s(4.08),246(4.40), 266(4.37)	5-1395-76
	M HCl	227(4.26),274(4.34), 309(3.86)	5-1395-76
$C_5H_8BrNO_2$			
2-Oxazolidinone, 3-bromo-4,4-dimethyl-	H_2O	274(2.30)	78-1097-76
2-Oxazolidinone, 3-bromo-5,5-dimethyl-	H_2O	275(2.35)	78-1097-76
$C_5H_8N_2$			
1H-Imidazole, 1-ethyl-	pH 1	211(3.71)	103-0691-76
	EtOH	202(3.71),217(3.71)	103-0691-76
2-Propenenitrile, 3-(dimethylamino)-	H_2O	265(4.32)	149-0217-76B
1H-Pyrazole, 1-ethyl-	pH 1	226(3.79)	103-0691-76
	EtOH	221(3.66)	103-0691-76
	dioxan	221(3.64)	103-0691-76

Compound	Solvent	$\lambda_{max}(\log \epsilon)$	Ref.
$C_5H_8N_2O$			
3(2H)-Pyridazinone, 4,5-dihydro-6-methyl-	pH 5.0	241(3.80)	39-1424-76C
$C_5H_8N_2OS$			
2-Thiazolamine, 4,5-dimethyl-, 3-oxide	MeOH	268(3.52)	64-0251-76B
2-Thiazolamine, 4-ethyl-, 3-oxide	MeOH	264(3.53)	64-0251-76B
$C_5H_8N_2O_2$			
1H-Pyrazole-3-carboxylic acid, 4,5-dihydro-, methyl ester	MeOH	291(4.02)	103-1026-76
$C_5H_8N_2O_5$			
2-Pentanone, 5,5-dinitro-, potassium salt	pH 13	379(4.23)	104-2433-76
$C_5H_8N_2O_6$			
Butanoic acid, 4,4-dinitro-, methyl ester, potassium salt	pH 13	379(4.23)	104-2433-76
$C_5H_8N_2S$			
1H-Imidazole, 1-methyl-2-(methylthio)-, hydriodide	EtOH	220-253(4.36-3.81)	28-0143-76B
	CH_2Cl_2	228-255(4.25-3.81)	28-0143-76B
2H-Imidazole-2-thione, 1,3-dihydro-1,3-dimethyl-	EtOH	260(4.186)	28-0143-76B
	CH_2Cl_2	267(4.339)	28-0143-76B
2H-Imidazole-2-thione, 1-ethyl-1,3-dihydro-	EtOH	261.5(4.120)	28-0143-76B
	CH_2Cl_2	270(3.704)	28-0143-76B
$C_5H_8N_3$			
Pyridazinium, 3-amino-1-methyl-, iodide	pH 10.0	213s(4.26),243(3.82), 320(3.31)	39-1424-76C
Pyridazinium, 3-amino-2-methyl-, iodide	pH 8.0	228(3.63),289(3.45)	39-1424-76C
Pyridazinium, 4-amino-1-methyl-, iodide	pH 11.0	277(4.18)	39-1424-76C
	pH 14.7	287(4.21)	39-1424-76C
Pyridazinium, 4-amino-2-methyl-, iodide	pH 10.0	212s(4.22),260(3.98), 312(3.56)	39-1424-76C
$C_5H_8N_4O$			
4(1H)-Pyridinone, 2,3,6-triamino-, dihydrochloride	pH 7	224(3.96),300(3.93)	44-3784-76
1H-1,2,3-Triazole-4-carboxamide, N,1-dimethyl-	H_2O	214(4.07)	44-1041-76
1H-1,2,4-Triazolium, 4-(acetylamino)-1-methyl-, hydroxide, inner salt	$CHCl_3$	267(3.77)	94-2568-76
$C_5H_8N_4O_2S$			
1,2,4-Thiadiazole-2(3H)-carboximidic acid, 5-ethoxy-3-imino-	MeOH	272(4.11)	18-3170-76
$C_5H_8N_4S$			
4(1H)-Pyridinethione, 2,3,6-triamino-	pH 7	293(4.03),342(3.95)	44-3784-76
$C_5H_8N_4S_2$			
1,2,4-Thiadiazol-3(2H)-imine, 2-(1-iminoethyl)-5-(methylthio)-	H_2O	243(4.36)	18-3165-76
C_5H_8O			
Cyclobutanecarboxaldehyde	C_6H_{12}	297(1.40)	39-0669-76B
	EtOH	293(1.79)	39-0669-76B
1-Penten-3-one	MeOH	212(3.77),322(1.38)	40-0166-76

Compound	Solvent	$\lambda_{max}(\log \epsilon)$	Ref.
3-Penten-2-one, cis	C_6H_{12}	223(3.94),330(1.43)	23-2127-76
	H_2O	231(3.98),303(1.78)	23-2127-76
3-Penten-2-one, trans	C_6H_{12}	215(4.06),325(1.46)	23-2127-76
	H_2O	226(4.16),300(1.79)	23-2127-76
$C_5H_8O_2$			
2-Butenoic acid, methyl ester	C_6H_{12}	202(4.17),242(2.63)	48-0745-76
	EtOH	206(4.16),236(2.23)	48-0745-76
	MeCN	204(4.18),238(2.30)	48-0745-76
2-Propenoic acid, ethyl ester	C_6H_{12}	193(4.15),247(1.85)	48-0745-76
	EtOH	196(4.16),240(1.98)	48-0745-76
	MeCN	195(4.02),243(1.95)	48-0745-76
2-Propenoic acid, 2-methyl-, methyl	C_6H_{12}	199(4.05),241(2.26)	48-0745-76
ester	EtOH	205(3.96),234(2.18)	48-0745-76
	MeCN	204(4.00),237(2.18)	48-0745-76
$C_5H_8O_3$			
Butanoic acid, 2-methyl-3-oxo-	hexane	266(3.31),292s(3.08)	104-1252-76
	H_2O	291(2.75)	104-1252-76
$C_5H_9ClO_2$			
Propanoic acid, 3-chloro-, ethyl ester	neat	271.9(-1.02)	44-3312-76
	hexane	212.8(1.77)	44-3312-76
	H_2O	202.6(1.82)	44-3312-76
	MeCN	211.2(1.76)	44-3312-76
C_5H_9NO			
3-Buten-2-one, 4-(methylamino)-, cis-	C_6H_{12}	300(4.18)	35-2826-76
s-cis	H_2O	294(4.42)	35-2826-76
	MeOH	292(4.32)	35-2826-76
2-Propenal, 3-(dimethylamino)-	C_6H_{12}	272(4.47)	35-2826-76
	H_2O	289(4.56)	35-2826-76
	MeOH	285(4.55)	35-2826-76
2-Propenal, 3-(ethylamino)-	H_2O	282(4.56)	35-2826-76
	MeOH	280(4.49)	35-2826-76
cis-s-cis	C_6H_{12}	306(4.15)	35-2826-76
trans	C_6H_{12}	265(4.45)	35-2826-76
$C_5H_9N_3O$			
1H-Imidazole-4-ethanol, 2-amino-	EtOH	216(3.91)	35-3049-76
$C_5H_9N_3O_2$			
2H-1,2,4-Oxadiazin-3(6H)-one, 5-(dimeth-	EtOH	245(4.21)	44-3128-76
ylamino)-			
$C_5H_9N_4$			
Pyridazinium, 3,4-diamino-1-methyl-,	pH 7.0	204s(4.15),229(3.91),	39-1424-76C
iodide		305(4.05)	
	pH 15.0	317(4.13)	39-1424-76C
Pyridazinium, 3,5-diamino-1-methyl-,	pH 9.0	229(4.57),300(3.57)	39-1424-76C
iodide			
Pyridazinium, 3,5-diamino-2-methyl-,	pH 7.0	226(4.45),278(3.78)	39-1424-76C
iodide			
Pyridazinium, 3,6-diamino-1-methyl-,	pH 7.0	231(4.23),331(3.32)	39-1424-76C
iodide			
Pyridazinium, 4,5-diamino-1-methyl-,	pH 7.0	224(4.39),290s(3.79),	39-1424-76C
iodide		319(4.02)	
$C_5H_9N_5OS$			
1,2,4-Thiadiazole-2(3H)-carboximidamide,	MeOH	242(3.85)	18-3170-76
5-ethoxy-3-imino-			

Compound	Solvent	$\lambda_{max}(\log \epsilon)$	Ref.
$C_5H_9N_5O_2$			
1H-1,2,4-Triazole-1-acetic acid, 3,5-di-amino-, methyl ester	EtOH	215(3.8)	23-3620-76
$C_5H_9N_5O_3S$			
1H-Imidazole-4-sulfonamide, 5-[(methyl-amino)carbonyl]amino]-	H_2O	235(4.23)	103-0924-76
$C_5H_{10}N_2O_2$			
Butanal, 4-(N-methyl-N-nitrosoamino)-	EtOH	235(3.89),347(1.88)	88-0593-76
2-Pyrrolidinemethanol, 1-nitroso-	$C_6H_{12}-1\%$ dioxan	358s(1.95),370(2.01), 384(1.91)	78-0847-76
$C_5H_{10}N_2S$			
1H-Imidazole, 4,5-dihydro-1-methyl-	EtOH	220.5(4.391)	28-0485-76A
2-(methylthio)-, hydriodide	CH_2Cl_2	236(4.376)	28-0485-76A
2-Imidazolidinethione, 1-ethyl-	EtOH	241.2(4.207)	28-0485-76A
	CH_2Cl_2	248(4.237)	28-0485-76A
Pyrimidine, 1,4,5,6-tetrahydro-2-(meth-ylthio)-, monohydriodide	EtOH	221(4.406)	28-0485-76A
	CH_2Cl_2	232(4.322)	28-0485-76A
2(1H)-Pyrimidinethione, tetrahydro-1-methyl-	EtOH	244.8(4.125)	28-0485-76A
	CH_2Cl_2	254.0(4.140)	28-0485-76A
$C_5H_{10}N_4O_2S$			
Methanesulfonamide, N-(1-ethyl-1H-1,2,3-triazol-5-yl)-	EtOH	253(3.91)	32-0001-76
$C_5H_{10}O$			
2-Butanone, 3-methyl-	C_6H_{12}	284(1.30)	112-0777-76
3-Pentanone	C_6H_{12}	280(1.20)	112-0777-76
$C_5H_{10}O_2$			
Acetic acid, 1-methylethyl ester	hexane	211.4(1.77)	44-3312-76
	H_2O	204.2(1.77)	44-3312-76
	MeCN	208.5(1.75)	44-3312-76
Acetic acid, propyl ester	neat	273.5(-1.33)	44-3312-76
	hexane	211.3(1.76)	44-3312-76
	H_2O	203.0(1.77)	44-3312-76
	MeCN	207.8(1.76)	44-3312-76
Butanoic acid, 2-methyl-	neat	277s(1.56)	44-3312-76
	hexane	206.4(1.94)	44-3312-76
	H_2O	208.2(1.89)	44-3312-76
	MeCN	213.4(1.78)	44-3312-76
Butanoic acid, 3-methyl-	neat	273.8(-1.34)	44-3312-76
	hexane	206.8(1.83)	44-3312-76
	H_2O	204.9(1.80)	44-3312-76
	MeCN	213.8(1.66)	44-3312-76
Pentanoic acid	hexane	206.5(1.81)	44-3312-76
	H_2O	205.0(1.79)	44-3312-76
	MeCN	211.7(1.68)	44-3312-76
Propanoic acid, ethyl ester	hexane	208.0(1.83)	44-3312-76
	H_2O	204.3(1.89)	44-3312-76
	MeCN	208.1(1.82)	44-3312-76
Propanoic acid, 2,2-dimethyl-	hexane	207.9(2.00)	44-3312-76
	H_2O	208.7(1.93)	44-3312-76
	MeCN	213.9(1.83)	44-3312-76
$C_5H_{11}NO$			
2-Propanamine, 2-methyl-N-methylene-, N-oxide	ether	250(3.86)	4-1001-76

Compound	Solvent	$\lambda_{max}(\log \epsilon)$	Ref.
$C_5H_{11}NO_2$			
Pentanal, 5-hydroxy-, oxime	EtOH	209(3.38)	104-2034-76
Propanamide, N-hydroxy-2,2-dimethyl-	heptane	214(2.95)	34-0504-76
	H_2O	193(3.76)	34-0504-76
	EtOH	203(3.45)	34-0504-76
$C_5H_{11}N_5$			
1H-1,2,4-Triazole, 3,5-diamino-1-propyl-	EtOH	213(3.8)	23-3620-76
$C_5H_{12}N_2$			
2-Propanone, dimethylhydrazone	hexane	266(2.78)	104-0957-76
$C_5H_{12}N_2O$			
Diazene, (1,1-dimethylethyl)methyl-, 1-oxide	EtOH	215(3.80),270s(3)	44-1135-76
$C_5H_{13}N_5O_2S$			
Methanesulfonamide, N-(1-methyl-1H-triazol-5-yl)-, methylamine salt	EtOH	253(3.93)	32-0001-76
$C_5H_{13}OPSe$			
Phosphinoselenoic acid, diethyl-, O-methyl ester	hexane	191s(4.09),217s(3.84)	70-0174-76
	EtOH	194(4.03),213s(3.90)	70-0174-76
$C_5H_{23}B_{18}CoO$			
Cobalt, [[10,10'-μ-methoxybis-[(7,8,9,10,11-η)decahydro-7,8-dicarbaundecaborato]](3-)]-	CH_2Cl_2	255(4.33),292(4.01), 408(3.14),510(3.01)	73-3509-76
$C_5H_{23}B_{18}CoSe$			
Cobalt, [[10,10'-[μ-(methaneselenol-ato)]bis[(7,8,9,10,11-η)decahydro-7,8-dicarbaundecaborato]](3-)]-	CH_2Cl_2	250s(--),296(4.43), 322s(--),450(2.70), 485(2.71)	73-3509-76
$C_5H_{23}B_{18}CoTe$			
Cobalt, [[10,10'-μ-(methanetelluro-lato)]bis[(7,8,9,10,11-η)decahydro-7,8-dicarbaundecaborato]](3-)]-	CH_2Cl_2	249s(--),305(4.30), 321s(--),473(2.64)	73-3509-76

Compound	Solvent	$\lambda_{max}(\log \epsilon)$	Ref.
$C_6Br_4O_2$ 2,5-Cyclohexadiene-1,4-dione, 2,3,5,6-tetrabromo-	EtOH	245(3.82),305(4.18)	115-0203-76
	ether	245(3.81),305(4.18)	115-0203-76
	$CHCl_3$	310(4.21)	115-0203-76
	CCl_4	302(4.23)	115-0203-76
C_6F_6 Benzene, hexafluoro-	EtOH	259(2.21)	59-0269-76
C_6HF_5 Benzene, pentafluoro-	EtOH	260(2.82)	59-0269-76
$C_6H_2Cl_2F_3N$ Pyridine, 2,6-dichloro-3-(trifluoromethyl)-	MeOH	268(3.58)	33-0229-76
$C_6H_2Cl_3NO$ 3-Pyridinecarboxaldehyde, 2,5,6-trichloro-	MeOH	225(3.44),280(3.83)	33-0211-76
$C_6H_2Cl_3NO_2$ 3-Pyridinecarboxylic acid, 2,5,6-trichloro-	MeOH	229(3.98),284(3.69)	33-0222-76
$C_6H_2Cl_5N$ Pyridine, 2,6-dichloro-3-(trichloromethyl)-	MeOH	230(3.64),272(3.96)	33-0190-76
Pyridine, 2,3,6-trichloro-5-(dichloromethyl)-	MeOH	231(4.05),286(3.72)	33-0190-76
$C_6H_2F_4$ Benzene, 1,2,3,4-tetrafluoro-	EtOH	260(2.80)	59-0269-76
Benzene, 1,2,3,5-tetrafluoro-	EtOH	259(2.84)	59-0269-76
Benzene, 1,2,4,5-tetrafluoro-	EtOH	264(3.30)	59-0269-76
$C_6H_3Br_2NO$ 3-Pyridinecarboxaldehyde, 2,6-dibromo-	MeOH	221(3.76),272(4.01)	33-0211-76
$C_6H_3Br_3ClNO$ Pyridine, 2,6-dibromo-3-(bromomethyl)-5-chloro-, N-oxide	MeOH	237(4.51),275(3.90)	33-0190-76
$C_6H_3ClNSSe$ 1,2,3-Benzothiaselenazol-2-ium, 5-chloro-, chloride	93% H_2SO_4	380(4.04),470(2.92)	103-1128-76
1,2,3-Benzothiaselenazol-2-ium, 6-chloro-, chloride	93% H_2SO_4	392(3.90),450(3.42)	103-1128-76
$C_6H_3ClNS_2$ 1,2,3-Benzodithiazol-2-ium, 5-chloro-, chloride	93% H_2SO_4	360(4.61),440(3.18)	103-1128-76
1,2,3-Benzodithiazol-2-ium, 6-chloro-, chloride	93% H_2SO_4	370(4.06),427(3.61)	103-1128-76
$C_6H_3ClN_2$ 2,4-Hexadienedinitrile, 3-chloro-, cis-cis	EtOH	259s(4.22),267(4.29), 274s(4.26)	44-1389-76
$C_6H_3Cl_2NO$ 3-Pyridinecarboxaldehyde, 2,6-dichloro-	MeOH	220(3.93),270(3.65)	33-0211-76

Compound	Solvent	$\lambda_{max}(\log \epsilon)$	Ref.
$C_6H_3Cl_2NO_2$ 3-Pyridinecarboxylic acid, 2,6-dichloro-	MeOH	229(3.67),274(3.92)	33-0222-76
$C_6H_3Cl_3N_2O$ 3-Pyridinecarboximidoyl chloride, 2,6-dichloro-N-hydroxy-	MeOH	218(3.56),266(3.62)	33-0211-76
$C_6H_3Cl_3N_2O_2$ Pyridine, 2,6-dichloro-3-(chloromethyl)- 5-nitro-	MeOH	250(3.66),282(3.61)	33-0190-76
$C_6H_3Cl_4N$ Methanamine, 1-(2,3,4,5-tetrachloro- 2,4-cyclopentadien-1-ylidene)-	C_6H_{12}	231(3.55),323(4.41), 370s(3.26)	24-3939-76
Pyridine, 2,6-dichloro-3-(dichloro- methyl)-	MeOH	273.5(3.68)	33-0190-76
Pyridine, 2,3,6-trichloro-5-(chloro- methyl)-	MeOH	228(4.01),283(3.70)	33-0179-76
$C_6H_3Cl_4NO$ Pyridine, 2,3,6-trichloro-5-(chloro- methyl)-, N-oxide	MeOH	237(4.47),273(3.91)	33-0190-76
$C_6H_3Cl_5N_2O_2$ 1H-Imidazole-1-carboxylic acid, 2,4,5- trichloro-, dichloroethyl ester	n.s.g.	207(4.36),282(3.23), 290(3.21)	24-1625-76
$C_6H_3F_3$ Benzene, 1,2,4-trifluoro-	EtOH	264(3.19)	59-0269-76
Benzene, 1,3,5-trifluoro-	EtOH	252(2.46)	59-0269-76
$C_6H_3N_3$ 1H-Pyrrole-2,3-dicarbonitrile	H_2O	233(3.84),240(3.85), 267(4.01)	33-2786-76
$C_6H_3N_3O_7$ Phenol, 2,4,6-trinitro-	H_2O	356(4.15)	23-2436-76
	H_2O-acid	338(3.73)	23-2436-76
	EtOH-HCl	280s(3.95),325s(3.69)	23-2436-76
	EtOH-NaOEt	359(4.22)	23-2436-76
$C_6H_3N_5O$ Pyrazolo[5,1-c][1,2,3]triazine-3-carbo- nitrile, 1,4-dihydro-4-oxo-	EtOH	269s(3.84),275(3.86), 287s(3.70),295(3.69), 345(3.97)	39-1496-76C
$C_6H_3N_5O_2$ 8H-[1,2,4]Oxadiazolo[2,3-i]purin-8-one	EtOH	260(4.03),277(4.05)	39-1496-76C
$C_6H_4BrCl_2N$ Pyridine, 3-(bromomethyl)-2,6-dichloro-	MeOH	236(3.97),279(3.75)	33-0190-76
$C_6H_4BrCl_2NO$ Pyridine, 3-(bromomethyl)-2,6-dichloro-, 1-oxide	MeOH	235(4.50),270(3.93)	33-0190-76
$C_6H_4BrNO_3$ Furan, 2-(2-bromoethenyl)-5-nitro-, cis	MeOH	240(3.93),260(3.77), 345(4.18)	73-0614-76

Compound	Solvent	$\lambda_{max}(\log \epsilon)$	Ref.
$C_6H_4Br_2N_2O_2$			
Dibromodiaminobenzoquinone	EtOH	245(3.52),338(4.05)	115-0203-76
	ether	245(3.68),335(4.07)	115-0203-76
	CHCl₃	338(3.75)	115-0203-76
2,4(1H,3H)-Pyrimidinedione, 5-(2,2-di-bromoethenyl)-	MeOH	234(4.03),244s(4.05), 295(4.02)	88-2427-76
$C_6H_4Br_3N$			
Pyridine, 2,6-dibromo-3-(bromomethyl)-	MeOH	236(3.97),279(3.75)	33-0179-76
$C_6H_4Br_3NO$			
Pyridine, 2,6-dibromo-3-(bromomethyl)-, N-oxide	MeOH	241(4.51),272(3.91)	33-0190-76
$C_6H_4ClF_2N$			
Pyridine, 3-chloro-2,6-difluoro-5-methyl-	MeOH	270.5(3.65)	33-0229-76
$C_6H_4ClNO_4$			
2H-Pyrimidine-5-carboxylic acid, 6-amino-4-chloro-2-oxo-	dioxan	271(4.05),330(4.05)	39-2462-76C
$C_6H_4ClN_2$			
Benzenediazonium, 4-chloro-, tetra-fluoroborate	pH 1	214(3.77),282(4.25)	33-1438-76
	H₂O	280(4.21)	33-1427-76
	DMSO	350s(3.35)	33-1427-76
$C_6H_4ClN_5O$			
Pyrido[4,3-e]-1,2,4-triazin-3-amine, 5-chloro-, 1-oxide	pH 1	268(3.45),312(4.00), 394(3.50)	114-0405-76C
$C_6H_4Cl_2IN$			
Pyridine, 2,6-dichloro-3-(iodomethyl)-	MeOH	233(3.76),280(3.99)	33-0190-76
$C_6H_4Cl_2N_2O$			
3-Pyridinecarboxaldehyde, 2,6-dichloro-, oxime	MeOH	254(3.03),293(3.53)	33-0211-76
$C_6H_4Cl_3N$			
Pyridine, 2,6-dichloro-3-(chloromethyl)-	MeOH	223(3.97),273(3.66)	33-0179-76
Pyridine, 2,3,6-trichloro-5-methyl-	MeOH	282(3.72)	33-0190-76
$C_6H_4Cl_3NO$			
Pyridine, 2,6-dichloro-3-(chloromethyl)-, 1-oxide	MeOH	230(4.49),268(3.95)	33-0190-76
3-Pyridinemethanol, 2,5,6-trichloro-	MeOH	225(4.00),280(3.73)	33-0211-76
$C_6H_4F_2$			
Benzene, 1,2-difluoro-	EtOH	260(3.17)	59-0269-76
Benzene, 1,3-difluoro-	EtOH	260(2.92)	59-0269-76
Benzene, 1,4-difluoro-	EtOH	266(3.40)	59-0269-76
$C_6H_4F_6NiO_2S_4$			
Nickel, bis[O-(2,2,2-trifluoroethyl)-carbonodithioato-S,S']-	CHCl₃	393s(3.01),429(3.39), 496(3.23),656(1.88)	70-1388-76
$C_6H_4F_6O_2PdS_4$			
Palladium, bis[O-(2,2,2-trifluoroethyl)-carbonodithioato-S,S']-	CHCl₃	286(4.70),396(3.69), 468(2.29)	70-1388-76

Compound	Solvent	λ_{max} (log ϵ)	Ref.
C_6H_4NSSe			
1,2,3-Benzothiaselenazol-2-ium, chloride	93% H_2SO_4	375(4.05),450(3.20)	103-1128-76
$C_6H_4NS_2$			
1,2,3-Benzodithiazol-2-ium, chloride	93% H_2SO_4	350(4.08),430(3.24)	103-1128-76
$C_6H_4N_2$			
2,4-Hexadienedinitrile, (Z,Z)-	EtOH	252s(4.35),260(4.44), 270(4.31)	44-1389-76
$C_6H_4N_2O$			
2,4-Pentadienedinitrile, N-oxide	H_2O	222(3.88),287(4.30)	33-2727-76
2,4-Pentadienenitrile, 5-isocyanato-	C_6H_{12}	257(4.32)	33-2727-76
$C_6H_4N_2OSe$			
2,1,3-Benzoselenadiazole, 1-oxide	EtOH	234(3.83),346(3.65), 413(3.45)	1-0675-76
$C_6H_4N_2O_2$			
Furo[2,3-d]pyrimidin-2(1H)-one	pH 6	243(4.01),320(3.64)	78-2795-76
	pH 13	244(4.11),310(3.81)	78-2795-76
2,4(1H,3H)-Pyrimidinedione, 5-ethynyl-	MeOH	225(4.02),284(3.98)	88-2427-76
$C_6H_4N_2S_2$			
Thiazolo[5,4-b]pyridine-2(1H)-thione	dioxan	335(4.54)	104-2381-76
$C_6H_4N_3O_2$			
Benzenediazonium, 4-nitro-, tetrafluoro- borate	pH 1	261(4.18),312(3.34)	33-1438-76
	H_2O	259(4.22),308(3.39)	33-1427-76
	DMSO	317s(3.57)	33-1427-76
$C_6H_4N_4$			
1-Propene-1,1,3-tricarbonitrile, 2-amino-	H_2O	273(4.23)	149-0217-76B
Pyrido[3,4-e]-1,2,4-triazine	EtOH	226(4.29),332(3.43), 475(2.37)	114-0285-76C
$C_6H_4N_6$			
Benzene, 1,2-diazido-	isooctane	242(3.3),263(3.1)	135-0074-76
Benzene, 1,3-diazido-	isooctane	242(4.4)	135-0074-76
Benzene, 1,4-diazido-	isooctane	272(4.5)	135-0074-76
$C_6H_4N_6O$			
[1,2,4]Triazolo[5,1-c][1,2,4]triazine- 3-carbonitrile, 1,4-dihydro-1-methyl- 4-oxo-	EtOH	215(4.22),253(3.98), 261s(3.88),320(4.31)	39-1496-76C
$C_6H_4O_2$			
2,5-Cyclohexadiene-1,4-dione	EtOH	232(3.98),243(4.09), 293(3.16)	115-0203-76
	ether	240(4.11),290(3.18)	115-0203-76
	$CHCl_3$	290(3.03)	115-0203-76
$C_6H_4O_3$			
2,5-Cyclohexadiene-1,4-dione, 2-hydroxy-	acid	246(4.04)	18-1171-76
7-Oxabicyclo[2.2.1]hept-5-ene-2,3-dione	hexane	486(1.94)	118-0256-76
$C_6H_4O_3S_2$			
1,4-Dithiino[2,3-c]furan-5,7-dione, 2,3-dihydro-	MeCN	207(4.07),271(3.76), 394(3.88)	56-1523-76

Compound	Solvent	$\lambda_{max}(\log \epsilon)$	Ref.
1,4-Dithiino[2,3-c]furan-5,7-dione, 2,3-dihydro-, methanol adduct	MeOH	224(3.84),318(3.82)	56-1523-76
$C_6H_4S_4$			
[1,4]Dithiino[2,3-b]-1,4-dithiin	C_6H_{12}	235(3.90),248s(3.79), 270s(3.70)	44-1484-76
1,3-Dithiole, 2-(1,3-dithiol-2-ylidene)-	THF	308(4.09),316(4.09), 357(--),446(2.37)	97-0360-76
$C_6H_5AsO_2$			
Arsenin-4-carboxylic acid	EtOH	285(4.38)	88-0665-76
$C_6H_5BrN_2O_2$			
2,4(1H,3H)-Pyrimidinedione, 5-(2-bromo-ethenyl)-, (E)-	pH 1	239(4.09),288(3.85)	78-2795-76
	pH 13	267(4.19),305(4.06)	78-2795-76
$C_6H_5BrN_2O_3$			
2,4(1H,3H)-Pyrimidinedione, 5-(bromo-acetyl)-	pH 1	285(4.15)	87-0194-76
$C_6H_5Br_2NO$			
3-Pyridinemethanol, 2,6-dibromo-	MeOH	220(3.94),274(3.74)	33-0211-76
C_6H_5Cl			
Benzene, chloro-	C_5H_{10}	265(2.42)	80-0193-76
	pentane	265(2.41)	80-0193-76
	C_6H_{12}	264.7(2.43)	80-0193-76
	hexane	265(2.40)	80-0193-76
	heptane	264.8(2.41)	80-0193-76
	octane	265(2.40)	80-0193-76
	decane	265(2.41)	80-0193-76
	dodecane	265(2.42)	80-0193-76
	MeOH	264(2.30)	80-0193-76
	EtOH	264(2.32)	80-0193-76
	PrOH	264.5(2.38)	80-0193-76
	isoPrOH	265(2.40)	80-0193-76
	BuOH	265(2.35)	80-0193-76
	ether	265(2.33)	80-0193-76
	dioxan	265(2.33)	80-0193-76
	MeOAc	264(2.35)	80-0193-76
	EtOAc	264(2.34)	80-0193-76
	$CHCl_3$	265.4(2.33)	80-0193-76
	CCl_4	267.8(2.44)	80-0193-76
(also other solvents)	MeCN	254.4(2.45)	80-0193-76
$C_6H_5ClN_2O_2$			
3-Pyridinecarboxylic acid, 2-amino-6-chloro-	MeOH	249(3.99),326(3.87)	33-0222-76
$C_6H_5ClN_2O_3$			
4-Pyridazineacetic acid, 3-chloro-1,6-dihydro-6-oxo-	MeOH	295(3.32)	4-1155-76
4-Pyridazineacetic acid, 6-chloro-2,3-dihydro-3-oxo-	MeOH	297(3.39)	4-1155-76
$C_6H_5Cl_2N$			
Pyridine, 2,6-dichloro-3-methyl-	MeOH	273(3.69)	33-0190-76
$C_6H_5Cl_2NO$			
3-Pyridinemethanol, 2,6-dichloro-	MeOH	220(3.97),272(3.67)	33-0211-76

Compound	Solvent	$\lambda_{max}(\log \epsilon)$	Ref.
$C_6H_5Cl_4NO$ 2(1H)-Pyridinone, 3,5,6-trichloro- 3-(chloromethyl)-3,4-dihydro-	MeOH	266.5(3.57)	33-0179-76
C_6H_5F Benzene, fluoro-	EtOH	260(3.00)	59-0269-76
$C_6H_5FN_2O_2$ Benzenamine, 3-fluoro-4-nitro-	H_2O	226(3.86),375(4.20)	35-7777-76
C_6H_5NO Benzene, nitroso-	EtOH	<u>280(3.9),303(3.9),</u> <u>770(1.7)</u>	78-0467-76
C_6H_5NOS 3-Pyridinecarboxaldehyde, 1,2-dihydro- 2-thioxo-	EtOH	216(4.02),293s(3.75), 320(4.02),375(3.82)	1-0904-76
$C_6H_5NO_2$ Benzene, nitro- 2,5-Cyclohexadiene-1,4-dione, monooxime	EtOH 5% H_2SO_4 45% H_2SO_4 95% H_2SO_4 102% H_2SO_4	260(3.92),300(3.00) 300<u>(4.2)</u> <u>300(4.1)</u>,380s(3.7) <u>300(3.8)</u>,390(4.1) <u>320(4.4)</u>	110-1170-76 104-1451-76 104-1451-76 104-1451-76 104-1451-76
$C_6H_5NO_2S_2$ 5H-1,4-Dithiino[2,3-c]pyrrole-5,7(6H)- dione, 2,3-dihydro-	EtOH	204(3.79),253(3.99), 405(3.51)	56-1523-76
$C_6H_5NO_3$ Phenol, 4-nitro-, anion	NaOH DMSO-NaOH DMSO-Et$_2$NH DMSO-BuNH$_2$ BuNH$_2$	402(4.26) 434(4.51) 436(4.54) 438(4.53) 405(4.26)	70-0986-76 70-0986-76 70-0986-76 70-0986-76 70-0986-76
$C_6H_5NO_4$ 1,2-Benzenediol, 4-nitro-	pH 2.0 pH 7.1	<u>241(3.8),345(3.8)</u> <u>268(3.7),425(4.0)</u>	46-0722-76 46-0722-76
$C_6H_5N_2$ Benzenediazonium, tetrafluoroborate	pH 1 H_2O DMSO	263(4.12),298(3.32) 290(3.29) 290(3.54)	33-1438-76 33-1427-76 33-1427-76
$C_6H_5N_3$ Benzene, azido- 1H-Benzotriazole	0.1N H_2SO_4 pH 1	<u>253(4.0)</u> 198(4.34),258(3.75), 262(3.77),273(3.75), 279(3.74)	104-0405-76 103-0691-76
	EtOH	199(4.49),254(3.87), 260(3.87),276(3.78)	103-0691-76
	dioxan	252(3.79),258(3.77), 277(3.69),281(3.69)	103-0691-76
5H-Pyrrolo[2,3-b]pyrazine	EtOH	218(4.07),309(3.80)	39-1361-76C
$C_6H_5N_3O$ Imidazo[1,2-c]pyrimidin-5(6H)-one	EtOH	268(4.04),276s(--), 288s(--)	35-7408-76

Compound	Solvent	λ_{max}(log ϵ)	Ref.
Imidazo[1,2-c]pyrimidin-5(6H)-one (cont.)	EtOH-HCl	245(3.70),251s(--), 286(4.02),301s(--)	35-7408-76
	EtOH-NaOH	282s(--),285(4.07), 292s(--)	35-7408-76
$C_6H_5N_3OS$ 5H-1,3,4-Thiadiazolo[3,2-a]pyrimidin-5-one, 7-methyl-	isoPrOH	213(4.03),217(4.02), 235s(3.65),255s(3.49), 306(3.90)	4-0291-76
7H-1,3,4-Thiadiazolo[3,2-a]pyrimidin-7-one, 5-methyl-	isoPrOH	212(4.39),269(4.08)	4-0291-76
$C_6H_5N_3O_3$ Diazene, hydroxy(4-nitrophenyl)-, anion	DMF	420(4.5)	104-0395-76
$C_6H_5N_3O_4$ Benzenamine, 3,5-dinitro-	H_2O	240(4.08),260(4.10), 377(3.30)	35-7777-76
$C_6H_5N_5$ 1H-Imidazo[1,5-b][1,2,4]triazole-7-carbonitrile, 1-methyl-	2M HCl pH 1 and 11	227(3.85),265(4.32) 235(4.02),260(4.09)	44-1889-76 44-1889-76
Pyrido[4,3-e]-1,2,4-triazin-3-amine	EtOH	251(3.49),377(2.50)	114-0327-76D
$C_6H_5N_5O$ Pyrido[4,3-e]-1,2,4-triazin-3-amine, 1-oxide	EtOH	285(3.48),302(4.20), 385(3.51)	114-0405-76C
1H-1,2,3-Triazolo[4,5-c]pyridine-1-carboxamide	pH 1	263(3.69)	114-0405-76C
$C_6H_5N_5O_3$ 1H-Imidazo[1,5-b][1,2,4]triazole-2-carb-oxylic acid, 7-(aminocarbonyl)-	pH 1 pH 10	240s(3.90),279(3.86) 263(3.87),310(4.14)	44-1889-76 44-1889-76
Triazolo[3,4-c][1,2,4]triazine-6-carb-oxylic acid, 4,7-dihydro-7-oxo-, methyl ester	EtOH	280s(3.72),295(3.61)	22-1178-76
C_6H_6 1,3-Cyclopentadiene, 5-methylene-	C_6H_{12}	242(4.1),362(2.33)	39-0048-76C
2,4-Hexadiyne	EtOH	219(2.48),227(2.56), 236(2.52),250(2.20)	88-2663-76
C_6H_6ClN Benzenamine, 4-chloro-	H_2O	239(4.02),292(3.11)	35-7777-76
$C_6H_6ClN_3O_4$ Pyrimidine, 4-chloro-2,6-dimethoxy-5-nitro-	MeOH	216(4.45),265(4.23), 337(3.28)	39-1327-76C
$C_6H_6ClN_5$ 1H-Purin-2-amine, 6-chloro-8-methyl-	pH 1 H_2O	237(3.90),314(3.92) 217(4.48),242(3.82), 308(3.91)	44-0568-76 44-0568-76
	pH 13	268(3.57),303(3.89)	44-0568-76
$C_6H_6ClN_5O_2$ Guanidine, (3-chloro-5-nitro-4-pyridin-yl)-	EtOH	263(4.00)	114-0405-76C

Compound	Solvent	$\lambda_{max}(\log \epsilon)$	Ref.
$C_6H_6Cl_2N_2$			
3-Pyridinamine, 2,6-dichloro-5-methyl-	MeOH	247(4.04),313(3.73)	33-0190-76
$C_6H_6N_2$			
5H-Pyrrolo[1,2-a]imidazole	EtOH	267(3.87)	4-0111-76
5H-Pyrrolo[1,2-c]imidazole	EtOH	273(3.88)	4-0111-76
$C_6H_6N_2O$			
1H-Pyrrole-2-carbonitrile, 3-methoxy-	H_2O	247(4.28)	33-2786-76
$C_6H_6N_2OS$			
4-Oxazolecarbonitrile, 2-methyl-5-(methylthio)-	$CHCl_3$	274(3.86)	94-0948-76
$C_6H_6N_2O_2$			
Benzenamine, 4-nitro-	H_2O	226(3.83),382(4.14)	35-7777-76
protonated	H_2O	258(3.87),375(3.85)	35-7777-76
1,3-Cyclohexanedione, 2-diazo-	hexane	224(4.22),258(3.90)	104-1245-76
	H_2O	230(4.23),259(3.89)	104-1245-76
	12M HCl	236(4.26),266(3.81), 315s(2.04)	104-1245-76
$C_6H_6N_2O_3$			
5-Pyrimidinecarboxaldehyde, 1,2,3,4-tetrahydro-6-methyl-2,4-dioxo-	H_2O	231(3.81),283(3.98)	44-1858-76
$C_6H_6N_2O_4$			
5-Pyrimidinecarboxylic acid, 1,2,3,4-tetrahydro-6-methyl-2,4-dioxo-	pH 1	222(4.05),272(4.04)	44-1858-76
	H_2O	267(3.92)	44-1858-76
	pH 13	288(3.83)	44-1858-76
$C_6H_6N_2O_4S$			
Hydroxylamine-O-sulfonic acid, (2-pyridinylmethylene)-	H_2O	239(4.16),277(3.51)	33-2786-76
Hydroxylamine-O-sulfonic acid, (3-pyridinylmethylene)-	H_2O	240(4.11),275(3.73)	33-2786-76
Hydroxylamine-O-sulfonic acid, (4-pyridinylmethylene)-	H_2O	238(3.98),275(3.87)	33-2786-76
$C_6H_6N_4$			
1H-1,2,3-Triazolo[4,5-c]pyridine, 1-methyl-	pH 1	275(3.66)	44-1449-76
	pH 11	264(3.78)	44-1449-76
	MeOH	265(3.76)	44-1449-76
$C_6H_6N_4O$			
1H-Imidazole-4-carboxamide, 5-(cyanomethyl)-	pH 1	210(4.15)	35-1492-76
	pH 7	242(4.03)	35-1492-76
	pH 11	257(4.06)	35-1492-76
3H-Imidazo[4,5-b]pyridin-7(4H)-one, 5-amino-, dihydrochloride	pH 13	262(4.03),276(3.97)	44-3784-76
4H-Imidazo[4,5-c]pyridin-4-one, 6-amino-1,5-dihydro-	pH 1	273(4.05),311(3.80)	35-1492-76
	pH 7	262(4.01),298(3.91)	35-1492-76
	pH 11	262(3.98),298(3.89)	35-1492-76
3(5H)-Imidazo[1,2-b]-as-triazinone, 2-methyl-	EtOH	212(4.22),246(3.82), 308(3.69)	114-0413-76A
1H-Tetrazole, 5-(5-methyl-2-furanyl)-	EtOH	265(4.35)	73-1692-76
3H-1,2,3-Triazolo[4,5-b]pyridine, 3-methoxy-	EtOH	<u>280(3.0)</u>	103-1172-76

Compound	Solvent	$\lambda_{max}(\log \epsilon)$	Ref.
$C_6H_6N_4OS$ 6H-Purin-6-one, 1,7-dihydro-8-(methyl-thio)-	neutral anion	274(4.135) 280(--)	39-0090-76C 39-0090-76C
$C_6H_6N_4O_2S$ 1,2,3-Thiadiazolo[4,5-d]pyrimidine-5,7(4H,6H)-dione, 4,6-dimethyl-	EtOH	214(4.45),242s(3.81), 327(3.79)	88-1129-76
1,2,3-Thiadiazolo[4,5-d]pyrimidinium, 6,7-dihydro-5-hydroxy-3,6-dimethyl-7-oxo-, hydroxide, inner salt	EtOH	210(4.05),242(4.22), 405(3.42)	88-1129-76
$C_6H_6N_4O_3$ [1,2,5]Oxadiazolo[3,4-d]pyrimidine, 7-methoxy-5-methyl-, 1-oxide	MeOH	210(4.03),245(3.75), 283(3.54),350(3.45)	39-1327-76C
1H-Purine-2,6-dione, 3,7-dihydro-7-hydroxy-1-methyl-	pH 3.1 pH 7.2	204(4.40),268(3.95) 222(4.18),255(3.93), 275(3.76)	39-1507-76C 39-1507-76C
	pH 12.0	226(4.32),295(3.84)	39-1507-76C
$C_6H_6N_4O_4$ [1,2,5]Oxadiazolo[3,4-d]pyrimidine, 5,7-dimethoxy-, 1-oxide	MeOH	215(4.50),281(4.33), 290(4.32),345(3.80)	39-1327-76C
$C_6H_6N_4S$ 7H-Imidazo[4,5-b]pyridine-7-thione, 5-amino-1,4-dihydro-	pH 13	280(3.77),317(3.96)	44-3784-76
Thiazolo[4,5-c]pyridine-4,6-diamine	pH 13	314(4.02)	44-3784-76
C_6H_6OS Ethanone, 1-(2-thienyl)-	EtOH	260(3.98),282(3.84)	59-0501-76
2-Thiophenecarboxaldehyde, 3-methyl-	1% EtOH	280(4.11)	44-1952-76
$C_6H_6O_2$ 4,8-Dioxabicyclo[5.1.0]octa-2,5-diene	MeCN	227s(2.99)	35-6350-76
2-Furancarboxaldehyde, 5-methyl-	1% EtOH	293(4.20)	44-1952-76
4H-Pyran-4-carboxaldehyde	MeCN	246(2.77),310(2.04)	35-6350-76
$C_6H_6O_3$ 2-Furancarboxaldehyde, 5-(hydroxy-methyl)-	1% EtOH	230(3.30),280(4.17)	44-1952-76
7-Oxabicyclo[2.2.1]heptane-2,3-dione	hexane	500(1.72)	118-0256-76
$C_6H_6O_4$ Ethanone, 1-(2-furanyl)-2,2-dihydroxy-	H_2O	230(3.45),282(4.16)	102-1753-76
$C_6H_6O_9S_3$ 1,3,5-Benzenetrisulfonic acid, sodium salt	H_2O	<u>210(4.0),270(2.4)</u>	104-0613-76
$C_6H_6O_{12}S_4$ 1,2,4,5-Benzenetetrasulfonic acid, sodium salt	H_2O	<u>230(4.0),272(3.3), 280(3.5),290(3.5)</u>	104-0613-76
$C_6H_6O_{15}S_5$ Benzenepentasulfonic acid, sodium salt	H_2O	<u>220(4.4),305(2.3)</u>	104-0613-76
$C_6H_6O_{18}S_6$ Benzenehexasulfonic acid, sodium salt	H_2O	<u>212(4.4),260(4.3), 320(2.4)</u>	104-0613-76

Compound	Solvent	$\lambda_{max}(\log \epsilon)$	Ref.
$C_6H_6S_3$			
4H-Cyclopenta-1,3-dithiole-2-thione, 5,6-dihydro-	hexane	232(3.79),268(3.07), 380(4.02)	44-0730-76
C_6H_7AsO			
Arsenin, 4-methoxy-	EtOH	230(3.77),310(3.83)	88-4143-76
$C_6H_7BrN_2O_3$			
1H-Pyrazole-3-carboxylic acid, 4-bromo-2,5-dihydro-2-methyl-5-oxo-, methyl ester	MeOH	235(4.00),281(3.70)	24-0268-76
1H-Pyrazole-3-carboxylic acid, 4-bromo-5-hydroxy-1-methyl-, methyl ester	MeOH	226(3.93),267s(3.42), 310(2.72)	24-0261-76
$C_6H_7BrO_2$			
2-Cyclohexen-1-one, 3-bromo-2-hydroxy-	MeOH	273(4.30)	104-0089-76
2(5H)-Furanone, 3-(bromomethyl)-5-methyl-	EtOH	217(3.77)	33-0100-76
$C_6H_7BrO_3$			
2H-Pyran-3(6H)-one, 4-bromo-6-methoxy-, (S)-	n.s.g.	250(3.62)	136-0093-76D
$C_6H_7ClN_2O_4S$			
1H-Imidazole-4-carboxylic acid, 5-(chlorosulfonyl)-, ethyl ester	H_2O	224(3.75),253(3.92)	103-0924-76
$C_6H_7ClN_4O_2$			
4-Pyridazineacetic acid, 6-chloro-2,3-dihydro-3-oxo-, hydrazide	MeOH	206(4.36),225s(3.68), 298(3.45)	4-1155-76
$C_6H_7ClO_3$			
2H-Pyran-3(6H)-one, 4-chloro-6-methoxy-, (S)-	n.s.g.	242(3.68)	136-0093-76D
C_6H_7N			
Benzenamine (aniline)	H_2O	231(3.89),280(3.12)	35-7777-76
C_6H_7NO			
Pyridine, 2-methoxy-	EtOH	271(3.65)	35-0171-76
Pyridine, 4-methoxy-	EtOH	218(4.01),235s(3.34)	35-0171-76
Pyridinium, 3-hydroxy-1-methyl-, hydroxide, inner salt	EtOH	333(3.74)	39-1829-76B
2(1H)-Pyridinone, 1-methyl-	EtOH	302(3.74)	35-0171-76
4(1H)-Pyridinone, 1-methyl-	EtOH	265(4.48)	35-0171-76
1H-Pyrrole-2-carboxaldehyde, 1-methyl-	1% EtOH	255(3.30),293(4.20)	44-1952-76
C_6H_7NOS			
Thiocyanic acid, 1,2-dimethyl-3-oxo-1-propenyl ester	EtOH	271(4.12)	97-0049-76
$C_6H_7NO_4$			
4-Isoxazoleacetic acid, 2,3-dihydro-5-methyl-3-oxo-	EtOH	211(3.77)	1-0567-76
4-Isoxazolecarboxylic acid, 3-methoxy-5-methyl-	MeOH	215(3.87)	1-0567-76
C_6H_7NS			
4H-Cyclopent[d]isothiazole, 5,6-dihydro-	EtOH	218(3.672),254(3.714)	48-0507-76
Pyridine, 2-(methylthio)-	EtOH	248(4.02),291(3.61)	35-0171-76

Compound	Solvent	$\lambda_{max}(\log \epsilon)$	Ref.
Pyridine, 4-(methylthio)-	EtOH	263(4.07)	35-0171-76
2(1H)-Pyridinethione, 1-methyl-	EtOH	287(4.11),357(3.80)	35-0171-76
4(1H)-Pyridinethione, 1-methyl-	EtOH	231(3.81),348(4.40)	35-0171-76
$C_6H_7N_3$			
1H-Pyrrole-3-carbonitrile, 2-amino- 4-methyl-	MeOH	262(3.77)	118-0051-76
$C_6H_7N_3O$			
Hydrazine, 1-nitroso-1-phenyl-	H_2O	<u>282(3.9)</u>	104-0405-76
	0.1N H_2SO_4	<u>253(4.0)</u>	104-0405-76
$C_6H_7N_3O_2S$			
Acetic acid, [(4-amino-2-pyrimidinyl)- thio]-	pH 2.0	241(4.41)	24-3615-76
	pH 12	225(4.24),250s(3.94), 285(3.73)	24-3615-76
Butanamide, 3-oxo-N-1,3,4-thiadiazol- 2-yl-	isoPrOH	251s(3.91),277(4.00)	4-0291-76
$C_6H_7N_3O_3$			
2,4(1H,3H)-Pyrimidinedione, 5-(aminoace- tyl)-, monohydrobromide hemihydrate	pH 1	231(4.00),285(4.10)	87-0194-76
$C_6H_7N_5O$			
1H-Imidazo[1,5-b][1,2,4]triazole-7-carb- oxamide, 1-methyl-	pH 1	232(3.84),273(4.18)	44-1889-76
	pH 7 and 12	276(4.23)	44-1889-76
Triazolo[3,2-c]triazin-7(3H)-one, 3,6-dimethyl-	EtOH	250(3.86),312(3.91)	22-1178-76
Triazolo[3,2-c]triazin-7(4H)-one, 2,4-dimethyl-	EtOH	238(3.50),296(3.77)	22-1178-76
Triazolo[3,2-c]triazin-7(4H)-one, 2,6-dimethyl-	EtOH	244(3.69),304(3.92)	22-1178-76
Triazolo[3,2-c]triazin-7(4H)-one, 4,6-dimethyl-	EtOH	242(3.64),297(3.77)	22-1178-76
Triazolo[3,4-c]triazin-7(4H)-one, 1,4-dimethyl-	EtOH	243(3.86),330(3.73)	22-1178-76
Triazolo[3,4-c]triazin-7(4H)-one, 4,6-dimethyl-	EtOH	243(3.87),329(3.76)	22-1178-76
Triazolo[4,3-b]triazin-7(1H)-one, 1,3-dimethyl-	EtOH	230s(3.70),291(3.58)	22-1178-76
Triazolo[4,3-b]triazin-7(1H)-one, 1,6-dimethyl-	EtOH	230s(3.73),295(3.52)	22-1178-76
Triazolo[4,3-b]triazin-7(1H)-one, 3,6-dimethyl-	EtOH	230(3.77),289(3.58)	22-1178-76
Triazolo[4,3-b]triazin-7(8H)-one, 6,8-dimethyl-	EtOH	249(3.50),294(3.07)	22-1178-76
$C_6H_7N_5S$			
1H-1,2,3-Triazolo[4,5-b]pyridin- 5-amine, 7-(methylthio)-	pH 13	288(4.17),306s(4.06)	44-3784-76
$C_6H_7N_7O_2$			
Acetic acid, cyano(1H-tetrazol-5-yl- hydrazono)-, ethyl ester	EtOH	368(3.99)	39-1496-76C
Tetrazolo[5,1-c][1,2,4]triazine-6-carb- oxylic acid, 7-amino-, ethyl ester	EtOH	236(4.28),329(4.28)	39-1496-76C
$C_6H_8BrNO_3$			
2H-Pyran-3(6H)-one, 4-bromo-6-methoxy-, oxime, (S)-	n.s.g.	239(4.05)	136-0093-76D

Compound	Solvent	$\lambda_{max}(\log \epsilon)$	Ref.
$C_6H_8ClNO_3$			
2H-Pyran-3(6H)-one, 4-chloro-6-methoxy-, oxime, (S)-	n.s.g.	231(4.02)	136-0093-76D
$C_6H_8ClN_3O$			
Ethanol, 2-[(6-chloro-4-pyrimidinyl)amino]-	pH 1	255s(4.24),258(4.26)	39-1847-76C
	pH 13	245(4.14),276(3.55)	39-1847-76C
2-Pyrimidinamine, 4-chloro-6-methoxy-5-methyl-	MeOH	233(4.16),284(3.76)	94-0507-76
4-Pyrimidinamine, 2-chloro-6-ethoxy-	MeOH	209(4.48),239(3.86)	94-0507-76
4-Pyrimidinamine, 2-chloro-6-methoxy-5-methyl-	MeOH	208(4.37),245(3.80), 261(3.79)	94-0507-76
$C_6H_8F_3N_5$			
1,3,5-Triazine-2,4-diamine, N,N-dimethyl-6-(trifluoromethyl)-	MeOH	213(4.44),278(3.53)	36-0525-76
C_6H_8NO			
Pyridinium, 1-methoxy-, perchlorate	DMF	354(4.54)	83-0592-76
	DMSO	356(4.52)	83-0592-76
	HMPT	350(4.62)	83-0592-76
$C_6H_8N_2$			
1-Cyclopenbene-1-carbonitrile, 2-amino-	H_2O	263(4.08)	149-0217-76B
Hydrazine, phenyl-	n.s.g.	241(3.19),283(3.96)	35-4236-76
1H-Imidazole, 1-ethenyl-2-methyl-	pH 1	222(3.98)	103-0691-76
	EtOH	233(4.04)	103-0691-76
	dioxan	237(4.07)	103-0691-76
$C_6H_8N_2OS$			
4(1H)-Pyrimidinone, 5-ethyl-2,3-dihydro-2-thioxo-	pH 1	277(4.03)	73-0311-76
	pH 13	260(3.96),302s(3.73)	73-0311-76
4(3H)-Pyrimidinone, 2-(ethylthio)-	EtOH	233(3.94),291(3.89)	95-1094-76
4(3H)-Pyrimidinone, 6-methyl-2-(methylthio)-	EtOH	233(3.98),285(3.84)	95-1094-76
$C_6H_8N_2O_2$			
4-Oxa-1,2-diazaspiro[4.4]non-1-en-3-one	n.s.g.	224(3.55),367(2.50)	4-0649-76
1H-Pyrazole-4-carboxylic acid, 1-methyl-, methyl ester	MeOH	223(4.08)	24-2154-76
2,4(1H,3H)-Pyrimidinedione, 1,3-dimethyl-	pH 2-11	266(3.92)	88-2321-76
2,4(1H,3H)-Pyrimidinedione, 5-ethyl-	EtOH	265(3.90)	78-2795-76
$C_6H_8N_2O_3$			
1H-Pyrazole-3-carboxylic acid, 2,5-dihydro-5-oxo-, ethyl ester	EtOH	220(4.0),270(3.3)	23-0488-76
1H-Pyrazole-3-carboxylic acid, 5-hydroxy-1-methyl-, methyl ester	MeOH	223(4.01),253s(3.48)	24-0253-76
4(1H)-Pyrimidinone, 2,6-dimethoxy-	MeOH	223(3.60),261(3.92)	94-0026-76
$C_6H_8N_2S$			
Thiopyrano[4,3-c]pyrazole, tetrahydro-	EtOH	222(3.56)	4-0225-76
$C_6H_8N_4OS$			
Acetamide, 2-[(4-amino-2-pyrimidinyl)-thio]-	pH 1.5	241(4.38)	24-3615-76
	pH 5.5 and 12	225(4.28),245s(3.97), 285(3.76)	24-3615-76

Compound	Solvent	$\lambda_{max}(\log \epsilon)$	Ref.
$C_6H_8N_4O_2$			
Acetamide, N-(2H-6-methyl-5-oxo-1,2,4-triazin-3-yl)-	EtOH	219(4.41),236s(4.02)	114-0413-76A
Imidazo[1,2-b]-as-triazin-3(5H)-one, 6,7-dihydro-7-hydroxy-2-methyl-	EtOH	208(4.38),250(3.84)	114-0413-76A
$C_6H_8N_4O_2S$			
2,6-Pyridinediamine, 4-(methylthio)-3-nitro-	pH 7	272(3.90),342(3.96), 398(4.09)	44-3784-76
$C_6H_8N_4O_2S_3$			
3-Thiazolidineacetic acid, 4-[(aminothioxomethyl)hydrazono]-2-thioxo-	MeOH	252(4.34),294(4.16)	103-0749-76
$C_6H_8N_4O_3S$			
1,2,3-Thiadiazolo[4,5-d]pyrimidine-5,7-dione, 2,3-dihydro-3,6-dimethyl-, S-oxide	EtOH	205(4.13),265(4.25), 343(4.24)	88-1129-76
$C_6H_8N_6$			
1H-Imidazole-4-carbonitrile, 5-(3,3-dimethyl-1-triazenyl)-	EtOH	224(4.02),328(4.11)	103-0465-76
1H-Imidazo[4,5-d]-1,2,3-triazin-4-amine, N,N-dimethyl-	H_2O	214(4.15),269(4.10), 316(3.84)	103-0465-76
$C_6H_8N_6O_3$			
1H-Tetrazole, 1-acetyl-5-(2-amino-1-nitro-1-propenyl)-	H_2O	208(4.06),341(3.88)	39-1327-76C
C_6H_8O			
Ethanone, 1-(methylenecyclopropyl)-	C_6H_{12}	218(2.43),283(1.45)	88-4183-76
2-Propanone, 1-cyclopropylidene-	C_6H_{12}	215(4.15),337(1.43)	88-4183-76
2-Propenal, 1-cyclopropyl-	EtOH	246(4.21)	70-0428-76
C_6H_8OSe			
3(2H)-Selenophenone, 2,5-dimethyl-	C_6H_{12}	323(3.7),370(3.0)	11-0126-76B
$C_6H_8O_2$			
1,4-Cyclohexanedione	EtOH	281(1.46)	88-4851-76
$C_6H_8O_2S$			
4-Thiopyranone, 3-hydroxymethylene-2,3,5,6-tetrahydro-	EtOH	280(3.71)	4-0225-76
$C_6H_8O_3$			
1-Cyclopentanecarboxylic acid, 2-oxo-	hexane	258(3.36)	104-1252-76
	H_2O	257(2.85),321s(2.11)	104-1252-76
$C_6H_8O_5$			
2H-Pyran-3(6H)-one, 2-(dihydroxymethyl)-2-hydroxy-	dioxan	230(3.91),345(1.48)	102-1753-76
$C_6H_9BrN_4$			
4,6-Pyrimidinediamine, 2-bromo-5-ethyl-	MeOH	219(4.50),268(3.93)	94-0026-76
$C_6H_9ClN_4$			
4,6-Pyrimidinediamine, 2-chloro-5-ethyl-	MeOH	219(4.49),268(3.92)	94-0026-76
C_6H_9NO			
4(1H)-Pyridinone, 2,3-dihydro-6-methyl-	EtOH	302(4.13)	4-0253-76

Compound	Solvent	$\lambda_{max}(\log \epsilon)$	Ref.
C_6H_9NOS			
2-Propanone, 1-(2-thiazolidinylidene)-	MeOH	251(3.52),313(4.47)	48-0298-76
$C_6H_9NO_2$			
3H-Pyrrol-3-one, 4,5-dihydro-5,5-dimeth-yl-, 1-oxide	EtOH	269(4.27)	12-2511-76
$C_6H_9NO_2S$			
Acetic acid, isothiocyanato-, 1-methyl-ethyl ester	pH 8.06	245(3.06)	73-2216-76
	dioxan	249(3.07)	73-2216-76
Acetic acid, isothiocyanato-, propyl ester	pH 8.06	244(3.06)	73-2216-76
	dioxan	248(3.09)	73-2216-76
Propanoic acid, 3-isothiocyanato-, ethyl ester	pH 8.06	246(3.09)	73-2216-76
	MeOH	249(2.96)	73-2216-76
	dioxan	248(3.07)	73-2216-76
$C_6H_9NO_4$			
2-Propenoic acid, 3-[(methoxycarbonyl)-amino]-, methyl ester, (Z)-	EtOH	264(4.12)	73-2059-76
C_6H_9NS			
Isothiazole, 5-ethyl-4-methyl-	EtOH	218(3.362),252(3.511)	48-0507-76
Thiazole, 2,4,5-trimethyl-	EtOH	253(3.60)	70-1353-76
$C_6H_9N_2$			
Pyridinium, 4-amino-1-methyl-, iodide	H_2O	225(4.12),268(4.33)	35-7777-76
$C_6H_9N_2O$			
Pyrimidinium, 2,3-dihydro-1,3-dimethyl-2-oxo-	pH 9.18	239(3.93),296(2.70)	23-2681-76
cation (hydrogen sulfate)	pH 0.29	316(3.93)	23-2681-76
$C_6H_9N_3OS$			
4(1H)-Pyrimidinone, 6-amino-2-(ethyl-thio)-	EtOH	231(4.36),276(3.98)	95-1094-76
$C_6H_9N_3O_2$			
4(1H)-Pyrimidinone, 6-amino-2-ethoxy-	MeOH	210(4.29),263(4.11)	94-0026-76
4(1H)-Pyrimidinone, 6-amino-2-methoxy-5-methyl-	MeOH	211(4.12),269(4.11)	94-0026-76
4(1H)-Pyrimidinone, 6-[(2-hydroxyethyl)-amino]-	pH 1	261(4.09)	39-1847-76C
	pH 13	256(3.66)	39-1847-76C
$C_6H_9N_3O_4S$			
1H-Imidazole-4-carboxylic acid, 5-(am-inosulfonyl)-, ethyl ester	H_2O	211(3.89),243(3.97)	103-0924-76
$C_6H_9N_7$			
4H-Imidazo[4,5-d]-1,2,3-triazin-4-one, 1,5-dihydro-, dimethylhydrazone	H_2O	215(4.09),269(4.02), 314(3.80)	103-0465-76
$C_6H_9S_3$			
Cyclopropenylium, tris(methylthio)-, tetrafluoroborate	MeCN	275(4.23)	35-4668-76
$C_6H_9Se_3$			
Cyclopropenylium, tris(methylseleno)-, tetrafluoroborate	MeCN	297(4.30)	35-4668-76

Compound	Solvent	$\lambda_{max}(\log \epsilon)$	Ref.
$C_6H_9Te_3$			
Cyclopropenylium, tris(methyltelluro)-, tetrafluoroborate	MeCN	342(4.15)	35-4668-76
$C_6H_{10}ClNO_4$			
Imidodicarbonic acid, chloro-, diethyl ester	MeOH	218(4.36)	117-0025-76
$C_6H_{10}N_2$			
1H-Pyrazole, 1,3,5-trimethyl-, iodine	CH_2Cl_2	400(3.21)	65-0664-76
complex	CCl_4	420(3.12)	65-0664-76
$C_6H_{10}N_2O$			
4H-Imidazol-4-one, 1,5-dihydro-1,2,5-trimethyl-	MeOH	267(3.68)	20-0579-76
dimer	MeOH	265(4.08)	20-0579-76
4H-Imidazol-4-one, 3,5-dihydro-2,3,5-trimethyl-	EtOH	232(3.60)	20-0573-76
Pyrazolidinium, 1-(1-methylethylidene)-3-oxo-, hydroxide, inner salt	MeOH	268(4.3)	48-0946-76
$C_6H_{10}N_2O_2$			
2-Butenedial, 2,3-dimethyl-, dioxime	MeOH	283(4.40)	104-2039-76
4-Isoxazolemethanamine, 3-methoxy-5-methyl-, hydrochloride	MeOH	210(3.77)	1-0567-76
3(2H)-Isoxazolone, 4-(1-aminoethyl)-5-methyl-, hydrobromide	MeOH	211(3.71)	1-0567-76
3(2H)-Isoxazolone, 4-(2-aminoethyl)-5-methyl- (zwitterion)	MeOH	213(3.77)	1-0567-76
2-Propenoic acid, 3-(1-methyl-2-methylenehydrazino)-, methyl ester, (2E)-	MeOH	290(4.47)	24-2154-76
2H-Pyrrole-2-carboxamide, 3,4-dihydro-2-methyl-, 1-oxide	n.s.g.	236(3.67)	39-1955-76C
$C_6H_{10}N_2O_3$			
Morpholine, N-(2-nitroethenyl)-	EtOH	245(3.51),360(4.39)	104-1985-76
Proline, N-nitroso-, methyl ester	C_6H_{12}-10% dioxan	345s(1.86),359(2.00), 371(2.06),384(1.98)	78-0847-76
$C_6H_{10}N_2O_4S$			
Thiourea, N-(carboxymethyl)-N'-(carbomethoxymethyl)-	base	265(3.76)	73-2216-76
$C_6H_{10}N_2S$			
1H-Imidazole, 1-ethyl-2-(methylthio)-, monohydriodide	EtOH	220-253(4.022-3.494)	28-0143-76B
	CH_2Cl_2	231-255(3.734-3.789)	28-0143-76B
2H-Imidazole-2-thione, 1,3-dihydro-1-(1-methylethyl)-	EtOH	260(4.197)	28-0143-76B
	CH_2Cl_2	270(4.192)	28-0143-76B
2H-Imidazole-2-thione, 1-ethyl-1,3-dihydro-3-methyl-	EtOH	262(4.136)	28-0143-76B
	CH_2Cl_2	266(4.177)	28-0143-76B
$C_6H_{10}N_4OS$			
1,2,4-Thiadiazol-3(2H)-imine, 5-ethoxy-2-(1-iminoethyl)-	H_2O	271(3.95)	18-3165-76
$C_6H_{10}N_4O_2$			
4(1H)-Pyrimidinone, 5-amino-6-[(2-hydroxyethyl)amino]-	pH 1	262(3.85)	39-1847-76C
	pH 13	234s(3.56)	39-1847-76C
1H-1,2,4-Triazolium, 4-[(ethoxycarbonyl)amino]-1-methyl-, hydroxide, inner salt	$CHCl_3$	265(3.75)	94-2568-76

Compound	Solvent	$\lambda_{max}(\log \epsilon)$	Ref.
$C_6H_{10}N_4O_2S$ 1,2,4-Thiadiazole-2(3H)-carboximidic acid, 5-ethoxy-3-imino-, methyl ester	MeOH	263(4.00)	18-3170-76
$C_6H_{10}N_4O_4S$ Carbamic acid, [5-(aminosulfonyl)-1H-imidazol-4-yl]-, ethyl ester	H_2O	230(4.12)	103-0924-76
$C_6H_{10}N_4S$ 2,3,6-Pyridinetriamine, 4-(methylthio)-	pH 7	222s(4.16),236(4.26), 324(3.78)	44-3784-76
$C_6H_{10}N_6O_2$ 4-Pyridazineacetic acid, 6-hydrazino-2,3-dihydro-3-oxo-, hydrazide	MeOH	232(4.22),330(3.27)	4-1155-76
$C_6H_{10}N_6O_4S$ Purinium, 2,8-diamino-6-methyl-, sulfate	H_2O	231(4.29),263(3.90), 309(3.91)	39-1414-76C
$C_6H_{10}O$ Cyclobutanone, 2,2-dimethyl-	C_6H_{12}	295(1.32)	112-0777-76
Cyclobutanone, 2,4-dimethyl-, cis	C_6H_{12}	295(1.49)	112-0777-76
Cyclobutanone, 2,4-dimethyl-, trans	C_6H_{12}	298(1.36)	112-0777-76
Cyclobutanone, 3,3-dimethyl-	C_6H_{12}	278(1.18),284(1.23), 293(1.20)	112-0777-76
Cyclohexanone	C_6H_{12}	280(1.42)	78-2827-76
	hexane	291(1.18)	35-3242-76
1-Hexen-3-one	MeOH	212(3.72),323(1.42)	40-0166-76
4-Hexen-3-one, trans	C_6H_{12}	217(3.99),320(1.40)	23-2127-76
	H_2O	227(4.06),297(1.75)	23-2127-76
1-Penten-3-one, 4-methyl-	MeOH	215(3.75),327(1.40)	40-0166-76
$C_6H_{10}OS_2$ 2-Propene-1-sulfinothioic acid, S-2-propenyl ester	hexane	264(3.92)	102-0187-76
$C_6H_{10}O_2$ 2-Butenoic acid, ethyl ester	C_6H_{12}	199(4.19),246(2.30)	48-0745-76
	EtOH	203(4.17),240(2.30)	48-0745-76
	MeCN	201(4.19),243(2.18)	48-0745-76
Cyclopentanecarboxylic acid	hexane	207.6(1.98)	44-3312-76
	H_2O	206.8(1.91)	44-3312-76
	MeCN	214.8(1.88)	44-3312-76
2-Propenoic acid, 2-methyl-, ethyl ester	C_6H_{12}	199(4.05),246(2.08)	48-0745-76
	EtOH	205(4.02),238(2.30)	48-0745-76
	MeCN	202(4.04),241(2.26)	48-0745-76
$C_6H_{10}O_2S_2$ 2-Propene-1-sulfonothioic acid, S-2-propenyl ester	hexane	266(2.76)	102-0187-76
$C_6H_{10}S_2$ Disulfide, di-2-propenyl	hexane	none	102-0187-76
	EtOH	213(2.8)	142-0035-76B
$C_6H_{11}IN_2S$ 1H-Imidazole, 1-ethyl-2-(methylthio)-, monohydriodide	EtOH	220-253(4.022-3.494)	28-0143-76B
	CH_2Cl_2	231-255(3.734-3.789)	28-0143-76B

Compound	Solvent	$\lambda_{max}(\log \epsilon)$	Ref.
$C_6H_{11}NO$			
2-Butenal, 3-(dimethylamino)-	C_6H_{12}	290(4.33)	35-2826-76
	MeOH	301(4.55)	35-2826-76
3-Buten-2-one, 4-(dimethylamino)-	C_6H_{12}	288(4.34)	35-2826-76
	H_2O	304(4.47)	35-2826-76
	MeOH	300(4.45)	35-2826-76
3-Buten-2-one, 4-(ethylamino)-	H_2O	296(4.43)	35-2826-76
	MeOH	296(4.35)	35-2826-76
cis-s-cis	C_6H_{12}	299(4.18)	35-2826-76
4-Piperidinone, 1-methyl-	hexane	285(1.18),293(1.18)	35-3242-76
2-Propenal, 3-[(1-methylethyl)amino]-	H_2O	285(4.59)	35-2826-76
	MeOH	281(4.55)	35-2826-76
cis-s-cis	C_6H_{12}	306(4.15)	35-2826-76
trans	C_6H_{12}	264(4.45)	35-2826-76
$C_6H_{11}NO_2$			
3-Buten-2-one, 4-methoxy-4-(methyl-amino)-	MeOH	286(4.31)	39-0783-76C
$C_6H_{11}N_2$			
1,4-Diazepinium, 2,3-dihydro-1-methyl-, perchlorate	MeOH	336(4.18)	78-2339-76
$C_6H_{11}N_2S$			
1H-Imidazolium, 1,3-dimethyl-2-(methyl-thio)-, iodide	EtOH	221-261(4.269-3.579)	28-0143-76B
	CH_2Cl_2	238(4.377)	28-0143-76B
$C_6H_{12}N_2$			
1,2-Diazocine, 3,4,5,6,7,8-hexahydro-, (E)-	MeOH	363(1.58)	24-0518-76
$C_6H_{12}N_2O$			
Piperidine, 3-methyl-1-nitroso-, (S)-	C_6H_{12}	342s(1.70),364(1.92), 365(2.06),377(2.03)	78-0847-76
2-Propenamide, N,N-dimethyl-3-(methyl-amino)-	C_6H_{12}	282(4.19)	78-1025-76
	MeOH	277(4.36)	78-1025-76
	dioxan	283(4.15)	78-1025-76
$C_6H_{12}N_2O_2$			
Propanamide, 2-(acetylamino)-N-methyl-	EtOH	220(2.63)	20-0573-76
2-Propenoic acid, 3-(1,2-dimethyl-hydrazino)-, methyl ester	MeOH	277(4.36)	24-2154-76
Pyrrolidine, 2-(methoxymethyl)-1-ni-troso-	C_6H_{12}-10% dioxan	359s(1.95),370(2.08), 385(1.99)	78-0847-76
$C_6H_{12}N_2O_4$			
Ethane, 1,2-bis(ethyl-aci-nitro)-	CH_2Cl_2	318(4.57)	104-2039-76
$C_6H_{12}N_2S$			
1H-Imidazole, 1-ethyl-4,5-dihydro-2-(methylthio)-, monohydriodide	EtOH	219.0(4.378)	28-0485-76A
	CH_2Cl_2	235(4.375)	28-0485-76A
2-Imidazolidinethione, 1-(1-methyl-ethyl)-	EtOH	241.2(4.198)	28-0485-76A
	CH_2Cl_2	248(4.228)	28-0485-76A
Pyrimidine, 1,4,5,6-tetrahydro-1-meth-yl-, monohydriodide	EtOH	220(4.462)	28-0485-76A
	CH_2Cl_2	233(4.362)	28-0485-76A
2(1H)-Pyrimidinethione, 1-ethyltetra-hydro-	EtOH	246.3(4.418)	28-0485-76A
	CH_2Cl_2	257.5(4.122)	28-0485-76A
$C_6H_{12}N_4$			
4H-1,2,4-Triazol-4-amine, N-ethyl-3,5-dimethyl-	H_2O	192(3.77)	103-0711-76

Compound	Solvent	$\lambda_{max}(\log \epsilon)$	Ref.
$C_6H_{12}N_4O_2$			
1H-1,2,4-Triazol-3-amine, 1-(2,2-di-methoxyethyl)-	EtOH	228(3.50)	78-0341-76
1H-1,2,4-Triazol-5-amine, 1-(2,2-di-methoxyethyl)-	EtOH	218(3.32)	78-0341-76
4H-1,2,4-Triazol-3-amine, 4-(2,2-di-methoxyethyl)-	EtOH	210(3.57)	78-0341-76
$C_6H_{12}O$			
2-Butanone, 3,3-dimethyl-	C_6H_{12}	285(1.30)	112-0777-76
3-Pentanone, 2-methyl-	C_6H_{12}	283(1.34)	112-0777-76
$C_6H_{12}O_2$			
Acetic acid, butyl ester	hexane	210.5(1.78)	44-3312-76
	H_2O	202.3(1.79)	44-3312-76
	MeCN	208.5(1.75)	44-3312-76
Acetic acid, 1,1-dimethylethyl ester	hexane	218.0(1.76)	44-3312-76
	H_2O	208.9(1.74)	44-3312-76
	MeCN	215.2(1.82)	44-3312-76
Acetic acid, 1-methylpropyl ester	neat	281s(--)	44-3312-76
	hexane	212.2(1.81)	44-3312-76
	H_2O	204.1(1.79)	44-3312-76
	MeCN	210.9(1.80)	44-3312-76
Acetic acid, 2-methylpropyl ester	hexane	211.1(1.78)	44-3312-76
	H_2O	201.3(1.77)	44-3312-76
	MeCN	207.0(1.82)	44-3312-76
Butanoic acid, ethyl ester	neat	281.7(-1.68)	44-3312-76
	hexane	213.3(1.82)	44-3312-76
	H_2O	206.2(1.87)	44-3312-76
	MeCN	212.4(1.80)	44-3312-76
Hexanoic acid	hexane	205.6(1.84)	44-3312-76
	H_2O	202.1(1.80)	44-3312-76
	MeCN	213.0(1.66)	44-3312-76
Propanoic acid, 2-methyl-, ethyl ester	neat	279.3(-1.50)	44-3312-76
	hexane	213.4(1.90)	44-3312-76
	H_2O	207.6(2.03)	44-3312-76
	MeCN	209.1(1.90)	44-3312-76
$C_6H_{12}S_3$			
Propane(dithioperoxo)thioic acid, 2-methyl-, ethyl ester	C_6H_{12}	312(3.38),468(1.00)	138-0441-76
$C_6H_{13}IN_2S$			
1H-Imidazole, 1-ethyl-4,5-dihydro-2-(methylthio)-, monohydriodide	EtOH	219.0(4.378)	28-0485-76A
	CH_2Cl_2	235(4.375)	28-0485-76A
Pyrimidine, 1,4,5,6-tetrahydro-1-methyl-, monohydriodide	EtOH	220(4.462)	28-0485-76A
	CH_2Cl_2	233(4.362)	28-0485-76A
$C_6H_{13}N$			
Piperidine, N-methyl-	hexane	199(3.69)	35-3242-76
$C_6H_{13}NO$			
4-Piperidinol, 1-methyl-	hexane	196(3.59)	35-3242-76
$C_6H_{13}N_2O_4P$			
Acetic acid, [(dimethoxymethylphosphor-anylidene)hydrazono]-, methyl ester	EtOH	272.0(4.08)	65-1424-76
$C_6H_{13}N_2O_5P$			
Acetic acid, [(trimethoxyphosphoranyli-dene)hydrazono]-, methyl ester	EtOH	268.5(4.18)	65-1424-76

Compound	Solvent	λ_{max}(log ϵ)	Ref.
$C_6H_{14}N_2$ Diazene, bis(1-methylethyl)-, (Z)-	EtOH	373(2.01)	44-1146-76
$C_6H_{14}N_2O$ Butanal, 4-hydroxy-, dimethylhydrazone	EtOH	239(3.91)	104-2034-76
Diazene, bis(1-methylethyl)-, 1-oxide, (E)-	EtOH	232(3.90)	44-1146-76
Pentanal, 5-hydroxy-, methylhydrazone	EtOH	232(3.68)	104-2034-76
$C_6H_{14}O_2S_2$ 1-Propanesulfonothioic acid, S-propyl ester	hexane	256(2.44)	102-0187-76
$C_6H_{14}S_2$ Disulfide, dipropyl	hexane	252(2.60)	102-0187-76
$C_6H_{15}ClNOPSe$ Phosphoramidochloridoselenoic acid, diethyl-, O-ethyl ester	hexane	194(4.20),247(3.24)	70-0174-76
$C_6H_{15}OPSe$ Phosphinoselenoic acid, diethyl-, O-ethyl ester	hexane EtOH	192s(4.05),217s(3.80) 194s(4.11),211s(3.90)	70-0174-76 70-0174-76
$C_6H_{15}O_2PSe$ Phosphonoselenoic acid, ethyl-, O,O-diethyl ester	hexane or EtOH	196s(4.06),217s(3.72)	70-0174-76
$C_6H_{15}O_3PSe$ Phosphoroselenoic acid, O,O,O-triethyl ester	hexane or EtOH	190(3.92),213(3.55)	70-0174-76
$C_6H_{15}PSSe$ Phosphinoselenothioic acid, diethyl-, S-ethyl ester	hexane EtOH MeCN	198s(4.12),265(3.60) 197s(3.87),262(3.30) 262(3.52)	70-0174-76 70-0174-76 70-0174-76
$C_6H_{16}NPSe$ Phosphinoselenoic amide, P,P-diethyl-N,N-dimethyl-	hexane	195s(4.19),233s(3.73)	70-0174-76
$C_6H_{16}N_4$ Diazene, [1-(2,2-dimethylhydrazino)-ethyl]ethyl-	EtOH	362(1.48)	104-2207-76
$C_6H_{17}N_2PSe$ Phosphonoselenoic diamide, P-ethyl-N,N,N',N'-tetramethyl-	hexane	194s(4.09),232s(3.70)	70-0174-76
$C_6H_{18}N_2Si_2$ Diazene, bis(trimethylsilyl)-	pentane	175(3.43),192(3.22), 250(2.42),784(0.7)	35-0109-76
$C_6H_{18}N_3PSe$ Phosphoroselenoic triamide, hexamethyl-	hexane EtOH	194(4.14),229(3.76) 200(4.03),225s(3.76)	70-0174-76 70-0174-76
$C_6H_{20}N_5PS_2Si_2$ 1,3,2,4,6,5-Dithia(3-S^{IV})triazaphosphorine, 5,5-dihydro-5,5-bis[(trimethylsilyl)amino]-	isooctane	570(6.03)	89-0696B-76

Compound	Solvent	$\lambda_{max}(\log \epsilon)$	Ref.
$C_6H_{25}B_{18}CoO$ Cobalt, [[10,10'-μ-ethoxybis-[(7,8,9,10,11-η)-decahydro-7,8-dicarbaundecaborato]](3-)]-	CH_2Cl_2	254(4.32),292(4.00), 408(3.12),510(2.99)	73-3509-76
$C_6H_{26}B_{18}CoN$ Cobalt, [[10,10'-[μ-(N-methylmethanami-nato)]bis[(7,8,9,10,11-η)-decahydro-7,8-dicarbaundecaborato]](3-)]-	CH_2Cl_2	259(4.47),293s(--), 413(3.09),498(2.96)	73-3509-76

Compound	Solvent	$\lambda_{max}(\log \epsilon)$	Ref.
$C_7Cl_6O_2$ Bicyclo[2.2.1]hept-5-ene-2,3-dione, 1,4,5,6,7,7-hexachloro-	hexane	443(2.08)	118-0256-76
C_7F_7I Benzene, 1,2,4,5-tetrafluoro-3-iodo- 6-(trifluoromethyl)-	EtOH	242(3.50)	78-1829-76
C_7HN_7 1H-Imidazo[1,5-b][1,2,4]triazole- 2,5,7-tricarbonitrile	EtOH	229(4.49),279(4.07), 313(4.24),325(4.11)	44-1889-76
	MeCN	219(4.43),262(4.10), 282(3.71),305(3.84)	44-1889-76
potassium derivative	H_2O	220(4.53),279(4.05), 313(4.22)	44-1889-76
$C_7H_4Br_2N_2O$ Pyrazolo[1,5-a]pyridin-2(1H)-one, 1,3-dibromo-	MeOH	249(4.18),309(3.62), 350(3.61)	18-1980-76
$C_7H_4ClNO_3$ Benzaldehyde, 2-chloro-5-nitro- Benzaldehyde, 4-chloro-3-nitro-	pH 7.0 pH 7.0	240(4.31) 245(4.25)	39-1594-76B 39-1594-76B
$C_7H_4Cl_2N_2$ 3-Pyridineacetonitrile, 2,6-dichloro-	MeOH	216(3.42),271(3.49)	33-0190-76
$C_7H_4Cl_2O$ Benzaldehyde, 2,4-dichloro- Benzaldehyde, 2,6-dichloro- Benzaldehyde, 3,4-dichloro-	pH 7.0 pH 7.0 pH 7.0	264(4.06) 255(3.76) 260(4.09)	39-1594-76B 39-1594-76B 39-1594-76B
$C_7H_4Cl_2O_2S$ 4-Cyclopentene-1,3-dione, 4,5-dichloro- 2-[(methylthio)methylene]-	CH_2Cl_2	256(4.22),327(4.37)	83-0034-76
$C_7H_4Cl_3NO$ 3-Pyridinecarboxaldehyde, 2,6-dichloro- 5-(chloromethyl)-	MeOH	224(4.00),275(4.66)	33-0211-76
$C_7H_4Cl_4$ 1,3-Cyclopentadiene, 1,2,3,4-tetra- chloro-5-ethylidene-	C_6H_{12}	274(4.21),400(2.58)	24-3929-76
$C_7H_4Cl_4O$ 1,3-Cyclopentadiene, 1,2,3,4-tetra- chloro-5-(methoxymethylene)-	C_6H_{12}	240(3.40),304(4.35), 393(2.90)	24-3939-76
$C_7H_4N_2O_2$ 2,5-Cycloheptadiene-1,4-dione, 7-diazo-	CH_2Cl_2	245(4.20),302(2.64), 311(2.62),375(3.85)	88-2339-76
$C_7H_4N_2O_2S_2$ 2(3H)-Benzothiazolethione, 6-nitro-	dioxan	295(4.14),330(4.04), 365(4.16)	104-2381-76
$C_7H_4N_4O$ 1H-Pyrazolo[4,3-b]pyridine-6-carbo- nitrile, 4,7-dihydro-7-oxo-	EtOH	226(4.31),232(4.27), 293(3.98)	39-0507-76C

Compound	Solvent	$\lambda_{max}(\log \epsilon)$	Ref.
$C_7H_4N_4O_4$			
Benzoic acid, 5-azido-2-nitro-	EtOH	209(4.06),234(3.87), 376(4.17)	64-0263-76C
$C_7H_4O_4$			
2H-Oxocin-2,5,8-trione	MeOH	250s(2.08)	88-2339-76
$C_7H_5BrN_2O$			
Pyrazolo[1,5-a]pyridin-2(1H)-one, 3-bromo-	MeOH	235(4.02)	18-1980-76
$C_7H_5BrN_2S$			
Isothiazolo[5,4-b]pyridine, 3-bromo-6-methyl-	EtOH	311s(3.75)	48-0779-76
C_7H_5BrO			
Benzaldehyde, 4-bromo-	1% EtOH-pH 10	256(3.94)	44-1957-76
C_7H_5Cl			
Bicyclo[4.1.0]hepta-1,3,5-triene, 2-chloro-	EtOH	242(3.27),250(3.34), 258(3.38),266(3.32), 270(3.24)	88-0863-76
Bicyclo[4.1.0]hepta-1,3,5-triene, 3-chloro-	C_6H_{12}	274(2.20),278(2.29), 286(2.21)	88-4177-76
$C_7H_5ClN_2S$			
2-Benzothiazolamine, 6-chloro-	EtOH	222(4.43),266(4.02), 295s(3.25)	104-2374-76
$C_7H_5ClN_4$			
1,2,4-Benzotriazin-3-amine, 6-chloro-	EtOH	211(4.25),242(4.30), 303(3.41)	107-0457-76
C_7H_5ClO			
Benzaldehyde, 2-chloro-	pH 7	253(3.96)	39-1594-76B
Benzaldehyde, 3-chloro-	pH 7	249(4.03)	39-1594-76B
Benzaldehyde, 4-chloro-	1% EtOH-pH 10	254(3.94)	44-1957-76
C_7H_5ClOSe			
Benzenecarboselenoic acid, 4-chloro-, potassium salt	EtOH	239(4.22),333(3.63)	138-0203-76
$C_7H_5ClO_3$			
Benzoic acid, 2-chloro-3-hydroxy-	MeOH	215(4.14),243s(3.38), 295(3.31)	12-2003-76
2,4,6-Cycloheptatrien-1-one, 6-chloro-2,5-dihydroxy-	MeOH	228s(4.24),245(4.39), 267s(3.74),320s(3.73), 340(3.94),380s(3.98), 391(3.99),422s(3.48)	88-2339-76
$C_7H_5Cl_4N$			
Methanamine, N-methyl-1-(2,3,4,5-tetra-chloro-2,4-cyclopentadien-1-ylidene)-	C_6H_{12}	233(3.48),336(4.45)	24-3939-76
C_7H_5FO			
Benzaldehyde, 4-fluoro-	1% EtOH-pH 10	246(4.09)	44-1957-76

Compound	Solvent	$\lambda_{max}(\log \epsilon)$	Ref.
C_7H_5IO			
Benzaldehyde, 4-iodo-	1% EtOH- pH 10	258(3.83)	44-1957-76
C_7H_5NOS			
2,1-Benzisothiazol-3(1H)-one	$CHCl_3$	244s(4.07),254s(3.89), 349(3.54)	48-0161-76
$C_7H_5NO_3$			
Benzaldehyde, 2-nitro-	pH 7.0	225(4.07)	39-1594-76B
Benzaldehyde, 3-nitro-	pH 7.0	233(4.33)	39-1594-76B
Benzaldehyde, 4-nitro-	pH 7.0	268(4.18)	39-1594-76B
	1% EtOH- pH 10	260(4.11)	44-1957-76
2-Furancarboxylic acid, 5-cyano-, methyl ester	EtOH	214(4.47),268(4.42)	73-1692-76
$C_7H_5NO_4$			
3,4-Pyridinedicarboxylic acid	neutral	262-267(3.70)	32-0019-76
	anion	262-267(3.70)	32-0019-76
	dianion	265-270(c.3.4)	32-0019-76
	cation	264-269(--)	32-0019-76
$C_7H_5NS_2$			
2(3H)-Benzothiazolethione	dioxan	325(4.49)	104-2381-76
$C_7H_5N_3$			
5-Azacinnoline	EtOH	262(3.59),271(3.51), 304(3.60),316(3.61), 370(2.06)	103-0808-76
$C_7H_5N_3O_2$			
Pyrazolo[1,5-a]pyridin-2(1H)-one, 3-nitroso-	MeOH	273(4.30),317(3.76)	18-1980-76
$C_7H_5N_3O_3$			
Pyrazolo[1,5-a]pyridin-2(1H)-one, 3-nitro-	MeOH	220(4.30),276(3.51), 354(4.16)	18-1980-76
Pyrido[2,3-d]pyrimidine-2,4(1H,3H)- dione, 8-oxide	pH 1	237(4.53),336(3.73)	44-3027-76
	pH 7	237(4.38),257(4.33), 363(3.74)	44-3027-76
	pH 11	257(4.42),280(4.06), 368(3.85)	44-3027-76
$C_7H_5N_5O_5$			
1H-Imidazo[1,5-b][1,2,4]triazole-2,5-di- carboxylic acid, 7-(aminocarbonyl)-	neutral	221(4.33),276(3.99), 311(4.09)	44-1889-76
	anion	220(4.30),270(3.88), 307(4.20)	44-1889-76
	dianion	204(4.30),238(4.13), 268(3.97),302(4.08)	44-1889-76
	trianion	237(4.33),283(3.79), 328(4.35)	44-1889-76
$C_7H_5N_7O_2$			
1H-Imidazo[1,5-b][1,2,4]triazole-2,5- dicarboxamide, monosodium salt	pH 12	234(4.47),290(3.94), 323(4.25)	44-1889-76
$C_7H_6Br_2O_2S$			
1,3-Dioxolane, 2-(4,5-dibromo-2-thien- yl)-	isooctane	241(3.94),298(2.37)	44-1320-76

Compound	Solvent	$\lambda_{max}(\log \epsilon)$	Ref.
$C_7H_6ClNO_4$ 3-Pyridinecarboxylic acid, 4-chloro- 1,2-dihydro-6-hydroxy-2-oxo-, methyl ester	dioxan	275(4.11),293s(4.06), 303s(4.05),317s(3.87)	39-2462-76C
$C_7H_6ClNO_5S$ Benzenesulfonic acid, 4-amino-5-carboxy- 2-chloro-	pH 13	265(4.25),323(3.73)	133-0304-76
$C_7H_6ClN_3O_2$ Methanehydrazonoyl chloride, N-(2-nitro- phenyl)-	EtOH	252(4.20),407(3.68)	104-1649-76
$C_7H_6Cl_2N_2O_2$ 4-Pyridazineacetic acid, 3,6-dichloro-, methyl ester	MeOH	273(3.16),295s(2.63)	4-1155-76
C_7H_6NOSSe 1,2,3-Benzothiaselenazol-2-ium, 6-methoxy-, chloride	93% H_2SO_4	400(3.86),475(3.92)	103-1128-76
$C_7H_6NOS_2$ 1,2,3-Benzodithiazol-2-ium, 6-methoxy-, chloride	93% H_2SO_4	376(3.84),447(3.87)	103-1128-76
C_7H_6NS Thiazolo[3,2-a]pyridinium bromide	EtOH	227(4.10),296(4.10), 310(4.27)	103-0717-76
$C_7H_6N_2$ 1H-Benzimidazole	pH 1	209(4.04),233(3.56), 241(3.58),247(3.56), 253(3.56),262(3.67), 267(3.85),274(3.86)	103-0691-76
	EtOH	201(4.67),245(3.74), 250(3.75),259(3.52), 267(3.61),273(3.78), 280(3.85)	103-0691-76
	dioxan	246(3.89),251(3.88), 260(3.62),269(3.67), 275(3.82),282(3.88)	103-0691-76
Benzonitrile, 4-amino- protonated	H_2O H_2O	213(4.16),270(4.28) 220(4.01),269(3.00), 278(3.00)	35-7777-76 35-7777-76
2,4-Hexadienedinitrile, 3-methyl-, cis-cis	EtOH	254s(4.28),265(4.35), 274(4.26)	44-1389-76
$C_7H_6N_2O$ 2,4-Hexadienedinitrile, 3-methoxy-, cis-cis	EtOH	235(4.08),248s(3.92), 256(3.85),287(3.97)	44-1389-76
Pyrazolo[1,5-a]pyridin-2(1H)-one	MeOH	232(4.59),281(3.09), 310(3.10)	18-1980-76
$C_7H_6N_2OS$ Thiazolo[5,4-b]pyridine, 2-methoxy-	MeOH-NaOMe	215(4.33),246(3.91), 285(3.83)	4-0491-76
$C_7H_6N_2O_2S$ 2H-1,2,4-Benzothiadiazine, 1,1-dioxide	dioxan	264(4.12)	103-0402-76

Compound	Solvent	$\lambda_{max}(\log \epsilon)$	Ref.
$C_7H_6N_2O_2S_2$ Thiazolo[5,4-b]pyridine, 2-(methyl- sulfonyl)-	MeOH-NaOMe	214(4.01),240(3.72), 260(3.78),292(3.69)	4-0491-76
$C_7H_6N_2O_3$ Benzamide, 4-nitro-	EtOH	262(4.09),305(3.30)	110-1173-76
$C_7H_6N_2O_4$ 2-Propenoic acid, 3-(1,2,3,4-tetrahydro- 2,4-dioxo-5-pyrimidinyl)-, (E)-	pH 1	268(4.14),297(4.21)	87-0194-76
$C_7H_6N_2S$ 2-Benzothiazolamine	EtOH	222(4.45),261(4.12), 295s(3.45)	104-2374-76
Thieno[2,3-d]pyrimidine, 2-methyl-	H_2O	192(3.83),233(4.29), 270(3.15)	22-0761-76
$C_7H_6N_2S_2$ 2(3H)-Benzothiazolethione, 6-amino-	dioxan	340(4.46)	104-2381-76
Thiazolo[5,4-b]pyridine, 2-(methylthio)-	dioxan	300(4.28)	104-2381-76
$C_7H_6N_4$ Pyrido[3,2-e]-1,2,4-triazine, 3-methyl-	EtOH	312(3.48),330(3.94), 476(2.27)	114-0301-76C
Pyrido[3,4-e]-1,2,4-triazine, 3-methyl-	EtOH	230(5.10),342(4.30), 475(2.38)	114-0285-76C
$C_7H_6N_4O$ Pteridine, 4-methoxy-	MeOH	224(4.26),259(3.47), 302(3.88)	24-3208-76
4(1H)-Pteridinone, 1-methyl-	MeOH	230(4.09),260s(3.48), 324(3.93)	24-3208-76
4(3H)-Pteridinone, 3-methyl-	MeOH	235(4.07),276(3.62), 311(3.78)	24-3208-76
$C_7H_6N_4O_2$ 2,4(1H,3H)-Pyrimidinedione, 5-(1H-imid- azol-2-yl)-	pH 1	240(3.90),294(3.99)	87-0194-76
2,4(1H,3H)-Pyrimidinedione, 5-(1H-imid- azol-4-yl)-	pH 1	233(4.05),283(3.95)	87-0194-76
[1,2,4]Triazolo[1,5-a]pyridine, 2-methyl-8-nitro-	H_2O	230(4.16),330(3.81)	44-3124-76
[1,2,4]Triazolo[1,5-a]pyridine, 3-methyl-8-nitro-	H_2O MeOH	225(4.15),360(3.63) 225(4.18),360(3.68)	44-3124-76 114-0301-76C
$C_7H_6N_4O_2S$ 2,4(1H,3H)-Pyrimidinedione, 5-(2-amino- 4-thiazolyl)-, hydrobromide	pH 1	246(4.10),303(3.71)	87-0194-76
$C_7H_6N_4O_3$ 2-Furancarboxylic acid, 5-(1H-tetrazol- 5-yl)-, methyl ester	EtOH	262(4.25)	73-1692-76
$C_7H_6N_4S$ Pyrido[2,3-e]-1,2,4-triazine, 3-(meth- ylthio)-	EtOH	<u>255(4.4),270s(4.2), 395(3.8)</u>	103-0942-76
	+ 20% mor- pholine	<u>243(4.3),347(4.0)</u>	103-0942-76

Compound	Solvent	$\lambda_{max}(\log \epsilon)$	Ref.
$C_7H_6N_6O$			
6H-Pyrrolo[3,2-e]-1,2,4-triazolo[4,3-b]pyridazin-7(8H)-one, 6-amino-	MeOH	230(4.35),285(3.84)	4-1155-76
C_7H_6O			
Benzaldehyde	1% EtOH-pH 10	250(4.11)	44-1957-76
2,4,6-Cycloheptatrien-1-one	EtOH	310(3.90)	33-0747-76
Tricyclo[4.1.0.0²,⁷]hept-4-en-3-one	EtOH	253(3.63),341(1.88)	88-2129-76
C_7H_6OS			
2,4,6-Cycloheptatriene-1-thione, 2-hydroxy-	C_6H_{12}	420(4.08)	33-0747-76
C_7H_6OSe			
Benzenecarboselenoic acid, potassium salt	EtOH	231(4.19),325(3.61)	138-0203-76
$C_7H_6O_2$			
Benzaldehyde, 4-hydroxy-	MeOH	221(3.589),284(4.155)	24-3379-76
Benzoic acid	75% H_2SO_4	238(4.1)	104-0107-76
	76.3% H_2SO_4	240(4.1)	104-0107-76
(also other percentages)	89.5% H_2SO_4	249(4.2)	104-0107-76
Bicyclo[2.2.1]hept-5-ene-2,3-dione	hexane	460(2.13)	118-0256-76
2,4,6-Cycloheptatrien-1-one, 2-hydroxy-	C_6H_{12}	374(3.74)	33-0747-76
2,5-Cyclohexadiene-1,4-dione, 2-methyl-	EtOH	235(3.91),248(4.09),285(3.26),297(3.20)	115-0203-76
	ether	243(4.26),295(3.08)	115-0203-76
	CHCl₃	295(2.78)	115-0203-76
$C_7H_6O_3$			
2,5-Cyclohexadiene-1,4-dione, 2-hydroxy-5-methyl-	pH 5.6	270(4.00),470(3.43)	39-0339-76C
	EtOH-acid	262(4.20),390(2.95)	39-0339-76C
	EtOH-base	268(4.08),495(3.34)	39-0339-76C
$C_7H_6O_4S_2Se$			
1,3-Dithiole-4,5-dicarboxylic acid, 2-selenoxo-, dimethyl ester	C_6H_{12}	210(4.33),290(--),295(3.17),392(4.30),540(2.51)	44-0879-76
1,3-Thiaselenole-4,5-dicarboxylic acid, 2-thioxo-, dimethyl ester	C_6H_{12}	250(3.75),305(--),360(--),365(3.97)	44-0879-76
$C_7H_6O_4Se_3$			
1,3-Diselenole-4,5-dicarboxylic acid, 2-selenoxo-, dimethyl ester	C_6H_{12}	217(4.02),235(3.97),252s(3.90),405(4.02),562(2.17)	44-0882-76
$C_7H_6O_5$			
2,4-Furandicarboxylic acid, 5-methyl-	EtOH	251(4.16)	94-2421-76
2,5-Heptadienedioic acid, 4-oxo-, (E,E)-	H_2O	236(4.13)	33-0626-76
	H_2O	243(4.13)	33-0626-76
C_7H_6S			
2,4,6-Cycloheptatriene-1-thione	toluene	378(4.19)	33-0747-76
	EtOH	255s(--),380(4.23)	33-0747-76
	CCl₄	258s(--),376(4.21)	33-0747-76
	95% H_2SO_4	256(4.25),360(4.26)	33-0747-76
$C_7H_7BrN_2O$			
Diazene, (bromomethyl)phenyl-, 2-oxide, (Z)-	EtOH	265(4.04)	44-1141-76

Compound	Solvent	$\lambda_{max}(\log \epsilon)$	Ref.
$C_7H_7BrN_4$ [1,2,3]Triazolo[1,5-a]pyrimidine, 3-bromo-5,7-dimethyl-	MeOH	223(4.43),274(3.59), 284(4.60),310(3.51)	44-0385-76
$C_7H_7BrN_4O_2S$ 2,4(1H,3H)-Pyrimidinedione, 5-(2-amino- 4-thiazolyl)-, hydrobromide	pH 1	246(4.10),303(3.71)	87-0194-76
C_7H_7BrOS 4H-Thiopyran-4-one, 2-bromo-3,5-di- methyl-	EtOH	244(3.99),297(4.16), 304(4.15)	88-2167-76
$C_7H_7BrO_2S$ 1,3-Dioxolane, 2-(4-bromo-2-thienyl)-	isooctane	229(3.76),247(3.64)	44-1320-76
1,3-Dioxolane, 2-(5-bromo-2-thienyl)-	isooctane	242(4.00),289(2.59)	44-1320-76
$C_7H_7BrO_3$ 1,2-Benzenediol, 5-bromo-3-methoxy-	MeOH	228(4.33),275(3.37), 282(3.35)	12-2003-76
2-Cyclopenten-1-one, 2-acetoxy-3-bromo-	MeOH	240(4.64),305(1.79)	104-0089-76
$C_7H_7ClN_2O$ Diazene, (chloromethyl)phenyl-, 2-oxide, (Z)-	EtOH	258(4.04)	44-1141-76
$C_7H_7ClN_2O_3$ 4-Pyridazineacetic acid, 3-chloro-1,6- dihydro-6-oxo-, methyl ester	MeOH	210(4.38),225s(3.69), 295(3.32)	4-1155-76
4-Pyridazineacetic acid, 6-chloro-2,3- dihydro-3-oxo-, methyl ester	MeOH	209(4.31),225s(3.67), 297(3.46)	4-1155-76
$C_7H_7ClN_2O_4S$ Benzoic acid, 2-amino-5-(aminosulfonyl)- 4-chloro-	pH 13	262(4.05),317(3.51)	133-0304-76
$C_7H_7ClO_2S$ 2-Thiophenecarboxylic acid, 4-chloro-, ethyl ester	MeOH	244(3.95),280(3.71)	4-0393-76
$C_7H_7FN_2O_3$ Benzenamine, 3-fluoro-N-(hydroxymethyl)- 4-nitro-	n.s.g.	225(--),370(4.20)	35-7777-76
$C_7H_7F_3N_2O_3$ 2,4(1H,3H)-Pyrimidinedione, 5-[(2,2,2- trifluoroethoxy)methyl]-	EtOH	260(3.91)	104-0643-76
C_7H_7NO Benzamide	EtOH	226(4.03),270(2.85)	110-1173-76
2,4,6-Cycloheptatrien-1-one, 2-amino-	C_6H_{12}	395(3.89)	33-0747-76
C_7H_7NOS Thiocyanic acid, 2-formyl-1-cyclopent- en-1-yl ester	EtOH	291(3.77)	97-0049-76
$C_7H_7NO_2$ Cyclopenta[c]pyrrole-1,3(2H,4H)-dione, 5,6-dihydro-	EtOH	223(4.03),233(4.04)	78-2379-76
Furo[2,3-b]pyridin-6-ol, 2,3-dihydro-	MeOH	229(3.74),297(3.90), 332(3.44)	24-1269-76

Compound	Solvent	$\lambda_{max}(\log \epsilon)$	Ref.
$C_7H_7NO_2S$			
Benzene, 1-(methylthio)-2-nitro-	C_6H_{12}	243(4.32),263(3.83), 362(3.60)	104-1735-76
	heptane	243(4.34),264(3.80), 360(3.61)	
	MeOH	244(4.27),372(3.55)	
	dioxan	245(4.30),371(3.57)	
	AmOAc	244(4.28),371(3.57)	
	CCl_4	247(4.31),365(3.62)	
	$CHCl_3$	246(4.28),376(3.57)	
	CH_2Cl_2	246(4.28),376(3.47)	
	DMF	380(3.53)	
	MeCN	244(4.27),375(3.57)	
Benzene, 1-(methylthio)-3-nitro-	C_6H_{12}	252(4.24),344(3.13)	104-1735-76
	heptane	250(4.31),342(3.10)	
	MeOH	253(4.31),349(3.06)	
	dioxan	253(4.29),349(3.07)	
	AmOAc	253(4.30),348(3.08)	
	CCl_4	255(4.29),347(3.10)	
	$CHCl_3$	256(4.36),352(3.08)	
	CH_2Cl_2	255(4.32),352(3.08)	
	DMF	256(4.24),355(3.04)	
	MeCN	253(4.33),350(3.06)	
Benzene, 1-(methylthio)-4-nitro-	heptane	215-240(3.74-3.68), 326(4.15)	104-1735-76
	MeOH	212-250(3.76-3.50), 339(4.11)	
	dioxan	215-240(3.72-3.66), 336(4.12)	
	CCl_4	331(4.18)	
	$CHCl_3$	220-245(3.85-3.65), 341(4.14)	
	DMF	346(4.15)	
	MeCN	213-250(3.70-3.46), 342(4.11)	
$C_7H_7NO_2S_2$			
5H-1,4-Dithiino[2,3-c]pyrrole-5,7(6H)- dione, 2,3-dihydro-2-methyl-	C_6H_{12}	204(3.66),260(4.05), 400f(3.45)	56-1523-76
	H_2O	265(4.03),431(3.36)	
	EtOH	204(3.80),263(4.05), 417(3.40)	
	dioxan	263(4.01),408(3.42)	
	MeCN	211(3.42),263(4.09), 417(3.38)	
$C_7H_7NO_2Se$			
Benzene, 1-(methylseleno)-2-nitro-	C_6H_{12}	251(4.26),271(3.91), 383(3.65)	104-1735-76
	heptane	250(4.19),271(3.84), 382(3.60)	
	MeOH	252(4.22),391(3.57)	
	dioxan	254(4.23),391(3.60)	
	AmOAc	252(4.20),390(3.59)	
	CCl_4	253(4.22),386(3.61)	
	$CHCl_3$	256(4.22),396(3.58)	
	CH_2Cl_2	254(4.26),397(3.59)	
	DMF	258(4.20),399(3.59)	
	MeCN	253(4.21),395(3.59)	
Benzene, 1-(methylseleno)-3-nitro-	C_6H_{12}	250(4.15),347(2.95)	104-1735-76

Compound	Solvent	$\lambda_{max}(\log \epsilon)$	Ref.
Benzene, 1-(methylseleno)-3-nitro- (cont.)	heptane	249(4.16),345(2.96)	104-1735-76
	MeOH	252(4.12),352(2.91)	
	dioxan	252(4.15),353(2.98)	
	AmOAc	251(4.14),351(2.93)	
	CCl_4	253(4.20),350(2.96)	
	$CHCl_3$	255(4.17),357(2.91)	
	CH_2Cl_2	255(4.21),356(2.90)	
	DMF	255(4.11),359(2.91)	
	MeCN	256(4.10),353(2.90)	
Benzene, 1-(methylseleno)-4-nitro-	heptane	210-225(4.04-3.83), 230-245(3.80-3.78), 336(4.22)	104-1735-76
	MeOH	215-225(3.99-3.96), 230-250(3.90-3.71), 350(4.11)	
	dioxan	218(3.93),238(3.8), 349(4.10)	
	AmOAc	349(4.12)	
	CCl_4	342(4.19)	
	$CHCl_3$	353(4.10)	
	DMF	357(4.07)	
	MeCN	220(3.8),243(3.7), 352(4.05)	

$C_7H_7NO_3$

Compound	Solvent	$\lambda_{max}(\log \epsilon)$	Ref.
Benzene, 1-methoxy-2-nitro-	C_6H_{12}	213(4.21),250(3.56), 307(3.42)	104-1735-76
	heptane	213(4.19),250(3.54), 306(3.40)	
	MeOH	212(4.18),258(3.55), 321(3.42)	
	dioxan	214(4.17),257(3.53), 319(3.40)	
	AmOAc	256(3.51),317(3.39)	
	CCl_4	254(3.57),313(3.46)	
	$CHCl_3$	261(3.59),325(3.46)	
	CH_2Cl_2	260(3.59),326(3.46)	
	DMF	263(3.53),328(3.40)	
	MeCN	213(4.10),260(3.50), 325(3.37)	
Benzene, 1-methoxy-3-nitro-	C_6H_{12}	212(4.16),225(4.14), 262(3.81),318(3.39)	104-1735-76
	heptane	212(4.14),224(4.12), 261(3.80),317(3.36)	
	MeOH	212(4.20),227(4.08), 268(3.81),326(3.36)	
	dioxan	212(4.19),227(4.07), 268(3.78),326(3.38)	
	AmOAc	267(3.79),325(3.37)	
	CCl_4	265(3.83),322(3.42)	
	$CHCl_3$	270(3.77),328(3.33)	
	CH_2Cl_2	270(3.80),328(3.34)	
	DMF	272(3.74),330(3.32)	
	MeCN	213(4.14),227(4.01), 269(3.76),327(3.32)	
Benzene, 1-methoxy-4-nitro-	heptane	220(3.98),226(3.98), 293(4.08)	104-1735-76
	MeOH	227(3.87),307(4.03)	
	dioxan	227(3.87),304(4.03)	
	AmOAc	303(4.05)	

Compound	Solvent	$\lambda_{max}(\log \epsilon)$	Ref.
Benzene, 1-methoxy-4-nitro- (cont.)	CCl$_4$	298(4.09)	104-1735-76
	CHCl$_3$	229(3.86),308(4.01)	
	DMF	312.5(4.06)	
	MeCN	223(3.87),227(3.88),	
		308(4.07)	
Phenol, 3-methyl-2-nitro-	MeOH	274(3.34),340(3.03)	12-2003-76
$C_7H_7NO_5$			
1,3-Dioxolane, 2-(5-nitro-2-furanyl)-	n.s.g.	305(3.98)	103-1100-76
2-Furancarboxylic acid, 5-nitro-, ethyl ester	n.s.g.	296(4.05)	103-1100-76
C_7H_7NS			
2,4,6-Cycloheptatriene-1-thione, 2-amino-	C_6H_{12}	448(4.12)	33-0747-76
$C_7H_7N_2O$			
Benzenediazonium, 4-methoxy-, tetrafluoroborate	pH 1	230(3.85),314(4.38)	33-1438-76
$C_7H_7N_3$			
1,2,4-Triazolo[4,3-a]pyridine, 3-methyl-	H$_2$O	267(3.52),286(3.52)	44-3124-76
$C_7H_7N_3O$			
1H-Benzotriazole, 1-methoxy-	EtOH	175(3.74),263(3.73),	4-0509-76
		283(3.60)	
	EtOH	330(3.1)	103-1172-76
$C_7H_7N_3O_2$			
1H-Imidazole-4-carboxylic acid, 5-(cyanomethyl)-, methyl ester	pH 1	222(4.03)	35-1492-76
	pH 11	251(4.03)	35-1492-76
Imidazo[1,2-a]pyrimidine-2,5(1H,3H)-di-one, 7-methyl-	pH 6.1	238(4.03),278(3.84)	114-0413-76A
	EtOH	228(3.98),280(3.83)	114-0413-76A
Imidazo[1,2-a]pyrimidine-2,7(1H,3H)-di-one, 5-methyl-	pH 4.1	263s(3.92)	114-0413-76A
	EtOH	217(4.36),248s(3.95)	114-0413-76A
	dioxan	211(4.36),260(4.06)	114-0413-76A
$C_7H_7N_3O_3$			
Diazene, methoxy(4-nitrophenyl)-	DMF	280(4.1)	104-0395-76
Formic acid, 2-(2-nitrophenyl)hydrazide	EtOH	227(4.27),276(3.72),	104-1649-76
		392(3.64)	
Formic acid, 2-(4-nitrophenyl)hydrazide	EtOH	227(3.93),352(4.16)	104-1649-76
Isoxazole, 3-methoxy-4-diazoacetyl-5-methyl-	MeOH	246(3.96),291(4.21)	1-0567-76
Methyl(4-nitrophenyl)nitrosoamine	DMF	315(4.3)	104-0395-76
$C_7H_7N_5$			
5-Azacinnoline, 4-hydrazino-	EtOH	256(3.83),262(3.84),	103-0808-76
		265(3.84),268(3.83),	
		283(3.80),385(3.79)	
$C_7H_7N_5O$			
Ethanone, 1-(4-methyl[1,2,4]triazolo-[5,1-c][1,2,4]triazin-3-yl)-	EtOH	208(3.67),319(3.78)	39-0421-76C
$C_7H_7N_5O_3$			
1H-Imidazo[1,5-b][1,2,4]triazole-2-carb-oxylic acid, 7-(aminocarbonyl)-1-meth-yl-	2M HCl	244(3.92),285(3.79)	44-1889-76
	pH 7	261(4.09),285s(3.90)	44-1889-76

Compound	Solvent	$\lambda_{max}(\log \epsilon)$	Ref.
Pyrimido[5,4-e]-1,2,4-triazine-5,7-(1H,6H)-dione, 1,6-dimethyl-, 4-oxide	dioxan	264(4.29),321(3.65), 412(3.37)	39-0713-76C
Triazolo[3,2-c]triazine-6-carboxylic acid, 4,7-dihydro-7-oxo-, ethyl ester	EtOH	270(3.21),315(3.72)	22-1178-76
	EtOH	208(4.06),220s(3.93), 261s(3.64),273(3.68), 318(4.00)	39-1496-76C
Triazolo[3,4-c][1,2,4]triazine-6-carboxylic acid, 4,7-dihydro-1-methyl-7-oxo-, methyl ester	EtOH	278(3.76),303(3.56)	22-1178-76
Triazolo[3,4-c][1,2,4]triazine-6-carboxylic acid, 4,7-dihydro-7-oxo-, ethyl ester	EtOH	281s(3.72),296(3.61)	22-1178-76
Triazolo[4,3-b][1,2,4]triazine-6-carboxylic acid, 1,7-dihydro-7-oxo-, ethyl ester	EtOH	245(3.79),305(3.65)	22-1178-76
Triazolo[5,1-c][1,2,4]triazine-6-carboxylic acid, 4,7-dihydro-7-oxo-, ethyl ester	EtOH	208(4.06),220s(3.93), 261s(3.64),273(3.68), 318(4.00)	39-0421-76C
$C_7H_7N_7O_3$ 1H-Imidazo[1,5-b][1,2,4]triazole-2,5,7-tricarboxamide	pH 13	238(4.39),295(3.92), 335(4.31)	44-1889-76
$C_7H_8Br_2N_2$ Pyrimidine, 2-(dibromomethyl)-4,6-dimethyl-	MeOH	225(3.83),250s(3.81)	44-0385-76
$C_7H_8ClIN_2$ Pyrimidine, 2-(chloroiodomethyl)-4,6-dimethyl-	MeOH	230(3.86),260s(3.83)	44-0385-76
C_7H_8ClN Benzenamine, 4-chloro-N-methyl-	H_2O	247(4.04),298(3.10)	35-7777-76
C_7H_8ClNO Benzenamine, 4-chloro-N-(hydroxymethyl)-	H_2O	247(4.08),295(3.18)	35-7777-76
$C_7H_8ClNO_2$ Pyridine, 4-chloro-2,6-dimethoxy-	EtOH	230(3.92),278(4.05)	39-2462-76C
2(1H)-Pyridinone, 4-chloro-6-methoxy-1-methyl-	EtOH	233(3.62),308(3.91)	39-2462-76C
$C_7H_8ClN_3O_2$ Acetamide, 2-chloro-N-(3,4-dihydro-6-methyl-4-oxo-2-pyrimidinyl)-	EtOH	226(3.94),286(3.91)	114-0413-76A
	dioxan	238(3.89),278(3.88)	114-0413-76A
$C_7H_8ClN_3S$ Hydrazinecarbothioamide, 2-(4-chlorophenyl)-	EtOH	245(4.33),285s(3.85)	104-2374-76
$C_7H_8Cl_2N_2$ Pyrimidine, 2-(dichloromethyl)-4,6-dimethyl-	MeOH	228(3.85),250s(3.79)	44-0385-76
$C_7H_8Cl_2O$ 2-Propenal, 3-(2,2-dichloro-3-methylcyclopropyl)-	EtOH	243(4.54)	70-0396-76
$C_7H_8NO_2$ Pyridinium, 3-carboxy-1-methyl-, iodide, sodium salt	EtOH	209(4.07),221(4.09)	39-0315-76C

Compound	Solvent	$\lambda_{max}(\log \epsilon)$	Ref.
$C_7H_8N_2OS$			
4(1H)-Pyrimidinone, 5-cyclopropyl-	pH 1	280(3.72)	73-0311-76
2,3-dihydro-2-thioxo-	pH 13	263(4.18),300s(3.90)	73-0311-76
4(1H)-Pyrimidinone, 6-cyclopropyl-	pH 1	272(4.27)	73-0311-76
2,3-dihydro-2-thioxo-	pH 13	259(4.27),300(3.89)	73-0311-76
4(1H)-Pyrimidinone, 2-(2-propenylthio)-	EtOH	227(3.94),289(3.89)	95-1094-76
$C_7H_8N_2O_2$			
Benzenamine, N-methyl-4-nitro-	H_2O	230(3.90),407(4.27)	35-7777-76
Methanol, (phenyl-ONN-azoxy)-, (Z)-	EtOH	246(4.05)	44-1141-76
2-Pyridinecarboxaldehyde, 1,6-dihydro-	H_2O	322(3.92)	83-0769-76
1-methyl-6-oxo-, 2-oxime, (E)-	pH 13	332.5(4.06)	83-0769-76
	MeOH	239(3.80),331(3.94)	83-0769-76
2,4(1H,3H)-Pyrimidinedione, 5-cyclo-	pH 1	268(3.95)	73-0311-76
propyl-	pH 13	290(3.80)	73-0311-76
2,4(1H,3H)-Pyrimidinedione, 6-cyclo-	pH 1	267(4.11)	73-0311-76
propyl-	pH 13	283(4.03)	73-0311-76
$C_7H_8N_2O_2S_2$			
1H-1,2-Thiazine-4-carbonitrile,	EtOH	230(4.48),256(3.75),	142-1875-76
5-hydroxy-1-methyl-3-(methyl-		317(3.97)	
thio)-, 1-oxide			
$C_7H_8N_2O_3$			
Aziridine, 1-[(3-methoxy-5-isoxazolyl)-	MeOH	231(4.06)	1-0281-76
carbonyl]-			
Benzenamine, N-(hydroxymethyl)-4-nitro-	H_2O	225(--),380(4.30)	35-7777-76
4-Oxazolecarboxamide, 2-acetyl-	MeOH	217(3.98),269(3.96)	35-8237-76
5-methyl-			
1H-Pyrazole-3-carboxylic acid, 1-acet-	MeOH	217(4.00)	103-1026-76
yl-, methyl ester			
4-Pyridazineacetic acid, 2,3-dihydro-	MeOH	212(4.14),224(3.47),	4-1155-76
3-oxo-, methyl ester		288(3.55)	
$C_7H_8N_2O_4$			
1(2H)-Pyrimidineacetic acid, 3,4-di-	EtOH	208(3.91),264(4.02)	114-0413-76A
hydro-6-methyl-2,4-dioxo-			
$C_7H_8N_2O_5$			
1H-Pyrazole-5-carboxylic acid, 3-(carb-	MeOH	223(3.99),258(3.51)	24-0268-76
oxymethoxy)-1-methyl-			
3,5-Pyrazoledicarboxylic acid, 4-hy-	EtOH	275(4.14)	78-0269-76
droxy-, dimethyl ester			
$C_7H_8N_2O_6$			
2-Furancarboxaldehyde, 3,4-dimethoxy-	SDA 30	353(4.10)	87-0729-76
5-nitro-, oxime			
$C_7H_8N_4$			
[1,2,4]Triazolo[1,5-a]pyridin-8-amine,	H_2O	270(4.02),294(3.83)	44-3124-76
2-methyl-			
[1,2,4]Triazolo[4,3-a]pyridin-8-amine,	H_2O	225(4.10),297(4.06)	44-3124-76
3-methyl-			
$C_7H_8N_4O$			
1H-Purine, 2-methoxy-1-methyl-	pH 9.0	276(3.91)	39-1176-76B
6H-Purin-6-one, 1,7-dihydro-1,8-di-	MeOH	253(3.98)	39-0239-76C
methyl-			
5H-1,2,3-Triazolo[4,5-b]pyridin-5-one,	MeOH	249(3.66),308(4.02)	23-1029-76
1,4-dihydro-1,4-dimethyl-			

Compound	Solvent	$\lambda_{max}(\log \epsilon)$	Ref.
5H-1,2,3-Triazolo[4,5-b]pyridin-5-one, 2,4-dihydro-2,4-dimethyl-	MeOH	238(3.67),302(4.19), 312(4.20)	23-1029-76
$C_7H_8N_4OS$			
6H-Purin-6-one, 1,7-dihydro-1-methyl- 8-(methylthio)	neutral	274(4.080)	39-0090-76C
	anion	280(--)	39-0090-76C
	cation	277(--)	39-0090-76C
6H-Purin-6-one, 1,7-dihydro-7-methyl- 8-(methylthio)-	neutral	278(3.963)	39-0090-76C
	anion	278(--)	39-0090-76C
	cation	277(--)	39-0090-76C
6H-Purin-6-one, 1,9-dihydro-9-methyl- 8-(methylthio)-	neutral	270(4.145)	39-0090-76C
	anion	277(--)	39-0090-76C
	cation	275(--)	39-0090-76C
6H-Purin-6-one, 3,7-dihydro-3-methyl- 8-(methylthio)-	neutral	287(4.324)	39-0090-76C
	anion	295(--)	39-0090-76C
	cation	287(--)	39-0090-76C
$C_7H_8N_4O_2$			
6H-Purin-6-one, 1,9-dihydro-9-(2-hy- droxyethyl)-	pH 1	250(4.05)	39-1847-76C
	pH 13	256(4.06),260s(4.01)	39-1847-76C
$C_7H_8N_4O_2S$			
Hydrazinecarbothioamide, 2-(4-nitro- phenyl)-	EtOH	243(4.27),343(4.19)	104-2374-76
$C_7H_8N_4O_3$			
Acetic acid, 2-(3-nitro-4-pyridinyl)hy- drazide	pH 1	231(4.23),264(4.03), 350(3.54)	114-0285-76C
1H-Purine-2,6,8(3H)-trione, 7,9-di- hydro-1,3-dimethyl-	EtOH	290.5(4.07)	95-1453-76
$C_7H_8N_4O_4$			
[1,2,5]Oxadiazolo[3,4-d]pyrimidine- 5,7(4H,6H)-dione, 4-ethyl-6-methyl-, 1-oxide	MeOH	269(3.99),317s(3.08)	39-1327-76C
4-Pyrimidineacetic acid, 2-amino- 1,6-dihydro-α-(hydroxyimino)-6- oxo-, methyl ester	EtOH	227(4.27),325(3.69)	44-2850-76
$C_7H_8N_4S$			
1H-Imidazo[4,5-b]pyridin-5-amine, 7-(methylthio)-	pH 13	233s(4.19),271s(3.85), 277(3.89),312(3.98)	44-3784-76
1H-Purine, 1-methyl-2-(methylthio)-	pH 8.0	247(4.46),278(3.77)	39-1176-76B
$C_7H_8N_6OS$			
Acetamide, N-[7-(methylthio)-1H-1,2,3- triazolo[4,5-d]pyrimidin-5-yl]-	EtOH-HCl	227(4.13),252(4.21), 277(4.03),302(4.13)	87-1186-76
	EtOH-pH 7	238(4.33),307(4.09)	87-1186-76
	EtOH-pH 13	237(4.26),308(4.05)	87-1186-76
$C_7H_8N_6O_2$			
Acetic acid, cyano(1H-1,2,4-triazol- 3-ylhydrazono)-, ethyl ester	EtOH	371(4.19)	39-1496-76C
[1,2,4]Triazolo[5,1-c][1,2,4]triazine- 3-carboxylic acid, 4-amino-, ethyl ester	EtOH	215(4.32),229(4.35), 282(4.17),325(4.29)	39-1496-76C
C_7H_8O			
Phenol, 4-methyl-	dioxan	282(3.35),288(3.26)	126-3481-76
Tricyclo[2.2.1.0^{2,6}]heptan-3-one	3-Mepentane	284(1.29)	144-0475-76

Compound	Solvent	$\lambda_{max}(\log \epsilon)$	Ref.
$C_7H_8O_2$			
Bicyclo[2.2.1]heptane-2,3-dione	hexane	484(1.67)	118-0256-76
1,3-Cyclopentadien-1-ol, acetate	heptane	245(3.43)	104-1002-76
2-Furanmethanol, 5-ethenyl-	EtOH	269(4.59)	70-0683-76
4H-Pyran-4-one, 2,6-dimethyl-	heptane	181(4.30),207(3.93), 239(4.07)	18-3685-76
	$C_3H_2F_6O$	183(4.36),203(3.89), 246(4.22)	18-3685-76
$C_7H_8O_2S$			
Benzenesulfinic acid, 4-methyl-, sodium salt	n.s.g.	219(4.11),254(3.85), 325(4.05)	78-1411-76
$C_7H_8O_3$			
2-Cyclopenten-1-one, 4-acetoxy-	MeOH	210(4.11)	78-1713-76
1,3-Dioxolane, 2-(2-furanyl)-	n.s.g.	208(3.23)	103-1100-76
2-Furancarboxylic acid, ethyl ester	n.s.g.	251(4.08)	103-1100-76
$C_7H_8O_3S$			
Benzenesulfonic acid, 4-methyl-, sodium salt	50% MeOH	260(2.35)	104-2433-76
$C_7H_8O_4$			
2-Furancarboxaldehyde, 3,4-dimethoxy-	SDA 30	284(4.24)	87-0729-76
$C_7H_8O_4S$			
Benzenesulfonic acid, 2-hydroxy-4-methyl-	H_2O	<u>220(3.8)</u>,280(3.47)	40-0497-76
Benzenesulfonic acid, 4-hydroxy-2-methyl-	H_2O	232(3.93),<u>280(3.3)</u>	40-0497-76
$C_7H_8O_5S$			
4H-Cyclopenta-1,3,2-dioxathiol-4-one, 3a,6a-dihydro-3a,5-dimethyl-	n.s.g.	223(3.93),339(1.63)	77-0823-76
$C_7H_8S_3$			
3H-1,2-Benzodithiole-3-thione, 4,5,6,7-tetrahydro-	hexane	234(3.84),273(3.15), 372(4.072)	44-0730-76
C_7H_9AsO			
4(1H)-Arseninone, 1-ethyl-	EtOH	230(3.88),314(3.93)	88-4143-76
$C_7H_9BrO_3$			
2H-Pyran-3(6H)-one, 4-bromo-2-methoxy-6-methyl-, (2R-trans)-	n.s.g.	255(3.89)	136-0093-76D
$C_7H_9ClO_3$			
2H-Pyran-3(6H)-one, 4-chloro-2-methoxy-6-methyl-, (2R-trans)-	n.s.g.	246(3.86)	136-0093-76D
$C_7H_9ClS_3$			
2,4,9-Trithiatricyclo[3.3.1.1^{3,7}]decane, 7-chloro-	hexane	247(3.37)	11-0220-76A
$C_7H_9Cl_3O_3$			
2,4-Pentanedione, 3-(2,2,2-trichloro-1-hydroxyethyl)-	$CHCl_3$	290(2.15)	18-1447-76
C_7H_9N			
Benzenamine, N-methyl-	H_2O	237(3.99),285(3.18)	35-7777-76

Compound	Solvent	λ_{max}(log ϵ)	Ref.
Benzenamine, 4-methyl-	H$_2$O	232(3.95),287(3.19)	35-7777-76
C$_7$H$_9$NO			
Benzenamine, 4-methoxy-	H$_2$O	233(3.88),295(3.26)	35-7777-76
Isoxazole, 4-ethenyl-3,5-dimethyl-	isooctane	230(4.07)	39-0570-76C
Methanol, (phenylamino)-	H$_2$O	238(4.01),283(3.14)	35-7777-76
Pyridinium, 5-hydroxy-1,2-dimethyl-, hydroxide, inner salt	MeOH	220(3.96),256(3.65), 336(3.45)	39-2285-76C
C$_7$H$_9$NOS			
Thiocyanic acid, 1-ethyl-2-methyl-3-oxo-1-propenyl ester, (E)-	EtOH	271(3.96)	97-0049-76
Thiocyanic acid, 2-formyl-1-methyl-1-butenyl ester, (E)-	EtOH	293(3.83)	97-0049-76C
C$_7$H$_9$NO$_2$			
Ethanone, 1-[4-(hydroxymethyl)-1H-pyrrol-3-yl]- (verrucarin E)	MeOH	249(3.94)	33-1698-76
C$_7$H$_9$NO$_3$			
Cyclohexanecarboxamide, 2,6-dioxo-	EtOH	257(4.25)	36-1794-76
	EtOH-base	257(4.26)	36-1794-76
C$_7$H$_9$NO$_4$			
2-Furancarboxaldehyde, 3,4-dimethoxy-, oxime	SDA 30	280(4.19)	87-0729-76
C$_7$H$_9$NS			
1,2-Benzisothiazole, 4,5,6,7-tetrahydro-	EtOH	218(3.587),254(3.679)	48-0507-76
C$_7$H$_9$NS$_3$			
5,6-Cyclopenta-1,4,2-dithiazine, 6,7-dihydro-3-(methylthio)-	C$_6$H$_{12}$	225s(3.8),250(4.06), 289(5.32),349(2.65)	48-0127-76
C$_7$H$_9$N$_3$OS			
4(1H)-Pyrimidinone, 6-amino-2-(2-propenylthio)-	EtOH	279(4.06)	95-1094-76
C$_7$H$_9$N$_3$O$_2$			
Acetamide, N-(1,4-dihydro-6-methyl-4-oxo-2-pyrimidinyl)-	dioxan	235(4.00),286(3.88)	114-0413-76A
C$_7$H$_9$N$_3$O$_2$S			
Acetic acid, [(4-amino-2-pyrimidinyl)-thio]-, methyl ester	pH 2.0	241(4.43)	24-3615-76
	pH 5.5 and 11	223(4.36),244s(4.01), 285(3.79)	24-3615-76
C$_7$H$_9$N$_3$O$_3$S			
Propanoic acid, 3-oxo-3-(1,3,4-thiadiazol-2-ylamino)-, ethyl ester	isoPrOH	250(3.94)	4-0291-76
C$_7$H$_9$N$_3$O$_4$			
1(2H)-Pyrimidinepropanoic acid, α-amino-3,4-dihydro-2,4-dioxo-	pH 1	264(3.93)	123-0598-76
C$_7$H$_9$N$_3$S			
Hydrazinecarbothioamide, N-phenyl-	EtOH	242(4.82),280s(3.70)	104-2374-76
C$_7$H$_9$N$_5$			
1H-Pyrazolo[3,4-d]pyrimidin-4-amine, N-ethyl-	pH 1	265(3.99)	87-0555-76

Compound	Solvent	λ_{max} (log ϵ)	Ref.
1H-Pyrazolo[3,4-d]pyrimidin-4-amine, N-ethyl- (cont.)	pH 13	268(3.98)	87-0555-76
	EtOH	278(4.10)	87-0555-76
$C_7H_9N_5O$			
1,2,4-Triazolo[3,2-c][1,2,4]triazin-7(4H)-one, 2,4,6-trimethyl-	EtOH	241(3.60),296(3.83)	22-1178-76
1,2,4-Triazolo[3,4-c][1,2,4]triazin-7(4H)-one, 1,4,6-trimethyl-	EtOH	245(3.92),333(3.84)	22-1178-76
1,2,4-Triazolo[4,3-b][1,2,4]triazin-7(1H)-one, 1,3,6-trimethyl-	EtOH	231s(3.74),292(3.53)	22-1178-76
$C_7H_9N_5O$			
2,3,4-Pentanetrione, 3-(1H-1,2,4-triazol-3-ylhydrazone)	EtOH	207(3.89),230s(3.65), 306(3.95)	39-0421-76C
C_7H_{10}			
1,3,5-Hexatriene, 2-methyl-	EtOH	248(4.54),257(4.65), 267(4.54)	39-1792-76C +118-0108-76
$C_7H_{10}BrNO_3$			
2H-Pyran-3(6H)-one, 4-bromo-2-methoxy-6-methyl-, oxime, (2R-trans)-	n.s.g.	242(4.04)	136-0093-76D
$C_7H_{10}BrN_3O$			
2-Pyrimidinamine, 4-bromo-6-propoxy-	MeOH	232(4.10),280(3.79)	94-0507-76
$C_7H_{10}ClNO_3$			
2H-Pyran-3(6H)-one, 4-chloro-2-methoxy-6-methyl-, oxime, (2R-trans)-	n.s.g.	241(4.12)	136-0093-76D
$C_7H_{10}ClN_3O$			
4-Pyrimidinamine, 2-chloro-6-(1-methylethoxy)-	MeOH	210(4.46),240(3.83)	94-0507-76
4-Pyrimidinamine, 2-chloro-6-propoxy-	MeOH	210(4.44),238(3.86)	94-0507-76
$C_7H_{10}Cl_2N_2S$			
2-Thiazolamine, 4,5-dichloro-N,N-diethyl-	EtOH	240s(3.45),280(3.96)	4-1297-76
$C_7H_{10}NO$			
Pyridinium, 1-methoxy-2-methyl-, perchlorate	DMF	353(4.18)	83-0592-76
	DMSO	356(4.17)	83-0592-76
	HMPT	350(4.35)	83-0592-76
Pyridinium, 1-methoxy-3-methyl-, methosulfate	DMF	355(4.37)	83-0592-76
	DMSO	363(4.39)	83-0592-76
	HMPT	356(4.50)	83-0592-76
Pyridinium, 1-methoxy-4-methyl-, methosulfate	DMF	354(4.32)	83-0592-76
	DMSO	362(4.23)	83-0592-76
	HMPT	355(4.30)	83-0592-76
$C_7H_{10}NO_2$			
8-Azabicyclo[3.2.1]oct-8-yloxy, 3-oxo-	CHCl$_3$	240(3.48),470(0.93)	78-0239-76
$C_7H_{10}N_2$			
2,4-Cyclopentadien-1-one, dimethylhydrazone	hexane	327(4.41)	39-0048-76C
1H-Pyrazole, 1-ethenyl-3,5-dimethyl-	pH 1	219(3.87),252(3.98)	103-0691-76
	EtOH	227(3.85),255(4.18)	103-0691-76
	dioxan	256(4.19)	103-0691-76
1H-Pyrrole-2-carboxamide, N,N-dimethyl-	MeOH	263(4.08)	44-3050-76

Compound	Solvent	$\lambda_{max}(\log \epsilon)$	Ref.
$C_7H_{10}N_2O$			
2(1H)-Pyrazinone, 3,5,6-trimethyl-	MeOH	228(3.93),332(3.87)	87-0165-76
	MeOH-HCl	227(3.97),326(3.76)	87-0165-76
	MeOH-NaOH	233(4.03),337(3.89)	87-0165-76
Pyrazolo[1,5-a]pyridin-2(1H)-one, 4,5,6,7-tetrahydro-	MeOH	229(3.72),260(3.48)	18-1980-76
2H-Pyrrole-5-carbonitrile, 3,4-dihydro-2,2-dimethyl-, 1-oxide	MeOH	271(4.02)	39-1951-76C
$C_7H_{10}N_2OS$			
2-Benzothiazolamine, 4,5,6,7-tetrahydro-, 3-oxide	MeOH	268(3.79)	64-0251-76B
4(1H)-Pyrimidinone, 2,3-dihydro-6-propyl-2-thioxo-	EtOH	276(4.24)	95-1094-76
4(1H)-Pyrimidinone, 2-(ethylthio)-6-methyl-	EtOH	234(3.94),287(3.84)	95-1094-76
4(1H)-Pyrimidinone, 2-(propylthio)-	EtOH	233(3.92),290(3.88)	95-1094-76
$C_7H_{10}N_2O_2$			
3,5-Heptanedione, 4-diazo-	hexane	228(4.22),239s(4.11)	104-1245-76
	pH 1	230(4.22),240s(4.19)	104-1245-76
	H_2O	230(4.22),240s(4.18)	104-1245-76
	NaOH	229(4.16),241s(4.11), 270(3.83)	104-1245-76
4-Oxa-1,2-diazaspiro[4.5]dec-1-en-3-one	n.s.g.	231(3.59),366(2.39)	4-0649-76
3-Pyridinemethanol, 4-amino-5-hydroxy-6-methyl-, hydrochloride	pH 1	276(4.04)	87-0999-76
	pH 7	283(3.78),308(3.91)	87-0999-76
	pH 13	298(3.79)	87-0999-76
$C_7H_{10}N_2O_3$			
1H-Pyrazole-3-carboxylic acid, 1-acetyl-4,5-dihydro-, methyl ester	MeOH	280(4.29)	103-1026-76
1H-Pyrazole-3-carboxylic acid, 2,5-dihydro-1,2-dimethyl-5-oxo-	MeOH	225(3.95),304(3.62)	24-2154-76
1H-Pyrazole-3-carboxylic acid, 2,5-dihydro-1-methyl-5-oxo-, ethyl ester	EtOH	<u>225(4.0),300(3.2)</u>	23-0488-76
1H-Pyrazole-3-carboxylic acid, 2,5-dihydro-4-methyl-5-oxo-, ethyl ester	EtOH	<u>230(4.1),280(3.5)</u>	23-0488-76
1H-Pyrazole-3-carboxylic acid, 4-(hydroxymethyl)-, ethyl ester	MeOH	221(3.91),238s(3.78)	103-1029-76
1H-Pyrazole-3-carboxylic acid, 5-(hydroxymethyl)-, ethyl ester	MeOH	221(3.98)	103-1029-76
1H-Pyrazole-3-carboxylic acid, 5-methoxy-1-methyl-, methyl ester	MeOH	223(4.03),250s(3.52), 303(2.60)	24-0261-76
1H-Pyrazole-5-carboxylic acid, 3-methoxy-1-methyl-, methyl ester	MeOH	225(4.08),269(3.58)	24-0268-76
4(1H)-Pyrimidinone, 2,6-dimethoxy-5-methyl-	MeOH	230(3.57),270(3.89)	94-0026-76
4(1H)-Pyrimidinone, 2-ethoxy-6-methoxy-	MeOH	224(3.61),262(3.91)	94-0026-76
4(1H)-Pyrimidinone, 6-ethoxy-2-methoxy-	MeOH	223(3.60),260(3.92)	94-0026-76
$C_7H_{10}N_2O_4$			
1H-Pyrazole-3-carboxylic acid, 5-hydroxy-4-(2-hydroxyethyl)-1-methyl-	MeOH	224(3.95),262(3.52)	24-0261-76
2,4(1H,3H)-Pyrimidinedione, 5-(dimethoxymethyl)-	EtOH	259(3.63)	87-0909-76
$C_7H_{10}N_4O$			
1H-Imidazole-4-carboxamide, 5-amino-1-(2-propenyl)-	EtOH	267(4.1)	2-0346-76

Compound	Solvent	λ_{max} (log ϵ)	Ref.
$C_7H_{10}N_4OS$			
1H-Imidazole-4-carboxamide, 5-amino-2,3-dihydro-1-(2-propenyl)-2-thioxo-	EtOH	275s(--),304(4.37)	2-0351-76
Propanamide, 3-[(4-amino-2-pyrimidinyl)-thio]-	pH 2.0	240(4.29)	24-3615-76
	pH 5.3	225(4.21),246s(4.07), 279(3.88)	24-3615-76
4-Thiazolecarboxamide, 5-amino-2-(2-propenylamino)-	EtOH	264(3.96),310(3.86)	2-0351-76
$C_7H_{10}N_4O_2$			
Lumazine, 5,6,7,8-tetrahydro-1-methyl-	pH 2.0	265(4.26)	24-3184-76
	pH 7.0	244(3.62),298(4.02)	24-3184-76
	pH 13.0	295(3.82)	24-3184-76
$C_7H_{10}N_4O_2S_3$			
3-Thiazolidinepropanoic acid, 4-[(aminothioxomethyl)hydrazono]-2-thioxo-	MeOH	255(4.27),295(4.12)	103-0749-76
$C_7H_{10}N_8O$			
1,2,4-Triazolo[4,3-b]pyridazine-7-acetic acid, 6-hydrazino-, hydrazide	MeOH	221(4.36),283(3.78)	4-1155-76
$C_7H_{10}N_8O_5S_2$			
1H-Imidazole-4-sulfonamide, 5,5'-(carbonyldiimino)bis-	H_2O	234(4.19)	103-0924-76
$C_7H_{10}O$			
Bicyclo[2.2.1]heptan-2-one	3-Mepentane	295(1.36)	144-0475-76
	MeCN	289(1.39)	144-0475-76
Bicyclo[3.2.0]heptan-3-one	isooctane	290(1.62),300(1.72), 311(1.68),323(1.39)	77-0338-76
2-Butenal, 4-cyclopropyl-, trans	EtOH	218(4.28)	138-1297-76
2-Propenal, 3-(2-methylcyclopropyl)-	EtOH	251(4.23)	70-0428-76
$C_7H_{10}OS_3$			
2,4,9-Trithiatricyclo[3.3.1.1^{3,7}]decan-7-ol	hexane	245(3.42)	11-0220-76A
$C_7H_{10}OSe$			
Selenophene, 3-methoxy-2,5-dimethyl-	C_6H_{12}	245(3.9),267s(3.7)	11-0126-76B
3(2H)-Selenophenone, 2,2,5-trimethyl-	C_6H_{12}	322(3.8)	11-0126-76B
$C_7H_{10}O_2$			
Cyclohexanone, 2-(hydroxymethylene)-	EtOH	280(4.28),370(2.09), 390(2.05)	65-2037-76
$C_7H_{10}O_3$			
DL-Hex-2-enopyranos-4-ulose, 5-C-methyl-2,3,6-trideoxy-	n.s.g.	207(3.86)	78-1051-76
$C_7H_{10}S_3$			
2,4,9-Trithiatricyclo[3.3.1.1^{3,7}]decane	hexane	248(3.39)	11-0220-76A
$C_7H_{11}NO$			
Cyclohexanone, 2-(aminomethylene)-	C_6H_{12}	302(4.00)	35-2826-76
	MeOH	316(4.23)	35-2826-76
4(1H)-Pyridinone, 2,3-dihydro-2,6-dimethyl-	EtOH	303(4.12)	4-0253-76

Compound	Solvent	$\lambda_{max}(\log \epsilon)$	Ref.
$C_7H_{11}NOS_2$			
Ethanethioic acid, S-(4,5-dihydro-2,5-dimethyl-5-thiazolyl) ester	EtOH	233(3.48)	70-1353-76
$C_7H_{11}NO_2$			
1H-Azepine-2,5-dione, tetrahydro-7-methyl-	n.s.g.	283(1.43)	44-0400-76
4-Isoxazolemethanol, $\alpha,3,5$-trimethyl-	H_2O	220(3.72)	39-0570-76C
3H-Pyrrol-3-one, 4,5-dihydro-2,5,5-trimethyl-, 1-oxide	EtOH	272(4.26)	12-2561-76
$C_7H_{11}NO_2S$			
Acetic acid, isothiocyanato-, butyl	pH 8.06	244(3.06)	73-2216-76
ester	dioxan	248(3.09)	73-2216-76
Acetic acid, isothiocyanato-, 1,1-dimethylethyl ester	pH 8.06	245(3.17)	73-2216-76
	dioxan	251(3.29)	73-2216-76
Butanoic acid, 2-isothiocyanato-,	pH 8.06	245(3.07)	73-2216-76
ethyl ester	MeOH	245(3.20)	73-2216-76
	dioxan	249(3.10)	73-2216-76
5-Thiazolecarboxylic acid, 2,3-dihydro-2,2-dimethyl-, methyl ester	EtOH	211(3.83),292(3.38), 335(3.76)	39-0584-76C
$C_7H_{11}NS$			
Isothiazole, 5-methyl-4-propyl-	EtOH	218(3.327),252(3.522)	48-0507-76
Thiazole, 4-ethyl-2,5-dimethyl-	EtOH	252(3.26)	70-1353-76
$C_7H_{11}N_2O$			
Pyrimidinium, 1-ethyl-2,3-dihydro-3-methyl-2-oxo-	pH 9.18	228(4.23),296(3.27)	23-2681-76
cation iodide	pH 0.29	225(4.16),316(3.93)	23-2681-76
$C_7H_{11}N_3$			
Pyrrolidine, 1-(2-cyano-1-iminoethyl)-	pH 7.2	210(3.98)	39-0125-76C
	pH 12.0	260(4.27)	39-0125-76C
$C_7H_{11}N_3OS$			
4(1H)-Pyrimidinone, 6-amino-2-(propylthio)-	EtOH	232(4.29),277(3.94)	95-1094-76
$C_7H_{11}N_3O_2$			
Ethanol, 2-[(6-methoxy-4-pyrimidinyl)-amino]-	pH 1	256(4.10)	39-1847-76C
	pH 13	246(4.01)	39-1847-76C
4(1H)-Pyrimidinone, 6-amino-5-ethyl-2-methoxy-	MeOH	211(4.20),270(4.16)	94-0026-76
$C_7H_{11}N_3O_5S$			
1,2,4-Thiadiazol-3(2H)-one, 5-amino-2-β-D-ribofuranosyl-	pH 1	220(4.17),273(3.70)	4-0169-76
	pH 7	220(4.14),255(3.80)	4-0169-76
	pH 11	230(4.05),278s(3.88)	4-0169-76
$C_7H_{11}N_4OS$			
Pyrimidinium, 4-amino-2-(carbamoylmethylthio)-1-methyl-, iodide	pH 5.3	238(4.51)	24-3615-76
$C_7H_{12}N_2$			
Propanenitrile, 2-[(1,1-dimethylethyl)-imino]-	C_6H_{12}	295(2.06)	35-1099-76
$C_7H_{12}N_2O$			
3-Cyclohexen-1-amine, N-methyl-N-nitroso-	MeOH	340(1.98)	23-2385-76

Compound	Solvent	$\lambda_{max}(\log \epsilon)$	Ref.
2-Cyclopentene-1-methanamine, N-methyl- N-nitroso-	MeOH-HCl	345(1.95)	23-2385-76
2(1H)-Pyrimidinone, 3,4-dihydro-4,4,6- trimethyl-	pH 5 MeOH	246(3.28) 236(3.51)	23-2681-76 23-2681-76
$C_7H_{12}N_2O_2$			
3(2H)-Isoxazolone, 4-(2-aminopropyl)- 5-methyl-, hydrobromide	MeOH	211(3.77)	1-0567-76
4-Oxazolemethanamine, 3-methoxy-α,5-di- dimethyl-, hydrochloride	MeOH	212(3.66)	1-0567-76
2H-Pyrrole-5-carboxaldehyde, 3,4-di- hydro-2,2-dimethyl-, oxime, 1-oxide	EtOH	292(4.09)	39-1951-76C
2H-Pyrrole-5-carboxaldehyde, 3,4-di- hydro-3,3-dimethyl-, oxime, 1-oxide	MeOH	292(4.07)	39-1951-76C
3H-Pyrrol-3-one, 4,5-dihydro-2,5,5-tri- methyl-, oxime, 1-oxide	EtOH	273(4.33)	12-2561-76
$C_7H_{12}N_2O_4$			
2-Butenedioic acid, 2-(1-methylhydra- zino)-, dimethyl ester, (E)-	MeOH	280(4.35)	24-0253-76
$C_7H_{12}N_2S$			
1H-Imidazole, 1-(1-methylethyl)-2-(meth- ylthio)-, monohydriodide	EtOH CH_2Cl_2	219-252(4.336-3.796) 231-255(4.383-3.851)	28-0143-76B 28-0143-76B
2H-Imidazole-2-thione, 1,3-dihydro- 1-methyl-3-(1-methylethyl)-	EtOH CH_2Cl_2	262(4.119) 266(4.127)	28-0143-76B 28-0143-76B
2H-Imidazole-2-thione, 1-(1,1-dimethyl- ethyl)-1,3-dihydro-	EtOH CH_2Cl_2	267(4.130) 275(4.165)	28-0143-76B 28-0143-76B
$C_7H_{12}N_2S_2$			
Methanaminium, N-(4,4-dimethyl-2-thioxo- 5-thiazolidinylidene)-N-methyl-, hydroxide, inner salt	EtOH $CHCl_3$	273(4.11),340s(2.69) 280(4.13),341s(2.82)	33-2566-76 33-2566-76
5(4H)-Thiazolethione, 2-(dimethylamino)- 4,4-dimethyl-	EtOH	293(4.05),330s(3.38), 478(1.23)	33-2566-76
$C_7H_{12}O$			
1-Penten-3-one, 4,4-dimethyl-	MeOH	219(3.78),329(1.40)	40-0166-76
$C_7H_{12}O_2$			
Cycloheptanone, 2-hydroxy-	MeOH	274(1.40)	104-2286-76
Cyclohexanecarboxylic acid	hexane H_2O MeCN	207.0(1.94) 208s(--) 218.8(1.82)	44-3312-76 44-3312-76 44-3312-76
Cyclopentanol, acetate	neat hexane MeCN	296.9(-1.07) 212.2(1.80) 210.4(1.79)	44-3312-76 44-3312-76 44-3312-76
$C_7H_{12}O_3$			
Pentanoic acid, 2-ethyl-3-oxo-	hexane H_2O	265(3.10),305s(2.26) 282(2.38)	104-1252-76 104-1252-76
$C_7H_{13}BrN_2O_2$			
3(2H)-Isoxazolone, 4-(2-aminopropyl)- 5-methyl-, hydrobromide	MeOH	211(3.77)	1-0567-76
$C_7H_{13}ClN_4O$			
Pyrimidinium, 4-amino-2-[(2-hydroxy- ethyl)amino]-1-methyl-, chloride	H_2O	216(4.45),240s(4.17), 274(3.82)	5-1395-76
	10M NaOH	240(4.44),307(3.38)	5-1395-76

Compound	Solvent	$\lambda_{max}(\log \epsilon)$	Ref.
$C_7H_{13}N$			
2H-Pyrrole, 2-ethyl-3,4-dihydro-5-methyl-	ether	228(2.00)	4-0405-76
$C_7H_{13}NO$			
3-Buten-2-one, 4-[(1-methylethyl)amino]-	H_2O	298(4.45)	35-2826-76
	MeOH	299(4.36)	35-2826-76
cis-s-cis	C_6H_{12}	300(4.23)	35-2826-76
1-Penten-3-one, 1-amino-4,4-dimethyl-	C_6H_{12}	282(4.02)	35-2826-76
	MeOH	294(4.13)	35-2826-76
1-Penten-3-one, 1-(dimethylamino)-	C_6H_{12}	288(4.33)	35-2826-76
	H_2O	306(4.47)	35-2826-76
	MeOH	303(4.40)	35-2826-76
1-Penten-3-one, 2-methyl-1-(methyl-amino)-	H_2O	304(4.40)	35-2826-76
	MeOH	300(4.37)	35-2826-76
cis-s-cis	C_6H_{12}	316(4.05)	35-2826-76
trans	C_6H_{12}	282(4.32)	35-2826-76
1-Penten-3-one, 4-methyl-1-(methyl-amino)-	H_2O	300(4.35)	35-2826-76
	MeOH	299(4.32)	35-2826-76
cis-s-cis	C_6H_{12}	299(4.24)	35-2826-76
3-Penten-2-one, 4-(dimethylamino)-	C_6H_{12}	293(4.38)	35-2826-76
	MeOH	310(4.46)	35-2826-76
3-Penten-2-one, 3-methyl-4-(methyl-amino)-	H_2O	328(4.00)	35-2826-76
	MeOH	330(4.16)	35-2826-76
cis-s-cis	C_6H_{12}	322(4.11)	35-2826-76
2-Propenal, 3-(1,1-dimethylethyl)amino-	H_2O	286(4.60)	35-2826-76
	MeOH	285(4.54)	35-2826-76
cis-s-cis	C_6H_{12}	305(4.18)	35-2826-76
trans	C_6H_{12}	269(4.48)	35-2826-76
2H-Pyrrole, 3,4-dihydro-2,2,5-trimethyl-, 1-oxide	EtOH	229(3.86)	44-0855-76
$C_7H_{13}NO_4$			
1,3-Dioxane, 2-(1-methylethyl)-5-nitro-, cis	heptane	214(3.39),278(1.52)	35-0956-76
trans	heptane	214(3.39),282(1.49)	35-0956-76
$C_7H_{13}N_2S$			
1H-Imidazolium, 1-ethyl-3-methyl-2-(methylthio)-, iodide	EtOH	220-260(4.243-3.522)	28-0143-76B
	CH_2Cl_2	243(4.070)	28-0143-76B
$C_7H_{13}N_3S$			
2H-Imidazole-2-thione, 4-(dimethylami-no)-1,5-dihydro-5,5-dimethyl-	EtOH	285(4.50)	33-2566-76
$C_7H_{13}N_4O$			
Pyrimidinium, 4-amino-2-[(2-hydroxy-ethyl)amino]-1-methyl-, chloride	H_2O	216(4.45),240s(4.17), 274(3.82)	5-1395-76
	10M NaOH	240(4.44),307(3.38)	5-1395-76
$C_7H_{14}ClNO$			
Oxazolidine, 3-chloro-2,2,4,4-tetra-methyl-	H_2O	265(2.36)	36-1733-76
$C_7H_{14}N_2O$			
Diazene, cyclohexylmethyl-, 1-oxide, (E)-	EtOH	232(3.77)	44-1146-76
(Z)-	EtOH	214(3.78),280s(1.30)	44-1146-76
Diazene, cyclohexylmethyl-, 2-oxide, (E)-	EtOH	233(3.90)	44-1146-76
(Z)-	EtOH	222(3.88)	44-1146-76

Compound	Solvent	$\lambda_{max}(\log \epsilon)$	Ref.
2-Propenamide, 3-(dimethylamino)-N,N-dimethyl-	C_6H_{12}	273(4.34)	78-1025-76
	MeOH	285(4.43)	78-1025-76
	dioxan	278(4.40)	78-1025-76
2-Propenamide, 3-(ethylamino)-N,N-dimethyl-	C_6H_{12}	283(4.35)	78-1025-76
	MeOH	279(4.42)	78-1025-76
	dioxan	283(4.25)	78-1025-76
$C_7H_{14}N_2O_2$			
Pyruvic acid, ethyl ester, dimethyl-hydrazone	hexane	311(3.53)	104-0957-76
	EtOH	308(3.59)	104-0957-76
$C_7H_{14}N_2S$			
1H-Imidazole, 4,5-dihydro-1-(1-methyl-ethyl)-2-(methylthio)-, hydriodide	EtOH	220(4.392)	28-0485-76A
	CH_2Cl_2	237(4.381)	28-0485-76A
2-Imidazolidinethione, 1-(1,1-dimethyl-ethyl)-	EtOH	244(4.118)	28-0485-76A
	CH_2Cl_2	255(4.102)	28-0485-76A
Pyrimidine, 1-ethyl-1,4,5,6-tetrahydro-2-(methylthio)-, hydriodide	EtOH	221(4.444)	28-0485-76A
	CH_2Cl_2	233(4.393)	28-0485-76A
2(1H)-Pyrimidinethione, tetrahydro-1-(1-methylethyl)-	EtOH	246.5(4.145)	28-0485-76A
	CH_2Cl_2	255.5(4.145)	28-0485-76A
$C_7H_{14}O$			
3-Pentanone, 2,2-dimethyl-	C_6H_{12}	284(1.48)	112-0777-76
3-Pentanone, 2,4-dimethyl-	C_6H_{12}	288(1.42)	112-0777-76
$C_7H_{14}O_2$			
Acetic acid, 2,2-dimethylpropyl ester	neat	277.6(-0.94)	44-3312-76
	hexane	208.7(1.82)	44-3312-76
	MeCN	209.1(1.81)	44-3312-76
Acetic acid, 2-methylbutyl ester	hexane	211.6(1.76)	44-3312-76
	MeCN	208.7(1.73)	44-3312-76
Acetic acid, 3-methylbutyl ester	neat	313.4(-0.97)	44-3312-76
	hexane	212.1(1.78)	44-3312-76
	H_2O	201.3(1.78)	44-3312-76
	MeCN	209.4(1.77)	44-3312-76
Acetic acid, pentyl ester	hexane	212.6(1.73)	44-3312-76
	H_2O	202.5(1.76)	44-3312-76
	MeCN	209.2(1.80)	44-3312-76
Acetic acid, tert-pentyl ester	hexane	215.9(1.80)	44-3312-76
	H_2O	210.0(1.76)	44-3312-76
	MeCN	213.7(1.83)	44-3312-76
Butanoic acid, 2-methyl-, ethyl ester	neat	275s(--)	44-3312-76
	hexane	214.0(1.89)	44-3312-76
	H_2O	207.9(2.01)	44-3312-76
	MeCN	212.2(1.95)	44-3312-76
Butanoic acid, 3-methyl-, ethyl ester	hexane	208.3(1.84)	44-3312-76
	H_2O	205.4(1.93)	44-3312-76
	MeCN	208.2(1.89)	44-3312-76
Pentanoic acid, ethyl ester	neat	271.8(-1.30)	44-3312-76
	hexane	212.4(1.81)	44-3312-76
	H_2O	205.9(1.86)	44-3312-76
	MeCN	211.8(1.83)	44-3312-76
Propanoic acid, 2,2-dimethyl-, ethyl ester	hexane	215.0(2.00)	44-3312-76
	H_2O	208.1(1.99)	44-3312-76
	MeCN	214.3(1.98)	44-3312-76
$C_7H_{15}N_2O_4P$			
Phosphorohydrazidic acid, (2-oxopropyli-dene)-, diethyl ester, (E)-	EtOH	261.5(4.15)	65-1670-76

Compound	Solvent	$\lambda_{max}(\log \epsilon)$	Ref.
$C_7H_{15}N_5$			
Pyrimidinium, 4-amino-2-(2-ammonioeth-ylamino)-1-methyl-, dichloride	H_2O	215(4.35),240s(4.04), 274(3.81)	5-1395-76
	10M NaOH	239(4.37),308(3.31)	5-1395-76
$C_7H_{16}NO_4P$			
Phosphonic acid, (1-methyl-1-nitroso-ethyl)-, diethyl ester	EtOH	677(1.23)	70-0465-76
$C_7H_{16}N_2O$			
Butanal, 4-hydroxy-, propylhydrazone	EtOH	234(3.64)	104-2034-76
Pentanal, 5-hydroxy-, dimethylhydrazone	EtOH	240(3.98)	104-2034-76
Pentanal, 5-hydroxy-, ethylhydrazone	EtOH	235(3.51)	104-2034-76
$C_7H_{16}OSn$			
2-Butanone, 4-(trimethylstannyl)-	C_6H_{12}	282(1.52)	35-0104-76
	CCl_4	242(2.24),288(1.66)	35-0104-76
$C_7H_{17}N_2OPSe$			
Phosphorodiamidoselenoic acid, tetra-methyl-, O-2-propenyl ester	hexane or EtOH	192s(4.13),225s(3.80)	70-0174-76
$C_7H_{17}N_2O_3P$			
Formaldehyde, (triethoxyphosphoranyli-dene)hydrazone	EtOH	228.0(3.46)	65-1424-76
$C_7H_{18}N_2Si$			
Diazene, (1,1-dimethylethyl)(trimethyl-silyl)-	heptane	200(3.03),500(0.95)	35-0109-76
$C_7H_{18}N_2Si_2$			
Diazomethane, bis(trimethylsilyl)-	n.s.g.	249(3.80),359(1.13)	64-1317-76B
Silanamine, N-isocyano-1,1,1-trimethyl-N-(trimethylsilyl)-	n.s.g.	239(3.49)	64-1317-76B
$C_7H_{18}Sn$			
Stannane, butyltrimethyl-	CCl_4	242(2.19)	35-0104-76
$C_7H_{19}N_6PS$			
2-Propanone, methyl(tetrahydro-2,4-di-methyl-1,2,4,5,3-tetraazaphosphorin-3(2H)-yl)hydrazone, P-sulfide	$CHCl_3$	242(3.00)	78-2633-76

Compound	Solvent	$\lambda_{max}(\log \epsilon)$	Ref.

$C_8Cl_2F_6$
  Bicyclo[4.2.0]octa-1,3,5-triene, 7,7-di-
  chloro-2,3,4,5,8,8-hexafluoro-
	heptane	256.4(2.43)	104-0818-76

$C_8Cl_4F_4$
  Bicyclo[4.2.0]octa-1,3,5-triene,
  2,3,4,5-tetrachloro-7,7,8,8-
  tetrafluoro-
	heptane	235(3.02),242(3.02),
		248(3.02),275(2.45),
		283(2.54),294s(2.20)
			104-0818-76

Bicyclo[4.2.0]octa-1,3,5-triene,
  7,7,8,8-tetrachloro-2,3,4,5-
  tetrafluoro-
	heptane	261(2.64)	104-0818-76

C_8Cl_7N 3H-Indole, 2,3,3,4,5,6,7-heptachloro-	C_6H_{12}	241(4.34),249(4.44), 257(4.43)	39-0162-76C

$C_8F_4O_2$
  Bicyclo[4.2.0]octa-1,3,5-triene-7,8-di-
  one, 2,3,4,5-tetrafluoro-
	CCl_4	274s(--),287(3.23),
		297(3.23),369(2.54),
		391s(2.38)
			104-0818-76

C_8F_6O
  Bicyclo[4.2.0]octa-1,3,5-trien-7-one,
  2,3,4,5,8,8-hexafluoro-
	heptane	241s(3.82),246(3.84),
		262(3.02),269(3.01),
		281(2.98)
			104-0818-76

C_8HCl_6N 1H-Indole, 2,3,4,5,6,7-hexachloro-	C_6H_{12}	293(4.00)	39-0162-76C

$C_8H_2CrO_5S_3$
  Chromium, pentacarbonyl(1,3-dithiole-
  2-thione-S^2)-, (OC-6-22)-
	benzene	327(3.85),363(3.74),
		408(3.51),495(3.90)
			101-0369-76Q

$C_8H_2FeO_6$
  Iron, tetracarbonyl(1,4-dioxo-2-butene-
  1,4-diyl)-, (OC-6-22)-
	EtOH	256(3.80),440(1.72)	101-0235-76H

$C_8H_2MoO_5S_3$
  Molybdenum, pentacarbonyl(1,3-dithiole-
  2-thione-S^2)-, (OC-6-22)-
	benzene	323(3.83),362(3.87),
		470(3.98)
			101-0369-76Q

$C_8H_3Cl_4N$ 1H-Indole, 3,5,6,7-tetrachloro-	C_6H_{12}	286(3.79),294(3.76), 302(3.67),307(3.62)	39-0162-76C
1H-Indole, 4,5,6,7-tetrachloro-	C_6H_{12}	283(3.91),295(3.80), 306(3.46)	39-0162-76C
1H-Isoindole, 4,5,6,7-tetrachloro-	CH_2Cl_2	234(4.65),268(3.26), 277(3.18),293(3.09), 332(3.60),344(3.70), 356(3.60)	88-1661-76

$C_8H_3F_3N_6O$
  Acetamide, N-(3-cyano-1H-pyrazolo[3,4-
  d]pyrimidinyl)-
	EtOH-pH 1	238(3.81),246(3.81),
		254(3.82),306(4.16),
		314s(--)
			69-1005-76

	EtOH-pH 7	255(3.83),305(4.14), 315s(--)	69-1005-76
	EtOH-pH 11	303(3.88)	69-1005-76

Compound	Solvent	$\lambda_{max}(\log \epsilon)$	Ref.
$C_8H_3N_7$			
1H-Imidazo[1,5-b][1,2,4]triazole-2,5,7-tricarbonitrile, 1-methyl-	EtOH	221(4.35),264(4.10), 298(3.82)	44-1889-76
3H-Imidazo[1,5-b][1,2,4]triazole-2,5,7-tricarbonitrile, 3-methyl-	EtOH	217(4.34),240(4.18), 248(4.22),271(3.49), 325(4.22),336(4.32)	44-1889-76
$C_8H_4ClF_3$			
Benzene, 1-chloro-3-fluoro-2-(2,2-difluoroethenyl)-	EtOH	231(3.92),274(3.03)	12-1435-76
$C_8H_4Cl_2N_2$			
Phthalazine, 1,4-dichloro-	dioxan	260(3.71)	103-0342-76
$C_8H_4Cl_3N$			
1H-Indole, 5,6,7-trichloro-	C_6H_{12}	279(3.86),292(3.75), 304(3.72)	39-0162-76C
$C_8H_4CrO_5S_3$			
Chromium, pentacarbonyl(1,3-dithiolane-2-thione-S^2)-, (OC-6-22)-	benzene	273(4.15),395(3.57), 490(3.95)	101-0369-76Q
$C_8H_4F_3NO_2$			
Benzene, 1-fluoro-2-(2,2-difluoroethenyl)-4-nitro-	EtOH	236(4.23),265s(3.90)	12-1435-76
$C_8H_4F_6O$			
7-Oxabicyclo[2.2.1]hepta-2,5-diene, 2,3-bis(trifluoromethyl)-	EtOH	229s(2.42),258s(2.08)	24-2823-76
7-Oxabicyclo[4.1.0]hepta-2,4-diene, 2,5-bis(trifluoromethyl)-	EtOH	257(3.69)	24-2823-76
$C_8H_4MoO_5S_3$			
Molybdenum, pentacarbonyl(1,3-dithiolane-S^2)-, (OC-6-22)-	benzene	294(4.00),376(3.57), 465(3.95)	101-0369-76Q
$C_8H_4N_2O$			
Propanedinitrile, (2-furanylmethylene)-	C_6H_{12}	338(4.23)	80-0781-76
	EtOH	341(4.25)	80-0781-76
	acetone	340(4.25)	80-0781-76
	CCl_4	340(4.25)	80-0781-76
	$CHCl_3$	345(4.37)	80-0781-76
non-ion	H_2O	345(4.38),476(3.79)	80-0781-76
ion	H_2O	277(3.97),578(3.91)	80-0781-76
$C_8H_5BrN_2$			
1,7-Naphthyridine, 3-bromo-	EtOH	217(4.18),234(4.50), 268(3.53),304(3.31), 317(3.27)	4-0961-76
1,7-Naphthyridine, 4-bromo-	EtOH	214(4.66),269(3.81), 307(3.45),318(3.39)	56-0451-76
$C_8H_5BrN_2O$			
4(1H)-Quinazolinone, 6-bromo-	pH 0.29	266(3.97),293(3.72), 302(3.56)	44-0838-76
$C_8H_5BrN_4O_3$			
Ethanone, 1-(3-azido-5-nitrophenyl)-2-bromo-	MeOH	245(4.37),333(3.36)	64-0389-76C

Compound	Solvent	λ_{max}(log ϵ)	Ref.
C$_8$H$_5$ClN$_2$			
1,7-Naphthyridine, 4-chloro-	EtOH	213(3.20),267(3.52), 307(3.20),318(3.18)	56-0451-76
Phthalazine, 1-chloro-	dioxan	266(3.64)	103-0342-76
C$_8$H$_5$ClN$_2$O			
1(2H)-Phthalazinone, 4-chloro-	dioxan	258(4.00),294(3.79)	103-0342-76
C$_8$H$_5$Cl$_2$F			
Benzene, (2,2-dichloro-1-fluoroethenyl)-	n.s.g.	207(4.10),256(4.13)	44-1487-76
C$_8$H$_5$Cl$_4$NO			
Acetamide, N-[(2,3,4,5-tetrachloro-2,4-cyclopentadien-1-ylidene)methyl]-	C$_6$H$_{12}$	247(3.31),328(4.49), 390s(2.97)	24-3939-76
C$_8$H$_5$F$_3$N$_4$			
1,2,4-Benzotriazin-3-amine, 6-(trifluoromethyl)-	EtOH	203(4.28),236(4.38), 252s(4.07),280s(3.33)	107-0457-76
C$_8$H$_5$F$_3$O			
Benzaldehyde, 4-(trifluoromethyl)-	1% EtOH-pH 10	246(3.94)	44-1957-76
C$_8$H$_5$NOS			
2-Benzothiazolecarboxaldehyde	1% EtOH	217(4.26),255(3.85), 296(3.30)	44-1952-76
C$_8$H$_5$NO$_2$			
2-Propynoic acid, 3-(3-pyridinyl)-	pH 1	204(4.27),245(4.02), 280(3.88)	104-0436-76
C$_8$H$_5$NO$_4$			
1(3H)-Isobenzofuranone, 6-nitro-	buffer	255(4.0)	130-0037-76
C$_8$H$_6$BrNO$_2$			
Benzene, (2-bromo-2-nitroethenyl)-, cis	EtOH	246(4.08),327(3.28)	44-2112-76
trans	EtOH	226(3.95),324(4.08)	44-2112-76
C$_8$H$_6$BrNO$_5$			
2-Propenoic acid, 3-(5-bromo-2-furanyl)-2-nitro-, methyl ester	n.s.g.	321(4.315)	73-2422-76
C$_8$H$_6$BrN$_3$			
Methanimidamide, N-(4-bromophenyl)-N'-cyano-	EtOH	283(4.37),304s(4.15)	48-0347-76
C$_8$H$_6$BrN$_3$S			
1,2,3-Thiadiazol-5-amine, N-(4-bromophenyl)-	dioxan	255(4.03),325(4.28)	73-1182-76
1H-1,2,3-Triazole-5-thiol, 1-(4-bromophenyl)-	dioxan	255(3.98)	73-1551-76
C$_8$H$_6$Br$_2$O$_2$			
3,5-Cyclooctadiene-1,2-dione, 3,7-dibromo-	C$_6$H$_{12}$	212(3.81),235(3.58), 299(3.88)	44-3760-76
C$_8$H$_6$ClNO$_2$			
Benzene, (2-chloro-2-nitroethenyl)-, cis	EtOH	223(4.11),320(3.60)	44-2112-76
trans	EtOH	226(4.11),320(4.28)	44-2112-76

Compound	Solvent	$\lambda_{max}(\log \epsilon)$	Ref.
$C_8H_6ClN_3$ Methanimidamide, N-(4-chlorophenyl)- N'-cyano-	EtOH	281(4.36),300s(4.20)	48-0347-76
$C_8H_6Cl_2O_2$ Ethanone, 1-(3,5-dichloro-2-hydroxy- phenyl)-	EtOH CCl$_4$ MeCN	260(3.73),347(3.48) 264(4.23),350(4.00) 255(4.00),340(3.81)	97-0365-76 97-0365-76 97-0365-76
$C_8H_6Cl_2O_2S_2$ 4-Cyclopentene-1,3-dione, 2-[bis(meth- ylthio)methylene]-4,5-dichloro-	CH$_2$Cl$_2$	253(4.13),330s(3.92), 360(4.12)	83-0034-76
$C_8H_6Cl_2O_4$ 4-Cyclopentene-1,3-dione, 4,5-dichloro- 2-(dimethoxymethylene)-	CH$_2$Cl$_2$	251(4.34),259(4.32), 274(4.26)	83-0034-76
$C_8H_6Cl_3NO_2$ 4-Cyclopentene-1,3-dione, 4,5-dichloro- 2-[chloro(dimethylamino)methylene]- 3-Pyridinemethanol, 2,5,6-trichloro-, acetate	CH$_2$Cl$_2$ MeOH	253(4.15),263(4.07), 320(4.30) 226(4.03),281(3.72)	83-0034-76 33-0211-76
$C_8H_6Cl_4$ 1,3-Cyclopentadiene, 1,2,3,4-tetra- chloro-5-(1-methylethylidene)-	C$_6$H$_{12}$	289(4.24),406(2.82)	24-3929-76
$C_8H_6Cl_4O$ 1,3-Cyclopentadiene, 1,2,3,4-tetra- chloro-5-(ethoxymethylene)-	C$_6$H$_{12}$	239(3.38),304(4.37), 391(2.91)	24-3939-76
$C_8H_6Cl_4O_2$ Benzene, 1,2,4,5-tetrachloro-3,6-di- methoxy-	MeOH	295(3.21)	2-0230-76
$C_8H_6Cl_4S$ 1,3-Cyclopentadiene, 1,2,3,4-tetra- chloro-5-[(ethylthio)methylene]-	C$_6$H$_{12}$	347(4.51),410s(2.86), 359(4.48)	24-3939-76
$C_8H_6Cl_5N$ Methanamine, 1-chloro-N,N-dimethyl- 1-(2,3,4,5-tetrachloro-2,4-cyclo- pentadien-1-ylidene)-	C$_6$H$_{12}$	235(3.51),378(4.36)	24-3939-76
$C_8H_6INO_5$ 2-Propenoic acid, 3-(5-iodo-2-furanyl)- 2-nitro-, methyl ester, (Z)-	n.s.g.	333(4.214)	73-2422-76
$C_8H_6N_2$ Phthalazine	dioxan	262(3.62)	103-0342-76
$C_8H_6N_2O$ 1(2H)-Phthalazinone 4(3H)-Quinazolinone 5-Quinoxalinol	dioxan pH 2.23 N H$_2$SO$_4$	254(3.95),286(3.90) 255(3.76),261(3.77), 283(3.68),291(3.65) 261(4.47),337(3.56), 415(2.90)	103-0342-76 44-0838-76 54-0285-76
$C_8H_6N_2OS$ 1,3,4-Oxadiazole-2(3H)-thione, 5-phenyl-	EtOH	221(4.08),253(3.96), 298(4.22)	106-0153-76

Compound	Solvent	$\lambda_{max}(\log \epsilon)$	Ref.
$C_8H_6N_2O_2$			
Benzeneacetonitrile, 4-nitro-, carbanion	DMSO	540(4.68)	70-0762-76
$C_8H_6N_2O_2S_2$			
Benzothiazole, 2-(methylthio)-6-nitro-	dioxan	330(4.24)	104-2381-76
$C_8H_6N_2O_4$			
[1,4]Dioxino[2,3-f]-2,1,3-benzoxadiazole, 6,7-dihydro-, 1-oxide	EtOH	218(4.25),350(3.70)	4-1327-76
$C_8H_6N_2O_5$			
Ethanone, 1-(3,5-dinitrophenyl)-	MeOH	219(4.39)	64-0389-76C
$C_8H_6N_2O_7$			
2-Propenoic acid, 2-nitro-3-(5-nitro-2-furanyl)-, methyl ester, (Z)-	n.s.g.	355(4.313)	73-2422-76
$C_8H_6N_2S$			
2H,5H-Thieno[2,3-b:4,5-b']dipyrrole	EtOH	214(4.17),230(4.16), 255(4.00)	23-1074-76
2H,7H-Thieno[2,3-b:5,4-b']dipyrrole	EtOH	208(4.36)	23-1074-76
4H,5H-Thieno[3,2-b:4,5-b']dipyrrole	EtOH	224(3.76),263(4.19), 272(4.25),282(4.16)	23-1074-76
$C_8H_6N_2S_2$			
2,3-Quinoxalinedithiol	DMF	272(3.21),279(3.19), 287(3.19),313(2.92), 324(2.80),365(2.75), 385(2.94),407(3.02), 429(2.98),456(2.72), 534(1.18),560(1.11)	59-1015-76
$C_8H_6N_4$			
Benzene, 1,3-bis(diazomethyl)-	MeOH	276(4.81),476(1.57)	12-1699-76
$C_8H_6N_4O_2$			
[4,5'-Bipyrimidine]-4',6(1H,1'H)-dione	0.25M HCl	233(3.95),307(3.93)	19-0029-76
	pH 12	235(4.12),288(3.90)	19-0029-76
Methanimidamide, N-cyano-N'-(4-nitrophenyl)-	DMF	339(4.14),419(3.84)	48-0347-76
4H-Pyrrolo[2,3-b]pyrazine, 7-(2-nitroethenyl)-	EtOH	228(4.15),296(4.08), 349(4.28)	103-0707-76
$C_8H_6N_4O_2S$			
1,2,3-Thiadiazol-5-amine, N-(4-nitrophenyl)-	dioxan	234s(--),296(3.70), 365(4.32)	73-1182-76
$C_8H_6N_4O_3$			
Ethanone, 1-(3-azido-5-nitrophenyl)-	MeOH	233(4.34),332(3.23)	64-0389-76C
$C_8H_6N_4O_5$			
2,4-Imidazolidinedione, 1-[[(5-nitro-2-furanyl)methylene]amino]-	H_2O	231(3.60),278(3.95), 390(c.4.2)	96-0986-76
anion	H_2O	231(3.60),278(3.95), 370+(4.2+)	96-0986-76
protonated	H_2O	230s(3.30),264(3.78), 365(3.95)	96-0986-76
diprotonated	H_2O	233(3.30),266(3.90), 368(4.16)	96-0986-76

Compound	Solvent	$\lambda_{max}(\log \epsilon)$	Ref.
$C_8H_6N_4S$			
1-Thia-3,5,8,8b-tetraazaacenaphthylene, 2-methyl-	EtOH	250(4.13),257(4.11), 282(3.69),293(3.54), 357(3.78),470s(2.77), 505(2.86),544(2.84), 590(2.59),656(1.95)	1-0468-76
$C_8H_6N_4Se$			
1-Selena-3,5,8,8b-tetraazaacenaphthylene, 2-methyl-	EtOH	236(4.21),244(4.20), 282(3.69),292(3.56), 366(3.86),378(3.87), 475(2.80),510(2.89), 551(2.86),603(2.60), 663(1.90)	1-0468-76
$C_8H_6N_6O$			
Acetamide, N-(3-cyano-1H-pyrazolo [3,4-d]pyrimidin-4-yl)-	EtOH-pH 1	278(4.04)	69-1005-76
	EtOH-pH 7	278(4.02)	69-1005-76
	EtOH-pH 11	312(4.00),340s(--)	69-1005-76
$C_8H_6O_2$			
Bicyclo[3.2.1]octa-3,6-diene-2,8-dione	EtOH	224(3.84),275(3.39), 310s(3.06)	77-0446-76
Bicyclo[4.2.0]octa-4,7-diene-2,3-dione	EtOH	233(3.72),328(3.15)	88-0839-76
$C_8H_6O_3$			
Benzoic acid, 4-formyl-	1% EtOH- pH 10	255(3.89)	44-1957-76
$C_8H_6S_2$			
2,2'-Bithiophene	heptane	250(3.81),300(4.03)	103-0981-76
2,3'-Bithiophene	heptane	236(3.96),285(4.10)	103-0981-76
3,3'-Bithiophene	heptane	2?2(4.30),262(4.06)	103-0981-76
$C_8H_6S_3$			
1,3-Benzodithiole-2-thione, 5-methyl-	benzene	290(3.51),363(4.37), 422(2.10)	101-0369-76Q
$C_8H_7BrN_2S$			
Isothiazolo[5,4-b]pyridine, 3-bromo- 4,6-dimethyl-	EtOH	311s(3.76)	48-0779-76
$C_8H_7BrN_4O_2S$			
Benzenesulfonamide, 4-bromo-N-(1H- 1,2,3-triazol-5-yl)-	EtOH	230(4.20),249(4.00)	32-0001-76
$C_8H_7BrO_3$			
Benzaldehyde, 6-bromo-2-hydroxy-3-methoxy-	MeOH	213(4.09),233(3.94), 273(3.74),360(3.34)	12-2003-76
1,3-Benzodioxole, 4-bromo-7-methoxy-	MeOH	209(4.60),274(2.94), 284s(2.72)	12-2003-76
$C_8H_7Br_2NO_2$			
3-Pyridinemethanol, 2,6-dibromo-, acetate	MeOH	225(3.85),275(3.71)	33-0211-76
$C_8H_7ClN_2O_3$			
Benzenecarboximidoyl chloride, N-methoxy-4-nitro-, (E)-	EtOH	227(4.00),296(4.15)	44-0252-76
(Z)-	EtOH	220(3.94),293(3.98)	44-0252-76

Compound	Solvent	$\lambda_{max}(\log \epsilon)$	Ref.
$C_8H_7ClN_4$			
1(2H)-Phthalazinone, 4-chloro-	isooctane	275(3.38),328(3.04)	103-0342-76
	MeOH	274(3.93),318(3.82)	103-0342-76
	dioxan	279(4.11),333(3.84)	103-0342-76
	MeCN	275(3.90),318(3.88)	103-0342-76
$C_8H_7ClN_6O_2$			
1,3,4-Oxadiazol-2-ol, 2-[(6-chloro-1,2,4-triazolo[4,3-b]pyridazin-7-yl)methyl]-2,3-dihydro-	MeOH	217(4.52),275(3.29),301(3.38)	4-1155-76
$C_8H_7ClO_3$			
Benzoic acid, 2-chloro-3-methoxy-	MeOH	243s(3.28),290(3.26)	12-2003-76
$C_8H_7Cl_2NO$			
Benzenecarboximidoyl chloride, 4-chloro-N-methoxy-, (E)-	EtOH	214s(4.14),220s(3.98),263(4.24),295s(3.19)	44-0252-76
(Z)-	EtOH	220s(3.88),260(4.04)	44-0252-76
$C_8H_7Cl_2NO_2$			
4-Cyclopentene-1,3-dione, 4,5-dichloro-2-[(dimethylamino)methylene]-	CH_2Cl_2	253(4.20),303(4.47)	83-0034-76
$C_8H_7Cl_4N$			
Methanamine, N,N-dimethyl-1-(2,3,4,5-tetrachloro-2,4-cyclopentadien-1-ylidene)-	C_6H_{12}	234(3.64),346(4.46)	24-3939-76
C_8H_7N			
Benzeneacetonitrile, carbanion	DMSO	343(4.48)	70-0762-76
C_8H_7NO			
Benzocyclobutenone, oxime	EtOH-acid	247(4.09),280s(3.60),288(3.74),296(3.80)	12-1685-76
	EtOH-base	278s(3.98),296(4.02),309s(3.97)	12-1685-76
C_8H_7NOS			
2,1-Benzisothiazol-3(1H)-one, 1-methyl-	$CHCl_3$	250s(4.05),260(4.03),374(3.74)	48-0161-76
Benzothiazole, 2-methoxy-	MeOH-NaOMe	215(4.61),247(3.91)	4-0491-76
$C_8H_7NO_2$			
Benzene, (2-nitroethenyl)-, cis	EtOH	223(4.00),306(3.78)	44-2112-76
Benzeneacetamide, α-oxo-	EtOH	256(3.95)	12-2459-76
$C_8H_7NO_2S$			
Benzene, 1-(ethenylthio)-4-nitro-	n.s.g.	228(3.92),240(4.05),254(3.76)	47-0845-76
$C_8H_7NO_2S_2$			
Benzothiazole, 2-(methylsulfonyl)-	MeOH-NaOMe	216(4.17),237(3.93),271(3.96)	4-0491-76
4H-1,3-Thiazin-4-one, 6-(2-furanyl)-tetrahydro-2-thioxo-	n.s.g.	257(4.11),313(4.17)	73-1388-76
$C_8H_7NO_3$			
2(3H)-Benzoxazolone, 6-methoxy-	EtOH	209(4.01),232(4.05),290(3.80)	102-1997-76

Compound	Solvent	λ_{max} (log ϵ)	Ref.
$C_8H_7NO_3S$ 3H,7H-Azeto[2,1-b]furo[3,4-d][1,3]thia-zine-1,7(4H)-dione, 5a,6-dihydro-	EtOH	257(3.97)	33-2294-76
$C_8H_7NO_4$ Benzoic acid, 4-nitro-, methyl ester	EtOH	258(4.09),300(3.26)	110-1170-76
$C_8H_7NO_4S$ Benzoic acid, 4-mercapto-3-nitro-, methyl ester	EtOH	257(4.18),278(4.07), 339(3.90)	33-0855-76
	EtOH-HOAc	257(4.30),278(4.17), 350(3.51)	33-0855-76
	EtOH-Et$_3$N	250(3.70),278(3.64), 337(4.27)	33-0855-76
$C_8H_7NO_5$ Benzoic acid, 3-methoxy-2-nitro-	MeOH	240s(3.51),298(3.42)	12-2003-76
2-Propenoic acid, 3-(2-furanyl)-2-nitro-, methyl ester	n.s.g.	313(4.204)	73-2422-76
$C_8H_7NS_2$ 2,1-Benzisothiazole-3(2H)-thione, 1-methyl-	EtOH	253(3.92),294(3.74), 359(3.39),438(3.63)	48-0161-76
Benzothiazole, 2-(methylthio)-	dioxan	275(4.21)	104-2381-76
$C_8H_7NS_3$ 1,4,2-Benzodithiazine, 3-(methylthio)-	C_6H_{12}	220(4.13),265(4.22), 332(3.41)	48-0127-76
$C_8H_7N_3$ 1H-Benzotriazole, 1-ethenyl-	pH 1	198(4.40),215(4.28), 218(4.28),227(4.14), 265(3.87),273(3.87), 297(3.80)	103-0691-76
	EtOH	199(4.40),215(4.22), 219(4.24),228(4.16), 263(3.87),271(3.84), 298(3.83)	103-0691-76
	dioxan	220(4.19),233(4.20), 262(3.85),270(3.82), 299(3.83)	103-0691-76
1,7-Naphthyridin-3-amine	EtOH	210(4.16),250(4.31), 293(3.82),344(3.70)	56-0451-76
1,7-Naphthyridin-4-amine	EtOH	208(4.68),225(4.47), 247(4.20),328(4.21)	56-0451-76
Pyrido[3,2-c]pyridazine, 6-methyl-	EtOH	264(3.76),300(3.77), 313(3.81),360(2.43)	103-0808-76
Pyrido[3,2-c]pyridazine, 7-methyl-	EtOH	252(3.51),258(3.54), 265s(3.46),310(3.63), 323(3.67),366(2.34)	103-0808-76
Pyrido[3,2-c]pyridazine, 8-methyl-	EtOH	265(3.62),271s(3.61), 305(3.59),317(3.57), 370(2.30)	103-0808-76
$C_8H_7N_3O$ Methanimidamide, N-cyano-N'-(2-hydroxy-phenyl)-	EtOH	286(4.30)	48-0347-76
$C_8H_7N_3O_3S$ 2H-1,2,4-Benzothiadiazine-3-carboxamide, 1,1-dioxide	dioxan	288(3.75)	103-0402-76

Compound	Solvent	$\lambda_{max}(\log \epsilon)$	Ref.
$C_8H_7N_3O_6$			
2-Oxazolidinone, 3-[[(4,5-dihydro-5-aci-nitro-4-oxo-2-furanyl)methylene]amino]-, hemihydrate	H_2O	415(4.38)	87-0729-76
dihydrate	H_2O	415(4.33)	87-0729-76
$C_8H_7N_3S$			
1,2,3-Thiadiazol-5-amine, N-phenyl-	dioxan	247(3.91),286s(--), 322(4.22)	73-1182-76
$C_8H_7N_3S_2$			
1,2,4-Triazolidine-3,5-dithione, 1-phenyl-	EtOH	220(4.14),290(3.84)	65-0174-76
$C_8H_7N_4O_7$			
Benzofurazan, 4,?-dihydro-4,4-dimethoxy-5,7-dinitro-, ion(1-)	MeOH	450(4.44)	28-0195-76A
$C_8H_7N_5O$			
9H-Imidazo[1,2-a]purin-9-one, 1,4-dihydro-4-methyl-	pH 1.0	223(4.53),228(4.53), 246(3.65),283(4.02)	69-0898-76
	pH 5.8	228(4.50),257(3.71), 307(3.86)	69-0898-76
	pH 11.0	229(4.53),269(3.78), 302(3.98)	69-0898-76
$C_8H_7N_5O_4S$			
Benzenesulfonamide, 4-nitro-N-(1H-1,2,3-triazol-5-yl)-	EtOH	257(4.13)	32-0001-76
$C_8H_7N_5O_5$			
1H-Imidazo[1,5-b][1,2,4]triazole-2,5-dicarboxylic acid, 7-(aminocarbonyl)-1-methyl-	2M HCl	224(4.25),245s(4.11), 278(4.00),313(4.09)	44-1889-76
	pH 3	247(4.10),272(3.82), 307(4.21)	44-1889-76
	pH 7 and 10	250(4.14),273s(3.95), 302(4.15)	44-1889-76
$C_8H_8BrN_2S$			
Isothiazolo[5,4-b]pyridinium, 3-bromo-6,7-dimethyl-, perchlorate	DMF	316(3.89)	48-0779-76
$C_8H_8Br_2O_2$			
5-Cyclooctene-1,2-dione, 3,8-dibromo-, trans	C_6H_{12}	224(2.87),280(2.48), 288(2.39)	44-3760-76
C_8H_8ClNO			
Benzenecarboximidoyl chloride, N-methoxy-, (E)-	C_6H_{12}	212s(4.03),218s(3.94), 257(4.08),290s(2.76)	44-0252-76
	EtOH	217s(3.96),257(4.11), 290s(2.88)	44-0252-76
Benzenecarboximidoyl chloride, N-methoxy-, (Z)-	C_6H_{12}	218s(3.91),256(3.88)	44-0252-76
	EtOH	217s(3.85),254(3.92)	44-0252-76
$C_8H_8ClNO_3S$			
Carbamothioic acid, (4-chloro-2-oxo-2H-pyran-6-yl)-, S-ethyl ester	EtOH	243(3.96),341(4.17), 380s(3.68)	39-2462-76C
Carbamothioic acid, O-(4-chloro-1,6-dihydro-6-oxo-2-pyridinyl)-, S-ethyl ester	MeCN	226(3.88),274(3.97)	39-2462-76C

Compound	Solvent	$\lambda_{max}(\log \epsilon)$	Ref.
2H-Pyran-5-carbothioic acid, 6-amino-4-chloro-2-oxo-, S-ethyl ester	MeCN	297(4.11),338(4.17)	39-2462-76C
Sulfamoyl chloride, 7-oxabicyclo[4.2.1]-nona-2,4-dien-8-ylidene-	EtOH	265(3.64)	44-3583-76
$C_8H_8ClNO_4$			
3-Pyridinecarboxylic acid, 4-chloro-1,2-dihydro-6-hydroxy-2-oxo-, ethyl ester	MeCN	273(4.08),292s(4.01), 305(4.00),315s(3.84)	39-2462-76C
$C_8H_8ClN_3O_2$			
Ethanehydrazonoyl chloride, N-(2-nitrophenyl)-	EtOH	260(4.09),407(3.62)	104-1649-76
Ethanehydrazonoyl chloride, N-(4-nitrophenyl)-	EtOH	247(3.95),375(4.17)	104-1649-76
$C_8H_8Cl_2Ge$			
9-Germabicyclo[4.2.1]nona-2,4,7-triene, 9,9-dichloro-	heptane	242(3.36),320(3.03)	70-0602-76
$C_8H_8Cl_2N_2O_2$			
Benzenamine, 5-chloro-N-(2-chloroethyl)-2-nitro-	EtOH	239(4.41),270s(3.75), 431(3.75)	78-0057-76
$C_8H_8F_4N_2O_3$			
2,4(1H,3H)-Pyrimidinedione, 5-[(2,2,3,3-tetrafluoropropoxy)methyl]-	EtOH	262(3.88)	104-0643-76
C_8H_8NOSSe			
1,2,3-Benzothiaselenazol-2-ium, 6-ethoxy-, chloride	93% H_2SO_4	400(3.85),475(3.96)	103-1128-76
$C_8H_8NOS_2$			
1,2,3-Benzodithiazol-2-ium, 6-ethoxy-, chloride	93% H_2SO_4	370(3.81),450(3.95)	103-1128-76
$C_8H_8N_2$			
1H-Benzimidazole, 1-methyl-, iodine complex	CH_2Cl_2 CCl_4	380(3.39) 410(3.18)	65-0664-76 65-0664-76
2,3'-Bi-1H-pyrrole	EtOH	205(4.21),213s(4.18), 232(4.04),257(4.13), 279s(3.81)	23-1083-76
3,3'-Bi-1H-pyrrole	EtOH	206(4.29),237(3.79)	23-1083-76
Diazabasketene	isooctane	373s(1.94),381s(2.20), 387(2.37),390(2.51), 394(2.38),403(2.68), 412s(1.43)	35-2398-76
2,4-Hexadienedinitrile, 3,4-dimethyl-, cis-cis	EtOH	204(4.11),222s(3.87)	44-1389-76
Imidazo[1,2-a]pyridine, 7-methyl-, compd. with hydrogen triiodide	EtOH	215(4.46),224s(4.26), 287(4.51),356(4.23)	44-3549-76
$C_8H_8N_2O$			
Benzonitrile, 4-[N-(hydroxymethyl)-amino]-	H_2O	216(--),275(4.07)	35-7777-76
Propanedinitrile, (tetrahydro-4H-pyran-4-ylidene)-	C_6H_{12}	214(3.36),251(4.11)	78-2827-76
Pyrazolo[1,5-a]pyridine, 2-methoxy-	MeOH	234(4.57),282(3.30), 307(3.31)	18-1980-76
Pyrazolo[1,5-a]pyridin-2(1H)-one, 1-methyl-	MeOH	242(4.47),281(3.89), 288(3.88),341(3.32)	18-1980-76

Compound	Solvent	$\lambda_{max}(\log \epsilon)$	Ref.
$C_8H_8N_2OS$			
2-Benzothiazolamine, 6-methoxy-	EtOH	223(4.72),270(4.31), 297s(3.63)	104-2374-76
7H-Thiazolo[3,2-a]pyrimidin-7-one, 5,6-dimethyl-	MeOH	222(4.53),232s(4.41), 239s(4.20),279(4.30), 316s(3.73)	39-1908-76C
$C_8H_8N_2O_2$			
Benzenamine, N-(2-nitroethenyl)-	EtOH	236(4.12),380(4.38)	104-1985-76
$C_8H_8N_2O_3$			
Ethanone, 1-(3-amino-5-nitrophenyl)-	MeOH	219(4.40),382(3.32)	64-0389-76C
Furo[2',3':4,5]furo[2,3-d]pyrimidin-2(1H)-one, 4b,6,7,7a-tetrahydro-, cis	H_2O	277(3.66)	4-0929-76
2,4(1H,3H)-Pyrimidinedione, 5-(3-oxo-1-butenyl)-	pH 1	309(4.14)	87-0194-76
	pH 13	287(3.68),355(3.33)	87-0194-76
$C_8H_8N_2O_4$			
2-Propenoic acid, 3-(1,2,3,4-tetra-hydro-6-methyl-2,4-dioxo-5-pyrim-idinyl)-, (E)-	H_2O	270s(4.11),293(4.15)	44-1858-76
$C_8H_8N_2O_6$			
3,6-Pyridazinedicarboxylic acid, 4,5-di-hydroxy-, dimethyl ester	EtOH	318(4.09)	78-0269-76
$C_8H_8N_2S$			
2-Benzothiazolamine, 6-methyl-	EtOH	223(4.50),263(4.16), 297s(3.38)	104-2374-76
Propanedinitrile, (tetrahydro-4H-thio-pyran-4-ylidene)-	C_6H_{12}	230(4.13),294(3.46)	78-2827-76
$C_8H_8N_2S_2$			
6-Benzothiazolamine, 2-(methylthio)-	dioxan	305(4.31)	104-2381-76
$C_8H_8N_4$			
1,2,4-Benzotriazin-3-amine, 6-methyl-	EtOH	208(4.22),237(4.24), 308(3.38)	107-0457-76
Phthalazine, 1-hydrazino-	neutral	273(3.8),305(3.7)	36-0274-76
	cation	260(4.0),290s(3.7), 302(3.7),315(3.6)	36-0274-76
	dication	300s(3.8),318(3.8)	36-0274-76
1(2H)-Phthalazinone, hydrazone	isooctane	273(3.70),332(3.48)	103-0342-76
	MeOH	273(4.11),318(3.88)	103-0342-76
	dioxan	275(4.15),340(3.66)	103-0342-76
	MeCN	273(4.11),320(3.68)	103-0342-76
$C_8H_8N_4O_2S$			
Benzenesulfonamide, N-(1H-1,2,3-triazol-5-yl)-	EtOH	213(4.01),249(3.92)	32-0001-76
$C_8H_8N_4O_3S$			
2H-1,2,4-Benzothiadiazine-3-carboxylic acid, hydrazide, 1,1-dioxide	dioxan	292(3.86)	103-0402-76
$C_8H_8N_4O_4$			
[1,2,5]Oxadiazolo[3,4-d]pyrimidine-5,7(4H,6H)-dione, 6-methyl-4-(2-propenyl)-, 1-oxide	MeOH	269(3.92),315(3.08)	39-1327-76C

Compound	Solvent	$\lambda_{max}(\log \epsilon)$	Ref.
$C_8H_8N_4O_5$			
[1,2,4]Oxadiazolo[3,4-d]pyrimidine-5,7(4H,6H)-dione, 6-methyl-4-(2-oxopropyl)-, 1-oxide	MeOH	267(4.00),315(3.08)	39-1327-76C
$C_8H_8N_4S$			
3H-1,2,4-Triazole-3-thione, 5-amino-1,2-dihydro-2-phenyl-	EtOH	228(4.07),302(3.72)	65-0174-76
$C_8H_8N_6O_3$			
Glycine, N-[(1H-pyrazolo[3,4-d]pyrimidin-4-ylamino)carbonyl]	pH 1	267(4.14)	87-0555-76
	pH 13	267s(--),271(4.06), 297(3.64)	87-0555-76
	EtOH	265(4.14),275s(--), 285s(--)	87-0555-76
C_8H_8O			
Benzaldehyde, 4-methyl-	1% EtOH-pH 10	260(4.07)	44-1957-76
9-Oxabicyclo[4.2.1]nona-2,4,7-triene	hexane	253(3.62),263(3.58), 273(3.38)	78-2239-76
C_8H_8OSe			
Benzenecarboselenoic acid, 4-methyl-, potassium salt	EtOH	240(4.17),324(3.65)	138-0203-76
$C_8H_8O_2$			
Benzaldehyde, 4-methoxy- (in 1% EtOH)	pH 10	257(4.02)	44-1957-76
Benzoic acid, methyl ester	EtOH	228(4.05),272(2.95)	110-1170-76
Bicyclo[2.2.2]oct-5-ene-2,3-dione	dioxan	451(2.05)	118-0256-76
Cyclopenta[b]pyran-4(5H)-one, 6,7-dihydro-	MeOH	255(4.0)	5-1689-76
2,3:5,6-Dimethanocyclohexane-1,4-dione, cis	EtOH	280(1.67)	88-4851-76
trans	EtOH	271(1.76)	88-4851-76
7-Oxabicyclo[4.2.1]nona-2,4-dien-8-one	EtOH	263(3.63)	44-3583-76
1(3aH)-Pentalenone, 4,6a-dihydro-2-hydroxy-	MeOH	259(3.78),313(2.28)	138-0215-76
1(3aH)-Pentalenone, 4,6a-dihydro-6a-hydroxy-	MeOH	222(3.82),338(2.08)	138-0215-76
1(3aH)-Pentalenone, 6,6a-dihydro-6a-hydroxy-	MeOH	220(3.79),331(1.61)	138-0215-76
$C_8H_8O_2S_2$			
2-Propanone, 1-(4-acetyl-1,3-dithiol-2-ylidene)-	n.s.g.	294(3.51)	104-1131-76
2-Propanone, 1,1'-(1,3-dithietane-2,4-diylidene)bis-	CHCl$_3$	322(5.48),357(5.49)	70-1913-76
	n.s.g.	322(5.48)	104-1131-76
$C_8H_8O_2Se$			
Benzenecarboselenoic acid, 4-methoxy-, potassium salt	EtOH	259(4.11),319(3.79)	138-0203-76
$C_8H_8O_3$			
Ethanone, 1-(3,4-dihydroxyphenyl)-	MeOH	275(3.78),305(3.63)	105-0099-76
$C_8H_8O_4$			
2(6H)-Benzofuranone, 7,7a-dihydro-6,7-dihydroxy-, (6α,7β,7aβ)-	MeOH	256(3.85)	100-0385-76
Benzoic acid, 3-hydroxy-4-methoxy-	EtOH	250(3.97)	105-0551-76

Compound	Solvent	λ_{max}(log ϵ)	Ref.
Benzoic acid, 3-hydroxy-4-methoxy-	HOAc	260(4.27)	105-0551-76
Benzoic acid, 4-hydroxy-3-methoxy-	EtOH	256(4.02)	105-0551-76
	HOAc	263(4.06)	105-0551-76
3,5-Cyclohexadiene-1,2-dione, 3,5-di-methoxy-	H_2O	300(3.51),350(3.77), 490(2.70)	5-1435-76
3,5-Cyclohexadiene-1,2-dione, 4,5-di-methoxy-	H_2O	290(4.11),430(2.74)	135-0717-76A
	MeOH	285(4.11),411(2.74)	135-0717-76A
	dioxan	280(4.11),400(2.74)	135-0717-76A
	$CHCl_3$	283(4.11),407(2.74)	135-0717-76A
Ethanone, 1-(2,4,6-trihydroxyphenyl)-	EtOH	228(4.16),288(4.21)	102-1785-76
2H-Pyran-2,4(3H)-dione, 3-acetyl-6-methyl-	n.s.g.	223(3.9),308(1.2)	31-1490-76
$C_8H_8O_5$			
2,3-Furandicarboxylic acid, dimethyl ester	EtOH	259(3.89)	88-3295-76
C_8H_8S			
Benzo[b]thiophene, 3a,7a-dihydro-	MeCN	237(3.70),260(3.34)	35-6405-76
$C_8H_8S_2Se_2$			
1,3-Thiaselenone, 5-methyl-2-(5-methyl-1,3-thiaselenol-2-ylidene)-	hexane	240(3.79),295(4.32), 370(3.40),490(2.30)	88-0423-76
$C_8H_9AsO_3$			
1(4H)-Arseninacetic acid, 4-oxo-, methyl ester	EtOH	226(3.82),306(3.79)	88-4143-76
$C_8H_9BrO_3$			
2-Cyclohexen-1-one, 2-acetoxy-3-bromo-	MeOH	247(4.19),308(1.92)	104-0089-76
$C_8H_9ClN_2O_2$			
Benzenamine, N-(2-chloroethyl)-2-nitro-	EtOH	232(4.32),279(3.66), 418(3.76)	78-0057-76
$C_8H_9Cl_2F$			
1-Octen-3-yne, 1,1-dichloro-2-fluoro-	n.s.g.	236(4.17),241s(--), 247(4.13)	44-1487-76
$C_8H_9F_2NO$			
Tyramine, 3,5-difluoro-	0.05M HCl	262(2.74)	44-2373-76
	0.05M NaOH	277(3.30)	44-2373-76
C_8H_9NO			
Acetamide, N-phenyl-	EtOH	240(4.10),282(2.70)	110-1173-76
Benzene, (2-nitrosoethyl)-, dimer	EtOH	290(3.95)	78-1267-76
C_8H_9NOS			
Thiocyanic acid, 2-formyl-1-cyclohexen-1-yl ester	EtOH	292(3.74)	97-0049-76
$C_8H_9NO_2$			
Benzamide, 2-hydroxy-N-methyl-	C_6H_{12}	306(3.70)	18-2679-76
	MeOH	300(3.64)	18-2679-76
	EtOH	301(3.64)	18-2679-76
	ether	305(3.71)	18-2679-76
	dioxan	304(3.70)	18-2679-76
	EtOAc	304(3.69)	18-2679-76
	MeCN	303(3.66)	18-2679-76
	$C_2H_4Cl_2$	304(3.69)	18-2679-76
Benzeneacetamide, α-hydroxy-	EtOH	258(2.51)	12-2459-76

Compound	Solvent	$\lambda_{max}(\log \epsilon)$	Ref.
Benzoic acid, 2-amino-, methyl ester	n.s.g.	219(4.45)	40-0462-76
Benzoic acid, 3-amino-, methyl ester	n.s.g.	223(4.45)	40-0462-76
Benzoic acid, 4-amino-, methyl ester	n.s.g.	295(4.34)	40-0462-76
Carbamic acid, phenyl-, methyl ester	n.s.g.	235(4.26)	40-0462-76
Furo[2,3-b]pyridin-6(2H)-one, 3,7-dihydro-7-methyl-	MeOH	238(3.78),331(4.00)	24-1269-76
$C_8H_9NO_2S_2$			
5H-1,4-Dithiino[2,3-c]pyrrole-5,7(6H)-dione, 6-ethyl-2,3-dihydro-	EtOH	204(3.89),263(4.07), 414(3.42)	56-1523-76
$C_8H_9NO_3$			
Benzoic acid, 2-amino-3-methoxy-	MeOH	225(3.06),248s(3.82)	12-2003-76
2,5-Cyclohexadiene-1,4-dione, 2-(2-aminoethyl)-5-hydroxy-	pH 5.6	270(4.04),495(3.34)	39-0339-76C
	EtOH-acid	262(4.15),380(--)	39-0339-76C
	EtOH-base	272(4.01),495(3.34)	39-0339-76C
2,5-Cyclohexadiene-1,4-dione, 2-methoxy-5-(methylamino)-	EtOH	242(3.64),282(3.20), 340(4.28)	115-0203-76
	ether	240(3.66),280(3.30), 330(4.28)	115-0203-76
	CCl_4	277(3.79),320(4.13), 335(4.26)	115-0203-76
$C_8H_9NO_4$			
4-Oxazolecarboxylic acid, 2-acetyl-5-methyl-, methyl ester	MeOH	204(3.88),265(3.96)	35-8237-76
$C_8H_9NO_5$			
1,3-Dioxolane, 4-methyl-2-(5-nitro-2-furanyl)-	n.s.g.	306(3.92)	103-1100-76
2-Furancarboxylic acid, 5-nitro-, propyl ester	n.s.g.	295(4.08)	103-1100-76
C_8H_9NS			
Benzenecarbothioamide, N-methyl-	MeOH	239(4.03),286(3.88)	78-0661-76
	MeCN	240(3.96),287(3.78)	78-0661-76
Ethanethioamide, N-phenyl-	MeOH	217(4.13),297(4.10)	78-0661-76
	MeCN	219(4.20),296(4.09)	78-0661-76
$C_8H_9N_3$			
1H-Benzotriazole, 1-ethyl-	pH 1	199(4.41),260(3.82), 273(3.87),279(3.76), 284(3.80)	103-0691-76
	EtOH	202(4.48),258(3.82), 263(3.84),273(3.78), 279(3.78),284(3.76)	103-0691-76
	dioxan	257(3.80),263(3.82), 273(3.72),279(3.76), 284(3.76)	103-0691-76
$C_8H_9N_3O$			
1H-Benzotriazole, 1-ethoxy-	EtOH	262(3.82),283(3.76)	4-0509-76
1H-Imidazo[4,5-c]pyridine, 2-methoxy-1-methyl-	pH 4.0	265(3.71)	39-1176-76B
	pH 10.0	267(3.65)	39-1176-76B
3H-Imidazo[4,5-c]pyridine, 2-methoxy-3-methyl-	pH 4.0	279.5(4.08)	39-1176-76B
	pH 10.0	248(3.78),268(3.71), 276(3.62)	39-1176-76B
$C_8H_9N_3O_2$			
Ethanimidamide, N-(4-nitrophenyl)-	neutral	357(4.12)	39-0211-76B

Compound	Solvent	$\lambda_{max}(\log \epsilon)$	Ref.
Ethanimidamide, N-(4-nitrophenyl)-	cation	282(4.16)	39-0211-76B
Imidazo[1,2-a]pyrimidine-2,5(1H,3H)-dione, 3,7-dimethyl-	pH 6.1	237(4.06),278(3.86)	114-0413-76A
Imidazo[1,2-a]pyrimidine-2,5(3H,8H)-dione, 7,8-dimethyl-	EtOH	210(3.91),246(3.96), 272(3.66)	114-0413-76A
	dioxan	242(4.08),274(3.71)	114-0413-76A
Imidazo[1,2-a]pyrimidine-2,7(1H,3H)-dione, 1,5-dimethyl-	EtOH	220(4.39),258(3.84)	114-0413-76A
	dioxan	224(4.26),253(3.81)	114-0413-76A
Imidazo[1,2-a]pyrimidine-2,7(3H,8H)-dione, 5,8-dimethyl-	EtOH	215(4.37),269(4.08)	114-0413-76A
	dioxan	214(4.35),262(4.08)	114-0413-76A
$C_8H_9N_3O_3$			
Acetic acid, 2-(2-nitrophenyl)hydrazone	EtOH	228(4.27),276(3.67), 392(3.59)	104-1649-76
Acetic acid, 2-(4-nitrophenyl)hydrazone	EtOH	223(3.83),352(4.11)	104-1649-76
$C_8H_9N_5$			
1H-Pyrazolo[3,4-d]pyrimidin-4-amine, N-2-propenyl-	pH 1	266(4.02)	87-0555-76
	pH 13	268(4.01)	87-0555-76
	EtOH	277.5(4.15)	87-0555-76
Pyrido[3,2-c]pyridazine, 4-hydrazino-6-methyl-	EtOH	260s(3.82),265(3.85), 268s(3.83),279(3.82), 292s(3.80),372(3.88)	103-0808-76
Pyrido[3,2-c]pyridazine, 4-hydrazino-7-methyl-	EtOH	254(3.76),294(3.53), 308(3.52),322(3.48), 370(3.59)	103-0808-76
$C_8H_9N_5O$			
Pyrido[4,3-e]-1,2,4-triazin-3-amine, N-(methoxymethyl)-	EtOH	252(3.55),369(2.40)	114-0327-76D
$C_8H_9N_5O_2$			
Acetic acid, cyano(1H-pyrazol-3-ylhydrazono)-, ethyl ester	EtOH	358(4.15)	39-1496-76C
Carbamic acid, 1H-pyrazolo[3,4-d]pyrimidin-4-yl-, ethyl ester	pH 1	265(4.03)	87-0555-76
	pH 13	290(4.01)	87-0555-76
	EtOH	260(4.09),280s(--)	87-0555-76
1H-Imidazo[1,5-b][1,2,4]triazole-7-carboxamide, N-acetyl-1-methyl-	pH 1	239(3.56),297(4.31)	44-1889-76
	pH 7	250(3.46),299(4.37)	44-1889-76
	pH 11	286s(4.20),298(4.24) (anom.)	44-1889-76
4,7(1H,8H)-Pteridinedione, 2-(dimethylamino)-	pH 5.0	221(4.31),293(4.04), 353(4.16)	24-3228-76
	pH 8.8	233(4.42),265s(3.96), 283s(3.90),340(4.09)	24-3228-76
	pH 13.0	226(4.52),258(4.14), 352(4.13)	24-3228-76
Pyrazolo[5,1-c][1,2,4]triazine-3-carboxylic acid, 4-amino-, ethyl ester	EtOH	234(4.14),310(4.05), 344(3.92)	39-1496-76C
Pyrimido[5,4-e]-1,2,4-triazine-5,7-(6H,8H)-dione, 3,6,8-trimethyl-	EtOH	240(4.32),350(3.64)	39-0713-76C
$C_8H_9N_5O_3$			
Pyrimido[5,4-e]-1,2,4-triazine-5,7-(1H,6H)-dione, 1,3,6-trimethyl-, 4-oxide	dioxan	264(4.35),324(3.65), 420(3.41)	39-0713-76C
Pyrimido[5,4-e]-1,2,4-triazine-5,7-(6H,8H)-dione, 3,6,8-trimethyl-, 4-oxide	dioxan	240(4.38),301(3.71), 361(3.56)	39-0713-76C

Compound	Solvent	$\lambda_{max}(\log \epsilon)$	Ref.
Triazolo[3,2-c]triazine-6-carboxylic acid, 4,7-dihydro-2-methyl-7-oxo-, ethyl ester	EtOH	275(3.08),313(3.61)	22-1178-76
Triazolo[3,4-c]triazine-6-carboxylic acid, 4,7-dihydro-1-methyl-7-oxo-, ethyl ester	EtOH	279(3.77),305(3.57)	22-1178-76
Triazolo[4,3-b]triazine-6-carboxylic acid, 7,8-dihydro-8-methyl-7-oxo-, ethyl ester	EtOH	259(3.62),310(2.98)	22-1178-76
$C_8H_9N_5O_4$ Urea, N,N''-(5-nitro-1,3-phenylene)bis-	pH 1	223(4.72),238(4.50), 300(3.62)	87-0426-76
C_8H_{10} Cyclohexene, 4,5-bis(methylene)-	EtOH	219(3.73)	44-1635-76
Cyclohexene, 4-ethynyl-	EtOH	245(4.04)	88-1265-76
1,3,5-Hexatriene, 3-ethenyl-	n.s.g.	214s(4.08),220(4.15), 227(4.11),275(4.50)	78-2681-76
2,4-Octadiyne	EtOH	226(2.69),237(2.46)	78-3041-76
1,2,5,7-Octatetraene	EtOH	225(4.43)	88-0887-76
$C_8H_{10}BrN_3O_5$ 1,2,4-Triazine-3,5(2H,4H)-dione, 2-(2-bromo-2-deoxy-β-D-ribofuranosyl)-	MeOH	261(3.80)	73-2110-76
$C_8H_{10}ClN$ Benzenamine, 4-chloro-N,N-dimethyl-	H_2O	260(4.2),311(3.3)	35-7777-76
$C_8H_{10}ClNO$ Benzenamine, 4-chloro-N-(hydroxymethyl)-N-methyl-	H_2O	250(4.07),298(3.1)	35-7777-76
$C_8H_{10}ClN_3O_2$ Acetamide, 2-chloro-N-(1,4-dimethyl-6-oxo-2-pyrimidinyl)-	EtOH dioxan	218(4.21),273(4.15) 224(4.15),272(4.13)	114-0413-76A 114-0413-76A
$C_8H_{10}ClN_3O_5$ 1,2,4-Triazine-3,5(2H,4H)-dione, 2-(2-chloro-2-deoxy-β-D-ribofuranosyl)-	MeOH	262(3.71)	73-2110-76
$C_8H_{10}Cl_3NO$ Pyrrolidine, 1-(3,4,4-trichloro-1-oxo-3-butenyl)-	heptane	209(4.15)	24-2159-76
$C_8H_{10}Cl_3NO_2$ Morpholine, 4-(3,4,4-trichloro-1-oxo-3-butenyl)-	heptane	208(4.01)	24-2159-76
$C_8H_{10}FNO$ Phenol, 4-(2-aminoethyl)-2-fluoro-	0.05M HCl 0.05M NaOH	271(3.17) 288(3.41)	44-2373-76 44-2373-76
$C_8H_{10}N_2$ Ethanimidamide, N-phenyl-	neutral cation	235(4.06) 231(4.02)	39-0211-76B 39-0211-76B
$C_8H_{10}N_2O$ 2-Buten-1-one, 1-(1H-imidazol-2-yl)-3-methyl-	EtOH	297(4.24)	33-2738-76

Compound	Solvent	$\lambda_{max}(\log \epsilon)$	Ref.
2-Buten-1-one, 1-(1H-imidazol-4-yl)-3-methyl-	EtOH	279(4.13)	33-2738-76
Diazene, methyl(2-methylphenyl)-, 1-oxide, (Z)-	EtOH	228(3.99),294(3.75)	44-1135-76
Diazene, methyl(2-methylphenyl)-, 2-oxide, (Z)-	EtOH	223(3.89),264(3.18)	44-1135-76
Diazene, methyl(phenylmethyl)-, 2-oxide, (Z)-	EtOH	273(2.54),278(2.35)	44-1135-76
1H-Imidazole, 1-(3-methyl-1-oxo-2-butenyl)-	THF	245(4.14)	33-2738-76
$C_8H_{10}N_2OS$			
4(1H)-Pyrimidinone, 6-methyl-2-(2-propenylthio)-	EtOH	231(4.02),286(3.92)	95-1094-76
$C_8H_{10}N_2O_2$			
1,3-Cyclohexanedione, 2-diazo-5,5-dimethyl-	hexane	226(4.22),257(3.94)	104-1245-76
	H_2O	232(4.23),259(3.93)	104-1245-76
Diazene, (methoxymethyl)phenyl-, 2-oxide	EtOH	246(4.27)	44-1141-76
2(1H)-Pyrazinone, 5-acetyl-3,6-dimethyl-	EtOH	285(4.14)	18-2805-76
2(1H)-Pyridinone, 6-[1-(hydroxyimino)-ethyl]-1-methyl-, (E)-	H_2O	305.5(3.87)	83-0769-76
	pH 13	310.5(3.96)	83-0769-76
	MeOH	230(3.78),311(3.91)	83-0769-76
2(1H)-Pyridinone, 6-[1-(hydroxyimino)-ethyl]-1-methyl-, (Z)-	H_2O	304(3.79)	83-0769-76
	pH 13	305(3.83)	83-0769-76
	MeOH	231(3.77),309(3.85)	83-0769-76
$C_8H_{10}N_2O_2S$			
4(1H)-Pyrimidinone, 2,3-dihydro-5-(4-hydroxy-1-butenyl)-2-thioxo-	H_2O	284(4.16),307(4.18)	4-0929-76
$C_8H_{10}N_2O_2S_2$			
1H-1,2-Thiazine-4-carbonitrile, 5-methoxy-1-methyl-3-(methylthio)-, 1-oxide	EtOH	231(4.32),254(3.67), 315(3.91)	142-1875-76
$C_8H_{10}N_2O_3$			
Ethanamine, 2-(3-nitrophenoxy)-	n.s.g.	225(4.00),265(3.75), 329(3.29)	20-0421-76
Ethanol, 2-[(2-nitrophenyl)amino]-	EtOH	232(4.29),280(3.60), 410(3.80)	78-0057-76
Pyrazinecarboxylic acid, 4,5-dihydro-3,6-dimethyl-5-oxo-, methyl ester	EtOH	264(4.14),307(3.82)	18-2805-76
2,4(1H,3H)-Pyrimidinedione, 5-(4-hydroxy-1-butenyl)-, (E)-	H_2O	267(3.69),290(3.86)	4-0929-76
(Z)-	H_2O	258(3.67),277(3.82)	4-0929-76
2,4(1H,3H)-Pyrimidinedione, 5-(tetrahydro-2-furanyl)-	H_2O	263(3.89)	4-0929-76
$C_8H_{10}N_2O_3S$			
4-Oxazolecarboxamide, N-acetyl-2-methyl-5-(methylthio)-	$CHCl_3$	298(4.09)	94-0948-76
$C_8H_{10}N_2O_3S_3$			
4-Thiazolidinone, 3-ethyl-5-[1-(methylthio)-2-nitroethylidene]-2-thioxo-	EtOH	387(4.46)	94-1671-76
$C_8H_{10}N_2O_4$			
1H-Pyrazole-3-carboxylic acid, 5-acetyl-4,5-dihydro-4-oxo-, ethyl ester	EtOH	249s(3.991),287(4.124)	23-1752-76

Compound	Solvent	$\lambda_{max}(\log \epsilon)$	Ref.
1H-Pyrazole-3-carboxylic acid, 5-(acet-yloxy)-1-methyl-, methyl ester	MeOH	215(4.10)	24-0261-76
2,4(1H,3H)-Pyrimidinedione, 5-(tetra-hydro-3-hydroxy-2-furanyl)-	H_2O	263(3.90)	4-0929-76
5-Pyrimidinepropanoic acid, 1,2,3,4-tetrahydro-6-methyl-2,4-dioxo-	H_2O	266(3.92)	44-1858-76
Sydnone, 4-(3-ethoxy-3-oxo-1-propenyl)-3-methyl-, (E)-	EtOH	337(4.35)	48-0823-76
$C_8H_{10}N_2O_5S$			
2-Thiophenemethanol, α-(1-methyl-1-nitroethyl)-5-nitro-	MeOH	322(4.00)	12-0327-76
$C_8H_{10}N_3$			
Pyridinium, 1-cyano-4-(dimethylamino)-, perchlorate	H_2O	301(4.41)	77-0021-76
$C_8H_{10}N_4O$			
2H-Pyrrolo[2,3-d]pyrimidin-2-one, 4-ami-no-1,3-dihydro-3,5-dimethyl-	DMSO	320(3.85)	24-2983-76
Urea, N-(3-cyano-4-methyl-1H-pyrrol-2-yl)-N'-methyl-	MeOH	266(4.10)	24-2983-76
$C_8H_{10}N_4OS$			
Hypoxanthine, 1,3-dimethyl-8-(methyl-thio)-	neutral	291(4.204)	39-0090-76C
	cation	288(--)	39-0090-76C
Hypoxanthine, 1,7-dimethyl-8-(methyl-thio)-	neutral	277.5(4.160)	39-0090-76C
	cation	279(--)	39-0090-76C
Hypoxanthine, 1,9-dimethyl-8-(methyl-thio)-	neutral	272(4.135)	39-0090-76C
	cation	>276(--)	39-0090-76C
Hypoxanthine, 3,7-dimethyl-8-(methyl-thio)-	neutral	288(4.383)	39-0090-76C
	cation	287(--)	39-0090-76C
$C_8H_{10}N_4O_2$			
1H-Purine-2,6-dione, 3,7-dihydro-1-propyl-	pH 0.0	230(3.86),261(3.97)	39-1507-76C
	pH 5.0	201(4.22),269(4.04)	39-1507-76C
	pH 10.5	240(3.95),279(3.94)	39-1507-76C
	pH 14.0	217(4.15),283(3.90)	39-1507-76C
3H-Purine-2,8-dione, 1,7-dihydro-1,3,7-trimethyl-	H_2O	223(4.13),257(3.60), 321(3.94)	39-1414-76C
Theobromine, 1-allyl-	pH 1	274(4.03)	18-3224-76
	pH 7	275(4.01)	18-3224-76
	pH 13	275(4.01)	18-3224-76
Theophylline, 7-allyl-	pH 1	273(3.98)	18-3224-76
	pH 7	275(3.94)	18-3224-76
	pH 13	275(3.94)	18-3224-76
$C_8H_{10}N_4O_3$			
1H-Purine-2,6-dione, 3,7-dihydro-7-hydroxy-2-propyl-	pH 3.0	204(4.33),270(4.05)	39-1507-76C
	pH 7.6	226(4.20),256(3.92), 277(3.80)	39-1507-76C
	pH 11.9	228(4.30),294(3.79)	39-1507-76C
$C_8H_{10}N_4O_5$			
Carbamic acid, (6-amino-1,4-dihydro-5-nitro-4-oxo-2-pyridinyl)-, ethyl ester	pH 7	255(4.07),333(3.99)	44-3784-76
$C_8H_{10}N_6O_2$			
9H-Purine-9-propanoic acid, α,6-diamino-	pH 1	260(4.15)	123-0598-76

Compound	Solvent	λ_{max}(log ϵ)	Ref.
$C_8H_{10}O$			
Benzene, 1-methoxy-2-methyl-	EtOH	270(3.29),277(3.27)	33-1763-76
Bicyclo[3.2.0]hept-2-en-7-one, 1-methyl-	EtOH	302(2.46)	33-1745-76
Bicyclo[3.2.0]hept-2-en-7-one, 2-methyl-	EtOH	302(2.49)	33-1745-76
Bicyclo[3.2.0]hept-2-en-7-one, 3-methyl-	EtOH	301(2.53)	33-1745-76
2,4-Cyclohexadiene-1-carboxaldehyde, 1-methyl-	EtOH	259(3.60)	33-1745-76
2,4-Cyclohexadien-1-one, 6,6-dimethyl-	4:1 isopen-tane-C_6H_{11}-Me	294(3.72),350s(1.91), 361(2.01),379(2.00), 395s(1.83),415(1.38)	24-1332-76
	EtOH	301(3.71),355s(1.31)	24-1332-76
	CF_3CH_2OH	305(3.71)	24-1332-76
2-Cyclohexen-1-one, 3-methyl-6-methyl-ene-	MeOH	255(4.13)	35-3393-76
2-Cyclopenten-1-one, 2-cyclopropyl-	EtOH	236(3.90)	39-0249-76C
2-Cyclopenten-1-one, 2-(2-propenyl)-	EtOH	227(3.97)	39-0249-76C
8-Oxabicyclo[5.1.0]octa-2,4-diene, 2-methyl-	EtOH	246(3.74)	33-1756-76
8-Oxabicyclo[5.1.0]octa-2,4-diene, 6-methyl-	EtOH	245(3.54)	33-1756-76
2-Propen-1-one, 1-(1-cyclopenten-1-yl)-	EtOH	260(4.04)	78-2533-76
$C_8H_{10}O_2$			
3(2H)-Benzofuranone, 4,5,6,7-tetrahydro-	EtOH	270(3.68)	22-1913-76
Bicyclo[2.2.2]octane-2,3-dione	hexane	478(1.86)	118-0256-76
5-Cyclooctene-1,2-dione	C_6H_{12}	230(2.00),281(1.55), 288(1.52),345(1.24)	44-3760-76
$C_8H_{10}O_2S_2$			
3-Cyclobutene-1,2-dione, 3,4-bis(ethyl-thio)-	C_6H_{12}	213(4.07),290s(4.17), 302s(4.29),311(4.36), 321s(4.22)	44-3904-76
$C_8H_{10}O_3$			
1,4-Benzenediol, 2-methoxy-6-methyl-	MeOH	208(4.10),289(3.59)	1-0368-76
2-Cyclopenten-1-one, 4,5-dihydroxy-3-(1-propenyl)- (terrein)	MeOH	271(4.41)	120-0120-76
1,3-Dioxolane, 2-(2-furanyl)-4-methyl-	n.s.g.	211(3.91)	103-1100-76
1,3-Dioxolane, 2-(5-methyl-2-furanyl)-	n.s.g.	219(3.94)	103-1100-76
2-Furancarboxylic acid, propyl ester	n.s.g.	254(4.14)	103-1100-76
2-Furancarboxylic acid, 5-methyl-, ethyl ester	n.s.g.	270(4.14)	103-1100-76
2,4-Heptadienoic acid, 5-methyl-6-oxo-	EtOH	282(4.42)	1-0064-76
2-Oxabicyclo[3.2.1]octane-4-carboxalde-hyde, 3-oxo-	MeOH	247(3.98)	39-0264-76C
	MeOH-NaOH	283(--)	39-0264-76C
2-Oxabicyclo[3.2.1]oct-3-ene-4-carbox-ylic acid	MeOH	246(4.10)	39-0264-76C
2H-Pyran-2-one, 3-ethyl-4-hydroxy-6-methyl-	EtOH	290(3.92)	78-0269-76
	EtOH-NaOH	289(3.98)	78-0269-76
$C_8H_{10}O_3S$			
3-Cyclobutene-1,2-dione, 3-ethoxy-4-(ethylthio)-	C_6H_{12}	258(3.98),285(4.34)	44-3904-76
2-Thiopheneacetic acid, α-hydroxy-, ethyl ester, (R)-	EtOH	224(3.62)	12-2459-76
$C_8H_{11}AsO$			
Arsenin, 4-propoxy-	EtOH	226(3.64),290(3.74)	88-4143-76
4(1H)-Arseninone, 1-propyl-	EtOH	213(3.85),316(3.90)	88-4143-76

Compound	Solvent	$\lambda_{max}(\log \epsilon)$	Ref.
$C_8H_{11}ClN_4O_4$ α-D-Xylofuranuronimidoyl chloride, 3-azido-3-deoxy-N-hydroxy-1,2-O- (1-methylethylidene)-	EtOH	207(3.36)	136-0119-76A
$C_8H_{11}FO$ 2-Cyclohexen-1-one, 6-fluoro-	C_6H_{12}	218(4.13)	78-2099-76
$C_8H_{11}IN_4O$ Hypoxanthinium, 1,3,8-trimethyl-, iodide	MeOH	266(3.88)	39-0239-76C
2H-Purinium, 3,7,8,9-tetrahydro-1,3,9- trimethyl-2,8-dioxo-, iodide	H_2O	226(4.36),340(4.03)	39-1414-76C
$C_8H_{11}IO$ Pyrylium, 2,4,6-trimethyl-, iodide	CH_2Cl_2	450(2.16)	59-1311-76
$C_8H_{11}N$ Benzenamine, N,4-dimethyl-	H_2O	238(4.02),290(3.22)	35-7777-76
protonated	H_2O	210(3.85),218s(3.66)	35-7777-76
Methanamine, 1-(2,4-cyclopentadien- 1-ylidene)-N,N-dimethyl-	hexane	316(4.49)	39-0048-76C
$C_8H_{11}NO$ Benzenamine, 4-methoxy-N-methyl-	H_2O	238(4.02),300(3.31)	35-7777-76
3H-Indol-3-one, 1,2,4,5,6,7-hexahydro-	MeOH	318(4.37)	5-0383-76
	$CHCl_3$	314(4.28)	89-0052-76
Methanol, methylphenylamino-	H_2O	239(3.97),285(3.18)	35-7777-76
Phenol, 3-(2-aminoethyl)-	EtOH	274(3.37)	95-1031-76
Pyrrole, 3-acetyl-2,5-dimethyl-	EtOH	246(3.96),294(3.77)	44-0351-76
$C_8H_{11}NOS$ Thiocyanic acid, 2-formyl-1-methyl- 1-pentenyl ester, (E)-	EtOH	295(3.80)	97-0049-76
$C_8H_{11}NOS_4$ 4-Thiazolinone, 5-[bis(methylthio)meth- ylene]-3-ethyl-2-thioxo-	EtOH	290(3.98),320(3.86), 420(4.42)	94-1671-76
$C_8H_{11}NO_2$ Benzenamine, N-(hydroxymethyl)-4-meth- oxy-	H_2O	238(3.97),298(3.25)	35-7777-76
Benzenemethanol, α-(aminomethyl)-3-hy- droxy-	EtOH	275.5(3.25)	95-1031-76
$C_8H_{11}NO_3$ 1,2,4-Benzenetriol, 5-(2-aminoethyl)-	pH 5.6	295(3.64)	39-0339-76C
2-Pyrrolidinone, N-(trans-3-methoxy- acryloyl)-	n.s.g.	271(4.34)	39-1241-76C
$C_8H_{11}NO_4$ 4-Isoxazoleacetic acid, 2,3-dihydro- 2,5-dimethyl-3-oxo-, methyl ester	MeOH	233(3.83)	1-0567-76
4-Isoxazoleacetic acid, 3-methoxy- 5-methyl-, methyl ester	MeOH	211(3.78)	1-0567-76
$C_8H_{11}NS$ 4H-Cyclohept[d]isothiazole, 5,6,7,8- tetrahydro-	EtOH	254(3.846)	48-0507-76

Compound	Solvent	$\lambda_{max}(\log \epsilon)$	Ref.
$C_8H_{11}NS_2$			
1,2-Benzisothiazole, 4,5,6,7-tetrahydro-3-(methylthio)-	EtOH	234s(3.4),279(4.03), 280(4.10)	48-0127-76
$C_8H_{11}NS_3$			
1,4,2-Benzodithiazine, 5,6,7,8-tetrahydro-3-(methylthio)-	C_6H_{12}	259(3.98),285s(3.3), 341(2.78)	48-0127-76
$C_8H_{11}N_2O$			
Pyridinium, 1-methyl-3-[(methylamino)-carbonyl]-, iodide	EtOH	280(4.26)	39-0315-76C
$C_8H_{11}N_2OS$			
Isothiouronium, S-[(Z)-1-phenyloxidodiazenylmethyl]-, bromide	EtOH	260(4.28)	44-1141-76
$C_8H_{11}N_3$			
3-Cinnolinamine, 5,6,7,8-tetrahydro-	MeOH	236(4.16),310(3.55)	48-0663-76
$C_8H_{11}N_3OS$			
Hydrazinecarbothioamide, 2-(4-methoxyphenyl)-	EtOH	244(4.22),295(3.25)	104-2374-76
$C_8H_{11}N_3O_2$			
Acetamide, N-(3,4-dihydro-3,6-dimethyl-4-oxo-2-pyrimidinyl)-	dioxan	221(4.18),270(4.14)	114-0413-76A
1,2-Ethanediamine, N-(2-nitrophenyl)-, hydrobromide	EtOH	230(4.30),280(3.66), 417(3.73)	78-0057-76
hydrochloride	EtOH	231(4.24),279(3.60), 416(3.86)	78-0057-76
$C_8H_{11}N_3O_2S$			
Acetic acid, [(4-amino-2-pyrimidinyl)-thio]-, ethyl ester	pH 2.0	240(4.49)	24-3615-76
	pH 5.5 and 11	222(4.41),245s(4.03), 284(3.81)	24-3615-76
2-Butenoic acid, 3-(1,3,4-thiadiazol-2-ylamino)-, ethyl ester	isoPrOH	250s(3.51),306(4.40)	4-0291-76
$C_8H_{11}N_3O_5$			
2'-Deoxy-6-azauridine	MeOH	265(3.71)	73-2110-76
$C_8H_{11}N_3S$			
Hydrazinecarbothioamide, 2-(4-methylphenyl)-	EtOH	243(4.33),280s(3.23)	104-2374-76
$C_8H_{11}N_4O$			
Hypoxanthinium, 1,3,8-trimethyl-, iodide	MeOH	266(3.88)	39-0239-76C
2H-Purinium, 3,7,8,9-tetrahydro-1,3,9-trimethyl-2,8-dioxo-, iodide	H_2O	226(4.36),340(4.03)	39-1414-76C
$C_8H_{11}N_5$			
Pyrazolo[3,4-d]pyrimidin-4-amine, N-(1-methylethyl)-	pH 1	265(4.03)	87-0555-76
	pH 13	269(4.09)	87-0555-76
	EtOH	279(4.19)	87-0555-76
$C_8H_{11}N_5O$			
[1,2,4]Triazolo[1,5-a]pyrimidine-5-methanol, 7-(ethylamino)-	MeOH	224(--),265(3.76), 294(4.22)	106-0051-76

Compound	Solvent	$\lambda_{max}(\log \epsilon)$	Ref.
$C_8H_{11}N_5O_2$			
Urea, N,N''-(5-amino-1,3-phenylene)bis-	MeOH	228(4.66),243(4.32), 278(3.11),290(3.00)	87-0426-76
$C_8H_{11}O$			
Pyrylium, 2,4,6-trimethyl-, iodide	CH_2Cl_2	450(2.16)	59-1311-76
thiocyanate	CH_2Cl_2	400(2.37)	59-1311-76
selenocyanate	CH_2Cl_2	445(2.35)	59-1311-76
C_8H_{12}			
Cyclohexane, 1,2-bis(methylene)-	EtOH	220(3.80)	44-1635-76
$C_8H_{12}BrN_3O_2$			
1,2-Ethanediamine, N-(2-nitrophenyl)-, hydrobromide	EtOH	230(4.30),280(3.66), 417(3.73)	78-0057-76
$C_8H_{12}ClN_3O$			
4-Pyrimidinamine, 6-butoxy-2-chloro-	MeOH	210(4.57),238(3.91)	94-0507-76
$C_8H_{12}ClN_3O_2$			
1,2-Ethanediamine, N-(2-nitrophenyl)-, hydrochloride	EtOH	231(4.24),279(3.60), 416(3.86)	94-0507-76
$C_8H_{12}Cl_3NO$			
3-Butenamide, 3,4,4-trichloro-N,N-di-ethyl-	heptane	209(4.11)	24-2159-76
$C_8H_{12}F_3N_5$			
1,3,5-Triazine-2,4-diamine, N-butyl-6-(trifluoromethyl)-	MeOH	213(4.53),273(3.57)	36-0525-76
$C_8H_{12}NO$			
Pyridinium, N-methoxy-2,3-dimethyl-,	DMF	350(2.53)	83-0592-76
methosulfate	DMSO	368(2.57)	83-0592-76
	HMPT	330(2.60)	83-0592-76
Pyridinium, N-methoxy-2,6-dimethyl-,	DMF	369(2.81)(changing)	83-0592-76
perchlorate	DMSO	374(2.86)(changing)	83-0592-76
	HMPT	359(2.57)(changing)	83-0592-76
Pyridinium, N-methoxy-3,5-dimethyl-,	DMF	356(3.74)(changing)	83-0592-76
perchlorate	DMSO	370(3.59)(changing)	83-0592-76
	HMPT	361(4.07)(changing)	83-0592-76
$C_8H_{12}N_2$			
1H-Benzimidazole, 4,5,6,7-tetrahydro-1-methyl-	EtOH	226(3.79)	44-0019-76
Cycloheptimidazole, 1,4,5,6,7,8-hexa-hydro-	MeOH	219(3.98)	44-0019-76
1-Cyclohexene-1-carbonitrile, 2-(methyl-amino)-	EtOH	276(4.12)	44-0019-76
$C_8H_{12}N_2O$			
1H-Imidazole, 1-(2,2-dimethyl-1-oxo-propyl)-	THF	245(3.74)	33-2738-76
1H-Imidazole, 2-(2,2-dimethyl-1-oxo-propyl)-	EtOH	278(4.12)	33-2738-76
1H-Imidazole, 4-(2,2-dimethyl-1-oxo-propyl)-	EtOH	257(4.09)	33-2738-76
3-Pyridinecarboxamide, 1,4-dihydro-N,1-dimethyl-	MeCN	347(3.74)	44-2976-76

Compound	Solvent	$\lambda_{max}(\log \epsilon)$	Ref.
$C_8H_{12}N_2OS$			
4(1H)-Pyrimidinone, 2-(butylthio)-	EtOH	232(3.90),289(3.85)	95-1094-76
4(1H)-Pyrimidinone, 6-methyl-2-(propylthio)-	EtOH	234(3.89),287(3.84)	95-1094-76
4(1H)-Pyrimidinone, 2-(methylthio)-6-propyl-	EtOH	233(3.99),285(3.90)	95-1094-76
$C_8H_{12}N_2O_2$			
2,4-Hexanedione, 3-diazo-5,5-dimethyl-	hexane	230(4.26),243s(4.18), 364(1.48)	104-1245-76
	pH 1	234(4.17),243(4.11), 355(1.77)	104-1245-76
	H_2O	233(4.21),244s(4.12), 345(1.75)	104-1245-76
4-Oxa-1,2-diazaspiro[4.6]undec-1-en-3-one	n.s.g.	225(3.56),372(2.49)	4-0649-76
$C_8H_{12}N_2O_3$			
Acetic acid, [3-(hydroxyimino)-2-piperidinylidene]-, methyl ester	n.s.g.	230(5.15),338(4.90)	39-1241-76C
3,5-Hexadien-2-one, 6-(dimethylamino)-3-nitro-	EtOH	325(4.00),417(4.72)	70-0577-76
4(1H)-Pyrimidinone, 5-ethyl-2,6-dimethoxy-	MeOH	230(3.69),270(4.00)	94-0026-76
$C_8H_{12}N_2O_3S_2$			
Sulfoximine, N-[2-cyano-3-methoxy-1-(methylthio)-3-oxo-1-propenyl]-S,S-dimethyl-	EtOH	246(3.78),318(4.27)	142-1875-76
$C_8H_{12}N_2O_4$			
2-Butenedioic acid, 2-(methylmethylenehydrazino)-, dimethyl ester, (E)-	MeOH	292(4.55)	24-2154-76
1H-Pyrazole-3-carboxylic acid, 5-hydroxy-4-(2-hydroxyethyl)-1-methyl-, methyl ester	MeOH	226(4.02),259s(3.52)	24-0261-76
$C_8H_{12}N_2O_7$			
2H-1,2,4-Oxadiazine-3,5(4H,6H)-dione, 2-β-D-ribofuranosyl-	EtOH	219(3.26)	44-3128-76
$C_8H_{12}N_2O_8$			
1,4-Butanediol, 2,3-dinitro-, diacetate	EtOH-NaOH	248(3.88)	104-2039-76
$C_8H_{12}N_4O_2$			
Acetamide, N-methyl-N-(2,6-dimethyl-5-oxo-1,2,3-triazin-3-yl)-	dioxan	246(4.10)	114-0413-76A
$C_8H_{12}N_4O_2S_3$			
3-Thiazolidinebutanoic acid, 4-[(aminothioxomethyl)hydrazono]-2-thioxo-	MeOH	256(4.27),295(4.15)	103-0749-76
$C_8H_{12}N_4O_3$			
Urea, N-methyl-N'-(1,2,3,4-tetrahydro-1,3-dimethyl-2,4-dioxo-5-pyrimidinyl)-	H_2O	208(3.97),277(3.81)	39-1414-76C
$C_8H_{12}N_4O_4$			
1H-1,2,3-Triazole-4-carboxamide, 1-(2-deoxy-β-D-ribofuranosyl)-	MeOH	210(4.08)	44-1041-76

Compound	Solvent	$\lambda_{max}(\log \epsilon)$	Ref.
$C_8H_{12}N_4O_5$			
1H-1,2,3-Triazole-4-carboxamide, 1-β-D-ribofuranosyl-	H_2O	210(4.09)	44-1041-76
1H-1,2,3-Triazole-5-carboxamide, 4-β-D-ribofuranosyl-	EtOH	210(4.00)	44-0084-76
$C_8H_{12}N_6$			
1H-Imidazole-4-carbonitrile, 5-(3,3-diethyl-1-triazenyl)-	EtOH	224(4.01),333(4.03)	103-0465-76
9H-Purine-9-propanamine, 6-amino-	pH 1	257(4.01)	70-0384-76
	pH 7	261(4.11)	70-0384-76
	pH 13	261(4.11)	70-0384-76
$C_8H_{12}N_6O$			
9H-Purine-9-ethanol, 6-amino-α-(aminomethyl)-	pH 1	257(4.09)	70-0384-76
	pH 7	261(4.11)	70-0384-76
	pH 13	261(4.12)	70-0384-76
$C_8H_{12}O$			
Bicyclo[2.2.1]heptan-2-one, 1-methyl-	3-Mepentane	290(1.33)	144-0475-76
	MeCN	288(1.37)	144-0475-76
Bicyclo[2.2.1]heptan-2-one, 3-methyl-	3-Mepentane	295(1.27)	144-0475-76
	MeCN	293(1.30)	144-0475-76
isomer	MeCN	291(1.43)	144-0475-76
2-Cyclohexen-1-one, 6,6-dimethyl-	EtOH	228(4.00)	33-2012-76
2-Cyclopenten-1-one, 3-(1-methylethyl)-	EtOH	226(4.17)	28-0761-76A
2-Cyclopenten-1-one, 3,5,5-trimethyl-	EtOH	229(4.10)	23-1449-76
Furan, 2-methyl-5-(1-methylethyl)-	n.s.g.	220(3.46)	33-1253-76
3,4-Heptadien-2-one, 6-methyl-	n.s.g.	219(3.98)	33-1253-76
2-Octyn-4-one	C_6H_{12}	310(1.43)	35-2683-76
2-Propenal, 3-cyclopentyl-	EtOH	229(4.21)	22-0849-76
$C_8H_{12}O_2$			
1,3-Cyclobutanedione, 2,2,4,4-tetramethyl-	C_6H_{12}	309(1.76),350(1.40)	59-1415-76
	EtOH	306(--),343(--)	59-1415-76
Cyclohexanone, 2-(hydroxymethylene)-6-methyl-	EtOH	280(4.28),350(2.05)	65-2037-76
2-Cyclohexen-1-one, 2-hydroxy-3,5-dimethyl-	EtOH	272(3.99)	118-0541-76
2-Cyclohexen-1-one, 2-hydroxy-5,5-dimethyl-	EtOH	268(3.82)	118-0541-76
2-Cyclohexen-1-one, 3-hydroxy-5,5-dimethyl-	MeOH	258(4.27),288s(3.83)	30-0366-76
sodium salt	MeOH	286(4.49)	30-0366-76
2,4-Heptadienoic acid, methyl ester, trans	EtOH	262(4.40)	70-0547-76
3,5-Heptadienoic acid, methyl ester, trans	EtOH	233(4.32)	70-0547-76
3,5-Heptadienoic acid, 6-methyl-	$C_6H_{11}Me$	238(4.38)	24-1332-76
	EtOH	238(4.38)	24-1332-76
1-Propanone, 1-(1-cyclopenten-1-yl)-3-hydroxy-	EtOH	240(3.84)	78-2533-76
4H-Pyran-4-one, 2,3-dihydro-6-propyl-	EtOH	266.5(3.98)	4-0253-76
$C_8H_{12}O_2S$			
Acetic acid, (tetrahydro-4H-thiopyran-4-ylidene)-, methyl ester	C_6H_{12}	210(4.20),254(3.46)	78-2827-76
$C_8H_{12}O_3$			
Cyclopentanecarboxylic acid, 4,4-dimethyl-2-oxo-	hexane	258(3.58)	104-1252-76
	H_2O	261(2.69),313s(2.15)	104-1252-76

$C_8H_{12}O_4-C_8H_{13}N_3$

Compound	Solvent	$\lambda_{max}(\log \epsilon)$	Ref.
Hex-2-eno-DL-pyranosid-4-ulose, 5-C-methyl-2,3,6-trideoxy-, methyl ester	n.s.g.	204(3.85),208(3.80), 214(3.73)	78-1051-76
$C_8H_{12}O_4$			
Griffonilide, tetrahydro-	MeOH	213(3.1)	100-0385-76
$C_8H_{12}S_2$			
1,3-Cyclobutanedithione, 2,2,4,4-tetramethyl-	C_6H_{12}	500(1.34)	59-1415-76
	EtOH	490(--)	59-1415-76
$C_8H_{12}S_3$			
2,4,9-Trithiatricyclo[3.3.1.1^{3,7}]decane, 1-methyl-	hexane	247(3.35)	11-0220-76A
2,4,9-Trithiatricyclo[3.3.1.1^{3,7}]decane, 7-methyl-	hexane	248(3.39)	11-0220-76A
$C_8H_{13}NO$			
2-Cyclohexen-1-one, 3-amino-5,5-dimethyl-	pH 1	276(4.33)	39-2207-76C
	H_2O	287(4.42)	39-2207-76C
	pH 13	288(4.42)	39-2207-76C
2-Cyclohexen-1-one, 3-(dimethylamino)-	MeOH	298(4.52)	35-2826-76
3,5-Heptadienamide, 6-methyl-, trans	EtOH	230(4.55)	104-0307-76
4(1H)-Pyridinone, 2,3-dihydro-6-propyl-	EtOH	304(4.03)	4-0253-76
4(1H)-Pyridinone, 2,3-dihydro-1,2,6-trimethyl-	EtOH	323(4.11)	4-0253-76
2-Pyrrolidinone, 1-ethyl-4-methyl-3-methylene-	n.s.g.	266(3.10)	48-0471-76
2H-Pyrrol-2-one, 3,4-diethyl-1,5-dihydro-	MeOH	217(4.21)	118-0335-76
Tropinone	hexane	241(2.70)	35-3242-76
$C_8H_{13}NO_2$			
Cyclohexanecarboxamide, N-methyl-2-oxo-	H_2O	256(2.00)	44-0013-76
	pH 13	284(4.02)	44-0013-76
3(4H)-Pyridinone, 5,6-dihydro-2,6,6-trimethyl-, 1-oxide	C_6H_{12}	288(4.25)	39-1951-76C
$C_8H_{13}NO_2S$			
Butanoic acid, 2-isothiocyanato-3-methyl-, ethyl ester	pH 8.06	245(3.06)	73-2216-76
	MeOH	247(3.06)	73-2216-76
	dioxan	248(3.09)	73-2216-76
$C_8H_{13}NS$			
Isothiazole, 4-butyl-5-methyl-	EtOH	217(3.558),252(3.726)	48-0507-76
Thiazole, 5-methyl-4-(1-methylpropyl)-	EtOH	252(3.44)	70-1353-76
$C_8H_{13}N_2O$			
Pyrimidinium, 1,3-diethyl-2,3-dihydro-2-oxo-, chloride	pH 0.29	317(3.91)	23-2681-76
pseudobase	pH 9.18	238(3.71),300(2.84)	23-2681-76
iodide	cation	223(4.0+),317(3.90)	23-2681-76
	pseudobase	225(4+),299(3.12)	23-2681-76
bisulfate	cation	317(3.92)	23-2681-76
	pseudobase	238(3.69),300(2.83)	23-2681-76
Pyrimidinium, 2,3-dihydro-1-methyl-3-(1-methylethyl)-2-oxo-	pH 0.29	317(3.92)	23-2681-76
	pH 9.18	238(3.76),298(3.02)	23-2681-76
$C_8H_{13}N_3$			
Piperidine, 1-(2-cyano-1-iminoethyl)-	pH 7.2	214(4.04)	39-0125-76C
	pH 12.0	265(4.21)	39-0125-76C

Compound	Solvent	$\lambda_{max}(\log \epsilon)$	Ref.
$C_8H_{13}N_3O$			
1H-Imidazole-1-carboxamide, N,N-diethyl-	EtOH	209(4.04)	33-2738-76
1H-Imidazole-2-carboxamide, N,N-diethyl-	EtOH	258(3.99)	33-2738-76
1H-Imidazole-4-carboxamide, N,N-diethyl-	EtOH	237(3.88)	33-2738-76
$C_8H_{13}N_3OS$			
4(1H)-Pyrimidinone, 6-amino-2-(butyl-thio)-	EtOH	232(4.25),278(4.01)	95-1094-76
$C_8H_{13}N_3OS_2$			
Thiourea, N-(4,5-dihydro-4,4-dimethyl-5-oxo-2-thiazolyl)-N,N'-dimethyl-	EtOH	237(4.05),269(4.12)	33-2768-76
$C_8H_{13}N_3O_2S$			
1H-Imidazole-5-carboxylic acid, 4-amino-1-methyl-2-(methylthio)-, ethyl ester	EtOH	221(4.08),296(4.23)	49-1413-76
C_8H_{14}			
2,4-Hexadiene, 2,3-dimethyl-	heptane	231(3.08)	70-0841-76
$C_8H_{14}Ge$			
Germa-3,5-cycloheptadiene, 1,1-dimethyl-	heptane	232(3.78),244(3.86), 249(3.88)	70-0602-76
$C_8H_{14}I_2N_6$			
2,8-Diamino-1,7,9-trimethyl-9H-purinium diiodide	H_2O	230(4.76),321(3.96)	39-1414-76C
$C_8H_{14}N_2O_2$			
Ethanone, 1-(3,4-dihydro-2,2-dimethyl-2H-pyrrol-5-yl)-, oxime, N-oxide	MeOH	279(4.13)	39-1951-76C
Ethanone, 1-(3,4-dihydro-3,3-dimethyl-2H-pyrrol-5-yl)-, oxime, N-oxide	MeOH	279(4.11)	39-1951-76C
2H-Pyrrole-5-carboxaldehyde, 3,4-dihydro-2,2,3-trimethyl-, oxime, 1-oxide	MeOH	293(4.11)	39-1951-76C
$C_8H_{14}N_2O_3$			
2-Propenoic acid, 3-(2-acetyl-1,2-di-methylhydrazino)-, methyl ester, (E)-	MeOH	266(4.34)	24-2154-76
$C_8H_{14}N_2O_4$			
2-Butene, 1,4-bis(ethyl-aci-nitro)-	CH_2Cl_2	240(3.67),338(4.65), 354(4.81)	104-2039-76
2-Butenedioic acid, 2-(1,2-dimethyl-hydrazino)-, dimethyl ester, (E)-	MeOH	277(4.34)	24-0253-76
$C_8H_{14}N_2S$			
1H-Imidazole, 1-(1,1-dimethylethyl)-	EtOH	220-253(4.290-3.817)	28-0143-76B
2-(methylthio)-, monohydriodide	CH_2Cl_2	232-255(4.257-3.830)	28-0143-76B
2H-Imidazole-2-thione, 1-(1,1-dimethyl-ethyl)-1,3-dihydro-3-methyl-	EtOH	265(4.052)	28-0143-76B
	CH_2Cl_2	273(4.074)	28-0143-76B
$C_8H_{14}N_4O_3$			
1,2-Cyclopentanediol, 3-(5-amino-1H-1,2,4-triazol-3-yl)-5-(hydroxy-methyl)-, (1α,2α,3β,5β)-	MeOH	214(3.40)	23-0861-76
$C_8H_{14}N_4O_3S$			
as-Triazin-5(2H)-one, 3-(2,2-dimethoxy-ethylamino)-6-methyl-	EtOH	210(4.38),244(3.84)	114-0413-76A

Compound	Solvent	$\lambda_{max}(\log \epsilon)$	Ref.
$C_8H_{14}O$			
Cyclobutanone, 2,2,4,4-tetramethyl-	C_6H_{12}	311(1.38)	112-0777-76
Cyclohexanone, 3-ethyl-	EtOH	285(1.32)	44-3076-76
4-Hexen-3-one, 2,5-dimethyl-	EtOH	242(3.40)	78-2725-76
4-Hexen-3-one, 4,5-dimethyl-, trans	C_6H_{12}	239(3.77),314(1.79)	23-2127-76
	H_2O	251(3.65)	23-2127-76
$C_8H_{14}O_2$			
Acetic acid, cyclohexyl ester	neat	324s(--)	44-3312-76
	hexane	211.7(1.81)	44-3312-76
	H_2O	208s(--)	44-3312-76
	MeCN	210.2(1.81)	44-3312-76
Hexanoic acid, ethyl ester	hexane	214.8(1.95)	44-3312-76
	H_2O	208.6(1.96)	44-3312-76
	MeCN	212.6(1.95)	44-3312-76
4-Hexen-3-one, 2-hydroxy-2,5-dimethyl-	n.s.g.	240(4.10)	23-1449-76
$C_8H_{14}O_2Si$			
1-Oxa-2-silacyclohexa-3,5-diene, 2-methoxy-2,3,6-trimethyl-	C_6H_{12}	273.0(3.93)	44-1799-76
$C_8H_{14}O_3Si$			
1,3-Dioxa-2-silacyclohepta-4,6-diene, 2-methoxy-2,4,7-trimethyl-	C_6H_{12}	248.0(4.02)	44-1799-76
$C_8H_{14}Si$			
Silacyclohepta-3,5-diene, 1,1-dimethyl-	C_6H_{12}	247.0(3.48)	44-1799-76
$C_8H_{15}NO$			
3-Buten-2-one, 4-(diethylamino)-	C_6H_{12}	292(4.33)	35-2826-76
	H_2O	306(4.48)	35-2826-76
	MeOH	305(4.44)	35-2826-76
3-Buten-2-one, 4-[(1,1-dimethylethyl)- amino]-	H_2O	305(4.47)	35-2826-76
	MeOH	305(4.39)	35-2826-76
cis-s-cis	C_6H_{12}	302(4.21)	35-2826-76
1-Hexen-3-one, 1-(dimethylamino)-	C_6H_{12}	289(4.39)	35-2826-76
	H_2O	307(4.46)	35-2826-76
	MeOH	305(4.43)	35-2826-76
1-Penten-3-one, 1-(dimethylamino)-4- methyl-	C_6H_{12}	290(4.34)	35-2826-76
	H_2O	310(4.47)	35-2826-76
	MeOH	306(4.42)	35-2826-76
1-Penten-3-one, 4,4-dimethyl-1-(methyl- amino)-	H_2O	308(4.33)	35-2826-76
	MeOH	303(4.29)	35-2826-76
cis-s-cis	C_6H_{12}	300(4.21)	35-2826-76
1-Penten-3-one, 1-(ethylamino)-4-methyl-	H_2O	304(4.35)	35-2826-76
	MeOH	304(4.30)	35-2826-76
cis-s-cis	C_6H_{12}	300(4.23)	35-2826-76
3-Penten-2-one, 4-(propylamino)-	H_2O	312(4.34)	35-2826-76
	MeOH	312(4.25)	35-2826-76
cis-s-cis	C_6H_{12}	306(4.19)	35-2826-76
2H-Pyrrole, 5-ethyl-3,4-dihydro-3,3-di- methyl-, 1-oxide	EtOH	231(3.90)	39-1951-76C
$(C_8H_{15}NO)_n$			
Poly(N-sec-butyl-N-methylacrylamide)	MeOH	200(3.84)	108-0039-76
	$(CF_3)_2CHOH$	199(3.84)	108-0039-76
$C_8H_{15}N_2S$			
1H-Imidazolium, 1-methyl-3-(1-methyl- ethyl)-2-(methylthio)-, iodide	EtOH	222-262(4.318-3.586)	28-0143-76B
	CH_2Cl_2	243(4.294)	28-0143-76B

Compound	Solvent	$\lambda_{max}(\log \epsilon)$	Ref.
$C_8H_{15}N_5OS$ 1,2,4-Thiadiazole-2(3H)-carboximidamide, 5-ethoxy-3-imino-N-propyl-	MeOH	272(4.04)	18-3170-76
$C_8H_{16}N_2O$ 2-Propenamide, N,N-diethyl-3-(methyl-amino)-	C_6H_{12}	282(4.31)	78-1025-76
	MeOH	279(4.31)	78-1025-76
	dioxan	280(4.14)	78-1025-76
2-Propenamide, N,N-dimethyl-3-[(1-methylethyl)amino]-	C_6H_{12}	284(4.31)	78-1025-76
	MeOH	283(4.40)	78-1025-76
	dioxan	284(4.29)	78-1025-76
$C_8H_{16}N_2O_2$ Butanal, 4-(butylnitrosoamino)-	EtOH	234(3.88),350(1.96)	88-0593-76
$C_8H_{16}N_2O_3$ 3-(Dimethylamino)propenylidenedimethyl-ammonium bicarbonate	EtOH	312(4.65)	70-0577-76
$C_8H_{16}N_2O_4$ Butane, 2,3-bis(ethyl-aci-nitro)-	CH_2Cl_2	252(3.30)	104-2039-76
$C_8H_{16}N_2S$ Pyrimidine, 1,4,5,6-tetrahydro-1-(1-methylethyl)-2-(methylthio)-, monohydriodide	EtOH	221(4.425)	28-0485-76A
	CH_2Cl_2	232(4.371)	28-0485-76A
$C_8H_{16}N_4O_2$ 4H-1,2,4-Triazol-3-amine, 4-(2,2-diethoxyethyl)-	EtOH	210(3.54)	78-0341-76
$C_8H_{16}N_6O$ 1,3,4-Triazahex-2-en-4-one, 1-(3,5-dimethyl-1,2,4-triazol-4-yl)-	H_2O	193(4.06)	103-0711-76
$C_8H_{16}O$ 3-Pentanone, 2,2,4-trimethyl-	C_6H_{12}	296(1.34)	112-0777-76
$C_8H_{16}O_2$ Hexanoic acid, ethyl ester	neat	284s(--)	44-3312-76
	hexane	211.6(1.86)	44-3312-76
	H_2O	204.4(--)	44-3312-76
	MeCN	209.2(1.87)	44-3312-76
$C_8H_{16}O_3S$ 1,3-Dioxane, 2-(1-methylethyl)-5-(methylsulfinyl)-, cis	C_6H_{12}	202.5(3.17)	35-0956-76
trans	C_6H_{12}	208(3.53)	35-0956-76
$C_8H_{16}O_6$ L-Sorbose, 5,6-di-O-methyl-	MeOH	283(1.54)	5-0269-76
$C_8H_{17}NO$ 2-Butanone, 4-(diethylamino)-	hexane	199(3.61)	35-3242-76
$C_8H_{17}N_2O_4P$ Acetic acid, [(diethoxymethylphosphoranylidene)hydrazono]-, methyl ester	EtOH	275.0(4.04)	65-1424-76

Compound	Solvent	$\lambda_{max}(\log \epsilon)$	Ref.
$C_8H_{18}ClO_2PSe$ Phosphorochloridoselenoic acid, 0,0- bis(2-methylpropyl) ester	hexane EtOH	190(4.08),244(3.26) 194(3.91),242(3.26)	70-0174-76 70-0174-76
$C_8H_{18}ClPSe$ Phosphinoselenoic chloride, dibutyl-	C_6H_{12}	201(3.43),265(3.89)	70-0174-76
$C_8H_{18}N_2$ Diazene, bis(1,1-dimethylethyl)-	gas	168(3.83),200(3.26), 368(1.08)	35-0109-76
$C_8H_{18}N_2O$ Pentanal, 5-hydroxy-, (1-methylethyl)- hydrazone Pentanal, 5-hydroxy-, propylhydrazone	EtOH EtOH	235(3.53) 237(3.51)	104-2034-76 104-2034-76
$C_8H_{18}OSn$ 2-Pentanone, 5-(trimethylstannyl)-	C_6H_{12} CCl_4	284(1.38) 242(2.15),285(1.51)	35-0104-76 35-0104-76
$(C_8H_{18}Sn)_n$ Poly(dibutylstannylene)	C_6H_{12}	210(2.8),300s(1.3)	18-2837-76
$C_8H_{19}N_4O_2P$ Acetic acid, [[bis(dimethylamino)phos- phoranylidene]hydrazono]-, methyl ester	EtOH	291.0(4.17)	65-1424-76
$C_8H_{20}ClN_2PSe$ Phosphorodiamidoselenoic chloride, tetraethyl-	hexane	198(4.20),252(3.38)	70-0174-76
$C_8H_{20}NO_2PSe$ Phosphoramidoselenoic acid, diethyl-, 0,0-diethyl ester	hexane or EtOH	192s(4.03),217s(3.69)	70-0174-76
$C_8H_{20}N_4$ Diazene, butyl[1-(2,2-dimethylhydra- zino)ethyl]- Diazene, [1-(2,2-diethylhydrazino)eth- yl]ethyl- Diazene, [1-(2,2-dimethylhydrazino)but- yl]ethyl- Diazene, [1-(2,2-dimethylhydrazino)- 2-methylpropyl]ethyl-	EtOH EtOH EtOH EtOH	365(1.51) 363(1.60) 362(1.57) 370(1.70)	104-2207-76 104-2207-76 104-2207-76 104-2207-76
$C_8H_{21}N_6PS$ Acetaldehyde, methyl(tetrahydro-2,4,6,6- tetramethyl-1,2,4,5,3-tetraazaphosph- orin-3(2H)-yl)hydrazone, P-sulfide	$CHCl_3$	243(3.00)	78-2633-76

Compound	Solvent	$\lambda_{max}(\log \epsilon)$	Ref.
C_9F_{10}			
Bicyclo[4.2.0]octa-1,3,5-triene, 2,3,4,5,7,7,8-heptafluoro-8-(trifluoromethyl)-	heptane	254.5(2.55)	104-0818-76
Bicyclo[4.2.0]octa-1,3,5-triene, 2,3,4,7,7,8,8-heptafluoro-5-(trifluoromethyl)-	heptane	267(3.11)	104-0818-76
$C_9H_3Cl_6N$			
1H-Indole, 2,3,4,5,6,7-hexachloro-1-methyl-	C_6H_{12}	303(3.98)	39-0162-76C
$C_9H_4Cl_4N_2O_2S$			
1H-Imidazole, 2,4,5-trichloro-1-[(4-chlorophenyl)sulfonyl]-	n.s.g.	200(4.54),230(4.24)	24-1625-76
$C_9H_4Cl_6N_2$			
3H-Indol-2-amine, 3,3,4,5,6,7-hexachloro-1-methyl-	C_6H_{12}	258(4.51),261(4.41)	39-0162-76C
$C_9H_4N_2OSe$			
8H-Indeno[1,2-d][1,2,3]selenadiazol-8-one	C_6H_{12}	209(4.25),250(4.1), 317(3.4),360(2.75)	24-1650-76
$C_9H_5BrN_2O_2$			
5,8-Quinolinedione, 6-amino-7-bromo-	pH 1	228(3.85),262(3.58)	4-1063-76
	pH 11	235(3.60),268(3.59)	4-1063-76
$C_9H_5Cl_2N$			
Quinoline, 3,4-dichloro-	MeOH	231(4.66),311(3.52), 325(3.51)	95-0725-76
$C_9H_5Cl_2NO$			
Isoxazole, 3,4-dichloro-5-phenyl-	HOAc	269(4.25)	103-0851-76
	+17M H_2SO_4	313(4.25)	103-0851-76
$C_9H_5Cl_4N$			
1H-Indole, 4,5,6,7-tetrachloro-1-methyl-	C_6H_{12}	292(3.81),312(3.67)	39-0162-76C
$C_9H_5F_2IO_2$			
2-Propenoic acid, 2,3-difluoro-3-(4-iodophenyl)-	EtOH	275(4.47)	104-1920-76
$C_9H_5F_2NO_4$			
2-Propenoic acid, 2,3-difluoro-3-(4-nitrophenyl)-	EtOH	309(4.35)	104-1920-76
$C_9H_5N_3OS$			
[1]Benzothieno[3,2-d]-1,2,3-triazin-4(1H)-one	EtOH	234(4.50),253(4.25), 320(3.80)	44-1733-76
$C_9H_5N_3O_2$			
[1]Benzopyrano[3,4-d][1,2,3]triazol-4(1H)-one	EtOH	255(4.09),265(4.07), 290(3.91),300(3.88)	39-1260-76C
$C_9H_5N_3S$			
[1]Benzothieno[3,2-d]-1,2,3-triazine	EtOH	225(4.41),265(4.17), 320s(--)	44-1733-76
3-Cinnolinecarbonitrile, 1,4-dihydro-4-thioxo-	EtOH	237(4.37),470(3.11)	39-0592-76C

Compound	Solvent	$\lambda_{max}(\log \epsilon)$	Ref.
$C_9H_5N_5O_2S$ Thiazole, 2-azido-4-(4-nitrophenyl)-	n.s.g.	236(4.19),330(4.20)	124-1298-76
C_9H_6BrNO Quinoline, 4-bromo-, 1-oxide	acid base	322(3.92) 344(3.97)	39-0456-76B 39-0456-76B
C_9H_6BrNOS 3H-Indol-3-one, 4-bromo-2-(methylthio)-	hexane	218(4.15),255(4.33), 262(4.32),280s(3.75), 333(3.00),347(3.08), 362(3.04),431(3.41), 452(3.42)	12-1023-76
C_9H_6ClN Acetonitrile, chloro-2,4,6-cyclohepta- trien-1-ylidene-	CH_2Cl_2	242(4.09),354(4.24)	88-0251-76
C_9H_6ClNO Isoxazole, 3-chloro-5-phenyl- Isoxazole, 4-chloro-5-phenyl- Isoxazole, 5-chloro-3-phenyl- Quinoline, 4-chloro-, 1-oxide	HOAc +17M H_2SO_4 HOAc +17M H_2SO_4 HOAc +17M H_2SO_4 acid base	261(4.31) 298(4.38) 268(4.24) 307(4.28) 250(4.01) 282(4.27) 322(3.90) 344(3.94)	103-0851-76 103-0851-76 103-0851-76 103-0851-76 103-0851-76 103-0851-76 39-0456-76B
C_9H_6ClNOS 4-Thiazolol, 5-(4-chlorophenyl)-	EtOH	224(4.33),247(4.29), 307(3.61)	118-0261-76
$C_9H_6ClNO_5$ 4H,5H-Pyrano[3,4-e]-1,3-oxazine-4,5-di- one, 7-chloro-2-ethoxy-	$CHCl_3$	275(4.08),283(4.06), 322(3.97)	39-2462-76C
C_9H_6ClNS Thiazole, 2-(4-chlorophenyl)-	MeCN	290(4.33)	39-1901-76C
$C_9H_6Cl_2N_2O_2$ 1,2,4-Oxadiazole, 5-(2,4-dichlorophen- yl)-3-methyl-, 4-oxide	EtOH	240(4.00),315(3.89)	44-2985-76
$C_9H_6Cl_6O$ Ethanone, 1-(1,4,5,6,7,7-hexachlorobi- cyclo[2.2.1]hept-5-en-2-yl)-, endo-	benzene	287(1.43)	44-2943-76
$C_9H_6F_2O_2$ 2-Propenoic acid, 2,3-difluoro-3-phen- yl-, trans	EtOH	260(4.30)	104-1920-76
$C_9H_6N_2O$ 3-Cinnolinecarboxaldehyde	1% EtOH	229(4.56),250(4.26), 285(3.48),325(3.60)	44-1952-76
$C_9H_6N_2O_3$ Isoxazole, 4-nitro-3-phenyl-	HOAc +17M H_2SO_4	252(3.78) 283(3.98)	103-0851-76 103-0851-76

Compound	Solvent	$\lambda_{max}(\log \epsilon)$	Ref.
$C_9H_6N_2O_3S$			
4-Thiazolol, 5-(4-nitrophenyl)-	EtOH	255(4.35),360(--)	118-0261-76
$C_9H_6N_2Se$			
4H-Indeno[2,1-d]-1,2,3-selenadiazole	C_6H_{12}	228(4.2),285(4.1), 320(3.8)	24-1650-76
$C_9H_6N_4$			
Propanedinitrile, (phenylazo)-	EtOH	244(3.91),359(4.30)	94-1331-76
$C_9H_6N_4O_3$			
7H-Furo[3,2-b]pyrimido[4,5-e][1,4]oxa-zin-7-one, 2-amino-5-methyl-	EtOH	338(3.82)	94-1189-76
$C_9H_6N_6$			
Propanedinitrile, (1-methyl-1H-pyrazolo-[3,4-d]pyrimidin-4-yl)-	EtOH	233(3.92),237(3.92), 251(3.62),297(3.46), 344(4.51)	95-1352-76
C_9H_6O			
Benzaldehyde, 2-ethynyl-	EtOH	250(3.94),255(3.96), 305(3.38),314(3.38)	18-2840-76
$C_9H_6OS_2$			
1,3-Dithiol-1-ium, 4-hydroxy-2-phenyl-, hydroxide, inner salt	MeOH	220(4.24),265(3.96), 285(3.89),325(3.46)	44-1724-76
	dioxan	222(3.99),238s(3.90), 266(3.92),290s(3.64), 483(3.78)	24-0740-76
Methanone, di-2-thienyl-	EtOH	215(3.64),270(4.00), 313(4.18)	59-0501-76
Methanone, 2-thienyl-3-thienyl-	EtOH	223(3.88),263(4.09), 294(4.03)	59-0501-76
$C_9H_6O_2$			
1H-2-Benzopyran-1-one	EtOH	228(4.17),250s(3.53), 280(2.9),320(3.57)	25-0954-76
1H-Indene-1,3(2H)-dione	n.s.g.	248(4.07),290(3.44), 300(3.40)	36-0134-76
$C_9H_6O_2S$			
Methanone, 2-furanyl-2-thienyl-	EtOH	212(3.70),277(3.94), 315(4.17)	59-0501-76
Methanone, 3-furanyl-2-thienyl-	EtOH	265(3.87),294(3.89)	59-0501-76
$C_9H_6O_3$			
2-Benzofurancarboxaldehyde, 2,3-dihydro-3-oxo-	EtOH	250(3.83),308(4.56)	111-0533-76
4H-1-Benzopyran-4-one, 5-hydroxy-	MeOH	223(4.30),255(4.09), 329(3.56)	4-0211-76
	MeOH-NaOMe	259(--),364(--)	4-0211-76
	MeOH-AlCl₃	229(--),246(--), 264(--),284s(--), 387(--)	4-0211-76
4H-1-Benzopyran-4-one, 6-hydroxy-	MeOH	228(4.34),237s(4.24), 248s(4.01),334(3.80)	4-0211-76
	MeOH-NaOMe	243(--),264s(--), 378(--)	4-0211-76
	MeOH-NaOAc	336(--),380(--)	4-0211-76
	MeOH-NaOAc-H₃BO₃	333(--)	4-0211-76

Compound	Solvent	$\lambda_{max}(\log \epsilon)$	Ref.
4H-1-Benzopyran-4-one, 7-hydroxy-	MeOH	236s(4.14),241(4.22), 248(4.26),287s(3.91), 300(3.99)	4-0211-76
4H-1-Benzopyran-4-one, 8-hydroxy-	MeOH	224(4.24),255(4.10), 315(3.63)	4-0211-76
Methanone, di-2-furanyl-	EtOH	232(3.46),281s(3.96), 313(4.28)	59-0501-76
$C_9H_6O_4$			
4H-1-Benzopyran-4-one, 5,6-dihydroxy-	MeOH	233(4.21),262(4.08), 360(3.54)	4-0211-76
4H-1-Benzopyran-4-one, 5,7-dihydroxy-	MeOH	224s(4.15),253s(4.26), 258(4.28),296(3.88), 320s(3.67)	4-0211-76
4H-1-Benzopyran-4-one, 5,8-dihydroxy-	MeOH	223(4.16),259(4.12), 306(3.33),360(3.47)	4-0211-76
4H-1-Benzopyran-4-one, 6,7-dihydroxy-	MeOH	227s(4.19),245s(3.90), 282(3.69),327(3.94)	4-0211-76
4H-1-Benzopyran-4-one, 6,8-dihydroxy-	MeOH	229s(4.08),239(4.09), 261(4.01),342(3.60)	4-0211-76
4H-1-Benzopyran-4-one, 7,8-dihydroxy-	MeOH	252s(4.31),258(4.37), 305(3.81)	4-0211-76
$C_9H_6S_2$			
4H-Cyclopenta[2,1-b:3,4-b']dithiophene (protonated)	CF_3COOH at -15°	313(3.89),360s(4.43), 376(4.54)	39-0323-76B
4H-Cyclopenta[2,1-b:3,4-c']dithiophene (protonated)	CF_3COOH at -15°	321(3.85),362(4.33), 445s(3.89),458(4.05), 472(4.05)	39-0323-76B
7H-Cyclopenta[1,2-b:3,4-b']dithiophene (protonated)	CF_3COOH at -15°	305s(3.92),340(4.39), 378s(4.15)	39-0323-76B
$C_9H_7BrN_2O$			
4(3H)-Quinazolinone, 6-bromo-3-methyl-	pH 2.10	264(3.85),298(3.37), 311(2.23)	44-0838-76
$C_9H_7BrO_4$			
1,3-Benzodioxole-5-carboxaldehyde, 7-bromo-4-methoxy-	MeOH	218(4.30),238(4.24), 292(3.90)	12-2003-76
$C_9H_7ClN_2$			
4-Quinolinamine, 3-chloro-	MeOH	218(4.52),244(4.42), 316(3.88)	95-0725-76
5-Quinolinamine, 6-chloro-	MeOH	217(4.09),262(4.35), 373(4.26)	95-0725-76
$C_9H_7ClN_2O$			
Phthalazine, 1-chloro-4-methoxy-	dioxan	273(3.88)	103-0342-76
1(2H)-Phthlazinone, 4-chloro-2-methyl-	dioxan	261(3.69),303(3.84)	103-0342-76
$C_9H_7ClN_2O_2$			
1,2,4-Oxadiazole, 5-(4-chlorophenyl)- 3-methyl-, 4-oxide	EtOH	228(3.84),242(4.10), 317(4.08)	44-2985-76
2(1H)-Quinolinone, 3-amino-6-chloro- 4-hydroxy-, hydrochloride	MeOH	218(4.54),238(4.70), 305(3.96),330(4.15), 342(4.19)	118-0805-76
$C_9H_7ClN_2S_2$			
1H-1,4-Benzodiazepine-2,5-dithione, 7-chloro-3,4-dihydro-	MeOH	210(4.36),230s(4.16), 303(4.32),342s(4.05)	44-2724-76

Compound	Solvent	$\lambda_{max}(\log \epsilon)$	Ref.
C₉H₇Cl₄NO			
Acetamide, N-methyl-N-[(2,3,4,5-tetra-chloro-2,4-cyclopentadien-1-ylidene)-methyl]-	C₆H₁₂	253(3.22),341(4.50)	24-3939-76
C₉H₇FN₂O₂			
1,2,4-Oxadiazole, 5-(4-fluorophenyl)-3-methyl-, 4-oxide	EtOH	226s(3.88),243(3.98), 310(3.99)	44-2985-76
C₉H₇F₃			
Styrene, 2,ß,ß-trifluoro-α-methyl-	EtOH	226(3.79),268(3.17)	12-1435-76
C₉H₇F₇N₂O₃			
2,4(1H,3H)-Pyrimidinedione, 5-[(2,2,3,3-4,4,4-heptafluorobutoxy)methyl]-	EtOH	261(3.86)	104-0643-76
C₉H₇N			
Quinoline	EtOH	225(4.48),270(3.59), 313(3.37)	65-1108-76
C₉H₇NO			
1H-Indole-2-carboxaldehyde	1% EtOH	235(4.11),310(4.36)	44-1952-76
iminate anion	+ NaOH	247(4.26),335(4.38)	44-1952-76
1H-Indole-3-carboxaldehyde	1% EtOH	244(4.11),261(4.11)	44-1952-76
iminate anion	+ NaOH	265(4.34),324(4.34)	44-1952-76
Isoxazole, 3-phenyl-	HOAc	251(3.88)	103-0851-76
	+17M H₂SO₄	275(4.17)	103-0851-76
Isoxazole, 5-phenyl-	HOAc	262(4.27)	103-0851-76
	+17M H₂SO₄	290(4.32)	103-0851-76
Quinoline, 1-oxide	acid	318(3.80)	39-0456-76B
	base	335(3.78)	39-0456-76B
8-Quinolinol	pH 1	251(4.71),308(3.30), 318(3.31),356(3.24)	65-1108-76
	pH 13	252(4.61),332(3.54), 350(3.49)	65-1108-76
	EtOH	241(4.63),313(3.40)	65-1108-76
4(1H)-Quinolinone	EtOH	317(4.08),330(4.12)	39-1428-76B
	CHCl₃	319(4.05),332(4.11)	39-1428-76B
	MeCN	317(4.08),330(4.12)	39-1428-76B
C₉H₇NOS			
4-Thiazolol, 5-phenyl-	EtOH	250(3.69),307(3.97)	118-0261-76
C₉H₇NO₂			
5(2H)-Oxazolone, 4-phenyl-	n.s.g.	260(4.03)	33-2149-76
C₉H₇NS			
Isothiazole, 4-phenyl-	EtOH	270(3.887)	48-0507-76
Isothiazole, 5-phenyl-	EtOH	204(4.091),268(4.184)	48-0507-76
8-Quinolinethiol	pH 1	240(4.32),315(3.72)	65-1108-76
	pH 13	260(4.35),367(3.63)	65-1108-76
	EtOH	250(4.35),324(3.61)	65-1108-76
Thiazole, 2-phenyl-	EtOH	286(4.23)	39-1901-76C
C₉H₇N₃O			
1H-1,2,3-Triazole-5-carboxaldehyde, 1-phenyl-	EtOH	223(3.90)	136-0081-76B
C₉H₇N₃O₄			
1,2,4-Oxadiazole, 3-methyl-5-(4-nitro-phenyl)-, 4-oxide	EtOH	254(4.07),347(3.95)	44-2985-76

Compound	Solvent	$\lambda_{max}(\log \epsilon)$	Ref.
$C_9H_7N_3O_5$			
2-Butynoic acid, 4-(6-amino-1,2,3,4-tetrahydro-2,4-dioxo-5-pyrimidinyl)-4-oxo-, methyl ester	pH 1	262s(--),276(4.16), 427(3.96)	44-1095-76
	pH 7, 11	273(4.00),333(4.37)	44-1095-76
Pyrido[2,3-d]pyrimidine-5-carboxylic acid, 1,2,3,4,7,8-hexahydro-2,4,7-trioxo-	pH 1	275(3.99),312(4.12)	44-1095-76
	pH 7	279(4.00),322(4.16)	44-1095-76
	pH 14	270(3.85),324(4.15)	44-1095-76
$C_9H_7N_5$			
1H-Pyrazolo[4,3-c]cinnolin-3-imine, 2,3-dihydro-	EtOH	234(4.39),287(3.65), 343(3.18)	39-0592-76C
$C_9H_7N_5O$			
8H-Imidazo[4,5-g]quinazolin-8-one, 6-amino-1,7-dihydro-, dihydrochloride	pH 1	228s(--),234(4.63), 240(4.65),256(3.95), 267s(--),278(3.56), 316(3.79),325s(--)	44-3529-76
	pH 2.8	235s(--),240(4.64), 266s(--),277s(--), 290s(--),317(3.76), 329s(--)	44-3529-76
	pH 6.8	233(4.64),254(4.35), 270(3.84),285(3.86), 324(3.77)	44-3529-76
	pH 10.3	234(4.61),270(3.86), 283(3.85),324(3.76)	44-3529-76
	pH 13	238(4.58),254(4.60), 280s(--),296s(--), 310(3.81),345(3.75), 374s(--)	44-3529-76
Pyrimido[1,2-a]purin-10(1H)-one, 7-methyl-	pH 1	218(4.35),248(4.37), 301s(3.69),312(3.75), 342(3.69)	44-0294-76
	pH 6.8	218(4.47),256(4.37), 309s(3.56),319(3.60), 355(3.62)	44-0294-76
	pH 10.1	228(4.59),250s(4.20), 265(4.26),315(3.50), 372(3.53)	44-0294-76
$C_9H_7N_5O_2$			
Pyrimido[1,2-a]purin-10(1H)-one, 7-methoxy-	pH 1	231(4.33),259(4.32), 267s(4.28),287(3.63), 300(3.61),312s(3.48), 355(3.58)	44-0294-76
	pH 6.8	227(4.45),265s(4.36), 271(4.38),300s(3.45), 365(3.58)	44-0294-76
	pH 10.1	270(4.32),317(3.49), 335(3.43),379(3.55)	44-0294-76
$C_9H_8BrN_3S$			
1H-1,2,3-Triazole, 1-(4-bromophenyl)-5-(methylthio)-	dioxan	242(4.12)	73-1551-76
$C_9H_8Br_2$			
Bicyclo[4.2.0]octa-1,3,5-triene, 7,8-dibromo-7-methyl-	ether	274(3.52),280(3.48)	22-0493-76

Compound	Solvent	$\lambda_{max}(\log \epsilon)$	Ref.
$C_9H_8ClNO_6$ Carbonic acid, 3-(aminocarbonyl)-6-chloro-2-oxo-2H-pyran-4-yl ethyl ester	MeCN	274(4.20),328(3.94)	39-2462-76C
C_9H_8ClNS Thiazole, 2-(4-chlorophenyl)-4,5-dihydro-	MeCN	248(4.27)	39-1901-76C
$C_9H_8ClN_3$ 1(2H)-Phthalazinimine, 4-chloro-2-methyl-	dioxan	275(4.00),320(3.78)	103-0342-76
$C_9H_8Cl_2N_2O_2$ 2(1H)-Quinolinone, 3-amino-6-chloro-4-hydroxy-, hydrochloride	MeOH	218(4.54),238(4.70), 305(3.96),330(4.15), 342(4.19)	118-0805-76
$C_9H_8Cl_4$ 1,3-Cyclopentadiene, 1,2,3,4-tetrachloro-5-(1-methylpropylidene)-	C_6H_{12}	290(4.28),405(2.80)	24-3929-76
1,3-Cyclopentadiene, 1,2,3,4-tetrachloro-5-(2-methylpropylidene)-	C_6H_{12}	276(4.27),402(2.78)	24-3929-76
$C_9H_8NO_5P$ 2(1H)-Quinolinone, 3-(phosphonooxy)-	EtOH	221(4.55),266(3.76), 274(3.86),283(3.80), 312(3.86),322(3.95), 335(3.80)	102-0029-76
$C_9H_8N_2$ 1H-Benzimidazole, 1-ethenyl-	pH 1	210(4.26),264(3.94), 269(3.96),275(3.89)	103-0691-76
	EtOH	202(4.49),227(4.22), 233(4.23),253(4.04), 260(3.97),281(3.71), 290(3.64)	103-0691-76
	dioxan	232(4.23),237(4.28), 252(4.14),262(3.97), 283(3.74),292(3.70)	103-0691-76
$C_9H_8N_2O$ 1,2,4-Oxadiazole, 5-methyl-3-phenyl-	EtOH	238(4.11)	98-0876-76
1(2H)-Phthalazinone, 2-methyl-	dioxan	255(3.87),294(3.90)	103-0342-76
4(1H)-Quinazolinone, 2-methyl-	EtOH	224(4.33),263(3.79)	115-0341-76
5-Quinolinamine, 1-oxide	MeOH	236(4.53),266(4.14), 294(3.82),341(3.59), 395(3.70)	95-0725-76
$C_9H_8N_2OS$ 1,3,4-Oxadiazole, 2-(methylthio)-5-phenyl-	EtOH	271(4.11)	106-0153-76
1,3,4-Thiadiazolium, 5-hydroxy-3-methyl-2-phenyl-, hydroxide, inner salt	MeOH	233(3.89),311(3.91)	78-0661-76
	MeCN	234(3.99),320(3.87)	78-0661-76
$C_9H_8N_2OSe$ 1,2,5-Selenadiazole, 3-methyl-4-phenyl-, 2-oxide	EtOH	233(4.11),299(3.99)	1-0675-76
$C_9H_8N_2O_2$ 2H-1-Benzopyran-2-one, 3,4-diamino-	EtOH	228s(4.18),252(4.13), 336(4.25)	103-0268-76

Compound	Solvent	$\lambda_{max}(\log \epsilon)$	Ref.
1H-Indazole-1-carboxylic acid, methyl ester	EtOH	246(3.96),252(3.91), 288(3.75),297(3.71)	87-0839-76
2H-Indazole-2-carboxylic acid, methyl ester	EtOH	278(3.93),289(3.97)	87-0839-76
1H-Indazole-4,7-dione, 3,6-dimethyl-	EtOH	225(4.10),254(4.22), 328(3.36)	94-1731-76
1H-Indole, 3-methyl-5-nitro-	EtOH	272(4.29),328(3.97)	30-0629-76
1H-Indole, 3-methyl-7-nitro-	EtOH	254(3.62),286(3.67), 373(3.48)	30-0629-76
Indolizine, 2-methyl-6-nitro-	EtOH	225(4.35),300(4.29), 434(3.16)	103-0766-76
Indolizine, 2-methyl-8-nitro-	EtOH	260(3.80),350(3.10), 475(2.90)	103-0766-76
2H-Indol-2-one, 1,3-dihydro-3-(methyl-imino)-, N-oxide	MeOH	211(4.17),259(4.15), 277(3.87),328(4.07), 407(3.07)	24-0200-76
1,2,4-Oxadiazole, 3-methyl-5-phenyl-, 4-oxide	EtOH	220s(3.84),243(3.95), 311(3.94)	44-2985-76
1,2,4-Oxadiazole-5-methanol, 3-phenyl-	EtOH	238(4.08)	98-0876-76
Phenol, 4-(5-methyl-1,2,4-oxadiazol-3-yl)-	EtOH	260(4.22)	98-0876-76
Pyrazolo[1,5-a]pyridin-2-ol, acetate	MeOH	223(4.60),286(3.59)	18-1980-76
2(1H)-Quinolinone, 3-amino-4-hydroxy-, hydrochloride	MeOH	217(4.51),229(4.49), 300(3.86),321(4.03), 333(4.03)	118-0805-76
$C_9H_8N_2O_3$ 1H-Indazole-4,7-dione, 5-hydroxy-3,6-di-methyl-	EtOH	236(4.10),270(4.15), 390(3.12)	94-1731-76
$C_9H_8N_2O_3S$ 5-Thiazolecarboxamide, 2-[5-(hydroxy-methyl)-2-furanyl]-	EtOH	231(4.19),315(4.26)	44-4074-76
7H-Thiazolo[3,2-a]pyrimidine-5-acetic acid, 7-oxo-, methyl ester	MeOH	218(4.37),228s(4.30), 276(4.20)	39-1908-76C
2H-Thiopyrano[2,3-d]pyrimidine-2,4,5(1H,3H)-trione, 1,3-dimethyl-	EtOH	218(4.34),294(4.15)	94-1390-76
$C_9H_8N_2O_4S$ Thiazolo[3,2-a]pyrimidine-5-carboxylic acid, 6-hydroxy-3-methyl-7-oxo-, methyl ester	MeOH	250(4.20),370(3.90)	64-0251-76B
$C_9H_8N_2O_5S$ 1,2,4-Oxadiazole-5-methanol, 3-phenyl-, hydrogen sulfate, potassium salt	EtOH	238(4.10)	98-0876-76
Phenol, 4-(5-methyl-1,2,4-oxadiazol-3-yl)-, hydrogen sulfate, potassium salt	EtOH	248(4.23)	98-0876-76
$C_9H_8N_2S$ 2H-Imidazole-2-thione, 1,3-dihydro-1-phenyl-	EtOH	283(3.635)	28-0143-76B
	CH_2Cl_2	292(3.737)	28-0143-76B
$C_9H_8N_2S_2$ 1,3,4-Thiadiazolium, 2,3-dihydro-4-methyl-5-phenyl-2-thioxo-, hydroxide, inner salt	MeOH	265(4.16),380(3.86)	78-0661-76
	MeCN	271(4.09),385(3.76)	78-0661-76
1,3,4-Thiadiazolium, 4,5-dihydro-2-methyl-3-phenyl-5-thioxo-, hydroxide, inner salt	M HCl	260(4.10),325(3.59)	4-1273-76
	H_2O	260(4.10),325(3.60)	4-1273-76
	MeOH	262(4.08),350(3.51)	78-0661-76

Compound	Solvent	$\lambda_{max}(\log \epsilon)$	Ref.
1,3,4-Thiadiazolium, 4,5-dihydro-2-methyl-3-phenyl-5-thioxo-, hydroxide, inner salt (cont.)	isoPrOH	263(4.12),362(3.52)	4-1273-76
	isoBuOH	270(4.12),369(3.46)	4-1273-76
	isoAmOH	364(3.49)	4-1273-76
	tert-AmOH	262(4.11),360(3.51)	4-1273-76
	acetone	377(3.63)	4-1273-76
	EtOAc	275(4.11),383(3.58)	4-1273-76
	$HCONH_2$	269(4.11),351(3.58)	4-1273-76
	MeCN	268(4.07),367(3.55)	4-1273-76
	MeCN	266(4.07),365(3.48)	78-0661-76
	pyridine	379(3.43)	4-1273-76
	$CHCl_3$	268(4.10),380(3.53)	4-1273-76
$C_9H_8N_2Se$			
1,2,5-Selenadiazole, 4-methyl-3-phenyl-	EtOH	223(3.85),305(4.00)	1-0675-76
$C_9H_8N_4$			
4,5'-Bipyrimidine, 5-methyl-	EtOH	208(4.20),223(4.11), 263(4.05)	94-1489-76
4,5'-Bipyrimidine, 6-methyl-	EtOH	206(4.38),236(4.34), 263(4.30)	94-1489-76
$C_9H_8N_4O_2$			
Pyrrolo[1,2-a]-1,3,5-triazine-8-carbo-nitrile, 1,2,3,4-tetrahydro-3,7-di-methyl-2,4-dioxo-	MeOH	267(3.93)	24-2983-76
$C_9H_8N_4O_3$			
Benzenamine, N-hydroxy-3-nitro-4-(1H-pyrazol-1-yl)-	EtOH	250(4.28),350(3.34)	22-0839-76
Benzenamine, N-hydroxy-5-nitro-2-(1H-pyrazol-1-yl)-	EtOH	220(4.18),255(4.12), 300(3.91),365(3.58)	22-0839-76
$C_9H_8N_6O$			
Urea, N-2-propynyl-N'-1H-pyrazolo-[3,4-d]pyrimidin-4-yl-	pH 1	266(4.16)	87-0555-76
	pH 13	272(4.05),297(3.70)	87-0555-76
	EtOH	264(4.18)	87-0555-76
$C_9H_8N_6O_2$			
3H-Pyridazino[4,3-b]pyrimido[4,5-e]-[1,4]oxazin-3-one, 8-amino-1,2-di-hydro-6-methyl-	EtOH	345(4.02)	94-1189-76
C_9H_8O			
Benzofuran, 2-methyl-	EtOH	217(3.97),246(4.05), 276(3.50),282(3.55)	39-0001-76C
	neutral	208(3.97),245(3.76)	78-1767-76
	cation	217(4.26),257(3.86)	78-1767-76
2-Propen-1-one, 1-phenyl-	MeOH	257(3.86),352(1.74)	40-0166-76
$C_9H_8O_2$			
Benzeneacetaldehyde, α-(hydroxymethyl-ene)-	pH 1	252(4.11)	44-0294-76
	pH 6.8	274(4.36)	44-0294-76
	pH 13	274(4.36)	44-0294-76
	90% H_2SO_4	247(4.16),305(3.69)	44-0294-76
1H-2-Benzopyran-4(3H)-one	MeCN	245(4.01),288(3.18)	1-0619-76
Tricyclo[3.2.2.0^{2,4}]non-8-ene-6,7-dione	hexane	450(1.93)	118-0256-76
$C_9H_8O_2S$			
1H-2-Benzothiopyran-4(3H)-one, 2-oxide	EtOH	253(3.99),294(3.20)	23-0455-76
4H-1-Benzothiopyran-4-one, 2,3-dihydro-, 1-oxide	EtOH	243(4.01),288(3.28), 333s(1.70)	23-0455-76

Compound	Solvent	$\lambda_{max}(\log \epsilon)$	Ref.
$C_9H_8O_3$			
Benzenepropanoic acid, α-oxo-	M HCl	289(4.40)	64-0001-76C
	MeOH	292(4.37)	64-0001-76C
	dioxan	287(4.21)	64-0001-76C
$C_9H_8O_4$			
1,3,6-Cycloheptatriene-1-carboxylic acid, 6-hydroxy-5-oxo-, methyl ester	EtOH	248(4.45),327(3.68)	39-2289-76C
C_9H_8S			
Benzo[b]thiophene, 2-methyl-	neutral	230(4.44)	78-1767-76
	cation	213(4.57)	78-1767-76
$C_9H_9BrN_2S$			
Isothiazolo[5,4-b]pyridine, 3-bromo-6,7-dihydro-4,7-dimethyl-6-methylene-	EtOH	314(3.93)	48-0779-76
$C_9H_9BrN_4O_2S$			
Benzenesulfonamide, 4-bromo-N-(1-methyl-1H-1,2,3-triazol-5-yl)-	EtOH	232(4.16),250(4.00)	32-0001-76
C_9H_9BrO			
Benzenemethanol, 3-bromo-α-ethenyl-	EtOH	260(2.98)	73-0262-76
Benzenemethanol, 4-bromo-α-ethenyl-	EtOH	259(2.99)	73-0262-76
2-Propen-1-ol, 3-(3-bromophenyl)-	EtOH	254(4.29)	73-0262-76
2-Propen-1-ol, 3-(4-bromophenyl)-	EtOH	258(4.38)	73-0262-76
$C_9H_9ClN_2O_2$			
2(1H)-Quinolinone, 3-amino-4-hydroxy-, hydrochloride	MeOH	217(4.51),229(4.49), 300(3.86),321(4.03), 333(4.03)	118-0805-76
$C_9H_9ClN_2O_4$			
Benzoic acid, 4-[(2-chloroethyl)amino]-3-nitro-	EtOH	263(4.35),286(4.28), 410(3.74)	78-0057-76
$C_9H_9ClN_4$			
Phthalazine, 1-chloro-4-(1-methylhydrazino)-	dioxan	315(3.90)	103-0342-76
1(2H)-Phthalazinone, 4-chloro-2-methyl-, hydrazone, (Z)-	dioxan	285(4.08),350(3.56)	103-0342-76
C_9H_9ClO			
Benzene, [(2-chloro-2-propenyl)oxy]-	EtOH	223(3.73),267(3.01), 272(3.15),279(3.08)	39-0001-76C
Benzenemethanol, 2-chloro-α-ethenyl-	EtOH	253(2.99)	73-0262-76
Benzenemethanol, 3-chloro-α-ethenyl-	EtOH	255.5(2.95)	73-0262-76
Benzenemethanol, 4-chloro-α-ethenyl-	EtOH	258(2.93)	73-0262-76
Phenol, 2-(2-chloro-2-propenyl)-	EtOH	222(3.75),278(3.41)	39-0001-76C
1-Propanone, 1-(4-chlorophenyl)-	MeOH	310s(2.0)	47-1901-76
2-Propen-1-ol, 3-(2-chlorophenyl)-	EtOH	250(4.26)	73-0262-76
2-Propen-1-ol, 3-(3-chlorophenyl)-	EtOH	255(4.29)	73-0262-76
2-Propen-1-ol, 3-(4-chlorophenyl)-	EtOH	256(4.34)	73-0262-76
C_9H_9ClOSe			
Benzeneethaneselenoic acid, 4-chloro-, O-methyl ester	MeOH	275(3.85)	44-0729-76
$C_9H_9ClO_3$			
Benzoic acid, 2-chloro-3-methoxy-, methyl ester	MeOH	243s(3.44),293(3.42)	12-2003-76

Compound	Solvent	$\lambda_{max}(\log \epsilon)$	Ref.
$C_9H_9Cl_4N$			
Ethanamine, N,N-dimethyl-1-(2,3,4,5-tetrachloro-2,4-cyclopentadien-1-ylidene)-	C_6H_{12}	235s(3.73),379(4.35)	24-3939-76
$C_9H_9Cl_4NO$			
Methanamine, 1-methoxy-N,N-dimethyl-1-(2,3,4,5-tetrachloro-2,4-cyclo-pentadien-1-ylidene)-	C_6H_{12}	344(4.23)	24-3939-76
$C_9H_9FN_2O_5$			
2,2'-Anhydro-1-β-D-arabinofuranosyl-5-fluorouracil	H_2O	223(3.85),254(3.93)	35-7381-76
C_9H_9FO			
Benzenemethanol, α-ethenyl-2-fluoro-	EtOH	251(2.81)	73-0262-76
Benzenemethanol, α-ethenyl-3-fluoro-	EtOH	251(2.70)	73-0262-76
Benzenemethanol, α-ethenyl-4-fluoro-	EtOH	252(2.85)	73-0262-76
2-Propen-1-ol, 3-(2-fluorophenyl)-	EtOH	249(4.23)	73-0262-76
2-Propen-1-ol, 3-(3-fluorophenyl)-	EtOH	250(4.24)	73-0262-76
2-Propen-1-ol, 3-(4-fluorophenyl)-	EtOH	250(4.25)	73-0262-76
C_9H_9NO			
2H-Azirine-2-methanol, 3-phenyl-	EtOH	245(4.15)	35-2605-76
1,4-Cyclohexadiene-1-carbonitrile, 6,6-dimethyl-3-oxo-	EtOH	239(4.22),334(1.59)	44-2950-76
	tert-BuOH	239(4.23),334(1.58)	44-2950-76
2-Propenamide, 3-phenyl-	EtOH	244(3.91),274(4.33)	110-0703-76
C_9H_9NOS			
2-Oxazolidinethione, 5-phenyl- (resedinine)	EtOH	246(4.29)	105-0247-76
$C_9H_9NO_2$			
Benzene, (1-methyl-2-nitroethenyl)-, trans	EtOH	226(4.04),293(4.11)	44-2112-76
Benzene, [1-(nitromethyl)ethenyl]-	EtOH	237(4.04)	44-2112-76
Benzene, (2-nitro-1-propenyl)-, cis	EtOH	225(4.15),282(3.48)	44-2112-76
trans	EtOH	226(4.00),305(4.08)	44-2112-76
Formamide, N-(2-acetylphenyl)-	C_6H_{12}	327(3.61)	46-1804-76
	$C_6H_{11}Me$	329(3.68)	46-1804-76
	H_2O	230(4.4),260(4.0), 315(3.49)	46-1804-76
	MeOH	320(3.60)	46-1804-76
	EtOH	320(3.60)	46-1804-76
2(3H)-Furanone, dihydro-5-(3-pyridinyl)-	MeOH	211(3.82),255(3.79), 260(3.84),266(3.69)	23-1262-76
2-Oxazolidinone, 5-phenyl- (resedine)	EtOH	253(2.23),258(2.33)	105-0245-76
$C_9H_9NO_2S$			
2-Furanmethanol, 5-(4-methyl-2-thiazo-lyl)-	EtOH	219(3.93),318(4.21)	44-4074-76
$C_9H_9NO_3$			
2(3H)-Benzoxazolone, 6-methoxy-3-methyl-	EtOH	210(3.64),234(3.78), 292(3.52)	102-1997-76
1-Propanone, 1-(4-nitrophenyl)-	MeOH	260(4.2),285s(3.2)	47-1901-76
$C_9H_9NO_3Se$			
Benzeneethaneselenoic acid, 4-nitro-, O-methyl ester	MeOH	275(4.23)	44-0729-76

Compound	Solvent	$\lambda_{max}(\log \epsilon)$	Ref.
$C_9H_9NO_4$			
Benzoic acid, 3-methoxy-2-nitro-, methyl ester	MeOH	227(3.00),252s(3.76), 340(3.67)	12-2003-76
Furo[2,3-b]pyridine-4-carboxylic acid, 2,3,6,7-tetrahydro-6-oxo-, methyl ester	MeOH	207(4.24),235s(--), 339(3.71),370s(--)	24-1269-76
2,3-Pyridinediol, diacetate	MeCN	227(3.15),260(3.49)	56-2193-76
2(1H)-Pyridinone, 1-acetyl-3-(acetyloxy)-	MeCN	228(3.32),308(3.72), 319(3.79),331(3.72), 348(3.36)	56-2193-76
1H-Pyrrole-1-acetic acid, α-ethylidene-2,5-dihydro-3-methyl-2,5-dioxo-, (Z)-	H_2O	215(4.28)	23-2862-76
$C_9H_9NO_5$			
1,3-Benzodioxole-5-carboxylic acid, 6-amino-4-hydroxy-, methyl ester	EtOH	212(4.18),244(4.33), 275(3.95),354(3.80)	24-0855-76
Glycine, N-(6-methoxy-3,4-dioxo-1,5-cyclohexadien-1-yl)-, sodium salt	H_2O	312(4.19),494(3.26)	135-0717-76A
	MeOH	305(4.19),475(3.26)	135-0717-76A
	dioxan	294(4.19),471(3.26)	135-0717-76A
2-Propenoic acid, 3-(5-methyl-2-furanyl)-2-nitro-, methyl ester, (Z)-	n.s.g.	325(4.311)	73-2422-76
$C_9H_9N_3$			
5-Azacinnoline, 7-ethyl-	EtOH	252(3.59),262(3.62), 267s(3.58),313(3.70), 325(3.75),365(2.41)	103-0808-76
Benzenamine, 4-(1H-pyrazol-1-yl)-	EtOH	272(4.21)	22-0195-76 +22-0839-76
Methanimidamide, N-cyano-N'-(3-methylphenyl)-	EtOH	280(4.25),292s(4.18)	48-0347-76
Methanimidamide, N-cyano-N'-(4-methylphenyl)-	EtOH	282(4.30),293s(4.23)	48-0347-76
1H-1,2,3-Triazole, 1-(4-methylphenyl)-	dioxan	250(4.12)	73-1551-76
1H-1,2,3-Triazole, 1-(phenylmethyl)-	dioxan	210(4.09)	73-1551-76
$C_9H_9N_3O$			
Benzenamine, N-hydroxy-4-(1H-pyrazol-1-yl)-	EtOH	275(4.18)	22-0195-76
5H-1,3,4-Benzotriazepin-5-one, 1,4-dihydro-4-methyl-	n.s.g.	237(4.40),257(4.15), 278(3.70),320(2.86)	44-2736-76
5H-1,3,4-Benzotriazepin-5-one, 3,4-dihydro-4-methyl-	EtOH	231(5.32),260s(4.33), 288s(4.47)	44-2732-76
	n.s.g.	233(4.47),251(4.28), 282(3.63),331(2.76)	44-2736-76
Methanimidamide, N-cyano-N'-(4-methoxyphenyl)-	EtOH	285(4.32)	48-0347-76
$C_9H_9N_3OS$			
1,2,3-Thiadiazol-5-amine, N-(4-methoxyphenyl)-	dioxan	251(4.14),276s(--), 323(3.76)	73-1182-76
$C_9H_9N_3O_2S$			
1,2-Benzenediol, 4-(2-amino-6H-1,3,4-thiadiazin-5-yl)-	EtOH	238s(--),331(4.23)	103-0868-76
1,4-Benzenediol, 2-(2-amino-6H-1,3,4-thiadiazin-5-yl)-	EtOH	250(4.08),367(4.00)	103-0868-76
$C_9H_9N_3O_3$			
Propanal, 2-oxo-, 1-[(4-nitrophenyl)hydrazone]	n.s.g.	372(4.44)	59-1593-76

Compound	Solvent	$\lambda_{max}(\log \epsilon)$	Ref.
1H-Pyrazolo[4,3-b]pyridine-6-carboxylic acid, 4,7-dihydro-7-oxo-, ethyl ester	EtOH	225(4.33),293(4.07)	39-0507-76C
Pyrido[2,3-d]pyrimidine-2,4(1H,3H)-dione, 1,3-dimethyl-, 8-oxide	pH 1	243(4.43),342(3.53)	44-3027-76
	pH 7	243(4.44),344(3.55)	44-3027-76
	pH 11	247(4.16)	44-3027-76
Pyrido[2,3-d]pyrimidine-2,4(1H,3H)-dione, 6-hydroxy-1,3-dimethyl-	pH 1	307(4.12)	44-3027-76
	pH 7	273(3.88),308(4.29), 321(4.25)	44-3027-76
	pH 11	273(3.85),308(4.28), 321(4.25)	44-3027-76
Pyrido[2,3-d]pyrimidine-2,4,7(1H,3H,8H)-trione, 1,3-dimethyl-	pH 1	265s(--),281s(--), 306s(--),311(4.14)	44-1095-76
	pH 1	248(3.95),338(3.78)	44-3027-76
	pH 7	274(3.89),308(4.26), 321(4.26)	44-1095-76
	pH 7	340(3.70)	44-3027-76
	pH 11	274(3.89),308(4.31), 321(4.27)	44-1095-76
	pH 11	268(4.05)	44-3027-76
$C_9H_9N_3O_3S$			
1,2,3-Benzenetriol, 4-(2-amino-6H-1,3,4-thiadiazin-5-yl)-	EtOH	217(4.30),344(4.20)	103-0868-76
2H-1,2,4-Benzothiadiazine-3-carboxamide, N-methyl-, 1,1-dioxide	dioxan	288(3.74)	103-0402-76
$C_9H_9N_3O_4S$			
1,4-Benzothiazine, 2,3-dihydro-4-methyl-5,7-dinitro-	MeOH	282(3.99),312s(3.70), 398(4.01)	88-4825-76
1,4-Benzothiazine, 2,3-dihydro-4-methyl-6,8-dinitro-	MeOH	265(4.16),315(3.92), 445(3.61)	88-4825-76
$C_9H_9N_3O_5$			
4-Pyrimidineacetic acid, 2-(acetyl-amino)-1,6-dihydro-α,6-dioxo-, methyl ester	EtOH	217(4.10),233(4.09), 287(3.83)	44-2850-76
	THF	244(4.04),335(3.65)	44-2850-76
$C_9H_9N_3O_6$			
2-Oxazolidinone, 3-[[[4,5-dihydro-5-methyl-aci-nitro-4-oxo-2-fur-anyl]methylene]amino]-	SDA 30	385(4.27)	87-0729-76
$C_9H_9N_3O_7$			
1,2,4-Triazine-3,5(2H,4H)-dione, 2-(2,3-O-carbonyl-β-D-ribofuranosyl)-	MeOH	259(3.74)	73-2110-76
$C_9H_9N_3S$			
1,2,3-Thiadiazol-5-amine, N-(2-methyl-phenyl)-	dioxan	247(3.84),285s(--), 320(4.13)	73-1182-76
1,2,3-Thiadiazol-5-amine, N-(3-methyl-phenyl)-	dioxan	248(3.90),287s(--), 324(4.23)	73-1182-76
1,2,3-Thiadiazol-5-amine, N-(4-methyl-phenyl)-	dioxan	249(3.95),287s(--), 325(4.21)	73-1182-76
1H-1,2,3-Triazole-5-thiol, 1-(4-methyl-phenyl)-	dioxan	257(3.95)	73-1551-76
$C_9H_9N_3S_2$			
1,2,3-Thiadiazol-5-amine, N-[4-(methyl-thio)phenyl]-	dioxan	265(3.99),331(4.30)	73-1182-76

Compound	Solvent	$\lambda_{max}(\log \epsilon)$	Ref.
$C_9H_9N_3Se$			
2,4-Selenazolediamine, 5-phenyl-, hydrobromide	EtOH	235(4.33),284(3.65)	78-0173-76
1H-1,2,4-Triazolium, 2,3-dihydro-1-methyl-4-phenyl-3-selenoxo-, hydroxide, inner salt	MeCN	204s(4.23),248(4.01), 274(3.89),318(3.25)	70-1162-76
$C_9H_9N_5O$			
9H-Imidazo[1,2-a]purin-9-one, 1,4-dihydro-4,6-dimethyl-	pH 1	227(4.57),231(4.57), 255(3.70),284(3.99)	69-0898-76
	pH 5.8	231(4.52),265(3.77), 306(3.81)	69-0898-76
	pH 11.0	231(4.56),275(3.88), 301(3.96)	69-0898-76
9H-Imidazo[1,2-a]purin-9-one, 1,4-dihydro-4,7-dimethyl-	pH 1.0	229(4.52),232(4.52), 251s(3.62),288(3.91)	69-0898-76
	pH 5.8	232(4.52),256(3.68), 313(3.72)	69-0898-76
	pH 11.0	233(4.54),264(3.72), 307(3.90)	69-0898-76
$C_9H_9N_5O_2$			
1H-1,2,4-Triazol-4-amine, 3-methyl-N-(2-nitrophenyl)-	EtOH	220(4.00),242(3.93), 288(4.06),338(3.76)	104-1535-76
$C_9H_9N_5O_3$			
Formazan, 1-acetyl-5-(2-nitrophenyl)-	EtOH	246(3.93),300(4.04), 345(3.90),400(3.70)	104-1535-76
$C_9H_9N_5O_4S$			
Benzenesulfonamide, N-(1-methyl-1H-1,2,3-triazol-5-yl)-	EtOH	255(4.18),350(3.36)	32-0001-76
$C_9H_9N_5O_5$			
9,5'-Cyclo-3-β-D-ribofuranosyl-8-azaxanthine	MeOH	240(3.91),257(4.03)	44-1100-76
C_9H_{10}			
Bicyclo[2.2.1]hept-2-ene, 5,6-bis(methylene)-	EtOH	241(3.98)	44-1635-76
$C_9H_{10}BrClN_2O_5$			
2,4(1H,3H)-Pyrimidinedione, 5-bromo-1-(3-chloro-3-deoxy-β-D-xylofuranosyl)-	pH 6.9	280(3.96)	44-2995-76
Uridine, 5-bromo-2'-chloro-2'-deoxy-	pH 6.9	279(3.99)	44-2995-76
$C_9H_{10}BrN_2S$			
Isothiazolo[5,4-b]pyridinium, 3-bromo-4,6,7-trimethyl-, perchlorate	DMF	320(3.88)	48-0779-76
$C_9H_{10}BrN_3O_3$			
Imidazo[1,2-a]pyrimidin-5-one, 1-acetyl-6-bromo-3-hydroxy-7-methyl-2,3-dihydro-	EtOH	244(3.92),296(4.08)	114-0413-76A
$C_9H_{10}BrN_3O_4$			
2,2'-Anhydro-1-β-D-arabinofuranosyl-5-bromocytidine, hydrochloride	pH 7.2	235s(3.93),280(3.98)	44-2995-76

Compound	Solvent	$\lambda_{max}(\log \epsilon)$	Ref.
$C_9H_{10}Br_2$ Bicyclo[6.1.0]nona-3,5-diene, 9,9-di- bromo-	hexane	233(3.64)	78-2153-76
$C_9H_{10}Br_2O_3$ Benzene, 1,3-dibromo-2,4,6-trimethoxy-	EtOH	235(4.27),292(3.56)	102-0767-76
$C_9H_{10}ClN$ Benzenamine, N-(2-chloro-2-propenyl)-	EtOH	209(3.72),245(4.06), 294(3.27)	39-0001-76C
$C_9H_{10}ClNO$ Benzenecarboximidoyl chloride, N-meth- oxy-2-methyl-, (E)-	EtOH	242(3.86)	44-0252-76
(Z)-	EtOH	232s(3.81),273s(2.98)	44-0252-76
Benzenecarboximidoyl chloride, N-meth- oxy-4-methyl-, (E)-	EtOH	212(4.15),219s(3.95), 262(4.18),293s(3.06)	44-0252-76
(Z)-	EtOH	220s(3.88),261(4.08)	44-0252-76
Propane, 1-(4-chlorophenyl)-2-nitroso-, dimer	EtOH	294(3.88)	78-1267-76
$C_9H_{10}ClNO_2$ Benzenecarboximidoyl chloride, N,4-di- methoxy-, (E)-	EtOH	271(4.29),300s(3.54)	44-0252-76
(Z)-	EtOH	215s(4.05),271(4.20)	44-0252-76
$C_9H_{10}ClNO_3S$ Carbamothioic acid, (4-chloro-2-oxo-2H- pyran-6-yl)methyl-, S-ethyl ester	EtOH	223s(3.97),336(4.07)	39-2462-76C
$C_9H_{10}ClN_3O$ 1H-Benzimidazole-1-ethanol, 2-amino- 5-chloro-	EtOH	217(4.58),252(3.78), 294(3.98)	78-0839-76
Imidazo[1,2-c]pyrimidin-5(6H)-one, 2-chloro-6-propyl-	pH 1	218(4.11),286(4.03), 296s(--),310s(--)	35-7408-76
	pH 7.0	217s(--),276(4.05), 282s(--),296s(--)	35-7408-76
	pH 13	214(4.11),276(4.03), 284s(--),296s(--)	35-7408-76
$C_9H_{10}ClN_3O_2$ Propanehydrazonoyl chloride, N-(4-nitro- phenyl)-	EtOH	240(3.89),370(4.20)	104-1649-76
$C_9H_{10}Cl_2O_2$ 2H-Pyran-2-one, 4-chloro-6-(2-chloro-1- methylethenyl)-5,6-dihydro-3-methyl-	MeOH	233(4.08)	88-4455-76
$C_9H_{10}N_2$ 6-Azaindolizine, 2,7-dimethyl-	EtOH	239(4.31),270s(3.44), 280(3.57),291(3.61)	44-0351-76
1H-Benzimidazole, 1-ethyl-	pH 1	250(3.59),256(3.65), 262(3.73),269(3.88), 276(3.86)	103-0691-76
	EtOH	204(4.67),249(3.78), 254(3.79),267(3.62), 275(3.68),282(3.71)	103-0691-76
	dioxan	252(3.78),256(3.80), 269(3.61),278(3.65), 285(3.68)	103-0691-76

Compound	Solvent	$\lambda_{max}(\log \epsilon)$	Ref.
3H-Indazole, 3,3-dimethyl-	isoPrOH	219(3.98),226s(3.81), 261(3.79),300s(2.92), 347(2.45)	4-0033-76
Propanedinitrile, cyclohexylidene-	C_6H_{12}	235(4.26)	78-2827-76
$C_9H_{10}N_2O$			
1H-Benzimidazole, 2-methoxy-1-methyl-	pH 9.0	238(3.71),275(3.74), 281(3.71)	39-1176-76B
Propanal, 2-oxo-, 1-(phenylhydrazone)	n.s.g.	343(3.99)	59-1593-76
Urea, (2-phenylethenyl)-, (E)-	EtOH	278(4.27)	95-0962-76
Urea, (2-phenylethenyl)-, (Z)-	EtOH	272(3.96)	95-0962-76
$C_9H_{10}N_2O_2$			
Benzenamine, N-methyl-N-(2-nitroethen-yl)-	EtOH	241(4.03),365(4.36)	104-1985-76
Benzonitrile, 2-amino-3,5-dimethoxy-	MeCN	245(3.91),342(3.80)	22-1993-76
Ethanone, 1-[4-(2-furanyl)-4,5-dihydro-1H-pyrazol-3-yl]-	EtOH	217(3.98),311(4.05)	104-2358-76
$C_9H_{10}N_2O_2S$			
1H-Benzimidazole, 1-methyl-2-(methyl-sulfonyl)-	pH 9.0	279(4.00)	39-1176-76B
$C_9H_{10}N_2O_3$			
Acetic acid, cyano(5-oxo-2-pyrrolidin-ylidene)-, ethyl ester	EtOH	276(4.35)	94-3011-76
potassium salt	EtOH	309(4.37)	94-3011-76
geometric isomer m. 182-4°	EtOH	276(4.37)	94-3011-76
Methanol, phenyl-ONN-azoxy-, acetate	EtOH	250(4.12)	44-1141-76
Propane, 1-(4-nitrophenyl)-2-nitroso-, dimer	EtOH	272(4.36)	78-1267-76
$C_9H_{10}N_2O_4$			
2-Propenoic acid, 3-(1,2,3,4-tetrahydro-2,4-dioxo-5-pyrimidinyl)-, ethyl ester, trans	pH 1	297(4.30)	87-0194-76
$C_9H_{10}N_2O_4S$			
2-Propenoic acid, 3-[(1,2,3,6-tetra-hydro-1,3-dimethyl-2,6-dioxo-4-pyr-imidinyl)thio]-, (E)-	EtOH	254(3.85),296(3.65)	94-1390-76
$C_9H_{10}N_2O_5$			
Benzoic acid, 4-[(2-hydroxyethyl)amino]-3-nitro-	EtOH	263(4.28),290(4.27), 417(3.72)	78-0057-76
$C_9H_{10}N_2O_6$			
6,5'(R)-Cyclouridine	pH 1	272(4.01)	44-3133-76
	pH 10.8	272(3.91)	44-3133-76
6,5'(S)-Cyclouridine	pH 1	268(4.03)	44-3133-76
	pH 10.8	268(3.93)	44-3133-76
$C_9H_{10}N_2S$			
1H-Benzimidazole, 1-methyl-2-(methyl-thio)-	pH 9.0	252(3.54),259(3.52), 285(3.86),292(3.87)	39-1176-76B
2-Imidazolidinethione, 1-phenyl-	EtOH	270(3.929)	28-0485-76A
	CH_2Cl_2	274.8(3.931)	28-0485-76A
$C_9H_{10}N_4$			
1(2H)-Phthalazinone, methylhydrazone	dioxan	279(4.15),344(3.70)	103-0342-76

Compound	Solvent	$\lambda_{max}(\log \epsilon)$	Ref.
1(2H)-Phthalazinone, 2-methyl-, hydrazone	dioxan	280(4.00),355(3.64)	103-0342-76
Quinoline, 3-amino-4-hydrazino-, dihydrochloride	EtOH-HCl	313(3.30),387(3.80)	24-2338-76
$C_9H_{10}N_4O_2$			
Ethanediamine, N-cyano-N'-(2-nitro-phenyl)-	EtOH	231(4.31),281(3.66), 425(3.74)	78-0057-76
$C_9H_{10}N_4O_2S$			
Benzenesulfonamide, N-(1-methyl-1H-1,2,3-triazol-5-yl)-	EtOH	219(4.05),243s(3.67)	32-0001-76
Benzenesulfonamide, 4-methyl-N-(1H-1,2,3-triazol-5-yl)-	EtOH	220(4.09),250(3.97)	32-0001-76
2H-1,3,4-Thiadiazin-2-one, 5-(2,5-di-hydroxyphenyl)-3,6-dihydro-, hydrazone	EtOH	238s(--),359(4.08)	103-0868-76
2H-1,3,4-Thiadiazin-2-one, 5-(3,4-di-hydroxyphenyl)-3,6-dihydro-, hydrazone	EtOH	222s(--),327(4.00)	103-0868-76
$C_9H_{10}N_4O_3S$			
2H-1,3,4-Thiadiazin-2-one, 3,6-dihydro-5-(2,3,4-trihydroxyphenyl)-, hydrazone	EtOH	216s(--),333(4.04)	103-0868-76
$C_9H_{10}N_4O_5$			
Furo[2,3-h]-1,3,7-oxadiazinine-2,4-(3H,5H)-dione, 5-diazo-7a,9,10,10a-tetrahydro-10-hydroxy-9-methyl-, [7aR-(7aα,9β,10α,10aα)]-	MeOH	249(4.48)	44-1041-76
4-Pyrimidineacetic acid, 2-(acetyl-amino)-1,6-dihydro-α-(hydroxyimino)-6-oxo-, methyl ester	EtOH	241(4.42),315(3.67)	44-2850-76
$C_9H_{10}N_4S$			
1H-1,2,4-Triazol-3-amine, 5-(methyl-thio)-1-phenyl-	EtOH	278(4.01)	65-0174-76
1H-1,2,4-Triazolium, 4-amino-3-mercapto-1-(phenylmethyl)-, hydroxide, inner salt	MeCN	248(4.57)	103-0361-76
$C_9H_{10}N_6O$			
Urea, N-2-propenyl-N'-1H-pyrazolo-[3,4-d]pyrimidin-4-yl-	pH 1	270(4.15)	87-0555-76
	pH 13	267s(--),273(4.02), 298(3.57)	87-0555-76
	EtOH	266(4.13),275s(--), 285s(--)	87-0555-76
$C_9H_{10}O$			
Benzenemethanol, α-ethenyl-	EtOH	252(2.74)	73-0262-76
1-Propanone, 1-phenyl-	MeOH	280(3.0),320s(1.8)	47-1901-76
2-Propen-1-ol, 3-phenyl-	EtOH	251(4.26)	73-0262-76
Tricyclo[3.3.0.0³,⁷]octan-2-one, 4-methylene, (-)-	isooctane	292(2.58),298s(2.61), 301(2.65),307s(2.56)	44-1229-76
$C_9H_{10}OS$			
Benzeneethanethioic acid, O-methyl ester	MeOH	240(3.85)	44-0729-76

Compound	Solvent	$\lambda_{max}(\log \epsilon)$	Ref.
$C_9H_{10}OSe$			
Benzeneethaneselenoic acid, O-methyl ester	MeOH	275(3.83)	44-0729-76
$C_9H_{10}O_2$			
Benzaldehyde, 2-ethoxy-	1% EtOH-pH 10	257(3.99)	44-1957-76
Benzeneacetic acid, methyl ester	MeOH	214(3.78)	44-0729-76
Ethanone, 1-[5-(1-hydroxyethylidene)-1,3-cyclopentadien-1-yl]-	heptane	248(4.44),328(4.25), 396(4.12)	30-0072-76
Fulvene, 6-acetoxy-6-methyl-	heptane	280(4.32),320(4.06)	30-0072-76
1-Propanone, 1-(4-hydroxyphenyl)-	MeOH	275(4.0),325s(2.3)	47-1901-76
Tricyclo[3.2.2.0^{2,4}]nonane-6,7-dione	hexane	460(1.91)	118-0256-76
$C_9H_{10}O_2S_2$			
Benzenecarbodithioic acid, 2-hydroxy-4-methoxy-, methyl ester	heptane-10% CH_2Cl_2	324(4.21),381(4.26), 450s(2.53)	1-0695-76
Benzenecarbodithioic acid, 4-hydroxy-2-methoxy-, methyl ester	heptane-10% CH_2Cl_2	325(4.12),486(2.30)	1-0695-76
$C_9H_{10}O_3$			
Benzoic acid, 3-hydroxy-, ethyl ester	MeOH	236(3.81),298(3.41)	95-0683-76
Furan, 3,5-diacetyl-2-methyl-	EtOH	277(4.42)	94-2421-76
2-Furanmethanol, 5-ethenyl-, acetate	EtOH	269(4.59)	70-0683-76
$C_9H_{10}O_4$			
Benzeneacetic acid, 3,5-dihydroxy-4-methyl-	EtOH	274(3.00),281(3.00)	35-7733-76
Benzoic acid, 2,4-dihydroxy-3,6-dimethyl-	EtOH	266(4.09),302(3.56)	35-7733-76
	EtOH	218(4.42),266(4.10), 302(3.57)	94-1853-76
Benzoic acid, 4,6-dihydroxy-2,3-dimethyl-	EtOH	262(4.01),307(3.70)	35-7733-76
2,5-Cyclohexadiene-1-acetic acid, 1-hydroxy-4-oxo-, methyl ester (jacaranone)	EtOH	225(4.16)	100-0255-76
4H-Pyran-2-acetic acid, 6-methyl-4-oxo-, methyl ester	EtOH	248(4.09)	35-7733-76
$C_9H_{10}O_5S$			
Benzaldehyde, 3-methoxy-4-[(methylsulfonyl)oxy]-	n.s.g.	255(3.96),307(3.52)	39-1056-76C
$C_9H_{10}S_3$			
Benzenecarbo(dithioperoxo)thioic acid, ethyl ester	C_6H_{12}	295(3.95),527(1.85)	138-0441-76
$C_9H_{11}BrClN_3O_4$			
2,2'-Anhydro-1-β-D-arabinofuranosyl-5-bromocytidine, hydrochloride	pH 7.2	235s(3.93),280(3.98)	44-2995-76
$C_9H_{11}BrN_2O$			
Benzaldehyde, 5-bromo-2-hydroxy-, dimethylhydrazone	benzene	300(4.15),330(4.12)	104-0625-76
	DMF	298(4.18),327(4.15)	104-0625-76
$C_9H_{11}BrN_2O_3$			
Acetamide, N,N'-[(5-bromo-2-furanyl)-methylene]bis-	EtOH	216(4.60)	103-0142-76

Compound	Solvent	$\lambda_{max}(\log \epsilon)$	Ref.
$C_9H_{11}BrN_2O_5$			
Uridine, 5-bromo-5'-deoxy-	MeOH	279(4.00)	44-1041-76
$C_9H_{11}BrO_3$			
1,3-Dioxepane, 2-(5-bromo-2-furanyl)-	EtOH	217(3.99)	103-0142-76
$C_9H_{11}ClN_2O_4S$			
1,2-Benzisothiazolium, 2-methyl-3-(methylamino)-, perchlorate	MeOH	235(4.32),327(3.84), 338s(--)	24-0659-76
$C_9H_{11}ClN_2O_5$			
2,4(1H,3H)-Pyrimidinedione, 1-(3-chloro-3-deoxy-β-D-xylofuranosyl)-	MeOH	262(4.03)	44-2995-76
Uridine, 2'-chloro-2'-deoxy-	H_2O	260(3.99)	44-2995-76
$C_9H_{11}ClN_2S_2$			
2,6(1H,3H)-Pyridinedithione, 5-chloro-3-[[(1-methylethyl)amino]methylene]-	MeOH	247(4.21),431(4.38)	33-0190-76
$C_9H_{11}ClO_2$			
2,4,6-Heptatrienoic acid, 7-chloro-, ethyl ester	EtOH	296(4.70)	70-0652-76
$C_9H_{11}FN_2O_5$			
Uridine, 2'-deoxy-5-fluoro-	acid	268(3.92)	35-7381-76
	base	268(3.82)	35-7381-76
$C_9H_{11}FN_2O_6$			
Uridine, 5-fluoro-	acid	268(3.96)	35-7381-76
	base	268(3.86)	35-7381-76
$C_9H_{11}IO_3$			
2,4-Cyclohexadien-1-one, 2-iodo-5,6-dimethoxy-6-methyl-	EtOH	388(3.75)	1-0064-76
1,3-Dioxepane, 2-(5-iodo-2-furanyl)-	EtOH	232(3.99)	103-0142-76
$C_9H_{11}IO_4$			
2,4-Cyclohexadien-1-one, 2-iodo-5,6,6-trimethoxy-	EtOH	394(3.73)	1-0064-76
$C_9H_{11}NO$			
Benzene, (2-nitrosopropyl)-, dimer	EtOH	292(3.91)	78-1267-76
meso	EtOH	294(3.91)	78-1267-76
1,3,5-Cycloheptatriene-1-carboxamide, 7-methyl-	EtOH	250(3.38)	44-3583-76
2,4,6-Cycloheptatrien-1-one, 4-(dimethylamino)-	EtOH	375(4.08)	39-1915-76C
Methanamine, 1-methoxy-N-(phenylmethylene)-	EtOH	246(3.93),280(2.93), 285(2.88)	35-2605-76
1-Propanone, 1-(4-aminophenyl)-	MeOH	312(4.1)	47-1901-76
1H-2-Pyrindin-1-one, 2,5,6,7-tetrahydro-3-methyl-	n.s.g.	237(3.75),306(3.95)	104-0681-76
$C_9H_{11}NOS$			
Benzenecarbothioamide, 2-hydroxy-N,N-dimethyl-	heptane-10% CH_2Cl_2	242(3.91),290(4.01), 377(2.90)	1-0695-76
Benzenecarbothioamide, 4-hydroxy-N,N-dimethyl-	heptane-10% CH_2Cl_2	259(4.06),276(4.13), 383(2.75)	1-0695-76
Thiocyanic acid, 2-formyl-1-cyclohepten-1-yl ester	EtOH	297(3.75)	97-0049-76

Compound	Solvent	$\lambda_{max}(\log \epsilon)$	Ref.
$C_9H_{11}NO_2$			
Furo[2,3-b]pyridin-6(2H)-one, 3,7-di-hydro-2,2-dimethyl-	MeOH	231(3.75),298(3.86), 331(3.58)	24-1269-76
2,5-Pyrrolidinedione, 3-(3-methyl-2-butenylidene)-	EtOH	289(4.31)	44-2950-76
$C_9H_{11}NO_2S_2$			
5H-1,4-Dithiino[2,3-c]pyrrole-5,7(6H)-dione, 2,3-dihydro-6-(1-methylethyl)-	EtOH	205(3.86),264(4.07), 414(3.42)	56-1523-76
$C_9H_{11}NO_3$			
3,5-Cyclohexadiene-1,2-dione, 4-(dimeth-ylamino)-5-methoxy-	H_2O	330(4.20),516(3.56)	135-0717-76A
	MeOH	325(4.20),495(3.56)	135-0717-76A
	dioxan	320(4.20),472(3.56)	135-0717-76A
	$CHCl_3$	323(4.20),484(3.56)	135-0717-76A
$C_9H_{11}NO_4S$			
1H,3H-Oxazolo[4,3-c][1,4]thiazine-6-carboxylic acid, 8,8a-dihydro-3-methyl-1-oxo-, methyl ester	EtOH	215(3.81),261(3.49), 312(4.03)	39-0584-76C
$C_9H_{11}NO_5$			
1,3-Dioxepane, 2-(5-nitro-2-furanyl)-	EtOH	306(3.96)	103-0142-76
2,6-Pyridinedicarboxylic acid, 1,2,3,4-tetrahydro-4-oxo-, dimethyl ester	EtOH	337(3.97)	33-0626-76
$C_9H_{11}NO_7$			
2H-1,3-Oxazine-2,6(3H)-dione, 3-β-D-ribofuranosyl-2,3-dihydro-	H_2O	265(3.86)	87-0643-76
$C_9H_{11}N_2S$			
1,2-Benzisothiazolium, 2-methyl-3-(methylamino)-, perchlorate	MeOH	235(4.32),327(3.84), 338s(--)	24-0659-76
$C_9H_{11}N_3$			
1H-Indole-3-carbonitrile, 2-amino-4,5,6,7-tetrahydro-	MeOH	269(3.80)	48-0663-76
$C_9H_{11}N_3O$			
1H-1,2,3-Benzotriazole, 1-propoxy-	EtOH	264(3.84),283(3.76)	4-0509-76
Imidazo[1,2-c]pyrimidin-5(6H)-one, 6-propyl-, hydrochloride	pH 1	255s(--),290(4.08)	35-7408-76
	pH 7	272(4.06)	35-7408-76
	pH 13	271(4.06)	35-7408-76
$C_9H_{11}N_3O_2$			
Benzamide, N-[(methylnitrosoamino)meth-yl]-	EtOH	233(4.23),353(1.90), 356s(1.90),372s(1.72)	94-0369-76
Imidazo[1,2-c]pyrimidine-2,5(3H,6H)-di-one, 6-propyl-	pH 1	237(3.68),305(4.21)	35-7408-76
	pH 7.0	218(3.68),309(4.17), 320s(--)	35-7408-76
	pH 13	223(4.03),275(3.88), 288s(--),300s(--)	35-7408-76
Imidazo[1,2-a]pyrimidine-2,7(3H,8H)-di-one, 3,5,8-trimethyl-	pH 2-9.85	269(4.11)	114-0413-76A
Imidazo[1,2-a]pyrimidin-5(1H)-one, 1-acetyl-2,3-dihydro-6-methyl-	EtOH or dioxan	240(3.86),282(4.04)	114-0413-76A
Imidazo[1,2-a]pyrimidin-7(1H)-one, 1-acetyl-2,3-dihydro-6-methyl-	dioxan	224(4.41),247s(4.03)	114-0413-76A
1H-Pyrrole-2-carboxylic acid, 5-amino-4-cyano-3-methyl-, ethyl ester	EtOH	222s(3.93),256(3.89), 302(4.29)	103-1379-76

Compound	Solvent	$\lambda_{max}(\log \epsilon)$	Ref.
$C_9H_{11}N_3O_3$			
Benzaldehyde, 2-hydroxy-5-nitro-,	benzene	300(4.30)	104-0625-76
dimethylhydrazone	DMF	305(4.34)	104-0625-76
Imidazo[1,2-a]pyrimidin-5(1H)-one, 1-acetyl-2,3-dihydro-3-hydroxy-7-methyl-	EtOH	234(3.88),283(4.00)	114-0413-76A
Propanoic acid, 2-(4-nitrophenyl)-,	EtOH	225(3.95),359(4.16)	104-1649-76
hydrazide	dioxan	347(4.22)	104-1649-76
2-Propenamide, N,N-dimethyl-3-(1,2,3,4-tetrahydro-2,4-dioxo-5-pyrimidinyl)-, (E)-	pH 1	278(4.13),297(4.20)	87-0194-76
$C_9H_{11}N_3O_4$			
2,2'-Anhydro-1-β-D-arabinofuranosyl-cyclocytosine, hydrochloride	MeOH-acid	232(3.98),262(4.04)	87-0663-76
2,2'-Anhydro-1-β-D-arabinofuranosyl-cytosine, hydrochloride	MeOH-acid	231(4.00),263(4.03)	87-0654-76
O^2-2'-Cyclo-5-amino-5'-deoxyuridine	MeOH	258(3.80),289(3.86)	44-1041-76
$C_9H_{11}N_3O_5$			
Morpholine, 4-[(hydroxyimino)(5-nitro-2-furanyl)methyl]-	EtOH	222(4.11),250(3.92), 317(4.08)	73-3085-76
$C_9H_{11}N_5$			
Pyrido[3,2-c]pyridazine, 7-ethyl-4-hydrazino-	EtOH	252(3.82),298(3.61), 312(3.60),327(3.56), 373(3.68)	103-0808-76
1H-1,2,4-Triazole-3,5-diamine, 1-(phenylmethyl)-	EtOH	210(4.1)	23-3620-76
$C_9H_{11}N_5O$			
2-Pteridinamine, 4-methoxy-N,N-dimethyl-	MeOH	237(4.31),277(4.16), 389(3.79)	24-3208-76
4(1H)-Pteridinone, 2-(dimethylamino)-8-methyl-	MeOH	235(3.74),275(4.17), 380s(3.75),422(3.94)	24-3208-76
$C_9H_{11}N_5O_2$			
4,7-Pteridinedione, 2-(dimethylamino)-tetrahydro-8-methyl-	pH 4.0	226(4.45),296(4.06), 353(4.21)	24-3228-76
	pH 11.0	233(4.00),264(4.08), 373(4.26)	24-3228-76
$C_9H_{11}N_5O_3$			
L-erythro-Biopterin	pH 13	254(4.37),365(3.86)	35-2301-76
$C_9H_{11}N_5O_4$			
4(1H)-Pteridinone, 2-amino-6-(1,2-di-hydroxypropyl)-, 8-oxide, [S-(R*,R*)]-	pH 13	265(4.54),295s(3.90), 385(3.95)	35-2301-76
C_9H_{12}			
1-Cyclohexene, 1-methyl-4,5-bis(methylene)-	EtOH	218(3.73)	44-1635-76
$C_9H_{12}BrN_3$			
Pyrido[2,3-b]pyrazine, 7-bromo-1,2,3,4-tetrahydro-2,3-dimethyl-	EtOH	272(3.85),338(3.92)	22-0251-76
$C_9H_{12}BrN_3O_4S$			
1H-Imidazole-4-carbothioamide, 5-bromo-1-β-D-ribofuranosyl-	pH 1	267(4.00),300(4.03)	87-1020-76
	pH 11	267(4.06),302(4.04)	87-1020-76

Compound	Solvent	λ_{max}(log ϵ)	Ref.
$C_9H_{12}BrN_3O_5$ 1H-Imidazole-4-carboxamide, 5-bromo- 1-β-D-ribofuranosyl-	pH 1 pH 11	239(3.90) 244(3.96)	87-1020-76 87-1020-76
$C_9H_{12}ClNO_4S$ 3-Pentenoic acid, 3-chloro-5-[[(ethyl-thio)carbonyl]amino]-5-oxo-, methyl ester	EtOH	229(4.32)	39-2462-76C
$C_9H_{12}ClN_3O_4S$ 1H-Imidazole-4-carbothioamide, 5-chloro- 1-β-D-ribofuranosyl-	pH 1 pH 11	262(4.05),300(4.07) 263(4.10),301(4.08)	87-1020-76 87-1020-76
$C_9H_{12}ClN_3O_5$ 1H-Imidazole-4-carboxamide, 5-chloro- 1-β-D-ribofuranosyl-	pH 1 pH 11	237(3.91) 241(3.93)	87-1020-76 87-1020-76
$C_9H_{12}Cl_2N_4$ Quinoline, 3-amino-4-hydrazino-, dihydrochloride	EtOH-HCl	313(3.30),387(3.80)	24-2338-76
$C_9H_{12}Cl_2O_2$ 2-Propenoic acid, 3-(2,2-dichloro-3-methylcyclopropyl)-, ethyl ester	EtOH	233(4.20)	70-0909-76
$C_9H_{12}Cl_2O_4$ α-D-erythro-Pentofuranose, 3-deoxy-3-C-(dichloromethylene)-1,2-O-iso-propylidene-	EtOH	217(3.65)	111-0489-76
$C_9H_{12}Cl_3NO$ Piperidine, 1-(3,4,4-trichloro-1-oxo-3-butenyl)-	heptane	208(4.11)	24-2159-76
$C_9H_{12}FN_2O_9P$ 5'-Uridylic acid, 4'-C-fluoro-, barium salt	H_2O	260(3.98)	44-3010-76
$C_9H_{12}FN_3O_4$ Cytidine, 2'-deoxy-5-fluoro-	acid base	291(4.10) 281(3.94)	35-7381-76 35-7381-76
$C_9H_{12}FN_3O_5$ Cytidine, 5-fluoro- Cytosine, 5-fluoro-1-β-D-arabino-furanosyl-	acid base acid base	289(4.04) 281(3.90) 290(4.07) 282(3.91)	35-7381-76 35-7381-76 35-7381-76 35-7381-76
$C_9H_{12}FN_3O_8S$ Uridine, 4'-C-fluoro-, 5'-sulfamate	MeOH	259(3.98)	44-3010-76
$C_9H_{12}F_3N$ Ethanamine, 2,2,2-trifluoro-N-(3-methyl-2-cyclohexen-1-ylidene)-	C_6H_{12} pH 2 pH 12	228(4.34) 268(4.27) 238(4.30)	35-4174-76 35-4174-76 35-4174-76
$C_9H_{12}IN_3O_5$ 1H-Imidazole-4-carboxamide, 5-iodo- 1-β-D-ribofuranosyl-	pH 1 pH 11	241(3.95) 247(4.00)	87-1020-76 87-1020-76

Compound	Solvent	$\lambda_{max}(\log \epsilon)$	Ref.
$C_9H_{12}N_2$			
Ethanimidamide, N-methyl-N-phenyl-	neutral	376(4.17)	39-0211-76B
	cation	234s(4.11)	39-0211-76B
$C_9H_{12}N_2O$			
Benzaldehyde, 2-hydroxy-, dimethylhydrazone	benzene	325(4.00)	104-0625-76
	EtOH	280(3.99),320(3.90)	104-0625-76
Diazene, methyl(1-phenylethyl)-, 2-oxide, (Z)-	EtOH	257(2.5),263(2.5), 268(2.5)	44-1135-76
Ethanimidamide, N-(4-methoxyphenyl)-	neutral	241(4.10)	39-0211-76B
	cation	237(4.12)	39-0211-76B
4H-Pyrido[1,2-a]pyrimidin-4-one, 5,6,7,8-tetrahydro-5-methyl-	pH 7.35	227(3.81),275(3.72)	145-0478-76
$C_9H_{12}N_2O_2$			
Cyclopent[e][1,3]oxazin-4(5H)-one, 2-(dimethylamino)-6,7-dihydro-	MeOH	260(3.85)	5-1689-76
Diazene, (dimethoxymethyl)phenyl-	EtOH	215(4.11),270(3.28)	44-1141-76
1,2-Diazepine-4-carboxylic acid, 1-methyl-, ethyl ester	MeOH	224(3.93),350s(2.42)	88-4859-76
1,2-Diazepine-6-carboxylic acid, 1-methyl-, ethyl ester	MeOH	224(3.90),330s(2.38)	88-4859-76
2(1H)-Pyridinone, 6-[1-(hydroxyimino)-propyl]-1-methyl-, (E)-	H_2O	304.5(3.87)	83-0769-76
	pH 13	308(3.95)	83-0769-76
	MeOH	230(3.80),310(3.91)	83-0769-76
2(1H)-Pyridinone, 6-[1-(hydroxyimino)-propyl]-1-methyl-, (Z)-	H_2O	304(3.81)	83-0769-76
	pH 13	305(3.85)	83-0769-76
	MeOH	230(3.77),308(3.86)	83-0769-76
$C_9H_{12}N_2O_2S_2$			
Pentanoic acid, 2,5-diisothiocyanato-, ethyl ester	pH 8.06	244(3.36)	73-2216-76
	MeOH	246(3.31)	73-2216-76
	dioxan	247(3.06)	73-2216-76
$C_9H_{12}N_2O_3$			
Acetamide, N,N'-(2-furanylmethylene)-bis-	EtOH	209(4.57)	103-0142-76
Pyrazinecarboxylic acid, 4,5-dihydro-3,6-dimethyl-5-oxo-, ethyl ester	EtOH	264(4.27),308(3.65)	18-2805-76
1H-Pyrazole-3-carboxylic acid, 5-hydroxy-1-methyl-4-(2-propenyl)-, methyl ester	MeOH	228(3.98),261s(3.48)	24-0261-76
1H-Pyrazole-3-carboxylic acid, 1-methyl-5-(2-propenyloxy)-, methyl ester	MeOH	222(4.00),253s(3.52)	24-0261-76
1H-Pyrazole-5-carboxylic acid, 3-hydroxy-1-methyl-4-(2-propenyl)-, methyl ester	MeOH	233(3.96),278(3.58)	24-0268-76
1H-Pyrazole-5-carboxylic acid, 1-methyl-3-(2-propenyloxy)-, methyl ester	MeOH	226(4.19),270(3.59)	24-0268-76
$C_9H_{12}N_2O_3S$			
1(2H)-Pyrimidineacetic acid, 6-methyl-4-oxo-2-thioxo-, ethyl ester	EtOH	219(4.16),274(4.13)	114-0413-76A
$C_9H_{12}N_2O_4$			
1H-Pyrazole-3-carboxylic acid, 4-(acetoxymethyl)-, ethyl ester	MeOH	221(3.92),238s(3.75)	103-1029-76
1H-Pyrazole-3-carboxylic acid, 5-(acetoxymethyl)-, ethyl ester	MeOH	221(4.01)	103-1029-76
2,4(1H,3H)-Pyrimidinedione, 5-(tetrahydro-3-methoxy-2-furanyl)-, trans	H_2O	264(3.90)	4-0929-76

Compound	Solvent	$\lambda_{max}(\log \epsilon)$	Ref.
$C_9H_{12}N_2O_5$ 1H-Pyrazole-3,5-dicarboxylic acid, 4-hydroxy-, diethyl ester	EtOH	233s(4.274),276(4.230)	23-1752-76
$C_9H_{12}N_2O_5S$ 4-Thiazolecarboxamide, 2-β-D-ribo- furanosyl-	EtOH	215(3.97),237(3.88)	44-4074-76
$C_9H_{12}N_2S$ Benzenecarboximidamide, 2-mercapto- N,N'-dimethyl-	MeOH	276(4.01)	24-0659-76
$C_9H_{12}N_2S_2$ 2,6(1H,3H)-Pyridinedithione, 3-[[(1- methylethyl)amino]methylene]-	MeOH	288(3.68),338(4.28), 419(4.58)	33-0190-76
$C_9H_{12}N_3O_6PS$ Cytidine, 2'-deoxy-2'-thio-, 2',3'- phosphorothioate, sodium salt	H$_2$O	272(3.79)	78-2991-76
$C_9H_{12}N_4$ 1H-Benzimidazole-1,2-diamine, 5,6-di- methyl-	EtOH EtOH-acid	243(3.86),287(4.01) 232s(4.09),280(4.04), 282(4.06),287(4.03)	107-0457-76 107-0457-76
Ethanediamine, N-(2-amino-1-benzimid- azolyl)-, dihydrobromide	EtOH	204(4.69),250(3.34), 276(3.90),282(3.88)	78-0839-76
1H-Imidazol-2-amine, 1-(2-aminophenyl)- 4,5-dihydro-, dihydrobromide	EtOH	211s(4.19),240(3.99), 296(3.53)	78-0057-76
$C_9H_{12}N_4O_2$ 1,3-Propanediol, 2-amino-1-(4-azidophen- yl)-, [R-(R*,R*)]-	MeOH	253(4.19),288(3.40)	130-0025-76
3H-Pyrimido[1,2-b]-as-triazin-3-one, 5-acetyl-5,6,7,8-tetrahydro-2-methyl-	dioxan	239(4.37)	114-0413-76A
2H-Pyrrolo[2,3-d]pyrimidin-2-one, 4-amino-1,3-dihydro-3-(methoxy- methyl)-5-methyl-	DMSO	328(3.89)	24-2983-76
Urea, N-(3-cyano-4-methyl-1H-pyrrol- 2-yl)-N'-(methoxymethyl)-	MeOH	226(4.00),265(4.12)	24-2983-76
$C_9H_{12}N_4O_2S$ Isothiazolo[3,4-d]pyrimidine-4,6(5H,7H)- dione, 3-(dimethylamino)-7-ethyl-	EtOH	233(<u>4.2</u>),276(<u>4.0</u>), 305(<u>3.9</u>)	94-0970-76
$C_9H_{12}N_4O_3$ Ethanol, 2-[[9-(methoxymethyl)-9H-purin- 6-yl]oxy]-	n.s.g.	248(4.04)	88-4427-76
Lumazine, 5-acetyl-5,6,7,8-tetrahydro- 1-methyl-	pH 8.0 pH 14.0	252(3.87),283(4.24) 255s(3.93),278(4.16)	24-3184-76 24-3184-76
1H-1,2,4-Triazole-5-acetic acid, α-[(acetylamino)methylene]-1- methyl-, methyl ester	EtOH	274(4.25)	94-2568-76
$C_9H_{12}N_4O_4$ 1H-Imidazole-4-carboxamide, 5-amino- 1-(2,3-anhydro-β-D-ribofuranosyl)-	pH 1 pH 7 pH 13	267(4.04) 265(4.07) 265(4.06)	94-2089-76 94-2089-76 94-2089-76
[1,2,5]Oxadiazolo[3,4-d]pyrimidine- 5,7(4H,6H)-dione, 4-butyl-6-methyl-, 1-oxide	MeOH	269(4.09),317s(3.19)	39-1327-76C

Compound	Solvent	$\lambda_{max}(\log \epsilon)$	Ref.
$C_9H_{12}N_4O_5$ 1H-1,2,4-Triazole-3-carboxamide, 1-(6-deoxy-β-D-erythro-hex-5-en- 2-ulofuranosyl)-	MeOH	207(4.08)	44-1836-76
$C_9H_{12}N_4O_7$ 1H-Pyrazole-3-carboxamide, 4-nitro- 1-α-D-ribofuranosyl-	MeOH	214(3.95),268(3.86)	104-1538-76
β-	MeOH	212(3.85),268(3.78)	104-1538-76
1H-Pyrazole-5-carboxamide, 4-nitro- 1-β-D-ribofuranosyl-	MeOH	210(3.56),268(3.78)	104-1538-76
$C_9H_{12}N_6O$ 4(1H)-Pteridinone, 2-hydrazino- 1,6,7-trimethyl-	H_2O	242(3.98),279s(3.27), 337(3.78)	12-0459-76
	pH 13	236(4.00),281(3.68), 402(3.51)	12-0459-76
Urea, N-(1-methylethyl)-N'-1H-pyrazolo- [3,4-d]pyrimidin-4-yl-	pH 1	271(4.20)	87-0555-76
	pH 13	267s(--),273(4.07), 298(3.58)	87-0555-76
	EtOH	266(4.17),275s(--), 285s(--)	87-0555-76
$C_9H_{12}N_6O_3$ 3H-Purine-3-propanoic acid, 8-amino- 2,7-dihydro-6-(hydroxymethyl)-2- imino-, (R)-	H_2O	229(4.11),252s(3.72), 335(4.18)	35-5414-76
$C_9H_{12}N_6O_4$ 3H-Purine-3-propanoic acid, 8-amino- 2,7-dihydro-α-hydroxy-6-(hydroxy- methyl)-2-imino-, (R)-	H_2O	232(4.08),252s(3.64), 337(4.20)	35-5414-76
$C_9H_{12}N_6O_4S$ [1,2,3]Thiadiazolo[5,4-d]pyrimidine- 5,7-diamine, N^7-β-D-ribofuranosyl-	pH 1	248(4.34),275(4.05), 285s(4.03)	87-1186-76
	pH 7	247(4.17),311(4.06)	87-1186-76
	pH 13	248(4.16),313(4.05)	87-1186-76
7H-1,2,3-Triazolo[4,5-d]pyrimidine- 7-thione, 5-amino-3,4-dihydro-3- β-D-ribofuranosyl-	MeOH-HCl	211(4.33),230s(4.07), 263(3.95),338(4.31)	87-1186-76
	MeOH-pH 7	212(4.26),258(4.02), 334(4.25)	87-1186-76
	MeOH-NaOH	257(4.04),292(3.89), 328(4.23)	87-1186-76
$C_9H_{12}O$ Bicyclo[3.2.1]oct-3-en-2-one, 4-methyl-	C_6H_{12}	228(4.06)	78-2805-76
	EtOH	241(3.86)	78-2805-76
Brexan-2-one, (-)-	isooctane	281s(1.25),290(1.33), 300(1.30)	44-3725-76
2,5-Cyclohexadien-1-one, 4-ethyl-4-meth- yl-	isooctane	227(4.142)	39-1349-76B
	EtOH	237(4.178)	39-1349-76B
2,3-Cyclononadien-1-one	EtOH	229(3.37),298(2.38)	88-0779-76
2-Cyclopenten-1-one, 2-cyclopropyl- 3-methyl-	EtOH	242(3.97)	39-0249-76C
4H-Inden-4-one, 1,2,3,5,6,7-hexahydro-	n.s.g.	250(4.23)	35-6752-76
$C_9H_{12}O_2$ Benzenemethanol, 4-methoxy-α-methyl-	EtOH	255(3.66),275(3.36), 282(3.30)	39-1466-76C

Compound	Solvent	$\lambda_{max}(\log \epsilon)$	Ref.
4H-1-Benzopyran-4-one, 2,3,5,6,7,8-hexa-hydro-	EtOH	275(3.95)	22-1913-76
Bicyclo[3.2.2]nonane-6,7-dione	hexane	532(1.62)	118-0256-76
2,4-Cyclohexadien-1-one, 6-hydroxy-2,5,6-trimethyl-	EtOH	317(3.46),362s(2.98)	1-0064-76
1,3-Cyclohexanedione, 2-(2-propenyl)-	EtOH	261(4.13)	39-0249-76C
$C_9H_{12}O_3$			
Bicyclo[2.2.1]heptane-7-carboxylic acid, 2-oxo-, methyl ester, syn	isooctane	284(1.23),290(1.24),300s(1.20)	44-3725-76
1,3-Dioxepane, 2-(2-furanyl)-	EtOH	208(3.95)	103-0142-76
2(5H)-Furanone, 4-methyl-5-(1-methyl-2-oxopropyl)-	EtOH	210(4.14)	1-0064-76
2-Oxabicyclo[3.2.1]oct-3-ene-4-carbox-ylic acid, methyl ester	MeOH	248(4.13)	39-0264-76C
$C_9H_{12}O_4$			
2,4-Cyclohexadien-1-one, 5,6,6-trimeth-oxy-	EtOH	358(3.66)	1-0064-76
$C_9H_{12}O_5$			
Octanoic acid, 3,5,7-trioxo-, methyl ester	EtOH	265(3.98),318s(3.62)	35-7733-76
$C_9H_{13}BrN_4O_5$			
1H-Imidazole-4-carboximidamide, 5-bromo-N-hydroxy-1-β-D-ribofuranosyl-	pH 1	251(3.93)	87-1020-76
	pH 11	231(3.85)	87-1020-76
$C_9H_{13}ClN_4O_4$			
1H-Imidazole-4-carboxamide, 5-amino-1-(3-chloro-3-deoxy-β-D-xylofuranosyl)-	pH 1	266(4.01)	94-2089-76
	pH 7	265(4.04)	94-2089-76
	pH 13	266(4.08)	94-2089-76
$C_9H_{13}ClN_4O_5$			
1H-Imidazole-4-carboximidamide, 5-chlo-ro-N-hydroxy-1-β-D-ribofuranosyl-	pH 1	248(4.01)	87-1020-76
	pH 11	232(3.88)	87-1020-76
$C_9H_{13}ClO_2$			
2,4-Hexadien-1-ol, 6-chloro-3-methyl-, acetate, (E,E)-	hexane	237(4.449)	33-0397-76
3,5-Hexadien-1-ol, 2-chloro-3-methyl-, acetate, (E)-	hexane	235(4.390)	33-0397-76
$C_9H_{13}Cl_3O_3$			
2,4-Hexanedione, 3-(2,2,2-trichloro-1-hydroxyethyl)-5-methyl-	$CHCl_3$	290(2.14)	18-2817-76
$C_9H_{13}NO_2$			
Acetamide, N-(7-oxo-1-cyclohepten-1-yl)-	MeOH	212(3.94),263(3.63)	104-2286-76
Ethanamine, 2-(3-methoxyphenyl)-	n.s.g.	210(3.82),220(3.85),275(3.28)	20-0421-76
$C_9H_{13}NO_2S$			
2-Pentenoic acid, 2-isothiocyanato-4-methyl-, ethyl ester, (Z)-	C_6H_{12}	233(3.90),279(4.06)	5-1850-76
$C_6H_{13}NO_3$			
Acetic acid, (4,4-dimethyl-6-oxo-2-pip-eridinylidene)-	EtOH	268(4.25)	94-3011-76
3.5-Heptadienoic acid, 3-(aminocarbo-nyl)-6-methyl-	EtOH	278(3.94)	44-2950-76

Compound	Solvent	$\lambda_{max}(\log \epsilon)$	Ref.
$C_9H_{13}NO_4S$			
Butanedioic acid, isothiocyanato-, diethyl ester	pH 8.06	249(3.09)	73-2216-76
	MeOH	249(--)	73-2216-76
	dioxan	245(3.12)	73-2216-76
2-Propenoic acid, 3-[[1-(methoxycarbo-nyl)ethenyl]amino]-2-(methylthio)-, methyl ester, (Z)-	EtOH	207(3.70),284(3.74), 319s(3.56)	39-0584-76C
D-Ribitol, 1,4-anhydro-1-C-(4-methyl-2-thiazolyl)-, (R)-	EtOH	250(3.83)	44-4074-76
$C_9H_{13}NS_2$			
4H-Cyclohept[d]isothiazole, 5,6,7,8-tetrahydro-3-(methylthio)-	EtOH	280(4.10)	48-0127-76
$C_9H_{13}NS_3$			
5H-Cyclohepta-1,4,2-dithiazine, 6,7,8,9-tetrahydro-3-(methylthio)-	C_6H_{12}	260(4.07),286(3.61), 343(2.88)	48-0127-76
$C_9H_{13}N_3$			
Pyrido[2,3-b]pyrazine, 1,2,3,4-tetra-hydro-2,3-dimethyl-	EtOH	265(3.68),328(3.88)	22-0251-76
$C_9H_{13}N_3O_2$			
Acetamide, N-methyl-N-(1,4-dimethyl-6-oxodihydro-2-pyrimidinyl)-	EtOH	224(3.83),278(3.70)	114-0413-76A
Acetamide, N-methyl-N-(1,6-dimethyl-4-oxodihydro-2-pyrimidinyl)-	EtOH	242(4.18)	114-0413-76A
	dioxan	242(4.16)	114-0413-76A
$C_9H_{13}N_3O_3$			
2,4(1H,3H)-Pyrimidinedione, 5-[3-(di-methylamino)-1-oxopropyl]-, hydro-chloride	pH 1	230(3.88),287(4.00)	87-0194-76
$C_9H_{13}N_3O_4$			
1H-Pyrazole-3-carboxamide, 1-(2-deoxy-α-D-erythro-pentofuranosyl)-	pH 1	198(4.10)	18-3552-76
	H_2O	212(4.11)	18-3552-76
	pH 13	218(4.07)	18-3552-76
1H-Pyrazole-3-carboxamide, 1-(2-deoxy-β-D-erythro-pentofuranosyl)-	pH 1	198(4.08)	18-3552-76
	H_2O	213(4.08)	18-3552-76
	pH 13	218(4.06)	18-3552-76
1H-Pyrazole-5-carboxamide, 1-(2-deoxy-α-D-erythro-pentofuranosyl)-	pH 1	200(3.91)	18-3552-76
	H_2O	219(3.92)	18-3552-76
	pH 13	220(3.95)	18-3552-76
1H-Pyrazole-5-carboxamide, 1-(2-deoxy-β-D-erythro-pentofuranosyl)-	pH 1	198(4.11)	18-3552-76
	H_2O	220(4.11)	18-3552-76
	pH 13	220(4.04)	18-3552-76
$C_9H_{13}N_3O_4S$			
1,2,4-Triazin-5(2H)-one, 6-[2,3-dihy-droxy-4-(hydroxymethyl)cyclopentyl]-3,4-dihydro-3-thioxo-, (1α,2α,3β,4α)-	MeOH	272(4.23)	23-2925-76
$C_9H_{13}N_3O_4S_2$			
1H-Imidazole-4-carbothioamide, 5-mercap-to-1-β-D-ribofuranosyl-	pH 1	278(3.81),355(4.29)	87-1020-76
	pH 11	260s(3.87),354(4.29)	87-1020-76
$C_9H_{13}N_3O_5$			
Cytosine, 1-β-D-arabinofuranosyl-	MeOH-acid	212(3.99),285(4.13)	87-0667-76
Isocytosine, 5-α-D-ribofuranosyl-	pH 1	222(4.00),262(3.87)	44-2793-76
	pH 7.2	270(3.59),290(3.60)	44-2793-76
	pH 13	232(3.97),277(3.86)	44-2793-76

Compound	Solvent	$\lambda_{max}(\log \epsilon)$	Ref.
Isocytosine, 5-β-D-ribofuranosyl-	pH 1	221(4.02),262(3.89)	44-2793-76
	pH 13	232(3.99),277(3.87)	44-2793-76
1H-Pyrazole-3-carboxamide, 1-α-D-arabinofuranosyl-	pH 1	198(4.11)	18-3552-76
	H_2O	231(4.11)	18-3552-76
	pH 13	219(4.07)	18-3552-76
1H-Pyrazole-3-carboxamide, 1-β-D-arabinofuranosyl-	pH 1	198(4.08)	18-3552-76
	H_2O	213(4.06)	18-3552-76
	pH 13	219(4.04)	18-3552-76
1H-Pyrazole-5-carboxamide, 1-α-D-arabinofuranosyl-	pH 1	198(4.15),222(4.03)	18-3552-76
	H_2O	219(4.14)	18-3552-76
	pH 13	220(4.00)	18-3552-76
Uracil, 1-(2-amino-2-deoxy-β-D-ribofuranosyl)-	MeOH	260(3.97)	44-3138-76
Uridine, 5-amino-5'-deoxy-	pH 1	265(4.00)	44-1041-76
	H_2O	293(3.90)	44-1041-76
hydrochloride	pH 1	264(4.04)	44-1041-76
	H_2O	294(3.89)	44-1041-76
$C_9H_{13}N_3O_6$			
1H-Pyrazole-5-carboxamide, 4-hydroxy-3-α-D-ribofuranosyl-	pH 13	304(3.92)	44-0287-76
	EtOH	225(3.90),276(3.83)	44-0287-76
1H-Pyrazole-5-carboxamide, 4-hydroxy-3-β-D-ribofuranosyl-	pH 12	233(3.68),307(3.92)	44-0287-76
	EtOH	226(3.88),267(3.78)	44-0287-76
1H-1,2,3-Triazole-4-carboxylic acid, 1-β-D-ribofuranosyl-, methyl ester	MeOH	214(4.04)	44-1041-76
1H-1,2,3-Triazole-4-carboxylic acid, 5-β-D-ribofuranosyl-, methyl ester	pH 13	235(3.84)	44-0084-76
	EtOH	218(3.88)	44-0084-76
$C_9H_{13}N_3O_6S$			
Acetamide, N-(2,3-dihydro-3-oxo-2-β-D-ribofuranosyl-1,2,4-thiadiazol-5-yl)-	pH 1	235(4.03),275(3.54)	4-0169-76
	pH 7	254(4.00),277s(3.73)	4-0169-76
	pH 11	254(4.10),277s(3.77)	4-0169-76
$C_9H_{13}N_4O$			
1H-Purinium, 6,7-dihydro-1,7,8,9-tetramethyl-6-oxo-, iodide	MeOH	254(3.98)	39-0239-76C
$C_9H_{13}N_4O_2$			
2H-Purinium, 3,7,8,9-tetrahydro-1,3,7,9-tetramethyl-2,8-dioxo-, iodide	H_2O	229(4.37),343(4.06)	39-1414-76C
$C_9H_{13}O$			
Pyrylium, 4-ethyl-2,6-dimethyl-, iodide	CH_2Cl_2	444(1.81)	59-1311-76
thiocyanate	CH_2Cl_2	408(2.40)	59-1311-76
selenocyanate	CH_2Cl_2	441(2.38)	59-1311-76
C_9H_{14}			
Cyclobutene, 3-isopropenyl-4,4-dimethyl-	EtOH	212(3.81)	78-0441-76
1,3-Cyclohexadiene, 5-ethyl-5-methyl-	EtOH	257(3.63)	118-0108-76
1,3-Cyclohexadiene, 1,5,5-trimethyl-	EtOH	264(3.78)	78-0441-76
1,3-Cyclooctadiene, 1-methyl-	EtOH	232(3.79)	70-0547-76
1,3-Cyclooctadiene, 2-methyl-	EtOH	229(3.74)	70-0547-76
1,3-Cyclopentadiene, 5-(1-methylpropyl)-	hexane	246(3.54)	104-1796-76
1,3,5-Heptatriene, 2,6-dimethyl-, (E)-	EtOH	262(4.43),272(4.56), 283(4.44)	78-0441-76
1,3,5-Heptatriene, 2,6-dimethyl-, (Z)-	EtOH	264s(4.13),273(4.18), 283s(4.05)	78-0441-76
2,3,5-Heptatriene, 2,6-dimethyl-	EtOH	230(4.32)	78-0441-76

Compound	Solvent	$\lambda_{max}(\log \epsilon)$	Ref.
$C_9H_{14}Br_2N_4$ Ethanediamine, N-(2-amino-1-benzimid-azolyl)-, dihydrobromide	EtOH	204(4.69),250(3.34), 276(3.90),282(3.88)	78-0839-76
$C_9H_{14}ClN_3O_3$ 2,4(1H,3H)-Pyrimidinedione, 5-[3-(di-methylamino)-1-oxopropyl]-, hydro-chloride	pH 1	230(3.88),287(4.00)	87-0194-76
$C_9H_{14}ClN_3O_5$ Uridine, 5-amino-5'-deoxy-, hydrochlor-ide	pH 1 H_2O	264(4.04) 294(3.89)	44-1041-76 44-1041-76
$C_9H_{14}ClN_4$ Methanaminium, N-[1-chloro-2-cyano-3-(dimethylamino)-2-propenylidene]-amino]methylene]-N-methyl-, perchlorate	HOAc	384(4.28)	73-1565-76
$C_9H_{14}F_3N_5$ 1,3,5-Triazine-2,4-diamine, N-pentyl-6-(trifluoromethyl)-	MeOH	213(4.52),273(3.57)	36-0525-76
$C_9H_{14}GeN_2$ Diazene, phenyl(trimethylgermyl)-	hexane	214(3.89),255(3.83), 562(1.42)	35-0109-76
$C_9H_{14}NO_2$ 8-Azabicyclo[3.2.1]oct-8-yloxy, 1,5-di-methyl-3-oxo-	C_6H_{12}	245(3.40),500(1.00)	78-0239-76
$C_9H_{14}N_2$ 1-Cyclohexene-1-carbonitrile, 2-(dimeth-ylamino)-	EtOH	285(4.14)	44-0019-76
3-Pyridinemethanamine, N-(1-methyleth-yl)-	MeOH	259(3.37)	33-0190-76
$C_9H_{14}N_2O$ Benzenemethanol, 2-hydrazino-α,α-di-methyl-	isoPrOH	244(3.99),291(3.38)	4-0033-76
2H-Pyrrole-5-carbonitrile, 3,4-dihydro-2,2,4,4-tetramethyl-, 1-oxide	EtOH	271(4.01)	39-1951-76C
$C_9H_{14}N_2OS$ 4(1H)-Pyrimidinone, 2-(butylthio)-6-methyl-	EtOH	235(3.97),287(3.89)	95-1094-76
4(1H)-Pyrimidinone, 2-(ethylthio)-6-propyl-	EtOH	236(3.91),287(3.86)	95-1094-76
4(1H)-Pyrimidinone, 2-[(3-methylbutyl)-thio]-	EtOH	232(3.91),290(3.89)	95-1094-76
$C_9H_{14}N_2OS_2$ 5(4H)-Thiazolethione, 4,4-dimethyl-2-morpholino-	EtOH	294(4.05),479(1.23)	33-2566-76
$C_9H_{14}N_2O_2$ 3,5-Heptanedione, 4-diazo-2,6-dimethyl-	hexane H_2O NaOH	230(4.23),241s(4.13) 234(4.25),242s(4.19) 233(4.17),241s(4.11), 270(3.75)	104-1245-76 104-1245-76 104-1245-76

Compound	Solvent	$\lambda_{max}(\log \epsilon)$	Ref.
4-Oxa-1,2-diazaspiro[4.7]dodec-1-en-3-one	n.s.g.	225(3.54),374(2.46)	4-0649-76
$C_9H_{14}N_2O_2S$			
4,6(1H,5H)-Pyrimidinedione, dihydro-5-(1-phenylbutyl)-2-thioxo-, (S)-	50%MeOH-HCl	287.5(4.43)	87-0521-76
	dioxan	287(4.35)	87-0521-76
	CH_2Cl_2	286(4.33)	87-0521-76
	MeCN	284(4.31)	87-0521-76
Thiourea, S-(2,6-dihydroxy-4,4-dimethyl-cyclohexyl)-, inner salt	MeOH	266(4.29)	44-0125-76
$C_9H_{14}N_2O_4$			
Carbamic acid, 2-[(2,3-dihydro-5-methyl-3-oxo-4-isoxazolyl)ethyl]-, ethyl ester	EtOH	213(3.76)	1-0567-76
$C_9H_{14}N_2O_5$			
2-Butenedioic acid, 2-(2-acetyl-1-methylhydrazino)-, dimethyl ester, (E)-	MeOH	271(4.28)	24-0253-76
1H-Imidazole, 1-α-D-galactopyranosyl-	H_2O	214(3.53)	136-0275-76D
1H-Imidazole, 1-β-D-galactopyranosyl-	H_2O	210(3.00)	136-0275-76D
1H-Imidazole, 1-α-D-mannopyranosyl-	H_2O	215(3.50)	136-0275-76D
1H-Imidazole, 1-β-D-mannopyranosyl-	H_2O	210(3.00)	136-0275-76D
$C_9H_{14}N_2Si$			
Diazene, phenyl(trimethylsilyl)-	hexane	215(4.26),220(4.19), 267(4.21),580(1.46)	35-0109-76
$C_9H_{14}N_3O_3$			
5-Pyrimidineethanaminium, 1,2,3,4-tetra-hydro-N,N,N-trimethyl- ,2,4-trioxo-, bromide	pH 1	286.5(4.12)	87-0194-76
$C_9H_{14}N_4O_2$			
Lumazine, 5,6,7,8-tetrahydro-1,6,7-tri-methyl-	pH 2.0	265(4.52)	24-3184-76
	pH 7.0	243(3.69),297(4.07)	24-3184-76
	pH 13.0	240s(3.83),290(4.04)	24-3184-76
$C_9H_{14}N_4O_3$			
Urea, N,N'-dimethyl-N-(1,2,3,4-tetra-hydro-1,3-dimethyl-2,4-dioxo-5-pyr-imidinyl)-	H_2O	206(3.90),275(3.76)	39-1414-76C
Urea, N-methyl-N'-(1,2,3,4-tetrahydro-1,3,6-trimethyl-2,4-dioxo-5-pyrimi-dinyl)-	H_2O	206(4.05),273(3.96)	39-1414-76C
$C_9H_{14}N_4O_3S$			
1H-Imidazole-4-carbothioamide, 5-amino-1-(3-deoxy-β-D-ribofuranosyl)-	pH 1	279(3.97),325(4.20)	94-2089-76
	pH 7	271(3.95),325(4.19)	94-2089-76
	pH 13	271(3.96),326(4.20)	94-2089-76
$C_9H_{14}N_4O_4$			
1H-Imidazole-4-carboxamide, 5-amino-1-(3-deoxy-β-D-ribofuranosyl)-	pH 1	267(4.06)	94-2089-76
	pH 7	266(4.08)	94-2089-76
	pH 13	267(4.09)	94-2089-76
1H-Imidazole-4-carboximidamide, 1-β-D-ribofuranosyl-, hydrobromide	pH 1	244(4.06)	87-1020-76
	pH 11	243(4.03)	87-1020-76
Uridine, 2',3'-diamino-2',3'-dideoxy-	MeOH	260(3.96)	44-3138-76

Compound	Solvent	$\lambda_{max}(\log \epsilon)$	Ref.
$C_9H_{14}N_4O_5$			
1H-Pyrazole-3-carboxamide, 4-amino-1-β-D-ribofuranosyl-	MeOH	214(3.89),287(3.51)	104-1538-76
1H-Pyrazole-3-carboxamide, 4-amino-5-β-D-ribofuranosyl-	H_2O	233s(3.65),282(3.68)	4-1359-76
	pH 11	233s(3.69),282(3.69)	4-1359-76
1H-Pyrazole-5-carboxamide, 4-amino-1-β-D-ribofuranosyl-	MeOH	213(3.75),229(3.66), 288(3.60)	104-1538-76
$C_9H_{14}N_4O_6$			
1H-1,2,4-Triazole-3-carboxamide, 1-β-D-psicofuranosyl-	pH 1	208(3.26)	44-1836-76
$C_9H_{14}N_6O$			
9H-Purine-9-ethanol, 6-amino-α-[(methylamino)methyl]-	pH 1	257(4.11)	70-0384-76
	pH 7	261(4.13)	70-0384-76
	pH 13	261(4.13)	70-0384-76
$C_9H_{14}O$			
Bicyclo[2.2.1]heptan-2-one, 3,3-dimethyl-	3-Mepentane	289(1.30)	144-0475-76
	MeCN	290(1.25)	144-0475-76
Bicyclo[2.2.1]heptan-2-one, 7,7-dimethyl-	3-Mepentane	295(1.43)	144-0475-76
	MeCN	292(1.46)	144-0475-76
Bicyclo[3.1.0]hexan-2-one, 1,4,4-trimethyl-	n.s.g.	288(1.45)	33-0082-76
Cyclobutanone, 3,3-dimethyl-2-(1-propenyl)-, (E)-	n.s.g.	295(1.89)	33-0082-76
2-Cyclohexen-1-one, 4-ethyl-4-methyl-	EtOH	228(4.053)	39-1349-76B
2-Cyclohexen-1-one, 2,5,5-trimethyl-	n.s.g.	232(3.69),335(1.46)	33-0082-76
2-Cyclohexen-1-one, 2,6,6-trimethyl-	EtOH	236(3.78)	33-2012-76
3-Cyclohexen-1-one, 2,5,5-trimethyl-	n.s.g.	255(2.01)	33-0082-76
2-Propen-1-one, 1-cyclohexyl-	MeOH	215(3.74),325(1.31)	40-0166-76
$C_9H_{14}O_2$			
Acetic acid, cyclohexylidene-	C_6H_{12}	218(4.19)	78-2827-76
2-Cyclohexen-1-one, 2-hydroxy-3-(1-methylethyl)-	EtOH	275.5(3.98)	18-2292-76
2-Cyclohexen-1-one, 2-hydroxy-3,5,5-trimethyl-	EtOH	274(4.00)	118-0541-76
2-Cyclopenten-1-one, 2-(2-methylpropoxy)-	EtOH	251(3.99)	39-0249-76C
2(5H)-Furanone, 5,5-dimethyl-4-(1-methylethyl)-	n.s.g.	206(4.23)	33-1253-76
2(5H)-Furanone, 5-pentyl-	n.s.g.	205(4.08)	70-0103-76
2,4-Heptadienoic acid, 6-methyl-, methyl ester, trans	EtOH	274(4.53)	70-0547-76 +104-0307-76
2,5-Heptadienoic acid, 6-methyl-, methyl ester, 2-cis	EtOH	208(4.10)	70-0547-76
trans	EtOH	208(4.13)	104-0307-76
	EtOH	206(4.19)	70-0547-76
3,5-Heptadienoic acid, 6-methyl-, methyl ester, trans	EtOH	234(4.36)	70-0547-76 +104-0307-76
2-Propanone, 1-(1-cyclohexen-1-yl)-1-hydroxy-	EtOH	233(3.81)	78-2533-76
$C_9H_{14}O_2S$			
Thiophene, 5-ethylidene-2,2,4-trimethyl-, 1,1-dioxide, (E)-	EtOH	242(4.18)	44-0242-76
$C_9H_{14}O_3$			
2-Butenoic acid, 3-(3,3-dimethyloxiranyl)-, methyl ester	EtOH	210(4.04)	104-0307-76

Compound	Solvent	$\lambda_{max}(\log \epsilon)$	Ref.
2(5H)-Furanone, 4-butyl-3-hydroxy-5-methyl-	EtOH	234(4.04)	107-0299-76
2(5H)-Furanone, 3-methoxy-5-methyl-4-propyl-	EtOH	226(3.93)	107-0299-76
2-Heptenoic acid, 6-methyl-5-oxo-, methyl ester	EtOH	208(4.08)	104-0307-76
2,4-Hexadiene-1,6-diol, 3-methyl-, 1-acetate, (E,E)-	hexane	232(4.408)	33-0397-76
$C_9H_{14}S_3$			
2,4,9-Trithiatricyclo[3.3.1.1^{3,7}]decane, 1,3-dimethyl-	hexane	244(3.34)	11-0220-76A
2,4,9-Trithiatricyclo[3.3.1.1^{3,7}]decane, 1,7-dimethyl-	hexane	245(3.37)	11-0220-76A
$C_9H_{15}BrN_2$			
Pyrimidine, 5-bromo-2,2-diethyl-1,2-dihydro-4-methyl-, hydrochloride	MeOH	378(3.56)	39-1784-76C
$C_9H_{15}BrN_4O_4$			
1H-Imidazole-4-carboximidamide, 1-β-D-ribofuranosyl-, hydrobromide	pH 1	244(4.06)	87-1020-76
	pH 11	243(4.03)	87-1020-76
$C_9H_{15}IN_2$			
Pyrimidine, 2,2-diethyl-1,2-dihydro-5-iodo-4-methyl-, hydrochloride	MeOH	384(3.54)	39-1784-76C
$C_9H_{15}NO$			
Cyclohexanone, 2-[(dimethylamino)methylene]-	C_6H_{12}	309(4.17)	35-2826-76
	H_2O	336(--)	35-2826-76
	MeOH	334(4.28)	35-2826-76
2-Cyclohexen-1-one, 5,5-dimethyl-3-(methylamino)-	pH 1	280(4.37)	39-2207-76C
	H_2O	291(4.46)	39-2207-76C
	pH 13	292(4.47)	39-2207-76C
4,6-Heptadien-3-one, 6-(dimethylamino)-	EtOH	370(4.78)	70-0577-76
3,5-Hexadien-2-one, 6-(dimethylamino)-3-methyl-	EtOH	248(3.81)	70-0577-76
2-Propenal, 3-(cyclohexylamino)-	H_2O	285(4.57)	35-2826-76
	MeOH	284(4.53)	35-2826-76
cis-s-cis	C_6H_{12}	308(4.16)	35-2826-76
trans	C_6H_{12}	265(4.46)	35-2826-76
3H-Pyrrol-3-one, 4-butyl-1,2-dihydro-5-methyl-	MeOH	317.5(3.84)	5-0383-76
$C_9H_{15}NO_2$			
Acetamide, N-(2-oxocycloheptyl)-	EtOH	242(3.18)	103-1379-76
$C_9H_{15}NO_2S$			
Hexanoic acid, 2-isothiocyanato-, ethyl ester	MeOH	245(3.32)	73-2216-76
Pentanoic acid, 2-isothiocyanato-3-methyl-, ethyl ester	MeOH	246(3.15)	73-2216-76
Pentanoic acid, 2-isothiocyanato-4-methyl-, ethyl ester	pH 8.06	246(3.15)	73-2216-76
	MeOH	245(3.26)	73-2216-76
	dioxan	249(3.08)	73-2216-76
$C_9H_{15}NO_3$			
2H-Pyrrole-4-carboxylic acid, 3,4-dihydro-2,2-dimethyl-, ethyl ester, 1-oxide	EtOH	235(2.24)	12-2511-76

Compound	Solvent	$\lambda_{max}(\log \epsilon)$	Ref.
$C_9H_{15}NO_3S$ Tetramethylammonium 3-(methylthio)-cyclobutenedion-4-olate	H_2O	257(4.11),310(4.30)	44-3904-76
$C_9H_{15}NO_4$ Butanoic acid, 4-[(1-methoxy-3-oxo-1-propenyl)amino]-, methyl ester	n.s.g.	292(4.31)	39-1241-76C
$C_9H_{15}NS$ Thiazole, 4,5-dimethyl-2-(2-methyl-propyl)-	EtOH	253(3.82)	70-1353-76
$C_9H_{15}NS_2$ Cyclopentanecarbodithioic acid, 2-[(1-methylethyl)imino]-	EtOH	304(4.06),398(4.39)	39-1706-76C
$C_9H_{15}N_2O$ Pyrimidinium, 1-(1,1-dimethylethyl)-2,3-dihydro-3-methyl-2-oxo-pseudobase	pH 0.29	316(3.92)	23-2681-76
	pH 9.18	236(3.61),308(3.22)	23-2681-76
$C_9H_{15}N_3OS$ 4(1H)-Pyrimidinone, 6-amino-2-[(3-meth-ylbutyl)thio]-	EtOH	232(4.25),275(3.89)	95-1094-76
$C_9H_{15}N_3O_3$ 2,4(1H,3H)-Pyrimidinedione, 5-[3-(di-methylamino)-1-hydroxypropyl]-, hydrochloride	pH 1	264(3.98)	87-0194-76
$C_9H_{15}N_5O_4$ D-Monapterin, 5,6,7,8-tetrahydro-, dihydrochloride	pH 4.3	267(3.98)	33-0248-76
4(1H)-Pteridinone, 2-amino-5,6,7,8-tetrahydro-6-(1,2,3-trihydroxy-propyl)-, dihydrochloride, L-	pH 4.3	267(4.00)	33-0248-76
$C_9H_{15}N_5O_5$ 1H-Pyrazole-3-carboximidamide, 4-amino-N-hydroxy-5-β-D-ribofuranosyl-	pH 11	237(3.80),270(3.79)	4-1359-76
1H-Pyrazole-3-carboxylic acid, 4-amino-5-β-D-ribofuranosyl-, hydrazide	pH 11 MeOH	240s(3.36),276(3.43) 231(3.40),279(3.42)	4-1359-76 4-1359-76
$C_9H_{16}N_2$ 1H-Pyrazole, 3,5-dimethyl-4-(1-methyl-propyl)-	EtOH	224(3.53)	44-2874-76
Pyrimidine, 2,2-diethyl-1,2-dihydro-4-methyl-, hydrobromide	MeOH	356(3.66)	39-1784-76C
$C_9H_{16}N_2O$ 3-Cyclohexene-1-ethanamine, N-methyl-N-nitroso-	MeOH-HCl	343(1.99)	23-2385-76
4H-Pyrazol-4-ol, 3,5-dimethyl-4-(1-meth-ylpropyl)-	EtOH	232(3.11)	44-2874-76
4H-Pyrazol-4-one, 5-butyl-1,5-dihydro-3,5-dimethyl-	EtOH base	204(3.48),330(3.68) 211(4.33),335(3.82), 357(3.85)	44-2874-76 44-2874-76
4H-Pyrazol-4-one, 1,5-dihydro-3,5-di-methyl-5-(1-methylpropyl)-	EtOH EtOH-base	202(3.32),330(3.60) 212(4.3+),356(3.71)	44-2874-76 44-2874-76
4H-Pyrazol-4-one, 5-(1,1-dimethylethyl)-1,5-dihydro-3,5-dimethyl-	EtOH	330(3.60)	44-2874-76

Compound	Solvent	$\lambda_{max}(\log \epsilon)$	Ref.
Quinoline, decahydro-1-nitroso-, trans	C_6H_{12}-10% dioxan	345s(1.60),357s(1.79), 369(1.92),382(1.92)	78-0847-76
$C_9H_{16}N_2O_2$			
Ethanone, 1-(3,4-dihydro-2,2,3-trimeth-yl-2H-pyrrol-5-yl)-, oxime, N'-oxide	MeOH	279(4.15)	39-1951-76C
4H-Pyrrolo[3,2-d]isoxazole, 3a,5,6,6a-tetrahydro-6-hydroxy-3,3a,5,6a-tetra-methyl-, isomer A	EtOH	210(3.52)(end abs.)	44-0855-76
isomer B	EtOH	210(3.41)(end abs.)	44-0855-76
$C_9H_{16}N_2O_3$			
Ethanone, 1-(2,5-dihydro-1-hydroxy-2,4,5,5-tetramethyl-1H-imidazol-2-yl)-, N-oxide	EtOH	236(3.93)	70-1997-76
$C_9H_{16}N_3O_3$			
1H-Imidazol-1-yloxy, 1,5-dihydro-4-[1-(hydroxyimino)ethyl]-2,2,5,5-tetramethyl-, 3-oxide	EtOH	236(4.09)	70-1997-76
$C_9H_{16}O_2$			
Cyclohexanecarboxylic acid, ethyl ester	hexane	218.0(1.96)	44-3312-76
	H_2O	207.0(--)	44-3312-76
	MeCN	215.7(1.96)	44-3312-76
2-Heptanone, 3-ethylidene-1-hydroxy-	EtOH	234(4.03)	107-0299-76
2,4-Pentanedione, 3-methyl-3-propyl-	EtOH	291(2.14)	44-0855-76
$C_9H_{16}O_3$			
Pentanoic acid, 4-methyl-2-(1-methyl-ethyl)-3-oxo-	hexane	272(1.98),290s(1.85)	104-1252-76
	H_2O	288(1.88)	104-1252-76
$C_9H_{16}Si$			
Silane, [1,3(and 4)-cyclopentadien-1-ylmethyl]trimethyl-	hexane	259(3.63)	65-1170-76
$C_9H_{17}N$			
1-Butanamine, N-cyclopentylidene-	C_6H_{12}	250(2.46)	59-1415-76
	EtOH	240(--)	59-1415-76
Quinoline, decahydro-, trans	heptane	294(1.54)	103-0902-76
	MeOH	290(1.85)	103-0902-76
	MeCN	290(1.60)	103-0902-76
hydrochloride	H_2O	245(1.70),278(1.95)	103-0902-76
	MeOH	203(2.35),280(1.30)	103-0902-76
methiodide	MeOH	220(4.20),300s(1.78)	103-0902-76
$C_9H_{17}NO$			
4-Heptanone, 3-[(methylamino)methylene]-	H_2O	306(4.39)	35-2826-76
	MeOH	300(4.38)	35-2826-76
cis-s-cis	C_6H_{12}	316(4.09)	35-2826-76
trans	C_6H_{12}	282(4.37)	35-2826-76
1-Penten-3-one, 1-(diethylamino)-	C_6H_{12}	292(4.36)	35-2826-76
	MeOH	307(4.42)	35-2826-76
1-Penten-3-one, 1-(dimethylamino)-4,4-dimethyl-	C_6H_{12}	292(4.37)	35-2826-76
	H_2O	317(4.46)	35-2826-76
	MeOH	311(4.42)	35-2826-76
1-Penten-3-one, 1-(ethylamino)-4,4-di-methyl-	H_2O	310(4.36)	35-2826-76
	MeOH	306(4.33)	35-2826-76
cis-s-cis	C_6H_{12}	300(4.24)	35-2826-76

Compound	Solvent	$\lambda_{max}(\log \epsilon)$	Ref.
1-Penten-3-one, 4-methyl-1-[(1-methyl-ethyl)amino]-	H_2O	304(--)	35-2826-76
	MeOH	304(4.30)	35-2826-76
cis-s-cis	C_6H_{12}	300(4.21)	35-2826-76
Propane, 1-cyclohexyl-2-nitroso-, dimer	EtOH	295(3.19)	78-1267-76
2-Propenal, 3-[bis(1-methylethyl)amino]-	C_6H_{12}	276(4.50)	35-2826-76
	H_2O	291(4.63)	35-2826-76
	MeOH	289(4.58)	35-2826-76
2-Propenal, 3-(dipropylamino)-	C_6H_{12}	276(4.51)	35-2826-76
	H_2O	292(4.65)	35-2826-76
	MeOH	288(4.61)	35-2826-76
$C_9H_{17}N_2$			
Butenylium, 2-amino-4-piperidino-, (picrate)	MeOH	315(4.72)	39-1784-76C
$C_9H_{17}N_2S$			
1H-Imidazolium, 1-(1,1-dimethylethyl)-3-methyl-2-(methylthio)-	EtOH	221-265(4.300-3.522)	28-0143-76B
	CH_2Cl_2	243(4.260)	28-0143-76B
$C_9H_{17}N_3O_3$			
Ethanone, 1-(2,5-dihydro-1-hydroxy-2,4,5,5-tetramethyl-1H-imidazol-2-yl)-, oxime, N-oxide	EtOH	235(3.96)	70-1997-76
$C_9H_{18}N_2O$			
2-Propenamide, 3-(diethylamino)-N,N-di-methyl-	C_6H_{12}	277(4.33)	78-1025-76
	MeOH	288(4.51)	78-1025-76
	dioxan	281(4.49)	78-1025-76
2-Propenamide, 3-[(1,1-dimethylamino)-N,N-dimethyl-	C_6H_{12}	284(4.32)	78-1025-76
	MeOH	289(4.45)	78-1025-76
	dioxan	286(4.31)	78-1025-76
$C_9H_{18}O$			
3-Pentanone, 2,2,4,4-tetramethyl-	C_6H_{12}	297(1.32)	112-0777-76
$C_9H_{18}OS$			
3-Pentanethione, 2,2,4,4-tetramethyl-, S-oxide	EtOH	264(3.95)	39-2079-76C
$C_9H_{18}O_3Si_2$			
2-Cyclopropen-1-one, 2,3-bis[(trimethyl-silyl)oxy]-	hexane	225s(3.26),275s(2.35)	35-3641-76
$C_9H_{18}Se$			
3-Pentaneselone, 2,2,4,4-tetramethyl-	C_6H_{12}	230(3.45),268(3.86), 710(1.32)	39-2079-76C
$C_9H_{19}NO$			
Isobutyramide, N-sec-butyl-N-methyl-, (+)-	C_6H_{12}	197.5(3.83)	108-0039-76
	MeOH	197.5(3.90)	108-0039-76
	MeCN	197.5(3.86)	108-0039-76
2-Pentanamine, 2,4,4-trimethyl-N-meth-ylene-, N-oxide	ether	251(3.85)	4-1001-76
$C_9H_{19}N_2O$			
1H-1,4-Diazepin-1-yloxy, hexahydro-2,2,7,7-tetramethyl-	C_6H_{12}	236(3.38),452(0.96)	77-0083-76
	MeOH	235(3.41),430(0.92)	77-0083-76
$C_9H_{19}N_2O_5P$			
Acetic acid, [[(triethoxyphosphoranyli-dene)hydrazono]-, methyl ester	EtOH	270.5(4.17)	65-1424-76

Compound	Solvent	$\lambda_{max}(\log \epsilon)$	Ref.
$C_9H_{20}N_2O$			
1-Pentanol, 5-(diethylazo)-	EtOH	247(3.66)	104-2034-76
1-Pentanol, 5-[(2-methylpropyl)azo]-	EtOH	236(3.62)	104-2034-76
$C_9H_{20}N_3O_4P$			
Acetic acid, [[(dimethylamino)diethoxy-phosphoranylidene]hydrazono]-, methyl ester	EtOH	277.0(4.17)	65-1424-76
$C_9H_{20}N_4$			
1-Piperidinamine, N-[1-(ethylazo)ethyl]-	hexane	370(1.53)	104-2207-76
$C_9H_{20}OSn$			
2-Hexanone, 6-(trimethylstannyl)-	C_6H_{12}	282(1.32)	35-0104-76
	CCl_4	242(2.15),283(1.43)	35-0104-76
$C_9H_{21}N_4O_3P$			
Acetic acid, [[bis(dimethylamino)ethoxy-phosphoranylidene]hydrazono]-, methyl ester	EtOH	285.5(4.31)	65-1424-76
$C_9H_{22}N_5O_2P$			
Acetic acid, [[tris(dimethylamino)phos-phoranylidene]hydrazono]-, methyl ester	EtOH	293.0(4.12)	65-1424-76

Compound	Solvent	$\lambda_{max}(\log \epsilon)$	Ref.
$C_{10}F_{10}$			
1H-Indene, 1,1,3,4,5,6,7-heptafluoro-2-(trifluoromethyl)-	heptane	278(3.59),285(3.59), 301s(3.50)	70-2141-76
$C_{10}F_{11}NO_2$			
1H-Indene, 1,1,3,3,4,5,6,7-octafluoro-2-nitro-2-(trifluoromethyl)-	heptane	264s(2.93),269(2.98)	70-2141-76
$C_{10}H_4N_4O_6$			
5H,10H-Dipyrrolo[1,2-a:1',2'-a]pyrazine-5,10-dione, 2,7-dinitro-	CHCl$_3$	266(4.54),288(4.45)	44-3050-76
$C_{10}H_5BrO_2$			
1,2-Naphthalenedione, 5-bromo-	EtOH	248s(4.21),255(4.24), 260s(4.23),402(3.38)	78-2693-76
1,4-Naphthalenedione, 5-bromo-	EtOH	241(4.30),357(3.47)	78-2693-76
$C_{10}H_5ClN_2$			
Propanedinitrile, [(2-chlorophenyl)methylene]-	C_6H_{12}	232(3.72),255(3.13), 299(4.05)	80-0781-76
	EtOH	237(3.74),255(3.60), 302(3.82)	80-0781-76
	CHCl$_3$	306(4.13)	80-0781-76
	CCl$_4$	301(4.25)	80-0781-76
non-ion	H$_2$O	305(4.14)	80-0781-76
ion	H$_2$O	250(4.11)	80-0781-76
Propanedinitrile, [(4-chlorophenyl)methylene]-	C_6H_{12}	231(4.04),255(3.22), 320(4.42)	80-0781-76
	EtOH	237(3.83),258(3.81), 324(4.11)	80-0781-76
	CHCl$_3$	324(4.40)	80-0781-76
	CCl$_4$	320(4.37)	80-0781-76
non-ion	H$_2$O	320(4.31)	80-0781-76
ion	H$_2$O	260(4.26)	80-0781-76
$C_{10}H_5ClO_2$			
1,2-Naphthalenedione, 5-chloro-	EtOH	216(4.40),240(4.54), 309(3.48),345(3.38)	78-2693-76
1,4-Naphthalenedione, 5-chloro-	EtOH	237(4.21),246s(4.20), 354(3.41)	78-2693-76
$C_{10}H_5Cl_2F$			
Benzene, (4,4-dichloro-3-fluoro-3-buten-1-ynyl)-	n.s.g.	207(4.20),217(4.16), 223(4.14),236(4.00), 273(4.28),284(4.36), 290s(--),301(4.28)	44-1487-76
$C_{10}H_5FO_2$			
1,2-Naphthalenedione, 5-fluoro-	EtOH	245(4.31),340(3.43), 398(3.48)	78-2693-76
1,4-Naphthalenedione, 5-fluoro-	EtOH	243(4.23),248(4.23), 343(3.55)	78-2693-76
$C_{10}H_5F_5O_2$			
2-Propenoic acid, 2,3-difluoro-3-[4-(trifluoromethyl)phenyl]-, (E)-	EtOH	265(4.34)	104-1920-76
$C_{10}H_5IO_2$			
1,2-Naphthalenedione, 5-iodo-	EtOH	212(4.27),247(4.32), 415(3.36)	78-2693-76

Compound	Solvent	$\lambda_{max}(\log \epsilon)$	Ref.
1,4-Naphthalenedione, 5-iodo-	EtOH	213(4.29),247(4.30), 377(3.38)	78-2693-76
$C_{10}H_5NO_4$			
1,2-Naphthalenedione, 5-nitro-	EtOH	213(4.28),265s(3.93), [400(2.97)(end abs.)]	78-2693-76
1,4-Naphthalenedione, 5-nitro-	EtOH	229s(4.12),250(4.19), 333(3.22)	78-2693-76
$C_{10}H_5N_3O_2$			
Propanedinitrile, [(2-nitrophenyl)methylene]-, non-ion	H_2O	257(4.01)	80-0781-76
ion	H_2O	270(3.86)	80-0781-76
Propanedinitrile, [(4-nitrophenyl)methylene]-	C_6H_{12}	238(3.57),257(3.60), 300(3.11)	80-0781-76
	EtOH	238(4.01),264(4.10), 310(3.49)	80-0781-76
	$CHCl_3$	307(4.34)	80-0781-76
	CCl_4	305(4.27)	80-0781-76
non-ion	H_2O	270(4.10),310(3.94)	80-0781-76
ion	H_2O	268(4.11)	80-0781-76
$C_{10}H_6ClNOS$			
2-Propenoyl isothiocyanate, 3-(4-chlorophenyl)-	C_6H_{12}	314(4.52)	73-1388-76
$C_{10}H_6Cl_6N_2$			
3H-Indol-2-amine, 3,3,4,5,6,7-hexachloro-N,N-dimethyl-	C_6H_{12}	259.5(4.50)	39-0162-76C
$C_{10}H_6FeO_6$			
Iron, tetracarbonyl(2,3-dimethyl-1,4-dioxo-2-butene-1,4-diyl)-, (OC-6-22)-	hexane	241(4.06),417(1.92)	101-0235-76H
$C_{10}H_6N_2$			
Benzonitrile, 2-(2-cyanoethenyl)-	EtOH	223(4.20),230(4.19), 237(4.06),271(4.16), 315s(2.83)	44-1389-76
Propanedinitrile, (phenylmethylene)-	C_6H_{12}	230(3.74),302(4.39)	80-0781-76
	EtOH	231(4.00),308(4.12)	80-0781-76
	EtOH	231(3.89),309(4.31)	118-0705-76
	$CHCl_3$	312(4.34)	80-0781-76
	CCl_4	305(4.28)	80-0781-76
non-ion	H_2O	310(4.31)	80-0781-76
ion	H_2O	250s(4.15)	80-0781-76
$C_{10}H_6N_2O$			
1(2H)-Naphthalenone, 2-diazo-	hexane	218(4.32),257(4.50), 292(3.88),303(3.88), 318(3.73),397(3.99)	99-0081-76
	EtOH	262(4.52),302(3.70), 313(3.55),401(3.92)	99-0081-76
	hexanol?	264(4.50),292(3.83), 304(3.81),317(3.68), 403(3.99)	99-0081-76
	$C_2H_4Cl_2$	265(4.54),299(3.85), 313(3.83),400(3.98)	99-0081-76
	DMF	264(4.42),303(3.67), 316(3.55),402(3.91)	99-0081-76
	DMSO	267(4.46),304(3.81), 317(3.60),402(3.92)	99-0081-76

Compound	Solvent	$\lambda_{max}(\log \epsilon)$	Ref.
1(2H)-Naphthalenone, 2-diazo- (cont.)	HCl	215(4.35),262(4.51), 303(3.92),313(4.04), 413(3.74)	99-0081-76
Propanedinitrile, [(2-hydroxyphenyl)-methylene]-	C_6H_{12}	261(3.80),278(3.65), 350(3.87)	80-0781-76
	EtOH	265(3.99),278(3.91), 360(3.48)	80-0781-76
	$CHCl_3$	280(4.01),355(3.60)	80-0781-76
	CCl_4	279(3.97),353(3.51)	80-0781-76
non-ion	H_2O	305(4.06)	80-0781-76
ion	H_2O	293(3.90)	80-0781-76
$C_{10}H_6N_2O_2$ 5H,10H-Dipyrrolo[1,2-a:1',2'-d]pyrazine-5,10-dione	MeOH	235(4.34),273(4.15), 303(4.10),315(4.11)	44-3050-76
$C_{10}H_6N_2S_2$ 1,3,4-Thiadiazolo[2,3-a]isoquinolin-4-ium, 2-mercapto-, hydroxide, inner salt	benzene	442(--)	2-0176-76
	MeOH	244(4.26),254(4.26), 280s(--),295(4.43), 352s(3.57),388(3.77)	2-0176-76
	$CHCl_3$	416(--)	2-0176-76
	MeCN	405(--)	2-0176-76
1,3,4-Thiadiazolo[3,2-a]quinolin-10-ium, 2-mercapto-, hydroxide, inner salt	benzene	458(3.72)	2-0176-76
	MeOH	223(4.25),254(3.66), 283(4.47),325(3.80), 338(3.93),400s(3.66)	2-0176-76
	EtOH	402s(3.65)	2-0176-76
	$CHCl_3$	432(3.79)	2-0176-76
	CH_2Cl_2	428(3.83)	2-0176-76
	MeCN	417(3.73)	2-0176-76
$C_{10}H_6N_4$ 1,3,5,7-Tetraazaphenanthrene	EtOH	222(4.46),253(4.34), 283(3.70),316(3.93), 328(3.59)	94-0591-76
1,2,4-Triazino[5,6-c]quinoline	EtOH	270(4.16),350(3.65)	24-2338-76
$C_{10}H_6N_4O$ Imidazo[5,1-c][1,2,4]benzotriazine-1-carboxaldehyde	EtOH	247(4.17),372(3.75)	44-0158-76
$C_{10}H_6N_4O_6$ 2,4(1H,3H)-Pyrimidinedione, 5-nitro-1-(4-nitrophenyl)-	pH 1	240(3.98),304(4.15)	87-1072-76
	pH 11	325(4.29)	87-1072-76
$C_{10}H_6OS$ Naphtho[1,8-bc]thiete, 1-oxide	EtOH	221(4.48),265(3.68), 276(3.65),288(3.66), 319(2.54)	35-6643-76
$C_{10}H_6OS_2$ 2-Cyclobuten-1-one, 2-mercapto-3-phenyl-4-thioxo-, potassium salt	EtOH	202(4.35),240(4.13), 254(4.12),372(4.58)	118-0445-76
$C_{10}H_6O_2$ 1,2-Naphthalenedione	heptane	195(--),245(--), 251(--),326(--), 338(--),382(--)	18-1401-76

$C_{10}H_6O_2S_2-C_{10}H_7ClO_2$

Compound	Solvent	$\lambda_{max}(\log \epsilon)$	Ref.
1,2-Naphthalenedione (cont.)	$CF_3CHOHCF_3$	170s(4.35),196(4.39), 249(4.44),255(4.44), 354(3.53),425(4.34)	18-1401-76
1,4-Naphthalenedione	EtOH	246(4.32),260(4.02), 334(3.52)	115-0203-76
	ether	225(4.18),243(4.26), 270(3.88),328(3.42)	115-0203-76
	$CHCl_3$	260(4.05),330(3.33)	115-0203-76
$C_{10}H_6O_2S_2$ Naphtho[1,8-cd]-1,2-dithiole, 1,1-dioxide	EtOH	243(4.20),325(3.56)	35-6643-76
$C_{10}H_6O_2Se$ 3-Cyclobutene-1,2-dione, 3-phenyl-	EtOH	202(4.30),247(4.22), 254s(4.20),260s(4.16), 268s(4.11),328(4.26), 383(3.96)	89-0704-76
$C_{10}H_6O_3$ Furano[6,7:2',3']coumaran-3-one	EtOH	230(3.50),245(3.26), 258(2.87),272(2.84), 325(2.78)	102-1553-76
$C_{10}H_6O_4$ 1H-2-Benzopyran-4-carboxylic acid, 1-oxo-	MeOH	225(4.35),290(3.19), 315(3.6)	25-0954-76
1,4-Naphthalenedione, 2,5-dihydroxy-	MeOH-HCl	283(3.84),411(3.36)	78-1353-76
	MeOH-NaOMe	261(4.15),386(3.20), 470(3.32)	78-1353-76
	$CHCl_3$	284(4.08),429(3.54)	78-1353-76
$C_{10}H_6S$ Naphtho[1,8-bc]thiete	EtOH	220(3.51),260(3.70), 280(3.45),290s(3.23)	35-6643-76
$C_{10}H_7Br$ Azulene, 2-bromo-	C_6H_{12}	281(4.81),289(4.88), 303(3.86),332(3.78), 346(3.86),359(3.74), 557(2.70),596(2.66), 655(2.25)	44-1811-76
Azulene, 6-bromo-	C_6H_{12}	278(4.81),282(4.85), 288(4.90),304(3.80), 334(3.63),340(3.58), 347(3.77),362(3.13), 521(2.17),539(2.32), 560(2.41),582(2.49), 607(2.45),636(2.46), 662(2.14),702(2.13)	44-1811-76
$C_{10}H_7BrN_2S_2$ [1,2]Dithiolo[5,1-e][1,2,3]thiadiazole-7-S^{IV}, 1-(4-bromophenyl)-	C_6H_{12}	204(4.36),231(4.38), 253(4.24),292(4.07), 484(4.17)	39-0228-76C
$C_{10}H_7ClO_2$ 8H-Cyclohepta[b]furan-8-one, 7-chloro-2-methyl-	EtOH	233(4.18),239(4.09), 274(4.51),305s(3.79), 317s(3.74),332(3.60), 347(3.64),368(3.51)	39-2403-76C

Compound	Solvent	$\lambda_{max}(\log \epsilon)$	Ref.
2,4,6-Cycloheptatrien-1-one, 2-chloro-7-(2-propynyloxy)-	EtOH	252(4.46),256(4.47), 325(3.90),355(3.85), 371s(3.71)	39-2403-76C
2,4,6-Cycloheptatrien-1-one, 3-chloro-2-(2-propynyloxy)-	EtOH	247(4.41),320s(3.86)	39-2403-76C
$C_{10}H_7IS$			
Thiophene, 2-iodo-3-phenyl-	isooctane	233(4.21),258(4.01)	44-1320-76
$C_{10}H_7NOS_2$			
4H-1,3-Thiazin-4-one, 2,3-dihydro-6-phenyl-2-thioxo-	MeOH	254(4.30),310(4.23)	73-1388-76
$C_{10}H_7NO_2$			
2-Naphthalenol, 1-nitroso-	isooctane	213(4.42),273(4.10), 370(3.78)	3-1915-76
	CHCl$_3$	275(4.15),380(3.77)	3-1915-76
$C_{10}H_7NO_2S$			
4-Oxazolecarboxaldehyde, 5-mercapto-2-phenyl-	heptane	265(4.26),370(3.78)	104-1575-76
5(4H)-Oxazolone, 4-(mercaptomethylene)-2-phenyl-	EtOH	405(4.35)	23-2089-76
4H-1,3-Thiazine-4,6(5H)-dione, 2-phenyl-, sodium salt	EtOH	247(4.47)	103-0862-76
$C_{10}H_7NO_3$			
Acetic acid, (1,2-dihydro-2-oxo-3H-indol-3-ylidene)-	MeOH	253(4.28),304(3.73)	94-2305-76
geometric isomer	MeOH	255(4.33),261(4.32), 304(3.73)	94-2305-76
1H-Inden-1-one, 2-methyl-4-nitro-	hexane	214(4.03),242(4.01), 274(4.08),328(3.23), 342(3.30),357(3.23)	44-3540-76
	MeOH	345(3.24)	44-3540-76
	CHCl$_3$	350(3.41)	44-3540-76
$C_{10}H_7NO_3S$			
Acetic acid, (1,2-dihydro-2-oxo-3H-indol-3-ylidene)mercapto-	MeOH	253(4.03),274(3.82), 378(4.05)	94-2305-76
$C_{10}H_7NO_4$			
1H-2-Benzopyran-1-one, 3-methyl-5-nitro-	n.s.g.	233s(4.13),298(3.73), 353(3.54)	39-1073-76C
$C_{10}H_7N_3$			
5H-Pyrazino[2,3-b]indole	EtOH	214(4.36),244(3.95), 261(4.00),319(3.92)	39-1361-76C
1,9,9b-Triazaphenalene	EtOH	246(3.98),264(4.11), 268s(4.08),365s(4.02), 378(4.13),385s(4.05), 399(3.76),452s(2.81), 483(2.83),503s(2.57), 602s(2.18),650(2.38)	1-0466-76
$C_{10}H_7N_3O$			
1,9-Diazaphenoxazine	MeOH	210(4.29),217(4.30), 338(4.12)	4-0107-76

Compound	Solvent	$\lambda_{max}(\log \epsilon)$	Ref.
$C_{10}H_7N_3O_2$			
Benzonitrile, 4-(3-methyl-1,2,4-oxadia-zol-5-yl)-, N-oxide	EtOH	250(4.16),332(4.03)	44-2985-76
$C_{10}H_7N_3O_2S_2$			
1H-[1,2]Dithiolo[5,1-e][1,2,3]thiadia-zole-7-S^{IV}, 1-(4-nitrophenyl)-	C_6H_{12}	205(4.38),226(4.44), 251(4.10),327(4.20), 339s(4.13),495(4.27)	39-0228-76C
$C_{10}H_7N_3O_3S$			
1-Naphthalenesulfonamide, 6-diazo-5,6-dihydro-5-oxo-	HCl	223(4.45),273(4.50), 302(4.03),406(3.74)	99-0081-76
	DMF	264(4.22),340(3.83), 402(3.81)	99-0081-76
	MeCN	228(4.39),262(4.29), 339(3.90),399(3.83)	99-0081-76
$C_{10}H_7N_3O_6$			
Benzeneacetic acid, α-cyano-2,4-dini-tro-, methyl ester	MeOH	474(4.07)	73-0590-76
	+20% DMSO	480(4.08)	73-0590-76
	+40% DMSO	486(4.09)	73-0590-76
	+60% DMSO	490(4.11)	73-0590-76
	+80% DMSO	494(4.11)	73-0590-76
$C_{10}H_7N_5$			
[1,2,4]Triazino[6,5-c]quinolin-2-amine	EtOH	271(4.52),382(3.72)	114-0395-76C
$C_{10}H_7N_5O$			
[1,2,4]Triazino[6,5-c]quinolin-2-amine, 4-oxide	EtOH	219(4.34),236(4.20), 271(4.47)	114-0395-76C
$C_{10}H_8$			
Naphthalene, lithium salt of dianion	ether	530(3.80)	35-5706-76
$C_{10}H_8BrNOS$			
Spiro[3H-indole-3,2'-oxiran], 4-bromo-2-(methylthio)-	hexane	232s(4.29),238(4.39), 244(4.32),280(3.72), 290(3.73),302(3.71), 322(3.79)	12-1023-76
Spiro[3H-indole-3,2'-oxiran], 6-bromo-2-(methylthio)-	EtOH	249(4.30),295(3.62), 306(3.67),317(3.69)	12-1023-76
$C_{10}H_8BrNOS_2$			
4H-1,3-Thiazin-4-one, 6-(4-bromophenyl)-tetrahydro-2-thioxo-	n.s.g.	260(4.54),312(4.22)	73-1388-76
$C_{10}H_8ClNOS$			
4-Thiazolol, 5-(4-chlorophenyl)-2-meth-yl-	EtOH	306(4.06)	118-0261-76
$C_{10}H_8ClNOS_2$			
4H-1,3-Thiazin-4-one, 6-(4-chlorophen-yl)tetrahydro-2-thioxo-	n.s.g.	260(4.13),312(4.21)	73-1388-76
$C_{10}H_8ClNO_2$			
7,8-Quinolinediol, 5-chloro-2-methyl-	MeOH	207(4.45),252(4.61), 305(3.26),350(3.48)	24-2259-76
$C_{10}H_8ClNO_2S$			
4-Thiazolecarboxylic acid, 2-(4-chloro-phenyl)-4,5-dihydro-	MeCN	251(4.01)	39-1901-76C

Compound	Solvent	$\lambda_{max}(\log \epsilon)$	Ref.
$C_{10}H_8ClN_3O$ 4(1H)-Pyrimidinone, 6-amino-2-chloro-5-phenyl-	MeOH	218(4.33),265(3.85)	94-0026-76
$C_{10}H_8Cl_4$ 1,3-Cyclopentadiene, 1,2,3,4-tetra-chloro-5-cyclopentylidene-	C_6H_{12}	292(4.28),303(4.27), 406(2.82)	24-3929-76
$C_{10}H_8F_2O$ 2-Propenoic acid, 2,3-difluoro-3-(4-methylphenyl)-	EtOH	266(4.37)	104-1920-76
$C_{10}H_8F_2O_3$ 2-Propenoic acid, 2,3-difluoro-3-(4-methoxyphenyl)-	EtOH	277(4.48)	104-1920-76
$C_{10}H_8F_8N_2O_3$ 2,4(1H,3H)-Pyrimidinedione, 5-[[(2,2,3-3,4,4,5,5-octafluoropentyl)oxy]meth-yl]-	EtOH	266(3.90)	104-0643-76
$C_{10}H_8FeN_6O_4S_2$ Ferrocene, 1,1'-bis(azidosulfonyl)-	EtOH	207(4.53),247(4.03), 307(3.35)	44-0491-76
$C_{10}H_8FeO_3$ Iron, dicarbonyl[(η^5-2,4-cyclopentadien-1-ylidene)(3-oxo-1,3-propanediyl)]-	hexane	246s(3.83),325(3.61)	24-1429-76
$C_{10}H_8MoO_3$ Molybdenum, tricarbonyl(1-ethylcyclo-pentadiene-1,2'-diyl)-	hexane	220(4.35),262s(3.94), 296s(3.60),426(2.59)	24-1429-76
$C_{10}H_8N_2$ Bipyridine	isooctane	236(4.07),281(4.16)	62-0497-76A
	H_2O	233(3.99),281(4.11)	62-0497-76A
	88% MeOH	235(4.07),282(4.19)	62-0497-76A
	88% isoPrOH	236(4.07),283(4.18)	62-0497-76A
	90% dioxan	238(4.07),283(4.19)	62-0497-76A
protonated	H_2O	241(3.86),301(4.18)	62-0497-76A
	88% isoPrOH	241(3.91),303(4.11)	62-0497-76A
	90% dioxan	243(3.95),303(4.22)	62-0497-76A
2,2'-Bipyridine, ferrous complex	MeOH	522(3.94)	118-0001-76
4,4'-Bipyridine	MeOH	224(4.54),272(4.45), 352(3.58)	18-1725-76
1H-Indole-2-carbonitrile, 1-methyl-	EtOH	284(4.504)	103-0669-76
$C_{10}H_8N_2O$ 11-Azabicyclo[4.4.1]undeca-1,3,5,7,9-pentaene, 11-nitroso-	MeOH	250(4.69),295(3.80), 376(2.38),388(2.58), 398(2.80),409(2.86)	89-0228-76
[2,2'-Bipyridin]-5-ol	EtOH	253(3.76),297(3.95)	64-0115-76B
1H-Furo[2,3-g]indazole, 7-methyl-	EtOH	272(3.81),288(3.75), 299(3.73)	32-1083-76
1H-Imidazole, 1-benzoyl-	EtOH	240(4.05),272s(3.46)	33-2738-76
Methanone, 1H-imidazol-2-ylphenyl-	EtOH	260(3.94),297(4.17)	33-2738-76
Methanone, 1H-imidazol-4-ylphenyl-	EtOH	257s(4.04),276(4.13)	33-2738-76
Pyridine, 2,2'-oxybis-	EtOH	262(3.85)	64-0122-76B

Compound	Solvent	$\lambda_{max}(\log \epsilon)$	Ref.
$C_{10}H_8N_2OS$			
5(4H)-Oxazolethione, 4-(aminomethylene)-2-phenyl-	heptane	270(3.90),317(3.69), 400(3.62)	104-1575-76
4(1H)-Pyrimidinone, 2,3-dihydro-6-phenyl-2-thioxo-	EtOH	228(4.28),279(4.51)	95-1094-76
$C_{10}H_8N_2O_2$			
6-Indolizinecarbonitrile, 3,5-dihydro-2-hydroxy-7-methyl-5-oxo-	EtOH	260(3.92),364s(4.22), 377(4.27),399s(3.73)	48-0313-76
3-Isoxazolecarbonitrile, 4,5-dihydro-5-phenyl-, 2-oxide	CCl₄	268(4.02)	104-1974-76
5H-Oxazolo[2,3-b]quinazolin-5-one, 2,3-dihydro-	MeOH	221(4.61),226s(4.55), 245(3.83),254(3.83), 262(3.79),308(3.56), 318s(--)	118-0469-76
4(1H)-Pyrimidinone, 5-hydroxy-2-phenyl-	pH 1	240(4.05),288(4.07)	39-2038-76C
	pH 11	227(4.06),310(4.10)	39-2038-76C
$C_{10}H_8N_2O_2S$			
Pyridine, 2,2'-sulfonylbis-	MeOH	206s(3.89),220(3.98), 261(3.83)	39-1845-76B
$C_{10}H_8N_2O_3$			
4H-Pyrido[1,2-a]pyrimidine-3-carboxylic acid, 6-methyl-4-oxo-	pH 7.35	255(3.97),322(3.82), 370(4.07)	145-0478-76
5,8-Quinolinedione, 7-amino-6-methoxy-	pH 1	232(4.21),271(4.14), 475(3.41)	4-1063-76
	pH 11	232(4.23),271(4.24), 475(3.45)	4-1063-76
$C_{10}H_8N_2O_3S$			
[2,2'-Bipyridine]-5-sulfonic acid	pH 6.0	240(3.80),288(3.93)	64-0115-76B
4-Thiazolol, 2-methyl-5-(4-nitrophenyl)-	EtOH	305(4.10),374(4.16), 471(4.52)	118-0261-76
$C_{10}H_8N_2O_4S_2$			
5-Isothiazolecarboxamide, 3-[(phenylsulfonyl)oxy]-	MeOH	222(4.39),262(3.91)	1-0781-76
$C_{10}H_8N_2O_5S$			
2-Furancarboximidothioic acid, N-hydroxy-5-nitro-, 2-furanylmethyl ester	EtOH	227(4.29),351(4.05)	73-3085-76
$C_{10}H_8N_2S$			
3aH-Cyclohepta[b]thiophene-3-carbonitrile, 2-amino-	EtOH	215(3.9)	33-0747-76
8H-Cyclohepta[b]thiophene-3-carbonitrile, 2-amino-	EtOH	338(4.02)	33-0747-76
$C_{10}H_8N_2S_2$			
1H-[1,2]Dithiolo[5,1-e][1,2,3]thiadiazole-7-S^{IV}, 1-phenyl-	C_6H_{12}	203(4.34),234(4.42), 250s(4.23),289(3.96), 483(4.20)	39-0228-76C
$C_{10}H_8N_2Se$			
Naphtho[1,2-d][1,2,3]selenadiazole, 4,5-dihydro-	C_6H_{12}	211(4.2),255(4.15), 335(3.2)	24-1650-76
Naphtho[2,1-d][1,2,3]selenadiazole, 4,5-dihydro-	C_6H_{12}	232(4.05),287(4.0), 332(3.8)	24-1650-76

Compound	Solvent	$\lambda_{max}(\log \epsilon)$	Ref.
$C_{10}H_8N_3O_3$			
Pyridinium, 3-hydroxy-1-(5-nitro-2-pyridinyl)-, chloride	H_2O	211(4.11),225(4.18), 250(4.16),280(4.01), 350(3.57)	39-2296-76C
$C_{10}H_8N_4$			
1,2,4-Triazolo[3,4-a]phthalazine, 3-methyl-	MeOH	234(4.49),240(4.62), 248(4.53)	94-1068-76
$C_{10}H_8N_4O$			
Imidazo[5,1-c][1,2,4]benzotriazine-1-methanol	EtOH	247(4.19),262s(4.10), 375(3.76)	44-0158-76
[1,2,5]Oxadiazolo[3,4-d]pyrimidine, 6,7-dihydro-5-phenyl-	EtOH	242(4.50),302(4.05)	94-0235-76
[1,2,4]Triazolo[1,5-b]isoquinolin-5(1H)-one, 1-amino-	EtOH	232(4.34),317(4.22), 375(3.88),392(3.76)	95-0700-76
$C_{10}H_8N_4O_3$			
3-Pyridinol, 2-[(3-nitro-2-pyridinyl)-amino]-	MeOH	210(4.23),240s(4.09), 312(3.91),355(3.89), 400(3.73)	4-0107-76
5H-Pyrrolo[2,3-b]pyrazine, 5-acetyl-7-(2-nitroethenyl)-	EtOH	226(4.19),250(4.12), 320(4.28)	103-0707-76
$C_{10}H_8N_4O_4$			
2-Butenal, 3-(5-nitro-2H-benzotriazol-2-yl)-, N-oxide, (E)-	EtOH	235(4.02),345(4.13)	22-0184-76
(Z)-	EtOH	229(4.00),275(3.95)	22-0184-76
Hydrazinecarboxylic acid, [cyano(4-nitrophenyl)methylene]-, methyl ester anion	EtOH	321(4.35)	142-0147-76S
	EtOH-NaOEt	410(4.35)	142-0147-76S
$C_{10}H_8N_4O_6$			
2,4(1H,3H)-Pyrimidinedione, dihydro-5-nitro-1-(4-nitrophenyl)-	pH 1	289(4.02)	87-1072-76
	pH 11	321(4.31)	87-1072-76
$C_{10}H_8N_6O_7$			
Acetic acid, α-diazopicrylimino-, ethyl ester	EtOH	233(4.33),255(4.34), 300(4.15)	32-0001-76
$C_{10}H_8O$			
Cyclobuta[b]benzofuran, 2a,7b-dihydro-	C_6H_{12}	217(3.71),285(3.56), 291(3.56)	88-4473-76
1H-Inden-1-one, 2-methyl-	hexane	392(2.74)	44-3540-76
	MeOH	238(4.46),243(4.64), 325(3.00),400(2.62)	44-3540-76
	$CHCl_3$	400(2.63)	44-3540-76
2,3,4-Metheno-2H-1-benzopyran, 3,4-dihydro-	C_6H_{12}	233(3.51),237s(3.49), 246s(3.22),279(3.29), 286(3.55)	88-4473-76
2-Naphthalenol	isooctane	224(4.85),263(--), 273(3.65),285(--), 328(3.37)	3-1915-76
	$CHCl_3$	266(--),275(3.65), 287(--),330(3.29)	3-1915-76
$C_{10}H_8OS_2$			
1,3-Dithiol-1-ium, 4-hydroxy-5-methyl-2-phenyl-, hydroxide, inner salt	dioxan	225s(3.92),252s(3.90), 270(3.98),294s(3.42), 343(4.35)	24-0740-76

Compound	Solvent	$\lambda_{max}(\log \epsilon)$	Ref.
Ethanone, 1-[2,2'-bithiophene]-5-yl-	ether	343(4.35)	102-1309-76
$C_{10}H_8O_2$			
1,4-Epidioxynaphthalene, 1,4-dihydro-	C_6H_{12}	205(4.42)	89-0228-76
$C_{10}H_8O_2S_2$			
1,3-Dithiol-1-ium, 4-hydroxy-2-(4-meth-oxyphenyl)-, hydroxide, inner salt	MeOH	230(4.21),255(3.83), 330(3.83)	44-1724-76
$C_{10}H_8O_3$			
Benzo[3,4]cyclobuta[1,2-c]furan-1,3-di-one, 3a,3b,7a,7b-tetrahydro-	MeCN	274.5(3.52)	89-0779B-76
4H-1-Benzopyran-4-one, 5-methoxy-	MeOH	220(4.32),252(4.08), 315(3.68)	4-0211-76
4H-1-Benzopyran-4-one, 6-methoxy-	MeOH	226(4.23),235s(4.17), 246s(3.90),327(3.78)	4-0211-76
4H-1-Benzopyran-4-one, 7-methoxy-	MeOH	236s(4.16),240(4.24), 246(4.26),281s(3.89), 297(4.00),303s(3.99)	4-0211-76
4H-1-Benzopyran-4-one, 8-methoxy-	MeOH	222(4.37),253(4.11), 312(3.72)	4-0211-76
1,3,8-Naphthalenetriol	MeOH	243(3.84),294(3.41), 305(3.47),330s(3.42), 339(3.43)	12-1865-76
	EtOH	230(4.64),292(3.69), 304(3.72),329(3.57), 340(3.59)	78-1353-76
	EtOH-NaOEt	337.5(3.90)	78-1353-76
$C_{10}H_8O_4$			
Anemonin	EtOH	220(3.99)	100-0178-76
4H-1-Benzopyran-4-one, 3,8-dihydroxy-2-methyl-	EtOH	238s(4.41),242(4.46), 331(3.85)	1-0705-76
	EtOH-NaOH	255(4.37)	1-0705-76
4H-1-Benzopyran-4-one, 5-hydroxy-7-methoxy-	MeOH	227(4.10),252s(4.24), 257(4.26),293(3.85), 310s(3.68)	4-0211-76
4H-1-Benzopyran-4-one, 6-hydroxy-5-methoxy-	MeOH	232(4.27),250s(4.00), 338(3.71)	4-0211-76
4H-1-Benzopyran-4-one, 8-hydroxy-7-methoxy-	MeOH	220(4.13),253s(4.30), 258(4.36),303(3.76)	4-0211-76
1,4:5,8-Diepoxynaphthalene-2,3-dione, 1,4,4a,5,8,8a-hexahydro-	dioxan	505(1.76)	118-0256-76
$C_{10}H_8O_4S_4$			
1,3-Dithiole-4-carboxylic acid, 2-[4-(methoxycarbonyl)-1,3-dithiol-2-yl-idene]-, methyl ester	CH_2Cl_2	290s(4.00),302(4.03), 315(4.04),444(3.39)	118-0489-76
$C_{10}H_8O_5$			
4H-1-Benzopyran-4-one, 3,5,6-trihydroxy-2-methyl-	EtOH	238(4.17),255(4.24), 295s(3.63),367(3.62)	1-0705-76
	EtOH-NaOH	235(4.07),500(3.25)	1-0705-76
4H-1-Benzopyran-4-one, 3,5,8-trihydroxy-2-methyl-	EtOH	248(4.35),370(3.65)	1-0705-76
	EtOH-NaOH	235s(3.23),365(3.79)	1-0705-76
$C_{10}H_8O_5S$			
2H-1-Benzopyran-2-one, 4-[(methylsulfon-yl)oxy]-	EtOH	271(3.96),280(4.00), 310(3.86)	80-1543-76

Compound	Solvent	$\lambda_{max}(\log \epsilon)$	Ref.
$C_{10}H_8S$			
Thiophene, 2-phenyl-	EtOH	220(3.92),264(3.81), 282(4.12)	18-2158-76
Thiophene, 3-phenyl-	EtOH	222(4.04),260(4.04), 275(3.80)	18-2158-76
$C_{10}H_8S_2Se_4$			
1,3-Diselenolo[4,5-c]thiophene, 2-(4,6-dihydro-1,3-diselenolo[4,5-c]thien-2-ylidene)-4,6-dihydro-	PhCN	317(3.92),438(2.70), 495(2.72)	88-1719-76
$C_{10}H_9BrN_2O$			
3H-Pyrazol-3-one, 4-bromo-1,2-dihydro-1-(phenylmethyl)-	MeOH	238(3.87)	24-0268-76
$C_{10}H_9BrN_2O_2S$			
Isothiazolo[5,4-b]pyridine-4-carboxylic acid, 3-bromo-6-methyl-, ethyl ester	EtOH	308(3.80)	48-0779-76
$C_{10}H_9BrN_4$			
4,6-Pyrimidinediamine, 2-bromo-5-phenyl-	MeOH	211(4.45),290(3.93)	94-0026-76
$C_{10}H_9BrN_4O_2$			
4,5'-Bipyrimidine, 5-bromo-2,2'-dimethoxy-	EtOH	232(4.11),253(4.05), 306(3.97)	19-0029-76
$C_{10}H_9BrO_5$			
3-Oxatetracyclo[3.2.0.0^{2,7}.0^{4,6}]heptane-6,7-dicarboxylic acid, 1-bromo-, dimethyl ester	EtOH	289(3.60)	24-2823-76
$C_{10}H_9ClN_2$			
4-Quinolinamine, 3-chloro-N-methyl-	MeOH	221(4.56),248(4.38), 328(4.04)	95-0725-76
4-Quinolinamine, 8-chloro-N-methyl-	MeOH	220(4.56),246(4.13), 332(4.14)	95-0725-76
$C_{10}H_9ClN_2O$			
Pyrazolidinium, 1-[(3-chlorophenyl)methylene]-3-oxo-, hydroxide, inner salt	MeOH	341(3.7)	48-0946-76
Pyrazolidinium, 1-[(4-chlorophenyl)methylene]-3-oxo-, hydroxide, inner salt	MeOH	345(4.4)	48-0946-76
$C_{10}H_9ClN_4$			
4,6-Pyrimidinediamine, 2-chloro-5-phenyl-	MeOH	211(4.42),289(3.93)	94-0026-76
$C_{10}H_9ClN_4S$			
1H-Imidazo[1,2-a][1,3,5]benzotriazepine-5(6H)-thione, 8-chloro-2,3-dihydro-	DMSO	272(4.44),324(4.26), 380s(3.00)	78-0057-76
$C_{10}H_9ClO$			
Benzofuran, 5-chloro-2,3-dimethyl-	MeOH	285(3.56),293(3.53)	87-1214-76
2-Propen-1-one, 1-(4-chlorophenyl)-2-methyl-	MeOH	335(2.1)	47-1901-76
$C_{10}H_9Cl_2F$			
1-Decene-3,5-diyne, 1,1-dichloro-2-fluoro-	n.s.g.	215(4.49),222(4.53), 248(3.84),262(4.15), 278(4.35),294(4.29)	44-1487-76

$C_{10}H_9Cl_3O_3-C_{10}H_9NO_2S$

Compound	Solvent	$\lambda_{max}(\log \epsilon)$	Ref.
$C_{10}H_9Cl_3O_3$			
3,5-Cyclohexadien-1-one, 2,2,4-trichlo-ro-3-methoxy-6-propanoyl-	n.s.g.	284(3.86),340(3.0)	120-0101-76
Tricyclo[2.2.1.0^{2,6}]heptane-3-carbox-ylic acid, 3,4,6-trichloro-2-methyl-5-oxo-, methyl ester	CHCl$_3$	265(1.60)	44-2943-76
$C_{10}H_9CoNO_2$			
Cobaltocenium, nitro-	MeOH	277(3.08),368(2.59), 425(2.16)	101-0189-76I
$C_{10}H_9F$			
Tricyclo[3.3.1.0^{2,8}]nona-3,6-diene, 9-(fluoromethylene)-	C$_6$H$_{12}$	230s(3.43)	88-1565-76
$C_{10}H_9F_4NO_3$			
Acetamide, 2,2,2-trifluoro-N-[2-(2-flu-oro-3,4-dihydroxyphenyl)ethyl]-	0.05M HCl	270(2.94)	44-2373-76
	pH 9.40	276(3.35)	44-2373-76
Acetamide, 2,2,2-trifluoro-N-[2-(2-flu-oro-4,5-dihydroxyphenyl)ethyl]-	0.05M HCl	281(3.50)	44-2373-76
	pH 9.40	290(3.72)	44-2373-76
Acetamide, 2,2,2-trifluoro-N-[2-(3-flu-oro-4,5-dihydroxyphenyl)ethyl]-	0.05M HCl	270(3.00)	44-2373-76
	pH 9.40	277(3.38)	44-2373-76
$C_{10}H_9NO$			
Isoxazole, 3-methyl-5-phenyl-	HOAc	262(4.27)	103-0851-76
	+17M H$_2$SO$_4$	287(4.37)	103-0851-76
Isoxazole, 5-methyl-3-phenyl-	HOAc	253(4.03)	103-0851-76
	+17M H$_2$SO$_4$	273(4.28)	103-0851-76
Quinoline, 4-methyl-, N-oxide	acid	316(3.84)	39-0456-76B
	base	338(3.85)	39-0456-76B
$C_{10}H_9NOS$			
Spiro[3H-indole-3,2'-oxirane], 2-(methylthio)-	EtOH	235(4.37),289(3.74), 320(3.86)	12-1023-76
4-Thiazolol, 2-methyl-5-phenyl-	EtOH	303(4.09)	118-0261-76
$C_{10}H_9NOSSe$			
4H-1,3-Selenazin-4-one, tetrahydro-6-phenyl-2-thioxo-	MeOH	260(4.17),311(4.27)	73-1388-76
$C_{10}H_9NOS_2$			
4H-1,3-Thiazin-4-one, tetrahydro-6-phenyl-2-thioxo-	n.s.g.	261(4.11),312(4.21)	73-1388-76
$C_{10}H_9NO_2$			
1H-2-Benzopyran-1-one, 5-amino-3-methyl-	n.s.g.	235(4.27),243(4.29), 274(3.97),364(3.65)	39-1073-76C
Furo[2,3-g]-1,2-benzisoxazole, 4,5-di-hydro-7-methyl-	EtOH	228(4.00),290(3.87)	32-1083-76
2H-Isoindole-1-carboxylic acid, methyl ester	EtOH	228(4.32),256(4.17), 262(4.23),336(4.22), 350(4.19)	32-0065-76
Naphtho[1,2-b:3,4-b']bisoxiren-2b,6a-imine, 1a,1b,2a,6b-tetrahydro-	MeCN	255(3.67)	89-0229-76
5(2H)-Oxazolone, 2-methyl-4-phenyl-	n.s.g.	262(4.05)	33-2149-76
Quinoline, 4-methoxy-, 1-oxide	acid	313(3.83)	39-0456-76B
	base	333(3.89)	39-0456-76B
$C_{10}H_9NO_2S$			
1-Benzothiepin-4,5-dione, 2,3-dihydro-, 4-oxime	EtOH	245(4.26),263(3.97), 343(3.27)	73-2771-76

Compound	Solvent	$\lambda_{max}(\log \epsilon)$	Ref.
4-Thiazolecarboxylic acid, 4,5-dihydro-2-phenyl-	MeCN	243(3.92)	39-1901-76C
4-Thiazolol, 5-(4-methoxyphenyl)-	EtOH	249(3.56),303(3.79)	118-0261-76
$C_{10}H_9NO_3$			
2-Propen-1-one, 2-methyl-1-(4-nitrophenyl)-	MeOH	<u>260(4.1),350s(2.8)</u>	47-1901-76
2(1H)-Quinolinone, 4-hydroxy-6-methoxy-	EtOH	232(4.44),275(3.68), 284(3.59),337(3.54)	105-0282-76
$C_{10}H_9NO_4$			
2H-1-Benzopyran, 8-methoxy-6-nitro-	C_6H_{12}	216(4.20),244s(4.18), 255s(4.09),275s(3.89), 282(3.94),336(3.79)	22-2039-76
	EtOH	216(4.12),244s(4.2), 250(4.16),260s?(4.04), 283(3.81),351(3.72)	22-2039-76
2(3H)-Benzoxazolone, 3-acetyl-6-methoxy-	EtOH	204(4.38),254(3.89), 281(2.87)	102-1997-76
$C_{10}H_9NO_7$			
2-Furancarboxylic acid, 5-(3-methoxy-2-nitro-3-oxo-1-propenyl)-, methyl ester, (Z)-	n.s.g.	320(4.371)	73-2422-76
$C_{10}H_9NS$			
Isothiazole, 4-methyl-3-phenyl-	EtOH	260(4.340)	48-0507-76
Thiazole, 5-methyl-2-phenyl-	EtOH	296(4.14)	70-1353-76
$C_{10}H_9NS_2$			
Isothiazole, 3-(methylthio)-4-phenyl-	EtOH	242(3.89),289(4.01)	48-0127-76
Isothiazole, 3-(methylthio)-5-phenyl-	EtOH	270(4.23)	48-0127-76
$C_{10}H_9NS_3$			
1,4,2-Dithiazine, 3-(methylthio)-5-phenyl-	C_6H_{12}	239(4.22),254(4.25), 304(3.82),357(3.60)	48-0127-76
1,4,2-Dithiazine, 3-(methylthio)-6-phenyl-	C_6H_{12}	230(4.27),269(4.30), 307s(3.75),351(3.00)	48-0127-76
$C_{10}H_9N_3O$			
3-Buten-2-one, 4-(1H-benzotriazol-1-yl)-, (E)-	EtOH	262(4.25),317(4.10)	22-0184-76
$C_{10}H_9N_3OS$			
Ethanone, 1-[4-(5-mercapto-1H-1,2,3-triazol-1-yl)phenyl]-	dioxan	260(4.15)	73-1551-76
Ethanone, 1-[4-(1,2,3-thiadiazol-5-ylamino)phenyl]-	dioxan	233(3.97),280(3.84), 342(4.50)	73-1182-76
$C_{10}H_9N_3OS_2$			
1,2,4-Triazolidine-3,5-dithione, 1-acetyl-2-phenyl-	EtOH	259(4.00),287(4.01)	65-0174-76
$C_{10}H_9N_3OS_4$			
4-Thiazolidinone, 3-ethyl-5-(2-methyl-5H-thiazolo[3,2-b]-1,3,4-thiadiazol-5-ylidene)-2-thioxo-	EtOH	465(5.1)	103-0650-76
$C_{10}H_9N_3O_2$			
3-Buten-2-one, 4-(1H-benzotriazol-1-yl)-, N-oxide, (E)-	EtOH	230s(4.14),273(4.17), 360(4.33),372s(4.21)	22-0184-76

Compound	Solvent	$\lambda_{max}(\log \epsilon)$	Ref.
$C_{10}H_9N_3O_2S$			
Benzoic acid, 4-(1,2,3-thiadiazol-5-yl-amino)-, methyl ester	dioxan	272(3.87),336(4.46)	73-1182-76
[2,2'-Bipyridine]-5-sulfonamide	EtOH	244(3.84),250(3.82), 289(4.00)	64-0115-76B
Phenol, 4-[(4-methyl-2-thiazolyl)azo]-, N-oxide	n.s.g.	440(4.18)	64-0981-76B
$C_{10}H_9N_3O_3$			
Pyrazolidinium, 1-[(3-nitrophenyl)meth-ylene]-3-oxo-, hydroxide, inner salt	MeOH	329(4.1)	48-0946-76
Pyrazolidinium, 1-[(4-nitrophenyl)meth-ylene]-3-oxo-, hydroxide, inner salt	MeOH	363(4.2)	48-0946-76
$C_{10}H_9N_3O_3S$			
4H-Pyran-3-carboxamide, 2,6-dimethyl-4-oxo-N-1,3,4-thiadiazol-2-yl-	isoPrOH	209(4.36),247(4.15), 272s(4.02)	4-0291-76
$C_{10}H_9N_3O_4S$			
4,5-Thiazoledicarboxamide, 2-[5-(hydr-oxymethyl)-2-furanyl]-	EtOH	239(4.19),337(4.23)	44-4074-76
$C_{10}H_9N_3O_5$			
Pyrido[2,3-d]pyrimidine-5-carboxylic acid, 1,2,3,4,7,8-hexahydro-1,3-dimethyl-2,4,7-trioxo-	pH 1	263(3.78),281s(--), 312(4.09)	44-1095-76
	pH 7	274(3.89),307(4.26), 316(4.25)	44-1095-76
	pH 11	274(3.89),307(4.26), 316(4.26)	44-1095-76
Pyrido[2,3-d]pyrimidine-5-carboxylic acid, 1,2,3,4,7,8-hexahydro-1-methyl-2,4,7-trioxo-, methyl ester	pH 1	263(3.84),283s(--), 314(4.08)	44-1095-76
	pH 7	275(3.94),312(4.21), 322(4.21)	44-1095-76
	pH 14	253(4.08),275(3.93), 328(4.13)	44-1095-76
Pyrido[2,3-d]pyrimidine-5-carboxylic acid, 1,2,3,4,7,8-hexahydro-3-methyl-2,4,7-trioxo-, methyl ester	pH 1	275(3.92),313(4.15)	44-1095-76
	pH 7	297(3.96),323(4.18)	44-1095-76
	pH 14	245(4.11),268(3.76), 326(4.19)	44-1095-76
Pyrido[2,3-d]pyrimidine-5-carboxylic acid, 1,2,3,4,7,8-hexahydro-8-methyl-2,4,7-trioxo-, methyl ester	pH 1	279(4.08),313(4.12)	44-1095-76
	pH 7	284(4.05),335(4.17)	44-1095-76
	pH 14	253(4.08),279(4.16), 345(4.26)	44-1095-76
$C_{10}H_9N_5$			
1H-Pyrazolo[4,3-c]cinnolin-3-amine, 1-methyl-	EtOH	234(4.24),290(3.74), 348(3.30)	39-0592-76C
$C_{10}H_9N_5O$			
9H-Purin-6-amine, 9-(5-methyl-2-furan-yl)-	MeOH	248(4.36),280s(3.87)	35-8213-76
$C_{10}H_9N_5O_2$			
Guanidine, (3-nitro-4-quinolinyl)-	H_2O	336(3.96)	114-0395-76C
	DMF	340(4.18)	114-0395-76C
hemihydrate	H_2O	370(3.47)	114-0395-76C
	DMF	440(3.65)	114-0395-76C
9H-Purin-6-amine, 9-[5-(hydroxymethyl)-2-furanyl]-	MeOH	247(4.36),280s(3.90)	35-8213-76

Compound	Solvent	$\lambda_{max}(\log \epsilon)$	Ref.
$C_{10}H_9OS$ 2-Benzothiopyrylium, 4-hydroxy-1-methyl-, perchlorate	MeCN	228(4.12),258(4.26), 293(3.34),312(3.33), 324(3.10),392(3.81)	1-0024-76
$C_{10}H_{10}$ 1,2:3,4-Dicyclobutabenzene	EtOH	266(3.13),269(3.14), 275(3.19)	35-0628-76
Tricyclo[5.3.0.0^{2,10}]deca-3,5-diene	dioxan	267(3.61)	89-0053B-76
$C_{10}H_{10}BrNOS_2$ 3H-Indol-3-one, 4-bromo-1,2-dihydro-2,2-bis(methylthio)-	EtOH	245(4.36),406(3.46)	12-1023-76
$C_{10}H_{10}BrN_2O$ Quinazolinium, 6-bromo-3,4-dihydro-1,3-dimethyl-4-oxo-	pH 2.10	274(3.96),294(3.80), 305(3.66)	44-0838-76
$C_{10}H_{10}ClNO_5S$ 4H,5H-Pyrano[3,4-e]-1,3-oxazine-4,5-dione, 7-chloro-2-(ethylthio)-2,3-dihydro-2-methoxy-	MeCN	310(4.06)	39-2462-76C
$C_{10}H_{10}ClN_3$ 1-Phthalazinamine, 4-chloro-N,N-dimethyl-	dioxan	320(3.76)	103-0342-76
$C_{10}H_{10}Co$ Cobaltocenium (cation)	MeOH	262(4.63),300(3.07), 403(2.34)	101-0189-76I
$C_{10}H_{10}F_3NO$ Propane, 1-(3-trifluoromethylphenyl)-2-nitroso-, dimer	EtOH	294(3.91)	78-1267-76
$C_{10}H_{10}F_3N_5O_4$ Uridine, 5'-azido-2',5'-dideoxy-5-(trifluoromethyl)-	EtOH	261(3.97)	87-0915-76
$C_{10}H_{10}Fe$ Ferrocene	gas	188(4.42),197(4.51), 202(4.11),237(3.45), 244(3.66),247(3.30), 265(3.30),325(2+), 440(2.0)	46-0717-76
$C_{10}H_{10}MoO_3$ Molybdenum, tricarbonylmethyl[(1,2,3,4-5-η)-1-methyl-2,4-cyclopentadien-1-yl]-	hexane	222s(4.18),258s(3.82), 314(3.33),353s(3.07)	24-1429-76
$C_{10}H_{10}N$ Isoquinolinium, N-methyl-	H_2O	282(3.8),322(3.5)	78-1773-76
$C_{10}H_{10}NS$ Thiazolium, 3-methyl-2-phenyl-, iodide	MeOH	278(4.22)	78-1773-76
$C_{10}H_{10}N_2$ 1H-Imidazole, 1-(phenylmethyl)-	EtOH	209(4.15)	33-2738-76
1H-Imidazole, 2-(phenylmethyl)-	EtOH	212(4.05)	33-2738-76

Compound	Solvent	$\lambda_{max}(\log \epsilon)$	Ref.
1H-Imidazole, 4-(phenylmethyl)-	EtOH	212(4.06)	33-2738-76
4-Quinolinamine, N-methyl-	MeOH	216(4.52),234(4.16), 241(4.14),326(4.08)	95-0725-76
$C_{10}H_{10}N_2O$			
2H-Furo[2,3-g]indazole, 4,5-dihydro-7-methyl-	EtOH	256(3.92)	32-1083-76
1H-Indole-2-carboxamide, 1-methyl-	EtOH	293(4.20)	103-1249-76
1(2H)-Isoquinolinone, 3-(methylamino)-	EtOH	229(4.37),244(4.02), 291(4.19),301(4.24), 373(3.66)	95-0154-76
Pyrazolidinium, 3-oxo-1-(phenylmethylene)-, hydroxide, inner salt	MeOH	347(4.4)	48-0946-76
Pyrrolo[1,2-c]pyrimidine-7-carboxaldehyde, 3,6-dimethyl-	EtOH	228(4.27),247(4.07), 253s(4.01),264(3.97), 360s(4.17),367(4.22)	44-0351-76
4-Quinolinamine, N-hydroxy-N-methyl-, hydrochloride	MeOH	220(4.52),243(4.18), 340(4.21),353(4.23)	95-0725-76
4-Quinolinamine, N-methyl-, 1-oxide, hydrochloride	MeOH	218(4.55),236(4.12), 247(4.11),342(4.14), 354(4.16)	95-0725-76
$C_{10}H_{10}N_2O_2$			
2H-1-Benzopyran-2-one, 3-amino-4-(methylamino)-	EtOH	239s(4.06),250(4.03), 335(4.15)	103-0268-76
2H-Indol-2-one, 1,3-dihydro-1-methyl-	MeOH	212(4.13),255(4.12), 284(4.02),327(4.16), 412(2.94)	24-0200-76
Pyridinium, 1-cyano-2-ethoxy-2-oxoethylide	H$_2$O	232(4.31),376(3.81)	78-2647-76
	EtOH	229(4.26),390(4.08)	78-2647-76
	dioxan	228(4.29),404(4.26)	78-2647-76
	MeCN	227(4.31),391(4.25)	78-2647-76
2(1H)-Quinolinone, 3-amino-4-hydroxy-6-methyl-, hydrochloride	MeOH	217(4.55),233(4.58), 299(3.91),325(4.07), 338(4.06)	118-0805-76
$C_{10}H_{10}N_2O_3$			
2-Butenamide, N-(2-nitrophenyl)-, trans	MeOH	245(5.21),345(4.51)	2-0323-76
1H-Pyrazole-3-carboxylic acid, 4-(2-furanyl)-1-methyl-, methyl ester	MeOH	220(4.15),263(4.05), 273s(4.00)	5-1531-76
1H-Pyrazole-4-carboxylic acid, 3-(2-furanyl)-1-methyl-, methyl ester	MeOH	216(4.12),250s(3.80), 256s(3.88),280(4.10)	5-1531-76
1H-Pyrazole-4-carboxylic acid, 5-(2-furanyl)-1-methyl-, methyl ester	MeOH	215(4.17),241(3.88), 278(3.99)	5-1531-76
1H-Pyrazole-5-carboxylic acid, 4-(2-furanyl)-1-methyl-, methyl ester	MeOH	223(4.11),246s(3.82), 265s(3.67),302(3.93)	5-1531-76
2(1H)-Quinolinone, 3-amino-4-hydroxy-6-methoxy-, hydrochloride	MeOH	216(4.50),236(4.57), 305(3.84),327(4.08), 347(4.06)	118-0805-76
4(1H)-Quinolinone, 2,3-dihydro-2-methyl-8-nitro-	MeOH	225(4.29),260(4.11), 380(3.88)	2-0323-76
$C_{10}H_{10}N_2O_4S$			
7H-Thiazolo[3,2-a]pyrimidine-5-carboxylic acid, 6-hydroxy-2,3-dimethyl-7-oxo-, methyl ester	MeOH	270(4.22),375(3.90)	64-0251-76B
$C_{10}H_{10}N_2S$			
1H-Imidazole, 2-(methylthio)-1-phenyl-, hydriodide	EtOH	219-261(4.047-3.735)	28-0143-76B
	CH$_2$Cl$_2$	230-263(4.304-3.859)	28-0143-76B

Compound	Solvent	$\lambda_{max}(\log \epsilon)$	Ref.
2H-Imidazole-2-thione, 1,3-dihydro-1-methyl-3-phenyl-	EtOH	260(4.307)	28-0143-76B
	CH_2Cl_2	267(4.346)	28-0143-76B
Pyrrolo[1,2-c]pyrimidine-7-carbothioaldehyde, 3,6-dimethyl-	EtOH	227(4.20),238s(4.02), 253s(3.91),300s(3.72), 308(3.76),370(3.65), 420s(3.98),433s(4.15), 439(4.20)	44-0351-76
$C_{10}H_{10}N_4$			
Butanenitrile, 3-imino-2-(phenylhydrazono)-	EtOH	236(1.58),355(4.15)	33-0551-76
1H-Indole, 2-(azidomethyl)-3-methyl-	EtOH	225(4.38),280(3.79), 284(3.80),292(3.73)	88-2347-76
$C_{10}H_{10}N_4O$			
1H-Imidazo[1,2-a][1,3,5]benzotriazepin-5(6H)-one, 2,3-dihydro-	EtOH	209(4.31),235(4.58), 273s(3.33)	78-0057-76
1,2,4-Triazin-5(4H)-one, 4-amino-3-methyl-6-phenyl-	MeOH	312(4.06)	5-2206-76
1H-1,2,4-Triazolium, 4-(benzoylamino)-1-methyl-, hydroxide, inner salt	$CHCl_3$	280(4.05),289(4.06)	94-2568-76
$C_{10}H_{10}N_4OS$			
Acetamide, N-(2,5-dihydro-1-phenyl-5-thioxo-1H-1,2,4-triazol-3-yl)-	EtOH	225(4.26),260(3.85), 294(3.97)	65-0174-76
Hydrazinecarbothioamide, 2-(1,3-dihydro-3-oxo-2H-indol-2-ylidene)-N$^{(\omega)}$-methyl-	EtOH	242(4.12),267(4.27), 340(4.13),460(3.99)	103-0076-76
Hydrazinecarbothioamide, 2-(1,3-dihydro-3-oxo-2H-indol-2-ylidene)-1-methyl-	EtOH	257(4.47),295(3.82), 340(3.83),456(3.85)	103-0076-76
Hydrazinecarbothioamide, 2-methyl-2-(3-oxo-3H-indol-2-yl)-	EtOH	267(4.38),344(4.11), 456(3.96)	103-0076-76
1,2,4-Triazin-3(2H)-one, 6-amino-4,5-dihydro-2-(phenylmethyl)-5-thioxo-	EtOH	249(3.79),316(3.89), 358(3.90)	94-2274-76
$C_{10}H_{10}N_4OS_2$			
Pyrimido[5,6-e]-1,3-thiazino[2,3-b]pyrimidin-6(2H)-one, 3,4-dihydro-9-(methylthio)-	n.s.g.	250(3.95),269(3.92)	2-0759-76
$C_{10}H_{10}N_4O_2$			
4,5'-Bipyrimidine, 4',6-dimethoxy-	EtOH	242(3.98),278(4.03)	19-0029-76
Pyridine, 2-(3,5-dimethyl-1H-pyrazol-1-yl)-3-nitro-	EtOH	235(3.97),272(3.94), 326s(3.18)	22-0839-76
$C_{10}H_{10}N_4O_3S$			
Pyrimido[4,5-c]pyridazine-3-carboxylic acid, 1,4-dihydro-7-(methylthio)-4-oxo-, ethyl ester	EtOH	270(4.48),345(4.10), 350s(4.09)	94-2637-76
$C_{10}H_{10}N_4O_5$			
8,2'-Anhydro-8-oxy-9-β-D-arabinofuranosylhypoxanthine	pH 1	251(4.11),286(3.59)	94-0672-76
	pH 7	251(4.11),286s(3.59)	94-0672-76
	pH 13	252.5(4.11)	94-0672-76
8,5'-Anhydro-8-oxyinosine	pH 1	251(4.14)	94-0672-76
	pH 7	252(4.13)	94-0672-76
	pH 13	255(4.18)	94-0672-76
8,3'-Anhydro-8-oxy-9-β-D-xylofuranosylhypoxanthine	pH 2 and 7	254(4.15)	94-0672-76
	pH 12	257(4.18)	94-0672-76

Compound	Solvent	$\lambda_{max}(\log \epsilon)$	Ref.
$C_{10}H_{10}N_4O_5S$			
Acetimidic acid, N-(4-nitrophenylsulfon-yl)-2-diazo-, O-ethyl ester	EtOH	284(4.43)	32-0001-76
$C_{10}H_{10}N_4S$			
1H-Imidazo[1,2-a][1,3,5]benzotriazepine-5(6H)-thione, 2,3-dihydro-	EtOH	204(4.38),264(4.40), 312(4.22),360s(3.20)	78-0057-76
1H-[1,3,5]Triazepino[1,2-a]benzimida-zole-2(3H)-thione, 4,5-dihydro-	EtOH	205(4.49),266(4.12), 298(4.37)	78-0839-76
1H-1,2,4-Triazolium, 2,3-dihydro-1-meth-yl-4-[(phenylmethylene)amino]-3-thi-oxo-, hydroxide, inner salt	MeCN	244(4.67),348(3.15)	103-0361-76
$C_{10}H_{10}N_4S_2$			
4,5'-Bipyrimidine, 4',6-bis(methylthio)-	EtOH	231(4.23),268(4.20)	19-0029-76
1,2,4-Thiadiazol-3(2H)-imine, 2-(imino-phenylmethyl)-5-(methylthio)-	H_2O	256(4.51)	18-3165-76
$C_{10}H_{10}N_6$			
1H-Cyclohepta[1,2-a:3,4-d']diimidazole-2,8-diamine, 4-methyl-	MeOH	250(3.93),297(4.38), 360(3.60),400(3.83)	35-3049-76
	MeOH-HCl	237(3.82),286(4.44), 340s(--),394(3.93)	35-3049-76
1H-Cyclohepta[1,2-a:4,5-d']diimidazole-2,6-diamine, 4-methyl-	MeOH	295(4.48),403(4.03)	35-3049-76
	MeOH-HCl	286(4.52),381(3.97)	35-3049-76
$C_{10}H_{10}O$			
2H-1-Benzopyran, 4-methyl-	EtOH	223(3.97),262(3.45), 306(3.34)	39-0001-76C
Methanone, cyclopropylphenyl-	EtOH	242(4.37),273(3.26)	44-3067-76
2-Propen-1-one, 2-methyl-1-phenyl-	MeOH	330(2.0)	47-1901-76
Tricyclo[4.3.1.0^{2,5}]deca-3,8-dien-7-one, 2,5-endo	pentane	225(3.9)	2-0462-76
$C_{10}H_{10}O_2$			
Benzofuran, 5-methoxy-2-methyl-	$CHCl_3$	248(4.07),292(3.64), 300(3.58)	102-0999-76
2-Propen-1-one, 1-(4-hydroxyphenyl)-2-methyl-	MeOH	287(4.1)	47-1901-76
$C_{10}H_{10}O_2S$			
4H-1-Benzothiopyran-4-one, 2,3-dihydro-3-methyl-, 1-oxide	EtOH	244(3.96),288(3.28), 295s(3.25),340(2.24)	23-0455-76
3-Buten-2-one, 1-(phenylsulfinyl)-	EtOH	216(4.16)	35-4577-76
$C_{10}H_{10}O_3$			
Benzenepropanoic acid, 2-hydroxy-α-meth-ylene-	MeOH	243(3.81),252(3.80), 263(3.67),295(2.67)	35-3555-76
Benzenepropanoic acid, α-oxo-, methyl ester	EtOH	224(3.73),230(3.64), 294(4.08),304(3.97)	12-2459-76
Benzo[b]furan-4-one, 5-(hydroxymethyl-ene)-2-methyl-4,5,6,7-tetrahydro-	EtOH	275(3.78),305(3.78)	32-1083-76
1H-2-Benzopyran-4(3H)-one, 7-methoxy-	95% MeOH	224(4.06),276(4.19)	13-0197-76B
1(2H)-Naphthalenone, 3,4-dihydro-3,8-di-hydroxy- (vermelone)	EtOH	259(4.03),334(3.60)	44-2468-76
	EtOH-NaOEt	266s(--),333s(--), 346(3.72),374(3.74)	44-2468-76
1(2H)-Naphthalenone, 3,4-dihydro-6,8-di-hydroxy-	MeOH	224(4.12),284(4.15), 317(3.84)	12-1865-76
2-Propenal, 3-hydroxy-3-(4-methoxyphen-yl)-	pH 1	234(4.24),265s(3.95)	44-0294-76
	pH 6.8	275(4.33)	44-0294-76

Compound	Solvent	$\lambda_{max}(\log \epsilon)$	Ref.
2-Propenal, 3-hydroxy-3-(4-methoxyphenyl)- (cont.)	pH 13	275(4.34)	44-0294-76
	90% H_2SO_4	247(4.37),292(3.76), 315s(3.61)	44-0294-76
2-Propenoic acid, 3-(2-hydroxyphenyl)-, methyl ester	MeOH	243(3.81),252(3.80), 263(3.67),295(2.67)	35-3555-76
Spiro[bicyclo[4.2.0]octa-4,7-diene-2,2'-[1,3]dioxolan]-3-one	EtOH	223(3.71),344(1.89)	88-0839-76
$(C_{10}H_{10}O_3)_n$ Benzoic acid, 5-ethenyl-2-hydroxy-, methyl ester, homopolymer	THF	236(3.83),316(3.54)	47-2725-76
$C_{10}H_{10}O_4$ 1H-2-Benzopyran-1-one, 3,4-dihydro-4,8-dihydroxy-3-methyl-, (3S-cis)-	EtOH	246(3.70),314(3.57)	102-0537-76
2(5H)-Furanone, 3-acetyl-5-(2-butenylidene)-4-hydroxy-	EtOH	246(4.03),304(4.29)	4-0521-76
1(2H)-Naphthalenone, 3,4-dihydro-3,6,8-trihydroxy-, (+)-	EtOH	218s(--),222(3.96), 232(3.79),239s(--), 285(4.00),317s(--)	78-1353-76
	EtOH-NaOEt	254(3.63),335(4.31)	78-1353-76
$C_{10}H_{10}O_5$ Ethanone, 1,1'-(2,4,6-trihydroxy-1,3-phenylene)bis-	EtOH	206(4.27),269(3.72)	102-1785-76
1(2H)-Naphthalenone, 3,4-dihydro-3,4,6,8-tetrahydroxy-, (-)-cis	EtOH	217(4.14),234(3.92), 284(4.12),312(3.81)	78-1353-76
	EtOH-NaOEt	256(3.76),337(4.45)	78-1353-76
$C_{10}H_{10}O_5S$ 2H-1-Benzopyran-2-one, 3,4-dihydro-7-[(methylsulfonyl)oxy]-	EtOH	276(3.18)	80-1543-76
$C_{10}H_{10}S$ Benzo[b]thiophene, 2,6-dimethyl-	EtOH	236(4.20),260(3.99), 292(3.38),296s(3.30), 302(3.44)	39-0001-76C
$C_{10}H_{11}BrN_4O_2S$ Benzenesulfonamide, 4-bromo-N-(1-ethyl-1H-1,2,3-triazol-5-yl)-	EtOH	235(4.17)	32-0001-76
$C_{10}H_{11}BrN_5O_7P$ Adenosine, 8-bromo-, cyclic 3',5'-(hydrogen phosphate), 1-oxide	pH 1	236(4.10),262(4.12)	87-0419-76
	pH 7 and 11	236(4.55),263(3.99), 293s(3.40)	87-0419-76
$C_{10}H_{11}ClN_2O$ 4-Quinolinamine, N-hydroxy-N-methyl-, hydrochloride	MeOH	220(4.52),243(4.18), 340(4.21),353(4.23)	95-0725-76
4-Quinolinamine, N-methyl-, 1-oxide, hydrochloride	MeOH	218(4.55),236(4.12), 247(4.11),342(4.14), 354(4.16)	95-0725-76
$C_{10}H_{11}ClN_2O_2$ 2(1H)-Quinolinone, 3-amino-4-hydroxy-6-methyl-, hydrochloride	MeOH	217(4.55),233(4.58), 299(3.91),325(4.07), 338(4.06)	118-0805-76
$C_{10}H_{11}ClN_2O_4$ Benzene, 1-(1-chloro-2-methyl-2-nitropropyl)-4-nitro-	C_6H_{12}	258(4.09)	12-2631-76

Compound	Solvent	$\lambda_{max}(\log \epsilon)$	Ref.
$C_{10}H_{11}ClN_2O_5$			
Quinazolinium, 3,4-dihydro-1,3-dimethyl-4-oxo-, perchlorate	pH 2.10	273(3.77),283(3.77), 294(3.67)	44-0838-76
$C_{10}H_{11}ClN_4O_4$			
1H-1,2,3-Triazolo[4,5-c]pyridine, 4-chloro-1-β-D-ribofuranosyl-	pH 1 H_2O pH 11	268(3.87) 267(3.86) 268(3.90)	44-1449-76 44-1449-76 44-1449-76
2H-1,2,3-Triazolo[4,5-c]pyridine, 4-chloro-2-β-D-ribofuranosyl-	pH 1 H_2O pH 11	261(3.81),292(3.83) 262(3.66),292(3.72) 260(3.79),292(3.80)	44-1449-76 44-1449-76 44-1449-76
3H-1,2,3-Triazolo[4,5-c]pyridine, 4-chloro-3-β-D-ribofuranosyl-	pH 1 pH 11	232(3.64),293(3.86) 236(3.63),295(3.86)	44-1449-76 44-1449-76
$C_{10}H_{11}ClO$			
Benzene, [(3-chloro-2-butenyl)oxy]-	EtOH	226(3.48),270(3.21), 278(3.14)	39-0001-76C
$C_{10}H_{11}ClS$			
Benzene, [(3-chloro-2-butenyl)thio]-	EtOH	224(3.63),254(3.60)	39-0001-76C
Benzene, 1-[(2-chloro-2-propenyl)thio]-4-methyl-	EtOH	228(3.71),253(3.66)	39-0001-76C
$C_{10}H_{11}Cl_3N_2S$			
Diazene, [(1,1-dimethylethyl)thio]-(2,4,6-trichlorophenyl)-, (E)-	benzene	332(3.90),403(2.92)	144-0399-76
$C_{10}H_{11}CoN$			
Cobaltocenium, amino-	MeOH	276(4.46),357(3.62), 417(3.10)	101-0189-76I
$C_{10}H_{11}N$			
1H-Indole, 2,3-dimethyl-	MeOH	225(4.57),280(3.90), 290(3.81)	117-0041-76
$C_{10}H_{11}NO$			
Acetamide, N-bicyclo[4.2.0]octa-1,3,5-trien-7-yl-	EtOH	253s(2.78),260(3.06), 266(3.23),272(3.21)	12-1685-76
1-Isoquinolinol, 1,2-dihydro-2-methyl-	aq DMSO- Me$_4$NOH	312(4.13)	78-1773-76
2-Propenal, 3-(methylphenylamino)-	H_2O	306(4.51)	69-0114-76
5(6H)-Quinolinone, 7,8-dihydro-3-methyl-	n.s.g.	234(3.90),291(3.70)	39-0975-76C
$C_{10}H_{11}NO_2$			
2-Butene, 2-(2-nitrophenyl)-	MeCN	245s(3.9)	104-1289-76
2-Propenoic acid, 3-(3-pyridinyl)-, ethyl ester, trans	MeOH	204(4.15),258(4.23), 282s(4.07)	104-0436-76
3-Pyridinemethanol, α-ethynyl-5-hydroxy-α,6-dimethyl-	MeOH	204(3.82),223(3.70), 284(3.84)	104-1544-76
$C_{10}H_{11}NO_3$			
Acetamide, N-(4-ethoxyphenyl)-2-oxo-, hydrate	EtOH	248(3.72)	65-0702-76
Furo[3,2-d]-2,1-benzisoxazole, 3,3a,4,5-tetrahydro-3-hydroxy-7-methyl-	EtOH	260(3.97)	32-1083-76
$C_{10}H_{11}NO_4$			
Benzeneacetic acid, 4-nitro-, ethyl ester	DMSO	550(4.63)	70-0762-76

Compound	Solvent	$\lambda_{max}(\log \epsilon)$	Ref.
$C_{10}H_{11}NO_4S$			
2-Thiophenemethanol, α-(1-methylethen-yl)-5-nitro-, acetate	MeOH	244(3.84),368(4.02)	12-0327-76
$C_{10}H_{11}NO_5$			
Benzoic acid, 2-(hydroxymethyl)-4-nitro-, ethyl ester	buffer	260(4.2)	130-0037-76
2-Furanmethanol, α-(1-methylethenyl)-5-nitro-, acetate	MeOH	235(4.00),360s(3.95)	12-0327-76
$C_{10}H_{11}NS$			
1(2H)-Isoquinolinethione, 3,4-dihydro-3-methyl-	isooctane	262(3.8),300(3.8)	103-0086-76
Isothiocyanic acid, α-benzylethyl ester, (+)-	isooctane	252(3.17)	103-0086-76
$C_{10}H_{11}N_2O$			
Quinazolinium, 3,4-dihydro-1,3-dimethyl-4-oxo-, perchlorate	pH 2.10	273(3.77),283(3.77), 294(3.67)	44-0838-76
$C_{10}H_{11}N_3$			
Benzenamine, 4-(3-methyl-1H-pyrazol-1-yl)-	EtOH	274(4.22)	22-0839-76
$C_{10}H_{11}N_3O$			
5H-1,3,4-Benzotriazepin-5-one, 1,4-di-hydro-1,4-dimethyl-	n.s.g.	237(4.40),257(4.15), 278(3.70),320(2.86)	44-2732-76
5H-1,3,4-Benzotriazepin-5-one, 1,4-di-hydro-2,4-dimethyl-	EtOH	228(5.25),252s(4.97), 288s(4.51)	44-2732-76
1H-1,2,3-Triazole, 1-(4-ethoxyphenyl)-	dioxan	257(4.12)	73-1551-76
$C_{10}H_{11}N_3OS$			
1,2,3-Thiadiazol-5-amine, N-(4-ethoxy-phenyl)-	dioxan	250(3.98),286s(--), 323(4.17)	73-1182-76
1H-1,2,3-Triazole-5-thiol, 1-(4-ethoxy-phenyl)-	dioxan	270(3.96)	73-1551-76
$C_{10}H_{11}N_3O_2$			
L-7-Azatryptophan	pH 13	291(3.87)	94-3149-76
Imidazo[2,1-b]quinazoline-2,5(1H,3H)-dione, 6,7,8,9-tetrahydro-	EtOH	244(4.12),288(3.88)	114-0413-76A
$C_{10}H_{11}N_3O_2S_3$			
Propanamide, 2-cyano-3-(3-ethyl-4-oxo-2-thioxo-5-thiazolidinylidene)-3-(methylthio)-	EtOH	292(3.95),384(4.10), 479(4.29)	94-1671-76
$C_{10}H_{11}N_3O_3$			
1H-Pyrazolo[4,3-b]pyridine-6-carboxylic acid, 4,7-dihydro-1-methyl-7-oxo-, ethyl ester	EtOH	227(4.29),298(4.06)	39-0507-76C
$C_{10}H_{11}N_3O_3S$			
Acetimidic acid, 2-diazo-N-(4-methyl-phenyl)-, O-methyl ester	EtOH	222(4.16),277(4.45)	32-0001-76
$C_{10}H_{11}N_3O_6$			
2-Butenedioic acid, 2-(6-amino-1,2,3,4-tetrahydro-2,4-dioxo-5-pyrimidinyl)-, dimethyl ester, (E)-	pH 1	267(4.16),328(3.87)	44-1095-76
	pH 7	267(4.11),338(3.81)	44-1095-76
	pH 11	271(4.22),388(3.67)	44-1095-76

Compound	Solvent	$\lambda_{max}(\log \epsilon)$	Ref.
2-Butenedioic acid, 2-(6-amino-1,2,3,4- tetrahydro-2,4-dioxo-5-pyrimidinyl)-, (Z)-	pH 1 pH 7 pH 11	269(4.05),312(4.06) 279(4.00),322(4.16) 228(4.24),273(3.92), 322(4.16)	44-1095-76 44-1095-76 44-1095-76
Glycine, N-(2,4-dinitrophenyl)-, ethyl ester	EtOH	286(3.73),383(3.95), 434(3.98),487(4.26)	78-1221-76
2-Oxazolidinone, 3-[[[5-(ethyl-aci-ni- tro)-4,5-dihydro-4-oxo-2-furanyl]- methylene]amino]-	SDA 30	385(4.32)	87-0729-76
$C_{10}H_{11}N_3O_7$ 2-Oxazolidinone, 3-[[(3,4-dimethoxy- 5-nitro-2-furanyl)methylene]amino]-	SDA 30	370(4.28)	87-0729-76
$C_{10}H_{11}N_3S$ 1H-1,2,3-Triazole, 1-(4-methylphenyl)- 5-(methylthio)-	dioxan	238(4.04)	73-1551-76
$C_{10}H_{11}N_3S_2$ 1,2,3-Thiadiazol-5-amine, N-[4-(ethyl- thio)phenyl]-	dioxan	266(3.92),330(4.32)	73-1182-76
1H-1,2,4-Triazole, 3,5-bis(methylthio)- 1-phenyl	EtOH	240(3.33),269(3.38)	65-0174-76
$C_{10}H_{11}N_5$ 2-Propanone, pyrido[3,2-d]pyrimidin- 4-ylhydrazone	EtOH	237(4.12),305(3.56), 320s(3.64),372(4.08)	103-0808-76

$C_{10}H_{11}N_5O$
9H-Imidazo[1,2-a]purin-9-one, 1,4-di-
hydro-1,4,6-trimethyl-
	pH 2.42
	
	pH 5.44
		230s(4.53),234(4.54),
		  288(3.91)
		233(4.48),265(3.72),
		  311(3.79)
			69-0898-76
			
			69-0898-76

$C_{10}H_{11}N_5O_2$ Adenine, 9-(2,3-dideoxy-β-D-glycero- pent-2-enofuranosyl)-	MeOH	258(4.19)	35-8213-76
$C_{10}H_{11}N_5O_3$ Adenine, 9-(2-deoxy-D-erythro-pent- 1-enofuranosyl)-	MeOH	250(4.22),281s(3.86), 290s(3.67)	35-8213-76
Adenine, 9-(3-deoxy-β-D-glycero-pent- 3-enofuranosyl)-	MeOH	258(4.18)	35-8213-76
Adenine, 9-(5-deoxy-β-D-erythro-pent- 4-enofuranosyl)-	MeOH	259(4.16)	44-1836-76
$C_{10}H_{11}N_5O_4S$ 8,2'-Anhydro-8-mercapto-9-β-D-arabino- furanosyladenine sulfoxide	pH 1 pH 7.4	273.5(4.20) 282(4.10)	78-0043-76 78-0043-76
8,3'-Anhydro-8-mercapto-9-β-D-xylo- furanosyladenine sulfoxide	pH 1 pH 7.4	223(--),274(4.10) 284(4.21)	78-0043-76 78-0043-76
8,5'-Anhydro-8-mercaptoadenosine sulf- oxide	pH 1 pH 7.4 pH 13	273(4.24) 281(4.14) 279(4.14)	78-0043-76 78-0043-76 78-0043-76
Benzenesulfonamide, N-(1-ethyl-1H-1,2,3- triazol-5-yl)-4-nitro-	EtOH	255(4.10),351(3.35)	32-0001-76
$C_{10}H_{11}OS$ 1,3-Benzoxathiol-1-ium, 2-propyl-, tetrafluoroborate	EtOH	244s(3.66),247(3.69), 252(3.65),298(3.89)	39-0323-76C

Compound	Solvent	$\lambda_{max}(\log \epsilon)$	Ref.
$C_{10}H_{12}$			
1-Butene, 3-phenyl-	hexane	245(2.98)	44-2485-76
Tricyclo[5.2.0.0^{2,9}]nona-3,5-diene, 9-methyl-	C_6H_{12}	276(3.58)	88-3903-76
$C_{10}H_{12}BrN_5O_3$			
9H-Purin-6-amine, 9-(3-bromo-3-deoxy-β-D-arabinofuranosyl)-	pH 7	259(4.18)	24-1395-76
9H-Purin-6-amine, 9-(2-bromo-2-deoxy-β-D-xylofuranosyl)-	pH 7	258(4.18)	24-1395-76
9H-Purin-6-amine, 9-(3-bromo-3-deoxy-β-D-xylofuranosyl)-	pH 7	258(4.17)	24-1395-76
$C_{10}H_{12}Br_2N_5S_2$			
Thiazolium, 5-bromo-2-[3-(5-bromo-3-ethyl-2(3H)-thiazolylidene)-1-triazenyl]-3-ethyl-, tetrafluoroborate	EtOH	486(4.45)	33-0155-76
$C_{10}H_{12}ClN$			
Benzenamine, N-(3-chloro-2-buten-1-yl)-	EtOH	210(3.69),247(4.00), 294(3.26)	39-0001-76C
$C_{10}H_{12}ClNO_2$			
Benzene, 1-(1-chloro-2-methylpropyl)-4-nitro-	hexane	263(4.09)	12-2621-76
$C_{10}H_{12}ClNO_4S$			
Carbamic acid, chloro[(4-methylphenyl)-sulfonyl]-, ethyl ester	MeOH	230(4.10)	117-0025-76
$C_{10}H_{12}ClN_3O_2$			
Butanehydrazonoyl chloride, N-(4-nitrophenyl)-	EtOH	245(3.90),372(4.24)	104-1649-76
$C_{10}H_{12}ClN_3O_2S$			
Diazene, (2-chloro-4-nitrophenyl)-[(1,1-dimethylethyl)thio]-, (E)-	EtOH	356(4.23)	144-0399-76
$C_{10}H_{12}ClN_5O_3$			
Adenosine, 2'-chloro-2'-deoxy-	pH 7	258(4.2)	24-0433-76
Adenosine, 3'-chloro-3'-deoxy-	pH 7	258(4.2)	24-0433-76
9H-Purin-6-amine, 9-(3-chloro-3-deoxy-β-D-arabinofuranosyl)-	pH 7	259(4.18)	24-1395-76
9H-Purin-6-amine, 9-(2-chloro-2-deoxy-β-D-xylofuranosyl)-	pH 7	259(4.18)	24-1395-76
9H-Purin-6-amine, 9-(3-chloro-3-deoxy-β-D-xylofuranosyl)-	pH 7	259(4.19)	24-1395-76
$C_{10}H_{12}Cl_2N_5S_2$			
Thiazolium, 5-chloro-2-[3-(5-chloro-3-ethyl-2(3H)-thiazolylidene)-1-triazenyl]-3-ethyl-, tetrafluoroborate	EtOH	483(4.41)	33-0155-76
$C_{10}H_{12}Cl_4N_2$			
Methanediamine, N,N,N',N'-tetramethyl-1-(2,3,4,5-tetrachloro-2,4-cyclopentadien-1-ylidene)-	dioxan	261(3.87),346(4.31)	24-3939-76
$C_{10}H_{12}CoN_2$			
Cobaltocenium, 1,1'-diamino-	MeOH	267(4.20),315(4.43), 375(3.74),423(3.53)	101-0189-76I

Compound	Solvent	$\lambda_{max}(\log \epsilon)$	Ref.
$C_{10}H_{12}F_3N_3O_4$			
Thymidine, 5'-amino-5'-deoxy-α,α,α-trifluoro-	pH 1	260(3.82)	87-0915-76
	pH 6.8	260(3.81)	87-0915-76
	pH 13	260(3.71)	87-0915-76
$C_{10}H_{12}IN_5O_3$			
9H-Purin-6-amine, 9-(2-deoxy-2-iodo-β-D-arabinofuranosyl)-	pH 7	259(4.18)	24-1395-76
9H-Purin-6-amine, 9-(3-deoxy-3-iodo-β-D-arabinofuranosyl)-	pH 7	259(4.18)	24-1395-76
9H-Purin-6-amine, 9-(3-deoxy-3-iodo-β-D-xylofuranosyl)-	pH 7	259(4.19)	24-1395-76
$C_{10}H_{12}N$			
Isoquinolinium, 3,4-dihydro-N-methyl-	pH 7	282(4.1)	78-1773-76
	0.5M NaOH	264(2.7),271(2.6)	78-1773-76
$C_{10}H_{12}NO_2$			
2,1-Benzisoxazolium, 3-ethyl-1,3-di-hydro-3-methyl-1-oxo-, hexafluoroantimonate	MeCN	<u>332(4.0)</u>	104-1289-76
Pyridinium, 3-(3-hydroxy-1-oxo-2-buten-yl)-1-methyl-, iodide	EtOH	270(3.90),320(4.19)	39-0315-76C
$C_{10}H_{12}NS$			
Thiazolium, 4,5-dihydro-3-methyl-2-phenyl-, iodide	MeOH-H_3BO_3	262(4.05)	78-1773-76
$C_{10}H_{12}N_2$			
1H-Pyrazole, 4,5-dihydro-1-methyl-3-phenyl-	benzene	309(4.15)	61-0288-76
1H-Pyrazole, 4,5-dihydro-3-methyl-1-phenyl-	benzene	284(4.15)	61-0288-76
Pyrrolo[1,2-c]pyrimidine, 1,3,6-tri-methyl-	EtOH	235(4.59),250s(3.97), 279(3.89),290(3.92), 350(3.26)	44-0351-76
$C_{10}H_{12}N_2O$			
Urea, (1-methyl-2-phenylethenyl)-, (E)-	EtOH	270(4.09)	95-0962-76
Urea, (1-methyl-2-phenylethenyl)-, (Z)-	EtOH	267(4.24)	95-0962-76
Urea, (1-phenyl-1-propenyl)-, (E)-	EtOH	251(3.79)	95-0962-76
Urea, (1-phenyl-1-propenyl)-, (Z)-	EtOH	254(4.63)	95-0962-76
$C_{10}H_{12}N_2O_2$			
Ethanone, 1,1'-(3,6-dimethyl-2,5-pyra-zinediyl)bis-	EtOH	234(3.72),298(3.93)	18-2805-76
$C_{10}H_{12}N_2O_3$			
Acetic acid, cyano(1-methyl-5-oxo-2-pyr-imidinylidene)-, ethyl ester	EtOH	283(4.47)	94-3011-76
4H-Pyrido[1,2-a]pyrimidine-3-carboxylic acid, 6,7,8,9-tetrahydro-6-methyl-4-oxo-	pH 7.35	232(3.75),290(--)	145-0478-76
$C_{10}H_{12}N_2O_4$			
Acetamide, N-[2-(4-nitrophenoxy)ethyl]-	EtOH	326(4.45),441(4.28), 488(4.37)	78-1221-76
2-Propenoic acid, 3-(1,2,3,4-tetrahydro-6-methyl-2,4-dioxo-5-pyrimidinyl)-, ethyl ester	EtOH	270s(3.97),303(4.25)	44-1858-76

Compound	Solvent	$\lambda_{max}(\log \epsilon)$	Ref.
1H-Pyrazole-3-carboxylic acid, 1-[1-(methoxycarbonyl)cyclopropyl]-, methyl ester	MeOH	221(4.11)	103-1026-76
$C_{10}H_{12}N_2O_4S$ 2-Propenoic acid, 3-[(1,2,3,6-tetrahydro-1,3-dimethyl-2,6-dioxo-4-pyrimidinyl)thio]-, methyl ester, (E)-	EtOH	264(4.14),298(4.14)	94-1390-76
$C_{10}H_{12}N_2O_5$ Benzenemethanol, α-(1-methyl-1-nitroethyl)-3-nitro-	MeOH	263(3.91)	12-0327-76
$C_{10}H_{12}N_2O_5S$ 2-Butenedioic acid, 2-hydroxy-3-(2-imino-4-methyl-3(2H)-thiazolyl)-, dimethyl ester	MeOH	259(4.16)	64-0251-76B
$C_{10}H_{12}N_2O_6$ 3,6-Pyridazinedicarboxylic acid, 1,4-dihydro-5-methoxy-1-methyl-4-oxo-, dimethyl ester	EtOH	292(4.14)	78-0269-76
$C_{10}H_{12}N_2O_6S$ 2-Thiophenemethanol, α-(1-methyl-1-nitroethyl)-5-nitro-, acetate	MeOH	315(4.00)	12-0327-76
$C_{10}H_{12}N_2S$ 1,2-Benzisothiazoline, 2-ethyl-3-(methylimino)- (isomer mixture)	heptane	308(3.62),347(3.52), 360s(--),380s(--)	24-0659-76
	MeOH	327(3.84),336s(--)	24-0659-76
Benzo[b]thiophene-3-carbonitrile, 2-amino-4,5,6,7-tetrahydro-4-methyl-	EtOH	221(4.39),293(3.77)	2-0357-76
1H-Imidazole, 4,5-dihydro-2-(methylthio)-1-phenyl-, hydriodide	EtOH	222(4.413)	28-0485-76A
	CH_2Cl_2	235(4.381)	28-0485-76A
$C_{10}H_{12}N_4O_2$ Benzeneacetic acid, α-(acetylhydrazono)-, hydrazide, anti	MeOH	221(4.05),255(4.05)	5-2206-76
syn	MeOH	219(4.15),286(4.23)	5-2206-76
1H-Pyrazolo[3,4-d]pyrimidine-4-acetic acid, 1-methyl-, ethyl ester	EtOH	243s(3.86),318(3.87), 359(3.88),378s(3.94), 394(4.10),413(4.18)	95-1352-76
$C_{10}H_{12}N_4O_2S$ Benzenesulfonamide, N-(1-ethyl-1H-1,2,3-triazol-5-yl)-	EtOH	220(4.05)	32-0001-76
Benzenesulfonamide, N-(1-methyl-1H-1,2,3-triazol-5-yl)-4-methyl-	EtOH	222(4.05),255(4.00)	32-0001-76
$C_{10}H_{12}N_4O_3S$ 6-Purinethiol, 8-(2-α-deoxyribosyl)-	pH 1	230(4.06),326(4.30)	111-0067-76
	H_2O	233(4.08),328(4.43)	111-0067-76
	pH 13	235(4.19),313(4.30)	111-0067-76
6-Purinethiol, 8-(2-β-deoxyribosyl)-	pH 1	230(4.02),326(4.29)	111-0067-76
	H_2O	233(4.06),328(4.34)	111-0067-76
	pH 13	235(4.23),313(4.29)	111-0067-76
$C_{10}H_{12}N_4O_4$ [1,1'-Bi-1H-pyrazole]-3,3'-dicarboxylic acid, 4,5-dihydro-, dimethyl ester	MeOH	213(4.10),253(4.06)	103-1026-76

Compound	Solvent	λ_{max}(log ϵ)	Ref.
Imidazo[1,2-b][1,2,4]triazin-7(1H)-one, 3-acetoxy-1-acetyl-2,3-dihydro-6-methyl-	EtOH	224(4.44)	114-0413-76A
Pyrazolo[3,4-b]pyrazine, 1-β-D-ribofuranosyl-	MeOH	217(4.1),277(3.8), 302s(3.8)	104-0901-76
1H-1,2,3-Triazolo[4,5-c]pyridine, 1-β-D-ribofuranosyl-	pH 1	266(3.73)	44-1449-76
	pH 11	259(3.82)	44-1449-76
	MeOH	256(3.83)	44-1449-76
2H-1,2,3-Triazolo[4,5-c]pyridine, 2-β-D-ribofuranosyl-	pH 1	263(3.96),296(3.81)	44-1449-76
	H_2O	262(3.91),279s(3.82)	44-1449-76
	pH 11	263(3.90),279(3.82)	44-1449-76
3H-1,2,3-Triazolo[4,5-c]pyridine, 3-β-D-ribofuranosyl-	pH 1	224(3.48),300(3.83)	44-1449-76
	H_2O	240(3.64),289(3.75)	44-1449-76
	pH 11	240(3.61),290(3.72)	44-1449-76
$C_{10}H_{12}N_4O_4S$ 6-Purinethiol, 8-arabinosyl-	pH 1	233(4.03),330(4.26)	111-0067-76
	pH 13	235(4.21),317(4.26)	111-0067-76
6-Purinethiol, 8-ribosyl-	pH 1	233(4.07),328(4.29)	111-0067-76
	pH 13	235(4.14),314(4.27)	111-0067-76
β-D-Ribofuranosylamine, N-1,2,3-thiadiazolo[5,4-b]pyridin-7-yl-	pH 1	252(4.08),275(3.92), 322(3.94)	44-1449-76
	pH 11	243(4.12),272(3.75), 338(3.77)	44-1449-76
	MeOH	234(4.14),270(3.72), 339(3.76)	44-1449-76
4H-1,2,3-Triazolo[4,5-c]pyridine-4-thione, 2,5-dihydro-2-β-D-ribofuranosyl-	pH 1	237(4.29),365(4.11)	44-1449-76
	pH 11	236(4.36),362(3.97)	44-1449-76
	MeOH	240(4.25),369(4.13)	44-1449-76
4H-1,2,3-Triazolo[4,5-c]pyridine-4-thione, 3,5-dihydro-3-β-D-ribofuranosyl-	pH 1	266(4.24),366(4.12)	44-1449-76
	H_2O	226(4.19),367(4.07)	44-1449-76
	pH 11	351(4.03)	44-1449-76
$C_{10}H_{12}N_4O_5$ Butanal, 4-hydroxy-, 2,4-dinitrophenylhydrazone	EtOH	222(4.17),354(4.26)	104-2034-76
Pyrazolo[3,4-b]pyrazine, 1-β-D-ribofuranosyl-, 4-oxide	MeOH	225(4.2),296(3.9), 327(3.8)	104-0901-76
Pyrazolo[1,5-a][1,3,5]triazin-4(1H)-one, 8-β-D-ribofuranosyl-	pH 1	262(3.96)	4-1305-76
	pH 7	269(3.99)	4-1305-76
	pH 13	269(4.01)	4-1305-76
$C_{10}H_{12}N_4O_6$ 3,6-Epoxy-2H,8H-pyrimido[6,1-b][1,3]-oxazocine-8,10(9H)-dione, 11-diazohexahydro-4,5-dihydroxy-9-methyl-, [3R-(3α,4α,5α,6α,11aα)]-	MeOH	265(4.23)	44-1041-76
Pyrazolo[1,5-a]-1,3,5-triazine-2,4-(1H,3H)-dione, 8-β-D-ribofuranosyl-	MeOH	231s(3.69),253(3.63)	44-3000-76
$C_{10}H_{12}N_4S$ 1,4-Benzenediamine, N,N-dimethyl-N-1,2,3-thiadiazol-5-yl-	dioxan	266(4.05),306s(--), 326(4.08)	73-1182-76
$C_{10}H_{12}N_5O_6P$ Guanosine, 2'-deoxy-, cyclic (3',5'-hydrogen phosphate)	pH 1	261(4.06),281s(3.88)	69-0217-76
	pH 11	263(4.05)	69-0217-76
$C_{10}H_{12}N_5O_7P$ Guanine, 9-β-D-xylofuranosyl-, cyclic 3',5'-(hydrogen phosphate)	pH 1	255(4.13),275s(3.97)	87-1026-76

Compound	Solvent	$\lambda_{max}(\log \epsilon)$	Ref.
$C_{10}H_{12}N_6O_3$ 9H-Purine-9-propanoic acid, α-(acetyl-amino)-6-amino-	pH 1	260(4.11)	123-0598-76
$C_{10}H_{12}N_6O_4$ L-Threonine, N-[(1H-pyrazolo[3,4-d]pyr-imidin-4-ylamino)carbonyl]-	pH 1	267(4.21)	87-0555-76
	pH 13	267s(--),272(4.14), 296(3.73)	87-0555-76
	EtOH	265(4.23),275s(--), 285s(--)	87-0555-76
$C_{10}H_{12}N_7O_4S_2$ Thiazolium, 3-ethyl-2-[3-(3-ethyl-5-ni-tro-2(3H)-thiazolylidene)-1-triazen-yl]-5-nitro-, perchlorate	HOAc	502(4.47)	33-0155-76
$C_{10}H_{12}N_8O_3$ Adenosine, 2'-azido-2'-deoxy-	pH 2	257(4.18)	88-4485-76
	pH 7	258(4.17)	24-0433-76
	pH 12	259(4.18)	88-4485-76
	50% EtOH	259(4.18)	88-4485-76
Adenosine, 3'-azido-3'-deoxy-	pH 7	259(4.18)	24-0433-76
9H-Purin-6-amine, 9-(2-azido-2-deoxy-β-D-arabinofuranosyl)-	pH 7	258(4.2)	24-0433-76
9H-Purin-6-amine, 9-(3-azido-3-deoxy-β-D-arabinofuranosyl)-	pH 7	259(4.19)	24-0433-76
$C_{10}H_{12}O$ Benzaldehyde, 4-(1-methylethyl)-	1% EtOH- pH 10	262(4.15)	44-1957-76
Benzene, 1-methoxy-2-(2-propenyl)-	EtOH	270(3.34),277(3.30)	33-1763-76
Benzenemethanol, α-ethenyl-2-methyl-	EtOH	252(2.88)	73-0262-76
Benzenemethanol, α-ethenyl-3-methyl-	EtOH	256(2.71)	73-0262-76
Benzenemethanol, α-ethenyl-4-methyl-	EtOH	257(2.90)	73-0262-76
2,4,6-Cyclooctatrien-1-one, 8-ethyl-	C_6H_{12}	216(4.11),247(3.65), 281(3.52),360(2.60)	39-1838-76C
2-Propen-1-ol, 3-(2-methylphenyl)-	EtOH	252(4.23)	73-0262-76
2-Propen-1-ol, 3-(3-methylphenyl)-	EtOH	254.5(4.24)	73-0262-76
2-Propen-1-ol, 3-(4-methylphenyl)-	EtOH	254.5(4.29)	73-0262-76
$C_{10}H_{12}OSe$ Benzeneethaneselenoic acid, O-ethyl ester	EtOH	277(3.89)	44-0729-76
$C_{10}H_{12}O_2$ Benzenemethanol, α-ethenyl-2-methoxy-	EtOH	256(3.00)	73-0262-76
Benzenemethanol, α-ethenyl-3-methoxy-	EtOH	259(2.88)	73-0262-76
Benzenemethanol, α-ethenyl-4-methoxy-	EtOH	262(3.23)	73-0262-76
Benzoic acid, 2,4,5-trimethyl-	EtOH	238(4.03),281(3.16), 288(3.12)	33-1763-76
Benzoic acid, 2,4,6-trimethyl-	71.7% H_2SO_4	none	104-0107-76
	99.9% H_2SO_4	282(4.3)	104-0107-76
Bicyclo[4.2.0]octa-1,3,5-triene, 2,4-di-methoxy-	hexane	275(3.20),277(3.20), 282(3.22)	35-6401-76
Dispiro[3.1.3.1]decane-5,10-dione	CH_2Cl_2	312(1.91),359(1.40)	18-0741-76
1H-Indene-1,5(6H)-dione, 2,3,7,7a-tetra-hydro-7a-methyl-	EtOH	236(4.01)	54-0223-76
1(3H)-Isobenzofuranone, 3a,4-dihydro-3,3-dimethyl-	EtOH	296(3.80)	39-0484-76C
1-Propanone, 2-hydroxy-2-methyl-1-phen-yl-	EtOH	244(3.86),325(1.80)	44-3067-76

Compound	Solvent	$\lambda_{max}(\log \epsilon)$	Ref.
2-Propen-1-ol, 3-(2-methoxyphenyl)-	EtOH	251(4.28)	73-0262-76
2-Propen-1-ol, 3-(3-methoxyphenyl)-	EtOH	256(4.32)	73-0262-76
2-Propen-1-ol, 3-(4-methoxyphenyl)-	EtOH	261(4.35)	73-0262-76
$C_{10}H_{12}O_2Se$			
Benzeneethaneselenoic acid, 4-methoxy-, O-methyl ester	MeOH	275(3.88)	44-0729-76
$C_{10}H_{12}O_3$			
Benzeneacetic acid, 2-hydroxy-α-methyl-, methyl ester	MeOH	215(3.83),275(3.40)	35-3555-76
Benzenepropanoic acid, α-hydroxy-, methyl ester	EtOH	258(2.56)	12-2459-76
Benzoic acid, 4-methoxy-2,6-dimethyl-	EtOH	245(3.65)	39-1466-76C
$C_{10}H_{12}O_4$			
Acetic acid, [(3-methoxyphenyl)methoxy]-	95% MeOH	222(3.69),274(3.24)	13-0197-76B
Benzeneacetic acid, 3,5-dihydroxy-4-methyl-, methyl ester	EtOH	274(3.01),282(3.01)	35-7733-76
2-Furancarboxylic acid, 4-acetyl-5-methyl-, ethyl ester	EtOH	259(4.22)	94-2421-76
3-Furancarboxylic acid, 5-acetyl-2-methyl-, ethyl ester	EtOH	275(4.20)	94-2421-76
$C_{10}H_{12}O_5$			
2-Oxabicyclo[3.2.0]hept-3-ene-1,5-dicarboxylic acid, dimethyl ester	EtOH	230s(4.06)	88-3295-76
7-Oxabicyclo[4.1.0]hept-4-ene-1,4-dicarboxylic acid, dimethyl ester	EtOH	235(3.96)	88-3295-76
$C_{10}H_{12}S_3$			
Propane(dithioperoxo)thioic acid, 2-methyl-, phenyl ester	C_6H_{12}	291(3.69),484(1.48)	138-0441-76
$C_{10}H_{13}BrS_4$			
2,5-Methano-7H-1,3-dithiolo[4,5-d][1,3]-dithiin, 7-(bromomethylene)dihydro-3a,5,8-trimethyl-	isooctane	227(3.86),245(3.88), 280(3.50)	70-2214-76
$C_{10}H_{13}ClN_2$			
Ethanimidamide, N'-(4-chlorophenyl)-N,N-dimethyl-	neutral	247(4.17)	39-0211-76B
	cation	240(4.16)	39-0211-76B
$C_{10}H_{13}ClN_2S$			
Diazene, (4-chlorophenyl)[(1,1-dimethylethyl)thio]-	EtOH	332(4.25)	144-0399-76
$C_{10}H_{13}ClN_2S_2$			
2,6(1H,3H)-Pyridinedithione, 5-chloro-3-[[(1,1-dimethylethyl)amino]methylene]-	MeOH	297(3.73),345(4.26), 430(4.53)	33-0190-76
$C_{10}H_{13}ClO_2$			
Benzene, 1-chloro-3-(1-ethoxyethoxy)-	ether	272(3.08)	39-1001-76B
Benzene, 1-chloro-4-(1-ethoxyethoxy)-	ether	284(3.10)	39-1001-76B
Benzene, 4-(2-chloroethyl)-1,2-dimethoxy-	n.s.g.	226(3.85),285(3.54)	20-0421-76
$C_{10}H_{13}FN_2O_6$			
Uridine, 5-fluoro-2'-O-methyl-	acid	269(4.02)	35-7381-76
	base	267(3.95)	35-7381-76

Compound	Solvent	$\lambda_{max}(\log \epsilon)$	Ref.
Uridine, 5-fluoro-3'-O-methyl-	acid	269(3.92)	35-7381-76
	base	269(3.81)	35-7381-76
$C_{10}H_{13}FN_6O_6S$			
Adenosine, 4'-C-fluoro-, 5'-sulfamate	MeOH	259(4.18)	35-3346-76
9H-Purin-6-amine, 9-[5-O-(aminosulfon-	pH 2	257(4.17)	35-3346-76
yl)-4-C-fluoro-α-L-lyxofuranosyl)-	pH 11	259(4.15)(changing)	35-3346-76
$C_{10}H_{13}IN_2O_5$			
2,4(1H,3H)-Pyrimidinedione, 1-(2,6-di-deoxy-6-iodo-β-D-arabinohexopyrano-syl)-	H_2O	259(4.12)	44-0600-76
$C_{10}H_{13}IN_2S$			
1H-Imidazole, 4,5-dihydro-2-(methyl-thio)-1-phenyl-, hydriodide	EtOH	222(4.413)	28-0485-76A
	CH_2Cl_2	235(4.381)	28-0485-76A
$C_{10}H_{13}NO$			
1H-2-Pyrindin-1-one, 2,5,6,7-tetrahydro-3,4-dimethyl-	n.s.g.	235(3.73),305(3.92)	104-0681-76
$C_{10}H_{13}NOS$			
Benzenecarbothioamide, 2-methoxy-N,N-dimethyl-	heptane-10% CH_2Cl_2	245(3.93),281(4.03),379(2.48)	1-0695-76
Benzenecarbothioamide, 4-methoxy-N,N-dimethyl-	heptane-10% CH_2Cl_2	253(4.18),276(4.16),389(2.73)	1-0695-76
$C_{10}H_{13}NO_2$			
Butanamide, 3-hydroxy-N-phenyl-, (+)-	EtOH	224(4.19)	12-2459-76
$C_{10}H_{13}NO_2S$			
Acetic acid, cyano(tetrahydro-4H-thio-pyran-4-ylidene)-, ethyl ester	C_6H_{12}	229(4.00),288(3.32)	78-2827-76
Acetic acid, cyclopentylideneisothio-cyanato-, ethyl ester	C_6H_{12}	236(3.98),282(4.15)	5-1850-76
Benzenecarbothioamide, 2-hydroxy-4-methoxy-N,N-dimethyl-	heptane-10% CH_2Cl_2	279(4.04),288(4.04),370(3.16)	1-0695-76
$C_{10}H_{13}NO_2S_2$			
5H-1,4-Dithiino[2,3-c]pyrrole-5,7(6H)-dione, 6-(1,1-dimethylethyl)-2,3-di-hydro-	EtOH	205(3.83),263(4.06),410(3.40)	56-1523-76
$C_{10}H_{13}NO_3$			
Acetic acid, cyano(tetrahydro-4H-pyran-4-ylidene)-, ethyl ester	C_6H_{12}	210(3.74),248(4.11)	78-2827-76
Benzenemethanol, α-(1-methylethyl)-4-nitro-	hexane	265(4.06)	12-2621-76
1H-Inden-1-one, 2,3,3a,4,5,6-hexahydro-7-methyl-4-nitro-	EtOH	250(3.996)	39-0410-76C
$C_{10}H_{13}NO_3S$			
Valine, N-(2-thenoyl)-, (±)-	EtOH	252(4.03)	12-0339-76
$C_{10}H_{13}NO_4$			
2,1-Benzisoxazole-4-carboxylic acid, 3,3a,4,5,6,7-hexahydro-3-oxo-, ethyl ester	n.s.g.	260(4.04)	128-0341-76
Butanoic acid, 3-oxo-2-(5-oxo-2-pyrrol-idinylidene)-, ethyl ester	EtOH	286.5(4.00)	94-3011-76

Compound	Solvent	λ_{max}(log ϵ)	Ref.
1H-Pyrrole-1-acetic acid, 2,5-dihydro-2,5-dioxo-, 1,1-dimethylethyl ester	EtOH	221(3.89)	73-1954-76
$C_{10}H_{13}NO_4S$			
Carbamic acid, [(4-methylphenyl)sulfonyl]-, ethyl ester	MeOH	228(4.01)	117-0025-76
2-Propenoic acid, 3-(2-methyl-4-methylene-5-oxooxazolidin-3-yl)-2-(methylthio)-, methyl ester, (Z)-	EtOH	208(4.51),223s(3.46), 291(3.67),331s(3.48)	39-0584-76C
2-Thiopheneacetic acid, 5-amino-4-(ethoxycarbonyl)-3-methyl-	EtOH	227(4.42),265(3.68), 305(3.76)	2-0357-76
$C_{10}H_{13}NO_5$			
2-Furanmethanol, α-(1-methyl-1-nitroethyl)-, acetate	MeOH	315(4.00)	12-0327-76
1H-Pyrrole-2,5-dione, 3-[2,3-dihydroxy-4-(hydroxymethyl)cyclopentyl]-	MeOH	228(3.92)	88-1063-76
$C_{10}H_{13}NO_5S$			
2H-1,4-Thiazine-3,6-dicarboxylic acid, 4-acetyl-3,4-dihydro-, dimethyl ester, (R)-	EtOH	207s(3.79),219(3.92), 245s(3.42),309(4.00)	39-2540-76C
$C_{10}H_{13}NO_6S$			
2H-1,4-Thiazine-3,6-dicarboxylic acid, 4-acetyl-3,4-dihydro-, 1-oxide, (1R-3R)-	EtOH	204(3.73),276(4.25)	39-2540-76C
(1S,3R)-	EtOH	205(3.61),276(4.19)	39-2540-76C
2,5-Thiazoledicarboxylic acid, 3-acetyl-2,3-dihydro-2-methyl-, dimethyl ester, (1R,2R)-	EtOH	206(3.82),224s(3.66), 283(4.08)	39-2540-76C
$C_{10}H_{13}NS_2$			
Carbamodithioic acid, dimethyl-, 4-methylphenyl ester	EtOH	251(4.13),276s(3.90), 343(1.99)	97-0355-76
$C_{10}H_{13}N_3$			
Cyclohepta[b]pyrrole-3-carbonitrile, 2-amino-1,4,5,6,7,8-hexahydro-	EtOH	270(3.82)	103-1379-76
1H-Indole-3-carbonitrile, 4,5,6,7-tetrahydro-2-(methylamino)-	MeOH	275(3.77)	48-0663-76
$C_{10}H_{13}N_3O$			
1H-Benzotriazole, 1-butoxy-	EtOH	264(3.88),283(3.81)	4-0509-76
$C_{10}H_{13}N_3O_2$			
Acetamide, N,N-(5-amino-1,3-phenylene)-bis-	MeOH	235(4.60),305(3.36)	87-0426-76
Acetamide, N-[[nitroso(phenylmethyl)amino]methyl]-	EtOH	237(3.95),359s(1.84), 366(1.86),374s(1.75)	94-0369-76
Benzamide, N-[(ethylnitrosoamino)methyl]-	EtOH	233(4.27),355(1.90), 364(1.90),374s(1.76)	94-0369-76
Butanal, 4-(methylnitrosoamino)-4-(3-pyridinyl)-	EtOH	230(3.81),258s(3.57), 351(1.84)	88-0593-76
Ethanimidamide, N,N-dimethyl-N-(4-nitrophenyl)-	neutral	362(4.01)	39-0211-76B
	cation	287(4.00)	39-0211-76B
Imidazo[1,2-a]pyrimidine-2,5(1H,3H)-dione, 6-ethyl-3,7-dimethyl-	pH 6.1	246(4.06),281(3.84)	114-0413-76A

Compound	Solvent	$\lambda_{max}(\log \epsilon)$	Ref.
$C_{10}H_{13}N_3O_3$			
Acetamide, N-[2-[(2-nitrophenyl)amino]-ethyl]-	EtOH	236(4.33),280(3.70), 423(3.80)	78-0839-76
Butanoic acid, 2-(4-nitrophenyl)hydrazide	EtOH	225(3.92),357(4.15)	104-1649-76
$C_{10}H_{13}N_3O_5$			
2(1H)-Pyrimidinone, 4-amino-1-(6-deoxy-β-D-erythro-hex-5-en-2-ulofuranosyl)-	pH 13	229(3.93),272(3.94)	44-1836-76
$C_{10}H_{13}N_3O_6$			
2,2'-Anhydro-1-β-D-arabinofuranosyl-cytosine formate	MeOH	233(4.09),264(4.11)	44-2995-76
$C_{10}H_{13}N_3O_6S$			
4,5-Thiazoledicarboxamide, 2-β-D-ribofuranosyl-	EtOH	220(4.15),269(3.93)	44-4074-76
$C_{10}H_{13}N_3O_8$			
1,4-Cyclohexadiene, 3-(ethyl-aci-nitro-6,6-dimethoxy-1,5-dinitro-	MeOH	366(4.25)	104-1798-76
$C_{10}H_{13}N_5O_3$			
Adenine, 8-(2-deoxy-α-ribosyl)-	pH 1	267(4.17)	111-0067-76
	H_2O	264(4.17)	111-0067-76
	pH 13	272(4.16)	111-0067-76
Adenine, 8-(2-deoxy-β-ribosyl)-	pH 1	267(4.19)	111-0067-76
	H_2O	264(4.17)	111-0067-76
	pH 13	272(4.16)	111-0067-76
Adenine, 9-(2-deoxy-α-D-threo-pento-furanosyl)-	EtOH	260(4.18)	136-0041-76C
Adenine, 9-(3-deoxy-α-L-threo-pento-furanosyl)-	MeOH	258(4.20)	35-8213-76
Adenosine, 2'-deoxy-	pH 7	258(4.17)	24-1395-76
Adenosine, 3'-deoxy-	pH 7	258(4.17)	24-1395-76
$C_{10}H_{13}N_5O_4$			
Adenine, 8-arabinosyl-	pH 1	269(4.13)	111-0067-76
	H_2O	267(4.22)	111-0067-76
	pH 13	275(4.20)	111-0067-76
Adenine, 8-ribosyl-	pH 1	266(4.13)	111-0067-76
	H_2O	265(4.18)	111-0067-76
	pH 13	272(4.16)	111-0067-76
D-Ribitol, 1-C-(4-aminopyrazolo[1,5-a]-[1,3,5]triazin-8-yl)-1,4-anhydro-	pH 1	255(3.76),290s(3.58)	4-1305-76
	pH 7	273(3.95)	4-1305-76
1H-1,2,3-Triazolo[4,5-c]pyridin-4-amine, 1-β-D-ribofuranosyl-	pH 1	274(4.08)	44-1449-76
	H_2O	290(3.89)	44-1449-76
	pH 11	295(3.88)	44-1449-76
$C_{10}H_{13}N_5O_5$			
Thymidine, α-azido-	M HCl	262(4.11)	87-0909-76
	H_2O	262(4.11)	87-0909-76
	M NaOH	261(4.00)	87-0909-76
7H-1,2,3-Triazolo[4,5-b]pyridin-7-one, 5-amino-3,4-dihydro-3-β-D-ribofuranosyl-	pH 1	250(3.90),296(4.25)	4-1365-76
	pH 11	284(4.26)	4-1365-76
	EtOH	272s(4.02),291(4.17)	4-1365-76
$C_{10}H_{13}N_7O$			
Piperidine, 1-[(4-amino[1,2,4]triazolo-[5,1-c][1,2,4]triazin-3-yl)carbonyl]-	EtOH	223(4.16),280(3.87), 332(4.01)	39-1496-76C

Compound	Solvent	$\lambda_{max}(\log \epsilon)$	Ref.
$C_{10}H_{14}$			
Benzene, 1,2,3,4-tetramethyl-	EtOH	267(2.48),272(2.39), 276(2.37)	35-0628-76
Benzene, 1,2,4,5-tetramethyl-	EtOH	268(2.85),273(2.82), 278(2.86)	35-0628-76
Bicyclo[4.1.1]octa-2,4-diene, 7,7-di-methyl-	hexane	270(3.38),280(3.50), 292(3.47),304(3.19)	88-4019-76
Bicyclo[2.2.2]octane, 2,3-bis(methyl-ene)-	EtOH	247(4.00)	44-1635-76
$C_{10}H_{14}BrClS_4$			
2,4,6,8-Tetrathiatricyclo[3.3.1.1^{3,7}]-decane, 1-(bromomethyl)-9-chloro-3,5,7-trimethyl-	isooctane	223(3.35),255(3.08)	70-2214-76
$C_{10}H_{14}Br_2NO_3P$			
Phosphonic acid, [(2,6-dibromo-3-pyri-dinyl)methyl]-, diethyl ester	MeOH	226.5(4.03),227(3.72) [sic]	33-0190-76
$C_{10}H_{14}ClN_3$			
4-Pyrimidinamine, 6-chloro-N-cyclohexyl-	pH 1	254(4.08),260s(4.06)	39-1847-76C
	pH 13	248(4.08),279(3.41)	39-1847-76C
$C_{10}H_{14}Cl_2N_2$			
3-Pyridinemethanamine, 2,6-dichloro-N,N-diethyl-	MeOH	272(3.66)	33-0190-76
$C_{10}H_{14}Cl_2O_5$			
α-D-ribo-Hexofuranose, 3-deoxy-3-(di-chloromethylene)-1,2-O-(1-methyleth-ylidene)-	EtOH	218s(--),235(2.00)	111-0489-76
$C_{10}H_{14}FN_3O_5$			
Cytidine, 5-fluoro-2'-O-methyl-	acid	290(4.13)	35-7381-76
	base	280(4.03)	35-7381-76
Cytidine, 5-fluoro-3'-O-methyl-	acid	290(4.12)	35-7381-76
	base	279(4.03)	35-7381-76
$C_{10}H_{14}N_2$			
Diazene, (1,1-dimethylethyl)phenyl-	hexane	210(4.08),260(3.92), 410(2.16)	35-0109-76
Ethanimidamide, N,N-dimethyl-N'-phenyl-	neutral	240(4.01)	39-0211-76B
	cation	237(4.00)	39-0211-76B
$C_{10}H_{14}N_2O$			
Benzaldehyde, 2-methoxy-, dimethylhydra-zone	benzene	320(4.07)	104-0625-76
	EtOH	285(3.90),315(3.99)	104-0625-76
1H-Imidazole, 1-(cyclohexylcarbonyl)-	THF	244(3.74)	33-2738-76
Methanone, cyclohexyl-1H-imidazol-2-yl-	EtOH	279(4.11)	33-2738-76
Methanone, cyclohexyl-1H-imidazol-4-yl-	EtOH	257(4.09)	33-2738-76
2(1H)-Pyridinone, 3-(1-methyl-2-pyrroli-dinyl)-	MeOH	230(3.84),303(3.58)	83-0197-76
2(1H)-Pyridinone, 5-(1-methyl-2-pyrroli-dinyl)-	MeOH	230(4.04),305(3.68)	83-0197-76
1H-Pyrrolo[1,2-c]imidazol-1-one, 2,3-di-hydro-2-methyl-3-(1-methylethyl)-	MeOH	235(3.76),242(3.74), 278(3.95)	44-3050-76
$C_{10}H_{14}N_2OS$			
Carbamothioic acid, methyl-, S-[4-(di-methylamino)phenyl] ester	MeOH	269(3.75),310(3.78)	83-0631-76

Compound	Solvent	$\lambda_{max}(\log \epsilon)$	Ref.
4(1H)-Pyrimidinone, 2-(2-propenylthio)-6-propyl-	EtOH	234(3.95),288(3.92)	95-1094-76
$C_{10}H_{14}N_2O_2$			
Benzenemethanol, α-[1-(methylnitroso-amino)ethyl]-	C_6H_{12}-10% dioxan	352s(1.96),364(2.07), 376s(2.00)	78-0847-76
Diazene, [(1-methylethoxy)methyl]phenyl-, 2-oxide, (Z)-	EtOH	248(4.04)	44-1141-76
Ethanone, 1-[6-methyl-2-(methylamino)-4-(methylimino)-4H-pyran-3-yl]-	EtOH	233(4.50),316(4.21)	44-0013-76
$C_{10}H_{14}N_2O_2S_2$			
Hexanoic acid, 2,6-diisothiocyanato-, ethyl ester	pH 8.06	245(3.29)	73-2216-76
	MeOH	248(3.29)	73-2216-76
	dioxan	247(3.30)	73-2216-76
$C_{10}H_{14}N_2O_3$			
Acetamide, N,N'-[(5-methyl-2-furanyl)-methylene]bis-	EtOH	217(4.59)	103-0142-76
$C_{10}H_{14}N_2O_4$			
Nitrous acid, 1,2,4,5-tetramethyl-4-nitro-2,5-cyclohexadien-1-yl ester	MeOH	199(3.15)	23-1795-76
2,4(1H,3H)-Pyrimidinedione, 1-[2-hydroxy-4-(hydroxymethyl)cyclopentyl]-, (1α,2β,4α)-(+)-	pH 1 and 7 pH 13	268(4.01) 266(3.88)	4-1015-76 4-1015-76
2,4(1H,3H)-Pyrimidinedione, 1-[3-hydroxy-4-(hydroxymethyl)cyclopentyl]-, (1α,3β,4α)-(+)-	pH 1 and 7 pH 13	268(4.02) 266(3.90)	4-1015-76 4-1015-76
$C_{10}H_{14}N_2O_5$			
1H-Pyrazole-5-carboxylic acid, 3-(2-ethoxy-2-oxoethoxy)-1-methyl-, methyl ester	MeOH	225(4.05),265(3.56)	24-0268-76
2,4(1H,3H)-Pyrimidinedione, 1-(2,6-dideoxy-β-D-arabino-hexopyranosyl)-	pH 2-7 pH 12	259(3.95) 259(3.82)	44-0600-76 44-0600-76
2,4(1H,3H)-Pyrimidinedione, 1-[2,3-dihydroxy-4-(hydroxymethyl)cyclopentyl]-, (1α,2β,3β,4α)-(+)-	pH 1 and 7 pH 2 pH 13	267(4.03) 262(4.01) 266(3.90)	4-1015-76 73-2096-76 4-1015-76
$C_{10}H_{14}N_2O_8$			
3-Hexene-1,6-diol, 2,5-dinitro-, diacetate	EtOH	253(3.88)	104-2039-76
$C_{10}H_{14}N_2S$			
Diazene, [(1,1-dimethylethyl)thio]phenyl-, (E)-	heptane MeOH	325(4.10),396(2.73) 326(4.11),390(2.70)	144-0399-76 144-0399-76
(Z)-	heptane	248(4.06),306(3.34), 382(2.68)	144-0399-76
$C_{10}H_{14}N_2S_2$			
2,6(1H,3H)-Pyridinedithione, 3-[(butylamino)methylene]-	MeOH	225(4.20),293(3.60), 342(4.28),420(4.55)	33-0190-76
2,6(1H,3H)-Pyridinedithione, 3-[[(1,1-dimethylethyl)amino]methylene]-	MeOH	226(4.19),250(3.58), 291(3.61),341(4.25), 421(4.60)	33-0190-76
$C_{10}H_{14}N_4O$			
1H-Imidazol-2-amine, 1-(2-amino-4-methoxyphenyl)-4,5-dihydro-, dihydrobromide	H_2O	206(4.65),230s(3.98), 287(3.47)	78-0057-76

Compound	Solvent	λ_{max}(log ϵ)	Ref.
$C_{10}H_{14}N_4O_2$			
Imidazo[1,2-b]-1,2,4-triazine-2,7(1H,3H)-dione, 1-butyl-6-methyl-	EtOH	214(4.35),233(4.15), 263s(3.85)	114-0413-76A
	dioxan	214(4.29),230s(4.10), 265s(3.86)	114-0413-76A
4H-Pyrido[1,2-a]pyrimidine-3-carboxylic acid, 6,7,8,9-tetrahydro-6-methyl-4-oxo-, hydrazide	pH 7.35	231(3.79),299s(--)	145-0478-76
5(2H)-as-Triazinone, 2-acetyl-6-methyl-3-pyrrolidino-	dioxan	222(4.04),244(4.05), 257(4.08)	114-0413-76A
$C_{10}H_{14}N_4O_2S$			
Isothiazolo[3,4-d]pyrimidine-4,6(5H,7H)-dione, 5,7-diethyl-3-(methylamino)-	EtOH	232(4.38),269(4.1), 291(3.8)	94-0970-76
$C_{10}H_{14}N_4O_3$			
6(1H)-Purinone, 9-(1,5-dihydroxy-3-pentyl)-	pH 1	250(3.74)	103-0461-76
	H_2O	250(3.81)	103-0461-76
	pH 13	257(3.81)	103-0461-76
$C_{10}H_{14}N_4O_4$			
Acetamide, N-(2-acetoxyethyl)-N-(6-methyl-5-oxo-3-triazinyl)-	EtOH	210(4.38),242s(3.90)	114-0413-76A
1H-1,2,4-Triazole-5-acetic acid, α-[[(ethoxycarbonyl)amino]methylene]-1-methyl-, methyl ester	EtOH	268(4.25)	94-2568-76
$C_{10}H_{10}N_4O_6$			
1H-Pyrazole-3,5-dicarboxamide, 1-α-D-arabinofuranosyl-	pH 1	207(4.03)	18-3552-76
	H_2O	205(4.08)	18-3552-76
	pH 13	219(3.95)	18-3552-76
$C_{10}H_{14}N_5O_8P$			
Guanine, 9-β-D-xylofuranosyl-, 5'-(hydrogen phosphate)	pH 1	255(4.10),274s(3.93)	87-1026-76
	pH 7	252(4.14),270s(3.99)	87-1026-76
	pH 11	256-267(4.08)	87-1026-76
5'-Thymidylic acid, α-azido-	pH 1	263(3.99)	87-0903-76
	pH 7	263(3.99)	87-0903-76
	pH 13	261(3.85)	87-0903-76
$C_{10}H_{14}N_5S_2$			
Thiazolium, 3-ethyl-2-[3-(3-ethyl-2(3H)-thiazolylidene)-1-triazenyl]-, tetrafluoroborate	EtOH	465(4.41)	33-0155-76
$C_{10}H_{14}N_6O$			
Urea, N-butyl-N'-1H-pyrazolo[3,4-d]pyrimidin-4-yl-	pH 1	270(4.16)	87-0555-76
	pH 13	272(4.06),298(3.62)	87-0555-76
	EtOH	264(4.15)	87-0555-76
Urea, N-(1,1-dimethylethyl)-N'-1H-pyrazolo[3,4-d]pyrimidin-4-yl-	pH 1	272(4.19)	87-0555-76
	pH 13	273(4.05),300(3.57)	87-0555-76
	EtOH	266(4.16)	87-0555-76
Urea, N-(2-methylpropyl)-N'-1H-pyrazolo[3,4-d]pyrimidin-4-yl-	pH 1	272(4.21)	87-0555-76
	pH 13	272(4.13),294(3.76)	87-0555-76
	EtOH	267(4.14)	87-0555-76
$C_{10}H_{14}N_6O_2$			
Urea, N-(2-ethoxyethyl)-N'-1H-pyrazolo[3,4-d]pyrimidin-4-yl-	pH 1	271(4.21)	87-0555-76
	pH 13	272(4.13),295(3.78)	87-0555-76
	EtOH	267(4.22)	87-0555-76

Compound	Solvent	λ_{max}(log ϵ)	Ref.
$C_{10}H_{14}N_6O_3$			
Adenosine, 2'-amino-2'-deoxy-	pH 2	256.5(4.17)	88-4485-76
	H_2O	259.5(4.17)	88-4485-76
	pH 12	259.5(4.16)	88-4485-76
$C_{10}H_{14}N_6O_4$			
Adenosine, 2-amino-	MeOH	217(4.36),257(3.96),	5-0745-76
		281(4.0)	
$C_{10}H_{14}N_6O_4S$			
Benzenesulfonamide, N-(1H-1-methyl-1,2,3-triazol-5-yl)-4-nitro-, anion, salt with methylamine	EtOH	255(4.22),348(3.38)	32-0001-76
3H-1,2,3-Triazolo[4,5-d]pyrimidin-5-amine, 7-(methylthio)-3-β-D-ribofuranosyl-	pH 1	221(4.15),247(3.99), 279(3.96),315(4.10)	87-1186-76
	pH 7	221(4.16),248(4.03), 277(3.94),316(4.10)	87-1186-76
	pH 13	221(4.12),244(4.03), 277(3.94),316(4.10)	87-1186-76
$C_{10}H_{14}N_6O_5$			
1H-1,2,3-Triazolo[4,5-d]pyrimidin-5-amine, 7-methoxy-1-β-D-ribofuranosyl-	EtOH-HCl	215s(4.28),245s(3.86), 294(3.82)	87-1186-76
	EtOH-pH 7	211(4.39),313(3.66)	87-1186-76
	EtOH-pH 13	310(3.68)	87-1186-76
2H-1,2,3-Triazolo[4,5-d]pyrimidin-5-amine, 7-methoxy-2-β-D-ribofuranosyl-	pH 1	204(4.36),243(3.80), 292(4.12)	87-1186-76
	pH 7	215(3.50),313(3.92)	87-1186-76
	pH 13	216(3.43),265s(3.65), 307(3.93)	87-1186-76
3H-1,2,3-Triazolo[4,5-d]pyrimidin-5-amine, 7-methoxy-3-β-D-ribofuranosyl-	pH 1	213(4.34),244(3.71), 283(4.03)	87-1186-76
	pH 7	214(4.35),246(3.73), 287(4.03)	87-1186-76
	pH 13	244(3.73),287(4.03)	87-1186-76
$C_{10}H_{14}O$			
Benzofuran, 4,5,6,7-tetrahydro-3,6-dimethyl-	EtOH	223(3.76)	42-0050-76
Bicyclo[3.1.1]hept-3-en-2-one, 4,6,6-trimethyl- (verbenone)	C_6H_{12}	241(3.81)	78-2805-76
Bicyclo[4.1.1]oct-3-en-2-one, 7,7-dimethyl-	hexane	221(3.90)	88-4019-76
2,5-Cyclohexadien-1-one, 4-ethyl-3,4-dimethyl-	H_2O	234.5(4.017)	39-1349-76B
2,5-Cyclohexadien-1-one, 4-methyl-4-propyl-	H_2O	243(4.17)	39-1349-76B
Cyclohexanol, 2-(1-buten-3-ynyl)-	C_6H_{12}	227(3.90)	88-1539-76
2-Cyclohexen-1-one, 3-methyl-4-(1-methylethenyl)-	EtOH	234(4.00),292s(2.72)	39-0484-76C
2-Cyclohexen-1-one, 3-methyl-4-(1-methylethylidene)-	EtOH	229s(3.62),299(3.94)	39-0484-76C
2-Cyclopenten-1-one, 2-(2-butenyl)-3-methyl-, (E)-	EtOH	236(4.01)	39-0249-76C
2-Cyclopenten-1-one, 2-(2,2-dimethylcyclopropyl)-	EtOH	239(3.69)	39-0249-76C
2-Cyclopenten-1-one, 2-(1,1-dimethyl-2-propenyl)-	EtOH	226(3.90)	39-0249-76C
2-Cyclopenten-1-one, 3-methyl-2-(2-methyl-2-propenyl)-	EtOH	235(4.05)	39-0249-76C

Compound	Solvent	λ_{max}(log ϵ)	Ref.
5H-Inden-5-one, 1,2,3,6,7,7a-hexahydro-1-methyl-	EtOH	240(4.1)	2-0238-76
Tricyclo[3.3.1.1^{3,7}]decanone	hexane	290(1.22)	35-3242-76
$C_{10}H_{14}O_2$			
Benzene, (1-ethoxyethoxy)-	ether	269(3.08)	39-1001-76B
1-Benzoxepin-5(2H)-one, 3,4,6,7,8,9-hexahydro-	EtOH	282(4.20)	22-1913-76
Bicyclo[3.1.0]hexane-6-carboxaldehyde, 1,4,4-trimethyl-2-oxo-	n.s.g.	293(1.85)	33-0082-76
2,4-Cyclohexadien-1-one, 6-ethoxy-5,6-dimethyl-	EtOH	316(3.66),385s(2.66)	1-0064-76
2,4-Cyclohexadien-1-one, 6-methoxy-2,5,6-trimethyl-	EtOH	321(3.68),390s(2.82)	1-0064-76
Cyclopenta[c]pyran-3(1H)-one, hexahydro-7-methyl-4-methylene-	MeOH	238(4.38)	12-1375-76
2-Cyclopenten-1-one, 2-cyclopropyl-4-methoxy-3-methyl-	EtOH	230(4.01)	39-0249-76C
8-Oxabicyclo[5.1.0]oct-5-en-2-one, 1,4,4-trimethyl-	n.s.g.	224(3.29),316(1.76)	33-0082-76
2-Propenal, 2-(2-formyl-3-methyl-1-cyclopentyl)-	H$_2$O	223(3.84)	12-1375-76
2H-Pyran-2-one, 5,6-dihydro-6-(2-pentenyl)-	EtOH	213.8(3.93)	88-1659-76
$C_{10}H_{14}O_3$			
Benzene, 1,2,5-trimethoxy-3-methyl-	n.s.g.	228(3.72),283(3.35)	39-2594-76C
5H-1-Benzopyran-5-one, 2,3,4,6,7,8-hexahydro-7-hydroxy-2-methyl-	MeOH	263(4.07)	77-0340-76
Bicyclo[2.2.1]heptane-7-acetic acid, 2-oxo-, methyl ester, (1R-syn)-	isooctane	275s(1.26),283(1.33), 296(1.35),305(1.31)	44-3725-76
2,5-Cyclohexadiene-1-carboxylic acid, 2-methoxy-1-methyl-, methyl ester	EtOH	272(1.72)	44-2401-76
2,4-Cyclohexadien-1-one, 6-ethoxy-5-methoxy-6-methyl-	EtOH	347(3.70),362s(3.60)	1-0064-76
2,5-Cyclohexadien-1-one, 2,4-dimethoxy-4,5-dimethyl-	EtOH	242(4.12),293(3.15)	5-1435-76
1-Cyclohexene-1-carboxylic acid, 3-hydroxy-6-(1-methylethenyl)-	EtOH	221(3.96)	39-0484-76C
1,3-Dioxepane, 2-(5-methyl-2-furanyl)-	EtOH	218(3.99)	103-0142-76
2H-Pyran-2-one, 5-butylidene-5,6-dihydro-6-(hydroxymethyl)-	EtOH	271(3.92)	102-0537-76
$C_{10}H_{14}O_4$			
Butanedioic acid, 2-butenylidene-, dimethyl ester	EtOH	272(4.39)	35-1204-76
2,5-Cyclohexadien-1-one, 2-hydroxy-4,4-dimethoxy-3,5-dimethyl-	EtOH	236(4.07),323(3.26)	5-1435-76
Cyclopropanecarboxylic acid, 1-(3-methoxy-3-oxo-1-propenyl)-, ethyl ester, (E)-	MeOH	224(3.93),283(3.85)	24-1269-76
Phenol, 2,6-dimethoxy-4-(methoxymethyl)-	EtOH	233s(3.88),273(3.26), 279s(3.22)	5-1435-76
2H-Pyran-2-one, 5-(1-hydroxybutyl)-6-(hydroxymethyl)-, (S)-	EtOH	298(3.76)	102-0537-76
$C_{10}H_{15}Cl_2N_3$			
1,3,5-Benzenetriamine, N,N-bis(2-chloroethyl)-, dihydrochloride	MeOH	235(4.36),281(4.27), 374(3.97)	87-0426-76

Compound	Solvent	$\lambda_{max}(\log \epsilon)$	Ref.
$C_{10}H_{15}DO$			
2,4-Cycloheptadien-1-d-1-ol, 2,6,6-tri-methyl-	n.s.g.	250(3.84),255s(3.82), 277s(3.53)	33-0082-76
3-Cyclohepten-1-one-5-d, 2,6,6-trimeth-yl-	n.s.g.	292(1.62),310s(1.54), 320s(1.28)	33-0082-76
$C_{10}H_{15}N$			
Butanamine, γ-phenyl-	isooctane	208(4.03),248(2.14), 252(2.27),258(2.34), 260(2.33),262(2.23), 268(2.22)	123-0714-76
	EtOH	208(3.98),242(2.07), 248(2.19),255(2.29), 258(2.35),260(2.33), 264(2.24),267(2.22)	123-0714-76
$C_{10}H_{15}NO$			
Benzenemethanol, α-[1-(methylamino)-ethyl]-	MeOH	283(4.37)	88-4621-76
Cyclopentanone, 2-[3-(dimethylamino)-2-propenylidene]-	EtOH	400(4.58)	70-0577-76
	EtOH-HCl	401(--)	70-0577-76
1(2H)-Naphthalenone, 8-amino-3,4,4a,5,6,7-hexahydro-cis-s-cis	MeOH	322(4.13)	35-2826-76
	C_6H_{12}	306(3.98)	35-2826-76
2-Propanamine, 1-(3-methylphenoxy)-	n.s.g.	225(3.91),275(3.20)	20-0421-76
$C_{10}H_{15}NO_2$			
2-Propenoic acid, 3-(1,2,5,6-tetrahydro-1-methyl-3-pyridinyl)-, methyl ester	EtOH	258(4.39)	88-3447-76
2-Propenoic acid, 3-(1,4,5,6-tetrahydro-1-methyl-3-pyridinyl)-, methyl ester	EtOH	357(4.62)	88-3447-76
3-Pyridinemethanol, α-ethyl-5-hydroxy-α,6-dimethyl-	MeOH	203(3.94),221(3.73), 284(3.83),326(2.45)	104-1544-76
$C_{10}H_{15}NO_3$			
2,5-Cyclohexadien-1-ol, 4-methyl-1-(1-methylethyl)-4-nitro-, cis	MeOH	192(3.31)	23-0423-76
Ethanamine, 2-(3,4-dimethoxyphenoxy)-	n.s.g.	226(3.85),285(3.54)	20-0421-76
Ethanamine, 2-(3,5-dimethoxyphenoxy)-	n.s.g.	210(4.20),265(2.71)	20-0421-76
$C_{10}H_{15}NO_4S$			
Pentanedioic acid, 2-isothiocyanato-, diethyl ester	pH 8.06	247(3.11)	73-2216-76
	MeOH	249(--)	73-2216-76
	dioxan	249(3.14)	73-2216-76
$C_{10}H_{15}NO_5S$			
2H-1,4-Thiazine-3,6-dicarboxylic acid, 3,4-dihydro-4-(methoxymethyl)-, dimethyl ester, (3R)-	EtOH	214(3.81),257(3.49), 315(3.97)	39-2540-76C
$C_{10}H_{15}NO_5S_2$			
9-Oxa-4-thia-2-azabicyclo[4.2.1]nonane-2-propanoic acid, 7,8-dihydroxy-3-thioxo-, methyl ester	n.s.g.	262(3.79),294(4.04)	33-2972-76
$C_{10}H_{15}NO_6S$			
4,9-Dioxa-2-azabicyclo[4.2.1]nonane-2-propanoic acid, 7,8-dihydroxy-3-thioxo-, methyl ester	n.s.g.	265(4.09)	33-2972-76

Compound	Solvent	$\lambda_{max}(\log \epsilon)$	Ref.
2H-1,4-Thiazine-3,6-dicarboxylic acid, 3,4-dihydro-4-(methoxymethyl)-, 1-oxide, (1R,3R)-	EtOH	207(3.83),275(4.13)	39-2540-76C
(1S,3R)-	EtOH	206(3.79),275(4.11)	39-2540-76C
2,5-Thiazoledicarboxylic acid, 2,3-dihydro-3-(methoxymethyl)-2-methyl-, dimethyl ester, 1-oxide, (1R,2R)-	EtOH	208s(3.78),214(3.79), 287(3.88)	39-2540-76C
$C_{10}H_{15}N_2O$			
4H-Pyrido[1,2-a]pyrimidinium, 6,7,8,9-tetrahydro-1,6-dimethyl-4-oxo-, methyl sulfate	pH 7.35	234(3.96),265(3.59)	145-0478-76
$C_{10}H_{15}N_2O_9P$			
5'-Thymidylic acid, α-hydroxy-	pH 1	264(4.03)	87-0903-76
	pH 7	264(4.03)	87-0903-76
	pH 13	264(3.91)	87-0903-76
$C_{10}H_{15}N_2PS$			
1H-1,3,2-Benzodiazaphosphole, 2-ethyl-2,3-dihydro-1,3-dimethyl-	EtOH	249(3.61),293(3.77)	65-0807-76
$C_{10}H_{15}N_3O$			
4(1H)-Pyrimidinone, 6-(cyclohexylamino)-	pH 1	262(4.08)	39-1847-76C
	pH 13	256(3.78)	39-1847-76C
$C_{10}H_{15}N_3O_2S$			
1H-Imidazole-5-carboxylic acid, 4-amino-2-(methylthio)-1-(2-propenyl)-, ethyl ester	EtOH	225(4.07),300(4.18)	49-1413-76
$C_{10}H_{15}N_3O_4$			
Hexanoic acid, 2-amino-6-(1-uracilyl)-	pH 1	266(4.00)	24-2615-76
Pentanoic acid, 2-amino-5-(1-thyminyl)-	pH 1	270(3.97)	24-2615-76
1H-Pyrazole-3-carboxamide, 4-[2,3-dihydroxy-4-(hydroxymethyl)cyclopentyl]-, (1α,2β,3β,4α)-(+)-	EtOH	220(3.85)	23-0861-76
2(1H)-Pyrimidinone, 4-amino-1-(2,5-dideoxy-β-D-erythro-hexofuranosyl)-	0.1N H_2SO_4	214(4.05),280(4.10)	28-0021-76B
	H_2O	228s(3.88),272(3.94)	28-0021-76B
2(1H)-Pyrimidinone, 4-amino-1-[2,3-dihydroxy-4-(hydroxymethyl)cyclopentyl]-, (1α,2β,3β,4α)-(+)-	pH 1	214(4.01),284(4.12)	4-1353-76
	pH 7	225s(--),274(3.97)	4-1353-76
$C_{10}H_{15}N_3O_5$			
1H-Pyrazole-3-carboxamide, 5-(2,3-dihydroxy-4-(hydroxymethyl)cyclopentyl]-4-hydroxy-	pH 1	226(3.81),270(3.63)	88-1063-76
	pH 13	235(3.71),311(3.93)	88-1063-76
$C_{10}H_{15}N_3O_6$			
Cytosine, 1-β-D-psicofuranosyl-	pH 13	226(3.96),273(3.97)	44-1836-76
1H-Pyrazole-3-carboxamide, 4-(hydroxymethyl)-1-α-D-ribofuranosyl-	MeOH	218(4.03)	65-1370-76
β-	MeOH	218(3.86)	65-1370-76
1H-Pyrazole-3-carboxamide, 5-(hydroxymethyl)-1-β-D-ribofuranosyl-	MeOH	218(3.90)	65-1370-76
1H-Pyrazole-5-carboxamide, 3-(hydroxymethyl)-1-β-D-ribofuranosyl-	MeOH	224(4.06)	65-1370-76
1H-Pyrazole-5-carboxamide, 4-(hydroxymethyl)-1-β-D-ribofuranosyl-	MeOH	227(3.92)	65-1370-76

Compound	Solvent	$\lambda_{max}(\log \epsilon)$	Ref.
$C_{10}H_{15}N_3O_9$			
Uracil, 1-N-(1-deoxy-D-mannitol-1-yl)- 5-nitro-, monohydrate	acid	242(3.94),309(4.09)	136-0049-76G
	H_2O	241(3.91),312(4.10)	136-0049-76G
	base	220(3.95),322(4.25)	136-0049-76G
$C_{10}H_{15}N_5$			
Imidodicarbonimidic diamide, N-(1-phenylethyl)-	n.s.g.	238(4.26)	2-0887-76
1H-Purin-6-amine, N-(3-methylbutyl)-	pH 1	269(3.99)	87-0555-76
	pH 13	268(4.04)	87-0555-76
	EtOH	278.5(4.17)	87-0555-76
$C_{10}H_{15}N_5O_2$			
1,5-Pentanediol, 3-(6-amino-9H-purin-9-yl)-	pH 1	260(3.40)	103-0461-76
	H_2O	267(3.30)	103-0461-76
	pH 13	265(3.18)	103-0461-76
$C_{10}H_{15}N_5O_3$			
D-erythro-Pentitol, 5-(6-amino-9H-purin-9-yl)-2,5-dideoxy-	pH 1	258(4.20)	78-2409-76
	pH 7	262(4.21)	78-2409-76
	pH 13	262(4.20)	78-2409-76
$C_{10}H_{15}N_5O_6$			
1H-Pyrazole-3,5-dicarboxamide, 4-amino-1-α-D-arabinofuranosyl-	pH 1	207(4.99),304(3.46)	18-3552-76
	H_2O	204(4.04)	18-3552-76
	pH 13	219(4.26),302(3.83)	18-3552-76
$(C_{10}H_{15}N_6O_7P)_n$			
Poly(2-aminoadenylic acid)	pH 8.0	214(4.24),257(3.87), 278(3.82)	69-3783-76
$C_{10}H_{15}O$			
Pyrylium, 2,6-diethyl-4-methyl-, iodide	CH_2Cl_2	453(2.06)	59-1311-76
selenocyanate	CH_2Cl_2	441(2.43)	59-1311-76
thiocyanate	CH_2Cl_2	405(2.31)	59-1311-76
Pyrylium, 2,6-dimethyl-4-(1-methylethyl)-, iodide	CH_2Cl_2	445(1.77)	59-1311-76
selenocyanate	CH_2Cl_2	435(2.29)	59-1311-76
thiocyanate	CH_2Cl_2	401(2.46)	59-1311-76
$C_{10}H_{15}S$			
Sulfonium, ethylmethyl(4-methylphenyl)-, bromide, (S)-(+)-	EtOH	218(4.04),229(4.30), 251(3.11),254(3.14), 266(3.21),272(3.02), 274(3.14)	44-3096-76
tetrafluoroborate	EtOH	215(4.34),230(4.53), 254(3.23),255(3.24), 263(3.23),267(3.28), 272(3.16),275(3.26)	44-3096-76
$C_{10}H_{16}$			
1,3-Cyclohexadiene, 1,3,5,5-tetramethyl-	EtOH	264(3.66)	78-0441-76
Cyclohexane, 2-butenylidene-, (E)-	EtOH	233(4.46),240(4.51), 247s(--)	39-2386-76C
Cyclohexane, 2-butenylidene-, (Z)-	EtOH	234s(--),241(4.44), 249s(--)	39-2386-76C
1-Cyclohexene, 1,5,5-trimethyl-3-methylene-	EtOH	231(4.04)	78-0441-76
1-Cyclohexene, 1,5,5-trimethyl-6-methylene-	pentane	229(4.11),236(--), 244(3.98)	33-0032-76

Compound	Solvent	$\lambda_{max}(\log \epsilon)$	Ref.
1,5-Heptadiene, 2,6-dimethyl-4-methyl-ene-	EtOH	235(4.00)	78-0441-76
2,5-Heptadiene, 2,6-dimethyl-4-methyl-ene-	EtOH	221(4.16),242(4.08)	78-0441-76
1,3,5-Heptatriene, 2,4,6-trimethyl-, (E)-	EtOH	261(4.10)	78-0441-76
(Z)-	EtOH	217(3.95),241(4.08)	78-0441-76
1,3,6-Heptatriene, 2,4,6-trimethyl-, (E)-	EtOH	236(4.00)	78-0441-76
(Z)-	EtOH	234(3.95)	78-0441-76
$C_{10}H_{16}N_2OS$			
4(1H)-Pyrimidinone, 6-methyl-2-[(3-methylbutyl)thio]-	EtOH	236(3.92),287(3.89)	95-1094-76
4(1H)-Pyrimidinone, 6-propyl-2-(propylthio)-	EtOH	236(3.97),288(3.92)	95-1094-76
1H-Pyrrole-3-carboxylic acid, 2-amino-4-methyl-, 1,1-dimethylethyl ester	MeOH	213(3.95),285(3.63)	118-0051-76
$C_{10}H_{16}N_2O_4$			
Carbamic acid, [2-(2,3-dihydro-5-methyl-3-oxo-4-isoxazolyl)-1-methylethyl]-, ethyl ester	MeOH	216(3.83)	1-0567-76
$C_{10}H_{16}N_2O_5$			
2-Butenedioic acid, 2-(2-acetyl-1,2-dimethylhydrazino)-, dimethyl ester, (E)-	MeOH	268(4.27)	24-2154-76
$C_{10}H_{16}N_3O_7P$			
D-erythro-Pentofuranoside, methyl 5-(4-amino-2-oxo-1(2H)-pyrimidinyl)-2,5-dideoxy-, 3-(dihydrogen phosphate)	pH 1	282(3.90)	78-2409-76
	pH 7	274(3.73)	78-2409-76
	pH 13	274(3.73)	78-2409-76
$C_{10}H_{16}N_4O$			
4(1H)-Pyrimidinone, 5-amino-6-(cyclohexylamino)-	pH 1	262(3.97)	39-1847-76C
	pH 13	234s(3.54)	39-1847-76C
$C_{10}H_{16}N_4OS_3$			
1-Thia-4,6,8,10-tetraazaspiro[4.5]decan-2-one, 3,3,6,8,10-pentamethyl-7,9-dithioxo-	EtOH	278(4.49)	33-2768-76
$C_{10}H_{16}N_4O_2S_3$			
3-Thiazolidinehexanoic acid, 4-[(aminothioxomethyl)hydrazono]-2-thioxo-	MeOH	255(4.23),295(4.13)	103-0749-76
$C_{10}H_{16}N_4O_4$			
Morpholine, 4,4'-(azodicarbonyl)bis-	H_2O	306(3.54),448(1.63)	35-3001-76
	MeCN	298(3.31),427(1.62)	35-3001-76
2-Propenoic acid, 3,3'-(1,4-dimethyl-2-tetrazene-1,4-diyl)bis-, dimethyl ester, (E,E,?)-	MeOH	256(4.08),352(4.66)	24-0253-76
$C_{10}H_{16}N_5O_6P$			
D-erythro-Pentitol, 5-(6-amino-9H-purin-9-yl)-2,5-dideoxy-, 1-(dihydrogen phosphate)	pH 1	260(4.25)	78-2409-76
	pH 7	262(4.26)	78-2409-76
	pH 13	262(4.26)	78-2409-76
D-erythro-Pentitol, 5-(6-amino-9H-purin-9-yl)-2,5-dideoxy-, 3-(dihydrogen phosphate)	pH 1	258(4.11)	78-2409-76
	pH 7 and 13	262(4.13)	78-2409-76

Compound	Solvent	$\lambda_{max}(\log \epsilon)$	Ref.

$C_{10}H_{16}O$

Bicyclo[2.2.1]heptan-2-one, 1,3,3-tri-methyl-	3-Mepentane	288(1.23)	144-0475-76
	MeCN	287(1.20)	144-0475-76
Bicyclo[2.2.1]heptan-2-one, 1,7,7-tri-methyl- (camphor)	3-Mepentane	292(1.41)	144-0475-76
	MeCN	290(1.44)	144-0475-76
3-Cyclohepten-1-one, 2,6,6-trimethyl-	n.s.g.	288(1.67)	33-0082-76
Cyclohexanone, 2-methyl-2-(2-propenyl)-	EtOH	288(1.65)	44-0855-76
2-Cyclohexen-1-one, 5-ethyl-2,4-dimeth-yl-	EtOH	238(4.00)	78-1699-76
2-Cyclohexen-1-one, 4,4,6,6-tetramethyl-	C_6H_{12}	220(4.20),342(1.73)	33-2244-76
Cyclopentanone, 4,4-dimethyl-2-(1-prop-enyl)-, (Z)-	n.s.g.	286(1.86)	33-0082-76
Ethanol, 1-(2-cyclooocten-1-ylidene)-	EtOH	255(c.4.0)	18-3673-76
Ethanone, 1-(1-cyclooocten-1-yl)-	MeOH	234(4.00),306(0.89)	18-3673-76
	MeCN	233(4.00)	18-3673-76
2,4,6-Octatrien-1-ol, 3,7-dimethyl-, (Z,E)-	EtOH	279(4.35)	78-0107-76

$C_{10}H_{16}O_2$

1-Cyclohexene-1-carboxaldehyde, 3-hydroxy-2,6,6-trimethyl-	MeOH	245(4.00)	35-3661-76
2-Cyclohexen-1-one, 2-hydroxy-3-methyl-6-(1-methylethyl)-	EtOH	274(3.98)	18-2292-76
2-Cyclohexen-1-one, 2-hydroxy-6-methyl-3-(1-methylethyl)-	EtOH	275(4.00)	18-2292-76

$C_{10}H_{16}O_2S$

Sulfonium, dimethyl-, 4,4-dimethyl-2,6-dioxocyclohexylide	MeOH	261(4.30)	30-0366-76

$C_{10}H_{16}O_3$

1-Cyclopentanecarboxylic acid, 2,2,4,4-tetramethyl-5-oxo-	hexane	259(3.95)	104-1252-76
	MeOH	252(3.67)	104-1252-76
2(5H)-Furanone, 4-butyl-3-methoxy-5-methyl-	EtOH	226(3.96)	107-0299-76

$C_{10}H_{16}O_4$

2,6-Oxecanedione, 4-hydroxy-10-methyl-	MeOH	290(3.95)	77-0340-76

$C_{10}H_{16}O_4S$

Sulfonium, dimethyl-, 1-acetyl-4-ethoxy-2,4-dioxobutylide	EtOH	232(3.91),272(4.06)	39-1688-76C

$C_{10}H_{16}S_3$

2,4,9-Trithiatricyclo[3.3.1.1^{3,7}]decane, 1,3,5-trimethyl-	hexane	242(3.35)	11-0220-76A
2,4,9-Trithiatricyclo[3.3.1.1^{3,7}]decane, 1,3,7-trimethyl-	hexane	243(3.36)	11-0220-76A

$C_{10}H_{16}Se$

Bicyclo[2.2.1]heptane-2-selone, 1,3,3-trimethyl-	C_6H_{12}	224(3.57),272(3.98), 616(1.62)	39-2079-76C

$C_{10}H_{17}BrN_2$

Pyrimidine, 5-bromo-2,2-diethyl-1,2-di-hydro-4,6-dimethyl-, hydrobromide	MeOH	368(3.64)	39-1784-76C

$C_{10}H_{17}Cl_4N_3$

1,3,5-Benzenetriamine, N,N-bis(2-chloro-ethyl)-, dihydrochloride	MeOH	235(4.36),281(4.27), 374(3.97)	87-0426-76

Compound	Solvent	$\lambda_{max}(\log \epsilon)$	Ref.
$C_{10}H_{17}NO$			
3-Buten-2-one, 4-(cyclohexylamino)-	H_2O	302(4.41)	35-2826-76
	MeOH	303(4.33)	35-2826-76
cis-s-cis	C_6H_{12}	303(4.21)	35-2826-76
Cyclohexanone, 2-methyl-2-(2-propenyl)-, oxime	EtOH	204(3.45)(end absorption)	44-0855-76
2-Cyclohexen-1-one, 3-(dimethylamino)- 5,5-dimethyl-	pH 1	286(4.36)	39-2207-76C
	H_2O	302(4.51)	39-2207-76C
	pH 13	302(4.53)	39-2207-76C
4,6-Heptadien-3-one, 7-(dimethylamino)- 4-methyl-	EtOH	370(4.79)	70-0577-76
$C_{10}H_{17}NOS$			
4H-Thiopyran-4-one, 3-[(diethylamino)- methylene]-2,3,5,6-tetrahydro-	EtOH	275s(3.54),330(4.19)	4-0225-76
$C_{10}H_{17}NO_2$			
Acetic acid, (5,5-dimethyl-2-pyrrolidin- ylidene)-, ethyl ester	EtOH	281(4.35)	12-1735-76
3H-Pyrrole, 2-acetyl-4,5-dihydro- 3,3,5,5-tetramethyl-, 1-oxide	EtOH	289(4.00)	39-1951-76C
3H-Pyrrol-3-one, 2-(1,1-dimethylethyl)- 4,5-dihydro-5,5-dimethyl-, 1-oxide	EtOH	277(4.22)	12-2561-76
$C_{10}H_{17}NO_3S$			
Tetramethylammonium 3-ethoxycyclobutene- dione-4-thiolate	H_2O	316(4.50)	44-3904-76
$C_{10}H_{17}NO_4$			
Butanoic acid, 4-[(1-methoxy-3-oxo- 1-butenyl)amino]-, methyl ester	n.s.g.	289.5(5.3)	39-1241-76C
$C_{10}H_{17}N_2O$			
Pyrimidinium, 2,3-dihydro-1,3-bis- (1-methylethyl)-2-oxo-, ethyl sulfate, cation	pH 0.29	319(3.95)	23-2681-76
pseudobase	pH 9.18	237(3.62),305(3.31)	23-2681-76
Pyrimidinium, 2,3-dihydro-2-oxo-1,3-di- propyl-, iodide, cation	pH 0.29	223(4.09)	23-2681-76
pseudobase	pH 9.18	227(4+),303(3.50)	23-2681-76
$C_{10}H_{17}N_3S_2$			
5(4H)-Thiazolethione, 4,4-dimethyl- 2-(N-methylpiperazino)-	EtOH	294(4.08),475(1.23)	33-2566-76
$C_{10}H_{18}ClNO$			
1-Oxa-4-azaspiro[4.5]decane, 4-chloro- 3,3-dimethyl-	H_2O	265(2.20)	36-1733-76
$C_{10}H_{18}N_2$			
1H-Pyrazole, 1,3,5-trimethyl-4-(1-meth- ylpropyl)-, (+)-	EtOH	230(3.64)	44-2874-76
Pyrimidine, 2,2-diethyl-1,2-dihydro- 1,4-dimethyl-, hydriodide	MeOH	374(3.71)	39-1784-76C
Pyrimidine, 2,2-diethyl-1,2-dihydro- 4,6-dimethyl-, picrate	MeOH	352(4.30)	39-1784-76C
$C_{10}H_{18}N_2O$			
d-Camphidine, N-nitroso-	C_6H_{12}	346s(1.78),357(1.97), 368(2.08),380(2.01)	78-0847-76

Compound	Solvent	$\lambda_{max}(\log \epsilon)$	Ref.
4H-Pyrazol-4-one, 1,5-dihydro-1,3,5-tri-methyl-5-(1-methylpropyl)-	EtOH	367(3.76)	44-2874-76
$C_{10}H_{18}N_2O_2$			
3H-Pyrrol-3-one, 2-(1,1-dimethylethyl)-4,5-dihydro-5,5-dimethyl-, oxime, 1-oxide	EtOH	277(4.29)	12-2561-76
$C_{10}H_{18}N_2O_4$			
Butanoic acid, 3,3'-azobis[3-methyl-	H_2O	360(1.43)	126-0395-76
2-Butene, 1,4-bis(ethyl-aci-nitro)-2,3-dimethyl-	CH_2Cl_2	248(--),343(4.49)	104-2039-76
3,4-Heptanedione, 2,2,6-trimethyl-6-nitro-, 4-oxime	EtOH	213s(--),231(3.90)	12-2561-76
$C_{10}H_{18}O$			
Cyclobutanone, 2,4-bis(1-methylethyl)-, cis	C_6H_{12}	304(1.68)	112-0777-76
trans	C_6H_{12}	312(1.38),322(1.38)	112-0777-76
4-Hepten-2-one, 3,5,6-trimethyl-	n.s.g.	280s(2.24),287(2.29), 294(2.31),303s(2.23), 314s(1.94)	33-1253-76
$C_{10}H_{18}O_2$			
2,4-Decanedione	benzene	280(3.95)	90-1357-76
	o-xylene	294(3.73)	90-1357-76
	$CHCl_3$	278(3.96)	90-1357-76
	CCl_4	277(4.01)	90-1357-76
$C_{10}H_{18}O_4Si_2$			
3-Cyclobutene-1,2-dione, 3,4-bis[(tri-methylsilyl)oxy]-	hexane	227(4.96),258(4.37), 275s(4.27),320s(3.77)	35-3641-76
$C_{10}H_{18}S_2$			
Disulfide, di-4-pentenyl	EtOH	250(2.6)	142-0035-76S
$C_{10}H_{19}ClN_2O$			
1-Oxa-4,8-diazaspiro[4.5]decane, 4-chloro-3,3,8-trimethyl-	H_2O	265(2.43)	36-1733-76
$C_{10}H_{19}I$			
2-Octene, 8-iodo-2,6-dimethyl-	heptane	256(2.65)	78-1391-76
$C_{10}H_{19}NO$			
3-Buten-2-one, 4-[bis(1-methylethyl)-amino]-	C_6H_{12}	296(4.34)	35-2826-76
	H_2O	312(4.48)	35-2826-76
	MeOH	309(4.45)	35-2826-76
1-Penten-3-one, 1-[(1,1-dimethylethyl)-amino]-4-methyl-	H_2O	312(4.36)	35-2826-76
	MeOH	312(4.36)	35-2826-76
cis-s-cis	C_6H_{12}	300(4.24)	35-2826-76
1-Penten-3-one, 1-(diethylamino)-2-meth-yl-	C_6H_{12}	294(4.37)	35-2826-76
	H_2O	314(4.46)	35-2826-76
	MeOH	306(4.43)	35-2826-76
1-Penten-3-one, 1-(diethylamino)-4-meth-yl-	C_6H_{12}	294(4.37)	35-2826-76
	H_2O	313(4.43)	35-2826-76
	MeOH	307(4.42)	35-2826-76
2H-Pyrrole, 5-(1,1-dimethylethyl)-3,4-dihydro-2,2-dimethyl-	EtOH	235(3.95)	12-2511-76

Compound	Solvent	$\lambda_{max}(\log \epsilon)$	Ref.
$C_{10}H_{19}NSe$ Pentane, 3-isoselenocyanato-2,2,4,4-tetramethyl-	C_6H_{12}	250(3.97)	39-2079-76C
$C_{10}H_{19}N_3OS$ 2(1H)-Pyrimidinone, 4-(1-aminobutyl)-tetrahydro-1,3-dimethyl-6-thioxo-diastereomer	n.s.g. n.s.g.	271(4.14) 269(4.14)	88-0297-76 88-0297-76
$C_{10}H_{19}N_5O$ 1H-Imidazole-4-carboxamide, 5-amino-1-[2-(diethylamino)ethyl]-	EtOH	270(4.17)	2-0346-76
$C_{10}H_{20}N_2O$ 2-Propenamide, 3-(methylamino)-N,N-bis-(1-methylethyl)-	C_6H_{12} MeOH dioxan	282(4.26) 279(4.27) 281(4.20)	78-1025-76 78-1025-76 78-1025-76
$C_{10}H_{20}N_4OS_2$ Propanamide, N,N,2-trimethyl-2-[[[methyl[(methylamino)thioxomethyl]amino]-thioxomethyl]amino]-	EtOH	256(3.98),276(3.98)	33-2768-76
$C_{10}H_{20}N_4S_2$ Thiourea, N-[5-(dimethylamino)-4,5-di-hydro-4,4-dimethyl-2-thiazolyl]-N,N'-dimethyl-	EtOH	266(4.29)	33-2768-76
$C_{10}H_{20}O_2$ 3,7-Nonanediol, 5-methylene-	EtOH	250(2.87),256(2.43)	78-1839-76
$C_{10}H_{21}N_2O_5P$ Acetic acid, [(triethoxyphosphoranyli-dene)hydrazono]-, ethyl ester	EtOH	272.0(4.18)	65-1424-76
$C_{10}H_{22}N_4$ 1H-1-Azepin-1-amine, N-[1-(ethylazo)eth-yl]hexahydro-	hexane	369(1.58)	104-2207-76
$C_{10}H_{24}N_4$ Diazene, [1-[2,2-bis(1-methylethyl)hy-drazino]ethyl]ethyl- Diazene, [1-(2,2-diethylhydrazino)-2-methylpropyl]ethyl-	EtOH C_6H_{12}	367(1.67) 370(1.53)	104-2207-76 104-2207-76
$C_{10}H_{24}Si_3$ Silacycloprop-2-ene, 1,1-dimethyl-2,3-bis(trimethylsilyl)-	pentane	205(3.54),215(3.46), 345(1.57)	35-6382-76
$C_{10}H_{25}N_2OPSe$ Phosphorus selenide, P,P-bis(diethyl-amino)-P-ethoxy-	hexane or EtOH	227(3.91)	70-0174-76

Compound	Solvent	$\lambda_{max}(\log \epsilon)$	Ref.

$C_{11}HN_6$
Pentadiene-1,1,2,4,5,5-hexacarbonitrile,   MeCN   540(4.91)   35-0558-76
   ion(1-)

$C_{11}H_4ClF_3N_4O_2$
Benzo[g]pteridine-2,4(1H,3H)-dione,   EtOH   250(4.57),260s(4.45),   103-0938-76
   8-chloro-7-(trifluoromethyl)-   308(3.93),375s(4.12),
                              390(4.18),410s(4.03)

$C_{11}H_5BrOS$
5H-Naphtho[1,8-bc]thiophen-5-one,   MeOH   236(3.94),265(4.11),   89-0775-76
   2-bromo-   275(4.07),369(4.22)

$C_{11}H_5Cl_3O_2$
2-Furancarbonyl chloride, 5-(3,4-dichlo-   dioxan   333(4.67)   73-2577-76
   rophenyl)-

$C_{11}H_5Cl_4NO$
Phenol, 4-[(2,3,4,5-tetrachloro-2,4-cy-   pet ether   483(4.02)   39-0048-76C
   clopentadien-1-ylidene)amino]-

$C_{11}H_5N_3$
Ethenetricarbonitrile, phenyl-   CHCl$_3$ 342(4.23)   35-0558-76

$C_{11}H_6BrCl_2NO_6S$
Furan, 2-bromo-5-[2-[(dichloromethyl)-   EtOH   232(3.80),309(4.40),   73-3391-76
   sulfonyl]-2-(5-nitro-2-furanyl)eth-   375s(4.01)
   enyl]-

$C_{11}H_6BrNO$
2-Furancarbonitrile, 5-(4-bromophenyl)-   EtOH   214(4.19),294(4.59)   73-1692-76

$C_{11}H_6ClNO$
2-Furancarbonitrile, 5-(4-chlorophenyl)-   EtOH   215(4.21),300(4.56)   73-1692-76

$C_{11}H_6ClNO_4$
2-Furancarbonyl chloride, 5-(2-nitro-   dioxan   317(4.24)   73-2577-76
   phenyl)-
2-Furancarbonyl chloride, 5-(3-nitro-   dioxan   322(4.44)   73-2577-76
   phenyl)-
2-Furancarbonyl chloride, 5-(4-nitro-   dioxan   345(4.52)   73-2577-76
   phenyl)-

$C_{11}H_6ClN_3OS$
Ethanone, 1-[2-(4-chlorophenyl)-4-thia-   pH 1   287(4.30)   80-0113-76
   zolyl]-2-diazo-   40% dioxan   303(4.54)   80-0113-76

$C_{11}H_6Cl_2INO_6S$
Furan, 2-[1-[(dichloromethyl)sulfonyl]-   EtOH   243(3.92),313(4.42),   73-3391-76
   2-(5-iodo-2-furanyl)ethenyl]-5-nitro-   366s(4.12)

$C_{11}H_6Cl_2N_2O_8S$
Furan, 2,2'-[1-[(dichloromethyl)sulfon-   EtOH   243(4.12),308(4.41),   73-3391-76
   yl]-1,2-ethenediyl]bis[5-nitro-   371(4.08)

$C_{11}H_6Cl_2O_2$
2-Furancarbonyl chloride, 5-(2-chloro-   dioxan   325(4.44)   73-2577-76
   phenyl)-
2-Furancarbonyl chloride, 5-(4-chloro-   dioxan   336(4.84)   73-2577-76
   phenyl)-

Compound	Solvent	$\lambda_{max}(\log \epsilon)$	Ref.

$C_{11}H_6Cl_3NO_6S$
  Furan, 2-chloro-5-[2-[(dichloromethyl)-
  sulfonyl]-2-(3-nitro-2-furanyl)eth-
  enyl]-
	EtOH	236(3.80),306(4.44),
		371s(4.04)
			73-3391-76

$C_{11}H_6Cl_4O_3$
  Spiro[1,3-benzodioxole-2,1'-cyclopent-
  an]-2'-one
	C_6H_{12}	298(3.43),305(3.52),
		340(2.23)
			5-0793-76

	benzene	334.5s(2.23)	5-0793-76

$C_{11}H_6F_5NO_3$
  2-Propenoic acid, 2,3-difluoro-3-[4-
  [(trifluoroacetyl)amino]phenyl]-, (E)-
	EtOH	287(4.54)	104-1920-76

$C_{11}H_6N_2O_2$
  5H-[1]Benzopyrano[4,3-d]pyrimidin-5-one
  Imidazo[1,2-b]isoquinoline-5,10-dione
	MeOH
	MeOH
	DMF
		267(4.16),305(3.78)
		290(4.15)
		322(3.60),365(3.31)
			118-0274-76
			44-0836-76
			44-0836-76

$C_{11}H_6N_2O_3$
  2-Furancarbonitrile, 5-(2-nitrophenyl)-
  2-Furancarbonitrile, 5-(3-nitrophenyl)-
	EtOH
	EtOH
		214(4.54),266(4.31)
		214(4.19),290(4.52)
			73-1692-76
			73-1692-76

$C_{11}H_6N_4$ 1,9,9b-Triazaphenalene-3-carbonitrile	EtOH	254(4.45),268s(4.29), 361(4.33),391(4.29), 412(4.26),452(2.93), 570s(2.36),623(2.56), 686(2.62)	1-0466-76

$C_{11}H_6N_4O_3S$
  Ethanone, 2-diazo-1-[2-(4-nitrophenyl)-
  4-thiazolyl]-
	pH 1
	40% dioxan
		290(4.15)
		312(4.42)
			80-0113-76
			80-0113-76

$C_{11}H_7BrN_2O_2S$
  4,6(1H,5H)-Pyrimidinedione, 5-[(4-bromo-
  phenyl)methylene]dihydro-2-thioxo-
	dioxan	250(4.01),317(4.29),
		363(4.45)
			104-1124-76

$C_{11}H_7BrN_4O$
  1H-Tetrazole, 5-[5-(4-bromophenyl)-
  2-furanyl]-
	EtOH	204(4.20),224(4.05),
		314(4.51)
			73-1692-76

$C_{11}H_7BrO_2$ 1-Azulenecarboxylic acid, 6-bromo-	EtOH	295(4.71),306(4.75), 345(3.76),354(3.80), 371(3.72),543(2.62), 590s(2.53),650s(2.00)	44-1822-76

$C_{11}H_7BrO_3S$
  2-Furancarboxylic acid, 5-[(4-bromo-
  phenyl)thio]-
	EtOH	208(3.84),256(4.25)	73-2571-76

$C_{11}H_7BrO_5S$
  2-Furancarboxylic acid, 5-[(4-bromo-
  phenyl)sulfonyl]-
	EtOH	205(4.16),235(4.22),
		266(4.30)
			73-2571-76

$C_{11}H_7ClN_2O$
  2-Oxa-3,7-diazabicyclo[3.2.0]hepta-
  3,5,7-triene, 6-(2-chlorophenyl)-
  4-methyl-
	EtOH	262(4.01)	88-1825-76

Compound	Solvent	$\lambda_{max}(\log \epsilon)$	Ref.
$C_{11}H_7ClN_2O_4$			
2-Furancarboximidoyl chloride, N-hydroxy-5-(2-nitrophenyl)-	EtOH	303(4.23)	73-3085-76
2-Furancarboximidoyl chloride, N-hydroxy-5-(3-nitrophenyl)-	EtOH	315(4.64)	73-3085-76
2-Furancarboximidoyl chloride, N-hydroxy-5-(4-nitrophenyl)-	EtOH	355(4.33)	73-3085-76
$C_{11}H_7ClN_4O$			
1H-Tetrazole, 5-[5-(4-chlorophenyl)-2-furanyl]-	EtOH	202(4.11),223(4.00), 313(4.42)	73-1692-76
$C_{11}H_7ClN_4OS$			
4H-1,3,4-Thiadiazolo[2,3-c][1,2,4]triazin-4-one, 7-(3-chlorophenyl)-3-methyl-	MeOH	263(3.86),332(3.88)	4-0117-76
4H-1,3,4-Thiadiazolo[2,3-c][1,2,4]triazin-4-one, 7-(4-chlorophenyl)-3-methyl-	MeOH	270(4.23),356(4.34)	4-0117-76
$C_{11}H_7ClO_2$			
5H-Benzocyclohepten-5-one, 8-chloro-7-hydroxy-	EtOH	266(4.48),275(4.60), 284(4.54),306(3.72), 331(3.78),350(3.78)	18-2230-76
$C_{11}H_7ClO_3$			
1,4-Naphthalenedione, 2-chloro-5-hydroxy-3-methyl-	MeOH	280(4.07),431(3.59)	39-0997-76C
	MeOH-base	283(4.13),541(3.71)	39-0997-76C
Naphth[2,3-b]oxirene-2,7-dione, 5-chloro-1a-methyl-	n.s.g.	232(4.44),267(3.86), 304(3.23)	33-0664-76
$C_{11}H_7ClO_3S$			
2-Furancarboxylic acid, 5-[(4-chlorophenyl)thio]-	EtOH	206(4.09),254(4.25), 293s(3.96)	73-2571-76
$C_{11}H_7ClO_5S$			
2-Furancarboxylic acid, 5-[(4-chlorophenyl)sulfonyl]-	EtOH	205(4.05),230(4.17), 266(4.22)	73-2571-76
$C_{11}H_7Cl_2NO$			
Pyridine, 2,6-dichloro-3-[2-(2-furanylethenyl]-	MeOH	335(4.42)	33-0190-76
$C_{11}H_7Cl_2NO_2$			
2-Furancarboxamide, 5-(3,4-dichlorophenyl)-	dioxan	308(4.45)	73-2577-76
$C_{11}H_7Cl_2NO_6S$			
Furan, 2-[1-[(dichloromethyl)sulfonyl]-2-(2-furanyl)ethenyl]-5-nitro-	EtOH	243(3.87),304(4.49), 377s(4.07)	73-3391-76
$C_{11}H_7Cl_6NO$			
3H-Indole, 3,3,4,5,6,7-hexachloro-2-(1-methylethoxy)-	C_6H_{12}	239(4.43),246(4.50), 254(4.43)	39-0162-76C
$C_{11}H_7IN_2O_2S$			
4,6(1H,5H)-Pyrimidinedione, dihydro-5-[(4-iodophenyl)methylene]-2-thioxo-	dioxan	232(4.18),370(4.51)	104-1124-76
$C_{11}H_7IO_2$			
1-Azulenecarboxylic acid, 2-iodo-	CH_2Cl_2	295(4.64),306(4.74), 316s(4.43),350s(3.84), 364s(3.72),515(2.83)	44-1822-76

Compound	Solvent	$\lambda_{max}(\log \epsilon)$	Ref.
$C_{11}H_7N$			
6-Azulenecarbonitrile	C_6H_{12}	283(4.01),333(3.65), 346(3.82),362(3.70), 584(2.40),606(2.48), 631(2.57),660(2.51), 696(2.54),721(2.20), 774(1.96)	44-1811-76
$C_{11}H_7NO$			
2-Furancarbonitrile, 5-phenyl-	EtOH	214(4.10),294(4.47)	73-1692-76
$C_{11}H_7NO_2S$			
4H-[1]Benzothiopyrano[3,4-d]oxazol- 4-one, 2-methyl-	MeOH	233(4.5),302(4.0)	24-3462-76
$C_{11}H_7NO_5S$			
2-Furancarboxylic acid, 5-[(4-nitrophen- yl)thio]-	EtOH	206(4.16),239(4.23), 317(4.17)	73-2571-76
$C_{11}H_7NO_7S$			
2-Furancarboxylic acid, 5-[(4-nitrophen- yl)sulfonyl]-	EtOH	206(4.14),250(4.17), 256(4.15),317(4.34)	73-2571-76
$C_{11}H_7N_3$			
[2,2'-Bipyridine]-5-carbonitrile	EtOH	246(3.82),251(3.85), 295(4.06)	64-0115-76B
$C_{11}H_7N_3OS$			
Ethanone, 2-diazo-1-(2-phenyl-4-thiazo- lyl)-	pH 1	280(4.25)	80-0113-76
	40% dioxan	300(4.44)	80-0113-76
Ethanone, 2-diazo-1-(2-phenyl-5-thiazo- lyl)-	pH 1	318(4.36)	80-0113-76
	40% dioxan	335(4.56)	80-0113-76
$C_{11}H_7N_3O_4S$			
4,6(1H,5H)-Pyrimidinedione, dihydro-5- [(2-nitrophenyl)methylene]-2-thioxo-	dioxan	270(4.27),340(4.30)	104-1124-76
4,6(1H,5H)-Pyrimidinedione, dihydro-5- [(3-nitrophenyl)methylene]-2-thioxo-	dioxan	235(4.20),270(4.11), 375(4.64)	104-1124-76
4,6(1H,5H)-Pyrimidinedione, dihydro-5- [(4-nitrophenyl)methylene]-2-thioxo-	dioxan	287(4.39),375(3.85)	104-1124-76
$C_{11}H_7N_5O_2$			
[1,2,4]Triazolo[5,1-c][1,2,4]triazin- 4(1H)-one, 3-benzoyl-	EtOH	210(4.20),257(4.06), 305(3.98)	39-0421-76C
$C_{11}H_7N_5O_3$			
1H-Tetrazole, 5-[5-(2-nitrophenyl)- 2-furanyl]-	EtOH	204(4.29),212(4.25), 291(4.28)	73-1692-76
1H-Tetrazole, 5-[5-(3-nitrophenyl)- 2-furanyl]-	EtOH	203(4.07),212(4.18), 308(4.32)	73-1692-76
$C_{11}H_7N_5O_6$			
2,5-Pyrrolidinedione, 1-[(5-azido-2-ni- trobenzoyl)oxy]-	EtOH	229(4.12),312(4.04)	64-0263-76C
$C_{11}H_8BrNOS$			
5-Thiazolecarboxaldehyde, 2-(4-bromo- phenyl)-4-methyl-	n.s.g.	240(3.85),326(4.14)	80-1073-76

Compound	Solvent	$\lambda_{max}(\log \epsilon)$	Ref.
$C_{11}H_8BrNO_3S$ Furan, 2-[[(4-bromophenyl)thio]methyl]- 5-nitro-	n.s.g.	212(4.22),230(4.13), 258(4.04),322(3.98)	78-1411-76
$C_{11}H_8BrNO_5S$ Furan, 2-[[(4-bromophenyl)sulfonyl]- methyl]-5-nitro-	n.s.g.	209(4.06),237(4.23), 312(3.98)	78-1411-76
$C_{11}H_8ClNO_2$ 2-Furancarboxamide, 5-(2-chlorophenyl)- 2-Furancarboxamide, 5-(4-chlorophenyl)-	dioxan dioxan	301(4.34) 308(4.42)	73-2577-76 73-2577-76
$C_{11}H_8ClNO_3S$ Furan, 2-[[(4-chlorophenyl)thio]methyl]- 5-nitro-	n.s.g.	214(4.04),258(3.84), 322(3.92)	78-1411-76
$C_{11}H_8ClNO_5S$ Furan, 2-[[(4-chlorophenyl)sulfonyl]- methyl]-5-nitro-	n.s.g.	209s(4.03),230(4.25), 309(4.00)	78-1411-76
$C_{11}H_8Cl_3NO_5S$ 2-Propenoic acid, 3-[[(2,5-dihydro-4- isocyanato-5-oxo-3-furanyl)methyl]- thio]-, 2,2,2-trichloroethyl ester	hexane	239(4.14),279(4.15)	33-1988-76
$C_{11}H_8F_4$ Indan, 4,7-difluoro-1-(difluoromethyl- ene)-3-methyl-	EtOH	246(4.12),252s(4.06), 280(3.14),290(3.07)	12-1435-76
$C_{11}H_8N_2$ Imidazo[2,1-a]isoquinoline	MeOH	243(4.59),252(4.74), 280(3.85),294(3.67)	39-0075-76C
	MeOH	214(3.90),243(4.58), 252(4.74),280(3.86), 294(3.67)	39-0075-76C
Propanedinitrile, (1-phenylethylidene)- 1H-Pyrrole-2-carbonitrile, 3-phenyl-	EtOH H_2O	232(3.88),292(4.11) 265(4.18)	118-0705-76 33-2786-76
$C_{11}H_8N_2O$ 3-Indolizinecarbonitrile, 1-acetyl-	MeOH	205(4.37),219(4.34), 242s(4.62),246(4.74), 277(4.65),316s(4.18), 336s(4.08),350s(3.96)	39-1908-76C
Methanone, di-2-pyridinyl-	hexane MeOH	235(3.89),268(3.82) 246s(3.83),269(4.01)	59-0351-76 59-0351-76
Methanone, di-3-pyridinyl-	hexane	204(4.21),234(4.01), 240(4.00),266(3.79)	59-0351-76
	MeOH	209(4.01),231(4.01), 238(4.01),267(3.88)	59-0351-76
Methanone, di-4-pyridinyl- Methanone, 2-pyridinyl-3-pyridinyl-	isoPrOH hexane	228(4.16),280(3.62) 202(4.22),242(4.05), 271(4.01)	59-0351-76 59-0351-76
	MeOH	210(4.05),244(4.03), 273(4.02)	59-0351-76
Methanone, 2-pyridinyl-4-pyridinyl-	hexane-4% MeOH	235(4.08),270(3.89)	59-0351-76
	MeOH	205(4.06),230(4.01), 268(3.89)	59-0351-76
Methanone, 3-pyridinyl-4-pyridinyl-	hexane MeOH	205(3.78),258(3.25) 212(3.68),257(3.48), 280s(3.09)	59-0351-76 59-0351-76

Compound	Solvent	$\lambda_{max}(\log \epsilon)$	Ref.
Propanedinitrile, [(2-methoxyphenyl)-methylene]-	C_6H_{12}	235(3.79),299(4.12), 359(3.98)	80-0781-76
	EtOH	240(3.88),297(3.98), 368(3.82)	80-0781-76
	$CHCl_3$	300(4.14),372(4.01)	80-0781-76
	CCl_4	300(4.23),362(4.15)	80-0781-76
non-ion	H_2O	305(4.06),370(3.96)	80-0781-76
ion	H_2O	256(4.04)	80-0781-76
Propanedinitrile, [(4-methoxyphenyl)-methylene]-	C_6H_{12}	247(3.85),280(3.65), 342(4.47)	80-0781-76
	EtOH	248(4.10),275(3.91), 349(4.41)	80-0781-76
	acetone	358(4.45)	80-0781-76
	$CHCl_3$	352(4.47)	80-0781-76
	CCl_4	280(3.65),345(4.41)	80-0781-76
non-ion	H_2O	350(4.13)	80-0781-76
ion	H_2O	285(3.92)	80-0781-76
$C_{11}H_8N_2OS$ 4-Oxazolecarbonitrile, 2-methyl-5-(phenylthio)-	$CHCl_3$	271(3.76)	94-0948-76
$C_{11}H_8N_2OS_2$ 1H-[1,2]Dithiolo[5,1-e][1,2,3]thiadiazole-7-S^{IV}-3-carboxaldehyde, 1-phenyl-	C_6H_{12}	234(4.38),243s(4.33), 312(4.02),346(3.97), 475(4.36)	39-0880-76C
$C_{11}H_8N_2O_2$ [1]Benzopyrano[3,4-c]pyrazol-4(3H)-one, 3-methyl-	EtOH	252(4.78),303(4.95)	39-1260-76C
5H-[1]Benzopyrano[4,3-d]pyrimidin-5-ol	MeOH	280(4.10),321(3.86)	118-0274-76
[2,2'-Bipyridine]-5-carboxylic acid	EtOH	242(3.76),248(3.75), 291(4.00)	64-0115-76B
1-Indolizinecarboxylic acid, 3-cyano-, methyl ester	MeOH	216(4.51),237s(4.63), 242(4.76),265s(4.18), 273(4.15),303s(4.16), 316(4.26),330s(4.10), 335s(4.02),345s(3.74)	39-1908-76C
Methanone, (2-hydroxyphenyl)-5-pyrimidinyl-	MeOH	267(4.00),340(3.58)	118-0274-76
$C_{11}H_8N_2O_2S$ 4,6(1H,5H)-Pyrimidinedione, dihydro-5-(phenylmethylene)-2-thioxo-	dioxan	260(3.84),310(3.90), 382(4.59)	104-1124-76
$C_{11}H_8N_2O_3$ Benzoic acid, 2-(1H-imidazol-2-ylcarbonyl)-	MeOH	287(4.12)	44-0836-76
$C_{11}H_8N_2O_4$ 2-Furancarboxamide, 5-(2-nitrophenyl)-	dioxan	290(4.22)	73-2577-76
2-Furancarboxamide, 5-(3-nitrophenyl)-	dioxan	302(4.35)	73-2577-76
2-Furancarboxamide, 5-(4-nitrophenyl)-	dioxan	350(4.43)	73-2577-76
$C_{11}H_8N_2O_4S$ 2-Furancarboximidothioic acid, N-hydroxy-5-nitro-, phenyl ester	EtOH	208(4.26),333(4.09)	73-3085-76
1-Naphthalenesulfonic acid, 6-diazo-5,6-dihydro-5-oxo-, methyl ester	HCl	223(4.22),270(4.32), 302(3.79),404(3.57)	99-0081-76
	dioxan	230(4.15),258(4.04), 343(3.74),400(3.71)	99-0081-76

Compound	Solvent	$\lambda_{max}(\log \epsilon)$	Ref.
1-Naphthalenesulfonic acid, 6-diazo-5,6-dihydro-5-oxo-, methyl ester	MeCN	228(4.18),260(4.11), 342(3.86),400(3.83)	99-0081-76
$C_{11}H_8N_2O_5S$ Furan, 2-nitro-5-[[(4-nitrophenyl)thio]-methyl]-	n.s.g.	214(4.22),231s(4.09), 327(4.39)	78-1411-76
$C_{11}H_8N_2O_7S$ Furan, 2-nitro-5-[[(4-nitrophenyl)sulfonyl]methyl]-	n.s.g.	214(4.06),248(4.20), 309(4.15)	78-1411-76
$C_{11}H_8N_2S$ 1H-Perimidine-2(3H)-thione	MeOH	233(4.77),282(4.64), 321(4.27)	83-0928-76
$C_{11}H_8N_2S_2$ [1]Benzothieno[3,2-d]pyrimidine, 2-(methylthio)-	MeOH	249(4.40),268(4.38), 298(3.88),307(3.85), 362(3.49)	73-2771-76
$C_{11}H_8N_4O$ 1(2H)-Phthalazinone, 4-(1H-imidazol-2-yl)-	CF_3COOH	292(4.01),300(3.98), 310(3.83)	44-0836-76
1H-Tetrazole, 5-(5-phenyl-2-furanyl)-	EtOH	203(4.17),217(4.18), 308(4.48)	73-1692-76
$C_{11}H_8N_4O_2$ Benzo[g]pteridine-2,4(3H,10H)-dione, 10-methyl-	EtOH	217(4.37),267(4.46), 333(3.77),436(3.93)	35-0830-76
	50% EtOH-5M HCl	213(4.33),259(4.38), 369(4.11)	35-0830-76
Pyrazolo[1,5-a]pyrimidin-5(4H)-one, 7-hydroxy-3-(3-pyridinyl)-	MeOH-pH 1	205(4.44),215(4.46), 242(4.31),280(4.29)	87-0291-76
	MeOH-pH 11	232(4.50),275(4.40)	87-0291-76
$C_{11}H_8N_4O_3$ Benzo[g]pteridine-2,4(3H,10H)-dione, 10-methyl-, 5-oxide	EtOH	213(3.96),265(4.33), 340(3.74),355s(3.64), 451(3.69)	35-0830-76
	50% EtOH-5M HCl	267(4.27),369(3.79), 395s(3.58)	35-0830-76
$C_{11}H_8N_6O$ Methanone, (4-amino-1,2,4-triazolo[5,1-c][1,2,4]triazin-3-yl)phenyl-	EtOH	216(4.22),243s(3.86), 300(4.07),331(4.15)	39-0421-76C
$C_{11}H_8N_6O_5$ 2,4,6-Cycloheptatrien-1-one, 2,5-diacetoxy-4,6-diazido-	CH_2Cl_2	267(4.29),292(4.56), 376s(3.97),399(4.03), 420(3.94)	88-2339-76
$C_{11}H_8OS$ Methanone, phenyl-2-thienyl-	EtOH	263(4.06),293(4.06)	59-0501-76
2-Thiophenecarboxaldehyde, 4-phenyl-	EtOH	252(4.45),314(3.46)	44-1320-76
2-Thiophenecarboxaldehyde, 5-phenyl-	isooctane	228(3.94),320(4.21)	44-1320-76
$C_{11}H_8O_2$ 1,2-Naphthalenedione, 5-methyl-	EtOH	257(4.40),372s(3.52), 415(3.54)	78-2693-76
1,4-Naphthalenedione, 5-methyl-	EtOH	247(4.26),350(3.53)	78-2693-76

Compound	Solvent	λ_{max}(log ϵ)	Ref.
$C_{11}H_8O_3$			
1,2-Naphthalenedione, 5-methoxy-	EtOH	255(4.19),376s(3.19), 459(3.57)	78-2693-76
1,4-Naphthalenedione, 5-hydroxy-6-methyl-	EtOH	219(4.20),252(4.13), 434(3.62)	1-0353-76
Naphth[2,3-b]oxirene-2,7-dione, 1a,7a-dihydro-1a-methyl-	n.s.g.	225(4.42),264(3.76), 302(4.22)	33-0664-76
$C_{11}H_8O_3S$			
2-Furancarboxylic acid, 5-(phenylthio)-	EtOH	207(4.23),250(4.29), 290s(3.93)	73-2571-76
$C_{11}H_8O_4$			
1H-2-Benzopyran-3-carboxylic acid, 1-oxo-, methyl ester	MeOH	231(4.34),251(4.12), 281(3.00),291(4.13), 315(3.80)	35-5581-76
1H-2-Benzopyran-4-carboxylic acid, 1-oxo-, methyl ester	MeOH	220(4.44),290(3.52), 315(3.67)	25-0954-76
$C_{11}H_8O_5$			
2(5H)-Furanone, 3-acetyl-5-(2-furanyl-methylene)-4-hydroxy-	EtOH	246(3.98),344(4.31)	4-0521-76
Oosponol (lenzitin)	EtOH	215(4.45),233s(--), 256(4.04),333(3.84)	102-0327-76
$C_{11}H_8O_5S$			
2-Furancarboxylic acid, 5-(phenylsul-fonyl)-	EtOH	205s(3.98),221(4.08), 262(4.26)	73-2571-76
$C_{11}H_9BS_2$			
4H-Borepino[3,2-b:6,7-b']dithiophene, 4-methyl-	C_6H_{12}	<u>250(4.5),400f(4.4)</u>	46-0287-76
$C_{11}H_9BrN_2O_2$			
5H-Oxazolo[2,3-b]quinazolin-5-one, 2-(bromomethyl)-2,3-dihydro-	MeOH	221(4.63),226s(4.58), 245(3.85),254(3.86), 262(3.82),307(3.54), 318s(--)	118-0469-76
$C_{11}H_9ClN_4O$			
5H-1,2,4-Triazolo[4,3-d][1,4]benzodiaz-epin-6(7H)-one, 10-chloro-7-methyl-	MeOH	231(4.55),250(4.10), 298(3.28)	88-1931-76
$C_{11}H_9ClN_4O_2$			
3H-1,2,4-Triazolo[4,3-d][1,4]benzodiaz-epine-3,6(5H)-dione, 10-chloro-2,7-dihydro-7-methyl-	MeOH	251(4.25),300(3.57)	88-1931-76
$C_{11}H_9ClO_3$			
Oxireno[c][1]benzoxepin-8(2H)-one, 6-chloro-1a,8a-dihydro-1a-methyl-	n.s.g.	251(3.83),325(3.42)	33-0664-76
$C_{11}H_9NO$			
6-Azabicyclo[3.1.0]hex-3-en-2-one, 6-phenyl-	EtOH	229(4.00)	39-2338-76C
2-Propenal, 3-(1H-indol-3-yl)-	MeCN	222(4.18),273(3.86), 336(4.30)	5-1039-76
Pyridinium, 3-hydroxy-1-phenyl-, hydroxide, inner salt	EtOH	342(3.73)	39-1829-76B

Compound	Solvent	$\lambda_{max}(\log \epsilon)$	Ref.
Quinoline, 8-(ethenyloxy)-	pH 1	249(4.59),307(3.42), 316(3.48),345(3.33)	65-1108-76
	EtOH	239(4.54),302(3.63), 314(3.59)	65-1108-76
hydrochloride	EtOH	243(4.43),254(4.50), 307(3.46),315(3.45), 360(3.06)	65-1108-76
methiodide	EtOH	255(4.57),310(3.39), 321(3.44),355(3.40)	65-1108-76
$C_{11}H_9NOS$ 5-Thiazolecarboxaldehyde, 4-methyl-2-phenyl-	n.s.g.	236(4.00),322(4.28)	80-1073-76
$C_{11}H_9NO_2$ 1,4-Methanonaphthalene, 1,4-dihydro-6-nitro-	EtOH	290(3.90)	44-2006-76
$C_{11}H_9NO_3$ Benzamide, N-(2,5-dihydro-5-oxo-2-furanyl)-	EtOH	227(3.18),272(1.91)	138-0793-76
Benzamide, N-(2-formyl-3-oxo-1-propenyl)-	EtOH	228(4.01),267(4.14), 312(4.03)	138-0793-76
4-Isoxazolecarboxylic acid, 5-methyl-3-phenyl-	EtOH	216(4.15)	33-2074-76
2-Propenoic acid, 2-cyano-3-hydroxy-3-phenyl-, methyl ester	EtOH	290(3.83)	22-0177-76
2H-Pyrrol-2-one, 1-benzoyl-1,5-dihydro-5-hydroxy-	EtOH	239.5(3.99)	138-0793-76
$C_{11}H_9NO_3S$ Acetamide, N-(4-hydroxy-2-oxo-2H-1-benzothiopyran-3-yl)-	MeOH	231(4.5),315(4.0)	24-3462-76
Acetic acid, (1,2-dihydro-2-oxo-3H-indol-3-ylidene)(methylthio)-, cis	MeOH	277(4.06),338(4.14)	94-2305-76
trans	MeOH	253(4.18),284(3.82), 332(4.17)	94-2305-76
Furan, 2-nitro-5-[(phenylthio)methyl]-	n.s.g.	218(4.02),244(3.84), 323(3.91)	78-1411-76
4H-1,3-Thiazine-4,6(5H)-dione, 2-(phenylmethoxy)-	EtOH	225(4.26),267(4.09)	103-0862-76
$C_{11}H_9NO_4$ Benzo[b]cyclopropa[d]pyran-2(1H)-one, 1a,7b-dihydro-1-methyl-1a-nitro-	EtOH	237(3.54),275s(3.03)	39-1260-76C
2H-1-Benzopyran-2-one, 4-ethyl-3-nitro-	EtOH	282(4.02),317(3.74)	39-1260-76C
$C_{11}H_9NO_5S$ Furan, 2-nitro-5-[(phenylsulfonyl)methyl]-	n.s.g.	221(4.07),260(3.34), 268(3.45),274(3.56), 314(4.03)	78-1411-76
Thieno[2,3-b]pyridine-5,6-dicarboxylic acid, 4,7-dihydro-4-oxo-, dimethyl ester	MeOH	221(4.28),240(4.52), 260s(3.75),281s(3.57), 318(3.71)	5-1972-76
$C_{11}H_9NO_7S$ 2H-1-Benzopyran-2-one, 4-methyl-7-[(methylsulfonyl)oxy]-8-nitro-	EtOH	272(4.14),314(4.01), 360s(3.61)	80-1543-76

Compound	Solvent	$\lambda_{max}(\log \epsilon)$	Ref.
$C_{11}H_9NS$			
Naphth[2,3-d]isothiazole, 4,9-dihydro-	EtOH	226(3.91),269(3.99), 277(4.00),287(4.05), 302(3.96)	48-0507-76
2(1H)-Pyridinethione, 1-phenyl-	EtOH	234s(3.81),292(4.01), 371(3.76)	1-0863-76
Quinoline, 8-(ethenylthio)-	pH 1	256(4.02),306(3.84), 313(3.85)	65-1108-76
	EtOH	251(4.37),338(3.78)	65-1108-76
$C_{11}H_9N_2S_2$			
1,3,4-Thiadiazolo[2,3-a]isoquinolin-4-ium, 2-(methylthio)-, iodide	MeOH	225(4.46),259(4.53), 320(4.00)	2-0176-76
1,3,4-Thiadiazolo[3,2-a]quinolin-4-ium, 2-(methylthio)-, iodide	MeOH	226(4.39),258(4.59), 320s(--),330(4.26)	2-0176-76
$C_{11}H_9N_3O_2$			
Benzamide, 2-(1H-imidazol-2-ylcarbonyl)-	MeOH	275(3.10)	44-0836-76
$C_{11}H_9N_3O_2Se_2$			
1H-[1,2]Diselenolo[5,1-e][1,2,3]selenadiazole-7-Se^{IV}, 5-methyl-1-(4-nitrophenyl)-	C_6H_{12}	203(4.50),220(4.36), 244(4.35),277s(4.00), 346(4.20),543(4.34)	39-0228-76C
$C_{11}H_9N_3S$			
[1,2,4]Triazolo[1,5-a]isoquinolinium, 2-mercapto-3-methyl-, hydroxide, inner salt	MeOH	233(4.23),270(4.62), 316(3.72),363(3.55)	2-0176-76
$C_{11}H_9N_5O$			
Pyridazino[3,2-b]quinazolin-10-one, 2-hydrazino-	n.s.g.	224(4.21),243(4.19), 288(4.49),302(4.30), 317(4.22)	2-0879-76
$C_{11}H_{10}BrNOS$			
Quinoline, 5-bromo-3-methoxy-2-(methylthio)-	hexane	220(4.46),259(4.37), 265(4.44),280s(3.71), 293(3.63),300(3.58), 306(3.60),313(3.71), 321(3.76),328(4.06), 335(3.91),345(4.21)	12-1023-76
Quinoline, 7-bromo-3-methoxy-2-(methylthio)-	EtOH	220(4.59),262(4.42), 287(3.61),300(3.49), 317(3.72),323(3.80), 331(4.07),338(3.97), 347(4.23)	12-1023-76
$C_{11}H_{10}BrN_3O$			
8-Azabicyclo[3.2.1]oct-3-ene-6-carbonitrile, 4-bromo-8-(2-cyanoethyl)-2-oxo-, exo	EtOH	250(3.95)	39-2334-76C
$C_{11}H_{10}BrN_3O_2$			
1H-Pyrazole, 4-bromo-3,5-dimethyl-1-(2-nitrophenyl)-	EtOH	250(4.04)	22-0184-76
$C_{11}H_{10}ClN$			
1H-Pyrrole, 2-(4-chlorophenyl)-5-methyl-	EtOH	305(4.35)	23-1827-76

Compound	Solvent	λ_{max}(log ϵ)	Ref.
$C_{11}H_{10}ClN_3$			
4-Pyrimidinamine, 6-chloro-N-(phenyl-	pH 1	258(4.06),261(4.06)	39-1847-76C
	pH 13	246(3.97),278s(3.31)	39-1847-76C
$C_{11}H_{10}ClN_3O$			
2-Pyrimidinamine, 4-chloro-6-methoxy-5-phenyl-	MeOH	245(4.22),284(3.87)	94-0507-76
4-Pyrimidinamine, 2-chloro-6-methoxy-5-phenyl-	MeOH	210(4.48),258(3.93)	94-0507-76
$C_{11}H_{10}Cl_4$			
Cyclohexane, (2,3,4,5-tetrachloro-2,4-cyclopentadien-1-ylidene)-	C_6H_{12}	299(4.31),405(2.83)	24-3929-76
$C_{11}H_{10}CoO_2$			
Cobaltocenium, carboxy-	MeOH	268(4.49),310(3.15), 410(2.44)	101-0189-76I
$C_{11}H_{10}NO_2$			
Pyridinium, 1-(3,4-dihydroxyphenyl)-, bromide	MeOH	249(3.92),295(4.1)	2-0516-76
Quinolizinium, 4-(carboxymethyl)-, perchlorate	EtOH	213(4.43),234(4.35), 289(3.63),318(4.08), 325(4.07),332(4.28)	39-0341-76C
$C_{11}H_{10}N_2$			
5-Azacycl[3.2.2]azine, 2,6-dimethyl-	EtOH	249(4.62),291(3.75), 306(3.54),433(3.54)	44-0351-76
Benzenamine, 2-(2-pyridinyl)-	C_6H_{12}	275(4.27),305s(3.85)	18-2770-76
	MeOH	273(4.29),313s(3.85)	18-2770-76
	MeCN	273(4.32),308s(3.88)	18-2770-76
3-Indolizinecarbonitrile, 1,2-dimethyl-	MeOH	207s(4.10),214s(4.31), 216s(4.34),220(4.40), 250(4.45),258(4.59), 324s(3.87),375(3.96)	39-1908-76C
$C_{11}H_{10}N_2O$			
2,2'-Bipyridine, 5-methoxy-	EtOH	240(3.47),306(3.79)	64-0115-76B
Cycloprop[cd]azulen-2(1H)-one, 1-diazo-2a,2b,6a,6b-tetrahydro-2a-methyl-	EtOH	250(4.08),298s(3.66), 400(1.5)	88-3903-76
Furo[2,3-g]indazole, 1,7-dimethyl-	EtOH	273(3.94),291(3.73), 303(3.57)	32-1083-76
Furo[2,3-g]indazole, 2,7-dimethyl-	EtOH	245(3.81),285s(4.00), 303(4.12),315s(3.95)	32-1083-76
Imidazo[1,2-b]isoquinolin-5-one, 2,3-dihydro-	CHCl$_3$	227(4.37)	117-0287-76
3(2H)-Pyridazinone, 6-methyl-4-phenyl-	C_6H_{12}	221s(3.75),270(3.63), 318(3.74)	59-0837-76
	EtOH	224s(3.65),270(3.55), 317(3.75)	59-0837-76
	H_2SO_4	237(3.77),342(3.89), 362s(3.84)	59-0837-76
$C_{11}H_{10}N_2OS$			
4(1H)-Pyrimidinone, 2-(methylthio)-6-phenyl-	EtOH	254(4.42),311(3.85)	95-1094-76
4(1H)-Pyrimidinone, 2-[(phenylmethyl)-thio]-	EtOH	289(3.93)	95-1094-76

Compound	Solvent	$\lambda_{max}(\log \epsilon)$	Ref.
$C_{11}H_{10}N_2OS_2$ 1H-[1,2]Dithiolo[5,1-e][1,2,3]thiadiazole-7-S^{IV}, 1-(4-methylphenyl)-	C_6H_{12}	216(4.22),234(4.41), 248s(4.27),297(3.97), 329s(3.52),486(4.14)	39-0228-76C
$C_{11}H_{10}N_2O_2$ 3-Isoxazolecarbonitrile, 4,5-dihydro-5-(4-methylphenyl)-, 2-oxide	CCl_4	268(4.01)	104-1974-76
5H-Oxazolo[2,3-b]quinazolin-5-one, 2,3-dihydro-3-methyl-	MeOH	221(4.65),226s(--), 245(3.89),254(3.90), 263(3.85),308(3.59), 318s(--)	118-0469-76
Thymine, 1-phenyl-	pH 1 and 7 pH 13	273(4.07) 273(3.98)	4-1041-76 4-1041-76
$C_{11}H_{10}N_2O_2S$ 4-Oxazolecarboxamide, 2-methyl-5-(phenylthio)-	$CHCl_3$	282(3.82)	94-0948-76
$C_{11}H_{10}N_2O_3$ 1,2,4-Oxadiazole-5-methanol, 3-phenyl-, acetate	EtOH	238(4.10)	98-0876-76
Phenol, 4-(5-methyl-1,2,4-oxadiazol-3-yl)-, acetate	EtOH	242(4.22)	98-0876-76
1H-Pyrazole-3-carboxylic acid, 5-hydroxy-1-(phenylmethyl)-	MeOH	207(4.16),220(4.12), 254s(3.56)	24-0261-76
1H-Pyrazole-5-carboxylic acid, 3-hydroxy-1-(phenylmethyl)-	MeOH	225(3.92),266(3.57)	24-0268-76
4H-Pyrido[1,2-a]pyrimidine-9-carboxylic acid, 2-methyl-4-oxo-, methyl ester	1% MeOH	237(3.94),254(4.00), 335(3.99)	4-0797-76
4H-Pyrido[1,2-a]pyrimidin-4-one, 9-acetoxy-2-methyl-	EtOH	249(3.96),256(3.95), 318s(3.91),338(3.96)	4-0797-76
$C_{11}H_{10}N_2O_4S_2$ 5-Isothiazolecarboxamide, 3-[[(4-methylphenyl)sulfonyl]oxy]-	MeOH	230(4.44),262(3.89)	1-0781-76
$C_{11}H_{10}N_2O_7$ Benzeneacetic acid, α-acetyl-2,4-dinitro-	MeOH +20% DMSO +40% DMSO +60% DMSO	455(3.72) 468(3.79) 481(3.85) 500(3.92)	73-0590-76 73-0590-76 73-0590-76 73-0590-76
$C_{11}H_{10}N_2O_8$ Propanedioic acid, (2,4-dinitrophenyl-, dimethyl ester	MeOH +20% DMSO +40% DMSO +60% DMSO +80% DMSO	480(3.93) 488(4.10) 503(4.15) 510(4.21) 518(4.26)	73-0590-76 73-0590-76 73-0590-76 73-0590-76 73-0590-76
$C_{11}H_{10}N_2Se$ 4H-Benzo[6,7]cyclohepta[1,2-d][1,2,3]-selenadiazole, 5,6-dihydro-	C_6H_{12}	224(3.95),249(3.85), 318(2.95)	24-1650-76
$C_{11}H_{10}N_2Se_2$ 1H-[1,2]Diselenolo[5,1-e][1,2,3]selenadiazole-7-Se^{IV}, 5-methyl-1-phenyl-	C_6H_{12}	202(4.42),212(4.41), 250(4.47),280(4.19), 340(3.54),529(4.17)	39-0228-76C

Compound	Solvent	λ_{max}(log ϵ)	Ref.
$C_{11}H_{10}N_4$			
1H-Imidazole-4-carbonitrile, 5-amino-1-(phenylmethyl)-	EtOH	246(4.07)	2-0346-76
Propanedinitrile, 2-[(4-methylphenyl)-azo]-2-methyl-	CHCl$_3$	300(4.18),383(2.54)	126-1357-76
$C_{11}H_{10}N_4O$			
6H-1,2,4,5-Tetrazino[1,6-b]isoquinolin-6-one, 1,2-dihydro-3-methyl-	EtOH	233(4.32),316(4.18), 373(3.85),390(3.73)	95-0700-76
	EtOH	316(4.2),373(3.9), <u>391(3.8)</u>	4-1343-76
[1,2,4]Triazolo[1,5-b]isoquinolin-5(1H)-one, 1-(methylamino)-	EtOH	234(4.29),316(4.17), 375(3.84),392(3.71)	95-0700-76
$C_{11}H_{10}N_4O_2S$			
1-Thia-3,5,8,8b-tetraazaacenaphthylene-6-carboxylic acid, 2-methyl-, ethyl ester	EtOH	251(4.10),279s(3.67), 292s(3.51),391(3.96), 469(2.88),501(2.89), 536(2.79),588(2.48), 646(1.74)	1-0468-76
$C_{11}H_{10}N_4O_2Se$			
1-Selena-3,5,8,8b-tetraazaacenaphthyl-ene-6-carboxylic acid, 2-methyl-, ethyl ester	EtOH	236(4.16),248s(4.12), 279s(3.70),290s(3.49), 397(3.90),467(2.99), 504(2.96),542(2.88), 590(2.57)	1-0468-76
$C_{11}H_{10}N_4O_3$			
Isoxazole, 3,5-dimethyl-4-[(4-nitrophen-yl)azo]-	EtOH	246s(4.06),300(3.88), 329s(3.70),447(4.82)	48-0359-76
$C_{11}H_{10}N_4O_4$			
3-Buten-2-one, 3-methyl-4-(5-nitro-2H-benzotriazol-2-yl)-, N-oxide, (E)-	EtOH	267(3.85),314(4.40), 405(3.60)	22-0184-76
(Z)-	EtOH	304(4.30),395(3.72)	22-0184-76
3-Penten-2-one, 4-(5-nitro-2H-benzotria-zol-2-yl)-, N-oxide, (Z)-	EtOH	274(4.26),300s(4.12), 385(3.74)	22-0184-76
$C_{11}H_{10}N_4O_5$			
2-Butenoic acid, 3-(5-nitro-2H-benzo-triazol-2-yl)-, methyl ester, N-oxide, (Z)-	EtOH	272(4.20),295s(4.06), 387(3.68)	22-0184-76
$C_{11}H_{10}O$			
Azulene, 6-methoxy-	C_6H_{12}	282(4.77),287(4.80), 293(4.82),310(3.93), 330(3.51),338(3.58), 344(3.59),353(3.67), 366(3.48),514(2.28), 533(2.34),551(2.29), 579(3.34),605(1.93), 635(1.80)	44-1811-76
Bicyclo[5.3.1]undeca-1,3,5,9-tetraen-8-one	C_6H_{12}	220(4.19),252s(3.77), 318(3.46),390s(2.41)	35-8277-76
1H-Inden-1-one, 2,4-dimethyl-	hexane	237(4.58),242(4.63), 315(3.08),337(3.18), 341(3.10),395(2.88)	44-3540-76
	MeOH	412(2.81)	44-3540-76
	CHCl$_3$	410(2.82)	44-3540-76

Compound	Solvent	$\lambda_{max}(\log \epsilon)$	Ref.
1H-Inden-1-one, 2,6-dimethyl-	hexane	237(4.58),242(4.62), 316(2.85),329(2.75), 405(2.75)	44-3540-76
	MeOH	420(2.59)	44-3540-76
	CHCl$_3$	419(2.59)	44-3540-76
1-Naphthalenol, 5-methyl-	EtOH	223(4.62),299(3.82), 312(3.70),326(3.53)	95-1161-76
$C_{11}H_{10}O_2$			
1-Benzoxepin-5(2H)-one, 3-methyl-	n.s.g.	262(4.06),313(3.23)	33-0664-76
1H-Inden-1-one, 6-methoxy-2-methyl-	hexane	245(4.58),252(3.51), 288(2.95),300(2.88), 435(2.82)	44-3540-76
1,4-Methanonaphthalene-5,8-diol, 1,4-di-hydro-	ether	299(3.54),303(3.57)	39-0550-76C
Tetracyclo[5.4.0.0^{2,4}.0^{3,6}]undeca-1(7),7,10-triene-8,11-diol	EtOH	295(3.53)	39-0550-76C
$C_{11}H_{10}O_3$			
1H-2-Benzopyran-4-ol, acetate	MeCN	244(3.91),280(3.77)	1-0619-76
2H-1-Benzopyran-2-one, 7-hydroxy-4,5-dimethyl-	n.s.g.	245(4.23),250(4.27), 289(4.04)	2-0994-76
4H-1-Benzopyran-4-one, 3-acetyl-2,3-dihydro-	EtOH	220(3.97),255(3.73), 306(3.88),351(3.88)	39-2444-76C
Cyclopenta[c]pyran-4-carboxylic acid, 7-methyl-, methyl ester	hexane	222(4.10),260(4.13), 340(3.38),433(2.81)	32-0733-76
	HOAc	258(4.10),263(4.18), 336(3.52),425(2.98)	32-0733-76
	HOAc-HClO$_4$	286(3.56),345(3.42), 591s(3.56),630(3.75)	32-0733-76
$C_{11}H_{10}O_3S$			
1H-2-Benzothiopyran-3-carboxylic acid, 3,4-dihydro-4-oxo-, methyl ester	MeOH	260(4.01),368(3.73)	35-5581-76
$C_{11}H_{10}O_4$			
1H-2-Benzopyran-3-carboxylic acid, 1-hydroxy-, methyl ester	MeOH	227(3.95),233(3.90), 296(4.21)	35-5581-76
2H-1-Benzopyran-4-carboxylic acid, 3,4-dihydro-3-oxo-, methyl ester	MeOH	233(3.94),293(3.67)	35-3555-76
2H-1-Benzopyran-2-one, 5,7-dimethoxy-	EtOH	220(4.1),250(3.8), 260(3.8),330(4.2)	32-0681-76
2H-1-Benzopyran-2-one, 6,7-dimethoxy-	EtOH	232(4.2),251(3.8), 295(3.8),346(4.1)	32-0681-76
2H-1-Benzopyran-2-one, 7,8-dimethoxy-	EtOH	230(4.0),250s(3.6), 300(4.04)	32-0681-76
4H-1-Benzopyran-4-one, 5,6-dimethoxy-	MeOH	232(4.36),251s(4.02), 335(3.69)	4-0211-76
4H-1-Benzopyran-4-one, 5,7-dimethoxy-	MeOH	227(4.16),247(4.25), 254(4.25),287(3.91)	4-0211-76
4H-1-Benzopyran-4-one, 5,8-dimethoxy-	MeOH	221(4.35),256(4.12), 338(3.62)	4-0211-76
4H-1-Benzopyran-4-one, 6,7-dimethoxy-	MeOH	235(4.27),246s(3.98), 277(3.78),318(4.02)	4-0211-76
4H-1-Benzopyran-4-one, 6,8-dimethoxy-	MeOH	228s(4.20),237(4.20), 258(4.03),332(3.71)	4-0211-76
4H-1-Benzopyran-4-one, 7,8-dimethoxy-	MeOH	219(4.26),245s(4.29), 250(4.34),298(3.99)	4-0211-76
1H-Indene-2-carboxylic acid, 2,3-di-hydro-2-hydroxy-1-oxo-, methyl ester	MeOH	249(4.05),294(3.36)	35-5581-76

Compound	Solvent	$\lambda_{max}(\log \epsilon)$	Ref.
$C_{11}H_{10}O_4S$			
1-Benzothiopyran-8-carboxylic acid, 2,3-dihydro-4-oxo-, S-oxide, methyl ester	EtOH	250s(3.93),289(2.98), 298s(2.92),342(2.63)	23-0455-76
2-Propenoic acid, 2-[(carboxymethyl)-thio]-3-phenyl-	EtOH	281(3.74)	42-0038-76
$C_{11}H_{10}O_5S$			
2H-1-Benzopyran-2-one, 4-methyl-6-[(methylsulfonyl)oxy]-	EtOH	270(4.09),278s(3.90), 319(3.70)	80-1543-76
2H-1-Benzopyran-2-one, 4-methyl-7-[(methylsulfonyl)oxy]-	EtOH	278s(4.08),310(3.99)	80-1543-76
2H-1-Benzopyran-2-one, 5-methyl-7-[(methylsulfonyl)oxy]-	EtOH	292(4.03),315(3.85)	80-1543-76
$C_{11}H_{10}S$			
Benzo[b]cyclobuta[d]thiophene, 2a,7b-dihydro-1-methyl-	C_6H_{12}	217(4.26),255s(3.89), 257(3.92),263s(3.83), 297(3.29),305(3.22)	18-0825-76
Benzo[b]cyclobuta[d]thiophene, 2a,7b-dihydro-2a-methyl-	C_6H_{12}	218(4.24),223s(4.06), 254(3.95),258(3.96), 296(3.29),298(3.27), 305(3.23)	18-0825-76
$C_{11}H_{11}BrO$			
Bicyclo[3.2.0]hepta-2,4,7-trien-6-one, 4-bromo-7-(1,1-dimethylethyl)-	EtOH	225(4.22),244(4.31), 280(3.63),288(3.67), 299(3.52),387s(2.43)	33-0021-76
$C_{11}H_{11}BrO_5$			
3-Oxatricyclo[5.1.0.0^{2,4}]oct-5-ene-6,7-dicarboxylic acid, 5-bromo-, dimethyl ester, (1α,2α,4α,7α)-	MeOH	241(3.76)	24-2823-76
4H-Oxocin-5,6-dicarboxylic acid, 7-bromo-, dimethyl ester	MeOH	289s(2.97)	24-2823-76
$C_{11}H_{11}ClO_2$			
1,4-Pentanedione, 1-(4-chlorophenyl)-	EtOH	252(3.64)	12-0339-76
$C_{11}H_{11}Cl_2N_5O_3$			
Adenosine, 3'-deoxy-3'-C-(dichloromethylene)-	EtOH	217(4.27),258(4.10)	111-0489-76
$C_{11}H_{11}Cl_3O_4$			
3,5-Cyclohexadien-1-one, 2,2,4-trichloro-3,5-dimethoxy-6-propanoyl-	n.s.g.	283(4.13),337(3.31)	120-0101-76
$C_{11}H_{11}F_2NO_2$			
2-Propenoic acid, 3-[4-(dimethylamino)-phenyl]-2,3-difluoro-, (E)-	EtOH	319(4.36)	104-1920-76
$C_{11}H_{11}F_3$			
Styrene, α-ethyl-β,β,2-trifluoro-5-methyl-	EtOH	226(3.73),270(3.28), 275(3.28)	12-1435-76
$C_{11}H_{11}N$			
1H-Pyrrole, 2-methyl-5-phenyl-	EtOH	295(4.27)	23-1827-76
$C_{11}H_{11}NO$			
2H-1-Benzazepin-2-one, 1,5-dihydro-4-methyl-	EtOH	232(4.24),255s(3.90), 290(3.08)	94-1544-76

Compound	Solvent	$\lambda_{max}(\log \epsilon)$	Ref.
2-Cyclopropen-1-one, 2-(dimethylamino)-3-phenyl-	MeOH	276(4.31),282(4.31)	88-2405-76
Ethanone, 1-(2-methyl-1H-indol-3-yl)-	MeOH	213(4.64),240(4.26), 265(4.15),299(4.15)	23-1020-76
2-Propanone, 1-(1H-indol-2-yl)-	MeOH	273(3.90),279(3.88), 290(3.78)	24-3282-76
Quinoline, 8-ethoxy-	pH 1	251(4.59),276(2.94), 307(3.23),317(3.25), 354(3.20)	65-1108-76
	EtOH	241(4.55),303(3.45)	65-1108-76
$C_{11}H_{11}NOS$			
Ethanone, phenyl-2-thiazolidinylidene-	MeOH	245(4.04),342(4.47)	48-0298-76
5(2H)-Isoxazolethione, 2-ethyl-3-phenyl-	EtOH	251(4.14),348(4.20)	97-0270-76
Quinoline, 3-methoxy-2-(methylthio)-	hexane	217(4.73),250(4.42), 255(4.43),260(4.50), 287(4.50),298(3.72), 305(3.63),312(3.60), 319(3.83),327(3.91), 333(4.06),343(4.28)	12-1023-76
$C_{11}H_{11}NOS_2$			
Isothiazole, 4-(4-methoxyphenyl)-3-(methylthio)-	EtOH	243(4.02),289(4.07)	48-0127-76
Isothiazole, 5-(4-methoxyphenyl)-3-(methylthio)-	EtOH	290(4.53)	48-0127-76
4H-1,3-Thiazin-4-one, tetrahydro-6-(4-methylphenyl)-2-thioxo-	n.s.g.	261(4.16),312(4.28)	73-1388-76
$C_{11}H_{11}NO_2$			
Benzoic acid, 2-(3-cyanopropyl)-	MeOH	229(3.89),278(3.51)	73-2771-76
2H-Isoindole-1-carboxylic acid, 2-methyl-, methyl ester	EtOH	231(4.33),258(4.16), 264(4.22),336(4.19), 350(4.17)	32-0065-76
1,4-Methanonaphthalene, 1,2,3,4-tetrahydro-5-nitro-	hexane	275(3.71),302(3.18)	44-2006-76
1,4-Methanonaphthalene, 1,2,3,4-tetrahydro-6-nitro-	hexane	271(3.93)	44-2006-76
5(2H)-Oxazolone, 2,2-dimethyl-4-phenyl-	n.s.g.	262(4.14)	33-2149-76
5(4H)-Oxazolone, 4,4-dimethyl-2-phenyl-	n.s.g.	241(4.40)	33-2149-76
$C_{11}H_{11}NO_2S$			
3H-3-Benzazepine, 3-(methylsulfonyl)-	n.s.g.	244(4.50),315(3.0)	138-0519-76
Benzo[b]thiophene-3-propanoic acid, α-amino-, L-	pH 13	261(3.62),290(3.45), 299(3.46)	94-3149-76
4-Thiazolol, 5-(4-methoxyphenyl)-2-methyl-	EtOH	306(4.10)	118-0261-76
$C_{11}H_{11}NO_2S_2$			
4H-1,3-Thiazin-4-one, tetrahydro-6-(4-methoxyphenyl)-2-thioxo-	n.s.g.	258(4.13),313(4.20)	73-1388-76
$C_{11}H_{11}NO_3$			
7-Azabicyclo[4.2.2]deca-2,4,9-triene-7-carboxylic acid, 8-oxo-, methyl ester	EtOH	254(3.54),264s(3.24)	35-3636-76
Glycine, N-(1-oxo-3-phenyl-2-propenyl)-	MeOH-acid	273(4.42)	20-0647-76
$C_{11}H_{11}NO_4$			
Glycine, N-[3-(4-hydroxyphenyl)-1-oxo-2-propenyl]-, (E)-	MeOH-acid	227(4.19),294(4.37), 310(4.37)	20-0647-76

Compound	Solvent	$\lambda_{max}(\log \epsilon)$	Ref.
Glycine, N-[3-(4-hydroxyphenyl)-1-oxo-2-propenyl]-, (E)- (cont.)	MeOH-base	237(4.14),313s(4.21), 347(4.49)	20-0647-76
$C_{11}H_{11}NO_4S$			
Benzo[b]thiophene-3-propanoic acid, α-amino-, 1,1-dioxide, hydrochloride, (S)-	pH 1 pH 13	307(3.30) 260(3.81),304(3.20)	94-3149-76 94-3149-76
$C_{11}H_{11}NO_5$			
Glycine, N-[3-(3,4-dihydroxyphenyl)-1-oxo-2-propenyl]-, (E)-	MeOH-acid	240(4.13),295(4.18), 323(4.26)	20-0647-76
	MeOH-base	252(4.09),297s(3.92), 307(3.97),358(4.34)	20-0647-76
$C_{11}H_{11}NS$			
2-Propynethioamide, N,N-dimethyl-3-phenyl-	C_6H_{12}	269s(4.21),273s(4.23), 283(4.31),290s(4.20), 300(4.18),327(3.89), 446(2.23)	97-0355-76
Quinoline, 8-(ethylthio)-	pH 1	239(4.37),308(3.73), 315(3.75)	65-1108-76
	EtOH	253(4.40),337(3.65)	65-1108-76
Thiazole, 4,5-dimethyl-2-phenyl-	EtOH	303(4.08)	70-1353-76
$C_{11}H_{11}NS_2$			
Isothiazole, 4-(4-methylphenyl)-3-(methylthio)-	EtOH	243(4.02),289(4.07)	48-0127-76
Isothiazole, 5-(4-methylphenyl)-3-(methylthio)-	EtOH	279(4.44)	48-0127-76
2(5H)-Thiazolethione, 5,5-dimethyl-4-phenyl-	C_6H_{12}	232(3.85),309(4.26), 370(3.81)	24-0139-76
$C_{11}H_{11}N_2O$			
Pyridinium, 1-methyl-2-(2-pyridinyloxy)-, iodide	H_2O	278(4.10)	64-0122-76B
Pyrimidinium, 2,3-dihydro-1-methyl-2-oxo-3-phenyl-, chloride	pH 0.29	326(3.91)	23-2681-76
pseudobase	pH 9.18	235(3.85),335(1.66)	23-2681-76
Pyrimidinium, 2,3-dihydro-1-methyl-2-oxo-3-phenyl-, iodide (or isomer)	pH 0.29	225(4.15),325(3.95)	23-2681-76
pseudobase	pH 9.18	227(4.33),338(2.00)	23-2681-76
$C_{11}H_{11}N_3$			
Pyrrolo[1,2-a]pyrimidine-8-carbonitrile, 2,4,7-trimethyl-	MeOH	230(4.59),249(4.35), 257(4.36),290(3.82), 301(3.82),344(3.72)	24-2983-76
$C_{11}H_{11}N_3O$			
Acetamide, N-[4-(1H-pyrazol-1-yl)phenyl]-	EtOH	276(4.22)	22-0195-76
8-Azabicyclo[3.2.1]oct-2-ene-8-propanenitrile, 7-cyano-4-oxo-, exo	EtOH	222(2.95)	39-2334-76C
Isoxazole, 3,5-dimethyl-4-(phenylazo)-	EtOH	229(4.06),231s(4.04), 311s(4.17),319(4.17), 423(2.91)	48-0359-76
Pyridinium, 1-(4,6-dimethyl-2-pyrimidinyl)-3-hydroxy-, hydroxide, inner salt	EtOH	367(3.64)	39-1829-76B
4(1H)-Pyrimidinone, 6-[(phenylmethyl)-amino]-	pH 1	262(4.02)	39-1847-76C
	pH 13	258(3.68)	39-1847-76C

Compound	Solvent	$\lambda_{max}(\log \epsilon)$	Ref.
$C_{11}H_{11}N_3OS$			
Ethanone, 1-[4-[5-(methylthio)-1H-1,2,3-triazol-1-yl]phenyl]-	dioxan	258(4.26)	73-1551-76
Isoxazole, 3-methyl-5-(methylthio)-4-(phenylazo)-	EtOH	230s(4.02),235(4.02), 420s(3.98),266(3.87), 361(3.86),428s(3.19)	48-0359-76
5(2H)-Isoxazolethione, 3-methyl-4-[(2-methylphenyl)azo]-	EtOH	251(4.26),342(3.66), 467(4.27),504s(4.06)	48-0359-76
5(2H)-Isoxazolethione, 3-methyl-4-[(4-methylphenyl)azo]-	EtOH	251(4.25),342(3.69), 470(4.27),504s(4.12)	48-0359-76
4(1H)-Pyrimidinone, 6-amino-2-[(phenyl-methyl)thio]-	EtOH	277(3.95)	95-1094-76
$C_{11}H_{11}N_3O_2$			
Compound, m. 147-8°	EtOH	223(4.40),260(4.24), 311(3.50),377(4.28)	95-0160-76
3,5-Heptadienedinitrile, 2-(1-hydroxy-ethylidene)-6-[(methoxyimino)methyl]-	MeOH	437(4.68)	5-1799-76
	$CHCl_3$	362(4.48)	5-1799-76
3-Penten-2-one, 4-(2H-benzotriazol-2-yl)-, N-oxide, (E)-	EtOH	275s(3.98),319(4.17), 360s(3.65)	22-0184-76
(Z)-	EtOH	305(3.86),342s(3.62)	22-0184-76
4(1H)-Pyrimidinone, 6-amino-2-methoxy-5-phenyl-	MeOH	210(4.38),267(4.11)	94-0026-76
1,2,4-Triazine, 5,6-dimethyl-3-phenyl-, 1,4-dioxide	MeOH	263(3.76),317(3.29)	5-0153-76
$C_{11}H_{11}N_3O_2S$			
Benzoic acid, 4-(1,2,3-thiadiazol-5-yl-amino)-, ethyl ester	dioxan	271(3.84),336(4.47)	73-1182-76
$C_{11}H_{11}N_3O_4$			
Pyrido[2,3-d]pyrimidine-5-carboxylic acid, 1,2,3,4-tetrahydro-1,3-di-methyl-2,4-dioxo-, methyl ester	pH 1	248s(--),315(3.80)	44-1095-76
	pH 7	248s(--),315(3.83)	44-1095-76
	pH 11	248s(--),315(3.79)	44-1095-76
Pyrido[2,3-d]pyrimidine-2,4(1H,3H)-di-one, 6-acetoxy-1,3-dimethyl-	pH 1	317(3.80)	44-3027-76
	pH 7	317(3.80)	44-3027-76
	pH 11	269(4.06),322(3.48)	44-3027-76
$C_{11}H_{11}N_3O_5$			
Pyrido[2,3-d]pyrimidine-5-carboxylic acid, 1,2,3,4,7,8-hexahydro-1,3-di-methyl-2,4,7-trioxo-, methyl ester	pH 1	262(4.05),278(3.99), 315(4.15)	44-1095-76
	pH 7	273(4.08),313(4.29), 323(4.28)	44-1095-76
	pH 11	273(4.08),312(4.29), 322(4.28)	44-1095-76
$C_{11}H_{11}N_5O_2$			
1H-Pyrazolo[3,4-d]pyrimidine-4-acetic acid, α-cyano-1-methyl-, ethyl ester	EtOH	240(3.88),325(4.45), 336(4.50)	95-1352-76
$C_{11}H_{12}$			
Tricyclo[3.3.1.0^{2,8}]nona-3,6-diene, 9-ethylidene-	C_6H_{12}	230(3.70)	88-1565-76
$C_{11}H_{12}BrN_3$			
Benzenamine, 4-(4-bromo-3,5-dimethyl-1H-pyrazol-1-yl)-	EtOH	258(4.22),351(3.20)	22-0839-76
Pyrido[2,3-b]pyrazine, 7-bromo-2,3-di-ethenyl-1,2,3,4-tetrahydro-	EtOH	272(3.70),340(3.93)	22-0251-76

Compound	Solvent	λ_{max}(log ϵ)	Ref.
$C_{11}H_{12}Br_2N_2O_5$ Uridine, 2'-deoxy-5-(2,2-dibromoethen-yl)-	MeOH	235(4.07),297(4.03)	88-2427-76
$C_{11}H_{12}ClNO$ 2-Pentanone, 4-[(3-chlorophenyl)imino]-	EtOH	208(4.33),230(3.88), 326(4.34)	40-0451-76
2-Pentanone, 4-[(4-chlorophenyl)imino]-	EtOH	212(3.87),232(3.86), 326(4.31)	40-0451-76
$C_{11}H_{12}ClNO_5S$ 2H-Pyran-3-carboxylic acid, 6-chloro-4-[[(ethylthio)carbonyl]amino]-2-oxo-, ethyl ester	MeCN	347.5(4.23)	39-2462-76C
4H,5H-Pyrano[3,4-e]-1,3-oxazine-4,5-di-one, 7-chloro-2-ethoxy-2-(ethylthio)-2,3-dihydro-	MeCN	312(4.11)	39-2462-76C
$C_{11}H_{12}ClNO_6$ Carbonimidic acid, [(6-chloro-4-hydroxy-2-oxo-2H-pyran-3-yl)carbonyl]-, diethyl ester	MeCN	269(3.99),337(4.14)	39-2462-76C
$C_{11}H_{12}ClN_2O_2$ Methanaminium, N-[3-chloro-3-(4-nitro-phenyl)-2-propenylidene]-N-methyl-, perchlorate	MeOH	264s(4.06),299(4.21)	48-0731-76
$C_{11}H_{12}ClN_5O_3$ Hydrazinecarboxamide, 2-[3-[(2-chloro-phenyl)amino]-2-(hydroxyimino)-1-methyl-3-oxopropylidene]-	EtOH	263(4.36)	42-0454-76
Hydrazinecarboxamide, 2-[3-[(3-chloro-phenyl)amino]-2-(hydroxyimino)-1-methyl-3-oxopropylidene]-	EtOH	264(4.37)	42-0454-76
Hydrazinecarboxamide, 2-[3-[(4-chloro-phenyl)amino]-2-(hydroxyimino)-1-methyl-3-oxopropylidene]-	EtOH	264(4.40)	42-0454-76
$C_{11}H_{12}Cl_2N_4O_3$ Acetamide, N-[2-(4-azidophenyl)-2-hy-droxy-1-(hydroxymethyl)ethyl]-2,2-dichloro-, [R-(R*,R*)]-	MeOH	253(4.21),288(3.40)	130-0025-76
$C_{11}H_{12}Cl_2O$ 2-Cyclopropen-1-one, 2,3-bis(2-chloro-1-methyl-1-propenyl)-	CH_2Cl_2	293(4.43)	44-2258-76
$C_{11}H_{12}Cl_3$ Cyclopropenylium, chlorobis(2-chloro-1-methyl-1-propenyl)-, hexachloro-antimonate	CH_2Cl_2	292(4.48)	44-2258-76
$C_{11}H_{12}Cl_4O_3$ Ethanone, 1-(1,4,5,6-tetrachloro-7,7-di-methoxybicyclo[2.2.1]hept-5-en-2-yl)-, endo	benzene	288(1.43)	44-2943-76
$C_{11}H_{12}FN_5O_6$ Uridine, 5'-azido-5'-deoxy-4'-C-fluoro-2',3'-O-(methoxymethylene)-	MeOH	256(3.99)	44-3010-76

Compound	Solvent	λ_{max} (log ϵ)	Ref.
$C_{11}H_{12}FeO$			
Ferrocene, (hydroxymethyl)-	50% MeOH	440(2.00)	104-2433-76
$C_{11}H_{12}FeO_2$			
Iron, [1,4-butanediyl(η^5-2,4-cyclopenta-dien-1-ylidene)]dicarbonyl-	hexane	251s(3.75),276s(3.59), 348(3.17)	24-1429-76
$C_{11}H_{12}N$			
Isoquinolinium, 1,2-dimethyl-, perchlorate	EtOH	325(3.68),330(3.65)	103-0683-76
methylene base	EtOH-NaOH	370(3.84)	103-0683-76
Quinolinium, 1,2-dimethyl-, perchlorate	EtOH	315(3.95)	103-0683-76
methylene base	EtOH-NaOH	390(3.56)	103-0683-76
$C_{11}H_{12}N_2O$			
1H-Indole-2-carboxamide, N,N-dimethyl-	MeOH	293(4.24)	44-3050-76
1(2H)-Isoquinolinone, 3-(ethylamino)-	EtOH	229(4.39),246(3.98), 292(4.25),302(4.32), 374(3.71)	95-0154-76
1H-Pyrazole, 3-methoxy-1-(phenylmethyl)-	MeOH	228(3.91)	24-0268-76
1H-Pyrazolium, 4,5-dihydro-3-hydroxy-1-[(4-methylphenyl)methylene]-, hydroxide, inner salt	MeOH	343(4.4)	48-0946-76
4H-Pyrazol-4-ol, 3,5-dimethyl-4-phenyl- (plus shoulder not listed)	EtOH	212(3.60)	44-2874-76
4H-Pyrazol-4-one, 1,5-dihydro-3,5-dimethyl-5-phenyl-	EtOH	207(4.00),331(3.71)	44-2874-76
	EtOH-NaOH	211(4.23),350(3.80), 366(3.85)	44-2874-76
Pyrrolo[1,2-c]pyrimidine-5-carboxaldehyde, 1,3,6-trimethyl-	EtOH	232(4.39),261s(3.60), 338(4.16)	44-0351-76
Pyrrolo[1,2-c]pyrimidine-5-carboxaldehyde, 3,6,7-trimethyl-	EtOH	240(4.41),264s(3.71), 274s(3.37),340(4.10)	44-0351-76
Pyrrolo[1,2-c]pyrimidine-7-carboxaldehyde, 1,3,6-trimethyl-	EtOH	228(4.26),246(4.06), 253s(4.03),260s(3.99), 355s(4.23),365(4.29)	44-0351-76
4-Quinolinamine, N,N-dimethyl-, 1-oxide, hydrochloride	MeOH	223(4.55),250(4.03), 369(4.13)	95-0725-76
$C_{11}H_{12}N_2O_2$			
5H-Indazol-5-one, 2-acetyl-2,3-dihydro-3,3-dimethyl-	isoPrOH	248(3.42),373(4.49)	4-0033-76
3H-Indol-3-one, 1,2-dihydro-2-[(1-methylethyl)imino]-, N-oxide	MeOH	223(4.01),270(4.20), 347(3.98),463(3.83)	24-0200-76
1,2,4-Oxadiazole, 3-(1-methylethoxy)-5-phenyl-	MeOH	255(4.14)	18-3607-76
1H-Pyrazolium, 4,5-dihydro-3-hydroxy-1-[(3-methoxyphenyl)methylene]-, hydroxide, inner salt	MeOH	347(3.7)	48-0946-76
1H-Pyrazolium, 4,5-dihydro-3-hydroxy-1-[(4-methoxyphenyl)methylene]-, hydroxide, inner salt	MeOH	350(4.4)	48-0946-76
$C_{11}H_{12}N_2O_2S$			
1H-Pyrazole-3-carboxylic acid, 1-methyl-4-(2-thienyl)-, ethyl ester	MeOH	217(4.14),274(3.91), 278s(3.90)	5-1531-76
1H-Pyrazole-4-carboxylic acid, 1-methyl-3-(2-thienyl)-, ethyl ester	MeOH	218(4.16),253s(3.86), 263s(3.90),283(3.99)	5-1531-76
1H-Pyrazole-4-carboxylic acid, 1-methyl-5-(2-thienyl)-, ethyl ester	MeOH	225(4.12),236s(4.10), 268s(3.73)	5-1531-76
1H-Pyrazole-5-carboxylic acid, 1-methyl-4-(2-thienyl)-, ethyl ester	MeOH	230(4.08),295(3.86)	5-1531-76

Compound	Solvent	$\lambda_{max}(\log \epsilon)$	Ref.
$C_{11}H_{12}N_2O_3$			
Benzeneacetic acid, α-(acetylhydra- zono)-, methyl ester, anti	MeOH	217(3.93),265(4.09)	5-2206-76
syn	MeOH	223(4.03),292(4.13)	5-2206-76
$C_{11}H_{12}N_2O_3S$			
4-Thiazolecarboxamide, 2-[5-(ethoxy- methyl)-2-furanyl]-	EtOH	231(4.18),313(4.26)	44-4074-76
$C_{11}H_{12}N_2O_3S_3$			
Propanoic acid, 2-cyano-3-(3-ethyl- 4-oxo-2-thioxo-5-thiazolidinyli- dene)-3-(methylthio)-, methyl ester	EtOH	242(4.52),282(4.17), 315(4.12),475(4.36)	94-1671-76
$C_{11}H_{12}N_2O_5$			
3aH-Oxireno[e]indazole-3,3a-dicarbox- ylic acid, 1a,6,6a,6b-tetrahydro-, dimethyl ester, (1aα,3aα,6aα,6bα)-	MeOH	322(2.26)	24-2823-76
Uridine, 2'-deoxy-5-ethynyl-	MeOH	225(3.98),287(4.04)	88-2427-76
$C_{11}H_{12}N_2O_6$			
2,2'-Anhydro-1-(3'-O-acetyl-β-D-arabino- furanosyl)uracil	MeOH	224(3.99),251(3.90)	44-2995-76
$C_{11}H_{12}N_2S$			
3-Quinolinamine, 4-methyl-2-(methyl- thio)-	EtOH	306(3.49),350(3.92), 360(4.56),385s(3.72), 393s(3.60)	48-0039-76
$C_{11}H_{12}N_3O$			
Pyridinium, 1-(4,6-dimethyl-2-pyrimidin- yl)-3-hydroxy-, chloride	H_2O	205(4.20),223(4.30), 240(4.28)	39-2296-76C
$C_{11}H_{12}N_4$			
Butanenitrile, 3-imino-2-[(2-methyl- phenyl)hydrazono]-	EtOH	240(1.58),350(4.31)	33-0551-76
Butanenitrile, 3-imino-2-[(3-methyl- phenyl)hydrazono]-	EtOH	238(1.68),350(4.34)	33-0551-76
1(2H)-Phthalazinone, (1-methylethyli- dene)hydrazone	MeOH	229(3.83),269(4.11), 278(4.28),345(3.98)	94-1068-76
$C_{11}H_{12}N_4O$			
Furo[2,3-b]indole, 3a-azido-8a-methyl-	EtOH	247(3.88),307(3.24)	88-2347-76
2,4-Pentadienenitrile, 5-[(2-cyano- 1-methylethenyl)amino]-2-[(meth- oxyimino)methyl]-, (E,E,E,Z)-	MeOH	403(4.57)	5-1799-76
4(1H)-Pyrimidinone, 5-amino-6-[(phenyl- methyl)amino]-	pH 1	262(4.05)	39-1847-76C
	pH 13	236s(3.65)	39-1847-76C
Urea, (6,7-dimethyl-2-quinoxalinyl)-	pH 1-8	348(4.11),356(4.08)	5-1276-76
	pH 13	375(4.03)	5-1276-76
	6M HCl	367(4.23)	5-1276-76
	H_2SO_4	408(4.30)	5-1276-76
$C_{11}H_{12}N_4OS$			
Hydrazinecarbothioamide, N,N-dimethyl- 2-(3-oxo-3H-indol-2-ylidene)-	EtOH	242(4.27),265(4.23), 337(4.06),460(3.61), 500(3.61)	103-0076-76
Hydrazinecarbothioamide, N,1-dimethyl- 2-(3-oxo-3H-indol-2-ylidene)-	EtOH	261(4.47),336(4.06), 476(3.69)	103-0076-76

Compound	Solvent	$\lambda_{max}(\log \epsilon)$	Ref.
Hydrazinecarboximidothioic acid, 2-methyl-2-(3-oxo-3H-indol-2-yl)-, methyl ester	EtOH	268(4.69),335(4.06), 410(3.48),478(3.22)	103-0076-76
1H-Imidazo[1,2-a][1,3,5]benzotriazepine-5(6H)-thione, 2,3-dihydro-8-methoxy-	DMSO	276(4.46),313(4.36)	78-0057-76
1H-Imidazole-4-carboxamide, 5-amino-2,3-dihydro-1-(phenylmethyl)-2-thioxo-	EtOH	276s(--),308(4.29)	2-0351-76
1,2,4-Thiadiazol-3(2H)-imine, 5-ethoxy-2-(iminophenylmethyl)-	H_2O	278(4.08)	18-3165-76
1,2,4-Triazin-5(2H)-one, 6-(2-aminophenyl)-2-methyl-3-(methylthio)-	EtOH	204(4.36),237(4.44), 294s(3.95),364(3.50)	78-1735-76
$C_{11}H_{12}N_4O_2$ Acetic acid, 2-(2-amino-1,2-dihydro-1-oxo-3-isoquinolinyl)hydrazide	EtOH	230(4.37),298(4.12), 356(3.61)	95-0700-76
Benzenamine, 2-(3,5-dimethyl-1H-pyrazol-1-yl)-3-nitro-	EtOH	235(4.27),360(3.28)	22-0839-76
Benzenamine, 4-(3,5-dimethyl-1H-pyrazol-1-yl)-3-nitro-	EtOH	246(4.37),365(3.04)	22-0839-76
1H-Purine-2,6-dione, 3,7-dihydro-3,7-di-2-propenyl-	pH 1	269(4.13)	18-3224-76
	pH 7	273(3.96)	18-3224-76
	pH 13	275(4.00)	18-3224-76
$C_{11}H_{12}N_4O_3$ Benzenamine, 4-[(4,6-dimethoxy-1,3,5-triazin-2-yl)oxy]-	C_6H_{12}	<u>230(4.1),287(3.5)</u>	144-0325-76
Benzenamine, 2-(3,5-dimethyl-1H-pyrazol-1-yl)-N-hydroxy-3-nitro-	EtOH	232(4.27),340(3.30)	22-0839-76
Benzenamine, 2-(3,5-dimethyl-1H-pyrazol-1-yl)-N-hydroxy-5-nitro-	EtOH	255(4.23),242(4.18), 300(3.73),360(3.48)	22-0839-76
Benzenamine, 4-(3,5-dimethyl-1H-pyrazol-1-yl)-N-hydroxy-3-nitro-	EtOH	248(4.33),353(3.28)	22-0839-76
$C_{11}H_{12}N_4O_4S$ 8,2'-Anhydro-6-(methylthio)-8-oxy-9-β-D-arabinofuranosylpurine	pH 1	228(4.08),293(4.28), 318s(3.04)	94-0672-76
	pH 7	228(4.08),293(4.31)	94-0672-76
	pH 13	227(4.13),295(4.24), 315s(3.75)	94-0672-76
8,3'-Anhydro-6-(methylthio)-8-oxy-9-β-D-xylofuranosylpurine	pH 2	231(4.08),295(4.28), 320s(3.28)	94-0672-76
	pH 7	230(4.09),295(4.31)	94-0672-76
	pH 12	232(4.05),296(4.32)	94-0672-76
$C_{11}H_{12}N_4O_5$ 4(3H)-Pteridinone, 3-β-D-ribofuranosyl-	pH -3.0	227s(3.87),295s(3.91), 303(3.99)	24-3208-76
	pH 3.0	236(4.07),268(3.69), 310(3.81)	24-3208-76
	MeOH	238(4.04),270(3.68), 310(3.77)	24-3208-76
$C_{11}H_{12}N_4O_6$ 4-Pyrimidineacetic acid, 2-(acetylamino)-α-(acetoxyimino)-1,6-dihydro-6-oxo-, methyl ester	EtOH	242(4.42),316(3.67)	44-2850-76
D-Ribose, 5-deoxy-5-(3,4-dihydro-2,4-dioxo-8(2H)-pteridinyl)-	pH -4.0	242(4.06),273(3.37), 350(4.01)	24-3175-76
	pH 6.0	257(4.18),275s(3.88), 400(3.96)	24-3175-76

Compound	Solvent	$\lambda_{max}(\log \epsilon)$	Ref.
D-Ribose, 5-deoxy-5-(3,4-dihydro-2,4-dioxo-8(2H)-pteridinyl)- (cont.)	pH 12.0	227(4.24),279(4.06), 306(3.93),400(2.94)	24-3175-76
s-Triazolo[1,5-a]pyridine, 2-β-D-ribofuranosyl-8-nitro-	H_2O	235(4.14),328(3.74)	44-3124-76
$C_{11}H_{12}N_4S$			
3H-[1,3,5]Triazepino[1,2-a]benzimidazole, 4,5-dihydro-2-(methylthio)-hydriodide	EtOH	214(4.36),242s(4.00), 280s(3.95),313(4.42)	78-0839-76
	EtOH	217(4.56),242(4.00), 312(4.44)	78-0839-76
$C_{11}H_{12}N_4S_2$			
1,2,4-Thiadiazol-3(2H)-imine, 2-[imino-(4-methylphenyl)methyl]-5-(methylthio)-	H_2O	263(4.54)	18-3165-76
1,2,4-Thiadiazol-3(2H)-imine, 2-(1-imino-2-phenylethyl)-5-(methylthio)-	H_2O	247(4.45)	18-3165-76
$C_{11}H_{12}N_5O_8P$			
9H-Purine-8-carboxylic acid, 6-amino-9-(3,5-O-phosphinico-β-D-ribofuranosyl)-	H_2O	262(4.11)	94-2052-76
$C_{11}H_{12}O$			
Bicyclo[3.2.0]hepta-2,4,7-trien-6-one, 7-(1,1-dimethylethyl)-	EtOH	233(4.33),275(3.65), 282(3.67),293(3.51), 375s(2.00)	33-0021-76
2-Buten-1-one, 3-methyl-1-phenyl-	EtOH	259(4.29),346(2.19)	44-3067-76
2,4,6-Cyclooctatrien-1-one, 8-(2-propenyl)-	C_6H_{12}	215(3.99),246(3.77), 282(3.56),361(2.60)	39-1838-76C
2,3,8-Methenocyclonona-4,6-dien-1-one, 2-methyl-	EtOH	260(3.14)	88-3903-76
1-Penten-3-one, 1-phenyl-	C_6H_{12}	279(4.38)	23-2127-76
$C_{11}H_{12}OS$			
3-Buten-2-one, 4-[4-(methylthio)phenyl]-, (E)-	MeOH	244(3.99),336(4.39)	87-1195-76
2-Thiophenecarboxaldehyde, 4-(1-cyclohexen-1-yl)-	isooctane	245(4.33),318(3.79)	44-1320-76
2-Thiophenecarboxaldehyde, 5-(1-cyclohexen-1-yl)-	MeCN	226(3.87),331(4.21)	44-1320-76
$C_{11}H_{12}O_2$			
1,3-Butanedione, 2-methyl-1-phenyl-	MeOH	246(4.03),285(3.23)	23-1197-76
1H-Dicyclopenta[b,e]pyran-8(5H)-one, 2,3,6,7-tetrahydro-	MeOH	256.5(4.15)	5-1689-76
1,2-Pentanedione, 1-phenyl-	MeCN	250(3.91),388(1.56)	35-8125-76
2-Propenoic acid, 2-methyl-, 4-methylphenyl ester	dioxan	266(3.13),273(3.01)	126-3481-76
$C_{11}H_{12}O_2S$			
2-Propenoic acid, 2-(ethylthio)-3-phenyl-	EtOH	298(4.08)	2-0547-76
2-Propenoic acid, 2-(methylthio)-3-phenyl-, methyl ester	75% dioxan	270(4.0),320s(3.8)	104-0149-76
$C_{11}H_{12}O_3$			
4H-1-Benzopyran-4-one, 2,3-dihydro-7-methoxy-5-methyl-	EtOH	273(4.17),309(3.70)	39-0499-76C
Carbonic acid, 1,3,5,7-cyclooctatetraen-1-yl ethyl ester	C_6H_{12}	279(3.20)	39-1838-76C

Compound	Solvent	$\lambda_{max}(\log \epsilon)$	Ref.
$C_{11}H_{12}O_5S$ 4H-1-Benzopyran-4-one, 2,3-dihydro- 2-methyl-7-[(methylsulfonyl)oxy]-	EtOH	256(3.99),318(3.61)	80-1543-76
$C_{11}H_{13}BrN_5O_6P$ Adenosine, 8-bromo-2-methyl-, cyclic 3',5'-(hydrogen phosphate)	pH 1 pH 11	262(4.26) 267(4.23)	87-0419-76 87-0419-76
$C_{11}H_{13}BrO_4S$ Acetic acid, bromo-, 2-[(4-methylphen- yl)sulfonyl]ethyl ester	EtOH	225(4.10)	39-1718-76C
$C_{11}H_{13}ClN$ Methanaminium, N-(3-chloro-3-phenyl-2- propenylidene)-N-methyl-, perchlorate	MeOH	249(4.02),317(3.71)	48-0731-76
$C_{11}H_{13}ClN_2O$ 4-Quinolinamine, N,N-dimethyl-, 1-oxide, hydrochloride	MeOH	223(4.55),250(4.03), 369(4.13)	95-0725-76
$C_{11}H_{13}ClN_4$ 3H-1,4-Benzodiazepine-2,5-diamine, 7-chloro-N,N'-dimethyl-	MeOH	218s(4.27),272(4.16), 319(3.46)	44-2724-76
1(2H)-Phthalazinone, 4-chloro-2-methyl-, dimethylhydrazone, (Z)-	dioxan	275(4.00),355(3.54)	103-0342-76
$C_{11}H_{13}ClO_2S$ Benzenepropanoic acid, β-chloro-α-(meth- ylthio)-, methyl ester	75% dioxan	none above 220 nm	104-0149-76
$C_{11}H_{13}Cl_4N$ 1-Propanamine, N,N,2-trimethyl- 1-(2,3,4,5-tetrachloro-2,4- cyclopentadien-1-ylidene)-	C_6H_{12}	240s(3.56),389(4.25)	24-3939-76
$C_{11}H_{13}FO_2$ 1,6(2H,7H)-Naphthalenedione, 7-fluoro- 3,4,8,8a-tetrahydro-8a-methyl-	C_6H_{12}	243(4.03)	78-2099-76
$C_{11}H_{13}IN_4S$ 3H-[1,3,5]Triazepino[1,2-a]benzimida- zole, 4,5-dihydro-2-(methylthio)-, hydriodide	EtOH	217(4.56),242(4.00), 312(4.44)	78-0839-76
$C_{11}H_{13}NO$ Benzeneacetonitrile, 4-methoxy-2,6-di- methyl-	EtOH	274(3.05),282(3.06)	39-1466-76C
Dicyclopenta[b,d]pyridin-5(1H)-one, 2,3,4,6,7,8-hexahydro-	n.s.g.	235(3.63),3?3(3.86)	104-0681-76
2-Furanamine, tetrahydro-N-(phenylmeth- ylene)-	MeOH	247(4.08)	35-2605-76
2-Pentanone, 4-(phenylimino)-	EtOH	204(4.03),228(3.87), 324(4.31)	40-0451-76
$C_{11}H_{13}NOS$ 2H-Indol-2-one, 3-(ethylthio)-1,3-di- hydro-3-methyl-	EtOH	253(3.85),287(3.11)	39-0745-76C
$C_{11}H_{13}NO_2$ 7-Azabicyclo[4.2.2]deca-2,4,9-triene- 7-carboxylic acid, methyl ester	EtOH	264(3.49)	35-3636-76

Compound	Solvent	$\lambda_{max}(\log \epsilon)$	Ref.
Benzeneacetaldehyde, 4-(dimethylamino)- α-(hydroxymethylene)-	pH 1	250(4.21)	44-0294-76
	pH 6.8	272(4.35)	44-0294-76
	pH 13	272(4.37)	44-0294-76
	90% H_2SO_4	243(4.39),249(4.39), 255(4.39),260s(4.31), 280s(4.14)	44-0294-76
2-Pentanone, 4-[(3-hydroxyphenyl)imino]-	EtOH	211(4.19),233(3.73), 326(4.26)	40-0451-76
2-Pentanone, 4-[(4-hydroxyphenyl)imino]-	EtOH	203(4.06),230(3.76), 320(4.26)	40-0451-76
$C_{11}H_{13}NO_2S$ 1H-Indole, 2-(ethylsulfonyl)-3-methyl-	EtOH	222(4.43),279(3.94), 296s(3.70),307s(3.43)	39-0745-76C
$C_{11}H_{13}NO_3$ Benzaldehyde, 2,3,4,5-tetramethyl- 6-nitro-	EtOH	215(4.06),265(4.26), 304(3.65)	104-1057-76
Benzoic acid, 2-[(ethoxymethylene)- amino]-	n.s.g.	227(4.62),253(4.14), 312(3.72),338(3.62)	44-2732-76
3,5-Cyclohexadiene-1,2-dione, 4-methyl- 5-morpholino-	EtOH	496(3.49)	18-2333-76
Gentiananine	n.s.g.	219(3.69),265(3.76)	105-0116-76
1-Isoquinolinecarboxylic acid, 1,2,3,4- tetrahydro-6-hydroxy-1-methyl-	EtOH	277(3.39)	95-1031-76
1-Propanone, 2,2-dimethyl-1-(4-nitro- phenyl)-	hexane	262(4.09)	12-2621-76
5H-2-Pyrindine- -carboxylic acid, 6,7-	EtOH	284(4.27)	102-0576B-76
dihydro-6-hydroxy-7-methyl-, methyl	EtOH-acid	284(4.27)	102-0576B-76
ester, cis (cantleyine)	EtOH-base	284(4.27)	102-0576B-76
$C_{11}H_{13}NO_3S$ 2H-Indol-2-one, 3-(ethylsulfonyl)-1,3- dihydro-3-methyl-	EtOH	256(3.70),266s(3.57), 295(3.15)	39-0745-76C
$C_{11}H_{13}NO_4$ Formamide, N-[2-(4-methoxy-1,3-benzodi- oxol-5-yl)ethyl]-	MeOH	232s(3.77),278(3.06), 283s(3.04)	12-2003-76
Furo[2,3-b]pyridine-4-carboxylic acid, 2,3,6,7-tetrahydro-2,2-dimethyl- 6-oxo-, methyl ester	MeOH	207(4.21),235s(--), 339(3.70),370s(--)	24-1269-76
1-Isoquinolinecarboxylic acid, 1,2,3,4- tetrahydro-4,6-dihydroxy-1-methyl-	EtOH	278(3.19)	95-1031-76
$C_{11}H_{13}NO_5S$ Benzenemethanol, α-(1-methylethylidene)- 4-nitro-, methanesulfonate	MeOH	298(3.96)	12-2631-76
$C_{11}H_{13}NO_7$ β-D-Ribofuranoside, 4-nitrophenyl	EtOH	302(4.06)	94-0394-76
$C_{11}H_{13}NO_9$ Methanediol, [4-(acetyloxy)-4,5-dihydro- 5-nitro-2-furanyl]-, diacetate	HOAc	284(2.01)	103-0502-76
$C_{11}H_{13}N_2S$ 1H-Imidazolium, 3-methyl-4-(methylthio)- 1-phenyl-, iodide	EtOH	221-266(4.360-3.599)	28-0143-76B
	CH_2Cl_2	242(4.276)	28-0143-76B

Compound	Solvent	$\lambda_{max}(\log \epsilon)$	Ref.
$C_{11}H_{13}N_3$			
Benzenamine, 2-(3,5-dimethyl-1H-pyrazol-1-yl)-	EtOH	231(4.06),298(3.33)	22-0839-76
Benzenamine, 4-(3,5-dimethyl-1H-pyrazol-1-yl)-	EtOH	256(4.17)	22-0195-76 +22-0839-76
Ethanimidamide, N'-(4-cyanophenyl)-N,N-dimethyl-	neutral	285(4.17)	39-0211-76B
	cation	262(4.16)	39-0211-76B
1H-Pyrazol-1-amine, N,N-dimethyl-3-phenyl-	MeOH	252(4.23)	94-3001-76
Pyrido[2,3-b]pyrazine, 2,3-diethenyl-1,2,3,4-tetrahydro-	EtOH	263(3.64),325(3.87)	22-0251-76
1-Pyrrolidineacetonitrile, 2-(3-pyridinyl)-	EtOH	260(3.27)	87-1168-76
$C_{11}H_{13}N_3O$			
Benzenamine, 2-(3,5-dimethyl-1H-pyrazol-1-yl)-N-hydroxy-	EtOH	232(4.24),290(3.43)	22-0839-76
5H-1,3,4-Benzotriazepin-5-one, 2-ethyl-1,4-dihydro-4-methyl-	EtOH	229(5.24),251s(4.98), 285s(4.41)	44-2732-76
1H-Benzotriazole, 1-(cyclopentyloxy)-	EtOH	264(3.78),283(3.71)	4-0509-76
$C_{11}H_{13}N_3OS$			
1H-1,2,3-Triazole, 1-(4-ethoxyphenyl)-5-(methylthio)-	dioxan	248(4.15)	73-1551-76
$C_{11}H_{13}N_3O_2S$			
Acetic acid, [4-amino-3-(1-propenyl)-2(3H)-thiazolidinylidene]cyano-, ethyl ester	MeOH	229(4.06),302(3.79), 349(3.95)	48-0343-76
1H-Furo[2,3-g]indazole-1-thiocarbox-amide, 3a,4,5,8b-tetrahydro-8b-hy-droxy-7-methyl-	EtOH	275(3.90)	32-1083-76
$C_{11}H_{13}N_3O_3$			
1H-Furo[2,3-g]indazole-1-carboxamide, 3a,4,5,8b-tetrahydro-8b-hydroxy-7-methyl-	EtOH	287(4.00)	32-1083-76
1H-Pyrazolo[4,3-b]pyridine-6-carboxylic acid, 4,7-dihydro-1,3-dimethyl-7-oxo-, ethyl ester	EtOH	227(4.41),303(4.13)	39-0507-76C
2H-Pyrazolo[4,3-b]pyridine-6-carboxylic acid, 4,7-dihydro-2,3-dimethyl-7-oxo-, ethyl ester	EtOH	233(4.32),313(4.05)	39-0507-76C
$C_{11}H_{13}N_3O_4$			
Methanone, 4-morpholinyl(4-nitrophenyl)-, oxime, (E)-	n.s.g.	264(4.10),340(3.08)	77-0862-76
(Z)-	n.s.g.	267(3.93),340(3.30)	77-0862-76
s-Triazolo[4,3-a]pyridine, 3-α-D-ribofuranosyl-	H_2O	261(3.56),272(3.65), 280(3.58)	44-3124-76
β-	H_2O	265(3.64),270(3.67), 280(3.60)	44-3124-76
$C_{11}H_{13}N_3O_5$			
Cytosine, 2,2'-anhydro-1-(3'-O-acetyl-β-D-arabinofuranosyl)-, hydrochloride	MeOH-acid	232(3.98),263(4.00)	87-0654-76
	EtOH	234(4.00),265(4.04)	44-2995-76
Imidazo[1,2-c]pyrimidin-5(1H)-one, 1-β-D-arabinofuranosyl-	pH 1	244(3.69),285s(4.04), 292(4.06)	87-0814-76
	pH 7	250(3.79),301(4.13), 309s(4.05)	87-0814-76

Compound	Solvent	$\lambda_{max}(\log \epsilon)$	Ref.
Imidazo[1,2-c]pyrimidin-5(1H)-one, 1-β-D-arabinofuranosyl- (cont.)	pH 11	250(3.77),302(4.16), 309s(4.09)	87-0814-76
$C_{11}H_{13}N_3S$ 9-Azathieno[2,3-b]bicyclo[3.2.1]octane-3-carbonitrile, 2-amino-9-methyl-	EtOH	221(4.36),289(3.84)	2-0357-76
Benzonitrile, 4-[[(1,1-dimethylethyl)-thio]azo]-, (E)-	heptane	338(4.33),406(2.95)	144-0399-76
	MeOH	338(4.26),399(2.95)	144-0399-76
(Z)-	heptane	264(4.01),382(2.78)	144-0399-76
	MeOH	265(4.02),377(2.85)	144-0399-76
$C_{11}H_{13}N_5$ 1,3,5-Triazine-2,4-diamine, N-(1-phenyl-ethyl)-	n.s.g.	262(3.59)	2-0887-76
$C_{11}H_{13}N_5O_3$ Hydrazinecarboxamide, 2-[2-(hydroxyimi-no)-1-methyl-3-oxo-3-(phenylamino)-propylidene]-	EtOH	260(4.37)	42-0454-76
$C_{11}H_{13}N_5O_4$ 9H-Purin-6-amine, 9-(1,6-anhydro-α-D-psicofuranosyl)-	MeOH-base	259(4.19)	44-1836-76
9H-Purin-6-amine, 9-(6-deoxy-α-D-erythro-hex-5-en-2-ulosyl)-	MeOH-base	259(4.14)	44-1836-76
9H-Purin-6-amine, 9-(6-deoxy-β-D-erythro-hex-5-en-2-ulosyl)-	MeOH-base	260(4.19)	44-1836-76
$C_{11}H_{13}N_6O_7P$ Adenosine, 8-(aminocarbonyl)-, cyclic 3',5'-(hydrogen phosphate)	H_2O	288(4.01),224(4.28)	94-2052-76
$C_{11}H_{13}N_8O_6P$ Adenosine, 8-azido-2-methyl-, cyclic 3',5'-(hydrogen phosphate)	pH 1	276(4.17)	87-0419-76
	pH 7	275(4.06)	87-0419-76
	pH 11	275(4.06)	87-0419-76
$C_{11}H_{13}OS$ 1,3-Benzoxathiol-1-ium, 2-(1,1-dimethyl-ethyl)-, tetrafluoroborate	96% H_2SO_4	244s(3.46),248(3.54), 252(3.45),300(3.84)	39-0323-76C
$C_{11}H_{14}$ Benzene, (1,2-dimethyl-2-propenyl)-	EtOH	250(2.38),240(2.45), 260(2.48),261(3.34), 270(2.26)	78-1839-76
Benzene, (2-methyl-1-butenyl)-	EtOH	247(4.09)	78-1839-76
Benzene, (2-methyl-2-butenyl)-	EtOH	250(2.34),255(2.38), 260(2.38),263(2.34), 271(2.26)	78-1839-76
Tricyclo[5.2.0.0²,⁹]nona-3,5-diene, 8,9-dimethyl-, 8-endo	C_6H_{12}	281(3.55)	88-3903-76
8-exo	C_6H_{12}	277(3.70)	88-3903-76
$C_{11}H_{14}Br_2O_3$ Benzene, 1,2-dibromo-3-(ethoxymethyl)-5,6-dimethoxy-	EtOH	227(3.93),292(3.15)	102-0767-76
2H-Pyran-2-one, 6-(1,2-dibromo-1-methyl-propyl)-4-methoxy-3-methyl-, (R*,R*)-	EtOH	312(3.76)	88-1903-76

Compound	Solvent	$\lambda_{max}(\log \epsilon)$	Ref.
$C_{11}H_{14}ClNO_2$			
Benzene, 1-(1-chloro-2,2-dimethylpropyl)-2-nitro-	hexane	252(3.60)	12-2621-76
Benzene, 1-(1-chloro-2,2-dimethylpropyl)-4-nitro-	hexane	264(4.09)	12-2621-76
$C_{11}H_{14}ClN_3O$			
Butanamide, 4-chloro-N-(3-cyano-4,5-dimethyl-1H-pyrrol-2-yl)-	MeOH	210(4.00),284(3.91)	36-0908-76
$C_{11}H_{14}Cl_2O_2$			
2,4-Pentadienoic acid, 5-(2,2-dichloro-3-methylcyclopropyl)-, ethyl ester	EtOH	208(3.85),271(4.60)	70-0909-76
$C_{11}H_{14}N$			
Isoquinolinium, 3,4-dihydro-1,2-dimethyl-, perchlorate	EtOH	278(4.11),315s(3.39)	103-0683-76
methylene base	EtOH-NaOH	236(4.03),300(3.45)	103-0683-76
$C_{11}H_{14}N_2$			
1H-Indole-3-methanamine, N,N-dimethyl- (gramine)	EtOH	218(3.4)	44-3441-76
Pyrrolo[1,2-a]pyrimidine, 2,4,6,7-tetramethyl-	EtOH	236s(4.27),252(4.48), 288(3.39),296(3.45), 310(3.24),384(3.33)	103-1379-76
Pyrrolo[1,2-c]pyrimidine, 1,3,6,7-tetramethyl-	EtOH	240(4.39),277s(3.71), 284(3.83),295(3.82), 358(3.15)	44-0351-76
$C_{11}H_{14}N_2O$			
Urea, (1-phenyl-1-butenyl)-, (E)-	EtOH	251(3.75)	95-0962-76
Urea, (1-phenyl-1-butenyl)-, (Z)-	EtOH	254(3.95)	95-0962-76
$C_{11}H_{14}N_2O_2$			
Butanal, 4-[nitroso(phenylmethyl)amino]-	EtOH	237(3.88),355(1.88)	88-0593-76
Donoxaridine	n.s.g.	240(3.82),290(3.42)	105-0499-76
2-Propenoic acid, 3-[1-(phenylmethyl)-hydrazino]-, methyl ester, (E)-	MeOH	278(4.38)	24-0253-76
2(1H)-Pyridinone, 1-methyl-3-(1-methyl-5-oxo-2-pyrrolidinyl)-, (S)-	MeOH	232(4.31),305(3.92)	106-0540-76
2(1H)-Pyridinone, 1-methyl-5-(1-methyl-5-oxo-2-pyrrolidinyl)-, (S)-	MeOH	231(3.77),308(3.84)	106-0540-76
$C_{11}H_{14}N_2O_3$			
4H-Pyrido[1,2-a]pyrimidine-3-carboxylic acid, 1,6,7,8-tetrahydro-1,6-dimethyl-4-oxo-	pH 7.35	238(3.84),281(3.73)	145-0478-76
4H-Pyrido[1,2-a]pyrimidine-3-carboxylic acid, 6,7,8,9-tetrahydro-4-oxo-, ethyl ester	pH 7.35	231(3.84),298(3.95)	145-0478-76
$C_{11}H_{14}N_2O_4$			
Benzene, 1-(1,2-dimethyl-2-nitropropyl)-4-nitro-	hexane	262(4.06)	12-2621-76
1H-1,2-Diazepine-1,4-dicarboxylic acid, diethyl ester	MeOH	222(3.99),300(2.95)	88-4859-76
Hydrazinecarboxylic acid, [(3,4-dimethoxyphenyl)methylene]-, methyl ester	MeOH	280s(4.11),286(4.13), 309(4.10)	33-0949-76

Compound	Solvent	$\lambda_{max}(\log \epsilon)$	Ref.
$C_{11}H_{14}N_2O_5S$ 2-Butenedioic acid, 2-(4-ethyl-2-imino-3(2H)-thiazolyl)-3-hydroxy-, dimethyl ester	MeOH	258(4.09)	64-0251-76B
$C_{11}H_{14}N_2O_6$ 2,4(1H,3H)-Pyrimidinedione, 5-acetyl-1-(2-deoxy-α-D-ribofuranosyl)-	M HCl	228(3.85),282(4.06)	87-0909-76
	H_2O	227(3.78),282(4.07)	87-0909-76
	M NaOH	231s(3.92),282(3.92)	87-0909-76
Uridine, 5-acetyl-2'-deoxy-	M HCl	229(3.98),282(4.09)	87-0909-76
	H_2O	230(4.00),282(4.11)	87-0909-76
	M NaOH	223s(--),274(3.99)	87-0909-76
$C_{11}H_{14}N_2O_7S$ Benzenemethanol, α-(1-methyl-1-nitro-ethyl)-4-nitro-, methanesulfonate	EtOH	262(4.04)	12-2631-76
$C_{11}H_{14}N_2S$ Ethanamine, N-(2-ethyl-1,2-benzisothia-zol-3(2H)-ylidene)-	heptane	309(3.54),348s(--), 364s(--),380s(--)	24-0659-76
	MeOH	327(3.86),340s(--)	24-0659-76
	MeOH-HCl	327(3.86),340s(--)	24-0659-76
Propanedinitrile, [3-(ethylthio)cyclo-hexylidene]-	C_6H_{12}	238(4.11),286(3.23)	78-2827-76
$C_{11}H_{14}N_4$ 1,3-Benzenediamine, 2-(3,5-dimethyl-1H-pyrazol-1-yl)-	EtOH	231(4.13),300(3.26)	22-0839-76
1,3-Benzenediamine, 4-(3,5-dimethyl-1H-pyrazol-1-yl)-	EtOH	223(4.56),248(3.96), 300(3.58)	22-0839-76
$C_{11}H_{14}N_4O$ Acetamide, N-[2-(2-amino-1H-benzimid-azol-1-yl)ethyl]-, hydrobromide	EtOH	208(4.46),250(3.46), 284(3.71)	78-0839-76
1,3-Benzenediamine, 4-(3,5-dimethyl-1H-pyrazol-1-yl)-N -hydroxy-	EtOH	225(4.49),254(4.03), 293(3.53)	22-0839-76
Hypoxanthine, 3-cyclohexyl-	pH 1	254(4.03)	39-1847-76C
	pH 13	265(4.06),274s(4.03)	39-1847-76C
$C_{11}H_{14}N_4O_2$ Imidazolidine, 1-(2-amino-4-carboxymeth-ylphenyl)-2-imino-, dihydrobromide	EtOH	222(4.26),330(3.47)	78-0057-76
$C_{11}H_{14}N_4O_2S$ Benzenesulfonamide, N-(1-ethyl-1H-1,2,3-triazol-5-yl)-4-methyl-	EtOH	221(4.06),255(4.00)	32-0001-76
$C_{11}H_{14}N_4O_3$ 7H-Pyrrolo[2,3-d]pyrimidin-4-amine, 7-(2-deoxy-β-D-erythro-pentofur-anosyl)- (2'-deoxytubercidin)	pH 1	227(4.42),272(4.11)	77-0269-76
	pH 13	270(4.13)	77-0269-76
$C_{11}H_{14}N_4O_4$ D-Ribitol, 1-C-(8-amino-1,2,4-triazolo-[1,5-a]pyridin-2-yl)-1,4-anhydro-, (S)-	H_2O	275(3.95),295(3.78)	44-3124-76
Theophylline, 7-N-(2-acetoxyethyl)-	EtOH	273(4.00)	136-0049-76G
$C_{11}H_{14}N_4O_4S$ 1H-Pyrazolo[3,4-d]pyrimidine, 4-(methyl-thio)-1-β-D-ribofuranosyl-	EtOH	224s(4.08),290(4.23), 298s(4.15)	104-1143-76

Compound	Solvent	$\lambda_{max}(\log \epsilon)$	Ref.
2H-Pyrazolo[4,3-d]pyrimidine, 7-(methyl-thio)-2-β-D-ribofuranosyl-	MeOH	218(4.31),261(3.79), 268(3.79),309(4.10), 320(4.16),330(4.02)	104-1141-76
1H-1,2,3-Triazolo[4,5-c]pyridine, 4-(methylthio)-1-β-D-ribofuranosyl-	pH 1	312(4.32)	44-1449-76
	pH 11	303(4.15)	44-1449-76
	MeOH	299(4.20)	44-1449-76
$C_{11}H_{14}N_4O_5$			
1H-Imidazole-4-carboxamide, 5-(cyano-methyl)-1-β-D-ribofuranosyl-	pH 1	217(3.94),231s(3.91)	35-1492-76
	pH 7	219s(3.91),232(3.94)	35-1492-76
	pH 11	234(3.94)	35-1492-76
1H-Imidazo[4,5-c]pyridin-4(5H)-one, 6-amino-1-β-D-ribofuranosyl-	pH 1	284(4.14),309s(3.82)	35-1492-76
	pH 7	270(4.00),298(3.91)	35-1492-76
	pH 11	272(4.00),295s(3.91)	35-1492-76
4H-Imidazo[4,5-c]pyridin-4-one, 6-amino-3,5-dihydro-3-β-D-ribofuranosyl-	pH 1	277(4.06),317(3.77)	35-1492-76
	pH 7	258(3.61),317(3.87)	35-1492-76
	pH 11	258(3.79),316(3.85)	35-1492-76
Imidazo[1,2-c]pyrimidin-5(1H)-one, 7-amino-1-β-D-arabinofuranosyl-	pH 1	270s(4.00),300(4.29)	87-0814-76
	pH 7	292(4.31)	87-0814-76
	pH 11	292(4.31)	87-0814-76
Pentanal, 5-hydroxy-, 2,4-dinitrophenyl-hydrazone	EtOH	226(4.25),354(4.39)	104-2034-76
4H-Pyrrolo[2,3-d]pyrimidin-4-one, 2-amino-1,7-dihydro-7-β-D-ribo-furanosyl-	pH 1	218(4.31),262(4.05)	4-1363-76
	pH 11	262(4.10)	4-1363-76
	EtOH	260(4.20),280s(4.05)	4-1363-76
5H-1,2,3-Triazolo[4,5-b]pyridin-5-one, 2,4-dihydro-2-methyl-4-β-D-ribo-furanosyl-	MeOH	235(3.90),303(4.22), 312(4.20)	23-1029-76
5H-1,2,3-Triazolo[4,5-b]pyridin-5-one, 2,4-dihydro-4-methyl-2-β-D-ribo-furanosyl-	MeOH	233(3.90),304(4.35), 312(4.34)	23-1029-76
$C_{11}H_{14}N_5O_6P$			
9H-Purin-6-amine, 9-(5,6-dideoxy-6-phosphono-β-D-ribohex-5-eno-furanosyl)-, (E)-	pH 2	257(4.17)	130-0031-76
$C_{11}H_{14}N_6O_2$			
2(1H)-Pyrimidinone, 1,1'-(1,3-propane-diyl)bis[4-amino-	pH 1	283.5(--)	56-1805-76
	pH 7	274.5(4.20)	56-1805-76
	pH 13	274.0(--)	56-1805-76
$C_{11}H_{14}N_6O_4S$			
1H-Pyrazolo[3,4-d]pyrimidine-3-carbo-thioamide, 4-amino-1-β-D-ribo-furanosyl-	pH 2	223(4.23),270(4.01)	69-1005-76
	pH 7	234(4.06),293(4.05)	69-1005-76
	pH 11	232(4.12),279(4.09)	69-1005-76
$C_{11}H_{14}N_6O_5$			
1H-Pyrazolo[3,4-d]pyrimidine-3-carbox-amide, 4-amino-1-β-D-ribofuranosyl-	pH 2	216s(--),270(3.96)	69-1005-76
	pH 7	223s(--),282(4.03)	69-1005-76
	pH 12	230s(--),280(4.05)	69-1005-76
$C_{11}H_{14}O$			
Benzene, 1-methoxy-4-methyl-2-(2-prop-enyl)-	EtOH	278(3.39),284(3.35)	33-1763-76
Benzenemethanol, α-cyclopropyl-α-methyl-	EtOH	247-267f(1.88-2.27)	44-3067-76
Ethanone, 1-(2,4,5-trimethylphenyl)-	EtOH	254(4.10),292(3.24)	33-1763-76
Tricyclo[5.2.0.0^{2,9}]nona-3,5-diene-8-methanol, 9-methyl-	EtOH	216(3.67),277(3.47)	88-3903-76
Tricyclo[3.3.0.0^{3,7}]octan-2-one, 4-(1-methylethylidene)-, (+)-	isooctane	290s(2.72),296s(2.78), 298(2.79),304s(2.73)	44-1229-76

Compound	Solvent	$\lambda_{max}(\log \epsilon)$	Ref.
$C_{11}H_{14}OS_2$			
Cyclopentanecarbodithioic acid, 3-cyclopentylidene-2-oxo-	EtOH	245s(3.63),252(3.66), 284s(2.12),296(2.20), 329s(3.63),340(3.70), 424(4.42)	39-1706-76C
$C_{11}H_{14}OSe$			
Benzeneethaneselenoic acid, O-(1-methylethyl) ester	isoProH	282(3.92)	44-0729-76
$C_{11}H_{14}O_2$			
Benzeneacetic acid, α,α-dimethyl-, methyl ester	EtOH	248-262f(2.61-2.92)	44-3067-76
Benzoic acid, 2-(1-methylethyl)-, methyl ester	EtOH	229(3.68),277(2.88)	39-0484-76C
2-Butanone, 3-methoxy-3-phenyl-	EtOH	255(2.91),291(2.41)	44-3067-76
1,3-Cyclohexadiene-1-carboxylic acid, 6-(1-methylethenyl)-, methyl ester, (R)-	EtOH	293(3.79)	39-0484-76C
1H-Indene-4-carboxaldehyde, 2,6,7,7a-tetrahydro-1-hydroxy-7a-methyl-	EtOH	216(4.24)	118-0307-76
Norcaradiene-7-carboxylic acid, 2,3,7-trimethyl-	EtOH	239(3.59),277(3.56)	33-1763-76
Oxirane, 2-methoxy-3,3-dimethyl-2-phenyl-	EtOH	244-281f(1.81-2.86)	44-3067-76
Phenol, 2-(2-butenyl)-3-methoxy-	MeOH	273(3.00),281(3.00)	77-0340-76
1-Propanone, 2-methoxy-2-methyl-1-phenyl-	EtOH	247(4.01),280s(3.01), 330(1.86)	44-3067-76
$C_{11}H_{14}O_3$			
Benzene, 1-ethenyl-2,4,5-trimethoxy- (same spectrum in acid and base)	EtOH	258(4.09),264s(4.05), 314(3.83)	102-0347-76
Benzeneacetic acid, 4-methoxy-2,6-dimethyl-	EtOH	276(3.02),284(3.04)	39-1466-76C
Benzoic acid, 5-(1,1-dimethylethyl)-2-hydroxy-	MeOH	208(4.46),230(3.95), 305(3.65)	39-0465-76C
Benzoic acid, 4-methoxy-2,6-dimethyl-, methyl ester	EtOH	250(3.78),282(3.00)	39-1466-76C
Bicyclo[3.1.1]hept-2-ene-2-carboxylic acid, 6,6-dimethyl-4-oxo-, methyl ester, (1R)-	EtOH	257(3.81)	39-0484-76C
2-Cyclopenten-1-one, 4-acetoxy-3-methyl-2-(2-propenyl)-	EtOH	228(4.12)	39-0249-76C
o-Mentha-1,8-dien-7-oic acid, 3-oxo-, methyl ester	EtOH	233(4.15)	39-0484-76C
Phenol, 2-methoxy-4,5-dimethyl-, acetate	EtOH	276s(3.06),279(3.07), 283s(3.03)	5-1435-76
$C_{11}H_{14}O_4$			
Ethanone, 1-(6-hydroxy-3,4-dimethoxy-2-methylphenyl)-	n.s.g.	220(4.02),234(3.93), 277(3.75),318(3.54)	39-2594-76C
Propanoic acid, 3-(3-methoxy-5-methylphenoxy)-	EtOH	271(3.35),278(3.37)	39-0499-76C
1H,3H-Pyrano[3,4-c]pyran-1-one, 5β-ethenyl-6α-methoxy-4,4aα,5,6-tetrahydro-	MeOH	244.5(3.93)	24-3626-76
Spiro[cyclopenta[b]pyran-5(2H),2'-[1,3]-dioxolan]-2-one, 3,4,4a,6-tetrahydro-4a-methyl-	EtOH	207(3.94)	28-0951-76A

$C_{11}H_{14}O_5-C_{11}H_{15}NO_2$

Compound	Solvent	$\lambda_{max}(\log \epsilon)$	Ref.
$C_{11}H_{14}O_5$			
1,4-Benzenediol, 2,3-dimethoxy-5-methyl-, 4-acetate	EtOH	280(3.25)	65-1350-76
2,8-Dioxabicyclo[3.3.1]non-3-ene-4-carboxylic acid, 9-ethenyl-7-hydroxy-, methyl ester	MeOH	234(3.94)	24-3640-76
2,5-Methano-4H,5H-pyrano[2,3-d]-1,3-dioxin-6-carboxylic acid, 4a,8a-dihydro-4-methyl-, methyl ester (sarracenin)	EtOH	232(3.99)	35-1569-76
2H-Pyran-5-carboxylic acid, 4α-(1-formyl-1-propenyl)-3,4-dihydro-2-hydroxy-, methyl ester, (E)-	MeOH	231(4.08)	24-3640-76
1H,3H-Pyrano[3,4-c]pyran-1-one, 5ß-ethenyl-4,4aα,5,6-tetrahydro-3-hydroxy-6α-methoxy-	MeOH	237(4.01)	24-3626-76
$C_{11}H_{14}O_6$			
2H-Pyran-4-acetic acid, 3-formyl-3,4-dihydro-5-(methoxycarbonyl)-2-methyl-	MeOH	237(3.94)	39-0160-76C
2H-Pyran-2,2-dicarboxylic acid, 3,4-dihydro-4-oxo-, diethyl ester	MeOH	257(3.78)	44-2850-76
$C_{11}H_{14}S_3$			
Benzenecarbo(dithioperoxo)thioic acid, 1,1-dimethylethyl ester	C_6H_{12}	298(3.96),537(1.88)	138-0441-76
$C_{11}H_{15}BrO_3$			
1-Cyclohexene-1-carboxylic acid, 2-(bromomethyl)-6,6-dimethyl-3-oxo-, methyl ester	MeOH	230(3.93),245s(3.89)	35-3661-76
$C_{11}H_{15}ClN_2S$			
Diazene, (4-chloro-2-methylphenyl)-[(1,1-dimethylethyl)thio]-	EtOH	334(4.13)	144-0399-76
$C_{11}H_{15}FN_2O_6$			
2,4(1H,3H)-Pyrimidinedione, 1-(6-deoxy-6-fluoro-β-D-galactopyranosyl)-5-methyl-	H_2O	265(3.94)	48-0079-76
2,4(1H,3H)-Pyrimidinedione, 1-(6-deoxy-6-fluoro-β-D-glucopyranosyl)-5-methyl-	H_2O	264(3.84)	48-0079-76
Uridine, 5-fluoro-2',3'-di-O-methyl-	acid	268(3.96)	35-7381-76
	base	267(3.98)	35-7381-76
$C_{11}H_{15}N$			
Pyridine, 2,6-dimethyl-3-(2-methyl-1-propenyl)-	EtOH	216(3.77),239(3.85), 280(3.49)	39-2103-76C
$C_{11}H_{15}NO$			
Benzaldimine, N-(1,1-dimethylethyl)-, N-oxide	EtOH	296(4.21)	4-1001-76
Oxaziridine, 2-(1,1-dimethylethyl)-3-phenyl-	EtOH	249(2.95)	4-1001-76
2-Propanamine, 2-methoxy-N-(phenylmethylene)-	EtOH	247(4.01),282(2.00), 285(1.90)	35-2605-76
$C_{11}H_{15}NO_2$			
Acetic acid, cyanocyclohexylidene-, ethyl ester	C_6H_{12}	235(4.17)	78-2827-76
1H-Indole-3-carboxylic acid, 4,5,6,7-tetrahydro-1-methyl-, methyl ester	MeOH	203(4.23),237(4.08), 262s(3.72)	24-2154-76

Compound	Solvent	$\lambda_{max}(\log \epsilon)$	Ref.
Propane, 1-(2,6-dimethylphenoxy)-2-nitroso-, dimer	EtOH	292(3.95)	78-1267-76
2-Propenoic acid, 3-(3,4-dimethyl-1H-pyrrol-2-yl)-, ethyl ester	EtOH	206(4.08),287(3.54), 349(4.35)	4-1145-76
2H-Pyrano[4,3-b]pyridin-2-one, 1,5,7,8-tetrahydro-4,5,7-trimethyl-	EtOH	308(3.81)	22-0995-76
2(1H)-Pyridinone, 5-cyclohexyl-4-hydroxy-	EtOH	285(3.57)	88-3827-76
$C_{11}H_{15}NO_2S$			
Benzenecarbothioamide, 2,4-dimethoxy-N,N-dimethyl-	heptane-10% CH_2Cl_2	255(4.09),283(4.11), 378(2.75)	1-0695-76
$C_{11}H_{15}NO_3$			
Benzene, 1,2-dimethoxy-4-(2-nitroso-propyl)-, dimer	EtOH	281(4.09)	78-1267-76
1,2-Benzenediol, 4-methyl-5-morpholino-	EtOH	263s(3.77),295(3.57)	18-2333-76
1H-Indole-3-acetic acid, 2,4,5,6,7,7a-hexahydro-1-methyl-2-oxo-	EtOH	215(4.13)	22-0776-76
1H-Indole-2-carboxylic acid, 4,5,6,7-tetrahydro-3-hydroxy-, ethyl ester	MeOH	278.5(4.43)	5-0383-76
$C_{11}H_{15}NO_3S$			
Morpholine, 4-[(4-methylphenyl)sulfonyl]-	EtOH	212(4.23),241(3.95)	118-0339-76
$C_{11}H_{15}NO_4S$			
1H,3H-Oxazolo[4,3-c][1,4]thiazine-6-carboxylic acid, 8,8a-dihydro-3,8,8-trimethyl-1-oxo-, methyl ester	EtOH	216(3.84),265(3.42), 277s(3.65),318(3.94)	39-0584-76C
2-Propenoic acid, 3-(2,2-dimethyl-4-methylene-5-oxo-3-oxazolidinyl)-2-(methylthio)-, methyl ester, (Z)-	EtOH	215(3.68),275(3.89), 320s(3.73)	39-0584-76C
3(2H)-Thiazoleacetic acid, 5-(methoxycarbonyl)-2,2-dimethyl-α-methylene-, methyl ester	EtOH	208(3.80),222s(3.70), 285(3.34),334(3.70), 345s(3.68)	39-0584-76C
3(2H)-Thiazoleacetic acid, 5-(methoxycarbonyl)-2-methyl-α-(1-methylethylidene)-	EtOH	214(3.95),298(3.45), 351(3.74)	39-0584-76C
$C_{11}H_{15}NO_5$			
3,5-Oxepindicarboxylic acid, 2-amino-6,7-dihydro-, 5-ethyl 3-methyl ester	MeOH	275(3.15),327(4.31)	24-1269-76
$C_{11}H_{15}NO_5S$			
Propanol, 2-methyl-2-nitro-1-phenyl-, methanesulfonate	MeOH	258(2.66),262(2.64), 268(2.56)	12-2631-76
3(2H)-Thiazoleacetic acid, 5-(methoxycarbonyl)-2,2-dimethyl-α-methylene-, methyl ester, 1-oxide, (+)-	EtOH	207(3.75),275s(3.52), 301(3.76)	39-2540-76C
2,5-Thiazoledicarboxylic acid, 2,3-dihydro-2-methyl-3-(1-methylethenyl)-, dimethyl ester, 1-oxide, trans	EtOH	215s(3.89),225(3.94), 309(4.15)	39-2540-76C
$C_{11}H_{15}N_2O_4P$			
Acetic acid, [(dimethoxyphenylphosphoranylidene)hydrazono]-, methyl ester	EtOH	271.5(4.22)	65-1424-76
$C_{11}H_{15}N_2O_9P$			
5'-Uridylic acid, 5-acetyl-2'-deoxy-	pH 1	228(4.01),281(4.09)	87-0903-76

$C_{11}H_{15}N_3O-C_{11}H_{15}N_5$

Compound	Solvent	$\lambda_{max}(\log \epsilon)$	Ref.
5'-Uridylic acid, 5-acetyl-2'-deoxy- (cont.)	pH 7 pH 13	228(4.09),281(4.11) 284(4.00)	87-0903-76 87-0903-76
$C_{11}H_{15}N_3O$ 1H-Benzotriazole, 1-(pentyloxy)-	EtOH	264(3.74),283(3.65)	4-0509-76
$C_{11}H_{15}N_3O_2$ Acetamide, N-[[nitroso(2-phenylethyl)- amino]methyl]-	EtOH	236(3.93),361s(1.89), 364(1.92),376s(1.74)	94-0369-76 94-0369-76
Benzamide, N-[[(1-methylethyl)nitroso- amino]methyl]-	EtOH	232(4.29),366(1.88)	94-0369-76
Imidazo[1,2-a]pyrimidine-2,5(1H,3H)-di- one, 1-butyl-7-methyl-	EtOH	230(4.07),242s(3.74), 280(3.94)	114-0413-76A
Imidazo[1,2-a]pyrimidine-2,5(1H,3H)-di- one, 6-ethyl-1,3,7-trimethyl-	pH 2.1-10.3	239(4.06),285(3.92)	114-0413-76A
2-Quinoxalinecarboxylic acid, 3-amino- 4,6,7,8-tetrahydro-4-methyl-, methyl ester	EtOH	225(4.04),319(4.52), 360(3.68)	88-2899-76
$C_{11}H_{15}N_3O_3$ 1,2-Diazabicyclo[5.2.0]nona-3,5-diene- 2-carboxylic acid, 8-amino-5-methyl- 9-oxo-, ethyl ester	MeOH	278(3.89)	142-0471-76S
$C_{11}H_{15}N_3O_4$ Benzoic acid, 4-(2-aminoethylamino)-3- nitro-, ethyl ester, hydrochloride	EtOH	263(4.32),287(4.31), 412(3.70)	78-0057-76
$C_{11}H_{15}N_3O_6$ 2(1H)-Pyrimidinone, 1-(3-O-acetyl-β-D- arabinofuranosyl)-4-amino-	MeOH-acid	212(3.97),279(4.14)	87-0667-76
1,2,4-Triazine-3,5(2H,4H)-dione, 6-cy- clopropyl-2-β-D-ribofuranosyl-	H_2O pH 13	276(3.79) 262(3.75)	73-3635-76 73-3635-76
$C_{11}H_{15}N_3O_7S$ Acetamide, N-[2-(5-O-acetyl-β-D-ribo- furanosyl)-2,3-dihydro-3-oxo-1,2,4- thiadiazol-5-yl]-	pH 1 pH 7 and 11	235(3.94),270s(3.54) 255(3.94),276s(3.67)	4-0169-76 4-0169-76
$C_{11}H_{15}N_3O_9$ β-D-Ribofuranuronic acid, 1-deoxy- 1-(tetrahydro-5-nitro-2,4-dioxo- 1(2H)-pyrimidinyl)-, ethyl ester	pH 11	322(4.15)	87-1072-76
$C_{11}H_{15}N_3S$ 4H-6,9-Ethanopyrido[3,4-d]thiazolo- [3,2-a]pyrimidine, 2,3,6a,7,8,9a- hexahydro-	n.s.g.	244(3.81)	2-0759-76
$C_{11}H_{15}N_4O_8P$ 1H-Imidazole-4-carboxamide, 5-(cyano- methyl)-1-(5-O-phosphono-β-D-ribo- furanosyl)-	pH 1 pH 7 pH 11	213(4.00) 235(3.94) 238(3.87)	35-1492-76 35-1492-76 35-1492-76
1H-Imidazo[4,5-c]pyridin-4(5H)-one, 6-amino-1-(5-O-phosphono-β-D-ribo- furanosyl)-	pH 1 pH 7 pH 11	287(3.92),306s(3.69) 272(3.92),303(3.80) 272(3.91),303(3.78)	35-1492-76 35-1492-76 35-1492-76
$C_{11}H_{15}N_5$ 3H-Purin-6-amine, 3-methyl-N-(3-methyl- 2-butenyl)-	pH 2 pH 7	285.2(4.35) 286.7(4.26)	88-3807-76 88-3807-76

Compound	Solvent	$\lambda_{max}(\log \epsilon)$	Ref.
3H-Purin-6-amine, 3-methyl-N-(3-methyl-2-butenyl)-	pH 12	287.7(4.25)	88-3807-76
	MeOH	291.7(4.21)	88-3807-76
$C_{11}H_{15}N_5O$			
6H-Purin-6-one, 2-amino-1,7-dihydro-8-methyl-7-(3-methyl-2-butenyl)-	pH 1	247(4.10),276(3.93)	44-0568-76
	H_2O	248s(3.83),283(3.93)	44-0568-76
	pH 13	278(3.93)	44-0568-76
6H-Purin-6-one, 2-amino-1,9-dihydro-8-methyl-9-(3-methyl-2-butenyl)-	pH 1	252(3.11),277(3.94)	44-0568-76
	H_2O	252(4.13),273s(3.99)	44-0568-76
	pH 13	255s(4.05),269(4.07)	44-0568-76
6H-Purin-6-one, 2-amino-3,7-dihydro-8-methyl-3-(3-methyl-2-butenyl)-	pH 1	245s(3.91),267(4.11)	44-0568-76
	H_2O	236(3.90),271(4.13)	44-0568-76
	pH 13	277(4.17)	44-0568-76
$C_{11}H_{15}N_5O_3$			
1,2-Cyclopentanediol, 3-(6-amino-9H-purin-9-yl)-5-(hydroxymethyl)-(aristeromycin)	pH 2	259(4.14)	73-2096-76
9H-Purin-6-amine, 9-(3-deoxy-3-methyl--D-ribofuranosyl)-	MeOH	260(4.18)	87-1265-76
9H-Purin-6-amine, 3,5'-(1'-O-methyl-2',5'-dideoxy-D-ribofuranosyl)-	EtOH	274(4.09)	102-1523-76
$C_{11}H_{15}N_5O_4$			
9H-Purin-6-amine, 9-(2-deoxy-α-D-arabinohexofuranosyl)-	EtOH	260(4.15)	136-0041-76C
β-	EtOH	260(4.15)	136-0041-76C
9H-Purin-6-amine, 9-(6-deoxy-β-D-gulofuranosyl)-	pH 1	257(4.14)	44-0306-76
	H_2O	259(4.15)	44-0306-76
	pH 13	259(4.16)	44-0306-76
9H-Purin-6-amine, 9-(6-deoxy-α-D-idofuranosyl)-	H_2O	259(4.15)	44-0306-76
	pH 11	260(4.15)	44-0306-76
$C_{11}H_{15}OP$			
1H-Phosphindoline, 1,3,3-trimethyl-, 1-oxide	EtOH	269(2.92),276(--)	39-2556-76C
$C_{11}H_{16}BrN_3$			
Pyrido[2,3-b]pyrazine, 7-bromo-2,3-diethyl-1,2,3,4-tetrahydro-	EtOH	270(3.68),340(3.86)	22-0251-76
$C_{11}H_{16}Br_2N_4O_2$			
Imidazolidine, 1-(2-amino-4-carboxymethylphenyl)-2-imino-, dihydrobromide	EtOH	222(4.26),330(3.47)	78-0057-76
$C_{11}H_{16}ClNO_2$			
2-Butanamine, 1-(4-chloro-3-methoxyphenoxy)-	n.s.g.	230(3.92),280(3.47)	20-0421-76
$C_{11}H_{16}FN_3O_5$			
Cytidine, 5-fluoro-2',3'-di-O-methyl-	pH 1	289(4.04)	35-7381-76
	pH 13	280(3.91)	35-7381-76
$C_{11}H_{16}N$			
Bicyclo[4.2.0]octa-1,3,5-trien-7-aminium, N,N,N-trimethyl-, bromide	EtOH	247s(2.38),253s(2.72), 260(3.01),266(3.21), 272(3.20)	12-1685-76
$C_{11}H_{16}N_2$			
Ethanimidamide, N,N-dimethyl-N'-(4-methylphenyl)-	neutral	245(4.04)	39-0211-76B
	cation	230(4.06)	39-0211-76B

Compound	Solvent	$\lambda_{max}(\log \epsilon)$	Ref.
$C_{11}H_{16}N_2O$			
Ethanimidamide, N'-(4-methoxyphenyl)- N,N-dimethyl-	neutral cation	245(4.08) 242(4.05)	39-0211-76B 39-0211-76B
2(1H)-Pyridinone, 1-methyl-3-(2-piperi- dinyl)-, (-)-	MeOH	230(3.75),305(3.80)	106-0540-76
2(1H)-Pyridinone, 1-methyl-5-(2-piperi- dinyl)-, (-)-	MeOH	232(4.11),305(3.81)	106-0540-76
2(1H)-Pyridinone, 1-methyl-3-(1-methyl- 2-pyrrolidinyl)-	MeOH	232(4.00),305(4.09)	83-0197-76
2(1H)-Pyridinone, 1-methyl-5-(1-methyl- 2-pyrrolidinyl)-	MeOH	232(4.07),307(3.75)	83-0197-76
$C_{11}H_{16}N_2OS$			
Benzo[b]thiophene-3-carboxamide, 2-ami- no-N-ethyl-4,5,6,7-tetrahydro-	EtOH	225(4.14),215(3.89), 263(3.81),317(3.86)	2-0357-76
$C_{11}H_{16}N_2O_2$			
Acetic acid, 2-[2-(1-hydroxy-1-methyl- ethyl)phenyl]hydrazide	isoPrOH	238(4.03),285(3.34)	4-0033-76
Acetic acid, cyano(1-methyl-4-piperi- dinylidene)-, ethyl ester	C_6H_{12}	240(4.15)	78-2827-76
$C_{11}H_{16}N_2O_3S$			
5-Pyrimidinepropanoic acid, 1,4-dihydro- 6-methyl-2-(methylthio)-4-oxo-, ethyl ester	EtOH	235(3.96),288(3.93)	44-1858-76
$C_{11}H_{16}N_2O_4$			
1H,3H-Pyrimidine-2,4-dione, 1-[3-hydr- oxy-4-(hydroxymethyl)cyclopentyl]- 5-methyl-, (1α,3β,4α)-	pH 1 or 7 pH 13	273(4.02) 272(3.91)	4-1041-76 4-1041-76
$C_{11}H_{16}N_2O_6$			
2-Butenedioic acid, 2-(2,2-diacetyl- 1-methylhydrazino)-, dimethyl ester, (E)-	MeOH	265(4.24)	24-0253-76
1H-Pyrazole-3-carboxylic acid, 4-β-D- ribofuranosyl-, ethyl ester	pH 13 EtOH	241(4.02) 222(4.06)	44-0084-76 44-0084-76
2,4(1H,3H)-Pyrimidinedione, 1-(2-deoxy- β-D-arabino-hexofuranosyl)-5-methyl-	EtOH	267(3.97)	136-0041-76C
Uridine, 2'-deoxy-5-(methoxymethyl)-	H_2O	208(4.00),261(4.04)	87-0909-76
$C_{11}H_{16}N_2O_8S$			
2,4(1H,3H)-Pyrimidinedione, 1-[2-deoxy- 6-O-(methylsulfonyl)-β-D-arabino- hexopyranosyl]-	MeOH	259(4.00)	44-0600-76
$C_{11}H_{16}N_2S$			
Benzenecarboximidamide, N,N'-diethyl- 2-mercapto-	MeOH	275(3.99)	24-0659-76
$C_{11}H_{16}N_2S_2$			
8H-Thiapyrano[4,3-d]thiazolo[3,2-a]- [1,3]diazocine, 1,2,3,4,5,10,11,12- octahydro-, hydrobromide	n.s.g.	278(3.92)	2-0773-76
$C_{11}H_{16}N_4O_3$			
Lumazine, 5-acetyl-5,6,7,8-tetrahydro- 1,6,7-trimethyl-	pH 8.0 pH 14.0	250(3.90),283(4.26) 255(3.93),279(4.16)	24-3184-76 24-3184-76
1H-Purine-2,6-dione, 1-hexyl-3,7-di- hydro-7-hydroxy-	pH 3.2	204(4.32),270(4.05)	39-1507-76C

Compound	Solvent	$\lambda_{max}(\log \epsilon)$	Ref.
1H-Purine-2,6-dione, 1-hexyl-3,7-di-hydro-7-hydroxy- (cont.)	pH 7.6	226(4.23),256(3.95), 278(3.84)	39-1507-76C
	pH 11.9	228(4.33),294(3.81)	39-1507-76C
$C_{11}H_{16}N_4O_6$ 2,4(1H,3H)-Pteridinedione, 1-β-D-ribo-furanosyl-	pH 2.0	268(4.20)	24-3184-76
	pH 7.0	245(3.50),304(3.98)	24-3184-76
$C_{11}H_{16}N_5O_6P$ D-erythro-Pentofuranoside, methyl 5-(6-amino-9H-purin-9-yl)-2,5-dideoxy-, 3-(dihydrogen phosphate)	pH 1	258(4.04)	78-2409-76
	pH 7	262(4.06)	78-2409-76
	pH 13	262(4.06)	78-2409-76
9H-Purin-6-amine, 9-(5,6-dideoxy-6-phos-phono-β-D-ribo-hexofuranosyl)-, disodium salt	pH 3	257(4.18)	130-0031-76
	pH 12	259(4.19)	130-0031-76
$C_{11}H_{16}N_6O$ Urea, N-(2,2-dimethylpropyl)-N'-1H-pyra-zolo[3,4-d]pyrimidin-4-yl-	pH 1	270(4.21)	87-0555-76
	pH 13	272(4.10),298(3.62)	87-0555-76
	EtOH	267(4.19)	87-0555-76
Urea, N-(3-methylbutyl)-N'-1H-pyrazolo-[3,4-d]pyrimidin-4-yl-	pH 1	270(4.20)	87-0555-76
	pH 13	272(4.08),297(3.61)	87-0555-76
	EtOH	265(4.18)	87-0555-76
Urea, N-pentyl-N'-1H-pyrazolo[3,4-d]pyr-imidin-4-yl-	pH 1	272(4.16)	87-0555-76
	pH 13	272(4.07),295(3.70)	87-0555-76
	EtOH	266(4.12)	87-0555-76
$C_{11}H_{16}N_6O_2$ Cyclopentanemethanol, 2-amino-4-(6-ami-no-9H-purin-9-yl)-3-hydroxy-, (1α,2β,3α,4α)-	pH 1	258(4.16)	88-3005-76
	pH 13	260(4.17)	88-3005-76
$C_{11}H_{16}N_6O_4$ N^1,5'-Anhydro-N^5-(2',3'-O-isopropyli-dene-β-D-ribofuranosyl)-4-carboxy-hydrazino-5-amino-1,2,3-triazole	MeOH	238(4.15),259s(4.00)	44-1100-76
$C_{11}H_{16}O$ 2,4-Cyclohexadien-1-one, 6-(1-methyl-ethyl)-3,6-dimethyl-	n.s.g.	295(3.75)	33-1253-76
2,5-Cyclohexadien-1-one, 3,4-dimethyl-4-(1-methylethyl)-	n.s.g.	230(4.22)	33-1253-76
2,5-Cyclohexadien-1-one, 4-ethyl-3,4,5-trimethyl-	hexane	232(4.197)	39-1349-76B
	H_2O	237.5(4.232)	39-1349-76B
2,5-Cyclohexadien-1-one, 2,3,4,4,5-pen-tamethyl-	EtOH	244(3.95),273s(3.54)	104-1277-76
2-Cyclohexen-1-ol, 3-ethynyl-2,4,4-tri-methyl-	EtOH	234(4.00)	23-2310-76
3,5-Hexadien-2-one, 6-cyclopentyl-	EtOH	279s(4.52),289s(4.45)	22-0849-76
1H-Inden-1-ol, 2,6,7,7a-tetrahydro-4,7a-dimethyl-	EtOH	236(4.34),240(4.34), 252(4.16)	118-0307-76
1H-Inden-1-one, 2,3,4,5,6,7-hexahydro-7,7-dimethyl-	EtOH	237(4.07)	39-0410-76C
1(2H)-Naphthalenone, 3,4,6,7,8,8a-hexa-hydro-8a-methyl-	C_6H_{12}	198(3.67),288(1.80), 296(1.83),305(1.81)	12-2207-76
1(2H)-Naphthalenone, 3,5,6,7,8,8a-hexa-hydro-8a-methyl-	C_6H_{12}	198(3.71),298(2.11), 304(2.11)	12-2207-76
$C_{11}H_{16}O_2$ Acetic acid, (2,6,6-trimethyl-2-cyclo-hexen-1-ylidene)-	EtOH	256(4.00)	23-2310-76

Compound	Solvent	$\lambda_{max}(\log \epsilon)$	Ref.
Benzene, 1-(1-ethoxyethoxy)-4-methyl (partial spectrum)	ether	278(3.12)	39-1001-76B
2(4H)-Benzofuranone, 5,6,7,7a-tetrahydro-4,4,7a-trimethyl-	EtOH	212(4.14)	39-0532-76C
6,7-Dioxabicyclo[3.2.2]non-8-ene, 1,4,4-trimethyl-2-methylene-	EtOH	300(2.04)	39-1796-76C
2,4,6-Octatrienoic acid, 3,7-dimethyl-, methyl ester, (Z,E)-	EtOH	315(4.55)	78-0107-76
4-Penten-1-yn-3-ol, 3-methyl-, α-tetrahydropyranyl ether	MeOH	202(2.62)	104-1544-76
$C_{11}H_{16}O_3$			
Benzene, 1-(1-ethoxyethoxy)-4-methoxy- (partial spectrum)	ether	286(3.35)	39-1001-76B
2(4H)-Benzofuranone, 5,6,7,7a-tetrahydro-6-hydroxy-4,4,7a-trimethyl- (loliolide)	MeOH	215(4.17)	106-0051B-76
	EtOH	216(4.17)	36-1549-76
	EtOH	214(4.03)	39-0296-76C
Bicyclo[3.1.1]hept-2-ene-2-carboxylic acid, 4-hydroxy-6,6-dimethyl-, methyl ester (methyl 4α-hydroxymyrtenate)	EtOH	237(3.66)	39-0484-76C
2,5-Cyclohexadien-1-one, 4,4-dimethoxy-2,3,5-trimethyl-	EtOH	230(4.17),278(3.28), 340s(1.88)	1-0064-76
1-Cyclohexene-1-carboxylic acid, 3-hydroxy-6-(1-methylethenyl)-, methyl ester	EtOH	217(4.06)	39-0484-76C
1-Cyclohexene-1-carboxylic acid, 2,6,6-trimethyl-3-oxo-, methyl ester	MeOH	232(4.01)	35-3661-76
$C_{11}H_{16}O_4$			
Butanedioic acid, 2-butenylidene-3-methyl-, dimethyl ester	EtOH	266(4.28)	35-1204-76
Butanedioic acid, (3-methyl-2-butenylidene)-, dimethyl ester	EtOH	275(4.33)	35-1204-76
$C_{11}H_{16}O_5$			
α-D-xylo-Hept-5-enodialdo-1,4-furanose, 5,6-dideoxy-3-O-methyl-1,2-O-(1-methylethylidene)-, (E)-	EtOH	226(3.84)	136-0127-76A
$C_{11}H_{16}O_5S_2$			
α-D-Glucofuranose, 3-O-methyl-1,2-O-(1-methylethylidene)-6-thio-, cyclic 5,6-carbonodithioate	MeOH	283(4.22)	118-0449-76
$C_{11}H_{16}O_6S$			
α-D-Glucofuranose, 3-O-methyl-1,2-O-(1-methylethylidene)-, cyclic carbonothioate	MeOH	235(4.11)	118-0449-76
$C_{11}H_{17}ClN_4O_2$			
Methanimidamide, N'-(5-chloro-1,3-diethyl-1,2,3,6-tetrahydro-2,6-dioxo-4-pyrimidinyl)-N,N-dimethyl-	EtOH	235(4.1),255s(4.0), 310(4.2)	94-0970-76
$C_{11}H_{17}ClO$			
Cyclohexanone, 2-(chloromethylene)-3-methyl-6-(1-methylethyl)-, cis	heptane	240(3.62)	104-2192-76
trans	heptane	240(3.88)	104-2192-76

Compound	Solvent	$\lambda_{max}(\log \epsilon)$	Ref.
$C_{11}H_{17}ClOSi$			
Silane, (4-chloro-3,5-dimethoxyphenoxy)-trimethyl-	dioxan	278(2.85),283(2.85)	56-0363-76
$C_{11}H_{17}NO$			
Cyclohexanone, 2-[3-(dimethylamino)-2-propenylidene]-	H_2O	430(--)	70-0577-76
	aq HCl	385(--)	70-0577-76
	EtOH	393(4.53)	70-0577-76
	EtOH-HCl	390(--)	70-0577-76
1(2H)-Naphthalenone, 3,4,4a,5,6,7-hexahydro-8-(methylamino)-cis-s-cis	H_2O	333(4.06)	35-2826-76
	MeOH	334(4.18)	35-2826-76
	C_6H_{12}	325(4.05)	35-2826-76
Tricyclo[3.3.1.1^{3,7}]decan-1-amine, N-methylene-, N-oxide	ether	249(3.94)	4-1001-76
$C_{11}H_{17}NOS$			
4H-Thiopyran-4-one, tetrahydro-3-(1-piperidinylmethylene)-, (E)-	EtOH	275s(3.57),334(4.21)	4-0225-76
$C_{11}H_{17}NO_2$			
5-Isoxazolol, 4,5-dihydro-3,5-dimethyl-4,4-di-2-propenyl-	EtOH	210(3.46)(end abs.)	44-0855-76
$C_{11}H_{17}NO_3$			
Acetic acid, (4,4-dimethyl-6-oxo-2-piperidinylidene)-, ethyl ester	EtOH	270.5(4.30)	94-3011-76
2-Propanamine, 1-(3,4-dimethoxyphenoxy)-	n.s.g.	210(3.98),230(3.86),285(3.55)	20-0421-76
$C_{11}H_{17}NO_4$			
1H-Pyrrole-1,3-dicarboxylic acid, 4,5-dihydro-2-methyl-, diethyl ester	MeOH	281(4.29)	1-0567-76
$C_{11}H_{17}NO_6S$			
2,5-Thiazoledicarboxylic acid, 3-(ethoxymethyl)-2,3-dihydro-2-methyl-, dimethyl ester, 1-oxide, trans-($\pm$)-	EtOH	214(3.91),287(4.03)	39-2540-76C
$C_{11}H_{17}N_2O$			
4H-Pyrido[1,2-a]pyrimidinium, 6,7,8,9-tetrahydro-1,3,6-trimethyl-4-oxo-, methyl sulfate	pH 7.35	240.8(3.96)	145-0478-76
Pyrimidinium, 1-cyclohexyl-2,3-dihydro-3-methyl-2-oxo-, iodide	pH 0.29	223(4+),319(3.96)	23-2681-76
pseudobase	pH 9.18	227(4.16),300(3.21)	23-2681-76
$C_{11}H_{17}N_2O_9P$			
5'-Thymidylic acid, α-methoxy-	pH 1	261(4.03)	87-0903-76
	pH 7	207(4.03),261(4.05)	87-0903-76
	pH 13	261(3.89)	87-0903-76
$C_{11}H_{17}N_3$			
Pyrido[2,3-b]pyrazine, 2,3-diethyl-1,2,3,4-tetrahydro-	EtOH	265(3.90),330(3.92)	22-0251-76
$C_{11}H_{17}N_3O$			
4-Pyrimidinamine, N-cyclohexyl-6-methoxy-	pH 1	255(4.10)	39-1847-76C
	pH 13	247(3.98)	39-1847-76C

Compound	Solvent	$\lambda_{max}(\log \epsilon)$	Ref.
$C_{11}H_{17}N_3O_2S$ 1(2H)-Pyrimidineacetamide, N-butyl-3,4- dihydro-6-methyl-4-oxo-2-thioxo-	EtOH	220(4.26),276(4.18)	114-0413-76A
$C_{11}H_{17}N_3O_4$ 5-Pyrimidineheptanoic acid, 2-amino- 1,4,5,6-tetrahydro-4,6-dioxo-	pH 1	266(3.98)	24-2615-76
1(2H)-Pyrimidinehexanoic acid, α-amino- 3,4-dihydro-5-methyl-2,4-dioxo-	pH 12	271(3.94)	24-2615-76
$C_{11}H_{17}N_4O_5S$ Pyrimidinium, 4-amino-2-[(2-amino-2-oxo- ethyl)thio]-1-β-D-ribofuranosyl-, chloride	H_2O	246(4.29),273s(3.85)	5-1395-76
$C_{11}H_{18}$ 1,3-Cycloheptadiene, 1-butyl-	C_6H_{12}	247.5(3.70)	22-0142-76
1,3-Cycloheptadiene, 2-butyl-	C_6H_{12}	252(4.05)	22-0142-76
Cyclohexane, (3-methyl-2-butenylidene)-	EtOH	239(4.26),247(4.31), 255(4.19)	39-2386-76C
Cyclohexene, 2-ethenyl-1,3,3-trimethyl-	pentane	233(3.60)	33-0032-76
$C_{11}H_{18}NO_2$ Spiro[bicyclo[2.2.1]heptane-2,2'-oxaz- olidin]-3'-yloxy, 4',4'-dimethyl-, (1α,2α,4α)-	MeOH	229(3.32),417(1.08)	131-0275-76B
(1α,2β,4α)-	MeOH	232(3.38),422(1.11)	131-0275-76B
Spiro[bicyclo[2.2.1]heptane-2,4'-oxaz- olidin]-3'-yloxy, 2',2'-dimethyl-	MeOH	240(3.32),404(1.00)	131-0275-76B
$C_{11}H_{18}N_2$ 1-Cyclohexene-1-carbonitrile, 2-[(1,1- dimethylethyl)amino]-	EtOH	278(4.21)	44-0019-76
$C_{11}H_{18}N_2O$ 1H-Imidazole, 1-(1-oxooctyl)-	THF	244(3.65)	33-2738-76
1-Octanone, 1-(1H-imidazol-2-yl)-	EtOH	277(4.12)	33-2738-76
2,4-Pentadienal, 3-methyl-5-(1-pyrroli- dinyl)-, O-methyloxime	H_2O	374(--)	83-0592-76
	MeOH	360(4.58)	83-0592-76
	$CHCl_3$	363(4.54)	83-0592-76
	DMF	357(4.58)	83-0592-76
$C_{11}H_{18}N_2OS$ 4(1H)-Pyrimidinone, 2-(butylthio)- 6-propyl-	EtOH	238(3.91),289(3.89)	95-1094-76
$C_{11}H_{18}N_2O_2$ 1(2H)-Pyridinecarboxylic acid, 3,4-di- hydro-5-(1-pyrrolidinyl)-, methyl ester	MeOH	250(3.96)	1-0884-76
Pyrimidine, 2-(diethoxymethyl)-4,6-di- methyl-	MeOH	250(3.93)	44-0385-76
$C_{11}H_{18}N_2O_2S$ 4,6(1H,5H)-Pyrimidinedione, 5-ethyldi- hydro-5-(1-methylbutyl)-2-thioxo-	50%MeOH-HCl	288(4.33)	87-0521-76
	dioxan	287(4.37)	87-0521-76
	$C_2H_4Cl_2$	284(4.38)	87-0521-76
	MeCN	284(4.42)	87-0521-76

Compound	Solvent	$\lambda_{max}(\log \epsilon)$	Ref.
$C_{11}H_{18}N_4O_2$			
Methanimidamide, N'-(1,3-diethyl-1,2,3,6-tetrahydro-2,6-dioxo-4-pyrimidinyl)-N,N-dimethyl-	EtOH	248(<u>4.2</u>),305(<u>4.3</u>)	94-0970-76
2,4(1H,3H)-Pteridinedione, 5,6,7,8-tetrahydro-1,3,5,6,7-pentamethyl-	M HCl	264(4.23)	78-2303-76
$C_{11}H_{18}N_4O_3$			
4H-Cyclopenta-1,3-dioxole-4-methanol, 6-(5-amino-1H-1,2,4-triazol-3-yl)-tetrahydro-2,2-dimethyl-, (3aα,4α,6α,6aα)-	MeOH	217(3.40)	23-0861-76
$C_{11}H_{18}O$			
3-Buten-2-one, 4-cycloheptyl-	EtOH	223.0(3.68)	22-0849-76
1,3-Cycloheptadiene, 7-methoxy-1,5,5-trimethyl-	n.s.g.	249(3.90)	33-0082-76
1,3-Cyclohexadiene-1-ethanol, 2,6,6-trimethyl-	EtOH	265(3.61)	23-2310-76
2-Cyclohexen-1-one, 6-ethyl-4,4,6-trimethyl-	C_6H_{12}	221(4.18),342(1.57)	33-2244-76
3-Cycloocten-1-one, 4,8,8-trimethyl-	EtOH	220s(2.90),278(1.89)	33-0001-76
$C_{11}H_{18}O_2$			
Cyclohexanone, 2-butyl-6-(hydroxymethylene)-	EtOH	285(4.40),350(2.04)	65-2037-76
Cyclohexanone, 4-(1,1-dimethylethyl)-2-(hydroxymethylene)-	EtOH	280(4.28),350(2.05)	65-2037-76
Cyclohexanone, 2-(hydroxymethylene)-3-methyl-6-(1-methylethyl)-	heptane	290(3.94),345s(2.4)	104-1399-76
	heptane	292(3.88),345s(--)	104-2193-76
	DMF	292(2.64),360(1.67)	104-2193-76
Cyclohexanone, 2-(hydroxymethylene)-6-methyl-6-(1-methylethyl)-	EtOH	280(4.28),350(2.05)	65-2037-76
Ethanone, 1-[2,5-dihydro-5,5-dimethyl-4-(1-methylethyl)-2-furanyl]-	n.s.g.	216(3.69),293(2.31)	33-1253-76
Ethanone, 1,1'-[3-methyl-3-(1-methethyl)-1,2-cyclopropanediyl]bis-, trans	n.s.g.	288(2.05)	33-1253-76
3,4-Heptadien-2-one, 6-hydroxy-6-methyl-5-(1-methylethyl)-	n.s.g.	225(4.14)	33-1253-76
3-Heptene-2,6-dione, 5-methyl-5-(1-methylethyl)-, (E)-	n.s.g.	232(4.02),295(2.84)	33-1253-76
(Z)-	n.s.g.	226(3.77)	33-1253-76
3-Hepten-2-one, 5,6-epoxy-6-methyl-5-(1-methylethyl)-, (E)-	n.s.g.	231(4.09),332(1.63)	33-1253-76
4-Hepten-2-one, 6-methyl-5-[(1-methylethenyl)oxy]-	n.s.g.	280(2.42)	33-1253-76
3-Hexenoic acid, 5-methyl-2-(1-methylethylidene)-, methyl ester	EtOH	240(4.08)	88-2441-76
3-Octene-2,5-dione, 6,6,7-trimethyl-, (E)-	n.s.g.	234(4.11)	33-1253-76
2,4-Pentanedione, 3-(2,3-dimethyl-1-butenyl)-	EtOH	217(3.78),290(3.93)	33-1253-76
	EtOH-KOH	246(3.74),316(4.20)	33-1253-76
$C_{11}H_{18}O_3$			
2(5H)-Furanone, 4-butyl-3-ethoxy-5-methyl-	EtOH	226(3.96)	107-0299-76
$C_{11}H_{18}S_3$			
2,4,9-Trithiatricyclo[3.3.1.1^{3,7}]decane, 1,3,5,7-tetramethyl-	hexane	241(3.35)	11-0220-76A

Compound	Solvent	$\lambda_{max}(\log \epsilon)$	Ref.
$C_{11}H_{19}ClN_3O_2$			
Methanaminium, N-[[[1-chloro-3-(dimeth-ylamino)-2-(ethoxycarbonyl)-2-propen-ylidene]amino]methylene]-N-methyl-	HOAc	388(4.13)	73-1565-76
$C_{11}H_{19}NO$			
Cyclohexanone, 2-[(butylamino)methyl-ene]-	H_2O	329(4.40)	35-2826-76
	MeOH	330(4.28)	35-2826-76
cis-s-cis	C_6H_{12}	325(4.11)	35-2826-76
Cyclohexanone, 2-[(diethylamino)methyl-ene]-	C_6H_{12}	311(4.24)	35-2826-76
	H_2O	336(--)	35-2826-76
	MeOH	334(4.33)	35-2826-76
3(1H)-Pyridinone, 2,6-dihydro-2,5-di-methyl-1-(1,1-dimethylethyl)-, hydrochloride	EtOH	242(4.08)	70-0455-76
3(1H)-Pyridinone, 2,6-dihydro-4,5-di-methyl-1-(1,1-dimethylethyl)-, hydrochloride	EtOH	246(4.02)	70-0455-76
$C_{11}H_{19}NO_5$			
α-D-xylo-Hexofuranose, 5,6-dideoxy-3-O-methyl-1,2-O-(1-methylethyli-dene)-5-(methylimino)-, N-oxide, (Z)-	EtOH	243(3.90)	136-0127-76A
α-L-lyxo-Hexopyranoside, methyl 4,6-di-deoxy-2,3-O-(1-methylethylidene)-4-(methylimino)-, N-oxide, (Z)-	EtOH	245(3.99)	136-0127-76A
$C_{11}H_{19}NO_6$			
2-Cyclohexen-1-one, 5-hydroxy-3-[[2-hy-droxy-1-(hydroxymethyl)ethyl]amino]-5-(hydroxymethyl)-2-methoxy-, (-)-, (mycosporine)	H_2O	310(4.44)	23-1105-76
$C_{11}H_{19}N_5$			
1H-Imidazole-4-carbonitrile, 5-amino-1-[3-(diethylamino)propyl]-	EtOH	241.5(4.08)	2-0346-76
$C_{11}H_{20}$			
2,4-Undecadiene	C_6H_{12}	230(4.29)	22-0142-76
$C_{11}H_{20}N_2O$			
2-Propenamide, 3-(1-aziridinyl)-N,N-bis-(1-methylethyl)-, cis	C_6H_{12}	222(3.93),250(3.78)	78-1025-76
	MeOH	219(3.82),251(3.98)	78-1025-76
	dioxan	223(3.93),252(3.84)	78-1025-76
trans	C_6H_{12}	247(4.23)	78-1025-76
	MeOH	257(4.15)	78-1025-76
	dioxan	252(4.24)	78-1025-76
$C_{11}H_{20}O$			
4,9-Decadien-2-ol, 2-methyl-	EtOH	238(4.26)	18-3646-76
$C_{11}H_{20}O_6$			
L-Sorbose, 5,6-di-O-methyl-3,4-O-(1-methylethylidene)-	MeOH	285(1.55)	5-0269-76
$C_{11}H_{21}ClN_4$			
Methanaminium, N-[[[1-chloro-3-(dimeth-ylamino)-2-[(dimethyliminio)methyl]-2-propenylidene]amino]methylene]-N-methyl-, diperchlorate	MeCN	370(4.10)	73-1565-76

Compound	Solvent	$\lambda_{max}(\log \epsilon)$	Ref.
$C_{11}H_{21}NO$ 1-Penten-3-one, 1-(diethylamino)-4,4-dimethyl-	C_6H_{12} H_2O MeOH	297(4.34) 320(4.45) 316(4.42)	35-2826-76 35-2826-76 35-2826-76
$C_{11}H_{21}N_5O$ 1H-Imidazole-4-carboxamide, 5-amino-1-[3-(diethylamino)propyl]-	EtOH	270(4.12)	2-0346-76
$C_{11}H_{21}N_5OS$ 1H-Imidazole-4-carboxamide, 5-amino-1-[3-(diethylamino)propyl]-2,3-di-hydro-2-thioxo-	EtOH	273s(--),303(4.31)	2-0351-76
$C_{11}H_{22}ClFN_2O_4S$ 1-Oxa-4,8-diazaspiro[4.5]decane, 4-chloro-3,3,8,8-tetramethyl-, fluorosulfonate	H_2O	265(2.42)	36-1733-76
$C_{11}H_{22}N_2O$ 2-Propenamide, 3-(ethylamino)-N,N-bis-(1-methylethyl)-	C_6H_{12} MeOH dioxan	284(4.32) 281(4.41) 283(4.24)	78-1025-76 78-1025-76 78-1025-76
$C_{11}H_{22}N_4OS_2$ Thiourea, N-[5-(dimethylamino)-4,5-di-hydro-5-methoxy-4,4-dimethyl-2-thia-zolyl]-N,N-dimethyl-	EtOH	264(4.21)	33-2768-76
$C_{11}H_{22}O_2Si$ 2-Cyclopenten-1-ol, 4-[[(1,1-dimethyl-ethyl)dimethylsilyl]oxy]-	MeOH	209(4.08)	78-1713-76

Compound	Solvent	$\lambda_{max}(\log \epsilon)$	Ref.
$C_{12}Cl_{10}NO$ Nitroxide, bis(pentachlorophenyl)	C_6H_{12}	222(4.65),243s(4.30), 266s(3.55),304s(3.59), 330s(3.79),339s(3.89), 400s(3.72),550s(1.83)	88-0945-76
$C_{12}HCl_{10}NO$ Benzenamine, 2,3,4,5,6-pentachloro- N-hydroxy-N-(pentachlorophenyl)-	C_6H_{12}	222(4.63),245s(4.33), 332(4.17)	88-0945-76
$C_{12}H_2Br_8$ 1,1'-Biphenyl, 2,2',3,3',5,5',6,6'-octa- bromo-	hexane or EtOH	<u>223</u>(4.48)	98-1062-76
$C_{12}H_2Cl_6O_2$ Dibenzo[b,e][1,4]dioxin, 1,2,3,4,7,8- hexachloro-	$CHCl_3$	313(3.61)	44-2435-76
$C_{12}H_3Cl_5O_2$ Dibenzo[b,e][1,4]dioxin, 1,2,3,7,8-pen- tachloro-	$CHCl_3$	308(3.79)	44-2435-76
Dibenzo[b,e][1,4]dioxin, 1,2,4,7,8-pen- tachloro-	$CHCl_3$	307(3.57)	44-2435-76
$C_{12}H_3N_6$ Pentadiene-1,1,2,4,5,5-hexacarbonitrile, 3-methyl-, ion(1-)-	MeCN	589(4.54)	35-0558-76
$C_{12}H_4Br_4$ Biphenylene, 2,3,6,7-tetrabromo-	EtOH	234(4.27),250s(4.62), 259(4.97),267(5.16), 358(4.15),377(4.21)	78-2571-76
$C_{12}H_4Cl_6N_2$ Pyridine, 3,3'-(1,2-dichloro-1,2-ethene- diyl)bis[2,6-dichloro-, (E)-	MeOH	274.5(4.11)	33-0190-76
$C_{12}H_4FeO_6$ Iron, tetracarbonyl(1,2-phenylenedicar- bonyl)-, (OC-6-22)-	hexane	230(4.69),292s(3.54), 303(3.51),402(2.00)	101-0235-76H
$C_{12}H_5Cl_2F$ Benzene, (6,6-dichloro-5-fluoro- 5-hexene-1,4-diynyl)-	n.s.g.	208(4.40),218(4.42), 231(4.44),253(4.53), 265(4.49),274s(--), 291(4.29),309(4.44), 330(4.39)	44-1487-76
$C_{12}H_5Cl_3O_2$ Dibenzo[b,e][1,4]dioxin, 2,3,7-tri- chloro-	$CHCl_3$	305(3.72)	44-2435-76
$C_{12}H_5Cl_5O_2$ Phenol, 4,5-dichloro-2-(2,4,5-trichloro- phenoxy)-	EtOH	290(4.04),297s(4.00)	1-0796-76
$C_{12}H_6$ Benzocyclooctene, 5,6,9,10-tetradehydro-	pentane	228(4.38),233(4.37), 238s(4.43),244(4.67), 249(4.61),257(4.86),	89-0117B-76

Compound	Solvent	$\lambda_{max}(\log \epsilon)$	Ref.
Benzocyclooctene, 5,6,9,10-tetradehydro- (cont.)		327(2.76),332(2.77), 341(2.88),345(2.91), 361(2.95),366s(2.88)	89-0117B-76
$C_{12}H_6BrF_3O$ Ethanone, 1-(2-bromo-1-azulenyl)- 2,2,2-trifluoro-	C_6H_{12}	270(4.28),276(4.34), 315(4.56),325(4.59), 352(3.87),513(2.80), 540(2.77),590(2.35)	44-1822-76
$C_{12}H_6BrNO_2$ 5H-[1]Benzopyrano[2,3-b]pyridin-5-one, 7-bromo-	96% H_2SO_4	340(4.544)	103-1022-76
$C_{12}H_6ClF_3N_4O_2$ Benzo[g]pteridine-2,4(3H,10H)-dione, 8- chloro-10-methyl-7-(trifluoromethyl)-	EtOH	273(4.57),332(3.78), 430(3.99)	103-0938-76
$C_{12}H_6ClF_3O$ Ethanone, 1-(2-chloro-1-azulenyl)- 2,2,2-trifluoro-	CH_2Cl_2	275(4.44),323(4.61), 376s(4.15),392s(4.13), 495(2.95)	44-1822-76
$C_{12}H_6ClNO_2$ 5H-[1]Benzopyrano[2,3-b]pyridin-5-one, 7-chloro-	96% H_2SO_4	340(4.580)	103-1022-76
$C_{12}H_6Cl_2N_2S$ Quinazoline, 4-(2-thienyl)-, product with PCl_5-$POCl_3$	MeOH	238(3.72),310(3.25), 359(3.59)	139-0105-76C
$C_{12}H_6Cl_4$ Benzene, [(2,3,4,5-tetrachloro-2,4-cy- clopentadien-1-ylidene)methyl]-	C_6H_{12}	245(3.84),335(4.32), 430(2.65)	24-3929-76
	C_6H_{12}	340(4.31)	39-0048-76C
$C_{12}H_6Cl_4O$ Benzene, [(2,3,4,5-tetrachloro-2,4-cy- clopentadien-1-ylidene)methoxy]-	C_6H_{12}	312(4.46),395(2.93)	24-3939-76
Phenol, 4-[(2,3,4,5-tetrachloro-2,4-cy- clopentadien-1-ylidene)methyl]-	C_6H_{12}	370(4.44)	39-0048-76C
$C_{12}H_6Cl_6O$ 4,8-Ethenoazulen-5(3H)-one, 4,6,7,8,9- 10-hexachloro-3a,4,8,8a-tetrahydro-, $(3a\alpha,4\alpha,8\alpha,8a\alpha)$-	$CHCl_3$	256(3.78)	78-2265-76
$(3a\alpha,4\beta,8\beta,8a\alpha)$-	$CHCl_3$	252(3.73)	78-2265-76
$C_{12}H_6FNO_2$ 5H-[1]Benzopyrano[2,3-b]pyridin-5-one, 7-fluoro-	96% H_2SO_4	338(4.483)	103-1022-76
$C_{12}H_6F_3IO$ Ethanone, 2,2,2-trifluoro-1-(2-iodo- 1-azulenyl)-	C_6H_{12}	274(4.23),323(4.50), 334(4.50),362s(3.93), 523(2.78),552s(2.75)	44-1822-76
$C_{12}H_6F_{12}P_2$ 1,4-Diphosphabicyclo[2.2.2]octa-2,5,7- triene, 2,3-dimethyl-5,6,7,8-tetra- kis(trifluoromethyl)-	MeOH	272(2.93),326(2.79)	88-3715-76

Compound	Solvent	λ_{max}(log ϵ)	Ref.
$C_{12}H_6N_2O_5S$			
Thiazolo[2,3-f][1,6]naphthyridin-4-ium,	pH 2	228(4.13),272(4.13)	35-0299-76
3,8-dicarboxy-6-hydroxy-, hydroxide,	pH 12	232(3.98),294(4.11)	35-0299-76
inner salt (berninamycinic acid)			+35-8237-76
$C_{12}H_7BrN_2$			
Propanedinitrile, [2-bromo-2-(2,4,6-cy-	CH_2Cl_2	273(3.95),473(4.44)	88-0251-76
cloheptatrien-1-ylidene)ethylidene]-			
$C_{12}H_7Br_3$			
1,1'-Biphenyl, 2,4',5-tribromo-	hexane or	230(4.08)	98-1062-76
	EtOH		
$C_{12}H_7ClN_2$			
Propanedinitrile, [2-chloro-2-(2,4,6-cy-	CH_2Cl_2	275(3.81),283(3.78),	88-0251-76
cloheptatrien-1-ylidene)ethylidene]-		477(4.43)	
Propanedinitrile, (3-chloro-3-phenyl-	CH_2Cl_2	348(4.38)	48-0705-76
2-propenylidene)-			
$C_{12}H_7ClN_2O$			
3-Pyridinecarbonitrile, 6-(4-chlorophen-	HOAc	354(4.21)	48-0705-76
yl)-1,2-dihydro-2-oxo-			
$C_{12}H_7ClN_2S$			
Quinazoline, x-chloro-4-(2-thienyl)-	MeOH	220(3.95),272(3.20),	139-0105-76C
		345(3.21)	
$C_{12}H_7Cl_2NO_2S$			
Benzenamine, 4-(2,4-dichlorophenoxy)-	n.s.g.	347(4.11)	97-0138B-76
N-sulfinyl-			
$C_{12}H_7Cl_3N_2$			
Pyridine, 2,3,6-trichloro-5-[2-(3-pyri-	MeOH	290(4.28)	33-0190-76
dinyl)ethenyl]-, (E)-			
$C_{12}H_7Cl_3N_2O_2$			
Pyridine, 2,6-dichloro-3-[(4-chloro-	MeOH	271.5(3.81)	33-0190-76
3-nitrophenyl)methyl]-			
Pyridine, 2,3,6-trichloro-5-[(4-nitro-	MeOH	279(4.15)	33-0190-76
phenyl)methyl]-			
$C_{12}H_7Cl_4N$			
Pyridine, 2,3,6-trichloro-5-[(x-chloro-	MeOH	282(3.76)	33-0190-76
phenyl)methyl]-			
$C_{12}H_7F_6N$			
Quinoline, 2-[2,2,2-trifluoro-1-(tri-	hexane	224(4.72),227(4.75),	70-0837-76
fluoromethyl)ethyl]-		231(4.71),275(3.69),	
		302(3.60),315(3.64)	
$C_{12}H_7F_6NO$			
4(1H)-Quinolinone, 2-[2,2,2-trifluoro-	MeOH	212(4),228(3.84),	70-0837-76
1-(trifluoromethyl)ethyl]-		318(3.54),331(3.57)	
$C_{12}H_7NO_2$			
1-Azulenecarboxylic acid, 2-cyano-	EtOH	261(4.04),294(4.52),	44-1822-76
		306(4.59),347(3.88),	
		360(3.58),545s(2.89),	
		568(2.90)	
5H-[1]Benzopyrano[2,3-b]pyridin-5-one	96% H_2SO_4	332(4.462)	103-1022-76

Compound	Solvent	$\lambda_{max}(\log \epsilon)$	Ref.
$C_{12}H_7NO_3$ Benzoic acid, 4-(5-cyano-2-furanyl)-	EtOH	214(4.44),307(4.61)	73-1692-76
$C_{12}H_7N_3O_2$ Propanedinitrile, [3-(2-nitrophenyl)- 2-propenylidene]-	MeOH	216(4.42),331(4.22)	65-1141-76
$C_{12}H_7N_3O_3$ Quinoxaline, 2-(2-furanyl)-8-nitro-	EtOH	270(3.81),292(3.81), 377(3.80)	103-0261-76
$C_{12}H_7N_5$ 1,2,4,5-Benzenetetracarbonitrile, 3-amino-6-ethyl-	EtOH	217(4.53),257(4.48), 422(3.94)	35-0558-76
$C_{12}H_8BrN_3O_2$ Pyrazolo[1,5-a]pyrimidin-5(4H)-one, 3-(4-bromophenyl)-7-hydroxy-	MeOH-pH 1 MeOH-pH 11	202(4.50),247(4.44) 234(4.62),308(4.57)	87-0291-76 87-0291-76
$C_{12}H_8BrN_3O_4$ Benzenamine, 5-bromo-2-nitro-N-(4-nitro- phenyl)-	n.s.g.	246(4.24),288(3.93), 345(4.12),400(4.14)	40-0138-76
$C_{12}H_8Br_2$ 1,1'-Biphenyl, 2,4-dibromo- 1,1'-Biphenyl, 4,4'-dibromo-	hexane hexane	<u>243</u>(3.49) <u>264</u>(4.54)	98-1062-76 98-1062-76
$C_{12}H_8Br_2N_2$ Pyridine, 2,6-dibromo-3-[2-(3-pyridin- yl)ethenyl]-, (E)-	MeOH	310(4.36)	33-0190-76
$C_{12}H_8ClHgIN_2$ Mercury, chloro[2-[(2-iodophenyl)azo]- phenyl]- Mercury, chloro[3-iodo-2-(phenylazo)- phenyl]-	EtOH EtOH	210(4.28),328(4.13), 443(2.90) 226(4.26),252s(3.85), 342(4.17),460(2.46)	101-0039-76M 101-0039-76M
$C_{12}H_8ClNO_2S_2$ 5H-1,4-Dithiino[2,3-c]pyrrole-5,7(6H)- dione, 6-(2-chlorophenyl)-2,3-dihydro- 5H-1,4-Dithiino[2,3-c]pyrrole-5,7(6H)- dione, 6-(3-chlorophenyl)-2,3-dihydro- 5H-1,4-Dithiino[2,3-c]pyrrole-5,7(6H)- dione, 6-(4-chlorophenyl)-2,3-dihydro-	EtOH EtOH EtOH	204(4.28),208(4.27), 242(3.94),265(3.98), 417(3.51) 204(4.37),248(4.19), 270(3.99),420(3.42) 202(4.20),221(3.99), 246(4.26),270(3.98), 420(3.38)	56-1523-76 56-1523-76 56-1523-76
$C_{12}H_8ClN_3OS$ Ethanone, 1-[2-(4-chlorophenyl)-4-meth- yl-5-thiazolyl]-2-diazo-	pH 1 40% dioxan	325(4.34) 337(4.50)	80-0113-76 80-0113-76
$C_{12}H_8ClN_3O_2$ Pyrazolo[1,5-a]pyrimidin-5(4H)-one, 3-(4-chlorophenyl)-7-hydroxy-	MeOH-pH 1 MeOH-pH 11	204(4.42),245(4.37) 233(4.38),305(4.33)	87-0291-76 87-0291-76
$C_{12}H_8ClN_3O_3$ 2,4-Pentadienamide, 5-chloro-2-cyano- 5-(4-nitrophenyl)-	CH_2Cl_2	332(4.12)	48-0705-76

Compound	Solvent	$\lambda_{max}(\log \epsilon)$	Ref.
$C_{12}H_8ClN_3O_4$			
Benzenamine, 2-chloro-4-nitro-N-(4-nitrophenyl)-	n.s.g.	392(4.49)	40-0138-76
Benzenamine, 3-chloro-4-nitro-N-(4-nitrophenyl)-	n.s.g.	243(4.09),395(4.57)	40-0138-76
Benzenamine, 5-chloro-2-nitro-N-(4-nitrophenyl)-	n.s.g.	245(4.51),344(4.36), 398(4.38)	40-0138-76
$C_{12}H_8ClO_2PS$			
Dibenzo[d,f][1,3,2]dioxaphosphepin, 6-chloro-, 6-sulfide	MeOH	243(4.23)	31-1111-76
$C_{12}H_8ClO_3P$			
Dibenzo[d,f][1,3,2]dioxaphosphepin, 6-chloro-, 6-oxide	MeOH	246(4.02)	31-1111-76
$C_{12}H_8Cl_2FN$			
Pyridine, 2,6-dichloro-3-[(4-fluorophenyl)methyl]-	MeOH	271(3.72)	33-0190-76
$C_{12}H_8Cl_2N_2$			
Diazene, bis(4-chlorophenyl)-	C_6H_{12}	331(4.42)	18-1381-76
Pyridine, 2,6-dichloro-3-[2-(3-pyridinyl)ethenyl]-, (E)-	MeOH	308(4.34)	33-0190-76
$C_{12}H_8Cl_2N_2O$			
2,4-Pentadienamide, 5-chloro-5-(4-chlorophenyl)-2-cyano-	CH_2Cl_2	334(4.41)	48-0705-76
$C_{12}H_8Cl_2N_2O_2$			
Pyridine, 2,6-dichloro-3-[(4-nitrophenyl)methyl]-	MeOH	272(4.20)	33-0190-76
Pyridine, 2,6-dichloro-3-nitro-5-(phenylmethyl)-	MeOH	287(3.63)	33-0190-76
$C_{12}H_8Cl_2O_3S$			
Phenol, 2-chloro-6-[(4-chlorophenyl)sulfonyl]-	C_6H_{12}	250(4.03),300(3.93)	42-0602-76
	EtOH	235(3.54),280(3.65), 295(3.65)	42-0602-76
$C_{12}H_8Cl_3N$			
Pyridine, 2,6-dichloro-3-[(4-chlorophenyl)methyl]-	MeOH	272.5(3.72)	33-0190-76
Pyridine, 2,5,6-trichloro-5-(phenylmethyl)-	MeOH	282(3.72)	33-0190-76
$C_{12}H_8Cl_6O$			
4,8-Ethenoazulen-5(1H)-one, 4,6,7,8,9-10-hexachloro-2,3,3a,4,8,8a-hexahydro-, (3aα,4β,8β,8aα)-	$CHCl_3$	259(3.79)	78-2265-76
$C_{12}H_8F_3NO_2$			
2-Furancarboxamide, 5-[3-(trifluoromethyl)phenyl]-	dioxan	302(4.41)	73-2577-76
$C_{12}H_8IN_3O_2$			
Diazene, (2-iodo-6-nitrophenyl)phenyl-	EtOH	222(4.33),304(4.13), 448(2.65)	101-0039-76M
Diazene, (2-iodophenyl)(2-nitrophenyl)-	EtOH	211(4.29),248s(4.02), 323(4.11),454(2.51)	101-0039-76M

Compound	Solvent	$\lambda_{max}(\log \epsilon)$	Ref.
$C_{12}H_8I_2N_2$			
Diazene, (2,6-diiodophenyl)phenyl-	EtOH	216(4.36),278(4.06), 310s(3.93),437(2.54)	101-0039-76M
$C_{12}H_8N_2$			
1,10-Phenanthroline, ferrous complex	MeOH	510(4.05)	118-0001-76
Propanedinitrile, (2,3-dihydro-1H-inden-1-ylidene)-	EtOH	230(3.97),308(4.18)	118-0705-76
Propanedinitrile, (3-phenyl-2-propenylidene)-	C_6H_{12}	246(3.77),265(3.20), 345(4.20)	80-0781-76
	EtOH	247(3.86),265(3.38), 349(4.45)	80-0781-76
	acetone	347(4.45)	80-0781-76
	$CHCl_3$	265(3.39),353(4.47)	80-0781-76
	CCl_4	349(4.47)	80-0781-76
non-ion	H_2O	350(4.43)	80-0781-76
ion	H_2O	292(4.27)	80-0781-76
$C_{12}H_8N_2O$			
3-Pyridinecarbonitrile, 1,2-dihydro-2-oxo-6-phenyl-	HOAc	353(4.27)	48-0705-76
$C_{12}H_8N_2OS_2$			
10H-[1]Benzothiopyrano[3,2-d]pyrimidin-10-one, 2-(methylthio)-	MeOH	231(4.12),261s(4.15), 293(4.51),418(3.60)	73-2771-76
5,11-Epoxy-5H,11H-[1,5]dithiocino[2,3-b:6,7-b']dipyridine dihydrochloride	EtOH	243(4.14),298(3.75)	1-0904-76
	EtOH	242(4.36),298(3.97)	1-0904-76
$C_{12}H_8N_2O_2$			
5-Quinoxalinol, 2-(2-furanyl)-	EtOH	260(4.31),304(4.63), 371(4.03)	103-0261-76
5-Quinoxalinol, 3-(2-furanyl)-	EtOH	260(4.31),304(4.62), 369(4.09)	103-0261-76
$C_{12}H_8N_2S$			
Quinazoline, 4-(2-thienyl)-	MeOH	229(4.04),301(3.79), 341(4.07)	139-0105-76C
$C_{12}H_8N_4$			
9H-Carbazole, 4-azido-	ether	250(4.36),292(3.86), 301(3.86),328(3.54), 343(3.60)	18-2495-76
1,1,2,2-Ethenetetracarbonitrile, 1-(2,4-cyclohexadien-1-yl)-	CH_2Cl_2	265(3.6)	33-2635-76
Pyrido[3,2-e]-1,2,4-triazine, 3-phenyl-	EtOH	210(4.43),269(4.54), 352(3.91)	114-0301-76C
Pyrido[3,4-e]-1,2,4-triazine, 3-phenyl-	EtOH	253(4.28),285(4.38), 490(2.34)	114-0285-76C
$C_{12}H_8N_4O$			
Phenol, 2-pyrido[3,2-e]-1,2,4-triazin-3-yl-	dioxan	262(4.40),309(4.05), 385(3.65)	114-0301-76C
Phenol, 2-pyrido[3,4-e]-1,2,4-triazin-3-yl-	EtOH	287(4.31),390(3.62)	114-0285-76C
Phenol, 4-pyrido[3,2-e]-1,2,4-triazin-3-yl-	EtOH	275(3.60),380(3.25)	114-0301-76C
Phenol, 4-pyrido[3,4-e]-1,2,4-triazin-3-yl-	EtOH	280(4.43),305(4.30), 390(3.74)	114-0285-76C

$C_{12}H_8N_4OS-C_{12}H_8O_5$

Compound	Solvent	λ_{max} (log ϵ)	Ref.
$C_{12}H_8N_4OS$ 2,1,3-Benzothiadiazolo[4,5-h][1,6]naphthyridin-6(7H)-one, 8,9-dihydro-	EtOH	212(4.10),247s(4.32), 252(4.34),295(4.34), 328s(3.66),382(3.61)	103-0056-76
$C_{12}H_8N_4O_3$ Benzoic acid, 4-[5-(1H-tetrazol-5-yl)-2-furanyl]-	EtOH	204(4.06),229(4.03), 322(4.48)	73-1692-76
$C_{12}H_8N_4O_3S$ Ethanone, 2-diazo-1-[4-methyl-2-(4-nitrophenyl)-5-thiazolyl]-	pH 1 40% dioxan	340(4.30) 352(4.47)	80-0113-76 80-0113-76
$C_{12}H_8N_4O_4$ Diazene, bis(4-nitrophenyl)-	C_6H_{12}	336(4.39)	18-1381-76
$C_{12}H_8N_4O_6$ Benzenamine, 3,4-dinitro-N-(4-nitrophenyl)-	n.s.g.	398(4.46)	40-0138-76
$C_{12}H_8N_4O_8$ Naphthalene, 2,3-dimethyl-1,4,5,7-tetranitro-	dioxan	254(4.35),310s(3.87)	12-2247-76
$C_{12}H_8N_4S$ Propanedinitrile, (4-nitro-3-phenyl-2(3H)-thiazolylidene)-	MeOH	231(3.99),299(3.77), 357(4.00)	48-0343-76
2,4-Thiophenedicarbonitrile, 3-amino-5-(phenylamino)-	MeOH	260(4.54),296(4.02), 340(4.00)	48-0343-76
$C_{12}H_8N_4S_2$ 1,3,4-Thiadiazolium, 5-[(dicyanomethyl)-thio]-3-methyl-2-phenyl-, hydroxide, inner salt	MeOH	206(4.46),256(4.01), 298(4.08),372s(3.90)	70-1163-76
$C_{12}H_8N_6$ 1,1'-Biphenyl, 2,2'-diazido-	ether	232(4.28),252(4.26)	18-2495-76
$C_{12}H_8OS$ Thieno[2,3-b]furan, 2-phenyl-	EtOH	222(4.53),298(4.59)	104-1550-76
$C_{12}H_8OSe$ Selenolo[2,3-b]furan, 2-phenyl-	EtOH	230(4.31),299(4.50)	104-1550-76
$C_{12}H_8O_3$ 1H,3H-Naphtho[1,8-cd]pyran-1-one, 3-hydroxy-	MeOH	310(3.86)	118-0249-76
Tricyclo[4.4.0.0^{2,5}]deca-3,7,9-triene-1,6-dicarboxylic acid anhydride	ether	276(3.32)	88-3591-76
$C_{12}H_8O_4$ 5H-Furo[3,2-g][1]benzopyran-5-one, 4-hydroxy-7-methyl- (norvisnagin)	EtOH	214(4.66),246s(4.86), 253(4.92),261s(4.76), 282(4.16),336(3.96)	94-0580-76
	EtOH-pH 11	220(4.66),263s(4.81), 269(4.82),296(4.13), 318(4.18),370(3.99)	94-0580-76
$C_{12}H_8O_5$ Psoralen, 8-hydroxy-5-methyl-	EtOH	248(4.18),265(4.13), 310(4.05)	2-0816-76

Compound	Solvent	$\lambda_{max}(\log \epsilon)$	Ref.
$C_{12}H_8SSe$ Selenolo[2,3-b]thiophene, 2-phenyl-	EtOH	255(4.34),305(4.34)	104-1550-76
$C_{12}H_8S_2$ Thieno[2,3-b]thiophene, 2-phenyl-	EtOH	246(3.90),300(3.85)	104-1550-76
$C_{12}H_9$ 1,1'-Biphenyl, anion	H_2O NH_3 THF	405(4.42) 403(4.49) 402(4.58)	46-0122-76 46-0122-76 46-0122-76
$C_{12}H_9AsO_2$ 4-Arseninol, benzoate	EtOH	222(4.30),274(4.15)	88-4143-76
$C_{12}H_9BrN_2O$ 1,2-Diazepine, 1-benzoyl-4-bromo- 1,2-Diazepine, 1-benzoyl-6-bromo-	MeOH MeOH	226(4.15),358(2.60) 227(4.08),360(2.61)	88-4859-76 88-4859-76
$C_{12}H_9BrN_2O_2S_2$ 4-Pyrimidinecarboxylic acid, 5-[(4-bromophenyl)thio]-2-(methylthio)-	MeOH	268(4.33),289(4.22)	73-2771-76
$C_{12}H_9BrN_4$ Propanedinitrile, [(4-bromophenyl)azo]-2-propenyl-	CHCl$_3$	305(4.22)	126-1357-76
$C_{12}H_9BrO_2$ 1-Azulenecarboxylic acid, 6-bromo-, methyl ester	C_6H_{12}	297(4.81),302(4.28), 309(4.89),347(3.79), 355(3.88),373(3.85), 547(2.58),592(2.50), 650(2.07)	44-1822-76
$C_{12}H_9ClHgN_2$ Mercury, chloro[2-(phenylazo)phenyl]-	EtOH	230(4.19),332(4.25), 449(2.63)	101-0039-76M
$C_{12}H_9ClN_2$ Diazene, (4-chlorophenyl)phenyl- 1H-Pyrido[2,3-b]indole, 2-chloro-1-methyl-	hexane n.s.g.	230(4.12),323(4.30), 442(2.89) 224(4.11),265(4.19), 273(4.21),282(4.23), 331(4.01)	61-0301-76 103-0329-76
$C_{12}H_9ClN_2O$ 1,2-Diazepine, 1-benzoyl-4-chloro- 1,2-Diazepine, 1-benzoyl-6-chloro- 2,4-Pentadienamide, 5-chloro-2-cyano-5-phenyl-	MeOH MeOH CH$_2$Cl$_2$	230(4.02),358(2.67) 227(4.11),358(2.48) 345(4.40)	88-4859-76 88-4859-76 48-0705-76
$C_{12}H_9ClN_2OS_2$ 4-Pyrimidinecarbonyl chloride, 2-(methylthio)-5-(phenylthio)-	MeOH	269(4.20),287(4.25), 365(3.29)	73-2771-76
$C_{12}H_9ClN_2O_2$ 2,1-Benzisoxazole, 5-chloro-3-(3,5-dimethyl-4-isoxazolyl)-	EtOH	337(4.07)	4-0661-76
$C_{12}H_9ClN_2O_5S_2$ Benzenesulfonimidoyl chloride, 4-nitro-N-(phenylsulfonyl)-	CHCl$_3$	256(4.29)	139-0087-76B

Compound	Solvent	λ_{max} (log ϵ)	Ref.
$C_{12}H_9ClN_2S$			
2-Pyridinecarbothioamide, N-(3-chloro-phenyl)-	20% EtOH	282(4.09)	94-1451-76
2-Pyridinecarbothioamide, N-(4-chloro-phenyl)-	20% EtOH	283(4.04)	94-1451-76
$C_{12}H_9ClN_4O_2$			
Benzo[g]pteridine-2,4(3H,10H)-dione, 8-chloro-3,10-dimethyl-	EtOH	224(4.67),269(4.71), 338(4.09),434(4.23)	35-0830-76
	50% EtOH-5M HCl	222(4.57),259(4.57), 389(4.46)	35-0830-76
$C_{12}H_9ClN_4O_4$			
Furo[2,3-c]pyridin-2(6H)-one, 3-chloro-6-(4,6-dimethoxy-1,3,5-triazin-2-yl)-	MeCN	217(3.56),378(4.36), 398(4.34)	88-2959-76
$C_{12}H_9ClN_4S$			
4H-Pyrimido[2,3-b]thiazolo[4,5-b]quin-oxaline, 8(or 9)-chloro-2,3-dihydro-, hydrochloride	n.s.g.	225(4.70),266(3.38)	2-0759-76
$C_{12}H_9ClN_6O$			
Urea, N-(2-chlorophenyl)-N'-1H-pyrazolo-[3,4-d]pyrimidin-4-yl-	pH 1	278(4.29)	87-0555-76
	pH 13	280(4.26),307(3.87)	87-0555-76
	EtOH	280(4.29)	87-0555-76
Urea, N-(3-chlorophenyl)-N'-1H-pyrazolo-[3,4-d]pyrimidin-4-yl-	pH 1	281(4.37)	87-0555-76
	pH 13	280(4.34),310(4.10)	87-0555-76
	EtOH	277(4.39)	87-0555-76
Urea, N-(4-chlorophenyl)-N'-1H-pyrazolo-[3,4-d]pyrimidin-4-yl-	pH 1	283(4.33)	87-0555-76
	pH 13	280(4.34),307(4.05)	87-0555-76
	EtOH	279(4.36)	87-0555-76
$C_{12}H_9ClO_2$			
1-Azulenecarboxylic acid, 2-chloro-, methyl ester	CH_2Cl_2	294(4.72),304(4.77), 340(3.81),350(3.84), 366(3.51),515(2.72), 538s(2.70),590s(2.28)	44-1822-76
5H-Benzocyclohepten-5-one, 6-chloro-9-methoxy-	EtOH	237(4.36),257(4.24), 332(3.81),382(3.92)	18-2230-76
5H-Benzocyclohepten-5-one, 8-chloro-1-methoxy-	EtOH	235(4.42),290(3.99), 348(4.01)	18-2230-76
5H-Benzocyclohepten-5-one, 8-chloro-2-methoxy-	EtOH	234(4.26),282(4.53), 340s(3.74)	18-2230-76
5H-Benzocyclohepten-5-one, 8-chloro-6-methoxy-	EtOH	237(4.38),260s(4.16), 330(3.90),368(3.14)	18-2230-76
5H-Benzocyclohepten-5-one, 8-chloro-7-methoxy-	EtOH	252(4.78),261(4.76), 270s(4.60),335(4.00), 368s(3.26)	18-2230-76
5H-Benzocyclohepten-5-one, 8-chloro-9-methoxy-	EtOH	231(4.39),312(3.92), 355(3.83)	18-2230-76
7H-Benzocyclohepten-7-one, 6-chloro-1-methoxy-	EtOH	228(4.41),235(4.40), 295(4.59)	18-2230-76
7H-Benzocyclohepten-7-one, 6-chloro-5-methoxy-	EtOH	238(4.39),275(4.59), 340(3.52),357(3.11)	18-2230-76
7H-Benzocyclohepten-7-one, 6-chloro-8-methoxy-	EtOH	244(4.54),250(4.56), 286(4.76),316(4.06)	18-2230-76
7H-Benzocyclohepten-7-one, 8-chloro-5-methoxy-	EtOH	256(4.58),263(4.60), 273(4.68),288s(4.22), 312s(3.92)	18-2230-76
2-Furancarbonyl chloride, 5-(4-methyl-phenyl)-	dioxan	335(4.44)	73-2577-76

Compound	Solvent	$\lambda_{max}(\log \epsilon)$	Ref.
$C_{12}H_9ClO_3$ 2-Furancarbonyl chloride, 5-(4-methoxy- phenyl)-	dioxan	351(4.52)	73-2577-76
$C_{12}H_9ClO_3S$ Phenol, 2-[(4-chlorophenyl)sulfonyl]-	C_6H_{12} EtOH	243(3.97),280(3.23) 235(4.40),275(3.87), 290(3.92)	42-0602-76 42-0602-76
Phenol, 4-[(4-chlorophenyl)sulfonyl]-	C_6H_{12} EtOH	250(3.92),300(3.24) 235(4.33),290(3.84)	42-0602-76 42-0602-76
$C_{12}H_9Cl_2N$ Pyridine, 2,6-dichloro-3-(phenylmethyl)-	MeOH	273(3.73)	33-0190-76
$C_{12}H_9Cl_2NO_6S$ Furan, 2-[1-[(dichloromethyl)sulfonyl]- 2-(5-methyl-2-furanyl)ethenyl]- 5-nitro-	EtOH	234(3.90),311(4.48), 393s(4.08)	73-3391-76
$C_{12}H_9Cl_2NO_6S_2$ Furan, 2-[1-(dichloromethyl)sulfonyl]- 2-[5-(methylthio)-2-furanyl]ethenyl]- 5-nitro-	EtOH	227(4.11),326(4.27), 402(4.10)	73-3391-76
$C_{12}H_9Cl_2N_2O_2$ Pyridinium, 1-[(2,6-dichlorophenyl)meth- yl]-3-nitro-, bromide	H_2O	270(3.60)	35-5689-76
$C_{12}H_9FN_2$ Diazene, (4-fluorophenyl)phenyl-	THF	230(4.13),318(4.27), 440(2.71)	5-0946-76
$C_{12}H_9FN_2O$ 1,2-Diazepine, 1-benzoyl-4-fluoro- 1,2-Diazepine, 1-benzoyl-6-fluoro-	MeOH MeOH	227(3.93),362(2.30) 225(4.02),360(2.40)	88-4859-76 88-4859-76
$C_{12}H_9FN_6O$ Urea, N-(2-fluorophenyl)-N'-1H-pyrazolo- [3,4-d]pyrimidin-4-yl-	pH 1 pH 13 EtOH	281(4.33) 280(4.34),307(4.01) 280(4.33)	87-0555-76 87-0555-76 87-0555-76
$C_{12}H_9IN_2$ Diazene, (2-iodophenyl)phenyl-	EtOH	222(4.22),250(3.97), 320(3.24),456(2.58)	101-0039-76M
Diazene, (4-iodophenyl)phenyl-	hexane	232(4.15),329(4.43), 441(2.92)	61-0301-76
$C_{12}H_9IN_2O$ 1,2-Diazepine, 1-benzoyl-4-iodo- 1,2-Diazepine, 1-benzoyl-6-iodo-	MeOH MeOH	227(4.13),363(2.84) 228(4.19),362(2.49)	88-4859-76 88-4859-76
$C_{12}H_9IO_2$ 1-Azulenecarboxylic acid, 2-iodo-, methyl ester	C_6H_{12}	304(4.71),317(4.73), 346(4.08),363(4.12), 532(2.71),565(2.68), 615s(2.30)	44-1822-76
$C_{12}H_9N$ 1-Azuleneacetonitrile	CH_2Cl_2	277(4.70),282(4.67), 286(4.62),341(3.66),	44-1811-76

Compound	Solvent	$\lambda_{max}(\log \epsilon)$	Ref.
1-Azuleneacetonitrile (cont.)		355(3.44),581(2.55), 616(2.51)	44-1811-76
Carbazole	hexane	232(4.60),292(4.12), 332(3.46)	135-1054-76
Pyrido[2,1,6-de]quinolizine	C_6H_{12}	254(4.09),275(4.08), 306(4.10),312(4.14), 410s(3.78),422(3.99), 431(4.10),446(4.19), 458(4.45)	39-0341-76C
$C_{12}H_9NO$ 2-Furancarbonitrile, 5-(4-methylphenyl)-	EtOH	214(4.52),296(4.55)	73-1692-76
1H-Indole, 2-(2-furanyl)-	EtOH	248(4.03),310(4.51)	2-0579-76
Methanone, phenyl-2-pyridinyl-	hexane	264(4.14),360(2.21)	39-0869-76B
	hexane	230s(3.94),266(4.21)	59-0351-76
	pH 1	252s(2.41),275(4.13)	39-0869-76B
	pH 7	267(4.14),335s(2.53)	39-0869-76B
	MeOH	263(4.08)	59-0351-76
	EtOH	265(4.14),349(2.38)	39-0869-76B
Methanone, phenyl-3-pyridinyl-	hexane	248(4.21),349(2.07)	39-0869-76B
	hexane	202(4.40),247(4.33), 275s(3.60)	59-0351-76
	pH 1	266(4.13),338s(2.30)	39-0869-76B
	pH 7	265(4.11),330s(2.34)	39-0869-76B
	MeOH	205(3.25),257(3.18)	59-0351-76
	EtOH	255(4.28),341(2.20)	39-0869-76B
Methanone, phenyl-4-pyridinyl-	hexane	252(4.08),347(2.00)	39-0869-76B
	hexane	225(3.95),254(4.02), 278s(3.58)	59-0351-76
	pH 1	270(4.03),355s(2.15)	39-0869-76B
	pH 7	265(4.10),335s(2.27)	39-0869-76B
	MeOH	220s(4.01),258(3.99), 286s(3.52)	59-0351-76
	EtOH	258(4.06),338(2.09)	39-0869-76B
3H-Pyrrolo[1,2-a]indol-3-one, 1-methyl-	EtOH	210(4.55),266(4.19), 273(4.16),348(4.03)	83-0131-76
4H-Pyrrolo[3,2,1-ij]quinolin-4-one, 6-methyl-	EtOH	204(4.31),223(4.12), 290(3.87),325(4.04)	83-0185-76
$C_{12}H_9NOS$ 3-Pyridinecarboxaldehyde, 1,2-dihydro-1-phenyl-2-thioxo-	EtOH	298s(3.78),321(3.99), 385(3.49)	1-0863-76
β-Styrylacryloyl isothiocyanate	C_6H_{12}	340(4.78)	73-1388-76
Thiocyanic acid, 2-formyl-3,4-dihydro-1-naphthalenyl ester ?	EtOH	229(3.85),240(3.80), 312(3.96)	97-0049-76
$C_{12}H_9NOS_2$ 2,7-Dithiabicyclo[2.2.1]heptane-6-carbonitrile, 3-oxo-1-phenyl-, endo	MeOH	200(4.33)	44-1724-76
$C_{12}H_9NO_2$ 2-Furancarbonitrile, 5-(4-methoxyphenyl)-	EtOH	214(4.49),304(4.58)	73-1692-76
3H-Naphth[1,2-e][1,3]oxazine, 2-oxide	MeOH	240(4.41),260(4.23), 335(4.03),360(4.01)	24-1106-76
3-Pyridinecarboxaldehyde, 1,2-dihydro-2-oxo-1-phenyl-	EtOH	238s(4.01),360(3.82)	1-0863-76
$C_{12}H_9NO_2S$ Benzenamine, 4-phenoxy-N-sulfinyl-	n.s.g.	353(3.94)	97-0318B-76

Compound	Solvent	$\lambda_{max}(\log \epsilon)$	Ref.
$C_{12}H_9NO_2S_2$			
5H-1,4-Dithiino[2,3-c]pyrrole-5,7(6H)-dione, 2,3-dihydro-2-phenyl-	EtOH	204(4.19),236(4.15), 270(3.92),420(3.38)	56-1523-76
$C_{12}H_9NO_3$			
Acetamide, N-(5,6-dihydro-5,6-dioxo-1-naphthalenyl)-	EtOH	250(4.26),420(3.36)	78-2693-76
Acetamide, N-(5,8-dihydro-5,8-dioxo-1-naphthalenyl)-	EtOH	213(4.48),257(4.32), 418(3.56)	78-2693-76
Furo[2,3-b]quinolin-8-ol, 4-methoxy- (robustine)	pH 13	260(4.52),370(3.72)	100-0134-76
$C_{12}H_9NO_3S$			
Methanone, (3-methyl-2-thienyl)(2-nitro-phenyl)-	MeOH	202(4.23),212(4.18)	139-0105-76C
$C_{12}H_9NO_4S$			
2-Thiophenecarboxylic acid, 5-(4-nitro-phenyl)-, methyl ester	MeOH	239s(3.84),271s(3.81), 327(4.39)	48-0731-76
$C_{12}H_9N_3$			
Methanimidamide, N-cyano-N'-1-naphtha-lenyl-	EtOH	225(4.76),250s(4.16), 303(4.11)	48-0347-76
Methanimidamide, N-cyano-N'-2-naphtha-lenyl-	EtOH	253s(4.23),262(4.47), 271(4.51),301(4.25), 307s(4.25)	48-0347-76
$C_{12}H_9N_3O$			
Pyridine, 2-(diazophenylmethyl)-, 1-oxide	C_6H_{12}	269(4.46),338(3.85), 465(2.05)	64-1009-76B
5-Quinoxalinamine, 3-(2-furanyl)-	EtOH	265(4.01),315(4.57), 370(3.68)	103-0261-76
$C_{12}H_9N_3OS$			
Ethanone, 2-diazo-1-[2-(4-methylphenyl)-4-thiazolyl]-	pH 1 40% dioxan	287(4.18) 302(4.42)	80-0113-76 80-0113-76
Ethanone, 2-diazo-1-(4-methyl-2-phenyl-5-thiazolyl)-	pH 1 40% dioxan	319(4.38) 335(4.56)	80-0113-76 80-0113-76
$C_{12}H_9N_3O_2$			
Diazene, (3-nitrophenyl)phenyl-	hexane	231(4.14),317(4.22), 445(2.65)	61-0301-76
Diazene, (4-nitrophenyl)phenyl-	hexane C_6H_{12}	329(4.38),452(2.83) 330(4.41)	61-0301-76 18-1381-76
Pyrazolo[1,5-a]pyrimidine-5,7-diol, 3-phenyl-	MeOH-pH 1 MeOH-pH 11	203(4.37),238(4.26) 234(4.29),297(4.22)	87-0291-76 87-0291-76
Pyrimido[4,5-b]quinoline-2,4(3H,10H)-dione, 10-methyl-	EtOH	220(4.29),262(4.44), 318(3.83),398(3.96)	39-1805-76C
	50% EtOH-HCl	215(4.29),257(4.40), 334(4.08)	39-1805-76C
$C_{12}H_9N_3O_2S$			
Ethanone, 2-diazo-1-[2-(4-methoxyphen-yl)-4-thiazolyl]-	pH 1 40% dioxan	300(4.45) 302(4.42)	80-0113-76 80-0113-76
Thieno[3.2-c]cinnoline-2-carboxylic acid, 2,3-dihydro-3-imino-, methyl ester	EtOH	242(4.24),252(3.28), 258s(4.27),317(4.19), 355s(3.48)	39-0592-76C
$C_{12}H_9N_3O_3$			
2,4-Pentadienamide, 2-cyano-5-(2-nitro-phenyl)-	MeOH	213(5.22),295(5.08)	65-1141-76

Compound	Solvent	$\lambda_{max}(\log \epsilon)$	Ref.

$C_{12}H_9N_3O_4$
   Pyridine, 2-[(2,4-dinitrophenyl)methyl]-
     photochromic form
	EtOH
	 EtOH
		250(4.22)
		 567(4.63)
			104-2361-76
			 104-2361-76

$C_{12}H_9N_3O_6$
   Naphthalene, 2,3-dimethyl-1,4,5-tri-
     nitro-
	dioxan	303(3.50),311s(3.49),
		   324s(3.46)
			12-2247-76

$C_{12}H_9N_3O_7$
   Phenol, 4-(2,6-dinitro-4-aci-nitro-
     2,5-cyclohexadien-1-yl)-
	H_2O
	 aq acid
	 EtOH
	 EtOH-HCl
	 DMSO
		470(4.32),550s(--)
		 366(4.16)
		 465(4.40),545(4.02)
		 368(4.25)
		 468(4.50),570(4.19)
			23-2436-76
			 23-2436-76
			 23-2436-76
			 23-2436-76
			 23-2436-76

$C_{12}H_9N_3S$
   1,2,3-Thiadiazol-5-amine, N-1-naphtha-
     lenyl-
   1,2,3-Thiadiazol-5-amine, N-2-naphtha-
     lenyl-
	dioxan
	
	 dioxan
		246(4.18),342(4.11)
		
		 251(3.44),278(4.03),
		   332(4.29),286(4.04)
			73-1182-76
			
			 73-1182-76

$C_{12}H_9N_5O_2$
   Propanedinitrile, [(4-nitrophenyl)azo]-
     2-propenyl-
   [1,2,4]Triazolo[5,1-c][1,2,4]triazin-
     4(1H)-one, 3-benzoyl-1-methyl-
	$CHCl_3$
	
	 EtOH
		283(4.27)
		
		 209(4.21),259(4.08),
		   312(4.05)
			126-1357-76
			
			 39-0421-76C

$C_{12}H_{10}$
   Benzene, 1,3-hexadiynyl- (neocapillen)
	EtOH	211(3.42),222(3.55),
		   244(2.63),257(3.13),
		   272(3.32),289(3.20)
			102-1987-76

$C_{12}H_{10}ClF_4N$
   Benzenamine, 4-(4-chloro-1,2,3,4-tetra-
     fluoro-1,3-butadienyl)-N,N-dimethyl-
	heptane	330(4.47)	104-1565-76

$C_{12}H_{10}ClNO_2$
   1,4-Naphthalenedione, 5-amino-6-chloro-
     2,3-dimethyl-
	EtOH	271(4.01),494(3.50)	12-2247-76

$C_{12}H_{10}ClNO_3S_2$
   Benzenesulfonimidoyl chloride, N-(phen-
     ylsulfonyl)-
	$CHCl_3$	246(3.99)	139-0087-76B

$C_{12}H_{10}Cl_2N_2$
   Pyridine, 2,6-dichloro-3-[2-(1-methyl-
     1H-pyrrol-2-yl)ethenyl]-, (E)-
   3-Pyridinemethanamine, 2,6-dichloro-
     N-phenyl-
	MeOH
	
	 MeOH
		265(4.04)
		
		 242(4.15),272(3.77)
			33-0190-76
			
			 33-0190-76

$C_{12}H_{10}Cl_2N_2OS_2$
   5,11-Epoxy-5H,11H-[1,5]dithiocino[2,3-
     b:6,7-b']dipyridine, dihydrochloride
	EtOH	242(4.36),298(3.97)	1-0904-76

$C_{12}H_{10}CoO_4$
   Cobaltocenium, 1,1'-dicarboxy-
	MeOH	274(4.32),322(3.15),
		   405(2.49)
			101-0189-76I

$C_{12}H_{10}F_6$
   1H-Indene, 2,3-dihydro-1-[2,2,2-triflu-
     oro-1-(trifluoromethyl)ethyl]-
	EtOH	261(3.78),267(3.95),
		   273(3.98)
			104-0339-76

Compound	Solvent	$\lambda_{max}(\log \epsilon)$	Ref.
1H-Indene, 3a,4,7,7a-tetrahydro-1-[2,2,2-trifluoro-1-(trifluoro-methyl)ethylidene]-	EtOH	251(3.82)	104-0339-76
$C_{12}H_{10}Fe$ Ferrocene, ethynyl-, monocation	CH_2Cl_2	480s(--),565(2.48), 700(2.49)	44-2700-76
$C_{12}H_{10}N_2$ Benzo[g]phthalazine, 1,4-dihydro-	n.s.g.	265(3.78),274(3.81), 285(3.80),295(3.63), 314(2.85),319(2.78), 360(2.45)	33-2273-76
Propanedinitrile, (1-phenylpropylidene)-	EtOH	232(3.92),292(4.11)	118-0705-76
1H-Pyrido[2,3-b]indole, 1-methyl-	n.s.g.	204(4.14),220(4.33), 240(3.90),269(4.32), 274(4.33),278(4.38), 321(4.08),402(3.38)	103-0329-76
$C_{12}H_{10}N_2O$ 5-Azacycl[3.2.2]azine-1-carboxaldehyde, 2,6-dimethyl-	EtOH	227(4.38),258(4.45), 267s(4.33),286s(4.03), 312(4.00),424(3.92)	44-0351-76
5-Azacycl[3.2.2]azine-4-carboxaldehyde, 2,6-dimethyl-	EtOH	224(4.33),249(4.47), 281(4.20),309(4.01), 315(4.01),428(3.99), 435s(3.98)	44-0351-76
Isoquinoline-4-carbonitrile, 3-ethoxy-	EtOH	232(4.73),275(3.33), 286(3.38),298(3.25), 355(3.66)	2-0964-76
5H-Pyrido[2,3-b]indol-2-ol, 4-methyl-	EtOH	220(4.31),268(3.83), 328(3.93)	103-1186-76
1H-Pyrrole-2-carbonitrile, 3-(phenyl-methoxy)-	H_2O	247(4.18)	33-2786-76
$C_{12}H_{10}N_2OS_2$ Ethanone, 1-[4-(1H-[1,2]dithiolo[5,1-e]-[1,2,3]thiadiazol-7-S^{IV}-1-yl)phenyl]-	C_6H_{12}	205(4.35),227(4.38), 248s(4.09),305(4.36), 492(4.22)	39-0228-76C
$C_{12}H_{10}N_2O_2$ [1]Benzopyrano[3,4-c]pyrazol-4(1H)-one, 1,1-dimethyl-	EtOH	237(4.88),245(4.92), 310(4.79),340(4.81)	39-1260-76C
[2,2'-Bipyridine]-5-carboxylic acid, methyl ester	pH 6.0	245(4.10),291(4.30)	64-0115-76B
[2,2'-Bipyridin]-5-ol, acetate	EtOH	241(4.02),286(4.15)	64-0115-76B
Isoquinoline-4-carbonitrile, 3-ethoxy-8-hydroxy-	EtOH	235(4.53),266(4.25), 306(3.33),390(3.76)	2-0964-76
$C_{12}H_{10}N_2O_2S$ 4,6(1H,5H)-Pyrimidinedione, dihydro-5-[(4-methylphenyl)methylene]-2-thioxo-	dioxan	241(4.05),368(4.60)	104-1124-76
$C_{12}H_{10}N_2O_2S_2$ 4-Pyrimidinecarboxylic acid, 2-(methyl-thio)-5-(phenylthio)-	MeOH	273(4.21),286(4.22), 372(3.29)	73-2771-76
$C_{12}H_{10}N_2O_3$ Acetamide, N-(7-hydroxy-8-nitroso-1-naphthalenyl)-	EtOH	273(4.15),420(3.69)	12-2499-76

Compound	Solvent	$\lambda_{max}(\log \epsilon)$	Ref.
Formamidine, N^2-(1,3-dihydroxy-2-naph-thoyl)-	EtOH	215(4.50),250(4.64), 260(4.58),317(4.18), 327(4.21),368(4.18), 380(4.27),398(4.05)	94-2585-76
Pyridinium, 3-hydroxy-1-methyl-5-(4-ni-trophenyl)-, hydroxide, inner salt	MeCN	215(4.1),242(3.9), 277(4.1),290(4.0), 372(3.6)	39-2329-76C
$C_{12}H_{10}N_2O_3S$ 4,6(1H,5H)-Pyrimidinedione, dihydro-5-[(4-methoxyphenyl)methylene]-2-thioxo-	dioxan	255(4.18),315(4.08), 390(4.66)	104-1124-76
$C_{12}H_{10}N_2O_4$ 2-Furancarboxamide, N-methyl-5-(2-nitro-phenyl)-	dioxan	288(4.17)	73-2577-76
2-Furancarboxamide, N-methyl-5-(3-nitro-phenyl)-	dioxan	304(4.40)	73-2577-76
2-Furancarboxamide, N-methyl-5-(4-nitro-phenyl)-	dioxan	351(4.36)	73-2577-76
$C_{12}H_{10}N_2O_4S$ 2-Furancarboximidothioic acid, N-hydr-oxy-5-nitro-, phenylmethyl ester	EtOH	209(4.25),352(4.06)	73-3085-76
$C_{12}H_{10}N_2O_6S$ 2-Furancarboximidic acid, N-hydroxy-5-nitro-, anhydride with 4-methyl-benzenesulfinic acid	EtOH	207(4.25),234(4.32), 324(4.15)	73-3085-76
$C_{12}H_{10}N_2S$ Benzenecarbothioamide, N-3-pyridinyl-	EtOH	285(4.26),320(4.11)	39-0315-76C
2-Pyridinecarbothioamide, N-phenyl-	20% EtOH	281(4.08)	94-1451-76
4-Pyridinecarbothioamide, N-phenyl-	20% EtOH	283(4.16)	94-1451-76
$C_{12}H_{10}N_2S_2$ [1]Benzothieno[3,2-d]pyrimidine, 8-methyl-2-(methylthio)-	MeOH	248(4.44),269(4.42), 304(3.92),313(3.93)	73-2771-76
$C_{12}H_{10}N_3O_4$ Pyridinium, 1-[(4-nitrophenyl)methyl]-3-nitro-, bromide	H_2O	266(4.15)	35-5689-76
$C_{12}H_{10}N_4$ Propanedinitrile, [(4-ethenylphenyl)-azo]methyl-	MeOH	322(4.26)	126-1357-76
Pyrido[3,4-e]-1,2,4-triazine, 1,2-di-hydro-3-phenyl-	pH 1	226(4.18),257(4.11), 400(2.79)	114-0285-76C
$C_{12}H_{10}N_4O$ Phenol, 2-(1,2-dihydropyrido[3,4-e]-1,2,4-triazin-3-yl)-, hydrochloride	EtOH	260(5.43),323(4.95)	114-0285-76C
Phenol, 4-(1,2-dihydropyrido[3,2-e]-1,2,4-triazin-3-yl)-, dihydrochloride	EtOH	269(4.13),394(3.93), 434s(--)	114-0301-76C
1(2H)-Phthalazinone, 4-(1H-imidazol-2-yl)-2-methyl-	CF_3COOH	302(4.08),313(4.02)	44-0836-76
4(1H)-Pyrimidinone, 6-amino-5-(1H-indol-3-yl)-	EtOH	203s(4.52),216(4.64), 278(4.02),288(3.94)	103-1246-76
1H-Tetrazole, 5-[5-(4-methylphenyl)-furanyl]-	EtOH	204(4.27),219(4.18), 311(4.46)	73-1692-76

Compound	Solvent	$\lambda_{max}(\log \epsilon)$	Ref.
$C_{12}H_{10}N_4OS$			
4(1H)-Pyrimidinone, 6-amino-2,3-dihydro-5-(1H-indol-3-yl)-2-thioxo-	EtOH	205(4.64),218s(4.55), 290(3.98)	103-1246-76
4H-1,3,4-Thiadiazolo[2,3-c][1,2,4]triazin-4-one, 1,3-dimethyl-7-(4-methylphenyl)-	MeOH	276(3.82),354(3.98)	4-0117-76
$C_{12}H_{10}N_4O_2$			
Acetamide, N-(5-oxo-1,2,4-triazolo[1,5-b]isoquinolin-1(5H)-yl)-	EtOH	234(4.35),312(4.18), 371(3.89),388(3.78)	95-0700-76
Benzo[g]pteridine-2,4(1H,3H)-dione, 1,3-dimethyl-	n.s.g.	326(3.93),382(3.85)	88-1389-76
Benzo[g]pteridine-2,4(3H,10H)-dione, 3,10-dimethyl-	EtOH	218(4.40),268(4.51), 334(3.82),438(3.92)	35-0830-76
	50% EtOH-5M HCl	214(4.30),259(4.40), 368(4.10)	35-0830-76
Benzo[g]pteridine-2,4(3H,10H)-dione, 10-ethyl-	EtOH	219(4.43),267(4.50), 333(3.81),436(3.98)	35-0830-76
	50% EtOH-5M HCl	213(4.34),259(4.37), 369(4.10)	35-0830-76
Imidazo[5,1-c]-1,2,4-benzotriazine, 1-(1,3-dioxolan-2-yl)-	EtOH	247(4.17),372(3.76)	44-0158-76
6H-1,2,4,5-Tetrazino[2,3-b]isoquinolin-6-one, 2-acetyl-1,2-dihydro-	EtOH	233(4.35),311(4.19), 370(3.91),388(3.79)	95-0700-76
1H-Tetrazole, 5-[(4-methoxyphenyl)-2-furanyl]-	EtOH	204(4.19),222(3.95), 314(4.42)	73-1692-76
$C_{12}H_{10}N_4O_2S$			
4H-1,3,4-Thiadiazolo[2,3-c][1,2,4]triazin-4-one, 7-(3-methoxyphenyl)-3-methyl-	MeOH	264(4.13),334(4.28)	4-0117-76
4H-1,3,4-Thiadiazolo[2,3-c][1,2,4]triazin-4-one, 7-(4-methoxyphenyl)-3-methyl-	MeOH	272(2.07),354(4.15)	4-0117-76
$C_{12}H_{10}N_4O_3$			
Benzoic acid, 2-(3-nitro-2-pyridinyl)-hydrazide	EtOH	222(4.08),262(3.74), 376(3.23)	114-0301-76C
Benzoic acid, 2-(3-nitro-4-pyridinyl)-hydrazide	pH 1	266(4.16),330(3.63)	114-0285-76C
Benzo[g]pteridine-2,4(3H,10H)-dione, 3,10-dimethyl-, 5-oxide	EtOH	213(4.02),271(4.49), 342(3.96),355s(3.88), 452(3.85)	35-0830-76
	50% EtOH-5M HCl	271(4.51),367(4.04), 397s(3.80)	35-0830-76
Benzo[g]pteridine-2,4(3H,10H)-dione, 10-ethyl-, 5-oxide	EtOH	212(3.95),265(4.26), 341(3.70),356s(3.65), 450(3.66)	35-0830-76
	50% EtOH-5M HCl	268(4.27),368(3.84), 395s(3.66)	35-0830-76
$C_{12}H_{10}N_4O_4$			
Benzoic acid, 2-hydroxy-, 2-(3-nitro-4-pyridinyl)hydrazide	pH 1	233(4.33),265(4.17), 306(3.83)	114-0285-76C
Benzoic acid, 4-hydroxy-, 2-(3-nitro-4-pyridinyl)hydrazide	pH 1	234(4.26),270(4.34), 327(3.61)	114-0285-76C
$C_{12}H_{10}N_4O_4S$			
Diazene, (2,5-dimethyl-3-thienyl)-(2,4-dinitrophenyl)-	n.s.g.	218(4.35),378(4.20)	39-1639-76C

Compound	Solvent	$\lambda_{max}(\log \epsilon)$	Ref.
Diazene, (3,5-dimethyl-2-thienyl)-(2,4-dinitrophenyl)-	n.s.g.	418(4.39)	39-1639-76C
2-Thiophenecarboxaldehyde, 4-methyl-, 2,4-dinitrophenylhydrazone	n.s.g.	393(4.39)	39-1639-76C
2-Thiophenecarboxaldehyde, 5-methyl-, 2,4-dinitrophenylhydrazone	n.s.g.	223(4.19),309(3.80), 395(4.43)	39-1639-76C
	n.s.g.	223(4.17),309(3.81), 394(4.44)	39-1639-76C
$C_{12}H_{10}N_4O_6$			
1,1-Hydrazinedicarboxylic acid, 2-[cyano(4-nitrophenyl)methylene]-, dimethyl ester	EtOH	275(4.07),323(3.91)	142-0147-76S
$C_{12}H_{10}N_6O$			
Urea, N-phenyl-N'-1H-pyrazolo[3,4-d]-pyrimidin-4-yl-	pH 1	283(4.30)	87-0555-76
	pH 13	279(4.34),305(3.93)	87-0555-76
	EtOH	280(4.32)	87-0555-76
$C_{12}H_{10}N_6O_3$			
Pyrimido[5,4-e]-1,2,4-triazine-5,7-(6H,8H)-dione, 1,6-dimethyl-3-(3-pyridinyl)-, 4-oxide	dioxan	291(4.46),431(3.48)	39-0713-76C
$C_{12}H_{10}OS_2$			
1-Naphthalenecarbodithioic acid, 2-hydroxy-, methyl ester	heptane-10% CH_2Cl_2	328(3.85),401(3.08), 478s(2.46)	1-0695-76
2-Naphthalenecarbodithioic acid, 1-hydroxy-, methyl ester	heptane-10% CH_2Cl_2	330(4.37),422(4.06), 439(4.04)	1-0695-76
$C_{12}H_{10}O_2$			
1-Azuleneacetic acid	CH_2Cl_2	277(4.72),283(4.70), 288(4.64),332(3.53), 341(3.67),356(3.43), 586(2.53),630(2.47), 691(2.05)	44-1811-76
1-Azulenecarboxylic acid, 3-methyl-	EtOH	291(4.64),296(4.63), 303(4.72),366(3.91), 380(3.99),564(2.63), 665s(2.40)	44-1822-76
5,10-Benzocyclooctenedione, 6,7-dihydro-	EtOH	245(3.95)	138-1011-76
1,2-Naphthalenedione, 5-ethyl-	EtOH	254(4.35),258(4.36), 370(3.43),414(3.47)	78-2693-76
1,4-Naphthalenedione, 5-ethyl-	EtOH	250(4.30),255(4.26), 261(4.05),352(3.52)	78-2693-76
$C_{12}H_{10}O_2S$			
2-Thiophenecarboxylic acid, 5-phenyl-, methyl ester	MeOH	244(4.00),313(4.28)	48-0731-76
$C_{12}H_{10}O_2S_2$			
1,4-Butanedione, 1,4-di-2-thienyl-	EtOH	262(4.27),287(4.20)	12-0339-76
$C_{12}H_{10}O_2Se$			
2-Selenophenecarboxylic acid, 5-phenyl-, methyl ester	MeCN	321(4.29)	118-0521-76
$C_{12}H_{10}O_3$			
1-Azulenecarboxylic acid, 3-methoxy-	CH_2Cl_2	301(4.46),308(4.46), 313(4.53),405s(3.89), 620(2.68)	44-1822-76

Compound	Solvent	$\lambda_{max}(\log \epsilon)$	Ref.
1-Azulenecarboxylic acid, 6-methoxy-	EtOH	302s(4.65),314(4.72), 346(3.92),357(3.90), 370s(3.63),485(2.72)	44-1822-76
2H-1-Benzopyran-2-one, 3-(2-oxopropyl)-	MeOH	276(4.10),310(3.89)	39-1773-76B
1H-Cyclopenta[b]benzofuran-3-one, 2,3-dihydro-7-methoxy-	EtOH	219(4.16),284(4.39), 289(4.38),325(3.75)	18-0737-76
Ethanone, 1-(3,4-dihydroxy-2-naphthalenyl)-	CHCl$_3$	263(4.59),300(4.33), 409(3.87)	18-2790-76
2(3H)-Furanone, 3-[(2-hydroxyphenyl)-methylene]-5-methyl-, cis	MeOH	279(4.08),368s(3.81)	39-1773-76B
trans	MeOH	278(3.84),370(4.04)	39-1773-76B
3(2H)-Furanone, 2-acetyl-5-phenyl-	EtOH	227(4.13),233(4.15), 244(4.11),329(4.40)	39-1688-76C
1,2-Naphthalenediol, monoacetate	CHCl$_3$	278(3.78),288(3.70), 329(3.36)	18-2790-76

$C_{12}H_{10}O_3S$

2-Furancarboxylic acid, 5-[(4-methyl-phenyl)thio]-	EtOH	207(4.17),251(4.19), 294(3.95)	73-2571-76

$C_{12}H_{10}O_4$

6H-Cyclohepta[b]furan-6-one, 3-acetyl-5-hydroxy-2-methyl-	MeOH	266(4.57),320s(3.65), 331s(3.71),346s(3.83), 362(4.00),376(4.04), 402s(3.09)	88-2339-76
1,4-Naphthalenedione, 5,7-dihydroxy-2,3-dimethyl-	EtOH	218(4.40),228s(3.94), 272(4.11),281s(4.02), 435(3.48)	12-2247-76
	Na$_2$HPO$_4$	290(--),512(--)	12-2247-76
1,2,4-Naphthalenetriol, 1-acetate	MeOH	239(4.37),269(3.93), 331(3.49)	12-1087-76
1,2,4-Naphthalenetriol, 2-acetate	MeOH	238(4.56),290(3.66), 301(3.66),332(3.46)	12-1087-76

$C_{12}H_{10}O_4S$

2-Furancarboxylic acid, 5-[(4-methoxy-phenyl)thio]-	EtOH	204(4.22),253(4.25), 297s(3.98)	73-2571-76

$C_{12}H_{10}O_5S$

2-Furancarboxylic acid, 5-[(4-methyl-phenyl)sulfonyl]-	EtOH	205(4.02),229(4.10), 263(4.24)	73-2571-76

$C_{12}H_{10}O_6$

4H,5H-Pyrano[4,3-b]pyran-2-carboxylic acid, 7-methyl-4,5-dioxo-, ethyl ester	CHCl$_3$	248(4.18),301(3.96)	128-0041-76
4H,5H-Pyrano[4,3-b]pyran-7-carboxylic acid, 2-methyl-4,5-dioxo-, ethyl ester	CHCl$_3$	245(4.06),305(3.97)	128-0041-76

$C_{12}H_{10}O_6S$

5,7-Benzo[b]thiophenedicarboxylic acid, 4,6-dihydroxy-, dimethyl ester	EtOH	252(4.60),260s(4.50), 269s(4.43),294(3.99), 302(3.99),341(3.80)	78-0269-76
2-Furancarboxylic acid, 5-[(4-methoxy-phenyl)sulfonyl]-	EtOH	206(4.16),239(4.21), 273(4.28)	73-2571-76
2-Propenoic acid, 3-(1,3-benzodioxol-5-yl)-2-[(carboxymethyl)thio]-	EtOH	323(4.35)	42-0038-76
2H-Pyran-3-carboxylic acid, 4-hydroxy-6-methoxy-2-oxo-5-(2-thienyl)-, methyl ester	EtOH	239(3.82),267(3.81), 315(3.60)	78-0269-76
	EtOH-NaOH	274(3.86)	78-0269-76

Compound	Solvent	$\lambda_{max}(\log \epsilon)$	Ref.
$C_{12}H_{10}S_2$ Thiophene, 2,2'-(1,3-butadiene-1,4-di- yl)bis-	DMF	345s(4.55),361(4.67), 379(4.55)	126-1857-76
$C_{12}H_{11}BrN_5O_6P$ 3H-Imidazo[2,1-i]purine, 5-bromo-3-(3,5- O-phosphinico-β-D-ribofuranosyl)-	pH 1	228(4.49),280(4.09)	94-1561-76
$C_{12}H_{11}BrO$ Naphthalene, 1-(bromomethyl)-6-methoxy-	MeOH	216(4.49),233(3.61), 284(3.76),291(3.74), 323(3.36),332(3.39), 336(3.36)	24-3025-76
$C_{12}H_{11}BrO_2$ 2(5H)-Furanone, 4-bromo-3-ethyl-5-phen- yl-	EtOH	205(4.04),227(3.95)	5-0979-76
$C_{12}H_{11}ClN_2O_2$ 5H-Oxazolo[2,3-b]quinazolin-5-one, 7-chloro-2,3-dihydro-3,3-dimethyl-	MeOH	223(4.51),259(4.04), 269(3.97),318(3.42), 332s(--)	118-0469-76
2,5-Piperazinedione, 3-(chlorophenyl- methylene)-1-methyl-, (E)-	EtOH	224(4.46),259(4.45)	39-0628-76C
$C_{12}H_{11}ClN_4O_2$ Benzoic acid, 2-chloro-, 2,4-diamino- 6-methyl-5-pyrimidinyl ester	EtOH	280(4.08)	94-2461-76
$C_{12}H_{11}ClN_4O_5$ 1H-Pyrrolo[2,3-d]pyrimidine-5-carbo- nitrile, 2-chloro-4,7-dihydro-4-oxo- 7-β-D-ribofuranosyl-	pH 1 pH 11	257(3.72) 265(3.77),277s(3.54)	35-7870-76 35-7870-76
$C_{12}H_{11}ClO$ Naphthalene, 4-chloro-1-methoxy-5-meth- yl-	EtOH	222(4.64),305(3.93), 319(3.83),332(3.67)	95-1161-76
$C_{12}H_{11}ClO_3$ Pentanoic acid, 3-[(4-chlorophenyl)meth- ylene]-	EtOH	222(4.22),282(4.09)	2-0620-76
$C_{12}H_{11}Cl_3N_2$ 2H-Imidazole, 2,2-dimethyl-4-phenyl- 5-(trichloromethyl)-	EtOH	262(3.41)	33-1018-76
$C_{12}H_{11}Cl_3O_3$ 1,3-Butanedione, 1-phenyl-2-(2,2,2-tri- chloro-1-hydroxyethyl)-	$CHCl_3$	252(2.50)	18-2817-76
$C_{12}H_{11}CoS_2$ Cobalt, (η^5-2,4-cyclopentadien-1-yl)- [4-methyl-1,2-benzenedithiolato(2-)- S,S']-	benzene	289(4.46),586(4.18)	101-0369-76Q
$C_{12}H_{11}I_2NO_3S$ Benzenepropanoic acid, 4-hydroxy-3,5-di- iodo-α-isothiocyanato-, ethyl ester	pH 8.06 dioxan	225(4.43),248(--), 311(3.71) 224(4.79),311(3.71)	73-2216-76 73-2216-76

Compound	Solvent	$\lambda_{max}(\log \epsilon)$	Ref.
$C_{12}H_{11}N$			
Carbazole, 1,4-dihydro-	MeOH	226(4.32),278(3.62), 287s(3.54)	117-0223-76
$C_{12}H_{11}NO$			
2-Propenal, 3-(1-methyl-1H-indol-3-yl)-	MeCN	222(4.18),273(3.86), 336(4.30)	5-1039-76
2-Propenal, 3-(2-methyl-1H-indol-3-yl)-	MeCN	227(4.31),268(3.89), 281(3.90),346(4.36)	5-1039-76
Pyridinium, 3-hydroxy-1-methyl-5-phenyl-, hydroxide, inner salt	EtOH	336(3.63)	39-1829-76B
	MeCN	226(4.18),259(4.04), 283(3.94),360(3.56)	39-2329-76C
Pyridinium, 3-hydroxy-1-(phenylmethyl)-, hydroxide, inner salt	EtOH	333(3.72)	39-1829-76B
$C_{12}H_{11}NOS$			
2(1H)-Pyridinethione, 3-(hydroxymethyl)-1-phenyl-	EtOH	238s(3.90),286(4.05), 368(3.85)	1-0863-76
5-Thiazolecarboxaldehyde, 4-methyl-2-(3-methylphenyl)-	n.s.g.	236(3.78),324(4.24)	80-1073-76
5-Thiazolecarboxaldehyde, 4-methyl-2-(4-methylphenyl)-	n.s.g.	238(3.94),330(4.32)	80-1073-76
$C_{12}H_{11}NOS_2$			
4H-1,3-Thiazin-4-one, tetrahydro-6-(2-phenylethenyl)-	MeOH	259(4.54),313(4.26)	73-1388-76
$C_{12}H_{11}NO_2$			
Acetic acid, 4H-quinolizin-4-ylidene-, methyl ester	C_6H_{12}	211(4.27),233s(3.91), 276(3.93),286(3.98), 311(4.11),322(4.18), 451(4.08)	39-0341-76C
	EtOH-HClO₄	235(4.33),290(3.64), 317(4.03),324s(4.0), 330(4.21)	39-0341-76C
Ethanone, 1,1'-(1,3-indolizinediyl)bis-	MeOH	228(4.40),250s(4.65), 254(4.72),284s(4.32), 291(4.41),341(4.66), 350(4.67)	39-1908-76C
Ethanone, 1-(5-methyl-3-phenyl-4-isoxazolyl)-	EtOH	220(4.15)	33-2074-76
Ethanone, 1-(5-methyl-2-phenyl-4-oxazolyl)-	EtOH	217s(4.06),269(4.36), 279s(4.29)	33-2074-76
2-Furancarboxamide, 5-(4-methylphenyl)-	dioxan	303(4.48)	73-2577-76
2H-Indol-2-one, 1,3-dihydro-3-(1-methyl-2-oxopropylidene)-	EtOH	201(4.17),256(4.43), 295(3.61),370(3.02)	94-0782-76
2,7-Methano-1H-cyclopropa[b]naphthalene, 1a,2,7,7a-tetrahydro-3-nitro-, endo	hexane	260(3.48)	44-2006-76
exo	hexane	259(3.88)	44-2006-76
2,7-Methano-1H-cyclopropa[b]naphthalene, 1a,2,7,7a-tetrahydro-4-nitro-, endo	hexane	272.5(3.92)	44-2006-76
exo	hexane	276(3.96)	44-2006-76
	EtOH	290(--)	44-2006-76
1,4-Naphthalenedione, 5-amino-2,3-dimethyl-	12M HCl	247(4.26),274(4.16), 333(3.44)	12-2247-76
	EtOH	278(4.14),500(3.67)	12-2247-76
1H-Naphth[1,2-e][1,3]oxazine, 2,3-dihydro-2-hydroxy-	MeOH	230(4.82),266(3.65), 276(3.73),287(3.62), 319(3.43),332(3.48)	24-1106-76
3H-Naphth[1,2-c]-1,2-oxazin-3-one, 4,4a,5,6-tetrahydro-	MeCN	261(4.29),290s(3.60), 299s(3.43)	33-2499-76

Compound	Solvent	$\lambda_{max}(\log \epsilon)$	Ref.
Spiro[cyclopropane-1,3'-[3H]indol]- 2'(1'H)-one, 2-acetyl-	EtOH	204(4.27),228(4.33), 261(3.76),297(3.32)	94-0782-76
$C_{12}H_{11}NO_2S$			
Ethanone, 2-hydroxy-1-(5-methyl-2-phen- yl-4-thiazolyl)-	40% dioxan	318(4.26)	80-0113-76
2-Propenoic acid, 2-isothiocyanato- 3-phenyl-, ethyl ester, (Z)-	C_6H_{12}	260(4.27),320(4.34)	5-1850-76
$C_{12}H_{11}NO_3$			
Acetic acid, (1,2-dihydro-2-oxo-3H-ind- ol-3-ylidene)-, ethyl ester	MeOH	254(4.34),316(3.86)	94-2305-76
2-Furancarboxamide, 5-(4-methoxyphenyl)-	dioxan	312(4.44)	73-2577-76
1H-Indole-4-carboxylic acid, 3-acetyl- 2-methyl-	n.s.g.	223(4.34),278(3.92), 303(3.91)	39-1073-76C
1-Indolizinecarboxylic acid, 3-acetyl-, methyl ester	MeOH	223s(4.45),230s(4.52), 234(4.54),246s(4.56), 251(4.60),269s(4.48), 276(4.55),342s(4.45), 347(4.48)	39-1908-76C
2-Propenoic acid, 2-cyano-3-hydroxy- 3-phenyl-, ethyl ester	EtOH	286(3.79)	22-0177-76
2-Propenoic acid, 2-cyano-3-methoxy- 3-phenyl-, methyl ester	EtOH	269(3.76)	22-0177-76
$C_{12}H_{11}NO_3S$			
Acetamide, N-(4-hydroxy-2-oxo-2H-1-ben- zothiopyran-3-yl)-N-methyl-	MeOH	231(4.5),307(4.0)	24-3462-76
Acetamide, N-(2-methoxy-4-oxo-4H-1-ben- zothiopyran-3-yl)-	MeOH	248(4.3),326(4.0)	24-3462-76
Acetamide, N-(4-methoxy-2-oxo-2H-1-ben- zothiopyran-3-yl)-	MeOH	223(4.5),302(4.0)	24-3462-76
1-Benzothiepin-4,5-dione, 2,3-dihydro-, 4-(O-acetyloxime)	EtOH	245(4.34),346(3.36)	73-2771-76
5-Thiazolecarboxylic acid, 2-(4-methoxy- phenyl)-, methyl ester	MeCN	327(4.40)	118-0403-76
$C_{12}H_{11}NO_4$			
Benzonitrile, 4-[bis(acetyloxy)methyl]-	MeOH	228(4.28),232(4.26), 237(4.24)	12-0327-76
$C_{12}H_{11}NO_4S$			
Carbamic acid, [2,5-dihydro-4-(mercapto- methyl)-2-oxo-3-furanyl]-, phenyl ester	EtOH	245(3.80)	33-1988-76
$C_{12}H_{11}NO_5S$			
Furan, 2-[[(4-methylphenyl)sulfonyl]- methyl]-5-nitro-	n.s.g.	228(4.30),274s(3.64), 314(4.15)	78-1411-76
$C_{12}H_{11}NO_6$			
Benzoic acid, 2-(1-acetyl-2-oxopropyl)- 3-nitro-	n.s.g.	278(3.88)	39-1073-76C
$C_{12}H_{11}NO_6S$			
Furan, 2-[[(4-methoxyphenyl)sulfonyl]- methyl]-5-nitro-	n.s.g.	211(3.94),242(4.22), 269s(3.56),315(3.97)	78-1411-76
$C_{12}H_{11}N_2OS$			
Thiopyrylium, 2-amino-3-(aminocarbonyl)- 6-phenyl-, perchlorate	HOAc	402(4.11)	48-0705-76

Compound	Solvent	$\lambda_{max}(\log \epsilon)$	Ref.
$C_{12}H_{11}N_2O_2$ Pyrylium, 2-amino-3-(aminocarbonyl)-6-phenyl-, perchlorate	HOAc	384(4.30)	48-0705-76
$C_{12}H_{11}N_2O_3$ Pyridinium, 3-hydroxy-1-methyl-5-(4-nitrophenyl)-, bromide	EtOH	213(4.3),260(4.2), 281(4.3),308(4.2), 350(3.6)	39-2329-76C
$C_{12}H_{11}N_2O_3P$ Phosphonic acid, (2,2-dicyano-1-phenylethenyl)-, dimethyl ester	EtOH	318(3.40)	22-0177-76
$C_{12}H_{11}N_3$ Benzenamine, 2-(phenylazo)-	hexane	256(4.00),313(4.20), 406(3.95)	61-0301-76
Benzenamine, 3-(phenylazo)-	hexane	227(4.27),314(4.16), 370(3.48),459(3.90)	61-0301-76
Benzenamine, 4-(phenylazo)-	hexane	240(4.10),362(4.50), 430s(3.25)	61-0301-76
1H-Pyrrole-3-carbonitrile, 2-amino-4-methyl-5-phenyl-	C_6H_{12} EtOH	363(4.46) 286s(4.05),309(4.20)	18-1381-76 48-0663-76
$C_{12}H_{11}N_3O$ Spiro[3H-indole-3,3'-[3H]pyrazol-2(1H)-one, 1,4'-dimethyl-	EtOH	202(4.03),228(4.60), 264(4.13),279(4.03), 311(4.12),324(4.15)	94-0782-76
$C_{12}H_{11}N_3O_2$ 1,4-Benzenediamine, 2-nitro-N^1-phenyl-	EtOH	494(3.74)	146-0131-76
$C_{12}H_{11}N_3O_2S_2$ 1H-[1,2]Dithiolo[5,1-e][1,2,3]thiadiazole-7-S^{IV}, 3,4-dimethyl-1-(4-nitrophenyl)-	C_6H_{12}	202(4.44),228(4.45), 258(4.12),339(4.19), 349s(4.11),513(4.31)	39-0228-76C
$C_{12}H_{11}N_3O_2Se_2$ 1H-[1,2]Diselenolo[5,1-e][1,2,3]selenadiazole-7-Se^{IV}, 3,4-dimethyl-1-(4-nitrophenyl)-	C_6H_{12}	203(4.40),242(4.27), 266s(4.06),286(4.01), 350(4.15),361s(4.11), 561(4.24)	39-0228-76C
$C_{12}H_{11}N_3O_3$ Benzoic acid, 2-[(3,5-dimethyl-4-isoxazolyl)azo]-	EtOH	315(4.11),419(2.78)	48-0359-76
$C_{12}H_{11}N_3O_3S$ 1H-Pyrimido[5,4-b][1,4]benzothiazine-2,4(3H,4aH)-dione, 4a-hydroxy-1,3-dimethyl-	EtOH	225(4.46),252(4.17), 311(3.91)	94-3135-76
$C_{12}H_{11}N_3O_4$ 1,1-Hydrazinedicarboxylic acid, 2-(cyanophenylmethylene)-, dimethyl ester	EtOH	275(3.91)	142-0147-76S
$C_{12}H_{11}N_5$ 5-Azacinnoline, 4-(3,5-dimethyl-1-pyrazolyl)-	EtOH	204(4.80),260(3.57), 268(3.51),316(3.91)	103-0808-76
1H-Pyrazolo[3,4-d]pyrimidin-4-amine, N-(phenylmethyl)-	pH 1	291.5(4.05)	87-0555-76

Compound	Solvent	$\lambda_{max}(\log \epsilon)$	Ref.
1H-Pyrazolo[3,4-d]pyrimidin-4-amine, N-(phenylmethyl)- (cont.)	pH 13 EtOH	268(4.05) 278(4.19)	87-0555-76 87-0555-76
$C_{12}H_{11}N_5O$ 4(1H)-Pyrimidinone, 2,6-diamino-5-(1H-indol-3-yl)-	EtOH	210(4.82),282(4.37)	103-1246-76
$C_{12}H_{11}N_5O_2$ Acetamide, N-(6,7-dihydro-5-phenyl-[1,2,5]oxadiazolo[3,4-d]pyrimidin-7-yl)-	EtOH	242(4.08),292(3.95)	94-0235-76
4H-1,2,4-Oxadiazol-5(2H)-one, 3-[5-amino-1-(phenylmethyl)-1H-imidazol-4-yl]-	EtOH	276(4.24)	23-2804-76
$C_{12}H_{11}N_5O_3S$ Imidazo[3',4':2,3]pyridazo[6,5-d]pyrimidin-3(9H)-one, 9-acetyl-4-hydroxy-1-methyl-7-(methylthio)-	EtOH	244(4.16),294(4.10), 330(4.25),344(4.35)	94-2637-76
$C_{12}H_{11}N_5O_4$ 5-Pyrimidinol, 2,4-diamino-6-methyl-, 2-nitrobenzoate	EtOH	281(4.10)	94-2461-76
$C_{12}H_{11}N_8O_6P$ 3H-Imidazo[2,1-i]purine, 2-azido-3-(3,5-O-phosphinico-β-D-ribo-furanosyl)-	pH 1 pH 13	234(4.35),287(4.24) 239(4.18),246s(4.14), 274(4.02),284(4.02), 292s(3.92),303s(3.83)	94-1561-76 94-1561-76
$C_{12}H_{12}$ Bicyclo[2.2.2]oct-2-ene, 5,6,7,8-tetra-kis(methylene)-	isooctane	221(3.97),228(4.01), 236(3.98),252(3.96), 260(3.95),296s(3.73)	88-4271-76
Spiro[cyclopropane-1,1'-[1H]indene], 2-methyl-	hexane	230(4.03),240(3.84), 260(3.76),285(3.10), 292s(2.98),297(2.46)	35-3910-76
$C_{12}H_{12}BrN_3O_2$ 5-Pyrimidinecarboxamide, 4-(4-bromophen-yl)-1,2,3,4-tetrahydro-6-methyl-2-oxo-	EtOH	203(4.18),228(4.15), 275(3.65)	103-0191-76
$C_{12}H_{12}BrN_4O_8P$ Inosine, 8-bromo-, cyclic 3',5'-(hydro-gen phosphate) 2'-acetate, triethyl-amine salt	pH 1 pH 7 pH 11	252(4.17) 252(4.17) 257(4.13)	69-0217-76 69-0217-76 69-0217-76
$C_{12}H_{12}ClN$ 2H-Pyrrole, 4-chloro-2,2-dimethyl-5-phenyl-	EtOH	224(3.83),247(3.91), 275s(3.72)	27-0453-76
$C_{12}H_{12}ClN_3O_2$ Glycine, N-(4-chlorophenyl)-N-[(cyano-imino)methyl]-, ethyl ester	EtOH	271(4.30)	49-1413-76
1H-Imidazole-5-carboxylic acid, 4-amino-1-(4-chlorophenyl)-, ethyl ester	EtOH	222(4.13),290(4.04)	49-1413-76
$C_{12}H_{12}CoO_2$ Cobaltocenium, (methoxycarbonyl)-	MeOH	270(4.56),310(3.30), 409(2.42)	101-0189-76I

Compound	Solvent	$\lambda_{max}(\log \epsilon)$	Ref.
$C_{12}H_{12}F_3NO_3$			
Oxazole, 2,5-dihydro-2,5-dimethoxy-4-phenyl-5-(trifluoromethyl)-	MeOH	252(4.19)	35-2605-76
$C_{12}H_{12}F_3N_5$			
1,3,5-Triazine-2,4-diamine, N-(2-phenylethyl)-6-(trifluoromethyl)-	MeOH	213(4.55),273(3.61)	36-0525-76
$C_{12}H_{12}MoO_3$			
Molybdenum, tricarbonyl(1-5-η-butylcyclopentadiene-1,4'-diyl)-	hexane	221s(4.27),254s(3.95), 313(3.38),361s(2.58)	24-1429-76
$C_{12}H_{12}NO$			
Pyridinium, 3-hydroxy-1-methyl-5-phenyl-, bromide	EtOH	208(4.4),235(4.4), 310(3.9),347(3.6)	39-2329-76C
Quinolinium, 8-(ethenyloxy)-1-methyl-, iodide	EtOH	255(4.57),310(3.39), 321(3.44),355(3.40)	65-1108-76
$C_{12}H_{12}N_2$			
Benzenamine, N-(1-methyl-2(1H)-pyridinylidene)-	C_6H_{12}	280(4.07),373(3.61)	18-2770-76
	H_2O	250(3.90),318(4.01)	18-2770-76
	MeOH	260(3.93),320(3.88)	18-2770-76
	$CHCl_3$	288(4.13),366(3.77)	18-2770-76
	CCl_4	283(4.07),374(3.62)	18-2770-76
	MeCN	285(4.05),366(3.68)	18-2770-76
	DMF	288(4.11),370(3.74)	18-2770-76
1,2-Benzenediacetonitrile, α,α'-dimethyl-	EtOH	275(2.73)	78-0487-76
1,3-Benzenediacetonitrile, α,α'-dimethyl-	EtOH	250(2.63),260(2.58), 267s(2.48)	78-0487-76
1,4-Benzenediacetonitrile, α,α'-dimethyl-	EtOH	254(--),260(--), 266(--)	78-0487-76
3-Indolizinecarbonitrile, 2-ethyl-1-methyl-	MeOH	207s(4.23),216s(4.41), 221(4.47),250(4.47), 258(4.59),323s(3.92), 331s(4.00),335(4.00)	39-1908-76C
1H-Pyrazole, 3-(1-methylethenyl)-1-phenyl-	MeOH	276(4.26)	18-0748-76
2-Pyridinamine, N-methyl-N-phenyl-	C_6H_{12}	243(4.05),293(4.04)	18-2770-76
	MeOH	241(3.95),282(3.90), 310(3.77)	18-2770-76
	MeCN	240(3.96),285(3.97), 313(3.81)	18-2770-76
3H-Pyrido[3,4-b]indole, 4,9-dihydro-1-methyl- (harmalan)	MeOH	237(3.89),319(3.86), 346s(3.62)	12-2023-76
	MeOH-acid	246(3.77),349(4.05)	12-2023-76
$C_{12}H_{12}N_2O$			
2,2'-Bipyridine, 5-ethoxy-	EtOH	255(3.79),293(3.97)	64-0115-76B
Dipyrido[1,2-a:2',1'-c]pyrazinediium, 6,7-dihydro-3-hydroxy-, dibromide	pH 6.0	264(3.80),313(3.65), 390(4.06)	64-0115-76B
1(2H)-Isoquinolinone, 3-(2-propenylamino)-	EtOH	228(4.50),244(4.15), 294(4.29),301(4.34), 368(3.80)	95-0154-76
Pyrazolidinium, 3-oxo-1-(3-phenyl-2-propenylidene)-, hydroxide, inner salt	MeOH	354(4.5)	48-0946-76
3(2H)-Pyridazinone, 4,5-dihydro-6-(2-phenylethenyl)-	C_6H_{12}	224(4.13),303s(4.35), 312(4.36),324s(4.27)	59-0837-76
	EtOH	227(4.04),306s(4.65), 316(4.69)	59-0837-76

Compound	Solvent	$\lambda_{max}(\log \epsilon)$	Ref.
3(2H)-Pyridazinone, 6-methyl-4-(4-methylphenyl)-	C_6H_{12}	227(3.98),319(4.05)	59-0837-76
	EtOH	230s(3.80),320(4.12)	59-0837-76
	H_2SO_4	243(3.92),403(4.11)	59-0837-76
3-Pyridinecarboxamide, 1,2-dihydro-1-phenyl-	MeOH	251(5.05),410(4.93)	39-0045-76C
3-Pyridinecarboxamide, 1,4-dihydro-1-phenyl-	MeOH	273(5.04),355(4.80)	39-0045-76C
3-Pyridinecarboxamide, 1,6-dihydro-1-phenyl-	MeOH	246(5.07),372(4.88)	39-0045-76C
$C_{12}H_{12}N_2OS$			
4(1H)-Pyrimidinone, 2-(ethylthio)-6-phenyl-	EtOH	257(4.43),313(3.88)	95-1094-76
4(1H)-Pyrimidinone, 6-methyl-2-[(phenylmethyl)thio]-	EtOH	286(3.93)	95-1094-76
$C_{12}H_{12}N_2OSe_2$			
1H-[1,2]Diselenolo[5,1-e][1,2,3]selenadiazole-7-Se^{IV}, 1-(4-methoxyphenyl)-5-methyl-	C_6H_{12}	216(4.36),249(4.46), 289(4.15),348(3.69), 541(4.23)	39-0228-76C
$C_{12}H_{12}N_2O_2$			
1,1'-Dimethylenedi-2-pyridone	n.s.g.	228(4.23),302(4.16)	33-2841-76
1,7-Ethano-5,8-etheno-1,2,4a,5,6,7,8,8a-octahydro-1,7-naphthyridine-2,6-dione	n.s.g.	265s(3.08)	33-2841-76
Furo[2,3-g]indazole, 2-acetyl-4,5-dihydro-7-methyl-	EtOH	254(3.89),303(3.58)	32-1083-76
4-Isoquinolinecarbonitrile, 3-ethoxy-5,6,7,8-tetrahydro-8-oxo-	EtOH	229(4.49),268(4.17)	2-0964-76
Norharmine, tetrahydroketo-	MeOH	249(4.41),314(4.29)	102-1559-76
5H-Oxazolo[2,3-b]quinazolin-5-one, 2,3-dihydro-3,3-dimethyl-	MeOH	222(4.57),227s(4.50), 245(3.82),255(3.83), 264(3.79),309(3.51), 320s(--)	118-0469-76
1,6-Phenazinedione, 2,3,4,7,8,9-hexahydro-	EtOH	230(4.14),317(4.19)	18-2805-76
3(2H)-Pyridazinone, 4-(4-methoxyphenyl)-6-methyl-	C_6H_{12}	221(4.08),326(4.05)	59-0837-76
	EtOH	228s(3.69),237(3.71), 329(4.11)	59-0837-76
	H_2SO_4	242(3.98),343(4.16), 401s(3.78)	59-0837-76
2,4-Pyrrolidinedione, 3-[1-(phenylamino)ethylidene]-	EtOH	235(4.00),319(4.33)	95-0927-76
3-Quinolinecarboxylic acid, 2-amino-, ethyl ester	CHCl$_3$	253(4.63),292(3.63), 300(3.63),382(3.45)	12-0357-76
$C_{12}H_{12}N_2O_3$			
1H-Imidazole-2-carboxylic acid, 1-phenyl-, 2-hydroxyethyl ester	EtOH	260(4.08)	44-0158-76
1H-Pyrazole-3-carboxylic acid, 2,5-dihydro-5-oxo-1-phenyl-, ethyl ester	EtOH	<u>257(4.1),320(3.3)</u>	23-0488-76
1H-Pyrazole-3-carboxylic acid, 5-hydroxy-1-(phenylmethyl)-, methyl ester	MeOH	223(4.14),258s(3.53), 300(2.60)	24-0261-76
4H-Pyrido[1,2-a]pyrimidine-3-carboxylic acid, 6-methyl-4-oxo-, ethyl ester	pH 7.35	258(4.06),303(3.71), 385s(4.24)	145-0478-76
4H-Pyrido[1,2-a]pyrimidine-9-carboxylic acid, 2-methyl-4-oxo-, ethyl ester	1% MeOH	236(3.95),254(4.00), 335(3.99)	4-0797-76
4(1H)-Pyrimidinone, 2,6-dimethoxy-5-phenyl-	MeOH	280(4.06)	94-0026-76

Compound	Solvent	$\lambda_{max}(\log \epsilon)$	Ref.
3-Quinolinecarboxylic acid, 2-amino-, ethyl ester, 1-oxide	CHCl$_3$	264(4.72),307(3.67), 318(3.62),404(3.54)	12-0357-76
$C_{12}H_{12}N_2O_3S$ Ethanone, 1-(1-methyl-1H-imidazol-5-yl)- 2-(phenylsulfonyl)-	MeOH acid	265(4.19) 246(4.08)	78-1085-76 78-1085-76
$C_{12}H_{12}N_2O_4$ Pyrazolo[1,5-a]pyridine-2-acetic acid, 3-(methoxycarbonyl)-, methyl ester	MeOH	221(4.69),224(3.76), 226s(4.51),242(4.39), 248(4.40),303(4.41), 324s(3.79)	39-1908-76C
$C_{12}H_{12}N_2O_4S$ 2H,6H-Cyclopenta[4,5]thiazolo[3,2-a]- pyrimidine-4-carboxylic acid, 7,8- dihydro-3-hydroxy-2-oxo-, ethyl ester	MeOH	266(4.16),375(3.93)	64-0251-76B
$C_{12}H_{12}N_2O_5S$ 2H-Thiopyrano[2,3-d]pyrimidine-6-carb- oxylic acid, 1,3,4,5-tetrahydro-1,3- dimethyl-2,4,5-trioxo-, ethyl ester	EtOH	243(4.13),314(3.59)	94-1390-76
$C_{12}H_{12}N_2S_2$ 1H-[1,2]Dithiolo[5,1-e][1,2,3]thiadia- zole-7-S^{IV}, 3,4-dimethyl-1-phenyl-	C$_6$H$_{12}$	203(4.39),235(4.41), 254s(4.18),293(4.00), 498(4.12)	39-0228-76C
$C_{12}H_{12}N_2Se$ Benzo[7,8]cycloocta[1,2-d][1,2,3]selena- diazole, 4,5,6,7-tetrahydro-	C$_6$H$_{12}$	225(4.05),248(3.8), 308(2.9)	24-1650-76
$C_{12}H_{12}N_2Se$ 1H-[1,2]Diselenolo[5,1-e][1,2,3]selena- diazole-7-Se^{IV}, 3,4-dimethyl-1-phenyl-	C$_6$H$_{12}$	202(4.48),251(4.46), 286(4.22),345s(3.39), 542(4.14)	39-0228-76C
$C_{12}H_{12}N_4$ Benzenamine, 2,2'-azobis-	benzene	316(4.02),329s(3.88), 420s(3.95),448(4.00), 468(4.01)	33-2906-76
1,3-Benzenediamine, 4-(phenylazo)-	hexane	267(3.95),335(4.10), 386(4.19)	61-0301-76
1H-Pyrrole-3-carbonitrile, 1,2-diamino- 4-methyl-5-phenyl-	MeOH	279(3.98),301(4.02)	48-0663-76
$C_{12}H_{12}N_4O$ Benzoic acid, 2-(3-amino-2-pyridinyl)- hydrazide	EtOH	307(3.57)	114-0301-76C
Benzoic acid, 2-(3-amino-4-pyridinyl)- hydrazide	EtOH	232(4.34),350(3.62)	114-0285-76C
6H-1,2,4,5-Tetrazino[1,6-b]isoquinolin- 6-one, 3-ethyl-1,2-dihydro-	EtOH	233(4.32),316(4.17), 373(3.84),390(3.72)	95-0700-76
[1,2,4]Triazolo[1,5-b]isoquinolin-5(1H)- one, 2-methyl-1-(methylamino)-	EtOH	235(4.33),315(4.18), 371(3.86),398(3.71)	95-0700-76
$C_{12}H_{12}N_4O_2$ Benzoic acid, 2-hydroxy-, 2-(3-amino- 2-pyridinyl)hydrazide	EtOH	210(3.79),242(3.54), 303(3.45)	114-0301-76C
Benzoic acid, 2-hydroxy-, 2-(3-amino- 4-pyridinyl)hydrazide	EtOH	233(4.39),311(4.14)	114-0285-76C

Compound	Solvent	λ_{max} (log ϵ)	Ref.
Benzoic acid, 4-hydroxy-, 2-(3-amino-2-pyridinyl)hydrazide	pH 1	260(4.28),318(3.90)	114-0301-76C
5-Pyrimidinol, 2,4-diamino-6-methyl-, benzoate	EtOH	280(4.09)	94-2461-76
$C_{12}H_{12}N_4O_3$			
Benzoic acid, 2-hydroxy-, 2,4-diamino-6-methyl-5-pyrimidinyl ester	EtOH	320(4.08)	94-2461-76
Benzoic acid, 4-hydroxy-, 2,4-diamino-6-methyl-5-pyrimidinyl ester	EtOH	282(4.16)	94-2461-76
1H-Imidazo[1,2-a][1,3,5]benzotriazepine-8-carboxylic acid, 2,3,5,6-tetrahydro-5-oxo-, methyl ester	EtOH	224(4.42),232(4.40), 293(3.62),318(3.60)	78-0057-76
Oxazolo[4,5-b]quinoxaline-9a(2H)-carboxamide, 3,9-dihydro-6,7-dimethyl-2-oxo-	6M HCl	372(4.17)	5-1276-76
	pH 1.4	340(3.88),370(3.90)	5-1276-76
	pH 7	315(3.73),363(3.90)	5-1276-76
1H-Pyrazole-3-carboxylic acid, 4,5-dihydro-1-methyl-5-oxo-4-(phenylhydrazono)-, methyl ester	MeOH	220(4.03),251(3.83), 258(3.84),372s(4.10), 430(4.30)	24-0261-76
5-Pyrazolecarboxylic acid, 3-hydroxy-1-methyl-4-(phenylazo)-, methyl ester	MeOH	222(4.05),240s(3.88), 250s(3.80),356(4.27)	24-0268-76
$C_{12}H_{12}N_4O_4$			
Glycine, N-[(cyanoimino)methyl]-N-(4-nitrophenyl)-, ethyl ester	EtOH	228(4.11),310(4.22), 377s(--)	49-1413-76
3-Penten-2-one, 3-methyl-4-(5-nitro-2H-benzotriazol-2-yl)-, N-oxide, (Z)-	EtOH	275(4.11),380(3.67)	22-0184-76
5-Pyrimidinecarboxamide, 1,2,3,4-tetrahydro-6-methyl-4-(3-nitrophenyl)-2-oxo-	EtOH	207(4.07),218s(3.97), 269(3.95)	103-0191-76
5-Pyrimidinecarboxamide, 1,2,3,4-tetrahydro-6-methyl-4-(4-nitrophenyl)-2-oxo-	EtOH	203(4.28),218s(4.10), 270(3.65)	103-0191-76
$C_{12}H_{12}N_4O_4Se$			
1H-Pyrrolo[2,3-d]pyrimidine-5-carbonitrile, 4,7-dihydro-7-β-D-ribofuranosyl-4-selenoxo-	pH 1	355(4.18)	4-0745-76
	H_2O	362(4.15)	4-0745-76
	pH 11	334(4.14)	4-0745-76
$C_{12}H_{12}N_4O_6$			
3H-Pyrrolo[2,3-d]pyrimidine-5-carbonitrile, 4,7-dihydro-3-hydroxy-4-oxo-7-β-D-ribofuranosyl-	MeOH	274(3.90),302s(3.67)	35-7870-76
$C_{12}H_{12}N_5O_6PS$			
5H-Imidazo[2,1-i]purine-5-thione, 3,4-dihydro-3-(3,5-O-phosphinico-β-D-ribofuranosyl)-	pH 1	241(4.24),311(4.33)	94-1561-76
	pH 13	249(4.16),280s(3.87), 292s(4.05),318(4.19)	94-1561-76
$C_{12}H_{12}N_5O_7P$			
5H-Imidazo[2,1-i]purin-5-one, 3,4-dihydro-3-(3,5-O-phosphinico-β-D-ribofuranosyl)-	pH 1	260(3.87),296(4.20), 306s(4.10)	94-1561-76
	pH 13	273s(4.15),282(4.21), 292(4.22),303(4.02)	94-1561-76
$C_{12}H_{12}N_6$			
1H-Imidazole-4-carbonitrile, 5-[[4-(dimethylamino)phenyl]azo]-	EtOH	275(3.84),480(4.55)	103-0465-76

Compound	Solvent	$\lambda_{max}(\log \epsilon)$	Ref.
$C_{12}H_{12}O$			
1-Azuleneethanol	CH_2Cl_2	278(4.70),283(4.67), 288(4.58),343(3.63), 359(3.32),596(2.46), 646(2.39),704(1.93)	44-1811-76
6-Azulenemethanol, 2-methyl-	MeOH	235(4.16),266s(4.44), 276(4.80),282(4.78), 286(4.87),304s(3.79), 332(3.63),348(3.73), 358s(3.08),555(2.49), 594s(2.44)	88-2045-76
Bicyclo[5.3.1]undeca-1,3,5,7,9-pentaene, 8-methoxy-	C_6H_{12}	244(4.0),290(4.1), 355(3.4),505(2.5)	35-8277-76
2-Cyclopenten-1-one, 3-methyl-2-phenyl-	EtOH	225(4.17),252s(3.96)	12-0339-76
Naphthalene, 1-methoxy-5-methyl-	EtOH	222(4.66),295(3.87), 309(3.70),323(3.40)	95-1161-76
1-Naphthalenol, 5-ethyl-	EtOH	222(4.58),300(3.83), 312s(3.72),326s(3.54)	95-1161-76
$C_{12}H_{12}OS_2$			
1H-Inden-1-one, 2-(1,3-dithiolan-2-yl)- 2,3-dihydro-	C_6H_{12}	241(4.13),248s(4.09), 278s(3.33),286(3.45), 296(3.44)	148-0435-76
$C_{12}H_{12}OS_3$			
Cyclopentanone, 2-(4,5,6,7-tetrahydro- 4-thioxocyclopenta-1,3-dithiin- 2-ylidene)-	EtOH	258(3.73),345(4.43), 429(3.54)	39-1706-76C
$C_{12}H_{12}O_2$			
1H-2-Benzopyran-1-one, 3-propyl-	ether	266(4.03),277(3.95), 320(3.57)	102-1318-76
1-Benzoxepin-5(2H)-one, 3,4-dihydro- 4-methyl-3-methylene-	n.s.g.	246(3.92),297(3.38)	33-0664-76
1-Benzoxepin-5(2H)-one, 3,4-dimethyl-	n.s.g.	271(3.97),325(3.41)	33-0664-76
1,4:5,8-Dimethanonaphthalene-2,3-dione, 1,4,4a,5,8,8a-hexahydro-	hexane	478(2.14)	118-0256-76
13,14-Dioxatricyclo[8.2.1.1^{4,7}]tetra- deca-4,6,10,12-tetraene	C_6H_{12}	223(4.18),231s(4.01), 238s(3.44),246s(3.09)	89-0442-76
1,4-Ethanonaphthalene-5,8-diol, 1,4-di- hydro-	EtOH	293(3.54)	39-0550-76C
2(5H)-Furanone, 3-ethyl-5-phenyl-	EtOH	205(3.80)	5-0979-76
2(5H)-Furanone, 5-ethyl-5-phenyl-	EtOH	204(4.27)	5-0979-76
1-Naphthalenemethanol, 6-methoxy-	MeOH	215(4.78),228(4.85), 256(3.49),265(3.63), 275(3.70),285(3.59), 304(2.96),316(3.21), 325(3.17),330(3.32)	24-3025-76
Tetracyclo[6.4.0.0^{2,4}.0^{3,7}]dodeca- 1(8),9,11-triene-9,12-diol	EtOH	293(3.36)	39-0550-76C
Tetracyclo[6.4.0.0^{3,11}.0^{6,10}]dodec- 4-ene-2,7-dione	EtOH	307(2.75)	138-1011-76
Tricyclo[5.3.0.0^{2,10}]deca-3,5-dien- 9-one, 8-(hydroxymethylene)-10-methyl-	EtOH	261(4.08),310(3.55)	88-3903-76
$C_{12}H_{12}O_2S_2$			
Cyclopentanone, 2,2'-(1,3-dithietane- 2,4-diylidene)bis-	EtOH	226(3.82),360s(4.31), 373(4.38)	39-1706-76C

Compound	Solvent	$\lambda_{max}(\log \epsilon)$	Ref.
$C_{12}H_{12}O_2S_3$			
Cyclopentanone, 2,2'-(1,2,4-trithiolane-3,5-diylidene)bis-	EtOH	250(3.95),342(4.33)	39-1706-76C
$C_{12}H_{12}O_3$			
4H-1-Benzopyran-4-one, 3-acetyl-2,3-dihydro-6-methyl-	EtOH	225(4.09),258(3.75), 310(3.92),362(3.88)	39-2444-76C
4H-1-Benzopyran-4-one, 3-(hydroxymethyl)-2,6-dimethyl-	EtOH	237(4.09),268s(3.69), 305(3.81)	39-2444-76C
2-Butenoic acid, 4-(2,4-dimethylphenyl)-4-oxo-	n.s.g.	280(3.90)	115-0621-76
2-Butenoic acid, 4-(2,5-dimethylphenyl)-4-oxo-	n.s.g.	215(4.19),275(3.76)	115-0621-76
2-Butenoic acid, 4-(3,4-dimethylphenyl)-4-oxo-	n.s.g.	235(4.08),285(4.03)	115-0621-76
2-Buten-1-one, 2-acetoxy-1-phenyl-	MeCN	222(4.04),295(2.21)	35-8125-76
1-Naphthalenecarboxylic acid, 1,2,3,4-tetrahydro-2-oxo-, methyl ester	MeOH	208(4.42),225(4.19), 275(4.21),302(3.94)	12-1393-76
2(1H)-Naphthalenone, 1-ethyl-1-hydroperoxy-	EtOH	233(4.13),239(4.14), 315(3.90)	39-2570-76C
Pentanoic acid, 4-oxo-3-(phenylmethylene)-	EtOH	278(4.30)	2-0620-76
$C_{12}H_{12}O_4$			
3-Benzofuranpropanoic acid, 5-methoxy-	EtOH	210(4.38),252(3.92), 304(3.61),302(3.57)	18-0737-76
Isochromene-3-carboxylic acid, 1-methoxy-, methyl ester	MeOH	225(3.85),232(3.79), 294(4.03),307s(3.89)	35-5581-76
$(C_{12}H_{12}O_4)_n$			
Benzoic acid, 2-acetoxy-5-ethenyl-, methyl ester, homopolymer	THF	285(3.17)	47-2725-76
$C_{12}H_{12}O_4S$			
2-Propenoic acid, 3-(1,3-benzodioxol-5-yl)-2-(ethylthio)-	EtOH	318(4.19)	2-0547-76
$C_{12}H_{12}O_5$			
Acetic acid, (2,3,5,6,7,8-hexahydro-2,5-dioxo-4H-1-benzopyran-4-ylidene)-, methyl ester	n.s.g.	261(4.06)	44-0743-76
$C_{12}H_{12}O_7$			
Benzeneacetic acid, 4-formyl-3,5-dihydroxy-2-(methoxycarbonyl)-, methyl ester	EtOH	238(4.28),261(4.22), 338(3.61)	35-7733-76
2-Butenedioic acid, 2-(3,4-dihydro-6-methyl-2,4-dioxo-2H-pyran-3-yl)-, dimethyl ester, (Z)-	EtOH	218(4.11),283(3.72)	39-2137-76C
$C_{12}H_{12}O_8S_2$			
2H-1-Benzopyran-2-one, 4-methyl-5,7-bis[(methylsulfonyl)oxy]-	EtOH	282(3.98),310(3.81)	80-1543-76
2H-1-Benzopyran-2-one, 4-methyl-7,8-bis[(methylsulfonyl)oxy]-	EtOH	280(4.00),317(3.80)	80-1543-76
$C_{12}H_{12}S$			
Benzo[b]cyclobuta[d]thiophene, 2a,7b-dihydro-1,2a-dimethyl-	C_6H_{12}	218(4.23),255s(3.91), 258(3.92),297(3.32), 305(3.27)	18-0825-76

Compound	Solvent	$\lambda_{max}(\log \epsilon)$	Ref.
Benzo[b]cyclobuta[d]thiophene, 2a,7b-dihydro-1,7b-dimethyl-	C_6H_{12}	214(4.31),235s(3.79), 256(3.95),263s(3.89), 294(3.30),303(3.25)	18-0825-76
Benzo[b]cyclobuta[d]thiophene, 2a,7b-dihydro-2a,7b-dimethyl-	C_6H_{12}	216(4.21),236(3.70), 255(3.96),293(3.30), 302(3.25)	18-0825-76
$C_{12}H_{12}S_4$ 4H-Cyclopenta-1,3-dithiole, 2-(5,6-di-hydro-4H-cyclopenta-1,3-dithiol-2-ylidene)-5,6-dihydro-	$C_2H_4Cl_2$	307(4.02),327s(3.97), 462(2.25)	44-0730-76
cation	THF	305(4.36),464(2.57)	97-0360-76
	MeCN	337(--),433(--), 460(--),497s(--), 701(--)	97-0360-76
dication	MeCN	438(4.04)	97-0360-76
$C_{12}H_{13}BrCuN_6S_2$ Copper, [[2,2'-[1-(4-bromophenyl)-1,2-ethanediylidene]bis[N-methylhydrazinecarbothioamidato]](2-)-$N^2,N^2{}',S,S']-$, (SP-4-4)-	EtOH	307(4.38),500(3.65)	87-0131-76
$C_{12}H_{13}Cl$ 1,4-Ethanonaphthalene, 1-chloro-1,2,3,4-tetrahydro-	$CHCl_3$	208(3.97),261(2.50)	88-4669-76
$C_{12}H_{13}ClCuN_6S_2$ Copper, [[2,2'-[1-(4-chlorophenyl)-1,2-ethanediylidene]bis[N-methylhydrazine-carbothioamidato]](2-)-$N^2,N^2{}',S,S']-$, (SP-4-4)-	EtOH	307(4.30),500(3.90)	87-0131-76
$C_{12}H_{13}ClN_2$ 4-Quinolinamine, 7-chloro-N-propyl-	pH 6.9	<u>330</u>(4.25),<u>343</u>(4.28)	35-0613-76
$C_{12}H_{13}ClN_2S_2$ Pyrimidine, 6-(chloromethyl)-1,4-dihy-dro-2-(methylthio)-5-(phenylthio)-, hydrochloride	MeOH	242(4.08),312(3.82)	73-2771-76
$C_{12}H_{13}ClO$ Naphthalene, 5-chloro-1,2-dihydro-8-methoxy-4-methyl-	EtOH	228(4.37),265(3.87), 303(3.14)	95-1161-76
$C_{12}H_{13}CuN_7O_2S_2$ Copper, [[2,2'-[1-(4-nitrophenyl)-1,2-ethanediylidene]bis[N-methylhydrazine-carbothioamidato]](2-)]-, (SP-4-4)-	3:1 $CHCl_3$-MeOH	307(4.18),463(3.66)	87-0131-76
$C_{12}H_{13}FN_2$ 2H-Imidazole, 4-(fluoromethyl)-2,2-di-methyl-5-phenyl-	EtOH	261(3.84)	33-1018-76
$C_{12}H_{13}FN_2O_5$ 2,5'-Anhydro-4'-fluoro-2',3'-O-isopro-pylideneuridine	MeOH	237(4.13)	44-3010-76
$C_{12}H_{13}N$ 3-Azabicyclo[3.1.0]hex-2-ene, 1-methyl-2-phenyl-	C_6H_{12}	239(4.04)	44-0180-76

Compound	Solvent	$\lambda_{max}(\log \epsilon)$	Ref.
1,2-Butanoindole	EtOH	226(4.52),284f(3.93)	20-0819-76
1H-Pyrrole, 2-methyl-5-(4-methylphenyl)-	EtOH	295(4.38)	23-1827-76
$C_{12}H_{13}NO$			
2H-1-Benzazepin-2-one, 1,5-dihydro-4,7-dimethyl-	EtOH	235(4.46),255s(3.95),295s(3.11)	94-1544-76
2H-1-Benzazepin-2-one, 1,5-dihydro-4,8-dimethyl-	EtOH	232(4.41),254s(3.92),290s(3.11)	94-1544-76
Benz[g]indole, tetrahydro-3a-hydroxy-	EtOH	210(4.18),250(4.14)	2-0819-76
1-Carbazolol, tetrahydro-	n.s.g.	229(4.22),284(3.70),293(3.62)	44-0102-76
Ethanone, 1-(1,2-dimethyl-3-indolizinyl)-	MeOH	205(4.24),227s(4.41),229(4.45),250s(4.17),255s(4.22),270s(4.46),276(4.54),374(4.29),385s(4.28)	39-1908-76C
s-Indacen-1(2H)-one, 3,5,6,7-tetrahydro-, oxime	MeOH	213(4.41),257(4.14),301(3.93),313(3.90)	73-2020-76
1H-Pyrrole, 2-(4-methoxyphenyl)-5-methyl-	EtOH	285(4.29)	23-1827-76
3H-Pyrrol-3-one, 4,5-dihydro-5,5-dimethyl-2-phenyl-	EtOH	217(3.90),269(3.87)	12-2511-76
$C_{12}H_{13}NOS$			
Ethanone, 1-(4-methylphenyl)-2-thiazolidinylidene-	MeOH	255(4.20),341(4.56)	48-0298-76
$C_{12}H_{13}NO_2$			
1,3-Benzodioxole-2-butanenitrile, 2-methyl-	MeOH	232(3.48),284(3.60)	33-0999-76
1H-Indole-2-acetic acid, ethyl ester	MeOH	219(4.55),272(3.93),280(3.91),289(3.79)	24-3282-76
3-Indolizinol, 2-methyl-, propanoate	EtOH	232(4.44),243s(4.31),272(3.48),281(3.51),292(3.49),344(3.33),353s(3.30),372s(2.94)	1-0198-76
2H-Pyrrol-2-one, 4-ethyl-1,5-dihydro-3-hydroxy-5-phenyl-	EtOH	213(4.21),241(3.74)	40-1100-76
3H-Pyrrol-3-one, 4,5-dihydro-5,5-dimethyl-2-phenyl-, 1-oxide	EtOH	240(4.23),305(4.13)	12-2561-76
$C_{12}H_{13}NO_2S$			
Benzenepropanoic acid, α-isothiocyanato-, ethyl ester	pH 8.06	248(3.10)	73-2216-76
	MeOH	248(3.07)	73-2216-76
	dioxan	251(3.15)	73-2216-76
$C_{12}H_{13}NO_3$			
Benzoic acid, 3-[(1-methyl-3-oxobutylidene)amino]-	EtOH	213(4.31),228(4.07),326(4.33)	40-0451-76
Benzoic acid, 4-[(1-methyl-3-oxobutylidene)amino]-	EtOH	207(3.41),232(3.88),338(4.29)	40-0451-76
1H-Indole-2-acetic acid, 1-hydroxy-, ethyl ester	MeOH	221(4.57),273(3.86),284(3.84),298s(--)	24-3282-76
2-Propenenitrile, 3-(2,4,5-trimethoxyphenyl)-	EtOH	240(4.11),289(4.15),350(4.17)	39-0499-76C
2(1H)-Quinolinone, 4,6-dimethoxy-1-methyl-	EtOH	232(4.64),273(3.85),282(3.82),341(3.80)	105-0282-76
$C_{12}H_{13}NO_3S$			
Benzenepropanoic acid, 4-hydroxy-α-isothiocyanato-, ethyl ester	pH 8.06	226(3.99),248(--)	73-2216-76

Compound	Solvent	$\lambda_{max}(\log \epsilon)$	Ref.
Benzenepropanoic acid, 4-hydroxy-α-iso- thiocyanato-, ethyl ester (cont.)	MeOH dioxan	228(4.19) 229(3.98),311(3.71)	73-2216-76 73-2216-76
$C_{12}H_{13}NO_4S$ Butanedioic acid, (1-methyl-2(1H)-pyri- dinylidene)thioxo-, dimethyl ester	EtOH	220(3.82),264(3.81), 332(4.26)	142-0705-76
$C_{12}H_{13}NO_{10}$ 1,1-Cyclopropanedicarboxylic acid, 2-[2-methoxy-1-(methoxycarbonyl)- 2-oxoethylidene]-3-aci-nitro-, dimethyl ester	EtOH	230(4.14),350(4.16)	104-0668-76
$C_{12}H_{13}N_3O$ 1(2H)-Isoquinolinone, 3-[(1-methyleth- ylidene)hydrazino]-	EtOH	232(4.46),311(4.49), 376(3.83)	95-0154-76
4-Pyrimidinamine, 6-methoxy-N-(phenyl- methyl)-	pH 1 pH 13	258(4.10) 244(4.02)	39-1847-76C 39-1847-76C
$C_{12}H_{13}N_3O_2$ Glycine, N-[(cyanoimino)methyl]-N-phen- yl-, ethyl ester	EtOH	264(4.20)	49-1413-76
1H-Imidazole-5-carboxylic acid, 4-amino- 1-phenyl-, ethyl ester	EtOH	247(3.76),291(4.02)	49-1413-76
5-Pyrimidinecarboxamide, 1,2,3,4-tetra- hydro-6-methyl-2-oxo-4-phenyl-	EtOH	203(4.14),218s(3.9), 275(3.76)	103-0191-76
$C_{12}H_{13}N_3O_4S$ Acetic acid, cyano(1,3-dimethyl-6-ura- cilylthiomethylene)-, ethyl ester	EtOH	220(5.13),299(4.74)	94-1390-76
$C_{12}H_{13}N_3O_5$ Pyrido[2,3-d]pyrimidine-5-carboxylic acid, 1,2,3,4,7,8-hexahydro-1,3-di- methyl-2,4,7-trioxo-, ethyl ester	pH 1 pH 7 pH 11	276(4.16),317(4.33) 283(4.26),331(4.40) 283(4.25),331(4.39)	44-1095-76 44-1095-76 44-1095-76
Pyrido[2,3-d]pyrimidine-5-carboxylic acid, 1,2,3,4-tetrahydro-7-methoxy- 1,3-dimethyl-2,4-dioxo-, methyl ester	pH 1	261(3.86),307(4.07)	44-1095-76
$C_{12}H_{13}N_5O$ 2,4-Pentanedione, pyrido[3,2-c]pyrida- zin-4-ylhydrazone	EtOH	235(4.30),282(4.13), 315(4.32),442(4.46), 468(4.55)	103-0808-76
$C_{12}H_{13}N_5O_2$ 4(1H)-Pyrimidinone, 6-[(2-hydroxyethyl)- amino]-5-(phenylazo)-	pH 13	243(4.28),385(4.27)	39-1847-76C
$C_{12}H_{13}N_5O_4S$ Pyrimido[4,5-c]pyridazine-3-carboxamide, 1,2-diacetyl-1,2-dihydro-4-hydroxy- 7-(methylthio)-	EtOH	277(4.20),340(4.34)	94-2637-76
$C_{12}H_{13}N_5O_5$ 9,5'-Cyclo-3-(2,3-O-isopropylidene- β-D-ribofuranosyl)-8-azaxanthine	MeOH	235(3.99),255(4.02)	44-1100-76
1H-Pyrrolo[2,3-d]pyrimidine-5-carbo- nitrile, 2-amino-4,7-dihydro-4-oxo- 7-β-D-ribofuranosyl-	pH 1 pH 11	227(4.24),268(3.93), 288(3.84) 226(4.30),268(3.81), 286(3.85)	35-7870-76 35-7870-76

Compound	Solvent	$\lambda_{max}(\log \epsilon)$	Ref.
$C_{12}H_{13}N_5O_7$ 1,4-Diazepinium, 2,3-dihydro-5-methyl-, picrate	MeOH	330(4.42)	39-1784-76C
$C_{12}H_{13}N_6O_6P$ 3H-Imidazo[2,1-i]purin-5-amine, 3-(3,5-O-phosphinico-β-D-ribofuranosyl)-	pH 1 pH 13	274(4.15),294(4.12) 277s(4.14),284(4.16), 294(4.13),306s(3.93)	94-1561-76 94-1561-76
$C_{12}H_{13}N_7O_2$ 2,4(1H,3H)-Pyrimidinedione, 1-[3-(6-amino-9H-purin-9-yl)propyl]-	H_2O	261(4.33)	19-0517-76
$C_{12}H_{14}$ Azulene, 4,5-dihydro-4,4-dimethyl- Benzene, 1,3-hexadienyl- Bicyclo[2.2.2]octane, 2,3,5,6-tetrakis-(methylene)-	C_6H_{12} EtOH isooctane	317(4.10) 285(4.30) 231(4.01),240(3.99), 252(3.97),258(3.96), 268s(3.73)	35-7095-76 44-0745-76 88-4271-76
$C_{12}H_{14}BrNO_4$ Benzenemethanol, 4-bromo-α-(1-methyl-1-nitroethyl)-, acetate	MeOH	265(2.83),271(2.73)	12-0327-76
$C_{12}H_{14}BrN_3O_7$ 1,2,4-Triazine-3,5(2H,4H)-dione, 2-(3,5-di-O-acetyl-2-bromo-2-deoxy-β-D-ribofuranosyl)-	MeOH	259(3.75)	73-2110-76
$C_{12}H_{14}BrN_5O_5$ Uridine, 5'-azido-5-bromo-5'-deoxy-2',3'-O-(1-methylethylidene)-	MeOH	272(4.01)	44-1100-76
$C_{12}H_{14}ClNO_3$ Valine, N-(4-chlorobenzoyl)-, (+)-	EtOH	238(4.16)	12-0339-76
$C_{12}H_{14}ClNO_4$ Benzenemethanol, 2-chloro-α-(1-methyl-1-nitroethyl)-, acetate Benzenemethanol, 4-chloro-α-(1-methyl-1-nitroethyl)-, acetate Benzoic acid, 4-(1-chloro-2-methyl-2-nitropropyl)-, methyl ester	MeOH MeOH MeOH	268(2.72),276(2.69) 224(4.10),257(2.83), 265(2.76) 237(4.27),277(3.11), 284(3.04)	12-0327-76 12-0327-76 12-2631-76
$C_{12}H_{14}ClN_3O_7$ 1,2,4-Triazine-3,5(2H,4H)-dione, 2-(3,5-di-O-acetyl-2-deoxy-2-chloro-β-D-ribofuranosyl)-	MeOH	259(3.72)	73-2110-76
$C_{12}H_{14}ClN_3O_7S$ 5-Thia-1-azabicyclo[4.2.0]oct-2-ene-2-carboxylic acid, 3-[[(aminocarbonyl)oxy]methyl]-7-[(chloroacetyl)-amino]-7-methoxy-8-oxo-, (6R-cis)-	pH 6.86	267(3.83)	94-2629-76
$C_{12}H_{14}Cl_4S_2$ 1,3-Cyclopentadiene, 5-[bis[(ethylthio)-methyl]methylene]-1,2,3,4-tetrachloro-	C_6H_{12}	298(4.22),350(3.62)	24-3929-76

Compound	Solvent	$\lambda_{max}(\log \epsilon)$	Ref.
$C_{12}H_{14}Co$ Cobaltocenium, 1,1'-dimethyl-	MeOH	270(4.59),314(3.17), 403(2.49)	101-0189-76I
$C_{12}H_{14}CuN_6OS_2$ Copper, [[2,2'-[1-(4-hydroxyphenyl)-1,2-ethanediylidene]bis[N-methylhydrazine-carbothioamidato]](2-)-N^2,$N^2{}'$,S,S']-, (SP-4-4)-	EtOH	307(4.45),505(3.81)	87-0131-76
$C_{12}H_{14}CuN_6S_2$ Copper, [[2,2'-(1-phenyl-1,2-ethanedi-ylidene)bis[N-methylhydrazinecarbo-thioamidato]](2-)-N^2,$N^2{}'$,S,S']-, (SP-4-4)-	EtOH	306(4.42),500(3.78)	87-0131-76
$C_{12}H_{14}FIN_2O_5$ Uridine, 5'-deoxy-4'-C-fluoro-5'-iodo-2',3'-O-(1-methylethylidene)-	MeOH	257(3.97)	44-3010-76
$C_{12}H_{14}FN_5O_5$ Uridine, 5'-azido-5'-deoxy-4'-C-fluoro-2',3'-O-(1-methylethylidene)-	MeOH	256(4.00)	44-3010-76
$C_{12}H_{14}N$ Quinolinium, 1,2,3-trimethyl-, perchlorate	EtOH	315(3.99)	103-0683-76
methylene base	EtOH	380(3.78)	103-0683-76
$C_{12}H_{14}NOS$ Isoxazolium, 2-ethyl-5-(methylthio)-3-phenyl-, iodide	EtOH	261(3.99),302(4.24)	97-0270-76
$C_{12}H_{14}NO_3$ 1,3-Dioxolo[4,5-g]isoquinolinium, 7,8-dihydro-9-methoxy-6-methyl-, iodide	MeOH	220(4.47),245s(4.21), 350(4.02)	12-2003-76
$C_{12}H_{14}N_2$ Propanedinitrile, [4-(1-methylethyli-dene)cyclohexylidene]-	C_6H_{12}	228(4.11),272(3.52)	78-2827-76
$C_{12}H_{14}N_2O$ 3-Pyridinecarboxamide, 1,2,5,6-tetra-hydro-1-phenyl-	MeOH	250(5.07),285s(4.36)	39-0045-76C
3-Pyridinecarboxamide, 1,4,5,6-tetra-hydro-1-phenyl-	MeOH	240(4.61),318(5.46)	39-0045-76C
Pyridinediium, 2,2'-oxybis[1-methyl-, difluorosulfate	pH 5.0	281(4.25)	64-0122-76B
2H-Pyrrol-2-one, 3-amino-4-ethyl-1,5-dihydro-5-phenyl-	EtOH	207(4.30),243(4.25), 340(3.18)	40-1100-76
Pyrrolo[1,2-a]pyrimidine-5-carboxalde-hyde, 1,3,6,7-tetramethyl-	EtOH	239(4.40),267s(3.60), 279s(3.30),349(4.14)	44-0351-76
Spiro[3H-indole-3,3'-pyrrolidin]-2(1H)-one, 1'-methyl-	MeOH	250(3.89),272s(3.20)	87-0892-76
$C_{12}H_{14}N_2OS_2$ 4-Pyrimidinemethanol, 1,6-dihydro-2-(methylthio)-5-(phenylthio)-	MeOH	236(4.01),290(3.78)	73-2771-76

Compound	Solvent	λ_{max}(log ϵ)	Ref.
$C_{12}H_{14}N_2O_2$			
2-Butenoic acid, 4-(2,4-dimethylphenyl)-4-oxo-, hydrazone	n.s.g.	287(4.25)	115-0621-76
4H-Imidazole, 2,4,4-trimethyl-5-phenyl-, 1,3-dioxide	EtOH	285(3.94)	103-1280-76
Indolizine, 2-(1,1-dimethylethyl)-6-nitro-	EtOH	228(4.30),303(4.23), 430(3.09)	103-0766-76
3H-Indol-3-one, 2-[(1,1-dimethylethyl)-imino]-1,2-dihydro-, N-oxide	MeOH	225(4.07),267(4.15), 347s(--),357(3.93), 471(3.81)	24-0200-76
1,2,4-Oxadiazole, 3-butoxy-5-phenyl-	MeOH	255(4.18)	18-3607-76
Propanoic acid, 2-(1H-indol-3-yl)-3-amino-, methyl ester, hydrochloride	EtOH	210(4.57),280(3.78), 288(3.73)	104-1087-76
1H-Pyrazole-5-carboxylic acid, 4,5-dihydro-3-(2,4-dimethylphenyl)-	n.s.g.	210(4.63)	115-0621-76
3H-Pyrrol-3-one, 4,5-dihydro-5,5-dimethyl-2-phenyl-, oxime, 1-oxide	EtOH	241(4.22),298(4.20)	12-2561-76
$C_{12}H_{14}N_2O_3$			
Furo[2,3-b]pyridine-2-carboxylic acid, 3-amino-4,6-dimethyl-, ethyl ester	EtOH	261(4.08),304(4.09), 330(3.94)	48-0313-76
3-Heptenenitrile, 5-(1-hydroxyethylidene)-2-[2-(methoxyimino)ethylidene]-6-oxo-	CHCl$_3$ MeOH	336(4.37) 375(--)	5-1799-76 5-1799-76
2,4-Octadienenitrile, 6-(1-hydroxyethylidene)-2-[(methoxyimino)methyl]-7-oxo-	MeOH	391(4.50)	5-1799-76
1-Pyridineacetic acid, 3-cyano-4,6-dimethyl-2-oxo-1,2-dihydro-, ethyl ester	MeOH	235(3.85),333(4.11)	48-0313-76
Pyrrolo[2,3-b]indole-1(2H)-carboxylic acid, 3,3a,8,8a-tetrahydro-3a-hydroxy-, methyl ester	EtOH	242(3.94),298(3.38)	35-0635-76
$C_{12}H_{14}N_2O_4$			
1,2-Pentanedione, 4-methyl-4-nitro-1-phenyl-, 2-oxime	EtOH	213(3.92),250(4.06)	12-2561-76
$C_{12}H_{14}N_2O_6$			
1-Propanol, 2-methyl-2-nitro-1-(2-nitrophenyl)-, acetate	MeOH	257(3.66)	12-0327-76
1-Propanol, 2-methyl-2-nitro-1-(3-nitrophenyl)-, acetate	MeOH	259(3.91)	12-0327-76
1-Propanol, 2-methyl-2-nitro-1-(4-nitrophenyl)-, acetate	MeOH	263(4.06)	12-0327-76
$C_{12}H_{14}N_2O_6S$			
2-Butenedioic acid, 2-[[(1,2,3,6-tetrahydro-1,3-dimethyl-2,6-dioxo-4-pyrimidinyl)thio]-, dimethyl ester, (E)-	EtOH	280(4.13)	94-1390-76
$C_{12}H_{14}N_2S_2$			
2,4-Imidazolidinedithione, 5,5-dimethyl-3-(phenylmethyl)-	EtOH	298(4.41),384(1.79)	33-2566-76
$C_{12}H_{14}N_3O_2$			
Isoxazolium, 5-methoxy-2,3-dimethyl-4-(phenylazo)-, perchlorate	HOAc	235(4.01),389(4.11)	48-0359-76
tetrafluoroborate	HOAc	233(4.01),386(4.10)	48-0359-76

Compound	Solvent	$\lambda_{max}(\log \epsilon)$	Ref.
$C_{12}H_{14}N_4OS$			
Benzenamine, N,N-dimethyl-4-[(4-methyl-2-thiazolyl)azo]-, N-oxide	n.s.g.	550(4.68)	64-0981-76B
Hydrazinecarbothioamide, 2-(1,2-dihydro-1-methyl-3-oxo-2H-indol-2-ylidene)-N,N-dimethyl-	EtOH	275(4.28),342(4.13), 522(3.79)	103-0076-76
Hydrazinecarbothioamide, 2-(1,2-dihydro-3-oxo-3H-indol-2-ylidene)-N,N,S-trimethyl-	EtOH	237(4.09),272(4.12), 358(4.07),481(4.15)	103-0076-76
$C_{12}H_{14}N_4O_2$			
Propanoic acid, 2-(2-amino-1,2-dihydro-1-oxo-3-isoquinolinyl)hydrazide	EtOH	231(4.33),300(4.01), 358(3.67)	95-0700-76
$C_{12}H_{14}N_4O_4$			
2,4(1H,3H)-Pyrimidinedione, 1-[3-(3,4-dihydro-2,4-dioxo-1(2H)-pyrimidinyl)-propyl]-5-methyl-	H_2O	269(4.22)	56-2041-76
$C_{12}H_{14}N_4O_5$			
Oxireno[5,6]benzo[2,1-c:3,4-c']dipyrazole-7a,7b-dicarboxylic acid, 3,3a,3b-4a,4b,5-hexahydro-, dimethyl ester, (3aα,3bα,4aα,4bα,7aα,7bα)-	MeOH	321(2.56)	24-2823-76
$C_{12}H_{14}N_5O_8P$			
Guanosine, cyclic 3',5'-(hydrogen phosphate), 2'-acetate	pH 1	256(4.15),278s(4.00)	69-0217-76
	pH 7	252(4.20),274s(4.03)	69-0217-76
	pH 11	257(4.16)	69-0217-76
$C_{12}H_{14}N_6O_4$			
7,11-Epoxy-1H,6H-8,10-dioxa-2,4,5,5a-11a-pentaazacyclopenta[5,6]cyclooct-[1,2,3-cd]inden-1-one, 3-amino-7,7a,10a,11-tetrahydro-9,9-dimethyl-, [7R-(7α,7aβ,10aβ,11α)]-	MeOH	223(4.15),253(3.81), 271(3.85)	44-1100-76
$C_{12}H_{14}O$			
2H-1-Benzopyran, 2,2,3-trimethyl-	C_6H_{12}	221(4.27),260s(3.55), 267(3.60),275s(3.55), 308(3.43),314s(3.37)	22-2039-76
	EtOH	219(4.34),260s(3.71), 265(3.72),272s(3.65), 306(3.50),316s(3.43)	22-2039-76
Methanone, (2,2-dimethylcyclopropyl)-phenyl-	EtOH	245(4.11)	44-3067-76
2(4aH)-Naphthalenone, 5,6-dihydro-4a,8-dimethyl-	EtOH	224(4.09),247(3.94), 300(3.04)	44-3632-76
$C_{12}H_{14}O_2$			
1,3-Butanedione, 2,2-dimethyl-1-phenyl-	MeOH	243(3.98),275(3.08)	23-1197-76
2-Butenoic acid, 3-phenyl-, ethyl ester, (E)-	EtOH	265(4.20)	18-3239-76
3-Butenoic acid, 3-phenyl-, ethyl ester	EtOH	243(4.03)	18-3239-76
Cyclopenta[b][1]benzopyran-9(1H)-one, 2,3,5,6,7,8-hexahydro-	MeOH	256.5(4.15)	5-1689-76
1,4:5,8-Dimethanonaphthalene-2,3-dione, octahydro-	hexane	473(1.78)	118-0256-76
1(2H)-Naphthalenone, 3,4-dihydro-6-methoxy-4-methyl-	EtOH	227(4.25),276(4.20)	2-0312-76

Compound	Solvent	λ_{max}(log ϵ)	Ref.
1,2-Pentanedione, 4-methyl-1-phenyl-	MeCN	215(3.26),280(2.70)	35-8125-76
Tricyclo[5.2.0.0^{2,9}]nona-3,5-diene-8-carboxylic acid, 9-methyl-, methyl ester	EtOH	233s(3.33),275(3.47), 314s(2.50)	88-3903-76
$C_{12}H_{14}O_3$			
1,3-Benzodioxole-2-butanal, 2-methyl-	MeOH	232(3.50),285(3.62)	33-0999-76
1H-2-Benzopyran-4-ol, 4-ethenyl-3,4-di-hydro-7-methoxy-	95% MeOH	217(3.78),227(3.83), 278(3.65)	13-0197-76B
2-Cyclohexen-1-one, 3-hydroxy-2-(3-oxo-1-cyclohexenyl)-	EtOH	236(4.10),261(4.24)	94-0591-76
1H-Cyclopenta[b]benzofuran-3-ol, 2,3-di-hydro-7-methoxy-	EtOH	204(4.16),230(3.62), 300(3.46)	18-0737-76
1,4-Pentanedione, 1-(4-methoxyphenyl)-	EtOH	218(4.21),272(4.24)	12-0339-76
1-Propanone, 2-acetoxy-2-methyl-1-phen-yl-	EtOH	249(4.12),269(3.00), 314(1.98)	44-3067-76
2-Propenoic acid, 3-(4-methoxyphenyl)-, ethyl ester	EtOH	273(4.32)	100-0218-76
$C_{12}H_{14}O_3S$			
2-Propenoic acid, 2-(ethylthio)-3-(4-methoxyphenyl)-	EtOH	295(4.14)	2-0547-76
$C_{12}H_{14}O_4$			
Benzenepropanoic acid, 2-(acetyloxy)-, methyl ester	n.s.g.	268(3.57)	39-0407-76C
3-Benzofuranpropanoic acid, 2,3-dihydro-5-methoxy-	EtOH	204(4.21),230(3.80), 299(3.68)	18-0737-76
6H-2-Benzopyran-6-one, 7,8-dihydro-7,8-dihydroxy-3,5,7-trimethyl-, (7R,8R)-	EtOH	231(3.84),352(4.30)	39-0204-76C
(7R,8S)-	EtOH	234(3.81),356(4.27)	39-0204-76C
Bicyclo[4.2.0]octa-2,4-diene-7,8-dicarb-oxylic acid, dimethyl ester, cis	MeCN	274(3.54),295s(--)	89-0779B-76
trans	MeCN	274(3.54),295s(--)	89-0779B-76
Bicyclo[4.2.0]octa-2,4-diene-7,8-diol, diacetate	MeCN	262s(--),267(3.58), 286s(--),291s(--)	89-0779B-76
at -10°	MeCN	275s(--),286s(--), 297(--),310(4.78), 323(--)	89-0779B-76
2,4,6,8-Decatetraenedioic acid, dimeth-yl ester, (all-E)-	MeCN	304(4.34),319s(4.64), 333(4.83),349(4.79)	89-0779B-76
3-Furancarboxylic acid, 5-(2-butenyli-dene)-4,5-dihydro-2-methyl-4-oxo-	EtOH	208(3.65),298(4.12), 337(4.15)	4-0521-76
$C_{12}H_{14}O_5$			
Benzeneacetaldehyde, α-(hydroxymethyl-ene)-3,4,5-trimethoxy-	pH 1	238(4.23)	44-0294-76
	pH 6.8	273(4.35)	44-0294-76
	pH 13	273(4.35)	44-0294-76
	90% H_2SO_4	252(4.43),300s(3.55)	44-0294-76
Methanediol, (3-methoxyphenyl)-, diace-tate	MeOH	275(3.36),282(3.32)	12-0327-76
3-Oxatetracyclo[6.1.0.0^{2,4}.0^{5,7}]nonane-7,8-dicarboxylic acid, dimethyl ester, cis	MeOH	229(2.30)(end abs.)	24-2823-76
trans	MeOH	230(2.40)	24-2823-76
3-Oxatricyclo[5.1.0.0^{2,4}]oct-5-ene-6,7-dicarboxylic acid, 5-methyl-, dimethyl ester	MeOH	228(3.79)	24-2823-76
4H-Oxocin-5,6-dicarboxylic acid, 7-methyl-, dimethyl ester	MeOH	330(2.70)	24-2823-76

Compound	Solvent	$\lambda_{max}(\log \epsilon)$	Ref.
5,6-Oxoninedicarboxylic acid, 4,7-di- hydro-, dimethyl ester	EtOH	240s(3.48)	24-2823-76
$C_{12}H_{14}O_5S$ 4H-1-Benzopyran-4-one, 2,3-dihydro-2,2- dimethyl-7-[(methylsulfonyl)oxy]-	EtOH	258(4.09),322(3.66)	80-1543-76
$C_{12}H_{14}O_6$ Benzeneacetic acid, 3,5-dihydroxy- 2-(methoxycarbonyl)-4-methyl-, methyl ester	EtOH	269(4.14),304(3.69)	35-7733-76
2-Propenoic acid, 3-(3,4-dihydroxyphen- yl)-, 1,2,3-propanetriyl ester	MeOH	220(3.76),248(3.71), 325(3.94)	100-0409-76
$C_{12}H_{14}O_7$ 2-Propenoic acid, 3-[3,5-bis(hydroxy- methyl)-4-methoxy-2-oxo-2H-pyran-6- yl]-, methyl ester, (E)- (rosellisin)	EtOH	228(4.34),333(4.10)	102-1090-76
$C_{12}H_{15}BrN_2O$ Phenol, 4-bromo-2-[(1-piperidinylimino)- methyl]-	benzene DMF	298(4.18),327(4.11) 295(4.20),325(4.13)	104-0625-76 104-0625-76
$C_{12}H_{15}BrN_6S_2$ Hydrazinecarbothioamide, 2,2'-[1-(4-bro- mophenyl)-1,2-ethanediylidene]- bis[N-methyl-	EtOH	337(4.53)	87-0131-76
$C_{12}H_{15}ClN$ Methanaminium, N-[3-chloro-3-(4-methyl- phenyl)-2-propenylidene]-N-methyl-, perchlorate	MeOH	250(4.05),323(3.91)	48-0731-76
$C_{12}H_{15}ClN_2S_2$ 2,6(1H,3H)-Pyridinedithione, 5-chloro- 3-[(cyclohexylamino)methylene]-	MeOH	297(3.73),364(4.26), 432(4.56)	33-0190-76
$C_{12}H_{15}ClN_4O_6$ 1,2,4-Triazin-3(2H)-one, 5-amino- 2-(3,5-di-O-acetyl-2-chloro-2-deoxy- β-D-ribofuranosyl)-	MeOH	264(3.92)	73-2110-76
$C_{12}H_{15}ClN_6S_2$ Hydrazinecarbothioamide, 2,2'-[1-(4- chlorophenyl)-1,2-ethanediylidene]- bis[N-methyl-	MeOH	337(4.42)	87-0131-76
$C_{12}H_{15}FN_2O_6$ Uridine, 4'-fluoro-2',3'-O-(1-methyl- ethylidene)-	MeOH	256(3.97)	44-3010-76
$C_{12}H_{15}F_3N_2O_7$ Uridine, 5-[(2,2,2-trifluoroethoxy)- methyl]-	pH 13 EtOH	263(3.80) 263(4.01)	104-0643-76 104-0643-76
$C_{12}H_{15}N$ 1H-Indole, 1-(1-methylpropyl)-, (+)-	EtOH	223(4.54),277(3.80), 284(3.86),295(3.69)	118-0414-76
2H-Pyrrole, 3,4-dihydro-2,2-dimethyl- 5-phenyl-	EtOH	243(4.15)	12-2511-76

Compound	Solvent	$\lambda_{max}(\log \epsilon)$	Ref.
$C_{12}H_{15}NO$			
2H-1-Benzazepin-2-one, 1,3,4,5-tetra-hydro-4,8-dimethyl-	EtOH	215(4.45),245(3.95), 276s(3.11),285(3.08)	94-1544-76
4H-Cyclopenta[c]quinolin-4-one, 1,2,3,5,6,7,8,9-octahydro-	n.s.g.	235(3.68),308(3.96)	104-0681-76
3H-Indol-3-ol, 2-methyl-3-(1-methyl-ethyl)-	MeOH	213s(4.28),218(4.32), 224s(4.18),255(3.51), 290(3.30)	23-1015-76
6-Oxa-1-azabicyclo[3.1.0]hexane, 2,2-dimethyl-3-phenyl-	EtOH	221(3.29),243(1.99), 249(2.12),254(2.25), 260(2.35),266(2.26), 269(2.12),279(1.67)	12-2511-76
2-Pentanone, 4-[(3-methylphenyl)imino]-	EtOH	206(4.19),230(3.85), 324(4.25)	40-0451-76
2-Pentanone, 4-[(4-methylphenyl)imino]-	EtOH	207(4.04),230(3.89), 322(4.33)	40-0451-76
2H-Pyrrole, 3,4-dihydro-2,2-dimethyl-5-phenyl-, 1-oxide	EtOH	225(3.92),287s(--), 293(4.18)	12-2511-76
$C_{12}H_{15}NOS$			
1H-Indole, 2-(ethylsulfinyl)-1,3-di-methyl-	EtOH	225(4.44),286(4.15), 305s(3.90)	39-0745-76C
$C_{12}H_{15}NO_2$			
Hydroperoxide, 2-methyl-3-(1-methyl-ethyl)-3H-indol-3-yl-	MeOH	213s(4.30),218(4.36), 224s(4.20),257(3.52), 290(3.36)	23-1015-76
Pentanamide, 2-methyl-3-oxo-N-phenyl-	EtOH	280(2.72)	36-0068-76
2-Pentanone, 4-[(3-methoxyphenyl)-imino]-	EtOH	210(4.33),234(3.67), 324(4.32)	40-0451-76
2-Pentanone, 4-[(4-methoxyphenyl)-imino]-	EtOH	204(4.15),228(3.90), 320(4.30)	40-0451-76
4aH-Phenoxazin-4a-ol, 1,2,3,4,10,10a-hexahydro-	MeOH	209(4.47),245(3.85), 295(3.58)	44-4026-76
	MeOH-NaOH	216(4.06),244(3.78), 295(3.49)	44-4026-76
$C_{12}H_{15}NO_2S$			
1H-Indole, 2-(ethylsulfonyl)-1,3-di-methyl-	EtOH	225(4.58),275s(4.01), 282(4.07),305(3.82), 315(3.68)	39-0745-76C
$C_{12}H_{15}NO_2S_2$			
5H-1,4-Dithiino[2,3-c]pyrrole-5,7(6H)-dione, 6-cyclohexyl-2,3-dihydro-	EtOH	205(3.90),270(4.08), 417(3.40)	56-1523-76
$C_{12}H_{15}NO_3$			
Acetamide, N-(4-butoxyphenyl)-2-oxo-(hydrate)	EtOH	212(3.56),254(3.47)	65-0702-76
Acetamide, N-[1-methyl-2-(4-methyl-3,6-dioxo-1,4-cyclohexadien-1-yl)ethyl]-	H_2O	256(4.23)	44-3627-76
Benzenepropanenitrile, 2,4,5-trimethoxy-	EtOH	209(4.11),232(3.94), 290(3.70)	39-0499-76C
$C_{12}H_{15}NO_3S$			
2H-Indol-2-one, 3-(ethylsulfonyl)-1,3-dihydro-1,3-dimethyl-	EtOH	259(3.75),269s(3.61), 293(3.08)	39-0745-76C
$C_{12}H_{15}NO_4$			
Benzenemethanol, α-(1-methylethyl)-2-nitro-, acetate	MeOH	258(2.54),262(2.49), 264(2.49),268(2.36)	12-0327-76

Compound	Solvent	$\lambda_{max}(\log \epsilon)$	Ref.
Butanedioic acid, methyl[(1-methyl-1H-pyrrol-2-yl)methylene]-, 4-methyl ester, (E)-	EtOH	215(3.97),240(3.88), 286(3.54)	48-0816-76
1-Oxa-5-azaspiro[2.5]oct-6-ene-4,8-dione, 6-methyl-2-(3-methyl-1-oxobutyl)-, (±)- (isoflavipucine)	MeOH	326(3.81)	77-0566-76
2,6-Piperidinedione, 4-[(3,4-dihydro-5-methyl-4-oxo-2H-pyran-2-yl)methyl]-	EtOH	269(4.02)	33-2393-76
Propanoic acid, 3-[(4-methoxyphenyl)-amino]-3-oxo-, ethyl ester	EtOH	253(4.20)	105-0282-76
2H-Pyrano[4,3-b]pyridine-3-carboxylic acid, 1,5,7,8-tetrahydro-4,5,7-trimethyl-2-oxo-	EtOH	330(3.83)	22-0995-76
Pyridinium, 2-ethoxy-1-(ethoxycarbonyl)-2-oxoethylide	benzene	441(3.59)	78-2647-76
	H_2O	249(4.43),370(3.10)	78-2647-76
	aq HCl	261(3.74)	78-2647-76
	EtOH	247(4.03),398(3.09)	78-2647-76
	dioxan	246(4.16),439(3.19)	78-2647-76
	MeCN	245(4.42),420(3.37)	78-2647-76
	DMF	420(3.37)	78-2647-76
	$CHCl_3$	430(3.46)	78-2647-76
$C_{12}H_{15}NO_5$			
Benzoic acid, 4-(1-hydroxy-2-methyl-2-nitropropyl)-, methyl ester	MeOH	245(4.16)	12-0327-76
2,6-Piperidinedione, 4-[(3,6-dihydro-4-hydroxy-5-methyl-6-oxo-2H-pyran-2-yl)methyl]-	EtOH-acid	248(4.14)	33-2393-76
	EtOH-base	280(4.41)	33-2393-76
$C_{12}H_{15}NO_5S_3$			
Propanedioic acid, [(3-ethyl-4-oxo-2-thioxo-5-thiazolidinylidene)-(methylthio)methyl]-, dimethyl ester	EtOH	294(3.90),384(4.34)	94-1671-76
$C_{12}H_{15}NO_{10}$			
1,1-Cyclopropanedicarboxylic acid, 2-[2-methoxy-1-(methoxycarbonyl)-2-oxoethyl]-3-nitro-, dimethyl ester	EtOH	none	104-0668-76
$C_{12}H_{15}NS$			
1H-Indole, 2-(ethylthio)-1,3-dimethyl-	EtOH	227(4.59),288(4.04), 295(4.04)	39-0745-76C
2-Propene-1-thione, 3-(dimethylamino)-1-(4-methylphenyl)-	MeCN	315(4.20),431(4.26), 526s(2.80)	118-0521-76
$C_{12}H_{15}NSe$			
2-Propene-1-selone, 3-(dimethylamino)-1-(4-methylphenyl)-	MeCN	329(4.10),470(3.39), 585s(2.78)	118-0521-76
$C_{12}H_{15}N_3$			
1H-Pyrazol-1-amine, N,N,3-trimethyl-4-phenyl-	MeOH	245(4.08)	94-3001-76
1H-Pyrazol-1-amine, N,N,4-trimethyl-3-phenyl-	MeOH	251(4.09)	94-3001-76
1H-Pyrazol-1-amine, N,N,5-trimethyl-3-phenyl-	MeOH	253(4.22)	94-3001-76
1H-Pyrazol-1-amine, N,N,5-trimethyl-4-phenyl-	MeOH	245(4.13)	94-3001-76

Compound	Solvent	$\lambda_{max}(\log \epsilon)$	Ref.
$C_{12}H_{15}N_3O_2$			
Ethanone, 1-(2,3,6,7,8,9-hexahydro-5-oxo-1H-imidazo[1,2-b]quinazolin-1-yl)-	EtOH	244(3.90),287(3.98)	114-0413-76A
$C_{12}H_{15}N_3O_3$			
Benz[f]imidazo[1,2-a]pyrimidine-2,5(1H,3H)-dione, 6,7,8,9-tetrahydro-1-(2-hydroxyethyl)-	EtOH	236(4.04),282(3.90)	114-0413-76A
Phenol, 4-nitro-2-[(1-piperidinyl-imino)methyl]-, (E)-	benzene	298(4.31)	104-0625-76
	DMF	300(4.26)	104-0625-76
1H-Pyrazolo[4,3-b]pyridine-6-carboxylic acid, 7-ethoxy-1-methyl-, ethyl ester	EtOH	229(4.51),277(3.71), 306(3.63)	39-0507-76C
1H-Pyrazolo[4,3-b]pyridine-6-carboxylic acid, 4-ethyl-4,7-dihydro-1-methyl-7-oxo-, ethyl ester	EtOH	228(4.26),304(4.13)	39-0507-76C
$C_{12}H_{15}N_3O_5$			
6H-Furo[2',3':4,5]oxazolo[3,2-a]pyrimid-ine-2-methanol, 2,3,3a,9a-tetrahydro-6-imino-3-(1-oxopropoxy)-, hydrochlor-ide, [2R-(2α,3β,3aβ,9aβ)]-	MeOH-acid	230(3.97),262(4.01)	87-0654-76
$C_{12}H_{15}N_3O_6$			
Carbamic acid, (5-nitro-1,3-phenylene)-bis-, diethyl ester	MeOH	240(4.43),297(3.54)	87-0426-76
$C_{12}H_{15}N_3O_7$			
1,2,4-Triazine-3,5(2H,4H)-dione, 2-(3,5-di-O-acetyl-2-deoxy-β-D-erythro-pentofuranosyl)-	MeOH	263(3.70)	73-2110-76
$C_{12}H_{15}N_5OS$			
1,2,4-Thiadiazole-2(3H)-carboximidamide, 5-ethoxy-3-imino-N-(phenylmethyl)-	MeOH	267(4.00)	18-3170-76
$C_{12}H_{15}N_5O_3$			
Hydrazinecarboxamide, 2-[2-(hydroxyimi-no)-1-methyl-3-[(2-methylphenyl)ami-no]-3-oxopropylidene]-	EtOH	264(4.32)	42-0454-76
Hydrazinecarboxamide, 2-[2-(hydroxyimi-no)-1-methyl-3-[(4-methylphenyl)ami-no]-3-oxopropylidene]-	EtOH	260(4.37)	42-0454-76
$C_{12}H_{15}N_5O_4$			
Hydrazinecarboxamide, 2-[2-(hydroxy-imino)-3-[(2-methoxyphenyl)amino]-1-methyl-3-oxopropylidene]-	EtOH	265(4.33)	42-0454-76
Hydrazinecarboxamide, 2-[2-(hydroxy-imino)-3-[(4-methoxyphenyl)amino]-1-methyl-3-oxopropylidene]-	EtOH	265(4.40)	42-0454-76
$C_{12}H_{17}N_7O_2S_2$			
Hydrazinecarbothioamide, 2,2'-[1-(4-ni-trophenyl)-1,2-ethanediylidene]bis-[N-methyl-	3:1 CHCl$_3$-MeOH	336(4.34)	87-0131-76
$C_{12}H_{16}$			
Benzene, (2-ethyl-1-butenyl)-	EtOH	247(4.09)	78-1839-76
1-Butene, 2-ethyl-3-phenyl-	EtOH	245(2.91),249(2.92),	78-1839-76

Compound	Solvent	$\lambda_{max}(\log \epsilon)$	Ref.
1-Butene, 2-ethyl-3-phenyl- (cont.)		253(2.91),260(2.83), 266(2.60),276(2.48)	78-1839-76
2-Pentene, 3-methyl-4-phenyl-	EtOH	250(3.00),255(2.98), 260(2.91),270(2.30), 272(2.67)	78-1839-76
$C_{12}H_{16}ClN_3O_2$ 1-Pyrrolidineethanamine, N-(4-chloro-2-nitrophenyl)-	EtOH	240(4.44),270s(3.80), 440(3.79)	78-0057-76
$C_{12}H_{16}ClN_3O_5$ 1H-Imidazo[1,2-c]pyrimidin-4-ium, 5,6-dihydro-1-methyl-5-oxo-6-β-D-ribofuranosyl-, chloride	pH 1 and 7 pH 13	250s(3.61),288(4.11), 293(4.12),307s(3.89) 275(4.00)	35-7408-76 35-7408-76
$C_{12}H_{16}ClN_3O_6$ Thymidine, α-[(chloroacetyl)amino]-	M HCl H_2O M NaOH	264(3.90) 263(3.86) 263(3.79)	87-0909-76 87-0909-76 87-0909-76
$C_{12}H_{16}ClN_5O_3S$ 9H-Purin-6-amine, 2-chloro-9-(2-S-ethyl-2-thio-α-D-lyxofuranosyl- β-	EtOH EtOH	265(4.18) 265(4.19)	136-0041-76C 136-0041-76C
$C_{12}H_{16}CuN_2O_2S_2$ Copper, bis[1-(4,5-dihydro-2-thiazolyl)-2-propanonato-N,O]-	$CHCl_3$	230(4.35),284(4.35), 309(4.27),400s(2.82), 570s(1.93)	48-0298-76
$C_{12}H_{16}FN_3O_8S$ Uridine, 4'-C-fluoro-2',3'-O-(1-methyl-ethylidene)-, 5'-sulfamate	MeOH	255(3.99)	44-3010-76
$C_{12}H_{16}NO_2Sb$ Antimony, hydroxytrimethyl(8-quinolin-olato-N^1,O^8)-	CH_2Cl_2	350(3.38)	18-0817-76
$C_{12}H_{16}N_2$ Phenazine, 1,2,3,4,6,7,8,9-octahydro-	EtOH	290(4.03)	18-2805-76
$C_{12}H_{16}N_2NiO_2S_2$ Nickel, bis[1-(4,5-dihydro-2-thiazolyl)-2-propanonato-N,O]-	$CHCl_3$	270(4.54),300s(4.07), 350(3.72),500(1.96), 613(1.67)	48-0298-76
$C_{12}H_{16}N_2O$ Benzo[g]quinoxalin-2(1H)-one, 5,5a,6,7-8,9,9a,10-octahydro-, racemic-trans	EtOH	227(3.96),340(3.86)	142-0299-76S
1H-Imidazole, 2,5-dihydro-1-hydroxy-2,5,5-trimethyl-4-phenyl-hydrochloride	EtOH EtOH	260(3.30) 236(3.79)	103-1280-76 103-1280-76
1H-Imidazole, 4,5-dihydro-1-hydroxy-2,5,5-trimethyl-4-phenyl-	EtOH	259(3.80)	103-1280-76
Phenol, 2-[(1-piperidinylimino)methyl]-	benzene DMF	287(4.33),315(4.19) 290(4.30),317(4.19)	104-0625-76 104-0625-76
3-Piperidinecarboxamide, 1-phenyl-	MeOH	252(4.87),290s(4.34)	39-0045-76C
$C_{12}H_{16}N_2O_2$ 1-Pyrazolidinecarboxylic acid, 2-phen-yl-, ethyl ester	EtOH	237(4.03),278(4.03)	44-1244-76

Compound	Solvent	$\lambda_{max}(\log \epsilon)$	Ref.
$C_{12}H_{16}N_2O_2PdS_2$ Palladium, bis[1-(4,5-dihydro-2-thiazolyl)-2-propanonato-N,O]-	$CHCl_3$	229(4.52),279(4.44), 330(4.09)	48-0298-76
$C_{12}H_{16}N_2O_2S_2Zn$ Zinc, bis[1-(4,5-dihydro-2-thiazolyl)-2-propanonato-N,O]-, (T-4)-	$CHCl_3$	245(3.96),308(4.64)	48-0298-76
$C_{12}H_{16}N_2O_3$ 1H-Pyrazole-3-carboxylic acid, 4,5-dihydro-1-methyl-5-oxo-4,4-di-2-propenyl-, methyl ester	MeOH	288(3.85)	24-0261-76
4H-Pyrido[1,2-a]pyrimidine-3-carboxylic acid, 6,7,8,9-tetrahydro-6-methyl-4-oxo-, ethyl ester	pH 7.35	231(3.79),299(3.95)	145-0478-76
4H-Pyrido[1,2-a]pyrimidine-9-carboxylic acid, 6,7,8,9-tetrahydro-2,9-dimethyl-4-oxo-, methyl ester	EtOH	229(4.00)	4-0797-76
$C_{12}H_{16}N_2O_4$ Benzene, 1-(1-ethyl-2-methyl-2-nitropropyl)-4-nitro-	hexane	262(4.09)	12-2621-76
$C_{12}H_{16}N_2O_4S$ 4,12-Epoxy-5H,8H-1,3-dioxolo[4,5-e]pyrimido[2,1-b][1,3]thiazocin-8-one, 3a,4,9,10,12,12a-hexahydro-2,2-dimethyl-, [3aR-(3aα,4β,12β,12aα)]-	n.s.g.	272.5(4.09)	33-2972-76
$C_{12}H_{16}N_2O_5S$ 2-Butenedioic acid, 2-hydroxy-3-(2-imino-4-methyl-3(2H)-thiazolyl)-, diethyl ester	MeOH	265(4.04)	64-0251-76B
4-Thiazolecarboxamide, 2-[2,3-O-(1-methylethylidene)-β-D-ribofuranosyl]-	EtOH	215(3.89),237(3.87)	44-4074-76
Thymidine, 4-thio-, 3'-acetate	n.s.g.	201(4.09),239(3.59), 334(4.25)	128-0351-76
$C_{12}H_{16}N_2O_5S_2$ Uridine, 2'-deoxy-5-(1,3-dithiolan-2-yl)-	M HCl H_2O M NaOH	271(4.20) 270(4.20) 270(4.11)	87-0909-76 87-0909-76 87-0909-76
$C_{12}H_{16}N_2O_6$ [5,5'-Bi-2H-pyrrole]-2,2'-dicarboxylic acid, 3,3',4,4'-tetrahydro-2,2'-dimethyl-, 1,1'-dioxide	n.s.g.	338(4.29)	39-1955-76C
1H-Cyclopentapyrimidine-2,4(3H,5H)-dione, 6,7-dihydro-1-β-D-ribofuranosyl-	H_2O pH 13	270(4.02) 271(3.91)	19-0037-76 19-0037-76
1H-Pyrazole-3-carboxylic acid, 5-acetoxy-4-(2-acetoxyethyl)-1-methyl-, methyl ester	MeOH	219(3.92),246s(3.60)	24-0261-76
Uridine, 5-cyclopropyl-	H_2O pH 13	272(3.89) 272(3.78)	73-3635-76 73-3635-76
$C_{12}H_{16}N_2S$ Benzo[g]quinoxaline-2(1H)-thione, 5,5a,6,7,8,9,9a,10-octahydro-, (5aS-trans)-	EtOH	286(4.18),409(3.93)	142-0299-76S

Compound	Solvent	$\lambda_{max}(\log \epsilon)$	Ref.
$C_{12}H_{16}N_2S_2$ 2,6(1H,3H)-Pyridinedithione, 3-[(cyclo-hexylamino)methylene]-	MeOH	226(4.21),251(3.47), 291(3.61),341(4.26), 421(4.61)	33-0190-76
$C_{12}H_{16}N_3O_5$ Imidazo[1,2-c]pyrimidin-4-ium, 5,6-di-hydro-1-methyl-5-oxo-6-β-D-ribo-furanosyl-, chloride	pH 1 and 7 pH 13	250s(3.61),288(4.11), 293(4.12),307s(3.89) 275(4.00)	35-7408-76 35-7408-76
$C_{12}H_{16}N_4O$ Pyrido[2,3-d]pyrimidin-4(1H)-one, 2-amino-6-pentyl-	pH 13	238(4.29),271(3.92), 335(3.64)	4-0439-76
$C_{12}H_{16}N_4O_2S_2$ Benzenesulfonamide, 4-(1-butyl-5-mer-capto-1H-1,3,4-triazol-2-yl)-	EtOH base	230(4.15),255(4.32), 300(3.77) 230(4.30),285(3.90)	2-0887-76 2-0887-76
$C_{12}H_{16}N_4O_4S_2$ 9H-Purine, 6,8-bis(methylthio)- 9-β-D-ribofuranosyl-	pH 1 pH 13 EtOH	249(4.19),305s(4.27), 312(4.31),332s(3.81) 248(4.23),305s(4.35), 312(4.36) 248(4.22),305s(4.35), 311(4.36)	94-0565-76 94-0565-76 94-0565-76
$C_{12}H_{16}N_4O_5$ 4-Pyrimidineacetic acid, 2-(acetylami-no)-1,6-dihydro-α-(hydroxyimino)- 6-oxo-, 1,1-dimethylethyl ester	EtOH	242(4.38),307(3.67)	44-2850-76
$C_{12}H_{16}N_4O_6$ Pyrazolo[1,5-a]-1,3,5-triazine- 2,4(1H,3H)-dione, 1,3-dimethyl- 8-β-D-ribofuranosyl-	MeOH	215(4.19),240(3.75), 254(3.76)	44-3000-76
1,2,4-Triazin-3(2H)-one, 5-amino- 2-(3,5-di-O-acetyl-2-deoxy- β-D-erythro-pentofuranosyl)-	MeOH	272(3.76)	73-2110-76
$C_{12}H_{16}N_6OS_2$ Hydrazinecarbothioamide, 2,2'-[1-(4-hy-droxyphenyl)-1,2-ethanediylidene]-bis[N-methyl-	EtOH	341(4.54)	87-0131-76
$C_{12}H_{16}N_6O_2$ [4,5'-Bipyrimidine]-4',6(1H,1'H)-dione, 2,2'-bis(dimethylamino)-	pH 12	243s(4.15),276(4.16), 346(4.31),361s(4.15)	19-0017-76
2(1H)-Pyrimidinone, 4-amino-1-[3-(4-am-ino-2-oxo-1(2H)-pyrimidinyl)propyl]- 5-methyl-	pH 1 pH 7	284.0(--) 275.5(4.20)	56-1805-76 56-1805-76
$C_{12}H_{16}N_6O_3$ 1H-Pyrrole-1-carboxamide, 2-[bis[[(meth-ylamino)carbonyl]amino]-3-cyano-N,4-dimethyl-	MeOH	239(4.05)	24-2983-76
$C_{12}H_{16}N_6S_2$ Hydrazinecarbothioamide, 2,2'-(1-phenyl-1,2-ethanediylidene)bis[N-methyl-	EtOH	337(4.42)	87-0131-76

Compound	Solvent	$\lambda_{max}(\log \epsilon)$	Ref.
$C_{12}H_{16}O$			
Bicyclo[5.3.0]deca-1,6-dien-3-one, 5,10-dimethyl-	EtOH	228(4.1)	2-0243-76
2,4-Cyclohexadien-1-one, 2,4,6-trimethyl-6-(2-propenyl)-	n.s.g.	321(3.60)	39-0634-76C
2-Cyclopenten-1-one, 2-(1-cyclopenten-1-ylmethyl)-3-methyl-	EtOH	233(3.98)	39-0249-76C
2(4aH)-Naphthalenone, 4a-ethyl-5,6,7,8-tetrahydro-	EtOH	243(4.12)	39-1349-76B
2(4aH)-Naphthalenone, 5,6,7,8-tetrahydro-4a,8-dimethyl-, cis	EtOH	239(4.06)	44-3632-76
Phenol, 4-cyclohexyl-	dioxan	279(3.33),286(3.24)	126-3481-76
Spiro[4.5]deca-3,6-dien-2-one, 6,10-dimethyl-	EtOH	223(3.98)	44-3632-76
$C_{12}H_{16}O_2$			
Dispiro[4.1.4.1]dodecane-6,12-dione	CH_2Cl_2	306(1.43),347(1.38)	18-0741-76
2(4aH)-Naphthalenone, 5,6,7,8-tetrahydro-3-hydroxy-4a,8-dimethyl-, trans	EtOH	250(4.15),288(3.53)	44-1539-76
Norcaradiene-7-carboxylic acid, 2,3,5,7-tetramethyl-	EtOH	246(3.74),277(3.69)	33-1763-76
Norcaradiene-7-carboxylic acid, 2,3,7-trimethyl-, methyl ester	EtOH	243(3.63),276(3.54)	33-1763-76
7-Oxabicyclo[4.2.1]nona-2,4-dien-8-one, 9-(1,1-dimethylethyl)-	EtOH	258(3.66),265(3.63)	44-3583-76
Phenol, 3-(2-hydroxycyclohexyl)-	EtOH	273(3.28)	95-1031-76
$C_{12}H_{16}O_3$			
Benzenepropanoic acid, 2-hydroxy-, propyl ester	n.s.g.	273(3.58)	39-0407-76C
Benzoic acid, 5-(1,1-dimethylethyl)-2-hydroxy-, methyl ester	MeOH	211(4.54),240(3.96), 315(3.65)	39-0465-76C
Bicyclo[3.1.1]hept-3-en-2-one, 4-(acetoxymethyl)-6,6-dimethyl- (4-oxomyrtenyl acetate)	C_6H_{12} EtOH EtOH	238(3.72) 218s(3.62),247(3.87) 247(3.87)	78-2805-76 39-0484-76C 78-2805-76
2-Cyclohexen-1-one, 3-(acetoxymethyl)-4-(1-methylethenyl)-	EtOH EtOH C_6H_{12}	228(4.03),301s(2.45) 228(4.03) 223(--)	39-0484-76C 78-2805-76 78-2805-76
2-Cyclopenten-1-one, 4-acetoxy-3-methyl-2-(2-butenyl)-, (E)-($\pm$)-	EtOH	227(3.93)	39-0249-76C
(Z)-	EtOH	228(4.00)	39-0249-76C
p-Mentha-1,4(8)-dien-7-ol, 3-oxo-, acetate	EtOH	232(3.93),278(3.73)	39-0008-76C
p-Mentha-1,8-dien-7-ol, 3-oxo-, acetate	EtOH	230(4.08)	39-0008-76C
p-Mentha-1,8-dien-7-ol, 6-oxo-, acetate	EtOH	229(3.68)	39-0008-76C
Spiro[1,3-dioxolan-2,4'-[4H]inden]-1'(2'H)-one, 3',3'a,5',6'-tetrahydro-7'-methyl-	EtOH	255(3.98)	39-0410-76C
$C_{12}H_{16}O_4$			
Acetic acid, [(3-methoxyphenyl)methoxy]-, ethyl ester	95% MeOH	274(3.24),280(3.19)	13-0197-76B
Butenedioic acid, 2,4-hexadienylidene-, dimethyl ester	EtOH	303(4.50)	35-1204-76
Ethanone, 1-(3,4,6-trimethoxy-2-methylphenyl)-	n.s.g.	224(4.03),268(3.60), 295(3.54)	39-2594-76C
Propanoic acid, 3-(3-methoxy-5-methylphenoxy)-, methyl ester	EtOH	223(3.96),273(3.33), 279(3.33)	39-0499-76C
2H-Pyran-2-one, 6-(acetoxymethyl)-5-butylidene-5,6-dihydro-	EtOH	267(3.11)	102-0537-76

Compound	Solvent	$\lambda_{max}(\log \epsilon)$	Ref.
$C_{12}H_{16}O_5$			
2H-Oxeto[2',3':4,5]cyclopenta[1,2-c]pyr-an-6-carboxylic acid, 2aβ,2bα,3,6aα-7,7aβ-hexahydro-3α-methoxy-, methyl ester	MeOH	237.5(4.02)	24-3626-76
2H-Pyran-5-carboxylic acid, 3β-ethenyl-3,4-dihydro-2α-methoxy-4β-(2-oxoeth-yl)-, methyl ester	MeOH	235(4.05)	24-3626-76
2H-Pyran-2,2-dicarboxylic acid, 3,4-di-hydro-4-methylene-, diethyl ester	MeOH	245(4.10)	44-2850-76
$C_{12}H_{16}O_7$			
2-Propenoic acid, 3-(tetrahydro-3-form-yl-5-(hydroxymethyl)-4-methoxy-6-oxo-2H-pyran-2-yl)-, methyl ester (dihy-drorosellisin)	EtOH	206(4.25),291(3.87)	102-1090-76
$C_{12}H_{16}S$			
1-Butanethione, 2,2-dimethyl-1-phenyl-	C_6H_{12}	228(4.02),297(3.60), 567(2.04)	35-6218-76
$C_{12}H_{17}Cl_2N_3S$			
Benzenamine, 2,6-dichloro-4-[[(1,1-di-methylethyl)thio]azo]-N,N-dimethyl-, (E)-	heptane MeOH	351(4.11) 351(4.12)	144-0399-76 144-0399-76
(Z)-	heptane MeOH	306(3.69) 307(3.67)	144-0399-76 144-0399-76
$C_{12}H_{17}Cl_2N_5O_2$			
Urea, N,N''-[5-[bis(2-chloroethyl)amino]-1,3-phenylene]bis-	MeOH	219(4.54),239(4.76), 307(3.42)	87-0426-76
$C_{12}H_{17}F_3O_2$			
Menthone, 2-(trifluoroacetyl)-	isooctane EtOH	285(3.72) 288(3.8)	104-1399-76 104-1399-76
$C_{12}H_{17}NO$			
Phenol, 3-(2-aminocyclohexyl)-	EtOH	274(3.29)	95-1031-76
$C_{12}H_{17}NO_2$			
Benzoic acid, 2-amino-3,4,5,6-tetrameth-yl-, methyl ester	heptane	224(4.34),250s(3.78), 331(4.49)	104-1057-76
Bicyclo[3.2.1]oct-3-en-2-one, 4-morpho-lino-	EtOH CHCl$_3$	299(c.4.4) 293(c.4.38)	22-0889-76 22-0889-76
$C_{12}H_{17}NO_2S$			
Piperidine, 1-[(4-methylphenyl)sulfon-yl]-	EtOH	207(4.04),240(3.89)	118-0339-76
$C_{12}H_{17}NO_4$			
2,5-Cyclohexadien-1-ol, 4-methyl-1-(1-methylethyl)-4-nitro-, acetate	MeOH	197(3.09)	23-0423-76
diastereomer	MeOH	196(3.10)	23-0423-76
2,5-Cyclohexadien-1-ol, 1,2,3,4-tetra-methyl-4-nitro-, acetate, cis	MeOH	198.5(3.49)	23-1795-76
trans	MeOH	198(3.66)	23-1795-76
2,5-Cyclohexadien-1-ol, 1,2,4,5-tetra-methyl-4-nitro-, acetate, cis	MeOH	195(3.35)	23-1795-76
trans	MeOH	196(3.60)	23-1795-76
2,5-Cyclohexadien-1-ol, 1,3,4,5-tetra-methyl-4-nitro-, acetate	MeOH	199(3.18)	23-1795-76

$C_{12}H_{17}NO_4S-C_{12}H_{17}N_3O_2$

Compound	Solvent	$\lambda_{max}(\log \epsilon)$	Ref.
2,5-Cyclohexadien-1-ol, 1,3,4,5-tetra-methyl-4-nitro-, acetate, diastereo-mer	MeOH	199(3.09)	23-1795-76
2,5-Cyclohexadien-1-ol, 2,3,4,5-tetra-methyl-4-nitro-, acetate	MeOH	196(3.72)	23-1795-76
diastereomer	MeOH	196(3.45)	23-1795-76
$C_{12}H_{17}NO_4S$			
3(2H)-Thiazoleacetic acid, 5-(methoxy-carbonyl)-2,2-dimethyl-α-(1-methyl-ethylidene)-	EtOH	214(3.97),285s(3.40), 349(3.85)	39-0584-76C
K salt	EtOH	209(3.98),310(3.63), 350(3.98)	39-0584-76C
3(2H)-Thiazoleacetic acid, 5-(methoxy-carbonyl)-2-methyl-α-(1-methylethyl-idene)-, methyl ester	EtOH	219(4.20),286(3.61), 311s(3.90),346(3.94)	39-0584-76C
$C_{12}H_{17}N_2$			
Methanaminium, N-methyl-N-[3-(methyl-phenylamino)-2-propenylidene]-, perchlorate	H_2O	327(4.65)	69-0114-76
$C_{12}H_{17}N_2OPS$			
1,3,2-Thiaazaphospholidin-4-one, 2-(di-ethylamino)-3-phenyl-	hexane ether	227(4.00) 203(4.45),225(4.13)	65-1642-76 65-1642-76
$C_{12}H_{17}N_2OPS_2$			
1,3,2-Thiaazaphospholidin-4-one, 2-(di-ethylamino)-3-phenyl-, 2-sulfide	hexane MeOH	295(2.54) 205(4.28),218(4.15), 300(2.56)	65-1642-76 65-1642-76
	ether	204(4.36),218(3.18), 297(2.48)	65-1642-76
$C_{12}H_{17}N_2O_8P$			
5'-Uridylic acid, 2'-deoxy-5-(2-propen-yl)-	pH 1 pH 7 pH 13	266(3.99) 266(3.99) 264(3.83)	87-0903-76 87-0903-76 87-0903-76
$C_{12}H_{17}N_2O_9P$			
5'-Thymidylic acid, α-oxiranyl-	pH 1 pH 7 pH 13	268(4.00) 268(4.01) 263(3.99)	87-0903-76 87-0903-76 87-0903-76
$C_{12}H_{17}N_2S$			
1,2-Benzisothiazolium, 2-(1,1-dimethyl-ethyl)-3-(methylamino)-, perchlorate	MeOH	331(3.87),344(3.86)	24-0659-76
1,2-Benzisothiazolium, 3-(ethylamino)-2-(1-methylethyl)-, perchlorate	MeOH	329(3.91),340(3.90)	24-0659-76
1,2-Benzisothiazolium, 3(2)-(ethylami-no)-2(3)-propyl-, perchlorate (isomer mixture)	MeOH	328(3.89),338s(--)	24-0659-76
$C_{12}H_{17}N_3O$			
1H-Benzotriazole, 1-(hexyloxy)-	EtOH	264(3.69),283(3.65)	4-0509-76
$C_{12}H_{17}N_3O_2$			
Benzamide, N-[(butylnitrosoamino)meth-yl]-	EtOH	233(4.31),356(1.88), 364(1.88),374(1.73)	94-0369-76
Benzamide, N-[[(1,1-dimethylethyl)-nitrosoamino]methyl]-	EtOH	229(4.24),368(1.64)	94-0369-76

Compound	Solvent	$\lambda_{max}(\log \epsilon)$	Ref.
$C_{12}H_{17}N_3O_6$			
Acetamide, N-[1-(2-deoxy-β-D-arabino-hexopyranosyl)-1,2-dihydro-2-oxo-4-pyrimidinyl]-	H_2O	249(4.22),297(3.91)	44-0600-76
1H-Pyrazole-3-carboxamide, 4-hydroxy-5-[2,3-O-(1-methylethylidene)-α-D-ribofuranosyl]-	pH 12	310(3.91)	44-0287-76
	EtOH	220(3.88),263(3.76)	44-0287-76
β-	pH 12	310(3.91)	44-0287-76
	EtOH	225(3.96),264(3.86)	44-0287-76
$C_{12}H_{17}N_5O_3$			
Adenosine, 3'-deoxy-3'-ethyl-	MeOH	260(4.18)	87-1265-76
Propanoic acid, 2,2-dimethyl-, (2-amino-1,6-dihydro-8-methyl-6-oxo-9H-purin-9-yl)methyl ester	H_2O	251(4.14),276(3.95)	44-0568-76
$C_{12}H_{17}N_5O_3S$			
9H-Purin-6-amine, 9-(2-S-ethyl-2-thio-α-D-lyxofuranosyl)-	EtOH	260(4.15)	136-0041-76C
β-	EtOH	261(4.15)	136-0041-76C
$C_{12}H_{17}N_6O_6P$			
Adenosine, N-(2-aminoethyl)-, cyclic 3',5'-(hydrogen phosphate)	pH 1	262(4.21)	69-3724-76
	pH 11	267(4.21)	69-3724-76
$C_{12}H_{18}$			
Bicyclo[2.2.1]heptane, 1,7,7-trimethyl-2,3-bis(methylene)-, (1R)-	hexane	240s(3.95),248(4.00), 258s(3.76)	78-0309-76
3,4-Nonadien-6-yne, 5-ethyl-3-methyl-	C_6H_{12}	215s(3.83),221(3.87)	35-8426-76
2,5-Octadiyne, 4,4-diethyl-	C_6H_{12}	215s(3.83),221(3.87)	35-8426-76
$C_{12}H_{18}Cl_2N_2O_6S$			
2-(N,N-Dichloroamino)-2-methyl-1-propyl-N-methylnicotinate methyl sulfate	H_2O	303(2.42)	36-1733-76
$C_{12}H_{18}Cl_2N_3O_2$			
Morpholinium, 4-[2,3-dichloro-3-[(4-morpholinylmethylene)amino]-2-propenylidene]-, perchlorate	HOAc	414(3.48)	73-1565-76
$C_{12}H_{18}N_2O_2S$			
1H-Thiazolo[3,2-a][1,3]diazocine-8-carboxylic acid, 2,3,4,5-tetrahydro-9-methyl-, ethyl ester, hydrochloride	n.s.g.	289(4.21)	2-0773-76
$C_{12}H_{18}N_2O_6$			
1H-Pyrazole-5-carboxylic acid, 1-methyl-4-β-D-ribofuranosyl-, ethyl ester	pH 13	236(3.71)	44-0084-76
	EtOH	233(3.94),246s(3.83)	44-0084-76
$C_{12}H_{18}N_2O_6S$			
Acetic acid, α-amino-α-cyano-, p-toluenesulfonic acid salt hydrate	H_2O	261(2.56)	104-0910-76
$C_{12}H_{18}N_2O_8$			
Propanedioic acid, 2,2'-[azobis[1-methylethylidene)]bis-	H_2O	360(1.43)	126-0395-76
$C_{12}H_{18}N_2O_9S$			
2,4(1H,3H)-Pyrimidinedione, 5-methyl-1-[6-O-(methylsulfonyl)-β-D-galactopyranosyl]-	MeOH	265(3.95)	48-0079-76

Compound	Solvent	$\lambda_{max}(\log \epsilon)$	Ref.
$C_{12}H_{18}N_2S$ Benzothiazolo[3,2-a][1,3]diazocine, 1,2,3,4,5,8,9,10,11,12-decahydro-, hydrochloride	n.s.g.	280(3.94)	2-0773-76
$C_{12}H_{18}N_4$ Benzo[g]quinoxalin-2(1H)-one, 5,5a,6,7- 8,9,9a,10-octahydro-, hydrazone, (5aS-trans)-	MeCN	236(3.93),330(3.75)	142-0299-76S
$C_{12}H_{18}N_4O_2$ [5,5'-Bi-4H-imidazole]-4,4'-dione, 3,3',5,5'-tetrahydro-2,2',3,3',5,5'- hexamethyl-, dl	EtOH	232(3.85)	20-0573-76
meso	EtOH	232(3.90)	20-0573-76
$C_{12}H_{18}N_4O_3$ 2,4(1H,3H)-Pyrimidinedione, 1,3-diethyl- 5-formyl-6-[(dimethylamino)methylene- amino]-	EtOH	251(4.42),280(<u>4.1</u>), 315(<u>3.9</u>)	94-0970-76
$C_{12}H_{18}N_4O_3S_2$ Hydrazinecarbothioamide, N-butyl- 1-(4-sulfonamidobenzoyl)-	EtOH base	240(4.06),330(3.58) 225(4.80),305(4.12)	2-0887-76 2-0887-76
$C_{12}H_{18}N_4O_5S$ 1H-Imidazole-4-carboxamide, 5-amino- 2,3-dihydro-1-[2,3-O-(1-methylethyl- idene)-β-D-ribofuranosyl]-2-thioxo-	pH 1 pH 7 pH 13	299(4.17) 298(4.18) 268s(--),298(3.97)	94-2089-76 94-2089-76 94-2089-76
$C_{12}H_{18}N_4O_7$ 2,4(1H,3H)-Pteridinedione, 1-β-D-gluco- pyranosyl-5,6,7,8-tetrahydro-	pH 2.0 pH 7.0 pH 13.0	268(4.21) 245(3.46),305(3.99) 307(3.91)	24-3184-76 24-3184-76 24-3184-76
$C_{12}H_{18}N_5O_4S_4$ Thiazolium, 3-ethyl-2-[3-[3-ethyl-5- (methylsulfonyl)-2(3H)-thiazolyli- dene]-1-triazenyl]-5-(methylsulfon- yl)-, tetrafluoroborate	EtOH	477(4.41)	33-0155-76
$C_{12}H_{18}N_5S_2$ Thiazolium, 3-ethyl-2-[3-(3-ethyl- 5-methyl-2(3H)-thiazolylidene]-1-tri- azenyl]-5-methyl-, tetrafluoroborate	EtOH	477(4.46)	33-0155-76
$C_{12}H_{18}N_6O_4S$ Benzenesulfonamide, 4-nitro-N-(1-ethyl- 1,2,3-triazol-5-yl)-, ethylamine salt	EtOH	255(4.18),350(3.38)	32-0001-76
$C_{12}H_{18}N_7O_6P_2$ Adenosine, 2-[(2-aminoethyl)amino]-, cyclic 3',5'-(hydrogen phosphate) Adenosine, 8-[(2-aminoethyl)amino]-, cyclic 3',5'-(hydrogen phosphate)	pH 1 pH 11 pH 1 pH 11	253(4.13),295(4.01) 258(4.10),285(3.94) 275(4.22) 276(4.27)	69-3724-76 69-3724-76 69-3724-76 69-3724-76
$C_{12}H_{18}O$ Benzenemethanol, α-(1,1-dimethylethyl)- α-methyl-	EtOH	247-254f(2.10-2.31)	44-3067-76
2-Cyclohexen-1-one, 4,4,6-trimethyl- 6-(2-propenyl)-	C_6H_{12}	222(4.08),343(1.53)	33-2244-76

Compound	Solvent	$\lambda_{max}(\log \epsilon)$	Ref.
1H-Inden-1-one, 2,4,5,6,7,7a-hexahydro-4,4,7a-trimethyl-	pentane	286(1.38)	33-0032-76
1H-Inden-1-one, 3a,4,5,6,7,7a-hexahydro-4,4,7a-trimethyl-	pentane	222(3.92)	33-0032-76
2(3H)-Naphthalenone, 4,4a,5,6,7,8-hexahydro-4a,8-dimethyl-, cis	EtOH	240(4.14)	44-3632-76
$C_{12}H_{18}OS_5$			
Ethanethioic acid, S-(dihydro-2a,3,5,7-tetramethyl-2,5-methano-7H-1,3-dithiolo[4,5-d][1,3]dithiin-7-yl) ester	isooctane	224(3.66),254(3.27)	70-2353-76
$C_{12}H_{18}O_2$			
Benzenemethanol, α-(1-methoxy-1-methylethyl)-α-methyl-	EtOH	247-264f(2.37-2.64)	44-3067-76
Bicyclo[2.1.0]pentan-2-one, 5-acetyl-1,3,3,4,5-pentamethyl-	MeOH	225(4.41)	44-3377-76
Ethanol, 1-(2-cycloocten-1-ylidene)-, acetate	EtOH	241(4.15)	18-3673-76
geometric isomer	EtOH	239(3.98)	18-3673-76
2(3H)-Furanone, 4-(1,2-dimethyl-1-propenyl)dihydro-3,4-dimethyl-5-methylene-	MeOH	225(3.12)	44-3377-76
2(5H)-Furanone, 4-(1,2-dimethyl-1-propenyl)-3,5,5-trimethyl-	MeOH	223(3.42),255(3.03)	44-3377-76
2(5H)-Furanone, 5-(1,2-dimethyl-1-propenyl)-3,4,5-trimethyl-	MeOH	230(3.49)	44-3377-76
2(3H)-Naphthalenone, 4,4a,5,6,7,8-hexahydro-5-hydroxy-1,4a-dimethyl-, (+)-	MeOH	250(4.17)	44-2663-76
(-)-	MeOH	250(4.11)	44-2663-76
4-Oxatricyclo[3.2.0.0^{2,7}]heptan-3-one, 1,2,5,6,6,7-hexamethyl-	EtOH	220(2.95)	44-3377-76
$C_{12}H_{18}O_3$			
Cyclopentaneacetaldehyde, 2-(1,3-dioxolan-2-yl)-3-methyl-α-methylene-, [1S-(1α,2α,3β)]-	H_2O	223(3.77)	12-1375-76
$C_{12}H_{18}O_4$			
Cyclopropanecarboxylic acid, 1-(3-methoxy-3-oxo-1-propenyl)-2,2-dimethyl-, ethyl ester	MeOH	224(3.94),283(3.86)	24-1269-76
$C_{12}H_{18}O_6$			
Cyclopenta[c]pyran-4-carboxylic acid, 1,4aα,5,6,7,7aα-hexahydro-6α-hydroxy-7α-(hydroxymethyl)-1α-methoxy-, methyl ester	MeOH	237(4.04)	24-3626-76
$C_{12}H_{19}NO$			
Cyclohexanone, 2-(1-piperidinylmethylene)-	C_6H_{12}	312(4.24)	35-2826-76
	MeOH	334(4.33)	35-2826-76
2-Cyclohexen-1-one, 5,5-dimethyl-3-(1-pyrrolidinyl)-	pH 1	288(4.39)	39-2207-76C
	H_2O	304(4.55)	39-2207-76C
	pH 13	304(4.55)	39-2207-76C
1(2H)-Naphthalenone, 8-(ethylamino)-3,4,4a,5,6,7-hexahydro-	H_2O	335(4.11)	35-2826-76
	MeOH	335(4.21)	35-2826-76
cis-s-cis	C_6H_{12}	327(4.01)	35-2826-76
$C_{12}H_{19}NO_2$			
2-Cyclohexen-1-one, 5,5-dimethyl-3-morpholino-	pH 1	295(4.43)	39-2207-76C

Compound	Solvent	$\lambda_{max}(\log \epsilon)$	Ref.
2-Cyclohexen-1-one, 5,5-dimethyl-3-morpholino- (cont.)	H_2O pH 13	307(4.55) 307(4.55)	39-2207-76C 39-2207-76C
$C_{12}H_{19}NO_3$ 2-Butanamine, 3-(3,4-dimethoxyphenoxy)-	n.s.g.	210(3.99),230(3.87), 285(3.51)	20-0421-76
2-Butanamine, 3-(3,5-dimethoxyphenoxy)-	n.s.g.	215(4.09),265(2.75)	20-0421-76
1H-Pyrrole-2-carboxylic acid, 4-butyl-3-hydroxy-5-methyl-, ethyl ester	MeOH	276.5(4.64)	5-0383-76
$C_{12}H_{19}NO_4$ 1H-Pyrrole-1,3-dicarboxylic acid, 4,5-dihydro-2,5-dimethyl-, diethyl ester	MeOH	281(4.24)	1-0567-76
$C_{12}H_{19}NO_5$ α-D-xylo-Hept-5-enofuranose, 5,6,7-trideoxy-3-O-methyl-1,2-O-(1-methylethylidene)-7-(methylimino)-, N-oxide	EtOH	284(4.09)	136-0127-76A
1H-Pyrrole-1,3-dicarboxylic acid, 5-ethyl-2,5-dihydro-4-hydroxy-, diethyl ester	EtOH	278(3.74)	4-0113-76
$C_{12}H_{19}NO_6S$ 2H-1,4-Thiazine-3,6-dicarboxylic acid, 3,4-dihydro-4-[(1-methylethoxy)methyl]-, dimethyl ester, 1-oxide, (1R,3R)-(1S,3R)-	EtOH EtOH	206(3.78),276(4.18) 206(3.87),275(4.12)	39-2540-76C 39-2540-76C
2,5-Thiazoledicarboxylic acid, 2,3-dihydro-2-methyl-3-[(1-methylethoxy)methyl]-, dimethyl ester, 1-oxide, trans-(±)-	EtOH	215(3.89),288(4.02)	39-2540-76C
$C_{12}H_{19}N_3O_2$ 3-Octanone, 1-(2-amino-4-hydroxy-6-pyrimidinyl)-	pH 13	230(3.57),275(3.51)	4-0439-76
$C_{12}H_{19}N_3O_4$ 1(2H)-Pyrimidineheptanoic acid, α-amino-3,4-dihydro-5-methyl-2,4-dioxo-	pH 1	271(3.95)	24-2615-76
$C_{12}H_{19}N_3O_5$ 4-Pyrimidineacetic acid, 2-amino-α,α-diethoxy-1,6-dihydro-6-oxo-, ethyl ester	EtOH	224(4.00),292(3.95)	44-2850-76
$C_{12}H_{19}N_3O_7$ 1H-Imidazole-4-carboxylic acid, 5-amino-1-β-D-glucopyranosyl-, ethyl ester	H_2O	266(4.05)	65-1850-76
$C_{12}H_{19}N_5O_2$ 1,5-Pentanediol, 3-[6-(dimethylamino)-9H-purin-9-yl]-, hydrochloride	pH 1 H_2O pH 13	277(4.25) 277(4.64) 280(3.98)	103-0461-76 103-0461-76 103-0461-76
$C_{12}H_{19}N_5O_4$ Urea, N,N''-[5-[bis(2-hydroxyethyl)amino]-1,3-phenylene]bis-	pH 1	221(4.35),238(4.55), 258(4.03),310(3.28)	87-0426-76
$C_{12}H_{19}O$ Pyrylium, 4-methyl-2,6-bis(1-methylethyl)-, iodide	CH_2Cl_2	453(2.11)	59-1311-76

Compound	Solvent	$\lambda_{max}(\log \epsilon)$	Ref.
Pyrylium, 4-methyl-2,6-bis(1-methylethyl)-, selenocyanate	CH_2Cl_2	443(2.27)	59-1311-76
thiocyanate	CH_2Cl_2	409(2.55)	59-1311-76
$C_{12}H_{20}$			
Cyclohexane, (3-methyl-2-pentenylidene)-, (E)-	EtOH	241(4.43),248(4.49), 257(4.36)	39-2386-76C
(Z)-	EtOH	241(4.42),248(4.47), 257(4.33)	39-2386-76C
3-Heptyne, 2,2,6,6-tetramethyl-5-methylene-	C_6H_{12}	222(3.09),232(3.02)	44-3191-76
$C_{12}H_{20}Cl_2N_2O$			
Diazene, bis(2-chlorocyclohexyl)-, 1-oxide	EtOH	226(3.97)	88-0153-76
$C_{12}H_{20}Cl_3N_2O_3$			
1H-1,4-Diazepin-1-yloxy, hexahydro-2,2,7,7-tetramethyl-4-[(2,2,2-trichloroethoxy)carbonyl]-	C_6H_{12}	228(3.45),245s(3.29), 465(0.91)	77-0083-76
	MeOH	227(3.45),245s(3.29), 444(0.91)	77-0083-76
$C_{12}H_{20}N_2OS$			
4(1H)-Pyrimidinone, 2-[(3-methylbutyl)thio]-6-propyl-	EtOH	235(3.92),288(3.90)	95-1094-76
$C_{12}H_{20}N_2O_2$			
1(2H)-Pyridinecarboxylic acid, 3,4-dihydro-5-(1-pyrrolidinyl)-, ethyl ester	MeOH	250(3.99)	1-0884-76
$C_{12}H_{20}N_2O_2S$			
4,6(1H,3H)-Pyrimidinedione, 5-(1-methylbutyl)-5-propyl-2-thioxo-	50%MeOH-HCl	287(4.39)	87-0521-76
	dioxan	288(4.39)	87-0521-76
	MeCN	285.5(4.44)	87-0521-76
	$C_2H_4Cl_2$	284(4.36)	87-0521-76
$C_{12}H_{20}N_2O_4$			
1H-Pyrrole-1,3-dicarboxylic acid, 4-amino-5-ethyl-2,5-dihydro-, diethyl ester	EtOH	274(4.23)	4-0113-76
$C_{12}H_{20}N_2O_5$			
2-Propenamide, 3-ethoxy-N-[[[2-hydroxy-4-(hydroxymethyl)cyclopentyl]amino]carbonyl]-, (1α,2β,4α)-(±)-	pH 1 or 7	252(4.33)	4-1015-76
2-Propenamide, 3-ethoxy-N-[[[3-hydroxy-4-(hydroxymethyl)cyclopentyl]amino]carbonyl]-, (1α,3β,4α)-(±)-	pH 1 or 7	252(4.31)	4-1015-76
$C_{12}H_{20}N_2O_6$			
2-Propenamide, 3-ethoxy-N-[[[2,3-dihydroxy-4-(hydroxymethyl)cyclopentyl]amino]carbonyl]-, (1α,2β,3β,4α)-(±)-	pH 1	252(4.32)	4-1015-76
	pH 7	252(4.33)	4-1015-76
$C_{12}H_{20}N_2S_2$			
3-Cyclobutene-1,2-dithione, 3,4-bis(diethylamino)-	MeOH	283(4.34),371s(4.44), 403(4.60)	44-3904-76

Compound	Solvent	$\lambda_{max}(\log \epsilon)$	Ref.
$C_{12}H_{20}N_3PS$			
2H-1,3,2-Benzodiazaphosphole-2-amine, N,N-diethyl-1,3-dihydro-1,3-dimethyl-, 2-sulfide	EtOH	248(3.66),292(3.75)	65-0807-76
$C_{12}H_{20}N_4O$			
1,3,5-Triazanaphthalene, 2-amino-5,6,7-8-tetrahydro-4-hydroxy-6-pentyl-, hydrochloride	pH 13	252(3.91),305(3.76)	4-0439-76
$C_{12}H_{20}O$			
Bicyclo[2.2.1]heptan-2-one, 1,3,3,7,7-pentamethyl-	3-Mepentane	297(1.30)	144-0475-76
	MeCN	295(1.35)	144-0475-76
2-Butanone, 1-(1-cyclohexen-1-yl)-3,3-dimethyl-	EtOH	293.5(1.85)	44-3076-76
2-Butanone, 1-cyclohexylidene-3,3-dimethyl-	EtOH	238(4.16),325(2.01)	44-3076-76
3-Buten-2-one, 4-cyclooctyl-	EtOH	223.0(3.73)	22-0849-76
1-Cyclohexene-1-propanal, 2,6,6-tri-methyl-	pentane	287(1.85)	33-0032-76
	isoPrOH	295(1.90)	33-0032-76
Furan, 2,5-bis(1,1-dimethylethyl)-, cation	n.s.g.	255(3.95)	78-1767-76
$C_{12}H_{20}OS_5$			
Ethanol, 2-[(dihydro-2,3a,5,7-tetramethyl-2,5-methano-7H-1,3-dithiolo[4,5-d]-[1,3]dithiin-7-yl)thio]-	isooctane	231(3.12),258(2.52), 310(2.11)	70-2353-76
$C_{12}H_{20}O_2$			
2-Butanone, 1-(1-hydroxy-2-cyclohexen-1-yl)-3,3-dimethyl-	EtOH	292.5(1.28)	44-3076-76
2-Cyclohexen-1-one, 6-(4-hydroxy-1-methylbutyl)-3-methyl-	EtOH	232(4.17)	33-0695-76
2-Cyclopenten-1-one, 2-(7-hydroxyheptyl)-	EtOH	227(3.99)	39-2550-76C
Menthone, 2-acetyl-	isooctane	293(3.76),350s(1.18)	104-1399-76
	EtOH	294(3.77)	104-1399-76
$C_{12}H_{20}O_3$			
2(5H)-Furanone, 3,5-bis(1,1-dimethylethyl)-5-hydroxy-	MeOH	206(4.10)	35-1510-76
2(5H)-Furanone, 4-hexyl-3-methoxy-5-methyl-	EtOH	226(3.94)	107-0299-76
$C_{12}H_{20}O_6S_2$			
α-D-Glucofuranose, 3-0-methyl-1,2-0-(1-methylethylidene)-, 6-(S-methyl carbonodithioate)	MeOH	223(3.88),281(4.08)	118-0449-76
$C_{12}H_{20}S$			
Thiophene, 2,5-bis(1,1-dimethylethyl)-	neutral	245(3.60)	78-1767-76
	cation	249(3.60),310(3.59)	78-1767-76
$C_{12}H_{20}Si$			
Silane, [(3,5-dimethylphenyl)methyl]-trimethyl-	EtOH	271(2.42)	78-0051-76
$C_{12}H_{21}N$			
1H-Pyrrole, 2,5-bis(1,1-dimethylethyl)-	neutral	217(4.07)	78-1767-76
	cation	218.5(3.90)	78-1767-76

Compound	Solvent	$\lambda_{max}(\log \epsilon)$	Ref.
$C_{12}H_{21}NO$			
2-Cyclohexen-1-one, 3-(butylamino)-6,6-dimethyl-	MeOH	291(4.52)	35-2826-76
1-Penten-3-one, 1-(cyclohexylamino)-4-methyl-	H_2O	310(--)	35-2826-76
	MeOH	311(4.33)	35-2826-76
cis-s-cis	C_6H_{12}	304(4.24)	35-2826-76
$C_{12}H_{21}NOS_4$			
Ethanol, 2-(dihydro-2a,3,5,7-tetramethyl-2,5-methano-7H-1,3-dithiolo[4,5-d]-[1,3]dithiin-7-yl)amino]-	isooctane	256(2.38),296(2.21)	70-2353-76
$C_{12}H_{21}NO_2$			
2-Butanone, 1-(3,4-dihydro-2,2-dimethyl-2H-pyrrol-5-yl)-3,3-dimethyl-, N-oxide	EtOH-HCl	233(3.86),308(3.34)	39-1944-76C
	EtOH-NaOH	326(4.39)	39-1944-76C
$C_{12}H_{21}N_2O$			
Pyrimidinium, 1,3-dibutyl-2,3-dihydro-2-oxo-, sulfate(1:1)	pH 0.29	319(3.95)	23-2681-76
pseudobase	pH 9.18	240(3.61),301(2.88)	23-2681-76
$C_{12}H_{21}N_3O_2S$			
Acetamide, N-[1-(hexahydro-1,3-dimethyl-2-oxo-6-thioxo-4-pyrimidinyl)butyl]-	n.s.g.	283(4.10)	88-0297-76
diastereomer	n.s.g.	285(4.10)	88-0297-76
$C_{12}H_{21}N_3S$			
Thiopyrano[4,3-c]pyrazole-2(4H)-ethan-amine, N,N-diethyl-6,7-dihydro-	EtOH	229.5(3.69)	4-0225-76
$C_{12}H_{21}N_5S_3$			
Propanethioamide, N,N,2-trimethyl-2-[tetrahydro-1,3,5-trimethyl-4,6-di-thioxo-1,3,5-triazin-2(1H)-ylidene-amino]-	EtOH	276(4.70)	33-2768-76
$C_{12}H_{22}N_6O_2$			
Piperazine, 1,1'-(azodicarbonyl)bis-[4-methyl-	H_2O	299(3.56),442(1.60)	35-3001-76
	MeCN	291(3.26),423(1.59)	35-3001-76
$C_{12}H_{22}O$			
Cyclobutanone, 2,4-bis(1,1-dimethyleth-yl)-, cis	C_6H_{12}	312(1.82)	112-0777-76
trans	C_6H_{12}	304(1.45)	112-0777-76
Cyclohexanone, 5-butyl-3,3-dimethyl-	n.s.g.	285(1.40)	44-2396-76
$C_{12}H_{22}O_2$			
2-Butanone, 1-(1-hydroxycyclohexyl)-3,3-dimethyl-	EtOH	290.5(1.57)	44-3076-76
$C_{12}H_{22}S_2$			
Disulfide, di-4-hexenyl-, (E,E)-	EtOH	253(2.8)	142-0035-76S
$C_{12}H_{23}NO$			
1-Penten-3-one, 1-[bis(1-methylethyl)-amino]-4-methyl-	C_6H_{12}	296(4.35)	35-2826-76
	H_2O	316(4.44)	35-2826-76
	MeOH	313(4.43)	35-2826-76
$C_{12}H_{23}NSi_2$			
Silanamine, 1,1,1-trimethyl-N-phenyl-N-(trimethylsilyl)-	C_6H_{12}	235(3.60),266(2.67),272(2.65),298s(2.00)	101-0299-76M

Compound	Solvent	$\lambda_{max}(\log \epsilon)$	Ref.
$C_{12}H_{24}Si_3$			
Silane, (1-methyl-1,2-pentadien-4-yne-1,3,5-triyl)tris[dimethyl-	C_6H_{12}	235s(4.05),241(4.08), 255(4.09)	35-8426-76
Silane, (5-methyl-1,3-pentadiyn-1-yl-5-ylidene)tris[dimethyl-	C_6H_{12}	240(3.70),252(3.71), 266(3.67),282(3.13)	35-8426-76
$C_{12}H_{25}ClN_2O_5S$			
1-Oxa-4,8-diazaspiro[4.5]decan-8-ium, 4-chloro-3,3,8,8-tetramethyl-, methyl sulfate	H_2O	268(2.42)	36-1733-76
$C_{12}H_{25}N_2O_5P$			
Acetic acid, [(tripropoxyphosphoranyli-dene)hydrazono]-, methyl ester	EtOH	273.0(4.09)	65-1424-76
Acetic acid, [[tris(1-methylethoxy)-phosphoranylidene]hydrazono]-, methyl ester	EtOH	273.0(4.14)	65-1424-76
$C_{12}H_{26}O$			
Udecane, 1-methoxy-	MeOH	206(4.98),230s(4.34), 350(3.28)	39-0465-76C
$C_{12}H_{26}O_2Si_2$			
Silane, [(1,4-dimethyl-1,3-butadiene-1,5-diyl)bis(oxy)]bis[trimethyl-	C_6H_{12}	252.0(4.32)	44-1799-76
isomer	C_6H_{12}	252.0(4.09)	44-1799-76
$C_{12}H_{27}PSe$			
Phosphine selenide, tributyl-	EtOH	195(4.18),208s(4.06), 238(3.53)	70-0174-76
$C_{12}N_6$			
1-Cyclopropene-1,2-diacetonitrile, α,α'-dicyano-3-(dicyanomethylene)-, ion(2-)	MeCN	222(4.55),285s(4.34), 315(4.52)	35-0610-76
Propanedinitrile, 2,2',2"-(1,2,3-cyclo-propanetriylidene)tris-, radical ion (1-), potassium	MeCN	320(4.48),598(4.11),	35-0611-76

Compound	Solvent	$\lambda_{max}(\log \epsilon)$	Ref.
$C_{13}ClN_6$ 1,2,4,5-Benzenetetracarbonitrile, 3-chloro-6-(dicyanomethyl)-, ion(1-), pyridinium salt	MeCN	382(4.20),600(3.32)	24-2469-76
$C_{13}H_3N_5O_2$ 1-Cyclopropene-1-acetic acid, 2-(dicya- nomethyl)-3-(dicyanomethylene)-, methyl ester, ion(2-)	MeCN	224(4.51),293s(4.39), 322(4.54)	35-0610-76
$C_{13}H_4F_2N_6O$ 9H-Fluoren-9-one, 1,3-diazido-2,4-di- fluoro-	EtOH	207(4.29),212s(4.29), 256(4.56),267s(4.51), 360(3.84)	104-0857-76
$C_{13}H_4F_4O$ 9H-Fluoren-9-one, 1,2,3,4-tetrahydro-	EtOH	245(4.72),253(4.81), 287(3.45),298(3.61), 312(3.47),326(3.41), 380(3.90)	104-0857-76
$C_{13}H_4F_{18}OP_2$ 1,4-Diphosphabicyclo[2.2.2]octa-2,5-di- ene, 7-methoxy-2,3,5,6,7,8-hexakis- (trifluoromethyl)-	MeOH	243(3.22),332(3.00)	88-3715-76
$C_{13}H_5N_6$ 1,3-Pentadiene-1,1,2,4,5,5-hexacarbo- nitrile, 3-ethyl-, ion(1-)	MeCN	592(4.43)	35-0558-76
$C_{13}H_6MoO_5S_3$ Molybdenum, pentacarbonyl(5-methylbenzo- 1,3-dithiole-2-thione)-	benzene	338(3.95),365(3.92), 482(4.08)	101-0369-76Q
$C_{13}H_7BrN_4Se$ 1-Selena-3,5,8,8b-tetraazaacenaphthyl- ene, 6-bromo-2-phenyl-	EtOH	232(4.08),296(3.56), 308(3.54),320(3.49), 339(3.50),395(3.93), 495(2.57),528(2.59), 573(2.50),630(2.20)	78-0173-76
$C_{13}H_7Cl_5O_2$ Benzene, 1,2-dichloro-4-methoxy- 5-(2,4,5-trichlorophenoxy)-	EtOH	289(4.10),296s(4.05)	1-0796-76
$C_{13}H_7N$ 1-Acenaphthylenecarbonitrile	ether	236(4.39),330(4.05), 349(3.73),356(3.82), 475(1.96)	22-0217-76
$C_{13}H_7NO_4$ 9H-Xanthen-9-one, 2-nitro-	n.s.g.	217(4.31),250(4.34), 272(4.18),280(4.17), 305(3.92),330(3.58)	39-2241-76C
$C_{13}H_8BrClN_4$ 2H-Tetrazole, 2-(4-bromophenyl)- 5-(4-chlorophenyl)-	EtOH	278(4.41)	18-0762-76

Compound	Solvent	$\lambda_{max}(\log \epsilon)$	Ref.
$C_{13}H_8BrN_5O_2$ 2H-Tetrazole, 2-(4-bromophenyl)- 5-(4-nitrophenyl)-	EtOH	213(4.34),298(4.48)	18-0762-76
$C_{13}H_8ClF_3N_4O_3$ Benzo[g]pteridine-2,4(3H,10H)-dione, 8-chloro-10-(2-hydroxyethyl)- 7-(trifluoromethyl)-	EtOH	273(4.49),333(3.74), 430(3.99)	103-0938-76
$C_{13}H_8ClNO$ Cyclohept[b]indol-6(5H)-one, 2-chloro-	MeOH	275(4.32),317s(4.19), 324(4.22),338(4.16), 384(3.72),402(3.84)	18-1101-76
$C_{13}H_8ClNO_2$ Furo[2,3-c]pyridin-2(6H)-one, 3-chloro- 6-phenyl-	MeCN	235(3.00),366(4.70)	88-2959-76
Methanone, (5-chloro-2-nitrophenyl)phen- yl-	isoPrOH	245(4.23),292(3.98), 319(3.97)	4-0907-76
$C_{13}H_8ClN_3O_4$ Benzenecarboximidoyl chloride, 4-nitro- N-(4-nitrophenyl)-	hexane	274(4.49),324(3.92)	104-0161-76
$C_{13}H_8ClN_5O_2$ 2H-Tetrazole, 2-(4-chlorophenyl)- 5-(4-nitrophenyl)-	EtOH	296(4.47)	18-0762-76
2H-Tetrazole, 5-(4-chlorophenyl)- 2-(4-nitrophenyl)-	EtOH	258(4.24),300(4.44)	18-0762-76
$C_{13}H_8ClOS$ 1,3-Benzoxathiol-1-ium, 2-(2-chlorophen- yl)-, tetrafluoroborate	96% H_2SO_4	236(3.61),275(3.77), 373(4.34)	39-0323-76C
$C_{13}H_8ClS_2$ 1,3-Benzodithiol-1-ium, 2-(4-chlorophen- yl)-, perchlorate	96% H_2SO_4	250s(3.69),298(3.55), 401(4.47)	118-0759-76
$C_{13}H_8Cl_2FNO_5S$ Furan, 2-[1-(dichloromethyl)sulfonyl]- 2-(2-fluorophenyl)ethenyl]-5-nitro-	EtOH	207(4.31),287(4.18)	73-3391-76
Furan, 2-[1-(dichloromethyl)sulfonyl]- 2-(4-fluorophenyl)ethenyl]-5-nitro-	EtOH	207(4.29),292(4.26)	73-3391-76
$C_{13}H_8Cl_2N_4$ 2H-Tetrazole, 2,5-bis(4-chlorophenyl)-	EtOH	277(4.42)	18-0762-76
$C_{13}H_8Cl_3N$ Pyridine, 2,5,6-trichloro-5-(2-phenyl- ethenyl)-	MeOH	296(4.33),325(4.29)	33-0190-76
$C_{13}H_8Cl_3NO_5S$ Furan, 2-[2-(2-chlorophenyl)-1-[(dichlo- romethyl)sulfonyl]ethenyl]-5-nitro-	EtOH	207(4.40),292(4.13)	73-3391-76
Furan, 2-[2-(4-chlorophenyl)-1-[(dichlo- romethyl)sulfonyl]ethenyl]-5-nitro-	EtOH	206(4.29),226s(4.21), 297(4.24)	73-3391-76
$C_{13}H_8Cl_4N_2$ Benzenemethanamine, α-[(2,3,4,5-tetra- chloro-2,4-cyclopentadien-1-ylidene)- methyl]-	dioxan	256(3.85),397(4.40)	24-3939-76

Compound	Solvent	$\lambda_{max}(\log \epsilon)$	Ref.
$C_{13}H_8FNO$			
9(10H)-Acridinone, 2-fluoro-	MeOH	213(4.11),247(4.55), 257(3.53),386(3.88), 405(3.88)	5-0946-76
6(5H)-Phenanthridinone, 8-fluoro-	MeOH	229(4.57),259(4.21), 326(3.86),341(3.76)	5-0946-76
$C_{13}H_8FN_3O$			
1,2,3-Benzotriazin-4(3H)-one, 6-fluoro-3-phenyl-	MeOH	225(4.45),290(3.86)	5-0946-76
$C_{13}H_8FN_3O_4$			
Benzenecarboximidoyl fluoride, 4-nitro-N-(4-nitrophenyl)-	hexane	279(4.31),300(4.27)	104-0161-76
$C_{13}H_8F_2N_2OS$			
Thieno[2,3-d]pyrimidin-2(3H)-one, 4-(2,6-difluorophenyl)-3-methyl-	EtOH	221(4.28),252(4.39), 292(3.43),374(3.72)	18-1395-76
$C_{13}H_8IN_3$			
Benzonitrile, 3-iodo-2-(phenylazo)-	EtOH	224s(4.26),260(3.94), 316(4.09),456(2.57)	101-0039-76M
$C_{13}H_8N_2$			
1H-Cyclopenta[gh]perimidine	MeOH	<u>220s(3.8),418(4.0)</u>	103-0450-76
N-anion	MeOH	<u>225(3.9),300(3.7), 315s(3.7),440(3.8)</u>	103-0450-76
$C_{13}H_8N_2O$			
9H-Xanthene, 9-diazo-	C_6H_{12}	237(4.10),244(4.11), 273(3.75),281(3.78), 321(3.65),335(3.65), 347(3.69),362(3.71)	22-1131-76
$C_{13}H_8N_2O_2$			
5H-Oxazolo[4,5-b]phenoxazine	benzene	325(3.99)	23-0037-76
4H-Pyrano[2,3-b]pyridin-4-one, 2-(2-pyridinyl)-	EtOH	297(4.06)	22-1916-76
4H-Pyrano[2,3-b]pyridin-4-one, 2-(3-pyridinyl)-	EtOH	220(4.11),292(4.24)	22-1916-76
4H-Pyrano[2,3-b]pyridin-4-one, 2-(4-pyridinyl)-	EtOH	220(4.00),285(4.10)	22-1916-76
4H-Pyrano[3,2-c]pyridin-4-one, 2-(3-pyridinyl)-	EtOH	236(4.31),300(4.25)	22-1916-76
4H-Pyrano[3,2-c]pyridin-4-one, 2-(4-pyridinyl)-	EtOH	238(4.26),280(4.02)	22-1916-76
$C_{13}H_8N_2O_3$			
Cyclohept[b]indol-6(5H)-one, 2-nitro-	MeOH	255(4.34),296(4.32), 305s(4.27),320(4.27), 376(4.01),397(4.02)	18-1101-76
Cyclohept[b]indol-6(5H)-one, 4-nitro-	MeOH	254(4.45),338(4.09), 366(4.13),386(4.15)	18-1101-76
1-Phenazinecarboxylic acid, 6-hydroxy-	EtOH	269(4.47),366(3.76), 373(3.79),456(3.21)	39-2248-76C
$C_{13}H_8N_4O$			
1H-Pyrazolo[4,3-b]pyridine-6-carbonitrile, 4,7-dihydro-7-oxo-1-phenyl-	EtOH	227(4.46),306(4.16)	39-0507-76C

Compound	Solvent	$\lambda_{max}(\log \epsilon)$	Ref.
$C_{13}H_8N_4O_3$ 1,2,3-Benzotriazin-4(3H)-one, 6-nitro-3-phenyl-	MeOH	222(4.48),332(3.94)	5-0946-76
$C_{13}H_8N_4S$ 1-Thia-3,5,8,8b-tetraazaacenaphthylene, 2-phenyl-	EtOH	228(4.37),281(3.79), 305(3.83),318(3.98), 334(4.09),381(4.19), 480(2.83),514(2.86), 553(2.79),606(2.51)	78-0173-76
$C_{13}H_8N_4Se$ 1-Selena-3,5,8,8b-tetraazaacenaphthylene, 2-phenyl-	EtOH	229(4.33),294(3.78), 305(3.81),319(3.82), 335(3.82),388(4.06), 481(2.78),516(2.81), 556(2.72),610(2.36)	78-0173-76
$C_{13}H_8N_6O_2$ 1-Butene-1,1,2-tricarbonitrile, 3-[(4-nitrophenyl)hydrazono]-	dioxan	462(4.27)	24-1787-76
$C_{13}H_8N_6O_4$ 2H-Tetrazole, 2,5-bis(4-nitrophenyl)-	EtOH	213(4.34),304(4.54)	18-0762-76
$C_{13}H_8O$ Bicyclo[3.2.0]hepta-2,4,7-trien-6-one, 7-phenyl-	EtOH	210(4.00),225(4.07), 267(4.06),325s(4.16), 335(4.23),347s(4.10), 415s(2.76)	33-0021-76
$C_{13}H_8OS_2$ Thieno[2,3-b]thiophene-2-carboxaldehyde, 5-phenyl-	EtOH	290(4.29),347(3.40)	104-1550-76
$C_{13}H_8O_2S$ Thieno[2,3-b]furan-5-carboxaldehyde, 2-phenyl-	EtOH	220(4.29),275(4.48), 347(4.07)	104-1550-76
$C_{13}H_8O_2S_2$ Thieno[2,3-b]thiophene-5-carboxylic acid, 2-phenyl-	EtOH	305(4.54)	104-1550-76
$C_{13}H_8O_5$ 9H-Xanthen-9-one, 1,3,7-trihydroxy-	n.s.g.	218(3.99),268(4.24), 360(4.32),312(3.93), 377(3.64)	39-1377-76C
$C_{13}H_8S_2$ 2,2'-Bithiophene, 5-(1,3-pentadiynyl)-	ether	348(4.42)	102-1309-76
$C_{13}H_9BrN_2S$ Isothiazolo[5,4-b]pyridine, 3-bromo-4-methyl-6-phenyl-	EtOH	321s(4.16)	48-0779-76
$C_{13}H_9BrN_4$ 2H-Tetrazole, 2-(4-bromophenyl)-5-phenyl-	EtOH	240(4.18),275(4.36)	18-0762-76

Compound	Solvent	$\lambda_{max}(\log \epsilon)$	Ref.
$C_{13}H_9Br_2N$			
Pyridine, 2,6-dibromo-3-(2-phenylethen-yl)-, (E)-	MeOH	301(4.40)	33-0190-76
$C_{13}H_9ClN_2$			
Propanedinitrile, [3-chloro-3-(4-methyl-phenyl)-2-propenylidene]-	CH_2Cl_2	367(4.43)	48-0705-76
$C_{13}H_9ClN_2O_2$			
Benzenecarboximidoyl chloride, N-(4-ni-trophenyl)-	hexane	258(4.18),285(4.29)	104-0161-76
Benzenecarboximidoyl chloride, 4-nitro-N-phenyl-	hexane	230s(--),268(4.29), 337(3.76),365s(3.48)	104-0161-76
$C_{13}H_9ClN_4$			
2H-Tetrazole, 2-(4-chlorophenyl)-5-phen-yl-	EtOH	241(4.17),274(4.32)	18-0762-76
2H-Tetrazole, 5-(4-chlorophenyl)-2-phen-yl-	EtOH	276(4.38)	18-0762-76
$C_{13}H_9ClO_4S$			
Benzoic acid, 4-[(4-chlorophenyl)sulfon-yl]-	EtOH	198(4.7),230(4.2), 250(4.6)	104-2133-76
$C_{13}H_9ClS_3$			
Benzenecarbo(dithioperoxo)thioic acid, 4-chloro-, phenyl ester	C_6H_{12}	306(4.13),537(1.95)	138-0441-76
$C_{13}H_9Cl_2N$			
Pyridine, 2,6-dichloro-3-(2-phenyleth-enyl)-, (E)-	MeOH	258(4.34)	33-0190-76
$C_{13}H_9Cl_2NO_3S$			
2H-Pyran-5-carbothioic acid, 6-amino-4-chloro-2-oxo-, S-[(4-chlorophenyl)-methyl] ester	$CHCl_3$	308s(4.27),331(4.36)	39-2462-76C
$C_{13}H_9Cl_2NO_5S$			
Furan, 2-[1-[(dichloromethyl)sulfonyl]-2-phenylethenyl]-5-nitro-	EtOH	207(4.41),219s(4.21), 287(4.26)	73-3391-76
Phenol, 2,6-dichloro-4-nitro-, 4-methylbenzenesulfonate	MeOH	268(3.99)	40-0631-76
$C_{13}H_9Cl_2NO_8S$			
2-Furancarboxylic acid, 5-[2-[(dichloro-methyl)sulfonyl]-2-(5-nitro-2-furan-yl)ethenyl]-, methyl ester	EtOH	213(4.18),303(4.48), 370s(3.97)	73-3391-76
$C_{13}H_9Cl_2N_5O_2$			
Pyrimido[5,4-e]-1,2,4-triazine-5,7(1H,6H)-dione, 3-(3,4-dichloro-phenyl)-1,6-dimethyl-	dioxan	301(4.48),434(3.44)	39-0713-76C
Pyrimido[5,4-e]-1,2,4-triazine-5,7(6H,8H)-dione, 3-(3,4-dichloro-phenyl)-6,8-dimethyl-	dioxan	288(4.48),377(3.57)	39-0713-76C
$C_{13}H_9Cl_2N_5O_3$			
Pyrimido[5,4-e]-1,2,4-triazine-5,7(1H,6H)-dione, 3-(3,4-dichloro-phenyl)-1,6-dimethyl-, 4-oxide	dioxan	295(4.51),432(3.47)	39-0713-76C

Compound	Solvent	$\lambda_{max}(\log \epsilon)$	Ref.
Pyrimido[5,4-e]-1,2,4-triazine-5,7(6H,8H)-dione, 3-(3,4-dichloro-phenyl)-6,8-dimethyl-, 4-oxide	dioxan	262(4.39),300(4.37), 389(3.37)	39-0713-76C
$C_{13}H_9Cl_3N_2$ Benzenecarbohydrazonoyl chloride, 4-chloro-N-(4-chlorophenyl)-	C_6H_{12}	243(4.25),256s(--), 298s(--),308(4.17), 338(4.39)	48-0881-76
$C_{13}H_9FN_2O_2$ Benzenecarboximidoyl fluoride, N-(4-ni-trophenyl)-	hexane	246(4.09),299(4.33)	104-0161-76
Benzenecarboximidoyl fluoride, 4-nitro-N-phenyl-	hexane	238(4.14),266(4.10), 306(4.09),318(4.12), 332(4.09)	104-0161-76
$C_{13}H_9F_3N_2S$ 2-Pyridinecarbothioamide, N-[3-(tri-fluoromethyl)phenyl]-	20% EtOH	282(4.04)	94-1451-76
$C_{13}H_9F_3O$ Ethanone, 2,2,2-trifluoro-1-(3-methyl-1-azulenyl)-	C_6H_{12}	272(4.20),298(4.37), 309(4.45),316(4.54), 392(4.05),415(4.11), 553(2.74),596(2.62), 632(2.20),658(2.11)	44-1822-76
$C_{13}H_9F_3O_2$ Ethanone, 2,2,2-trifluoro-1-(3-methoxy-1-azulenyl)-	C_6H_{12}	282(4.19),313(4.30), 319(4.32),327(4.43), 423(3.99),450(4.04), 576(2.67),610(2.72), 623(2.73),657(2.62), 682(2.56),735(2.16), 765(2.00)	44-1822-76
$C_{13}H_9N$ Acridine	pH 7.0	249(5.05)	95-1334-76
$C_{13}H_9NO$ 9(10H)-Acridinone	pH 1	212(4.32),253(4.75), 385(3.70)	56-1267-76
	MeOH	251(4.55),380(3.62), 399(3.62)	100-0134-76
	EtOH	255(4.73),265s(4.56), 268s(4.23),294(3.37), 307(3.15),344s(3.27), 363s(3.67),368(3.93), 398(3.95)	35-0171-76
9H-Indeno[2,1-b]pyridin-9-one, 4-methyl-(onychine)	EtOH	253(4.62),279(3.85), 289(3.88),308(3.30)	102-1186-76
	EtOH-HCl	252(4.42),292s(4.00), 298(4.04),320s(3.60), 331s(3.48)	102-1186-76
$C_{13}H_9NOS$ 2,1-Benzisothiazol-3(1H)-one, 1-phenyl-	CHCl$_3$	265(4.04),290s(3.62), 374(3.83)	48-0161-76

Compound	Solvent	$\lambda_{max}(\log \epsilon)$	Ref.
$C_{13}H_9NO_2$			
1-Azulenecarboxylic acid, 2-cyano-, methyl ester	EtOH	261(4.18),295(4.72), 306(4.82),347(4.03), 361(3.75),545s(2.96), 568(2.99)	44-1822-76
5H-[1]Benzopyrano[2,3-b]pyridin-5-one, 7-methyl-	96% H_2SO_4	341(4.532)	103-1022-76
Furo[2,3-c]pyridin-2(6H)-one, 6-phenyl-	MeCN	245(3.00),367(4.60)	88-2959-76
1H-Indole-3-carboxaldehyde, 2-(2-furan-yl)-	EtOH	228(4.17),259(4.47), 282s(3.88)	2-0579-76
$C_{13}H_9NO_2S$			
2H-Pyran-3-carbonitrile, 4-(methylthio)-2-oxo-6-phenyl-	EtOH	239(3.08),255(3.42), 330(3.50),370(3.08)	142-1493-76
$C_{13}H_9NO_3$			
9(10H)-Acridinone, 1,3-dihydroxy-	EtOH	210(4.07),222(4.18), 244(4.44),260(4.29), 269(4.68),294(4.12), 313s(3.72),328(3.88), 392(3.90)	39-2089-76C
$C_{13}H_9NO_3S$			
Benzonitrile, 2-hydroxy-5-(phenylsulfon-yl)-	EtOH	254(4.03),292(4.03)	56-0973-76
Benzonitrile, 2-[(phenylsulfonyl)oxy]-	EtOH	267(3.28),274(3.32), 283(3.09)	56-0973-76
$C_{13}H_9NO_4$			
Benzaldehyde, 2-(4-nitrophenoxy)-	n.s.g.	211(4.36),227(4.11), 245(3.96),302(4.17)	39-2241-76C
[1,1'-Biphenyl]-4-carboxaldehyde, 3-hydroxy-2-nitro-	octane	255(4.4),390(3.4)	103-0174-76
[1,1'-Biphenyl]-4-carboxaldehyde, 5-hydroxy-2-nitro-	octane	270(4.2),330(3.7)	103-0174-76
$C_{13}H_9NO_6$			
Methanone, (2-nitrophenyl)(2,4,6-tri-hydroxyphenyl)-	EtOH	213(4.51),228s(4.34), 295(4.31)	39-2089-76C
$C_{13}H_9NO_7$			
1,3-Dioxolo[4,5-g]quinoline-6,7-dicarb-oxylic acid, 5,8-dihydro-8-oxo-, 7-methyl ester (DMF solvate)	MeOH	205(4.35),222(4.38), 248(4.42),266(4.34), 276s(4.11),324s(3.89), 336(3.99),350(3.97), 367(3.76)	5-1972-76
$C_{13}H_9NS$			
Cyclohept[b]indole-6-thiol	MeOH	234(4.30),300(4.41), 403(3.94),461(4.17)	18-1101-76
$C_{13}H_9N_3$			
Benzonitrile, 2-(phenylazo)-	EtOH	217s(4.20),234(4.09), 243s(4.00),249s(3.86), 324(4.28),337s(4.26), 354s(3.98),452(2.62)	101-0039-76M
$C_{13}H_9N_3O$			
1,2,3-Benzotriazin-4(3H)-one, 3-phenyl-	MeOH	225(4.53),290(3.85), 314s(3.80)	5-0946-76

Compound	Solvent	$\lambda_{max}(\log \epsilon)$	Ref.
1,2-Diazepine-4-carbonitrile, 1-benzoyl-Pyrido[2,1-f][1,2,4]triazin-9-ium, 1,4-dihydro-4-oxo-2-phenyl-, hydroxide, inner salt	MeOH EtOH	232(3.90),360(2.70) 253(4.43),343(4.15)	88-4859-76 138-0413-76
$C_{13}H_9N_3O_2$ 9-Acridinamine, 1-nitro-	pH 1 pH 6.8	268(4.47),423(3.70) 267(4.56),422(3.78)	56-1603-76 56-1603-76
$C_{13}H_9N_3O_3$ Diazene, benzoyl(4-nitrophenyl)-	dioxan	237(4.13),288(4.16), 448(2.08)	65-1358-76
Diazene, (4-nitrobenzoyl)phenyl-	dioxan	257(4.08),297s(3.51), 360s(2.60)	65-1358-76
$C_{13}H_9N_3O_4$ Benzenamine, 4-nitro-N-[(4-nitrophenyl)-methylene]-	$C_2H_4Cl_2$	298(4.37),335(4.24)	104-0161-76
Pyrazolo[1,5-a]pyrimidin-5(4H)-one, 3-(1,3-benzodioxol-5-yl)-7-hydroxy-	MeOH-pH 1 MeOH-pH 11	295(4.02),305(4.48) 230(4.24),275(4.15)	87-0291-76 87-0291-76
$C_{13}H_9N_5$ 1,2,4,5-Benzenetetracarbonitrile, 3-amino-6-propyl-	EtOH	220(4.52),259(4.52), 423(3.93)	35-0558-76
1-Butene-1,1,2-tricarbonitrile, 3-(phenylhydrazono)-	dioxan	475(4.38)	24-1787-76
$C_{13}H_9N_5O_2$ 8H-1,2,4-Oxadiazolo[2,3-i]purin-2(7H)-one, 7-(phenylmethyl)-	EtOH	260(4.00),270(4.20), 277(4.18)	23-2804-76
2H-Tetrazole, 2-(4-nitrophenyl)-5-phen-yl-	EtOH	241(4.27),304(4.36)	18-0762-76
2H-Tetrazole, 5-(4-nitrophenyl)-2-phen-yl-	EtOH	212(4.31),294(4.42)	18-0762-76
$C_{13}H_{10}$ 1H-Cyclobuta[b]cyclopropa[g]naphthalene, 4,5-dihydro-	n.s.g.	227(4.92),256(3.49), 267(3.68),288(3.75), 297(3.28),304(3.34), 310(3.56),324(3.73)	88-2815-76
Trideca-1,3,11-triene-5,7,9-triyne, (3Z,11E)-	ether	224(4.52),234(4.48), 243(4.44),274(4.73), 289(4.72),299(4.10), 310(4.00),319(4.25), 331(4.04),342(4.39), 354(3.92),366(4.26), 382(3.48)	88-0109-76
$C_{13}H_{10}BrNO$ Pyridinium, 2-(4-bromophenyl)-2-oxoeth-ylide	benzene EtOH	453(4.38) 247(4.02),260(4.03), 424(4.19)	78-2647-76 78-2647-76
	dioxan	246(4.01),263(3.89), 451(4.50)	78-2647-76
	MeCN	246(4.05),256(3.99), 437(4.44)	78-2647-76
$C_{13}H_{10}BrN_3O$ Diazene, [(3-bromophenyl)(hydroxyimino)-methyl]phenyl-	EtOH 50% EtOH-pH 12.5	307(4.23),436(2.49) 240(4.03),275(3.81), 350(4.42)	23-3260-76 23-3260-76

Compound	Solvent	$\lambda_{max}(\log \epsilon)$	Ref.
Diazene, [(4-bromophenyl)(hydroxyimino)-methyl]phenyl-	EtOH	305(4.24),436(2.51), 350(4.40)	23-3260-76
$C_{13}H_{10}BrOS$			
Naphtho[1,8-bc]thiolium, 2-bromo-5-eth-oxy-, tetrafluoroborate	CH_2Cl_2	434(3.96)	89-0775-76
$C_{13}H_{10}ClNO_2$			
Benzoic acid, 2-[(2-chlorophenyl)amino]-, β-cyclodextrin complex	pH 7.0	290(4.16),330(3.83)	133-0343-76
Benzoic acid, 2-[(3-chlorophenyl)amino]-, β-cyclodextrin complex	pH 7.0	285(4.08),325(3.68)	133-0343-76
2,4-Pentadienoic acid, 5-chloro-2-cyano-5-(4-methylphenyl)-	CH_2Cl_2	349(4.35)	48-0705-76
$C_{13}H_{10}ClNO_3S$			
Carbamothioic acid, N-(4-chloro-2-oxo-2H-pyran-6-yl)-, S-(phenylmethyl) ester	EtOH	241(3.38),341(4.18), 378s(3.95)	39-2462-76C
2H-Pyran-5-carbothioic acid, 6-amino-4-chloro-2-oxo-, S-(phenylmethyl) ester	MeCN	303(4.15),338(4.25)	39-2462-76C
$C_{13}H_{10}ClNO_5S$			
Phenol, 2-chloro-4-nitro-, 4-methylben-zenesulfonate	MeOH	264(3.96)	40-0631-76
Phenol, 4-chloro-2-nitro-, 4-methylben-zenesulfonate	MeOH	253(3.65)	40-0631-76
$C_{13}H_{10}ClN_3O$			
Diazene, [(2-chlorophenyl)(hydroxyimi-no)methyl]phenyl-	EtOH	310(4.46),430(2.61)	23-3260-76
	50% EtOH-pH 12.5	238(4.08),275(3.67), 346(4.46)	23-3260-76
Diazene, [(3-chlorophenyl)(hydroxyimi-no)methyl]phenyl-	EtOH	307(4.36),436(2.62)	23-3260-76
	50% EtOH-pH 12.5	240(4.13),275(3.95), 350(4.54)	23-3260-76
Diazene, [(4-chlorophenyl)(hydroxyimi-no)methyl]phenyl-	EtOH	307(4.24),434(2.55)	23-3260-76
	50% EtOH-pH 12.5	240(4.03),275(3.90), 350(4.42)	23-3260-76
$C_{13}H_{10}ClN_3O_2$			
Furo[2,3-c]pyridin-2(6H)-one, 3-chloro-6-(4,6-dimethyl-2-pyrimidinyl)-	MeCN	237(3.08),374(4.06), 396(4.46)	88-2959-76
$C_{13}H_{10}ClN_5O_2$			
Pyrimido[5,4-e][1,2,4]triazine-5,7(1H,6H)-dione, 3-(4-chloro-phenyl)-1,6-dimethyl-	dioxan	300(4.54),436(3.49)	39-0713-76C
Pyrimido[5,4-e][1,2,4]triazine-5,7(6H,8H)-dione, 3-(4-chloro-phenyl)-6,8-dimethyl-	dioxan	288(4.39),377(3.44)	39-0713-76C
$C_{13}H_{10}ClN_5O_3$			
Pyrimido[5,4-e][1,2,4]triazine-5,7(1H,6H)-dione, 3-(4-chloro-phenyl)-1,6-dimethyl-, 4-oxide	dioxan	298(4.53),435(3.49)	39-0713-76C
Pyrimido[5,4-e][1,2,4]triazine-5,7(6H,8H)-dione, 3-(4-chloro-phenyl)-6,8-dimethyl-, 4-oxide	dioxan	265(4.37),292(4.44), 385(3.47)	39-0713-76C

Compound	Solvent	$\lambda_{max}(\log \epsilon)$	Ref.
$C_{13}H_{10}Cl_2N_2$			
Benzenecarbohydrazonoyl chloride, 4-chloro-N-phenyl-	C_6H_{12}	231s(--),312s(--), 336(4.25)	48-0881-76
Benzenecarbohydrazonoyl chloride, N-(4-chlorophenyl)-	C_6H_{12}	229(3.87),238s(--), 290s(--),297(3.72), 328(4.38),345s(--)	48-0881-76
$C_{13}H_{10}Cl_2N_2O_2$			
Benzoic acid, 2-[[(2,6-dichloro-3-pyridinyl)methyl]amino]-	MeOH	250(4.04),268(3.72), 340(3.77)	33-0190-76
$C_{13}H_{10}Cl_2O_3S$			
Phenol, 2,4-dichloro-, 4-methylbenzenesulfonate	MeOH	268(3.99)	40-0631-76
$C_{13}H_{10}Cl_3N$			
Pyridine, 2,3,6-trichloro-5-[(x-methylphenyl)methyl]-	MeOH	282(3.75)	33-0190-76
$C_{13}H_{10}Cl_4N_2$			
1,4-Benzenediamine, N,N-dimethyl-N'-(2,3,4,5-tetrachloro-2,4-cyclopentadien-1-ylidene)-	C_6H_{12}	575(4.45)	39-0048-76C
$C_{13}H_{10}F_3NO_3$			
4-Isoxazolecarboxylic acid, 5-methyl-3-phenyl-, 2,2,2-trifluoroethyl ester	EtOH	218(4.17)	33-2074-76
4-Oxazolecarboxylic acid, 5-methyl-2-phenyl-, 2,2,2-trifluoroethyl ester	EtOH	266(4.33),270s(4.30), 276s(4.26),289s(3.94)	33-2074-76
$C_{13}H_{10}F_7N$			
Benzenamine, N,N-dimethyl-4-[1,2,3,4-tetrafluoro-4-(trifluoromethyl)-1,3-butadienyl]-	heptane	344(4.44)	104-1565-76
$C_{13}H_{10}N_2$			
1-Acridinamine	pH 7.0	238(4.33),262(4.35)	95-1334-76
3-Acridinamine	pH 7.0	238(4.46),262(4.49)	95-1334-76
4-Acridinamine	pH 7.0	240(4.54),262(4.61)	95-1334-76
9-Acridinamine	pH 1	220(4.48),260(4.82), 400(3.85),420(3.80)	56-1267-76
	pH 7.0	250s(4.62),260(4.82)	95-1334-76
Propanedinitrile, (3,4-dihydro-1(2H)-naphthalenylidene)-	EtOH	231(3.76),313(4.12)	118-0705-76
$C_{13}H_{10}N_2O$			
3-Pyridinecarbonitrile, 1,2-dihydro-2-oxo-6-(4-methylphenyl)-	HOAc	360(4.34)	48-0705-76
$C_{13}H_{10}N_2O_2$			
Benzenamine, N-[(4-nitrophenyl)methylene]-	$C_2H_4Cl_2$	230s(--),292(4.21), 347(4.06)	104-0161-76
Benzenamine, 4-nitro-N-(phenylmethylene)-	$C_2H_4Cl_2$	226(4.20),260s(3.91), 332(4.25)	104-0161-76
Imidazo[2,1-a]isoquinoline-2-carboxylic acid, methyl ester	MeOH	248(4.75),251s(4.65), 256s(4.61)	39-0075-76C
3-Indolizinecarbonitrile, 1,2-diacetyl-	MeOH	210s(4.46),215s(4.49), 220s(4.56),223s(4.58), 246(4.60),290s(4.09), 327(4.28),346s(4.17)	39-1908-76C

Compound	Solvent	$\lambda_{max}(\log \epsilon)$	Ref.
$C_{13}H_{10}N_2O_4$			
Benzoic acid, 2-[(2-nitrophenyl)amino]-, β-cyclodextrin complex	pH 7.0	268(4.15),285(3.82), 445(4.11)	133-0343-76
Benzoic acid, 2-[(3-nitrophenyl)amino]-, β-cyclodextrin complex	pH 7.0	283(4.24),325(3.79)	133-0343-76
Benzoic acid, 2-[(4-nitrophenyl)amino]-, β-cyclodextrin complex	pH 7.0	265(4.01),415(4.26)	133-0343-76
4,5-Pyrimidinedicarboxylic acid, 6-methyl-2-phenyl-	EtOH	267(4.25)	5-1809-76
$C_{13}H_{10}N_2O_4S$			
Sulfonium, diphenyl-, dinitromethylide	MeCN	273(4.03),324(3.83)	70-1906-76
	DMSO	273(--),333(3.86)	70-1906-76
Thiazolo[5,4-f]quinoline-8-carboxylic acid, 6-ethyl-6,9-dihydro-2-hydroxy-9-oxo-	MeOH	261.0(4.42)	94-0147-76
$C_{13}H_{10}N_2O_7$			
2,3-Quinolinedicarboxylic acid, 1,4-di-hydro-7-nitro-4-oxo-, dimethyl ester	MeOH	222(4.40),267(4.30), 320s(3.77),329(3.79)	5-1972-76
2,8-Quinolinedicarboxylic acid, 1,4-di-hydro-5-nitro-4-oxo-, dimethyl ester	MeOH	205(4.51),217(4.45), 251(4.24),254s(4.23), 291(3.29),305(3.21), 343s(3.68),364s(4.00), 376(4.08)	5-1972-76
$C_{13}H_{10}N_2O_7S$			
Phenol, 2,4-dinitro-, 4-methylbenzene-sulfonate	MeOH	254(4.07)	40-0531-76
$C_{13}H_{10}N_2S$			
Quinazoline, 4-(3-methyl-2-thienyl)-	MeOH	225(4.49),265(3.67), 335(3.92)	139-0105-76C
$C_{13}H_{10}N_4$			
Pyrido[3,2-e]-1,2,4-triazine, 3-(phen-ylmethyl)-	dioxan	265(3.63),317(3.90), 489(2.33)	114-0301-76C
Pyrido[3,4-e]-1,2,4-triazine, 3-(phen-ylmethyl)-	EtOH	234(4.27),335(4.32)	114-0285-76C
2H-Tetrazole, 2,5-diphenyl-	EtOH	240(4.14),272(4.26)	18-0762-76
$C_{13}H_{10}N_4O$			
Pyrido[3,4-e]-1,2,4-triazine, 3-(4-meth-oxyphenyl)-	EtOH	270(4.09),313(4.45), 400(3.59)	114-0285-76C
$C_{13}H_{10}N_4O_2$			
[1,2,4]Triazolo[1,5-a]pyridine, 8-nitro-2-(phenylmethyl)-	H_2O	230(4.21),326(3.78)	44-3124-76
$C_{13}H_{10}N_4O_3$			
8-Azabicyclo[3.2.1]oct-3-ene-6-carbo-nitrile, 8-(5-nitro-2-pyridinyl)-2-oxo-, endo	EtOH	210(4.10),228(4.27), 356(4.26)	39-2307-76C
exo	EtOH	208(4.12),225(4.29), 350(4.26)	39-2307-76C
Diazene, [(hydroxyimino)(2-nitrophenyl)-methyl]phenyl-	EtOH	310(4.72),436(2.59)	23-3260-76
	50% EtOH-pH 12.5	240(4.20),265(4.06), 344(4.50)	23-3260-76
Diazene, [(hydroxyimino)(3-nitrophenyl)-methyl]phenyl-	EtOH	307(4.32),434(2.53)	23-3260-76
	50% EtOH pH 12.5	270(4.17),349(4.52)	23-3260-76

Compound	Solvent	λ_{max}(log ϵ)	Ref.
Diazene, [(hydroxyimino)(4-nitrophenyl)-methyl]phenyl-	EtOH	305(4.40),435(2.54)	23-3260-76
	50% EtOH-pH 12.5	270(4.10),349(4.48)	23-3260-76
$C_{13}H_{10}N_4O_4S$			
[1,2,5]Thiadiazolo[3,4-h]quinoline-7-carboxylic acid, 6-[(2-carboxyethyl)amino]-	EtOH	228(4.07),251s(4.55), 258(4.75),290(4.43), 405(3.83)	103-0056-76
$C_{13}H_{10}N_4O_5$			
[1,2,5]Oxadiazolo[3,4-d]pyrimidine-5,7(4H,6H)-dione, 6-methyl-4-(2-oxo-2-phenylethyl)-, 1-oxide	MeOH	248(4.24),265s(4.11), 322(3.42)	39-1327-76C
$C_{13}H_{10}N_6O_4$			
Pyrimido[5,4-e]-1,2,4-triazine-5,7(6H,8H)-dione, 6,8-dimethyl-3-(4-nitrophenyl)-	dioxan	309(4.38),362s(3.09)	39-0713-76C
$C_{13}H_{10}N_6O_5$			
Pyrimido[5,4-e]-1,2,4-triazine-5,7(6H,8H)-dione, 6,8-dimethyl-3-(4-nitrophenyl)-, 4-oxide	dioxan	271(4.24),308(4.31), 375s(3.54)	39-0713-76C
$C_{13}H_{10}O$			
Methanone, diphenyl-	hexane	246(4.32),346(2.06)	39-0869-76B
	pH 1 and 7	260(4.25),326s(2.42)	39-0869-76B
	EtOH	250(4.30),333(2.21)	39-0869-76B
$C_{13}H_{10}OS$			
2-Propen-1-one, 1-phenyl-3-(2-thienyl)-	n.s.g.	345(4.28)	39-0380-76C
2-Propen-1-one, 3-phenyl-1-(2-thienyl)-	n.s.g.	320(4.29)	39-0380-76C
$C_{13}H_{10}OS_2$			
Thieno[2,3-b]thiophene-2-methanol, 5-phenyl-	EtOH	255(4.26),302(4.16)	104-1550-76
$C_{13}H_{10}OS_3$			
2,2':5',2"-Terthiophene, 3'-methoxy-	ether	370(4.33)	24-0901-76
$C_{13}H_{10}O_2$			
2,8-Nonadiene-4,6-diyn-1-ol, 9-(2-furanyl)-, (E,E)- (atractylodinol)	EtOH	259(4.12),273(4.17), 337(4.66),356(4.56)	95-1322-76
2-Propen-1-one, 1-(2-furanyl)-3-phenyl-	n.s.g.	324(4.03)	39-0380-76C
2-Propen-1-one, 3-(2-furanyl)-1-phenyl-	n.s.g.	344(4.43)	39-0380-76C
$C_{13}H_{10}O_2S$			
Benzoic acid, 2-[2-(2-thienyl)ethenyl]-	EtOH	230(4.18)	4-0083-76
1H-2-Benzopyran-1-one, 3,4-dihydro-3-(2-thienyl)-	EtOH	230(4.29),275(3.32)	4-0083-76
2-Propenoic acid, 3-(4-phenyl-2-thienyl)-, (E)-	MeOH	263(4.48),322(4.00)	44-1320-76
2-Propenoic acid, 3-(5-phenyl-2-thienyl)-, (E)-	EtOH	243(4.00),346(4.44)	44-1320-76
Thieno[2,3-b]furan-5-methanol, 2-phenyl-	EtOH	227(4.16),300(4.21)	104-1550-76
$C_{13}H_{10}O_3$			
2(5H)-Furanone, 4-methyl-5-(2-oxo-2-phenylethylidene)-	EtOH	249(4.45)	23-0415-76

Compound	Solvent	$\lambda_{max}(\log \epsilon)$	Ref.
$C_{13}H_{10}O_4$			
2-Furancarboxylic acid, 4-acetyl-5-phen- yl-	EtOH	252(4.30)	94-2421-76
2(5H)-Furanone, 3-acetyl-4-hydroxy- 5-(phenylmethylene)-	EtOH	232(4.07),312(4.43)	4-0521-76
2H-Pyran-2-one, 5-benzoyl-4-hydroxy-	EtOH	272(4.17)	78-0269-76
6-methyl-	EtOH-base	265(4.19)	78-0269-76
$C_{13}H_{10}O_5S$			
Benzoic acid, 2-hydroxy-5-(phenylsulfon- yl)-	EtOH	256(4.13)	56-0973-76
2,4,6-Cycloheptatrien-1-one, 2,5-di- hydroxy-6-(phenylsulfonyl)-	MeOH	235s(4.35),247(4.36), 342s(3.95),357(4.01), 415(3.94),444s(3.78)	88-2339-76
2,4,6-Cycloheptatrien-1-one, 2,5-di- hydroxy-7-(phenylsulfonyl)-	MeOH	230(4.30),250(4.29), 344(3.91),360s(3.83), 404s(3.80),427(3.89), 454s(3.71)	88-2339-76
$C_{13}H_{10}S_3$			
Benzenecarbo(dithioperoxo)thioic acid, phenyl ester	C_6H_{12}	293(4.11),537(1.86)	138-0441-76
2,2':5',2"-Terthiophene, 5-methyl-	ether	357(4.35)	24-0901-76
$C_{13}H_{11}AsO_2$			
Benzeneacetic acid, 4-arseninyl ester	EtOH	216(3.95),274(3.85)	88-4143-76
$C_{13}H_{11}BrN_2O_3$			
Pyridinium, 3-nitro-1-(2-oxo-2-phenyl- ethyl)-, bromide	H_2O	253(4.30)	35-5689-76
$C_{13}H_{11}ClHgN_2$			
Mercury, chloro[3-methyl-2-(phenylazo)- phenyl]-	EtOH	233(4.08),342(4.17), 452(2.67)	101-0039-76M
$C_{13}H_{11}ClHgN_2O$			
Mercury, chloro[2-[(2-methoxyphenyl)- azo]phenyl]-	EtOH	234(4.07),326(3.91), 368(3.89),450s(2.99)	101-0039-76M
$C_{13}H_{11}ClN_2$			
Benzenecarbohydrazonoyl chloride, N-phenyl-	C_6H_{12}	229(3.84),238s(--), 290s(--),297(3.72), 328(4.38),345s(--)	48-0881-76
Pyridinium, 3-cyano-1-(phenylmethyl)-, chloride	MeOH	268(3.59)	44-2976-76
1H-Pyrido[2,3-b]indole, 2-chloro-1,4-di- methyl-	n.s.g.	207(4.19),223(4.08), 242(3.75),272(4.29), 282(4.35),326(4.03)	103-0329-76
$C_{13}H_{11}ClN_2O$			
2,4-Pentadienamide, 5-chloro-2-cyano- 5-(4-methylphenyl)-	CH_2Cl_2	344(4.37)	48-0705-76
$C_{13}H_{11}ClN_2O_2$			
2,4-Pentadienamide, 5-chloro-2-cyano- 5-(4-methoxyphenyl)-	CH_2Cl_2	370(4.30)	48-0705-76
$C_{13}H_{11}ClN_2O_2S_2$			
5-Pyrimidineacetic acid, 4-[(4-chloro- phenyl)thio]-2-(methylthio)-	MeOH	257(4.35),308(3.87)	73-2771-76

Compound	Solvent	λ_{max} (log ϵ)	Ref.
$C_{13}H_{11}ClN_2O_3$			
2,4,6(1H,3H,5H)-Pyrimidinetrione, 5-[(3-chlorophenyl)methylene]-1,3-dimethyl-	EtOH	236(3.97),318(4.04)	94-0607-76
2,4,6(1H,3H,5H)-Pyrimidinetrione, 5-[(4-chlorophenyl)methylene]-1,3-dimethyl-	EtOH	235(4.03),239(4.03), 256s(3.88),333(4.22)	94-0607-76
$C_{13}H_{11}ClN_2S$			
2,4-Pentadienethioamide, 5-chloro-2-cyano-5-(4-methylphenyl)-	CH_2Cl_2	369(4.05)	48-0705-76
$C_{13}H_{11}ClN_4$			
1H-Benzotriazole-1-methanamine, N-(2-chlorophenyl)-	n.s.g.	255(3.74),276(3.66)	2-0718-76
1H-Benzotriazole-1-methanamine, N-(3-chlorophenyl)-	n.s.g.	255(3.72),276(3.64)	2-0718-76
$C_{13}H_{11}ClO_2S$			
Benzene, 1-chloro-4-[(4-methylphenyl)-sulfonyl]-	EtOH	<u>200(4.7),220s(4.1), 247(4.3)</u>	104-2133-76
$C_{13}H_{11}ClO_3$			
5H-Benzocyclohepten-5-one, 8-chloro-6,7-dimethoxy-	EtOH	250s(4.95),265(4.38), 340(3.83)	18-2230-76
$C_{13}H_{11}ClO_3S$			
Phenol, 2-[(4-chlorophenyl)sulfonyl]-6-methyl-	C_6H_{12}	245(4.22),300(3.75)	42-0602-76
	EtOH	232(4.02),242(3.95), 285(3.47)	42-0602-76
Phenol, 4-[(4-chlorophenyl)sulfonyl]-3-methyl-	C_6H_{12}	230(4.53),280(3.82), 300(3.99)	42-0602-76
	EtOH	242(4.21),280(3.54), 295(3.75)	42-0602-76
$C_{13}H_{11}Cl_2N$			
Pyridine, 2,6-dichloro-3-[(4-methylphenyl)methyl]-	MeOH	272.5(3.70)	33-0190-76
$C_{13}H_{11}Cl_2NO$			
Pyridine, 2,6-dichloro-3-[(x-methoxyphenyl)methyl]-	MeOH	272.5(3.84)	33-0190-76
$C_{13}H_{11}Cl_2N_3O$			
Imidazo[1,2-a]pyrimidin-5(3H)-one, 8-(2,6-dichlorophenyl)-2,8-dihydro-7-methyl-	EtOH EtOH-HCl	220(4.48),280(3.57) 211(4.47),256(3.81)	33-1203-76 33-1203-76
Imidazo[1,2-a]pyrimidin-7(3H)-one, 8-(2,6-dichlorophenyl)-2,8-dihydro-5-methyl-	EtOH EtOH-HCl	217(4.28),308(3.79) 217(4.12),279(3.90)	33-1203-76 33-1203-76
$C_{13}H_{11}IN_2$			
Diazene, (2-iodo-6-methylphenyl)phenyl-	EtOH	220(4.26),278s(3.94), 310(4.08),456(2.58)	101-0039-76M
$C_{13}H_{11}IN_2O$			
Diazene, (2-iodophenyl)(2-methoxyphenyl)-	EtOH	213(4.30),240(3.98), 321(4.03),367(4.09), 454(2.85)	101-0039-76M
$C_{13}H_{11}N$			
Benzenamine, N-(phenylmethylene)-	hexane	235(4.03),263(4.25), 312(3.86)	104-0161-76

Compound	Solvent	$\lambda_{max}(\log \epsilon)$	Ref.
$C_{13}H_{11}NO$			
Benzenamine, N-(phenylmethylene)-, N-oxide	EtOH	322(4.29)	4-1001-76
9H-Carbazol-2-ol, 3-methyl-	EtOH	240(4.56),258(4.38), 304(4.31),325(3.89)	2-0329-76
9H-Carbazol-2-ol, 6-methyl-	EtOH	240(4.72),260(4.55), 303(4.41),325(3.96)	2-0329-76
Ethanone, 1-(2-methylpyrrolo[2,1,5-cd]-indolizin-1-yl)-	EtOH	240(4.05),257(4.23), 277s(3.80),313(3.78), 395s(3.77),403(3.86), 411(3.89)	1-0198-76
Onychine, dihydro-	EtOH	283(3.92),298(3.90), 310(4.04)	102-1186-76
	EtOH-HCl	296(3.88),326(4.26)	102-1186-76
Pyridinium, 2-oxo-2-phenylethylide	benzene	454(4.31)	78-2647-76
	EtOH	246(4.10),420(4.00)	78-2647-76
	dioxan	247(4.00),450(4.05)	78-2647-76
	MeCN	246(4.03),437(4.22)	78-2647-76
3H-Pyrrolo[1,2-a]indol-3-one, 1,9-di-methyl-	CHCl$_3$	241(4.85?),271(4.10), 353(4.09)	83-0131-76
$C_{13}H_{11}NO_2$			
Benzoic acid, 2-(phenylamino)-, β-cyclodextrin complex	pH 7.0	291(4.10),332(3.72)	133-0343-76
Ethanone, 1-phenyl-2-(2-pyridinyl)-, N-oxide	EtOH	245(4.26)	39-1944-76C
	EtOH-NaOH	243(4.14),264s(4.05), 399(4.34)	39-1944-76C
Phenol, 2-[[(2-hydroxyphenyl)imino]meth-yl]-	MeOH	226s(--),269(4.00), 347(4.08),452s(3.28)	114-0053-76A
	BuOH?	270(3.98),357(4.03)	114-0053-76A
Phenol, 2-[[(3-hydroxyphenyl)imino]meth-yl]-	MeOH	225s(--),267(3.98), 338(3.98),425s(2.50)	114-0053-76A
	BuOH?	268(4.04),342(4.07)	114-0053-76A
Phenol, 2-[[(4-hydroxyphenyl)imino]meth-yl]-	MeOH	230s(--),270(3.91), 348(4.20),430s(2.60)	114-0053-76A
	BuOH?	268(3.93),350(4.14)	114-0053-76A
$C_{13}H_{11}NO_2S$			
Cyclopenta[e][1,3]oxazin-4(5H)-one, 2,3,6,7-tetrahydro-3-phenyl-2-thioxo-	MeOH	220(4.20),278(4.15)	5-1689-76
$C_{13}H_{11}NO_2S_2$			
5H-1,4-Dithiino[2,3-c]pyrrole-5,7(6H)-dione, 2,3-dihydro-6-(2-methylphenyl)-	EtOH	207(4.26),231(4.02), 267(3.97),417(3.45)	56-1523-76
5H-1,4-Dithiino[2,3-c]pyrrole-5,7(6H)-dione, 2,3-dihydro-6-(3-methylphenyl)-	EtOH	203(4.32),237(4.22), 270(3.96),420(3.42)	56-1523-76
5H-1,4-Dithiino[2,3-c]pyrrole-5,7(6H)-dione, 2,3-dihydro-6-(4-methylphenyl)-	EtOH	204(4.24),236(4.23), 269(3.91),420(3.37)	56-1523-76
5H-1,4-Dithiino[2,3-c]pyrrole-5,7(6H)-dione, 2,3-dihydro-6-(phenylmethyl)-	EtOH	206(4.17),265(4.01), 417(3.42)	56-1523-76
$C_{13}H_{11}NO_3$			
3,5-Cyclohexadiene-1,2-dione, 4-methoxy-5-(phenylamino)-	H$_2$O	292(3.90),328(4.13), 510(3.51)	135-0717-76A
	MeOH	292(3.90),322(4.13), 488(3.51)	135-0717-76A
	dioxan	292(3.90),307(4.13), 460(3.51)	135-0717-76A
	CHCl$_3$	292(3.90),321(4.13), 476(3.51)	135-0717-76A

Compound	Solvent	$\lambda_{max}(\log \epsilon)$	Ref.
$C_{13}H_{11}NO_3S_2$			
5H-1,4-Dithiino[2,3-c]pyrrole-5,7(6H)-dione, 2,3-dihydro-6-(2-methoxyphenyl)-	EtOH	204(4.27),214(4.21), 232(3.98),272(4.06), 413(3.45)	56-1523-76
5H-1,4-Dithiino[2,3-c]pyrrole-5,7(6H)-dione, 2,3-dihydro-6-(3-methoxyphenyl)-	EtOH	204(4.38),217(4.30), 239(4.13),275(4.04), 420(3.38)	56-1523-76
5H-1,4-Dithiino[2,3-c]pyrrole-5,7(6H)-dione, 2,3-dihydro-6-(4-methoxyphenyl)-	EtOH	202(4.27),237(4.31), 268(3.93),321(3.32), 422(3.32)	56-1523-76
$C_{13}H_{11}NO_4$			
Furo[2,3-b]quinolin-8-ol, 4,7-dimethoxy-	EtOH	250(4.67),325(3.73)	105-0709-76
$C_{13}H_{11}NO_4S$			
Benzamide, 2-hydroxy-5-(phenylsulfonyl)-	EtOH	255(4.03),291(3.91)	56-0973-76
2-Furancarboxylic acid, 5-[[4-(acetylamino)phenyl]thio]-	EtOH	207(4.30),268(4.39), 305s(3.78)	73-2571-76
$C_{13}H_{11}NO_5$			
2,3-Quinolinedicarboxylic acid, 1,4-dihydro-4-oxo-, dimethyl ester	MeOH	217(4.52),238s(4.19), 250(4.27),332(3.89), 345(3.92),360s(3.73)	5-1972-76
$C_{13}H_{11}NO_5S$			
Benzenesulfonic acid, 4-methyl-, 2-nitrophenyl ester	MeOH	256(3.78)	40-0631-76
Benzenesulfonic acid, 4-methyl-, 4-nitrophenyl ester	MeOH	263(4.03)	40-0631-76
$C_{13}H_{11}NO_6S$			
2-Furancarboxylic acid, 5-[[4-(acetylamino)phenyl]sulfonyl]-	EtOH	206(4.23),244s(4.24), 250(4.28),284(4.41)	73-2571-76
$C_{13}H_{11}NS$			
Benzenecarbothioamide, N-phenyl-	MeOH	236(4.22),266(4.13), 314(3.93)	78-0661-76
	MeCN	238(4.21),266(4.12), 326(3.93)	78-0661-76
6H-Thieno[2,3-b]pyrrole, 6-methyl-5-phenyl-	EtOH	245(4.03),295(4.04)	104-1550-76
$C_{13}H_{11}NSe$			
6H-Selenolo[2,3-b]pyrrole, 6-methyl-5-phenyl-	EtOH	248(4.21),299(4.16)	104-1550-76
$C_{13}H_{11}N_2$			
Pyridinium, 3-cyano-1-(phenylmethyl)-, chloride	MeOH	268(3.59)	44-2976-76
$C_{13}H_{11}N_2O$			
Pyrylium, 2-amino-3-cyano-6-(4-methylphenyl)-, perchlorate	HOAc	389(4.40)	48-0705-76
$C_{13}H_{11}N_2O_3$			
Pyridinium, 3-nitro-1-(2-oxo-2-phenylethyl)-, bromide	H_2O	253(4.30)	35-5689-76
$C_{13}H_{11}N_2S$			
Thiopyrylium, 2-amino-3-cyano-6-(4-methylphenyl)-, perchlorate	HOAc	422(4.24)	48-0705-76

Compound	Solvent	$\lambda_{max}(\log \epsilon)$	Ref.
$C_{13}H_{11}N_3$			
Dipyrido[1,2-a:2',1'-c]pyrazinediium, 3-cyano-6,7-dihydro-, dibromide	pH 6.0	272(3.53),315(3.84), 328(3.83)	64-0115-76B
Proflavine	pH 7.0	250s(4.53),261(4.63), 444(4.51)	95-1334-76
Propanedinitrile, [3-amino-3-(4-methylphenyl)-2-propenylidene]-	CH_2Cl_2	383(4.46)	48-0705-76
3-Pyridinecarbonitrile, 2-amino-6-(4-methylphenyl)-	CH_2Cl_2	343(4.22)	48-0705-76
Pyrrolo[1,2-a]pyrimidin-2-amine, 7-phenyl-	MeOH	226(4.2),259(4.63), 285s(4.05),302s(3.94), 313s(3.86),373(3.38)	39-1991-76C
Pyrrolo[1,2-c]pyrimidin-3-amine, 6-phenyl-	MeOH	262(4.7),360s(3.2), 390s(3.1)	39-1991-76C
Quinoline, 2-(1-methyl-1H-imidazol-5-yl)- (isomacrorine)	EtOH	210(4.45),234(4.12), 267(4.37),326(3.99), 341(3.98)	78-1085-76
	EtOH-acid	213(4.45),251(4.52), 256(4.51),318(3.79), 332(3.73)	78-1085-76
1,2,4-Triazolo[1,5-a]pyridine, 7-methyl-2-phenyl-	EtOH	252(4.43),344(4.21)	138-0413-76
1,2,4-Triazolo[4,3-a]pyridine, 3-(phenylmethyl)-	H_2O	268(3.63),287(3.56)	44-3124-76
$C_{13}H_{11}N_3O$			
Diazene, [(hydroxyimino)phenylmethyl]phenyl]-	C_6H_{12}	305(4.27)	23-3260-76
	EtOH	307(4.23),436(2.55)	23-3260-76
	50% EtOH-pH 12.5	237(4.01),350(4.40)	23-3260-76
	dioxan	302(4.21)	23-3260-76
	HOAc	310(4.19)	23-3260-76
	DMF	305(4.23)	23-3260-76
	DMSO	302(4.23)	23-3260-76
	$CHCl_3$	310(4.24)	23-3260-76
6H-[1,2,4]Triazino[2,3-b]isoquinolin-6-one, 2,3-dimethyl-	EtOH	298s(3.9),357(3.9), 365(3.9),433(3.4)	4-1343-76
$C_{13}H_{11}N_3OS$			
Ethanone, 2-diazo-1-[4-methyl-2-(4-methylphenyl)-5-thiazolyl]-	pH 1	330(4.51)	80-0113-76
	40% dioxan	337(4.65)	80-0113-76
$C_{13}H_{11}N_3O_2$			
Benzaldehyde, α-nitro-, phenylhydrazone	EtOH	403(3.89)	44-2981-76
Diazene, (2-methyl-4-nitrophenyl)phenyl-	C_6H_{12}	336(4.39)	18-1381-76
Diazene, (2-methylphenyl)(4-nitrophenyl)-	C_6H_{12}	341(4.38)	18-1381-76
1,2-Diazepine-4-carboxamide, 1-benzoyl-	MeOH	229(4.23),370(3.34)	88-4859-76
Pyrazolo[1,5-a]pyrimidin-5(4H)-one, 7-hydroxy-3-(3-methylphenyl)-	MeOH-pH 1	207(4.47),240(4.31)	87-0291-76
	MeOH-pH 11	230(4.42),286(4.27)	87-0291-76
Pyrazolo[1,5-a]pyrimidin-5(4H)-one, 7-hydroxy-3-(4-methylphenyl)-	MeOH-pH 1	203(4.49),243(4.31)	87-0291-76
9H-Pyrido[3,4-b]indole-1-carboxamide, 7-methoxy-	$CHCl_3$	255(4.54),279(4.44), 321(4.35)	102-1559-76
Pyrimido[4,5-b]quinoline-2,4(3H,10H)-dione, 3,10-dimethyl-	EtOH	220(4.53),263(4.66), 319(4.00),398(4.08)	39-1805-76C
	50% EtOH-HCl	216(4.41),256(4.54), 333(4.21)	39-1805-76C
Pyrimido[4,5-b]quinoline-2,4(3H,10H)-dione, 10-ethyl-	EtOH	221(4.50),262(4.51), 319(3.89),396(4.03)	39-1805-76C

Compound	Solvent	$\lambda_{max}(\log \epsilon)$	Ref.
Pyrimido[4,5-b]quinoline-2,4(3H,10H)-dione, 10-ethyl- (cont.)	50% EtOH-HCl	216(4.42),257(4.53), 334(4.21)	39-1805-76C
1H-Pyrrolo[2,3-d]pyrimidine-2,4(3H,7H)-dione, 3-(phenylmethyl)-	MeOH	240(3.87),272(3.84)	24-2983-76
1,9,9b-Triazaphenalene-3-carboxylic acid, ethyl ester	EtOH	256(4.37),282s(3.88), 354(3.97),384s(3.87), 399(4.03),420(4.05), 564s(2.11),612(2.32), 672(2.34)	1-0466-76

$C_{13}H_{11}N_3O_2S$

Compound	Solvent	$\lambda_{max}(\log \epsilon)$	Ref.
Ethanone, 2-diazo-1-[2-(4-methoxyphenyl)-4-methyl-5-thiazolyl]-	pH 1	343(4.35)	80-0113-76
	40% dioxan	352(4.44)	80-0113-76

$C_{13}H_{11}N_3O_4$

Compound	Solvent	$\lambda_{max}(\log \epsilon)$	Ref.
Benzenemethanamine, 4-nitro-N-(2-nitrophenyl)-	MeOH	232(4.43),273(4.25), 415(3.90)	39-0394-76C

$C_{13}H_{11}N_3O_5$

Compound	Solvent	$\lambda_{max}(\log \epsilon)$	Ref.
2,4,6(1H,3H,5H)-Pyrimidinetrione, 1,3-dimethyl-5-[(3-nitrophenyl)methylene]-	EtOH	260(4.80),300s(4.29)	94-0607-76

$C_{13}H_{11}N_4$

Compound	Solvent	$\lambda_{max}(\log \epsilon)$	Ref.
3H-Imidazo[3,2-a]pyridin-4-ium, 3-(phenylhydrazono)-, perchlorate	n.s.g.	225(5),275(3.61), 360(3.87),385(3.82)	104-0929-76

$C_{13}H_{11}N_5O_2$

Compound	Solvent	$\lambda_{max}(\log \epsilon)$	Ref.
Pyrimido[5,4-e]-1,2,4-triazine-5,7-(1H,6H)-dione, 1,6-dimethyl-3-phenyl-	dioxan	295(4.49),433(3.47)	39-0713-76C
Pyrimido[5,4-e]-1,2,4-triazine-5,7-(6H,8H)-dione, 6,8-dimethyl-3-phenyl-	dioxan	278(4.45),375(3.55)	39-0713-76C

$C_{13}H_{11}N_5O_3$

Compound	Solvent	$\lambda_{max}(\log \epsilon)$	Ref.
Pyrimido[5,4-e]-1,2,4-triazine-5,7-(1H,6H)-dione, 1,6-dimethyl-3-phenyl-, 4-oxide	dioxan	290(4.48),431(3.47)	39-0713-76C
Pyrimido[5,4-e]-1,2,4-triazine-5,7-(6H,8H)-dione, 6,8-dimethyl-3-phenyl-, 4-oxide	dioxan	268(4.34),283(4.36), 380(3.47)	39-0713-76C

$C_{13}H_{11}N_5O_4S_3$

Compound	Solvent	$\lambda_{max}(\log \epsilon)$	Ref.
3-Thiazolidineacetic acid, 4-[(aminothioxomethyl)hydrazono]-5-[(4-nitrophenyl)methylene]-2-thioxo-	MeOH	245(4.27),278(4.17), 375(4.46)	103-0749-76

$C_{13}H_{12}$

Compound	Solvent	$\lambda_{max}(\log \epsilon)$	Ref.
Benzo[3,4]tetracyclo[6.1.1.0^{2,7}.0^{5,9}]-non-3-ene	hexane	250(2.35),256(2.50), 258(2.49),261s(--), 262(2.62),266(2.44), 269(2.62)	35-2600-76
Benzo[8,9]tricyclo[5.2.0.0^{2,5}]nona-3,8-diene	hexane	249s(2.69),256s(3.00), 260(3.23),267(3.42), 273(3.42)	35-2600-76

$C_{13}H_{12}ClN_3O_3S$

Compound	Solvent	$\lambda_{max}(\log \epsilon)$	Ref.
Butanoic acid, 2-[(4-chlorophenyl)hydrazono]-3-oxo-4-thiocyanato-, ethyl ester	n.s.g.	255(3.99),300(3.73), 390(4.17)	124-1162-76

Compound	Solvent	$\lambda_{max}(\log \epsilon)$	Ref.
$C_{13}H_{12}Cl_2$			
1H-Cyclobuta[b]cyclopropa[g]naphthalene, 1,1-dichloro-1a,2,4,5,7,7a-hexahydro-	hexane	220(3.71),266(3.30), 271(3.44),275(3.53), 281(3.53)	88-2815-76
$C_{13}H_{12}F_3NO_2$			
4-Oxa-7-azaspiro[2.4]hept-6-ene, 5-methoxy-6-phenyl-5-(trifluoromethyl)-	MeOH	255(4.00)	35-2605-76
6-Oxa-4-azaspiro[2.4]hept-4-ene, 7-methoxy-5-phenyl-7-(trifluoromethyl)-	MeOH	252(4.17)	35-2605-76
$C_{13}H_{12}MoO_3$			
Molybdenum, [η^6-5,6-bis(methylene)bicyclo[2.2.2]oct-2-ene]tricarbonyl-	C_6H_{12}	218(4.47),334(4.06)	23-1958-76
$C_{13}H_{12}N_2$			
Diazene, (4-methylphenyl)phenyl-	hexane	231(4.13),323(4.31), 444(2.83)	61-0301-76
Propanedinitrile, (2-methyl-1-phenylpropylidene)-	EtOH	232(4.04),283(3.73)	118-0705-76
Propanedinitrile, (1-phenylbutylidene)-	EtOH	233(3.88),292(4.06)	118-0705-76
3-Pyridinecarbonitrile, 1,4-dihydro-1-(phenylmethyl)-	MeCN	340(3.75)	44-2976-76
1H-Pyrido[2,3-b]indole, 1,4-dimethyl-	n.s.g.	207(4.38),219(4.32), 240(4.04),272(4.38), 278(4.44),316(4.09), 324(4.06)	103-0329-76
2H-Pyrrole-3-carbonitrile, 2,2-dimethyl-5-phenyl-	EtOH	237(4.17),276s(3.76)	33-1018-76
$C_{13}H_{12}N_2O$			
Diazene, (4-methoxyphenyl)phenyl-	C_6H_{12}	342(4.33)	18-1381-76
4-Isoquinolinecarbonitrile, 3-ethoxy-1-methyl-	EtOH	235(4.75),276(3.48), 286(3.54),298(3.40), 360(3.77)	2-0964-76
Pyridinium, 3-[(hydroxyphenylmethylene)-amino]-1-methyl-, hydroxide, inner salt	EtOH	275(4.00),298(3.98)	39-0315-76C
$C_{13}H_{12}N_2OS$			
2-Pyridinecarbothioamide, N-(4-methoxyphenyl)-	20% EtOH	284(4.09)	94-1451-76
4(1H)-Pyrimidinone, 6-phenyl-2-(2-propenylthio)-	EtOH	256(4.41),311(3.87)	95-1094-76
2H-Thiopyran-3-carboxamide, 2-imino-6-(4-methylphenyl)-	CH_2Cl_2	411(3.93)	48-0705-76
$C_{13}H_{12}N_2O_2$			
Acetic acid, cyano(1,3-dihydro-2H-indol-2-ylidene)-, ethyl ester	MeOH	224(4.10),293(4.04), 323(4.37)	24-3282-76
Benzenamine, 4-methyl-2-nitro-N-phenyl-	EtOH	438(3.86)	146-0131-76
[1]Benzopyrano[3,4-c]pyrazol-4(3H)-one, 3-ethyl-1-methyl-	EtOH	252(4.73),306(4.72)	39-1260-76C
[2,2'-Bipyridin]-5-ol, propanoate	EtOH	242(4.14),286(4.26)	64-0115-76B
Dipyrido[1,2-a:2',1'-c]pyrazinediium, 3-carboxy-6,7-dihydro-, dibromide	pH 6.0	244(3.89),312(4.27)	64-0115-76B
1H-Indole-1-carboxylic acid, 2-(cyanomethyl)-, ethyl ester	EtOH	225(4.31),253(4.09), 261(4.07),281(3.50), 291(3.54)	39-2587-76C
4-Isoquinolinecarbonitrile, 3-ethoxy-8-hydroxy-1-methyl-	EtOH	236(4.48),265(4.21), 385(3.72)	2-0964-76

Compound	Solvent	λ_{max}(log ϵ)	Ref.
Pyridine, 2,5-dimethyl-4-(4-nitrophenyl)-	EtOH	276(4.14)	103-0312-76
2(1H)-Pyridinone, 6-[(hydroxyimino)phenylmethyl]-1-methyl-, (E)-	MeOH	235(4.19),314(4.01)	83-0769-76
(Z)-	MeOH	235(4.09),255(4.06), 310(3.90)	83-0769-76
$C_{13}H_{12}N_2O_2S$ Thiocyanic acid, 3,3a,8,8a-tetrahydro-3a,8-dimethyl-2-oxo-2H-furo[2,3-b]-indol-5-yl ester	MeOH	245(3.40),270(4.18)	83-0631-76
$C_{13}H_{12}N_2O_3$ Benzenamine, 4-methoxy-2-nitro-N-phenyl-	EtOH	465(3.91)	146-0131-76
Benzoic acid, 2-(1H-imidazol-2-ylcarbonyl)-, ethyl ester	EtOH	214(4.14),290(4.12)	44-0836-76
$C_{13}H_{12}N_2O_3S$ 4-Oxazolecarboxamide, N-acetyl-2-methyl-5-(phenylthio)-	CHCl$_3$	300(4.03)	94-0948-76
$C_{13}H_{12}N_2O_4$ 1,3-Dioxolo[4,5-g]quinolinium, 8-amino-7-carboxy-5-ethyl-, hydroxide, inner salt	MeOH	265(4.46),274(4.36), 317(3.89),332(3.94), 347(3.98)	142-1347-76
	EtOH	271(4.4),275(4.3), 343(3.9),350(3.9)	4-0765-76
2-Furancarboxamide, N,N-dimethyl-5-(2-nitrophenyl)-	dioxan	294(4.17)	73-2577-76
2-Furancarboxamide, N,N-dimethyl-5-(3-nitrophenyl)-	dioxan	303(4.43)	73-2577-76
2-Furancarboxamide, N,N-dimethyl-5-(4-nitrophenyl)-	dioxan	352(4.38)	73-2577-76
2-Furancarboxamide, N-ethyl-5-(2-nitrophenyl)-	dioxan	286(4.15)	73-2577-76
2-Furancarboxamide, N-ethyl-5-(3-nitrophenyl)-	dioxan	303(4.43)	73-2577-76
2-Furancarboxamide, N-ethyl-5-(4-nitrophenyl)-	dioxan	353(4.28)	73-2577-76
1H-Pyrrole-3-carboxylic acid, 5-benzoyl-4,5-dihydro-4-oxo-, ethyl ester	EtOH	229(4.52),257(4.47), 332s(3.96)	23-1752-76
$C_{13}H_{12}N_2O_4S$ Acetamide, N-[4-[[(5-nitro-2-furanyl)-methyl]thio]phenyl]-	n.s.g.	214(4.16),268(4.27), 321(4.01)	78-1411-76
4,6(1H,5H)-Pyrimidinedione, 5-[(3,4-dimethoxyphenyl)methylene]dihydro-2-thioxo-	dioxan	245(4.20),330(3.84), 420(4.51)	104-1124-76
$C_{13}H_{12}N_2S$ 2-Pyridinecarbothioamide, N-(3-methylphenyl)-	20% EtOH	282(4.05)	94-1451-76
2-Pyridinecarbothioamide, N-(4-methylphenyl)-	20% EtOH	283(4.10)	94-1451-76
Pyridinium, 3-[(mercaptophenylmethylene)amino]-1-methyl-, hydroxide, inner salt	EtOH	272(3.98),325(4.03)	39-0315-76C
$C_{13}H_{12}N_2S_2$ [1,2]Dithiolo[4,5,1-hi][1,2,3]benzothiadiazole-3-S^{IV}, 2,6,7,8-tetrahydro-2-phenyl-	C_6H_{12}	203(4.34),230(4.41), 254s(4.21),295(4.02), 504(4.17)	39-0228-76C

Compound	Solvent	$\lambda_{max}(\log \epsilon)$	Ref.
$C_{13}H_{12}N_4$			
1H-Benzotriazole-1-methanamine, N-phenyl-	n.s.g.	244(4.14),279(4.76)	2-0718-76
Phthalazine, 1-(3,5-dimethyl-1H-pyrazol-1-yl)-	MeOH	220(4.54),285(3.86), 349(3.98)	94-1068-76
Pyridine, 2-(4,5-dihydro-4-phenyl-1H-1,2,3-triazol-5-yl)-	MeOH	245(3.9),280(4.0), 325(4.01),385(4.02)	2-0793B-76
1,2,4-Triazolo[3,4-a]phthalazine, 3-(1-methyl-1-propenyl)-	MeOH	252(4.43),259s(4.40), 270(4.30)	94-1068-76
1,2,4-Triazolo[1,5-a]pyridin-8-amine, 2-(phenylmethyl)-	H_2O	272(4.02),294(3.86)	44-3124-76
$C_{13}H_{12}N_4O$			
6H-Purin-6-one, 3,7-dihydro-8-methyl-3-(phenylmethyl)-	MeOH	266(4.12)	39-0239-76C
4(1H)-Pyrimidinone, 6-amino-5-(1H-indol-3-yl)-2-methyl-	EtOH	203s(4.58),216(4.64), 278(4.04),288(4.03)	103-1246-76
$C_{13}H_{12}N_4O_2$			
Benzo[g]pteridine-2,4(3H,10H)-dione, 10-ethyl-3-methyl-	EtOH	219(4.44),269(4.58), 335(3.90),438(4.01)	35-0830-76
	50% EtOH-5M HCl	215(4.34),259(4.44), 368(4.16)	35-0830-76
Benzo[g]pteridine-2,4(3H,10H)-dione, 10-propyl-	EtOH	219(4.26),267(4.36), 334(3.67),437(3.84)	35-0830-76
	50% EtOH-5M HCl	215(4.18),259(4.24), 369(3.96)	35-0830-76
Imidazo[4,5-d]pyridazine-4,7-diol, 3-methyl-1-(phenylmethyl)-	pH 2	273(3.53)	44-2303-76
	pH 7	299(3.54)	44-2303-76
	pH 12	307(3.58)	44-2303-76
6H-1,2,4,5-Tetrazino[1,6-b]isoquinolin-6-one, 2-acetyl-1,2-dihydro-1-methyl-	EtOH	236(4.33),310(4.13), 367(3.87),384(3.74)	95-0700-76
6H-1,2,4,5-Tetrazino[1,6-b]isoquinolin-6-one, 2-acetyl-1,2-dihydro-3-methyl-	EtOH	234(4.35),310(4.15), 366(3.88),383(3.76)	95-0700-76
6H-1,2,4,5-Tetrazino[1,6-b]isoquinolin-6-one, 1,2-dihydro-2-(1-oxopropyl)-	EtOH	234(4.32),311(4.16), 370(3.88),388(3.76)	95-0700-76
$C_{13}H_{12}N_4O_3$			
Benzeneacetic acid, 2-(3-nitro-4-pyridinyl)hydrazide	EtOH	232(4.25),352(3.56)	114-0285-76C
Benzo[g]pteridine-2,4(3H,10H)-dione, 10-ethyl-3-methyl-, 5-oxide	EtOH	213(4.07),271(4.58), 342(3.85),356s(3.78), 451(3.75)	35-0830-76
	EtOH-5M HCl	270(4.36),366(3.89), 398s(3.64)	35-0830-76
Benzo[g]pteridine-2,4(3H,10H)-dione, 10-propyl-, 5-oxide	EtOH	213(4.34),266(4.67), 340(4.10),356s(4.04), 452(4.04)	35-0830-76
	50% EtOH-5M HCl	266(4.34),368(4.19), 395s(4.01)	35-0830-76
$C_{13}H_{12}N_4O_4$			
Benzoic acid, 2-(3-nitro-4-pyridinyl)hydrazide	isoPrOH	237(4.45),260(4.46), 349(4.45)	114-0285-76C
$C_{13}H_{12}N_4O_4S$			
2-Thiophenecarboxaldehyde, 3,5-dimethyl-, 2,4-dinitrophenylhydrazone	n.s.g.	402(4.43)	39-1639-76C
2-Thiophenecarboxaldehyde, 4,5-dimethyl-, 2,4-dinitrophenylhydrazone	n.s.g.	396(4.51)	39-1639-76C

Compound	Solvent	$\lambda_{max}(\log \epsilon)$	Ref.
3-Thiophenecarboxaldehyde, 2,5-di-methyl-, 2,4-dinitrophenylhydrazone	n.s.g.	388(4.28)	39-1639-76C
$C_{13}H_{12}N_5O$ 3H-Imidazo[1,2-a]pyrimidin-4-ium, 6-methoxy-3-(phenylhydrazono)-, perchlorate	n.s.g.	310(4.14),320(4.13), 380(4.28)	104-0929-76
$C_{13}H_{12}O$ 1-Propanone, 1-(2-naphthalenyl)-	MeOH	210(3.2),246(3.9), 284(4.65),337(3.84)	19-0175-76
$C_{13}H_{12}OSe$ 2-Naphthaleneethaneselenoic acid, O-methyl ester	MeOH	276(4.11)	44-0729-76
$C_{13}H_{12}O_2$ 1,2-Naphthalenedione, 5-(1-methylethyl)-	EtOH	258(4.41),370(3.42), 414(3.45)	78-2693-76
9H-Xanthen-9-one, 1,2,3,4-tetrahydro-	EtOH	228(4.26),270(3.72)	22-1913-76
$C_{13}H_{12}O_2S$ 2-Thiophenecarboxylic acid, 5-(4-methyl-phenyl)-, methyl ester	MeOH MeCN	318(4.37) 317(4.40)	48-0731-76 118-0521-76
2-Thiophenepropanoic acid, 4-phenyl-	EtOH	232(4.19),264(4.34)	44-1320-76
$C_{13}H_{12}O_2Se$ 2-Selenophenecarboxylic acid, 5-(4-meth-ylphenyl)-, methyl ester	MeCN	328(4.34)	118-0521-76
$C_{13}H_{12}O_3$ 1-Azulenecarboxylic acid, 3-methoxy-, methyl ester	C_6H_{12}	287(4.39),291(4.50), 298(4.61),305(4.57), 311(4.67),382(3.96), 409(4.07),623(2.59), 640(2.62),675(2.54), 706(2.54),755(2.14)	44-1822-76
Bicyclo[5.3.1]undeca-1,3,5,8-tetraene-8-carboxylic acid, 10-oxo-, methyl ester	C_6H_{12}	204(4.35),235(4.16), 336(2.86),406s(1.91), 430s(1.51)	35-8277-76
Ethanone, 1-[6-hydroxy-2-(1-methylethen-yl)-5-benzofuranyl]- (euparin)	EtOH	262(3.59),290(3.19), 359(2.79)	73-2047-76
1-Naphthalenecarboxylic acid, 6-meth-oxy-, methyl ester	MeOH	212(4.55),236(4.58), 253(4.02),295(3.69), 334(3.45)	24-3025-76
2H-Pyran-2-one, 5,6-dihydro-4-hydroxy-6-(2-phenylethenyl)-, (E)-	n.s.g.	256(4.24),263(4.24)	1-0613-76
$C_{13}H_{12}O_4$ 2-Furanacetic acid, 2,3-dihydro-2-meth-yl-3-oxo-5-phenyl-	EtOH	242(4.02),295(4.32), 304(4.33)	23-0415-76
2-Furancarboxylic acid, 2,3-dihydro-3-oxo-5-phenyl-, ethyl ester	EtOH	223(4.19),242(3.85), 310(4.36)	39-1688-76C
2(5H)-Furanone, 5-hydroxy-4-methyl-5-(2-oxo-2-phenylethyl)-	MeCN	243(4.03),277s(3.34), 338(3.48)	23-0415-76
1,4-Naphthalenedione, 5,7-dimethoxy-2-methyl-	n.s.g.	265(4.28),401(3.63)	33-0664-76
$C_{13}H_{12}O_5$ 2,5-Benzodioxocin-3-carboxylic acid,	MeOH	238(3.89)	35-5581-76

Compound	Solvent	$\lambda_{max}(\log \epsilon)$	Ref.
1,6-dihydro-4-methyl-6-oxo-, methyl ester (cont.)			35-5581-76
1H-2-Benzopyran-3-carboxylic acid, 4-acetoxy-, methyl ester	MeOH	230(3.86),238(3.83), 313(3.68)	35-5581-76
2H-1-Benzopyran-4-carboxylic acid, 3-acetoxy-, methyl ester	MeOH	220(3.93),275(3.52), 305(3.30)	35-3555-76
2,5-Epoxy-1,3-benzodioxepin-5(4H)-carboxylic acid, 2-methyl-4-methylene-	MeOH	275(3.45),283(3.44)	35-3555-76
3-Furancarboxylic acid, 5-(2-furanylmethylene)-4,5-dihydro-2-methyl-4-oxo-, ethyl ester	EtOH	220(3.05),374(4.31)	4-0521-76
1,4-Naphthalenedione, 3-hydroxy-5,7-dimethoxy-2-methyl-	n.s.g.	225s(4.18),268(4.19), 301(4.16),373(3.65)	33-0664-76
Pentanoic acid, 3-(1,3-benzodioxol-5-ylmethylene)-4-oxo-	EtOH	322(4.30)	2-0620-76
$C_{13}H_{12}O_6$			
2-Propenoic acid, 3-(3,4-diacetoxyphenyl)-	MeOH	216(4.3),272(4.3)	102-1305-76
$C_{13}H_{13}BrN_4O_2S$			
Thiourea, N-[4-[(5-bromo-1-uracilyl)-methyl]phenyl]-N'-methyl-	pH 9.2	272.5(4.28)	106-0598-76
$C_{13}H_{13}BrO_2$			
2(5H)-Furanone, 4-bromo-3-ethyl-5-methyl-5-phenyl-	EtOH	206(4.15),228(4.01)	5-0979-76
$C_{13}H_{13}BrO_7$			
4,7-Ethanoisobenzofuran-1,3,8-trione, 7a-bromo-3a,4,7,7a-tetrahydro-4,9,9-trimethoxy-	EtOH	320(1.97)	1-0403-76
$C_{13}H_{13}ClD_2N_4O_4$			
9H-Purine, 9-(2-O-acetyl-4,6-dideoxy-β-L-xylo-hexopyranosyl-2,3-C-d_2)-6-chloro-	MeOH	264(3.90)	44-3827-76
$C_{13}H_{13}ClN_2O$			
Benzenemethanol, 5-chloro-2-hydrazino-α-phenyl-	isoPrOH	252(4.05),302(3.40)	4-0033-76
Quinoline, 2-chloro-4-morpholino-	MeOH	233(4.56),314(3.96)	95-0725-76
Quinoline, 3-chloro-4-morpholino-	MeOH	232(4.54),325(3.77)	95-0725-76
$C_{13}H_{13}ClN_2O_6$			
Pyrylium, 2-amino-3-(aminocarbonyl)-6-(4-methylphenyl)-, perchlorate	HOAc	391(4.37)	48-0705-76
$C_{13}H_{13}ClN_3O$			
Pyridinium, 2-amino-3-(aminocarbonyl)-6-(4-chlorophenyl)-1-methyl-, perchlorate	HOAc	341(4.10)	48-0705-76
$C_{13}H_{13}ClN_4O_2S$			
Thiourea, N-[4-[(5-chloro-1-uracilyl)-methyl]phenyl]-N'-methyl-	pH 9.2	268.0(4.17)	106-0598-76
$C_{13}H_{13}ClO_3$			
2-Benzofurancarboxylic acid, 5-chloro-2,3-dihydro-2-methyl-3-methylene-, ethyl ester	MeOH	322(3.87),335(3.87)	87-1214-76

Compound	Solvent	λ_{max}(log ϵ)	Ref.
$C_{13}H_{13}FN_4O_2S$ Thiourea, N-[4-[(5-fluoro-1-uracilyl)- methyl]phenyl]-N'-methyl-	pH 9.2	270.0(4.25)	·106-0598-76
$C_{13}H_{13}IN_2O$ Pyridinium, 3-(benzoylamino)-1-methyl-, iodide	MeOH	275(4.08),300(3.95)	39-0315-76C
$C_{13}H_{13}IO$ Pyrylium, 2,4-dimethyl-6-phenyl-, iodide	CH_2Cl_2	475(1.72)	59-1311-76
Pyrylium, 2,6-dimethyl-4-phenyl-, iodide	CH_2Cl_2	487(2.09)	59-1311-76
$C_{13}H_{13}N$ Aziridine, 1-methyl-2-(2-naphthalenyl)-	hexane	267(3.69),276(3.71), 286s(3.53),321(2.38)	95-1426-76
	heptane	269(3.73),278(3.76), 288(3.52),308(2.68), 322(2.60)	95-1426-76
	EtOH	269(3.73),278(3.75), 287s(3.58),308(2.64), 321(2.48)	95-1426-76
	BuOH	269(3.72),278(3.75), 287s(3.58),308(2.64), 321(2.48)	95-1426-76
	HOAc	269(3.67),276(3.69), 285s(3.54)	95-1426-76
	$HCONH_2$	270(3.69),278(3.70), 287s(3.56),304s(2.70)	95-1426-76
	MeCN	271(3.72),278(3.75), 287s(3.58),308(2.62), 322(2.56)	95-1426-76
	pyridine	301(2.89),309(2.79), 323(2.66)	95-1426-76
	DMSO	273(3.75),278(3.78), 290s(3.63),309(2.78), 323(2.68)	95-1426-76
	$CHCl_3$	274(3.74),280(3.76), 290s(3.60),309(2.75), 323(2.60)	95-1426-76
Benzenamine, N-methyl-N-phenyl-	n.s.g.	243(3.88),288(4.14)	65-0435-76
$C_{13}H_{13}NO$ 1H-Cyclopenta[b]quinoline, 2,3-dihydro- 7-methoxy-	EtOH	234(4.52),280(3.48), 310(3.59),323(3.64)	95-0968-76
Ethanone, 1-(2-methyl-5-phenyl-1H-pyr- rol-3-yl)-	EtOH	229s(4.22),236s(4.31), 242(4.34),288(4.31), 310(4.06)	44-0351-76
Isoxazole, 5-methyl-4-(1-methylethenyl)- 3-phenyl-	EtOH	234(4.04)	33-2074-76
Oxazole, 5-methyl-4-(1-methylethenyl)- 2-phenyl-	EtOH	218(4.16),282(4.12)	33-2074-76
2-Propenal, 3-(1,2-dimethyl-1H-indol- 3-yl)-	MeCN	229(4.32),276(3.93), 354(4.36)	5-1039-76
1H-Pyrrole-3-carboxaldehyde, 2,5-dimeth- yl-1-phenyl-	1% EtOH	260(3.93),303(3.70)	44-1952-76
2H-Pyrrol-2-one, 1,5-dihydro-3-methyl- 5-[(4-methylphenyl)methylene]-, (E)-	EtOH	242(3.85),250(3.85), 254(3.73),340(4.32)	49-0831-76

Compound	Solvent	$\lambda_{max}(\log \epsilon)$	Ref.
2H-Pyrrol-2-one, 1,5-dihydro-3-methyl-5-[(4-methylphenyl)methylene]-, (Z)-	EtOH	238(3.78),246(3.74), 254(3.34),325s(4.37), 339(4.44),357s(4.26)	49-0831-76
$C_{13}H_{13}NOS$			
1-Naphthalenecarbothioamide, 2-hydroxy-N,N-dimethyl-	heptane-10% CH_2Cl_2	285(4.06),334(3.56), 382(3.07)	1-0695-76
2-Naphthalenecarbothioamide, 1-hydroxy-N,N-dimethyl-	heptane-10% CH_2Cl_2	278(4.21),289s(4.13), 346(3.49),373(3.46)	1-0695-76
Sulfilimine, S-(2-methoxyphenyl)-S-phenyl-, (-)-	MeOH	225(4.08),283(3.59)	138-0363-76
$C_{13}H_{13}NO_2$			
L-Alanine, 3-(1-naphthalenyl)-	pH 13	273(3.82),283(3.89), 294s(3.72)	94-3149-76
L-Alanine, 3-(2-naphthalenyl)-	pH 13	269(3.62),276(3.63), 286s(3.48)	94-3149-76
2-Naphthalenecarbonitrile, 1,2,3,4-tetrahydro-4-methoxy-4-methyl-1-oxo-	EtOH	231(3.89),289(3.99)	2-0312-76
Naphth[2,1-d]isoxazole, 4,5-dihydro-7-methoxy-5-methyl-	EtOH	224(4.05),291(4.20), 298(4.20)	2-0312-76
Pyridinium, 3-hydroxy-5-(4-methoxyphenyl)-1-methyl-, hydroxide, inner salt	MeCN	207(3.9),220(3.7), 250(3.7),271(3.8), 360(3.4)	39-2329-76C
Spiro[cyclopropane-1,4'(1'H)-isoquinoline]-1',3'(2'H)-dione, 2,2'-dimethyl-	EtOH	244s(4.02),295(3.40)	95-0160-76
$C_{13}H_{13}NO_2S$			
Benzenesulfonamide, 4-methyl-N-phenyl-	EtOH	207(4.04),238(3.98)	118-0339-76
Ethanone, 2-hydroxy-1-[4-methyl-2-(4-methylphenyl)-5-thiazolyl]-	40% dioxan	330(4.37)	80-0113-76
Sulfoximine, S-(2-methoxyphenyl)-S-phenyl-, (R)-	isooctane	284(3.49)	78-3003-76
	MeOH	219(4.12),287(3.58)	78-3003-76
	dioxan	286(3.59)	78-3003-76
	MeCN	220(4.13),286(3.59)	78-3003-76
5-Thiazolecarboxaldehyde, 2-(4-ethoxyphenyl)-4-methyl-	n.s.g.	248(3.91),346(4.40)	80-1073-76
$C_{13}H_{13}NO_2S_2$			
4H-1,3-Thiazine-4,6(5H)-dione, 5,5-dimethyl-2-[(phenylmethyl)thio]-	EtOH	270(4.3)	103-0862-76
$C_{13}H_{13}NO_3$			
1-Indolizinecarboxylic acid, 3-acetyl-2-methyl-, methyl ester	MeOH	204(3.84),230s(3.93), 238(4.00),260(3.86), 274s(3.92),279(3.98), 336(3.90),350(3.92)	39-1908-76C
4-Isoxazolecarboxylic acid, 5-methyl-3-phenyl-, ethyl ester	EtOH	218(4.15)	33-2074-76
2,4-Pentadienal, 2,4-dimethyl-5-(4-nitrophenyl)-, (E,E)-	EtOH	220(4.04),267s(3.93), 331(4.32)	39-0404-76C
$C_{13}H_{13}NO_3S$			
Ethanone, 2-hydroxy-1-[2-(4-methoxyphenyl)-4-methyl-5-thiazolyl]-	40% dioxan	343(4.38)	80-0113-76
5-Thiazolecarboxylic acid, 2-(4-methoxyphenyl)-, ethyl ester	MeCN	327(4.35)	118-0403-76
$C_{13}H_{13}NO_4$			
Cyclopropa[c][1]benzoxepin-2-one, 1,1a-8,8a-tetrahydro-1,8-dimethyl-1a-nitro-	EtOH	232(3.49)	39-1260-76C

Compound	Solvent	$\lambda_{max}(\log \epsilon)$	Ref.
1H-Indole-3,4-dicarboxylic acid, 2-methyl-, 3-ethyl ester	n.s.g.	219(4.49),258(3.92), 293(3.87)	39-1073-76C
1,4-Naphthalenedione, 5,8-dihydroxy-2,3-dimethyl-6-(methylamino)-	EtOH	248s(4.17),258(4.33), 329(3.95),478s(3.91), 504(3.97),540s(3.88)	12-2247-76

$C_{13}H_{13}NO_5S$
| 2-Furancarboxylic acid, 5-[[4-(dimethylamino)phenyl]sulfonyl]- | EtOH | 206(4.09),243(4.30), 280(4.16) | 73-2571-76 |

$C_{13}H_{13}NS$
| 1-Naphthalenecarbothioamide, N,N-dimethyl- | heptane-10% CH_2Cl_2 | 275(4.13),304s(3.89), 379(2.58) | 1-0695-76 |
| 2-Naphthalenecarbothioamide, N,N-dimethyl- | heptane-10% CH_2Cl_2 | 275(4.11),286(4.13), 295s(4.03),384(2.84) | 1-0695-76 |

$C_{13}H_{13}N_2O$
| Pyridinium, 3-(benzoylamino)-1-methyl-, iodide | MeOH | 275(4.08),300(3.95) | 39-0315-76C |

$C_{13}H_{13}N_2OS$
| Thiopyrrylium, 2-amino-3-(aminocarbonyl)-6-(4-methylphenyl)-, perchlorate | HOAc | 410(4.16) | 48-0705-76 |

$C_{13}H_{13}N_2O_2$
| Pyrylium, 2-amino-3-(aminocarbonyl)-6-(4-methylphenyl)-, perchlorate | HOAc | 391(4.37) | 48-0705-76 |

$C_{13}H_{13}N_2PS$
| 1H-1,3,2-Benzodiazaphosphole, 2,3-dihydro-1-methyl-2-phenyl-, 2-sulfide | EtOH | 245(4.00),293(3.77), 318s(2.80) | 65-0807-76 |

$C_{13}H_{13}N_2S$
| Pyridinium, 1-methyl-3-[(phenylthioxomethyl)amino]-, iodide | EtOH | 278(4.10) | 39-0315-76C |

$C_{13}H_{13}N_3$
| Benzenamine, N-methyl-4-(phenylazo)- | hexane | 244(4.01),386(4.48), 455s(3.26) | 61-0301-76 |
| Benzenamine, 4-[(4-methylphenyl)azo]- | benzene | 363(4.45) | 18-1381-76 |

$C_{13}H_{13}N_3OS$
| Imidazo[1,2-a]pyridin-2(3H)-one, 3-(3,4,5-trimethyl-2(3H)-thiazolylidene)- | EtOH | 427(4.32) | 124-0961-76 |

$C_{13}H_{13}N_3O_2$
| 1H-Pyrrolo[2,3-d]pyrimidine-2,4(3H,5H)-dione, 6,7-dihydro-3-(phenylmethyl)- | DMSO | 281(4.28) | 24-2983-76 |

$C_{13}H_{13}N_3O_2S$
| 4,6(1H,5H)-Pyrimidinedione, 5-[[4-(dimethylamino)phenyl]methylene]dihydro-2-thioxo- | dioxan | 245(4.23),275(4.06), 320(3.94),473(5.03) | 104-1124-76 |

$C_{13}H_{13}N_3O_3$
| L-Alanine, N-[(3-methyl-2-quinoxalinyl)carbonyl]- | EtOH | 201(4.57),240(4.47), 319(3.87),326(3.83) | 78-2931-76 |

Compound	Solvent	$\lambda_{max}(\log \epsilon)$	Ref.
$C_{13}H_{13}N_3O_3S$			
Butanoic acid, 3-oxo-2-(phenylhydra-zono)-4-thiocyanato-, ethyl ester	n.s.g.	245(4.09),360(4.26)	124-1162-76
1H-Pyrimido[5,4-b][1,4]benzothiazine-2,4(3H,4aH)-dione, 4α-methoxy-1,3-dimethyl-	EtOH	225(4.55),251(4.46), 313(4.32)	94-3135-76
$C_{13}H_{13}N_3O_6$			
2H-Pyran-2-carboxylic acid, 2-[2-(acet-ylamino)-1,6-dihydro-6-oxo-4-pyrimid-inyl]-3,4-dihydro-4-oxo-, methyl ester	MeOH	236(3.79),258(3.60), 288(3.49)	44-2850-76
$C_{13}H_{13}N_3S$			
[1,2,4]Triazolo[1,5-a]isoquinolinium, 2-mercapto-3-propyl-, hydroxide, inner salt	MeOH	234(4.23),270(4.60), 317(3.72),365(3.53)	2-0176-76
$C_{13}H_{13}N_4O_3P$			
Phosphonic acid, (diazophenylmethyl)-, bis(2-cyanoethyl) ester	EtOH	213(4.11),264(4.10), 410(2.48)	35-7327-76
$C_{13}H_{13}N_5O$			
Methanimidamide, N,N-dimethyl-N'-(5-oxo-1,2,4-triazolo[1,5-b]isoquinolin-1(5H)-yl)-	EtOH	232(4.36),325(4.26), 384(3.90),402(3.80)	95-0700-76
$C_{13}H_{13}N_5O_3S$			
2,4-Pyrimidinediamine, N^2-acetyl-6-(benzylthio)-5-nitro-	EtOH-pH 1	259(4.31),345(4.11)	87-1186-76
	EtOH-pH 7	215(4.31),256(4.31), 348(4.13)	87-1186-76
	EtOH-pH 13	256(4.10),351(4.24)	87-1186-76
$C_{13}H_{13}O$			
Pyrylium, 2,4-dimethyl-6-phenyl-, iodide	CH_2Cl_2	475(1.72)	59-1311-76
selenocyanate	CH_2Cl_2	474(1.90)	59-1311-76
thiocyanate	CH_2Cl_2	441(2.29)	59-1311-76
Pyrylium, 2,6-dimethyl-4-phenyl-, iodide	CH_2Cl_2	487(2.09)	59-1311-76
selenocyanate	CH_2Cl_2	436(2.30)	59-1311-76
thiocyanate	CH_2Cl_2	416(2.53)	59-1311-76
$C_{13}H_{14}$			
Naphthalene, 1,2,6-trimethyl-	hexane	278(3.75),282(3.75), 290s(3.67),313(3.18), 320(3.04),327(3.27)	44-0066-76
$C_{13}H_{14}ClN_3O_5$			
Pyridinium, 2-amino-3-(aminocarbonyl)-1-methyl-6-phenyl-, perchlorate	HOAc	341(4.13)	48-0705-76
$C_{13}H_{14}Cl_3NO_4$			
Methanaminium, N-[3-chloro-5-(2-chloro-phenyl)-2,4-pentadienylidene]-N-methyl-, perchlorate	MeOH	230s(4.10),239(4.02), 268s(4.13),286(4.20), 309s(4.00),317s(3.84), 360(3.86)	48-0731-76
Methanaminium, N-[3-chloro-5-(4-chloro-phenyl)-2,4-pentadienylidene]-N-methyl-, perchlorate	MeOH	226s(4.07),233s(3.97), 271s(4.23),284s(4.31), 293(4.33),302s(4.30)	48-0731-76

Compound	Solvent	$\lambda_{max}(\log \epsilon)$	Ref.
$C_{13}H_{14}Cl_6FN_2O_9P$			
5'-Uridylic acid, 4'-C-fluoro-, bis-(2,2,2-trichloroethyl) ester	MeOH	257(3.98)	44-3010-76
$C_{13}H_{14}F_3N_5O$			
1,3,5-Triazine-2,4-diamine, N-[2-(4-methoxyphenyl)ethyl]-6-(trifluoro-methyl)-	MeOH	215(4.55),277(3.75), 284(3.67)	36-0525-76
$C_{13}H_{14}FeN_2O_4$			
Ferrocene, (2,2-dinitropropyl)-	MeOH	375(2.37)	104-2433-76
$C_{13}H_{14}NO$			
Pyrylium, 4-[4-(dimethylamino)phenyl]-, perchlorate	CH_2Cl_2	253(3.95),283s(3.73), 516(4.78)	4-1089-76
	MeCN	258s(3.93),280s(3.74), 500(4.73)	4-1089-76
$C_{13}H_{14}NOS_2$			
Morpholinium, 4-(5-phenyl-3H-1,2-dithi-ol-3-ylidene)-, perchlorate	MeOH	220(4.16),360(4.04)	48-0221-76
$C_{13}H_{14}NO_2$			
Pyridinium, 3-hydroxy-5-(4-methoxyphen-yl)-1-methyl-, bromide	EtOH	210(4.4),235(4.1), 262(4.3),290(3.9), 335(3.9)	39-2329-76C
$C_{13}H_{14}NO_5P$			
2-Propenoic acid, 2-cyano-3-(dimethoxy-phosphinyl)-3-phenyl-, methyl ester	EtOH	289(3.53)	22-0177-76
$C_{13}H_{14}NS$			
Thiopyrylium, 4-[4-(dimethylamino)phen-yl]-, perchlorate	CH_2Cl_2	265(3.86),275s(3.80), 302(3.75),558(4.77)	4-1089-76
	MeCN	265(3.66),276(3.66), 298(3.62),536(4.65)	4-1089-76
$C_{13}H_{14}N_2$			
1,2-Benzenediacetonitrile, α,α,α'-tri-methyl-	EtOH	259(2.49),266(2.48), 271(2.34)	78-0487-76
6H-Cyclohepta[b]quinoxaline, 7,8,9,10-tetrahydro-	MeOH	273(4.53),315(2.95)	104-2286-76
3H-Pyrido[3,4-b]indole, 4,9-dihydro-1,4-dimethyl-	n.s.g.	247(4.19),300(4.55), 368(3.82)	103-0329-76
$C_{13}H_{14}N_2O$			
2,2'-Bipyridine, 5-propoxy-	EtOH	256(3.66),292(3.84)	64-0115-76B
6H-Dipyrido[1,2-a:2',1'-c][1,4]diaze-pinediium, 7,8-dihydro-3-hydroxy-, dibromide	pH 6.0	255(4.00),314(3.78), 364(4.14)	64-0115-76B
6H-Dipyrido[1,2-a:2',1'-c]pyrazinediium, 6,7-dihydro-3-methoxy-, dibromide	pH 6.0	279(3.65),340(4.06)	64-0115-76B
1(2H)-Isoquinolinone, 3-pyrrolidino-	EtOH	228(4.40),244(4.05), 294(4.24),308(4.39), 380(3.68)	95-0154-76
1-Naphthalenecarboxaldehyde, 2-hydroxy-, dimethylhydrazone	benzene	322(4.01),360(4.03)	104-0625-76
	EtOH	318(4.07),358(4.07)	104-0625-76
	DMF	320(3.99),360(4.03)	104-0625-76
1H-Pyrazole, 1-(phenylmethyl)-3-(2-prop-enyloxy)-	MeOH	228(3.90),266(3.53)	24-0268-76

Compound	Solvent	$\lambda_{max}(\log \epsilon)$	Ref.
3(2H)-Pyridazinone, 4,5-dihydro-6-[2-(4-methylphenyl)ethenyl]-	C_6H_{12}	230(3.86),306s(4.22), 316(4.42),329s(4.12)	59-0837-76
	EtOH	230(3.94),308s(4.09), 319(4.59)	59-0837-76
$C_{13}H_{14}N_2OS$			
Ethanone, 1-[2-[4-(dimethylamino)phenyl]-5-thiazolyl]-	MeCN	396(4.47)	118-0403-76
2H-Imidazole-2-thione, 1-acetyl-1,5-dihydro-5,5-dimethyl-4-phenyl-	C_6H_{12}	225(3.8),295(4.3), 330s(4.0)	24-0154-76
4(1H)-Pyrimidinone, 6-phenyl-2-(propylthio)-	EtOH	257(4.41),312(3.88)	95-1094-76
$C_{13}H_{14}N_2OS_2$			
1H-[1,2]Dithiolo[5,1-e][1,2,3]thiadiazole-7-S^{IV}, 1-(4-methoxyphenyl)-3,4-dimethyl-	C_6H_{12}	207(4.27),235(4.44), 256s(4.23),301(4.05), 335s(3.49),505(4.16)	39-0228-76C
$C_{13}H_{14}N_2O_2$			
1H-3a,9a-Diazacyclohepta[def]biphenylene-4,9-dione, 2,3,6a,6b,9b,9c-hexahydro-	n.s.g.	263(3.34)	33-2841-76
1(2H)-Isoquinolinone, 3-morpholino-	EtOH	230(4.30),247(3.84), 294(4.13),352(3.64)	95-0154-76
5(4H)-Oxazolone, 4-[(dimethylamino)methylene]-2-(phenylmethyl)-	$CHCl_3$	323(4.36)	5-2185-76
2,5-Piperazinedione, 1,4-dimethyl-3-(phenylmethylene)-, (Z)-	EtOH	224(4.32),289(4.46)	39-0628-76C
2,5-Piperazinedione, 1,6-dimethyl-3-(phenylmethylene)-, (Z)-	EtOH	224(4.20),296(4.38)	39-0628-76C
2(1H)-Pyrazinone, 3,6-dihydro-5-methoxy-1-methyl-3-(phenylmethylene)-, (Z)-	EtOH	220(4.06),224(4.13), 231(4.08),295(4.43), 307(4.51),320(4.36)	39-0628-76C
3(2H)-Pyridazinone, 4,5-dihydro-6-[2-(4-methoxyphenyl)ethenyl]-	C_6H_{12}	226(3.68),249(3.36), 254(3.35),260(3.35), 323(4.22),335s(4.09)	59-0837-76
	EtOH	229(3.81),317s(3.94), 324(4.50)	59-0837-76
2(1H)-Pyridinone, 1,1'-(1,3-propanediyl)bis-	n.s.g.	228(4.12),301(4.03)	33-2841-76
Quinoline, 4-morpholinyl-, 1-oxide, hydrochloride	MeOH	225(4.54),365(4.08)	95-0725-76
$C_{13}H_{14}N_2O_2S$			
5-Thiazolecarboxylic acid, 2-[4-(dimethylamino)phenyl]-, methyl ester	MeCN	383(4.46)	118-0403-76
$C_{13}H_{14}N_2O_3$			
1H-Pyrazole-3-carboxylic acid, 4,5-dihydro-4-oxo-5-(phenylmethyl)-	EtOH	229(4.326),275(4.125)	23-1752-76
$C_{13}H_{14}N_2O_4$			
Benzonitrile, 4-[1-(acetyloxy)-2-methyl-2-nitropropyl]-	MeOH	231(4.26),238(4.20), 273(2.99),281(2.97)	12-0327-76
Quinolinium, 4-amino-3-carboxy-1-ethyl-6-hydroxy-7-methoxy-, hydroxide, inner salt	MeOH	227(4.53),273(4.47), 335(4.00),348(4.62)	142-1347-76
$C_{13}H_{14}N_2O_4S$			
2H-Pyrimido[2,1-b]benzothiazole-4-carb-	MeOH	262(4.13),372(3.89)	64-0251-76B

Compound	Solvent	λ_{max}(log ϵ)	Ref.
oxylic acid, 6,7,8,9-tetrahydro- 3-hydroxy-2-oxo-, ethyl ester (cont.)			64-0251-76B
$C_{13}H_{14}N_2O_5S$ Benzenamine, N,N-dimethyl-4-[[(5-nitro- 2-furanyl)methyl]sulfonyl]-	n.s.g.	217(4.19),290(4.46)	78-1411-76
$C_{13}H_{14}N_2O_7$ 2,2'-Anhydro-1-(3',5'-di-O-acetyl-β-D- arabinofuranosyl)uracil	pH 6.8	224(3.92),251(3.91)	44-2995-76
$C_{13}H_{14}N_3O$ Isoxazolo[2,3-a]pyridinium, 4,5,6,7- tetrahydro-4-(phenylhydrazono)-, bromide	n.s.g.	245(4.2),300s(4.2), 380(4.2)	39-1241-76C
Pyridinium, 2-amino-3-(aminocarbonyl)- 1-methyl-6-phenyl-, perchlorate	HOAc	341(4.13)	48-0705-76
$C_{13}H_{14}N_4$ 2-Butenal, 2-methyl-, 1-phthalazinyl- hydrazone	MeOH	240(3.89),279s(4.38), 288(4.53),355(4.32)	94-1068-76
Pyrido[3,2-e]-1,2,4-triazine, 1,2-di- hydro-3-(phenylmethyl)-, dihydro- chloride	EtOH	383(3.70)	114-0301-76C
$C_{13}H_{14}N_4O$ Benzeneacetic acid, 2-(3-amino-4-pyri- dinyl)hydrazide	EtOH	245(4.26),339(3.45), 450(4.03)	114-0285-76C
1,2,4-Triazolo[1,5-b]isoquinolin-5(1H)- one, 2-ethyl-1-(methylamino)-	EtOH	235(4.39),315(4.23), 370(3.90),388(3.78)	95-0700-76
$C_{13}H_{14}N_4O_2$ Benzoic acid, 2-methyl-, 2,4-diamino- 6-methyl-5-pyrimidinyl ester	EtOH	281(4.17)	94-2461-76
Benzoic acid, 3-methyl-, 2,4-diamino- 6-methyl-5-pyrimidinyl ester	EtOH	280(4.17)	94-2461-76
Benzoic acid, 4-methyl-, 2,4-diamino- 6-methyl-5-pyrimidinyl ester	EtOH	280(4.07)	94-2461-76
1H-Imidazo[4,5-d]pyridazine-4,7-dione, 2,3,5,6-tetrahydro-1-methyl-3-(phen- ylmethyl)-	pH 7 pH 10	340(3.60) 342(3.61)	44-2303-76 44-2303-76
$C_{13}H_{14}N_4O_2S$ Thiourea, N-methyl-N'-[[(1-uracilyl)- methyl]phenyl]-	pH 9.2	263.0(4.26)	106-0598-76
$C_{13}H_{14}N_4O_3$ Acetic acid, 2-[2-(acetylamino)-1,2-di- hydro-1-oxo-3-isoquinolinyl]hydrazide	EtOH	231(4.28),300(4.00), 362(3.60)	95-0700-76
Oxazolo[4,5-b]quinoxaline-9a(2H)-carb- oxamide, 3,9-dihydro-N,6,7-trimethyl- 2-oxo-	6M HCl pH 1.4 pH 7	372(4.15) 340(3.88),375(4.87) 315(3.74),365(3.92)	5-1276-76 5-1276-76 5-1276-76
$C_{13}H_{14}N_4O_4$ 1H-Pyrazole-4-carboxylic acid, 5-amino- 3-methyl-1-(4-nitrophenyl)-, ethyl ester	EtOH	295(4.15)	4-1137-76
$C_{13}H_{14}N_4O_4Se$ 7H-Pyrrolo[2,3-d]pyrimidine-5-carbo-	pH 1	315(4.20)	4-0745-76

Compound	Solvent	$\lambda_{max}(\log \epsilon)$	Ref.
nitrile, 4-(methylseleno)-7-β-D-ribo-furanosyl- (cont.)	H_2O pH 11	312(4.27) 312(4.23)	4-0745-76 4-0745-76
$C_{13}H_{14}N_5O_6P$ 3H-Imidazo[2,1-i]purine, 5-methyl- 3-(3,5-O-phosphinico-β-D-ribo- furanosyl)-	pH 1 pH 13	274(4.08) 255s(3.71),265(3.83), 275(3.88),293(3.71)	94-1561-76 94-1561-76
$C_{13}H_{14}N_5O_6PS$ 3H-Imidazo[2,1-i]purine, 5-(methylthio)- 3-(3,5-O-phosphinico-β-D-ribofurano- syl)-	pH 1 pH 13	237(4.33),292(4.28) 240(4.33),244s(4.32), 273(3.92),283(4.01), 303(3.97)	94-1561-76 94-1561-76
$C_{13}H_{14}N_5O_7P$ 3H-Imidazo[2,1-i]purine, 5-methoxy- 3-(3,5-O-phosphinico-β-D-ribofurano- syl)-	pH 1 pH 13	262s(4.05),281(4.20), 291s(4.06) 273(4.13)	94-1561-76 94-1561-76
$C_{13}H_{14}O$ Ethanone, 1-(1,2-dihydro-2-methyl- 2-naphthalenyl)-	MeOH	215(4.2),272(3.7)	19-0175-76
Ethanone, 1-(1,4-dihydro-1-methyl- 1-naphthalenyl)-	MeOH	206(4.19),264(2.80)	19-0175-76
Naphthalene, 1-ethyl-5-methoxy-	EtOH	222(4.65),295(3.90), 309(3.73),323(3.44)	95-1161-76
1-Naphthalenol, 5-(1-methylethyl)-	EtOH	223(4.58),299(3.84), 312s(3.74),325s(3.56)	95-1161-76
2-Naphthalenol, 1-(1-methylethyl)-	EtOH	232(4.78),271(3.60), 281(3.69),292(3.61), 325(3.38),337(3.41)	39-2570-76C
$C_{13}H_{14}OS_2$ 1(2H)-Naphthalenone, 2-(1,3-dithiolan- 2-yl)-3,4-dihydro-	C_6H_{12}	247(4.13),253s(4.02), 289(3.27),298(3.20)	148-0435-76
$C_{13}H_{14}O_2$ Acetic acid, tricyclo[3.3.1.0^{2,8}]nona- 3,6-dien-9-ylidene-, ethyl ester	C_6H_{12}	220(4.11),260(3.90)	88-1565-76
Bicyclo[5.3.1]undeca-1,3,5,8-tetraene- 8-carboxylic acid, methyl ester	C_6H_{12}	206(4.39),255(3.32)	35-8277-76
Bicyclo[5.3.1]undeca-3,5,8,10-tetraene- 8-carboxylic acid, methyl ester	C_6H_{12}	230(4.01),304(3.74)	35-8277-76
Cyclohexanone, 2-[(2-hydroxyphenyl)meth- ylene]-	EtOH	279(3.46),315(2.54)	18-0245-76
3-Cyclohexene-1-carboxylic acid, 4-phenyl-	EtOH	247(4.19)	39-1669-76C
2(5H)-Furanone, 3-ethyl-5-methyl- 5-phenyl-	EtOH	206(4.25)	5-0979-76
2(5H)-Furanone, 5-ethyl-3-methyl- 5-phenyl-	EtOH	206(4.36)	5-0979-76
6-Heptene-2,5-dione, 7-phenyl-, (E)-	EtOH	222(4.09),227(4.07), 288(4.36)	12-0339-76
2,6-Methano-1-benzooxonin-2(3H)-ol, 4,5-dihydro-	EtOH	235(4.15),277(3.46)	18-0245-76
2(1H)-Naphthalenone, 1-hydroxy- 1-(1-methylethyl)-	EtOH	237(4.09),313(3.82)	39-2570-76C
1,5,11-Tridecatriene-7,9-diyne-3,4-diol, 1-trans-5-trans	EtOH	230(4.47),237(4.47), 248(4.33),261(3.92), 277(4.20),293(4.37), 312(4.27)	39-0735-76C

Compound	Solvent	λ_{max}(log ϵ)	Ref.
$C_{13}H_{14}O_2S$			
2-Propenoic acid, 3-[4-(1-cyclohexen-1-yl)-2-thienyl]-, (E)-	EtOH	252(4.28),287(4.19), 330(3.88)	44-1320-76
$C_{13}H_{14}O_3$			
Bicyclo[5.3.1]undeca-1,3,5,8-tetraene-8-carboxylic acid, 10-hydroxy-, methyl ester	C_6H_{12}	251s(4.1),266(4.2), 295(4.4),364(3.9), 480(2.8)	35-8277-76
2(5H)-Furanone, 3-methoxy-5-methyl-4-(phenylmethyl)-	EtOH	229(4.03)	107-0299-76
Naphthalene, 1,3,8-trimethoxy-	EtOH	230(4.67),285(3.72), 294(3.73),319(3.51), 331(3.53)	78-1353-76
1(2H)-Naphthalenone, 3,4-dihydro-2-(hydroxymethylene)-6-methoxy-4-methyl-	EtOH	230(3.94),256(3.64), 283(3.89),360(4.05)	2-0312-76
2(1H)-Naphthalenone, 1-hydroperoxy-1-(1-methylethyl)-	EtOH	234(4.17),240(4.19), 313(3.92)	39-2570-76C
Pentanoic acid, 3-[(4-methylphenyl)methylene]-4-oxo-	EtOH	212(4.14),290(4.01)	2-0620-76
$C_{13}H_{14}O_4$			
1-Benzoxepin-5(2H)-one, 6,8-dimethoxy-3-methyl-	n.s.g.	239(4.07),308(3.86)	33-0664-76
1H-Indene-2-acetic acid, 2,3-dihydro-5-methoxy-1-oxo-, methyl ester	EtOH	267(4.17),286(4.07)	78-0065-76
1-Naphthalenecarboxylic acid, 1,2,3,4-tetrahydro-7-methoxy-2-oxo-, methyl ester	MeOH	217(4.46),232(4.32), 258(4.14),270(4.18), 297(4.16)	12-1393-76
1-Naphthalenecarboxylic acid, 1,2,3,4-tetrahydro-8-methoxy-2-oxo-, methyl ester	MeOH	230(2.74),277(2.74), 330(2.43)	12-1393-76
Pentanoic acid, 3-[(4-methoxyphenyl)-methylene]-4-oxo-	EtOH	214(4.00),302(3.63)	2-0620-76
$C_{13}H_{14}O_5$			
4,7-Ethanoisobenzofuran-1,3,8-trione, 3a,4,7,7a-tetrahydro-9-hydroxy-4,6,9-trimethyl-	EtOH	205(3.61),308(2.11)	1-0403-76
$C_{13}H_{14}O_6S$			
2-Propenoic acid, 2-[(carboxymethyl)-thio]-3-(3,4-dimethoxyphenyl)-	EtOH	308(4.46)	42-0038-76
$C_{13}H_{14}O_8$			
1,3-Cyclopentadiene-1,2,3,4-tetracarboxylic acid, tetramethyl ester, sodium deriv.	MeOH	215(3.80),267(4.46), 302(4.16)	24-1928-76
$C_{13}H_{15}BrN_2O_3$			
Pyridinium, 1,2,3,6-tetrahydro-1,1-dimethyl-5-(4-nitrophenyl)-3-oxo-, bromide	EtOH	217(4.0),294(4.3)	39-2329-76C
$C_{13}H_{15}ClN_2O_2$			
Quinoline, 4-morpholinyl-, 1-oxide, hydrochloride	MeOH	225(4.54),365(4.08)	95-0725-76
$C_{13}H_{15}ClN_4O_4$			
9H-Purine, 9-(2-O-acetyl-4,6-dideoxy-β-L-xylohexopyranosyl)-6-chloro-	MeOH	264(3.81)	44-3827-76

Compound	Solvent	$\lambda_{max}(\log \epsilon)$	Ref.
1H-1,2,3-Triazolo[4,5-c]pyridine, 4-chloro-1-[2,3-0-(1-methylethylidene)-β-D-ribofuranosyl]-	pH 1 pH 11 MeOH	270(3.98) 268(3.96) 267(3.96)	44-1449-76 44-1449-76 44-1449-76
$C_{13}H_{15}ClN_6O_5S_2$ 5-Thia-1-azabicyclo[4.2.0]oct-2-ene-2-carboxylic acid, 7-[(chloroacetyl)-amino]-7-methoxy-3-[[(1-methyl-1H-tetrazol-5-yl)thio]methyl]-8-oxo-, (6R-cis)-	pH 6.86	271(3.96)	94-2629-76
$C_{13}H_{15}ClO$ Naphthalene, 5-chloro-4-ethyl-1,2-di-hydro-8-methoxy- Naphthalene, 8-chloro-1-ethylidene-1,2,3,4-tetrahydro-5-methoxy-	EtOH EtOH	229(4.37),265(3.88), 303(3.14) 227(4.45),295(3.22)	95-1161-76 95-1161-76
$C_{13}H_{15}Cl_2NO_4$ Methanaminium, N-(3-chloro-5-phenyl-2,4-pentadienylidene)-N-methyl-, perchlorate	MeOH	234s(3.88),266(4.06), 291(4.03),301s(3.99), 310s(3.91),380(4.26)	48-0731-76
$C_{13}H_{15}Cl_2NO_5$ Methanaminium, N-[3-chloro-5-(4-hydroxy-phenyl)-2,4-pentadienylidene]-N-meth-yl-, perchlorate	MeOH	279(4.05),294s(3.94), 301s(3.93),307s(3.92), 324s(3.81),439(4.54)	48-0731-76
$C_{13}H_{15}Cl_2NO_6$ Methanaminium, N-[3-chloro-3-[4-(meth-oxycarbonyl)phenyl]-2-propenylidene]-N-methyl-, perchlorate	MeOH	269(4.21),314s(3.58), 331s(3.52)	48-0731-76
$C_{13}H_{15}FN_8O_3$ Adenosine, 5'-azido-5'-deoxy-4'-C-flu-oro-2',3'-0-(1-methylethylidene)- 9H-Purin-6-amine, 9-[5-azido-5-deoxy-4-C-fluoro-2,3-0-(1-methylethylidene)-α-L-lyxo-furanosyl]-	MeOH MeOH	258(4.12) 258(4.15)	35-3346-76 35-3346-76
$C_{13}H_{15}N$ 6H-Azepino[1,2-a]indole, 7,8,9,10-tetra-hydro-	EtOH	226(4.54),284f(3.88)	20-0819-76
$C_{13}H_{15}NO$ Benzo[f]quinolin-4a-ol, 2,3,4,4a,9,10-hexahydro- 4(1H)-Pyridinone, 2,3-dihydro-2,6-di-methyl-1-phenyl-	EtOH EtOH	215(4.20),250(4.15) 322(4.21)	2-0819-76 4-0253-76
$C_{13}H_{15}NOS$ 4H-Thiopyran-4-one, tetrahydro-3-[(meth-ylphenylamino)methylene]-	EtOH	230s(3.77),343(4.17)	4-0225-76
$C_{13}H_{15}NO_2$ Benzo[f]quinoline-4a-hydroperoxide, 2,3,4,4a,9,10-hexahydro- 2-Propen-1-one, 3-morpholino-1-phenyl- 2H-Pyrrol-2-one, 1,5-dihydro-3-hydroxy-5-phenyl-4-propyl-	EtOH EtOH EtOH	206(4.18),235(3.34) 242(4.06),342(4.39) 216(4.14),240(3.72)	2-0819-76 32-0205-76 40-1100-76

Compound	Solvent	$\lambda_{max}(\log \epsilon)$	Ref.
$C_{13}H_{15}NO_3$			
Butanamide, 3-oxo-N-[2-(2-oxopropyl)-phenyl]-	EtOH	210(4.23),232(3.85)	94-1544-76
Cyclopropanecarboxamide, N-[2-(1,3-benzodioxol-5-yl)ethyl]-	n.s.g.	233(3.63),287(3.67)	39-0146-76C
Glycine, N-(1-oxo-3-phenyl-2-propenyl)-, ethyl ester	MeOH-acid	274(4.33)	20-0647-76
1H-Indole, 1-(2-deoxy-α-D-erythro-pento-furanosyl)-	EtOH	299(3.65)	28-0227-76B
β-	EtOH	299(3.72)	28-0227-76B
$C_{13}H_{15}NO_3S$			
4-Thiazolidinecarboxylic acid, 3-acetyl-2-(4-methylphenyl)-, (2R-cis)-	MeOH	258(3.15)	87-1002-76
$C_{13}H_{15}NO_4S$			
Butanedioic acid, (1,3-dimethyl-2(1H)-pyridinylidene)thioxo-, dimethyl ester	EtOH	220(3.94),272(3.89), 334(4.33)	142-0705-76
Butanedioic acid, (1,5-dimethyl-2(1H)-pyridinylidene)thioxo-, dimethyl ester	EtOH	220(4.03),272(3.87), 332(4.28)	142-0705-76
Butanedioic acid, (1,6-dimethyl-2(1H)-pyridinylidene)thioxo-, dimethyl ester	EtOH	220(3.80),272(3.92), 330(4.28)	142-0705-76
$C_{13}H_{15}NO_5$			
4-Azabicyclo[5.1.0]octa-2,5-diene-1,7-dicarboxylic acid, 4-acetyl-, dimethyl ester, cis	MeOH	248(4.14)	24-3505-76
Cyclopenta[b]pyrrole-3a,6a(1H,4H)-dicarboxylic acid, 1-acetyl-, dimethyl ester	MeOH	252(3.08),271(2.70)	24-3505-76
$C_{13}H_{15}NO_6$			
4-Azabicyclo[5.1.0]octa-2,5-diene-1,4,7-tricarboxylic acid, trimethyl ester, cis	EtOH	230(4.08)	24-3505-76
Cyclopenta[b]pyrrole-1,3a,6a(4H)-tricarboxylic acid, trimethyl ester	EtOH	234s(3.36)	24-3505-76
$C_{13}H_{15}NO_6S$			
2-Butenedioic acid, 2-[[3-(ethoxycarbonyl)-2-thienyl]amino]-, dimethyl ester, (Z)-	MeOH	219(4.20),232(4.13), 257(3.91),353(--)	5-1972-76
$C_{13}H_{15}NO_8$			
Cyclopentadienetetracarboxylic acid, 5-amino-, tetramethyl ester, sodium derivative	MeOH	226(4.52),268(3.85)	24-1928-76
$C_{13}H_{15}N_2O_3$			
Pyridinium, 1,2,3,6-tetrahydro-1,1-dimethyl-5-(4-nitrophenyl)-3-oxo-, bromide	EtOH	217(4.0),294(4.3)	39-2329-76C
$C_{13}H_{15}N_3$			
Pyrrolo[1,2-a]pyrimidine-8-carbonitrile, 6-ethyl-2,4,7-trimethyl-	EtOH	222(4.41),235s(4.34), 240(4.35),257(4.31), 267(4.34),285s(3.54),	103-1379-76

Compound	Solvent	$\lambda_{max}(\log \epsilon)$	Ref.
Pyrrolo[1,2-a]pyrimidine-8-carbonitrile, 6-ethyl-2,4,7-trimethyl- (cont.)		297(3.67),308(3.68), 364(3.38)	103-1379-76
Pyrrolo[1,2-a]pyrimidine-8-carbonitrile, 7-ethyl-2,4,6-trimethyl-	EtOH	220(4.33),235(4.28), 238(4.28),257(4.29), 266(4.32),286s(3.54), 309(3.65),366(3.52)	103-1379-76
Quinoline, 1,2,3,4-tetrahydro-2-(1-methyl-1H-imidazol-5-yl)- (tetrahydroisomacrorine)	MeOH	205(4.30),248(3.82), 299(3.29)	78-1085-76

$C_{13}H_{15}N_3O$

Acetamide, N-[4-(3,5-dimethyl-1H-pyrazol-1-yl)phenyl]-	EtOH	261(4.18)	22-0195-76
Ethanone, 1-[2-amino-5-(3,5-dimethyl-1H-pyrazol-1-yl)phenyl]-	EtOH	237(4.20),267(3.90)	22-0195-76
Morpholine, 4-(3-phenyl-1H-pyrazol-1-yl)-	MeOH	252.5(4.24)	94-3001-76
Piperidine, 1-(5-phenyl-1,2,4-oxadiazol-3-yl)-	MeOH	234(3.99),278(4.21)	18-3607-76

$C_{13}H_{15}N_3O_2$

1H-Indazole-4,7-dione, 3,6-dimethyl-5-pyrrolidino-	EtOH	224(4.11),274(4.15), 494(3.65)	94-1731-76

$C_{13}H_{15}N_3O_2S$

1H-Imidazole-5-carboxylic acid, 4-amino-2-(methylthio)-1-phenyl-, ethyl ester	EtOH	226(4.13),303(4.13)	49-1413-76

$C_{13}H_{15}N_3O_3$

1H-Indazole-4,7-dione, 3,6-dimethyl-5-morpholino-	EtOH	221(4.22),265(4.06), 320(3.71),491(3.55)	94-1731-76
Propanoic acid, 2-(3-indolyl)-3-ureido-, methyl ester	EtOH	219(4.54),280(3.83), 288(3.78)	104-1087-76

$C_{13}H_{15}N_3O_3S_2$

1,2,4-Triazine-3,5(2H,4H)-dithione, 6-[(3,4,5-trimethoxyphenyl)methyl]-	MeOH	274(4.78)	39-2521-76C

$C_{13}H_{15}N_3O_4S$

1,2,4-Triazin-5(2H)-one, 3,4-dihydro-3-thioxo-6-[(3,4,5-trimethoxyphenyl)methyl]-	MeOH	272(4.62)	39-2521-76C

$C_{13}H_{15}N_3O_5$

1,2,4-Triazine-3,5(2H,4H)-dione, 6-[(3,4,5-trimethoxyphenyl)methyl]-	MeOH	268(4.56)	39-2521-76C

$C_{13}H_{15}N_3O_6$

2,2'-Anhydro-1-(3',5'-diacetyl-β-D-arabinofuranosyl)cytosine, hydrochloride	MeOH-acid EtOH	234(3.98),263(4.00) 236(4.01),264(4.06)	87-0663-76 44-2995-76

$C_{13}H_{15}N_5$

Cyclohexanone, pyrido[3,2-c]pyridazin-4-ylhydrazone	EtOH	259(4.10),305(3.45), 380(4.04)	103-0808-76

$C_{13}H_{15}N_5O_3$

Adenosine, 4',5'-didehydro-5'-deoxy-2',3'-O-(1-methylethylidene)-	MeOH	259(4.17)	35-3346-76
1H-Purine-2,6-dione, 3,7-dihydro-8-[(5-hydroxy-2,4-pentadienyli-	pH 12	455(4.86)	142-1257-76

Compound	Solvent	$\lambda_{max}(\log \epsilon)$	Ref.
dene)amino]-1,3,7-trimethyl-, mono-sodium salt, (E,E,E)- (cont.)			142-1257-76
$C_{13}H_{15}N_5O_5$ 9H-Imidazo[1,2-a]purin-9-one, 3,4-di-hydro-6-methyl-3-β-D-ribofuranosyl-	pH 1.0	227(4.40),240s(4.35), 277(3.92),300(3.89)	69-0898-76
	pH 7.5	231(4.55),285(4.07)	69-0898-76
	pH 11.3	239(4.51),284s(3.71), 308(3.85)	69-0898-76
$C_{13}H_{15}N_5O_6$ D-glycero-β-D-allo-Octofuranuronic acid, 1-(6-amino-9H-purin-9-yl)-3,7-anhydro-1,6-dideoxy-	0.05M HCl 0.05M NaOH	257(4.10) 260(4.11)	88-1687-76 88-1687-76
$C_{13}H_{15}N_5O_7$ 1,4-Diazepinium, 2,3-dihydro-1,5-dimeth-yl-, picrate	MeOH	338(4.45)	39-1784-76C
$C_{13}H_{15}N_7O_2$ 9,10-Propanopyrimido[4',5':3,4]azeto-[2,1-f]purine-6,8(5aH,7H)-dione, 4-amino-9a,9b-dihydro-5a-methyl-	H_2O	227(4.26),284(3.84)	19-0517-76
2,4(1H,3H)-Pyrimidinedione, 1-[3-(6-ami-no-9H-purin-9-yl)propyl]-5-methyl-	H_2O	261(4.30)	19-0517-76
$C_{13}H_{15}OS$ 1,3-Benzoxathiol-1-ium, 2-cyclohexyl-, tetrafluoroborate	96% H_2SO_4	244s(3.63),248(3.67), 252(3.61),300(3.92)	39-0323-76C
$C_{13}H_{15}S_2$ 1,3-Benzodithiol-1-ium, 2-cyclohexyl-, perchlorate	96% H_2SO_4	225s(3.89),259(3.86), 263(3.88),316(3.89), 370s(3.37)	118-0759-76
$C_{13}H_{16}$ Acenaphthene, 2a,3,4,5-tetrahydro-6-methyl-	EtOH	254(4.13),312(3.54)	22-1599-76
Acenaphthene, 2a,3,4,5-tetrahydro-8-methyl-	EtOH	258(4.16),300(3.44)	22-1599-76
Naphthalene, 1,2-dihydro-2,5,6-trimeth-yl-	MeOH	223(4.53),267(4.08)	44-0066-76
$C_{13}H_{16}ClN_3$ 2,4-Quinolinediamine, 6-chloro-N,N,N',N'-tetramethyl-	EtOH	244(4.43),274(4.42), 355(3.74)	1-0133-76
$C_{13}H_{16}ClN_3O_2$ Butanamide, N-(1-acetyl-3-cyano-4,5-di-methyl-1H-pyrrol-2-yl)-4-chloro-	MeOH	209(4.12),284(3.99)	36-0908-76
$C_{13}H_{16}ClN_5O_6$ Carbamic acid, (7-chloro-3-β-D-ribo-furanosyl-3H-1,2,3-triazolo[4,5-b]-pyridin-5-yl)-, ethyl ester	pH 1	265(4.01),274(4.04), 302(4.32),310s(4.23)	4-1365-76
	pH 11	266(4.00),274(3.99), 302(4.27),310s(4.23)	4-1365-76
$C_{13}H_{16}CuN_6OS_2$ Copper, [[2,2'-[1-(4-methoxyphenyl)-1,2-ethanediylidene]bis[N-methylhydrazine-	EtOH	307(4.40),498(3.73)	87-0131-76

Compound	Solvent	$\lambda_{max}(\log \epsilon)$	Ref.
carbothioamidato]](2-)-$N^2,N^{2'},S,S'$]-, (SP-4-4)- (cont.)			87-0131-76
$C_{13}H_{16}CuN_6S_2$ Copper, [[2,2'-[1-(4-methylphenyl)-1,2-ethanediylidene]bis[N-methylhydrazine-carbothioamidato]](2-)-$N^2,N^{2'},S,S'$]-, (SP-4-4)-	EtOH	305(4.34),500(3.70)	87-0131-76
$C_{13}H_{16}FN_5O_4$ Adenosine, 4'-C-fluoro-2',3'-O-(1-methylethylidene)-	MeOH	258(4.12)	35-3346-76
9H-Purin-6-amine, 9-(4-C-fluoro-2,3-O-(1-methylethylidene)-α-L-lyxofuranosyl]-	MeOH	258(4.16)	35-3346-76
$C_{13}H_{16}F_4N_2O_7$ Uridine, 5-[(2,2,3,3-tetrafluoropropoxy)methyl]-	pH 13 EtOH	266(3.82) 266(3.97)	104-0643-76 104-0643-76
$C_{13}H_{16}NO$ Pyridinium, 1,2,3,6-tetrahydro-1,1-dimethyl-3-oxo-5-phenyl-, bromide	EtOH	239(3.9),334(4.3)	39-2329-76C
$C_{13}H_{16}N_2$ Pyrimido[1,2-a]indole, 6,7,8,9-tetrahydro-2,4-dimethyl-	EtOH	235s(4.2),255(4.43), 285s(3.37),298(3.44), 310(3.25),380(3.27)	103-1379-76
$C_{13}H_{16}N_2O$ 2H-Pyrrol-2-one, 3-amino-1,5-dihydro-5-phenyl-4-propyl-	EtOH	209(4.04),230(4.06), 251(4.06)	40-1100-76
Spiro[3H-indole-3,2'-piperidin]-2(1H)-one, 1'-methyl-	MeOH	248(3.85),272s(3.20)	87-0892-76
Spiro[3H-indole-3,2'-pyrrolidin]-2(1H)-one, 1'-ethyl-	MeOH	249(3.81),272s(3.16)	87-0892-76
Spiro[3H-indole-3,3'-pyrrolidin]-2(1H)-one, 1'-ethyl-	MeOH	252(3.71),274s(3.02)	87-0892-76
$C_{13}H_{16}N_2O_2$ 2H-1-Benzopyran-2-one, 3-amino-4-(diethylamino)-	EtOH	250(3.91),265s(3.88), 330(4.18)	103-0268-76
Spiro[3H-indole-3,6'-[6H-1,3]oxazin]-2(1H)-one, 2',3',4',5'-tetrahydro-2',3'-dimethyl- (donaxarine)	n.s.g.	246(3.60),297(3.11)	105-0499-76
$C_{13}H_{16}N_2O_2S$ Ethanone, 1-[3-amino-4-ethyl-5-methyl-6-(methylthio)furo[2,3-b]pyridin-2-yl]-	EtOH	257(3.96),272(3.86), 341(4.37),360s(4.27)	48-0313-76
3-Pyridinecarbonitrile, 4-ethyl-5-methyl-6-(methylthio)-2-(2-oxopropoxy)-	EtOH	229(3.97),264(4.03), 320(4.30)	48-0313-76
$C_{13}H_{16}N_2O_4$ 2-Butenedioic acid, 2-[1-(phenylmethyl)-hydrazino]-, dimethyl ester, (E)-	MeOH	282(4.40)	24-0261-76
D-Erythrose, 2,4-O-ethylidene-, benzoyl-hydrazone	EtOH	211s(3.94),233(4.18)	136-0185-76D
3,5-Heptadienoic acid, 5-cyano-2-(1-hydroxyethylidene)-7-(methoxyimino)-, ethyl ester, (E,E,Z,E)-	MeOH CHCl$_3$	422(--) 348(4.39)	5-1799-76 5-1799-76

Compound	Solvent	$\lambda_{max}(\log \epsilon)$	Ref.
3,5-Heptadienoic acid, 6-cyano-2-(1-hy-droxyethylidene)-7-(methoxyimino)-, ethyl ester, (E,E,Z,E)-	MeOH CHCl$_3$	440(4.38) 347(4.42)	5-1799-76 5-1799-76
Pyrrolo[3,2-b][1,4]benzoxazine-2-carbox-ylic acid, 1,2,3,3a,9,9a-hexahydro-3a-methoxy-, methyl ester	EtOH	239(3.64),289(3.27)	142-0053-76S
$C_{13}H_{16}N_2O_5$			
4,6-Ethano-3H-oxireno[f]indazole-3a,6a-dicarboxylic acid, 4,4a,5a,6-tetra-hydro-, dimethyl ester	MeOH	323(2.28)	24-2823-76
3H-Oxonino[5,6-c]pyrazole-3a,10a(4,10H)-dicarboxylic acid, dimethyl ester	MeOH	322(2.30)	24-2823-76
$C_{13}H_{16}N_2O_8$			
2-Propenoic acid, 3-(1,2,3,4-tetrahydro-2,4-dioxo-1-β-D-ribofuranosyl-5-pyrim-idinyl)-, methyl ester, (E)-	H$_2$O	300(4.30)	35-1587-76
Uridine, 3',5'-diacetate	EtOH	260(3.95)	44-2995-76
$C_{13}H_{16}N_2S$			
2(1H)-Pyrimidinethione, 3,4-dihydro-4,4,6-trimethyl-1-phenyl-	EtOH	275(4.10)	42-1223-76
Pyrrolo[2,3-b]indole-5-thiol, 2,3,3a,9-tetrahydro-1,3a,8-trimethyl-	MeOH	270(3.88),312(3.77)	83-0631-76
4H-1,3-Thiazin-2-amine, 4,4,6-trimethyl-N-phenyl-	EtOH	304(4.34)	42-1223-76
$C_{13}H_{16}N_3O$			
Isoxazolium, 2-ethyl-3,5-dimethyl-4-(phenylazo)-	HOAc	230(4.03),317(4.16), 429(2.81)	48-0359-76
$C_{13}H_{16}N_3OS$			
Isoxazolium, 2-ethyl-3-methyl-5-(meth-ylthio)-4-(phenylazo)-	HOAc	292(4.18),351(4.09), 363s(4.05),417s(3.06)	48-0359-76
$C_{13}H_{16}N_4O$			
Benzenamine, 4-[(3,5-dimethyl-4-isoxa-zolyl)azo]-N,N-dimethyl-	EtOH	263(4.06),302s(3.60), 317(3.61),329(3.58), 397(4.42),420s(4.39)	48-0359-76
2-Quinoxalinecarboxamide, N,6,7-trimeth-yl-3-(methylamino)-	6M HCl pH 1.4 pH 6-13	348(3.97),400(3.99) 343(3.94),397(3.99) 315(3.76),411(3.85)	5-1276-76 5-1276-76 5-1276-76
$C_{13}H_{16}N_4OS$			
Benzenamine, 4-[(4,5-dimethyl-2-thiazo-lyl)azo]-N,N-dimethyl-, N-oxide	n.s.g.	550(4.67)	64-0981-76B
Carbamohydrazonothioic acid, N'-(1,3-di-hydro-1-methyl-3-oxo-2H-indol-2-yli-dene)-N,N-dimethyl-, methyl ester	EtOH	236(4.16),279(4.14), 360(4.05),507(3.96)	103-0076-76
$C_{13}H_{16}N_4O_4$			
Propanedioic acid, (1-methyl-1H-pyra-zolo[3,4-d]pyrimidin-4-yl)-, diethyl ester	EtOH	241(4.03),329(4.43), 336(4.43)	95-1352-76
1,2,4-Triazin-5(2H)-one, 3-amino-6-[(3,4,5-trimethoxyphenyl)methyl]-	MeOH	266(4.58)	39-2521-76C
$C_{13}H_{16}N_4O_5$			
6H-Purin-6-one, 1,7-dihydro-1-[2,3-O-(1-methylethylidene)-α-D-ribofuranosyl)-	pH 1	248(3.98)	88-0045-76

Compound	Solvent	$\lambda_{max}(\log \epsilon)$	Ref.
6H-Purin-6-one, 1,7-dihydro-1-[2,3-O-(1-methylethylidene)-β-D-ribofuranosyl)-	pH 1	249(3.99)	88-0045-76
$C_{13}H_{16}O$			
Bicyclo[6.4.1]trideca-8,10,12-trien-13-one	hexane	202(4.01),238(4.00), 284(3.62)	78-2381-76
Ethanone, 1-(1,2,3,4-tetrahydro-2-methyl-2-naphthalenyl)-	MeOH	205(3.95),263(3.09), 274(3.03)	19-0175-76
1-Hexen-3-one, 5-methyl-1-phenyl-, (E)-	EtOH	283(4.34)	2-0472-76
1(2H)-Naphthalenone, 3,4-dihydro-3,7,8-trimethyl-	MeOH	247(3.91),298(3.31)	44-0066-76
$C_{13}H_{16}O_2$			
2-Penten-1-one, 2-hydroxy-4,4-dimethyl-1-phenyl-	CCl_4	282(3.85)	35-8125-76
	MeCN	252(3.85),279(3.88)	35-8125-76
Tricyclo[4.4.0.0^{5,9}]deca-3,7-dien-2-one, 5-hydroxy-2,8,9-trimethyl-	MeOH	244(3.79),325s(1.71)	35-7040-76
Tricyclo[4.4.0.0^{5,9}]deca-3,7-dien-2-one, 5-hydroxy-3,8,9-trimethyl-	MeOH	244(3.79),325(1.76)	35-7040-76
Tricyclo[4.4.0.0^{3,7}]dec-8-ene-2,5-dione, 3,8,9-trimethyl-	MeOH	286(1.89)	35-7040-76
Tricyclo[4.4.0.0^{3,7}]dec-8-ene-2,5-dione, 4,8,9-trimethyl-	MeOH	283(1.89)	35-7040-76
Tricyclo[4.4.0.0^{3,9}]dec-7-ene-2,5-dione, 3,8,9-trimethyl-	MeOH	294(2.68),307s(2.63)	35-7040-76
$C_{13}H_{16}O_3$			
2H-Indeno[1,2-b]furan-2,5(3H)-dione, 3a,4,6,7,8,8b-hexahydro-8,8-dimethyl-	EtOH	245(4.016)	39-0410-76C
Pterosin C	EtOH	220(4.47),261(4.17), 295(3.30)	2-0817-76
$C_{13}H_{16}O_4$			
Benzoic acid, 2-acetoxy-5-(1,1-dimethylethyl)-	MeOH	212(4.14),230s(3.79)	39-0465-76C
1H-Indene-2-acetic acid, 2,3,4,5,6,7-hexahydro-7,7-dimethyl-1,4-dioxo-	EtOH	260(4.12)	35-3661-76
	EtOH	260(4.08)	39-0410-76C
1H-Indene-2-carboxylic acid, 2,3,4,5,6-7-hexahydro-7,7-dimethyl-1,4-dioxo-, methyl ester	EtOH	222(3.96),256(3.88), 325(3.81)	44-0563-76
$C_{13}H_{16}O_4S$			
2-Propenoic acid, 3-(3,4-dimethoxyphenyl)-2-(ethylthio)-	EtOH	310(4.09)	2-0547-76
$C_{13}H_{16}O_5$			
Benzenemethanol, 4-acetoxy-5-methoxy-2-methyl-, acetate	EtOH	278(3.11)	5-1435-76
5-Oxabicyclo[7.1.0]deca-3,6-diene-1,9-dicarboxylic acid, dimethyl ester	EtOH	275s(2.51)	24-2823-76
$C_{13}H_{16}O_5S$			
4H-1-Benzopyran-4-one, 2,3-dihydro-2,2,5-trimethyl-7-[(methylsulfonyl)-oxy]-	EtOH	261(3.96),323(3.60)	80-1543-76
$C_{13}H_{16}O_6$			
1,4-Benzenediol, 2,3-dimethoxy-5-methyl-, diacetate	EtOH	269(2.74)	65-1350-76
Cyclopenta[c]pyran-4-carboxylic acid,	MeOH	240(3.97)	94-2644-76

Compound	Solvent	$\lambda_{max}(\log \epsilon)$	Ref.
7-(acetoxymethyl)-1,4a,5,7a-tetrahy-dro-1-hydroxy-, methyl ester, [1R-(1α,4aα,7aα)]- (10-acetylgenipin)			94-2644-76
2H-Pyran-5-carboxylic acid, 2-acetoxy-4α-(1-formyl-1-propenyl)-3,4-dihydro-, methyl ester, (E)-	MeOH	230(4.05)	24-3640-76
$C_{13}H_{16}O_8S_2$			
4H-1-Benzopyran-4-one, 2,3-dihydro-2,2-dimethyl-5,7-bis[(methylsulfonyl)oxy]-	EtOH	258(3.97),323(3.70)	80-1543-76
$C_{13}H_{17}BF_2O_4$			
1-Oxa-3-oxonia-2-boratanaphthalene, 2,2-difluoro-1,2-dihydro-5,7-di-methoxy-8-methyl-4-(1-methylethyl)-	MeCN	215(4.1),320(4.4), 385(3.5)	5-1514-76
$C_{13}H_{17}Br$			
Bicyclo[6.4.1]trideca-8,10,12-triene, 13-bromo-, syn	C_6H_{12}	214(4.23),271(3.62)	78-2381-76
$C_{13}H_{17}ClN_2O_4$			
Benzoic acid, 4-chloro-3-nitro-, 2-(diethylamino)ethyl ester, hydrochloride	EtOH	228(4.32),294(3.04), 330s(2.5)	78-0057-76
$C_{13}H_{17}ClN_4O_2$			
Cyanamide, butyl[2-[(4-chloro-2-nitro-phenyl)amino]ethyl]-	EtOH	238(4.44),270s(3.80), 430(3.77)	78-0057-76
$C_{13}H_{17}FN_6O_3$			
9H-Purin-6-amine, 9-[5-amino-5-deoxy-4-C-fluoro-2,3-O-(1-methylethylid-ene)-α-L-lyxofuranosyl]-	MeOH	257(4.17)	35-3346-76
$C_{13}H_{17}FN_6O_6S$			
Adenosine, 4'-C-fluoro-2',3'-O-(1-meth-ylethylidene)-, 5'-sulfamate	MeOH	259(4.13)	35-3346-76
9H-Purin-6-amine, 9-[5-O-(aminosulfon-yl)-4-C-fluoro-2,3-O-(1-methylethyl-idene)-α-L-lyxofuranosyl]-	MeOH	259(4.19)	35-3346-76
$C_{13}H_{17}N$			
Azete, 2,3-dihydro-2,2,3,3-tetramethyl-4-phenyl-	MeOH	245(4.20),287s(3.87)	77-0729-76
1H-Indole, 2-(2,2-dimethylpropyl)-	MeOH	220(4.62),280(4.00), 288(3.93)	23-1020-76
2H-Pyrrole, 2,2,3-trimethyl-5-phenyl-	EtOH	244(4.14),277s(3.5), 285(3.3)	27-0453-76
$C_{13}H_{17}NO$			
Bicyclo[3.2.0]hepta-2,4,7-trien-6-one, 7-[bis(1-methylethyl)amino]-	EtOH	252(4.52),304(4.15), 350(3.08)	33-0021-76
Cyclohepta[b]cyclopenta[d]pyridin-4(1H)-one, 2,3,5,6,7,8,9,10-octahydro-	n.s.g.	240(3.74),308(3.91)	104-0681-76
Ethanone, 1-phenyl-2-piperidino-	n.s.g.	243(3.99)	36-0134-76
3H-Indol-3-ol, 3-(1,1-dimethylethyl)-2-methyl-	MeOH	220(4.36),255(3.51), 285(3.45)	23-1015-76
6(5H)-Phenanthridinone, 1,2,3,4,7,8,9-10-octahydro-	n.s.g.	235(3.68),308(3.96)	104-2200-76

Compound	Solvent	$\lambda_{max}(\log \epsilon)$	Ref.
$C_{13}H_{17}NO_4$			
Benzenemethanol, 4-methyl-α-(1-methyl-1-nitroethyl)-, acetate	MeOH	256(2.86),263(2.75), 269(2.63),273(2.55)	12-0327-76
Valine, N-(4-methoxybenzoyl)-, (±)-	EtOH	254(4.27)	12-0339-76
$C_{13}H_{17}NO_5$			
2,6-Piperidinedione, 4-[(3,4-dihydro-6-methoxy-5-methyl-4-oxo-2H-pyran-2-yl)methyl]-	EtOH	276(4.19)	33-2393-76
2,6-Piperidinedione, 4-[(3,6-dihydro-4-methoxy-5-methyl-6-oxo-2H-pyran-2-yl)methyl]-	EtOH	252(4.13)	33-2393-76
$C_{13}H_{17}NO_5S$			
4H-1-Benzopyran-4-one, 2,3-dihydro-7-methoxy-2,2-dimethyl-, O-(methyl-sulfonyl)oxime	EtOH	276(4.16),314(4.01)	80-1543-76
$C_{13}H_{17}NO_6S$			
1H-Azonine-5,6-dicarboxylic acid, 4,7-dihydro-1-(methylsulfonyl)-, dimethyl ester	EtOH	230(3.72)(end abs.)	24-3505-76
3-Azatetracyclo[6.1.0.0^{2,4}.0^{5,7}]nonane-7,8-dicarboxylic acid, dimethyl ester, (1α,2α,4α,5α,7α,8α)-	EtOH	230(2.60)(end abs.)	24-3505-76
3-Azatricyclo[5.1.0.0^{2,4}]oct-5-ene-6,7-dicarboxylic acid, 5-methyl-3-(methylsulfonyl)-, dimethyl ester, (1α,2α,4α,7α)-	EtOH	224(3.89)	24-3505-76
$C_{13}H_{17}NO_7$			
Furo[2,3-b]pyridine-3a,4(6H)-dicarboxylic acid, 2,3,7,7a-tetrahydro-7a-methoxy-6-oxo-, 3a-ethyl 4-methyl ester	MeOH	220(4.17),250s(--)	24-1269-76
$C_{13}H_{17}NO_7S$			
Benzoic acid, 4-[2-methyl-1-(methylsul-fonoxy)-2-nitropropyl]-, methyl ester	C_6H_{12}	236(4.25),276(3.06), 284(3.01)	12-2631-76
$C_{13}H_{17}N_2$			
Cyclopropenylium, bis(dimethylamino)-phenyl-, chloride	MeOH	285(4.21),292s(4.19)	88-2405-76
$C_{13}H_{17}N_3$			
1H-Pyrazol-1-amine, N,N,3,5-tetramethyl-4-phenyl-	MeOH	241(4.05)	94-3001-76
1H-Pyrazol-1-amine, N,N,4,5-tetramethyl-3-phenyl-	MeOH	250(4.07)	94-3001-76
$C_{13}H_{17}N_3O$			
1H-Imidazole-5-methanol, α-[2-(2-amino-phenyl)ethyl]-1-methyl-	MeOH	233s(3.83),286(3.28)	78-1085-76
$C_{13}H_{17}N_3O_3$			
Cycloheptanone, 2-hydroxy-, (4-nitro-phenyl)hydrazone	MeOH	248(4.38),384(4.70)	104-2286-76
1H-Pyrazolo[4,3-b]pyridine-6-carboxylic acid, 7-ethoxy-1,3-dimethyl-, ethyl ester	EtOH	234(4.54),276(3.70), 316(3.68)	39-0507-76C

Compound	Solvent	$\lambda_{max}(\log \epsilon)$	Ref.
2H-Pyrazolo[4,3-b]pyridine-6-carboxylic acid, 7-ethoxy-2,3-dimethyl-, ethyl ester	EtOH	237(4.61),283(3.66), 318(3.83)	39-0507-76C
1H-Pyrazolo[4,3-b]pyridine-6-carboxylic acid, 4-ethyl-4,7-dihydro-1,3-dimethyl-7-oxo-, ethyl ester	EtOH	227(4.38),308(4.2)	39-0507-76C
2H-Pyrazolo[4,3-b]pyridine-6-carboxylic acid, 4-ethyl-4,7-dihydro-2,3-dimethyl-7-oxo-, ethyl ester	EtOH	233(4.37),318(4.20)	39-0507-76C
$C_{13}H_{17}N_3O_4$			
Benzoic acid, 3-nitro-4-[[2-(1-pyrrolidinyl)ethyl]amino]-, hydrochloride	EtOH	261(4.30),285(4.19), 402(3.64)	78-0057-76
$C_{13}H_{17}N_3O_5$			
Butanoic acid, 2,3,3a,9a-tetrahydro-2-(hydroxymethyl)-6-imino-6H-furo[2',3':4,5]oxazolo[3,2-a]pyrimidin-3-yl ester, hydrochloride	MeOH-acid	230(3.97),264(4.03)	87-0654-76
Propanoic acid, 2-methyl-, 2,3,3a,9a-tetrahydro-2-(hydroxymethyl)-6-imino-6H-furo[2',3':4,5]oxazolo[3,2-a]pyrimidin-3-yl ester, hydrochloride	MeOH-acid	230(3.98),263(4.04)	87-0654-76
$C_{13}H_{17}N_3O_7$			
2(1H)-Pyrimidinone, 4-amino-1-(3,5-di-O-acetyl-β-D-arabinofuranosyl)-, hydrochloride	MeOH-acid	211(4.01),282(4.16)	87-0667-76
$C_{13}H_{17}N_3S$			
Piperidine, 1-[(3-cyano-4,6-dimethyl-2-pyridinyl)thio]-	EtOH	224(4.43),270(4.00), 309(3.66),414(1.85)	48-0779-76
$C_{13}H_{17}N_5O_3$			
Hydrazinecarboxamide, 2-[3-[(2,4-dimethylphenyl)amino]-2-(hydroxyimino)-1-methyl-3-oxopropylidene]-	EtOH	265(4.31)	42-0454-76
1,2,4-Triazine-3,5-diamine, 6-[(3,4,5-trimethoxyphenyl)methyl]-	MeOH	224s(4.34),300(3.74)	39-2521-76C
$C_{13}H_{17}N_5O_4$			
Hydrazinecarboxamide, 2-[3-[(4-ethoxyphenyl)amino]-2-(hydroxyimino)-1-methyl-3-oxopropylidene]-	EtOH	260(4.43)	42-0454-76
6H-Purin-6-imine, 1,7-dihydro-1-[2,3-O-(1-methylethylidene)-α-D-ribofuranosyl]-	pH 1	259(4.01)	88-0045-76
β-	pH 1	261(3.98)	88-0045-76
$C_{13}H_{17}N_5O_5$			
Hydrazinecarboxamide, 2-[3-(2,4-dimethoxyphenyl)amino]-2-(hydroxyimino)-1-methyl-3-oxopropylidene]-	EtOH	265(4.40)	42-0454-76
$C_{13}H_{17}N_5O_6$			
4,7(1H,8H)-Pteridinedione, 2-(dimethylamino)-8-β-D-ribofuranosyl-	pH 5.0	221(4.43),298(4.03), 359(4.18)	24-3228-76
	pH 11.0	230(4.07),267(4.02), 380(4.20)	24-3228-76

Compound	Solvent	$\lambda_{max}(\log \epsilon)$	Ref.
$C_{13}H_{17}N_5O_6S$ 3H-Azirino[5,6]benzo[2,1-c:3,4-c']di-pyrazole-7a,7b-dicarboxylic acid, 3a,3b,4,4a,4b,5-hexahydro-4-(methyl-sulfonyl)-, dimethyl ester, (3aα,3bα,4aα,4bα,7aα,7bα)-	EtOH	322(2.46)	24-3505-76
$C_{13}H_{18}$ Bicyclo[6.4.1]trideca-8,10,12-triene	hexane	216(4.11),272(3.54)	78-2381-76
1-Butene, 2-ethyl-3-methyl-1-phenyl-	EtOH	248(4.02)	78-1839-76
1-Butene, 2-ethyl-3-methyl-3-phenyl-	EtOH	250(2.15),254(2.30), 260(2.34),262(2.26), 270(2.15)	78-1839-76
1-Butene, 2-(1-methylethyl)-3-phenyl-	EtOH	250(2.85),255(2.83), 260(2.78),267(2.62), 270(2.53)	78-1839-76
Cyclohexene, 1-(2-cyclohexen-1-ylidene-methyl)-, (E)-	hexane	265(4.24),273(4.25)	35-4867-76
(Z)-	hexane	265(4.35),273(4.38)	35-4867-76
1H-Fluorene, 2,3,4,4aα,4bα,5,6,7-octa-hydro-	hexane	245(4.19)	35-4867-76
1H-Fluorene, 2,3,4,4aα,4bβ,5,6,7-octa-hydro-	hexane	248(4.21)	35-4867-76
1H-Fluorene, 2,3,4,5,6,7,8,9-octahydro-	hexane	266(3.58)	35-4867-76
3H-Fluorene, 4,4aα,4bβ,5,6,7,8,8a-octa-hydro-	hexane	243(4.24)	35-4867-76
4,7-Methano-2H-indene, 4,5,6,7-tetra-hydro-4,8,8-trimethyl-, (4R)-	hexane	241(3.72)	78-0309-76
2-Pentene, 3-ethyl-2-phenyl-	EtOH	232(3.70)	78-1839-76
2-Pentene, 3-ethyl-4-phenyl-	EtOH	270(2.66),275(2.30)	78-1839-76
$C_{13}H_{18}BrN_3O_7S$ Cytidine, 5-bromo-2',3'-O-(1-methyleth-ylidene)-, 5'-methanesulfonate	MeOH	284(3.83)	44-1100-76
$C_{13}H_{18}ClFO_5$ α-D-ribo-Hexofuranose, 3-(chlorofluoro-methylene)-3-deoxy-1,2:5,6-bis-O-(1-methylethylidene)-, cis	EtOH	234(2.51)	136-0009-76A
$C_{13}H_{18}ClN_5O_4S$ 9H-Purin-6-amine, 2-chloro-9-(2-S-ethyl-2-thio-α-D-mannofuranosyl)-	EtOH	265(4.13)	136-0041-76C
$C_{13}H_{18}Cl_2N_2O$ Benzamide, 3,5-dichloro-N-[2-(diethyl-amino)ethyl]-, hydrochloride	H_2O	208(4.68),238s(3.85), 285(2.98)	133-0314-76
$C_{13}H_{18}F_2O_5$ α-D-ribo-Hexofuranose, 3-deoxy-3-(di-fluoromethylene)-1,2:5,6-bis-O-(1-methylethylidene)-	EtOH	226(1.68)	136-0009-76A
α-D-xylo-Hexofuranose, 3-deoxy-3-(di-fluoromethylene)-1,2:5,6-bis-O-(1-methylethylidene)-	EtOH	212(2.10),225s(--), 255(2.06)	136-0009-76A
$C_{13}H_{18}N_2$ 1H-Indole-2-carbonitrile, 4,5,6,7-tetra-hydro-1-(2-methylpropyl)-	EtOH	269(4.17)	20-0819-76

Compound	Solvent	λ_{max}(log ϵ)	Ref.
$C_{13}H_{18}N_2O$			
1-Piperidinamine, N-[(2-methoxyphenyl)-	benzene	300(4.09),321(4.24)	104-0625-76
methylene]-, (E)-	DMF	285(4.10),320(4.21)	104-9625-76
2-Propenal, 3,3-bis(dimethylamino)-	H_2O	221(3.94),272(4.14),	88-2405-76
2-phenyl-		320(4.00)	
Spiro[4.5]dec-1-ene-2-carbonitrile,	EtOH	264(4.08)	44-0010-76
1-amino-3,3-dimethyl-6-oxo-			
$C_{13}H_{18}N_2O_2$			
Benzoic acid, 4-[[1-(dimethylamino)eth-	neutral	290(4.17)	39-0211-76B
ylidene]amino]-, ethyl ester	cation	270(4.16)	39-0211-76B
$C_{13}H_{18}N_2O_3$			
4H-Pyrido[1,2-a]pyrimidine-3-acetic	pH 7.35	229(3.78),277(3.84)	145-0478-76
acid, 6,7,8,9-tetrahydro-6-methyl-			
4-oxo-, ethyl ester			
$C_{13}H_{18}N_2O_5$			
Benzimidazole, N-(1-deoxy-D-mannitol-	H_2O	248(3.76),254(3.76),	136-0049-76G
1-yl)-, hydrochloride		263(3.72),269(3.78),	
		276(3.74)	
	base	247(3.81),253(3.81),	136-0049-76G
		265(3.66),273(3.66),	
		280(3.57)	
$C_{13}H_{18}N_2O_7$			
Uridine, 2',3'-di-O-methyl-, 5'-acetate	H_2O	264(3.99)	35-7381-76
$C_{13}H_{18}N_2S$			
Benzo[g]quinoxaline, 5,5a,6,7,8,9,9a,10-	MeCN	250(3.81),304s(3.54),	142-0299-76S
octahydro-2-(methylthio)-, (5aS-trans)-		327(3.76)	
$C_{13}H_{18}N_4O_2S$			
Benzenesulfonamide, N-(1-butyl-1H-1,2,3-	EtOH	229(4.21),250(4.05)	32-0001-76
triazol-5-yl)-4-methyl-			
$C_{13}H_{18}N_4O_5$			
1H-Purine-2,6-dione, 7-(4,6-dideoxy-	MeOH	275(3.86)	44-3827-76
β-L-xylo-hexopyranosyl)-3,7-dihydro-			
1,3-dimethyl-			
$C_{13}H_{18}N_4O_5S$			
1H-Imidazole-4-carbothioamide, 5-amino-	pH 1	278(3.98),324(4.22)	94-2089-76
1-(2,5-di-O-acetyl-3-deoxy-β-D-ery-	pH 7	271(3.99),325(4.19)	94-2089-76
thro-pentofuranosyl)-	pH 13	270(3.95),326(4.18)	94-2089-76
$C_{13}H_{18}N_4O_6$			
1H-Imidazole-4-carboxamide, 5-amino-1-	pH 1	268(4.02)	94-2089-76
(2,5-di-O-acetyl-3-deoxy-β-D-erythro-	pH 7	266(4.08)	94-2089-76
pentofuranosyl)-	pH 13	267(4.08)	94-2089-76
$C_{13}H_{18}N_5O_6PS$			
Adenosine, 8-(propylthio)-, cyclic	pH 2.0	284(4.29)	69-1408-76
3',5'-(hydrogen phosphate)	pH 13.0	282(--)	69-1408-76
$C_{13}H_{18}N_6OS_2$			
Hydrazinecarbothioamide, 2,2'-[1-(4-	MeOH	338(4.48)	87-0131-76
methoxyphenyl)-1,2-ethanediylidene]-			
bis[N-methyl-			

Compound	Solvent	λ_{max} (log ϵ)	Ref.
$C_{13}H_{18}N_6O_2$			
2(1H)-Pyrimidinone, 4-amino-1-[3-(4-amino-2-oxo-1(2H)-pyrimidinyl)propyl]-5-ethyl-	pH 1 pH 7	285.0(--) 276.0(4.19)	56-1805-76 56-1805-76
2(1H)-Pyrimidinone, 1,1'-(1,3-propanediyl)bis[4-amino-5-methyl-	pH 1.0 pH 7.0	289.5(--) 278.0(4.17)	56-1805-76 56-1805-76
$C_{13}H_{18}N_6S_2$			
Hydrazinecarbothioamide, 2,2'-[1-(4-methylphenyl)-1,2-ethanediylidene]-bis[N-methyl-	MeOH	338(4.42)	87-0131-76
$C_{13}H_{18}O$			
Benzenemethanol, α-(2,2-dimethylcyclopropyl)-α-methyl-	EtOH	240-266f(1.96-2.41)	44-3067-76
1-Naphthalenol, 5,6,7,8-tetrahydro-2,3,4-trimethyl-	EtOH	214(4.11),280s(3.09), 287(3.12)	70-0917-76
1(5H)-Naphthalenone, 6,7,8,8a-tetrahydro-2,4,8a-trimethyl-	n.s.g.	335(3.73)	39-0634-76C
1-Pentanone, 3,3-dimethyl-1-phenyl-	EtOH	241(4.01),277(3.08)	44-3067-76
$C_{13}H_{18}OS_2$			
[1,2]Dithiolo[1,5-b][1,2]benzoxathiole-9-S^{IV}, 2-(1,1-dimethylethyl)-4,5,6,7-tetrahydro-	C_6H_{12}	227(4.29),420(4.13), 434s(4.11)	39-0880-76C
$C_{13}H_{18}O_2$			
3H-2-Benzopyran-3-one, 1,5,6,7,8,8a-hexahydro-5,5,8a-trimethyl-1-methylene-	n.s.g.	224(4.05)	39-0532-76C
Bicyclo[4.1.0]hepta-2,4-diene-7-carboxylic acid, 1,2,3,7-tetramethyl-, methyl ester, (1α,6α,7α)-	EtOH	242(3.58),277(3.50)	33-1763-76
Bicyclo[4.1.0]hepta-2,4-diene-7-carboxylic acid, 2,3,5,7-tetramethyl-, methyl ester, (1α,6α,7α)-	EtOH	248(3.79),275(3.70)	33-1763-76
2-Cyclohexen-1-ol, 3-ethynyl-2,4,4-trimethyl-	EtOH	231(4.06)	23-2310-76
Dispiro[4.1.5.1]tridecane-6,13-dione	CH_2Cl_2	309(1.54),345(1.30)	18-0741-76
1,3(2H,5H)-Naphthalenedione, 6,7,8,8a-tetrahydro-5,5,8a-trimethyl-	EtOH KOH	243(4.00),291(3.71) 329(--)	39-0532-76C 39-0532-76C
2(4aH)-Naphthalenone, 5,6,7,8-tetrahydro-3-methoxy-4a,8-dimethyl-, trans	EtOH	251(4.05),285(3.52)	44-1539-76
3-Penten-1-ol, 4-(2-methoxy-4-methylphenyl)-	MeOH	282(4.06)	33-0695-76
Spiro[4.5]deca-3,6-dien-2-one, 3-methoxy-6,10-dimethyl-, cis-(+)-	EtOH	256(3.90)	44-1539-76
$C_{13}H_{18}O_3$			
Acetic acid, (4,7,7-trimethyl-3-oxobicyclo[2.2.1]hept-2-ylidene)-, methyl ester, (1S)-	hexane	243(4.13)	78-0309-76
2-Cyclopenten-1-one, 4-acetoxy-3-methyl-2-(2-pentenyl)-, (E)-	EtOH	227(4.04)	39-0249-76C
(Z)-	EtOH	227(4.04)	39-0249-76C
$C_{13}H_{18}O_3S$			
Cyclohexanol, 1-[5-(1,3-dioxolan-2-yl)-2-thienyl]-	THF	266(4.06),291(4.09)	44-1320-76
Cyclohexanol, 1-[5-(1,3-dioxolan-2-yl)-3-thienyl]-	dioxan	232(3.87),293(2.07)	44-1320-76

Compound	Solvent	$\lambda_{max}(\log \epsilon)$	Ref.
$C_{13}H_{18}O_4$			
1,3-Cyclohexadiene-1-carboxylic acid, 6-(1-acetoxy-1-methylethyl)-, methyl ester, (S)-	EtOH	290(3.54)	39-0484-76C
1-Cyclohexene-1-carboxylic acid, 3-acetoxy-6-(1-methylethenyl)-, methyl ester	EtOH	215(3.96)	39-0484-76C
Loliolide acetate	EtOH	214(4.28)	36-1549-76
Myrtenic acid, 4α-acetoxy-, methyl ester	EtOH	234(3.80)	39-0484-76C
Phenol, 2,6-dimethoxy-4-propyl-, acetate	MeOH	227(3.79),275(2.90)	18-1940-76
1-Propanone, 1-(2-hydroxy-4,6-dimethoxy-3-methylphenyl)-2-methyl- (baeckeol)	MeCN	211(4.2),290(4.3), 332s(--)	5-1514-76
$C_{13}H_{18}O_6$			
D-Glucitol, 2,4(S)-O-(phenylmethylene)-	MeCN-DMSO	257(2.31)	12-1859-76
$C_{13}H_{18}O_7$			
β-D-Glucopyranoside, 4-(hydroxymethyl)-phenyl	MeOH	271(3.04),278(2.96)	102-1986-76
$C_{13}H_{18}S$			
1-Pentanethione, 2,2-dimethyl-1-phenyl-	C_6H_{12}	230(3.99),298(3.59), 567(2.04)	35-6218-76
$C_{13}H_{19}ClN_4$			
Benzenamine, 2-(3-butyl-2-imino-1-imidazolidinyl)-5-chloro-, hydrobromide	EtOH	210(4.53),246(4.02), 304(3.64)	78-0057-76
dihydrobromide	EtOH	210(4.55),245(4.04), 305(3.64)	78-0057-76
1H-Benzimidazole-1-ethanamine, 2-amino-N-butyl-5-chloro-	EtOH	218(4.66),253(3.80), 295(3.99)	78-0839-76
	EtOH-HCl	210(4.69),229s(4.00), 286(3.92)	78-0839-76
dihydrobromide	EtOH	213(4.62),254(3.55), 292(3.88)	78-0839-76
$C_{13}H_{19}FO_5$			
α-D-ribo-Hexofuranose, 3-deoxy-3-(fluoromethylene)-1,2:5,6-bis-O-(1-methylethylidene)-, cis	EtOH	233(1.65)	136-0009-76A
trans	EtOH	228.5(2.18)	136-0009-76A
α-D-xylo-Hexofuranose, 3-deoxy-3-(fluoromethylene)-1,2:5,6-bis-O-(1-methylethylidene)-, cis	EtOH	215(2.27),227s(--)	136-0009-76A
trans	EtOH	232s(--),257(2.65)	136-0009-76A
$C_{13}H_{19}NO$			
6(5H)-Phenanthridinone, 1,2,3,4,4a,7,8-9,10,10b-decahydro-	MeOH	248(3.48)	95-0919-76
$C_{13}H_{19}NO_2$			
1,3-Cyclopentanedione, 2-(4,4-dimethyl-2-pyrrrolidinylidene)-4,4-dimethyl-	EtOH	238(4.08),295(4.23)	39-1944-76C
3,5-Heptadienoic acid, 3-cyano-6-methyl-, 1,1-dimethylethyl ester	EtOH	271(4.45)	44-2950-76
$C_{13}H_{19}NO_2S$			
Acetic acid, cyano[3-(ethylthio)cyclohexylidene]-, ethyl ester	C_6H_{12}	234(4.14),278(3.33)	78-2827-76

Compound	Solvent	$\lambda_{max}(\log \epsilon)$	Ref.
$C_{13}H_{19}NO_3$ 1H-Indole-3-acetic acid, 2,4,5,6,7,7a- hexahydro-1-(1-methylethyl)-2-oxo-	EtOH	214.5(4.18)	22-0776-76
$C_{13}H_{19}NO_4$ Butanoic acid, 2-(4,4-dimethyl-6-oxo- 2-piperidinylidene)-3-oxo-, ethyl ester	EtOH	296(4.15)	94-3011-76
2(1H)-Pyridinone, 4-hydroxy-3-(1-meth- oxy-4-methyl-2-oxopentyl)-6-methyl-	MeOH	289(3.84)	77-0566-76
$C_{13}H_{19}NO_4S$ 3(2H)-Thiazoleacetic acid, 5-(methoxy- carbonyl)-2,2-dimethyl-α-(1-methyl- ethylidene)-, methyl ester	EtOH	217(4.10),285s(3.75), 319(3.87),337(3.87)	39-0584-76C
$C_{13}H_{19}NO_6S$ 4,9-Epoxy-4H-1,3-dioxolo[4,5-e][1,3]ox- azocine-5(6H)-propanoic acid, tetra- hydro-2,2-dimethyl-6-thioxo-, methyl ester, [3aS-(3aα,4α,9α,9aα)]-	n.s.g.	265.5(4.22)	33-2972-76
Propanedioic acid, [4-(ethoxycarbonyl)- 5,5-dimethyl-2-thiazolidinylidene]-, dimethyl ester	C_6H_{12}	286(4.33)	5-2185-76
$C_{13}H_{19}N_2O_3$ 4H-Pyrido[1,2-a]pyrimidinium, 3-(ethoxy- carbonyl)-6,7,8,9-tetrahydro-1,6-di- methyl-4-oxo-, methyl sulfate	pH 7.35	259.7(4.13)	145-0478-76
$C_{13}H_{19}N_3OS$ Pyrido[4,3-d]thiazolo[3,2-a][1,3]diazo- cine, 8-acetyl-1,2,3,4,5,8,9,10,11,12- decahydro-, hydrobromide	n.s.g.	275(4.04)	2-0773-76
$C_{13}H_{19}N_3O_3$ 2-Quinoxalinecarboxylic acid, 3-amino- 4,6,7,8-tetrahydro-4-(2-methoxyeth- yl)-, methyl ester	EtOH	230(3.94),312(4.50), 320(4.50),371(3.63)	88-2899-76
$C_{13}H_{19}N_3O_5$ 4H-Cyclopenta-1,3-dioxole-4-methanol, 6-(5-amino-1,3,4-oxadiazol-2-yl)- tetrahydro-2,2-dimethyl-, acetate, (3aα,4α,6α,6aα)-($\pm$)-	EtOH	222(3.98)	23-0861-76
$C_{13}H_{19}N_3O_6$ 2(1H)-Pyrimidinone, 4-amino-1-[3-O-(1- oxobutyl)-β-D-arabinofuranosyl]-	MeOH-acid	212(3.97),283(4.13)	87-0667-76
$C_{13}H_{19}N_3O_8$ 1(2H)-Pyrimidinebutanoic acid, α-amino- 3,6-dihydro-2,6-dioxo-3-α-D-ribo- furanosyl-, hydrochloride, (S)-	pH 7.0	263(3.91)	24-0082-76
$C_{13}H_{19}N_3O_8S_2$ Sparsomycin peroxide oxidation product	H_2O	270s(4.15),301(4.35)	44-1858-76
$C_{13}H_{19}N_3S_3$ Hydrazinecarbodithioic acid, 2-[(butyl-	MeOH	240(4.32),280(3.98)	78-0661-76

Compound	Solvent	$\lambda_{max}(\log \epsilon)$	Ref.
amino)thioxomethyl]-1-methyl-, phenyl ester (cont.)	MeCN	243(4.29),285(3.87)	78-0661-76
$C_{13}H_{19}N_5O_4S$ 9H-Purin-6-amine, 9-(2-S-ethyl-2-thio-α-D-mannofuranosyl)-	H_2O	260(4.16)	136-0041-76C
β-	H_2O	260(4.18)	136-0041-76C
$C_{13}H_{19}N_6O_6P$ Adenosine, 8-(propylamino)-, cyclic 3',5'-(hydrogen phosphate)	pH 2.0 pH 13.0	277(4.13) 277(--)	69-1408-76 69-1408-76
$C_{13}H_{20}$ Cyclohexene, 1-(cyclohexylidenemethyl)- Naphthalene, 1,2,3,4,7,8-hexahydro-1,1,6-trimethyl- (contains 1/7 isomer)	EtOH n.s.g.	234(4.01) 267(3.85)	35-4867-76 88-4521-76
$C_{13}H_{20}NO$ Piperidinium, 1,1-dimethyl-3-phenoxy-, bromide	H_2O	264s(2.94),269(3.04), 276(2.94)	87-0692-76
$C_{13}H_{20}N_2O$ Benzamide, N-[2-(diethylamino)ethyl]-, hydrochloride	H_2O	200(--),228(4.00), 270s(2.74)	133-0314-76
1,3,6,8-Nonatetraen-5-one, 1,9-bis(dimethylamino)-	EtOH	258(4.00),296(3.73), 470(4.81)	70-0577-76
2-Pyrrolidinone, 5-[(3,4-dihydro-2,2-dimethyl-2H-pyrrol-5-yl)methylene]-4,4-dimethyl-	EtOH EtOH-HCl EtOH-NaOH	284(3.79),349(4.14) 310(4.40) 290(3.86),350(4.18)	39-1944-76C 39-1944-76C 39-1944-76C
$C_{13}H_{20}N_2O_2$ 2-Pyrrolidinone, 5-[(3,4-dihydro-2,2-dimethyl-2H-pyrrol-5-yl)methylene]-4,4-dimethyl-, N-oxide	EtOH EtOH-HCl EtOH-NaOH	233(3.89),314(4.15) 310(4.51) 255(3.59),340(4.55)	39-1944-76C 39-1944-76C 39-1944-76C
$C_{13}H_{20}N_2O_3$ 4H-Pyrido[1,2-a]pyrimidine-3-carboxylic acid, 1,6,7,8,9,9a-hexahydro-1,6-dimethyl-4-oxo-, ethyl ester	pH 7.35	241(4.35),318(3.83)	145-0478-76
$C_{13}H_{20}N_2O_5S$ 4,9-Epoxy-4H-1,3-dioxolo[4,5-e][1,3]thiazocine-5(6H)-propanoic acid, tetrahydro-6-imino-2,2-dimethyl-, methyl ester, hydrochloride	n.s.g.	226(3.36),278s(2.69)	33-2972-76
$C_{13}H_{20}N_2O_6$ 4(1H)-Pyrimidinone, 5,6-dihydro-2-methoxy-1-[2,3-O-(1-methylethylidene)-β-D-ribofuranosyl]-	n.s.g.	244(4.05)	33-2972-76
$C_{13}H_{20}N_2O_6S$ 2,4(1H,3H)-Pyrimidinedione, 1-(2-S-ethyl-2-thio-β-D-mannofuranosyl)-5-methyl-	EtOH	268(3.99)	136-0041-76C
$C_{13}H_{20}N_4$ Benzenamine, 2-(3-butyl-2-imino-1-imidazolidinyl)-, dihydrobromide	EtOH	216s(4.22),238(4.04), 296(3.59)	78-0057-76
1H-Benzimidazole-1-ethanamine, 2-amino-N-butyl-	EtOH	212(4.71),250(3.93), 285(3.94),334(2.31)	78-0839-76

Compound	Solvent	$\lambda_{max}(\log \epsilon)$	Ref.
1H-Benzimidazole-1-ethanamine, 2-amino-N-butyl- (cont.)	EtOH-HCl	204(4.72),222s(4.18), 276(3.98),282(3.94), 328(2.40),338(2.40)	78-0839-76
$C_{13}H_{20}N_4O_6$ 2,4(1H,3H)-Pteridinedione, 5,6,7,8-tetrahydro-6,7-dimethyl-1-β-D-ribofuranosyl-	pH 1.0 pH 7.0	262(4.27) 240s(3.71),295(4.01)	24-3184-76 24-3184-76
$C_{13}H_{20}N_4O_7$ D-Mannitol, 1-deoxy-1-(1,2,3,6-tetrahydro-1,3-dimethyl-2,6-dioxo-7H-purin-7-yl)-	H_2O	273(3.80)	136-0049-76G
$C_{13}H_{20}N_5O$ 1-Piperidinyloxy, 2,2,6,6-tetramethyl-4-(2-pyrimidinylhydrazono)-	EtOH	264(4.36)	64-0101-76C
$C_{13}H_{20}O$ Bicyclo[7.2.0]undec-5-en-2-one, 10,10-dimethyl-	EtOH	260(2.30)	118-0461-76
2-Butanone, 1-bicyclo[4.1.0]hept-2-ylidene-3,3-dimethyl- (isomer mixture)	EtOH	264(4.08),330s(2.23)	44-3076-76
isomer 16a	EtOH	263(4.17)	44-3076-76
isomer 16b	EtOH	264(3.97)	44-3076-76
2-Cyclohexen-1-one, 4,4,6-trimethyl-6-(2-methyl-2-propenyl)-	C_6H_{12}	222(4.09),343(1.60)	33-2244-76
$C_{13}H_{20}OS$ 1-Nonanone, 1-(2-thienyl)-	heptane	258(3.97),275(3.86)	34-0233-76
$C_{13}H_{20}OS_2$ Bicyclo[2.2.1]heptan-2-one, 3-(1,3-dithiolan-2-yl)-1,7,7-trimethyl-	C_6H_{12}	244s(3.75),290s(4.00), 302(4.10)	148-0435-76
Bicyclo[3.1.1]heptan-3-one, 2-(1,3-dithiolan-2-yl)-4,6,6-trimethyl-	C_6H_{12}	303(3.33),313(3.36)	148-0435-76
$C_{13}H_{20}OSi$ Cyclohexanol, 2-[4-(trimethylsilyl)-1,3-butadiynyl]-, trans	C_6H_{12}	217(2.60),229(2.78), 241(2.95),254(3.00), 261s(2.60),269(2.85)	88-1539-76
$C_{13}H_{20}O_2$ Acetic acid, (2,6,6-trimethyl-2-cyclohexen-1-ylidene)-, ethyl ester, cis	EtOH	257(3.92)	23-2310-76
trans	EtOH	257(3.85)	23-2310-76
Bicyclo[4.2.2]dec-9-en-7-one, 1-hydroxy-2,2,6-trimethyl-, (+)-	EtOH	292(1.40)	33-0001-76
1,3-Cyclohexadiene-1-ethanol, 2,6,6-trimethyl-, acetate	EtOH	265(3.63)	23-2310-76
2-Cyclononene-1,4-dione, 5,5,9,9-tetramethyl-	hexane	217(3.77),224(3.65), 326(2.56)	33-0001-76
2-Cycloocten-1-one, 4-acetyl-4,8,8-trimethyl-, cis	hexane	217(3.51),293s(2.37), 301s(2.42),310s(2.42), 317s(2.30)	33-0001-76
	EtOH	220(3.38),301(2.44)	33-0001-76
trans, m. 81-2°	hexane	212(3.63),294(1.90)	33-0001-76
	EtOH	208(3.60),290(2.04)	33-0001-76
trans, m. 93-5°	hexane	209(3.70),290(1.95)	33-0001-76
	EtOH	210(3.60),290(2.06)	33-0001-76

Compound	Solvent	$\lambda_{max}(\log \epsilon)$	Ref.
2-Furanone, 3,5-bis(1,1-dimethylethyl)-	MeOH	211(4.11)	35-1510-76
2H-Inden-2-one, 1,4,5,6,7,7a-hexahydro-1-hydroxy-1,4,4,7a-tetramethyl-	EtOH	237(4.09)	39-0532-76C
2-Pentene-1,4-dione, 1-(1,2,2-trimethyl-cyclopentyl)-	hexane	306(2.16)	33-0001-76
	EtOH	220(3.82),290(2.16)	33-0001-76
$C_{13}H_{20}O_3$			
3H-2-Benzopyran-3-one, 1,5,6,7,8,8a-hexahydro-1-hydroxy-1,5,5,8a-tetra-methyl-	EtOH	227(4.06)	39-0532-76
3-Cyclohexene-1-propanoic acid, β,4-di-methyl-2-oxo-, ethyl ester	MeOH	234(4.17)	33-0695-76
2-Cyclohexen-1-one, 6-(3-acetoxy-1-meth-ylpropyl)-3-methyl-	EtOH	233(3.95)	33-0695-76
Spiro[4.5]dec-3-en-2-one, 6-hydroxy-3-methoxy-6,10-dimethyl-	EtOH	258(3.92)	44-1539-76
$C_{13}H_{20}O_4$			
3,6-Dioxabicyclo[3.1.0]hexane-2-carbox-aldehyde, 2,5-bis(1,1-dimethylethyl)-4-oxo-	MeOH	199(3.00),300(1.38)	35-1510-76
$C_{13}H_{20}O_8$			
Crassinodine, trans	MeOH	215.5(4.0)	1-0297-76
$C_{13}H_{21}NO$			
Benzenamine, 4-(3-hydroxy-3-methylbut-yl)-N,N-dimethyl-	MeOH	252(4.09),301(3.18)	35-6711-76
2-Cyclohexen-1-one, 5,5-dimethyl-3-pip-eridino-	pH 1	290(4.47)	39-2207-76C
	H_2O	308(4.60)	39-2207-76C
	pH 13	308(4.62)	39-2207-76C
4(1H)-Pyridinone, 2,6-bis(1,1-dimethyl-ethyl)-	C_6H_{12}	255(3.35)	39-1428-76B
	+1% EtOH	251(3.38)	39-1428-76B
	$CHCl_3$	263(4.13)	39-1428-76B
$C_{13}H_{21}NO_3$			
Pentanoic acid, 5-(4,4-dimethyl-2-pyrro-lidinylidene)-3,3-dimethyl-4-oxo-	EtOH	306(4.44)	39-1944-76C
$C_{13}H_{21}NO_4$			
1-Cyclohexene-1-carboxylic acid, 2-[(2-ethoxy-2-oxoethyl)amino]-, ethyl ester	MeOH	296.5(4.20)	5-0383-76
2H-Pyrrole-5-pentanoic acid, 3,4-dihy-dro-β,β,2,2-tetramethyl-γ-oxo-, 1-oxide	EtOH	233(3.95),312(3.18)	39-1944-76C
	EtOH-HCl	228(3.72),295(3.28)	39-1944-76C
	EtOH-NaOH	326(4.48)	39-1944-76C
2H-Pyrrole-5-pentanoic acid, 3,4-dihy-dro-β,β,3,3-tetramethyl-γ-oxo-, 1-oxide	EtOH	239(3.95),312(3.30)	39-1944-76C
	EtOH-NaOH	331(4.48)	39-1944-76C
$C_{13}H_{21}NO_4S$			
Acetamide, N-[2-(acetyloxy)-1-[(1,1-di-methylethyl)thio]-5-oxo-3-pentenyl]-, [R*,S*-(E)]-	MeOH	220(3.86),332(2.88)	4-0321-76
$C_{13}H_{21}NO_5S$			
1H-Pyrrole-1,3-dicarboxylic acid, 2,5-dihydro-4-hydroxy-5-[2-(methylthio)-ethyl]-, diethyl ester	EtOH	247(3.82)	4-0113-76

Compound	Solvent	$\lambda_{max}(\log \epsilon)$	Ref.
$C_{13}H_{21}NO_6$ α-D-xylo-Hexofuranose, 3-deoxy-1,2:5,6-bis-O-(1-methylethylidene)-3-(methyl-imino)-, N-oxide, (E)-	EtOH	250(3.87)	136-0127-76A
$C_{13}H_{21}NO_6S$ 2H-1,4-Thiazine-3,6-dicarboxylic acid, 4-[(1,1-dimethylethoxy)methyl]-3,4-dihydro-, dimethyl ester, 1-oxide, (1R,3R)-	EtOH	209(3.80),276(4.21)	39-2540-76C
(1S,3R)-	EtOH	207(3.82),276(4.15)	39-2540-76C
2,5-Thiazoledicarboxylic acid, 3-[(1,1-dimethylethoxy)methyl]-2,3-dihydro-2-methyl-, dimethyl ester, 1-oxide, trans-(±)-	EtOH	207(3.92),214s(3.89), 288(4.00)	39-2540-76C
$C_{13}H_{21}N_7O_4$ 3H-1,2,3-Triazolo[4,5-d]pyrimidine-5,7-diamine, N^7-butyl-3-β-D-ribofuranosyl-	pH 1	213(4.34),261(4.12), 285(4.03)	87-1186-76
	pH 7	233(4.31),291(4.15)	87-1186-76
	pH 13	232(4.33),291(4.16)	87-1186-76
$C_{13}H_{21}S$ Sulfonium, butylethyl(4-methylphenyl)-, tetrafluoroborate	EtOH	230(3.95),253(3.21), 263(3.12),266(3.13), 273(2.98),275(3.04)	44-3096-76
$C_{13}H_{22}$ 3,5,8-Undecatriene, 4,5-dimethyl-	heptane	232(4.13)	70-0841-76
$C_{13}H_{22}N_2O_4S$ 1,3-Pyrroledicarboxylic acid, 4-amino-2,5-dihydro-5-[2-(methylthio)ethyl]-, diethyl ester	EtOH	275.5(4.15)	4-0113-76
$C_{13}H_{22}O$ 1-Cyclohexene-1-butanal, 2,6,6-trimethyl-	pentane	294(1.43)	33-0032-76
2-Cyclopentene-1-butanal, γ,γ,2,3-tetra-methyl-	pentane	294(1.42)	33-0032-76
$C_{13}H_{22}OS_2$ Cyclohexanone, 2-(1,3-dithiolan-2-yl)-3-methyl-6-(1-methylethyl)-	C_6H_{12}	240(2.67),304(2.26)	148-0435-76
Furan, 3-[(2-methylpropyl)thio]-2-[[(2-methylpropyl)thio]methyl]-	EtOH	225(3.85),245(2.63)	136-0209-76D
Furan, 4-[(2-methylpropyl)thio]-2-[[(2-methylpropyl)thio]methyl]-	EtOH	225(3.92)	136-0209-76D
$C_{13}H_{22}OSi$ Cyclohexanol, 2-[4-(trimethylsilyl)-1-buten-3-ynyl]-, [1α,2β(Z)]-	C_6H_{12}	231s(3.94),239(4.06), 249(3.97)	88-1539-76
$C_{13}H_{22}O_2$ 2-Butanone, 1-(2-hydroxybicyclo[4.1.0]-hept-2-yl)-3,3-dimethyl-	EtOH	294(1.53)	44-3076-76
stereoisomer	EtOH	292(1.48)	44-3076-76
Cyclohexanone, 2-(ethoxymethylene)-3-methyl-6-(1-methylethyl)-, trans	heptane	209(4.08),262(4.53), 345(3.04)	104-0458-76
	EtOH	209(3.89),273(4.49), 330(3.08)	104-0458-76

Compound	Solvent	$\lambda_{max}(\log \epsilon)$	Ref.
Cyclooctanone, 6-acetyl-2,2,6-trimethyl-	EtOH	290(1.72)	33-0001-76
$C_{13}H_{22}O_4$ 2-Pentenedioic acid, 2,4-bis(1,1-dimeth- ylethyl)-	MeOH	205.5(3.82)	35-1510-76
$C_{13}H_{22}O_5S_3$ α-D-Glucofuranose, 5-0,6-S-[bis(methyl- thio)methylene]-3-0-methyl-1,2-0-(1- methylethylidene)-6-thio-	MeOH EtOH	210(2.99) 207(3.30),227(2.78)	118-0449-76 39-2112-76C
$C_{13}H_{22}S$ Thiophene, 2-nonyl-	heptane	234(3.92)	34-0233-76
$C_{13}H_{23}NO$ 1-Penten-3-one, 1-(cyclohexylamino)- 4,4-dimethyl- cis-s-cis	H$_2$O MeOH C$_6$H$_{12}$	315(4.38) 312(4.32) 303(4.20)	35-2826-76 35-2826-76 35-2826-76
$C_{13}H_{23}NO_3$ 2H-Pyrrole-4-carboxylic acid, 5-(1,1-di- methylethyl)-3,4-dihydro-2,2-dimeth- yl-, ethyl ester, 1-oxide	EtOH	235(3.26),276(2.76)	12-2511-76
$C_{13}H_{24}O$ 2-Butanone, 3,3-dimethyl-1-(1-methyl- cyclohexyl)- 4,9-Decadien-2-ol, 2,4,9-trimethyl- 4,9-Decadien-2-ol, 2,5,9-trimethyl- Ionone, tetrahydro-, cis 5,10-Undecadien-3-ol, 2,3-dimethyl-	EtOH EtOH EtOH pentane EtOH	295(1.48) 239(4.40) 239(4.40) 285(1.60) 231(4.32)	44-3076-76 18-3646-76 18-3646-76 33-0032-76 18-3646-76
$C_{13}H_{25}NO$ 1-Penten-3-one, 1-[bis(1-methylethyl)- amino]-4,4-dimethyl-, (E)-	C$_6$H$_{12}$ H$_2$O MeOH	297(4.35) 320(4.47) 316(4.45)	35-2826-76 35-2826-76 35-2826-76
$C_{13}H_{25}N_6O_2$ Piperazinium, 1,1-dimethyl-4-[[[(4-meth- yl-1-piperazinyl)carbonyl]azo]carbo- nyl]-	H$_2$O MeCN	293(3.49) 287(3.24),426(1.58)	35-3001-76 35-3001-76

Compound	Solvent	$\lambda_{max}(\log \epsilon)$	Ref.
$C_{14}H_4N_6O_{10}$ Pyrido[2,3,4,5-lmn]phenanthridine-5,10-dione, 4,9-dihydro-1,3,6,8-tetranitro-	DMF	306(4.18),370(3.99), 490(4.11)	103-0834-76
$C_{14}H_5Cl_2F$ Benzene, (8,8-dichloro-7-fluoro-7-octene-1,3,5-triynyl)-	n.s.g.	205(4.45),224(4.41), 251(4.65),264(4.86), 274(4.78),291(4.77), 300s(--),320(4.36), 343(4.49),369(4.35)	44-1487-76
$C_{14}H_6ClN_3O_2$ 1H-Anthra[1,2-d]triazole-6,11-dione, 4-chloro-	EtOH	350(3.66)	104-0865-76
$C_{14}H_6N_4O_4$ 1-Cyclopropene-1,2-diacetic acid, α,α'-dicyano-3-(dicyanomethylene)-, dimethyl ester, ion(2-)	MeCN	227(4.48),302s(4.38), 326(4.57)	35-0610-76
Propanedioic acid, [2,3-bis(dicyanomethyl)-2-cyclopropen-1-ylidene]-, dimethyl ester, ion(2-)	MeCN	223(4.44),269(4.24), 322(4.52)	35-0610-76
$C_{14}H_7N_3$ Pyrido[2,1,6-de]quinolizine-1,3-dicarbonitrile	EtOH	239s(3.83),249s(3.88), 287(4.65),402(4.20), 425(4.23),449(4.36)	39-0341-76C
$C_{14}H_7N_6$ 1,3-Pentadiene-1,1,2,4,5,5-hexacarbonitrile, 3-(1-methylethyl)-, ion(1-)	MeCN	600(4.27)	35-0558-76
1,3-Pentadiene-1,1,2,4,5,5-hexacarbonitrile, 3-propyl-, ion(1-)	MeCN	593(4.40)	35-0558-76
$C_{14}H_8BrN_5$ Benzonitrile, 4-[2-(4-bromophenyl)-2H-tetrazol-5-yl]-	EtOH	282(4.47)	18-0762-76
$C_{14}H_8Br_2$ Phenanthrene, 2,7-dibromo-	heptane	260(3.90),276s(3.43), 290(3.32)	104-2355-76
$C_{14}H_8ClFOS$ Dibenzo[b,f]thiepin-10(11H)-one, 8-chloro-7-fluoro-	MeOH	243(4.27),267s(4.03), 273s(4.02),329(3.50)	73-0881-76
$C_{14}H_8ClFS$ Dibenzo[b,f]thiepin, 3-chloro-2-fluoro-	MeOH	223(4.42),263(4.32), 295(3.80)	73-0881-76
$C_{14}H_8ClNO_2$ 9,10-Anthracenedione, 2-amino-1-chloro- 1H-2,3-Benzoxazin-1-one, 4-(4-chlorophenyl)-	CCl_4 H_2SO_4	406(3.69) 295(3.94)	104-2499-76 104-1797-76
$C_{14}H_8ClN_3O_6$ Benzene, 2-[2-(4-chlorophenyl)ethenyl]-1,3,5-trinitro-, (E)-	EtOH	250(4.26),350(4.21)	80-1127-76

Compound	Solvent	$\lambda_{max}(\log \epsilon)$	Ref.
$C_{14}H_8ClN_5$ Benzonitrile, 4-[2-(4-chlorophenyl)- 2H-tetrazol-5-yl]-	EtOH	281(4.20)	18-0762-76
$C_{14}H_8ClN_5O$ 1,2,4,5-Benzenetetracarbonitrile, 3-chloro-6-morpholino-	MeCN	323(3.98),464(3.45)	24-2469-76
$C_{14}H_8Cl_2N_2$ Benzeneacetonitrile, α-[(2,6-dichloro- 3-pyridinyl)methylene]-	MeOH	309.5(4.30)	33-0211-76
$C_{14}H_8Cl_2OS$ Dibenzo[b,f]thiepin-10(11H)-one, 7,8-di- chloro-	MeOH	251(4.39),345(3.65)	73-0881-76
$C_{14}H_8Cl_2S$ Dibenzo[b,f]thiepin, 2,3-dichloro-	MeOH	220(4.48),266(4.33), 297(3.78)	73-0881-76
$C_{14}H_8Cl_2S_3$ Benzenecarbodithioic acid, 4-chloro-, anhydrosulfide	C_6H_{12}	323(4.22),548(2.19)	89-0766-76
$C_{14}H_8Cl_4$ Benzene, [3-(2,3,4,5-tetrachloro-2,4-cy- clopentadien-1-ylidene)-1-propenyl]-	C_6H_{12}	252(3.70),379(4.69)	24-3929-76
$C_{14}H_8CsN$ 9H-Fluorene-9-carbonitrile, cesium deriv.	$(MeOCH_2)_2$	416(3.35),439(3.34)	104-2220-76
$C_{14}H_8F_2OS$ Dibenzo[b,f]thiepin-10(11H)-one, 7,8-di- fluoro-	MeOH	225(4.20),240(4.17), 328(3.50)	73-0881-76
$C_{14}H_8F_2S$ Dibenzo[b,f]thiepin, 2,3-difluoro-	MeOH	259(4.36),290(3.67)	73-0881-76
$C_{14}H_8F_3NO_2$ Acetamide, N-(3-dibenzofuranyl)-2,2,2- trifluoro-	C_6H_{12}	298(4.52),313(4.52)	40-0133-76
$C_{14}H_8F_3N_3O_4$ Benzenamine, 4-nitro-N-[2,2,2-trifluoro- 1-(4-nitrophenyl)ethylidene]-	hexane	270(4.27),335(3.51)	104-0161-76
$C_{14}H_8I_2$ Phenanthrene, 2,7-diiodo-	heptane	267(3.89),277s(3.65), 294(3.49),302(3.22)	104-2355-76
$C_{14}H_8LiN$ 9H-Fluorene-9-carbonitrile, lithium deriv.	$(MeOCH_2)_2$	406(3.31),428(3.31)	104-2220-76
$C_{14}H_8NNa$ 9H-Fluorene-9-carbonitrile, sodium deriv.	$(MeOCH_2)_2$	411(3.36),434(3.34)	104-2220-76

Compound	Solvent	$\lambda_{max}(\log \epsilon)$	Ref.
$C_{14}H_8N_2$			
Indeno[1,2,3-ij][2,7]naphthyridine (canagine)	EtOH	228(4.30),235(4.31), 278(4.21),287(4.18), 295s(3.95),335s(3.54), 349(3.77),369(3.76)	100-0459-76
Indolo[3,2-b]indole, 5,10-dihydro-, ion(2-)	DMF	290(4.43),347(4.19), 400(3.58)	5-1090-76
semiquinone	DMF	264(4.52),291(4.84), 303(4.77),348s(4.13), 372s(3.68),416s(3.66), 504(3.70),528s(3.75), 541(3.81),568(3.75), 585(3.78),617(3.46), 638(3.46)	5-1090-76
oxidized form	DMF	259(4.54),266(4.82), 274(5.00),403(3.71), 423(3.80),448(3.59)	5-1090-76
$C_{14}H_8N_2O$			
6H-Indolo[3,2,1-de][1,5]naphthyridin-6-one (6-canthinone)	MeOH	251(4.07),260(4.05), 268(4.03),300(3.87), 347(3.89),362(4.09), 380(4.05)	24-0705-76
	EtOH	259(4.50),270(4.38), 300(4.29),362(4.55), 380(4.52)	94-1532-76
	dioxan	261(4.04),270(4.03), 292(3.87),299(3.88), 348(3.89),364(4.12), 382(4.09)	24-0705-76
$C_{14}H_8N_2OSe$			
Phenanthro[9,10-c][1,2,5]selenadiazole, 1-oxide	EtOH	238(4.50),302(4.15), 377(4.05)	1-0675-76
$C_{14}H_8N_2O_2$			
6H-Indolo[3,2,1-de][1,5]naphthyridin-6-one, 3-oxide	EtOH	247(4.07),280(4.38), 370(4.13)	94-1532-76
$C_{14}H_8N_2O_4$			
Benzene, 1,1'-(1,2-ethynediyl)bis[4-nitro-	n.s.g.	332(4.54)	2-0101-76
Phenanthrene, 2,5-dinitro-	EtOH	270(4.20),313(3.71)	104-2355-76
Phenanthrene, 2,7-dinitro-	EtOH	266(4.54),274(4.59), 320(4.26)	104-2355-76
$C_{14}H_8N_2O_6$			
[1,1'-Biphenyl]-2,2'-dicarboxaldehyde, 4,4'-dinitro-	EtOH	285(4.19)	104-2355-76
[1,1'-Biphenyl]-2,2'-dicarboxaldehyde, 4.6'-dinitro-	EtOH	270(4.10)	104-2355-76
$C_{14}H_8N_2O_8$			
[1,1'-Biphenyl]-2,2'-dicarboxylic acid, 4,4'-dinitro-	EtOH	274(4.18)	104-2355-76
[1,1'-Biphenyl]-2,2'-dicarboxylic acid, 4,6'-dinitro-	EtOH	267(4.11)	104-2355-76
$C_{14}H_8N_4$			
1,3,5,7-Tetraazatriphenylene	EtOH	209(4.35),253(4.60),	94-2585-76

Compound	Solvent	$\lambda_{max}(\log \epsilon)$	Ref.
1,3,5,7-Tetraazatriphenylene (cont.)		274(4.21),307(3.67), 322(3.54),337(3.41)	94-2585-76
1,3,6,8-Tetraazatriphenylene	EtOH	250(4.45),280(3.80), 324(3.47),338(3.55)	94-2585-76
1,3,10,12-Tetraazatriphenylene	EtOH	220(4.18),253(4.18), 260(4.16),325(3.62)	94-2585-76
$C_{14}H_8N_4O_8$ Benzene, 1,3,5-trinitro-2-[2-(4-nitro-phenyl)ethenyl]-, (E)-	EtOH	310(4.30)	80-1127-76
$C_{14}H_8N_6$ Propanedinitrile, (1-phenyl-1H-pyrazolo-[3,4-d]pyrimidin-4-yl)-	EtOH	230s(4.14),250(4.28), 346(4.47)	95-1352-76
$C_{14}H_8N_6O_2$ Benzonitrile, 4-[2-(4-nitrophenyl)-2H-tetrazol-5-yl]-	EtOH	257(4.23),300(4.44)	18-0762-76
$C_{14}H_8N_8O_6S_2$ Benzenesulfonyl azide, 4,4'-(2,4-dioxo-1,3-diazetidine-1,3-diyl)bis-	MeCN	297(4.77)	93-0163-76
$C_{14}H_8O_2$ 1,2-Phenanthrenedione	EtOH	218(4.47),232(4.42), 291(4.31),370(3.38), 485(3.00)	78-2693-76
1,4-Phenanthrenedione	EtOH	224(4.68),262(4.17), 279(4.13),289(4.13), 370(3.49)	78-2693-76
$C_{14}H_8O_3S$ 1H-2-Benzopyran-1,3(4H)-dione, 4-(2-thi-enylmethylene)-, (E)-	EtOH	220(4.26),385(4.22)	4-0083-76
$C_{14}H_8O_4$ 1,2-Anthracenedione, 3,9-dihydroxy-	CHCl$_3$	263(4.30),305(4.45), 450(3.68),481(3.75), 543(3.90)	12-1535-76
1,4-Anthracenedione, 2,10-dihydroxy-	MeOH-HCOOH	285(4.32),460(3.87)	12-1535-76
1H-2-Benzopyran-1,3(4H)-dione, 4-(2-fur-anylmethylene)-, (E)-	EtOH	220(4.32),380(4.22)	4-0083-76
$C_{14}H_8O_5$ 6H-Dibenzo[b,d]pyran-6-one, 2-hydroxy-8,9-(methylenedioxy)-	EtOH	224(4.38),247(4.45), 259(4.47),277s(--), 320(3.97),350s(--)	24-0855-76
$C_{14}H_8O_6$ 2H,5H-Pyrano[3,2-c][1]benzopyran-3-carb-oxylic acid, 2,5-dioxo-, methyl ester	n.s.g.	361(4.31)	64-0514-76B
$C_{14}H_8O_7$ 6H-Dibenzo[b,d]pyran-6-one, 2,3,4-tri-hydroxy-8,9-(methylenedioxy)-	EtOH	236(4.39),274(4.27), 310(4.17),357s(--)	24-0855-76
$C_{14}H_8S_4$ 1,3-Benzodithiole, 2-(1,3-benzodithiol-2-ylidene)-	THF	255(4.30),289(4.28), 310(4.30),436(2.49)	97-0360-76
cation	MeCN	424(--),442(--), 631(--)	97-0360-76

Compound	Solvent	$\lambda_{max}(\log \epsilon)$	Ref.
$C_{14}H_9BrCl_2$			
Benzene, 1-[2-(3-bromophenyl)ethenyl]-2,4-dichloro-, (E)-	EtOH	315(4.42)	39-1731-76B
$C_{14}H_9BrN_4O_4$			
1H-Benzimidazole, 5-bromo-2-[(2,4-dinitrophenyl)methyl]-	0.1N H_2SO_4	240(4.34),278(4.20), 285(4.14)	104-1097-76
	EtOH	245(4.33),283(4.09), 290(4.01)	104-1097-76
	EtOH-KOH	492(4.14),650(3.60)	104-1097-76
photochromic form	EtOH	523(4.70)	104-2361-76
$C_{14}H_9Br_2NO$			
9(10H)-Acridinone, 2,7-dibromo-10-methyl-	EtOH	255(4.55),401(3.81), 420(3.83)	48-0515-76
$C_{14}H_9ClN_2$			
Propanedinitrile, (3-chloro-5-phenyl-2,4-pentadienylidene)-	CH_2Cl_2	404(4.69)	48-0705-76
$C_{14}H_9ClN_2O$			
Cinnoline, 6-chloro-4-phenyl-, 1-oxide	isoPrOH	235(4.44),283(3.89), 295s(3.82),336s(3.88), 367(4.08)	4-0907-76
Cinnoline, 6-chloro-4-phenyl-, 2-oxide	isoPrOH	217(4.44),270(4.49), 315(3.91),365(3.69), 380s(3.66)	4-0907-76
$C_{14}H_9ClN_2O_4$			
Benzene, 1,1'-(1-chloro-1,2-ethenediyl)-bis[4-nitro-, cis	n.s.g.	325(4.17)	2-0101-76
trans	n.s.g.	328(4.38)	2-0101-76
$C_{14}H_9ClN_2S_2$			
1,3,4-Thiadiazolium, 2-(4-chlorophenyl)-5-mercapto-3-phenyl-, hydroxide, inner salt	M HCl	272(4.30),366(3.78)	4-1273-76
	H_2O	275(4.30),366(3.77)	4-1273-76
	MeOH	279(4.34),400(3.79)	4-1273-76
	EtOH	280(4.33),405(3.73)	4-1273-76
	PrOH	281(4.33),414(3.76)	4-1273-76
	isoPrOH	279(4.20),408(3.66)	4-1273-76
	BuOH	280(4.34),414(3.76)	4-1273-76
	isoBuOH	280(4.32),419(3.68)	4-1273-76
	sec-BuOH	282(4.32),414(3.75)	4-1273-76
	isoAmOH	282(4.31),411(3.75)	4-1273-76
	tert-AmOH	279(4.32),410(3.73)	4-1273-76
	$PhCH_2OH$	394(3.68)	4-1273-76
	acetone	427(3.76)	4-1273-76
	$HCONH_2$	282(4.31),400(3.77)	4-1273-76
	MeCN	280(4.36),415(3.80)	4-1273-76
	$MeNO_2$	413(3.70)	4-1273-76
	pyridine	433(3.68)	4-1273-76
	EtOAc	261(4.35),440(3.75)	4-1273-76
	$CHCl_3$	263(4.41),432(3.73)	4-1273-76
$C_{14}H_9ClN_4O_4$			
1H-Benzimidazole, 5-chloro-2-[(2,4-dinitrophenyl)methyl]-	0.1N H_2SO_4	240(4.33),270(4.21), 278(4.22)	104-1097-76
	EtOH	247(4.32),283(4.07), 290(4.01)	104-1097-76
	EtOH-KOH	494(4.09),650(3.57)	104-1097-76
photochromic form	EtOH	522(4.71)	104-2361-76

Compound	Solvent	$\lambda_{max}(\log \epsilon)$	Ref.
$C_{14}H_9Cl_2NO$ 9(10H)-Acridinone, 2,7-dichloro-10-methyl-	EtOH	255(4.95),388(4.14), 406(4.17)	48-0515-76
$C_{14}H_9Cl_5N_2$ 2-Azabicyclo[4.1.0]hept-3-ene-4-carbonitrile, 5,7,7-trichloro-2-[(2,6-dichlorophenyl)methyl]-	CHCl$_3$	260(3.7),318(3.6)	64-0807-76B
3-Azabicyclo[4.1.0]hept-4-ene-5-carbonitrile, 2,7,7-trichloro-3-[(2,6-dichlorophenyl)methyl]-	CHCl$_3$	260(3.7),334(3.5)	64-0807-76B
3-Pyridinecarbonitrile, 1-[(2,6-dichlorophenyl)methyl]-2-(trichloromethyl)-	CHCl$_3$	378(4.6)	64-0807-76B
$C_{14}H_9FN_2S_2$ 1,3,4-Thiadiazolium, 2-(4-fluorophenyl)-5-mercapto-3-phenyl-, hydroxide, inner salt	M HCl	275(4.26),363(3.79)	4-1273-76
	H$_2$O	275(4.28),363(3.78)	4-1273-76
	MeOH	276(4.31),397(3.74)	4-1273-76
	EtOH	278(4.32),404(3.78)	4-1273-76
	PrOH	277(4.28),406(3.76)	4-1273-76
	isoPrOH	277(4.21),405(3.73)	4-1273-76
	BuOH	277(4.32),409(3.76)	4-1273-76
	isoBuOH	279(4.30),413(3.73)	4-1273-76
	sec-BuOH	279(4.31),409(3.70)	4-1273-76
	isoAmOH	278(4.29),404(3.66)	4-1273-76
	tert-AmOH	277(4.29),404(3.70)	4-1273-76
	PhCH$_2$OH	388(3.61)	4-1273-76
	acetone	422(3.76)	4-1273-76
	HCONH$_2$	279(4.35),394(3.78)	4-1273-76
	MeCN	278(4.30),411(3.71)	4-1273-76
	MeNO$_2$	405(3.64)	4-1273-76
	pyridine	425(3.68)	4-1273-76
	EtOAc	280(4.29),4.4(3.69)	4-1273-76
	CHCl$_3$	283(4.32),421(3.71)	4-1273-76
$C_{14}H_9F_3N_2O_2$ Benzenamine, 4-nitro-N-(2,2,2-trifluoro-1-phenylethylidene)-	hexane	265(3.95),280(4.06)	104-0161-76
Benzenamine, N-[2,2,2-trifluoro-1-(4-nitrophenyl)ethylidene]-	hexane	250s(4.12),274s(4.12), 340(3.26)	104-0161-76
$C_{14}H_9N$ 9H-Fluorene-9-carbonitrile, anion	diglyme	430(3.40)	70-0762-76
$C_{14}H_9NO_2$ 1,2-Anthracenedione, 1-oxime	hexane	224(4.28),249(4.3), 260s(4.16),316(4.08), 380(3.42),393(3.42), 416(3.46),438(3.54), 462(3.42)	110-1711-76
	EtOH	238(4.56),250(4.46), 288s(4.4),299(4.54), 315s(4.25),330s(3.8), 376(3.84),411(3.88), 438(3.84),473(3.4)	110-1711-76
	1:4 dioxanhexane	242(4.54),260s(4.4), 316(4.42),335s(3.7), 376(3.69),408(3.74), 434(3.78),462(3.68)	110-1711-76

Compound	Solvent	$\lambda_{max}(\log \epsilon)$	Ref.
1,2-Anthracenedione, 2-oxime	hexane	219(4.36),243s(4.04), 250(4.08),272s(4.02), 295(4.34),310s(4.18), 436(3.42),458(3.47), 490(3.35)	110-1711-76
	EtOH	267s(4.16),290(4.64), 299(4.64),325s(4.15), 404s(3.75),425(3.9), 446(3.84),480s(3.24)	110-1711-76
	1:4 dioxan-hexane	230(4.08),243(4.04), 266s(--),294(4.5), 310s(4.26),399(3.52), 413s(3.64),435(3.63), 450s(3.6),484(3.04)	110-1711-76
Benzene, 1-nitro-4-(phenylethynyl)-	n.s.g.	327(4.53)	2-0101-76
2H-1,4-Benzoxazin-2-one, 3-phenyl-	EtOH	235(3.84),334(4.17)	78-2033-76
1H-2,3-Benzoxazin-1-one, 4-phenyl-	H_2SO_4	295(3.98)	104-1797-76
$C_{14}H_9NO_2S_2$ 1,3-Benzodithiole, 5-nitro-2-(phenyl-methylene)-	$CHCl_3$	295(4.34),316(4.30), 343(3.61)	39-0692-76C
$C_{14}H_9NO_3$ 1H-Indole-3-carboxaldehyde, 2-(5-formyl-2-furanyl)-	EtOH	252(4.17),271s(4.07)	2-0579-76
$C_{14}H_9NO_4$ 6H-Dibenzo[b,d]pyran-6-one, 2-amino-8,9-(methylenedioxy)-	EtOH	250(4.69),289(4.17), 325(4.09),353s(--)	24-0855-76
$C_{14}H_9N_3$ Indazolo[2,3-a]quinazoline	EtOH	242(4.64),298(4.09), 370(3.97)	4-1325-76
$C_{14}H_9N_3O_4$ Benzoic acid, 2-(1H-benzimidazol-1-yl)-5-nitro-	EtOH	246(4.21),270s(3.98), 317(3.82)	2-0001-76
$C_{14}H_9N_3O_6$ Benzene, 2,4-dinitro-1-[2-(2-nitrophen-yl)ethenyl]-, (E)-	EtOH	240(4.33),340(4.44)	80-1127-76
Benzene, 2,4-dinitro-1-[2-(3-nitrophen-yl)ethenyl]-, (E)-	EtOH	245(4.44),340(4.46)	80-1127-76
Benzene, 2,4-dinitro-1-[2-(4-nitrophen-yl)ethenyl]-, (E)-	EtOH	352(4.47)	80-1127-76
Benzene, 1,3,5-trinitro-2-(2-phenyleth-enyl)-	EtOH	350(4.16)	80-1127-76
$C_{14}H_9N_5$ Benzonitrile, 4-(2-phenyl-2H-tetrazol-5-yl)-	EtOH	279(4.47)	18-0762-76
$C_{14}H_9N_5O$ 1H-Imidazole-4-carbonitrile, 5-[(2-hy-droxy-1-naphthalenyl)azo]-	EtOH	279(3.93),476(4.24)	103-0465-76
Pyrimido[1,2-a]purin-10(9H)-one, 7-phenyl-	pH 1	248(4.35),289(4.28), 315s(3.95),360s(3.48)	44-0294-76
	pH 6.8	244(4.42),285(4.34), 320s(3.77),370(3.47)	44-0294-76
	pH 10.1	245(4.50),287(4.42), 323s(3.75),375(3.42)	44-0294-76

Compound	Solvent	$\lambda_{max}(\log \epsilon)$	Ref.
$C_{14}H_9N_5O_6$			
1H-Benzimidazole, 2-[(2,4-dinitrophenyl)methyl]-5-nitro-	0.1N H_2SO_4	230(4.50),265(4.27)	104-1097-76
	EtOH	238(4.52),305(4.08)	104-1097-76
	EtOH-KOH	370(4.06),516(3.90)	104-1097-76
photochromic form	EtOH	546(4.56)	104-2361-76
$C_{14}H_{10}$			
Anthracene, lithium salt of dianion	ether	554(4.26)	35-5706-76
Naphthalene, 2-(1-buten-3-ynyl)-, cis	MeOH	213(4.57),235s(4.17),	39-1104-76B
		243(4.24),259(4.50),	
		268(4.53),288(4.15),	
		298(4.34),311(4.34),	
		323s(3.72),332s(3.32),	
		350(2.95)	
Naphthalene, 2-(1-buten-3-ynyl)-, trans	MeOH	212(4.29),230s(4.13),	39-1104-76B
		236(4.29),244(4.37),	
		261(4.57),270(4.61),	
		286(4.13),297(4.49),	
		311(4.52),325s(3.70),	
		334(3.50),351(3.10)	
$C_{14}H_{10}BrClN_2O$			
Benzeneethanehydrazonoyl bromide, N-(4-chlorophenyl)-α-oxo-	EtOH	229(4.33),257(4.33), 372(4.21)	18-0321-76
$C_{14}H_{10}BrN_3$			
Pyrido[2,3-b]pyrazine, 7-bromo-6-methyl-3-phenyl-	EtOH	248(4.11),348(4.07)	22-0251-76
Pyrido[2,3-b]pyrazine, 7-bromo-8-methyl-3-phenyl-	EtOH	250(4.26),350(4.16)	22-0251-76
$C_{14}H_{10}BrN_3O_3$			
Benzeneethanehydrazonoyl bromide, 4-nitro-α-oxo-N-phenyl-	EtOH	229(4.36),270(4.20), 320(3.91),384(4.01)	18-0321-76
$C_{14}H_{10}BrN_3O_4$			
Methanone, [4-(5-bromo-2-furanyl)-4,5-dihydro-1H-pyrazol-3-yl](4-nitrophenyl)-	EtOH	217(4.21),269(4.21), 351(4.02)	104-2358-76
$C_{14}H_{10}BrN_5O_2S$			
Isothiazolo[5,4-b]pyridine, 3-bromo-6,7-dihydro-7-methyl-6-[[(4-nitrophenyl)azo]methylene]-	DMF	547(4.49)	48-0779-76
$C_{14}H_{10}BrN_5O_3S_3$			
3-Thiazolidineacetic acid, 4-[(aminothioxomethyl)hydrazono]-5-(5-bromo-1,2-dihydro-2-oxo-3H-indol-3-ylidene)-2-thioxo-	MeOH	249(4.41),290s(--), 399(4.39),415s(--)	103-0749-76
$C_{14}H_{10}Br_2N_2O$			
Benzeneethanehydrazonoyl bromide, 4-bromo-α-oxo-N-phenyl-	EtOH	227(4.30),261(4.27), 374(4.18)	18-0321-76
$C_{14}H_{10}Br_2O_4S$			
2(3H)-Furanone, 5-(3,5-dibromo-2,4-dihydroxyphenyl)dihydro-5-(2-thienyl)-	EtOH-pH 8.4	550(2.94)	18-2027-76

Compound	Solvent	$\lambda_{max}(\log \epsilon)$	Ref.
$C_{14}H_{10}ClF_3N_2OS_2$ 1-Propen-2-ol, 1-[4-[(4-chlorophenyl)-thio]-2-(methylthio)-5-pyrimidinyl]-3,3,3-trifluoro-	MeOH	257(4.34),307(3.89)	73-2771-76
$C_{14}H_{10}ClNO_2$ Benzamide, N-benzoyl-N-chloro-	MeOH	244(4.36)	117-0025-76
Benzene, 1-(2-chloro-2-phenylethenyl)-4-nitro-, trans	n.s.g.	328(4.24)	2-0101-76
Furo[2,3-c]pyridin-2(6H)-one, 3-chloro-6-(phenylmethyl)-	MeCN	265(3.80),396s(4.37), 408(4.47)	88-2959-76
1H-Indole-4,7-dione, 2-(4-chlorophenyl)-5,6-dihydro-	MeOH	217(4.19),261(4.39), 287(4.33),335(4.11)	111-0241-76
$C_{14}H_{10}ClN_3O$ 4(1H)-Quinazolinone, 2-[(2-chlorophenyl)amino]-	EtOH	217(4.43),258(4.25), 302(3.99)	114-0341-76D
4(1H)-Quinazolinone, 2-[(4-chlorophenyl)amino]-	EtOH	213(4.37),237(4.39), 273(4.25),306(4.13)	114-0341-76D
$C_{14}H_{10}ClN_3O_3S$ 2H-1,2,4-Benzothiadiazine-3-carboxamide, N-(3-chlorophenyl)-, 1,1-dioxide	dioxan	290(4.20)	103-0402-76
$C_{14}H_{10}Cl_2N_2O_2$ 4a,8a-Naphthalenedicarbonitrile, 2,3-dichloro-1,4,5,8-tetrahydro-6,7-dimethyl-1,4-dioxo-, cis	MeOH	270(3.78),340s(2.60)	35-7040-76
$C_{14}H_{10}Cl_2OS$ Dibenzo[b,e]thiepin-11-ol, 7,8-dichloro-10,11-dihydro-	MeOH	272(3.99)	73-0881-76
$C_{14}H_{10}Cl_4O_2$ 1H-Cyclopenta[b][1,4]benzodioxin, 5,6,7,8-tetrachloro-8a,9a-dihydro-1-(1-methylethylidene)-	C_6H_{12}	217(5.42),300(3.23)	78-0147-76
$C_{14}H_{10}F_3NO$ Acetamide, N-[1,1'-biphenyl]-4-yl-2,2,2-trifluoro-	C_6H_{12} MeOH	273(4.48) 274(4.49)	40-0133-76 40-0133-76
$C_{14}H_{10}F_3NO_2$ Benzoic acid, 2-[[3-(trifluoromethyl)-phenyl]amino]-, β-cyclodextrin complex	pH 7.0	292(4.16),320(3.93)	133-0343-76
$C_{14}H_{10}N_2$ Cycloocta[b]quinoxaline	C_6H_{12}	205(4.44),245(4.36), 284(3.61),334(3.71), 354(4.49)	44-3760-76
1H-Cyclopenta[gh]perimidine, 1-methyl-	MeOH	<u>218s(3.8)</u>,416(<u>3.9</u>)	103-0450-76
$C_{14}H_{10}N_2O$ Benzeneacetonitrile, α-[4-(hydroxyimino)-2,5-cyclohexadien-1-ylidene]-	EtOH	371(4.58)	40-0144-76
Methanone, 1H-indol-3-yl-3-pyridinyl-	EtOH	260s(4.16),269(4.17), 322(4.14)	39-1155-76C
3-Pyridinecarbonitrile, 1,2-dihydro-2-oxo-6-(2-phenylethenyl)-	HOAc	376(4.34)	48-0705-76

Compound	Solvent	$\lambda_{max}(\log \epsilon)$	Ref.
Pyridinium, 1-(1-cyano-2-hydroxy-2-phen-ylethenyl)-, hydroxide, inner salt	EtOH	240(4.07),279(3.77), 334(4.10)	35-5671-76
Pyridinium, 1-cyano-2-oxo-1-phenyleth-ylide	benzene	415(4.11)	78-2647-76
	aq. HCl	231(4.04),258(3.95), 365(3.41)	78-2647-76
	H_2O	239(4.09),264(3.95), 370(3.70)	78-2647-76
	EtOH	240(4.09),272(3.84), 391(3.90)	78-2647-76
	dioxan	241(4.06),276(3.66), 410(4.05)	78-2647-76
	DMF	401(3.96)	78-2647-76
	MeCN	238(4.13),279(3.65), 296(4.05)	78-2647-76
	$CHCl_3$	272(3.74),408(4.03)	78-2647-76
$C_{14}H_{10}N_2OS$			
Benzothiazole, 6-nitroso-7-(phenylmeth-yl)-	$CHCl_3$	760(1.59)	118-0270-76
Phenol, 4-(benzo[b]thien-5-ylazo)-	C_6H_{12}	247(4.02),349(4.11)	62-0249-76A
	EtOH	251(4.33),285s(3.81), 360(4.42)	62-0249-76A
	acetone	352(4.59)	62-0249-76A
	$CHCl_3$	252(4.57),285s(4.15), 358(4.66)	62-0249-76A
	CCl_4	285s(4.02),353(4.51)	62-0249-76A
$C_{14}H_{10}N_2OSe$			
1,2,5-Selenadiazole, 3,4-diphenyl-, 2-oxide	EtOH	243(4.30),316(4.00)	1-0675-76
$C_{14}H_{10}N_2O_2$			
1,2-Anthracenedione, dioxime	EtOH	220(4.7),245s(4.3), 302(4.60),313(4.58), 350(3.97),365(3.87), 388(3.6),400(3.58), 434(3.37)	110-1711-76
	1:4 dioxan-hexane	220(4.7),242s(4.3), 294s(4.48),303(4.58), 313(4.58),350(3.97), 365(3.88),390(3.6), 408(3.62),430(3.34)	110-1711-76
9,10-Anthracenedione, 1,2-diamino-	EtOH	535(3.92)	104-2503-76
9,10-Anthracenedione, 2,3-diamino-	EtOH	503(3.38)	104-2503-76
Benzeneacetonitrile, 4-nitro-α-phenyl-	EtOH	266(4.08)	40-0144-76
1,3-Diazetidine-2,4-dione, 1,3-diphenyl-	THF	250(4.36)	95-0962-76
1H-Indole, 5-nitro-3-phenyl-	EtOH	265(4.33),327(3.91)	30-0629-76
1H-Indole, 7-nitro-3-phenyl-	EtOH	281(4.35),372(3.78)	30-0629-76
Indolizine, 6-nitro-2-phenyl-	EtOH	240(4.35),312(4.26), 350-450d(3.9-3.4)	103-0766-76
Indolizine, 8-nitro-2-phenyl-	EtOH	248(4.62),375(--), 450-500s(3.4)	103-0766-76
2H-Indol-2-one, 1,3-dihydro-3-(phenyl-imino)-, N-oxide	MeOH	210(4.12),261(3.99), 338(3.92)	24-0200-76
3H-Indol-3-one, 1,2-dihydro-2-(phenyl-imino)-, N-oxide	MeOH	219(4.16),270(4.20), 348(4.05),471(3.71)	24-0200-76
$C_{14}H_{10}N_2O_3$			
1-Phenazinecarboxylic acid, 6-methoxy-	EtOH	264(4.67),365(3.94), 428(3.40)	39-2248-76C

Compound	Solvent	$\lambda_{max}(\log \epsilon)$	Ref.
1-Phenazinecarboxylic acid, 8-methoxy-	EtOH	258(4.57),365(3.49), 400(3.38)	39-2248-76C
1-Phenazinecarboxylic acid, 9-methoxy-	EtOH	263(4.52),366(3.86), 420s(3.42)	39-2248-76C
$C_{14}H_{10}N_2O_3S$ Phenol, 4-(benzo[b]thien-5-ylazo)-, S,S-dioxide	acetone	366(4.53)	62-0249-76A
	EtOAc	253(4.33),295s(3.95), 374(4.39)	62-0249-76A
	CHCl$_3$	252(4.39),290s(4.08), 364(4.55)	62-0249-76A
$C_{14}H_{10}N_2O_4$ Benzene, 2,4-dinitro-1-(2-phenylethen-yl)-	EtOH	230(4.34),356(4.36)	80-1127-76
Imidazo[2,1-a]isoquinoline-2,3-dicarb-oxylic acid, 2-methyl ester	MeOH	256(4.79),318(3.65)	39-0075-76C
$C_{14}H_{10}N_2S_2$ 2-Benzothiazolecarbothioamide, N-phenyl-	20% EtOH	327(4.15)	94-1451-76
1,3,4-Thiadiazolium, 2-mercapto-4,5-di-phenyl-, hydroxide, inner salt	M HCl	275(4.26),363(3.79)	4-1273-76
	H$_2$O	271(4.28),362(3.78)	4-1273-76
	MeOH	273(4.29),395(3.74)	4-1273-76
	MeOH	281(4.26),398(3.72)	78-0661-76
	EtOH	275(4.26),400(3.65)	4-1273-76
	PrOH	277(4.32),408(3.73)	4-1273-76
	isoPrOH	277(4.18),406(3.67)	4-1273-76
	BuOH	277(4.32),409(3.73)	4-1273-76
	isoBuOH	278(4.28),415(3.64)	4-1273-76
	sec-BuOH	278(4.28),411(3.67)	4-1273-76
	isoAmOH	278(4.29),405(3.71)	4-1273-76
	PhCH$_2$OH	390(3.71)	4-1273-76
	acetone	422(3.76)	4-1273-76
	EtOAc	280(4.28),435(3.72)	4-1273-76
	HCONH$_2$	278(4.34),395(3.75)	4-1273-76
	MeCN	278(4.30),410(3.77)	4-1273-76
	MeCN	276(4.29),410(3.74)	78-0661-76
	MeNO$_2$	406(3.74)	4-1273-76
	pyridine	427(3.77)	4-1273-76
	CHCl$_3$	283(4.32),421(3.71)	4-1273-76
$C_{14}H_{10}N_4$ Indazolo[2,3-a]quinazolin-5-amine	EtOH	239(4.69),295(4.33), 372(3.94)	4-1325-76
1,2,4,5-Tetrazine, 3,6-diphenyl-	EtOH	294(4.48)	103-0533-76
$C_{14}H_{10}N_4O_2$ 4-Cinnolineacetic acid, α,3-dicyano-, ethyl ester	EtOH	240(4.77),408(2.70)	39-0592-76C
Phenol, 4-pyrido[3,4-e]-1,2,4-triazin-3-yl-, acetate	dioxan	263(4.57),353(4.12), 484(2.49)	114-0301-76C
$C_{14}H_{10}N_4O_3$ 4(3H)-Quinazolinone, 3-amino-2-(4-nitro-phenyl)	EtOH	215(4.46),227(4.53), 257(4.31),310(4.21)	114-0341-76D
$C_{14}H_{10}N_4O_4$ 1H-Benzimidazole, 2-[(2,4-dinitrophen-yl)methyl]-	0.1N H$_2$SO$_4$	241(4.28),270(4.24), 276(4.19)	104-1097-76
	EtOH	243(4.34),274(4.14), 281(4.08)	104-1097-76

Compound	Solvent	$\lambda_{max}(\log \epsilon)$	Ref.
1H-Benzimidazole, 2-[(2,4-dinitrophen-yl)methyl]- (cont.)	EtOH-KOH	490(4.34),645(3.84)	104-1097-76
photochromic form	EtOH	515(4.54)	104-2361-76
$C_{14}H_{10}O$			
1-Anthracenol, hyamine salt	DMF	480(3.60)	112-0487-76
2-Anthracenol, hyamine salt	DMF	446(3.60)	112-0487-76
9-Anthracenol, hyamine salt	DMF	507(3.64)	112-0487-76
$C_{14}H_{10}OS$			
Benzo[b]thiophen-3(2H)-one, 2-phenyl-	EtOH	220(4.05),235(4.07), 265(3.63)	111-0533-76
$C_{14}H_{10}OSSe$			
Ethanone, 1-(2-phenylselenolo[2,3-b]thi-en-5-yl)-	EtOH	295(4.47),350s(--)	104-1550-76
$C_{14}H_{10}OS_2$			
Ethanone, 1-(5-phenylthieno[2,3-b]thien-2-yl)-	EtOH	287(4.66),345(3.85)	104-1550-76
$C_{14}H_{10}O_2$			
3(2H)-Benzofuranone, 2-phenyl-	EtOH	220(3.70),267(3.00), 280(2.90)	111-0533-76
9H-Xanthen-9-one, 2-methyl-	n.s.g.	233(4.52),241(4.53), 261(4.06),290(3.54), 340(3.76)	39-2241-76C
$C_{14}H_{10}O_2S$			
Dibenzo[b,f]thiepin-10(11H)-one, 7-hy-droxy-	MeOH	257(4.36),288s(4.06), 315(3.64)	73-3607-76
Ethanone, 1-(2-phenylthieno[2,3-b]furan-5-yl)-	EtOH	277(4.51),345(4.16)	104-1550-76
$C_{14}H_{10}O_2Se$			
Ethanone, 1-(2-phenylselenolo[2,3-b]fur-an-5-yl)-	EtOH	290(4.69),368(4.14)	104-1550-76
$C_{14}H_{10}O_3$			
Benzoic acid, 2-benzoyl-[14]C-	MeOH	247(4.14)	56-0499-76
9H-Xanthen-9-one, 2-methoxy-	n.s.g.	234(4.52),248(4.45), 302(3.54),358(3.76)	39-2241-76C
9H-Xanthen-9-one, 4-methoxy-	n.s.g.	212(4.02),232(4.38), 246(4.46),275(3.69), 288(3.51),345(3.66)	39-2241-76C
$C_{14}H_{10}O_4$			
9(10H)-Anthracenone, 1,3,8-trihydroxy-	MeOH-1% HCOOH	258(3.95),264(3.89), 302s(3.91),355(4.21)	12-1523-76
$C_{14}H_{10}O_4S$			
Benzeneacetic acid, 2-carboxy-α-(2-thi-enylmethylene)-, (E)-	EtOH	300(4.14)	4-0083-76
$C_{14}H_{10}O_4S_4$			
1,3-Dithiole-4,5-dicarboxylic acid, 2-(1,3-benzodithiol-2-ylidene)-, dimethyl ester	$C_2H_4Cl_2$	255(4.28),313(4.24), 416(2.00)	77-0966-76

Compound	Solvent	$\lambda_{max}(\log \epsilon)$	Ref.
$C_{14}H_{10}O_5$ Benzeneacetic acid, 2-carboxy-α-(2-furanylmethylene)-, (E)-	EtOH	300(4.25)	4-0083-76
$C_{14}H_{10}O_6$ 1,4-Dibenzofurandione, 8-hydroxy-3,7-dimethoxy-	EtOH	250(4.44),280(4.20), 325(4.01),490(3.71)	115-0185-76
$C_{14}H_{10}S$ Benzo[b]thiophene, 3-phenyl-	EtOH	232(4.3),263(3.8), 293(3.6),302(3.6)	12-0883-76
Dibenzo[b,d]thiepin	C_6H_{12}	213s(4.39),229(4.48), 260s(4.15),300s(3.06)	89-0611-76
$C_{14}H_{10}S_3$ Benzenecarbodithioic acid anhydrosulfide	C_6H_{12}	307(4.16),548(2.16)	89-0766-76
$C_{14}H_{11}BrN_2O$ Benzeneethanehydrazonoyl bromide, α-oxo-N-phenyl-	EtOH	253(4.257),372(4.169)	18-0321-76
Salicylaldehyde, [(4-bromophenyl)methylene]hydrazone	EtOH	224(4.27),234s(4.10), 248s(3.92),254s(3.89), 301(4.49),312s(4.47), 345(4.33)	95-0044-76
$C_{14}H_{11}BrN_2O_2$ Methanone, [4-(5-bromo-2-furanyl)-4,5-dihydro-1H-pyrazol-3-yl]phenyl-	EtOH	223(4.16),258(3.93), 333(4.10)	104-2358-76
Methanone, (4-bromophenyl)[4-(2-furanyl)-4,5-dihydro-1H-pyrazol-3-yl]-	EtOH	217(4.16),266(3.99), 338(4.07)	104-2358-76
$C_{14}H_{11}BrN_4$ 2H-Tetrazole, 2-(4-bromophenyl)-5-(4-methylphenyl)-	EtOH	245(4.27),277(4.33)	18-0762-76
$C_{14}H_{11}BrN_4O$ 2H-Tetrazole, 2-(4-bromophenyl)-5-(4-methoxyphenyl)-	EtOH	255(4.33),282(4.34)	18-0762-76
$C_{14}H_{11}BrO_2$ 2,4,6-Cycloheptatrien-1-one, 2-bromo-7-(phenylmethoxy)-	EtOH	256(4.47),328(3.98), 345(3.89),361(3.94), 371s(3.78)	39-2403-76C
2,4,6-Cycloheptatrien-1-one, 3-bromo-2-(phenylmethoxy)-	EtOH	253(4.28),322(3.82)	39-2403-76C
$C_{14}H_{11}BrO_4S$ 2H-Pyran-3-carboxylic acid, 6-(4-bromophenyl)-4-(methylthio)-2-oxo-, methyl ester	EtOH	245(3.82),338(4.08)	142-1493-76
$C_{14}H_{11}Cl$ Ethene, 2-chloro-1,1-diphenyl-	EtOH	227(4.2),257(4.0), 292(3.0),303(2.8)	12-0883-76
$C_{14}H_{11}ClN_2$ 3H-Indazole, 5-chloro-3-methyl-3-phenyl-	isoPrOH	233(4.22),270(3.89), 350(2.38)	4-0033-76

Compound	Solvent	$\lambda_{max}(\log \epsilon)$	Ref.
$C_{14}H_{11}ClN_2O$			
Benzaldehyde, 2-hydroxy-, [(4-chloro-phenyl)methylene]hydrazone]	EtOH	224(4.29),245s(3.93), 253(3.88),300(4.52), 344(4.31)	95-0044-76
2,4,6-Heptatrienamide, 5-chloro-2-cyano-7-phenyl-	CH_2Cl_2	385(4.57)	48-0705-76
1H-Indazole, 5-chloro-1-methyl-3-phen-yl-, 2-oxide	isoPrOH	232(4.43),247(4.39), 311(4.18)	4-0033-76
$C_{14}H_{11}ClN_2O_2$			
Methanone, [5-chloro-2-(methyl-ONN-az-oxy)phenyl]phenyl-	isoPrOH	250(4.34),290s(3.53)	4-0907-76
geometric isomer	isoPrOH	236(4.27),290s(3.92)	4-0907-76
$C_{14}H_{11}ClN_2O_3$			
Benzenecarboximidoyl chloride, 4-meth-oxy-N-(4-nitrophenyl)-	hexane	250s(3.81),298(4.41)	104-0161-76
Benzenecarboximidoyl chloride, N-(4-methoxyphenyl)-4-nitro-	hexane	260(4.22),345s(3.70), 365(3.81),375s(3.78)	104-0161-76
$C_{14}H_{11}ClN_4$			
2H-Tetrazole, 2-(4-chlorophenyl)-5-(4-methylphenyl)-	EtOH	244(4.28),276(4.33)	18-0762-76
2H-Tetrazole, 5-(4-chlorophenyl)-2-(4-methylphenyl)-	EtOH	277(4.37)	18-0762-76
$C_{14}H_{11}ClN_4O$			
2H-Tetrazole, 2-(4-chlorophenyl)-5-(4-methoxyphenyl)-	EtOH	254(4.34),281(4.34)	18-0762-76
$C_{14}H_{11}ClN_4S$			
1H-1,2,4-Triazolium, 4-amino-5-(4-chlo-rophenyl)-2,3-dihydro-1-phenyl-3-thi-oxo-, hydroxide, inner salt	MeCN	254(4.72),334(4.48)	103-0361-76
$C_{14}H_{11}ClO_2S$			
2-Thiophenecarboxylic acid, 5-[2-(2-chlorophenyl)ethenyl]-, methyl ester	MeOH	244(4.04),250(4.02), 344(4.40),367s(4.17)	48-0731-76
2-Thiophenecarboxylic acid, 5-[2-(4-chlorophenyl)ethenyl]-, methyl ester	MeOH	227s(4.08),234(4.09), 243(4.06),250s(4.04), 335s(4.55),348(4.60), 366s(4.49)	48-0731-76
$C_{14}H_{11}ClO_3$			
Benzoic acid, 2-chloro-3-(phenylmethoxy)-	MeOH	290(3.28)	12-2003-76
$C_{14}H_{11}ClO_3S$			
Benzoic acid, 5-chloro-2-[(3-methoxy-phenyl)thio]-	MeOH	223(4.39),258(4.05), 280s(3.89),288(3.83)	73-3607-76
$C_{14}H_{11}ClO_4S$			
3,4-Thiophenedicarboxylic acid, 2-(4-chlorophenyl)-, dimethyl ester	MeOH	233(4.36),275(4.03)	44-1724-76
$C_{14}H_{11}ClO_6$			
1,3-Naphthalenedicarboxylic acid, 6-chloro-2,4-dihydroxy-, dimethyl ester	EtOH	257(4.55),297(3.89), 309(3.78),354(3.72)	78-0269-76
2H-Pyran-3-carboxylic acid, 5-(4-chloro-phenyl)-4-hydroxy-6-methoxy-2-oxo-, methyl ester	EtOH EtOH-base	246(4.12),314(4.00) 275s(3.98)	78-0269-76 78-0269-76

Compound	Solvent	$\lambda_{max}(\log \epsilon)$	Ref.
$C_{14}H_{11}Cl_2NO_2$ Benzoic acid, 2-[(2,6-dichloro-3-methyl-phenyl)amino]-, β-cyclodextrin complex	pH 7.0	285(3.68),320(3.63)	133-0343-76
$C_{14}H_{11}Cl_2NO_5S$ Furan, 2-[1-[(dichloromethyl)sulfonyl]-2-(4-methylphenyl)ethenyl]-5-nitro-	EtOH	205(4.29),226s(4.19), 298(4.38)	73-3391-76
$C_{14}H_{11}Cl_2NO_6S$ Furan, 2-[1-[(dichloromethyl)sulfonyl]-2-(3-methoxyphenyl)ethenyl]-5-nitro-	EtOH	206(4.47),289(4.25)	73-3391-76
$C_{14}H_{11}Cl_4N$ Benzenamine, N,N-dimethyl-4-[(2,3,4,5-tetrachloro-2,4-cyclopentadien-1-yl-idene)methyl]-	C_6H_{12}	461(4.69)	39-0048-76C
Benzenemethanamine, N,N-dimethyl-α-(2,3,4,5-tetrachloro-2,4-cyclo-pentadien-1-ylidene)-	C_6H_{12}	266(4.08),404(4.19)	24-3939-76
$C_{14}H_{11}Cl_4NO$ Methanamine, N,N-dimethyl-1-phenoxy-1-(2,3,4,5-tetrachloro-2,4-cyclo-pentadien-1-ylidene)-	dioxan	358(4.30)	24-3939-76
$C_{14}H_{11}FN_2O_3$ Benzenecarboximidoyl fluoride, 4-meth-oxy-N-(4-nitrophenyl)-	hexane	265s(3.85),310(4.36)	104-0161-76
Benzenecarboximidoyl fluoride, N-(4-methoxyphenyl)-4-nitro-	hexane	250(4.29),266s(4.14), 345(4.26),361(4.27), 385s(4.03)	104-0161-76
$C_{14}H_{11}IO_3$ Benzoic acid, 2-iodo-3-(phenylmethoxy)-	MeOH	278(2.45),285(2.48)	12-2003-76
$C_{14}H_{11}Li$ Lithium, (9-methyl-9H-fluoren-9-yl)-	$(MeOCH_2)_2$	386(4.01),455s(--), 482(2.92),516(3.05), 556(3.93)	104-2411-76
$C_{14}H_{11}N$ 1-Anthracenamine	DMF	402(3.53)	112-0487-76
2-Anthracenamine	DMF	413(3.55)	112-0487-76
9-Anthracenamine	DMF	417(3.72)	112-0487-76
$C_{14}H_{11}NO$ Acridine, 9-methoxy-	EtOH	253(5.19),323s(3.42), 337(3.72),348s(3.86), 354(3.54),370s(3.76), 385(3.94)	35-0171-76
9(10H)-Acridinone, 10-methyl-	EtOH	254(4.80),388(3.94), 405(4.02)	48-0515-76
(unchanged in acid or base)	EtOH	254(4.66),293(3.32), 307(2.71),385(--), 403(3.91)	100-0134-76
9H-Carbazole, 9-acetyl-	hexane	313(3.90)	135-1054-76
Cyclohept[b]indol-6(5H)-one, 2-methyl-	MeOH	278(4.34),313(4.24), 326(4.23),338(4.19), 392s(3.68),409(3.82)	18-1101-76

Compound	Solvent	λ_{max} (log ϵ)	Ref.
Cyclohept[b]indol-6(5H)-one, 4-methyl-	MeOH	281(4.32),309(4.13), 323(4.11),337(4.02), 390s(3.72),407(3.83)	18-1101-76
2-Propen-1-one, 1-phenyl-3-(2-pyridinyl)-	n.s.g.	303(4.02)	39-0380-76C
2-Propen-1-one, 3-phenyl-1-(2-pyridinyl)-	n.s.g.	318(4.16)	39-0380-76C
C$_{14}$H$_{11}$NOS			
2,1-Benzisothiazol-3(1H)-one, 1-(phenyl-methyl)-	CHCl$_3$	250s(4.07),261(4.05), 371(3.76)	48-0161-76
6H-Thieno[2,3-b]pyrrole-2-carboxalde-hyde, 6-methyl-5-phenyl-	EtOH	295(4.53)	104-1550-76
C$_{14}$H$_{11}$NO$_2$			
Benzamide, N-benzoyl-	MeOH	243(5.29)	117-0025-76
1H-Benz[f]isoindole-1,3(2H)-dione, 2,4-dimethyl-	ether	218(4.38),252(4.66), 259(4.76),281(3.86), 292(3.90),305(3.73), 324(3.20),339(3.54), 357(3.74)	22-0493-76
2H-1,4-Benzoxazin-2-ol, 3-phenyl-	EtOH	233(3.96),242(3.94), 288(4.12),320(4.05)	78-2033-76
9H-Carbazole-1-carboxaldehyde, 2-hy-droxy-3-methyl-	EtOH	225(4.62),234(4.60), 287(4.29),295(4.24), 390(3.87)	2-0329-76
9H-Carbazole-1-carboxylic acid, methyl ester	n.s.g.	225(4.36),243(4.06), 250(4.07),279(4.20), 303(3.83),359(3.78)	44-0102-76
3-Dibenzofuranamine, N-acetyl-	C$_6$H$_{12}$	296(4.56),315(4.58)	40-0133-76
Pyrido[2,1,6-de]quinolizine-1-carbox-ylic acid, methyl ester	C$_6$H$_{12}$	247(4.10),282(4.39), 404(3.96),451(4.21), 467(4.28),476(4.49)	39-0341-76C
C$_{14}$H$_{11}$NO$_3$			
8H-1,3-Dioxolo[4,5-h]pyrrolo[2,1-b]-[3]benzazepin-8-one, 5,6-dihydro-	EtOH	263(3.91),375(4.18)	77-0505-76
11H-1,3-Dioxolo[4,5-h]pyrrolo[2,1-b]-[3]benzazepin-11-one, 5,6-dihydro-	MeCN	236(4.51),277(4.20), 240(4.58)[sic]	44-0875-76
1H-Indole-4,7-dione, 5,6-dihydro-2-(2-hydroxyphenyl)-	MeOH	218(4.35),257(4.32), 273(4.26),286(4.21), 308(4.11),351(4.07)	111-0241-76
1H-Indole-4,7-dione, 5,6-dihydro-2-(3-hydroxyphenyl)-	MeOH	222(4.34),258(4.37), 342(4.07)	111-0241-76
Oxirane, 2-(4-nitrophenyl)-3-phenyl-	EtOH	218(3.24),281(3.14)	44-3820-76
C$_{14}$H$_{11}$NO$_3$S			
Benzonitrile, 2-hydroxy-5-[(4-methyl-phenyl)sulfonyl]-	EtOH	289(4.36)	56-0973-76
Benzonitrile, 2-[[(4-methylphenyl)sul-fonyl]oxy]-	EtOH	269(3.18),275(3.24), 283(3.06)	56-0973-76
2H-Pyran-3-carbonitrile, 6-(4-methoxy-phenyl)-4-(methylthio)-2-oxo-	EtOH	250(3.42),342(3.42), 395(3.83)	142-1493-76
C$_{14}$H$_{11}$NO$_3$S$_2$			
5H-1,4-Dithiino[2,3-c]pyrrole-5,7(6H)-dione, 6-(2-acetylphenyl)-2,3-dihydro-	EtOH	205(4.40),228(4.18), 269(4.04),414(3.43)	56-1523-76
5H-1,4-Dithiino[2,3-c]pyrrole-5,7(6H)-dione, 6-(3-acetylphenyl)-2,3-dihydro-	EtOH	206(4.40),236(4.38), 270(4.00),420(3.42)	56-1523-76
5H-1,4-Dithiino[2,3-c]pyrrole-5,7(6H)-dione, 6-(4-acetylphenyl)-2,3-dihydro-	EtOH	204(4.32),281(4.35), 422(3.40)	56-1523-76

Compound	Solvent	$\lambda_{max}(\log \epsilon)$	Ref.
$C_{14}H_{11}NO_4$			
9-Acridanone, 1,3-dihydroxy-5-methoxy-	EtOH	253(4.57),266(4.34), 275(4.40),294s(4.08), 395(3.81)	100-0399-76
Benzoic acid, 2-[(2-carboxyphenyl)amino]-, β-cyclodextrin complex	pH 7.0	289(4.07),338(3.86)	133-0343-76
Benzoic acid, 2-[(3-carboxyphenyl)amino]-, β-cyclodextrin complex	pH 7.0	288(4.11),320(3.81)	133-0343-76
Benzoic acid, 2-[(4-carboxyphenyl)amino]-, β-cyclodextrin complex	pH 7.0	320(4.27)	133-0343-76
Cyclopenta[e][1,3]oxazine-2,4(3H,5H)-dione, 3-benzoyl-	MeOH	249(4.2)	5-1689-76
1H-Indole-4,7-dione, 2-(2,4-dihydroxyphenyl)-5,6-dihydro-	MeOH	221(4.31),262(4.33), 288(4.23),311(4.17), 368(4.05)	111-0241-76
1H-Indole-4,7-dione, 2-(2,5-dihydroxyphenyl)-5,6-dihydro-	MeOH	220(4.32),258(4.28), 281(4.13),331(4.05)	111-0241-76
1H-Indole-4,7-dione, 2-(3,4-dihydroxyphenyl)-5,6-dihydro-	MeOH	217(4.25),262(4.33), 360(3.98),455(2.78)	111-0241-76
$C_{14}H_{11}NO_4S$			
2,5-Furandione, 3-(4,5-dihydro-4-methyl-5-oxo-2-phenyl-4-thiazolyl)dihydro-, (R*,S*)-	MeOH	209(4.08),243(4.17), 270s(3.57)	78-0571-76
3H-Thiopyrano[3,4-c]furan-4,7-imine-1,3,6-trione, tetrahydro-7-methyl-4-phenyl-, (3aα,4β,7β,7aα)-	MeOH	213(3.98),248(3.94)	78-0571-76
$C_{14}H_{11}NO_5$			
Benzoic acid, 4-[(6-methoxy-3,4-dioxo-1,5-cyclohexadien-1-yl)amino]-,	H_2O	270(4.40),335(4.18), 509(3.42)	135-0717-76A
sodium salt	MeOH	279(4.40),334(4.18), 490(3.42)	135-0717-76A
	dioxan	280(4.40),328(4.18), 476(3.42)	135-0717-76A
1,3-Dioxolo[4,5-g]furo[2,3-b]quinoline, 4,9-dimethoxy- (flindersiamine)	EtOH	253(4.53),311(3.76)	102-0743-76
4H-Pyrrolo[2,1-c][1,4]benzoxazine-4-acetic acid, 4-carboxy-	MeOH	264(3.93),293(3.82)	4-0311-76
$C_{14}H_{11}NO_5S$			
Benzoic acid, 2-[(3-methoxyphenyl)thio]-5-nitro-	MeOH	260s(3.88),343(4.17)	73-3607-76
$C_{14}H_{11}NO_7$			
1,3-Dioxolo[4,5-g]quinoline-6,7-dicarboxylic acid, 5,8-dihydro-8-oxo-, dimethyl ester	MeOH	210(4.26),223(4.38), 247(4.39),267(4.27), 276s(4.20),327s(3.83), 344s(3.94),355(3.96), 371s(3.83)	5-1972-76
$C_{14}H_{11}N_3O$			
4(3H)-Quinazolinone, 3-amino-2-phenyl-	EtOH	215(4.31),232(4.30), 270(4.20),305(4.01)	114-0341-76D
$C_{14}H_{11}N_3OS$			
5(2H)-Isoxazolethione, 3-methyl-4-(2-naphthalenylazo)-	EtOH	243s(4.47),288s(3.87), 300(3.83),311(3.79), 341(3.71),475(4.26)	48-0359-76

Compound	Solvent	$\lambda_{max}(\log \epsilon)$	Ref.
$C_{14}H_{11}N_3O_2$			
9-Acridinamine, N-methyl-1-nitro-	pH 1	272(4.43),324(3.52), 438(3.79)	56-1603-76
	pH 6.8	273(4.46),322(3.49), 439(3.76)	56-1603-76
1,2,3-Benzotriazin-4(3H)-one, 6-methoxy-3-phenyl-	MeOH	232(4.46),310(4.08)	5-0946-76
3-Cinnolinecarbonitrile, 4-(2,4-dioxopentyl)-	EtOH	243(4.76),288(4.37)	39-0592-76C
Pyrido[3,4-c]cinnolin-4(3H)-one, 1-acetyl-2-methyl-	EtOH	244(4.23),363(3.70)	39-0592-76C
2H-Pyrido[1,2-a]-1,3,5-triazinium, 3,4-dihydro-2,4-dioxo-1-(phenylmethyl)-, hydroxide, inner salt	MeOH	220s(3.92),235(3.97), 285s(3.39),315(3.65)	24-3661-76
$C_{14}H_{11}N_3O_3$			
Diazene, [1,3-benzodioxol-5-yl(hydroxyimino)methyl]phenyl-	EtOH	302(4.32),436(2.58)	23-3260-76
	50% EtOH-pH 12.5	238(4.11),290(4.00), 350(4.36)	23-3260-76
$C_{14}H_{11}N_3O_4$			
Methanone, [4-(2-furanyl)-4,5-dihydro-1H-pyrazol-3-yl](4-nitrophenyl)-	EtOH	213(4.17),268(4.15), 348(3.97)	104-2358-76
$C_{14}H_{11}N_3O_5$			
Methanone, (2,5-dimethyl-4-pyridinyl)-(2,4-dinitrophenyl)-	EtOH	232(4.12),300(3.56)	103-0312-76
$C_{14}H_{11}N_5$			
1-Butene-1,1,2-tricarbonitrile, 3-(methylphenylhydrazono)-	EtOH	220(3.97),265(3.93), 485(4.27)	24-1787-76
1H-Pyrazolo[3,4-d]pyrimidine-4-acetonitrile, 1-methyl-α-phenyl-	EtOH	332(4.25)	95-1352-76
$C_{14}H_{11}N_5OS$			
1H-Pyrrolo[2,3-d]pyrimidine-5-carbonitrile, 4-amino-2,3-dihydro-6-mercapto-2-oxo-3-(phenylmethyl)-	5% NaOH	328(3.95)	24-2983-76
$C_{14}H_{11}N_5O_2$			
2H-Tetrazole, 2-(4-methylphenyl)-5-(4-nitrophenyl)-	EtOH	299(4.44)	18-0762-76
2H-Tetrazole, 5-(4-methylphenyl)-2-(4-nitrophenyl)-	EtOH	247(4.32),312(4.30)	18-0762-76
4H-1,2,4-Triazol-4-amine, N-(2-nitrophenyl)-3-phenyl-	EtOH	242(4.06),288(4.13), 347(3.93)	104-1535-76
$C_{14}H_{11}N_5O_3$			
2H-Tetrazole, 5-(4-methoxyphenyl)-2-(4-nitrophenyl)-	EtOH	259(4.39),326(4.22)	18-0762-76
$C_{14}H_{11}N_6O_4$			
Diazenium, 1-(2-carboxyethenyl)-2-(2,4-dihydroxyphenyl)-1-(1H-purin-6-yl)-, (Z)-, sulfate(2:1)	EtOH-pH 1.5	270(4.14),328(4.10), 478(4.35)	103-0817-76
	EtOH-pH 13	282(4.16),353(--), 511(4.40)	103-0817-76
$C_{14}H_{11}OS_2$			
1,3-Benzodithiol-1-ium, 2-(4-methoxyphenyl)-, perchlorate	96% H_2SO_4	230f(4.22),260s(3.58), 299)3.39),428(4.50)	118-0759-76

Compound	Solvent	λ_{max}(log ϵ)	Ref.
$C_{14}H_{12}$ Azulene, 1,3-diethenyl-	hexane	258(4.09),292(4.23), 302(3.60),391(3.60), 413(3.48),610s(2.28), 661(2.37),702(2.28), 734(2.25),793(1.92)	117-0169-76
9H-Fluorene, 9-methyl-, lithium salt	$(MeOCH_2)_2$	386(4.01),455s(--), 482(2.92),516(3.05), 556(3.93)	104-2411-76
5H-4a,5,8-Methenobenzo[2,3]cyclopropa- [1,2-a]pentalene, 4b,4c,7a,8-tetra- hydro-	n.s.g.	210(3.91),275(3.57)	35-1988-76
Pentaleno[2,1,6-ija]azulene, 2a,8a,8b,8c-tetrahydro-	n.s.g.	323(3.60),349(3.65)	35-1988-76
$C_{14}H_{12}BrN_2S$ Isothiazolo[5,4-b]pyridinium, 3-bromo- 4,7-dimethyl-6-phenyl-, perchlorate	DMF	321(4.04)	48-0779-76
$C_{14}H_{12}BrN_3$ Pyrido[2,3-b]pyrazine, 7-bromo-4,8-di- hydro-8-methyl-3-phenyl-	EtOH	249(4.26),272(4.23), 400(3.99)	22-0251-76
$C_{14}H_{12}ClNO$ Benzenecarboximidoyl chloride, N-(4-methoxyphenyl)-	hexane	240(4.33),285s(3.89), 330(3.81)	104-0161-76
Benzenecarboximidoyl chloride, 4-methoxy-N-phenyl-	hexane	274(4.33),282(4.23), 315(3.74)	104-0161-76
$C_{14}H_{12}ClNO_2$ 2,4-Pentadienoic acid, 5-chloro-2-cyano- 5-phenyl-, ethyl ester	CH_2Cl_2	343(4.48)	48-0705-76
$C_{14}H_{12}ClNO_3S$ Carbamothioic acid, (4-chloro-2-oxo-2H- pyran-6-yl)methyl-, S-(phenylmethyl) ester	EtOH	337(3.95)	39-2462-76C
$C_{14}H_{12}ClN_3$ 1H-Indazol-4-amine, 3-(4-chlorophenyl)- 1-methyl-	MeOH	241s(4.13),320(4.02)	24-1898-76
$C_{14}H_{12}ClN_3O_2$ 1H-Pyrazole, 1-acetyl-4-(5-chloro-2,1- benzisoxazol-3-yl)-3,5-dimethyl-	EtOH	261(3.86),342(4.06)	94-1106-76
$C_{14}H_{12}Cl_2N_2$ Benzenecarbohydrazonoyl chloride, 4-chloro-N-(4-methylphenyl)-	C_6H_{12}	339(3.70),340(3.92)	48-0881-76
$C_{14}H_{12}Cl_2N_2O$ Benzenecarbohydrazonoyl chloride, N-(4-chlorophenyl)-4-methoxy-	C_6H_{12}	241(4.27),249s(--), 292s(--),305(4.02), 342(4.34)	48-0881-76
$C_{14}H_{12}Cl_4O_2$ 1H-Cyclopenta[b][1,4]benzodioxin, 5,6,7,8-tetrachloro-2,3,3a,9a-tetra- hydro-1-(1-methylethylidene)-, cis	C_6H_{12}	219(4.83),232s(4.16), 293s(3.28),302(3.38)	78-0147-76

Compound	Solvent	$\lambda_{max}(\log \epsilon)$	Ref.
$C_{14}H_{12}FNO$			
Benzenecarboximidoyl fluoride, N-(4-methoxyphenyl)-	hexane	231(4.22),285s(4.08), 304(4.17),317(4.16)	104-0161-76
Benzenecarboximidoyl fluoride, 4-methoxy-N-phenyl-	hexane	269(4.33),285(4.29)	104-0161-76
$C_{14}H_{12}FNO_2$			
Benzoic acid, 5-fluoro-2-(phenylamino)-, methyl ester	MeOH	239(4.27),283(4.10), 366(3.83)	5-0946-76
$C_{14}H_{12}MoO_3$			
Molybdenum, [η^6-8,9-bis(methylene)tricyclo[3.2.2.0^{2,4}]non-6-ene]tricarbonyl-	C_6H_{12}	218(4.49),331(4.06)	23-1958-76
$C_{14}H_{12}N$			
Acridinium, N-methyl-, triiodide	CH_2Cl_2	260(5.16),292(4.47), 360(4.49)	104-1531-76
$C_{14}H_{12}NO_8$			
1,3-Cyclopentadiene-1,2,3,4-tetracarboxylic acid, 5-cyano-, tetramethyl ester, ion(1-), sodium deriv.	MeOH	212(3.96),263(4.63), 300(4.14)	24-1928-76
$C_{14}H_{12}NS$			
Benzothiazolium, 3-methyl-2-phenyl-	MeOH-H_3BO_3	300(4.14)	78-1773-76
$C_{14}H_{12}NS_2$			
Benzenaminium, N-3H-1,2-benzodithiol-3-ylidene-N-methyl-, iodide	EtOH	231(4.25),247(4.25), 285(4.01),391(3.83)	48-0221-76
$C_{14}H_{12}N_2$			
9-Acridinamine, N-methyl-	pH 1	220(4.40),260(4.73), 400(4.08),420(4.00)	56-1267-76
Benzo[c]cinnoline, 1,10-dimethyl-	MeOH	252(4.56),331(3.95)	18-2495-76
3H-Indazole, 3-methyl-3-phenyl-	isoPrOH	220(4.26),262(3.81)	4-0033-76
1H-Indole, 3-(3-pyridinylmethyl)-	EtOH	270(3.93),282(3.87), 291(3.81)	39-1155-76C
2,7-Phenanthrenediamine	EtOH	288(4.20),326(3.86), 386(3.36)	104-2355-76
1,10-Phenanthroline, 2,9-dimethyl-, Cu(I) complex	MeOH	457(3.90)	118-0001-76
Propanedinitrile, 1-benzocycloheptylidene-	EtOH	231(4.00),297(3.98)	118-0705-76
$C_{14}H_{12}N_2O$			
Ethanone, 1-[4-(phenylazo)phenyl]-	hexane	230(4.03),325(4.38), 452(2.83)	61-0301-76
3H-Indazol-5-ol, 3-methyl-3-phenyl-	isoPrOH	237(4.04),308(3.97)	4-0033-76
1H-Indole-3-methanol, α-3-pyridinyl-	EtOH	268(3.90),280s(3.86), 288(3.79)	39-1155-76C
$C_{14}H_{12}N_2O_2$			
2,3a-Epoxy-3,7-methano-3aH-indene-7,7a(1H,6H)-dicarbonitrile, 2,3-dihydro-1,8-dimethyl-6-oxo-	MeOH	258(3.51),330(1.82)	35-7040-76
Ethanone, 1-(7-methoxy-9H-pyrido-[3,4-b]indol-1-yl)-	CHCl$_3$	260(4.18),289(4.37), 330(4.11)	102-1559-76
Methanone, [4-(2-furanyl)-4,5-dihydro-1H-pyrazol-3-yl]phenyl-	EtOH	217(4.12),253(3.86), 330(4.11)	104-2358-76

Compound	Solvent	$\lambda_{max}(\log \epsilon)$	Ref.
4a,8a-Naphthalenedicarbonitrile, 1,4,5,8-tetrahydro-1,4-dimethyl-5,8-dioxo-, (1α,4α,4aβ,8aβ)-	MeOH	240(3.79),352(1.81)	35-7040-76
Tricyclo[4.4.0.0^{5,9}]deca-3,7-dien-2-one, 1,6-dicyano-5-hydroxy-8,9-dimethyl-	MeOH	239(3.58)	35-7040-76
$C_{14}H_{12}N_2O_3$			
Benzenamine, 4-methoxy-N-[(4-nitrophenyl)methylene]-	hexane	257(4.02),291(3.92), 365(3.97),375(3.98)	104-0161-76
Benzenamine, N-[(4-methoxyphenyl)methylene]-4-nitro-	hexane	270s(3.85),347(4.18)	104-0161-76
9H-Pyrido[3,4-b]indole-1-carboxylic acid, 4-methoxy-, methyl ester	EtOH	236(4.32),247(4.29), 267(4.41),278(4.31), 303(3.94),363(3.87)	94-1532-76
$C_{14}H_{12}N_2O_4$			
Benzoic acid, 5-nitro-2-(phenylamino)-, methyl ester	MeOH	215(4.32),257(3.89), 364(4.30)	5-0946-76
2-Furancarboxamide, 5-(2-nitrophenyl)-N-2-propenyl-	dioxan	290(4.19)	73-2577-76
2-Furancarboxamide, 5-(3-nitrophenyl)-N-2-propenyl-	dioxan	303(4.37)	73-2577-76
2-Furancarboxamide, 5-(4-nitrophenyl)-N-2-propenyl-	dioxan	351(4.30)	73-2577-76
2-Indolizineacetic acid, 3-cyano-1-(methoxycarbonyl)-, methyl ester	MeOH	207s(4.07),210s(4.16), 218s(4.38),220s(4.39), 222(4.40),240s(4.45), 246(4.55),261s(3.90), 266s(3.99),274(4.12), 302s(3.97),317(4.09), 329s(3.95),340s(3.75)	39-1908-76C
2,4-Pentadienoic acid, 2-cyano-5-(2-nitrophenyl)-, ethyl ester	MeOH	225(4.05),330(4.39)	65-1141-76
5H-1-Pyrindine-4-carboxylic acid, 3-cyano-6,7-dihydro-2,6-dimethyl-5,7-dioxo-, ethyl ester (enol)	H_2O 50% EtOH	337(3.78),385-420s(2.9) 337(3.81),385-420s(3.04)	103-0220-76 103-0220-76
anion	H_2O 50% EtOH	362(3.85),460(3.28) 366(3.86),463(3.28)	103-0220-76 103-0220-76
protonated enol	H_2O	356(3.79)	103-0220-76
$C_{14}H_{12}N_2O_4S$			
Thiazolo[5,4-f]quinoline-8-carboxylic acid, 6-ethyl-6,9-dihydro-2-methoxy-9-oxo-	MeOH	259(4.47),268(4.46)	94-0147-76
Thiazolo[5,4-f]quinoline-8-carboxylic acid, 6-ethyl-2,3,6,9-tetrahydro-3-methyl-2,9-dioxo-	MeOH	263.0(4.63)	94-0147-76
$C_{14}H_{12}N_2S$			
4,1,2-Benzothiadiazine, 1-methyl-3-phenyl-	EtOH	229(4.20),265(4.31), 316(3.56),380(3.53)	39-0038-76C
$C_{14}H_{12}N_4$			
Propanedinitrile, [(4-ethenylphenyl)azo]-2-propenyl-	CHCl$_3$	325(4.15)	126-1357-76
Pyrido[3,4-e]-1,2,4-triazine, 3-(2-phenylethyl)-	EtOH	233(4.41),341(3.48)	114-0285-76C
1,2,4,5-Tetrazine, 1,2-dihydro-3,6-diphenyl-	EtOH	248(4.45)	103-0533-76
2H-Tetrazole, 2-(4-methylphenyl)-5-phenyl-	EtOH	242(4.10),273(4.30)	18-0762-76

Compound	Solvent	$\lambda_{max}(\log \epsilon)$	Ref.
2H-Tetrazole, 5-(4-methylphenyl)-2-phenyl-	EtOH	244(4.23),275(4.27)	18-0762-76
$C_{14}H_{12}N_4O$			
2-Propanone, 1-(1-phenyl-1H-pyrazolo-[3,4-d]pyrimidin-4-yl)-	EtOH	265(4.20),344(4.53), 357(4.52)	95-1352-76
2H-Tetrazole, 5-(4-methoxyphenyl)-2-phenyl-	EtOH	256(4.26),279(4.30)	18-0762-76
$C_{14}H_{12}N_4O_2$			
Pyrido[3,4-e]-1,2,4-triazine, 3-(3,5-dimethoxyphenyl)-	EtOH	243(4.29),296(4.28), 370(3.60)	114-0285-76C
$C_{14}H_{12}N_4O_3$			
Acetamide, N-[4-(4,5-dihydro-7-hydroxy-5-oxopyrazolo[1,5-a]pyrimidin-3-yl)-phenyl]-	MeOH-pH 1 MeOH-pH 11	204(4.58),262(4.49) 232(4.56),311(4.51)	87-0291-76 87-0291-76
$C_{14}H_{12}N_4O_3S$			
Acetic acid, [[4-amino-1,6-dihydro-5-(1H-indol-3-yl)-6-oxo-2-pyrimidinyl]thio]-	EtOH	203s(4.48),220(4.57), 278(4.08),288(4.08)	103-1246-76
$C_{14}H_{12}N_4O_5$			
Acetamide, N-[3-(acetyloxy)-2-pyridinyl]-N-(3-nitro-2-pyridinyl)-	MeOH	208(4.26),225(4.29), 270(3.99)	4-0107-76
$C_{14}H_{12}N_4S$			
4H-1,2,4-Triazolium, 4-amino-2,3-dihydro-1,5-diphenyl-3-thioxo-, hydroxide, inner salt	MeCN	252(4.29),328(3.48)	103-0361-76
$C_{14}H_{12}N_6$			
1,1'-Biphenyl, 2,2'-diazido-6,6'-dimethyl-	hexane	254(4.30),288s(3.40)	18-2495-76
$C_{14}H_{12}O$			
Indeno[2,1-b]pyran, 2,4-dimethyl-	dioxan	291(4.14),324(3.92), 474(3.24)	78-2219-76
Indeno[2,1-b]pyran, 2-ethyl-	dioxan	228(4.04),263(3.90), 270(4.00),290(4.09), 339(3.61),360(3.29), 492(2.99)	78-2219-76
$C_{14}H_{12}OS$			
Benzene, [6-(methylsulfinyl)-5-heptene-1,3-diynyl]-	ether	300(4.20)	24-1964-76
trans isomer	ether	284(4.22),302(4.28), 321(4.22)	24-1964-76
$C_{14}H_{12}OSe$			
2,4,6-Cyclooctatrien-1-one, 8-(phenylseleno)-	EtOH	240s(4.05),280(3.74)	39-1838-76C
$C_{14}H_{12}O_2$			
1,2-Anthracenediol, 1,2-dihydro-, cis	MeOH	244(4.75),276(4.13), 287(4.23),298(4.21)	35-5988-76
trans	MeOH	244(4.59),276(3.97), 286(4.09),298(4.08)	35-5988-76

Compound	Solvent	$\lambda_{max}(\log \epsilon)$	Ref.
2-Fluorenol, 7-methoxy-	EtOH	275(4.37)	78-0065-76
	EtOH-NaOEt	297(4.42)	78-0065-76
7-Oxabicyclo[4.2.1]nona-2,4-dien-8-one	EtOH	255(3.64),265(3.61)	44-3583-76
1,2-Phenanthrenediol, 1,2-dihydro-, trans	MeOH	242(4.30),250(4.57), 259(4.68)	35-5988-76
3,4-Phenanthrenediol, 3,4-dihydro-, cis	MeOH	216(4.46),242(4.46), 250(4.72),258(4.81), 310(3.76)	35-5988-76
$C_{14}H_{12}O_2S$			
Benzeneacetic acid, α-(phenylthio)-	EtOH	215(4.00)	111-0533-76
Benzoic acid, 2-[2-(5-methyl-2-thienyl)-ethenyl]-	EtOH	230(4.13)	4-0083-76
1H-2-Benzopyran-1-one, 3,4-dihydro-3-(5-methyl-2-thienyl)-	EtOH	282(3.31)	4-0083-76
2-Propenoic acid, 3-(4-phenyl-2-thien-yl)-, methyl ester, (E)-	EtOH	265(4.38),362(4.02)	44-1320-76
2-Propenoic acid, 3-(5-phenyl-2-thien-yl)-, methyl ester, (E)-	MeOH	245(3.99),351(4.47)	44-1320-76
2-Thiophenecarboxylic acid, 5-(2-phenyl-ethenyl)-, methyl ester	MeOH	233(4.06),239s(4.04), 329s(4.37),344(4.45), 361s(4.30)	48-0731-76
$C_{14}H_{12}O_2S_2$			
5H,7H-Dibenzo[b,g][1,5]dithiocin, 12,12-dioxide	EtOH	250(3.86)	39-0913-76C
$C_{14}H_{12}O_3$			
Benzaldehyde, 2-(4-methoxyphenoxy)-	n.s.g.	210(4.31),220(4.18), 230(4.06),250(3.98), 274(3.34),282(3.28), 315(3.54)	39-2241-76C
Benzeneacetic acid, α-phenoxy-	EtOH	222(3.98),263(3.15), 276(3.06)	111-0533-76
2H,8H-Benzo[1,2-b:3,4-b']dipyran-2-one, 8,8-dimethyl- (seselin)	EtOH	219(4.45),285(4.04), 295(4.09),333(4.12)	39-1857-76C
Ethanone, 1-(5-benzoyl-2-methyl-3-furan-yl)-	EtOH	259(4.22)	94-2421-76
Ethanone, 1,1'-(5-phenyl-2,4-furandiyl)-bis-	EtOH	257(4.20),278(4.28)	94-2421-76
2H-Furo[2,3-h][1]benzopyran-2-one, 8,9-dihydro-8-(1-methylethenyl)- (angenomalin)	n.s.g.	250(3.50),262(4.0), 328(4.11)	105-0335-76
2H-Furo[2,3-h][1]benzopyran-2-one, 8-(1-methylethyl)-	EtOH	206(4.35),253(4.42), 305(4.03)	39-1857-76C
4H-Naphtho[1,2-b]pyran-5,6-dione, 3-methyl-	EtOH	256(4.42),280s(3.72), 330(3.20),427(3.24)	39-0600-76C
4H-Naphtho[2,3-b]pyran-5,10-dione, 3-methyl-	EtOH	250(4.37),281(4.13), 333(3.45),375s(3.07)	39-0600-76C
2H-Pyran-2-one, 4-methoxy-6-(2-phenyl-ethenyl)-	EtOH	230(4.16),255(4.13), 340(4.44),360s(4.36)	2-0127-76
4H,5H-Pyrano[3,2-c][1]benzopyran-4-one, 2,9-dimethyl-	EtOH	232(4.09),268(3.99), 279(4.11),289(4.07), 346(3.86)	39-2444-76C
$C_{14}H_{12}O_3S$			
5H,7H-Dibenz[b,g][1,5]oxathiocin, S,S-dioxide	EtOH	220(4.39)	39-0913-76C
Ethanone, 1-phenyl-2-(phenylsulfonyl)-	MeOH	257(4.20),274s(3.66), 288s(3.24)	30-0366-76

Compound	Solvent	$\lambda_{max}(\log \epsilon)$	Ref.
Ethanone, 1-phenyl-2-(phenylsulfonyl)- (cont.)	isoBuOH	255(4.17),273s(3.68), 283s(3.32)	30-0366-76
	ether	254(4.24),275(3.50), 283s(3.22)	30-0366-76
	CH_2Cl_2	257(4.26),273s(3.81), 283s(3.73)	30-0366-76
sodium deriv.	MeOH	220s(4.26),306(4.16)	30-0366-76
2-Furancarbothioic acid, 4-acetyl-5-phenyl-, S-methyl ester	EtOH	253(3.99)	94-2421-76
2(3H)-Furanone, dihydro-5-(4-hydroxyphenyl)-5-(2-thienyl)-	EtOH-pH 9	420(3.72)	18-2027-76
2-Thiophenecarboxylic acid, 5-[2-(4-hydroxyphenyl)ethenyl]-, methyl ester	MeOH	232s(4.02),258(4.00), 261s(3.99),288s(3.80), 362(4.61)	48-0731-76
$C_{14}H_{12}O_4$			
2(5H)-Furanone, 3-acetyl-4-hydroxy-5-[(4-methylphenyl)methylene]-	EtOH	235(4.18),316(4.50)	4-0521-76
2(5H)-Furanone, 3-acetyl-4-hydroxy-5-(1-phenylethylidene)-	EtOH	250(4.10),302(4.33)	4-0521-76
1,8-Naphthalenedicarboxylic acid, dimethyl ester	EtOH	299(3.54)	44-2401-76
2H-Pyran-2-one, 6-[2-(4-hydroxyphenyl)ethenyl]-4-methoxy-	EtOH	230(4.0),288(3.52), 302(3.67),350(4.02)	2-0300-76
	EtOH-base	236(3.96),288(3.73), 302(3.67),415(4.21)	2-0300-76
$C_{14}H_{12}O_4S$			
2,4,6-Cycloheptatrien-1-one, 4-[[(4-methylphenyl)sulfonyl]oxy]-	EtOH	303(3.81),313(3.78)	39-1915-76C
2(3H)-Furanone, 5-(2,4-dihydroxyphenyl)-dihydro-5-(2-thienyl)-	EtOH-pH 8.8	500(3.50)	18-2027-76
2-Thiophenecarboxylic acid, 5-[4-(methoxycarbonyl)phenyl]-, methyl ester	MeOH	228(3.97),311s(4.44), 321(4.47),339s(4.24)	48-0731-76
3,4-Thiophenedicarboxylic acid, 2-phenyl-, dimethyl ester	MeOH	230(4.28),275(3.91)	44-1724-76
$C_{14}H_{12}O_4S_2$			
5H,7H-Dibenzo[b,g][1,5]dithiocin, 6,6,12,12-tetraoxide	EtOH	220(4.26),262(3.70)	39-0913-76C
$C_{14}H_{12}O_5$			
6,12-Dioxatricyclo[6.2.2.0^{1,9}]dodec-3-ene-2,5,11-trione, 8-methyl-4-(1-propynyl)-, [1S-(1R*,8S*,9S*)]-	MeOH	273.5(4.11)	33-1809-76
2(5H)-Furanone, 3-acetyl-4-hydroxy-5-[(4-methoxyphenyl)methylene]-	EtOH	243(4.19),334(4.51)	4-0521-76
$C_{14}H_{12}O_5S$			
Benzoic acid, 2-hydroxy-5-[(4-methylphenyl)sulfonyl]-	EtOH	262(4.22)	56-0973-76
2(3H)-Furanone, dihydro-5-(2-thienyl)-5-(2,4,6-trihydroxyphenyl)-	EtOH-pH 9	480(3.30)	18-2027-76
$C_{14}H_{12}O_6$			
1,3-Naphthalenedicarboxylic acid, 2,4-dihydroxy-, dimethyl ester	EtOH	253(4.61),298(3.89), 310(3.87),364(3.71)	78-0269-76
2,3-Naphthalenedicarboxylic acid, 1,4-dihydroxy-, dimethyl ester	EtOH	220(4.99),264(4.84), 364(4.39)	78-0269-76

Compound	Solvent	$\lambda_{max}(\log \epsilon)$	Ref.
1,4-Naphthalenedione, 3-acetyl-2-hydroxy-5,7-dimethoxy-	MeOH	231(4.17),257(4.19), 304(4.26),426(3.34)	12-1087-76
2H-Pyran-3-carboxylic acid, 4-hydroxy-6-methoxy-2-oxo-5-phenyl-, methyl ester	EtOH EtOH-base	237(4.12),315(4.06) 273s(3.94)	78-0269-76 78-0269-76
$C_{14}H_{12}O_7$ 6H-1,3-Benzodioxolo[5,6-c][1]benzopyran-6-one, 1,2,3,4-tetrahydro-2,3,4-tri-hydroxy-, (2α,3α,4α)-(+)-	EtOH	242(4.58),258(4.30), 274(3.73),286(3.80), 297(3.77),333(3.69), 348s(--)	24-0855-76
$C_{14}H_{12}O_8$ 6H-1,3-Benzodioxolo[5,6-c][1]benzopyran-6-one, 1,2,3,4-tetrahydro-2,3,4,7-tetrahydroxy-, (2α,3α,4α)-(+)-	EtOH	249(4.64),287(3.50), 340(3.63),355s(--)	24-0855-76
$C_{14}H_{12}O_8Se_4$ 1,3-Diselenole-4,5-dicarboxylic acid, 2-[4,5-bis(methoxycarbonyl)-1,3-di-selenol-2-ylidene]-, dimethyl ester	C_6H_{12}	212(4.16),260(4.39), 285(4.43),328s(3.70), 422(4.00)	44-0882-76
$C_{14}H_{12}S_3$ Benzenecarbo(dithioperoxo)thioic acid, 4-methyl-, phenyl ester	C_6H_{12}	310(4.13),534(1.99)	138-0441-76
$C_{14}H_{13}BrN_2O$ Benzaldehyde, 5-bromo-2-hydroxy-, methylphenylhydrazone, (E)-	benzene DMF	305(4.10),350(4.38) 305(3.96),353(4.36)	104-0625-76 104-0625-76
$C_{14}H_{13}BrN_4O_4S$ Thiourea, N-(carboxymethyl)-N'-[[4-(5-bromo-1-uracilyl)methyl]phenyl]-	pH 9.2	272.5(4.28)	106-0598-76
$C_{14}H_{13}BrO_4$ 1,4-Naphthalenedione, 6-bromo-7-ethoxy-5-hydroxy-2,3-dimethyl-	EtOH	222(4.49),263s(4.16), 270(4.20),295(4.01), 417(3.65)	12-2247-76
$C_{14}H_{13}ClHgN_2$ Mercury, chloro[2-[(2,6-dimethylphenyl)-azo]phenyl]-	EtOH	226s(4.16),320(4.06), 460(2.88)	101-0039-76M
Mercury, chloro[3-methyl-2-[(2-methyl-phenyl)azo]phenyl]-	EtOH	233(4.08),342(4.14), 453(2.90)	101-0039-76M
$C_{14}H_{13}ClNO_2S$ Thiopyrylium, 2-amino-6-(4-chlorophen-yl)-3-(ethoxycarbonyl)-, perchlorate	HOAc	411(4.27)	48-0705-76
$C_{14}H_{13}ClN_2$ Benzenecarbohydrazonoyl chloride, N-(4-methylphenyl)-	C_6H_{12}	240(3.81),285(4.00), 336(4.01),345s(--)	48-0881-76
$C_{14}H_{13}ClN_2O$ Benzenecarbohydrazonoyl chloride, 4-methoxy-N-phenyl-	C_6H_{12}	238(3.86),306s(--), 339(4.23)	48-0881-76
$C_{14}H_{13}ClN_2O_2$ Benzenemethanol, 5-chloro-2-(methyl-nitrosoamino)-α-phenyl-	isoPrOH	219s(4.32),253s(3.88)	4-0033-76

Compound	Solvent	$\lambda_{max}(\log \epsilon)$	Ref.
$C_{14}H_{13}ClN_4O_4S$ Thiourea, N-(carboxymethyl)-N'-[[4-(5-chloro-1-uracilyl)methyl]phenyl]-	pH 9.2	268.0(4.20)	106-0598-76
$C_{14}H_{13}ClO_5$ 6,12-Dioxatricyclo[6.2.2.0^{1,9}]dodec-3-ene-2,5,11-trione, 4-(1-chloro-1-propenyl)-8-methyl-, [1S-[1R*,4(Z),8s*,9s*]]-	MeOH	218(3.84),275(4.15)	33-1809-76
$C_{14}H_{13}Cl_2N$ Pyridine, 2,6-dichloro-3-[(2,5-dimethyl-phenyl)methyl]-	MeOH	279(3.72)	33-0190-76
$C_{14}H_{13}FN_4O_4S$ Thiourea, N-(carboxymethyl)-N'-[[4-(5-fluoro-1-uracilyl)methyl]phenyl]-	pH 9.2	270(4.24)	106-0598-76
$C_{14}H_{13}IN_2$ Diazene, (2,6-dimethylphenyl)(2-iodo-phenyl)-	EtOH	224s(4.13),239s(4.00), 324(4.09),370s(3.61), 462(2.69)	101-0039-76M
Diazene, (2-iodo-6-methylphenyl)(2-meth-ylphenyl)-	EtOH	220(4.22),320(4.03), 457(2.56)	101-0039-76M
$C_{14}H_{13}N$ 9H-Carbazole, 9-ethyl-	hexane	344(3.67)	135-1054-76
Methanamine, N-(diphenylmethylene)-	C_6H_{12}	246(4.15)	101-0295-76G
$C_{14}H_{13}NO$ Acetamide, N-[1,1'-biphenyl]-4-yl-	C_6H_{12} MeOH	271(4.54) 272(4.54)	40-0133-76 40-0133-76
7-Azabicyclo[4.2.1]nona-2,4-dien-8-one, 4-phenyl-	EtOH	232(4.03),265s(3.66)	44-3583-76
7-Azabicyclo[4.2.1]nona-2,4-dien-8-one, 6-phenyl-	EtOH	265(3.68),273(3.67)	44-3583-76
Benzenamine, N-(1-phenylethylidene)-, N-oxide	EtOH	288(3.54)	39-0076-76B
9H-Carbazole, 2-methoxy-3-methyl-	EtOH	235(4.37),252(3.8), 299(3.9),330(3.32)	2-0329-76
9H-Carbazole, 2-methoxy-6-methyl-	EtOH	235(4.37),257(3.77), 299(3.89),330(3.32)	2-0329-76
$C_{14}H_{13}NOS$ Benzenecarbothioamide, 4-methoxy-N-phen-yl-	MeOH MeCN	222(4.29),290(4.29) 225(4.33),292(4.33)	78-0661-76 78-0661-76
3-Pyridinecarboxaldehyde, 1,2-dihydro-1-(1-phenylethyl)-2-thioxo-, (+)-	EtOH	294s(3.77),318(4.06), 388(3.49)	1-0904-76
3-Pyridinecarboxaldehyde, 1,2-dihydro-1-(2-phenylethyl)-2-thioxo-	EtOH	294s(3.72),320(4.01), 377(3.36)	1-0904-76
6H-Thieno[2,3-b]pyrrole-2-methanol, 6-methyl-5-phenyl-	EtOH	247(3.75),290(3.71)	104-1550-76
$C_{14}H_{13}NO_2$ 1H-Azepine-2-carboxylic acid, 7-phenyl-, methyl ester	EtOH	388(4.47)	44-0543-76
Benzamide, 2-hydroxy-N-(phenylmethyl)-	MeOH	301(3.65)	18-2679-76
Benzoic acid, 2-[(2-methylphenyl)ami-no]-, β-cyclodextrin complex	pH 7.0	283(3.95),330(3.69)	133-0343-76
Benzoic acid, 2-[(4-methylphenyl)ami-no]-, β-cyclodextrin complex	pH 7.0	288(4.16),330(3.80)	133-0343-76

Compound	Solvent	$\lambda_{max}(\log \epsilon)$	Ref.
Methanone, (2-aminophenyl)(2-methoxyphenyl)-	MeOH	207(4.32),227(4.35), 262(3.85),382(3.74)	39-2089-76C
3-Pyridinecarboxylic acid, 6-phenyl-, ethyl ester	EtOH	290(4.46)	44-0543-76

$C_{14}H_{13}NO_2S$

Compound	Solvent	$\lambda_{max}(\log \epsilon)$	Ref.
2,4-Pentadienoic acid, 2-isothiocyanato-5-phenyl-, ethyl ester	C_6H_{12}	271(4.35),346(4.54)	5-1850-76
2(1H)-Pyridinethione, 3-(1,3-dioxolan-2-yl)-1-phenyl-	EtOH	236s(4.05),293(4.33), 378(3.94)	1-0863-76
6H-1,3-Thiazin-6-one, 2,3-dihydro-4-methyl-2-(2-oxopropylidene)-3-phenyl-	EtOH	223(4.26),240(4.04), 288(3.93),351(4.65), 356(4.64)	103-0831-76

$C_{14}H_{13}NO_2S_2$

Compound	Solvent	$\lambda_{max}(\log \epsilon)$	Ref.
7H-1,2,4-Dithiazepin-7-one, 3,4-dihydro-5-methyl-3-(2-oxopropylidene)-4-phenyl-	EtOH	343(4.80)	103-0831-76
5H-1,4-Dithiino[2,3-c]pyrrole-5,7(6H)-dione, 6-(2,6-dimethylphenyl)-2,3-dihydro-	EtOH	207(4.31),265(4.03), 272s(3.99),417(3.49)	56-1523-76

$C_{14}H_{13}NO_3$

Compound	Solvent	$\lambda_{max}(\log \epsilon)$	Ref.
Benzene, 1-methyl-2-nitro-3-(phenylmethoxy)-	MeOH	247(3.19),253(3.16), 259(3.16),265(3.16), 269(3.16),278(3.11), 300(2.93),350(2.58)	12-2003-76
Benzoic acid, 2-amino-3-(phenylmethoxy)-	MeOH	237(4.41),248s(3.90), 335(3.65),345s(3.62), 434(2.90)	12-2003-76
Benzoic acid, 2-[(2-methoxyphenyl)amino]-, β-cyclodextrin complex	pH 7.0	282(4.05),300(3.99), 330(3.83)	133-0343-76
Benzoic acid, 2-[(3-methoxyphenyl)amino]-, β-cyclodextrin complex	pH 7.0	288(4.09),295(4.08), 330(3.77)	133-0343-76
Benzoic acid, 2-[(4-methoxyphenyl)amino]-, cyclodextrin complex	pH 7.0	288(4.10),330(3.76)	133-0343-76
[1,3]Dioxolo[4,5-j]phenanthridin-6(5H)-one, 3,4,4a,11b-tetrahydro-, trans	EtOH	224(4.64),263(3.77), 271s(--),306(3.90)	94-2969-76
Ethanone, 1-(2,3-dihydro-4-methoxyfuro[2,3-b]quinolin-2-yl)- (dubinidinone)	EtOH	230(4.57),237(4.40), 265s(3.70),274(3.79), 283s(3.68),308(3.42), 321(3.57)	105-0193-76
Ethanone, 1,1',1"-(1,2,3-indolizinetriyl)tris-	MeOH	206(4.11),229(4.41), 252(4.60),284s(4.01), 292(4.11),319(4.31), 338(4.49),350(4.67)	39-1908-76C

$C_{14}H_{13}NO_3S$

Compound	Solvent	$\lambda_{max}(\log \epsilon)$	Ref.
Benzoic acid, 5-amino-2-[(3-methoxyphenyl)thio]-	MeOH	263(4.16),345s(3.44)	73-3607-76

$C_{14}H_{13}NO_4$

Compound	Solvent	$\lambda_{max}(\log \epsilon)$	Ref.
[1,3]Dioxolo[4,5-j]phenanthridin-6(2H)-one, 1,3,4,5-tetrahydro-2-hydroxy-	EtOH	228(4.29),249(4.34), 264(4.27),247s(--), 280(3.75),297(3.78), 322s(--),337(3.57), 352(3.39)	94-2969-76
[1,3]Dioxolo[4,5-j]phenanthridin-6(2H)-one, 3,4,4a,5-tetrahydro-2-hydroxy-, trans	EtOH	241(4.50),280s(--), 303(3.80)	94-2969-76

Compound	Solvent	$\lambda_{max}(\log \epsilon)$	Ref.
Furo[2,3-b]quinoline, 4,7,8-trimethoxy-(skimmianine)	EtOH	248(5.07),320(4.22), 333(4.04)	100-0134-76
2,4-Pentanedione, 3-[3-(2-nitrophenyl)-2-propenylidene]-	MeOH	225(4.22),305(4.52)	65-1141-76
$C_{14}H_{13}NO_4S$			
Benzamide, 2-hydroxy-5-[(4-methylphenyl)sulfonyl]-	EtOH	289(4.24)	56-0973-76
$C_{14}H_{13}NO_5S$			
Cyclopent[e][1,3]oxazine-2,4(3H,5H)-dione, 6,7-dihydro-3-[(4-methylphenyl)-sulfonyl]-	MeOH	231.5(4.16)	5-1689-76
$C_{14}H_{13}NO_6$			
(+)-Lycoricidine	EtOH	244(4.48),305(3.95)	94-2977-76
6-Phenanthridinone, 1,2,3,4-tetrahydro-2r,3c,4c-trihydroxy-8,9-(methylenedioxy)-	EtOH	229(4.38),247s(--), 252(4.32),263(4.25), 275s(--),287(3.80), 298(3.87),318s(--), 333(3.54),349s(--)	24-0855-76
$C_{14}H_{13}NO_8$			
1,3-Cyclopentadiene-1,2,3,4-tetracarboxylic acid, 5-cyano-, tetramethyl ester, ion(1-), sodium deriv.	MeOH	212(3.96),263(4.63), 300(4.14)	24-1928-76
protonated cyano group	MeOH	207(3.90),263(4.52), 300(3.99)	24-1928-76
$C_{14}H_{13}N_3$			
1H-Indazol-4-amine, 1-methyl-3-phenyl-	MeOH	241s(4.04),320(3.92)	24-1898-76
3-Pyridinecarbonitrile, 1,2-dihydro-2-imino-1-methyl-6-(4-methylphenyl)-	CH_2Cl_2	262(4.15)	48-0705-76
perchlorate	HOAc	347(4.23)	48-0705-76
Pyrrolo[1,2-a]pyrimidin-2-amine, 4-methyl-7-phenyl-	MeOH	225(4.17),259(4.66), 283s(4.1),298s(3.98), 310s(3.9),361(3.49)	39-1991-76C
$C_{14}H_{13}N_3O$			
Acetamide, N-[4-(phenylazo)phenyl]-	EtOH	240(4.10),347(4.28), 436(3.18)	61-0301-76
Diazene, [(hydroxyimino)(4-methylphenyl)methyl]phenyl-	C_6H_{12}	300(4.22)	23-3260-76
	EtOH	300(4.21),437(2.53)	23-3260-76
	50% EtOH-pH 12.5	240(4.25),352(4.56)	23-3260-76
	HOAc	300(4.23)	23-3260-76
	dioxan	300(4.19)	23-3260-76
	DMF	300(4.25)	23-3260-76
	DMSO	305(4.20)	23-3260-76
	$CHCl_3$	305(4.21)	23-3260-76
$C_{14}H_{13}N_3O_2$			
Diazene, [(hydroxyimino)(4-methoxyphenyl)methyl]phenyl-	EtOH	290(4.23),430(2.55)	23-3260-76
	50% EtOH-pH 12.5	228(4.26),354(4.38)	23-3260-76
Diazene, (2-methyl-4-nitrophenyl)-(2-methylphenyl)-	C_6H_{12}	345(4.32)	18-1381-76
Ethanimidamide, 2-aci-nitro-N,N'-diphenyl-	n.s.g.	230(4.23),342(4.36)	104-0914-76
1,2-Pyridazinedicarboximide, hexahydro-4,5-bis(methylene)-N-phenyl-	n.s.g.	218(4.41)	35-1875-76

Compound	Solvent	$\lambda_{max}(\log \epsilon)$	Ref.
Pyrimido[4,5-b]quinoline-2,4(3H,10H)-di-one, 10-ethyl-3-methyl-	EtOH	221(4.49),263(4.62), 320(3.93),398(4.02)	39-1805-76C
	50% EtOH-HCl	216(4.35),256(4.49), 333(4.16)	39-1805-76C
Pyrimido[4,5-b]quinoline-2,4(3H,10H)-di-one, 10-ethyl-7-methyl-	EtOH	224(4.43),261(4.44), 323(3.87),404(3.89)	39-1805-76C
	50% EtOH-HCl	218(4.34),262(4.45), 334(4.10)	39-1805-76C
Pyrimido[4,5-b]quinoline-2,4(3H,10H)-di-one, 7,8,10-trimethyl-	EtOH	225(4.32),261(4.25), 324(3.95),402(3.69)	39-1805-76C
	50% EtOH-HCl	220(4.23),262(4.32), 347(4.03)	39-1805-76C
$C_{14}H_{13}N_3O_2S$			
Acetic acid, (4-amino-3-phenyl-2(3H)-thiazolylidene)cyano-, ethyl ester	MeOH	228(4.12),301(3.64), 356(4.06)	48-0343-76
2-Thiophenecarboxylic acid, 3-amino-4-cyano-5-(phenylamino)-, ethyl ester	MeOH	269(4.30),300(4.05), 341(4.14)	48-0343-76
3-Thiophenecarboxylic acid, 4-amino-5-cyano-2-(phenylamino)-, ethyl ester	MeOH	267(4.40),299(4.11), 349(3.97)	48-0343-76
$C_{14}H_{13}N_3O_3$			
Ethanimidamide, N-hydroxy-N'-(2-nitro-phenyl)-N-phenyl-	EtOH	235(4.02),339(3.79)	12-0357-76
$C_{14}H_{13}N_3O_4$			
8-Azabicyclo[3.2.1]oct-3-ene-6-carbox-aldehyde, 6-methyl-8-(5-nitro-2-pyr-idinyl)-2-oxo-	EtOH	210(4.14),227(4.29), 354(4.30)	39-2307-76C
8-Azabicyclo[3.2.1]oct-3-en-2-one, 6-acetyl-8-(5-nitro-2-pyridinyl)-, endo	EtOH	209(4.28),225(4.40), 348(4.38)	39-2307-76C
8-Azabicyclo[3.2.1]oct-3-en-2-one, 6-acetyl-8-(5-nitro-2-pyridinyl)-, exo	EtOH	210(4.27),228(4.39), 352(4.38)	39-2307-76C
$C_{14}H_{13}N_3O_4S$			
2,4-Imidazolidinedione, 5-acetoxy-5-(2-benzothiazolyl)-1,3-diphenyl-	EtOH	219(4.27),263(3.98), 290(3.43),300(3.26)	94-3135-76
$C_{14}H_{13}N_3O_5$			
8-Azabicyclo[3.2.1]oct-3-ene-6-carbox-ylic acid, 8-(5-nitro-2-pyridinyl)-2-oxo-, endo exo	EtOH	210(4.11),227(4.26), 356(4.27)	39-2307-76C
	EtOH	210(4.11),228(4.26), 359(4.28)	39-2307-76C
2,4-Cyclohexadien-1-one, 4-(dimethyl-amino)-6-(2-nitro-4-aci-nitro-2,5-cyclohexadien-1-ylidene)-	EtOH	495(4.16)	104-2333-76
	dioxan	495(4.11)	104-2333-76
	DMF	428(4.36)	104-2333-76
	DMSO	625(3.82)	104-2333-76
$C_{14}H_{13}N_3O_6S$			
Methanesulfonamide, N-(2-nitrophenyl)-N-[(4-nitrophenyl)methyl]-	MeOH	218(4.46),266(4.48)	39-0394-76C
$C_{14}H_{13}N_5O_2$			
Formazan, 3-methyl-5-(4-nitrophenyl)-1-phenyl-	EtOH	216(4.06),265(3.76), 360(4.54),460(2.39)	104-1649-76

Compound	Solvent	$\lambda_{max}(\log \epsilon)$	Ref.
$C_{14}H_{13}N_5O_3$			
Acetamide, N-(4,7-dihydro-3-methoxy-7-oxo-2-phenyl-2H-pyrazolo[4,3-d]-pyrimidin-5-yl)-	pH 13	241(4.12),317(3.57)	4-0439-76
Pyrimido[5,4-e]-1,2,4-triazine-5,7-(1H,6H)-dione, 3-(4-methoxyphenyl)-1,6-dimethyl-	dioxan	308(4.54),456(3.40)	39-0713-76C
Pyrimido[5,4-e]-1,2,4-triazine-5,7-(6H,8H)-dione, 3-(4-methoxyphenyl)-6,8-dimethyl-	dioxan	303(4.51),397(3.54)	39-0713-76C
$C_{14}H_{13}N_5O_4$			
Pyrimido[5,4-e]-1,2,4-triazine-5,7-(1H,6H)-dione, 3-(4-methoxyphenyl)-1,6-dimethyl-, 4-oxide	dioxan	304(4.55),450(3.44)	39-0713-76C
Pyrimido[5,4-e]-1,2,4-triazine-5,7-(6H,8H)-dione, 3-(4-methoxyphenyl)-6,8-dimethyl-, 4-oxide	dioxan	305(4.49),401(3.37)	39-0713-76C
$C_{14}H_{13}N_5O_4S$			
Imidazo[1,5-b]pyrimido[5,4-e]pyridazin-6(10H)-one, 5-acetoxy-10-acetyl-8-methyl-2-(methylthio)-	EtOH	245(3.97),296(3.94),332(4.10),345(4.15)	94-2637-76
$C_{14}H_{13}N_5O_4S_3$			
3-Thiazolidinepropanoic acid, 4-[(amino-thioxomethyl)hydrazono]-5-[(4-nitro-phenyl)methylene]-2-thioxo-	MeOH	245(4.27),278(4.01),376(4.42)	103-0749-76
$C_{14}H_{13}O_2P$			
Acetic acid, (diphenylphosphino)-	50% dioxan	262(3.06),268(3.18),274(3.10)	90-0125-76
$C_{14}H_{14}$			
2H-Fluorene, 4a,9-dihydro-4a-methyl-	MeOH	249(3.38),256(3.42),262(3.39),278(2.74),291(2.58)	44-2706-76
$C_{14}H_{14}BrNO$			
2H-Pyrrol-2-one, 5-[(3-bromo-4-methyl-phenyl)methylene]-1,5-dihydro-3,4-dimethyl-, (E)-	EtOH	253s(3.95),305(4.14)	49-0307-76
(Z)-	EtOH	233(3.89),243(3.91),251(3.89),332(4.36)	49-0307-76
2H-Pyrrol-2-one, 5-[(4-bromo-3-methyl-phenyl)methylene]-1,5-dihydro-3,4-dimethyl-, (E)-	EtOH	245s(3.88),310(4.23)	49-0307-76
(Z)-	EtOH	233(4.00),245s(3.85),335(4.41)	49-0307-76
$C_{14}H_{14}ClNO_7$			
Pyrylium, 2-amino-3-(ethoxycarbonyl)-6-phenyl-, perchlorate	HOAc	383(4.30)	48-0705-76
$C_{14}H_{14}ClN_3$			
Benzenamine, 4-[(4-chlorophenyl)azo]-N,N-dimethyl-	C_6H_{12}	410(4.53)	18-1381-76
$C_{14}H_{14}ClN_3O$			
4H-Indazol-4-one, 3-(4-chlorophenyl)-1,5,6,7-tetrahydro-1-methyl-, oxime	MeOH	230(4.43),265s(4.01)	24-1898-76

Compound	Solvent	$\lambda_{max}(\log \epsilon)$	Ref.
2,4-Pentadienamide, 5-(4-chlorophenyl)-2-cyano-5-(dimethylamino)-	CH_2Cl_2	400(4.54)	48-0705-76
Pyrazolo[4,3-b]azepin-5(1H)-one, 3-(4-chlorophenyl)-4,6,7,8-tetrahydro-1-methyl-	MeOH	230(4.08),257(4.04)	24-1898-76
Pyrazolo[4,3-c]azepin-4(1H)-one, 3-(4-chlorophenyl)-5,6,7,8-tetrahydro-1-methyl-	MeOH	250s(4.10)	24-1898-76
3-Pyridinecarboxamide, 6-(4-chlorophenyl)-1-ethyl-1,2-dihydro-2-imino-perchlorate	CH_2Cl_2	250(4.10)	48-0705-76
	HOAc	338(4.04)	48-0705-76
$C_{14}H_{14}CoO_4$			
Cobaltocene, 1,1'-bis(methoxycarbonyl)-, (cation)	MeOH	276(4.57),310(3.23), 422(2.44)	101-0189-76I
$C_{14}H_{14}N_2$			
Diazene, bis(3-methylphenyl)-	hexane	235(4.06),326(4.24), 442(2.40)	61-0301-76
Diazene, bis(4-methylphenyl)-	C_6H_{12}	330(4.43)	18-1381-76
	hexane	237(4.16),329(4.31), 443(2.87)	61-0301-76
Diazene, (2,6-dimethylphenyl)phenyl-	EtOH	312(4.07),453(2.78)	101-0039-76M
1,4-Ethanonaphtho[1,8-ef]-1,4-diazepine, 2,3-dihydro-	isooctane	323(3.25)	77-0173-76
Ethanone, 1-phenyl-, phenylhydrazone	EtOH	233(4.08),303(4.09), 331(4.23)	39-0456-76C
Perimidine, 2,4,5-trimethyl-	n.s.g.	237(4.44),332(4.02), 345(3.81),380(2.93)	12-2499-76
Perimidine, 2,4,8-trimethyl-	n.s.g.	240(4.57),333(4.04), 345(3.91),380(3.21)	12-2499-76
Propanedinitrile, (2,2-dimethyl-1-phenylpropylidene)-	EtOH	236(4.31),282(2.79)	118-0705-76
Propanedinitrile, (1-phenylpentylidene)-	EtOH	233(3.90),293(4.06)	118-0705-76
$C_{14}H_{14}N_2O$			
Acetylharmalan	MeOH	222s(4.38),231(4.39), 248s(4.10),306(4.32), 330s(3.76)	12-2023-76
	MeOH-acid	237(4.21),313(4.28), 347s(3.71)	12-2023-76
Benzaldehyde, 2-hydroxy-, methylphenylhydrazone	benzene	298(4.23),345(4.34)	104-0625-76
	DMF	345(4.36)	104-0625-76
Benzenecarboximidamide, N-hydroxy-N'-methyl-N-phenyl-	EtOH	301(3.78)	12-0357-76
$C_{14}H_{14}N_2OS$			
Benzenecarbothioic acid, 4-methoxy-, 2'-phenylhydrazide	EtOH	282(4.08)	39-0038-76C
2-Pyridinecarbothioamide, N-(4-ethoxyphenyl)-	20% EtOH	284(4.09)	94-1451-76
$C_{14}H_{14}N_2OS_2$			
[1,2]Dithiolo[4,5,1-hi][1,2,3]benzothiadiazole-3-S^{IV}, 2,6,7,8-tetrahydro-2-(4-methoxyphenyl)-	C_6H_{12}	205(4.36),230(4.42), 253s(4.22),302(4.05), 338(3.45),515(4.19)	39-0228-76C
Ethanone, 1-[4-(3,4-dimethyl-1H-[1,2]-dithiolo[5,1-e][1,2,3]thiadiazole-7-S^{IV}-1-yl)phenyl]-	C_6H_{12}	205(4.37),229(4.38), 254(4.11),314(4.38), 509(4.25)	39-0228-76C

Compound	Solvent	$\lambda_{max}(\log \epsilon)$	Ref.
$C_{14}H_{14}N_2O_2$			
Diazene, bis(4-methoxyphenyl)-	C_6H_{12}	353(4.42)	18-1381-76
Dipyrido[1,2-a:2',1'-c]pyrazinediium, 6,7-dihydro-3-(methoxycarbonyl)-, dibromide	pH 6.0	252(3.72),314(4.37), 323(4.35)	64-0115-76B
1H-Indole-1-carboxylic acid, 2-(cyanomethyl)-3-methyl-, ethyl ester	EtOH	229(4.32),263(4.06), 286(3.37),294(3.37)	39-2587-76C
1H-Indole-1-carboxylic acid, 2-(cyanomethyl)-5-methyl-, ethyl ester	EtOH	232(4.39),256(4.14), 263s(4.11),286(3.53), 290(3.56),297(3.56)	39-2587-76C
1H-Indole-1-carboxylic acid, 2-(cyanomethyl)-6-methyl-, ethyl ester	EtOH	227(4.15),259(3.95), 263s(3.94),283s(3.20), 288s(3.08),294(3.08)	39-2587-76C
4-Isoquinolinecarbonitrile, 1,3-diethoxy-	EtOH	237(4.63),274(3.71), 282(3.92),293(3.96), 359(3.86)	2-0964-76
2(1H)-Pyridinone, 6-[1-(hydroxyimino)-2-phenylethyl]-1-methyl-, (E)-	H_2O	305.5(3.88)	83-0769-76
	pH 13	310(3.96)	83-0769-76
	MeOH	231(3.83),314(3.90)	83-0769-76
(Z)-	H_2O	305(3.82)	83-0769-76
	pH 13	305(3.85)	83-0769-76
	MeOH	229(3.89),310(3.86)	83-0769-76
$C_{14}H_{14}N_2O_2S$			
1H-Pyrazole, 5-methyl-1-[(4-methylphenyl)sulfonyl]-3-(1-propynyl)-	EtOH	226(4.24),255(4.27)	78-1293-76
$C_{14}H_{14}N_2O_3$			
1H-Indole-1-carboxylic acid, 2-(cyanomethyl)-5-methoxy-, ethyl ester	EtOH	240(4.48),261(4.23), 297(3.71),307(3.69)	39-2587-76C
$C_{14}H_{14}N_2O_3S$			
Morpholine, 4-(2-nitro-5-phenyl-3-thienyl)-	MeCN	231(4.18),353(4.21), 430(3.69)	48-0221-76
$C_{14}H_{14}N_2O_4$			
2-Furancarboxamide, N-(1-methylethyl)-5-(2-nitrophenyl)-	dioxan	290(4.22)	73-2577-76
2-Furancarboxamide, N-(1-methylethyl)-5-(3-nitrophenyl)-	dioxan	305(4.43)	73-2577-76
2-Furancarboxamide, N-(1-methylethyl)-5-(4-nitrophenyl)-	dioxan	354(4.38)	73-2577-76
2-Furancarboxamide, 5-(3-nitrophenyl)-N-propyl-	dioxan	303(4.43)	73-2577-76
2-Furancarboxamide, 5-(4-nitrophenyl)-N-propyl-	dioxan	352(4.41)	73-2577-76
2,4,6(1H,3H,5H)-Pyrimidinetrione, 5-[(3-methoxyphenyl)methylene]-1,3-dimethyl-	EtOH	219(4.19),252(4.02), 324(4.09)	94-0607-76
$C_{14}H_{14}N_2O_5$			
1,3-Benzodiazocine-4-carboxylic acid, 3-acetyl-1,2,3,4,5,6-hexahydro-2,6-dioxo-, methyl ester	EtOH	226(4.18),244(4.00), 319(3.16)	142-0053-76S
Sydnone, 4-(3-ethoxy-3-oxo-1-propenyl)-3-(4-methoxyphenyl)-, (E)-	EtOH	345(4.28)	48-0823-76
$C_{14}H_{14}N_2O_6$			
5H-Benz[e]inden-5-one, 1,2,3,3a,4,9b-hexahydro-3a-methyl-7-nitro-3-(nitrooxy)-, (3α,3aα,9bβ)-	EtOH	240(3.32),270(3.98)	104-0796-76

Compound	Solvent	$\lambda_{max}(\log \epsilon)$	Ref.
$C_{14}H_{14}N_2O_6S$ Pyridinium, 4-[(4-nitrophenyl)methyl]-1-[2-(sulfooxy)ethyl]-, hydroxide, inner salt	H_2O	265(4.26),273s(--)	4-0517-76
$C_{14}H_{14}N_4$ [2,4'-Bipyridine]-3',5-dicarbonitrile, 1,1',2,4'-tetrahydro-1,1'-dimethyl-	MeOH	247(4.04),349(3.81)	44-3967-76
diastereomer	MeOH	247(4.05),349(3.81)	44-3967-76
[4,4'-Bipyridine]-3,3'-dicarbonitrile, 1,1',4,4'-tetrahydro-1,1'-dimethyl-	MeOH	235s(--),342(3.87)	44-3967-76
diastereomer	MeOH	235s(--),342(3.87)	44-3967-76
$C_{14}H_{14}N_4O$ 3H-Pyrrolo[2,3-d]pyrimidin-2(7H)-one, 4-amino-3-benzyl-5-methyl-	DMSO	322(3.92)	24-2983-76
Urea, N-(3-cyano-4-methyl-1H-pyrrol-2-yl)-N'-(phenylmethyl)-	MeOH	227(4.02),266(4.10)	24-2983-76
$C_{14}H_{14}N_4O_2$ Benzenamine, N,N-dimethyl-4-[(3-nitrophenyl)azo]-	C_6H_{12}	420(4.50)	18-1381-76
Benzenamine, N,N-dimethyl-4-[(4-nitrophenyl)azo]-	n.s.g.	445(4.52)	18-1381-76
Benzo[g]pteridine-2,4(1H,3H)-dione, 1,3,7,8-tetramethyl-	n.s.g.	343(3.96),388(3.94)	88-1389-76
Benzo[g]pteridine-2,4(3H,10H)-dione, 10-butyl-	EtOH	219(4.31),267(4.47), 338(3.82),439(3.92)	35-0830-76
	50% EtOH-5M HCl	214(4.26),262(4.37), 369(4.07)	35-0830-76
Benzo[g]pteridine-2,4(3H,10H)-dione, 3-methyl-10-propyl-	EtOH	219(4.44),269(4.59), 335(3.91),440(4.01)	35-0830-76
	50% EtOH-5M HCl	215(4.32),259(4.41), 368(4.13)	35-0830-76
Benzo[g]pteridine-2,4(3H,10H)-dione, 3,7,8,10-tetramethyl-	EtOH	223(4.45),271(4.53), 349(3.90),447(4.02)	35-0830-76
	50% EtOH-5M HCl	221(4.35),265(4.47), 393(4.23)	35-0830-76
6H-1,2,4,5-Tetrazino[1,6-b]isoquinolin-6-one, 2-acetyl-1,2-dihydro-1,3-dimethyl-	EtOH	235(4.37),319(4.15), 363(3.90),380(3.77)	95-0700-76
6H-1,2,4,5-Tetrazino[1,6-b]isoquinolin-6-one, 1,2-dihydro-1-methyl-2-(1-oxopropyl)-	EtOH	236(4.37),310(4.17), 367(3.90),385(3.77)	95-0700-76
$C_{14}H_{14}N_4O_2S$ Isothiazolo[3,4-d]pyrimidine-4,6(5H,7H)-dione, 3-(dimethylamino)-7-(phenylmethyl)-	EtOH	233(_4.2_),277(_4.0_), 306(_3.9_)	94-0970-76
$C_{14}H_{14}N_4O_3$ Benzenepropanoic acid, 2-(3-nitro-4-pyridinyl)hydrazide	EtOH	234(4.19),350(3.58)	114-0285-76C
Benzo[g]pteridine-2,4(3H,10H)-dione, 10-butyl-, 5-oxide	EtOH	212(4.14),266(4.49), 341(3.91),354s(3.85), 452(3.85)	35-0830-76
	50% EtOH-5M HCl	268(4.44),368(3.98), 396s(3.77)	35-0830-76
Benzo[g]pteridine-2,4(3H,10H)-dione, 3-methyl-10-propyl-, 5-oxide	EtOH	212(4.10),271(4.48), 343(3.92),256s(3.86), 453(3.82)	35-0830-76

Compound	Solvent	λ_{max}(log ϵ)	Ref.
Benzo[g]pteridine-2,4(3H,10H)-dione, 3-methyl-10-propyl-, 5-oxide (cont.)	EtOH-5M HCl	270(4.60),367(3.99), 395s(3.76)	35-0830-76
Benzo[g]pteridine-2,4(3H,10H)-dione, 3,7,8,10-tetramethyl-, 5-oxide	EtOH	217(4.34),275(4.61), 348(3.93),362s(3.92), 461(3.93)	35-0830-76
	50% EtOH-5M HCl	273(4.68),388(4.12)	35-0830-76
2,4-Pentadienamide, 2-cyano-5-(dimethyl-amino)-5-(4-nitrophenyl)-	CH_2Cl_2	374(4.27)	48-0705-76
1H-1,2,4-Triazole-5-acetic acid, α-[(benzoylamino)methylene]-1-methyl-, methyl ester	EtOH	240(3.98),300(4.27)	94-2568-76
$C_{14}H_{14}N_4O_4S$			
Diazene, [5-(1,1-dimethylethyl)-2-thien-yl](2,4-dinitrophenyl)-	n.s.g.	330(3.88),412(4.40)	39-1639-76C
2-Thiophenecarboxaldehyde, 3,4,5-tri-methyl-, 2,4-dinitrophenylhydrazone	n.s.g.	406(4.54)	39-1639-76C
3-Thiophenecarboxaldehyde, 2,4,5-tri-methyl-, 2,4-dinitrophenylhydrazone	n.s.g.	390(4.33)	39-1639-76C
Thiourea, N-(carboxymethyl)-	pH 9.2	263.0(4.26)	106-0598-76
$C_{14}H_{14}N_4O_5$			
Benzoic acid, 3,5-dimethoxy-2-(3-nitro-4-pyridinyl)hydrazide	EtOH	258(4.23),327(3.74)	114-0285-76C
4,14-Epoxy-5H,8H-1,3-dioxolo[5,6][1,3]-oxazocino[3,2-a]pteridin-8-one, 3a,4,14,14a-tetrahydro-2,2-dimethyl-, [3aR-(3aα,4β,14β,14aα)]-	MeOH	229(4.21),260s(3.61), 316(3.96)	24-1359-76
$C_{14}H_{14}N_6$			
1,2,4,5-Tetrazine, 3-hydrazino-1,4-di-hydro-1,6-diphenyl-	MeOH	254(4.26),356(3.26)	103-0600-76
$C_{14}H_{14}O$			
1H-Benz[e]inden-1-one, 2,3,4,5-tetra-hydro-3-methyl-	EtOH	230(4.1),292(3.7)	39-0722-76C
3,4-Benzotricyclo[4.3.1.0^{1,6}]dec-3-en-2-one	EtOH	244(4.06),286(3.10)	44-2162-76
Bicyclo[3.2.1]oct-2-en-6-one, 2-phenyl-	EtOH	247(4.12)	39-1669-76C
3(2H)-Phenanthrenone, 1,9,10,10a-tetra-hydro-	EtOH	225(4.19),296(4.25)	78-0073-76
1-Propanone, 2-methyl-1-(2-naphthal-enyl)-	MeOH	210(2.14),248(3.88), 283(4.62),334(4.24)	19-0175-76
$C_{14}H_{14}O_2$			
2-Azuleneethanol, acetate	CH_2Cl_2	278(4.73),283(4.70), 288(4.60),343(3.67), 357(3.38),593(2.49), 643(2.41)	44-1811-76
2H-Benz[e]inden-2-one, 3,3a,4,5-tetra-hydro-7-methoxy-, (+)-	EtOH	248(4.20),290(3.95)	88-2997-76
5,10-Benzocyclooctenedione, 6,7-dihydro-2,3-dimethyl-	EtOH	234(4.00),272(3.77)	138-1011-76
4,6,12-Tetradecatriene-8,10-diyn-3-one, 1-hydroxy-, (E,E,E)-	EtOH	212(4.30),263(4.29), 274(4.28),303(4.24), 322s(4.47),339(4.55), 359s(4.42)	39-0735-76C

Compound	Solvent	$\lambda_{max}(\log \epsilon)$	Ref.
$C_{14}H_{14}O_2S$			
2-Thiophenecarboxylic acid, 5-(4-methylphenyl)-, ethyl ester	MeOH	229(4.08),320(4.38)	48-0731-76
$C_{14}H_{14}O_3$			
1,2-Benzenediol, 3-[2-(4-hydroxyphenyl)-ethyl]-	MeOH	215(3.20),226(3.30), 280(2.62)	78-0109-76
Bicyclo[5.3.1]undeca-1,3,5,7,9-pentaene-8-carboxylic acid, 10-methoxy-, methyl ester	C_6H_{12}	248(4.8),294(4.6), 375(4.1),500s(3.1)	35-8277-76
$C_{14}H_{14}O_3S$			
Sulfoxonium, dimethyl-, 4-acetyl-1,2-dihydro-1-oxo-2-naphthalenylide	EtOH	231(4.45),259(4.27), 264(4.27),330(4.27), 349(4.27),365(4.12)	94-2421-76
$C_{14}H_{14}O_3S_2$			
2,7-Dithiabicyclo[2.2.1]heptane-6-carboxylic acid, 3-oxo-1-phenyl-, ethyl ester, endo	MeOH	200(4.60)	44-1724-76
$C_{14}H_{14}O_4$			
4H-1-Benzopyran-4-one, 3-(1-acetoxyethylidene)-2,3-dihydro-6-methyl-, (E)-	EtOH	226(4.16),274(4.18), 354(3.62)	39-2444-76C
4H-1-Benzopyran-4-one, 3-(acetoxymethyl)-2,6-dimethyl-	EtOH	233(4.28),304(3.80)	39-2444-76C
Ethanone, 1-(4-acetoxy-6-methyl-2H-1-benzopyran-3-yl)-	EtOH	227(4.08),245(4.04), 291(4.00),367(3.69)	39-2444-76C
Euparin, 7-methoxy-	EtOH	237(3.64),270(4.11), 363(3.15)	73-2047-76
1,4-Naphthalenedione, 7-ethoxy-5-hydroxy-2,3-dimethyl-	EtOH	269(4.17),288(4.07), 428(3.61)	12-2247-76
$C_{14}H_{14}O_5$			
3-Benzofuranpropanoic acid, 2-acetyl-5-methoxy-	EtOH	205(4.33),296(4.25)	18-0737-76
6H-2-Benzopyran-6,8(7H)-dione, 7-acetoxy-3,5,7-trimethyl-, (S)-	EtOH	221(4.06),338(4.09)	39-0204-76C
6,12-Dioxatricyclo[6.2.2.0^{1,9}]dodec-3-ene-2,5,11-trione, 8-methyl-4-(1-propenyl)-, [1S-[1R*,4(E),8S*,9S*]]-	MeOH	278.5(4.25)	33-1809-76
4,2-(Epoxymethano)-1H,7H-cyclopropa[c]-benzofuran-7-one, 1a,2,3a,4-tetrahydro-3a,4-dihydroxy-2-methyl-5-(1-propynyl)-, [1aR-(1aα,2α,3aβ,4α,7aS*)]-	MeOH	278(4.17)	33-1809-76
1(2H)-Naphthalenone, 6,8-diacetoxy-3,4-dihydro-	MeOH	217(4.16),255(4.12), 290(3.55)	12-1865-76
$C_{14}H_{14}O_6S$			
2H-Pyran-3-carboxylic acid, 5-(2,5-dimethyl-3-thienyl)-4-hydroxy-6-methoxy-2-oxo-, methyl ester	EtOH EtOH-NaOH	238(4.26),316(4.09) 278(3.99)	78-0269-76 78-0269-76
$C_{14}H_{14}S_2$			
1,3-Butadiene, 1,4-bis(5-methyl-2-thienyl)-	DMF	355s(4.52),372(4.66), 392(4.54)	126-1857-76
$C_{14}H_{15}BrN_2S_2$			
1H-[1,2]Dithiolo[5,1-e][1,2,3]thiadiazole-7-S^{IV}, 1-(4-bromophenyl)-5-(1,1-dimethylethyl)-	C_6H_{12}	204(4.42),231(4.32), 267(4.21),289s(4.05), 485(4.28)	39-0228-76C

Compound	Solvent	$\lambda_{max}(\log \epsilon)$	Ref.
$C_{14}H_{15}BrO_2$			
2(5H)-Furanone, 4-bromo-3,5-diethyl-5-phenyl-	EtOH	206(4.17),229(4.05)	5-0979-76
$C_{14}H_{15}ClN_2O$			
Benzenemethanol, 5-chloro-2-hydrazino-α-methyl-α-phenyl-	isoPrOH	253(4.12),305(3.40)	4-0033-76
$C_{14}H_{15}ClO_5$			
6,12-Dioxatricyclo[6.2.2.0^{1,9}]dodecane-2,5,11-trione, 4-(1-chloro-1-propenyl)-8-methyl-	MeOH	293(1.66)	33-1809-76
6,12-Dioxatricyclo[6.2.2.0^{1,9}]dodec-3-ene-5,11-dione, 4-(1-chloro-1-propenyl)-2-hydroxy-8-methyl-	MeOH	234(4.12)	33-1809-76
Microline	MeOH	227(3.95),294(4.06)	33-1809-76
$C_{14}H_{15}Cl_2N_3O_5$			
3-Pyridinecarboxamide, 6-(4-chlorophenyl)-1-ethyl-1,2-dihydro-2-imino-, perchlorate	HOAc	338(4.04)	48-0705-76
$C_{14}H_{15}IN_4O_5$			
2,4(1H,3H)-Pteridinedione, 1-[5-deoxy-5-iodo-2,3-O-(1-methylethylidene)-β-D-ribofuranosyl]-	pH 6.0	228(4.12),314(3.83),330s(3.71)	24-3159-76
	pH 11.0	237(4.18),275s(3.60),323(3.85),340s(3.74)	24-3159-76
	MeOH	231(4.13),315(3.81),337s(3.57)	24-3159-76
$C_{14}H_{15}N$			
Benzenamine, 4-(2,4-cyclopentadien-1-ylidenemethyl)-N,N-dimethyl-	heptane	390(4.52)	39-0048-76C
Quinoline, 2-methyl-3-(2-methyl-1-propenyl)-	EtOH	216(4.52),230(4.52),250(4.31),278(3.75),319(3.63)	39-2103-76C
Quinoline, 4-methyl-3-(2-methyl-1-propenyl)-	EtOH	227(4.60),280(3.83),318(3.52)	39-2103-76C
$C_{14}H_{15}NO$			
Benzenamine, 4-methoxy-N-(4-methylphenyl)-	EtOH	232(4.32),263(4.60),272(4.58)	42-0321-76
2-Cyclohexen-1-one, 3-[(2-phenylethenyl)amino]-	EtOH	254(3.623),289(3.760),349(4.013)	39-2484-76C
Ethanamine, 1-methoxy-N-(2-naphthylmethylene)-	EtOH	224(4.54),244(4.46),248(4.55),275(3.71),283(3.85),292(3.81)	35-2605-76
1H-Inden-1-one, 2,3-dihydro-2-(pyrrolidinomethylene)-	EtOH	256(4.02),290(3.47),298(3.49),367(4.45)	4-1201-76
$C_{14}H_{15}NOS$			
4H-1-Benzothiopyran-4-one, 2,3-dihydro-3-(pyrrolidinomethylene)-	EtOH	249(4.28),260s(4.19),376(4.16)	4-0225-76
$C_{14}H_{15}NO_2$			
Ethanone, 1-[4-[(phenylmethoxy)methyl]-1H-pyrrol-3-yl]-	MeOH	245(3.96)	33-1698-76
1H-Inden-1-one, 2,3-dihydro-2-(4-morpholinomethylene)-	EtOH	254(3.96),289s(3.39),298(3.47),359(4.41)	4-1201-76
1H-Indole-3-carboxylic acid, N-(1,1-dimethyl-2-propenyl)-	EtOH	218(4.49),288(4.10)	94-1853-76

Compound	Solvent	$\lambda_{max}(\log \epsilon)$	Ref.
2-Naphthalenecarbonitrile, 1,2,3,4-tetrahydro-6-methoxy-2,4-dimethyl-1-oxo-	EtOH	231(3.99),289(4.12)	2-0312-76
2-Propenoic acid, 2-methyl-3-(3-phenyl-2H-azirin-2-yl)-, ethyl ester	C_6H_{12}	243(4.44)	44-0543-76
1H-Pyrrole-3-carboxylic acid, 2-methyl-5-phenyl-, ethyl ester	EtOH	268(4.30)	44-0543-76
$C_{14}H_{15}NO_2S$			
Benzenesulfonamide, 4-methyl-N-(phenylmethyl)-	EtOH	208(3.60),241(3.28)	118-0339-76
$C_{14}H_{15}NO_3$			
1H-Carbazole-1-carboxylic acid, tetrahydro-1-hydroxy-, methyl ester	n.s.g.	230(4.23),276s(3.88), 283(3.90),292(3.81)	44-0102-76
1H-Indole-2-acetic acid, α-acetyl-, ethyl ester	MeOH	219(4.49),282(3.93), 292(3.88),333(3.80)	24-3282-76
	CH_2Cl_2	235(4.16),337(4.41)	24-3282-76
$C_{14}H_{15}NO_4$			
5H-Benz[e]inden-5-one, 1,2,3,3a,4,9b-hexahydro-3-hydroxy-3a-methyl-7-nitro-, (3α,3aα,9bβ)-	EtOH	240(3.25),274(3.90)	104-0796-76
1H-Indole-6-carboxylic acid, 4-acetoxy-1,5-dimethyl-, methyl ester	EtOH	227(4.32),240(4.36), 287(3.99),320(3.77)	48-0816-76
1,4-Naphthalenedione, 5-(dimethylamino)-6,8-dihydroxy-2,3-dimethyl-	EtOH	224(4.30),252(4.14), 307s(3.84),318(3.85), 510s(3.67),547(3.84), 587(3.86),614s(3.48)	12-2247-76
9H-Pyrrolo[1,2-a]indol-9-one, 3,9a-dihydro-3-hydroxy-7,9a-dimethoxy-6-methyl-, trans	MeOH	235(4.13),269s(3.74), 337(3.18)	142-1637-76
$C_{14}H_{15}NO_6S$			
1,2,3-Butanetricarboxylic acid, 3-[(phenylthioxomethyl)amino]-, (R*,S*)-	$CHCl_3$	248(4.12),285(3.73), 392(2.24)	78-0571-76
$C_{14}H_{15}N_2O_3$			
Pyridinium, 1-(methoxymethyl)-4-[(4-nitrophenyl)methyl]-, chloride	EtOH	261(4.15)	4-0517-76
$C_{14}H_{15}N_2PS$			
1H-1,3,2-Benzodiazaphosphole, 2,3-dihydro-1,3-dimethyl-2-phenyl-, 2-sulfide	EtOH	250(4.04),294(3.81), 316s(2.80)	65-0807-76
$C_{14}H_{15}N_3$			
Benzenamine, N,N-dimethyl-4-(phenylazo)-	C_6H_{12} hexane	390(4.49) 254(4.04),398(4.38), 465s(3.42)	18-1381-76 61-0301-76
Benzenamine, N-[2-(phenylazo)ethyl]-	EtOH	244(4.26),248(4.30), 254(4.26),260(4.17)	12-1853-76
Pyrimido[1,2-a]indole-10-carbonitrile, 6,7,8,9-tetrahydro-2,4-dimethyl-	EtOH	221(4.33),235(4.3), 249(4.32),259(4.3), 268(4.34),298(3.7), 309(3.68),364(3.57)	103-1379-76
$C_{14}H_{15}N_3O$			
4H-Indazol-4-one, 1,5,6,7-tetrahydro-1-methyl-3-phenyl-	MeOH	225(4.12),257s(3.90)	24-1898-76
2,4-Pentadienamide, 2-cyano-5-(dimethylamino)-5-phenyl-	CH_2Cl_2	412(4.26)	48-0705-76

Compound	Solvent	$\lambda_{max}(\log \epsilon)$	Ref.
Pyrazolo[4,3-b]azepin-5(1H)-one, 4,6,7,8-tetrahydro-1-methyl-3-phenyl-	MeOH	228(4.11),250(4.10)	24-1898-76
Pyrazolo[4,3-c]azepin-4(1H)-one, 5,6,7,8-tetrahydro-1-methyl-3-phenyl-	MeOH	240s(4.06)	24-1898-76
3-Pyridinecarboxamide, 1,2-dihydro-2-imino-1-methyl-6-(4-methylphenyl)-, monoperchlorate	HOAc	343(4.14)	48-0705-76

$C_{14}H_{15}N_3OS$

Imidazo[1,2-a]pyridin-2(3H)-one, 3-(3-ethyl-4,5-dimethyl-2(3H)-thiazolylidene)-, monoperchlorate	EtOH	420(4.30)	124-0961-76

$C_{14}H_{15}N_3O_2$

3-Pyridinecarboxamide, 1,2-dihydro-2-imino-6-(4-methoxyphenyl)-1-methyl-	CH$_2$Cl$_2$	267(4.20)	48-0705-76
hydrobromide	HOAc	353(4.17)	48-0705-76
monoperchlorate	HOAc	354(4.13)	48-0705-76
Pyrrolo[1,2-a]pyrimidine-6-carboxylic acid, 8-cyano-2,4,7-trimethyl-, ethyl ester	EtOH	231(4.49),257(4.52), 265(4.49),318(3.83), 345(3.54)	103-1379-76

$C_{14}H_{15}N_3O_2S_2$

1H-[1,2]Dithiolo[5,1-e][1,2,3]thiadiazole-7-S^{IV}, 5-(1,1-dimethylethyl)-1-(4-methoxyphenyl)-	C$_6$H$_{12}$	205(4.43),225(4.41), 255(4.07),331(4.17), 343s(4.10),498(4.35)	39-0228-76C

$C_{14}H_{15}N_3O_3$

L-Alanine, N-[(3-methyl-2-quinoxalinyl)-carbonyl]-, methyl ester	EtOH	200(4.56),239(4.47), 319(3.85),326(3.81)	78-2931-76
1-Pyrrolidinecarboxylic acid, 2-(diazoacetyl)-, phenylmethyl ester, (S)-	EtOH	248(4.05),270s(3.90)	33-1917-76

$C_{14}H_{15}N_3O_3S$

Butanoic acid, 2-[(4-methylphenyl)hydrazono]-3-oxo-4-thiocyanato-, ethyl ester	n.s.g.	248(4.07),370(4.21)	124-1162-76

$C_{14}H_{15}N_3O_4$

8-Azabicyclo[3.2.1]oct-3-en-2-one, 6-ethoxy-8-(5-nitro-2-pyridinyl)-, endo	EtOH	209(4.11),225(4.24), 350(4.28)	39-2307-76C
1-Naphthalenamine, N-butyl-4,5-dinitro-	benzene	420(4.02)	44-0044-76

$C_{14}H_{15}N_3O_4S$

Butanoic acid, 2-[(4-methoxyphenyl)hydrazono]-3-oxo-4-thiocyanato-, ethyl ester	n.s.g.	245(4.10),367(4.25)	124-1162-76
L-Cysteine, N-acetyl-S-[(3-phenyl-1,2,4-oxadiazol-5-yl)methyl]-	EtOH	238(4.11)	98-0876-76

$C_{14}H_{15}N_3O_5$

L-Alanine, N-[(3-methyl-2-quinoxalinyl)-carbonyl]-, methyl ester, N',N''-dioxide	EtOH	205(4.12),231(4.33), 260s(4.28),266(4.42), 346(3.77),367(4.03), 383(4.10)	78-2931-76
	MeCN	206(4.01),228(4.31), 260s(4.20),268(4.40), 360s(3.77),377(3.99), 394(4.02)	78-2931-76

Compound	Solvent	$\lambda_{max}(\log \epsilon)$	Ref.
2-Cyclohexen-1-one, 5,5-dimethyl-2-nitro-3-[(4-nitrophenyl)amino]-	12M HCl	287(4.28)	39-2207-76C
	M HCl	276(4.11),339(4.11)	39-2207-76C
	H_2O	274(4.08),338(4.08)	39-2207-76C
	pH 13	348(4.15)	39-2207-76C
	M NaOH	346(4.15)	39-2207-76C
2-Cyclohexen-1-one, 3-[(3,5-dinitrophenyl)amino]-5,5-dimethyl-	12M HCl	308(4.33)	39-2207-76C
	M HCl	305(4.26)	39-2207-76C
	pH 1	302(4.25)	39-2207-76C
	H_2O	301(4.26)	39-2207-76C
	pH 13	305(4.30)	39-2207-76C
	M NaOH	305(4.29)	39-2207-76C
2H-Pyran-2-carboxylic acid, 2-[2-(acetylamino)-1,6-dihydro-6-oxo-4-pyrimidinyl]-3,4-dihydro-4-methylene-, methyl ester	MeOH	237(4.34),286(3.88)	44-2850-76
4-Pyrimidineacetic acid, 2-(acetylamino)-1,6-dihydro-α-(3-methyl-2-furanyl)-6-oxo-, methyl ester	pH 12	335(--)	44-2850-76
	EtOH	214(4.23),236(4.20), 285(3.83),335(3.53)	44-2850-76

$C_{14}H_{15}N_3O_6$

Compound	Solvent	$\lambda_{max}(\log \epsilon)$	Ref.
L-Threonine, N-[(3-methyl-2-quinoxalinyl)carbonyl]-, N',N''-dioxide	EtOH	205(4.12),232(4.32), 260s(4.28),267(4.43), 369(4.05),383(4.10)	78-2931-76
	MeCN	204(4.47),225(4.23), 260s(3.86),268(3.96), 280s(3.54),380(3.42), 392(3.46)	78-2931-76

$C_{14}H_{15}N_3S$

Compound	Solvent	$\lambda_{max}(\log \epsilon)$	Ref.
[1,2,4]Triazolo[5,1-a]isoquinolinium, 1-butyl-2-mercapto-, hydroxide, inner salt	MeOH	240(4.53),255s(4.43), 280(4.44),342(3.68)	2-0176-76
	benzene	388(--)	2-0176-76
	MeCN	358(--)	2-0176-76
	$CHCl_3$	365(--)	2-0176-76
[1,2,4]Triazolo[5,1-a]isoquinolinium, 3-butyl-2-mercapto-, hydroxide, inner salt	benzene	432(3.58)	2-0176-76
	MeOH	236(4.19),272(4.59), 317(3.70),362s(3.48)	2-0176-76
	MeCN	387(3.51)	2-0176-76
	$CHCl_3$	400(3.55)	2-0176-76

$C_{14}H_{15}N_4O$

Compound	Solvent	$\lambda_{max}(\log \epsilon)$	Ref.
Hypoxanthinium, 1-benzyl-3,8-dimethyl-, bromide	pH 1	259(4.17)	39-0239-76C
	pH 8	255(4.12)	39-0239-76C

$C_{14}H_{15}N_5$

Compound	Solvent	$\lambda_{max}(\log \epsilon)$	Ref.
1H-Imidazole-4-carbonitrile, 5-(dimethylaminomethyleneamino)-1-(phenylmethyl)-	EtOH	302(4.03)	2-0346-76
1H-Imidazo[4,5-f]quinazolin-9-amine, N-(3-methyl-2-butenyl)-	pH 1	231(4.13),276(4.24), 309s(--),321(4.02), 332s(--)	102-0609-76
	pH 7	230s(--),238(4.11), 262(4.12),311s(--), 322(3.98),337(3.92)	102-0609-76
	pH 13	273(4.42),297s(--), 331(3.93),341s(--)	102-0609-76
1H-Imidazo[4,5-g]quinazolin-8-amine, N-(3-methyl-2-butenyl)-	EtOH	228(4.60),239s(--), 245(4.08),261s(--), 297(3.76),310s(--), 324s(--),335(4.15), 350(4.06)	102-0609-76

Compound	Solvent	$\lambda_{max}(\log \epsilon)$	Ref.
1H-Imidazo[4,5-g]quinazolin-8-amine, N-(3-methyl-2-butenyl)- (cont.)	EtOH-HCl	218(4.59),237s(--), 263(3.95),288s(--), 326s(--),338(3.59), 354(3.61)	102-0609-76
	EtOH-NaOH	244(4.64),249(4.61), 270(4.41),278(4.46), 315(3.98),326(4.10), 348(4.00),365(3.92)	102-0609-76
1H-Imidazo[4,5-h]quinazolin-6-amine, N-(3-methyl-2-butenyl)-	EtOH	227(4.45),253(4.56), 301(4.17),314(4.13), 316s(--)	102-0609-76
	EtOH-HCl	222(4.32),233s(--), 261(4.54),315(4.18), 328s(--)	102-0609-76
	EtOH-NaOH	266(4.92),312(4.09)	102-0609-76
$C_{14}H_{15}N_5OS$ Isothiazolo[3,4-d]pyrimidin-6(7H)-one, 4-amino-3-(dimethylamino)-7-(phenyl-	n.s.g. acid	262(3.7),299(4.2) 296(--),339(--)	94-0970-76 94-0970-76
$C_{14}H_{15}N_5O_4$ 1H-Imidazo[4,5-g]quinazolin-8-amine, 1-β-D-ribofuranosyl-	pH 1	221(4.51),232s(--), 259(4.08),267(4.06), 300(3.80),320s(--), 332(3.05),347(4.01)	35-3987-76
	pH 7	223(4.45),237(4.49), 263(4.23),290(3.74), 304s(--),324(3.80), 339(3.94),354(3.84)	35-3987-76
	pH 12	239(4.52),260(4.24), 292(3.75),305(3.67), 326(3.78),339(3.93), 346(3.83)	35-3987-76
1H-Imidazo[4,5-g]quinazolin-8-amine, 3-β-D-ribofuranosyl-	pH 1	226(4.50),234s(--), 259(4.14),267(4.06), 290s(--),308s(--), 320s(--),332(4.09), 347(4.04)	35-3987-76
	pH 7	231(4.63),259(4.25), 316(3.89),330(3.99), 345(3.86)	35-3987-76
	pH 12	231(4.63),258(4.22), 306s(--),317(3.88), 330(3.99),346(3.85)	35-3987-76
$C_{14}H_{15}N_7O_2$ Formazan, 1-(4,6-dimethyl-2-pyrimidin-yl)-3-methyl-5-(4-nitrophenyl)-	EtOH EtOH-NaOH benzene dioxan	406(4.36) 578(--) 407(--) 405(--)	103-1165-76 103-1165-76 103-1165-76 103-1165-76
$C_{14}H_{15}O$ Pyrylium, 2,4,6-trimethyl-3-phenyl-, iodide	CH_2Cl_2	462(2.41)	59-1311-76
selenocyanate	CH_2Cl_2	443(2.44)	59-1311-76
thiocyanate	CH_2Cl_2	412(2.52)	59-1311-76
$C_{14}H_{15}O_8$ 1,3-Cyclopentadiene-1,2,3,4-tetracarb-oxylic acid, 5-methyl-, tetramethyl ester, ion(1-)	MeOH	216(3.85),270(4.50), 306(4.19)	24-1928-76

Compound	Solvent	$\lambda_{max}(\log \epsilon)$	Ref.
$C_{14}H_{15}O_8S$ 1,3-Cyclopentadiene-1,2,3,4-tetracarb-oxylic acid, 5-(methylthio)-, tetra-methyl ester, ion(1-)	MeOH	207(3.83),264(4.55), 299s(4.09)	24-1928-76
$C_{14}H_{16}$ 2H-Benz[e]indene, 3,3a,4,5-tetrahydro-3a-methyl-	C_6H_{12}	258(3.88)	35-6218-76
Cyclopent[a]indene, 1,2,8,8a-tetrahydro-3,8a-dimethyl-	C_6H_{12}	259(3.90)	35-6218-76
$C_{14}H_{16}BrN_3OS$ Isothiazolo[5,4-c]isoquinoline, 1-bromo-6,7,8,9-tetrahydro-5-morpholino-	EtOH	321s(4.25)	48-0779-76
$C_{14}H_{16}BrN_5O_8P$ Guanosine, 8-bromo-2'-O-butyryl-, cyc-lic 3',5'-phosphate, triethylamine salt	pH 1 pH 11	260(4.19),275s(4.11) 270(4.13)	69-0217-76 69-0217-76
$C_{14}H_{16}ClNO$ 2-Cyclohexen-1-one, 3-[(2-chlorophenyl)-amino]-5,5-dimethyl-	pH 1 H_2O pH 13	294(4.34) 299(4.39) 301(4.40)	39-2207-76C 39-2207-76C 39-2207-76C
2-Cyclohexen-1-one, 3-[(3-chlorophenyl)-amino]-5,5-dimethyl-	pH 1 H_2O pH 13	301(4.34) 312(4.38) 310(4.36)	39-2207-76C 39-2207-76C 39-2207-76C
2-Cyclohexen-1-one, 3-[(4-chlorophenyl)-amino]-5,5-dimethyl-	pH 1 H_2O pH 13	301(4.34) 312(4.38) 311(4.37)	39-2207-76C 39-2207-76C 39-2207-76C
2-Cyclohexen-1-one, 3-[(2-chloro-2-phen-ylethyl)amino]-	EtOH	288(4.328)	39-2484-76C
$C_{14}H_{16}ClN_3O_5$ 3-Pyridinecarboxamide, 1,2-dihydro-2-imino-1-methyl-6-(4-methylphenyl)-, monoperchlorate	HOAc	343(4.14)	48-0705-76
$C_{14}H_{16}ClN_3O_6$ 3-Pyridinecarboxamide, 1,2-dihydro-2-imino-1-methyl-6-(4-methoxyphenyl)-, monoperchlorate	HOAc	354(4.13)	48-0705-76
$C_{14}H_{16}ClN_5O$ Methanimidamide, N'-[6-amino-5-chloro-2,3-dihydro-2-oxo-3-(phenylmethyl)-4-pyrimidinyl]-N,N-dimethyl-	n.s.g.	311(4.2)	94-0970-76
$C_{14}H_{16}FeN_2O_4$ Ferrocene, (2,2-dinitrobutyl)-	MeOH	430(2.20)	104-2433-76
$C_{14}H_{16}IN_3S$ [1,2,4]Triazolo[1,5-a]quinolinium, 2-(methylthio)-3-propyl-, iodide	MeOH	247(4.67),300(4.12)	2-0176-76
$C_{14}H_{16}NO_2$ Pyridinium, 1-[(3,4-dimethoxyphenyl)-methyl]-, bromide	MeOH	232(4.18),284(3.78)	2-0516-76

Compound	Solvent	$\lambda_{max}(\log \epsilon)$	Ref.
$C_{14}H_{16}N_2$			
1,3-Benzenediacetonitrile, $\alpha,\alpha,\alpha',\alpha'$-tetramethyl-	EtOH	260(2.38)	78-0487-76
1,4-Benzenediacetonitrile, $\alpha,\alpha,\alpha',\alpha'$-tetramethyl-	EtOH	255(2.40),262(2.45), 267(2.38)	78-0487-76
1H-Indene-1-carbonitrile, 2,3-dihydro-1,3,3-trimethyl-2-(methylimino)-	EtOH	256(2.81),263(2.92), 270(3.00)	78-0487-76
Pyridinium, 3,3'-(1,2-ethenediyl)bis-[1-methyl-, bis(tetrafluoroborate)	MeCN	287(4.18),317s(4.01)	35-6233-76
$C_{14}H_{16}N_2O$			
Benzenemethanol, 2-hydrazino-α-methyl-α-phenyl-	isoPrOH	206(4.57),245(3.94), 292(3.39)	4-0033-76
Cyclopenta[b]pyrimido[1,2-a]indol-7(6H)-one, 1,2,3,4,5,12b-hexahydro-, monohydrochloride	EtOH	251(4.11),278(3.52), 286(3.46)	44-3775-76
3H-2a,5a-Diazacyclopenta[jk]fluoren-2(1H)-one, 4,5,9b,9c-tetrahydro-9b-methyl-, cis	n.s.g.	246(3.96),296(3.41)	33-2704-76
[1,5]Diazocino[1,2-a]indol-2(1H)-one, 3,4,5,6-tetrahydro-12-methyl-	n.s.g.	229(4.49),288(3.81)	33-2704-76
6H-Dipyrido[1,2-a:2',1'-c]diazepinediium, 7,8-dihydro-3-methoxy-, dibromide	EtOH	282(3.86),318(4.07)	64-0115-76B
Dipyrido[1,2-a:2',1'-c]pyrazinediium, 3-ethoxy-6,7-dihydro-, dibromide	EtOH	280(3.62),343(3.95)	64-0115-76B
1,11-Ethenopyrrolo[1,2-e][1,5]diazecin-9(10H)-one, 5,6,7,8-tetrahydro-2-methyl-	n.s.g.	230(4.42),293(3.78)	33-2704-76
1H-Imidazo[1,5-a]indol-1-one, 2,3-dihydro-2-methyl-3-(1-methylethyl)-	MeOH	236(4.07),290(4.06), 298(4.11),312(3.99)	44-3050-76
1(2H)-Isoquinolinone, 3-piperidino-	EtOH	229(4.36),247(3.87), 294(4.21),350(3.64)	95-0154-76
1-Naphthalenecarboxaldehyde, 2-methoxy-, dimethylhydrazone, (E)-	EtOH	310(3.97),358(4.06)	104-0625-76
Oxazolo[3',2':1,2]pyrido[3,4-b]indole, 2,3,5,6,11,11b-hexahydro-11b-methyl-	MeOH	226(4.18),248(3.79), 275(3.36),283(3.38), 291(3.42),357(4.00)	23-1262-76
	MeOH-HCl	211(4.15),249(3.99), 357(4.32)	23-1262-76
3H-Pyrido[3,4-b]indole-1-propanol, 4,9-dihydro-	MeOH	215(4.10),236(4.07), 318(4.08)	23-1262-76
	MeOH-HCl	214(4.09),246(3.92), 353(4.29)	23-1262-76
	MeOH-KOH	216(4.11),236(4.11), 314(4.12)	23-1262-76
$C_{14}H_{16}N_2OS$			
4(1H)-Pyrimidinone, 2-(butylthio)-6-phenyl-	EtOH	257(4.41),312(3.89)	95-1094-76
4(1H)-Pyrimidinone, 2-[(phenylmethyl)-thio]-6-propyl-	EtOH	287(3.90)	95-1094-76
$C_{14}H_{16}N_2O_2$			
2H-1-Benzopyran-2-one, 3-amino-4-piperidino-	EtOH	261(3.97),332(4.08)	103-0268-76
2,7-Butanocyclobuta[1,2-c:4,3-c']dipyridine-1,8-dione, 4a,4b,8a,8b-tetrahydro-, (4aα,4bβ,8aα,8bα)-	n.s.g.	264(3.38)	33-2841-76
1-Naphthalenamine, N-butyl-4-nitro-	benzene	409(4.20)	44-0044-76

Compound	Solvent	$\lambda_{max}(\log \epsilon)$	Ref.
2(1H)-Pyridinone, 1,1'-(1,4-butanediyl)-bis-	n.s.g.	229(4.14),301(4.04)	33-2841-76
$C_{14}H_{16}N_2O_3$			
2-Cyclohexen-1-one, 5,5-dimethyl-3-[(2-nitrophenyl)amino]-	pH 1	298(4.33)	39-2207-76C
	H_2O	305(4.31)	39-2207-76C
	pH 13	304(4.35)	39-2207-76C
2-Cyclohexen-1-one, 5,5-dimethyl-3-[(3-nitrophenyl)amino]-	pH 1	302(4.36)	39-2207-76C
	H_2O	312(4.39)	39-2207-76C
	pH 13	307(4.37)	39-2207-76C
2-Cyclohexen-1-one, 5,5-dimethyl-3-[(4-nitrophenyl)amino]-	12M HCl	323(4.24)	39-2207-76C
	M HCl	288(4.03),330(4.05)	39-2207-76C
	pH 1	288(4.16),346(4.13)	39-2207-76C
	H_2O	286(4.18),370(4.25)	39-2207-76C
	pH 13	296(4.27),370(4.00)	39-2207-76C
	M NaOH	298(4.09),408(3.43)	39-2207-76C
4-Isoquinolinecarbonitrile, 1,3-diethoxy-5,6,7,8-tetrahydro-8-oxo-	EtOH	225(4.47),276(4.13), 304(4.16)	2-0964-76
$C_{14}H_{16}N_2O_4$			
Cycloocta-1,3,2-dioxazole, 3a,4,5,8,9-9a-hexahydro-2-(4-nitrophenyl)-, trans	hexane	306(3.95)	89-0372-76
1H-Indole-3-acetic acid, α-(aminomethyl)-1-(methoxycarbonyl)-, methyl ester, sulfate (1:1)	EtOH	227(4.39),260(4.05), 285(3.80),293(3.85)	104-1087-76
$C_{14}H_{16}N_2O_7$			
Cyclopentaneacetic acid, 1-methyl-2-(nitrooxy)-5-(4-nitrophenyl)-, (1α,2β,5β)-	EtOH	265(3.99)	104-0049-76
$C_{14}H_{16}N_2S_2$			
1H-[1,2]Dithiolo[5,1-e][1,2,3]thiadiazole-7-S^{IV}, 5-(1,1-dimethylethyl)-1-phenyl-	C_6H_{12}	202(4.43),234(4.39), 260s(4.15),285s(3.96), 483(4.23)	39-0228-76C
$C_{14}H_{16}N_3OS$			
Imidazo[1,2-a]pyridinium, 2,3-dihydro-1-methyl-2-oxo-3-(3,4,5-trimethyl-2(3H)-thiazolylidene)-, perchlorate	EtOH	409(4.28)	124-0961-76
Imidazo[1,2-a]pyridinium, 2,3-dihydro-5-methyl-2-oxo-3-(3,4,5-trimethyl-2(3H)-thiazolylidene)-, perchlorate	EtOH	416(4.30)	124-0961-76
$C_{14}H_{16}N_3PS$			
1H-1,3,2-Benzodiazaphosphol-2-amine, 1,3-dihydro-1,3-dimethyl-N-phenyl-, 2-sulfide	EtOH	237(4.40),290(3.76)	65-0807-76
$C_{14}H_{16}N_3S$			
[1,2,4]Triazolo[1,5-a]quinolinium, 2-(methylthio)-3-propyl-, iodide	MeOH	247(4.67),300(4.12)	2-0176-76
$C_{14}H_{16}N_4$			
[1,2,4]Triazepino[3,4-a]phthalazine, 4,5-dihydro-3,5,5-trimethyl-	MeOH	240s(3.98),267s(3.93), 278s(3.74),341(4.18)	94-1068-76
$C_{14}H_{16}N_4O$			
Benzenepropanoic acid, 2-(3-amino-4-pyridinyl)hydrazide	pH 1	228(4.04),280(3.73)	114-0285-76C

Compound	Solvent	λ_{max}(log ϵ)	Ref.
$C_{14}H_{16}N_4O_3$			
Benzoic acid, 3,5-dimethoxy-, 2-(3-amino-4-pyridinyl)hydrazide	pH 1	249(4.28),290(4.03)	114-0285-76C
2-Quinoxalinecarboxamide, 3-(methoxycarbonylamino)-N,6,7-trimethyl-	6M HCl	372(4.08)	5-1276-76
	pH 1.2	320s(3.81),336(3.81), 375(3.74)	5-1276-76
$C_{14}H_{16}N_4O_4$			
Hydrazinecarboxylic acid, 2-(3-nitro-4-quinolinyl)-, 1,1-dimethylethyl ester	DMF-EtOH	275(4.20),298(4.15), 390(3.62)	24-2338-76
2-Propenal, 3-cyclopentyl-, 2,4-dinitrophenylhydrazone	CHCl$_3$	382.0(4.40)	22-0849-76
$C_{14}H_{16}N_4O_6$			
2,4(1H,3H)-Pteridinedione, 1-[2,3-O-(1-methylethylidene)-β-D-ribofuranosyl]-	pH 6.0	227(4.14),315(3.88)	24-3159-76
	pH 11.0	236(4.18),280(3.64), 320(4.09)	24-3159-76
	MeOH	231(4.14),315(3.85)	24-3159-76
$C_{14}H_{16}N_4O_7S$			
Azirino[2',3':3,4]pyrrolo[1,2-a]indole-8a(7H)-sulfonic acid, 6-amino-8-[[(aminocarbonyl)oxy]methyl]-1,1a,2,4,8,8b-hexahydro-5-methyl-4,7-dioxo-, monosodium salt, [1aR-(1aα,8β,8aα,8bα)]-	H$_2$O	372(4.15)	35-7069-76
$C_{14}H_{16}N_5OP$			
Phosphorodiamidic azide, N,N-dimethyl-N,N'-diphenyl-	MeOH	207(4.37),228(4.23), 262(2.90),300(1.51)	139-0207-76A
$C_{14}H_{16}N_5O_6PS$			
3H-Imidazo[2,1-i]purine, 5-(ethylthio)-3-(3,5-O-phosphinico-β-D-ribofuranosyl)-	pH 1	238(4.30),293(4.28)	94-1561-76
	pH 13	240(4.32),244s(4.31), 274(3.90),283(3.99), 304(3.97)	94-1561-76
$C_{14}H_{16}N_6$			
Formazan, 1-(4,6-dimethyl-2-pyrimidinyl)-3-methyl-5-phenyl-	EtOH	376(4.22)	103-1165-76
	EtOH-NaOH	490(--)	103-1165-76
	benzene	390(--)	103-1165-76
	dioxan	376(--)	103-1165-76
$C_{14}H_{16}O$			
2-Cyclopenten-1-one, 4,4-dimethyl-3-(4-methylphenyl)-	MeOH	223(3.79)	107-0011-76
6,10-Methano-6H-benzocyclononen-6-ol, 5,7,8,9-tetrahydro-	EtOH	255(4.08)	44-2162-76
Naphthalene, 1-methoxy-5-(1-methylethyl)-	EtOH	222(4.65),295(3.91), 309s(3.73),322(3.41)	95-1161-76
4,6,12-Tetradecatriene-8,10-diyn-1-ol, (E,E,E)-	MeOH	252(4.50),267(4.46), 281(4.20),296(4.46), 316(4.59),337(4.46)	39-0735-76C
$C_{14}H_{16}O_2$			
5H-Benz[e]inden-5-one, 1,2,3,3a,4,9b-hexahydro-3-hydroxy-3a-methyl-, (3α,3aα,9bβ)-	EtOH	254(4.03),296(3.30)	104-0796-76
Bicyclo[3.2.1]octan-6-one, 2-hydroxy-2-phenyl-, exo-(±)-	EtOH	250(1.96),256(2.1), 262(1.94)	39-1669-76C

Compound	Solvent	$\lambda_{max}(\log \epsilon)$	Ref.
3-Cyclohexene-1-carboxylic acid, 1-methyl-4-phenyl-	EtOH	247(4.2)	39-1669-76C
3-Cyclohexene-1-carboxylic acid, 4-phenyl-, methyl ester	EtOH	247(4.1)	39-1669-76C
2H-Cyclopenta[b]furan-2-one, hexahydro-3a-methyl-4-phenyl-, (3aα,4β,6aβ)-	EtOH	244(1.95),250(2.08), 254(2.23),257(2.20), 259(2.32),264(2.15), 265(2.20)	104-0049-76
1,5-Ethanonaphthalene-4,9(1H)-dione, 4a,5,6,8a-tetrahydro-7,8-dimethyl-	EtOH	244(3.93)	138-1011-76
9H-Fluoren-9-one, 1,2,3,4,4a,9a-hexahydro-6-methoxy-	EtOH	218(4.21),264(4.23), 288(3.90),302(3.88)	12-2683-76
9H-Fluoren-9-one, 1,2,3,4,4a,9a-hexahydro-7-methoxy-	EtOH	220(4.34),249(3.98), 319(3.63)	12-2683-76
2(5H)-Furanone, 3,5-diethyl-5-phenyl-	EtOH	206(4.38)	5-0979-76
1H-Inden-1-one, 6-(2-hydroxyethyl)-2,5,7-trimethyl-	MeOH	217(4.78),249(4.93), 255(4.99),335(3.83)	94-0173-76
	EtOH	216(3.94),252(4.61), 334(3.40),411(2.70)	102-0995-76
4,6,12-Tetradecatriene-8,10-diyne-1,3-diol, (E,E,E)-	EtOH	250(4.52),266(4.47), 279(4.23),296(4.48), 315(4.61),336(4.47)	39-0735-76C
Tetraspiro[2.0.2.1.2.0.2.1]tetradecane-7,14-dione	EtOH	271(2.13)	88-4851-76
$C_{14}H_{16}O_3$			
3-Cyclohexene-1-carboxylic acid, 4-(4-methoxyphenyl)-	EtOH	254(4.29)	39-1669-76C
1-Heptene-3,6-dione, 1-(4-methoxyphenyl)-, (E)-	EtOH	233(4.03),320(4.38)	12-0339-76
1-Hexen-3-one, 6-(2-methyl-1,3-benzodioxol-2-yl)-	MeOH	212(4.01),231(3.61), 284(3.63)	33-0999-76
$C_{14}H_{16}O_4$			
4H-1-Benzopyran-4-one, 5,7-dimethoxy-2-propyl-	EtOH	230(4.32),244(4.31), 252(4.30),283(3.99)	12-1087-76
Ethanone, 1-(5-hydroxy-8-methoxy-2,2-dimethyl-2H-1-benzopyran-6-yl)-	MeOH	255(4.31),320(3.81), 340(3.74)	102-1795-76
Evodionol	MeOH	229s(3.96),237s(4.03), 264(4.59),290s(3.97), 352(3.64)	12-2023-76
2,4-Pentadienoic acid, 2-methoxy-5-(3-methoxyphenyl)-, methyl ester, (E,E)-	MeOH	213(4.25),236(3.89), 246(3.83),261(3.68), 321(4.41)	24-3025-76
(Z,E)-	MeOH	208(4.20),239(3.87), 246(3.87),254(3.77), 318(4.52)	24-3025-76
$C_{14}H_{16}O_5$			
Armin	EtOH	263(4.0323),322(4.1798)	105-0262-76
	EtOH-base	280(--),330(--), 375s(--)	105-0262-76
Dechloromicroline	MeOH	288.5(3.83)	33-1809-76
2(4aH)-Naphthalenone, 4a-(acetoxymethyl)-5,8-dihydro-1-hydroxy-4-methoxy-	EtOH	250(4.02),297(3.40)	44-3468-76
Pentanoic acid, 3-[(3,4-dimethoxyphenyl)methylene]-4-oxo-	EtOH	282(3.87),313(3.77)	2-0620-76
$C_{14}H_{16}S_4$			
1,3-Benzodithiole, 4,5,6,7-tetrahydro-	MeCN	297(4.025),323s(3.98),	44-0730-76

Compound	Solvent	$\lambda_{max}(\log \epsilon)$	Ref.
2-(4,5,6,7-tetrahydro-1,3-benzodi-thiol-2-ylidene)- (cont.)	THF	472(2.283) 259(4.03),301(4.29), 477(2.51)	44-0730-76 97-0360-76
cation	MeCN	268(--),344(--), 435(--),459(--), 513s(--),664(--)	97-0360-76
dication	MeCN	287(4.35),430(3.45)	97-0360-76
$C_{14}H_{17}AsO$ 4(1H)-Arseninone, 1,3,5-tri-2-propenyl-	EtOH	234s(3.86),316(3.78)	89-0690-76
$C_{14}H_{17}BrN_2O_4$ 1,3-Dioxane-4-carboxaldehyde, 5-acetoxy-2-methyl-, α-(4-bromophenyl)hydrazone	EtOH	211(4.07),242(3.82), 287(4.34),310s(3.99)	136-0185-76D
$C_{14}H_{17}ClN$ Methanaminium, N-[3-chloro-5-(4-methyl-phenyl)-2,4-pentadienylidene]-N-meth-yl-, perchlorate	MeOH	227s(4.04),253s(3.95), 271(4.18),293(4.25), 300s(4.23),312s(4.18), 380(4.20)	48-0731-76
$C_{14}H_{17}ClNO$ Methanaminium, N-[3-chloro-5-(4-methoxy-phenyl)-2,4-pentadienylidene]-N-meth-yl-, perchlorate	MeOH	277(4.03),307s(3.84), 429(4.56)	48-0731-76
$C_{14}H_{17}ClN_2O$ Acetamide, 2-chloro-N-[3-(3-methyl-1H-indol-1-yl)propyl]-	n.s.g.	227(4.50),291(3.72)	33-2704-76
$C_{14}H_{17}ClN_4$ 3H-[1,3,5]Triazepino[1,2-a]benzimida-zole, 3-butyl-9-chloro-4,5-dihydro-	EtOH	208(4.43),220s(4.27), 274(3.92),319(4.36)	78-0839-76
$C_{14}H_{17}ClN_4O$ 1H-Imidazo[1,2-a][1,3,5]benzotriazepin-5(6H)-one, 3-butyl-8-chloro-2,3-di-hydro-	EtOH	218s(4.30),239(4.51), 305(3.40)	78-0057-76
1H-Imidazo[1,2-a][1,3,5]benzotriazepin-5(6H)-one, 3-butyl-9-chloro-2,3-di-hydro-	EtOH	241(4.66),304(3.38)	78-0057-76
$C_{14}H_{17}ClN_4S$ 1H-Imidazo[1,2-a][1,3,5]benzotriazepine-5(6H)-thione, 3-butyl-8-chloro-2,3-di-hydro-	EtOH	275(4.50),315(4.30), 370s(3.20)	78-0057-76
1H-Imidazo[1,2-a][1,3,5]benzotriazepine-5(6H)-thione, 3-butyl-9-chloro-2,3-di-hydro-	EtOH	212(4.28),275(4.46), 314(4.32),365s(3.30)	78-0057-76
1H-[1,3,5]Triazepino[1,2-a]benzimida-zole-2(3H)-thione, 3-butyl-9-chloro-4,5-dihydro-	EtOH	211(4.52),222s(4.37), 266(4.26),298(4.30)	78-0839-76
$C_{14}H_{17}ClO_3$ 3-Hexanone, 1-chloro-6-(2-methyl-1,3-benzodioxol-2-yl)-	MeOH	232(3.50),285(3.61)	33-0999-76
$C_{14}H_{17}ClO_5$ 4,2-(Epoxymethano)-1H,7H-cyclopropa[c]-benzofuran-3a,4,7-triol, 5-(1-chloro-1-propenyl)-1a,2-dihydro-2-methyl-	MeOH	224(3.75),262(4.1)	33-1809-76

Compound	Solvent	λ_{max}(log ϵ)	Ref.
C$_{14}$H$_{17}$F$_2$N$_5$O$_7$S			
Serotonin, 4,6-difluoro-, creatinine	pH 6	265s(3.66)	4-1253-76
sulfate monohydrate	0.05M NaOH	305(3.43)	4-1253-76
C$_{14}$H$_{17}$N			
6-Azulenemethanamine, N,N,2-trimethyl-	MeOH	238(4.23),267s(4.56),	88-2045-76
		276(4.88),281s(4.86),	
		285(4.95),305(3.82),	
		333(3.73),348(3.82),	
		360s(2.95),565(2.46),	
		610s(2.38)	
Benzo[h]quinoline, 2,3,4,4a,5,6-hexa-	EtOH	216(4.18),252(4.10)	2-0819-76
hydro-4a-methyl-			
C$_{14}$H$_{17}$NO			
2-Cyclohexen-1-one, 5,5-dimethyl-	pH 1	299(4.33)	39-2207-76C
3-(phenylamino)-	H$_2$O	309(4.38)	39-2207-76C
	pH 13	308(4.37)	39-2207-76C
2-Cyclohexen-1-one, 3-[(2-phenylethyl)-	EtOH	289(4.305)	39-2484-76C
amino]-			
Ethanone, 2-(4,4-dimethyl-2-pyrrolidin-	EtOH	242(4.03),336(4.33)	39-1944-76C
ylidene)-1-phenyl-			
1H-Inden-1-one, 2-[(diethylamino)methyl-	EtOH	253(4.02),289s(3.43),	4-1201-76
ene]-2,3-dihydro-		297(3.47),361(4.42)	
C$_{14}$H$_{17}$NOS			
4H-1-Benzothiopyran-4-one, 3-[(diethyl-	EtOH	248(4.30),260s(4.15),	4-0225-76
amino)methylene]-2,3-dihydro-, (E)-		368(4.06)	
C$_{14}$H$_{17}$NOSe			
2-Propene-1-selone, 1-(4-methoxyphenyl)-	MeCN	350(4.15),482(3.88),	118-0521-76
3-pyrrolidino-		598s(2.86)	
C$_{14}$H$_{17}$NO$_2$			
9(2H)-Acridinone, 1,3,4,4a,9a,10-hexa-	EtOH	236(4.42),264(3.88),	39-0481-76C
hydro-9a-hydroxy-10-methyl-		403(3.66)	
Benzamide, N-(3-cyclohexen-1-ylmethyl)-	EtOH	235(3.66)	44-0863-76
N-hydroxy-			
5H-Benz[e]inden-5-one, 7-amino-1,2,3,3a-	EtOH	238(3.08),354(3.23)	104-0796-76
4,9b-hexahydro-3-hydroxy-3a-methyl-			
Benzo[h]quinolin-4a(2H)-ol, 3,4,5,6-	EtOH	220(4.20),270(4.16)	2-0819-76
tetrahydro-8-methoxy-			
3-Buten-2-one, 4,4'-(3,4-dimethyl-1H-	CHCl$_3$	283(4.07),317(3.80),	4-1145-76
pyrrole-2,5-diyl)bis-		425(4.37)	
2-Cyclohexen-1-one, 3-[(2-hydroxyphen-	pH 1	291(4.37)	39-2207-76C
yl)amino]-5,5-dimethyl-	H$_2$O	301(4.38)	39-2207-76C
	pH 13	296(4.27)	39-2207-76C
2-Cyclohexen-1-one, 3-[(3-hydroxyphen-	pH 1	299(4.29)	39-2207-76C
yl)amino]-5,5-dimethyl-	H$_2$O	312(4.37)	39-2207-76C
	pH 13	310(4.34)	39-2207-76C
2-Cyclohexen-1-one, 3-[(4-hydroxyphen-	pH 1	302(4.24)	39-2207-76C
yl)amino]-5,5-dimethyl-	H$_2$O	306(4.42)	39-2207-76C
	pH 13	307(4.25)	39-2207-76C
2-Cyclohexen-1-one, 3-[(2-hydroxy-	EtOH	290(4.389)	39-2484-76C
2-phenylethyl)amino]-			
Ethanone, 2-(3,4-dihydro-2,2-dimethyl-	EtOH-HCl	237(4.08),331(4.08)	39-1944-76C
2H-pyrrol-5-yl)-1-phenyl-, N-oxide	EtOH-NaOH	234(4.09),264(3.96),	39-1944-76C
		368(4.33)	

Compound	Solvent	λ_{max} (log ϵ)	Ref.
$C_{14}H_{17}NO_3$			
Butanamide, N-[4-methyl-2-(2-oxopropyl)-phenyl]-3-oxo-	EtOH	210(4.45),240(3.86)	94-1544-76
Butanamide, N-[5-methyl-2-(2-oxopropyl)-phenyl]-3-oxo-	EtOH	209(4.42),238(3.83)	94-1544-76
6-Phenanthridinecarboxylic acid, 1,2,3-4,4a,5,6,10b-octahydro-9-hydroxy-	EtOH	279(3.27)	95-1031-76
$C_{14}H_{17}NO_3S$			
Sulfonium, dimethyl-, 1-acetyl-2,4-di-oxo-4-(phenylamino)butylide	EtOH	248(4.33)	39-1688-76C
$C_{14}H_{17}NO_4$			
1H-Benz[e]indene-3,5-diol, 2,3,3a,4,5-9b-hexahydro-3a-methyl-7-nitro-, (3α,3aα,5α,9bβ)-	EtOH	282(4.00)	104-0796-76
5'-Pyridoxylideneacetic acid, 3,4'-O-isopropylidene-, methyl ester	MeOH	205(4.18),242(4.32), 272(4.18),309(4.08)	104-0440-76
$C_{14}H_{17}NO_5$			
1H-Pyrrole-2-carboxylic acid, 5-methyl-, 5,6-dihydro-6-methoxy-2,2-dimethyl-5-oxo-2H-pyran-4-yl ester, (R)-	EtOH	237(4.06),283(4.32)	78-2057-76
$C_{14}H_{17}NO_6$			
Benzoic acid, 4-(1-acetoxy-2-methyl-2-nitropropyl)-, methyl ester	MeOH	236(4.28),275(3.16), 283(3.10)	12-0327-76
$C_{14}H_{17}NO_7$			
Benzeneacetonitrile, α-(β-D-glucopyrano-syloxy)-4-hydroxy- (taxiphyllin)	MeOH	232(4.049),274(3.167), 281(3.079)	24-3379-76
D-Mannitol, 1-deoxy-1-(1,3-dihydro-1,3-dioxo-2H-isoindol-2-yl)-	H_2O	298(3.43)	136-0049-76G
$C_{14}H_{17}NSe$			
2-Propene-1-selone, 1-(4-methylphenyl)-3-pyrrolidino-	MeCN	335(4.13),481(3.83), 597s(2.80)	118-0521-76
2-Propene-1-selone, 1-phenyl-3-piperi-dino-	MeCN	325(4.00),475(3.60), 588s(2.77)	118-0521-76
$C_{14}H_{17}N_3$			
Pyrrolo[1,2-a]pyrimidine-8-carbonitrile, 2,4,7-trimethyl-6-propyl-	EtOH	221(4.38),235(4.30), 241(4.31),253(4.25), 265(4.31),297(3.65), 310(3.65),364(3.55)	103-1379-76
$C_{14}H_{17}N_3O$			
Morpholine, 4-(3-methyl-4-phenyl-1H-py-razol-1-yl)-	MeOH	245(4.10)	94-3001-76
Morpholine, 4-(4-methyl-3-phenyl-1H-py-razol-1-yl)-	MeOH	251(4.13)	94-3001-76
$C_{14}H_{17}N_3O_2$			
Acetic acid, [3-(phenylhydrazino)-2-piperidinylidene]-, methyl ester	n.s.g.	242(4.15),330(4.22)	39-1241-76C
2H-1-Benzopyran-2-one, 3-amino-4-(4-methyl-1-piperazinyl)-	EtOH	253(3.94),327(4.16)	103-0268-76
1H-Indazole-4,7-dione, 3,6-dimethyl-5-piperidino-	EtOH	267(4.00),504(3.49)	94-1731-76

Compound	Solvent	λ_{max}(log ϵ)	Ref.
$C_{14}H_{17}N_3O_3S$ 2H-1,2,4-Benzothiadiazine-3-carboxamide, N-cyclohexyl-	dioxan	262(4.58)	103-0402-76
$C_{14}H_{17}N_3O_3S_2$ 1,2,4-Triazine-5(2H)-thione, 3-(methyl- thio)-6-[(3,4,5-trimethoxyphenyl)- methyl]-	MeOH	268(4.75)	39-2521-76C
$C_{14}H_{17}N_3O_4S$ 1,2,4-Triazin-5(2H)-one, 3-(methylthio)- 6-[(3,4,5-trimethoxyphenyl)methyl]-	MeOH	270(5.20)	39-2521-76C
$C_{14}H_{17}N_3O_7$ 2H-Pyran-2-carboxylic acid, 2-[2-(acet- ylamino)-1,6-dihydro-6-oxo-4-pyrimi- dinyl]tetrahydro-6-methoxy-4-oxo-, methyl ester	MeOH	235(4.16),287(3.92)	44-2850-76
$C_{14}H_{17}N_3S$ Thiocyanic acid, 1,2,3,3a,8,8a-hexahy- dro-1,3a,8-trimethylpyrrolo[2,3-b]- indol-5-yl-	MeOH	278(4.13)	83-0631-76
$C_{14}H_{17}N_4O_8P$ Inosine, cyclic 3',5'-(hydrogen phos- phate) 2'-butanoate	pH 1 pH 7 pH 11	247(4.06) 247(4.06) 251(4.08)	69-0217-76 69-0217-76 69-0217-76
$C_{14}H_{17}N_5O$ Guanidine, N-(1,4-dihydro-6-methyl-4- oxo-2-pyrimidinyl)-N'-(2-phenylethyl)-	EtOH anion cation	265(4.20) 270(4.00) 260(4.30)	2-0887-76 2-0887-76 2-0887-76
$C_{14}H_{17}N_5O_5$ 9H-Imidazo[1,2-a]purin-9-one, 1,4-dihy- dro-4,6-dimethyl-1-D-ribofuranosyl- 9H-Imidazo[1,2-a]purin-9-one, 3,5-dihy- dro-5,6-dimethyl-3-β-D-ribofuranosyl-	pH 1.0 pH 7.0 pH 0.95 pH 6.6	230(4.48),234(4.49), 253s(3.64),288(3.89) 232(4.41),263(3.67), 313(3.73) 228(4.35),242s(4.29), 279(3.90),305(3.83) 233(4.53),289(4.05)	69-0898-76 69-0898-76 69-0898-76 69-0898-76
$C_{14}H_{17}N_5O_7$ 1,4-Diazepinium, 2,3-dihydro-1,4,5-tri- methyl-, picrate	MeOH	343(4.46)	39-1784-76C
$C_{14}H_{17}N_6O_6P$ 3H-Imidazo[2,1-i]purin-5-amine, N,N-di- methyl-3-(3,5-O-phosphinico-β-D-ribo- furanosyl-	pH 1 pH 13	230(4.28),288(4.25) 288(4.13)	94-1561-76 94-1561-76
$C_{14}H_{17}N_7O_2$ 9,10-Propanopyrimido[4',5':3,4]azeto- [2,1-f]purine-6,8(5aH,7H)-dione, 4- amino-5a-ethyl-9a,9b-dihydro- 2,4(1H,3H)-Pyrimidinedione, 1-[3-(6-am- ino-9H-purin-9-yl)propyl]-5-ethyl-	H_2O H_2O	230(4.26),286(3.83) 263(4.28)	19-0517-76 19-0517-76

Compound	Solvent	$\lambda_{max}(\log \epsilon)$	Ref.
$C_{14}H_{18}$			
Acenaphthene, 2a,3,4,5-tetrahydro-6,8-dimethyl-	EtOH	260(4.23),308(3.53)	22-1599-76
1H-Benz[cd]azulene, 2,2a,3,4,5,6-hexahydro-9-methyl-	EtOH	250(4.04),303(3.46)	22-1599-76
Benzene, (2,5,5-trimethyl-1-cyclopenten-1-yl)-	C_6H_{12}	235(3.66)	35-6218-76
Tetracyclo[6.3.1.0^{1,5}.0^{6,9}]dodeca-2,4-diene	C_6H_{12}	256(3.51)	33-2902-76
$C_{14}H_{18}BrNO_2$			
Pyridinium, 1,2,3,6-tetrahydro-5-(4-methoxyphenyl)-1,1-dimethyl-3-oxo-, bromide	EtOH	239(4.0),329(4.4)	39-2329-76C
$C_{14}H_{18}ClN_5$			
3H-[1,3,5]Triazepino[1,2-a]benzimidazol-2-amine, 3-butyl-9-chloro-4,5-dihydro-	EtOH	218(4.46),245s(4.15),277(3.98),314(4.44)	78-0839-76
$C_{14}H_{18}FN_3O_6$			
Uridine, 5'-(acetylamino)-5'-deoxy-4'-C-fluoro-2',3'-O-(1-methylethylidene)-	MeOH	256(3.97)	44-3010-76
$C_{14}H_{18}FN_5O_2$			
Serotonin, 6-fluoro-, creatinine sulfate monohydrate	pH 6	292(3.76)	4-1253-76
	0.05M NaOH	315(3.70)	4-1253-76
$C_{14}H_{18}N$			
1-Azulenemethanaminium, N,N,N-trimethyl-, iodide	EtOH	277(4.76),282(4.71),287(4.27),328(3.63),336(3.74),351(3.58),554(2.62),591(2.55),649(2.12)	44-1811-76
$C_{14}H_{18}NO_2$			
Pyridinium, 1,2,3,6-tetrahydro-5-(4-methoxyphenyl)-1,1-dimethyl-3-oxo-, bromide	EtOH	239(4.0),329(4.4)	39-2329-76C
$C_{14}H_{18}N_2O$			
2-Cyclohexen-1-one, 3-[(2-aminophenyl)-amino]-5,5-dimethyl-	pH 1	294(4.37)	39-2207-76C
	H$_2$O	298(4.39)	39-2207-76C
	pH 13	297(4.39)	39-2207-76C
2-Cyclohexen-1-one, 3-[(3-aminophenyl)-amino]-5,5-dimethyl-	pH 1	303(4.30)	39-2207-76C
	H$_2$O	310(4.32)	39-2207-76C
	pH 13	310(4.32)	39-2207-76C
2-Cyclohexen-1-one, 3-[(4-aminophenyl)-amino]-5,5-dimethyl-	pH 1	302(4.37)	39-2207-76C
	H$_2$O	307(4.36)	39-2207-76C
	pH 13	307(4.35)	39-2207-76C
Spiro[3H-indole-3,4'-piperidin]-2(1H)-one, 1'-ethyl-	MeOH	248(3.94),272s(3.30)	87-0892-76
$C_{14}H_{18}N_2O_2$			
Butanamide, 4-hydroxy-N-[2-(1H-indol-3-yl)ethyl]-	MeOH	229(4.00),275(3.69),282(3.72),291(3.65)	23-1262-76
$C_{14}H_{18}N_2O_3$			
Ethanone, 1-(2,5-dihydro-1-hydroxy-2,5,5-trimethyl-4-phenyl-1H-imidazol-2-yl)-, N-oxide	EtOH	282(4.08)	70-1997-76

Compound	Solvent	λ_{max}(log ϵ)	Ref.
$C_{14}H_{18}N_2O_6$ 2,4(1H,3H)-Pyrimidinedione, 1-[6-deoxy-2,3-O-(1-methylethylidene)-α-L-lyxo-hexopyranos-4-ulos-1-yl]-5-methyl-	MeOH	265(4.01)	28-0757-76A
$C_{14}H_{18}N_2O_8$ 3,6,9,16-Tetraoxa-12,13-diazabicyclo-[9.3.3]heptadeca-12,14-diene-2,10-dione, 11-acetyl-14-methyl-	MeCN	236(3.62),294(4.05)	77-0964-76
$C_{14}H_{18}N_2O_8$ Uridine, 2'-O-methyl-, 3',5'-diacetate	H_2O	260(4.04)	35-7381-76
$C_{14}H_{18}N_2O_{10}$ 6H-1,2,4-Oxadiazine-3,5(2H,4H)-dione, 2-(2,3,5-tri-O-acetyl-β-D-ribo-furanosyl)-	EtOH	217(3.22)	44-3128-76
$C_{14}H_{18}N_3O$ Isoxazolium, 2-ethyl-3,5-dimethyl-4-[(2-methylphenyl)azo]-, tetra-fluoroborate	HOAc	231(4.05),322(4.01), 436(2.86)	48-0359-76
$C_{14}H_{18}N_3O_3$ 1H-Imidazol-1-yloxy, 2,5-dihydro-2-[1-(hydroxyimino)ethyl]-2,5,5-trimethyl-4-phenyl-, 3-oxide	EtOH	288(4.11)	70-1997-76
$C_{14}H_{18}N_4O$ 1H-Imidazo[1,2-a][1,3,5]benzotriazepin-5(6H)-one, 3-butyl-2,3-dihydro-	EtOH	212s(4.14),239(4.51), 292(3.19)	78-0057-76
$C_{14}H_{18}N_4O_3$ Benzenamine, 2-[5-(1,1-dimethylethyl)-3-methyl-1H-pyrazol-1-yl]-N-hydroxy-5-nitro-	EtOH	230(4.18),240(4.19), 355(3.26)	22-0195-76
Benzenamine, 4-[5-(1,1-dimethylethyl)-3-methyl-1H-pyrazol-1-yl]-N-hydroxy-5-nitro-	EtOH	248(4.27),348(3.45)	22-0195-76
$C_{14}H_{18}N_4O_3S$ 1,2,4-Triazin-5-amine, 3-(methylthio)-6-[(3,4,5-trimethoxyphenyl)methyl]-	MeOH	230(4.44),266(4.68)	39-2521-76C
$C_{14}H_{18}N_4O_4$ 2,4(1H,3H)-Pyrimidinedione, 1-[3-(3,4-dihydro-5-methyl-2,4-dioxo-1(2H)-py-rimidinyl)propyl]-5-ethyl-	H_2O	272(4.26)	56-2041-76
$C_{14}H_{18}N_4O_5$ 1H-Imidazole-4-carboxamide, 5-(cyano-methyl)-1-[2,3-O-(1-methylethylid-ene)-β-D-ribofuranosyl]-	pH 1 pH 7 pH 11	217(3.99) 233(3.99) 239(3.99)	35-1492-76 35-1492-76 35-1492-76
$C_{14}H_{18}N_4O_8$ 1H-1,2,3-Triazole-4-carboxamide, 5-(2,3,5-tri-O-acetyl-β-D-ribo-furanosyl)-	pH 13 EtOH	238(3.87) 209(3.94)	44-0084-76 44-0084-76

Compound	Solvent	$\lambda_{max}(\log \epsilon)$	Ref.
$C_{14}H_{18}N_4S$			
1H-Imidazo[1,2-a][1,3,5]benzotriazepine-5(6H)-thione, 3-butyl-2,3-dihydro-	EtOH	270(4.50),312(4.34), 375s(3.10)	78-0057-76
$C_{14}H_{18}N_5O_7PS$			
Adenosine, 7,8-dihydro-N-(1-oxobutyl)-8-thioxo-, cyclic 3',5'-(hydrogen phosphate)	pH 7	243(4.22),320(4.36)	87-0899-76
Adenosine, 7,8-dihydro-8-thioxo-, cyclic 3',5'-(hydrogen phosphate) 2'-butanoate	pH 7	227(4.28),297(4.38)	87-0899-76
$C_{14}H_{18}N_5O_8P$			
Guanosine, cyclic 3',5'-(hydrogen phosphate) 2'-butanoate, sodium salt	pH 1	257(4.07),282s(3.92)	69-0217-76
	pH 7	253(4.12),277s(3.94)	69-0217-76
	pH 11	260(4.07)	69-0217-76
$C_{14}H_{18}N_6O_6$			
Guanosine, N-acetyl-2'-(acetylamino)-2'-deoxy-	pH 1	258(4.15)	94-2955-76
	H_2O	253(4.19)	94-2955-76
	pH 13	265(4.13)	94-2955-76
$C_{14}H_{18}O$			
Benzaldehyde, 2-methyl-4-(4-methyl-3-pentenyl)-	EtOH	263(4.20)	33-2261-76
Naphthalene, 1,2-dihydro-8-methoxy-4-(1-methylethyl)-	EtOH	221(4.37),264(3.90)	95-1161-76
Tetraspiro[2.0.2.1.2.0.2.1]tetradecan-7-one	EtOH	279(2.23)	88-4851-76
$C_{14}H_{18}O_2$			
Pterosin B	EtOH	220(4.47),260(4.15), 305(3.47)	2-0817-76
	EtOH	218(4.50),260(4.17), 305(3.36)	102-0995-76
$C_{14}H_{18}O_3$			
Cyclopentaneacetic acid, 2-hydroxy-1-methyl-5-phenyl-, (1α,2β,5β)-	EtOH	244(2.11),249(2.26), 251(2.24),255(2.34), 257(2.30),260(2.36), 264(2.23),265(2.26)	104-0049-76
1-Hexen-3-ol, 6-(2-methyl-1,3-benzodioxol-2-yl)-	MeOH	232(3.50),285(3.62)	33-0999-76
13-Nor-4-oxocolorat-8-enolide	EtOH	219(3.96)	39-0850-76C
$C_{14}H_{18}O_4$			
10-Oxabicyclo[7.2.1]dodeca-1(12),4-diene-4-carboxylic acid, 5-methyl-11-oxo-, methyl ester, (E)-	EtOH	215(4.11)	88-4409-76
2-Propenoic acid, 3-(4-hydroxy-3-methoxyphenyl)-, 1,1-dimethyl ethyl ester, (E)-	MeOH	237(4.10),295(4.19), 323(4.29)	20-0657-76
	MeOH-NaOH	257(4.05),297(3.76), 308(3.76),376(4.42)	20-0657-76
$C_{14}H_{18}O_4Zn$			
Zinc, bis(2-oxocyclohexanecarboxaldehydato-O,O')-, (T-4)-	$CHCl_3$	308(4.2)	65-2037-76
$C_{14}H_{18}O_5$			
6,12-Dioxatricyclo[6.2.2.0^{1,9}]dodecane-2,5,11-trione, 8-methyl-4-propyl-	MeOH	293(1.62)	33-1809-76

Compound	Solvent	$\lambda_{max}(\log \epsilon)$	Ref.
6,12-Dioxatricyclo[6.2.2.0^{1,9}]dodecane-2,5,11-trione, 8-methyl-4-propyl-, isomer	MeOH	289(1.75)	33-1809-76
$C_{14}H_{18}O_6$ 2H-Pyran-2-one, 5-(1-hydroxybutyl)-6-(hydroxymethyl)-, diacetate, (S)-	EtOH	293(3.73)	102-0537-76
$C_{14}H_{18}O_7$ Ethanone, 1-[4-(β-D-glucopyranosyloxy)-phenyl]-	MeOH	265(3.97)	105-0099-76
$C_{14}H_{18}O_8$ Ethanone, 1-[3-(β-D-glucopyranosyloxy)-4-hydroxyphenyl]-	MeOH	272(4.09)	105-0099-76
$C_{14}H_{18}S$ 1H-Indene-1-thione, 2-butyl-2,3-dihydro-2-methyl-	C_6H_{12}	296s(4.05),309(4.13), 324(4.14),548(1.26), 587(1.23)	35-6218-76
1(2H)-Naphthalenethione, 3,4-dihydro-2-methyl-2-propyl-	C_6H_{12}	230(3.91),235s(3.87), 250s(3.43),315(4.08), 588(1.64)	35-6218-76
$C_{14}H_{19}BF_2O_4$ 1-Oxa-3-oxonia-2-boratanaphthalene, 2,2-difluoro-1,2-dihydro-5,7-di-methoxy-8-methyl-4-(2-methylpropyl)-	MeCN	216(4.3),319(4.6), 386(3.6)	5-1514-76
$C_{14}H_{19}BrN_5O_6P$ Adenosine, 8-bromo-2-butyl-, cyclic 3',5'-(hydrogen phosphate)	pH 1 pH 7 pH 11	263(4.27) 267(4.24) 267(4.23)	87-0419-76 87-0419-76 87-0419-76
$C_{14}H_{19}ClN_3$ Methanaminium, N-[3-chloro-3-[[(dimeth-ylamino)methylene]amino]-2-phenyl-2-propenylidene]-N-methyl-, per-chlorate	HOAc	400(4.44)	73-1565-76
$C_{14}H_{19}Cl_2N_3O_2$ Acetamide, N,N'-[5-[bis(2-chloroethyl)-amino]-1,3-phenylene]bis-	MeOH	226(4.42),243(4.61)	87-0426-76
$C_{14}H_{19}NO$ Benzamide, N-(1,1-dimethyl-4-pentenyl)-	EtOH	222(4.04)	44-0855-76
$C_{14}H_{19}NO_2$ Benzamide, N-hydroxy-N-(1,2,2-trimethyl-3-buten-1-yl)-	EtOH	219s(3.84),240s(3.74)	44-0863-76
Pyrrolidine, 1-(benzoyloxy)-2,2,5-tri-methyl-	EtOH	228(4.08),273(3.00), 279(3.91)	44-0855-76
Pyrrolidine, 1-(benzoyloxy)-3,3,5-tri-methyl-	EtOH	225(4.15),273(2.93), 280(2.86)	44-0863-76
$C_{14}H_{19}NO_3$ Butanamide, 3-hydroxy-N-[5-methyl-2-(2-oxopropyl)phenyl]-	EtOH	209(4.38),240s(3.81)	94-1544-76
3,5-Cyclohexadiene-1,2-dione, 4-(1,1-di-methylethyl)-5-morpholino-	CHCl$_3$	272(4.01),330(3.34), 510(3.46)	18-2333-76

Compound	Solvent	$\lambda_{max}(\log \epsilon)$	Ref.
$C_{14}H_{19}NO_8$ Griffonin	MeOH	262(4.49)	100-0385-76
$C_{14}H_{19}NO_8S$ 4,5-Thiazoledicarboxylic acid, 2-β-D- ribofuranosyl-, diethyl ester	EtOH	213(4.14),263(4.14)	44-4074-76
$C_{14}H_{19}N_3$ 2-Pyrimidinamine, 1,4-dihydro-N,4,4,6- tetramethyl-1-phenyl-	n.s.g.	262(4.00)	103-0471-76
2,4-Quinolinediamine, N,N,N',N',6-penta- methyl-	EtOH	247(4.52),302(3.75), 348(3.76)	1-0133-76
$C_{14}H_{19}N_3O$ Benzenamine, 4-[5-(1,1-dimethylethyl)- 3-methyl-1H-pyrazol-1-yl]-N-hydroxy-	EtOH	253(4.15)	22-0195-76
2,4-Quinolinediamine, 6-methoxy- N,N,N',N'-tetramethyl-	EtOH	240(4.58),267(4.45), 360(3.67)	1-0133-76
2,4-Quinolinediamine, 8-methoxy- N,N,N',N'-tetramethyl-	EtOH	263(4.49),346(3.67)	1-0133-76
$C_{14}H_{19}N_3OS$ Benzo[b]thiophene-3-carbonitrile, 2-am- ino-4,5,6,7-tetrahydro-4-(morpholino- methyl)-	EtOH	215(4.32),293(3.78)	2-0357-76
$C_{14}H_{19}N_3O_2$ Benzamide, N-[(cyclohexylnitrosoamino)- methyl]-	EtOH	231(4.18),367(1.76)	94-0369-76
$C_{14}H_{19}N_3O_3$ Ethanone, 1-(2,5-dihydro-1-hydroxy- 2,5,5-trimethyl-4-phenyl-1H-imid- azol-2-yl)-, oxime, N-oxide	EtOH	286(3.95)	70-1997-76
4-Pyrimidinol, 2-acetamido-6-(3-oxo- 1-octenyl)-	pH 13	258(3.98),328(3.41)	4-0439-76
$C_{14}H_{19}N_3O_4$ L-Phenylalanine, N-(N-acetyl-D-alanyl)- 4-amino-	EtOH	240(4.26),291(3.08)	33-2421-76
$C_{14}H_{19}N_3O_6S$ 4,6-Ethanoazirino[2,3-f]indazole- 3a,6a(3H,4H)-dicarboxylic acid, 4a,5,5a,6-tetrahydro-5-(methyl- sulfonyl)-, dimethyl ester	EtOH	324(2.30)	24-3505-76
$C_{14}H_{19}N_5O_5$ 9H-Purine, 6-morpholino-9-β-D-ribo- furanosyl-	MeOH	217(4.22),278(4.35)	5-0745-76
$C_{14}H_{19}N_5O_6$ 3-Hexanone, 5,5-dimethyl-6-nitro-, 2,4-dinitrophenylhydrazone	EtOH	226(4.22),359(4.34)	39-1951-76C
$C_{14}H_{19}N_7O_6$ Adenosine, 5'-deoxy-5'-[(3-amino-3-carb- oxy-1-oxopropyl)amino]-, (S)-	H_2O	259(4.18)	87-0684-76

Compound	Solvent	$\lambda_{max}(\log \epsilon)$	Ref.
$C_{14}H_{20}$			
Bicyclo[2.2.2]oct-5-ene, 2,3-dimethyl-ene-5-methyl-8-(1-methylethyl)-	EtOH	246(3.92)	44-1635-76
Bicyclo[6.4.1]trideca-8,10,12-triene, 13-methyl-	C_6H_{12}	220(4.17),283(3.56)	78-2381-76
1,3-Cyclohexadiene, 2-ethenyl-1,3,4,5,5-pentamethyl-6-methylene-	C_6H_{12}	219(4.17),319(3.82)	5-1103-76
$C_{14}H_{20}ClFN_2O_2$			
Benzamide, 3-chloro-N-[2-(diethylamino)-ethyl]-5-fluoro-2-methoxy-, hydrochloride	H_2O	201(4.52),231(3.98), 285(3.26)	133-0314-76
$C_{14}H_{20}N_2O$			
Benzo[g]quinoxaline, 2-ethoxy-5,5a,6,7-8,9,9a,10-octahydro-, (5aS-trans)-	MeCN	217(3.92),287s(3.80), 304(3.91)	142-0299-76S
$C_{14}H_{20}N_2OS$			
1H-Indole-3-ethanamine, 2-(ethylsulfin-yl)-N,N-dimethyl-	EtOH	222(4.42),284(4.16), 298s(3.97)	39-0745-76C
$C_{14}H_{20}N_2O_3$			
Ethanone, 1,1'-[1,4-dihydro-3,6-dimeth-yl-1-(tetrahydro-2-furanyl)-2,5-pyra-inediyl]bis-	EtOH	234(3.95),289(4.14)	18-2805-76
Isoxazole, 4,4'-(oxydiethylidene)bis-[3,5-dimethyl-	isooctane	216(3.89)	39-0570-76C
$C_{14}H_{20}N_2O_4$			
4H-Pyrido[1,2-a]pyrimidine-3-carboxylic acid, 1-acetyl-1,6,7,8,9,9a-hexahydro-6-methyl-4-oxo-, ethyl ester	pH 7.35	219(4.06),249(4.04), 310(3.77)	145-0478-76
$C_{14}H_{20}N_2O_5$			
2,5-Pyrazinedicarboxylic acid, 1,4-di-hydro-3,6-dimethyl-1-(tetrahydro-2-furanyl)-, dimethyl ester	EtOH	267(4.15),331(3.59)	18-2505-76
1H-Pyrrole-1,3-dicarboxylic acid, 4-(acetylamino)-5-ethyl-, diethyl ester	EtOH	215(4.46)	4-0113-76
$C_{14}H_{20}N_2O_6$			
1H-Pyrazole-3-carboxylic acid, 4-[4-[(acetyloxy)methyl]-2,3-dihydroxy-1-cyclopentyl]-, ethyl ester, (1α,2β,3β,4α)-(±)-	EtOH	222(3.85)	23-0861-76
2,4(1H,3H)-Pyrimidinedione, 1-[2,3-O-(1-methylethylidene)-α-L-rhamnopyrano-syl]-	MeOH	263(4.05)	28-0757-76A
$C_{14}H_{20}N_2O_7$			
1H-Pyrazole-3-carboxylic acid, 4-hydr-oxy-5-[2,3-O-(1-methylethylidene)-α-D-ribofuranosyl]-, ethyl ester	pH 12 EtOH	317(3.90) 227(3.86),271(3.72)	44-0287-76 44-0287-76
$C_{14}H_{20}N_2S$			
1H-Indole-3-ethanamine, 2-(ethylthio)-N,N-dimethyl-	EtOH	224(4.51),248s(3.71), 283s(4.06),291(4.12), 300(4.04)	39-0745-76C

Compound	Solvent	$\lambda_{max}(\log \epsilon)$	Ref.
$C_{14}H_{20}N_4$			
Pyrido[3,2-e]-1,2,4-triazine, 3-octyl-	EtOH	233(4.33),341(3.48)	114-0285-76C
$C_{14}H_{20}N_4O_6S$			
1H-Imidazole-4-carboxamide, 1-[5-O-acet-	pH 1	298(4.18)	94-2089-76
yl-2,3-O-(1-methylethylidene)-β-D-	pH 7	296(4.24)	94-2089-76
ribofuranosyl]-5-amino-2,3-dihydro-	pH 13	270s(--),298(3.93)	94-2089-76
2-thioxo-			
$C_{14}H_{20}N_4O_8$			
2-Butenedioic acid, 2,2'-(1,4-dimethyl-	MeOH	256(3.48+),352(4.58+)	24-0253-76
2-tetrazene-1,4-diyl)bis-, tetrameth-			
yl ester, (E,E,?)-			
$C_{14}H_{20}N_5O_6PS$			
Adenosine, 8-(butylthio)-, cyclic	pH 2.0	284(4.30)	69-1408-76
3',5'-(hydrogen phosphate)	pH 13.0	282(--)	69-1408-76
$C_{14}H_{20}N_6O_2$			
[4,5'-Bipyrimidine]-2,2'-diamine, 4',6-	EtOH	253(4.24),283(4.35),	19-0017-76
dimethoxy-N,N,N',N'-tetramethyl-		321(4.24)	
3H-Purine-3-butanoic acid, α-amino-	pH 2	287.0(4.35)	88-3807-76
6-[(3-methyl-2-butenyl)amino]-	pH 7	288.9(4.24)	88-3807-76
(discadenine)	pH 12	290.0(4.24)	88-3807-76
	MeOH	292.7(4.20)	88-3807-76
2(1H)-Pyrimidinone, 4-amino-1-[3-(4-am-	pH 1	292.0(--)	56-1805-76
ino-5-ethyl-2-oxo-1(2H)-pyrimidinyl)-	pH 7	280.0(4.16)	56-1805-76
propyl]-5-methyl-	pH 13	279.5(--)	56-1805-76
$C_{14}H_{20}N_7O$			
1-Piperidinyloxy, 2,2,6,6-tetramethyl-	EtOH	224(4.23),298(4.43)	64-0101-76C
4-(1H-purin-6-ylhydrazono)-			
$C_{14}H_{20}O$			
1H-Indene-5-ethanol, 2,3-dihydro-2,4,6-	EtOH	220(4.12),272(3.23)	102-0995-76
trimethyl-			
$C_{14}H_{20}O_2$			
3,5-Cyclohexadiene-1,2-dione, 3,5-bis-	CHCl$_3$	268(4.21)	44-3627-76
(1,1-dimethylethyl)-			
Dispiro[5.1.5.1]tetradecane-7,14-dione	CH$_2$Cl$_2$	302(1.69),313(1.63),	18-0741-76
		342(1.34)	
$C_{14}H_{20}O$			
2H-1-Benzopyran-6-ol, 3,4-dihydro-	EtOH	220(4.02),288(3.46)	33-0290-76
2-methoxy-2,5,7,8-tetramethyl-			
Cuspidiol	EtOH	279(3.42),286(3.35)	95-1458-76
Cyclohexanone, 5-(3-hydroxy-1-methyl-	EtOH	257(3.60)	138-0991-76
2-oxo-3-cyclopenten-1-yl)-2,5-di-			
methyl-			
1H-Indene-2-acetic acid, 2,3,4,5,6,7-	EtOH	238(4.31)	95-1458-76
hexahydro-7,7-dimethyl-1-oxo-,			
methyl ester			
2(3H)-Naphthalenone, 3-acetoxy-4,4a,5,6-	EtOH	242(4.10)	44-1539-76
7,8-hexahydro-4a,8-dimethyl-			
13-Norcolorat-8-enolide, 4 β-hydroxy-	EtOH	220(3.98)	39-0850-76C
$C_{14}H_{20}O_4$			
1-Butanone, 1-(2-hydroxy-4,6-dimethoxy-	MeCN	212(4.1),231s(--),	5-1514-76
3-methylphenyl)-3-methyl-		291(4.2),331s(--)	

Compound	Solvent	$\lambda_{max}(\log \epsilon)$	Ref.
Tetronic acid, 2-(3-methylbutanoyl)- 4-(3-methyl-2-butenyl)-	MeOH-HCl MeOH-NaOH	268(3.99) 228(4.06),260(4.13)	78-2923-76 78-2923-76
$C_{14}H_{20}O_5$ 1-Cyclohexene-1-propanoic acid, 2-(meth- oxycarbonyl)-3,3-dimethyl-6-oxo-, methyl ester	EtOH	237(4.09)	44-0563-76
$C_{14}H_{20}O_6$ 2,5-Cyclohexadiene-1,4-dione, 2,3,5,6- tetraethoxy-	CHCl$_3$	304(4.03)	12-0179-76
$C_{14}H_{21}BrN_2O_2$ Benzamide, 5-bromo-N-[2-(diethylamino)- ethyl]-2-methoxy-, hydrochloride	H$_2$O	210(4.54),230(4.06), 304(3.47)	133-0314-76
$C_{14}H_{21}BrN_2O_5S_2$ Choline salt of 6-bromo-2-(methylthio)- indoxyl sulfate	EtOH	227(4.60),304(4.24)	88-1233-76
$C_{14}H_{21}ClN_2O_2$ Benzamide, 5-chloro-N-[2-(diethylamino)- ethyl]-2-methoxy-, hydrochloride	H$_2$O	208(4.51),228(4.02), 304(3.49)	133-0314-76
$C_{14}H_{21}FN_2O_2$ Benzamide, 5-fluoro-N-[2-(diethylamino)- ethyl]-2-methoxy-, hydrochloride	H$_2$O	202(4.42),232(3.91), 300(3.59)	133-0314-76
$C_{14}H_{21}NO_2$ 12-Azabicyclo[9.2.1]tetradeca-11(14),13- diene-13-carboxylic acid	pH 13 EtOH	266(4.13) 277(4.19)	78-1863-76 78-1863-76
$C_{14}H_{21}NO_3$ 1,2-Benzenediol, 4-(1,1-dimethylethyl)- 5-morpholino-	EtOH	220s(3.90),285(3.53)	18-2333-76
Butanamide, 3-hydroxy-N-[2-(2-hydroxy- propyl)-5-methylphenyl]-	EtOH	212(4.45),245(3.99)	94-1544-76
1H-Indole-3-acetic acid, 2,4,5,6,7,7a- hexahydro-1-(1-methylethyl)-2-oxo-, methyl ester	EtOH	213(4.27)	22-0776-76
$C_{14}H_{21}N_2O_3$ 4H-Pyrido[1,2-a]pyrimidinium, 3-(2-eth- oxy-2-oxoethyl)-6,7,8,9-tetrahydro- 1,6-dimethyl-4-oxo-	pH 7.35	240(3.87),268(3.73)	145-0478-76
$C_{14}H_{21}N_3O_2$ Carbamic acid, (2,3-di-1-pyrrolidinyl- 2-cyclopropen-1-ylidene)-, ethyl ester	MeOH	238(3.59),317(2.76)	138-1215-76
$C_{14}H_{21}N_3O_3$ 4-Pyrimidinol, 2-acetamido-6-(3-oxooct- yl)-	pH 13	239(3.66),272(3.49)	4-0439-76
$C_{14}H_{21}N_3O_3S_2$ 4-Thiazolidinone, 5-(dimorpholinometh- ylene)-3-ethyl-2-thioxo-	EtOH	262(4.04),380(4.37)	94-1671-76

Compound	Solvent	$\lambda_{max}(\log \epsilon)$	Ref.
$C_{14}H_{21}N_3O_4$			
Acetamide, N,N'-[5-[bis(2-hydroxyethyl)-amino]-1,3-phenylene]bis-	MeOH	230(4.46),245(4.61), 320(3.49)	87-0426-76
$C_{14}H_{21}N_5O$			
Benzamide, 3-amino-4-(2-amino-4,5-di-hydro-1H-imidazol-1-yl)-N-butyl-, dihydrobromide	EtOH	220s(4.20),318(3.58)	78-0057-76
$C_{14}H_{21}N_5O_3$			
Adenosine, 3'-butyl-3'-deoxy-	MeOH	260(4.18)	87-1265-76
$C_{14}H_{21}N_6O_6P$			
Adenosine, N-(4-aminobutyl)-, cyclic 2',5'-(hydrogen phosphate)	pH 1	263(4.21)	69-3724-76
	pH 11	267(4.21)	69-3724-76
Adenosine, 8-(butylamino)-, cyclic 3',5'-(hydrogen phosphate)	pH 2.0	277(4.10)	69-1408-76
	pH 13.0	277(--)	69-1408-76
$C_{14}H_{21}N_7O_5$			
Adenosine, 5'-[(3-amino-3-carboxyprop-yl)amino]-5'-deoxy-, (S)-	H_2O	259(4.23)	87-0684-76
$C_{14}H_{22}$			
Cyclohexane, 1,1'-(1,2-ethanediylidene)-bis-	EtOH	242(4.50),250(4.58), 259(4.42)	39-2386-76C
Cyclohexene, 1-[(2-methyl-1-cyclohexyli-dene)methyl]-, (E)-	hexane	233(3.92)	35-4867-76
(Z)-	hexane	233(3.93)	35-4867-76
5,6-Decadien-3-yne, 5,7-diethyl-	C_6H_{12}	222(3.78)	35-8426-76
$C_{14}H_{22}BrN_3O_2$			
Benzamide, 4-amino-5-bromo-N-[2-(dieth-ylamino)ethyl]-2-methoxy-, hydro-chloride	H_2O	213(4.41),230s(4.15), 272(4.11),308(4.05)	133-0314-76
$C_{14}H_{22}ClN_3O_2$			
Benzamide, 4-amino-5-chloro-N-[2-(dieth-ylamino)ethyl]-2-methoxy-, hydro-chloride	H_2O	212(4.42),228s(4.11), 272(4.14),309(4.08)	133-0314-76
$C_{14}H_{22}N_2O$			
1,3,6,8-Nonatetraen-5-one, 1,9-bis(di-methylamino)-4-methyl-	EtOH	270(4.09),370(4.23), 450(4.86)	70-0577-76
$C_{14}H_{22}N_2O_2$			
Benzamide, N-[2-(diethylamino)ethyl]-2-methoxy-, monophosphate	H_2O	203(4.52),236(3.94), 292(3.52)	133-0314-76
$C_{14}H_{22}N_2O_3$			
Ethanone, 1,1'-[1-(1-ethoxyethyl)-1,4-dihydro-3,6-dimethyl-2,5-pyrazinedi-yl]bis-	EtOH	234(3.73),288(3.90)	18-2805-76
4H-Pyrido[1,2-a]pyrimidine-3-carboxylic acid, 1-ethyl-1,6,7,8,9,9a-hexahydro-6-methyl-4-oxo-, ethyl ester	pH 7.35	241(4.36),319(3.86)	145-0478-76
$C_{14}H_{22}N_2O_5$			
2,5-Pyrazinedicarboxylic acid, 1-(1-eth-oxyethyl)-1,4-dihydro-3,6-dimethyl-, dimethyl ester	EtOH	266(4.15),322(3.72)	18-2805-76

Compound	Solvent	$\lambda_{max}(\log \epsilon)$	Ref.
1H-Pyrrole-1,3-dicarboxylic acid, 4-(acetylamino)-5-ethyl-2,5-di-hydro-, diethyl ester	EtOH	276(4.16)	4-0113-76
$C_{14}H_{22}N_4O_3$ Nonanoic acid, 2-(3-nitro-4-pyridinyl)-hydrazide	EtOH	350(3.61)	114-0285-76C
$C_{14}H_{22}N_4O_4$ Benzamide, 4-amino-N-[2-(diethylamino)-ethyl]-2-methoxy-5-nitro-, sulfate	H_2O	195(4.60),217(4.37), 268(4.70),310(4.02), 390(4.07)	133-0314-76
$C_{14}H_{22}N_4O_5$ 1H-Purine-2,6-dione, 7-(3,3-diethoxy-2-hydroxypropyl)-3,7-dihydro-1,3-di-methyl-	H_2O	209(4.31),273(4.09)	111-0527-76
$C_{14}H_{22}N_6O$ Urea, N-octyl-N'-1H-pyrazolo[3,4-d]py-rimidin-4-yl-	pH 1	269(4.22)	87-0555-76
	pH 13	272(4.12),297(3.63)	87-0555-76
	EtOH	265(4.22)	87-0555-76
$C_{14}H_{22}O$ 3-Buten-2-one, 3-methyl-1-(2,6,6-tri-methyl-1-cyclohexen-1-yl)-	isoPrOH	203(4.07),215s(4.05)	33-0567-76
1H-Indene-5-carboxaldehyde, 2,3,3a,6,7-7a-hexahydro-1,3a,4,7a-tetramethyl-	n.s.g.	246(3.62)	88-3619-76
2(1H)-Naphthalenone, 3,4,4a,5,6,7-hexa-hydro-7-methyl-4a-(1-methylethyl)-, (+)-	dioxan	300(1.89)	22-0957-76
(-)-	dioxan	300(1.80)	22-0957-76
2(3H)-Naphthalenone, 4,4a,5,6,7,8-hexa-hydro-5-methyl-8-(1-methylethyl)-	MeOH	240(4.42)	2-0901-76
2(3H)-Naphthalenone, 4,4a,5,6,7,8-hexa-hydro-7-methyl-4a-(1-methylethyl)-, (+)-(4aS,7R)-	dioxan	325(1.78)	22-0957-76
(-)-	dioxan	325(1.70)	22-0957-76
7-Oxabicyclo[4.1.0]heptane, 2,2,6-tri-methyl-1-(3-methyl-1,3-butadienyl)-, (E)-	n.s.g.	233.5(4.40)	33-0907-76
$C_{14}H_{22}O_2$ 2H-1-Benzopyran-2-ol, 3,5,6,7,8,8a-hexa-hydro-2,5,5,8a-tetramethyl-3-methyl-ene-	pentane	240(4.10)	33-0727-76
2-Cyclohexen-1-one, 3-methyl-6-(1-meth-ylethyl)-2-(3-oxobutyl)-	MeOH	243(4.01)	18-3137-76
2,4,9-Undecatrienoic acid, 6,10-dimeth-yl-, methyl ester, (E,E)-	EtOH	283(4.14)	70-0547-76
2,5,9-Undecatrienoic acid, 6,10-dimethyl-, methyl ester, (E,E)-	EtOH	202(4.31)	70-0547-76
(Z,E)-	EtOH	203(4.34)	70-0547-76
3,5,9-Undecatrienoic acid, 6,10-dimeth-yl-, methyl ester, (E,E)-	EtOH	235(4.13)	70-0547-76
$C_{14}H_{22}O_4$ 1-Cyclopentene-1-heptanoic acid, 4-hy-droxy-3,3-dimethyl-2-oxo-	EtOH	223(3.86)	88-0211-76

Compound	Solvent	$\lambda_{max}(\log \epsilon)$	Ref.
2-Furanacetic acid, 2,4-bis(1,1-dimeth-ylethyl)-2,5-dihydro-5-oxo-	MeOH	208.5(4.06)	35-1510-76
Tetronic acid, 2-(3-methylbutanoyl)-4-(3-methylbutyl)- (author's name)	MeOH-HCl	228(4.01),265(4.10)	78-2923-76
	MeOH-NaOH	230(4.09),262(4.17)	78-2923-76
$C_{14}H_{22}O_5$			
2-Furanacetic acid, 2,4-bis(1,1-dimeth-ylethyl)-2,5-dihydro-α-hydroxy-5-oxo-	MeOH	208(4.08)	35-1510-76
3-Furanheptanoic acid, 2,5-dihydro-4-methoxy-2-methyl-5-oxo-, methyl ester	EtOH	226(3.93)	107-0299-76
$C_{14}H_{22}O_6S_2$			
α-D-Galactopyranose, 1,2:3,4-bis-O-(1-methylethylidene)-, S-methyl carbo-nodithioate	EtOH	221(3.84),278(4.01)	39-2112-76C
$C_{14}H_{22}O_6S_4$			
α-D-Glucofuranose, 3-O-methyl-1,2-O-(1-methylethylidene)-, bis(S-methyl carbonodithioate)	EtOH	222(4.12),277(4.32)	39-2112-76C
$C_{14}H_{23}NO$			
3,5-Heptadienamide, N-cyclohexyl-6-meth-yl-, (E)-	ether	240(4.41)	24-1332-76
(Z)-	ether	241(4.41)	24-1332-76
1(2H)-Naphthalenone, 8-(butylamino)-3,4,4a,5,6,7-hexahydro-	H$_2$O	337(4.18)	35-2826-76
cis-s-cis	MeOH	337(4.17)	35-2826-76
	C$_6$H$_{12}$	328(4.07)	35-2826-76
$C_{14}H_{23}NO_2$			
Acetic acid, (3-piperidinocyclohexyli-dene)-, methyl ester	C$_6$H$_{12}$	213(4.18)	78-2827-76
$C_{14}H_{23}NO_5$			
1H-Pyrrole-1,3-dicarboxylic acid, 2,5-dihydro-4-hydroxy-5-(2-methylpropyl)-, diethyl ester	EtOH	250(3.80)	4-0113-76
$C_{14}H_{23}N_3O_2$			
Benzamide, 4-amino-N-[2-(diethylamino)-ethyl]-2-methoxy-, monohydrochloride	H$_2$O	205(4.31),223s(4.04), 273(4.11),299(4.12)	133-0314-76
$C_{14}H_{23}N_3O_4S$			
Benzamide, 5-(aminosulfonyl)-N-[2-(di-ethylamino)ethyl]-2-methoxy-, hydro-chloride	H$_2$O	211(4.57),234s(4.11), 292(3.42)	133-0314-76
$C_{14}H_{23}O$			
Pyrylium, 2,6-bis(1,1-dimethylethyl)-4-methyl-, iodide	CH$_2$Cl$_2$	454(2.00)	59-1311-76
selenocyanate	CH$_2$Cl$_2$	442(2.38)	59-1311-76
thiocyanate	CH$_2$Cl$_2$	405(2.42)	59-1311-76
$C_{14}H_{23}O_4P$			
Phosphonic acid, [(4-methoxy-2,6-dimeth-ylphenyl)methyl]-, diethyl ester	EtOH	230(4.09),275(3.16), 283(3.18)	39-1466-76C
$C_{14}H_{24}N_2$			
4H-1,2-Diazepine, 3,7-bis(1,1-dimethyl-ethyl)-5-methyl-	EtOH	237(3.83)	80-0241-76

Compound	Solvent	$\lambda_{max}(\log \epsilon)$	Ref.
$C_{14}H_{24}N_2O_4$ 1H-Pyrrole-1,3-dicarboxylic acid, 4-amino-2,5-dihydro-5-(2-methylpropyl)-, diethyl ester	EtOH	276(4.18)	4-0113-76
$C_{14}H_{24}N_4O$ Nonanoic acid, 2-(3-amino-4-pyridinyl)-hydrazide	EtOH	247(4.25),340(3.50), 415(3.20)	114-0285-76C
$C_{14}H_{24}N_4O_4S$ Benzamide, 5-(aminosulfonyl)-4-amino-N-[2-(diethylamino)ethyl]-2-methoxy-, hydrochloride	H_2O	225(4.56),279(4.26), 305s(3.86)	133-0314-76
$C_{14}H_{24}O$ 2-Butanone, 1-(3-ethyl-1-cyclohexen-1-yl)-3,3-dimethyl-	EtOH	295(1.91)	44-3076-76
2-Butanone, 3-methyl-4-(2,6,6-trimethyl-1-cyclohexen-1-yl)-	pentane	287(1.63)	33-0032-76
2-Cyclododecen-1-one, 4-ethyl-	EtOH	231(4.05)	78-1867-76
2-Cyclododecen-1-one, 12-ethyl-	EtOH	230(4.01)	78-1867-76
2(1H)-Naphthalenone, octahydro-7-methyl-4a-(1-methylethyl)-, (4aR,7R,8aR)-(+)-	dioxan	282(1.60)	22-0957-76
(4aS,7R,8aR)-(+)-	dioxan	282(1.68)	22-0957-76
2-Pentanone, 4-(2,6,6-trimethyl-2-cyclohexen-1-yl)-	pentane	283(1.43)	33-0032-76
2-Pentanone, 5-(2,6,6-trimethyl-1-cyclohexen-1-yl)-	pentane	281(1.45)	33-0032-76
$C_{14}H_{24}O_2$ 2-Cyclobuten-1-one, 3-ethoxy-4,4-dimethyl-2-hexyl-	EtOH	242(5.082)	78-1189-76
1,3-Cyclohexanedione, 5-methyl-2-(1-methylethyl)-5-(2-methylpropyl)-	MeOH-HCl MeOH-NaOH	266(4.16) 295(4.36)	2-0263-76 2-0263-76
$C_{14}H_{24}O_4$ 2-Pentenoic acid, 3-(2,2-dimethyl-1-oxopropoxy)-4,4-dimethyl-, ethyl ester	MeOH	216(4.11)	35-8204-76
$C_{14}H_{25}NO_4$ Hexanoic acid, 2-[1-[(2-ethoxy-2-oxoethyl)amino]ethylidene]-, ethyl ester	MeOH	297(4.17)	5-0383-76
$C_{14}H_{26}N_8O_4S_2$ 1,2,4,5-Tetrazine-3,6-dicarboximidamide, N,N,N'',N''-tetraethyl-N',N'''-bis(methylsulfonyl)-	MeCN	225(4.52),272(3.63), 525(2.62)	32-0001-76
$C_{14}H_{27}NO_3$ 4H-Azepin-4-one, hexahydro-5-hydroxy-1-(2-hydroxy-1,1-dimethylethyl)-3,3,6,6-tetramethyl-	EtOH	248s(2.30),290(1.73)	44-1768-76
$C_{14}H_{28}N_6O_2$ Piperazinium, 4,4'-(azodicarbonyl)bis[1,1-dimethyl-	H_2O MeCN	291(3.52),447(1.59) 284(--),425(--)	35-3001-76 35-3001-76
$C_{14}H_{29}NO_2$ Hexanamide, 2,2-dibutyl-N-hydroxy-	heptane EtOH	215(3.33) 204(3.62)	34-0504-76 34-0504-76

Compound	Solvent	$\lambda_{max}(\log \epsilon)$	Ref.
$C_{14}H_{30}N_2$ Diazene, bis(1,1,2,2-tetramethylpropyl)-	toluene	376(1.28)	73-1557-76
$C_{14}H_{30}Si_4$ Silane, 3,4-hexadien-1-yne-1,3,5,6-tetrayltetrakis[dimethyl-	C_6H_{12}	213(4.34),235s(4.17), 243(4.22),253s(4.19)	35-8426-76
$C_{14}H_{34}N_{12}P_2S_2$ Formaldehyde, 2,4,11,13-tetramethyl-1,2,4,5,10,11,13,14-octaaza-3,12-diphosphadispiro[5.2.5.2]hexadecane-3,12-diylidenebis(methylhydrazone), P,P'-disulfide	CHCl$_3$	242(3.08)	78-2633-76
$C_{14}H_{38}GeSi_4$ Germylene, bis[bis(trimethylsilyl)methyl]-	hexane	227(4.29),312(3.16), 414(2.99)	77-0261-76

Compound	Solvent	$\lambda_{max}(\log \epsilon)$	Ref.
$C_{15}H_5N_5$ 1,1,2,2,3-Azulenepentacarbonitrile	CH_2Cl_2	366(4.73),382s(3.60)	88-0251-76
$C_{15}H_6Br_2N_2O_4$ 5,8-Quinolinedione, 6,7-dibromo-2-(2-nitrophenyl)-	MeCN	300(3.64),345(3.42)	23-2563-76
$C_{15}H_6Cl_2N_2O_2$ 2,5-Cyclohexadiene-1,4-dione, 2,3-dichloro-5-cyano-6-(3-indolyl)-	EtOH	213(4.57),277(4.08), 302(4.11),600(3.67)	1-0853-76
$C_{15}H_6Cl_2N_2O_4$ 5,8-Quinolinedione, 6,7-dichloro-2-(2-nitrophenyl)-	MeCN	290(4.38),340(3.51)	23-2563-76
$C_{15}H_6F_{12}$ Benzene, [4,4-bis(trifluoromethyl)-3-[2,2,2-trifluoro-1-(trifluoromethyl)-ethylidene]-1-cyclobuten-1-yl]-	n.s.g.	302(5.61)	70-1068-76
$C_{15}H_8BrN_3O_4$ 5,8-Quinolinedione, 6-amino-7-bromo-2-(2-nitrophenyl)-	MeCN	300(5.23),450(4.30)	23-2563-76
$C_{15}H_8ClN_3O_2$ 1H-Anthra[1,2-d]triazole-6,11-dione, 4-chloro-1-methyl-	EtOH	340(3.80),370s(3.71)	104-0865-76
2H-Anthra[1,2-d]triazole-6,11-dione, 4-chloro-2-methyl-	EtOH	338(3.85),370s(3.71)	104-0865-76
3H-Anthra[1,2-d]triazole-6,11-dione, 4-chloro-3-methyl-	EtOH	338(3.82)	104-0865-76
$C_{15}H_8Cl_2N_2O_2$ Benzonitrile, 3,4-dichloro-2,5-dihydroxy-6-(3-indolyl)-	EtOH	219(4.81),278(3.95), 354(3.98)	1-0853-76
$C_{15}H_8Cl_3NO_2S$ Acridinium, 9-chloro-, 2-chloro-1-(chlorosulfinyl)-2-oxoethylide	MeCN	208(4.31),260(4.60), 350s(3.83),365(4.03), 388(3.90),455s(3.32)	44-3406-76
$C_{15}H_8Cl_4O_2$ 2-Propen-1-one, 1-(3,5-dichloro-2-hydroxyphenyl)-3-(2,4-dichlorophenyl)-, (E)-	EtOH MeCN CCl_4	235(4.03),325(4.21) 232(4.32),325(4.15) 255(3.81),325(4.26)	97-0365-76 97-0365-76 97-0365-76
$C_{15}H_8Cl_6N_2$ 3-Azabicyclo[4.1.0]hept-4-ene-5-carbonitrile, 7,7-dichloro-2-(dichloromethylene)-3-[(2,6-dichlorophenyl)-methyl]-	$CHCl_3$	329(4.3)	64-0807-76B
$C_{15}H_8N_2O_2$ Benzimidazo[1,2-b]isoquinoline-6,11-dione	DMF	283(4.31),418(3.34)	44-0836-76
$C_{15}H_8N_8$ 4-Cyclohexene-1,1,2,2,3,4,5-heptacarbonitrile, 3-amino-6,6-dimethyl-	MeCN	268(3.93)	35-0558-76

Compound	Solvent	$\lambda_{max}(\log \epsilon)$	Ref.
$C_{15}H_8O_3$ 11H-Benzofuro[3,2-b][1]benzopyran- 11-one	EtOH	212(4.23),244(4.20), 302(4.38)	22-1975-76
$C_{15}H_8O_4$ 6H-Benzofuro[3,2-c][1]benzopyran-6-one, 8-hydroxy-	dioxan	225(4.34),290s(4.11), 330(4.35)	2-0861-76
$C_{15}H_8O_4S$ 11H-[1]Benzothieno[3,2-b][1]benzopyran- 11-one, 1,3-dihydroxy-	MeOH	235(4.18),266(4.46), 284(4.05),322(4.17), 357(3.86)	83-0018-76
$C_{15}H_8O_5$ Coumarin, 3-(p-benzoquinonyl)-4-hydroxy-	MeOH	210(4.48),280s(4.11), 300(4.08)	2-0861-76
$C_{15}H_9BrOS_2$ 1,3-Dithiol-1-ium, 2-(4-bromophenyl)- 4-hydroxy-5-phenyl-, hydroxide, inner salt	dioxan	235(4.03),286(4.23), 562(4.11)	24-0740-76
1,3-Dithiol-1-ium, 4-(4-bromophenyl)- 5-hydroxy-2-phenyl-, hydroxide, inner salt	dioxan	245(4.10),277(4.22), 301(4.23),557(4.19)	24-0740-76
$C_{15}H_9ClN_2O_2S$ 1,3,4-Thiadiazol-2(3H)-one, 5-benzoyl- 3-(4-chlorophenyl)-	EtOH	250s(4.21),322(4.05)	4-0045-76
$C_{15}H_9ClN_4O_2S$ Methanone, [4-(4-chlorophenyl)-4,5-di- hydro-5-(nitrosoimino)-1,3,4-thiadi- azol-2-yl]phenyl-	EtOH	280(4.48),360(4.37), 480(1.84)	4-0045-76
$C_{15}H_9ClO_3$ 9,10-Anthracenedione, 2-(chloromethyl)- 1-hydroxy-	benzene	412(4.06)	2-0054-76
$C_{15}H_9ClO_4$ 9,10-Anthracenedione, 2-(chloromethyl)- 1,4-dihydroxy-	benzene	475(3.87)	2-0054-76
$C_{15}H_9Cl_2NO_2$ 9,10-Anthracenedione, 1-amino-4-chloro- 2-(chloromethyl)-	benzene	480(3.84)	146-0001-76
$C_{15}H_9Cl_2NO_4S$ Chlorosulfurous acid, (1-amino-4-chloro- 9,10-dihydro-9,10-dioxo-2-anthracen- yl)methyl ester	benzene	470(3.52)	2-0054-76
$C_{15}H_9Cl_3O_2$ 2-Propen-1-one, 3-(2-chlorophenyl)- 1-(3,5-dichloro-2-hydroxyphenyl)-	EtOH MeCN CCl₄	230(4.02),320(3.97) 230(4.11),320(4.11) 320(4.19)	97-0365-76 97-0365-76 97-0365-76
$C_{15}H_9Cl_7N_2$ 2-Azabicyclo[4.1.0]hept-3-ene-4-carbo- nitrile, 5,7,7-trichloro-5-(dichloro- methyl)-2-[(2,6-dichlorophenyl)methyl]-	CHCl₃	290(4.1)	64-0807-76B

Compound	Solvent	$\lambda_{max}(\log \epsilon)$	Ref.
$C_{15}H_9F_3O_2$ 1(3H)-Isobenzofuranone, 3-[4-(trifluoro-methyl)phenyl]-	EtOH	222(4.08),278(3.40)	111-0533-76
$C_{15}H_9NO_3S_2$ 1,3-Dithiol-1-ium, 4-hydroxy-2-(4-nitro-phenyl)-5-phenyl-, hydroxide, inner salt	dioxan	250s(4.08),279(4.16), 325s(3.88),580(4.11)	24-0740-76
	CH_2Cl_2	252s(4.01),282(4.11), 322s(3.93),576(4.25)	24-0740-76
1,3-Dithiol-1-ium, 4-hydroxy-5-(4-nitro-phenyl)-2-phenyl-, hydroxide, inner salt	CH_2Cl_2	249(4.19),276s(4.06), 305(3.80),387(4.02), 550(4.25)	24-0740-76
$C_{15}H_9NO_4$ 1H-2-Benzopyran-1-one, 5-nitro-3-phenyl-	n.s.g.	251(4.12),329(4.08)	39-1073-76C
$C_{15}H_9NO_4S$ Methanone, (3-hydroxybenzo[b]thien-2-yl)(2-nitrophenyl)-	dioxan	260(4.38),312(4.24), 364(3.88)	83-0018-76
Methanone, (3-hydroxybenzo[b]thien-2-yl)(4-nitrophenyl)-	dioxan	268(3.43),300(4.17), 390(3.88)	83-0018-76
$C_{15}H_9N_3O_2$ 1H-Anthra[1,2-d]triazole-6,11-dione, 1-methyl-	EtOH	344(3.72)	104-0865-76
2H-Anthra[1,2-d]triazole-6,11-dione, 2-methyl-	EtOH	342(3.85),380(3.58)	104-0865-76
3H-Anthra[1,2-d]triazole-6,11-dione, 3-methyl-	EtOH	339(3.81)	104-0865-76
$C_{15}H_9N_3O_4$ 5,8-Quinolinedione, 6-amino-2-(2-nitro-phenyl)-	MeCN	280(4.65),420(3.34)	23-2563-76
$C_{15}H_9N_3O_4S$ 1,3,4-Thiadiazol-2(3H)-one, 5-benzoyl-3-(4-nitrophenyl)-	EtOH	262(4.61),322(4.30)	4-0045-76
$C_{15}H_9N_3O_6$ 1-Cyclopropene-1,2-diacetic acid, α,α'-dicyano-3-(1-cyano-2-methoxy-2-oxoeth-ylidene)-, dimethyl ester, ion(2-)	MeCN	232(4.51),305s(4.47), 328(4.59)	35-0610-76
$C_{15}H_9N_5O_4S$ Methanone, [4,5-dihydro-4-(4-nitrophen-yl)-5-(nitrosoimino)-1,3,4-thiadiazol-2-yl]phenyl-	EtOH	276(4.40),360(4.27), 480(1.86)	4-0045-76
$C_{15}H_9OS$ [1]Benzothieno[3,2-b][1]benzopyrylium perchlorate	n.s.g.	266(4.32),280(4.35), 408(4.20),471(3.74), 650(3.07)	103-0977-76
$C_{15}H_{10}$ 2H-Cyclopenta[jk]fluorene	hexane	230(4.49),247(4.62), 255(4.80),272(3.90), 278(3.90),292(3.80), 296(3.80),306(3.87), 321(3.86),340(2.86)	24-2596-76

Compound	Solvent	$\lambda_{max}(\log \epsilon)$	Ref.
$C_{15}H_{10}BrNO$			
Isoxazole, 4-bromo-3,5-diphenyl-	HOAc	273(4.27)	103-0851-76
	+17M H_2SO_4	319(4.39)	103-0851-76
$C_{15}H_{10}BrN_3$			
11H-Indolo[3,2-c][1,8]naphthyridine, 8-bromo-3-methyl-	EtOH	244(4.68),260s(4.21), 282(4.72)	4-0097-76
$C_{15}H_{10}BrN_3O$			
Benzenepropanenitrile, 4-bromo-β-oxo-α-(phenylhydrazono)-	EtOH	245(4.376),280(4.165), 370(4.375)	18-0321-76
$C_{15}H_{10}BrN_3OS$			
Methanone, [4-(3-bromophenyl)-4,5-dihydro-5-imino-1,3,4-thiadiazol-2-yl]-phenyl-	EtOH	223(4.40),265(4.17), 366(3.96)	4-0045-76
Methanone, [4-(4-bromophenyl)-4,5-dihydro-5-imino-1,3,4-thiadiazol-2-yl]-phenyl-	EtOH	220(4.28),268(4.26), 370(4.02)	4-0045-76
$C_{15}H_{10}Br_2O_2$			
Benzo[b]benzo[3,4]cyclobuta[1,2-e][1,4]-dioxin, 7,8-dibromo-4b,10a-dihydro-4b-methyl-	ether	258(3.30),265(3.40), 272(3.43),296(3.57)	22-0493-76
$C_{15}H_{10}ClF_3N_2O_3S_2$			
3-Pyrimidineacetic acid, 4-[(4-chlorophenyl)thio]-2-(methylthio)-α-(2,2,2-trifluoro-1-hydroxyethylidene)-, monosodium salt	MeOH	226(4.17),257(4.32), 307(3.93)	73-2771-76
$C_{15}H_{10}ClN$			
Cyclohepta[b]pyrrole, 2-chloro-4-phenyl-	MeOH	228(3.76),275(4.68), 308s(3.73),337(3.75), 470(3.37)	142-0221-76
Cyclohepta[b]pyrrole, 2-chloro-6-phenyl-	MeOH	226(3.24),273(4.58), 307(4.39),346(4.12), 361(4.15),476(3.36)	142-0221-76
Cyclohepta[b]pyrrole, 2-chloro-8-phenyl-	MeOH	223(3.80),268(4.51), 292(4.24),338(3.74), 357s(3.46)	142-0221-76
Quinoline, 6-chloro-2-phenyl-	EtOH	261(4.57),327(3.90)	94-2409-76
Quinoline, 7-chloro-2-phenyl-	EtOH	259(4.71),327(3.98)	94-2409-76
$C_{15}H_{10}ClNO$			
Isoxazole, 4-chloro-3,5-diphenyl-	HOAc	274(4.31)	103-0851-76
	+17M H_2SO_4	319(4.38)	103-0851-76
$C_{15}H_{10}ClNOS$			
4-Thiazolol, 5-(4-chlorophenyl)-2-phenyl-	EtOH	275(1.91),364(4.29)	118-0261-76
$C_{15}H_{10}ClNO_2$			
9,10-Anthracenedione, 1-amino-2-(chloromethyl)-	benzene	475(3.92)	146-0001-76
9,10-Anthracenedione, 1-chloro-2-(methylamino)-	CCl_4	431(3.79)	104-2499-76
$C_{15}H_{10}ClNO_4S$			
Chlorosulfurous acid, (1-amino-9,10-di-	benzene	470(3.46)	2-0054-76

Compound	Solvent	$\lambda_{max}(\log \epsilon)$	Ref.
hydro-9,10-dioxo-2-anthracenyl)methyl ester (cont.)			2-0054-76
$C_{15}H_{10}ClN_3$			
11H-Indolo[3,2-c][1,8]naphthyridine, 8-chloro-3-methyl-	EtOH	242(4.68),260s(4.18), 281(4.70)	4-0097-76
11H-Indolo[3,2-c][1,8]naphthyridine, 10-chloro-3-methyl-	EtOH	238(4.64),256s(4.21), 278(4.71)	4-0097-76
$C_{15}H_{10}ClN_3O$			
Benzenepropanenitrile, α-[(4-chlorophenyl)azo]-β-oxo-	EtOH	298(4.153),280(3.869), 378(4.390)	18-0321-76
$C_{15}H_{10}ClN_3OS$			
Methanone, [4-(3-chlorophenyl)-4,5-dihydro-5-imino-1,3,4-thiadiazol-2-yl)-phenyl-	EtOH	217(4.27),265(4.25), 368(4.03)	4-0045-76
Methanone, [4-(4-chlorophenyl)-4,5-dihydro-5-imino-1,3,4-thiadiazol-2-yl)-phenyl-	EtOH	220(4.26),268(4.27), 366(4.05)	4-0045-76
$C_{15}H_{10}Cl_2O_2$			
2-Propen-1-one, 1-(3,5-dichloro-2-hydroxyphenyl)-3-phenyl-	EtOH	235(3.92),325(4.26)	97-0365-76
	MeCN	232(4.34),325(4.43)	97-0365-76
	CCl₄	325(4.38)	97-0365-76
$C_{15}H_{10}Cl_3N$			
Pyridine, 2,3,6-trichloro-5-(4-phenyl-1,3-butadienyl)-, (E,X)-	MeOH	345(4.50)	33-0190-76
$C_{15}H_{10}FN_3$			
11H-Indolo[3,2-c][1,8]naphthyridine, 8-fluoro-3-methyl-	EtOH	230(4.61),258s(4.22), 280(4.71)	4-0097-76
$C_{15}H_{10}F_3NO$			
Acetamide, N-9H-fluoren-2-yl-2,2,2-trifluoro-	C_6H_{12}	290(4.54),315(4.51)	40-0133-76
	MeOH	290(4.52),312(4.47)	40-0133-76
$C_{15}H_{10}F_6O$			
Benzene, 1,1'-[1,1,2-trifluoro-2-(trifluoromethoxy)-1,2-ethanediyl]bis-	MeOH	235(3.52),254(3.18), 260(3.15),267(3.04)	39-0101-76C
$C_{15}H_{10}N_2$			
Benzimidazo[2,1-a]isoquinoline	MeOH	258(4.38),268(4.55), 279(4.69),316(3.93), 333(3.88),349(3.79)	39-0075-76C
Benzimidazo[1,2-a]quinoline	MeOH	236(4.46),249(4.51), 257(4.45),266(4.39), 330(4.09),350(4.01), 366(3.78)	39-0075-76C
$C_{15}H_{10}N_2O$			
Pyrrolo[2,1-a]isoquinoline-3-carbonitrile, 1-acetyl-	MeOH	207(4.70),225(4.75), 246(4.69),275s(4.96), 277(4.98),315s(4.21), 324(4.33),338(4.42), 355(4.46)	39-1908-76C
$C_{15}H_{10}N_2O_2$			
Benzonitrile, 3-nitro-4-(2-phenylethenyl)-, (E)-	EtOH	314(4.34)	80-1127-76

Compound	Solvent	$\lambda_{max}(\log \epsilon)$	Ref.
Benzonitrile, 5-nitro-2-(2-phenylethen-yl)-, (E)-	EtOH	250(4.20),340(4.31)	80-1127-76
Canthin-6-one, 1-methoxy-	EtOH	280(4.09),330(3.87), 362(4.24),370(4.01), 379(4.28)	94-1532-76
Indolo[2,1-b]quinazolin-6(12H)-one, 12-hydroxy-	MeOH	201(4.32),228(4.13), 253(4.21),260(4.21), 270(4.16),320(3.83), 450(3.73)	39-2004-76C
Pyrrolo[2,1-a]isoquinoline-1-carboxylic acid, 3-cyano-, methyl ester	MeOH	207s(4.13),219s(4.35), 223(4.40),237s(4.30), 242(4.35),265s(4.63), 273(4.71),279(4.26), 308s(3.88),319(3.98), 334(4.00),350(4.02)	39-1908-76C
$C_{15}H_{10}N_2O_2S$ 1,3,4-Thiadiazol-2(3H)-one, 5-benzoyl-3-phenyl-	EtOH	246(4.03),270s(3.93), 322(3.99)	4-0045-76
$C_{15}H_{10}N_2O_3$ Benzoic acid, 2-(1H-benzimidazol-2-yl-carbonyl)-	MeOH	240(4.01),310(4.18)	44-0836-76
Benzonitrile, 2-[2-(4-hydroxyphenyl)-ethenyl]-5-nitro-, (E)-	EtOH	265(4.16),395(4.47)	80-1127-76
Benzonitrile, 4-[2-(4-hydroxyphenyl)-ethenyl]-3-nitro-, (E)-	EtOH	345(4.45)	80-1127-76
5,8-Quinazolinedione, 6-hydroxy-4-meth-yl-2-phenyl-	M NaOH	247(4.14),312(4.54), 469(3.40)	5-1809-76
	EtOH	235s(--),257(4.18), 296(4.31),320s(--), 388s(--)	5-1809-76
$C_{15}H_{10}N_2O_3S$ 4-Thiazolol, 5-(4-nitrophenyl)-2-phenyl-	EtOH	260(3.99),403(4.35), 491(1.59)	118-0261-76
$C_{15}H_{10}N_4O$ 1(2H)-Phthalazinone, 4-(1H-benzimidazol-2-yl)-	DMF	281(4.15),324(4.26)	44-0836-76
$C_{15}H_{10}N_4OS$ 1,2,4-Triazolo[3,4-b][1,3,4]thiadiazol-ium, 5,6-dihydro-6-oxo-2,3-diphenyl-, hydroxide, inner salt	MeCN	201s(4.72),278(4.13)	30-0104-76
$C_{15}H_{10}N_4O_2S$ Methanone, [4,5-dihydro-5-(nitrosoimi-no)-4-phenyl-1,3,4-thiadiazol-2-yl]-phenyl-	EtOH	277(4.36),360(4.29), 480(1.74)	4-0045-76
$C_{15}H_{10}N_4O_3$ Benzenepropanenitrile, 4-nitro-β-oxo-α-(phenylhydrazono)-	EtOH	265(4.200),376(4.467)	18-0321-76
$C_{15}H_{10}N_4O_3S$ Methanone, [4,5-dihydro-5-imino-4-(3-ni-trophenyl)-1,3,4-thiadiazol-2-yl]phen-yl-	EtOH	214(4.19),262(4.32), 360(3.96)	4-0045-76
Methanone, [4,5-dihydro-5-imino-4-(4-ni-	EtOH	230(4.24),267(4.24),	4-0045-76

Compound	Solvent	$\lambda_{max}(\log \epsilon)$	Ref.
trophenyl)-1,3;4-thiadiazol-2-yl]-phenyl- (cont.)		372(4.21)	4-0045-76
$C_{15}H_{10}N_4O_4S$			
Benzo[b]thiophene, 2-[(2,4-dinitrophenyl)azo]-3-methyl-	n.s.g.	217(4.56),277(4.02), 406(4.49)	39-1639-76C
Benzo[b]thiophene, 3-[(2,4-dinitrophenyl)azo]-2-methyl-	n.s.g.	222(4.38),288(4.17), 408(4.23)	39-1639-76C
$C_{15}H_{10}N_6S$			
1-Butene-1,1,2-tricarbonitrile, 3-[(3-methyl-2(3H)-benzothiazolylidene)-hydrazono]-	EtOH	292(4.59),535(4.29)	24-1787-76
$C_{15}H_{10}O$			
9(10H)-Anthracenone, 10-methylene-(qualitative spectra in other solvents)	isooctane	273(4.23)	73-1676-76
6H-Cyclohepta[b]naphthalen-6-one	MeOH	224s(4.28),229(4.26), 244(4.34),252(4.39), 298(3.95),307(3.87), 326s(3.50),360s(3.29), 398s(3.16)	88-1117-76
	70% H_2SO_4	222(4.12),247(4.33), 260(4.37),275s(4.25), 313(4.36),340s(4.17), 380(3.76),399s(3.34), 520(3.46)	88-1117-76
$C_{15}H_{10}OS$			
4H-1-Benzothiopyran-4-one, 2-phenyl-	EtOH	225(4.06),265(4.18), 345(3.88)	114-0309-76A
	2% H_2SO_4	265(4.30),350(3.90)	114-0309-76A
	97% H_2SO_4	265(4.58),380(3.96)	114-0309-76A
$C_{15}H_{10}OS_2$			
1,3-Dithiol-1-ium, 4-hydroxy-2,5-diphenyl-, hydroxide, inner salt	dioxan	<u>240(4.1),280(4.2),</u> 553(4.11)	24-0740-76
	benzene	558.7(--)	24-0740-76
	MeOH	530(--)	24-0740-76
	HOAc	522.5(--)	24-0740-76
	DMF	546.5(--)	24-0740-76
	MeCN	542(--)	24-0740-76
	DMSO	545(--)	24-0740-76
	CCl_4	565(--)	24-0740-76
$C_{15}H_{10}O_2$			
11H-Benzofuro[3,2-b][1]benzopyran	EtOH	226(4.28),244s(--), 278(3.85)	22-1967-76
Flavone	EtOH	215s(4.23),252(4.24), 298(4.33)	114-0309-76A
	2% H_2SO_4	253(4.34),302(4.37)	114-0309-76A
	97% H_2SO_4	250(4.30),348(4.48)	114-0309-76A
$C_{15}H_{10}O_3$			
9,10-Anthracenedione, 1-hydroxy-2-methyl-	MeOH	254(4.21),272s(3.87), 326(3.18),408(3.53)	64-0403-76C
Phenanthro[2,3-b]oxirene-7,9-dione, 7a,8a-dihydro-7a-methyl-	n.s.g.	223(4.35),269(4.54), 325(3.59),351(3.55)	33-0664-76

Compound	Solvent	$\lambda_{max}(\log \epsilon)$	Ref.
$C_{15}H_{10}O_3S$			
1H-2-Benzopyran-1,3(4H)-dione, 4-[(5-methyl-2-thienyl)methylene]-, (E)-	EtOH	220(4.14),400(4.12)	4-0083-76
$C_{15}H_{10}O_4$			
Benzo[d][1,3]dioxolo[4,5-h][2]benzoxepin-5(7H)-one	EtOH	243(4.37),285(3.90)	94-2191-76
Benzo[d][1,3]dioxolo[4,5-h][2]benzoxepin-7(5H)-one	EtOH	270(3.76),305(3.84)	94-2191-76
Nordalbergin	EtOH	262(2.55)	78-2407-76
9H-Xanthene-1-carboxaldehyde, 8-hydroxy-3-methyl-9-oxo-	EtOH	234(4.49),256(4.35), 294(4.10),369(3.68)	23-1703-76
$C_{15}H_{10}O_4S$			
Methanone, (2,4-dihydroxyphenyl)(3-hydroxybenzo[b]thien-2-yl)-	MeOH	262(4.13),304(3.98), 376(4.19),427(4.13)	83-0018-76
$C_{15}H_{10}O_5$			
1,4-Anthracenedione, 2,5,10-trihydroxy-7-methyl-	MeOH-DMSO-HCOOH	292(4.11),303(4.12), 349s(3.60),474s(3.89), 507(4.01),545s(3.94), 599s(3.54)	12-1535-76
1,4-Anthracenedione, 5,7,10-trihydroxy-2-methyl-	MeOH-DMSO-HCOOH	279s(4.00),338(3.98), 503s(3.69),530(3.93), 577s(3.69)	12-1535-76
9,10-Anthracenedione, 1,5-dihydroxy-3-(hydroxymethyl)-	MeOH	254(4.37),279s(4.03), 289(4.04),418(4.06), 428(4.06)	102-0317-76
Apigenin	EtOH	269(4.27),300s(4.13), 340(4.32)	112-0865-76
	EtOH-NaOEt	277(4.34),330(4.11), 400(4.50)	112-0865-76
10H-Benz[b]indeno[2,1-d]furan-10-one, 5a,10a-dihydro-3,5,10a-trihydroxy-	MeOH	215s(3.23),241(4.13), 289(3.68),340s(3.18)	83-0081-76
2H-1-Benzopyran-2-one, 4-hydroxy-3-(2,5-dihydroxyphenyl)-	MeOH	210(4.45),280s(3.97), 310(4.09)	2-0861-76
2H-1-Benzopyran-2,5,8-trione, 3,4-dihydro-7-hydroxy-4-phenyl-	n.s.g.	245(4.18),288(4.59)	39-0407-76C
9H-Xanthene-1-carboxylic acid, 8-hydroxy-6-methyl-9-oxo-	MeOH	235(4.47),250(4.45), 300(4.07),360(3.69)	2-0336-76
9H-Xanthene-4-carboxylic acid, 8-hydroxy-6-methyl-9-oxo-	MeOH	235(4.33),250(4.34), 303(3.87),360(3.66)	2-0336-76
$C_{15}H_{10}O_5S$			
Methanone, (3-hydroxybenzo[b]thien-2-yl)(2,3,4-trihydroxyphenyl)-	MeOH	263(4.10),304(3.86), 350(4.14),391(4.23)	83-0018-76
Methanone, (3-hydroxybenzo[b]thien-2-yl)(2,4,5-trihydroxyphenyl)-	MeOH	266(4.17),328(4.00), 413(4.14)	83-0018-76
$C_{15}H_{10}O_6$			
2-Anthracenecarboxylic acid, 9,10-dihydro-4,5,7-trihydroxy-10-oxo-	MeOH-1% HCOOH	264(4.12),270(4.12), 309s(3.51),366(4.22)	12-1509-76
9,10-Anthracenedione, 1,2,3,8-tetrahydroxy-6-methyl-	MeOH-HOAc	286(4.42),432(3.99)	12-2231-76
9,10-Anthracenedione, 1,2,4,5-tetrahydroxy-3-methyl-	MeOH	259(4.49),295s(4.09), 402s(3.59),466s(3.99), 491(4.10),518s(4.00), 527(4.01)	64-0403-76C
9,10-Anthracenedione, 1,2,4,5-tetrahydroxy-7-methyl-	dioxan-1% HOAc	259(4.54),306(4.05), 380(3.39),474s(4.00), 493(4.09),515s(3.98)	12-2231-76

Compound	Solvent	$\lambda_{max}(\log \epsilon)$	Ref.
9,10-Anthracenedione, 1,3,8-trihydroxy-6-(hydroxymethyl)-	MeOH-HOAc	250(4.36),266(4.38), 288(4.42),439(4.17)	12-2231-76
1,3-Benzodioxole-5-carboxylic acid, 6-(3,6-dioxo-1,4-cyclohexadien-1-yl)-, methyl ester	EtOH	227(4.57),252s(--), 300s(--),398(3.15)	24-0855-76
$C_{15}H_{10}O_7$			
1,3-Benzodioxole-5-carboxylic acid, 6-(3,6-dioxo-1,4-cyclohexadien-1-yl)-4-hydroxy-, methyl ester	EtOH	240(4.59),256s(--), 283s(--),317s(--), 347(3.57),362s(--), 410s(--)	24-0855-76
2H,5H-Pyrano[3,2-c][1]benzopyran-3-carboxylic acid, 8-hydroxy-2,5-dioxo-, ethyl ester	pH 4.0	225(4.05),282(3.80), 389(4.35)	49-0783-76
	pH 9.0	255(4.02),330(3.93), 438(4.40)	49-0783-76
$C_{15}H_{10}O_9$			
Chiodectonic acid	MeOH	287(4.33),510(4.01), 538(4.00)	102-0799-76
$C_{15}H_{11}Cl$			
Anthracene, 2-(chloromethyl)- (all log ϵ approximate)	hexane	256(5.00),325(2.95), 339(3.15),353(3.28), 363(3.20)	88-2815-76
1H-Indene, 2-chloro-3-phenyl-	EtOH	213(4.15),232(4.15), 265(3.86),290s(3.02), 300(2.96)	12-0883-76
$C_{15}H_{11}ClN_2O$			
1H-Indazole, 1-acetyl-5-chloro-3-phenyl-	isoPrOH	229(4.54),239s(4.46), 315(4.17)	4-0033-76
4H-Pyrazol-4-one, 5-(4-chlorophenyl)-	EtOH	260(4.498),302s(4.093)	23-1752-76
$C_{15}H_{11}ClN_2O_2$			
9,10-Anthracenedione, 1,4-diamino-2-(chloromethyl)-	benzene	592(3.79)	146-0001-76
$C_{15}H_{11}ClN_2O_4S$			
Chlorosulfurous acid, (1,4-diamino-9,10-dihydro-9,10-dioxoanthracenyl)-methyl ester	benzene	590(3.56)	2-0054-76
$C_{15}H_{11}ClN_4O_2$			
Urea, N-[2-(4-chlorophenyl)-3,4-dihydro-4-oxo-3-quinazolinyl]-	EtOH	215s(4.45),234(4.50), 280(4.23),312s(3.99)	114-0341-76D
$C_{15}H_{11}ClN_4S$			
2-Thiazolamine, 5-[(4-chlorophenyl)azo]-4-phenyl-	EtOH	268(4.324),444(4.180)	4-0045-76
$C_{15}H_{11}ClOS$			
Dibenzo[b,f]thiepin, 2-chloro-8-methoxy-	MeOH	227(4.58),265(4.43), 295(3.78)	73-3607-76
2-Propene-1-thione, 3-(3-chlorophenyl)-3-hydroxy-1-phenyl-	EtOH	261(4.03),333(4.18), 408(4.27),503s(2.33)	4-0691-76
2-Propene-1-thione, 3-(4-chlorophenyl)-3-hydroxy-1-phenyl-	EtOH	269(4.05),332(4.15), 412(4.31),498s(2.35)	4-0691-76
$C_{15}H_{11}ClO_2S$			
Dibenzo[b,f]thiepin-10(11H)-one, 2-chloro-7-methoxy-	MeOH	258(4.36),264(4.36), 312s(3.57)	73-3607-76

Compound	Solvent	$\lambda_{max}(\log \epsilon)$	Ref.
Dibenzo[b,f]thiepin-10(11H)-one, 2-chloro-8-methoxy-	MeOH	234(4.48),256s(4.05), 348(3.59)	73-3607-76
$C_{15}H_{11}Cl_2N$ Pyridine, 2,6-dichloro-3-(4-phenyl-1,3-butadienyl)-, (E,X)-	MeOH	341(4.66)	33-0190-76
$C_{15}H_{11}F_3O_3$ Benzeneacetic acid, α-[3-(trifluoromethyl)phenoxy]-	EtOH	211(4.21),221(4.22), 275(3.42),282(3.36)	111-0533-76
$C_{15}H_{11}LiO_2$ 9H-Fluorene-9-carboxylic acid, methyl ester, lithium salt	$(MeOCH_2)_2$	375s(--),389(3.75), 409(3.68)	104-2411-76
$C_{15}H_{11}N$ Quinoline, 2-phenyl-	EtOH	258(4.62),323(3.86)	94-2409-76
$C_{15}H_{11}NO$ Isoquinoline, 5-phenoxy-	EtOH	208(3.26),274(2.97), 284(2.86),312(2.61), 324(2.60)	39-2285-76C
3-Isoquinolinol, 1-phenyl-	EtOH	350(3.76),433(3.36)	103-0910-76
	ether	352(3.99)	103-0910-76
	$CHCl_3$	355(3.73),440(3.34)	103-0910-76
	CCl_4	353(3.85),438(3.15)	103-0910-76
Isoxazole, 3,5-diphenyl-	HOAc	268(4.39)	103-0851-76
	+17M H_2SO_4	312(4.47)	103-0851-76
4(1H)-Quinolinone, 2-phenyl-	pH 2	255(3.53),320(4.08)	114-0309-76A
	EtOH	210(4.55),256(4.58), 333(3.97)	114-0309-76A
	97% H_2SO_4	265(4.48),315(4.26)	114-0309-76A
$C_{15}H_{11}NOS$ 4-Thiazolol, 2,5-diphenyl-	EtOH	267(3.88),363(4.28)	118-0261-76
$C_{15}H_{11}NOS_2$ Pyridinium, 2-oxo-4-phenyl-3-thioxo-cyclobutene-1-thiolate	CH_2Cl_2	251(4.25),256(4.25), 374(4.50)	118-0445-76
$C_{15}H_{11}NO_2$ 1,2-Anthracenedione, 1-(O-methyloxime)	hexane	259(4.20),285s(4.35), 294(4.57),340(3.61), 356(3.52),375s(3.52), 396(3.68),412(3.58)	110-1711-76
	EtOH	240(4.37),258s(4.27), 284s(4.26),300(4.55), 300(4.55),358(3.61), 413(3.83),433s(--)	110-1711-76
	1:4 dioxan-hexane	254(4.28),284(4.36), 297s(4.26),340(3.64), 380s(3.60),395(3.71), 416(3.64)	110-1711-76
1,2-Anthracenedione, 2-(O-methyloxime)	hexane	220(--),274(4.18), 286s(4.4),297(4.46), 330s(3.8),388(3.48), 408(3.63),430(3.34), 450s(2.6)	110-1711-76
	EtOH	220(--),274s(4.18), 294(4.3),301(4.35),	110-1711-76

Compound	Solvent	$\lambda_{max}(\log \epsilon)$	Ref.
1,2-Anthracenedione, 2-(O-methyloxime) (cont.)		320s(3.8),396(3.54), 431(3.6),453(3.54)	110-1711-76
	1:4 dioxan-hexane	220(--),289(4.37), 298(4.44),320s(3.8), 397(3.44),413(3.62), 430(3.5)	110-1711-76
9,10-Anthracenedione, 1-amino-2-methyl-	CCl_4	457(3.84)	104-2499-76
1H-2-Benzopyran-1-one, 5-amino-3-phenyl-	n.s.g.	259(4.30),303(4.06), 312s(4.05),390(4.05)	39-1073-76C
4H-1-Benzopyran-4-one, 2-phenyl-, oxime	MeOH	239(4.40),271(4.21), 321(4.14)	56-1067-76
1H-2,3-Benzoxazin-1-one, 4-(4-methylphenyl)-	H_2SO_4	280(3.82)	104-1797-76
5(2H)-Oxazolone, 2,4-diphenyl-	n.s.g.	263(4.09)	33-2149-76
5(4H)-Oxazolone, 2,4-diphenyl-	n.s.g.	242(4.18)	33-2149-76
Phenol, 2-(3-phenyl-5-isoxazolyl)-	MeOH	245(4.32),264(4.27), 307(4.13)	56-1067-76
Pyridinium, 1-methyl-, 1,3-dihydro-1,3-dioxo-2H-inden-2-ylide	EtOH	235(4.38),293(3.99), 390(4.32)	73-1363-76
$C_{15}H_{11}NO_2S$			
Methanone, (2-aminophenyl)(3-hydroxybenzo[b]thien-2-yl)-	dioxan	241(4.38),266(4.31), 318(4.10),406(4.10)	83-0018-76
$C_{15}H_{11}NO_3$			
1H-2-Benzopyran-1,3(4H)-dione, 4-[(1-methyl-1H-pyrrol-2-yl)-methylene]-, (Z)-	EtOH	225(3.99),430(4.11)	4-0083-76
2-Propen-1-one, 1-(4-nitrophenyl)-3-phenyl-	n.s.g.	313(4.27)	39-0380-76C
2-Propen-1-one, 3-(4-nitrophenyl)-1-phenyl-	n.s.g.	306(4.32)	39-0380-76C
$C_{15}H_{11}NO_4$			
Benzeneacetic acid, 2-carboxy-α-(3-pyridinylmethylene)-	EtOH	287(4.18)	39-0315-76C
$C_{15}H_{11}NO_4S_2$			
Benzeneacetic acid, 4-nitro-α-[(phenylthioxomethyl)thio]-	dioxan	301(4.29),500(2.02)	24-0740-76
$C_{15}H_{11}NO_6$			
1,4-Benzenedicarboxylic acid, 2-(2-carboxyphenylamino)-, β-cyclodextrin complex	pH 7.0	298(4.12),330(3.88)	133-0343-76
1,3-Benzodioxole-5-carboxylic acid, 6-[3-(hydroxyimino)-6-oxo-1,4-cyclohexadien-1-yl]-, methyl ester	EtOH	225(4.36),302(4.26)	24-0855-76
$C_{15}H_{11}N_3$			
11H-Indolo[3,2-c][1,8]naphthyridine, 3-methyl-	EtOH	234(4.63),257s(4.23), 280(4.73)	4-0097-76
2,2':6',2"-Terpyridine, ferrous complex	MeOH	552(4.10)	118-0001-76
$C_{15}H_{11}N_3O$			
Benzenepropanenitrile, β-oxo-α-(phenylhydrazono)-	EtOH	248(3.893),280(3.605), 373(3.732)	18-0321-76
Benzonitrile, 4-[[[(2-hydroxyphenyl)-methylene]hydrazono]methyl]-	EtOH	224(4.26),248(3.95), 255(4.00),304(4.51), 314(4.44),351(4.28)	95-0044-76

Compound	Solvent	$\lambda_{max}(\log \epsilon)$	Ref.
Phenol, 4-(6-quinolinylazo)-	C_6H_{12}	280(3.82),350(3.44)	115-0379-76
	EtOH	250(4.13),290(3.76), 363(4.32)	115-0379-76
	acetone	358(4.38)	115-0379-76
	$CHCl_3$	248(4.26),358(4.50)	115-0379-76
	CCl_4	352(4.09)	115-0379-76
2-Propen-1-one, 3-(1H-benzotriazol-1-yl)-1-phenyl-, (E)-	EtOH	272(4.24),328(4.09)	22-0184-76
$C_{15}H_{11}N_3OS$			
5(2H)-Isoxazolethione, 3-phenyl-4-(phenylazo)-	EtOH	238(4.32),258s(4.20), 360(3.84),466(4.22), 496s(4.08)	48-0359-76
Methanone, (4,5-dihydro-5-imino-4-phenyl-1,3,4-thiadiazol-2-yl)phenyl-	EtOH	215(4.00),265(4.04), 368(3.82)	4-0045-76
	$CHCl_3$	280(4.40),368(4.39), 500(1.73)	4-0045-76
$C_{15}H_{11}N_3O_2$			
4-Oxazolecarboxamide, 2-phenyl-N-(2-pyridinyl)-	EtOH	285(4.12)	39-0315-76C
hydrochloride	EtOH	283(4.11)	39-0315-76C
2-Propen-1-one, 3-(1H-benzotriazol-1-yl)-1-phenyl-, N-oxide, (E)-	EtOH	235s(4.02),280(4.14), 298s(4.02),372(4.32), 382s(4.24)	22-0184-76
$C_{15}H_{11}N_3O_3$			
4H-Pyrazol-4-one, 1,5-dihydro-5-(4-nitrophenyl)-3-phenyl-	EtOH	262(4.672),333(4.686)	23-1752-76
4-Quinolinol, 3-nitro-2-(phenylamino)-	$CHCl_3$	279(3.96),303s(3.66), 319(3.68),369(4.20)	48-0039-76
$C_{15}H_{11}N_3O_4$			
Benzoic acid, 2-(1H-benzimidazol-1-yl)-5-nitro-, methyl ester	EtOH	247(4.22),275s(3.90), 317(3.77)	2-0001-76
11H-Dibenzo[b,e][1,4]diazepin-11-one, 5-acetyl-5,10-dihydro-2-nitro-	EtOH	267(4.09)	2-0001-76
$C_{15}H_{11}N_3O_7$			
Benzene, 2-[2-(4-methoxyphenyl)ethenyl]-1,3,5-trinitro-, (E)-	EtOH	280(4.19),400(4.17)	80-1127-76
$C_{15}H_{11}N_5$			
Benzonitrile, 4-[2-(4-methylphenyl)-2H-tetrazol-5-yl]-	EtOH	281(4.48)	18-0762-76
3H-Pyrazolo[4,3-c]cinnolin-3-imine, 1,2-dihydro-2-phenyl-	EtOH	231(4.27),294(4.22), 352(3.48)	39-0592-76C
$C_{15}H_{11}N_5O_2$			
Pyrimido[1,2-a]purin-10(1H)-one, 7-(4-methoxyphenyl)-	pH 1	244(4.26),258(4.27), 306(4.40),370s(3.42)	44-0294-76
	pH 6.8	253(4.41),256s(4.40), 304(4.41),370(3.49)	44-0294-76
	pH 10.0	254(4.44),301(4.42), 335(3.79),380(3.42)	44-0294-76
$C_{15}H_{11}N_5O_2S$			
2-Thiazolamine, 5-[(4-nitrophenyl)azo]-4-phenyl-	EtOH	272(4.301),440(4.301)	4-0045-76

Compound	Solvent	$\lambda_{max}(\log \epsilon)$	Ref.
$C_{15}H_{11}N_5O_4$ Urea, N-[2-(4-nitrophenyl)-4-oxodihydro-3-quinazolinyl]-	EtOH	225(4.50),280s(4.25), 313(4.17)	114-0341-76D
$C_{15}H_{12}$ 9H-Fluorene, 9-ethylidene-	isooctane	229(4.73),247(4.53), 256(4.77),270(4.24), 280(4.26),296(4.14), 311(4.11),340s(2.6)	24-2596-76
Naphthalene, 2-(1-penten-3-ynyl)-, cis	MeOH	214(4.50),222s(4.33), 235(4.26),243(4.28), 253s(4.31),261(4.44), 271(4.46),287(4.19), 299(4.36),313(4.37), 325s(3.71),332s(3.46), 350(2.89)	39-1104-76B
trans	MeOH	230s(4.18),237(4.32), 245(4.39),254s(4.41), 262(4.58),272(4.62), 287(4.36),298(4.57), 313(4.60),333s(3.65), 351(3.09)	39-1104-76B
$C_{15}H_{12}BrN$ 1H-Indole, 2-(4-bromophenyl)-1-methyl-	MeOH	300(4.29)	12-2747-76
$C_{15}H_{12}BrNO$ Phenol, 2-[3-[(4-bromophenyl)imino]-1-propenyl]-	MeOH dioxan DMSO DMSO-NaOMe	298(4.11),358(4.23) 298(4.16),354(4.28) 300(4.23),360(4.33) 350(4.18),495(4.30)	104-0840-76 104-0840-76 104-0840-76 104-0840-76
$C_{15}H_{12}BrNO_2$ Pyridinium 1-(4-bromobenzoyl)-2-oxopropylide	EtOH dioxan MeCN	237(4.16),285(4.18), 363(3.16) 245(4.20),267(4.02), 417(3.07) 237(4.24),284(4.10), 403(3.21)	78-2647-76 78-2647-76 78-2647-76
$C_{15}H_{12}BrN_3$ 11H-Indolo[3,2-c][1,8]naphthyridine, 8-bromo-5,6-dihydro-3-methyl-	EtOH	222(4.62),270(4.12), 379(4.33)	4-0097-76
$C_{15}H_{12}BrN_5O_2S$ Isothiazolo[5,4-b]pyridine, 3-bromo-6,7-dihydro-4,7-dimethyl-6-[[(4-nitrophenyl)azo]methylene]-	DMF	559(4.56)	48-0779-76
$C_{15}H_{12}BrN_5O_3S_3$ 3-Thiazolidinepropanoic acid, 4-[aminothioxomethyl)hydrazono]-5-(5-bromo-1,2-dihydro-2-oxo-3H-indol-3-ylidene)-2-thioxo-	MeOH	248(4.46),291(3.82), 390s(--),415(--)	103-0749-76
$C_{15}H_{12}ClN$ 1H-Indole, 2-(4-chlorophenyl)-1-methyl-	MeOH	299(4.30)	12-2747-76
$C_{15}H_{12}ClN_3$ 11H-Indolo[3,2-c][1,8]naphthyridine, 8-chloro-5,6-dihydro-3-methyl-	EtOH	222(4.60),268(4.09), 378(4.31)	4-0097-76

Compound	Solvent	$\lambda_{max}(\log \epsilon)$	Ref.
11H-Indolo[3,2-c][1,8]naphthyridine, 10-chloro-5,6-dihydro-3-methyl-	EtOH	220(4.55),274s(3.76), 374(4.30)	4-0097-76
$C_{15}H_{12}ClN_3O_2$ 4-Oxazolecarboxamide, 2-phenyl-N-(2-pyridinyl)-, hydrochloride	EtOH	283(4.11)	39-0315-76C
$C_{15}H_{12}Cl_2$ Benzene, 2,4-dichloro-1-[2-(3-methylphenyl)ethenyl]-, (E)-	EtOH	335(4.40)	39-1731-76B
$C_{15}H_{12}Cl_2N_2O$ 1H-Indazole, 5-chloro-2-(chloroacetyl)-2,3-dihydro-3-phenyl-	isoPrOH	248(3.99),294s(3.36)	4-0033-76
$C_{15}H_{12}Cl_2N_2O_6S$ Acetamide, N-[4-[2-[(dichloromethyl)sulfonyl]-2-(5-nitro-2-furanyl)ethenyl]phenyl]-	EtOH	230(4.19),310(4.49)	73-3391-76
$C_{15}H_{12}Cl_2O$ Methanone, bis[3-(chloromethyl)phenyl]-	MeOH	253(4.25)	44-2509-76
$C_{15}H_{12}Cl_3N_3O$ 1H-1,2,3-Triazole, 4,5-dihydro-1-(4-phenoxyphenyl)-5-(trichloromethyl)-	dioxan	246(4.23),273(3.93)	97-0102-76
$C_{15}H_{12}FN_3$ 11H-Indolo[3,2-c][1,8]naphthyridine, 8-fluoro-5,6-dihydro-3-methyl-	EtOH	215(4.55),270(4.08), 376(4.31)	4-0097-76
$C_{15}H_{12}F_2N_2OS$ 5H-Thieno[2,3-c]pyrrole-6-carboxamide, 4-(2,6-difluorophenyl)-N,5-dimethyl-	EtOH	221(4.21),251(4.20), 321(4.28)	18-1395-76
$C_{15}H_{12}F_3NO_2$ Benzoic acid, 2-[[3-(trifluoromethyl)phenyl]methyl]amino-, β-cyclodextrin complex	pH 7.0	252(4.10),285(3.87)	133-0343-76
$C_{15}H_{12}F_3NO_2S$ Sulfilimine, S-(2-methoxyphenyl)-S-phenyl-N-(trifluoroacetyl)-	MeOH dioxan	225s(4.14),288(3.57) 225s(4.17),287(3.58)	78-3003-76 78-3003-76
$C_{15}H_{12}FeN_2O_3$ Ferrocene, [[(5-nitro-2-furanyl)methylene]amino]-	n.s.g.	574(3.49)	65-2512-76
$C_{15}H_{12}NO_2$ Pyridinium, 1-[(2-oxo-2H-1-benzopyran-6-yl)methyl]-, bromide	MeOH	268(3.22),312(4.02)	2-0516-76
$C_{15}H_{12}NO_5P$ 2(1H)-Quinolinone, 4-phenyl-3-(phosphonooxy)-	EtOH	224(4.51),285(3.85), 321(3.91)	102-0029-76
$C_{15}H_{12}NSe$ Isoselenazolium, 2,5-diphenyl-, perchlorate	HOAc	300s(3.96),340(4.20)	118-0273-76

Compound	Solvent	$\lambda_{max}(\log \epsilon)$	Ref.
$C_{15}H_{12}N_2$			
Pyrrolo[2,1-a]isoquinoline-3-carbo-nitrile, 1,2-dimethyl-	MeOH	209(4.20),225s(4.04), 230s(4.01),242(4.00), 270s(4.40),275(4.44), 282s(4.21),315s(3.74), 331(3.76),349s(3.74), 364s(3.56)	39-1908-76C
$C_{15}H_{12}N_2O$			
2H-Cyclopenta[gh]perimidin-2-one, 1,3-dihydro-1,3-dimethyl-	n.s.g.	385(3.96)	103-0455-76
Indolo[2,3-a]quinolizin-4(6H)-one, 7,12-dihydro-	MeOH	254(3.91),260(3.94), 275(3.77),300(3.56), 370(4.39),390(4.33)	24-0705-76
1(2H)-Isoquinolinone, 3-(phenylamino)-	EtOH	232(4.27),248(4.21), 311(4.25),366(3.75)	95-0154-76
3(2H)-Isoquinolinone, 2-amino-1-phenyl-	EtOH	245(2),430(--)	111-0083-76
4H-Pyrazol-4-one, 1,5-dihydro-3,5-di-phenyl-	EtOH	252(4.355),276(4.230), 295(4.000)	23-1752-76
3(2H)-Pyridazinone, 6-methyl-4-(1-naph-thalenyl)-	C_6H_{12}	223(4.81),261(3.65), 272(3.68),282(3.72), 293(3.72),316s(3.75)	59-0837-76
	EtOH	225(4.49),282(3.82), 293(3.82),316(3.82)	59-0837-76
	H_2SO_4	235s(4.86),279(3.86), 322(3.54),329(3.48), 352(3.54)	59-0837-76
Pyrido[1,2-b]pyridazin-9-ium, 4-hydroxy-8-methyl-3-phenyl-, hydroxide, inner salt	EtOH	282(4.67),332(4.11)	44-1570-76
2-Quinolinamine, 3-phenyl-, 1-oxide	$CHCl_3$	253(4.70),304(3.71), 312s(3.70),364(3.74)	12-0357-76
$C_{15}H_{12}N_2OS$			
Benzothiazole, 2-methyl-6-nitroso-7-(phenylmethyl)-	$CHCl_3$	760(1.60)	118-0270-76
Benzothiazole, 6-nitroso-7-(2-phenyl-ethyl)-	$CHCl_3$	760(1.60)	118-0270-76
Phenol, 4-(benzo[b]thien-5-ylazo)-2-methyl-	C_6H_{12}	250(4.33),255s(4.30), 352(4.39)	62-0249-76A
	EtOH	253(4.37),365(4.43)	62-0249-76A
	acetone	363(4.49)	62-0249-76A
	$CHCl_3$	254(4.44),362(4.50)	62-0249-76A
	CCl_4	357(4.56)	62-0249-76A
$C_{15}H_{12}N_2OS_2$			
1,3,4-Thiadiazolium, 5-mercapto-2-(4-methoxyphenyl)-3-phenyl-, hydroxide, inner salt	MeOH	280(4.01),395(3.96)	78-0661-76
	MeCN	286(4.24),408(3.85)	78-0661-76
$C_{15}H_{12}N_2O_2$			
9,10-Anthracenedione, 1-amino-2-(methyl-amino)-	EtOH	547(3.99)	104-2503-76
9,10-Anthracenedione, 2-amino-1-(methyl-amino)-	EtOH	517(3.73)	104-2503-76
9,10-Anthracenedione, 2-amino-3-(methyl-amino)-	EtOH	521(3.37)	104-2503-76
2H-1-Benzopyran-2-one, 3-amino-4-(phen-ylamino)-	EtOH	241(3.85),267(3.94), 344(3.86)	103-0268-76
Ethanone, 1-(3-amino-6-phenylfuro[2,3-b]pyridin-2-yl)-	EtOH	271(4.20),331(4.34), 371(4.08)	48-0313-76

$C_{15}H_{12}N_2O_3-C_{15}H_{12}N_4O_3$

Compound	Solvent	$\lambda_{max}(\log \epsilon)$	Ref.
1H-Indazole-1-carboxylic acid, phenyl-methyl ester	EtOH	246(4.06),253(4.02), 288(3.86),298(3.83)	87-0839-76
2H-Indazole-2-carboxylic acid, phenyl-methyl ester	EtOH	279(3.93),290(3.96)	87-0839-76
3-Pyridinecarbonitrile, 2-(2-oxoprop-oxy)-6-phenyl-	EtOH	264(4.08),320(4.38)	48-0313-76
$C_{15}H_{12}N_2O_3$			
Benzamide, 3-nitro-4-(2-phenylethenyl)-	EtOH	274(4.16),310(4.32)	80-1127-76
1-Phenazinecarboxylic acid, 6-ethoxy-	EtOH	264(4.54),370(3.94), 425(3.48)	39-2248-76C
1-Phenazinecarboxylic acid, 8-ethoxy-	EtOH	265(4.49),368(3.74), 415(3.66)	39-2248-76C
1-Phenazinecarboxylic acid, 2-methoxy-, methyl ester	EtOH	215(4.47),257(4.87), 363(3.94),398(3.75)	39-2248-76C
1-Phenazinecarboxylic acid, 6-methoxy-, methyl ester	EtOH	264(4.69),348(3.74), 366(3.93),415(3.42)	39-2248-76C
1-Phenazinecarboxylic acid, 8-methoxy-, methyl ester	EtOH	254(4.84),356(3.95), 397(3.90)	39-2248-76C
1-Phenazinecarboxylic acid, 9-methoxy-, methyl ester	EtOH	263(4.82),346(3.95), 364(4.09),410(3.73)	39-2248-76C
Pyrrolo[2,1-a]isoquinoline-1-carboxylic acid, 3-(aminocarbonyl)-, methyl ester	MeOH	223(4.57),245(4.48), 264s(4.66),272(4.83), 295s(4.45),323(4.14), 339(4.19),359(4.22)	39-1908-76C
$C_{15}H_{12}N_2O_3S$			
Phenol, 2-(benzo[b]thien-5-ylazo)-4-methyl-, S,S-dioxide	C_6H_{12}	245(3.88),339(3.95), 420(3.54)	62-0249-76A
	EtOH	238(4.20),333(4.26), 414(3.93)	62-0249-76A
	$CHCl_3$	337(4.29),424(3.93)	62-0249-76A
	CCl_4	338(4.36),424(4.02)	62-0249-76A
Phenol, 4-(benzo[b]thien-5-ylazo)-2-methyl-, S,S-dioxide	C_6H_{12}	280(4.24),374(3.60)	62-0249-76A
	EtOH	254(4.22),381(4.31)	62-0249-76A
	acetone	375(4.18)	62-0249-76A
	$CHCl_3$	225s(4.35),265(4.36), 374(4.04)	62-0249-76A
	CCl_4	220s(4.43),279(4.44), 372(4.21)	62-0249-76A
$C_{15}H_{12}N_2O_4$			
Imidazo[2,1-a]isoquinoline-2,3-dicarb-oxylic acid, dimethyl ester	MeOH	245(4.80),259s(4.53), 304(3.56),318(3.66), 333(3.64)	39-0075-76C
$C_{15}H_{12}N_2O_5$			
Benzene, 1,1'-(1-methoxy-1,2-ethenedi-yl)bis[4-nitro-	n.s.g.	350(4.37)	2-0101-76
$C_{15}H_{12}N_4$			
Benzenepropanenitrile, β-imino-α-(phen-ylhydrazono)-	EtOH	240(3.39),370(4.34)	33-0551-76
$C_{15}H_{12}N_4O_2$			
Urea, (4-oxo-2-phenyl-3(4H)-quinazolin-yl)-	EtOH	216s(4.43),233(4.50), 275(4.21),307(4.03)	114-0341-76D
$C_{15}H_{12}N_4O_3$			
4-Cinnolinamine, 3-methoxy-4-(2-nitro-phenyl)-	$CHCl_3$	409(4.03)	39-0399-76C

Compound	Solvent	$\lambda_{max}(\log \epsilon)$	Ref.
$C_{15}H_{12}N_4O_4$			
1H-Benzimidazole, 2-[1-(2,4-dinitrophen-yl)ethyl]-	0.1N H_2SO_4	242(4.29),271(4.26), 278(4.21)	104-1097-76
	EtOH	245(4.29),274(4.11), 280(4.03)	104-1097-76
	EtOH-KOH	344(4.38),495(3.36)	104-1097-76
photochromic form	EtOH	580(4.55)	104-2361-76
1H-Benzimidazole, 2-[(2,4-dinitrophen-yl)methyl]-1-methyl-	0.1N H_2SO_4	242(4.32),271(4.24), 278(4.18)	104-1097-76
	EtOH	242(4.30),272(4.08), 278(4.00)	104-1097-76
	EtOH-KOH	495(4.48),646(3.98)	104-1097-76
photochromic form	EtOH	520(4.66)	104-2361-76
1H-Benzimidazole, 2-[(2,4-dinitrophen-yl)methyl]-5-methyl-	0.1N H_2SO_4	240(4.33),273(4.25), 280(4.26)	104-1097-76
	EtOH	245(4.33),279(4.10), 286(4.03)	104-1097-76
	EtOH-KOH	492(4.38),650(3.85)	104-1097-76
photochromic form	EtOH	516(4.67)	104-2361-76
4-Cinnolinamine, 3-methoxy-N-(2-nitro-phenyl)-, 1-oxide	$CHCl_3$	424(4.14)	39-0399-76C
$C_{15}H_{12}N_4O_5$			
1H-Benzimidazole, 2-[(2,4-dinitrophen-yl)methyl]-5-methoxy-	0.1N H_2SO_4	232(4.33),276(4.11), 285(4.14)	104-1097-76
	EtOH	245(4.32),288(4.07), 294(4.02)	104-1097-76
	EtOH-KOH	494(4.35),650(3.80)	104-1097-76
photochromic form	EtOH	520(4.68)	104-2361-76
$C_{15}H_{12}N_4S$			
Benzothiazole, 2-(1H-2,5-dihydro-4-phen-yl-1,2,3-triazol-2-yl)-	MeOH	270(3.9),325(4.03), 380(4.04)	2-0793B-76
2-Thiazolamine, 4-phenyl-5-(phenylazo)-	EtOH	263(4.312),426(4.457)	4-0045-76
$C_{15}H_{12}O$			
Anthracene, 9-methoxy-	EtOH	248(4.98),256(5.27), 318(2.91),333(3.32), 349(3.66),368(3.86), 388(3.81)	35-0171-76
Methanone, (2-ethenylphenyl)phenyl-	MeOH	244(4.38),270(3.88), 290(3.41),330(2.63)	35-0541-76
2-Propen-1-one, 1-[1,1'-biphenyl]-4-yl-	EtOH	300(4.3)	104-1968-76
$C_{15}H_{12}OS$			
4H-1-Benzothiopyran-4-one, 2,3-dihydro-2-phenyl-	EtOH	240(4.48),256s(3.90), 348(3.53)	114-0309-76A
	2% H_2SO_4	235(4.46),265s(3.95), 355(3.53)	114-0309-76A
	97% H_2SO_4	255(4.26),307(3.95), 480(3.60)	114-0309-76A
$C_{15}H_{12}O_2$			
Benzo[b]benzo[3,4]cyclobuta[1,2-e][1,4]-dioxin, 4b,10a-dihydro-4b-methyl-, cis	ether	259(3.50),265(3.69), 272(3.78)	22-0493-76
Benzofuran, 2-(2-methoxyphenyl)-	C_6H_{12}	234(4.08),278(4.16), 290(4.26),298(4.28), 312(4.53),327(4.52)	18-2560-76
Benzofuran, 2-(4-methoxyphenyl)-	C_6H_{12}	247(4.67),270(4.34),	18-2560-76

Compound	Solvent	$\lambda_{max}(\log \epsilon)$	Ref.
Benzofuran, 2-(4-methoxyphenyl)- (cont.)		280(4.20),303(4.49), 310(4.53),325(4.31)	18-2560-76
3(2H)-Benzofuranone, 5-methyl-2-phenyl-	EtOH	220(3.94),275(3.18), 280(3.16)	111-0533-76
1H-2-Benzopyran-4(3H)-one, 3-phenyl-	MeOH	247(4.02),290(3.42)	35-5581-76
4H-1-Benzopyran-4-one, 2,3-dihydro-	EtOH	214(4.49),252(3.97), 320(3.56)	114-0309-76A
	2% H_2SO_4	255(4.04),325(3.58)	114-0309-76A
	97% H_2SO_4	295(4.26),410(3.56)	114-0309-76A
9H-Fluorene-9-carboxylic acid, methyl ester	$(MeOCH_2)_2$	375s(--),389(3.75), 409(3.68)	104-2411-76
anion	DMSO	410(3.63)	70-0762-76
1H-Inden-1-one, 2,3-dihydro-2-hydroxy-2-phenyl-	MeOH	249(3.99),285(3.23)	35-5581-76
Indeno[2,1-b]pyran-9-carboxaldehyde, 2,4-dimethyl-	dioxan	252(4.46),317(4.37), 346(4.05),470(3.23)	78-2225-76
Methanone, (4-ethenylphenyl)(4-hydroxy-phenyl)-, homopolymer	THF	285(4.23)	126-3089-76
$C_{15}H_{12}O_2S$ 2-Propenoic acid, 3-phenyl-2-(phenyl-thio)-	75% dioxan	260(4.2),278(4.2)	104-0149-76
$C_{15}H_{12}O_3$ Benzo[d][1,3]dioxolo[4,5-h][2]benzoxe-pin, 5,7-dihydro-	EtOH	268(4.31),298(4.21)	94-0716-76
Benzoic acid, 4-benzoyl-, methyl ester	HOAc	254(4.38),341(2.28)	35-2928-76
	CCl_4	256(4.40),336s(2.26), 346(2.30),355s(2.26)	35-2928-76
	Freon 113	252(4.49),283s(4.01), 345(2.21)	35-2928-76
2,8-Nonadiene-4,6-diyn-1-ol, 9-(2-furan-yl)-, acetate, (E,E)-	EtOH	260(4.13),273(4.16), 339(4.64),358(4.53)	95-1322-76
9H-Xanthen-9-one, 1-hydroxy-6,8-dimeth-yl-	EtOH	234(4.48),253(4.40), 260s(4.28),288(3.98), 302(3.95),362(3.70)	23-1703-76
	EtOH-NaOH	235(4.53),257(4.29), 263(4.29),301s(3.90), 310(3.94),389(3.75)	23-1703-76
$C_{15}H_{12}O_4$ 7,10-Epoxy-10aH-benzo[b]cyclohepta[d]-furan-10a-carboxylic acid, 7,10-di-hydro-, methyl ester	MeOH	220(4.32),260(3.93), 276(3.86),383(3.77)	35-3555-76
2(5H)-Furanone, 3-acetyl-4-hydroxy-5-(3-phenyl-2-propenylidene)-	EtOH	246(4.18),340(4.64)	4-0521-76
9H-Xanthen-9-one, 1-hydroxy-8-(hydroxy-methyl)-6-methyl-	EtOH	233(4.23),254(4.19), 262s(4.10),292(3.83), 300s(3.81),363(3.48)	23-1703-76
	EtOH-NaOH	235(4.29),255(4.08), 264(4.10),303(3.76), 310(3.76),399(3.43)	23-1703-76
9H-Xanthen-9-one, 1-hydroxy-6-methoxy-8-methyl-	EtOH	234(4.46),248(4.32), 266(4.16),301(4.21), 349(3.78)	142-0167-76S
$C_{15}H_{12}O_4S$ Benzeneacetic acid, 2-carboxy-α-[(5-methyl-2-thienyl)methylene]-, (E)-	EtOH	307(4.15)	4-0083-76
Benzeneacetic acid, 2-carboxy-α-(2-thi-enylmethylene)-, α-methyl ester, (E)-	EtOH	313(4.20)	4-0083-76

Compound	Solvent	$\lambda_{max}(\log \epsilon)$	Ref.
Butanedioic acid, phenyl(2-thienylmethylene)-, (E)-	EtOH	207(4.28),302(4.39)	4-0285-76
$C_{15}H_{12}O_5$			
9(10H)-Anthracenone, 1,3,4,8-tetrahydroxy-6-methyl-	MeOH-HCOOH	272(4.13),311(3.87), 363(4.16)	12-1535-76
9(10H)-Anthracenone, 1,4,6,8-tetrahydroxy-3-methyl-	MeOH-HCOOH	273s(3.97),306s(3.90), 355(4.05)	12-1535-76
9(10H)-Anthracenone, 1,3,8-trihydroxy-6-(hydroxymethyl)-	MeOH-HCOOH	270(3.81),315s(3.72), 358(4.05)	12-1523-76
Benzeneacetic acid, 2-carboxy-α-(2-furanylmethylene)-, α-methyl ester, (E)-	EtOH	310(4.28)	4-0083-76
1,3-Propanedione, 1-(2,4-dihydroxyphenyl)-3-(4-hydroxyphenyl)- (licodione)	MeOH	285(4.3),376(4.6)	88-2539-76
	MeOH-NaOMe	242(4.1),342(4.7), 415(4.2)	88-2539-76
9H-Xanthen-9-one, 1,4-dihydroxy-6-methoxy-8-methyl-	EtOH	236(4.39),256(4.13), 279(4.33),300(4.13), 379(3.61)	142-0167-76S
9H-Xanthen-9-one, 3,8-dihydroxy-6-methoxy-1-methyl- (griseoxanthone C)	EtOH	240(4.58),264s(4.03), 310(4.31),340s(4.06)	35-5380-76
$C_{15}H_{12}O_6$			
4H-1-Benzopyran-4-one, 2,3-dihydro-5,6,7-trihydroxy-2-(4-hydroxyphenyl)-	EtOH	294(4.18),358(3.64)	95-0381-76
	EtOH-NaOAc	299(4.28),380(3.83)	95-0381-76
	EtOH-AlCl₃	314(4.14),420(3.53)	95-0381-76
4H-1-Benzopyran-4-one, 2,3-dihydro-5,7,8-trihydroxy-2-(4-hydroxyphenyl)-(carthamidin)	EtOH	297(4.13),330(3.74), 372(3.52)	95-0381-76
	EtOH-NaOAc	300(4.23),375(3.79)	95-0381-76
	EtOH-AlCl₃	314(4.22),394(3.28)	95-0381-76
9H-Xanthen-9-one, 1,7-dihydroxy-3,6-dimethoxy-	EtOH	238(4.36),256(4.24), 315(3.77),372(4.18)	102-2016-76
	EtOH-AlCl₃	231(4.32),263(4.29), 333(4.13),396(3.86)	102-2016-76
$C_{15}H_{12}O_6S$			
2-Propenoic acid, 3-[3,4-dihydroxy-5-(phenylsulfonyl)phenyl]-	EtOH	240(4.32),295(4.14), 332(4.08),372(3.71)	88-0313-76
$C_{15}H_{12}O_7$			
Taxifolin, (+)-	MeOH	290(4.30),335s(--)	64-0876-76B
$C_{15}H_{12}O_8$			
1,3-Benzodioxole-5-carboxylic acid, 6-(4,5-dihydroxy-3,6-dioxo-1-cyclohexen-1-yl)-, methyl ester, cis-(±)-	EtOH	225(4.40),267s(--), 304(3.68),338(3.52)	24-0855-76
$C_{15}H_{12}O_9$			
1,3-Benzodioxole-5-carboxylic acid, 6-(4,5-dihydroxy-3,6-dioxo-1-cyclohexen-1-yl)-4-hydroxy-, methyl ester, cis-(±)-	EtOH	250(4.09),275s(--), 370(3.80)	24-0855-76
$C_{15}H_{12}S$			
Benzo[b]thiophene, 2-methyl-3-phenyl-	EtOH	232(4.5),264(3.8), 292(3.5),302(3.5)	12-0883-76
$C_{15}H_{12}S_2$			
Benzo[b]thiophene, 3-[2-(methylthio)phenyl]-	C_6H_{12}	255(4.14),290(3.72), 300(3.72)	138-0389-76

Compound	Solvent	$\lambda_{max}(\log \epsilon)$	Ref.
$C_{15}H_{13}BrN_2$			
1H-Pyrazole, 1-(4-bromophenyl)-4,5-dihydro-3-phenyl-	benzene	365(4.35)	64-1248-76B
1H-Pyrazole, 3-(4-bromophenyl)-4,5-dihydro-1-phenyl-	benzene	372(4.34)	64-1248-76B
$C_{15}H_{13}BrN_2O$			
Benzeneethanehydrazonoyl bromide, 4-methyl-α-oxo-N-phenyl-	EtOH	229(4.268),257(4.214), 368(4.195)	18-0321-76
Benzeneethanehydrazonoyl bromide, N-(4-methylphenyl)-α-oxo-	EtOH	253(4.258),378(4.198)	18-0321-76
$C_{15}H_{13}BrN_2O_2$			
Benzeneethanehydrazonoyl bromide, N-(4-methoxyphenyl)-α-oxo-	EtOH	253(4.329),310(3.727), 386(4.223)	18-0321-76
$C_{15}H_{13}Cl$			
1-Propene, 2-chloro-1,1-diphenyl-	EtOH	209(4.3),241(4.2)	12-0883-76
$C_{15}H_{13}ClN_2$			
1H-Indazole, 5-chloro-1-ethyl-3-phenyl-	isoPrOH	204(4.48),224(4.52), 254(4.07),281s(3.70), 320(3.99)	4-0033-76
2H-Indazole, 5-chloro-2-ethyl-3-phenyl-	isoPrOH	220(4.54),263(3.81), 292s(3.83),316(4.01)	4-0033-76
1H-Pyrazole, 1-(4-chlorophenyl)-4,5-dihydro-3-phenyl-	benzene	365(4.32)	64-1248-76B
1H-Pyrazole, 3-(4-chlorophenyl)-4,5-dihydro-1-phenyl-	benzene	371(4.32)	64-1248-76B
$C_{15}H_{13}ClN_2O_3$			
Benzoic acid, 2-[(6-chloro-3-pyridazinyl)oxy]-3-(2-propenyl)-, methyl ester	EtOH	318(3.79)	94-1588-76
$C_{15}H_{13}ClO_2S$			
Benzenepropanoic acid, β-chloro-α-(phenylthio)-	75% dioxan	260(3.5)	104-0149-76
$C_{15}H_{13}Cl_3N_2O_2$			
2(1H)-Quinolinone, 4-morpholino-3-(trichloroethenyl)-	MeOH	229(4.33),263(4.51), 306(4.53)	24-2159-76
$C_{15}H_{13}Cl_4NO$			
Benzenemethanamine, N-(4-chlorophenyl)-4-methoxy-α-(trichloromethyl)-	n.s.g.	248(4.33),280(3.57), 297(3.37)	98-0724-76
$C_{15}H_{13}FN_2$			
1H-Pyrazole, 1-(4-fluorophenyl)-4,5-dihydro-3-phenyl-	benzene	361(4.23)	64-1248-76B
1H-Pyrazole, 3-(4-fluorophenyl)-4,5-dihydro-1-phenyl-	benzene	359(4.25)	64-1248-76B
$C_{15}H_{13}FN_2OS$			
5H-Thieno[2,3-c]pyrrole-6-carboxamide, N,5-dimethyl-4-(2-fluorophenyl)-	EtOH	224(4.17),251(4.20), 324(4.28)	18-1395-76
$C_{15}H_{13}F_3O$			
Ethanone, 2,2,2-trifluoro-1-(4,6,8-trimethyl-2-azulenyl)-	benzene	496(3.11),590s(2.11), 620(3.17),670s(2.96)	70-1558-76

Compound	Solvent	$\lambda_{max}(\log \epsilon)$	Ref.
$C_{15}H_{13}F_3O_4$ Benzeneacetic acid, α-methoxy-α-(tri-fluoromethyl)-4-oxo-2-cyclopenten-1-yl ester	MeOH	210(4.15)	78-1713-76
$C_{15}H_{13}FeNO$ Ferrocene, [(2-furanylmethylene)amino]-	n.s.g.	391(3.34),496(3.37)	65-2512-76
$C_{15}H_{13}IN_2$ 1H-Pyrazole, 4,5-dihydro-3-(4-iodophen-yl)-1-phenyl-	benzene	376(4.38)	64-1248-76B
$C_{15}H_{13}N$ Dibenzo[b,f]cycloprop[d]azepine, 1,1a,6,10b-tetrahydro-	EtOH	280(3.85)	94-2751-76
1H-Indole, 1-methyl-2-phenyl-	MeOH	294(4.25)	12-2747-76
$C_{15}H_{13}NO$ Acetamide, N-9H-fluoren-2-yl-	C_6H_{12}	288(4.60),317(4.52)	40-0133-76
	MeOH	288(4.57),314(4.48)	40-0133-76
9(10H)-Acridinone, 10-ethyl-	EtOH	255(4.98),390(3.99),405(4.10)	48-0515-76
2H-Azirine-2-methanol, α,3-diphenyl-, erythro-	EtOH	236(4.13)	35-2605-76
9H-Carbazole-3-carboxaldehyde, N-ethyl-	1% EtOH	233(4.32),242(4.26),276(4.38),293(4.26)	44-1952-76
geminal diol anion	1% EtOH-NaOH	248(3.6),275(3.6),300(4.08),350(4.00)	44-1952-76
Cyclohept[b]indol-6(5H)-one, 2,4-dimeth-yl-	MeOH	282(4.35),316(4.21),326(4.22),338(4.14),396s(3.81),409(3.88)	18-1101-76
Cyclohept[b]indol-6(5H)-one, 3,4-dimeth-yl-	MeOH	286(4.36),310(4.30),326s(4.21),337s(4.11),390s(3.80),408(3.90)	18-1101-76
Phenol, 2-[3-(phenylimino)-1-propenyl]-	MeOH	287(4.31),345(4.27)	104-0840-76
	MeOH-NaOMe	296(4.29),402(4.25)	104-0840-76
	dioxan	294(4.22),345(4.25)	104-0840-76
	DMSO	297(4.25),352(4.37)	104-0840-76
	DMSO-NaOMe	345(4.23),485(4.42)	104-0840-76
2-Propenamide, N,3-diphenyl-	EtOH	243(3.67),267(4.04)	110-0703-76
4(1H)-Quinolinone, 2,3-dihydro-2-phenyl-	EtOH	235(4.42),260s(3.90),373(3.54)	114-0309-76A
	2% H_2SO_4	235(4.40),260s(3.90),375(3.60)	114-0309-76A
	97% H_2SO_4	250(4.15),285(3.30)	114-0309-76A
$C_{15}H_{13}NOS$ Ethanone, 1-(6-methyl-5-phenyl-6H-thi-eno[2,3-b]pyrrol-2-yl)-	EtOH	305(4.19),320s(4.15)	104-1550-76
$C_{15}H_{13}NO_2$ 1,4-Butanedione, 1-phenyl-4-(3-pyridin-yl)-	EtOH	243(4.28),270s(3.70),278s(3.59)	44-3438-76
4a,9a-Methano-9H-carbazole-9-carboxylic acid, methyl ester	n.s.g.	244(3.93),252(3.94),282s(3.51),325(3.03),340(3.03)	107-0399-76
Pyridinium 1-benzoyl-2-oxopropylide	benzene	425(3.31)	78-2647-76
	H_2O	285(4.24)	78-2647-76
	aq HCl	255(4.20)	78-2647-76
	EtOH	283(4.20),364(3.10)	78-2647-76

Compound	Solvent	$\lambda_{max}(\log \epsilon)$	Ref.
Pyridinium 1-benzoyl-2-oxopropylide (cont.)	dioxan	238(4.05),246(4.04), 270(3.84),423(2.89)	78-2647-76
	DMF	283(4.05),411(3.18)	78-2647-76
	MeCN	226(4.05),281(4.12), 404(3.18)	78-2647-76
$C_{15}H_{13}NO_3$			
Acetamide, N-[2-(2-hydroxybenzoyl)phenyl]-	MeOH	221(4.28),235(4.28), 264(4.04),326(3.70)	39-2089-76C
Benzeneacetic acid, α-(benzoylamino)-	n.s.g.	244(4.26)	33-2149-76
1H-Benz[de]isoquinoline-1,3(2H)-dione, 2-ethyl-5-methoxy-	EtOH	225(4.42),242(4.62), 337(4.04),380(3.88)	56-1257-76
1H-Benz[de]isoquinoline-1,3(2H)-dione, 2-ethyl-6-methoxy-	EtOH	228(4.26),242(4.53), 252(4.49),368(4.15)	56-1257-76
$C_{15}H_{13}NO_4$			
Benzeneacetic acid, 2-carboxy-α-[(1-methyl-1H-pyrrol-2-yl)methylene]-, (Z)-	EtOH	330(4.21)	4-0083-76
$C_{15}H_{13}NO_5$			
Benzeneacetic acid, α-(3-nitrobenzoyloxy)-, methyl ester	EtOH	220(4.14),227(4.14), 265(3.82)	111-0533-76
Benzeneacetic acid, α-(4-nitrobenzoyloxy)-, methyl ester	EtOH	220(4.01),300(4.08)	111-0533-76
1H-Carbazole-1,4(9H)-dione, 6-hydroxy-2,7-dimethoxy-9-methyl-	EtOH	250(4.26),280(4.03), 325(3.85),500(3.59)	115-0185-76
$C_{15}H_{13}NO_6$			
Methanediol, (5-nitro-1-naphthalenyl)-, diacetate	MeOH	219(4.73),240s(3.92), 326(3.65)	12-0327-76
Methanone, (2-hydroxy-4,6-dimethoxyphenyl)(2-nitrophenyl)-	MeOH	216(4.14),295(4.06)	39-2089-76C
$C_{15}H_{13}NO_6S$			
2-Propenoic acid, 3-[[[2,5-dihydro-5-oxo-4-[(phenoxycarbonyl)amino]-3-furanyl]methyl]thio]-, cis	EtOH	259(4.18)	33-1988-76
trans	EtOH	255(4.23)	33-1988-76
$C_{15}H_{13}NO_8$			
1,3-Benzodioxole-5-carboxylic acid, 6-[4,5-dihydro-3-(hydroxyimino)-6-oxo-1-cyclohexen-1-yl]-, methyl ester, cis-(+)-	EtOH	225(4.39),266(4.17), 310s(--)	24-0855-76
4H-Quinolizine-1,2,3,4-tetracarboxylic acid, 1,2-dimethyl ester, disodium salt	MeOH	208(4.12),245(3.89), 313(3.98),328s(3.85), 440(3.80)	39-1911-76C
	MeOH-acid	208(4.11),270(3.66), 314(3.04)	39-1911-76C
$C_{15}H_{13}NO_8S$			
Glycine, N-[(9-methoxy-7-oxo-7H-furo[3,2-g][1]benzopyran-4-yl)sulfonyl]-, methyl ester	EtOH	227(4.44),255(4.37), 307(4.13)	56-0347-76
$C_{15}H_{13}N_3$			
11H-Indolo[3,2-c][1,8]naphthyridine, 5,6-dihydro-3-methyl-	EtOH	218(4.60),268(4.05), 374(4.30)	4-0097-76

Compound	Solvent	$\lambda_{max}(\log \epsilon)$	Ref.
$C_{15}H_{13}N_3O$			
5H-1,3,4-Benzotriazepin-5-one, 1,4-di-hydro-4-methyl-2-phenyl-	EtOH	229(5.34),250s(4.25), 297s(4.81)	44-2732-76
Isoxazole, 3,5-dimethyl-4-(1-naphtha-lenylazo)-	EtOH	256s(4.10),263(4.12), 272s(4.09),296s(3.89), 365(4.00),384s(3.87), 444s(3.14)	48-0359-76
Isoxazole, 3,5-dimethyl-4-(2-naphtha-lenylazo)-	EtOH	242s(4.20),264(4.29), 275(4.29),286(4.27), 311s(4.27),324(4.35), 336s(4.29),366s(4.01), 428(3.15)	48-0359-76
4(3H)-Quinazolinone, 3-amino-2-(2-meth-ylphenyl)-	EtOH	216(4.35),230s(4.28), 260(4.24),303(3.96)	114-0341-76D
4(3H)-Quinazolinone, 3-amino-2-(4-meth-ylphenyl)-	EtOH	214(4.25),237(4.21), 277(4.12),308(3.92)	114-0341-76D
1,2,4-Triazol-3-one, 5-methyl-1,4-di-phenyl- (mesoionic)	EtOH	216(4.11),272(3.79)	39-0863-76C
$C_{15}H_{13}N_3O_2$			
9-Acridinamine, N,N-dimethyl-1-nitro-	pH 1	273(4.45),440(3.78)	56-1603-76
	pH 6.8	271(4.42),439(3.73)	56-1603-76
1H-Pyrazole, 4,5-dihydro-1-(4-nitro-phenyl)-3-phenyl-	benzene	414(4.55)	64-1248-76B
1H-Pyrazole, 4,5-dihydro-3-(4-nitro-phenyl)-1-phenyl-	benzene	449(4.31)	64-1248-76B
4(3H)-Quinazolinone, 3-amino-2-(4-meth-oxyphenyl)-	EtOH	218(4.40),235s(4.32), 277(4.31),306(4.17)	114-0341-76D
$C_{15}H_{13}N_3O_2S$			
2-Naphthalenol, 1-[(4,5-dimethyl-2-thi-azolyl)azo]-, N-oxide	n.s.g.	545(4.20)	64-0981-76B
1H-1,2,4-Triazole, 3-[(4-methylphenyl)-sulfonyl]-1-phenyl-	EtOH	249(4.32)	44-0711-76
1H-1,2,4-Triazole, 5-[(4-methylphenyl)-sulfonyl]-1-phenyl-	EtOH	241(4.24)	44-0711-76
$C_{15}H_{13}N_3O_3$			
7-Azatricyclo[4.3.1.1^{2,5}]undeca-3,8-di-en-10-one, 7-(5-nitro-2-pyridinyl)-	CHCl$_3$	252(4.04),385(4.38)	39-2307-76C
11-Azatricyclo[5.3.1.0^{2,6}]undeca-3,9-di-en-8-one, 11-(5-nitro-2-pyridinyl)-, (1RS,2RS,6SR,7RS)-	EtOH	207(4.00),228(4.19), 355(4.25)	39-2307-76C
1H-Pyrazolo[4,3-b]pyridine-6-carboxylic acid, 4,7-dihydro-7-oxo-1-phenyl-, ethyl ester	EtOH	227(4.44),301(4.19)	39-0507-76C
$C_{15}H_{13}N_3O_3S$			
2H-1,2,4-Benzothiadiazine-3-carboxamide, N-(4-methylphenyl)-, 1,1-dioxide	dioxan	302(3.91)	103-0402-76
$C_{15}H_{13}N_3O_4S$			
2H-1,2,4-Benzothiadiazine-3-carboxamide, N-(3-methoxyphenyl)-, 1,1-dioxide	dioxan	297(4.50)	103-0402-76
$C_{15}H_{13}N_3O_5$			
Acetamide, N-(2-nitrophenyl)-N-[(4-ni-trophenyl)methyl]-	MeOH	208(3.91),271(3.79)	39-0394-76C

Compound	Solvent	λ_{max}(log ϵ)	Ref.
$C_{15}H_{13}N_3O_6$ Ethanol, 2-[(2,4-dinitrophenyl)amino]-, benzoate	EtOH	250(3.84),286(4.24), 384(4.03),433(4.35), 492(4.21)	78-1221-76
$C_{15}H_{13}N_3Se$ 1H-1,2,4-Triazolium, 2,3-dihydro-4-phenyl-1-(phenylmethyl)-3-selenoxo-, hydroxide, inner salt	MeCN	204s(4.56),247s(4.1), 269s(4.02),328s(3.23)	70-1162-76
$C_{15}H_{13}N_6O_3$ Diazenium, 1-(2-carboxyethenyl)-2-(4-hydroxy-2-methylphenyl)-1-(1H-purin-6-yl)-, (Z)-, sulfate(2:1)	EtOH-pH 1.5 EtOH-pH 13	272(4.18),327(4.14), 443(4.22) 280(4.11),347(4.14), 490(4.37)	103-0817-76 103-0817-76
$C_{15}H_{13}N_6O_4$ Diazenium, 1-(2-carboxyethenyl)-2-(2,4-dihydroxy-6-methylphenyl)-1-(1H-purin-6-yl)-, (Z)-, sulfate(2:1)	EtOH-pH 1 EtOH-pH 13	269(4.12),328(4.04), 471(4.31) 282(4.04),354(4.07), 520(4.42)	103-0817-76 103-0817-76
$C_{15}H_{13}OS$ [1]Benzothieno[3,2-b][1]benzopyrylium, 1,2,3,4-tetrahydro-, perchlorate	n.s.g.	256(4.44),291(4.07), 390(4.41)	103-0977-76
$C_{15}H_{14}$ 9H-Fluorene, 1,8-dimethyl- 1-Propene, 1,1-diphenyl-	EtOH n.s.g.	266(4.26) 245(4.15)	78-0065-76 35-7818-76
$C_{15}H_{14}ClNO_2$ Benzenecarboximidoyl chloride, 4-methoxy-N-(4-methoxyphenyl)- 2,4-Pentadienoic acid, 5-chloro-2-cyano-5-(4-methylphenyl)-, ethyl ester	hexane CH_2Cl_2	233s(3.96),274(4.22), 324(3.80) 346(4.42)	104-0161-76 48-0705-76
$C_{15}H_{14}Cl_2N_2O_2$ Acetic acid, chloro-, 2-[4-chloro-2-(hydroxyphenylmethyl)phenyl]hydrazide	isoPrOH	245(3.65),295(3.00)	4-0033-76
$C_{15}H_{14}FNO_2$ Benzenecarboximidoyl fluoride, 4-methoxy-N-(4-methoxyphenyl)-	hexane	233(3.88),266(4.17), 298s(4.29),308(4.32), 318s(4.29)	104-0161-76
$C_{15}H_{14}F_3N_5O_4$ Benzo[g]pteridine-2,4(3H,10H)-dione, 10-(2-hydroxyethyl)-8-[(2-hydroxyethyl)amino]-7-(trifluoromethyl)-	EtOH	257(4.84),318(4.18), 462(4.75)	103-0938-76
$C_{15}H_{14}NO_2$ Pyridinium, 3-(3-hydroxy-1-oxo-3-phenyl-2-propenyl)-1-methyl-, iodide	EtOH	275(3.93),350(4.27)	39-0315-76C
$C_{15}H_{14}NO_2S_2$ Thiazolo[2,3-a]isoquinolinium, 3-(ethoxycarbonyl)-2-(methylthio)-, sulfate(1:1)	EtOH	220(4.33),224(4.35), 265(4.53),354(4.18)	94-1299-76

Compound	Solvent	$\lambda_{max}(\log \epsilon)$	Ref.
$C_{15}H_{14}N_2$			
Cyclohept[b]indol-6-amine, N,N-dimethyl-	C_6H_{12}	236(4.20),268(4.16), 301(4.45),311(4.49), 363(4.13),382(4.22), 400s(3.95),507(3.68)	18-1101-76
[1,4]Diazepino[6,7,1-jk]carbazole, 1,2,3,4-tetrahydro-, hydrochloride	EtOH	237(4.59),262(4.22), 291(4.17)	4-1187-76
1H-Pyrazole, 4,5-dihydro-1,3-diphenyl-	benzene	363(4.31)	61-0288-76 +64-1248-76B
	isooctane	355(4.31)	61-0288-76
	MeOH	353(4.28)	61-0288-76
	dioxan	358(4.30)	61-0288-76
	acetone	356(4.30)	61-0288-76
3H-Pyrazole, 4,5-dihydro-3,5-dimethyl-, cis	n.s.g.	323(2.54)	78-0619-76
trans	n.s.g.	326(2.62)	78-0619-76
Pyrrolo[1,2-a]pyrimidine, 2,4-dimethyl-7-phenyl-	EtOH	259(4.63),315(3.88), 360(3.49)	103-1379-76
Pyrrolo[1,2-c]pyrimidine, 1,3-dimethyl-6-phenyl-	EtOH	213(4.19),257(4.61), 290s(3.81)	44-0351-76
$C_{15}H_{14}N_2O$			
8-Azabicyclo[3.2.1]oct-3-ene-6-carbonitrile, 8-methyl-2-oxo-4-phenyl-, exo	EtOH	225(3.9),289(4.2)	39-2329-76C
8-Azabicyclo[3.2.1]oct-3-ene-6-carbonitrile, 2-oxo-8-(phenylmethyl)-, isomeric mixture	EtOH	218(4.049)	39-2334-76C
Benzaldehyde, 2-hydroxy-, [(4-methylphenyl)methylene]hydrazone	EtOH	224(4.27),234(4.13), 301(4.48),310(4.44), 342(4.32)	95-0044-76
Indolo[2,3-a]quinolizin-4(1H)-one, 6,7,12,12b-tetrahydro-	MeOH	224(4.55),272(3.91), 290(3.80),280s(--)	24-0705-76
Phenol, 4-(4,5-dihydro-1-phenyl-1H-pyrazol-3-yl)-	benzene	357(4.28)	64-1248-76B
2H-Pyrido[1,2-a]pyrimidin-2-one, 3,4-dihydro-3-methyl-4-phenyl-, cis	EtOH	265(4.07),346(3.89)	44-3546-76
$C_{15}H_{14}N_2OS$			
2-Benzothiazolamine, N-[(4-methoxyphenyl)methyl]-	EtOH	270(4.25),298s(3.58)	94-2532-76
$C_{15}H_{14}N_2OS_2$			
Ethanone, 1-[4-(7,8-dihydro[1,2]dithiolo[4,5,1-hi][1,2,3]benzothiadiazol-3-S^{IV}-2(6H)-yl)phenyl]-	C_6H_{12}	205(4.41),225(4.40), 257(4.10),315(4.39), 519(4.29)	39-0228-76C
$C_{15}H_{14}N_2O_2$			
2-Benzoxazolamine, N-[(4-methoxyphenyl)-methyl]-	EtOH	245(4.27),282(4.02)	94-2532-76
2-Pyridineacetic acid, α-(2-pyridinylmethylene)-, ethyl ester, (E)-	MeOH	257(4.04),291(4.05)	44-2536-76
(Z)-	MeOH	263(4.09),311(4.37)	44-2536-76
$C_{15}H_{14}N_2O_2S$			
Benzoic acid, 3-[(2-pyridinylthioxomethyl)amino]-, ethyl ester	20% EtOH	282(4.12)	94-1451-76
Benzoic acid, 4-[(2-pyridinylthioxomethyl)amino]-, ethyl ester	20% EtOH	283(4.18)	94-1451-76
Imidazo[2,1-a]isoquinoline-3-carboxylic acid, 2-(methylthio)-, ethyl ester	EtOH	271(4.78),316(3.91)	142-0939-76

Compound	Solvent	$\lambda_{max}(\log \epsilon)$	Ref.
$C_{15}H_{14}N_2O_3$			
Benzenamine, 2-methoxy-5-[2-(4-nitro-phenyl)ethenyl]-	MeOH	234(4.28),384(4.30)	56-0967-76
Methanone, [4-(2-furanyl)-4,5-dihydro-1H-pyrazol-3-yl](4-methoxyphenyl)-	EtOH	223(4.17),294(4.04), 336(4.17)	104-2358-76
$C_{15}H_{14}N_2O_3S$			
Phenol, 2-[(2,3-dihydrobenzo[b]thien-5-yl)azo]-4-methyl-, S,S-dioxide	C_6H_{12}	251(3.72),332(3.88), 415(3.44)	62-0249-76A
	EtOH	246s(4.06),329(4.42), 406(4.03)	62-0249-76A
	acetone	326(4.38),405(4.00)	62-0249-76A
	CHCl$_3$	252(4.15),333(4.53), 418(4.10)	62-0249-76A
	CCl$_4$	326(4.20),420(3.78)	62-0249-76A
Phenol, 4-[(2,3-dihydrobenzo[b]thien-5-yl)azo]-2-methyl-, S,S-dioxide	EtOH	256(4.06),374(4.34)	62-0249-76A
	acetone	368(4.55)	62-0249-76A
	CHCl$_3$	254(3.86),366(4.15)	62-0249-76A
	CCl$_4$	362(4.15)	62-0249-76A
$C_{15}H_{14}N_2O_4$			
4,5-Pyrimidinedicarboxylic acid, 6-meth-yl-2-phenyl-, dimethyl ester	EtOH	270(4.32)	5-1809-76
Pyrimido[2,1,6-cd]pyrrolizine-1,2-dicar-boxylic acid, 4,7-dimethyl-, dimethyl ester	EtOH	239(4.51),263(4.36), 315(4.07),438(3.93)	44-0351-76
$C_{15}H_{14}N_2O_4S$			
Sulfonium, bis(phenylmethyl)-, dinitro-methylide	H$_2$O	333(4.07)	70-1906-76
	MeCN	325(4.03)	70-1906-76
	DMSO	330(4.04)	70-1906-76
Thiazolo[5,4-f]quinoline-8-carboxylic acid, 3,6-diethyl-2,3,6,9-tetrahydro-2,9-dioxo-	MeOH	263.0(4.69)	94-0147-76
Thiazolo[5,4-f]quinoline-8-carboxylic acid, 2-ethoxy-6-ethyl-6,9-dihydro-9-oxo-	MeOH	259(4.34),268(4.34)	94-0147-76
$C_{15}H_{14}N_2O_5$			
Acetic acid, [(6,7-dihydro-4-oxo-4H-pyr-azolo[5,1-c][1,4]oxazin-2-yl)oxy]-, phenylmethyl ester	MeOH	229(4.00),272(3.38)	24-0268-76
$C_{15}H_{14}N_3S_3$			
Benzothiazolium, 3-ethyl-2-[(2-methyl-5H-thiazolo[4,3-b]-1,3,4-thiadiazol-5-ylidene)methyl]-, iodide	EtOH	452(2.6)	103-0650-76
$C_{15}H_{14}N_4$			
2H-Tetrazole, 2,5-bis(4-methylphenyl)-	EtOH	246(4.23),275(4.34)	18-0762-76
$C_{15}H_{14}N_4O$			
2-Butanone, 3-(1-phenyl-1H-pyrazolo-[3,4-d]pyrimidin-4-yl)-	EtOH	262(4.08),344(4.61), 359(4.64)	95-1352-76
2H-Tetrazole, 5-(4-methoxyphenyl)-2-(4-methylphenyl)-	EtOH	257(4.28),279(4.33)	18-0762-76
$C_{15}H_{14}N_4O_2$			
1H-Pyrazolo[3,4-d]pyrimidine-4-acetic acid, 1-phenyl-, ethyl ester	EtOH	247(4.30),260(4.15), 320(4.35)	95-1352-76

Compound	Solvent	$\lambda_{max}(\log \epsilon)$	Ref.
$C_{15}H_{14}N_4O_3$			
Pyrido[3,2-e]-1,2,4-triazine, 3-(3,4,5-trimethoxyphenyl)-	EtOH	222(4.46),277(4.44), 370(3.82)	114-0301-76C
Pyrido[3,4-e]-1,2,4-triazine, 3-(3,4,5-trimethoxyphenyl)-	EtOH	282(4.33),384(3.75)	114-0285-76C
1H-Pyrimido[4,5-b][1,4]diazepine-2,7,9-(3H,6H,8H)-trione, 6,8-dimethyl-4-phenyl-	EtOH	226(4.26),261(4.26), 374(3.90)	95-1453-76
3H-Pyrimido[4,5-b][1,4]diazepine-4,7,9-(5H,6H,8H)-trione, 6,8-dimethyl-2-phenyl-	EtOH	240s(4.29),257s(4.16), 309(4.15)	95-1453-76
$C_{15}H_{14}N_6O_2$			
Imidazo[1,2-c]pyrimidin-5(6H)-one, 6,6'-(1,3-propanediyl)bis-	pH 1	290(4.38)	35-7408-76
	pH 7	272(4.33)	35-7408-76
	pH 13	272(4.33)	35-7408-76
$C_{15}H_{14}N_6O_4$			
Propanal, 2-[(4-nitrophenyl)hydrazono]-, (4-nitrophenyl)hydrazone	n.s.g.	438(4.77)	59-1593-76
$C_{15}H_{14}O$			
Benzene, 1-methoxy-4-(2-phenylethenyl)-	hexane	307(4.45),318(4.45), 335(4.43)	104-0161-76
2,4,6-Cycloheptatrien-1-one, 8-(phenylmethyl)-	C_6H_{12}	220(4.06),248(3.76), 284(3.54),361(2.60)	39-1838-76C
Fluorene, 7-methoxy-1-methyl-	EtOH	272(4.29),301(3.76), 310(3.71)	78-0065-76
Indeno[1,2-b]pyran, 2,3,4-trimethyl-	dioxan	228(4.34),300(4.31), 355(3.98),480(3.43)	78-2219-76
Indeno[2,1-b]pyran, 2-(1-methylethyl)-	dioxan	228(4.10),287(4.17), 356(3.37),492(3.03)	78-2219-76
Indeno[2,1-b]pyran, 2-propyl-	dioxan	230(3.93),254(3.69), 292(4.03),338(3.53), 489(2.92)	78-2219-76
Indeno[2,1-b]pyran, 2,4,9-trimethyl-	dioxan	226(4.25),244(4.06), 250(3.98),295(4.36), 335(3.86),480(3.37)	78-2219-76
6,11-Methanobenzocyclodecen-13-one, 5,6,11,12-tetrahydro-	MeOH	239(3.45),248(3.52), 256(3.58),265(3.53), 292(2.58)	88-1117-76
$C_{15}H_{14}OS_2$			
Methanone, bis[3-(mercaptomethyl)phenyl]-	MeOH	225(4.20)	44-2509-76
$C_{15}H_{14}O_2$			
1H-Benz[e]inden-1-one, 2,3-dihydro-7-methoxy-3-methyl-	EtOH	243(4.6),300(3.9)	39-0722-76C
9H-Fluoren-2-ol, 7-methoxy-1-methyl-	EtOH	277(4.40)	78-0065-76
	EtOH-NaOEt	301(4.45)	78-0065-76
Methanone, [2-(2-hydroxyethyl)phenyl]phenyl-	MeCN	249(4.06),280(3.38), 290(3.16),330(1.89)	35-0541-76
1-Propanone, 1-(4-hydroxyphenyl)-3-phenyl-	EtOH	284(3.81)	114-0381-76C
	EtOH-NaOEt	332(4.32)	114-0381-76C
$C_{15}H_{14}O_2S$			
Benzene, [(1-methyl-2-phenylethenyl)sulfonyl]-	EtOH	263(4.20)	22-0519-76
Benzeneacetic acid, α-(phenylthio)-, methyl ester	EtOH	227(4.05),265(3.48)	111-0533-76

Compound	Solvent	$\lambda_{max}(\log \epsilon)$	Ref.
2-Thiophenecarboxylic acid, 5-[2-(4-methylphenyl)ethenyl]-, methyl ester	MeOH	238(4.08),244(4.08), 335s(4.43),350(4.49), 369s(4.37)	48-0731-76
$C_{15}H_{14}O_2S_2$ 3-Butyn-1-ol, 4-(5'-methyl[2,2'-bithiophene]-5-yl)-, acetate	ether	335(4.34)	24-0901-76
$C_{15}H_{14}O_3$ Benzeneacetic acid, α-(3-methylphenoxy)-	EtOH	220(3.86),267(2.85), 277(2.78)	111-0533-76
Benzeneacetic acid, α-phenoxy-, methyl ester	EtOH	222(4.07),265(3.26), 272(3.15)	111-0533-76
1H-Fluorene-1-carboxylic acid, 2,3,4,9-tetrahydro-2-oxo-, methyl ester	EtOH EtOH-NaOEt	240(4.18),272(4.16) 304(4.32)	78-0065-76 78-0065-76
9H-Fluorene-1-methanol, 2-hydroxy-7-methoxy-	EtOH EtOH-NaOEt	276(4.35),315(3.66) 297(4.42)	78-0065-76 78-0065-76
$C_{15}H_{14}O_3S$ 2-Thiophenecarboxylic acid, 5-[2-(4-methoxyphenyl)ethenyl]-, methyl ester	MeOH	257(4.01),262s(4.00), 281s(3.84),351(4.28)	48-0731-76
$C_{15}H_{14}O_4$ 1(2H)-Anthracenone, 3,4-dihydro-3,8,9-trihydroxy-6-methyl- (germichrysone)	dioxan	271(4.77),300(3.64), 311(3.66),402(3.94)	102-1295-76
Ethanone, 1-(5-benzoyl-2-ethoxy-3-furanyl)-	EtOH	261(4.15),321(4.30)	94-2421-76
3-Furancarboxylic acid, 5-benzoyl-2-methyl-, ethyl ester	EtOH	259(4.00),293(4.29)	94-2421-76
3-Furancarboxylic acid, 4,5-dihydro-2-methyl-4-oxo-5-(phenylmethylene)-, ethyl ester	EtOH	206(4.18),232(4.00), 342(4.37)	4-0521-76
1,4-Methanonaphthalene-5,8-diol, 1,4-dihydro-, diacetate	ether	266(2.48)	39-0550-76C
2H-Naphtho[2,3-b]pyran-5,10-dione, 3,4-dihydro-4-hydroxy-2,2-dimethyl-, (S)-	EtOH	245(4.62),250(4.63), 280(4.47),330(3.72)	102-0570-76
	EtOH-NaOH	250(4.37),274(4.60)	102-0570-76
Tetracyclo[5.4.0^{2,4}.0^{3,6}]undeca-1(7),8,10-triene-8,11-diol, diacetate	EtOH	265(2.44)	39-0550-76C
$C_{15}H_{14}O_4S$ 2(3H)-Furanone, 5-(2,4-dihydroxy-6-methylphenyl)dihydro-5-(2-thienyl)-	EtOH-pH 8.8	500(3.56)	18-2027-76
$C_{15}H_{14}O_5$ 1,3-Butanedione, 1-(3-acetyl-3,4-dihydro-4-oxo-2H-1-benzopyran-6-yl)-	EtOH	230s(3.95),253(4.08), 314(4.41)	39-2444-76C
2-Propenoic acid, 3-[5-[3-(3-ethyloxiranyl)-1-oxo-2-propynyl]-2-furanyl]-, methyl ester, [2α(E),3α]-	EtOH	238(4.16),347(4.45)	102-1119-76
$C_{15}H_{14}O_5S$ 2H-Pyran-3-carboxylic acid, 6-(4-methoxyphenyl)-4-(methylthio)-2-oxo-, methyl ester	EtOH	245(3.96),340(4.02), 380(4.22)	142-1493-76
3,4-Thiophenedicarboxylic acid, 2-(4-methoxyphenyl)-, dimethyl ester	MeOH	240(4.41),275(4.83)	44-1724-76
$C_{15}H_{14}O_6$ 2(5H)-Furanone, 3-acetyl-5-[(3,4-dimethoxyphenyl)methylene]-4-hydroxy-	EtOH	245(4.24),340(4.48)	4-0521-76

Compound	Solvent	$\lambda_{max}(\log \epsilon)$	Ref.
Methanone, (2,6-dihydroxy-4-methoxyphenyl)(2,4-dihydroxy-6-methylphenyl)-	EtOH	294(4.22),340(3.78)	35-5380-76
Methanone, (4-hydroxy-2-methoxy-6-methylphenyl)(2,4,6-trihydroxyphenyl)-	EtOH	226s(4.24),294(4.36), 333s(3.86)	35-5380-76
1,3-Naphthalenedicarboxylic acid, 2,4-dihydroxy-6-methyl-, dimethyl ester	EtOH	253(4.60),299(3.94), 311(3.88),369(3.72)	78-0269-76
1,4-Naphthalenedione, 3-acetyl-2,5,7-trimethoxy-	MeOH	216(4.44),262(4.19), 296(3.98),353(3.45), 416(3.38)	12-1087-76
2H-Pyran-3-carboxylic acid, 4-hydroxy-6-methoxy-5-(4-methylphenyl)-2-oxo-, methyl ester	EtOH EtOH-base	242(4.14),315(4.12) 275s(4.00)	78-0269-76 78-0269-76

$C_{15}H_{14}O_7$
| 1,3-Naphthalenedicarboxylic acid, 2,4-dihydroxy-6-methoxy-, dimethyl ester | EtOH | 242(4.66),269s(4.48), 279s(4.40),298(4.05), 380(3.76) | 78-0269-76 |
| 2H-Pyran-3-carboxylic acid, 4-hydroxy-6-methoxy-5-(4-methoxyphenyl)-2-oxo-, methyl ester | EtOH EtOH-base | 254(4.30),315(4.23) 276s(4.27) | 78-0269-76 78-0269-76 |

$C_{15}H_{14}O_8$
| 1,3-Benzodioxole-5-carboxylic acid, 6-(3,4,5-trihydroxy-6-oxo-1-cyclohexen-1-yl)-, methyl ester, $(3\alpha,4\alpha,5\alpha)-(\pm)-$ | EtOH | 225(4.49),265s(--), 300(3.82) | 24-0855-76 |

$C_{15}H_{15}Br_2N_5$
| Pyridinium, 4,4'-(1,3,5-triazine-2,4-diyl)bis[1-methyl-, dibromide | pH 7 | 247(4.64),282s(4.00) | 78-0615-76 |

$C_{15}H_{15}ClN_2O$
| Acetaldehyde, 2-[4-chloro-2-(hydroxyphenylmethyl)phenyl]hydrazone | isoPrOH | 281(4.33) | 4-0033-76 |
| 6H-Perimidin-6-one, 9-chloro-1,2-dihydro-2,2,4,5-tetramethyl- | EtOH | 285(4.32),500(3.34) | 12-2247-76 |

$C_{15}H_{15}ClN_2O_2$
| Acetic acid, 2-[4-chloro-2-(hydroxyphenylmethyl)phenyl]hydrazide | isoPrOH | 248(4.08),297(3.35) | 4-0033-76 |

$C_{15}H_{15}Cl_2NO_4$
| Methanaminium, N-[3-chloro-3-(2-naphthalenyl)-2-propenylidene]-N-methyl-, perchlorate | MeOH | 280(4.10),290(4.17), 351(4.36) | 48-0731-76 |

$C_{15}H_{15}Cl_2N_5$
| Pyridinium, 4,4'-(1,3,5-triazine-2,4-diyl)bis[1-methyl-, dichloride | pH 7 | 247(4.56),283s(3.97) | 78-0615-76 |

$C_{15}H_{15}Cl_3N_2O$
| 2(1H)-Quinolinone, 4-(diethylamino)-3-(trichloroethenyl)- | MeOH | 228(4.30),261(4.40), 307(4.12) | 24-2159-76 |

$C_{15}H_{15}FeN_3O_4$
| Ferrocene, (4-cyano-2,2-dinitrobutyl)- | MeOH | 380(2.41) | 104-2433-76 |

$C_{15}H_{15}IN_2O$
| Diazene, [2-iodo-6-(1-methylethoxy)phenyl]phenyl- | EtOH | 220(4.39),274(3.99), 314(3.95),456(2.73) | 101-0039-76M |

Compound	Solvent	$\lambda_{max}(\log \epsilon)$	Ref.
$C_{15}H_{15}NO$			
Benzenemethanamine, 4-methoxy-N-(phenyl-methylene)-	EtOH	247(4.19),280(3.20), 285(3.15)	35-2605-76
Cyclohexanone, 2-(1-isoquinolinyl)-	EtOH	218(4.57),280(3.94), 440(3.86)	118-0764-76
$C_{15}H_{15}NO_2$			
Benzoic acid, 2-[(2,3-dimethylphenyl)-amino]-, β-cyclodextrin complex	pH 7.0	283(3.85),336(3.66)	133-0343-76
[1,1'-Biphenyl]-4-propanoic acid, α-amino-, hydrochloride, (S)-	pH 13	253.5(4.18)	94-3149-76
9H-Carbazole, 2,3-dimethoxy-6-methyl-	EtOH	236(4.66),265(4.18), 303(4.32)	42-0861-76
9H-Carbazole, 2,4-dimethoxy-6-methyl-	EtOH	233(4.47),237(4.47), 260(4.08),309(4.20)	42-0861-76
1H-Indeno[4,5,6-ij]quinolizine-1,9(5H)-dione, 2,3,6,7,10,11-hexahydro-	MeOH	245(4.18),270(4.13), 320(4.48),380(3.85)	42-0861-76
Propanamide, 3-hydroxy-N,3-diphenyl-	EtOH	210(4.927),244(5.034)	23-1205-76
$C_{15}H_{15}NO_2S$			
2-Cyclohexen-1-one, 2-(2-benzothiazo-1yl)-3-hydroxy-5,5-dimethyl-	MeOH	238(3.95),260(4.14), 271s(4.00),314s(--), 331s(4.42),344(4.52)	44-0125-76
	CH_2Cl_2	239s(4.03),259(4.15), 271s(4.00),334s(4.43), 347(4.55)	44-0125-76
Sulfilimine, N-acetyl-S-(2-methoxyphen-yl)-S-phenyl-, (S)-	MeOH	225s(4.16),282(3.56)	78-3003-76
	dioxan	282(3.63)	78-3003-76
2H-Thiopyran-3-carboxylic acid, 2-imino-6-(4-methylphenyl)-, ethyl ester, hydrobromide	HOAc	419(4.18)	48-0705-76
perchlorate	HOAc	417(4.14)	48-0705-76
$C_{15}H_{15}NO_3$			
2-Butenoic acid, 3-(5-methyl-3-phenyl-4-isoxazolyl)-, methyl ester, (E)-	EtOH	230(4.20)	33-2074-76
(Z)-	EtOH	226(4.13)	33-2074-76
2-Butenoic acid, 3-(5-methyl-2-phenyl-4-oxazolyl)-, methyl ester, (E)-	EtOH	219(4.23),289(4.35)	33-2074-76
(Z)-	EtOH	218(4.18),283(4.19)	33-2074-76
4-Isoxazolepropanoic acid, 5-methyl-β-methylene-3-phenyl-, methyl ester	EtOH	228s(4.10)	33-2074-76
Methanone, (2-aminophenyl)(2,4-dimeth-oxyphenyl)-	EtOH	209(4.45),233(4.52), 261(4.09),377(3.92)	39-2089-76C
2,4-Pentadienoic acid, 2-cyano-5-hydr-oxy-5-(4-methylphenyl)-, ethyl ester	HOAc	260(4.23)	48-0705-76
2H-Pyran-3-carboxylic acid, 2-imino-6-(4-methylphenyl)-, perchlorate	HOAc	395(4.30)	48-0705-76
$C_{15}H_{15}NO_4$			
3,7-Methano-1H-2-benzazonin-1-one, 3-(hydroxymethyl)tetrahydro-9,10-(methylenedioxy)-	n.s.g.	224(4.49),266(3.90), 305(3.88)	95-0631-76
LiAlH₄ reduction product	n.s.g.+HCl	236(2.38),290(2.46)	95-0631-76
Methanone, (2-aminophenyl)(2-hydroxy-4,6-dimethoxyphenyl)-	MeOH	212(4.54),231(4.45), 262(3.96),374(3.88)	39-2089-76C
6(5H)-Phenanthridinone, 3,4,4a,10b-tet-rahydro-4-(hydroxymethyl)-8,9-(meth-ylenedioxy)-	n.s.g.	224(4.53),264(3.61), 305(3.77)	95-0631-76
LiAlH₄ reduction product	n.s.g.+HCl	235(2.37),296(2.47)	95-0631-76

Compound	Solvent	$\lambda_{max}(\log \epsilon)$	Ref.
$C_{15}H_{15}NO_5$ 2-Indolizineacetic acid, 3-acetyl-1-(methoxycarbonyl)-, methyl ester	MeOH	206(3.93),208s(3.89), 227s(4.25),235s(4.42), 239(4.46),247s(4.33), 252s(4.31),256s(4.29), 272s(4.27),278(4.34), 323s(4.15),347(4.20), 352s(4.18)	39-1908-76C
$C_{15}H_{15}NO_6$ 2H-Quinolizine-1,2,3-tricarboxylic acid, trimethyl ester	MeOH	208(4.18),264(4.38), 296(3.78),465(3.77)	39-1911-76C
	MeOH-acid	208(4.15),283(4.11)	39-1911-76C
4H-Quinolizine-1,2,3-tricarboxylic acid, trimethyl ester	MeOH	211(4.15),260(3.90), 312(4.20),346(4.04), 444(4.00)	39-1911-76C
	MeOH-acid	214(4.09),252s(--), 313(4.12),340s(--), 442(3.58)	39-1911-76C
$C_{15}H_{15}NO_6S$ 1H-Pyrrole-3,4-dicarboxylic acid, 1-[(4-methylphenyl)sulfonyl]-, dimethyl ester	MeOH	237(4.35),275s(2.98)	18-3314-76
$C_{15}H_{15}NO_8$ 1H-Indole-1,5,7-tricarboxylic acid, 1-ethyl 5,7-dimethyl ester	EtOH	253(4.58),341(3.51)	78-0269-76
	EtOH-NaOH	257(4.47),279(4.47), 305s(4.37),354(3.76)	78-0269-76
1H-Pyrrole-1-carboxylic acid, 2-[4-hydroxy-6-methoxy-3-(methoxycarbonyl)-2-oxo-2H-pyran-5-yl]-, ethyl ester	EtOH	221(4.43),248s(4.10), 312(4.03)	78-0269-76
	EtOH-NaOH	268(4.15)	78-0269-76
$C_{15}H_{15}NSe$ 2-Propene-1-selone, 3-(dimethylamino)-1-(2-naphthalenyl)-	MeCN	330(4.12),456(3.78), 602s(2.76)	118-0521-76
$C_{15}H_{15}N_3$ Acridine Yellow	pH 7.0	246(4.44),266s(4.60), 436(4.21)	95-1334-76
Propanedinitrile, [3-(dimethylamino)-3-(4-methylphenyl)-2-propenylidene]-	CH_2Cl_2	407(4.21)	48-0705-76
$C_{15}H_{15}N_3O$ Acetamide, N-methyl-N-[4-(phenylazo)phenyl]-	hexane	231(4.13),330(4.28), 443(2.88)	61-0301-76
1H-Indazol-4-amine, 3-(4-methoxyphenyl)-1-methyl-	MeOH	221(4.40),250(4.05), 321(4.00)	24-1898-76
3-Pyridinecarboxamide, 1,2-dihydro-2-imino-1-methyl-6-(2-phenylethenyl)-, perchlorate	HOAc	364(4.05)	48-0705-76
$C_{15}H_{15}N_3O_2$ Benzamide, N-[(nitroso(phenylmethyl)amino]methyl]-	EtOH	231(4.32),360s(1.85), 364(1.85),374s(1.73)	94-0369-76
Benzenamine, N-acetoxy-N-methyl-4-(phenylazo)-	EtOH	353(4.23)	94-1485-76
Benzenamine, N,N-dimethyl-4-[[(4-nitrophenyl)imino]methyl]-	$C_2H_4Cl_2$	225(4.27),312(4.04), 405(4.44)	104-0161-76
1,4-Benzenediamine, N,N-dimethyl-N'-[(4-nitrophenyl)methylene]-	$C_2H_4Cl_2$	282(4.28),450(4.27)	104-0161-76

Compound	Solvent	$\lambda_{max}(\log \epsilon)$	Ref.
Pyridinium, 1-[[[(ethoxycarbonyl)amino]-phenylmethylene]amino]-, hydroxide, inner salt	EtOH	251(4.22),338(3.37)	104-0161-76
Pyrimido[5,4-b]quinoline-2,4(3H,10H)-di-one, 10-butyl-	EtOH	221(4.49),263(4.52), 320(3.92),398(4.05)	39-1805-76C
	50% EtOH-HCl	216(4.39),257(4.50), 334(4.22)	39-1805-76C
Pyrimido[5,4-b]quinoline-2,4(3H,10H)-di-one, 10-ethyl-3,7-dimethyl-	EtOH	224(4.53),264(4.69), 324(4.01),405(4.06)	39-1805-76C
	50% EtOH-HCl	220(4.39),259(4.60), 334(4.23)	39-1805-76C
Pyrimido[5,4-b]quinoline-2,4(3H,10H)-di-one, 3-methyl-10-propyl-	EtOH	221(4.57),263(4.71), 320(4.03),399(4.11)	39-1805-76C
	50% EtOH-HCl	216(4.45),257(4.58), 333(4.26)	39-1805-76C
$C_{15}H_{15}N_3O_3$			
7-Azabicyclo[4.3.1]deca-3,8-dien-10-one, 2-methyl-7-(5-nitro-2-pyridinyl)- (isomeric mixture)	EtOH	209(4.08),250(4.08), 385(4.34)	39-2307-76C
Benzenamine, 4-[(4-acetoxyphenyl)azo]-N-hydroxy-N-methyl-	EtOH	407(4.30)	94-1485-76
Benzenamine, N-[[4-(dimethylamino)phen-yl]methylene]-3-nitro-, N-oxide	benzene	392(4.44)	47-2983-76
	EtOAc	388(4.41)	47-2983-76
	CHCl$_3$	400(4.49)	47-2983-76
	polystyrene	395(4.41)	47-2983-76
in poly(methyl methacrylate)		392(4.43)	47-2983-76
1(4H)-Pyridinecarboxamide, N,N-dimethyl-4-[(4-nitrophenyl)methylene]-	EtOH	236(4.04),309(4.08), 467(4.18)	4-0517-76
$C_{15}H_{15}N_3O_3S_2$			
1H-[1,2]Dithiolo[5,1-e][1,2,3]thiadia-zole-7-S^{IV}-3-carboxaldehyde, 5-(1,1-dimethylethyl)-1-(4-nitrophenyl)-	C$_6$H$_{12}$	219(4.38),246s(4.19), 282(4.04),315(4.01), 361(4.26),487(4.34)	39-0880-76C
$C_{15}H_{15}N_3O_4$			
Piperidine, 1-(4,5-dinitro-1-naphthalen-yl)-	benzene	390(3.88)	44-0044-76
$C_{15}H_{15}N_3O_5$			
Morpholine, 4-[(hydroxyimino)[5-(2-ni-trophenyl)-2-furanyl]methyl]-	EtOH	210(4.24),292(4.26)	73-3085-76
Morpholine, 4-[(hydroxyimino)[5-(3-ni-trophenyl)-2-furanyl]methyl]-	EtOH	213(4.30),307(4.40)	73-3085-76
Morpholine, 4-[(hydroxyimino)[5-(4-ni-trophenyl)-2-furanyl]methyl]-	EtOH	207(4.27),269(4.09), 369(4.44)	73-3085-76
$C_{15}H_{15}N_4OS$			
Methanaminium, N-[[[5-cyano-2-(4-meth-oxyphenyl)-6H-1,3-thiazin-6-ylidene]-amino]methylene]-N-methyl-, perchlor-ate	HOAc	310(4.16),450(3.17)	97-0268-76
$C_{15}H_{15}N_5$			
Pyridinium, 4,4'-(1,3,5-triazine-2,4-diyl)bis[1-methyl-, dichloride	pH 7	247(4.56),283s(3.97)	78-0615-76
$C_{15}H_{15}N_5O_2$			
Carbamic acid, [1-(phenylmethyl)-1H-py-razolo[3,4-d]pyrimidin-4-yl]-, ethyl ester	pH 1	266(4.15)	87-0555-76
	pH 13	298(4.37),306s(--)	87-0555-76

Compound	Solvent	$\lambda_{max}(\log \epsilon)$	Ref.
Carbamic acid, [1-(phenylmethyl)-1H-py-razolo[3,4-d]pyrimidin-4-yl]-, ethyl ester (cont.)	50% EtOH	257(4.01),263(4.04), 283(3.88)	87-0555-76
7(1H)-Pteridinone, 2-(dimethylamino)-4-(phenylmethoxy)-	pH -2.0	222s(4.27),338(4.27)	24-3228-76
	pH 3.0	241(4.22),284(3.72), 362(4.25)	24-3228-76
	pH 11.0	228(4.68),274(3.94), 350(4.22)	24-3228-76
$C_{15}H_{15}N_5O_4S_3$ 3-Thiazolidinebutanoic acid, 4-[(amino-thioxomethyl)hydrazono]-5-[(4-nitro-phenyl)methylene]-2-thioxo-	MeOH	244(4.34),278(4.07), 378(4.47)	103-0749-76
$C_{15}H_{15}O_{10}$ 1,3-Cyclopentadiene-1,2,3,4,5-pentacarb-oxylic acid, pentamethyl ester, ion(1-), sodium	MeOH	212(3.82),262(4.68), 297(4.20)	24-1928-76
$C_{15}H_{16}$ 1H-Fluorene, 4,9-dihydro-9,9-dimethyl-	MeOH	262(4.09),272(3.97), 282(3.53),288(3.35), 299(3.38)	44-2706-76
$C_{15}H_{16}BrNO_2S$ 2H-Thiopyran-3-carboxylic acid, 2-imino-6-(4-methylphenyl)-, ethyl ester, hydrobromide	HOAc	419(4.18)	48-0705-76
$C_{15}H_{16}BrNO_7$ β-D-Ribofuranosyl bromide, 2,3-O-(1-methylethylidene)-, 4-nitrobenzoate	ether	255(4.12)	44-0287-76
$C_{15}H_{16}BrN_3O_7$ 1H-Imidazole-4-carbonitrile, 5-bromo-1-(2,3,5-tri-O-acetyl-β-D-ribo-furanosyl)-	pH 1	228(4.02)	87-1020-76
	pH 11	235(3.99)	87-1020-76
$C_{15}H_{16}Br_2O_2$ 2,4-Methano-1,3-dioxacyclopenta[cd]pen-talene, 6,6a-dibromo-5-(1-buten-3-yn-yl)-6-ethyloctahydro- (isomaneonene A)	EtOH	229(4.16)	88-4227-76
isomer B	EtOH	228(4.13)	88-4227-76
$C_{15}H_{16}ClN_3$ Benzenamine, 4-[(4-chloro-2-methylphen-yl)azo]-N,N-dimethyl-	C_6H_{12}	413(4.51)	18-1381-76
Benzenamine, 4-[(4-chlorophenyl)azo]-N,N,3-trimethyl-	C_6H_{12}	415(4.52)	18-1381-76
$C_{15}H_{16}ClN_3O_5$ 3-Pyridinecarboxamide, 1,2-dihydro-2-im-ino-1-methyl-6-(2-phenylethenyl)-, perchlorate	HOAc	364(4.05)	48-0705-76
$C_{15}H_{16}ClN_3O_7$ 1H-Imidazole-4-carbonitrile, 5-chloro-1-(2,3,5-tri-O-acetyl-β-D-ribofurano-syl)-	pH 1	227(4.06)	87-1020-76
	pH 11	232(4.00)	87-1020-76

Compound	Solvent	$\lambda_{max}(\log \epsilon)$	Ref.
$C_{15}H_{16}IN_3O_7$			
1H-Imidazole-4-carbonitrile, 5-iodo-1-(2,3,5-tri-O-acetyl-β-D-ribofuranosyl)-	pH 1 pH 11	238(4.05) 239(4.04)	87-1020-76 87-1020-76
$C_{15}H_{16}N_2$			
Benzenamine, N,N-dimethyl-4-[(phenylimino)methyl]-	$C_2H_4Cl_2$	240s(4.15),320s(4.26), 355(4.55)	104-0161-76
Benzenediamine, N,N-dimethyl-N'-(phenylmethylene)-	$C_2H_4Cl_2$	255(4.27),375(4.26)	104-0161-76
Propanedinitrile, (1-phenylhexylidene)-	EtOH	232(3.95),292(4.15)	118-0705-76
1-Propanone, 1-phenyl-, phenylhydrazone	EtOH	235(4.12),303(4.13), 334(4.30)	39-0456-76C
$C_{15}H_{16}N_2O$			
Benzaldehyde, 2-methoxy-, methylphenylhydrazone	benzene EtOH	310(3.88),350(4.37) 305(3.99),350(4.41)	104-0625-76 104-0625-76
Benzenecarboximidamide, N'-ethyl-N-hydroxy-N-phenyl-	EtOH	301(3.80)	12-0357-76
Benzenemethanamine, α-methyl-N-nitroso-N-(phenylmethyl)-, (S)-	C_6H_{12}	346s(1.57),358s(1.78), 370(1.90),383(1.80)	78-0847-76
Diazene, [2-(1-methylethoxy)phenyl]phenyl-	EtOH	220s(4.03),319(4.13), 350s(4.02),440(2.94)	101-0039-76M
6H-Perimidin-6-one, 1,2-dihydro-2,2,4,5-tetramethyl-	EtOH 3M HCl	284(4.23),509(3.23) 227(4.40),306(4.02), 583(3.49)	12-2247-76 12-2247-76
$C_{15}H_{16}N_2OS$			
Benzo[b]thiophene-3-carboxamide, 2-amino-4,5,6,7-tetrahydro-N-phenyl-	EtOH	257(4.25),319(3.87)	2-0357-76
$C_{15}H_{16}N_2OS_2$			
1H-[1,2]Dithiolo[5,1-e][1,2,3]thiadiazole-7-S^{IV}-3-carboxaldehyde, 5-(1,1-dimethylethyl)-1-phenyl-	C_6H_{12}	235(4.40),273(4.08), 319(3.95),348(4.04), 474(4.13)	39-0880-76C
$C_{15}H_{16}N_2O_2$			
Benzonitrile, 4-(3a,4,5,8,9,9a-hexahydrocycloocta-1,3,2-dioxazol-2-yl)-, trans	hexane	266(4.12)	89-0372-76
2,4,6-Heptatrienamide, 4-formyl-7-(methylamino)-N-phenyl-	EtOH	285(4.19)	39-0315-76C
Piperidine, 1-(4-nitro-1-naphthalenyl)-	benzene	395(3.95)	44-0044-76
$C_{15}H_{16}N_2O_4$			
2-Furancarboxamide, N,N-diethyl-5-(4-nitrophenyl)-	dioxan	354(4.37)	73-2577-76
3-Pyridinecarboxylic acid, 1,2-dihydro-6-hydroxy-2-oxo-4-[(phenylmethyl)amino]-, ethyl ester	MeCN	273(4.15),305(4.12), 320(4.09)	39-2462-76C
$C_{15}H_{16}N_2O_5S$			
Pyridinium, 4-[(4-nitrophenyl)methyl]-1-(3-sulfopropyl)-, hydroxide, inner salt	H_2O	265(4.28),275s(--)	4-0517-76
$C_{15}H_{16}N_4$			
Propanal, 2-(phenylhydrazono)-, phenylhydrazone	n.s.g.	362(4.67)	59-1593-76

Compound	Solvent	$\lambda_{max}(\log \epsilon)$	Ref.
$C_{15}H_{16}N_4O_2$			
Benzenamine, N,N,2-trimethyl-4-(4-nitro-phenylazo)-	n.s.g.	410(4.36)	18-1381-76
Benzo[g]pteridine-2,4(3H,10H)-dione, 10-butyl-3-methyl-	EtOH	218(4.43),270(4.55), 336(3.86),438(3.95)	35-0830-76
	50% EtOH-HCl	216(4.31),260(4.42), 368(4.13)	35-0830-76
6H-1,2,4,5-Tetrazino[1,6-b]isoquinolin-6-one, 3-ethyl-1,2-dihydro-2-(1-oxo-propyl)-	EtOH	234(4.41),311(4.20), 368(3.92),385(3.81)	95-0700-76
$C_{15}H_{16}N_4O_3$			
Benzo[g]pteridine-2,4(3H,10H)-dione, 10-butyl-3-methyl-, 5-oxide	EtOH	213(4.26),271(4.55), 343(3.98),358s(3.92), 453(3.87)	35-0830-76
	50% EtOH-HCl	270(4.52),367(4.06), 397s(3.81)	35-0830-76
Pyrido[3,2-e]-1,2,4-triazine, 1,2-di-hydro-3-(3,4,5-trimethoxyphenyl)-	MeOH	220(4.13),269(3.95), 376(3.70)	114-0301-76C
$C_{15}H_{16}N_4O_4Se$			
7H-Pyrrolo[2,3-d]pyrimidine-5-carbo-nitrile, 4-(2-propenylseleno)-7-β-ribofuranosyl-	pH 1	314(4.20)	4-0745-76
	H_2O	312(4.31)	4-0745-76
	pH 11	313(4.25)	4-0745-76
$C_{15}H_{16}N_4O_6$			
Benzoic acid, 3,4,5-trimethoxy-, 2-(3-nitro-2-pyridinyl)hydrazide	EtOH	219(4.53),265(3.98), 377(3.43)	114-0301-76C
Benzoic acid, 3,4,5-trimethoxy-, 2-(3-nitro-4-pyridinyl)hydrazide	EtOH	260(4.20),345(3.75)	114-0285-76C
1H-Purine-2,6-dione, 7-(4-acetoxy-3,6-dihydro-6-methyl-3-oxo-2H-pyran-2-yl)-3,7-dihydro-1,3-dimethyl-	MeOH	273(4.00)	44-3827-76
1H-Pyrazole-3,4-dicarboxylic acid, 5-amino-1-(4-nitrophenyl)-, diethyl ester	EtOH	270(4.26)	4-1137-76
$C_{15}H_{16}N_4O_9$			
3-Pyrazolecarbonitrile, 4-nitro-1-(2,3,5-tri-O-acetyl-β-D-ribofuranosyl)-	MeOH	209(4.18),260(3.83)	104-1538-76
5-Pyrazolecarbonitrile, 4-nitro-1-(2,3,5-tri-O-acetyl-β-D-ribofuranosyl)-	MeOH	207(3.72),263(3.58)	104-1538-76
$C_{15}H_{16}N_6O$			
Urea, N-(3-phenylpropyl)-N'-1H-pyrazolo-[3,4-a]pyrimidin-4-yl-	pH 1	270(4.21)	87-0555-76
	pH 13	272(4.08),299(3.60)	87-0555-76
	EtOH	266(4.18)	87-0555-76
$C_{15}H_{16}N_6O_2$			
Pyrimido[5,4-e]-1,2,4-triazine-5,7-(1H,6H)-dione, 3-[4-(dimethylamino)-phenyl]-1,6-dimethyl-	dioxan	263(3.85),359(4.55)	39-0713-76C
Pyrimido[5,4-e]-1,2,4-triazine-5,7-(6H,8H)-dione, 3-[4-(dimethylamino)-phenyl]-6,8-dimethyl-	dioxan	353(4.61),437(3.21)	39-0713-76C
$C_{15}H_{16}N_6O_3$			
Pyrimido[5,4-e]-1,2,4-triazine-5,7-(1H,6H)-dione, 3-[4-(dimethylamino)-phenyl]-1,6-dimethyl-, 4-oxide	dioxan	266(3.20),357(4.53)	39-0713-76C

Compound	Solvent	$\lambda_{max}(\log \epsilon)$	Ref.
Pyrimido[5,4-e]-1,2,4-triazine-5,7-(6H,8H)-dione, 3-[4-(dimethylamino)-phenyl]-6,8-dimethyl-, 4-oxide	dioxan	358(4.55),439(3.39)	39-0713-76C
$C_{15}H_{16}N_8O_5S_4$			
5-Thia-1-azabicyclo[4.2.0]oct-2-ene-2-carboxylic acid, 7-methoxy-3-[[(1-methyl-1H-tetrazol-5-yl)thio]methyl]-8-oxo-7-[[(1,3,4-thiadiazol-2-ylthio)-acetyl]amino]-, (6R-cis)-	pH 6.86	266(4.06)	94-2629-76
bis(cyclohexylamine) salt	pH 6.86	267(4.15)	94-2629-76
$C_{15}H_{16}O$			
Bicyclo[3.2.1]oct-2-en-6-one, 5-methyl-2-phenyl-	EtOH	247(4.10)	39-1669-76C
1(2H)-Phenanthrenone, 3,4,9,10-tetra-hydro-2-methyl-	EtOH	230(4.01),236(4.01),253(3.64),298(4.08)	42-0812-76
1-Propanone, 2,2-dimethyl-1-(1-naphtha-lenyl)-	EtOH	220(4.81),271s(3.71),282(3.75),293s(3.71)	44-3067-76
1-Propanone, 2,2-dimethyl-1-(2-naphtha-lenyl)-	EtOH	213(4.47),222(4.43),242(4.53),248(4.51),283(3.84)	44-3067-76
$C_{15}H_{16}O_2$			
1H-Benz[e]inden-1-one, 2,3,4,5-tetra-hydro-7-methoxy-3-methyl-	EtOH	245(4.38),290(3.99)	39-0722-76C
Bicyclo[3.2.1]oct-2-en-6-one, 2-(4-meth-oxyphenyl)-	EtOH	258(4.2)	39-1669-76C
Diketone from catalponol isomer oxida-tion	MeOH	224(4.33),248(3.90),257s(3.81),302(3.05)	83-0829-76
$C_{15}H_{16}O_3$			
1,2-Benzenediol, 3-[2-(4-methoxyphenyl)-ethyl]-	MeOH	227(4.10),280(3.48)	78-0109-76
	MeOH-NaOH	228(4.04),245s(3.78),282(3.48),300s(3.08)	78-0109-76
Cannabispirenone	EtOH	213(4.26),232(4.29),284(3.46)	78-2939-76
Indeno[5,6-b]furan-5,7-dione, 2,3-di-hydro-4,6,6,8-tetramethyl-	CHCl$_3$	242(3.83),259(3.66),287(3.78)	39-0876-76C
Indeno[2,1-c]pyran-1-acetic acid, 1,3,4,9-tetrahydro-1-methyl-	MeOH	258(4.04)	111-0571-76
1,2-Naphthalenedione, 3-(1-hydroxy-1-methylethyl)-5,8-dimethyl-	EtOH	265(4.61),437(3.84)	102-1267-76
2(1H)-Naphthalenone, 1-acetoxy-1-(1-methylethyl)-	EtOH	233(4.11),238(4.14),309(3.93)	39-2570-76C
Naphtho[2,3-b]furan-4,8-dione, 6,7,8a,9-tetrahydro-3,5,8a-trimethyl-, (S)-	EtOH	218(4.06),256(3.90),290(3.69)	56-1931-76
$C_{15}H_{16}O_4$			
2H-1-Benzopyran-8-propanal, 7-methoxy-β-methyl-2-oxo-	EtOH	250(3.69),258(3.79),322(4.17)	25-0410-76
2H-Indeno[4,5-b]furan-2,8(4H)-dione, 8b-acetyl-5,5a,8a,8b-tetrahydro-3,5a-dimethyl-, [5aS-(5aα,8aα,8bα)]-	n.s.g.	225(4.23)	39-0433-76C
Isoauraptone	EtOH	248(3.52),255(3.57),324(4.22)	25-0410-76
1-Naphthalenepropanoic acid, 2-(carboxy-methyl)-3,4-dihydro-	EtOH	267(4.05)	78-0073-76
Photosantonene, 4-hydroxy-	n.s.g.	249(3.85)	39-0433-76C
Spiro[6H-cyclopropa[1,5]cyclopenta[1,2-	n.s.g.	257(4.01)	39-0433-76C

Compound	Solvent	$\lambda_{max}(\log \epsilon)$	Ref.
b]furan-6,6'-[2]oxabicyclo[3.1.0]hexane-2,3'(4H)-dione, 5,5a-dihydro-1',3,5a-trimethyl- (cont.)			39-0433-76C
$C_{15}H_{16}O_5$			
2H-1-Benzopyran-2-one, 6-(1,2-dihydroxy-3-methyl-3-butenyl)-7-methoxy- (thamnosmonin)	EtOH	207(4.38),222(4.28), 243(3.77),252(3.66), 330(4.19)	100-0134-76
Ethanone, 1-(1-hydroxy-3,6,8-trimethoxy-2-naphthalenyl)-	EtOH	231(4.44),275(4.53), 312(3.90),324(3.95), 389(3.76)	12-1087-76
1H-Indene-3-propanoic acid, 2-(carboxymethyl)-6-methoxy-	EtOH	267(4.24)	78-0065-76
Isokhellactone, methyl-, cis	EtOH	220(4.15),247(3.45), 257(3.60),300s(4.24), 324(4.38)	102-1293-76
trans	EtOH	220(4.08),247(3.35), 257(3.52),300s(4.17), 324(4.30)	102-1293-76
2-Propenoic acid, 3-[5-[3-(3-ethyloxiranyl)-1-hydroxy-2-propynyl]-2-furanyl]-, methyl ester (wyerol epoxide)	EtOH	310(4.30)	102-1119-76
$C_{15}H_{16}O_8$			
2-Propenoic acid, 3-[3,4-bis[(ethoxycarbonyl)oxy]phenyl]-, (E)-	MeOH-acid	267(4.42)	20-0647-76
$C_{15}H_{16}O_{10}$			
1,3-Cyclopentadiene-1,2,3,4,5-pentacarboxylic acid, pentamethyl ester, ion(1-), sodium	MeOH	212(3.82),262(4.68), 297(4.20)	24-1928-76
$C_{15}H_{17}BrO_3$			
2,5-Methano-1-benzoxepin-2(3H)-carboxylic acid, 7-bromo-4,5-dihydro-5,8,10-trimethyl-, (2α,5β,10S*)-(-)-	hexane	234(3.90),284(3.37), 292(3.39)	12-2533-76
$C_{15}H_{17}ClN_2O$			
Benzenemethanol, 5-chloro-2-(2-ethylhydrazino)-α-phenyl-	isoPrOH	253(4.02),303(3.35)	4-0033-76
4H-Pyrrolo[2,1-c][1,4]benzoxazine-4-methanamine, 1-chloro-N,N,4-trimethyl-, hydrochloride	MeOH	257(4.04),283(3.74), 288(3.71)	4-0311-76
4H-Pyrrolo[2,1-c][1,4]benzoxazine-4-methanamine, 8-chloro-N,N,4-trimethyl-, hydrochloride	MeOH	228(4.47),266(3.95), 303(3.85)	4-0311-76
$C_{15}H_{17}ClN_4O_6$			
Alloxazinium, 1,3,5,7,8-pentamethyl-, perchlorate	6M HCl	220(4.43),270(4.69), 402(4.19),460(3.79)	78-2303-76
$C_{15}H_{17}Cl_2NO_4$			
Methanaminium, N-(3-chloro-7-phenyl-2,4,6-heptatrienylidene)-N-methyl-, perchlorate	MeOH	231(4.05),243s(3.96), 285(4.11),306(4.13), 319(4.18),334(4.11), 421(4.40)	48-0731-76
$C_{15}H_{17}IN_2O_4$			
1,3-Dioxolo[4,5-g]quinolinium, 8-amino-7-(ethoxycarbonyl)-5-ethyl-, iodide	EtOH	272(4.52),277(4.49), 318s(3.95),343s(3.99), 350(4.00)	4-0765-76

Compound	Solvent	$\lambda_{max}(\log \epsilon)$	Ref.
$C_{15}H_{17}N$			
3H-Indole, 2,3-dimethyl-3-(3-methyl-1,2-butadienyl)-	EtOH	212(4.32),225s(4.19), 260(3.76)	39-2103-76C
Quinoline, 2,4-dimethyl-3-(2-methyl-1-propenyl)-	EtOH	228(4.71),276(3.83), 306(3.60),319(3.64)	39-2103-76C
Quinoline, 4-methyl-3-(1-pentenyl)-, cis	EtOH	230(4.61),282(3.80), 320(3.50)	39-2103-76C
trans	EtOH	245(4.56),286(3.99), 326(3.44)	39-2103-76C
$C_{15}H_{17}NO$			
Benzo[3,4]cyclobuta[1,2-c]quinolin-6(5H)-one, 6a,6b,7,8,9,10,10a,10b-octahydro-, (6aα,6bα,10aα,10bα)-	EtOH	258(2.92)	1-0189-76
1H-Inden-1-one, 2,3-dihydro-2-(1-piperidinylmethylene)-	EtOH	253(4.04),258s(4.01), 290(3.47),297(3.51), 360(4.44)	4-1201-76
2H-Pyrrole, 5-ethoxy-4-methyl-2-[(4-methylphenyl)methylene]-, (E)-	EtOH	238(3.89),330(4.39)	49-0831-76
$C_{15}H_{17}NOS$			
4H-1-Benzothiopyran-4-one, 2,3-dihydro-3-(1-piperidinylmethylene)-	EtOH	247(4.28),260s(4.12), 375(4.02)	4-0225-76
$C_{15}H_{17}NO_2$			
Pentanoic acid, 4-cyano-4-methyl-2-methylene-, 4-methylphenyl ester	dioxan	266(3.07),272(2.99)	126-3481-76
$C_{15}H_{17}NO_3$			
Araliopsine, (+)- (same spectra in acid or base)	EtOH	232(4.54),286(3.87), 296(3.94),319(3.88), 331(3.80)	102-0743-76
1H-Carbazole-1-carboxylic acid, 2,3,4,9-tetrahydro-1-methoxy-, methyl ester	n.s.g.	227(4.23),278s(3.80), 284(3.82),292(3.73)	44-0102-76
Haplamine, dihydro-	EtOH	217(4.48),233(4.51), 278(3.92),287(3.89), 334(3.91)	105-0282-76
Isoplatydesmine, (+)-	EtOH and EtOH-NaOH	216(4.39),236(4.35), 250s(4.13),298(3.89), 308(4.01),319(3.97)	102-0743-76
	EtOH-HCl	217(4.43),235(4.52), 293(4.05),300s(4.03), 312s(3.82)	102-0743-76
Ribalinine, (-)-	EtOH and EtOH-NaOH	238(4.41),316(3.97), 328(3.95)	102-0743-76
	EtOH-HCl	237(4.58),299(4.02)	102-0743-76
$C_{15}H_{17}NO_3S_2$			
Pyrrolidine, 1-(3,4-dihydro-5H-[1]benzothiopyrano[3,4-e]-1,2-oxathiin-4-yl)-, S^2,S^2-dioxide	EtOH	220(4.06),248(4.28), 285(3.49),295s(3.35), 332(3.20)	4-0225-76
$C_{15}H_{17}NO_4$			
2,4-Pentadienoic acid, 2,4-dimethyl-, 5-(4-nitrophenyl)-, ethyl ester, (E,E)-	EtOH	265(3.96),327(4.25)	39-0404-76C
Propanedioic acid, 1H-indol-2-yl-, diethyl ester	MeOH	271(3.97),280(3.95), 290(3.82)	24-3282-76
	MeOH-NaOH	281(3.99),320(3.68)	24-3282-76

Compound	Solvent	λ_{max}(log ϵ)	Ref.
C$_{15}$H$_{17}$NO$_5$			
Propanedioic acid, (1-hydroxy-1H-indol-2-yl)-, diethyl ester	MeOH	274(3.87),286s(3.80)	24-3282-76
C$_{15}$H$_{17}$NO$_{10}$			
2H-1,3-Oxazine-2,6(3H)-dione, 3-(2,3,5-tri-O-acetyl-β-D-ribofuranosyl)-	dioxan	262(3.82)	87-0643-76
C$_{15}$H$_{17}$N$_2$O$_4$			
1,3-Dioxolo[4,5-g]quinolinium, 8-amino-7-(ethoxycarbonyl)-5-ethyl-, iodide	EtOH	272(4.52),277(4.49), 318s(3.95),343s(3.99), 350(4.00)	4-0765-76
C$_{15}$H$_{17}$N$_3$			
Benzenamine, N,N-dimethyl-4-[(2-methylphenyl)azo]-	C$_6$H$_{12}$	394(4.48)	18-1381-76
Benzenamine, N,N-dimethyl-4-[(4-methylphenyl)azo]-	C$_6$H$_{12}$	400(4.50)	18-1381-76
Benzenamine, N,N,2-trimethyl-4-(phenylazo)-	C$_6$H$_{12}$	405(4.48)	18-1381-76
6H-Cyclohepta[4,5]pyrrolo[1,2-a]pyrimidine-11-carbonitrile, 7,8,9,10-tetrahydro-2,4-dimethyl-	EtOH	223(4.35),240(4.3), 242(4.3),258(4.28), 267(4.31),292(3.58), 299(3.67),360(3.57)	103-1379-76
C$_{15}$H$_{17}$N$_3$O			
2,4-Pentadienamide, 2-cyano-5-(dimethylamino)-5-(4-methylphenyl)-	CH$_2$Cl$_2$	428(4.15)	48-0705-76
2,4-Pentadienamide, 2-cyano-5-(ethylamino)-5-(4-methylphenyl)-	CH$_2$Cl$_2$	383(4.54)	48-0705-76
C$_{15}$H$_{17}$N$_3$OS			
Imidazo[1,2-a]pyridin-2(3H)-one, 3-(3-ethyl-4,5-dimethyl-2(3H)-thiazolylidene)-7-methyl-, perchlorate	EtOH	420(4.37)	124-0961-76
C$_{15}$H$_{17}$N$_3$O$_2$			
4H-Indazol-4-one, 1,5,6,7-tetrahydro-3-(4-methoxyphenyl)-1-methyl-, oxime	MeOH	235(4.36),268(4.01)	24-1898-76
2,4-Pentadienamide, 2-cyano-5-(dimethylamino)-5-(4-methoxyphenyl)-	CH$_2$Cl$_2$	402(4.58)	48-0705-76
Pyrazolo[4,3-b]azepin-5(1H)-one, 4,6,7,8-tetrahydro-3-(4-methoxyphenyl)-1-methyl-	MeOH	240s(4.20),260(4.31)	24-1898-76
Pyrazolo[4,3-c]azepin-4(1H)-one, 5,6,7,8-tetrahydro-3-(4-methoxyphenyl)-1-methyl-	MeOH	254s(4.06)	24-1898-76
3-Pyridinecarboxamide, 1-ethyl-1,2-dihydro-2-imino-6-(4-methoxyphenyl)-perchlorate	CH$_2$Cl$_2$	266(3.95)	48-0705-76
	HOAc	344(4.10)	48-0705-76
C$_{15}$H$_{17}$N$_3$O$_2$S			
4,6(1H,5H)-Pyrimidinedione, 5-[[4-(diethylamino)phenyl]methylene]dihydro-2-thioxo-	dioxan	289(3.94),315(3.87), 335(3.75),480(5.06)	104-1124-76
C$_{15}$H$_{17}$N$_3$O$_3$			
7-Azabicyclo[4.3.1]deca-3,8-dien-10-ol, 5-methyl-7-(5-nitro-2-pyridinyl)-	EtOH	208(4.17),253(4.17), 392(4.42)	39-2307-76C
4H-Pyrrolo[2,1-c][1,4]benzoxazine-4-	MeOH	325(4.02)	4-0311-76

Compound	Solvent	$\lambda_{max}(\log \epsilon)$	Ref.
methanamine, N,N,4-trimethyl- 1-nitro-, hydrochloride (cont.)			4-0311-76
$C_{15}H_{17}N_3O_4$ Spiro[10H-[1]benzoxepino[3,4-c]pyrazole- 10,2'-oxirane], 3,3a,4,10a-tetrahydro- 3,3',4-trimethyl-10a-nitro-, [3α,3aα,4α,10β(R*),10aα]-	EtOH	260(3.24),266(3.25), 273(3.15),327(2.49)	39-1260-76C
$C_{15}H_{17}N_3O_5$ 1H-Indole-3-acetic acid, α-[[(aminocarb- onyl)amino]methyl]-1-(methoxycarbo- nyl)-, methyl ester	EtOH	227(4.34),260(3.98), 283(3.63),291(3.76)	104-1087-76
$C_{15}H_{17}N_3O_5S$ 1H-Pyrrolo[2,3-d]pyrimidine-2,4(3H,5H)- dione, 6,7-dihydro-3-(methoxymethyl)- 7-[(4-methylphenyl)sulfonyl]-	DMSO	287(4.46)	24-2983-76
$C_{15}H_{17}N_3S$ 2,4-Pentadienethioamide, 2-cyano-5-(di- methylamino)-5-(4-methylphenyl)-	CH_2Cl_2	447(4.45)	48-0705-76
$C_{15}H_{17}N_4OS$ Methanaminium, N-[[[2-(acetylamino)- 5-phenyl-6H-1,3-thiazin-6-ylidene]- amino]methylene]-N-methyl-, perchlor- ate	HOAc	315(4.16),425(4.16)	97-0268-76
$C_{15}H_{17}N_4O_2$ Benzo[g]pteridinium, 1,2,3,4-tetrahydro- 1,3,5,7,8-pentamethyl-2,4-dioxo-, perchlorate	6M HCl	220(4.43),270(4.69), 402(4.19),460(3.79)	78-2303-76
$C_{15}H_{17}N_5O_2$ Propanamide, 2,2-dimethyl-N-[9-(5-meth- yl-2-furanyl)-9H-purin-6-yl]-	MeOH	210(4.36),262(4.42)	35-8213-76
$C_{15}H_{17}N_5O_4$ 1H-Purine-2,6-dione, 8-[(5-acetoxy-2,4- pentadienylidene)amino]-3,7-dihydro- 1,3,7-trimethyl-, (E,E,E)-	$CHCl_3$	320(4.34),382(4.42), 395(4.43)	142-1257-76
$C_{15}H_{17}O_8$ 1,3-Cyclopentadiene-1,2,3,4-tetracarb- oxylic acid, 5-ethyl-, tetramethyl ester, ion(1-), sodium	MeOH	224(3.88),270(4.50), 306(4.15)	24-1928-76
$C_{15}H_{18}$ Bicyclo[9.2.2]pentadeca-2,9,11,13,14- pentaene	n.s.g.	244(3.75)	88-1243-76
1H-Cyclopenta[jk]fluorene, 2,2a,3,4,5,5a,9b,9c-octahydro-	hexane	218(4.92),261(2.98), 267(3.18),274(3.27)	24-2956-76
$C_{15}H_{18}Br_2O_3$ 2-Naphthaleneacetic acid, 1-bromo- 8-(bromomethyl)-1,2,3,4,4a,7-hexa- hydro-α,4a-dimethyl-7-oxo-, [1R- [1α,2α(S*),4aα]]-	EtOH	255(4.03)	44-1256-76

Compound	Solvent	$\lambda_{max}(\log \epsilon)$	Ref.
$C_{15}H_{18}ClN_3O_3$ 4H-Pyrrolo[2,1-c][1,4]benzoxazine- 4-methanamine, N,N,4-trimethyl- 1-nitro-, hydrochloride	MeOH	325(4.02)	4-0311-76
$C_{15}H_{18}ClN_5O_2$ 3H-1,2,4-Triazolo[4,3-d][1,4]benzodi- azepine-3,6(5H)-dione, 10-chloro-2- [2-(dimethylamino)ethyl]-2,7-dihydro- 7-methyl-, monohydrochloride	MeOH	253(4.20),302(3.59)	88-1931-76
$C_{15}H_{18}Cl_2N_2O$ 4H-Pyrrolo[2,1-c][1,4]benzoxazine- 4-methanamine, 1-chloro-N,N,4- trimethyl-, hydrochloride	MeOH	257(4.04),283(3.74), 288(3.71)	4-0311-76
$C_{15}H_{18}Cl_2N_2O_5$ Propanedioic acid, (acetylamino)[(2,6- dichloro-3-pyridinyl)methyl]-, diethyl ester	MeOH	221(4.00),275(3.68)	33-0190-76
$C_{15}H_{18}FN_5O_5$ Adenosine, 4'-C-fluoro-2',3'-O-(1-meth- ylethylidene)-, 5'-acetate	MeOH	259(4.16)	35-3346-76
$C_{15}H_{18}NO$ Pyrylium, 4-[4-(dimethylamino)phenyl]- 2,6-dimethyl-, perchlorate	MeCN CH_2Cl_2	282(4.09),485(4.80) 255s(3.56),285(4.13), 500(4.92)	4-1089-76 4-1089-76
$C_{15}H_{18}NO_8$ 1,3-Cyclopentadiene-1,2,3,4-tetracarbox- ylic acid, 5-(dimethylamino)-, tetra- methyl ester, ion(1-), sodium	MeOH	215(3.66),257(4.47), 291(4.14)	24-1928-76
$C_{15}H_{18}NS$ Thiopyrylium, 4-[4-(dimethylamino)phen- yl]-2,6-dimethyl-, perchlorate	MeCN CH_2Cl_2	260(3.73),288(3.89), 525(4.71) 262(3.75),295(3.92), 547(4.80)	4-1089-76 4-1089-76
$C_{15}H_{18}N_2$ 1H-Imidazole, 1-cyclohexyl-4-phenyl- Indolo[2,3-a]quinolizine, 1,2,3,4,6,7- 12,12b-octahydro- Perimidine, 2,3-dihydro-2,2,4,5-tetra- methyl- 3H-Pyrido[2,3-b]indole, 4,9-dihydro- 1-methyl-4-propyl-	EtOH EtOH n.s.g. n.s.g.	255(4.32) 274s(3.72),279s(3.74), 282(3.75),289(3.66) 238(4.72),335(4.07), 348(4.09) 248(2.83),355(3.45)	88-4753-76 130-0283-76 12-2499-76 103-0329-76
$C_{15}H_{18}N_2O$ Cyclopenta[b]pyrimido[1,2-a]indol-7(4H)- one, 1,2,3,5,6,12b-hexahydro-5-methyl- Dipyrido[1,2-a:2',1'-c]pyrazinediium, 6,7-dihydro-3-propoxy-, dibromide 1(2H)-Isoquinolinone, 3-(cyclohexyl- amino)- 3-Pyridinecarboxamide, 1,4-dihydro- N,N-dimethyl-1-(phenylmethyl)-	EtOH EtOH EtOH EtOH	251(4.10),278(3.51), 286(3.45) 279(3.63),342(3.92) 230(4.39),244(4.07), 294(4.27),303(4.34), 378(3.71) 345(3.63)	44-3775-76 64-0115-76B 95-0154-76 44-2976-76

Compound	Solvent	$\lambda_{max}(\log \epsilon)$	Ref.
4H-Pyrrolo[2,1-c][1,4]benzoxazine-4-methanamine, N,N,4-trimethyl-, hydrochloride	MeOH	218(4.40),264(4.02), 290(3.85)	4-0311-76
2H-Pyrrol-2-one, 5-[[3-(dimethylamino)-phenyl]methylene]-1,5-dihydro-3,4-di-methyl-, (E)-	EtOH	248(3.80),280(3.75), 333s(3.59)	49-0831-76
(Z)-	EtOH	<u>255(4.1),290(4.2), 325(4.3)</u>	49-0831-76
2H-Pyrrol-2-one, 5-[[4-(dimethylamino)-phenyl]methylene]-1,5-dihydro-3,4-di-methyl-, (E)-	EtOH	258(4.06),392(4.37)	49-0831-76
(Z)-	EtOH	<u>263(4.7),395(4.4)</u>	49-0831-76
$C_{15}H_{18}N_2OS$ 4(1H)-Pyrimidinone, 2-[(3-methylbutyl)-thio]-6-phenyl-	EtOH	257(4.47),312(3.97)	95-1094-76
$C_{15}H_{18}N_2OS_2$ 1H-[1,2]Dithiolo[5,1-e][1,2,3]thiadia-zole-7-S^{IV}, 5-(1,1-dimethylethyl)-1-(4-methoxyphenyl)-	C_6H_{12}	209(4.31),233(4.39), 254s(4.19),296(3.94), 488(4.22)	39-0228-76C
$C_{15}H_{18}N_2O_2$ 9-Acridinecarbonitrile, 1,2,3,4,4a,9,9a-10-octahydro-4a-hydroxy-9a-methoxy-	EtOH	247(4.16),298(3.47)	95-0968-76
4-Isoquinolinecarbonitrile, 3-ethoxy-8-ethylidene-5,6,7,8-tetrahydro-1-methyl-	EtOH	239(3.85),301(3.85)	2-0731-76
2,7-Pentanocyclobuta[1,2-c:4,3-c']di-pyridine-1,8-dione, 4a,4b,8a,8b-tetrahydro-	n.s.g.	264(3.79)	33-2841-76
$C_{15}H_{18}N_2O_3$ 3-Butenenitrile, 4-(2-hydroxy-4,4-di-methyl-6-oxo-1-cyclohexen-1-yl)-2-[2-(methoxyimino)ethylidene]-, (E,Z,E)-	MeOH CHCl$_3$	411(4.40) 350(4.39)	5-1799-76 5-1799-76
3,5-Hexadienenitrile, 6-(2-hydroxy-4,4-dimethyl-6-oxo-1-cyclohexen-1-yl)-2-(methoxyimino)-, (Z,E,E)-	MeOH	414(4.50)	5-1799-76
2,4-Pentadienenitrile, 5-(2-hydroxy-4,4-dimethyl-6-oxo-1-cyclohexen-1-yl)-2-[(methoxyimino)methyl]-, (E,Z,E)-	MeOH	423(4.62)	5-1799-76
$C_{15}H_{18}N_2O_4$ Pyrrolo[1,2-a]pyrimidine-7,8-dicarbox-ylic acid, 2,4-dimethyl-, diethyl ester	MeOH	221(4.52),239(4.52), 245(4.59),288(3.77), 299(3.73),343(3.69)	24-2983-76
$C_{15}H_{18}N_2O_5$ Benzoic acid, [(5-acetoxy-2-methyl-1,3-dioxan-4-yl)methylene]hydrazide	EtOH	223(4.23),291(4.24)	136-0185-76D
$C_{15}H_{18}N_2O_6$ 2-Butenedioic acid, 2-[3-cyano-5-(meth-oxyimino)-1,3-pentadienyl]-3-hydroxy-, diethyl ester, (E,E,Z,E)-	MeOH CHCl$_3$	408(4.36) 363(4.37)	5-1799-76 5-1799-76
2-Butenedioic acid, 2-[4-cyano-5-(meth-oxyimino)-1,3-pentadienyl]-3-hydroxy-, diethyl ester, (E,E,Z,E)-	MeOH CHCl$_3$	421(4.54) 378(4.45)	5-1799-76 5-1799-76

Compound	Solvent	$\lambda_{max}(\log \epsilon)$	Ref.
2-Butenedioic acid, 2-[5-cyano-5-(meth-oxyimino)-1,3-pentadienyl]-3-hydroxy-, diethyl ester, (Z,E,E,E)-	MeOH CHCl$_3$	407(4.40) 366(4.35)	5-1799-76 5-1799-76
$C_{15}H_{18}N_2S_2$ 1H-[1,2]Dithiolo[5,1-e][1,2,3]thiadia-zole-7-S^{IV}, 5-(1,1-dimethylethyl)-1-(4-methylphenyl)-	C$_6$H$_{12}$	206(4.38),234(4.38), 257(4.18),288s(3.96), 481(4.21)	39-0228-76C
$C_{15}H_{18}N_3S$ [1,2,4]Triazolo[5,1-a]isoquinolinium, 1-butyl-2-(methylthio)-, iodide	MeOH	247(4.67),310(4.09)	2-0176-76
[1,2,4]Triazolo[1,5-a]quinolinium, 3-butyl-2-(methylthio)-, iodide	MeOH	244(4.67),280(3.80), 314(3.43),329(3.50)	2-0176-76
$C_{15}H_{18}N_4O_2$ Benzo[g]pteridine-2,4(1H,3H)-dione, 5,10-dihydro-1,3,5,7,8-pentamethyl-, hydrochloride	6M HCl	222(4.36),250(4.05), 285(4.02),306(4.01)	78-2303-76
$C_{15}H_{18}N_4O_2S$ 2,6-Pyridinediamine, 3-nitro-4-[[(2,4,6-trimethylphenyl)methyl]thio]-	pH 1	224(4.39),296(3.88), 375(4.19)	44-3784-76
$C_{15}H_{18}N_4O_3$ Propanoic acid, 2-[1,2-dihydro-1-oxo-2-[(1-oxopropyl)amino]-3-isoquino-linyl]hydrazide	EtOH	231(4.27),300(4.00), 362(3.59)	95-0700-76
$C_{15}H_{18}N_4O_4$ Benzoic acid, 3,4,5-trimethoxy-, 2-(3-amino-2-pyridinyl)hydrazide	EtOH	214(3.72),252(3.35), 295(3.18)	114-0301-76C
Benzoic acid, 3,4,5-trimethoxy-, 2-(3-amino-4-pyridinyl)hydrazide	pH 1	258(4.21),455(3.10)	114-0285-76C
1H-Pyrazole-4-carboxylic acid, 5-amino-3-(1-methylethyl)-1-(4-nitrophenyl)-, ethyl ester	EtOH	300(4.68)	4-1137-76
$C_{15}H_{18}N_4O_6$ 4,11-Epoxy-1,3-dioxolo[4,5-f]imidazo-[1,5-a][1,3]diazocine-7-carboxamide, N-acetyl-6-formyl-3a,4,5,6,11,11a-hexahydro-2,2-dimethyl-, [3aR-(3aα,4β,11β,11aα)]-	MeOH	217(4.41),266(4.06)	35-3346-76
β-D-Ribofuranoside, methyl 5-deoxy-5-(3,4-dihydro-2,4-dioxo-8(2H)-pteri-dinyl)-2,3-O-(1-methylethylidene)-	pH -4.0	242(4.08),270(3.36), 348(4.02)	24-3175-76
	pH 6.0	258(4.17),275s(3.90), 398(3.97)	24-3175-76
	pH 12.0	227(4.25),278(4.10), 306(3.97),398(2.84)	24-3175-76
$C_{15}H_{18}N_4O_8S$ 2,4(1H,3H)-Pteridinedione, 1-[2,3-O-(1-methylethylidene)-5-O-(methylsulfon-yl)-β-D-ribofuranosyl]-	pH 6.0 pH 11.0	228(4.12),313(3.87) 236(4.17),275(3.56), 321(3.86)	24-3159-76 24-3159-76
	MeOH	231(4.15),316(3.84), 335s(3.65)	24-3159-76
$C_{15}H_{18}N_5O_6PS$ 3H-Imidazo[2,1-i]purine, 5-[(1-methyl-	pH 1	238(4.29),293(4.28)	94-1561-76

Compound	Solvent	$\lambda_{max}(\log \epsilon)$	Ref.
ethyl)thio]-3-(3,5-O-phosphinico- β-D-ribofuranosyl)- (cont.)	pH 13	240(4.32),244s(4.30), 274(3.88),283(3.97), 306(3.96)	94-1561-76
3H-Imidazo[2,1-i]purine, 3-(3,5-O-phos- phinico-β-D-ribofuranosyl)-5-(propyl- thio)-	pH 1	238(4.29),293(4.28), 274(3.90),284(3.99), 304(3.98)	94-1561-76
$C_{15}H_{18}N_6$ Formazan, 1-(4,6-dimethyl-2-pyrimidin- yl)-3-methyl-5-(4-methylphenyl)-	EtOH EtOH-NaOH benzene dioxan	382(4.32) 490(--) 392(--) 380(--)	103-1165-76 103-1165-76 103-1165-76 103-1165-76
$C_{15}H_{18}O$ 1H-Benz[cd]azulen-6-one, 2,2a,3,4,5,6- hexahydro-7,9-dimethyl-	EtOH	260(3.97),300(3.35)	22-1599-76
Bicyclo[3.2.1]octan-6-one, 5-methyl- 2-phenyl-, endo	EtOH	258(2.45)	39-1669-76C
$C_{15}H_{18}OS$ Azulene, 1-(ethylsulfinyl)-4,6,8-tri- methyl-	hexane	540(2.83)	104-1966-76
$C_{15}H_{18}O_2$ Bicyclo[3.2.1]octan-6-one, 1-hydroxy- 5-methyl-2-phenyl-, exo-(±)-	EtOH	250(2.86)	39-1669-76C
3-Cyclohexene-1-carboxylic acid, 1-meth- yl-4-phenyl-, methyl ester	EtOH	247(4.15)	39-1669-76C
9H-Fluoren-9-one, 6-methoxy-4a-methyl- 1,2,3,4,4a,9a-hexahydro-	EtOH	224(4.15),269(4.16), 288(4.04)	12-2683-76
2-Naphthalenemethanol, 3-hydroxy- α,α,5,8-tetramethyl-	EtOH	233s(4.72),243(4.73), 277s(3.78),289(3.87), 301(3.81),326s(3.51), 340(3.57)	102-1267-76
1-Naphthalenone, 1,2,3,4-tetrahydro- 4-hydroxy-3-(2-isopentenyl)-, (3R,4S)- (catalponol isomer)	MeOH	243(3.85),251(3.68), 290(3.19)	83-0829-76
$C_{15}H_{18}O_2S$ Sulfonium, methylphenyl-, 4,4-dimethyl- 2,6-dioxocyclohexylide	MeOH	230(4.07),259(4.24)	30-0366-76
$C_{15}H_{18}O_3$ Cannabispirone	EtOH	213(4.23),225(4.06), 279(3.30),283(3.34)	78-2939-76
Colorata-5,8-dienolide, 7-oxo-	EtOH	249.0(4.13)	39-0850-76C
3-Cyclohexene-1-carboxylic acid, 4-(4- methoxyphenyl)-, methyl ester	EtOH	254(4.2)	39-1669-76C
3-Cyclohexene-1-carboxylic acid, 4-(4- methoxyphenyl)-1-methyl-	EtOH	255(4.28)	39-1669-76C
Furanoeremophilan-6-one, 1,10β-epoxy-	ether	206(4.19),259(3.52)	24-0819-76
2-Naphthaleneacetic acid, 3,4,4a,7- tetrahydro-α,4a,8-trimethyl-7-oxo-, [S-(R*,R*)]-	EtOH	232(4.11),316(4.11)	44-1256-76
Naphtho[2,3-b]furan-4(6H)-one, 7,8,8a,9- tetrahydro-8-hydroxy-3,5,8a-trimeth- yl-, (8S-cis)- (nehipetol)	EtOH	219(4.07),263(3.80), 294(3.75)	56-1931-76
Spiro[furan-2(5H),2'(1'H)-naphthalen]-5- one, 3,3',4,4'-tetrahydro-6'-methoxy- 3-methyl-	EtOH	278(3.36)	39-0722-76C

Compound	Solvent	$\lambda_{max}(\log \epsilon)$	Ref.
$C_{15}H_{18}O_4$			
1H-Dibenzo[b,d]pyran-6-carboxylic acid, 2,3,4,4a,6,10b-hexahydro-9-hydroxy-6-methyl-	EtOH	277(3.50)	95-1031-76
Eremophil-7(11)-ene-12,18α:14β,6α-diolide	EtOH	217(4.29)	18-3196-76 +94-0360-76
Helenalin	EtOH	219.5(4.08)	83-0333-76
$C_{15}H_{18}O_5$			
Eremophil-7(11)-ene-12,8α:14β,6α-diolide, 8β-hydroxy-	EtOH	214(4.09)	18-3196-76 +94-0360-76
$C_{15}H_{18}O_6$			
2-Propenoic acid, 3-[5-(1,4,5-trihydroxy-2-heptynyl)-2-furanyl]-, methyl ester	EtOH	310(4.26)	102-1119-76
$C_{15}H_{18}O_6S$			
2-Propenoic acid, 2-[(carboxymethyl)-thio]-3-(3,4-dimethoxyphenyl)-, 1-ethyl ester	EtOH	334(4.1)	42-0038-76
$C_{15}H_{18}O_8$			
Bicyclo[2.2.2]octa-2,5-diene-2,3-dicarboxylic acid, 1,7,7-trimethoxy-8-oxo-, dimethyl ester	EtOH	240s(3.42),305s(2.40)	1-0403-76
1,3-Cyclopentadiene-1,2,3,4-tetracarboxylic acid, 5-ethyl-, tetramethyl ester, ion(1-), sodium	MeOH	224(3.88),270(4.50), 306(4.15)	24-1928-76
$C_{15}H_{18}S$			
Azulene, 1-(ethylthio)-4,6,8-trimethyl-	hexane	580(2.93)	104-1966-76
$C_{15}H_{19}BrN_4O_8$			
1H-Imidazole-4-carboximidamide, 5-bromo-N-hydroxy-1-(2,3,5-tri-O-acetyl-β-D-ribofuranosyl)-	pH 1 pH 11	249(3.94) 232(3.91)	87-1020-76 87-1020-76
$C_{15}H_{19}BrO$			
2,5-Methano-1-benzoxepin, 7-bromo-2,3,4,5-tetrahydro-2,5,8,10-tetramethyl-, (2α,5α,10S*)-(-)- (filiformin)	hexane	237(3.96),287(3.43), 290(3.43),296(3.45)	12-2533-76
Phenol, 4-bromo-2-(1,2-dimethyl-3-methylenecyclopentyl)-5-methyl-, cis-(+)- (allolaurinterol)	hexane	283(3.33),289(3.33)	12-2533-76
$C_{15}H_{19}BrO_2$			
2,5-Methano-1-benzoxepin-2(3H)-methanol, 7-bromo-4,5-dihydro-5,8,10-trimethyl-, (2α,5β,10S*)-(-)- (filiforminol)	hexane	235(3.92),286(3.45), 289(3.44),295(3.46)	12-2533-76
$C_{15}H_{19}BrO_3$			
2-Naphthaleneacetic acid, 1-bromo-1,2,3,4,4a,7-hexahydro-α,4a,8-trimethyl-7-oxo-, [1R-[1α,2α(S*),4aα]]-	EtOH	251(4.06)	44-1256-76
2-Naphthaleneacetic acid, 8-(bromomethyl)-1,2,3,4,4a,7-hexahydro-α,4a-dimethyl-7-oxo-, [2R-[2α(S*),4aα]]-	EtOH	241(4.00)	44-1256-76
Naphtho[1,2-b]furan-2,8(3H,4H)-dione, 7-bromo-3a,5,5a,9,9a,9b-hexahydro-	EtOH	254(3.83)	94-2810-76

Compound	Solvent	$\lambda_{max}(\log \epsilon)$	Ref.
3,5a,9-trimethyl-, [3S-(3α,3aα,5aβ-9α,9aα,9bβ)]- (cont.)			94-2810-76
$C_{15}H_{19}Br_2ClO$			
2H-Pyran, 3-bromo-2-(3-bromo-2-penten-yl)-5-chlorotetrahydro-6-(2-penten-4-ynyl)- (isodactylyne)	isooctane	224(4.19),233s(--)	44-3480-76
$C_{15}H_{19}ClN_4O_8$			
Imidazole-4-carboximidamide, 5-chloro-N-hydroxy-1-(2,3,5-tri-O-acetyl-β-D-ribofuranosyl)-	pH 1	247(4.02)	87-1020-76
	pH 11	236(3.89)	87-1020-76
$C_{15}H_{19}ClO_3$			
Naphtho[1,2-b]furan-2,8(3H,4H)-dione, 7-chloro-3a,5,5a,9,9a,9b-hexahydro-3,5a,9-trimethyl-, [3S-(3α,3aα,5aβ-9α,9aα,9bβ)]-	EtOH	245(3.86)	94-2810-76
$C_{15}H_{19}N$			
1H-Benz[e]indole, 2,3,3a,9b-tetrahydro-3,3a,9b-trimethyl-, cis-(±)-	EtOH	212(4.33),217(4.33), 222s(4.15),259(3.86)	88-0767-76
$C_{15}H_{19}NO$			
2-Cyclohexen-1-one, 5,5-dimethyl-3-[(2-methylphenyl)amino]-	pH 1	292(4.38)	39-2207-76C
	H_2O	299(4.45)	39-2207-76C
	pH 13	299(4.43)	39-2207-76C
2-Cyclohexen-1-one, 5,5-dimethyl-3-[(3-methylphenyl)amino]-	pH 1	299(4.34)	39-2207-76C
	H_2O	310(4.39)	39-2207-76C
	pH 13	309(4.38)	39-2207-76C
2-Cyclohexen-1-one, 5,5-dimethyl-3-[(4-methylphenyl)amino]-	pH 1	299(4.35)	39-2207-76C
	H_2O	309(4.41)	39-2207-76C
	pH 13	308(4.41)	39-2207-76C
2-Cyclopropen-1-one, 2-[bis(1-methyl-ethyl)amino]-3-phenyl-	MeOH	278(4.25)	88-2405-76
$C_{15}H_{19}NO_2$			
Acetic acid, (5,5-dimethyl-4-phenyl-2-pyrrolidinylidene)-, methyl ester	EtOH	209(3.91),281(4.40)	12-1735-76
2-Cyclohexen-1-one, 3-[(2-methoxyphen-yl)amino]-5,5-dimethyl-	pH 1	292(4.30)	39-2207-76C
	H_2O	304(4.41)	39-2207-76C
	pH 13	304(4.42)	39-2207-76C
2-Cyclohexen-1-one, 3-[(3-methoxyphen-yl)amino]-5,5-dimethyl-	pH 1	298(4.31)	39-2207-76C
	H_2O	312(4.36)	39-2207-76C
	pH 13	310(4.35)	39-2207-76C
2-Cyclohexen-1-one, 3-[(4-methoxyphen-yl)amino]-5,5-dimethyl-	pH 1	302(4.27)	39-2207-76C
	H_2O	306(4.42)	39-2207-76C
	pH 13	306(4.38)	39-2207-76C
$C_{15}H_{19}NO_3$			
Cocculolidine	MeOH	214(4.15)	36-0132-76
6-Phenanthridinecarboxylic acid, 1,2,3,4,4a,5,6,10b-octahydro-9-hydroxy-6-methyl-	EtOH	277.5(3.21)	95-1031-76
2H-Pyrrole-4-carboxylic acid, 3,4-di-hydro-2,2-dimethyl-5-phenyl-, ethyl ester, 1-oxide	EtOH	245(4.23)	12-2511-76
$C_{15}H_{19}NO_3S$			
3H-Indeno[2,1-e]-1,2-oxathiin-4-amine, N,N-diethyl-4,5-dihydro-, 2,2-dioxide	EtOH	216s(4.17),224s(4.03), 259(4.10)	4-1201-76

Compound	Solvent	$\lambda_{max}(\log \epsilon)$	Ref.
$C_{15}H_{19}NO_3S_2$ 5H-[1]Benzothiopyrano[3,4-e]-1,2-oxathiin-4-amine, N,N-diethyl-3,4-dihydro-, 2,2-dioxide	EtOH	220(4.09),248(4.27), 285(3.52),295s(3.39), 332(3.22)	4-0225-76
$C_{15}H_{19}NO_4$ 2-Propenoic acid, 3-(2,2,8-trimethyl-4H-1,3-dioxino[4,5-c]pyridin-5-yl)-, ethyl ester, (E)-	MeOH	205(4.14),242(4.27), 274(4.07),311(4.04)	104-0440-76
$C_{15}H_{19}NO_5$ 3-Azatetracyclo[7.1.0.0^{2,4}.0^{5,7}]decane-7,9-dicarboxylic acid, 3-acetyl-, dimethyl ester, cis	MeOH	240s(2.58)	24-3505-76
Carbamic acid, (1,4-dioxo-4-phenylbutoxy)-, 1,1-dimethylethyl ester	MeCN	241(4.16),277(3.20)	33-2499-76
$C_{15}H_{19}NO_6$ 3,6,9,12-Tetraoxa-16-azabicyclo[12.3.1]-octadeca-1(18),14,16-triene-2,13-dione, 15,17-dimethyl-	MeCN	234(4.09),273(3.57), 280(3.48)	77-0964-76
$C_{15}H_{19}NO_7$ Benzeneacetonitrile, α-(β-D-glucopyranosyloxy)-4-methoxy-, (R)-	MeOH	232(4.108),273(3.121), 280(3.056)	24-3379-76
$C_{15}H_{19}NO_8$ 1,3-Cyclopentadiene-1,2,3,4-tetracarboxylic acid, 5-(dimethylamino)-, tetramethyl ester, ion(1-), sodium	MeOH	215(3.66),257(4.47), 291(4.14)	24-1928-76
N-protonated	MeOH	217(3.70),257(4.46), 292(4.14)	24-1928-76
$C_{15}H_{19}NS$ 2-Cyclopropene-1-thione, 2-[bis(1-methylethyl)amino]-3-phenyl-	MeOH	240(3.63)	88-2405-76
$C_{15}H_{19}N_3$ Piperidine, 1-(3-methyl-4-phenyl-1H-pyrazol-1-yl)-	MeOH	246(4.11)	94-3001-76
Piperidine, 1-(4-methyl-3-phenyl-1H-pyrazol-1-yl)-	MeOH	252(4.14)	94-3001-76
$C_{15}H_{19}N_3O_3S_2$ 1,2,4-Triazine, 3,5-bis(methylthio)-6-[(3,4,5-trimethoxyphenyl)methyl]-	MeOH	220(4.27),332(3.77)	39-2521-76C
$C_{15}H_{19}N_3O_5$ 3,10,11-Triazatetracyclo[7.3.0.0^{2,4}-0^{5,7}]dodec-10-ene-7,9-dicarboxylic acid, 3-acetyl-, dimethyl ester, cis	CH_2Cl_2	325(2.34)	24-3505-76
$C_{15}H_{19}N_3O_6$ 8a(1H)-Quinazolinecarboxylic acid, 8-acetoxy-2-(acetylamino)-4,4a,5,8-tetrahydro-6-methyl-4-oxo-, methyl ester, isomer A	EtOH	254(3.81)	44-2124-76
isomer C	EtOH	254(3.98)	44-2124-76
8a(1H)-Quinazolinecarboxylic acid, 8-acetoxy-2-(acetylamino)-4,4a,5,8-tetrahydro-7-methyl-4-oxo-, methyl ester	EtOH	207(4.16),255(3.81)	44-2124-76

Compound	Solvent	$\lambda_{max}(\log \epsilon)$	Ref.
$C_{15}H_{19}N_3O_9S$ Acetamide, N-[2,3-dihydro-3-oxo-2-(2,3,5- tri-O-acetyl-β-D-ribofuranosyl)-1,2,4- thiadiazol-5-yl]-	pH 1 pH 7 and 11	236(4.08),275s(3.60) 256(4.11),278s(3.81)	4-0169-76 4-0169-76
$C_{15}H_{19}N_3S$ [1]Benzothiopyrano[4,3-c]pyrazole-1(4H)- propanamine, N,N-dimethyl-	EtOH	225(4.07),245s(4.19), 254(4.24),280s(3.80), 318(3.49)	4-0225-76
$C_{15}H_{19}N_4S$ Methanaminium, N-[[[2-(dimethylamino)- 5-phenyl-6H-1,3-thiazin-6-ylidene]- amino]methylene]-N-methyl-, perchlor- ate	HOAc	301(4.24),451(4.44)	97-0268-76
$C_{15}H_{19}N_5O$ Methanimidamide, N,N-dimethyl-N'-[3- [(1-methylethylidene)hydrazino]- 1-oxo-2(1H)-isoquinolinyl]-	EtOH	229(4.41),315(4.18), 374(3.73)	95-0700-76
$C_{15}H_{19}N_5OS$ 1H-Imidazo[1,2-a][1,3,5]benzotriazepine- 8-carboxamide, N-butyl-2,3,5,6-tetra- hydro-5-thioxo-	DMSO	263(4.52),284(4.55), 312s(4.12)	78-0057-76
$C_{15}H_{19}N_5O_5$ 3,7,8,12,13-Pentaazatetracyclo[9.3.0- $0^{2,4}.0^{5,9}$]tetradeca-7,12-diene-9,11- dicarboxylic acid, 3-acetyl-, dimeth- yl ester, cis	MeOH	323(2.52)	24-3505-76
4(1H)-Pteridinone, 1-[2,3-O-(1-methyl- ethylidene)-β-D-ribofuranosyl]- 2-(methylamino)-	pH -1.0 pH 4.0 2M NaOH	220(4.21),233(4.17), 313(3.90) 237(4.15),280s(3.72), 320(3.94) 223(4.30),260(3.85), 363(3.82)	24-3159-76 24-3159-76 24-3159-76
$C_{15}H_{19}N_5O_6$ 3,7,8,12,13-Pentaazatetracyclo[9.3.0- $0^{2,4}.0^{5,9}$]tetradeca-7,12-diene- 3,9,11-tricarboxylic acid, trimethyl ester, cis	EtOH	323(2.54)	24-3505-76
$C_{15}H_{19}N_7O_2$ 9,10-Propanopyrimido[4',5':3,4]azeto- [2,1-f]purine-6,8(5aH,7H)-dione, 4-amino-9a,9b-dihydro-5a-propyl-	H_2O	228(4.26),285(3.83)	19-0517-76
2,4(1H,3H)-Pyrimidinedione, 1-[3-(6-am- ino-9H-purin-9-yl)propyl]-5-propyl-	H_2O	264(4.28)	19-0517-76
$C_{15}H_{20}$ Benzene, (3-ethyl-5,5-dimethyl-1-cyclo- penten-1-yl)-	C_6H_{12}	247(3.98)	35-6218-76
Benzene, (2,3,5,5-tetramethyl-1-cyclo- penten-1-yl)-	C_6H_{12}	235(3.65)	35-6218-76
$C_{15}H_{20}NO$ 1-Azulenemethanaminium, 3-methoxy- N,N,N-trimethyl-, iodide	EtOH	286(4.61),360(3.62), 377(3.62),697(2.88)	44-1811-76

Compound	Solvent	$\lambda_{max}(\log \epsilon)$	Ref.
$C_{15}H_{20}N_2$			
Cyclopenta[b]pyrimido[1,2-a]indole, 1,2,3,4,5,6,7,12b-octahydro-5-methyl-	EtOH	248(4.07),303(3.40)	44-3775-76
1H-Indole-2-carbonitrile, 1-(5-hexenyl)-4,5,6,7-tetrahydro-	EtOH	270(4.13)	20-0819-76
$C_{15}H_{20}N_2O$			
Acetamide, N-[2-(2,2-dimethylpropyl)-1H-indol-3-yl]-	MeOH	222(4.52),281(3.93), 288(3.88)	23-1020-76
Spiro[3H-indole-3,3'-piperidin]-2(1H)-one, 1'-propyl-	MeOH	250(3.82),268s(3.17)	87-0892-76
$C_{15}H_{20}N_2O_2$			
Pyrrolo[1,2-a]pyrimidine-8-carboxylic acid, 2,4,7-trimethyl-, 1,1-dimethylethyl ester	MeOH	232(4.46),258(4.24), 266(4.27),296(3.84), 307(3.88),347(3.72)	24-2983-76
$C_{15}H_{20}N_2O_4$			
2,6-Piperidinedione, 4-[2-amino-2-(4,4-dimethyl-2,6-dioxocyclohexylidene)-ethyl]-	EtOH	341(4.16),399(4.08)	30-0139-76
$C_{15}H_{20}N_2O_4S$			
7-Methoxyisochromanylideneethylisothiouronium acetate	95% MeOH	221(4.09),258s(3.93), 273(4.19),298s(3.67)	13-0197-76B
$C_{15}H_{20}N_4O_6$			
1H-Purine-2,6-dione, 7-(3-O-acetyl-4,6-dideoxy-α-L-ribo-hexopyranosyl)-3,7-dihydro-1,3-dimethyl-	MeOH	274(3.87)	44-3827-76
$C_{15}H_{20}N_4S$			
2,3,6-Pyridinetriamine, 4-[[(2,4,6-trimethylphenyl)methyl]thio]-	pH 1	242(4.36),270(4.18), 333(3.98)	44-3784-76
$C_{15}H_{20}N_5O_7PS$			
Adenosine, 8-(methylthio)-, cyclic 3',5'-(hydrogen phosphate) 2'-butanoate	pH 7	278(4.27)	87-0899-76
Adenosine, 8-(methylthio)-N-(1-oxobutyl)-, cyclic 3',5'-(hydrogen phosphate), monosodium salt	pH 7	295(4.28)	87-0899-76
$C_{15}H_{20}O$			
Cyclohexanone, 3-methyl-2-[(2-methylphenyl)methyl]-	EtOH	264(2.53),271(2.49)	78-0065-76
2-Naphthalenol, 5,6-dihydro-3,8-dimethyl-5-(1-methylethyl)-	ether	221(4.42),265(3.89), 273s(3.81),304(3.68), 314s(3.63)	24-2021-76
$C_{15}H_{20}O_2$			
Anhydrolactarorufin N	EtOH	211(3.91),285(3.17)	56-2095-76
2(3H)-Benzofuranone, 6-ethenylhexahydro-6-methyl-3-methylene-7-(1-methylethenyl)-	EtOH	211.5(3.94)	78-0765-76
2-Butenoic acid, 3-(4-methylphenyl)-, 2-methylpropyl ester	EtOH	272(4.17)	2-0472-76
2,6-Naphthalenedione, 3,4,4a,5-tetrahydro-1,4a-dimethyl-7-(1-methylethyl)-(8-oxo-β-cyperone)	ether	304(4.23)	102-1075-76

Compound	Solvent	$\lambda_{max}(\log \epsilon)$	Ref.
Naphtho[1,2-c]furan-1(3H)-one, 4,5,5a,6-7,8,9,9a-octahydro-7,9a-dimethyl-6-methylene- (colorata-4(13),8-dienolide)	EtOH	219.5(4.23)	39-0850-76C
Naphtho[2,3-b]furan-2(3H)-one, 3a,4,4a-5,6,7,9,9a-octahydro-4a,5-dimethyl-3-methylene-	MeOH	205(4.03)	94-1419-76
$C_{15}H_{20}O_3$			
Azuleno[4,5-b]furan-2,8(3H,4H)-dione, 3a,5,6,9,9a,9b-hexahydro-3,6,9-trimethyl- (hypochaerin)	EtOH	229(4.2)	102-0991-76
Colorat-8-enolide, 4α,13-epoxy-	EtOH	219(4.07)	39-0850-76C
Colorat-8-enolide, 7-oxo-	EtOH	248.5(4.07)	39-0850-76C
Cyclopentaneacetic acid, 2-hydroxy-1-methyl-5-phenyl-, methyl ester, (1α,2β,5β)-	EtOH	244(2.08),249(2.18), 254(2.28),256(2.26), 260(2.34),264(2.20), 266(2.23)	104-0049-76
2H-Cyclopropa[f]benzofuran-2-one, octahydro-5a-methyl-3-methylene-5-(3-oxobutyl)- (carabron)	EtOH	211(3.91)	83-0333-76
2-Naphthaleneacetic acid, 1,2,3,4,4a,7-hexahydro-α,4a,8-trimethyl-, [2R-[2α(S*),4aα]]-	EtOH	241(3.99)	44-1256-76
1(2H)-Naphthalenone, 3,4-dihydro-2-hydroxy-3-(1-hydroxy-1-methylethyl)-5,8-dimethyl-, (2R-trans)-	EtOH	259(4.05),309(3.45)	102-1267-76
Nehipetol, dihydro-	EtOH	207(4.22),270(3.55)	56-1931-76
A/B cis	EtOH	207(4.21),269(3.69)	56-1931-76
isomer 7	EtOH	207(4.21),269(3.26)	56-1931-76
$C_{15}H_{20}O_3S$			
Bicyclo[3.2.1]octan-3-ol, 4-methylbenzenesulfonate, endo	EtOH	257(2.69),262(2.78), 267(2.73),273(2.68)	39-1506-76B
exo	EtOH	257(2.69),262(2.79), 267(2.74),273(2.69)	39-1506-76B
Sulfonium, dimethyl-, 1-(ethoxycarbonyl)-2-oxo-4-phenylbutylide	EtOH	228(3.89),261(4.12)	39-1688-76C
$C_{15}H_{20}O_4$			
2H-1-Benzopyran-2-acetic acid, 3,4-dihydro-6-hydroxy-2,5,7,8-tetramethyl-, (±)-	EtOH	225(4.05),292(3.50)	33-0290-76
2H-1-Benzopyran-2,6-diol, 3,4-dihydro-2,5,7,8-tetramethyl-, 6-acetate, (±)-	EtOH	223s(4.00),276(3.18), 283(3.24)	33-0290-76
Budlein B	MeOH	224(3.86)	102-0525-76
5αH-4,6,11β-Eudesman-6,13-olide, 2,3-dioxo-	EtOH	281.5(3.96)	2-0657-76
Granilin	EtOH	210(3.97)	102-1531-76
Nehipediol	EtOH	207(4.20),271(3.45)	56-1931-76
Pterosin W	MeOH	217(4.49),260(4.00), 303(3.12)	94-1961-76
5α-Santanolide, 1α,2α-epoxy-3-oxo-	EtOH	299(1.51)	94-2810-76
5α-Santanolide, 1β,2β-epoxy-3-oxo-	EtOH	296(1.51)	94-2810-76
$C_{15}H_{20}O_5$			
4,10-Cyclodecadiene-1,4-dicarboxylic acid, 5-methyl-9-oxo-, dimethyl ester, (E,Z)-	EtOH	225(3.96)	88-4409-76

Compound	Solvent	$\lambda_{max}(\log \epsilon)$	Ref.
1-Propanone, 1-(2-acetoxy-4,6-dimethoxy-3-methylphenyl)-2-methyl- (baeckeol acetate	MeCN	216(4.2),261(3.7), 286(3.6)	5-1514-76
$C_{15}H_{20}O_6$ Benzenemethanol, 4-acetoxy-α-ethyl-3,5-dimethoxy-, acetate	MeOH	226(3.98),276(2.99)	18-1940-76
$C_{15}H_{20}O_8$ Ethanone, 1-[4-(β-D-glucopyranosyloxy)-3-methoxyphenyl]-	MeOH	268(3.95),300(3.8)	105-0099-76
$C_{15}H_{21}Br_3O_2$ 1,3-Benzenediol, 2,4,6-tribromo-5-nonyl-	MeOH	262(3.21),295(3.35)	12-1989-76
$C_{15}H_{21}Cl_6Sb$ Cyclopropenylium, tris(1-methyl-1-propenyl)-, hexachloroantimonate	CH_2Cl_2	263(4.40),283(3.46)	44-2258-76
Cyclopropenylium, tris(2-methyl-1-propenyl)-, hexachloroantimonate	CH_2Cl_2	248(4.26),306(4.56), 315(4.56)	44-2258-76
$C_{15}H_{21}NO$ 1H-Indole-2-carboxaldehyde, 1-(5-hexenyl)-4,5,6,7-tetrahydro-	EtOH	264(3.91),312(4.24)	20-0819-76
$C_{15}H_{21}NO_2$ 2,5-Cyclohexadiene-1,4-dione, 2,3,5-trimethyl-6-(1-piperidinylmethyl)-, nitrate	H_2O	261(4.21),350(2.51)	12-1163-76
2-Propanol, 1-(1H-inden-4-yloxy)-3-[(1-methylethyl)amino]-, hydrochloride	H_2O	251(4.05),259(3.93), 292(3.35),302(3.29)	94-0552-76
2-Propanol, 1-(1H-inden-7-yloxy)-3-[(1-methylethyl)amino]-, hydrochloride	H_2O	249(3.93),287(3.05), 296(2.98)	94-0552-76
$C_{15}H_{21}NO_3$ Acetamide, N-[4-(heptyloxy)phenyl]-2-oxo- (hydrate)	EtOH	254(3.12)	65-0702-76
$C_{15}H_{21}NO_4$ 1-Isoquinolineacetic acid, 1,2,3,4-tetrahydro-6,7-dimethoxy-, ethyl ester	n.s.g.	225(4.20),285(3.05)	2-0784-76
$C_{15}H_{21}NO_6$ β-D-Glucopyranoside, ethyl 2-benzoylamino-2-deoxy-	H_2O	230(3.81)	44-4038-76
2H-Pyran-3(4H)-one, 5,6-dihydro-2,5α-dimethoxy-6,6-dimethyl-4-(5-methyl-2-pyrrolylcarbonyloxy)-5β-	EtOH	230(3.55),278(4.32)	78-2057-76
5β-	EtOH	229(3.57),279(4.21)	78-2057-76
$C_{15}H_{21}N_3$ 2,4-Quinolinediamine, 3-ethyl-N,N,N',N'-tetramethyl-	EtOH	252(4.37),327(3.85)	1-0133-76
2,4-Quinolinediamine, N,N,N',N',7,8-hexamethyl-	EtOH	255(4.58),345(3.71)	1-0133-76
$C_{15}H_{21}N_3OS$ Carbamothioic acid, methyl-, S-(1,2,3-3a,8,8a-hexahydro-1,3a,8-trimethylpyrrolo[2,3-b]indol-5-yl) ester (thioeserin)	MeOH	269(3.73),308(3.74)	83-0631-76

Compound	Solvent	$\lambda_{max}(\log \epsilon)$	Ref.
$C_{15}H_{21}N_3O_4$ Benzoic acid, 3-nitro-4-[[2-(1-pyrrolidinyl)ethyl]amino]-, ethyl ester, hydrochloride	EtOH	262(4.26),286(4.24), 401(3.67)	78-0057-76
$C_{15}H_{21}N_3O_5$ 2,2'-Anhydro-1-(3'-hexanoyl-β-D-arabinofuranosyl)cytosine, hydrochloride	MeOH-acid	230(3.98),263(4.06)	87-0654-76
$C_{15}H_{21}N_3O_7$ Cytidine, N-acetyl-2',3'-di-O-methyl-, 5'-acetate	H_2O	214(4.29),247(4.26), 297(3.96)	35-7381-76
$C_{15}H_{21}N_7O_6$ Adenosine, 5'-[(4-amino-4-carboxy-1-oxobutyl)amino]-5'-deoxy-, (S)-	H_2O	259(4.15)	87-0684-76
$C_{15}H_{22}$ Cyclohexene, 1,5,5-trimethyl-6-(3-methyl-2,4-pentadienylidene)-	EtOH	300(4.71),312(4.79), 325(4.68)	33-0387-76
Naphthalene, 1,2,3,4,4a,8a-hexahydro-4a,8-dimethyl-2-(1-methylethenyl)-, [2R-(2α,4aβ,8aα)]-	MeOH	267(3.56)	44-3847-76
Spiro[4.5]deca-1,6-diene, 6,10-dimethyl-2-(1-methylethenyl)-, cis-(±)-	EtOH	238(4.31)	44-1539-76
Spiro[4.5]deca-2,6-diene, 6,10-dimethyl-3-(1-methylethenyl)-, cis-(±)-	EtOH	237(4.21)	44-1539-76
$C_{15}H_{22}ClN_4O_4P$ Benzamide, 2-amino-5-chloro-N-(di-4-morpholinylphosphinyl)-	$CHCl_3$	239(3.86),249(3.95), 354(3.69)	44-2720-76
$C_{15}H_{22}N_2O$ Cyclopentanone, 2,5-bis[3-(dimethylamino)-2-propenylidene]-	EtOH	255(4.10),505(4.91)	70-0577-76
$C_{15}H_{22}N_2O_2$ 2(1H)-Pyridinone, 6-[octahydro-1-(hydroxymethyl)-2H-quinolizin-3-yl]-, (1 ,3 ,9a)-(±)- (mamanine)	MeOH MeOH-acid	233(3.86),308(3.84) 219(--),289(--)	78-0919-76 78-0919-76
$C_{15}H_{22}N_2O_3$ Imidazolidinetrione, dicyclohexyl-	MeOH	223(3.18),263s(2.70)	44-1836-76
$C_{15}H_{22}N_2O_5$ 2,5-Cyclohexadiene-1,4-dione, 2,3,5-trimethyl-6-(1-piperidinylmethyl)-, nitrate	H_2O	261(4.21),350(2.51)	12-1163-76
D-Mannitol, 1-deoxy-1-(5,6-dimethyl-1H-benzimidazol-1-yl)-, monohydrochloride	H_2O	257s(3.76),271s(3.86), 277(3.95),286(3.93)	136-0049-76G
$C_{15}H_{22}N_2O_5S$ 1H-Pyrrole-1,3-dicarboxylic acid, 4-(acetylamino)-5-[2-(methylthio)-ethyl]-, diethyl ester	EtOH	214.5(4.47)	4-0113-76
$C_{15}H_{22}N_2O_7S$ Uridine, 2'-O-(tetrahydro-4-methoxy-2H-pyran-4-yl)-4-thio-	EtOH	247(3.56),329(4.26)	54-0108-76

Compound	Solvent	$\lambda_{max}(\log \epsilon)$	Ref.
$C_{15}H_{22}N_3O_2$			
1-Piperidinyloxy, 2,2,6,6-tetramethyl-4-[(3-pyridinylcarbonyl)amino]-	H_2O pH 13	262(3.75) 270(3.75)	103-0300-76 103-0300-76
$C_{15}H_{22}N_4$			
Pyrido[3,2-e]-1,2,4-triazine, 3-nonyl-	C_6H_{12}	263(3.19),311(3.53), 323(3.45),498(2.29)	114-0301-76C
Pyrido[3,4-e]-1,2,4-triazine, 3-nonyl-	EtOH	233(4.32),341(3.45)	114-0285-76C
$C_{15}H_{22}N_4O_2$			
Benzamide, 4-amino-5-cyano-2-methoxy-N-[2-(diethylamino)ethyl]-, hydrochloride	H_2O	235(4.66),282(4.24), 311s(3.73)	133-0314-76
$C_{15}H_{22}N_4O_5S_2$			
D-Ribose, 5-deoxy-5-(3,4-dihydro-2,4-dioxo-8(2H)-pteridinyl)-, diethyl mercaptal	pH -3.0	243(4.09),275(3.50), 348(3.98)	24-3175-76
	pH 6.0	257(4.17),275s(3.92), 398(3.94)	24-3175-76
	pH 12.0	231(4.23),278(4.13), 313(3.90)	24-3175-76
$C_{15}H_{22}N_5O_6PS$			
Adenosine, 8-(pentylthio)-, cyclic 3',5'-(hydrogen phosphate)	pH 2.0 pH 13	284(4.29) 282(--)	69-1408-76 69-1408-76
$C_{15}H_{22}N_6O_2$			
2(1H)-Pyrimidinone, 1,1'-(1,3-propanediyl)bis[4-amino-5-ethyl-	pH 1 pH 7.0	291.5(--) 280.0(4.16)	56-1805-76 56-1805-76
$C_{15}H_{22}N_6O_4$			
9H-Purine, 6-(4-methyl-1-piperazinyl)-9-β-D-ribofuranosyl-	MeOH	214(4.27),277(4.44)	5-0745-76
$C_{15}H_{22}O$			
Bicyclo[7.2.0]undec-4-en-3-one, 4,11,11-trimethyl-8-methylene-, (4E)-	MeOH	235(3.67)	88-3717-76
(4Z)-	MeOH	236(4.04)	88-3717-76
Germazone	EtOH	254(3.98)	88-3833-76
2,5-Heptadien-4-one, 2-methyl-6-(4-methyl-3-cyclohexen-1-yl)-, (E)-	EtOH	270(4.23)	107-0515-76
Occidol, (+)-	EtOH	224s(4.16),266(2.95)	102-1267-76
$C_{15}H_{22}OS$			
1-Hexanethione, 1-(4-methoxyphenyl)-2,2-dimethyl-	C_6H_{12}	225(4.05),236(4.03), 334(3.96),568(2.32)	35-6218-76
$C_{15}H_{22}O_2$			
Bicyclo[7.2.0]undec-4-en-3-one, 2-hydroxy-4,11,11-trimethyl-8-methylene-	MeOH	237(3.97)	88-3717-76
2H-Cyclopenta[a]pentalen-2-one, 1,3,3a,3b,4,5,6,6a,7,7a-decahydro-3a-hydroxy-4,4,6a-trimethyl-3-methylene-	MeOH	226(3.79)	78-1171-76
2H-Cyclopenta[a]pentalen-2-one, 1,3b,4,5,6,6a,7,7a-octahydro-5-hydroxy-3,4,4,6a-tetramethyl-	MeOH	243(4.16)	78-1171-76
2H-Cyclopenta[a]pentalen-2-one, 1,3b,4,5,6,6a,7,7a-octahydro-6-hydroxy-3,4,4,6a-tetramethyl-	MeOH	242.5(4.09)	78-1171-76

Compound	Solvent	$\lambda_{max}(\log \epsilon)$	Ref.
2H-Cyclopenta[a]pentalen-2-one, 1,3b,4,5,6,6a,7,7a-octahydro-7-hydroxy-3,4,4,6a-tetramethyl-	MeOH	244(4.10)	78-1171-76
5αH,4,6,11βH-Eudesm-2-en-6,13-olide	EtOH	204(2.86)	2-0157-76
2-Naphthalenemethanol, 1,2,3,4-tetra-hydro-3-hydroxy-α,α,5,8-tetramethyl-	EtOH	223(4.05),272(2.70)	102-1267-76
Naphtho[1,2-c]furan-1(3H)-one, 4,5,5a,6,7,8,9,9a-octahydro-6,6,9a-trimethyl- (isodrimenin)	EtOH	218.0(4.00)	39-0850-76C
Naphtho[1,2-c]furan-1(3H)-one, 4,5,5a,6,7,8,9,9a-octahydro-6,7,9a-trimethyl- (colorat-8-enolide)	EtOH	217.0(4.15)	39-0850-76C
Naphtho[2,3-b]furan-2(4H)-one, 4a,5,6,7,8,8a,9,9a-octahydro-3,4a,5-trimethyl-	MeOH	222(5.36)	94-1419-76
2-Penten-4-ynal, 5-(1-hydroxy-2,2,6-trimethylcyclohexyl)-3-methyl-	isoPrOH	277(4.33)	33-0567-76
Spiro[furan-3(2H),2'-[2H]inden]-2-one, decahydro-3'a,4'-dimethyl-4-methyl-ene- (bakkenolide A)	EtOH	212(3.01)	73-2047-76
5,7,9,11-Tetradecatetraenoic acid, methyl ester	ether	289(4.53),303(4.72), 318(4.68)	102-1318-76
$C_{15}H_{22}O_3$			
Blennin A	EtOH	222(3.85)	102-1953-76
Capnell-$\Delta^{9(12)}$-ene-2ξ,10α-diol-8-one	MeOH	225(3.74)	78-1171-76
Capnell-$\Delta^{9(12)}$-ene-5α,10α-diol-8-one	MeOH	225(3.71)	78-1171-76
Coloratanolide, 7-oxo-	EtOH	283.0(1.51)	39-0850-76C
Eremophil-7(11)-en-12,8α-olide, 6β-hydroxy-	EtOH	218(4.16)	18-3196-76 +94-0360-76
Lactororufin N	EtOH	227(3.89)	56-2095-76
Nehipetol, tetrahydro-	EtOH	274(4.00)	56-1931-76
2-Oxa-1,2,3,5a,6,7,8,9a,9b-decahydro-dibenzofuran-3-one, 1,1,9,9a-tetra-methyl-	EtOH	249(4.04),331(2.06)	83-0638-76
$C_{15}H_{22}O_4$			
2H-Cyclopenta[b]furan-2-one, hexahydro-4-hydroxy-5-(3-oxo-1-octenyl)-	n.s.g.	227(4.14)	88-1753-76
Eremophil-7(11)-en-12,8α-olide, 6β,8β-dihydroxy-	EtOH	219(4.07)	18-3196-76
Naphtho[2,3-b]furan-2(4H)-one, 4a,5,6,7,8,8a,9,9a-octahydro-4,9a-dihydroxy-3,4a,5-trimethyl-	EtOH	219(4.07)	94-0360-76
5α-Santanolide, 1α-hydroxy-3-oxo-	EtOH	285(1.36)	94-2810-76
5α-Santanolide, 1β-hydroxy-3-oxo-	EtOH	286.5(1.51)	94-2810-76
$C_{15}H_{22}O_5$			
4-Cyclodecene-1,4-dicarboxylic acid, 5-methyl-9-oxo-, dimethyl ester, (Z)-	EtOH	212(3.74)	88-4409-76
isomer	EtOH	215(3.71)	88-4409-76
$C_{15}H_{22}O_7S$			
Spiro[2H-pyran-2,2'(3'H)-thieno[3,2-g]-benzofuran]-3'-one, dodecahydro-3'a,4,4',6'-tetrahydroxy-5-methyl- (breynolide)	MeOH MeOH-NaOMe	none 349(3.85)	94-0114-76 94-0114-76
$C_{15}H_{22}O_8$			
β-D-Glucopyranoside, 1,4a,5,7a-tetra-	EtOH	209(3.2)	32-0725-76

Compound	Solvent	$\lambda_{max}(\log \epsilon)$	Ref.
hydro-7-(hydroxymethyl)cyclopenta[c]-pyran-1-yl (bartsioside) (cont.)			32-0725-76
$C_{15}H_{23}NO$			
Oxaziridine, 3-phenyl-2-(1,1,3,3-tetra-methylbutyl)-	EtOH	250(2.97)	4-1001-76
2-Pentanamine, 2,4,4-trimethyl-N-(phen-ylmethylene)-, N-oxide	EtOH	297(4.21)	4-1001-76
$C_{15}H_{23}NO_6$			
α-L-lyxo-Hexopyranoside, methyl 6-deoxy-5-C-methyl-4-O-methyl-, 3-(5-methyl-1H-pyrrole-2-carboxylate)	EtOH	230(3.58),279(4.33)	78-2057-76
β-	EtOH	230(3.54),279(4.32)	78-2057-76
α-L-xylo-Hexopyranoside, methyl 6-deoxy-5-C-methyl-4-O-methyl-, 3-(5-methyl-1H-pyrrole-2-carboxylate)	EtOH	232(3.51),278(3.27)	78-2057-76
β-	EtOH	231(3.56),278(4.33)	78-2057-76
$C_{15}H_{23}N_3O_4S$			
Benzamide, 5-(aminosulfonyl)-N-[(1-eth-yl-2-pyrrolidinyl)methyl]-2-methoxy-, monohydrochloride	H_2O	224(4.55),279(4.24), 303s(3.88)	133-0314-76
$C_{15}H_{23}N_3O_6$			
2(1H)-Pyrimidinone, 4-amino-1-[3-O-(1-oxohexyl)-β-D-arabinofuranosyl]-	MeOH-acid	213(3.99),283(4.16)	87-0667-76
$C_{15}H_{23}N_4O_3$			
1-Piperidinyloxy, 4-hydroxy-2,2,6,6-tetramethyl-4-[2-(3-pyridinylcarbo-nyl)hydrazino]-	H_2O pH 13	262(4.10) 270(3.96)	103-0300-76 103-0300-76
$C_{15}H_{23}N_4O_4P$			
Benzamide, 2-amino-N-(di-4-morpholinyl-phosphinyl)-	M HCl MeCN	230(4.07),270s(3.24) 219(4.42),250(3.90), 338(3.72)	44-2720-76 44-2720-76
$C_{15}H_{23}N_7O_4$			
9H-Purin-2-amine, 6-(4-methyl-1-pipera-zinyl)-9-β-D-ribofuranosyl-	MeOH	232(4.4),255(4.03), 261(4.02),288(4.2)	5-0745-76
$C_{15}H_{24}$			
Epizonarene, (+)-	MeOH	245(4.57)	2-0901-76
1H-Indene, 5-ethenyl-2,3,3a,4,7,7a-hexa-hydro-1,3a,4,7a-tetramethyl-	n.s.g.	243(4.32)	88-3619-76
1H-Indene, 2,3,3a,4-tetrahydro-3,3a,6-trimethyl-1-(1-methylethyl)- (casca-rillediene)	n.s.g.	264(3.83)	22-0088-76
Naphthalene, decahydro-1,1,4a-trimethyl-5,6-bis(methylene)-, (4S-trans)-	EtOH	216(3.78)	33-0075-76
$C_{15}H_{24}BrNO$			
Cyclohexanaminium, N,N,N-trimethyl-2-phenoxy-, bromide, cis	H_2O	263s(2.98),269(3.08), 276(2.99)	87-0692-76
trans	H_2O	264s(2.99),270(3.09), 276(3.00)	87-0692-76
$C_{15}H_{24}N_2O_4S$			
Benzamide, N-[2-(diethylamino)ethyl]-	H_2O	211(4.59),235(4.13),	133-0314-76

Compound	Solvent	λ_{max}(log ϵ)	Ref.
2-methoxy-5-(methylsulfonyl)-, hydro- chloride (cont.)		288(3.35)	133-0314-76
$C_{15}H_{24}N_2O_5S$ 1H-Pyrrole-1,3-dicarboxylic acid, 4-(acetylamino)-2,5-dihydro-5- [2-(methylthio)ethyl]-, diethyl ester	EtOH	275(4.15)	4-0113-76
$C_{15}H_{24}N_3$ Methanaminium, N-[1,3-bis(dimethylami- no)-2-phenyl-2-propenylidene)-N- methyl-, chloride	H_2O	223(3.93),275(4.24), 345(4.10)	88-2405-76
$C_{15}H_{24}N_4$ Pyrido[3,2-e]-1,2,4-triazine, 1,2-di- hydro-3-nonyl-, hydrochloride	MeOH	383(3.39),397s(--)	114-0301-76C
$C_{15}H_{24}N_4O_3$ Decanoic acid, 2-(3-nitro-2-pyridinyl)-, hydrazide	EtOH	256(3.70),376(3.69)	114-0301-76C
Decanoic acid, 2-(3-nitro-4-pyridinyl)-, hydrazide	EtOH	260(4.10),336(3.60)	114-0285-76C
$C_{15}H_{24}N_5O_6P$ D-erythro-Pentofuranoside, methyl 5-(6- amino-3H-purin-3-yl)-2,5-dideoxy-, 3-(diethyl phosphate)	EtOH	274(4.26)	102-1523-76
$C_{15}H_{24}O$ Acoragermacrone	MeOH	242(3.82)	100-0412-76
Acorenone B	MeOH	241(3.86)	24-2805-76
4-Azulenol, 1,2,3,3a,4,5,6,8a-octahydro- 4,8-dimethyl-2-(1-methylethenyl)-	hexane	248(4.36),283(4.58), 291(4.63),310(3.62), 336(3.48),350(3.56)	77-0034-76
4(1H)-Azulenone, 2,5,6,7,8,8a-hexahydro- 3,8-dimethyl-5-(1-methylethyl)-, [5S- (5α,8α,8aβ)]-	MeOH	256(3.92)	18-3148-76
2-Cyclohexen-1-one, 4,4,6-trimethyl- 6-(3-methyl-2-methylenebutyl)-	C_6H_{12}	222(4.08),343(1.76)	33-2244-76
2-Cyclopenten-1-one, 2-[[2-methyl-5-(1- methylethyl)cyclopentyl]methyl]-	n.s.g.	235(3.97)	39-1762-76C
4-Epiacorenone	MeOH	241(3.80)	24-2805-76
1-Naphthalenol, 1,2,3,7,8,8a-hexahydro- 1,6-dimethyl-4-(1-methylethyl)-	MeOH	241(4.04),247(4.06)	18-3148-76
stereoisomer	MeOH	241(4.20),246(4.21)	18-3148-76
2(1H)-Naphthalenone, 3,4,4a,5,6,7-hexa- hydro-3,7-dimethyl-4a-(1-methylethyl)-, (4aR,7R)-(-)-	dioxan	297(2.00)	22-0957-76
(4aS,7R)-(+)-	dioxan	298(1.93)	22-0957-76
2(3H)-Naphthalenone, 4,4a,5,6,7,8a-hexa- hydro-3,7-dimethyl-4a-(1-methylethyl)-, (3R,4aS,7R)-(+)-	dioxan	325(1.99)	22-0957-76
(3S,4aR,7R)-(-)-	dioxan	326(1.95)	22-0957-76
7-Oxabicyclo[4.1.0]heptane, 1-(2,3-di- methyl-1,3-butadienyl)-2,2,6-tri- methyl-, (E)-	n.s.g.	234(4.40)	33-0907-76
Spiro[4.5]dec-8-en-7-one, 4,8-dimethyl- 1-(1-methylethyl)- (4-epiacorenone B)	MeOH	240(3.83)	24-0041-76
Tetradeca-4,8,11,13-tetraen-2-ol, 2-methyl-	EtOH	229(4.93)	18-3646-76

Compound	Solvent	$\lambda_{max}(\log \epsilon)$	Ref.
$C_{15}H_{24}O_2$			
1,3-Benzenediol, 5-nonyl-	MeOH	223(3.82),276(3.58), 281(3.59)	12-1989-76
2(3H)-Benzofuranone, 6-ethylhexahydro-6-methyl-3-methylene-7-(1-methylethyl)-, [3aS-(3aα,6α,7β,7aβ)]-	EtOH	213(3.80)	78-0765-76
α-Bisabololone	EtOH	242(4.01)	107-0515-76
Butanoic acid, 3-methyl-, 3,7-dimethyl-2,4,6-octatrienyl ester, (Z,E)-	EtOH	280(4.44)	78-0107-76
Cyclohexanol, 1-(5-hydroxy-3-methyl-3-penten-1-ynyl)-2,2,6-trimethyl-, (E)-	isoPrOH	228(4.15)	33-0567-76
diastereomer	isoPrOH	228(4.18)	33-0567-76
Ethanone, 1-[4,5-dihydro-2,5,5-trimethyl-4-(3-methyl-4-pentenyl)-3-furanyl]-	n.s.g.	275(3.25)	39-1762-76C
Isocentdarone	n.s.g.	245(3.79)	102-0557-76
1(2H)-Naphthalenone, 3,4,4a,5,6,7-hexahydro-5-hydroxy-4a,8-dimethyl-2-(1-methylethyl)-	MeOH	248(3.66)	18-3137-76
2-Oxabicyclo[4.4.0]deca-3,5-diene, 1,3,7,7-tetramethyl-4-(methoxymethyl)-	pentane	286(3.66)	33-0727-76
11-Oxabicyclo[5.3.1]undeca-4,6-dien-2-ol, 1,5-dimethyl-8-(1-methylethyl)-, (1α,2α,8α)-	MeOH	243(3.88)	18-3137-76
11-Oxabicyclo[5.3.1]undec-6-en-2-ol, 1-methyl-5-methylene-8-(1-methylethyl)-, (1α,2α,8α)-	MeOH	240(4.21)	18-3137-76
11-Oxabicyclo[8.1.0]undec-6-en-5-one, 1,7-dimethyl-4-(1-methylethyl)-, [1R-(1α,4α,10β)]-	MeOH	243(3.81)	18-3137-76
2H-Pyran-2-one, 5-(1,2-dimethylcyclohexyl)-5,6-dihydro-6,6-dimethyl-	EtOH	312(2.32)	83-0638-76
2,4,5-Tetradecatrienoic acid, methyl ester	hexane	252.5(4.36)	88-0275-76
$C_{15}H_{24}O_2S$			
2-Thiophenecarboxylic acid, 5-nonyl-, methyl ester	EtOH	254(3.91),275(4.03)	34-0233-76
$C_{15}H_{24}O_4$			
3H-Pyrano[4,3-b]benzofuran-3-one, decahydro-4a-hydroxy-1,1,9,9a-tetramethyl-	EtOH	277(2.20)	83-0638-76
$C_{15}H_{24}O_7S$			
Breynolide, dihydro-	MeOH	none	94-0114-76
	MeOH-NaOH	none	94-0114-76
$C_{15}H_{24}O_9$			
β-D-Glucopyranoside, 1,4a,5,6,7,7a-hexahydro-5,7-dihydroxy-4-methylcyclopenta[c]pyran-1-yl	MeOH	218(2.97)	32-0057-76
$C_{15}H_{25}ClN_4$			
Pyrido[3,2-e]-1,2,4-triazine, 1,2-dihydro-3-nonyl-, hydrochloride	MeOH	383(3.39),397s(--)	114-0301-76C
$C_{15}H_{25}NO$			
Cyclopentanone, 2-(6-pentyl-2-piperidinylidene)-	n.s.g.	331.0(4.26)	142-0445-76S

Compound	Solvent	$\lambda_{max}(\log \epsilon)$	Ref.
$C_{15}H_{25}N_6O_7P$ 5'-Adenylic acid, mono(2-aminopentyl) ester, (S)-	H_2O	267(4.23)	87-1279-76
$C_{15}H_{25}O_5P$ Phosphonic acid, [3-methyl-3-[(tetra-hydro-2H-pyran-2-yl)oxy]-4-penten-1-ynyl]-, diethyl ester	MeOH	205(2.99),236(2.35)	104-1544-76
$C_{15}H_{25}O_6P$ Phosphonic acid, [4-(1-hydroxy-1-methyl-2-propenyl)-2-propoxy-3-furanyl]-, diethyl ester	MeOH	223(3.90)	104-1544-76
$C_{15}H_{26}N_2O_4S$ tert-Butylammonium α-(5-methoxycarbonyl-2-methyl-4-thiazolin-3-yl)-β,β-dimethylacrylate	EtOH	210(4.06),302s(3.42), 353(4.06)	39-0584-76C
$C_{15}H_{26}N_4O$ Decanoic acid, 2-(3-nitro-2-pyridinyl)-hydrazide	MeOH	288(3.56)	114-0301-76C
$C_{15}H_{26}O$ 2(1H)-Naphthalenone, octahydro-3,7-di-methyl-4a-(1-methylethyl)-, [3R-(3α,4aβ,7β,8aβ)]-	dioxan	283(1.67)	22-0957-76
[3S-(3α,4aβ,7α,8aβ)]-	dioxan	283(1.73)	22-0957-76
2-Propenal, 3-cyclododecyl-	EtOH	231.0(4.00)	22-0849-76
$C_{15}H_{26}O_2$ 1,3-Cyclohexanedione, 5-nonyl-	MeOH	258(4.12),278(3.94)	12-1989-76
Nerolidol, (-)	MeOH	235s(2.38)	12-2023-76
$C_{15}H_{26}O_3$ 2-Cyclodecen-1-one, 6,7-dihydroxy-3,7-dimethyl-10-(1-methylethyl)-	MeOH	238(3.62)	18-3137-76
Cyclooctanone, 6-acetyl-7-ethoxy-2,2,6-trimethyl-	EtOH	288(1.74)	33-0001-76
2H-Pyran-2-one, 5-(1,2-dimethylcyclo-hexyl)tetrahydro-4-hydroxy-6,6-di-methyl-	EtOH	278(1.40)	83-0638-76
$C_{15}H_{27}NO_3S$ 2,6-Pyridinediol, 1-acetyl-4-(1,1-di-methylethyl)-3-[(1,1-dimethylethyl)-thio]-1,2,3,6-tetrahydro-	MeOH	209(3.86)	4-0861-76
$C_{15}H_{28}Si_2$ Silane, [(3,5-dimethylphenyl)methylene]-bis[trimethyl-	EtOH	266(2.64)	78-0051-76
Silane, [(5-methyl-1,3-phenylene)bis-(methylene)]-	EtOH	274(2.56)	78-0051-76
$C_{15}H_{30}Ge_3$ Germane, (1-methyl-1,2-pentadien-4-yne-1,3,5-triyl)tris[trimethyl-	C_6H_{12}	245(4.16),255(4.16)	35-8426-76
$C_{15}H_{30}Si_3$ Silane, (1-methyl-1,2-pentadien-4-yne-1,3,5-triyl)tris[trimethyl-	EtOH	235s(4.04),241(4.08), 252(4.10)	35-8426-76

Compound	Solvent	$\lambda_{max}(\log \epsilon)$	Ref.
$C_{15}H_{31}N_2O_5P$ Acetic acid, [(tributoxyphosphoranyli- dene)hydrazono]-, methyl ester	EtOH	272.0(4.11)	65-1424-76
$C_{15}H_{34}N_5O_2P$ Acetic acid, [[tris(diethylamino)phos- phoranylidene]hydrazono]-, methyl ester	EtOH	296.5(4.05)	65-1424-76

Compound	Solvent	$\lambda_{max}(\log \epsilon)$	Ref.

$C_{16}H_2N_8$
  1,2,4,5-Benzenetetracarbonitrile,
  3,6-bis(dicyanomethyl)-, compound
  with pyridine (1:2)
	MeCN	412(4.27),710(3.23)	24-2469-76

$C_{16}H_5Cl_2F$
  Benzene, (10,10-dichloro-9-fluoro-
  9-decene-1,3,5,7-tetraynyl)-
	n.s.g.	209(4.52),277(4.89),
		292(5.06),301s(--),
		315(4.88),325s(--),
		349(4.32),376(4.37),
		407(4.16)
			44-1487-76

$C_{16}H_7ClN_2$
  9,10-Phenanthrenedicarbonitrile,
  3-chloro-
	50% EtOH	219(4.35),243(4.66),
		254(4.58),263(4.61),
		274(4.40),332(4.18),
		343(4.24),383(3.30)
			18-2224-76

$C_{16}H_7ClO_3$ Anthra[9,1-bc]pyran-2,7-dione, 3-chloro-	n.s.g.	342(3.60),532(3.95), 565(4.00),602s(3.75)	104-1988-76
Anthra[9,1-bc]pyran-2,7-dione, 6-chloro-	n.s.g.	352(3.58),543(4.03), 580(4.09),630(3.84)	104-1988-76

$C_{16}H_7Cl_2N_3O_2$
  2,4-Cyclohexadien-1-one, 2,3-dichloro-
  5,6-dicyano-4-hydroxy-6-(3-indolyl)-
	EtOH	215(4.48),262(4.31),
		359(3.81)
			1-0853-76

Indole complex with 2,3-dichloro-5,6-di- cyanoquinone (1:1)	CH_2Cl_2	592(2.02)	1-0853-76

$C_{16}H_8Cl_2N_2O_2$
  Quinone, 5,6-dichloro-3-cyano-2-(N-meth-
  yl-3-indolyl)-
	EtOH	216(4.23),302(3.77),
		381(2.66),602(3.45)
			1-0853-76

$C_{16}H_8F_{12}$
  4,7-Methano-1H-indene, 3a,4,7,7a-tetra-
  hydro-1,8-bis[2,2,2-trifluoro-1-(tri-
  fluoromethyl)ethylidene]-
	MeOH	252(3.85),302(2.57)	104-0339-76

$C_{16}H_8N_2$ 3,10-Phenanthrenedicarbonitrile	50% EtOH	232(4.65),250(4.67), 281(4.00),313(4.18), 324(4.17),347(3.28), 368(3.23)	18-2224-76
9,10-Phenanthrenedicarbonitrile	50% EtOH	238(4.54),252(4.54), 262(4.58),269s(4.44), 325(4.10),339(4.20)	18-2224-76

$C_{16}H_8N_2S_2$
  [1]Benzothieno[3,2-h]isoquinoline-
  6-carbonitrile
	MeOH	240(4.15),248(4.15),
		290(4.13),310s(3.33),
		378(3.22)
			4-0141-76

$C_{16}H_8O_4$
  [2]Benzopyrano[4,3-b][1]benzopyran-
  5,7-dione
	EtOH	218(4.32),232(4.28),
		252(4.32),305(4.20),
		315(4.18),332(4.15),
		348(4.04)
			22-1975-76

Compound	Solvent	$\lambda_{max}(\log \epsilon)$	Ref.
$C_{16}H_9BrN_2O_5$ 5,8-Quinolinedione, 7-bromo-6-methoxy- 2-(2-nitrophenyl)-	MeCN	294(4.11),390(3.26)	23-2563-76
$C_{16}H_9BrN_4OS$ 4H-[1,3,4]Thiadiazolo[2,3-c][1,2,4]tria- zin-4-one, 7-(4-bromophenyl)-3-phenyl-	MeOH	264(4.03),356(4.27)	4-0117-76
$C_{16}H_9ClN_2$ 2-Butenedinitrile, 2-(4-chlorophenyl)- 3-phenyl-, (E)-	50% EtOH	222(4.14),237s(3.96), 337(4.22)	18-2224-76
9,10-Phenanthrenedicarbonitrile, 3-chloro-9,10-dihydro-	50% EtOH	272(4.22)	18-2224-76
$C_{16}H_9ClN_2O_5$ 5,8-Quinolinedione, 7-chloro-6-methoxy- 2-(2-nitrophenyl)-	MeCN	290(4.26),340(3.30)	23-2563-76
$C_{16}H_9ClN_4OS$ 4H-[1,3,4]Thiadiazolo[2,3-c][1,2,4]tria- zin-4-one, 7-(2-chlorophenyl)-3-phen- yl-	MeOH	250(3.78),359(4.13)	4-0117-76
4H-[1,3,4]Thiadiazolo[2,3-c][1,2,4]tria- zin-4-one, 7-(3-chlorophenyl)-3-phen- yl-	MeOH	254(4.15),360(4.26)	4-0117-76
4H-[1,3,4]Thiadiazolo[2,3-c][1,2,4]tria- zin-4-one, 7-(4-chlorophenyl)-3-phen- yl-	MeOH	258(4.37),358(4.39)	4-0117-76
$C_{16}H_9Cl_3N_2O_3$ 2-Cyclobuten-1-one, 2,4,4-trichloro-3- [4-[(4-nitrophenyl)methylene]-1(4H)- pyridinyl]-	EtOH	262(4.11),436(4.36)	4-0517-76
$C_{16}H_9NO_2$ 5H-Benzo[b]carbazole-6,11-dione	EtOH	290(4.00),334(3.58), 462(3.63)	115-0185-76
$C_{16}H_9NO_3$ 5H-Benzo[b]carbazole-6,11-dione, 2-hy- droxy-	EtOH	270(4.39),290(4.12), 332(3.59),475(3.72)	115-0185-76
$C_{16}H_9NO_4$ 1H,3H-Naphtho[1,8-cd]pyran-1,3-dione, 6-(5-methyl-2-oxazolyl)-	toluene	375(4.28),440(3.46)	103-0736-76
$C_{16}H_{10}$ Acephenanthrylene	MeOH	227(4.49),232(4.50), 252(4.47),258(4.46), 269s(4.15),286(3.88), 298(3.98),316(3.78), 328(3.83),344(3.82), 364(3.88)	54-0165-76
$C_{16}H_{10}BrNO_3$ 1H-Inden-1-one, 3-bromo-2,3-dihydro- 6-nitro-2-(phenylmethylene)-	MeCN	254(4.37),333(4.40)	44-3540-76
$C_{16}H_{10}Br_2O_4S_2$ 1,4-Dithiin, 2,5-bis(4-bromophenyl)-, 1,1,4,4-tetraoxide	n.s.g.	283(3.66)	4-0057-76

$$C_{16}H_{10}ClN_3OS-C_{16}H_{10}N_2O_4$$

Compound	Solvent	λ_{max}(log ϵ)	Ref.
$C_{16}H_{10}ClN_3OS$ Quinazoline, 6-chloro-4-phenyl-2-(thio-cyanatomethyl)-, 3-oxide	EtOH	235(4.40),269(4.54), 310s(3.76),370s(3.69)	4-0433-76
$C_{16}H_{10}Cl_2O_2$ 1H-Inden-1-one, 2-(2,4-dichlorobenzoyl)-2,3-dihydro-	MeOH	251(3.95),319(4.23)	83-0356-76
$C_{16}H_{10}Cl_2O_4S_2$ 1,4-Dithiin, 2,5-bis(4-chlorophenyl)-, 1,1,4,4-tetraoxide	n.s.g.	280(3.83)	4-0057-76
$C_{16}H_{10}Cl_2S$ Thiophene, 3,4-bis(4-chlorophenyl)-	EtOH	240(4.54)	42-0490-76
$C_{16}H_{10}F_2O_4S_2$ 1,4-Dithiin, 2,5-bis(4-fluorophenyl)-, 1,1,4,4-tetraoxide	n.s.g.	278(3.91)	4-0057-76
$C_{16}H_{10}N_2$ 3,3'-Bi-1H-indole	DMF	258(4.79),322(4.08), 360s(3.85)	5-1060-76
semiquinone (radical ion)	DMF	268(4.64),345(4.21), 361(4.17),398s(3.71), 416s(3.64),485(3.51), 526s(3.66),610(3.92)	5-1060-76
ion(2-)	DMF	267(4.30),294s(4.14), 409(4.21),428(4.13)	5-1060-76
2-Butenedinitrile, 2,3-diphenyl-, trans	50% EtOH	215s(4.03),238(3.86), 331(4.16)	18-2224-76
9,10-Phenanthrenedicarbonitrile, 9,10-dihydro-	50% EtOH	270(4.23)	18-2224-76
Propanedinitrile, (diphenylmethylene)-	EtOH	227(4.16),320(3.28)	118-0705-76
$C_{16}H_{10}N_2O$ Benz[4,5]oxepino[2,3-b]quinoxaline	C_6H_{12}	228(4.51),247(4.47), 322(3.93),354(3.93)	32-0781-76
Benz[6,7]oxepino[2,3-b]quinoxaline	C_6H_{12}	250(4.9),311(3.95), 360(4.04)	32-0871-76
3-Pyridinecarbonitrile, 1,2-dihydro-6-(2-naphthalenyl)-2-oxo-	HOAc	367(4.32)	48-0705-76
$C_{16}H_{10}N_2O_2$ 8H-Pyrazolo[5,1-a]isoindol-8-one, 2-(4-hydroxyphenyl)-	THF	235(4.38),240(4.40), 265(4.46),286(4.30), 297(4.19),332(4.05), 347(4.04),390(4.31)	44-0110-76
$C_{16}H_{10}N_2O_3S$ Acetic acid, (10-oxothiazolo[3,2-a]per-imidin-9(10H)-ylidene)-, methyl ester	MeOH	225(4.64),271(4.10), 332(4.14),443(4.22)	83-0928-76
$C_{16}H_{10}N_2O_4$ Benzonitrile, 4-[2-(1,3-benzodioxol-5-yl)ethenyl]-3-nitro-, (E)-	EtOH	255(4.19),375(4.33)	80-1127-76
Indolo[2,1-b]quinazoline-2-carboxylic acid, 6,12-dihydro-12-hydroxy-6-oxo-	MeOH	201(4.38),232(4.19), 255(4.29),272s(3.86), 318(3.83),450(3.74)	39-2004-76C

Compound	Solvent	$\lambda_{max}(\log \epsilon)$	Ref.
$C_{16}H_{10}N_2O_4S_2$ 1,3-Dithiole, 4-(4-nitrophenyl)-2-[(4-nitrophenyl)methylene]-	$CHCl_3$	278(4.15),443(4.21)	39-1404-76B
$C_{16}H_{10}N_2O_5$ 5,8-Quinolinedione, 6-methoxy-2-(2-nitrophenyl)-	MeCN	290(4.73),340(3.32)	23-2563-76
$C_{16}H_{10}N_2S$ Benzo[b]thiophene-3-acetonitrile, α-(2-pyridinylmethylene)-, cis	MeOH	231(4.13),258(3.99), 284(3.67),330(3.24)	4-0141-76
	MeOH-HCl	228(--),262(--), 280(--),343(--)	4-0141-76
trans	MeOH	260(4.17),295s(4.14), 338(4.16)	4-0141-76
	MeOH-HCl	262(--),280s(--), 343(--)	4-0141-76
Benzo[b]thiophene-3-acetonitrile, α-(3-pyridinylmethylene)-	MeOH	260(4.15),300(3.37), 340(3.92)	4-0141-76
Benzo[b]thiophene-3-acetonitrile, α-(4-pyridinylmethylene)-	MeOH	260(4.15),292(3.67), 330(4.14)	4-0141-76
	MeOH-HCl	264(--),355(--)	4-0141-76
$C_{16}H_{10}N_2S_2$ Thioindigo, cis	n.s.g.	484(4.14)	104-1804-76
Thioindigo, trans	n.s.g.	543(4.23)	104-1804-76
$C_{16}H_{10}N_2Se_2$ Selenoindigo, cis	n.s.g.	485(4.02)	104-1804-76
Selenoindigo, trans	n.s.g.	562(4.15)	104-1804-76
$C_{16}H_{10}N_4O$ 2-Naphthalenol, 3-pyrido[3,4-e]-1,2,4-triazin-3-yl-	EtOH	226(4.36),266(4.37), 324(4.30),388(3.62)	114-0285-76C
$C_{16}H_{10}N_4OS$ 4H-[1,3,4]Thiadiazolo[2,3-c][1,2,4]-triazin-4-one, 3,7-diphenyl-	MeOH	276(3.98),356(4.39)	4-0117-76
$C_{16}H_{10}N_4O_4S$ Diazene, (2,4-dinitrophenyl)(5-phenyl-2-thienyl)-	n.s.g.	453(4.39)	39-1639-76C
$C_{16}H_{10}O_2$ Benzo[1,2:4,5]dicycloheptene-3,9-dione	$CHCl_3$	243(4.30),305(4.98), 330s(4.38),400s(3.27)	22-0914-76
Dibenzo[a,c]cyclooctene-5,8-dione	EtOH	225s(4.32),285s(3.41), 365s(2.30)	35-3627-76
1,2-Naphthalenedione, 5-phenyl-	EtOH	253(4.41),415(3.35)	78-2693-76
$C_{16}H_{10}O_3$ [2]Benzopyrano[4,3-b][1]benzopyran-7(5H)-one	EtOH	222(4.52),238s(--), 286(4.00),336(3.54)	22-1967-76
$C_{16}H_{10}O_4$ Benzo[1,2:4,5]dicycloheptene-3,9-dione, 2,8-dihydroxy-	$CHCl_3$	260(4.17),315(4.95), 327s(4.72),347s(4.13), 363s(4.01),383s(3.94), 415(3.59),442(3.67)	22-0914-76
Benzo[1,2:4,5]dicycloheptene-3,9-dione,	$CHCl_3$	255(4.14),260s(4.13),	22-0914-76

Compound	Solvent	λ_{max}(log ϵ)	Ref.
2,10-dihydroxy- (cont.)		322(4.98),347s(4.28), 363s(4.05),418(3.77), 445(3.78)	22-0914-76
6H-Benzofuro[3,2-c][1]benzopyran-6-one, 9-methoxy-	MeOH	242(4.54),334(4.49), 345(4.43)	18-1955-76
$C_{16}H_{10}O_5$ Acacetin	EtOH	269s(4.31),298s(4.21), 330(4.32)	112-0865-76
	EtOH-NaOEt	278(4.51),295s(4.32), 376(4.15)	112-0865-76
Coumarin, 3-(p-benzoquinonyl)-4-methoxy-	MeOH	210(4.51),285s(4.07), 325(4.24)	2-0861-76
$C_{16}H_{10}O_6$ 6H-[1,3]Benzodioxolo[5,6-c][1]benzopyran-6-one, 2-acetoxy-	EtOH	247(4.60),254(4.62), 310(4.08),317s(--), 338s(--)	24-0855-76
6H-[1,3]Benzodioxolo[5,6-c][1]benzopyran-6-one, 4-acetoxy-	EtOH	249(4.56),256(4.59), 290s(--),305(4.02), 315s(--),324s(--), 337s(--)	24-0855-76
4H-1-Benzopyran-4-one, 3-(1,3-benzodioxol-5-yl)-5,7-dihydroxy-	EtOH	265(4.51),294s(4.25), 336s(3.80)	95-0254-76
	EtOH-NaOEt	275(4.53),300s(4.24), 338(4.00)	95-0254-76
	EtOH-NaOAc	275(4.52),296s(4.28), 330(4.02)	95-0254-76
	EtOH-AlCl$_3$	275(4.51),296s(4.25), 381(3.61)	95-0254-76
8H-1,3-Dioxolo[4,5-g][1]benzopyran-8-one, 9-hydroxy-6-(4-hydroxyphenyl)-	EtOH	284(4.14),336(4.33)	94-1609-76
	EtOH-NaOAc	283(4.10),337(4.29)	94-1609-76
	EtOH-AlCl$_3$	289s(4.14),301(4.19), 354(4.33)	94-1609-76
	EtOH-NaOAc-H$_3$BO$_3$	281(4.14),334(4.31)	94-1609-76
$C_{16}H_{10}O_8$ 2-Anthracenecarboxylic acid, 9,10-dihydro-3,6,7,8-tetrahydroxy-1-methyl-9,10-dioxo- (ceroalbolinic acid)	MeOH-HOAc	293(4.45),418(3.68)	12-2225-76
[1]Benzopyrano[5,4,3-cde][1]benzopyran-5,10-dione, 2,7-dihydroxy-3,8-dimethoxy- (3,4-di-O-methylellagic acid)	EtOH	255(4.38),277s(4.33), 343(3.88),362(3.95)	95-0894-76
	EtOH-NaOEt	256s(4.66),269(4.73), 295(4.64),375(4.20)	95-0894-76
	EtOH-NaOAc	255(4.38),274(4.32), 330(3.87),354(3.92)	95-0894-76
	EtOH-NaOAc-H$_3$BO$_3$	260(4.47)	95-0894-76
$C_{16}H_{11}BrN_2O_2$ 2H-Indol-2-one, 5-bromo-1,3-dihydro-3-hydroxy-3-(1H-indol-3-yl)-	MeOH	219(3.36),266(2.83), 290(2.51)	103-0178-76
$C_{16}H_{11}BrN_2O_4$ Benzene, 1,1'-[3-(bromonitromethylene)-2-nitro-1-propene-1,3-diyl]bis-, cis-cis	C_6H_{12}	231(4.18),291(3.95)	44-2112-76
cis-trans	C_6H_{12}	225(4.20),308(4.20)	44-2112-76
trans-trans	C_6H_{12}	228(4.26),310(4.18)	44-2112-76

Compound	Solvent	$\lambda_{max}(\log \epsilon)$	Ref.
$C_{16}H_{11}BrO$ 1H-Inden-1-one, 3-bromo-2,3-dihydro- 2-(phenylmethylene)-	hexane	229(4.21),234s(4.18), 254s(4.02),319(4.37), 330(4.28)	44-3540-76
$C_{16}H_{11}BrO_2$ 1H-Inden-1-one, 2-(4-bromobenzoyl)- 2,3-dihydro-	MeOH	256(4.32),353(4.31)	83-0356-76
$C_{16}H_{11}Br_2NO$ 9(10H)-Acridinone, 2,7-dibromo-10-(2- propenyl)-	EtOH	255(4.58),396(3.79), 415(3.86)	48-0515-76
$C_{16}H_{11}ClN_2O$ 2,4-Pentadienamide, 5-chloro-2-cyano- 5-(2-naphthalenyl)-	CH_2Cl_2	345(4.23)	48-0705-76
$C_{16}H_{11}ClN_2O_3$ 5,8-Quinolinedione, 2-(2-aminophenyl)- 7-chloro-6-methoxy-	MeCN	290(3.95),310(3.79), 380(3.57),490(2.95)	23-2563-76
$C_{16}H_{11}ClN_2O_4$ Benzene, 1,1'-[3-(chloronitromethylene)- 2-nitro-1-propene-1,3-diyl]bis-, cis-trans	C_6H_{12}	226(4.20),309(4.18)	44-2112-76
trans-trans	C_6H_{12}	227(4.28),309(4.18)	44-2112-76
$C_{16}H_{11}ClN_2Se$ Benzenamine, N-[6-(4-chlorophenyl)-2H- 1,3-selenazin-2-ylidene]-, monoper- chlorate	HOAc	424(3.76)	88-2005-76
$C_{16}H_{11}ClO$ 1H-Inden-1-one, 2-[(3-chlorophenyl)meth- ylene]-2,3-dihydro-	isooctane	230(--),304(4.45), 318(4.32),353(2.15), 374(2.18),390(2.08), 410s(--)	65-2057-76
	EtOH	230(3.98),315(4.42)	65-2057-76
1H-Inden-1-one, 2-[(4-chlorophenyl)meth- ylene]-2,3-dihydro-	isooctane	230(4.09),315(4.44), 328(4.34),390(2.42), 410(2.04)	65-2057-76
	EtOH	230(4.14),324(4.47)	65-2057-76
$C_{16}H_{11}ClOS$ Benzo[b]thiophen-3(2H)-one, 2-[(4-chlo- rophenyl)methylene]-5-methyl-, (Z)-	octane	329(4.32),437(4.09)	103-0854-76
$C_{16}H_{11}ClOSe$ Benzo[b]selenophen-3(2H)-one, 2-[(4- chlorophenyl)methylene]-5-methyl-, (Z)-	octane	332(4.34),442(4.03)	103-0854-76
$C_{16}H_{11}ClO_2$ 3(2H)-Benzofuranone, 2-[(4-chlorophen- yl)methylene]-5-methyl-, (Z)-	octane	324(4.49),379(4.31)	103-0854-76
4H-1-Benzopyran-4-one, 3-[(4-chlorophen- yl)methylene]-2,3-dihydro-	isooctane	235s(--),297(4.28), 340(3.90)	65-2057-76
	EtOH	227(4.19),305(4.19), 356(3.90)	65-2057-76
1H-Inden-1-one, 2-(4-chlorobenzoyl)- 2,3-dihydro-	MeOH	255(3.88),319(4.23)	83-0356-76

Compound	Solvent	λ_{max}(log ϵ)	Ref.
$C_{16}H_{11}ClO_3S$ 1-Naphthalenol, 4-[(4-chlorophenyl)sulfonyl]-	C_6H_{12} EtOH	220(4.14),275(3.50) 228(4.42),240(4.11), 285(3.88)	42-0602-76 42-0602-76
$C_{16}H_{11}ClO_5$ 9,10-Anthracenedione, 1-chloro-4,5-dihydroxy-7-methoxy-2-methyl-	EtOH	226(4.47),252(4.19), 269(4.24),288(4.12), 440(4.04)	39-1852-76C
9,10-Anthracenedione, 1-chloro-4,7-dihydroxy-5-methoxy-2-methyl-	EtOH	222(4.49),251(4.09), 273(4.20),286(4.24), 434(3.90)	39-1852-76C
$C_{16}H_{11}Cl_2N_3O_2$ 2-Pyrimidinamine, 4,6-bis(4-chlorophenoxy)-	EtOH	230(4.64),270(4.21)	42-0913-76
4-Pyrimidinamine, 2,6-bis(4-chlorophenoxy)-	EtOH	230(4.62),250(3.92)	42-0913-76
$C_{16}H_{11}FOS$ Benzo[b]thiophen-3(2H)-one, 2-[(4-fluorophenyl)methylene]-5-methyl-, (Z)-	octane	324(4.26),433(4.06)	103-0854-76
$C_{16}H_{11}FO_2$ 3(2H)-Benzofuranone, 2-[(4-fluorophenyl)methylene]-5-methyl-, (Z)-	octane	322(4.38),375(4.20)	103-0854-76
$C_{16}H_{11}NOS$ Cyclobutenediylium, 1-hydroxy-3-mercapto-2-phenyl-4-(phenylamino)-, dihydroxide, bis(inner salt)	MeOH	202(4.30),229(4.22), 283(4.10),327(4.37), 442(3.43)	89-0704-76
$C_{16}H_{11}NOSe$ Cyclobutenediylium, 1-hydroxy-2-phenyl-4-(phenylamino)-3-selenyl-, dihydroxide, bis(inner salt)	MeOH	202(4.37),233(4.29), 285(3.96),347(4.40), 475(3.29)	89-0704-76
$C_{16}H_{11}NO_2$ Cyclobutenediylium, 1,3-dihydroxy-2-phenyl-4-(phenylamino)-, dihydroxide, bis(inner salt)	MeOH	202(4.31),215(4.14), 257(4.13),268s(4.29), 282s(4.19),393(3.71)	89-0704-76
$C_{16}H_{11}NO_2S$ Thiophene, 2-(4-nitrophenyl)-5-phenyl-	MeOH	285(3.91),384(4.36)	48-0731-76
$C_{16}H_{11}NO_3$ 4H-Anthra[1,2-d][1,3]oxazine-7,12-dione, 1,2-dihydro- (generalized spectra for several related compounds)	MeOH	477-498(3.84-4.21)	2-0504-76
Benz[b]indeno[2,1-e][1,4]oxazin-11(4bH)-one, 4b-methoxy-	MeOH	219(4.46),237(4.51), 279(4.26),294(4.27), 320s(4.15)	83-0081-76
Benzo[d][1,3]dioxolo[4,5-h][2]benzoxepin-5-carbonitrile, 5,7-dihydro-	EtOH	270(4.19),300(4.19)	94-0716-76
1H-Inden-1-one, 2,3-dihydro-2-[(4-nitrophenyl)methylene]-	isooctane EtOH	275s(--),315(4.35) 270(3.88),338(4.24)	65-2057-76 65-2057-76
1H-Inden-1-one, 2,3-dihydro-6-nitro-2-(phenylmethylene)-	MeCN	245(4.29),255(4.31), 327(4.42)	44-3540-76

Compound	Solvent	$\lambda_{max}(\log \epsilon)$	Ref.
$C_{16}H_{11}NO_3S$ Benzo[b]thiophen-3(2H)-one, 5-methyl-2-[(4-nitrophenyl)methylene]-, (Z)-	octane	324(4.31),447(4.01)	103-0854-76
$C_{16}H_{11}NO_3Se$ Benzo[b]selenophen-3(2H)-one, 5-methyl-2-[(4-nitrophenyl)methylene]-, (Z)-	octane	322(4.18),459(3.88)	103-0854-76
$C_{16}H_{11}NO_4$ 3(2H)-Benzofuranone, 5-methyl-2-[(4-nitrophenyl)methylene]-, (Z)-	octane	327(4.25),383(4.10)	103-0854-76
4H-1-Benzopyran-4-one, 2,3-dihydro-3-[(4-nitrophenyl)methylene]-	isooctane EtOH	297(3.89),340(3.54) 265s(--),303(4.21), 365(3.65)	65-2057-76 65-2057-76
1H-Inden-1-one, 2,3-dihydro-2-(2-nitrobenzoyl)-	MeOH	248(4.03),331(4.13)	83-0356-76
1H-Inden-1-one, 2,3-dihydro-2-(4-nitrobenzoyl)-	MeOH	257(4.13),359(4.23)	83-0356-76
$C_{16}H_{11}NO_4S$ 10(9H)-Acridineacetic acid, 9-oxo-α-sulfinyl-, methyl ester	isoPrOH	214(4.32),252(4.63), 285s(3.81),365s(3.81), 378(3.86),395s(3.49), 445(2.69)	44-3406-76
$C_{16}H_{11}NO_6$ 6H-[1,3]Benzodioxolo[5,6-c][1]benzopyran-6-one, 4-acetoxy-2-amino-	EtOH	208(4.41),249(4.70), 291(4.12),330s(--), 360s(--)	24-0855-76
$C_{16}H_{11}NS$ [1]Benzothieno[3,2-h]isoquinoline, 5-methyl-	MeOH	255(4.40),280(4.44), 323(4.16)	4-0141-76
	MeOH-HCl	248(--),265(--), 303(--),390(--), 398(--)	4-0141-76
$C_{16}H_{11}N_3O$ 1,7-Naphthyridine-8-carbonitrile, 7-benzoyl-7,8-dihydro-	EtOH	293(4.05)	94-1813-76
$C_{16}H_{11}N_3OS_4$ 4-Thiazolidinone, 5-(2-phenyl-5H-thiazolo[4,3-b][1,3,4]thiadiazol-5-ylidene)-3-(2-propenyl)-2-thioxo-	EtOH	488(5.05)	103-0650-76
$C_{16}H_{11}N_3O_2$ 1H-Anthra[1,2-d]triazole-6,11-dione, 1-ethyl-	EtOH	345(3.73)	104-0865-76
1H-Anthra[1,2-d]triazole-6,11-dione, 2-ethyl-	EtOH	342(3.88),370s(3.48)	104-0865-76
1H-Anthra[1,2-d]triazole-6,11-dione, 3-ethyl-	EtOH	338(3.73)	104-0865-76
Pyrazolo[1,5-a]pyrimidine-5,7-diol, 3-(1-naphthalenyl)-	MeOH-pH 1	205(4.51),256(4.20), 290(4.08)	87-0291-76
	MeOH-pH 11	230(4.29),275(4.20),	87-0291-76
13H-Quino[2,3-b][1,4]benzodiazepin-13-one, 5,12-dihydro-11-hydroxy-	DMF	272(4.06),283(4.00), 301s(3.59),313(3.60), 322s(3.54)	48-0039-76

Compound	Solvent	λ_{max}(log ϵ)	Ref.
$C_{16}H_{11}N_3O_3S$ Benzoic acid, 4-(5-benzoyl-2-imino- 1,3,4-thiadiazol-3(2H)-yl)-	EtOH	217(4.15),273(4.26), 370(4.03)	4-0045-76
$C_{16}H_{11}N_5$ 1,2,4-Triazolo[5,1-c][1,2,4]triazine, 6,7-diphenyl-	EtOH	248(4.32),349(3.84)	39-1492-76C
$C_{16}H_{11}N_5O_6$ 4-Oxazolecarboxylic acid, 2-phenyl-, 2-(2,4-dinitrophenyl)hydrazide	EtOH	275(4.14),300s(3.83), 430(3.95)	39-0315-76C
$C_{16}H_{12}$ Acephenanthrene	MeOH	220(4.32),250s(4.81), 252(4.90),265s(4.36), 278(4.03),290(4.00), 302(4.09),320(2.87), 334(3.10),345(2.82), 352(3.24)	54-0165-76
Benzene, 1,1'-(1-buten-3-yne-1,4-diyl)- bis-, (E)-	MeOH	223(4.23),227s(4.20), 248s(4.01),303s(4.57), 312(4.59),331s(4.42)	39-1104-76B
9bH-Cyclopenta[jk]fluorene, 9b-methyl-	hexane	223s(4.17),268(4.26), 273(4.26),283s(4.17), 293s(4.00),321(3.15), 334(3.03)	24-2596-76
Dicyclobuta[b,h]biphenylene, 1,2,5,6- tetrahydro-	EtOH	245(4.8),255(5.0), 365(4.0),387(4.0)	35-3579-76
Naphthalene, 2-phenyl-	EtOH	210(4.81),249(4.63), 285(4.00)	22-1829-76
$C_{16}H_{12}BrNO$ 1H-Indole-3-carboxaldehyde, 2-(4-bromo- phenyl)-1-methyl-	MeOH	257(4.44),307(4.17)	12-2747-76
$C_{16}H_{12}BrN_3OS_2$ Thiourea, [3-(4-bromophenyl)-4-oxo- 2-thiazolidinylidene]phenyl-	dioxan	412(3.94)	103-0751-76
$C_{16}H_{12}Br_2O$ 9(10H)-Phenanthrenone, 2,7-dibromo- 10,10-dimethyl-	hexane	246(4.37),254(4.32), 276s(--),284(4.36), 292(4.21),304(4.19), 336(3.67),346s(--)	104-1298-76
9(10H)-Phenanthrenone, 3,6-dibromo- 10,10-dimethyl-	hexane	249(4.61),255(4.63), 275(4.02),285(4.02), 303(3.72),311s(--), 331s(--)	104-1298-76
$C_{16}H_{12}Br_2O_4$ 2(3H)-Furanone, 5-(3,5-dibromo-2,4-di- hydroxyphenyl)dihydro-5-phenyl-	EtOH-pH 9.1	530(3.55)	18-2027-76
$C_{16}H_{12}ClNO$ 1H-Indole-3-carboxaldehyde, 2-(4-chloro- phenyl)-1-methyl-	MeOH	257(4.46),308(4.19)	12-2747-76
$C_{16}H_{12}ClNO_2$ 9,10-Anthracenedione, 1-chloro-2-(di- methylamino)-	CCl$_4$	416(3.69)	104-2499-76

Compound	Solvent	$\lambda_{max}(\log \epsilon)$	Ref.
$C_{16}H_{12}ClN_3$			
2-Pyrimidinamine, 4-(3-chlorophenyl)-6-phenyl-	EtOH	252(4.58),334(4.15)	4-0257-76
2-Pyrimidinamine, 4-(4-chlorophenyl)-6-phenyl-	EtOH	256(4.56),334(4.16)	4-0257-76
$C_{16}H_{12}ClN_3OS_2$			
Thiourea, [3-(3-chlorophenyl)-4-oxo-2-thiazolidinylidene]phenyl-	dioxan	336(3.96)	103-0751-76
$C_{16}H_{12}Cl_2N_2O_4Se$			
Benzenamine, N-[6-(4-chlorophenyl)-2H-1,3-selenazin-2-ylidene]-, monoperchlorate	HOAc	424(3.76)	88-2005-76
$C_{16}H_{12}Cl_2O_3$			
2-Propen-1-one, 3-(3,5-dichloro-2-hy-droxyphenyl)-1-(4-methoxyphenyl)-	EtOH	250(3.73),370(3.91)	97-0365-76
	MeCN	248(4.41),370(4.65)	97-0365-76
	CCl_4	260(4.03),375(4.36)	97-0365-76
$C_{16}H_{12}Cl_2O_5$			
1,4-Naphthalenedione, 5-acetoxy-8-(di-chloroacetyl)-2,7-dimethyl- (mollisin acetate)	$CHCl_3$	255(4.35),272s(4.17), 354(3.43)	35-2636-76
$C_{16}H_{12}N_2$			
2,3'-Bi-1H-indole	MeCN	213(4.44),222(4.47), 227(4.49),292s(4.11), 311(4.26)	5-1039-76
$C_{16}H_{12}N_2O$			
12H-Isoquinolino[1,2-b]quinazolin-12-one, 1,2-dihydro-	n.s.g.	220(3.74),242(3.64), 306(3.54)	2-0879-76
4(1H)-Pyridazinone, 3,5-diphenyl-	glyme	329(4.17)	88-0521-76
Pyrrolo[3,4-b]indol-3(2H)-one, 1,4-di-hydro-1-phenyl-	EtOH	300(4.091)	103-1162-76
2H-Pyrrol-2-one, 1,5-dihydro-1-phenyl-5-(phenylimino)-	EtOH	250(4.33),377(3.54)	44-2946-76
$C_{16}H_{12}N_2O_2$			
1H-Indole, 1-acetyl-3-(3-pyridinylcarbo-nyl)-	EtOH	210(4.29),228(4.42), 252(4.16),309(4.05)	39-1155-76C
6H-Indolo[2,3-a]quinolizin-5-ium, 3-car-boxy-7,12-dihydro-, hydroxide, inner salt	MeOH	251(3.87),324(4.11), 400(4.23)	24-0705-76
Naphtho[2,3-f]quinoxaline-7,12-dione, 1,2,3,4-tetrahydro-	EtOH and EtOH-NaOH	574(4.02)	104-0172-76
	$MeNO_2$	561(3.89)	104-0172-76
	$MeNO_2-HClO_4$	478(3.54)	104-0172-76
5H-Oxazolo[2,3-b]quinazolin-5-one, 2,3-dihydro-3-phenyl-	MeOH	222(4.64),228s(4.60), 257(3.89),264(3.81), 307(3.49)	118-0469-76
8H-Pyrazolo[5,1-a]isoindol-8-one, 3,3a-dihydro-2-(4-hydroxyphenyl)-	DMF	324(4.25)	44-0110-76
$C_{16}H_{12}N_2O_2S$			
1,3,4-Thiadiazol-2(3H)-one, 5-benzoyl-3-(4-methylphenyl)-	EtOH	248(4.13),272s(3.96), 328(3.99)	4-0045-76

Compound	Solvent	$\lambda_{max}(\log \epsilon)$	Ref.
$C_{16}H_{12}N_2O_2S_2$			
4-Pyrimidinecarboxylic acid, 2-(methyl-thio)-5-(2-naphthalenylthio)-	MeOH	265(4.39)	73-2771-76
$C_{16}H_{12}N_2O_3$			
Acetic acid, (1,2-dihydro-2-oxo-3H-ind-ol-3-ylidene)(phenylamino)-, cis	MeOH	276(4.20),369(4.40)	94-2305-76
Benzamide, 5-nitro-2-(2-phenylethenyl)-, trans	EtOH	336(4.27)	80-1127-76
Benzoic acid, 2-[5-(4-hydroxyphenyl)-1H-pyrazol-3-yl]-	EtOH	261(4.48)	44-0110-76
Benzonitrile, 2-[2-(4-methoxyphenyl)-ethenyl]-5-nitro-, (E)-	EtOH	265(4.23),390(4.52)	80-1127-76
Benzonitrile, 4-[2-(4-methoxyphenyl)-ethenyl]-3-nitro-, (E)-	EtOH	240(4.50),340(4.49)	80-1127-76
5,8-Quinazolinedione, 6-methoxy-4-meth-yl-2-phenyl-	dioxan	290(4.43)	5-1809-76
$C_{16}H_{12}N_2O_3S$			
Benzene, 2-isothiocyanato-1-methoxy-4-[2-(4-nitrophenyl)ethenyl]-	MeOH	288(4.30),365(4.45)	56-0967-76
1,3,4-Thiadiazol-2(3H)-one, 5-benzoyl-3-(4-methoxyphenyl)-	EtOH	252(4.20),270s(4.07), 335(3.90)	4-0045-76
$C_{16}H_{12}N_2O_4$			
Benzene, 1,1'-[2-nitro-3-nitromethyl-ene)-1-propene-1,3-diyl]bis-, cis-cis	C_6H_{12}	232(4.08),322(4.26)	44-2112-76
cis-trans	C_6H_{12}	228(4.30),308(4.20)	44-2112-76
trans-trans	C_6H_{12}	226(4.18),307(4.38)	44-2112-76
Benzonitrile, 4-[2-(4-hydroxy-3-methoxy-phenyl)ethenyl]-3-nitro-, (E)-	EtOH	250(4.18),366(4.33)	80-1127-76
Cyclobutene, 3,4-dinitro-1,2-diphenyl-	n.s.g.	303(4.18)	104-0022-76
$C_{16}H_{12}N_2O_5S$			
2-Furancarboximidothioic acid, N-hydr-oxy-5-(2-nitrophenyl)-, 2-furanyl-methyl ester	EtOH	221(4.44),299(4.28)	73-3085-76
2-Furancarboximidothioic acid, N-hydr-oxy-5-(3-nitrophenyl)-, 2-furanyl-methyl ester	EtOH	219(4.44),318(4.41)	73-3085-76
2-Furancarboximidothioic acid N-hydr-oxy-5-(4-nitrophenyl)-, 2-furanyl-methyl ester	EtOH	220(4.34),372(4.38)	73-3085-76
$C_{16}H_{12}N_2O_8$			
1-Cyclopropene-1,2-diacetic acid, α-cy-ano-3-(1-cyano-2-methoxy-2-oxoethyli-dene)-α'-(methoxycarbonyl)-, dimethyl ester, ion(2-)	MeCN	234(4.41),305s(4.50), 320(4.54)	35-0610-76
1-Cyclopropene-1,2-diacetic acid, 3-(di-cyanomethylene)-α,α'-bis(methoxycarbo-nyl)-, dimethyl ester, ion(2-)	MeCN	268(4.36),315(4.51)	35-0610-76
$C_{16}H_{12}N_2O_8S_2$			
Benzoic acid, 4,4'-dithiobis[3-nitro-, dimethyl ester	n.s.g.	255(4.56),277s(4.41), 350(3.95)	33-0855-76
$C_{16}H_{12}N_2S$			
1H-Indole, 2-(1H-indol-3-ylthio)-	MeCN	222(4.12),239(4.02), 260(3.89),270(3.91),	5-1039-76

Compound	Solvent	λ_{max} (log ϵ)	Ref.
1H-Indole, 2-(1H-indol-3-ylthio)- (cont.)		296s(4.23),302(4.35), 349(3.77),415(4.10), 429s(4.06)	5-1039-76
2-Quinolinecarbothioamide, N-phenyl-	20% EtOH	320(4.10)	94-1451-76
$C_{16}H_{12}N_2S_2$ 1H-[1,2]Dithiolo[5,1-e][1,2,3]thiadia- zole-7-S^{IV}, 1,4-diphenyl-	C_6H_{12}	205(4.44),233(4.43), 257s(4.21),280(4.13), 299(4.11),491(4.14)	39-0228-76C
1H-[1,2]Dithiolo[5,1-e][1,2,3]thiadia- zole-7-S^{IV}, 1,5-diphenyl-	C_6H_{12}	204(4.63),237(4.57), 251s(4.45),302(4.13), 504(4.29)	39-0228-76C
$C_{16}H_{12}N_2Se$ Benzenamine, N-(6-phenyl-2H-1,3-selena- zin-2-ylidene)-, perchlorate	HOAc	423(3.85)	88-2005-76
$C_{16}H_{12}N_4$ 3H-1,2,3-Triazolo[4,5-b]quinoline, 9-methyl-3-phenyl-	DMF	313s(3.79),325(3.91), 370(3.75)	48-0039-76
$C_{16}H_{12}N_4O$ 1(2H)-Phthalazinone, 4-(1H-benzimidazol- 2-yl)-2-methyl-	DMF	281(4.15),332(4.25), 401(2.92)	44-0836-76
$C_{16}H_{12}N_4O_2S$ Methanone, [4,5-dihydro-4-(4-methylphen- yl)-5-(nitrosoimino)-1,3,4-thiadiazol- 2-yl]phenyl-	EtOH	275(3.97),358(3.97), 480(1.75)	4-0045-76
1-Thia-3,5,8,8b-tetraazaacenaphthylene- 6-carboxylic acid, 2-phenyl-, ethyl ester	EtOH	233(4.43),292(3.83), 304(3.89),318(3.99), 334(4.00),410(4.16), 510(2.97),550(2.81), 607(2.40)	78-0173-76
$C_{16}H_{12}N_4O_2Se$ 1-Selena-3,5,8,8b-tetraazaacenaphthyl- ene-6-carboxylic acid, 2-phenyl-, ethyl ester	EtOH	233(4.47),295(3.91), 305(3.93),318(3.89), 335(3.83),419(4.17), 520(3.14),564(2.96), 616(2.59)	78-0173-76
$C_{16}H_{12}N_4O_3S$ Methanone, [4,5-dihydro-4-(4-methoxy- phenyl)-5-(nitrosoimino)-1,3,4-thi- adiazol-2-yl]phenyl-	EtOH	288(3.79),358(3.72), 480(1.80)	4-0045-76
$C_{16}H_{12}N_4O_3S_2$ Thiourea, [3-(3-nitrophenyl)-4-oxo- 2-thiazolidinylidene]phenyl-	dioxan	300(4.25),403(4.08)	103-0751-76
Thiourea, [3-(4-nitrophenyl)-4-oxo- 2-thiazolidinylidene]phenyl-	dioxan	270(4.20),370(4.10)	103-0751-76
$C_{16}H_{12}N_4O_4$ 3-Buten-2-one, 4-(5-nitro-2H-benzotria- zol-2-yl)-4-phenyl-, N-oxide, (Z)-	EtOH	274(4.50),304s(4.08), 380(3.67)	22-0184-76
2-Propen-1-one, 2-methyl-3-(5-nitro-2H- benzotriazol-2-yl)-1-phenyl-, N-oxide, (E)-	EtOH	270(4.01),317(4.39), 350(4.01),392(3.68)	22-0184-76
(Z)-	EtOH	302(4.32),347(3.85), 385(3.71)	22-0184-76

Compound	Solvent	$\lambda_{max}(\log \epsilon)$	Ref.
$C_{16}H_{12}N_6$			
1,2,4-Triazolo[4,3-b][1,2,4]triazin-3-amine, 6,7-diphenyl-	EtOH	243(4.17),284(4.17), 335s(3.80),420(3.38)	39-1492-76C
1,2,4-Triazolo[5,1-c][1,2,4]triazin-7-amine, 3,4-diphenyl-	EtOH	260(4.43),350(3.89)	39-1492-76C
$C_{16}H_{12}O$			
10H,11H-Benzo[b]indeno[2,1-e]pyran	EtOH	210(4.11),228(4.15), 270(3.60)	22-1967-76
1H-Inden-1-one, 2,3-dihydro-2-(phenyl-methylene)-	isooctane	225(4.05),298(4.36), 312(4.48),325(4.37), 362(2.40),375(2.38), 390(2.28),410(1.95)	65-2057-76
	EtOH	225(3.89),320(4.34)	65-2057-76
1-Naphthalenol, 5-phenyl-	EtOH	240(4.54),307(3.97), 314(3.96),327(3.88)	95-1161-76
$C_{16}H_{12}OS$			
Benzo[b]thiophen-3(2H)-one, 5-methyl-2-(phenylmethylene)-, (Z)-	octane	324(4.25),434(4.07)	103-0854-76
$C_{16}H_{12}OS_2$			
1,3-Dithiol-1-ium, 4-hydroxy-2-(4-meth-ylphenyl)-5-phenyl-, hydroxide, inner salt	dioxan	236(4.05),243s(4.02), 283(4.28),295s(4.21), 556(4.18)	24-0740-76
1,3-Dithiol-1-ium, 4-hydroxy-5-(4-meth-ylphenyl)-2-phenyl-, hydroxide, inner salt	dioxan	236(4.07),244s(4.08), 284(4.28),295s(4.21), 558(4.17)	24-0740-76
$C_{16}H_{12}OSe$			
Benzo[b]selenophen-3(2H)-one, 5-methyl-2-(phenylmethylene)-, (Z)-	octane	326(4.27),440(3.96)	103-0854-76
$C_{16}H_{12}O_2$			
9,10-Anthracenedione, 2,3-dimethyl-	EtOH	209(4.34),263(4.66), 278s(4.24),334(3.78)	12-2257-76
3(2H)-Benzofuranone, 5-methyl-2-(phen-ylmethylene)-	octane	322(4.37),376(4.20)	103-0854-76
6H,7H-[1]Benzopyrano[4,3-b][1]benzopyr-an	EtOH	212(4.26),225(4.26), 282(3.54),310(3.65)	22-1967-76
4H-1-Benzopyran-4-one, 2,3-dihydro-3-(phenylmethylene)-	isooctane	224(3.49),292(4.33), 340(3.92),390s(--), 410s(--)	65-2057-76
	EtOH	225(4.02),305(4.03), 354(3.72)	65-2057-76
	EtOH	222(4.20),298(4.23), 340s(--)	22-1967-76
6H-Cyclohepta[b]furan-8-one, 2-methyl-7-phenyl-	EtOH	277(4.48),329(3.89)	39-2403-76C
2,4,6-Cycloheptatrien-1-one, 2-phenyl-7-(2-propynyloxy)-	EtOH	236(4.32),340(4.02)	39-2403-76C
2,4,6-Cycloheptatrien-1-one, 3-phenyl-2-(2-propynyloxy)-	EtOH	222(4.20),274(4.08), 324(3.86)	39-2403-76C
Dibenzo[a,c]cyclooctene-5,8-dione, 6,7-dihydro-	EtOH	245s(4.01),290(3.32)	35-3627-76
1,4-Phenanthrenedione, 2-ethyl-	n.s.g.	223(4.60),275(4.25), 284s(4.14),357(3.40), 401s(3.20)	33-0664-76
Tricyclo[3.2.2.0^{2,8}]non-6-ene-3,4-dione, 9-methylene-2-phenyl-	C_6H_{12}	222(3.98),254(3.30), 268s(3.24),275(3.20), 427(1.90)	39-2403-76C

Compound	Solvent	$\lambda_{max}(\log \epsilon)$	Ref.
$C_{16}H_{12}O_2S$			
Benzo[b]thiophen-3(2H)-one, 2-[(4-hydroxyphenyl)methylene]-5-methyl-, (Z)-	octane	338(4.31),436(4.28)	103-0854-76
2-Thiophenecarboxylic acid, 5-(2-naphthalenyl)-, methyl ester	MeOH	227(4.67),249s(4.07), 275(4.29),287(4.28), 331(4.42),347s(4.49)	48-0731-76
$C_{16}H_{12}O_2S_2$			
1,3-Dithiol-1-ium, 4-hydroxy-2-(4-methoxyphenyl)-5-phenyl-, hydroxide, inner salt	dioxan	239(4.11),290(4.22), 326(3.94),563(4.20)	24-0740-76
1,3-Dithiol-1-ium, 4-hydroxy-5-(4-methoxyphenyl)-2-phenyl-, hydroxide, inner salt	dioxan	247(4.09),278(4.27), 302s(4.10),570(4.15)	24-0740-76
$C_{16}H_{12}O_2Se$			
Benzo[b]selenophen-3(2H)-one, 2-[(4-hydroxyphenyl)methylene]-5-methyl-, (Z)-	octane	340(4.33),441(4.14)	103-0854-76
$C_{16}H_{12}O_3$			
9,10-Anthracenedione, 4-ethyl-1-hydroxy-	EtOH	204(3.76),250(3.60), 387(3.57)	56-0759-76
3(2H)-Benzofuranone, 2-[(4-hydroxyphenyl)methylene]-5-methyl-, (Z)-	octane	332(4.31),385(4.34)	103-0854-76
[2]Benzopyrano[4,3-b][1]benzopyran-5-one, 5,6a,7,12a-tetrahydro-	EtOH	224(4.11),273(3.48), 280(3.48)	22-1967-76
4H-1-Benzopyran-4-one, 2,3-dihydro-3-[(4-hydroxyphenyl)methylene]-	EtOH	332(4.20),360(4.20)	22-1967-76
4H-1-Benzopyran-4-one, 3-[(2-hydroxyphenyl)methyl]-	EtOH	224(4.40),244s(--), 302(3.78)	22-1967-76
3(2H)-Furanone, 2-hydroxy-2,4-diphenyl-	MeOH	244(4.27),292(3.80)	44-0388-76
1H-Inden-1-one, 2,3-dihydro-2-(2-hydroxybenzoyl)-	MeOH	248(4.06),284(3.99), 371(4.39)	83-0356-76
$C_{16}H_{12}O_3S$			
Dibenzo[b,f]thiepin-10(11H)-one, 7-acetoxy-	MeOH	247(4.34),323(3.57)	73-3607-76
2-Propene-1-thione, 3-(1,3-benzodioxol-5-yl)-3-hydroxy-1-phenyl-	EtOH	258(3.97),323(4.10), 349s(4.01),434(4.37)	4-0691-76
$C_{16}H_{12}O_4$			
Benzoic acid, 2-(2-hydroxy-1-oxo-3-phenyl-2-propenyl)-	ether	315(4.37)	44-0388-76
4H-1-Benzopyran-4-one, 5,7-dihydroxy-6-methyl-2-phenyl-	EtOH	210(4.33),273(4.33)	2-0009-76
4H-1-Benzopyran-4-one, 5,7-dihydroxy-8-methyl-2-phenyl-	EtOH	249(4.13),273(4.36) 318(4.05)	2-0009-76
2-Butenedioic acid, 2,3-diphenyl-, (E)-	dioxan	261(4.041)	24-0576-76
Dalbergin	EtOH	238(3.91),304(3.49)	78-2407-76
$C_{16}H_{12}O_4S_2$			
1,4-Dithiin, 2,5-diphenyl-, 1,1,4,4-tetraoxide	n.s.g.	276(4.16)	4-0057-76
$C_{16}H_{12}O_5$			
9,10-Anthracenedione, 1,6-dihydroxy-8-methoxy-3-methyl- (questin)	EtOH	224(4.55),248(4.14), 285(4.35),430(3.95)	39-1852-76C
9,10-Anthracenedione, 1,8-dihydroxy-4-methoxy-2-methyl-	MeOH	231(4.45),249s(4.07), 284s(3.82),472(3.87)	102-0317-76
9,10-Anthracenedione, 2,5-dihydroxy-7-methoxy-3-methyl-	MeOH	254s(4.21),272s(4.32), 280(4.34),412(3.85)	102-0317-76

Compound	Solvent	$\lambda_{max}(\log \epsilon)$	Ref.
4H-1-Benzopyran-4-one, 5-hydroxy-2-(4-hydroxyphenyl)-7-methoxy- (genkwanin)	EtOH	268(4.33),336(4.43)	112-0865-76
	EtOH-NaOEt	273(4.33),300s(4.08), 360s(4.46),382(4.54)	112-0865-76
	EtOH-NaOAc	344s(4.49),368(4.32), 396(4.22)	112-0865-76
Biochanin A	EtOH	264(4.51),330s(3.70)	95-0254-76
	EtOH-NaOEt	275(4.52),338(3.92)	95-0254-76
	EtOH-NaOAc	275(4.50),333(3.90)	95-0254-76
	EtOH-AlCl$_3$	274(4.51),308s(3.81), 377(3.46)	95-0254-76

$C_{16}H_{12}O_6$

Compound	Solvent	$\lambda_{max}(\log \epsilon)$	Ref.
9,10-Anthracenedione, 1,4,5-trihydroxy-2-methoxy-7-methyl- (xanthorin)	dioxan-1% HOAc	257(4.58),306(4.02), 465s(4.05),492(4.18), 511s(4.03),524s(4.01)	12-2231-76
9,10-Anthracenedione, 2,4,5-trihydroxy-1-methoxy-3-methyl-	MeOH	257(4.14),296s(3.76), 460(3.48),488(3.51)	64-0403-76C
4H-1-Benzopyran-4-one, 5,7-dihydroxy-3-(2-hydroxy-4-methoxyphenyl)-	EtOH	221(4.26),263(4.37), 290(4.06)	102-1029-76
	EtOH-NaOH	235(4.27),273(4.30), 327(4.01)	102-1029-76
	EtOH-NaOAc	225(4.26),275(4.35), 336(3.76)	102-1029-76
	EtOH-AlCl$_3$	221(4.30),275(4.38), 378(3.08)	102-1029-76
4H-1-Benzopyran-4-one, 3,5,7-trihydroxy-6-methoxy-2-phenyl-	EtOH	270(4.25),325(4.25), 350(4.20)	28-0683-76A
	EtOH-NaOAc	272(--),328(--), 368(--)	28-0683-76A
	EtOH-AlCl$_3$	278(--),350(--), 415(--)	28-0683-76A
2H,5H-Pyrano[3,2-c][1]benzopyran-3-carboxylic acid, 7-methyl-2,5-dioxo-, ethyl ester	n.s.g.	365(4.42)	64-0514-76B

$C_{16}H_{12}O_6S$

Compound	Solvent	$\lambda_{max}(\log \epsilon)$	Ref.
1,3-Dibenzothiophenedicarboxylic acid, 2,4-dihydroxy-, dimethyl ester	EtOH	258(4.37),272(4.30), 286(4.26),308(4.14), 378(3.56),396(3.54)	78-0269-76
2H-Pyran-3-carboxylic acid, 5-(3-benzo[b]thienyl)-4-hydroxy-6-methoxy-2-oxo-, methyl ester	EtOH	225(4.49),268(3.95), 292(3.95),300(3.97), 315(3.86)	78-0269-76
	EtOH-NaOH	273(4.04)	78-0269-76

$C_{16}H_{12}O_7$

Compound	Solvent	$\lambda_{max}(\log \epsilon)$	Ref.
2,4-Dibenzofurandicarboxylic acid, 1,3-dihydroxy-, dimethyl ester	EtOH	256(4.36),316(3.78), 327s(3.70),380(3.50)	78-0269-76
2H-Pyran-3-carboxylic acid, 5-(2-benzofuranyl)-4-hydroxy-6-methoxy-2-oxo-, methyl ester	EtOH	246(4.27),312(3.81)	78-0269-76

$C_{16}H_{12}S$

Compound	Solvent	$\lambda_{max}(\log \epsilon)$	Ref.
Benzo[b]naphtho[1,2-d]thiophene, 5,6-dihydro-	EtOH	220(4.94),241(4.80), 249(4.81),265(3.84), 292(3.88),302(3.88), 315(3.74)	12-0883-76
Thiophene, 2,4-diphenyl-	EtOH	222(4.16),256(4.32), 312(4.12)	18-2158-76
	EtOH	258(4.20),318(4.20)	42-0490-76
Thiophene, 2,5-diphenyl-	EtOH	220(4.15),255(4.37), 294(4.12)	18-2158-76

Compound	Solvent	$\lambda_{max}(\log \epsilon)$	Ref.
Thiophene, 3,4-diphenyl-	EtOH	230(3.89),255(3.72)	18-2158-76
	EtOH	234(4.35),257s(4.09)	42-0490-76
$C_{16}H_{12}S_2$			
1,3-Dithietane, 2,4-bis(phenylmethylene)-	CHCl$_3$	305s(3.84),315(3.83), 365(3.73)	39-0692-76C
$C_{16}H_{13}BrN_2O_3$			
1H-Pyrazole, 1-acetyl-3-benzoyl-4-(5-bromo-2-furanyl)-4,5-dihydro-	EtOH	223(4.18),269(4.00), 305(4.21)	104-2358-76
1H-Pyrazole, 1-acetyl-3-(4-bromobenzoyl)-4-(2-furanyl)-4,5-dihydro-	EtOH	217(4.23),268(4.16), 311(4.20)	104-2358-76
$C_{16}H_{13}ClN_2$			
3-Quinolinamine, 6-chloro-2-methyl-4-phenyl-	EtOH	206(4.40),219(4.43), 249(4.63),279s(3.73), 290s(3.63),302s(3.46), 349(3.84)	44-1743-76
$C_{16}H_{13}ClN_2O$			
4H-2,3-Benzodiazepin-4-one, 8-chloro-3,5-dihydro-3-methyl-1-phenyl-	EtOH	238(4.9)	111-0083-76
$C_{16}H_{13}ClN_2O_2$			
1H-Indazole-1-methanol, 5-chloro-3-phenyl-, acetate	isoPrOH	220(4.57),251s(4.00), 257s(3.93),270s(3.78), 309(3.98),313s(3.98)	4-0033-76
$C_{16}H_{13}ClN_2S$			
3-Quinolinamine, 6-chloro-2-(methylthio)-4-phenyl-	EtOH	229(4.66),260(4.62), 297s(3.63),309s(3.57), 359(4.05)	48-0039-76
$C_{16}H_{13}ClN_4O_2$			
Propanamide, 2-azido-3-benzoyl-N-(4-chlorophenyl)-	EtOH	247(4.45)	115-0001-76
Propanamide, 2-azido-3-(4-chlorobenzoyl)-N-phenyl-	EtOH	251(4.44)	115-0001-76
$C_{16}H_{13}ClN_4O_4$			
4-Cinnolinamine, 6-chloro-3-methoxy-N-(4-methyl-2-nitrophenyl)-, 1-oxide	CHCl$_3$	440(4.03)	39-0399-76C
$C_{16}H_{13}ClN_4O_5$			
4-Cinnolinamine, 6-chloro-3-methoxy-N-(4-methoxy-2-nitrophenyl)-, 1-oxide	CHCl$_3$	444(3.96)	39-0399-76C
$C_{16}H_{13}ClO_2$			
1,4-Butanedione, 1-(4-chlorophenyl)-4-phenyl-	EtOH	222s(4.20),248s(4.20), 272(4.31)	12-0339-76
$C_{16}H_{13}ClO_2S$			
Dibenzo[b,f]thiepin, 8-chloro-2,3-dimethoxy-	MeOH	226(4.57),245s(4.41), 266(4.39),305s(3.67), 340s(3.35)	73-1396-76
$C_{16}H_{13}ClO_3$			
9H-Xanthene-2-carboxylic acid, 7-chloro-, ethyl ester	EtOH	273(4.40)	94-1588-76

Compound	Solvent	$\lambda_{max}(\log \epsilon)$	Ref.
$C_{16}H_{13}ClO_3S$ Dibenzo[b,f]thiepin-10(11H)-one, 8-chloro-2,3-dimethoxy-	MeOH	233(4.46),265s(4.09), 342(3.65)	73-1396-76
$C_{16}H_{13}ClO_6$ Spiro[benzofuran-2(3H),1'-[2,5]cyclo- hexadiene]-3,4'-dione, 7-chloro-4- hydroxy-2',6-dimethoxy-6'-methyl-	EtOH	223(4.49),292(4.50), 330(3.63),370(3.30)	35-5380-76
$C_{16}H_{13}Cl_2NO_3$ Butanoic acid, 4-[(2,6-dichloro-3-pyri- dinylmethyl)phenyl]-4-oxo-	MeOH	248.5(4.35)	33-0190-76
$C_{16}H_{13}F_3N_2$ 1H-Pyrazole, 4,5-dihydro-1-phenyl- 3-[4-(trifluoromethyl)phenyl]-	benzene	382(4.32)	64-1248-76B
$C_{16}H_{13}F_3O_3$ Benzeneacetic acid, α-[3-(trifluorometh- yl)phenoxy]-, methyl ester	EtOH	221(4.10),273(3.30), 280(3.24)	111-0533-76
$C_{16}H_{13}I$ 1H-Indene, 2,3-dihydro-1-[(2-iodophen- yl)methylene]-, cis	MeOH	265(4.07),293(4.01)	54-0165-76
trans	MeOH	277s(4.11),312(4.33)	54-0165-76
$C_{16}H_{13}N$ 2H-Azirine, 2-phenyl-3-(2-phenylethen- yl)-, (E)-	EtOH	293(4.42)	44-0543-76
Isoquinoline, 3-methyl-1-phenyl-, [1-^{14}C]	pH 2 H_2O	268(3.95),350(3.93) 275(3.78),333(3.79)	56-0499-76 56-0499-76
Quinoline, 6-methyl-2-phenyl-	EtOH EtOH	258(4.63),326(3.91) 257(4.66),325(3.95)	94-2409-76 94-2409-76
$C_{16}H_{13}NO$ 9(10H)-Acridinone, 10-(2-propenyl)-	EtOH	253(4.85),380(4.00), 400(4.07)	48-0515-76
1H-Indole-3-carboxaldehyde, N-methyl- 2-phenyl-	MeOH	254(4.43),307(4.17)	12-2747-76
1H-Indole-3-carboxaldehyde, 1-(phenyl- methyl)-	1% EtOH	248(3.85),263(c.3.6), 305(3.90)	44-1952-76
3(2H)-Isoquinolinone, 2-methyl-1-phenyl-	n.s.g.	438(3.72)	103-0910-76
3(2H)-Isoquinolinone, 1-(3-methylphen- yl)-	EtOH ether CHCl$_3$ CCl$_4$	350(3.83),433(3.65) 352(3.86) 355(3.58),437(3.38) 352(3.82),440(3.12)	103-0910-76 103-0910-76 103-0910-76 103-0910-76
3(2H)-Isoquinolinone, 1-(4-methylphen- yl)-	EtOH ether CHCl$_3$ CCl$_4$	350(3.77),433(3.46) 353(3.94) 355(3.86),440(3.70) 353(3.85),440(3.01)	103-0910-76 103-0910-76 103-0910-76 103-0910-76
Isoxazole, 4-methyl-3,5-diphenyl-	HOAc +17M H_2SO_4	267(4.29) 304(4.42)	103-0851-76 103-0851-76
2H-Pyrrol-2-one, 1,5-dihydro-3,4-diphen- yl-	EtOH	286(3.99)	22-1491-76
Quinoline, 6-methoxy-2-phenyl-	EtOH EtOH	263(4.50),334(3.85) 217(4.47),260(4.59), 335(3.94)	94-2409-76 95-0968-76
$C_{16}H_{13}NOS$ Benzo[b]thiophen-3(2H)-one, 2-[(4-amino- phenyl)methylene-5-methyl-, (Z)-	octane	350(4.27),448(4.43)	103-0854-76

Compound	Solvent	$\lambda_{max}(\log \epsilon)$	Ref.
$C_{16}H_{13}NOS_2$ 2-Methylpyridinium 2-oxo-4-phenyl-3-thi-oxocyclobutene-1-thiolate	CH_2Cl_2	260(4.30),376(4.57)	118-0445-76
$C_{16}H_{13}NO_2$ 9,10-Anthracenedione, 2-methyl-1-(meth-ylamino)-	CCl_4	496(3.76)	104-2499-76
9,10-Anthracenedione, 2-methyl-3-(meth-ylamino)-	CCl_4	432(3.64)	104-2499-76
3(2H)-Benzofuranone, 2-[(4-aminophenyl)-methylene]-5-methyl-, (Z)-	octane	340(4.06),402(4.30)	103-0854-76
3-Indolizinol, 2-phenyl-, acetate	EtOH	252(4.68),284s(3.99), 294s(3.88),307s(3.62), 352(3.69),365(3.65), 384s(3.54)	1-0198-76
2H-Isoindole-1-carboxylic acid, 2-phen-yl-, methyl ester	EtOH	237(4.35),256(4.25), 305(3.66),339(4.05), 349(4.04)	32-0065-76
1H-Isoindol-1-one, 2-(2-acetylphenyl)-2,3-dihydro-	MeOH	230(4.34),262s(3.85), 269s(3.84),277s(3.80)	83-0356-76
1H-Isoindol-1-one, 2-(4-acetylphenyl)-2,3-dihydro-	MeOH	223(4.15),307(4.42)	83-0356-76
3-Isoquinolinol, 1-(3-methoxyphenyl)-	EtOH	350(3.72),420(3.37)	103-0910-76
	ether	353(3.90)	103-0910-76
	$CHCl_3$	355(3.26),438(3.06)	103-0910-76
	CCl_4	353(3.72),440(2.97)	103-0910-76
3-Isoquinolinol, 1-(4-methoxyphenyl)-	EtOH	350(3.75),433(3.45)	103-0910-76
	ether	353(3.91)	103-0910-76
	$CHCl_3$	355(3.74),440(3.52)	103-0910-76
	CCl_4	353(3.64),440(2.94)	103-0910-76
2H-1,3-Oxazin-2-one, 3,4-dihydro-4,6-di-phenyl-	MeOH	221(4.0),256(4.1)	2-0477-76
2H-Pyrrol-2-one, 1,5-dihydro-3-hydroxy-4,5-diphenyl-	EtOH	210(4.32),291(4.26)	40-1100-76
2H-Pyrrol-2-one, 1,5-dihydro-5-hydroxy-3,4-diphenyl-	EtOH	226(4.30),251s(3.92), 295(3.94)	22-1491-76
$C_{16}H_{13}NO_2S$ 4-Thiazolol, 5-(4-methoxyphenyl)-2-phen-yl-	EtOH	268(3.95),270(4.17)	118-0261-76
$C_{16}H_{13}NO_3$ Benzoic acid, 2-(1,4-dihydro-1-methyl-4-oxocyclopenta[b]pyrrol-5-yl)-, methyl ester	EtOH	265(4.29),348(4.16)	4-0083-76
Pyrido[2,1,6-de]quinolizine-1-carbox-ylic acid, 3-formyl-, ethyl ester	EtOH	239(4.04),281(4.42), 293s(4.2),323(3.58), 341(3.72),372(4.07), 390(4.38),431s(4.21), 452(4.38)	39-0341-76C
4(1H)-Quinolinone, 3-hydroxy-6-methoxy-2-phenyl-	EtOH	264(4.30),322(3.72), 363(3.74)	142-1089-76
$C_{16}H_{13}NO_3S$ Benzenepropanethioic acid, 2-(aminocarb-onyl)-β-oxo-, S-phenyl ester	MeOH	288.5(4.0)	149-0383-76B
$C_{16}H_{13}NO_4$ 4(1H)-Quinolinone, 1,3-dihydroxy-6-meth-oxy-2-phenyl-	EtOH	265(3.38),321(3.78), 367(3.84)	142-1089-76

Compound	Solvent	$\lambda_{max}(\log \epsilon)$	Ref.
$C_{16}H_{13}NO_5$			
Isoxazolo[4,5-b]dibenzofuran-6-one, 6,6a-dihydro-7,9-dihydroxy-5,6a,10-trimethyl-	MeCN	222(4.18),301(3.40)	23-3721-76
Methanone, [3-(5-methoxy-2-nitrophenyl)-oxiranyl]phenyl-	EtOH	246(4.59),304(4.25)	142-1089-76
$C_{16}H_{13}NO_7$			
1,3-Benzodioxole-5-carboxylic acid, 6-nitro-4-(phenylmethoxy)-, methyl ester	EtOH	207(4.50),248s(--), 337(3.91)	24-0855-76
$C_{16}H_{13}NS$			
Pyridine, 4-(2-benzo[b]thien-3-yl-1-methylethenyl)-, (E)-	MeOH	280(3.51),310(4.13)	4-0141-76
	MeOH-HCl	270(--),368(--)	4-0141-76
$C_{16}H_{13}NS_2$			
Isothiazole, 3-(methylthio)-4,5-diphenyl-	EtOH	241(4.15),276(4.04)	48-0127-76
$C_{16}H_{13}NS_3$			
1,4,2-Dithiazine, 3-(methylthio)-5,6-diphenyl-	MeOH	229(4.06),269(3.94), 309(3.53),362(3.22)	48-0127-76
$C_{16}H_{13}N_2Se$			
Benzenaminium, N-(6-phenyl-2H-1,3-selenazin-2-ylidene)-, perchlorate	HOAc	423(3.85)	88-2005-76
$C_{16}H_{13}N_3$			
Benzonitrile, 4-(4,5-dihydro-1-phenyl-1H-pyrazol-3-yl)-	benzene	403(4.39)	64-1248-76B
Benzonitrile, 4-(4,5-dihydro-3-phenyl-1H-pyrazol-1-yl)-	benzene	368(3.31)	64-1248-76B
Pyrrolo[1,2-a]pyrimidine-8-carbonitrile, 2,4-dimethyl-7-phenyl-	EtOH	221(4.26),260(4.55), 315(3.84),350(3.72)	103-1379-76
$C_{16}H_{13}N_3O$			
Benzenepropanenitrile, 4-methyl-β-oxo-α-(phenylhydrazono)-	EtOH	247(4.131),285(3.909), 374(4.356)	18-0321-76
Benzenepropanenitrile, α-[(4-methylphenyl)hydrazono]-β-oxo-	EtOH	250(4.037),280s(3.801), 384(4.301)	18-0321-76
Benzimidazo[1,2-c]quinazolin-6(7H)-one, 2,3-dimethyl-	n.s.g.	228(4.53),248(4.11), 278(3.96),290(4.08), 307(4.15)	2-0879-76
Cyanamide, [[(2-oxo-2-phenylethyl)phenylamino]methylene]-	EtOH	248(4.51),270s(4.33)	49-1413-76
11H-Indolo[3,2-c][1,8]naphthyridine, 8-methoxy-3-methyl-	EtOH	236(4.68),264(4.44), 286(4.60)	4-0097-76
Methanone, (4-amino-1-phenyl-1H-imidazol-5-yl)phenyl-	EtOH	243(4.05),341(4.00)	49-1413-76
Phenol, 2-methyl-4-(6-quinolinylazo)-	C_6H_{12}	250(--),353(3.47)	115-0379-76
	EtOH	255(4.02),290(3.77), 372(4.30)	115-0379-76
	acetone	364(4.40)	115-0379-76
	CHCl$_3$	260(4.41),284(4.28), 363(4.44)	115-0379-76
	CCl$_4$	287(3.91),356(4.18)	115-0379-76
Phenol, 4-methyl-2-(6-quinolinylazo)-	C_6H_{12}	228(3.79),255(3.61), 332(3.90),414(3.58)	115-0379-76
	EtOH	230(4.03),255(3.94), 330(4.17),400(3.86)	115-0379-76

Compound	Solvent	$\lambda_{max}(\log \epsilon)$	Ref.
Phenol, 4-methyl-2-(6-quinolinylazo)- (cont.)	acetone	290(4.11),333(4.09), 411(3.81)	115-0379-76
	CHCl$_3$	259(4.02),335(4.15), 400(3.90)	115-0379-76
Spiro[3H-indole-3,3'-[3H]pyrazol-2(1H)- one, 2',4'-dihydro-5'-phenyl-	EtOH	210(4.32),259(4.12), 296(4.20)	94-0782-76
$C_{16}H_{13}N_3OS$ Methanone, [4,5-dihydro-5-imino-4-(3- methylphenyl)-1,3,4-thiadiazol- 2-yl]phenyl-	EtOH	216(4.25),265(4.21), 368(4.03)	4-0045-76
Methanone, [4,5-dihydro-5-imino-4-(4- methylphenyl)-1,3,4-thiadiazol- 2-yl]phenyl-	EtOH	219(4.13),265(4.15), 370(3.99)	4-0045-76
$C_{16}H_{13}N_3OS_2$ Thiourea, (4-oxo-3-phenyl-2-thiazolidin- ylidene)phenyl-	dioxan	245(4.15),405(4.02)	103-0751-76
$C_{16}H_{13}N_3O_2$ Benzenepropanenitrile, α-[(4-methoxy- phenyl)hydrazono]-β-oxo-	EtOH	250(3.914),290(3.693), 400(4.185)	18-0321-76
2-Buten-1-one, 3-(2H-benzotriazol-2-yl)- 1-phenyl-, N-oxide	EtOH	255(4.09),295(4.00), 350s(3.56)	22-0184-76
4-Oxazolecarboxamide, N-(2-aminophenyl)- 2-phenyl-	EtOH	280(4.16)	39-0315-76C
4-Oxazolecarboxylic acid, 2-phenyl-, 2-phenylhydrazide	EtOH	285(4.26)	39-0315-76C
2-Pyrimidinamine, 4,6-diphenoxy-	EtOH	234(4.69),270(4.68)	42-0913-76
Pyrrolo[2,1-a]isoquinoline-3-carboxylic acid, 2-amino-1-cyano-, ethyl ester	EtOH	220(4.17),270(4.67), 295(4.58),336(4.01), 367(3.76)	94-1299-76
2-Quinolinamine, 4-methyl-3-nitro-N- phenyl-	CHCl$_3$	267(4.50),294(4.18), 308(4.13),414s(3.14)	48-0039-76
5,8-Quinolinedione, 2,4-dimethyl- 6-(phenylamino)-	EtOH	226(4.22),273(4.29), 283s(--),477(3.70)	5-1809-76
$C_{16}H_{13}N_3O_2S$ Methanone, [4,5-dihydro-5-imino-4-(4- methoxyphenyl)-1,3,4-thiadiazol- 2-yl]phenyl-	EtOH	225(4.25),265(4.14), 370(3.95)	4-0045-76
$C_{16}H_{13}N_3O_3$ 5,8-Quinolinedione, 7-amino-2-(2-amino- phenyl)-6-methoxy-	MeCN	320(4.32),380(4.15), 500(3.48)	23-2563-76
$C_{16}H_{13}N_3O_4$ 1,3-Butanedione, 2-[(4-nitrophenyl)azo]- 1-phenyl-	n.s.g.	242(4.18),277(4.92), 390(4.54)	65-2619-76
$C_{16}H_{13}N_5$ 1H-Imidazo[4,5-f]quinazolin-9-amine, N-(phenylmethyl)-	pH 1	232(4.12),276(4.31), 307s(--),312s(--), 323(4.02),329s(--)	102-0609-76
	pH 7	238(4.09),265(4.25), 277s(--),311s(--), 323(3.98),336(3.93)	102-0609-76
	pH 13	272(4.43),296s(--), 332(3.93),343s(--)	102-0609-76
1H-Imidazo[4,5-g]quinazolin-8-amine, N-(phenylmethyl)-	EtOH	227(4.58),236s(--), 264(4.11),272(4.09),	102-0609-76

Compound	Solvent	$\lambda_{max}(\log \epsilon)$	Ref.
1H-Imidazo[4,5-g]quinazolin-8-amine, N-(phenylmethyl)- (cont.)		322s(--),334(4.15), 351(4.05)	102-0609-76
	EtOH-HCl	217(4.54),237s(--), 265(3.89),324s(--), 338(4.30),354(4.32)	102-0609-76
	EtOH-NaOH	244(4.63),248s(--), 270s(--),277(4.50), 324(4.10),349(3.98), 365(3.90)	102-0609-76
1H-Imidazo[4,5-h]quinazolin-6-amine, N-(phenylmethyl)-	EtOH	228(4.42),253(4.62), 300(4.15),314(4.10)	102-0609-76
	EtOH-HCl	222(4.31),261(4.53), 314(4.17)	102-0609-76
	EtOH-NaOH	266(4.81),314(4.05)	102-0609-76
$C_{16}H_{13}N_5O_2$ 1H-Pyrazolo[3,4-d]pyrimidine-4-acetic acid, α-cyano-1-phenyl-, ethyl ester	EtOH	265(4.03),325(4.39), 337(4.44)	95-1352-76
$C_{16}H_{13}N_5O_2S$ 1,2,4-Triazolo[3,4-b][1,3,4]thiadiazol-ium, 1-(phenylamino)-6-[(phenylsul-fonyl)amino]-, hydroxide, inner salt	MeCN	272(4.20)	30-0104-76
$C_{16}H_{13}N_5O_5$ Uridine, 2'-azido-2',3'-didehydro-2',3'-dideoxy-, 5'-benzoate	MeOH	230(4.33),258s(4.13)	44-3138-76
$C_{16}H_{13}O_3S$ 1,3-Benzoxathiol-1-ium, 2-[4-(ethoxy-carbonyl)phenyl]-, tetrafluoroborate	96% H_2SO_4	275(4.17),373(4.31)	39-0323-76C
$C_{16}H_{14}$ 1,3-Methano-1H-indene, 2,3-dihydro-2-phenyl-, endo	MeOH	256s(2.77),262(2.90), 269(2.98),275(2.93)	35-1052-76
exo-	MeOH	254s(2.99),260(3.09), 267(3.15),274(3.13)	35-1052-76
Tricyclo[3.3.1.0^{2,8}]nona-3,6-diene, 9-(phenylmethylene)-	C_6H_{12}	263(4.15)	88-1565-76
$C_{16}H_{14}BrN$ 5H-Dibenz[b,f]azepine, 10-bromo-5-ethyl-	MeOH	258(4.44),283(3.81)	87-1345-76
$C_{16}H_{14}BrN_2$ Pyridinium, 1-[(6-bromo-1-methyl-2(1H)-quinolinylidene)methyl]-, perchlorate	EtOH	238(4.05),256(4.15), 455(3.06)	104-0444-76
$C_{16}H_{14}BrN_5O_3S_3$ 3-Thiazolidinebutanoic acid, 4-[(amino-thioxomethyl)hydrazono]-5-(5-bromo-1,2-dihydro-2-oxo-3H-indol-3-ylidene)-2-thioxo-	MeOH	245(4.45),290(3.68), 297s(--),420s(--)	103-0749-76
$C_{16}H_{14}BrO$ 1H-Benzo[c]furanium, 3-(4-bromophenyl)-1,1-dimethyl-	H_2SO_4	365(4.56)	104-2523-76
$C_{16}H_{14}BrS$ 1H-Benzo[c]thiolium, 3-(4-bromophenyl)-1,1-dimethyl-	H_2SO_4	382(4.37)	104-2523-76

Compound	Solvent	$\lambda_{max}(\log \epsilon)$	Ref.
$C_{16}H_{14}ClF_3N_4O_6$			
Riboflavin, 7-chloro-7-demethyl-$\alpha^8,\alpha^8,\alpha^8$-trifluoro-	EtOH	221(4.49),270(4.57), 321(3.81),450(3.98)	103-1386-76
Riboflavin, 8-chloro-8-demethyl-$\alpha^7,\alpha^7,\alpha^7$-trifluoro-	EtOH	270(4.50),335(3.82), 430(3.96)	103-0938-76
	EtOH	228(4.52),272(4.55), 335(3.89),430(4.06)	103-1386-76
$C_{16}H_{14}ClF_3O_6$			
Acetic acid, trifluoro-, monoester with [1aR-[1aα,2α,3aβ,4α,5(Z),7aS*]]-5-(1-chloro-1-propenyl)-1a,2,3a,4-tetrahydro-3a,4-dihydroxy-2-methyl-4,2-(epoxymethano)-1H,7H-cyclopropa[c]benzofuran-7-one	MeOH	228(3.93),293(4.05)	33-1809-76
$C_{16}H_{14}ClNO_2$			
1,4-Benzoxazepine, 7-chloro-2,3-dihydro-8-methoxy-5-phenyl-	n.s.g.	210(4.36),255(4.02), 305(4.18)	20-0787-76
2,4,6-Heptatrienoic acid, 5-chloro-2-cyano-7-phenyl-, ethyl ester	CH_2Cl_2	389(4.64)	48-0705-76
1H-Indole, 2-(4-chlorophenyl)-4,7-dimethoxy-	MeOH	212(4.42),258(4.44), 301(4.40)	111-0241-76
$C_{16}H_{14}ClNO_5S$			
2H-Pyran-3-carboxylic acid, 6-chloro-2-oxo-4-[[[(phenylmethyl)thio]carbonyl]amino]-, ethyl ester	MeCN	372(4.25)	39-2462-76C
$C_{16}H_{14}ClN_2$			
Pyridinium, 1-[(6-chloro-1-methyl-2(1H)-quinolinylidene)methyl]-, perchlorate	EtOH	237(3.74),259(3.84), 460(4.31)	104-0444-76
$C_{16}H_{14}ClN_3O_4$			
Glycine, N-[4-[[[(4-chlorophenyl)amino]-carbonyl]amino]benzoyl]-	EtOH	212(4.59),280(5.02)	80-1345-76
$C_{16}H_{14}ClO$			
1H-Benzo[c]furanium, 3-(4-chlorophenyl)-1,1-dimethyl-	H_2SO_4	359(4.53)	104-2523-76
$C_{16}H_{14}ClS$			
1H-Benzo[c]thiolium, 3-(4-chlorophenyl)-1,1-dimethyl-	H_2SO_4	380(4.35)	104-2523-76
$C_{16}H_{14}Cl_4N_2$			
Diazene, bis[1-chloro-1-(3-chlorophenyl)ethyl]-, meso	benzene	355(1.66)	39-1249-76B
Diazene, bis[1-chloro-1-(4-chlorophenyl)ethyl]-, meso	CH_2Cl_2	353(1.76)	39-1249-76B
$C_{16}H_{14}Cl_5NO$			
Benzenemethanamine, N-(2,4-dichlorophenyl)-4-ethoxy-	n.s.g.	247(4.25),282(3.27), 303(3.33)	98-0724-76
$C_{16}H_{14}F_3NO_3$			
2-Butenoic acid, 3-(5-methyl-3-phenyl-4-isoxazolyl)-, 2,2,2-trifluoroethyl ester, (E)-	EtOH	235(4.18)	33-2074-76
(Z)-	EtOH	225(4.17)	33-2074-76

Compound	Solvent	$\lambda_{max}(\log \epsilon)$	Ref.
2-Butenoic acid, 3-(5-methyl-2-phenyl-4-oxazolyl)-, 2,2,2-trifluoroethyl ester, (E)-	EtOH	218(4.17),289(4.31)	33-2074-76
(Z)-	EtOH	215(4.25),282(4.28)	33-2074-76
4-Isoxazolepropanoic acid, 5-methyl-β-methylene-3-phenyl-, 2,2,2-tri-fluoroethyl ester	EtOH	230(4.08)	33-2074-76
$C_{16}H_{14}F_3N_3O_2$			
Benzenamine, N,N-dimethyl-4-(2,2,2-tri-fluoro-1-[(4-nitrophenyl)imino]ethyl]-	hexane	223(4.09),304(4.07), 365(4.12)	104-0161-76
1,4-Benzenediamine, N,N-dimethyl-N'-[2,2,2-trifluoro-1-(4-nitrophenyl)-ethylidene]-	hexane	254(4.13),320(3.80), 412(3.63)	104-0161-76
$C_{16}H_{14}IN_2$			
Pyridinium, 1-[(6-iodo-1-methyl-2(1H)-quinolinylidene)methyl]-, perchlorate	EtOH	235(4.23),260(4.11), 453(3.40)	104-0444-76
$C_{16}H_{14}NS_2$			
Benzenaminium, N-methyl-N-(5-phenyl-3H-1,2-dithiol-3-ylidene)-, iodide	MeOH	312(4.24),361(4.09)	48-0221-76
$C_{16}H_{14}NSe$			
Isoselenazolium, 5-(4-methylphenyl)-2-phenyl-, perchlorate	HOAc	299s(3.73),357(4.34)	118-0273-76
$C_{16}H_{14}N_2$			
Indolo[3,2-b]indole, 5,10-dihydro-5,10-dimethyl-	MeCN	240s(4.16),264(4.79), 312s(4.16),326(4.26), 355(3.79)	5-1090-76
semiquinone	MeCN	221(4.41),231(4.34), 248s(4.32),270(4.55), 282(4.39),303(4.20), 317s(4.12),357(3.26), 368(3.26),396s(3.42), 411(3.80),442(3.63), 464s(3.27),500(3.23), 616(3.29),655(3.34), 710s(3.20)	5-1090-76
Pyrrolo[2,1-a]isoquinoline-3-carbo-nitrile, 2-ethyl-1-methyl-	MeOH	211(4.44),223s(4.24), 228s(4.19),241(4.28), 270s(4.68),276(4.71), 284s(4.45),316s(4.21), 330(4.26),347(4.25), 353s(4.18),364(4.15)	39-1908-76C
2-Quinolinamine, 4-methyl-N-phenyl-	hexane	345(4.03),360(3.97)	18-2770-76
	MeOH	349(3.94),355(3.92)	18-2770-76
$C_{16}H_{14}N_2O$			
Acetamide, N-(2-phenyl-1H-indol-3-yl)-	MeOH	222(4.52),280(3.94), 288(3.90)	23-1020-76
1,3,6-Cycloheptatriene-1-carbonitrile, 6-(dimethylamino)-5-oxo-3-phenyl-	MeCN	280(4.2),362(3.3)	39-2329-76C
Dibenzo[b,f]cycloprop[d]azepine-6(1H)-carboxamide, 1a,10b-dihydro-	EtOH	271(2.96)	94-2751-76
Furo[2,3-g]indazole, 4,5-dihydro-7-methyl-1-phenyl-	EtOH	254(4.08)	32-1083-76
1H-Indole-2-carboxamide, 1-methyl-N-phenyl-	EtOH	304(4.41)	103-1249-76

Compound	Solvent	$\lambda_{max}(\log \epsilon)$	Ref.
1(2H)-Isoquinolinone, 3-[(phenylmethyl)-amino]-	EtOH	229(4.29),244(3.97), 292(4.14),302(4.20), 372(3.68)	95-0154-76
1H-Pyrrole, 3-methoxy-5-phenyl-2-[(2H-pyrrol-2-ylidene)methyl]-	EtOH-HCl	484(4.81)	4-0447-76
	EtOH-NaOH	436(4.46)	4-0447-76
Pyrrolo[1,2-c]pyrimidine-5-carboxalde-hyde, 1,3-dimethyl-6-phenyl-	EtOH	242(4.50),282s(3.79), 348(4.19)	44-0351-76

$C_{16}H_{14}N_2OS$

Compound	Solvent	$\lambda_{max}(\log \epsilon)$	Ref.
Benzothiazole, 2-methyl-6-nitroso-7-(2-phenylethyl)-	$CHCl_3$	760(1.81)	118-0270-76

$C_{16}H_{14}N_2O_2$

Compound	Solvent	$\lambda_{max}(\log \epsilon)$	Ref.
1,2-Anthracenedione, bis(O-methyloxime)	hexane	220(4.7),293(4.38), 310s(4.26),338(3.82), 358s(3.89),385s(3.36), 404(3.15)	110-1711-76
	EtOH	223(4.6),295(4.62), 316s(4.4),344(4.0), 362s(3.84),390s(3.69), 412s(3.64)	110-1711-76
	1:4 dioxan-hexane	220(4.6),298(4.48), 314s(4.22),342(3.95), 390(3.48),410(3.55)	110-1711-76
9,10-Anthracenedione, 1-amino-2-(dimeth-ylamino)-	EtOH	500(3.77)	104-2503-76
9,10-Anthracenedione, 2-amino-1-(dimeth-ylamino)-	EtOH	483(3.69)	104-2503-76
9,10-Anthracenedione, 2-amino-3-(dimeth-ylamino)-	EtOH	469(3.49)	104-2503-76
9,10-Anthracenedione, 1,2-bis(methylam-ino)-	EtOH	523(3.85)	104-2503-76
9,10-Anthracenedione, 2,3-bis(methylam-ino)-	EtOH	533(3.75)	104-2503-76
2H-1-Benzopyran-2-one, 3-amino-4-[(phen-ylmethyl)amino]-	EtOH	253(4.09),341(4.09)	103-0268-76
3H-Indazol-5-ol, 3-methyl-3-phenyl-, acetate	isoPrOH	220(4.26),268(3.87), 346(2.44)	4-0033-76
5H-Indazol-5-one, 2-acetyl-2,3-dihydro-3-methyl-3-phenyl-	isoPrOH	252(3.51),374(4.46)	4-0033-76
Methanone, (3-amino-4,6-dimethylfuro-[2,3-b]pyridin-2-yl)phenyl-	EtOH	256(4.17),314(4.09), 386(4.20)	48-0313-76
3,5-Pyrazolidinedione, 4-methyl-1,2-di-phenyl-	hexane	238(4.15),279(3.34)	28-0343-76B
	pH 13	248(4.18),262(4.24)	28-0343-76B
	dioxan	240(4.16),277(3.43)	28-0343-76B
4H-Pyrazol-4-one, 1,5-dihydro-5-(4-meth-oxyphenyl)-3-phenyl-	EtOH	259(4.477)	23-1752-76
2H-Pyrrol-2-one, 1,5-dihydro-5-(hydroxy-phenylamino)-1-phenyl-	EtOH	230(4.08),280(3.68)	44-2496-76
4(3H)-Quinazolinone, 3-ethyl-2-(2-furan-ylethenyl)-	EtOH	220(4.60),312(4.38)	115-0341-76

$C_{16}H_{14}N_2O_3$

Compound	Solvent	$\lambda_{max}(\log \epsilon)$	Ref.
Acetamide, N-(2-ethyl-2,3-dihydro-1,3-dioxo-1H-benz[de]isoquinolin-5-yl)-	EtOH	209(4.23),256(4.69), 342(4.02),375(3.74)	56-1257-76
Acetamide, N-(2-ethyl-2,3-dihydro-1,3-dioxo-1H-benz[de]isoquinolin-6-yl)-	EtOH	214(4.54),241(4.54), 367(4.16)	56-1257-76
Acetic acid, [(3-cyano-6-phenyl-2-pyri-dinyl)oxy]-, ethyl ester	EtOH	264(4.10),320(4.40)	48-0313-76
Benzoic acid, 2-[3-(4-hydroxyphenyl)-1H-pyrazol-5-yl]-	DMF	282(4.24)	44-0110-76

Compound	Solvent	$\lambda_{max}(\log \epsilon)$	Ref.
Furo[2,3-b]pyridine-2-carboxylic acid, 3-amino-6-phenyl-, ethyl ester	EtOH	267(4.27),324(4.35), 357s(4.02)	48-0313-76
1H-Pyrazole, 1-acetyl-3-benzoyl-4-(2-furanyl)-4,5-dihydro-	EtOH	219(4.18),268(4.03), 305(4.20)	104-2358-76
3-Pyridinecarboxylic acid, 1,6-dihydro-1-[2-(1H-indol-3-yl)ethyl]-6-oxo-	MeOH	221(4.53),260(4.22), 290(3.95)	24-0705-76
5,8-Quinolinediol, 2-(2-aminophenyl)-6-methoxy-	MeCN	290(4.30),360(3.49), 470(4.00)	23-2563-76

$C_{16}H_{14}N_2O_4S$

Compound	Solvent	$\lambda_{max}(\log \epsilon)$	Ref.
Benzeneacetamide, N-(1,4,5a,6-tetrahydro-1,7-dioxo-3H,7H-azeto[2,1-b]furo[3,4-d][1,3]thiazin-6-yl)-, (5aR-cis)-	EtOH	259(4.00)	33-2298-76
(5aR-trans)-	EtOH	262(3.99)	33-2298-76
1H-Indazole-4,7-dione, 3,6-dimethyl-1-[(4-methylphenyl)sulfonyl]-	EtOH	236(4.28),259(4.16), 319(3.59)	94-1731-76
Pyrrolo[2,1-a]isoquinoline-3-carboxylic acid, 2-(methylthio)-1-nitro-, ethyl ester	EtOH	220(4.42),257(4.45), 279(4.55),340(3.89), 390(3.58)	94-1299-76

$C_{16}H_{14}N_2O_5$

Compound	Solvent	$\lambda_{max}(\log \epsilon)$	Ref.
Benzene, 1,1'-(1-ethoxy-1,2-ethenediyl)-bis[4-nitro-	n.s.g.	352(4.44)	2-0101-76

$C_{16}H_{14}N_2O_6$

Compound	Solvent	$\lambda_{max}(\log \epsilon)$	Ref.
2,2'-Anhydro-1-(5-O-benzoyl-β-D-arabinofuranosyl)uracil	MeOH	227(4.34),251s(3.93)	44-3138-76

$C_{16}H_{14}N_2S$

Compound	Solvent	$\lambda_{max}(\log \epsilon)$	Ref.
3-Quinolinamine, 2-(methylthio)-4-phenyl-	CHCl$_3$	258(4.51),298(3.56), 312(3.55),355(3.93)	48-0039-76

$C_{16}H_{14}N_3O_2$

Compound	Solvent	$\lambda_{max}(\log \epsilon)$	Ref.
Pyridinium, 1-[(1-methyl-6-nitro-2(1H)-quinolinylidene)methyl]-, perchlorate	EtOH	245(4.12),260(3.95), 435(3.39)	104-0444-76

$C_{16}H_{14}N_4$

Compound	Solvent	$\lambda_{max}(\log \epsilon)$	Ref.
Benzenepropanenitrile, β-imino-α-[(2-methylphenyl)hydrazono]-	EtOH	235(3.32),375(4.36)	33-0551-76
Benzenepropanenitrile, β-imino-α-[(3-methylphenyl)hydrazono]-	EtOH	235(3.42),373(4.46)	33-0551-76
Benzenepropanenitrile, β-imino-α-[(4-methylphenyl)hydrazono]-	EtOH	240(3.45),370(4.44)	33-0551-76

$C_{16}H_{14}N_4O$

Compound	Solvent	$\lambda_{max}(\log \epsilon)$	Ref.
Benzaldehyde, (2-amino-1,2-dihydro-1-oxo-3-isoquinolinyl)-	EtOH	229(4.39),259(4.27), 294(4.03),365(4.29)	95-0700-76
Cyclopentanone, 2-(1-phenyl-1H-pyrazolo[3,4-d]pyrimidin-4-yl)-	EtOH	227s(3.68),237s(3.66), 268(3.88),358(4.47), 373(4.52)	95-1352-76

$C_{16}H_{14}N_4OS$

Compound	Solvent	$\lambda_{max}(\log \epsilon)$	Ref.
2-Thiazolamine, 5-[(4-methoxyphenyl)azo]-4-phenyl-	EtOH	270(4.362),446(4.452)	4-0045-76

$C_{16}H_{14}N_4O_2$

Compound	Solvent	$\lambda_{max}(\log \epsilon)$	Ref.
Benzenebutanamide, α-azido-γ-oxo-N-phenyl-	EtOH	244(4.40)	115-0001-76
Benzoic acid, 2-(2-amino-1,2-dihydro-1-oxo-3-isoquinolinyl)hydrazide	EtOH	230(4.51),298(4.15), 358(3.71)	95-0700-76
Urea, N-[3,4-dihydro-2-(4-methylphenyl)-4-oxo-3-quinazolinyl]-	EtOH	215s(4.43),232(4.46), 275(4.23),310(4.02)	114-0341-76D

Compound	Solvent	$\lambda_{max}(\log \epsilon)$	Ref.
$C_{16}H_{14}N_4O_3$			
Benzenamine, N-hydroxy-4-(3-methyl-5-phenyl-1H-pyrazol-1-yl)-3-nitro-	EtOH	248(4.35),352(3.31)	22-0839-76
Benzenamine, N-hydroxy-4-(4-methyl-5-phenyl-1H-pyrazol-1-yl)-3-nitro-	EtOH	248(4.34),357(3.32)	22-0839-76
Urea, N-[3,4-dihydro-2-(4-methoxyphenyl)-4-oxo-3-quinazolinyl]-	EtOH	218(4.43),237(4.29), 277(4.31),308(4.12)	114-0341-76D
$C_{16}H_{14}N_4O_4$			
1H-Benzimidazole, 2-[1-(2,4-dinitrophenyl)ethyl]-1-methyl-	EtOH	246(4.27),275(4.03), 282(3.93)	104-2361-76
photochromic form	EtOH	520(4.31)	104-2361-76
anion	50% EtOH	545(4.53)	104-2361-76
1H-Benzimidazole, 2-[(2,4-dinitrophenyl)methylene]-2,3-dihydro-1,3-dimethyl-	benzene	525(4.62)	104-1097-76
	EtOH	504(4.36)	104-1097-76
	DMF	506(4.59)	104-1097-76
	$CHCl_3$	524(4.56)	104-1097-76
4-Cinnolinamine, 3-ethoxy-N-(2-nitrophenyl)-, 1-oxide	$CHCl_3$	440(4.01)	39-0399-76C
4-Cinnolinamine, 3-methoxy-N-(2-methyl-6-nitrophenyl)-, 1-oxide	$CHCl_3$	448(3.93)	39-0399-76C
4-Cinnolinamine, 3-methoxy-N-(4-methyl-2-nitrophenyl)-, 1-oxide	$CHCl_3$	438(4.24)	39-0399-76C
$C_{16}H_{14}N_4O_5$			
4-Cinnolinamine, 3-methoxy-N-(4-methoxy-2-nitrophenyl)-, 1-oxide	$CHCl_3$	445(4.05)	39-0399-76C
$C_{16}H_{14}N_4S$			
2-Thiazolamine, 5-[(4-methylphenyl)azo]-4-phenyl-	EtOH	270(4.298),430(4.395)	4-0045-76
1H-1,2,4-Triazolium, 3-mercapto-1-(phenylmethyl)-4-[(phenylmethylene)amino]-, hydroxide, inner salt	MeCN	266(4.52),342(3.41)	103-0361-76
$C_{16}H_{14}N_4Se$			
Imidazo[1,2-a]pyridine, 3,3'-selenobis-[7-methyl-, 1:1 compound with hydrogen triiodide	EtOH	220(4.62),227(4.61), 234(4.59),287(4.56), 356(4.11)	44-3549-76
$C_{16}H_{14}N_6$			
Formazan, 3-methyl-1-phenyl-5-(4-quinazolinyl)-	C_6H_{12}	290(4.15),355(4.21), 385(3.90),470(3.56)	103-0596-76
	EtOH	290(4.20),355(4.18)	103-0596-76
	EtOH-KOH	583(3.83)	103-0596-76
	dioxan	365(4.00),393(3.96), 465(3.82)	103-0596-76
nickel complex	EtOH-pH 7	652(3.99)	103-0596-76
1H-Indazole, 3,3'-azobis[1-methyl-	MeCN	210(4.51),254s(3.99), 261(4.00),273s(3.95), 395(4.46),424s(4.24)	5-1039-76
$C_{16}H_{14}N_6O$			
Pyrimido[1,2-a]purin-10(1H)-one, 7-[4-(dimethylamino)phenyl]-	pH 1	245(4.35),263s(4.32), 317(4.06),360s(3.46)	44-0294-76
	pH 6.8	222(4.35),257(4.30), 329(4.41)	44-0294-76
	pH 10.0	231(4.42),270(4.30), 324(4.42)	44-0294-76

Compound	Solvent	$\lambda_{max}(\log \epsilon)$	Ref.
$C_{16}H_{14}N_8O_5$			
Uridine, 2',3'-diazido-2',3'-dideoxy-, 5'-benzoate	MeOH	228(4.23),259(4.05)	44-3138-76
$C_{16}H_{14}O$			
9(10H)-Anthracenone, 2,3-dimethyl-	EtOH	216(4.29),264(4.21), 275s(4.20)	12-2257-76
9(10H)-Anthracenone, 10,10-dimethyl-	EtOH	268(4.19),290s(3.78), 300s(3.62)	35-0171-76
3-Buten-2-one, 3,4-diphenyl- (4-^{14}C)	MeOH	226(4.15),296(4.25)	56-0499-76
Methanone, phenyl(2-phenylcyclopropyl)-	EtOH	244(4.31)	44-3067-76
2-Propen-1-one, 1-(4-methylphenyl)-3-phenyl-	n.s.g.	312(4.34)	39-0380-76C
2-Propen-1-one, 3-(4-methylphenyl)-1-phenyl-	n.s.g.	334(4.32)	39-0380-76C
$C_{16}H_{14}OS$			
Benzo[b]thiophene, 3-(4-methoxyphenyl)-2-methyl-	C_6H_{12}	235(4.61),263(4.10), 292(3.81),302(3.76)	138-0389-76
2-Propene-1-thione, 3-hydroxy-3-(4-methylphenyl)-1-phenyl-	EtOH	270(3.93),327(4.10), 414(4.24),505s(2.36)	4-0691-76
$C_{16}H_{14}O_2$			
3(2H)-Benzofuranone, 5,6-dimethyl-2-phenyl-	EtOH	220(3.86),275(3.15), 283(3.06)	111-0533-76
2H-1-Benzopyran, 3-methoxy-2-phenyl-	EtOH	222(4.30),276(3.04), 298(3.00)	22-1967-76
2H-1-Benzopyran, 3-methoxy-4-phenyl-	MeOH	276(3.78),301(3.68)	35-3555-76
Indeno[1,2-b]pyran-9-carboxaldehyde, 2,3,4-trimethyl-	dioxan	244(4.41),281(3.85), 320(4.36),348(4.04), 476(3.82)	78-2225-76
Naphth[2,1-b]oxepin-1(2H)-one, 3,4-dihydro-4-methyl-3-methylene-	n.s.g.	218(4.59),240s(4.23), 307(3.81)	33-0664-76
Naphth[2,1-b]oxepin-1(2H)-one, 3,4-dimethyl-	n.s.g.	263(4.63),312(3.68), 350(3.48)	33-0664-76
Naphth[2,1-b]oxepin-1(2H)-one, 3-ethylidene-3,4-dihydro-	n.s.g.	221(4.52),236s(4.34), 264(3.94),311(3.89), 344(3.68)	33-0664-76
Naphth[2,1-b]oxepin-1(4H)-one, 3,4-dimethyl-	n.s.g.	220(4.75),250(4.17), 319(3.71),338s(3.62)	33-0664-76
2-Oxabenzobicyclo[3.1.0]hexene, 1-methoxy-5-phenyl-	MeOH	277(3.36),284(3.34)	35-3555-76
4(1H)-Phenanthrenone, 1-hydroxy-1,2-dimethyl-	n.s.g.	218(4.56),250(4.27), 325s(3.87),332(3.89)	33-0664-76
2-Propen-1-one, 1-(3-methoxyphenyl)-3-phenyl-	n.s.g.	300(4.34)	39-0380-76C
2-Propen-1-one, 1-(4-methoxyphenyl)-3-phenyl-	n.s.g.	312(4.41)	39-0380-76C
2-Propen-1-one, 3-(3-methoxyphenyl)-1-phenyl-	n.s.g.	302(4.28)	39-0380-76C
2-Propen-1-one, 3-(4-methoxyphenyl)-1-phenyl-	n.s.g.	334(4.43)	39-0380-76C
$(C_{16}H_{14}O_2)_n$			
Benzaldehyde, 4-[(ethenylphenyl)methoxy]-, homopolymer	THF	273(4.31),282s(4.25), 328s(2.70)	47-1661-76 +116-0221-76
$C_{16}H_{14}O_2S$			
Dibenzo[b,f]thiepin, 2,3-dimethoxy-	MeOH	248(4.37),265(4.41), 300s(3.67),340s(3.37)	73-1396-76

Compound	Solvent	$\lambda_{max}(\log \epsilon)$	Ref.
2-Propenoic acid, 3-phenyl-2-(phenyl-thio)-, methyl ester	75% dioxan	260(4.2),278(4.2)	104-0149-76
2-Thiophenecarboxylic acid, 5-(4-phenyl-1,3-butadienyl)-, methyl ester	MeOH	231s(4.02),240(4.00), 249(3.98),265s(4.18), 271(4.21),349s(4.51), 367(4.60),381(4.49)	48-0731-76
$C_{16}H_{14}O_2S_2$			
Cycloocta[1,2-c:3,4-c']dithiophene-4,7-dione, 1,3,8,10-tetramethyl-	EtOH	232s(4.42),295(3.47), 360(3.08)	35-3627-76
$C_{16}H_{14}O_2S_3$			
Benzenecarbodithioic acid, 4-methoxy-, anhydrosulfide	C_6H_{12}	355(4.55),559(2.56)	89-0766-76
$C_{16}H_{14}O_3$			
9H-Fluorene-1-carboxylic acid, 2-hydr-oxy-8-methyl-, methyl ester	EtOH EtOH-NaOEt	240(4.18),270(4.21) 300(4.32)	78-0065-76 78-0065-76
2(3H)-Furanone, dihydro-5-(4-hydroxy-phenyl)-5-phenyl-	EtOH-pH 9.2	412(3.66)	18-2027-76
1-Phenanthrenecarboxylic acid, 9,10-di-hydro-2-hydroxy-, methyl ester	EtOH EtOH-NaOH	245(3.94),274(4.0) 304(4.20)	78-0073-76 78-0073-76
Psoralen, 3-(α,α-dimethylallyl)-	EtOH	246(4.46),293(4.14), 330(4.0)	32-0681-76
9H-Xanthene-2-carboxylic acid, ethyl ester	EtOH	275(4.20)	94-1588-76
$C_{16}H_{14}O_3S$			
Dibenzo[b,f]thiepin-10(11H)-one, 2,3-di-methoxy-	MeOH	233(4.40),251(4.20), 290s(3.49),336(3.52)	73-1396-76
$C_{16}H_{14}O_3S_2$			
Benzeneacetic acid, α-[[(4-methoxyphen-yl)thioxomethyl]thio]-	dioxan	233(4.27),286s(4.06), 305(4.83),500(2.10)	24-0740-76
$C_{16}H_{14}O_4$			
Benzoic acid, 2-(5-ethyl-2-hydroxybenz-oyl)-	EtOH	260(4.16),340(3.75)	56-0759-76
Dibenz[c,e]oxepin-5(7H)-one, 1,2-dimeth-oxy-	EtOH	242(4.52),295(3.47)	94-2191-76
Dibenz[c,e]oxepin-5(7H)-one, 2,3-dimeth-oxy-	EtOH	246(4.41),282(4.02)	94-2191-76
Dibenz[c,e]oxepin-5(7H)-one, 9,10-di-methoxy-	EtOH	268(3.95),301(3.83)	94-2191-76
Dibenz[c,e]oxepin-5(7H)-one, 10,11-di-methoxy-	EtOH	220(4.57),287(3.67)	94-2191-76
9H-Fluorene-1-carboxylic acid, 2-hydr-oxy-7-methoxy-, methyl ester	EtOH EtOH-NaOEt	248(4.24),277(4.38) 248(4.16),293(4.46)	78-0065-76 78-0065-76
2(3H)-Furanone, 5-(2,4-dihydroxyphenyl)-dihydro-5-phenyl-	EtOH-pH 9.1	500(3.52)	18-2027-76
1,2-Heptalenedicarboxylic acid, dimeth-yl ester	hexane	204(4.36),266(4.29), 337(3.63)	89-0104-76
$C_{16}H_{14}O_4S$			
Benzeneacetic acid, 2-carboxy-α-[(5-methyl-2-thienyl)methylene]-, α-methyl ester, (E)-	EtOH	320(4.29)	4-0083-76
Butanedioic acid, [(5-methyl-2-thienyl)-methylene]phenyl-, (E)-	EtOH	205(4.11),265(3.72), 310(4.24)	4-0285-76
Butanedioic acid, phenyl(2-thienylmeth-ylene)-, 1-methyl ester, (E)-	EtOH	212(3.99),270(4.06), 315(4.26)	4-0285-76

Compound	Solvent	λ_{max}(log ϵ)	Ref.
3-Furancarboxylic acid, tetrahydro-5-oxo-4-phenyl-2-(2-thienyl)-, methyl ester	EtOH	219(4.02),233(4.10)	4-0285-76
$C_{16}H_{14}O_5$			
1,3-Azulenedicarboxylic acid, 6-formyl-2-methyl-, dimethyl ester	$CHCl_3$	252(4.50),309(4.73), 320(4.86),350s(3.86), 367(3.99),570(2.59)	88-2045-76
1,3-Benzenediol, 3-(7-hydroxy-2H-1-benzopyran-3-yl)-6-methoxy- (sepiol)	EtOH	323(4.44)	88-1741-76
	EtOH-NaOAc-H_3BO_3	338(--)	88-1741-76
4H-1-Benzopyran-4-one, 2,3-dihydro-5-hydroxy-2-(2-hydroxyphenyl)-7-methoxy-	EtOH	286(4.30),326s(3.52)	2-0644-76
Butanedioic acid, (2-furanylmethylene)-phenyl-, 1-methyl ester, (E)-	EtOH	211(4.20),301(4.49)	48-0816-76
(Z)-	EtOH	207(4.29),301(4.03)	48-0816-76
3-Furancarboxylic acid, 2-(2-furanyl)-tetrahydro-5-oxo-4-phenyl-, methyl ester	EtOH	209(4.10),278(3.90)	48-0816-76
2(3H)-Furanone, dihydro-5-phenyl-5-(2,4,6-trihydroxyphenyl)-	EtOH-pH 8.2	475(3.45)	18-2027-76
9H-Xanthen-9-one, 1-hydroxy-3,6-dimethoxy-8-methyl- (lichexanthone)	MeOH	242(4.37),252s(4.18), 269s(3.88),306(4.09), 340s(3.63)	102-1093-76
	MeOH-NaOH	239(4.54),270(4.22), 308(4.12),347(3.73)	102-1093-76
9H-Xanthen-9-one, 1-hydroxy-3,6-dimethoxy-8-methyl-	MeOH	257(3.8),270(3.78), 307(3.90),340(3.65)	102-1799-76
9H-Xanthen-9-one, 1-hydroxy-4,6-dimethoxy-8-methyl-	EtOH	235(4.38),253(4.06), 277(4.20),298(4.05), 371(3.54)	142-0167-76S
9H-Xanthen-9-one, 1,3,7-trimethoxy-	n.s.g.	209(4.21),242(4.48), 255(4.60),303(4.13), 359(3.85)	39-1377-76C
$C_{16}H_{14}O_6$			
Dibenzofuran-1(9bH)-one, 2-acetyl-6,9b -dimethyl-3,7,9-trihydroxy-	MeOH	231(4.04),267(3.82), 325(3.22)	23-3721-76
$C_{16}H_{14}O_6S$			
2-Propenoic acid, 3-[3,4-dihydroxy-5-(phenylsulfonyl)phenyl]-, methy⁻ ester	EtOH	242(4.32),301(3.94), 347(4.16),370(4.15)	88-0313-76
$C_{16}H_{14}S$			
Benzo[b]thiophene, 2-ethyl-3-phenyl-	EtOH	234(4.49),265(3.97), 292(3.57),302(3.57)	12-0883-76
$C_{16}H_{14}S_3$			
Benzenecarbodithioic acid, 4-methyl-, anhydrosulfide	C_6H_{12}	334(4.40),568(2.18)	89-0766-76
$C_{16}H_{15}BrN_2$			
Quinolinium, 6-bromo-1-methyl-2-(pyridiniomethyl)-, diperchlorate	EtOH	259(4.56),319(3.80), 448(3.59)	104-0444-76
$C_{16}H_{15}BrS_2$			
Benzene, 1,1'-(bromoethenylidene)bis-[2-(methylthio)-	C_6H_{12}	229(4.36),245(4.32)	138-0389-76

Compound	Solvent	$\lambda_{max}(\log \epsilon)$	Ref.
$C_{16}H_{15}Cl$ 1-Butene, 2-chloro-1,1-diphenyl-	EtOH	213(4.03),240(3.95)	12-0883-76
$C_{16}H_{15}ClN_2$ Quinolinium, 6-chloro-1-methyl-2-(pyri- diniomethyl)-, diperchlorate	EtOH	260(4.69),320(3.58), 449(3.31)	104-0444-76
$C_{16}H_{15}ClN_2O$ 1H-Indazole, 2-acetyl-5-chloro-2,3-di- hydro-3-methyl-3-phenyl-	isoPrOH	257(4.06),300s(3.31)	4-0033-76
$C_{16}H_{15}ClN_2O_3$ Benzoic acid, 4-[(6-chloro-3-pyridazin- yl)oxy]-3-(2-propenyl)-, ethyl ester	EtOH	230(3.97),311(4.24)	94-1588-76
$C_{16}H_{15}ClO_2S$ Benzenepropanoic acid, β-chloro- α-(phenylthio)-, methyl ester	75% dioxan	260(3.5)	104-0149-76
Dibenzo[b,f]thiepin, 10-chloro-10,11-di- hydro-2,3-dimethoxy-	MeOH	259(3.91),270(3.90), 283s(3.79)	73-1396-76
$C_{16}H_{15}ClO_3$ 1H-Xanthene-7-carboxylic acid, 2-chloro- 4,9-dihydro-, ethyl ester	EtOH	225(4.03),277(3.93), 297(3.82)	94-1588-76
$C_{16}H_{15}Cl_2N$ Aziridine, 2,2-dichloro-3-methyl- 1-(3-methylphenyl)-3-phenyl-	C_6H_{12}	236(4.11)	18-1893-76
$C_{16}H_{15}Cl_3N_2O$ 2(1H)-Quinolinone, 4-piperidino-3-(tri- chloroethenyl)-	MeOH	228(4.18),264(4.38), 308(4.08)	24-2159-76
$C_{16}H_{15}Cl_3O_3$ Naphtho[1,2-b]furan-2,8(3H,4H)-dione, 5,5a-dihydro-3,5a,9-trimethyl-3-(tri- chloromethyl)-, (3R-cis)-	n.s.g.	227(3.87),250(3.94), 313(4.09)	39-0442-76C
(3S-trans)-	n.s.g.	225(4.00),248(4.00), 310(4.03)	39-0442-76C
$C_{16}H_{15}Cl_4N_5O_8$ 1,2-Hydrazinedicarboxylic acid, 1,1'- (4,5,6,7-tetrachloro-2H-isoindole- 1,3-diyl)bis-, tetramethyl ester	MeCN	239(4.72),268(3.25), 302(3.10),350(3.81), 359(3.78)	88-1661-76
$C_{16}H_{15}F_3N_2$ 1,4-Benzenediamine, N,N-dimethyl-N'- (2,2,2-trifluoro-1-phenylethylidene)-	hexane	248s(4.04),252(4.06), 324s(3.75),358(3.94)	104-0161-76
Benzenamine, N,N-dimethyl-4-[2,2,2-tri- fluoro-1-(phenylimino)ethyl]-	hexane	243s(4.05),332(4.40)	104-0161-76
$C_{16}H_{15}IN_2$ Quinolinium, 6-iodo-1-methyl-2-(pyri- diniomethyl)-, diperchlorate	EtOH	259(4.57),318(3.72), 447(3.42)	104-0444-76
$C_{16}H_{15}N$ Dibenzo[b,f]cycloprop[d]azepine, 1,1a,6,10b-tetrahydro-6-methyl-	EtOH	258(3.77),281(3.67)	94-2751-76
$C_{16}H_{15}NO$ 9(10H)-Acridinone, 10-(1-methylethyl)-	EtOH	253(4.83),385(3.97), 405(4.01)	48-0515-76

Compound	Solvent	$\lambda_{max}(\log \epsilon)$	Ref.
9(10H)-Acridinone, 10-propyl-	EtOH	254(4.83),387(3.93), 406(4.02)	48-0515-76
3-Buten-2-one, 3,4-diphenyl-, oxime	EtOH	230(4.20),288(4.27)	56-0499-76
1H-Indole, 2-(4-methoxyphenyl)-1-methyl-	MeOH	295(4.31)	12-2747-76
Phenol, 2-[3-[(phenylmethyl)imino]- 1-propenyl]-	MeOH	285(4.22),333(4.11)	104-0840-76
	dioxan	280(4.40),325(4.15)	104-0840-76
	DMSO	284(4.23),332(4.04)	104-0840-76
	DMSO-NaOMe	293(4.17),428(4.05)	104-0840-76
6H-Pyrido[3,2,1-jk]carbazol-6-one, 8,9,10,11-tetrahydro-4-methyl-	EtOH	206(4.49),212(4.47), 333(4.09)	83-0185-76
$C_{16}H_{15}NOS$			
1H-Indole, 2-(ethylsulfinyl)-3-phenyl-	EtOH	230s(4.42),235(4.43), 289(4.16)	39-0745-76C
2-Propenamide, 2-(methylthio)-N,3-di- phenyl-	75% dioxan	228(4.2),301(4.3)	104-0149-76
$C_{16}H_{15}NO_2$			
Benzeneacetic acid, α-(2-pyridinylmeth- ylene)-, ethyl ester, (E)-	MeOH	250(4.00),292(4.07)	44-2536-76
(Z)-	MeOH	223(4.04),304(4.22)	44-2536-76
2-Pyridineacetic acid, α-(phenylmethyl- ene)-, ethyl ester, (E)-	MeOH	281(4.20)	44-2536-76
(Z)-	MeOH	220(4.06),299(4.27)	44-2536-76
Pyridinium, 1,3-dioxo-1-phenyl-2-pentyl- ide	EtOH	283(4.21),367(3.12)	78-2647-76
	dioxan	229(4.14),281(4.05), 425(3.26)	78-2647-76
	MeCN	227(4.07),282(4.12), 413(3.23)	78-2647-76
Pyridinium, 1-(4-methylphenyl)-1,3-di- oxo-2-butylide	aq HCl	262.5(4.28)	78-2647-76
	H_2O	287(4.23)	78-2647-76
	EtOH	285(4.20),365(3.12)	78-2647-76
	dioxan	244(4.07),263(4.00), 422(2.97)	78-2647-76
	MeCN	235(4.11),282(4.11), 410(3.20)	78-2647-76
$C_{16}H_{15}NO_2S$			
1H-Indole, 2-(ethylsulfonyl)-3-phenyl-	EtOH	225(4.53),285(4.17)	39-0745-76C
Pyrrolo[2,1-a]isoquinoline-1-carboxylic acid, 2-(methylthio)-, ethyl ester	EtOH	220(3.88),289(4.42), 318(3.90),330(4.87), 350(3.73),369(3.81)	94-1299-76
Pyrrolo[2,1-a]isoquinoline-3-carboxylic acid, 2-(methylthio)-, ethyl ester	EtOH	220(3.86),288(4.41), 318(3.91),330(3.86), 351(3.71),369(3.79)	94-1299-76
$C_{16}H_{15}NO_3$			
Acetamide, N-[2-(2-methoxybenzoyl)phen- yl]-	MeOH	211s(4.20),233(4.33), 262(3.95),269(3.94), 326(3.61)	39-2089-76C
2H-Pyran-3-carboxylic acid, 2-imino- 6-(2-phenylethenyl)-, ethyl ester, perchlorate	HOAc	425(4.31)	48-0705-76
$C_{16}H_{15}NO_3S$			
2H-Indol-2-one, 3-(ethylsulfonyl)-1,3- dihydro-3-phenyl-	EtOH	216(4.40),259(3.74), 269s(3.62),300(3.30)	39-0745-76C
$C_{16}H_{15}NO_4$			
Benzeneacetic acid, 2-carboxy-α-[(1-	EtOH	335(4.35)	4-0083-76

Compound	Solvent	$\lambda_{max}(\log \epsilon)$	Ref.
methyl-1H-pyrrol-2-yl)methylene]-, α-methyl ester, (Z)- (cont.)			4-0083-76
1H-2-Benzopyran-4-carboxylic acid, 3,4-dihydro-3-(1-methyl-1H-pyrrol-2-yl)-1-oxo-, methyl ester, cis	EtOH	222(4.12),338(4.25)	4-0083-76
$C_{16}H_{15}NO_5$			
1,2-Benzenedicarboxylic acid, 3-[(2-hydroxyphenyl)amino]-, dimethyl ester	MeCN	242(3.85),287(3.95)	4-0311-76
6(5H)-Phenanthridinone, 2-acetoxy-2,3,4,4a-tetrahydro-8,9-methylenedioxy)-	EtOH	243(4.46),280s(--), 306(3.87)	94-2969-76
$C_{16}H_{15}NO_6$			
Isoxazolo[4,5-b]dibenzofuran-6-one, 1a,2,6,6a-tetrahydro-1aα,7,9-trihydroxy-5,6aα,10-trimethyl-	MeOH	221(3.84),262(3.20), 286(3.08)	23-3721-76
Isoxazolo[4,5-b]dibenzofuran-6-one, 1a,2,6,6a-tetrahydro-1aβ,7,9-trihydroxy-5,6aα,10-trimethyl-	MeOH	223(4.06),260(2.30), 329(2.35)	23-3721-76
$C_{16}H_{15}NO_8$			
4H-Quinolizine-1,2,3,4-tetracarboxylic acid, 1,2,3-trimethyl ester, sodium salt	MeOH	258(3.40),314(3.52), 348(3.60),435(3.48)	39-1911-76C
$C_{16}H_{15}NO_8S$			
L-Alanine, N-[(9-methoxy-7-oxo-7H-furo-[3,2-g][1]benzopyran-4-yl)sulfonyl]-, methyl ester	EtOH	227(4.44),255(4.37), 307(4.13)	56-0347-76
$C_{16}H_{15}NS$			
1H-Indole, 2-(ethylthio)-3-phenyl-	EtOH	226(4.46),253(4.09), 284s(4.14),292(4.17), 301s(4.14)	39-0745-76C
$C_{16}H_{15}N_2$			
Pyridinium, 1-[(1-methyl-2(1H)-quinolinylidene)methyl]-, perchlorate	EtOH	238(4.06),260(4.17), 465(3.42)	104-0444-76
$C_{16}H_{15}N_2O$			
Pyridinium, 3-formyl-1-[2-(1H-indol-3-yl)ethyl]-, bromide	MeOH	218(4.56),263(3.86), 282(3.66),288(3.54)	35-3645-76
$C_{16}H_{15}N_3$			
3-Azabicyclo[3.2.0]hept-2-ene-1,5-dicarbonitrile, 4,4-dimethyl-2-phenyl-	EtOH	250(4.24)	33-1018-76
Benzenamine, 2-(3-methyl-5-phenyl-1H-pyrazol-1-yl)-	EtOH	241(4.32),296(3.52)	22-0839-76
Pyridine, 2-(2,2-dimethyl-5-phenyl-2H-imidazol-4-yl)-	EtOH	218(4.00),263(3.96)	33-1018-76
Pyridine, 4-(2,2-dimethyl-5-phenyl-2H-imidazol-4-yl)-	EtOH	225(4.08),267(3.80)	33-1018-76
2,3-Quinolinediamine, 4-methyl-N^2-phenyl-	DMF	276(4.62),354(4.20), 367(4.20)	48-0039-76
$C_{16}H_{15}N_3O$			
Benzamide, 2-(1H-benzimidazol-2-yl)-N,N-dimethyl-	EtOH	245(4.15),272s(3.67), 280s(3.58)	2-0001-76
Benzamide, 4-(4,5-dihydro-3-phenyl-1H-pyrazol-1-yl)-	benzene	372(4.34)	64-1248-76B

Compound	Solvent	$\lambda_{max}(\log \epsilon)$	Ref.
11H-Indolo[3,2-c][1,8]naphthyridine, 5,6-dihydro-8-methoxy-3-methyl-	EtOH	220(4.58),270(4.02), 376(4.30)	4-0097-76
$C_{16}H_{15}N_3O_2$ Quinolinium, 1-methyl-6-nitro-2-(pyridiniomethyl)-, diperchlorate	EtOH	256(4.52),317(3.76), 428(3.38)	104-0444-76
$C_{16}H_{15}N_3O_2S$ Isoquinolinium, 2-(cyanoimino)-1-(ethoxycarbonyl)-2-(methylthio)ethylide	EtOH	233(4.75),268(4.13), 310(4.41),425(3.48)	142-0939-76
$C_{16}H_{15}N_3O_3S$ Glycine, N-[4-[[(phenylamino)thioxomethyl]amino]benzoyl]-	EtOH	222(5.34),275(5.22)	80-1345-76
$C_{16}H_{15}N_3O_4$ Benzenamine, 4-[2-(2,4-dinitrophenyl)-ethenyl]-N,N-dimethyl-, (E)-	EtOH	300(4.18),476(4.42)	80-1127-76
Glycine, N-[4-[[(phenylamino)carbonyl]-amino]benzoyl]-	EtOH	209(5.64),280(5.89)	80-1349-76
$C_{16}H_{15}N_3O_5$ 2,2'-Anhydro-1-(3'-benzoyl-β-D-arabino-furanosyl)cytosine	pH 1	232(4.23),264s(3.83)	44-1886-76
	MeOH	217(4.51),227s(4.47), 262(3.64)	44-1886-76
hydrochloride	MeOH-acid	233(4.37),263(4.08)	87-0654-76
$C_{16}H_{15}N_3O_7$ 8-Azabicyclo[3.2.1]oct-2-ene-6,7-dicarboxylic acid, 8-(5-nitro-2-pyridinyl)-4-oxo-, methyl ester, (6-endo,7-exo)-	EtOH	210(4.22),225(4.34), 348(4.35)	39-2307-76C
(6-exo,7-exo)-	EtOH	210(4.25),226(4.36), 348(4.36)	39-2307-76C
$C_{16}H_{15}N_5O_4$ 1H-Benzimidazolium, 2-[(2,4-dinitrophenyl)methyl]-1,3-dimethyl-, perchlorate	0.1N H_2SO_4	267(4.48),273(4.49), 280(4.40)	104-1097-76
	EtOH-KOH	504(4.37)	104-1097-76
$C_{16}H_{15}N_5$ Benzaldehyde, (7-ethylpyrido[3,2-c]pyridazin-4-yl)hydrazone	EtOH	213(4.55),273(4.34), 300s(4.48),330(4.38), 386(4.52),480(3.66)	103-0808-76
$C_{16}H_{15}N_5O_6$ 2,4(1H,3H)-Pyrimidinedione, 1-(3-azido-5-O-benzoyl-3-deoxy-β-D-arabino-furanosyl)-	MeOH	226(4.19),260(4.01)	44-3138-76
Uridine, 2'-azido-2'-deoxy-, 5'-benzoate	MeOH	231(4.15),261(3.94)	44-3138-76
$C_{16}H_{15}O$ 1H-Isobenzofurylium, 1,1-dimethyl-3-phenyl-	H_2SO_4	340(4.42)	104-2523-76
$C_{16}H_{15}S$ 1H-Benzo[c]thiolium, 1,1-dimethyl-3-phenyl-	H_2SO_4	360(4.49)	104-2523-76
$C_{16}H_{16}$ Butene, 1,3-diphenyl-	EtOH	213(3.93),252(3.93)	1-0904-76

Compound	Solvent	$\lambda_{max}(\log \epsilon)$	Ref.
$C_{16}H_{16}BrClO_5$ Cyclobuta[b]naphthalene-3,8-dione, 8a-bromo-4-chloro-1,2,2a,8a-tetrahydro-1,1,7-trimethoxy-5-methyl-	CHCl$_3$	344(3.76)	39-0613-76C
$C_{16}H_{16}Br_2N_4O_2$ Butanal, 2-[(4-bromophenyl)hydrazono]-3,4-dihydroxy-, (4-bromophenyl)hydrazone, (S)- (5λ,4ε)	EtOH	210(4.18),230(4.13), 264(4.12),318(4.28), 406(?)	136-0185-76D
$C_{16}H_{16}ClNOS$ Benzenepropanamide, β-chloro-α-(methylthio)-N-phenyl-	75% dioxan	<u>252(4.2)</u>	104-0149-76
$C_{16}H_{16}ClNO_2$ 1,4-Benzoxazepine, 7-chloro-2,3,4,5-tetrahydro-8-methoxy-5-phenyl-	n.s.g.	215(4.30),235(3.97), 280(3.30)	20-0898-76
$C_{16}H_{16}ClNO_3$ Benzamide, N-[2-(4-chloro-3-methoxyphenoxy)ethyl]-	n.s.g.	230(4.28),280(3.53)	20-0787-76
$C_{16}H_{16}ClNO_7$ 2H-Pyran-3-carboxylic acid, 2-imino-6-(2-phenylethenyl)-, ethyl ester, perchlorate	HOAc	425(4.31)	48-0705-76
$C_{16}H_{16}ClN_3$ 1H-Indazol-4-amine, 3-(4-chlorophenyl)-1,5,6-trimethyl-	MeOH	250s(4.17),326(3.98)	24-1898-76
Pyrrolidine, 1-[4-[[(4-chlorophenyl)azo]-phenyl]-	C$_6$H$_{12}$	417(4.54)	18-1381-76
$C_{16}H_{16}Cl_2N_2$ Diazene, bis(1-chloro-1-phenylethyl)-, meso	C$_6$H$_{12}$	360(1.72)	39-1249-76B
$C_{16}H_{16}Cl_2N_2O$ Diazene, bis(2-chloro-2-phenylethyl)-, 1-oxide	EtOH	220(4.33)	88-0153-76
$C_{16}H_{16}Cl_2N_4O_2$ Butanal, 2-[(4-chlorophenyl)hydrazono]-3,4-dihydroxy-, (4-chlorophenyl)hydrazone, (S)-	EtOH	210(4.16),228(4.10), 264(4.34),318(4.10), 404(4.10)	136-0185-76D
$C_{16}H_{16}Cl_2O_5$ α-D-erythro-Pentofuranose, 5-O-benzoyloxy-3-deoxy-3-C-(dichloromethylene)-1,2-O-isopropylidene-	EtOH	222(3.64)	111-0489-76
$C_{16}H_{16}Cl_3NO$ Benzenemethanamine, 4-methoxy-N-(4-methylphenyl)-α-(trichloromethyl)-	n.s.g.	272s(3.55),281(3.52), 292s(3.38)	98-0724-76
$C_{16}H_{16}Cl_3NO_2$ Benzenemethanamine, 4-methoxy-N-(4-methoxyphenyl)-α-(trichloromethyl)-	n.s.g.	233(4.27),280(3.31), 306(3.23)	98-0724-76

Compound	Solvent	λ_{max}(log ϵ)	Ref.
$C_{16}H_{16}N_2$			
3H-Indol-2-amine, 3-ethyl-3-phenyl-, benzene complex	MeOH	272(3.95)	56-0857-76
1H-Pyrazole, 3-[1,1'-biphenyl]-4-yl-4,5-dihydro-1-methyl-	EtOH	315(4.3)	104-1968-76
1H-Pyrazole, 4,5-dihydro-1-(4-methylphenyl)-3-phenyl-	benzene	367(4.25)	64-1248-76B
1H-Pyrazole, 4,5-dihydro-3-(4-methylphenyl)-1-phenyl-	benzene	360(4.31)	64-1248-76B
1H-Pyrazole, 4,5-dihydro-3-phenyl-1-(phenylmethyl)-	benzene	310(4.21)	61-0288-76
Quinolinium, 1-methyl-2-(pyridiniomethyl)-, diiodide	EtOH	255(4.21),360(3.53), 460(3.30)	104-0444-76
$C_{16}H_{16}N_2O$			
1H-Indazole, 2-acetyl-2,3-dihydro-3-methyl-3-phenyl-	isoPrOH	248(3.96),290s(3.35)	4-0033-76
1H-Pyrazole, 4,5-dihydro-1-(4-methoxyphenyl)-3-phenyl-	benzene	370(4.21)	64-1248-76B
1H-Pyrazole, 4,5-dihydro-3-(4-methoxyphenyl)-1-phenyl-	benzene	358(4.31)	64-1248-76B
2H-Pyrido[1,2-a]pyrimidin-2-one, 3,4-dihydro-3,8-dimethyl-4-phenyl-, cis	EtOH	268(4.03),340(3.91)	44-3546-76
2-Pyrrolidinone, 1-phenyl-5-(phenylamino)-	EtOH	245(4.10),287(3.23)	44-2496-76
$C_{16}H_{16}N_2OS$			
Sulfilimine, N-(2-cyanoethyl)-S-(2-methoxyphenyl)-S-phenyl-, (S)-	MeOH	287(3.64)	78-3003-76
$C_{16}H_{16}N_2OS_2Se$			
4-Thiazolidinone, 3-ethyl-5-[(7-ethylselenolo[2,3-b]pyridin-4(7H)-ylidene)-ethylidene]-2-thioxo-	EtOH	578(5.03)	103-0052-76
$C_{16}H_{16}N_2O_2$			
8-Azabicyclo[3.2.1]oct-3-ene-6-carbonitrile, 4-(4-methoxyphenyl)-8-methyl-2-oxo-, exo	EtOH	235(3.7),325(4.0)	39-2329-76C
Benzenamine, N,N-dimethyl-4-[2-(4-nitrophenyl)ethenyl]-	$C_2H_4Cl_2$	438(4.51)	104-0161-76
1H-Furo[2,3-g]indazole, 3a,4,5,8b-tetrahydro-8b-hydroxy-7-methyl-1-phenyl-	EtOH	248(4.23)	32-1083-76
Indolo[2,3-a]quinolizine-3-carboxylic acid, 1,4,6,7,12,12b-hexahydro-	MeOH	224(4.55),274(3.84), 281(3.85),290(3.74)	24-0705-76
1H-Naphtho[2,3-b][1,4]diazepine-6,11-dione, 2,5-dihydro-2,2,4-trimethyl-	EtOH	610(3.2)(changing)	103-0125-76
	CHCl$_3$	308s(4.21),322(4.26), 510(3.08),615(3.06)	103-0125-76
	CCl$_4$	284s(4.13),308s(4.22), 319(4.30),600(3.33)	
hydrochloride	EtOH-HCl	294(4.06),326(3.97), 431(3.43)	103-0125-76
$C_{16}H_{16}N_2O_2S$			
1H-Pyrazole, 4,5-dihydro-3-[(4-methylsulfonyl)phenyl]-1-phenyl-	benzene	397(4.36)	64-1248-76B
$C_{16}H_{16}N_2O_2S_2$			
4-Thiazolidinone, 3-ethyl-5-[(7-ethylfuro[2,3-b]pyridin-4(7H)-ylidene)ethylidene]-2-thioxo-	EtOH	571(5.10)	103-0052-76

Compound	Solvent	$\lambda_{max}(\log \epsilon)$	Ref.
$C_{16}H_{16}N_2O_2S_3$ 2H-Indol-2-one, 3-[(3-ethyl-4-oxo-2-thi-oxo-5-thiazolidinylidene)(methylthio)-methyl]-1,3-dihydro-1-methyl-	EtOH	258(4.14),289(4.09), 386(4.33),525(3.74)	94-1671-76
$C_{16}H_{16}N_2O_3$ Acetamide, N-(4-methoxyphenyl)-2-[(4-methoxyphenyl)imino]-	EtOH	231(3.51)	65-0702-76
Benzoic acid, 2-[(2-aminobenzoyl)meth-ylamino]-, methyl ester	dioxan	255(4.32),360(4.10)	83-0018-76
$C_{16}H_{16}N_2O_4$ Acetamide, N-(4-methoxyphenyl)-2-[(4-methoxyphenyl)imino]-, N-oxide	EtOH	225(4.2),295(4.1), 340(4.2)	65-0702-76
Benzamide, N,N'-1,2-ethanediylbis[2-hydroxy-	MeOH	302(3.94)	18-2679-76
$C_{16}H_{16}N_2O_4S_2$ Spiro[benzothiazole-2(3H),2'-[2H]pyr-role]-3',4'-dicarboxylic acid, 3-methyl-5'-(methylthio)-, dimethyl ester	EtOH	220(4.23),254(4.23), 286(4.42),300(4.29), 405(3.78)	142-0705-76
$C_{16}H_{16}N_2O_6$ 1-Naphthalenemethanol, α-(1-methyl-1-nitroethyl)-5-nitro-, acetate	MeOH	220(4.73),330(3.65)	12-0327-76
$C_{16}H_{16}N_2O_6S$ 1,6-Naphthyridine-6(5H)-acetic acid, 8-methoxy-2-(methoxycarbonyl)-α-[(methylthio)methylene]-, methyl ester	EtOH	275(4.18)	35-0299-76 +35-8237-76
$C_{16}H_{16}N_2S$ 1H-Pyrazole, 4,5-dihydro-3-[4-(methyl-thio)phenyl]-1-phenyl-	benzene	373(4.42)	64-1248-76B
$C_{16}H_{16}N_4$ Benzonitrile, 4-[[(1,2,3,4-tetrahydro-2-quinoxalinyl)methyl]amino]-	pH 7.8	290(4.45)	35-3678-76
$C_{16}H_{16}N_4O$ 2-Pentanone, 3-(1-phenyl-1H-pyrazolo-[3,4-d]pyrimidin-4-yl)-	EtOH	264(4.08),360(4.44), 372(4.48)	95-1352-76
$C_{16}H_{16}N_4O_2$ Benzo[g]pteridine-2,4(1H,3H)-dione, 3,7,8-trimethyl-1-(2-propenyl)-	n.s.g.	343(4.04),388(4.00)	88-1389-76
1H-Indazol-4-amine, 1,5,6-trimethyl-3-(4-nitrophenyl)-	MeOH	221(4.41),280(4.06)	24-1898-76
$C_{16}H_{16}N_4O_4$ 1,5-Cyclohexadiene-1,3-dicarboxylic acid, 5-amino-4,4,6-tricyano-3-methyl-, diethyl ester	n.s.g.	220(4.09),265(3.81), 340(3.68)	20-0141-76
Isoxazole, 3-(6-azidotetrahydro-2,2-di-methylfuro[2,3-d]-1,3-dioxol-5-yl)-5-phenyl-, [3aR-(3aα,5α,6α,6aα)]-	EtOH	209(--),262(3.91)	136-0119-76A

Compound	Solvent	$\lambda_{max}(\log \epsilon)$	Ref.
$C_{16}H_{16}N_4O_5$			
1H-Imidazole-4-carbonitrile, 5-amino-2-(5-O-benzoyl-α-D-arabinofuranosyl)-	H_2O	237(4.26),253s(4.05)	111-0067-76
$C_{16}H_{16}O$			
Cyclopent[cd]azulen-2(1H)-one, 4-methyl-6-(1-methylethyl)-	C_6H_{12}	235(4.44),242(4.44), 258(4.26),268(4.35), 300s(4.36),304(4.47), 312(4.51),318(4.62), 350s(3.86),362(3.98), 375s(3.96),381(4.03), 524s(2.63),544s(2.73), 559(2.82),584(2.79), 609(2.84),640s(2.50), 674(2.51)	18-1650-76
Indeno[2,1-b]pyran, 3-ethyl-2,4-dimethyl-	dioxan	245(4.05),253(3.96), 300(4.36),342(3.98), 357(3.74),481(3.46)	78-2219-76
Methanone, phenyl(2,4,6-trimethylphenyl)-	EtOH	248(3.88),285s(3.08)	44-3083-76
$C_{16}H_{16}O_2$			
Benzeneacetic acid, α-phenyl-, ethyl ester, carbanion	DMSO	348(4.35)	70-0762-76
1,4:9,12-Diepoxycyclobuta[1,2:3,4]dicyclooctene, 1,4,6a,6b,9,12,12a,12b-octahydro-, (1α,4α,6aα,6bβ,9α,12α-12aβ,12bα)-	hexane	210(3.70)	78-2239-76
9H-Fluorene, 2,7-dimethoxy-1-methyl-	EtOH	277(5.05),310(4.25)	78-0065-76
Spiro[3.5]nona-5,8-diene-7-carboxylic acid, 7-phenyl-	EtOH	269(3.10)	35-7835-76
$C_{16}H_{16}O_3$			
Benzeneacetic acid, α-(3-methylphenoxy)-, methyl ester	EtOH	220(3.16),229(4.08), 272(3.08),280(3.08)	111-0533-76
Benzeneacetic acid, α-(4-methylphenoxy)-, methyl ester	EtOH	220(3.16),229(4.08), 272(3.08),280(3.08)	111-0533-76
9bH-Benz[e]inden-9bα-acetic acid, 1,3aα,4,5-tetrahydro-3-methyl-1-oxo-	EtOH	229(4.3)	39-0722-76C
1H-Fluorene-1-carboxylic acid, 8-methyl-2-oxo-, methyl ester	EtOH EtOH-NaOEt	238(4.26),269(4.24) 300(4.27)	78-0065-76 78-0065-76
1-Phenanthrenecarboxylic acid, 2,3,4,4a,9,10-hexahydro-2-oxo-, methyl ester	EtOH	227(4.14)	78-0073-76
4-Phenanthrenecarboxylic acid, 1,2,3,4,9,10-hexahydro-3-oxo-, methyl ester	EtOH NaOH	263(3.96),271(3.95) 245(4.40),299(4.23)	78-0073-76 78-0073-76
4-Phenanthrenecarboxylic acid, 1,2,3,9,10,10a-hexahydro-3-oxo-, methyl ester	EtOH NaOH	228(3.90),299(4.17) 297(5.02)	78-0073-76 78-0073-76
1-Propanone, 1-(2,4-dihydroxy-6-methyl-phenyl)-3-phenyl-	EtOH	214(4.07),282(3.94), 308(3.73)	39-0499-76C
$C_{16}H_{16}O_3S_2$			
Sulfonium, dimethyl-, 2-oxo-2-phenyl-1-(phenylsulfonyl)ethylide	MeOH	218s(4.34),266(4.10)	30-0366-76
$C_{16}H_{16}O_4$			
Benzoic acid, 2-hydroxy-6-methyl-4-(phenylmethoxy)-, methyl ester	EtOH	216(4.46),264(4.30), 290s(3.87)	35-5380-76
Benzoic acid, 2-methoxy-6-methyl-4-(phenylmethoxy)-	EtOH	276(3.88)	35-5380-76

Compound	Solvent	$\lambda_{max}(\log \epsilon)$	Ref.
Cyclopentane-2,5-diacetic acid, 1,1-di-hydroxy-2-methyl-3-phenyl-, dilactone	EtOH	250(2.26),252(2.34), 255(2.18),256(2.30), 260(2.38),263(2.23), 268(2.27)	104-0049-76
1H-Fluoreno[1,9-cd]pyran-1,3(6bH)-dione, 7,8,9,10,10a,10b-hexahydro-4-methoxy-, (6bα,10aα,10bβ)-	ether	321(3.69)	44-2401-76
3-Furancarboxylic acid, 4,5-dihydro-2-methyl-5-[(4-methylphenyl)meth-ylene]-4-oxo-, ethyl ester	EtOH	210(4.22),320(4.18), 352(4.28)	4-0521-76
3-Furancarboxylic acid, 4,5-dihydro-2-methyl-4-oxo-5-(1-phenylethyli-dene)-, ethyl ester	EtOH	206(4.20),234(4.05), 328(4.23)	4-0521-76
1,4-Naphthalenedione, 5,8-dihydroxy-2-(4-methyl-3-pentenyl)-	n.s.g.	275(2.99),493(2.81), 526(2.87),568(2.48)	105-0652-76
2H-Naphtho[1,8-bc]furan-2-one, 4-hydr-oxy-3-methoxy-7-methyl-5-(1-methyl-ethyl)-	MeOH	260(4.03),362(3.58)	2-0616-76
$C_{16}H_{16}O_5$			
Ethanone, 1-(2,4-dihydroxy-3-methoxy-phenyl)-2-(4-methoxyphenyl)-	EtOH	238s(4.01),287(4.19), 321s(3.89)	102-1019-76
3-Furancarboxylic acid, 4,5-dihydro-5-[(4-methoxyphenyl)methylene]-2-methyl-4-oxo-, ethyl ester	EtOH	221(3.87),248(3.88), 352(4.31)	4-0521-76
$C_{16}H_{16}O_5S$			
1H-Naphtho[2,1-b]pyran-1-one, 2,3-di-hydro-3,3-dimethyl-6-[(methylsul-fonyl)oxy]-	EtOH	240s(4.40),263s(3.87), 344(3.62),354s(3.61)	80-1543-76
1H-Naphtho[2,1-b]pyran-1-one, 2,3-di-hydro-3,3-dimethyl-9-[(methylsul-fonyl)oxy]-	EtOH	265s(4.08),316(3.91), 360(3.79)	80-1543-76
$C_{16}H_{16}O_6$			
Ethanone, 1-(1-acetoxy-3-hydroxy-6,8-di-methoxy-2-naphthalenyl)-	EtOH	228(4.49),250(4.42), 323(3.78)	12-1087-76
Ethanone, 1-(2,4-dihydroxy-3-methoxy-phenyl)-2-(3-hydroxy-4-methoxyphenyl)-	EtOH	231s(4.16),289(4.21), 315s(3.91)	102-1019-76
$C_{16}H_{16}O_7$			
Butanedioic acid, [2-(carboxymethyl)-2,3-dihydro-5-methoxy-1H-inden-1-ylidene]-	EtOH	288(4.12),317(4.24)	78-0065-76
Ethanone, 1-(4-acetoxy-1,3-dihydroxy-6,8-dimethoxy-2-naphthalenyl)-	MeOH	233(4.43),279(4.48), 381(3.60)	12-1087-76
$C_{16}H_{17}Br_2N_5$			
Pyridinium, 4,4'-(6-methyl-1,3,5-triaz-ine-2,4-diyl)bis[1-methyl-, dibromide	pH 7	246(4.48),282s(3.90)	78-0615-76
$C_{16}H_{17}ClN_2O$			
Propanimidamide, 2-chloro-N-hydroxy-2-methyl-N,N'-diphenyl-	EtOH	235(4.00),283(3.95), 372(3.37)	12-0357-76
2-Propanone, 2-[4-chloro-2-(hydroxy-phenylmethyl)phenyl]hydrazone	isoPrOH	286(4.33)	4-0033-76
$C_{16}H_{17}ClN_2O_2$			
Acetic acid, 2-[4-chloro-2-(1-hydroxy-1-phenylethyl)phenyl]hydrazide	isoPrOH	248(4.14),298(3.41)	4-0033-76

Compound	Solvent	$\lambda_{max}(\log \epsilon)$	Ref.
$C_{16}H_{17}ClN_2O_4S$ 1H-2,1-Benzisothiazol-1-ium, 3-[4-(di-methylamino)phenyl]-1-methyl-, per-chlorate	EtOH	291(4.27),335s(3.36), 520(4.44)	48-0161-76
$C_{16}H_{17}ClN_3O_2S$ Morpholinium, 4-[[[5-chloro-2-(4-meth-oxyphenyl)-6H-1,3-thiazin-6-ylidene]-amino]methylene]-, perchlorate	HOAc	299(3.95),420(4.16)	97-0268-76
$C_{16}H_{17}ClO_3S$ Benzeneethanol, 2-[(4-chlorophenyl)-thio]-4,5-dimethoxy-	MeOH	254(4.31),280(4.00)	73-1396-76
$C_{16}H_{17}ClSi$ Silane, chloro(1,2-diphenylethenyl)di-methyl-, (E)-	n.s.g.	242(4.35),272(4.40), 298(4.20),311(2.65), 322(2.32)	65-1044-76
$C_{16}H_{17}NO$ Cycloheptanone, 2-(2-quinolinyl)-	EtOH	220(4.63),331(4.08), 445(4.12)	118-0764-76
$C_{16}H_{17}NO_3$ 8-Azabicyclo[3.2.1]oct-3-ene-6-carbox-ylic acid, 8-methyl-2-oxo-4-phenyl-, methyl ester, endo	EtOH	225(3.9),287(4.2)	39-2329-76C
exo	EtOH	225(3.9),290(4.1)	39-2329-76C
1,3-Cycloheptadiene-1-carboxylic acid, 6-(methylphenylamino)-5-oxo-, methyl ester	EtOH	209(4.32),251(4.18), 293(4.02)	39-2289-76C
2H-Pyran-2-one, 6-[2-[4-(dimethylamino)-phenyl]ethenyl]-4-methoxy-	MeOH	232(5.2),245(5.21), 408(5.58)	44-4070-76
$C_{16}H_{17}NO_4$ Hamayne (O-demethylcrinamine)	EtOH	240(3.64),296(3.85)	18-3363-76
Methanone, (2-aminophenyl)(2,4,6-tri-methoxyphenyl)-	EtOH	207(4.70),229(4.40), 260(3.87),361(3.79)	39-2089-76C
Propanedioic acid, 4H-quinolizin-4-yli-dene-, diethyl ester	EtOH	211(4.54),250(4.39), 310(3.89),422(3.73)	39-0341-76C
	EtOH-HClO₄	213(4.44),236(4.34), 290(3.53),316(3.97), 322s(3.95),330(4.17)	39-0341-76C
2H-Pyrrole-3,4-dicarboxylic acid, 2,2-dimethyl-5-phenyl-, dimethyl ester	n.s.g.	231(3.87),248s(3.74), 283s(3.19)	33-2149-76
$C_{16}H_{17}NO_4S_2$ Butanedioic acid, [(4-methyl-2-phenyl-5-thiazolyl)thio]-, dimethyl ester	MeOH	216(4.06),310(4.24)	78-0583-76
$C_{16}H_{17}NO_5$ Spiro[3H-benz[e]indene-3,2'-[1,3]dioxol-an]-5(2H)-one, 1,3a,4,9b-tetrahydro-3a-methyl-7-nitro-, trans	EtOH	240(4.00),274(3.96)	104-0796-76
$C_{16}H_{17}NO_6$ 2H-Quinolizine-1,2,3-tricarboxylic acid, 6-methyl-, trimethyl ester	MeOH	213(4.00),263(4.07), 305(3.85),265(3.76)	39-1911-76C
	MeOH-acid	208(4.15),287(3.95)	39-1911-76C

Compound	Solvent	$\lambda_{max}(\log \epsilon)$	Ref.
4H-Quinolizine-1,2,3-tricarboxylic acid, 6-methyl-, trimethyl ester	MeOH	212(4.04),264(3.90), 314(4.02),344(4.12), 461(3.97)	39-1911-76C
	MeOH-acid	217(4.13),317(4.06), 450(3.63)	39-1911-76C
$C_{16}H_{17}NO_8S$ 8aH-Thiazolo[3,2-a]pyridine-6,7,8,8a-tetracarboxylic acid, 2-methyl-, tetramethyl ester	MeOH and MeOH-acid	228(4.33),286(4.49), 443(3.81)	39-1269-76C
$C_{16}H_{17}N_2O$ Pyridinium, 1-methyl-3-[1-oxo-3-(phenylamino)-2-butenyl]-, iodide	EtOH	275(3.97),350(4.24)	39-0315-76C
$C_{16}H_{17}N_2S$ 1H-2,1-Benzisothiazol-1-ium, 3-[4-(dimethylamino)phenyl]-1-methyl-, perchlorate	EtOH	291(4.27),335s(3.36), 520(4.44)	48-0161-76
$C_{16}H_{17}N_3$ 1H-Indazol-4-amine, 1,5,6-trimethyl-3-phenyl-	MeOH	249s(4.08),324(3.93)	24-1898-76
Pyrrolidine, 1-[4-(phenylazo)phenyl]-	C_6H_{12} THF	407(4.52) 259(4.03),317s(3.61), 330s(3.48),415(4.48)	18-1381-76 5-0946-76
$C_{16}H_{17}N_3O$ 7-Azatricyclo[4.3.1.1^{2,5}]undeca-3,8-dien-10-one, 7-(4,6-dimethyl-2-pyrimidinyl)-, (1α,2β,5β,6α)-	CHCl$_3$	278(4.51)	39-2307-76C
Benzoic acid, 2-hydroxy-, [[4-(dimethylamino)phenyl]methylene]hydrazone	EtOH	244(4.17),277(4.06), 325s(3.99),389(4.61)	95-0044-76
2,4,6-Heptatrienamide, 2-cyano-5-(dimethylamino)-7-phenyl-	MeCN	400(4.51)	48-0705-76
1H-Indole-3-carbonitrile, 2-amino-4,5,6,7-tetrahydro-1-(4-methoxyphenyl)-	MeOH	270(3.81)	48-0663-76
3H-Pyrrole-3,3-dicarbonitrile, 4-ethoxy-2,4-dihydro-2,2-dimethyl-5-phenyl-	EtOH	249(4.17)	33-1018-76
$C_{16}H_{17}N_3O_2$ Acetic acid, cyano[[(1,3-dihydro-1,3,3-trimethyl-2H-indol-2-ylidene)methyl]-imino]-, methyl ester	EtOH	458(4.76)	78-3031-76
1H-Indazol-4-amine, 3-(3,4-dimethoxyphenyl)-1-methyl-	MeOH	284s(3.94),322(4.04)	24-1898-76
Pyrazolo[3,4,5-kl]acridine, 2,3,4,5-tetrahydro-8,9-dimethoxy-2-methyl-	MeOH	250s(4.56),257(4.61), 263s(4.52),296s(3.88), 303(3.92),316(3.88), 331(3.70)	24-1898-76
Pyridinium, 1-[[[[(ethoxycarbonyl)amino]phenylmethylene]amino]-4-methyl-, hydroxide, inner salt	EtOH	249(4.24),330(3.40)	138-0413-76
Pyrimido[4,5-b]quinoline-2,4(3H,10H)-dione, 10-butyl-3-methyl-	EtOH	220(4.42),263(4.56), 320(3.89),399(3.97)	39-1805-76C
	50% EtOH-HCl	216(4.28),256(4.43), 333(4.11)	39-1805-76C

Compound	Solvent	$\lambda_{max}(\log \epsilon)$	Ref.
$C_{16}H_{17}N_3O_3$			
7-Azabicyclo[4.3.1]deca-3,8-dien-10-one, 3,4-dimethyl-7-(5-nitro-2-pyridinyl)-	EtOH	255(4.11),392(4.41)	39-2307-76C
Carbamic acid, [[(phenylmethyl)-2-pyridinyl]amino]carbonyl]-, ethyl ester	MeOH	215s(4.08),240(4.09), 280(3.68),320(3.29)	24-3661-76
5,7-Ethano-1H,5H-cyclopropa[d][1,2,4]-triazolo[1,2-a]pyridazine-1,3(2H)-dione, tetrahydro-2-(4-methoxyphenyl)-	n.s.g.	231(3.24)	2-0883-76
$C_{16}H_{17}N_3O_4$			
Ethanimidamide, N,N'-bis(4-methoxyphenyl)-2-aci-nitro-	n.s.g.	230(4.42),338(4.42)	104-0914-76
$C_{16}H_{17}N_3O_5$			
2,4-Cyclohexadien-1-one, 4-(diethylamino)-6-(2-nitro-4-aci-nitro-2,5-cyclohexadien-1-ylidene)-	EtOH	495(4.45)	104-2333-76
	dioxan	495(4.11)	104-2333-76
	DMF	428(4.52)	104-2333-76
	DMSO	428(4.52)	104-2333-76
	$CHCl_3$	495(4.23)	104-2333-76
Cytidine, N-benzoyl-2'-deoxy-	EtOH	259(4.32),304(--)	138-0019-76
$C_{16}H_{17}N_3O_6$			
Cytidine, N-benzoyl-	EtOH	259(4.35),303(--)	138-0019-76
2,4(1H,3H)-Pyrimidinedione, 1-(3-amino-5-O-benzoyl-3-deoxy-β-D-arabinofuranosyl)-	MeOH	227(4.32),262(4.16)	44-3138-76
$C_{16}H_{17}N_5$			
Pyridinium, 4,4'-(6-methyl-1,3,5-triazine-2,4-diyl)bis[1-methyl-, dibromide	pH 7	246(4.48),282s(3.90)	78-0615-76
$C_{16}H_{17}N_5O_2$			
7(8H)-Pteridinone, 2-(dimethylamino)-4-methoxy-8-(phenylmethyl)-	MeOH	242(4.17),288(3.68), 361(4.21)	24-3228-76
$C_{16}H_{17}O_{10}$			
1,3-Cyclopentadiene-1,2,3,4-tetracarboxylic acid, 5-(2-methoxy-2-oxoethyl)-, tetramethyl ester, ion(1-), sodium	MeOH	215(3.83),266(4.59), 302(4.18)	24-1928-76
$C_{16}H_{18}$			
1H-Fluorene, 4,9-dihydro-1,9,9-trimethyl-	MeOH	266(4.04),293(3.14)	44-2706-76
2H-Fluorene, 4a,9-dihydro-4a,9,9-trimethyl-	MeOH	259(2.50),264(2.80), 271(3.30),289(--)	44-2706-76
$C_{16}H_{18}ClN_3$			
Benzenamine, 4-[(4-chloro-2-methylphenyl)azo]-N,N,3-trimethyl-	C_6H_{12}	411(4.49)	18-1381-76
$C_{16}H_{18}ClN_3O$			
4H-Indazol-4-one, 3-(4-chlorophenyl)-1,5,6,7-tetrahydro-1,6,6-trimethyl-, oxime, (E)-	MeOH	232(4.42),270s(4.23)	24-1898-76
Pyrazolo[4,3-b]azepin-5(1H)-one, 3-(4-chlorophenyl)-4,6,7,8-tetrahydro-1,7,7-trimethyl-	MeOH	235(4.08),260(4.11)	24-1898-76
Pyrazolo[4,3-c]azepin-4(1H)-one, 3-(4-chlorophenyl)-5,6,7,8-tetrahydro-1,7,7-trimethyl-	MeOH	256(4.15)	24-1898-76

Compound	Solvent	$\lambda_{max}(\log \epsilon)$	Ref.
$C_{16}H_{18}ClRhS_3$ Rhodium, chloro[(1,2,5,6-η)-1,5-cyclo- octadiene](5-methyl-1,3-benzodithiole- 2-thione-S^2)-	$C_2H_4Cl_2$	373(4.21),400(3.96)	101-0369-76Q
$C_{16}H_{18}Cl_6FN_2O_9P$ 5'-Uridylic acid, 4'-C-fluoro-2',3'-O- (1-methylethylidene)-, bis(2,2,2-tri- chloroethyl) ester	MeOH	255(3.99)	44-3010-76
$C_{16}H_{18}FN_5O_7$ Adenosine, 8-fluoro-, 2',3',5'-triacet- ate	THF acid base	249(4.12) 251(4.13) 250(4.15)	77-0430-76 77-0430-76 77-0430-76
$C_{16}H_{18}FeN_2O_5$ Ferrocene, (2,2-dinitro-5-oxohexyl)-	MeOH	420(2.29)	104-2433-76
$C_{16}H_{18}FeN_2O_6$ Ferrocene, (5-methoxy-2,2-dinitro- 5-oxopentyl)-	MeOH	425(2.26)	104-2433-76
$C_{16}H_{18}N_2$ 9H-Carbazole-2-ethanamine, N,1-dimethyl-	EtOH	239(4.67),249(4.55), 260(4.31),295(4.26), 326(3.61),339(3.60)	88-1873-76
Propanedinitrile, (2-phenylheptylidene)-	EtOH	232(3.99),292(4.15)	118-0705-76
2H-1,5-Propano-1H-naphtho[1,8-bc]-1,5- diazocine, 3,4-dihydro-	isooctane	380(2.35)	77-0173-76
1-Propanone, 2-methyl-1-phenyl-, phenylhydrazone, (E)-	EtOH	260(4.02),283(4.07)	39-0456-76C
Pyridine, 3-(1-methyl-5-phenyl-2-pyrro- lidinyl)-, (2S-cis)-	EtOH	258(3.57),262s(3.59), 268s(3.47)	44-3438-76
	EtOH-acid	260(3.70)	44-3438-76
(2S-trans)-	EtOH	257(3.55),262(3.55), 268s(3.43)	44-3438-76
	EtOH-acid	260(3.69)	44-3438-76
$C_{16}H_{18}N_2O$ Benzenecarboximidamide, N-hydroxy-N'-(1- methylethyl)-N-phenyl-, hydrochloride	EtOH	300(3.92)	12-0357-76
$C_{16}H_{18}N_2OS_2$ Ethanone, 1-[4-[5-(1,1-dimethylethyl)- 1H-[1,2]dithiolo[5,1-e][1,2,3]thia- diazol-7-S^{IV}-1-yl]phenyl]-	C_6H_{12}	209(4.33),227(4.32), 252(4.07),307(4.37), 490(4.31)	39-0228-76C
$C_{16}H_{18}N_2O_2$ Acetic acid, 2-[2-(1-hydroxy-1-phenyl- ethyl)phenyl]hydrazide	isoPrOH	239(4.00),288(3.40)	4-0033-76
3H-Indol-3-one, 2-(1,3,4,5,6,7-hexa- hydro-3-oxo-2H-indol-2-ylidene)- 1,2,4,5,6,7-hexahydro-	CHCl$_3$	315(4.38),520(4.11)	89-0052-76
1H-Naphtho[2,3-b][1,4]diazepine-6,11- dione, 2,3,4,5-tetrahydro-2,2,4-tri- methyl-	EtOH	234(4.12),268(4.11), 302(4.28),584(3.26)	103-0125-76
hydrochloride	EtOH	266(4.20),296(4.00), 440(3.27)	103-0125-76
2,4-Pentadienoic acid, 2-cyano-5-(di- methylamino)-5-phenyl-, ethyl ester	CH$_2$Cl$_2$	404(4.18)	48-0705-76

Compound	Solvent	$\lambda_{max}(\log \epsilon)$	Ref.
$C_{16}H_{18}N_2O_2S$			
Acetamide, N-(3-morpholino-5-phenyl-2-thiazolyl)-	MeOH	239s(4.08),252s(4.03), 329(4.18)	48-0221-76
$C_{16}H_{18}N_2O_4$			
Isoxazole, 3-(3-amino-3-deoxy-1,2-O-isopropylidene-α-D-xylo-tetrafuranos-4-yl)-5-phenyl-	EtOH	208(3.74),263(4.02)	136-0119-76A
β-D-ribo-Pentodialdo-1,4-furanoside, methyl 5-C-1H-imidazol-3-yl-2,3-O-(1-methylethylidene)-	EtOH	215(4.05),237(3.96), 242(3.94),302(4.04)	4-1241-76
$C_{16}H_{18}N_2O_7$			
Spiro[5H-benz[e]indene-5,2'-[1,3]dioxolan]-3-ol, 1,2,3,3a,4,9b-hexahydro-3a-methyl-7-nitro-, nitrate, (3α,3aα,9bβ)-	EtOH	275(4.03)	104-0796-76
$C_{16}H_{18}N_3O_5$			
Piperidine, 1-(1,4-dihydro-1-methoxy-2,4-dinitro-1-naphthalenyl)-, ion(1-), piperidinium salt	DMSO	364(4.20),519(4.43)	18-1521-76
$C_{16}H_{18}N_3O_8P$			
5'-Cytidylic acid, N-benzoyl-2'-deoxy-	H_2O	259(4.27),304(--)	138-0019-76
$C_{16}H_{18}N_3O_9P$			
5'-Cytidylic acid, N-benzoyl-	H_2O	259(4.27),303(--)	138-0019-76
$C_{16}H_{18}N_4OS$			
4-Thiazolecarboxamide, 2-(1-aminoethyl)-N-[2-(1H-indol-3-yl)ethyl]-, (S)-	MeOH	223(4.76),275(4.02), 282(4.02),291(3.92)	94-0092-76
$C_{16}H_{18}N_4O_2$			
Benzenamine, N,N-diethyl-4-[(4-nitrophenyl)azo]-	n.s.g.	458(4.57)	18-1381-76
Benzenamine, N,N,3-trimethyl-4-[(2-methyl-5-nitrophenyl)azo]-	C_6H_{12}	425(4.47)	18-1381-76
Butanal, 3,4-dihydroxy-, 2-(phenylhydrazono)-, phenylhydrazone, (S)-	EtOH	209(4.20),230s(--), 256(4.35),310(4.06), 400(4.33)	136-0185-76D
6H-1,2,4,5-Tetrazino[1,6-b]isoquinolin-6-one, 3-ethyl-1,2-dihydro-1-methyl-2-(1-oxopropyl)-	EtOH	234(4.41),319(4.19), 364(3.93),381(3.81)	95-0700-76
$C_{16}H_{18}N_4O_3$			
4H-Indazol-4-one, 1,5,6,7-tetrahydro-1,6,6-trimethyl-3-(4-nitrophenyl)-, oxime, (E)-	MeOH	216(4.36),242s(4.13), 314(4.02)	24-1898-76
Pyrazolo[4,3-b]azepin-5(1H)-one, 4,6,7,8-tetrahydro-1,7,7-trimethyl-3-(4-nitrophenyl)-	MeOH	224s(4.29),322(4.09)	24-1898-76
Pyrazolo[4,3-c]azepin-4(1H)-one, 5,6,7,8-tetrahydro-1,7,7-trimethyl-3-(4-nitrophenyl)-	MeOH	308(4.13)	24-1898-76
$C_{16}H_{18}N_4O_3S$			
1H-Pyrimido[5,4-b][1,4]benzothiazine-2,4(3H,4aH)-dione, 1,3-dimethyl-4a-morpholino-	EtOH	225(4.51),258(4.30), 301(3.93),312(3.92)	94-3135-76

Compound	Solvent	λ_{max} (log ϵ)	Ref.
$C_{16}H_{18}N_4O_5$			
4,14-Epoxy-5H,8H-1,3-dioxolo[5,6][1,3]-oxazocino[3,2-a]pteridin-8-one, 3a,4-14,14a-tetrahydro-2,2,10,11-tetramethyl-, [3aR-(3aα,4β,14β,14aα)]-	MeOH	227(4.45),252s(4.24), 271(4.19),361(4.15)	24-3159-76
Uridine, 2',3'-diamino-2',3'-dideoxy-, 5'-benzoate	MeOH	224(4.21),261(4.04)	44-3138-76
$C_{16}H_{18}N_6O$			
Urea, N-(4-phenylbutyl)-N'-1H-pyrazolo-[3,4-d]pyrimidin-4-yl-	pH 1	269(4.20)	87-0555-76
	pH 13	272(4.10),294(3.64)	87-0555-76
	EtOH	265(4.19)	87-0555-76
$C_{16}H_{18}N_6O_4S$			
3H-1,2,3-Triazolo[4,5-d]pyrimidin-5-amine, 7-[(phenylmethyl)thio]-3-β-D-ribofuranosyl-	EtOH-HCl	216(4.28),249(3.97), 280(3.96),317(4.13)	87-1186-76
	EtOH-pH 7	216(4.29),249(3.99), 278(3.95),317(4.13)	87-1186-76
$C_{16}H_{18}O$			
Benzenemethanol, 2,4,6-trimethyl-α-phenyl-	EtOH	216s(4.24),257-274f(<2.6)	44-3083-76
$C_{16}H_{18}O_2$			
Bicyclo[3.2.1]oct-2-en-6-one, 2-(4-methoxyphenyl)-5-methyl-	EtOH	260(4.29)	39-1669-76C
Furan, tetrahydro-2-methoxy-5-(1,3,9-undecatriene-5,7-diynyl)-	ether	251(4.54),267(4.50), 280(4.25),297(4.47), 316(4.60),337(4.47)	24-2291-76
2(1H)-Naphthalenone, 1-cyclohexyl-1-hydroxy-	EtOH	235(4.29),313(3.75)	39-2570-76C
$C_{16}H_{18}O_3$			
2H-1-Benzopyran-2-one, 3-(2,2-dimethyl-1-oxopropyl)-4-ethyl-	EtOH	225(3.63),278(3.98), 315(3.83)	39-1260-76C
1-Benzoxepin-2(5H)-one, 3-(2,2-dimethyl-1-oxopropyl)-5-methyl-	EtOH	235(3.44),272s(3.19)	39-1260-76C
1H-Fluorene-8-carboxylic acid, 2,3,4,9-tetrahydro-7-methoxy-, methyl ester	EtOH	268(4.17),275s(4.08), 329(3.28)	44-2401-76
7H-Furo[3,2-g][1]benzopyran-7-one, 6-(1,1-dimethylpropyl)-2,3-dihydro-	EtOH	209(4.3),226(4.09), 301s(3.93),334(4.33)	32-0681-76
1H-Inden-1-one, 6-(2-acetoxyethyl)-2,5,7-trimethyl-	EtOH	216(3.92),251(4.67), 333(3.37),406(2.72)	102-0995-76
2(1H)-Naphthalenone, 1-cyclohexyl-1-hydroperoxy-	EtOH	236(4.12),241(4.14), 316(3.89)	39-2570-76C
Naphtho[1,2-b]furan-2,8(3H,4H)-dione, 5,5a-dihydro-3,3,5a,9-tetramethyl-, (S)-	n.s.g.	238(4.06),317(4.01)	39-0442-76C
Phenol, 2-methoxy-6-[2-(4-methoxyphenyl)ethyl]-	MeOH	210(2.93),226(3.00), 278(2.26)	78-0109-76
Santonene, 4-methyl-	n.s.g.	223(4.08),287(4.28)	39-0442-76C
photoisomer	n.s.g.	255(4.22)	39-0442-76C
$C_{16}H_{18}O_4$			
1,4-Ethanonaphthalene-5,6,10-trione, 1,4,4a,8a-tetrahydro-9-hydroxy-4,7,8,9-tetramethyl-	EtOH	273(3.79),308s(3.29), 418(1.87)	1-0064-76
1,4-Ethanonaphthalene-5,8,9-trione, 1,4,4a,8a-tetrahydro-10-hydroxy-1,6,7,10-tetramethyl-	EtOH	203(3.88),252(4.01), 315s(2.49),360(2.10)	1-0064-76

Compound	Solvent	$\lambda_{max}(\log \epsilon)$	Ref.
4H,9H-Furo[2',3',4':4,5]naphtho[2,1-c]-pyran-4,9-dione, 1,2,3,3a,5a,7,10b-10c-octahydro-3a,10b-dimethyl-	EtOH	258.5(4.13)	39-2407-76C
$C_{16}H_{18}O_5$ 2H-1-Benzopyran-2-one, 5,7-dimethoxy-6-(3-methyl-2-oxobutyl)- (puberulin)	EtOH	210(4.55),228s(4.26), 298(4.02),342(3.86)	102-1080-76
$C_{16}H_{18}O_6$ Ethanone, 1-(1-hydroxy-3,4,6,8-tetra-methoxy-2-naphthalenyl)-	MeOH	236(4.48),271(4.45), 316(3.76),397(3.69)	12-1087-76
Tricyclo[4.4.0.0^{5,9}]deca-3,7-diene-1,6-dicarboxylic acid, 5-hydroxy-8,9-dimethyl-2-oxo-, dimethyl ester	MeOH	241(3.56),325s(1.70)	35-7040-76
Tricyclo[4.4.0.0^{3,9}]dec-4-ene-1,6-di-carboxylic acid, 3,4-dimethyl-7,10-dioxo-, dimethyl ester	MeOH	300(2.22),313(2.13)	35-7040-76
$C_{16}H_{18}O_8$ Cyclohexanecarboxylic acid, 1,3,4-tri-hydroxy-5-[[3-(2-hydroxyphenyl)-1-oxo-2-propenyl)oxy]-	MeOH-acid	226(4.03),277(4.25), 328(4.00)	20-0663-76
	MeOH-base	234(4.16),288(4.11), 387(4.06)	20-0663-76
Cyclohexanecarboxylic acid, 1,3,5-tri-hydroxy-4-[[3-(2-hydroxyphenyl)-1-oxo-2-propenyl)oxy]-	MeOH-acid	213(4.13),277(4.21), 326(3.95)	20-0663-76
	MeOH-base	235(4.15),288(3.98), 394(3.90)	20-0663-76
$C_{16}H_{18}O_9$ 9-Oxabicyclo[4.2.1]nona-4,7-diene-1,4,7,8-tetracarboxylic acid, tetramethyl ester	MeCN	230s(3.86)	88-3303-76
$C_{16}H_{19}BrO_2$ 2H-1-Benzopyran-6-ol, 7-bromo-2-methyl-2-(4-methyl-3-pentenyl)- (cymopo-chromenol)	EtOH	235(4.13),267(3.48), 335(3.53)	39-1696-76C
$C_{16}H_{19}BrO_3$ 2,5-Methano-1-benzoxepin-2(3H)-carbox-ylic acid, 7-bromo-4,5-dihydro-5,8,10-trimethyl-, methyl ester, (2α,5β,10S*)-	hexane	237(3.90),287(3.40), 295(3.43)	12-2533-76
2,6-Octadien-1-one, 1-(4-bromo-2,5-di-hydroxyphenyl)-3,7-dimethyl-, cis	EtOH	283(4.22),375(3.22)	39-1696-76C
trans (cymopolone)	EtOH	280(4.30),378(3.83)	39-1696-76C
$C_{16}H_{19}ClO$ 1-Propanone, 3-chloro-1-(1,2-dihydro-1,2,6-trimethyl-1-naphthalenyl)-	MeOH	222(4.36),264(3.82)	44-0066-76
$C_{16}H_{19}FN_2OS$ Methanone, (4-fluorophenyl)(3,4,7,8,9-10-hexahydro-2H,6H-[1,3]thiazino[3,2-a][1,3]diazocin-2-yl)-, hydrochloride	n.s.g.	219(4.33)	2-0773-76
$C_{16}H_{19}IN_4O_5$ 2,4(1H,3H)-Pteridinedione, 1-[5-deoxy-5-iodo-2,3-O-(1-methylethylidene)-β-D-ribofuranosyl]-6,7-dimethyl-	pH 6.0	245(3.97),323(3.95), 335s(3.87)	24-3159-76
	pH 11.0	241(4.21),332(4.01), 345s(3.89)	24-3159-76

Compound	Solvent	$\lambda_{max}(\log \epsilon)$	Ref.
2,4(1H,3H)-Pteridinedione, 1-[5-deoxy-5-iodo-2,3-O-(1-methylethylidene)-β-D-ribofuranosyl]-6,7-dimethyl- (cont.)	MeOH	230(4.10),247s(4.02), 325(3.94),338s(3.83)	24-3159-76
$C_{16}H_{19}N$			
3H-Indole, 3-(1-ethynylbutyl)-2,3-di-methyl-	EtOH	214(4.20),220(4.20), 226s(4.07),260(3.66)	39-2103-76C
3H-Indole, 3-(1,2-hexadienyl)-2,3-di-methyl-	EtOH	215(4.24),220(4.22), 228s(4.06),260(3.68)	39-2103-76C
Quinoline, 2,4-dimethyl-3-(1-pentenyl)-, cis	EtOH	212(4.29),229(4.42), 272(3.55),306(3.27), 319(4.27)	39-2103-76C
$C_{16}H_{19}NO$			
Benzeneethanol, 4-(dimethylamino)-α-phenyl-	MeOH	256(4.20),301(3.24)	35-6711-76
Benzo[3,4]cyclobuta[1,2-c]quinolin-6(5H)-one, 6a,6b,7,8,9,10,10a,10b-octahydro-10b-methyl-, (6aα,6bα,10aα,10bα)-	EtOH	256(2.91)	1-0189-76
1-Butanone, 4-(dimethylamino)-1-(2-naph-thalenyl)-	heptane	238(4.46),246(3.54), 281(3.71),324(3.08), 340(3.15)	35-4515-76
4H-1-Pyrindin-4-one, 1,2,3,5,6,7-hexa-hydro-1-(1-phenylethyl)-	heptane	310(4.16)	103-0428-76
	MeOH	340(4.32)	103-0428-76
	CF_3COOH	320(4.00)	103-0428-76
2H-Pyrrol-2-one, 1,5-dihydro-3,4-dimeth-yl-5-[(2,4,6-trimethylphenyl)methyl-ene]-, (Z)-	EtOH	255s(4.04),302(4.15)	49-0907-76
4(1H)-Quinolinone, 2-(1-heptenyl)-	EtOH	258(4.62),267(4.61), 308(3.98),337(4.09)	70-1115-76
	EtOH-HCl	258(4.64),267(4.68), 318(4.21)	70-1115-76
	EtOH-NaOH	260(4.55),267s(4.51), 323(3.86)	70-1115-76
$C_{16}H_{19}NO_2$			
Eleocarpine	EtOH	253(3.94),320(3.51)	42-0531-76
Isoeleocarpine	EtOH	258(3.98),326(3.49)	42-0531-76
$C_{16}H_{19}NO_3$			
Spiro[3H-benz[e]indene-3,2'-[1,3]dioxol-an]-5(2H)-one, 7-amino-1,3a,4,9b-tetrahydro-3a-methyl-, trans	EtOH	238(4.33),358(3.30)	104-0796-76
$C_{16}H_{19}NO_3S_2$			
Piperidine, 1-(3,4-dihydro-5H-[1]benzo-thiopyrano[3,4-e]-1,2-oxathiin-4-yl)-, S^2,S^2-dioxide	EtOH	220(4.09),248(4.29), 285(3.54),295s(3.41), 332(3.26)	4-0225-76
$C_{16}H_{19}NO_6$			
Glycine, N-[3-[4-[(ethoxycarbonyl)oxy]-phenyl]-1-oxo-2-propenyl]-, ethyl ester	MeOH-acid	277(4.45)	20-0647-76
1H-1-Pyrindine-3,4-dicarboxylic acid, 4,5,6,7-tetrahydro-2,6-dimethyl-5,7-dioxo-, diethyl ester	EtOH-HCl	251(4.39),391(3.81)	103-0220-76
	EtOH-KOH	264(4.42),380(3.77), 494(3.30)	103-0220-76

Compound	Solvent	$\lambda_{max}(\log \epsilon)$	Ref.
$C_{16}H_{19}N_3$			
Benzenamine, N,N-diethyl-4-(phenylazo)-	C_6H_{12}	407(4.51)	18-1381-76
Benzenamine, 4-[(2,4-dimethylphenyl)-azo]-N,N-dimethyl-	C_6H_{12}	393(4.49)	18-1381-76
Benzenamine, 4-[(2,6-dimethylphenyl)-azo]-N,N-dimethyl-	toluene	389(4.36)	18-1381-76
Benzenamine, N,N,3,5-tetramethyl-4-(phenylazo)-	toluene	391(4.30)	18-1381-76
Benzenamine, N,N,3-trimethyl-4-[(2-methylphenyl)azo]-	C_6H_{12}	400(4.46)	18-1381-76
Benzenamine, N,N,3-trimethyl-4-[(4-methylphenyl)azo]-	C_6H_{12}	405(4.48)	18-1381-76
$C_{16}H_{19}N_3O$			
4H-Indazol-4-one, 1,5,6,7-tetrahydro-1,6,6-trimethyl-3-phenyl-, oxime, (E)-	MeOH	226(4.31),258s(3.89)	24-1898-76
Pyrazolo[4,3-b]azepin-5(1H)-one, 4,6,7-8-tetrahydro-1,7,7-trimethyl-3-phenyl-	MeOH	229(4.11),250(4.15)	24-1898-76
Pyrazolo[4,3-c]azepin-4(1H)-one, 5,6,7-8-tetrahydro-1,7,7-trimethyl-3-phenyl-	MeOH	246s(4.04)	24-1898-76
$C_{16}H_{19}N_3O_3$			
4H-Indazol-4-one, 3-(3,4-dimethoxyphenyl)-1,5,6,7-tetrahydro-1-methyl-, oxime, (E)-	MeOH	238s(4.33),274(3.99)	24-1898-76
L-Isoleucine, N-[(3-methyl-2-quinoxalinyl)carbonyl]-	EtOH	203(4.60),242(4.55), 320(3.85)	78-2931-76
L-Leucine, N-[(3-methyl-2-quinoxalinyl)carbonyl]-	EtOH	203(4.59),241(4.50), 320(3.84)	78-2931-76
Pyrazolo[4,3-c]azepin-4(1H)-one, 3-(3,4-dimethoxyphenyl)-5,6,7,8-tetrahydro-1-methyl-	MeOH	250s(4.02),285s(3.89)	24-1898-76
$C_{16}H_{19}N_3O_4$			
Ethanol, 2,2'-[[3-nitro-4-(phenylamino)-phenyl]imino]bis-	EtOH	524(3.71)	146-0131-76
$C_{16}H_{19}N_3O_5$			
Cytidine, N-(phenylmethyl)-	pH 2	285(4.21)	44-1597-76
	pH 7	238(--),273(--)	44-1597-76
L-Isoleucine, N-[(3-methyl-2-quinoxalinyl)carbonyl]-, N,N'-dioxide	EtOH	205(4.12),232(4.31), 260s(4.25),267(4.01), 370s(4.03),384(4.08)	78-2931-76
	MeCN	205(4.01),228(4.27), 260s(4.13),268(4.33), 360s(3.72),379(3.92), 393(3.99)	78-2931-76
L-Leucine, N-[(3-methyl-2-quinoxalinyl)-carbonyl]-, N,N'-dioxide	EtOH	205(4.14),232(4.34), 260s(4.29),267(4.43), 370s(4.05),384(4.11)	78-2931-76
	MeCN	205(4.21),228(4.30), 260s(4.14),268(4.31), 360s(3.69),380s(3.89), 393(3.95)	78-2931-76
$C_{16}H_{19}N_4O_3P$			
Diazenecarboxylic acid, [bis(methylphenylamino)phosphinyl]-, methyl ester	MeOH	206(4.27),232(4.11), 270(3.18),464(1.53)	139-0207-76A

Compound	Solvent	$\lambda_{max}(\log \epsilon)$	Ref.
$C_{16}H_{19}N_5O$ 4(1H)-Pyrimidinone, 6-(cyclohexylamino)- 5-(phenylazo)-	pH 13	244(4.34),386(4.32)	39-1847-76C
$C_{16}H_{19}O_8$ 1,3-Cyclopentadiene-1,2,3,4-tetracarbox- ylic acid, 5-(1-methylethyl)-, tetra- methyl ester, ion(1-), sodium	MeOH	209(4.04),267(4.16), 306(4.03)	24-1928-76
$C_{16}H_{20}BrN_3O_2$ 5-Pyrimidinecarboxamide, 4-(4-bromophen- yl)-N,N-diethyl-1,2,3,4-tetrahydro- 6-methyl-2-oxo-	EtOH	203(4.43),228(4.19), 269(3.59)	103-0191-76
$C_{16}H_{20}ClN_4O_6$ 1-Piperidinyloxy, 4-[(5-chloro-2,4-di- nitrobenzoyl)amino]-2,2,6,6-tetra- methyl-	EtOH	229s(--),252(4.10)	64-0328-76C
$C_{16}H_{20}N_2$ [1,1'-Biphenyl]-4,4'-diamine, N,N,N',N'- tetramethyl-	MeOH	207(4.56),307(4.51)	139-0151-76A
$C_{16}H_{20}N_2O$ 9H-Carbazole-9-propanamine, 1,2,3,4- tetrahydro-α-methyl-γ-oxo-, hydro- chloride	EtOH	246(4.23),266(3.98), 300(3.70)	44-3775-76
4H-Pyrrolo[2,1-c][1,4]benzoxazine-4-eth- anamine, N,N,4-trimethyl-, hydrochlor- ide	MeOH	218(4.38),265(4.01), 291(3.86)	4-0311-76
4H-Pyrrolo[2,1-c][1,4]benzoxazine- 1-methanamine, N,N,4-tetramethyl-, hydrochloride	MeOH	264(4.05),291(3.84)	4-0311-76
$C_{16}H_{20}N_2O_2$ 9-Acridinecarbonitrile, 9a-ethoxy- 1,2,3,4,4a,9,9a,10-octahydro- 4a-hydroxy-	EtOH	247(4.16),300(3.71)	95-0968-76
1,6-Phenazinedione, 2,3,4,7,8,9-hexa- hydro-3,3,8,8-tetramethyl-	EtOH	233(3.97),317(3.97)	18-2805-76
$C_{16}H_{20}N_2O_3$ Butanamide, 4-acetoxy-N-[2-(1H-indol- 3-yl)ethyl]-	MeOH	229(4.09),275(3.68), 282(3.72),291(3.64)	23-1262-76
4H-Pyrrolo[3,2-d]isoxazole, 6-(benzoyl- oxy)-3a,5,6,6a-tetrahydro-3,3a,5,6a- tetramethyl-	EtOH	229(4.16),278(3.02), 280(2.92)	44-0855-76
$C_{16}H_{20}N_4$ Benzenamine, 4,4'-azobis[N,N-dimethyl-	C_6H_{12}	410(4.57)	18-1381-76
$C_{16}H_{20}N_4O_4$ 5-Pyrimidinecarboxamide, N,N-diethyl- 1,2,3,4-tetrahydro-6-methyl-4-(3- nitrophenyl)-2-oxo-	EtOH	202(4.53),211s(4.37), 266(4.08)	103-0191-76
5-Pyrimidinecarboxamide, N,N-diethyl- 1,2,3,4-tetrahydro-6-methyl-4-(4- nitrophenyl)-2-oxo-	EtOH	203(4.53),215s(4.32), 269(4.07)	103-0191-76

Compound	Solvent	$\lambda_{max}(\log \epsilon)$	Ref.
$C_{16}H_{20}N_4O_6$			
α-D-xylo-Pentodialdo-1,4-furanose, 3-(acetylamino)-3-deoxy-1,2-O-(1-methylethylidene)-, 5-(4-nitrophenyl)-hydrazone	EtOH	207(4.09),248(3.98), 378(4.39)	136-0119-76A
$C_{16}H_{20}N_5O_6PS$			
3H-Imidazo[2,1-a]purine, 5-(butylthio)-3-(3,5-O-phosphinico-β-D-ribofuranosyl)-	pH 1	238(4.29),293(4.29)	94-1561-76
	pH 13	241(4.32),245s(4.31), 274(3.91),284(4.00), 305(3.99)	94-1561-76
3H-Imidazo[2,1-a]purine, 5-(isobutylthio)-3-(3,5-O-phosphinico-β-D-ribofuranosyl)-	pH 1	238(4.28),293(4.29)	94-1561-76
	pH 13	240(4.30),245s(4.29), 274(3.89),283(3.98), 306(3.98)	94-1561-76
$C_{16}H_{20}N_6O$			
Urea, N-1H-pyrazolo[3,4-d]pyrimidin-4-yl-N'-tricyclo[3.3.1.1^{3,7}]dec-2-yl-	pH 1	272(4.22)	87-0555-76
	pH 13	272(4.10),296(3.61)	87-0555-76
	EtOH	267(4.20)	87-0555-76
$C_{16}H_{20}O$			
6,8,14-Hexadecatriene-10,12-diyn-1-ol, (E,E,E)-	MeOH	252(4.51),267(4.44), 281(4.23),296(4.45), 316(4.58),337(4.45)	39-0735-76C
1-Naphthalenemethanol, α-(1,1-dimethylethyl)-α-methyl-	ether	224(4.90),2-0-295f(3.68-3.93)	44-3067-76
2-Naphthalenemethanol, α-(1,1-dimethylethyl)-α-methyl-	ether	224(4.97),248-288f(3.51-3.71)	44-3067-76
$C_{16}H_{20}O_2$			
2H-Benz[e]inden-3-ol, 3,3a,4,5-tetrahydro-7-methoxy-3a,5-dimethyl-, (3α,3aα,5β)	EtOH	269(4.06)	2-0312-76
3H-Benz[e]inden-3-one, 1,2,3a,4,5,9b-hexahydro-7-methoxy-3a,5-dimethyl-, cis	EtOH	282(3.40),289(3.38)	2-0312-76
trans	EtOH	282(3.45),289(3.40)	2-0312-76
Cyclohexanecarboxylic acid, 1-methyl-3-(phenylmethylene)-, methyl ester	EtOH	250(4.30)	88-4633-76
2-Propenoic acid, 2-methyl-, 4-cyclohexylphenyl ester	dioxan	270.5(3.09)	126-3481-76
$C_{16}H_{20}O_3$			
Eudesma-4,6-dien-12,6-olactone, 11-methyl-3-oxo-	n.s.g.	305(4.4)	39-0442-76C
2(5H)-Furanone, 4-butyl-5-methyl-3-(phenylmethoxy)-	EtOH	226(3.91)	107-0299-76
2-Naphthaleneacetic acid, 3,4,4a,7-tetrahydro-α,4a,8-trimethyl-7-oxo-, methyl ester, [S-(R*,R*)]-	EtOH	235(4.08),310(4.09)	44-1256-76
2-Naphthalenepropanoic acid, 1,2,3,4-tetrahydro-β-methyl-1-oxo-, ethyl ester	EtOH	249(4.1),292(3.3)	39-0722-76C
2(1H)-Naphthalenone, 3-acetoxy-3,4-dihydro-1,1,4,4-tetramethyl-	EtOH	285(1.58),300(1.36)	18-3132-76
2(1H)-Naphthalenone, 3-acetyl-3,4-dihydro-1,1,4,4-tetramethyl-3-hydroxy-	EtOH	315(2.18)	18-3132-76
1H-Naphtho[2,1-b]pyran-1,10(4aH)-dione, 2,3,7,8,9,10b-hexahydro-4a,8,10b-trimethyl-	MeOH	300(3.86)	44-0066-76
	MeOH-NaOH	305(3.98)	44-0066-76

Compound	Solvent	$\lambda_{max}(\log \epsilon)$	Ref.
1H-Naphtho[2,1-b]pyran-1-one, 2,3,4a,5-6,10b-hexahydro-10-hydroxy-4a,8,10b-trimethyl-, (4aR-cis)-	MeOH	215(3.73),278(3.23), 284(3.26)	44-0066-76
	MeOH-NaOH	293(3.48)	44-0066-76
2,6-Octadienal, 8-(2,5-dihydroxyphenyl)-2,6-dimethyl-, (E,E)- (alliodorin)	EtOH	227(4.33),295(3.61)	78-0665-76
Santonene, 1,2-dihydro-4-methyl-	n.s.g.	287(4.35)	39-0442-76C
Santonene, 1,2-dihydro-6α-methyl-	n.s.g.	237(4.3)	39-0442-76C

$C_{16}H_{20}O_4$

Compound	Solvent	$\lambda_{max}(\log \epsilon)$	Ref.
1H-Dibenzo[b,d]pyran-6-acetic acid, 2,3,4,4a,6,10b-hexahydro-9-hydroxy-6-methyl-	EtOH	278.5(3.38)	95-1031-76
1H-Dibenzo[b,d]pyran-6-carboxylic acid, 2,3,4,4a,6,10b-hexahydro-9-hydroxy-6-methyl-, methyl ester	EtOH	277(3.27)	95-1031-76
1-Naphthalenecarboxylic acid, 1,2,3,4-tetrahydro-7-methoxy-6-(1-methylethyl)-2-oxo-, methyl ester	MeOH	212(4.67),248(4.18), 255(4.09),292(3.80), 300(3.79),315(3.57)	18-1985-76
8-Nonenoic acid, 9-(3,4-methylenedioxyphenyl)-	MeOH	214(4.37),269(4.09), 286(3.86)	102-2018-76

$C_{16}H_{20}O_4S$

Compound	Solvent	$\lambda_{max}(\log \epsilon)$	Ref.
Spiro[1,3-benzoxathiole-2,1'-cyclohexane]-2',4',6'(5H)-trione, 6,7-dihydro-4',4',6,6-tetramethyl-	MeOH	289s(3.57),306(3.62), 380s(2.18)	44-0125-76
	CH_2Cl_2	246s(4.24),292(3.66), 312s(3.61),380s(2.08)	44-0125-76

$C_{16}H_{20}O_5$

Compound	Solvent	$\lambda_{max}(\log \epsilon)$	Ref.
Butanedioic acid, [1-(2-methoxy-4-methylphenyl)ethylidene]-, 1-ethyl ester	MeOH	270(4.10)	33-0695-76
Eremophil-7(11)-ene-12,8α:14β,6α-diolide, 8β-methoxy-	EtOH	216(4.17)	18-3196-76

$C_{16}H_{20}O_6$

Compound	Solvent	$\lambda_{max}(\log \epsilon)$	Ref.
1,9-Dioxacyclohexadeca-3,11-diene-2,5,10,13-tetrone, 8,16-dimethyl-, (E,E)- (pyrenophorin)	EtOH	211(4.36)	39-1718-76C

$C_{16}H_{20}O_8$

Compound	Solvent	$\lambda_{max}(\log \epsilon)$	Ref.
1,3-Cyclopentadiene-1,2,3,4-tetracarboxylic acid, 5-(1-methylethyl)-, tetramethyl ester, ion(1-), sodium	MeOH	209(4.04),267(4.16), 306(4.03)	24-1928-76

$C_{16}H_{20}O_{11}$

Compound	Solvent	$\lambda_{max}(\log \epsilon)$	Ref.
Ixoside	MeOH	219(4.16)	94-1216-76

$C_{16}H_{20}S_4$

Compound	Solvent	$\lambda_{max}(\log \epsilon)$	Ref.
4H-Cyclohepta-1,3-dithiole, 5,6,7,8-tetrahydro-2-(5,6,7,8-tetrahydro-4H-cyclohepta-1,3-dithiol-2-ylidene)-	THF	258(4.46),291(4.37), 500(2.37)	97-0360-76
cation	MeCN	463(--),661(--)	97-0360-76
dication	MeCN	275(3.90),422(4.08)	97-0360-76

$C_{16}H_{21}BF_2O_5$

Compound	Solvent	$\lambda_{max}(\log \epsilon)$	Ref.
1-Oxa-3-oxonia-2-boratanaphthalene, 2,2-difluoro-1,2-dihydro-8-isobutyryl-4-isopropyl-5,7-dimethoxy-	MeCN	213(4.2),236(3.9), 315(4.5),366(3.7)	5-1514-76

$C_{16}H_{21}BrO_2$

Compound	Solvent	$\lambda_{max}(\log \epsilon)$	Ref.
1,4-Benzenediol, 2-bromo-5-(3,7-dimethyl-2,6-octadienyl)-, (E)- (cymopol)	EtOH	299(3.88)	39-1696-76C

Compound	Solvent	$\lambda_{max}(\log \epsilon)$	Ref.
$C_{16}H_{21}ClN_3O$			
Morpholinium, 4-[3-chloro-3-[(dimethyl-amino)methylene]amino]-2-phenyl-2-propenylidene]-, perchlorate	HOAc	400(4.52)	73-1565-76
$C_{16}H_{21}NO$			
1H-Benz[e]indole, 2,3,3a,9b-tetrahydro-8-methoxy-3,3a,9b-trimethyl-, cis-($\pm$)-	EtOH	215(4.43),272(4.21)	88-0767-76
2-Cyclohexen-1-one, 3-[(2,4-dimethyl-phenyl)amino]-5,5-dimethyl-	pH 1	293(4.36)	39-2207-76C
	H_2O	299(4.44)	39-2207-76C
	pH 13	299(4.44)	39-2207-76C
2-Cyclohexen-1-one, 3-[(2,6-dimethyl-phenyl)amino]-5,5-dimethyl-	pH 1	290(4.38)	39-2207-76C
	H_2O	296(4.45)	39-2207-76C
	pH 13	296(4.44)	39-2207-76C
1-Naphthaleneethanamine, 3,4-dihydro-6-methoxy-N,N-dimethyl-β-methylene-	MeOH	273(4.03)	44-0531-76
$C_{16}H_{21}NOS$			
4H-1-Benzothiopyran-4-one, 3-[[bis(1-methylethyl)amino]methylene]-2,3-dihydro-	EtOH	247(4.29),260s(4.18), 380(4.16)	4-0225-76
3-Buten-2-one, 4-[4-(methylthio)phenyl]-1-(2-piperidinyl)-, hydrochloride	MeOH	247(3.99),344(4.40)	87-1195-76
$C_{16}H_{21}NO_3$			
$\Delta^{\alpha,\beta}$-Piperlonguminine, dihydro-	MeOH	208(4.20),212s(4.15), 240s(3.81),290(3.58), 339(3.44)	102-0822-76
2H-Pyran-2-one, 6-[2-[4-(dimethylamino)-phenyl]ethyl]-5,6-dihydro-4-methoxy-	MeOH	246(5.26),301(4.26)	44-4070-76
$C_{16}H_{21}NO_6$			
Propanedioic acid, (3-formyl-1,2-di-hydro-1-methyl-2-oxo-4-pyridinyl)-, 1,1-dimethylethyl ethyl ester	MeOH	234(3.87),315(3.40), 366(3.77)	24-1389-76
2(1H)-Pyridinone, 4-acetoxy-3-(1-acet-oxy-4-methyl-2-oxopentyl)-6-methyl-	MeOH	230(3.74),309(3.94)	77-0566-76
$C_{16}H_{21}NS_4$			
2,5-Methano-7H-1,3-dithiolo[4,5-d]-[1,3]dithiin-7-amine, dihydro-2a,3,5,7-tetramethyl-N-phenyl-	isooctane	241(3.82),255(3.70), 292(2.74)	70-2353-76
$C_{16}H_{21}N_2O_3P$			
Phosphonic acid, (3a,4,5,6,7,7a-hexa-hydro-3-phenyl-4,7-methano-3H-inda-zol-3-yl)-, dimethyl ester	MeOH	208(4.1),338(2.4)	23-0044-76
$C_{16}H_{21}N_3O_2$			
5-Pyrimidinecarboxamide, N,N-diethyl-1,2,3,4-tetrahydro-6-methyl-2-oxo-4-phenyl-	EtOH	205(4.40),218s(4.09), 265(3.38)	103-0191-76
$C_{16}H_{21}N_3O_6$			
Propanoic acid, 2,2-dimethyl-, (3-acet-oxy-2,3,3a,9a-tetrahydro-6-imino-6H-furo[2',3':4,5]oxazolo[3,2-a]pyrimi-din-2-yl)methyl ester, hydrochloride	MeOH-acid	235(4.00),262(4.07)	87-0663-76

Compound	Solvent	$\lambda_{max}(\log \epsilon)$	Ref.
$C_{16}H_{21}N_3O_8$ Cytidine, N-acetyl-2'-O-methyl-, 3',5'-diacetate	H_2O	213(4.36),247(4.34), 300(4.05)	35-7381-76
$C_{16}H_{21}N_3S$ [1]Benzothiopyrano[4,3-c]pyrazole- 1(4H)-ethanamine, N,N-diethyl-	EtOH	225(4.09),245s(4.20), 254(4.25),280s(3.83), 318(3.53)	4-0225-76

$C_{16}H_{21}N_5O$
Methanimidamide, N,N-dimethyl-N'-[3-
  [(1-methylpropylidene)hydrazino]-
  1-oxo-2(1H)-isoquinolinyl]-
	EtOH	230(4.44),316(4.21),
		375(3.74)
			95-0700-76

$C_{16}H_{21}N_7O_2$
9,10-Propanopyrimido[4',5':3,4]azeto-
  [2,1-f]purine-6,8(5aH,7H)-dione,
  4-amino-5a-butyl-9a,9b-dihydro-
	H_2O	230(4.24),288(3.82)	19-0517-76

2,4(1H,3H)-Pyrimidinedione, 1-[3-(6-am- ino-9H-purin-9-yl)propyl]-5-butyl-	H_2O	264(4.28)	19-0517-76

$C_{16}H_{21}O_3P$
Phosphonic acid, (3-phenyltricyclo-
  [3.2.1.0^{2,4}]oct-3-yl)-, dimethyl
  ester, syn
	MeOH	214(3.1),337(2.3)	23-0044-76

$C_{16}H_{22}$ Bicyclo[10.2.2]hexadeca-2,12,14,15-tet- raene	C_6H_{12}	202(4.27),239(3.88)	94-1724-76
$C_{16}H_{22}Co$ Cobaltocenium, 1,1'-bis(1-methylethyl)-	MeOH	272(4.46),310(3.21), 403(2.60)	101-0189-76I

$C_{16}H_{22}INO$
Benzenaminium, N,N,N-trimethyl-4-[(2-
  oxocyclohexylidene)methyl]-,
  iodide, (E)-
	EtOH	219(4.14),280(4.13)	36-0538-76

$C_{16}H_{22}N_2O_3$
1H-Indole-3-carboxylic acid, 6-[(dimeth-
  ylamino)methyl]-5-methoxy-2-methyl-,
  ethyl ester, hydrochloride
	n.s.g.	218(4.52),248(4.43),
		286(4.02),296(4.00)
			103-0245-76

$C_{16}H_{22}N_2O_9$ Thymine, 1-(2',3',4'-tri-O-acetyl-α- L-rhamnosyl)-	MeOH-pH 7	263(4.00)	28-0757-76A

$C_{16}H_{22}N_2S_4$
1,4-Benzenediamine, N-(dihydro-2a,3,5,7-
  tetramethyl-2,5-methano-7H-1,3-dithi-
  olo[4,5-d][1,3]dithiin-7-yl)-
	isooctane	246(3.78),292(2.92)	70-2353-76

$C_{16}H_{22}N_2Si_2$ Cyclodisilazane, 2,2,4,4-tetramethyl- 1,3-diphenyl-	heptane	258(4.65),282(3.59), 287(3.59),299(3.40)	101-0299-76M
$C_{16}H_{22}N_4O$ 1H-Pyrazol-1-amine, N,N-dimethyl- 5-(4-morpholinylmethyl)-3-phenyl-	MeOH	253(4.25)	94-3001-76

Compound	Solvent	$\lambda_{max}(\log \epsilon)$	Ref.
1H-Pyrazol-1-amine, N,N-dimethyl- 5-(4-morpholinylmethyl)-4-phenyl-	MeOH	246(4.12)	94-3001-76
$C_{16}H_{22}N_5O_4S_2$ Thiazolium, 5-(ethoxycarbonyl)-2-[3-[5- (ethoxycarbonyl)-3-ethyl-2(3H)-thia- zolylidene]-1-triazenyl]-3-ethyl-, tetrafluoroborate	EtOH	487(4.47)	33-0155-76
$C_{16}H_{22}N_5O_6$ 1-Piperidinyloxy, 4-[[5-(aminocarbonyl)- 2,4-dinitrophenyl]amino]-2,2,6,6- tetramethyl-	EtOH	227(4.03),271(3.99), 349(4.09),400s(--)	64-0328-76C
$C_{16}H_{22}O$ 4,9-Decadien-2-ol, 2-phenyl- 3-Oxatricyclo[11.2.2.0^{2,4}]heptadeca- 13,15,16-triene	EtOH heptane	225(4.38) 198(4.55),200(4.55), 228(3.99),254s(2.58), 261(2.52),269(2.51), 278(2.36)	18-3646-76 94-1724-76
$C_{16}H_{22}O_2$ 1H-Benz[e]inden-3-ol, 2,3,3a,4,5,9b-hex- ahydro-7-methoxy-3a,5-dimethyl-, cis	EtOH	282(3.30),289(3.25)	2-0312-76
trans	EtOH	282(3.30),289(3.25)	2-0312-76
1,4-Naphthalenedione, 4a,5,8,8a-tetra- hydro-2,3,4aß,5α,8α,8aß-hexamethyl-	benzene	350(1.92)	35-7040-76
1,4-Naphthalenedione, 4a,5,8,8a-tetra- hydro-2,3,4aß,5ß,8ß,8aß-hexamethyl-	MeOH	251(3.94),340(1.85)	35-7040-76
1,4-Naphthalenedione, 4a,7,8,8a-tetra- hydro-2,3,4aß,5,8ß,8aß-hexamethyl-	MeOH	254(4.04),352(1.85)	35-7040-76
11-Oxatetracyclo[6.2.1^{1,7}.0^{5,10}]undec- 2-en-4-one, 2,3,5,6,9,10-hexamethyl-	MeOH	266(3.79),325(2.38)	35-7040-76
Tetracyclo[4.4.0.0^{3,9}.0^{4,8}]decane-2,5- dione, 1,3,4,6,7,10-hexamethyl-	MeOH	227(2.52),299(1.60), 315(1.52)	35-7040-76
Tricyclo[4.4.0.0^{3,10}]dec-8-ene-2,5-di- one, 1,3,4,6,7,10-hexamethyl-	MeOH	290(1.83)	35-7040-76
$C_{16}H_{22}O_3$ 2H-1-Benzopyran-2-propanal, 3,4-dihydro- 6-hydroxy-2,5,7,8-tetramethyl-, (S)-	EtOH	225s(3.97),291(3.51)	33-0290-76
2H-Indeno[4,5-b]furan-2,8(4H)-dione, 5,5a,6,7,8a,8b-hexahydro-3,5a-dimeth- yl-8b-(1-methylethyl)-, [5aS- (5aα,8aα,8bß)]-	n.s.g.	223(4.04)	39-0442-76C
1H-Naphtho[2,1-b]pyran-1-one, 2,3,4a,8,9,10,10a,10b-octahydro- 10-hydroxy-4a,8,10b-trimethyl-	MeOH	242(4.30)	44-0066-76
$C_{16}H_{22}O_4$ 2H-1-Benzopyran-2-acetic acid, 3,4-di- hydro-6-hydroxy-2,5,7,8-tetramethyl-, methyl ester, (S)-	EtOH	223s(4.00),290(3.50)	33-0290-76
2H-1-Benzopyran-6-ol, 3,4-dihydro-2- methoxy-2,5,7,8-tetramethyl-, acetate	EtOH	223(3.99),273(3.15), 280(3.19),283(3.19)	33-0290-76
Ligolide	n.s.g.	225(4.38),265s(--)	105-0666-76
2-Naphthaleneacetic acid, 1,2,3,4,4a,7- hexahydro-8-(hydroxymethyl)-α,4a-di- methyl-7-oxo-, methyl ester	EtOH	242(4.06)	44-1256-76

Compound	Solvent	$\lambda_{max}(\log \epsilon)$	Ref.
4a(2H)-Naphthalenecarboxylic acid, 6-formyl-1,3,4,7,8,8a-hexahydro-1,1-dimethyl-7-oxo-, ethyl ester, trans-(+)-	EtOH	245(3.71)	44-1005-76
Naphtho[1,2-b]furan-2,7-dione, 3,3a,4,5-5a,6,9a,9b-octahydro-8-methoxy-3,5a,9-trimethyl-, [3S-(3α,3aα,5aβ,9aα,9bβ)]-	EtOH	261(3.95)	2-0657-76
$C_{16}H_{22}O_4Zn$			
Zinc, bis(3-methyl-2-oxocyclohexanecarboxaldehydato-O,O')-, (T-4)-	CHCl$_3$	295(3.9)	65-2037-76
$C_{16}H_{22}O_9$			
Sweroside	MeOH	245(3.90)	95-0683-76
	MeOH	242(4.2)	102-1305-76
Tarennoside	MeOH	250(4.10)	94-1216-76
$C_{16}H_{22}O_{10}$			
Geniposidic acid	MeOH	237(3.64)	94-1216-76
$C_{16}H_{22}S_2$			
1,3-Dithietane, 2-[2,2-dimethyl-1-(1-methylethyl)propylidene]-4-phenyl-	hexane	240(4.5),360s(--)	24-0906-76
$C_{16}H_{23}NO_2$			
2-Piperidineethanol, α-[2-(4-methoxyphenyl)ethenyl]-, fumarate, (E)-	EtOH	263(4.68),270s(--), 293s(--),305s(--)	87-1195-76
$C_{16}H_{23}NO_3$			
1H-Indole-3-acetic acid, 1-cyclohexyl-2,4,5,6,7,7a-hexahydro-2-oxo-	EtOH	215(4.09)	22-0776-76
Piperlonguminine, tetrahydro-	MeOH	210(3.77),237s(3.51), 290(3.48)	102-0822-76
$C_{16}H_{23}NO_4$			
1H-Pyrrole-2-propanoic acid, 5-(3-ethoxy-3-oxo-1-propenyl)-3,4-dimethyl-, ethyl ester	EtOH	202(4.09),292(3.38), 356(4.34)	4-1145-76
Streptimidone	MeOH	291(2.90)	33-2393-76
$C_{16}H_{23}N_3O_9$			
Uracil, 1-N-(1-deoxy-3,4:5,6-di-O-isopropylidene-D-mannitol-1-yl)-5-nitro-	10% EtOH + base	238(3.86),308(4.00) 244s(3.81),324(3.99)	136-0049-76G 136-0049-76G
$C_{16}H_{24}$			
Bicyclo[3.2.1]octa-3,6-diene, 1,3,4,6-7,8,8-heptamethyl-2-methylene-	EtOH	244(4.14)	35-1551-76
Bicyclo[3.2.1]octa-3,6-diene, 1,3,5,6-7,8,8-heptamethyl-2-methylene-	EtOH	244(4.09)	35-1551-76
Bicyclo[3.2.1]octa-3,6-diene, 1,4,5,6-7,8,8-heptatmethyl-2-methylene-	EtOH	243(4.11)	35-1551-76
Bicyclo[3.2.1]octa-3,6-diene, 3,4,5,6-7,8,8-heptamethyl-2-methylene-	EtOH	245(4.10)	35-1551-76
Bicyclo[8.2.2]tetradeca-10,12,13-triene	isooctane	234(3.85),283(2.56), 288s(2.53)	44-4081-76
Tricyclo[3.2.1.0^{2,7}]oct-3-ene, 1,3,4,5-7,8,8-heptamethyl-6-methylene-	EtOH	242s(3.49)	35-1551-76
$C_{16}H_{24}Br_2O_2$			
1,3-Benzenediol, 4,6-dibromo-2-methyl-5-nonyl-	MeOH	268(3.14)	12-1989-76

Compound	Solvent	$\lambda_{max}(\log \epsilon)$	Ref.
$C_{16}H_{24}IN_6O_9P$ 5'-Adenylic acid, N-[2-[2-[(iodoacetyl)-aminolethoxylethyl]-, disodium salt	H_2O	267(4.22)	87-1279-76
$C_{16}H_{24}NO_5P$ Phosphonic acid, [2-(2,2,8-trimethyl-4H-1,3-dioxino[4,5-c]pyridin-5-yl)-ethenyl]-, diethyl ester, cis	EtOH	218(4.11),294(3.75)	104-0440-76
trans	EtOH	236(4.26),307(3.91)	104-0440-76
$C_{16}H_{24}N_2O$ 2,5-Cyclohexadien-1-one, 4-(1-aziridin-ylimino)-2,6-bis(1,1-dimethylethyl)-	n.s.g.	295(3.24)	70-2007-76
Cyclohexanone, 2,6-bis[3-(dimethylamino)-2-propenylidene]-	EtOH	259(4.04),490(4.78)	70-0577-76
$C_{16}H_{24}N_2O_3S$ Benzo[b]thiophene-3-carboxylic acid, 2-amino-4,5,6,7-tetrahydro-4-(mor-pholinomethyl)-, ethyl ester	EtOH	229(4.42),310(3.77)	2-0357-76
$C_{16}H_{24}N_2O_5$ 1H-Pyrrole-1,3-dicarboxylic acid, 4-(acetylamino)-5-(2-methylpropyl)-, diethyl ester	EtOH	217(4.69)	4-0113-76
$C_{16}H_{24}N_2O_6S_2Zn$ Zinc bis(monothiosquarate), bis(tetra-methylammonium) salt	H_2O	269(4.34),324(4.81)	44-3904-76
$C_{16}H_{24}N_4O_6$ Lumazine, 1-(2,3-O-isopropylidene-β-D-ribofuranosyl)-6,7-dimethyl-	pH 6.0	227s(4.05),244s(3.94), 323(3.94),337s(3.83)	24-3159-76
	pH 11.0	239(4.16),275(3.43), 330(3.97)	24-3159-76
	MeOH	228(4.06),245s(3.97), 323(3.80),337s(3.79)	24-3159-76
$C_{16}H_{24}N_5O_6PS$ Adenosine, 8-(hexylthio)-, cyclic 3',5'-(hydrogen phosphate)	pH 2.0	285(4.31)	69-1408-76
	pH 13	282(--)	69-1408-76
$C_{16}H_{24}N_6O_5S$ L-Homocysteine, N[6]-dimethyladenosyl-, (S)-	H_2O	275(4.37)	87-1094-76
$C_{16}H_{24}O$ 1H-Cyclopentacyclododecen-1-one, 2,3,4,5,6,7,8,9,10,11,12,13-dodeca-hydro-2-methylene-	EtOH	244(3.99),264s(3.90)	33-1226-76
$C_{16}H_{24}O_2$ Dispiro[6.1.6.1]hexadecane-8,16-dione	CH_2Cl_2	312(1.76),333(1.43)	18-0741-76
$C_{16}H_{24}O_3$ 2-Cyclopenten-1-one, 4-hydroxy-3-(2-methoxyethyl)-2-(1-octynyl)-	MeOH	259(4.22)	20-0503-76
1H-Naphtho[2,1-b]pyran-1,10-diol, 2,3,4a,8,9,10,10a,10b-octahydro-4a,8,10b-trimethyl-	MeOH	232s(4.25),239(4.31), 247(4.17)	44-0066-76

Compound	Solvent	$\lambda_{max}(\log \epsilon)$	Ref.
$C_{16}H_{24}O_4$			
4a(2H)-Naphthalenecarboxylic acid, octa-hydro-2-(hydroxymethylene)-8,8-dimeth-yl-2-oxo-, ethyl ester, trans-($\pm$)-	EtOH base	272(3.21) 312(3.64)	44-1005-76 44-1005-76
$C_{16}H_{24}O_5$			
2-Epilaksholic acid, methyl ester	MeOH	229(3.76)	39-0967-76C
Laksholic acid, methyl ester	MeOH	230(3.76)	39-0967-76C
$C_{16}H_{25}BrO$			
2,5-Cyclohexadien-1-one, 4-(bromometh-yl)-2,6-bis(1,1-dimethylethyl)-4-methyl-	hexane	236.5(3.924)	70-1780-76
$C_{16}H_{25}ClO_2S$			
2-Thiophenecarboxylic acid, 4-(chloro-methyl)-5-nonyl-, methyl ester	heptane	255(3.93),277(3.97)	34-0233-76
$C_{16}H_{25}NO_2$			
12-Azabicyclo[9.2.1]tetradeca-11(14),13-diene-13-carboxylic acid, ethyl ester	EtOH	280(4.23)	78-1863-76
Benzene, 1,3-bis(1,1-dimethylethyl)-5-ethyl-2-nitro-	hexane	247(2.98),349(2.26)	18-1142-76
$C_{16}H_{25}NO_4$			
Streptimidol L	MeOH	230(4.37)	33-2393-76
$C_{16}H_{25}NO_6$			
α-D-ribo-Hexofuranose, 3-deoxy-1,2:5,6-bis-O-(1-methylethylidene)-3-[2-(meth-ylimino)propylidene]-, N-oxide, (E,Z)-	EtOH	295(3.95)	136-0127-76A
(Z,Z)-	EtOH	291(3.76)	136-0127-76A
$C_{16}H_{25}NS$			
1-Hexanethione, 1-[4-(dimethylamino)-	C_6H_{12}	257(4.01),310(3.04), 322(3.08),395(4.26), 570(2.74)	35-6218-76
	MeCN	257(4.01),313(3.04), 325(3.04),420(4.26), 557(2.98)	35-6218-76
$C_{16}H_{25}N_2O$			
Pyrimidinium, 1,3-dicyclohexyl-2,3-di-hydro-2-oxo-, ethyl sulfate	EtOH-pH 0.3	322(4.02)	103-2681-76
pseudobase	EtOH-pH 9	239(3.73)	103-2681-76
$C_{16}H_{25}N_3O_2$			
Benzamide, N-[(nitrosooctylamino)meth-yl]-	EtOH	233(4.29),356(1.86), 362(1.86),374s(1.71)	94-0369-76
Carbamic acid, (2,3-di-1-piperidinyl-2-cyclopropen-1-ylidene)-, ethyl ester	MeOH	233(3.98),315(3.68)	138-1215-76
$C_{16}H_{25}N_3O_7$			
Cytidine, N-methyl-2'-O-(tetrahydro-4-methoxy-2H-pyran-4-yl)-	EtOH	270(4.03)	54-0108-76
$C_{16}H_{25}N_5O_3$			
Adenosine, 3'-deoxy-3'-hexyl-	MeOH	260(4.22)	87-1265-76

Compound	Solvent	$\lambda_{max}(\log \epsilon)$	Ref.
$C_{16}H_{25}N_6O_6P$			
Adenosine, N-(6-aminohexyl)-, cyclic	pH 1	262(4.21)	69-3724-76
2',5'-(hydrogen phosphate)	pH 11	267(4.21)	69-3724-76
Adenosine, 8-(hexylamino)-, cyclic	pH 2.0	277(4.13)	69-1408-76
3',5'-(hydrogen phosphate)	pH 13.0	277(--)	69-1408-76
$C_{16}H_{26}$			
Cyclohexane, 1,1'-(1,2-ethanediylidene)-	EtOH	233s(4.36),242(4.50),	35-3732-76
bis[4-methyl-		250(4.59),259(4.44)	
$C_{16}H_{26}BrN_5O_2S$			
Benzenesulfonamide, 4-bromo-N-(1-butyl-	EtOH	229(4.17),250(4.03)	32-0001-76
1H-1,2,3-triazol-5-yl)-, anion,			
tetrabutylammonium salt			
$C_{16}H_{26}N_2$			
Pyrrolidine, 1-[[3-methylene-2-(1-pyrro-	n.s.g.	239(4.18),295(3.01)	94-2494-76
lidinyl)-1-cyclohexen-1-yl]methyl]-			
$C_{16}H_{26}N_2O$			
Morpholine, 4-[6-methylene-2-(1-pyrro-	n.s.g.	238(4.01),287(3.14)	94-2494-76
lidinylmethyl)-1-cyclohexen-1-yl]-			
$C_{16}H_{26}N_2O_2$			
Morpholine, 4-[[3-methylene-2-(4-morpho-	n.s.g.	238(4.26),286(3.30)	94-2494-76
linyl)-1-cyclohexen-1-yl]methyl]-			
$C_{16}H_{26}N_2O_4S$			
Benzamide, 5-(ethylsulfonyl)-2-methoxy-	H_2O	211(4.55),230(4.16),	133-0314-76
N-[2-(diethylamino)ethyl]-, hydro-		285(3.31)	
chloride			
$C_{16}H_{26}N_2O_5$			
1H-Pyrrole-1,3-dicarboxylic acid, 4-	EtOH	275(4.19)	4-0113-76
(acetylamino)-2,5-dihydro-5-(2-methyl-			
propyl)-, diethyl ester			
$C_{16}H_{26}O_2$			
1,3-Benzenediol, 2-methyl-5-nonyl-	MeOH	276(2.97),281(2.96)	12-1989-76
$C_{16}H_{26}O_2S$			
2-Thiophenecarboxylic acid, 4-methyl-	heptane	254(3.94),284(3.96)	34-0233-76
5-nonyl-, methyl ester			
$C_{16}H_{26}O_4S$			
1(2H)-Naphthalenone, 3,4,4a,5,6,7-hexa-	MeOH	246(3.62)	18-3137-76
hydro-4a,8-dimethyl-2-(1-methylethyl)-			
5-[(phenylsulfonyl)oxy]-, [2S-			
(2α,4aα,5α)]-			
$C_{16}H_{26}O_8$			
1-Cyclohexene-1-carboxylic acid,	H_2O	194(3.74)	94-2644-76
4-(β-D-glucopyranosyloxy)-2,6,6-tri-			
methyl-, (R)- (picrocrocinic acid)			
$C_{16}H_{27}B$			
Borane, bis(3-methyl-1,3-butadienyl)-	ether	227(--),268(4.52)	101-0303-76N
(1,1,2-trimethylpropyl)-, (E,E)-			

Compound	Solvent	$\lambda_{max}(\log \epsilon)$	Ref.
$C_{16}H_{27}N$ Ethanamine, N,N-dimethyl-2-(2,3,4,4,5,6-hexamethyl-2,5-cyclohexadien-1-ylidene)-	C_6H_{12}	260(4.29)	5-1103-76
$C_{16}H_{27}NO_4$ Streptimidone, 6,7,8,9-tetrahydro-	MeOH	227s(2.45),237s(2.26), 251s(2.02),282(1.78)	33-2393-76
$C_{16}H_{27}N_4O_3$ Morpholinium, 4-[3-(4-morpholinyl)-2-[(4-morpholinylmethylene)amino]-2-propenylidene]-, perchlorate, monoperchlorate	MeCN	390(4.42)	73-1565-76
$C_{16}H_{27}N_6O_7P$ 5'-Adenylic acid, N-(6-aminohexyl)-	H_2O	267(4.24)	87-1279-76
$C_{16}H_{27}N_6O_9P$ 5'-Adenylic acid, N-[2-[2-(2-aminoethoxy)ethoxy]ethyl]-	H_2O	267(4.23)	87-1279-76
$C_{16}H_{28}O$ 3-Buten-2-one, 4-cyclododecyl-	EtOH	224.0(3.86)	22-0849-76
$C_{16}H_{28}O_2$ 1,3-Cyclohexanedione, 2-methyl-5-nonyl-	MeOH	266(3.91),287(3.79)	12-1989-76
$C_{16}H_{29}NS$ Butanethioamide, N-1-cyclohexen-1-yl-N,3,3-trimethyl-2-(1-methylethyl)-	hexane	282(3.9),367(1.5)	24-0906-76
$C_{16}H_{30}N_2$ 2-Pyridinamine, 3,5-bis(1,1-dimethylethyl)-5,6-dihydro-N,N,5-trimethyl-	hexane MeOH	272(3.67) 244(3.78)	20-0147-76 20-0147-76
$C_{16}H_{32}N_6O_2$ Piperazinium, 4,4'-(azodicarbonyl)bis[1-ethyl-1-methyl-	H_2O	291(4.53),442(1.62)	35-3001-76
$C_{16}H_{34}N_2$ Diazene, bis(1,1,3,3-tetramethylbutyl)-	toluene	372(1.29)	73-1557-76

Compound	Solvent	$\lambda_{max}(\log \epsilon)$	Ref.
$C_{17}H_5N_6$ 1,3-Pentadiene-1,1,2,4,5,5-hexacarbo-nitrile, 3-phenyl-, ion(1-)	MeCN	594(4.40)	35-0558-76
$C_{17}H_6Br_2N_2O_3$ 2H-Anthra[1,9,8-cdef][2,7]naphthyridine-1,6,11(10H)-trione, 5,7-dibromo-	H_2SO_4	245(4.45),280(4.30), 335(4.15),360(3.30), 410(4.20)	103-1352-76
$C_{17}H_6Cl_2N_2O_3$ 2H-Anthra[1,9,8-cdef][2,7]naphthyridine-1,6,11(10H)-trione, 5,7-dichloro-	H_2SO_4	256(4.55),338(4.26), 425(3.84)	103-1352-76
$C_{17}H_6N_4O_7$ 2H-Anthra[1,9,8-cdef][2,7]naphthyridine-1,6,11(10H)-trione, 5,7-dinitro-	H_2SO_4	255(4.35),334(4.25), 415(4.35)	103-1352-76
$C_{17}H_7N_3$ 3,9,10-Phenanthrenetricarbonitrile	50% EtOH	215(4.33),248(4.67), 260(4.57),333(4.07), 344(4.14),370(3.38), 390(3.43)	18-2224-76
$C_{17}H_8Cl_2N_2O_2S$ Benzenamine, 2,6-dichloro-N-2H-naphtho-[1,8-bc]thien-2-ylidene-4-nitro-	dioxan	378(4.10)	103-0857-76
$C_{17}H_9ClN_2O_2S$ Benzenamine, 2-chloro-N-2H-naphtho-[1,8-bc]thien-2-ylidene-4-nitro-	dioxan	385(4.13)	103-0857-76
$C_{17}H_9Cl_2N_3O_2$ 2,4-Cyclohexadiene-1,2-dicarbonitrile, 4,5-dichloro-3-hydroxy-1-(1-methyl-1H-indol-3-yl)-6-oxo-	EtOH	219(4.65),263(4.38), 380(3.36)	1-0853-76
2,4-Cyclohexadiene-1,2-dicarbonitrile, 4,5-dichloro-1-(1H-indol-3-yl)-3-methoxy-6-oxo-	EtOH	214(4.49),342(4.56), 445(2.91)	1-0853-76
$C_{17}H_9Cl_4NO_6S$ Furan, 2-[1-[(dichloromethyl)sulfonyl]-2-[5-(3,4-dichlorophenyl)-2-furanyl]-ethenyl]-5-nitro-	EtOH	216(4.20),248(4.21), 324(4.33),397(4.20)	73-3391-76
$C_{17}H_9NO_2$ Benz[c]acridine-5,6-dione	$CHCl_3$	258(4.32),322(4.52), 430s(3.02)	4-0619-76
$C_{17}H_9NO_3$ 8H-Benzo[g]-1,3-benzodioxolo[6,5,4-de]-quinolin-8-one (liriodenine) ($9\lambda,8\epsilon$)	C_6H_{12}	244(4.57),259(4.50), 261(4.48),268(4.13), 278(3.60),335(4.10), 377(4.16),398(2.90), 475(?)	102-0547-76
	EtOH	222s(4.43),249(4.45), 269(4.37),311(3.86), 400s(4.04),416(4.07)	95-1458-76
$C_{17}H_9N_3$ 2-Butenedinitrile, 2-(4-cyanophenyl)-3-phenyl-, (E)-	50% EtOH	330(4.24)	18-2224-76

Compound	Solvent	λ_{max}(log ϵ)	Ref.
3,9,10-Phenanthrenetricarbonitrile, 9,10-dihydro-	50% EtOH	233(4.53),239(4.39), 271(4.20)	18-2224-76
$C_{17}H_9N_3O_4S$			
Benzenamine, N-2H-naphtho[1,8-bc]thien-2-ylidene-2,4-dinitro-	dioxan	392(4.14)	103-0857-76
$C_{17}H_{10}BrCl_2NO_6S$			
Furan, 2-(4-bromophenyl)-5-[2-[(dichloromethyl)sulfonyl]-2-(5-nitro-2-furanyl)ethenyl]-	EtOH	205(4.31),257(4.26), 328(4.39),405(4.27)	73-3391-76
$C_{17}H_{10}ClNO$			
Benzeneacetonitrile, 3-chloro-α-(1-oxo-3-phenyl-2-propynyl)-	pH 13	230(4.23),251s(4.19), 349(4.16)	34-0115-76
	EtOH	218s(--),227(4.12), 245(4.1),265s(3.97), 342(4.3)	34-0115-76
Benzeneacetonitrile, 4-chloro-α-(1-oxo-3-phenyl-2-propynyl)-	pH 13	235(4.35),265s(4.2), 353(4.51)	34-0115-76
	EtOH	213(4.14),236(4.18), 266s(4.04),354(4.32)	34-0115-76
Benzeneacetonitrile, α-[3-(2-chlorophenyl)-1-oxo-2-propynyl]-	pH 13	222(4.16),252(4.05), 356(4.12)	34-0115-76
	EtOH	218s(--),225(3.98), 260(3.88),342(4.08)	34-0115-76
Benzeneacetonitrile, α-[3-(4-chlorophenyl)-1-oxo-2-propynyl]-	pH 13	234(4.14),256(4.19), 349(4.21)	34-0115-76
	EtOH	217s(--),247(4.06), 265(3.96),344(4.23)	34-0115-76
$C_{17}H_{10}Cl_2N_2O_8S$			
Furan, 2-[2-(dichloromethyl)sulfonyl]-2-(5-nitro-2-furanyl)ethenyl]-5-(2-nitrophenyl)-	EtOH	211s(4.17),244(4.26), 315(4.42),383(4.18)	73-3391-76
Furan, 2-[2-(dichloromethyl)sulfonyl]-2-(5-nitro-2-furanyl)ethenyl]-5-(3-nitrophenyl)-	EtOH	212s(4.07),244(4.37), 323(4.38),389(4.18)	73-3391-76
Furan, 2-[2-(dichloromethyl)sulfonyl]-2-(5-nitro-2-furanyl)ethenyl]-5-(4-nitrophenyl)-	EtOH	204(4.31),251(4.29), 328(4.37),405(4.26)	73-3391-76
$C_{17}H_{10}Cl_3NO_6S$			
Furan, 2-(2-chlorophenyl)-5-[2-(dichloromethyl)sulfonyl]-2-(5-nitro-2-furanyl)ethenyl]-	EtOH	214(4.12),251(4.27), 332(4.39),426(4.25)	73-3391-76
Furan, 2-(4-chlorophenyl)-5-[2-(dichloromethyl)sulfonyl]-2-(5-nitro-2-furanyl)ethenyl]-	EtOH	215(4.14),280(4.29), 341(4.28),427(4.26)	73-3391-76
$C_{17}H_{10}Cl_3O_2P$			
Hydroxybenzophenalenium salt with POCl$_3$	n.s.g.	482(4.49)	104-0180-76
$C_{17}H_{10}Cl_4O_5$			
1H-2-Benzopyran-4-ol, 1-(2,3,4,5-tetrachloro-6-hydroxyphenoxy)-, 4-acetate	MeCN	217(4.35),263(4.01), 298(3.44)	1-0619-76
$C_{17}H_{10}FNO$			
Benzeneacetonitrile, 2-fluoro-α-(1-oxo-3-phenyl-2-propynyl)-	pH 13	230(4.14),262(4.13), 348(4.21)	34-0115-76

Compound	Solvent	$\lambda_{max}(\log \epsilon)$	Ref.
Benzeneacetonitrile, 2-fluoro-α-(1-oxo-3-phenyl-2-propynyl)- (cont.)	EtOH	215(4.4),260(4.02), 340(4.29)	34-0115-76
Benzeneacetonitrile, 3-fluoro-α-(1-oxo-3-phenyl-2-propynyl)-	pH 13	230(4.21),250(4.16), 260s(4.13),348(4.28)	34-0115-76
	EtOH	226(4.04),248s(3.97), 263(3.88),342(4.19)	34-0115-76
Benzeneacetonitrile, 4-fluoro-α-(1-oxo-3-phenyl-2-propynyl)-	pH 13	229(4.14),250(4.08), 347(4.23)	34-0115-76
	EtOH	214(4.09),230(4.09), 250(4.08),347(4.26)	34-0115-76
$C_{17}H_{10}N_2O$ 9,10-Phenanthrenedicarbonitrile, 3-methoxy-	50% EtOH	219(4.37),254(4.66), 263s(4.53),284(4.29), 352(4.16),362(4.17), 380(4.07)	18-2224-76
$C_{17}H_{10}N_2O_2$ Imidazo[1,2-b]isoquinoline-5,10-dione, 2(or 3)-phenyl-	DMF	274(4.38),402(3.38)	44-0836-76
$C_{17}H_{10}N_2O_2S$ Benzenamine, N-2H-naphtho[1,8-bc]thien-2-ylidene-4-nitro-	dioxan	392(4.20)	103-0857-76
$C_{17}H_{10}N_4$ 4-Cinnolineacetonitrile, 3-cyano-α-phenyl-	EtOH	232(4.70),282(4.15), 410(4.38)	39-0592-76C
$C_{17}H_{10}N_4S_2$ 1,3,4-Thiadiazolium, 5-[(dicyanomethyl)-thio]-2,3-diphenyl-, hydroxide, inner salt	MeOH	212(4.55),241(4.26), 266s(4.11),308(4.04), 389s(3.91)	70-1163-76
$C_{17}H_{10}O$ Benzanthrone	hexane	425s(1.3)	39-1850-76B
$C_{17}H_{10}OS$ 5H-Naphtho[1,8-bc]thiophen-5-one, 2-phenyl-	MeOH	388(4.10)	89-0775-76
$C_{17}H_{10}O_2$ Benzo[c]xanthone	EtOH	260(4.63),270(4.72), 360(3.65)	22-1975-76
$C_{17}H_{10}O_3$ Benzo[1,2-b:3,4-b']difuran-3(2H)-one, 2-(phenylmethylene)-	EtOH	222(3.93),267(3.87), 295(3.71)	102-1553-76
$C_{17}H_{10}O_4$ Benzo[1,2-b:3,4-b']difuran-3(2H)-one, 2-benzoyl-	EtOH	220(5.22),235(4.84), 317(4.55),375(5.22)	102-1553-76
Benzo[1,2-b:3,4-b']difuran-3(2H)-one, 4-hydroxy-2-(phenylmethylene)-	EtOH	224(4.67),252(4.46), 282(4.82),355(3.51)	102-1553-76
Spiro[furan-2(3H),1'(3'H)-isobenzofuran]-3,3'-dione, 4-phenyl-	ether	235(4.41),276(3.60), 284(3.61),306(3.58)	44-0388-76
$C_{17}H_{11}$ Cyclohept[a]acenaphthylenylium tetrafluoroborate	MeCN	263(4.31),333(4.44), 438(4.23)	89-0240-76

Compound	Solvent	$\lambda_{max}(\log \epsilon)$	Ref.
$C_{17}H_{11}BrO$ 2(1H)-Naphthalenone, 3-bromo-1-(phenyl-methylene)-	EtOH	240(3.6),281(3.9), 327(3.1)	39-0695-76C
$C_{17}H_{11}Cl_2NO_6S$ Furan, 2-[2-[(dichloromethyl)sulfonyl]-2-(5-nitro-2-furanyl)ethenyl]-5-phen-yl-	EtOH	210s(3.99),246(4.28), 328(4.47),398(4.27)	73-3391-76
$C_{17}H_{11}N$ Benzo[4,5]cyclohept[1,2,3-ij]isoquino-line, hydrobromide	EtOH	238(4.58),278(3.94), 301(3.95),313(3.84), 322(3.71)	4-1049-76
$C_{17}H_{11}NO_2S$ 1H-Indene-1,3(2H)-dione, 2-(3-methyl-2(3H)-benzothiazolylidene)-	EtOH	235(4.50),304(3.91), 380(4.70)	73-1363-76
$C_{17}H_{11}NO_2Se$ 1H-Indene-1,3(2H)-dione, 2-(3-methyl-2(3H)-benzoselenazolylidene)-	EtOH	235(4.52),305(4.06), 385(4.69)	73-1363-76
$C_{17}H_{11}NO_3$ 1H-Indene-1,3(2H)-dione, 2-hydroxy-2-(1H-indol-3-yl)-	EtOH	226(4.79),248(3.90), 316(3.20)	83-0081-76
$C_{17}H_{11}NO_7$ 2,3-Furandione, 4-(1,3-benzodioxol-5-yl)dihydro-5-(2-nitrophenyl)-	EtOH	208(4.53),253(4.23), 296(4.21),322(4.27)	22-1541-76
2,3-Furandione, 4-(1,3-benzodioxol-5-yl)dihydro-5-(3-nitrophenyl)-	EtOH	207(4.47),255(4.18), 295(4.15),321(4.21)	22-1541-76
2,3-Furandione, 4-(1,3-benzodioxol-5-yl)dihydro-5-(4-nitrophenyl)-	EtOH	241(4.09),297(4.18), 323(4.17)	22-1541-76
$C_{17}H_{11}N_3O_3S_2$ 1H-[1,2]Dithiolo[5,1-e][1,2,3]thiadia-zole-7-S^{IV}-3-carboxaldehyde, 1-(4-ni-trophenyl)-5-phenyl-	C_6H_{12}	204(4.54),221s(4.43), 328(4.33),370(4.12), 506(4.36)	39-0880-76C
$C_{17}H_{11}N_5O$ Methanone, phenyl(4-phenyl-1,2,4-tria-zolo[5,1-c][1,2,4]triazin-3-yl)-	EtOH	213(4.31),258(4.29), 331(3.91)	39-0421-76C
$C_{17}H_{11}OS$ 1,3-Benzoxathiol-1-ium, 2-(2-naphtha-lenyl)-, tetrafluoroborate	96% H_2SO_4	248(4.08),261(4.04), 278(4.00),293(3.65), 398(4.46),430s(4.18)	39-0323-76C
$C_{17}H_{12}$ 4H-Cyclobuta[1]cyclopenta[def]phenan-threne, 7b,9a-dihydro-	C_6H_{12}	231(4.21),238(4.09), 244(4.27),254(4.03), 263s(4.02),276(4.06), 284(3.97),300s(3.48)	88-1223-76
4H-Cycloocta[def]fluorene	C_6H_{12}	255(4.47),273s(4.33), 300(3.66),346s(3.13)	88-1223-76
$C_{17}H_{12}BrN_3$ 1H-Pyrrole-3-carbonitrile, 2-amino-1-(4-bromophenyl)-4-phenyl-	EtOH	242(4.61),289(3.95)	48-0663-76

Compound	Solvent	$\lambda_{max}(\log \epsilon)$	Ref.
$C_{17}H_{12}BrN_3O$ 1H-Pyrazole-5-carbonitrile, 3-(4-bromo-benzoyl)-4,5-dihydro-1-phenyl-	EtOH	270(4.20),380(4.20)	18-0321-76
$C_{17}H_{12}Br_2O$ 1,4-Pentadien-3-one, 2,4-dibromo-1,5-di-phenyl-, cis-trans	EtOH	255(3.95),311(4.10)	39-0695-76C
trans-trans	EtOH	257(4.10),312(4.24)	39-0695-76C
$C_{17}H_{12}ClNO$ Ethanone, 1-(6-chloro-4-phenyl-3-quino-linyl)-	isoPrOH	243(4.68),285(3.84), 325(3.43)	4-0131-76
$C_{17}H_{12}ClNO_2$ 3H-Furo[3,4-b]indol-3-one, 1-(2 and 4-chlorophenyl)-1,4-dihydro-4-methyl-	n.s.g.	300(4.16)	103-0947-76
2,6(1H,3H)-Pyridinedione, 3-(3-chloro-phenyl)-4-phenyl-	pH 13	223(4.00),227(3.99), 282(3.99)	34-0120-76
	EtOH	221(4.07),288(4.18)	34-0120-76
	CHCl$_3$	231(3.62),293(4.28)	34-0120-76
2,6(1H,3H)-Pyridinedione, 4-(2-chloro-phenyl)-3-phenyl-	pH 13	226(3.90),234s(3.85), 282(3.99)	34-0120-76
	EtOH	219(3.79),225s(3.75), 282(3.81)	34-0120-76
	CHCl$_3$	234(3.73),286(4.03)	34-0120-76
2,6(1H,3H)-Pyridinedione, 4-(4-chloro-phenyl)-3-phenyl-	pH 13	225(4.00),229s(3.97), 295(4.23)	34-0120-76
	EtOH	223(4.16),293(4.41)	34-0120-76
	CHCl$_3$	234(3.58),300(4.37)	34-0120-76
$C_{17}H_{12}ClN_3O$ 1H-Pyrazole-5-carbonitrile, 3-benzoyl-1-(4-chlorophenyl)-4,5-dihydro-	EtOH	254(4.448),374(4.442)	18-0321-76
$C_{17}H_{12}ClN_3O_2S$ Acetamide, N-[5-benzoyl-3-(4-chlorophen-yl)-1,3,4-thiadiazol-2(3H)-ylidene]-	EtOH	260(4.32),323(4.58)	4-0045-76
$C_{17}H_{12}ClN_5O$ 4H-[1,2,4]Triazolo[4,3-a][1,4]benzodia-zepine, 6-(2-chlorophenyl)-1-methyl-8-nitroso-	isoPrOH	215s(4.41),247(4.17), 291(4.03),315(4.00)	87-1378-76
$C_{17}H_{12}Cl_2N_4O$ 4H-1,2,4-Triazolo[4,3-a][1,5]benzodia-zepin-5(6H)-one, 8-chloro-1-(chloro-methyl)-6-phenyl-	EtOH	229(4.53),283(3.29), 291s(3.22)	87-0192-76
$C_{17}H_{12}FNO_2$ 2,6(1H,3H)-Pyridinedione, 3-(2-fluoro-phenyl)-4-phenyl-	pH 13	224(4.04),289(4.25)	34-0120-76
	EtOH	220(4.07),226(4.00), 286(4.28)	34-0120-76
	CHCl$_3$	233(3.67),292(4.16)	34-0120-76
2,6(1H,3H)-Pyridinedione, 3-(3-fluoro-phenyl)-4-phenyl-	pH 13	226(3.87),290(4.11)	34-0120-76
	EtOH	220(4.12),226(4.05), 288(4.31)	34-0120-76
	CHCl$_3$	232(3.69),293(4.23)	34-0120-76
$C_{17}H_{12}F_3NO_2$ Spiro[4.5]deca-3,6,9-triene-2,8-dione, 4-amino-3-[3-(trifluoromethyl)phenyl]-	EtOH	227(4.26),258(4.40)	88-4125-76

Compound	Solvent	$\lambda_{max}(\log \epsilon)$	Ref.
$C_{17}H_{12}N_2$ 11-Demethylellipticine	85% EtOH-HCl	238(4.43),248s(4.40), 270s(4.30),307(4.86), 350(3.85)	39-1155-76C
$C_{17}H_{12}N_2O$ 5-Azacycl[3.2.2]azine-1-carboxaldehyde, 6-methyl-2-phenyl-	EtOH	225(4.40),233s(4.33), 284s(4.42),297(4.46), 315s(4.27),438(3.97)	44-0351-76
5-Azacycl[3.2.2]azine-4-carboxaldehyde, 6-methyl-2-phenyl-	EtOH	230s(4.46),245(4.59), 342(4.36),454(3.98)	44-0351-76
2-Butenedinitrile, 2-(4-methoxyphenyl)-3-phenyl-, (E)-	50% EtOH	228(4.12),243s(3.96), 290(3.53),365(4.22)	18-2224-76
5H-Dibenz[b,f]azepine-10-carbonitrile, 5-acetyl-	MeOH	228(4.31),243(4.10), 297(4.13)	78-1081-76
9,10-Phenanthrenedicarbonitrile, 9,10-dihydro-3-methoxy-	50% EtOH	234(4.48),241(4.51), 272(4.21)	18-2224-76
$C_{17}H_{12}N_2OS_2$ 1H-[1,2]Dithiolo[5,1-e][1,2,3]thiadiazole-7-S^{IV}-3-carboxaldehyde, 1,5-diphenyl-	C_6H_{12}	205(4.52),239(4.54), 315(4.30),356s(3.87), 492(4.24)	39-0880-76C
$C_{17}H_{12}N_2O_2$ 1H-Benzimidazolium, 1-(3-hydroxy-1-oxo-1H-inden-2-yl)-3-methyl-, hydroxide, inner salt	MeOH	253(4.55),277(3.88), 296s(3.59),302(3.60), 313s(3.57),325(3.56), 402(3.21)	83-0092-76
Pyrrolo[2,1-a]isoquinoline-3-carbonitrile, 1,2-diacetyl-	MeOH	217(4.48),256s(4.82), 264(4.85),276s(4.73), 340(4.01),354(4.12), 373(4.09)	39-1908-76C
$C_{17}H_{12}N_2O_3$ Benzoic acid, 2-[(4-phenyl-1H-imidazol-2-yl)carbonyl]-	MeOH	250(4.10),323(4.21)	44-0836-76
$C_{17}H_{12}N_2O_4S$ 2-Furancarboximidothioic acid, N-hydroxy-5-(2-nitrophenyl)-, phenyl ester	EtOH	209(4.41),305(4.37)	73-3085-76
2-Furancarboximidothioic acid, N-hydroxy-5-(3-nitrophenyl)-, phenyl ester	EtOH	210(4.30),314(4.20)	73-3085-76
2-Furancarboximidothioic acid, N-hydroxy-5-(4-nitrophenyl)-, phenyl ester	EtOH	207(4.19),372(4.38)	73-3085-76
$C_{17}H_{12}N_4O$ 1(2H)-Phthalazinone, 4-(1H-imidazol-2-yl)-2-phenyl-	DMF	284(3.51),443(3.28)	44-0836-76
[1,2,4]Triazolo[1,5-b]isoquinolin-5(1H)-one, 1-[(phenylmethylene)amino]-	EtOH	232(4.21),314(4.01), 402(4.09),420(4.00)	95-0700-76
$C_{17}H_{12}N_4OS$ 4H-1,3,4-Thiadiazolo[2,3-c][1,2,4]triazin-4-one, 7-(4-methylphenyl)-3-phenyl-	MeOH	256(4.10),358(4.09)	4-0117-76
$C_{17}H_{12}N_4O_2$ Benzo[g]pteridine-2,4(3H,10H)-dione, 10-methyl-3-phenyl-	EtOH	216(4.49),266(4.54), 335(3.89),435(3.98)	88-2899-76

Compound	Solvent	$\lambda_{max}(\log \epsilon)$	Ref.
6H-1,2,4,5-Tetrazino[1,6-b]isoquinolin-6-one, 2-benzoyl-1,2-dihydro-	EtOH	232(4.44),312(4.13), 370(3.89),388(3.80)	95-0700-76
$C_{17}H_{12}N_4O_2S$ 4H-1,3,4-Thiadiazolo[2,3-c][1,2,4]triazin-4-one, 7-(3-methoxyphenyl)-3-phenyl-	MeOH	282(3.78),358(4.07)	4-0117-76
4H-1,3,4-Thiadiazolo[2,3-c][1,2,4]triazin-4-one, 7-(4-methoxyphenyl)-3-phenyl-	MeOH	276(3.87),352(4.15)	4-0117-76
$C_{17}H_{12}N_4O_4S$ Acetamide, N-[5-benzoyl-3-(4-nitrophenyl)-1,3,4-thiadiazol-2(3H)-ylidene]-	EtOH	265(4.26),320(4.40)	4-0045-76
$C_{17}H_{12}O$ 2-Propen-1-one, 3-(2-ethynylphenyl)-1-phenyl-	EtOH	249(4.21),254(4.23), 308(4.26)	18-2840-76
$C_{17}H_{12}OS$ Indeno[2,1-b]pyran, 4-methyl-2-(2-thienyl)-	dioxan	227(4.63),273(4.48), 375(4.49),518(3.83)	78-2219-76
2H-Thiopyran-2-one, 3,6-diphenyl-	MeOH	222(4.39),275(3.94), 379(4.13)	39-2565-76C
$C_{17}H_{12}OSe$ Methanone, phenyl(5-phenylselenophene-2-yl)-	MeCN	347(4.33)	118-0521-76
$C_{17}H_{12}O_2$ Benzo[c]xanthone, 5,6-dihydro-	EtOH	212(4.23),252(4.18), 312(4.15)	22-1975-76
2,4,6-Cycloheptatrien-1-one, 2-hydroxy-4-(1-naphthalenyl)-	MeOH	221(4.79),248(4.42), 319(4.16),386(3.83)	18-0831-76
2,4,6-Cycloheptatrien-1-one, 2-hydroxy-5-(1-naphthalenyl)-	MeOH	219(4.80),346(4.24)	18-0831-76
2H-Cyclopenta[jk]fluorene-1-carboxylic acid, methyl ester	EtOH	223(4.40),256(4.55), 263(4.75),278(3.07), 289(3.89),317(4.09), 327(4.13),340s(3.96)	24-2596-76
	DMSO-EtOH-NaOEt	530(3.70),555s(3.68)	24-2956-76
2H-Cyclopenta[jk]fluorene-2-carboxylic acid, methyl ester	dioxan	223(4.52),248(4.63), 257(4.80),278(3.93), 306(3.93),320(3.81), 340(2.99)	24-2956-76
	DMSO-EtOH-NaOEt	450(3.85),470s(3.79)	24-2956-76
2H-Pyran-2-one, 4,6-diphenyl-	EtOH	229(4.04),236(4.05), 262(4.36),344(4.01)	18-3239-76
2,2'-Spirobi[2H-1-benzopyran]	C_6H_{12}	212(4.62),220(4.60), 246(4.32),255(4.30), 246s(4.23),300(3.70)	22-2039-76
	EtOH	213(4.60),222(4.58), 250(4.23),262(4.17), 298(3.61)	22-2039-76
$C_{17}H_{12}O_2S$ Benzoic acid, 2-(3-phenyl-2-thienyl)-	MeOH	225(4.29)	44-1320-76
Benzoic acid, 2-(4-phenyl-2-thienyl)-	MeOH	257(4.48)	44-1320-76

Compound	Solvent	λ_{max}(log ϵ)	Ref.
Benzoic acid, 2-(5-phenyl-2-thienyl)-	MeOH	315(4.32)	44-1320-76
$C_{17}H_{12}O_3$			
2(3H)-Furanone, 3-(hydroxyphenylmethyl-ene)-5-phenyl-	EtOH	217(4.02),227(4.01), 245(4.14),257(4.12), 287(3.99),385(4.24)	12-0339-76
3(2H)-Furanone, 2-benzoyl-5-phenyl-	EtOH	261(4.27),358(4.34)	39-1688-76C
1H-Indene-5-carboxylic acid, 2,3-di-hydro-3-oxo-2-phenylmethylene-	MeCN	234(4.48),318(4.42)	44-3540-76
$C_{17}H_{12}O_3S$			
Benzoic acid, 3-hydroxy-2-(4-phenyl-2-thienyl)-	MeOH	253(4.47)	44-1320-76
$C_{17}H_{12}O_4$			
7-Methoxy-[2]benzopyrano[4,3-b][1]benzo-pyran-5-one	EtOH	224(4.58),284(3.78), 324(3.48)	22-1975-76
$C_{17}H_{12}O_5$			
Anhydropisatin	EtOH	234s(4.01),251s(3.88), 290s(3.62),324s(4.06), 340(4.28),358(4.28)	95-0254-76
9,10-Anthracenedione, 1,3,8-trihydroxy-6-(1-propenyl)-, (E)-	EtOH	233(4.25),267(4.14), 288(4.20),296(4.19), 325(3.82),450(3.91)	39-0613-76C
6H-Benzofuro[3,2-c][1]benzopyran-6-one, 2,9-dimethoxy-	MeOH	235(4.58),320(4.09), 345(4.31)	18-1955-76
6H-Benzofuro[3,2-c][1]benzopyran-6-one, 4,9-dimethoxy-	MeOH	242(4.46),335(4.30)	18-1955-76
Isounonal	EtOH	225(4.37),252(4.42), 295(4.37)	2-0009-76
Unonal	EtOH	267s(4.36),290(4.46), 335(3.92)	2-0009-76
$C_{17}H_{12}O_6$			
9,10-Anthracenedione, 2-acetyl-1,6,8-trihydroxy-3-methyl-	MeOH-HOAc	248(4.33),265(4.29), 289(4.34),439(4.07)	12-2231-76
$C_{17}H_{12}O_7$			
9,10-Anthracenedione, 2-acetyl-1,5,6,8-tetrahydroxy-3-methyl-	dioxan-1% HOAc	260(4.53),307(4.07), 385(3.46),480s(4.09), 497(4.14),520s(4.02)	12-2231-76
5H-[1,3]Benzodioxolo[5,6-c]-1,3-dioxolo-[4,5-h][1]benzopyran-5-one, 12-hydr-oxy-2,2-dimethyl-	EtOH	208(4.49),239(4.78), 277(4.29),316(4.25), 355s(--)	24-0855-76
8H-1,3-Dioxolo[4,5-g][1]benzopyran-8-one, 4,9-dihydroxy-6-(4-methoxy-phenyl)-	EtOH	235(4.05),280(4.07), 342(4.27)	94-0815-76
	EtOH-AlCl$_3$	241s(4.09),294(4.04), 365(4.23)	94-0815-76
	EtOH-NaOAc- H$_3$BO$_3$	235s(4.05),279(4.10), 341(4.25)	94-0815-76
$C_{17}H_{12}O_8$			
1,3-Benzodioxole-5-carboxylic acid, 6-(5-acetoxy-3,6-dioxo-1,4-cyclo-hexadien-1-yl)-, methyl ester	EtOH	226(4.72),257(4.42), 300s(--),420s(--)	24-0855-76
1,3-Benzodioxole-5-carboxylic acid, 4-acetoxy-6-(3,6-dioxo-1,4-cyclo-hexadien-1-yl)-, methyl ester	EtOH	299(4.39),400(3.19)	24-0855-76
Ellagic acid, 3,3',4-tri-O-methyl-	EtOH	250(4.49),285s(3.81),	95-0984-76

Compound	Solvent	$\lambda_{max}(\log \epsilon)$	Ref.
Ellagic acid, 3,3',4-tri-O-methyl- (cont.)		355(3.84),371(3.90)	95-0984-76
	EtOH-NaOEt	254(4.34),299(4.28), 315(3.77),409(3.86)	95-0984-76
	EtOH-NaOAc	251(4.33),282(4.13), 354(3.55),408(3.64)	95-0984-76
$C_{17}H_{12}O_{17}S_3$ 4H-1-Benzopyran-4-one, 3-acetoxy- 2-[3,4-bis(sulfooxy)phenyl]-5-hydroxy- 7-(sulfooxy)-, tripotassium salt	50% EtOH	241(4.22),266(4.34), 344(4.29)	100-0253-76
	+ NaOMe	237(4.30),272(4.24), 300s(--),383(4.38)	100-0253-76
	+ NaOAc	265(4.25),400(4.35)	100-0253-76
	+ AlCl$_3$	254s(--),275(4.34), 296s(--),340(4.27), 389(4.17)	100-0253-76
$C_{17}H_{13}BrN_2O_2$ [3,3'-Bi-1H-indol]-2(3H)-one, 5-bromo- 3-hydroxy-1'-methyl-	MeOH	214(3.33),266(2.81), 290(2.53)	103-0178-76
$C_{17}H_{13}ClN_2O_3$ Ethanone, 1-(5-chloro-2,4-dihydroxyphen- yl)-2-(1-phenyl-1H-pyrazol-4-yl)-	EtOH	265(4.42),340(4.26)	103-0914-76
$C_{17}H_{13}ClO$ Naphthalene, 4-chloro-1-methoxy-5-phen- yl-	EtOH	225s(4.54),242s(4.39), 310(4.01),319s(3.98), 333(3.88)	95-1161-76
1(2H)-Naphthalenone, 2-[(3-chlorophen- yl)methylene]-3,4-dihydro-	isooctane	282(4.30),300s(--), 364(2.28),375(2.28), 395s(--)	65-2057-76
	EtOH	298(4.22)	65-2057-76
1(2H)-Naphthalenone, 2-[(4-chlorophen- yl)methylene]-3,4-dihydro-	isooctane	298(4.20),315s(--), 362(2.46),375(2.43), 400s(--)	65-2057-76
	EtOH	306(4.20),320s(--)	65-2057-76
$C_{17}H_{13}ClO_5$ 9,10-Anthracenedione, 1-chloro-7-hydr- oxy-4,5-dimethoxy-2-methyl-	EtOH	221(4.48),268(4.25), 280(4.28),400(3.80)	39-0613-76C
$C_{17}H_{13}Cl_2N_3O_2$ 1H-Pyrazole-4-carboxamide, 3-(4-chloro- benzoyl)-N-(4-chlorophenyl)-4,5-di- hydro-	EtOH	253(4.54),339(4.10)	115-0001-76
$C_{17}H_{13}N$ Benzo[4,5]cyclohept[1,2,3-ij]isoquino- line, 7,8-dihydro-, hydrochloride	EtOH	228(4.51),290(3.84), 302(3.81),331(3.89), 340(3.88)	4-1049-76
6H-Dibenzo[de,g]quinoline, 6-methyl-	MeOH	228(4.46),275(4.20), 300(3.41),339(4.03), 352(4.09),404(3.77), 427(3.75)	35-7108-76
$C_{17}H_{13}NO$ 6H-Benzocyclohepten-5,9-imin-6-one, 5,9-dihydro-10-phenyl-	EtOH	208(3.30),240(3.38)	39-2285-76C

Compound	Solvent	$\lambda_{max}(\log \epsilon)$	Ref.
$C_{17}H_{13}NO_2$			
Ethanone, 1-(3-benzoyl-1-indolizinyl)-	MeOH	206(4.44),232(4.51), 258(4.52),287s(4.34), 293(4.37),327s(4.42), 345s(4.55),355(4.56)	39-1908-76C
2H-Indol-2-one, 1,3-dihydro-3-(1-methyl-2-oxo-2-phenylethylidene)-	EtOH	204(4.46),253(4.60), 293(3.88),370(3.02)	94-0782-76
2H-Pyran-2-one, 3-methyl-6-phenyl-4-(2-pyridinyl)-	EtOH	231s(3.90),240s(4.01), 248(4.12),253s(4.04), 265(3.98),336(3.96)	1-0512-76
Spiro[cyclopropane-1,3'-[3H]indol]-2'(1'H)-one, 2-benzoyl-	EtOH	209(4.57),250(4.38), 282s(3.62),313s(3.30)	94-0782-76
$C_{17}H_{13}NO_2S$			
Thiophene, 2-(4-methylphenyl)-5-(4-nitrophenyl)-	MeOH	290(3.98),391(4.43)	48-0731-76
$C_{17}H_{13}NO_3$			
Fuseine	EtOH	235(4.48),273(4.55), 317(3.91)	102-1187-76
1(2H)-Naphthalenone, 3,4-dihydro-2-[(4-nitrophenyl)methylene]-	isooctane EtOH	300(4.31) 310(4.28)	65-2057-76 65-2057-76
$C_{17}H_{13}NO_6$			
Benzoic acid, 2-(1-benzoyl-2-oxopropyl)-3-nitro-	n.s.g.	304(3.88)	39-1073-76C
$C_{17}H_{13}NS$			
[1]Benzothieno[3,2-h]isoquinoline, 5-ethyl-	MeOH	239(4.29),255(4.26), 280(4.24),330(3.90)	4-0141-76
$C_{17}H_{13}N_3$			
3H-Imidazo[4,5-b]quinoline, 9-methyl-3-phenyl-	DMF	303s(3.76),318(3.97), 331(3.85),343s(3.67)	48-0039-76
1H-Pyrrole-3-carbonitrile, 2-amino-1,4-diphenyl-	EtOH	247(4.59),293(3.96)	48-0663-76
$C_{17}H_{13}N_3O$			
1H-Pyrazole-5-carbonitrile, 3-benzoyl-4,5-dihydro-1-phenyl-	EtOH	240(4.142),376(4.214)	18-0321-76
$C_{17}H_{13}N_3O_2$			
4-Oxazolecarboxylic acid, 2-phenyl-, (phenylmethylene)hydrazide	EtOH	287(4.27),300s(4.04)	39-0315-76C
2-Pyrimidinamine, 4-(1,3-benzodioxol-5-yl)-6-phenyl-	EtOH	239(4.48),341(4.31)	4-0257-76
$C_{17}H_{13}N_3O_2S$			
Acetamide, N-(5-benzoyl-3-phenyl-1,3,4-thiadiazol-2(3H)-ylidene)-	EtOH	275(4.23),333(4.08)	4-0045-76
Isoquinolinium 3,3-dicyano-1-(methoxycarbonyl)-2-(methylthio)-2-propenyl-ide	EtOH	233(4.71),362(4.28), 465(3.79)	142-0939-76
$C_{17}H_{13}N_3O_3$			
Acetamide, N-[4-[2-(4-cyano-2-nitrophenyl)ethenyl]phenyl]-	EtOH	260(4.08),370(4.31)	80-1127-76
4-Oxazolecarboxylic acid, 2-phenyl-, 2-benzoylhydrazide	EtOH	282(4.16)	39-0315-76C

Compound	Solvent	$\lambda_{max}(\log \epsilon)$	Ref.
$C_{17}H_{13}N_3O_5$ Benzoic acid, 2-[(1,4-dihydro-3-nitro-4-oxo-2-quinolinyl)amino]-, methyl ester	$CHCl_3$	282(3.97),318s(3.71), 370(4.41)	48-0039-76
$C_{17}H_{13}N_3O_6S$ 3H-Pyrazole, 4-(2-furanyl)-4,5-dihydro-3-(5-nitro-2-furanyl)-3-(phenylsul-fonyl)-	EtOH	208(4.45),220s(4.46), 311(4.18)	73-3371-76
$C_{17}H_{13}N_5O$ 3H-Imidazo[4,5-e]-1,2,4-triazin-3-one, 1,2-dihydro-6-phenyl-2-(phenylmethyl)-	EtOH	256(4.11),316(4.42)	94-2274-76
$C_{17}H_{13}N_5OS$ 1,2,4-Triazolo[3,4-b][1,3,4]thiadiazol-ium, 6-(benzoylamino)-1-(phenylmeth-yl)-, hydroxide, inner salt	MeCN	202(4.74),232(4.27), 305(4.30)	30-0104-76
1,2,4-Triazolo[3,4-b][1,3,4]thiadiazol-ium, 6-(benzoylamino)-2-(phenylmeth-yl)-, hydroxide, inner salt	MeCN	199s(4.90),240(4.34), 312(4.59)	30-0104-76
$C_{17}H_{13}O$ Pyrylium, 2,6-diphenyl-, iodide	CH_2Cl_2	546(2.32)	59-1311-76
selenocyanate	CH_2Cl_2	505(2.59)	59-1311-76
thiocyanate	CH_2Cl_2	473(2.62)	59-1311-76
$C_{17}H_{14}$ 4H-Benz[de]anthracene, 5,6-dihydro-	EtOH	214(4.50),225(4.22), 251(4.68),258(4.76), 271(4.15),279(4.08), 295(4.05),302(4.12)	117-0133-76
$C_{17}H_{14}Br_2$ Phenanthrene, 2,7-dibromo-9,10-dihydro-9,9-dimethyl-10-methylene-	hexane	222(4.41),246s(--), 251(4.39),295(4.35)	104-1298-76
	H_2SO_4	286s(--),294(4.40), 337(3.91),396s(--), 577(3.80)	104-1298-76
Phenanthrene, 3,6-dibromo-9,10-dihydro-9,9-dimethyl-10-methylene-	hexane	230s(--),243(4.54), 251(4.52),285(4.05), 319s(--)	104-1298-76
	H_2SO_4	271(4.36),295s(--), 373(4.59),528(3.98)	104-1298-76
$C_{17}H_{14}ClN$ Benzo[4,5]cyclohept[1,2,3-ij]isoquino-line, 7,8-dihydro-, hydrochloride	EtOH	228(4.51),290(3.84), 302(3.81),331(3.89), 340(3.88)	4-1049-76
$C_{17}H_{14}ClNO$ Ethanone, 1-(6-chloro-1,2-dihydro-4-phenyl-3-quinolinyl)-	isoPrOH	250(4.47),305s(3.65), 425(3.49)	4-0131-76
$C_{17}H_{14}ClN_2Se$ Benzenaminium, N-[6-(4-chlorophenyl)-2H-1,3-selenazin-2-ylidene]-N-methyl-	HOAc	356(4.37),418s(3.58)	88-2005-76
$C_{17}H_{14}ClN_3O_2$ 1H-Pyrazole-4-carboxamide, 3-benzoyl-N-(4-chlorophenyl)-4,5-dihydro-	EtOH	252(4.48),334(4.12)	115-0001-76

Compound	Solvent	$\lambda_{max}(\log \epsilon)$	Ref.
1H-Pyrazole-4-carboxamide, 3-(4-chloro-benzoyl)-4,5-dihydro-N-phenyl-	EtOH	246(4.29),340(4.12)	115-0001-76
$C_{17}H_{14}ClN_5O$ 4H-[1,2,4]Triazolo[4,3-a][1,4]benzodi-azepin-8-amine, 6-(2-chlorophenyl)-N-hydroxy-1-methyl-	isoProH	210(4.47),243(4.37), 270s(4.07),330(3.30)	87-1378-76
$C_{17}H_{14}Cl_2N_2O$ 3-Pyrazolidinone, 1-[2-(2,6-dichloro-phenyl)ethenyl]-2-phenyl-, (E)-	EtOH	273(4.22)	44-3775-76
Pyrimido[1,2-a]indol-4(3H)-one, 10-(2,6-dichlorophenyl)-1,2,10,10a-tetrahydro-	EtOH	265(4.16),301(3.51)	44-3775-76
hydrochloride	EtOH	249(4.14),279(3.51), 288(3.44)	44-3775-76
$C_{17}H_{14}Cl_2O_5$ 11H-Dibenzo[b,e][1,4]dioxepin-11-one, 2,7-dichloro-3,8-dihydroxy-1,4,6,9-tetramethyl- (norvicanicin)	90% EtOH	230(4.27),270(3.97), 323s(2.90),330s(2.85)	12-2263-76
$C_{17}H_{14}Cl_3NO_2$ 1,4-Benzoxazepine, 7-chloro-5-(2,4-di-chlorophenyl)-2,3-dihydro-8-methoxy-3-methyl-	n.s.g.	220(4.44),248(4.08), 313(4.16)	20-0787-76
$C_{17}H_{14}Cl_4F_3NO$ Benzenemethanamine, N-[4-chloro-2-(tri-chloromethyl)phenyl]-4-ethoxy-α-(tri-chloromethyl)-	n.s.g.	249(4.42),281(3.39), 311(3.64)	98-0724-76
$C_{17}H_{14}Cl_4O_2$ 1H-Cyclopenta[b][1,4]benzodioxin, 5,6,7,8-tetrachloro-1-cyclohex-ylidene-3a,9a-dihydro-, cis	C_6H_{12}	217(5.07),293s(3.47), 300(3.55)	78-0147-76
$C_{17}H_{14}D_2O_5$ Eucomin-3,6-d_2, 3,9-dihydro-	EtOH	211(4.43),222(4.36), 291(4.25),325(3.84)	33-2048-76
$C_{17}H_{14}F_6N_4O_6$ Riboflavin,$\alpha^7,\alpha^7,\alpha^7,\alpha^8,\alpha^8,\alpha^8$-hexafluoro-	EtOH	221(4.28),276(4.49), 320(3.54),432(3.91)	103-1386-76
$C_{17}H_{14}FeN_2O_2$ Ferrocene, [[(4-nitrophenyl)methylene]-amino]-	n.s.g.	546(3.53)	65-2512-76
$C_{17}H_{14}N$ Dibenzo[a,g]quinolizinium, 5,6-dihydro-, chloride	EtOH	211(4.33),224s(4.31), 228(4.33),261s(4.53), 265(4.55),303(4.00), 315(4.08),366(3.76),	73-3157-76
iodide	EtOH	222(4.55),260s(4.57), 265(4.58),303(4.05), 315(4.13),365(3.81)	73-3157-76
$C_{17}H_{14}N_2$ 5H-Dibenz[b,f]azepine-10-carbonitrile, 5-ethyl-	MeOH	220(4.29),258(4.44), 296(3.90)	87-1345-76

Compound	Solvent	$\lambda_{max}(\log \epsilon)$	Ref.
$C_{17}H_{14}N_2O$			
7H,13H-Benz-3-azepino[2,3-b]quinazoline, 1,2-dihydro-13-oxo-	n.s.g.	225(4.50),268(3.96), 274(3.83),306(3.59), 317(3.50)	2-0879-76
5H-Dibenz[b,f]azepine, 5-acetyl-10-cyano-10,11-dihydro-	MeOH	232s(3.92)	78-1081-76
$C_{17}H_{14}N_2OS$			
3H-Pyrazol-3-one, 1,2-dihydro-5-methyl-2-phenyl-1-(phenylthioxomethyl)-	EtOH	386(4.23)	39-0038-76C
4(1H)-Pyrimidinone, 6-phenyl-2-[(phenylmethyl)thio]-	EtOH	257(4.42),312(3.93)	95-1094-76
$C_{17}H_{14}N_2O_2$			
1H-Anthra[1,2-d]imidazole-6,11-dione, 2,3-dihydro-2,2-dimethyl-	EtOH	570(4.16)	104-0172-76
1H-Anthra[2,3-d]imidazole-5,10-dione, 2,3-dihydro-2,2-dimethyl-	EtOH	595s(3.26)	104-0172-76
Benzaldehyde, 2-[[3-[(2-formylphenyl)-amino]-2-propenylidene]amino]-, (Z,E)-	EtOH	217(4.30),238(4.35), 254(4.34),333(4.22), 347s(--),399(4.38)	12-2271-76
hydrochloride	EtOH	253s(--),313(4.07), 367(4.19)	12-2271-76
Benzoic acid, 2-(1-methyl-3-phenyl-1H-pyrazol-5-yl)-	EtOH	248(4.37)	44-0110-76
[3,3'-Bi-1H-indol]-2(3H)-one, 3-hydroxy-1-methyl-	MeOH	219(4.76),266(4.06), 280(3.96),290(3.88)	103-0178-76
[3,3'-Bi-1H-indol]-2(3H)-one, 3-hydroxy-2'-methyl-	MeOH	213(3.38),220(3.37), 290(2.60)	103-0178-76
Cycloprop[a]inden-1a(1H)-amine, 6,6a-dihydro-N-[(4-nitrophenyl)methylene]-	C_6H_{12}	306(3.99)	35-1048-76
1H-Indene-1,3(2H)-dione, 2-[[4-(dimethylamino)phenyl]imino]-	MeOH	259(4.16),279(3.97), 478(4.20),558(4.42)	83-0092-76
1H-Pyrrole-2,5-dione, 1-phenyl-3-[(phenylmethyl)amino]-	EtOH	234(4.26),263(3.95), 365(3.50)	78-0431-76
$C_{17}H_{12}N_2O_2S$			
3-Thiophenamine, N-methyl-2-nitro-N,5-diphenyl-	MeCN	260(4.25),303(3.90), 361(4.22),452(3.94)	48-0221-76
$C_{17}H_{12}N_2O_3$			
Benzoic acid, 2-[5-(4-methoxyphenyl)-1H-pyrazol-3-yl]-	DMF	266(4.55)	44-0110-76
4H-Quinolizine-1-carboxylic acid, 4-oxo-3-(2-pyridinyl)-, ethyl ester	EtOH	265(4.3)	23-0097-76
$C_{17}H_{14}N_2O_6$			
1H-Isoindole-1,3(2H)-dione, 2-[2-hydroxy-1-(hydroxymethyl)-2-(4-nitrophenyl)ethyl]-, [R-(R*,R*)]-	MeOH	228(4.55),272(4.03)	130-0025-76
$C_{17}H_{14}N_2S$			
2-Quinolinecarbothioamide, N-(3-methylphenyl)-	20% EtOH	322(4.10)	94-1451-76
$C_{17}H_{14}N_2S_2$			
1H-[1,2]Dithiolo[5,1-e][1,2,3]thiadiazole-7-S^{IV}, 3-methyl-1,5-diphenyl-	C_6H_{12}	204(4.57),238(4.53), 254s(4.41),306(4.10), 507(4.29)	39-0228-76C

Compound	Solvent	$\lambda_{max}(\log \epsilon)$	Ref.
$C_{17}H_{14}N_4$			
Isoquinoline, 1-(1H-4-phenyl-1,2,3-triazol-2-yl)-	MeOH	240(3.8),285(4.0), 330(4.05)	2-0793B-76
Quinoline, 2-(1H-4-phenyl-1,2,3-triazol-2-yl)-	MeOH	275(4.02),324(4.2), 370(4.25)	2-0793B-76
$C_{17}H_{14}N_4O_2S$			
1-Thia-3,5,8,8b-tetraazaacenaphthylene-6-carboxylic acid, 4-methyl-2-phenyl-, ethyl ester	EtOH	230(4.42),305(3.91), 318(4.00),333(4.03), 407(4.14),500(3.09), 540(2.93)	78-0173-76
$C_{17}H_{14}N_4O_2Se$			
1-Selena-3,5,8,8b-tetraazaacenaphthylene-6-carboxylic acid, 4-methyl-2-phenyl-, ethyl ester	EtOH	229(4.21),305(3.81), 317(3.82),333(3.81), 409(3.79),510(2.90), 550(2.72)	78-0173-76
$C_{17}H_{14}N_4O_3$			
8-Azabicyclo[3.2.1]oct-3-en-2-one, 8-(5-nitro-2-pyridinyl)-6-(4-pyridinyl)-, endo	EtOH	212(4.41),225(4.36), 340(4.30)	39-2307-76C
Naphtho[1,2-g]pteridine-9,11(7H,10H)-dione, 7-(2-hydroxyethyl)-10-methyl-	5% DMF	303(4.28),456(4.22), 467(4.21)	12-2753-76
Naphtho[2,1-g]pteridine-8,10(9H,12H)-dione, 12-(2-hydroxyethyl)-9-methyl-	5% DMF	303(4.46),374(3.89), 480(4.06),498(4.06)	12-2753-76
$C_{17}H_{14}N_4O_4$			
1H-Isoindole-1,3(2H)-dione, 2-[2-[(4-azidophenyl)-2-hydroxy-1-(hydroxymethyl)ethyl]-, [R-(R*,R*)]-	MeOH	243(4.31),250(4.29), 290(3.62)	130-0025-76
$C_{17}H_{14}N_6O_7$			
2,4-Cyclohexadien-1-one, 2,6-dimethyl-6-[[1-(2,4,6-trinitrophenyl)-1H-1,2,3-triazol-4-yl]methyl]-	n.s.g.	310(3.65)	44-0678-76
$C_{17}H_{14}O$			
Benzene, 1-methoxy-4-(4-phenyl-1-buten-3-ynyl)-, (E)-	MeOH	225(4.22),231s(4.19), 249s(4.06),312s(4.56), 324(4.61),345s(4.42)	39-1104-76B
1H-Inden-1-one, 2,3-dihydro-2-[(4-methylphenyl)methylene]-	isooctane	230(4.10),265(3.92), 305s(--),315(4.51), 330(4.46),390(2.38), 410s(--)	65-2057-76
	EtOH	235(3.93),270s(--), 330(4.35)	65-2057-76
1H-Inden-1-one, 2,3-dihydro-6-methyl-2-(phenylmethylene)-	MeOH	235(3.90),325(4.45)	44-3540-76
Naphthalene, 1-methoxy-5-phenyl-	EtOH	237(4.54),301(4.04), 310s(4.00),323s(3.82)	95-1161-76
1(2H)-Naphthalenone, 3,4-dihydro-2-(phenylmethylene)-	isooctane	226(4.05),297(4.20), 310s(--),362(2.42), 375(2.36),390(2.26), 412s(--)	65-2057-76
	EtOH	228(4.10),306(4.24), 316s(--)	65-2057-76
$C_{17}H_{14}OS$			
Benzo[b]thiophen-3(2H)-one, 5-methyl-2-[(4-methylphenyl)methylene]-, (Z)-	octane	330(4.36),436(4.24)	103-0854-76

Compound	Solvent	$\lambda_{max}(\log \epsilon)$	Ref.
$C_{17}H_{14}OS_3$			
Ethane(dithioic) acid, (methylthio)-9H-xanthen-9-ylidene-, methyl ester	CCl_4	278(3.76),332(3.92), 428(3.22),520s(2.20)	88-0209-76
$C_{17}H_{14}O_2$			
3(2H)-Benzofuranone, 5-methyl-2-[(4-methylphenyl)methylene]-, (Z)-	octane	326(4.36),380(4.33)	103-0854-76
4H-1-Benzopyran-4-one, 2,3-dihydro-3-[(4-methoxyphenyl)methylene]-	isooctane	275s(--),305(4.31), 342(4.02)	65-2057-76
	EtOH	275s(--),318(4.19), 355(3.96)	65-2057-76
1H-Inden-1-one, 2,3-dihydro-2-[(4-methoxyphenyl)methylene]-	isooctane	245(4.11),260s(--), 335(4.51),350(4.45), 348(4.41),405s(--)	65-2057-76
	EtOH	250(4.14),270s(--), 355(4.45)	65-2057-76
1H-Inden-1-one, 2,3-dihydro-2-(2-methylbenzoyl)-	MeOH	266(4.05),325(4.23)	83-0356-76
1H-Inden-1-one, 2,3-dihydro-2-(3-methylbenzoyl-	MeOH	250(4.09),353(4.27)	83-0356-76
1H-Inden-1-one, 2,3-dihydro-2-(4-methylbenzoyl-	MeOH	253(4.12),359(4.31)	83-0356-76
1(2H)-Naphthalenone, 2-[(2-hydroxyphenyl)methylene]-	EtOH	214(4.42),263(4.04), 296(3.95),340(4.00)	22-1975-76
$C_{17}H_{14}O_2S$			
Benzo[b]thiophen-3(2H)-one, 2-[(4-methoxyphenyl)methylene]-5-methyl-, (Z)-	octane	341(4.37),438(4.32)	103-0854-76
Dibenzo[b,d]thiepin-6-carboxylic acid, ethyl ester	C_6H_{12}	213(4.33),245(4.41), 260s(4.29),346(3.41)	89-0611-76
$C_{17}H_{14}O_2Se$			
Benzo[b]selenophen-3(2H)-one, 2-[(4-methoxyphenyl)methylene]-5-methyl-, (Z)-	octane	344(4.35),442(4.21)	103-0854-76
$C_{17}H_{14}O_3$			
3(2H)-Benzofuranone, 2-[(4-methoxyphenyl)methylene]-5-methyl-, (Z)-	octane	335(4.39),393(4.52)	103-0854-76
4H-1-Benzopyran-4-one, 2,3-dihydro-3-[(4-methoxyphenyl)methylene]-	isooctane	227(4.23),237s(--), 274(3.93),323(4.33), 347(4.21)	65-2057-76
	EtOH	245(4.13),270s(--), 335(4.14),365(4.02)	65-2057-76
7H-1-Benzopyran-7-one, 5-methoxy-8-methyl-2-phenyl-	EtOH	272(4.04),320(3.93), 330s(3.86),378(3.65), 500(3.83)	39-1570-76C
2,5-Epoxy-1,3-benzodioxepin, 4,5-dihydro-2-methyl-4-methylene-5-phenyl-	MeOH	276(3.25),283(3.21)	35-3555-76
2(5H)-Furanone, 5-hydroxy-5-methyl-3,4-diphenyl-	dioxan	220(4.25),285(4.05)	24-0576-76
3(2H)-Furanone, 2-methoxy-2,4-diphenyl-	MeOH	241(4.27),307(3.54)	44-0388-76
1H-Inden-1-one, 2,3-dihydro-2-(2-methoxybenzoyl)-	MeOH	248(3.99),341(4.21)	83-0356-76
1H-Inden-1-one, 2,3-dihydro-2-(4-methoxybenzoyl)-	MeOH	269(4.23),368(4.34)	83-0356-76
$C_{17}H_{14}O_4$			
2H-1-Benzopyran-2-one, 8-methoxy-3-(4-methoxyphenyl)-	MeOH	215(4.42),248(4.18), 330(4.30)	78-0109-76

Compound	Solvent	$\lambda_{max}(\log \epsilon)$	Ref.
4H-1-Benzopyran-4-one, 5-hydroxy-7-methoxy-6-methyl-2-phenyl-	EtOH	207(4.54),273(4.52)	2-0009-76
9H-Xanthen-9-one, 8-acetoxy-1,3-dimethyl-	EtOH	236s(4.54),241(4.56), 273(4.11),277s(4.10), 340(3.79)	23-1703-76

$C_{17}H_{14}O_5$

9,10-Anthracenedione, 1,3,8-trihydroxy-6-propyl-	EtOH	222(4.38),253(4.09), 266(4.08),290(4.15), 445(3.92)	39-0613-76C
9(10H)-Anthracenone, 2-acetyl-1,6,8-trihydroxy-3-methyl-	MeOH-HOAc	253(4.30),270s(4.21), 310s(4.18),356(4.34)	12-2225-76
4H-1-Benzopyran-4-one, 7-hydroxy-8-methoxy-3-(4-methoxyphenyl)-	EtOH	257(4.49),306s(3.90)	102-1019-76
	EtOH-NaOMe	269(--),324s(--), 348(--)	102-1019-76
4H-1-Benzopyran-4-one, 2,3-dihydro-5,7-dihydroxy-3-[(4-methoxyphenyl)methylene]-, (E)- (eucomin)	MeOH	356(4.479)	33-2048-76
	EtOH	212(4.47),359(4.481)	33-2048-76
(Z)-	EtOH	214(4.47),363(4.394)	33-2048-76
Indano[1,2-b]benzofuran-5-one, 4b,9b-dihydro-9b-hydroxy-2,4b-dimethoxy-	MeOH	221s(4.29),239s(4.14), 291(3.67),340s(3.09)	83-0081-76
1H-Indene-1,3(2H)-dione, 2-(2,4-dimethoxyphenyl)-2-hydroxy-	MeOH	226(4.75),245s(4.17), 280s(3.56)	83-0081-76

$C_{17}H_{14}O_6$

9,10-Anthracenedione, 1,3,8-trihydroxy-6-(1-hydroxypropyl)-	EtOH	222(4.46),250(4.19), 266(4.18),290(4.22), 435(4.01)	39-0613-76C
Benzaldehyde, 3-(1,3-dioxo-3-phenylpropyl)-2,4,6-trihydroxy-5-methyl-	EtOH	272(4.6),342(3.6)	2-0009-76
4H-1-Benzopyran-4-one, 5,7-dihydroxy-3,6-dimethoxy-2-phenyl-	EtOH	245(3.22),270(4.23), 322(3.96)	28-0683-76
	EtOH-NaOAc	245(--),274(--), 330(--)	28-0683-76
	EtOH-AlCl₃	282(--),338(--)	28-0683-76
4H-1-Benzopyran-4-one, 3-(2,4-dimethoxyphenyl)-5,7-dihydroxy-	EtOH	218(4.28),262(4.36), 322(3.40)	102-1029-76
	EtOH-NaOH	235(4.28),275(4.42), 335(4.02)	102-1029-76
	EtOH-NaOAc	225(4.24),265(4.33), 333(3.67)	102-1029-76
	EtOH-AlCl₃	219(4.31),275(4.37), 378(2.97)	102-1029-76
4H-1-Benzopyran-4-one, 5-hydroxy-2-(4-hydroxyphenyl)-3,7-dimethoxy- (kumatakenin)	EtOH	269(4.28),302s(4.04), 353(4.29)	94-1242-76
Irisolidone	EtOH	271(4.48),340s(3.82)	95-0254-76
	EtOH-NaOEt	250(4.28),275(4.49), 346(4.04)	95-0254-76
	EtOH-NaOAc	275(4.51),344(4.10)	95-0254-76
	EtOH-AlCl₃	278(4.45),318(3.87), 378(3.44)	95-0254-76
9H-Xanthen-9-one, 1-(hydroperoxymethyl)-8-acetoxy-3-methyl-	EtOH	236s(4.40),241(4.42), 272(4.01),277s(3.97), 340(3.63)	23-1703-76

$C_{17}H_{14}O_7$

4H-1-Benzopyran-4-one, 5,6-dihydroxy-2-(4-hydroxy-3-methoxyphenyl)-7-methoxy-	EtOH	284(4.48),344(4.51)	102-0838-76
	EtOH-NaOAc	287(4.49),344(4.52), 400(4.45)	102-0838-76

Compound	Solvent	$\lambda_{max}(\log \epsilon)$	Ref.
4H-1-Benzopyran-4-one, 5,6-dihydroxy-2-(4-hydroxy-3-methoxyphenyl)-7-methoxy- (cont.)	EtOH-AlCl$_3$	296(4.45),308s(4.30), 369(4.46)	102-0838-76
4H-1-Benzopyran-4-one, 5,6-dihydroxy-2-(4-hydroxyphenyl)-3,7-dimethoxy-	MeOH	215(4.48),235s(--), 281(4.29),340(4.38)	88-0067-76
4H-1-Benzopyran-4-one, 5,8-dihydroxy-2-(4-hydroxyphenyl)-3,7-dimethoxy-	MeOH	223(4.16),278(4.19), 305(4.12),327(4.07)	88-0067-76
4H-1-Benzopyran-4-one, 3,5,7-trihydroxy-6-methoxy-2-(4-methoxyphenyl)-	EtOH	255(4.21),270(4.26), 340s(--),364(4.32)	88-0067-76
2H-Pyran-3-carboxylic acid, 4-hydroxy-6-methoxy-5-(3-methyl-2-benzofuranyl)-2-oxo-, methyl ester	EtOH	263(4.42),280s(4.34), 285s(4.31)	78-0269-76
$C_{17}H_{14}O_8$ 4H-1-Benzopyran-4-one, 2-(3,4-dihydroxyphenyl)-5,6-dihydroxy-3,7-dimethoxy-	MeOH	239(4.17),258(4.17), 280(4.18),351(4.34)	88-0067-76
4H-1-Benzopyran-4-one, 2-(3,4-dihydroxyphenyl)-5,8-dihydroxy-3,7-dimethoxy-	MeOH	260s(--),278(4.30), 303s(--),342(4.15), 380s(--)	88-0067-76
4H-1-Benzopyran-4-one, 3,5,6-trihydroxy-2-(4-hydroxy-3-methoxyphenyl)-7-methoxy-	MeOH	240(4.28),258(4.27), 275s(--),358(4.39)	88-0067-76
4H-1-Benzopyran-4-one, 5,6,7-trihydroxy-2-(3-hydroxy-4-methoxyphenyl)-3-methoxy-	MeOH	250(4.22),278(4.24), 348(4.37)	88-0067-76
4H-1-Benzopyran-4-one, 5,6,7-trihydroxy-2-(4-hydroxy-3-methoxyphenyl)-3-methoxy-	MeOH	223(4.49),246(4.43), 265s(--),280s(--), 350(4.34)	88-0067-76
$C_{17}H_{15}BrN_4O_2$ 5-Pyrimidinecarboxamide, 4-(4-bromophenyl)-1,2,3,4-tetrahydro-6-methyl-2-oxo-N-2-pyridinyl-	EtOH	203(4.31),232(4.17), 294(4.13)	103-0191-76
$C_{17}H_{15}ClN_2O_2$ Benzaldehyde, 2-[[3-[(2-formylphenyl)-amino]-2-propenylidene]amino]-, (Z,E)-, hydrochloride	EtOH	253s(--),313(4.07), 367(4.19)	12-2271-76
Ethanol, 2-[(7-chloro-5-phenyl-3H-1,4-benzodiazepin-2-yl)oxy]-	isoPrOH	218(4.52),255s(4.17), 322(3.35)	44-2724-76
$C_{17}H_{15}ClN_2O_4$ 1,4-Benzoxazepine, 7-chloro-2,3-dihydro-8-methoxy-3-methyl-5-(4-nitrophenyl)-	n.s.g.	215(4.38),245(4.21), 313(4.18)	20-0787-76
$C_{17}H_{15}ClN_2O_4S$ 1,3-Thiazinium, 6-(4-methylphenyl)-2-(phenylamino)-, perchlorate	HOAc	424(3.97)	88-2005-76
$C_{17}H_{15}ClN_4O_2$ Hydrazinecarboxylic acid, 2-(7-chloro-5-phenyl-3H-1,4-benzodiazepin-2-yl)-, methyl ester	isoPrOH	210s(4.27),230s(4.18), 258(4.12),337(2.95)	44-2724-76
$C_{17}H_{15}ClO$ Naphthalene, 5-chloro-1,2-dihydro-8-methoxy-4-phenyl-	EtOH	227(4.52),265s(3.82), 310s(3.07)	95-1161-76
$C_{17}H_{15}Cl_2FN_2$ 1H-1,4-Benzodiazepine, 7-chloro-2-(chlo-	MeOH-base	232s(--),239(4.34),	111-0501-76

Compound	Solvent	$\lambda_{max}(\log \epsilon)$	Ref.
romethyl)-5-(2-fluorophenyl)-2,3-di-hydro-1-methyl-, hydrochloride, (cont.)		268s(--),368(3.34)	111-0501-76
1,5-Benzodiazocine, 3,8-dichloro-6-(2-fluorophenyl)-1,2,3,4-tetrahydro-1-methyl-, hydrochloride	EtOH-base	227(4.33),267(4.13), 368(3.43)	111-0501-76
$C_{17}H_{15}Cl_2NO_2$			
1,4-Benzoxazepine, 7-chloro-5-(2-chloro-phenyl)-2,3-dihydro-8-methoxy-3-meth-yl-	n.s.g.	215(4.36),245(4.07), 310(4.28)	20-0787-76
$C_{17}H_{15}Cl_3N_2$			
1H-1,4-Benzodiazepine, 7-chloro-2-(chloro-methyl)-5-(2-chlorophenyl)-2,3-di-hydro-1-methyl-	EtOH	234(4.34),266s(--), 366(3.38)	111-0501-76
hydrochloride	EtOH	252(4.30),285s(--), 460(3.69)	111-0501-76
1,5-Benzodiazocine, 3,8-dichloro-6-(2-chlorophenyl)-1,2,3,4-tetrahydro-1-methyl-	EtOH	226s(--),270(4.08), 373(3.43)	111-0501-76
$C_{17}H_{15}FeN$			
Ferrocene, [(phenylmethylene)amino]-	n.s.g.	392(3.23),434(3.20)	65-2512-76
$C_{17}H_{15}NO$			
1H-Inden-1-one, 2,3-dihydro-2-[(methyl-phenylamino)methylene]-	EtOH	258(4.10),290s(3.55), 298(3.56),371(4.44)	4-1201-76
3(2H)-Isoquinolinone, 1-(4-ethylphenyl)-	EtOH	357(3.87),422(3.60)	103-0910-76
	ether	353(3.93)	103-0910-76
	CHCl$_3$	355(3.72),440(3.52)	103-0910-76
	CCl$_4$	353(3.85),440(3.02)	103-0910-76
$C_{17}H_{15}NOS$			
4H-1-Benzothiopyran-4-one, 2,3-dihydro-3-[(methylphenylamino)methylene]-, (E)-	EtOH	248(4.39),265s(4.13), 382(4.27)	4-0225-76
$C_{17}H_{15}NOS_2$			
1,3-Dithiol-1-ium, 2-[4-(dimethylamino)-phenyl]-4-hydroxy-5-phenyl-, hydrox-ide, inner salt	dioxan	236(4.05),250s(4.05), 302(4.20),398(4.14), 599(4.36)	24-0740-76
$C_{17}H_{15}NO_2$			
9,10-Anthracenedione, 1-(dimethylamino)-2-methyl-	CCl$_4$	477(3.33)	104-2499-76
9,10-Anthracenedione, 2-(dimethylamino)-3-methyl-	CCl$_4$	415(3.48)	104-2499-76
1H-Indole-3-carboxaldehyde, 2-(4-meth-oxyphenyl)-1-methyl-	MeOH	257(4.53),307(4.24)	12-2747-76
2H-Isoindole-1-carboxylic acid, 2-(phen-ylmethyl)-, methyl ester	EtOH	237(4.27),257(4.04), 264(4.22),338(3.93), 352(3.92)	32-0065-76
Quinoline, 3,4-dimethoxy-2-phenyl-	EtOH	249(4.35),288(3.42), 318(3.20)	142-1089-76
4(1H)-Quinolinone, 3-methoxy-1-methyl-2-phenyl-	EtOH	251(4.39),338(3.85), 346(3.92)	142-1089-76
$C_{17}H_{15}NO_3$			
1H-Indene-1,3(2H)-dione, 2-[4-(dimethyl-amino)phenyl]-2-hydroxy-	EtOH	229(4.67),258(4.55), 305(3.91)	83-0081-76

Compound	Solvent	$\lambda_{max}(\log \epsilon)$	Ref.
$C_{17}H_{15}NO_4$			
10(9H)-Acridineacetic acid, 9-oxo-, 2-hydroxyethyl ester	isoPrOH	214(4.42),253(4.72), 290(3.42),355s(3.59), 374(3.95),394(4.04)	44-3406-76
$C_{17}H_{15}NO_7$			
9H-Carbazole-9-acetic acid, 1,4-dihydro-6-hydroxy-2,7-dimethoxy-α-methyl-1,4-dioxo-	EtOH	220(4.35),260(4.16), 273(4.12),290(4.08), 315(4.09),475(3.54)	146-0667-76
$C_{17}H_{15}NO_{10}$			
1-Cyclopropene-1,2-diacetic acid, α-cyano-α'-(methoxycarbonyl)-3-[2-methoxy-1-(methoxycarbonyl)-2-oxoeth-ylidene]-, dimethyl ester, ion(2-)	MeCN	240(4.31+),300s(4.54), 314(4.59)	35-0610-76
bis(N,N,N-tributyl-1-butanaminium) salt	MeCN-Bu₄NOH	239(4.35),267(4.34), 310(4.56)	35-0610-76
$C_{17}H_{15}NS$			
Pyridine, 4-[(1-benzo[b]thien-3-ylmeth-ylene)propyl]-, (E)-	MeOH	270(4.13),310(4.16)	4-0141-76
$C_{17}H_{15}N_2OSe$			
1,3-Selenazinium, 6-(4-methoxyphenyl)-2-(phenylamino)-, perchlorate	HOAc	461(4.18)	88-2005-76
$C_{17}H_{15}N_2S$			
1,3-Thiazinium, 6-(4-methylphenyl)-2-(phenylamino)-, perchlorate	HOAc	424(3.97)	88-2005-76
$C_{17}H_{15}N_2Se$			
1,3-Selenazinium, 6-(4-methylphenyl)-2-(phenylamino)-, perchlorate	HOAc	430(3.97)	88-2005-76
1,3-Selenazinium, 2-(N-methyl-N-phenyl-amino)-6-phenyl-, perchlorate	HOAc	343(4.27),408s(3.52)	88-2005-76
$C_{17}H_{15}N_3$			
2-Pyrimidinamine, 4-(4-methylphenyl)-6-phenyl-	EtOH	255(4.46),333(4.15)	4-0257-76
$C_{17}H_{15}N_3O$			
2-Pyrimidinamine, 4-(4-methoxyphenyl)-6-phenyl-	EtOH	252(4.42),334(4.30), 285s(4.23)	4-0257-76
Spiro[3H-indole-3,3'-[3H]pyrazol]-2(1H)-one, 2',4'-dihydro-1-methyl-5'-phenyl-	EtOH	210(4.73),263(4.33), 295(4.38)	94-0782-76
$C_{17}H_{15}N_3OS$			
Carbamimidothioic acid, N'-cyano-N-(2-oxo-2-phenylethyl)-N-phenyl-, methyl ester	EtOH	285(4.88),292(4.40)	49-1413-76
Isoxazole, 5-(ethylthio)-3-phenyl-4-(phenylazo)-	EtOH	236(4.28),281(4.17), 365(3.91),438s(3.21)	48-0359-76
Methanone, [4-amino-2-(methylthio)-1-phenyl-1H-imidazol-5-yl]phenyl-	EtOH	245(4.09),345(4.02)	49-1413-76
$C_{17}H_{15}N_3OS_2$			
Thiourea, [3-(4-methylphenyl)-4-oxo-2-thiazolidinylidene]phenyl-	dioxan	278(4.16),413(4.07)	103-0751-76

Compound	Solvent	$\lambda_{max}(\log \epsilon)$	Ref.
$C_{17}H_{15}N_3O_2$			
Benzamide, 2-[5-(4-methoxyphenyl)-1H-pyrazol-3-yl]-	EtOH	261(4.50)	44-0110-76
Benzonitrile, 2-[2-[4-(dimethylamino)-phenyl]ethenyl]-5-nitro-, (E)-	EtOH	300(3.95),464(4.38)	80-1127-76
Benzonitrile, 4-[2-[4-(dimethylamino)-phenyl]ethenyl]-3-nitro-, (E)-	EtOH	260(4.45),430(4.59)	80-1127-76
Cyanamide, [[(4-methoxyphenyl)(2-oxo-2-phenylethyl)amino]methylene]-	EtOH	249(4.80),277s(4.28)	49-1413-76
Methanone, [4-amino-1-(4-methoxyphenyl)-1H-imidazol-5-yl]-	EtOH	230(4.28),341(3.99)	49-1413-76
1H-Pyrazole-4-carboxamide, 3-benzoyl-4,5-dihydro-N-phenyl-	EtOH	247(4.39),336(4.13)	115-0001-76
5,8-Quinazolinedione, 2-ethyl-4-methyl-6-(phenylamino)-	EtOH	228(4.20),275(4.29), 284s(--),475(3.69)	5-1809-76
$C_{17}H_{15}N_3O_2S_2$			
Thiourea, [3-(4-methoxyphenyl)-4-oxo-2-thiazolidinylidene]phenyl-	dioxan	264(4.23),420(4.02)	103-0751-76
$C_{17}H_{15}N_3O_3$			
4(1H)-Pyrimidinone, 2-amino-6-(diphenoxymethyl)-	EtOH	292(3.83)	44-2850-76
$C_{17}H_{15}N_3O_3S$			
[1]Benzothieno[2,3-b]pyridin-4(1H)-one, 5,6,7,8-tetrahydro-3-nitro-2-(phenylamino)-	CHCl$_3$	324(4.00),332s(3.99), 392s(2.39)	48-0039-76
$C_{17}H_{15}N_3O_4$			
1H-1,4-Benzodiazepine-2,6,9(3H)-trione, 7-amino-1-(methoxymethyl)-5-phenyl-	isoPrOH	215(4.43),257(4.18), 323(3.85),510(3.15)	87-1378-76
$C_{17}H_{15}N_5O$			
4(1H)-Pyrimidinone, 5-(phenylazo)-6-[(phenylmethyl)amino]-	pH 13	244(4.32),386(4.24)	39-1847-76C
[1,2,4]Triazolo[5,1-c][1,2,4]triazine, 1,4-dihydro-4-methoxy-3,4-diphenyl-	EtOH	297(4.16)	39-1492-76C
$C_{17}H_{15}N_5O_3$			
1H-Pyrazole-4-carboxamide, 5-amino-3-methyl-1-(4-nitrophenyl)-N-phenyl-	EtOH	282(4.415)	4-1137-76
$C_{17}H_{15}N_5O_4$			
5-Pyrimidinecarboxamide, 1,2,3,4-tetra-hydro-6-methyl-4-(3-nitrophenyl)-2-oxo-N-2-pyridinyl-	EtOH	205(4.19),250(4.00), 285(4.26)	103-0191-76
5-Pyrimidinecarboxamide, 1,2,3,4-tetra-hydro-6-methyl-4-(4-nitrophenyl)-2-oxo-N-2-pyridinyl-	EtOH	203(4.23),249(4.15), 287(4.37)	103-0191-76
Pyrimido[1,2-a]purin-10(1H)-one, 7-(3,4,5-trimethoxyphenyl)-	pH 1	243(4.36),307(4.33), 370s(3.43)	44-0294-76
	pH 6.8	253(4.39),256s(4.39), 304(4.31),370(3.46)	44-0294-76
	pH 10.1	255(4.42),277(4.40), 295(4.38),380(3.45)	44-0294-76
$C_{17}H_{15}N_5O_6$			
7,11-Epoxy-1H,6H-8,10-dioxa-2,4,5,5a,11a-pentaazacyclopenta[5,6]cyclooct[1,2,3-	MeOH	225(4.39),255(4.12)	44-1100-76

Compound	Solvent	$\lambda_{max}(\log \epsilon)$	Ref.
cd]indene-1,3(2H)-dione, 7,7a,10a,11- tetrahydro-9-(4-methoxyphenyl)- (cont.)			44-1100-76
$C_{17}H_{16}$ 4H-Benz[de]anthracene, 5,6,6a,7-tetra- hydro-	EtOH	261(4.72),283(4.12), 294(4.12),305(4.20), 323(2.64),333(3.08), 354(3.12)	104-0180-76
Phenanthrene, 9,10-dihydro-9,9-dimethyl- 10-methylene-	H_2SO_4	266(4.56),3??s(--), 337(4.03),537(3.85)	104-1298-76
$C_{17}H_{16}BrNO_3$ 2-Propen-1-one, 1-(3-bromo-2-hydroxy- 5-methylphenyl)-3-(4-methoxyphenyl)-, oxime	isoBuOH	224(4.45),270(4.16), 322(3.76)	42-1067-76
$C_{17}H_{16}Br_2N_2O$ 9(10H)-Acridinone, 2,7-dibromo-10-[2- (dimethylamino)ethyl]-	EtOH	253(5.01),415(4.24)	48-0515-76
$C_{17}H_{16}ClFN_2O$ 1H-1,4-Benzodiazepine-2-methanol, 7-chloro-5-(2-fluorophenyl)- 2,3-dihydro-	EtOH	231s(--),237(4.33), 268s(--),374(3.34)	111-0501-76
$C_{17}H_{16}ClF_3N_4O_6$ Mannitol, 1-[8-chloro-3,4-dihydro-2,4- dioxo-7-(trifluoromethyl)benzo[g]- pteridin-10(2H)-yl]-1,6-dideoxy-	EtOH	272(4.56),335(3.84), 430(4.02)	103-0938-76
Rhamnitol, 1-[7-chloro-3,4-dihydro-2,4- dioxo-8-(trifluoromethyl)benzo[g]- pteridin-10(2H)-yl]-	EtOH	222(4.50),271(4.61), 322(3.79),451(3.95)	103-1386-76
$C_{17}H_{16}ClF_3N_4O_7$ Galactitol, 1-[7-chloro-3,4-dihydro-2,4- dioxo-8-(trifluoromethyl)benzo[g]pter- idin-10(2H)-yl]-1-deoxy-	EtOH	220(4.41),270(4.44), 321(3.75),450(3.90)	103-1386-76
Galactitol, 1-[8-chloro-3,4-dihydro-2,4- dioxo-7-(trifluoromethyl)benzo[g]pter- idin-10(2H)-yl]-1-deoxy-	EtOH	272(4.48),334(3.81), 430(3.97)	103-0938-76
D-Glucitol, 1-[8-chloro-3,4-dihydro-2,4- dioxo-8-(trifluoromethyl)benzo[g]pter- idin-10(2H)-yl]-1-deoxy-	EtOH	219(4.35),269(4.43), 322(3.73),449(3.99)	103-1386-76
D-Glucitol, 1-[8-chloro-3,4-dihydro-2,4- dioxo-7-(trifluoromethyl)benzo[g]pter- idin-10(2H)-yl]-1-deoxy-	EtOH	271(4.52),335(3.84), 430(3.98)	103-0938-76
$C_{17}H_{16}ClN$ Acridine, 9-chloro-2-(1,1-dimethyleth- yl)-	MeOH	215(4.20),258(4.92), 345(3.73),365(3.88), 390(3.58)	39-0465-76C
$C_{17}H_{16}ClNO_2$ Ethanone, 1-(6-chloro-1,2,3,4-tetrahy- dro-4-hydroxy-4-phenyl-3-quinolinyl)-	isoPrOH	255(4.09),318(3.47)	4-0131-76
3-Quinolinecarboxaldehyde, 6-chloro- 1,2,3,4-tetrahydro-4-hydroxy-1-methyl- 4-phenyl-	isoPrOH	262(4.21),321(3.46)	4-0131-76

Compound	Solvent	$\lambda_{max}(\log \epsilon)$	Ref.
$C_{17}H_{16}ClNO_3$ 1,4-Benzoxazepine, 5-(4-chlorophenyl)- 2,3-dihydro-7,8-dimethoxy-	n.s.g.	210(4.43),270(4.07), 310(4.14)	20-0787-76
$C_{17}H_{16}ClNO_5$ 1,3-Azulenedicarboxylic acid, 2-chloro- 6-[(hydroxyimino)methyl]-, diethyl ester	MeOH	239(4.45),269(4.11), 326(4.76),360(4.21), 383(4.15),520(2.83)	88-2045-76
$C_{17}H_{16}ClN_3O_2S$ Propanoic acid, 3-[(4-chlorophenyl)ami- no]-2-(phenylhydrazono)-3-thioxo-, ethyl ester, (Z)-	EtOH	246(4.20),290(4.11), 360(4.32)	104-1912-76
$C_{17}H_{16}ClN_7$ 9H-Purine-9-propanamine, 6-amino- N-(7-chloro-4-quinolinyl)-	pH 6.9	<u>333</u>(4.09),<u>346</u>(4.11)	35-0613-76
$C_{17}H_{16}Cl_2N_2$ 1,5-Benzodiazocine, 3-chloro-6-(2-chlo- rophenyl)-1,2,3,4-tetrahydro-1-methyl-	EtOH	225s(--),235s(--), 260s(--),362(3.48)	111-0501-76
1,5-Benzodiazocine, 3,8-dichloro- 1,2,3,4-tetrahydro-1-methyl- 6-phenyl-, hydrochloride	EtOH-base	229(4.32),254(4.30), 364(3.40)	111-0501-76
$C_{17}H_{16}Cl_2N_2O$ 1H-1,4-Benzodiazepine-2-methanol, 7-chloro-5-(2-chlorophenyl)- 2,3-dihydro-1-methyl-	EtOH	228s(--),272(3.86), 374(3.30)	111-0501-76
1,5-Benzodiazocin-3-ol, 8-chloro- 6-(2-chlorophenyl)-1,2,3,4-tetra- hydro-1-methyl-, hydrochloride	EtOH-base	225s(--),273(4.03), 376(3.45)	111-0501-76
$C_{17}H_{16}Cl_2N_2O_2$ Benzeneethanol, 4-chloro-α-[(4-chloro- phenyl)azo]-α-methyl-, acetate	C_6H_{12}	280(4.32),400(2.35)	22-0930-76
$C_{17}H_{16}Cl_2O_7$ α-D-erythro-Pentofuranose, 1,2-di-O- acetyl-5-O-benzoyl-3-deoxy-3-C-(di- chloromethylene)-	EtOH	223(3.83)	111-0489-76
β-	EtOH	220(3.70)	111-0489-76
$C_{17}H_{16}Cl_3NO_3$ Benzamide, 2,4-dichloro-N-2-(4-chloro- 3-methoxyphenoxy)-1-methylethyl]-	n.s.g.	210(4.39),280(3.53)	20-0787-76
$C_{17}H_{16}Cl_4O_2$ 1H-Cyclopenta[b][1,4]benzodioxin, 5,6,7,8-tetrachloro-1-cyclohexyl- idene-2,3,3a,9a-tetrahydro-, cis	C_6H_{12}	219(4.85),232(4.22), 293(3.30),301(3.40)	78-0147-76
$C_{17}H_{16}NOS_2$ Benzenaminium, 4-methoxy-N-methyl- N-(5-phenyl-3H-1,2-dithiol-3-yl- idene)-, bromide	MeOH	232(4.20),281(4.26), 377(3.89)	48-0221-76
$C_{17}H_{16}NOSe$ Isoselenazolium, 5-(4-methoxyphenyl)- 2-(4-methylphenyl)-, perchlorate	HOAc	290(3.67),391(4.24)	118-0273-76

Compound	Solvent	$\lambda_{max}(\log \epsilon)$	Ref.
$C_{17}H_{16}NO_2Se$			
Isoselenazolium, 2,5-bis(4-methoxyphenyl)-, perchlorate	HOAc	288(3.66),389(4.32)	118-0273-76
$C_{17}H_{16}NS_2$			
Benzenaminium, N-ethyl-N-(5-phenyl-3H-1,2-dithiol-3-ylidene)-, bromide	MeOH	318(4.16),364(3.95)	48-0221-76
$C_{17}H_{16}N_2$			
Benzenamine, N-(1,4-dimethyl-2(1H)-quinolinylidene)-	hexane	383(3.95)	18-2770-76
	MeOH	370(4.03)	18-2770-76
	CHCl$_3$	378(3.97)	18-2770-76
	CCl$_4$	383(3.89)	18-2770-76
5H-Dibenz[b,f]azepine-10-carbonitrile, 5-ethyl-10,11-dihydro-	MeOH	253.5(4.00)	87-1345-76
2-Quinolinamine, N,4-dimethyl-N-phenyl-	hexane	346(3.96),358s(3.91)	18-2770-76
	MeOH	347(3.86),356(3.84)	18-2770-76
	CHCl$_3$	350(3.93),360s(3.90)	18-2770-76
$C_{17}H_{16}N_2O$			
Cyclohept[b]indole, 6-morpholino-	C_6H_{12}	237(4.29),279(4.27), 305(4.48),315(4.50), 370(4.11),389(4.22), 408(4.03),513(3.55)	18-1101-76
[1,4]Diazepino[6,7,1-jk]carbazole, 3-acetyl-1,2,3,4-tetrahydro-	EtOH	236(4.61),262(4.22), 291(4.17)	4-1187-76
Ethanone, 1-[4-(4,5-dihydro-1-phenyl-1H-pyrazol-3-yl)phenyl]-	benzene	409(4.38)	64-1248-76B
Ethanone, 1-[4-(4,5-dihydro-3-phenyl-1H-pyrazol-1-yl)phenyl]-	benzene	378(4.59)	64-1248-76B
Imidazo[1,2-b]isoquinolin-5(1H)-one, 2,3,10,10a-tetrahydro-10a-phenyl-	n.s.g.	237(4.08),252(4.00), 263(3.88)	39-1073-76C
1-Propanone, 1-[2-methyl-3-(2-pyridinyl-1-indolizinyl]-	EtOH	231(4.20),265s(4.09), 270(4.15),325(4.04), 338(4.06),352s(4.01), 370s(3.79)	1-0198-76
$C_{17}H_{16}N_2O_2$			
9,10-Anthracenedione, 1-(dimethylamino)-2-(methylamino)-	EtOH	488(3.78)	104-2503-76
9,10-Anthracenedione, 2-(dimethylamino)-1-(methylamino)-	EtOH	540(3.83)	104-2503-76
9,10-Anthracenedione, 2-(dimethylamino)-3-(methylamino)-	EtOH	490(3.55)	104-2503-76
Benzoic acid, 4-(4,5-dihydro-1-phenyl-1H-pyrazol-3-yl)-, methyl ester	benzene	400(4.35)	64-1248-76B
Isoquinoline, 6,7-dimethoxy-1-(2-pyridinylmethyl)-	MeOH	238(4.85),263(4.13), 312(3.75),326(3.88)	83-0072-76
3,5-Pyrazolidinedione, 4,4-dimethyl-1,2-diphenyl-	hexane	238(4.26),279(3.48)	28-0343-76B
	dioxan	239(4.26),277(3.46)	28-0343-76B
$C_{17}H_{16}N_2O_2S_2$			
Acetamide, N-[2-[(6-methoxy-2-methyl-2-benzothiazolyl)thio]phenyl]-	MeOH	242s(4.30),297s(3.78), 308s(3.70)	88-0851-76
$C_{17}H_{16}N_2O_3$			
9(10H)-Acridinone, 2-(1-methylpropyl)-6-nitro-	MeOH	239(4.58),270(4.65), 390(3.71),430(3.77)	39-0465-76C
Indolo[2,3-a]quinolizin-4(1H)-one, 3-acetyl-6,7,12,12b-tetrahydro-2-hydroxy-	MeOH	225(4.46),260(4.05), 285(4.20),292(4.14)	24-3825-76

Compound	Solvent	$\lambda_{max}(\log \epsilon)$	Ref.
3-Pyridinecarboxylic acid, 1,6-dihydro-1-[2-(1H-indol-3-yl)ethyl]-6-oxo-, methyl ester	MeOH	221(4.53),260(4.22), 290(3.95)	24-0705-76
$C_{17}H_{16}N_2O_4$			
Acetamide, N-[2-methoxy-5-[2-(4-nitrophenyl)ethenyl]phenyl]-	MeOH	249(4.28),374(4.45)	56-0967-76
Benzoic acid, 2-[2-[4-(dimethylamino)phenyl]ethenyl]-5-nitro-, (E)-	EtOH	295(4.09),444(4.33)	80-1127-76
1,4-Benzoxazepine, 2,3-dihydro-8-methoxy-3-methyl-5-(2-nitrophenyl)-	n.s.g.	215(4.36),240(4.09), 315(4.31)	20-0787-76
1H-Isoindole-1,3(2H)-dione, 2-[2-(4-aminophenyl)-2-hydroxy-1-(hydroxymethyl)ethyl]-, [R-(R*,R*)]-	MeOH	221(4.53),236(4.33), 242(4.34),292(3.54)	130-0025-76
$C_{17}H_{16}N_2O_5$			
Benzonitrile, 2-[(2,6-dimethoxy-4-oxo-2,5-cyclohexadien-1-ylidene)amino]-3,5-dimethoxy-	MeCN	218(4.54),295(4.36), 500(3.44)	22-1993-76
Benzonitrile, 4-[(2,6-dimethoxy-4-oxo-2,5-cyclohexadien-1-ylidene)amino]-3,5-dimethoxy-	MeCN	220(4.53),298(4.43), 480(3.20)	22-1993-76
$C_{17}H_{16}N_4O$			
1,2,3-Benzotriazin-4(3H)-one, 3-phenyl-6-pyrrolidino-	MeOH	234(4.42),374(4.43)	5-0946-76
Cyclohexanone, 2-(1-phenyl-1H-pyrazolo[3,4-d]pyrimidin-4-yl)-	EtOH	244(3.90),255(3.89), 265(3.90),362(4.60), 375(4.61)	95-1352-76
$C_{17}H_{16}N_4OS$			
1,2,4-Triazin-3(2H)-one, 4,5-dihydro-2-(phenylmethyl)-6-[(phenylmethyl)amino]-5-thioxo-	EtOH	249(3.96),309(3.93), 384(3.75)	94-2274-76
$C_{17}H_{16}N_4O_2$			
Benzenebutanamide, α-azido-4-methyl-γ-oxo-N-phenyl-	EtOH	252(4.44)	115-0001-76
Benzenebutanamide, α-azido-N-(4-methylphenyl)-γ-phenyl-	EtOH	245(4.46)	115-0001-76
Benzo[g]pteridine-2,4(3H,10H)-dione, 6,7,8,9-tetrahydro-10-methyl-3-phenyl-	EtOH	266(4.24),318(3.30), 416(4.10)	88-2899-76
1,2,3-Benzotriazin-4(3H)-one, 6-morpholino-3-phenyl-	MeOH	225(4.34),354(4.28)	5-0946-76
3-Butenenitrile, 4-(5-hydroxy-3-methyl-1-phenyl-1H-pyrazol-4-yl)-2-[2-(methoxyimino)ethylidene]-, (E,Z,E)-	MeOH CHCl₃	435(4.52) 378(4.50)	5-1799-76 5-1799-76
Glycine, N-(dicyanomethylene)-1,3-dihydro-2-(1,3,3-trimethyl-2H-indol-2-ylidene)-, methyl ester	EtOH	449(4.64)	78-3031-76
3,5-Hexadienenitrile, 6-(5-hydroxy-3-methyl-1-phenyl-1H-pyrazol-4-yl)-2-methoxyimino-, (Z,E,E)-	MeOH CHCl₃	436(4.59) 375(4.49)	5-1799-76 5-1799-76
2,4-Pentadienenitrile, 5-(5-hydroxy-3-methyl-1-phenyl-1H-pyrazol-4-yl)-2-[(methoxyimino)methyl]-, (E,Z,E)-	MeOH CHCl₃	441(4.70) 388(4.49)	5-1799-76 5-1799-76
5-Pyrimidinecarboxamide, 1,2,3,4-tetrahydro-6-methyl-2-oxo-4-phenyl-N-2-pyridinyl-	EtOH	203(4.20),242(3.10), 294(3.24)	103-0191-76

Compound	Solvent	λ_{max}(log ϵ)	Ref.
$C_{17}H_{16}N_4O_6$			
Uridine, 2',3'-(carbonyldiimino)- 2',3'-dideoxy-, 5'-benzoate	MeOH	226(4.17),259(4.04)	44-3138-76
$C_{17}H_{16}O$			
4H-Benz[de]anthracen-7-ol, 5,6,6a,7- tetrahydro-	EtOH	262(4.84),304(4.04), 357(3.12),377(3.16)	104-0180-76
7H-Cyclohept[b]indeno[1,2-e]pyran, 8,9,10,11-tetrahydro-	dioxan	228(4.32),257(3.91), 287(4.33),292(4.35), 302(4.27),342(4.12), 499(3.40)	78-2219-76
5H-Dibenzo[a,c]cyclobuta[e]cyclohepten- 5-ol, 6,7-dihydro-	EtOH	250(4.13)	80-1497-76
Dibenzo[a,c]cyclopropa[e]cycloheptene- 4b(5H)-methanol, 5a,6-dihydro-	EtOH	252(4.14)	80-1497-76
1,7,9-Heptadecatriene-11,13,15-triyn- 6-one, (E,E)-	EtOH	272(4.47),284(4.63), 306(4.15),325(4.32), 341(4.51),365(4.45)	39-0735-76C
[2.2]Metacyclophane-4-carboxaldehyde	EtOH	330s(2.30)	49-0195-76
Spiro[5H-6,7-dihydrodibenzo[a,c]cyclo- hepten-6-ol-5,1'-cyclopropane]	EtOH	249(4.16)	80-1497-76
$C_{17}H_{16}O_2$			
Benzeneacetic acid, α-(phenylmethylene)-, ethyl ester, (E)-	MeOH	219(4.21),284(4.17)	44-2536-76
(Z)-	MeOH	222(4.16),287(4.32)	44-2536-76
3-Phenanthrenol, 2-methoxy-7,8-dimethyl-	MeOH	248s(4.44),257(4.59), 280(4.26)	102-0844-76
	MeOH-NaOH	248s(4.41),256(4.54), 266s(4.39),295(4.19)	102-0844-76
1,3-Propanedione, 2,2-dimethyl-1,3-di- phenyl-	MeOH	248(4.26),280(3.34)	23-1197-76
$C_{17}H_{16}O_2S$			
3-Cyclohexene-1-carboxylic acid, 6-(5- phenyl-2-thienyl)-, trans	MeOH	292(4.23),345(2.34)	44-1320-76
$C_{17}H_{16}O_3$			
Benzeneacetic acid, 2-(4-methylbenzoyl)-, methyl ester	n.s.g.	260(4.16)	103-0910-76
Benzo[b]cyclohepta[d]pyran-11a(6H)-carb- oxylic acid, 7-methyl-, methyl ester	EtOH	221(4.11),273(3.67), 281s(3.60)	33-1763-76
Benzo[b]cyclohepta[d]pyran-11a(11H)- carboxylic acid, 7-methyl-, methyl ester	EtOH	223(4.02),251(3.96), 279(3.92),306(3.88)	33-1763-76
2H-1-Benzopyran, 2,3-dimethoxy-2-phenyl-	EtOH	218(3.60),264(3.90), 292(4.34)	22-1975-76
1,4-Butanedione, 1-(4-methoxyphenyl)- 4-phenyl-	EtOH	222s(4.17),248s(4.20), 272(4.32)	12-0339-76
1,3-Pentanedione, 1-(2-hydroxyphenyl)- 5-phenyl-	MeOH	211(3.92),250(3.62), 313(4.48),346s(3.28)	12-2023-76
	MeOH-base	251(3.85),345(4.26)	12-2023-76
1-Phenanthrenecarboxylic acid, 9,10-di- hydro-2-methoxy-, methyl ester	EtOH	274(4.33)	78-0073-76
1-Propanone, 1-(2-acetoxyphenyl)-2-phen- yl-	EtOH	256(3.98),329(3.58)	114-0381-76C
$C_{17}H_{16}O_4$			
Benzoic acid, 2-(5-ethyl-2-hydroxybenz- oyl)-, methyl ester	EtOH	200(4.42),216(4.38), 256(4.13)	56-0759-76

Compound	Solvent	λ_{max}(log ϵ)	Ref.
2H-1-Benzopyran-8-carboxaldehyde, 3,4-dihydro-7-hydroxy-5-methoxy-2-phenyl-, (S)-	EtOH	296(4.29),330s(3.58)	39-1570-76C
4H-1-Benzopyran-4-one, 2,3-dihydro-7-methoxy-2-(4-methoxyphenyl)-	EtOH	228(4.45),279(4.38),320(4.05)	102-0773-76
	EtOH-NaOH	243(4.33),279(4.19),356(3.97)	102-0773-76
9H-Fluorene-1-carboxylic acid, 2,7-dimethoxy-, methyl ester	EtOH	247(4.00),276(4.39),330(3.67)	78-0065-76
3-Furancarboxylic acid, 4,5-dihydro-2-methyl-4-oxo-5-(3-phenyl-2-propenylidene)-, ethyl ester	EtOH	210(4.17),373(4.55)	4-0521-76
2(3H)-Furanone, 5-(2,4-dihydroxy-6-methylphenyl)dihydro-5-phenyl-	EtOH-pH 9.5	495(3.33)	18-2027-76
2H-Naphtho[1,2-b]pyran-5-carboxylic acid, 6-hydroxy-2,2-dimethyl-(mollugin)	n.s.g.	283(4.46),392(3.86)	5-1295-76
C$_{17}$H$_{16}$O$_4$S			
Butanedioic acid, [(5-methyl-2-thienyl)-methylene]phenyl-, 1-methyl ester (E)-	EtOH	210(4.24),270(3.91),315(4.42)	4-0285-76
C$_{17}$H$_{16}$O$_5$			
2,3-Benzobicyclo[2.2.2]octatetraene-5,6-dicarboxylic acid, 10-methoxy-, dimethyl ester	C$_6$H$_{12}$	229s(3.77),280(3.34),285(3.30),304s(2.81)	33-1469-76
5,10-Benzocyclooctenedicarboxylic acid, 2-methoxy-, dimethyl ester	C$_6$H$_{12}$	224s(4.16),271s(3.59)	33-1469-76
Benzo[a]cyclopropa[cd]pentalene-2a,6c-dicarboxylic acid, 2b,6b-dihydro-4-methoxy-, dimethyl ester	C$_6$H$_{12}$	245s(3.88),287(3.36),293(3.34)	33-1469-76
Benzo[a]cyclopropa[cd]pentalene-2a,6c-dicarboxylic acid, 2b,6b-dihydro-5-methoxy-, dimethyl ester	C$_6$H$_{12}$	223s(4.13),252s(3.76),282(3.45),292(3.30)	33-1469-76
1,4-Dibenzofurandione, 2,3-diethoxy-7-methyl-	n.s.g.	253(4.38),329(3.97),376(3.45)	12-0179-76
1,4-Dibenzofurandione, 2,3-diethoxy-8-methyl-	n.s.g.	253(4.47),329(4.07),376s(3.36)	12-0179-76
2H-Naphtho[2,3-b]pyran-5,10-dione, 4-acetoxy-3,4-dihydro-2,2-dimethyl-	EtOH	223s(4.36),245(4.57),251(4.59),283(4.44)	102-0570-76
C$_{17}$H$_{16}$O$_6$			
Benzenepropanoic acid, β-(2-hydroxy-6-methoxy-3,4-dioxo-1,5-cyclohexadien-1-yl)-, methyl ester	n.s.g.	399(2.98)	39-0407-76C
1,4-Dibenzofurandione, 2,3-diethoxy-6-methoxy-	n.s.g.	246(3.80),268(3.49),274s(3.47),330(3.28),395(2.67)	12-0179-76
1,4-Dibenzofurandione, 2,3-diethoxy-7-methoxy-	n.s.g.	254(4.48),310(3.87),500(3.15)	12-0179-76
1,4-Dibenzofurandione, 2,3-diethoxy-8-methoxy-	n.s.g.	256(4.20),308(4.04),433(3.34)	12-0179-76
Ethanone, 2-(3,4-dimethoxyphenyl)-1-(6-hydroxy-1,3-benzodioxol-5-yl)-	EtOH	236(4.19),281(3.90),349(3.89)	102-1019-76
Eucomin, 3,9-dihydro-, (+)-	MeOH	289.5(4.315)	33-2048-76
	EtOH	212(4.40),222(4.33),291(4.23),326(3.87)	33-2048-76
C$_{17}$H$_{16}$O$_7$			
5H-[1,3]Benzodioxolo[5,6-c]-1,3-dioxolo-	EtOH	242(4.53),258s(--),	24-0855-76

Compound	Solvent	$\lambda_{max}(\log \epsilon)$	Ref.
[4,5-h][1]benzopyran-5-one, 3a,11,12-12a-tetrahydro-12-hydroxy-2,2-dimeth-yl-, (3aα,12β,12aα)-(+)- (cont.)		287(3.78),297(3.78), 330(3.66),350s(--)	24-0855-76
6H-[1,3]Benzodioxolo[5,6-c]-1,3-dioxolo-[4,5-g][1]benzopyran-6-one, 3a,4,12-12a-tetrahydro-4-hydroxy-2,2-dimeth-yl-, (3aα,4β,12aα)-	EtOH	235s(--),243(4.39), 258(4.20),275s(--), 287(3.70),298(3.67), 335(3.61)	24-0855-76
Ethanone, 2-(1,3-benzodioxol-5-yl)-1-(2,4-dihydroxy-3,6-dimethoxyphenyl)-	MeOH	292(4.4),320s(--)	2-0584-76
$C_{17}H_{16}S_3$			
1(2H)-Naphthalenethione, 2-(hexahydro-1,3-benzodithiol-2-ylidene)-	CH_2Cl_2	323(4.39),353(4.03), 392s(3.41),540(4.07)	88-3815-76
$C_{17}H_{17}$			
Tricyclo[8.3.2.2^{4,7}]heptadeca-4,6,10,12-14,16-hexaen-1-ylium, (deloc-1,10,11-12,13,14,15)-, tetrafluoroborate	MeCN	238s(3.90),278(3.33), 296(3.51),323(3.35), 353(3.20),400s(2.72)	35-8446-76
	CH_2Cl_2	226(4.2),235s(4.1), 295(3.8),325(3.7), 355(3.6)	88-3899-76
$C_{17}H_{17}Br$			
Tricyclo[8.3.2.2^{4,7}]heptadeca-1(14),4,6-10(15),11,16-hexaene, 13-bromo-	EtOH	220s(4.26),238s(3.80), 283(3.43)	35-8446-76
$C_{17}H_{17}BrN_2O$			
Pyridinium, 3-acetyl-1-[2-(1H-indol-3-yl)ethyl]-, bromide	MeOH	222(4.52),302(4.57)	35-3645-76
$C_{17}H_{17}ClN$			
Methanaminium, N-[3-[1,1'-biphenyl]-4-yl-3-chloro-2-propenylidene]-N-methyl-, perchlorate	MeOH	280(4.27),326(3.97)	48-0731-76
Methanaminium, N-[3-chloro-5-(2-naphtha-lenyl)-2,4-pentadienylidene]-N-meth-yl-, perchlorate	MeOH	247s(4.47),253s(4.46), 277(4.33),296s(4.27), 305(4.30),319(4.21), 383(4.31)	48-0731-76
$C_{17}H_{17}ClN_2$			
1H-Indole-2-methanamine, 3-[(2-chloro-phenyl)methyl]-1-methyl-, hydrochlor-ide	EtOH	250(4.398)	103-0669-76
$C_{17}H_{17}ClN_2O$			
1,5-Benzodiazocin-3-ol, 8-chloro-1,2,3-4-tetrahydro-1-methyl-6-phenyl-	EtOH	224(4.35),250(4.28), 261s(--),366(3.51)	111-0501-76
1,5-Benzodiazocin-3-ol, 6-(2-chlorophen-yl)-1,2,3,4-tetrahydro-1-methyl-	EtOH	225s(--),237s(--), 264s(--),366(3.51)	111-0501-76
4H-3,1-Benzoxazin-1(4H)-amine, 6-chloro-N-ethylidene-2-methyl-4-phenyl-	isoProH	286(4.21)	4-0033-76
isomer m. 112-115°	isoProH	259(3.80),280s(3.74)	4-0033-76
$C_{17}H_{17}ClN_2O_3S$			
3H-Phenothiazine-2,3(10H)-dione, 8-chlo-ro-10-[3-(dimethylamino)-2-hydroxy-propyl]-, hydrochloride	EtOH	228(4.37),260(4.68), 500(4.20)	4-1067-76
3H-Phenothiazine-2,3(10H)-dione, 8-chlo-ro-10-[3-(dimethylamino)propyl]-1-hy-droxy-, hydrochloride	EtOH	220(3.87),265(4.39), 542(3.98)	4-1067-76

Compound	Solvent	$\lambda_{max}(\log \epsilon)$	Ref.
$C_{17}H_{17}ClN_2O_5$			
Benzamide, N-[2-(4-chloro-3-methoxyphen-oxy)-1-methylethyl]-4-nitro-	n.s.g.	210(4.38),285(3.54)	20-0787-76
$C_{17}H_{17}ClN_5O_6PS$			
Adenosine, 8-[(4-chlorophenyl)thio]-2-methyl-, cyclic 3',5'-(hydrogen phosphate)	pH 1 pH 7 pH 11	278(4.29) 283(4.25) 283(4.25)	87-0419-76 87-0419-76 87-0419-76
$C_{17}H_{17}ClO_3$			
1H-Xanthene-7-carboxylic acid, 2-chloro-9,9a-dihydro-9a-methyl-, ethyl ester	EtOH	232(3.91),315(3.94), 327(3.90)	94-1581-76
$C_{17}H_{17}Cl_2NO_3$			
Benzamide, 2-chloro-N-[2-(4-chloro-3-methoxyphenoxy)-1-methylethyl]-	n.s.g.	210(4.33),280(3.44)	20-0787-76
$C_{17}H_{17}Cl_3N_2$			
1,5-Benzodiazocine, 3,8-dichloro-1,2,3,4-tetrahydro-1-methyl-6-phenyl-, hydrochloride	EtOH-base	229(4.32),254(4.30), 364(3.40)	111-0501-76
$C_{17}H_{17}FN_2O$			
Pyrrolidine, 1-[5-fluoro-2-(phenylami-no)benzoyl]-	MeOH	210(4.35),295(4.19), 340s(3.36)	5-0946-76
$C_{17}H_{17}FN_2O_2$			
Morpholine, 4-[5-fluoro-2-(phenylamino)-benzoyl]-	MeOH	240(3.98),284(4.12)	5-0946-76
$C_{17}H_{17}F_3N_4O_6$			
Riboflavin, $\alpha^7,\alpha^7,\alpha^7$-trifluoro-	EtOH	225(4.36),270(4.48), 339(3.82),430(4.01)	103-1386-76
Riboflavin, $\alpha^8,\alpha^8,\alpha^8$-trifluoro-	EtOH	218(4.32),270(4.48), 325(3.78),450(3.89)	103-1386-76
$C_{17}H_{17}NO$			
9(10H)-Acridinone, 10-butyl-	EtOH	255(4.78),385(3.99), 405(4.05)	48-0515-76
1H-Indole, 2-(4-methoxy-3-methylphenyl)-1-methyl-	MeOH	295(4.29)	12-2747-76
$C_{17}H_{17}NO_2$			
Benzamide, 2-[1-(hydroxymethyl)-2-phen-ylethenyl]-N-methyl-	EtOH	273s(3.16),285s(3.02)	4-0597-76
Benzenamine, N-[3-(2-methoxymethoxy)-phenyl]-2-propen-1-ylidene]-	MeOH dioxan DMSO	280(4.22),335(4.16) 284(4.15),340(4.08) 282(4.26),340(4.23)	104-0840-76 104-0840-76 104-0840-76
4H-1-Benzopyran-4-one, 2,3-dihydro-2-(2-phenylethyl)-, oxime, (-)-	MeOH	210(4.47),252(4.00), 302(3.69),312s(3.61)	12-2023-76
$C_{17}H_{17}NO_2S_2$			
Benzeneacetic acid, α-[[[4-(dimethylami-no)phenyl]thioxomethyl]thio]-, ($\pm$)-	dioxan	238(4.06),250s(3.98), 272s(3.82),342(4.11), 425(4.23),480s(3.31)	24-0740-76
$C_{17}H_{17}NO_3$			
1,3,6-Cycloheptatriene-1-carboxylic acid, 6-(dimethylamino)-5-oxo-3-phenyl-, methyl ester	MeCN	277(3.8),360(3.3)	39-2329-76C

Compound	Solvent	$\lambda_{max}(\log \epsilon)$	Ref.
Ethanone, 1-[2-(1,3-dihydro-1-hydroxy-3-methoxy-2H-isoindol-2-yl)phenyl]-	MeOH	228(4.50),255(3.98), 356(3.75)	83-0356-76
1H-Indole, 4,7-dimethoxy-2-(2-methoxyphenyl)-	MeOH	211(4.60),256(4.30), 314(4.18)	111-0241-76
1H-Indole, 4,7-dimethoxy-2-(3-methoxyphenyl)-	MeOH	213(4.58),256(4.44), 306(4.33)	111-0241-76
1H-Indole, 4,7-dimethoxy-2-(4-methoxyphenyl)-	MeOH	212(4.42),258(4.43), 301(4.40)	111-0241-76
1-Isoquinolinecarboxylic acid, 1,2,3,4-tetrahydro-6-hydroxy-1-(phenylmethyl)-	EtOH	277(3.31)	95-1031-76
2-Propen-1-one, 1-(2-hydroxy-5-methylphenyl)-3-(4-methoxyphenyl)-, oxime	isoBuOH	222(4.37),266(4.15), 316(3.73)	42-1067-76

$C_{17}H_{17}NO_4$

Compound	Solvent	$\lambda_{max}(\log \epsilon)$	Ref.
Acetamide, N-[2-(2,4-dimethoxybenzoyl)-phenyl]-	EtOH	209(4.32),231s(4.33), 236(4.35),263(3.95), 270(3.95),323(3.82)	39-2089-76C
2-Azabicyclo[3.2.0]heptane-5-carboxylic acid, 7-ethenyl-3,4-dioxo-1-phenyl-, ethyl ester, (1α,5α,7α)-	dioxan	258(3.58)	142-1229-76
Benzamide, N-[2-(4-methoxy-3,6-dioxo-1,4-cyclohexadien-1-yl)-1-methylethyl]-	CHCl$_3$	263(4.18)	44-3627-76
1,4-Benzoxazepin-7-ol, 2,3-dihydro-5-(4-hydroxyphenyl)-8-methoxy-3-methyl-	n.s.g.	262(4.02),320(4.25), 387(4.07)	20-0898-76
Butanedioic acid, [(1-methyl-1H-pyrrol-2-yl)methylene]phenyl-, 1-methyl ester	EtOH	207(4.24),326(3.84)	48-0816-76
geometric isomer	EtOH	212(4.44),227(4.28), 327(4.59)	48-0816-76
1-Isoquinolinecarboxylic acid, 1,2,3,4-tetrahydro-6-hydroxy-1-[(4-hydroxyphenyl)methyl]-	EtOH	277(3.48)	95-1031-76
2-Propenoic acid, 3-(1,5-dihydrospiro-[benz[cd]indole-4(3H),2'-[1,3]dioxolan]-5-yl)-2-methyl-, (E)-	MeOH	222(4.64),278(3.84), 285(3.87),293(3.84)	24-2126-76

$C_{17}H_{17}NO_5S$

Compound	Solvent	$\lambda_{max}(\log \epsilon)$	Ref.
1,3-Dioxolane-4-carboxaldehyde, 2,2-dimethyl-4-[[[(5-oxo-2-phenyl-4(5H)-oxazolylidene)methyl]thio]methyl]-	MeOH	361(4.63)	23-2089-76

$C_{17}H_{17}NO_6$

Compound	Solvent	$\lambda_{max}(\log \epsilon)$	Ref.
Bis[1,3]dioxolo[4,5-c:4',5'-j]phenan-thridin-5(3aH)-one, 4,11,12,12a-tetrahydro-12-hydroxy-2,2-dimethyl-, (3aα,12β,12aα)-(+)-	EtOH	231(4.33),247s(--), 251(4.41),263(4.35), 275s(--),288(3.84), 298(3.91),332(3.70), 350s(--)	24-0855-76

$C_{17}H_{17}NO_7$

Compound	Solvent	$\lambda_{max}(\log \epsilon)$	Ref.
Bis[1,3]dioxolo[4,5-c:4',5'-j]phenan-thridin-5(3aH)-one, 4,11,12,12a-tetrahydro-6,12-dihydroxy-2,2-dimethyl-, (3aα,12β,12aα)-(+)-	EtOH	251(4.69),275s(--), 289(3.67),299(3.79), 333(3.82),348(3.82)	24-0855-76

$C_{17}H_{17}N_2$

Compound	Solvent	$\lambda_{max}(\log \epsilon)$	Ref.
Pyridinium, 1-[(1,6-dimethyl-2(1H)-quin-olinylidene)methyl]-, perchlorate	EtOH	234(4.04),255(4.13), 471(3.54)	104-0444-76

Compound	Solvent	$\lambda_{max}(\log \epsilon)$	Ref.
$C_{17}H_{17}N_2O$			
Pyridinium, 3-acetyl-1-[2-(1H-indol-3-yl)ethyl]-, bromide	MeOH	222(4.52),302(4.57)	35-3645-76
Pyridinium, 1-[(6-methoxy-1-methyl-2(1H)-quinolinylidene)methyl]-, perchlorate	EtOH	235(4.10),255(3.00), 470(3.37)	104-0444-76
$C_{17}H_{17}N_3$			
Benzenamine, 2-(3,5-dimethyl-1-phenyl-1H-pyrazol-4-yl)-	MeOH	241(4.28),295(3.64)	23-1020-76
1H-Pyridazino[4,5-b]indole, 2,3,4,5-tetrahydro-5-methyl-1-phenyl-, hydrochloride	EtOH	285(4.20)	103-1008-76
2,4-Quinolinediamine, N^2,N^2-dimethyl-N^4-phenyl-	EtOH	243(4.51),320(4.08)	1-0133-76
$C_{17}H_{17}N_3O$			
Acetamide, N-[4-(4,5-dihydro-1-phenyl-1H-pyrazol-3-yl)phenyl]-	benzene	371(4.40)	64-1248-76B
10(9H)-Acridineacetamide, N-methyl-9-(methylimino)-	EtOH	220(4.45),245s(4.35), 257(4.40),268(4.39), 285s(3.99),377(3.86), 410(3.62),430(3.60)	44-3406-76
Phenazine, 2-methyl-3-morpholino-	EtOH	235(4.47),256(4.66), 382(4.08),410(3.94)	18-2333-76
$C_{17}H_{17}N_3OS_3$			
4-Thiazolidinone, 5-(4-amino-3-ethyl-5-phenyl-2(3H)-thiazolylidene)-3-(2-propenyl)-2-thioxo-	n.s.g.	452(4.58)	103-0648-76
$C_{17}H_{17}N_3O_2$			
9-Acridinamine, N-butyl-1-nitro-	pH 1	269(4.43),440(3.69)	56-1603-76
	pH 6.8	273(4.45),440(3.78)	56-1603-76
9-Acridinamine, 2-(1-methylpropyl)-6-nitro-	MeOH	210(4.40),252(4.67), 285(4.63),325(4.59), 400(3.96)	39-0465-76C
Imidazo[1,2-b]quinazoline-2,5-dione, 1,2,3,4,6,7,8,9-octahydro-1-(phenylmethyl)-	EtOH	236(4.00),282(3.92)	114-0413-76A
$C_{17}H_{17}N_3O_2S$			
Propanoic acid, 3-(phenylamino)-2-(phenylhydrazono)-3-thioxo-, ethyl ester, (Z)-	EtOH	290(3.87),355(4.07)	104-1912-76
$C_{17}H_{17}N_3O_3$			
Anthranilic acid pyrrolidide, 5-nitro-N-phenyl-	MeOH	210(4.34),256(4.09), 377(4.26)	5-0946-76
Benzamide, 4-[2-[4-(dimethylamino)ethenyl]-3-nitro-	EtOH	240(4.33),400(4.37)	80-1127-76
2H-1,4-Benzodiazepin-2-one, 7-amino-1,3-dihydro-6-hydroxy-1-(methoxymethyl)-5-phenyl-	isoPrOH	248(4.40),345(3.36)	87-1378-76
2H-1,4-Benzodiazepin-2-one, 1,3-dihydro-7-(hydroxyamino)-1-(methoxymethyl)-5-phenyl-	isoPrOH	245(4.46),338(3.32)	87-1378-76
1,4-Methano-6H-benzocyclohepten-5,9-imin-6-one, 1,2,3,4,4a,5,9,9a-octahydro-10-(5-nitro-2-pyridinyl)-	EtOH	209(4.34),227(4.50), 362(4.54)	39-2307-76C

Compound	Solvent	$\lambda_{max}(\log \epsilon)$	Ref.
2,5-Pyrrolidinedione, 1-[2-[3-(1-methyl-1H-imidazol-5-yl)-3-oxopropyl]phenyl]-(dehydroisolongistrobine)	EtOH acid	258(4.23) 235(4.16)	78-1085-76 78-1085-76
$C_{17}H_{17}N_3O_4$ Benzoic acid, 4-[4-(N-acetoxy-N-methyl-amino)phenyl]azo]-, methyl ester	EtOH	369(4.46)	94-1485-76
$C_{17}H_{17}N_3O_4S$ 1H-Benzimidazole-2-butanoic acid, α-[(phenylsulfonyl)amino]-, (S)- hydrochloride	EtOH H_2O	236(3.72),275(3.77), 281(3.72) 231(3.78),270(3.82), 276(3.80)	42-0274-76 42-0274-76
sodium salt	H_2O	236(3.76),273(3.77), 280(3.74)	42-0274-76
$C_{17}H_{17}N_3O_5$ Benzeneacetic acid, 2,3,3a,9a-tetrahy-dro-2-(hydroxymethyl)-6-imino-6H-furo-[2',3':4,5]oxazolo[3,2-a]pyrimidin-3-yl ester, hydrochloride, [2R-(2α,3β,3aβ,9aβ)]-	MeOH-acid	232(4.04),264(4.05)	87-0654-76
$C_{17}H_{17}N_5O_3$ Pyrrolidine, 1-[5-nitro-2-(3-phenyl-1-triazenyl)benzoyl]-	MeOH	207(4.45),252s(4.15), 386(4.49)	5-0946-76
$C_{17}H_{17}N_5O_5$ Adenine, 8-(5'-O-benzoyl-α-D-arabino-furanosyl)-	pH 1	212(4.15),234(4.10), 268(4.16)	111-0067-76
$C_{17}H_{17}N_5O_8S$ 2,4(1H,3H)-Pyrimidinedione, 1-[3-azido-5-O-benzoyl-3-deoxy-2-O-(methylsul-fonyl)-β-D-arabinofuranosyl]-	MeOH	231(4.20),259(4.01)	44-3138-76
Uridine, 2'-azido-2'-deoxy-, 5'-benzo-ate 3'-methanesulfonate	MeOH	229(4.16),257(4.00)	44-3138-76
$C_{17}H_{17}O$ 1H-Benzo[c]furylium, 1,1-dimethyl-3-(3-methylphenyl)-	H_2SO_4	342(4.26)	104-2523-76
1H-Benzo[c]furylium, 1,1-dimethyl-3-(4-methylphenyl)-	H_2SO_4	362(4.45)	104-2523-76
$C_{17}H_{17}OS$ 1H-Benzo[c]thiolium, 3-(4-methoxyphen-yl)-1,1-dimethyl-	H_2SO_4	<u>350(4.1)</u>,439(4.44)	104-2523-76
$C_{17}H_{17}O_2$ 1H-Benzo[c]furylium, 3-(4-methoxyphen-yl)-1,1-dimethyl-	H_2SO_4	<u>290(3.8)</u>,390(4.54)	104-2523-76
$C_{17}H_{17}S$ 1H-Benzo[c]thiolium, 1,1-dimethyl-3-(3-methylphenyl)-	H_2SO_4	364(4.27)	104-2523-76
1H-Benzo[c]thiolium, 1,1-dimethyl-3-(4-methylphenyl)-	H_2SO_4	396(4.28)	104-2523-76
$C_{17}H_{18}$ Benzene, 1,3,5-trimethyl-2-(1-phenyleth-enyl)-	EtOH	247(3.81)	44-3083-76

Compound	Solvent	$\lambda_{max}(\log \epsilon)$	Ref.
9H-Fluorene, 1,2,3,4-tetramethyl-	EtOH	211(4.63),263s(4.27), 271(4.34),280s(4.19), 303(3.64)	117-0163-76
[2.2]Metacyclophane, 4-methyl-	EtOH	272(2.74),276s(2.73)	49-0195-76
Tricyclo[8.3.2.2^{4,7}]heptadeca-1(14),4-6,10(15),12,16-hexaene	EtOH	220s(4.12),238s(3.69), 280(3.18)	35-8446-76
$C_{17}H_{18}BrNO_4$			
4(5H)-Oxazolone, 5-(2-acetoxy-1,1-di-methylethyl)-2-[2-(4-bromophenyl)-ethenyl]-, perchlorate	n.s.g.	333(4.73)	104-1146-76
$C_{17}H_{18}Br_2$			
Tetracyclo[8.3.2.2^{4,7}.0^{10,12}]heptadeca-4,6,10(15),16-tetraene	EtOH	226s(4.01),238s(3.85), 282(2.76)	35-8446-76
$C_{17}H_{18}ClNO_3$			
1,4-Benzoxazepine, 5-(4-chlorophenyl)-2,3,4,5-tetrahydro-7,8-dimethoxy-	n.s.g.	220(4.29),285(3.52)	20-0898-76
$C_{17}H_{18}ClNO_4$			
Benzamide, 4-chloro-N-[2-(3,4-dimethoxy-phenoxy)ethyl]-	n.s.g.	210(4.13),282(3.54)	20-0787-76
$C_{17}H_{18}ClN_3$			
1H-Pyridazino[4,5-b]indole, 2,3,4,5-tetrahydro-5-methyl-1-phenyl-, hydrochloride	EtOH	285(4.20)	1-0133-76
$C_{17}H_{18}Cl_2N_2O_3S$			
3H-Phenothiazine-2,3(10H)-dione, 8-chlo-ro-10-[3-(dimethylamino)-2-hydroxy-propyl]-, hydrochloride	EtOH	228(4.37),260(4.68), 500(4.20)	4-1067-76
3H-Phenothiazine-2,3(10H)-dione, 8-chlo-ro-10-[3-(dimethylamino)propyl]-1-hydroxy-, hydrochloride	EtOH	220(3.87),265(4.39), 542(3.98)	4-1067-76
$C_{17}H_{18}N_2$			
3-Acridinamine, 7-(1-methylpropyl)-, hydrochloride	MeOH	236(3.44),276(3.51), 352(3.96),367(4.02), 456(3.99)	39-0465-76C
9-Acridinamine, 2-(1,1-dimethylethyl)-, hydrochloride	MeOH	226(3.75),265(4.90), 382(3.61),402(3.80), 424(3.70)	39-0465-76C
9-Acridinamine, N-(1-methylpropyl)-, hydrochloride	MeOH	222(4.35),258(4.67), 268(4.78),314(3.30), 327(3.08),394(3.82), 413(3.04),434(3.04)	39-0465-76C
9-Acridinamine, 2-(1-methylpropyl)-, hydrochloride	MeOH	226(3.76),365(4.90), 382(3.61),402(3.80), 424(3.71)	39-0465-76C
Benzenamine, N,N-dimethyl-4-(1-methyl-1H-indol-2-yl)-	MeOH	266(4.18),307(4.45)	12-2747-76
Quinolinium, 1,6-dimethyl-2-(pyridinio-methyl)-, diperchlorate	EtOH	264(4.13),325(3.56), 460(3.29)	104-0444-76
$C_{17}H_{18}N_2O$			
9(10H)-Acridinone, 10-[2-(dimethylami-no)ethyl]-	EtOH	255(4.75),285(3.97), 400(4.05)	48-0515-76
1,5-Benzodiazocin-3-ol, 1,2,3,4-tetra-	EtOH-base	223(4.38),236(4.36),	111-0501-76

Compound	Solvent	$\lambda_{max}(\log \epsilon)$	Ref.
hydro-1-methyl-6-phenyl-, Z-2-butene-dioate (cont.)		360(3.42)	111-0501-76
Phenol, 2-[3-[[4-(dimethylamino)phenyl]-imino]-1-propenyl]-	MeOH	320(4.05),420(4.29)	104-0840-76
	dioxan	325(4.43),418(4.25)	104-0840-76
	DMSO	323(4.08),430(4.33)	104-0840-76
	DMSO-NaOMe	336(4.09),500(4.30)	104-0840-76
2-Propenamide, N-[4-(dimethylamino)phen-yl]-3-phenyl-	EtOH	247(4.06),275(4.35), 300(4.42)	110-0703-76
Quinolinium, 6-methoxy-1-methyl-2-(pyri-diniomethyl)-, diperchlorate	EtOH	262(4.54),322(4.00), 458(3.44)	104-0444-76
$C_{17}H_{18}N_2O_2$			
Indolo[2,3-a]quinolizine-3-carboxylic acid, 1,4,6,7,12,12b-hexahydro-, methyl ester	94% EtOH	205(4.47),224(4.61), 284(3.96),292(3.92)	1-0251-76
2-Propenoic acid, 3-(3-cyano-3,4-di-hydro-5-phenyl-2H-pyrrol-2-yl)-2-methyl-, ethyl ester, cis	EtOH	247(4.17)	44-0543-76
trans	EtOH	247(4.17)	44-0543-76
1H-Pyrazole, 3-(2,6-dimethoxyphenyl)-4,5-dihydro-1-phenyl-	benzene	297(4.13)	61-0288-76
Spiro[3H-indole-3,1'(5'H)-indolizin]-3'(2'H)-one, N-formyl-4',6',7',7'a-tetrahydro-	EtOH	213(4.18),251(4.03), 260s(3.90),283s(3.51), 289(3.53)	130-0283-76
isomer II	EtOH	213(4.21),251(4.05), 260s(3.93),283(3.51), 289(3.53)	130-0283-76
$C_{17}H_{18}N_2O_3$			
Acetamide, N-(4-ethoxyphenyl)-2-[(4-methoxyphenyl)imino]-	EtOH	236(3.29)	65-0702-76
Acetamide, 2-[(4-ethoxyphenyl)imino]-N-(4-methoxyphenyl)-	EtOH	243(3.36)	65-0702-76
$C_{17}H_{18}N_2O_4$			
Acetamide, 2-[(4-ethoxyphenyl)imino]-N-(4-methoxyphenyl)-, N-oxide	EtOH	228(4.2),295(4.1), 325(4.1),345(4.1)	65-0702-76
Acetamide, N-(4-ethoxyphenyl)-2-[(4-methoxyphenyl)imino]-, N-oxide	EtOH	228(4.3),300(4.2), 325(4.2),340(4.18), 345(4.12)	65-0702-76
Benzamide, N,N'-1,3-propanediylbis[2-hy-droxy-	MeOH	302(3.95)	18-2679-76
Benzoic acid, 2-[[4-(1-methylpropyl)-phenyl]amino]-4-nitro-	MeOH	210(4.30),270(4.20)	39-0465-76C
2-Butenoic acid, 3-methyl-2-[[[2-(phen-ylmethyl)-4-oxazolyl]carbonyl]amino]-, methyl ester	MeOH	215(4.30)	88-0375-76
$C_{17}H_{18}N_2O_4S$			
Benzamide, N-methyl-N-[2-[[(4-methyl-phenyl)sulfonyl]amino]-2-oxoethyl]-	MeOH	226(4.32)	44-0813-76
Thiazolo[5,4-f]quinoline-8-carboxylic acid, 2-butoxy-6-ethyl-6,9-dihydro-9-oxo-	MeOH	260(4.45),269(4.45)	94-0147-76
Thiazolo[5,4-f]quinoline-8-carboxylic acid, 3-butyl-6-ethyl-2,3,6,9-tetra-hydro-2,9-dioxo-	MeOH	263.5(4.53)	94-0147-76
$C_{17}H_{18}N_2O_5$			
Benzamide, N-[2-(3-methoxyphenoxy)-1-methylethyl]-2-nitro-	n.s.g.	210(4.33),265(3.81)	20-0787-76

Compound	Solvent	$\lambda_{max}(\log \epsilon)$	Ref.
$C_{17}H_{18}N_3O$			
Isoxazolium, 2-ethyl-3,5-dimethyl-4-(1-naphthalenylazo)-, tetrafluoroborate	HOAc	270(4.12),277s(4.11), 386(3.84)	48-0359-76
Isoxazolium, 2-ethyl-3,5-dimethyl-4-(2-naphthalenylazo)-, tetrafluoroborate	HOAc	267(4.18),273s(4.17), 284s(4.11),329(4.14), 377s(3.69)	48-0359-76
$C_{17}H_{18}N_4$			
Benzonitrile, 2-[[4-(diethylamino)phenyl]azo]-	benzene	455(--)	39-0042-76C
	EtOH	462(4.48)	39-0042-76C
Benzonitrile, 3-[[4-(diethylamino)phenyl]azo]-	benzene	430(--)	39-0042-76C
	EtOH	446(4.45)	39-0042-76C
Benzonitrile, 4-[[4-(diethylamino)phenyl]azo]-	benzene	452(--)	39-0042-76C
	EtOH	466(4.51)	39-0042-76C
Propanenitrile, 3-[ethyl[4-(phenylazo)phenyl]amino]-	hexane	238(3.97),377(4.46), 390(4.46),455s(3.20)	61-0301-76
$C_{17}H_{18}N_4O_3S$			
1H-Benzimidazole-2-butanamide, α-[(phenylsulfonyl)amino]-, (S)-	EtOH	227(3.77),272(3.83), 276(3.83)	42-0274-76
hydrochloride	EtOH	238(3.90),275(3.89), 282(3.90)	42-0274-76
$C_{17}H_{18}N_4O_5$			
1H-Indene-4-carboxaldehyde, 2,6,7,7a-tetrahydro-1-hydroxy-7a-methyl-, 2,4-dinitrophenylhydrazone	$CHCl_3$	385(4.16)	118-0307-76
$C_{17}H_{18}N_4O_6$			
1H-Pyrazole-4-carboxamide, 1-(4-nitrophenyl)-3-(tetrahydro-2,2-dimethylfuro[3,4-d]-1,3-dioxol-4-yl)-, [3aS-(3aα,4α,6aα)]-	$CHCl_3$	311(4.22)	136-0019-76A
1H-Pyrazole-5-carboxamide, 1-(4-nitrophenyl)-3-(tetrahydro-2,2-dimethylfuro[3,4-d]-1,3-dioxol-4-yl)-, [3aS-(3aα,4α,6aα)-	$CHCl_3$	300(4.10)	136-0019-76A
$C_{17}H_{18}N_4O_9$			
β-D-Ribofuranose, 5-deoxy-5-(3,4-dihydro-2,4-dioxo-8(2H)-pteridinyl)-, 1,2,3-triacetate	pH -4.9	242(4.07),270(3.22), 349(4.02)	24-3175-76
	pH 6.0	259(4.18),275s(3.90), 402(3.94)	24-3175-76
	pH 11.0	230(4.14),278(4.02), 308(3.90),402(2.62)	24-3175-76
$C_{17}H_{18}N_8O_3$			
Phenol, 2-[5-amino-2-[(2,4-diamino-6-methyl-5-pyrimidinyl)oxy]-2,3-dihydro-7-methyloxazolo[4,5-d]pyrimidin-2-yl]-	EtOH	302(4.85),340(4.38)	94-2461-76
$C_{17}H_{18}O$			
Benzenemethanol, α-methyl-α-(2-phenylcyclopropyl)-	EtOH	249-272f(2.64-2.95)	44-3067-76
1,7,9-Heptadecatriene-11,13,15-triyn-6-ol, (E,E)-	EtOH	247(4.39),259(4.72), 269(5.00),288(4.19), 306(4.43),325(4.61), 348(4.54)	39-0735-76C
1-Pentanone, 1,3-diphenyl-	EtOH	243(4.13)	44-3067-76
3-Penten-1-ol, 1,4-diphenyl-, (E)-	C_6H_{12}	247(4.19)	44-2485-76

Compound	Solvent	$\lambda_{max}(\log \epsilon)$	Ref.
$C_{17}H_{18}O_2$			
1H-Benz[e]inden-1-one, 3aα,4,5,9b-tetra-hydro-3-methyl-9bα-(2-oxopropyl)-	EtOH	231(4.21)	39-0722-76C
2H-1-Benzopyran-4-ol, 3,4-dihydro-2-(2-phenylethyl)-	MeOH	223s(3.92),275(3.33), 283(3.33)	12-2023-76
19,20-Cyclopodocarpa-8,11,13-triene-7,19-dione, (±)-	EtOH	248(3.98),287(3.15)	44-1089-76
2H-1,4aβ-Ethanophenanthrene-3,11(4H)-dione, 1,9,10,10aα-tetrahydro-1α-methyl-	EtOH	310(2.57)	39-0722-76C
1-Naphthalenecarboxylic acid, cyclohexyl ester	EtOH	222(4.73),298(3.96)	12-2683-76
3-Phenanthrenol, 9,10-dihydro-2-methoxy-7,8-dimethyl-	MeOH	218(4.67),276(4.22), 291(4.02),313(4.03)	102-0844-76
	MeOH-NaOH	221(4.37),254(4.24), 277(3.93),325(3.73)	102-0844-76
$C_{17}H_{18}O_2S$			
Cyclohexanecarboxylic acid, 2-(4-phenyl-2-thienyl)-, trans	MeOH	233(4.33),263(4.10)	44-1320-76
$C_{17}H_{18}O_2S_2$			
Butanoic acid, 3-methyl-, 4-[2,2'-bi-thiophene]-5-yl-3-butynyl ester	ether	333(4.31)	102-1309-76
$C_{17}H_{18}O_3$			
Benzene, 1,2-dimethoxy-3-[2-(4-methoxy-phenyl)ethenyl]-, (E)-	MeOH	220(3.30),238s(3.18), 250s(3.00),307(3.48), 320(3.36)	78-0109-76
Benzofuran, 2,3-dihydro-2,3-dimethoxy-2-methyl-3-phenyl-	MeOH	278(3.46),285(3.39)	35-3555-76
1,4a-Ethano-4aH-fluorene-9,11(2H)-dione, 1,3,4,9a-tetrahydro-7-methoxy-1-meth-yl-, (1α,4aα,9aα)-(±)-	EtOH	222(4.28),252(3.75), 325(3.20)	35-3721-76
$C_{17}H_{18}O_3S_2$			
Butanoic acid, 3-methyl-, 4-[2,2'-bi-thiophen]-5-yl-2-hydroxy-3-butynyl ester	ether	332(4.34)	24-0901-76
$C_{17}H_{18}O_4$			
Angolensin, 2-O-methyl-	n.s.g.	270(4.1),315(3.98)	39-0186-76C
Benzoic acid, 2-methoxy-6-methyl-4-(phenylmethoxy)-, methyl ester	EtOH	256(3.93),275s(3.88)	35-5380-76
1H-Fluorene-1-carboxylic acid, 2,3,4,9-tetrahydro-7-methoxy-1-methyl-2-oxo-, methyl ester	EtOH	269(4.15)	78-0065-76
4a(2H)-Phenanthrenecarboxylic acid, 3,4,9,10-tetrahydro-5-methoxy-2-oxo-, methyl ester	MeOH	223(4.22),240(4.11), 272(3.42),279(3.38), 338(2.70)	12-2087-76
4a(2H)-Phenanthrenecarboxylic acid, 3,4,9,10-tetrahydro-6-methoxy-2-oxo-, methyl ester	MeOH	230(4.32),245(4.18), 288(3.51)	12-2087-76
Phenol, 3-[2-(3,4-dimethoxyphenyl)ethen-yl]-5-methoxy-, (E)-	MeOH	305s(4.41),323(4.51)	102-2006-76
1,5,11-Tridecatriene-7,9-diyne-3,4-diol, diacetate, 1-trans-5-trans	EtOH	230(4.49),238(4.49), 248(4.35),262(3.90), 277(4.16),293(4.33), 313(4.26)	39-0735-76C

Compound	Solvent	$\lambda_{max}(\log \epsilon)$	Ref.
$C_{17}H_{18}O_5$			
5,10-Benzocyclooctenedicarboxylic acid, 7,8-dihydro-2-methoxy-, dimethyl ester	MeOH	218s(4.28),276s(3.45)	33-1469-76
4βH-Photosantonene, 4-acetoxy-	n.s.g.	254(4.15)	39-0442-76C
$C_{17}H_{18}O_6$			
Ethanone, 1-(1-acetoxy-3,6,8-trimethoxy-2-naphthalenyl)-	EtOH	229(4.54),248(4.51), 303(3.72)	12-1087-76
3-Furancarboxylic acid, 5-[(3,4-dimethoxyphenyl)methylene]-4,5-dihydro-2-methyl-4-oxo-, ethyl ester	EtOH	210(4.08),260(3.85), 387(4.27)	4-0521-76
Isokhellactone, trans-acetylmethyl-	EtOH	220(4.05),246(3.26), 257(3.30),300s(4.00), 324(4.22)	102-1293-76
2H-Pyran-3-carboxylic acid, 4-hydroxy-6-methoxy-2-oxo-5-(2,4,6-trimethylphenyl)-, methyl ester	EtOH EtOH-base	221(4.26),313(3.97) 276(3.86)	78-0269-76 78-0269-76
2H-Pyran-2-one, 4-methoxy-6-[2-(3,4,5-trimethoxyphenyl)ethenyl]-	MeOH	224(4.29),362(4.36), 345(4.38)	44-4070-76
$C_{17}H_{18}O_7$			
Ethanone, 1-(4-acetoxy-1-hydroxy-3,6,8-trimethoxy-2-naphthalenyl)-	MeOH	236(4.48),270(4.45), 314(3.77),388(3.70)	12-1087-76
7H-Furo[3,2-g][1]benzopyran-7-one, 9-methoxy-4-(1,2,3-trihydroxy-3-methylbutyl)- (thamontanin)	EtOH	220(4.49),250(4.36), 264(4.29),308(4.14)	100-0134-76
Isokhellactone diacetate	EtOH	220(4.05),246(3.28), 257(3.36),300s(4.00), 324(4.22)	102-1293-76
$C_{17}H_{19}ClN_2O_2S$			
10H-Phenothiazine-2,7-diol, 8-chloro-10-[3-(dimethylamino)propyl]-	EtOH	244(4.34),309(3.76)	4-1067-76
10H-Phenothiazine-3,7-diol, 2-chloro-10-[3-(dimethylamino)propyl]-	EtOH	216(4.34),251(4.42), 311(3.71)	4-1067-76
$C_{17}H_{19}ClN_2O_3S$			
10H-Phenothiazine-2,3-diol, 8-chloro-10-[3-(dimethylamino)-2-hydroxypropyl]-, hydrochloride	EtOH	216(4.31),237(4.42), 280(4.26),316(3.78)	4-1067-76
10H-Phenothiazine-2,3,7-triol, 8-chloro-10-[3-(dimethylamino)propyl]-, hydrochloride	EtOH	219(4.16),241(4.37), 320(3.98)	4-1067-76
$C_{17}H_{19}ClN_2S$			
10H-Phenothiazine-10-propanamine, 2-chloro-N,N-dimethyl- (chlorpromazine)	n.s.g.	254.5(4.52)	133-0065-76
$C_{17}H_{19}Cl_2NO_6$			
β-D-Glucopyranoside, 5,7-dichloro-2-methyl-8-quinolinyl 4-O-methyl-	MeOH	209(4.59),244(4.65), 294s(3.70),300(3.72), 312(3.67),326(3.51)	24-2259-76
$C_{17}H_{19}N$			
Azete, 2,3-dihydro-2,2,3,3-tetramethyl-4-(1-naphthalenyl)-	C_6H_{12}	243(4.28),293(3.85), 305(3.91),315(3.76), 323(3.72)	77-0729-76
Butanamine, N-benzylidene-γ-phenyl-, (+)-	isooctane	245(4.33),275(3.07), 285(2.92)	123-0714-76

Compound	Solvent	$\lambda_{max}(\log \epsilon)$	Ref.
$C_{17}H_{19}NO$			
Compound 36	MeOH	206(4.48),227(4.30)	39-0465-76C
Cyclobutanone, 2-(2-quinolinyl)-	EtOH	220(4.55),320(3.91), 444(3.95)	118-0764-76
$C_{17}H_{19}NO_2$			
Benzoic acid, 2-[[4-(1-methylpropyl)-phenyl]amino]-	MeOH	210(4.23),295(4.01), 340(3.43)	39-0465-76C
$C_{17}H_{19}NO_2S$			
Sulfilimine, S-(2-methoxyphenyl)-N-(2-methyl-1-oxopropyl)-S-phenyl-	MeOH	225s(4.01),287(3.61)	78-3003-76
	dioxan	283(3.63)	78-3003-76
$C_{17}H_{19}NO_3$			
7H-Benzo[h]cyclopenta[c]quinolin-7-one, 6,6a,8,9,9a,9b,10,11-octahydro-9b-hydroxy-2-methoxy-, (6aα,9aβ,9bα)-(±)-	EtOH	209(4.24),225(4.10), 275(4.06)	39-1889-76C
1,4-Benzoxazepin-7-ol, 2,3,4,5-tetrahydro-8-methoxy-3-methyl-5-phenyl-, hydrochloride	n.s.g.	215(4.23),240(3.72), 287(3.60),385(1.98)	20-0898-76
$C_{17}H_{19}NO_5$			
2-Azabicyclo[3.2.0]heptane-5-carboxylic acid, 7-ethoxy-3,4-dioxo-1-phenyl-, ethyl ester, (1α,5α,7α)-	dioxan	225(3.71),255s(3.52)	142-1229-76
4,6-Heptadienoic acid, 4,6-dimethyl-7-(4-nitrophenyl)-3-oxo-, ethyl ester, (E,E)-	EtOH	220(3.92),334(4.13)	39-0404-76C
	EtOH-NaOH	218(4.30),351(4.05)	39-0404-76C
$C_{17}H_{19}NO_6$			
Benzenamine, 2,4,6-trimethoxy-N-(2,6-dimethoxy-4-oxo-2,5-cyclohexadien-1-ylidene)-	MeCN	238s(4.10),297(4.31), 544(3.68)	22-1993-76
$C_{17}H_{19}NO_8S$			
8aH-Thiazolo[3,2-a]pyridine-6,7,8,8a-tetracarboxylic acid, 2,5-dimethyl-, tetramethyl ester	MeOH and MeOH-HClO$_4$	229(4.25),287(4.46), 442(3.81)	39-1269-76C
$C_{17}H_{19}N_3$			
Acridine Orange	pH 7.0	230(4.19),268(4.58), 290(4.24),470(4.46), 492(4.54)	95-1334-76
Pyrrolidine, 1-[4-[(4-methylphenyl)azo]-phenyl]-	C_6H_{12}	407(4.52)	18-1381-76
$C_{17}H_{19}N_3O$			
1H-Indazol-4-amine, 3-(4-methoxyphenyl)-1,5,6-trimethyl-	MeOH	221(4.48),250(4.18), 324(3.99)	24-1898-76
3H-Pyrrole-3,3-dicarbonitrile, 4-ethoxy-2,4-dihydro-2,2,4-trimethyl-5-phenyl-	EtOH	247(4.12)	33-1018-76
$C_{17}H_{19}N_3O_3$			
2-Pyrrolidinone, 5-hydroxy-1-[2-[3-(1-methyl-1H-imidazol-5-yl)-3-oxopropyl]phenyl]-	EtOH	257(4.18)	78-1085-76
$C_{17}H_{19}N_3O_4S$			
1,2-Diazabicyclo[5.2.0]nona-3,5-diene-2-carboxylic acid, 5-methyl-9-oxo-8-	MeOH	274(3.96)	142-0471-76S

Compound	Solvent	$\lambda_{max}(\log \epsilon)$	Ref.
[(2-thienylacetyl)amino]-, ethyl ester, cis (cont.)			142-0471-76S
$C_{17}H_{19}N_3O_6$ 2(1H)-Pyrimidinone, 4-amino-1-[3-O-(phenylacetyl)-β-D-arabinofurano-syl]-, hydrochloride	MeOH-acid	284(4.17)	87-0667-76
$C_{17}H_{19}N_5O_4S_3$ 3-Thiazolidinehexanoic acid, 4-[(amino-thioxomethyl)hydrazono]-5-[(4-nitro-phenyl)methylene]-2-thioxo-	MeOH	243(4.33),278(4.05), 378(4.51)	103-0749-76
$C_{17}H_{20}ClNO_3$ 1,4-Benzoxazepin-7-ol, 2,3,4,5-tetra-hydro-8-methoxy-3-methyl-5-phenyl-, hydrochloride	n.s.g.	215(4.23),240(3.72), 287(3.60),385(1.98)	20-0898-76
$C_{17}H_{20}ClNO_7$ β-D-Glucopyranoside, 5-chloro-7-hydroxy-2-methyl-8-quinolinyl 4-O-methyl-	MeOH	209(4.41),223s(4.25), 242(4.38),279(3.53), 306(3.40),333(3.51)	24-2259-76
	MeOH-NaOH	260(4.37),292(3.62), 377(3.54)	24-2259-76
β-D-Glucopyranoside, 5-chloro-8-hydroxy-2-methyl-7-quinolinyl 4-O-methyl-	MeOH	251(4.55),323(3.38), 335(3.38)	24-2259-76
	MeOH-NaOH	264(4.47),343(4.44), 373(3.41)	24-2259-76
$C_{17}H_{20}Cl_2N_2O_3S$ 10H-Phenothiazine-2,3-diol, 8-chloro-10-[3-(dimethylamino)-2-hydroxy-propyl]-, hydrochloride	EtOH	216(4.31),237(4.42), 280(4.26),316(3.78)	4-1067-76
10H-Phenothiazine-2,3,7-triol, 8-chloro-10-[3-(dimethylamino)propyl]-, hydro-chloride	EtOH	219(4.16),241(4.37), 320(3.98)	4-1067-76
$C_{17}H_{20}N_2$ 1-Propanone, 2,2-dimethyl-1-phenyl-, phenylhydrazone, (E)-	EtOH	267(4.19)	39-0456-76C
$C_{17}H_{20}N_2O$ Benzamide, N-(1,1-dimethylethyl)-2-(phenylamino)-	MeOH	210(4.36),235(3.63), 290(4.13)	39-0465-76C
[1,4]Diazepino[6,7,1-jk]carbazole, 3-acetyl-1,2,3,4,8,9,10,11-octahydro-	EtOH	234(4.49),287(3.84), 297s(3.76)	4-1187-76
$C_{17}H_{20}N_2O_2$ 2,4-Pentadienoic acid, 2-cyano-5-(di-methylamino)-5-(4-methylphenyl)-, ethyl ester	CH$_2$Cl$_2$	403(4.30)	48-0705-76
2,4-Pentadienoic acid, 2-cyano-5-(ethyl-amino)-5-(4-methylphenyl)-, ethyl ester	CH$_2$Cl$_2$	390(4.30)	48-0705-76
$C_{17}H_{20}N_2O_5$ Benzo[f]cinnoline-3,4-dicarboxylic acid, 2,4a,5,6-tetrahydro-8-meth-oxy-, dimethyl ester	EtOH	262(4.24)	4-1333-76

Compound	Solvent	$\lambda_{max}(\log \epsilon)$	Ref.
$C_{17}H_{20}N_2S$			
Promazine	n.s.g.	252(4.49)	133-0065-76
$C_{17}H_{20}N_3O_3S$			
Methanaminium, N-[[[5-(ethoxycarbonyl)-2-(4-methoxyphenyl)-6H-1,3-thiazin-6-ylidene]amino]methylene]-N-methyl-, perchlorate	HOAc	312(4.19),385s(4.02), 440(4.18)	97-0268-76
$C_{17}H_{20}N_3O_5$			
Piperidine, 1-(1-ethoxy-1,2-dihydro-2,4-dinitro-1-naphthalenyl)-, ion(1-)	DMSO	357(4.23),522(4.38)	18-1521-76
$C_{17}H_{20}N_4O_6$			
Riboflavin	H_2O	224(4.41),268(4.46), 373(3.96),447(4.03)	35-0830-76
	5M HCl	222(4.31),267(4.47), 396(4.29)	35-0830-76
	EtOH	223(4.43),270(4.49), 362(3.93),449(4.0)	103-1386-76
$C_{17}H_{20}N_4O_7$			
Riboflavin, 5-oxide	H_2O	217(4.23),272(4.37), 370(3.71),461(3.72)	35-0830-76
	5M HCl	272(4.27),393(3.88)	35-0830-76
$C_{17}H_{20}N_4O_7S$			
1H-Pyrazolo[3,4-d]pyrimidine, 4-(methyl-thio)-1-(2,3,5-tri-O-acetyl-β-D-ribo-furanosyl)-	EtOH	224s(4.08),290(4.23), 298s(4.15)	104-1143-76
2H-Pyrazolo[4,3-d]pyrimidine, 7-(methyl-thio)-2-(2,3,5-tri-O-acetyl-β-D-ribo-furanosyl)-	MeOH	218(4.23),261(3.63), 267(3.62),311(4.00), 320(4.06),334(3.91)	104-1141-76
$C_{17}H_{20}N_4O_8$			
5H-1,2,3-Triazolo[4,5-b]pyridin-5-one, 2,4-dihydro-4-methyl-2-(2,3,4-tri-O-acetyl-β-D-ribofuranosyl)-	MeOH	245(3.60),302(4.18), 311(4.18)	23-1029-76
5H-1,2,3-Triazolo[4,5-b]pyridin-5-one, 2,4-dihydro-4-methyl-2-(2,3,5-tri-O-acetyl-β-D-ribofuranosyl)-	MeOH	239(3.87),304(4.32), 313(4.30)	23-1029-76
$C_{17}H_{20}O$			
Benzenemethanol, α,2,4,6-tetramethyl-α-phenyl-	EtOH	245-260f(c.2.6)	44-3083-76
19,20-Cyclopodocarpa-8,11,13-trien-19-one, (+)-	EtOH	266(2.65),274(2.64)	44-1089-76
1,4-Ethano-2H-benz[f]inden-2-one, 1,3-3a,4,9,9a-hexahydro-1,3a-dimethyl-	EtOH	261(2.71),266(2.81), 273(2.85)	39-1975-76C
3,5-Ethano-2H-benz[e]inden-2-one, 1,3-3a,4,5,9b-hexahydro-3,4-dimethyl-, (3α,3aα,4α,5α,9bβ)-(+)-	EtOH	252(3.35),273(3.10)	39-1975-76C
$C_{17}H_{20}O_2$			
1,3-Dioxan, 2,2-dimethyl-4-(1,3,9-nona-triene-5,7-diynyl)-, (E,E,E)-	EtOH	233(4.35),250(4.61), 266(4.45),279(4.31), 291(4.56),314(4.69), 336(4.56)	39-0735-76C
1,3-Pentanediol, 1,5-diphenyl-, (R*,S*)-(+)-	MeOH	257(2.64)	12-2023-76

Compound	Solvent	$\lambda_{max}(\log \epsilon)$	Ref.
$C_{17}H_{20}O_3$			
Benzene, 1,2-dimethoxy-3-[2-(4-methoxy-phenyl)ethyl]-	MeOH	212(3.36),225(3.40), 284(2.85),310(2.85)	78-0109-76
1(2H)-Naphthalenone, 2-acetoxy-3,4-di-hydro-5,8-dimethyl-3-(1-methylethen-yl)-, (2R-trans)-	EtOH	258(4.29),308(3.76)	102-1267-76
2-Oxaestra-1(10),4-diene-3,17-dione	MeOH	300(3.78)	44-2864-76
2-Oxaestra-4,9-diene-3,17-dione	MeOH	285(4.23)	44-2864-76
$C_{17}H_{20}O_4$			
2H-Benz[e]indene-1-carboxylic acid, 3,3a,4,5-tetrahydro-3-hydroxy-7-meth-oxy-3a,5-dimethyl-, (3α,3aα,5β)-	EtOH	277(4.37)	2-0312-76
1-Naphthalenepropanoic acid, 3,4-di-hydro-2-(2-methoxy-2-oxoethyl)-, methyl ester	EtOH	263(4.08)	78-0073-76
2-Naphthalenepropanoic acid, 3,4-di-hydro-1-(2-methoxy-2-oxoethyl)-, methyl ester	EtOH	264(4.07)	78-0073-76
2-Pentenedioic acid, 2-(1-phenylethen-yl)-, ethyl ester	EtOH	238(4.10)	18-3243-76
2-Pentenedioic acid, 4-(1-phenylethyl-idene)-, diethyl ester, (E,Z)-	EtOH	234(4.05),285(4.17)	18-3243-76
Phenol, 5-[3-(4-hydroxy-2-methoxyphen-yl)propyl]-2-methoxy-	EtOH EtOH-NaOH	225s(4.29),280(3.89) 242(4.28),294(4.04)	102-0567-76 102-0567-76
$C_{17}H_{20}O_4S$			
Sulfonium, dimethyl-, 3-benzoyl-1-(eth-oxycarbonyl)-4-oxo-2-pentenylide	EtOH	253(4.19),342(4.28), 380(4.07)	94-2421-76
$C_{17}H_{20}O_5$			
1H-Indene-3-propanoic acid, 6-methoxy-2-(2-methoxy-2-oxoethyl)-, methyl ester	EtOH	267(4.15)	78-0065-76
Photosantone, 4-acetoxy-1,2-dihydro-	n.s.g.	256(4.14)	39-0442-76C
$C_{17}H_{20}O_6$			
Oxireno[9,10]cyclodeca[1,2-b]furan-4,9(1aH,5H)-dione, 7-acetoxyocta-hydro-1a-methyl-5,8-bis(methylene)-, [1aR-(1aR*,7R*,7aR*,10aS*,10bR*)]-(anhydroperoxyferolide)	MeOH	212(4.24),323(1.48)	77-0402-76
$C_{17}H_{20}O_9$			
Quinic acid, 3-O-ferulyl-, D-(-)-	MeOH-acid MeOH-base	298s(4.15),327(4.32) 300s(3.58),310(3.67), 378(4.48)	20-0663-76 20-0663-76
Quinic acid, 3-O-isoferulyl-, D-(-)-	MeOH-acid	244(4.11),296(4.23), 327(4.27)	20-0663-76
	MeOH-base	267(4.29),308(4.21), 366(4.00)	20-0663-76
$C_{17}H_{21}BrO_2$			
Phenol, 4-bromo-2-(1,2-dimethyl-3-meth-ylenecyclopentyl)-5-methyl-, acetate, cis-(+)- (allolaurinterol acetate)	hexane	273(2.86),281(2.87)	12-2533-76
$C_{17}H_{21}BrO_3$			
2,5-Methano-1-benzoxepin-2(3H)-methanol, 7-bromo-4,5-dihydro-5,8,10-trimethyl-, acetate, (2α,5β,10S*)-(-)-	hexane	235(3.98),285(3.42), 288(3.42),294(3.42)	12-2533-76

Compound	Solvent	$\lambda_{max}(\log \epsilon)$	Ref.
$C_{17}H_{21}ClO$			
Ethanone, 2-chloro-1-(1,2,3,4,4a,9,10-10a-octahydro-1-methyl-1-phenanthren-yl)-, (1α,4aα,10aβ)-(±)-	EtOH	266(2.72),274(2.69)	44-1089-76
$C_{17}H_{21}FN_2O_9$			
2,4(1H,3H)-Pyrimidinedione, 5-methyl-1-(2,3,4-tri-O-acetyl-6-deoxy-6-fluoro-β-D-galactopyranosyl)-	MeOH	265(3.94)	48-0079-76
$C_{17}H_{21}NO$			
Benzenepropanol, 4-(dimethylamino)-α-phenyl-	MeOH	254(4.15),302(3.19)	35-6711-76
Benzo[3,4]cyclobuta[1,2-c]quinolin-6(5H)-one, 6a,6b,7,8,9,10,10a,10b-octahydro-6a,10b-dimethyl-, (6aα,6bα,10aα,10bα)-	EtOH	256(3.03)	1-0189-76
4H-1-Pyrindin-4-one, 1,2,3,5,6,7-hexa-hydro-1-(1-methyl-2-phenylethyl)-, (S)-	heptane	318(4.12)	103-0428-76
	MeOH	340(4.26)	103-0428-76
	CF$_3$COOH	315(4.03)	103-0428-76
4(1H)-Quinolinone, 2,3,5,6,7,8-hexahy-dro-1-(1-phenylethyl)-, (S)-	heptane	315(4.20)	103-0428-76
	MeOH	338(4.28)	103-0428-76
	CF$_3$COOH	328(4.15)	103-0428-76
$C_{17}H_{21}NO_2$			
2-Azaestra-1(10),4-diene-3,17-dione	MeOH	306(3.72)	44-2864-76
Cocculine	MeOH	211(3.76),223s(3.65), 286(3.19)	36-0132-76
	MeOH-KOH	214(3.84),230s(3.62), 248s(3.44),291(3.18), 307(2.94)	36-0132-76
$C_{17}H_{21}NO_3$			
2-Propanamine, 1-[3-methoxy-4-(phenyl-methoxy)phenoxy]-	n.s.g.	215(4.16),285(3.54)	20-0421-76
$C_{17}H_{21}NO_5$			
Carbamic acid, [[(1,2,3,4-tetrahydro-1-oxo-2-naphthalenyl)acetyl]oxy]-, 1,1-dimethylethyl ester	MeCN	246(4.10),289(3.30)	33-2499-76
1H-Pyrrole-1,3-dicarboxylic acid, 2,5-dihydro-4-hydroxy-5-(phenyl-methyl)-, diethyl ester	EtOH	247(3.63)	4-0113-76
$C_{17}H_{21}NSi$			
Methanamine, N-(diphenylmethylene)-1-(trimethylsilyl)-	C_6H_{12}	253(4.23)	101-0295-76G
$C_{17}H_{21}NSn$			
Methanamine, N-(diphenylmethylene)-1-(trimethylstannyl)-	C_6H_{12}	267(4.22)	101-0295-76G
$C_{17}H_{21}N_2O_9P$			
Uridine, 2'-deoxy-5-[(phenylmethoxy)-methyl]-, 5'-phosphate	pH 1	264(3.99)	87-0903-76
	pH 7	264(3.99)	87-0903-76
	pH 13	263(3.88)	87-0903-76
$C_{17}H_{21}N_3$			
Benzenamine, 4-[(2,4-dimethylphenyl)-azo]-N,N,3-trimethyl-	C_6H_{12}	399(4.30)	18-1381-76

Compound	Solvent	$\lambda_{max}(\log \epsilon)$	Ref.
Benzenamine, 4-[(2,6-dimethylphenyl)-azo]-N,N,3-trimethyl-	toluene	377(4.42)	18-1381-76
Benzenamine, N,N,3,5-tetramethyl-4-[(2-methylphenyl)azo]-	toluene	395(4.39)	18-1381-76
Benzenamine, N,N,3,5-tetramethyl-4-[(4-methylphenyl)azo]-	toluene	395(4.39)	18-1381-76
$C_{17}H_{21}N_3O$			
7-Azabicyclo[4.3.1]deca-3,8-dien-10-one, 7-(4,6-dimethyl-2-pyrimidinyl)-3,4-dimethyl-	EtOH	206(3.98),226(3.86), 274(4.51)	39-2307-76C
$C_{17}H_{21}N_3O_2$			
4H-Indazol-4-one, 1,5,6,7-tetrahydro-3-(4-methoxyphenyl)-1,6,6-trimethyl-, oxime, (E)-	MeOH	235(4.42),268(4.06)	24-1898-76C
Pyrazolo[4,3-b]azepin-5(1H)-one, 4,6,7,8-tetrahydro-3-(4-methoxy-phenyl)-1,7,7-trimethyl-	MeOH	240s(4.16),257(4.23)	24-1898-76
Pyrazolo[4,3-c]azepin-4(1H)-one, 5,6,7,8-tetrahydro-3-(4-methoxy-phenyl)-1,7,7-trimethyl-	MeOH	260(4.12)	24-1898-76
$C_{17}H_{21}N_3O_3$			
L-Isoleucine, N-[(3-methyl-2-quinoxalin-yl)carbonyl]-, methyl ester	EtOH	203(4.59),241(4.53), 320(3.83)	78-2931-76
L-Leucine, N-[(3-methyl-2-quinoxalinyl)-carbonyl]-, methyl ester	EtOH	203(4.59),241(4.52), 320(3.84)	78-2931-76
$C_{17}H_{21}N_3O_4S$			
L-Cysteine, N-acetyl-S-[(3-phenyl-1,2,4-oxadiazol-5-yl)methyl]-, 1-methyl-ethyl ester	EtOH	238(4.07)	98-0876-76
$C_{17}H_{21}N_3O_5$			
Ethanol, 2,2'-[[4-[(4-methoxyphenyl)ami-no]-3-nitrophenyl]imino]bis-	EtOH	529(3.73)	146-0131-76
L-Isoleucine, N-[(3-methyl-2-quinoxalin-yl)carbonyl]-, methyl ester, N,N'-di-oxide	EtOH	206(4.18),232(4.33), 260s(4.25),267(4.40), 349s(3.79),369(4.01), 383(4.06)	78-2931-76
	MeCN	205(4.20),260s(4.16), 268(4.36),358s(3.72), 378(3.94),395(4.01)	78-2931-76
L-Leucine, N-[(3-methyl-2-quinoxalinyl)-carbonyl]-, methyl ester, N,N'-diox-ide	EtOH	206(4.10),233(4.32), 260s(4.28),267(4.41), 350s(3.79),370(4.03), 385(4.09)	78-2931-76
	MeCN	206(4.12),227(4.33), 261s(4.23),267(4.41), 359s(3.77),378(3.97), 394(4.07)	78-2931-76
$C_{17}H_{21}N_3O_6$			
Cytidine, N,4-didehydro-3,4-dihydro-3-[(4-methoxyphenyl)methyl]-, hydrobromide	pH 2 pH 11.5	280(4.13) 268(--)	44-1597-76 44-1597-76
Cytidine, N-[(4-methoxyphenyl)methyl]-	pH 2 pH 7	283(4.22) 273(--)	44-1597-76 44-1597-76

Compound	Solvent	$\lambda_{max}(\log \epsilon)$	Ref.
$C_{17}H_{21}N_3O_6S$			
D-Ribitol, 1-deoxy-1-(1,2,3,4-tetrahy-dro-7,8-dimethyl-2,4-dioxo-10H-pyri-mido[5,4-b][1,4]benzothiazin-10-yl)-	pH 2	252(4.27),280s(3.98), 345(3.30)	89-0443-76
	pH 13	262(4.44),298s(3.98), 316s(3.30)	89-0443-76
$C_{17}H_{21}N_3O_{11}$			
1H-Pyrazole-3-carboxylic acid, 4-nitro-1-(2,3,5-tri-O-acetyl-α-D-ribofurano-syl)-, ethyl ester	MeOH	207(4.06),264(3.85)	104-1538-76
β-	MeOH	207(4.02),264(3.78)	104-1538-76
1H-Pyrazole-5-carboxylic acid, 4-nitro-1-(2,3,5-tri-O-acetyl-α-D-ribofurano-syl)-, ethyl ester	MeOH	218(3.61),265(3.89)	104-1538-76
β-	MeOH	205(3.98),260(3.90)	104-1538-76
$C_{17}H_{21}N_4OS$			
Methanaminium, N-methyl-N-[[[2-(4-mor-pholinyl)-5-phenyl-6H-1,3-thiazin-6-ylidene]amino]methylene]-, per-chlorate	HOAc	304(4.24),454(4.42)	97-0268-76
$C_{17}H_{21}N_4O_3P$			
Diazenecarboxylic acid, [bis(methylphen-ylamino)phosphinyl]-, ethyl ester	MeOH	212(3.92),230(4.01), 271(3.15),464(1.67)	139-0207-76A
$C_{17}H_{22}$			
Bicyclo[11.2.2]heptadeca-2,11,13,15,16-pentaene	n.s.g.	251(4.03)	88-1243-76
$C_{17}H_{22}Br_2O_2$			
Phenol, 5-bromo-2-[(3-bromo-2,2-dimeth-yl-6-methylenecyclohexyl)methyl]-4-methoxy-, (1S-cis)-	EtOH	220s(4.10),297(3.61)	39-1696-76C
$C_{17}H_{22}ClN_4O_5P$			
1H-1,4-Benzodiazepine-2,5-dione, 7-chloro-4-(di-4-morpholinyl-phosphinyl)-3,4-dihydro-	EtOH	220(4.55),250(4.14), 310(3.45)	44-2720-76
$C_{17}H_{22}N_2O$			
Cyclohept[b]indole-5(6H)-propanamine, 7,8,9,10-tetrahydro-α-methyl-γ-oxo-, hydrochloride	EtOH	249(4.17),303(3.76)	44-3775-76
5H-Cyclooct[b]indole-5(6H)-propanamine, 6,7,8,9,10,11-hexahydro-γ-oxo-, hydrochloride	EtOH	248(4.19),271(3.98), 298(3.69),304(3.71)	44-3775-76
$C_{17}H_{22}N_2O_3S$			
Proline, 3,3-dimethyl-4-[(phenylacetyl)-amino]-5-thioxo-, ethyl ester, trans	C_6H_{12}	270(4.11)	5-2185-76
$C_{17}H_{22}N_2O_4$			
1H-Pyrrole-1,3-dicarboxylic acid, 4-ami-no-2,5-dihydro-5-(phenylmethyl)-, diethyl ester	EtOH	276.5(4.08)	4-0113-76
$C_{17}H_{22}N_2O_9$			
1H-Imidazole, 1-(2,3,4,6-tetra-O-acetyl-α-D-galactopyranosyl)-	MeOH	225(2.96)	136-0275-76D

Compound	Solvent	$\lambda_{max}(\log \epsilon)$	Ref.
1H-Imidazole, 1-(2,3,4,6-tetra-O-acetyl-α-D-mannopyranosyl)-	EtOH	225(3.15)	136-0275-76D
β-	EtOH	225(3.25)	136-0275-76D
$C_{17}H_{22}N_4O_2$			
1H-Benzo[g]pyrrolo[2,1-e]pteridine-4,6(5H,7H)-dione, 2,3,7a,8-tetrahydro-5,8,10,11-tetramethyl-	KCl	248s(3.72),305(3.56)	69-1791-76
$C_{17}H_{22}N_4O_7$			
1H-Purine-2,6-dione, 7-(2,3-di-O-acetyl-4,6-dideoxy-α-L-ribo-hexopyranosyl)-3,7-dihydro-1,3-dimethyl-	MeOH	274(3.90)	44-3827-76
1H-Purine-2,6-dione, 7-(2,3-di-O-acetyl-4,6-dideoxy-β-L-xylo-hexopyranosyl)-3,7-dihydro-1,3-dimethyl-	MeOH	274(3.88)	44-3827-76
$C_{17}H_{22}N_4O_8S$			
2,4(1H,3H)-Pteridinedione, 6,7-dimethyl-1-[2,3-O-(1-methylethylidene)-5-O-(methanesulfonyl)-β-D-ribo-furanosyl]-	pH 6.0	245(3.94),324(3.97),336s(3.90)	24-3159-76
	pH 11.0	240(4.19),330(3.99),345s(3.87)	24-3159-76
	MeOH	230(4.09),245s(4.00),324(3.94),337s(3.87)	24-3159-76
$C_{17}H_{22}O$			
Ethanone, 1-(1,4-dihydro-1-pentyl-1-naphthalenyl)-	MeOH	204(3.82),260(3.02)	19-0175-76
$C_{17}H_{22}O_3$			
2-Dehydroxyemmotin F, acetate	EtOH	258(4.36),311(3.74)	102-1267-76
Naphtho[2,3-b]furan-6-ol, 4,4a,5,6,7,8-8a,9-octahydro-3,8a-dimethyl-5-methyl-ene-, acetate, [4aR-(4aα,6β,8aβ)]-	EtOH	220(3.84)	95-1089-76
$C_{17}H_{22}O_4$			
2H-1-Benzopyran-2-acetaldehyde, 6-acet-oxy-3,4-dihydro-2,5,7,8-tetramethyl-	EtOH	225(3.98),278(3.24),284(3.29)	33-0290-76
$C_{17}H_{22}O_5$			
2H-1-Benzopyran-2-acetic acid, 6-acet-oxy-3,4-dihydro-2,5,7,8-tetramethyl-,(S)-	EtOH	223s(4.03),276(3.22),283(3.27)	33-0290-76
Naphtho[1,2-b]furan-2,7-dione, 8-acet-oxy-3,5a,9-trimethyl-	EtOH	246(4.15)	2-0657-76
Naphtho[1,2-b]furan-2,8(3H,4H)-dione, 7-acetoxy-3a,6,7,9,9a,9b-hexahydro-3,6,9-trimethyl-	EtOH	209(3.4117)	2-0657-76
$C_{17}H_{22}O_7$			
β-D-Glucopyranoside, (1-oxo-6-tetralin-yl) 4-O-methyl-	MeOH	218(4.20),268(4.20)	24-2259-76
$C_{17}H_{22}O_8$			
2-Butenedioic acid, 2-[3-(2,2,5-trimeth-yl-4,6-dioxo-1,3-dioxan-5-yl)-1-buten-yl]-, dimethyl ester	EtOH	267(4.14)	35-1204-76
$C_{17}H_{22}S_2$			
Azulene, 1,3-bis(ethylthio)-4,6,8-tri-methyl-	hexane	600(2.71)	104-1966-76

Compound	Solvent	$\lambda_{max}(\log \epsilon)$	Ref.
$C_{17}H_{23}AsO$ 4(1H)-Arseninone, 1-(2-butenyl)-3,5-bis- (1-methyl-2-propenyl)-	EtOH	232s(3.83),326(3.65)	89-0690-76
$C_{17}H_{23}Cl_2N_4O_5P$ Benzamide, 5-chloro-2-[(chloroacetyl)- amino]-N-(di-4-morpholinylphosphinyl)-	isoPrOH	223(4.38),257(4.16), 312(3.58)	44-2720-76
$C_{17}H_{23}NO$ Benzo[g]pyrrolo[1,2-a]quinolin-5(1H)- one, 2,3,6,6a,7,8,9,10,10a,11-deca- hydro-10a-methyl-, cis	EtOH	266(4.09)	23-1512-76
$C_{17}H_{23}NO_2S$ Acetamide, N-[5-oxo-4-(tricyclo[3.3.1- $1^{3,7}$]dec-1-ylthio)-1,3-pentadienyl]-, (Z,E)-	MeOH	336(4.43)	4-0321-76
$C_{17}H_{23}NO_3$ 2(1H)-Quinolinone, 3-ethyl-8-hydroxy- 4-methoxy-7-(3-methylbutyl)-	EtOH	220(4.50),235s(4.29), 253(4.07),261(4.11), 282s(3.86),295s(3.90), 325(4.15),339(4.12)	105-0709-76
$C_{17}H_{23}NO_3S_2$ 5H-[1]Benzothiopyrano[3,4-e]-1,2-oxa- thiin-4-amine, 3,4-dihydro-N,N-bis- (1-methylethyl)-, 2,2-dioxide	EtOH	217(4.11),249(4.29), 287(3.66),297s(3.61)	4-0225-76
$C_{17}H_{23}N_2O_2$ 1-Pyrrolidinyloxy, 3-[[(3-ethenylphen- yl)amino]carbonyl]-2,2,5,5-tetrameth- yl-	EtOH	242(4.95)	23-3545-76
$C_{17}H_{23}N_2O_3$ 1-Pyrrolidinyloxy, 2,2,5,5-tetramethyl- 3-[[(3-oxiranylphenyl)amino]carbonyl]-	EtOH	240(3.86)	23-3545-76
$C_{17}H_{23}N_2O_4P$ Phosphonic acid, [3a,4,5,6,7,7a-hexahy- dro-3-(4-methoxyphenyl)-4,7-methano- 3H-indazol-3-yl]-, dimethyl ester, (3α,3aα,4β,7β,7aα)-	MeOH	206(3.4),230(4.3), 278(3.0),286(3.0), 337(2.4)	23-0044-76
$C_{17}H_{23}N_3O_2$ 4-Hepten-3-one, 5-(2H-benzotriazol-2- yl)-2,2,6,6-tetramethyl-, N-oxide, (E)-	EtOH	302(3.83),340(3.62)	22-0184-76
1H-Pyrazole, 3,5-bis(1,1-dimethylethyl)- 1-(2-nitrophenyl)-	EtOH EtOH	235(4.15),280(3.46) 234(3.96),290s(3.26)	22-0184-76 22-0839-76
1H-Pyrazole, 3,5-bis(1,1-dimethylethyl)- 1-(4-nitrophenyl)-	EtOH	254(4.15),295(3.26)	22-0839-76
$C_{17}H_{23}N_3O_6$ 2,2'-Anhydro-1-(3',5'-dibutyryl-β-D-ara- binofuranosyl)cytosine, hydrochloride	MeOH-acid	234(4.01),263(4.06)	87-0663-76
$C_{17}H_{23}N_5O_5$ 4(1H)-Pteridinone, 6,7-dimethyl-2-(meth- ylamino)-1-[2,3-0-(1-methylethylid-	pH 0.0	223(4.26),250s(3.96), 320(3.97)	24-3159-76

Compound	Solvent	$\lambda_{max}(\log \epsilon)$	Ref.
ene)-β-D-ribofuranosyl]- (cont.)	pH 5.0	240(4.21),275(3.57), 330(4.02)	24-3159-76
	2M NaOH	230(4.31),263(3.89), 365(3.88)	24-3159-76
$C_{17}H_{23}N_8O_9P$			
5'-Adenylic acid, mono[2-(4-amino-2-oxo-	pH 1	267(4.28)	78-2409-76
1(2H)-pyrimidinyl)-1-(hydroxymethyl)-	pH 7	262(4.30)	78-2409-76
ethyl] ester, ammonium salt, (S)-	pH 13	262(4.28)	78-2409-76
5'-Adenylic acid, mono[3-(4-amino-2-oxo-	pH 1	267(4.29)	78-2409-76
1(2H)-pyrimidinyl)-2-hydroxypropyl]	pH 7	262(4.30)	78-2409-76
ester, ammonium salt, (S)-	pH 13	262(4.29)	78-2409-76
$C_{17}H_{23}O_4P$			
Phosphonic acid, [3-(4-methoxyphenyl)-	MeOH	207(3.3),235(4.1),	23-0044-76
tricyclo[3.2.1.0^{2,4}]oct-3-yl]-, di-		281(3.1),288(3.0)	
methyl ester, (1α,2β,3β,4β,5α)-			
$C_{17}H_{24}NO$			
1-Naphthaleneethanaminium, 3,4-dihydro-	MeOH	218(4.44),268(4.00)	44-0531-76
6-methoxy-N,N,N-trimethyl-β-methyl-			
ene-, iodide			
$C_{17}H_{24}N_2O$			
2H-Pyrrol-2-one, 3,4-diethyl-5-[(4-eth-	MeOH	417(4.56)	118-0335-76
yl-3,5-dimethyl-1H-pyrrol-2-yl)meth-			
ylene]-1,5-dihydro-			
$C_{17}H_{24}N_2O_2$			
Carbamic acid, N-phenyl-, 3-pyrrolidino-	H_2O	233(4.18),265(2.81)	106-0024-76
cyclohexyl ester			
1H-Pyrano[3,2-c:5,6-c']dipyridinium,	MeOH	261(3.88)	83-0425-76
2,3,4,6,7,8,9,10-octahydro-2,2,8,8-			
tetramethyl-4,6-bis(methylene)-			
10-oxo-, dichloride			
$C_{17}H_{24}N_2O_3$			
1H-Indole-3-carboxylic acid, 6-[(dimeth-	n.s.g.	222(4.46),252(4.44),	103-0245-76
ylamino)methyl]-5-methoxy-1,2-dimeth-		290(4.01),298(3.98)	
yl-, ethyl ester, hydrochloride			
2H-Pyrrol-2-one, 5-[(3,4-diethyl-1,5-di-	MeOH	277(3.79)	31-1107-76
hydro-5-oxo-2H-pyrrol-2-ylidene)meth-			
yl]-3-ethyl-1,5-dihydro-5-methoxy-			
4-methyl-			
$C_{17}H_{24}N_4O$			
1H-Pyrazol-1-amine, N,N,3-trimethyl-	MeOH	243(4.01)	94-3001-76
5-(4-morpholinylmethyl)-4-phenyl-,			
hydrochloride			
$C_{17}H_{24}N_4O_3$			
Benzenamine, 2-[3,5-bis(1,1-dimethyleth-	EtOH	240(4.18),360(3.26)	22-0195-76
yl)-1H-pyrazol-1-yl]-N-hydroxy-5-nitro-			
Benzenamine, 4-[3,5-bis(1,1-dimethyleth-	EtOH	246(3.93),360(3.04)	22-0195-76
yl)-1H-pyrazol-1-yl]-N-hydroxy-3-nitro-			+22-0195-76
$C_{17}H_{24}O$			
4,6,8-Undecatrien-10-yn-3-one, 9-(1,1-	EtOH	230s(3.74),239(3.82),	18-2306-76
dimethylethyl)-2,2-dimethyl-		336(4.58)	

Compound	Solvent	$\lambda_{max}(\log \epsilon)$	Ref.
$C_{17}H_{24}O_2$			
2-Naphthalenemethanol, 1,2,3,4-tetra-hydro-α,α,5,8-tetramethyl-, acetate, (R)- (occidol acetate)	EtOH	223s(4.12),261(4.11)	102-1267-76
$C_{17}H_{24}O_3$			
Benzeneacetic acid, α-hydroxy-, 3,3,5-trimethylcyclohexyl ester (cyclo-spasmol)	MeOH	252(--),258(<u>2.3</u>), 265(--),268s(--)	106-0363-76
	CHCl$_3$	253(--),259(<u>2.3</u>), 265(--),269(--)	106-0363-76
Bicyclo[7.2.0]undec-4-en-3-one, 2-acet-oxy-4,11,11-trimethyl-8-methylene-(buddledin A)	MeOH	238(3.93)	88-3717-76
1H-Naphtho[2,1-b]pyran-1-one, 2,3,4a,8-9,10,10a,10b-octahydro-10-hydroxy-2,4a,8,10b-tetramethyl-	MeOH	242(4.34)	44-0066-76
$C_{17}H_{24}O_4$			
Lactarorufin N, acetate 2-Naphthaleneacetic acid, 1,2,3,4,4a,7-hexahydro-8-(methoxymethyl)- ,4a-di-methyl-7-oxo-, methyl ester, [2R-[2α(S*),4aα]]-	EtOH	228(3.88)	56-2095-76
$C_{17}H_{24}O_5$			
Arsanin acetate	EtOH	284(1.45)	94-2810-76
Arsantin acetate	EtOH	283(1.34)	94-2810-76
$C_{17}H_{24}O_{12}$			
Secogalioside	EtOH	238(4.05)	1-0743-76
$C_{17}H_{25}NO$			
Benzo[g]pyrrolo[1,2-a]quinolin-5(1H)-one, 2,3,3a,4,6,6a,7,8,9,10,10a,11-dodecahydro-10a-methyl-	EtOH	340(4.17)	23-1512-76
$C_{17}H_{25}NO_3$			
1H-Indole-3-acetic acid, 1-cyclohexyl-2,4,5,6,7,7a-hexahydro-5-methyl-2-oxo-	EtOH	216.5(4.10)	22-0776-76
$C_{17}H_{25}NO_4$			
2,6-Piperidinedione, 4-(2-hydroxy-5,7-dimethyl-4-oxo-6,8-decadienyl)-	EtOH	235(4.00),289(2.80)	12-0673-76
$C_{17}H_{25}NS$			
2-Azetidinethione, 3-(1,1-dimethyleth-yl)-1-methyl-3-(1-methylethyl)-4-phenyl-	hexane	245(4.0),268(4.1), 349(1.8)	24-0906-76
$C_{17}H_{25}N_3$			
Benzenamine, 4-[3,5-bis(1,1-dimethyleth-yl)-1H-pyrazol-1-yl]-	EtOH	261(4.20)	22-0195-76
Benzo[g]quinoxaline, 5,5a,6,7,8,9,9a,10-octahydro-2-piperidino-, (5aS-trans)-	MeCN	260(4.15),301(3.30), 346(3.83)	142-0299-76S
$C_{17}H_{25}N_3O$			
Benzenamine, 2-[3,5-bis(1,1-dimethyleth-yl)-1H-pyrazol-1-yl]-N-hydroxy-	EtOH	283(3.76)	22-0839-76

Compound	Solvent	$\lambda_{max}(\log \epsilon)$	Ref.
$C_{17}H_{25}N_3O_5$ 2,2'-Anhydro-1-(3-octanoyl-β-D-arabino- furanosyl)cytosine, hydrochloride	MeOH	231(3.99),263(4.03)	87-0654-76
$C_{17}H_{25}N_3O_7$ 2(1H)-Pyrimidinone, 4-amino-1-[3,5-bis- O-(1-oxobutyl)-β-D-arabinofuranosyl]-	MeOH-acid	213(4.00),283(4.17)	87-0667-76
$C_{17}H_{26}$ Pentalene, 1,3a,4,6a-tetrahydro-1,1,2,3- 3a,5,6,6a-octamethyl-4-methylene-, cis	EtOH	244(4.23)	35-1545-76
Pentalene, 1,3a,6,6a-tetrahydro-1,1,2,3- 3a,4,5,6a-octamethyl-6-methylene-, cis	EtOH	247(4.15)	35-1545-76
$C_{17}H_{26}IN_6O_8P$ 5'-Adenylic acid, N-[5-[(iodoacetyl)- amino]pentyl]-, disodium salt	H_2O	267(4.23)	87-1279-76
$C_{17}H_{26}N_2O_6$ 1H-Pyrazole-3-carboxylic acid, 4-[6- (acetyloxy)methyl]tetrahydro-2,2- dimethyl-4H-cyclopenta-1,3-dioxol- 4-yl]-4,5-dihydro-, ethyl ester	EtOH	285(3.59)	23-0861-76
1H-Pyrrole-1,3-dicarboxylic acid, 4-(acetylamino)-5-(4-methoxy- butyl)-, diethyl ester	EtOH	216(4.54)	4-0113-76
$C_{17}H_{26}N_2O_7$ Spiro[furo[2,3-d]-1,3-dioxole-6(5H),3'- [3H]pyrazole-4'-carboxylic acid, 5-(2,2-dimethyl-1,3-dioxolan-4- yl)-3a,4',5',6a-tetrahydro-2,2- dimethyl-, methyl ester	ether	291(3.02)	136-0079-76G
Spiro[furo[2,3-d]-1,3-dioxole-6(5H),4'- [3H]pyrazole-3'-carboxylic acid, 5-(2,2-dimethyl-1,3-dioxolan-4- yl)-3a,4',5',6a-tetrahydro-2,2- dimethyl-, methyl ester	ether	298(2.90)	136-0079-76G
$C_{17}H_{26}N_4$ 1,3-Benzenediamine, 4-[3,5-bis(1,1-di- methylethyl)-1H-pyrazol-1-yl]-	EtOH	231(4.38),248(3.94), 300(3.30)	22-0839-76
$C_{17}H_{26}N_5O_6PS$ Adenosine, 8-(heptylthio)-, cyclic 3',5'-(hydrogen phosphate)	pH 2.0 pH 13.0	285(4.31) 282(--)	69-1408-76 69-1408-76
$C_{17}H_{26}O$ 2,5-Cyclohexadien-1-one, 2,6-bis(1,1-di- methylethyl)-4-(1-methylethylidene)-	isooctane	317(4.45)	73-1676-76
$C_{17}H_{26}O_2$ 1-Naphthalenol, 1,2,3,7,8,8a-hexahydro- 1,6-dimethyl-4-(1-methylethyl)-, acetate, (1R-cis)-	MeOH	241(4.01),247(4.02)	18-3148-76
2,4-Pentadien-1-ol, 3-methyl-5-(2,6,6- trimethyl-1-cyclohexen-1-yl)-, acetate, (E,E)-	EtOH	230(4.01),267(3.89)	33-0387-76

Compound	Solvent	$\lambda_{max}(\log \epsilon)$	Ref.
$C_{17}H_{26}O_3$			
1(2H)-Naphthalenone, 5-acetoxy-3,4,4a,5-6,7-hexahydro-4a,8-dimethyl-2-(1-methylethyl)-, (2α,4aα,5α,8aα)-	MeOH	246(3.68)	18-3137-76
3-Penten-2-one, 5-acetoxy-3-methyl-1-(2,6,6-trimethyl-1-cyclohexen-1-yl)-, (E)-	MeOH	225(3.92)	33-0567-76
$C_{17}H_{27}NO_2$			
Benzene, 1,3-bis(1,1-dimethylethyl)-5-(1-methylethyl)-2-nitro-	hexane	246s(3.24),340(2.54)	18-1142-76
$C_{17}H_{27}NO_4$			
2-Propenoic acid, 3-[1,2,5,6-tetrahydro-1-[4-(2-methyl-1,3-dioxolan-2-yl)butyl]-3-pyridinyl]-, methyl ester	EtOH	266(2.67)	88-3447-76
$C_{17}H_{27}N_3O_6$			
2(1H)-Pyrimidinone, 4-amino-1-[3-O-(1-oxooctyl)-β-D-arabinofuranosyl]-	MeOH-acid	212(4.01),283(4.16)	87-0667-76
$C_{17}H_{27}N_3O_7$			
Cytidine, N,N-dimethyl-2'-O-(tetrahydro-4-methoxy-2H-pyran-4-yl)-	EtOH	278(4.12)	54-0108-76
$C_{17}H_{28}NO$			
Cyclohexanaminium, 2-(2,6-dimethylphenoxy)-N,N,N-trimethyl-, bromide, cis	H_2O	267(2.67),274s(2.58)	87-0692-76
$C_{17}H_{28}N_2$			
Piperidine, 1-[[3-methylene-2-(1-pyrrolidinyl)-1-cyclohexen-1-yl]methyl]-	n.s.g.	240(4.02),296(2.97)	94-2494-76
$C_{17}H_{28}N_2O$			
Morpholine, 4-[[3-methylene-2-(1-piperidinyl)-1-cyclohexen-1-yl]methyl]-	n.s.g.	238(4.21),291(3.22)	94-2494-76
Morpholine, 4-[6-methylene-2-(1-piperidinylmethyl)-1-cyclohexen-1-yl]-	n.s.g.	238(4.22),285(3.27)	94-2494-76
$C_{17}H_{28}N_2O_2$			
1H-Pyrano[3,2-c:5,6-c']dipyridinium, 2,3,4,6,7,8,9,10-octahydro-2,2,4,6,8,8-hexamethyl-10-oxo-	MeOH	250(4.06)	83-0425-76
$C_{17}H_{28}N_4O_3$			
Dodecanoic acid, 2-(3-nitro-4-pyridinyl)hydrazide	EtOH	227(4.23),342(3.58)	114-0285-76C
$C_{17}H_{28}O$			
Cyclohexene, 1-[2-(1,1-dimethylethoxy)-cyclohexylidenemethyl]-, (E)-	hexane	232(3.96)	35-4867-76
(Z)-	hexane	234(4.05)	35-4867-76
$C_{17}H_{28}OS$			
1-Tridecanone, 1-(2-thienyl)-	heptane	258(4.04),274(3.89)	34-0233-76
$C_{17}H_{29}B$			
Borane, dicyclohexyl(3-methyl-1,3-butadienyl)-, (E)-	octane	245(4.28)	101-0303-76N
	ether	244(4.30)	101-0303-76N
	THF	244(4.26)	101-0303-76N

Compound	Solvent	$\lambda_{max}(\log \epsilon)$	Ref.
$C_{17}H_{29}NO_4S$ 2,5-Pyridinediol, 1-acetyl-4-(1,1-di- methylethyl)-6-[(1,1-dimethylethyl)- thio]-1,2,5,6-tetrahydro-, 5-acetate	MeOH	212(3.75)	4-0861-76
$C_{17}H_{29}N_6O_7P$ 5'-Adenylic acid, N-(7-aminoheptyl)-	H_2O	266(4.23)	87-1279-76
$C_{17}H_{30}N_4O$ Dodecanoic acid, 2-(3-amino-4-pyridin- yl)hydrazide	EtOH	340(3.59),445(3.14)	114-0285-76C
$C_{17}H_{30}O$ 2-Cyclopenten-1-one, 2,4,5-tris(1,1-di- methylethyl)-	EtOH	233(4.08)	118-0604-76
$C_{17}H_{30}O_2$ 2-Cyclopenten-1-one, 2,4,5-tris(1,1-di- methylethyl)-3-hydroxy-	EtOH	262(4.13)	118-0604-76
$C_{17}H_{30}S$ Thiophene, 2-tridecyl-	heptane	234(3.91)	34-0233-76

Compound	Solvent	λ_{max}(log ϵ)	Ref.
$C_{18}H_8Br_2O_3$			
Naphtho[2,3-c]furan-1,3-dione, 5-bromo-4-(3-bromophenyl)-	C_6H_{12}	260(4.81),268(4.91), 297s(3.95)	42-1053-76
Naphtho[2,3-c]furan-1,3-dione, 7-bromo-4-(3-bromophenyl)-	C_6H_{12}	263(4.73),326(3.81)	42-1053-76
$C_{18}H_8Cl_2N_6$			
Propanedinitrile, [3-amino-4-cyano-2-(3,5-dichlorophenyl)-8(2H)-cin-nolinylidene]-	MeOH	295(4.32),350(3.98), 735(3.68)	64-0371-76B
$C_{18}H_8Cl_2O$			
Dibenzo[a,e]cyclobuta[c]cycloocten-1(2H)-one, 2,2-dichloro-7,8-didehydro-	hexane	250(4.51),275(4.90), 283(4.99),310(3.44), 326(3.34),362(3.85)	88-2715-76
$C_{18}H_8Cl_2O_3$			
7H-Benzo[c]fluorene-6-carboxylic acid, 1,10-dichloro-7-oxo-	HOAc	298(4.90),308(4.88), 360s(3.24),380s(2.74), 434(2.72)	42-1053-76
7H-Benzo[c]fluorene-6-carboxylic acid, 3,10-dichloro-7-oxo-	HOAc	301(4.62),362s(3.38), 450(2.77)	42-1053-76
Naphtho[2,3-c]furan-1,3-dione, 5-chloro-4-(3-chlorophenyl)-	C_6H_{12}	257(4.73),267(4.81), 303(3.90),314s(3.85)	42-1053-76
Naphtho[2,3-c]furan-1,3-dione, 7-chloro-4-(3-chlorophenyl)-	C_6H_{12}	258s(4.59),264(4.68), 323(3.71)	42-1053-76
$C_{18}H_8I_2O_3$			
Naphtho[2,3-c]furan-1,3-dione, 7-iodo-4-(3-iodophenyl)-	C_6H_{12}	269(4.78),285s(3.94), 315(3.93)	42-1053-76
$C_{18}H_8N_6$			
[4,4'-Bicinnoline]-3,3'-dicarbonitrile	EtOH	253(4.30),315(3.60)	39-0592-76C
$C_{18}H_8O_2$			
Dibenzo[a,e]cyclobuta[c]cyclooctene-1,2-dione, 7,8-didehydro- (unstable)	EtOH	272(4.38),282(4.50), 363(3.43),382(3.43)	88-2715-76
$C_{18}H_9BrN_4O_2$			
1H-Benz[de]isoquinoline-1,3(2H)-dione, 2-(1H-benzotriazol-1-yl)-5-bromo-	EtOH	210(4.38),237(4.65), 333(4.13),360(3.95)	56-1257-76
1H-Benz[de]isoquinoline-1,3(2H)-dione, 2-(1H-benzotriazol-1-yl)-6-bromo-	EtOH	210(4.70),234(4.80), 342(4.08),358(4.30)	56-1257-76
1H-Benz[de]isoquinoline-1,3(2H)-dione, 2-(2H-benzotriazol-2-yl)-5-bromo-	EtOH	210(4.46),240(4.54), 337(4.10),364(3.85)	56-1257-76
1H-Benz[de]isoquinoline-1,3(2H)-dione, 2-(2H-benzotriazol-2-yl)-6-bromo-	EtOH	215(4.50),234(4.62), 342(4.24),358(4.15)	56-1257-76
$C_{18}H_9ClN_4O_2$			
1H-Benz[de]isoquinoline-1,3(2H)-dione, 2-(1H-benzotriazol-1-yl)-6-chloro-	EtOH	212(4.43),233(4.63), 340(4.26),360(4.11)	56-1257-76
1H-Benz[de]isoquinoline-1,3(2H)-dione, 2-(2H-benzotriazol-2-yl)-6-chloro-	EtOH	216(4.57),233(4.72), 340(4.27),360(4.15)	56-1257-76
1H-Benz[de]isoquinoline-1,3(2H)-dione, 2-(5-chloro-1H-benzotriazol-1-yl)-	EtOH	212(4.76),229(4.66), 335(4.07)	56-1257-76
$C_{18}H_9ClN_6$			
Propanedinitrile, [3-amino-2-(4-chloro-phenyl)-4-cyano-8(2H)-cinnolinylid-ene]-	MeOH	295(4.32),350(4.00), 735(3.69)	64-0371-76B

Compound	Solvent	$\lambda_{max}(\log \epsilon)$	Ref.
$C_{18}H_9Cl_2N_3O_3$ 2,4-Cyclohexadiene-1,2-dicarbonitrile, 3-acetoxy-4,5-dichloro-1-(1H-indol- 3-yl)-6-oxo-	EtOH	217(4.58),264(4.06), 282(4.11),462(2.99)	1-0853-76
$C_{18}H_9FO_2$ Benz[a]anthracene-7,12-dione, 6-fluoro-	EtOH	283(4.5),360(3.5), 405(3.5)	24-1038-76
$C_{18}H_9N_3O_2S$ Benzonitrile, 2-(2H-naphtho[1,8-bc]thi- en-2-ylidenimino)-5-nitro-	dioxan	390(4.17)	103-0857-76
$C_{18}H_{10}Br_2S_4$ 1,3-Dithiole, 4-(4-bromophenyl)-2-[4-(4- bromophenyl)-1,3-dithiol-2-ylidene]-	THF	272(4.55),331(4.29), 400(3.78)	97-0360-76
cation	MeCN	460(--),683(--)	97-0360-76
dication	MeCN	266(4.59),478(3.87)	97-0360-76
$C_{18}H_{10}Cl_2O$ Dibenzo[a,e]cyclobuta[c]cycloocten- 1(2H)-one, 2,2-dichloro-	C_6H_{12}	230s(4.37),259(4.53), 310(3.82)	88-2715-76
$C_{18}H_{10}Cl_2O_4S$ 2,5-Thiophenedicarboxylic acid, 3,4-bis- (2-chlorophenyl)-	EtOH	237s(4.32),278(4.12)	42-0490-76
2,5-Thiophenedicarboxylic acid, 3,4-bis- (3-chlorophenyl)-	EtOH	242(4.37),284(4.06)	42-0490-76
2,5-Thiophenedicarboxylic acid, 3,4-bis- (4-chlorophenyl)-	EtOH	246(4.40),283(4.00)	42-0490-76
$C_{18}H_{10}Cl_2S_4$ 1,3-Dithiole, 4-(4-chlorophenyl)-2-[4- (4-chlorophenyl)-1,3-dithiol-2-ylid- ene]-	THF	269(4.52),330(4.25), 424(3.80)	97-0360-76
cation	MeCN	459(--),681(--)	97-0360-76
dication	MeCN	260(--),484(4.06)	97-0360-76
$C_{18}H_{10}Cl_4O_2$ Benzene, [phenoxy(2,3,4,5-tetrachloro- 2,4-cyclopentadien-1-ylidene)methoxy]-	C_6H_{12}	230(3.85),312(4.49), 385(3.05)	24-3939-76
$C_{18}H_{10}F_6$ Benzene, 1,1'-(1,2,3,4,5,6-hexafluoro- 1,3,5-hexatriene-1,6-diyl)bis-	heptane	314(4.51)	104-1565-76
$C_{18}H_{10}N_2$ Indolo[3,2-b]carbazole	DMF	274(4.53),313(4.72), 335(4.44),386(4.44), 394s(3.60),462(3.48), 490(3.38)	5-1090-76
$C_{18}H_{10}N_2O_2$ 6H,13H-Pyrazino[1,2-a:4,5-a']diindole- 6,13-dione	$CHCl_3$	263(4.34),276(4.54), 308(4.26),380(4.30)	44-3050-76
$C_{18}H_{10}N_4O_2$ 1H-Benz[de]isoquinoline-1,3(2H)-dione, 2-(1H-benzotriazol-1-yl)-	EtOH	211(4.56),230(4.61), 340(4.00)	56-1257-76
1H-Benz[de]isoquinoline-1,3(2H)-dione, 2-(2H-benzotriazol-2-yl)-	EtOH	212(4.71),230(4.87), 333(4.38)	56-1257-76

Compound	Solvent	λ_{max}(log ϵ)	Ref.
$C_{18}H_{10}N_6$			
Propanedinitrile, (3-amino-4-cyano-2-phenyl-8(2H)-cinnolinylidene)-	MeOH	295(4.24),350(3.89), 735(3.59)	64-0371-76B
$C_{18}H_{10}O_2$			
Dibenzo[a,e]cyclobuta[c]cyclooctene-1,2-dione	EtOH	265(4.13),315s(3.43)	88-2715-76
$C_{18}H_{10}O_3$			
Benz[a]anthracene-7,12-dione, 1-hydroxy-	MeOH	251(4.35),311(4.31), 485(3.09)	39-0997-76C
	MeOH-base	594(2.98)	39-0997-76C
Benz[a]anthracene-7,12-dione, 3-hydroxy-	MeOH	225(4.35),245s(4.16), 306(4.28),390(3.59)	39-0997-76C
	MeOH-base	241(4.34),337(4.13), 420(3.38),538(3.59)	39-0997-76C
5,12-Naphthacenedione, 6-hydroxy-	CHCl$_3$	255(4.35),275(3.96), 283(3.95),298(3.82), 312(3.42),442(3.60)	39-0503-76C
Naphtho[2,3-b]furan-4,9-dione, 2-phenyl-	CHCl$_3$	270(4.38),290(4.45), 421(3.63)	18-3713-76
$C_{18}H_{10}O_4$			
5,6,11,12-Naphthacenetetrone, 1,4-dihydro-	MeOH	213(4.32),230s(4.22), 252(4.17),315(3.61)	44-2296-76
$C_{18}H_{10}O_5$			
Benzo[1,2-b:3,4-b']difuran-3(2H)-one, 2-(1,3-benzodioxol-5-ylmethylene)-	EtOH	224(4.65),245(4.00), 330(4.46)	102-1553-76
6H-[1,3]Dioxolo[5,6]benzofuro[3,2-c]-furo[3,2-g][1]benzopyran (neoduleen)	EtOH	224(4.30),252(4.16), 294(3.84),346(4.28), 364(4.18)	102-1283-76
1H,3H-Naphtho[1,8-cd]pyran-1,3-dione, 5,6-dihydroxy-6-phenyl-	MeOH	266(4.27),345(3.72)	102-1413-76
$C_{18}H_{11}BrO$			
Indeno[2,1-b]pyran, 2-(4-bromophenyl)-	dioxan	226(4.35),267(4.26), 281(4.25),293(4.30), 313(4.19),327(4.20), 344(4.29),365(4.37), 520(3.54)	78-2219-76
$C_{18}H_{11}BrO_3$			
12H-Benzo[a]xanthen-12-one, 4-bromo-3-methoxy-	n.s.g.	220(4.33),238(4.40), 255(4.35),262(4.34), 275(4.11),302(3.93), 325(3.94),372(3.59)	39-2241-76C
$C_{18}H_{11}ClO$			
Indeno[2,1-b]pyran, 2-(4-chlorophenyl)-	dioxan	226(4.26),268(4.19), 282(4.21),291(4.23), 344(4.25),360(4.32), 529(3.47)	78-2219-76
$C_{18}H_{11}Cl_2N_3O_2$			
2,4-Cyclohexadiene-1,2-dicarbonitrile, 4,5-dichloro-3-methoxy-1-(1-methyl-1H-indol-3-yl)-6-oxo-	EtOH	217(4.62),283(4.10)	1-0853-76

Compound	Solvent	$\lambda_{max}(\log \epsilon)$	Ref.
$C_{18}H_{11}F$			
Benz[a]anthracene, 6-fluoro-	EtOH	230(4.5),258(4.4), 270(4.6),279(4.8), 290(4.9),302s(--), 318(3.6),332(3.8), 344(3.8),364(3.7), 390(2.8)	24-1038-76
$C_{18}H_{11}FO$			
Benz[a]anthracen-12-ol, 6-fluoro-	EtOH	282(4.8),290(4.8), 305(4.3),375(3.8), 396(3.7)	24-1038-76
Benz[a]anthracen-12(7H)-one, 6-fluoro-	EtOH	224(4.5),260s(--), 347(3.9)	24-1038-76
$C_{18}H_{11}NO_3$			
1H-2-Benzopyran-1,3(4H)-dione, 4-(1H- indol-3-ylmethylene)-, (Z)-	EtOH	220(4.54),275(4.18), 430(4.49)	4-0083-76
$C_{18}H_{11}NO_4$			
5H-Benzo[g]-1,3-benzodioxolo[6,5,4-de]- quinoline-5,6(7H)-dione, 7-methyl- (cepharadione A)	MeOH	239(3.98),278(3.98), 301(4.17),313(4.20), 433(4.11)	102-1323-76
$C_{18}H_{11}N_3O_6$			
1,1':2',1"-Terphenyl, 4,4',4"-trinitro-	EtOH	285(4.49)	22-0901-76
1,1':3',1"-Terphenyl, 2,2',4"-trinitro-	EtOH	267(4.29)	22-0901-76
1,1':3',1"-Terphenyl, 2',4,4"-trinitro-	EtOH	274(4.39)	22-0901-76
1,1':3',1"-Terphenyl, 4,4',4"-trinitro-	EtOH	293(4.44)	22-0901-76
1,1':4',1"-Terphenyl, 2,2',4"-trinitro-	EtOH	220(4.46),268(4.30)	22-0901-76
1,1':4',1"-Terphenyl, 2',4,4"-trinitro-	EtOH	219(4.39),303(4.46)	22-0901-76
$C_{18}H_{12}$			
Benzene, 1,1'-(1,2-diethynyl-1,2-ethene- diyl)bis-	n.s.g.	265(4.42),268(4.43), 280(4.54),288(4.40), 296(4.49)	104-0223-76
$C_{18}H_{12}ClNO$			
Benzeneacetonitrile, 4-chloro-α-(1-meth- oxy-3-phenyl-2-propynylidene)-	EtOH	210(4.02),228(4.27), 343(4.48)	24-0115-76
$C_{18}H_{12}Cl_2N_2O_5$			
Benzenepropanoic acid, 5-(2,3-dichloro- 5,6-dicyano-4-hydroxyphenoxy)-2-hy- droxy-, methyl ester	n.s.g.	252(4.53),387(3.97)	39-0407-76C
$C_{18}H_{12}Cl_2N_3$			
1,2,4-Triazolo[4,3-a]pyridinium, 1,3- bis(4-chlorophenyl)-, tetrafluoro- borate	MeOH	234(4.29),250(4.27), 276s(--),303s(--)	48-0881-76
$C_{18}H_{12}Cl_2O_3$			
1H-Inden-1-one, 2-[(acetyloxy)(2,4-di- chlorophenyl)methylene]-2,3-dihydro-	MeOH	239s(3.92),280(4.24)	83-0376-76
$C_{18}H_{12}FNO$			
Benzeneacetonitrile, 4-fluoro-α-(1-meth- oxy-3-phenyl-2-propynylidene)-	EtOH	208(3.86),224(4.14), 336(4.45)	34-0115-76

Compound	Solvent	$\lambda_{max}(\log \epsilon)$	Ref.
$C_{18}H_{12}F_6O$			
9(10H)-Phenanthrenone, 10,10-dimethyl-3,6-bis(trifluoromethyl)-	hexane	245s(--),267s(--), 274(4.02),292s(--), 321(3.53)	104-1298-76
$C_{18}H_{12}N_2$			
Benzo[1,2:3,4]dicycloheptene-6,7-dicarbonitrile	CH_2Cl_2	240(4.51),294(4.06), 409(4.31)	88-0251-76
$C_{18}H_{12}N_2O$			
Benzeneacetonitrile, α,α'-(oxydimethylidyne)bis-, (E,E)-	EtOH	294(4.26)	28-0319-76A
(E,Z)-	EtOH	302(4.40)	28-0319-76A
(Z,Z)-	EtOH	305(4.40)	28-0319-76A
$C_{18}H_{12}N_2O_2$			
Benzo[a]phenazin-5-ol, acetate	MeOH	223(4.62),278(4.68), 358s(3.84),385(4.02), 404(4.09)	49-0879-76
2H-1,4-Benzoxazine, 2,2'-(1,2-ethanediylidene)bis-	MeOH	267(4.44),465(4.52)	78-1407-76
	MeOH-acid	267(4.36),350s(3.84), 544(4.49)	78-1407-76
Methanone, 4,5-pyridazinediylbis[phenyl-	$CHCl_3$	266(4.26)	24-3886-76
$C_{18}H_{12}N_2O_2S$			
Benzenamine, 2-methyl-N-2H-naphtho-[1,8-bc]thien-2-ylidene-4-nitro-	dioxan	380(4.00)	103-0857-76
$C_{18}H_{12}N_2O_3$			
Benzoic acid, 2-[(4-cyano-1,2-dihydro-1-oxo-3-isoquinolinyl)methyl]-	EtOH	253(4.11),295(4.13)	95-0154-76
4H-1-Benzopyran-4-one, 7-hydroxy-3-(1-phenyl-1H-pyrazol-4-yl)-	EtOH	270(4.61)	103-0914-76
$C_{18}H_{12}N_2O_4$			
1,1':2',1"-Terphenyl, 2,4"-dinitro-	EtOH	294(4.12)	22-0901-76
1,1':2',1"-Terphenyl, 4,4'-dinitro-	EtOH	287.5(4.35)	22-0901-76
1,1':2',1"-Terphenyl, 4,4"-dinitro-	EtOH	290.5(4.37)	22-0901-76
1,1':2',1"-Terphenyl, 4',4"-dinitro-	EtOH	297.5(4.23)	22-0901-76
1,1':3',1"-Terphenyl, 2',4-dinitro-	EtOH	none	22-0901-76
1,1':3',1"-Terphenyl, 4,4'-dinitro-	EtOH	300(4.20)	22-0901-76
1,1':3',1"-Terphenyl, 4,4"-dinitro-	EtOH	305(4.50)	22-0901-76
1,1':3',1"-Terphenyl, 4',4"-dinitro-	EtOH	297(4.35)	22-0901-76
1,1':4',1"-Terphenyl, 2,4"-dinitro-	EtOH	220(4.32),314(4.35)	22-0901-76
1,1':4',1"-Terphenyl, 4,4"-dinitro-	EtOH	223(4.30),334(4.55)	22-0901-76
$C_{18}H_{12}N_2S$			
Thieno[3,4-d]pyridazine, 5,7-diphenyl-	CH_2Cl_2	388(4.0)	28-0607-76A
$C_{18}H_{12}N_2S_2$			
2H-1,4-Benzothiazine, 2,2'-(1,2-ethanediylidene)bis-	MeOH	283(4.40),330s(4.08), 507(4.45)	78-1407-76
	MeOH-acid	285(4.34),360s(3.94), 620(4.46)	78-1407-76
$C_{18}H_{12}N_2Te$			
Tellurolo[3,4-d]pyridazine, 5,7-diphenyl-	$CHCl_3$	276(4.30),350(3.98), 417(4.01)	24-3886-76

Compound	Solvent	$\lambda_{max}(\log \epsilon)$	Ref.
$C_{18}H_{12}O$			
Indeno[1,2-b]pyran, 2-phenyl-	dioxan	227(4.34),258(4.28), 291(4.26),367(4.32), 512(3.49)	78-2219-76
$C_{18}H_{12}O_2$			
5,7,13,15-Cyclohexadecatetraene-2,9,11- triyne-1,4-dione, 8,13-dimethyl-	$CHCl_3$	266(4.38),277s(4.32), 312(4.40),343s(4.16), 398(4.10),422s(4.02)	88-3841-76
Naphtho[2,3-c]furan-1(3H)-one, 4-phenyl-	EtOH	241(4.65),291(3.72), 301(3.74),331s(3.41), 342(3.49)	4-0741-76
$C_{18}H_{12}O_2S$			
Thieno[3,2-b]furan-2(5H)-one, 3,6-di- phenyl-	CH_2Cl_2	350(4.25)	28-0607-76A
3,4-Thiophenedicarboxaldehyde, 2,5-di- phenyl-	CH_2Cl_2	311(4.0)	28-0607-76A
Tribenzo[b,d,f]thiepin 9,9-dioxide	EtOH	222(4.49),227(4.56), 234(4.67),241(4.67), 277(3.95),290(3.83), 323(2.75)	39-1577-76C
$C_{18}H_{12}O_3$			
12H-Benzo[a]xanthen-12-one, 3-methoxy-	n.s.g.	212(4.42),236(4.40), 260(4.40),300(3.94), 318(3.92),368(3.58)	39-2241-76C
$C_{18}H_{12}O_4$			
Benzo[1,2-b:3,4-b']difuran-3(2H)-one, 4-methoxy-2-(phenylmethylene)-	EtOH	221(4.48),268(4.66), 306(4.29),337(3.93)	102-1553-76
4a,9a-[2]Butenoanthracene-1,4,9,10- tetrone	MeOH	227(4.65),245s(4.48), 296(3.80)	44-2296-76
Derriobtusone A	EtOH	258(6.02),263(5.99), 327(6.06)	102-1553-76
5,13:6,12-Diepoxydibenzo[a,f]cyclodec- ene-7,14-dione, 5,6,12,13-tetrahydro-, anti	MeCN	240(4.45),291(3.42)	1-0619-76
syn	MeCN	250(4.28),296(3.34)	1-0619-76
5,6,11,12-Naphthacenetetrone, 1,4,4a,12a-tetrahydro-	MeOH	213(4.32),230s(4.22), 253(4.17),315(3.61)	44-2296-76
$C_{18}H_{12}O_4S$			
2,5-Thiophenedicarboxylic acid, 3,4-di- phenyl-	EtOH	242(4.32),285(4.01)	42-0490-76
$C_{18}H_{12}O_5$			
2,6,11(1H)-Naphthacenetrione, 3,4-di- hydro-5,12-dihydroxy-	MeOH	252(4.44),256(4.43), 288(3.85),457(3.81), 482(3.86),514(3.72)	44-2296-76
$C_{18}H_{12}O_8$			
6H-[1,3]Benzodioxolo[5,6-c][1]benzo- pyran-6-one, 2,3-diacetoxy-	EtOH	224(4.46),238s(--), 249(4.67),255(4.71), 281s(--),310(4.21)	24-0855-76
$C_{18}H_{12}S_4$			
1,3-Dithiole, 4-phenyl-2-(4-phenyl- 1,3-dithiol-2-ylidene)-	THF	258(4.56),331(4.36), 416(3.77)	97-0360-76
cation	MeCN	330(--),460(--), 680(--)	97-0360-76

Compound	Solvent	$\lambda_{max}(\log \epsilon)$	Ref.
1,3-Dithiole, 4-phenyl-2-(4-phenyl-1,3-dithiol-2-ylidene)-, dication	MeCN	279(4.63),484(4.04)	97-0360-76
$C_{18}H_{13}BrN_2O$ 1-Naphthalenecarboxaldehyde, 2-hydroxy-, [(4-bromophenyl)methylene]hydrazone	EtOH	231(4.65),276(4.33), 290s(4.18),323s(4.25), 333(4.29),384(4.40)	95-0044-76
2-Naphthalenecarboxaldehyde, 1-hydroxy-, [(4-bromophenyl)methylene]hydrazone	EtOH	220(4.39),252s(3.90), 259(3.96),288s(4.43), 296(4.47),321s(4.26), 330(4.29),387(4.14)	95-0044-76
$C_{18}H_{13}BrO_3$ Methanone, (3-acetoxy-1H-inden-2-yl)-(4-bromophenyl)-	MeOH	228(4.05),312(4.40)	83-0376-76
$C_{18}H_{13}ClN_2$ 1H-Pyrido[2,3-b]indole, 2-chloro-4-methyl-1-phenyl-	n.s.g.	210(4.16),224(4.17), 243(4.06),273(4.36), 282(4.39),329(4.10)	103-0329-76
$C_{18}H_{13}ClN_2O$ 1-Naphthalenecarboxaldehyde, 2-hydroxy-, [(4-chlorophenyl)methylene]hydrazone	EtOH	231(4.70),275(4.33), 289s(4.16),324s(4.24), 332(4.29),383(4.38), 400s(4.30)	95-0044-76
2-Naphthalenecarboxaldehyde, 1-hydroxy-, [(4-chlorophenyl)methylene]hydrazone	EtOH	220(4.52),250s(4.03), 259(4.08),289s(4.58), 295(4.61),320s(4.38), 329(4.41),388(4.28)	95-0044-76
$C_{18}H_{13}ClN_2O_4$ 2-Oxa-6,7-diazabicyclo[2.2.1]heptane-3,5-dione, 6-(4-chlorobenzoyl)-7-methyl-1-phenyl-	MeOH	246(4.42),263s(4.21), 330(3.76)	44-0813-76
$C_{18}H_{13}ClN_3$ 1,2,4-Triazolo[4,3-a]pyridinium, 1-(4-chlorophenyl)-3-phenyl-, tetrafluoroborate	MeOH	233(4.23),244s(--), 278s(--),304(3.99)	48-0881-76
1,2,4-Triazolo[4,3-a]pyridinium, 3-(4-chlorophenyl)-1-phenyl-, tetrafluoroborate	MeOH	226(4.31),247(4.30), 277s(--),306s(--)	48-0881-76
$C_{18}H_{13}ClO_2$ 1,4-Naphthalenedione, 2-chloro-3-(2-phenylethyl)-	MeOH	246(4.20),252(4.22), 274(4.15),339(3.26)	39-0997-76C
$C_{18}H_{13}ClO_7$ 2-Anthracenecarboxylic acid, 6-chloro-9,10-dihydro-5,8-dihydroxy-1-methoxy-3-methyl-	EtOH	234(4.36),257(4.32), 294(3.87),350(3.31), 470(3.88)	44-3018-76
$C_{18}H_{13}Cl_2NO_6S$ Furan, 2-[1-[(dichloromethyl)sulfonyl]-2-[5-(4-methylphenyl)-2-furanyl]-ethenyl]-5-nitro-	EtOH	211(4.08),250(4.19), 330(4.29),412(4.19)	73-3391-76
$C_{18}H_{13}Cl_2N_3O$ Imidazo[1,2-a]pyrimidin-5(8H)-one,	EtOH	210(4.49),304(3.59)	33-1203-76

Compound	Solvent	λ_{max}(log ϵ)	Ref.
8-(2,6-dichlorophenyl)-2,3-dihydro-7-phenyl- (cont.)	EtOH-HCl	210(4.54),269(3.92)	33-1203-76
Imidazo[1,2-a]pyrimidin-7(3H)-one, 8-(2,6-dichlorophenyl)-2,8-dihydro-5-phenyl-	EtOH EtOH-HCl	234(4.32),325(3.79) 243(4.20),281(4.00)	33-1203-76 33-1203-76

$C_{18}H_{13}NO$

Compound	Solvent	λ_{max}(log ϵ)	Ref.
Benzeneacetonitrile, 4-methyl-α-(1-oxo-3-phenyl-2-propynyl)-	pH 13	220(4.10),234(4.24), 264s(4.08),352(4.32)	34-0115-76
	EtOH	211(4.25),232(4.22), 346(4.41)	34-0115-76
1H-Benz[f]isoindol-1-one, 2,3-dihydro-4-phenyl-	EtOH	238(4.85),288(3.93), 298(3.95),318(3.47), 324(3.55)	44-2571-76
1H-Benz[f]isoindol-1-one, 2,3-dihydro-9-phenyl-	EtOH	241(4.82),278s(3.71), 289(3.88),300(3.88), 323(3.39),336(3.49)	44-2571-76
2-Propynamide, 3-phenyl-N-(3-phenyl-2-propynyl)-	EtOH	242(4.51),252(4.51)	44-2571-76

$C_{18}H_{13}NOS$

Compound	Solvent	λ_{max}(log ϵ)	Ref.
Benzoxazole, 2-[5-(4-methylphenyl)-2-thienyl]-	CH_2Cl_2	257s(3.79),349(4.45), 360s(4.44),410s(3.02)	48-0731-76

$C_{18}H_{13}NO_2$

Compound	Solvent	λ_{max}(log ϵ)	Ref.
Benzeneacetonitrile, α-[3-(4-methoxyphenyl)-1-oxo-2-propynyl]-	pH 13	224(4.07),263(4.02), 348(4.18)	34-0115-76
	EtOH	217(4.06),236(4.11), 255(4.04),344(4.37)	34-0115-76
1,1':2',1"-Terphenyl, 2-nitro-	EtOH	235(4.29)	22-0901-76
1,1':2',1"-Terphenyl, 4-nitro-	EtOH	235(4.14),298(4.02)	22-0901-76
1,1':2',1"-Terphenyl, 4'-nitro-	EtOH	243(4.16),297(3.99)	22-0901-76
1,1':3',1"-Terphenyl, 2'-nitro-	EtOH	232(4.45)	22-0901-76
1,1':3',1"-Terphenyl, 4-nitro-	EtOH	227(4.37),302(4.21)	22-0901-76
1,1':3',1"-Terphenyl, 4'-nitro-	EtOH	232(4.33)	22-0901-76
1,1':4',1"-Terphenyl, 2-nitro-	EtOH	258(4.45)	22-0901-76
1,1':4',1"-Terphenyl, 2'-nitro-	EtOH	257(4.44)	22-0901-76
1,1':4',1"-Terphenyl, 4-nitro-	EtOH	327(4.35)	22-0901-76

$C_{18}H_{13}NO_2S$

Compound	Solvent	λ_{max}(log ϵ)	Ref.
Thiophene, 2-(4-nitrophenyl)-5-(2-phenylethenyl)-	MeOH	337(4.16),411(4.53)	48-0731-76

$C_{18}H_{13}NO_3$

Compound	Solvent	λ_{max}(log ϵ)	Ref.
1-Naphthalenecarboxylic acid, 4-(aminocarbonyl)-3-phenyl-	EtOH	246(4.32),303(3.60)	22-1829-76

$C_{18}H_{13}NO_4$

Compound	Solvent	λ_{max}(log ϵ)	Ref.
Benzeneacetic acid, 2-carboxy-α-(1H-indol-3-ylmethylene)-, (Z)-	EtOH	280(3.97),327(4.09)	4-0083-76

$C_{18}H_{13}NO_4S$

Compound	Solvent	λ_{max}(log ϵ)	Ref.
Benzoic acid, 4-[5-(4-nitrophenyl)-2-thienyl]-, methyl ester	MeOH	290s(3.81),311s(3.94), 382(4.48)	48-0731-76

$C_{18}H_{13}NO_5$

Compound	Solvent	λ_{max}(log ϵ)	Ref.
1H-Inden-1-one, 2-[(acetyloxy)(4-nitrophenyl)methylene]-2,3-dihydro-, (Z)-	MeOH	231(4.07),306(4.29)	83-0376-76
Methanone, (3-acetoxy-1H-inden-2-yl)-(4-nitrophenyl)-	MeOH	311(4.37)	83-0376-76

Compound	Solvent	$\lambda_{max}(\log \epsilon)$	Ref.
$C_{18}H_{13}NO_6$			
2,3-Anthracenedicarboxylic acid, 1-amino-9,10-dihydro-9,10-dioxo-, dimethyl ester	CH_2Cl_2	<u>465(3.9)</u>	5-1873-76
2,3-Anthracenedicarboxylic acid, 5-amino-9,10-dihydro-9,10-dioxo-, dimethyl ester	CH_2Cl_2	<u>475(3.8)</u>	5-1873-76
2,3-Anthracenedicarboxylic acid, 6-amino-9,10-dihydro-9,10-dioxo-, dimethyl ester	CH_2Cl_2	<u>300s(4.3),420(3.6)</u>	5-1873-76
$C_{18}H_{13}NS$			
Benzonitrile, 4-[5-(4-methylphenyl)-2-thienyl]-	MeOH	248(4.08),354(4.51)	48-0731-76
$C_{18}H_{13}N_3$			
Propanedinitrile, [3-phenyl-3-(phenylamino)-2-propenylidene]-	CH_2Cl_2	405(4.57)	48-0705-76
$C_{18}H_{13}OS$			
[1]Benzothieno[3,2-b]pyrylium, 2-methyl-4-phenyl-, perchlorate	$C_2H_4Cl_2$	255(3.70),292(3.70), 350(3.90),370(3.94), 387(3.98),427(--)	103-0977-76
$C_{18}H_{14}$			
1,1':2',1"-Terphenyl	EtOH	none	22-0901-76
	EtOH	233(4.42)	33-2012-76
1,1':3',1"-Terphenyl	CHCl$_3$	252(4.64)	22-0901-76
1,1':4',1"-Terphenyl	hexane	275(4.54)	22-0901-76
	CHCl$_3$	280(4.40)	22-0901-76
$C_{18}H_{14}BF_4N_3$			
1,2,4-Triazolo[4,3-a]pyridinium, 1,3-diphenyl-, tetrafluoroborate	MeOH	231(4.11),245(4.10), 275(3.92),303(3.90)	48-0881-76
$C_{18}H_{14}BrNO_5S$			
1H-Pyrrolo[2,1-c][1,4]thiazine-7-carboxylic acid, 1-(4-bromophenyl)-1-[(2-carboxyethenyl)oxy]-4-methyl-	MeOH	260(4.29),287(4.38)	44-0187-76
$C_{18}H_{14}BrN_3O_3$			
8-Azabicyclo[3.2.1]oct-3-en-2-one, 6-(4-bromophenyl)-8-(5-nitro-2-pyridinyl)-, endo	EtOH	212(4.36),225(4.36), 353(4.30)	39-2307-76C
2,9-Methano-3H-indeno[2,1-b]pyridin-3-one, 6-bromo-1,2,4,4a,9,9a-hexahydro-1-(5-nitro-2-pyridinyl)-	CHCl$_3$	363(4.40)	118-0105-76
$C_{18}H_{14}BrN_3O_7$			
1-Azuleneethanol, 2-bromo-, trinitrobenzene complex	CH_2Cl_2	285(4.79),295(4.80), 335(3.69),349(3.76), 362(3.13),568(2.58), 605(2.53),665(2.14)	44-1811-76
$C_{18}H_{14}Br_2N_4O$			
Acetic acid, (4-bromophenyl)[[1-(4-bromophenyl)-1H-pyrazol-3-yl]methylene]-hydrazide	EtOH	218(4.56),301(4.78)	136-0185-76D

Compound	Solvent	$\lambda_{max}(\log \epsilon)$	Ref.
$C_{18}H_{14}Br_2S$ Thiophene, 3,4-bis(bromomethyl)-2,5-di-phenyl-	CH_2Cl_2	296(4.2)	28-0607-76A
$C_{18}H_{14}ClNO_2$ Cyclohepta[b]pyrrole-3-carboxylic acid, 2-chloro-4-phenyl-, ethyl ester	MeOH	230(4.32),283(4.68), 337(3.89),468(3.44)	142-0221-76
Cyclohepta[b]pyrrole-3-carboxylic acid, 2-chloro-6-phenyl-, ethyl ester	MeOH	283s(4.49),298(4.55), 313(4.57),345(4.34), 361(4.36),443(3.34)	142-0221-76
Cyclohepta[b]pyrrole-3-carboxylic acid, 2-chloro-8-phenyl-, ethyl ester	MeOH	279(4.56),300s(4.42), 338s(4.01),447(3.20)	142-0221-76
$C_{18}H_{14}ClNO_4$ 1H-Indole-4,7-diol, 2-(4-chlorophenyl)-, diacetate	MeOH	211(4.53),250(4.41), 308(4.42)	111-0241-76
$C_{18}H_{14}ClN_3O$ Acetamide, N-[4-(3-chlorophenyl)-6-phenyl-2-pyrimidinyl]-	EtOH	254(4.57),318(4.21), 291s(4.05)	4-0257-76
Acetamide, N-[4-(4-chlorophenyl)-6-phenyl-2-pyrimidinyl]-	EtOH	258(4.53),318(4.23), 292(4.03)	4-0257-76
$C_{18}H_{14}ClN_3O_3$ 8-Azabicyclo[3.2.1]oct-3-en-2-one, 6-(4-chlorophenyl)-8-(5-nitro-2-pyridinyl)-, endo	EtOH	215(4.33),227(4.33), 349(4.29)	39-2307-76C
2,9-Methano-3H-indeno[2,1-b]pyridin-3-one, 6-chloro-1,2,4,4a,9,9a-hexa-hydro-1-(5-nitro-2-pyridinyl)-	$CHCl_3$	363(4.34)	118-0105-76
$C_{18}H_{14}ClN_3O_4$ Pyridinium, 1-[4,4-dicyano-1-(4-methyl-phenyl)-1,3-butadienyl]-, perchlorate	HOAc	361(4.39)	48-0705-76
$C_{18}H_{14}ClN_3O_7$ 1-Azuleneethanol, 2-chloro-, trinitro-benzene complex	CH_2Cl_2	283(4.85),291(4.86), 304(3.90),324(3.55), 334(3.72),348(3.81), 565(2.56),600(2.52), 660(2.11)	44-1811-76
$C_{18}H_{14}Cl_5NO_5$ L-Threonine, N-[(phenylmethoxy)carbo-nyl]-, pentachlorophenyl ester	EtOH	252s(3.65),301s(2.81), 325(3.03)	136-0001-76E
	NaOH	256s(4.10),311s(3.70), 324(3.80)	136-0001-76E
$C_{18}H_{14}Cl_9NO_{13}$ 2H-1,3-Oxazine-2,6(3H)-dione, 3-[2,3,5-tris-O-[(2,2,2-trichloroethoxy)carbo-nyl]-β-D-ribofuranosyl]-	dioxan	260(3.78)	87-0643-76
$C_{18}H_{14}F_2N_3S_2$ Benzothiazolium, 2-[1,2-difluoro-2-[(3-methyl-2(3H)-benzothiazolylidene)ami-no]ethenyl]-3-methyl-, tetrafluoro-borate	CH_2Cl_2	477(4.70)	124-0204-76

Compound	Solvent	$\lambda_{max}(\log \epsilon)$	Ref.
$C_{18}H_{14}N_2$			
1H-Pyrido[2,3-b]indole, 4-methyl-1-phenyl-	n.s.g.	205(4.61),221(4.61), 241(4.19),265(4.32), 281(4.44),323(4.06), 409(3.56)	103-0329-76
$C_{18}H_{14}N_2O$			
1-Naphthalenecarboxaldehyde, 2-hydroxy-, (phenylmethylene)hydrazone	toluene	290(4.06),324(4.19), 334(4.26),383(4.34), 400(4.26)	95-0044-76
	EtOH	231(4.66),266s(4.31), 272(4.34),286s(4.10), 322s(4.21),330(4.27), 380(4.34)	95-0044-76
$C_{18}H_{14}N_2OS$			
2-Thiazolepropanenitrile, 2,5-dihydro-5-oxo-2,4-diphenyl-	MeOH	208(4.34),254(4.32), 274(4.30)	78-0571-76
$C_{18}H_{14}N_2O_2$			
3H-Benz[f]indazol-3-one, 1,2-dihydro-4-hydroxy-1-methyl-2-phenyl-	EtOH	262(4.51),269(5.51), 384(4.17)	94-0596-76
1,3-Diazetidine-2,4-dione, 1,3-bis-(2-phenylethenyl)-, (E,E)-	THF	309(4.58)	95-0962-76
Ellipticine, 8,9-(methylenedioxy)-	EtOH	213(4.60),229(4.58), 249(4.46),273s(4.66), 283(4.76),316(4.86), 346(4.19)	39-1155-76C
1H-Indene-1,3(2H)-dione, 2-(1,3-dihydro-1,3-dimethyl-2H-benzimidazol-2-ylidene)-	EtOH	252(4.25),299(4.30), 340(4.50),390s(3.35)	73-1363-76
1H-Pyrano[3,4-c]pyridin-1-one, 5-ethyl-3-(1H-indol-2-yl)-	EtOH	230(4.22),269(3.86), 382(4.39)	88-1085-76
Vincarpine	pH 1	225(4.17),247s(3.81), 332(3.89),402(3.70)	88-4887-76
	pH 13	292(3.81),318(3.68), 390(3.86)	88-4887-76
	EtOH	222s(4.22),250s(3.77), 315(3.81),395(3.73)	88-4887-76
$C_{18}H_{14}N_2O_2S$			
1H-[1]Benzothiepino[3,4-c]pyrazole-3,4-(2H,10H)-dione, 1-methyl-2-phenyl-	EtOH	240(4.17),325(3.99)	94-0596-76
	EtOH-NaOH	325(4.08)	94-0596-76
	+ NaOH	325(4.20)	94-0596-76
5H-Thiazolo[3,4-b]indazol-4-ium, 1-(ethoxycarbonyl)-3-phenyl-, hydroxide, inner salt	CHCl$_3$	269(4.23),306(4.39), 322s(4.35),363(3.90), 375s(3.86),542(3.92)	44-0129-76
$C_{18}H_{14}N_2O_2Se$			
Benzenamine, N-[6-[4-(methoxycarbonyl)-phenyl]-2H-1,3-selenazin-2-ylidene]-, perchlorate	HOAc	418(3.69)	88-2005-76
$C_{18}H_{14}N_2O_4$			
2-Oxa-6,7-diazabicyclo[2.2.1]heptane-3,5-dione, 6-benzoyl-7-methyl-1-phenyl-	MeOH	240(4.42),259s(4.05), 323(3.75)	44-0813-76
Pyrrolo[2,1-a]isoquinoline-2-acetic acid, 3-cyano-1-(methoxycarbonyl)-, methyl ester	MeOH	206(3.95),220s(3.93), 224(3.94),239s(3.92), 243(3.95),255s(3.96),	39-1908-76C

Compound	Solvent	$\lambda_{max}(\log \epsilon)$	Ref.
Pyrrolo[2,1-a]isoquinoline-2-acetic acid, 3-cyano-1-(methoxycarbonyl)-, methyl ester (cont.)		264(4.19),270s(4.21), 273(4.24),318(3.52), 334(3.52),351(3.56)	39-1908-76C
$C_{18}H_{14}N_2O_4S$			
2-Furancarboximidothioic acid, N-hydroxy-5-(2-nitrophenyl)-, phenylmethyl ester	EtOH	211(4.42),298(4.23)	73-3085-76
2-Furancarboximidothioic acid, N-hydroxy-5-(3-nitrophenyl)-, phenylmethyl ester	EtOH	211(4.26),320(4.19)	73-3085-76
2-Furancarboximidothioic acid, N-hydroxy-5-(4-nitrophenyl)-, phenylmethyl ester	EtOH	208(4.40),373(4.42)	73-3085-76
$C_{18}H_{14}N_2O_6S$			
2-Furancarboximidic acid, N-hydroxy-5-(2-nitrophenyl)-, anhydride with 4-methylbenzenesulfinic acid	EtOH	208(4.42),229(4.41), 310(4.38)	73-3085-76
2-Furancarboximidic acid, N-hydroxy-5-(3-nitrophenyl)-, anhydride with 4-methylbenzenesulfinic acid	EtOH	207(4.26),228(4.27), 321(4.37)	73-3085-76
2-Furancarboximidic acid, N-hydroxy-5-(4-nitrophenyl)-, anhydride with 4-methylbenzenesulfinic acid	EtOH	206(4.40),231(4.40), 364(4.54)	73-3085-76
$C_{18}H_{14}N_2S$			
1H-Benzimidazole, 2-[5-(4-methylphenyl)-2-thienyl]-	CH_2Cl_2	262(3.99),358(4.52)	48-0731-76
$C_{18}H_{14}N_3$			
Pyridinium, 1-[4,4-dicyano-1-(4-methylphenyl)-1,3-butadienyl]-, perchlorate	HOAc	361(4.39)	48-0705-76
1,2,4-Triazolo[4,3-a]pyridinium, 1,3-diphenyl-, tetrafluoroborate	MeOH	231(4.11),245(4.10), 275(3.92),303(3.90)	48-0881-76
$C_{18}H_{14}N_4$			
1H-Indole-3-carboxaldehyde, (1H-indol-3-ylmethylene)hydrazone	DMF	275(4.15),338s(4.53), 350(4.58),364(4.42)	5-1039-76
	DMF-tert-BuOK	384s(4.45),403(4.58), 415s(4.55)	5-1039-76
$C_{18}H_{14}N_4O$			
1(2H)-Phthalazinone, 2-methyl-4-(4-phenyl-1H-imidazol-2-yl)-	DMF	339(4.08),413(3.38)	44-0836-76
4(1H)-Pyrimidinone, 6-amino-5-(1H-indol-3-yl)-2-phenyl-	EtOH	205(4.66),222(4.60), 278(3.93),288(3.89)	103-1246-76
$C_{18}H_{14}N_4O_2$			
6H-1,2,4,5-Tetrazino[1,6-b]isoquinolin-6-one, 2-benzoyl-1,2-dihydro-1-methyl-	EtOH	232(4.44),312(4.12), 368(3.88),386(3.77)	95-0700-76
$C_{18}H_{14}N_6O$			
1H-Pyrazole, 1,1'-(azoxydi-4,1-phenylene)bis-	EtOH	244(3.81),365(4.18)	22-0195-76
$C_{18}H_{14}O$			
7H-Benz[de]anthracen-7-ol, 7-methyl-	EtOH	222(4.2),229(4.34), 247(4.15),312(4.11), 328(4.28),343(4.25)	42-0286-76

Compound	Solvent	$\lambda_{max}(\log \epsilon)$	Ref.
7H-Benz[de]anthracen-7-ol, 7-methyl-, cation	H_2SO_4	256(4.9),283(5.14), 357(4.0),408(4.54), 475(4.14),565(3.95)	42-0286-76
$C_{18}H_{14}OS$			
Indeno[2,1-b]pyran, 4,9-dimethyl-2-(2-thienyl)-	dioxan	232(4.35),270(4.14), 306(4.18),341(4.25), 356(4.26),523(3.52)	78-2219-76
1-Naphthaleneethanethioic acid, S-phenyl ester	MeOH	273(3.8),283(3.8), 294(3.7)	149-0383-76B
$C_{18}H_{14}OSe$			
Methanone, [5-(4-methylphenyl)seleno-phene-2-yl]phenyl-	MeCN	353(4.34)	118-0521-76
$C_{18}H_{14}O_2$			
Benzo[1,2:4,5]dicycloheptene-3,9-dione, 2,8(or 10)-dimethyl-	$CHCl_3$	253(4.40),302(5.11), 324s(4.46),340s(4.34), 354s(4.03),385(3.70), 407(3.63)	22-0914-76
1,4-Naphthalenedione, 2-methyl-3-(phen-ylmethyl)-	n.s.g.	247(4.26),260s(4.18), 268(4.19),327(3.46)	33-0664-76
Spiro[9H-fluorene-9,2'(5'H)-furan]-5'-one, 3',4'-dimethyl-	n.s.g.	227(4.54),234(4.45), 274(4.08),285(4.00)	35-3627-76
$C_{18}H_{14}O_2S$			
Benzoic acid, 2-(3-phenyl-2-thienyl)-, methyl ester	MeOH	225(4.31)	44-1320-76
Benzoic acid, 2-(5-phenyl-2-thienyl)-, methyl ester	MeOH	313(4.21)	44-1320-76
3-Thiophenecarboxaldehyde, 4-(hydroxy-methyl)-2,5-diphenyl-	CH_2Cl_2	261(4.34),310s(3.95)	28-0607-76A
2-Thiophenecarboxylic acid, 5-[2-(2-naphthalenyl)ethenyl]-, methyl ester	MeOH	244(4.54),271s(4.21), 283(4.25),297(4.24), 333s(4.36),345(4.40), 354s(4.39),379s(4.21)	48-0731-76
$C_{18}H_{14}O_3$			
Benzaldehyde, 2-[(6-methoxy-2-naphtha-lenyl)oxy]-	n.s.g.	210(4.54),219(4.83), 268(3.94),278(3.74), 320(3.74)	39-2241-76C
2H-1-Benzopyran-2-one, 3-benzoyl-4-eth-yl-	EtOH	254(3.93),280(3.95), 312(3.73)	39-1260-76C
1-Benzoxepin-2(3H)-one, 3-benzoyl-5-methyl-	EtOH	247(4.34)	39-1260-76C
1H-Indene-3-carboxylic acid, 2,3-di-hydro-1-oxo-2-(phenylmethylene)-, methyl ester	MeOH	230(4.09),250(4.08)	44-3540-76
1H-Indene-5-carboxylic acid, 2,3-di-hydro-3-oxo-2-(phenylmethylene)-, methyl ester	MeCN	234(3.48),318(4.41)	44-3540-76
$C_{18}H_{14}O_4$			
9,10-Anthracenedione, 1-acetoxy-4-ethyl-	EtOH	206(4.42),252(4.33), 338(3.46),408(3.27)	56-0759-76
Benzo[1,2:4,5]dicycloheptene-3,9-dione, 2,8-dimethoxy-	$CHCl_3$	259(4.19),267(4.20), 312(5.03),327s(4.55), 348s(4.24),368(4.24), 414(3.65),440(3.57)	22-0914-76

Compound	Solvent	$\lambda_{max}(\log \epsilon)$	Ref.
Benzo[1,2:4,5]dicycloheptene-3,9-dione, 2,10-dimethoxy-	$CHCl_3$	260(4.20),312(4.93), 365s(4.14),398(3.68), 424(3.57)	22-0914-76
1,4-Naphthalenedione, 2-hydroxy-3-[2-(3-hydroxyphenyl)ethyl]-	MeOH	255(4.36),277(4.39), 342s(2.75)	39-0997-76C
	MeOH-base	235(4.40),277(4.52), 496(3.53)	39-0997-76C
$C_{18}H_{14}O_4S$			
Benzo[b]thiophene-6-carboxylic acid, 4-acetoxy-5-phenyl-, methyl ester	EtOH	211(4.41),240(4.61), 268(3.99),315(3.82)	4-0285-76
2-Propenoic acid, 3-phenyl-2-[(1-carboxy-2-phenylethenyl)thio]-	EtOH	272(4.09)	42-0038-76
$C_{18}H_{14}O_5$			
2,3-Furandione, 4-(1,2-benzodioxol-5-yl)dihydro-5-(4-methylphenyl)-	EtOH	298(4.18),321(4.23)	22-1541-76
5,12-Naphthacenedione, 7,8,9,10-tetrahydro-6,8,11-trihydroxy-	MeOH	252(4.60),256s(4.59), 287(3.98),457(3.94), 481(3.99),514(3.78)	44-2296-76
Unonal 7-methyl ether	EtOH	283(4.66),330(4.12)	2-0009-76
$C_{18}H_{14}O_6$			
6H-Benzofuro[3,2-c][1]benzopyran-6-one, 2,3,9-trimethoxy-	MeOH	235(4.52),353(4.47)	18-1955-76
2H-1-Benzopyran-2-one, 3-(1-acetoxy-4-methoxy-6-oxo-2,4-cyclohexadien-1-yl)-	MeOH	285(4.15),318(4.03)	18-1955-76
8H-1,3-Dioxolo[4,5-g][1]benzopyran-8-one, 7-(3,4-dimethoxyphenyl)-	EtOH	264(4.35),286s(4.13), 323(4.05),330s(4.02)	102-1019-76
2H-Pyran-3-carboxylic acid, 4-hydroxy-6-methoxy-5-(1-naphthalenyl)-2-oxo-, methyl ester	EtOH	246(4.46),315(4.13)	78-0269-76
$C_{18}H_{14}O_7$			
4H-1-Benzopyran-4-one, 3-(1,3-benzodioxol-5-yl)-7-hydroxy-5,8-dimethoxy-	MeOH MeOH-NaOAc	260(4.24),292s(--) 274(--)	2-0584-76 2-0584-76
$C_{18}H_{14}O_8$			
11H-Dibenzo[b,e][1,4]dioxepin-6-carboxylic acid, 4-formyl-3-hydroxy-8-methoxy-1,9-dimethyl-11-oxo-	n.s.g.	241(4.43),273s(3.94), 302(3.56)	12-1059-76
$C_{18}H_{14}S$			
4H-Thiopyran, 4-(diphenylmethylene)-	CH_2Cl_2	344(4.32)	44-2279-76
$C_{18}H_{15}BrN_2O_2$			
[3,3'-Bi-1H-indol]-2(3H)-one, 5-bromo-3-hydroxy-1',2'-dimethyl-	MeOH	215(3.63),220(3.63), 268s(3.06),294(2.93)	103-0178-76
$C_{18}H_{15}BrO_5$			
9,10-Anthracenedione, 3-(bromomethyl)-1,6,8-trimethoxy-	EtOH	223(4.59),279(4.43), 398(3.95)	39-0613-76C
2-Propenoic acid, 3-[4-[2-(4-bromophenyl)-2-oxoethoxy]-3-methoxyphenyl]-, (E)-	MeOH-NaOH	271(4.92)	20-0657-76
$C_{18}H_{15}ClN_2O$			
Acetamide, N-(6-chloro-2-methyl-4-phenyl-3-quinolinyl)-	EtOH	210s(4.50),233(4.66), 282(3.77),296s(3.70), 311(3.61),326(3.63)	44-1743-76

Compound	Solvent	$\lambda_{max}(\log \epsilon)$	Ref.
Acetamide, N-[(6-chloro-4-phenyl-2-quin-olinyl)methyl]-	EtOH	234(4.70),290(3.90), 310s(3.73),325(3.65)	44-1743-76
4(3H)-Quinazolinone, 2-[2-(2-chlorophen-yl)ethenyl]-3-ethyl-	EtOH	312(4.38)	115-0341-76
$C_{18}H_{15}ClN_2O_2$			
Acetamide, N-(6-chloro-2-methyl-4-phen-yl-3-quinolinyl)-, N-oxide	EtOH	224s(4.42),236(4.51), 252(4.55),331(4.04)	44-1743-76
1H-Pyrazole-1-carboxylic acid, 5-(4-chlorophenyl)-3-phenyl-, ethyl ester	EtOH	245s(4.39),266(4.46)	4-0257-76
$C_{18}H_{15}ClN_2O_3$			
Ethanone, 1-(5-chloro-2-hydroxy-4-meth-oxyphenyl)-2-(1-phenyl-1H-pyrazol-4-yl)-	EtOH	270(4.52),324(4.08)	103-0914-76
$C_{18}H_{15}ClN_2O_5S_2$			
7-Thia-2,6-diazabicyclo[2.2.1]heptane-2-sulfonic acid, 1-(4-chlorophenyl)-3,5-dioxo-6-phenyl-, ethyl ester	MeOH	260(4.15),285s(3.86), 411(4.13)	44-0813-76
$C_{18}H_{15}ClN_4O_4$			
1H-Pyrazole-4-carboxylic acid, 5-amino-3-(2-chlorophenyl)-1-(4-nitrophenyl)-, ethyl ester	EtOH	300(4.19)	4-1137-76
1H-Pyrazole-4-carboxylic acid, 5-amino-3-(4-chlorophenyl)-1-(4-nitrophenyl)-, ethyl ester	EtOH	305(4.24)	4-1137-76
$C_{18}H_{15}ClN_6O$			
4H-1,2,4-Triazolo[4,3-a][1,4]benzodiaz-epine, 6-(2-chlorophenyl)-1-methyl-8-(methyl-ONN-azoxy)-	isoPrOH	234(4.47),265s(4.14)	87-1378-76
$C_{18}H_{15}Cl_2N_3$			
1H-1,4-Benzodiazepine, 7-chloro-5-(2-chlorophenyl)-2-(cyanomethyl)-2,3-dihydro-1-methyl-	EtOH	227s(--),265s(--), 364(3.30)	111-0501-76
$C_{18}H_{15}FN_3S_2$			
Benzothiazolium, 2-[2-fluoro-2-[(3-meth-yl-2(3H)-benzothiazolylidene)amino]-ethenyl]-3-methyl-, tetrafluoroborate	MeNO$_2$	451(4.76)	124-0204-76
$C_{18}H_{15}F_4N$			
Benzenamine, N,N-dimethyl-4-(1,2,3,4-tetrafluoro-4-phenyl-1,3-butadienyl)-	heptane	350(4.30)	104-1565-76
$C_{18}H_{15}N$			
Isoquinoline, 1-[(ethenylphenyl)meth-yl]-, homopolymer	THF	<u>275(3.8),310(3.5), 320(3.6)</u>	116-0010-76
picrate	DMSO	283s(4.00),287s(3.87), 310s(3.80),332(3.92), 380(4.22)	116-0010-76
Isoquinoline, 1-[(4-ethenylphenyl)meth-yl]-, homopolymer	THF	267(3.67),272(3.71), 280s(3.61),308(3.40), 316s(3.36),321(3.51)	116-0010-76
$C_{18}H_{15}NO$			
1H-Benz[f]isoindol-1-one, 2,3,3a,4-tet-rahydro-4-phenyl-	EtOH	227(4.38),233(4.40), 293(4.16)	44-2571-76

Compound	Solvent	$\lambda_{max}(\log \epsilon)$	Ref.
1H-Benz[f]isoindol-1-one, 2,3,3a,4-tet-rahydro-9-phenyl-	EtOH	235(4.45),295(4.08)	44-2571-76
6H-Benzocyclohepten-5,9-imin-6-one, 5,9-dihydro-9-methyl-10-phenyl-	EtOH	204(3.34),239(3.29)	39-2285-76C
6,13-Methano-5H-dibenzo[b,g]azonin-14-one, 6,13-dihydro-5-methyl-	EtOH	248(4.31),275(4.33)	39-2285-76C
2-Propenamide, 3-phenyl-N-(3-phenyl-2-propynyl)-, cis	EtOH	223(4.20),250(4.42), 240(4.44)	44-2571-76
trans	EtOH	242(4.35),251(4.39), 279s(4.43),273(4.44)	44-2571-76
2-Propynamide, 4-phenyl-N-(3-phenyl-2-propenyl)-, cis	EtOH	249(4.52),258s(4.45)	44-2571-76
trans	EtOH	255(4.55)	44-2571-76
$C_{18}H_{15}NO_2$			
Murrayacine	EtOH	226(4.60),282(4.57), 301(4.58)	102-0356-76
2-Quinolinemethanol, 3-methyl-, benzo-ate	EtOH	233(4.90),260(3.77), 306(3.66),319(3.66)	95-0968-76
$C_{18}H_{15}NO_2S$			
3(2H)-Benzothiophenone, 2-(2-isopropyl-ideneamino-α-hydroxyphenylmethylene)-	dioxan	241(4.73),376(3.74)	83-0018-76
2-Propenoic acid, 2-isothiocyanato-3,3-diphenyl-, ethyl ester		235(4.25),320(4.22)	5-1850-76
$C_{18}H_{15}NO_3$			
Acetamide, N-[2-[(2,3-dihydro-1-oxo-1H-inden-2-yl)carbonyl]phenyl]-	MeOH	235(4.30),349(4.08)	83-0356-76
Acetamide, N-[4-[(2,3-dihydro-1-oxo-1H-inden-2-yl)carbonyl]phenyl]-	MeOH	266(4.08),291(4.09), 370(4.40)	83-0356-76
2-Furancarboxylic acid, 1,2,3,4-tetra-hydro-4-acridinyl ester	EtOH	231(4.65),236(4.64), 252(4.27),308(3.73), 323(3.78)	95-0968-76
1-Indolizinecarboxylic acid, 3-acetyl-2-phenyl-, methyl ester	MeOH	206(4.18),229(4.19), 248s(4.18),274s(4.06), 277(4.10),342s(3.86), 351(3.88)	39-1908-76C
2,6(1H,3H)-Pyridinedione, 4-(4-methoxy-phenyl)-3-phenyl-	pH 13	231(4.03),315(4.28)	34-0120-76
	EtOH	220(3.99),230(4.02), 318(4.33)	34-0120-76
	CHCl₃	231(3.84),323(4.34)	34-0120-76
$C_{18}H_{15}NO_4$			
2,3-Anthracenedicarboxylic acid, 5-amino-, dimethyl ester	CH₂Cl₂	<u>295(4.7),395(3.7)</u>	5-1873-76
2,3-Anthracenedicarboxylic acid, 6-amino-, dimethyl ester	CH₂Cl₂	<u>295(5.0),335(4.2), 415(4.0)</u>	5-1873-76
Crystamidine	EtOH	235(4.13),267(4.15), 357(3.31)	94-0052-76
$C_{18}H_{15}NO_5$			
1H-Inden-1-one, 3-ethoxy-2,3-dihydro-2-(4-nitrobenzoyl)-	MeOH	259(4.30),355(4.17)	83-0356-76
$C_{18}H_{15}NO_5S$			
1H-Pyrrolo[2,1-c][1,4]thiazine-7-carbox-ylic acid, 1-[(2-carboxyethenyl)oxy]-4-methyl-1-phenyl-, (Z)-	MeOH	235(4.21),290(4.33)	44-0187-76

Compound	Solvent	$\lambda_{max}(\log \epsilon)$	Ref.
$C_{18}H_{15}NO_6$			
Isoxazolo[4,5-b]dibenzofuran-6-one, 10-acetyl-6,6a-dihydro-7,9-dihydroxy-5,6aα,8-trimethyl-	MeOH	219(4.63),281(3.76), 329(3.69),374(2.66)	23-3713-76
Isoxazolo[5,4-a]dibenzofuran-4-one, 7-acetyl-4,10b-dihydro-8,10-dihydroxy-3,9,10bα-trimethyl-	MeOH	226(4.48),258(4.11), 282(4.23),336(3.73), 390(3.34)	23-3713-76
2H-[1,2]Oxazocino[4,5,6,7,8-amlk]dibenzofuran-4,11-dione, 1-acetyl-3a,4,10b-11-tetrahydro-8-hydroxy-3,9,10bα-trimethyl-	MeOH	233(4.45),281(4.24), 326(3.83)	23-3713-76
Pyrido[2,1,6-de]quinolizine-1,2,3-tricarboxylic acid, trimethyl ester	EtOH	250(4.16),288(4.47), 407(4.29),450(4.35)	39-0341-76C
$C_{18}H_{15}NO_9$			
Butanedioic acid, (1,4-dihydro-6-hydroxy-2,7-dimethoxy-1,4-dioxo-9H-carbazol-9-yl)-, (S)-	EtOH	225(4.27),250(4.19), 273(4.06),285(4.02), 310(3.97),352(4.03), 458(3.92)	146-0667-76
$C_{18}H_{15}N_2PS$			
1H-1,3,2-Benzodiazaphosphole, 2,3-dihydro-1,2-diphenyl-, 2-sulfide	EtOH	256(4.06),298(3.78), 324s(2.80)	65-0807-76
$C_{18}H_{15}N_3$			
Benzenamine, N-phenyl-2-(phenylazo)-	EtOH	285(4.40),316(4.49), 450(4.00)	23-1599-76
3H-Imidazo[4,5-b]quinoline, 2,9-dimethyl-3-phenyl-	EtOH	245(4.79),314(4.16), 327(4.13),339s(3.87)	48-0039-76
$C_{18}H_{15}N_3O$			
Acetamide, N-(4,6-diphenyl-2-pyrimidinyl)-	EtOH	254(4.56),290(4.29), 315(4.21)	4-0257-76
Imidazo[1,2-a]pyrimidin-5(8H)-one, 2,3-dihydro-7,8-diphenyl-	EtOH EtOH-HCl	210(4.40),311(3.61) 213(4.46),271(3.94)	33-1203-76 33-1203-76
Imidazo[1,2-a]pyrimidin-7(3H)-one, 2,8-dihydro-5,8-diphenyl-	EtOH EtOH-HCl	235(4.28),326(3.58) 239(4.26),286(4.00)	33-1203-76 33-1203-76
1H-Pyrazole-5-carbonitrile, 3-benzoyl-4,5-dihydro-1-(4-methylphenyl)-	EtOH	248(4.134),386(4.196)	18-0321-76
1H-Pyrrole-3-carbonitrile, 2-amino-1-(4-methoxyphenyl)-4-phenyl-	EtOH	241(4.67),289(4.07)	48-0663-76
Rutaecarpine, dihydro-	EtOH	275(3.89),292(3.76), 304s(3.45),316(3.24)	102-0578-76
$C_{18}H_{15}N_3O_2$			
1H-2,3a-Diazapentalene, 2,3-dihydro-5-nitro-2,3-diphenyl-	MeOH	256(4.20),303(4.04)	44-3050-76
1H-Pyrazole-5-carbonitrile, 3-benzoyl-4,5-dihydro-1-(4-methoxyphenyl)-	EtOH	244(4.181),396(4.214)	18-0321-76
4H-Pyridazino[4,5-b]indol-4-one, 2-acetyl-1,2,3,5-tetrahydro-1-phenyl-	EtOH	302(4.21)	103-1008-76
$C_{18}H_{15}N_3O_2S$			
Acetamide, N-[5-benzoyl-3-(4-methylphenyl)-1,3,4-thiadiazol-2(3H)-ylidene]-	EtOH	275(4.27),335(4.07)	4-0045-76
Isoquinolinium, 3,3-dicyano-1-(ethoxycarbonyl)-2-(methylthio)-2-propenylide	EtOH	232(4.50),364(4.15), 470(3.58)	142-0939-76

Compound	Solvent	$\lambda_{max}(\log \epsilon)$	Ref.
$C_{18}H_{15}N_3O_3$			
8-Azabicyclo[3.2.1]oct-3-en-2-one, 8-(5-nitro-2-pyridinyl)-6-phenyl-, endo	EtOH	212(4.33),225(4.34), 352(4.32)	39-2307-76C
Benzeneacetic acid, α-[[(3-methyl-2-quinoxalinyl)carbonyl]amino]-	EtOH	203(4.67),242(4.55), 320(3.85)	78-2931-76
	EtOH-KOH	200(4.92),241(4.56), 321(3.83)	78-2931-76
2,9-Methano-3H-indeno[2,1-b]pyridin-3-one, 1,2,4,4a,9,9a-hexahydro-1-(5-nitro-2-pyridinyl)-, (±)-	CHCl$_3$	365(4.37)	118-0105-76
Spiro[3H-indole-3,2'(1'H)-quinazoline]-4'-carboxylic acid, 1,2-dihydro-1,1'-dimethyl-2-oxo-	MeOH	206(4.56),237(4.54), 412(3.18)	39-2004-76C
Unidentified compound VII	EtOH	264(4.15),305s(3.87), 394(3.80)	95-0160-76
$C_{18}H_{15}N_3O_3S$			
Acetamide, N-[5-benzoyl-3-(4-methoxyphenyl)-1,3,4-thiadiazol-2(3H)-ylidene]-	EtOH	272(4.25),335(3.97)	4-0045-76
$C_{18}H_{15}N_3O_5$			
Benzeneacetic acid, α-[[(3-methyl-2-quinoxalinyl)carbonyl]amino]-, N,N'-dioxide, (S)-	EtOH	205(4.33),231(4.29), 259s(4.24),267(4.37), 370(3.97),382(4.03)	78-2931-76
	MeCN	205(4.32),232(4.26), 267(4.34),360s(3.75), 380(3.92),396(3.96)	78-2931-76
Benzoic acid, 2-[(1,4-dihydro-3-nitro-4-oxo-2-quinolinyl)amino]-, ethyl ester	CHCl$_3$	281(3.99),323(3.78), 369(4.49)	48-0039-76
$C_{18}H_{15}N_3O_8$			
Pyrido[2,1,6-de]quinolizine-1,3-dicarboxylic acid, 4,6-dinitro-, diethyl ester	EtOH	267(4.28),293s(4.13), 366(3.64),462(4.23), 532(4.31)	39-0341-76C
Pyrido[2,1,6-de]quinolizine-1,3-dicarboxylic acid, 4,7-dinitro-, diethyl ester	EtOH	275s(4.20),292(4.24), 330s(3.8),516s(4.30), 548(4.32)	39-0341-76C
Pyrido[2,1,6-de]quinolizine-1,3-dicarboxylic acid, 4,9-dinitro-, diethyl ester	EtOH	289(4.28),360(3.87), 563(4.34)	39-0341-76C
Pyrido[2,1,6-de]quinolizine-1,3-dicarboxylic acid, 6,7-dinitro-, diethyl ester	EtOH	284(4.29),349s(3.84), 504(4.50)	39-0341-76C
$C_{18}H_{15}N_3S$			
1,2,4-Triazolo[5,1-a]isoquinoline-2-thiol, 3-(2-phenylethyl)-, inner salt	MeOH	232(4.25),271(4.62), 316(3.77),364s(3.56)	2-0176-76
$C_{18}H_{15}O$			
Pyrylium, 2-methyl-4,6-diphenyl-, iodide	CH$_2$Cl$_2$	530(2.14)	59-1311-76
selenocyanate	CH$_2$Cl$_2$	488(1.88)	59-1311-76
thiocyanate	CH$_2$Cl$_2$	463(2.61)	59-1311-76
Pyrylium, 4-methyl-2,6-diphenyl-, iodide	CH$_2$Cl$_2$	510(2.41)	59-1311-76
selenocyanate	CH$_2$Cl$_2$	465(2.22)	59-1311-76
thiocyanate	CH$_2$Cl$_2$	452(2.79)	59-1311-76

Compound	Solvent	λ_{max}(log ϵ)	Ref.
$C_{18}H_{16}$			
Benzene, 1,3-dimethyl-5-(4-phenyl-1-but-en-3-ynyl)-, cis	MeOH	236(4.26),241s(4.23), 316(4.36),335(4.19)	39-1104-76B
Benzocyclotetradecene	C_6H_{12}	305(4.703),360s(3.666)	89-0365-76
Benzocyclotetradecene, 9,10,11,12-tetra-dehydro-7,8,13,14-tetrahydro-, (E,E)-	EtOH	234(4.348),255s(4.076)	89-0365-76
Dibenzotricyclo[4.3.1.0^{3,7}]deca-4,8-di-ene	MeOH	239s(3.75),246s(4.00), 251(4.19),258(4.29), 264(4.19)	35-1052-76
Naphthalene, 1,3-dimethyl-8-phenyl-	MeOH	232(4.61),288(3.86)	39-1115-76B
$C_{18}H_{16}BF_4N_3O$			
s-Triazolo[4,3-a]pyridinium, 3-(4-meth-oxyphenyl)-1-phenyl-, tetrafluoro-borate	MeOH	238s(--),258(4.31), 279s(--),311s(--)	48-0881-76
$C_{18}H_{16}BrN_3O_2$			
5-Pyrimidinecarboxamide, 4-(4-bromophen-yl)-1,2,3,4-tetrahydro-6-methyl-2-oxo-N-phenyl-	EtOH	204(4.46),232(4.14), 284(4.01)	103-0191-76
$C_{18}H_{16}ClN$			
1H-Carbazole, 9-(4-chlorophenyl)-2,3,4,9-tetrahydro-	EtOH	226(4.57),269(4.15), 287s(4.02),298s(3.90)	103-0880-76
1H-Carbazole, 6-chloro-2,3,4,9-tetra-hydro-9-phenyl-	EtOH	231(4.58),266(4.19), 259(4.00),302s(3.92)	103-0880-76
$C_{18}H_{16}ClNO$			
Ethanone, 1-(6-chloro-1,2-dihydro-1-methyl-4-phenyl-3-quinolinyl)-	isoPrOH	262(4.51),303s(3.62), 341(3.48)	4-0131-76
$C_{18}H_{16}ClNO_4$			
Phenol, 2-chloro-3-[(7,8-dihydro-9-meth-oxy-1,3-dioxolo[4,5-g]isoquinolin-5-yl)methyl]-, hydrochloride	MeOH	225(4.39),250s(4.00), 285(3.61),350(3.92)	12-2003-76
$C_{18}H_{16}ClN_3O_2$			
1H-Pyrazole-4-carboxamide, 3-benzoyl-N-(4-chlorophenyl)-4,5-dihydro-5-methyl-	EtOH	248(4.39),332(4.06)	115-0001-76
1H-Pyrazole-4-carboxamide, 3-(4-chloro-benzoyl)-4,5-dihydro-N-(4-methylphen-yl)-	EtOH	253(4.52),339(4.22)	115-0001-76
1H-Pyrazole-4-carboxamide, N-(4-chloro-phenyl)-4,5-dihydro-3-(4-methylbenz-oyl)-	EtOH	253(4.52),339(4.17)	115-0001-76
$C_{18}H_{16}ClN_4O_8PS$			
Inosine, 8-[(4-chlorophenyl)thio]-, cyclic 3',5'-(hydrogen phosphate), 2'-acetate	pH 1 pH 7 pH 11	256(4.21),275s(4.15) 256(4.21),275s(4.16) 244(4.11),282(4.19)	69-0217-76 69-0217-76 69-0217-76
$C_{18}H_{16}Cl_2N_4O_7$			
7H-Pyrrolo[2,3-d]pyrimidine-5-carboni-trile, 2,4-dichloro-7-(2,3,5-tri-O-acetyl-β-D-ribofuranosyl)-	MeOH	285(3.79)	35-7870-76
$C_{18}H_{16}NS_2$			
Isoquinolinium, 1,2,3,4-tetrahydro-2-(5-phenyl-3H-1,2-dithiol-3-yli-dene)-, perchlorate	MeOH	240s(4.28),251s(4.09), 323(4.19),397(3.99)	48-0221-76

Compound	Solvent	$\lambda_{max}(\log \epsilon)$	Ref.
$C_{18}H_{16}N_2$			
Benzimidazo[1,2-a]quinoline, 2,9,10-trimethyl-	MeOH	245(4.56),253(4.54), 283s(3.80),338(4.17), 355(4.11),373(3.87)	39-0075-76C
3,3'-Bi-1H-indole, 1,1'-dimethyl-	MeCN	239(4.65),281(3.78), 310(3.97)	5-1060-76
semiquinone	MeCN	248(4.26),303(3.87), 312s(3.90),317(4.12), 342(3.86),426(3.69), 441(3.72),720(3.87)	5-1060-76
1H-Pyrrolo[1,2-a]imidazole, 2,3-dihydro-2,3-diphenyl-	MeOH	232(3.89),283(4.00)	44-3050-76
$C_{18}H_{16}N_2O$			
8-Azabicyclo[3.2.1]oct-3-en-2-one, 8-phenyl-6-(4-pyridinyl)-, endo	EtOH	209(4.13),237(4.15)	39-2289-76C
$C_{18}H_{16}N_2OSe$			
Benzenamine, N-[6-(4-methoxyphenyl)-2H-1,3-selenazin-2-ylidene]-4-methyl-, perchlorate	HOAc	460(4.19)	88-2005-76
$C_{18}H_{16}N_2O_2$			
1H-Anthra[1,2-d]imidazole-6,11-dione, 2-ethyl-2,3-dihydro-2-methyl-	EtOH	570(4.18)	104-0172-76
1H-Anthra[2,3-d]imidazole-5,10-dione, 2-ethyl-2,3-dihydro-2-methyl-	EtOH	595s(3.21)	104-0172-65
Benzoic acid, 2-(5-phenyl-1H-pyrazol-3-yl)-, ethyl ester	EtOH	252(4.40)	44-0110-76
[3,3'-Bi-1H-indol]-2(3H)-one, 3-hydroxy-1',2'-dimethyl-	MeOH	214(2.85),223(2.84), 258s(2.23),294(2.18)	103-0178-76
2,3-Diazabicyclo[3.1.0]hex-2-ene-6-carboxylic acid, 1,5-diphenyl-, methyl ester, (1α,5α,6α)-	MeOH	331(2.66)	104-0802-76
1H-Pyrazole-1-carboxylic acid, 3,5-diphenyl-, ethyl ester	EtOH	238s(4.44),263(4.51)	4-0257-76
4-Pyridazinecarboxylic acid, 1,4-dihydro-3,5-diphenyl-, methyl ester	MeOH	242(4.19),370(3.94)	104-0802-76
$C_{18}H_{16}N_2O_2S$			
3-Thiophenamine, N-ethyl-2-nitro-N,5-diphenyl-	MeCN	261(4.23),307(3.71), 360(4.18),458(3.87)	48-0221-76
$C_{18}H_{16}N_2O_3$			
Acetamide, N-(2,4-dioxo-3,5-diphenyl-3-pyrrolidinyl)-, cis	n.s.g.	255s(3.13),262s(2.97), 269s(2.76),309(2.57)	33-2149-76
Benzoic acid, 2-[5-(4-methoxyphenyl)-1H-pyrazol-3-yl]-, methyl ester	EtOH	261(4.45)	44-0110-76
hydrochloride	EtOH	261(4.45)	44-0110-76
Ethanone, 1-(2-hydroxy-4-methoxyphenyl)-2-(1-phenyl-1H-pyrazol-4-yl)-	EtOH	275(4.48),315(4.07)	103-0914-76
1H-Pyrazole-3-carboxylic acid, 2,5-dihydro-5-oxo-1,4-diphenyl-, ethyl ester	EtOH	250(4.3),340(3.4)	23-0488-76
$C_{18}H_{16}N_2O_5S$			
2-Oxa-6,7-diazabicyclo[2.2.1]heptane-3,5-dione, 7-methyl-6-[(4-methylphenyl)sulfonyl]-1-phenyl-	MeOH	230(4.23)	44-0813-76

Compound	Solvent	$\lambda_{max}(\log \epsilon)$	Ref.
$C_{18}H_{16}N_2O_6$			
Pyrido[2,1,6-de]quinolizine-1,3-dicarb-oxylic acid, 4-nitro-, diethyl ester	EtOH	279(4.42),350(3.90), 413(4.28),581(4.27)	39-0341-76C
Pyrido[2,1,6-de]quinolizine-1,3-dicarb-oxylic acid, 6-nitro-, diethyl ester	EtOH	273(4.29),360(3.68), 400(3.81),574(4.59)	39-0341-76C
$C_{18}H_{16}N_2S$			
5H-Pyrrolo[1,2-c]imidazolium, 6,7-di-hydro-1-mercapto-2,3-diphenyl-, hydroxide, inner salt	MeOH	210(4.41),230(4.31), 348(3.94)	78-0579-76
$C_{18}H_{16}N_3O$			
s-Triazolo[4,3-a]pyridinium, 3-(4-meth-oxyphenyl)-1-phenyl-, tetrafluoro-borate	MeOH	238s(--),258(4.31), 279s(--),311s(--)	48-0881-76
$C_{18}H_{16}N_4$			
Bisbenzimidazo[1,2-a:1',2'-e][1,5]diazo-cine, 6,7,14,15-tetrahydro-	EtOH	248(4.12),254(4.13), 268s(3.95),275(4.03), 283(4.04)	39-0312-76C
1H-Indole, 3,3'-azobis[2-methyl-	DMF	268(4.53),290s(4.21), 400s(4.00),446s(4.44), 466(4.59),493(4.53)	5-1060-76
oxidized form	DMF	237(4.44),270s(4.28), 277(4.27),337(4.02), 362(3.99)	5-1060-76
Pyrazolo[4,3-c]pyrazole, 1,4-dihydro-1,4-dimethyl-3,6-diphenyl-	MeCN	212(4.31),222(4.38), 226(4.37),236s(4.27), 313(4.37)	5-1090-76
1H-Pyrrole-3-carbonitrile, 2-amino-4-methyl-5-phenyl-1-(phenylamino)-	MeOH	231(4.33),283(4.12), 305(4.06)	48-0663-76
$C_{18}H_{16}N_4O$			
Acetic acid, phenyl[(1-phenyl-1H-pyra-zol-3-yl)methylene]hydrazide	EtOH	212(4.34),297(4.61)	136-0185-76D
Pyrazinecarboxamide, 3-amino-N-methyl-5,6-diphenyl-	MeOH	222(4.26),278(4.24), 378(4.09)	24-3194-76
$C_{18}H_{16}N_4O_2$			
2,4(1H,3H)-Pteridinedione, 5,6,7,8-tet-rahydro-6,7-diphenyl-	pH 0.0	263(4.29)	24-3184-76
	pH 6.0	240s(3.83),290(4.04)	24-3184-76
$C_{18}H_{16}N_4O_2S$			
2,6-Pyridinediamine, 4-[(diphenylmeth-yl)thio]-3-nitro-	pH 1	255(4.43),297(4.00), 377(4.29)	44-3784-76
$C_{18}H_{16}N_4O_4$			
Morpholine, 4-[2-(1H-benzimidazol-1-yl)-5-nitrobenzoyl]-	EtOH	243(4.22),270s(3.94), 307(3.85)	2-0001-76
2-Propenoic acid, 2-cyano-3-[[3-nitro-4-(phenylamino)phenyl]amino]-, ethyl ester	EtOH	462(3.76)	146-0131-76
1H-Pyrazole-4-carboxylic acid, 5-amino-1-(4-nitrophenyl)-3-phenyl-, ethyl ester	EtOH	310(4.05)	4-1137-76
5-Pyrimidinecarboxamide, 1,2,3,4-tetra-hydro-6-methyl-4-(4-nitrophenyl)-2-oxo-N-phenyl-	EtOH	204(4.32),275(4.22)	103-0191-76

Compound	Solvent	$\lambda_{max}(\log \epsilon)$	Ref.
$C_{18}H_{16}N_4O_7$			
D-Ribitol, 1,4-anhydro-1-C-(8-nitro-[1,2,4]triazolo[1,5-a]pyridin-2-yl)-, 5-benzoate, (S)-	H_2O	243(3.99),328(3.72)	44-3124-76
$C_{18}H_{16}O$			
5H-Benzocyclohepten-5-one, 6,7,8,9-tetrahydro-6-(phenylmethylene)-	isooctane	233(4.08),262(4.03), 297(4.15),356(2.45), 370(2.45),388s(--), 410s(--)	65-2057-76
5H-Dibenzo[a,c]cyclohepten-5-one, 7-ethyl-6-methyl-	EtOH	237(4.51),325(3.47)	35-3627-76
1(2H)-Naphthalenone, 3,4-dihydro-2-[(4-methylphenyl)methylene]-	isooctane	225(4.18),260s(--), 300s(--),316(4.25), 375s(--),415s(--)	65-2057-76
	EtOH	230(4.11),270s(--), 322(4.27)	65-2057-76
2-Propyne, 1-ethoxy-1,1-diphenyl-	EtOH	248s(2.61),252(2.64), 258(2.66),263s(2.54)	56-0203-76
$C_{18}H_{16}O_2$			
1-Benzoxepin-5(2H)-one, 3-methyl-4-(phenylmethyl)-	n.s.g.	265(3.99),326(3.42)	33-0664-76
Bicyclo[1.1.0]butane-2-carboxylic acid, 1,3-diphenyl-, methyl ester, (1α,2β,3α)-	MeOH	273(3.95)	104-0802-76
5H-Dibenzo[a,c]cyclohepten-5-one, 7-(1-hydroxyethyl)-6-methyl-	EtOH	233(4.51),320(3.39)	35-3627-76
2(5H)-Furanone, 5,5-dimethyl-3,4-diphenyl-	dioxan	263(2.975)	24-0576-76
1(2H)-Naphthalenone, 3,4-dihydro-2-[(4-methoxyphenyl)methylene]-	isooctane	242(4.16),330(4.34), 390s(--),410s(--)	65-2057-76
	EtOH	246(4.06),344(4.25)	65-2057-76
Spiro[5H-dibenzo[a,c]cyclohepten-5,2'-oxiran]-7(6H)-one, 3',6-dimethyl-	EtOH	235(4.39),255s(3.98), 295(3.18)	35-3627-76
Tricyclo[9.3.1.1^4,^{8}]hexadeca-1(15),4,6-8(16),11,13-hexaene-5,14-dicarboxaldehyde	EtOH	330s(2.30)	49-0195-76
$C_{18}H_{16}O_2S$			
2(5H)-Furanone, 5-methyl-5-(methylthio)-3,4-diphenyl-	dioxan	223(4.276),278(4.000)	24-0576-76
Thiophene, 2,4-bis(4-methoxyphenyl)-	EtOH	265(4.56),315s(4.18)	42-0490-76
Thiophene, 3,4-bis(4-methoxyphenyl)-	EtOH	234(4.41),261s(4.18)	42-0490-76
3,4-Thiophenedimethanol, 2,5-diphenyl-	CH_2Cl_2	295(4.2)	28-0607-76A
$C_{18}H_{16}O_3$			
2(5H)-Furanone, 5-methoxy-5-methyl-3,4-diphenyl-	dioxan	222(4.247),290(4.021)	24-0576-76
$C_{18}H_{16}O_3S$			
1H-2-Benzopyran-3-carboxylic acid, 1-(phenylmethylthio)-, methyl ester	MeOH	235(4.40),240(4.31), 311(4.10)	35-5581-76
$C_{18}H_{16}O_4$			
1,4-Benzenediacetic acid, cyclic diester with 1,4-benzenedimethanol	CH_2Cl_2	240(3.53),264(2.72), 268(2.72),273(2.65)	88-3665-76
5-Benzofuranpropanol, 2-(1,3-benzodioxol-5-yl)-	EtOH	214(4.55),317(4.51), 331(4.42)	39-1514-76C
1H-2-Benzopyran-3-carboxylic acid, 1-(phenylmethoxy)-, methyl ester	MeOH	227(3.93),234(3.86), 295(4.05),307s(3.92)	35-5581-76

Compound	Solvent	$\lambda_{max}(\log \epsilon)$	Ref.
4H-1-Benzopyran-4-one, 5,7-dimethoxy-8-methyl-2-phenyl-	EtOH	212(4.64),262(4.38), 307(4.33)	2-0009-76
$C_{18}H_{16}O_4S$			
2H-1-Benzopyran-2-one, 4-ethyl-3-[(4-methylphenyl)sulfonyl]-	EtOH	284(3.82),327(3.11)	39-1260-76C
1-Benzoxepin-2(3H)-one, 5-methyl-3-[(4-methylphenyl)sulfonyl]-	EtOH	248(3.85)	39-1260-76C
$C_{18}H_{16}O_5$			
2H-1-Benzopyran-2-one, 3-(2,5-dimethoxyphenyl)-4-methoxy-	MeOH	210(4.68),285(4.12), 325(4.31)	2-0861-76
4H-1-Benzopyran-4-one, 2,5-dihydro-5-hydroxy-7-methoxy-3-[(4-methoxyphenyl)-methylene]- (7-O-methyleucomin)	EtOH EtOH	212(4.53),358(4.45) 212(4.49),358(4.508)	33-2048-76 33-2048-76
$C_{18}H_{16}O_6$			
Benzaldehyde, 3-(1,3-dioxo-3-phenylpropyl)-2,4-dihydroxy-6-methoxy-5-methyl-	EtOH	275(4.25),360(4.39)	2-0009-76
4H-1-Benzopyran-4-one, 3-(2,4-dimethoxyphenyl)-5-hydroxy-7-methoxy-	EtOH	217(4.12),261(4.24), 277s(3.91),319(3.27)	102-1029-76
	EtOH-NaOH	235(4.09),270(4.18), 358(3.13)	102-1029-76
4H-1-Benzopyran-4-one, 3-(3,4-dimethoxyphenyl)-7-hydroxy-8-methoxy-	EtOH	256(4.48),290s(4.15), 304s(4.02)	102-1019-76
	EtOH-NaOAc	264(--),344s(--)	102-1019-76
4H-1-Benzopyran-4-one, 3-(3-hydroxy-4-methoxyphenyl)-7,8-dimethoxy-	EtOH	224(4.47),254(4.53), 290(4.23)	102-1019-76
	EtOH-NaOMe	250(--),298(--)	102-1019-76
4H-1-Benzopyran-4-one, 2-(4-methoxyphenyl)-5,7-dimethoxy-	MeOH	260(4.28),266s(4.27), 310s(3.96),358(4.31), 420s(3.53)	95-1217-76
	MeOH-NaOMe	260(4.28),268s(4.25), 310s(3.92),362(4.26), 410s(3.78)	95-1217-76
Spiro[benzofuran-2(3H),1'-[3,5]cyclohexadiene]-2',3-dione, 6-acetoxy-4'-methoxy-4,6'-dimethyl-	EtOH	269(4.28),324(3.96)	39-0147-76C
$C_{18}H_{16}O_7$			
4H-1-Benzopyran-4-one, 3,5-dihydroxy-7,8-dimethoxy-2-(4-methylphenyl)-(tambulin)	EtOH	270(4.32),324(4.14), 380(4.21)	2-0233-76
	EtOH-base	265(4.44),330(4.04)	2-0233-76
	EtOH-AlCl$_3$	270(4.30),313(3.92), 350(4.14),440(4.28)	2-0233-76
1(9bH)-Dibenzofuranone, 2,8-diacetyl-3,7,9-trihydroxy-6,9b-dimethyl-	MeOH	231(4.12),285(4.08), 328(3.61)	23-2795-76
Isousnic acid, (+)-	MeOH	232(4.17),285(4.21), 336(3.65)	23-3721-76
$C_{18}H_{16}O_8$			
6H-[1,3]Benzodioxolo[5,6-c][1]benzopyran-6-one, 2,4-diacetoxy-1,2,3,4-tetrahydro-	EtOH	242(4.64),258s(--), 287(3.84),298(3.84), 328(3.77),348s(--)	24-0855-76
[5,5'-Bi-1,3-benzodioxole]-6-carboxylic acid, 3'a,4',7',7'a-tetrahydro-2',2'-dimethyl-4',7'-dioxo-, methyl ester, cis-(+)-	EtOH	226(4.50),265s(--), 303(3.77),345s(--)	24-0855-76
$C_{18}H_{16}S$			
Thiophene, 2,4-bis(4-methylphenyl)-	EtOH	260(4.43),312(4.09)	42-0490-76

Compound	Solvent	$\lambda_{max}(\log \epsilon)$	Ref.
Thiophene, 3,4-dimethyl-2,5-diphenyl-	CH_2Cl_2	305(4.3)	28-0607-76A
$C_{18}H_{17}ClN_2O_3$			
Benzoic acid, 2-[5-(4-methoxyphenyl)-1H-pyrazol-3-yl]-, methyl ester, hydrochloride	EtOH	261(4.45)	44-0110-76
Hydrazinecarboxylic acid, [3-(3-chlorophenyl)-3-oxo-1-phenylpropylidene]-, ethyl ester	EtOH	226s(4.18),279(4 37)	4-0257-76
Hydrazinecarboxylic acid, [3-(4-chlorophenyl)-3-oxo-1-phenylpropylidene]-, ethyl ester	EtOH	219(4.42),281(4.39)	4-0257-76
$C_{18}H_{17}ClN_2O_6$			
Benzimidazo[3,2-b]isoquinolinium, 9,10-dimethoxy-7-methyl-, perchlorate	MeOH	250(4.87),315(4.16), 365(3.78)	103-0205-76
$C_{18}H_{17}ClOS$			
Dibenzo[b,f]thiepin-10(11H)-one, 11-butyl-8-chloro-	MeOH	230(4.31),242(4.30), 263s(4.10),338(3.70)	73-3420-76
$C_{18}H_{17}Cl_2NO_3$			
1,4-Benzoxazepine, 5-(2,6-dichlorophenyl)-2,3-dihydro-7,8-dimethoxy-3-methyl-	n.s.g.	215(4.34),250(4.06), 320(4.27)	20-0787-76
$C_{18}H_{17}F_3N_2O_5$			
Acetamide, 2,2,2-trifluoro-N-[tetrahydro-2,2-dimethyl-5-(5-phenyl-3-isoxazolyl)furo[2,3-d]-1,3-dioxol-6-yl]-, [3aR-(3aα,5α,6α,6aα)]-	EtOH	207(3.94),262(4.29)	136-0119-76A
$C_{18}H_{17}NO$			
1H-Inden-1-one, 2-[4-[(dimethylamino)phenyl]methylene]-2,3-dihydro-	isooctane	260(4.32),305(3.72), 382(4.51),400(4.56)	65-2057-76
	EtOH	270(4.26),315s(--), 430(4.49)	65-2057-76
3(2H)-Isoquinolinone, 1-[4-(1-methylethyl)phenyl]-	EtOH	350(3.73),433(3.53)	103-0910-76
	ether	353(3.91)	103-0910-76
	$CHCl_3$	365(3.91),440(3.82)	103-0910-76
	CCl_4	353(3.81),439(3.12)	103-0910-76
2-Propenamide, 3-phenyl-N-(3-phenyl-2-propenyl)-, cis-cis	EtOH	248(4.35)	44-2571-76
cis-trans	EtOH	252(4.45),282s(3.96), 291s(3.77)	44-2571-76
trans-cis	EtOH	222(4.33),271(4.49)	44-2571-76
trans-trans	EtOH	216(4.25),222s(4.26), 268(4.52)	44-2571-76
$C_{18}H_{17}NOSe$			
Benzo[b]selenophen-3(2H)-one, 2-[[4-(dimethylamino)phenyl]methylene]-5-methyl-, (E)-	octane	375(4.32),461(4.59)	103-0854-76
$C_{18}H_{17}NO_2$			
3(2H)-Benzofuranone, 2-[[4-(dimethylamino)phenyl]methylene]-5-methyl-, (E)-	octane	345(3.96),437(4.68)	103-0854-76
4H-1-Benzopyran-4-one, 3-[[4-(dimethylamino)phenyl]methylene]-2,3-dihydro-	isooctane	220s(--),265(4.38), 342(4.23),390(4.57)	65-2057-76
	EtOH	277(4.20),365(3.78), 416(4.40)	65-2057-76

Compound	Solvent	λ_{max}(log ϵ)	Ref.
4H-1-Benzopyran-4-one, 3-[[4-(dimethyl-amino)phenyl]methylene]-2,3-dihydro-, perchlorate	EtOH	272(4.00),426(4.26)	22-1967-76
3-Butene-1,2-dione, 4-(dimethylamino)-1,3-diphenyl-	MeOH	250(4.09),300(4.12)	44-0388-76
1H-Indole-3-carboxaldehyde, 2-(4-meth-oxy-3-methylphenyl)-1-methyl-	MeOH	257(4.46),307(4.22)	12-2747-76
$C_{18}H_{17}NO_2S$			
2-Butanone, 4-(1H-indol-1-yl)-1-(phenyl-sulfinyl)-	EtOH	222(4.54),284(3.92), 292(3.80)	35-4577-76
$C_{18}H_{17}NO_3$			
Benzeneacetic acid, α-[2-(dimethylami-no)-2-oxo-1-phenylethylidene]-, (E)-	dioxan	261(4.223)	24-0576-76
Pyrrolo[2,1-a]isoquinoline-3-carboxylic acid, 1-acetyl-2-methyl-, ethyl ester	EtOH	260(4.12),278(4.73), 331(3.92),348(4.07), 366(4.13)	94-1299-76
Quinoline, 3,4,6-trimethoxy-2-phenyl-	EtOH	254(4.23),332(3.30)	142-1089-76
4(1H)-Quinolinone, 3,6-dimethoxy-1-methyl-2-phenyl-	EtOH	252(4.25),342s(3.83), 357(4.02)	142-1089-76
$C_{18}H_{17}NO_3S$			
Pyrrolo[2,1-a]isoquinoline-3-carboxylic acid, 1-acetyl-2-(methylthio)-, ethyl ester	EtOH	220(4.28),278(4.64), 331(3.95),348(4.07), 365(4.10)	94-1299-76
$C_{18}H_{17}NO_3S_2$			
5H-[1]Benzothiopyrano[3,4-e]-1,2-oxathi-in-4-amine, 3,4-dihydro-N-methyl-N-phenyl-, 2,2-dioxide	EtOH	252(4.59),287(3.79), 295s(3.75),330(3.43)	4-0225-76
$C_{18}H_{17}NO_4$			
Litseferine	EtOH	282(4.10),310(4.12)	2-0150-76
Phenanthro[4,5-def][1,3]dioxepin-3,7-diol, 1-[2-(methylamino)ethyl]-, hydrobromide	EtOH-HCl	244(4.49),262(4.71), 298(4.00),311(4.13), 324(4.21),345(3.38), 363(3.58),382(3.67)	33-2551-76
	EtOH-NaOH	252(4.65),271(4.64), 340s(4.21),351(4.26), 385(3.75)	33-2551-76
2-Propenoic acid, 2-[[(phenylmethoxy)-carbonyl]amino]-, phenylmethyl ester	EtOH	245(3.82)	44-3634-76
Pyrido[2,1,6-de]quinolizine-1,3-dicarb-oxylic acid, diethyl ester	EtOH	240(4.00),285(4.53), 316(3.81),331(3.81), 399(4.40),430s(4.31), 453(4.48)	39-0341-76C
$C_{18}H_{17}NO_5$			
1H-Isoindole-1,3(2H)-dione, 2-[2-(3,4-dimethoxyphenoxy)ethyl]-	n.s.g.	225(4.30),286(3.69)	20-0421-76
$C_{18}H_{17}NO_5S$			
1H-Pyrrolo[2,1-c][1,4]thiazine-7,8-di-carboxylic acid, 1-hydroxy-4-methyl-1-phenyl-, dimethyl ester	MeOH	214(4.38)	44-0187-76
$C_{18}H_{17}NO_6$			
1,3(2H,9bH)-Dibenzofurandione, 6-acetyl-2-(1-aminoethylidene)-7,9-dihydroxy-8,9b-dimethyl-, (-)-	MeCN	291(4.18),333(3.38)	23-2795-76

Compound	Solvent	$\lambda_{max}(\log \epsilon)$	Ref.
Pyrido[2,1,6-de]quinolizine-1,2,3-tri-carboxylic acid, 1,2-dihydro-, trimethyl ester	EtOH	212(4.41),235s(4.00), 283s(3.98),292(3.04), 326(3.97),452(4.14)	39-0341-76C
	EtOH-HClO$_4$	243(4.39),296(3.62), 324(3.95),332s(3.94), 337(4.19)	39-0341-76C
$C_{18}H_{17}NO_7$			
Benzofuro[3,2-f]-1,2-benzisoxazol-4(4aH)-one, 8-acetyl-9a,10-dihydro-5,7,9-trihydroxy-3,4a,6-trimethyl-, cis-(+)-	MeOH	223(3.91),286(3.79), 330(3.27)	23-3721-76
7,9-Epoxy-2H-1,2-benzoxazecine-4-carbox-ylic acid, 10-acetyl-5,6-dihydro-11-hydroxy-3,8,12-trimethyl-5-oxo-	MeOH	237(3.84),287(3.83), 356(3.07)	23-3713-76
4-Isoxazolecarboxylic acid, 5-[(7-acet-yl-4,6-dihydroxy-3,5-dimethyl-2-benzo-furanyl)methyl]-3-methyl-	MeOH	242(4.02),302(3.85), 350(3.33)	23-3713-76 +23-3721-76
$C_{18}H_{17}N_2O$			
Pyrimidinium, 2,3-dihydro-2-oxo-1,3-bis-(phenylmethyl)-, sulfate (1:1)	EtOH-pH 0.29	321(3.99)	103-2681-76
$C_{18}H_{17}N_2O_2$			
Benzimidazo[3,2-b]isoquinolinium, 9,10-dimethoxy-7-methyl-, perchlorate	MeOH	250(4.87),315(4.16), 365(3.78)	103-0205-76
$C_{18}H_{17}N_3O$			
8-Azabicyclo[3.2.1]octane-8-propane-nitrile, 6-cyano-2-oxo-3-(phenyl-methylene)-, endo	EtOH	209(3.69),225(3.68), 300(3.99)	39-2334-76C
2,4-Pentadienamide, 2-cyano-5-(dimethyl-amino)-5-(2-naphthalenyl)-	CH$_2$Cl$_2$	399(4.68)	48-0705-76
Spiro[5H-1,4-diazepine-5,3'-[3H]indol]-2'(1'H)-one, 2,3,4,6-tetrahydro-7-phenyl-	MeOH	209(4.47),272(4.11), 278(4.21),349(4.22)	103-0665-76
$C_{18}H_{17}N_3O_2$			
1H-Pyrazole-4-carboxamide, 3-benzoyl-N-(4-methylphenyl)-	EtOH	251(4.45),239(4.22)	115-0001-76
1H-Pyrazole-4-carboxamide, 3-(4-methyl-benzoyl)-N-phenyl-	EtOH	247(4.44),335(4.18)	115-0001-76
2-Pyrimidinamine, 4,6-bis(3-methylphen-oxy)-	EtOH	216(4.42),262(4.01)	42-0913-76
2-Pyrimidinamine, 4,6-bis(4-methylphen-oxy)-	EtOH	222(4.68),250(4.56), 260(4.51)	42-0913-76
4-Pyrimidinamine, 2,6-bis(3-methylphen-oxy)-	EtOH	230(4.65),260(4.52)	42-0913-76
4-Pyrimidinamine, 2,6-bis(4-methylphen-oxy)-	EtOH	222(4.45),238(4.16), 360(4.03)	42-0913-76
5-Pyrimidinecarboxamide, 1,2,3,4-tetra-hydro-6-methyl-2-oxo-N,4-diphenyl-	EtOH	203(4.30),284(4.10)	103-0191-76
$C_{18}H_{17}N_3O_2S_2$			
Thiourea, [3-(4-ethoxyphenyl)-4-oxo-2-thiazolidinylidene]phenyl-	dioxan	261(4.28),414(4.04)	103-0751-76
$C_{18}H_{17}N_3O_3S$			
4H-Cyclohepta[4,5]thieno[2,3-b]pyridin-4-one, 1,5,6,7,8,9-hexahydro-3-nitro-2-(phenylamino)-	CHCl$_3$	279(3.65),330(4.05), 391s(2.24)	48-0039-76

Compound	Solvent	λ_{max}(log ϵ)	Ref.
$C_{18}H_{17}N_3O_4$			
2-Pyrimidinamine, 4,6-bis(4-methoxyphen-oxy)-	EtOH	230(4.67),270(4.13)	42-0913-76
4-Pyrimidinamine, 2,6-bis(4-methoxyphen-oxy)-	EtOH	230(4.49),260(4.12)	42-0913-76
$C_{18}H_{17}N_3O_5$			
D-Ribitol, 1,4-anhydro-1-C-1,2,4-triaz-olo[4,3-a]pyridin-3-yl-, 5-benzoate, α-	H_2O	265(3.61),271(3.68), 280(3.59),296(3.48)	44-3124-76
β-	H_2O	263(3.66),271(3.72), 280(3.65),295(3.53)	44-3124-76
$C_{18}H_{17}N_5$			
1,2-Benzenedicarbonitrile, 3-[[4-(dieth-ylamino)phenyl]azo]-	EtOH	490(4.54)	39-0042-76C
	benzene	485(--)	39-0042-76C
1,2-Benzenedicarbonitrile, 4-[[4-(dieth-ylamino)phenyl]azo]-	EtOH	500(4.59)	39-0042-76C
	benzene	486(--)	39-0042-76C
1,3-Benzenedicarbonitrile, 2-[[4-(dieth-ylamino)phenyl]azo]-	EtOH	503(4.52)	39-0042-76C
	benzene	495(--)	39-0042-76C
1,3-Benzenedicarbonitrile, 4-[[4-(dieth-ylamino)phenyl]azo]-	EtOH	514(4.60)	39-0042-76C
	benzene	505(--)	39-0042-76C
1,3-Benzenedicarbonitrile, 5-[[4-(dieth-ylamino)phenyl]azo]-	EtOH	478(4.53)	39-0042-76C
	benzene	466(--)	39-0042-76C
1,4-Benzenedicarbonitrile, 2-[[4-(dieth-ylamino)phenyl]azo]-	EtOH	495(4.56)	39-0042-76C
	benzene	488(--)	39-0042-76C
$C_{18}H_{17}N_5O$			
[1,2,4]Triazolo[5,1-c][1,2,4]triazine, 4-ethoxy-1,4-dihydro-3,4-diphenyl-	EtOH	298(4.16)	39-1492-76C
$C_{18}H_{17}N_5O_3$			
1H-Pyrazole-4-carboxamide, 5-amino-3-ethyl-1-(4-nitrophenyl)-N-phenyl-	EtOH	310(4.29)	4-1137-76
$C_{18}H_{18}$			
Benzene, 1,2-di-1-hexen-5-ynyl-, (E,E)-	n.s.g.	234(4.391),260(4.257)	89-0365-76
Benzene, 1,1'-(1,2-ethynediyl)bis-[2,4-dimethyl-	C_6H_{12}	241(4.272),256(4.184), 331(4.267),405(4.175)	101-0385-76N
Benzene, 1,1'-(1,2-ethynediyl)bis-[2,5-dimethyl-	C_6H_{12}	288(4.411),301(4.275), 312(4.362)	101-0385-76N
3a,5a-Ethenopyrene, 1,2,3,6,7,8-hexa-hydro-	C_6H_{12}	213s(4.56),218(4.58), 262s(2.86),271(2.72), 283(2.42)	88-4559-76
Phenanthrene, 2-methyl-1-propyl-	EtOH	258(4.73),281(4.01), 289(3.89),302(4.00), 320(2.51),336(2.60), 352(2.43)	42-0812-76
$C_{18}H_{18}BrN_5O_3S_3$			
3-Thiazolidinehexanoic acid, 4-[(amino-thioxomethyl)hydrazono]-5-(5-bromo-1,2-dihydro-2-oxo-3H-indol-3-ylidene)-2-thioxo-	MeOH	244(4.41),290(3.61), 398(4.30),418s(--)	103-0749-76
$C_{18}H_{18}ClFN_2O$			
1H-1,4-Benzodiazepine, 7-chloro-5-(2-fluorophenyl)-2,3-dihydro-2-(methoxy-methyl)-1-methyl-, hydrochloride	EtOH-base	229s(--),235(4.31), 265s(--),368(3.32)	111-0501-76

Compound	Solvent	$\lambda_{max}(\log \epsilon)$	Ref.
$C_{18}H_{18}ClN$ Dibenzo[b,f]cycloprop[d]azepine, 6-(3-chloropropyl)-1,1a,6,10b-tetrahydro-	EtOH	256(3.76)	94-2751-76
$C_{18}H_{18}ClNOS$ Dibenzo[b,f]thiepin-10(11H)-one, 8-chloro-11-[2-(dimethylamino)ethyl]-, hydrochloride	MeOH	223(4.28),241(4.30), 265s(4.03),340(3.59)	73-3420-76
$C_{18}H_{18}ClNO_2$ Ethanone, 1-(6-chloro-1,2,3,4-tetrahydro-4-hydroxy-1-methyl-4-phenyl-3-quinolinyl)-, cis	isoPrOH	260(4.17),319(3.47)	4-0131-76
$C_{18}H_{18}ClNS$ Dibenzo[b,f]thiepin-10-ethanamine, 2-chloro-N,N-dimethyl-, hydrochloride	MeOH	223(4.41),260(4.27), 283s(3.73)	73-3420-76
$C_{18}H_{18}ClNS_2$ Benzenaminium, N-ethyl-4-methyl-N-(5-phenyl-3H-1,2-dithiol-3-ylidene)-, chloride	MeOH	318(4.21),367(4.04)	48-0221-76
$C_{18}H_{18}ClN_3O_2$ 1H-Pyrazole-4-carboxamide, 3-(4-chlorobenzoyl)-5-methyl-N-phenyl-	EtOH	249(4.31),340(4.11)	115-0001-76
$C_{18}H_{18}ClN_3O_2S$ Propanoic acid, 2-[(3-chloro-2-methylphenyl)hydrazono]-3-(phenylamino)-3-thioxo-, ethyl ester	EtOH	260(4.25),276(4.13), 402(4.48)	104-0544-76
Propanoic acid, 2-[(5-chloro-2-methylphenyl)hydrazono]-3-(phenylamino)-3-thioxo-, ethyl ester	EtOH	258(4.08),276(4.03), 402(4.43)	104-0544-76
(Z)-	EtOH	250(4.17),370(4.21)	104-1912-76
Propanoic acid, 2-[(4-chlorophenyl)hydrazono]-3-[(4-methylphenyl)amino]-3-thioxo-, ethyl ester	EtOH	255(4.10),280(4.17), 400(4.54)	104-0544-76
$C_{18}H_{18}Cl_2N_2O$ 1H-1,4-Benzodiazepine, 7-chloro-5-(2-chlorophenyl)-2,3-dihydro-2-(methoxymethyl)-1-methyl-, hydrochloride	MeOH-base	229s(--),264s(--), 372(3.34)	111-0501-76
$C_{18}H_{18}CuN_6S_2$ Cobalt, [[2,2'-(1-[1,1'-biphenyl]-4-yl-1,2-ethanediylidene)bis[N-methylhydrazinecarbothioamidato]](2-)-N^2,N^2,S,S']-	EtOH	313(4.22),500(3.53)	87-0131-76
$C_{18}H_{18}FNO_3$ 1,4-Benzoxazepine, 5-(3-fluorophenyl)-2,3-dihydro-7,8-dimethoxy-3-methyl-	n.s.g.	210(4.38),260(4.09), 315(4.15)	20-0787-76
$C_{18}H_{18}NO_3$ 1,3-Dioxolo[4,5-g]isoquinolinium, 7,8-dihydro-9-methoxy-6-(phenylmethyl)-, bromide	MeOH	248(4.21),360(4.07)	12-2003-76

Compound	Solvent	$\lambda_{max}(\log \epsilon)$	Ref.
$C_{18}H_{18}N_2$			
1H-Benzimidazole, 5,6-dimethyl-1-[2-(4-methylphenyl)ethenyl]-, (E)-	MeOH	208(4.56),238s(4.01), 259s(4.01),268(4.10), 344(4.59),360s(4.38)	39-0075-76C
1H-Imidazole, 4,5-dimethyl-2-phenyl-1-(phenylmethyl)-	n.s.g.	275(4.05)	33-2149-76
$C_{18}H_{18}N_2O$			
1H-Indole-3-carboxaldehyde, 2-[4-(dimethylamino)phenyl]-1-methyl-	MeOH	264(4.37),301(4.23), 341(4.12)	12-2747-76
3(2H)-Pyridazinone, 6-(2,4-dimethylphenyl)-4,5-dihydro-2-phenyl-	n.s.g.	278(4.22)	115-0621-76
3(2H)-Pyridazinone, 6-(2,5-dimethylphenyl)-4,5-dihydro-2-phenyl-	n.s.g.	277(4.19)	115-0621-76
3(2H)-Pyridazinone, 6-(3,4-dimethylphenyl)-4,5-dihydro-2-phenyl-	n.s.g.	255(4.11),300(4.26)	115-0621-76
$C_{18}H_{18}N_2O_2$			
9,10-Anthracenedione, 1,2-bis(dimethylamino)-	EtOH	522(3.68)	104-2503-76
9,10-Anthracenedione, 2,3-bis(dimethylamino)-	EtOH	490(3.48)	104-2503-76
2-Butenoic acid, 4-(2,4-dimethylphenyl)-4-(phenylhydrazono)-, methyl ester	n.s.g.	235(4.15),300(3.86), 345(4.46)	115-0621-76
2-Butenoic acid, 4-(2,5-dimethylphenyl)-4-(phenylhydrazono)-, methyl ester	n.s.g.	235(4.13),345(4.45)	115-0621-76
2-Butenoic acid, 4-(3,4-dimethylphenyl)-4-(phenylhydrazono)-, methyl ester	n.s.g.	235(4.12),300(3.90), 350(4.43)	115-0621-76
1,3-Diazetidine-2,4-dione, 1,3-bis-(2-phenylethyl)-	THF	260(2.46)	95-0962-76
Ethanone, 2-[3-ethyl-5-(hydroxymethyl)-4-pyridinyl]-1-(1H-indol-2-yl)-	EtOH	224(4.1),236(4.01), 273(3.68),310(4.27)	88-1085-76
Vincarpine, tetrahydro-	EtOH	225(4.6),274(3.8)	88-4887-76
$C_{18}H_{18}N_2O_2S$			
Methanone, [3-amino-4-ethyl-5-methyl-6-(methylthio)furo[2,3-b]pyridin-2-yl]phenyl-	EtOH	258(4.15),350(4.29), 390(4.36)	48-0313-76
3-Pyridinecarbonitrile, 4-ethyl-5-methyl-6-(methylthio)-2-(2-oxo-2-phenylethoxy)-	EtOH	244(4.29),261s(4.13), 321(4.27)	48-0313-76
$C_{18}H_{18}N_2O_3$			
Hydrazinecarboxylic acid, (3-oxo-1,3-diphenylpropylidene)-, ethyl ester	EtOH	226s(4.10),279(4.37)	4-0257-76
$C_{18}H_{18}N_2O_3S_2$			
Acetamide, N-[4-methoxy-2-[(6-methoxy-2-methyl-7-benzothiazolyl)thio]phenyl]-	MeOH	246s(4.30),297s(3.78), 308s(3.70)	88-0851-76
$C_{18}H_{18}N_2O_4$			
2-Furancarboxylic acid, 4-cyano-2-ethyl-1,2,3,4-tetrahydro-2-hydroxy-3-methyl-3-quinolinyl ester	EtOH	245(4.40),298(3.52)	95-0968-76
Pyrrolo[2,1-a]isoquinoline-1,3-dicarboxylic acid, 2-amino-, diethyl ester	EtOH	220(4.28),279(4.66), 306(4.55),342(4.06)	94-1299-76
$C_{18}H_{18}N_2O_7$			
L-Threonine, N-[(phenylmethoxy)carbonyl]-, 2-nitrophenyl ester	EtOH	258(3.70),343(3.01)	136-0001-76E
	NaOH	281(3.65)	136-0001-76E

Compound	Solvent	$\lambda_{max}(\log \epsilon)$	Ref.
$C_{18}H_{18}N_4$ Tricyclo[3.3.0.0^{2,6}]oct-7-ene-3,3,4,4- tetracarbonitrile, 1,2,5,6,7,8-hexa- methyl-	C_6H_{12}	223(3.70)	88-1987-76
$C_{18}H_{18}N_4O$ 1,2,3-Benzotriazin-4(3H)-one, 3-phenyl- 6-piperidino-	MeOH	233(4.33),368(4.34)	5-0946-76
$C_{18}H_{18}N_4O_3$ 2H-1,4-Benzodiazepin-2-one, 1,3-dihydro- 1-(methoxymethyl)-7-(N-methyl-ONN- azoxy)-5-phenyl-	isoPrOH	225(4.45),249(4.40), 310s(3.57)	87-1378-76
$C_{18}H_{18}N_4O_4$ Propanedioic acid, (1-phenyl-1H-pyraz- olo[3,4-d]pyrimidin-4-yl)-, diethyl ester	EtOH	259(4.24),328(4.48), 335(4.46)	95-1352-76
$C_{18}H_{18}N_4O_4S$ Propanoic acid, 2-[(2-methyl-5-nitro- phenyl)hydrazono]-3-(phenylamino)- 3-thioxo-, ethyl ester, (Z)-	EtOH	280(4.16),355(4.26)	104-1912-76
$C_{18}H_{18}N_4O_5$ D-Ribitol, 1-C-(8-amino-1,2,4-triazolo- [1,5-a]pyridin-2-yl)-1,4-anhydro-, 5-benzoate, (S)-	H_2O	280(3.99),299(3.84)	44-3124-76
$C_{18}H_{18}N_4O_5S$ Propanoic acid, 2-[(2-methoxy-5-nitro- phenyl)hydrazono]-3-(phenylamino)- 3-thioxo-, ethyl ester, (Z)-	EtOH	295(4.27),355(4.31)	104-1912-76
$C_{18}H_{18}N_4S$ 2,3,6-Pyridinetriamine, 4-[(diphenyl- methyl)thio]-, dihydrochloride	pH 7	337(3.76)	44-3784-76
$C_{18}H_{18}O$ Ethanone, 1-tricyclo[9.3.1.1^{4,8}]hexa- deca-1(15),4,6,8(16),11,13-hexaen- 5-yl-, (+)-	EtOH	330s(2.30)	49-0195-76
9(10H)-Phenanthrenone, 3,6,10,10-tetra- methyl-	hexane	245(4.54),252s(--), 272(4.01),281(4.05), 296s(--),323(3.49)	104-1298-76
$C_{18}H_{18}OS_2$ 9,13-Dithiatricyclo[13.3.1.1^{3,7}]eicosa- 1(19),3,5,7(20),15,17-hexaen-2-one	MeOH	253(4.15)	44-2509-76
$C_{18}H_{18}O_2$ 3,5,11,13-Hexadecatetraene-7,9-diyne- 2,15-dione, 6,11-dimethyl-, (E,E,Z,Z)-	EtOH	248s(4.10),260(4.25), 298(4.44),346s(4.49), 364(4.49),388(4.39)	18-0292-76
Phenanthrene, 5,6-dimethoxy-1,2-dimeth- yl-	CCl_4	261(4.66),282s(4.41)	102-0844-76
2H-Pyran-2-one, 5,6-dihydro-6-[2-(2- methyl-1-naphthalenyl)ethyl]-	EtOH	228(4.94),276(3.70), 285(3.74),293(3.60), 308(2.90),323(2.70)	39-1165-76C

Compound	Solvent	$\lambda_{max}(\log \epsilon)$	Ref.
Tricyclo[9.3.1.1^{4,8}]hexadeca-1(15),4,6-8(16),11,13-hexaene-5-carboxylic acid, methyl ester, (-)-	EtOH	295(3.08)	49-0195-76
$C_{18}H_{18}O_2S$			
3-Cyclohexene-1-carboxylic acid, 6-(4-phenyl-2-thienyl)-, methyl ester, trans	MeOH	232(4.34),263(4.09), 330(2.29)	44-1320-76
3-Cyclohexene-1-carboxylic acid, 6-(5-phenyl-2-thienyl)-, methyl ester, trans	MeOH	290(4.24),347(2.38)	44-1320-76
$C_{18}H_{18}O_2S_2$			
1-Propanone, 2,2'-dithiobis[1-phenyl-	EtOH	242(4.28)	32-0205-76
$C_{18}H_{18}O_3$			
Benzeneacetic acid, 3-methyl-, anhydride	EtOH	210(4.34),252(2.85), 257(2.85),265(2.88), 273(2.79)	78-2967-76
Benzeneacetic acid, 4-methyl-, anhydride	EtOH	251(2.99),256(2.97), 264(2.97),273(2.83)	78-2967-76
Benz[d]indeno[4,5-b]pyran-1(2H)-one, 5,10,11,11a-tetrahydro-7-methoxy-11a-methyl-, (+)	95% MeOH	221(4.09),297(3.96), 320(4.12)	13-0197-76B
2-Buten-1-one, 1,3-bis(4-methoxyphenyl)-	EtOH	325(4.28)	39-1466-76C
$C_{18}H_{18}O_4$			
Aurentiacin	EtOH	343(6.8)	102-0229-76
	EtOH-NaOH	344(--)	102-0229-76
2H-1-Benzopyran-8-carboxaldehyde, 3,4-dihydro-5,7-dimethoxy-2-phenyl-, (S)-	EtOH	229(4.10),290(4.09), 320s(3.60)	39-1570-76C
Isoaurentiacin	EtOH	288(4.3),320s(3.5)	102-0229-76
1,3-Pentanedione, 1-(2-hydroxy-3-methoxyphenyl)-5-phenyl-	MeOH	211(4.45),224s(4.26), 261(3.89),322(3.76), 345s(3.54)	12-2023-76
	MeOH-base	250(3.91),343(4.28)	12-2023-76
$C_{18}H_{18}O_5$			
Benzeneacetic acid, 2-methoxy-, anhydride	MeCN	217s(4.19),272(3.63), 278(3.61)	78-2967-76
Benzeneacetic acid, 3-methoxy-, anhydride	MeCN	274(3.55),280(3.52)	78-2967-76
9,10-Benzobicyclo[3.3.2]deca-2,9-dien-7-one, 3,8β-diacetoxy-	EtOH	218(3.67),262s(2.57), 265(2.59),266s(2.59), 272(2.57),290(1.00)	33-0765-76
4H-1-Benzopyran-4-one, 2,3-dihydro-5-hydroxy-7-methoxy-3-[(4-methoxyphenyl)methyl]-, (+)-	EtOH	215(4.44),225(4.32), 289(4.21)	33-2048-76
	EtOH	288.5(4.299)	33-2048-76
(+)-	EtOH	214(4.45),223(4.40), 289(4.31)	33-2048-76
$C_{18}H_{18}O_6$			
1,3-Azulenedicarboxylic acid, 6-(1,3-dioxolan-2-yl)-2-methyl-, dimethyl ester	MeOH	238(4.49),271(4.33), 297(4.68),308(4.79), 344(3.89),355s(3.79), 370(3.77),503(2.65)	88-2045-76
2,3-Benzobicyclo[2.2.2]octatetraene-5,6-dicarboxylic acid, 10,11-dimethoxy-, dimethyl ester	C_6H_{12}	212(4.54),242(3.72), 288(3.59),292(3.59), 297s(3.48),319s(2.93)	33-1469-76

Compound	Solvent	$\lambda_{max}(\log \epsilon)$	Ref.
Benzo[a]cyclopropa[cd]pentalene-1,6c-(2aH)-dicarboxylic acid, 2b,6b-di-hydro-4,5-dimethoxy-, dimethyl ester	C_6H_{12}	247s(3.79),287s(3.72), 292(3.79),295(3.80), 300s(3.70)	33-1469-76
4H-1-Benzopyran-4-one, 2,3-dihydro-3,5-dihydroxy-7-methoxy-3-[(4-methoxyphenyl)methyl]- (7-O-methyleucomol)	EtOH	216(4.46),291(4.302)	33-2048-76
3,4-Benzotricyclo[3.3.0.0^{2,8}]octa-3,6-diene-1,6-dicarboxylic acid, 10,11-dimethoxy-, dimethyl ester	C_6H_{12}	235(4.10),299(3.56)	33-1469-76
1,4-Naphthalenedione, 2-(1-acetoxy-4-methyl-3-pentenyl)-5,8-dihydroxy-	EtOH	275(3.80),493(3.75), 526(3.80),568(3.62)	105-0652-76
$C_{18}H_{18}O_7$ 1,3-Azulenedicarboxylic acid, 6-(1,3-dioxolan-2-yl)-2-methoxy-, dimethyl ester	MeOH	237(4.44),265(4.24), 297s(4.67),309(4.78), 345(3.85),362s(3.78), 476(2.58)	88-2045-76
Barbatic acid, 4-O-demethyl-	EtOH	278(4.42),308(4.06)	94-1853-76
1,4-Dibenzofurandione, 2,3-diethoxy-7,8-dimethoxy-	n.s.g.	255(4.25),279(3.96), 330(3.67),395(3.60), 565(3.32)	12-0179-76
1(4H)-Dibenzofuranone, 2,6-diacetyl-4a,9b-dihydro-3,7,9-trihydroxy-8,9b-dimethyl-	MeOH	227(4.19),283(4.38), 337(3.44)	23-2795-76
Ethanone, 2-(1,3-benzodioxol-5-yl)-1-(2-hydroxy-3,4,6-trimethoxyphenyl)-	MeOH	290(4.3)	2-0584-76
Ethanone, 2-(1,3-benzodioxol-5-yl)-1-(6-hydroxy-2,3,4-trimethoxyphenyl)-	MeOH	285(4.03),335(3.52)	2-0222-76
Ethanone, 2-(3,4-dimethoxyphenyl)-1-(6-hydroxy-4-methoxy-1,3-benzodioxol-5-yl)-	EtOH	225(4.39),289(4.03), 350s(3.85)	102-1019-76
4,2-(Epoxymethano)-1H,7H-cyclopropa[c]-benzofuran-7-one, 3a,4-diacetoxy-1a,2,3a,4-tetrahydro-2-methyl-5-(1-propynyl)- [1aR-(1aα,2α,3aβ,4α,7aS*)]-	MeOH	216(4.1),299(4.05)	33-1809-76
$C_{18}H_{18}O_{12}$ 1-Cyclopropene-1,2-diacetic acid, α,α'-bis(methoxycarbonyl)-3-[2-methoxy-1-(methoxycarbonyl)-2-oxoethylidene]-, dimethyl ester, ion(2-)	MeCN MeCN-Bu$_4$NOH	264(4.59),309(4.61) 265(4.51),305(4.59)	35-0610-76 35-0610-76
Propanedioic acid, 2,2',2''-(1,2,3-cyclo-propanetriylidene)tris-, hexamethyl ester	MeCN	223(4.14),375(4.32), 400s(4.21)	35-0611-76
$C_{18}H_{19}BrN_3S$ Isothiazolo[5,4-b]pyridinium, 3-bromo-6-[2-[4-(dimethylamino)phenyl]ethenyl]-4,7-dimethyl-, perchlorate	DMF	553(4.58)	48-0779-76
$C_{18}H_{19}ClN_2O_2$ Ethanone, 1-(6-chloro-1,2,3,4-tetra-hydro-4-hydroxy-1-methyl-4-phenyl-3-quinolinyl)-, oxime	isoPrOH	266(4.18),322(3.48)	4-0131-76
$C_{18}H_{19}ClO_4S$ Benzeneacetic acid, 2-[(4-chlorophenyl)-thio]-4,5-dimethoxy-, ethyl ester	MeOH	254(4.33),280(4.01)	73-1396-76

Compound	Solvent	$\lambda_{max}(\log \epsilon)$	Ref.
$C_{18}H_{19}ClO_7$ Microline, diacetate	MeOH	227(4.07),301(3.63)	33-1809-76
$C_{18}H_{19}Cl_2NO_3$ 1,4-Benzoxazepine, 5-(3,4-dichlorophen-yl)-2,3,4,5-tetrahydro-7,8-dimethoxy-3-methyl-	n.s.g.	220(4.28),285(3.58)	20-0898-76
$C_{18}H_{19}Cl_2NO_4$ Benzamide, 2,6-dichloro-N-[2-(3,4-di-methoxyphenoxy)-1-methylethyl]-	n.s.g.	210(4.34),285(3.55)	20-0787-76
$C_{18}H_{19}FN_2O$ Piperidine, 1-[5-fluoro-2-(phenylamino)-benzoyl]-	MeOH	240(3.98),283(4.14)	5-0946-76
$C_{18}H_{19}F_3N_2S$ 10H-Phenothiazine-10-propanamine, N,N-dimethyl-2-(trifluoromethyl)-	n.s.g.	255.5(4.49)	133-0065-76
$C_{18}H_{19}NO$ 6H-Dibenzo[a,h]quinolizine, 5,8,9,13b-tetrahydro-3-methoxy-	EtOH	218(4.21),279(3.38), 288(3.31)	118-0719-76
$C_{18}H_{19}NO_3$ Benzamide, N-(benzoyloxy)-N-(1,1-dimeth-ylethyl)-	EtOH	233(4.23)	44-0855-76
Ethanone, 1-[4-(1,3-dihydro-1,3-dimeth-oxy-2H-isoindol-2-yl)phenyl]-	MeOH	229(3.65),309(4.17)	83-0356-76
5H-Isoindolo[2,1-a][3,1]benzoxazine, 6a,11-dihydro-5,11-dimethoxy-5-methyl-	MeOH	248(4.09),264s(3.86), 270s(3.81)	83-0356-76
Tricyclo[8.2.2.2^{4,7}]hexadeca-4(16),6,10-12,13-pentaen-5-one, 6,16-dimethyl-15-nitro-	THF	239(4.01),290s(3.16), 320s(2.77),380s(1.74)	88-1509-76
$C_{18}H_{19}NO_4$ 1H-Indole, 2-(2,4-dimethoxyphenyl)-4,7-dimethoxy-	MeOH	214(4.59),259(4.27), 294(4.20),313(4.24)	111-0241-76
1H-Indole, 2-(2,5-dimethoxyphenyl)-4,7-dimethoxy-	MeOH	216(4.53),257(4.32), 297(4.08),329(4.16)	111-0241-76
1H-Indole, 2-(3,4-dimethoxyphenyl)-4,7-dimethoxy-	MeOH	215(4.54),258(4.35), 308(4.37)	111-0241-76
Norpallidine	EtOH	238(3.93),281(3.75)	102-1802-76
Pyrido[2,1,6-de]quinolizine-1,3-dicarb-oxylic acid, 3a,4-dihydro-, diethyl ester	EtOH	247s(4.0),302(4.47), 398(4.05),453(3.65), 530(3.81)	39-0341-76C
$C_{18}H_{19}NO_5$ Acetamide, N-[2-(2,4,6-trimethoxybenz-oyl)phenyl]-	MeOH	212(4.44),234(4.45), 266(4.09),270(4.07), 333(3.75)	39-2089-76C
Epimacronine	n.s.g.	232(4.67),272(3.99), 312(3.94)	105-0747-76
$C_{18}H_{19}NO_6$ 2-Azabicyclo[3.2.0]heptane-5-carboxylic acid, 7-acetoxy-7-methyl-3,4-dioxo-1-phenyl-, ethyl ester, (1α,5α,7α)-	dioxan	258(3.53)	142-1229-76

Compound	Solvent	$\lambda_{max}(\log \epsilon)$	Ref.
$C_{18}H_{19}N_2O_3$			
Isoquinolinium, 2-(2-aminophenyl)-3-hydroxy-6,7-dimethoxy-1-methyl-, perchlorate	MeOH	255(4.73),320(3.81), 420(3.63)	103-0205-76
$C_{18}H_{19}N_3O_2S$			
Propanoic acid, 3-[(4-methylphenyl)amino]-2-(phenylhydrazono)-3-thioxo-, ethyl ester, cis	EtOH	407(4.55)	104-1912-76
	EtOH	230(4.3),285(4.2), 405(4.5)	104-0544-76
trans	EtOH	247(4.23),385(4.48)	104-1912-76
	EtOH	250(4.3),393(4.5)	104-0544-76
Propanoic acid, 2-[(4-methylphenyl)hydrazono]-3-(phenylamino)-3-thioxo-, ethyl ester, cis	EtOH	236(4.25),254(4.10), 278(4.16),407(4.49)	104-0544-76
trans	EtOH	245(4.17),380(4.08)	104-1912-76
$C_{18}H_{19}N_3O_3$			
2H-1,4-Benzodiazepin-2-one, 1,3-dihydro-7-(methoxyamino)-1-(methoxymethyl)-5-phenyl-	isoPrOH	242(4.47),265s(4.19), 335(3.40)	87-1378-76
Piperidine, 1-[5-nitro-2-(phenylamino)-benzoyl]-	MeOH	211(4.29),254(4.02), 378(4.21)	5-0946-76
$C_{18}H_{19}N_3O_3S$			
Propanoic acid, 2-[(4-methoxyphenyl)hydrazono]-3-(phenylamino)-3-thioxo-, ethyl ester, cis	EtOH	228(4.23),286(4.47), 415(4.45)	104-0544-76
trans	EtOH	248(4.23),395(4.29)	104-1912-76
$C_{18}H_{19}N_3O_5$			
1H-Indole-2,3-diacetic acid, α^2-(aminocarbonyl)-α^3-cyano-, diethyl ester	MeOH	274(4.00),282(4.01), 290(3.92)	24-3282-76
$C_{18}H_{19}N_3O_6$			
1,1-Cyclobutanedicarbonitrile, 2,2,3,3-tetramethoxy-4-[2-(4-nitrophenyl)ethenyl]-	CHCl$_3$	309(4.26)	39-1538-76C
$C_{18}H_{19}N_3O_7$			
1H-Pyrazole-4-carboxylic acid, 1-(4-nitrophenyl)-3-(tetrahydro-2,2-dimethylfuro[3,4-d]-1,3-dioxol-4-yl)-, methyl ester, [3aS-(3aα,4α,6aα)]-	CHCl$_3$	310(4.26)	136-0019-76A
$C_{18}H_{19}N_3S$			
10H-Phenothiazine-2-carbonitrile, 10-[3-(dimethylamino)propyl]-	n.s.g.	268(4.44)	133-0065-76
$C_{18}H_{19}N_5OS$			
2H-Pyrrolo[2,3-d]pyrimidine-5-carbonitrile, 4-(dimethylamino)-3,7-dihydro-7-methyl-6-(methylthio)-2-oxo-3-(phenylmethyl)-	DMSO	326(4.50)	24-2983-76
$C_{18}H_{19}N_5O_3S$			
Glycine, N-[[6-[(phenylmethyl)thio]-9H-purin-9-yl]acetyl]-, ethyl ester	pH 10.5	294(4.28)	48-0291-76
$C_{18}H_{19}N_5O_4$			
D-erythro-Pentofuranoside, methyl 5-[6-	pH 1	252(4.09),291(4.49)	78-2409-76

Compound	Solvent	$\lambda_{max}(\log \epsilon)$	Ref.
(benzoylamino)-9H-purin-9-yl]-2,5-di- deoxy- (cont.)	pH 7 pH 13	282(4.39) 304(4.21)	78-2409-76 78-2409-76
$C_{18}H_{19}N_5O_7$ 1H-Pyrazole-4,5-dicarboxamide, 1-(4-ni- trophenyl)-3-(tetrahydro-2,2-dimeth- ylfuro[3,4-d]-1,3-dioxol-4-yl)-, [3aS-(3aα,4α,6aα)]-	$CHCl_3$	326(4.21)	136-0019-76A
$C_{18}H_{19}OS$ 1H-Benzo[c]thiolium, 3-(4-ethoxyphenyl)- 1,1-dimethyl-	H_2SO_4	441(4.44)	104-2523-76
$C_{18}H_{19}O_2$ 1H-Isobenzofurylium, 3-(4-ethoxyphenyl)- 1,1-dimethyl-	H_2SO_4	392(4.58)	104-2523-76
$C_{18}H_{20}$ Azulene, 1,3-di-1-butenyl-	C_6H_{12}	263(4.49),292(4.62), 395(4.08),669(2.49)	117-0169-76
Tricyclo[9.3.1.1^{4,8}]hexadeca-1(15),4,6- 8(16),11,13-hexaene, 5,14-dimethyl-, (+)-	EtOH	276(2.81)	49-0195-76
Tricyclo[10.2.2.2^{5,8}]octadeca-5,7,12,14- 15,17-hexaene	n.s.g.	271(2.67),283(2.43)	88-4095-76
Tricyclo[11.3.1.1^{5,9}]octadeca-1(17),5- 7,9(18),13,15-hexaene	C_6H_{12}	216s(4.00),226s(3.80), 256s(2.67),261s(2.53), 266s(2.38),275s(2.17), 282s(2.01)	138-1405-76
$C_{18}H_{20}ClNO_3$ 1,4-Benzoxazepine, 5-(2-chlorophenyl)- 2,3,4,5-tetrahydro-7,8-dimethoxy- 3-methyl-	n.s.g.	215(4.30),285(3.56)	20-0898-76
$C_{18}H_{20}ClN_3O$ Ethanone, 1-(6-chloro-1,2,3,4-tetra- hydro-4-hydroxy-1-methyl-4-phenyl- 3-quinolinyl)-, hydrazone	isoPrOH	256(4.19),320(3.48)	4-0131-76
$C_{18}H_{20}ClN_3O_5$ 1H-Imidazo[1,2-c]pyrimidin-4-ium, 5,6- dihydro-5-oxo-6-β-D-ribofuranosyl- 1-(phenylmethyl)-, chloride	pH 1 pH 7 pH 13	253s(3.63),287(4.11), 294(4.14),313(3.95) 253s(3.63),287(4.11), 294(4.14),313(3.95) 278(4.00)	35-7408-76 35-7408-76 35-7408-76
$C_{18}H_{20}Cl_2N_2$ Diazene, bis[1-chloro-1-(4-methylphen- yl)ethyl]-, meso	CH_2Cl_2	360(1.65)	39-1249-76B
$C_{18}H_{20}Cl_2N_2O$ Diazene, bis(2-chloro-1-methyl-2-phenyl- ethyl)-, 1-oxide	EtOH	224(4.53)	88-0153-76
$C_{18}H_{20}FNO_3$ 1,4-Benzoxazepine, 5-(3-fluorophenyl)- 2,3,4,5-tetrahydro-7,8-dimethoxy- 3-methyl-	n.s.g.	215(4.29),240(3.85), 285(3.58)	20-0898-76

Compound	Solvent	$\lambda_{max}(\log \epsilon)$	Ref.
$C_{18}H_{20}FNO_4$			
Benzamide, N-[2-(3,4-dimethoxyphenoxy)-1-methylethyl]-3-fluoro-	n.s.g.	225(4.23),280(3.62)	20-0787-76
$C_{18}H_{20}F_3N_3$			
1,4-Benzenediamine, N'-[1-[4-(dimethylamino)phenyl]-2,2,2-trifluoroethylidene]-N,N-dimethyl-	hexane	254(4.30),324(4.23), 370(3.94)	104-0161-76
$C_{18}H_{20}F_3N_5O_7$			
Ribitol, 1-deoxy-1-[3,4-dihydro-8-[(2-hydroxyethyl)amino]-2,4-dioxo-7-(trifluoromethyl)benzo[g]pteridin-10(2H)-yl]-	EtOH	258(4.32),318(3.95), 463(4.53)	103-0938-76
$C_{18}H_{20}FeN_2O_2S$			
Pyridinium, 1-[(ferrocenylsulfonyl)amino]-2,4,6-trimethyl-, hydroxide, inner salt	EtOH	212(4.57),249(4.13), 350(3.86),426(2.31)	44-0491-76
$C_{18}H_{20}N_2$			
2,7-Anthracenediamine, N,N,N',N'-tetramethyl-	C_6H_{12}	215(4.79),256s(5.11), 262(5.15),302s(5.34), 309(4.38),385s(4.46), 400(4.56),430s(3.94)	12-1613-76
	10% H_2SO_4	220(4.68),246s(4.62), 254(5.99),324(3.99), 338(4.23),355(4.36), 373(4.26)	12-1613-76
1,2-Diazocine, 3,4,5,6,7,8-hexahydro-3,8-diphenyl-, cis-(E)-	$CHCl_3$	371(1.83)	24-0518-76
cis-(Z)-	$CHCl_3$	381(2.04)	24-0518-76
trans-(E)-	$CHCl_3$	390(1.79)	24-0518-76
1H-Pyrazole, 4,5-dihydro-1-phenyl-3-(2,4,6-trimethylphenyl)-	benzene	295(4.15)	61-0288-76
$C_{18}H_{20}N_2O_2$			
2,4,6-Heptatrienoic acid, 2-cyano-5-(dimethylamino)-7-phenyl-, ethyl ester	CH_2Cl_2	401(4.49)	48-0705-76
Isoquinoline, 1,2-dihydro-6,7-dimethoxy-2-methyl-1-(2-pyridinylmethyl)-, diperchlorate	MeOH	255(4.75),315(4.20)	83-0072-76
Pyridinium, 1-(3-ethoxy-3-oxo-2-phenyl-1-propenyl)amino]-2,6-dimethyl-, hydroxide, inner salt, (E)-	EtOH	282(4.38)	44-1570-76
Vincarpine, hexahydro-	EtOH	225(4.7),278(4.1)	88-4887-76
$C_{18}H_{20}N_2O_3$			
Acetamide, N-(4-ethoxyphenyl)-2-[(4-ethoxyphenyl)imino]-	EtOH	234(3.12)	65-0702-76
Benzenamine, 4-(2,3-dihydro-7,8-dimethoxy-3-methyl-1,4-benzoxazepin-5-yl)-	n.s.g.	230(4.36),310(4.26)	20-0898-76
2,4-Pentadienoic acid, 2-cyano-5-(4-morpholinyl)-5-phenyl-, ethyl ester	CH_2Cl_2	419(4.14)	48-0705-76
$C_{18}H_{20}N_2O_4$			
Acetamide, N-(4-ethoxyphenyl)-2-[(4-ethoxyphenyl)imino]-, N-oxide	EtOH	225(4.2),288(4.1), 320(4.1),345(4.1), 380(4.0)	65-0702-76
Acetamide, 2-[(4-methoxyphenyl)imino]-N-(4-propoxyphenyl)-, N-oxide	EtOH	225(4.5),290(4.2), 320(4.2)	65-0702-76

Compound	Solvent	$\lambda_{max}(\log \epsilon)$	Ref.
1,1-Cyclobutanedicarbonitrile, 2,2,3,3-tetramethoxy-4-(2-phenylethenyl)-	$CHCl_3$	259(4.32)	39-1538-76C
$C_{18}H_{20}N_2O_4S$ 3-Oxa-1,7-diazaspiro[4.4]non-1-ene-8-carboxylic acid, 9,9-dimethyl-4-oxo-2-(phenylmethyl)-6-thioxo-, ethyl ester	C_6H_{12}	276(4.05)	5-2185-76
isomer	C_6H_{12}	274(4.15)	5-2185-76
$C_{18}H_{20}N_2O_5$ Acetamide, N-[tetrahydro-2,2-dimethyl-5-(5-phenyl-3-isoxazolyl)furo[2,3-d]-1,3-dioxol-6-yl]-, [3aR-3aα,5α,6β,6aα)]-	EtOH	208(4.08),261(4.31)	136-0119-76A
$C_{18}H_{20}N_2O_5S$ 2H-Pyran-3-carboxylic acid, 4-[[(ethyl-thio)carbonyl]amino]-2-oxo-6-[(phen-ylmethyl)amino]-, ethyl ester	MeCN	245(4.08),287(3.80), 321(4.12)	39-2462-76C
$C_{18}H_{20}N_2O_8P_2$ Phosphinedicarboxylic acid, [(diphen-oxyphosphinyl)azo]-, diethyl ester, 1-oxide	MeOH	272(4.34),278(3.24), 468(1.36)	139-0207-76A
$C_{18}H_{20}N_3O_5$ 1H-Imidazo[1,2-c]pyrimidin-4-ium, 5,6-dihydro-5-oxo-β-D-ribofuranosyl-1-(phenylmethyl)-, chloride	pH 1 and 7 pH 13	253s(3.63),287(4.11), 294(4.14),313(3.95) 278(4.00)	35-7408-76 35-7408-76
$C_{18}H_{20}N_4O_3$ 2-Quinoxalinecarboxylic acid, 3,4,5,6,7-8-hexahydro-4-methyl-3-[[(phenylami-no)carbonyl]imino]-, methyl ester	EtOH	236(4.03),299(4.27), 400(4.00)	88-2899-76
$C_{18}H_{20}N_5O_6PS$ Adenosine, 1-methyl-8-[(phenylmethyl)-thio]-, cyclic 3',5'-(hydrogen phosphate	pH 1 pH 7 pH 11	285(4.27) 285(4.26) 285(4.10)	87-0419-76 87-0419-76 87-0419-76
Adenosine, 2-methyl-8-[(phenylmethyl)-thio]-, cyclic 3',5'-(hydrogen phosphate)	pH 1 pH 7 pH 11	283(4.27) 283(4.23) 282(4.21)	87-0419-76 87-0419-76 87-0419-76
$C_{18}H_{20}N_6S_2$ Hydrazinecarbothioamide, 2,2'-(1'-[1,1'-biphenyl]-4-yl-1,2-ethanediylidene)-bis[N-methyl-	EtOH	337(4.61)	87-0131-76
$C_{18}H_{20}N_{10}O_2$ 2,2'-Hydrazobis[1,6,7-trimethylpteridin-4(1H)-one]	HOAc	317(4.51),398(3.90)	12-0459-76
$C_{18}H_{20}O$ Tricyclo[9.2.2.1⁴,⁸]hexadeca-6,8(16),11-13,14-pentaen-5-one, 4,7-dimethyl-	THF	230s(3.80),276s(3.08), 320(3.63),380s(2.52), 400s(2.28)	88-1509-76
$C_{18}H_{20}O_2$ Benzene, 5-methoxy-2-[2-(4-methoxyphen-yl)ethenyl]-1,3-dimethyl-, (E)-	EtOH	289(4.36)	39-1466-76C
(Z)-	EtOH	271(4.16)	39-1466-76C

Compound	Solvent	$\lambda_{max}(\log \epsilon)$	Ref.
3,6-Phenanthrenedione, 9,10-diethyl-4,4a,4b,5-tetrahydro-	MeOH	221(3.95),287(4.36), 406(4.29)	35-3262-76
$C_{18}H_{20}O_2S$			
Cyclohexanecarboxylic acid, 2-(4-phenyl-2-thienyl)-, methyl ester, trans	MeOH	232(4.36),262(4.10)	44-1320-76
$C_{18}H_{20}O_3$			
9,10-Anthracenedione, 1,2,3,4-tetrahydro-5-methoxy-7-propyl-	EtOH	212(4.59),249(4.18), 255s(4.17),275(4.16), 390(3.62)	39-0613-76C
9bH-Benz[e]indene-9b-acetic acid, tetrahydro-3-methyl-1-oxo-, ethyl ester	EtOH	230(3.9),290(3.4)	39-0722-76C
9bH-Benz[e]indene-9bα-acetic acid, 3a,4,5,9b-tetrahydro-3-methyl-1-oxo-,,ethyl ester	EtOH	228(4.2)	39-0722-76C
1H-Benz[e]inden-1-one, 3a,4,5,9b-tetrahydro-7-methoxy-3-methyl-9b-(2-oxopropyl)-, cis-(±)-	EtOH	229(4.28)	39-0722-76C
Benz[d]indeno[4,5-b]pyran-1-ol, 1,2,5,10,11,11a-hexahydro-7-methoxy-11a-methyl-, cis-(±)-	95% MeOH	225(4.00),282s(3.68), 321(4.25)	13-0197-76B
Ethanone, 1-(2-methoxy-4,6-dimethylphenyl)-2-(4-methoxyphenyl)-	EtOH	258s(3.68),277(3.64), 284(3.62)	39-1466-76C
Ethanone, 1-(4-methoxy-2,6-dimethylphenyl)-2-(4-methoxyphenyl)-	EtOH	273(4.32)	39-1466-76C
$C_{18}H_{20}O_3S$			
Thiophenium, tetrahydro-, 1-acetyl-3-benzoyl-4-oxo-2-pentenylide	EtOH	256(4.09),349(4.29)	94-2421-76
Tricyclo[5.2.0.0^{2,9}]nona-3,5-diene-8-methanol, 9-methyl-, 4-methylbenzenesulfonate	EtOH	225(4.08),274(3.49)	88-3903-76
$C_{18}H_{20}O_4$			
Benzene, 4-[2-(3,5-dimethoxyphenyl)ethenyl]-1,2-dimethoxy-	MeOH	302s(4.31),323(4.37)	102-2006-76
2H-Benz[e]indene-1-carboxylic acid, 3,3a,4,5-tetrahydro-7-methoxy-3a,5-dimethyl-3-oxo-, methyl ester, trans	EtOH	304(4.12)	2-0312-76
1,3-Cyclopentanedione, 2-[2-(7-methoxy-1H-2-benzopyran-4(3H)-ylidene)ethyl]-2-methyl-	95% MeOH	221(4.24),258s(4.24), 268(4.34),297s(3.82)	13-0197-76B
6H,15H-Dibenzo[b,i][1,4,8,11]tetraoxacyclotetradecin, 7,8,16,17-tetrahydro-	MeOH	275(3.30)	64-0330-76B
sodium derivative picrate	MeOH	274(3.97),278s(3.97)	64-0330-76B
4,6,12-Tetradecatriene-8,10-diyne-1,3-diol, diacetate, (E,E,E)-	EtOH	247(4.54),266(4.48), 279(4.22),296(4.47), 315(4.61),336(4.48)	39-0735-76C
$C_{18}H_{20}O_5$			
2-Oxaandrost-4-ene-1,3,17-trione, 6β,19-epoxy-	MeOH	223(3.93)	44-2864-76
$C_{18}H_{20}O_7$			
Ethanone, 1-(2,4-dihydroxy-5-methoxyphenyl)-2-(2,4,5-trimethoxyphenyl)-	MeOH	235(4.31),285(3.90), 342(3.54)	2-0951-76
$C_{18}H_{21}BrN_4O_4S$			
Hexanoic acid, ε-amino-N-[4-(5-bromo-1-uracilylmethyl)phenylaminothio-	pH 9.2	272.5(4.28)	106-0598-76

Compound	Solvent	$\lambda_{max}(\log \epsilon)$	Ref.
carbonyl]- (cont.)			106-0598-76
$C_{18}H_{21}ClN_4O_4S$ Hexanoic acid, ε-amino-N-[4-(5-chloro-1-uracilylmethyl)phenylaminothiocarbonyl]-	pH 9.2	268.0(4.19)	106-0598-76
$C_{18}H_{21}FN_4O_4S$ Hexanoic acid, ε-amino-N-[4-(5-fluoro-1-uracilylmethyl)phenylaminothiocarbonyl]-	pH 9.2	270.0(4.28)	106-0598-76
$C_{18}H_{21}F_3N_2O_{10}$ Uridine, 5-[(2,2,2-trifluoroethoxy)methyl]-, 2',3',5'-triacetate	pH 13 EtOH	264(3.85) 260(3.97)	104-0643-76 104-0643-76
$C_{18}H_{21}N$ 1-Propanamine, N-(diphenylmethylene)-2,2-dimethyl-	C_6H_{12}	247.5(4.20)	101-0295-76G
$C_{18}H_{21}NO$ 7-Azabicyclo[4.3.1]deca-3,8-dien-10-one, 3,4,7-trimethyl-9-phenyl-	EtOH	215(3.7),269s(3.0)	39-2329-76C
$C_{18}H_{21}NO_2$ 11-Azaestra-1,3,5(10),9(11)-tetraen-17-one, 3-methoxy-	EtOH	260(4.03),270(4.06)	39-1889-76C
$C_{18}H_{21}NO_3$ Benzamide, 2-(2-hydroxyethyl)-N-[2-(3-methoxyphenyl)ethyl]-	EtOH	220(4.16),273(3.42), 280(3.36)	118-0719-76
Benzamide, 2-(2-hydroxyethyl)-N-[2-(4-methoxyphenyl)ethyl]-	EtOH	227(4.25),279(3.16), 285(3.11)	118-0719-76
Erysodine	EtOH	233(4.14),284(3.81)	94-0052-76
$C_{18}H_{21}NO_3S$ Thiopyrano[4,3-b]benzo[g]quinolizin-13-one, 6,7,11b,12-tetrahydro-9,10-dimethoxy-	n.s.g.	262(3.51),287(3.81), 330(2.99)	2-0784-76
$C_{18}H_{21}NO_4$ 1,3-Azulenedicarboxylic acid, 6-[(dimethylamino)methyl]-2-methyl-, methyl ester	MeOH	237(4.48),272(4.35), 298(4.67),309(4.78), 346(3.92),355(3.81), 371(3.79),494(2.70)	88-2045-76
N-Nororientaline	EtOH	216(4.03),230(4.03), 286(3.81)	94-0052-76
Pyrido[2,1,6-de]quinolizine-1,3-dicarboxylic acid, 3a,4,5,6-tetrahydro-, diethyl ester	EtOH	265(4.02),272(4.02), 314(4.39),362(3.96), 507(3.95)	39-0341-76C
1H-Pyrrole-3-carboxylic acid, 5-[4-(2-methoxy-2-oxoethyl)phenyl]-2,4-dimethyl-, ethyl ester, cis	EtOH	248(4.36)	44-0543-76
trans	EtOH	248(4.31)	44-0543-76
Wisanine	EtOH	250(4.00),304(4.16), 309(4.12),378(4.32)	88-3049-76
$C_{18}H_{21}NO_5$ α-D-xylo-Heptofuranose, 6,6,7,7-tetradehydro-5,6,7-trideoxy-3-O-methyl-	EtOH	307(4.21),322(4.16)	136-0127-76A

Compound	Solvent	$\lambda_{max}(\log \epsilon)$	Ref.
1,2-O-(1-methylethylidene)-5-(methyl-imino)-7-phenyl-, N-oxide (cont.)			136-0127-76A
4(5H)-Oxazolone, 5-(2-acetoxy-1,1-di-methylethyl)-2-[2-(4-methoxyphenyl)-ethenyl]-, perchlorate	n.s.g.	360(4.60)	104-1146-76
$C_{18}H_{21}N_3$			
1,3-Propanediamine, N'-9-acridinyl-N,N-dimethyl-	pH 1	220(4.11),274(4.60), 415(3.96),425(3.85)	56-1267-76
$C_{18}H_{21}N_3O$			
Benzamide, N-(2,3-di-1-pyrrolidinyl-2-cyclopropen-1-ylidene)-	MeOH	234(3.90),315(3.25)	138-1215-76
$C_{18}H_{21}N_3O_2$			
1H-Indazol-4-amine, 3-(3,4-dimethoxy-phenyl)-1,5,6-trimethyl-	MeOH	252s(4.10),322(3.95)	24-1898-76
4H-Indazol-4-one, 1,5,6,7-tetrahydro-1,6,6-trimethyl-3-phenyl-, O-acet-yloxime, (E)-	MeOH	231(4.40),266(4.00)	24-1898-76
4-Pentenamide, N-[2-[3-(1-methyl-1H-imidazol-5-yl)-3-oxopropyl]phenyl]-	MeOH MeOH-acid	255(4.04) 232(4.06)	78-1085-76 78-1085-76
Pyrazolo[3,4,5-kl]acridine, 2,3,4,5-tetrahydro-8,9-dimethoxy-2,4,4-tri-methyl-	MeOH	250s(4.58),257(4.62), 264(4.54),298s(3.90), 305(3.94),316(3.93), 331(3.75)	24-1898-76
$C_{18}H_{21}N_3O_4$			
Butanoic acid, 4-[[2-[3-(1-methyl-1H-imidazol-5-yl)-3-oxopropyl]phenyl]-amino]-4-oxo-, methyl ester	MeOH	254(4.21)	78-1085-76
5(4H)-Oxazolone, 4-(di-4-morpholinyl-methylene)-2-phenyl-	EtOH	234(3.89),278(3.99), 355(4.08)	104-2170-76
$C_{18}H_{21}N_3O_5S$			
Cephalsporanic acid, 7α-ethoxyformamido-7β-phenoxyacetamidodeacetoxy-	EtOH	251(4.10),265(3.86), 272(3.79)	39-1918-76C
$C_{18}H_{21}N_3O_6$			
D-Ribitol, 1-deoxy-1-(2,3,4,10-tetra-hydro-7,8-dimethyl-2,4-dioxopyrimi-do[5,4-b]quinolin-10-yl)-	EtOH 50% EtOH-HCl	227(4.52),264(4.45), 332(4.02),402(4.07) 224(4.38),263(4.53), 355(4.27)	39-1805-76C 39-1805-76C
$C_{18}H_{21}N_3O_9$			
1H-Imidazole-4-carboxylic acid, 5-(cya-nomethyl)-1-(2,3,5-tri-O-acetyl-α-D-ribofuranosyl)-, methyl ester	pH 1 pH 7 pH 11	226(4.00) 236(4.02) 236(4.02)	35-1492-76 35-1492-76 35-1492-76
β-	pH 1 pH 7 pH 11	227(3.92) 235(3.94) 237(3.92)	35-1492-76 35-1492-76 35-1492-76
1H-Imidazole-5-carboxylic acid, 4-(cya-nomethyl)-1-(2,3,5-tri-O-acetyl-α-D-ribofuranosyl)-, methyl ester	pH 1 pH 7 pH 11	238(3.99) 247(4.06) 247(4.06)	35-1492-76 35-1492-76 35-1492-76
β-	pH 1 pH 7 pH 11	235(3.99) 242(4.06) 242(4.06)	35-1492-76 35-1492-76 35-1492-76
$C_{18}H_{22}Cl_2N_3O_2$			
Morpholinium, 4-[3-chloro-2-(4-chloro-	HOAc	409(4.41)	73-1565-76

$C_{18}H_{22}Cl_2O_2Si-C_{18}H_{22}N_6O_8$

Compound	Solvent	λ_{max}(log ϵ)	Ref.
phenyl)-3-[(4-morpholinylmethylene)-amino]-2-propenylidene]-, perchlorate			73-1565-76
$C_{18}H_{22}Cl_2O_2Si$ Silane, bis(4-chloro-3,5-dimethylphen-oxy)dimethyl-	dioxan	275(3.15),283(3.15)	56-0363-76
$C_{18}H_{22}N_2$ 3-Epiuleine	MeOH	213(4.38),307(4.28), 315(4.24)	102-1093-76
Phenazine, 5,10-dihydro-5,10-dipropyl-	EtOH	256(4.78),360(3.98)	118-0748-76
Uleine	MeOH	213(4.38),307(4.28), 315(4.24)	102-1093-76
$C_{18}H_{22}N_2O$ 2(1H)-Pyridinone, 3-(1-methyl-2-pyrrol-idinyl)-1-(2-phenylethyl)-, (S)-	MeOH	231s(3.60),307(3.63)	106-0603-76
2(1H)-Pyridinone, 5-(1-methyl-2-pyrrol-idinyl)-1-(2-phenylethyl)-, (S)-	MeOH	233(3.84),307(3.54)	106-0603-76
Tricyclo[3.3.1.1^{3,7}]decane-1-acetalde-hyde, α-(phenylhydrazono)-	CCl_4	383(4.16)	24-3735-76
$C_{18}H_{22}N_2OS$ Methopromazine	n.s.g.	251(4.43)	133-0065-76
$C_{18}H_{22}N_2O_2$ Isofumigaclavine A	EtOH	226(4.42),277s(3.77), 283(3.81),293(3.75)	31-0140-76
$C_{18}H_{22}N_2O_3$ 2H-Pyrrole-3-carboxylic acid, 3-cyano-4-ethoxy-3,4-dihydro-2,2-dimethyl-5-phenyl-, ethyl ester	EtOH	246(4.02)	33-1018-76
4H-Pyrrolo[3,2-d]isoxazole, 6-(benzoyl-oxy)-3a,5,6,6a-tetrahydro-3,5,6a-tri-methyl-3a-(2-propenyl)-	EtOH	225(4.10),271(2.99)	44-0855-76
$C_{18}H_{22}N_2O_5$ 1-Cyclobutanecarboxamide, 1-cyano-2,2,3,3-tetramethoxy-4-(2-phenyl-ethenyl)-	$CHCl_3$	257(4.35)	39-1538-76C
$C_{18}H_{22}N_2O_9S$ 2,4(1H,3H)-Pyrimidinedione, 5-methyl-1-[6-O-[(4-methylphenyl)sulfonyl]-β-D-galactopyranosyl]-	MeOH	223(4.22),265(4.04)	48-0079-76
$C_{18}H_{22}N_2S$ 10H-Phenothiazine-10-propanamine, N,N,β-trimethyl-	n.s.g.	252.5(4.46)	133-0065-76
$C_{18}H_{22}N_4O_4S$ Hexanoic acid, ε-amino-N-[4-(1-uracilyl-methyl)phenylaminothiocarbonyl]-	pH 9.2	263.0(4.28)	106-0598-76
$C_{18}H_{22}N_6O_8$ Guanosine, N-acetyl-2'-(acetylamino)-2'-deoxy-, 3',5'-diacetate	pH 1 H_2O pH 13	261(4.21) 258(4.21) 262(4.07)	94-2955-76 94-2955-76 94-2955-76

Compound	Solvent	$\lambda_{max}(\log \epsilon)$	Ref.
$C_{18}H_{22}O_2$			
19,20-Cyclopodocarpa-8,11,13-trien-19-one, 13-methoxy-, (+)-	EtOH	276(3.55)	44-1089-76
2H-Pyran-2-one, 5,6-dihydro-6-[2-(5,6,7,8-tetrahydro-2-methyl-1-naphthalenyl)ethyl]-	EtOH	228(4.94),276(3.70), 285(3.74),293(3.60)	39-1165-76C
Spiro[5.5]undec-8-en-1-one, 9-(4-methoxyphenyl)-	n.s.g.	257(3.78)	111-0279-76
$C_{18}H_{22}O_3$			
1-Benzoxepin-2(5H)-one, 3-(2,2-dimethyl-1-oxopropyl)-4-ethyl-5-methyl-	EtOH	242(3.49),267(3.34), 273(3.35),315s(2.78)	39-1260-76C
2H-1-Benzoxocin-2-one, 3-(2,2-dimethyl-1-oxopropyl)-5,6-dihydro-5,6-dimethyl-	EtOH	233(3.48),262(2.99), 270(2.91)	39-1260-76C
1H-Cyclodec[e]indene-3,6,9(2H)-trione, 11,12-didehydro-3a,4,5,5a,7,8,12,13-13a,13b-decahydro-3a-methyl-	MeCN	222(4.08)	35-5038-76
isomer	MeCN	223(3.97)	35-5038-76
2-Cyclohexene-1-carboxylic acid, 2-methyl-3-[(2-methylphenyl)methyl]-4-oxo-, ethyl ester	EtOH	242(4.08)	78-0065-76
7-Oxaestra-1,3,5(10),8-tetraen-17β-ol, 3-methoxy-	95% MeOH	221(3.99),286(4.00)	13-0197-76B
$C_{18}H_{22}O_4$			
Benzene, 1,1'-(1,2-dimethoxy-1,2-ethanediyl)bis[4-methoxy-	hexane	197(4.88),229(4.45), 275(3.58),281(3.56)	33-1925-76
Benzene, 1,1'-(2,2-dimethoxyethylidene)-bis[4-methoxy-	hexane	196(4.81),230(4.35), 276(3.59),283(3.51)	33-1925-76
2H-Benz[e]indene-1-carboxylic acid, 3,3a,4,5-tetrahydro-3-hydroxy-7-methoxy-3a,5-dimethyl-, methyl ester, (3α,3aα,5β)-	EtOH	210(3.98),226(3.96), 305(4.03)	2-0312-76
1,4-Ethanonaphthalene-5,8,9-trione, 10-ethoxy-1,4,4a,8a-tetrahydro-1,6,7,10-tetramethyl-	EtOH	203(3.88),252(3.94), 315s(2.23),365(2.02)	1-0064-76
$C_{18}H_{22}O_{10}$			
D-Quinic acid, 3-O-sinapyl-, (-)-	MeOH-acid	330(4.35)	20-0663-76
	MeOH-base	395(4.48)	20-0663-76
$C_{18}H_{23}ClN_3O_2$			
Morpholinium, 4-[3-chloro-3-[(4-morpholinylmethylene)amino]-2-phenyl-2-propenylidene]-, perchlorate	HOAc	408(4.38)	73-1565-76
$C_{18}H_{23}NO$			
4H-Cyclohepta[b]pyridin-4-one, 1,2,3,5,6,7,8,9-octahydro-1-(1-phenylethyl)-, (S)-	heptane	319(4.22)	103-0428-76
	MeOH	340(4.34)	103-0428-76
	CF_3COOH	320(4.08)	103-0428-76
Hasubanan, 9,10-didehydro-3-methoxy-17-methyl-, dl-	EtOH	215(4.36),273(4.12)	88-0767-76
4(1H)-Quinolinone, 2,3,5,6,7,8-hexahydro-1-(1-methyl-2-phenylethyl)-	heptane	315(4.09)	103-0428-76
	MeOH	338(4.37)	103-0428-76
	CF_3COOH	324(4.28)	103-0428-76
$C_{18}H_{23}NO_2$			
2-Azaestra-1,3,5(10)-trien-17-one, 3-methoxy-	MeOH	276(3.59)	44-2864-76
Cocculidine	MeOH	210(4.09),230s(3.91), 282(3.37),288s(3.36)	36-0132-76

Compound	Solvent	$\lambda_{max}(\log \epsilon)$	Ref.
5H-Inden-5-one, 1,2,3,6,7,7a-hexahydro-1-hydroxy-7a-methyl-4-[2-(6-methylpyridin-2-yl)ethyl]-, (1S-cis)-	EtOH	210(4.00),250(4.08)	35-4975-76
$C_{18}H_{23}NSi$ Ethanamine, N-(diphenylmethylene)-2-(trimethylsilyl)-	C_6H_{12}	247.5(4.21)	101-0295-76G
$C_{18}H_{23}N_2O_5P$ Benzoic acid, 4-[3-(dimethoxyphosphinyl)-3a,4,5,6,7,7a-hexahydro-4,7-methano-3H-indazol-3-yl]-, methyl ester, (3α,3aα,4β,7β,7aα)-	MeOH	210(4.0),243(4.2), 337(2.5)	23-0044-76
$C_{18}H_{23}N_3O_3$ 4H-Indazol-4-one, 3-(3,4-dimethoxyphenyl)-1,5,6,7-tetrahydro-1,6,6-trimethyl-, oxime, (E)-	MeOH	238s(4.32),271(3.99)	24-1898-76
Pyrazolo[4,3-c]azepin-4(1H)-one, 3-(3,4-dimethoxyphenyl)-5,6,7,8-tetrahydro-1,7,7-trimethyl-	MeOH	260(4.06),264s(4.06), 291s(3.90)	24-1898-76
1H-Pyrrole-3-carboxylic acid, 2-amino-4-methyl-5-[[(phenylmethyl)amino]carbonyl]-, 1,1-dimethylethyl ester	MeOH	270(4.10),307(4.35)	24-2983-76
$C_{18}H_{23}N_3O_4$ Imidazo[1,2-a][1,3]diazocine, 5,6,7,8-9,10-hexahydro-10-(3,4,5-trimethoxybenzoyl)-	n.s.g.	214(4.60),265(3.80)	2-0773-76
$C_{18}H_{23}N_5O$ Methanimidamide, N'-[3-(cyclohexylidenehydrazino)-1-oxo-2(1H)-isoquinolinyl]-N,N-dimethyl-	EtOH	224(4.40),316(4.10), 374(3.69)	95-0700-76
$C_{18}H_{23}N_5O_5$ Purine, 6-(benzylamino)-3-N-(1-deoxy-D-mannitol-1-yl)-, hydrochloride	aq acid H_2O aq base	286(4.37) 288(4.31) 289(4.30)	136-0049-76G 136-0049-76G 136-0049-76G
$C_{18}H_{23}N_5O_8$ Adenosine, 7,8-dihydro-N,N-dimethyl-8-oxo-, 2',3',5'-triacetate	pH 1 pH 13 EtOH	277(4.14),305s(3.96) 295(4.28) 282(4.18)	94-0565-76 94-0565-76 94-0565-76
$C_{18}H_{23}O_5P$ Benzoic acid, 4-[3-(dimethoxyphosphinyl)tricyclo[3.2.1.0^{2,4}]oct-3-yl]-, methyl ester, (1α,2β,3β,4β,5α)-	MeOH	212(2.30),254(3.5)	23-0044-76
$C_{18}H_{24}$ Naphthalene, octamethyl-	EtOH	252(4.79),309(3.76)	104-1057-76
$C_{18}H_{24}N_2O$ 5H-Cyclooct[b]indole-5-propanamine, 6,7,8,9,10,11-hexahydro-α-methyl-γ-oxo-, hydrochloride	EtOH	248(4.20),271(3.99), 304(3.73)	44-3775-76
Uleine, 1,13-dihydro-13-hydroxy-	MeOH	219(4.56),283(3.90), 290(3.85)	102-1093-76

Compound	Solvent	$\lambda_{max}(\log \epsilon)$	Ref.
$C_{18}H_{24}N_2O_2$			
1H-Pyrido[3,4-b]indole-1-propanoic acid, 2,3,4,9-tetrahydro-α-(1-methylpropyl)-(decahydrovincarpine)	EtOH	227(4.38),275(3.8)	88-4887-76
$C_{18}H_{24}N_2O_9S_2$			
2H-1,4-Thiazine-3,6-dicarboxylic acid, 4,4'-[oxybis(methylene)]bis[3,4-dihydro-, tetramethyl ester, [R-(R*,R*)]-	EtOH	212(3.90),256(3.56), 309(4.03)	39-2540-76C
$C_{18}H_{24}N_2O_{12}S$			
2,4(1H,3H)-Pyrimidinedione, 5-methyl-1-[2,3,4-tri-O-acetyl-6-O-(methylsulfonyl)-β-D-galactopyranosyl]-	MeOH	265(4.04)	48-0079-76
2,4(1H,3H)-Pyrimidinedione, 5-methyl-1-[2,3,4-tri-O-acetyl-6-O-(methylsulfonyl)-β-D-glucopyranosyl]-	MeOH	265(4.10)	48-0079-76
$C_{18}H_{24}N_5O_6$			
1-Piperidinyloxy, 4-[[5-(1-aziridinyl)-2,4-dinitrobenzoyl]amino]-2,2,6,6-tetramethyl-	EtOH	235(4.12),271(4.05), 329(4.04)	64-0328-76C
$C_{18}H_{24}N_5O_8PS$			
Adenosine, 7,8-dihydro-N-(1-oxobutyl)-8-thioxo-, cyclic 3',5'-(hydrogen phosphate) 2'-butanoate	pH 7	243(4.30),322(4.47)	87-0899-76
$C_{18}H_{24}N_5O_9P$			
Adenosine, 7,8-dihydro-8-oxo-N-(1-oxobutyl)-, cyclic 3',5'-(hydrogen phosphate) 2'-butanoate	pH 7	288(4.16)	87-0899-76
$C_{18}H_{24}N_6O_2$			
[4,5'-Bipyrimidine]-4',6(1H,1'H)-dione, 2,2'-dipiperidino-	pH 12	251(4.24),279(4.19), 365s(4.02),399(4.28)	19-0017-76
$C_{18}H_{24}O$			
1H-Inden-1-ol, 2,4,5,6,7,7a-hexahydro-4,4,7a-trimethyl-1-phenyl-, trans	pentane	259f(2.30)	33-0032-76
1-Oxaspiro[4.5]decane, 6,6-dimethyl-10-methylene-2-phenyl-	pentane	258f(2.57)	33-0032-76
2-Oxatricyclo[4.4.0.0^{3,7}]decane, 1,6,7-trimethyl-3-phenyl-	pentane	220(3.88),258f(2.41)	33-0032-76
10β-Oxatricyclo[5.2.1.0^{1,6}]decane, 2,2,6α-trimethyl-7-phenyl-	pentane	227(4.08),264f(2.46)	33-0032-76
1-Propanone, 1-phenyl-3-(2,6,6-trimethyl-1-cyclohexen-1-yl)-	pentane	239(4.18),279(3.02), 287(2.91),317(1.83), 320s(1.81),334s(1.77), 347s(1.58),359s(0.90)	33-0032-76
Spiro[cyclohexane-1,2'(1'H)-naphthalen]-1'-one, 3',4'-dihydro-2-propyl-	EtOH	230(4.15),266(2.36)	78-2731-76
Spiro[cyclopentane-1,2'(1'H)-naphthalen]-1'-one, 3',4'-dihydro-3-methyl-2-propyl-	EtOH	232(3.92),249(3.98), 298(3.16)	42-0812-76
$C_{18}H_{24}O_2$			
4-Cyclooctene-1-acetic acid, 5-phenyl-, ethyl ester	EtOH	248.5(3.95)	39-2457-76C
Estra-4,9-dien-3-one, 17-hydroxy-	EtOH	303(4.30)	35-4975-76

Compound	Solvent	$\lambda_{max}(\log \epsilon)$	Ref.
$C_{18}H_{24}O_3$			
1H-Benz[e]inden-3-ol, 2,3,3a,4,5,9b-hexahydro-7-methoxy-3a,5-dimethyl-, acetate, cis	EtOH	282(3.28),289(3.24)	2-0312-76
trans	EtOH	282(3.32),289(3.26)	2-0312-76
Estra-1,4-dien-3-one, 10α,17β-dihydroxy-	MeOH	234(4.11)	22-2021-76
Estra-1,4-dien-3-one, 10β,17β-dihydroxy-	MeOH	234(4.15)	22-2021-76
$C_{18}H_{24}O_4$			
1,4-Ethanonaphthalene-6,10(4H)-dione, 1,4a,5,8a-tetrahydro-5,9-dihydroxy-1,2,5,7,8,9-hexamethyl-	EtOH	203(3.80),253(3.86), 310s(2.36)	1-0064-76
1H-Naphtho[2,1-b]pyran-1-one, 10-acetoxy-2,3,4a,8,9,10,10a,10b-octahydro-4a,8,10b-trimethyl-, [4aR-(4aα,8α,10β,10aα,10bα)]-	MeOH	241(4.25)	44-0066-76
$C_{18}H_{24}O_5$			
1H-Fluorene-1,9-dicarboxylic acid, 2,3,4,4b,5,6,7,8,8a,9-decahydro-1-methyl-2-oxo-, dimethyl ester	EtOH	295(1.96)	44-2401-76
isomer	EtOH	310(1.84)	44-2401-76
2-Oxaandrost-4-en-3-one, 6β,17-epoxy-1,17-dihydroxy-	MeOH	226(3.94)	44-2864-76
$C_{18}H_{25}ClN_5O_4P$			
5H-1,4-Benzodiazepin-5-one, 7-chloro-4-(di-4-morpholinylphosphinyl)-3,4-dihydro-2-(methylamino)-	MeCN	227(4.35),274(4.20), 333(3.63)	44-2720-76
$C_{18}H_{25}NO$			
10-Aza-9-oxabicyclo[6.3.0]undec-10-ene, 11-(2,4,6-trimethylphenyl)-, cis	EtOH	260(3.29)	39-1030-76B
trans	EtOH	260(3.50)	39-1030-76B
6H-Naphtho[2,3-c]quinolizin-6-one, 1,2,3,4,7,7a,8,9,10,11,11a,12-dodecahydro-11a-methyl-, cis	EtOH	266(4.05)	23-1512-76
$C_{18}H_{25}NO_3$			
11,4a-(Epoxymethano)-4aH-indeno[4,5-h]-isoquinolin-2(4bH)-one, 3,4,5,6,6a,7-8,9,9a,9b,10,11-dodecahydro-7-hydroxy-6a-methyl-, [4aS-(4aα,4bβ,6aα,7α,9aβ-9bα,11α)]-	MeOH	220(4.05)	44-2864-76
$C_{18}H_{25}NO_5$			
Retroisosenine	EtOH	220(3.44)	73-2952-76
$C_{18}H_{26}Br_2O_4$			
Benzoic acid, 3,5-dibromo-2,4-dihydroxy-6-nonyl-, ethyl ester	MeOH	227(4.45),269(3.87), 320(3.65)	12-1989-76
$C_{18}H_{26}Co$			
Cobaltocenium, 1,1'-bis(1-methylpropyl)-	MeOH	273(4.52),310(3.26), 407(2.52)	101-0189-76I
$C_{18}H_{26}NO_7$			
3,6,9,12,15-Pentaoxa-19-azoniabicyclo-[15.3.1]heneicosa-1(21),17,19-triene, 18,19,20-trimethyl-2,16-dioxo-, perchlorate	MeCN	228(3.94),280(3.81), 288(3.80),310(2.95)	77-0964-76

Compound	Solvent	$\lambda_{max}(\log \epsilon)$	Ref.
$C_{18}H_{26}N_2O_2$ Carbamic acid, N-phenyl-, 3-(1-piperi-dinyl)cyclohexyl ester	H_2O	233(4.19),265(2.83)	106-0024-76
$C_{18}H_{26}N_2O_2SSi_2$ 1H-Pyrazole, 1-[(4-methylphenyl)sulfon-yl]-5-(trimethylsilyl)-3-[(trimethyl-silyl)ethynyl]-	EtOH	229(4.28),259(4.24)	78-1293-76
$C_{18}H_{26}N_2O_3$ 1H-Indole-3-acetamide, N-(5,6-dihydroxy-6-methylheptyl)-	EtOH	274(3.71),281(3.73), 289(3.66)	130-0283-76
$C_{18}H_{26}O$ 1-Propanone, 3-[2-methyl-5-(1-methyl-ethyl)cyclopentyl]-1-phenyl-	n.s.g.	241(3.09)	39-1762-76C
$C_{18}H_{26}O_3$ Estr-4-en-3-one, 10α,17β-dihydroxy- 1H-Naphtho[2,1-b]pyran-1-one, 2,3,4,4a,8-9,10,10a,10b-octahydro-10-hydroxy-2,2,4a,8,10b-pentamethyl-, [4aR-(4aα,8α,10β,10aα,10bα)]-	EtOH MeOH	235(4.11) 240(4.42)	22-2021-76 44-0066-76
$C_{18}H_{26}O_4$ 1,3-Cyclohexanedione, 5,5'-(1,6-hexane-diyl)bis-	MeOH	257(4.44)	12-1989-76
$C_{18}H_{26}O_8$ 1,3-Heptadiene-1,2,6,6-tetracarboxylic acid, 5-methyl-, 6,6-diethyl 1,2-di-methyl ester	EtOH	267(4.08)	35-1204-76
$C_{18}H_{27}NO$ Dispiro[4.1.4.1]dodecan-6-one, 12-(cy-clohexylimino)-	MeOH	317(1.49)	77-0787-76
6H-Naphtho[2,3-c]quinolizin-6-one, 1,2,3,4,4a,5,7,7a,8,9,10,11,11a,12-tetradecahydro-11a-methyl- isomer	EtOH EtOH	330(4.15) 334(4.15)	23-1512-76 23-1512-76
$C_{18}H_{27}NO_5$ Bulgarsenine	EtOH	222(3.93)	73-2952-76
$C_{18}H_{27}N_3O_5$ 2,2'-Anhydro-1-(3'-nonanoyl-β-D-arabino-furanosyl)cytosine, hydrochloride	MeOH	232(4.03),263(4.06)	87-0654-76
$C_{18}H_{27}N_3S_2$ 1,2,3-Thiadiazol-5-amine, N-[4-(decyl-thio)phenyl]-	dioxan	268(3.92),331(4.30)	73-1182-76
$C_{18}H_{28}$ Bicyclo[10.2.2]hexadeca-12,14,15-triene, 13,15-dimethyl-	isooctane	263(2.48),273(2.75), 282(2.78)	44-4081-76
$C_{18}H_{28}BrN_5O$ Benzamide, 3-amino-4-[3-(4-bromobutyl)-2-imino-1-imidazolidinyl]-N-butyl-, dihydrochloride	EtOH	211(4.42),225s(4.36), 320(3.57)	78-0057-76

Compound	Solvent	$\lambda_{max}(\log \epsilon)$	Ref.
$C_{18}H_{28}N_2O_2$ 2,3-Heptanediol, 7-[[2-(1H-indol-3-yl)-ethyl]amino]-2-methyl-, (±)-	EtOH	274(3.68),281(3.72), 289(3.65)	130-0283-76
$C_{18}H_{28}N_2S_4$ 2-Propanamine, N,N'-[dithiobis(carbono-thioyl-2-cyclopentyl-1-ylidene)]bis-	benzene	313(3.88),414(4.74)	39-1706-76C
$C_{18}H_{28}N_4O_8$ 1,1'-[(1,4-Benzoquinone-2,5-diyl)dimeth-yl]bispiperidinium dinitrate	H_2O	250(4.25),306(2.91)	12-1163-76
$C_{18}H_{28}N_5O_6PS$ Adenosine, 8-(octylthio)-, cyclic 3',5'-(hydrogen phosphate)	pH 2.0 pH 13.0	285(4.26) 282(--)	69-1408-76 69-1408-76
$C_{18}H_{28}N_6O_4S$ Alanine, 4-[[(4-amino-2-methyl-5-pyrimi-dinyl)methyl]formylamino]-3-[(2-amino-1-oxopropyl)thio]-3-pentenyl ester, trihydrochloride, (S)-	EtOH	246(4.08)	94-0852-76
$C_{18}H_{28}O$ 2-Butanone, 4-(3,4,4a,5,6,7,8,8a-octa-hydro-5,5,8a-trimethyl-2-methylene-naphthylidene)-	isooctane	213(3.85)	31-0966-76
$C_{18}H_{28}O_4$ Benzoic acid, 2,4-dihydroxy-6-nonyl-, ethyl ester	MeOH	265(4.14),301(3.73)	12-1989-76
$C_{18}H_{29}N$ 2-Buten-1-amine, 4-(2,3,4,4,5,6-hexa-methyl-2,5-cyclohexadien-1-ylidene)-N,N-dimethyl-	C_6H_{12}	293(4.55),304s(--)	5-1103-76
$C_{18}H_{29}N_3O_2$ Carbamic acid, [2,3-bis(hexahydro-1H-azepin-1-yl)-2-cyclopropen-1-yli-dene]-, ethyl ester	MeOH	235(3.59),315(3.37)	138-1215-76
$C_{18}H_{29}N_3O_6$ Cytosine, 1-(3'-nonanoyl-β-D-arabino-furanosyl-1,2-dihydro-2-oxo-4-pyri-midinyl)-	MeOH-acid	213(4.01),283(4.17)	87-0667-76
$C_{18}H_{29}N_5O$ Benzamide, 3-amino-N-butyl-4-(3-butyl-2-imino-1-imidazolidinyl)-, dihydro-bromide	EtOH	210s(4.30),319(3.51)	78-0057-76
$C_{18}H_{30}N_2$ Piperidine, 1-[[3-methylene-2-(1-piperi-dinyl)-1-cyclohexen-1-yl]methyl]-	n.s.g.	237(4.24),290(3.26)	94-2494-76
$C_{18}H_{30}O$ 3,5-Hexadien-2-one, 6-cyclododecyl-	EtOH	285.0(4.44)	22-0849-76
$C_{18}H_{30}O_2S$ 2-Thiophenecarboxylic acid, 5-tridecyl-	heptane	254(4.24),274(4.03)	34-0233-76

Compound	Solvent	$\lambda_{max}(\log \epsilon)$	Ref.
$C_{18}H_{30}O_3$ 2,4-Nonadienoic acid, 3,7-dimethyl- 9-(tetrahydro-2-furanyl)-, 1-methyl- ethyl ester	EtOH	264(4.27)	73-1225-76
$C_{18}H_{30}O_4$ Cyclohexanecarboxylic acid, 2-nonyl- 4,6-dioxo-, ethyl ester	MeOH EtOH-base	259(4.09),280(4.05) 282(4.43)	12-1989-76 12-1989-76
$C_{18}H_{30}O_6$ 1-Cyclopentene-1-heptanoic acid, 2,4,4-triethoxy-5-oxo-	MeOH	265(4.17)	20-0503-76
$C_{18}H_{31}NOS$ 1-Dodecanesulfinamide, N-phenyl-	EtOH	205(4.15),235(3.96)	118-0339-76
$C_{18}H_{31}NO_5$ 4H-Azepin-4-one, 5-acetoxy-1-(2-acetoxy- 1,1-dimethylethyl)hexahydro-3,3,6,6- tetramethyl-	EtOH	244s(2.43),290(1.74)	44-1768-76
$C_{18}H_{31}N_6O_7P$ 5'-Adenylic acid, N-(8-aminooctyl)-	H_2O	267(4.24)	87-1279-76
$C_{18}H_{32}O$ Indan-4-one, 1-(1,5-dimethylhexyl)hexa- hydro-7a-methyl-	EtOH	283(1.46)	73-2788-76
$C_{18}H_{32}O_2$ 2-Cyclopenten-1-one, 2,4,5-tris(1,1-di- methylethyl)-3-methoxy- Oxacyclotridec-10-en-2-one, 13-hexyl-, [R-(E)]-	EtOH pentane	253(4.11) 225s(1.70)	118-0604-76 33-0755-76
$C_{18}H_{32}S$ Thiophene, 3-methyl-2-tridecyl-	heptane	236(3.81)	34-0233-76
$C_{18}H_{32}S_2$ 1,3-Dithietane, 2,4-bis[2,2-dimethyl- 1-(1-methylethyl)propylidene]-	C_6H_{12}	260s(--),265(4.5), 367(2.1)	24-0906-76
$C_{18}H_{32}S_3$ 1,2,4-Trithiolane, 3,5-bis[2,2-dimethyl- 1-(1-methylethyl)propylidene]-	C_6H_{12}	268(4.1),320s(--)	24-0906-76
$C_{18}H_{36}N_2$ 3-Pentanone, 2,2,4,4-tetramethyl-, [1-(1,1-dimethylethyl)-2,2-di- methylpropylidene]hydrazone	C_6H_{12}	242(3.77),260(2.78)	39-2079-76C
$C_{18}H_{36}OSe$ Octadecaneselenoic acid, potassium salt	EtOH	274(3.70)	138-0203-76
$C_{18}H_{36}Si_3$ Silane, [1,3,5-benzenetriyltris(methyl- ene)tris[trimethyl-	C_6H_{12}	233s(4.14),241(4.18), 255(4.20)	35-8426-76

Compound	Solvent	$\lambda_{max}(\log \epsilon)$	Ref.
$C_{19}H_8Cl_8N_2$ Benzenecarboximidamide, N,N'-bis-[(2,3,4,5-tetrachloro-2,4-cyclopentadien-1-ylidene)methyl]-	dioxan	319(4.30),436(4.43)	24-3939-76
$C_{19}H_{10}Br_2N_2O_3$ 2H-Anthra[1,9,8-cdef][2,7]naphthyridine-1,6,11(10H)-trione, 5,7-dibromo-2,10-dimethyl-	H_2SO_4	244(4.25),290(4.10), 400(3.81)	103-1352-76
$C_{19}H_{10}Br_2O_8S$ Dibromopyrogallol Red	15M H_2SO_4 5M H_2SO_4 pH 2.8 pH 6.8 pH 10.8 M NaOH	476(4.58) 477(4.52) 436(4.19),528(4.16) 561(4.76) 557(4.36) 598(4.44)	140-1691-76 140-1691-76 140-1691-76 140-1691-76 140-1691-76 140-1691-76
$C_{19}H_{10}ClF_3N_2O_3$ 4H-1-Benzopyran-4-one, 6-chloro-7-hydroxy-3-(1-phenyl-1H-pyrazol-4-yl)-2-(trifluoromethyl)-	EtOH	270(4.39),330(3.76), 363(3.85)	103-0914-76
$C_{19}H_{10}Cl_2N_2O_5$ 1,2-Benzenedicarbonitrile, 3-[(4-acetoxy-1H-2-benzopyran-1-yl)oxy]-4,5-dichloro-6-hydroxy-	MeCN	220(4.5),255(4.1), 347(3.8)	1-0619-76
$C_{19}H_{10}N_2O_3$ 2H-Pyrano[3,2-c]quinoline-3-carbonitrile, 5,6-dihydro-2,5-dioxo-6-phenyl-	n.s.g.	367(4.36),385(4.32)	64-0514-76B
$C_{19}H_{10}N_4O_7$ 2H-Anthra[1,9,8-cdef][2,7]naphthyridine-1,6,11(10H)-trione, 2,10-dimethyl-5,7-dinitro-	H_2SO_4	260(4.30),390(4.10), 430(4.00)	103-1352-76
$C_{19}H_{11}BrO_2$ Benz[a]anthracene-7,12-dione, 5-bromo-2-methyl-	$CHCl_3$	251(4.42),290(4.49), 333(3.53),364s(3.40), 414(3.38)	18-3713-76
$C_{19}H_{11}ClN_2O_2$ 5H-Oxazolo[4,5-b]phenoxazine, 2-(2-chlorophenyl)-	benzene	385(4.09)	23-0037-76
5H-Oxazolo[4,5-b]phenoxazine, 2-(3-chlorophenyl)-	benzene	382(4.37)	23-0037-76
5H-Oxazolo[4,5-b]phenoxazine, 2-(4-chlorophenyl)-	benzene	385(4.13)	23-0037-76
$C_{19}H_{11}Cl_2N_3O_3$ 2,4-Cyclohexadiene-1,2-dicarbonitrile, 3-acetoxy-4,5-dichloro-1-(1-methyl-1H-indol-3-yl)-6-oxo-	EtOH	217(4.58),264(4.06), 282(4.07),462(2.99)	1-0853-76
$C_{19}H_{11}N$ Acephenanthryleno[5,4-c]pyridine	C_6H_{12}	220(4.49),224(4.58), 230(4.54),231(4.54), 244(4.40),242(4.41), 255(4.47),264(4.29), 267(4.31),272(4.30),	73-1208-76

Compound	Solvent	$\lambda_{max}(\log \epsilon)$	Ref.
Acephenanthryleno[5,4-c]pyridine (cont.)		278(4.38),282(4.31), 285(4.34),289(4.35), 293(4.19),297(4.30), 310(3.72),330(3.93), 340(4.03),348(4.03), 354(4.09),365(3.89), 372(4.09),377(4.49)	73-1208-76
Azuleno[1,2-f]azulene-5-carbonitrile	CH_2Cl_2	259(4.37),296s(4.14), 336(4.49),350s(4.51), 362(4.56),376s(4.45), 407(4.63),434(4.79), 666s(2.93)	118-0673-76
Benz[h]indeno[1,2,3-de]quinoline	C_6H_{12}	215(4.42),224(4.46), 234(4.31),252(4.57), 258(4.44),268(4.26), 275(4.38),285(4.53), 293(4.29),298(4.36), 315(3.65),325(3.74), 340(3.31),342(3.84), 359(4.05),365(3.83), 373(3.91),378(4.16), 384(3.70)	73-1208-76
$C_{19}H_{11}N_3O_4$ 5H-Oxazolo[4,5-b]phenoxazine, 2-(2-ni- trophenyl)-	benzene	415(4.39)	23-0037-76
5H-Oxazolo[4,5-b]phenoxazine, 2-(3-ni- trophenyl)-	benzene	298(4.39)	23-0037-76
5H-Oxazolo[4,5-b]phenoxazine, 2-(4-ni- trophenyl)-	benzene	288(4.55),448(4.22)	23-0037-76
$C_{19}H_{11}OS$ [1]Benzothieno[3,2-b]naphtho[1,2-e]pyr- ylium, perchlorate	$C_2H_4Cl_2$	255(4.42),305(4.47), 345(4.07),480(4.53)	103-0977-76
$C_{19}H_{12}$ Fluoradene	n.s.g.	220(4.63),275(4.49)	44-2120-76
$C_{19}H_{12}Br_2O$ 2,5-Cyclohexadien-1-one, 2,6-dibromo- 4-(diphenylmethylene)-	isooctane	389(4.42)	73-1676-76
$C_{19}H_{12}Br_4O_3$ Cyclopentanone, 2,5-bis[(3,5-dibromo- 2-hydroxyphenyl)methylene]-	EtOH	220(4.76),258(4.37), 310(3.83)	114-0381-76B
	ether	215(--),258(--), 310(--)	114-0381-76B
	$CHCl_3$	310(3.74)	114-0381-76B
	CCl_4	310(3.75)	114-0381-76B
ion	n.s.g.	405(4.68)	114-0381-76B
non-ion	n.s.g.	310(4.63)	114-0381-76B
$C_{19}H_{12}ClNOS$ Benzoxazole, 4-chloro-2-[5-(2-phenyl- ethenyl)-2-thienyl]-	CH_2Cl_2	250s(4.03),263(3.97), 271(3.97),371s(4.57), 386(4.62),404s(4.48)	48-0731-76
$C_{19}H_{12}ClNO_2$ 1H-Indole, 1-[5-(2-chlorophenyl)- 1,3-dioxo-4-pentynyl]-	EtOH	218(4.19),248(4.09), 281(4.01),340(4.28)	12-1583-76

Compound	Solvent	$\lambda_{max}(\log \epsilon)$	Ref.
1H-Indole, 1-[5-(4-chlorophenyl)-1,3-dioxo-4-pentynyl]-	EtOH	217(4.12),248s(4.13), 254(4.15),344(4.3)	12-1583-76
$C_{19}H_{12}ClNO_3S_2$			
4,7-Epithiopyrano[3,4-c]pyrrole-1,3,6-(2H)-trione, 4-(4-chlorophenyl)tetrahydro-2-phenyl-, $(3a\alpha,4\alpha,7\alpha,7a\alpha)$-	MeOH	200(4.61),223(4.34)	44-1724-76
$C_{19}H_{12}N_2O_2$			
Benzo[g]quinazoline-5,10-dione, 4-methyl-2-phenyl-	EtOH	252(4.27),293(4.43), 336(3.78)	5-1809-76
5H-Oxazolo[4,5-b]phenoxazine, 2-phenyl-	benzene	372(4.15)	23-0037-76
$C_{19}H_{12}N_2O_3$			
Phenol, 2-(5H-oxazolo[4,5-b]phenoxazin-2-yl)-	benzene	385(4.20)	23-0037-76
Phenol, 4-(5H-oxazolo[4,5-b]phenoxazin-2-yl)-	benzene	288(4.46),378(4.23)	23-0037-76
$C_{19}H_{12}N_4O_2$			
1H-Benz[de]isoquinoline-1,3(2H)-dione, 2-(5-methyl-1H-benzotriazol-1-yl)-	EtOH	210(4.66),230(4.64), 340(4.15)	56-1257-76
5H-Oxazolo[4,5-b]phenoxazine, 2-(phenylazo)-	benzene	455(3.98)	23-0037-76
$C_{19}H_{12}N_4O_3$			
1H-Benz[de]isoquinoline-1,3(2H)-dione, 2-(1H-benzotriazol-1-yl)-5-methoxy-	EtOH	210(4.45),236(4.59), 334(4.11),378(3.98)	56-1257-76
1H-Benz[de]isoquinoline-1,3(2H)-dione, 2-(1H-benzotriazol-1-yl)-6-methoxy-	EtOH	212(4.45),240(4.42), 280(3.85),370(4.12)	56-1257-76
1H-Benz[de]isoquinoline-1,3(2H)-dione, 2-(2H-benzotriazol-2-yl)-5-methoxy-	EtOH	218(4.51),234(4.65), 332(4.06),356(3.88)	56-1257-76
1H-Benz[de]isoquinoline-1,3(2H)-dione, 2-(2H-benzotriazol-2-yl)-6-methoxy-	EtOH	212(4.55),235(4.36), 280(4.26),370(4.19)	56-1257-76
1H-Benz[de]isoquinoline-1,3(2H)-dione, 2-(5-methoxy-1H-benzotriazol-1-yl)-	EtOH	212(4.64),230(4.57), 340(4.16)	56-1257-76
$C_{19}H_{12}O_2$			
13H-Benzofuro[3,2-b]naphtho[1,2-e]pyran	EtOH	227(4.59),264(4.04)	22-1967-76
$C_{19}H_{12}O_3$			
Naphtho[2,3-b]furan-4,9-dione, 2-(4-methylphenyl)-	$CHCl_3$	273(4.40),295(4.51), 437(3.72)	18-3713-76
$C_{19}H_{12}O_4$			
Benz[a]anthracene-7,12-dione, 1,8-dihydroxy-3-methyl-	MeOH	227(4.58),249s(4.24), 318(4.26),435(3.72)	39-0997-76C
	MeOH-base	310(4.28),527(3.73)	39-0997-76C
Benz[a]anthracene-7,12-dione, 3,8-dihydroxy-1-methyl-	MeOH	226(4.40),262s(4.17), 312(4.34),413(3.75)	39-0997-76C
	MeOH-base	237(4.36),332(4.26), 504(3.69)	39-0997-76C
$C_{19}H_{12}O_6$			
Derriobtusone B	EtOH	224(4.43),240(4.39), 267(4.19),315(4.27) 340(4.41)	102-1553-76
$C_{19}H_{12}O_8S$			
Pyrogallol Red	pH 0.3	466(4.42)	140-1691-76
	pH 4.6	432(4.09),511(4.15)	140-1691-76

Compound	Solvent	λ_{max} (log ϵ)	Ref.
Pyrogallol Red (cont.)	pH 8.5	547(4.64)	140-1691-76
	pH 11.0	542(4.29)	140-1691-76
	M NaOH	584(4.35)	140-1691-76
$C_{19}H_{13}ClN_2O_2$			
2,6(1H,3H)-Pyridinedione, 4-(2-chloro- phenyl)-3-(1H-indol-1-yl)-	EtOH	222(4.30),273(4.14)	12-1583-76
1H-Pyrrolo[3,2-b]quinoline-3-carboxylic acid, 7-chloro-2-methyl-9-phenyl-	EtOH	204(4.50),236s(4.60), 251(4.75),332(4.11), 358(3.95),372s(3.91)	44-1743-76
$C_{19}H_{13}ClO$			
2-Propen-1-one, 3-(4-chlorophenyl)- 1-(1-naphthalenyl)-, cis	EtOH	310(4.33)	32-0617-76
trans	EtOH	317.5(4.38)	32-0617-76
2-Propen-1-one, 3-(4-chlorophenyl)- 1-(2-naphthalenyl)-, cis	EtOH	313.4(4.24)	32-0617-76
trans	EtOH	316.5(4.49)	32-0617-76
$C_{19}H_{13}F$			
Benz[a]anthracene, 6-fluoro-12-methyl-	EtOH	221(4.4),261(4.5), 271s(--),280(4.8), 288(4.8),301s(--), 325s(--),338(3.8), 352(3.9),365(3.7), 390(2.9)	24-1038-76
$C_{19}H_{13}Li$			
Lithium, (9-phenyl-9H-fluoren-9-yl)-	(MeOCH$_2$)$_2$	373(3.97),409(4.07), 455s(--),489(3.09), 525(2.97)	104-2411-76
$C_{19}H_{13}N$			
Acridine, 9-phenyl-	MeOH	255(5.5),355(4.0)	46-2614-76
	MeOH-HCl	260(5.1),340(4.0), 355(4.2)	46-2614-76
Benzenamine, N-9H-fluoren-9-ylidene-	C$_6$H$_{12}$	394(4.20)	39-0048-76C
$C_{19}H_{13}NO$			
Phenol, 4-(9H-fluoren-9-ylideneamino)-	C$_6$H$_{12}$	416(4.24)	39-0048-76C
$C_{19}H_{13}NOS$			
Benzoxazole, 2-[5-(2-phenylethenyl)- 2-thienyl]-	CH$_2$Cl$_2$	248s(4.24),260(4.02), 267(4.03),365s(4.60), 381(4.65),398s(4.49)	48-0731-76
$C_{19}H_{13}NO_2$			
1H-Benz[f]isoindole-1,3(2H)-dione, 2-methyl-4-phenyl-	ether	257(4.73),294(3.96), 341(3.49),357(3.64)	22-0493-76
1H-Indene-1,3(2H)-dione, 2-(1-methyl- 2(1H)-quinolinylidene)-	EtOH	223(4.56),238(4.49), 291(4.21),441(4.53)	73-1363-76
1H-Indole, 1-(1,3-dioxo-5-phenyl- 4-pentynyl)-	EtOH	218(4.12),247(4.09), 253(4.1),278s(3.99), 341(4.32)	12-1583-76
$C_{19}H_{13}NO_3$			
Benzoic acid, 2-(1-oxo-2-cyclopenteno- [b]indolyl)-, methyl ester	EtOH	282(4.17),325(3.93)	4-0083-76
2-Propen-1-one, 1-(1-naphthalenyl)- 3-(3-nitrophenyl)-, cis	EtOH	253(4.31)	32-0617-76
trans	EtOH	274(4.49)	32-0617-76

Compound	Solvent	$\lambda_{max}(\log \epsilon)$	Ref.
2-Propen-1-one, 1-(1-naphthalenyl)- 3-(4-nitrophenyl)-, cis	EtOH	285(4.20)	32-0617-76
trans	EtOH	300(4.30)	32-0617-76
2-Propen-1-one, 1-(2-naphthalenyl)- 3-(3-nitrophenyl)-, cis	EtOH	253(4.31)	32-0617-76
trans	EtOH	274(4.49)	32-0617-76
2-Propen-1-one, 1-(2-naphthalenyl)- 3-(4-nitrophenyl)-, cis	EtOH	285(4.20)	32-0617-76
trans	EtOH	300(4.30)	32-0617-76
$C_{19}H_{13}NO_3S_2$ 4,7-Epithiothiopyrano[3,4-c]pyrrole- 1,3,6(2H)-trione, tetrahydro-3,4- diphenyl-, (3aα,4α,7α,7aα)-	MeOH	200(4.91),215s(4.27)	44-1724-76
$C_{19}H_{13}NO_4$ Decarine	EtOH	216(4.34),249(4.57), 258(4.60),277(4.72), 327(4.22),368(3.49), 388(3.50)	95-1458-76
	EtOH-NaOH	222(4.46),256(4.53), 299(4.71),331(4.32), 414(3.55)	95-1458-76
$C_{19}H_{13}NO_4S$ 2,5-Furandione, 3-(4,5-dihydro-5-oxo- 2,4-diphenyl-4-thiazolyl)dihydro-	n.s.g.	210(4.17),245(4.25), 270s(3.68)	78-0571-76
$C_{19}H_{13}NS_2$ Benzothiazole, 2-[5-(2-phenylethenyl)- 2-thienyl]-	CH_2Cl_2	269(4.04),275s(4.03), 304s(3.80),371s(4.50), 391(4.55),413s(4.39), 450s(3.22)	48-0731-76
$C_{19}H_{13}N_3$ Pyrido[2,3-b]pyrazine, 3,7-diphenyl-	EtOH	258(4.45),360(4.34)	22-0251-76
$C_{19}H_{13}N_3O$ Benzonitrile, 4-[[[(1-hydroxy-2-naphtha- lenyl)methylene]hydrazono]methyl]-	EtOH	221(4.49),286(4.55), 295(4.60),333(4.44), 399(4.29)	95-0044-76
Benzonitrile, 4-[[[(2-hydroxy-1-naphtha- lenyl)methylene]hydrazono]methyl]-	EtOH	231(4.69),273s(4.37), 278(4.38),292s(4.19), 330s(4.21),336(4.23), 393(4.39)	95-0044-76
2-Naphthalenol, 1-(6-quinolinylazo)-	C_6H_{12}	238(4.09),340(3.82), 420(3.56)	115-0379-76
	EtOH	333(4.19),475(4.22)	115-0379-76
	acetone	474(3.90)	115-0379-76
	$CHCl_3$	340(4.27),480(3.92)	115-0379-76
	CCl_4	448(4.32)	115-0379-76
$C_{19}H_{13}N_3O_2$ Benzenamine, 2-(5H-oxazolo[4,5-b]phen- oxazin-2-yl)-	benzene	387(4.27)	23-0037-76
Benzenamine, 3-(5H-oxazolo[4,5-b]phen- oxazin-2-yl)-	benzene	382(4.13)	23-0037-76
Benzenamine, 4-(5H-oxazolo[4,5-b]phen- oxazin-2-yl)-	benzene	302(4.38),376(4.23)	23-0037-76
Cyclohept[b]indol-6-amine, N-(4-nitro- phenyl)-	MeOH	277(4.31),323(4.20), 421(4.31)	18-1101-76

Compound	Solvent	$\lambda_{max}(\log \epsilon)$	Ref.
$C_{19}H_{13}N_5$			
1-Butene-1,1,2-tricarbonitrile, 3-(di-phenylhydrazono)-	EtOH	220(3.94),270(4.16), 485(4.35)	24-1787-76
1H-Pyrazolo[3,4-d]pyrimidine-4-aceto-nitrile, α,1-diphenyl-	EtOH	268(4.20),326(4.33)	95-1352-76
$C_{19}H_{14}ClN_3O_2$			
1H-Pyrazole, 1-benzoyl-4-(5-chloro-2,1-benzisoxazol-3-yl)-3,5-dimethyl-	EtOH	248(4.07),342(4.07)	94-1106-76
$C_{19}H_{14}Cl_2N_2O$			
Benzenecarboximidamide, 4-chloro-N'-(4-chlorophenyl)-N-hydroxy-N-phenyl-	EtOH	266(4.07),322(3.96)	12-0357-76
$C_{19}H_{14}Cl_2N_2O_5$			
Benzenepropanoic acid, 5-(2,3-dichloro-5,6-dicyano-4-hydroxyphenoxy)-2-hy-droxy-, ethyl ester	n.s.g.	254(4.48),390(3.94)	39-0407-76C
$C_{19}H_{14}Cl_2O$			
Cyclopentanone, 2,5-bis[(4-chlorophen-yl)methylene]-	EtOH	238(4.27),358(4.64)	114-0381-76B
	ether	237(4.18),345(4.53)	114-0381-76B
	$CHCl_3$	355(4.85)	114-0381-76B
	CCl_4	348(4.53)	114-0381-76B
$C_{19}H_{14}FNO_5S$			
Furan, 2-[2-(4-fluorophenyl)-1-(phenyl-sulfonyl)-1-propenyl]-5-nitro-	EtOH	211(4.60),223(4.58), 317(4.26)	73-3371-76
$C_{19}H_{14}FN_3O_5S$			
3H-Pyrazole, 4-(4-fluorophenyl)-4,5-di-hydro-3-(5-nitro-2-furanyl)-3-(phen-ylsulfonyl)-	EtOH	210(4.62),220(4.60), 312(4.25)	73-3371-76
$C_{19}H_{14}FN_4S_2$			
Benzothiazolium, 2-[2-cyano-1-fluoro-2-[(3-methyl-2(3H)-benzothiazolylidene)-amino]ethenyl]-3-methyl-, tetrafluoro-borate	$MeNO_2$	514(4.57)	124-0204-76
$C_{19}H_{14}F_6$			
Phenanthrene, 9,9-dihydro-9,9-dimethyl-10-methylene-3,6-bis(trifluoromethyl)-	hexane	230s(--),240s(--), 246(4.39),281(4.04)	104-1298-76
	H_2SO_4	265(4.36),310(3.70), 336(3.81),527(3.52)	104-1298-76
$C_{19}H_{14}NO_4$			
Bis[1,3]benzodioxolo[5,6-a:5',6'-g]quin-olizinium, 5,6-dihydro-, chloride (coptisine chloride)	EtOH	229(4.37),241(4.37), 268(4.32),354s(4.35), 363(4.36),467(3.70)	73-3157-76
	EtOH	268(4.03),362(4.04), 465(3.38)	102-0545-76
Deoxythalidastine chloride	MeOH	230(3.78),265(3.78), 275(3.70),348(3.64), 430(3.08)	100-0065-76
	MeOH-NaOH	230(--),245(--), 288(--),378(--)	100-0065-76
Pseudocoptisine chloride	EtOH	220s(4.54),231(4.46), 266(4.39),289(4.50), 317(4.31),346(4.14), 380(3.76)	73-3157-76

Compound	Solvent	$\lambda_{max}(\log \epsilon)$	Ref.
$C_{19}H_{14}N_2$			
Cyclohept[b]indol-6-amine, N-phenyl-	MeOH	277(4.22),305(4.32), 372(4.18)	18-1101-76
$C_{19}H_{14}N_2O$			
Benzimidazolone, 1,3-diphenyl-	MeOH	246(4.0),286(3.63)	83-0316-76
11H-Dibenzo[b,e][1,4]diazepin-11-one, 5,10-dihydro-5-phenyl-	MeOH	204(4.38),220s(--), 236(3.96),258s(--), 280(3.60)	83-0316-76
3H-Indazol-3-one, 1,2-dihydro-1,2-di-phenyl-	MeOH	222(4.6),254(4.4), 320(4.07)	83-0316-76
Isoxazolo[5,4-b]pyridine, 6-(4-methyl-phenyl)-3-phenyl-	EtOH	247(3.96),320(4.12)	18-3339-76
6H-Perimidin-6-one, 4,5-dimethyl-2-phen-yl-	CHCl$_3$	275(4.49),322(3.99), 335(3.95),410(3.55)	12-2247-76
2-Propen-1-one, 1-phenyl-2,3-di-2-pyri-dinyl-, (E)-	MeOH	259(4.15),293(4.09)	44-2536-76
(Z)-	MeOH	254(4.34),310(4.33)	44-2536-76
$C_{19}H_{14}N_2O_2$			
Isoxazolo[5,4-b]pyridine, 6-(4-methoxy-phenyl)-3-phenyl-	EtOH	243(4.12),332(4.43)	18-3339-76
2,6(1H,3H)-Pyridinedione, 3-(1H-indol-1-yl)-4-phenyl-	EtOH	224(4.29),283(4.23)	12-1583-76
$C_{19}H_{14}N_2O_2S$			
1H-Thieno[3,4-d]imidazole-4-carboxylic acid, 2,6-diphenyl-, methyl ester	MeOH	249(4.33),328(4.25), 339(4.26),371(4.33)	49-0299-76
$C_{19}H_{14}N_2O_3$			
Benzoic acid, 2-[(4-cyano-1,2-dihydro-1-oxo-3-isoquinolinyl)methyl]-, methyl ester	EtOH	254(4.07),298(4.20)	95-0154-76
4H-1-Benzopyran-4-one, 7-hydroxy-5-meth-yl-3-(1-phenyl-1H-pyrazol-4-yl)-	EtOH	273(4.44)	103-0914-76
4H-1-Benzopyran-4-one, 7-methoxy-3-(1-phenyl-1H-pyrazol-4-yl)-	EtOH	270(4.12)	103-0914-76
1H-1,2-Diazepin-4-ol, 1-benzoyl-, benz-oate	MeOH	235(4.36),350s(2.76)	88-4859-76
1H-1,2-Diazepin-6-ol, 1-benzoyl-, benz-oate	MeOH	233(4.13),350s(2.58)	88-4859-76
$C_{19}H_{14}N_2O_7S$			
Furan, 2-nitro-5-[2-(4-nitrophenyl)-1-(phenylsulfonyl)-1-propenyl]-, (E)-	EtOH	208(4.47),218(4.46), 304(4.18),278(4.26)	73-3371-76
$C_{19}H_{14}N_4O$			
Ethanone, 1-phenyl-2-(1-phenyl-1H-pyra-zolo[3,4-d]pyrimidin-4-yl)-	EtOH	262(4.15),375(4.59)	95-1352-76
2H-Indazole, 2-acetyl-5-(dicyanomethyl-ene)-3,5-dihydro-3-methyl-3-phenyl-	isoPrOH	220s(4.05),253(3.72), 450(4.54)	4-0033-76
2(1H)-Pyrimidinone, 4-phenyl-6-(4-quin-azolinylmethyl)-	EtOH	428(4.99),445(5.05)	4-0383-76
$C_{19}H_{14}N_4OS$			
1H-1,2,4-Triazolium, 4-[(2-furanylmeth-ylene)amino]-3-mercapto-1,5-diphenyl-, hydroxide, inner salt	MeCN	258(4.68),296(4.57)	103-0361-76

Compound	Solvent	$\lambda_{max}(\log \epsilon)$	Ref.
$C_{19}H_{14}N_4O_3$			
2H-Anthra[1,9,8-cdef][2,7]naphthyridine-1,6,11(10H)-trione, 5,7-bis(methylamino)-	DMF	288(4.35),420(3.28), 450(3.55),525(4.29), 560(4.30)	103-1352-76
$C_{19}H_{14}N_4O_7$			
6-Azulenecarbonitrile, 1-(2-hydroxyethyl)-, trinitrobenzene complex	CH_2Cl_2	281s(4.85),290(5.03), 338(3.68),353(3.80), 657(2.53),725(2.44)	44-1811-76
$C_{19}H_{14}N_4O_7S$			
3H-Pyrazole, 4,5-dihydro-3-(5-nitro-2-furanyl)-4-(4-nitrophenyl)-3-(phenylsulfonyl)-	EtOH	207(4.64),218(4.62), 276(4.25),303(4.13)	73-3371-76
$C_{19}H_{14}O$			
2,5-Cyclohexadien-1-one, 4-(diphenylmethylene)-	isooctane	358(4.45)	73-1676-76
Indeno[2,1-b]pyran, 2-methyl-4-phenyl-	dioxan	242(4.27),264(4.14), 302(4.46),313(4.46), 338(3.86),533(3.45)	78-2225-76
Indeno[2,1-b]pyran, 4-methyl-2-phenyl-	dioxan	255(4.16),307(4.30), 360(4.24),510(3.51)	78-2219-76
Indeno[2,1-b]pyran, 2-(4-methylphenyl)-	dioxan	225(4.33),259(4.17), 292(4.18),310(4.13), 364(4.32),518(3.55)	78-2219-76
2-Propen-1-one, 1-(1-naphthalenyl)-3-phenyl-, cis	EtOH	312(4.47)	32-0617-76
trans	EtOH	317(4.51)	32-0617-76
2-Propen-1-one, 1-(2-naphthalenyl)-3-phenyl-, cis	EtOH	312(4.32)	32-0617-76
trans	EtOH	313.5(4.46)	32-0617-76
$C_{19}H_{14}O_2$			
5H-Oxeto[2',3':3,4]naphtho[1,2-b]furan, 2,3-dihydro-2-phenyl-	$CHCl_3$	242(3.84),266(3.69), 274(3.79),280(3.80), 291(3.68),307(2.94), 313(2.76),322(2.82)	18-2596-76
$C_{19}H_{14}O_3$			
Ethanone, 1-(5-benzoyl-2-phenyl-3-furanyl)-	EtOH	320(4.29)	94-2421-76
$C_{19}H_{14}O_3S$			
4H-1-Benzopyran-4-thione, 3-(2-benzofuranyl)-7-methoxy-2-methyl-	EtOH	285(4.40),375(4.46)	103-1214-76
$C_{19}H_{14}O_4$			
Benz[a]anthracene-7,12-dione, 5,6-dihydro-1,8-dihydroxy-3-methyl-	MeOH	269(4.25),390s(3.48), 458(3.86)	39-0997-76C
	MeOH-base	246s(4.37),314s(4.04), 561(3.70)	39-0997-76C
Benz[a]anthracene-7,12-dione, 5,6-dihydro-3,8-dihydroxy-1-methyl-	MeOH	273(4.35),313s(3.95), 478(3.82)	39-0997-76C
	MeOH-base	299(4.34),552(3.96)	39-0997-76C
12H-Benzo[b]xanthen-12-one, 6-hydroxy-3-methoxy-1-methyl-	EtOH	207(4.26),232(4.12), 236(4.10),264(4.59), 287(4.30),313(3.92), 332(3.71),405(3.45)	39-0499-76C

Compound	Solvent	$\lambda_{max}(\log \epsilon)$	Ref.
2H-Cyclopenta[jk]fluorene-1,2-dicarbox-ylic acid, dimethyl ester	EtOH	224(4.40),254(4.57), 264(4.81),279(3.90), 292(3.91),319(4.11), 328(4.14),340s(3.98)	24-2596-76
	DMSO-EtOH-NaOEt	478(4.05)	24-2596-76
$C_{19}H_{14}O_5$ 4a,9a-[2]Butenoanthracene-1,4,9,10-tet-rone, 12-methoxy-	EtOH	227(4.57),253(4.05), 298(3.38)	44-2296-76
$C_{19}H_{14}O_5S$ 11bH-Thiopyrano[4,3,2-kl]xanthene-2,3-dicarboxylic acid, dimethyl ester	dioxan	252(4.15),288(3.82), 353(3.87)	88-3563-76
$C_{19}H_{14}O_6$ 2-Propen-1-one, 3-(1,3-benzodioxol-5-yl)-1-(7-hydroxy-4-methoxy-5-benzo-furanyl)-	EtOH	237(4.53),362(4.53)	102-1553-76
$C_{19}H_{14}S$ Naphthalene, 2-(1H-inden-3-ylthio)-	MeOH	251(4.5),285(3.9), 299(3.8),320(3.2), 336(3.0)	35-3564-76
$C_{19}H_{15}$ Methyl, triphenyl- (cation)	EtOH	564(5.01)	44-0221-76
	8% DMSO	404(4.18),584(4.84)	44-0221-76
hexachloroantimonate benzene complex	CH_2Cl_2	501.8(2.40)	97-0192-76
hexachloroantimonate toluene complex	CH_2Cl_2	503.0(2.57)	97-0192-76
hexachloroantimonate mesitylene com-plex	CH_2Cl_2	507.1(2.53)	97-0192-76
hexachloroantimonate hexamethylbenz-ene complex (also other complexes)	CH_2Cl_2	614.3(2.27)	97-0192-76
$C_{19}H_{15}BrClNO$ Ethanone, 1-(4-bromophenyl)-2-[1-(4-chlorophenyl)-1,4-dihydro-4-pyrid-inyl]-	MeOH	286(4.20),435(4.92)	104-1105-76
$C_{19}H_{15}BrN_2O$ Benzaldehyde, 5-bromo-2-hydroxy-, diphenylhydrazone, (E)-	benzene	310(4.17),360(4.43)	104-0625-76
	DMF	307(3.99),360(4.36)	104-0625-76
Benzenecarboximidamide, N'-(4-bromo-phenyl)-N-hydroxy-N-phenyl-	EtOH	261(3.92),317(3.96)	12-0357-76
$C_{19}H_{15}BrN_2O_2S$ Thiazolium, 3-[1-(4-bromobenzoyl)-2-oxo-2-(phenylamino)ethyl]-4-methyl-, hydroxide, inner salt	MeOH	248(4.28),300(4.28)	44-0187-76
$C_{19}H_{15}Br_2NO$ Quinolinium, 2-(2-hydroxy-3,5-dibromo-styryl)-1,3-dimethyl-, hydroxide, inner salt	benzene	325(3.72),430s(2.47), 530(2.23),690(2.65)	103-0683-76
	EtOH	243(4.51),258s(4.37), 305s(3.75),355(4.08), 530(4.06)	103-0683-76
$C_{19}H_{15}ClN_2O$ Benzenecarboximidamide, 2-chloro-N-hydroxy-N,N'-diphenyl-	EtOH	266(4.10),309(4.05)	12-0357-76

Compound	Solvent	$\lambda_{max}(\log \epsilon)$	Ref.
Benzenecarboximidamide, 4-chloro-N-hydroxy-N,N'-diphenyl-	EtOH	256(4.00),321(3.83)	12-0357-76
Benzenecarboximidamide, N'-(4-chloro-phenyl)-N-hydroxy-N-phenyl-	EtOH	260(3.89),317(3.91)	12-0357-76
$C_{19}H_{15}ClN_2O_2$ Ethanone, 1-[3-benzoyl-1-(4-chlorophen-yl)-5-methyl-1H-pyrazol-4-yl]-	EtOH	250(4.393)	4-0989-76
$C_{19}H_{15}ClN_2O_3$ Ethanone, 2-[1-(4-chlorophenyl)-1,4-di-hydro-4-pyridinyl)-1-(4-nitrophenyl)-	MeOH	266(4.19),450(4.65)	104-1105-76
$C_{19}H_{15}ClN_3$ 1,2,4-Triazolo[4,3-a]pyridinium, 3-(4-chlorophenyl)-1-(4-methyl-phenyl)-, tetrafluoroborate	MeOH	232(4.33),248s(--),278s(--),306s(--)	48-0881-76
$C_{19}H_{15}ClN_3O$ 1,2,4-Triazolo[4,3-a]pyridinium, 1-(4-chlorophenyl)-3-(4-methoxy-phenyl)-, tetrafluoroborate	MeOH	243s(--),259(4.36),279s(--),309s(--)	48-0881-76
$C_{19}H_{15}ClO_3$ Benzeneacetic acid, 3-chloro-α-(1-oxo-3-phenyl-2-propynyl)-, ethyl ester	pH 13	221(3.67),250(3.77),334(3.69)	34-0118-76
	EtOH	218(4.08),242(3.89),319(4.16)	34-0118-76
	CHCl$_3$	251(3.87),323(4.21)	34-0118-76
Benzeneacetic acid, α-[3-(2-chlorophen-yl)-1-oxo-2-propynyl]-, ethyl ester	pH 13	223(3.87),255(3.91),338(3.82)	34-0118-76
	EtOH	217(4.05),230(4.05),282s(3.94),322(4.13)	34-0118-76
	CHCl$_3$	248(3.65),327(4.07)	34-0118-76
$C_{19}H_{15}ClO_4$ 1,4-Naphthalenedione, 2-chloro-5-hydr-oxy-3-[2-(3-hydroxy-5-methylphenyl)-ethyl]-	MeOH	281(4.01),433(3.40)	39-0997-76C
	MeOH-base	286(4.04),547(3.53)	39-0997-76C
$C_{19}H_{15}Cl_2NO_2$ Indeno[1,2-b]pyran-2(3H)-one, 3,3-di-chloro-4,5-dihydro-4-(methylphenyl-amino)-	EtOH	250(4.33),267s(4.22)	4-1201-76
$C_{19}H_{15}FO_3$ Benzeneacetic acid, 2-fluoro-α-(1-oxo-3-phenyl-2-propynyl)-, ethyl ester	pH 13	222(4.01),250(4.20),330(4.18)	34-0118-76
	EtOH	221(4.00),245s(3.80),272s(3.94),315(4.31)	34-0118-76
	CHCl$_3$	250(3.67),318(4.18)	34-0118-76
Benzeneacetic acid, 3-fluoro-α-(1-oxo-3-phenyl-2-propynyl)-, ethyl ester	pH 13	224(4.08),251(4.24),331(4.20)	34-0118-76
	EtOH	223(3.94),247(3.85),318(4.19)	34-0118-76
	CHCl$_3$	249(3.83),321(4.23)	34-0118-76
$C_{19}H_{15}N$ Benz[c]acridine, 7,10-dimethyl-	EtOH	281(3.73)	39-2277-76C

Compound	Solvent	$\lambda_{max}(\log \epsilon)$	Ref.
$C_{19}H_{15}NO$			
8-Azabicyclo[3.2.1]octa-3,6-dien-2-one, 6,8-diphenyl-	EtOH	208(4.42),246(4.43)	39-2289-76C
Benzeneacetonitrile, 3,4-dimethyl-α-(1-oxo-3-phenyl-2-propynyl)-	pH 13	235(4.16),350(4.12)	34-0115-76
	EtOH	216(4.16),250(4.07), 344(4.29)	34-0115-76
Benzeneacetonitrile, 4-ethyl-α-(1-oxo-3-phenyl-2-propynyl)-	pH 13	235(4.27),265s(4.11), 353(4.34)	34-0115-76
	EtOH	211(4.02),234(4.01), 348(4.18)	34-0115-76
Benzeneacetonitrile, α-(1-methoxy-3-phenyl-2-propynylidene)-4-methyl-	EtOH	210(4.14),228(4.32), 242(4.53)	34-0115-76
$C_{19}H_{15}NO_2$			
Acetamide, N-[3-(1-naphthalenyl)-7-oxo-1,3,5-cycloheptatrien-1-yl]-	MeOH	221(4.84),253(4.34), 324(4.17),368(3.97), 388s(3.81)	18-0831-76
Acetamide, N-[4-(1-naphthalenyl)-7-oxo-1,3,5-cycloheptatrien-1-yl]-	MeOH	219(4.80),253(4.31), 280s(4.00),344(4.20), 390s(4.01)	18-0831-76
Acetamide, N-[5-(1-naphthalenyl)-7-oxo-1,3,5-cycloheptatrien-1-yl]-	MeOH	221(4.83),261(4.33), 290s(3.98),326s(4.14), 370s(3.97)	18-0831-76
Benzoic acid, 2-([1,1'-biphenyl]-3-yl-amino)-, β-cyclodextrin complex	pH 7.0	293(4.15),335(3.86)	133-0343-76
1H-Cyclopenta[b]quinolin-3-ol, 2,3-dihydro-, benzoate	EtOH	232(4.79),280(3.63), 300(3.72),320(3.85)	95-0968-76
$C_{19}H_{15}NO_4$			
Benzeneacetic acid, 2-carboxy-α-(1H-indol-3-ylmethylene)-, α-methyl ester, (Z)-	EtOH	335(4.23)	4-0083-76
1H-2-Benzopyran-4-carboxylic acid, 3,4-dihydro-3-(1H-indol-3-yl)-1-oxo-, methyl ester, cis	EtOH	250(4.21),295(3.86)	4-0083-76
Berberrubine	EtOH	240(4.55),278(4.44), 328(4.06),396(4.14), 521(3.81)	73-3654-76
$C_{19}H_{15}NO_5$			
8H-Benzo[g]-1,3-benzodioxolo[5,6-a]quinolizin-8-one, 5,6-dihydro-9-hydroxy-10-methoxy-	EtOH	226(4.60),348(4.36), 370(4.33),388(4.20)	73-3654-76
$C_{19}H_{15}NO_5S$			
Furan, 2-nitro-5-[2-phenyl-1-(phenyl-sulfonyl)-1-propenyl]-, (E)-	EtOH	211(4.46),220(4.64), 315(4.20)	73-3371-76
$C_{19}H_{15}NO_9$			
1,3-Benzodioxole-5-carboxylic acid, 6-[5-acetoxy-3-(acetoxyimino)-6-oxo-1,4-cyclohexadien-1-yl]-, methyl ester	EtOH	226(4.49),274s(--), 297(4.28),420(3.46)	24-0855-76
$C_{19}H_{15}N_3$			
Benzenamine, 4-(phenylazo)-N-(phenyl-methylene)-	hexane	233(3.97),356(4.20), 440s(2.94)	61-0301-76
Propanedinitrile, [3-(4-methylphenyl)-3-(phenylamino)-2-propenylidene]-	CH_2Cl_2	408(4.32)	48-0705-76
Pyrido[2,3-b]pyrazine, 1,2-dihydro-2,3-diphenyl-	EtOH	244(4.30),268(4.31), 403(3.67)	22-0251-76

Compound	Solvent	$\lambda_{max}(\log \epsilon)$	Ref.
3-Pyrrolecarbonitrile, 4-methyl-5-phenyl-2-(phenylmethyleneamino)-	MeOH	272(4.23),393(4.20)	48-0663-76
$C_{19}H_{15}N_3O_2$ 5H-Pyridazino[4,5-b]indol-4-ol, 5-methyl-1-phenyl-, acetate	EtOH	234(4.65),253(4.48), 295(3.97),325(4.03)	103-1008-76
$C_{19}H_{15}N_3O_3$ Acetamide, N-[4-(1,3-benzodioxol-5-yl)-6-phenyl-2-pyrimidinyl]-	EtOH	232(4.51),333(4.32)	4-0257-76
Benzenecarboximidamide, N-hydroxy-4-nitro-N,N'-diphenyl-	EtOH	284(4.05),372(3.44)	12-0357-76
Benzenecarboximidamide, N-hydroxy-N'-(2-nitrophenyl)-N-phenyl-	EtOH	251(4.04),314(3.93), 390(3.52)	12-0357-76
Benzenecarboximidamide, N-hydroxy-N'-(3-nitrophenyl)-N-phenyl-	EtOH	258(4.10),312(4.02)	12-0357-76
Euxylophoricine F	MeOH	247(4.50),337(4.48), 347(4.51),364(4.43)	102-1095-76
	MeOH-HCl	372(--)	102-1095-76
	MeOH-NaOH	304(--)	102-1095-76
$C_{19}H_{15}N_3O_4$ Ethanone, 1-[3-benzoyl-5-methyl-1-(4-nitrophenyl)-1H-pyrazol-4-yl]-	EtOH	260(4.47)	4-0989-76
Ethanone, 1-[5-methyl-3-(4-nitrobenzoyl)-1-phenyl-1H-pyrazol-4-yl]-	EtOH	250(4.37)	4-0989-76
$C_{19}H_{15}N_3O_5$ 2,9-Methano-3H-indeno[2,1-b]pyridin-3-one, 1-(2,4-dinitrophenyl)-1,2,4,4a,9,9a-hexahydro-, (±)-	CHCl$_3$	360(4.19)	118-0105-76
$C_{19}H_{15}N_3O_5S$ 3H-Pyrazole, 4,5-dihydro-3-(5-nitro-2-furanyl)-4-phenyl-3-(phenylsulfonyl)-	EtOH	210(4.58),221(4.57), 313(4.23)	73-3371-76
$C_{19}H_{15}OS$ Naphtho[1,8-bc]thiolium, 5-ethoxy-2-phenyl-, tetrafluoroborate	CH$_2$Cl$_2$	456(4.11)	89-0775-76
$C_{19}H_{16}$ Cyclobuta[b]naphthalene, 1,2-dihydro-8-methyl-1-phenyl-	CHCl$_3$	243(3.90),276(3.70), 307(3.23),322(3.15)	18-2596-76
Pyrene, 4,5-dihydro-4,4-dimethyl-5-methylene-	hexane	227(4.77),236s(--), 270s(--),275(4.54), 299s(--)	104-1298-76
	H$_2$SO$_4$	293(4.14),306s(--), 318(4.08),353s(--), 501(3.86),634(3.44)	104-1298-76
	CF$_3$COOH	276s(--),284s(--), 300s(--),315s(--), 351s(--),365s(--), 456s(--),497(3.32), 626(2.84)	104-1298-76
$C_{19}H_{16}BrNO$ 8-Azabicyclo[3.2.1]oct-3-en-2-one, 6-(4-bromophenyl)-8-phenyl-, endo	EtOH	209(4.40),232(4.48)	39-2289-76C
Ethanone, 1-(4-bromophenyl)-2-(1,4-dihydro-1-phenyl-4-pyridinyl)-	MeOH	265(4.19),440(4.81)	104-1105-76

Compound	Solvent	λ_{max}(log ϵ)	Ref.
Ethanone, 2-[1-(4-bromophenyl)-1,4-di-hydro-4-pyridinyl]-1-phenyl-	MeOH	260(4.15),440(4.81)	104-1105-76
Spiro[2H-1-benzopyran-2,2'(1'H)-quino-line], 6-bromo-1',3'-dimethyl-	octane	230(4.66),252s(4.50),270s(4.12),320(3.67)	103-0683-76
	EtOH	243(4.54),288s(3.39),340(4.05),376s(4.05),540(3.68)	103-0683-76
$C_{19}H_{16}BrN_3$			
Pyrido[2,3-b]pyrazine, 7-bromo-1,2,3,4-tetrahydro-2,3-diphenyl-	EtOH	270(3.90),340(4.05)	22-0251-76
$C_{19}H_{16}ClNO$			
8-Azabicyclo[3.2.1]oct-3-en-2-one, 6-(4-chlorophenyl)-8-phenyl-, endo	EtOH	209(4.32),236(4.26)	39-2289-76C
Ethanone, 1-(4-chlorophenyl)-2-(1,4-di-hydro-1-phenyl-4-pyridinyl)-	MeOH	250(3.96),435(4.85)	104-1105-76
$C_{19}H_{16}ClNO_2$			
2-Propenoic acid, 3-(4-chlorophenyl)-3-(1H-indol-1-yl)-, ethyl ester	EtOH	273(4.02),338(3.94)	12-1583-76
$C_{19}H_{16}ClNO_3S_2$			
Sulfilimine, N-[(4-chlorophenyl)sulfon-yl]-S-(2-methoxyphenyl)-S-phenyl-	MeOH	231(4.35),285(3.59)	78-3003-76
	dioxan	232(4.34),285(3.61)	78-3003-76
$C_{19}H_{16}ClN_3O_3$			
1H-Pyrazole-4-carboxamide, 1-acetyl-3-benzoyl-N-(4-chlorophenyl)-4,5-dihydro-	EtOH	252(4.30),305(4.12)	115-0001-76
$C_{19}H_{16}CrN_2O_3$			
Chromium, tricarbonyl[1-(η^6-phenyl)eth-anone)(1-phenylethylidene)hydrazone]-	MeOH	217(4.52),260(4.41),295s(4.17),323(4.09),410(3.54)	23-0448-76
$C_{19}H_{16}F_3N_3O_4$			
Acetamide, N-[2,3-dihydro-1-(methoxy-methyl)-2-oxo-5-phenyl-1H-1,4-benz-odiazepin-7-yl]-2,2,2-trifluoro-N-hydroxy-	isoPrOH	219(4.43),255(4.36),275s(4.27),320s(3.35)	87-1378-76
$C_{19}H_{16}NOP$			
Phenophosphazine, 5,10-dihydro-5-methyl-10-phenyl-, 10-oxide	n.s.g.	275(4.15),306(3.81),336(3.80)	65-0435-76
$C_{19}H_{16}NO_4$			
Berberinium, 9-demethoxy-9-hydroxy-, iodide	EtOH	234(4.42),275(4.38),338(--),353(4.31),455(3.73)	73-3654-76
Berberrubine chloride	EtOH	234(4.42),275(4.37),356(4.31),455(3.73)	73-3157-76
$C_{19}H_{16}NP$			
Phenophosphazine, 5,10-dihydro-5-methyl-10-phenyl-	n.s.g.	284(4.11),315s(3.83)	65-0435-76
$C_{19}H_{16}N_2$			
Pyridazine, 3-(4-methylphenyl)-6-(2-phenylethenyl)-, (E)-	C_6H_{12}	224s(4.28),313(4.70)	59-0837-76
	EtOH	227(3.83),319(4.66)	

Compound	Solvent	$\lambda_{max}(\log \epsilon)$	Ref.
Pyridazine, 3-(4-methylphenyl)-6-(2-phenylethenyl)-, (E)- (cont.)	H_2SO_4	229s(4.21),280(4.00), 398(4.61)	59-0837-76
$C_{19}H_{16}N_2O$			
Benzaldehyde, 2-hydroxy-, diphenylhydrazone, (E)-	benzene	310(4.06),350(4.26)	104-0625-76
	EtOH	310(4.02),345(4.24)	104-0625-76
Benzenecarboximidamide, N-hydroxy-N,N'-diphenyl-	EtOH	255(3.91),316(3.90)	12-0357-76
1-Naphthalenecarboxaldehyde, 2-hydroxy-, [(4-methylphenyl)methylene]hydrazone	EtOH	231(4.670),274(4.28), 290(4.14),322(4.27), 332(4.33),382(4.40), 395(4.33)	95-0044-76
2-Naphthalenecarboxaldehyde, 1-hydroxy-, [(4-methylphenyl)methylene]hydrazone	EtOH	220(4.48),250s(4.56), 256(4.07),290(4.56), 298(4.59),330(4.38), 340s(4.27),383(4.26)	95-0044-76
1H-Pyrazole, 3-(2-furanyl)-1,5-diphenyl-	n.s.g.	359(4.34)	124-0383-76
Pyridazine, 3-(4-methoxyphenyl)-6-(2-phenylethenyl)-	C_6H_{12}	225(4.40),319(4.63)	59-0837-76
	EtOH	225(3.96),326(4.49)	59-0837-76
	H_2SO_4	227s(4.30),333(4.64), 362(4.66)	59-0837-76
$C_{19}H_{16}N_2OS$			
14H-Naphtho[1,2-d]thiazolo[3,2-b]quinazoline, 7,8-dihydro-10-methoxy-, hydrobromide	n.s.g.	303(4.20),340(4.11)	2-0759-76
$C_{19}H_{16}N_2OS_2$			
Thiazolium, 3-[1-benzoyl-2-(phenylamino)-2-thioxoethyl]-4-methyl-, hydroxide, inner salt	MeOH	210(4.31),326(4.43)	44-0187-76
$C_{19}H_{16}N_2O_2$			
1-Naphthalenecarboxaldehyde, 2-hydroxy-, [(4-methoxyphenyl)methylene]hydrazone	EtOH	232(4.63),276(4.10), 298(4.12),325s(4.32), 334(4.39),382(4.46), 397s(4.37)	95-0044-76
2-Naphthalenecarboxaldehyde, 1-hydroxy-, [(4-methoxyphenyl)methylene]hydrazone	EtOH	220(4.48),254(4.15), 260(4.16),294(4.44), 304(4.48),334(4.43), 344(4.36),382(4.35), 390(4.33)	95-0044-76
3-Pyridinecarboxylic acid, 1,2,3,4-tetrahydro-4-acridinyl ester	EtOH	231(4.65),262(3.80), 308(3.70),322(3.74)	95-0968-76
2,4-Pyrrolidinedione, 3-[1-(phenylamino)ethylidene]-5-(phenylmethylene)-	EtOH	281(4.29),334(4.47)	95-0927-76
2,4-Pyrrolidinedione, 3-[3-phenyl-1-(phenylamino)-2-propenylidene]-	EtOH	276(4.27),354(4.41)	95-0927-76
Spiro[2H-anthra[1,2-d]imidazole-2',1-cyclopentane]-6,11-dione, 1,3-dihydro-	EtOH	572(4.19)	104-0172-76
$C_{19}H_{16}N_2O_2S$			
Isoquinoline, 1,2,3,4-tetrahydro-2-(2-nitro-5-phenyl-3-thienyl)-	MeCN	262(4.37),309s(3.78), 357(4.22),474(3.97)	48-0221-76
Spiro[2H-1-benzothiopyran-2,2'-imidazolidine]-4',5'-dione, 1',3'-dimethyl-3-phenyl-	dioxan	243(4.60),348(3.94)	88-3563-76
Thiazolium, 3-[1-benzoyl-2-oxo-2-(phenylamino)ethyl]-4-methyl-, hydroxide, inner salt	MeOH	242(4.24),296(3.95)	44-0187-76

Compound	Solvent	$\lambda_{max}(\log \epsilon)$	Ref.
$C_{19}H_{16}N_2O_3$ Ethanone, 2-(1,4-dihydro-1-phenyl- 4-pyridinyl)-1-(4-nitrophenyl)-	MeOH	270(4.07),455(4.57)	104-1105-76
$C_{19}H_{16}N_2O_4S$ Thiazolo[2,3-a]isoquinolinium, 2-(1-cy- ano-2-ethoxy-2-oxoethyl)-3-(ethoxy- carbonyl)-, hydroxide, inner salt	EtOH	224(4.47),252(4.38), 311(4.40),345(4.14), 440(4.09)	94-1299-76
$C_{19}H_{16}N_2S$ 1H-Pyrazole, 4,5-dihydro-1,5-diphenyl- 3-(2-thienyl)-	n.s.g.	373(4.47)	124-0383-76
$C_{19}H_{16}N_3$ 1,2,4-Triazolo[4,3-a]pyridinium, 1-(4- methylphenyl)-3-phenyl-, tetrafluoro- borate	MeOH	227(4.16),245s(--), 275s(--),304(3.92)	48-0881-76
$C_{19}H_{16}N_4O_2$ 1H-Anthra[1,2-d]triazole-6,11-dione, 4-piperidino-	EtOH	520(3.97)	104-0865-76
Benzo[g]pteridine-2,4(1H,3H)-dione, 7,8-dimethyl-3-(phenylmethyl)-	6M HCl pH 7 H_2SO_4	390(4.24) 353(4.14),388(4.01) 440(4.44)	5-1276-76 5-1276-76 5-1276-76
1H-Imidazo[4,5-d]pyridazinium, 4,5,6,7- tetrahydro-4,7-dioxo-1,3-bis(phenyl- methyl)-, hydroxide, inner salt	pH 2 pH 7 pH 12	264(3.51),268(3.53), 277(3.53) 299(3.52) 306(3.53)	44-2303-76 44-2303-76 44-2303-76
$C_{19}H_{16}N_4O_3$ Benzo[g]pteridine-2,4(3H,10H)-dione, 10-(2-methoxyethyl)-3-phenyl-	EtOH	267(4.48),337(3.88), 434(3.94)	88-2899-76
Oxazolo[4,5-d]pyrimidin-2-ol, 5-amino- 3-benzoyl-2,3-dihydro-7-methyl- 2-phenyl-	EtOH	308(4.02)	94-2461-76
$C_{19}H_{16}N_4S$ 1H-Imidazo[4,5-b]pyridin-5-amine, 7-[(diphenylmethyl)thio]-	pH 13	273s(3.92),279(3.96), 318(4.00)	44-3784-76
$C_{19}H_{16}N_6$ 1,3,5-Benzenetricarbonitrile, 2-[[4-(di- ethylamino)phenyl]azo]-	EtOH benzene	562(4.67) 552(--)	39-0042-76C 39-0042-76C
$C_{19}H_{16}O$ 2-Cyclobuten-1-one, 4-(1-methylethyli- dene)-2,3-diphenyl-	$CHCl_3$	<u>280(4.4),360(3.7)</u>	24-1506-76
Cyclopentanone, 2,5-bis(phenylmethyl- ene)-	EtOH ether $CHCl_3$ CCl_4	230(4.22),354(4.64) 230(4.38),338(4.67) 350(4.55) 343(4.62)	114-0381-76B 114-0381-76B 114-0381-76B 114-0381-76B
$C_{19}H_{16}O_2$ 10H-Benzo[a]fluoren-10-one, 10a,11-di- hydro-3-methoxy-10a-methyl-, (+)-	MeOH	224(3.75),246(4.61), 264(4.16),286(3.97), 299(3.92),352(3.80), 368(3.86),411(4.00)	24-3025-76
2,2'-Spirobi[2H-1-benzopyran], 3,3'-di- methyl-	C_6H_{12}	212(4.79),225s(4.75), 252(4.46),265(4.36), 298(3.84),308s(3.78)	22-2039-76

Compound	Solvent	$\lambda_{max}(\log \epsilon)$	Ref.
2,2'-Spirobi[2H-1-benzopyran], 3,3'-di-methyl- (cont.)	EtOH	213(4.49),224(4.61), 252(4.32),265(4.23), 297(3.81),304s(3.76)	22-2039-76
$C_{19}H_{16}O_3$			
Cyclobuta[b]naphthalene-3,8-dione, 1,2,2a,8a-tetrahydro-8a-methoxy-1-phenyl-, (1α,2aα,8aα)-	MeCN	301(3.42),310(3.38), 339(2.65)	18-2596-76
	CCl₄	300(3.45),310(3.42), 343(2.65)	18-2596-76
Cyclobuta[2,3]naphth[2,1-b]oxet-5(3H)-one, 1,4,4a,9b-tetrahydro-9b-hydroxy-3-phenyl-	CHCl₃	290(3.48)	18-2596-76
Cyclopentanone, 2,5-bis[(2-hydroxyphen-yl)methylene]-	EtOH	238(4.22),260(4.11), 332(4.27),390(4.64)	114-0381-76B
	ether	234(--),255(--), 320(--),367(--)	114-0381-76B
	CHCl₃	325(--),380(--)	114-0381-76B
	CCl₄	315(--),366(--)	114-0381-76B
ion	n.s.g.	490(4.50)	114-0381-76B
non-ion	n.s.g.	395(4.49)	114-0381-76B
Cyclopentanone, 2,5-bis[(4-hydroxyphen-yl)methylene]-	EtOH	245(4.95),395(4.45)	114-0381-76B
	ether	235(--),375(--)	114-0381-76B
	CHCl₃	385(--)	114-0381-76B
Methanone, (3-acetoxy-1H-inden-2-yl)-(3-methylphenyl)-	MeOH	228(3.93),309(4.35)	83-0376-76
Methanone, (3-acetoxy-1H-inden-2-yl)-(4-methylphenyl)-	MeOH	230(3.99),318(4.38)	83-0376-76
$C_{19}H_{16}O_4$			
1,3-Cyclopentadiene-1-carboxylic acid, 3-hydroxy-5-oxo-2,3-diphenyl-, methyl ester	ether	282(4.14)	22-1491-76
2-Furancarboxylic acid, 5-methoxy-3,4-diphenyl-, methyl ester	ether	235s(4.16),300(4.06)	22-1491-76
1H-Inden-1-one, 2-[acetoxy(2-methoxy-phenyl)methylene]-2,3-dihydro-	MeOH	241(3.82),300(4.19)	83-0376-76
Methanone, (3-acetoxy-1H-inden-2-yl)-(3-methoxyphenyl)-	MeOH	309(4.33)	83-0376-76
Methanone, (3-acetoxy-1H-inden-2-yl)-(4-methoxyphenyl)-	MeOH	232(4.08),329(4.32)	83-0376-76
$C_{19}H_{16}O_4S$			
Benzo[b]thiophene-6-carboxylic acid, 4-acetoxy-2-methyl-5-phenyl-, methyl ester	EtOH	212(4.38),242(4.58), 278(4.07),318(3.85)	4-0285-76
$C_{19}H_{16}O_5$			
5-Benzofuranpropanoic acid, 2-benzoyl-5-methoxy-	EtOH	207(4.52),257(3.97), 317(4.32)	18-0737-76
4H-1-Benzopyran-6-carboxaldehyde, 5,7-dimethoxy-8-methyl-4-oxo-2-phenyl-	EtOH	223(4.43),260(4.32), 303(4.32)	2-0009-76
4H-1-Benzopyran-2-carboxylic acid, 7-methoxy-5-methyl-4-oxo-3-(phen-ylmethyl)-	EtOH	218(4.33),226(4.30), 238(4.24),253(4.09), 263(3.94),302(3.94)	39-0499-76C
2H-Furo[2,3-h]-1-benzopyran-8-carbox-ylic acid, 3,4-dihydro-5-methoxy-2-phenyl-	EtOH	288(4.12),299s(3.15)	39-1570-76C
5,12-Naphthacenedione, 7,8,9,10-tetra-hydro-6,11-dihydroxy-1-methoxy-	CHCl₃	292(3.95),370(3.57), 475(4.05),501(4.13), 537(3.94)	88-1637-76

Compound	Solvent	$\lambda_{max}(\log \epsilon)$	Ref.
$C_{19}H_{16}O_6$			
4H-1-Benzopyran-4-one, 7-acetoxy-8-methoxy-3-(4-methoxyphenyl)-	EtOH	256(4.52),310s(3.80)	102-1019-76
$C_{19}H_{16}O_7$			
2H-1-Benzopyran-2-one, 3-(1-acetoxy-4,5-dimethoxy-6-oxo-2,4-cyclohexadien-1-yl)-	MeOH	350(4.10),390s(3.99)	18-1955-76
2H-1-Benzopyran-2-one, 3-(1-acetoxy-4-methoxy-6-oxo-2,4-cyclohexadien-1-yl)-6-methoxy-	MeOH	284(4.31),344(4.04)	18-1955-76
2H-1-Benzopyran-2-one, 3-(1-acetoxy-4-methoxy-6-oxo-2,4-cyclohexadien-1-yl)-8-methoxy-	MeOH	257(4.19),301(4.13)	18-1955-76
4H-1-Benzopyran-4-one, 3-(1,3-benzodioxol-5-yl)-5,6,7-trimethoxy-	MeOH	260(4.33),290(4.02)	2-0222-76
4H-1-Benzopyran-4-one, 3-(3-hydroxy-4-methoxyphenyl)-7,8-dimethoxy-, monoacetate	EtOH EtOH-NaOMe EtOH-NaOAc	255(4.54),305s(4.04) 274(--),354(--) 273(--),352(--)	102-1019-76 102-1019-76 102-1019-76
$C_{19}H_{16}O_8$			
11H-Dibenzo[b,e][1,4]dioxepin-6-carboxylic acid, 4-formyl-3-hydroxy-8-methoxy-1,9-dimethyl-11-oxo-, methyl ester	n.s.g.	241(4.43),273s(3.90), 302(3.57)	12-1059-76
$C_{19}H_{16}S$			
Naphthalene, 2-[(1-methyl-2-phenylethenyl)thio]-	EtOH	260(4.46),277(4.40), 290(4.36),300(4.31), 336(3.45)	35-3564-76
Naphthalene, 2-[(1-phenyl-1-propenyl)thio]-	MeOH	252(5.18),288(4.58), 300(4.49),322(3.78), 337(3.61)	35-3564-76
$C_{19}H_{17}BrN_2OS$			
14H-Naphtho[1,2-d]thiazolo[3,2-b]quinazoline, 7,8-dihydro-10-methoxy-, hydrobromide	n.s.g.	303(4.20),340(4.11)	2-0759-76
$C_{19}H_{17}BrN_2O_3$			
1H-Pyrazole-5-carboxylic acid, 3-(4-bromobenzoyl)-4,5-dihydro-1-phenyl-, ethyl ester	EtOH	260(4.126),275s(4.084), 398(4.283)	18-0321-76
$C_{19}H_{17}Br_2N_4S_2$			
Isothiazolo[5,4-b]pyridinium, 3-bromo-6-[3-(3-bromo-4,7-dimethylisothiazolo-[5,4-b]pyridin-6(7H)-ylidene)-1-propenyl]-4,7-dimethyl-, perchlorate	DMF	663(4.18)	48-0779-76
$C_{19}H_{17}ClN_2O_3$			
1H-Pyrazole-5-carboxylic acid, 3-benzoyl-1-(4-chlorophenyl)-4,5-dihydro-, ethyl ester	EtOH	260(4.179),310s(3.518), 390(4.330)	18-0321-76
$C_{19}H_{17}ClN_6$			
Formazan, 3-(4-chlorophenyl)-1-(4,6-dimethyl-2-pyrimidinyl)-5-phenyl-	EtOH EtOH-NaOH benzene dioxan	440(4.16) 496(--) 458(--) 449(--)	103-1165-76 103-1165-76 103-1165-76 103-1165-76

Compound	Solvent	$\lambda_{max}(\log \epsilon)$	Ref.
$C_{19}H_{17}CuN_4$			
Copper(1+), (16,17-dihydro-7H-dibenzo-[f,m][1,4,8,12]tetraazacyclopenta-decinato-N^5,N^9,N^{15},N^{18})-, (SP-4-2)-, perchlorate	EtOH	237(4.49),251(4.52), 277(4.49),334(3.76), 395(3.81),463(4.27)	12-2271-76
$C_{19}H_{17}FeN$			
Ferrocene, [(3-phenyl-2-propenylidene)-amino]-	n.s.g.	406(3.4),562(3.5)	65-2512-76
$C_{19}H_{17}NO$			
8-Azabicyclo[3.2.1]oct-3-en-2-one, 6,8-diphenyl-, endo	EtOH	209(4.34),238(4.27)	39-2289-76C
Ethanone, 2-(1,4-dihydro-1-phenyl-4-pyridinyl)-1-phenyl-	MeOH	265(3.95),435(4.63)	104-1105-76
6,13-Methano-5H-dibenz[b,g]azonin-14-one, 6,13-dihydro-5,11-dimethyl-	EtOH	247(4.30),275(4.33)	39-2285-76C
2,9-Methano-3H-indeno[2,1-b]pyridin-3-one, 1,2,4,4a,9,9a-hexahydro-1-phenyl-	$CHCl_3$	256(4.28)	118-0105-76
$C_{19}H_{17}NOS$			
Acetamide, N-[4-[5-(4-methylphenyl)-2-thienyl]phenyl]-	MeOH	340(4.52)	48-0731-76
$C_{19}H_{17}NO_2$			
Benz[c]acridine-5,6-diol, 5,6-dihydro-7,10-dimethyl-, cis	EtOH	264(4.55),271(4.58), 333(3.99),348(4.03)	4-0619-76
6H-Dibenzo[de,g]quinoline, 10,11-di-methoxy-6-methyl-	MeOH	238(4.51),269(4.11), 278(4.14),340(4.09), 353(4.18),395(3.69), 417(3.74),445(3.71)	35-7108-76
Ethanone, 1,1'-(6-methyl-7-phenyl-1,3-indolizinediyl)bis-	EtOH	234(4.18),264(4.44), 294(4.16),352(4.40)	103-0424-76
2-Propenoic acid, 3-(1H-indol-1-yl)-3-phenyl-, ethyl ester	EtOH	271(4.03),335(4.17)	12-1583-76
2,6(1H,3H)-Pyridinedione, 3-(2,5-dimeth-ylphenyl)-4-phenyl-	pH 13	226(4.11),288(4.25)	34-0120-76
	EtOH	221(4.29),225(4.25), 287(4.39)	34-0120-76
	$CHCl_3$	232(3.50),293(4.24)	34-0120-76
2,6(1H,3H)-Pyridinedione, 3-(3,4-dimeth-ylphenyl)-4-phenyl-	pH 13	227(3.94),290(4.07)	34-0120-76
	EtOH	222(4.37),226(4.39), 288(4.53)	34-0120-76
	$CHCl_3$	234(3.73),292(4.24)	34-0120-76
$C_{19}H_{17}NO_3$			
Anonaine, N-acetyl-, (-)-	MeOH	217(4.24),269(4.12), 312(3.49)	102-1169-76
$C_{19}H_{17}NO_3S_2$			
Sulfilimine, S-(2-methoxyphenyl)-S-phen-yl-N-(phenylsulfonyl)-, (S)-(-)-	MeOH	225s(4.16),288(3.59)	78-3003-76
	dioxan	225s(4.26),286(3.62)	78-3003-76
$C_{19}H_{17}NO_4$			
Acronycine, 6,12-demethyl-11-methoxy-	MeOH	257(3.47),272(3.43), 283(3.43),295s(3.30), 315s(3.09),404(2.70)	100-0399-76
6H-Bis[1,3]benzodioxolo[5,6-a:5',6'-g]-quinolizine, 5,8,13,13a-tetrahydro-(stylopine)	EtOH	238(3.93),290(3.91)	73-3157-76

Compound	Solvent	λ_{max}(log ϵ)	Ref.
1H-Indole-6-carboxylic acid, 4-acetoxy-1-methyl-5-phenyl-, methyl ester	EtOH	208(4.40),244(4.63), 285(3.99),318(3.91)	48-0816-76
Methanone, (4-ethyl-4,5-dihydro-5-nitro-2-phenyl-3-furanyl)phenyl-, trans	$C_2H_4Cl_2$	261(4.38)	104-0640-76
7H-Pyrano[2,3-c]acridin-7-one, 3,12-dihydro-6-hydroxy-11-methoxy-3,3-dimethyl-	EtOH	252(4.57),266(4.34), 275(4.4),295(4.08), 392(3.67)	100-0399-76
1H-Pyrrolium, 2-(2-carboxyphenyl)-2,5-dihydro-3-hydroxy-1,1-dimethyl-5-oxo-4-phenyl-, hydroxide, inner salt	pH 13	262(4.22)	44-0390-76
$C_{19}H_{17}NO_5$			
Pyrido[2,1,6-de]quinolizine-1,3-dicarboxylic acid, 4-formyl-, diethyl ester	EtOH	253s(4.08),286(4.40), 326s(3.76),412(4.30), 506(4.24)	39-0341-76C
Pyrido[2,1,6-de]quinolizine-1,3-dicarboxylic acid, 6-formyl-, diethyl ester	EtOH	285(4.45),337s(3.57), 382s(3.84),400(4.11), 521(4.73)	39-0341-76C
$C_{19}H_{17}NO_6S$			
1,2,3-Propanetricarboxylic acid, 1-phenyl-1-[(phenylthioxomethyl)amino]-, (R*,R*)-	$CHCl_3$	250(4.04),285s(3.81), 395(2.30)	78-0571-76
$C_{19}H_{17}NO_8S$			
9H-Pyrido[2,1-b]benzothiazole-6,7,8,9-tetracarboxylic acid, tetramethyl ester	MeOH	229(4.36),256(4.12), 295s(4.52),309(4.28), 418(4.29)	39-1269-76C
$C_{19}H_{17}NO_9$			
Pentanedioic acid, 2-(1,4-dihydro-6-hydroxy-2,7-dimethoxy-1,4-dioxo-9H-carbazol-9-yl)-, (S)-	EtOH	217(4.34),258(4.24), 273(4.21),290(4.18), 312(4.17),355(4.23), 460(4.14)	146-0667-76
$C_{19}H_{17}N_3$			
Pyrido[2,3-b]pyrazine, 1,2,3,4-tetrahydro-2,3-diphenyl-	EtOH	260(3.77),330(4.01)	22-0251-76
$C_{19}H_{17}N_3O$			
Acetamide, N-[4-(4-methylphenyl)-6-phenyl-2-pyrimidinyl]-	EtOH	256(4.45),316(4.18), 282(4.01)	4-0257-76
2,4-Pentadienamide, 2-cyano-5-(4-methylphenyl)-5-(phenylamino)-	CH_2Cl_2	399(4.43)	48-0705-76
Spiro[3H-indole-3,3'-[3H]pyrazol]-2(1H)-one, 2',4'-dihydro-5'-phenyl-1-(2-propenyl)-	EtOH	211(4.60),262(4.20), 295(4.24)	94-0782-76
$C_{19}H_{17}N_3O_2$			
Acetamide, N-[4-(4-methoxyphenyl)-6-phenyl-2-pyrimidinyl]-	EtOH	259(4.33),325(4.28), 285(4.14)	4-0257-76
3-Pyridinecarboxylic acid, 5-[(4,9-dihydro-3H-pyrido[3,4-b]indol-1-yl)-methyl]-, methyl ester	MeOH	245s(--),324(4.04), 352s(--)	23-1262-76
$C_{19}H_{17}N_3O_3$			
8-Azabicyclo[3.2.1]oct-3-en-2-one, 6-methyl-8-(5-nitro-2-pyridinyl)-6-phenyl-, endo-exo	EtOH	210(4.36),227(4.38), 360(4.38)	39-2307-76C
exo-endo	EtOH	210(4.36),227(4.37), 360(4.38)	39-2307-76C

Compound	Solvent	λ_{max}(log ϵ)	Ref.
Benzeneacetic acid, α-[[(3-methyl-2-quinoxalinyl)carbonyl]amino]-, methyl ester, (S)-	EtOH	203(4.67),242(4.55), 320(3.85)	78-2931-76
2,9-Methano-3H-indeno[2,1-b]pyridin-3-one, 1,2,4,4a,9,9a-hexahydro-9-methyl-1-(5-nitro-2-pyridinyl)-, (±)-	EtOH	210(4.50),225(4.43), 362(4.62)	118-0105-76
L-Phenylalanine, N-[(3-methyl-2-quinoxalinyl)carbonyl]-	EtOH	203(4.63),242(4.50), 320(3.81)	78-2931-76
	EtOH-KOH	202(4.68),242(4.48), 321(3.76)	78-2931-76
1H-Pyrazole-4-carboxamide, 1-acetyl-3-benzoyl-4,5-dihydro-N-phenyl-	EtOH	244(4.30),306(4.23)	115-0001-76
Spiro[3H-indole-3,2'(1'H)-quinazoline]-4'-carboxylic acid, 1,2-dihydro-1,1'-dimethyl-2-oxo-, methyl ester, (+)-	MeOH	207(4.46),237(4.48), 416(3.15)	39-2004-76C
$C_{19}H_{17}N_3O_4$			
8-Azabicyclo[3.2.1]oct-3-en-2-one, 6-(4-methoxyphenyl)-8-(5-nitro-2-pyridinyl)-	EtOH	215(4.37),226(4.37), 350(4.33)	39-2307-76C
2,9-Methano-3H-indeno[2,1-b]pyridin-3-one, 1,2,4,4a,9,9a-hexahydro-6-methoxy-1-(5-nitro-2-pyridinyl)-, (±)-	CHCl	368(4.37)	118-0105-76
$C_{19}H_{17}N_3O_4S$			
4-Imidazolidinone, 3-[2-methoxy-5-[2-(4-nitrophenyl)ethenyl]phenyl]-5-methyl-2-thioxo-	MeOH	267(4.38),366(4.45)	56-0967-76
1H-Pyrrolo[2,3-d]pyrimidine-2,4(3H,7H)-dione, 6,7-dihydro-7-[(4-methylphenyl)sulfonyl]-3-phenyl-	DMSO	292(4.14)	24-2983-76
$C_{19}H_{17}N_3O_5$			
Benzeneacetic acid, α-[[(3-methyl-2-quinoxalinyl)carbonyl]amino]-, N,N'-dioxide, methyl ester, (S)-	EtOH	205(4.36),235(4.34), 259s(4.26),267(4.39), 350s(3.80),370(4.00), 390(4.04)	78-2931-76
	MeCN	205(4.36),230(4.34), 268(4.36),347s(3.60), 377(3.92),394(3.99)	78-2931-76
1,2-Diazabicyclo[5.2.0]nona-3,5-diene-2-carboxylic acid, 8-(1,3-dihydro-1,3-dioxo-2H-isoindol-2-yl)-5-methyl-9-oxo-, ethyl ester, cis	EtOH	277(4.00)	142-0471-76S
1,2-Diazabicyclo[5.2.0]nona-3,5-diene-2-carboxylic acid, 5-methyl-9-oxo-8-[(3-oxo-1(3H)-isobenzofuranylidene)-amino]-, ethyl ester, cis	MeOH	276(4.10)	142-0471-76S
L-Phenylalanine, N-[(3-methyl-2-quinoxalinyl)carbonyl]-, N,N'-dioxide	EtOH	205(4.36),232(4.35), 260s(4.29),265(4.44), 350s(3.83),370(4.06), 384(4.11)	78-2931-76
	MeCN	205(4.36),268(4.38), 360s(3.77),380(3.97), 393(4.03)	78-2931-76
$C_{19}H_{17}N_3O_5S$			
1-Propanone, 1-(1-methyl-1H-imidazol-5-yl)-3-(2-nitrophenyl)-2-(phenylsulfonyl)-, (±)-	MeOH	271(4.22)	78-1085-76

Compound	Solvent	$\lambda_{max}(\log \epsilon)$	Ref.
$C_{19}H_{17}N_3O_7$			
1-Azuleneethanol, 2-methyl-, trinitro-benzene complex	CH_2Cl_2	281(4.68),289(4.76), 304(3.74),334(3.50), 349(3.61),576(2.36), 615(2.30),687(1.83)	44-1811-76
1-Azuleneethanol, 3-methyl-, trinitro-benzene complex	EtOH	281(4.76),334s(3.55), 349(3.76),365(3.63), 625(2.50),660s(2.43), 685(2.40),765(1.95)	44-1811-76
$C_{19}H_{17}N_3O_8$			
1-Azuleneethanol, 2-methoxy-, trinitro-benzene complex	CH_2Cl_2	283(4.72),293(4.81), 318(3.89),345(3.61), 361(3.69),375(3.51), 535(2.35),570(2.14), 625(1.69)	44-1811-76
1-Azuleneethanol, 3-methoxy-, trinitro-benzene complex	CH_2Cl_2	291(4.62),362(3.68), 378(3.64),695(2.52), 770s(2.44)	44-1811-76
$C_{19}H_{17}N_4Ni$			
Nickel(1+), 16,17-dihydro-7H-dibenzo-[f,m][1,4,8,12]tetraazacyclopenta-decinato-N^5,N^9,N^{15},N^{18})-, chloride, (SP-4-2)-	EtOH	249(4.55),268(4.52), 330(4.12),445(4.16), 573(3.53)	12-2271-76
$C_{19}H_{17}N_7O_2$			
Formazan, 1-(4,6-dimethyl-2-pyrimidin-yl)-5-(4-nitrophenyl)-3-phenyl-	EtOH	413(4.02),471(4.05)	103-1165-76
	EtOH-NaOH	584(--)	103-1165-76
	benzene	490(--)	103-1165-76
	dioxan	425(--),482(--)	103-1165-76
$C_{19}H_{17}O$			
Pyrylium, 3,4-dimethyl-2,6-diphenyl-, iodide	CH_2Cl_2	500(2.19)	59-1311-76
selenocyanate	CH_2Cl_2	474(1.97)	59-1311-76
$C_{19}H_{17}O_8$			
1,3-Cyclopentadiene-1,2,3,4-tetracarb-oxylic acid, 5-phenyl-, tetramethyl ester, ion(1-), sodium	MeOH	214(4.02),269(4.19), 302(4.18)	24-1928-76
$C_{19}H_{18}ClFN_2S$			
Piperazine, 1-(7-chloro-8-fluorodibenzo-[b,f]thiepin-10-yl)-4-methyl-	MeOH	235(4.35),264s(4.13), 310(3.88)	73-0881-76
$C_{19}H_{18}ClN_3O$			
3H-1,4-Benzodiazepine, 7-chloro-2-(4-morpholinyl)-5-phenyl-	isoPrOH	233(4.44),272(4.27), 285s(4.24),351(3.49)	44-2724-76
3-Quinolinamine, 6-chloro-2-(4-morpho-linyl)-4-phenyl-	isoPrOH	232(4.52),255(4.57), 353(4.05)	44-2724-76
$C_{19}H_{18}ClN_3O_3S$			
Butanoic acid, 2-[(4-chlorophenyl)azo]-3-oxo-2-[(phenylamino)thioxomethyl]-, ethyl ester	EtOH	305(3.98),330(3.97)	104-1912-76
$C_{19}H_{18}ClN_5O$			
4H-[1,2,4]Triazolo[4,3-a][1,5]benzodi-azepin-5(6H)-one, 8-chloro-1-[(di-methylamino)methyl]-6-phenyl-	EtOH	228(4.58),282s(3.32), 290s(3.27)	87-0192-76

Compound	Solvent	$\lambda_{max}(\log \epsilon)$	Ref.
$C_{19}H_{18}ClN_5O_3S$ Benzenesulfonamide, 4-[[[(4-chlorophen-yl)amino]carbonyl]amino]-N-(4,6-di-methyl-2-pyrimidinyl)-	EtOH	218(5.06),273(5.31)	80-1345-76
$C_{19}H_{18}Cl_2N_2S$ Piperazine, 1-(7,8-dichlorodibenzo[b,f]-thiepin-10-yl)-4-methyl-	MeOH	219(4.58),240s(4.29), 270s(4.12),315(3.86)	73-0881-76
$C_{19}H_{18}F_3NO_3$ 1,4-Benzoxazepine, 2,3-dihydro-7,8-di-methoxy-3-methyl-5-[3-(trifluoro-methyl)phenyl]-	n.s.g.	210(4.37),260(4.13), 315(4.14)	20-0787-76
$C_{19}H_{18}IN$ Isoquinolinium, 2-methyl-1-(vinylbenz-yl)-, iodide, polymer	DMSO	292(4.08),348(3.80)	116-0010-76
$C_{19}H_{18}NO_2S_2$ Benzenaminium, N-(2-ethoxy-2-oxoethyl)-N-(5-phenyl-3H-1,2-dithiol-3-yli-dene)-, bromide	MeOH	321(4.19),393(4.11)	48-0221-76
$C_{19}H_{18}N_2$ 1H-Perimidine, 2,3-dihydro-4,5-dimethyl-2-phenyl-	n.s.g.	239(4.65),335(4.04), 346(4.06)	12-2499-76
$C_{19}H_{18}N_2O$ 2,2'-Bipyridine, 4-(4-methoxyphenyl)-6,6'-dimethyl-, Cu(I) complex	EtOH	465(4.04)	118-0001-76
Pseudoyohimbone, 18,19,20,21-tetra-dehydro-, dl-	MeOH	225(3.76),275(3.15), 400(3.59)	35-3645-76
2H-Pyrrol-2-one, 4-ethyl-1,5-dihydro-5-phenyl-3-[(phenylmethylene)amino]-	EtOH	211(4.36),269(4.25), 280(4.26)	40-1100-76
$C_{19}H_{18}N_2OS$ Acetamide, N-[3-(methylphenylamino)-5-phenyl-2-thienyl]-	MeOH	239s(4.40),283s(4.20), 299(4.20),325s(4.12)	48-0221-76
$C_{19}H_{18}N_2O_2$ Benzoic acid, 2-(1 or 2-methyl-3-phenyl-pyrazol-5-yl)-, ethyl ester, hydro-chloride	EtOH	244(4.35)	44-0110-76
2,3-Diazabicyclo[3.1.0]hex-2-ene-6-carb-oxylic acid, 1,5-diphenyl-, ethyl ester, (1α,5α,6α)-	MeOH	332(2.65)	104-0802-76
Pyrano[2,3-c]pyrazole, 1,4,5,6-tetra-hydro-6-methoxy-1,4-diphenyl-, cis	EtOH	245(4.17)	35-2947-76
trans	EtOH	244.5(4.17)	35-2947-76
$C_{19}H_{18}N_2O_2S$ 3-Thiophenamine, N-ethyl-N-(4-methyl-phenyl)-2-nitro-5-phenyl-	MeCN	238s(4.11),261(4.18), 308(3.85),359(4.14), 454(3.90)	48-0221-76
$C_{19}H_{18}N_2O_3$ Benzoic acid, 2-[5-(4-methoxyphenyl)-1H-pyrazol-3-yl]-, ethyl ester hydrochloride	EtOH	260(4.45)	44-0110-76
	EtOH	262(4.45)	44-0110-76

Compound	Solvent	$\lambda_{max}(\log \epsilon)$	Ref.
Ethanone, 1-(2,4-dimethoxyphenyl)-2-(1-phenyl-1H-pyrazol-4-yl)-	EtOH	270(4.33),302(3.88)	103-0914-76
Ethanone, 1-(2-hydroxy-4-methoxy-6-methylphenyl)-2-(1-phenyl-1H-pyrazol-4-yl)-	EtOH	265(4.32)	103-0914-76
1H-Pyrazole-1-carboxylic acid, 5-(4-methoxyphenyl)-3-phenyl-, ethyl ester	EtOH	229(4.37),245s(4.42), 267(4.49)	4-0257-76
1H-Pyrazole-5-carboxylic acid, 3-benzoyl-4,5-dihydro-1-phenyl-, ethyl ester	EtOH	254(4.244),390(4.374)	18-0321-76
$C_{19}H_{18}N_2O_4$			
Benzoic acid, 2-[5-(4-methoxyphenyl)-1H-pyrazol-3-yl]-, 2-hydroxyethyl ester	EtOH	261(4.45)	44-0110-76
3-Quinolinecarboxylic acid, 1-ethyl-1,4-dihydro-6-hydroxy-7-methoxy-4-(phenylimino)-	MeOH	242(3.45),260(4.41), 279(4.42)	142-1347-76
$C_{19}H_{18}N_2O_5$			
Hydrazinecarboxylic acid, [3-(1,3-benzodioxol-5-yl)-3-oxo-1-phenylpropylidene]-, ethyl ester	EtOH	225s(4.15),283(4.36)	4-0257-76
$C_{19}H_{18}N_2O_8$			
9H-Carbazole-9-propanoic acid, 2-[(2-carboxyethyl)amino]-1,4-dihydro-6-hydroxy-7-methoxy-1,4-dioxo-	EtOH	382(3.81),482(3.77)	146-0667-76
sodium salt	EtOH	386(3.94),475(3.89)	146-0667-76
$C_{19}H_{18}N_3S$			
Methanaminium, N-[[(2,5-diphenyl-6H-1,3-thiazin-6-ylidene)amino]methylene]-N-methyl-, perchlorate	HOAc	271(4.26),321(4.13), 429(4.11)	97-0268-76
$C_{19}H_{18}N_4O$			
Pyrazinecarboxamide, N-methyl-3-(methylamino)-5,6-diphenyl-	MeOH	228(4.24),286(4.29), 391(4.03)	24-3194-76
$C_{19}H_{18}N_4O_2$			
2-Butenoic acid, 2-(dicyanomethyleneamino)-4-(2,3-dihydro-1,3,3-trimethyl-indol-2-ylidene)-, methyl ester	EtOH	557(4.80)	78-3031-76
geometric isomer	EtOH	545(4.79)	78-3031-76
Lumazine, 5,6,7,8-tetrahydro-1-methyl-6,7-diphenyl-	pH 0.0	266(4.26)	24-3184-76
	pH 6.0	242s(3.77),293(4.01)	24-3184-76
$C_{19}H_{18}N_4O_3$			
2(5H)-Furanone, 2-(4-amidinobenzyl)-3-(4-amidinophenyl)-4-hydroxy-, dihydrochloride	pH 1	298(4.33)	106-0279-76
	pH 13	340(4.31)	106-0279-76
	MeOH	238(4.42),365(4.33)	106-0279-76
Oxazolo[4,5-b]quinoxaline-9a(2H)-carboxamide, 3,9-dihydro-6,7-dimethyl-2-oxo-N-(phenylmethyl)-	6M HCl	372(4.18)	5-1276-76
	pH 1.4	330(3.90),376(3.85)	5-1276-76
	pH 6	314(3.76),366(3.95)	5-1276-76
1H-Pyrazole-3-carboxylic acid, 5-amino-1-phenyl-4-[(phenylamino)carbonyl]-, ethyl ester	EtOH	307(4.11)	4-1137-76
$C_{19}H_{18}N_4O_4$			
1H-Pyrazole-4-carboxylic acid, 5-amino-3-(4-methylphenyl)-1-(4-nitrophenyl)-, ethyl ester	EtOH	265(4.31)	4-1137-76

Compound	Solvent	$\lambda_{max}(\log \epsilon)$	Ref.
$C_{19}H_{18}N_4S$			
1H-Imidazole-5-carbonitrile, 4-[(phenyl-methyl)amino]-1-[2-(phenylthio)ethyl]-	EtOH	256(4.03),275(4.025)	2-0346-76
$C_{19}H_{18}N_5OP$			
Phosphinic amide, N-methyl-N-phenyl-P,P-bis(phenylazo)-	MeOH	205(3.76),222(3.63), 306(3.66),507(2.35)	139-0207-76A
$C_{19}H_{18}N_6$			
Formazan, 1-(4,6-dimethyl-2-pyrimidin-yl)-3,5-diphenyl-	EtOH	440(4.01)	103-1165-76
	EtOH-NaOH	492(--)	103-1165-76
	benzene	457(--)	103-1165-76
	dioxan	448(--)	103-1165-76
$C_{19}H_{18}O$			
7H-Benzo[c]xanthene, 5,6-dihydro-5,5-di-methyl-	EtOH	210(4.20),235(4.28), 288(3.70)	22-1967-76
2-Cyclohexen-1-one, 2-methyl-5,5-diphen-yl-	EtOH	234(3.87)	33-2012-76
Cyclopropane, 1-methoxy-1-methyl-2-phen-yl-3-(phenylethynyl)-	hexane	252(4.31)	44-3931-76
isomer	EtOH	252(4.33)	44-3931-76
2-Cyclopropen-1-one, 2,3-bis(2,4-dimeth-ylphenyl)-	MeCN	298(4.320),311(4.379), 323(4.293)	101-0385-76N
2-Cyclopropen-1-one, 2,5-bis(2,5-dimeth-ylphenyl)-	MeCN	287(4.288),300(4.287), 311(4.281),323s(4.14)	101-0385-76N
1-Penten-4-yne, 3-methoxy-3-methyl-1,5-diphenyl-, trans	EtOH	252(4.48)	44-3931-76
$C_{19}H_{18}O_2$			
10H-Benzo[a]fluoren-10-one, 8,9,10a,11-tetrahydro-3-methoxy-10a-methyl-, ($\pm$)-	MeOH	216(4.12),236(4.40), 244(4.56),252(4.73), 261(4.77),283(4.10), 294(4.28),306(4.29), 323(2.20),339(3.28), 356(3.23)	24-3025-76
1H-Inden-1-one, 2,3-dihydro-2-(2,4,6-trimethylbenzoyl)-	MeOH	257(3.92),314(4.27)	83-0356-76
$C_{19}H_{18}O_3$			
2(5H)-Furanone, 5-ethoxy-5-methyl-3,4-diphenyl-	dioxan	222(4.199),292(3.995)	24-0576-76
1,3-Naphthalenediol, 2-(2-hydroxy-2-phenylethyl)-4-methyl-	$CHCl_3$	246(3.86),275(3.64), 283(3.58),321(3.08)	18-2596-76
2-Pentenoic acid, 5-oxo-3,5-diphenyl-, ethyl ester, (E)-	EtOH	246(4.34),268(4.22)	18-3239-76
$C_{19}H_{18}O_4$			
3-Benzofuranpropanoic acid, 5-methoxy-2-(phenylmethyl)-	EtOH	210(4.50),255(4.15), 292(3.75),302(3.70)	18-0737-76
2H-Cyclopenta[b]furan-2,4(3H)-dione, tetrahydro-6a-(6-methoxy-2-naphtha-lenyl)-3a-methyl-, (3aS-cis)-	EtOH	231(4.94),262(3.75), 270(3.78),314(3.20), 329(3.29)	78-0079-76
Dipetalolactone	EtOH	222(4.17),244s(4.37), 250(4.47),294s(4.37), 297(4.38),307s(4.28), 344(4.04)	102-0313-76
Furo[3,2-c]naphth[2,1-e]oxepin-10,12-dione, 6,7,8,9-tetrahydro-1,6,6-trimethyl-	EtOH	237(4.00),270(3.80), 306(3.97),325(3.93)	39-1716-76C

Compound	Solvent	$\lambda_{max}(\log \epsilon)$	Ref.
Hortiline	EtOH	251(4.36),290(4.32), 340(4.03)	32-0681-76
Hortiolone	pH 13	252(4.36),317(4.5), 408(3.78)	32-0681-76
	EtOH	235s(4.11),255(4.02), 260(4.08),285s(4.46), 303(4.55),315(4.53), 400s(3.5)	32-0681-76
	EtOH-NaOH	214(4.47),256(4.42), 323(4.66),415(3.89)	32-0681-76
Propanedioic acid, (1,2-diphenylethen- yl)-, dimethyl ester	hexane	233(4.91),241(4.95), 325(4.66),343(4.46)	104-0230-76
Propanedioic acid, (1,2-diphenylethyli- dene)-, dimethyl ester	hexane	221(4.82),265(4.86), 340(4.02)	104-0230-76
$C_{19}H_{18}O_5$ 4H-1-Benzopyran-4-one, 2,3-dihydro- 5,7-dimethoxy-3-[(4-methoxyphenyl)- methylene]-, (E)-	EtOH	212(4.57),339(4.467)	33-2048-76
3-Furancarboxylic acid, 4-methyl- 2-(6,7,8,8a-tetrahydro-6,6-dimethyl- 2-oxo-2H-naphtho[1,8-bc]furan-3-yl)-	EtOH	239(4.14),298(3.79), 325(3.92)	39-1716-76C
1H-Inden-1-one, 2,3-dihydro-2-(2,4,6- trimethoxybenzoyl)-	MeOH	341(4.20)	83-0356-76
$C_{19}H_{18}O_6$ Machicendiol	n.s.g.	218(4.3),318(4.56)	2-0613-76
Ulugbekic acid, methyl ester	EtOH	236s(--),250s(--), 292(4.07),336(4.10)	105-0539-76
$C_{19}H_{18}O_7$ Isoflavone, 7-hydroxy-2',4',5',6-tetra- methoxy-	MeOH	252(4.2),300(4.07), 327(3.99)	2-0951-76
	MeOH-NaOAc	253(4.2),300(3.78), 344(4.1)	2-0951-76
Isoflavone, 5,7,8-trimethoxy-3',4'- (methylenedioxy)-	MeOH	260(4.2),289s(--)	2-0584-76
$C_{19}H_{18}O_8$ 1,3-Cyclopentadiene-1,2,3,4-tetracarb- oxylic acid, 5-phenyl-, tetramethyl ester, ion(1-), sodium	MeOH	214(4.02),269(4.19), 302(4.18)	24-1928-76
Dibenzo[b,d]pyran-6-one, 2r-acetoxy- 3,4c-isopropylidenedioxy-8,9-(meth- ylenedioxy)-1,2,3,4-tetrahydro-	EtOH	241(4.53),258s(--), 287(3.71),330(3.65), 348s(--)	24-0855-76
Dibenzo[b,d]pyran-6-one, 4c-acetoxy- 2,3r-isopropylidenedioxy-8,9-(meth- ylenedioxy)-1,2,3,4-tetrahydro-	EtOH	242(4.68),278(4.13), 298(4.16),332(3.97)	24-0855-76
$C_{19}H_{18}P$ Phosphonium, methyltriphenyl-	MeOH	225s(4.64),263(3.66), 269(3.71),277(3.62)	30-0405-76
methosulfate	MeOH	226(4.48),257s(3.24), 264(3.40),277(3.46)	30-0405-76
$C_{19}H_{19}BrN_2O_7$ Uridine, 5-bromo-2',3'-O-(1-methyleth- ylidene)-, 5'-benzoate	MeOH	225(4.13),274(3.92)	44-1100-76

Compound	Solvent	$\lambda_{max}(\log \epsilon)$	Ref.
$C_{19}H_{19}ClN_2O_3$ Benzoic acid, 2-[5-(4-methoxyphenyl)- 1H-pyrazol-3-yl]-, ethyl ester, hydrochloride	EtOH	262(4.45)	44-0110-76
$C_{19}H_{19}ClN_2O_6$ 5H-Benzimidazo[3,2-b]isoquinolinium, 7- ethyl-9,10-dimethoxy-, perchlorate	MeOH	280(4.30),290(4.25), 300(4.20)	103-0205-76
$C_{19}H_{19}FO$ Furan, 2-(2-fluoro-3,3-diphenyl-2-prop- enyl)tetrahydro-	MeOH	246(3.95)	44-0940-76
$C_{19}H_{19}NO$ 1H-Carbazole, 2,3,4,9-tetrahydro-6-meth- oxy-9-phenyl-	isoPrOH	224(4.20),227(3.98), 299s(3.74)	103-0880-76
1H-Carbazole, 2,3,4,9-tetrahydro- 9-(4-methoxyphenyl)-	EtOH	225(4.48),261(4.08), 285(3.92),290s(3.92)	103-0880-76
1(2H)-Naphthalenone, 2-[[4-(dimethylami- no)phenyl]methylene]-3,4-dihydro-	isooctane	260(4.26),305(3.78), 384(4.30)	65-2057-76
	EtOH	272(4.26),418(4.33)	65-2057-76
$C_{19}H_{19}NOS$ 2-Furanamine, N,N-dimethyl-5-(methyl- thio)-3,4-diphenyl-	dioxan	232(4.262),296(3.977)	24-0576-76
$C_{19}H_{19}NO_2$ Benzeneacetamide, N,N-dimethyl-α-(2-oxo- 1-phenylpropylidene)-, cis	dioxan	225s(4.243),282(3.987)	24-0576-76
trans	dioxan	262(4.158)	24-0576-76
$C_{19}H_{19}NO_2S$ Benzeneethanethioic acid, α-[2-(dimeth- ylamino)-2-oxo-1-phenylethylidene]-, S-methyl ester	dioxan	226s(4.231),295(4.140)	24-0576-76
$C_{19}H_{19}NO_3$ 2-Pentenoic acid, 5-oxo-3-phenyl- 5-(phenylamino)-, ethyl ester	EtOH	224(4.14),247(4.35), 270s(4.20)	18-3239-76
$C_{19}H_{19}NO_4$ Aporphin-2-ol, 10-methoxy-1,11-(methyl- enedioxy)-, (6aS)-, hydrochloride	EtOH	274(4.06),305(3.68)	33-2551-76
	EtOH-NaOH	252(4.22),282(4.06), 336s(3.53)	33-2551-76
Berbine, 3-hydroxy-2-methoxy-10,11- (methylenedioxy)-	EtOH	230s(4.03),290(3.92)	73-1219-76
	EtOH-NaOH	240s(4.00),296(4.01)	73-1219-76
Cheilanthifoline, (+)-	EtOH	232s(3.9),285(3.7)	102-0545-76
	EtOH-base	293(--)	102-0545-76
Groenlandicine, tetrahydro-	EtOH	230s(3.99),287(3.83)	73-1219-76
	EtOH-NaOH	245(4.07),292(3.84)	73-1219-76
Oliveridine	EtOH	222(4.22),238s(4.11), 284(4.20),318s(3.64)	100-0350-76
$C_{19}H_{19}NO_4S$ Pyrrolo[2,1-a]isoquinoline-1,3-dicarbox- ylic acid, 2-(methylthio)-, diethyl ester	EtOH	220(4.24),271(4.64), 328(3.93),345(4.01), 362(4.03)	94-1299-76
$C_{19}H_{19}NO_5$ Luteoreticulin	EtOH	226(4.37),256s(4.05), 368(4.26)	39-0404-76C

Compound	Solvent	$\lambda_{max}(\log \epsilon)$	Ref.
N-Oxyoliveridine	EtOH	222(4.42),240s(4.15), 287(4.28),320s(3.88)	100-0350-76
DL-Phenylalanine, N-(α-methylcinnamoyl)- β-3,4-dihydroxy-, (E)-	EtOH	264(4.29)	44-1466-76
$C_{19}H_{19}NO_5S$ 1H-Indole-3-carboxylic acid, 5-hydroxy- 1,2-dimethyl-4-(phenylsulfonyl)-, ethyl ester	EtOH	222(4.48),324(4.17)	103-0044-76
$C_{19}H_{19}NO_6$ 1,3(2H,9bH)-Dibenzofurandione, 6-acetyl- 7,9-dihydroxy-8,9b-dimethyl-2-[(1- (methylamino)ethylidene]-, (-)-	MeOH	223(4.07),293(4.19), 345(3.39)	23-2795-76
Gravacridrontriol	EtOH	227(4.15),250(4.44), 265s(4.51),272(4.60), 300(4.21),332(3.88), 398(3.73)	102-0240-76
$C_{19}H_{19}NO_7$ Bis[1,3]dioxolo[4,5-c:4',5'-j]phenan- thridin-5(3aH)-one, 12-acetoxy- 3b,4,12,12a-tetrahydro-2,2-di- methyl-, (3aα,3bβ,12α,12α)-(+)-	EtOH	248(4.45),308(4.04)	94-2977-76
4-Isoxazolecarboxylic acid, 5-[(7-acet- yl-4,6-dihydroxy-3,5-dimethyl-2-benzo- furanyl)methyl]-3-methyl-, methyl ester	MeOH	242(4.04),303(3.85), 349(3.35)	23-3713-76
$C_{19}H_{19}NS$ 2-Thiophenamine, N,N,5-trimethyl-3,4-di- phenyl)-	dioxan	246(4.387)	24-0576-76
$C_{19}H_{19}N_2O_2$ 5H-Benzimidazo[3,2-b]isoquinolinium, 7- ethyl-9,10-dimethoxy-, perchlorate	MeOH	280(4.30),290(4.25), 300(4.20)	103-0205-76
$C_{19}H_{19}N_3O$ 8-Azabicyclo[3.2.1]oct-3-en-2-one, 8-(4,6-dimethyl-2-pyrimidinyl)- 6-phenyl-, endo	EtOH	213(4.23),242(4.37), 295(3.54)	39-2307-76C
Oxazolo[3',2':1,2]pyrido[3,4-b]indole, 2,3,5,6,11,11b-hexahydro-11b-methyl- 2-(3-pyridinyl)- (naucleonine)	MeOH	224(4.33),252(3.88), 268(3.73),281(3.60), 291(3.51),360(4.02)	23-1262-76
	MeOH-HCl	213(4.22),251(4.08), 363(4.38)	23-1262-76
	MeOH-KOH	227(4.47),269(3.94), 282(3.88),291(3.81)	23-1262-76
1H-Pyridazino[4,5-b]indole, 2-acetyl- 2,3,4,5-tetrahydro-5-methyl-1-phenyl-	EtOH	285(4.21)	103-1008-76
3H-Pyrido[3,4-b]indole-1-propanol, 4,9-dihydro-α-3-pyridinyl-	MeOH	213(4.25),239(4.10), 243(4.10),260s(--), 268s(--),319(4.08)	23-1262-76
	MeOH-HCl	213(4.25),250(4.00), 356(4.26)	23-1262-76
	MeOH-KOH	213(4.35),239(4.16), 243(4.16),260s(--), 268s(--),318(4.13)	23-1262-76
Spiro[5H-1,4-diazepine-5,3'-[3H]indol]- 2'(1'H)-one, 2,3,4,6-tetrahydro-1'- methyl-7-phenyl-	MeOH	211(4.42),270(4.15), 273(4.18),281(4.23), 345(4.22)	103-0665-76

Compound	Solvent	$\lambda_{max}(\log \epsilon)$	Ref.
$C_{19}H_{19}N_3O_2$			
1H-Pyrazole-4-carboxamide, 3-benzoyl-4,5-dihydro-5-methyl-N-(4-methyl-phenyl)-	EtOH	252(4.35),336(4.09)	115-0001-76
1H-Pyrazole-4-carboxamide, 4,5-dihydro-5-methyl-4-(4-methylbenzoyl)-N-phenyl-	EtOH	246(4.36),333(4.20)	115-0001-76
1H-Pyrazole-4-carboxamide, 4,5-dihydro-3-(4-methylbenzoyl)-N-(4-methylphenyl)-	EtOH	250(4.43),334(4.16)	115-0001-76
3-Pyridinecarboxylic acid, 5-[(2,3,4,9-tetrahydro-1H-pyrido[3,4-b]indol-1-yl)methyl]-, methyl ester	MeOH	275(3.84),291s(--)	23-1262-76
$C_{19}H_{19}N_3O_2S_3$			
Acetamide, N-[3-ethyl-2,3-dihydro-2-[4-oxo-3-(2-propenyl)-2-thioxo-5-thiazo-linylidene]-5-phenyl-4-thiazolyl]-	n.s.g.	437(4.72)	103-0648-76
$C_{19}H_{19}N_3O_3$			
3-Pyridinecarboxylic acid, 5-[2-[[2-(1H-indol-3-yl)ethyl]amino]-2-oxoethyl]-, methyl ester	EtOH	243(4.15),260s(3.55),267s(3.45),300(3.47)	95-0968-76
5-Pyrimidinecarboxamide, 1,2,3,4-tetra-hydro-4-(3-methoxyphenyl)-6-methyl-2-oxo-N-phenyl-	EtOH	204(4.67),238(4.10),286(4.20)	103-0191-76
$C_{19}H_{19}N_3O_3S$			
Butanoic acid, 3-oxo-2-[(phenylamino)-thioxomethyl]-2-(phenylazo)-, ethyl ester	EtOH	236(3.68),314(4.02),335(3.96)	104-1912-76
$C_{19}H_{19}N_3O_4$			
Acetamide, N-[2,3-dihydro-1-(methoxy-methyl)-2-oxo-5-phenyl-1H-1,4-benzo-diazepin-7-yl]-N-hydroxy-	isoPrOH	215s(4.43),253(4.42),317(3.37)	87-1378-76
$C_{19}H_{19}N_3O_5S$			
L-Alanine, N-[[[2-methoxy-5-[2-(4-nitro-phenyl)ethenyl]phenyl]amino]thioxo-methyl]-	MeOH	252(4.30),370(4.36)	56-0967-76
$C_{19}H_{19}N_3O_6$			
1,2-Diazabicyclo[5.2.0]nona-3,5-diene-2-carboxylic acid, 8-[(2-carboxyben-zoyl)amino]-5-methyl-9-oxo-, 2-ethyl ester, cis	MeOH	275(4.00)	142-0471-76S
$C_{19}H_{19}N_3O_9$			
1H-Pyrazole-4,5-dicarboxylic acid, 1-(4-nitrophenyl)-3-(tetrahydro-2,2-dimeth-ylfuro[3,4-d]-1,3-dioxol-4-yl)-, 4-methyl ester, [3aS-(3aα,4aα,6aα)]-	CHCl$_3$	310(4.27)	136-0019-76A
$C_{19}H_{19}N_4S$			
Methanaminium, N-methyl-N-[[[5-phenyl-2-(phenylamino)-6H-1,3-thiazin-6-yl-idene]amino]methylene]-, perchlorate	HOAc HOAc	457(4.26) 261(4.17),312(4.30),457(4.26)	88-2005-76 97-0268-76
$C_{19}H_{19}N_4Se$			
Methanaminium, N-methyl-N-[[[5-phenyl-2-(phenylamino)-6H-1,3-selenazin-6-yl-idene]amino]methylene]-, perchlorate	HOAc	461(4.24)	88-2005-76

Compound	Solvent	$\lambda_{max}(\log \epsilon)$	Ref.
$C_{19}H_{19}N_5O$ Methanimidamide, N,N-dimethyl-N'-[1-oxo-3-[(phenylmethylene)hydrazino]-2(1H)-isoquinolinyl]-	EtOH	231(4.43),276(4.28), 363(4.20)	95-0700-76
$C_{19}H_{19}N_5O_2S_2$ Benzenesulfonamide, N-(4,6-dimethyl-2-pyrimidinyl)-4-[[(phenylamino)-thioxomethyl]amino]-	EtOH	222(5.34),275(5.22)	80-1345-76
$C_{19}H_{19}N_5O_3S$ Benzenesulfonamide, N-(4,6-dimethyl-2-pyrimidinyl)-4-[[(phenylamino)-carbonyl]amino]-	EtOH	208(4.85),222s(4.70), 275(4.99)	80-1345-76
$C_{19}H_{20}$ 1-Cyclopentene, 3,3-dimethyl-1,2-diphenyl-	C_6H_{12}	259(4.12)	35-6218-76
Ethyne, (2,4-dimethylphenyl)(2,4,6-trimethylphenyl)-	C_6H_{12}	279(4.39),284(4.43), 294(4.53),301(4.43), 313(4.48)	101-0385-76N
Phenanthrene, 9,10-dihydro-3,6,9,9-tetramethyl-10-methylene-	hexane	225(4.45),243(4.42), 249(4.42),279s(--), 288(4.07)	104-1298-76
	H_2SO_4	269(4.41),285s(--), 347(4.33),528(3.90)	104-1298-76
	CF_3COOH	268(4.45),286s(--), 353(4.33),529(3.92)	104-1298-76
Phenanthrene, 2,6-dimethyl-1-propyl-	EtOH	260(4.70),282(3.56), 292(3.87),304(4.02), 322(3.56),338(3.70), 354(3.66)	42-0812-76
$C_{19}H_{20}BrNO_4$ Scoulerine, 12-bromo-	EtOH EtOH-NaOH	235s(4.11),286(3.84) 249(4.19),302(3.99)	2-0841-76 2-0841-76
$C_{19}H_{20}Br_2N_2O$ 9(10H)-Acridinone, 2,7-dibromo-10-[2-(diethylamino)ethyl]-	EtOH	257(5.05),420(5.35)	48-0515-76
$C_{19}H_{20}ClNOS$ Dibenzo[b,f]thiepin-10(11H)-one, 8-chloro-11-[3-(dimethylamino)-propyl]-, hydrochloride	MeOH	227(4.31),242(4.32), 265s(4.07),341(3.65)	73-3420-76
$C_{19}H_{20}ClNO_2S$ 2H-Pyrrole, 4-chloro-3,4-dihydro-2,2-dimethyl-3-[(4-methylphenyl)sulfonyl]-5-phenyl-, trans	EtOH	210(4.13),230(4.29), 245(4.19),273s(3.52)	27-0453-76
$C_{19}H_{20}ClNS$ Dibenzo[b,f]thiepin-10-propanamine, 2-chloro-N,N-dimethyl-	MeOH	223(4.47),259(4.36), 280s(3.82)	73-3420-76
$C_{19}H_{20}ClN_3O_4$ Oxazolo[3,2-d][1,4]benzodiazepin-6(5H)-one, 11b-(2-chlorophenyl)-2,3,7,11b-tetrahydro-10-(hydroxyamino)-7-(methoxymethyl)-	isoPrOH	254(4.06),305s(3.34)	87-1378-76

Compound	Solvent	$\lambda_{max}(\log \epsilon)$	Ref.
$C_{19}H_{20}Cl_2N_2O$ 9(10H)-Acridinone, 2,7-dichloro-10-[2-(diethylamino)ethyl]-	EtOH	253(5.01),400(4.07), 420(4.25)	48-0515-76
$C_{19}H_{20}F_3NO_3$ 1,4-Benzoxazepine, 2,3,4,5-tetrahydro-7,8-dimethoxy-3-methyl-5-[3-(trifluoromethyl)phenyl]-	n.s.g.	215(4.28),285(3.56)	20-0898-76
$C_{19}H_{20}F_3NO_4$ Benzamide, N-[2-(3,4-dimethoxyphenoxy)-1-methylethyl]-3-(trifluoromethyl)-	n.s.g.	225(4.23),285(3.57)	20-0787-76
$C_{19}H_{20}FeN_2$ Ferrocene, [[[4-(dimethylamino)phenyl]-methylene]amino]-	n.s.g.	464(3.23)	65-2512-76
$C_{19}H_{20}N_2O$ Δ^3-14-Dehydrovincamone	n.s.g.	243(4.25),268(3.96), 296(--),304(3.62)	88-0435-76
$C_{19}H_{20}N_2O_3$ Hydrazinecarboxylic acid, [3-(4-methylphenyl)-3-oxo-1-phenylpropylidene]-, ethyl ester	EtOH	225s(4.25),281(4.39)	4-0257-76
3(2H)-Isoquinolinone, 2-(2-aminophenyl)-1-ethyl-6,7-dimethoxy-, monoperchlorate	MeOH	260(4.73),410(--)	103-0205-76
Pentanamide, 2-acetyl-4-oxo-N-phenyl-3-(phenylamino)-	MeOH	235(4.02)	83-0467-76
$C_{19}H_{20}N_2O_4$ Hydrazinecarboxylic acid, [3-(4-methoxyphenyl)-3-oxo-1-phenylpropylidene]-, ethyl ester	C_6H_{12}	223(4.33),229(4.31), 278(4.41)	4-0257-76
	EtOH	226s(4.40),280(4.42)	4-0257-76
$C_{19}H_{20}N_2O_6$ 1,4-Benzoxazepine, 2,3-dihydro-7,8-dimethoxy-5-(3-methoxy-4-nitrophenyl)-3-methyl-	n.s.g.	250(4.14),320(4.18)	20-0787-76
$C_{19}H_{20}N_2O_6S$ Propanoic acid, 2-(benzoylmethylamino)-3-[[(4-methylphenyl)sulfonyl]amino]-3-oxo-, methyl ester	MeOH	226(4.45)	44-0813-76
$C_{19}H_{20}N_4O$ 1,2,3-Benzotriazin-4-one, 6-(cyclohexylamino)-3,4-dihydro-3-phenyl-	MeOH	238(4.34),365(4.34)	5-0946-76
$C_{19}H_{20}N_4OS$ Ethanimidamide, 2-cyano-2-[formyl-[2-(phenylthio)ethyl]amino]-N-(phenylmethyl)-	EtOH	222(3.98),255(4.27), 260(4.25)	2-0346-76
1H-Imidazole-5-carboxamide, 4-[(phenylmethyl)amino]-1-[2-(phenylthio)ethyl]-	EtOH	250(4.03),290(3.95)	2-0346-76
$C_{19}H_{20}N_4O_3$ Benzo[g]pteridine-2,4(3H,10H)-dione, 6,7,8,9-tetrahydro-10-(2-methoxyethyl)-3-phenyl-	EtOH	267(4.24),320(3.15), 420(4.10)	88-2899-76

Compound	Solvent	$\lambda_{max}(\log \epsilon)$	Ref.
$C_{19}H_{20}N_4O_6$ 1,2-Diazabicyclo[5.2.0]nona-3,5-diene- 2-carboxylic acid, 5-methyl-8-[[(4- nitrophenyl)acetyl]amino]-9-oxo-, ethyl ester, cis	MeOH	272(4.21)	142-0471-76S
$C_{19}H_{20}N_4O_8$ 1H-Pyrazole-4-carboxylic acid, 5-(amino- carbonyl)-1-(4-nitrophenyl)-3-(tetra- hydro-2,2-dimethylfuro[3,4-d]-1,3-di- oxol-4-yl)-, methyl ester, [3aS- (3aα,4α,6aα)]-	CHCl$_3$	245(3.81),291(4.12)	136-0019-76A
$C_{19}H_{20}O_2$ D-Homo-C-norestra-1,3,5(10),6,8,14-hexa- en-17aα-ol, 3-methoxy-, (±)-	MeOH	214(4.13),234(4.37), 243(4.53),250(4.73), 259(4.80),281(4.07), 292(4.23),303(4.27), 327(3.11),342(3.26), 358(3.22)	24-3025-76
D-Homo-C-norestra-1,3,5(10),6,8,14-hexa- en-17aβ-ol, 3-methoxy-, (±)-	MeOH	216(4.12),234(4.35), 243(4.51),250(4.70), 259(4.77),280(4.03), 292(4.20),303(4.23), 326(3.08),341(3.23), 357(3.19)	24-3025-76
D-Homo-C-norestra-1,3,5(10),6,8-pentaen- 17a-one, 3-methoxy-, (±)-(14α)-	MeOH	232(4.80),258(3.61), 268(3.70),278(3.72), 289(3.50),313(3.06), 326(3.32),335(3.32), 341(3.41)	24-3025-76
D-Homo-C-norestra-1,3,5(10),6,8-pentaen- 17a-one, 3-methoxy-, (±)-(14β)-	MeOH	228(4.77),242(4.41), 266(3.77),277(3.76), 285(3.53),304(3.25), 310(3.14),324(3.34), 339(3.44)	24-3025-76
$C_{19}H_{20}O_3$ Benzenepentanoic acid, δ-oxo-β-phenyl-, ethyl ester	EtOH	243(4.11)	18-3239-76
1H-Benz[e]indene-3-butanoic acid, 7-methoxy-2-methyl-	MeOH	210(4.49),254(4.58), 264(4.50),288(3.75), 299(3.88),311(3.87), 335(3.15),350(3.24), 366(3.15)	24-3025-76
1,3-Cyclohexanedione, 2-[(6-methoxy- 1-naphthalenyl)methyl]-2-methyl-	MeOH	230(4.63),270(3.62), 279(3.69),290(3.61), 318(3.22),327(3.19), 332(3.32)	24-3025-76
$C_{19}H_{20}O_4$ 5-Benzofuranpropanol, 2-(3,4-dimethoxy- phenyl)-	EtOH	215(4.52),303s(4.40), 317(4.56),331(4.45)	39-1514-76C
Benzoic acid, 4,4'-methylenebis[2-meth- yl-, dimethyl ester	MeOH	237(4.31),284(3.45)	44-2509-76
2H-1-Benzopyran-2-one, 7-[(3,7-dimethyl- 5-oxo-2,6-octadienyl)oxy]-	ether	232(4.22),320(4.17)	24-1584-76
2H-Cyclopenta[b]furan-2,4(3H)-dione, 6a-(3,4-dihydro-6-methoxy-2-naphtha- lenyl)tetrahydro-3a-methyl-, (3aS-cis)-	EtOH	278(4.25)	78-0079-76

Compound	Solvent	$\lambda_{max}(\log \epsilon)$	Ref.
1,2-Naphthalenediol, 5,8-dimethyl-3-(1-methylethenyl)-, diacetate	EtOH	243(4.54),299(3.76)	102-1267-76
$C_{19}H_{20}O_5$			
4H-1-Benzopyran-4-one, 2,3-dihydro-5,7-dimethoxy-3-[(4-methoxyphenyl)methyl]-	EtOH	212(4.40),227(4.41), 284(4.28)	33-2048-76
2-Naphthaleneheptanoic acid, 6-methoxy-β-methyl-γ,ζ-dioxo-	EtOH	238(4.58),242s(4.54), 255(4.47),308(4.13)	78-0079-76
4H-Naphtho[1,2-b]pyran-4-one, 5,8,10-trimethoxy-2-propyl-	EtOH	238(4.58),275(4.53), 357(3.80)	12-1087-76
$C_{19}H_{20}O_6$			
1,3-Benzodioxole, 4-methoxy-6-[2-(3,4,5-trimethoxyphenyl)ethenyl]-, cis	EtOH	310(3.68)	102-1057-76
trans	EtOH	312(4.00),331(4.04), 346(3.91)	102-1057-76
4H-1-Benzopyran-4-one, 2,3-dihydro-3-hydroxy-5,7-dimethoxy-3-[(4-methoxyphenyl)methyl]-	EtOH	214(4.46),223(4.46), 283(4.29)	33-2048-76
15-Deoxygoyazensolide	n.s.g.	266(3.75)	102-1775-76
$C_{19}H_{20}O_7$			
Goyazensolide	n.s.g.	205(4.28),268(3.93)	102-0191-76
$C_{19}H_{21}BrN_2O_8S$			
Uridine, 5-bromo-2',3'-(1-methylethylidene)-, 5'-(4-methylbenzenesulfonate)	MeOH	220(4.27),270(4.01)	44-1100-76
$C_{19}H_{21}ClN_2O_7$			
3(2H)-Isoquinolinone, 2-(2-aminophenyl)-1-ethyl-6,7-dimethoxy-, monoperchlorate	MeOH	260(4.73),410(--)	103-0205-76
$C_{19}H_{21}N$			
Benzo[4,5]cyclohept[1,2,3-ij]isoquinoline, 1-ethyl-1,2,3,7,8,12b-hexahydro-, hydrochloride	EtOH	264(2.79),270(2.72)	4-1049-76
$C_{19}H_{21}NO$			
6H-Dibenzo[a,h]quinolizine, 5,8,9,13b-tetrahydro-2,3-dimethoxy-	EtOH	216(4.40),236s(3.84), 287(3.56)	118-0719-76
$C_{19}H_{21}NO_2$			
1,4-Benzoxazepine, 2,3-dihydro-7,8-dimethoxy-5-(4-methoxyphenyl)-3-phenyl-	n.s.g.	210(4.42),260(4.01), 317(4.28)	20-0787-76
Corytuberine, (S)-	EtOH-HCl	225(4.59),268(4.14), 276s(4.11),304(3.85)	33-2551-76
	EtOH-NaOH	228s(4.58),272(3.99), 280(3.99),318(3.93)	33-2551-76
Isoboldine	EtOH	219(4.54),273s(4.03), 281(4.11),305(4.16), 311s(4.14)	36-0294-76
	EtOH-NaOH	232(4.20),299s(3.95), 326(4.07)	36-0294-76
$C_{19}H_{21}NO_4$			
Benzene, 1,2-dimethoxy-4-[1-[(4-methoxyphenyl)methylene]propyl]-5-nitro-, (E)-	MeOH	213(4.54),254(4.44), 340(3.83)	33-2201-76
1,4-Benzoxazepine, 5-(1,3-benzodioxol-5-yl)-2,3,4,5-tetrahydro-7,8-dimethoxy-3-methyl-	n.s.g.	215(4.30),240(4.06), 285(3.91)	20-0898-76

Compound	Solvent	λ_{max} (log ϵ)	Ref.
$C_{19}H_{21}NO_6$			
1-Butanone, 2-(4,5-dimethoxy-2-nitro-phenyl)-1-(4-methoxyphenyl)-	MeOH	218(4.28),255(4.10), 283(4.25),346(3.61)	33-2201-76
$C_{19}H_{21}NO_6S$			
3-Azatetracyclo[6.1.0.0^{2,4}.0^{5,7}]nonane-7,8-dicarboxylic acid, 3-[(4-methyl-phenyl)sulfonyl]-, dimethyl ester, (1α,2α,4α,5α,7α,8α)-	MeOH	267(3.00),273(3.00)	24-3505-76
3-Azatricyclo[3.2.2.0^{2,4}]non-6-ene-6,7-dicarboxylic acid, 3-[(4-methylphen-yl)sulfonyl]-, dimethyl ester, (1α,2α,4α,5α)-	MeOH	261(3.20),267(3.08), 273(2.98)	24-3505-76
3-Azatricyclo[5.1.0.0^{2,4}]oct-5-ene-6,7-dicarboxylic acid, 5-methyl-3-[(4-methylphenyl)sulfonyl]-, dimethyl ester, (1α,2α,4α,7α)-	MeOH	227(4.30),266(2.95), 273(2.78)	24-3505-76
1H-Azonine-5,6-dicarboxylic acid, 4,7-dihydro-1-[(4-methylphenyl)sulfonyl]-, dimethyl ester	ether	230(4.15),245(3.85)	24-3505-76
$C_{19}H_{21}N_2O$			
Furo[2,3-b]pyridinium, 4-[2-[4-(dimeth-ylamino)phenyl]ethenyl]-7-ethyl-, iodide	EtOH	500(5.12)	103-0052-76
$C_{19}H_{21}N_2S$			
Thieno[2,3-b]pyridinium, 4-[2-[4-(di-methylamino)phenyl]ethenyl]-7-ethyl-, iodide	EtOH	508(5.08)	103-0052-76
$C_{19}H_{21}N_2Se$			
Selenolo[2,3-b]pyridinium, 4-[2-[4-(di-methylamino)phenyl]ethenyl]-7-ethyl-, iodide	EtOH	496(4.96)	103-0052-76
$C_{19}H_{21}N_3$			
Benzenamine, 2-[3-methyl-5-(1-methyleth-yl)-1-phenyl-1H-pyrazol-4-yl]-	MeOH	231(4.34),290(3.66)	23-1020-76
$C_{19}H_{21}N_3O_2$			
Quinoline, 6-methoxy-4-(4-methoxyphenyl-amino)-2-(dimethylamino)-	EtOH	240(4.52),315(3.99), 340s(--)	1-0133-76
$C_{19}H_{21}N_3O_3$			
Benzenecarboximidamide, N'-cyclohexyl-N-hydroxy-4-nitro-N-phenyl-	EtOH	260(4.24),366(3.38)	12-0357-76
3-Pyridinecarboxamide, 1,4,5,6-tetrahy-dro-6-hydroxy-5-(hydroxy-2-pyridinyl-methyl)-1-(phenylmethyl)-	H$_2$O MeCN	290(4.31) 285(4.16)	88-2541-76 88-2541-76
$C_{19}H_{21}N_3O_4$			
1,2-Diazabicyclo[5.2.0]nona-3,5-diene-2-carboxylic acid, 5-methyl-9-oxo-8-[(phenylacetyl)amino]-, ethyl ester	MeOH	275(3.90)	142-0471-76S
$C_{19}H_{21}N_3O_4S$			
1H-Benzimidazole-2-butanoic acid, α-[(phenylsulfonyl)amino]-, ethyl ester, hydrochloride, (S)-	H$_2$O	271(4.91),277(4.93)	42-0274-76

Compound	Solvent	$\lambda_{max}(\log \epsilon)$	Ref.
$C_{19}H_{21}N_3O_5$ 1,2-Diazabicyclo[5.2.0]nona-3,5-diene- 2-carboxylic acid, 5-methyl-9-oxo- 8-[(phenoxyacetyl)amino]-, ethyl ester, cis	MeOH	269(4.01),275(4.02)	142-0471-76S
$C_{19}H_{21}N_5O_2$ 5-Pyrimidinecarboxamide, 4-[4-(dimethyl- amino)phenyl]-1,2,3,4-tetrahydro- 6-methyl-2-oxo-N-2-pyridinyl-	EtOH	205(4.31),264(4.11), 294(4.09)	103-0191-76
$C_{19}H_{21}N_5O_6S$ 3,7,8,11,12-Pentaazatetracyclo[8.3.0- $0^{2,4}.0^{5,9}$]trideca-7,11-diene-9,10- dicarboxylic acid, 3-[(4-methylphen- yl)sulfonyl]-, dimethyl ester	MeOH	254(2.83),262(2.87), 272(2.71),322(2.61)	24-3505-76
$C_{19}H_{22}ClN$ Benzo[4,5]cyclohept[1,2,3-ij]isoquino- line, 1-ethyl-1,2,3,7,8,12b-hexahy- dro-, hydrochloride	EtOH	264(2.79),270(2.72)	4-1049-76
$C_{19}H_{22}ClN_4O_2S$ Morpholinium, 4-[[[5-(4-chlorophenyl)- 2-(4-morpholinyl)-6H-1,3-thiazin-6-yl- idene]amino]methylene]-, perchlorate	HOAc	308(4.34),459(4.37)	97-0268-76
$C_{19}H_{22}Cl_2N_4O_7S$ 9H-Purine, 2,6-dichloro-9-(3,5,6-tri- O-acetyl-2-S-ethyl-2-thio-α-D-manno- furanosyl)-	EtOH	255(3.76),274(3.97)	136-0041-76C
$C_{19}H_{22}F_4N_2O_{10}$ Uridine, 5-[(2,2,3,3-tetrafluoroprop- oxy)methyl]-, 2',3',5'-triacetate	pH 13 EtOH	263(3.82) 261(3.97)	104-0643-76 104-0643-76
$C_{19}H_{22}NO_3$ Benzo[d][1,3]dioxolo[4,5-h][2]benzoxe- pin-5-methanaminium, 5,7-dihydro- N,N,N-trimethyl-, iodide	EtOH	272(4.01),300(3.98)	94-0716-76
$C_{19}H_{22}N_2O$ 9(10H)-Acridinone, 10-[2-(diethylamino)- ethyl]-	EtOH	255(4.66),385(3.95), 400(4.01)	48-0515-76
Benzenecarboximidamide, N'-cyclohexyl- N-hydroxy-N-phenyl-	EtOH	302(4.01)	12-0357-76
Cyclohexanecarboximidamide, N-hydroxy- N,N'-diphenyl-	EtOH	270(4.01)	12-0357-76
Rhazinilam	EtOH	220s(4.31),274s(3.21), 314s(1.94)	18-2000-76
$C_{19}H_{22}N_2O_2$ 1,4-Benzenediamine, N'-[3-[2-(methoxy- methoxy)phenyl]-2-propenylidene]- N,N-dimethyl-	MeOH dioxan DMSO	335(4.02),420(4.16) 335(4.03),418(4.09) 336(4.08),422(4.03)	104-0840-76 104-0840-76 104-0840-76
2,5-Cyclohexadiene-1-carbonitrile, 1-methoxy-4-[(4-methoxy-2,6-dimethyl- phenyl)imino]-3,5-dimethyl-	MeCN	245(4.58),426(3.20)	22-1993-76
Oxayohimban-17-one, 19-methyl-	MeOH	224(4.58),283(3.86), 290(3.78)	35-3645-76

Compound	Solvent	λ_{max}(log ϵ)	Ref.
1H-Pyrazole, 3-(2,5-dimethoxyphenyl)-1-(2,6-dimethylphenyl)-4,5-dihydro-	benzene	343(4.24)	61-0288-76
1H-Pyrazole, 3-(2,6-dimethoxyphenyl)-1-(2,6-dimethylphenyl)-4,5-dihydro-	benzene	284(3.98)	61-0288-76
$C_{19}H_{22}N_2O_4$			
Acetamide, N-(4-butoxyphenyl)-2-[(4-methoxyphenyl)imino]-, N-oxide	EtOH	230(4.2),290(4.0),330(4.1)	65-0702-76
$C_{19}H_{22}N_2O_4S$			
3-Oxa-1,7-diazaspiro[4.4]nona-1,6-diene-8-carboxylic acid, 9,9-dimethyl-6-(methylthio)-4-oxo-2-(phenylmethyl)-, ethyl ester	CHCl$_3$	243(3.75)	5-2185-76
isomer b	CHCl$_3$	242(3.74)	5-2185-76
$C_{19}H_{22}N_2O_5$			
Benzo[f]cinnoline-3,4-dicarboxylic acid, 1,2-dihydro-8-methoxy-, diethyl ester	EtOH	240(4.76),267(3.86),278(3.90),327(3.20),341(3.26)	4-1333-76
1,1-Cyclobutanedicarbonitrile, 2,2,3,3-tetramethoxy-4-[2-(4-methoxyphenyl)-ethenyl]-	CHCl$_3$	274(4.38)	39-1538-76C
1H-Pyrrole-1,3-dicarboxylic acid, 4-(acetylamino)-5-(phenylmethyl)-, diethyl ester	EtOH	212(4.58)	4-0113-76
$C_{19}H_{22}N_2O_6$			
2,4-Cyclohexadiene-1-carbonitrile, 1,3,5-trimethoxy-6-[(2,4,6-tri-methoxyphenyl)imino]-	MeCN	204(4.83),247(4.21),411(3.52)	22-1993-76
$C_{19}H_{22}N_2O_7$			
Benzamide, N-[2-(3,4-dimethoxyphenoxy)-1-methylethyl]-3-methoxy-4-nitro-	n.s.g.	225(4.23),265(3.85),285(3.85)	20-0787-76
$C_{19}H_{22}N_4O_3S$			
1H-Benzimidazole-2-butanamide, N,N-di-methyl-α-[(phenylsulfonyl)amino]-, hydrochloride, (S)-	H$_2$O	238(4.75),274(4.75),281(4.71)	42-0274-76
$C_{19}H_{22}O$			
Indeno[2,1-b]pyran, 2,4-dimethyl-3-(3-methylbutyl)-	dioxan	236(4.10),306(4.02),326(3.96),507(3.15)	78-2219-76
$C_{19}H_{22}O_2$			
Androsta-1,4,6-triene-3,17-dione, (14β)-	EtOH	297(4.15)	78-1375-76
Benzene, 5-methoxy-2-[2-(4-methoxyphen-yl)-1-propenyl]-1,3-dimethyl-, (E)-	EtOH	265(4.25)	39-1466-76C
(Z)-	EtOH	268(3.94)	39-1476-76C
8αH-Estra-1,3,5(10),6-tetraen-17-one, 3-hydroxy-1-methyl-	CHCl$_3$	268(3.90)	22-1889-76
8αH-Estra-1,3,5(10),8-tetraen-17-one, 3-hydroxy-1-methyl-	CHCl$_3$	268(4.04)	22-1889-76
14βH-Estra-1,3,5(10),8-tetraen-17-one, 1-hydroxy-4-methyl-	CHCl$_3$	272(4.02)	22-1889-76
14βH-Estra-1,3,5(10),8-tetraen-17-one, 3-hydroxy-1-methyl-	CHCl$_3$	272(4.12)	22-1889-76
D-Homo-C-nor-14α-estra-1,3,5(10),6,8-pentaen-17aα-ol, 3-methoxy-, ($\pm$)-	MeOH	222(4.81),258(3.53),268(3.65),278(3.68),	24-3025-76

Compound	Solvent	$\lambda_{max}(\log \epsilon)$	Ref.
D-Homo-C-nor-14α-estra-1,3,5(10),6,8-pentaen-17aα-ol, 3-methoxy-, (±)-(cont.)		289(3.47),315(3.07), 328(3.35),338(3.36), 343(3.45)	24-3025-76
14β-	MeOH	229(4.78),236(4.70), 243(4.35),257(3.53), 266(3.65),277(3.67), 287(3.45),313(3.02), 320(3.16),325(3.32), 336(3.31),340(3.42)	24-3025-76
Spiro[2.5]octa-4,7-diene-6-carboxylic acid, 1,1,2,2-tetramethyl-6-phenyl-	EtOH	218(4.34),269s(2.58)	35-7835-76
$C_{19}H_{22}O_3$			
Acerogenin A	EtOH	278(3.38)	77-0338-76
Androsta-1,4-diene-3,12,17-trione	MeOH	244(4.17)	78-0089-76
Ethanone, 2-(4-methoxy-2,6-dimethylphenyl)-1-(4-methoxy-2-methylphenyl)-	EtOH	270(4.25)	39-1466-76C
$C_{19}H_{22}O_4$			
7H-Dibenzo[b,d]pyran-7-one, 1-acetoxy-6,8,9,10-tetrahydro-3,6,6,9-tetramethyl-	EtOH	300(4.07),350(3.81)	4-1101-76
Dipetalolactone, tetrahydro-	EtOH	211(4.55),225s(4.13), 253(3.80),262(3.85), 338(4.16)	102-0313-76
Hortiline, tetrahydro-	EtOH	228s(3.98),253s(4.01), 262(4.06),336(4.02)	32-0681-76
Hortiolone, tetrahydro-	EtOH	228(4.42),255(4.15), 276(4.34),295(4.15), 323(4.14)	32-0681-76
2-Naphthalenemethanol, 3-acetoxy-α,α,5-8-tetramethyl-, acetate	EtOH	242(3.96),286s(4.00), 297(4.03)	102-1267-76
$C_{19}H_{22}O_4S$			
Thiophenium, tetrahydro-, 3-benzoyl-1-(ethoxycarbonyl)-4-oxo-2-pentenylide	EtOH	254(4.17),344(4.33), 390(4.07)	94-2421-76
$C_{19}H_{22}O_5$			
Androst-2-ene-1,4,17-trione, 5,6-epoxy-14-hydroxy-, (5β,6α)-	EtOH	225.0(4.10)	94-1403-76
2H-1-Benzopyran, 3,4-dihydro-7-methoxy-3-(2,3,4-trimethoxyphenyl)-	EtOH	280(3.70),288(3.60)	88-1741-76
Podocarpa-8,11,13-triene-19,20-dioic acid, 7-oxo-, dimethyl ester, (±)-	EtOH	248(4.05),285(3.28)	44-1089-76
$C_{19}H_{22}O_6$			
2,5-Cyclohexadiene-1,4-dione, 2,3,5-triethoxy-6-(3-methylphenoxy)-	CHCl$_3$	301(4.54)	12-0179-76
2'-Epistrigol, (±)-	EtOH	238(4.20)	39-0410-76C
2H-Indeno[1,2-b]furan-2-one, 3-[[(2,5-dihydro-4-methyl-5-oxo-2-furanyl)oxy]-methylene]-3,3a,4,5,6,7,8,8b-octahydro-5-hydroxy-8,8-dimethyl- (strigol)	EtOH	238(4.20)	39-0410-76C
5-epimer	EtOH	237(4.19)	39-0410-76C
1,2-Naphthalenediacetic acid, 3,4-di-hydro-α¹-(2-methoxy-2-oxoethyl)-, dimethyl ester	EtOH	264(4.08)	78-0073-76
1-Propanone, 1-(2,4-dihydroxy-6-methyl-phenyl)-3-(2,4,5-trimethoxyphenyl)-	EtOH	208(4.39),223(4.26), 232(4.17),288(4.04), 322(3.58)	39-0499-76C

Compound	Solvent	λ_{max}(log ϵ)	Ref.
Vernopectolide B	EtOH	207(4.36),235s(3.91)	78-2545-76
$C_{19}H_{22}O_7$			
Butanedioic acid, [2,3-dihydro-5-meth-oxy-2-(2-methoxy-2-oxoethyl)-1H-inden-1-ylidene]-, dimethyl ester	EtOH	299(4.11),315(4.22)	78-0065-76
Ethanone, 1-[3,5-dimethoxy-4-(3,4,5-tri-methoxyphenoxy)phenyl]-	MeOH	231(4.2),274(3.86), 340(3.71)	2-0951-76
$C_{19}H_{22}S$			
1-Pentanethione, 2,2-dimethyl-1,5-di-phenyl-	C_6H_{12}	228(4.00),295(3.61), 367(2.04)	35-6218-76
$C_{19}H_{23}BrO_3$			
Androst-4-ene-3,6,17-trione, 16α-bromo-	EtOH	250(4.08)	69-4730-76
$C_{19}H_{23}ClN_2O$			
Imipramine, 3-chloro-8-hydroxy-	EtOH	222(4.23),252(3.94), 271(3.96)	4-0269-76
$C_{19}H_{23}ClO$			
2,4-Cyclopentadien-1-one, 3-(4-chloro-phenyl)-2,5-bis(1,1-dimethylethyl)-	C_6H_{12}	414(2.77)	89-0160-76
$C_{19}H_{23}NO$			
Benzenecarboximidic acid, N-[4-(1,1-di-methylethyl)phenyl]-, ethyl ester	MeOH	212(4.47),235(4.45), 265(3.77)	39-0465-76C
1-Propanone, 1-[1,1'-biphenyl]-4-yl-3-(butylamino)-	EtOH	285(4.3)	104-1968-76
$C_{19}H_{23}NO_3$			
Laurifine	MeOH	221(4.34),284(3.96)	39-2197-76C
Laurifinine (same spectrum from deuter-ated derivative)	MeOH	223(4.24),284(3.84)	39-2197-76C
$C_{19}H_{23}NO_4$			
Benzamide, N-[2-(3,4-dimethoxyphenyl)-ethyl]-2-(2-hydroxyethyl)-	EtOH	225(4.36),278(3.57)	118-0719-76
1,4-Benzoxazepine, 5-(3,4-dimethoxyphen-yl)-2,3,4,5-tetrahydro-8-methoxy-3-methyl-, hydrochloride	n.s.g.	215(4.30),235(4.30), 280(3.67)	20-0898-76
1,4-Benzoxazepine, 2,3,4,5-tetrahydro-7,8-dimethoxy-5-(4-methoxyphenyl)-3-methyl-	n.s.g.	215(4.30),230(4.26), 280(3.65)	20-0898-76
14-Episinomenine	MeOH	211(4.41),272(3.87)	88-1631-76
Indeno[1,2-j]isoquinoline-3,10-diol, 3,5,6,7,7a,8-hexahydro-2,11-dimethoxy-7-methyl-	EtOH	235s(3.94),291(3.74)	142-0235-76
Ocobotrine	MeOH	210(4.52),264(4.01)	88-1631-76
$C_{19}H_{23}NO_5$			
Benzamide, N-[2-(3,4-dimethoxyphenoxy)-1-methylethyl]-4-methoxy-	n.s.g.	210(4.36),250(4.22)	20-0787-76
$C_{19}H_{23}NO_6$			
5H-1-Pyrindine-3,4-dicarboxylic acid, 6,7-dihydro-2,6-dimethyl-5,7-dioxo-, 4-ethyl 3-pentyl ester (enol)	H_2O	336(3.66),406s(2.90)	103-0220-76
	50% EtOH	323(3.77),332(3.77), 406s(3.0)	103-0220-76
anion	H_2O	352(3.86),460(3.25)	103-0220-76
	50% EtOH	357(3.87),460(3.25)	103-0220-76

Compound	Solvent	$\lambda_{max}(\log \epsilon)$	Ref.
5H-1-Pyrindine-3,4-dicarboxylic acid, 6,7-dihydro-2,6-dimethyl-5,7-dioxo-, 4-ethyl 3-pentyl ester, protonated enol	H_2O 50% EtOH	350(3.90),406s(3.23) 351(3.91),410s(3.3)	103-0220-76 103-0220-76
$C_{19}H_{23}N_3O_2$ Benzoic acid, 4-[methyl[(1,2,3,4-tetrahydro-2-quinoxalinyl)methyl]amino]-, ethyl ester	EtOH	222(4.56),314(4.50)	35-3678-76
Benzoic acid, 4-[[(1,2,3,4-tetrahydro-1-methyl-2-quinoxalinyl)methyl]amino]-, ethyl ester	50% dioxan	312(4.39)	35-3678-76
$C_{19}H_{23}N_3O_5S$ L-Phenylalanine, 3'-ester with 4-thiothymidine	n.s.g.	201(4.37),240(3.74), 331(4.30)	128-0351-76
$C_{19}H_{23}N_3O_6$ D-Ribitol, 1-C-(3,4-dihydro-3,7,8-trimethyl-2,4-dioxopyrimido[4,5-b]quinolin-10(2H)-yl)-	EtOH	227(4.44),264(4.49), 333(3.98),403(4.00)	39-1805-76C
	50% EtOH-HCl	225(4.33),262(4.50), 355(4.23)	39-1805-76C
$C_{19}H_{23}N_3O_7$ 1H-Benzotriazole, 5,6-dimethyl-1-(2,3,5-tri-O-acetyl-α-D-ribofuranosyl)-	MeOH	266(3.81),280s(3.68)	22-1983-76
β-	MeOH	265(3.81),279s(3.66)	22-1983-76
2H-Benzotriazole, 5,6-dimethyl-2-(2,3,5-tri-O-acetyl-β-D-ribofuranosyl)-	MeOH	284(4.10),293s(4.07)	22-1983-76
$C_{19}H_{23}N_4O_2S$ Morpholinium, 4-[[[2-(4-morpholinyl)-5-phenyl-6H-1,3-thiazin-6-ylidene]-amino]methylene]-, perchlorate	HOAc	306(4.07),479(4.20)	97-0268-76
$C_{19}H_{23}N_5O_2$ Benzo[g]pteridinium, 5-[3-(dimethylamino)-2-propenylidene]-1,2,3,4,5,10-hexahydro-3,7,8,10-tetramethyl-2,4-dioxo-, hydroxide, inner salt	pH 2 pH 7	375(4.42) 390(4.38)	69-0114-76 69-0114-76
$C_{19}H_{24}NO_3$ Oblongine	EtOH	224s(4.47),284(3.78)	105-0111-76
$C_{19}H_{24}N_2O$ Cinchonamine	MeOH	284(3.93)	33-2268-76
Rhazinine	EtOH	227(4.54),281(3.88), 289(3.86)	18-2000-76
	EtOH-HCl	221(4.62),274(3.92), 284(3.90)	18-2000-76
$C_{19}H_{24}N_2O_2$ Isoretuline, N_a-deacetyl-18-hydroxy-	n.s.g.	244(3.85),296(3.40)	102-0321-76
Pentanoic acid, 4-cyano-2-(1-cyano-1-methylethyl)-2,4-dimethyl-, 4-methylphenyl ester	dioxan	266(2.73),273(2.67)	126-3481-76
Pentanoic acid, 4-cyano-2-(2-cyano-2-methylpropyl)-4-methyl-, 4-methylphenyl ester	dioxan	266(2.80),272(2.73)	126-3481-76

Compound	Solvent	λ_{max}(log ϵ)	Ref.
$C_{19}H_{24}N_2O_5$			
Benzo[f]cinnoline-3,4-dicarboxylic acid, 2,4a,5,6-tetrahydro-8-methoxy-, diethyl ester	EtOH	264(4.19)	4-1333-76
2-Butenoic acid, 2-[[3-ethoxy-1-oxo-2-[(phenylacetyl)amino]-2-propenyl]-amino]-3-methyl-, methyl ester	MeOH	242(4.30)	88-0375-76
1H-Pyrrole-1,3-dicarboxylic acid, 4-(acetylamino)-2,5-dihydro-5-(phenylmethyl)-, diethyl ester	EtOH	275(4.22)	4-0113-76
$C_{19}H_{24}N_2O_7S_2$			
Uridine, 5,6-dihydro-2',3'-O-(1-methyl-ethylidene)-2-thio-, 5'-p-toluenesul-fonate	n.s.g.	233(4.21),275(4.13)	33-2972-76
$C_{19}H_{24}N_2O_{11}$			
1H-Pyrazole-3-carboxylic acid, 5-(acet-oxymethyl)-1-(2,3,5-tri-O-acetyl-β-D-ribofuranosyl)-, methyl ester	MeOH	220(4.00)	65-1378-76
1H-Pyrazole-3-carboxylic acid, 1-(2,3,4,6-tetra-O-acetyl-α-D-glucopyranosyl)-, methyl ester	MeOH	220(3.82)	65-1370-76
β-	MeOH	216(4.05)	65-1370-76
1H-Pyrazole-5-carboxylic acid, 3-(acet-oxymethyl)-1-(2,3,5-tri-O-acetyl-α-D-ribofuranosyl)-, methyl ester	MeOH	227(4.03)	65-1378-76
β-	MeOH	227(4.01)	65-1378-76
1H-Pyrazole-5-carboxylic acid, 1-(2,3,4,6-tetra-O-acetyl-α-D-glucopyranosyl)-, methyl ester	MeOH	227(4.03)	65-1370-76
β-	MeOH	227(3.91)	65-1370-76
$C_{19}H_{24}N_4O$			
1-Phenanthrenepropanenitrile, 2-azido-1,2,3,4,4a,9,10,10a-octahydro-7-meth-oxy-2-methyl-, [1S-(1α,2β,4aβ,10aα)]-	EtOH	275(3.25)	2-0618-76
$C_{19}H_{24}N_4O_{10}$			
2,4(1H,3H)-Pteridinedione, 5-acetyl-5,6,7,8-tetrahydro-1-(2,3,5-tri-O-acetyl-β-D-ribofuranosyl-	pH 7.0	245(3.73),286(4.18)	24-3184-76
	pH 12.0	257(3.85),286(4.09)	24-3184-76
	MeOH	240(3.76),286(4.15)	24-3184-76
$C_{19}H_{24}O$			
2,4-Cyclopentadien-1-one, 2,5-bis(1,1-dimethylethyl)-3-phenyl-	C_6H_{12}	414(2.72)	89-0160-76
Gona-1,3,5(10),8-tetraene, 3-methoxy-17β-methyl-, (13ξ,14ξ)-	MeOH	273(4.20)	94-0181-76
$C_{19}H_{24}O_2$			
Androsta-4,6-diene-3,17-dione, (14β)-	CHCl$_3$	282(4.23)	78-1375-76
C,18-Dinorpregna-4(13),17-diene-3,20-dione	EtOH	247(4.30)	18-0499-76
Estra-4,6-diene-3,17-dione, 7β-methyl-, (14β)-	n.s.g.	292(4.41)	22-1240-76
D-Homo-C,18-dinorandrosta-4,13(17a)-di-ene-3,11-dione, 17a-methyl-	EtOH	247(4.11)	138-0975-76
$C_{19}H_{24}O_3$			
Androsta-1,4-diene-3,17-dione, 12α-hy-droxy-	MeOH	244(4.20)	78-0089-76

Compound	Solvent	λ_{max}(log ϵ)	Ref.
Androsta-1,4-diene-3,17-dione, 12β-hydroxy-	MeOH	244(4.28)	78-0089-76
Androsta-1,4-dien-3-one, 6β,7β-epoxy-17β-hydroxy-	EtOH	249(4.27)	56-0097-76
Cyclobuta[b]naphthalene-3,8-dione, 1-hexyl-1,2,2a,8a-tetrahydro-8a-methoxy-, (1α,2aα,8aα)-	MeCN	300(3.18),308(3.15), 349(2.23)	18-2596-76
	CCl$_4$	300(3.15),309(3.11), 352(2.11)	18-2596-76
C$_{19}$H$_{24}$O$_4$			
Cyclopentaneacetic acid, 1-methyl-2-oxo-5-(1,2,3,4-tetrahydro-6-methoxy-2-naphthalenyl)-, (1S-trans)-	EtOH	278(3.36),287(3.33)	78-0079-76
C$_{19}$H$_{24}$O$_5$			
1(2H)-Naphthalenone, 2-acetoxy-3-(1-acetoxy-1-methylethyl)-3,4-dihydro-5,8-dimethyl-, (2R-trans)-	EtOH	259(4.29),311(3.67)	102-1267-76
Podolide	EtOH	218(4.11)	142-0595-76
C$_{19}$H$_{24}$O$_6$			
1,6-Cyclohexanecarboxylic acid, 2-benzoyl-, diethyl ester, c-2-c-6-	n.s.g.	232(4.07)	128-0341-76
c-2-t-6-	n.s.g.	232(4.07)	128-0341-76
t-2-c-6-	n.s.g.	232(4.06)	128-0341-76
t-2-t-6-	n.s.g.	232(4.09)	128-0341-76
4a,1-(Epoxymethano)-4aH-fluorene-9-carboxylic acid, 1,2,3,4,4b,5,6,8a,9,9a-decahydro-2-hydroxy-1,7-dimethyl-11-oxo-8a-(2-oxoethyl)-, [1S-(1α,2β,4aα,4bβ,8aβ,9β,9aβ)]-	n.s.g.	274(2.68)	88-3989-76
Pectorolide	EtOH	217(4.33)	78-2545-76
Propanedioic acid, [(3-methoxyphenyl)methyl](2-oxo-3-butenyl)-, diethyl ester	EtOH	215(4.18)	35-4577-76
Tagitinin C	n.s.g.	212(4.21),248(4.04)	36-0918-76
Tagitinin F	n.s.g.	215(4.09)	36-0918-76
C$_{19}$H$_{24}$O$_8$S			
Cyclopenta[c]pyran-4-carboxylic acid, 1,4aα,5,6,7,7aα-hexahydro-6α-hydroxy-1α-methoxy-7α-[[[(4-methylphenyl)sulfonyl]oxy]methyl]-	MeOH	227(4.29),237s(--)	24-3626-76
Cyclopenta[c]pyran-4-carboxylic acid, 1,4aα,5,6,7,7aα-hexahydro-6β-hydroxy-1α-methoxy-7α-[[[(4-methylphenyl)sulfonyl]oxy]methyl]-	MeOH	228(4.29),237s(--)	24-3626-76
C$_{19}$H$_{25}$BrO$_2$			
Androst-4-ene-3,17-dione, 6α-bromo-	EtOH	237.5(4.13)	69-4730-76
Androst-4-ene-3,17-dione, 6β-bromo-	EtOH	247(4.16)	69-4730-76
C$_{19}$H$_{25}$ClN$_3$O$_3$			
Morpholinium, 4-[3-chloro-2-(4-methoxyphenyl)-3-[(4-morpholinylmethylene)-amino]-2-propenylidene]-, perchlorate	HOAc	409(4.39)	73-1565-76
C$_{19}$H$_{25}$NO			
4H-Cyclohepta[b]pyridin-4-one, 1,2,3,5-6,7,8,9-octahydro-1-(1-methyl-2-phenylethyl)-, (S)-	heptane	310(4.26)	103-0428-76
	MeOH	332(4.22)	103-0428-76
	CF3COOH	315(4.11)	103-0428-76

Compound	Solvent	$\lambda_{max}(\log \epsilon)$	Ref.
$C_{19}H_{25}NO_3$			
2-Azaestra-3,5(10)-diene-1,17-dione, 3-methoxy-2-methyl-	MeOH	235(3.72),305(4.00)	44-2864-76
2-Azaestra-1(10),4-dien-3-one, 17-acetoxy-, (17β)-	MeOH	231(3.90),306(3.74)	44-2864-76
2-Azaestra-1,3,5(10)-trien-17-one, 1,3-dimethoxy-	MeOH	230(3.94),281(3.85)	44-2864-76
$C_{19}H_{25}NO_5$			
1,6-Cyclohexanedicarboxylic acid, 2-(benzoylamino)-, diethyl ester, c-2-c-6-	n.s.g.	228(4.08)	128-0341-76
c-2-t-6-	n.s.g.	228(4.07)	128-0341-76
t-2-c-6-	n.s.g.	228(4.06)	128-0341-76
t-2-t-6-	n.s.g.	228(4.10)	128-0341-76
$C_{19}H_{25}N_3O$			
Methanone, [3-[(dimethylamino)methyl]-1-methyl-1H-indol-2-yl](1,2,3,6-tetrahydro-1-methyl-4-pyridinyl)-	EtOH	243s(3.89),313(3.84), 350s(3.60)	77-0818-76
$C_{19}H_{25}N_4O_3P$			
Diazenecarboxylic acid, [bis(ethylphenylamino)phosphinyl]-, ethyl ester	MeOH	208(4.23),229(3.97), 280(3.28),470(1.72)	139-0207-76A
$C_{19}H_{25}N_5NaO_8PS$			
Adenosine, 8-(methylthio)-N-(1-oxobutyl)-, cyclic 3',5'-(hydrogen phosphate) 2'-butanoate, sodium salt	pH 7	296(4.31)	87-0899-76
$C_{19}H_{25}N_5O_7S$			
Adenosine, N,N-dimethyl-8-(methylthio)-, 2',3',5'-triacetate	pH 1	290(4.33)	94-0565-76
	pH 13	294(4.31)	94-0565-76
	EtOH	231(4.24),292(4.29)	94-0565-76
9H-Purin-6-amine, 9-(3,5,6-tri-O-acetyl-2-S-ethyl-2-thio-α-D-mannofuranosyl)-	EtOH	260(4.16)	136-0041-76C
β-	EtOH	260(4.17)	136-0041-76C
$C_{19}H_{26}N_2O_4$			
2,6-Piperidinedione, 4-[2-(4,4-dimethyl-2,6-dioxocyclohexylidene)-2-[(1-pyrrolidinyl)ethyl]-	EtOH	295(3.70),374(4.21)	30-0139-76
$C_{19}H_{26}N_2O_5$			
Benzimidazole, 1-(1-deoxy-3,4:5,6-di-O-isopropylidene-D-mannitol-1-yl)-	EtOH	248(3.85),253(3.85), 265(3.64),274(3.68), 281(3.70)	136-0049-76G
$C_{19}H_{26}N_2O_8S_2$			
2,4(1H,3H)-Pyrimidinedione, 1-(3,5-di-O-acetyl-6-S-acetyl-2-S-ethyl-2,6-dithio-β-D-mannofuranosyl)-	EtOH	266(3.99)	136-0041-76C
$C_{19}H_{26}N_2O_9S$			
2,4(1H,3H)-Pyrimidinedione, 5-methyl-1-(3,5,6-tri-O-acetyl-2-S-ethyl-2-thio-β-D-mannofuranosyl)-	EtOH	265(3.95)	136-0041-76C
$C_{19}H_{26}N_2O_{10}$			
Uridine, 2'-O-(tetrahydro-4-methoxy-2H-pyran-4-yl)-, 3',5'-diacetate	EtOH	262(4.03)	54-0108-76

Compound	Solvent	$\lambda_{max}(\log \epsilon)$	Ref.
$C_{19}H_{26}N_5O_8PS$			
Adenosine, 8-(methylthio)-N-(1-oxobut-yl)-, cyclic 3',5'-(hydrogen phosphate) 2'-butanoate, sodium salt	pH 7	296(4.31)	87-0899-76
$C_{19}H_{26}O$			
Spiro[cyclopentane-1,2'(1'H)-naphthalen]-1'-one, 3',4'-dihydro-3,7'-dimethyl-2-propyl-	EtOH	234(3.85),250(4.06), 300(3.30)	42-0812-76
$C_{19}H_{26}O_2$			
Androst-4-ene-3,17-dione	n.s.g.	242(4.26)	22-1240-76
Estr-4-ene-3,17-dione, 7α-methyl-	n.s.g.	242(4.24)	22-1240-76
$C_{19}H_{26}O_3$			
8,14-Secogona-1,3,5(10),9-tetraene-14α,17α-diol, 3-methoxy-13β-methyl-	EtOH	265(4.28)	114-0081-76D
14α,17β-	EtOH	265(4.30)	114-0081-76D
14β,17β-	EtOH	265(4.32)	114-0081-76D
$C_{19}H_{26}O_4$			
Androst-1-ene-3,11-dione, 2,17-dihydroxy-	EtOH	270.5(4.20)	95-0863-76
Androst-3-ene-2,11-dione, 3,17-dihydroxy-	EtOH	269(3.92)	95-0863-76
Androst-4-ene-3,11-dione, 4,17-dihydroxy-	EtOH	277.5(4.07)	95-0863-76
1,4-Ethanonaphthalene-6,10(4H)-dione, 1,4a,5,8a-tetrahydro-5-hydroxy-9-methoxy-1,2,5,7,8,9-hexamethyl-	EtOH	208(3.82),252(3.91), 305(2.86)	1-0064-76
$C_{19}H_{26}O_5$			
1H-Fluorene-1,9-dicarboxylic acid, 4,4b,5,6,7,8,8a,9-octahydro-2-methoxy-1-methyl-, dimethyl ester	EtOH	210(3.60)(end abs.)	44-2401-76
$C_{19}H_{26}O_6$			
Nagilactone E	EtOH	219(4.11)	142-0595-76
Orientin	EtOH	201(4.27)	105-0346-76
Pectorolide, dihydro-	EtOH	212(4.00)	78-2545-76
Tagitinin E	n.s.g.	213(3.98)	36-0918-76
$C_{19}H_{26}O_7$			
Tagitinin B	EtOH	214(4.11)	2-0077-76
	n.s.g.	214(4.11)	36-0918-76
$C_{19}H_{26}O_{11}$			
Geniposide, 10-acetyl-	MeOH	239(4.06)	94-2644-76
$C_{19}H_{27}BrN_2O_6S_2$			
Tyrindoxyl sulfate I, β,β-dimethylacrylylcholine III salt	EtOH	227(4.71),304(4.24)	88-1233-76
$C_{19}H_{27}DO$			
Androst-4-en-3-one-6β-d	dioxan	236(4.18)	20-0333-76
$C_{19}H_{27}NO$			
2H-Azeto[2,1-a]isoquinolin-2-one, 1,1-bis(1,1-dimethylethyl)-1,4,5,9b-tetrahydro-	hexane	270(0.6),277(0.6)	24-0906-76

Compound	Solvent	λ_{max}(log ϵ)	Ref.
$C_{19}H_{27}NO_4S$			
Acetamide, N-[2-(acetyloxy)-5-oxo-1-(tricyclo[3.3.1.1^{3,7}]dec-1-yl-thio)-3-pentenyl]-, [R*,S*-(E)]-	MeOH	222(4.07)	4-0321-76
2,5-Pyridinediol, 1-acetyl-1,2,5,6-tet-rahydro-6-(tricyclo[3.3.1.1^{3,7}]dec-1-ylthio)-, 5-acetate, (2α,5β,6α)-	MeOH	213(3.62)	4-0321-76
$C_{19}H_{27}NS$			
2H-Azeto[2,1-a]isoquinoline-2-thione, 1,1-bis(1,1-dimethylethyl)-1,4,5,9b-tetrahydro-	hexane	245s(--),275(4.2), 331(1.9)	24-0906-76
	EtOH	273(4.2),305s(--)	24-0906-76
2-Pentanethione, 1-(3,4-dihydro-1(2H)-isoquinolinylidene)-4,4-dimethyl-3-(1-methylethyl)-	hexane	260(4.0),275s(--), 426(4.3)	24-0906-76
$C_{19}H_{27}N_3O_5$			
2-Decenoic acid, 2,3,3a,9a-tetrahydro-2-(hydroxymethyl)-6-imino-6H-furo-[2',3':4,5]oxazolo[3,2-a]pyrimi-din-3-yl ester, hydrochloride, [2R-(2α,3β,3aβ,9aβ)]-	MeOH-acid	212(4.31),263(4.05)	87-0654-76
$C_{19}H_{27}N_5OS$			
1H-Imidazo[1,2-a][1,3,5]benzotriazepine-8-carboxamide, N,3-dibutyl-2,3,5,6-tetrahydro-5-thioxo-	EtOH	220(4.43),279(4.63), 316(4.30),389s(3.20)	78-0057-76
$C_{19}H_{27}N_5O_2$			
1H-Imidazo[1,2-a][1,3,5]benzotriazepine-8-carboxamide, N,3-dibutyl-2,3,5,6-tetrahydro-5-oxo-	EtOH	242(4.36),261(4.42), 320(3.40)	78-0057-76
$C_{19}H_{28}NO_3$			
Pyrrolidinium, 3-[(cyclopentylhydroxy-phenylacetyl)oxy]-1,1-dimethyl-, bromide	pH 1	251(--),258(2.3), 264(--),268s(--)	106-0363-76
	H_2O	251(--),258(2.3), 264(--),268s(--)	106-0363-76
	MeOH	252(--),258(2.3), 264(--),268s(--)	106-0363-76
$C_{19}H_{28}N_2O$			
Cyclohexanecarboximidamide, N'-cyclo-hexyl-N-hydroxy-N-phenyl-	EtOH	245(3.93),276(3.69)	12-0357-76
2(1H)-Quinolinone, 4-(decylamino)-	MeOH	223(4.65),298(4.11)	111-0555-76
$C_{19}H_{28}N_2O_3$			
1H-Indole-3-acetamide, N-(5,5-diethoxy-pentyl)-	EtOH	273(3.81),280(3.84), 289(3.78)	130-0283-76
$C_{19}H_{28}N_4$			
3-Aza-A-homo-4α-androsteno[3,4-d]tetra-zole	EtOH	243(4.22)	2-0618-76
$C_{19}H_{28}N_4O_7$			
D-Mannitol, 1-deoxy-3,4:5,6-bis-O-(1-methylethylidene)-1-(1,2,3,6-tetra-hydro-1,3-dimethyl-2,6-dioxo-7H-purin-7-yl)-	EtOH	273(3.97)	136-0049-76G

Compound	Solvent	$\lambda_{max}(\log \epsilon)$	Ref.
$C_{19}H_{28}O_2$ Estr-4-en-3-one, 17-hydroxy-12-methyl-, $(8\alpha,9\beta,10\alpha,12\alpha,14\beta,17\alpha)$-	n.s.g.	241(4.18)	39-1643-76C
$C_{19}H_{28}O_4$ 1,3-Benzenediol, 5-nonyl-, diacetate	MeOH	267(3.14)	12-1989-76
$C_{19}H_{28}O_5$ 2H-4a,1-(Epoxymethano)naphthalene-3,4- diol, 1-[2-(3-furanyl)-2-hydroxyeth- yl]octahydro-2,5-dimethyl-	n.s.g.	210.5(3.88)	105-0675-76
$C_{19}H_{28}O_6$ Tagitinin A, dihydro-, acetic anhydride product	EtOH	232(3.96)	2-0259-76
Tagitinin D	n.s.g.	218(4.28)	36-0918-76
$C_{19}H_{28}O_7$ Tagitinin A	EtOH n.s.g.	215(4.06) 215(4.06)	2-0259-76 36-0918-76
$C_{19}H_{29}NO$ 17a-Aza-D-homoandrosta-5,7-dien-3β-ol	EtOH	258(3.85),268(3.95), 278(3.95),288(3.70)	13-0225-76A
$C_{19}H_{29}NS$ 2-Azetidinethione, 3-(1,1-dimethyleth- yl)-1,3-bis(1-methylethyl)-4-phenyl-	hexane	247(4.0),271(4.1), 340s(--)	24-0906-76
$C_{19}H_{29}N_3O_5$ Decanoic acid, 2,3,9,9a-tetrahydro- 2-(hydroxymethyl)-6-imino-6H-furo- [2',3':4,5]oxazolo[3,2-a]pyrimidin- 3-yl ester, hydrochloride, [2R- $(2\alpha,3\beta,3a\beta,9a\beta)$]-	MeOH-acid	232(3.97),263(4.03)	87-0654-76
$C_{19}H_{29}N_3O_6$ 2(1H)-Pyrimidinone, 4-amino-1-[3-O-(1- oxo-2-decenyl)-β-D-arabinofuranosyl)-, hydrochloride	MeOH-acid	211(4.33),282 4.17)	87-0667-76
$C_{19}H_{30}$ Phenanthrene, 1,2,3,4,7,8,8a,9,10,10a- decahydro-3,3,8,8a,10a-pentamethyl-	hexane	244(3.67)	44-2670-76
$C_{19}H_{30}N_2O_2$ 1H-Indole-3-ethanamine, N-(5,5-diethoxy- pentyl)-	EtOH	274(3.77),281(3.79), 289(3.74)	130-0283-76
$C_{19}H_{30}N_4$ Pyrido[3,2-e]-1,2,4-triazine, 3-tridec- yl-	EtOH	234(4.31),341(3.47)	114-0285-76C
$C_{19}H_{30}O$ 18-Norcembra-2,7,11-trien-4-one	EtOH	230(3.85)	105-0167-76
$C_{19}H_{30}O_4$ 2,5-Cyclohexadiene-1,4-dione, 2,5-di- hydroxy-3-tridecyl- (rapanone)	MeOH	292(4.32)	12-1979-76

Compound	Solvent	$\lambda_{max}(\log \epsilon)$	Ref.
$C_{19}H_{31}BrO$			
2,5-Cyclohexadien-1-one, 4-(bromometh-yl)-2,4,6-tris(1,1-dimethylethyl)-	hexane	240(4.023)	70-1780-76
$C_{19}H_{31}NO$			
Bicyclo[10.2.2]hexadeca-12,14,15-trien-2-ol, 3-[(1-methylethyl)amino]-, [S-(R*,S*)]-, hydrochloride	MeOH	196(4.54),198(4.56), 215(3.80),223(3.87), 227s(3.80),260(2.27), 266(2.29),275(2.09)	94-1724-76
$C_{19}H_{31}N_3O_6$			
2(1H)-Pyrimidinone, 4-amino-1-[3-O-(1-oxodecyl)-β-D-arabinofuranosyl]-	MeOH-acid	213(3.98),284(4.14)	87-0667-76
2(1H)-Pyrimidinone, 4-amino-1-[5-O-(1-oxodecyl)-β-D-arabinofuranosyl]-	MeOH-acid	213(4.00),285(4.13)	87-0667-76
$C_{19}H_{31}N_6O_6P$			
Adenosine, N-(9-aminononyl)-, cyclic 2',5'-(hydrogen phosphate)	pH 1	263(4.21)	69-3724-76
	pH 11	269(4.21)	69-3724-76
$C_{19}H_{32}N_4O_3$			
Tetradecanoic acid, 2-(3-nitro-4-pyri-dinyl)hydrazide	pH 1	225(4.23),259(4.06), 327(3.57)	114-0285-76C
$C_{19}H_{32}OS$			
1-Pentadecanone, 1-(2-thienyl)-	heptane	258(3.98),275(3.88)	34-0233-76
$C_{19}H_{32}O_2S$			
2-Thiophenecarboxylic acid, 4-methyl-5-tridecyl-	EtOH	253(3.94),283(3.98)	34-0233-76
2-Thiophenecarboxylic acid, 5-tridecyl-, methyl ester	heptane	254(3.98),275(4.06)	34-0233-76
$C_{19}H_{32}O_3$			
2-Cyclopenten-1-one, 3-acetoxy-2,4,5-tris(1,1-dimethylethyl)-	EtOH	235(4.14)	118-0604-76
Oxirane-2-carboxaldehyde, 3-methyl-3-[7-(1-methylethyl)-4-methylundec-3-en-1-yl]-, (E)-	n.s.g.	228(4.22)	105-0278-76
$C_{19}H_{33}Cl_2N_{11}O_8$			
Tuberactinamine N	pH 1	268(4.43)	88-2343-76
	H_2O	268(4.43)	88-2343-76
	pH 13	285(4.23)	88-2343-76
$C_{19}H_{33}NOS$			
Dodecanesulfinamide, N-(phenylmethyl)-	EtOH	211(4.26),241(4.04)	118-0339-76
$C_{19}H_{34}S$			
Thiophene, 2-pentadecyl-	heptane	234(3.87)	34-0233-76
$C_{19}H_{35}NO_2S_2$			
2-Pyridinol, 1-acetyl-4-(1,1-dimethyl-ethyl)-3,6-bis[(1,1-dimethylethyl)-thio]-1,2,3,6-tetrahydro-, (2α,3β,6α)-	EtOH	212(4.26)	4-0861-76
$C_{19}H_{36}OS$			
Thiirane, [(1-hexadecenyloxy)methyl]-, cis	hexane	259(2.21)	104-0964-76
trans	hexane	258(1.72)	104-0964-76

Compound	Solvent	$\lambda_{max}(\log \epsilon)$	Ref.
$C_{19}H_{36}O_2$			
Oxirane, [(1-hexadecenyloxy)methyl]-, cis	hexane	258(2.48)	104-0964-76
trans	hexane	258(1.68)	104-0964-76
$C_{19}H_{38}O_3$			
1,2-Propanediol, 3-(1-hexadecenyloxy)-	hexane	258(1.88)	104-0964-76

Compound	Solvent	λ_{max} (log ϵ)	Ref.
$C_{20}H_9N_5$			
1H-Indole-3-acetonitrile, 2-cyano-α-	MeCN	513(4.3)	103-1343-76
[4-(dicyanomethylene)-2,5-cyclohexa-dien-1-ylidene]-	MeCN-base	725(4.4),775(4.6)	103-1343-76
$C_{20}H_{10}N_2S$			
Propanedinitrile, (2-phenyl-5H-naphtho-[1,8-bc]thien-5-ylidene)-	MeOH	250(4.02),312(3.95), 485(4.36)	88-4713-76
$C_{20}H_{10}O_2S$			
Triphenyleno[2,3-b]thiophene-9,13-dione	EtOH	250(4.66),297(4.10), 307(4.11),325s(3.86), 426(3.49)	118-0675-76
$C_{20}H_{10}O_2SSe$			
Naphtho[1,8-bc]thiopyran-3(2H)-one, 2-(3-oxobenzo[b]selenophene-2(3H)-ylidene)-, cis	n.s.g.	484(3.92)	104-1804-76
trans	n.s.g.	608(4.20)	104-1804-76
$C_{20}H_{10}O_2S_2$			
Naphtho[1,8-bc]thiopyran-3(2H)-one, 2-(3-oxobenzo[b]thien-2(3H)-yli-dene)-, cis	n.s.g.	480(4.22)	104-1804-76
trans	n.s.g.	592(4.51)	104-1804-76
$C_{20}H_{10}O_6$			
Benzo[j]fluoranthene-3,8-dione, 2,4,7,9-tetrahydroxy- (bulgarhodin)	H_2SO_4	258(4.45),324(3.97), 348(3.91),410(4.03), 450s(3.80),670(4.00)	39-2149-76C
	$CHCl_3$	254(--),294(--), 312(--),338s(--), 359(--),402(--), 514s(--),542(--), 670(--)	39-2149-76C
$C_{20}H_{11}F_{15}N_2$			
Benzenamine, N-[[4-(dimethylamino)phen-yl]methylene]-2,3,4,5,6-pentakis(tri-fluoromethyl)-	EtOH	360(4.31)	104-0467-76
$C_{20}H_{11}NO$			
Azuleno[1,2-f]azulene-1-carboxaldehyde, 5-cyano-	CH_2Cl_2	258(4.39),275(4.37), 284(4.36),316s(4.25), 326(4.30),368s(4.30), 375(4.33),388(4.45), 436s(4.54),455(4.70), 567s(3.52),660s(2.86)	118-0673-76
$C_{20}H_{11}NO_2S$			
Triphenyleno[2,3-b]thiophen-13(9)-one, 9(13)-(hydroxyimino)-	EtOH	251(4.64),257s(4.63), 295(4.09),307(4.05), 333(4.02),386(3.74)	118-0675-76
$C_{20}H_{11}N_5$			
1,4-Benzenediacetonitrile, α,α'-dicyano-α-1H-indol-3-yl-	EtOH	320(4.5)	103-1343-76
$C_{20}H_{12}$			
Azuleno[1,2,3-cd]phenalene	C_6H_{12}	265(4.3),351(4.8),	138-0543-76

Compound	Solvent	$\lambda_{max}(\log \epsilon)$	Ref.
Azuleno[1,2,3-cd]phenalene (cont.)		390(4.0),422(4.0), 449(4.4),475(4.7)	138-0543-76
	MeOH	680s(2.2),750(2.3), 840(2.3),950s(1.7)	138-0543-76
Benzo[j]fluoranthene	EtOH	245(4.64),283(4.26), 295(4.37),310(4.56), 334(4.25),351(3.78), 367(4.04),377(4.11), 386(4.16),414(3.25), 438(3.00)	39-2149-76C
Benzo[k]fluoranthene	EtOH	249(4.74),269(4.39), 285(4.41),298(4.68), 309(4.83),325(4.03), 338(3.91),363(3.93), 382(4.13),404(4.15)	39-2149-76C

$C_{20}H_{12}B_2OS_4$
| 4H-Borepino[3,2-b:6,7-b']dithiophene, 4,4'-oxybis- | C_6H_{12} | 240(4.8),260(4.5), 400f(4.8) | 46-0287-76 |

$C_{20}H_{12}Br_2O_2$
| Benzo[b]benzo[3,4]cyclobuta[1,2-e][1,4]-dioxin, 7,8-dibromo-4b,10a-dihydro-4b-phenyl-, cis | ether | 258(3.36),265(3.42), 272(3.40),295(3.58) | 22-0493-76 |

$C_{20}H_{12}Br_2O_3$
| Naphtho[2,3-c]furan-1,3-dione, 7-bromo-4-(3-bromo-4-methylphenyl)-6-methyl- | C_6H_{12} | 269(4.84),314s(3.82), 324(3.86) | 42-1053-76 |

$C_{20}H_{12}ClNO_5$
| 2H-Pyrano[3,2-c]quinoline-3-carboxylic acid, 9-chloro-5,6-dihydro-2,5-dioxo-6-phenyl- | n.s.g. | 375(4.30) | 64-0514-76B |

$C_{20}H_{12}F_2O_5$
| Naphtho[2,3-c]furan-1,3-dione, 5-fluoro-4-(3-fluoro-4-methoxyphenyl)-6-meth-oxy- | C_6H_{12} | 273(4.68) | 42-1053-76 |
| Naphtho[2,3-c]furan-1,3-dione, 7-fluoro-4-(3-fluoro-4-methoxyphenyl)-6-meth-oxy- | C_6H_{12} | 271(4.74),318s(3.96) | 42-1053-76 |

$C_{20}H_{12}I_2O_3$
| Naphtho[2,3-c]furan-1,3-dione, 5-iodo-4-(3-iodo-4-methylphenyl)-6-methyl- | C_6H_{12} | 254s(4.34),274s(4.62), 281(4.69),308s(3.96) | 42-1053-76 |
| Naphtho[2,3-c]furan-1,3-dione, 7-iodo-4-(3-iodo-4-methylphenyl)-6-methyl- | C_6H_{12} | 270(4.67),333(3.86) | 42-1053-76 |

$C_{20}H_{12}N_6O_{10}$
| 1,3-Benzenedicarboxamide, 4,6-dinitro-N,N'-bis(2-nitrophenyl)- | H_2SO_4 | 231(4.7505) | 116-0626-76 |

$C_{20}H_{12}O$
Benzo[a]pyren-1-ol (only most intense log ε given in these spectra)	MeOH	258(--),266(4.64), 277(--),298(--), 398(--)	44-0977-76
	90% MeOH-NaOH	262(4.61),310(--), 425(--)	44-0977-76
Benzo[a]pyren-2-ol	MeOH	287(4.67),301s(--), 382(--)	44-0977-76

Compound	Solvent	$\lambda_{max}(\log \epsilon)$	Ref.
Benzo[a]pyren-2-ol (cont.)	90% MeOH-NaOH	264(4.77),302(--), 387(--)	44-0977-76
Benzo[a]pyren-3-ol	MeOH	258(4.67),307(--), 380(--)	44-0977-76
	90% MeOH-NaOH	237(4.68),314(--), 394(--)	44-0977-76
Benzo[a]pyren-4-ol	MeOH	267(4.84),299(--), 375(--)	44-0977-76
	90% MeOH-NaOH	271(5.02),311(--), 390(--)	44-0977-76
Benzo[a]pyren-5-ol	MeOH	262(4.70),302(--), 381(--)	44-0977-76
	90% MeOH-NaOH	257(4.82),317(--), 405(--)	44-0977-76
Benzo[a]pyren-6-ol	MeOH	256(--),303(4.70), 391(--)	44-0977-76
	90% MeOH-NaOH	252(4.39),313(--), 433(--)	44-0977-76
Benzo[a]pyren-7-ol	MeOH	268(--),304(4.71), 399(--)	44-0977-76
	90% MeOH-NaOH	251(4.73),318(--), 438(--)	44-0977-76
Benzo[a]pyren-8-ol	MeOH	279(4.75),306s(--), 384(--)	44-0977-76
	90% MeOH-NaOH	264(--),296(4.82), 386(--)(changing)	44-0977-76
Benzo[a]pyren-9-ol	MeOH	250(--),267(4.75), 302(--),379(--)	44-0977-76
	90% MeOH-NaOH	269(4.75),292(--), 396(--)	44-0977-76
Benzo[a]pyren-10-ol	MeOH	256(4.57),302(--), 379(--)	44-0977-76
	90% MeOH-NaOH	251(4.61),325(4.61), 402(--)	44-0977-76
Benzo[a]pyren-11-ol	MeOH	268(4.80),301(--), 382(--)	44-0977-76
	90% MeOH-NaOH	276(4.80),308s(--), 395(--)	44-0977-76
Benzo[a]pyren-12-ol	MeOH	255(--),295(4.65), 380(--)	44-0977-76
	90% MeOH-NaOH	260(4.62),308(--), 418(--)	44-0977-76
Dibenzo[d,j]oxacyclotridecin, 14,15,16,17-tetradehydro-, (E,E)-	EtOH	237(4.49),258s(4.19), 277(4.30),311(3.78), 352(3.86),353(3.83)	18-3709-76
$C_{20}H_{12}O_2S$ Triphenyleno[2,3-b]thiophene, 10,11-dioxide	EtOH	244(4.51),279(4.57), 285(4.57),308(3.81), 320(3.80),337s(3.66), 353(3.48),372(3.30)	118-0675-76
$C_{20}H_{12}O_2S_2$ 1,3-Benzoxathiol-1-ium, 2,2'-(1,4-phenylene)bis-, bis(tetrafluoroborate)	96% H_2SO_4	278(4.10),434(4.55), 450s(4.47)	39-0323-76C
$C_{20}H_{12}O_3$ [2]Benzopyrano[4,3-b]naphtho[1,2-e]pyran-5-one, 5,7-dihydro-	EtOH	225(4.28),332(3.70)	22-1967-76

Compound	Solvent	$\lambda_{max}(\log \epsilon)$	Ref.
$C_{20}H_{12}O_6$			
Benzo[1,2-b:4,5-b']bisbenzofuran-6,12-dione, 2,8-dimethoxy-	$CHCl_3$	259(4.59),287s(3.79), 348(3.79),475(3.74)	12-0179-76
Benzo[1,2-b:4,5-b']bisbenzofuran-6,12-dione, 3,9-dimethoxy-	dioxan	262(4.00),292(3.82), 402(3.43)	12-0179-76
Benzo[1,2-b:5,4-b']bisbenzofuran-6,12-dione, 3,9-dimethoxy-	$CHCl_3$	262(4.18),375(3.46), 558(3.28)	12-0179-76
$C_{20}H_{12}O_8S$			
2,5-Thiophenedicarboxylic acid, 3,4-di-1,3-benzodioxol-5-yl-	EtOH	237s(4.32),288(4.22)	42-0490-76
$C_{20}H_{12}S$			
Triphenyleno[2,3-b]thiophene	EtOH	227(4.23),248s(4.51), 269(4.82),279(4.95), 302s(4.05),309(4.04), 325(3.81),342(3.16), 350(2.84),358(2.97)	118-0675-76
$C_{20}H_{13}Br$			
Naphthalene, 1-bromo-2-(4-phenyl-1-buten-3-ynyl)-, cis	MeOH	239(4.58),282s(4.18), 295(4.26),323(4.44), 345s(4.21)	39-1104-76B
trans	MeOH	235(4.32),256(4.17), 281(4.47),291(4.50), 330(4.64),350s(4.45)	39-1104-76B
$C_{20}H_{13}Cl_2P$			
Phosphine, bis(4-chlorophenyl)(phenylethynyl)-	MeOH	204(4.54),241(4.34), 262(4.31),286(3.94)	139-0169-76A
$C_{20}H_{13}Cl_4N_3O$			
1H-1,2,3-Triazole, 1-[4-(2,4-dichlorophenoxy)phenyl]-5-(2,4-dichlorophenyl)-4,5-dihydro-	dioxan	287(4.15)	97-0102-76
$C_{20}H_{13}F_2P$			
Phosphine, bis(4-fluorophenyl)(phenylethynyl)-	MeOH	250(4.10),263(4.04)	139-0169-76A
$C_{20}H_{13}I$			
Phenanthrene, 2-iodo-1-phenyl-	MeOH	280s(4.52),289s(4.40), 298s(4.16),329(2.47), 337(2.54),345(2.50)	39-1115-76B
Phenanthrene, 3-iodo-4-phenyl-	MeOH	280s(4.24),293(4.20), 305(4.26),337(2.76), 345(2.53),354(2.73)	39-1115-76B
$C_{20}H_{13}NO_2S$			
Thiophene, 2-(2-naphthalenyl)-5-(4-nitrophenyl)-	MeOH	280(4.20),292s(4.19), 305s(4.13),328(4.04), 398(4.49)	48-0731-76
$C_{20}H_{13}N_3$			
1,3-Benzenedicarbonitrile, 2-amino-4,6-diphenyl-	EtOH	240(4.55),253(4.41), 280(4.01),360(4.07)	104-0688-76
$C_{20}H_{13}N_3O_4$			
3-Pyridinecarbonitrile, 2-[2-(4-nitrophenyl)-2-oxoethoxy]-6-phenyl-	EtOH	275(4.42),320(4.32)	48-0313-76

Compound	Solvent	λ_{max} (log ϵ)	Ref.
$C_{20}H_{13}N_5$			
5H-Pyridazino[4,5-b]indole, 1,4-di-2-py-ridinyl-	MeOH	270s(4.45),286(4.52), 358(3.85)	18-1725-76
$C_{20}H_{13}N_5O_3$			
Acetamide, N-[2-(1H-benzotriazol-1-yl)-2,3-dihydro-1,3-dioxo-1H-benz[de]-isoquinolin-5-yl]-	EtOH	208(4.37),249(4.63), 340(4.10),370(3.80)	56-1257-76
Acetamide, N-[2-(1H-benzotriazol-1-yl)-2,3-dihydro-1,3-dioxo-1H-benz[de]-isoquinolin-6-yl]-	EtOH	210(4.60),238(4.53), 280(4.02),372(4.20)	56-1257-76
Acetamide, N-[2-(2H-benzotriazol-2-yl)-2,3-dihydro-1,3-dioxo-1H-benz[de]-isoquinolin-5-yl]-	EtOH	217(4.57),232(4.56), 335(4.11),375(3.95)	56-1257-76
Acetamide, N-[2-(2H-benzotriazol-2-yl)-2,3-dihydro-1,3-dioxo-1H-benz[de]-isoquinolin-6-yl]-	EtOH	212(4.65),233(4.40), 278(4.26),370(4.20)	56-1257-76
$C_{20}H_{14}$			
9H-Fluorene, 9-(phenylmethylene)-	dioxan	260(4.48),326(4.17)	39-0048-76C
Naphthalene, 1-(4-phenyl-1-buten-3-yn-yl)-, cis	MeOH	228(4.54),270(4.22), 279(4.29),328(4.20)	39-1104-76B
trans	MeOH	233(4.52),274(4.10), 281(4.16),331(4.46)	39-1104-76B
Naphthalene, 1-(4-phenyl-3-buten-1-yn-yl)-, trans	CH_2Cl_2	268s(3.94),277(4.07), 287(4.16),325s(4.41), 340(4.51),363(4.41)	39-1104-76B
Naphthalene, 2-(4-phenyl-1-buten-3-yn-yl)-, cis	MeOH	230s(4.51),236(4.61), 275(4.24),287(4.31), 295s(4.13),325(4.40), 340(4.27),356s(3.90)	39-1104-76B
trans	MeOH	232s(4.39),236(4.43), 247(4.36),256s(4.40), 265s(4.51),275(4.60), 286(4.62),317s(4.63), 325(4.67),345(4.55), 357s(4.05)	39-1104-76B
Phenanthrene, 1-phenyl-	MeOH	275s(4.27),284s(4.14), 298(4.16),333(2.56), 341(2.35),350(2.43)	39-1115-76B
Phenanthrene, 3-phenyl-	EtOH	261(4.73)	2-0038-76
Phenanthrene, 4-phenyl-	MeOH	275s(4.36),282s(4.27), 296(4.13),334(2.62), 349(2.58)	39-1115-76B
$C_{20}H_{14}BrNO_2$			
Pyridinium, 1-(4-bromophenyl)-3-phenyl-1,3-dioxo-2-propylide	EtOH	231(4.31),317(4.07)	78-2647-76
	dioxan	320(4.94),414(3.32)	78-2647-76
	MeCN	316(3.99),390(3.29)	78-2647-76
$C_{20}H_{14}BrN_3$			
Pyrido[2,3-b]pyrazine, 7-bromo-6-methyl-2,3-diphenyl-	EtOH	358(4.26)	22-0251-76
$C_{20}H_{14}BrN_5O_2S$			
Isothiazolo[5,4-b]pyridine, 3-bromo-4,7-dihydro-7-methyl-4-[[(4-nitro-phenyl)azo]methylene]-6-phenyl-	DMF	571(4.33)	48-0779-76

Compound	Solvent	$\lambda_{max}(\log \epsilon)$	Ref.
$C_{20}H_{14}ClN_2O_2$ Pyridinium, 1-[(1-amino-4-chloro-9,10-dihydro-9,10-dioxo-2-anthracenyl)-methyl]-, chloride	MeOH	462(3.87)	2-0054-76
$C_{20}H_{14}Cl_2N_4O_3$ 1H-1,2,3-Triazole, 1-[3-chloro-4-(2-chlorophenoxy)phenyl]-4,5-dihydro-5-(4-nitrophenyl)-	dioxan	283(4.29)	97-0102-76
$C_{20}H_{14}Cl_3N_3O$ 1H-1,2,3-Triazole, 1-[3-chloro-4-(2,4-dichlorophenoxy)phenyl]-4,5-dihydro-5-phenyl-	dioxan	292(4.17)	97-0102-76
$C_{20}H_{14}NO_3$ Pyridinium, 1-[(9,10-dihydro-1-hydroxy-9,10-dioxo-2-anthracenyl)methyl]-, chloride	MeOH	410(3.86)	2-0054-76
$C_{20}H_{14}NO_4$ Pyridinium, 1-[(9,10-dihydro-1,4-di-hydroxy-9,10-dioxo-2-anthracenyl)-methyl]-, chloride	MeOH	490(4.01)	2-0054-76
$C_{20}H_{14}N_2O_2$ Methanone, (3-amino-6-phenylfuro[2,3-b]-pyridin-2-yl)phenyl-	EtOH	271(4.28),338(4.37), 398(4.27)	48-0313-76
5H-Oxazolo[4,5-b]phenoxazine, 2-(2-meth-ylphenyl)-	benzene	375(4.21)	23-0037-76
5H-Oxazolo[4,5-b]phenoxazine, 2-(3-meth-ylphenyl)-	benzene	374(4.32)	23-0037-76
5H-Oxazolo[4,5-b]phenoxazine, 2-(4-meth-ylphenyl)-	benzene	378(4.18)	23-0037-76
Phthalazine, 1,4-diphenoxy-	dioxan	275(4.53)	103-0342-76
3-Pyridinecarbonitrile, 2-(2-oxo-2-phen-ylethoxy)-6-phenyl-	EtOH	251(4.34),320(4.33)	48-0313-76
$C_{20}H_{14}N_2O_3$ 5H-Oxazolo[4,5-b]phenoxazine, 2-(4-meth-oxyphenyl)-	benzene	376(4.25)	23-0037-76
$C_{20}H_{14}N_2O_3S$ 1H-Thieno[3,4-d]imidazole-4-carboxylic acid, 2-benzoyl-6-phenyl-, methyl ester	MeOH	311(4.48),400(4.07)	49-0299-76
$C_{20}H_{14}N_2O_4$ 4H-1-Benzopyran-4-one, 7-acetoxy-3-(1-phenyl-1H-pyrazol-4-yl)-	EtOH	270(4.42)	103-0914-76
$C_{20}H_{14}N_4$ 3H-Indole, 3-azido-2,3-diphenyl-	EtOH	220(4.19),249(4.13), 318(4.05)	88-2347-76
Pyridine, tri-2-pyridinyl-, ferrous complex	MeOH	570(4.40)	118-0001-76
$C_{20}H_{14}N_4O_4$ 4-Oxazolecarboxamide, 2-phenyl-, 2-[(2-phenyl-4-oxazolyl)carbonyl]hydrazide	EtOH	285(4.27)	39-0315-76C

Compound	Solvent	$\lambda_{max}(\log \epsilon)$	Ref.
$C_{20}H_{14}N_4O_5S$ Benzenesulfonamide, 4-[2-(4-nitrophenyl)-4-oxo-3(4H)-quinazolinyl]-	EtOH-NaOH	208(4.34),248(4.06), 285s(4.02)	106-0574-76
$C_{20}H_{14}N_6O_6$ 3,11-Diazatricyclo[5.3.1.1^{2,6}]dodeca-4,8-diene-10,12-dione, 3,11-bis(5-nitro-2-pyridinyl)-, (1α,2β,6β,7α)-	EtOH	210(4.10),227(4.16), 250(4.00),354(4.20)	39-2296-76C
$C_{20}H_{14}O$ Benzofuran, 2,3-diphenyl-	C_6H_{12}	237(4.35),305(4.36)	18-2560-76
Indeno[1,2-b]naphtho[1,2-e]pyran, 12,13-dihydro-	EtOH	218(4.58),242(4.63), 270s(--)	22-1967-76
Indeno[2,1-b]naphtho[2,1-e]pyran, 5,6-dihydro-	dioxan	234(4.23),271(4.25), 291(3.99),370(3.98), 536(3.28)	78-2219-76
$C_{20}H_{14}OS$ Benzothiophene, 6-hydroxy-2,3-diphenyl-	C_6H_{12}	307(4.19)	118-0451-76
$C_{20}H_{14}O_2$ Benzo[b]benzo[3,4]cyclobuta[1,2-e][1,4]-dioxin, 4b,10a-dihydro-4b-phenyl-	ether	259(3.50),265(3.49), 272(3.56)	22-0493-76
Indeno[2,1-b]pyran-9-carboxaldehyde, 4-methyl-2-phenyl-	dioxan	238(4.31),266(4.50), 328(4.26),340(4.29), 378(4.29),500(3.64)	78-2225-76
Methanone, 1,3-phenylenebis[phenyl-	EtOH	249(4.5),330(2.4)	47-0319-76
Methanone, 1,4-phenylenebis[phenyl-	EtOH	264(4.5),333(2.5)	47-0319-76
$C_{20}H_{14}O_3S$ 2-Propen-1-one, 3-(benzoyloxy)-3-phenyl-1-(2-thienyl)-	n.s.g.	314(4.22)	39-0380-76C
$C_{20}H_{14}O_4$ 5,12-Naphthacenedione, 6,11-dimethoxy-	MeOH	249(4.63),281s(4.30), 293(4.27),403(4.85)	44-2296-76
2-Propen-1-one, 3-(benzoyloxy)-1-(2-furanyl)-3-phenyl-	n.s.g.	324(4.13)	39-0380-76C
$C_{20}H_{14}O_6$ 4a,9a-[2]Butenoanthracene-1,4,9,10-tetrone, 11-acetoxy-	MeOH	228(4.55),310(3.34)	44-2296-76
5,12-Naphthacenedione, 7-acetoxy-7,10-dihydro-6,11-dihydroxy-	MeOH	251(4.57),256(4.56), 281s(3.98)	44-2296-76
5,6,11,12-Naphthacenetetrone, 1-acetoxy-1,4,4a,12a-tetrahydro-	MeOH	217(4.24),239(4.25), 257(4.24),278s(4.05), 348(3.62)	44-2296-76
$C_{20}H_{14}O_7$ Dehydroaverufine	MeOH	223(4.46),256(4.20), 270(4.27),292(4.47), 321(3.98),455(4.04)	20-0161-76
4H-Furo[2,3-h]-1-benzopyran-4-one, 3-methoxy-2-(7-methoxy-1,3-benzodioxol-5-yl)-	MeOH	245(3.98),335(3.83)	2-0229-76
$C_{20}H_{14}O_8S$ 2-Propenoic acid, 2,2'-thiobis[3-(1,3-benzodioxol-5-yl)-	EtOH	317(4.15)	42-0038-76

Compound	Solvent	$\lambda_{max}(\log \epsilon)$	Ref.
$C_{20}H_{14}O_{10}$ 6H-[1,3]Benzodioxolo[5,6-c][1]benzo-pyran-6-one, 2,3,7-triacetoxy-	EtOH	256(4.66),309(4.13), 317s(--),343s(--)	24-0855-76
$C_{20}H_{14}S$ Triphenyleno[2,3-b]thiophene, 9,13-di-hydro-	EtOH	225(4.36),246(4.67), 254(4.70),277(4.10), 284(3.96),296(3.93), 316(2.66),323(2.58), 331(2.62),338(2.39), 346(2.51)	118-0675-76
$C_{20}H_{14}S_2$ 2,2'-Bithiophene, 3,3'-diphenyl-	isooctane	232(4.34),251(4.32), 300(3.82)	44-1320-76
$C_{20}H_{15}BrN_2OS$ Benzeneethanehydrazonothioic acid, 4-bromo-α-oxo-N-phenyl-, phenyl ester	EtOH	242(4.221),270(4.171), 372(4.185)	18-0321-76
$C_{20}H_{15}ClN_2O$ Ethanone, 1-(7-chloro-2-methyl-9-phenyl-1H-pyrrolo[3,2-b]quinolin-3-yl)-	EtOH	209(4.44),248(4.80), 280(4.35),319s(3.88), 333(3.08),355(4.00), 369(3.95)	44-1743-76
$C_{20}H_{15}ClN_2OS$ Benzeneethanehydrazonothioic acid, N-(4-chlorophenyl)-α-oxo-, phenyl ester	EtOH	250(4.54),290(3.30), 370(3.43)	18-0321-76
$C_{20}H_{15}ClN_2O_5$ 5-Camptothecinol, 7-chloro-	MeOH	223(4.38),249(4.25), 257(4.25),298(3.75), 345s(3.99),361(4.04)	24-1389-76
$C_{20}H_{15}ClN_4O_3$ 1H-1,2,3-Triazole, 1-(3-chloro-4-phen-oxyphenyl)-4,5-dihydro-5-(4-nitro-phenyl)-	dioxan	281(4.34)	97-0102-76
$C_{20}H_{15}ClO_4$ 1,4-Naphthalenedione, 2-[2-(3-acetoxy-phenyl)ethyl]-3-chloro-	MeOH	246(4.17),252(4.20), 274(4.13),341(3.45)	39-0997-76C
$C_{20}H_{15}ClO_8$ 2,3-Anthracenedicarboxylic acid, 5-chlo-ro-9,10-dihydro-8-hydroxy-1-methoxy-6-methyl-9,10-dioxo-, dimethyl ester	EtOH	247(4.45),282s(3.96), 345(3.55),420(3.84)	44-3018-76
$C_{20}H_{15}Cl_2NO_8S$ Benzoic acid, 4-[5-[2-(dichloromethyl)-sulfonyl]-2-(5-nitro-2-furanyl)ethen-yl]-2-furanyl]-, ethyl ester	EtOH	210s(4.01),256(4.29), 326(4.38),402(4.27)	73-3391-76
$C_{20}H_{15}F$ Benz[a]anthracene, 6-fluoro-7,12-dimeth-yl-	EtOH	267(4.4),279s(--), 288(4.6),298(4.6), 333s(--),350(3.6), 369(3.7),385(3.6)	24-1038-76

Compound	Solvent	λ_{max}(log ϵ)	Ref.
C$_{20}$H$_{15}$NO			
9(10H)-Acridinone, 10-(phenylmethyl)-	EtOH	255(4.74),380(3.98), 400(4.08)	48-0515-76
2-Propen-1-one, 1,2-diphenyl-3-(2-pyridinyl)-, (Z)-	MeOH	256(4.34),303(4.28)	44-2536-76
2-Propen-1-one, 1,3-diphenyl-2-(2-pyridinyl)-, (E)-	MeOH	268(4.13),297(4.20)	44-2536-76
(Z)-	MeOH	258(4.33),297(4.33)	44-2536-76
C$_{20}$H$_{15}$NO$_2$			
1H-Benz[f]isoindol-1-one, 2-acetyl-2,3-dihydro-4-phenyl-	dioxan	250(4.85),282s(3.81), 293(4.02),304(4.04), 333(3.59),347(3.69)	44-2571-76
1H-Benz[f]isoindol-1-one, 2-acetyl-2,3-dihydro-9-phenyl-	EtOH	249(4.84),293(4.00), 304(4.00),338s(3.58), 347(3.63)	44-2571-76
Pyridinium dibenzoylmethylide	benzene	322(3.95),429(3.39)	78-2647-76
	H$_2$O	227(4.20),317(4.09)	78-2647-76
	aq acid	255(4.25)	78-2647-76
	EtOH	316(4.09)	78-2647-76
	EtOH-HCl	257(4.34)	78-2647-76
	dioxan	247(4.07),320(3.97), 420(3.32)	78-2647-76
	DMF	317(3.98),390(3.28)	78-2647-76
	MeCN	246(4.10),313(3.95), 390(3.25)	78-2647-76
	CHCl$_3$	317(4.03),402(3.32)	78-2647-76
C$_{20}$H$_{15}$NO$_2$S			
Benzoxazole, 2-[5-[2-(4-methoxyphenyl)-ethenyl]-2-thienyl]-	CH$_2$Cl$_2$	249(4.08),259s(4.06), 309s(4.00),352s(4.26), 371(4.30),460s(2.75)	48-0731-76
Thiophene, 2-(4-nitrophenyl)-5-(4-phenyl-1,3-butadienyl)-	MeOH	250s(4.06),260s(4.08), 269(4.07),348(4.25), 425(4.52)	48-0731-76
C$_{20}$H$_{15}$NO$_3$			
1H-Benz[f]isoindole-1,3(2H)-dione, 4-(2-hydroxyphenyl)-2,9-dimethyl-	ether	262(4.65),295(3.92), 344(3.56),362(3.66)	22-0493-76
Benzoic acid, 2-[(3-benzoylphenyl)-amino]-, β-cyclodextrin complex	pH 7.0	255(4.27),290(4.23), 320(3.90)	133-0343-76
C$_{20}$H$_{15}$NO$_4$			
Sanguinarine, dihydro-	EtOH	237(4.44),285(4.44), 322(4.09)	95-0527-76
C$_{20}$H$_{15}$N$_2$O$_2$			
Pyridinium, 1-[(1-amino-9,10-dihydro-9,10-dioxo-2-anthracenyl)methyl]-, chloride	MeOH	460(3.95)	2-0054-76
C$_{20}$H$_{15}$N$_3$			
Propanedinitrile, [5-phenyl-3-(phenyl-amino)-2,4-pentadienylidene]-	CH$_2$Cl$_2$	431(4.32)	48-0705-76
3-Pyridinecarbonitrile, 1,2-dihydro-2-imino-1-phenyl-6-(2-phenylethen-yl)-, monoperchlorate	HOAc	384(4.19)	48-0705-76
Pyrido[2,3-b]pyrazine, 6-methyl-2,3-di-phenyl-	EtOH	350(4.25)	22-0251-76

Compound	Solvent	$\lambda_{max}(\log \epsilon)$	Ref.
$C_{20}H_{15}N_3O_5$ Benzamide, N-(2-nitrophenyl)-N-[(4-nitrophenyl)methyl]-	MeOH	211(4.35),263(4.44)	39-0394-76C
$C_{20}H_{15}N_3O_6$ Carbamic acid, (5-nitro-1,3-phenylene)-bis-, diphenyl ester	MeCN	233(4.73),247(4.47), 303(3.54),350(3.23)	87-0426-76
$C_{20}H_{15}N_3Se$ 1H-1,2,4-Triazolium, 2,3-dihydro-1,4,5-triphenyl-3-selenoxo-, hydroxide, inner salt	MeCN	204s(4.71),238(4.34), 268s(4.18),348(3.23)	70-1162-76
$C_{20}H_{15}N_5O$ Pyridinium, 2-oxo-2-phenyl-1-(1-phenyl-1H-tetrazol-5-yl)ethylide	EtOH	240(4.13),320(3.80), 419(3.71)	39-0909-76C
	dioxan	333(3.75),454(3.97)	39-0909-76C
Pyridinium, 2-oxo-2-phenyl-1-(2-phenyl-2H-tetrazol-5-yl)ethylide	EtOH	243(--),330(--), 428(--)	39-0909-76C
	dioxan	362(3.89),478(3.74)	39-0909-76C
$C_{20}H_{16}$ Benzene, 1,1',1"-(1-ethenyl-2-ylidene)-tris-	MeOH	228(4.25),297(4.34)	23-3038-76
Tribenzo[a,c,f]cyclooctene, 9,14-dihydro-	C_6H_{12}	237s(3.95),265s(3.17), 273(3.05)	35-8268-76
$C_{20}H_{16}BrN_3$ Pyrido[2,3-b]pyrazine, 7-bromo-4,6-dihydro-6-methyl-2,3-diphenyl-	EtOH	318(3.93),395(4.08)	22-0251-76
$C_{20}H_{16}BrN_3O_8$ 1-Azuleneethanol, 2-bromo-, acetate, trinitrobenzene complex	CH_2Cl_2	285(4.81),295(4.82), 334(3.69),349(3.78), 361(3.17),566(2.59), 605(2.55),660(2.21)	44-1811-76
1-Azuleneethanol, 3-bromo-, acetate, trinitrobenzene complex	CH_2Cl_2	285(4.78),295(4.73), 340(3.74),348(3.84), 364(3.80),609(2.58)	44-1811-76
$C_{20}H_{16}ClN_3O_4$ 3-Pyridinecarbonitrile, 1,2-dihydro-2-imino-1-phenyl-6-(2-phenylethenyl)-, monoperchlorate	HOAc	384(4.19)	48-0705-76
$C_{20}H_{16}Cl_2N_2O_2$ Spiro[2H-anthra[1,2-d]imidazole-2,1'-cyclohexane]-6,11-dione, 4,5-dichloro-1,3-dihydro-	EtOH	534(4.09)	104-0172-76
$C_{20}H_{16}Cl_2O$ Cyclohexanone, 2,6-bis[(4-chlorophenyl)methyl]-	EtOH	238(4.28),332(4.56)	114-0381-76B
	ether	234(4.30),325(4.01)	114-0381-76B
	$CHCl_3$	334(4.54)	114-0381-76B
	CCl_4	330(4.52)	114-0381-76B
$C_{20}H_{16}Cl_2O_2$ Benzofuran, 5-chloro-2-[(5-chloro-2-methyl-3-benzofuranyl)methyl]-2,3-dihydro-2-methyl-3-methylene-	MeOH	285(4.58),293(3.61), 326(3.81),339(3.83)	87-1214-76

Compound	Solvent	$\lambda_{max}(\log \epsilon)$	Ref.
Benzofuran, 3,3'-(1,2-ethanediyl)bis-[5-chloro-2-methyl-	MeOH	287(3.87),294(3.86)	87-1214-76
$C_{20}H_{16}Cl_2Si$ Silane, dichloro(1,2-diphenylethenyl)-phenyl-	n.s.g.	232(4.51),267(4.50)	87-1214-76
$C_{20}H_{16}Fe_2$ Biferrocenylene	$CHCl_3$	315(3.26),465(2.65)	35-3181-76
$C_{20}H_{16}NO_4$ Corysamine chloride	EtOH	230(4.41),240s(4.34),268(4.37),344s(4.28),352(4.29),455(3.72)	73-3157-76
$C_{20}H_{16}N_2$ Cyclohept[b]indol-6-amine, N-(methyl-phenyl)-	MeOH	281(4.20),304(4.36),473(4.20)	18-1101-76
Indolo[3,2-b]carbazole, 5,11-dihydro-5,11-dimethyl-	MeCN	218(4.51),244(4.43),252(4.43),262(4.41),280(4.72),307s(4.02),315s(4.20),322(4.46),333(4.49),338(4.77),376s(3.39),392(3.63),416(3.76)	5-1090-76
$C_{20}H_{16}N_2O$ 1-Isoquinolinecarbonitrile, 2-benzoyl-1,2-dihydro-1-(2-propenyl)-	EtOH	228(4.48),287s(4.00),297(4.00),321s(4.00)	116-0010-76
$C_{20}H_{16}N_2OS$ Benzeneethanehydrazonothioic acid, α-oxo-N-phenyl-, phenyl ester	EtOH	243(4.306),275(4.057),370(4.278)	18-0321-76
$C_{20}H_{16}N_2O_2$ 2,2'-Bipyridine, 4,4'-bis(2-furanyl)-6,6'-dimethyl-, Cu(I) complex	EtOH	498(4.23)	118-0001-76
Isoxazolo[5,4-b]pyridine, 6-(4-methoxy-phenyl)-5-methyl-3-phenyl-	EtOH	236(4.19),321(4.15)	18-3339-76
Methanone, (3,6-dimethyl-2,5-pyrazine-diyl)bis[phenyl-	EtOH	256(4.37),290(4.32)	18-2805-76
Pyridinium, benzoyl(phenylaminocarbo-nyl)methylide	dioxan	252(4.16),298(4.09),440(3.46)	78-2647-76
	DMF	295(4.14),415(3.45)	78-2647-76
	MeCN	249(4.13),293(4.09),412(4.34)	78-2647-76
$C_{20}H_{16}N_2O_2S$ 1H-Thieno[3,4-d]imidazole-4-carboxylic acid, 6-phenyl-2-(phenylmethyl)-, methyl ester	MeOH	238(4.28),305(4.01),320(4.05),357(4.42)	49-0299-76
$C_{20}H_{16}N_2O_3$ 1H-Indole-2-carboxylic acid, 1-methyl-, anhydride	EtOH	314(4.356)	103-0669-76
$C_{20}H_{16}N_2O_3S$ 1H-Thieno[3,4-d]imidazole-4-carboxylic acid, 2-(hydroxyphenylmethyl)-6-phen-yl-, methyl ester	MeOH	239(4.27),319(4.02),306(3.99),357(4.39)	49-0299-76

Compound	Solvent	$\lambda_{max}(\log \epsilon)$	Ref.
$C_{20}H_{16}N_2O_4$ 1H-Pyrano[3',4':6,7]indolizino[1,2-b]- quinoline-3,14(4H,12H)-dione, 4-ethyl- 4-hydroxy-, (S)- (camptothecin)	EtOH	219(4.58),253(4.46), 290(3.70),370(4.28)	98-1085-76
$C_{20}H_{16}N_2O_4S$ Thiazolo[5,4-f]quinoline-8-carboxylic acid, 6-ethyl-6,9-dihydro-9-oxo- 2-(phenylmethoxy)-	MeOH	260(4.47),269(4.47)	94-0147-76
Thiazolo[5,4-f]quinoline-8-carboxylic acid, 6-ethyl-2,3,6,9-tetrahydro- 2,9-dioxo-3-(phenylmethyl)-	MeOH	263.5(4.38)	94-0147-76
$C_{20}H_{16}N_2O_5$ 2,3-Benzodiazocine-4,5-dicarboxylic acid, 1,2-dihydro-1-oxo-2-phenyl-, dimethyl ester	EtOH	210(4.36),245(4.15), 285s(3.84),350(3.00)	39-2281-76C
Cyclohexanone, 2,6-bis[(3-nitrophenyl)- methylene]-	EtOH	225(4.32),278(4.26), 315(4.44)	114-0381-76B
	ether	220(--),272(--), 310(--)	114-0381-76B
	CHCl$_3$	275(4.30),314(4.49)	114-0381-76B
	CCl$_4$	270(4.34),313(4.59)	114-0381-76B
Indeno[1,2-c]pyrazole-3,3a(1H)-dicarb- oxylic acid, 4,8b-dihydro-4-oxo- 1-phenyl-, dimethyl ester	EtOH	218(4.51),240(4.08), 243(4.30),290(3.76)	39-2281-76C
$C_{20}H_{16}N_3O_2$ Pyridinium, 1-[1,4-diamino-9,10-dihydro- 9,10-dioxo-2-anthracenyl)methyl]-, chloride	MeOH	565(3.96),607(4.06)	2-0054-76
$C_{20}H_{16}N_3S_3$ Benzothiazolium, 3-ethyl-2-[(2-phenyl- 5H-thiazolo[4,3-b]-1,3,4-thiadiazol- 5-ylidene)methyl]-, iodide	EtOH	478(2.55)	103-0650-76
$C_{20}H_{16}N_4O$ 1-Propanone, 2-(1-phenyl-1H-pyrazolo- [3,4-d]pyrimidin-4-yl)-1-phenyl-	EtOH	258(4.07),382(4.50)	95-1352-76
2(1H)-Pyrimidinone, 4-methyl-5-phenyl- 6-(4-quinazolinylmethyl)-	EtOH	400(4.38),420(4.63), 442(4.66)	4-0383-76
Tribenzo[c,g,k][1,2,5,10]tetraazacyclo- dodecin-12-ol, 11,12-dihydro-	benzene	316(3.98),432(3.80), 470s(3.69)	33-2906-76
$C_{20}H_{16}N_4O_2$ 2,4(1H,3H)-Pteridinedione, 1,3-dimethyl- 6,7-diphenyl-	MeOH	227(4.32),274(4.10), 362(4.07)	24-3194-76
$C_{20}H_{16}N_4O_3$ 1H-1,2,3-Triazole, 4,5-dihydro-5-(4-ni- trophenyl)-1-(4-phenoxyphenyl)-	dioxan	258(4.20),279(4.29)	97-0102-76
$C_{20}H_{16}N_4O_4S$ Thiazolo[3,2-a]benzimidazole-3-acetic acid, α-[(4-carboxyphenyl)hydraz- ono]-, 3-ethyl ester, hydrobromide	n.s.g.	250(4.27),290(4.19), 370(4.41)	124-1162-76
$C_{20}H_{16}N_6$ 1,3-Benzenediamine, 4,6-bis(1H-benzimid- azol-2-yl)-	H$_2$SO$_4$	305(4.5198)	116-0626-76

Compound	Solvent	$\lambda_{max}(\log \epsilon)$	Ref.
$C_{20}H_{16}N_6O$			
Pyridinium, 2-oxo-2-(phenylamino)-1-(1-phenyl-1H-tetrazol-5-yl)ethylide	EtOH	232(4.21),277(4.13), 314(4.32),469(3.55)	39-0909-76C
Pyridinium, 2-oxo-2-(phenylamino)-1-(2-phenyl-2H-tetrazol-5-yl)ethylide	EtOH	241(4.39),288(4.36), 375(3.72),486(3.51)	39-0909-76C
$C_{20}H_{16}N_6O_2$			
1,2,4-Triazolo[4,3-b][1,2,4]triazin-3-amine, 6,7-diphenyl-, diacetyl deriv.	EtOH	229(4.41),340(3.84)	39-1492-76C
1,2,4-Triazolo[5,1-c][1,2,4]triazin-2-amine, 6,7-diphenyl-, diacetyl deriv.	EtOH	257(4.45),325(3.94)	39-1492-76C
$C_{20}H_{16}N_6O_7$			
3,11-Diazatricyclo[5.3.1.1^{2,6}]dodec-4-ene-10,12-dione, 8-hydroxy-3,11-bis(5-nitro-2-pyridinyl)-	EtOH	208(4.31),227(4.28), 244(4.23),368(4.55)	39-2296-76C
$C_{20}H_{16}N_6S$			
Pyridinium, 2-(phenylamino)-1-(1-phenyl-1H-tetrazol-5-yl)-2-thioxoethylide	EtOH	280(4.01),338(4.35), 466(3.02)	39-0909-76C
Pyridinium, 2-(phenylamino)-1-(2-phenyl-2H-tetrazol-5-yl)-2-thioxoethylide	EtOH	255(4.34),316(4.33), 375s(4.01),480(2.87)	39-0909-76C
$C_{20}H_{16}O$			
Indeno[2,1-b]pyran, 2,4-dimethyl-3-phenyl-	dioxan	300(4.40),340(3.94), 485(3.42)	78-2219-76
Indeno[2,1-b]pyran, 2,4-dimethyl-9-phenyl-	dioxan	244(4.11),250(4.15), 308(4.20),509(3.21)	78-2219-76
Indeno[2,1-b]pyran, 4,9-dimethyl-2-phenyl-	dioxan	230(4.27),250(4.13), 311(4.33),334(4.21), 360(4.08),520(3.43)	78-2219-76
Indeno[2,1-b]pyran, 4-methyl-2-(4-methylphenyl)-	dioxan	256(4.45),262(4.48), 269(4.40),341(4.11), 360(4.19),372(4.17), 513(3.29)	78-2219-76
2-Propen-1-one, 3-(4-methylphenyl)-1-(1-naphthalenyl)-, cis	EtOH	317.5(4.41)	32-0617-76
trans	EtOH	322.5(4.47)	32-0617-76
2-Propen-1-one, 3-(4-methylphenyl)-1-(2-naphthalenyl)-, cis	EtOH	324(4.24)	32-0617-76
trans	EtOH	325.5(4.33)	32-0617-76
$C_{20}H_{16}OSe_2$			
Ethanone, 1-phenyl-2,2-bis(phenylseleno)-	EtOH	246(4.27)	39-1838-76C
$C_{20}H_{16}O_2$			
2-Propen-1-one, 3-(4-methoxyphenyl)-1-(1-naphthalenyl)-, cis	EtOH	290(3.59),330(3.52)	32-0617-76
trans	EtOH	343(4.16)	32-0617-76
2-Propen-1-one, 3-(4-methoxyphenyl)-1-(2-naphthalenyl)-, cis	EtOH	284(4.36)	32-0617-76
trans	EtOH	297.5(4.40)	32-0617-76
$C_{20}H_{16}O_2S_2$			
5,13:6,12-Diepithiodibenzo[a,f]cyclodecene-7,14-dione, 5,6,12,13-tetrahydro-5,12-dimethyl-, anti	MeCN	230(4.30),249(4.32), 301(3.55),360(2.37), 378(2.82),395(2.72)	1-0024-76

Compound	Solvent	$\lambda_{max}(\log \epsilon)$	Ref.
$C_{20}H_{16}O_2S_4$			
1,3-Dithiole, 4-(4-methoxyphenyl)- 2-[4-(4-methoxyphenyl)-1,3-dithiol- 2-ylidene]- cation	THF	271(4.59),334(4.28), 404(3.78)	97-0360-76
	MeCN	283(--),406(--), 467(--),750(--)	97-0360-76
dication	MeCN	298(4.42),598(4.00)	97-0360-76
$C_{20}H_{16}O_4$			
4H,8H-Benzo[1,2-b:3,4-b']dipyran-4-one, 5-hydroxy-8,8-dimethyl-2-phenyl-	EtOH	237(4.32),279(4.55), 362(3.60)	102-1553-76
5,12-Naphthacenedione, 1,4-dihydro- 6,11-dimethoxy-	MeOH	233(4.70),284(4.21), 295(4.19),412(3.82)	44-2296-76
5,12-Naphthacenedione, 7,10-dihydro- 6,11-dimethoxy-	MeOH	261(4.47),376(3.92)	44-2296-76
1,4-Naphthalenedione, 2-[(2,4-dimethyl- 3,6-dioxo-1,4-cyclohexadien-1-yl)- phenyl]-3-methyl-	EtOH	252(4.52),326(3.51)	39-2067-76C
1,4-Naphthalenedione, 2-[[2-(1,3-dioxo- lan-2-yl)phenyl]methyl]-	$CHCl_3$	252(4.29),260(4.18), 266(4.15),335(3.48)	39-0503-76C
2-Propenoic acid, 3-(4-methoxyphenyl)- 2-(2-naphthalenyloxy)-	EtOH	281(4.39)	2-0548-76
$C_{20}H_{16}O_4S$			
2,5-Thiophenedicarboxylic acid, 3,4-bis- (4-methylphenyl)-	EtOH	230(4.40),293(4.04)	42-0490-76
$C_{20}H_{16}O_5$			
4H-1-Benzopyran-4-one, 7-hydroxy-3-(5- hydroxy-2,2-dimethyl-2H-1-benzopyran- 3-yl)- (glabrone)	EtOH	243s(4.49),252(4.54), 257s(4.52),301s(4.19)	94-0991-76
	EtOH-NaOAc	264(--),330(--)	94-0991-76
2,5-Cyclohexadiene-1,4-dione, 3-hydroxy- 2,5-bis(4-methoxyphenyl)-	EtOH	256(4.49),388(3.87)	94-0613-76
2,3-Naphthalenedicarboxylic acid, 1-hy- droxy-4-phenyl-, dimethyl ester	EtOH	283(3.68),294(3.73), 306(3.72),345(3.81), 360(3.77)	39-0336-76C
$C_{20}H_{16}O_6$			
2,4-Furandicarboxylic acid, 4,5-dihydro- 5-oxo-3,4-diphenyl-, dimethyl ester	ether	260(3.81),277(3.78)	22-1491-76
5,12-Naphthacenedione, 1-acetoxy- 1,4,4a,12a-tetrahydro-6,11-di- hydroxy-	MeOH	237(4.43),252(4.36), 277(4.34),285(4.31), 397(4.11),417(4.09)	44-2296-76
$C_{20}H_{16}O_6S$			
2,5-Thiophenedicarboxylic acid, 3,4-bis- (4-methoxyphenyl)-	EtOH	228(4.41),284(4.05)	42-0490-76
$C_{20}H_{16}O_7$			
Deoxyaverufinone	MeOH	210(4.34),223(4.48), 255(4.13),266(4.16), 292(4.38),321(3.85), 455(3.84)	20-0161-76
$C_{20}H_{16}O_9$			
2-Naphthacenecarboxylic acid, 1,2,3,4,6- 11-hexahydro-2,4,5,12-tetrahydroxy- 7-methoxy-6,11-dioxo-, (2S-cis)-	MeOH	233(4.57),252(4.43), 288(3.94),479(4.08), 495(4.09),530(3.83)	87-0395-76

Compound	Solvent	$\lambda_{max}(\log \epsilon)$	Ref.
$C_{20}H_{16}S_3$			
Thiophene, 2,5-bis[4-(2-thienyl)-1,3-butadienyl]-	DMF	418s(4.65),440(4.72), 467(4.69)	126-1857-76
$C_{20}H_{16}S_4$			
1,3-Dithiole, 4-methyl-2-(4-methyl-5-phenyl-1,3-dithiol-2-ylidene)-5-phenyl-	THF	290(4.36),381(3.60), 484(2.65)	97-0360-76
cation	MeCN	464(--),669(--)	97-0360-76
dication	MeCN	463(4.04)	97-0360-76
1,3-Dithiole, 4-(4-methylphenyl)-2-[4-(4-methylphenyl)-1,3-dithiol-2-ylidene]-	THF	262(4.55),332(4.31), 411(3.76)	97-0360-76
cation	MeCN	260(--),395s(--), 443s(--),463(--), 701(--)	97-0360-76
dication	MeCN	274(4.45),532(4.13)	97-0360-76
$C_{20}H_{17}BrN_2O_2$			
Spiro[2H-anthra[1,2-d]imidazole-2,1'-cyclohexane]-6,11-dione, 4-bromo-1,3-dihydro-	EtOH	550(4.11)	104-0172-76
$C_{20}H_{17}ClN_2O_2$			
Acetamide, N-acetyl-N-(6-chloro-2-methyl-4-phenyl-3-quinolinyl)-	EtOH	216s(4.60),235(4.73), 281(3.77),300s(3.63), 314(3.59),328(3.65)	44-1743-76
3-Penten-2-one, 3-(7-chloro-5-phenyl-3H-1,4-benzodiazepin-2-yl)-4-hydroxy-	EtOH	218(4.45),258s(4.15), 276s(4.01),343(4.41)	44-1743-76
Spiro[2H-anthra[1,2-d]imidazole-2,1'-cyclohexane]-6,11-dione, 4-chloro-1,3-dihydro-	EtOH	552(4.15)	104-0172-76
Spiro[2H-anthra[1,2-d]imidazole-2,1'-cyclohexane]-6,11-dione, 5-chloro-1,3-dihydro-	EtOH	536(4.10)	104-0172-76
$C_{20}H_{17}NO$			
Benzeneacetonitrile, 4-ethyl-α-(1-methoxy-3-phenyl-2-propynylidene)-	EtOH	210(4.08),228(4.21), 344(4.49)	34-0115-76
$C_{20}H_{17}NO_2$			
4-Acridinol, 1,2,3,4-tetrahydro-, benzoate	EtOH	238(4.72),277(3.60), 310(3.63),323(3.62)	95-0968-76
1H-Benz[f]isoindol-1-one, 2-acetyl-2,3,3a,4-tetrahydro-9-phenyl-	EtOH	240(4.35),308(4.27)	44-2571-76
$C_{20}H_{17}NO_2S$			
Sulfilimine, N-benzoyl-S-(2-methoxyphenyl)-S-phenyl-, (S)-	MeOH	232(4.28),257(4.11), 275s(4.00)	78-3003-76
$C_{20}H_{17}NO_3$			
4,9[1',2']-Benzeno-1H-benz[f]isoindole-1,3(2H)-dione, 3a,4,9,9a-tetrahydro-4-hydroxy-2,9-dimethyl-	ether	254(3.09),264(3.11), 271(3.07)	88-1089-76
Tricholein	MeOH	269(4.17)	102-2018-76
$C_{20}H_{17}NO_4$			
4,11[1',2']-Benzeno-1H-[1,6]benzodioxocino[3,4-c]pyrrole-1,3(2H)-dione, 3a,4,11,11a-tetrahydro-2,4-dimethyl-, (3aα,4β,11β,11aβ)-	ether	264(3.07),270(3.14), 276(3.11)	22-0493-76

Compound	Solvent	$\lambda_{max}(\log \epsilon)$	Ref.
4,9-Epoxy-1H-benz[f]isoindole-1,3(2H)-dione, 3a,4,9,9a-tetrahydro-4-(2-hydroxyphenyl)-2,9-dimethyl-	ether	276(3.51),283(3.48)	22-0493-76
$C_{20}H_{17}NO_5$			
8H-Benzo[g]-1,3-benzodioxolo[5,6-a]quinolizin-8-one, 5,6-dihydro-9,10-dimethoxy-	EtOH	226(4.65),316s(4.15), 344(4.38),368s(4.20), 386s(4.04)	73-3654-76
7H-Dibenzo[de,g]quinolin-7-one, 1,2,9,10-tetramethoxy- (O-methylatheroline)	C_6H_{12}	241(4.54),259(4.46), 267(4.43),278(4.25), 288(4.13),329s(3.68), 375(4.07),395(4.11), 473(2.73)	102-0547-76
	H_2SO_4	250(4.59),275s(4.34), 330(3.66),379(3.72), 455(4.15),539(4.18)	102-0547-76
2,2'-Spirobi[2H-1-benzopyran], 6-methoxy-3,3'-dimethyl-8-nitro-	C_6H_{12}	221(4.33),249s(4.09), 253(4.12),294(3.49), 306(3.36),357(3.27)	22-2039-76
	EtOH	221(4.34),252(4.14), 290(3.76),304(3.66), 366(3.34)	22-2039-76
2,2'-Spirobi[2H-1-benzopyran], 8-methoxy-3,3'-dimethyl-6-nitro-	C_6H_{12}	218(4.47),244s(4.60), 250(4.64),260s(4.67), 280(4.54),310(3.97), 320s(3.94)	22-2039-76
	EtOH	217(4.54),243s(4.53), 251(4.56),258s(4.51), 308(3.96),338(4.00)	22-2039-76
4H-Quinolizine-1,2-dicarboxylic acid, 4-oxo-3-(phenylmethyl)-, dimethyl ester	MeOH	250(3.98),278(3.83), 383(4.09)	24-3668-76
$C_{20}H_{17}NO_6$			
Corydalispirone	EtOH	207(4.56),239(4.46), 290(4.38),317(4.35)	102-0577-76
	EtOH-HCl	245(--),298(--), 337(--),380s(--)	102-0577-76
1H-Indole-4,7-diol, 2-(2-acetoxyphenyl)-, diacetate	MeOH	210(4.56),244(4.35), 299(4.26)	111-0241-76
1H-Indole-4,7-diol, 2-(3-acetoxyphenyl)-, diacetate	MeOH	209(4.55),247(4.40), 305(4.39)	111-0241-76
1H-Indole-4,7-diol, 2-(4-acetoxyphenyl)-, diacetate	MeOH	210(4.55),247(4.43), 306(4.44)	111-0241-76
$C_{20}H_{17}NO_7$			
Isoxazolo[4,5-b]dibenzofuran-6-one, 7,9-diacetoxy-6,6a-dihydro-5,6a,10-trimethyl-	MeOH	216(4.21),245(3.75), 300(3.41)	23-3721-76
$C_{20}H_{17}N_3$			
Cyclohept[b]indole, 6-(2-methyl-2-phenylhydrazino)-	MeOH	249(4.44),303(4.25), 363(4.07)	18-1101-76
Pyrazino[2,3-b]quinoline, 3,4-dihydro-2,10-dimethyl-3-methylene-4-phenyl-	DMF	279(4.56),285s(4.51), 303(4.16),313(4.13), 337(3.79),353(3.68), 387(3.91),408(4.07), 423(3.99)	48-0039-76
Pyrido[2,3-b]pyrazine, 5,6-dihydro-6-methyl-2,3-diphenyl-	EtOH	315(3.91),398(4.10)	22-0251-76

$C_{20}H_{17}N_3O-C_{20}H_{18}Br_2N_4S_2$

Compound	Solvent	$\lambda_{max}(\log \epsilon)$	Ref.
$C_{20}H_{17}N_3O$			
2H-[1,4]Diazepino[2,3-b]quinolin-2-one, 1,5-dihydro-4,11-dimethyl-5-phenyl-	DMF	273s(4.12),283(4.18), 320s(4.00),332(4.09), 345(4.14),357(4.17), 364(4.20),382(3.85)	48-0039-76
1H-1,2,3-Triazole, 4,5-dihydro-1-(4-phenoxyphenyl)-5-phenyl-	dioxan	244(4.07),295(4.05)	97-0102-76
$C_{20}H_{17}N_3O_2$			
Benzenamine, N-(benzoyloxy)-N-methyl-4-(phenylazo)-	EtOH	357(4.32)	94-1485-76
$C_{20}H_{17}N_3O_3$			
1H-Pyridazino[4,5-b]indol-4-ol, 2-acetyl-2,5-dihydro-1-phenyl-, acetate	EtOH	315(4.32)	103-1008-76
$C_{20}H_{17}N_3O_4$			
Spiro[2H-anthra[1,2-d]imidazole-2,1'-cyclohexane]-6,11-dione, 1,3-dihydro-4-nitro-	EtOH	512(4.12)	104-0172-76
$C_{20}H_{17}N_3O_6S$			
Benzenesulfonamide, 4-methyl-N-(2-nitrophenyl)-N-[(4-nitrophenyl)methyl]-	MeOH	220(4.86),265(4.73)	39-0394-76C
$C_{20}H_{17}N_5O$			
4(1H)-Pteridinone, 2-[bis(phenylmethyl)-amino]-	pH −1.0	242(4.08),283(4.24), 320s(3.60),403(3.84)	24-3208-76
	pH 5.0	227s(4.21),284(4.31), 345(3.75),360s(3.68)	24-3208-76
	pH 10.0	230s(4.11),270(4.43), 377(3.88)	24-3208-76
$C_{20}H_{17}N_7O_6$			
3,11-Diazatricyclo[5.3.1.1^{2,6}]dodec-4-ene-10,12-dione, 8-amino-3,11-bis(5-nitro-2-pyridinyl)-	EtOH	209(4.34),227(4.36), 245(4.30),367(4.67)	39-2296-76C
$C_{20}H_{17}O_2P$			
6-Phosphanthridinol, 5,6-dihydro-5-methyl-6-phenyl-, 5-oxide	EtOH	212(4.60),265(3.86), 272(3.89)	39-2050-76C
$C_{20}H_{18}$			
1,1'-Bicyclopenta[cd]pentalene, 2a,2'a-4a,4'a,6a,6'a,6b,6'b-octahydro-, dl-meso-	C_6H_{12}	245(4.08),253(4.24), 262(4.08)	44-3524-76
	C_6H_{12}	245(4.10),253(4.24), 262(4.12)	44-3524-76
1,1':2',1''-Terphenyl, 4'-ethyl-	EtOH	235(4.41)	33-2012-76
$C_{20}H_{18}BrNO_2$			
Ethanone, 1-(4-bromophenyl)-2-[1,4-dihydro-1-(4-methoxyphenyl)-4-pyridinyl]-	MeOH	262(4.00),442(4.72)	104-1105-76
$C_{20}H_{18}Br_2N_4S_2$			
1H-[1,2]Dithiolo[5,1-e][1,2,3]thiadiazole-7-S^{IV}, 1-(4-bromophenyl)-3-[(4-bromophenyl)azo]-5-(1,1-dimethylethyl)-	C_6H_{12}	201(4.55),222(4.24), 260(4.41),286s(4.17), 310(4.13),433(4.44), 490s(4.29)	39-0228-76C

Compound	Solvent	$\lambda_{max}(\log \epsilon)$	Ref.
$C_{20}H_{18}ClNO$ Ethanone, 1-(4-chlorophenyl)-2-[1,4-di- hydro-1-(4-methylphenyl)-4-pyridinyl]-	MeOH	264(3.92),440(4.76)	104-1105-76
$C_{20}H_{18}ClNO_2$ Ethanone, 1-(4-chlorophenyl)-2-[1,4-di- hydro-1-(4-methoxyphenyl)-4-pyridin- yl]-	MeOH	256(4.29),436(4.79)	104-1105-76
Ethanone, 2-[1-(4-chlorophenyl)-1,4-di- hydro-4-pyridinyl]-1-(4-methoxyphen- yl)-	MeOH	295(3.62),450(4.40)	104-1105-76
$C_{20}H_{18}ClNO_5$ 1,3-Dioxolo[4,5-g]isoquinoline, 6-acet- yl-5-[(2-chloro-3-hydroxyphenyl)meth- ylene]-5,6,7,8-tetrahydro-9-methoxy-, (Z)-	MeOH	310(4.20)	12-2003-76
$C_{20}H_{18}Cl_2F_{12}N_2Pd$ Palladium, dichloro[2,4,6-trimethyl- 3,5-bis(trifluoromethyl)-1-azabicyclo- [2.2.0]hexa-2,5-diene][2,4,6-trimeth- yl-3,5-bis(trifluoromethyl)pyridinyl]-	CH_2Cl_2	245(4.37),280(3.73)	94-2219-76
$C_{20}H_{18}Cl_2N_2$ 2-Propenal, 3-chloro-3-(4-methylphenyl)-, [3-chloro-3-(4-methylphenyl)-2-propen- ylidene]hydrazone	CH_2Cl_2	358(4.52)	48-0705-76
$C_{20}H_{18}F_3N_3O_4$ Acetamide, N-[2,3-dihydro-1-(methoxy- methyl)-2-oxo-5-phenyl-1H-1,4-benzo- diazepin-7-yl]-2,2,2-trifluoro-N- methoxy-	isoPrOH	220(4.45),254(4.36), 315s(3.34)	87-1378-76
$C_{20}H_{18}NO_4$ Berberine, chloride	EtOH	230(4.43),267(4.41), 344s(4.36),352(4.36), 432(3.70)	73-3157-76 +73-3654-76
iodide	EtOH	226(4.59),267(4.46), 344s(4.42),353(4.43), 433(3.76)	73-3157-76 +73-3654-76
Pseudoepiberberine, iodide	EtOH	220(4.46),240(4.36), 264(4.34),290(4.62), 310(4.48),341(4.26), 380s(3.83)	73-3157-76
$C_{20}H_{18}N_2$ 1H-Indole, 3,3'-(1,2-ethenediyl)bis- [1-methyl-	MeCN	210(4.49),238(4.44), 278(4.14),337(4.33)	5-1060-76
semiquinone (radical ion)	MeCN	500(4.03),532(4.05), 758(3.85),845(3.96)	5-1060-76
oxidized form	MeCN	262(4.20),307(4.19), 486(4.44)	5-1060-76
1H-Indole, 3,3'-ethenylidenebis- [2-methyl-	MeOH	223(4.88),270(4.40)	23-1020-76
$C_{20}H_{18}N_2O$ Benzaldehyde, 2-methoxy-, diphenylhydra- zone	benzene EtOH	310(4.01),360(4.28) 305(3.91),350(4.27)	104-0625-76 104-0625-76

Compound	Solvent	$\lambda_{max}(\log \epsilon)$	Ref.
Benzenecarboximidamide, N-hydroxy-2-methyl-N,N'-diphenyl-	EtOH	260(3.98),309(4.07)	12-0357-76
Benzenecarboximidamide, N-hydroxy-4-methyl-N,N'-diphenyl-	EtOH	254(3.97),315(3.94)	12-0357-76
Benzenecarboximidamide, N-hydroxy-N'-(2-methylphenyl)-N-phenyl-	EtOH	257(3.80),313(3.80)	12-0357-76
Benzenecarboximidamide, N-hydroxy-N'-(4-methylphenyl)-N-phenyl-	EtOH	255(4.15),317(4.11)	12-0357-76
Benzenecarboximidamide, N-hydroxy-N-phenyl-N'-(phenylmethyl)-	EtOH	301(3.80)	12-0357-76
Benzo[b][1,6]naphthyridine, 2-acetyl-1,2,3,4-tetrahydro-10-phenyl-	EtOH	234(4.64),296(3.81), 308(3.78)	103-1144-76

$C_{20}H_{18}N_2O_2$

Compound	Solvent	$\lambda_{max}(\log \epsilon)$	Ref.
1H-Anthra[1,2-b][1,4]diazepine-8,13-di-one, 2,3-dihydro-2,2,4-trimethyl-	EtOH	257(3.57),324s(3.72), 530(3.86)	103-0125-76
1H-Anthra[2,3-b][1,4]diazepine-7,12-di-one, 2,3-dihydro-2,2,4-trimethyl-hydrochloride	EtOH	246(4.37),270s(4.21), 323(4.38),445(5.53)	103-0125-76
	EtOH-HCl	246(4.36),294(4.35), 373(3.86)	103-0125-76
Benzenecarboximidamide, N-hydroxy-2-methoxy-N,N'-diphenyl-	EtOH	261(4.03),294(4.13)	12-0357-76
Benzenecarboximidamide, N-hydroxy-4-methoxy-N,N'-diphenyl-	EtOH	253(4.17),317(4.13)	12-0357-76
Benzenecarboximidamide, N-hydroxy-N'-(2-methoxyphenyl)-N-phenyl-	EtOH	253(4.10),322(4.15)	12-0357-76
Benzenecarboximidamide, N-hydroxy-N'-(4-methoxyphenyl)-N-phenyl-	EtOH	253(3.94),316(3.85)	12-0357-76
Benzo[b][1,6]naphthyridine, 2-acetyl-1,2,3,4-tetrahydro-10-phenyl-, 5-oxide	EtOH	244(4.69),334(4.04)	103-1144-76
Ethanone, 1-[5-methyl-3-(4-methylbenz-oyl)-1-phenyl-1H-pyrazol-4-yl]-	EtOH	265(4.33)	4-0989-76
2,4-Pentadienoic acid, 2-cyano-5-phenyl-5-(phenylamino)-, ethyl ester	CH$_2$Cl$_2$	400(4.45)	48-0705-76
Spiro[2H-anthra[1,2-d]imidazole-2,1'-cy-clohexane]-6,11-dione, 1,3-dihydro-	EtOH	570(4.18)	104-0172-76
	EtOH-NaOH	640(3.90)	104-0172-76
	MeNO$_2$	550(3.97)	104-0172-76
	MeNO$_2$-HClO$_4$	598(4.04)	104-0172-76
Spiro[2H-anthra[2,3-d]imidazole-2,1'-cy-clohexane]-5,10-dione, 1,3-dihydro-	EtOH	600s(3.26)	104-0172-76
	EtOH-NaOH	630(3.64)	104-0172-76
	MeNO$_2$	534(3.23)	104-0172-76
	MeNO$_2$-HClO$_4$	403(3.40)	104-0172-76

$C_{20}H_{18}N_2O_3$

Compound	Solvent	$\lambda_{max}(\log \epsilon)$	Ref.
1H-Cyclopenta[b]quinoline-9-carboni-trile, 9a-(benzoyloxy)-2,3,3a,4,9,9a-hexahydro-3a-hydroxy-	EtOH	235(4.27),275s(3.31), 283(3.37),293(3.37)	95-0968-76
3H-Dibenz[f,ij]isoquinoline-2,7-dione, 1-[(3-methoxyphenyl)amino]-	CHCl$_3$	340(--),444(4.25)	104-1114-76
Ethanone, 2-[1,4-dihydro-1-(4-methyl-phenyl)-4-pyridinyl]-1-(4-nitro-phenyl)-	MeOH	266(4.31),452(4.77)	104-1105-76
Indolizino[1,2-b]quinoline-7-acetic acid, α-ethyl-9,11-dihydro-9-oxo-, methyl ester, (±)-	EtOH	219(4.63),249(4.41), 254(4.49),287(3.79), 321(3.87),366(4.24)	44-0699-76

$C_{20}H_{18}N_2O_4$

Compound	Solvent	$\lambda_{max}(\log \epsilon)$	Ref.
Acetamide, N-(1-acetyl-2,4-dioxo-3,5-diphenyl-3-pyrrolidinyl)-	n.s.g.	257s(3.25),264s(3.16), 269s(3.09),309(2.54)	33-2149-76

Compound	Solvent	$\lambda_{max}(\log \epsilon)$	Ref.
Benzoic acid, 2-[1-acetyl-5-(4-methoxy-phenyl)-1H-pyrazol-3-yl]-, methyl ester	EtOH	233(4.33),282(4.32), 300(4.16)	44-0110-76
$C_{20}H_{18}N_2O_4S$ Glycine, N-(2-nitro-5-phenyl-3-thienyl)-N-phenyl-, ethyl ester	MeCN	252(4.22),296(3.86), 363(4.18),443(3.92)	48-0221-76
1-Naphthalenesulfonic acid, 6-diazo-5,6-dihydro-5-oxo-, 4-(1,1-dimethylethyl)-phenyl ester	toluene $C_2H_4Cl_2$ DMF MeCN DMSO	345(3.88),401(3.85) 345(3.88),399(3.81) 349(3.93),399(3.85) 345(3.96),397(3.88) 351(3.86),401(3.78)	99-0081-76 99-0081-76 99-0081-76 99-0081-76 99-0081-76
$C_{20}H_{18}N_2O_5$ Indeno[1,2-c]pyrazole-3,3a(1H)-dicarb-oxylic acid, 2,3,4,8b-tetrahydro-4-oxo-1-phenyl-, dimethyl ester	EtOH	207(3.77),237(3.36), 300s(3.04),345(3.61)	39-2281-76C
$C_{20}H_{18}N_3O$ 1,2,4-Triazolo[4,3-a]pyridinium, 3-(4-methoxyphenyl)-1-(4-methyl-phenyl)-, tetrafluoroborate	MeOH	240s(--),258(4.32), 276s(--),308s(--)	48-0881-76
$C_{20}H_{18}N_4$ 1H-Indole-3-carboxaldehyde, 1-methyl-, [(1-methyl-1H-indol-3-yl)methylene]-hydrazone	DMF	275(4.18),345s(4.16), 357(4.66),372(4.50)	5-1039-76
$C_{20}H_{18}N_4O_2$ 1H-Anthra[1,2-d]triazole-6,11-dione, 1-methyl-4-piperidino-	EtOH	512(3.96)	104-0865-76
2H-Anthra[1,2-d]triazole-6,11-dione, 2-methyl-4-piperidino-	EtOH	522(4.00)	104-0865-76
3H-Anthra[1,2-d]triazole-6,11-dione, 3-methyl-4-piperidino-	EtOH	438(3.75)	104-0865-76
Benzo[g]pteridine-2,4(1H,3H)-dione, 3,7,8-trimethyl-1-(phenylmethyl)-	n.s.g.	340(3.93),389(3.89)	88-1389-76
$C_{20}H_{18}N_4O_2S$ Thiazolo[3,2-a]benzimidazole-3-acetic acid, α-[(4-methylphenyl)hydrazono]-, ethyl ester	n.s.g.	248(4.25),292(4.16), 370(4.31)	124-1162-76
$C_{20}H_{18}N_4O_3$ 2,4(1H,3H)-Pteridinedione, 5-acetyl-5,6,7,8-tetrahydro-6,7-diphenyl-	pH 4.0 pH 9.0	250(3.88),284(4.22) 217(4.51),283(4.16)	24-3184-76 24-3184-76
$C_{20}H_{18}N_4O_3S$ Thiazolo[3,2-a]benzimidazole-3-acetic acid, α-[(4-methoxyphenyl)hydraz-ono]-, ethyl ester, hydrobromide	n.s.g.	250(4.20),290(4.15), 386(4.20)	124-1162-76
$C_{20}H_{18}O$ Cyclohexanone, 2,6-bis(phenylmethylene)-	EtOH ether CHCl$_3$ CCl$_4$	230(3.82),330(3.97) 230(4.29),320(4.56) 322(4.39) 332(4.57)	114-0381-76B 114-0381-76B 114-0381-76B 114-0381-76B
Xanthene, 1,9a-dihydro-9a-methyl-2-phenyl-	EtOH	236(4.06),288(3.94), 341(4.28)	94-1581-76

Compound	Solvent	$\lambda_{max}(\log \epsilon)$	Ref.
$C_{20}H_{18}O_2$			
1H-Benz[de]anthracene-3-propanoic acid, 2,7-dihydro-	EtOH	218(4.33),262(3.88)	44-0977-76
Benzo[1,2:4,5]dicycloheptene-3,9-dione, 2,4,8,10-tetramethyl-	$CHCl_3$	250s(4.15),259(4.26), 308(5.17),325s(4.54), 344s(4.39),368s(3.57), 391(3.72),415(3.67)	22-0914-76
$C_{20}H_{18}O_3$			
Cyclohexanone, 2,6-bis[(4-hydroxyphenyl)methylene]-	EtOH	245(4.28),375(4.57)	114-0381-76B
	ether	242(--),350(--)	114-0381-76B
	$CHCl_3$	360(--)	114-0381-76B
	CCl_4	355(--)	114-0381-76B
ion	n.s.g.	460(4.56)	114-0381-76B
non-ion	n.s.g.	357(4.63)	114-0381-76B
1-Naphthalenol, 2-[[2-(1,3-dioxolan-2-yl)phenyl]methyl]-	$CHCl_3$	237(4.00),267(4.16), 274(3.92),287(3.96), 313(3.81),327(3.68)	39-0503-76C
1(2H)-Naphthalenone, 2-[[2-(1,3-dioxolan-2-yl)phenyl]methylene]-3,4-dihydro-	EtOH	225(4.00),285(4.16)	39-0503-76C
$C_{20}H_{18}O_4$			
Benzeneacetic acid, α-[3-(4-methoxyphenyl)-1-oxo-2-propynyl]-, ethyl ester	pH 13	222(3.61),258(3.75), 338(3.79)	34-0118-76
	EtOH	230(3.94),254s(3.76), 331(4.33)	34-0118-76
	$CHCl_3$	248s(3.95),253(3.97), 300s(4.16),337(4.51)	34-0118-76
1,3-Benzodioxole, 5-[4,5(or 6,7)-dimethoxy-1H-inden-2-yl]-6-ethenyl-	EtOH	224(4.25),265(3.88), 301(4.02),313(3.99)	78-1973-76
5H-Benzo[c]fluorene-5,8,11-trione, 7,7a,11a,11b-tetrahydro-11b-hydroxy-6,7a,9-trimethyl-, (7aα,11aα,11bβ)-	EtOH	248(4.45),280(3.95)	39-2067-76C
[3,6'-Bi-2H-1-benzopyran]-5',7-diol, 2',2'-dimethyl- (glabrene)	EtOH	248(4.25),281(4.26), 297(4.28),324(4.32)	94-0991-76
2(3H)-Furanone, 3-[(3,4-dimethoxyphenyl)methylene]dihydro-4-(phenylmethylene)-, (E,E)-	EtOH	279(3.70)	142-1481-76
5,12-Naphthacenediol, 1,4-dihydro-6,11-dimethoxy-	MeOH	218(4.67),272(4.45), 413(3.75)	44-2296-76
5,11-Naphthacenedione, 5a,6,11a,12-tetrahydro-1,3-dimethoxy-	MeOH	255(4.21),335(3.45)	2-0088-76
geometric isomer	MeOH	253(4.15),300(3.29), 335(3.39)	2-0088-76
5,12-Naphthacenedione, 1,4,4a,12a-tetrahydro-6,11-dimethoxy-	MeOH	235(4.44),265(4.56), 296(3.83),368(3.70)	44-2296-76
Naphtho[1,2-d]-1,3-dioxole, 9-(1,3-benzodioxol-5-yl)-8,9-dihydro-7,8-dimethyl-, (8R-trans)-	n.s.g.	241(4.14),274(4.15), 281(4.19)	31-0828-76
[1,1':4',1"-Terphenyl]-2',4-diol, 3',6'-dimethoxy-	EtOH	225(4.48),277(4.30)	94-0613-76
$C_{20}H_{18}O_5$			
9,10-Anthracenedione, 1,3,8-trimethoxy-6-(1-propenyl)-, (E)-	EtOH	231(4.52),287(4.50), 430(3.91)	39-0613-76C
5-Benzofuranpropanol, 2-(1,3-benzodioxol-5-yl)-, acetate	EtOH	214(4.52),305s(4.36), 320(4.50),335s(4.41)	39-1514-76C
4H-1-Benzopyran-2-carboxylic acid, 7-hydroxy-5-methyl-3-(phenylmethyl)-, ethyl ester	EtOH	226(4.33),236(4.26), 258(4.10),265(4.01), 311(3.99)	39-0499-76C

Compound	Solvent	$\lambda_{max}(\log \epsilon)$	Ref.
5,12-Naphthacenedione, 7,8,9,10-tetra-hydro-8-hydroxy-6,11-dimethoxy-	MeOH	223(4.29),260(4.54), 370(3.72)	44-2296-76
Naphtho[1,2-d]-1,3-dioxol-6(7H)-one, 9-(1,3-benzodioxol-5-yl)-8,9-dihydro-7,8-dimethyl-, [7R-(7α,8β,9α)]-(otobanone)	n.s.g.	236(4.35),288(4.00)	31-0828-76
[1,1':4',1"-Terphenyl]-2',4,4"-triol, 3',6'-dimethoxy-	EtOH	222(4.50),275(4.46)	94-0613-76

$C_{20}H_{18}O_6$

Compound	Solvent	$\lambda_{max}(\log \epsilon)$	Ref.
9,10-Anthracenedione, 1,3,8-trimethoxy-6-(3-methyl-2-oxiranyl)-	EtOH	223(4.55),279(4.33), 400(3.76)	39-0613-76C
Asarinin, (+)-	EtOH	207(4.49),233(4.95), 288(3.86)	2-0389-76
1,2-Benzenedicarboxylic acid, 3-methoxy-1-(5,8-dihydro-4-hydroxy-1-naphthal-enyl)-, 1-methyl ester	MeOH	283(3.58),302(3.63)	88-1637-76
1,2-Benzenedicarboxylic acid, 3-methoxy-2-(5,8-dihydro-4-hydroxy-1-naphthal-enyl)-, 1-methyl ester	MeOH	285(3.56),298(3.66)	88-1637-76
4H-1-Benzopyran-4-one, 3,5,7-trihydroxy-2-(4-hydroxyphenyl)-6-(3-methyl-2-but-enyl)- (licoflavanol)	EtOH	256s(4.07),271(4.15), 271(4.15),295s(3.86), 304s(3.87),340s(4.06), 369(4.13)	94-1242-76
1(3H)-Isobenzofuranone, 3-acetoxy-3-(2-acetoxy-5-ethylphenyl)-	EtOH	205(4.63),278(3.32)	56-0759-76
5,12-Naphthacenedione, 1-acetoxy-1,2,3,4,4a,12a-hexahydro-6,11-dihydroxy-	MeOH	237(4.43),252(4.37), 277(4.34),285(4.31), 398(4.12),417(4.10)	44-2296-76

$C_{20}H_{18}O_7$

Compound	Solvent	$\lambda_{max}(\log \epsilon)$	Ref.
2-Anthracenecarboxylic acid, 9,10-di-hydro-1,6,8-trimethoxy-3-methyl-9,10-dioxo-, methyl ester	EtOH	227(4.40),241(4.33), 279(4.44),343(3.63), 395(3.75)	44-3018-76
9,10-Anthracenedione, 1-acetyl-2,4,5,7-tetramethoxy-	EtOH	224(4.54),284(4.41), 415(3.77)	39-1852-76C
2,5-Furandicarboxylic acid, 2,5-di-hydro-2,5-dihydroxy-3,4-diphenyl-, dimethyl ester	ether	228(4.19),263(3.99)	22-1491-76
Isoflavone, 7-acetoxy-3',4',8-trimeth-oxy-	EtOH	257(4.39),285s(4.08)	102-1019-76

$C_{20}H_{18}O_8$

Compound	Solvent	$\lambda_{max}(\log \epsilon)$	Ref.
2H-1-Benzopyran-2-one, 3-(1-acetoxy-4,5-dimethoxy-6-oxo-2,4-cyclohexa-dien-1-yl)-6-methoxy-	MeOH	376(4.10)	18-1955-76
2H-1-Benzopyran-2-one, 3-(1-acetoxy-4,5-dimethoxy-6-oxo-2,4-cyclohexa-dien-1-yl)-7-methoxy-	MeOH	376(4.10)	18-1955-76
2H-1-Benzopyran-2-one, 3-(1-acetoxy-4,5-dimethoxy-6-oxo-2,4-cyclohexa-dien-1-yl)-8-methoxy-	MeOH	350(4.05),390s(3.96)	18-1955-76
11H-Dibenzo[b,e][1,4]dioxepin-6-carbox-ylic acid, 4-formyl-3,8-dimethoxy-1,9-dimethyl-11-oxo-, methyl ester	n.s.g.	241(4.43),273s(4.00), 302(3.69)	12-1059-76
Diferulic acid, trans-trans	pH 13	270(4.34),320(4.46), 329s(4.45)	102-1157-76
	M NaOH	244s(4.26),322s(4.32), 358(4.58)	102-1157-76
	MeOH	216s(4.33),243(4.35), 311s(4.44),324(4.47)	102-1157-76

Compound	Solvent	$\lambda_{max}(\log \epsilon)$	Ref.
Diferulic acid, trans-trans (cont.)	8% MeOH	218(4.37),241(4.38), 306s(4.43),318(4.44)	102-1157-76
Isousnic acid, 8-deacetyl-, diacetate, (+)-	MeOH	219(4.11),259(3.93), 280(3.75),325(3.39)	23-3721-76
$C_{20}H_{18}O_{10}$ 6H-[1,3]Benzodioxolo[5,6-c][1]benzopyran-6-one, 2,3,4-triacetoxy-1,2,3,4-tetrahydro-, $(2\alpha,3\alpha,4\alpha)-(\underline{+})-$	EtOH	242(4.56),258s(--), 287(3.74),298(3.75), 328(3.66),347s(--)	24-0855-76
$C_{20}H_{19}ClN_2O_2S$ Benzenesulfonamide, 4-chloro-N-(6,7,8,9-tetrahydro-5-methylcyclohept[b]indol-10(5H)-ylidene)-	EtOH	217(4.56),263(4.03), 275s(3.98),348(4.35)	39-0481-76C
$C_{20}H_{19}ClN_4O_8$ Microline, 2,4-dinitrophenylhydrazone	n.s.g.	228(3.93),293(4.04)	33-1809-76
$C_{20}H_{19}NO$ 8-Azabicyclo[3.2.1]oct-3-en-2-one, 8-methyl-4,6-diphenyl-, endo	EtOH	210(4.3),222(4.2), 285(4.2),345(3.5)	39-2329-76C
exo	EtOH	210(4.3),222(4.1), 285(4.1),345(3.3)	39-2329-76C
8-Azabicyclo[3.2.1]oct-3-en-2-one, 6-phenyl-8-(phenylmethyl)-, endo	EtOH	218(4.017)	39-2334-76C
exo	EtOH	218(3.97)	39-2334-76C
Ethanone, 2-[1,4-dihydro-1-(4-methylphenyl)-4-pyridinyl]-1-phenyl-	MeOH	252(4.12),435(4.83)	104-1105-76
$C_{20}H_{19}NO_2$ 1H-Benz[f]isoindol-1-one, 2-acetyl-2,3,3a,4,9,9a-hexahydro-4-phenyl-	EtOH	252s(3.05),306(2.70)	44-2571-76
Ethanone, 2-[1,4-dihydro-1-(4-methoxyphenyl)-4-pyridinyl]-1-phenyl-	MeOH	250(3.91),430(4.62)	104-1105-76
Ethanone, 1,1'-[6-methyl-7-(phenylmethyl)-1,3-indolizinediyl]bis-	EtOH	230(4.20),260(4.50), 293(4.10),345(4.39)	103-0424-76
2H-Pyrrole-3-carboxylic acid, 3,4-dihydro-2-phenyl-5-(2-phenylethenyl)-, methyl ester, cis	EtOH	287(4.34)	44-0543-76
Spiro[2H-1-benzopyran-2,2'(1'H)-quinoline], 8-methoxy-1',3'-dimethyl-	octane	240(4.47),251s(4.39), 267s(4.08),320(3.57)	103-0683-76
	EtOH	244(4.44),305(3.83), 325s(3.91),380(4.05), 580(3.42)	103-0683-76
$C_{20}H_{19}NO_3S_2$ Sulfilimine, S-(2-methoxyphenyl)-N-[(2-methylphenyl)sulfonyl]-S-phenyl-, S-(-)-	MeOH	225(4.29),288(3.59)	78-3003-76
Sulfilimine, S-(2-methoxyphenyl)-N-[(4-methylphenyl)sulfonyl]-S-phenyl-, S-(-)-	MeOH	229(4.19),288(3.59)	78-3003-76
$C_{20}H_{19}NO_4$ Corysamine, tetrahydro-	EtOH	238(3.95),291(3.85)	73-3157-76
2H-Quinolizine-1,3-dicarboxylic acid, 2-phenyl-, 1-ethyl 3-methyl ester	MeOH	208(4.36),268(4.49), 310(3.80),470(3.78)	39-1911-76C
	MeOH-acid	208(3.86),280(3.85)	39-1911-76C

Compound	Solvent	$\lambda_{max}(\log \epsilon)$	Ref.
$C_{20}H_{19}NO_4S_2$			
Sulfilimine, S-(2-methoxyphenyl)-N-[(4-methoxyphenyl)sulfonyl]-S-phenyl-, (S)-(-)-	MeOH	235(4.37),282(3.68)	78-3003-76
Sulfoximine, S-(2-methoxyphenyl)-N-[(4-methylphenyl)sulfonyl]-S-phenyl-, (S)-(-)-	MeOH	225(4.27),292(3.41)	78-3003-76
$C_{20}H_{19}NO_5$			
Elmerrilicine, N-acetyl-	MeOH	221(4.60),241s(4.27), 271s(4.14),277(4.17), 298(4.01)	12-2003-76
$C_{20}H_{19}NO_6$			
Pyrido[2,1,6-de]quinolizine-1,3,5-tricarboxylic acid, 1,3-diethyl 5-methyl ester	EtOH	244(4.13),298(4.45), 330(4.01),411(4.25), 452s(4.2),474(4.38)	39-0341-76C
$C_{20}H_{19}NO_7$			
1,3(2H,9bH)-Dibenzofurandione, 7,9-diacetoxy-2-(1-aminoethylidene)-6,9b-dimethyl-	MeOH	204(4.16),268(3.95), 284(3.85),328(3.42)	23-3721-76
4,9-Epoxy-4H-cycloocta[c]pyrrole-4,7-dicarboxylic acid, 1,2,3,3a,5,6,9,9c-octahydro-1,3-dioxo-2-phenyl-, dimethyl ester, (3aα,4β,9β,9aα)-	MeCN	218(4.25)	88-3303-76
$C_{20}H_{19}NO_9$			
[1,3]Dioxolo[4,5-j]phenanthridin-6(2H)-one, 2,3,4-triacetoxy-1,3,4,5-tetrahydro-	EtOH	232(4.45),250(4.48), 263s(--),287s(--), 298(3.99),315(3.75), 329(3.78),343(3.64)	24-0855-76
Lycoricidine, triacetate, (±)-	EtOH	245(4.49),306(3.93)	94-2977-76
$C_{20}H_{19}N_3O$			
4H-Indazol-4-one, 1,5,6,7-tetrahydro-3-phenyl-1-(phenylmethyl)-, 4-oxime	MeOH	227(4.44),266s(3.85)	24-1898-76
2H-Indol-2-one, 3-[[2-(1,1-dimethylethyl)-1H-indol-3-yl]imino]-1,3-dihydro-	MeOH	230(4.74),253(4.32), 290(4.08),510(3.85)	23-1020-76
1-Naphthalenecarboxaldehyde, 2-hydroxy-, [[4-(dimethylamino)phenyl]methylene]-hydrazone	EtOH	251s(4.38),258s(4.30), 282(3.82),322s(4.00), 352s(4.17),407(4.62)	95-0044-76
2-Naphthalenecarboxaldehyde, 1-hydroxy-, [[4-(dimethylamino)phenyl]methylene]-hydrazone	EtOH	224(4.46),255(4.27), 267(4.29),276(4.36), 300(4.13),318(4.04), 405(4.65)	95-0044-76
Pyrazolo[4,3-b]azepin-5(1H)-one, 4,6,7,8-tetrahydro-3-phenyl-1-(phenylmethyl)-	MeOH	230s(4.22),250s(4.18)	24-1898-76
Pyrazolo[4,3-c]azepin-4(1H)-one, 5,6,7,8-tetrahydro-3-phenyl-1-(phenylmethyl)-	MeOH	243s(4.11)	24-1898-76
$C_{20}H_{19}N_3O_2$			
2,4-Pentadienamide, 2-cyano-5-(4-methoxyphenyl)-5-[(4-methylphenyl)amino]-	CH_2Cl_2	404(4.49)	48-0705-76
4-Pyrazolecarboxamide, 1-acetyl-3-benzoyl-4,5-dihydro-5-methyl-	EtOH	243(4.27),305(4.19)	115-0001-76

Compound	Solvent	λ_{max}(log ϵ)	Ref.
$C_{20}H_{19}N_3O_3$			
15-Azapentacyclo[7.5.1.1^{3,6}.0^2,7^{10,14}]-hexadeca-4,12-dien-8-one, 15-(5-nitro-2-pyridinyl)-	EtOH	207(4.18),228(4.29), 370(4.05)	39-2307-76C
L-Phenylalanine, N-[(3-methyl-2-quinoxalinyl)carbonyl]-, methyl ester	EtOH	203(4.65),242(4.52), 320(3.83)	78-2931-76
3-Pyridinecarboxylic acid, 9-cyano-2,3,4,4a,9,10-hexahydro-4a-hydroxy-	EtOH	244(4.19),260s(3.66), 268s(3.55),300s(3.55)	95-0968-76
$C_{20}H_{19}N_3O_4S$			
1H-Pyrrolo[2,3-d]pyrimidine-2,4(3H,5H)-dione, 6,7-dihydro-7-[(4-methylphenyl)sulfonyl]-3-(phenylmethyl)-	DMSO	290(4.13)	24-2983-76
$C_{20}H_{19}N_3O_5$			
L-Phenylalanine, N-[(3-methyl-2-quinoxalinyl)carbonyl]-, N,N'-dioxide, methyl ester	EtOH	205(4.32),232(4.30), 260s(4.23),266(4.37), 349s(3.75),369(3.98), 384(4.03)	78-2931-76
	MeCN	205(4.35),228(4.28), 260s(4.15),268(4.33), 360s(3.70),378(3.94), 394(3.97)	78-2931-76
4-Quinolinecarbonitrile, 2-ethyl-1,2,3,4-tetrahydro-2-hydroxy-3-methyl-3-[(4-nitrobenzoyl)oxy]-	EtOH	246(4.32),265s(4.11), 295s(3.73)	95-0968-76
$C_{20}H_{19}N_4Ni$			
Nickel(1+), (16,17-dihydro-16-methyl-7H-dibenzo[f,m][1,4,8,12]tetraazacyclopentadecinato-N^5,N^9,N^{15},N^{18})-, iodide, (SP-4-4)-	EtOH	249(4.51),268(4.48), 332(4.11),446(4.13), 575(3.53)	12-2271-76
Nickel(1+), (7,8,9,18-tetrahydrodibenzo-[b,k][1,4,9,13]tetraazacyclohexadecinato-N^6,N^{10}N^{16},N^{20})-, iodide, (SP-4-2)-	EtOH	248(4.57),267s(--), 331(4.11),406s(--), 446(4.13),574(3.53)	12-2271-76
$C_{20}H_{19}N_5O_4$			
Adenosine, N-benzoyl-4',5'-didehydro-5'-deoxy-2',3'-O-(1-methylethylidene)-	MeOH	230(4.14),279(4.33)	35-3346-76
$C_{20}H_{20}Br_2N_8O_4S_2$			
1,2,4,5-Tetrazine-3,6-dicarboximidamide, N',N'''-bis[(4-bromophenyl)sulfonyl]-N,N,N'',N''-tetramethyl-	MeCN	250(4.57),525(2.65)	32-0001-76
$C_{20}H_{20}ClFN_2OS$			
1-Piperazineethanol, 4-(7-chloro-8-fluorodibenzo[b,f]thiepin-10-yl)-	MeOH	236(4.34),264s(4.13), 310(3.90)	73-0881-76
$C_{20}H_{20}ClNO_6$			
Acetamide, N-[2-[6-[2-(2-chloro-3-hydroxyphenyl)acetyl]-4-methoxy-1,3-benzodioxol-5-yl]ethyl]-	MeOH	286(3.96)	12-2003-76
$C_{20}H_{20}ClN_2S$			
Benzothiazolium, 2-[2-chloro-4-[4-(dimethylamino)phenyl]-1,3-butadienyl]-3-methyl-, perchlorate	MeOH	582(4.74)	124-1174-76
Benzothiazolium, 2-[4-chloro-4-[4-(dimethylamino)phenyl]-1,3-butadienyl]-3-methyl-, perchlorate	MeOH	562(4.70)	124-1174-76

Compound	Solvent	$\lambda_{max}(\log \epsilon)$	Ref.
$C_{20}H_{20}FN_2S$ Benzothiazolium, 2-[4-[4-(dimethylamino)phenyl]-1-fluoro-1,3-butadienyl]-3-methyl-, perchlorate	$MeNO_2$	570(4.64)	124-1174-76
$C_{20}H_{20}NO_4$ Jatrorrhizine, chloride	EtOH	228(4.37),241s(4.32), 267(4.35),275s(4.32), 352(4.40),440(3.72)	73-3157-76
$C_{20}H_{20}NO_5$ 1,3-Dioxolo[4,5-g]isoquinolinium, 7,8-dihydro-5-[[4-(hydroxymethyl)-1,3-benzodioxol-5-yl]methyl]-6-methyl-, chloride (pseudohypecorine chloride)	EtOH	247(4.22),295(3.94), 365(3.96)	105-0432-76
$C_{20}H_{20}N_2$ 3,3'-Biindolizine, 1,1',2,2'-tetramethyl-	MeCN	238(4.62),254(4.45), 263s(4.38),310(3.99), 364(3.59)	5-0317-76
$C_{20}H_{20}N_2O$ Oxazolo[3',2':1,2]pyrido[3,4-b]indole, 2,3,5,6,11,11b-hexahydro-11b-methyl-2-phenyl-	MeOH	225(4.31),250(3.79), 274(3.58),282(3.58), 291(3.52),360(4.06)	23-1262-76
	MeOH-HCl	214(4.22),250(4.02), 362(4.37)	23-1262-76
	MeOH-KOH	229(4.40),275(3.92), 282(3.92),291(3.85)	23-1262-76
3H-Pyrido[3,4-b]indole-1-propanol, 4,9-dihydro-α-phenyl-	MeOH	216(4.28),248(4.05), 289(3.81),318(3.93)	23-1262-76
	MeOH-HCl	214(4.26),305(3.67), 352(4.10)	23-1262-76
	MeOH-KOH	217(4.34),289(3.89), 316(4.01)	23-1262-76
2H-Pyrrol-2-one, 1,5-dihydro-5-phenyl-3-[(phenylmethylene)amino]-4-propyl-	EtOH	206(4.53),267(4.30), 278(4.30)	40-1100-76
$C_{20}H_{20}N_2OS$ Acetamide, N-[3-(ethylphenylamino)-5-phenyl-2-thienyl]-	MeOH	250(4.27),281s(4.20), 304(4.24),327s(4.16)	48-0221-76
4-Imidazolidinone, 3-(1,1-dimethylethyl)-5-(diphenylmethylene)-2-thioxo-	$CHCl_3$	260(4.06),381(4.34)	5-1997-76
$C_{20}H_{20}N_2O_2$ Benzimidazo[3,2-b]isoquinoline, 8,9-dimethoxy-11-propyl-, monoperchlorate	MeOH	278(4.88),340(3.46), 360(3.88)	103-0205-76
Pyrano[2,3-c]pyrazole, 6-ethoxy-1,4,5,6-tetrahydro-1,4-diphenyl-, cis	EtOH	245.5(4.23)	35-2947-76
trans	EtOH	244.5(4.21)	35-2947-76
$C_{20}H_{20}N_2O_2S$ Acetamide, N-[3-[(4-methoxyphenyl)methylamino]-5-phenyl-2-thienyl]-	MeOH	225(4.27),244(4.25), 281s(4.09),317(4.20), 340s(4.04)	48-0221-76
$C_{20}H_{20}N_2O_3$ Benzoic acid, 2-[5-(4-methoxyphenyl)-1H-pyrazol-3-yl]-, propyl ester	EtOH	261(4.46)	44-0110-76
1H-Inden-1-one, 2-[[(1,1-dimethylethyl)-amino]phenylmethyl]-4-nitro-	hexane	240(4.12),277(4.04), 345(3.65),395(3.40)	44-3540-76

Compound	Solvent	$\lambda_{max}(\log \epsilon)$	Ref.
1H-Inden-1-one, 2-[[(1,1-dimethylethyl)-amino]phenylmethyl]-5-nitro-	hexane	225(4.18),242(4.11), 253(4.08),275(4.08), 315(3.43),350(3.30), 405(3.23)	44-3540-76
	CHCl$_3$	405(2.93)	44-3540-76
	MeCN	405(2.70)	44-3540-76
1H-Inden-1-one, 2-[[(1,1-dimethylethyl)-amino]phenylmethyl]-6-nitro-	hexane	228(4.60),244(4.31), 330(4.04),385(3.74)	44-3540-76
	CHCl$_3$	395(3.36)	44-3540-76
	MeCN	397(3.29)	44-3540-76
2-Naphthalenecarboxamide, N-[4-(dimeth-ylamino)phenyl]-N-hydroxy-3-methoxy-, V(V) complex	CHCl$_3$-5M HCl	570(4.08)	3-0714-76
	CHCl$_3$-M HCl	535(3.95)	3-0714-76
	0.1M HCl	530(3.93)	3-0714-76
1H-Pyrazole-5-carboxylic acid, 3-benz-oyl-4,5-dihydro-1-(4-methylphenyl)-, ethyl ester	EtOH	256(4.187),400(4.299)	18-0321-76
1H-Pyrazole-5-carboxylic acid, 4,5-di-hydro-3-(4-methylbenzoyl)-1-phenyl-, ethyl ester	EtOH	253(4.189),290s(4.003), 392(4.375)	18-0321-76
4-Quinolinecarbonitrile, 3-(benzoyloxy)-2-ethyl-1,2,3,4-tetrahydro-2-hydroxy-3-methyl-	EtOH	237(4.31),275s(3.33), 283s(3.39),298(3.48)	95-0968-76
4-Quinolinecarbonitrile, 3-(benzoyloxy)-1,2,3,4-tetrahydro-2-hydroxy-2-(1-methylethyl)-	EtOH	237(4.29),275s(3.23), 287s(3.34),298(3.46)	95-0968-76
4-Quinolinecarbonitrile, 3-(benzoyloxy)-1,2,3,4-tetrahydro-2-hydroxy-2-propyl-	EtOH	235(4.29),275s(3.24), 285s(3.34),300(3.47)	95-0968-76
$C_{20}H_{20}N_2O_4$ 2-Propenoic acid, 3-(1-acetyl-4-cyano-1,2,2a,3-tetrahydro-2-hydroxybenz-[cd]indol-5-yl)-2-methyl-, ethyl ester, (E)-	MeOH	240(4.45),296(4.05)	24-2126-76
$C_{20}H_{20}N_2O_5$ L-Phenylalanine, N-[N-[3-(4-hydroxyphen-yl)-1-oxo-2-propenyl]glycyl]-	MeOH-acid	227(4.19),294(4.39), 311(4.37)	20-0647-76
	MeOH-base	238(4.11),312s(4.11), 352(4.46)	20-0647-76
$C_{20}H_{20}N_2O_6$ L-Phenylalanine, N-[N-[3-(3,4-dihydroxy-phenyl)-1-oxo-2-propenyl]glycyl]-, (E)-	MeOH-acid	238(4.32),297(4.35), 324(4.37)	20-0647-76
	MeOH-base	252(4.29),298(4.09), 308(4.13),363(4.47)	20-0647-76
$C_{20}H_{20}N_3OS$ 1,3-Thiazin-1-ium, 6-[[(dimethylamino)-methylene]amino]-2-(4-methoxyphenyl)-5-phenyl-, perchlorate	HOAc	295(4.13),335(4.20), 456(4.17)	97-0268-76
$C_{20}H_{20}N_4$ 1H-Indole, 3,3'-azobis[1,2-dimethyl-	CH$_2$Cl$_2$	214(4.38),242(4.21), 250(4.17),277s(4.11), 282(4.15),380(4.04), 403(4.34),424(4.44)	5-1060-76
oxidized form (dication)	MeCN	218(4.48),242s(4.40), 247(4.42),278(4.56), 332(3.97),400(3.71)	5-1060-76

Compound	Solvent	$\lambda_{max}(\log \epsilon)$	Ref.
Pyrazolo[4,3-c]pyrazole, 1,4-diethyl-1,4-dihydro-3,6-diphenyl-	MeCN	212(4.25),222(4.32), 225(4.31),236s(4.23), 310(4.35)	5-1090-76
$C_{20}H_{20}N_4OSe$ Methanimidamide, N'-[5-(4-methoxyphenyl)-2-(phenylimino)-2H-1,3-selenazin-6-yl]-N,N-dimethyl-, monoperchlorate	HOAc	473(4.17)	88-2005-76
$C_{20}H_{20}N_4O_4$ 1H-Pyrazole-3-carboxylic acid, 5-amino-1-(4-methoxyphenyl)-4-[(phenylamino)-carbonyl]-, ethyl ester	EtOH	307(4.19)	4-1137-76
$C_{20}H_{20}N_4O_5$ 9H-Fluoren-9-one, 1,2,3,4,4a,9a-hexahydro-7-methoxy-, 2,4-dinitrophenylhydrazone	EtOH CHCl$_3$	378(4.48) 394(4.51)	12-2683-76 12-2683-76
$C_{20}H_{20}N_4S_2$ 1H-[1,2]Dithiolo[5,1-e][1,2,3]thiadiazole-7-S^{IV}, 5-(1,1-dimethylethyl)-1-phenyl-3-(phenylazo)-	C$_6$H$_{12}$	201(4.52),224s(4.34), 238(4.38),250(4.37), 285(4.12),302s(4.05), 424(4.34),488(4.20)	39-0228-76C
$C_{20}H_{20}N_4Se$ Methanimidamide, N,N-dimethyl-N'-[2-[(4-methylphenyl)imino]-5-phenyl-2H-1,3-selenazin-6-yl]-, monoperchlorate	HOAc	461(4.19)	88-2005-76
$C_{20}H_{20}N_6$ Formazan, 1-(4,6-dimethyl-2-pyrimidinyl)-5-(4-methylphenyl)-3-phenyl-	EtOH EtOH-NaOH benzene dioxan	439(4.13) 473(--) 457(--) 448(--)	103-1165-76 103-1165-76 103-1165-76 103-1165-76
$C_{20}H_{20}O$ 2-Cyclohexen-1-one, 2-ethyl-5,5-diphenyl-	EtOH	234s(3.84)	33-2012-76
$C_{20}H_{20}O_2$ Estra-1,3,5(10),8,14-pentaen-17-one, 3-methoxy-11-methylene-, (±)-	MeOH	246(4.18),255(4.10), 313s(4.31),325(4.37), 338s(4.29)	44-0531-76
Ethanone, 1,1'-tricyclo[9.3.1.1^{4,8}]hexadeca-1(15),4,6,8(16),11,13-hexaene-5,14-diylbis-	EtOH	330s(2.30)	49-0195-76
$C_{20}H_{20}O_2S$ Sulfonium, diphenyl-, 4,4-dimethyl-2,6-dioxocyclohexylide	MeOH	225s(4.10),255(4.16)	30-0366-76
$C_{20}H_{20}O_3$ 2-Propen-1-one, 1-[2,4-dihydroxy-5-(3-methyl-2-butenyl)phenyl]-3-phenyl-, (E)-	EtOH	215(4.06),356(4.02)	2-0339-76
$C_{20}H_{20}O_4$ 2H,6H-Benzo[1,2-b:5,4-b']dipyran-6-one, 5-hydroxy-2,2,8-trimethyl-10-(3-methyl-1,3-butadienyl)-	MeOH	209(4.21),249(4.26), 290(4.50),340s(3.65)	24-2963-76

Compound	Solvent	$\lambda_{max}(\log \epsilon)$	Ref.
2-Cyclopentene-1-acetic acid, 2-(6-methoxy-2-naphthalenyl)-1-methyl-5-oxo-, methyl ester, (S)-	EtOH	235(4.78),292(3.78), 300(3.78)	78-0079-76
Glabridin	EtOH	282(4.18),288s(4.13), 312s(3.47)	94-0752-76
2-Propen-1-one, 1-(3,4-dihydro-3,5-dihydroxy-2,2-dimethyl-2H-1-benzopyran-6-yl)-3-phenyl- (flemistrictin C)	EtOH	215(4.19),345(4.45)	2-0339-76
2-Propen-1-one, 1-[2,3-dihydro-4-hydroxy-2-(1-hydroxy-1-methylethyl)-5-benzofuranyl]-3-phenyl-	EtOH	215(4.08),352(4.42)	2-0339-76
Tricyclo[8.2.2.2^{4,7}]hexadeca-4,6,10,12-13,15-hexaene-5,6-dicarboxylic acid, dimethyl ester	C_6H_{12}	325(<u>2.7</u>)	5-0140-76
Tricyclo[9.3.1.1^{4,7}]hexadeca-1(15),4,6-8(16),11,13-hexaene-5,14-dicarboxylic acid, dimethyl ester	EtOH	288(3.36)	49-0195-76
$C_{20}H_{20}O_4S$ Thiophene, 3,4-bis(3,4-dimethoxyphenyl)-	EtOH	238(4.48),264s(4.33)	42-0490-76
$C_{20}H_{20}O_4S_2$ 3,10-Dithiatricyclo[10.2.2.2^{5,8}]octadeca-5,7,12,14,15,17-hexaene-6,7-dicarboxylic acid, dimethyl ester	C_6H_{12}	<u>310s(2.9)</u>	5-0140-76
$C_{20}H_{20}O_5$ Bicyclo[3.2.1]oct-3-ene-2,8-dione, 7-(1,3-benzodioxol-5-yl)-3-methoxy-6-methyl-5-(2-propenyl)-, (6-exo-7-endo)	MeOH	230(3.72),276(3.68)	102-1033-76
$C_{20}H_{20}O_6$ 9,10-Anthracenedione, 1-ethyl-2,4,5,7-tetramethoxy-	EtOH	224(4.64),265s(4.32), 286(4.47),410(3.86)	39-1852-76C
$C_{20}H_{20}O_7$ Isoflavone, 2',4',5',6,7-pentamethoxy-	MeOH	254(4.27),300(4.2)	2-0951-76
$C_{20}H_{20}O_9$ Flavone, 5,7-dihydroxy-3,3',4',5',8-pentamethoxy-	EtOH EtOH-AlCl$_3$	279(4.20),360(3.90) 285(4.34),310(4.20), 350(4.25),415(3.72)	2-0849-76 2-0849-76
$C_{20}H_{21}BrN_2O_9$ D-Ribose, 1-C-(5-bromo-1H-imidazol-3-yl)-, 2,3,4,5-tetraacetate	EtOH	222(4.16),241(4.00), 248(4.00),292s(3.86), 301(3.96)	4-1241-76
D-Ribose, 1-C-(6-bromo-1H-imidazol-3-yl)-, 2,3,4,5-tetraacetate	EtOH	222(4.15),244(4.15), 249(4.15),301(4.07)	4-1241-76
$C_{20}H_{21}ClN_2O_3S$ Benzenesulfonamide, 4-chloro-N-(1,2,3-4,5,6-hexahydro-1-methyl-12-oxo-2,7-methano-7H-1-benzazonin-7-yl)-	EtOH	233(4.24),253s(4.05), 304(3.32)	39-0481-76C
$C_{20}H_{21}ClN_2O_6$ Benzimidazo[3,2-b]isoquinoline, 8,9-dimethoxy-11-propyl-, monoperchlorate	MeOH	278(4.88),340(3.46), 360(3.88)	103-0205-76

Compound	Solvent	$\lambda_{max}(\log \epsilon)$	Ref.
$C_{20}H_{21}NO$ 1H-Inden-1-one, 2-[[(1,1-dimethylethyl)-amino]phenylmethyl]-	hexane	238(4.60),244(4.62), 303(3.04),314(3.08), 327(2.95),389(3.00)	44-3540-76
	CHCl$_3$	390(2.60)	44-3540-76
$C_{20}H_{21}NOS$ 2-Azoniabicyclo[3.1.0]hex-3-ene, 3-hydroxy-2,2-dimethyl-1-(methylthio)-4,5-diphenyl-, hydroxide, inner salt	dioxan	221s(4.201),300(4.238)	24-0562-76
Benzeneacetamide, N,N-dimethyl-α-[2-(methylthio)-1-phenyl-2-propenylidene]-, cis	EtOH	225(4.312),280(4.000)	24-0576-76
trans	EtOH	227s(4.260),276(4.127)	24-0576-76
Propanethioamide, N,N-diethyl-2-(9H-xanthen-9-ylidene)-	CCl$_4$	282(4.28),304s(4.15), 325(4.11),375(2.97)	88-0209-76
Unknown compound II, m. 221-3°	EtOH	284(4.454)	24-0576-76
	dioxan	281(4.467)	24-0576-76
	MeCN	283(4.477)	24-0576-76
$C_{20}H_{21}NO_3$ Benzeneacetic acid, α-[2-(dimethylamino)-2-oxo-1-phenylethylidene]-, ethyl ester, (Z)-	dioxan	222(4.294),272(4.049)	24-0576-76
Nornuciferine, N-acetyl-, (±)-	MeOH	211(4.09),268(3.78), 300s(3.10)	102-1169-76
$C_{20}H_{21}NO_3S$ Benzeneacetamide, N,N-dimethyl- -[[2-(methylsulfinyl)oxiranyl]phenyl]-methylene]-, (Z)-	dioxan	229(4.238),282(3.939)	24-0576-76
$C_{20}H_{21}NO_4$ Aporphine, 2,10-dimethoxy-1,11-(methylenedioxy)-, (6aS)-	EtOH	276(4.16),305(3.78)	33-2551-76
Canadine	EtOH	229(4.13),288(3.75)	73-3157-76
Canadine, (±)-	EtOH	229s(4.13),287(3.75)	73-3654-76
9H-Carbazole, 9-[2,3-O-(1-methylethylidene)-α-D-ribofuranosyl]-	EtOH	235(4.58),248(4.28), 258(4.11),292(4.18), 323(3.60),332(3.69)	136-0043-76A
9H-Carbazole, 9-[2,3-O-(1-methylethylidene)-β-D-ribofuranosyl]-	EtOH	234(4.63),248(4.29), 257(4.01),290(4.22), 323(3.69),332(3.69)	136-0043-76A
1(3H)-Isobenzofuranone, 6,7-dimethoxy-3-(1,2,3,4-tetrahydro-2-methyl-1-isoquinolinyl)-, [S-(R*,S*)]-	EtOH	209(4.56),263s(3.05), 272(3.06),310(3.62)	44-1657-76
Isoquinoline, 1-[(2,3-dimethoxyphenyl)-methyl]-6,7-dimethoxy-, hydrochloride	EtOH	239(4.78),272(3.77), 312(3.68),326(3.71)	87-0882-76
Oliverine	EtOH	223(4.24),239s(4.10), 284(4.23),319s(3.75)	100-0350-76
Pseudoepiberberine, tetrahydro-	EtOH	233(3.94),290(3.82)	73-3157-76
Sinactine	EtOH	232s(4.13),287(3.86)	73-3157-76
$C_{20}H_{21}NO_5$ 1,3-Benzodioxole-4-methanol, 5-[(5,6,7,8-tetrahydro-6-methyl-1,3-dioxolo[4,5-g]isoquinolin-5-yl)methyl]-, (+)-	EtOH	240(4.93),294(4.98)	105-0432-76
13-Berbancarboxylic acid, 12,13-didehydro-7,8-(methylenedioxy)-14-oxo-, methyl ester	MeOH	230s(3.8),292(3.7)	24-1724-76

Compound	Solvent	$\lambda_{max}(\log \epsilon)$	Ref.
13-Berbancarboxylic acid, 12,17-didehydro-7,8-(methylenedioxy)-14-oxo-, methyl ester	MeOH	225(4.1),288(3.7)	24-1724-76
N-Oxyoliverine	EtOH	223(4.38),238s(4.03), 283(4.18),317s(3.70)	100-0350-76
$C_{20}H_{21}NO_6$			
2-Demethylcolchiceine	EtOH-NaOH	236(4.78),280s(4.10), 343(4.49),366s(4.39), 380(4.35)	73-3146-76
3-Demethylcolchiceine	EtOH-NaOH	242(4.71),258s(4.64), 266s(4.49),302(3.94), 368(4.46),397(4.45)	73-3146-76
1,3(2H,9bH)-Dibenzofurandione, 6-acetyl-2-(1-ethylaminoethylidene)-7,9-dihydroxy-8,9b-dimethyl-, (+)-	MeCN	294(4.38),348(3.56)	23-2795-76
Formamide, N-(5,6,7,9-tetrahydro-2-hydroxy-1,3,10-trimethoxy-9-oxobenzo[a]heptalen-7-yl)-, (S)-	EtOH	243(4.46),293s(3.54), 355(4.20)	73-3146-76
	EtOH	247(4.45),290(3.86), 356(4.19)	73-3146-76
Formamide, N-(5,6,7,9-tetrahydro-3-hydroxy-1,2,10-trimethoxy-9-oxobenzo[a]heptalen-7-yl)-, (S)-	EtOH	232(4.48),245s(4.44), 357(4.19)	73-3146-76
	EtOH-NaOH	225(4.52),245(4.51), 279(3.99),374(4.00)	73-3146-76
$C_{20}H_{21}NO_7$			
Benzofuro[3,2-f]-1,2-benzisoxazol-4(4aH)-one, 8-acetyl-9a-ethoxy-9a,10-dihydro-5,7-dihydroxy-3,4a,6-trimethyl-, cis-(+)-	MeOH	223(4.00),285(3.93), 330(3.54)	23-3721-76
$C_{20}H_{21}NS_2$			
2-Azoniabicyclo[3.1.0]hex-3-ene, 3-mercapto-2,2-dimethyl-1-(methylthio)-4,5-diphenyl-, hydroxide, inner salt	EtOH	222(4.362),325(4.093)	24-0562-76
Benzeneethanethioamide, N,N-dimethyl-α-[2-(methylthio)-1-phenyl-2-propenylidene]-, cis	EtOH	244(4.318),298s(4.158), 375s(2.982)	24-0576-76
trans	EtOH	249(4.276),286s(4.161), 370s(2.955)	24-0576-76
Unknown compound III	EtOH	245(4.356),285s(3.819), 385(2.415)	24-0576-76
$C_{20}H_{21}N_2$			
Isoquinolinium, 1-[2-[4-(dimethylamino)phenyl]ethenyl]-2-methyl-, perchlorate	EtOH	235(4.45),315(4.00), 480(4.05)	103-0683-76
Quinolinium, 2-[2-[4-(dimethylamino)phenyl]ethenyl]-, perchlorate	EtOH	230(4.04),280(3.70), 330(3.78),530(4.46)	103-0683-76
$C_{20}H_{21}N_2O_3$			
Benzoic acid, 2-[5-(4-methoxyphenyl)-1,2-dimethyl-1H-pyrazolium-3-yl)-, methyl ester, iodide	EtOH	218(4.57),277(4.29)	44-0110-76
$C_{20}H_{21}N_2S_2$			
Benzothiazolium, 2-[4-[4-(dimethylamino)phenyl]-1,3-butadienyl]-3-methyl-, methyl sulfate	$MeNO_2$	572(4.81)	124-1174-76

Compound	Solvent	$\lambda_{max}(\log \epsilon)$	Ref.
$C_{20}H_{21}N_3O$ 1,4-Diazepino[2,1-b]quinazolin-11-one, 1,2,3,4,5,11-hexahydro-3-(2-phenylethyl)-, dihydrochloride	n.s.g.	267(3.95),274(3.95), 305(3.64),316(3.55)	2-0879-76
$C_{20}H_{21}N_3OS_2$ 3-Thiazolidinecarbothioamide, 5,5-dimethyl-4-oxo-N-(phenylmethyl)-2-[(phenylmethyl)imino]-	EtOH	240s(4.11),276(4.09)	33-2768-76
4-Thiazolidinone, 5-[bis[(phenylmethyl)amino]methylene]-3-ethyl-2-thioxo-	EtOH	262(4.04),380(4.37)	94-1671-76
$C_{20}H_{21}N_3O_4$ Acetamide, N-[2,3-dihydro-1-(methoxymethyl)-2-oxo-5-phenyl-1H-1,4-benzodiazepin-7-yl]-N-methoxy-	isoPrOH	215s(4.49),249(4.42), 314(3.34)	87-1378-76
1,2-Hydrazinedicarboxylic acid, 1-(2-phenyl-3-indolizinyl)-, diethyl ester	MeOH	221(4.42),252(4.73), 307(3.94)	2-0057-76
$C_{20}H_{21}N_3O_4S$ 1H-Indazole-4,7-dione, 3,6-dimethyl-1-[(4-methylphenyl)sulfonyl]-5-pyrrolidino-	EtOH	236(4.38),540(3.63)	94-1731-76
$C_{20}H_{21}N_3O_5S$ 1H-Indazole-4,7-dione, 3,6-dimethyl-1-[(4-methylphenyl)sulfonyl]-5-morpholino-	EtOH	235(4.35),318(3.79), 537(3.47)	94-1731-76
$C_{20}H_{21}N_3O_6$ Propanedioic acid, [[[3-nitro-4-(phenylamino)phenyl]amino]methylene]-, diethyl ester	EtOH	457(3.75)	146-0131-76
$C_{20}H_{21}N_3O_{11}$ D-Ribose, 1-C-(6-nitro-1H-indazol-3-yl)-, 2,3,4,5-tetraacetate	EtOH	225s(3.87),272(4.35), 335s(3.48)	4-1241-76
$C_{20}H_{21}N_4OP$ Phosphinic diamide, N,N'-dimethyl-N,N'-diphenyl-P-(phenylazo)-	MeOH	207(4.33),226(4.30), 299(4.16),503(2.10)	139-0207-76A
$C_{20}H_{21}N_9O_4$ Pyrido[2,3-b]pyrazine, 1,2,3,5-tetrahydro-5-methyl-2,3-bis[2-(4-nitrophenyl)hydrazino]-, conjugate monoacid	n.s.g.	264(3.90),374(4.49)	103-0948-76
$C_{20}H_{22}$ Benzene, 1,1'-(1,2-ethynediyl)bis[2,4,6-trimethyl-	C_6H_{12}	280(4.35),286(4.39), 296(4.53),303(4.43), 315(4.52)	101-0385-76N
$C_{20}H_{22}ClNOS$ Dibenzo[b,f]thiepin-10(11H)-one, 8-chloro-11-[3-(dimethylamino)-2-methylpropyl]-, hydrochloride	MeOH	224(4.32),242(4.28), 263s(4.11),338(3.64)	73-3420-76
$C_{20}H_{22}ClNO_4$ Isoquinoline, 1-[(2,3-dimethoxyphenyl)methyl]-6,7-dimethoxy-, hydrochloride	EtOH	239(4.78),272(3.77), 312(3.68),326(3.71)	87-0882-76

Compound	Solvent	λ_{max} (log ϵ)	Ref.

$C_{20}H_{22}N_2O$
3(2H)-Indolizinone, hexahydro-2-phenyl-
  1-(phenylamino)-

| | EtOH | 248(4.27),295(3.36) | 39-0005-76C |
| | EtOH-HCl | 255(2.34) | 39-0005-76C |

$C_{20}H_{22}N_2O_2$
2-Butenediamide, N,N,N',N'-tetramethyl-
  2,3-diphenyl-, (E)-

| | dioxan | 226(4.346),267(4.037) | 24-0576-76 |

(Z)-

| | dioxan | 259(4.149) | 24-0576-76 |

6,11-Methanoindolo[3,2-b]quinolizine-
  6(5H)-carboxylic acid, 11-ethyl-
  8,11,11a,12-tetrahydro-, methyl
  ester, [6R-(6α,11α,11aβ)]-

| | EtOH | 225(4.55),281(3.83),
289(3.77) | 33-1213-76 |

1-Piperazinecarboxylic acid, 4-(9H-fluo-
  ren-2-yl)-, ethyl ester

| | MeOH | 293(4.38) | 73-0906-76 |

Quino[2,1-c][1,4]benzodiazepine-11-carb-
  oxylic acid, 5,6,6a,7,8,13-hexahydro-,
  ethyl ester

| | EtOH | 225(4.11),259(4.21),
302(4.34) | 35-3678-76 |

$C_{20}H_{22}N_2O_3$
Indolo[2,3-a]quinolizine-2-acetic acid,
  3-ethylidene-1,2,3,4,6,7,12,12b-octa-
  hydro-4-oxo-, methyl ester, (Z)-

| | MeOH | 224(4.62),274(3.97),
281(3.96),292(3.83) | 24-3825-76 |

3(2H)-Isoquinolinone, 2-(2-aminophenyl)-
  6,7-dimethoxy-1-propyl-, monoper-
  chlorate

| | MeOH | 255(4.65),405(3.57) | 103-0205-76 |

$C_{20}H_{22}N_2O_4$
Benzenemethanol, α,α'-bis[α-methyl-,
  diacetate, (R*,S*)-

| | CH_2Cl_2 | 357(1.60) | 39-1249-76B |

L-Phenylalanine, N-cinnamylglycyl-

| | MeOH-acid | 274(4.39) | 20-0647-76 |

$C_{20}H_{22}N_2O_5$
1,3(2H,9bH)-Dibenzofurandione, 7,9-di-
  hydroxy-8,9b-dimethyl-2-[1-(methyl-
  amino)ethylidene]-6-[1-(methylimino)-
  ethyl]-

| | MeCN | 234(4.05),250(3.96),
300(4.13) | 23-2795-76 |

$C_{20}H_{22}N_2O_6$
1,4-Benzoxazepine, 3-ethyl-2,3-dihydro-
  7,8-dimethoxy-5-(4-methoxy-3-nitro-
  phenyl)-

| | n.s.g. | 215(4.43),255(4.11),
310(4.35) | 20-0787-76 |

$C_{20}H_{22}N_2O_7$
9H-Carbazole-9-hexanoic acid, α-amino-
  1,4-dihydro-6-hydroxy-2,7-dimethoxy-
  1,4-dioxo-, (S)-

| | EtOH | 225(4.17),258(4.07),
273(4.00),290(3.94),
317(3.85),490(3.47) | 146-0667-76 |

4,1-(Iminomethano)benzofuro[3,2-c]pyrid-
  ine-1,4-dicarboxylic acid, 2,3,4a,9b-
  tetrahydro-2,10-dimethyl-3,11-dioxo-,
  diethyl ester, (1α,4α,4aα,9bα)-

| | EtOH | 273s(3.36),280(3.52),
288(3.52) | 78-0507-76 |

(1α,4β,4aα,9bα)-

| | EtOH | 273s(3.38),281(3.52),
288(3.53) | 78-0507-76 |

$C_{20}H_{22}N_2O_9$
1,2,3,4-Butanetetrol, 1-(5-phenyl-1,3,4-
  oxadiazol-2-yl)-, tetraacetate, [1R-
  (1R*,2S*,3S*)]-

| | EtOH | 251(4.42) | 136-0049-76D |

D-Ribose, 1-C-1H-indazol-3-yl-, 2,3,4,5-
  tetraacetate

| | EtOH | 237(3.91),242(3.90),
302(4.00) | 4-1241-76 |

Compound	Solvent	$\lambda_{max}(\log \epsilon)$	Ref.
$C_{20}H_{22}N_2S$ 1H-Imidazolium, 1-butyl-5-mercapto-3-methyl-2,4-diphenyl-, hydroxide, inner salt	MeOH	306(3.84)	2-0382B-76
$C_{20}H_{22}N_4O_2$ 5-Pyrimidinecarboxamide, 4-[4-(dimethylamino)phenyl]-1,2,3,4-tetrahydro-6-methyl-2-oxo-N-phenyl-	EtOH	208(4.33),264(4.12), 275(4.01)	103-0191-76
$C_{20}H_{22}N_4O_5$ 5-Pyrimidinecarboxamide, 1,2,3,4-tetrahydro-6-methyl-2-oxo-N-2-pyridinyl-4-(3,4,5-trimethoxyphenyl)-	EtOH	206(4.36),242(3.90), 290(3.91)	103-0191-76
$C_{20}H_{22}N_4O_7$ Uridine, 2',3'-bis(acetylamino)-2',3'-dideoxy-, 5'-benzoate	MeOH	224(4.22),259(4.05)	44-3138-76
$C_{20}H_{22}N_4O_9$ 1H-Pyrazole-4-carboxylic acid, 5-[(methoxycarbonyl)amino]-1-(4-nitrophenyl)-3-(tetrahydro-2,2-dimethylfuro[3,4-d]-1,3-dioxol-4-yl)-, methyl ester, [3aS-(3aα,4α,6aα)]-	CHCl	243(3.77),297(4.14)	136-0019-76A
$C_{20}H_{22}N_8O_4S_2$ 1,2,4,5-Tetrazine-3,6-dicarboximidamide, N,N,N'',N''-tetramethyl-N,N'''-bis(phenylsulfonyl)-	MeCN	240(4.61),525(2.65)	32-0001-76
$C_{20}H_{22}OS_2$ 9,13-Dithiatricyclo[13.3.1.1^{3,7}]eicosa-1(19),3,5,7(20),15,17-hexaen-2-one, 11,11-dimethyl-	MeOH	249(4.17)	44-2509-76
$C_{20}H_{22}O_2$ 6H-Cyclopenta[a]phenanthren-17-ol, 7,11,12,13,16,17-hexahydro-3-methoxy-13-methyl-11-methylene-, cis-($\pm$)-	MeOH	247(4.13),257(4.04), 312(4.30),324(4.37), 339(4.27)	44-0531-76
17H-Cyclopenta[a]phenanthren-17-one, 6,7,11,12,13,16-hexahydro-3-methoxy-12,13-dimethyl-, (12R-cis)-	n.s.g.	312(4.47)	39-1643-76C
(12S-trans)-	n.s.g.	312(4.46)	39-1643-76C
1,4-Naphthalenedione, 2,3-bis(3-methyl-2-butenyl)-	EtOH	243s(4.26),245(4.25), 260(4.19),268(4.20), 327(3.25)	102-0225-76
Tricyclo[10.2.2.2^{4,7}]octadeca-4,6,12,14-15,17-hexaene-8-carboxylic acid, methyl ester	n.s.g.	251(2.59),271(2.59), 282s(2.43),292s(1.90)	88-4095-76
Tricyclo[10.2.2.2^{4,7}]octadeca-4,6,12,14-15,17-hexaene-9-carboxylic acid, methyl ester	n.s.g.	271(2.65),283(2.49)	88-4095-76
$C_{20}H_{22}O_3$ 1,3-Cyclopentanedione, 2-[2-(3,4-dihydro-6-methoxy-1-naphthalenyl)-2-propenyl]-2-methyl-	MeOH	274(4.02)	44-0531-76

Compound	Solvent	$\lambda_{max}(\log \epsilon)$	Ref.
$C_{20}H_{22}O_4$			
Cyclopentaneacetic acid, 2-(6-methoxy-2-naphthalenyl)-1-methyl-5-oxo-, methyl ester	EtOH	230(4.17),232(4.19), 312(4.47)	78-0079-76
Galeon	MeOH	280(3.62)	88-3069-76
Glabrene, tetrahydro-	EtOH	282s(3.74),284(3.76), 288s(3.59)	94-0991-76
Isootabaphenol, (-)-	EtOH	237(4.09),292(3.79)	102-0773-76
Phenol, 4-[2-(1,3-benzodioxol-5-yl)-1-methylethyl]-5-methoxy-2-(2-propenyl)-	EtOH	225s(4.13),286(3.93)	88-4371-76
$C_{20}H_{22}O_4S_2$			
1-Propanone, 3,3'-dithiobis[1-(2-hydroxy-3-methylphenyl)-	MeOH	259(4.15),330(3.76)	23-0455-76
$C_{20}H_{22}O_5$			
3H-Benz[e]inden-3-one, 5-ethyl-1,2-dihydro-4,8-dihydroxy-7,9-dimethoxy-1-propylidene-	MeOH	226(4.30),289(4.45), 329(4.04),415(3.70)	18-1937-76
2H,6H-Benzo[1,2-b:5,4-b']dipyran-6-one, 5-hydroxy-10-(3-hydroxy-3-methyl-1-butenyl)-2,2,8-trimethyl-	MeOH	246(4.36),275(4.45), 320s(3.62)	24-2963-76
4H,8H-Benzo[1,2-b:3,4-b']dipyran-4-one, 5-hydroxy-6-(3-hydroxy-3-methyl-1-butenyl)-2,8,8-trimethyl-	MeOH	215(4.18),240(4.35), 275(4.38),350s(3.48)	24-2963-76
Hydroxygaleon	MeOH	279(3.88)	88-3069-76
Icetexone	EtOH	213(4.14),260s(--), 315(3.90),430(2.88)	88-2501-76
$C_{20}H_{22}O_6$			
Benzaldehyde, 2-(3,4-dihydro-6,7-dimethoxy-1H-2-benzopyran-3-yl)-4,5-dimethoxy-	dioxan	236(4.63),283(4.15), 312(3.86)	33-0949-76
D-Glucitol, 1,3:2,4-bis-O-(phenylmethylene)-, [1(R),2(S)]-	1:1 MeCN-DMSO	257(2.59)	12-1859-76
Matairesinol	EtOH	232(4.06),283(3.74)	102-1789B-76
4H-Naphtho[2,3-b]pyran-4-one, 5,6,8,10-tetramethoxy-2-propyl-	n.s.g.	227(4.45),253s(4.47), 270(4.67),316(3.35), 327(3.50),346(3.53), 383(3.99)	12-1087-76
Pinoresinol	EtOH	208(4.26),230(4.24), 283(3.66)	102-1789B-76
$C_{20}H_{22}O_7$			
Budlein A	EtOH	215(4.32),266(4.00)	102-0525-76
Spiro[furan-3(2H),6'(7'H)-[1H-3,9a]methano[2]benzoxepin-10'-carboxylic acid, 5-(3-furanyl)octahydro-7'-methyl-1',2-dioxo-, [3'R-(3'α,5'aα,6'β-7'β,9'aα,10'(S*)]-	EtOH	209(3.82)	78-1881-76
$C_{20}H_{22}O_8$			
β-D-Xylopyranoside, 3,4-dihydro-7-hydroxy-2-(4-hydroxyphenyl)-2H-1-benzopyran-5-yl, (S)-	n.s.g.	208(4.70),227s(4.38), 275(3.21)	102-1185-76
$C_{20}H_{22}O_9$			
Astringin	MeOH	221(4.37),305s(4.33), 327(4.43)	102-2006-76

Compound	Solvent	$\lambda_{max}(\log \epsilon)$	Ref.
$C_{20}H_{22}O_{10}$ Cinnamic acid, 3,4-dihydroxy-, glyceryl ester, tetraacetate	MeOH	218(4.23),283(4.36)	100-0409-76
$C_{20}H_{23}ClN_5O_6PS$ Adenosine, 2-butyl-8-[(4-chlorophenyl)-thio]-, cyclic 3',5'-(hydrogen phosphate)	pH 1 pH 7 pH 11	278(4.32) 282(4.26) 282(4.26)	87-0419-76 87-0419-76 87-0419-76
$C_{20}H_{23}Cl_2N_3O$ 1,4-Diazepino[2,1-b]quinazolin-11-one, 1,2,3,4,5,11-hexahydro-3-(2-phenyl-ethyl)-, dihydrochloride	n.s.g.	267(3.95),274(3.95), 305(3.64),316(3.55)	2-0879-76
$C_{20}H_{23}Cl_4N_5O_8$ 1,2-Hydrazinedicarboxylic acid, 1,1'-(4,5,6,7-tetrachloro-2H-isoindole-1,3-diyl)bis-, tetraethyl ester	CH_2Cl_2	240(4.75),269(3.29), 300(3.15),338(3.77), 349(3.87),360(3.82)	88-1661-76
$C_{20}H_{23}NO$ Methanone, phenyl[2-[2-(1-piperidinyl)-ethyl]phenyl]-	MeOH	250(4.18),280(3.55), 290(3.34),330(2.26)	35-0541-76
$C_{20}H_{23}NO_3$ Ethanone, 1-[4-(1,3-diethoxy-1,3-di-hydro-2H-isoindol-2-yl)phenyl]-	EtOH	230(3.92),309(4.47)	83-0356-76
$C_{20}H_{23}NO_4$ Corytenchine	EtOH	287(3.58)	39-0063-76C
mol. compd. with corytenchirine	EtOH	287(3.82)	39-0063-76C
Dibenzo[d,f]indol-2(5H)-one, 6,7,7a,8-tetrahydro-3,10,11-trimethoxy-7-meth-yl-, 1:1 compd. with borane, (4aR*,7aR*)-(+)-	EtOH	234s(4.29),262(4.15), 290(4.09),356(3.92)	44-3210-76
Indeno[1,2-j]isoquinolin-3(5H)-one, 6,7,7a,8-tetrahydro-2,10,11-trimeth-oxy-7-methyl-, 1:1 compd. with borane	EtOH	244(4.27),286(3.87)	44-3210-76
Norglaucine	EtOH	217(4.49),280(4.18), 300(4.09)	2-0134-76
Predicentrine	EtOH	218(4.58),236s(4.36), 273s(4.09),282(4.16), 304(4.14),314s(4.09)	36-0294-76
	EtOH-NaOH	233(4.11),305s(4.08), 330(4.23)	36-0294-76
Sebiferine	EtOH	238(4.25),282(3.89)	2-0150-76
$C_{20}H_{23}NO_5$ Colchicine, N-deacetyl-2-demethyl-N-methyl-	EtOH EtOH-NaOH	241(4.49),355(4.19) 244(4.60),287(3.96), 355(4.13)	73-3146-76 73-3146-76
Colchicine, N-deacetyl-3-demethyl-N-methyl-	EtOH EtOH-NaOH	244(4.47),356(4.20) 235(4.44),250(4.47), 291(3.87),318s(3.82), 394(4.01)	73-3146-76 73-3146-76
5-Isoxazolidinecarboxylic acid, 3-(2,5-dimethoxyphenyl)-5-methyl-2-phenyl-, methyl ester, cis	n.s.g.	250(3.92),293(3.70), 429(2.34)	142-0109-76S
$C_{20}H_{23}NO_6S$ 3-Azatetracyclo[7.1.0.0^{2,4}.0^{5,7}]decane-	EtOH	255(2.89),261(2.91),	24-3505-76

Compound	Solvent	$\lambda_{max}(\log \epsilon)$	Ref.
7,9-dicarboxylic acid, 3-[(4-methyl-phenyl)sulfonyl]-, dimethyl ester, $(1\alpha,2\alpha,4\alpha,5\alpha,7\alpha,9\alpha)$- (cont.)		266s(2.85),272(2.72)	24-3505-76
$C_{20}H_{23}NO_{12}$			
α-D-Glucopyranosiduronic acid, (4-nitro-phenyl)methyl methyl ester, 2,3,4-tri-acetate	EtOH	267(4.11)	94-0394-76
β-	EtOH	267(4.11)	94-0394-76
$C_{20}H_{23}N_2$			
Isoquinolinium, 1-[2-[4-(dimethylamino)-phenyl]ethenyl]-3,4-dihydro-2-methyl-, perchlorate	EtOH	300(3.65),494(4.05)	103-0683-76
$C_{20}H_{23}N_3$			
Benzenamine, 2-[3-methyl-5-(2-methylpro-pyl)-1-phenyl-1H-pyrazol-4-yl]-	MeOH	240(4.26),295(3.58)	23-1020-76
$C_{20}H_{23}N_3O$			
Phenazine, 2-(1,1-dimethylethyl)-3-mor-pholino-	EtOH	256(5.07),372(4.16)	18-2333-76
$C_{20}H_{23}N_3O_2$			
Acetamide, N-acetyl-N-(6,7-dihydro-1,6,6-trimethyl-3-phenyl-1H-inda-zol-4-yl)-	MeOH	235-258(4.04)	24-1898-76
Benzoic acid, 4-(5,6-dihydro-6-methyl-2H-1,5-methano-1,3,6-benzotriazocin-3(4H)-yl)-, ethyl ester	50% dioxan	258(4.05),311(4.34)	35-3678-76
2H-1,14-Ethanopyrido[3',2':6,7][1,3]di-azocino[1,8-a]indole-6,8(5H,7H)-dione, 4a-ethyl-3,4,4a,14b-tetrahydro-, trans-(±)-	MeOH	220(4.08),246(4.18), 261s(4.02),293(3.62), 301(3.59)	22-1961-76
$C_{20}H_{23}N_3O_3$			
Acetamide, N-[3-(3,4-dimethoxyphenyl)-1,5,6-trimethyl-1H-indazol-4-yl]-	MeOH	213(4.63),245s(4.07), 278(3.83),309(4.01)	24-1898-76
2H-1,4-Benzodiazepin-2-one, 1,3-dihydro-1-(methoxymethyl)-7-(methoxymethyl-amino)-3-methyl-5-phenyl-	isoPrOH	240(4.44),330(3.40)	87-1378-76
$C_{20}H_{23}N_3O_4S$			
Benzenesulfonic acid, 4-methyl-, [1-(4-methyl-3,6-dioxo-2-(1-pyrrolidinyl)-1,4-cyclohexadien-1-yl]ethylidene]-hydrazide	EtOH	235(4.45),275(4.00), 515(3.46)	94-1731-76
$C_{20}H_{23}N_3O_5S$			
Benzenesulfonic acid, 4-methyl-, [1-[4-methyl-2-(4-morpholinyl)-3,6-dioxo-1,4-cyclohexadien-1-yl]ethylidene]-hydrazide	EtOH	230(4.43),508(3.41)	94-1731-76
$C_{20}H_{23}N_3O_6$			
Ethanol, 2,2'-[[3-nitro-4-(phenylamino)-phenyl]imino]bis-, diacetate	EtOH	512(3.68)	146-0131-76
4,1-(Iminomethano)-1H-pyrido[4,3-b]in-dole-1,4(4aH)-dicarboxylic acid, 2,3,5,9b-tetrahydro-2,10-dimethyl-	EtOH	248(3.83),309(3.46)	78-0507-76

Compound	Solvent	$\lambda_{max}(\log \epsilon)$	Ref.
3,11-dioxo-, diethyl ester, (1α,4aα,9bα)- (cont.)			78-0507-76
(1α,4β,4aα,9bα)-	EtOH	247(3.79),307(3.43)	78-0507-76
$C_{20}H_{23}N_3O_7S$ 5-Thia-1-azabicyclo[4.2.0]oct-3-ene-2-carboxylic acid, 2-[(ethoxycarbonyl)-amino]-3-methyl-8-oxo-7-[(phenoxy-acetyl)amino]-, methyl ester	n.s.g.	266(3.63)	88-2909-76
$C_{20}H_{23}N_5O_6$ 7(8H)-Pteridinone, 2-(dimethylamino)-4-(phenylmethoxy)-8-β-D-ribofuranosyl-	MeOH	242(4.30),292(3.71), 363(4.28)	24-3228-76
$C_{20}H_{23}N_5O_6S$ Azirino[6,7]cyclohepta[2,1-c:4,5-c']di-pyrazole-4a,5a(1H,5H)-dicarboxylic acid, 1a,1b,2,8,8a,8b-hexahydro-1-[(4-methylphenyl)sulfonyl]-, dimethyl ester	MeOH	256(2.73),262(2.81), 267(2.76),274(3.70), 323(2.52)	24-3505-76
$C_{20}H_{24}$ 1,3,7,9-Cyclododecatetrayne, 5,5,6,6-11,11,12,12-octamethyl-	EtOH	236(2.86),248(2.89), 263(2.71)	88-2663-76
[2.2]Paracyclophane, 4,7,12,15-tetra-methyl-	isooctane	226(4.26),250s(3.57), 308s(2.10)	44-4081-76
$C_{20}H_{24}BrNO_4$ Isoquinoline, 1-[(2-bromo-4,5-dimethoxy-phenyl)methyl]-1,2,3,4-tetrahydro-6,7-dimethoxy-	EtOH	228(4.23),283(3.81)	2-0134-76
$C_{20}H_{24}N_2$ 3,3'-Bipyrrolidine, 1,1'-diphenyl-, (R*,S*)-	EtOH	203(4.08),257(4.02), 304(1.04)	104-2062-76
$C_{20}H_{24}N_2O$ Cycloaffinisine	MeOH	228(4.50),283(3.81), 292(3.77)	24-3527-76
$C_{20}H_{24}N_2OS$ 1H-Pyrrole, 3-methoxy-2-[(5-methyl-4-pentyl)-2H-pyrrol-2-ylidene)-methyl]-5-(2-thienyl)-	EtOH-HCl EtOH-NaOH	530(4.97) 495(4.42)	4-0497-76 4-0497-76
$C_{20}H_{24}N_2O_2$ Ibophyllidine	EtOH	227(4.08),302(3.95), 333(3.78)	78-2539-76
$C_{20}H_{24}N_2O_3$ Acetamide, N-(4-butoxyphenyl)-2-[(4-eth-oxyphenyl)imino]-	EtOH	235(3.03)	65-0702-76
$C_{20}H_{24}N_2O_4$ Acetamide, N-(4-butoxyphenyl)-2-[(4-eth-oxyphenyl)imino]-, N-oxide	EtOH	235(4.1),288(4.0), 340(4.0)	65-0702-76
Benzamide, N,N'-1,6-hexanediylbis[2-hy-droxy-	MeOH	301(3.94)	18-2679-76
7,11-Methano-2H-pyrido[1,2-a]azocine-9-carboxylic acid, 1,3,4,6,7,8,11,11a-octahydro-10-hydroxy-6-oxo-7-(2-pyrid-inyl)-, ethyl ester, (7α,11β,11aα)-(±)-	EtOH	254(4.1)	23-0097-76

Compound	Solvent	$\lambda_{max}(\log \epsilon)$	Ref.
$C_{20}H_{24}N_2O_4S_2$ 5(4H)-Oxazolone, 4-[[[[2,2-dimethyl- 4-(3-methyl-2-thiazolidinyl)-1,3-di- oxolan-4-yl]methyl]thio]methylene]- 2-phenyl-	EtOH	361(4.65)	23-2089-76
$C_{20}H_{24}N_2O_5S$ 5(4H)-Oxazolone, 4-[[[[2,2-dimethyl- 4-(3-methyl-2-oxazolidinyl)-1,3-diox- olan-4-yl]methyl]thio]methylene]- 2-phenyl-	MeOH	361(4.66)	23-2089-76
$C_{20}H_{24}N_2O_7$ Benzamide, N-[1-[(3,4-dimethoxyphenoxy)- methyl]propyl]-4-methoxy-3-nitro-	n.s.g.	230(4.42),285(3.66)	20-0787-76
$C_{20}H_{24}N_4O_4$ 2-Quinoxalinecarboxylic acid, 3,4,5,6- 7,8-hexahydro-4-(2-methoxyethyl)-3- [[(phenylamino)carbonyl]imino]-, methyl ester	EtOH	233(4.01),298(4.24), 402(3.97)	88-2899-76
$C_{20}H_{24}N_6O_3$ Adenosine, 5'-deoxy-2',3'-O-(1-methyl- ethylidene)-5'-[(phenylmethyl)amino]-	EtOH	259(4.27)	87-0684-76
$C_{20}H_{24}N_7$ Pyrido[2,3-b]pyrazinium, 1,2,3,4-tetra- hydro-2,3-bis(phenylhydrazino)- 5-methyl-	n.s.g.	272(4.38),319(4.12), 346(4.20)	103-0948-76
$C_{20}H_{24}N_8O_4S_2$ 1,2,4,5-Tetrazine-3,6-dicarboximidamide, 1,2-dihydro-N,N,N'',N''-tetramethyl- N',N'''-bis(phenylsulfonyl)-	MeCN	252(4.48)	32-0001-76
$C_{20}H_{24}O_2$ Benzene, 5-methoxy-2-[2-(4-methoxy-2,6- dimethylphenyl)ethenyl]-1,3-dimethyl-	EtOH	273(4.31)	39-1466-76C
Benzene, 5-methoxy-2-[2-(4-methoxy- 2-methylphenyl)-1-propenyl]-1,3- dimethyl-, (E)-	EtOH	235(4.16),283(3.65)	39-1466-76C
Benzene, 5-methoxy-2-[2-(4-methoxyphen- yl)-1-butenyl]-1,3-dimethyl-, (E)-	EtOH	256(4.23)	39-1466-76C
(Z)-	EtOH	270(4.04)	39-1476-76C
Estra-1,3,5(10),8(14),9(11)-pentaen- 17β-ol, 3-methoxy-11-methyl-, (±)-	MeOH	244(4.32),250s(4.31), 285(3.88)	44-0531-76
8α-Estra-1,3,5(10),8-tetraen-17-one, 3-methoxy-12β-methyl-	n.s.g.	279(4.20)	39-1643-76C
14β-Estra-1,3,5(10),8-tetraen-17-one, 3-methoxy-12α-methyl-	n.s.g.	273(4.18)	39-1643-76C
$C_{20}H_{24}O_3$ 5H-Inden-5-one, 1,2,3,6,7,7a-hexahydro- 1-hydroxy-4-[2-(3-methoxyphenyl)-2- propenyl]-7a-methyl-, (1S-cis)-	EtOH	211(4.37),244(4.18)	54-0223-76
$C_{20}H_{24}O_4$ Dehydrooopodin	EtOH	224(4.33),232(4.32)	105-0344-76

Compound	Solvent	$\lambda_{max}(\log \epsilon)$	Ref.
4a(2H)-Phenanthrenecarboxylic acid, 3,4,9,10-tetrahydro-6-methoxy-7-(1-methylethyl)-2-oxo-, methyl ester	MeOH	230(4.36),282(3.48)	12-2087-76 +18-1985-76
$C_{20}H_{24}O_5$			
Badkhysin	EtOH	252(4.12)	105-0344-76
Badkhyzinin	EtOH	216(3.75)	105-0344-76
2H,8H-Benzo[1,2-b:3,4-b']dipyran-6-propanoic acid, 5-methoxy-2,2,8,8-tetramethyl- (eriostemoic acid)	EtOH	206(4.29),254(4.33), 258(4.33),280(4.13), 331(3.39),347(3.29)	18-1653-76
$C_{20}H_{24}O_6$			
Benzene, 1,1'-(1,2-ethenediyl)bis[3,4,5-trimethoxy-, cis	EtOH	306(4.08)	102-1057-76
trans	EtOH	316(4.53),327(4.57), 344(4.38)	102-1057-76
Dibenzo[b,k][1,4,7,10,13,16]hexaoxacyclooctadecin, 6,7,9,10,17,18,20,21-octahydro-	MeOH	274(3.71)	64-0330-76B
calcium perchlorate complex	MeOH	274(3.72)	64-0330-76B
potassium perchlorate complex	MeOH	273(3.75),279(3.70)	64-0330-76B
sodium perchlorate complex	MeOH	274(3.72),279(3.69)	64-0330-76B
Spiro[furan-3(2H),6'(7'H)-[1H-3,9a]methano[2]benzoxepin]-1',2'-dione, 5-(3-furanyl)octahydro-10'-(hydroxymethyl)-7'-methyl-	EtOH	210(3.75)	78-1881-76
$C_{20}H_{25}NO_2$			
2-Aza-19-norpregna-1,3,5(10)-trien-20-yn-17-ol, 3-methoxy-, (17α)-	MeOH	277(3.57)	44-2864-76
Pentanoic acid, 4-cyano-4-methyl-2-methylene-, 4-cyclohexylphenyl ester	dioxan	264(3.16),270(3.08)	126-3481-76
$C_{20}H_{25}NO_3$			
Laurifonine	MeOH	221(4.30),283(3.90)	39-2197-76C
$C_{20}H_{25}NO_4$			
Erybidine	EtOH	223(5.24),281(3.90)	94-0052-76
$C_{20}H_{25}NO_5$			
1-Propanone, 1-[2-methyl-7-(1-oxopropoxy)-1-[1-(1-oxopropoxy)ethyl]-3-indolizinyl]-	EtOH	228(4.14),244s(3.93), 253s(3.99),263s(4.18), 267(4.21),349s(3.94), 360(4.00),408s(2.69)	1-0198-76
$C_{20}H_{25}NO_6$			
Cyclobutanecarboxylic acid, 1-cyano-2,2,3,3-tetramethyl-4-(2-phenylethenyl)-	CHCl$_3$	257(4.38)	39-1538-76C
$C_{20}H_{25}NO_7$			
D-Mannitol, 1-deoxy-1-(1,3-dihydro-1,3-dioxo-2H-isoindol-2-yl)-3,4:5,6-bis-O-(1-methylethylidene)-	EtOH	222(4.09),240s(3.60), 292(3.30),299s(3.28)	136-0049-76G
$C_{20}H_{25}N_3O$			
Benzamide, N-(2,3-di-1-piperidinyl-2-cyclopropen-1-ylidene)-	MeOH	234(4.02),317(3.54)	138-1215-76

Compound	Solvent	$\lambda_{max}(\log \epsilon)$	Ref.
$C_{20}H_{25}N_3O_4S_2$ 1H-[1,2]Dithiolo[5,1-e][1,2,3]thiadia-zole-7-S^{IV}-3-pentanoic acid, 5-(1,1-dimethylethyl)-1-(4-nitrophenyl)-, methyl ester	C_6H_{12}	225(4.49),258(4.03), 338(4.19),346s(4.16), 506(4.40)	39-0880-76C
$C_{20}H_{25}N_3O_5$ 2,4-Cyclohexadien-1-one, 4-(dibutyl-amino)-6-(2-nitro-4-aci-nitro-2,5-cyclohexadien-1-ylidene)-	benzene	488(4.32)	104-2333-76
	EtOH	500(4.40)	104-2333-76
	CHCl$_3$	495(4.18)	104-2333-76
	dioxan	495(4.15)	104-2333-76
	acetone	500(4.02)	104-2333-76
	THF	513(4.01)	104-2333-76
	EtOAc	488(4.08),530s(--)	104-2333-76
	$C_6H_5NO_2$	480(4.11)	104-2333-76
	pyridine	530(3.85),575s(--)	104-2333-76
$C_{20}H_{25}N_4O_3S$ Morpholinium, 4-[[[5-(4-methoxyphenyl)-2-(4-morpholinyl)-6H-1,3-thiazin-6-yl-idene]amino]methylene]-, perchlorate	HOAc	317(4.23),468(4.23)	97-0268-76
$C_{20}H_{25}N_5O_4$ Propanoic acid, 2,2-dimethyl-, [5-[6-[(2,2-dimethyl-1-oxopropyl)amino]-9H-purin-9-yl]-2-furanyl]methyl ester	MeOH	212(4.28),261(4.42)	35-8213-76
$C_{20}H_{26}$ 1,1'-Bicyclopenta[cd]pentalene, 2a,2'a-3,3',4,4',4a,4'a,5,5',6,6',6a,6'a-6b,6'b-hexadecahydro-stereoisomer	isooctane	237(4.28),245(4.37), 255(4.15)	44-3524-76
	isooctane	237(4.18),245(4.29), 255(4.12)	44-3524-76
$C_{20}H_{26}N_2O_2S$ 1H-Pyrazole, 3-(3,3-dimethyl-1-butynyl)-5-(1,1-dimethylethyl)-1-[(4-methyl-phenyl)sulfonyl]-	EtOH	227(4.26),252(4.18)	78-1293-76
$C_{20}H_{26}O$ 2,4-Cyclopentadien-1-one, 2,5-bis(1,1-dimethylethyl)-3-(2-methylphenyl)-	C_6H_{12}	410(2.71)	89-0160-76
2,4-Cyclopentadien-1-one, 2,5-bis(1,1-dimethylethyl)-3-(3-methylphenyl)-	C_6H_{12}	413(2.75)	89-0160-76
2,4-Cyclopentadien-1-one, 2,5-bis(1,1-dimethylethyl)-3-(4-methylphenyl)-	C_6H_{12}	414(2.82)	89-0160-76
Gona-1,3,5(10),8-tetraene, 3-methoxy-17,17-dimethyl-, (13ξ,14ξ)-	EtOH	273(4.24)	94-0181-76
$C_{20}H_{26}O_2$ 2,4-Cyclopentadien-1-one, 2,5-bis(1,1-dimethylethyl)-3-(2-methoxyphenyl)-	C_6H_{12}	410(2.75)	89-0160-76
2,4-Cyclopentadien-1-one, 2,5-bis(1,1-dimethylethyl)-3-(3-methoxyphenyl)-	C_6H_{12}	413(2.75)	89-0160-76
2,4-Cyclopentadien-1-one, 2,5-bis(1,1-dimethylethyl)-3-(4-methoxyphenyl)-	C_6H_{12}	417(2.87)	89-0160-76
Estra-1,3,5(10),8-tetraen-17β-ol, 3-methoxy-11β-methyl-, ($\pm$)-	MeOH	270s(4.13),278(4.14)	44-0531-76
Estra-1,3,5(10)-trien-17-one, 3-methoxy-11β-methyl-, ($\pm$)-	MeOH	278(3.30),287(3.29)	44-0531-76

Compound	Solvent	$\lambda_{max}(\log \epsilon)$	Ref.
8α-Estra-1,3,5(10)-trien-17-one, 3-methoxy-12β-methyl-	n.s.g.	278(3.43),287(3.40)	39-1643-76C
2(5H)-Furanone, 4-methyl-5-[5-(2,6,6-trimethyl-1-cyclohexenyl)-3-methyl-2(E),4(E)-pentadienylidene]-, (E)-	isoPrOH	388(4.59)	44-4108-76
(Z)-	isoPrOH	242(3.90),385(4.43)	44-4108-76
2,4,6,8,10-Undecapentaenoic acid, 11-(2,6,6-trimethyl-1-cyclohexen-1-yl)-, (all-E)-	EtOH	375(4.69)	36-0580-76
$C_{20}H_{26}O_3$			
Benzene, 1,1'-[oxybis(methylene)bis-[4-methoxy-2,6-dimethyl-	EtOH	274(3.02),283(3.03)	39-1466-76C
Jatrophatrione	EtOH	223(4.04),244s(3.95), 280(3.49)	44-1855-76
1(4H)-Naphthalenone, 4-[2-(3-furanyl)-ethenyl]-4a,5,6,7,8,8a-hexahydro-2-hydroxy-3,4a,8,8-tetramethyl-, [4R-[4α(E),4aα,8aβ)]]-	MeOH	215(4.11),230(4.16), 278(4.15)	102-0827-76
Potamogetonin	C_6H_{12}	201(3.92)	44-0593-76
$C_{20}H_{26}O_4$			
4H,8H-Benzo[1,2-b:3,4-b']dipyran-4-one, 9,10-dihydro-5-hydroxy-2,8,8-trimethyl-6-(3-methylbutyl)-	MeOH	209(4.31),233(4.17), 263(4.21),302(3.89), 335s(3.48)	24-2963-76
Kaur-16-en-18-oic acid, 11,15-dioxo-, (4α)-	n.s.g.	230(4.00)	44-1021-76
Stemolide	EtOH	218(4.10)	88-2849-76
$C_{20}H_{26}O_5$			
4H,8H-Benzo[1,2-b:3,4-b']dipyran-4-one, 9,10-dihydro-5-hydroxy-6-(3-hydroxy-3-methylbutyl)-2,8,8-trimethyl-	MeOH	213(4.39),233(4.28), 263(4.33),303(3.99), 328s(3.71)	24-2963-76
Kaur-16-en-18-oic acid, 6,7,9-trihydroxy-15-oxo-, γ-lactone, (4α,6α,7β)-	MeOH	232(3.87)	94-0549-76
Rastronol G, deacetyl-7,20-anhydro-	EtOH	232.5(3.97)	33-0772-76
$C_{20}H_{26}O_6$			
Conacytone	EtOH	204(4.38),272(4.10), 380(2.76)	88-2501-76
Peuarin	EtOH	222(4.45),247s(3.69), 258(3.59),300s(4.09), 325(4.28)	102-0209-76
$C_{20}H_{26}O_7$			
Budlein A, tetrahydro-	EtOH	220(3.85),259(4.00)	102-0525-76
$C_{20}H_{26}O_8$			
3-Heptenedioic acid, 4-methyl-3-[[2,4,6-trihydroxy-5-methyl-3-(2-methyl-1-oxopropyl)phenyl]methyl]-(mixture of double bond isomers)	n.s.g.	293(4.18)	30-0004-76
$C_{20}H_{26}Si$			
Silane, (1,2-diphenylethenyl)triethyl-	n.s.g.	235(4.46),265(4.50)	65-1044-76
$C_{20}H_{27}NO_3$			
2-Azaestra-1(10),4-dien-3-one, 17β-acetoxy-2-methyl-	MeOH	231(3.74),308(3.69)	44-2864-76
2-Azaestra-1,3,5(10)-trien-17β-ol, 3-methoxy-, acetate	MeOH	277(3.57)	44-2864-76

Compound	Solvent	$\lambda_{max}(\log \epsilon)$	Ref.
Cyclopenta[b]pyrrol-2(1H)-one, hexahydro-6a-methoxy-3a-methyl-4-(1,2,3,4-tetrahydro-6-methoxy-2-naphthalenyl)-	EtOH	278(3.38),289(3.36)	78-0079-76
$C_{20}H_{27}NO_4$			
Luteicine	MeOH	215(4.7),237s(4.1), 287(3.5)	105-0313-76
Regeline	n.s.g.	216(4.04),225(3.96), 290(3.37)	105-0702-76
$C_{20}H_{27}NO_5$			
3H-Pyrrole-3,3-dicarboxylic acid, 4-ethoxy-2,4-dihydro-2,2-dimethyl-5-phenyl-, diethyl ester	EtOH	242(4.19)	33-1018-76
$C_{20}H_{27}N_3O_8$			
Butanoic acid, L-threonyl-trans-4-hydroxy-L-prolyl-γ-(4-hydroxyphenyl)-γ-oxo-α-amino-, 1-methyl ester	H_2O	281(4.14)	33-2021-76
$C_{20}H_{27}N_3O_{11}S$			
Butanoic acid, L-threonyl-trans-4-hydroxy-L-prolyl-γ-(4-hydroxy-3-sulfophenyl)-γ-oxo-α-amino-, 1-methyl ester	H_2O	274(3.89)	33-2021-76
$C_{20}H_{27}N_5O_5$			
Adenosine, 1',2'-didehydro-2'-deoxy-N-(2,2-dimethyl-1-oxopropyl)-, 5'-(2,2-dimethylpropanoate)	MeOH	216s(4.20),248(4.28), 264(4.27)	35-8213-76
$C_{20}H_{28}IN_5O_5$			
Propanoic acid, 2,2-dimethyl-, 5-ester with N-[5-(2-deoxy-2-iodo-β-D-arabinofuranosyl)-9H-purin-6-yl]-2,2-dimethylpropanamide	MeOH	211(4.28),272(4.24)	35-8213-76
Propanoic acid, 2,2-dimethyl-, 5-ester with N-[9-(3-deoxy-3-iodo-β-D-xylofuranosyl)-9H-purin-6-yl]-2,2-dimethylpropanamide	MeOH	272(4.26)	35-8213-76
$C_{20}H_{28}N_2O$			
Benzenemethanamine, N-methyl-N-[[3-methylene-2-(4-morpholinyl)-1-cyclohexen-1-yl]methyl]-	n.s.g.	238(4.22),285(3.26)	94-2494-76
$C_{20}H_{28}N_2O_2$			
2(1H)-Pyridinone, 1,1'-(1,10-decanediyl)bis-	n.s.g.	229(4.14),303(4.04)	33-2841-76
$C_{20}H_{28}N_4$			
Benzenamine, 4,4'-azobis[N,N-diethyl-	C_6H_{12}	431(4.59)	18-1381-76
$C_{20}H_{28}N_6O_2$			
4,5'-Bipyrimidine, 4',6-dimethoxy-2,2'-dipiperidino-	EtOH	257(4.30),287(4.40), 319(4.30)	19-0017-76
$C_{20}H_{28}O_2$			
Androst-4-ene-3,17-dione, 7α-methyl-	n.s.g.	242(4.21)	13-0759-76A
8α,9β,14β-Estra-2,5(10)-dien-17-one, 3-methoxy-12α-methyl-	n.s.g.	237(4.17)	39-1643-76C

Compound	Solvent	$\lambda_{max}(\log \epsilon)$	Ref.
Estra-1,3,5(10)-trien-17-ol, 3-methoxy-11β-methyl-	MeOH	279(4.28),288(4.29)	44-0531-76
8α,9β,14β-Estra-1,3,5(10)-trien-17α-ol, 3-methoxy-12α-methyl-	n.s.g.	278(3.29),287(3.27)	39-1643-76C
$C_{20}H_{28}O_3$			
1,3-Cyclopentanediol, 2-[2-(3,4-dihydro-6-methoxy-1(2H)-naphthalenylidene)-ethyl]-2-ethyl-	EtOH	265(4.31)	114-0081-76D
1(2H)-Naphthalenone, 4-[2-(3-furanyl)-ethenyl]octahydro-2-hydroxy-3,4a,8,8-tetramethyl- (7-hydroxyhedychanone)	MeOH	210(4.09),232(4.15)	102-0827-76
1(4H)-Naphthalenone, 4-[2-(3-furanyl)-ethyl]-4a,5,6,7,8,8a-hexahydro-2-hydroxy-3,4a,8,8-tetramethyl-	MeOH	220(3.74),279(3.84)	102-0827-76
$C_{20}H_{28}O_4$			
17,19-Dinorkaur-4-en-6-one, 3β,20-epoxy-3,16α-dimethoxy-	MeOH	250(3.84)	39-2144-76C
1,4-Ethanonaphthalene-6,10(4H)-dione, 1,4a,5,8a-tetrahydro-5,9-dimethoxy-1,2,5,7,8,9-hexamethyl-	EtOH	206(3.79),252(3.91), 311(2.50)	1-0064-76
Kaur-16-en-18-oic acid, 11α-hydroxy-15-oxo-	n.s.g. MeOH	237(3.90) 238(3.74)	44-1021-76 94-0549-76 +94-1040-76
1,4-Phenanthrenedione, 4b,5,6,7,8,8a,9-10-octahydro-3,9-dihydroxy-4b,8,8-trimethyl-2-(1-methylethyl)- (horminone)	EtOH	277(4.02),410(2.92)	32-0119-76
Salvin	EtOH	212(4.33),233(3.56), 284(3.23)	105-0623-76
$C_{20}H_{28}O_5$			
11-Oxapregn-4-ene-3,20-dione, 17,21-di-hydroxy-	EtOH	239(4.2)	13-0717-76A
$C_{20}H_{28}O_6$			
Rastronol E	EtOH	233(3.79)	33-0772-76
$C_{20}H_{28}O_8$			
1,4-Ethanonaphthalene-6,10(4H)-dione, 1,4a,5,8a-tetrahydro-2,5,5,8,9,9-hexamethoxy-1,7-dimethyl-	EtOH	270(4.03),315s(3.08)	1-0064-76
Isopteroside	MeOH	217(4.36),263(3.96), 302(3.30)	94-0173-76
Wallichoside	EtOH	219(4.58),258(4.21), 300(3.30)	102-0995-76
$C_{20}H_{29}B$			
Borane, dicyclohexyl(2-phenylethenyl)-(and aromatic bands not listed)	ether	264(4.34)	101-0303-76N
$C_{20}H_{29}BrO_2$			
Sphaerococcenol A	MeOH	228(3.88)	88-0731-76
$C_{20}H_{29}NO_4$			
Valine, 2-(2-benzoylethyl)-N-hexanoyl-, (±)-	EtOH	243(4.11),281(3.02),	12-0339-76
Valine, 2-(2-benzoylethyl)-N-(4-methyl-pentanoyl)-, (±)-	EtOH	244(4.10),280(3.02), 288s(2.88)	12-0339-76

Compound	Solvent	$\lambda_{max}(\log \epsilon)$	Ref.
$C_{20}H_{29}N_3O_5$			
10-Undecenoic acid, 2,3,3a,9a-tetrahydro-2-(hydroxymethyl)-6-imino-6H-furo[2',3':4,5]oxazolo[3,2-a]pyrimidin-3-yl ester, hydrochloride, [2R-(2α,3β,3aα,9aβ)]-	MeOH-acid	232(4.00),264(4.03)	87-0654-76
$C_{20}H_{30}N_2O_3$			
Carbamic acid, (3-propoxyphenyl)-, 3-(1-pyrrolidinyl)cyclohexyl ester	H_2O	236(4.12),277(3.46)	106-0024-76
Carbamic acid, (4-propoxyphenyl)-, 3-(1-pyrrolidinyl)cyclohexyl ester	H_2O	239(4.26),281(3.25)	106-002 -76
$C_{20}H_{30}O$			
1,3-Cyclohexadiene-1-carboxaldehyde, 6-methyl-4,6-bis(4-methyl-3-pentenyl)-	EtOH	319(4.05)	33-2261-76
1,4-Cyclohexadiene-1-carboxaldehyde, 6-methyl-4,6-bis(4-methyl-3-pentenyl)-	EtOH	230s(3.95)	33-2261-76
$C_{20}H_{30}O_2$			
Androst-4-en-3-one, 17-hydroxy-7α-methyl-	n.s.g.	243(4.20)	13-0759-76A
β-	n.s.g.	240(4.20)	13-0759-76A
Cyclohexanol, 1-(7-hydroxy-3,7-dimethyl-3,5,8-nonatrien-1-ynyl)-2,2,6-trimethyl-	isoPrOH	270(4.42)	33-0567-76
2(5H)-Furanone, 4-[2-(1,2,3,4,4a,7,8,8a-octahydro-1,2,4a,5-tetramethyl-1-naphthalenyl)ethyl]-	EtOH	222(3.93)	94-0294-76
Linaradial	EtOH	235(4.05)	94-0294-76
8,11a-Methano-11aH-cyclohepta[a]naphthalen-3(4H)-one, 4a,5,6,6a,7,8,9,10,11-11b-decahydro-9-hydroxy-4,4,9,11b-tetramethyl-, [4aR-(4aα,6aβ,8α,9α-11aα,11bβ)]	MeOH	230(3.91)	36-0778-76
2(1H)-Naphthalenone, 5-[2-(3-furanyl)-ethyl]octahydro-1,5,6,8a-tetramethyl- (cascarillone)	n.s.g.	226(3.78),280s(--)	22-0088-76
$C_{20}H_{30}O_4$			
1,3-Benzenediol, 2-methyl-5-nonyl-, diacetate	MeOH	235(3.73),272(3.40)	12-1989-76
Prost-8(12)-en-13-yn-1-oic acid, 9-hydroxy-11-oxo-	MeOH	263(4.20)	20-0503-76
$C_{20}H_{30}O_5$			
4,14-Dioxatricyclo[11.3.2.0^{3,5}]octadec-8-en-15-one, 7,12-dihydroxy-5,9,13-trimethyl-16-methylene- (6ξ-hydroxy-sinulariolide)	MeOH	213(3.70)	20-0707-76
$C_{20}H_{31}NO$			
Dispiro[5.1.5.1]tetradecan-7-one, 14-(cyclohexylimino)-	MeOH	314(1.38),326(1.40)	77-0787-76
$C_{20}H_{31}NO_2$			
2,6-Methano-3-benzazocine-3(2H)-propanol, 1,4,5,6-tetrahydro-8-methoxy-α,α,6,11-tetramethyl-	MeOH	227(3.90),280(3.33), 287(3.29)	24-2657-76

Compound	Solvent	$\lambda_{max}(\log \epsilon)$	Ref.
$C_{20}H_{31}N_3O_3$			
Androst-5-en-17-one, 3β,16β-dihydroxy-, (aminocarbonyl)hydrazone	n.s.g.	222(4.09)	13-0845-76A
$C_{20}H_{31}N_3O_5$			
Undecanoic acid, 2,3,3a,9a-tetrahydro-2-(hydroxymethyl)-6-imino-6H-furo-[2',3':4,5]oxazolo[3,2-a]pyrimidin-3-yl ester, hydrochloride, [2R-(2α,3β,3aβ,9aβ)]-	MeOH-acid	232(4.00),264(4.05)	87-0654-76
$C_{20}H_{31}N_3O_6$			
1-Piperidinyloxy, 2,2,6,6-tetramethyl-4-[[(1-β-D-ribofuranosylpyridinium-3-yl)carbonyl]amino]-, bromide	H_2O	262(3.78)	103-0300-76
2(1H)-Pyrimidinone, 4-amino-1-[3-0-(1-oxo-10-undecenyl)-β-D-arabinofuranosyl]-	MeOH-acid	212(3.98),283(4.16)	87-0667-76
$C_{20}H_{31}N_3O_8S$			
Cyclopentaneacetic acid, α-[(aminothioxomethyl)hydrazono]-4-carboxy-2,3-bis(2,2-dimethyl-1-oxopropoxy)-, 1-methyl ester, (1α,2α,3β,4α)-	THF	326(4.10)	23-2925-76
$C_{20}H_{32}$			
Bicyclo[4.2.0]octa-1,3,5-triene, 7,8-bis(1,1-dimethylethyl)-2,3,4,5-tetramethyl-	hexane	237(4.43),243(4.44), 305s(2.37),312(2.42), 335s(2.37),372(2.43)	88-3513-76
$C_{20}H_{32}N_2O_3$			
Carbamic acid, (3-propoxyphenyl)-, 3-(diethylamino)cyclohexyl ester	H_2O	236(4.10),277(3.42)	106-0024-76
Carbamic acid, (4-propoxyphenyl)-, 3-(diethylamino)cyclohexyl ester	H_2O	239(4.21),281(3.22)	106-0024-76
$C_{20}H_{32}N_2S_4$			
2-Butanamine, N,N'-[dithiobis(carbonothioyl-2-cyclopentyl-1-ylidene]bis-	EtOH	307(3.95),407(4.83)	39-1706-76C
$C_{20}H_{32}N_4O_8$			
1,1'-[(3.6-Dimethyl-1,4-benzoquinone-2,5-diyl)methyl]bispiperidinium dinitrate	H_2O	262(4.29),320(2.62)	12-1163-76
$C_{20}H_{32}N_5O_6PS$			
Adenosine, 8-(decylthio)-, cyclic 3',5'-(hydrogen phosphate)	pH 2.0 pH 13.0	285(4.25) 282(--)	69-1408-76 69-1408-76
$C_{20}H_{32}O$			
Cembra-2,7,11-trien-5-one	n.s.g.	297(2.31)	105-0278-76
Cembra-2,7,11-trien-5-one, epimer	n.s.g.	297(2.32)	105-0278-76
1,3-Cyclohexadiene-1-methanol, 6-methyl-4,6-bis(4-methyl-3-pentenyl)-	EtOH	270(3.92)	33-2261-76
1-Cyclohexene-1-carboxaldehyde, 6-methyl-4,6-bis(4-methyl-3-pentenyl)-	EtOH	232(4.08)	33-2261-76
5,9-Cyclotetradecadien-1-one, 5,9-dimethyl-2-(1-methylethenyl)-	MeOH	244(3.86)	138-0219-76
5,9-Cyclotetradecadien-1-one, 5,9-dimethyl-2-(1-methylethylidene)-	MeOH	254(3.96)	138-0219-76

Compound	Solvent	$\lambda_{max}(\log \epsilon)$	Ref.
2,6,10-Cyclotetradecatrien-1-one, 3,7,11-trimethyl-14-(1-methylethyl)-	EtOH	242(3.97)	105-0100-76
$C_{20}H_{32}O_2$			
Cyclohexanone, 3-methyl-6-(1-methyleth- ylidene)-2-[1-methyl-1-(4-methyl-2- oxocyclohexyl)ethyl]-, (1R,5R,8R,9S)-	EtOH	257(3.92),335(2.20)	22-0978-76
(1R,5S,8R,9S)-	EtOH	256(3.95),333(2.19)	22-0978-76
$C_{20}H_{32}O_3$			
Prosta-10,14-dien-1-oic acid, 9-oxo-, (14E)-($\pm$)-	EtOH	219(3.97)	44-0986-76
$C_{20}H_{32}O_4$			
2,5-Cyclohexadiene-1,4-dione, 2-hydroxy- 5-methoxy-3-tridecyl-	MeOH	289(4.32)	12-1979-76
4a(2H)-Naphthalenecarboxylic acid, 3- [1,1-dimethylethoxy)methylene]octa- hydro-8,8-dimethyl-2-oxo-, ethyl ester, trans-($\pm$)-	EtOH or base acid + base	277(4.19) 273(4.00) 312(4.05)	44-1005-76 44-1005-76 44-1005-76
Prost-13-en-1-oic acid, 9,15-dioxo-	EtOH	228(4.08)	39-2550-76C
$C_{20}H_{33}B$			
Borane, [2-(1-cyclohexen-1-yl)ethenyl]- dicyclohexyl-, (E)-	ether	244(4.23)	101-0303-76N
$C_{20}H_{33}ClO_2S$			
2-Thiophenecarboxylic acid, 4-(chloro- methyl)-5-tridecyl-, methyl ester	heptane	255(3.92),278(3.97)	34-0233-76
$C_{20}H_{33}NO_2$			
Prosta-10,14-dien-1-amide, 9-oxo-, (14E)-($\pm$)-	EtOH	219(4.00)	44-0986-76
$C_{20}H_{33}NO_3$			
5H-Benzo[d]isoindolo[1,2-b]-1,3-oxazine, 5,11-diethoxy-6a,11-dihydro-5-methyl-	EtOH	247(4.07),264(3.86), 269(3.83)	83-0356-76
$C_{20}H_{33}N_3O_6$			
2(1H)-Pyrimidinone, 4-amino-1-[3-0-(1- oxoundecyl)-β-D-arabinofuranosyl]-	MeOH-acid	213(3.99),283(4.16)	87-0667-76
$C_{20}H_{33}N_6O_6P$			
Adenosine, 8-(decylamino)-, cyclic 3',5'-(hydrogen phosphate)	pH 2.0	277(4.03)	69-1408-76
$C_{20}H_{34}N_2$			
Cyclohexanamine, N,N'-(2,2,4,4-tetra- methyl-1,3-cyclobutanediylidene)bis-	C_6H_{12} EtOH	263(2.20),273(2.16) 249(--),265(--)	59-1415-76 59-1415-76
$C_{20}H_{34}N_2O_6Si$			
1H-Pyrazole-3-carboxylic acid, 5-[6- [[[(1,1-dimethylethyl)dimethylsilyl]- oxy]methyl]tetrahydro-2,2-dimethyl- 4H-cyclopenta-1,3-dioxol-4-yl]- 4-hydroxy-, methyl ester	pH 1 pH 13	230(3.85),274(3.70) 240(3.80),320(3.93)	88-1063-76 88-1063-76
$C_{20}H_{34}N_2O_8$			
Propanedioic acid, 2,2'-[azobis(1-meth- ylethylidene)]bis-, tetraethyl ester	isooctane	360(1.43)	126-0395-76

Compound	Solvent	$\lambda_{max}(\log \epsilon)$	Ref.
$C_{20}H_{34}O$ 1,3-Cyclohexadiene-1-carboxaldehyde, 6-methyl-4,6-bis(4-methylpentyl)-	EtOH	225(3.93)	33-2261-76
$C_{20}H_{34}O_2S$ 2-Thiophenecarboxylic acid, 4-methyl-5-tridecyl-, methyl ester	heptane	254(3.93),284(3.96)	34-0233-76
$C_{20}H_{34}O_6$ 1-Cyclopentene-1-heptanoic acid, 2,4,4-triethoxy-5-oxo-, ethyl ester	MeOH	265(4.19)	20-0503-76
$C_{20}H_{36}S$ Thiophene, 3-methyl-2-pentadecyl-	heptane	236(3.80)	34-0233-76

Compound	Solvent	$\lambda_{max}(\log \epsilon)$	Ref.
$C_{21}H_9ClN_6$ Propanedinitrile, [2-(4-chlorophenyl)- 4-cyano-2,3-dihydro-3-imino-9H-ind- eno[2,1-c]pyridazin-9-ylidene]-	MeCN	310(4.52)	64-0371-76B
$C_{21}H_{10}BrNO_3S$ 8H-Naphtho[2,3-c]thioxanthene-5,8,14- trione, 6-amino-10-bromo-	DMF	479.5(3.47)	80-1227-76
$C_{21}H_{10}BrNO_4$ 8H-Naphtho[2,3-c]xanthene-5,8,14-trione, 6-amino-10-bromo-	DMF	498(3.80),532(3.68), 569(3.53)	80-1227-76
$C_{21}H_{10}N_6$ Propanedinitrile, (4-cyano-2,3-dihydro- 3-imino-2-phenyl-9H-indeno[2,1-c]pyr- idazin-9-ylidene)-	MeCN	305(4.50)	64-0371-76B
$C_{21}H_{11}NO_3S$ 8H-Naphtho[2,3-c]thioxanthene-5,8,14- trione, 6-amino-	DMF	473(3.41),519(3.31)	80-1227-76
$C_{21}H_{11}NO_4$ 8H-Naphtho[2,3-c]xanthene-5,8,14-trione, 6-amino-	DMF	494(3.66),530(3.54), 569(3.37)	80-1227-76
$C_{21}H_{12}ClNO$ Benzonitrile, 3-chloro-4-[2-(3-dibenzo- furanyl)ethenyl]-	DMF	356(4.59)	33-0819-76
$C_{21}H_{12}O_3$ 10H-Benzo[d]phenanthro[9,10-b]pyran- 10-one, 13-hydroxy-	MeOH	255s(4.41),261(4.43), 290s(4.16),302(4.08), 341(3.75),358(3.87), 377(3.89),395(3.84)	78-1331-76
$C_{21}H_{13}ClN_4$ 3H-1,2,3-Triazolo[4,5-b]quinoline, 7-chloro-3,9-diphenyl-	DMF	321s(3.91),331(3.95), 386(3.86)	48-0039-76
$C_{21}H_{13}Cl_2N_3O_2$ 5,8-Quinazolinedione, 2-(3,4-dichloro- phenyl)-4-methyl-6-(phenylamino)-	EtOH	270(4.24),318(4.52), 480(3.54)	5-1809-76
$C_{21}H_{13}NO$ Benzonitrile, 2-[2-(3-dibenzofuranyl)- ethenyl]-	DMF	345(4.65)	33-0819-76
Benzonitrile, 4-[2-(3-dibenzofuranyl)- ethenyl]-	DMF	350(4.75)	33-0819-76
$C_{21}H_{13}NOS$ Benzoxazole, 2-[5-(2-naphthalenyl)- 2-thienyl]-	CH_2Cl_2	253s(4.17),274(4.05), 290(4.02),302(4.10), 360(4.58),387s(4.30)	48-0731-76
$C_{21}H_{13}NO_3$ 13H-Dibenzo[a,h]carbazole-7,12-dione, 5-hydroxy-13-methyl-	EtOH	240(3.75),280(3.95), 336(3.38),510(3.56)	115-0185-76

Compound	Solvent	$\lambda_{max}(\log \epsilon)$	Ref.
$C_{21}H_{13}NO_4S$ Benzoic acid, 2-[(4-amino-9,10-dihydro-9,10-dioxo-1-anthracenyl)thio]-	DMF	458(3.41),481(3.48)	80-1227-76
$C_{21}H_{13}NO_5$ Benzoic acid, 2-[(4-amino-9,10-dihydro-9,10-dioxo-1-anthracenyl)oxy]-	DMF	495(3.82),524(3.76), 568(3.60)	80-1227-76
$C_{21}H_{14}$ 1H-Dibenz[a,de]anthracene	EtOH	258(4.44),266(4.48), 276(4.30),287(4.06), 301(3.92),311(4.06), 325(4.24),341(4.26), 350(3.58),359(2.96), 370(3.20)	24-1991-76
$C_{21}H_{14}BrNO_2$ 9,10-Anthracenedione, 1-amino-4-bromo-2-(phenylmethyl)-	n.s.g.	478(3.91)	146-0001-76
$C_{21}H_{14}BrNO_3$ 9,10-Anthracenedione, 1-amino-4-bromo-2-[(4-hydroxyphenyl)methyl]-	n.s.g.	478(3.92)	146-0001-76
$C_{21}H_{14}BrN_3O_2$ 5,8-Quinazolinedione, 2-(3-bromophenyl)-4-methyl-6-(phenylamino)-	EtOH	233s(--),312(4.47), 485(3.57)	5-1809-76
2-Quinolinamine, 6-bromo-3-nitro-N,4-diphenyl-	CHCl$_3$	269(4.46),297s(4.16), 319s(4.12),416(3.20)	48-0039-76
$C_{21}H_{14}ClN$ Benzonitrile, 4-(2-[1,1'-biphenyl]-4-ylethenyl)-2-chloro-	DMF	345(4.63)	33-0819-76
$C_{21}H_{14}ClN_3$ 2,2',2"-Tripyridine, 4'-(4-chlorophenyl)-, ferrous complex	MeOH	567(4.40)	118-0001-76
$C_{21}H_{14}ClN_3O_2$ 2-Quinolinamine, 6-chloro-3-nitro-N,4-diphenyl-	CHCl$_3$	267(4.52),269s(4.27), 319s(4.12),417s(3.16)	48-0039-76
$C_{21}H_{14}Cl_2$ Azulene, 1,1'-methylenebis[2-chloro-	C_6H_{12}	279(4.90),292(4.85), 335(3.88),351(4.02), 364(3.44),573(2.78), 615(2.73),670(2.33), 685(2.28)	44-1811-76
$C_{21}H_{14}N_2OS_2$ 11a,5a-([1,2]Benzenomethano)-11H,12H-[1,4]benzothiazino[2,3-b][1,4]benzothiazin-19-one	CHCl$_3$	245(4.44),293(4.31)	83-0081-76
$C_{21}H_{14}N_4$ 3H-1,2,3-Triazolo[4,5-b]quinoline, 3,9-diphenyl-	DMF	318s(3.82),330(3.91), 380(3.88)	48-0039-76
$C_{21}H_{14}N_4O$ Pyrimido[5,4-a]phenazin-5-ol, 1-methyl-	dioxan	232s(--),252s(--),	5-1809-76

Compound	Solvent	$\lambda_{max}(\log \epsilon)$	Ref.
3-phenyl- (cont.)		315(4.67),368(4.01), 390(4.14),413(4.17)	5-1809-76
$C_{21}H_{14}N_4O_4$			
2-Propen-1-one, 3-(5-nitro-2H-benzotria- zol-2-yl)-1,3-diphenyl-, N-oxide, (Z)-	EtOH	277(4.49),315(4.17), 385(3.71)	22-0184-76
5,8-Quinazolinedione, 4-methyl-2-(4-ni- trophenyl)-6-(phenylamino)-	EtOH	251(4.19),319(4.59), 485(3.59)	5-1809-76
$C_{21}H_{14}N_4O_5$			
4H-1-Benzopyran-4-one, 2-phenyl-, 2,4-dinitrophenylhydrazone	dioxan	425(4.42)	114-0309-76A
$C_{21}H_{14}OSe$			
Methanone, [5-(2-naphthalenyl)seleno- phene-2-yl]phenyl-	MeCN	361(4.39)	118-0521-76
$C_{21}H_{14}O_3$			
9,10-Anthracenedione, 1-hydroxy- 2-(phenylazo)-	n.s.g.	410(3.91)	146-0001-76
1H-2-Benzopyran-1-one, 6-hydroxy- 3,4-diphenyl-	MeOH	248s(4.69),254(4.73), 289(4.24),300(4.19)	78-1331-76
4H-1-Benzopyran-4-one, 7-hydroxy- 2,3-diphenyl-	MeOH	241(4.77),337(4.88)	78-1331-76
$C_{21}H_{14}O_4$			
9,10-Anthracenedione, 1,4-dihydroxy- 2-(phenylmethyl)-	n.s.g.	480(4.06)	146-0001-76
9,10-Anthracenedione, 1-hydroxy- 2-[(4-hydroxyphenyl)methyl]-	n.s.g.	412(4.01)	146-0001-76
Benzoic acid, 2-(9-hydroxy-10-phenan- threnyl)-4-hydroxy-	MeOH	262(4.83),280(4.55), 290(4.53),303(4.44), 340(4.19),355(4.29), 380(4.28),393(4.27)	78-1331-76
$C_{21}H_{15}ClN_2O_5$			
4H-1-Benzopyran-2-carboxylic acid, 6-chloro-7-hydroxy-4-oxo-3-(1- phenyl-1H-pyrazol-4-yl)-, ethyl ester	EtOH	275(4.73),350(4.15)	103-0914-76
$C_{21}H_{15}ClOS$			
Dibenzo[b,f]thiepin-10(11H)-one, 8-chloro-11-(phenylmethyl)-	MeOH	262(4.09),338(3.65)	73-3420-76
$C_{21}H_{15}ClS$			
Dibenzo[b,f]thiepin, 2-chloro-10-(phen- ylmethyl)-	MeOH	259(4.31),283(4.28)	73-3420-76
$C_{21}H_{15}Cl_2NS$			
1,5-Benzothiazepine, 2-(2,4-dichloro- phenyl)-2,3-dihydro-4-phenyl-	dioxan	236(4.18),265(4.18), 340(3.56)	114-0293-76A
1,5-Benzothiazepine, 2-(2,6-dichloro- phenyl)-2,3-dihydro-4-phenyl-	EtOH	262(4.22),335(3.57)	114-0293-76A
1,5-Benzothiazepine, 2-(3,4-dichloro- phenyl)-2,3-dihydro-4-phenyl-		236(4.25),264(4.25), 340(3.62)	114-0293-76A
$C_{21}H_{15}CrO_6$			
Chromium, tris(2-hydroxybenzaldehydato- 0,0')-	EtOH	273(4.55),413(3.91)	24-2691-76

Compound	Solvent	$\lambda_{max}(\log \epsilon)$	Ref.
$C_{21}H_{15}I$			
Phenanthrene, 2-iodo-9-methyl-1-phenyl-	MeOH	282s(4.53),290s(4.41), 302(4.15),323(2.54), 339(2.67),348(2.47), 356(2.62)	39-1115-76B
$C_{21}H_{15}N$			
Benzonitrile, 2-(2-[1,1'-biphenyl]-4-ylethenyl)-	DMF	337(4.59)	33-0819-76
Benzonitrile, 4-(2-[1,1'-biphenyl]-4-ylethenyl)-	DMF	342(4.70)	33-0819-76
9H-Dibenzo[a,c]carbazole, 9-methyl-	EtOH	249(4.03),298(3.74)	142-1275-76
$C_{21}H_{15}NO$			
Dibenzo[a,d]cyclohepten-5,10-imin-11-one, 5,10-dihydro-12-phenyl-	EtOH	208(3.47),227(3.43) 240(3.37)	39-2285-76C
Isoxazole, 3,4,5-triphenyl-	HOAc	275(4.23)	103-0851-76
	HOAc-H_2SO_4	318(4.26)	103-0851-76
$C_{21}H_{15}NOS$			
Benzoxazole, 2-[5-(4-phenyl-1,3-butadienyl)-2-thienyl]-	CH_2Cl_2	250(4.08),258s(4.06), 275(3.99),283s(3.98), 299s(3.97),311(4.05), 359s(4.31),380(4.42), 399(4.45),420s(4.29), 467s(2.81)	48-0731-76
$C_{21}H_{15}NO_2$			
9,10-Anthracenedione, 1-amino-2-(phenylmethyl)-	n.s.g.	470(3.72)	146-0001-76
2H-Isoindole-5,7-dione, 2-methyl-1,3-diphenyl-	MeOH	229(4.42),276(4.13), 418(3.64)	18-3314-76
$C_{21}H_{15}NO_3$			
9,10-Anthracenedione, 1-amino-2-[(4-hydroxyphenyl)methyl]-	n.s.g.	475(3.83)	146-0001-76
2-Propen-1-one, 3-(benzoyloxy)-1-phenyl-3-(2-pyridinyl)-	n.s.g.	302(4.08)	39-0380-76C
2-Propen-1-one, 2-(4-nitrophenyl)-1,3-diphenyl-, (E)-	MeOH	266(4.26),289(4.26)	44-2536-76
(Z)-	MeOH	251(4.35),327(4.35)	44-2536-76
2-Propen-1-one, 3-(4-nitrophenyl)-1,2-diphenyl-, (E)-	EtOH	257(4.21),318(4.21)	44-2536-76
(Z)-	EtOH	251(4.33),328(4.31)	44-2536-76
$C_{21}H_{15}NS$			
Thiazole, 2,4,5-triphenyl-	MeOH	244(4.3),322(4.2)	24-0906-76
$C_{21}H_{15}NS_2$			
2(5H)-Thiazolethione, 4,5,5-triphenyl-	C_6H_{12}	255(4.12),320(4.25), 387(3.77)	24-0139-76
$C_{21}H_{15}N_3$			
2,2',2''-Tripyridine, 4'-phenyl-, ferrous complex	MeOH	566(4.35)	118-0001-76
$C_{21}H_{15}N_3O_2$			
5,8-Quinazolinedione, 4-methyl-2-phenyl-6-(phenylamino)-	EtOH	233s(--),267(4.22), 317(4.47),488(3.60)	5-1809-76
2-Quinolinamine, 3-nitro-N,4-diphenyl-	$CHCl_3$	269(4.57),311s(4.24), 419s(3.14)	48-0039-76

Compound	Solvent	$\lambda_{max}(\log \epsilon)$	Ref.
$C_{21}H_{15}N_3O_3$ 1,3,5-Triazine-2,4,6(1H,3H,5H)-trione, 1,3,5-triphenyl-	THF	259(3.49)	95-0962-76
$C_{21}H_{15}N_5O_2S_2$ 1,2,4-Triazolo[3,4-b][1,3,4]thiadiazolium, 2,3-diphenyl-6-[(phenylsulfonyl)amino]-, hydroxide, inner salt	MeCN	278(4.29)	30-0104-76
$C_{21}H_{15}N_7O_2$ Formazan, 1-(4-nitrophenyl)-3-phenyl-5-(4-quinazolinyl)-	C_6H_{12}	305(3.90),405(4.20), 420(4.16)	103-0596-76
	EtOH	310(3.92),410(4.38)	103-0596-76
	EtOH-KOH	507(4.24)	103-0596-76
	dioxan	293(3.96),420(4.25)	103-0596-76
nickel complex	EtOH-pH 7	592(4.03)	103-0596-76
$C_{21}H_{16}$ 4H-Dibenz[a,kl]anthracene, 5,6-dihydro-	EtOH	200(4.56),221(4.60), 232(4.38),248(4.00), 259(4.16),269(4.47), 277(4.76),287(4.92), 298(4.20),308(4.05), 321(4.05),334(3.83)	117-0125-76 +117-0133-76
Naphthalene, 1-methyl-4-(4-phenyl-1-buten-3-ynyl)-, cis	MeOH	233(4.54),273(4.23), 282(4.30),335(4.21)	39-1104-76B
Naphthalene, 1-methyl-4-(4-phenyl-1-buten-3-ynyl)-, trans	MeOH	236(4.55),275s(4.25), 282(4.28),336(4.49)	39-1104-76B
Naphthalene, 2-methyl-1-(4-phenyl-1-buten-3-ynyl)-, cis?	MeOH	224(4.65),268s(4.27), 275(4.33),289s(4.16), 325s(3.83)	39-1104-76B
Naphthalene, 2-methyl-1-(4-phenyl-1-buten-3-ynyl)-, trans	MeOH	225(4.70),277(4.26), 311(4.31)	39-1104-76B
Phenanthrene, 1-methyl-3-phenyl-	EtOH	263(4.5771)	2-0038-76
Phenanthrene, 9-methyl-1-phenyl-	MeOH	277s(4.32),290s(4.16), 302(4.16),336(2.71), 341(2.52),352(2.66)	39-1115-76B
$C_{21}H_{16}BrN_3$ 2,3-Quinolinediamine, 6-bromo-N^2,4-diphenyl-	DMF	275(4.63),281s(4.59), 365(4.27),379(4.26)	48-0039-76
$C_{21}H_{16}Br_2ClN_3O_2$ 1H-1,2,3-Triazole, 1-(3-chloro-4-phenoxyphenyl)-5-(3,5-dibromo-2-methoxyphenyl)-4,5-dihydro-	dioxan	292(4.11)	97-0102-76
$C_{21}H_{16}ClNO_3$ 1H-Indole-1-acetic acid, α-[3-(2-chlorophenyl)-1-oxo-2-propynyl]-, ethyl ester	EtOH	227(4.69),283(4.63), 290(4.65),310(4.67)	12-1583-76
1H-Indole-1-acetic acid, α-[3-(4-chlorophenyl)-1-oxo-2-propynyl]-, ethyl ester	EtOH	223(4.34),267s(4.12), 280(4.13),290(4.12), 312(4.18)	12-1583-76
$C_{21}H_{16}ClNS$ 1,5-Benzothiazepine, 2-(2-chlorophenyl)-2,3-dihydro-4-phenyl-	EtOH	262(4.24),335(3.66)	114-0293-76A
1,5-Benzothiazepine, 2-(3-chlorophenyl)-2,3-dihydro-4-phenyl-	EtOH	261(4.29),335(3.64)	114-0293-76A

Compound	Solvent	$\lambda_{max}(\log \epsilon)$	Ref.
1,5-Benzothiazepine, 2-(4-chlorophenyl)-2,3-dihydro-4-phenyl-	EtOH	262(4.30),340(3.64)	114-0293-76A
$C_{21}H_{16}ClN_2O_2$			
Pyridinium, 1-[(1-amino-4-chloro-9,10-dihydro-9,10-dioxo-2-anthracenyl)-methyl]-2-methyl-	MeOH	465(3.83)	2-0054-76
Pyridinium, 1-[(1-amino-4-chloro-9,10-dihydro-9,10-dioxo-2-anthracenyl)-methyl]-3-methyl-	MeOH	464(3.81)	2-0054-76
Pyridinium, 1-[(1-amino-4-chloro-9,10-dihydro-9,10-dioxo-2-anthracenyl)-methyl]-4-methyl-	MeOH	465(3.85)	2-0054-76
$C_{21}H_{16}ClN_3$			
2,3-Quinolinediamine, 6-chloro-N^2,4-di-phenyl-	DMF	275(4.61),363(4.23), 375(4.22)	48-0039-76
$C_{21}H_{16}NO_3$			
Pyridinium, 1-[(9,10-dihydro-1-hydroxy-9,10-dioxo-2-anthracenyl)methyl]-2-methyl-, chloride	MeOH	408(3.91)	2-0054-76
$C_{21}H_{16}NO_4$			
Pyridinium, 1-[(9,10-dihydro-1,4-di-hydroxy-9,10-dioxo-2-anthracenyl)-methyl]-2-methyl-, chloride	MeOH	488(3.88)	2-0054-76
$C_{21}H_{16}NS_2$			
1,2-Dithiol-1-ium, 3-(diphenylamino)-5-phenyl-	MeOH	235s(4.24),323(4.32), 374(4.14)	48-0221-76
$C_{21}H_{16}N_2$			
Benzeneacetonitrile, α-[phenyl(phenyl-amino)methylene]-	EtOH	352(4.05)	40-0144-76
12H-Diindolo[1,2,3-ef:3',2',1'-jk][1,5]-benzodiazepine, 13,14-dihydro-	DMF	268(4.52),286(4.25), 327(4.19),347(4.24), 364(4.11),384(4.29)	5-1090-76
dianion	DMF-tert-BuOK	264(4.59),286(4.54), 320(4.59),346s(4.24), 390s(4.05),406(4.23), 428(4.27)	5-1090-76
$C_{21}H_{16}N_2O$			
2-Aceperimidylenone, 1,3-dihydro-1-methyl-3-(phenylmethyl)-	n.s.g.	380(4.06)	103-0455-76
Acetaldehyde, 9(10H)-acridinylidene-(phenylamino)-	EtOH	252(4.9),362(3.8), 380(3.9),399(3.8), 523(3.1)	18-1138-76
Benzimidazo[1,2-b]isoquinolin-11(5H)-one, 5a,6-dihydro-6a-phenyl-	n.s.g.	240(4.33),350(3.86)	39-1073-76C
$C_{21}H_{16}N_2O_2$			
9,10-Anthracenedione, 1,4-diamino-2-(phenylmethyl)-	n.s.g.	550(4.15),590(4.16)	146-0001-76
9,10-Anthracenedione, 1,8-diamino-2-(phenylmethyl)-	n.s.g.	510(4.09)	146-0001-76
2H-Indol-2-one, 3-([1,1'-biphenyl]-2-yl-imino)-1,3-dihydro-1-methyl-, N-oxide	MeOH	214(4.39),248(4.28), 339(4.11)	24-0200-76
Methanone, [3-amino-6-(4-methylphenyl)-furo[2,3-b]pyridin-2-yl]phenyl-	EtOH	276(4.25),344(4.32), 397(4.27)	48-0313-76

Compound	Solvent	$\lambda_{max}(\log \epsilon)$	Ref.
3-Pyridinecarbonitrile, 6-(4-methyl-phenyl)-2-(2-oxo-2-phenylethoxy)-	EtOH	250(4.29),324(4.44)	48-0313-76
$C_{21}H_{16}N_2O_2S$ 1,5-Benzothiazepine, 2,3-dihydro-2-(3-nitrophenyl)-4-phenyl-	EtOH	262(4.39),342(3.66)	114-0293-76A
1,5-Benzothiazepine, 2,3-dihydro-2-(4-nitrophenyl)-4-phenyl-	dioxan	265(4.46),323(3.96)	114-0293-76A
$C_{21}H_{16}N_2O_3$ 9,10-Anthracenedione, 1,4-diamino-2-[(4-hydroxyphenyl)methyl]-	n.s.g.	550(4.18),590(4.19)	146-0001-76
9,10-Anthracenedione, 1,8-diamino-2-[(4-hydroxyphenyl)methyl]-	n.s.g.	512(4.29)	146-0001-76
$C_{21}H_{16}N_2O_4$ 4H-1-Benzopyran-4-one, 7-acetoxy-2-methyl-3-(1-phenyl-1H-pyrazol-4-yl)-	EtOH	270(4.31)	103-0914-76
$C_{21}H_{16}N_4OS$ 1H-1,2,4-Triazolium, 4-[[(2-hydroxyphenyl)methylene]amino]-3-mercapto-1,5-diphenyl-, hydroxide, inner salt	MeCN	256(4.40),332(3.84)	103-0361-76
$C_{21}H_{16}N_4O_3$ Benzenamine, 4-(3,5-diphenyl-1H-pyrazol-1-yl)-N-hydroxy-3-nitro-	EtOH	254(4.61),366(3.36)	22-0839-76
$C_{21}H_{16}N_4O_4S$ 4H-1-Benzothiopyran-4-one, 2,3-dihydro-2-phenyl-, 2,4-dinitrophenylhydrazone	dioxan	385(4.39)	114-0309-76A
$C_{21}H_{16}N_4O_5$ 4H-1-Benzopyran-4-one, 2,3-dihydro-2-phenyl-, 2,4-dinitrophenylhydrazone	dioxan	383(4.44)	114-0309-76A
$C_{21}H_{16}N_4O_8$ 2-Azulenecarbonitrile, 1-(2-acetoxyethyl)-, trinitrobenzene complex	CH_2Cl_2	256s(4.47),289(4.69), 300(4.64),334(3.78), 342(3.75),347(3.81), 357(3.48),606(2.69), 648(2.70),710s(2.45)	44-1811-76
6-Azulenecarbonitrile, 1-(2-acetoxyethyl)-, trinitrobenzene complex	CH_2Cl_2	280s(4.88),290(5.07), 306s(3.95),337(3.80), 354(3.88),654(2.53), 722(2.46)	44-1811-76
$C_{21}H_{16}N_4S$ 1H-1,2,4-Triazolium, 3-mercapto-1,5-diphenyl-4-[(phenylmethylene)amino]-, hydroxide, inner salt	MeCN	258(4.49),350(3.46)	103-0361-76
$C_{21}H_{16}N_6$ Formazan, 1,3-diphenyl-5-(4-quinazolinyl)-	C_6H_{12}	290(4.40),450(4.16)	103-0596-76
	EtOH	290(4.29),360(4.32), 370s(4.30)	103-0596-76
	EtOH-KOH	515(4.00)	103-0596-76
	dioxan	290(3.89),380(3.91), 420s(3.79)	103-0596-76
	H_2SO_4	620(3.64)	103-0596-76
nickel complex	EtOH-pH 7	613(3.95)	103-0596-76

Compound	Solvent	$\lambda_{max}(\log \epsilon)$	Ref.
$C_{21}H_{16}N_6O_5S$			
Benzenesulfonamide, N-(aminoiminometh-yl)-4-[2-(4-nitrophenyl)-4-oxo-3(4H)-quinazolinyl]-	EtOH-NaOH	210(4.48),275(4.06), 300s(3.97)	106-0574-76
$C_{21}H_{16}O$			
Anthracene, 9-methyl-10-phenoxy-	ether	258(5.16),340(3.47), 358(3.81),377(4.04), 398(4.02)	88-1089-76
5H-Benzo[6,7]cyclohept[1,2-b]indeno-[1,2-e]pyran, 6,7-dihydro-	dioxan	228(4.36),262(4.15), 294(4.27),358(4.29), 520(3.47)	78-2219-76
2,5-Cyclohexadien-1-one, 4-(9H-fluoren-9-ylidene)-2,6-dimethyl-	isooctane	244(4.49),273(4.40), 281(4.39),425(4.58)	44-0214-76
14H-Dibenzo[a,h]xanthene, 12,13-dihydro-	EtOH	218(4.57),242(4.57), 275s(--)	22-1967-76
9H-Fluorene, 9-[(4-methoxyphenyl)methyl-ene]-	dioxan	348(4.29)	39-0048-76C
2-Propen-1-one, 1,2,3-triphenyl-, (E)-	MeOH	255(4.22),295(4.16)	44-2536-76
2-Propen-1-one, 1,2,3-triphenyl-, (Z)-	MeOH	254(4.37),282(4.32)	44-2536-76
$C_{21}H_{16}O_2$			
Benzofuran, 3-(2-methoxyphenyl)-2-phen-yl-	C_6H_{12}	240(4.25),305(4.33)	18-2560-76
6,11-Epoxydibenz[b,e]oxepin, 6,11-di-hydro-6-methyl-11-phenyl-	ether	279(3.47),286(3.47)	88-1089-76
Indeno[2,1-b]pyran-9-carboxaldehyde, 4-methyl-2-(4-methylphenyl)-	dioxan	241(4.22),268(4.37), 325(4.11),339(4.16), 377(4.23),505(3.65)	78-2225-76
1,3-Propanedione, 1,2,3-triphenyl-	C_6H_{12}	246(4.52),286s(3.40), 334(3.45),355(3.43)	78-1549-76
$C_{21}H_{16}O_3$			
9,10-Epidioxyanthracene, 9,10-dihydro-9-methyl-10-phenoxy-	ether	260(3.23),266(3.25), 273(3.18)	88-1089-76
$C_{21}H_{16}O_4S$			
Butanedioic acid, phenyl(phenyl-2-thien-ylmethylene)-, (E)-	EtOH	207(4.56),245(4.26), 285(4.21)	4-0285-76
$C_{21}H_{16}O_5$			
1,3-Cyclopentadiene-1,4-dicarboxylic acid, 5-oxo-2,3-diphenyl-, dimethyl ester	ether	242s(4.03),334(3.86), 420s(2.88)	22-1491-76
$C_{21}H_{17}BrN_2OS$			
Benzeneethanehydrazonothioic acid, 4-bromo-α-oxo-N-phenyl-, 4-methyl-phenyl ester	EtOH	245(4.246),270(4.189), 374(4.189)	18-0321-76
$C_{21}H_{17}BrN_2O_3S$			
Ethanone, 1-(4-bromophenyl)-2-[(4-meth-ylphenyl)sulfonyl]-2-(phenylhydrazono)-	EtOH	231(4.356),255(4.274), 275(4.267),369(4.301)	18-0321-76
$C_{21}H_{17}BrO$			
Benzene, 1-(2-bromo-1,2-diphenylethen-yl)-4-methoxy-, cis	MeOH	233(4.37),257s(4.04), 292(3.95)	36-1545-76
trans	MeOH	241(4.34),302(3.96)	36-1545-76

Compound	Solvent	λ_{max}(log ϵ)	Ref.
$C_{21}H_{17}ClN_2OS$ Benzeneethanehydrazonothioic acid, N-(4-chlorophenyl)-α-oxo-, 4-methyl- phenyl ester	EtOH	245(4.348),275(4.155), 374(4.306)	18-0321-76
$C_{21}H_{17}ClN_2O_2$ 1H-Pyrrolo[3,2-b]quinoline-3-carboxylic acid, 7-chloro-2-methyl-9-phenyl-, ethyl ester	EtOH	238s(4.61),249(4.73), 334(4.10),354(3.98), 368s(3.92)	44-1743-76
$C_{21}H_{17}ClN_2O_3S$ Ethanone, 2-[(4-chlorophenyl)hydrazono]- 2-[(4-methylphenyl)sulfonyl]-1-phenyl-	EtOH	247(4.355),272(4.218), 365(4.360)	18-0321-76
$C_{21}H_{17}ClO_5$ 1,4-Naphthalenedione, 3-[2-(3-acetoxy- 5-methylphenyl)ethyl]-2-chloro-5-hy- droxy-	MeOH MeOH-base	281(3.95),293s(3.60), 435(3.27) 285(4.06),300s(3.91), 545(3.42)	39-0997-76C 39-0997-76C
$C_{21}H_{17}Cl_2NOS$ 1-Propanone, 3-[(2-aminophenyl)thio]- 3-(2,6-dichlorophenyl)-1-phenyl-	EtOH	243(4.38),320(3.62)	114-0293-76A
$C_{21}H_{17}Cl_2N_3O$ 1H-1,2,3-Triazole, 5-(2,4-dichlorophen- yl)-4,5-dihydro-1-[4-(4-methylphen- oxy)phenyl]-	dioxan	292(4.15)	97-0102-76
$C_{21}H_{17}CoN_6$ Cobalt, bis(cyano-C)(16,17-dihydro-7H- dibenzo[f,m][1,4,8,12]tetraazacyclo- decinato-N^5,N^9,N^{15},N^{18})-, (OC-6-22)-	EtOH	230(4.59),255s(--), 300(4.08),374(3.71), 425s(--),490(3.89), 626(3.48)	12-2271-76
$C_{21}H_{17}CoN_6O_2$ Cobalt, (16,17-dihydro-7H-dibenzo[f,m]- [1,4,8,12]tetraazacyclopentadecinato- N^5,N^9,N^{15},N^{18})bis(thiocyanato-N)-, (OC-6-22)-	EtOH	247(4.58),281(4.38), 368(3.86),396(3.68), 435(3.68),468(3.75), 491(3.82),625(3.34)	12-2271-76
$C_{21}H_{17}NOS$ 2-Propenamide, N,3-diphenyl-2-(phenyl- thio)-	75% dioxan	247(4.2),290(4.3)	104-0149-76
$C_{21}H_{17}NO_2$ Pyridinium, 1-benzoyl-2-oxo-2-(4-methyl- phenyl)ethylide	EtOH dioxan MeCN	227(4.29),250(4.08), 318(4.09) 281(3.91),320(3.98), 420(3.34) 315(4.02),395(3.31)	78-2647-76 78-2647-76 78-2647-76
$C_{21}H_{17}NO_2S$ Phenanthrene, 9,10-dihydro-9,10-tosyl- imino-	EtOH	232s(4.49),267(4.16), 275s(4.19),276(4.21), 290(4.03)	94-1013-76
$C_{21}H_{17}NO_3$ 1H-Indole-1-acetic acid, α-(1-oxo-3- phenyl-2-propynyl)-, ethyl ester	EtOH	225(4.35),265s(4.14), 283(4.21),290(4.24), 309(4.28)	12-1583-76

Compound	Solvent	$\lambda_{max}(\log \epsilon)$	Ref.
$C_{21}H_{17}NS$ 1,5-Benzothiazepine, 2,3-dihydro-2,4-di- phenyl-	EtOH	260(4.27),336(3.63)	114-0293-76A
$C_{21}H_{17}N_2O_2$ Pyridinium, 1-[(1-amino-9,10-dihydro- 9,10-dioxo-2-anthracenyl)methyl]- 2-methyl-	MeOH	460(3.94)	2-0054-76
Pyridinium, 1-[(1-amino-9,10-dihydro- 9,10-dioxo-2-anthracenyl)methyl]- 3-methyl-	MeOH	464(3.85)	2-0054-76
Pyridinium, 1-[(1-amino-9,10-dihydro- 9,10-dioxo-2-anthracenyl)methyl]- 4-methyl-	MeOH	462(3.89)	2-0054-76
$C_{21}H_{17}N_3$ 2,3-Quinolinediamine, N^2,4-diphenyl-	DMF	360(4.22),368s(4.22)	48-0039-76
$C_{21}H_{17}N_3O_2S$ 1H-Pyrazole, 4,5-dihydro-3-[2-(5-nitro- 2-thienyl)ethenyl]-1,5-diphenyl-	n.s.g.	512(5.17)	124-0383-76
$C_{21}H_{17}N_3O_3$ 1H-Pyrazole, 4,5-dihydro-3-[2-(5-nitro- 2-furanyl)ethenyl]-1,5-diphenyl-	n.s.g.	478(5.13)	124-0383-76
$C_{21}H_{17}N_3O_5S$ Ethanone, 2-[(4-methylphenyl)sulfonyl]- 1-(4-nitrophenyl)-2-(phenylhydrazono)-	EtOH	235(4.197),255(4.210), 270(4.190),370(4.156)	18-0321-76
$C_{21}H_{17}N_5O_5$ 4(1H)-Quinolinone, 2,3-dihydro-2-phen- yl-, 2,4-dinitrophenylhydrazone	dioxan	408(4.33)	114-0309-76A
$C_{21}H_{17}N_5S$ Propanedinitrile, [5-[[4-(dimethylami- no)phenyl]methylene]-4-imino-3-phenyl- 2-thiazolidinylidene]-, monoperchlor- ate	MeOH	330(4.07),370s(3.78), 408(3.54),552(4.62)	48-0343-76
$C_{21}H_{18}BrN_3$ Pyrido[2,3-b]pyrazine, 7-bromo-6-ethyl- 5,6-dihydro-2,3-diphenyl-	EtOH	315(3.97),401(4.11)	22-0251-76
$C_{21}H_{18}ClNOS$ Benzenepropanamide, β-chloro-N-phenyl- α-(phenylthio)-	75% dioxan	250(4.2)	104-0149-76
1-Propanone, 3-[(2-aminophenyl)thio]- 3-(2-chlorophenyl)-1-phenyl-	EtOH	242(4.38),310(3.79)	114-0293-76A
1-Propanone, 3-[(2-aminophenyl)thio]- 3-(3-chlorophenyl)-1-phenyl-	EtOH	244(4.30),308(3.82)	114-0293-76A
1-Propanone, 3-[(2-aminophenyl)thio]- 3-(4-chlorophenyl)-1-phenyl-	EtOH	228(4.51),312(3.98)	114-0293-76A
$C_{21}H_{18}Cl_3NO_4$ 6H-Benzo[g]-1,3-benzodioxolo[5,6-a]quin- olizine, 5,8-dihydro-9,10-dimethoxy- 8-(trichloromethyl)-	$CHCl_3$	283(3.97),362(4.01), 380(3.98)	100-0204-76
	$CHCl_3$-HCl	255(--),293(--), 312(--),385(--)	100-0204-76

Compound	Solvent	$\lambda_{max}(\log \epsilon)$	Ref.
$C_{21}H_{18}D_5NO_3$			
8H-Dibenzo[a,g]quinolizin-8-one-9,10,11,12-d_4, 5,6,13,13a-13a-d-2,3-dimethoxy-13,13-dimethyl-	n.s.g.	230(4.30),255(3.81), 266(3.76),280(3.74), 289(3.65)	44-2201-76
Isoquinoline, 2-(benzoyl-d_5)-1,2,3,4-tetrahydro-6,7-dimethoxy-1-(1-methylethylidene)-	n.s.g.	252(4.23),290(3.80)	44-2201-76
$C_{21}H_{18}N_2$			
1,4-Benzenediamine, N'-9H-fluoren-9-ylidene-N,N-dimethyl-	C_6H_{12}	476(4.65)	39-0048-76C
Benzo[a]phenazine, 1,4-dimethyl-6-(1-methylethenyl)-	EtOH	249(4.45),262(4.47), 291(4.44),406(3.95), 425(4.08)	102-1267-76
1H-Pyrazole, 3-([1,1'-biphenyl]-4-yl)-4,5-dihydro-1-phenyl-	benzene	379(4.43)	64-1248-76B
1H-Pyrazole, 4,5-dihydro-1,3,5-triphenyl-	benzene n.s.g.	362(4.31) 356(4.30)	61-0288-76 124-0383-76
$C_{21}H_{18}N_2O$			
1H-Pyrazole, 4,5-dihydro-3-(4-phenoxyphenyl)-1-phenyl-	benzene	362(4.33)	64-1248-76B
1H-Pyrazole, 3-[2-(2-furanyl)ethenyl]-4,5-dihydro-1,5-diphenyl-	n.s.g.	384(4.47)	124-0383-76
$C_{21}H_{18}N_2OS$			
Benzeneethanehydrazonothioic acid, N-(4-methylphenyl)-α-oxo-, phenyl ester	EtOH	245(4.275),278(3.986), 376(4.258)	18-0321-76
Benzeneethanehydrazonothioic acid, α-oxo-N-phenyl-, 4-methylphenyl ester	EtOH	242(4.522),280(4.262), 368(4.443)	18-0321-76
$C_{21}H_{18}N_2O_2$			
Isoxazolo[5,4-b]pyridine, 5-ethyl-6-(4-methoxyphenyl)-3-phenyl-	EtOH	234(4.19),318(4.08)	18-3339-76
$C_{21}H_{18}N_2O_2S$			
Benzeneethanehydrazonothioic acid, N-(4-methoxyphenyl)-α-oxo-, phenyl ester	EtOH	242(4.506),310(3.945), 386(4.476)	18-0321-76
$C_{21}H_{18}N_2O_3S$			
Ethanone, 2-[(4-methylphenyl)sulfonyl]-1-phenyl-2-(phenylhydrazono)-	EtOH	249(4.229),270(4.123), 352(4.262)	18-0321-76
$C_{21}H_{18}N_2O_8$			
Spiro-2,2'-bis[2H-1-benzopyran], 8,8'-dimethoxy-3,3'-dimethyl-6,6'-dinitro-	EtOH	213(4.30),251(4.64), 269s(4.10),343(4.07)	22-2039-76
$C_{21}H_{18}N_2S$			
1H-Pyrazole, 4,5-dihydro-1,5-diphenyl-3-[2-(2-thienyl)ethenyl]-	n.s.g.	398(4.59)	124-0383-76
1H-Pyrazole, 4,5-dihydro-1-phenyl-3-[4-(phenylthio)phenyl]-	benzene	382(4.38)	64-1248-76B
$C_{21}H_{18}N_3O_2$			
Pyridinium, 1-[(1,4-diamino-9,10-dihydro-9,10-dioxo-2-anthracenyl)methyl]-2-methyl-, chloride	MeOH	570(3.90),605(4.12)	2-0054-76
Pyridinium, 1-[(1,4-diamino-9,10-dihydro-9,10-dioxo-2-anthracenyl)methyl]-3-methyl-, chloride	MeOH	565(3.99),609(4.18)	2-0054-76

Compound	Solvent	$\lambda_{max}(\log \epsilon)$	Ref.
Pyridinium, 1-[(1,4-diamino-9,10-di-hydro-9,10-dioxo-2-anthracenyl)-methyl]-4-methyl-, chloride	MeOH	568(4.07),610(4.20)	2-0054-76
$C_{21}H_{18}N_4$ 1,3-Benzenediamine, 4-(3,5-diphenyl-1H-pyrazol-1-yl)-	EtOH	222(4.70),253(4.64), 299(3.89)	22-0839-76
$C_{21}H_{18}N_4O_3$ L-Tryptophan, N-[(3-methyl-2-quinoxa-linyl)carbonyl]-	EtOH	203(4.75),220(4.68), 242(4.52),274(3.92), 281(3.94),290(3.91), 321(3.80)	78-2931-76
	EtOH-KOH	201(4.80),222(4.67), 241(4.51),275(3.92), 283(3.94),291(3.91), 323(3.81)	78-2931-76
$C_{21}H_{18}N_4O_5$ L-Tryptophan, N-[(3-methyl-2-quinoxa-linyl)carbonyl]-, N,N'-dioxide	EtOH	199(4.49),222(4.69), 232s(4.36),259s(4.32), 267(4.48),283s(3.97), 290(3.83),370(4.03), 384(4.08)	78-2931-76
	MeCN	223(4.69),260s(4.28), 268(4.47),289(3.83), 360s(3.79),378(3.98), 393(4.04)	78-2931-76
$C_{21}H_{18}N_6$ Benzaldehyde, (1,2-dihydro-1,6-diphenyl-1,2,4,5-tetrazin-3-yl)hydrazone	MeOH	280(4.43)	103-0600-76
$C_{21}H_{18}N_6O$ Cyclohexanone, 2,6-bis[(4-azidophenyl)-methylene]-4-methyl-	benzene	354(4.26)	93-0163-76
$C_{21}H_{18}N_6O_7$ 3,11-Diazatricyclo[5.3.1.1^{2,6}]dodec-4-ene-10,12-dione, 8-methoxy-3,11-bis(5-nitro-2-pyridinyl)-	EtOH	208(4.30),226(4.28), 244(4.23),369(4.55)	39-2296-76C
$C_{21}H_{18}O$ 2,5-Cyclohexadien-1-one, 4-(diphenyl-methylene)-2,6-dimethyl-	isooctane	366(4.45)	73-1676-76
Indeno[2,1-b]pyran, 4,9-dimethyl-2-(4-methylphenyl)-	dioxan	231(4.35),256(4.23), 315(4.35),337(4.28), 362(4.21),523(3.53)	78-2219-76
Indeno[2,1-b]pyran, 2,4-dimethyl-3-(phenylmethyl)-	dioxan	296(4.27),340(3.94), 479(3.34)	78-2219-76
Indeno[2,1-b]pyran, 4-phenyl-2-propyl-	dioxan	237(4.28),264(4.07), 312(4.29),356(3.71), 540(3.34)	78-2225-76
$C_{21}H_{18}O_2$ 1(2H)-Acenaphthylenone, 2-hydroxy-2-(2,4,6-trimethylphenyl)-	hexane	219(4.77),257(4.11), 314(3.79),340(3.66)	35-1860-76
$C_{21}H_{18}O_3$ 2,5-Cyclohexadien-1-one, 4-(diphenyl-methylene)-2,6-dimethoxy-	isooctane	380(4.42)	73-1676-76

Compound	Solvent	$\lambda_{max}(\log \epsilon)$	Ref.
$C_{21}H_{18}O_3S_2$ Sulfonium, methylphenyl-, 2-oxo-2-phen- yl-1-(phenylsulfonyl)ethylide	MeOH	218s(4.43),263(4.05)	30-0366-76
$C_{21}H_{18}O_5$ 2,5-Cyclohexadiene-1,4-dione, 3-meth- oxy-2,5-bis(4-methoxyphenyl)-	dioxan	254(4.56),385(3.97)	94-0613-76
3-Cyclopentene-1,3-dicarboxylic acid, 2-oxo-4,5-diphenyl-, dimethyl ester	EtOH EtOH-KOH	292(4.18) 370(--)	104-0085-76 104-0085-76
Ovalichromene B	MeOH	265(4.2)	102-2011-76
$C_{21}H_{18}O_7$ Compd., m. 169°	n.s.g.	210(4.69),267(4.24), 274(4.23),310(4.28), 334(4.24)	39-0186-76C
$C_{21}H_{18}O_8$ 7-Deoxysteffimycinone	MeOH	213(4.39),235(4.42), 282(4.35),435(4.18)	69-4139-76
$C_{21}H_{18}O_9$ 2-Naphthacenecarboxylic acid, 1,2,3,4- 6,11-hexahydro-2,4,5,12-tetrahydroxy- 7-methoxy-6,11-dioxo-, methyl ester, (2S-cis)-	MeOH	233(4.57),251(4.40), 288(3.94),478(4.06), 494(4.07),529(3.81)	87-0395-76
$C_{21}H_{18}S$ Azulene, 4,6,8-trimethyl-1-[(phenyleth- ynyl)thio]-	hexane	560(2.66)	104-1966-76
$C_{21}H_{19}ClN_2O_2$ Spiro[2H-anthra[1,2-d]imidazole-2,1'-cy- clohexane]-6,11-dione, 4-chloro-1,3- dihydro-3-methyl-	EtOH	562(4.08)	104-0172-76
$C_{21}H_{19}ClN_3S$ Benzothiazolium, 2-[4-chloro-1-cyano- 4-[4-(dimethylamino)phenyl]-1,3-buta- dienyl]-3-methyl-, perchlorate	acetone	660(4.65)	124-1174-76
$C_{21}H_{19}ClSi$ Silane, chloro(1,2-diphenylethenyl)meth- ylphenyl-, (E)-	n.s.g.	236(4.56),268(4.57)	65-1044-76
$C_{21}H_{19}IN_4S$ 1,3,4-Triazoline-2-thione, 1-(benzyli- deneamino)-S-methyl-4,5-diphenyl-, iodide	MeCN	246(4.73)	103-0361-76
$C_{21}H_{19}NO$ Acridine, 9,10-dihydro-9-[(4-methoxy- phenyl)methyl]-	EtOH	287(4.24)	77-0198-76
Pyridinium, 2,5-dimethyl-4-phenyl-, 2-oxo-2-phenylethylide	EtOH	252(4.16),282(4.21), 434(3.46)	103-0424-76
$C_{21}H_{19}NOS$ 1-Propanone, 3-[(2-aminophenyl)thio]- 1,3-diphenyl-	EtOH	242(4.31),310(3.75)	114-0293-76A

Compound	Solvent	$\lambda_{max}(\log \epsilon)$	Ref.
$C_{21}H_{19}NO_2S$			
5H-Dibenzo[b,g][1,5]thiazocine, 6,7-dihydro-6-(phenylmethyl)-, 12,12-dioxide	EtOH	245(4.18)	39-0913-76C
$C_{21}H_{19}NO_3$			
Benzoic acid, 4-methoxy-, 1,2,3,4-tetrahydro-4-acridinyl ester	EtOH	231(4.63),255(4.37), 308(3.89),323(3.83)	95-0968-76
Vinyl acetate, 2-(2-acetyl-3-methyl-1-indolizinyl)-1-phenyl-, (Z)-	EtOH	231s(4.04),251(4.23), 267s(3.97),331(3.73), 394(3.64)	1-0512-76
$C_{21}H_{19}NO_4$			
2-Azabicyclo[3.2.0]heptane-5-carboxylic acid, 3,4-dioxo-1,7-diphenyl-, ethyl ester	dioxan	227(3.88),260s(3.59)	142-1229-76
2H-Azepine-4-carboxylic acid, 5,6-dihydro-3-hydroxy-2-oxo-6,7-diphenyl-, ethyl ester	dioxan	284(4.00)	142-1229-76
1H-Pyrrole-3,4-dicarboxylic acid, 1-methyl-2,5-diphenyl-, dimethyl ester	n.s.g.	215s(4.39),272(4.16)	39-2565-76C
2H-Pyrrol-2-one, 1-acetyl-3-(acetyloxy)-1,5-dihydro-5-phenyl-4-(phenylmethyl)-	EtOH	210(4.29),235(4.00), 283(4.25)	40-1100-76
$C_{21}H_{19}NO_5$			
Thaliglucinone	MeOH	238(4.46),256(4.56), 264(4.70),287(3.67), 312(4.16),390(3.80)	100-0065-76
$C_{21}H_{19}NO_6$			
Arnottianamide	EtOH	236(4.73),280s(4.01), 321s(3.63),324(3.65), 332(3.81)	95-1458-76
Benzo[g]-1,3-benzodioxolo[5,6-a]quinolizinium, 5,6-dihydro-13-hydroxy-8,9,10-trimethoxy-, hydroxide, inner salt	EtOH	230(4.50),262(4.11), 313(4.09),359(3.81), 374(3.80),455(3.83)	35-6714-76
Isoarnottianamide	MeOH	238(4.73),290(4.00), 332(3.86)	95-1458-76
1(3H)-Isobenzofuranone, 3-(6,7-dimethoxy-1-isoquinolinyl)-5,6-dimethoxy-, (±)-	EtOH	233s(3.79),248(3.93), 300(3.10),335(2.84)	78-0211-76
Spiro[7H-indeno[4,5-d]-1,3-dioxole-7,1'(2'H)-isoquinoline]-6,8-dione, 3',4'-dihydro-6',7'-dimethoxy-2'-methyl-	EtOH	242(4.59),300(3.92), 350(3.82)	39-0063-76C
$C_{21}H_{19}NS$			
Benzenamine, N,N-dimethyl-4-(9H-thioxanthen-9-yl)-	EtOH	214(4.67),270(4.51)	88-4609-76
Benzenemethanamine, N-(phenylmethylene)-2-[(phenylmethyl)thio]-	EtOH	252(4.30)	35-2605-76
$C_{21}H_{19}N_3O_3$			
1H-Pyridazino[4,5-b]indol-4-ol, 2-acetyl-2,5-dihydro-5-methyl-1-phenyl-, acetate	EtOH	315(4.34)	103-1008-76
$C_{21}H_{19}N_3O_4$			
1H-Pyridazino[4,5-b]indol-4-ol, 2-acet-	EtOH	315(4.31)	103-1008-76

Compound	Solvent	$\lambda_{max}(\log \epsilon)$	Ref.
yl-2,5-dihydro-1-(4-methoxyphenyl)-, acetate (cont.)			103-1008-76
Spiro[2H-anthra[1,2-d]imidazole-2,1'-cyclohexane]-6,11-dione, 1,3-dihydro-3-methyl-4-nitro-	EtOH	520(4.19)	104-0172-76
$C_{21}H_{19}N_3O_5$ 9-Acridinecarbonitrile, 1,2,3,4,4a,9,9a-10-octahydro-4a-hydroxy-9a-[(4-nitrobenzoyl)oxy]-	EtOH	246(4.33),265s(4.09), 298s(3.90)	95-0968-76
$C_{21}H_{19}N_4S$ 1,3,4-Triazoline-2-thione, 1-(benzylideneamino)-S-methyl-4,5-diphenyl-, iodide	MeCN	246(4.73)	103-0361-76
$C_{21}H_{19}O_2P$ 6-Phosphanthridinol, 5-ethyl-5,6-dihydro-6-phenyl-, 5-oxide	EtOH	213(4.64),235(4.26), 265(3.90),272(3.93), 286(3.81)	39-2050-76C
$C_{21}H_{19}O_3$ Flavylium, 4-(4-ethoxybutadienyl)-7-hydroxy-, perchlorate	n.s.g.	735(4.36)	103-0619-76
$C_{21}H_{20}ClN_3O_3S$ 4H-Indazol-4-one, 3-(4-chlorophenyl)-1,5,6,7-tetrahydro-1-methyl-, O-[(4-methylphenyl)sulfonyl]oxime, (E)-	MeOH	229(4.56),263(4.00)	24-1898-76
$C_{21}H_{20}ClN_4OSe$ 1,3-Selenazinium, 5-(4-chlorophenyl)-6-(morpholinomethyleneamino)-2-(phenylamino)-, perchlorate	HOAc	465(3.86)	88-2005-76
$C_{21}H_{20}NO$ Pyridinium, 2,5-dimethyl-1-(2-oxo-2-phenylethyl)-4-phenyl-, bromide	EtOH	250(4.36),282(4.36), 430(3.10)	103-0424-76
$C_{21}H_{20}NO_4$ Benzo[g]-1,3-benzodioxolo[5,6-a]quinolizinium, 5,6-dihydro-9,10-dimethoxy-13-methyl-, chloride	EtOH	233(4.43),266(4.44), 344(4.32),427(3.72)	73-3654-76
iodide	EtOH	228(4.58),266(4.48), 344(4.36),430(3.75)	73-3654-76
Benzo[g]-1,3-benzodioxolo[5,6-a]quinolizinium, 9-ethoxy-5,6-dihydro-10-methoxy-, iodide	EtOH	226(4.55),269(4.41), 345s(4.37),352(4.39), 435(3.73)	73-3654-76
Dibenzo[a,g]quinolizinium, 2,3,10,11-tetramethoxy-, chloride (norcoralynium chloride)	EtOH	219(4.35),237(4.33), 244s(4.30),282(4.58), 303(4.67),314(4.69), 327(4.55),365(3.85), 410(4.08),429(4.20)	73-3654-76
$C_{21}H_{20}NO_5$ Benzo[g]-1,3-benzodioxolo[5,6-a]quinolizinium, 5,6-dihydro-9,10,13-trimethoxy-, chloride	EtOH	235(4.50),267(4.47), 342s(4.39),352(4.41), 433(3.88)	73-3654-76
iodide	EtOH	227(4.54),268(4.45), 340s(4.37),351(4.39), 433(3.86)	73-3654-76

Compound	Solvent	$\lambda_{max}(\log \epsilon)$	Ref.
$C_{21}H_{20}N_2$			
1H-Indole, 1,2-dimethyl-3-[2-(1-methyl-1H-indol-3-yl)ethenyl]-	MeCN	208(4.58),241(4.42), 283(4.17),333(4.34)	5-1060-76
semiquinone (radical ion)	MeCN	225(4.57),260(4.17), 274s(4.07),283(4.05), 362(3.91),504(4.14), 534(4.09),758(3.91), 838(4.00)	5-1060-76
oxidized form	MeCN	224s(4.23),262(4.12), 305(4.20),478(4.34)	5-1060-76
$C_{21}H_{20}N_2O$			
Benzenecarboximidamide, N'-(2,6-dimethylphenyl)-N-hydroxy-N-phenyl-	EtOH	250(3.85),312(3.82)	12-0357-76
Benzenecarboximidamide, N-hydroxy-2-methyl-N'-(2-methylphenyl)-N-phenyl-	EtOH	252(3.81),308(3.93)	12-0357-76
Benzenecarboximidamide, N-hydroxy-4-methyl-N'-(4-methylphenyl)-N-phenyl-	EtOH	254(3.83),317(3.79)	12-0357-76
Benzo[a]phenazine-6-methanol, α,α,1,4-tetramethyl-	EtOH	241(4.49),263(4.52), 292(4.52),406(3.98), 425(4.00)	102-1267-76
1H-Indene-5-carbonitrile, 2-[[(1,1-dimethylethyl)amino]phenylmethyl]-1-oxo-	hexane	246(4.49),254(4.51), 310(3.57),320(3.58), 333(3.60),405(3.30)	44-3540-76
	CHCl₃	405(3.04)	44-3540-76
$C_{21}H_{20}N_2OS$			
Acetamide, N-[3-(3,4-dihydro-2(1H)-isoquinolinyl)-5-phenyl-2-thienyl]-	MeOH	254(4.15),279s(4.19), 301(4.23),329s(4.08)	48-0221-76
$C_{21}H_{20}N_2O_2$			
[2]Benzopyrano[4,3-b]pyrrol-5(1H)-one, 2-(dimethylamino)-1-ethyl-3-phenyl-	MeOH	233(4.44),313(4.27), 390(3.79)	44-0390-76
2,4-Pentadienoic acid, 2-cyano-5-(4-methylphenyl)-5-(phenylamino)-, ethyl ester	CH₂Cl₂	407(4.40)	48-0705-76
2H-Pyrrol-2-one, 1-acetyl-4-ethyl-1,5-dihydro-5-phenyl-3-[(phenylmethylene)amino]-	EtOH	206(4.42),221(4.00), 279(4.37)	40-1100-76
Spiro[2H-anthra[1,2-d]imidazole-2,1'-cycloheptane]-6,11-dione, 1,3-dihydro-	EtOH	568(4.17)	104-0172-76
Spiro[2H-anthra[1,2-d]imidazole-2,1'-cyclohexane]-6,11-dione, 1,3-dihydro-1-methyl-	EtOH	570(4.22)	104-0172-76
	EtOH-NaOH	570(4.22)	104-0172-76
	MeNO₂	565(4.08)	104-0172-76
	MeNO₂-HClO₄	593(4.46)	104-0172-76
Spiro[2H-anthra[1,2-d]imidazole-2,1'-cyclohexane]-6,11-dione, 1,3-dihydro-3-methyl-	EtOH	569(4.13)	104-0172-76
	EtOH-NaOH	630(4.26)	104-0172-76
	MeNO₂	560(3.88)	104-0172-76
	MeNO₂-HClO₄	410(3.49)	104-0172-76
$C_{21}H_{20}N_2O_3$			
9-Acridinecarbonitrile, 9a-(benzoyloxy)-1,2,3,4,4a,9,9a,10-octahydro-4a-hydroxy-	EtOH	237(4.28),277s(3.23), 290s(3.34),300(3.44)	95-0968-76
Benzenecarboximidamide, N-hydroxy-2-methoxy-N'-(2-methoxyphenyl)-N-phenyl-	EtOH	255(3.96),315(4.10)	12-0357-76
1H-Cyclopenta[b]quinoline-9-carbonitrile, 9a-(benzoyloxy)-2,3,3a,4,9,9a-hexahydro-3a-methoxy-	EtOH	235(4.28),275s(3.19), 282s(3.29),296(3.35)	95-0968-76
3H-Dibenz[f,ij]isoquinoline-2,7-dione, 1-[(3-methoxypropyl)amino]-3-methyl-	CHCl₃	340(--),449(4.31)	104-1114-76

Compound	Solvent	$\lambda_{max}(\log \epsilon)$	Ref.
$C_{21}H_{20}N_2O_4$			
1H-Benz[g]indolo[2,3-a]quinolizin-6-ium, 1-carboxy-2,3,4,7,8,13-hexahydro-5-hydroxy-11-methoxy-, hydroxide, inner salt	n.s.g.	218(4.32),267(3.79), 335(4.14),455(4.09)	64-0533-76B
	+ base	220(4.54),280(3.96), 335(3.98),403(4.12)	64-0533-76B
Dibenzo[c,g]azecine-7-carbonitrile, 5,6,7,8-tetrahydro-9,10-dimethoxy-2,3-(methylenedioxy)-	EtOH	286(3.89)	35-4301-76
1,3-Dioxolo[4,5-g]quinoline-7-carboxylic acid, 5-ethyl-5,8-dihydro-8-(phenylimino)-, ethyl ester	MeOH	243(4.42),265(4.52), 285(4.46)	142-1347-76
$C_{21}H_{20}N_3OS$			
Morpholinium, 4-[[(2,5-diphenyl-6H-1,3-thiazin-6-ylidene)amino]methylene]-, perchlorate	HOAc	275(4.15),348(4.05), 421s(3.92)	97-0268-76
$C_{21}H_{20}N_3S$			
Benzothiazolium, 2-[1-cyano-4-[4-(dimethylamino)phenyl]-1,3-butadienyl]-3-methyl-, perchlorate	MeOH	647(4.77)	124-1174-76
$C_{21}H_{20}N_4O_2$			
2H-Anthra[1,2-d]triazole-6,11-dione, 2-ethyl-4-piperidino-	EtOH	522(3.98)	104-0865-76
3H-Anthra[1,2-d]triazole-6,11-dione, 3-ethyl-4-piperidino-	EtOH	440(3.65)	104-0865-76
$C_{21}H_{20}N_4O_3$			
2,4(1H,3H)-Pteridinedione, 5-acetyl-5,6,7,8-tetrahydro-1-methyl-6,7-diphenyl-	pH 7.0	250s(3.94),288(4.18)	24-3184-76
	pH 13.0	283(4.12)	24-3184-76
$C_{21}H_{20}N_4O_3S$			
Thiazolo[3,2-a]benzimidazole-3-acetic acid, α-[(4-ethoxyphenyl)hydrazono]-, ethyl ester, hydrobromide	n.s.g.	250(4.23),290(4.19), 392(4.25)	124-1162-76
$C_{21}H_{20}O_2$			
1-Azulenecarboxylic acid, 4-methyl-7-(1-methylethyl)-, phenyl ester	C_6H_{12}	241(4.43),298(4.62), 304(4.64),337(3.72), 344(3.75),364(3.99), 377(4.09),510s(2.83), 532(2.92),570(2.87), 623(2.48)	18-1650-76
$C_{21}H_{20}O_3$			
Benzeneacetic acid, 2,5-dimethyl-α-(1-oxo-3-phenyl-2-propynyl)-, ethyl ester	pH 13	222(3.86),250(3.96), 332(3.90)	34-0118-76
	EtOH	218(4.10),314(4.25)	34-0118-76
	CHCl$_3$	249(3.43),317(4.11)	34-0118-76
Cyclobuta[b]naphthalene-3,8-dione, 1,2,2a,8a-tetrahydro-8a-(1-methylethoxy)-1-phenyl-	CHCl$_3$	300(3.36),310(3.26), 338(2.52)	18-2596-76
Cyclopentanone, 2,5-bis[(4-methoxyphenyl)methylene]-	EtOH	243(4.42),394(4.65)	114-0381-76B
	ether	240(--),370(--)	114-0381-76B
	CCl$_4$	373(4.67)	114-0381-76B
1H-Inden-1-one, 2-[(acetyloxy)(2,4,6-trimethylphenyl)methylene]-2,3-dihydro-	MeOH	238(3.79),302(4.24)	83-0376-76

Compound	Solvent	$\lambda_{max}(\log \epsilon)$	Ref.
$C_{21}H_{20}O_3S$			
Sulfonium, dimethyl-, 1,3-dibenzoyl-4-oxo-2-pentenylide	EtOH	254(4.29),342(4.09)	94-2421-76
$C_{21}H_{20}O_4$			
4H,8H-Benzo[1,2-b:3,4-b']dipyran-4-one, 2,3-dihydro-6-methoxy-8,8-dimethyl-2-phenyl (ovalichromene)	MeOH	255(4.19),315(3.64), 345(3.56)	102-1795-76
5,11-Naphthacenedione, 5a,6,11a,12-tetrahydro-1,3-dimethoxy-8-methyl-	MeOH	262(4.446),330(3.343)	2-0088-76
5,11-Naphthacenedione, 5a,6,11a,12-tetrahydro-1,3-dimethoxy-10-methyl-	MeOH	225(4.19),305(4.00), 330(3.14)	2-0088-76
2-Propen-1-one, 1-(5-hydroxy-8-methoxy-2,2-dimethyl-2H-1-benzopyran-6-yl)-3-phenyl-	MeOH	300(4.02),355(3.89)	102-1795-76
$C_{21}H_{20}O_4S$			
Sulfoxonium, dimethyl-, 1,3-dibenzoylpentenylide	EtOH	252(4.20),284(4.03), 388(4.03)	94-2421-76
$C_{21}H_{20}O_5$			
4H-1-Benzopyran-2-carboxylic acid, 7-methoxy-5-methyl-4-oxo-3-(phenylmethyl)-, ethyl ester	EtOH	227(4.40),237(4.27), 254(4.09),262(4.02), 306(4.01)	39-0499-76C
4H-1-Benzopyran-4-one, 3-[[2-(1,3-dioxolan-2-yl)phenyl]methyl]-7-methoxy-5-methyl-	EtOH	222(4.26),238(4.28), 246(4.21),271(3.89), 283(3.78),299(3.78)	39-0499-76C
1,3-Cyclopentanedicarboxylic acid, 2-oxo-4,5-diphenyl-, dimethyl ester	EtOH	253(3.76)	104-0085-76
	EtOH-KOH	283s(4.01),289(4.03)	104-0085-76
Cyclopentanone, 2,5-bis[(4-hydroxy-3-methoxyphenyl)methylene]-	EtOH	248(4.21),410(4.67)	114-0381-76B
	ether	250(--),395(--)	114-0381-76B
	$CHCl_3$	395(--)	114-0381-76B
	CCl_4	390(--)	114-0381-76B
ion	n.s.g.	505(4.74)	114-0381-76B
non-ion	n.s.g.	410(4.62)	114-0381-76B
4-Cyclopentene-1,3-dicarboxylic acid, 2-hydroxy-4,5-diphenyl-, dimethyl ester	EtOH	228(4.00),270(3.99)	104-0085-76
	EtOH-KOH	228(4.00),270(3.99)	104-0085-76
2-Naphthalenecarboxylic acid, 1,2,3,4-tetrahydro-6,8-dimethoxy-3-[(3-methylphenyl)methylene]-4-oxo-	MeOH	310(4.2416)	2-0088-76
Spiro[12H-benzo[b]xanthene-6(5aH),2'-[1,3]dioxolan]-12-one, 11,11a-dihydro-3-methoxy-1-methyl-, trans	EtOH	241(3.69),248(3.67), 282(3.29)	39-0499-76C
$C_{21}H_{20}O_6$			
Methanone, [3-(acetyloxy)-1H-inden-2-yl](2,4,6-trimethoxyphenyl)-	MeOH	270(4.07),310(4.10)	83-0376-76
$C_{21}H_{20}O_8$			
2-Anthracenecarboxylic acid, 9,10-dihydro-1,5,6,8-tetramethoxy-3-methyl-9,10-dihydro-, methyl ester	EtOH	230(4.37),248(4.34), 282(4.17),410(3.75)	44-3018-76
Pachybasin, 1-O-β-D-glucoside	MeOH-1% HOAc	215(4.20),259(4.40), 278s(4.15),345(3.49), 378(3.54)	12-2231-76
$C_{21}H_{20}O_{10}$			
Cosmosiin	MeOH	268(4.56),333(4.56)	95-1180-76
Emodin, 8-O-β-D-glucoside	MeOH-1% HOAc	222(4.52),284(4.36), 416(2.92)	12-2231-76

Compound	Solvent	$\lambda_{max}(\log \epsilon)$	Ref.
Vitexin	MeOH	270(4.11),306s(4.21), 336(4.15)	95-1180-76
$C_{21}H_{20}O_{11}$ 1,3-Benzodioxole-5-carboxylic acid, 6-(3,4,5-triacetoxy-6-oxo-1-cyclohexen-1-yl)-, methyl ester, (3α,4α,5α)-(+)-	EtOH	224(4.51),264(3.81), 298s(--)	24-0855-76
Coptiside II	MeOH	258(4.18),364(4.16)	94-0407-76
	MeOH-NaOAc	264(--),366(--)	94-0407-76
	MeOH-AlCl₃	270(--),405(--)	94-0407-76
Emodin, 7-hydroxy-, 8-0-β-D-glucoside	9:1 dioxan-EtOH	253(4.00),290(4.24), 418(3.70)	12-2231-76
$C_{21}H_{21}BrN_2O_2$ Pyrano[2,3-c]pyrazole, 4-(4-bromophenyl)-1,4,5,6-tetrahydro-6-(1-methylethoxy)-1-phenyl-, cis	EtOH	235(4.33)	35-2947-76
trans	EtOH	234(4.34)	35-2947-76
$C_{21}H_{21}ClN_2O_2$ Ethanone, 1-[6-chloro-2-[2-(dimethylamino)ethoxy]-4-phenyl-3-quinolinyl]-	EtOH	209(4.52),239(4.62), 282(3.79),330s(3.62), 345(3.76),358s(3.69)	44-1743-76
Pyrano[2,3-c]pyrazole, 4-(4-chlorophenyl)-1,4,5,6-tetrahydro-6-(1-methylethoxy)-1-phenyl-, cis	EtOH	241(4.27)	35-2947-76
trans	EtOH	241(4.27)	35-2947-76
$C_{21}H_{21}ClN_2O_2S$ Benzenesulfonamide, 4-chloro-N-(4,5,9-10-tetrahydro-7H-pyrrolo[3,2,1-de]-phenanthridin-7a(8H)-yl)-	EtOH	235(4.42),270s(3.98), 345(3.72)	39-2254-76C
$C_{21}H_{21}ClO_5$ 9,10-Anthracenedione, 1-butyl-8-chloro-5-hydroxy-2,4-dimethoxy-7-methyl-	EtOH	229(4.51),255(4.29), 264(4.32),285(4.16), 425(4.08)	39-1852-76C
$C_{21}H_{21}FN_2O_2$ Pyrano[2,3-c]pyrazole, 4-(4-fluorophenyl)-1,4,5,6-tetrahydro-6-(1-methylethoxy)-1-phenyl-, cis	EtOH	245(4.22)	35-2947-76
trans	EtOH	245(4.21)	35-2947-76
$C_{21}H_{21}NO$ 2-Cyclohexen-1-one, 3-[2-phenyl-3-(phenylmethyl)-1-aziridinyl]-	EtOH	279(4.356)	39-2484-76C
4H-Indol-4-one, 1,2,3,5,6,7-hexahydro-3-phenyl-2-(phenylmethyl)-	EtOH	300(4.029)	39-2484-76C
$C_{21}H_{21}NOS$ 2(1H)-Azocinone, 7,8-dihydro-1-methyl-5-(methylthio)-3,4-diphenyl-	EtOH	228(4.318),277(4.041)	24-0576-76
2H-Cyclopropa[gh]pyrrolizinium, 1,5a,5b-5c-tetrahydro-4-hydroxy-3-methyl-5c-(methylthio)-5,5a-diphenyl-, hydroxide, inner salt	EtOH	219s(4.152),294(4.086)	24-0562-76
$C_{21}H_{21}NO_3$ Phenanthrene, 5,6,7,8,9,10-hexahydro-2-methoxy-7-nitro-8-phenyl-	MeOH	220(4.19),275(4.17)	2-0799-76

Compound	Solvent	$\lambda_{max}(\log \epsilon)$	Ref.
Phenanthrene, 6,7,8,8a,9,10-hexahydro-2-methoxy-7-nitro-8-phenyl-	MeOH	216(4.31),264(4.26)	2-0799-76
Phenanthrene, 6,7,8,8a,9,10-hexahydro-2-methoxy-8-nitro-7-phenyl-	MeOH	218(4.38),265(4.38)	2-0799-76
$C_{21}H_{21}NO_4$			
Dibenzo[a,g]quinolizinium, 5,6-dihydro-9-hydroxy-2,3,10-trimethoxy-13-methyl-, hydroxide, inner salt	EtOH	238(4.47),279(4.41), 384(4.10),502(3.81)	88-1595-76
$C_{21}H_{21}NO_5$			
8H-Benzo[a][1,3]benzodioxolo[5,6-g]quinolizin-8-one, 5,6,14,14a-tetrahydro-2,3-dimethoxy-14-methyl-	MeOH	222(4.61),272(3.90), 291(3.93),303(3.86)	44-2201-76
8H-Dibenzo[a,g]quinolizin-8-one, 5,6-dihydro-2,3,10,11-tetramethoxy-	EtOH	232(3.90),265(3.73), 338(3.59),350s(3.46)	78-0211-76
Isoquinoline, 2-(1,3-benzodioxol-5-ylcarbonyl)-1-ethylidene-1,2,3,4-tetrahydro-6,7-dimethoxy-, (Z)-	MeOH	242(4.13),261(4.30), 282(3.95),297(4.08)	44-2201-76
$C_{21}H_{21}NO_6$			
Benzoic acid, 2-[(6,7-dimethoxy-1-isoquinolinyl)methyl]-4,5-dimethoxy-	EtOH	241(4.84),292(3.87), 311s(3.75),327(3.74)	88-1415-76
Peshawarine	EtOH	228(4.96),245(4.32), 293(3.78),333(3.69)	142-0041-76S
Spiro[7H-indeno[4,5-d]-1,3-dioxole-7,1'(2'H)-isoquinolin]-6(8H)-one, 3',4'-dihydro-8-hydroxy-6',7'-dimethoxy-2'-methyl-	EtOH	204(4.80),240(3.94), 291(3.91),313(3.99)	105-0118-76
Yenhusomidine	EtOH	238(4.21),290(3.74), 314(3.72)	39-0063-76C
$C_{21}H_{21}NO_6S$			
Thiazolo[2,3-a]isoquinolinium, 3-(ethoxycarbonyl)-3-[2-ethoxy-1-(ethoxycarbonyl)-2-oxoethyl]-, hydroxide, inner salt	EtOH	224(4.38),256(4.48), 314(4.36),360(3.96), 390(4.02),456(4.07)	94-1299-76
$C_{21}H_{21}NO_7$			
Benzoic acid, 6-[(7,8-dihydro-1,3-dioxolo[4,5-g]isoquinolin-5(6H)-ylidene)-hydroxymethyl]-2,3-dimethoxy-, methyl ester, hydrochloride	EtOH	210(4.50),228(4.37), 280(4.20),308(4.19)	35-6714-76
$C_{21}H_{21}NS_2$			
2(1H)-Azocinethione, 7,8-dihydro-1-methyl-5-(methylthio)-3,4-diphenyl-	dioxan	221s(4.385),265(4.309), 367s(2.799)	24-0576-76
2H-Cyclopropa[gh]pyrrolizinium, 1,5a,5b,5c-tetrahydro-4-mercapto-3-methyl-5c-(methylthio)-5,5a-diphenyl-, hydroxide, inner salt	EtOH	224(4.329),325(4.068)	24-0562-76
$C_{21}H_{21}N_2OS$			
Furo[2,3-b]pyridinium, 7-ethyl-4-[3-(3-ethyl-2(3H)-benzothiazolylidene)-1-propenyl]-, iodide	EtOH	569(5.16)	103-0052-76
$C_{21}H_{21}N_2O_2$			
Furo[2,3-b]pyridinium, 7-ethyl-4-[3-(7-ethylfuro[2,3-b]pyridin-6(7H)-ylidene)-1-propenyl]-, iodide	EtOH	616(5.26)	103-0052-76

Compound	Solvent	$\lambda_{max}(\log \epsilon)$	Ref.
$C_{21}H_{21}N_2O_4$ 1,3-Dioxolo[4,5-g]quinolinium, 7-(eth-oxycarbonyl)-5-ethyl-8-(phenylamino)-, iodide	MeOH	265(4.68),285(4.62), 330(4.28)	142-1347-76
$C_{21}H_{21}N_2SSe$ Benzothiazolium, 3-ethyl-2-[3-(7-ethyl-selenolo[2,3-b]pyridin-4(7H)-ylidene)-1-propenyl]-, iodide	EtOH	571(5.09)	103-0052-76
$C_{21}H_{21}N_2Se_2$ Selenolo[2,3-b]pyridinium, 7-ethyl-4-[3-(7-ethylselenolo[2,3-b]pyridin-6(7H)-ylidene)-1-propenyl]-, iodide	EtOH	624(5.08)	103-0052-76
$C_{21}H_{21}N_3O_2$ Pyrrolidine, 1-[2-[5-(4-methoxyphenyl)-1H-pyrazol-3-yl]benzoyl]-	EtOH	261(4.51)	44-0110-76
$C_{21}H_{21}N_3O_3$ Alkaloid ND-363B	MeOH	222(4.78),270(4.14), 279s(--),289s(--), 361(3.92)	23-1262-76
	MeOH-HCl	247(4.29),265s(--), 362(4.55)	23-1262-76
	MeOH-KOH	223(4.85),273(4.19), 280s(--),289s(--)	23-1262-76
3-Pyridinecarboxylic acid, 5-(2,3,5,6-11,11b-hexahydro-11b-methyloxazolo[3',2':1,2]pyrido[3,4-b]indol-2-yl)-, methyl ester (naucleonidine)	MeOH	224(4.45),272(3.82), 291(3.65),360(3.77)	23-1262-76
	MeOH-HCl	212(4.28),250(4.04), 364(4.33)	23-1262-76
	MeOH-KOH	225(4.55),273(4.00), 291(3.86)	23-1262-76
Pyrrolidine, 1-[[3-[[(aminocarbonyl)-oxy]methyl]-1-phenyl-1H-indol-2-yl]carbonyl]-	EtOH	262(4.51)	44-0110-76
13,14,16-Triazagona-1,3,5-triene-15,17-dione, 3-methoxy-16-phenyl-	EtOH	260(4.27)	4-1333-76
$C_{21}H_{21}N_3O_3S$ 4H-Indazol-4-one, 1,5,6,7-tetrahydro-1-methyl-3-phenyl-, O-[(4-methyl-phenyl)sulfonyl]oxime, (E)-	MeOH	226(4.51),256(3.98)	24-1898-76
$C_{21}H_{21}N_3O_4$ Pyrano[2,3-c]pyrazole, 1,4,5,6-tetrahy-dro-6-(1-methylethoxy)-4-(4-nitro-phenyl)-1-phenyl-, cis	EtOH	251(4.30)	35-2947-76
trans	EtOH	252(4.32)	35-2947-76
$C_{21}H_{21}N_3O_8$ Azulene, 1,3-bis(2-hydroxyethyl)-2-meth-yl-, trinitrobenzene complex	CH_2Cl_2	287(4.70),294(4.75), 307(3.90),338(3.54), 353(3.68),369(3.12), 595(2.40),635(2.34), 710(1.90)	44-1811-76
$C_{21}H_{21}N_4OS$ Morpholinium, 4-[[[5-phenyl-2-(phenyl-amino)-6H-1,3-thiazin-6-ylidene]-amino]methylene]-, perchlorate	HOAc	262(4.16),313(4.27), 462(4.29)	97-0268-76

Compound	Solvent	$\lambda_{max}(\log \epsilon)$	Ref.
$C_{21}H_{22}BrN_5O_5$			
Adenosine, N-benzoyl-5'-bromo-5'-deoxy-4'-C-methoxy-2',3'-O-(1-methylethylidene)-	MeOH	230(4.10),279(4.30)	35-3346-76
Benzamide, N-[9-[5-bromo-5-deoxy-4-C-methoxy-2,3-O-(1-methylethylidene)-α-L-lyxofuranosyl]-9H-purin-6-yl]-	MeOH	239(4.21),312(4.23)	35-3346-76
$C_{21}H_{22}ClNO_3$			
2-Butanone, 4-(3-acetyl-6-chloro-3,4-dihydro-4-hydroxy-4-phenyl-1(2H)-quinolinyl)-	isoPrOH	265(4.21),322(3.52)	4-0131-76
$C_{21}H_{22}ClN_3O_3$			
Quinazoline, 6-chloro-2-(2-ethyl-3-nitropentyl)-4-phenyl-, 3-oxide	isoPrOH	231(4.42),266(4.43), 310s(3.74),370(3.65)	4-0433-76
$C_{21}H_{22}Cl_3N_3O_7S$			
5-Thia-1-azabicyclo[4.2.0]oct-2-ene-2-carboxylic acid, 7-[(ethoxycarbonyl)amino]-3-methyl-8-oxo-7-[(phenoxyacetyl)amino]-, 2,2,2-trichloroethyl ester, (6R-cis)-	EtOH	265(3.83),272(3.80)	39-1918-76C
$C_{21}H_{22}NO_4$			
Palmatine, chloride	EtOH	228(4.39),240s(4.32), 268(4.38),276s(4.38), 343s(4.39),350(4.40), 433(3.71)	73-3157-76
Pseudopalmatine, chloride	EtOH	242(4.23),265(4.25), 288(4.57),310s(4.42), 342(4.20),380s(3.78)	73-3157-76
Thalphenine, chloride	MeOH	221(4.32),230s(4.20), 288(2.81),315(3.90), 328s(3.87)	100-0065-76
$C_{21}H_{22}N_2OS$			
Acetamide, N-[3-[ethyl(4-methylphenyl)-amino]-5-phenyl-2-thienyl]-	MeOH	224(4.31),252(4.20), 277s(4.08),314(4.20), 331s(4.11)	48-0221-76
$C_{21}H_{22}N_2O_2$			
Aspidospermidine-3-carboxylic acid, 2,3,6,7,20,21-hexadehydro-, methyl ester, hydrochloride, (5α,12β,19α)-	EtOH	227(3.99),301(3.95), 330(4.15)	78-2839-76
Isoapogeissoschizine	MeOH	222(4.43),275(4.31), 322(4.28)	24-3817-76
Pyrano[2,3-c]pyrazole, 1,4,5,6-tetrahydro-6-(1-methylethoxy)-1,4-diphenyl-, cis	EtOH	245(4.21)	35-2947-76
trans	EtOH	244.5(4.22)	35-2947-76
4-Pyridinamine, 2-methyl-3-(phenylmethoxy)-5-[(phenylmethoxy)methyl]-, hydrochloride	EtOH	277(4.02)	87-0999-76
	pH 7	272(--)	87-0999-76
$C_{21}H_{22}N_2O_3$			
Benzoic acid, 2-[5-(dimethylamino)-1-ethyl-2,3-dihydro-3-oxo-4-phenyl-1H-pyrrol-2-yl]-	MeOH	243(4.21),302(4.14)	44-0390-76
Benzoic acid, 2-[5-(4-methoxyphenyl)-1H-pyrazol-3-yl]-, butyl ester	EtOH	260(4.46)	44-0110-76

Compound	Solvent	$\lambda_{max}(\log \epsilon)$	Ref.
1H-Inden-1-one, 2-[[(1,1-dimethylethyl)-methylamino]phenylmethyl]-6-nitro-	hexane	225(4.42),273(4.20), 322(3.54),335(3.45), 392(3.27),415(3.23), 440(3.10),475(3.01)	44-3540-76
	$CHCl_3$	390(3.32),470(2.96)	44-3540-76
	MeCN	387(3.34),470(2.95)	44-3540-76
4-Quinolinecarbonitrile, 3-(benzoyloxy)-1,2,3,4-tetrahydro-2-hydroxy-2-(2-methylpropyl)-	EtOH	232(4.31),275s(3.19), 282s(3.30),300(3.45)	95-0968-76
Rhazinaline	EtOH	266(3.78)	18-2000-76
Yohimbinone, 15,16-didehydro-	MeOH	226(4.63),274(3.96), 280(3.98),290(3.88)	24-1737-76
hydrochloride	EtOH	227(4.66),271(3.97), 280(3.96),288(3.88)	24-1737-76
Yohimbinone, 15,20-didehydro-, hydrochloride	EtOH	225(4.63),271(3.95), 280(3.94),288(3.88)	24-1737-76
$C_{21}H_{22}N_2O_4$			
Benzoic acid, 4-methoxy-, 4-cyano-2-ethyl-1,2,3,4-tetrahydro-2-hydroxy-3-methyl-3-quinolinyl ester	EtOH	246(4.40),270s(4.09), 297(3.53)	95-0968-76
$C_{21}H_{22}N_2O_5$			
2,2'-Bipyridine, 4'-(2,4,6-trimethoxy-phenyl)-6,6'-dimethyl-, Cu(I) complex	EtOH	460(3.69)	118-0001-76
2,2'-Bipyridine, 4'-(3,4,5-trimethoxy-phenyl)-6,6'-dimethyl-, Cu(I) complex	EtOH	460(4.02)	118-0001-76
Propanedioic acid, [3-formyl-1,2,6,7,12-12b-hexahydroindolo[2,3-a]quinolizin-2-yl]-, dimethyl ester, cis	MeOH	224(4.51),294(4.63)	35-3645-76
$C_{21}H_{22}N_3S_2$			
Benzothiazolium, 2-[(4-amino-3-ethyl-5-phenyl-2(3H)-thiazolylidene)-3-ethyl-, iodide	n.s.g.	452(4.58)	103-0648-76
$C_{21}H_{22}O$			
2-Cyclopropen-1-one, 2,3-bis(2,4,6-tri-methylphenyl)-	MeCN	279(4.306),294s(4.192), 314(4.009),332(3.855)	101-0385-76N
$C_{21}H_{22}O_2$			
17H-Cyclopenta[a]phenanthren-17-one, 13-ethyl-6,7,11,12,13,16-hexahydro-3-methoxy-11-methylene-, (+)-	MeOH	248(4.11),257(4.04), 312(4.23),325(4.28), 339(4.19)	44-0531-76
$C_{21}H_{22}O_3$			
8H-Benzo[a]fluoren-10-ol, 9,10,10a,11-tetrahydro-3-methoxy-10a-methyl-, acetate, trans-(+)-	MeOH	215(4.21),235(4.41), 242(4.50),250(4.67), 259(4.75),281(4.02), 291(4.18),303(4.21), 326(3.11),340(3.24), 357(3.16)	24-3025-76
$C_{21}H_{22}O_4$			
Benzene, 2,3-dimethoxy-1-[2-(3,4-meth-ylenedioxyphenyl)-1-methylethenyl]-6-(2-propenyl)-	EtOH	266(4.12),296s(3.87)	88-4371-76
D(17a)-Homo-18-norpregna-2,13,15,17-tetraene-1,20-dione, 6,7-epoxy-5-hydroxy-, (5α,6α,7α)-	EtOH	212(4.53),265(4.27), 285s(3.34)	39-0304-76C

Compound	Solvent	$\lambda_{max}(\log \epsilon)$	Ref.
$C_{21}H_{22}O_5$			
2,5-Cyclohexadien-1-one, 4-[2-(1,3-ben-zodioxol-5-yl)-1-methylethenyl]-4,5-dimethoxy-2-(2-propenyl)- (isofuto-quinol B)	EtOH	237(4.09),288(3.80)	88-4371-76
2-Naphthalenecarboxylic acid, 1,2,3,4-tetrahydro-6,8-dimethoxy-3-[(3-meth-ylphenyl)methyl]-4-oxo-	MeOH	265(3.88),325(3.45)	2-0088-76
Tricyclo[4.2.0.0^{2,8}]oct-3-en-5-one, 7-(1,3-benzodioxol-5-yl)-2,3-di-methoxy-8-methyl-6-(2-propenyl)- (isofutoquinol A)	EtOH	233s(3.92),277(3.98)	88-4371-76
$C_{21}H_{22}O_8$			
2-Propenoic acid, 2-methyl-, 5-acetoxy-2,3,3a,4,5,6,9,10,11,11a-decahydro-10-methyl-3,6-bis(methylene)-2,9-dioxo-7,10-epoxycyclodeca[b]furan-4-yl ester, [3aR-(3aR*,4R*,5r*,10R*,11aS*)]-	n.s.g.	285(3.89)	102-0191-76
$C_{21}H_{23}ClN_2O_2$			
Aspidospermidine-3-carboxylic acid, 3-chloro-1,2,6,7-tetrahydro-, methyl ester, (5α,12β,13α)-	EtOH	227(4.21),282(3.80)	33-1213-76
$C_{21}H_{23}Cl_2N_3O$			
1H-1,4-Benzodiazepine, 7-chloro-5-(2-chlorophenyl)-2,3-dihydro-1-methyl-2-(morpholinomethyl)-	EtOH	228s(--),269(3.92), 374(3.36)	111-0501-76
$C_{21}H_{23}IN_2O_2$			
Aspidospermidine-3-carboxylic acid, 2,3,6,7-tetrahydro-20-iodo-, methyl ester, (5α,12β,19α,20R)-	EtOH	228(4.04),298(4.08), 330(4.15)	78-2839-76
epimer	EtOH	228(4.08),298(4.08), 330(4.15)	78-2839-76
$C_{21}H_{23}NO$			
1H-Inden-1-one, 2-[[(1,1-dimethylethyl)-amino]phenylmethyl]-6-methyl-	hexane	245(4.64),252(4.66), 306(2.85),320(3.08), 335(3.04),393(3.04)	44-3540-76
1H-Inden-1-one, 2-[[(1,1-dimethylethyl)-methylamino]phenylmethyl]-	hexane	390(2.90),407s(3.04), 430(3.08)	44-3540-76
	CHCl$_3$	390(2.83)	44-3540-76
	MeCN	390(2.85)	44-3540-76
$C_{21}H_{23}NOS$			
2-Azoniabicyclo[3.1.0]hex-3-ene, 1-eth-oxy-3-mercapto-2,2-dimethyl-4,5-di-phenyl-, hydroxide, inner salt	EtOH	221(4.346),325(4.110)	24-0562-76
Benzeneethanethioamide, α-(2-ethoxy-1-phenyl-2-propenylidene)-N,N-di-methyl-, (Z)-	EtOH	263(4.245),289s(4.161)	24-0576-76
$C_{21}H_{23}NO_2$			
Benzeneacetamide, α-(2-ethoxy-1-phenyl-2-propenylidene)-N,N-dimethyl-, (Z)-	EtOH	227s(4.219),284(4.082)	24-0576-76
$C_{21}H_{23}NO_3$			
Benzamide, N-(benzoyloxy)-N-(1,1-dimeth-yl-4-pentenyl)-	EtOH	232(4.18)	44-0855-76

Compound	Solvent	$\lambda_{max}(\log \epsilon)$	Ref.
8H-Dibenzo[a,g]quinolizin-8-one, 5,6,13,13a-tetrahydro-2,3-dimethoxy-13,13-dimethyl-	n.s.g.	230(4.31),254(3.88), 268(3.80),280(3.78), 289s(3.70)	44-2201-76
2,6-Methano-3-benzazocine-1,8-diol, 3-benzoyl-1,2,3,4,5,6-hexahydro-6,11-dimethyl-, (1α,2β,6β,11S*)-(+)-	MeOH	230(4.18),275(3.32), 283(3.24)	24-2657-76
2,6-Methano-3-benzazocine-4,8-diol, 3-benzoyl-1,2,3,4,5,6-hexahydro-6,11-dimethyl-, (2α,4α,6α,11R*)-(+)-	MeOH	217s(4.14),228s(4.06), 281(3.33),287s(3.28)	24-2657-76
2,6-Methano-3-benzazocin-8-ol, 1,2,3,4-5,6-hexahydro-6,11-dimethyl-3-(7-oxabicyclo[4.1.0]hepta-2,4-dien-3-ylcarbonyl)-	MeOH	219s(4.05),226s(3.97), 276(3.59),287s(3.48)	24-2657-76
2,6-Methano-3-benzazocin-8-ol, 1,2,3,4-5,6-hexahydro-3-(4-hydroxybenzoyl)-6,11-dimethyl-, (2α,6α,11R*)-(+)-	MeOH	225(4.16),245(3.96), 274(3.67),283(3.57)	24-2657-76
6-Phenanthridinecarboxylic acid, 1,2,3,4,4a,5,6,10b-octahydro-9-hydroxy-6-(phenylmethyl)-	EtOH	279.5(3.24)	95-1031-76
$C_{21}H_{23}NO_4$			
Benzo[e][1,3]dioxolo[4,5-k][3]benzazecine, 5,6,7,8-tetrahydro-3,4-dimethoxy-6-methyl-, cis	C_6H_{12}	292(3.83)	35-4301-76
Ochotensane, 9,10-dimethoxy-2,3-(methylenedioxy)-	EtOH	230s(4.24),287(3.77)	78-1973-76
6-Phenanthridinecarboxylic acid, 1,2,3,4,4a,5,6,10b-octahydro-9-hydroxy-6-[(4-hydroxyphenyl)methyl]-	EtOH	277.5(3.29)	95-1031-76
Spiro[5H-indeno[5,6-d]-1,3-dioxole-5,3'(2'H)-isoquinoline], 1',4',6,7-tetrahydro-7',8'-dimethoxy-2'-methyl-	EtOH	228s(4.07),288(3.75), 294(3.77)	78-1973-76
Thalictricavine	EtOH	227s(4.18),291(3.75)	73-3157-76
Tuliferoline, (-)-	MeOH	224(4.06),273(4.10)	102-1169-76
$C_{21}H_{23}NO_5$			
2-Furancarboxylic acid, tetrahydro-4-[2,4,6-trimethyl-7-(4-nitrophenyl)-2,4,6-heptatrienylidene]-, (Z,E,E,E)-(spectinabilic acid)	EtOH	248(4.15),283(4.18), 358(4.08)	78-0217-76
Pavinan, 2,3,7-trimethoxy-8,9-(methylenedioxy)-	EtOH	287(3.84)	105-0110-76
$C_{21}H_{23}NO_6$			
1,3-Benzodioxole-4-carboxylic acid, 5-[2-[6-[2-(dimethylamino)ethyl]-1,3-benzodioxol-5-yl]ethyl]-	EtOH	234s(3.38),291(3.22)	142-0041-76S
Berbane, 13-hydroxy-2,3,10,11-tetramethoxy-8-oxo-, cis-(+)-	EtOH	212(3.93),230(3.96), 275(3.34),297(3.30), 306s(3.11)	78-0211-76
trans-(+)-	EtOH	213(4.28),230(4.38), 270(3.81),295(3.43), 307s(3.22)	78-0211-76
Colchiceine	EtOH-NaOH	241(4.49),259s(4.40), 352(4.24),367s(4.29), 403(4.04)	73-3146-76
Colchicine, 2-demethyl-	EtOH	243(4.46),355(4.20)	73-3146-76
	EtOH-NaOH	247(4.45),290s(3.86), 356(4.19)	73-3146-76
Colchicine, 3-demethyl-	EtOH	232s(4.41),243(4.46), 355(4.19)	73-3146-76

Compound	Solvent	$\lambda_{max}(\log \epsilon)$	Ref.
Cochicine, 3-demethyl- (cont.)	EtOH-NaOH	251(4.46),286(3.86), 340s(3.86),395(4.06)	73-3146-76
1(3H)-Isobenzofuranone, 5,6-dimethoxy-3-(1,2,3,4-tetrahydro-6,7-dimethoxy-1-isoquinolinyl)-, (R*,R*)-	EtOH	212(3.88),227(3.82), 262(3.27),295(3.19), 310s(2.98)	78-0211-76
(R*,S*)-	EtOH	215(4.03),228(4.08), 263(3.58),295(3.57), 308s(3.38)	78-0211-76
Norboldine, N-(ethoxycarbonyl)-, (±)-	EtOH	216(4.61),284(4.15), 304(4.15)	77-0091-76
Yenhusomine	EtOH	241(4.11),288(3.85)	39-0063-76C

$C_{21}H_{23}NO_7$

4,5-Isoxazolidinedicarboxylic acid, 3-(2,5-dimethoxyphenyl)-2-phenyl-, dimethyl ester	n.s.g.	249(3.90),289(3.71), 429(2.37)	142-0109-76S

$C_{21}H_{23}N_2$

Quinolinium, 2-[2-[4-(dimethylamino)-phenyl]ethenyl]-1,3-dimethyl-, iodide	EtOH	307(4.39),505(4.05)	103-0683-76

$C_{21}H_{23}N_3O_3$

3-Pyridinebutanamide, γ-acetoxy-N-[2-(1H-indol-3-yl)ethyl]-	MeOH	225(4.36),262(3.70), 268(3.72),282(3.67), 291(3.60)	23-1262-76

$C_{21}H_{23}N_3O_4$

1,2-Hydrazinedicarboxylic acid, 1-(1-methyl-2-phenyl-3-indolizinyl)-, diethyl ester	MeOH	222(4.54),250(4.77), 312(4.19)	2-0057-76
1,2-Hydrazinedicarboxylic acid, 1-(3-methyl-2-phenyl-1-indolizinyl)-, diethyl ester	MeOH	224(4.60),253(4.95), 306(4.16)	2-0057-76

$C_{21}H_{23}N_3O_4S$

1H-Indazole-4,7-dione, 3,6-dimethyl-1-[(4-methylphenyl)sulfonyl]-5-piperidino-	EtOH	236(4.42),318(3.83), 550(3.56)	94-1731-76

$C_{21}H_{23}N_3O_6$

2,4-Hexadienoic acid, [2,3,3a,9a-tetra-hydro-6-imino-3-[(1-oxo-2,4-hexadien-yl)oxy]-6H-furo[2',3':4,5]oxazolo-[3,2-a]pyrimidin-2-yl]methyl ester, hydrochloride, [2R-(2α,3β,3aβ,9aβ)]-	MeOH-acid	234(4.33),263(4.70)	87-0663-76
L-Phenylalanine, N-(N-acetyl-D-alanyl)-, 4-nitrophenylmethyl ester	EtOH	269(4.01)	33-2421-76

$C_{21}H_{23}N_3O_{10}$

1H-Pyrazole-3-carboxylic acid, 4-hydr-oxy-5-[2,3-O-(1-methylethylidene)-5-O-(4-nitrobenzoyl)-α-D-ribofuranosyl]-, ethyl ester	EtOH	227(4.10),261(4.22)	44-0287-76

$C_{21}H_{23}N_4S$

Methanaminium, N-[[[2-[4-(dimethylami-no)phenyl]-5-phenyl-6H-1,3-thiazin-6-ylidene]amino]methylene]-N-methyl-, perchlorate	HOAc	317(4.07),389(4.47), 556(4.52)	97-0268-76

Compound	Solvent	$\lambda_{max}(\log \epsilon)$	Ref.
$C_{21}H_{24}ClN$			
Acridine, 9-chloro-2,7-bis(1,1-dimethyl-ethyl)-	MeOH	204(4.11),224(4.04), 260(5.15),348(3.81), 366(4.01)	39-0465-76C
$C_{21}H_{24}NO_4$			
Canadinium, N-methyl-	EtOH	234(4.07),287(3.76)	73-3654-76
$C_{21}H_{24}N_2O$			
Benzeneacetamide, α-[2-(dimethylamino)-1-phenyl-2-propenylidene]-N,N-dimethyl-, (Z)-	EtOH	227(4.428),275s(4041)	24-0576-76
3(2H)-Indolizinone, hexahydro-1-[(4-methylphenyl)amino]-2-phenyl-	EtOH EtOH-acid	249(4.21),305(3.34) 254(2.46)	39-0005-76C 39-0005-76C
$C_{21}H_{24}N_2O_2$			
3(2H)-Indolizinone, hexahydro-1-[(4-methoxyphenyl)amino]-2-phenyl-	EtOH EtOH-acid	249(4.24),295(3.28) 254(2.48)	39-0005-76C 39-0005-76C
$C_{21}H_{24}N_2O_2S$			
Piperazine, 1-(2,3-dimethoxydibenzo-[b,f]thiepin-10-yl)-4-methyl-	MeOH	237(4.41),274(4.17), 333s(3.84)	73-1396-76
$C_{21}H_{24}N_2O_3$			
Azacycloundecino[5,4-b]indole-9-carboxylic acid, 7-ethylidene-3-formyl-1,2,3,4,7,8,9,10-octahydro-, methyl ester, (Z,?)-($\pm$)-	EtOH	220(4.28),285(3.90), 293(3.90)	78-2839-76
Catharanthine, N-oxide	EtOH or MeOH	224(4.5),275s(3.8), 283(3.9),291(3.8)	33-2858-76
rearrangement product	EtOH or MeOH	226(4.5),276s(3.8), 285(3.9),293(3.8)	33-2858-76
15R,20S-Epoxycatharanthine	MeOH	226(4.62),285(3.97), 293(3.91)	88-3945-76
Geissoschizine	EtOH	221(4.32),267(3.88), 280s(3.57)	18-2000-76
	EtOH-NaOH	280(4.20)	18-2000-76
Rhazinalinol	EtOH	265(3.78)	18-2000-76
8,14-Seco-2-azaestra-1,3,5(10),9-tetraene-4-carbonitrile, 3-ethoxy-1-methyl-14,17-dioxo-	EtOH	258(3.95),312(3.53)	2-0731-76
$C_{21}H_{24}N_2O_4$			
Cathovalinine	EtOH	230(3.69),299(3.73), 329(3.90)	78-1899-76
Oxayohimban-16-carboxylic acid, 19-methyl-17-oxo-, methyl ester	MeOH	224(4.52),279(3.88), 290(3.75)	35-3645-76
epimer	MeOH	224(4.47),282(3.82), 290(3.74)	35-3645-76
2,6-Piperidinedione, 4-[2-(4,4-dimethyl-2,6-dioxocyclohexylidene)-2-(phenylamino)ethyl]-	EtOH	341(3.89),394(3.73)	30-0139-76
$C_{21}H_{24}N_2O_4S_2$			
Benzothiazolium, 2-[4-[4-(dimethylamino)phenyl]-1,3-butadienyl]-3-methyl-, methyl sulfate	MeNO$_2$	572(4.81)	124-1174-76
$C_{21}H_{24}N_2S$			
2-Azoniabicyclo[3.1.0]hex-3-ene, 1-(di-	EtOH	222(4.33+),323(4.04+)	24-0562-76

Compound	Solvent	$\lambda_{max}(\log \epsilon)$	Ref.
methylamino)-3-mercapto-2,2-dimethyl-4,5-diphenyl-, hydroxide, inner salt			24-0562-76
$C_{21}H_{24}N_4$ Pyrrolo[1,2-c]pyrimidine, 5,5'-methyl-enebis[3,6,7-trimethyl-	EtOH	243(4.53),277s(3.83), 287(3.91),299(3.91), 376(3.22)	44-0351-76
$C_{21}H_{24}N_5O_7PS$ Adenosine, N-(1-oxobutyl)-8-[(phenyl-methyl)thio]-, cyclic 3',5'-(hydrogen phosphate)	pH 7	298(4.24)	87-0899-76
$C_{21}H_{24}N_6O_5S$ Acetamide, N-[3-[2,3-O-(1-methylethylidene)-β-D-ribofuranosyl]-7-[(phenylmethyl)thio]-3H-1,2,3-triazolo[4,5-d]pyrimidin-5-yl]-	EtOH-HCl EtOH-pH 7	231(4.16),257(4.22), 283s(4.05),307(4.22) 231(4.16),257(4.23), 283s(4.05),307(4.22)	87-1186-76 87-1186-76
$C_{21}H_{24}OS_2$ 9,13-Dithiatricyclo[13.2.2.1^{3,7}]eicosa-3,5,7(20),15,17,18-hexaen-2-one, 11,11,16-trimethyl-	MeOH	251(4.11)	44-2509-76
$C_{21}H_{24}O_2$ 6H-Cyclopenta[a]phenanthren-17β-ol, 13-ethyl-7,11,12,13,16,17-hexahydro-3-methoxy-11-methylene-, (±)- D-Homo-C,18-dinorpregna-4,13,15,17-tetraene-3,20-dione, 17a-methyl-	MeOH EtOH	248(4.14),257(4.05), 312(4.29),324(4.35), 338(4.25) 239(4.11),295(3.57)	44-0531-76 138-0975-76
$C_{21}H_{24}O_3$ 1,3-Cyclopentanedione, 2-[2-(3,4-dihydro-6-methoxy-1-naphthalenyl)-2-propenyl]-2-ethyl- 14βH-Estra-1,3,5(10),8-tetraen-17-one, 1-hydroxy-3-methyl-, acetate	MeOH CHCl$_3$	273(4.04) 274(4.23)	44-0531-76 22-1889-76
$C_{21}H_{24}O_4$ Benzofuran, 2-(3,4-dimethoxyphenyl)-2,3-dihydro-7-methoxy-3-methyl-5-(1-propenyl)-, [2α,3β,5(E)]-(±)- Cyclohexanone, 3-acetoxy-2-[(6-methoxy-1-naphthalenyl)methyl]-2-methyl-, cis-(±)-	MeOH MeOH	275(3.99) 215(4.53),230(4.21), 244(4.17),259(3.52), 268(3.66),278(3.75), 289(3.67),303(3.05), 317(3.26),326(3.22), 331(3.38)	102-1033-76 24-3025-76
$C_{21}H_{24}O_5$ 3H-Benz[e]inden-3-one, 5-ethyl-1,2-dihydro-4-hydroxy-7,8,9-trimethoxy-1-propylidene- 2H,6H-Benzo[1,2-b:5,4-b']dipyran-6-one, 5-hydroxy-10-(3-methoxy-3-methyl-1-butenyl)-2,2,8-trimethyl- 4H,8H-Benzo[1,2-b:3,4-b']dipyran-4-one, 5-hydroxy-6-(3-methoxy-3-methyl-1-butenyl)-2,8,8-trimethyl-, (E)-	MeOH MeOH MeOH	225(4.29),291(4.48), 331(3.95),416(3.66) 247(4.42),276(4.56), 330s(3.61) 210(4.18),242(4.40), 275(4.55),350s(3.45)	18-1937-76 24-2963-76 24-2963-76

Compound	Solvent	$\lambda_{max}(\log \epsilon)$	Ref.
6(2H)-Benzofuranone, 2-(3,4-dimethoxy-phenyl)-3,3a-dihydro-5-methoxy-3-methyl-3a-(2-propenyl)-, [2S-(2α,3β,3aβ)]-	MeOH	235(4.18),260(4.29), 285s(4.03)	102-1033-76
6(2H)-Benzofuranone, 2-(3,4-dimethoxy-phenyl)-3,5-dihydro-5-methoxy-3-methyl-5-(2-propenyl)-, [2S-(2α,3β,5α)]-	MeOH	230(4.38),278(4.36), 313(4.28)	102-1033-76
1H-2-Benzopyran, 3-(2-ethenyl-4,5-dimethoxyphenyl)-3,4-dihydro-6,7-dimethoxy-, (±)-	MeOH	263(3.93),286s(3.71)	33-0949-76
2,5-Cyclohexadien-1-one, 4-[2-(1,3-benzodioxol-5-yl)-1-methylethyl]-4,5-dimethoxy-2-(2-propenyl)- diastereomer	EtOH	236(4.19),287(3.91)	88-4371-76
	EtOH	236(4.22),287(3.94)	88-4371-76
Tricyclo[4.2.0.0^{2,8}]oct-3-en-5-ol, 7-(1,3-benzodioxol-5-yl)-2,3-dimethoxy-8-methyl-6-(2-propenyl)-	EtOH	238s(3.78),290(3.61)	88-4371-76
$C_{21}H_{24}O_6$ Arctigenin	EtOH	230(4.16),281(3.76)	102-1789B-76
Bicyclo[3.2.1]oct-3-en-2-one, 7-(1,3-benzodioxol-5-yl)-8-hydroxy-1,3-dimethoxy-6-methyl-5-(2-propenyl)-, (6-exo,7-endo,8-anti)	MeOH	234(3.96),263(3.82)	102-1033-76
$C_{21}H_{24}O_7$ Butanoic acid, 3-hydroxy-3-methyl-, 1-(1,4-dihydro-5,8-dihydroxy-1,4-dioxo-2-naphthalenyl)-4-methyl-3-pentenyl ester	EtOH	275(3.86),493(3.71), 526(3.78),568(3.57)	105-0652-76
Phthalan, 1-(2-acetyl-4,5-dimethoxybenzyl)-3-oxo-5,6-dimethoxy-	MeOH	228(4.59),261(4.14), 280(4.06),294(4.05)	33-0949-76
Spiro[furan-3(2H),6'(7'H)-[1H-3,9a]methano[2]benzoxepin]-10'-carboxylic acid, 5-(3-furanyl)octahydro-7'-methyl-1',2-dioxo-, methyl ester, [3'R-(3'α,5'a,6'β,7'β,9'aα,10'(S*)]-	EtOH	210(3.78)	78-1881-76
$C_{21}H_{25}ClN_2S$ 3-Pyrrolidinamine, 1-(8-chloro-10,11-dihydrodibenzo[b,f]thiepin-10-yl)-N-ethyl-N-methyl-, (8R,10R)-	MeOH	258(4.15),277(4.22), 295(4.10)	87-0040-76
(8R,10S)-	MeOH	259(4.14),277(4.21), 295(4.09)	87-0040-76
(8S,10R)-	MeOH	257(4.27),274(4.32), 295(4.11)	87-0040-76
(8S,10S)-	MeOH	257(4.20),274(4.25), 295(4.05)	87-0040-76
$C_{21}H_{25}ClO_5$ Spiro[androst-4-ene-17,4'-[1,3]dioxolan]-2',3,11-trione, 5'-chloro-, (5'R,17α)-	n.s.g.	238(4.20)	44-0078-76
(5'S,17α)-	n.s.g.	238(4.21)	44-0078-76
$C_{21}H_{25}NO_2$ Benzeneacetamide, α-(2-ethoxy-1-phenyl-propylidene)-N,N-dimethyl-, (Z)-	dioxan	247s(4.017)	24-0576-76
$C_{21}H_{25}NO_4$ d-Corybulbine	EtOH	225s(4.20),283(3.66)	95-0527-76

Compound	Solvent	$\lambda_{max}(\log \epsilon)$	Ref.
4H-Dibenzo[de,g]quinoline, 5,6,6a,7- tetrahydro-1,2,10,11-tetramethoxy- 6-methyl-, hydrochloride	EtOH	224(4.61),270(4.16), 300s(3.71)	33-2551-76
6H-Dibenzo[a,h]quinolizine, 5,8,9,13b- tetrahydro-2,3,11,12-tetramethoxy-	EtOH	220(4.36),234s(4.15), 286(3.84)	118-0719-76
Palmatine, tetrahydro-	EtOH	227s(4.28),282(3.75)	73-3157-76

$C_{21}H_{25}NO_5$

Compound	Solvent	$\lambda_{max}(\log \epsilon)$	Ref.
Alloberban-13-carboxylic acid, 12,13-di- dehydro-7,8-dimethoxy-14-oxo-, methyl ester	MeOH	231(3.3),282(2.6)	24-1724-76
1,4-Benzoxazepine, 5-(3,4-dimethoxy- phenyl)-2,3-dihydro-7,8-dimethoxy- 2,3-dimethyl-	n.s.g.	210(4.48),257(4.19), 315(4.12)	20-0787-76
Berban-13-carboxylic acid, 12,13-di- dehydro-7,8-dimethoxy-14-oxo-, methyl ester	MeOH	229(4.2),282(3.6)	24-1724-76
Berban-13-carboxylic acid, 12,17-di- dehydro-7,8-dimethoxy-14-oxo-, methyl ester	MeOH	225s(4.1),285(3.7), 306s(3.3)	24-1724-76
Berbane, 2,3,10,11-tetramethoxy- 13-hydroxy-, cis-(+)-	EtOH	213(4.07),233s(3.56), 288(3.17)	78-0211-76
trans-(+)-	EtOH	212(4.17),230(3.93), 290(3.48)	78-0211-76
1-Capaurine	EtOH	225s(4.31),275(3.41)	95-0527-76
Protothalipine	MeOH	232(4.60),282(3.91)	100-0065-76
Triketone from oxidation of anopterine	EtOH	250(3.25),438(2.18)	12-1295-76

$C_{21}H_{25}NO_6$

Compound	Solvent	$\lambda_{max}(\log \epsilon)$	Ref.
1,3-Benzodioxole-5-ethanol, α-[6-[2-(di- methylamino)ethyl]-1,3-benzodioxol- 5-yl]-4-(hydroxymethyl)-	EtOH	225(4.21),288(3.63)	142-0041-76S
Nokoensine	EtOH	227s(4.17),276(3.33)	95-0527-76

$C_{21}H_{25}NO_8$

Compound	Solvent	$\lambda_{max}(\log \epsilon)$	Ref.
Pentanedioic acid, 2,4-diacetyl-3-[2-(2- nitrophenyl)ethenyl]-, diethyl ester	MeOH	208(4.34),220(4.29), 239(4.37),308(3.56)	65-1141-76

$C_{21}H_{25}N_3OS$

Compound	Solvent	$\lambda_{max}(\log \epsilon)$	Ref.
Spiro[[1]benzothieno[2,3-d]pyrimidine- 2(1H),1'-cyclohexan]-4(3H)-one, 5,6,7,8-tetrahydro-3-(phenylamino)-	EtOH	234(4.30),328(3.29)	2-0357-76

$C_{21}H_{25}N_3O_2$

Compound	Solvent	$\lambda_{max}(\log \epsilon)$	Ref.
Corynan-17-oic acid, 16-cyano-, methyl ester, (+)-	EtOH	282(3.88),290(3.80)	78-1019-76

$C_{21}H_{25}N_3O_4S$

Compound	Solvent	$\lambda_{max}(\log \epsilon)$	Ref.
Benzenesulfonic acid, 4-methyl-, [1-[4- methyl-3,6-dioxo-2-(1-piperidinyl)- 1,4-cyclohexadien-1-yl]ethylidene]- hydrazide	EtOH	231(4.43),521(3.37)	94-1731-76

$C_{21}H_{25}N_3O_5$

Compound	Solvent	$\lambda_{max}(\log \epsilon)$	Ref.
1H,5H-Imidazo[1',2':1,2]pyrido[2,3-b]in- dole-2,3,5-trione, 7a-(1,1-dimethyl- propyl)-7a,12a-dihydro-6,12-dimethoxy- 1-methyl-, (7aS-trans)-	MeOH	233(4.21),285(3.79)	78-2625-76

Compound	Solvent	$\lambda_{max}(\log \epsilon)$	Ref.
$C_{21}H_{25}N_3O_6$			
Bicyclo[2.2.2]oct-2-ene-1-carboxylic acid, 4-methyl-, (3-acetoxy-2,3,3a,9a-tetrahydro-6-imino-6H-furo[2',3':4,5]-oxazolo[3,2-a]pyrimidin-2-yl)methyl ester, hydrochloride	MeOH-acid	235(3.99),262(4.06)	87-0663-76
4,1-(Iminomethano)-1H-pyrido[4,3-b]-indole-1,4(4aH)-dicarboxylic acid, 2,3,5,9b-tetrahydro-2,9b,10-trimethyl-3,11-dioxo-, diethyl ester, (1α,4α,4aα,9bα)-	EtOH	244(3.85),308(3.46)	78-0507-76
(1α,4β,4aα,9bα)-	EtOH	245(3.79),305(3.42)	78-0507-76
$C_{21}H_{25}N_3O_7$			
Ethanol, 2,2'-[[4-[(4-methoxyphenyl)ami-no]-3-nitrophenyl]imino]bis-, diacet-ate	EtOH	520(3.71)	146-0131-76
$C_{21}H_{25}N_5O_7S$			
Acetamide, N-[4-[[2,3-O-(1-methylethyl-idene)-β-D-ribofuranosyl]amino]-5-ni-tro-6-[(phenylmethyl)thio]-2-pyrimi-dinyl]-	EtOH-HCl	219(4.37),261(4.43), 352(4.13)	87-1186-76
	EtOH-NaOH	260(4.20),357(4.24)	87-1186-76
$C_{21}H_{25}N_9O_4$			
L-Alanine, N-[N-(5-oxo-L-prolyl)-L-his-tidyl]-3-(1H-pyrrolo[2,3-b]pyridin-3-yl)hydrazide	HOAc	293(3.92)	94-3149-76
$C_{21}H_{26}BrNO_{10}$			
Pyridinium, 3-acetyl-1-(2,3,4,6-tetra-O-acetyl-β-D-glucopyranosyl)-, bromide	H_2O	268(3.62)	35-5689-76
$C_{21}H_{26}ClN_3$			
1-Piperazinamine, N-[(4-chlorophenyl)-methylene]-4-[(2,4,6-trimethylphen-yl)methyl]-	MeOH	297.5(4.30)	73-1035-76
$C_{21}H_{26}N_2$			
9-Acridinamine, 2,7-bis(1,1-dimethyl-ethyl)-, hydrochloride	MeOH	226(3.63),265(4.85), 402(3.76),424(3.68)	39-0465-76C
3,3'-Bipyrrolidine, 4-methyl-1,1'-di-phenyl-	EtOH	203(4.649),256(4.597), 307(3.746)	104-2062-76
diastereomer	EtOH	201(4.576),256(4.708), 302(3.703)	104-2062-76
$C_{21}H_{26}N_2O$			
Benzeneacetamide, α-[2-(dimethylamino)-1-phenylpropylidene]-N,N-dimethyl-, (Z)-	dioxan	255s(4.021)	24-0576-76
$C_{21}H_{26}N_2O_2$			
1,5-Methanoazocino[1',2':1,2]pyrido[3,4-b]indol-3(4H)-one, 1-ethyl-1,2,5,6,8-9,14,14b-octahydro-11-methoxy-, [1S-(1α,5α,14bα)]- (voaketone)	n.s.g.	226(4.70),279(4.19), 293s(4.09)	33-0532-76
Reflexine	EtOH	250(3.78),295(3.66)	31-1236-76
$C_{21}H_{26}N_2O_3$			
Epiheyneanine	EtOH	226(4.43),280(3.74), 285(3.79),293(3.72)	102-0551-76

Compound	Solvent	$\lambda_{max}(\log \epsilon)$	Ref.
16-Epiisositsirikine	EtOH	226(4.50),285(3.90), 293(3.80)	102-0543-76
Geissoschizine, dihydro-	EtOH	224(4.26),272(3.77), 282(3.72),286(3.69)	18-2000-76
Isoretuline, N_a-deacetyl-17-O-acetyl-18-hydroxy-	n.s.g.	245(3.80),295(3.40)	102-0321-76
Isoretuline, 18-hydroxy-	n.s.g.	250(4.10),280s(3.50), 288(3.48)	102-0321-76
Strychnosplendine, N_a-acetyl-	EtOH	250(4.08),278(3.41), 286s(3.31)	32-0773-76
$C_{21}H_{26}N_2O_5$ 12H-5,7-Ethanopyrido[2,1-c][1,4]benzodi-azonine-16-carboxylic acid, 7-ethyl-6,6a,7,8,9,10,13,14-octahydro-16-hy-droxy-6,14-dioxo-, methyl ester, [6aS-(6aR*,7R*,16R*)]-	EtOH	235(4.34),312(3.42)	88-0435-76
$C_{21}H_{26}N_2O_3$ Iboxyphylline	EtOH	227(4.08),295(3.90), 328(3.87)	78-2539-76
$C_{21}H_{26}N_5O_6PS$ Adenosine, 2-butyl-8-[(phenylmethyl)-thio]-, cyclic 3',5'-(hydrogen phosphate	pH 1 pH 7 pH 11	283(4.27) 285(4.20) 285(4.20)	87-0419-76 87-0419-76 87-0419-76
$C_{21}H_{26}O_2$ 19-Norpregna-1,3,5,17(20)-tetraen-21-al, 3-methoxy-	EtOH	234(4.30),242(4.30)	118-0025-76
17α-Pregna-1,4-dien-20-yn-3-one, 17β-hydroxy-	EtOH	245(4.18)	94-0828-76
$C_{21}H_{26}O_3$ Pregna-4,6,16-triene-3,20-dione, 12β-hy-droxy- (neridienone A)	MeOH	245(4.11),283(4.38)	102-1745-76
8,14-Seco-1,3,5(10),9-gonatetraene-14,17-dione, 3-methoxy-13-propyl-	EtOH	265(4.23)	114-0081-76D
$C_{21}H_{26}O_4$ 1,3-Benzodioxole, 5-[(1,2-dimethoxy-7-methyl-5-propylbicyclo[4.1.0]-hepta-2,4-dien-7-yl)methyl]-	EtOH	231s(3.90),284(3.97)	88-4371-76
12,17-Cyclo-18-nor-13,17-secopregn-4-ene-3,11,13,20-tetrone, 12-methyl-, (17α)-	EtOH	234(4.15)	138-0975-76
D(17a)-Homo-C,18-dinorpregn-4-ene-3,11,20-trione, 13,17a-epoxy-17a-methyl-, (13α,17α,17aα)-	EtOH	245(4.15)	138-0975-76
$C_{21}H_{26}O_5$ 8α-Estra-1,3,5(10)-trien-11-one, 17β-acetoxy-14α-hydroxy-3-methoxy-	EtOH	221(3.96),229(3.84), 278(3.21),285(3.16)	44-0707-76
8α,9β-Estra-1,3,5(10)-trien-11-one, 17β-acetoxy-14α-hydroxy-3-methoxy-	EtOH	221(3.92),227(3.89), 278(3.23),285(3.20)	44-0707-76
Kaur-16-en-18-oic acid, 11,12,15-tri-oxo-, methyl ester, (4α)-	n.s.g.	232(3.78)	44-1021-76
Methyl barbascoate, (-)-	MeOH	212(4.20),286(2.70)	35-3669-76

Compound	Solvent	$\lambda_{max}(\log \epsilon)$	Ref.
$C_{21}H_{26}O_6$			
10b,4-(Epoxymethano)-5a,8-methanocyclo-buta[5,6]benz[1,2-a]azulene-5-carbox-ylic acid, tetradecahydro-3-hydroxy-4,8-dimethyl-7,12-dioxo-, (α)-	MeOH	207(3.75),286(1.65)	78-2021-76
(β)-	MeOH	205(3.79),290(1.65)	78-2021-76
$C_{21}H_{26}O_7$			
Propanoic acid, 2-methyl-, 6-acetoxy-2,3,3a,4,5,6,9,11a-octahydro-6,10-di-methyl-3-methylene-2,9-dioxocyclodeca-[b]furan-4-yl ester	EtOH	247(4.02)	2-0077-76
$C_{21}H_{26}O_8$			
Vernopectolide A	EtOH	211(4.38)	78-2545-76
$C_{21}H_{26}O_9S$			
Cyclopenta[c]pyran-4-carboxylic acid, 6β-acetoxy-7α-[[[(4-methylphenyl)sul-fonyl]oxy]methyl]-1,4aα,5,6,7,7a-hexahydro-1α-methoxy-	MeOH	228(4.29),237s(--)	24-3626-76
$C_{21}H_{26}O_{10}$			
D-Glucitol, 2,4-O-(phenylmethylene)-, tetraacetate, (S)-	EtOH	256(2.38)	12-1859-76
$C_{21}H_{27}ClO_3$			
Androsta-1,4-dien-3-one, 17β-acetoxy-4-chloro-	EtOH	245(4.25)	56-2087-76
Pregn-4-en-18-oic acid, 21-chloro-20β-hydroxy-3-oxo-, γ-lactone	MeOH	240(4.22)	44-3532-76
$C_{21}H_{27}ClO_4$			
Spiro[androst-4-ene-17,4'-[1,3]dioxol-an]-2',3-dione, 5'-chloro-, (5'R,17α)-	n.s.g.	240(4.22)	44-0078-76
(5'S,17α)-	n.s.g.	240(4.22)	44-0078-76
$C_{21}H_{27}EuO_6$			
Europium, tris(2-oxocyclohexanecarbox-aldehydato-O,O')-	CHCl$_3$	305(4.4)	65-2037-76
$C_{21}H_{27}FO_2$			
Pregna-4,16-diene-3,20-dione, 21-fluoro-	EtOH	242(4.42)	33-1027-76
$C_{21}H_{27}FO_4$			
17α-Pregn-4-ene-3,20-dione, 16β,17-ep-oxy-9-fluoro-	MeOH	238(4.20)	24-0185-76
$C_{21}H_{27}NO$			
2H-Pyrrol-2-one, 1,5-dihydro-3,4-dimeth-yl-5-[(5,6,7,8-tetrahydro-2,3,4,8-tetramethyl-1-naphthalenyl)methyl-ene]-, (Z)-	EtOH	<u>265(4.0)</u>	49-0907-76
$C_{21}H_{27}NO_2$			
Benzoic acid, 5-(1,1-dimethylethyl)-2-[[4-(1,1-dimethylethyl)phenyl]amino]-	MeOH	209(4.54),295(4.32), 345(3.86)	39-0465-76C
$C_{21}H_{27}NO_3$			
2-Aza-19-norpregna-1,3,5(10)-trien-20-yn-17α-ol, 1,3-dimethoxy-	MeOH	230(3.96),281(3.87)	44-2864-76

Compound	Solvent	$\lambda_{max}(\log \epsilon)$	Ref.
Pregn-4-ene-18-nitrile, 21-hydroxy- 3,20-dioxo-	MeOH	238(4.18)	44-3532-76
$C_{21}H_{27}NO_5$ Diketone oxidation product from anop- terine	EtOH	230(3.49)	12-1295-76
Ketone, m. 250-252°	EtOH	320(2.72)	12-1295-76
$C_{21}H_{27}NO_6$ Benzamide, N-[2-(3,4-dimethoxyphenoxy)- 1-methylpropyl]-3,4-dimethoxy-	n.s.g.	220(--),257(4.10), 285(3.96)	20-0787-76
Benzamide, N-[2-(3,4-dimethoxyphenyl)- ethyl]-2-(2-hydroxyethyl)-4,5-di- methoxy-	EtOH	217(4.49),252s(3.89), 282(3.75)	118-0719-76
1,4-Benzoxazepine, 2,3,4,5-tetrahydro- 7,8-dimethoxy-3-methyl-5-(3,4,5-tri- methoxyphenyl)-	n.s.g.	217(4.30),280(3.61)	20-0898-76
$C_{21}H_{27}N_3$ 1-Piperazinamine, N-(phenylmethylene)- 4-[(2,4,6-trimethylphenyl)methyl]-	MeOH	292(4.23)	73-1035-76
$C_{21}H_{27}N_3O$ Phenol, 2-[[[4-[(2,4,6-trimethylphenyl)- methyl]-1-piperazinyl]imino]methyl]-	MeOH	285(4.21),312(4.17)	73-1035-76
$C_{21}H_{27}N_5O_5S$ Acetamide, N-[5-amino-4-[[2,3-O-(1-meth- ylethylidene)-β-D-ribofuranosyl]- amino]-6-[(phenylmethyl)thio]-2- pyrimidinyl]-	EtOH-HCl	218(4.27),243(4.32), 290(3.90),330(3.95)	87-1186-76
	EtOH-pH 7	213(4.42),273(3.98), 310(4.05)	87-1186-76
	EtOH-pH 13	310(4.03)	87-1186-76
$C_{21}H_{28}BrFO_4$ Pregn-4-ene-3,20-dione, 17-bromo- 9-fluoro-11β,16β-dihydroxy-	MeOH	237(4.20)	24-0185-76
$C_{21}H_{28}ClNO_3$ Pregn-4-ene-3,20-dione, 21-chloro- 18-nitroso-	MeOH	240(4.20),298(3.60)	44-3532-76
$C_{21}H_{28}F_2O_2$ Pregna-5,16-dien-20-one, 21,21-difluoro- 3β-hydroxy-	EtOH	250(3.95)	33-1027-76
$C_{21}H_{28}N_2O_2$ Pregna-5,16-dien-20-one, 21-diazo- 3β-hydroxy-	EtOH	253(4.02),300(4.03)	33-1027-76
$C_{21}H_{28}N_2O_3$ Isovelbanamine, 16-(methoxycarbonyl)-	MeOH	228(4.46),287(3.92), 294(3.88)	28-0759-76B
3,7-Secoepipandoline	EtOH	227(4.36),288(3.79), 294(3.79)	88-3567-76
3,7-Secopandoline A	EtOH	226(4.35),286(3.77), 292(3.74)	88-3567-76
3,7-Secopandoline B	EtOH	228(4.42),286(3.85), 294(3.83)	88-3567-76
Velbanamine, 16-(methoxycarbonyl)-	MeOH	228(4.47),286(3.95), 293(3.93)	28-0759-76B

Compound	Solvent	$\lambda_{max}(\log \epsilon)$	Ref.
$C_{21}H_{28}N_2O_8$			
Blastmycin, deisovaleryl-	MeOH	226(4.44),322(3.71)	138-0701-76
	MeOH-HCl	238(3.96),302(3.67)	138-0701-76
	MeOH-NaOH	343(3.94)	138-0701-76
$C_{21}H_{28}N_4O_{10}$			
2,4(1H,3H)-Pteridinedione, 5-acetyl-	pH 8.0	245(3.77),287(4.19)	24-3184-76
5,6,7,8-tetrahydro-6,7-dimethyl-	pH 12.0	258(3.84),287(4.11)	24-3184-76
1-(2,3,5-tri-O-acetyl-β-D-ribo-	MeOH	243(3.76),287(4.17)	24-3184-76
furanosyl)-			
$C_{21}H_{28}N_5O_8PS$			
3'-Adenylic acid, 2'-deoxy-, diethyl	EtOH	274(4.13)	102-1523-76
ester, 5'-(4-methylbenzenesulfonate)			
$C_{21}H_{28}O$			
Cycloheptanone, 3-methyl-2-(4-methyl-	EtOH	280(4.2)	2-0243-76
3-pentenyl)-7-(phenylmethylene)-			
$C_{21}H_{28}O_2$			
1-Dibenzofuranol, 5a,8,9,9a-tetrahydro-	EtOH	271(2.97),283(2.94)	39-0008-76C
6-methyl-9-(1-methylethenyl)-3-pentyl-,			
[5aS-(5aα,9α,9aα)]-			
6H-Dibenzo[b,d]pyran-1-ol, 6a,9,10,10a-	MeOH	228(4.34),276(3.28),	24-3606-76
tetrahydro-6,6-dimethyl-9-methylene-		282(3.28)	
3-pentyl-, (6aR-trans)-			
Gona-1,3,5(10),8-tetraen-17β-ol, 13-eth-	MeOH	271s(4.24),278(4.25)	44-0531-76
yl-3-methoxy-11β-methyl-, (+)-			
Gona-1,3,5(10)-trien-17-one, 13-ethyl-	MeOH	220(3.92),278(3.29),	44-0531-76
3-methoxy-11β-methyl-, (+)-		287(3.28)	
D-Homo-C,18-dinorpregna-4,13(17a)-diene-	EtOH	238(4.06)	18-0499-76
3,20-dione, 17a-methyl-, (17α)-			
Pregna-4,17(20)-dien-21-al, 3-oxo-	EtOH	242(4.52)	118-0025-76
17α-Pregna-1,4,20-trien-3-one, 17β-hy-	EtOH	245(4.19)	94-0828-76
droxy-			
$C_{21}H_{28}O_2S$			
Benzene, [[3-methyl-5-(2,6,6-trimethyl-	EtOH	272(4.20)	22-0513-76
1-cyclohexen-1-yl)-2,4-pentadienyl]-	EtOH	227(4.22),240(4.13),	33-0387-76
sulfonyl]-, (E,E)-		265(4.17),272(4.19)	
$C_{21}H_{28}O_3$			
6H-Dibenzo[b,d]pyran-1-ol, 10,10a-di-	MeOH	227(4.33),278(3.34),	24-3606-76
hydro-3-(2-hydroxypentyl)-6,6-dimeth-		284(3.36)	
yl-9-methylene-			
6H-Dibenzo[b,d]pyran-1-ol, 10,10a-di-	MeOH	227(4.26),276(3.37),	24-3606-76
hydro-3-(3-hydroxypentyl)-6,6-dimeth-		283(3.37)	
yl-9-methylene-			
6H-Dibenzo[b,d]pyran-1-ol, 10,10a-di-	MeOH	227(4.26),275(3.48),	24-3606-76
hydro-3-(4-hydroxypentyl)-6,6-dimeth-		282(3.46)	
yl-9-methylene-			
14βH-Estra-3,5-dien-17-one, 3-acetoxy-	n.s.g.	234(4.34)	22-1240-76
7β-methyl-			
Pregn-4-en-21-al, 3,20-dioxo-	EtOH	241(4.19)	44-2477-76
17α-Pregn-4-ene-3,20-dione, 16β,17-ep-	MeOH	240(4.21)	24-0185-76
oxy-			
8,14-Seco-1,3,5(10),9-gonatetraen-14-	EtOH	266(4.30)	114-0081-76D
one, 17β-hydroxy-3-methoxy-13β-propyl-			

Compound	Solvent	$\lambda_{max}(\log \epsilon)$	Ref.
$C_{21}H_{28}O_3S$			
2H,5H-Thiopyrano[3,4-c][1]benzopyran-5-one, 8-(1,2-dimethylheptyl)-1,4-dihydro-10-hydroxy-	EtOH	310(3.996)	87-0549-76
$C_{21}H_{28}O_4$			
D(17a)-Homo-C,18-dinorpregna-4,17-diene-3,20-dione, 11,13-dihydroxy-17a-methyl-, (11α,13α)-	EtOH	245(4.23)	138-0975-76
D(1 a)-Homo-C,18-dinorpregn-4-ene-3,11,20-trione, 17a-hydroxy-17a-methyl-, (17α,17aα)-	MeOH	236(4.15)	138-0975-76
Neridienone B	EtOH	280(4.51)	102-1745-76
Pregn-4-en-21-al, 11β-hydroxy-3,20-dioxo-	EtOH	241(4.18)	44-2477-76
Pregn-4-en-21-al, 17-hydroxy-3,20-dioxo-	EtOH	241(4.19)	44-2477-76
Pregn-4-ene-3,11-dione, 12,20-epoxy-20-hydroxy-, (12β,13α)-	EtOH	237(4.00)	138-0975-76
Taxodione, 14-methoxy-	EtOH	328(4.38),340(4.39), 420(3.33)	32-0119-76
$C_{21}H_{28}O_5$			
Androsta-1,4-diene-2-carboxylic acid, 11α,17β-dihydroxy-17α-methyl-3-oxo-	MeOH	252(4.10)	145-0537-76
Androst-1-ene-3,11-dione, 17-acetoxy-2-hydroxy-	EtOH	269.5(3.93)	95-0863-76
Cyclopentaneacetic acid, 2-acetoxy-1-methyl-5-(1,2,3,4-tetrahydro-6-methoxy-2-naphthalenyl)-	EtOH	278(3.48),286(3.42)	78-0079-76
Pregn-4-en-21-al, 11β,17-dihydroxy-3,20-dioxo-	EtOH	241(4.20)	44-2477-76
$C_{21}H_{28}O_{12}$			
2H-Pyran-2-one, 5,6-dihydro-4,6-dimethyl-5-[(2,3,4,6-tetra-O-acetyl-β-D-glucopyranosyl)oxy]-, (5R-cis)-	MeOH	214.5(4.04)	1-0297-76
(5S-trans)-	MeOH	213.5(4.06)	1-0297-76
$C_{21}H_{29}BrO_3$			
Pregn-4-ene-3,20-dione, 17-bromo-16β-hydroxy-	MeOH	240(4.23)	24-0185-76
$C_{21}H_{29}FO_2$			
Pregna-5,16-dien-20-one, 21-fluoro-3β-hydroxy-	EtOH	242(3.97)	33-1027-76
$C_{21}H_{29}FO_3$			
D(17a)-Homo-C,18-dinorpregn-4-ene-3,20-dione, 17a-hydroxy-17a-methyl-	EtOH	237(4.00)	138-0975-76
$C_{21}H_{29}FO_4$			
D(17a)-Homo-C,18-dinorpregn-4-ene-3,20-dione, 17a-fluoro-11,13-dihydroxy-17a-methyl-, (11α,13α,17α,17aβ)-	EtOH	239(4.00)	138-0975-76
Pregn-4-ene-3,20-dione, 21-fluoro-16α,17-dihydroxy-	EtOH	242(4.23)	33-1027-76
$C_{21}H_{29}FO_4S$			
Pregn-4-ene-3,20-dione, 9-fluoro-11β,16β-dihydroxy-17-mercapto-	MeOH	239(4.19)	24-0185-76

Compound	Solvent	$\lambda_{max}(\log \epsilon)$	Ref.
$C_{21}H_{29}NO_3$			
17a-Aza-D-homoandrosta-5,7-dien-17-one, 3-acetoxy-, (3β)-	EtOH	258(3.86),268(3.99), 278(3.99),288(3.74)	13-0225-76A
2,6-Methano-3-benzazocin-8-ol, 1,2,3,4-5,6-hexahydro-3-[(3-hydroxycyclohex-yl)carbonyl]-6,11-dimethyl-	MeOH	218s(4.09),227s(3.95), 281(3.35)	24-2657-76
9(11)-Secoestra-1,3,5(10)-trien-11-oic acid, 17α-amino-17β-ethoxy-3-methoxy-, lactam	EtOH	279(3.32),287(3.28)	78-0079-76
$C_{21}H_{29}NO_4$			
4H-Cyclopenta-1,3-dioxole-4-methanol, 6-[4,5-dihydro-3-(2,4,6-trimethyl-phenyl)-5-isoxazolyl]tetrahydro-2,2-dimethyl-	EtOH	212(4.00)	23-0861-76
$C_{21}H_{30}$			
Cycloheptane, 1-methyl-2-(4-methyl-3-pentenyl)-4-(phenylmethylene)-	EtOH	250(4.2)	2-0243-76
$C_{21}H_{30}ClNO_3$			
Pregn-4-en-18-al, 21-chloro-20β-hydroxy-3-oxo-, 18-oxime	MeOH	240(4.41)	44-3532-76
Pregn-4-en-3-one, 21-chloro-20β-hydroxy-18-nitroso-	MeOH	240(4.22),300(3.48)	44-3532-76
$C_{21}H_{30}N_2$			
Benzenemethanamine, N-methyl-N-[6-meth-ylene-2-(1-piperidinylmethyl)-1-cyclo-hexen-1-yl]-	n.s.g.	238(4.16),286(3.26)	94-2494-76
$C_{21}H_{30}N_2O_5$			
D-Mannitol, 1-deoxy-1-(5,6-dimethyl-1H-benzimidazol-1-yl)-3,4:5,6-bis-O-(1-methylethylidene)-	EtOH	251(3.92),255s(3.91), 280(3.84),289(3.84)	136-0049-76G
$C_{21}H_{30}N_4O_4$			
2-Propenal, 3-cyclododecyl-, 2,4-dini-trophenylhydrazone	$CHCl_3$	364.0(4.40)	22-0849-76
$C_{21}H_{30}N_4O_7$			
Butanoic acid, 4-[[(3a,4,13,13a-tetra-hydro-2,2-dimethyl-4,13-epoxy-5H-1,3-dioxolo[4,5-g]imidazo[5,1-d][1,3,5]-oxadiazecin-9-yl)carbonyl]amino]-, 1,1-dimethylethyl ester, [3aR-(3aα,4β,13β,13aα)]-	pH 1 pH 13 EtOH	277(--),288(--) 268(4.38) 277(4.32),288s(4.17)	87-0684-76 87-0684-76 87-0684-76
$C_{21}H_{30}N_5O_6$			
1-Piperidinyloxy, 4-[[2,4-dinitro-5-(1-piperidinyl)benzoyl]amino]-2,2,6,6-tetramethyl-	EtOH	236(4.08),255s(--), 380(4.06)	64-0328-76C
$C_{21}H_{30}O$			
Pregna-5,16(17)-dien-20-one	EtOH	240(3.99)	2-0073-76
$C_{21}H_{30}O_2$			
Gona-1,3,5(10)-trien-17β-ol, 13-ethyl-3-methoxy-11β-methyl-, (±)-	MeOH	220(3.92),278(3.29), 287(3.28)	44-0531-76
D-Homo-5,7-androstadien-3β-ol, formate	EtOH	270(4.04),280(4.06), 290(3.85)	13-0225-76A

Compound	Solvent	$\lambda_{max}(\log \epsilon)$	Ref.
$C_{21}H_{30}O_3$			
1,3-Benzenediol, 2-[4-hydroxy-3-methyl-6-(1-methylethenyl)-2-cyclohexen-1-yl]-5-pentyl-	EtOH	275s(2.96),282(2.96)	39-0008-76C
18,20-Cyclopregn-4-en-3-one, 20,21-di-hydroxy-	EtOH	242(4.23)	44-2552-76
6H-Dibenzo[b,d]pyran-1,7-diol, 6a,7,10,10a-tetrahydro-6,6,9-trimethyl-3-pentyl-	MeOH	228s(3.99),275(3.11), 282(3.12)	24-3606-76
A-Norandrost-3(5)-en-2-one, 17β-acetoxy-1β-methyl-	EtOH	233(4.21)	39-2363-76C
	EtOH	232(4.36)	39-2363-76C
Pregn-4-en-21-al, 20-hydroxy-3-oxo-	MeOH	241(4.21)	44-2477-76
Pregn-4-en-21-al, 20-hydroxy-3-oxo-, 3-methyl-2-benzothiazolonehydrazone derivative	MeOH	312(4.43)	44-2477-76
8,14-Seco-1,3,5(10),9-gonatetraene-14α,17α-diol, 3-methoxy-13β-propyl-	EtOH	265(4.27)	114-0081-76D
14α,17β-	EtOH	265(4.24)	114-0081-76D
14β,17β-	EtOH	265(4.30)	114-0081-76D
$C_{21}H_{30}O_3S$			
Pregn-4-ene-3,20-dione, 16β-hydroxy-17-mercapto-	MeOH	241(4.19)	24-0185-76
$C_{21}H_{30}O_4$			
Androsta-1,4-dien-3-one, 11,17-dihydr-oxy-2-(hydroxymethyl)-17-methyl-, (11α,17β)-	MeOH	250(4.20)	145-0537-76
Cannabitriol	EtOH	231(4.40),279(4.14)	31-0283-76
6H-Dibenzo[b,d]pyran-1,7-diol, 6a,7,10-10a-tetrahydro-3-(2-hydroxypentyl)-6,6,9-trimethyl-	MeOH	227s(3.99),275(3.10), 282(3.11)	24-3606-76
6H-Dibenzo[b,d]pyran-1,7-diol, 6a,7,10-10a-tetrahydro-3-(3-hydroxypentyl)-6,6,9-trimethyl-	MeOH	227s(3.98),276(3.19), 282(3.18)	24-3606-76
6H-Dibenzo[b,d]pyran-1,7-diol, 6a,7,10-10a-tetrahydro-3-(4-hydroxypentyl)-6,6,9-trimethyl-	MeOH	227s(3.97),275(3.14), 282(3.14)	24-3606-76
Pregn-4-en-21-al, 11β,20-dihydroxy-3-oxo-	MeOH	241(4.18)	44-2477-76
3-methyl-2-benzothiazolonehydrazone derivative	MeOH	312(4.24)	44-2477-76
Pregn-4-en-21-al, 17,20-dihydroxy-3-oxo-	MeOH	241(4.20)	44-2477-76
3-methyl-2-benzothiazolonehydrazone derivative	MeOH	312(4.40)	44-2477-76
Pregn-4-en-21-oic acid, 20-hydroxy-3-oxo-	EtOH	238(4.12)	69-0576-76
Prost-14-en-5-yn-1-oic acid, 9,13-di-oxo-, methyl ester, (14E)-	MeOH	227(4.13)	44-2477-76
$C_{21}H_{30}O_5$			
Pregn-4-en-21-al, 11β,17,20-trihydroxy-3-oxo-	MeOH	241(4.17)	44-2477-76
3-methyl-2-benzothiazolonehydrazone derivative	MeOH	312(4.41)	44-2477-76
$C_{21}H_{30}O_9$			
Pterosin, 2'-O-β-D-glucoside, (2R,3R)-	EtOH	218(4.61),261(4.20), 294(3.53),303(3.43)	94-2241-76

Compound	Solvent	$\lambda_{max}(\log \epsilon)$	Ref.
Pterosin, 2'-O-β-D-glucoside, (2S,3R)-	EtOH	218(4.62),261(4.18), 294(3.58),303(3.52)	94-2241-76
$C_{21}H_{30}O_{10}$ Penstemide	n.s.g.	214(4.33)	88-4119-76
$C_{21}H_{30}O_{14}$ Hyrcanoside	n.s.g.	224s(2.02),285(2.15)	105-0092-76
$C_{21}H_{31}ClO_2$ Pregn-4-en-3-one, 21-chloro-20β-hydroxy-	MeOH	242(4.15)	44-3532-76
$C_{21}H_{31}FO_3$ 6H-Dibenzo[b,d]pyran-1,10-diol, 9-fluoro-6a,7,8,9,10,10a-hexahydro-6,6,9-trimethyl-3-pentyl-, [6aR-(6aα,9α,10β,10aβ)]-	EtOH EtOH-NaOH	282(3.25) 286s(3.43),294(3.50)	33-1963-76 33-1963-76
$C_{21}H_{31}NO$ 5α-Pregn-17-ene-21-nitrile, 3α-hydroxy- 4(1H)-Quinolinone, 3-decyl-2,8-dimethyl-	EtOH C_6H_{12} EtOH MeCN $CHCl_3$	220(4.13) 321(3.95),335(3.99) 323(4.13),336(4.13) 322(4.10),335(4.11) 324(4.03),338(4.08)	13-0431-76A 39-1428-76B 39-1428-76B 39-1428-76B 39-1428-76B
$C_{21}H_{31}N_3O_6$ 2,2'-Anhydro-1-(3',5'-dihexanoyl-β-D-arabinofuranosyl)cytosine, hydrochloride	MeOH	234(3.99),263(4.03)	87-0663-76
$C_{21}H_{31}N_3O_7S$ Propanoic acid, 2,2-dimethyl-, 3-[(acetyloxy)methyl]-5-(2,3,4,5-tetrahydro-5-oxo-3-thioxo-1,2,4-triazin-6-yl)-1,2-cyclopentanediyl ester, (1α,2β,3α,5α)-	MeOH	272(4.30)	23-2925-76
$C_{21}H_{31}N_5O_6$ Butanoic acid, 4-[[9-[2,3-O-(1-methylethylidene)-β-D-ribofuranosyl]-9H-purin-6-yl]amino]-, 1,1-dimethylethyl ester	EtOH	268(4.23)	87-0684-76
$C_{21}H_{32}$ Benzene, (5,5-dimethyl-2-octyl-1-cyclopenten-1-yl)-	C_6H_{12}	235(3.70)	35-6218-76
$C_{21}H_{32}N_2O_3$ Carbamic acid, (3-butoxyphenyl)-, 3-(1-pyrrolidinyl)cyclohexyl ester Carbamic acid, (4-butoxyphenyl)-, 3-(1-pyrrolidinyl)cyclohexyl ester Carbamic acid, (3-propoxyphenyl)-, 3-(1-piperidinyl)cyclohexyl ester Carbamic acid, (4-propoxyphenyl)-, 3-(1-piperidinyl)cyclohexyl ester	H_2O H_2O H_2O H_2O	236(4.07),277(3.39) 239(4.20),281(3.17) 236(4.06),277(3.38) 239(4.20),282(3.18)	106-0024-76 106-0024-76 106-0024-76 106-0024-76
$C_{21}H_{32}N_6O_5$ Adenosine, 5'-deoxy-5'-[[4-(1,1-dimethylethoxy)-4-oxobutyl]amino]-2',3'-O-(1-methylethylidene)-	EtOH	259(4.12)	87-0684-76

Compound	Solvent	$\lambda_{max}(\log \epsilon)$	Ref.
$C_{21}H_{32}O_2$			
Androst-4-en-3-one, 7α-ethyl-17-hydroxy-	n.s.g.	242(4.21)	13-0759-76A
Androst-4-en-3-one, 7β-ethyl-17-hydroxy-	n.s.g.	243(4.20)	13-0759-76A
6H-Dibenzo[b,d]pyran-1,10-diol, 6a,7,8-9,10,10a-hexahydro-6,6,9-trimethyl-3-pentyl-, [6aR-(6aα,9β,10α,10aβ)]-	EtOH	277(3.13),284(3.13)	33-1963-76
	EtOH-NaOH	288s(3.39),295(3.43)	33-1963-76
3-Pentenoic acid, 3-formyl-5-(1,2,3,4-4a,7,8,8a-octahydro-1,2,4a,5-tetramethyl-1-naphthalenyl)-, methyl ester, [1S-[1α(E),2β,4aα,8aα]]-	ether	234.5(4.14)	94-0294-76
5α-Pregn-17-en-21-oic acid, 3α-hydroxy-	EtOH	222(4.13)	13-0431-76A
$C_{21}H_{32}O_7$			
Acetic acid, [decahydro-8-(1-hydroxy-1-methylethyl)-4a,10a-dimethyl-3-oxo-benzo[1,2-b:4,3-b']dipyran-2(3H)-yli-dene]methoxy-, methyl ester, [4aR-(4aα,6aβ,8α,10aα,10bβ)]-	MeOH	245(4.02)	32-0147-76
$C_{21}H_{33}NO$			
D-Homoandrosta-5,7-dien-3β-ol, 17a-(methylamino)-	EtOH	258(3.81),268(3.95), 278(3.95),288(3.70)	13-0225-76A
$C_{21}H_{33}N_2$			
Cyclopropenylium, bis[bis(1-methylethyl)amino]phenyl-, perchlorate	MeOH	242(3.84)	88-2405-76
$C_{21}H_{33}N_3O_2$			
Benzenamine, 4-[3,5-bis(1,1-dimethylethyl)-1H-pyrazol-1-yl]-N-(1,1-dimethylethoxy)-N-hydroxy-	EtOH	260(3.20)	22-0195-76
$C_{21}H_{33}N_3O_5$			
Dodecanoic acid, 2,3,3a,9a-tetrahydro-2-(hydroxymethyl)-6-imino-6H-furo-[2',3':4,5]oxazolo[3,2-a]pyrimidin-3-yl ester, hydrochloride, [2R-(2α,3β,3aβ,9aβ)]-	MeOH-acid	232(3.98),263(4.03)	87-0654-76
$C_{21}H_{33}N_3O_7$			
2(1H)-Pyrimidinone, 4-amino-1-[3,5-bis-O-(1-oxohexyl)-β-D-arabinofuranosyl]-	MeOH	213(3.99),283(4.17)	87-0667-76
$C_{21}H_{33}N_3O_8S$			
Cyclopentaneacetic acid, α-[(aminothi-oxomethyl)hydrazono]-2,3-bis(2,2-di-methyl-1-oxopropoxy)-4-(methoxycarb-onyl)-, methyl ester, (1α,2α,3β,4α)-	THF	328(4.09)	23-2925-76
$C_{21}H_{34}N_2O_3$			
Carbamic acid, (3-butoxyphenyl)-, 3-(diethylamino)cyclohexyl ester	H_2O	236(4.04),277(3.35)	106-0024-76
Carbamic acid, (4-butoxyphenyl)-, 3-(diethylamino)cyclohexyl ester	H_2O	239(4.17),281(3.21)	106-0024-76
$C_{21}H_{34}O_3$			
1-Naphthalenepentanoic acid, β-hydroxy-β,2,5,5,8a-pentamethyl-4a,5,6,7,8,8a-hexahydro-, methyl ester	EtOH	267(3.64)	32-0625-76

Compound	Solvent	$\lambda_{max}(\log \epsilon)$	Ref.
$C_{21}H_{35}N_3O_6$			
Cytosine, 1-(3'-dodecanoyl-β-D-arabino-furanosyl)-	MeOH-acid	212(3.99),283(4.16)	87-0667-76
$C_{21}H_{35}N_6O_6P$			
Adenosine, 8-(undecylamino)-, cyclic 3',5'-(hydrogen phosphate)	pH 2.0	277(4.03)	69-1408-76
	pH 13.0	277(--)	69-1408-76
$C_{21}H_{36}OS$			
1-Heptadecanone, 1-(2-thienyl)-	heptane	258(3.98),275(3.87)	34-0233-76
$C_{21}H_{36}O_2S$			
2-Thiophenecarboxylic acid, 4-methyl-5-pentadecyl-	heptane	252(3.92),284(3.99)	34-0233-76
2-Thiophenecarboxylic acid, 5-penta-decyl-, methyl ester	heptane	254(3.97),275(4.05)	34-0233-76
$C_{21}H_{38}N_2O$			
Imidazole, 2-(1-oxooctadecyl)-	THF	279(4.14)	33-2753-76
Imidazole, 4(or 5)-(1-oxooctadecyl)-	THF	260(4.12)	33-2753-76
$C_{21}H_{38}S$			
Thiophene, 2-heptadecyl-	heptane	235(3.87)	34-0233-76

Compound	Solvent	$\lambda_{max}(\log \epsilon)$	Ref.
$C_{22}H_{10}BrN_3O_2$ 9H-Benzo[e]quino[2,3-i]perimidine- 9,15(14H)-dione, 11-bromo-	DMF	270(3.61),368(3.15), 534(3.85),582(3.00), 630(3.08)	80-0257-76
$C_{22}H_{10}N_4O_4$ 9H-Benzo[e]quino[2,3-i]perimidine- 9,15(14H)-dione, 12-nitro-	DMF	270(4.31),366(4.01), 526(3.65)	80-0257-76
$C_{22}H_{11}N_3O_2$ 9H-Benzo[e]quino[2,3-i]perimidine- 9,15(14H)-dione	DMF	270(3.80),472(3.42)	80-0257-76
$C_{22}H_{12}$ Dibenzo[def,mno]chrysene	EtOH	209(4.79),213(4.78), 232(5.16),242s(4.79), 247s(4.70),255(4.71), 259(4.79),271(3.91), 282(4.31),294(4.80), 307(5.16),320s(4.05), 363(3.71),380(4.29), 384(4.29),396s(4.44), 401(4.69),406(4.71), 421(4.78),430(5.04)	117-0133-76
$C_{22}H_{12}BrNO_4S$ 8H-Naphtho[2,3-c]thioxanthene-5,8,14- trione, 6-amino-10-bromo-7-methoxy-	DMF	487(3.21),515(3.27), 548(3.24),589(3.04)	80-1227-76
$C_{22}H_{12}BrNO_5$ 8H-Naphtho[2,3-c]xanthene-5,8,14-trione, 6-amino-10-bromo-7-methoxy-	DMF	496(4.00),516(4.02), 549(3.93)	80-1227-76
$C_{22}H_{12}F_6O_3S$ Cyclopropenylium, tris(4-fluorophenyl)-, trifluoromethanesulfonate	MeCN	255(4.14),307(4.66), 322(3.60)	44-2258-76
$C_{22}H_{12}N_6$ Propanedinitrile, [4-cyano-2,3-dihydro- 3-imino-2-(4-methylphenyl)-9H-indeno- [2,1-c]pyridazin-9-ylidene]-	MeCN	310(4.52)	64-0371-76B
$C_{22}H_{12}N_6O$ Propanedinitrile, [4-cyano-2,3-dihydro- 3-imino-2-(2-methoxyphenyl)-9H-indeno- [2,1-c]pyridazin-9-ylidene]-	MeCN	310(4.50)	64-0371-76B
$C_{22}H_{12}O_6$ Dinaphtho[1,2-b:2',3'-d]furan-1,4,7,12- tetrone, 11-hydroxy-6,9-dimethyl-	$CHCl_3$	265s(4.58),271(4.61), 314s(3.90),322(3.92), 370s(3.86),405s(3.87), 448(3.92)	39-2155-76C
$C_{22}H_{13}ClO_2$ 1(3H)-Isobenzofuranone, 3-(10-chloro- 9(10H)-anthracenylidene)-	dioxan	355(4.26)	104-0168-76
$C_{22}H_{13}ClO_6$ [2,2'-Binaphthalene]-1,4,5',8'-tetrone, 6'-chloro-1',5-dihydroxy-3',7-dimeth- yl-	n.s.g.	252(4.26),288s(4.02), 439(3.85)	39-2155-76C

Compound	Solvent	$\lambda_{max}(\log \epsilon)$	Ref.
[2,2'-Binaphthalene]-1,4,5',8'-tetrone, 7'-chloro-1',5-dihydroxy-3',7-dimethyl-	n.s.g.	255(4.27),276s(4.14), 441(3.91)	39-2155-76C
$C_{22}H_{13}ClO_7$ [2,2'-Binaphthalene]-1,4,5',8'-tetrone, 7'-chloro-1',5,6'-trihydroxy-3',7-dimethyl-	n.s.g.	234s(4.13),254(4.10), 274s(4.01),296s(3.90), 438(3.78),556s(2.97)	39-2155-76C
$C_{22}H_{13}F_6P$ Phosphine, (phenylethynyl)bis[4-(trifluoromethyl)phenyl]-	MeOH	210(4.48),226(4.34), 254(4.35),264(4.34), 288(4.03)	139-0169-76A
$C_{22}H_{13}NO_4S$ 8H-Naphtho[2,3-c]thioxanthene-5,8,14-trione, 6-amino-7-methoxy-	DMF	482(3.35),518(3.45), 551(3.44),590(3.31)	80-1227-76
$C_{22}H_{13}NO_5$ 8H-Naphtho[2,3-c]xanthene-5,8,14-trione, 6-amino-7-methoxy-	DMF	495(3.92),545(3.87)	80-1227-76
$C_{22}H_{13}N_3O_2$ 7H-Benzimidazo[2,1-a]benz[de]isoquinolin-7-one, 3(or 4)-(5-methyl-2-oxazolyl)-	toluene	300(4.24),422(4.43), 515(3.56)	103-0736-76
$C_{22}H_{14}$ Acephenanthrylene, 5-phenyl-	MeOH	228(4.73),247s(4.56), 264(4.70),273s(4.35), 297(4.14),308(4.27), 323(4.00),335(4.06), 351(4.05),370(4.08)	54-0165-76
2H-Cyclopenta[jk]fluorene, 2-(phenylmethylene)-	hexane	223(4.59),257(4.33), 273(4.30),299(4.41), 306(4.41),323(4.34), 349(4.30)	24-2956-76
$C_{22}H_{14}BrNO_2$ 6-Quinolinol, 3-bromo-2-phenyl-, benzoate	EtOH	240(4.74),255s(4.59), 325(3.89)	95-0968-76
8-Quinolinol, 3-bromo-2-phenyl-, benzoate	EtOH	238(4.59),258(4.48), 305s(3.73)	95-0968-76
$C_{22}H_{14}Br_4N_6O_2$ 6-Quinoxalinamine, complex with bromanil(2:1)	CHCl$_3$	242(4.52),257(4.69), 312(4.18),382(4.04), 455(3.40)	56-0073-76
$C_{22}H_{14}ClN_2O_2S$ Benzothiazolium, 3-[(1-amino-4-chloro-9,10-dihydro-9,10-dioxo-2-anthracenyl)methyl]-, chloride	MeOH	490(3.91)	2-0054-76
$C_{22}H_{14}ClN_3$ 3H-Imidazo[4,5-b]quinoline, 7-chloro-3,9-diphenyl-	DMF	326(4.02),345(3.87), 356s(3.84)	48-0039-76
$C_{22}H_{14}ClN_3O_2$ 9,10-Anthracenedione, 1-amino-2-(1H-	MeOH	490(3.89)	2-0054-76

Compound	Solvent	λ_{max}(log ϵ)	Ref.
benzimidazol-1-ylmethyl)-4-chloro-, hydrochloride			2-0054-76
$C_{22}H_{14}ClN_3O_2S$ Benzamide, N-[5-benzoyl-3-(4-chloro-phenyl)-1,3,4-thiadiazol-2(3H)-ylidene]-	EtOH	275s(4.24),340(4.34)	4-0045-76
$C_{22}H_{14}ClN_7O_5$ Benzamide, N-[4-(4-chlorophenyl)amino]-6-(4-nitrophenyl)-1,3,5-triazin-2-yl]-4-nitro-	MeOCH$_2$CH$_2$OH	268(4.43)	40-0488-76
$C_{22}H_{14}Cl_2N_3$ 1,2,4-Triazolo[3,4-a]isoquinolinium, 1,3-bis(4-chlorophenyl)-, tetra-fluoroborate	MeOH	230s(--),252(4.39), 303(3.56),318(3.71), 331(3.73)	48-0881-76
1,2,4-Triazolo[4,3-a]quinolinium, 1,3-bis(4-chlorophenyl)-, tetra-fluoroborate	MeOH	246(4.28),304(4.07), 319(4.04),323(4.05)	48-0881-76
$C_{22}H_{14}Cl_4$ Benzene, 1,4-dichloro-2,5-bis[2-(4-chlo-rophenyl)ethenyl]-	DMF	366(4.74)	33-0802-76
$C_{22}H_{14}NO_3S$ Benzothiazolium, 3-[(9,10-dihydro-1-hy-droxy-9,10-dioxo-2-anthracenyl)meth-yl]-, chloride	MeOH	410(3.80)	2-0054-76
$C_{22}H_{14}NO_4S$ Benzothiazolium, 3-[(9,10-dihydro-1,4-dihydroxy-9,10-dioxo-2-anthracenyl)-methyl]-, chloride	MeOH	488(3.81)	2-0054-76
$C_{22}H_{14}N_2O$ 2H-Quinolizine-4-carbonitrile, 2-oxo-1,3-diphenyl-	EtOH	234(4.34),298s(4.23), 408s(3.80),604(4.03)	77-1045-76
$C_{22}H_{14}N_2O_3$ 1H-Benz[de]isoquinoline-1,3(2H)-dione, 6-(5-methyl-2-oxazolyl)-2-phenyl-	toluene	375(4.40),440(3.34)	103-0736-76
$C_{22}H_{14}N_3$ 1H-Indole, 3-(3H-indol-3-ylidene-4-pyri-dinylmethyl)-, ion(1-)	MeOH	521(4.56)	44-0870-76
$C_{22}H_{14}N_4O_4S$ Benzamide, N-[5-benzoyl-3-(4-nitrophen-yl)-1,3,4-thiadiazol-2(3H)-ylidene]-	EtOH	265(4.30),335(4.39)	4-0045-76
$C_{22}H_{14}N_8O_4$ 1H-1,2,4-Triazole, 3,3'-(4,6-dinitro-1,3-phenylene)bis[5-phenyl-	H$_2$SO$_4$	247(4.7324)	116-0626-76
$C_{22}H_{14}O_3$ Benzoic acid, 2-(9-anthracenylcarbonyl)-	EtOH	257(5.08),333(3.51), 348(3.72),366(3.88), 386(3.83)	104-0168-76
10H-Benzo[d]phenanthro[9,10-b]pyran-10-one, 13-methoxy-	MeOH	257(4.54),278s(4.28), 288s(4.17),300s(3.98),	78-1331-76

Compound	Solvent	$\lambda_{max}(\log \epsilon)$	Ref.
(cont.)		330(2.88),351(2.32)	78-1331-76
$C_{22}H_{14}O_6$			
[2,2'-Binaphthalene]-1,4,5',8'-tetrone, 1',8-dihydroxy-3,6'-dimethyl-	EtOH	245s(3.89),431(3.41)	39-2546-76C
Chitranone	dioxan	257(4.25),437(4.22)	102-0237-76
$C_{22}H_{14}O_7$			
4H-1-Benzopyran-5-carboxylic acid, 8-(1,4-dihydro-5-hydroxy-7-methyl-1,4-dioxo-2-naphthalenyl)-7-methyl-4-oxo-	n.s.g.	250s(4.14),306(3.74), 444(3.50)	39-2155-76C
Naphthalic anhydride, 3,4-diacetoxy-5-phenyl-	MeOH	239(4.27),310(3.72)	102-1413-76
$C_{22}H_{15}BrN_2O$			
2,3-Benzodiazocin-1(2H)-one, 6-bromo-2,4-diphenyl-	EtOH	207(4.56),247s(4.33), 280s(4.06)	39-2281-76C
$C_{22}H_{15}BrN_2OSe$			
Benzamide, N-[4-(4-bromophenyl)-3-phenyl-2(3H)-selenazolylidene]-	MeCN	235(4.58),283s(3.46), 345(4.27)	97-0018B-76
$C_{22}H_{15}ClN_2O$			
4(3H)-Quinazolinone, 2-[2-(3-chlorophenyl)ethenyl]-3-phenyl-	EtOH	325(4.51)	115-0341-76
$C_{22}H_{15}ClN_2OSe$			
Benzamide, N-[4-(4-chlorophenyl)-3-phenyl-2(3H)-selenazolylidene]-	MeCN	235s(4.44),346(4.27)	97-0018B-76
$C_{22}H_{15}ClN_2O_2$			
1H-Pyrazole-1-carboxylic acid, 5-(4-chlorophenyl)-3-phenyl-, phenyl ester	EtOH	241s(4.30),269(4.41)	4-0257-76
$C_{22}H_{15}ClN_3$			
1,2,4-Triazolo[3,4-a]isoquinolinium, 1-(4-chlorophenyl)-3-phenyl-, tetrafluoroborate	MeOH	227(4.51),254(4.49), 310s(--),318(3.64), 331(3.68)	48-0881-76
1,2,4-Triazolo[3,4-a]isoquinolinium, 3-(4-chlorophenyl)-1-phenyl-, tetrafluoroborate	MeOH	227(4.44),254(4.56), 304(3.47),318(3.63), 332(3.65)	48-0881-76
1,2,4-Triazolo[4,3-a]quinolinium, 1-(4-chlorophenyl)-3-phenyl-, tetrafluoroborate	MeOH	242(4.27),303(4.13), 323(4.09)	48-0881-76
1,2,4-Triazolo[4,3-a]quinolinium, 3-(4-chlorophenyl)-1-phenyl-, tetrafluoroborate	MeOH	242(4.27),304(4.07), 315(4.06),323(4.06)	48-0881-76
$C_{22}H_{15}ClN_6$			
2H-Benzotriazole, 2-[3-chloro-4-[2-(2-phenyl-2H-1,2,3-triazol-4-yl)ethenyl]phenyl]-	DMF	357(4.71)	33-2469-76
$C_{22}H_{15}Cl_2N_3O_2$			
9,10-Anthracenedione, 1-amino-2(1H-benzimidazol-1-ylmethyl)-4-chloro-, hydrochloride	MeOH	490(3.89)	2-0054-76

Compound	Solvent	$\lambda_{max}(\log \epsilon)$	Ref.
$C_{22}H_{15}N$ Pyrrolo[3,2,1-hi]indole, 1,5-diphenyl-	EtOH	245(4.52),295(4.70), 305(4.58)	49-0663-76
$C_{22}H_{15}NO$ Benzonitrile, 4-[2-(3-dibenzofuranyl)- ethenyl]-3-methyl-	DMF	350(4.69)	33-0819-76
$C_{22}H_{15}NO_2$ Benzamide, 2-(9-anthracenylcarbonyl)-	dioxan	335(3.51),350(3.70), 367(3.86),388(3.83)	104-0168-76
$C_{22}H_{15}NO_3$ Dinaphtho[1,2-b:2',3'-d]pyrrole-5,13- dione, 6-ethyl-11-hydroxy-	EtOH	240(4.08),276(4.31), 338(3.60),500(3.73)	115-0185-76
$C_{22}H_{15}NO_5$ 2-Propen-1-one, 3-(benzoyloxy)-3-(4-ni- trophenyl)-1-phenyl-	n.s.g.	306(4.28)	39-0380-76C
$C_{22}H_{15}NO_5S$ Benzoic acid, 2-[(4-amino-9,10-dihydro- 3-methoxy-9,10-dioxo-1-anthracenyl)- thio]-	DMF	490(3.83),517(3.98), 550(3.88)	80-1227-76
$C_{22}H_{15}NO_6$ Benzoic acid, 2-[(4-amino-9,10-dihydro- 3-methoxy-9,10-dioxo-1-anthracenyl)- oxy]-	DMF	490(3.75),551(3.74)	80-1227-76
$C_{22}H_{15}N_2O_2S$ Benzothiazolium, 3-[(1-amino-9,10-di- hydro-9,10-dioxo-2-anthracenyl)meth- yl]-, chloride	MeOH	490(3.98)	2-0054-76
$C_{22}H_{15}N_3$ 3H-Imidazo[4,5-b]quinoline, 3,9-diphen- yl-	EtOH	251(4.70),321(4.10), 440s(3.97)	48-0039-76
1H-Indole, 3-(3H-indol-3-ylidene-4-pyri- dinylmethyl)-	MeOH	450(4.42)	44-0870-76
dihydrochloride	MeOH	575(4.15)	44-0870-76
$C_{22}H_{15}N_3OS$ Isoquinolinium, 1-benzoyl-3,3-dicyano- 2-(methylthio)-2-propenylide	EtOH	232(4.73),320(4.09), 407(4.24)	142-0939-76
$C_{22}H_{15}N_3O_2$ 9,10-Anthracenedione, 1-amino-2-(1H- benzimidazol-1-ylmethyl)-, mono- hydrochloride	MeOH	491(3.96)	2-0054-76
$C_{22}H_{15}N_3O_2S$ Benzamide, N-(5-benzoyl-3-phenyl- 1,3,4-thiadiazol-2(3H)-ylidene)-	EtOH	280s(4.30),342(4.41)	4-0045-76
$C_{22}H_{15}N_3O_3$ 4(3H)-Quinazolinone, 2-[2-(2-nitrophen- yl)ethenyl]-3-phenyl-	EtOH	315(4.12)	115-0341-76
4(3H)-Quinazolinone, 2-[2-(3-nitrophen- yl)ethenyl]-3-phenyl-	EtOH	325(4.37)	115-0341-76

Compound	Solvent	$\lambda_{max}(\log \epsilon)$	Ref.
$C_{22}H_{15}N_3O_4$			
1H-Indene-1,3(2H)-dione, 2-[(2-nitro-phenyl)azo]-2-(phenylmethyl)-	EtOH	232(4.51),288s(4.03)	104-1440-76
1H-Indene-1,3(2H)-dione, 2-[(3-nitro-phenyl)azo]-2-(phenylmethyl)-	EtOH	232(4.54),254(4.44)	104-1440-76
1H-Indene-1,3(2H)-dione, 2-[(4-nitro-phenyl)azo]-2-(phenylmethyl)-	EtOH	228(4.52),288(4.26)	104-1440-76
	EtOH	228(4.54),287(4.28) (changing)	104-1440-76
	EtOH-HCl	228(4.56),286(4.30) (changing)	104-1440-76
	EtOH-NaOH	290s(4.11),370(3.93), 518(4.57)	104-1440-76
	dioxan	233(4.36),292(4.29)	104-1440-76
$C_{22}H_{15}N_5OS$			
1,2,4-Triazolo[3,4-b][1,3,4]thiadiazol-ium, 6-(benzoylamino)-2,3-diphenyl-, hydroxide, inner salt	MeCN	219s(4.54),280(4.28), 330(4.55)	30-0104-76
$C_{22}H_{15}N_7O_5$			
Benzamide, 4-nitro-N-[4-(4-nitrophenyl)-6-(phenylamino)-1,3,5-triazin-2-yl]-	MeOCH$_2$CH$_2$OH	269(4.54)	40-0488-76
$C_{22}H_{16}BrNO_2$			
9,10-Anthracenedione, 1-amino-4-bromo-2-[(4-methylphenyl)methyl]-	n.s.g.	478(3.74)	146-0001-76
$C_{22}H_{16}Br_2O_4$			
2(3H)-Furanone, 5-(3,5-dibromo-2,4-di-hydroxyphenyl)-5-(1,2-dihydro-3-ace-naphthylenyl)dihydro-	EtOH-pH 9.0	590(3.39)	18-2027-76
$C_{22}H_{16}ClN_3O$			
4(3H)-Quinazolinone, 3-[(2-chlorophen-yl)methyleneamino]-2-(4-methylphenyl)-	EtOH	215(4.47),230s(4.36), 283(4.39),322(4.38)	114-0341-76D
4(3H)-Quinazolinone, 3-[(4-chlorophen-yl)methyleneamino]-2-(4-methylphenyl)-	EtOH	215(4.42),230s(4.37), 280(4.33),310(4.37)	114-0341-76D
$C_{22}H_{16}ClN_3O_2$			
9,10-Anthracenedione, 1-amino-2-(1H-benzimidazol-1-ylmethyl)-, mono-hydrochloride	MeOH	491(3.96)	2-0054-76
$C_{22}H_{16}ClN_5O$			
Benzamide, N-[4-[(4-chlorophenyl)amino]-6-phenyl-1,3,5-triazin-2-yl]-	MeOCH$_2$CH$_2$OH	261(4.67),285s(4.52)	40-0488-76
$C_{22}H_{16}ClN_5O_3$			
1H-Pyrazole-4-carboxamide, 5-amino-3-(4-chlorophenyl)-1-(4-nitro-phenyl)-N-phenyl-	EtOH	288(4.49)	4-1137-76
$C_{22}H_{16}Cl_2$			
Benzene, 1,4-dichloro-2,5-bis(2-phenyl-ethenyl)-	DMF	362(4.69)	33-0802-76
$C_{22}H_{16}Cl_2N_2O_4$			
1,2-Benzenedicarbonitrile, 4,5-dichloro-3-hydroxy-6-[(3,3a,4,5-tetrahydro-3,7-dihydroxy-3a-methyl-2H-benz-[e]inden-1-yl)oxy]-, cis	EtOH	269(4.21),339(3.34)	25-1030-76

Compound	Solvent	$\lambda_{max}(\log \epsilon)$	Ref.
$C_{22}H_{16}F_2N_6$ 2a,4a-Diazacyclopenta[cd]pentalene- 1,6-dicarbonitrile, 2,5-diamino- 6a,6b-bis(4-fluorophenyl)-3,4,6a,6b- tetrahydro-	EtOH	212(4.18),257(4.30)	44-3389-76
$C_{22}H_{16}N_2$ Pyridazine, 3-(1-naphthalenyl)-6-(2- phenylethenyl)-	C_6H_{12} EtOH H_2SO_4	222(4.56),315(4.23) 222(4.73),317(4.54) 226(4.69),327s(4.46), 342(4.51)	59-0837-76 59-0837-76 59-0837-76
$C_{22}H_{16}N_2O$ Benzenemethanol, α-[(2-phenyl-4-quinazo- linyl)methylene]-, (Z)-	CDCl$_3$	260(4.5),395(4.4), 420(4.5)	5-2007-76
2,3-Benzodiazocin-1(2H)-one, 2,4-diphen- yl-	EtOH	215(4.82),228(4.75), 300(3.56)	39-2281-76C
2,5-Imino-2H-2-benzazepin-1(5H)-one, 3,10-diphenyl-	EtOH	207(4.72),245(4.68)	39-2281-76C
4(3H)-Quinazolinone, 3-ethyl-2-[2-(2- naphthalenyl)ethenyl]-	EtOH	330(4.18)	115-0341-76
$C_{22}H_{16}N_2OS$ Benzamide, N-(3,4-diphenyl-2(3H)-thia- zolylidene)-	MeCN	237(4.29),280(3.70), 286s(3.69),341(4.28)	97-0018B-76
$C_{22}H_{16}N_2OS_2$ 7-Thia-2,6-diazabicyclo[2.2.1]heptan- 3-one, 1,2,6-triphenyl-5-thioxo-	MeOH	203(4.65),277(4.10), 306s(3.85),424(4.06)	44-0813-76
$C_{22}H_{16}N_2OSe$ Benzamide, N-(3,4-diphenyl-2(3H)-selen- azolylidene)-	MeCN	237s(4.46),279s(3.82), 287s(3.75),345(4.25)	97-0018B-76
$C_{22}H_{16}N_2O_2$ 1H-Anthra[1,2-d]imidazole-6,11-dione, 2,3-dihydro-2-methyl-2-phenyl-	EtOH	552(4.08)	104-0172-76
1H-Indene-1,3(2H)-dione, 2-(phenylazo)- 2-(phenylmethyl)-	EtOH	230(4.50),258(4.17), 284(4.16)	104-1440-76
4(3H)-Quinazolinone, 2-(2-hydroxyphen- yl)-3-phenyl-	EtOH	345(4.00)	115-0341-76
$C_{22}H_{16}N_2O_2S$ 7-Thia-2,6-diazabicyclo[2.2.1]heptane- 3,5-dione, 1,2,6-triphenyl-	MeOH	203(4.58),234(4.20), 268(4.15),400(4.20)	44-0813-76
5-Thiazolol, 2,4-diphenyl-, phenylcarb- amate	ether	230(4.36),270(4.02), 300(4.00),320(4.04)	78-0571-76
3-Thiophenamine, 2-nitro-N,N,5-triphen- yl-	MeCN	259(4.50),277(4.49), 363(4.10),469(3.79)	48-0221-76
$C_{22}H_{16}N_2S_2$ 1H-[1,2]Dithiolo[5,1-e][1,2,3]thiadia- zole-7-S^{IV}, 1,3,4-triphenyl-	C_6H_{12}	206(4.54),229(4.44), 250s(4.37),293(4.23), 323s(3.97),500(4.11)	39-0228-76C
$C_{22}H_{16}N_3$ 1,2,4-Triazolo[3,4-a]isoquinolinium, 1,3-diphenyl-, tetrafluoroborate	MeOH	227(4.41),253(4.53), 304(3.57),317(3.87), 331(3.89)	48-0881-76

Compound	Solvent	λ_{max} (log ϵ)	Ref.
1,2,4-Triazolo[4,3-a]quinolinium, 1,3-diphenyl-, tetrafluoroborate	MeOH	239(4.31),274(3.67), 304(4.11),317(4.05), 323(4.08)	48-0881-76
$C_{22}H_{16}N_3O_2S$ Benzothiazolium, 3-[(1,4-diamino-9,10-dihydro-9,10-dioxo-2-anthracenyl)-methyl-, chloride	MeOH	565(3.90),617(4.14)	2-0054-76
$C_{22}H_{16}N_4$ 1,2-Benzenediamine, N-benzo[a]phenazin-5-yl-	dioxan	242(4.58),270(4.50), 311(4.42),322(4.41), 447(4.05)	49-0879-76
	50% dioxan	233(4.60),271(4.45), 316(4.43),464(4.06)	49-0879-76
$C_{22}H_{16}N_4O_2$ 9,10-Anthracenedione, 1,4-diamino-2-(1H-benzimidazol-1-ylmethyl)-, monohydrochloride	MeOH	565(4.01),622(4.21)	2-0054-76
$C_{22}H_{16}N_4O_2S_3$ 3H-Pyrazol-3-one, 4.4'-(1,2,4-trithio-lane-3,5-diylidene)bis[2,4-dihydro-5-methyl-2-phenyl-	benzene	277(3.64),363(3.94)	39-1706-76C
$C_{22}H_{16}N_4O_3$ 4(3H)-Quinazolinone, 2-(4-methylphenyl)-3-[(2-nitrophenyl)methyleneamino]-	EtOH	215(4.51),245s(4.46), 268(4.47),330s(4.24)	114-0341-76D
4(3H)-Quinazolinone, 2-(4-methylphenyl)-3-[(4-nitrophenyl)methyleneamino]-	EtOH	215(4.52),235s(4.45), 270(4.35),335(4.39)	114-0341-76D
$C_{22}H_{16}N_4O_5$ 9H-Xanthen-9-one, 1-hydroxy-6-methoxy-3-[(1-phenyl-1H-tetrazol-5-yl)oxy]-	EtOH	239(4.55),245(4.56), 303(4.26),339(3.88)	142-0167-76S
$C_{22}H_{16}N_6$ 2H-Benzotriazole, 2-[4-[2-(2-phenyl-2H-1,2,3-triazol-4-yl)ethenyl]phenyl]-	DMF	353(4.78)	33-2469-76
$C_{22}H_{16}N_6O_3$ Benzamide, N-[4-(3-nitrophenyl)amino]-6-phenyl-1,3,5-triazin-2-yl]-	MeOCH$_2$CH$_2$OH	261(4.59)	40-0488-76
Benzamide, N-[4-(4-nitrophenyl)amino]-6-phenyl-1,3,5-triazin-2-yl]-	MeOCH$_2$CH$_2$OH	258(4.61),350(4.36)	40-0488-76
$C_{22}H_{16}O_2$ 6H,7H-[1]Benzopyrano[4,3-b][1]benzopyr-an, 6-phenyl-	EtOH	226(4.34),286(3.54), 312(3.60),324s(--)	22-1967-76
$C_{22}H_{16}O_3$ 9,10-Anthracenedione, 1-hydroxy-2-[(4-methylphenyl)methyl]-	n.s.g.	410(3.94)	146-0001-76
1H-2-Benzopyran-1-one, 6-methoxy-3,4-diphenyl-	MeOH	267(4.67),300(3.39)	78-1331-76
4H-1-Benzopyran-4-one, 7-methoxy-2,3-diphenyl-	MeOH	243(4.75),270(4.91), 333(4.59)	78-1331-76
$C_{22}H_{16}O_4$ 9,10-Anthracenedione, 1,4-dihydroxy-2-[(4-methylphenyl)methyl]-	n.s.g.	480(3.90)	146-0001-76

Compound	Solvent	$\lambda_{max}(\log \epsilon)$	Ref.
Benzoic acid, 2-(10-hydroxy-9-phenan-threnyl)-4-methoxy-	MeOH	260(4.87),280(4.64), 290(4.61),302(4.52), 337(4.24),355(4.37), 375(4.36),390(4.38)	78-1331-76
$C_{22}H_{16}O_6$			
Benzo[1,2-b:5,4-b']bisbenzofuran-6,12-dione, 3,9-dimethoxy-1,11-dimethyl-	EtOH	250s(4.58),265(4.73), 395(3.87),558(3.77)	12-0179-76
[1,3]Dioxolo[6,7][1]benzopyrano[3,4-b]-furo[2,3-h][1]benzopyran-12(5H)-one, 2,3-dihydro-2-(1-methylethenyl)-, (R)-	EtOH	210(4.50),237(4.13), 295(4.18)	102-1283-76
5,12-Methano-5H-benzo[4,5]cyclohepta-[1,2-b]naphthalene-6,11,13(12H)-tri-one, 1,5,7-trihydroxy-3,9-dimethyl-	EtOH	266(4.18),294s(3.83), 324s(3.68),440(3.66)	77-0241-76
$C_{22}H_{17}ClN_2O_2S$			
Benzenesulfonamide, N-[3-[(4-chlorophen-yl)methylene]-3H-indol-2-yl]-4-methyl-	CHCl$_3$	280(4.18),350(4.24), 410s(3.94)	103-0292-76
$C_{22}H_{17}ClN_2O_3$			
Hydrazinecarboxylic acid, [3-(3-chloro-phenyl)-3-oxo-1-phenylpropylidene]-, phenyl ester	EtOH	226s(4.36),277(4.39)	4-0257-76
Hydrazinecarboxylic acid, [3-(4-chloro-phenyl)-3-oxo-1-phenylpropylidene]-, phenyl ester	EtOH	225s(4.31),279(4.45)	4-0257-76
$C_{22}H_{17}Cl_2N_3$			
1H-Indole, 3-(3H-indol-3-ylidene-4-pyri-dinylmethyl)-, dihydrochloride	MeOH	575(4.15)	44-0870-76
$C_{22}H_{17}I$			
Phenanthrene, 4-(3,5-dimethylphenyl)-3-iodo-	MeOH	280s(4.23),292(4.18), 306(4.22),338(2.78), 354(2.82)	39-1115-76B
$C_{22}H_{17}N$			
Benzonitrile, 4-(2-[1,1'-biphenyl]-4-ylethenyl)-3-methyl-	DMF	340(4.56)	33-0819-76
$C_{22}H_{17}NO$			
1H-Inden-1-one, 2-[(diphenylamino)meth-ylene]-2,3-dihydro-	EtOH	259(4.18),290s(3.84), 379(4.44)	4-1201-76
$C_{22}H_{17}NOS$			
4H-1-Benzothiopyran-4-one, 3-[(diphenyl-amino)methylene]-2,3-dihydro-	EtOH	248(4.34),265(4.29), 392(4.27)	4-0225-76
$C_{22}H_{17}NO_2$			
9,10-Anthracenedione, 1-amino-2-[(4-methylphenyl)methyl]-	n.s.g.	474(3.57)	146-0001-76
Benzeneacetamide, N-benzoyl-α-(phenyl-methylene)-	EtOH	228(4.25),300(4.20)	39-0989-76C
Pyridinium, 1,3-dioxo-1,5-diphenyl-4-penten-2-ylide	EtOH	243(4.19),296(4.13), 308(4.13),343(4.25)	78-2647-76
	dioxan	273(4.15),290(4.17), 343(4.09),413(3.52)	78-2647-76
	MeCN	223(4.30),285(4.17), 342(4.10)	78-2647-76
5(6H)-Quinolinone, 4-benzoyl-7,8-di-hydro-2-phenyl-	EtOH	233(4.26),252(4.30), 272(4.20),313(4.41)	94-1160-76

Compound	Solvent	$\lambda_{max}(\log \epsilon)$	Ref.
$C_{22}H_{17}NO_2S$ 1,5-Benzothiazepine, 2-(1,3-benzodioxol-5-yl)-2,3-dihydro-4-phenyl-	EtOH	261(4.36),296s(4.09), 330(3.75)	114-0293-76A
$C_{22}H_{17}NO_3$ Pyrido[2,1,6-de]quinolizine-1-carbox-ylic acid, 3-benzoyl-, ethyl ester	EtOH	244(4.12),283(4.43), 398(4.37),435s(4.2), 458(4.42)	39-0341-76C
$C_{22}H_{17}NO_4$ Pyrido[2,1,6-de]quinolizine-1,3-dicarb-oxylic acid, 2-phenyl-, dimethyl ester	EtOH	258(4.18),294(4.49), 416(4.15),472(4.39)	39-0341-76C
$C_{22}H_{17}NO_5$ Methanone, [2-[2-[(4-nitrobenzoyl)oxy]-ethyl]phenyl]phenyl-	MeOH	252(4.44),280(3.99), 290(3.73),340(2.66)	35-0541-76
$C_{22}H_{17}N_3$ 1H-Indole, 3-(3H-indol-3-ylidene-4-pyri-dinylmethyl)-, dihydrochloride	MeOH	543(4.30)	44-0870-76
2,2',2"-Terpyridine, 4'-(4-methylphen-yl)-, ferrous complex	MeOH	567(4.40)	118-0001-76
$C_{22}H_{17}N_3O$ 4(3H)-Quinazolinone, 3-(phenylmethylene-amino)-2-(4-methylphenyl)-	EtOH	219(4.48),239s(4.45), 285(4.49),308(4.48)	114-0341-76D
2,2',2"-Terpyridine, 4'-(4-methoxyphen-yl)-, ferrous complex	MeOH	568(4.45)	118-0001-76
$C_{22}H_{17}N_3OS$ 3-Thiophenecarbonitrile, 4-amino-5-benz-oyl-2-(1,2-dimethyl-1H-indol-3-yl)-	EtOH	270(4.14),352(4.03), 400(4.32)	142-0009-76
$C_{22}H_{17}N_3O_2$ 5,8-Quinazolinedione, 4-methyl-2-(3-methylphenyl)-	EtOH	237s(--),269(4.22), 315(4.42),480(3.81)	5-1809-76
4(3H)-Quinazolinone, 3-[(2-hydroxyphen-yl)methyleneamino]-2-(4-methylphenyl)-	EtOH	215(4.53),236(4.46), 283(4.40),323(4.32)	114-0341-76D
4(3H)-Quinazolinone, 3-[(4-hydroxyphen-yl)methyleneamino]-2-(4-methylphenyl)-	EtOH	217(4.42),232s(4.40), 270(4.26),322(4.40)	114-0341-76D
2-Quinolinamine, 4-(4-methylphenyl)-3-nitro-N-phenyl-	CHCl₃	269(4.56),311s(4.15), 420s(3.14)	48-0039-76
$C_{22}H_{17}N_3O_3$ 12-Azatricyclo[6.3.1.0²,⁶]dodeca-2(6),4-9-trien-11-one, 12-(5-nitro-2-pyridin-yl)-7-phenyl-, (1RS,7RS,8RS)-4-endo 4-exo	EtOH	209(4.36),227(4.30), 360(4.23)	39-2307-76C
	EtOH	210(4.35),228(4.29), 360(4.21)	39-2307-76C
$C_{22}H_{17}N_3O_4S$ Benzenesulfonamide, 4-methyl-N-[3-[(4-nitrophenyl)methylene]-3H-indol-2-yl]-	CHCl₃	287(4.68),354(4.26), 412s(3.85)	103-0292-76
$C_{22}H_{17}N_5O$ Benzamide, N-[4-phenyl-6-(phenylamino)-1,3,5-triazin-2-yl]-	MeOCH₂CH₂OH	260(4.66),277s(4.51)	40-0488-76

Compound	Solvent	λ_{max}(log ϵ)	Ref.
C$_{22}$H$_{17}$N$_5$O$_3$ 1H-Pyrazole-4-carboxamide, 5-amino- 1-(4-nitrophenyl)-N,3-diphenyl-	EtOH	288(4.45)	4-1137-76
C$_{22}$H$_{18}$ Bicyclo[4.2.0]octa-3,7-diene, 2,5-di- 2,4,6-cycloheptatrien-1-ylidene-	CH$_2$Cl$_2$	237(4.35),434(4.64)	89-0487B-76
1,3,6-Cyclooctatriene, 5,8-di-2,4,6- cycloheptatrien-1-ylidene-	C$_6$H$_{12}$	226(4.27),414(4.18)	89-0487B-76
1,2,3,4,5,7,9,15,17-Cyclooctadecanona- ene-11,13-diyne, 1,6,10,15-tetra- methyl-, (E,E,Z,Z,Z)-	THF	218(4.89),237(4.18), 249(4.17),259s(3.71), 266s(3.62),272s(3.57), 290s(3.68),306s(4.02), 335(4.55),371(5.36), 440s(3.33),452(3.51), 490(4.06),521(4.87), 582s(2.27),596(2.09), 622(2.08),639s(2.11), 652(2.22),682(2.50), 705(2.79),740s(1.57)	18-0292-76
Dibenzo[def,mno]chrysene, 1,2,3,7,8,9- hexahydro-	EtOH	205(4.60),213(4.27), 230(4.21),237(4.58), 247(4.86),258(4.13), 269(4.43),280(4.74), 304(3.60),317(4.00), 332(4.43),344(4.42), 349(4.61),364(3.77), 377(3.12),384(3.77)	117-0133-76
Naphthalene, 2-[4-(3,5-dimethylphenyl)- 1-buten-3-ynyl]-, cis	MeOH	238(4.56),276(4.21), 287(4.26),321s(4.37), 327(4.39),342s(4.27), 356s(3.98)	39-1104-76B
trans	MeOH	237(4.38),276(4.37), 287(4.44),327(4.66), 347(4.54),320s(4.62)	39-1115-76B
Phenanthrene, 4-(3,5-dimethylphenyl)-	MeOH	276s(4.28),282s(4.20), 296(4.12),333(2.72), 349(2.71)	39-1115-76B
C$_{22}$H$_{18}$BrN$_3$O$_3$ Spiro[2H-1-benzopyran-2,2'-[2H]pyrrolo- [3,2-h]quinoline, 8-bromo-1',3'-di- hydro-1',3',3'-trimethyl-6-nitro-	EtOH	252(4.61),341(3.81)	103-0680-76
C$_{22}$H$_{18}$ClN$_5$O$_3$S Benzenesulfonamide, 4-[[[(4-chlorophen- yl)amino]carbonyl]amino]-N-(1-phenyl- 1H-pyrazol-5-yl)-	EtOH	208(4.91),264(5.08)	80-1345-76
C$_{22}$H$_{18}$Fe$_2$ Ferrocene, 1,1"-(1,2-ethynediyl)bis-, monocation	CH$_2$Cl$_2$	545(3.32),720s(--), 1560(2.83)	44-2700-76
dication	CH$_2$Cl$_2$	500s(--),720(3.30)	44-2700-76
C$_{22}$H$_{18}$HgO$_2$ Mercury, (acetato-O)(triphenylethenyl)-	MeOH	235(4.33)	23-3038-76
C$_{22}$H$_{18}$NS$_2$ 1,2-Dithiol-1-ium, 3-phenyl-5-[phenyl- (phenylmethyl)amino]-	MeOH	319(4.41),367(4.16)	48-0221-76

Compound	Solvent	$\lambda_{max}(\log \epsilon)$	Ref.
$C_{22}H_{18}N_2$			
Pyridine, 3-[5-phenyl-1-(phenylmethyl)-1H-pyrrol-2-yl]-	EtOH	304(4.13)	44-3438-76
$C_{22}H_{18}N_2O$			
1-Azaspiro[4,5]deca-1,3,6,9-tetraen-8-one, 2-(methylphenylamino)-4-phenyl-	EtOH	253(4.41)	24-1643-76
Methanone, [4-(4,5-dihydro-1-phenyl-1H-pyrazol-3-yl)phenyl]phenyl-	benzene	414(4.37)	64-1248-76B
$C_{22}H_{18}N_2O_2$			
9,10-Anthracenedione, 1,4-diamino-2-[(4-methylphenyl)methyl]-	n.s.g.	550(4.09),590(4.10)	146-0001-76
9,10-Anthracenedione, 1,8-diamino-2-[(4-methylphenyl)methyl]-	n.s.g.	510(4.15)	146-0001-76
4,12-Diazatetracyclo[6.3.1.0^{2,7}.0^{3,5}]-dodec-9-ene-6,11-dione, 4,12-diphenyl-, (2RS,3RS,5SR,7SR,8RS)-	EtOH	236(4.35)	39-2338-76C
(1RS,2SR,3SR,5RS,7RS,8RS)-	EtOH	236(4.33)	39-2338-76C
3,8-Diazatricyclo[5.3.1.1^{2,6}]dodeca-4,9-diene-11,12-dione, 3,8-diphenyl-, (1α,2β,6β,7α)-	EtOH	222(4.35),268(4.35),341(3.86)	39-2338-76C
1H-Indene-1,3(2H)-dione, 2-[(1-ethyl-3-methyl-2(1H)-quinoxalinylidene)-ethylidene]-	CH_2Cl_2	545(4.75),578(4.65)	33-0692-76
$C_{22}H_{18}N_2O_2S$			
Benzenesulfonamide, 4-methyl-N-[3-(phenylmethylene)-3H-indol-2-yl]-	CHCl$_3$	272(4.44),346(4.13),410s(3.83)	103-0292-76 +123-0095-76
$C_{22}H_{18}N_2O_3$			
Hydrazinecarboxylic acid, (3-oxo-1,3-diphenylpropylidene)-, phenyl ester	EtOH	225s(4.16),279(4.36)	4-0257-76
4,8-Iminocyclohepta[c]pyrrole-1,3,5(2H)-trione, 3a,4,8,8a-tetrahydro-2-phenyl-9-(phenylmethyl)-, (3aα,4β,8β,8aα)-	EtOH	218(4.006)	39-2334-76C
$C_{22}H_{18}N_2O_3S$			
5H-Pyrrolo[3',4':3,4]pyrrolo[2,1-b]thiazole-6,8(5aH,7H)-dione, 5-benzoyl-8a,8b-dihydro-3-methyl-7-phenyl-	MeOH	200(4.38),245(4.22)	44-0187-76
$C_{22}H_{18}N_2O_4$			
2-Furancarboxylic acid, 4-cyano-1,2,3,4-tetrahydro-2-hydroxy-3-methyl-2-phenyl-3-quinolinyl ester	EtOH	245(4.37),296(3.44)	95-0968-76
$C_{22}H_{18}N_2S$			
1H-Imidazolium, 4-mercapto-1-methyl-2,3,5-triphenyl-, hydroxide, inner salt	MeOH	315(3.89)	2-0382B-76
$C_{22}H_{18}N_4$			
2-Propenal, 3-(1H-indol-3-yl)-, [3-(1H-indol-3-yl)-2-propenylidene]hydrazone	MeCN	209(4.39),233(4.41),272(4.06),397(4.06)	5-1039-76
	DMF-tert-BuOK	384(4.29),500(4.86),520s(4.83)	5-1039-76
$C_{22}H_{18}N_4O_2S$			
1H-1,2,4-Triazolium, 4-[[(4-hydroxy-3-methoxyphenyl)methylene]amino]-	MeCN	234(4.66),292(4.41),318(4.50)	103-0361-76

Compound	Solvent	$\lambda_{max}(\log \epsilon)$	Ref.
3-mercapto-1,5-diphenyl-, hydroxide, inner salt (cont.)			103-0361-76
$C_{22}H_{18}N_4O_6$			
Benzoic acid, 4-[[4-[methyl[(4-nitrobenzoyl)oxy]amino]phenyl]azo]-, methyl ester	EtOH	370(4.30)	94-1485-76
$C_{22}H_{18}N_6$			
2a,4a-Diazacyclopenta[cd]pentalene-1,6-dicarbonitrile, 2,5-diamino-3,4,6a,6b-tetrahydro-6a,6b-diphenyl-	EtOH	214(4.02),256(4.34)	44-3389-76
Formazan, 1-(4-methylphenyl)-3-phenyl-5-(4-quinazolinyl)-	C_6H_{12}	385(4.25),455(4.16)	103-0596-76
	EtOH	285(4.28),370(4.34)	103-0596-76
	EtOH-KOH	517(3.65)	103-0596-76
	dioxan	290(3.98),405(4.15), 435(4.12)	103-0596-76
nickel complex	EtOH-pH 7	612(4.04)	103-0596-76
$C_{22}H_{18}N_6O$			
Formazan, 1-(4-methoxyphenyl)-3-phenyl-5-(4-quinazolinyl)-	C_6H_{12}	335(4.22),450(4.26)	103-0596-76
	EtOH	285(4.21),360(4.48), 400s(4.33)	103-0596-76
	EtOH-KOH	518(4.14)	103-0596-76
	dioxan	335(4.02),350(4.02), 400(4.03),425(4.23), 440(4.20)	103-0596-76
nickel complex	EtOH-pH 7	618(4.12)	103-0596-76
$C_{22}H_{18}N_6O_2S$			
3,4-Pyrroledicarbonitrile, 2-benzyl-carbamoylthio-5-(3-benzylureido)-	DMSO	308(4.00),333(4.03)	24-2983-76
$C_{22}H_{18}N_8$			
1,3-Benzenediamine, 4,6-bis(5-phenyl-1H-1,2,4-triazol-3-yl)-	H_2SO_4	275(4.7243)	116-0626-76
Ethanedione, di-2-pyridinyl-, bis(2-pyridinylhydrazone)	EtOH	235s(--),331(4.62)	12-1153-76
copper chelate	EtOH	354(4.40),471(3.95)	12-1153-76
$C_{22}H_{18}N_8O_6$			
1,3-Benzenedicarboxylic acid, 4,6-dinitro-, bis[2-(iminophenylmethyl)-hydrazide]	H_2SO_4	247(4.6021)	116-0626-76
$C_{22}H_{18}O_2$			
Benzofuran, 3-(2-methoxyphenyl)-2-(4-methylphenyl)-	C_6H_{12}	238(4.24),307(4.39)	18-2560-76
	80% EtOH	240s(4.24),306(4.43)	18-2560-76
1,4-Butanedione, 1,2,4-triphenyl-	EtOH	258(4.37)	115-0761-76
Indeno[2,1-b]pyran-9-carboxaldehyde, 2,4-dimethyl-3-(phenylmethyl)-	dioxan	246(4.48),273(3.95), 281(3.92),320(4.44), 348(4.11),475(3.85)	78-2225-76
2(1H)-Naphthalenone, 1-methyl-1-[(1-methyl-2-naphthalenyl)oxy]-	EtOH	232(4.89),283(3.94), 294(4.02),312(3.99)	39-2570-76C
Spiro[2H-1-benzopyran-3,3'-trimethylene-2,2'-2'H-7',8'-benzoxocine]	hexane	212(4.71),258(4.51), 296(3.98)	103-1227-76
	dioxan	230(4.71),258(4.68), 296(4.10)	103-1227-76
	DMSO	263(4.67),296(4.19)	103-1227-76

Compound	Solvent	$\lambda_{max}(\log \epsilon)$	Ref.
$C_{22}H_{18}O_3$			
Benzofuran, 3-(2-methoxyphenyl)-2-(4-methoxyphenyl)-	C_6H_{12}	250(4.24),316(4.44)	18-2560-76
Benzo[a]pyrene-5a,6(5H)-diol, 4,6-dihydro-, 6-acetate, cis	MeOH	260(4.67),270(4.82), 300(4.01),311(4.08), 324(3.96)	88-1557-76
2(3H)-Furanone, 5-(1,2-dihydro-3-acenaphthylenyl)-4,5-dihydro-5-(4-hydroxyphenyl)-	EtOH-pH 9.8	430(3.43)	18-2027-76
$C_{22}H_{18}O_4$			
2(3H)-Furanone, 5-(1,2-dihydro-3-acenaphthylenyl)-5-(2,4-dihydroxyphenyl)dihydro-	EtOH-pH 9.5	580(3.02)	18-2027-76
$C_{22}H_{18}O_4S$			
Butanedioic acid, phenyl(phenyl-2-thienylmethylene)-, 1-methyl ester, (E)-	EtOH	212(4.04),252(3.37), 280(3.27)	4-0285-76
$C_{22}H_{18}O_5$			
2(3H)-Furanone, 5-(1,2-dihydro-3-acenaphthylenyl)dihydro-5-(2,4,6-trihydroxyphenyl)-	EtOH-pH 9.2	430(--),450(3.64)	18-2027-76
$C_{22}H_{18}O_7$			
4H-1-Benzopyran-3-carboxaldehyde, 2,3-dihydro-7-methoxy-3-[(7-methoxy-4-oxo-4H-1-benzopyran-3-yl)methyl]-4-oxo-	$CHCl_3$	246(4.35),252(4.30), 284(4.36),309(4.23)	2-0823-76
$C_{22}H_{19}BrO_2$			
Benzene, 1,1'-(2-bromo-2-phenylethenylidene)bis[2-methoxy-	C_6H_{12}	284(4.02)	18-2560-76
$C_{22}H_{19}ClN_2O_4$			
Camptothecin, 7-chloro-5-ethyl-	MeOH	224(4.46),249(4.42), 256(4.45),296(3.83), 342s(4.08),364(4.24)	24-1389-76
$C_{22}H_{19}ClO$			
Benzene, 1-chloro-4-(2-methoxy-2,3-diphenylcyclopropyl)-, (1α,2α,3β)-	C_6H_{12}	230(4.26)	12-2445-76
(1α,2β,3β)-	C_6H_{12}	229(4.30)	12-2445-76
Benzene, 1-chloro-4-(1-methoxy-1,3-diphenyl-2-propenyl)-, (E)-	C_6H_{12}	240s(4.02),253(4.05)	12-2445-76
Benzene, 1-chloro-4-(3-methoxy-3,3-diphenyl-1-propenyl)-, (E)-	C_6H_{12}	221(3.88),262(4.31)	12-2445-76
$C_{22}H_{19}FO$			
Benzene, 1-fluoro-4-(1-methoxy-1,3-diphenyl-2-propenyl)-, (E)-	C_6H_{12}	257(4.16)	12-2445-76
Benzene, 1-fluoro-4-(3-methoxy-3,3-diphenyl-1-propenyl)-, (E)-	C_6H_{12}	222(3.84),255(4.24)	12-2445-76
$C_{22}H_{19}N$			
Benzenamine, 4-(9H-fluoren-9-ylidenemethyl)-N,N-dimethyl-	dioxan	398(4.49)	39-0048-76C
Indolizine, 6-methyl-2-phenyl-7-(phenylmethyl)-	EtOH	264(4.60),350(3.46)	103-0424-76

Compound	Solvent	$\lambda_{max}(\log \epsilon)$	Ref.
$C_{22}H_{29}NO$			
Benzo[f]quinoline, 1-ethyl-3-(4-methoxy-phenyl)-	EtOH	217(4.33),227(4.31), 268(4.46),285(4.57), 347(3.88),364(3.89)	103-1012-76
perchlorate	EtOH	237(4.56),282(4.38), 313(4.41),389(4.48)	103-1012-76
picrate	EtOH	217(4.49),231(4.46), 284(4.56),347(4.25), 364(4.28),391(3.99)	103-1012-76
Benzo[f]quinoline, 3-(4-methoxyphenyl)-1,2-dimethyl-	EtOH	219(4.39),258(4.54), 273(4.47),328(3.50), 341(3.70),356(3.74)	103-1012-76
perchlorate	EtOH	220(4.50),256(4.57), 278(4.56),342(3.84), 359(3.94),377(3.72)	103-1012-76
picrate	EtOH	219(4.62),247(4.63), 276(4.53),343(4.25), 358(4.35),391(3.90)	103-1012-76
Dibenz[c,f]azocin-12(5H)-one, 6,7-di-hydro-6-(phenylmethyl)-	EtOH	221(4.39),265(3.11)	39-0913-76C
Ethanamine, N-salicylidene-1-(9-fluoren-yl)-	EtOH	240s(3.96),258(4.33), 280s(3.99),291(3.72), 302(3.82),316(3.53)	83-0512-76
$C_{22}H_{19}NOS$			
1,5-Benzothiazepine, 2,3-dihydro-2-(2-methoxyphenyl)-4-phenyl-	dioxan	264(4.35),335(3.70)	114-0293-76A
1,5-Benzothiazepine, 2,3-dihydro-2-(3-methoxyphenyl)-4-phenyl-	EtOH	262(4.27),338(3.62)	114-0293-76A
1,5-Benzothiazepine, 2,3-dihydro-2-(4-methoxyphenyl)-4-phenyl-	EtOH	231(4.33),261(4.34), 335(3.75)	114-0293-76A
$C_{22}H_{19}NO_2$			
4H-Indol-4-one, 1,5,6,7-tetrahydro-3-(2-oxo-2-phenylethyl)-2-phenyl-	EtOH	243(4.30),273(3.99)	94-1160-76
5(1H)-Quinolinone, 4-benzoyl-4,6,7,8-tetrahydro-2-phenyl-	EtOH	244(4.32),309(4.17)	94-1160-76
$C_{22}H_{19}NO_3$			
Benzoic acid, 2-[benzoyl(2-methylphen-yl)amino]-, methyl ester	MeOH	206(4.23),275(3.57)	39-0465-76C
Benzoic acid, 2-[[(2-methylphenyl)imi-no]phenylmethoxy]-, methyl ester	MeOH	208(4.57),230(4.39), 270(3.57)	39-0465-76C
$C_{22}H_{19}NO_3S$			
1-Propanone, 3-[(2-aminophenyl)thio]-3-(1,3-benzodioxol-5-yl)-1-phenyl-	EtOH	204(4.77),242(4.38), 294(3.88)	114-0293-76A
$C_{22}H_{19}NO_5S$			
Acetamide, N-[2-[2-[[(4-methylphenyl)-sulfonyl]oxy]benzoyl]phenyl]-	MeOH	205(4.64),229(4.58), 266(4.08),274(4.07), 333(3.64)	39-2089-76C
$C_{22}H_{19}NO_8$			
1H-Indol-4,7-diol, 2-(3,4-diacetoxy-phenyl)-, diacetate	MeOH	211(4.59),248(4.41), 308(4.43)	111-0241-76
$C_{22}H_{19}NS$			
1,5-Benzothiazepine, 2,3-dihydro-2-(4-methylphenyl)-4-phenyl-	EtOH	261(4.32),336(3.72)	114-0293-76A

Compound	Solvent	$\lambda_{max}(\log \epsilon)$	Ref.
$C_{22}H_{19}N_3$			
1H-Imidazo[4,5-b]quinoline, 5,6,7,8-tetrahydro-1,9-dimethyl-	EtOH	263(3.91),300(4.05)	49-1413-76
2,3-Quinolinediamine, 4-(4-methylphenyl)-N²-phenyl-	DMF	276(4.58),360(4.21), 367(4.21)	48-0039-76
$C_{22}H_{19}N_3O_3$			
Spiro[2H-1-benzopyran-2,2'-[2H]pyrrolo-[3,2-h]quinoline, 1',3'-dihydro-1',3',3'-trimethyl-6-nitro-	EtOH	245(4.09),255(4.70)	103-0680-76
$C_{22}H_{19}N_3O_4$			
Benzenemethanamine, 4-[(4-acetoxyphenyl)azo]-N-(benzoyloxy)-	EtOH	370(4.38)	94-1485-76
$C_{22}H_{19}N_5O_2S$			
Benzenesulfonamide, N-[1,2-bis(1H-benzimidazol-2-yl)ethyl]-	EtOH	243(4.13),275(4.19), 282(4.23)	42-0331-76
$C_{22}H_{19}N_5O_2S_2$			
Benzenesulfonamide, 4-[[(phenylamino)-thioxomethyl]amino]-N-(1-phenyl-1H-pyrazol-5-yl)-	EtOH	214(3.69),250s(3.68), 278(3.77)	80-1345-76
$C_{22}H_{19}N_5O_3S$			
Benzenesulfonamide, 4-[[(phenylamino)-carbonyl]amino]-N-(1-phenyl-1H-pyrazol-5-yl)-	EtOH	204(4.93),263(5.00)	80-1345-76
$C_{22}H_{19}O_2P$			
Phosphine, bis(2-methoxyphenyl)(phenylethynyl)-	MeOH	200(4.50),254(4.18), 265(4.12)	139-0169-76A
Phosphine, bis(4-methoxyphenyl)(phenylethynyl)-	MeOH	204(4.68),250(4.46)	139-0169-76A
$C_{22}H_{19}P$			
Phosphine, bis(2-methylphenyl)(phenylethynyl)-	MeOH	200(4.50),254(4.18), 265(4.12)	139-0169-76A
Phosphine, bis(4-methylphenyl)(phenylethynyl)-	MeOH	204(4.57),241(4.28), 265(4.18),288(3.88)	139-0169-76A
$C_{22}H_{20}ClNO_5$			
Benzo[f]quinoline, 1-ethyl-3-(4-methoxyphenyl)-, perchlorate	EtOH	237(4.56),282(4.38), 313(4.41),389(4.48)	103-1012-76
Benzo[f]quinoline, 3-(4-methoxyphenyl)-1,2-dimethyl-, perchlorate	EtOH	220(4.50),256(4.57), 278(4.56),342(3.84), 359(3.94),377(3.72)	103-1012-76
$C_{22}H_{20}ClNO_6$			
1,3-Dioxolo[4,5-g]isoquinoline, 6-acetyl-5-[(3-acetoxy-2-chlorophenyl)methylene]-5,6,7,8-tetrahydro-9-methoxy-, (Z)-	MeOH	308(4.10)	12-2003-76
$C_{22}H_{20}CuN_2O_2S_2$			
Copper, bis[2-(4,5-dihydro-2-thiazolyl)-1-phenylethanonato-N,O]-	CHCl₃	240(4.50),296(4.49), 336(4.46),620s(--)	48-0298-76
$C_{22}H_{20}F_3N_2OS$			
Benzothiazolium, 2-[4-[4-(dimethylamino)phenyl-1-(trifluoroacetyl)-1,3-	MeNO₂	580(4.76)	124-1174-76

Compound	Solvent	$\lambda_{max}(\log \epsilon)$	Ref.
$C_{22}H_{20}F_6N_2$			
Benzenamine, 4,4'-(1,2,3,4,5,6-hexafluoro-1,3,5-hexatriene-1,6-diyl)bis-[N,N-dimethyl-	heptane	369(4.70)	104-1565-76
$C_{22}H_{20}N_2$			
1H-Indole, 3,3'-(1,3-butadiene-1,4-diyl)bis[1-methyl-	MeCN	209(4.58),241(4.47), 284(4.10),359(4.64)	5-1060-76
semiquinone	MeCN	425(--),530s(--), 610(--),660(--), 685s(--),825(--), 930(--)	5-1060-76
oxidized form	MeCN	251(4.08),274(3.94), 338(4.17),501(4.70), 522s(4.66)	5-1060-76
4H-Pyrazino[3,2,1-jk]carbazole, 5,6-dihydro-8-methyl-4-(phenylmethyl)-	EtOH	264(4.75),330(3.64), 346(3.58),410(3.49), 430(3.43)	103-0662-76
$C_{22}H_{20}N_2NiO_2S_2$			
Nickel, bis[2-(4,5-dihydro-2-thiazolyl)-1-phenylethanonato-N,0]-	$CHCl_3$	267(4.95),293(4.58), 371s(4.00),488s(2.07), 627(1.79)	48-0298-76
$C_{22}H_{20}N_2O$			
2,3-Benzodiazocin-1(2H)-one, 3,4,5,6-tetrahydro-2,4-diphenyl-	EtOH	214(4.58),265(4.20)	39-2281-76C
$C_{22}H_{20}N_2OS$			
Benzeneethanehydrazonothioic acid, N-(4-methylphenyl)-α-oxo-, 4-methylphenyl ester	EtOH	244(4.390),280(4.089), 378(4.335)	18-0321-76
$C_{22}H_{20}N_2O_2$			
1,1'-Biisoquinoline, 2,2'-diacetyl-1,1',2,2'-tetrahydro-, meso	EtOH	240(4.46),300s(4.05), 312(4.07)	142-0427-76S
racemic	EtOH	227(4.43),304(4.14)	142-0427-76S
3H-Dibenz[f,ij]isoquinoline-2,7-dione, 1-(cyclohexylamino)-	$CHCl_3$	342(--),448(4.39)	104-1114-76
2,4,6-Heptatrienoic acid, 2-cyano-7-phenyl-5-(phenylamino)-, ethyl ester	CH_2Cl_2	439(4.28)	48-0705-76
$C_{22}H_{20}N_2O_2PdS_2$			
Palladium, bis[2-(4,5-dihydro-2-thiazolyl)-1-phenylethanonato-N,0]-	$CHCl_3$	248(4.71),289(4.52), 360(4.24)	48-0298-76
$C_{22}H_{20}N_2O_2S$			
Benzeneethanehydrazonothioic acid, N-(4-methoxyphenyl)-α-oxo-, 4-methylphenyl ester	EtOH	244(4.158),285(3.811), 386(4.091)	18-0321-76
2-Propyn-1-one, 3,3'-[thiobis(2,1-ethanediylimino)]bis[1-phenyl-	MeOH	259(4.17),303(4.17), 388(4.35)	48-0298-76
$C_{22}H_{20}N_2O_2S_2Zn$			
Zinc, bis[2-(4,5-dihydro-2-thiazolyl)-1-phenylethanonato-N,0]-, (T-4)-	$CHCl_3$	245(4.42),341(4.70)	48-0298-76
$C_{22}H_{20}N_2O_3$			
Spiro[2H-anthra[1,2-d]imidazole-2,1'-cyclohexane]-6,11-dione, 3-acetyl-1,3-dihydro-	EtOH	518(4.00)	104-0172-76

Compound	Solvent	$\lambda_{max}(\log \epsilon)$	Ref.
$C_{22}H_{20}N_2O_3S$			
Ethanone, 2-[(4-methylphenyl)hydrazono]-2-[(4-methylphenyl)sulfonyl]-1-phenyl-	EtOH	253(4.326),273(4.163), 374(4.357)	18-0321-76
Ethanone, 1-(4-methylphenyl)-2-[(4-methylphenyl)sulfonyl]-2-(phenylhydrazono)-	EtOH	230(4.364),275(4.201), 364(4.308)	18-0321-76
$C_{22}H_{20}N_2O_4S$			
Ethanone, 2-[(4-methoxyphenyl)hydrazono]-2-[(4-methylphenyl)sulfonyl]-1-phenyl-	EtOH	229(4.399),255(4.366), 390(4.399)	18-0321-76
$C_{22}H_{20}N_2O_5$			
Benzenamine, N-[(2-nitrophenyl)(2,4,6-trimethoxyphenyl)methylene]-	MeOH	207(4.68),260(3.82), 270(3.68),288(3.67), 305(3.48)	39-2089-76C
$C_{22}H_{20}N_2O_6$			
2-Naphthacenecarbonitrile, 4-(dimethylamino)-1,4,4a,5,12,12a-hexahydro-3,10,11,12a-tetrahydroxy-6-methyl-1,12-dioxo-, [4S-(4α,4aα,12aα)]-	MeCN	267(4.65),426(3.84)	36-1794-76
$C_{22}H_{20}N_2O_{11}$			
α-D-Ribofuranose, 2,3-O-(1-methylethylidene)-, bis(4-nitrobenzoate)	EtOH	258(4.43),300s(3.62)	44-0287-76
β-D-Ribofuranose, 2,3-O-(1-methylethylidene)-, bis(4-nitrobenzoate)	EtOH	257(4.42),300s(3.60)	44-0287-76
$C_{22}H_{20}N_3$			
3H-Indolizinium, 3-[[4-(dimethylamino)phenyl]imino]-2-phenyl-, perchlorate	Ac_2O	695(3.7012)	2-0551-76
$C_{22}H_{20}N_4O_2$			
4,12-Diazatetracyclo[6.3.1.0²,⁷.0³,⁵]-dodec-9-ene-6,11-dione, 12-(4,6-dimethyl-2-pyrimidinyl)-4-phenyl-, (1RS,2RS,3RS,5SR,7SR,8RS)-	EtOH	235(4.41)	39-2338-76C
(1RS,2SR,3SR,5RS,7RS,8RS)-	EtOH	235(4.44)	39-2338-76C
$C_{22}H_{20}N_4O_3$			
L-Tryptophan, N-[(3-methyl-2-quinoxalinyl)carbonyl]-, methyl ester	EtOH	203(4.77),220(4.70), 242(4.55),274(3.95), 282(3.97),290(3.94), 321(3.83)	78-2931-76
$C_{22}H_{20}N_4O_5$			
L-Tryptophan, N-[(3-methyl-2-quinoxalinyl)carbonyl]-, methyl ester, N,N'-dioxide	EtOH	222(4.63),267(4.30), 280s(3.96),290(3.89), 370(3.77),383(3.80)	78-2931-76
	MeCN	205(4.63),222(4.69), 260s(4.19),268(4.34), 280s(4.02),289(3.86), 352s(3.51),377(3.73), 395(3.81)	78-2931-76
$C_{22}H_{20}O$			
Acenaphthylene, 1-methoxy-2-(2,4,6-trimethylphenyl)-	n.s.g.	205(4.86),230(4.76), 419(2.94)	44-3599-76
Benzene, 1,1',1"-(1-methoxy-1,2,3-cyclopropanetriyl)tris-, (1α,2β,3α)-	C_6H_{12}	235(3.97)	12-2445-76

Compound	Solvent	$\lambda_{max}(\log \epsilon)$	Ref.
Benzene, 1,1',1"-(3-methoxy-1-propen-1-yl-3-ylidene)tris-	C_6H_{12}	233(3.87),258(4.28)	12-2445-76
Indeno[2,1-b]pyran, 4-butyl-2-phenyl-	dioxan	227(4.21),257(4.16), 306(4.14),333(4.08), 357(4.06),520(3.05)	78-2225-76
$C_{22}H_{20}O_2$			
Benzene, 1,1'-(2-phenylethenylidene)bis-[2-methoxy-	C_6H_{12}	250(4.10),284(4.20)	18-2560-76
Benzeneacetic acid, α-[2-(1,1-dimethyl-ethyl)-4-phenyl-1-buten-3-ynylidene]-	$CHCl_3$	274(4.50)	18-2515-76
[1,1'-Biphenyl]-4-acetic acid, α-phen-yl-, ethyl ester, carbanion	DMSO	440(4.30)	70-0762-76
7,9,15,17-Cyclooctadecatetraene-2,4,11,13-tetrayne-1,6-diol, 1,6,10,15-tetramethyl-	EtOH	245s(4.46),254(4.69), 264(4.79),292(3.71), 308s(3.82),327(4.06), 344(4.27),368(4.30)	18-0292-76
$C_{22}H_{20}O_4$			
4H-1-Benzopyran-4-one, 3-[(2,6-dimethyl-4-oxo-4H-1-benzopyran-3-yl)methyl]-2,3-dihydro-6-methyl-	EtOH	234(4.36),251(4.23), 312(3.93),335(3.53)	39-2444-76C
$C_{22}H_{20}O_4S$			
Methanone, [2-[2-[[(4-methylphenyl)sul-fonyl]oxy]ethyl]phenyl]phenyl-	MeOH	250(4.15),272(3.68), 280(3.48),320(2.26)	35-0541-76
3,4-Thiophenedimethanol, 2,5-diphenyl-	CH_2Cl_2	295(4.0)	28-0607-76A
$C_{22}H_{20}O_5$			
Aspulvinone I	MeOH	240(4.07),326(4.24), 370(4.05)	88-1013-76
2H,6H-Benzo[1,2-b:5,4-b']dipyran-6-one, 5-methoxy-8-(4-methoxyphenyl)-2,2-di-methyl-	MeOH	225(4.52),267(4.37), 327(4.47)	12-2023-76
4H,8H-Benzo[1,2-b:3,4-b']dipyran-4-one, 3,6-dimethoxy-8,8-dimethyl-2-phenyl-	EtOH	235(3.82),250(3.81), 280(3.31),342(3.30)	102-1553-76
Benzoic acid, 4-methoxy-2,6-bis(phenyl-methoxy)-	EtOH	270(4.22)	35-5380-76
$C_{22}H_{20}O_6$			
4H-1-Benzopyran-4-one, 3-[2-(2,4-dimeth-oxybenzoyl)-1-propenyl]-7-methoxy-	$CHCl_3$	238(3.98),287(4.63), 440(4.37)	2-0823-76
Ovalichromene A	MeOH	225(4.55),320(3.92), 345(3.96)	102-2011-76
$C_{22}H_{20}O_7$			
12H-Benzo[b]xanthen-12-one, 6-hydroxy-3,7,8,10-tetramethoxy-1-methyl-	$CHCl_3$	260(5.06),302(4.91), 340(4.26),414(4.27)	39-0499-76C
$C_{22}H_{20}O_8$			
2(3H)-Furanone, 4-(acetoxy-1,3-benzodi-oxol-5-ylmethyl)-3-(1,3-benzodioxol-5-ylmethyl)dihydro-	MeOH	236(3.89),286(3.88)	18-3359-76
2(3H)-Furanone, 4-(acetoxymethyl)-5-(1,3-benzodioxol-5-yl)-3-(1,3-benzodioxol-5-ylmethyl)dihydro-	MeOH	237(3.89),287(3.86)	18-3359-76
Nidirufin, 6,8-di-O-methyl-	EtOH	224(4.68),251(4.28), 288(4.49),314s(3.93), 444(3.94)	102-1037-76

Compound	Solvent	λ_{max}(log ϵ)	Ref.
$C_{22}H_{20}O_{10}$ Vulgamycin I	MeOH	250(4.23),283(4.05)	88-4367-76
$C_{22}H_{20}O_{11}$ 1,3,5-Benzenetriol, 2-[3,5-bis(acetyl-oxy)phenoxy]-, triacetate	MeCN	230s(4.20),267(3.32)	102-1082-76
$C_{22}H_{20}O_{12}$ 6H-[1,3]Benzodioxolo[5,6-c][1]benzopyr-an-6-one, 2,3,4,7-tetraacetoxy-1,2,3,4-tetrahydro-, (2α,3α,4α)-	EtOH	248(4.68),282(3.77), 297(3.75),330(3.77)	24-0855-76
$C_{22}H_{20}O_{13}$ Lagertannin	EtOH	251(4.87),344(4.19), 358(4.21)	95-0984-76
	EtOH-NaOEt	256s(4.66),269(4.73), 295(4.64),375(4.20)	95-0984-76
	EtOH-NaOAc	254(4.71),275(4.58), 315(4.28),355(4.21)	95-0984-76
	EtOH-NaOAc- H_3BO_3	251(4.87),344(4.19), 358(4.21)	95-0984-76
$C_{22}H_{20}S_3$ Thiophene, 2,5-bis[4-(5-methyl-2-thien-yl)-1,3-butadienyl]-	DMF	425s(4.66),448(4.78), 477(4.66)	126-1857-76
$C_{22}H_{21}ClN_3O_2S$ Morpholinium, 4-[[[5-(4-chlorophenyl)-2-(4-methoxyphenyl)-6H-1,3-thiazin-6-ylidene]amino]methylene]-, perchlorate	HOAc	300s(4.05),341(4.26), 454(3.96)	97-0268-76
$C_{22}H_{21}NOS$ 1-Propanone, 3-[(2-aminophenyl)thio]-3-(4-methylphenyl)-1-phenyl-	EtOH	242(4.38),310(3.70)	114-0293-76A
$C_{22}H_{21}NO_2S$ 1-Propanone, 3-[(2-aminophenyl)thio]-3-(2-methoxyphenyl)-1-phenyl-	EtOH	242(4.37),282(3.81), 308(3.68)	114-0293-76A
1-Propanone, 3-[(2-aminophenyl)thio]-3-(3-methoxyphenyl)-1-phenyl-	EtOH	242(4.29),284(3.73), 308(3.69)	114-0293-76A
1-Propanone, 3-[(2-aminophenyl)thio]-3-(4-methoxyphenyl)-1-phenyl-	EtOH	238(4.45),310(3.66)	114-0293-76A
$C_{22}H_{21}NO_3$ Benzoic acid, 4-[4-(2-oxo-2-phenyleth-yl)-1(4H)-pyridinyl]-, ethyl ester	MeOH	250(4.22),440(4.88)	104-1105-76
$C_{22}H_{21}NO_4$ Spiro[4.5]dec-2-ene, 1-(2-carboxyphen-yl)-2-hydroxy-4-oxo-3-phenyl-, hydr-oxide, inner salt	pH 13	263(4.23)	44-0390-76
$C_{22}H_{21}NO_4S_2$ Butanedioic acid, [[2-phenyl-4-(phenyl-methyl)-5-thiazolyl]thio]-, dimethyl ester	MeOH	216(3.22),310(3.22)	78-0583-76
$C_{22}H_{21}NO_6$ Aporphine-2,10-diol, 1,11-(methylenedi-oxy)-, diacetate, (6aS)-	EtOH	275(3.96),313s(3.42)	33-2551-76

Compound	Solvent	λ_{max}(log ϵ)	Ref.
Aporphine-2,10-diol, 1,11-(methylenedi-oxy)-, diacetate, (6aS)- (cont.)	EtOH-NaOH	248(4.25),281(3.90), 325(3.51)	33-2551-76
Elmerrillicine, N,O-diacetyl-synthetic-, (±)-	MeOH	243(4.22),276(4.15)	12-2003-76
	MeOH	243s(3.96),277(3.94)	12-2003-76
2H-Quinolizine-1,3,4-tricarboxylic acid, 2-phenyl-, 1-ethyl 3,4-dimethyl ester	MeOH	208(4.31),271(4.30), 312s(3.83),465(3.61)	39-1911-76C
	MeOH-acid	208(4.35),278(3.99)	39-1911-76C
4H-Quinolizine-1,3,4-tricarboxylic acid, 2-phenyl-, 1-ethyl 3,4-dimethyl ester	MeOH	214(4.11),280(4.05), 335(3.66),348s(3.53), 410(3.79)	39-1911-76C
	MeOH-HClO$_4$	224(4.39),262s(3.93), 290(4.06)	39-1911-76C

C$_{22}$H$_{21}$NO$_9$
| Pyrido[2,1,6-de]quinolizine-1,2,7,9-tetracarboxylic acid, 7,9-diethyl 1,2-dimethyl ester | EtOH | 292(4.45),328s(3.89), 413(4.27),488(4.28) | 39-0341-76C |

C$_{22}$H$_{21}$NO$_9$
| Isoxazolo[4,5-b]dibenzofuran-6-one, 1aα,7,9-triacetoxy-1a,2,6,6a-tetra-hydro-5,6aα,10-trimethyl- | MeCN | 220(3.88),238(3.67), 276(2.96) | 23-3721-76 |

C$_{22}$H$_{21}$NO$_{11}$
| 2-Epiisonarciclasin tetraacetate | EtOH | 233(4.51),250(4.61), 264s(--),275s(--), 288(4.00),298(4.06), 320s(--),333(3.93), 350s(--) | 24-0855-76 |

C$_{22}$H$_{21}$N$_2$
| Acridinium, 9-[4-(dimethylamino)phenyl]-10-methyl-, (triiodide) | CH$_2$Cl$_2$ | 260(5.07),292(4.73), 360(4.61),607(4.14) | 104-1531-76 |

C$_{22}$H$_{21}$N$_2$S
| 1H-2,1-Benzoisothiazol-2-ium, 3-[4-(di-methylamino)phenyl]-1-(phenylmethyl)-, perchlorate | EtOH | 290(4.34),340s(3.44), 530(4.50) | 48-0161-76 |

C$_{22}$H$_{21}$N$_3$O$_4$
| 1H-Pyridazino[4,5-b]indol-4-ol, 2-acet-yl-2,5-dihydro-1-(4-methoxyphenyl)-5-methyl-, acetate | EtOH | 315(4.34) | 103-1008-76 |

C$_{22}$H$_{21}$N$_3$S
| Thieno[2,3-c]pyridine-3-carbonitrile, 2-amino-4,5,6,7-tetrahydro-4,6-di-methyl-5,7-diphenyl- | EtOH | 215(4.0),290(3.78) | 2-0357-76 |

C$_{22}$H$_{21}$N$_5$O$_2$
| 7(8H)-Pteridinone, 2-(dimethylamino)-4-(phenylmethoxy)-8-(phenylmethyl)- | MeOH | 243(4.21),290(3.66), 362(4.23) | 24-3228-76 |

C$_{22}$H$_{21}$O$_2$P
| 6-Phosphanthridinol, 5,6-dihydro-5-(1-methylethyl)-6-phenyl-, 5-oxide | EtOH | 214(4.62),237(4.24), 265(3.84),273(3.87), 286(3.76) | 39-2050-76C |

C$_{22}$H$_{22}$
| 1,1':2',1"-Terphenyl, 4'-butyl- | EtOH | 235(4.44) | 33-2012-76 |

Compound	Solvent	$\lambda_{max}(\log \epsilon)$	Ref.
$C_{22}H_{22}BrNO_5S$			
1H-Pyrrolo[2,1-c][1,4]thiazine-7-carboxylic acid, 1-(4-bromophenyl)-1-[(3-ethoxy-3-oxo-1-propenyl)oxy]-4-methyl-, ethyl ester, (Z)-	MeOH	293(4.46)	44-0187-76
$C_{22}H_{22}NO$			
Pyridinium, 2,5-dimethyl-1-(2-oxo-2-phenylethyl)-4-(phenylmethyl)-, bromide	EtOH	250(4.28),420(2.78)	103-0424-76
$C_{22}H_{22}NO_4$			
Benzo[g]-1,3-benzodioxolo[5,6-a]quinolizinium, 13-ethyl-5,6-dihydro-9,10-dimethoxy-, iodide	EtOH	227(4.54),267(4.47), 344(4.35),425(3.76)	73-3654-76
Coralynium, chloride	EtOH	221(4.37),235(4.34), 243s(4.30),287s(4.54), 304s(4.66),314(4.71), 329(4.61),362(3.85), 406s(4.08),424(4.18)	73-3654-76
Dibenzo[a,g]quinolizinium, 2,3,10,11-tetramethoxy-6-methyl-, chloride	EtOH	240(4.44),281(4.68), 299(4.81),310(4.82), 330(4.66),360(3.98), 410(4.22),425(4.27)	87-0882-76
$C_{22}H_{22}NO_5$			
Benzo[g]-1,3-benzodioxolo[5,6-a]quinolizinium, 13-ethoxy-5,6-dihydro-9,10-dimethoxy-, iodide	EtOH	227(4.52),269(4.40), 342s(4.31),353(4.33), 432(3.81)	73-3654-76
$C_{22}H_{22}N_2$			
1H-Indole, 3,3'-(1,2-ethenediyl)bis[1,2-dimethyl-	MeCN	209(4.57),236(4.38), 285(4.14),330(4.29)	5-1060-76
semiquinone	MeCN	240s(4.22),254s(4.18), 261(4.16),284s(3.89), 311(3.83),350s(3.89), 366(3.92),482s(4.18), 507(4.35),752(4.00), 834(4.07)	5-1060-76
oxidized form (dication)	MeCN	224(4.19),261(4.14), 303(4.00),470(4.28), 496s(4.23)	5-1060-76
Indolizine, 1,1'-(1,2-ethenediyl)bis-[2,3-dimethyl-	CH$_2$Cl$_2$	255(4.62),347s(4.27), 362(4.30),413(4.14)	5-0317-76
semiquinone	1:1 CH$_2$Cl$_2$-MeCN	334(3.91),448(3.97), 612(4.48),895(4.04), 1010(4.13)	5-0317-76
oxidized form (dication)	MeCN	255(4.24),303(3.89), 458(4.54)	5-0317-76
Indolizine, 3,3'-(1,2-ethenediyl)bis-[1,2-dimethyl-	MeCN	290(4.27),300(4.30), 408(4.41)	5-0317-76
semiquinone	MeCN	642(4.69),982(4.18)	5-0317-76
oxidized form (dication)	MeCN	290(4.24),362(4.33), 512(4.32)	5-0317-76
$C_{22}H_{22}N_2O_2$			
2(5H)-Pyrrolone, 1-acetyl-3-(phenylmethyleneamino)-5-phenyl-4-propyl-	EtOH	209(4.37),269(4.37), 280(4.37)	40-1100-76
Spiro[2H-anthra[1,2-d]imidazole-2,1'-cyclohexane]-6,11-dione, 1,3-dihydro-1,3-dimethyl-	EtOH and EtOH-NaOH	574(4.10)	104-0172-76
	MeNO$_2$	562(4.05)	104-0172-76

Compound	Solvent	$\lambda_{max}(\log \epsilon)$	Ref.
(cont.)	MeNO$_2$–HClO$_4$	450(3.49)	104-0172-76
$C_{22}H_{22}N_2O_3$			
9-Acridinecarbonitrile, 9a-(benzoyloxy)-1,2,3,4,4a,9,9a,10-octahydro-4a-methoxy-	EtOH	232(4.15),280s(3.30), 297(3.46)	95-0968-76
Benzophenone imine, 2-amino-2',4',6'-trimethoxy-N-phenyl-	MeOH	208(4.68),234(4.61), 262s(3.99),351(3.86)	39-2089-76C
4a,9a-(Epoxymethanoxy)acridine-9-carbonitrile, 1,2,3,4,9,10-hexahydro-12-methoxy-12-phenyl-	EtOH	238(4.01),267s(2.53), 293(3.38)	95-0968-76
4-Pyridinecarboxamide, 2-methyl-3-(phenylmethoxy)-5-[(phenylmethoxy)methyl]-	EtOH	278(3.70)	87-0999-76
$C_{22}H_{22}N_2O_3S$			
Glycine, N-[2-(acetylamino)-5-phenyl-3-thienyl]-N-phenyl-, ethyl ester	MeOH	236s(4.46),275(4.06), 307s(4.14),323(4.15)	48-0221-76
$C_{22}H_{22}N_2O_4$			
Benzoic acid, 2-[1-acetyl-5-(4-methoxyphenyl)-1H-pyrazol-3-yl]-, propyl ester	EtOH	235(4.36),288(4.33), 305(4.10)	44-0110-76
$C_{22}H_{22}N_2O_4S$			
Pyridinium, 3-(3-anilino-3-oxoprop-1-enyl)-1-methyl-, tosylate	MeOH	280(4.23),310(4.25)	39-0315-76C
$C_{22}H_{22}N_2O_7$			
Anhydrotetracycline, hydrochloride	n.s.g.	269(4.77),430(3.99)	36-1794-76
$C_{22}H_{22}N_2O_8$			
3,6,14,17-Tetraoxa-10,21-diazatricyclo-[17.3.1.1^{8,12}]tetracosa-1(23),8,10-12(24),19,21-hexaene-2,7,13,18-tetrone, 9,11,20,22-tetramethyl-	MeCN	234(4.12),273(3.64), 282(3.54)	77-0964-76
$C_{22}H_{22}N_3O_2S$			
Morpholinium, 4-[[[2-(4-methoxyphenyl)-5-phenyl-6H-1,3-thiazin-6-ylidene]-amino]methylene]-, perchlorate	HOAc	301s(4.15),340(4.26), 455(4.14)	97-0268-76
$C_{22}H_{22}N_4$			
1-Indolizinecarboxaldehyde, 2,3-dimethyl-, [(2,3-dimethyl-1-indolizinyl)-methylene]hydrazone	CH$_2$Cl$_2$	245(4.47),320(3.87), 420s(4.53),442(4.54), 470(4.28)	5-0317-76
3-Indolizinecarboxaldehyde, 1,2-dimethyl-, [(1,2-dimethyl-3-indolizinyl)-methylene]hydrazone	CH$_2$Cl$_2$	304(4.36),312(4.45), 434(4.50),456(4.70), 484(4.64)	5-0317-76
oxidized form	CH$_2$Cl$_2$	332(4.21),556(4.63)	5-0317-76
$C_{22}H_{22}N_4O_3$			
Carbamic acid, methyl[3-[(methylamino)-carbonyl]-5,6-diphenylpyrazinyl]-, ethyl ester	MeOH	224(4.33),240s(4.23), 293(4.19),330(4.05)	24-3194-76
$C_{22}H_{22}N_4O_4$			
Benzo[g]pteridine-5(1H)-acetic acid, 2,3,4,10-tetrahydro-7,8,10-trimethyl-2,4-dioxo-3-(phenylmethyl)-	20% MeOH–pH 7	307(4.16),356(3.85)	5-2037-76

Compound	Solvent	$\lambda_{max}(\log \epsilon)$	Ref.
$C_{22}H_{22}N_4O_4S$			
Acetic acid, [[3,4,5,10-tetrahydro-7,8,10-trimethyl-2,4-dioxo-3-(phenyl-methyl)benzo[g]pteridin-4a(2H)-yl]-thio]-	50% MeOH–pH –0.5	392(3.87)	5-2037-76
	50% MeOH pH 7 and 10	300s(--),360(3.87)	5-2037-76
$C_{22}H_{22}N_6O_2$			
3,12-Diazatricyclo[5.3.1.1^{2,6}]dodeca-4,8(and 9)-diene-10(and 8),11-dione, 3,12-bis(4,6-dimethyl-2-pyrimidinyl)-(2:1 mixture)	EtOH	208(4.11),230(4.38), 245(4.43),368(3.81)	39-2296-76C
$C_{22}H_{22}O_2$			
1-Azulenecarboxylic acid, 3,4-dimethyl-7-(1-methylethyl)-, phenyl ester	C_6H_{12}	248(4.48),302(4.64), 309(4.66),314s(4.57), 377(4.05),389(4.14), 562(2.95),602s(2.86), 665(2.41)	18-1650-76
1-Azulenecarboxylic acid, 3,8-dimethyl-5-(1-methylethyl)-, phenyl ester	C_6H_{12}	249(4.40),305(4.53), 315s(4.46),386(4.02), 482s(2.28),567(2.72), 650s(2.61)	18-1650-76
$C_{22}H_{22}O_3$			
1-Azulenecarboxylic acid, 4-methyl-7-(1-methylethyl)-, 4-methoxyphenyl ester	C_6H_{12}	241(4.45),299(4.59), 304(4.61),309s(4.51), 364(3.91),377(4.03), 513s(2.87),535(2.95), 569(2.85),623(2.48)	18-1650-76
Benzenepropenoic acid, (1,2,3,4-tetra-hydro-6-methoxy-1-naphthalenylidene)-ethyl ester	MeOH	217(4.47),223(4.45), 276(4.53)	2-0799-76
Cyclohexanone, 2,6-bis[(4-methoxyphen-yl)methylene]-	EtOH	243(4.37),364(4.62)	114-0381-76B
	ether	238(4.21),347(4.48)	114-0381-76B
	CHCl$_3$	332(4.16)	114-0381-76B
	CCl$_4$	350(4.60)	114-0381-76B
$C_{22}H_{22}O_4$			
Naphthacene, 1,4-dihydro-5,6,11,12-tet-ramethoxy-	MeOH	240(4.44),269(5.03), 359(3.63),377(3.89), 397(3.90),420(3.81)	44-2296-76
1,7-Octadiene-3,6-dione, 1,8-bis(4-meth-oxyphenyl)-, (E,E)-	dioxan	303s(3.60),308(3.71)	12-0339-76
$C_{22}H_{22}O_5$			
Cyclohexanone, 2,6-bis[(4-hydroxy-3-methoxyphenyl)methylene]-	EtOH	255(4.05),390(4.51)	114-0381-76B
	ether	252(--),365(--)	114-0381-76B
	CHCl$_3$	370(4.46)	114-0381-76B
	CCl$_4$	365(4.25)	114-0381-76B
ion	n.s.g.	490(4.74)	114-0381-76B
non-ion	n.s.g.	395(4.55)	114-0381-76B
2(1H)-Naphthacenone, 3,4-dihydro-5,6,11,12-tetramethoxy-	MeOH	227(4.30),266(4.86), 325(3.77),390(3.79), 413(3.71)	44-2296-76
$C_{22}H_{22}O_7$			
2-Anthracenecarboxylic acid, 9,10-di-hydro-1,6,8-trimethoxy-3-propyl-, methyl ester	EtOH	227(4.38),243(4.32), 281(4.41),345(3.63), 390(3.73)	44-3018-76
9,10-Anthracenedione, 2,4,5,7-tetrameth-oxy-1-(1-oxobutyl)-	EtOH	224(4.62),284(4.38), 342(3.63),415(3.79)	39-1852-76C

Compound	Solvent	$\lambda_{max}(\log \epsilon)$	Ref.
Naphtho[2,3-c]furan-1,3-dione, 4-(3,4-dimethoxyphenyl)-3a,4,9,9a-tetrahydro-6,7-dimethoxy-, [3aS-(3aα,4α,9aβ)]-	MeOH	283(3.69)	2-0014-76
$C_{22}H_{22}O_8$			
4H-1-Benzopyran-2-carboxylic acid, 7-methoxy-5-methyl-4-oxo-3-[(2,4,5-trimethoxyphenyl)methyl]-	CHCl$_3$	244(4.74),286(4.62), 296(4.58)	39-0499-76C
2,3-Naphthalenedicarboxylic acid, 1,2-dihydro-7-hydroxy-1-(4-hydroxy-3-methoxyphenyl)-6-methoxy-, dimethyl ester	MeOH	250(4.41),310(4.20), 340(4.12)	2-0014-76
$C_{22}H_{22}O_{13}$			
4H-1-Benzopyran-4-one, 2-[3-(β-D-glucopyranosyloxy)-4,5-dihydroxyphenyl]-5,7-dihydroxy-3-methoxy-	MeOH	252(4.07),302(3.82), 356(4.03)	2-0384-76
$C_{22}H_{23}BrO_5$			
2-Propenoic acid, 3-[4-[2-(4-bromophenyl)-2-oxoethoxy)-3-methoxyphenyl]-, 1,1-dimethylethyl ester, (E)-	MeOH	259(4.39),283(4.37), 318(4.31)	20-0657-76
$C_{22}H_{23}Cl_3O_3$			
Androsta-4,6-dieno[16,17-c]furan-3-one, 6-chloro-5'-(dichloromethylene)-16,17-epoxy-2',5',16,17-tetrahydro-, (16α,17α)-	MeOH MeCN	223(4.16),283(4.32) 223(4.14),278(4.33)	44-3940-76 44-3940-76
D-Dihomo-17-oxaandrosta-4,6-diene-3,17b-dione, 6-chloro-17a-(dichloromethyl)-16-methylene-	MeOH	283(4.41)	44-3940-76
$C_{22}H_{23}NOS$			
2H-Azonin-2-one, 1-methyl-5-(methylthio)-3,4-diphenyl-	EtOH	229(4.393),255s(4.079), 278(4.064)	24-0576-76
Cycloprop[hi]indolizinium, 4,5,6,6a,6b,6c-hexahydro-2-hydroxy-3-methyl-6c-(methylthio)-1,6b-diphenyl-, hydroxide, inner salt	dioxan	224s(4.187),302(4.204)	24-0562-76
$C_{22}H_{23}NO_3$			
1H-Indene-3-carboxylic acid, 2-[[(1,1-dimethylethyl)amino]phenylmethyl]-1-oxo-, methyl ester	hexane CHCl$_3$	244(4.44),248(4.49), 325(3.38),420(3.02) 425(2.93)	44-3540-76 44-3540-76
1H-Indene-6-carboxylic acid, 2-[[(1,1-dimethylethyl)amino]phenylmethyl]-1-oxo-, methyl ester	hexane MeCN	240(4.45),315(3.85), 327(3.88),388(3.76) 388(3.45)	44-3540-76 44-3540-76
$C_{22}H_{23}NO_4$			
Tylophorinidine	MeOH	258(4.7),287(4.43), 310(3.9),340(3.26), 355(2.8)	102-1561-76
	MeOH-KOH	258(4.64),297(4.45), 329(3.99),353(3.79), 368(3.47)	102-1561-76
$C_{22}H_{23}NO_5$			
8H-Benzo[a][1,3]benzodioxolo[5,6-g]quinolizin-8-one, 5,6,14,14a-tetrahydro-2,3-dimethoxy-14,14-dimethyl-	n.s.g.	221(4.53),274(3.90), 289(3.93),304(3.85)	44-2201-76
Isoquinoline, 2-(1,3-benzodioxol-5-yl-	MeOH	254(4.24),292(4.06)	44-2201-76

Compound	Solvent	$\lambda_{max}(\log \epsilon)$	Ref.
carbonyl)-1,2,3,4-tetrahydro-6,7-di- methoxy-1-(1-methylethylidene)- (cont.)			44-2201-76
$C_{22}H_{23}NO_5S$ 1H-Pyrrolo[2,1-c][1,4]thiazine-7-carbox- ylic acid, 1-[(3-ethoxy-3-oxo-1-prop- enyl)oxy]-4-methyl-1-phenyl-, ethyl ester, (Z)-	MeOH	208(4.33),233(4.29), 292(4.43)	44-0187-76
$C_{22}H_{23}NS_2$ Cycloprop[hi]indolizinium, 4,5,6,6a,6b- 6c-hexahydro-2-mercapto-3-methyl- 6c-(methylthio)-1,6b-diphenyl-, hydroxide, inner salt	EtOH	223(4.365),323(4.083)	24-0562-76
$C_{22}H_{23}N_2O_2S$ Sulfonium, [4-(diethylamino)-3-nitro- phenyl]diphenyl-, tetrafluoroborate	EtOH	320(4.34),400(3.45)	30-0235-76
Sulfonium, [4-[(1,1-dimethylethyl)ami- no]-3-nitrophenyl]diphenyl-, tetra- fluoroborate	EtOH	300(4.27),400(3.56)	30-0235-76
$C_{22}H_{23}N_3$ 1H-Pyrazole, 1-[(cyclohexylimino)phenyl- methyl]-3-phenyl-	EtOH	275(4.18),280(4.13), 292(3.88)	88-4753-76
4H-1,2,4-Triazepine, 4-cyclohexyl- 3,7-diphenyl-	EtOH	205(4.41),259(4.41), 259(4.30),362(2.94)	88-2459-76
$C_{22}H_{23}N_3O$ 4H-Indazol-4-one, 1,5,6,7-tetrahydro- 6,6-dimethyl-3-phenyl-1-(phenyl- methyl)-, oxime, (E)-	MeOH	228(4.48),266s(3.98)	24-1898-76
Pyrazolo[4,3-b]azepin-5(1H)-one, 4,6,7,8-tetrahydro-7,7-dimethyl- 3-phenyl-1-(phenylmethyl)-	MeOH	230s(4.06),254s(4.02)	24-1898-76
Pyrazolo[4,3-c]azepin-4(1H)-one, 5,6,7,8-tetrahydro-7,7-dimethyl- 3-phenyl-1-(phenylmethyl)-	MeOH	250s(4.13)	24-1898-76
$C_{22}H_{23}N_3O_4S$ 4H-Indazol-4-one, 1,5,6,7-tetrahydro- 3-(4-methoxyphenyl)-1-methyl-, O-[(4- methylphenyl)sulfonyl]oxime, (E)-	MeOH	235(4.52),262s(4.04)	24-1898-76
$C_{22}H_{23}N_5O_2$ Roquefortine C	EtOH	209(4.47),240(4.21), 328(4.43)	31-0140-76
$C_{22}H_{24}ClNO_5$ 6,12-Dioxatricyclo[6.2.2.0^{1,9}]dodecane- 2,5,11-trione, 4-[1-chloro-2-[(2-phen- ylethyl)amino]propylidene]-8-methyl-	MeOH	208(4.21),250(3.38)	33-1809-76
$C_{22}H_{24}Cl_4O_3$ Pregna-4,6,20-trien-3-one, 6,21,21-tri- chloro-16α-(chloromethyl)-16β,20- epoxy-17α-hydroxy-	MeOH MeCN	283(4.36) 278(4.38)	44-3940-76 44-3940-76
$C_{22}H_{24}NO_4$ Dibenzo[a,g]quinolizinium, 5,6-dihydro-	EtOH	231(4.46),267(4.52),	88-1595-76

Compound	Solvent	$\lambda_{max}(\log \epsilon)$	Ref.
2,3,10,11-tetramethoxy-8-methyl-, chloride (cont.)		340(4.43),425(3.81)	88-1595-76
$C_{22}H_{24}N_2O_2$			
Pyrano[2,3-c]pyrazole, 6-butoxy-1,4,5,6-tetrahydro-1,4-diphenyl-, cis	EtOH	245(4.23)	35-2947-76
trans	EtOH	244(4.20)	35-2947-76
Pyrano[2,3-c]pyrazole, 6-(1,1-dimethylethoxy)-1,4,5,6-tetrahydro-1,4-diphenyl-, cis	EtOH	245.5(4.17)	35-2947-76
trans	EtOH	244.5(4.21)	35-2947-76
Pyrano[2,3-c]pyrazole, 1,4,5,6-tetrahydro-6-(1-methylethoxy)-4-(4-methylphenyl)-1-phenyl-, cis	EtOH	245(4.24)	35-2947-76
trans	EtOH	244(4.22)	35-2947-76
Pyrano[2,3-c]pyrazole, 1,4,5,6-tetrahydro-6-(2-methylpropoxy)-1,4-diphenyl-, cis	EtOH	245.5(4.22)	35-2947-76
trans	EtOH	244(4.18)	35-2947-76
$C_{22}H_{24}N_2O_3$			
Benzenebutanamide, γ-acetoxy-N-[2-(1H-indol-3-yl)ethyl]-	MeOH	226(4.48),275(3.79), 282(3.82),291(3.76)	23-1262-76
Benzoic acid, 2-[5-(4-methoxyphenyl)-1H-pyrazol-3-yl]-, pentyl ester	EtOH	261(4.45)	44-0110-76
Pyrano[2,3-c]pyrazole, 1,4,5,6-tetrahydro-4-(4-methoxyphenyl)-6-(1-methylethoxy)-1-phenyl-, cis	EtOH	244(4.26)	35-2947-76
trans	EtOH	243(4.24)	35-2947-76
$C_{22}H_{24}N_2O_4$			
Hydrazine, bis(β-p-toluoylpropanoyl)-	EtOH	250(4.47)	115-0761-76
$C_{22}H_{24}N_2O_5$			
Propanedioic acid, (3-acetyl-1,2,6,7,12-12b-hexahydroindolo[2,3-a]quinolizin-2-yl)-, dimethyl ester, cis-(±)-	MeOH	222(4.52),302(4.57)	35-3645-76
Propanedioic acid, (3-ethylidene-1,2,3-4,6,7,12,12b-octahydro-4-oxoindolo-[2,3-a]quinolizin-2-yl)-, dimethyl ester, (E)-	MeOH	222(4.62),273(4.00), 290(3.89)	24-3825-76
$C_{22}H_{24}N_2O_6$			
Propanedioic acid, [1,2,6,7,12,12b-hexahydro-3-(methoxycarbonyl)indolo[2,3-a]quinolizin-2-yl]-, dimethyl ester, cis-(±)-	MeOH	224(4.50),282(4.54), 291(4.43)	35-3645-76
3H-Pyrrolizin-3-one, 6-(4,5-dimethoxy-2-nitrophenyl)hexahydro-5-(4-methoxyphenyl)-, cis	MeOH	226(4.40),247(4.00), 277(3.67),283(3.67), 340(3.60)	33-2201-76
trans	MeOH	228(4.38),246s(4.12), 278(3.66),286(3.64), 342(3.67)	33-2201-76
$C_{22}H_{24}N_2O_7$			
Androsta-5,8-diene-2,3,7,11,16,17-hexone, 6-hydroxy-4,4,14α-trimethyl-, 2,16-dioxime	dioxan	233(4.24),265s(3.75), 320s(3.26)	5-1214-76
2-Pyrrolidinone, 5-[[[2-(4,5-dimethoxy-2-nitrophenyl)-1-(4-methoxyphenyl)-ethenyl]oxy]methyl]-, [S-(Z)]-	MeOH	282(4.37),362(3.83)	33-2201-76

Compound	Solvent	$\lambda_{max}(\log \epsilon)$	Ref.
2-Pyrrolidinone, 5-[2-(4,5-dimethoxy-2-nitrophenyl)-3-(4-methoxyphenyl)-3-oxopropyl]-	MeOH	252(4.26),274(4.20)	33-2201-76
$C_{22}H_{24}N_2O_{10}S$ β-D-Glucopyranoside, 5-phenyl-1,3,4-oxadiazol-2-yl 1-thio-, 2,3,4,6-tetraacetate	EtOH	267(4.29)	106-0153-76
1,3,4-Oxadiazole-2(3H)-thione, 5-phenyl-3-(2,3,4,6-tetra-O-acetyl-β-D-glucopyranosyl)-	EtOH	234(4.03),264(4.13), 289(4.27)	106-0153-76
$C_{22}H_{24}N_4$ Bisbenzimidazo[1,2-a:1',2'-e][1,5]diazocine, 6,7,14,15-tetrahydro-2,3,10,11-tetramethyl-	EtOH	252(4.07),257(4.06), 277s(3.96),282(4.06), 285(4.05),291(4.07)	39-0312-76C
$C_{22}H_{24}N_4O_2S_2$ 1H-[1,2]Dithiolo[5,1-e][1,2,3]thiadiazole-7-S^{IV}, 5-(1,1-dimethylethyl)-1-(4-methoxyphenyl)-3-[(4-methoxyphenyl)azo]-	C_6H_{12}	207(4.42),230(4.41), 246s(4.37),332(4.16), 434(4.41),444s(4.39), 489(4.29)	39-0228-76C
$C_{22}H_{24}N_4O_4$ 2-Propenoic acid, 3,3'-[1,4-bis(phenylmethyl)-2-tetrazene-1,4-diyl]bis-, dimethyl ester, (E,E,?)-	MeOH	264(3.76+),354(4.28+)	24-0253-76
$C_{22}H_{24}N_4O_4S_2$ Diazene, [5,5'-bis(1,1-dimethylethyl)-[2,3'-bithiophene]-2'-yl](2,4-dinitrophenyl)-	n.s.g.	526(5.32)	39-1639-76C
$C_{22}H_{24}N_4O_5$ D-glycero-Tetrulose, N-acetyl-di-O-acetyl-, phenylosazone	EtOH	220s(3.90),243(4.1), 268s(3.7),390(4.02)	136-0185-76D
$C_{22}H_{24}N_4S_2$ 1H-[1,2]Dithiolo[5,1-e][1,2,3]thiadiazole-7-S^{IV}, 5-(1,1-dimethylethyl)-1-(4-methylphenyl)-3-[(4-methylphenyl)azo]-	C_6H_{12}	204(4.49),224(4.38), 242s(4.38),251(4.39), 287(4.11),312(4.12), 426(4.39),487s(4.24)	39-0228-76C
$C_{22}H_{24}O$ 2-Cyclohexen-1-one, 2-butyl-3,5-diphenyl-	EtOH	232s(3.93)	33-2012-76
$C_{22}H_{24}O_2S$ Sulfonium, bis(phenylmethyl)-, 4,4-dimethyl-2,6-dioxocyclohexylide	MeOH	222s(4.36),260(4.27)	30-0366-76
$C_{22}H_{24}O_2S_3$ Cyclopentanone, 2,2'-(1,2,4-trithiolane-3,5-diylidene)bis[5-cyclopentylidene-	EtOH	232(3.50),239(3.48), 250s(3.40),288(3.40), 296(3.40),384(3.97)	39-1706-76C
$C_{22}H_{24}O_3$ 11H-Cyclopenta[a]phenanthren-17-ol, 12,13,16,17-tetrahydro-3-methoxy-6,13-dimethyl-, acetate, (13R-cis)-	EtOH	252(4.55),258(4.70), 268(4.72),299(4.19)	54-0223-76

Compound	Solvent	$\lambda_{max}(\log \epsilon)$	Ref.
Glabridin dimethyl ether	EtOH	280(4.13),290s(3.99), 313s(3.44)	94-0752-76
$C_{22}H_{24}O_4$ 4aH-Xanthen-4a-ol, 9-(2-furanyl)-5-(2-furanylmethylene)-1,2,3,4,5,6,7,8,9-9a-decahydro-	EtOH	211(3.99),296s(4.50), 308(4.61),323(4.47)	114-0051-76C
$C_{22}H_{24}O_5$ 2-Cyclopentene-1-acetic acid, 2-(methoxycarbonyl)-5-(6-methoxy-2-naphthalenyl)-1-methyl-, methyl ester, (1S-trans)-	EtOH	230(4.93),261(3.80), 272(3.79),317(3.20), 331(3.31)	78-0079-76
2-Naphthacenol, 1,2,3,4-tetrahydro-5,6,11,12-tetramethoxy-	MeOH	359(3.59),378(3.89), 396(3.90),430(3.79)	44-2296-76
Psoralen, 8-geranyloxy-5-methoxy-	EtOH	223(4.44),242(4.17), 249(4.16),269(4.27), 274s(4.26),315(4.09)	105-0150-76
$C_{22}H_{24}O_6$ 9,10-Anthracenedione, 1-butyl-2,4,5,7-tetramethoxy-	EtOH	226(4.50),270(4.19), 288(4.33),420(3.76)	39-1852-76C
$C_{22}H_{24}O_7$ Bicyclo[3.2.1]oct-3-en-2-one, 8-acetoxy-7-(1,3-benzodioxol-5-yl)-3-hydroxy-1-methoxy-6-methyl-5-(2-propenyl)-, (6-exo,7-endo,8-anti)	MeOH	233(3.75),274(3.81)	102-1033-76
	MeOH-NaOH	233(3.75),294(3.81), 319(3.81)	102-1033-76
	MeOH-NaOH-HCl added	233(3.75),274(3.81)	102-1033-76
	MeOH-AlCl$_3$	235(3.69),280(3.64), 324(3.38)	102-1033-76
$C_{22}H_{24}O_8$ 2,3-Naphthalenedicarboxylic acid, 1-(3,4-dimethoxyphenyl)-1,2,3,4-tetrahydro-6,7-dimethoxy-	MeOH	238(4.18),280(3.8)	2-0014-76
$C_{22}H_{24}O_{10}$ Haplanthin	EtOH	287(4.27),325s(3.55)	2-0644-76
2-Propen-1-one, 1-[4-(D-glucopyranosyloxy)-2-hydroxy-6-(methoxyphenyl)-3-(4-hydroxyphenyl)- (helichrysin)	MeOH	236(4.18),272(4.12), 296s(4.05),360(4.31)	39-1819-76C
	MeOH-NaOMe	242s(4.26),269(4.29), 302s(3.98),382(4.30)	39-1819-76C
	MeOH-NaOAc	270(4.48),302s(4.37), 322s(4.36),374(4.44), 446s(4.29)	39-1819-76C
	+ H$_3$BO$_3$	270(4.28),286(4.20), 350(4.28)	39-1819-76C
	MeOH-AlCl$_3$	246s(4.43),274s(4.34), 294(4.30),322s(4.29), 350s(4.32),370s(4.38), 396(4.42)	39-1819-76C
	MeOH-AlCl$_3$-HCl	250s(4.46),274s(4.38), 294(4.36),322s(4.34), 342s(4.37),362s(4.40), 386(4.44)	39-1819-76C
$C_{22}H_{25}ClO_3$ Androsta-4,6-dieno[16,17-c]furan-3-one, 6-chloro-16,17-epoxy-2',5',16,17-tetrahydro-5'-methylene-, (16α,17α)-	MeOH	207(4.14),283(4.31)	44-3940-76
	MeCN	207(4.14),283(4.34)	44-3940-76

Compound	Solvent	λ_{max}(log ϵ)	Ref.
$C_{22}H_{25}Cl_3O_4$			
Androsta-4,6-diene-3,17-dione, 15-acet-yl-6-chloro-16-(dichloromethyl)-16-hydroxy-, (15α,16α)-	MeOH	283(4.36)	44-3940-76
$C_{22}H_{25}FN_3S$			
Benzothiazolium, 2-[4-(dimethylamino)-4-[4-(dimethylamino)phenyl]-1-fluoro-1,3-butadienyl]-3-methyl-, perchlorate	MeOH	523(4.93)	124-1174-76
$C_{22}H_{25}NO_4$			
1,3-Benzodioxole-5-ethanamine, 6-[4,5-(or 6,7)-dimethoxy-1H-inden-2-yl]-N,N-dimethyl-	EtOH	298(4.24)	78-1973-76
Benzo[e][1,3]dioxolo[4,5-k][3]benzazecine, 5,6,7,8-tetrahydro-3,4-dimethoxy-6,15-dimethyl-, (E)-	EtOH	234(4.18),294(3.94)	78-1973-76
Isoquinoline, 3-(6-ethenyl-1,3-benzodioxol-5-yl)-1,2,3,4-tetrahydro-7,8-dimethoxy-2,4-dimethyl-, cis-(±)-	EtOH	264(4.05),300(3.69)	78-1973-76
Ochotensane, 2,3-(methylenedioxy)-9,10-dimethoxy-13-methyl-, picrate	EtOH	230s(4.20),286(3.80), 294(3.74)	78-1973-76
3H-Pyrrolizin-3-one, 6-(3,4-dimethoxyphenyl)hexahydro-5-(4-methoxyphenyl)-, (7aS)-	MeOH	227(4.31),278(3.68)	33-2201-76
stereoisomer	MeOH	229(4.22),278(3.56)	33-2201-76
Valine, 2-(2-benzoylethyl)-N-(phenylacetyl)-, (±)-	EtOH	243(4.13)	12-0339-76
$C_{22}H_{25}NO_6$			
1H-2-Benzopyran-1-one, 3-[6-[2-(dimethylamino)ethyl]-1,3-benzodioxol-5-yl]-3,4-dihydro-7,8-dimethoxy-	EtOH	230(4.38),247(4.04), 292(3.60),317(3.31)	142-0041-76S
1(3H)-Isobenzofuranone, 5,6-dimethoxy-3-(1,2,3,4-tetrahydro-6,7-dimethoxy-2-methyl-1-isoquinolinyl)-, (R*,R*)-(±)-	EtOH	218s(4.20),230(4.25), 262(3.95),295(3.89), 307s(3.73)	78-0211-76
(R*,S*)-	EtOH	230(4.13),262(3.97), 296(3.89),307s(3.78)	78-0211-76
$C_{22}H_{25}NO_8$			
1-Oxa-7-azaspiro[4.4]non-2-ene-4,6-dione, 8-benzoyl-2-(1,2-dihydroxy-3-hexenyl)-9-hydroxy-8-methoxy-3-methyl-, [5S-[2-(1R*,2R*,3Z),5α,8β,9β]]-	EtOH	253(4.22),280(4.05)	33-0133-76
$C_{22}H_{25}N_3O$			
Acetamide, N-[2-[3-methyl-5-(2-methylpropyl)-1-phenyl-1H-pyrazol-4-yl]-phenyl]-	MeOH	246(4.32)	23-1020-76
$C_{22}H_{25}N_3O_7$			
5α-Androst-8-ene-2,3,6,7,11,16,17-heptone, 4,4,14α-trimethyl-, 2,6,16-trioxime	dioxan	238(4.12),270s(3.74), 300s(3.56)	5-1214-76
$C_{22}H_{25}N_4O_4P$			
Phosphinic acid, di-4-morpholinyl-, 2-phenyl-4-quinazolinyl ester	isoPrOH	207(4.61),255s(4.53), 259(4.54),285(4.22), 330s(3.41)	44-2720-76

Compound	Solvent	$\lambda_{max}(\log \epsilon)$	Ref.
$C_{22}H_{25}N_7O_4S$ L-Alanine, 3-benzo[b]thien-3-yl-N- [N-(5-oxo-L-prolyl)-L-histidyl]-, hydrazide	HOAc	259(3.72),290(3.45), 299(3.48)	94-3149-76
$C_{22}H_{25}N_7O_5$ L-Alanine, N-[N-(5-oxo-L-prolyl)-L-his- tidyl]-3-(1H-pyrrolo[2,3-b]pyridin- 3-yl)-, methyl ester	HOAc	293.3(3.87)	94-3149-76
$C_{22}H_{25}N_7O_6S$ L-Alanine, 3-benzo[b]thien-3-yl-N- [N-(5-oxo-L-prolyl)-L-histidyl]-, hydrazide, S,S-dioxide	HOAc	262(3.93)	94-3149-76
$C_{22}H_{26}$ Benzene, 1,1'-(1,2-ethynediyl)bis- [4-(1,1-dimethylethyl)-	C_6H_{12}	272(4.41),280(4.44), 288(4.55),297(4.40), 307(4.50)	101-0385-76N
$C_{22}H_{26}Cl_2N_7$ 1H-Benzimidazolium, 5-chloro-2-[3-(5- chloro-1,3-diethyl-1,3-dihydro-2H- benzimidazol-2-ylidene)-1-triazenyl]- 1,3-diethyl-, tetrafluoroborate	MeOH	436(3.41)	33-0148-76
$C_{22}H_{26}Cl_2O_3$ Pregna-4,6-diene-3,20-dione, 6-chloro- 16β-(chloromethyl)-16,17-epoxy-	MeOH MeCN	283(4.33) 278(4.34)	44-3940-76 44-3940-76
$C_{22}H_{26}NOS$ 2-Azoniabicyclo[3.1.0]hex-3-ene, 3-eth- oxy-2,2-dimethyl-1-(methylthio)-4,5- diphenyl-, tetrafluoroborate	dioxan	256(4.042)	24-0562-76
$C_{22}H_{26}NS_2$ 2-Azoniabicyclo[3.1.0]hex-3-ene, 3-(eth- ylthio)-2,2-dimethyl-1-(methylthio)- 4,5-diphenyl-, tetrafluoroborate	dioxan	250(4.053)	24-0562-76
$C_{22}H_{26}N_2$ 1H-Imidazole, 2-(1-methylethyl)-1-(2- methylpropyl)-4,5-diphenyl-	n.s.g.	253(4.08),270s(4.02)	33-2149-76
$C_{22}H_{26}N_2OS_2$ 1-Piperazinepropanol, 4-[8-(methylthio)- dibenzo[b,f]thiepin-10-yl]-	MeOH	225(4.41),265(4.37), 275(4.38),315s(3.96)	73-0443-76
$C_{22}H_{26}N_2O_2$ Apovincaminic acid, ethyl ester	EtOH	229(4.45),275(4.08), 315(3.85)	145-1907-76
$C_{22}H_{26}N_2O_4$ Acetamide, N-[4-(2-ethyl-2,3-dihydro- 7,8-dimethoxy-3-methyl-1,4-benzoxa- zepin-5-yl)phenyl]-, hydrochloride	n.s.g.	245(4.24),320(4.29), 380(4.13)	20-0898-76
Benzenemethanol, α,α'-azobis[α-ethyl-, diacetate, meso	CH_2Cl_2	352(1.60)	39-1249-76B
Benzenemethanol, α,α'-azobis[α-methyl-, dipropanoate, meso	CH_2Cl_2	358(1.56)	39-1249-76B

Compound	Solvent	λ_{max}(log ϵ)	Ref.
3H-Pyrrolizin-3-one, 6-(2-amino-4,5-dimethoxyphenyl)hexahydro-5-(4-methoxyphenyl)-, (7aS)-	MeOH	285(3.64),300(3.69)	33-2201-76
isomer	MeOH	286(3.71),300(3.81)	33-2201-76
$C_{22}H_{26}N_2O_5$			
Cabucine oxindole A	MeOH	212(4.36),245(4.10), 295(3.30)	22-1473-76
Cabucine oxindole B	MeOH	212(4.35),245(4.08), 295(3.28)	22-1473-76
2-Pentenoic acid, 5-(1,3-dihydro-2H-pyrrolo[3,4-b]quinolin-2-yl)-3-(dimethoxymethyl)-2-ethyl-5-oxo-, methyl ester	EtOH	232(4.68),282(3.60), 288(3.62),295(3.68), 301(3.69),308(3.83), 314(3.78),322(3.96)	44-0699-76
3-Pentenoic acid, 5-(1,3-dihydro-2H-pyrrolo[3,4-b]quinolin-2-yl)-3-(dimethoxymethyl)-2-ethyl-5-oxo-, methyl ester	EtOH	234(4.61),288(3.61), 294(3.63),301(3.62), 307(3.75),314(3.69), 321(3.89)	44-0699-76
Tetraphylline oxindole A	MeOH	220(4.45),288(3.43), 295(3.28)	22-1473-76
Tetraphylline oxindole B	MeOH	220(4.45),288(3.43), 295(3.28)	22-1473-76
$C_{22}H_{26}N_3S$			
Benzothiazolium, 2-[4-(dimethylamino)-4-[4-(dimethylamino)phenyl]-1,3-butadienyl]-3-methyl-, perchlorate	MeOH	506(4.93)	124-1174-76
$C_{22}H_{26}N_4O_7$			
Acetamide, N-[4-[[4-[bis[2-(acetyloxy)-ethyl]amino]-2-nitrophenyl]amino]-phenyl]-	EtOH	522(3.73)	146-0131-76
$C_{22}H_{26}N_4S_2$			
Methanaminium, N-[4,4-dimethyl-2-[(phenylmethyl)[[(phenylmethyl)amino]thioxomethyl]amino]-5(4H)-thiazolylidene]-N-methyl-, hydroxide, inner salt	MeCN	249(4.37),348s(3.23)	33-2768-76
$C_{22}H_{26}N_8O_4S_2$			
1,2,4,5-Tetrazine-3,6-dicarboximidamide, N,N,N'',N''-tetramethyl-N',N'''-bis-[(4-methylphenyl)sulfonyl]-	MeCN	221(4.54),243(4.64), 525(2.68)	32-0001-76
$C_{22}H_{26}N_9O_4$			
1H-Benzimidazolium, 2-[3-(1,3-diethyl-1,3-dihydro-5-nitro-2H-benzimidazol-2-ylidene)-1-triazenyl]-1,3-diethyl-5-nitro-, tetrafluoroborate	MeOH	443(3.90)	33-0148-76
$C_{22}H_{26}O_2Si_2$			
Silane, [indeno[2,1-a]indene-5,10-diyl-bis(oxy)]bis[trimethyl-	ether	232(4.24),250(4.38), 262(4.44),267(4.47), 293(4.51),303(4.52), 366(3.83),387(4.16), 411(4.29)	44-4041-76
$C_{22}H_{26}O_3$			
2H-1-Benzopyran-2-acetaldehyde, 3,4-dihydro-2,5,7,8-tetramethyl-6-(phenylmethoxy)-	EtOH	227s(4.06),357s(3.03), 364s(3.03),368s(3.08), 381s(3.33),388(3.38)	33-0290-76

Compound	Solvent	$\lambda_{max}(\log \epsilon)$	Ref.
$C_{22}H_{26}O_4$			
Phenol, 3-[2-(3,4-dimethoxyphenyl)ethen-yl]-5-methoxy-2-(3-methyl-2-butenyl)-, (E)-	MeOH	321(4.43)	102-2006-76
Phenol, 3-[2-(3,4-dimethoxyphenyl)ethen-yl]-5-methoxy-6-(3-methyl-2-butenyl)-, (E)-	MeOH	320(4.38)	102-2006-76
$C_{22}H_{26}O_5$			
1H-2-Benzopyran, 3-(2-ethenyl-4,5-di-methoxyphenyl)-3,4-dihydro-6,7-di-methoxy-1-methyl-	MeOH	264(3.78),288s(3.52)	33-0949-76
2H-1-Benzopyran-7-ol, 3-[5-(1,1-dimeth-yl-2-propenyl)-3-hydroxy-2,4-dimeth-oxyphenyl]-3,4-dihydro-	EtOH	286(3.7),293s(3.5)	88-2391-76
Cyclopentaneacetic acid, 2-(methoxycarb-onyl)-5-(6-methoxy-2-naphthalenyl)-1-methyl-, methyl ester	EtOH	230(4.93),261(3.78), 271(3.79),316(3.20), 331(3.31)	78-0079-76
Pregna-4,20(21)-diene-3,11-dione, 17,20-[carbonylbis(oxy)]-	n.s.g.	237(4.20)	44-0078-76
$C_{22}H_{26}O_6$			
6(2H)-Benzofuranone, 2-(3,4-dimethoxy-phenyl)-3,3a-dihydro-5,7-dimethoxy-3-methyl-3a-(2-propenyl)-, [2S-(2α,3β,3aβ)]-	MeOH	229(4.30),271(4.32), 303(3.86)	102-1033-76
$C_{22}H_{26}O_7$			
Kaur-16-en-20-oic acid, 11-acetoxy-7,14-dihydroxy-15,18-dioxo-, 20,7-lactone, (4α,7α,11β,14R)-	EtOH	230(4.01)	33-0772-76
Spiro[furan-3(2H),6'(7'H)-[1H-3,9a]meth-ano[2]benzoxepin]-1',2-dione, 10'-(acetoxymethyl)-5-(3-furanyl)octa-hydro-7'-methyl-, [3'R-(3'α,5'aα,6'β-7'β,9'aα,10'R*)]-	EtOH	210(3.79)	78-1881-76
Spiro[furan-3(2H),1'(2'H)-naphthalene-4'a,5'(7'H)-dicarboxylic acid, 5-(3-furanyl)-3',4,4',5,8',8'a-hexahydro-2'-methyl-2-oxo-, dimethyl ester, [1'R-[1'α(R*),2'α,4'aα,8'aβ]]-	EtOH	209(4.02)	78-1881-76
Wikstromol dimethyl ether	EtOH	231(4.51),261(3.73)	102-1789B-76
$C_{22}H_{26}O_8$			
Corylifuran	EtOH	210(3.82)	78-1881-76
$C_{22}H_{26}O_9S$			
Breynogenin	MeOH	258(4.16)	94-0114-76
	MeOH-NaOH	288(4.24),352(3.93), 393(4.01)	94-0114-76
Isobreynogenin	MeOH	260(4.18)	94-0114-76
	MeOH-NaOH	289(4.25),348(3.94)	94-0114-76
$C_{22}H_{26}S_2$			
1,3-Dithietane, 4-[2,2-dimethyl-1-(1-methylethyl)propylidene]-2,2-diphenyl-	C_6H_{12}	219(4.4),235s(--), 295s(--),360s(--)	24-0906-76
$C_{22}H_{27}B$			
Borane, bis(2-phenylethenyl)(1,1,2-tri-methylpropyl)-, (E,E)- (also aroma-tic bands not listed)	ether	217(--),292(4.58)	101-0303-76N

Compound	Solvent	$\lambda_{max}(\log \epsilon)$	Ref.
$C_{22}H_{27}ClN_2O_4$			
Acetamide, N-[4-(2-ethyl-2,3-dihydro-7,8-dimethoxy-3-methyl-1,4-benzoxazepin-5-yl)phenyl]-, hydrochloride	n.s.g.	245(4.24),320(4.29), 380(4.13)	20-0898-76
$C_{22}H_{27}ClO_3$			
Androsta-1,4,6-trien-3-one, 4-chloro-17-(1-oxopropoxy)-, (17β)-	EtOH	229(4.15),260(3.86), 312(4.11)	56-2087-76
Pregna-4,6-diene-3,20-dione, 6-chloro-17α-hydroxy-16-methylene-	MeOH	285(4.35)	44-3940-76
$C_{22}H_{27}ClO_4$			
Androsta-1,4-diene-3,6-dione, 7α-chloro-17β-(1-oxopropoxy)-	EtOH	251(4.20)	56-0097-76
$C_{22}H_{27}ClO_5$			
Pregn-4-ene-3,11-dione, 17,20-[carbonyl-bis(oxy)]-20β-chloro-	n.s.g.	238(4.21)	44-0078-76
$C_{22}H_{27}ClO_6$			
Pregn-4-ene-3,11-dione, 17,20-[carbonyl-bis(oxy)]-20-chloro-21-hydroxy-	n.s.g.	238(4.19)	44-0078-76
$C_{22}H_{27}Cl_3O_4$			
Androsta-4,6-dien-3-one, 15-acetyl-6-chloro-16-(dichloromethyl)-16,17-dihydroxy-, (15α,16α,17β)-	MeOH	284(4.35)	44-3940-76
$C_{22}H_{27}NO_3$			
19-Nor-17α-pregn-4-ene-21-carboxylic acid, 7α-cyano-17-hydroxy-3-oxo-, γ-lactone	MeOH	233(4.23)	87-0975-76
$C_{22}H_{27}NO_4$			
1,3-Benzodioxole-5-ethanamine, 6-(2,3-dihydro-4,5-dimethoxy-1H-inden-2-yl)-N,N-dimethyl-, (±)-	EtOH	230(4.07),285(3.71), 292(3.67)	78-1973-76
Coralydine	EtOH	227s(4.22),287(3.88)	73-3157-76
Corydaline	EtOH	227s(4.31),282(3.75)	73-3157-76
$C_{22}H_{27}NO_5$			
1,4-Benzoxazepine, 5-(3,4-dimethoxyphenyl)-2,3-dihydro-7,8-dimethoxy-3-(1-methylethyl)-	n.s.g.	215(4.40),255(4.20), 315(4.17)	20-0787-76
$C_{22}H_{27}NO_6$			
Androsta-5,8-diene-7,11,16,17-tetrone, 3β,6-dihydroxy-4,4,14 -trimethyl-, 16-oxime	dioxan	236(3.75),275(3.24), 362(2.93)	5-1214-76
$C_{22}H_{27}N_3O_6$			
Tricyclo[3.3.1.1³,⁷]decane-1-carboxylic acid, (3-acetoxy-2,3,3a,9a-tetrahydro-6-imino-6H-furo[2',3':4,5]oxazolo[3,2-a]pyrimidin-2-yl)methyl ester, hydrochloride	MeOH-acid	235(4.05),262(4.09)	87-0663-76
$C_{22}H_{27}N_3O_7$			
Benzeneethanol, 4-[[4-[bis(2-acetoxyethyl)amino]-2-nitrophenyl]amino]-	EtOH	514(3.69)	146-0131-76

Compound	Solvent	$\lambda_{max}(\log \epsilon)$	Ref.
$C_{22}H_{27}N_4O_3P$			
2,3-Diazabicyclo[2.2.1]hept-5-ene-2-car-boxylic acid, 3-[bis(methylphenylami-no)phosphinyl]-, ethyl ester	MeOH	206(4.20),235(4.08), 270(3.08)	139-0207-76A
$C_{22}H_{28}Cl_2O_4$			
Androst-4-ene-3,6-dione, 4,7α-dichloro-17β-(1-oxopropoxy)-	EtOH	271(3.91)	56-1919-76
Androst-4-ene-3,7-dione, 4,6α-dichloro-17β-(1-oxopropoxy)-	EtOH	264(3.96)	56-1919-76
Androst-4-ene-3,7-dione, 4,6β-dichloro-17β-(1-oxopropoxy)-	EtOH	260(4.01)	56-1919-76
$C_{22}H_{28}N_2O_3$			
Acetamide, N-[4-(heptyloxy)phenyl]-2-[(4-methoxyphenyl)imino]-	EtOH	235(3.18)	65-0702-76
Strychnofendlerine	EtOH	248(4.16),278(3.53), 286s(3.45)	32-0773-76
$C_{22}H_{28}N_2O_4$			
Acetamide, N-[4-(2-ethyl-2,3,4,5-tetra-hydro-7,8-dimethoxy-3-methyl-1,4-ben-zoxazepin-5-yl)phenyl]-	n.s.g.	215(4.47),247(4.38), 285(3.63)	20-0898-76
Acetamide, N-[4-(heptyloxy)phenyl]-2-[(4-methoxyphenyl)imino]-	EtOH	235(4.1),288(4.0), 340(4.0)	65-0702-76
Cyclobutane-1,1-dicarbonitrile, 2,2,3,3-tetraethoxy-4-(2-phenylethenyl)-	CHCl$_3$	260(4.31)	39-1538-76C
$C_{22}H_{28}N_2O_5$			
Strychnosplendine, N_a-acetyl-12-hydroxy-11-methoxy-	EtOH	226(4.42),254(3.87), 296s(3.30)	32-0773-76
	EtOH-base	306(3.74)	32-0773-76
$C_{22}H_{28}N_2O_6$			
5α-Androst-8-ene-6,7,11,16,17-pentone, 3β-hydroxy-4,4,14α-trimethyl-, 6,16-dioxime	dioxan	234(3.94),272(3.75)	5-1214-76
12H-5,7-Ethanopyrido[2,1-c][1,4]benzodi-azonine-16-carboxylic acid, 6,6a,7,8-9,10,13,14-octahydro-16-hydroxy-3-methoxy-6,14-dioxo-, methyl ester, [6aS-(6aR*,7R*,16R*)]-	EtOH	250(4.43),281(4.08)	88-0435-76
$C_{22}H_{28}N_2Se$			
1,3,4-Selenadiazole, 2,2-bis(1,1-dimeth-yl)-2,5-dihydro-5,5-diphenyl-	C_6H_{12}	226(4.01),325(3.48), 356(3.45)(changing)	39-2079-76C
$C_{22}H_{28}N_4OS_2$			
Propanamide, N,N,2-trimethyl-2-[[[(phen-ylmethyl)[[(phenylmethyl)amino]thioxo-methyl]amino]thioxomethyl]amino]-	EtOH	267(4.22)	33-2768-76
$C_{22}H_{28}N_4O_{12}$			
2,4(1H,3H)-Pteridinedione, 5-acetyl-5,6,7,8-tetrahydro-1-(2,3,4,6-tetra-O-acetyl-β-D-glucopyranosyl)-	pH 7.0	240(3.81),287(4.22)	24-3184-76
	pH 12.0	257(3.90),287(4.13)	24-3184-76
	MeOH	240(3.80),287(4.19)	24-3184-76
$C_{22}H_{28}N_4S_2$			
Thiourea, N-[5-(dimethylamino)-4,5-di-hydro-4,4-dimethyl-2-thiazolyl]-N,N'-bis(phenylmethyl)-	EtOH	271(4.22)	33-2768-76

Compound	Solvent	$\lambda_{max}(\log \epsilon)$	Ref.
$C_{22}H_{28}N_7$			
1H-Benzimidazolium, 2-[3-(1,3-diethyl-1,3-dihydro-2H-benzimidazol-2-ylidene)-1-triazenyl]-1,3-diethyl-, tetrafluoroborate	MeOH	430(3.04)	33-0148-76
$C_{22}H_{28}O_2$			
2,4,6,8,10,12-Tridecahexaenoic acid, 13-(2,6,6-trimethyl-1-cyclohexen-1-yl)-, (all-E)-	EtOH	396(4.87)	36-0580-76
$C_{22}H_{28}O_4$			
5α-Androst-8-ene-3,7,11,17-tetrone, 4,4,14α-trimethyl-	dioxan	265(3.96)	5-1214-76
Hedychenone, 7-acetoxy-	MeOH	217(3.94),241(4.13)	102-0827-76
Jatrophone ethylene ketal	EtOH	255(4.30)	35-2295-76
Pregna-4,20-dien-3-one, 17,20-[carbonylbis(oxy)]-	n.s.g.	240(4.23)	44-0078-76
$C_{22}H_{28}O_4S$			
Pregn-4-ene-3,20-dione, 16,17-(oxycarbonylthio)-, (16β,17α)-	MeOH	239(4.25)	24-0185-76
$C_{22}H_{28}O_4S_2$			
8,12-Dithiatricyclo[12.2.2.2^{3,6}]eicosa-3,5,14,16,17,19-hexaene, 5,10,15,20-tetramethyl-, 8,8,12,12-tetraoxide, anti	MeOH	251(3.86)	44-2509-76
syn	MeOH	252(3.86)	44-2509-76
$C_{22}H_{28}O_5$			
Androsta-5,8-diene-7,11,17-trione, 3β,6-dihydroxy-4,4,14α-trimethyl-	dioxan	232(4.16),280(3.82), 355(3.53)	5-1206-76
$C_{22}H_{28}O_6$			
7,20-Anhydrorastronol G	EtOH	231(3.98)	33-0772-76
Corylifuran	EtOH	210(3.82)	78-1881-76
1H-Cyclopropa[3,4]benz[1,2-e]azulene-3-carboxaldehyde, 9a-acetoxy-1a,2,4,4a-5,7a,7b,8,9,9a-decahydro-4,7b-dihydroxy-1,1,6,8-tetramethyl-5-oxo-, [1aR-(1aα,1bβ,4aβ,7α,7bα,8α,9aα)]-	EtOH	240(4.19)	88-1737-76
11-Oxapregna-1,4-diene-3,20-dione, 21-acetoxy-17-hydroxy-	EtOH	242(4.05)	13-0717-76A
$C_{22}H_{28}O_7$			
Rastronol H	EtOH	231(3.93)	33-0772-76
$C_{22}H_{28}O_8$			
Lyoniresinol, (-)-	EtOH	279(3.57)	94-2102-76
Spiro[furan-3(2H),1'(2'H)-naphthalene]-4'a,5'(5'H)-dicarboxylic acid, 5-(3-furanyl)octahydro-6'-hydroxy-2'-methyl-2-oxo-, dimethyl ester, [1'R-[1'α(R*),2'α,4'aα,5'α,6'α,8'aβ]]-	EtOH	210(3.79)	78-1881-76
$C_{22}H_{28}O_9$			
Dethiobreynogenin	MeOH	258(4.29)	94-0114-76
	MeOH-NaOH	299(4.42),350(3.89)	94-0114-76

Compound	Solvent	$\lambda_{max}(\log \epsilon)$	Ref.
$C_{22}H_{28}O_9S$			
Breynogenin, dihydro-	MeOH	258(4.04)	94-0114-76
	MeOH-NaOH	300(4.23)	94-0114-76
$C_{22}H_{28}S_2$			
8,12-Dithiatricyclo[12.2.2.2^{3,6}]eicosa-3,5,14,16,17,19-hexaene, 5,10,15,20-tetramethyl-	MeOH	248(2.83)	44-2509-76
$C_{22}H_{29}ClO_4$			
Androsta-1,4-dien-3-one, 7α-chloro-6β-hydroxy-17β-(1-oxopropoxy)-	EtOH	246(4.20)	56-0097-76
Androst-4-en-3-one, 4-chloro-6α,7α-epoxy-17β-(1-oxopropoxy)-	EtOH	256(4.36)	56-1919-76
6β,17β-	EtOH	262(4.16)	56-1919-76
Pregn-4-en-3-one, 17,20-[carbonylbis(oxy)]-20α-chloro-	n.s.g.	240(4.21)	44-0078-76
Pregn-4-en-3-one, 17,20-[carbonylbis(oxy)]-20β-chloro-	n.s.g.	240(4.19)	44-0078-76
$C_{22}H_{29}ClO_5$			
Pregn-4-en-3-one, 17,20α-[carbonylbis(oxy)]-20β-chloro-21-hydroxy-	n.s.g.	241(4.21)	44-0078-76
$C_{22}H_{29}NO_2$			
2-Azaestra-1,3,5(10)-trien-17-one, 3-(cyclopentyloxy)-	MeOH	279(3.57)	44-2864-76
$C_{22}H_{29}NO_2S$			
2H-Quinolizin-4-ol, 6-(3-furanyl)octahydro-3,9-dimethyl-3-[(4-methylphenyl)thio]-	EtOH	242(3.36),248(3.40), 254(3.47),262(3.46), 268s(3.31)	44-2514-76
	EtOH-HClO₄	237s(3.67),242s(3.65), 247s(3.59),254s(3.51), 267s(3.41),275s(3.45)	44-2514-76
$C_{22}H_{29}NO_3S$			
Pregn-4-ene-3,20-dione, 16β-hydroxy-17α-thiocyanato-	MeOH	240(4.23)	24-0185-76
Pregn-4-ene-3,20-dione, 16β-hydroxy-17β-thiocyanato-	MeOH	239(4.24)	24-0185-76
$C_{22}H_{29}NO_6$			
Benzamide, N-[1-[(3,4-dimethoxyphenoxy)methyl]-2-methylpropyl]-3,4-dimethoxy-	n.s.g.	220(--),255(4.09), 285(3.96)	20-0787-76
$C_{22}H_{29}N_3O$			
Benzamide, N-[2,3-bis(hexahydro-1H-azepin-1-yl)-2-cyclopropen-1-ylidene]-	MeCN	234(3.94),339(3.58)	138-1215-76
$C_{22}H_{30}$			
1H-Indene, octahydro-1,3a-dimethyl-4-methylene-5-(1-methylethylidene)-1-(phenylmethyl)-, (1α,3aβ,7aα)-(+)-	C_6H_{12}	220(3.66)	35-5597-76
$C_{22}H_{30}Cl_2O_4$			
Androst-4-en-3-one, 4,6α-dichloro-7β-hydroxy-17β-(1-oxopropoxy)-	EtOH	266(4.05)	56-1919-76
Androst-4-en-3-one, 4,6β-dichloro-7α-hydroxy-17β-(1-oxopropoxy)-	EtOH	262(4.03)	56-1919-76

Compound	Solvent	$\lambda_{max}(\log \epsilon)$	Ref.
Androst-4-en-3-one, 4,7α-dichloro-6β-hydroxy-17β-(1-oxopropoxy)-	EtOH	253(4.12)	56-1919-76
$C_{22}H_{30}N_2O_2$ Propane, 1-(3,4-dimethylphenyl)-2-nitroso-, dimer	EtOH	294(3.89)	78-1267-76
$C_{22}H_{30}N_4S_2$ Benzenecarboximidamide, 2,2'-dithiobis-[N,N'-diethyl-	heptane MeOH MeOH-HCl CHCl$_3$	none 236(4.43),275(4.07), 328(3.95),340s(--) none 287(4.04)	24-0659-76 24-0659-76 24-0659-76 24-0659-76
$C_{22}H_{30}O$ Methanone, [octahydro-4,4,7a-trimethyl-5-(1-methylethylidene)-1H-inden-1-yl]phenyl-, (1α,3aα,7aβ)-(+)-	EtOH	244.5(4.06)	35-5597-76
$C_{22}H_{30}O_2$ 18,19-Dinor-17α-pregn-4-en-20-yn-3-one, 13-ethyl-17-hydroxy-11β-methyl-, (+)-	MeOH	240.5(4.22)	44-0531-76
18,19-Dinor-17α-pregn-5-en-20-yn-3-one, 13-ethyl-17-hydroxy-11β-methyl-, (+)-	MeOH	288(2.05)	44-0531-76
D-Homo-C,18-dinorpregna-4,13(17a)-diene-20-carboxaldehyde, 17a-methyl-3-oxo-, (17α)-	EtOH	238(4.02)	18-0499-76
Pregn-4-en-20-yn-3-one, 17-hydroxy-7-methyl-, (7α,17α)-	n.s.g.	243(4.20)	13-0759-76A
$C_{22}H_{30}O_4$ Androst-4-en-3-one, 6β,7β-epoxy-17β-(1-oxopropoxy)-	EtOH	245(4.23)	56-1919-76
5α-Androst-8-ene-7,11,17-trione, 3β-hydroxy-4,4,14α-trimethyl-	dioxan	268(3.93)	5-1206-76
Carnosol, dimethyl ether, (+)-	EtOH	272(2.97)	44-1005-76
Juferin	n.s.g.	214(4.3),256(4.5)	105-0491-76
17-Norabieta-8,11,13-trien-7-one, 10-(ethoxycarbonyl)-12-hydroxy-	EtOH EtOH-base	238(4.30),290(4.14) 257(4.12),359(4.48)	44-1005-76 44-1005-76
Pregna-4,6-diene-3,20-dione, 17β-hydroxy-16α-methoxy-	MeOH	282(4.21)	102-1745-76
$C_{22}H_{30}O_5$ Estr-5(10)-en-6-one, 3β,17β-diacetoxy-	MeOH	245(4.04)	44-0313-76
17-Norabieta-8,11,13-trien-7-one, 10-(ethoxycarbonyl)-11,12-dihydroxy-	EtOH EtOH-base	237(4.24),290(4.05) 262(4.03),366(4.39)	44-1005-76 44-1005-76
$C_{22}H_{30}O_6$ 5H-Cyclopropa[3,4]benz[1,2-e]azulen-5-one, 9a-acetoxy-1,1a,1b,4,4a,7a,7b,8-9,9a-decahydro-4a,7b-dihydroxy-3-(hydroxymethyl)-1,1,6,8-tetramethyl-(prostratin)	EtOH	201(4.05),235(3.71), 260s(3.53)	88-1737-76
11-Oxapregn-4-ene-3,20-dione, 21-acetoxy-17-hydroxy-	EtOH	238(4.15)	13-0717-76A
$C_{22}H_{30}O_7$ Rastronol G	EtOH	235(3.93)	33-0772-76

Compound	Solvent	$\lambda_{max}(\log \epsilon)$	Ref.
$C_{22}H_{30}O_8$			
Budlein A, hexahydro-, acetate	EtOH	258(4.09)	102-0525-76
2,4,8,10-Dodecatetraene-5,6,7,8-tetra-carboxylic acid, 2,11-dimethyl-, tetramethyl ester	EtOH	279(3.96)	35-1204-76
$C_{22}H_{30}O_9$			
Dethiodihydrobreynogenin	MeOH	259(4.01)	94-0114-76
	MeOH-NaOH	300(4.17)	94-0114-76
$C_{22}H_{31}ClO_4$			
Androst-4-en-3-one, 7α-chloro-6β-hydr-oxy-17β-(1-oxopropoxy)-	EtOH	237(4.16)	56-1919-76
$C_{22}H_{31}N_6O_9P$			
Adenosine, 8-amino-$N^6,N^6,O^{2'}$-tributyr-yl-, cyclic 3',5'-(hydrogen phosphate)	pH 7	255(3.86),293(4.10)	87-0899-76
$C_{22}H_{32}Br_2O_3$			
2(1H)-Phenanthrenone, 5-acetoxy-8-bromo-10a-(bromomethyl)-4a,4b,5,6,7,8,8a,9-10,10a-decahydro-5,8a-dimethyl-1-(1-methylethyl)-	n.s.g.	234(3.83)	32-0779-76
4(1H)-Phenanthrenone, 5-acetoxy-8-bromo-10a-(bromomethyl)-4a,4b,5,6,7,8,8a,9-10,10a-decahydro-5,8a-dimethyl-1-(1-methylethyl)-	n.s.g.	232(3.67)	32-0779-76
$C_{22}H_{32}N_2O$			
5H-Cyclododec[b]indole-5-propanamine, 6,7,8,9,10,11,12,13,14,15-decahydro-α-methyl-γ-oxo-, hydrochloride	EtOH	247(4.21),295(3.74), 302(3.74)	44-3775-76
$C_{22}H_{32}N_6O_4S$			
L-Proline, 4-[[(4-amino-2-methyl-5-pyri-midinyl)methyl]formylamino]-3-[(2-pyr-rolidinylcarbonyl)thio]-3-pentenyl ester, trihydrochloride, (S)-	EtOH	246(4.09)	94-0852-76
$C_{22}H_{32}O_2$			
Androst-4-en-3-one, 17-hydroxy-7-(2-pro-penyl)-, (7α,17β)-	n.s.g.	243(4.19)	13-0759-76A
2,5,7-Nonatrien-1-ol, 3,7-dimethyl-9-(2,6,6-trimethyl-2-cyclohexen-1-ylidene)-, acetate	EtOH	316(4.42)	33-0387-76
$C_{22}H_{32}O_2S$			
2H-Thieno[2,3-c][1]benzopyran-9-ol, 7-(1,2-dimethylheptyl)-1,4-dihydro-4,4-dimethyl-	EtOH	320(3.951)	87-0549-76
3H-Thieno[3,4-c][1]benzopyran-9-ol, 7-(1,2-dimethylheptyl)-1,4-dihydro-4,4-dimethyl-	EtOH	284(4.157)	87-0549-76
$C_{22}H_{32}O_3$			
Cyclohexanol, 1-(9-acetoxy-3,7-dimethyl-nona-3,5,7-trien-1-ynyl)-2,2,6-tri-methyl-	isoPrOH	292s(4.43),303(4.54), 316(4.45)	33-0567-76
2,4,6-Nonatrienoic acid, 3,7-dimethyl-8-oxo-9-(2,6,6-trimethylcyclohex-1-enyl)-, ethyl ester	isoPrOH	319(4.60)	33-0567-76

Compound	Solvent	$\lambda_{max}(\log \epsilon)$	Ref.
3,5,7-Nonatrien-2-one, 9-acetoxy-3,7-di-methyl-1-(2,2,6-trimethyl-1-cyclohex-en-1-yl)-, (E)-	isoPrOH	313(4.42)	33-0567-76
(Z)-	isoPrOH	310(4.36)	33-0567-76
2,4,6-Nonatrien-8-ynoic acid, 9-(1-hy-droxy-2,2,6-trimethylcyclohexyl)-	isoPrOH	332(4.61)	33-0567-76
17-Norabieta-8,11,13-triene-10-carbox-ylic acid, 12-hydroxy-, ethyl ester	EtOH	286(3.54)	44-1005-76
	base	305(3.66)	44-1005-76
4a(2H)-Phenanthrenecarboxylic acid, 1,3,4,9,10,10a-hexahydro-6-methoxy-1,1-dimethyl-7-(1-methylethyl)-, methyl ester, trans-($\pm$)-	MeOH	230(3.76),280(3.38), 287(3.37)	18-1985-76
2H-Pyran-2-one, 5-[3-[1-ethyl-2-(4-oxo-cyclohexyl)butyl]cyclopentyl]-, threo	EtOH	297(3.71)	5-1222-76
$C_{22}H_{32}O_4$			
Cannabitriol, methyl ether	n.s.g.	227(4.44),277(4.17)	31-0283-76
17-Nor-5α,8β,9β-abiet-13-ene-7,12-di-one, 10-(ethoxycarbonyl)-	EtOH	232(3.95)	44-1005-76
	base	255(3.88),440(4.30)	44-1005-76
4a(2H)-Phenanthrenecarboxylic acid, 1,3,4,9,10,10a-hexahydro-5,6-dimeth-oxy-1,1-dimethyl-7-(1-methylethyl)-, trans-($\pm$)- (carnosic acid dimethyl ether)	EtOH	276(2.85)	44-1005-76
$C_{22}H_{32}O_4S$			
Propanoic acid, 3-[[(15β,17β)-17-hydr-oxy-3-oxoandrost-4-en-15-yl]thio]-	MeOH	241(4.22)	13-0101-76B
$C_{22}H_{32}O_4S_4$			
10,23-Dioxa-1,5,15,19-tetrathiadispiro-[5.7.5.7]hexacosa-7,20-diene-9,22-di-one, 11,24-dimethyl-, (E,E)-	EtOH	212(4.21)	39-1718-76C
$C_{22}H_{32}O_6$			
Rastronol F	EtOH	235(3.93)	33-0772-76
$C_{22}H_{34}$			
Tricyclo[10.10.0.0^{3,14}]docosa-1,3(14),12-triene, (S)-	isooctane	255(3.80),291s(2.54)	44-4081-76
$C_{22}H_{34}Br_2O_4$			
Benzoic acid, 3,5-dibromo-2,4-dihydroxy-6-tridecyl-, ethyl ester	MeOH	225(4.45),269(3.86), 317(3.65)	12-1989-76
$C_{22}H_{34}N_2O_3$			
Carbamic acid, (3-butoxyphenyl)-, 3-(1-piperidinyl)cyclohexyl ester	H_2O	236(4.02),277(3.33)	106-0024-76
Carbamic acid, (4-butoxyphenyl)-, 3-(1-piperidinyl)cyclohexyl ester	H_2O	239(4.19),282(3.16)	106-0024-76
Carbamic acid, [3-(pentyloxy)phenyl]-, 3-(1-pyrrolidinyl)cyclohexyl ester	H_2O	236(4.06),277(3.39)	106-0024-76
Carbamic acid, [4-(pentyloxy)phenyl]-, 3-(1-pyrrolidinyl)cyclohexyl ester	H_2O	239(4.21),281(3.18)	106-0024-76
$C_{22}H_{34}O_2$			
Androst-4-en-3-one, 17-hydroxy-7β-(1-methylethyl)-	n.s.g.	245(4.18)	13-0759-76A
Androst-4-en-3-one, 17-hydroxy-7α-propyl-	n.s.g.	243(4.18)	13-0759-76A
7β-	n.s.g.	243(4.23)	13-0759-76A

Compound	Solvent	$\lambda_{max}(\log \epsilon)$	Ref.
$C_{22}H_{34}O_3$			
2H-Pyran-2-one, 5-[3-[1-ethyl-2-(4-hy-droxycyclohexyl)butyl]cyclopentyl]-, axial hydroxyl	EtOH	299(3.72)	5-1222-76
equatorial hydroxyl	EtOH	299(3.75)	5-1222-76
$C_{22}H_{34}O_5$			
Vitexilactone	EtOH	210(4.38)	94-1668-76
$C_{22}H_{34}O_5S_4$			
2-Propenoic acid, 3-[2-(3-hydroxybutyl)-1,3-dithian-2-yl]-, 3-[2-(2-carboxy-ethenyl)-1,3-dithian-2-yl]-1-methyl-propyl ester, [R*,R*-(E,E)]-(±)-	EtOH	211(4.20)	39-1718-76C
$C_{22}H_{34}O_8$			
1,3-Heptadiene-1,2,6,6-tetracarboxylic acid, 5-methyl-, 6,6-bis(1,1-dimethyl-ethyl) 1,2-dimethyl ester, (E,E)-	EtOH	269(4.01)	35-1204-76
Leuconolide A_3, 5,18-hemiacetal	EtOH	232(4.41)	94-1749-76
Leuconolide A_3, 9-dehydro-18-dihydro-	EtOH	279(4.00)	94-1749-76
$C_{22}H_{35}B$			
Borane, bis[2-(1-cyclohexen-1-yl)ethen-yl](1,1,2-trimethylpropyl)-, (E,E)-	THF-ether	278(4.40)	101-0303-76N
$C_{22}H_{35}OS$			
1,3-Benzoxathiol-1-ium, 2-pentadecyl-, tetrafluoroborate	96% H_2SO_4	244s(3.47),248(3.54), 252(3.45),299(3.85)	39-0323-76C
$C_{22}H_{36}N_2O_3$			
Carbamic acid, [3-(pentyloxy)phenyl]-, 3-(diethylamino)cyclohexyl ester	H_2O	236(4.03),277(3.35)	106-0024-76
Carbamic acid, [4-(pentyloxy)phenyl]-, 3-(diethylamino)cyclohexyl ester	H_2O	239(4.23),281(3.20)	106-0024-76
$C_{22}H_{36}N_5O_6PS$			
Adenosine, 8-(dodecylthio)-, cyclic 3',5'-(hydrogen phosphate)	pH 2.0	285(4.25)	69-1408-76
	pH 13.0	282(--)	69-1408-76
$C_{22}H_{36}N_6O_4S$			
L-Valine, 3-[(2-amino-3-methyl-1-oxo-butyl)thio]-4-[[(4-amino-2-methyl-5-pyrimidinyl)methyl]formylamino]-3-pentenyl ester, trihydrochloride, (S)-	EtOH	246(4.10)	94-0852-76
$C_{22}H_{36}O_3$			
Cyclopentaneacetaldehyde, 3-[2-(1,4-di-oxaspiro[4.5]dec-8-yl)-1-ethylbutyl]-α-methylene-	EtOH	222(3.96)	5-1222-76
$C_{22}H_{36}O_4$			
Benzoic acid, 2,4-dihydroxy-6-tridecyl-, ethyl ester	MeOH	266(4.12),302(3.73)	12-1989-76
2(5H)-Furanone, 4-[4-[(1R,2R)-1-ethyl-2-(4-hydroxycyclohexyl)butyl]-1-hydroxycyclohexyl]-	n.s.g.	210(4.24)	24-3318-76
2(5H)-Furanone, 4-[4-[(1R,2R)-1-ethyl-2-(4-hydroxycyclohexyl)butyl]-4-hydroxycyclohexyl]-	n.s.g.	210(4.23)	24-3318-76

Compound	Solvent	$\lambda_{max}(\log \epsilon)$	Ref.
5β-Pregn-20-en-22-al, 3β,14β-dihydroxy-, 17β-trans	EtOH	232(4.32)	87-1330-76
$C_{22}H_{36}O_5$			
2(3H)-Furanone, 4-[2-(4-acetoxydecahydro-1-hydroxy-2,5,5,8a-tetramethyl-1-naphthalenyl)ethyl]dihydro-(dihydrovitexilactone)	EtOH	205(3.46)	94-1668-76
2(5H)-Furanone, 4-[4-[(1R,2R)-1-ethyl-2-(4-hydroxycyclohexyl)butyl]-1,4-dihydroxycyclohexyl]-	n.s.g.	210(4.14)	24-3318-76
$C_{22}H_{36}Si_4$			
Silane, [3,4-bis(trimethylsilyl)ethynyl]-3-hexene-2,5-diyne-1,6-diyl]bis[trimethyl-	EtOH	245(4.00),250(4.00), 258(4.21),273s(3.66), 318s(4.08),327s(4.18), 337(4.43),362(4.54)	78-1293-76
Silane, [1-[4-(trimethylsilyl)-1,3-butadiynyl]-2-[(trimethylsilyl)ethynyl]-1-buten-3-yne-1,4-diyl]bis[trimethyl-	EtOH	213(4.11),223(4.23), 259s(3.81),268(3.95), 283(3.91),312s(4.11), 325(4.34),334(4.43), 349(4.55)	78-1293-76
$C_{22}H_{37}ClO_2S$			
2-Thiophenecarboxylic acid, 4-(chloromethyl)-5-pentadecyl-, methyl ester	heptane	256(3.93),278(3.98)	34-0233-76
$C_{22}H_{37}N_6O_6P$			
Adenosine, N-(12-aminododecyl)-, cyclic 2',5'-(hydrogen phosphate)	pH 1 pH 11	263(4.21) 268(4.21)	69-3724-76 69-3724-76
Adenosine, 8-(dodecylamino)-, cyclic 3',5'-(hydrogen phosphate)	pH 2.0 pH 13.0	277(4.05) 277(--)	69-1408-76 69-1408-76
$C_{22}H_{38}N_2$			
Cycloheptanamine, N,N'-(2,2,4,4-tetramethyl-1,3-cyclobutanediylidene)bis-	C_6H_{12} EtOH	260(2.15),274(2.16) 247(--),268(--)	59-1415-76 59-1415-76
$C_{22}H_{38}O_2S$			
2-Thiophenecarboxylic acid, 5-heptadecyl-	heptane	254(4.27),275(4.00)	34-0233-76
2-Thiophenecarboxylic acid, 4-methyl-5-pentadecyl-, methyl ester	heptane	254(3.94),283(3.96)	34-0233-76
$C_{22}H_{38}O_4$			
Cyclohexanecarboxylic acid, 2,4-dioxo-6-tridecyl-, ethyl ester	MeOH	257(4.17),283(3.73)	12-1989-76
Prost-13-ene-9,15-dione, 1-hydroxy-, cyclic 9-(1,2-ethanediyl acetal)-, (13E)-(±)-	EtOH	230(4.19)	39-2550-76C
$C_{22}H_{39}N_3O$			
Morpholine, 4-[2,6-bis(piperidinomethyl)-1-cyclohexen-1-yl]-	hexane	243s(4.07)	94-2494-76
$C_{22}H_{40}N_2O$			
Imidazole, 2-(4-methylstearoyl)-	THF	278(4.13)	33-2753-76
Imidazole, 4(or 5)-(4-methylstearoyl)-	THF	259(4.14)	33-2753-76
$C_{22}H_{40}S$			
Thiophene, 2-heptadecyl-3-methyl-	heptane	236(3.79)	34-0233-76

Compound	Solvent	$\lambda_{max}(\log \epsilon)$	Ref.

$C_{22}H_{41}NO$
  2,4-Octadecadienamide, N-(2-methylprop-
  yl)-, (E,E)-
	MeOH	260(4.47)	100-0060-76

$C_{22}H_{43}B$
  Borane, di-1-octenyl(1,1,2-trimethyl-
  propyl)-
	ether	226(4.28)	101-0303-76N

$C_{22}H_{46}Si_4$
  Silane, 3,4-hexadien-1-yne-1,3,5,6-
  tetrayltetrakis[ethyldimethyl-
	C_6H_{12}	235s(4.06),243(4.11),
		  255(4.09)
			35-8426-76

Compound	Solvent	$\lambda_{max}(\log \epsilon)$	Ref.
$C_{23}H_{12}N_4O_3$ 9H-Benzo[e]quino[2,3-i]perimidine-8-car-boxamide, 14,15-dihydro-9,15-dioxo-	DMF	270(4.123),365(3.820), 494(3.608),526(3.651)	80-0257-76
$C_{23}H_{12}N_4O_5$ 9H-Benzo[e]quino[2,3-i]perimidine-9,15-(14H)-dione, 8-methoxy-12-nitro-	DMF	270(4.146),366(3.643), 526(3.204)	80-0257-76
$C_{23}H_{13}NS_2$ Bis[1]benzothieno[3,2-b:2',3'-e]pyri-dine, 6-phenyl-	$C_2H_4Cl_2$	243(4.38),254(4.45), 285(4.88),336(3.75), 350(3.78)	103-0977-76
$C_{23}H_{13}N_3O_3$ 9H-Benzo[e]quino[2,3-i]perimidine-9,15(14H)-dione, 8-methoxy-	DMF	270(4.265),362(3.924), 490(3.832),522(3.934)	80-0257-76
$C_{23}H_{13}OS_2$ Bis[1]benzothieno[3,2-b:2',3'-e]pyryl-ium, 6-phenyl-, perchlorate	$C_2H_4Cl_2$	256(4.49),306(4.30), 377(4.20),424(4.23), 480(4.23)	103-0977-76
$C_{23}H_{14}ClNO$ Benzonitrile, 4-[2-[4-(2-benzofuranyl)-phenyl]ethenyl]-3-chloro-	DMF	369(4.69)	33-0819-76
Benzonitrile, 3-chloro-4-[2-(2-phenyl-6-benzofuranyl)ethenyl]-	DMF	372(4.68)	33-0819-76
$C_{23}H_{14}Cl_2N_2O_3S$ 7-Thia-2,6-diazabicyclo[2.2.1]heptane-3,5-dione, 2-(4-chlorobenzoyl)-1-(4-chlorophenyl)-6-phenyl-	MeOH	232(4.32),254(4.32), 399(4.16)	44-0813-76
$C_{23}H_{14}Cl_2O$ 2-Cyclobuten-1-one, 4-[(2,6-dichloro-phenyl)methylene]-2,3-diphenyl-, (Z)-	$CHCl_3$	292(4.42),352s(3.94)	24-1506-76
$C_{23}H_{14}N_2$ Phenanthro[9',10':4,5]imidazo[2,1-a]iso-quinoline	MeOH	252s(4.56),260(4.75), 281(4.91),318(4.15), 345(4.13),362(3.81)	39-2038-76C
$C_{23}H_{14}N_2O_2S$ Benzenamine, 4-nitro-N-(2-phenyl-5H-naphtho[1,8-bc]thien-5-ylidene)-	MeOH	296(4.19),417(4.30)	88-4713-76
$C_{23}H_{15}ClN_2$ 1H-Pyrido[2,3-b]indole, 2-chloro-1,4-diphenyl-	n.s.g.	204(4.57),228(4.33), 277(4.36),284(4.32), 346(4.05)	103-0329-76
$C_{23}H_{15}ClN_2O_3S$ 7-Thia-2,6-diazabicyclo[2.2.1]heptane-3,5-dione, 2-benzoyl-1-(4-chloro-phenyl)-6-phenyl-	MeOH	225(4.33),245s(4.25), 399(4.04)	44-0813-76
$C_{23}H_{15}ClN_4$ Benzonitrile, 2-chloro-4-[2-[4-(2-phen-yl-2H-1,2,3-triazol-4-yl)phenyl]ethen-yl]-	DMF	351(4.72)	33-0819-76

Compound	Solvent	$\lambda_{max}(\log \epsilon)$	Ref.
$C_{23}H_{15}ClN_4O$			
Benzoxazole, 2-[4-[2-[2-(4-chlorophenyl)-2H-1,2,3-triazol-4-yl]ethenyl]-phenyl-	DMF	352(4.85)	33-2469-76
Benzoxazole, 5-chloro-2-[4-[2-(2-phenyl-2H-1,2,3-triazol-4-yl)ethenyl]phenyl]-	DMF	352(4.82)	33-2469-76
Benzoxazole, 5-chloro-2-[2-[4-(2-phenyl-2H-1,2,3-triazol-4-yl)phenyl]ethenyl]-	DMF	359(4.79)	33-2469-76
Benzoxazole, 5-chloro-2-[2-[4-(4-phenyl-2H-1,2,3-triazol-2-yl)phenyl]ethenyl]-	DMF	362(4.79)	33-2469-76
$C_{23}H_{15}ClO$			
2-Cyclobuten-1-one, 4-[(4-chlorophenyl)-methylene]-2,3-diphenyl-, (Z)-	CHCl$_3$	310(4.50),330(4.58), 350s(4.4),380s(3.84)	24-1506-76
$C_{23}H_{15}ClO_2$			
2H-Pyran-2-one, 6-(3-chlorophenyl)-3,4-diphenyl-	EtOH	247s(4.15),255(4.17), 354(4.12)	4-0195-76
2H-Pyran-2-one, 6-(4-chlorophenyl)-3,4-diphenyl-	EtOH	238s(4.23),260(4.39), 358(4.23)	4-0195-76
$C_{23}H_{15}ClS_2$			
2H-Thiopyran-2-thione, 6-(4-chlorophenyl)-3,4-diphenyl-	MeOH	250(4.45),312(4.23), 338s(4.07),477(3.90)	44-0818-76
$C_{23}H_{15}NO$			
Benzonitrile, 2-[2-[4-(2-benzofuranyl)-phenyl]ethenyl]-	DMF	363(4.72)	33-0819-76
Benzonitrile, 4-[2-[4-(2-benzofuranyl)-phenyl]ethenyl]-	DMF	368(4.79)	33-0819-76
Benzonitrile, 2-[2-(2-phenyl-6-benzofuranyl)ethenyl]-	DMF	358(4.70)	33-0819-76
Benzonitrile, 4-[2-(2-phenyl-6-benzofuranyl)ethenyl]-	DMF	365(4.76)	33-0819-76
$C_{23}H_{15}NOS$			
Benzoxazole, 2-(5-[1,1'-biphenyl]-4-yl-2-thienyl)-	CH$_2$Cl$_2$	251s(4.08),263s(4.03), 294s(3.80),361(4.58)	48-0731-76
$C_{23}H_{15}NS$			
Benzenamine, N-(2-phenyl-5H-naphtho-[1,8-bc]thien-5-ylidene)-	MeOH	385(4.33)	88-4713-76
$C_{23}H_{15}NS_2$			
Thieno[3,4-c]isothiazole-5-S^{IV}, 3,4,6-triphenyl-	CH$_2$Cl$_2$	242(4.23),280(4.34), 529(4.13)	88-2163-76
$C_{23}H_{15}N_3OS$			
2-Pyridinecarboxylic acid, (2-phenyl-5H-naphtho[1,8-bc]thien-5-ylidene)-hydrazide	MeOH	296(4.21),420(4.52)	88-4713-76
3-Pyridinecarboxylic acid, (2-phenyl-5H-naphtho[1,8-bc]thien-5-ylidene)-hydrazide	MeOH	290(4.12),491(4.40)	88-4713-76
$C_{23}H_{15}N_3O_2S$			
5H-Naphtho[1,8-bc]thiophen-5-one, 2-phenyl-, (4-nitrophenyl)hydrazone	MeOH	300(4.20),489(4.70)	88-4713-76

Compound	Solvent	$\lambda_{max}(\log \epsilon)$	Ref.
$C_{23}H_{15}N_3O_4S_2$ Benzenesulfonic acid, 3-nitro-, (2-phen-yl-5H-naphtho[1,8-bc]thien-5-ylidene)-hydrazide	MeOH	292(4.11),398(4.27)	88-4713-76
$C_{23}H_{15}N_5O_5S_2$ Benzenesulfonamide, 4-[2-(4-nitrophen-yl)-4-oxo-3(4H)-quinazolinyl]-N-2-thiazolyl-	EtOH-NaOH	230(4.46),280(4.25), 335s(4.00)	106-0574-76
$C_{23}H_{15}OS$ [1]Benzothieno[3,2-b]pyrylium, 2,4-di-phenyl-, perchlorate	$C_2H_4Cl_2$	255(3.83),280(3.83), 314(4.30),370(4.82), 421(4.90)	103-0977-76
$C_{23}H_{16}$ Dibenzo[b,g]phenanthrene, 5-methyl-	C_6H_{12}	228(5.07),242(4.81), 247(4.99),254(4.56), 288(4.79),298(5.09), 311(5.18),390(4.05)	2-0939-76
$C_{23}H_{16}ClN$ Benzonitrile, 2-chloro-6-[2-[4-(2-phen-ylethenyl)phenyl]ethenyl]-	DMF	368(4.68)	33-0819-76
Benzonitrile, 3-chloro-4-[2-[4-(2-phen-ylethenyl)phenyl]ethenyl]-	DMF	369(4.74)	33-0819-76
$C_{23}H_{16}ClNO$ 2(1H)-Pyridinone, 6-(3-chlorophenyl)-3,4-diphenyl-	EtOH	244(4.39),296s(4.04), 333(4.21)	4-0195-76
$C_{23}H_{16}ClN_2O_2S$ Benzothiazolium, 3-[(1-amino-4-chloro-9,10-dihydro-9,10-dioxo-2-anthracen-yl)methyl]-2-methyl-, chloride	MeOH	489(3.90)	2-0054-76
$C_{23}H_{16}ClN_2O_3$ Benzoxazolium, 3-[(1-amino-4-chloro-9,10-dihydro-2-anthracenyl)methyl]-2-methyl-, chloride	MeOH	490(3.84)	2-0054-76
$C_{23}H_{16}ClN_3$ 3H-Imidazo[4,5-b]quinoline, 7-chloro-2-methyl-3,9-diphenyl-	DMF	330(4.17),351(4.01)	48-0039-76
$C_{23}H_{16}Cl_2N_6$ 2H-Benzotriazole, 2-[3-chloro-4-[2-[2-(4-chlorophenyl)-5-methyl-2H-1,2,3-triazol-4-yl]ethenyl]phenyl]-	DMF	367(4.75)	33-2469-76
$C_{23}H_{16}N$ Pyridinium, benz[a]anthracen-7-yl-, perchlorate	MeOH	274(4.68),281(4.74), 292(4.73),303(4.23), 323(3.74),337(3.87), 353(3.92),369(3.81), 389(3.47)	44-2679-76
$C_{23}H_{16}NO_4$ Benzoxazolium, 3-[(9,10-dihydro-1-hydr-oxy-9,10-dioxo-2-anthracenyl)methyl]-2-methyl-, chloride	MeOH	409(3.83)	2-0054-76

Compound	Solvent	$\lambda_{max}(\log \epsilon)$	Ref.
$C_{23}H_{16}NO_5$			
Benzoxazolium, 3-[(9,10-dihydro-1,4-di-hydroxy-9,10-dioxo-2-anthracenyl)-methyl]-2-methyl-, chloride	MeOH	488(3.80)	2-0054-76
$C_{23}H_{16}N_2$			
Imidazo[2,1-a]isoquinoline, 2,3-diphen-yl-	MeOH	266(4.78),310s(4.12)	39-2038-76C
1H-Phenanthro[9,10-d]imidazole, 1-(2-phenylethenyl)-, (E)-	MeOH	254(4.89),350(3.23)	39-2038-76C
1H-Pyrido[2,3-b]indole, 1,4-diphenyl-	n.s.g.	225(4.38),253(4.36), 273(4.35),288(4.17), 329(4.14)	103-0329-76
$C_{23}H_{16}N_2O_2S_2$			
Benzenesulfonic acid, (2-phenyl-5H-naph-tho[1,8-bc]thien-5-ylidene)hydrazide	MeOH	294(3.92),394(4.02)	88-4713-76
$C_{23}H_{16}N_2O_3S$			
2,4,6(1H,3H,5H)-Pyrimidinetrione, 1,3-dimethyl-5-(2-phenyl-5H-naphtho[1,8-bc]thien-5-ylidene)-	MeOH	334(3.93)	88-4713-76
$C_{23}H_{16}N_2O_6$			
Methanone, [4,5-dihydro-5-nitro-4-(3-nitrophenyl)-2-phenyl-3-furanyl]-phenyl-	$C_2H_4Cl_2$	234(4.32),260(4.13), 290(4.04)	104-0640-76
Methanone, [4,5-dihydro-5-nitro-4-(4-nitrophenyl)-2-phenyl-3-furanyl]-phenyl-, trans	$C_2H_4Cl_2$	232(4.30),269(4.38)	104-0640-76
$C_{23}H_{16}N_4$			
Benzonitrile, 3-[2-[4-(2-phenyl-2H-1,2,3-triazol-4-yl)phenyl]ethenyl]-	DMF	345(4.72)	33-0819-76
Benzonitrile, 4-[2-[4-(2-phenyl-2H-1,2,3-triazol-4-yl)phenyl]ethenyl]-	DMF	348(4.80)	33-0819-76
$C_{23}H_{16}N_4O$			
Benzoxazole, 2-[4-[2-(2-phenyl-2H-1,2,3-triazol-4-yl)phenyl]-	DMF	351(4.78)	33-2469-76
Benzoxazole, 2-[2-[4-(2-phenyl-2H-1,2,3-triazol-4-yl)phenyl]ethenyl]-	DMF	347(4.77),358(4.77)	33-2469-76
Benzoxazole, 2-[2-[4-(4-phenyl-2H-1,2,3-triazol-2-yl)phenyl]ethenyl]-	DMF	360(4.79)	33-2469-76
$C_{23}H_{16}O$			
Indeno[2,1-b]pyran, 4-methyl-2-(2-naph-thalenyl)-	dioxan	252(4.48),289(4.36), 333(4.19),367(4.33), 512(3.55)	78-2219-76
$C_{23}H_{16}OS$			
Indeno[2,1-b]pyran, 4-methyl-9-phenyl-2-(2-thienyl)-	dioxan	233(4.36),280(4.40), 367(4.29),551(3.52)	78-2219-76
$C_{23}H_{16}O_2$			
9-Anthraceneacetaldehyde, α-benzoyl-	pentane	220(4.37),254(5.2), 320(4.2),332(4.17), 349(4.1),368(4.12), 388(4.1)	44-2784-76
2H-Pyran-2-one, 3,4,6-triphenyl-	EtOH	243s(4.16),255(4.29), 356(4.21)	4-0195-76

Compound	Solvent	λ_{max}(log ϵ)	Ref.
Pyrylium, 3-hydroxy-2,4,6-triphenyl-, hydroxide, inner salt	EtOH	230(4.17),270(4.14), 309(4.6),490(4.47)	4-0237-76
$C_{23}H_{16}S_2$ 2H-Thiopyran-2-thione, 3,4,6-triphenyl-	MeOH	245(4.63),308(4.44), 330s(4.30),480(4.14)	44-0818-76
$C_{23}H_{17}BrN_2OSe$ Methanone, (4-bromophenyl)[2-(methyl-phenylamino)-4-phenyl-5-selenazolyl]-	MeCN	237(4.29),271(4.24), 387(4.11)	97-0018B-76
$C_{23}H_{17}ClN_2O$ 2(1H)-Pyridinone, 1-amino-6-(3-chloro-phenyl)-3,4-diphenyl-	EtOH	238s(4.30),258s(4.20), 288(3.85),340(4.12)	4-0195-76
2(1H)-Pyridinone, 1-amino-6-(4-chloro-phenyl)-3,4-diphenyl-	EtOH	238s(4.35),261s(4.22), 290(3.80),340(4.09)	4-0195-76
4(3H)-Quinazolinone, 2-[2-(2-chloro-phenyl)ethenyl]-3-(phenylmethyl)-	EtOH	325(4.28)	115-0341-76
$C_{23}H_{17}ClN_2OSe$ Methanone, (4-chlorophenyl)[2-(methyl-phenylamino)-4-phenyl-5-selenazolyl]-	MeCN	237(4.29),256(4.26), 385(4.08)	97-0018B-76
$C_{23}H_{17}ClN_2O_4S_2$ 7-Thia-2,6-diazabicyclo[2.2.1]heptane-3,5-dione, 1-(4-chlorophenyl)-2-[(4-methylphenyl)sulfonyl]-6-phenyl-	MeOH	223(3.82),259(3.79), 395(3.75)	44-0813-76
$C_{23}H_{17}ClN_3$ 1,2,4-Triazolo[3,4-a]isoquinolinium, 3-(4-chlorophenyl)-1-(4-methylphen-yl)-, tetrafluoroborate	MeOH	226(4.43),253(4.57), 304(3.74),318(3.64), 331(3.66)	48-0881-76
1,2,4-Triazolo[4,3-a]quinolinium, 3-(4-chlorophenyl)-1-(4-methyl-phenyl)-, tetrafluoroborate	MeOH	250(4.35),276(4.23), 303(4.16),317(4.12), 326(4.14)	48-0881-76
$C_{23}H_{17}ClN_3O$ 1,2,4-Triazolo[3,4-a]isoquinolinium, 1-(4-chlorophenyl)-3-(4-methoxy-phenyl)-, tetrafluoroborate	MeOH	228(4.34),263(4.51), 319(3.70),332(3.71)	48-0881-76
1,2,4-Triazolo[4,3-a]quinolinium, 1-(4-chlorophenyl)-3-(4-methoxy-phenyl)-, tetrafluoroborate	MeOH	247(4.33),304(4.12), 319(4.09),323(4.10)	48-0881-76
$C_{23}H_{17}ClN_6$ 2H-Benzotriazole, 2-[3-chloro-4-[2-(5-methyl-2-phenyl-2H-1,2,3-triazol-4-yl)ethenyl]phenyl]-	DMF	361(4.73)	33-2469-76
$C_{23}H_{17}ClN_6O$ 2H-Benzotriazole, 2-[3-chloro-4-[2-[2-(4-methoxyphenyl)-2H-1,2,3-triazol-4-yl]ethenyl]phenyl]-	DMF	360(4.73)	33-2469-76
$C_{23}H_{17}ClO_3$ Benzeneacetic acid, α-[3-(4-chlorophen-yl)-3-oxo-1-phenylpropylidene]-, (E)-	EtOH	253(4.39)	4-0195-76
$C_{23}H_{17}N$ Benzonitrile, 2-[2-[4-(2-phenylethenyl)-phenyl]ethenyl]-	DMF	362(4.76)	33-0819-76

Compound	Solvent	$\lambda_{max}(\log \epsilon)$	Ref.
Benzonitrile, 4-[2-[4-(2-phenylethenyl)-phenyl]ethenyl]-	DMF	368(4.81)	33-0819-76
$C_{23}H_{17}NO$			
Benzenamine, N-(4,6-diphenyl-2H-pyran-2-ylidene)-	EtOH	215(3.79),275(4.12), 395(3.60)	77-0899-76
2(1H)-Pyridinone, 1,4,6-triphenyl-	EtOH	220(4.06),250(4.13), 337(3.68)	77-0899-76
2(1H)-Pyridinone, 3,4,6-triphenyl-	EtOH	252(4.38),345(4.28)	4-0195-76
$C_{23}H_{17}NOS$			
1,2-Thiazepin-3(2H)-one, 2,4,7-triphenyl- (or isomer)	MeOH	265(4.32),314(4.27)	39-2565-76C
$C_{23}H_{17}NO_2$			
Benzamide, 2-(9-anthracenylcarbonyl)-N-methyl-	dioxan	335(3.51),349(3.72), 367(3.87),386(3.83)	104-0168-76
1-Isoquinolinemethanol, α-phenyl-, benzoate	THF	265(3.58),273(3.61), 282(3.53),309(3.28), 321(3.38)	47-1661-76
2,6(1H,3H)-Pyridinedione, 1,4,5-triphenyl-	MeOH	246s(4.19),356(4.07)	44-0818-76
1H-Pyrrole, 3-(benzoyloxy)-2,5-diphenyl-	EtOH	232(4.40),321(4.46)	39-0794-76C
6-Quinolinol, 3-methyl-2-phenyl-, benzoate	EtOH	230(4.42),260(4.51), 308(3.83),315s(3.80), 335s(3.65)	95-0968-76
$C_{23}H_{17}NO_4$			
Methanone, (4,5-dihydro-5-nitro-2,4-diphenyl-3-furanyl)phenyl-, cis	$C_2H_4Cl_2$	233(4.36),263(4.29)	104-0640-76
Methanone, (4,5-dihydro-5-nitro-2,4-diphenyl-3-furanyl)phenyl-, trans	$C_2H_4Cl_2$	233(4.28),258(4.11), 293(3.95)	104-0640-76
$C_{23}H_{17}N_2O_2S$			
Benzothiazolium, 3-[(1-amino-9,10-dihydro-9,10-dioxo-2-anthracenyl)-methyl]-2-methyl-, chloride	MeOH	490(3.97)	2-0054-76
$C_{23}H_{17}N_2O_3$			
Benzoxazolium, 3-[(1-amino-9,10-dihydro-9,10-dioxo-2-anthracenyl)methyl]-2-methyl-, chloride	MeOH	490(3.88)	2-0054-76
$C_{23}H_{17}N_3$			
3H-Imidazo[4,5-b]quinoline, 2-methyl-3,9-diphenyl-	EtOH	249(4.77),326(4.18)	48-0039-76
$C_{23}H_{17}N_3O_2S$			
Benzamide, N-[5-benzoyl-3-(4-methyl-phenyl)-1,3,4-thiadiazol-2(3H)-ylidene]-	EtOH	275(4.27),345(4.30)	4-0045-76
$C_{23}H_{17}N_3O_3$			
4(3H)-Quinazolinone, 2-[2-(3-nitrophenyl)ethenyl]-3-(phenylmethyl)-	EtOH	254(4.39),325(4.26)	115-0341-76
$C_{23}H_{17}N_3O_3S$			
Benzamide, N-[5-benzoyl-3-(4-methoxy-phenyl)-1,3,4-thiadiazol-2(3H)-ylidene]-	EtOH	265(4.24),340(4.17)	4-0045-76

Compound	Solvent	$\lambda_{max}(\log \epsilon)$	Ref.
$C_{23}H_{17}N_3O_5$			
1H-Indene-1,3(2H)-dione, 2-[(4-methoxy-phenyl)methyl]-2-[(2-nitrophenyl)-azo]-	EtOH	230(4.58),288s(4.06)	104-1440-76
1H-Indene-1,3(2H)-dione, 2-[(4-methoxy-phenyl)methyl]-2-[(3-nitrophenyl)-azo]-	EtOH	230(4.55),254(4.40)	104-1440-76
1H-Indene-1,3(2H)-dione, 2-[(4-methoxy-phenyl)methyl]-2-[(4-nitrophenyl)-azo]-	EtOH	228(4.65),288(4.34)	104-1440-76
$C_{23}H_{17}N_4Ni$			
Nickel(1+), (13H-tribenzo[b,f,m]-[1,4,8,12]tetraazacyclopentadecinato-N^5,N^{11},N^{15},N^{21})-, iodide, (SP-4-2)-	EtOH	236s(--),299(4.52), 335(4.48),361s(--), 428(3.92),461s(--), 648(3.30)	12-2271-76
$C_{23}H_{17}N_7O_4$			
1H-Pyrido[3,2-e][1,2,4]triazolo[1,2-a]-pyridazine-1,3(2H)-dione, 6-(3,5-di-oxo-4-phenyl-1,2,4-triazolidin-1-yl)-5,6-dihydro-2-phenyl-	12M HCl	260(4.28),315(3.54)	103-0590-76
$C_{23}H_{17}O$			
Pyrylium, 2,4,6-triphenyl-, iodide	CH_2Cl_2	564(2.02)	59-1311-76
selenocyanate	CH_2Cl_2	531(2.39)	59-1311-76
thiocyanate	CH_2Cl_2	488(2.66)	59-1311-76
$C_{23}H_{18}$			
Tribenzo[a,c,g]cycloundeca-5,10-diene, cis-trans	MeCN	240(4.49),297(4.23)	88-4529-76
trans-trans	MeCN	250(4.64),295s(4.52)	88-4529-76
Tricyclo[5.4.0.0^{1,6}]undeca-2,4,8,10-tetraene, 2,5-diphenyl-	n.s.g.	357(3.90)	88-3649-76
$C_{23}H_{18}BrFN_2O_7$			
Uridine, 2'-bromo-2'-deoxy-5-fluoro-, 3',5'-dibenzoate	pH 7	264(4.04)	73-3335-76
	pH 13	269(3.77)	73-3335-76
$C_{23}H_{18}ClN_3O_7$			
2,4(1H,3H)-Pyrimidinedione, 1-[5-O-benz-oyl-2-O-(4-chlorobenzoyl)-3-deoxy-2,3-imino-β-D-arabinofuranosyl]-	MeOH	237(4.50),261(4.18)	44-3138-76
$C_{23}H_{18}N_2$			
Pyrrolo[3,4-b]indole, 2,4-dihydro-4-phenyl-2-(phenylmethyl)-, picrate	MeOH	250(4.065),270(3.807), 282(3.817),303(3.678)	44-2031-76
$C_{23}H_{18}N_2O$			
Benzenecarboximidamide, N-hydroxy-N'-1-naphthalenyl-N-phenyl-	EtOH	320(4.18)	12-0357-76
1-Naphthalenecarboxaldehyde, 2-hydroxy-, diphenylhydrazone, (E)-	benzene	340(3.96),386(4.30)	104-0625-76
	EtOH	340(3.95),382(4.05)	104-0625-76
	DMF	340(4.00),386(4.30)	104-0625-76
1H-Phenanthro[9,10-d]imidazole-1-ethan-ol, α-phenyl-	MeOH	238s(4.55),247(4.81), 254(4.95),280(4.15), 300(3.91),333(3.26), 350(3.32)	39-2038-76C
2(1H)-Pyridinone, 1-amino-3,4,6-triphen-yl-	EtOH	230s(4.38),252s(4.23), 288(3.81),340(4.09)	4-0195-76

Compound	Solvent	$\lambda_{max}(\log \epsilon)$	Ref.
Pyrrolo[3,4-b]indol-3(2H)-one, 1,4-di- hydro-4-phenyl-2-(phenylmethyl)-	MeOH	246(4.265),299(4.141)	44-2031-76
$C_{23}H_{18}N_2OS$ Methanone, [2-(methylphenylamino)- 4-phenyl-5-thiazolyl]phenyl-	MeCN	258(4.24),374(4.01)	97-0018B-76
$C_{23}H_{18}N_2OSe$ Methanone, [2-(methylphenylamino)- 4-phenyl-5-selenazolyl]phenyl-	MeCN	231s(4.44),258s(4.32), 381(4.08)	97-0018-76B
$C_{23}H_{18}N_2O_2$ Benzimidazo[1,2-b]isoquinoline, 8,9-di- methoxy-11-phenyl-, perchlorate	MeOH	280(4.77),310(4.58)	103-0205-76
$C_{23}H_{18}N_2O_2S$ Cyclobuta[b]quinoxaline, 1,2-dihydro-1- [(4-methylphenyl)sulfonyl]-2-phenyl-	CHCl$_3$	236(4.59),328(3.84)	5-0284-76
$C_{23}H_{18}N_2O_3$ 1H-Indene-1,3(2H)-dione, 2-[(4-methoxy- phenyl)azo]-2-(phenylmethyl)-	EtOH	230(4.52),320(4.21)	104-1440-76
4-Quinolinecarbonitrile, 3-(benzoyloxy)- 1,2,3,4-tetrahydro-2-hydroxy-2-phenyl-	EtOH	236(4.34),272s(3.33), 282s(3.36),297(3.43)	95-0968-76
$C_{23}H_{18}N_2O_7$ 2,2'-Anhydro-1-(3,5-di-O-benzoyl- β-D-arabinofuranosyl)uracil	MeOH	227(4.63),252s(4.04)	44-3138-76
$C_{23}H_{18}N_3$ 1,2,4-Triazolo[3,4-a]isoquinolinium, 1-(4-methylphenyl)-3-phenyl-, tetrafluoroborate	MeOH	227(4.39),254(4.51), 303(3.38),317(3.68), 330(3.71)	48-0881-76
1,2,4-Triazolo[4,3-a]quinolinium, 1-(4- methylphenyl)-3-phenyl-, tetrafluoro- borate	MeOH	275(4.77),308(4.22), 315(4.02),320(4.11)	48-0881-76
$C_{23}H_{18}N_3O$ 1,2,4-Triazolo[3,4-a]isoquinolinium, 3-(4-methoxyphenyl)-1-phenyl-, tetrafluoroborate	MeOH	230(4.49),264(4.49), 320(3.66),332(3.64)	48-0881-76
1,2,4-Triazolo[4,3-a]quinolinium, 3-(4-methoxyphenyl)-1-phenyl-, tetrafluoroborate	MeOH	235(4.27),276(3.94), 304(4.10),318(4.08), 323(4.09)	48-0881-76
$C_{23}H_{18}N_3O_2S$ Benzothiazolium, 3-[(1,4-diamino-9,10- dihydro-9,10-dioxo-2-anthracenyl)- methyl]-2-methyl-, chloride	MeOH	570(3.85),620(4.06)	2-0054-76
$C_{23}H_{18}N_3O_3$ Benzoxazolium, 3-[(1,4-diamino-9,10-di- hydro-9,10-dioxo-2-anthracenyl)meth- yl]-2-methyl-, chloride	MeOH	562(3.96),615(4.11)	2-0054-76
$C_{23}H_{18}N_4O_3$ Benzoic acid, 2-[2-(benzoylamino)-1,2- dihydro-1-oxo-3-isoquinolinyl]hydra- zide	EtOH	233(5.07),294(4.22), 358(3.69)	95-0700-76

Compound	Solvent	$\lambda_{max}(\log \epsilon)$	Ref.
$C_{23}H_{18}N_4O_5$			
5H-Furo[2',3':4,5]oxazolo[3,2-a]pteridin-5-one, 7a,8,9,10a-tetrahydro-8-hydroxy-9-(hydroxymethyl)-2,3-diphenyl-, [7aS-(7aα,8α,9β,10aα)]-	MeOH	227(4.43),264(4.17), 358(4.15)	24-3217-76
$C_{23}H_{18}O$			
6H-Benzo[c]xanthene, 5,7-dihydro-5-phenyl-	EtOH	270(3.68),295s(3.60)	94-1588-76
Indeno[2,1-b]pyran, 2-ethyl-9-(1H-inden-2-yl)-	dioxan	253(4.06),260(4.17), 315(4.44),342(4.45), 552(3.42)	78-2225-76
Indeno[2,1-b]pyran, 9-(1H-inden-2-yl)-2,4-dimethyl-	dioxan	257(4.22),313(4.46), 343(4.50),535(3.51)	78-2219-76
2,4-Pentadien-1-one, 1,3,5-triphenyl-	EtOH	268(4.20),345(4.39)	1-0635-76
$C_{23}H_{18}O_2$			
1H-Indene-1-carboxylic acid, 2,3-diphenyl-, methyl ester	hexane	237(4.41),304(4.22)	104-0685-76
$C_{23}H_{18}O_3$			
Benzeneacetic acid, α-(3-oxo-1,3-diphenylpropylidene)-, (E)-	EtOH	244(4.24)	4-0195-76
5H-1-Benzopyran-5-one, 2,6,7,8-tetrahydro-2-(2-oxo-2-phenylethylidene)-4-phenyl-	MeOH	366(4.33),403(4.34)	24-1549-76
5H-1-Benzopyran-5-one, 4,6,7,8-tetrahydro-4-(2-oxo-2-phenylethylidene)-2-phenyl-	MeOH	274(4.37),386(4.31)	24-1549-76
2-Propen-1-one, 3-(benzoyloxy)-1-(4-methylphenyl)-3-phenyl-	n.s.g.	313(4.31)	39-0380-76C
Spiro[isobenzofuran-1(3H),9'-[9H]xanthen]-3-one, 2'-ethyl-7'-methyl-	EtOH	217(4.69),297(3.72), 302(3.76)	56-0759-76
$C_{23}H_{18}O_4$			
2-Propen-1-one, 3-(benzoyloxy)-1-(4-methoxyphenyl)-3-phenyl-	n.s.g.	313(4.35)	39-0380-76C
2-Propen-1-one, 3-(benzoyloxy)-3-(3-methoxyphenyl)-1-phenyl-	n.s.g.	300(4.26)	39-0380-76C
$C_{23}H_{18}O_7$			
Rotenone	MeOH	217(4.48),265(4.29), 295(4.19)	102-1283-76
$C_{23}H_{18}O_8$			
9,10-Anthracenedione, 1,3,8-triacetoxy-6-(1-propenyl)-, (E)-	EtOH	211(4.51),261(4.60), 277s(4.25),342(3.73)	39-0613-76C
$C_{23}H_{18}O_9$			
9,10-Anthracenedione, 1,3,8-triacetoxy-6-(3-methoxyoxiranyl)-	EtOH	213(4.42),262(4.53), 278s(4.07),343(3.69)	39-0613-76C
$C_{23}H_{18}O_{13}$			
Chiodectonic acid, tetraacetyl-	MeOH	214(4.33),290(4.55), 400(3.72)	102-0799-76
$C_{23}H_{19}BrN_2O_2$			
9,10-Anthracenedione, 1-amino-4-bromo-2-[[4-(dimethylamino)phenyl]methyl]-	n.s.g.	475(3.95)	146-0001-76

Compound	Solvent	$\lambda_{max}(\log \epsilon)$	Ref.
$C_{23}H_{19}BrO_8$ 9,10-Anthracenedione, 1,3,8-triacetoxy- 6-(1-bromopropyl)-	EtOH	212(4.50),260(4.59), 278s(4.15),341(3.76)	39-0613-76C
$C_{23}H_{19}ClO_6$ 1,4-Naphthalenedione, 5-acetoxy-3-[2-(3- acetoxy-5-methylphenyl)ethyl]-2-chloro-	MeOH	245(4.02),250s(4.00), 274(3.98),338(2.84)	39-0997-76C
$C_{23}H_{19}FN_2O_7$ Uridine, 2'-deoxy-5-fluoro-, 3',5'-di- benzoate	pH 7 pH 13	267(3.95) 268(3.83)	73-3335-76 73-3335-76
$C_{23}H_{19}HgI_2P$ Mercuric iodide, tetraphenylphosphonium cyclopentadienylide complex	THF	240(4.24),250(4.28), 264s(4.12),289(3.60)	35-7823-76
$C_{23}H_{19}N$ 1H-Pyrrole, 2,3-diphenyl-1-(phenylmeth- yl)-	EtOH	247(4.17),281(3.86)	88-2459-76
$C_{23}H_{19}NO_2$ 3-Buten-2-one, 3,4-diphenyl-, O-benzoyl- oxime	EtOH	233(4.32),293(4.20)	56-0499-76
5H-Oxazolo[3,2-b]isoquinolin-5-one, 2,3,10,10a-tetrahydro-2,10a-diphenyl-	n.s.g.	235(4.06),250(3.98), 263(3.78)	39-1073-76C
5(6H)-Quinolinone, 4-benzoyl-7,8-di- hydro-7-methyl-2-phenyl-	EtOH	232(4.16),251(4.21), 270(4.10),313(4.32)	94-1160-76
$C_{23}H_{19}NO_3$ 9,10-Anthracenedione, 2-[[4-(dimethyl- amino)phenyl]methyl]-1-hydroxy-	n.s.g.	410(3.85)	146-0001-76
Phenol, 2-methoxy-6-(1-methylbenzo[f]- quinolin-3-yl)-, acetate	EtOH	215(4.48),261(4.62), 273(4.61),339(3.60), 354(3.65)	65-0390-76
$C_{23}H_{19}NO_6$ 5(4H)-Oxazolone, 4-[(3,4-diacetoxyphen- yl)methylene]-2-(1-methyl-2-phenyl- ethenyl)-, (Z,E)-	CH_2Cl_2	291(4.08),377(4.66)	44-1466-76
$C_{23}H_{19}N_3$ 1H-Indole, 3-[1H-indol-3-yl(1-methyl- 4(1H)-pyridinylidene)methyl]- diprotonated	MeOH MeOH	367(4.18) 547(4.32)	44-0870-76 44-0870-76
1H-Pyrazole, 3-phenyl-1-[phenyl[(phen- ylmethyl)imino]methyl]-	EtOH	275(4.22),281(4.18), 292(4.04)	88-4753-76
4H-1,2,4-Triazepine, 3,7-diphenyl- 4-(phenylmethyl)-	n.s.g.	207(4.36),257(4.27), 365(2.92)	88-2459-76
$C_{23}H_{19}N_3O_2$ 1H-Pyrazole, 4,5-dihydro-3-[2-(4-nitro- phenyl)ethenyl]-1,5-diphenyl-	n.s.g.	452(5.43)	124-0383-76
4(3H)-Quinazolinone, 3-[[(4-methoxyphen- yl)methylene]amino]-2-(4-methylphen- yl)-	EtOH	222(4.51),232s(4.49), 282(4.38),322(4.32)	114-0341-76D
2,2',2"-Terpyridine, 4'-(2,4-dimethoxy- phenyl)-, ferrous complex	MeOH	567(4.38)	118-0001-76
$C_{23}H_{19}N_3O_2S$ 1H-Pyrazole, 4,5-dihydro-3-[4-(5-nitro-	n.s.g.	530(5.37)	124-0383-76

Compound	Solvent	λ_{max}(log ϵ)	Ref.
2-thienyl)-1,3-butadienyl]-1,5-diphenyl- (cont.)			124-0383-76
$C_{23}H_{19}N_3O_3$			
1H-Pyrazole, 4,5-dihydro-3-[4-(5-nitro-2-furanyl)-1,3-butadienyl]-1,5-diphenyl-	n.s.g.	490(5.41)	124-0383-76
3-Pyridinecarboxylic acid, 4-cyano-1,2,3,4-tetrahydro-2-hydroxy-3-methyl-2-phenyl-3-quinolinyl ester	EtOH	247(4.36),280s(4.05), 315(3.42)	95-0968-76
5,8-Quinazolinedione, 2-(2-ethoxyphenyl)-4-methyl-6-(phenylamino)-	EtOH	270s(--),305(4.36), 362(3.84),471(3.61)	5-1809-76
$C_{23}H_{19}N_3O_4$			
5H-Cyclopentacycloocten-4,8-imin-5-one, 3,4,5,9-tetrahydro-9-(4-methoxyphenyl)-10-(5-nitro-2-pyridinyl)-, (4α,8α,9α)-(+)-	CHCl$_3$	368(4.29)	39-2307-76C
5,8-Quinazolinedione, 2-(3,5-dimethoxyphenyl)-4-methyl-6-(phenylamino)-	EtOH	215s(--),272(4.28), 321(4.34),483(3.45)	5-1809-76
$C_{23}H_{19}N_3O_5$			
Benzoic acid, 2-[2-[(2-nitrophenyl)hydrazono]-1-oxo-3-phenylpropyl]-, methyl ester	EtOH	280(3.95),322(4.09), 404(3.97)	104-1440-76
Benzoic acid, 2-[2-[(3-nitrophenyl)hydrazono]-1-oxo-3-phenylpropyl]-, methyl ester	EtOH	282s(3.88),334(4.32)	104-1440-76
Benzoic acid, 2-[2-[(4-nitrophenyl)hydrazono]-1-oxo-3-phenylpropyl]-, methyl ester	EtOH	286(3.66),380(4.51)	104-1440-76
$C_{23}H_{19}N_5$			
Pyrido[2,3-b]pyrazine, 1,2,3,4-tetrahydro-2,3-di-1H-indol-3-yl-, conjugate monoacid	n.s.g.	272(4.27),291(4.19), 332(4.13)	103-0948-76
$C_{23}H_{19}N_5O$			
Benzamide, N-[4-[(4-methylphenyl)amino]-6-phenyl-1,3,5-triazin-2-yl]-	MeOCH$_2$CH$_2$OH	261(4.58)	40-0488-76
$C_{23}H_{19}N_5O_2$			
Benzamide, N-[4-[(4-methoxyphenyl)amino]-6-phenyl-1,3,5-triazin-2-yl]-	MeOCH$_2$CH$_2$OH	259(4.64),291s(4.32)	40-0488-76
1H-Pyrazole-3,4-dicarboxamide, 5-amino-N,N',1-triphenyl-	EtOH	295(4.335)	4-1137-76
$C_{23}H_{19}N_5O_7$			
Uridine, 2'-azido-2'-deoxy-, 3',5'-dibenzoate	MeOH	234(4.48),258(4.10)	44-3138-76
$C_{23}H_{19}OP$			
Phosphine oxide, (1-naphthalenemethyl)-diphenyl-, carbanion	DMSO	490(4.08)	70-0762-76
$C_{23}H_{20}N_2$			
1H-Pyrazole, 4,5-dihydro-1,5-diphenyl-3-(2-phenylethenyl)-	n.s.g.	382(4.50)	124-0383-76
Pyrrolo[3,4-b]indole, 1,2,3,4-tetrahydro-5-methyl-1,2-diphenyl-, hydrochloride	EtOH	225(4.59),252(4.25), 273(4.23)	103-1249-76

Compound	Solvent	$\lambda_{max}(\log \epsilon)$	Ref.
$C_{23}H_{20}N_2O$			
1H-Pyrazole, 3-[4-(2-furanyl)-1,3-buta-dienyl]-4,5-dihydro-1,5-diphenyl-	n.s.g.	406(4.56)	124-0383-76
$C_{23}H_{20}N_2OS$			
1H-Imidazolium, 4-mercapto-3-(4-methoxy-phenyl)-1-methyl-2,5-diphenyl-, hydroxide, inner salt	MeOH	286(3.88),310(3.92)	2-0382B-76
$C_{23}H_{20}N_2O_2$			
9,10-Anthracenedione, 1-amino-2-[[4-(di-methylamino)phenyl]methyl]-	n.s.g.	474(3.96)	146-0001-76
Phenol, 4-[4-(3,4-dihydro-2,2-dimethyl-2H-1-benzopyran-6-yl)-2,3-diisocyano-1,3-butadienyl]- (xanthoascin)	MeOH	244(4.18),302(4.12), 365(4.73),385(4.65)	94-2317-76
	CHCl$_3$	302(4.12),372(4.75), 390(4.67)	94-2317-76
$C_{23}H_{20}N_2O_2S$			
Benzenesulfonamide, N-[1,3-dihydro-1-methyl-3-(phenylmethylene)-2H-indol-2-ylidene]-4-methyl-	CHCl$_3$	282(5.18),373(2.25), 410(2.24)	103-0292-76
Benzenesulfonamide, N,4-dimethyl-N-[3-(phenylmethylene)-3H-indol-2-yl]-	CHCl$_3$	278(5.04),340(4.40), 380s(3.83)	103-0292-76
$C_{23}H_{20}N_2O_3$			
Benzoic acid, 2-[1-oxo-3-phenyl-2-(phen-ylhydrazono)propyl]-, methyl ester	EtOH	288(3.77),354(4.28)	104-1440-76
Hydrazinecarboxylic acid, [3-(4-methyl-phenyl)-3-oxo-1-phenylpropylidene]-, phenyl ester	EtOH	225s(4.26),279(4.38)	4-0257-76
3(2H)-Isoquinolinone, 2-(2-aminophenyl)-6,7-dimethoxy-1-phenyl-, perchlorate	MeOH	280(4.87),310(4.72), 410(3.53)	103-0205-76
$C_{23}H_{20}N_2O_3S$			
Benzenesulfonamide, N-[3-[(4-methoxy-phenyl)methylene]-3H-indol-2-yl]-4-methyl-	CHCl$_3$	270(4.10),373(4.24), 424s(4.04)	103-0292-76
$C_{23}H_{20}N_2O_4$			
Hydrazinecarboxylic acid, [3-(4-methoxy-phenyl)-3-oxo-1-phenylpropylidene]-, phenyl ester	EtOH	226s(4.36),280(4.38)	4-0257-76
$C_{23}H_{20}N_2O_5$			
3aH-Oxireno[e]indazole-3,3a-dicarbox-ylic acid, 1a,6,6a,6b-tetrahydro-2,6a-diphenyl-, dimethyl ester, (1aα,3aα,6aα,6bα)-	MeCN	265(3.86),329s(2.74)	24-2823-76
$C_{23}H_{20}N_2S$			
1H-Imidazolium, 4-mercapto-1-methyl-3-(4-methylphenyl)-2,5-diphenyl-, hydroxide, inner salt	MeOH	312(3.91)	2-0382B-76
1H-Pyrazole, 4,5-dihydro-1,5-diphenyl-3-[4-(2-thienyl)-1,3-butadienyl]-	n.s.g.	418(4.70)	124-0383-76
$C_{23}H_{20}N_4$			
2,2',2''-Terpyridine, 4'-[4-(dimethyl-amino)phenyl]-, ferrous complex	MeOH	620(4.39)	118-0001-76

Compound	Solvent	λ_{max}(log ϵ)	Ref.
$C_{23}H_{20}N_4O_6$			
2,4(1H,3H)-Pteridinedione, 1-α-D-ara-binofuranosyl-6,7-diphenyl-	pH 5.0	220(4.42),276(4.19), 360(4.14)	24-3217-76
	pH 11.0	220s(4.38),266(4.29), 360(4.18)	24-3217-76
2,4(1H,3H)-Pteridinedione, 1-β-D-ara-binofuranosyl-6,7-diphenyl-	pH 5.0	220s(4.42),274(4.18), 358(4.12)	24-3217-76
	pH 11.0	220s(4.36),264(4.29), 357(4.16)	24-3217-76
2,4(1H,3H)-Pteridinedione, 3-α-D-ara-binofuranosyl-6,7-diphenyl-	pH 5.0	220s(4.43),271(4.17), 363(4.17)	24-3217-76
	pH 11.0	240s(4.26),294(4.38), 392(4.07)	24-3217-76
2,4(1H,3H)-Pteridinedione, 6,7-diphenyl-1-β-D-ribofuranosyl-	pH 4.0	221(4.42),273(4.19), 358(4.14)	24-3217-76
	pH 11.0	220s(4.38),265(4.29), 358(4.19)	24-3217-76
2,4(1H,3H)-Pteridinedione, 6,7-diphenyl-3-β-D-ribofuranosyl-	pH 1.0	220s(4.43),272(4.19), 363(4.19)	24-3217-76
	pH 11.0	220s(4.41),240s(4.28), 292(4.39),388(4.07)	24-3217-76
$C_{23}H_{20}O$			
14H-Dibenzo[a,h]xanthene, 12,13-dihydro-12,12-dimethyl-	EtOH	213(4.90),244(4.83), 276s(--)	22-1967-76
1-Naphthalenemethanol, 4-[4-(3,5-dimeth-ylphenyl)-1-buten-3-yn-1-yl]-, trans	MeOH	236(4.50),277s(4.04), 283(4.04),336(4.49)	39-1104-76B
9-Phenanthrenemethanol, 1-(3,5-dimethyl-phenyl)-	MeOH	280s(4.21),290s(4.10), 301(4.17),335(2.74), 351(2.70)	39-1115-76B
$C_{23}H_{20}O_5$			
3-Oxatricyclo[5.1.0.0^{2,4}]oct-5-ene-6,7-dicarboxylic acid, 1,5-diphenyl-, dimethyl ester, cis	MeCN	255(3.95)	24-2823-76
4H-Oxocin-5,6-dicarboxylic acid, 3,7-diphenyl-, dimethyl ester	MeCN	247(4.44)	24-2823-76
$C_{23}H_{20}O_6$			
2-Propenoic acid, 3-(4-hydroxyphenyl)-, 9,10-dihydro-8,8-dimethyl-2-oxo-2H,8H-benzo[1,2-b:3,4-b']dipyran-9-yl ester, cis	EtOH	302s(4.41),318(4.50)	102-1049-76
trans	EtOH	302s(4.43),318(4.54)	102-1049-76
$C_{23}H_{20}O_7$			
α-Toxicarol, 6a,12a-dehydro-	MeOH	255(4.22),274(4.28), 277s(3.31),282(4.28), 312(3.87),328(3.88)	102-0234-76
Villosol	EtOH	209(4.69),279(4.70), 324(4.38)	2-0152-76
	EtOH-AlCl$_3$	232(4.78),289(4.77), 367(4.29)	2-0152-76
$C_{23}H_{21}BrO_2$			
Benzene, 1,1'-[2-bromo-2-(4-methylphen-yl)ethenylidene]bis[2-methoxy-	C_6H_{12}	235s(4.30),283(4.06)	18-2560-76
$C_{23}H_{21}BrO_3$			
Benzene, 1,1'-[2-bromo-2-(4-methoxyphen-yl)ethenylidene]bis[2-methoxy-	C_6H_{12}	240(4.23),285(4.13)	18-2560-76

Compound	Solvent	$\lambda_{max}(\log \epsilon)$	Ref.
$C_{23}H_{21}ClN_2O_2$ 1H-Pyrrolo[3,2-b]quinoline-3-carboxylic acid, 7-chloro-2-methyl-9-phenyl-, 1,1-dimethylethyl ester	EtOH	205(4.54),238s(4.59), 252(4.72),320s(3.87), 333(4.09),354(3.97), 367s(3.91)	44-1743-76
$C_{23}H_{21}ClO_3$ Benzene, 1,1'-[2-chloro-2-(4-methoxyphenyl)ethenylidene]bis[2-methoxy-	C_6H_{12}	240(4.21),284(4.19)	18-2560-76
$C_{23}H_{21}FeNO_3S$ Ferrocenylmethylpyridinium p-toluenesulfonate	50% MeOH	400(2.55)	104-2433-76
$C_{23}H_{21}NO_2$ Benzeneacetic acid, α-[phenyl[(phenylmethyl)amino]methylene]-, methyl ester	EtOH	241s(3.87),306(4.05)	78-0431-76
4H-Indol-4-one, 1,5,6,7-tetrahydro-6-methyl-3-(2-oxo-2-phenylethyl)-2-phenyl-	EtOH	243(4.36),273(4.05)	94-1160-76
$C_{23}H_{21}NO_2S$ 1,5-Benzothiazepine, 2-(3,4-dimethoxyphenyl)-2,3-dihydro-4-phenyl-	EtOH	237(4.29),261(4.32), 291s(4.01),335(3.67)	114-0293-76A
$C_{23}H_{21}NO_7S$ 3-[2-Carboxy-2-(2-carboxyphenyl)ethenyl]-1-methylpyridinium tosylate	EtOH	290(4.18)	39-0315-76C
$C_{23}H_{21}N_2$ 3H-Indolizinium, 3-[[4-(dimethylamino)phenyl]methylene]-2-phenyl-, bromide	Ac₂O	605(3.6878)	2-0551-76
perchlorate	Ac₂O	605(3.6970)	2-0551-76
$C_{23}H_{21}N_3$ 4H-1,2,4-Triazepine, 6,7-dihydro-3,7-diphenyl-5-(phenylmethyl)-	EtOH	241(4.00),324(3.68)	88-2459-76
$C_{23}H_{21}N_3O_2$ 9,10-Anthracenedione, 1,4-diamino-2-[[4-(dimethylamino)phenyl]methyl]-	n.s.g.	550(4.25),592(4.20)	146-0001-76
$C_{23}H_{21}N_3O_3$ 2,7-Naphthyridine-3-carboxylic acid, 5-ethenyl-1,2-dihydro-2-[2-(1H-indol-3-yl)ethyl]-1-oxo-, ethyl ester	EtOH	273(3.02),280s(3.01), 290s(2.93),335(2.95)	44-2542-76
$C_{23}H_{21}N_3O_4$ Azobenzene-4-carboxylic acid, 4'-[N-methyl-N-(4-methylbenzoyloxy)amino]-, methyl ester	EtOH	380(4.34)	94-1485-76
$C_{23}H_{21}N_5O_2S$ Benzenesulfonamide, N-[1,3-bis(1H-benzimidazol-2-yl)propyl]-	EtOH	245(4.00),275(4.10), 283(4.15)	42-0331-76
dihydrochloride	EtOH	226(4.08),272(4.13), 278(4.20)	42-0274-76

Compound	Solvent	λ_{max}(log ϵ)	Ref.
$C_{23}H_{21}N_5S$			
1H-1,2,4-Triazolium, 4-[[[4-(dimethyl-amino)phenyl]methylene]amino]-3-mer-capto-1,5-diphenyl-, hydroxide, inner salt	MeCN	248(4.76),348(4.07)	103-0361-76
$C_{23}H_{21}O_2$			
Pyrylium, 2-(4-ethoxy-1,3-butadienyl)-4,6-diphenyl-, perchlorate	n.s.g.	750(4.26)	103-0619-76
Pyrylium, 4-(4-ethoxy-1,3-butadienyl)-2,6-diphenyl-, perchlorate	n.s.g.	680(4.19)	103-0619-76
$C_{23}H_{22}NO$			
Benzo[f]quinolinium, 1-ethyl-3-(4-meth-oxyphenyl)-4-methyl-, iodide	EtOH	218(4.66),230(4.56), 284(4.50),348(3.97), 365(4.06),386(3.87)	103-1012-76
methosulfate	EtOH	218(4.35),232(4.37), 284(4.60),347(3.91), 364(3.96),388(3.48)	103-1012-76
perchlorate	EtOH	218(4.33),232(4.33), 285(4.61),347(3.90), 364(3.93),384(3.27)	103-1012-76
Benzo[f]quinolinium, 3-(4-methoxyphen-yl)-1,2,4-trimethyl-, iodide	EtOH	221(4.81),241(4.64), 281(4.66),318(3.98), 365(4.19),379(4.28)	103-1012-76
methosulfate	EtOH	218(4.36),259(4.47), 342(3.17),357(3.74), 384(3.15)	103-1012-76
perchlorate	EtOH	220(4.41),257(4.50), 278(4.49),343(3.70), 347(3.84),377(3.63)	103-1012-76
$C_{23}H_{22}N_2O_3$			
2-Propenoic acid, 2-(aminomethyl)-3-(1-benzoyl-1,2,2a,3-tetrahydrobenz[cd]in-dol-5-yl)-, methyl ester, (E)-($\pm$)-, p-toluenesulfonate	EtOH	234(4.23),255(4.27)	88-4311-76
$C_{23}H_{22}N_4O_2S$			
4-Thiazolecarboxamide, 2-[1-[[(2-hydr-oxyphenyl)methylene]amino]ethyl]-N-[2-(1H-indol-3-yl)ethyl]-, (S)-	EtOH	223(5.02),259(4.46), 283(4.11),292(4.08), 322(3.96)	94-0092-76
$C_{23}H_{22}N_8O_4$			
D-erythro-Pentos-2-ulose, 5-deoxy-5-(3,4-dihydro-2,4-dioxo-8(2H)-pteridinyl)-, 1,2-bis(phenylhydrazone)	MeOH	274(4.55),390(3.98), 457(3.81)	24-3175-76
$C_{23}H_{22}O$			
Benzene, 1-(2-methoxy-2,3-diphenylcyclo-propyl)-4-methyl-	C_6H_{12}	236(4.21)	12-2445-76
Benzene, 1-(1-methoxy-1,3-diphenyl-2-propenyl)-4-methyl-, (E)-	C_6H_{12}	224(3.98),259(4.25)	12-2445-76
Benzene, 1-(3-methoxy-3,3-diphenyl-1-propenyl)-4-methyl-, (E)-	C_6H_{12}	223(4.00),261(4.30)	12-2445-76
2,5-Cyclohexadien-1-one, 4-diphenyl-methylene-2,6-diethyl-	isooctane	364(4.42)	73-1676-76
$C_{23}H_{22}O_2$			
Benzene, 1-methoxy-4-(1-methoxy-2,3-di-phenylcyclopropyl)-, (1α,2α,3β)-	C_6H_{12}	240(3.87)	12-2445-76

Compound	Solvent	$\lambda_{max}(\log \epsilon)$	Ref.
Benzene, 1-methoxy-4-(1-methoxy-1,3-di-phenyl-2-propenyl)-, (E)-	C_6H_{12}	235(3.40),268(3.51)	12-2445-76
Benzene, 1-methoxy-4-(3-methoxy-3,3-di-phenyl-1-propenyl)-, (E)-	C_6H_{12}	227(3.73),268(4.03)	12-2445-76
Benzene, 1,1'-[2-(4-methylphenyl)ethen-ylidene]bis[2-methoxy-	C_6H_{12}	220(4.43),283(4.25)	18-2560-76
1,4-Pentadiyn-3-ol, 3-(1,1-dimethyleth-yl)-1,5-diphenyl-, acetate	$CHCl_3$	254(4.55),279(3.16)	18-2515-76

$C_{23}H_{22}O_3$

Benzene, 1,1'-[2-(4-methoxyphenyl)ethen-ylidene]bis[2-methoxy-	C_6H_{12}	208(4.57),285(4.28)	18-2560-76

$C_{23}H_{22}O_3S$

Thiophenium, tetrahydro-, 1,3-dibenzoyl-4-oxo-2-pentenylide	EtOH	251(4.24),350(4.27)	94-2421-76

$C_{23}H_{22}O_4$

Benzoic acid, 2-methyl-4,6-bis(phenyl-methoxy)-, methyl ester	EtOH	249s(3.78),282s(3.38)	35-5380-76

$C_{23}H_{22}O_5$

Benzoic acid, 4-methoxy-2,6-bis(phenyl-methoxy)-, methyl ester	EtOH	270(4.36)	35-5380-76
1-Propanone, 1-[2,4-dihydroxy-3-[[(2-hy-droxyphenyl)methyl]-6-methoxyphenyl]-3-phenyl- (uvaretin)	EtOH	254s(3.82),284s(4.11), 330(4.48)	44-1852-76

$C_{23}H_{22}O_8$

2-Propenoic acid, 3-(4-hydroxyphenyl)-, 2-acetoxy-1,3-propanediyl ester	n.s.g.	231(4.27),318(4.55)	102-0811-76
Villosinol	EtOH	235(4.35),293(4.43), 335(3.65)	2-0152-76
	EtOH-AlCl$_3$	241(4.67),311(4.64), 364(4.26)	2-0152-76

$C_{23}H_{22}O_{11}$

9,10-Anthracenedione, 2-acetyl-8-(β-D-glucopyranosyloxy)-1,6-dihydroxy-3-methyl-	dioxan	242s(4.07),283(4.36), 419(3.92)	12-2231-76

$C_{23}H_{22}O_{12}$

Acetylcynaroside	EtOH	257(4.09),268s(4.04), 355(4.09)	105-0228-76

$C_{23}H_{23}NO$

Benzenecarboximidic acid, N-[2-(1,1-di-methylethyl)phenyl]-, phenyl ester	MeOH	215(4.48),241(4.41), 310(4.02)	39-0465-76C
Benzenecarboximidic acid, N-[4-(1,1-di-methylethyl)phenyl]-, phenyl ester	MeOH	220s(4.28),240(4.33), 306(3.88)	39-0465-76C

$C_{23}H_{23}NOS_2Sn$

Morpholine, 4-[[thioxo(triphenylstann-yl)thio]methyl]-	$CHCl_3$	257(4.07),294(2.34)	39-0465-76C

$C_{23}H_{23}NO_3$

15H-Benzo[a]naphtho[2,1-f]quinolizin-15-one, 5,6,8,9,13b,14-hexahydro-11,12-dimethoxy-	n.s.g.	283(4.50),353(4.64)	2-0784-76

Compound	Solvent	$\lambda_{max}(\log \epsilon)$	Ref.
$C_{23}H_{23}NO_3S$ 1-Propanone, 3-[(2-aminophenyl)thio]- 3-(3,4-dimethoxyphenyl)-1-phenyl-	EtOH	242(4.42),282(3.81), 308(3.66)	114-0293-76A
$C_{23}H_{23}NO_5$ 12H-7,11-Nitrilo-6H-dibenzo[b,k]- [1,4,7,10,13]pentaoxacycloeicosin, 19,20,22,24-tetrahydro-	MeCN	272(3.87)	24-1803-76
NaSCN complex	MeCN	271(3.94)	24-1803-76
KSCN complex (also other complexes)	MeCN	271(3.93)	24-1803-76
$C_{23}H_{23}NO_6$ 8,13a-Propano-13aH-benzo[g]-1,3-benzo- dioxolo[5,6-a]quinolizin-6-one, 5,6,8,13-tetrahydro-13-hydroxy- 9,10-dimethoxy-	EtOH	230s(4.15),287(3.75)	73-3654-76
$C_{23}H_{23}NO_7S$ 1H-Pyrrolo[2,1-c][1,4]thiazine-7,8-di- carboxylic acid, 1-[(3-ethoxy-3-oxo- 1-propenyl)oxy]-4-methyl-1-phenyl-, dimethyl ester, (Z)-	MeOH	238(4.22),290(4.38)	44-0187-76
$C_{23}H_{23}NS_2Sn$ Pyrrolidine, 1-[[thioxo(triphenylstann- yl)thio]methyl]-	CHCl$_3$	250(4.17),280(2.92)	101-0055-76D
$C_{23}H_{23}N_2O$ Quinolinium, 1-ethyl-4-[3-(7-ethylfuro- [2,3-b]pyridin-6(7H)-ylidene)-1-prop- enyl]-, iodide	EtOH	663(5.28)	103-0052-76
$C_{23}H_{23}N_2S$ Quinolinium, 1-ethyl-2-[3-(7-ethylthi- eno[2,3-b]pyridin-4(7H)-ylidene)-1- propenyl]-, perchlorate	EtOH	674(5.24)	103-0052-76
$C_{23}H_{23}N_2Se$ Quinolinium, 1-ethyl-2-[3-(7-ethylselen- olo[2,3-b]pyridin-4(7H)-ylidene)-1- propenyl]-, iodide	EtOH	661(5.11)	103-0052-76
$C_{23}H_{23}N_3O_9$ Nocardicin A keto derivative	EtOH	300(4.19)	35-3023-76
$C_{23}H_{24}ClN_2$ Methanaminium, N-[4-[(2-chlorophenyl)- [4-(dimethylamino)phenyl]methylene]- 2,5-cyclohexadien-1-ylidene]-N-methyl-	EtOH 8% DMSO	640(5.10) 642(4.78)	44-0221-76 44-0221-76
$C_{23}H_{24}ClN_3O_3S$ 4H-Indazol-4-one, 3-(4-chlorophenyl)-1- 5,6,7-tetrahydro-1,6,6-trimethyl-, O- [(4-methylphenyl)sulfonyl]oxime, (E)-	MeOH	232(4.55),266(3.95)	24-1898-76
$C_{23}H_{24}ClN_4OS$ Morpholinium, 4-[[[5-(4-chlorophenyl)- 2-[4-(dimethylamino)phenyl]-6H-1,3- thiazin-6-ylidene]amino]methylene]-, perchlorate	HOAc	320(3.97),432(4.30), 569(4.37)	97-0268-76

Compound	Solvent	$\lambda_{max}(\log \epsilon)$	Ref.
$C_{23}H_{24}Cl_2N_2O_{11}$ D-Galactose, 1-C-(5,7-dichloro-1H-inda-zol-3-yl)-, 2,3,4,5,6-pentaacetate	EtOH	217(4.24),244(3.95), 250(3.96),295s(3.76), 316(3.91)	4-1241-76
$C_{23}H_{24}N_2O_3$ 1-Pyrrolidinecarbonitrile, 2-[(3,5,6-trimethoxy-9-phenanthrenyl)methyl]-, (S)-	MeOH	258(4.56),316(3.94), 353(3.17)	33-2201-76
1-Pyrrolidinecarbonitrile, 2-[(3,6,7-trimethoxy-9-phenanthrenyl)methyl]-, (S)-	MeOH	257(4.73),284(4.44), 311(3.90),342(2.89), 358(2.31)	33-2201-76
$C_{23}H_{24}N_2O_4$ Benzoic acid, 2-[1-acetyl-5-(4-methoxy-phenyl)-1H-pyrazol-3-yl]-, butyl ester	EtOH	235(4.33),288(4.32), 305(4.09)	44-0110-76
Benzoic acid, 4-[1,4,5,6-tetrahydro-6-(1-methylethoxy)-1-phenylpyrano[2,3-c]pyrazol-4-yl]-, methyl ester, cis	EtOH	244(4.46)	35-2947-76
trans	EtOH	243(4.40)	35-2947-76
Carbamic acid, [2-methyl-3-(phenylmeth-oxy)-5-[(phenylmethoxy)methyl]-4-pyri-dinyl]-, methyl ester	EtOH	267(3.59)	87-0999-76
$C_{23}H_{24}N_2O_5S$ 4-Thia-1-azabicyclo[3.2.0]heptane-2-carboxylic acid, 3,3-dimethyl-7-oxo-6-[(phenoxyacetyl)amino]-, phenylmethyl ester	EtOH	220(3.88),245(3.87), 328(4.41)	88-3979-76
$C_{23}H_{24}N_4O_5$ Pyrazinecarboxamide, N-methyl-5,6-di-phenyl-3-(α-D-ribopyranosylamino)-	MeOH	230(4.29),281(4.31), 374(4.06)	24-3194-76
$C_{23}H_{24}N_4O_5S$ 4H-Indazol-4-one, 1,5,6,7-tetrahydro-1,6,6-trimethyl-3-(4-nitrophenyl)-, O-[(4-methylphenyl)sulfonyl]oxime, (E)-	MeOH	220(4.48),254(4.08), 306(4.05)	24-1898-76
$C_{23}H_{24}N_4O_6$ 2,4(1H,3H)-Pteridinedione, 5,6,7,8-tetrahydro-6,7-diphenyl-1-β-D-ribofuranosyl-	pH 0.0 pH 6.0	267(4.20) 248s(3.80),287(4.05)	24-3184-76 24-3184-76
$C_{23}H_{24}N_4O_9$ Nocardicin A	EtOH EtOH-NaOH	220(4.32),272(4.20) 245(4.37),285(4.05)	35-3023-76 35-3023-76
$C_{23}H_{24}O_3$ 1-Azulenecarboxylic acid, 3,4-dimethyl-7-(1-methylethyl)-, 4-methoxyphenyl ester	C_6H_{12}	249(4.57),289s(4.53), 300(4.71),308(4.73), 314s(4.66),373(4.14), 388(4.21),526s(2.88), 562(3.00),604s(2.92), 666(2.48)	18-1650-76
1-Azulenecarboxylic acid, 3,8-dimethyl-5-(1-methylethyl)-, 4-methoxyphenyl ester	C_6H_{12}	247(4.85),303(4.89), 315s(4.54),387(4.22), 531s(2.77),569(2.86), 621s(2.74),684s(2.29)	18-1650-76

Compound	Solvent	$\lambda_{max}(\log \epsilon)$	Ref.
$C_{23}H_{24}O_8$			
4H-1-Benzopyran-2-carboxylic acid, 7-hy-droxy-5-methyl-4-oxo-3-[(2,4,5-tri-methoxyphenyl)methyl]-, ethyl ester	EtOH	209(4.33),217(4.34), 228(4.26),256(4.00), 300(3.94)	39-0499-76C
$C_{23}H_{24}O_{12}$			
Flavone, 4',5-dihydroxy-3',7-dimethoxy-, 6-glucoside	EtOH	272(4.20),344(4.35)	102-0838-76
	EtOH-NaOAc	269(4.30),351(4.38), 385(4.36)	102-0838-76
	EtOH-AlCl$_3$	302s(4.26),286(4.28), 370(4.38)	102-0838-76
$C_{23}H_{25}BrN_2O_{11}$			
D-Galactose, 1-C-(5-bromo-1H-indazol-3-yl)-, 2,3,4,5,6-pentaacetate	EtOH	222(4.10),243(4.10), 249(4.10),301(4.02)	4-1241-76
D-Galactose, 1-C-(6-bromo-1H-indazol-3-yl)-, 2,3,4,5,6-pentaacetate	EtOH	220(4.21),240(3.98), 247(3.96),290s(3.83), 320(3.92)	4-1241-76
$C_{23}H_{25}ClN_5O_6P$			
Phosphinic acid, di-4-morpholinyl-, 5-(2-chlorophenyl)-7-nitro-3H-1,4-benzodiazepin-2-yl ester	MeCN	215s(4.49),245s(4.24), 313(4.05)	44-2720-76
$C_{23}H_{25}Cl_2N_4O_4P$			
Phosphinic acid, di-4-morpholinyl-, 7-chloro-5-(2-chlorophenyl)-3H-1,4-benzodiazepin-2-yl ester	MeCN	216(4.62),270s(3.90), 315(3.32)	44-2720-76
$C_{23}H_{25}NO$			
Curryangine	MeOH	245(4.53),265(4.12), 305(4.04),340(3.64)	2-0430-76
Curryanine	MeOH	242(4.83),256(4.61), 305(4.41),330(3.87)	2-0430-76
$C_{23}H_{25}NOS$			
2(1H)-Azecinone, 7,8,9,10-tetrahydro-1-methyl-5-(methylthio)-3,4-diphenyl-	EtOH	227(4.403),256s(4.107), 273s(4.061)	24-0576-76
$C_{23}H_{25}NO_3$			
Desoxypergularinine	MeOH	257(4.2),286(3.9), 311(3.4),340(3.0), 360(2.6)	102-1561-76
Dibenzo[f,h]pyrrolo[1,2-b]isoquinoline, 9,11,12,13,13a,14-hexahydro-3,4,6-trimethoxy-, (S)-	MeOH	253(4.68),261(4.72), 288(3.95),303(4.01), 327(2.95),354(3.27), 372(3.34)	33-2201-76
1H-Indene-6-carboxylic acid, 2-[[(1,1-dimethylethyl)methylamino]phenyl-methyl]-1-oxo-, methyl ester	hexane	245(4.48),307(3.18), 320(3.18),335(3.08), 405(3.14),425(3.11), 455(3.02)	44-3540-76
	CHCl$_3$	425s(3.07),460(2.97)	44-3540-76
	MeCN	403(3.14),450(2.98)	44-3540-76
Phenanthroindolizidine, 3,6,7-trimeth-oxy-	MeOH	258(4.3),286(3.9), 311(3.5),341(3.1), 360(2.6)	102-1561-76
$C_{23}H_{25}NO_4$			
Atalaphyllidine	EtOH	255(4.39),285(4.34), 305(4.32),398(3.74)	102-1303-76

Compound	Solvent	$\lambda_{max}(\log \epsilon)$	Ref.
Pergularinine	MeOH	260(4.7),287(4.4), 313(3.9),341(3.2), 357(2.8)	102-1561-76
Triphyopeltine	EtOH	232(4.68),309(3.93), 323(3.89),377(3.82)	102-0817-76
Tylophorinidine, O-methyl-	MeOH	260(4.7),287(4.4), 313(3.8),340(3.3), 357(2.7)	102-1561-76
$C_{23}H_{25}NO_7$			
Isoquinoline, 2-acetyl-1-(1,3-dihydro-5,6-dimethoxy-3-oxo-1-isobenzofuran-yl)-1,2,3,4-tetrahydro-6,7-dimethoxy-, erythro	EtOH	215(4.46),230(4.43), 265(3.90),295(3.86), 305s(3.73)	78-0211-76
threo	EtOH	215(4.53),228(4.48), 265(3.91),295(3.86)	78-0211-76
$C_{23}H_{25}N_2$			
Malachite Green (cation)	EtOH	622(5.03)	44-0221-76
	8% DMSO	431(4.26),632(4.98)	44-0221-76
$C_{23}H_{25}N_3O_3S$			
4H-Indazol-4-one, 1,5,6,7-tetrahydro-1,6,6-trimethyl-3-phenyl-, O-[(4-methylphenyl)sulfonyl]oxime, (E)-	MeOH	227(4.52),262(3.98)	24-1898-76
$C_{23}H_{25}N_3O_5S$			
4H-Indazol-4-one, 3-(3,4-dimethoxyphen-yl)-1,5,6,7-tetrahydro-1-methyl-, O-[(4-methylphenyl)sulfonyl]oxime, (E)-	MeOH	218(4.49),261s(3.95), 271s(3.89),290s(3.76)	24-1898-76
$C_{23}H_{25}N_3O_{13}$			
D-Galactose, 1-C-(5-nitro-1H-indazol-3-yl)-, 2,3,4,5,6-pentaacetate	EtOH	234(4.04),265(4.03), 310(3.81)	4-1241-76
D-Galactose, 1-C-(6-nitro-1H-indazol-3-yl)-, 2,3,4,5,6-pentaacetate	EtOH	229s(3.85),273(4.37), 340s(3.48)	4-1241-76
$C_{23}H_{25}N_4OS$			
Morpholinium, 4-[[[2-[4-(dimethylamino)-phenyl]-5-phenyl-6H-1,3-thiazin-6-yli-dene]amino]methylene]-, perchlorate	HOAc	319(4.02),390(4.38), 565(4.47)	97-0268-76
$C_{23}H_{25}N_4S$			
Benzothiazolium, 2-[1-cyano-4-(dimethyl-amino)-4-[4-(dimethylamino)phenyl]-1,3-butadienyl]-3-methyl-, perchlorate	MeOH	478(4.73)	124-1174-76
$C_{23}H_{25}N_5O_5S$			
L-Alanine, 3-benzo[b]thien-3-yl-N-[N-(5-oxo-L-prolyl)-L-histidyl]-, methyl ester	HOAc	261(3.83),290(3.54), 299(3.60)	94-3149-76
$C_{23}H_{25}N_5O_7S$			
L-Alanine, 3-benzo[b]thien-3-yl-N-[N-(5-oxo-L-prolyl)-L-histidyl]-, methyl ester, S,S-dioxide	HOAc	302(3.36)	94-3149-76
$C_{23}H_{26}$			
Tricyclo[15.2.2.2^{3,6}]tricosa-3,5,7,15,17,19,20,22-octaene	n.s.g.	235(4.11),255(3.97)	88-1243-76

Compound	Solvent	$\lambda_{max}(\log \epsilon)$	Ref.
$C_{23}H_{26}ClN_4O_4P$			
Phosphinic acid, di-4-morpholinyl-, 3-amino-6-chloro-4-phenyl-2-quinolinyl ester	isoPrOH	249(4.62),344(3.93)	44-2720-76
Phosphinic acid, di-4-morpholinyl-, 7-chloro-5-phenyl-3H-1,4-benzo-diazepin-2-yl ester	isoPrOH	255s(4.16),317(3.31)	44-2720-76
$C_{23}H_{26}Cl_2O_5$			
Androsta-4,6-diene-17-carboxylic acid, 17-acetoxy-6-chloro-16-(chloromethyl)-16-hydroxy-3-oxo-, β-lactone, (16α,17α)-	MeOH	282(4.24)	44-3940-76
$C_{23}H_{26}N_2O$			
Cyclopentanone, 2,5-bis[[4-(dimethyl-amino)phenyl]methylene]-	EtOH	270(--),330(--)	114-0381-76B
	ether	270(4.17),325(3.60), 363(4.69)	114-0381-76B
ion	n.s.g.	495(4.72)	114-0381-76B
non-ion	n.s.g.	405(4.61)	114-0381-76B
2,5-Etheno-5H-cyclopenta[a]phenanthrene-3,4-dicarbonitrile, 1,2,6,7,8,9,10,11-12,13,14,15,16,17-tetradecahydro-10,13-dimethyl-17-oxo-, [2R-(2α,5α,8α,9β,10α,13α,14β)]-	EtOH	213(3.83),243(3.88)	23-3508-76
$C_{23}H_{26}N_2O_3$			
Benzoic acid, 2-[5-(4-methoxyphenyl)-1H-pyrazol-3-yl]-, hexyl ester	EtOH	262(4.42)	44-0110-76
Ethanone, 1-(5-hexyl-2,4-dihydroxyphen-yl)-2-(1-phenyl-1H-pyrazol-4-yl)-	EtOH	276(4.35),330(4.01)	103-0914-76
1-Pyrrolidinecarbonitrile, 2-[2-(3,4-di-methoxyphenyl)-3-(4-methoxyphenyl)-2-propenyl]-, [S-(Z)]-	MeOH	282(4.21)	33-2201-76
$C_{23}H_{26}N_2O_5$			
16,19-Secostrychnidine-10,16-dione, 21,22-epoxy-21,22-dihydro-4-meth-oxy-19-methyl-, (21α,22α)-	EtOH	217(4.29),256(4.11), 295(3.85)	102-1973-76
$C_{23}H_{26}N_2O_6$			
16,19-Secostrychnidine-10,16-dione, 21,22-epoxy-21,22-dihydro-14-hydroxy-4-methoxy-19-methyl-, (21α,22α)-	EtOH	217(4.29),256(4.11), 295(3.58)	102-1973-76
$C_{23}H_{26}N_2O_7$			
16,19-Secostrychnidine-10,16-dione, 21,22-epoxy-21,22-dihydro-4,14-dihy-droxy-3-methoxy-19-methyl-, (21α,22α)-	EtOH	237(4.21),270(3.80)	102-1973-76
$C_{23}H_{26}N_2O_{11}$			
D-Galactose, 1-C-1H-indazol-3-yl-, 2,3,4,5,6-pentaacetate	EtOH	236(4.02),242(4.00), 302(4.10)	4-1241-76
1,3,4-Oxadiazole, 2-(D-gluco-1,2,3,4,5-pentaacetoxypentyl)-5-phenyl-	EtOH	249(4.15)	136-0049-76D
$C_{23}H_{26}O_6$			
D-Glucitol, 1,3:2,4-di-O-benzylidene-5,6-O-(1-methylethylidene)-	MeCN	256(2.62)	12-1859-76

Compound	Solvent	$\lambda_{max}(\log \epsilon)$	Ref.
$C_{23}H_{26}O_7$ Bicyclo[3.2.1]oct-3-en-2-one, 8-acetoxy- 7-(1,3-benzodioxol-5-yl)-1,3-dimeth- oxy-6-methyl-5-(2-propenyl)-, (6-exo,7-endo,8-anti)-	MeOH	234(3.50),268(3.48)	102-1033-76
$C_{23}H_{26}O_8$ Glycerol, 1,3-bis-O-(p-coumaryl)-2-O- acetyl-, tetrahydro deriv.	n.s.g.	226(4.11),279(3.45), 286(3.36)	102-0811-76
$C_{23}H_{26}O_{10}$ Haplanthin, 5-O-methyl-	EtOH	283(4.28)	2-0644-76
$C_{23}H_{26}O_{11}$ Centapicrin, deacetyl- Lindleyin	MeOH EtOH	238(4.13),300(3.41) 230(4.38),282(4.33)	95-0683-76 102-0344-76
$C_{23}H_{27}ClO_7$ Pregn-4-en-21-oic acid, 17,20α-[carbo- nylbis(oxy)]-20β-chloro-3,11-dioxo-, methyl ester	n.s.g.	238(4.19)	44-0078-76
$C_{23}H_{27}NO$ Curryanine, dihydro-	MeOH	240(4.64),256(4.36), 305(4.18),330(3.61)	2-0430-76
$C_{23}H_{27}NO_3$ 17α-Pregna-1,4-diene-21-carboxylic acid, 7α-cyano-17-hydroxy-3-oxo-, γ-lactone	MeOH	241(4.25)	87-0975-76
$C_{23}H_{27}N_2O_4$ Dibenzo[a,g]quinolizinium, 5,6-dihydro- 9-[(2-hydroxyethyl)amino]-2,3,10-tri- methoxy-13-methyl-	EtOH	232(4.50),283(4.47), 346(4.17),480(3.86)	88-1595-76
$C_{23}H_{27}N_3OS_2$ Thiourea, [3-(4-heptylphenyl)-4-oxo- 2-thiazolidinylidene]phenyl-	dioxan	266(4.21),414(3.98)	103-0751-76
$C_{23}H_{27}N_3O_2$ Benzenamine, N,N-dimethyl-4-[1,4,5,6- tetrahydro-6-(1-methylethoxy)-1-phen- ylpyrano[2,3-c]pyrazol-4-yl]-, cis trans	EtOH EtOH	258.5(4.49) 257.5(4.43)	35-2947-76 35-2947-76
$C_{23}H_{28}N_2O$ 2,5-Ethano-5H-cyclopenta[a]phenanthrene- 3,4-dicarbonitrile, 1,2,6,7,8,9,10,11- 12,13,14,15,16,17-tetradecahydro- 10,13-dimethyl-17-oxo-, α- β-	EtOH EtOH	242(3.94) 242(3.90)	23-3508-76 23-3508-76
$C_{23}H_{28}N_2O_3$ 1H-Indole-3-carboxylic acid, 6-[(dimeth- ylamino)methyl]-5-methoxy-2-methyl- 1-(phenylmethyl)-, ethyl ester, hydrochloride	n.s.g.	221(4.53),250(4.44), 289(4.06)	103-0245-76
$C_{23}H_{28}N_2O_4$ 1-Pyrrolidinecarbonitrile, 2-[2-(3,4-	MeOH	218(4.22),277(3.62)	33-2201-76

Compound	Solvent	$\lambda_{max}(\log \epsilon)$	Ref.
dimethoxyphenyl)-3-hydroxy-3-(4-methoxyphenyl)propyl]- (cont.)			33-2201-76
$C_{23}H_{28}N_2O_6$			
Benz[cd]indole-1(3H)-carboxylic acid, 4-[(ethoxycarbonyl)amino]-5-(3-ethoxy-2-methyl-3-oxo-1-propenyl)-4,5-dihydro-, ethyl ester, [4α,5β(E)]-(+)-	MeOH	228(4.43),262(3.93), 268(3.93),288(3.76), 298(3.81)	24-2140-76
Herboxine	n.s.g.	224(4.39),250s(--)	105-0201-76
$C_{23}H_{28}O_3$			
14α-Carda-3,5,20(22)-trienolide, 7-oxo-	MeOH	214(4.21),267(4.31)	24-3304-76
4H-Indeno[5,4-b]furan, 6-(1,1-dimethylethoxy)-5,5a,6,7-tetrahydro-2-(3-methoxyphenyl)-5a-methyl-, (5aS-cis)-	MeOH	222(4.48),297s(4.26), 308(4.28),317(4.19)	24-2948-76
$C_{23}H_{28}O_4$			
2H-1-Benzopyran-2-acetic acid, 3,4-dihydro-2,5,7,8-tetramethyl-6-(phenylmethoxy)-, methyl ester, (S)-	EtOH	227s(4.08),258(2.94), 280(3.28),287(3.33)	33-0290-76
Pregna-1,4,16-triene-3,20-dione, 11β-acetoxy-	MeOH	239(4.39)	24-0185-76
$C_{23}H_{28}O_5$			
17α-Pregna-1,4-diene-3,20-dione, 11β-acetoxy-16β,17-epoxy-	MeOH	242(4.17)	24-0185-76
$C_{23}H_{28}O_{10}$			
Leucanthinin	MeOH	226(3.75)	44-3956-76
$C_{23}H_{29}BrO_5$			
Pregna-1,4-diene-3,20-dione, 11β-acetoxy-17-bromo-16β-hydroxy-	MeOH	241(4.18)	24-0185-76
$C_{23}H_{29}ClN_2O_9S$			
4-Thiouridine, 2'-O-(tetrahydromethoxypyranyl)-5'-O-(4-chlorophenoxyacetyl)-	EtOH	328(4.26)	54-0108-76
$C_{23}H_{29}FO_5$			
19-Norpregna-1,3,5(10)-trien-3-ol, 11α-fluoro-9β-methyl-17α,20:20,21-bis[methylenebis(oxy)]-	MeOH	281(3.13)	39-0101-76C
$C_{23}H_{29}NO_3$			
17α-Pregn-4-ene-21-carboxylic acid, 7α-cyano-17α-hydroxy-3-oxo-, γ-lactone	MeOH	235(4.19)	87-0975-76
$C_{23}H_{29}N_3O_3$			
Voaketone, O-acetyloxime	n.s.g.	225(4.35),280(3.91), 293s(3.85)	33-0532-76
$C_{23}H_{30}ClNO_4$			
Pregn-4-en-18-al, 21-chloro-3,20-dioxo-, 18-(O-acetyloxime)	MeOH	240(4.20)	44-3532-76
$C_{23}H_{30}N_2O_3$			
Acetamide, 2-[(4-ethoxyphenyl)imino]-N-[4-(heptyloxy)phenyl]-	EtOH	245(3.85)	65-0702-76

Compound	Solvent	$\lambda_{max}(\log \epsilon)$	Ref.
$C_{23}H_{30}N_2O_4$			
Acetamide, 2-[(4-ethoxyphenyl)imino]-N-[4-(heptyloxy)phenyl]-, N-oxide	EtOH	230(4.2),290(4.0), 310(4.0),340(4.0)	65-0702-76
$C_{23}H_{30}N_2O_5$			
Strychnofendlerine, 12-hydroxy-11-methoxy-	EtOH	226(4.41),254(3.90), 290s(3.28)	32-0773-76
	EtOH-base	308(3.77)	32-0773-76
$C_{23}H_{30}N_2O_{13}$			
1H-Pyrazole-3-carboxylic acid, 5-(acetoxymethyl)-1-(2,3,4,6-tetra-O-acetyl-β-D-glucopyranosyl)-, ethyl ester	MeOH	220(4.02)	65-1370-76
1H-Pyrazole-5-carboxylic acid, 3-(acetoxymethyl)-1-(2,3,4,6-tetra-O-acetyl-β-D-glucopyranosyl)-, ethyl ester	MeOH	228(4.06)	65-1370-76
$C_{23}H_{30}O_3$			
Benzoic acid, 3-[2-[octahydro-4a-methyl-5-(1-methylethenyl)-1-oxo-1H-inden-4-yl]ethyl]-, methyl ester, [3aS-(3aα,4α,5β,7aβ)]-	MeOH	232(4.03),279(3.26), 286(3.23)	23-3508-76
$C_{23}H_{30}O_4$			
14α-Carda-4,20(22)-dienolide, 11α-hydroxy-3-oxo-	MeOH	222(4.24)	24-3304-76
14α-Carda-5,20(22)-dienolide, 3β-hydroxy-7-oxo-	MeOH	220(3.94)	24-3304-76
17β,18-Cyclo-17α-pregn-4-ene-3,20-dione, 21-acetoxy-	n.s.g.	240(4.26)	39-1064-76C
5H-Inden-5-one, 1-(1,1-dimethylethoxy)-1,2,3,6,7,7a-hexahydro-4-[2-(3-methoxyphenyl)-2-oxoethyl]-7a-methyl-, (1S-cis)-	MeOH	217(4.43),250(4.33), 306(3.43)	24-2948-76
Pregna-4,6-diene-3,20-dione, 12β-acetoxy-	MeOH	283(4.42)	102-1745-76
$C_{23}H_{30}O_5$			
14α-Carda-4,20(22)-dienolide, 6β,11α-dihydroxy-3-oxo-	MeOH	221(3.99)	24-3304-76
14α-Carda-5,20(22)-dienolide, 3β,11α-dihydroxy-7-oxo-	MeOH	221(3.71)	24-3304-76
Carda-5,20(22)-dienolide, 3β,14α-dihydroxy-7-oxo-	MeOH	222(4.32)	24-3304-76
14α-Carda-5,20(22)-dienolide, 3β,17α-dihydroxy-7-oxo-	MeOH	224(4.29)	24-3304-76
Estra-1,3,5(10)-triene-3,17β-diol, 3-O-methyl-17α-(α-acetoxyacetyl)-	EtOH	278(3.97),287(3.93)	44-2312-76
Pregna-1,4-diene-3,20-dione, 17α,21-dihydroxy-, 21-acetate	EtOH	245(4.37)	94-0828-76
Pregn-4-ene-3,11,20-trione, 17-acetoxy-	n.s.g.	238(4.21)	44-0078-76
$C_{23}H_{30}O_5S$			
Pregna-1,4-diene-3,20-dione, 11β-acetoxy-16β-hydroxy-17-mercapto-	MeOH	241(4.19)	24-0185-76
$C_{23}H_{30}O_6$			
Androst-1-ene-3,11-dione, 2,17-diacetoxy-, (5α,17β)-	EtOH	235(3.98)	95-0863-76
Androst-3-ene-2,11-dione, 3,17-diacetoxy-, (5α,17β)-	EtOH	237(4.21)	95-0863-76

Compound	Solvent	$\lambda_{max}(\log \epsilon)$	Ref.
Benzene, 1,3-dimethoxy-5-[1-methyl-2-(3,4,5-trimethoxyphenyl)ethyl]-2-(2-propenyloxy)- (aurein)	EtOH	235s(4.10),270(4.08)	102-1289-76
Pregn-4-ene-3,11-dione, 17,20α-[carbonylbis(oxy)]-20β-methoxy-	n.s.g.	238(4.20)	44-0078-76
Pregn-4-ene-3,11-dione, 17,20β-[carbonylbis(oxy)]-20α-methoxy-	n.s.g.	238(4.20)	44-0078-76
$C_{23}H_{30}O_9$ Melampolidin	MeOH	222(4.08)	44-3956-76
$C_{23}H_{31}NO$ Polyalthenol	EtOH	226(4.52),285(3.76), 292(3.73)	88-3559-76
$C_{23}H_{31}NO_4$ 17α-Pregn-4-ene-21-carboxylic acid, 7α-cyano-17-hydroxy-3-oxo-, K salt	MeOH	241(4.17)	87-0975-76
$C_{23}H_{32}ClNO_4$ Pregn-4-en-18-al, 21-chloro-20β-hydroxy-3-oxo-, 18-(O-acetyloxime), (20S)-	MeOH	240(4.20)	44-3532-76
$C_{23}H_{32}O_2$ 9β,10α-Pregna-5,7,20-trien-3-one, cyclic 1,2-ethanediyl acetal	EtOH	270(4.007)	33-1850-76
$C_{23}H_{32}O_3$ Dzhamirone	n.s.g.	230(3.90),275(4.16), 312(3.85)	105-0736-76
$C_{23}H_{32}O_4$ 1,3-Benzenediol, 2-[3-(acetoxymethyl)-6-(1-methylethenyl)-2-cyclohexen-1-yl]-5-pentyl-, trans (7-acetoxycannabidiol)	EtOH	282(3.28)	39-0008-76C
14α-Card-20(22)-enolide, 5,6α-epoxy-3β-hydroxy-	MeOH	215(4.11)	24-3304-76
6H-Dibenzo[b,d]pyran-6-one, 3-(1,2-dimethylheptyl)-7,8,9,10-tetrahydro-9-hydroxy-1-methoxy-	EtOH	275s(3.35),283(3.42)	33-1963-76
17-Norabieta-8,11,13-trien-7-one, 10-(ethoxycarbonyl)-12-methoxy-	EtOH	236(4.41),281(4.23)	44-1005-76
Pregna-1,4-diene-20-carboxylic acid, 12α-hydroxy-3-oxo-, methyl ester	MeOH	244(4.23)	78-0089-76
$C_{23}H_{32}O_5$ 14α-Card-20(22)-enolide, 5,6α-epoxy-3β,11α-dihydroxy-	MeOH	214(4.20)	24-3304-76
14α-Card-20(22)-enolide, 5,6α-epoxy-3β,17α-dihydroxy-	MeOH	214(4.06)	24-3304-76
Juniferin	n.s.g.	211(4.0),260(4.3)	105-0491-76
Juniferinin	n.s.g.	211(4.0),260(4.3)	105-0491-76
Pregna-1,4-dien-3-one, 21-acetoxy-17α,20-dihydroxy-, (20R)-	EtOH	245(4.21)	94-0828-76
Pregn-4-en-3-one, 17,20-[carbonylbis(oxy)]-20-methoxy-	n.s.g.	240(4.23)	44-0078-76
$C_{23}H_{32}O_6$ Androst-4-en-3-one, 6-(3-carboxy-1-oxopropoxy)-17-hydroxy-, (6α,17β)-	pH 8.7	244(4.01)	95-0199-76

Compound	Solvent	λ_{max} (log ϵ)	Ref.
Androst-4-en-3-one, 6-(3-carboxy-1-oxo-propoxy)-17-hydroxy-, $(6\beta,17\beta)$-	pH 8.7	243(4.00)	95-0199-76
24-Norchol-20(22)-ene-19,23-dioic acid, 3,14,21-trihydroxy-, γ-lactone, $(3\beta,5\alpha,14\beta)$-	MeOH	218(4.31)	94-2886-76
$C_{23}H_{32}Si$ Silane, (1,2-diphenylethenyl)tripropyl-	n.s.g.	238(4.22),267(4.28)	65-1044-76
$C_{23}H_{33}FO_4$ 6H-Dibenzo[b,d]pyran-1,10-diol, 9-fluoro-6a,7,8,9,10,10a-hexahydro-6,6,9-trimethyl-3-pentyl-, 10-acetate, [6aR-(6α,9α,10β,10aβ)]-	EtOH	283(3.26)	33-1963-76
$C_{23}H_{33}NO_2$ Androsta-5,7-dien-3-one, 17-(2-aziridinyl)-, cyclic 1,2-ethanediyl acetal	EtOH	271(3.988)	33-1850-76
$C_{23}H_{34}O_2$ Pregn-4-ene-3,20-dione, 16,17-dimethyl-	EtOH	240(4.19)	39-1558-76C
$C_{23}H_{34}O_2S$ Thiopyrano[2,3-c][1]benzopyran-10-ol, 8-(1,2-dimethylheptyl)-1,2,3,5-tetrahydro-5,5-dimethyl-	EtOH	305(4.2625)	87-0549-76
2H,4H-Thiopyrano[3,4-c][1]benzopyran-10-ol, 8-(1,2-dimethylheptyl)-1,5-dihydro-5,5-dimethyl-	EtOH	275(3.6)	87-0549-76
$C_{23}H_{34}O_3$ 2H-Indeno[5,4-b]furan, 6-(1,1-dimethylethoxy)decahydro-2-(3-methoxyphenyl)-5a-methyl-, (3aS,5aS,8aS,8bR)-(+)-	MeOH	217(3.92),273(3.32), 280(3.28)	24-2948-76
5H-Inden-5-one, 1-(1,1-dimethylethoxy-decahydro-4-[2-(3-methoxyphenyl)ethyl]-7a-methyl-, (1S,3aS,4R,7aS)-(+)-	MeOH	217(3.92),273(3.31), 279(3.26)	24-2948-76
(1S,3aS,4S,7aS)-(+)-	MeOH	216(3.92),272(3.31), 280(3.25)	24-2948-76
4a(2H)-Phenanthrenecarboxylic acid, 1,3,4,9,10,10a-hexahydro-6-methoxy-1,1-dimethyl-7-(1-methylethyl)-, ethyl ester, trans-($\pm$)-	EtOH	240(3.86),281(3.58), 288(3.57)	44-1005-76
Pregna-5,16-dien-20-one, 21-ethoxy-3β-hydroxy-	EtOH	242(3.93)	33-1027-76
$C_{23}H_{34}O_3S$ Pregn-4-ene-3,20-dione, 21-[(2-hydroxyethyl)thio]-	MeOH	240(4.30)	44-2292-76
$C_{23}H_{34}O_4$ 9(5H)-Phenanthrenone, 4b,6,7,8,8a,10-hexahydro-1,3,4-trimethoxy-4b,8,8-trimethyl-2-(1-methylethyl)-, (4bS-trans)-	EtOH	234(3.34),282(2.91)	32-0119-76
Pregn-4-en-3-one, 18-acetoxy-20β-hydroxy-	MeOH	240(4.22)	44-3532-76
$C_{23}H_{34}O_5$ 14α-Card-20(22)-enolide, 3β,5α,6α-trihydroxy-	MeOH	215(3.93)	24-3304-76

Compound	Solvent	λ_{max}(log ϵ)	Ref.
14α-Card-20(22)-enolide, 3β,5α,6β-tri-hydroxy-	MeOH	216(4.18)	24-3304-76
Compactin	EtOH	230(4.28),237(4.30), 246(4.11)	39-1165-76C
Dibenzo[b,d]pyran-1,9α,10β-triol, 6a,7,8,9,10,10a-hexahydro-6,6,9β-trimethyl-3-pentyl-, 10-acetate	EtOH EtOH-NaOH	277(3.25),283(3.25) 286s(3.47),294(3.49)	33-1963-76 33-1963-76
Prost-14-en-5-yn-1-oic acid, 9,9-[1,2-ethanediylbis(oxy)]-13-oxo-, methyl ester, (14E)-	MeOH	227(4.11)	107-0039-76
$C_{23}H_{35}NO_3$ 4a(2H)-Phenanthrenecarboxylic acid, 5-amino-1,3,4,9,10,10a-hexahydro-6-methoxy-1,1-dimethyl-7-(1-methyl-ethyl)-, ethyl ester, trans-(±)-	EtOH	297(3.46)	44-1005-76
$C_{23}H_{36}N_2O_3$ Carbamic acid, [3-(hexyloxy)phenyl]-, 3-(1-pyrrolidinyl)cyclohexyl ester	H_2O	237(4.08),277(3.34)	106-0024-76
Carbamic acid, [4-(hexyloxy)phenyl]-, 3-(1-pyrrolidinyl)cyclohexyl ester	H_2O	239(4.19),282(3.18)	106-0024-76
Carbamic acid, [3-(pentyloxy)phenyl]-, 3-(1-piperidinyl)cyclohexyl ester	H_2O	236(4.07),277(3.39)	106-0024-76
Carbamic acid, [4-(pentyloxy)phenyl]-, 3-(1-piperidinyl)cyclohexyl ester	H_2O	239(4.20),282(3.16)	106-0024-76
$C_{23}H_{36}O_2$ Androst-4-en-3-one, 7α-butyl-17β-hydr-oxy-	n.s.g.	243(4.19)	13-0759-76A
Androst-4-en-3-one, 7β-butyl-17β-hydr-oxy-	n.s.g.	243(4.30)	13-0759-76A
$C_{23}H_{36}O_3$ 1H-Inden-5-ol, 1-(1,1-dimethylethoxy)-octahydro-4-[2-(3-methoxyphenyl)eth-yl]-7a-methyl-, (1S,3aS,4R,5S,7aS)-(+)-	MeOH	217(3.87),267s(3.14), 272(3.27),279(3.24)	24-2948-76
Phenanthrene, 1,2,3,4,4a,9,10,10a-octa-hydro-5,6,8-trimethoxy-1,1,4a-trimeth-yl-7-(1-methylethyl)-, (4aS-trans)-	EtOH	281(3.87)	32-0119-76
$C_{23}H_{36}O_4$ 9-Phenanthrenol, 4b,5,6,7,8,8a,9,10-octahydro-1,3,4-trimethoxy-4b,8,8-trimethyl-2-(1-methylethyl)-, [4bS-(4bα,8aβ,9α)]-	EtOH	280(2.99)	32-0119-76
$C_{23}H_{36}O_6S$ 4a(2H)-Naphthalenecarboxylic acid, octa-hydro-3-(hydroxymethylene)-8,8-dimeth-yl-4-[4-methyl-1-(methylsulfinyl)-2-oxopentyl]-2-oxo-, ethyl ester	EtOH	277(3.93)	44-1005-76
$C_{23}H_{37}NO_2$ 2H-[1]Benzopyrano[3,4-d]pyridin-10-ol, 8-(1,2-dimethylheptyl)-1,3,4,4a,5,10b-hexahydro-5,5-dimethyl-, cis	EtOH	280(2.63)	87-0445-76

Compound	Solvent	$\lambda_{max}(\log \epsilon)$	Ref.
$C_{23}H_{37}N_3O_2$ Pregn-5-en-20-one, 3 -hydroxy-16 -methyl-, semicarbazone	EtOH	228(4.16)	39-1558-76C
$C_{23}H_{37}N_3O_5$ Tetradecanoic acid, 2,3,3a,9a-tetrahydro-2-(hydroxymethyl)-6-imino-6H-furo-[2',3':4,5]oxazolo[3,2-a]pyrimidin-3-yl ester, hydrochloride, [2R-(2α,3β,3aβ,9aβ)]-	MeOH-acid	232(3.98),264(4.02)	87-0654-76
$C_{23}H_{38}N_2O_3$ Carbamic acid, [3-(hexyloxy)phenyl]-, 3-(diethylamino)cyclohexyl ester	H_2O	236(4.04),277(3.35)	106-0024-76
Carbamic acid, [4-(hexyloxy)phenyl]-, 3-(diethylamino)cyclohexyl ester	H_2O	239(4.20),281(3.15)	106-0024-76
$C_{23}H_{39}N_3O_6$ 2(1H)-Pyrimidinone, 4-amino-1-[3-O-(1-oxotetradecyl)-β-D-arabinofuranosyl]-	MeOH-acid	212(4.02),284(4.17)	87-0667-76
$C_{23}H_{40}OS$ 1-Nonadecanone, 1-(2-thienyl)-	heptane	258(3.98),275(3.87)	34-0233-76
$C_{23}H_{40}O_2S$ 2-Thiophenecarboxylic acid, 5-heptadecyl-, methyl ester	heptane	254(3.95),275(4.06)	34-0233-76
2-Thiophenecarboxylic acid, 5-heptadecyl-4-methyl-	heptane	253(3.92),284(3.96)	34-0233-76
$C_{23}H_{41}NO$ 2,4-Cyclohexadien-1-one, 2,4,6-tris(1,1-dimethylethyl)-6-methyl-, O-(1,1-dimethylethyl)oxime	hexane	314(3.81)	18-1142-76
$C_{23}H_{42}S$ Thiophene, 2-nonadecyl-	heptane	235(3.90)	34-0233-76
$C_{23}H_{54}Si_6$ Silane, 2,4-cyclopentadiene-2,3,4,5-tetrayl-1-ylidenehexakis[trimethyl-	heptane	214(3.85)	65-2630-76

Compound	Solvent	$\lambda_{max}(\log \epsilon)$	Ref.
$C_{24}H_{12}O_2$ 3,4:9,10-Dibenzopyrene-5,8-dione	EtOH	222(4.71),240(5.00), 268(4.65),284(4.78), 296(4.99),309(4.71), 329(4.38),351(4.35), 370(4.72),392(4.88), 420(3.85),445(3.88)	25-0029-76
$C_{24}H_{13}BrO_2$ Benz[a]anthracene-7,12-dione, 2-bromo- 5-phenyl-	CHCl$_3$	249(4.53),290(4.62), 367(3.53),416(3.69)	18-3361-76
$C_{24}H_{13}ClO_2$ Benz[a]anthracene-7,12-dione, 2-chloro- 5-phenyl-	CHCl$_3$	250(4.19),290(4.30), 366(3.21),416(3.36)	18-3361-76
$C_{24}H_{14}ClNO_2$ 2-Penten-4-ynenitrile, 3-(benzoyloxy)- 2-(4-chlorophenyl)-5-phenyl-	EtOH	213(4.23),237(4.40), 340(4.48)	34-0115-76
$C_{24}H_{14}Cl_2N_2$ Benzonitrile, 2,2'-(1,4-phenylenedi- 2,1-ethenediyl)bis[4-chloro-	DMF	370(4.74)	33-0819-76
Benzonitrile, 2,2'-(1,4-phenylenedi- 2,1-ethenediyl)bis[6-chloro-	DMF	374(4.71)	33-0819-76
Benzonitrile, 4,4'-(1,4-phenylenedi- 2,1-ethenediyl)bis[3-chloro-	DMF	378(4.82)	33-0819-76
$C_{24}H_{14}Cl_4N_2O_2S$ Sulfur diimide, bis[4-(2,4-dichlorophen- oxy)phenyl]-	n.s.g.	442(3.97)	97-0318B-76
$C_{24}H_{14}Cl_4O_2$ 1H-Cyclopenta[b][1,4]benzodioxin, 5,6,7,8-tetrachloro-1-(diphenyl- methylene)-3a,9a-dihydro-, cis	C$_6$H$_{12}$	217(5.15),232s(4.71), 291(4.61),299s(4.57)	78-0147-76
4,7-Ethano-1H-indene-8,9-dione, 4,5,6,7- tetrachloro-1-(diphenylmethylene)- 3a,9a-dihydro-, (3aα,4β,7β,7aα)-	C$_6$H$_{12}$	290(4.15),350s(3.33), 450s(2.35)	78-0147-76
Spiro[1,3-benzodioxole-2,1'-[3]cyclo- pentene], 4,5,6,7-tetrachloro- 2'-(diphenylmethylene)-	C$_6$H$_{12}$	218(4.69),283(4.176)	5-0793-76
$C_{24}H_{14}FNO_2$ Benzeneacetonitrile, α-[1-(benzoyloxy)- 3-phenyl-2-propynylidene]-4-fluoro-	EtOH	210(4.02),236(4.31), 329(4.37)	34-0115-76
$C_{24}H_{14}N_2O_2S$ Benzoxazole, 2-[5-[4-(2-benzoxazolyl)- 4-(2-benzoxazolyl)phenyl]-2-thienyl]-	DMF	279(3.90),299(3.85), 313(3.96),368(4.71), 380(4.74),401s(4.50)	48-0731-76
$C_{24}H_{14}O_2$ Benz[a]anthracene-7,12-dione, 5-phenyl-	CHCl$_3$	249(4.43),291(4.59), 334(3.61),368(3.48), 417(3.64)	18-3361-76
$C_{24}H_{14}O_3$ 5,12-Naphthacenedione, 6-hydroxy-11- phenyl-	EtOH	257(4.75),284(4.30), 299(4.20),438(4.00)	39-0336-76C

Compound	Solvent	$\lambda_{max}(\log \epsilon)$	Ref.
$C_{24}H_{15}Cl_2N_5O$ 1,3,4-Oxadiazole, 2-[3-chloro-4-[2-[2-(4-chlorophenyl)-2H-1,2,3-triazol-4-yl]ethenyl]phenyl]-5-phenyl-	DMF	348(4.74)	33-2469-76
$C_{24}H_{15}I$ Benzo[c]phenanthrene, 2-iodo-1-phenyl-	MeOH	314s(4.19),326s(4.02), 337s(3.83),368(2.66), 385(2.39)	39-1115-76B
Phenanthrene, 3-iodo-4-(1-naphthalenyl)-	MeOH	280(4.44),292(4.40), 305(4.31),337(2.87), 345(2.56),354(2.80)	39-1115-76B
$C_{24}H_{15}NOS_2$ Bis[1]benzothieno[3,2-b:2',3'-e]pyridine, 6-(4-methoxyphenyl)-	$C_2H_4Cl_2$	256(4.30),294(4.50), 307(4.50),374(3.30)	103-0977-76
$C_{24}H_{15}NO_5$ 2H-Benzo[f]pyrano[3,2-c]quinoline-3-carboxylic acid, 5,6-dihydro-2,5-dioxo-6-phenyl-, methyl ester	n.s.g.	382(4.40)	64-0514-76B
$C_{24}H_{16}$ Benzo[c]phenanthrene, 1-phenyl-	MeOH	307s(4.05),318s(3.93), 340s(3.68)	39-1115-76B
Naphthalene, 1,1'-(1-buten-3-yne-1,4-diyl)bis-, (E)-	CH_2Cl_2	267s(3.83),352(4.48)	39-1104-76B
Naphthalene, 1-[4-(2-naphthalenyl)-3-buten-1-ynyl]-, (E)-	CH_2Cl_2	266(4.27),276(4.34), 291(4.20),299(4.24), 354(4.63),350s(4.62), 374(4.55)	39-1104-76B
[2.2](2,7)-Naphthalenophane-1,11-diene	EtOH	239(4.99),283(4.31), 324(3.33),363s(3.20)	12-0163-76
Phenanthrene, 1-(1-naphthalenyl)-	MeOH	277(4.48),298(4.37)	39-1115-76B
Phenanthrene, 4-(1-naphthalenyl)-	MeOH	276(4.29),285(4.30), 297(4.28),333(2.84), 349(2.76)	39-1115-76B
Phenanthrene, 3-(4-phenyl-1-buten-3-ynyl)-, cis	MeOH	238(4.60),245(4.62), 252(4.68),272(4.41), 283(4.36),328s(4.39), 339(4.44),349s(4.33), 360s(4.23),367s(4.00)	39-1104-76B
trans	MeOH	226s(4.37),241s(4.45), 248(4.49),252(4.51), 274(4.47),285(4.46), 331s(4.64),341(4.70), 352s(4.51),368s(4.24)	39-1104-76B
$C_{24}H_{16}As$ 5,5'-Spirobi[5H-dibenzoarsolium], bromide	MeOH	237(4.45),242s(4.36), 273(3.96),330(3.48)	139-0151-76A
$C_{24}H_{16}BrN_3O$ [3,2':3',3''-Ter-1H-indol]-2''(3''H)-one, 5''-bromo-	MeOH	219(4.87),259(4.26), 289s(4.16),292(4.09)	103-0178-76
$C_{24}H_{16}Br_2N_2O$ Spiro[2H-1-benzopyran-2,5'(4'H)-pyrido-[2,3,4,5-lmn]phenanthridine], 6,8-dibromo-4',10'-dimethyl-	EtOH	230(4.85),255s(4.62), 315(3.98),370(3.85)	103-0273-76
	$C_2H_4Cl_2$	255(4.5),285s(4.24),	103-0273-76

Compound	Solvent	$\lambda_{max}(\log \epsilon)$	Ref.
(cont.)		305s(3.93),320s(3.9), 330s(3.87),340s(3.86), 360(3.8),375(3.79)	103-0273-76
$C_{24}H_{16}Br_2N_6O_2$ 2,5-Cyclohexadiene-1,4-dione, 2,5-di-bromo-3,6-bis[(2-methyl-6-quinoxal-inyl)amino]-	MeOH	214(3.80),234(3.70), 259(3.57)	56-0073-76
$C_{24}H_{16}ClNO_2$ Indeno[1,2-b]pyran-2(5H)-one, 3-chloro-4-(diphenylamino)-	EtOH	246(4.21),275(4.26), 360(4.36)	4-1201-76
$C_{24}H_{16}ClN_2O_2$ Isoquinolinium, 2-[(1-amino-4-chloro-9,10-dihydro-9,10-dioxo-2-anthra-cenyl)methyl]-, chloride	MeOH	470(3.82)	2-0054-76
Quinolinium, 1-[(1-amino-4-chloro-9,10-dihydro-9,10-dioxo-2-anthracenyl)-methyl]-, chloride	MeOH	470(3.80)	2-0054-76
$C_{24}H_{16}ClN_5O$ 1,2,4-Oxadiazole, 3-(4-chlorophenyl)-5-[4-[2-(2-phenyl-2H-1,2,3-triazol-4-yl)ethenyl]phenyl]-	DMF	340(4.71)	33-2469-76
1,2,4-Oxadiazole, 5-[4-[2-[2-(4-chloro-phenyl)-2H-1,2,3-triazol-4-yl]ethen-yl]phenyl]-3-phenyl-	DMF	341(4.76)	33-2469-76
1,3,4-Oxadiazole, 2-(4-chlorophenyl)-5-[4-[2-(2-phenyl-2H-1,2,3-triazol-4-yl)ethenyl]phenyl]-	DMF	344(4.76)	33-2469-76
1,3,4-Oxadiazole, 2-[3-chloro-4-[2-(2-phenyl-2H-1,2,3-triazol-4-yl)ethen-yl]phenyl]-5-phenyl-	DMF	348(4.70)	33-2469-76
$C_{24}H_{16}Cl_4O_2$ 1H-Cyclopenta[b][1,4]benzodioxin, 5,6,7,8-tetrachloro-1-(diphenyl-methylene)-2,3,3a,9a-tetrahydro-, cis	C_6H_{12}	217(4.97),231(4.69), 254s(4.34),289s(3.59)	78-0147-76
Spiro[1,3-benzodioxole-2,1'-cyclopent-ane], 4,5,6,7-tetrachloro-2'-(di-phenylmethylene)-	C_6H_{12}	219(4.48),240s(4.10), 295s(3.36),305(4.48)	5-0793-76
$C_{24}H_{16}N$ 9,9'-Spirobi[9H-carbazolium], chloride	MeOH	223(4.52),230(4.52), 266(4.52),275s(4.38), 290s(3.57),301(3.19)	139-0151-76A
$C_{24}H_{16}NO_3$ Quinolinium, 1-[(9,10-dihydro-1-hydroxy-9,10-dioxo-2-anthracenyl)methyl]-, chloride	MeOH	412(3.76)	2-0054-76
$C_{24}H_{16}N_2$ 1,10-Phenanthroline, 4,7-diphenyl-, ferrous complex	MeOH	???(4.36)	118-0001-76
$C_{24}H_{16}N_2O_2$ Benzo[1,2-d:4,5-d']bisoxazole, 2,6-bis(2-phenylethenyl)-	$CHCl_3$	364(4.76),380(4.80), 402s(4.58)	103-1331-76

Compound	Solvent	$\lambda_{max}(\log \epsilon)$	Ref.
Benzo[1,2-d:5,4-d']bisoxazole, 2,6-bis-(2-phenylethenyl)-	CHCl$_3$	332(4.69)	103-1331-76
$C_{24}H_{16}N_2Te$ 10H-Tellurolo[3',4':3,4]pyrrolo[1,2-a]-benzimidazole, 1,3-diphenyl-	CHCl$_3$	305(4.29),360(4.39)	24-3886-76
$C_{24}H_{16}N_4$ Benzo[a]phenazine, 5-(2-methyl-1H-benz-imidazol-1-yl)-	dioxan	226(4.69),279(4.68), 358s(3.83),379(4.07), 390(3.99),399(4.13)	49-0879-76
	50% dioxan	223(4.71),280(4.71), 364s(3.87),381(4.09), 390(4.03),401(4.16)	49-0879-76
$C_{24}H_{16}N_6$ Carbazole, 4-[(2'-azido[1,1'-biphenyl]-2-yl)azo]-	ether	250(4.55),306(3.99), 435(4.00)	18-2495-76
$C_{24}H_{16}N_8$ Diazene, bis(2'-azido[1,1'-biphenyl]-2-yl)-	ether	325(3.91),460(2.40)	18-2495-76
$C_{24}H_{16}O$ Indeno[2,1-b]pyran, 2,4-diphenyl-	dioxan	223(4.31),273(4.40), 375(4.23),526(3.47)	78-2219-76
$C_{24}H_{16}O_2$ 2-Cyclobuten-1-one, 4-(2-oxo-2-phenyl-ethylidene)-2,3-diphenyl-	CHCl$_3$	<u>290(4.2),330(4.4)</u>	24-1506-76
$C_{24}H_{16}O_2S$ Methanone, (2-phenyl-3,4-thiophenediyl)-bis[phenyl-	MeOH	255(4.62)	44-1724-76
$C_{24}H_{16}O_4$ 2H-Pyran-2-one, 6-(1,3-benzodioxol-5-yl)-3,4-diphenyl-	EtOH	236(4.38),265(4.29), 375(4.41)	4-0195-76
$C_{24}H_{16}O_4S_2$ 1,4-Dithiin, 2,5-di-2-naphthalenyl-, 1,1,4,4-tetraoxide	n.s.g.	208(4.11)	4-0057-76
$C_{24}H_{16}O_6$ 5a,11a-[2]Butenonaphthacene-5,6,11,12-tetrone, 13-acetoxy-	MeOH	232(4.60),262(4.21), 313(3.74)	44-2296-76
2H-Pyran-3-carboxylic acid, 4-hydroxy-6-methoxy-2-oxo-5-(1-pyrenyl)-, methyl ester	EtOH	241(4.97),265(4.67), 275(4.84),324(4.60), 239[sic](4.70)	78-0269-76
$C_{24}H_{16}P$ 5,5'-Spirobi[5H-dibenzophospholium], iodide	MeOH	241(4.99),273s(3.96), 346(3.55)	139-0151-76A
$C_{24}H_{17}BrN_2O$ Spiro[2H-1-benzopyran-2,5'(4'H)-pyrido-[2,3,4,5-lmn]phenanthridine], 6-bromo-4',10'-dimethyl-	EtOH	228(4.78),253(4.58), 285s(4.22),311s(3.95), 378(3.80)	103-0273-76
	$C_2H_4Cl_2$	256(4.56),285s(4.27), 305s(3.98),365s(3.8), 375(3.81)	103-0273-76

Compound	Solvent	$\lambda_{max}(\log \epsilon)$	Ref.
$C_{24}H_{17}BrN_2O_3S$ 2H-Pyridine-3-carboxamide, 1-(4-bromo-phenyl)-6-hydroxy-2-oxo-N-phenyl-4-(phenylthio)-	MeOH	336(4.55)	120-0103-76
$C_{24}H_{17}ClN_2O_2$ Ethanone, 1-[3-benzoyl-1-(4-chlorophen-yl)-5-phenyl-1H-pyrazol-4-yl]-	EtOH	251(4.50)	4-0989-76
$C_{24}H_{17}ClN_2O_3S$ 2H-Pyridine-3-carboxamide, 1-(4-chloro-phenyl)-6-hydroxy-2-oxo-N-phenyl-4-(phenylthio)-	MeOH	339.5(4.50)	120-0103-76
$C_{24}H_{17}ClN_4O$ Benzoxazole, 5-chloro-2-[4-[2-(5-methyl-2-phenyl-2H-1,2,3-triazol-4-yl)ethen-yl]phenyl]-	DMF	363(4.80)	33-2469-76
Benzoxazole, 2-[4-[2-[2-(4-chlorophenyl)-5-methyl-2H-1,2,3-triazol-4-yl]ethen-yl]phenyl]-	DMF	356(4.81)	33-2469-76
$C_{24}H_{17}Cl_2NO_2$ Indeno[1,2-b]pyran-2(3H)-one, 3,3-di-chloro-4-(diphenylamino)-4,5-dihydro-	EtOH	243(4.30),275(4.20)	4-1201-76
$C_{24}H_{17}NO$ Benzonitrile, 4-[2-[4-(2-benzofuranyl)-phenyl]ethenyl]-3-methyl-	DMF	364(4.76)	33-0819-76
$C_{24}H_{17}NOS$ [1]Benzothieno[3,2-b]pyridine, 4-(4-methoxyphenyl)-2-phenyl-	$C_2H_4Cl_2$	255(5.34),290(5.36), 345(4.71)	103-0977-76
$C_{24}H_{17}N_2O_2$ Isoquinolinium, 2-[(1-amino-9,10-di-hydro-9,10-dioxo-2-anthracenyl)-methyl]-, chloride	MeOH	472(3.95)	2-0054-76
Quinolinium, 1-[(1-amino-9,10-dihydro-9,10-dioxo-2-anthracenyl)methyl]-,chloride	MeOH	470(3.92)	2-0054-76
$C_{24}H_{17}N_3O_3$ Spiro[2H-1-benzopyran-2,5'(4'H)-pyrido-[2,3,4,5-lmn]phenanthridine], 4',10'-dimethyl-6-nitro-	EtOH	250s(4.6),340(4.18), 520(2.53)	103-0273-76
	$C_2H_4Cl_2$	250(4.57),355(4.17), 370s(4.09)	103-0273-76
$C_{24}H_{17}N_3O_4$ Ethanone, 1-[3-benzoyl-1-(4-nitrophen-yl)-5-phenyl-1H-pyrazol-4-yl]-	EtOH	263(4.50)	4-0989-76
$C_{24}H_{17}N_5O$ 1,2,4-Oxadiazole, 3-phenyl-5-[4-[2-(2-phenyl-2H-1,2,3-triazol-4-yl)ethen-yl]phenyl]-	DMF	340(4.69)	33-2469-76
1,3,4-Oxadiazole, 2-phenyl-5-[4-[2-(2-phenyl-2H-1,2,3-triazol-4-yl)ethen-yl]phenyl]-	DMF	343(4.75)	33-2469-76

Compound	Solvent	$\lambda_{max}(\log \epsilon)$	Ref.
$C_{24}H_{17}O_2S$ [1]Benzothieno[3,2-b]pyrylium, 4-(4-methoxyphenyl)-2-phenyl-, perchlorate	$C_2H_4Cl_2$	255(3.70),280(3.78), 318(3.73),420(4.60)	103-0977-76
$C_{24}H_{18}$ Dibenz[a,e]aceanthrylene, 1,2,3,4-tetra-hydro-	EtOH	286(4.52),297(4.48), 303(4.48),343(4.05), 356(4.07),377(3.88)	24-1991-76
Dibenzo[b,g]phenanthrene, 5,8-dimethyl-	C_6H_{12}	230(5.28),318(5.08), 289(5.03),300(5.22), 312(5.42),390(4.28)[sic]	2-0939-76
$C_{24}H_{18}Br_4N_6O_2$ Bromanil, 6-amino-2-methylquinoxaline complex (1:2)	CHCl₃	239(5.02),251(4.96), 311(4.54),377(4.13), 455(3.65)	56-0073-76
$C_{24}H_{18}ClNS$ 4(1H)-Pyridinethione, 2-(4-chlorophen-yl)-1-methyl-3,5-diphenyl-	MeOH	247(4.27),287(3.95), 359(4.01)	44-0818-76
$C_{24}H_{18}ClN_3O_2$ 1H-Pyrazole-4-carboxamide, 3-benzoyl-1-(4-chlorophenyl)-5-methyl-N-phenyl-	EtOH	257(4.624)	4-0989-76
$C_{24}H_{18}Fe_2$ Ferrocene, 1,1"-(1,3-butadiyne-1,4-diyl)-bis-, monocation	CH_2Cl_2	510(3.35),763(2.83), 1180(2.76)	44-2700-76
dication	CH_2Cl_2	580s(--),760(3.10)	44-2700-76
$C_{24}H_{18}N$ Pyridinium, 1-(benz[a]anthracen-7-yl-methyl)-, perchlorate	MeOH	230(4.51),254(4.48), 271(4.59),281(4.83), 292(4.91),326(3.67), 341(3.86),356(3.96), 372(3.85),390(3.47)	44-2679-76
Pyridinium, 1-(12-methylbenz[a]anthra-cen-7-yl)-, perchlorate	MeOH	222(4.52),278(4.66), 294(4.53),305(4.22), 346(3.80),360(3.88), 374(3.77),394(3.37)	44-2679-76
$C_{24}H_{18}N_2$ 1H-Perimidine, 1-methyl-2,4-diphenyl-	n.s.g.	360(4.20),433(3.24)	103-0586-76
Phenazine, 5,10-dihydro-5,10-diphenyl-, radical cation	MeCN	260(4.65),312(3.26), 375(3.68),442(3.86), 470(4.07),677(3.37), 750(3.32)	22-1567-76
$C_{24}H_{18}N_2O_2$ Ethanone, (3-benzoyl-1,5-diphenyl-1H-pyrazol-4-yl)-	EtOH	255(4.50)	4-0989-76
2(1H)-Indolone, 3-[benzoyl(3-indolyl)-methyl]-	EtOH	218(4.66),245(4.36), 283(3.94),291(3.88)	142-1675-76
2(1H)-Indolone, 3-(3-indolyl)-3-(2-oxo-2-phenylethyl)-	EtOH	221(4.65),246(4.36), 281(3.98),291(3.91)	142-1675-76
$C_{24}H_{18}N_2O_2S$ 1H-Pyrazole, 1-[(4-methylphenyl)sulfon-yl]-5-phenyl-3-(phenylethynyl)-	EtOH	235(4.42),246(4.42), 275(4.45),288s(4.44)	78-1293-76
Sulfur diimide, bis(4-phenoxyphenyl)-	n.s.g.	444(3.69)	97-0318-76B

Compound	Solvent	$\lambda_{max}(\log \epsilon)$	Ref.
$C_{24}H_{18}N_2O_2S_2$ Benzenesulfonic acid, 4-methyl-, (2-phenyl-5H-naphtho[1,8-bc]-thien-5-ylidene)hydrazide	MeOH	298(4.12),398(4.34)	88-4713-76
$C_{24}H_{18}N_2O_3$ 2(1H)-Pyridinone, 1-amino-6-(1,3-benzo-dioxol-5-yl)-3,4-diphenyl-	EtOH	225s(4.47),282s(3.96), 342(4.19)	4-0195-76
$C_{24}H_{18}N_2O_3S$ 2(1H)-Pyridine-3-carboxamide, 6-hydroxy-2-oxo-N,1-diphenyl-4-(phenylthio)-	MeOH	339(4.51)	120-0103-76
$C_{24}H_{18}N_2O_5S$ 1H-[1]Benzothieno[3,2-e]indazole-4,5-di-carboxylic acid, 2,3-dihydro-3-methyl-1-oxo-2-phenyl-, dimethyl ester	EtOH	231(4.20),276(4.03), 357(3.80)	94-0596-76
$C_{24}H_{18}N_2O_6S$ 8H-[2]Benzothiopyrano[8,1-fg]indazole-2,3-dicarboxylic acid, 9,10-dihydro-7-hydroxy-10-methyl-8-oxo-9-phenyl-, dimethyl ester	EtOH	266(4.11),300(3.82), 438(3.59)	94-0596-76
$C_{24}H_{18}N_3O_2$ Isoquinolinium, 2-[(1,4-diamino-9,10-di-hydro-9,10-dioxo-2-anthracenyl)-methyl]-, chloride	MeOH	568(4.03),610(4.15)	2-0054-76
Quinolinium, 1-[(1,4-diamino-9,10-di-hydro-9,10-dioxo-2-anthracenyl)-methyl]-, chloride	MeOH	565(3.91),609(4.11)	2-0054-76
$C_{24}H_{18}N_4O$ Acetamide, N-[2-(benzo[a]phenazin-5-yl-amino)phenyl]-	dioxan	232(4.60),270(4.56), 323(4.40),451(4.11)	49-0879-76
	50% dioxan	233(4.56),270(4.47), 326(4.38),468(4.09)	49-0879-76
Benzoxazole, 2-[4-[2-(5-methyl-2-phenyl-2H-1,2,3-triazol-4-yl)ethenyl]phenyl]-	DMF	360(4.73)	33-2469-76
Benzoxazole, 5-methyl-2-[4-[2-(2-phenyl-2H-1,2,3-triazol-4-yl)ethenyl]phenyl]-	DMF	353(4.82)	33-2469-76
Benzoxazole, 6-methyl-2-[4-[2-(2-phenyl-2H-1,2,3-triazol-4-yl)ethenyl]phenyl]-	DMF	354(4.79)	33-2469-76
Benzoxazole, 5-methyl-2-[2-[4-(2-phenyl-2H-1,2,3-triazol-4-yl)phenyl]ethenyl]-	DMF	350(4.76),362(4.78)	33-2469-76
Benzoxazole, 5-methyl-2-[4-[2-(4-phenyl-2H-1,2,3-triazol-2-yl)phenyl]ethenyl]-	DMF	363(4.78)	33-2469-76
Benzoxazole, 6-methyl-2-[2-[4-(2-phenyl-2H-1,2,3-triazol-4-yl)phenyl]ethenyl]-	DMF	350(4.77),362(4.77)	33-2469-76
Benzoxazole, 6-methyl-2-[4-[2-(4-phenyl-2H-1,2,3-triazol-2-yl)phenyl]ethenyl]-	DMF	354(4.78),364(4.78)	33-2469-76
$C_{24}H_{18}N_4O_2$ Benzoxazole, 2-[4-[2-[2-(4-methoxyphen-yl)-2H-1,2,3-triazol-4-yl]ethenyl]-phenyl]-	DMF	356(4.79)	33-2469-76
Benzoxazole, 5-methoxy-2-[4-[2-(2-phen-yl-2H-1,2,3-triazol-4-yl)ethenyl]-phenyl]-	DMF	358(4.76)	33-2469-76
Benzoxazole, 5-methoxy-2-[2-[4-(2-phen-yl-2H-1,2,3-triazol-4-yl)phenyl]ethenyl]-	DMF	357(4.73),367(4.73)	33-2469-76

Compound	Solvent	$\lambda_{max}(\log \epsilon)$	Ref.
Benzoxazole, 5-methoxy-2-[2-[4-(4-phenyl-2H-1,2,3-triazol-2-yl)phenyl]ethenyl]-	DMF	368(4.74)	33-2469-76
$C_{24}H_{18}N_4O_3S$ 5H-1,3,4-Thiadiazolo[3,2-a]pyrimidin-5-one, 6,7-dihydro-2-methyl-7-(4-nitrophenyl)-6,6-diphenyl-	EtOH	263(4.11)	94-2532-76
$C_{24}H_{18}O$ Indeno[2,1-b]pyran, 2,4-dimethyl-9-(2-naphthalenyl)-	dioxan	253(4.48),280(4.39), 308(4.42),321(4.46), 499(3.48)	78-2225-76
Indeno[2,1-b]pyran, 4,9-dimethyl-2-(2-naphthalenyl)-	dioxan	247(4.62),290(4.31), 318(4.25),337(4.23), 362(4.24),526(3.42)	78-2219-76
$C_{24}H_{18}O_2$ 2H-Pyran-2-one, 6-(4-methylphenyl)-3,4-diphenyl-	EtOH	245s(4.18),260(4.34), 360(4.25)	4-0195-76
$C_{24}H_{18}O_3$ 2H-Pyran-2-one, 4-hydroxy-3,5-diphenyl-6-(phenylmethyl)-	EtOH EtOH-NaOH	310(4.07) 307(4.05)	78-0269-76 78-0269-76
2H-Pyran-2-one, 6-(4-methoxyphenyl)-3,4-diphenyl-	EtOH	244(4.22),268(4.34), 372(4.35)	4-0195-76
$C_{24}H_{18}O_4S$ Benzo[b]thiophene-6-carboxylic acid, 4-acetoxy-5,7-diphenyl-, methyl ester	EtOH	205(4.46),238(4.56), 275(3.82),304(3.74)	4-0285-76
$C_{24}H_{18}O_7$ 4H-1-Benzopyran-5-carboxylic acid, 8-(1,4-dihydro-5-hydroxy-7-methyl-1,4-dioxo-2-naphthalenyl)-7-methyl-4-oxo-, ethyl ester	n.s.g.	248s(4.28),296(3.95), 304s(3.93),440(3.66)	39-2155-76C
$C_{24}H_{19}IO$ Pyrylium, 4-methyl-2,3,6-triphenyl-, iodide	CH_2Cl_2	486(2.37)	59-1311-76
$C_{24}H_{19}N$ Benzonitrile, 3-methyl-4-[2-[4-(2-phenylethenyl)phenyl]ethenyl]-	DMF	365(4.77)	33-0819-76
$C_{24}H_{19}NO$ Pyridine, 2-methoxy-3,4,6-triphenyl-	EtOH	253(4.41),313(4.32)	4-0195-76
$C_{24}H_{19}NOS_2$ 7,8-Dithia-2-azatricyclo[3.2.1.0²,⁴]octan-6-one, 3-methyl-1,4,5-triphenyl-	MeOH	243(4.09)	44-1724-76
$C_{24}H_{19}NO_2$ Benzenemethanol, α-[(3-phenyl-1-indolizinyl)methylene]-, acetate, (Z)-	EtOH	346(4.41),359s(4.36)	1-0512-76
2(1H)-Quinolinone, 3-(3-oxo-3-phenylpropyl)-4-phenyl-	dioxan	<u>225(4.3),335(3.2),</u> <u>450(3.3)</u>	56-0625-76
$C_{24}H_{19}NO_3$ Dinaphtho[1,2-b:2',3'-d]pyrrole-5,13-dione, N-butyl-11-hydroxy-	EtOH	240(3.92),280(4.18), 334(3.42),492(3.56)	115-0185-76

Compound	Solvent	$\lambda_{max}(\log \epsilon)$	Ref.
$C_{24}H_{19}NO_5$ Methanone, [4,5-dihydro-4-(4-methoxy-phenyl)-5-nitro-2-phenyl-3-furanyl]-phenyl-, trans	$C_2H_4Cl_2$	230(4.37),260(4.10), 286(4.10)	104-0640-76
$C_{24}H_{19}NS$ Benzenamine, N-(4,6-diphenyl-2H-thiopyr-an-2-ylidene)-4-methyl-	EtOH	225(4.03),275(4.11), 400(3.34)	77-0899-76
4(1H)-Pyridinethione, 1-methyl-2,3,5-triphenyl-	MeOH	226(4.13),267(3.71), 360(3.96)	44-0818-76
2(1H)-Pyridinethione, 1-(4-methylphen-yl)-4,6-diphenyl-	EtOH	220(4.01),280(4.08), 375(3.58)	77-0899-76
$C_{24}H_{19}N_2O_2S$ Sulfonium, [3-nitro-4-(phenylamino)phen-yl]diphenyl-, tetrafluoroborate	EtOH	302(4.29),400(3.67)	30-0235-76
$C_{24}H_{19}N_3O$ 7H-Pyridazino[4,5-d]azonine, 7-acetyl-1,4-diphenyl-	MeCN	255(4.47),396(3.78)	77-0313-76
$C_{24}H_{19}N_3O_2$ 1H-Pyrazole-4-carboxamide, 3-benzoyl-5-methyl-N,1-diphenyl-	EtOH	256(4.56)	4-0989-76
$C_{24}H_{19}N_3O_2S$ 5H-1,3,4-Thiadiazolo[3,2-a]pyrimidin-5-one, 6,7-dihydro-7-(4-methoxy-phenyl)-6,6-diphenyl-	EtOH	253.5(3.90)	94-2532-76
$C_{24}H_{19}N_3O_5$ 4-Quinolinecarbonitrile, 1,2,3,4-tetra-hydro-2-hydroxy-3-methyl-3-[(4-nitro-benzoyl)oxy]-2-phenyl-	EtOH	245(4.34),295s(3.76), 320s(3.00)	95-0968-76
$C_{24}H_{19}N_5O$ 3H-Imidazo[4,5-e]-1,2,4-triazin-3-one, 2,4-dihydro-6-phenyl-2,4-bis(phenyl-methyl)-	EtOH	248(3.93),333(4.17), 346(4.16)	94-2274-76
3H-Imidazo[4,5-e]-1,2,4-triazin-3-one, 2,5-dihydro-6-phenyl-2,5-bis(phenyl-methyl)-	EtOH	262(4.26),290(4.18), 369(3.98)	94-2274-76
3H-Imidazo[4,5-e]-1,2,4-triazin-3-one, 2,7-dihydro-6-phenyl-2,7-bis(phenyl-methyl)-	EtOH	257s(3.66),315(4.27)	94-2274-76
$C_{24}H_{19}N_7O_4$ 1H-Pyrido[3,2-c][1,2,4]triazolo[1,2-a]-pyridazine-1,3(2H)-dione, 6-(3,5-di-oxo-4-phenyl-1,2,4-triazolidin-1-yl)-5,6-dihydro-8-methyl-	12M HCl	265(4.30),317(3.57)	103-0590-76
1H-Pyrido[3,2-c][1,2,4]triazolo[1,2-a]-pyridazine-1,3(2H)-dione, 6-(3,5-di-oxo-4-phenyl-1,2,4-triazolidin-1-yl)-5,6-dihydro-9-methyl-	12M HCl	252(4.27),265s(4.18), 314(3.57)	103-0590-76
1H-Pyrido[3,2-c][1,2,4]triazolo[1,2-a]-pyridazine-1,3(2H)-dione, 6-(3,5-di-oxo-4-phenyl-1,2,4-triazolidin-1-yl)-5,6-dihydro-10-methyl-	12M HCl	268(4.26),316(3.53)	103-0590-76

Compound	Solvent	$\lambda_{max}(\log \epsilon)$	Ref.
$C_{24}H_{19}O$			
Pyrylium, 4-methyl-2,3,6-triphenyl-, iodide	CH_2Cl_2	486(2.37)	59-1311-76
selenocyanate	CH_2Cl_2	478(2.38)	59-1311-76
thiocyanate	CH_2Cl_2	476(2.03)	59-1311-76
$C_{24}H_{20}$			
[2.2](2,6)-Azulenophane, anti	$CHCl_3$	277(4.84),605(2.62)	88-1043-76
	$CHCl_3$	278(4.89),310s(4.02), 345s(3.50),367(3.48), 606(2.65)	88-2045-76
Benzo[c]phenanthrene, 1,2,3,4-tetrahydro-6-phenyl-	MeOH	300s(4.12),307(4.17), 341(2.94),358(2.79)	39-1115-76B
Naphthalene, 1-[2-(phenylethynyl)-1-cyclohexen-1-yl]-	MeOH	222(4.86),258s(4.18), 277(4.39),296(4.26)	39-1104-76B
[2.2](1,4)-Naphthalenophane, anti	C_6H_{12}	238(4.59),246(4.57), 273s(3.79),283s(3.75), 302s(3.74),310(3.78), 320s(3.69),327s(3.57), 333s(3.38)	89-0441-76
$C_{24}H_{20}Br_2O_2$			
1,3-Benzenedimethanol, α,α'-bis[2-(4-bromophenyl)ethenyl]-	n.s.g.	262(4.66),288(3.81), 298(3.53)	124-0842-76
Benzenemethanol, α,α'-1,3-phenylenedi-2,1-ethenediyl)bis[4-bromo-	n.s.g.	254(4.82)	124-0842-76
$C_{24}H_{20}Cl_2O_2$			
Benzene, 1,4-dichloro-2,5-bis[2-(4-methoxyphenyl)ethenyl]-	DMF	378(4.74)	33-0802-76
1,3-Benzenedimethanol, α,α'-bis[2-(4-chlorophenyl)ethenyl]-	n.s.g.	260(4.59),287(3.66), 299(3.42)	124-0842-76
Benzenemethanol, α,α'-(1,3-phenylenedi-2,1-ethenediyl)bis[4-chloro-	n.s.g.	252(4.78)	124-0842-76
$C_{24}H_{20}NO$			
Pyrylium, 4-(methylphenylamino)-2,6-diphenyl-, perchlorate	MeCN	260(4.31),323(4.24)	4-0073-76
$C_{24}H_{20}N_2$			
2,2'-Bipyridine, 6,6'-dimethyl-4,4'-diphenyl-, Cu(I) complex	EtOH	455(4.12)	118-0001-76
$C_{24}H_{20}N_2O$			
1-Naphthalenecarboxaldehyde, 2-methoxy-, diphenylhydrazone, (E)-	benzene	390(4.14)	104-0625-76
	EtOH	386(4.18)	104-0625-76
2(1H)-Pyridinone, 1-amino-6-(4-methylphenyl)-3,4-diphenyl-	EtOH	237s(4.34),254s(4.24), 289(3.79),341(4.09)	4-0195-76
$C_{24}H_{20}N_2OS$			
Acetamide, N-[3-(diphenylamino)-5-phenyl-2-thienyl]-	MeOH	290(4.55),340(4.95)	48-0221-76
$C_{24}H_{20}N_2O_2$			
2,3-Diazabicyclo[3.1.0]hex-2-ene-5-carboxylic acid, 1,6,6-triphenyl-, methyl ester	MeOH	323(2.67)	104-0802-76
Ethanone, 1-[4-[[2-(4-acetylphenyl)-2,3-dihydro-1H-isoindol-1-ylidene]amino]-phenyl]-	MeOH	238(4.28),330(4.46)	83-0356-76

Compound	Solvent	$\lambda_{max}(\log \epsilon)$	Ref.
4-Pyridazinecarboxylic acid, 2,5-dihydro-5,5,6-triphenyl-, methyl ester	MeOH	237(4.16),362(3.79)	104-0802-76
2(1H)-Pyridinone, 1-amino-6-(4-methoxyphenyl)-3,4-diphenyl-	EtOH	223s(4.51),253s(4.33), 343(4.23)	4-0195-76
$C_{24}H_{20}N_2O_3$ 4-Quinolinecarbonitrile, 3-(benzoyloxy)-1,2,3,4-tetrahydro-2-hydroxy-3-methyl-2-phenyl-	EtOH	235(4.35),275s(3.23), 282s(3.37),297(3.43)	95-0968-76
$C_{24}H_{20}N_2O_5$ 14H-Benzo[b]benzofuro[3,2-h][1,4]benzodiazepin-14-one, 4-acetyl-12,14a-dihydro-1,3-dihydroxy-2,13,14a-trimethyl-	MeCN	226(4.10),288(4.13), 305(4.13),370(3.09)	23-2795-76
$C_{24}H_{20}N_3O$ 1,2,4-Triazolo[3,4-a]isoquinolinium, 3-(4-methoxyphenyl)-1-(4-methylphenyl)-, tetrafluoroborate	MeOH	228(4.49),267(4.58), 319(3.75),332(3.79)	48-0881-76
1,2,4-Triazolo[4,3-a]quinolinium, 1-(4-methoxyphenyl)-3-(4-methylphenyl)-, tetrafluoroborate	MeOH	248(4.26),272(4.60), 302(3.99),318(3.92), 325(3.97)	48-0881-76
$C_{24}H_{20}N_4$ Benzenamine, 2,2'-azobis[N-phenyl-	EtOH	279(4.51),310(4.20)	23-1599-76
$C_{24}H_{20}N_4O_4S$ Spiro[2H-1-benzopyran-2,2'(3'H)-benzothiazole], 8-methoxy-3,3'-dimethyl-6-[(4-nitrophenyl)azo]-	EtOH	297(3.95),353s(4.28), 375(4.24),385(4.60)	28-0691-76A
$C_{24}H_{20}N_4O_5S$ Spiro[2H-1-benzopyran-2,2'(3'H)-benzothiazole], 3,8-dimethoxy-3'-methyl-6-[(4-nitrophenyl)azo]-	EtOH	302(4.04),350s(3.99), 375(4.40),384(4.75)	28-0691-76A
$C_{24}H_{20}O$ 7H-Benzo[c]xanthene, 6a,12a-dihydro-6a-methyl-5-phenyl-	EtOH	252(3.32),258(3.63), 266s(3.99),272(4.03), 280(3.97)	94-1581-76
2-Cyclohexen-1-one, 2,5,5-triphenyl-	EtOH	265(3.71)	33-2012-76
Indeno[2,1-b]pyran, 9-(3,4-dihydro-2-naphthalenyl)-2,4-dimethyl-	dioxan	240(4.24),305(4.16), 501(3.24)	78-2225-76
Indeno[2,1-b]pyran, 9-(1H-inden-2-yl)-2-(1-methylethyl)-	dioxan	251(4.25),317(4.46), 342(4.50),559(3.47)	78-2225-76
Indeno[2,1-b]pyran, 9-(1H-inden-2-yl)-2,3,4-trimethyl-	dioxan	227(4.51),259(4.39), 314(4.35),343(4.31), 551(3.55)	78-2225-76
$C_{24}H_{20}O_2$ Benzenepropanoic acid, α-(diphenylmethylene)-β-methylene-, methyl ester	MeOH	244(4.23)	104-0802-76
2-Hexene-1,6-dione, 1,4,6-triphenyl-, trans	EtOH	250(4.42)	12-0339-76
1-Naphthalenecarboxylic acid, 4-[4-(3,5-dimethylphenyl)-1-buten-3-ynyl]-, methyl ester	MeOH	238(4.45),281s(4.12), 286(4.13),351(4.50)	39-1104-76B

Compound	Solvent	$\lambda_{max}(\log \epsilon)$	Ref.
$C_{24}H_{20}O_3$ Cyclohepta[b]pyran-5(2H)-one, 6,7,8,9- tetrahydro-2-(2-oxophenylethylidene)- 4-phenyl-	MeOH	355(4.20),422(4.32)	24-1549-76
Spiro[isobenzofuran-1(3H),9'-xanthen]- 3-one, 2',7'-diethyl-	EtOH	218(4.67),292(3.71), 302(3.74)	56-0759-76
$C_{24}H_{20}O_4$ Benzeneacetic acid, α-[3-(4-methoxyphen- yl)-3-oxo-1-phenylpropylidene]-, (E)-	EtOH	223s(4.30),273(4.24)	4-0195-76
$C_{24}H_{20}O_7$ Carpachromene, diacetyl derivative	MeOH	222(4.51),265(4.46), 317(4.35)	12-2023-76
Glabrone, diacetate	EtOH	233s(4.55),247(4.58), 302(3.77)	94-0991-76
$C_{24}H_{20}O_8$ Isomilletone, 12a-acetoxy-	EtOH	210(4.42),240(4.08), 300(4.01)	102-1283-76
$C_{24}H_{21}BrN_2O_8$ Uridine, 5-bromo-2',3'-O-[(4-methoxy- phenyl)ethylidene]-, 5'-benzoate	MeOH	224(4.54),272(4.13)	44-1100-76
$C_{24}H_{21}ClN_2$ Pyrrolo[3,4-b]indole, 1-(2-chlorophen- yl)-1,2,3,4-tetrahydro-4-methyl-2-(4- methylphenyl)-	EtOH	225(4.70),255(4.49), 270(4.37)	103-1249-76
$C_{24}H_{21}ClN_2O_2$ 1,5-Benzodiazocin-3-ol, 8-chloro- 1,2,3,4-tetrahydro-1-methyl- 6-phenyl-, benzoate	EtOH	230(4.43),250s(--), 366(3.40)	111-0501-76
$C_{24}H_{21}NO_5$ 1H-Benz[de]isoquinoline-2(6H)-acetic acid, 5-hydroxy- -(1-methylpropyl)- 1,6-dioxo-7-phenyl-	MeOH	240(4.44),267(3.12), 325(4.03)	102-1413-76
7aH-Indole-7a-carboxylic acid, 3a-benz- oyl-1,2,3,3a,4,7-hexahydro-2,3-dioxo- 1-phenyl-, ethyl ester, cis	n.s.g.	245(4.29),349(3.28)	142-1355-76
Pyrano[4,3-b]pyrrole-7a(1H)-carboxylic acid, 6-ethenyl-2,3,6,7-tetrahydro- 2,3-dioxo-1,4-diphenyl-, ethyl ester, trans	n.s.g.	320(4.20)	142-1355-76
$C_{24}H_{21}N_3O_3$ Benzamide, 2-[5-(4-methoxyphenyl)-1H- pyrazol-3-yl]-N-(phenylmethoxy)-	EtOH	261(4.50)	44-0110-76
$C_{24}H_{21}N_3O_6$ Benzoic acid, 2-[3-(4-methoxyphenyl)-2- [(2-nitrophenyl)hydrazono]-1-oxoprop- yl]-, methyl ester	EtOH	280(4.04),322(4.11), 404(3.96)	104-1440-76
Benzoic acid, 2-[3-(4-methoxyphenyl)-2- [(3-nitrophenyl)hydrazono]-1-oxoprop- yl]-, methyl ester	EtOH	277(3.98),334(4.33)	104-1440-76
Benzoic acid, 2-[3-(4-methoxyphenyl)-2- [(4-nitrophenyl)hydrazono]-1-oxoprop- yl]-, methyl ester	EtOH	281(3.80),374(4.49)	104-1440-76

Compound	Solvent	$\lambda_{max}(\log \epsilon)$	Ref.
$C_{24}H_{21}N_3S_3$ 1,3,5-Triazine-2,4,6(1H,3H,5H)-trithi- one, 1,3,5-tris(phenylmethyl)-	EtOH	298(4.57)	33-2768-76
$C_{24}H_{21}N_5O_2$ 1H-Pyrazole-3,4-dicarboxamide, 5-amino- 1-(4-methylphenyl)-N,N'-diphenyl-	EtOH	295(4.405)	4-1137-76
$C_{24}H_{22}$ Benzo[rst]pentaphene, 1,2,3,4,9,10,11,12- octahydro-	EtOH	241(4.79),250(5.00), 262(4.25),272(4.61), 284(4.76),310(3.72), 321(4.17),338(4.57), 353(4.72)	25-0029-76
1H-Cyclopentachrysene, 3-ethyl-2,3-di- hydro-8-methyl-	EtOH	224(4.55),229(4.52), 263(4.91),273(5.14), 286(4.14),300(4.14), 313(4.18),326(4.52), 347(3.23)	78-0257-76
$C_{24}H_{22}NO_3$ Benz[f]quinolinium, 3-(2-acetoxy- 3-methoxyphenyl)-1,4-dimethyl-	EtOH	216(4.71),281(4.55), 357(4.08),377(4.18)	65-0390-76
$C_{24}H_{22}N_2$ 1H-Indole, 3,3'-(1,3,5-hexatriene-1,6- diyl)bis[1-methyl-	MeCN	210(4.61),243(4.47), 270s(4.13),328s(4.17), 385(4.76)	5-1060-76
oxidized form (two electrons lost)	MeCN	260(4.21),299(3.91), 308s(3.89),350s(3.99), 363(4.05),380s(3.98), 535(4.85)	5-1060-76
8H-Pyrido[1,2-c]quinazoline, 9,10,11,11a- tetrahydro-6,11a-diphenyl-	EtOH	235(4.31),313(3.84)	44-0497-76
Pyrrolo[3,4-b]indole, 1,2,3,4-tetra- hydro-4-methyl-2-(4-methylphenyl)- 1-phenyl-	EtOH	225(4.73),255(4.59), 272(4.29)	103-1249-76
Quinazoline, 7-(1,1-dimethylethyl)- 2,4-diphenyl-	EtOH	267(4.75),306s(3.88), 329(3.79)	44-0497-76
$C_{24}H_{22}N_2O_2$ 1,5-Benzodiazocin-3-ol, 1,2,3,4-tetra- hydro-1-methyl-6-phenyl-, benzoate	EtOH	232(4.47),357(3.42)	111-0501-76
$C_{24}H_{22}N_2O_4$ Benzoic acid, 2-[2-[(4-methoxyphenyl)- hydrazono]-1-oxo-3-phenylpropyl]-, methyl ester	EtOH	308(3.76),372(4.21)	104-1440-76
$C_{24}H_{22}N_2O_6$ 1,3(2H,9bH)-Dibenzofurandione, 6-acetyl- 2-[1-[(2-aminophenyl)amino]ethylidene]- 7,9-dihydroxy-8,9b-dimethyl-, (+)-	MeCN	228(4.24),293(4.20), 349(3.76)	23-2795-76
1,4-Naphthalenedicarbonitrile, 6,7,8- trimethoxy-2-(3,4,5-trimethoxyphenyl)-	EtOH	208(4.03),227(4.22), 274(4.19),372(3.53)	22-1829-76
$C_{24}H_{22}N_2O_6S$ 2(1H)-Pyrimidinone, 1-(3,4-di-O-benzoyl- 2,6-dideoxy-β-D-arabinohexopyranosyl)- 3,4-dihydro-4-thioxo-	EtOH	328(3.95)	44-0600-76

Compound	Solvent	$\lambda_{max}(\log \epsilon)$	Ref.
$C_{24}H_{22}N_2O_6S_2$ Dimer, m. 229-230°	MeOH	280s(3.48)	94-2250-76
$C_{24}H_{22}N_2O_7$ 2,4(1H,3H)-Pyrimidinedione, 1-(3,4-di- O-benzoyl-2,6-dideoxy- -D-arabino- hexopyranosyl)-	EtOH	235(4.39),252s(4.27)	44-0600-76
$C_{24}H_{22}N_4$ 2-Propenal, 3-(1-methyl-1H-indol-3-yl)-, [3-(1-methyl-1H-indol-3-yl)-2-propen- ylidene]hydrazone	MeCN	209(4.41),237(4.40), 272(4.08),407(4.77)	5-1039-76
$C_{24}H_{22}N_4O_6$ 2,4(1H,3H)-Pteridinedione, 1-methyl- 6,7-diphenyl-3-β-D-ribofuranosyl-	MeOH	223(4.39),277(4.22), 366(4.14)	24-3194-76
2,4(1H,3H)-Pteridinedione, 3-methyl- 6,7-diphenyl-1-β-D-ribofuranosyl-	MeOH	224(4.41),270(4.21), 358(4.14)	24-3194-76
$C_{24}H_{22}N_4O_8$ 1,4-Benzenediacetic acid, α,α'-bis(eth- oxycarbonyl)-, diethyl ester	MeCN	305(3.29),315(3.39), 408(2.30)	24-2469-76
$C_{24}H_{22}O$ Azulene, 6,6'-[oxybis(methylene)]bis[2- methyl-	$CHCl_3$	279s(4.85),286s(4.93), 297(5.07),335(3.99), 350(4.10),362s(3.58), 556(2.79),596s(2.74)	88-2045-76
$C_{24}H_{22}O_2$ 1,3-Benzenedimethanol, α,α'-bis(2-phen- ylethenyl)-	n.s.g.	254(4.49),282(3.70), 292(3.62)	124-0842-76
Benzenemethanol, α,α'-(1,3-phenylenedi- 2,1-ethenediyl)bis-	n.s.g.	252(4.53)	124-0842-76
2(1H)-Naphthalenone, 1-ethyl-1-[(1-eth- yl-2-naphthalenyl)oxy]-	EtOH	232(4.93),283(3.97), 294(4.04),313(4.01)	39-2570-76C
$C_{24}H_{22}O_5$ 4H-1-Benzopyran-4-one, 3-acetyl-3- [(2,6-dimethyl-4-oxo-4H-1-benzo- pyran-3-yl)methyl]-2,3-dihydro- 6-methyl-	EtOH	231(4.44),253(4.25), 310(3.90),335(3.50)	39-2444-76C
$C_{24}H_{22}O_6$ 3,5-Cyclopentadiene-1,3-dicarboxylic acid, 2-hydroxy-2-(2-oxopropyl)- 4,5-diphenyl-, dimethyl ester	ether	240s(4.09),289(3.82), 315s(3.75)	22-1491-76
Glabrene, diacetate	EtOH	242(4.35),275(4.21), 283(4.21),325(4.16)	94-0991-76
$C_{24}H_{22}O_{10}$ 2H-Benzo[a]xanthen-12-one, 9-(β-D-gluco- pyranosyloxy)-11-hydroxy-2-methoxy- (dillanoside)	EtOH	219(4.65),237s(4.47), 266(4.58),302s(3.67), 352(4.24)	94-0220-76
	EtOH-$AlCl_3$	222(4.71),240s(4.45), 285(4.52),320s(3.37), 376(4.20)	94-0220-76
$C_{24}H_{22}O_{11}$ Griseusin A, monoacetate	MeOH	213(4.56),253(4.01), 431(3.59)	78-2207-76

Compound	Solvent	$\lambda_{max}(\log \epsilon)$	Ref.
$C_{24}H_{23}ClN_2O$			
Benzamide, 2-[(4-chlorophenyl)(phenyl-imino)methyl]-N-(1,1-dimethylethyl)-1H-Isoindol-1-one, 3-(4-chlorophenyl)-2-(1,1-dimethylethyl)-2,3-dihydro-3-(phenylamino)-	EtOH EtOH	263(4.23),335(3.48) 233(4.49)	103-0981-76 +104-0237-76 104-0237-76
$C_{24}H_{23}ClN_2O_6$			
Indolizino[1,2-b]quinoline-7-acetic acid, 12-chloro-α-ethyl-9,11-dihydro-11-hydroxy-8-(methoxycarbonyl)-9-oxo-, 1-methylethyl ester	MeOH	222(4.59),255(4.41), 261(4.46),298(3.93), 362(4.26)	24-1389-76
$C_{24}H_{23}NO_2$			
Benzenamine, 4-(7-methoxy-3-phenyl-2H-1-benzopyran-2-yl)-N,N-dimethyl-4H-Inden-4-one, 1,5,6,7-tetrahydro-6,6-dimethyl-3-(2-oxo-2-phenylethyl)-2-phenyl-5(1H)-Quinolinone, 4-benzoyl-4,6,7,8-tetrahydro-7,7-dimethyl-2-phenyl-	n.s.g. EtOH EtOH	209(4.54),257(4.67), 335(4.27) 273(4.18) 251(4.37),330(3.80)	39-1389-76C 94-1160-76 94-1160-76
$C_{24}H_{23}NO_3$			
Benzoic acid, 2-[benzoyl[2-(1,1-dimeth-ylethyl)phenyl]amino]-	MeOH	205(4.20),270(3.59)	39-0465-76C
$C_{24}H_{23}NO_3S$			
1,5-Benzothiazepine, 2,3-dihydro-4-phen-yl-2-(3,4,5-trimethoxyphenyl)-	EtOH	261(4.32),330(3.65)	114-0293-76A
$C_{24}H_{23}NO_7$			
Acetamide, N-[2-[3,7-bis(acetyloxy)phen-anthro[4,5-def][1,3]dioxepin-1-yl]-ethyl]-N-methyl-	EtOH	255(4.64),283(4.31), 302(4.16),316(4.21), 330s(3.35),347(3.49), 365(3.51)	33-2551-76
$C_{24}H_{23}NO_8$			
1H-1,3a-Ethenopyrido[2,1,6-de]quinoli-zine-2,3,4,6-tetracarboxylic acid, 4,6-diethyl 2,3-dimethyl ester	EtOH	270s(4.0),281(4.06), 290s(4.2),307(4.34), 364(4.01),515(3.85)	39-0341-76C
$C_{24}H_{23}N_3O_2S$			
Benzenesulfonamide, N-[3-[[4-(dimethyl-amino)phenyl]methylene]-3H-indol-2-yl]-4-methyl-	CHCl$_3$	264(4.28),329(3.66), 475(4.32)	103-0292-76
$C_{24}H_{23}N_3O_4$			
Austamide, 12,13-dihydro-12-hydroxy-, (2S,9S,12R)-	MeOH	231(4.48),255(4.04), 390(3.47)	102-0355-76
$C_{24}H_{23}O_4P$			
Phosphine, bis(2,6-dimethoxyphenyl)-(phenylethynyl)-	MeOH	195(4.34),249(3.99), 258(3.87),278(3.75)	139-0169-76A
$C_{24}H_{24}$			
Tricyclo[10.2.2.2^{4,7}]octadeca-4,6,12,14-15,17-hexaene, 8-phenyl-Tricyclo[10.2.2.2^{4,7}]octadeca-4,6,12,14-15,17-hexaene, 9-phenyl-	n.s.g. n.s.g.	254(2.72),263(2.82), 270(2.85),283(2.56) 255(2.60),262(2.76), 268(2.83),283(2.49)	88-4095-76 88-4095-76

Compound	Solvent	$\lambda_{max}(\log \epsilon)$	Ref.
$C_{24}H_{24}CuN_2O_2S_2$ Copper, bis[2-(4,5-dihydro-2-thiazolyl)- 1-(4-methylphenyl)ethanonato-N,O]-	CHCl$_3$	242(4.18),297(4.52), 338(4.50),590s(1.99)	48-0298-76
$C_{24}H_{24}F_3N_2$ Methanaminium, N-[4-[[4-(dimethylamino)- phenyl][3-(trifluoromethyl)phenyl]- methylene]-2,5-cyclohexadien-1-yli- dene]-N-methyl-	EtOH 8% DMSO	634(4.99) 428(3.98),641(4.66)	44-0221-76 44-0221-76
Methanaminium, N-[4-[[4-(dimethylamino)- phenyl][4-(trifluoromethyl)phenyl]- methylene]-2,5-cyclohexadien-1-yli- dene]-N-methyl-	EtOH 8% DMSO	636(5.00) 427(4.06),647(4.87)	44-0221-76 44-0221-76
$C_{24}H_{24}N_2$ 1H-Indole, 3,3'-(1,3-butadiene-1,4-di- yl)bis[1,2-dimethyl-	MeCN	224s(4.39),236(4.37), 278(3.99),344s(4.61), 366(4.67),374s(4.52)	5-1060-76
semiquinone (radical ion)	MeCN	365(--),383(--), 514(--),598(4.58), 670(4.13),820(4.05), 930(4.26)	5-1060-76
oxidized form	MeCN	242(4.20),328s(4.20), 334(4.20),348s(4.07), 468s(4.55),487(4.63), 510s(4.55)	5-1060-76
Indolizine, 1,1'-(1,3-butadiene-1,4-di- yl)bis[2,3-dimethyl-	CH$_2$Cl$_2$	266(4.45),324(3.99), 400(4.50),430s(4.46)	5-0317-76
semiquinone	1:1 CH$_2$Cl$_2$- MeCN	360(--),500(--), 680(--),780(--), 975(--),1118(--)	5-0317-76
oxidized form	MeCN	273(4.04),335(3.81), 497(4.75)	5-0317-76
Indolizine, 3,3'-(1,3-butadiene-1,4-di- yl)bis[1,2-dimethyl-	CH$_2$Cl$_2$	308(4.27),317(4.27), 452(4.66)	5-0317-76
semiquinone	1:1 CH$_2$Cl$_2$- MeCN	590(4.52),726(4.75), 885(4.11),1112(4.21)	5-0317-76
oxidized form (diperchlorate)	MeCN	273(4.02),311(4.13), 396(4.37),409(4.38), 435(4.33),515(4.56)	5-0317-76
Quinazoline, 4-(1,1-dimethylethyl)- 1,4-dihydro-2,4-diphenyl-	EtOH	236(4.30),265(3.85), 320(3.75)	44-0497-76
Quinazoline, 7-(1,1-dimethylethyl)- 7,8-dihydro-2,4-diphenyl-	EtOH	239(4.23),253(4.30), 301(4.39)	44-0497-76
$C_{24}H_{24}N_2NiO_2S_2$ Nickel, bis[2-(4,5-dihydro-2-thiazolyl)- 1-(4-methylphenyl)ethanonato-N,O]-	CHCl$_3$	270(4.65),295(4.56), 390(4.00),495(2.32), 630(2.21)	48-0298-76
$C_{24}H_{24}N_2O_2$ 1,1'-Biisoquinoline, 2,2'-diacetyl- 1,1',2,2'-tetrahydro-3,3'-dimethyl-, meso	n.s.g.	233(4.49),287(4.19), 308s(4.12)	142-0427-76S
racemic	n.s.g.	228(4.43),287(4.22)	142-0427-76S
Piperidine, 1-benzoyl-3-ethenyl-4-[2- (1H-indol-2-yl)-2-oxoethyl]-, (3aR- cis)-	MeOH	308(4.37)	33-2268-76

Compound	Solvent	$\lambda_{max}(\log \epsilon)$	Ref.
$C_{24}H_{24}N_2O_2PdS_2$ Palladium, bis[2-(4,5-dihydro-2-thiazolyl)-1-(4-methylphenyl)ethanonato-N,O]-	CHCl$_3$	257(4.64),281(4.57), 362(4.27)	48-0298-76
$C_{24}H_{24}N_2O_2S$ 2-Propyn-1-one, 3,3'-[thiobis(2,1-ethanediylimino)]bis[1-(4-methylphenyl)-	MeOH	276(4.19),304(4.19), 389(4.35)	48-0298-76
$C_{24}H_{24}N_2O_3$ 4H-1-Benzopyran-4-one, 6-hexyl-7-hydroxy-3-(1-phenyl-1H-pyrazol-4-yl)-	EtOH	270(4.47)	103-0914-76
Ergoline-8-carboxylic acid, 1-benzoyl-9,10-didehydro-2,3-dihydro-6-methyl-, methyl ester	EtOH	253(4.59),307(3.89)	88-4311-76
stereoisomer	EtOH	242(4.30),305(3.67)	88-4311-76
$C_{24}H_{24}N_2O_6$ 1,3(2H,4H)-Dibenzofurandione, 6-acetyl-2-[1-[(2-aminophenyl)amino]ethylidene]-4a,9b-dihydro-7,9-dihydroxy-8,9b-dimethyl-, cis-(-)-	MeCN	230(4.15),293(4.21), 342(3.76)	23-2795-76
$C_{24}H_{24}N_4O_2S_2$ Ethanone, 1-[4-[3-[(4-acetylphenyl)azo]-5-(1,1-dimethylethyl)-1H-[1,2]dithiolo[5,1-e][1,2,3]thiadiazol-7-S^{IV}-1-yl]phenyl]-	C$_6$H$_{12}$	205(4.43),218s(4.35), 251(4.26),288(4.42), 444(4.47)	39-0228-76C
$C_{24}H_{24}N_6O_8$ D-glycero-β-D-allo-Octofuranuronamide, 3,7-anhydro-1-[6-(benzoylamino)-9H-purin-9-yl]-1,6-dideoxy-, 2,5-diacetate	EtOH	231(4.13),279(4.28)	88-1687-76
$C_{24}H_{24}O_2S_2Zn$ Zinc, bis[2-(4,5-dihydro-2-thiazolyl)-1-(4-methylphenyl)ethanonato-N,O]-	CHCl$_3$	253(4.36),340(4.62)	48-0298-76
$C_{24}H_{24}O_3$ 1-Azulenecarboxylic acid, 3,8-dimethyl-5-(1-methylethyl)-, 4-acetylphenyl ester	C$_6$H$_{12}$	251(4.70),305(4.70), 316(4.69),388(4.36), 524s(2.79),562(2.88), 613s(2.76),673s(2.24)	18-1650-76
$C_{24}H_{24}O_5$ Uvaretin, monomethyl ether	EtOH	290(4.40)	44-1852-76
$C_{24}H_{24}O_6$ Glabridin, diacetate	EtOH	280(4.10),291s(4.03), 313s(3.47)	94-0752-76
2H,6H-Pyrano[3,2-b]xanthen-6-one, 5,8-dihydroxy-9-methoxy-2,2-dimethyl-10-(3-methyl-2-butenyl)- (manglexanthone)	MeOH	235s(4.32),250s(4.36), 286(4.85),336(4.30), 376(4.09)	32-0651-76
$C_{24}H_{24}O_{11}$ Flavone, 5,7-diacetoxy-3,3',4',5',8-pentamethoxy-	EtOH	220(4.38),275(4.15), 320(4.08)	2-0849-76
Trifolirhizin, monoacetate	EtOH	279s(3.60),284(3.67), 311(3.94)	102-1089-76

Compound	Solvent	$\lambda_{max}(\log \epsilon)$	Ref.
$C_{24}H_{25}BrO_7$ Naphtho[2,3-c]furan-1(3H)-one, 9-(2-bromo-3,4,5-trimethoxyphenyl)-3,4-dihydro-6,7-dimethoxy-3a-methyl-	MeOH	240(4.67),285(4.34), 320(4.17)	2-0084-76
$C_{24}H_{25}BrO_9$ Naphtho[2,3-c]furan-1(3H)-one, 9-(2-bromo-3,4,5-trimethoxyphenyl)-3a,4-didihydro-4-hydroxy-3a-(hydroxymethyl)-6,7-dimethoxy-	MeOH	239(4.44),285(4.09), 320(3.92)	2-0084-76
$C_{24}H_{25}NO_3$ 1,4-Benzoxazepine, 2,3,4,5-tetrahydro-8-methoxy-3-methyl-5-phenyl-7-(phenylmethoxy)-	n.s.g.	215(4.30),240(4.19), 285(3.52)	20-0898-76
$C_{24}H_{25}NO_4S$ 1-Propanone, 3-[(2-aminophenyl)thio]-1-phenyl-3-(3,4,5-trimethoxyphenyl)-	EtOH	204(4.77),242(4.39), 312(3.67)	114-0293-76A
$C_{24}H_{25}NO_6$ 5H-1-Pyrindine-3,4-dicarboxylic acid, 6,7-dihydro-2-methyl-5,7-dioxo-6-phenyl-, 4-ethyl 3-pentyl ester (enol)	50% EtOH	346(4.1),451(3.56)	103-0220-76
anion	H_2O	378(4.15),470s(3.42)	103-0220-76
	50% EtOH	382(4.19),470(3.45)	103-0220-76
protonated enol	50% EtOH	385(4.02)	103-0220-76
$C_{24}H_{25}NO_7$ Acetamide, N-[2-[6-[2-[4-(acetoxymethyl)-1,3-benzodioxol-5-yl]ethenyl]-1,3-benzodioxol-5-yl]ethyl]-N-methyl-, (E)-	EtOH	225s(4.46),313(4.13)	105-0432-76
$C_{24}H_{25}NO_8$ 1H-1,3a-Ethanopyrido[2,1,6-de]quinolizine-2,3,4,6-tetracarboxylic acid, 4,6-diethyl 2,3-dimethyl ester	EtOH	267s(4.06),275(4.11), 309(4.36),362(3.88), 516(3.84)	39-0341-76C
$C_{24}H_{25}NO_9S$ Dibenzo[a,g]quinolizinium, 2,3,9,10-tetramethoxy-5-methyl-, acetosulfate	MeOH	246(4.14),270(4.17), 300s(4.46),311(4.83), 332(4.57),347(4.56), 403(3.83),426(4.15)	87-0882-76
Dibenzo[a,g]quinolizinium, 3,4,10,11-tetramethoxy-8-methyl-, acetosulfate	EtOH	235(4.24),317(4.76), 352(4.19),376(4.00), 400(3.95),420(3.89)	87-0882-76
Dibenzo[b,g]quinolizinium, 2,3,9,10-tetramethoxy-5-methyl-, acetosulfate	MeOH	246(4.14),270(4.17), 300s(4.46),311(4.83), 332(4.57),347(4.56), 403(3.83),426(4.15)	87-0882-76
$C_{24}H_{25}NO_{12}S$ 2-Azabicyclo[6.2.0]deca-3,5,9-triene-4,5,9,10-tetracarboxylic acid, 7-[3-methoxy-1-(methoxycarbonyl)-3-oxo-2-thioxopropylidene]-2-methyl-, tetramethyl ester	EtOH	220(4.54),298(3.80)	142-0705-76

Compound	Solvent	λ_{max}(log ϵ)	Ref.
$C_{24}H_{25}N_2$ Acridinium, 9-[4-(diethylamino)phenyl]- 10-methyl-, triiodide	CH_2Cl_2	260(5.04),292(4.72), 360(4.54),630(4.18)	104-1531-76
$C_{24}H_{25}N_2P$ Benzenamine, 4,4'-[(phenylethynyl)phos- phinidene]bis[N,N-dimethyl-	MeOH	211(4.58),298(4.61)	139-0169-76A
$C_{24}H_{25}N_5O_4$ 1H,5H-Imidazo[1',2':1,2]pyrido[2,3-b]in- dole-2,5(3H)-dione, 7a-(1,1-dimethyl- 2-propenyl)-7a,12-dihydro-3-(1H-imid- azol-4-ylmethylene)-6,12-dimethoxy-, [7aS-(3E,7aR*,12aS*)]- (oxaline)	MeOH	228(4.32),347(4.39)	78-2625-76
$C_{24}H_{26}BrN_5O_4$ Oxaline, bromodihydro-	MeOH	228(4.36),282(2.86), 351(4.44)	78-2625-76
$C_{24}H_{26}Br_4O_8$ Benzoic acid, 2,2'-(1,6-hexanediyl)bis- [3,5-dibromo-4,6-dihydroxy-, diethyl ester	MeOH	226(4.79),270(4.18), 316(3.84)	12-1989-76
$C_{24}H_{26}ClN_6O_4P$ 1H-1,2,4-Triazolo[4,3-a][1,4]benzodiaz- epin-1-one, 8-chloro-2-(di-4-morpho- linylphosphinyl)-2,4-dihydro-6-phenyl-	MeCN	208(4.62),245(4.30), 306(2.85)	44-2720-76
$C_{24}H_{26}Cl_4O_4$ Pregna-4,6,20-trien-3-one, 17α-acetoxy- 6,21,22-trichloro-16α-(chloromethyl)- 16β,20-epoxy-	MeOH	282.5(4.36)	44-3940-76
$C_{24}H_{26}F_6N_7$ 1H-Benzimidazolium, 2-[3-[1,3-diethyl- 1,3-dihydro-5-(trifluoromethyl)-2H- benzimidazol-2-ylidene]-1-triazenyl]- 1,3-diethyl-5-(trifluoromethyl)-, tetrafluoroborate	MeOH	427(2.96)	33-0148-76
$C_{24}H_{26}N_2O_2$ Pyrrolo[1,2-a:4,5-b']diindol-11-ol, 5a,12-diethyl-5a,10,10a,11-tetrahydro- 10a-methyl-, acetate, (5aα,10aα,11α)-	MeOH	228(4.60),295(3.98), 305(3.93)	23-1015-76
$C_{24}H_{26}N_4$ 1-Piperazinamine, 4-(diphenylmethyl)-N- [(6-methyl-2-pyridinyl)methylene]-	20% MeOH	223(4.28),310(4.28)	96-0469-76
anion	20% MeOH	242(4.27),335(4.30)	96-0469-76
diprotonated	20% MeOH	223(4.21),270(3.95), 349(4.35)	96-0469-76
triprotonated	20% MeOH	223(4.21),270(3.95), 353(4.33)	96-0469-76
$C_{24}H_{26}N_4O_2S$ Benzo[g]pteridine-2,4(1H,3H)-dione, 5,10-dihydro-7,8,10-trimethyl-3-(phen- ylmethyl)-5-(tetrahydro-2-thienyl)-	CHCl₃ CF₃COOH	246(4.26),335(3.75) 247(4.65),295s(--), 310(3.93)	5-2037-76 5-2037-76

Compound	Solvent	$\lambda_{max}(\log \epsilon)$	Ref.
$C_{24}H_{26}O$			
Benzo[a]biphenylen-6(5H)-one, 10b-(1,1-dimethylethyl)-6a,10b-dihydro-5-methyl-5-(1-methylethenyl)-, (5α,6aβ,10bβ)-	EtOH	267(3.325),273(3.236)	24-1301-76
Indeno[1,2-a]inden-9(4bH)-one, 4b-(1,1-dimethylethyl)-9a,10-dihydro-10-methyl-10-(1-methylethenyl)-, (4bα,9aα,10α)-	EtOH	249(4.079),272s(3.187), 296(3.309)	24-1294-76
$C_{24}H_{26}O_8$			
4H-1-Benzopyran-2-carboxylic acid, 7-methoxy-5-methyl-4-oxo-3-[(2,4,5-trimethoxyphenyl)methyl]-	EtOH	210(4.38),216(4.40), 227(4.35),254(4.09), 295(4.05)	39-0499-76C
Bicyclo[3.2.1]oct-3-en-2-one, 3,8-diacetoxy-7-(1,3-benzodioxol-5-yl)-1-methoxy-6-methyl-5-(2-propenyl)-, (6-exo,7-endo,8-anti)-	MeOH	235(3.54),279(3.26), 311(2.28)	102-1033-76
D-Glucitol, 1,3:2,4-bis-O-(phenylmethylene)-, diacetate, [1(R),2(S)]-	MeCN	256(2.63)	12-1859-76
$C_{24}H_{26}O_9$			
Naphtho[2,3-c]furan-1,4-dione, 3,3a,9,9a-tetrahydro-3a-(hydroxymethyl)-6,7-dimethoxy-9-(3,4,5-trimethoxyphenyl)-, (3aα,9α,9aα)-	MeOH	240(4.37),280(3.99), 320(3.86)	2-0084-76
$C_{24}H_{27}ClO_9$			
Peuchlorinin	EtOH	220(4.18),246(3.50), 257(3.47),300s(4.06), 324(4.27)	102-0209-76
$C_{24}H_{27}Cl_3O_5$			
Androsta-4,6-diene-3,17-dione, 15-acetyl-16-acetoxy-6-chloro-16-(dichloromethyl)-, (15α,16α)-	MeOH	282(4.25)	44-3940-76
$C_{24}H_{27}FO_6$			
Pregna-1,4-diene-3,11,20-trione, 21-acetoxy-16α,17α-epoxy-16β-(fluoromethyl)-	MeOH	238(4.23)	39-0101-76C
$C_{24}H_{27}NO_3$			
Majoritaire	EtOH	232(4.59),307(3.87), 321(3.82),335s(3.77)	102-0817-76
$C_{24}H_{27}NO_4$			
Tylophorinine	MeOH	259(4.6),291(4.4), 340(3.2),357(2.8)	102-1561-76
$C_{24}H_{27}NO_5$			
Alkaloid D from Pergularia pallida	MeOH	258(4.4),287(4.2), 302(3.6),339(3.1), 355(2.9)	102-1561-76
$C_{24}H_{27}N_2$			
Methanaminium, N-[4-[[4-(dimethylamino)-phenyl](4-methylphenyl)methylene]-2,5-cyclohexadien-1-ylidene]-N-methyl-	EtOH	618(5.03)	44-0221-76

Compound	Solvent	$\lambda_{max}(\log \epsilon)$	Ref.
$C_{24}H_{27}N_2O$			
Methanaminium, N-[4-[[4-(dimethylamino)-phenyl]-(4-methoxyphenyl)methylene]-2,5-cyclohexadien-1-ylidene]-N-methyl-	EtOH	610.5(4.99)	44-0221-76
Methylium, bis[4-(dimethylamino)phenyl]-(3-methoxyphenyl)-	EtOH	637(4.99)	44-0221-76
	8% DMSO	437(4.15),637(4.94)	44-0221-76
$C_{24}H_{27}N_3$			
1H-Indol-3-amine, 2-(1,1-dimethylethyl)-N-[2-(1,1-dimethylethyl)-3H-indol-3-ylidene]-	MeOH	231(4.64),260(4.08), 340(3.88),550(4.00)	23-1020-76
	MeOH-HCl	218(4.56),285(4.15), 298(4.08),615(4.26)	23-1020-76
$C_{24}H_{27}N_3O_2$			
Isoechinulin B	MeOH	228(4.45),272(4.27), 285s(4.22),370(4.02)	88-1601-76
$C_{24}H_{27}N_3O_3$			
Isoechinulin C	MeOH	229(4.43),272(4.26), 286s(4.20),371(3.99)	88-1601-76
$C_{24}H_{27}N_3O_4S$			
4H-Indazol-4-one, 1,5,6,7-tetrahydro-3-(4-methoxyphenyl)-1,6,6-trimethyl-, O-[(4-methylphenyl)sulfonyl]oxime, (E)-	MeOH	235(4.51),262s(4.04)	24-1898-76
$C_{24}H_{27}N_3O_{12}$			
1H-Pyrazole-3,5-dicarboxylic acid, 4,5-dihydro-5-[2,3-O-(1-methylethylidene)-5-O-(4-nitrobenzoyl)-α-D-ribofurano-syl]-4-oxo-, diethyl ester	EtOH	245(4.20),260s(4.16), 330(3.83)	44-0287-76
$C_{24}H_{27}N_5O_4$			
Oxaline dihydro-	MeOH	228(4.29),347(4.37)	78-2625-76
$C_{24}H_{27}N_7O_4$			
L-Alanine, 3-(1-naphthalenyl)-N-[N-(5-oxo-L-prolyl)-L-histidyl]-, hydrazide	HOAc	273(3.79),283(3.87), 294(3.71)	94-3149-76
L-Alanine, 3-(2-naphthalenyl)-N-[N-(5-oxo-L-prolyl)-L-histidyl]-, hydrazide	HOAc	268(3.64),276(3.65), 286(3.48)	94-3149-76
$C_{24}H_{28}$			
1H-Indene, 3-(1,1-dimethylethyl)-1-(2,2-dimethyl-1-phenylpropylidene)-	EtOH	235s(4.06),242s(4.09), 265(4.36),312(3.91), 322(3.86)	56-1301-76
$C_{24}H_{28}Br_2N_8O_4S_2$			
1,2,4,5-Tetrazine-3,6-dicarboximidamide, N',N'''-bis[(4-bromophenyl)sulfonyl]-N,N,N'',N''-tetraethyl-	MeCN	250(4.59),585(2.63)	32-0001-76
$C_{24}H_{28}Cl_2O_9$			
Peuchloridin	EtOH	220(4.28),246(3.60), 257(3.60),300s(4.20), 324(4.38)	102-0209-76
$C_{24}H_{28}NO_4$			
Pergularinine, methiodide	MeOH	258(4.7),285(4.4), 310(3.9),341(3.2), 357(2.8)	102-1561-76

Compound	Solvent	$\lambda_{max}(\log \epsilon)$	Ref.
$C_{24}H_{28}N_2O$			
Cyclohexanone, 2,6-bis[[4-(4-dimethyl-amino)phenyl]methylene]-, (E,E)-	EtOH	270(4.03),330(3.32), 447(4.59)	114-0381-76B
	ether	270(--),325(--), 407(--)	114-0381-76B
	$CHCl_3$	272(4.20),320(3.60), 435(4.47)	114-0381-76B
	CCl_4	268(4.28),320(3.89), 410(4.61)	114-0381-76B
$C_{24}H_{28}N_2O_3$			
Ethanone, 1-(5-hexyl-2-hydroxy-4-meth-oxyphenyl)-2-(1-phenyl-1H-pyrazol-4-yl)-	EtOH	270(4.40),330(3.79)	103-0914-76
$C_{24}H_{28}N_2O_6$			
L-Phenylalanine, N-[N-[3-(3,4-dihydroxy-phenyl)-1-oxo-2-propenyl]glycyl]-,	MeOH-acid	243(4.16),297(4.22), 327(4.29)	20-0647-76
1,1-dimethylethyl ester, (E)-	MeOH-base	258(4.11),298s(--), 308(3.96),364(4.38)	20-0647-76
16,19-Secostrychnidine-10,16-dione, 21,22-epoxy-21,22-dihydro-3,4-di-methoxy-19-methyl-, (21α,22α)-	EtOH	226(4.30),256(4.11), 295(3.58)	102-1973-76
$C_{24}H_{28}N_2O_{12}S$			
2,4(1H,3H)-Pyrimidinedione, 5-methyl-1-[2,3,4-tri-O-acetyl-6-O-[(4-methyl-phenyl)sulfonyl]-β-D-galactopyrano-syl]-	MeOH	222(4.21),265(4.03)	48-0079-76
$C_{24}H_{28}N_4$			
Cyclobuta[1,2-b:3,4-b']bis[1,4]diaze-pinium, 2,3,4,6,7,8,9,10-octahydro-1,6-dimethyl-5,10-diphenyl-, bis(tetrafluoroborate)	MeCN	253(4.47),300s(--), 341(4.54),388(4.74)	88-3515-76
$C_{24}H_{28}O$			
Benzo[a]biphenylen-5-ol, 5,10b-bis(1,1-dimethylethyl)-5,10b-dihydro-, endo	C_6H_{12}	251(3.984),290(3.724)	24-1287-76
Benzo[a]biphenylen-6-ol, 10b-(1,1-di-methylethyl)-5,6,6a,10b-tetrahydro-5-exo-methyl-5-endo-(1-methylethen-yl)-	EtOH	266(3.935),272(3.853)	24-1301-76
Benzo[a]biphenylen-6a(10bH)-ol, 5,10b-bis(1,1-dimethylethyl)-6a,10b-dihydro-	C_6H_{12}	231(4.230),244s(4.013), 272(3.739),279(3.702)	24-1287-76
Benzo[a]biphenylen-6(5H)-one, 5,10b-bis-(1,1-dimethylethyl)-6a,10b-dihydro-, (5α,6aα,10bα)-	EtOH	268(3.365),273(3.304)	24-1301-76
Indeno[1,2-a]inden-9(4bH)-one, 4b,10-bis(1,1-dimethylethyl)-9a,10-dihydro-, (4bα,9aα,10β)-	EtOH	247(4.076),268(3.404), 276(3.411),295(3.396)	24-1287-76
$C_{24}H_{28}O_2S_2$			
1,3-Benzoxathiol-1-ium, 2,2'-(1,10-dec-anediyl)bis-, bis(tetrafluoroborate)	96% H_2SO_4	244s(4.03),248(4.06), 252(4.02),299(4.24)	39-0323-76C
$C_{24}H_{28}O_8$			
2,3-Naphthalenedicarboxylic acid, 1-(3,4-dimethoxyphenyl)-1,2,3,4-tetra-hydro-6,7-dimethoxy-, dimethyl ester	MeOH	281(3.69)	2-0014-76

Compound	Solvent	$\lambda_{max}(\log \epsilon)$	Ref.
Naphtho[2,3-c]furan-1(3H)-one, 3a,4,9,9a- tetrahydro-3a-(hydroxymethyl)-6,7-di- methoxy-9-(3,4,5-trimethoxyphenyl)-	MeOH	225(4.68),280(4.03)	2-0084-76
$C_{24}H_{28}O_9$ Naphtho[2,3-c]furan-1(3H)-one, 3a,4,9,9a- tetrahydro-4-hydroxy-3a-(hydroxymeth- yl)-6,7-dimethoxy-9-(3,4,5-trimethoxy- phenyl)-, (3aα,4β,9α,9aα)-	MeOH	230(3.22),280(2.67)	2-0084-76
$C_{24}H_{29}ClO_5$ Pregna-4,6-diene-3,20-dione, 16α-(acet- oxymethyl)-6-chloro-16,17-epoxy-	MeOH	282(4.34)	44-3940-76
$C_{24}H_{29}ClO_7$ Pregn-4-ene-3,11-dione, 21-acetoxy- 17,20-[carbonylbis(oxy)]-20α-chloro-	n.s.g.	238(4.20)	44-0078-76
Pregn-4-ene-3,11-dione, 21-acetoxy- 17,20-[carbonylbis(oxy)]-20β-chloro-	n.s.g.	238(4.21)	44-0078-76
$C_{24}H_{29}Cl_3O_5$ Androsta-4,6-dien-3-one, 15-acetyl- 17-acetoxy-6-chloro-16-(dichloro- methyl)-16-hydroxy-, (15α,16α,17β)-	MeOH	283(4.23)	44-3940-76
$C_{24}H_{29}NO_6$ Floripavidine	n.s.g.	229s(4.37),273(4.25), 310(3.45)	105-0716-76
$C_{24}H_{29}NO_{12}$ Pentanedioic acid, 2-[2,3-O-(1-methyl- ethylidene)-5-O-(4-nitrobenzoyl)-α-D- ribofuranosyl]-3-oxo-, diethyl ester	pH 13 EtOH	279(4.45) 258(4.05)	44-0287-76 44-0287-76
2-Pentenedioic acid, 3-[[2,3-O-(1-meth- ylethylidene)-5-O-(4-nitrobenzoyl)- α-D-ribofuranosyl]oxy]-, diethyl ester	EtOH	234(4.21),260s(4.08)	44-0287-76
β-	EtOH	233(4.27),255s(4.15)	44-0287-76
$C_{24}H_{29}N_3O_2$ Isoechinulin A	EtOH	227(4.50),289(3.95), 341(3.97)	88-1601-76
$C_{24}H_{30}N_2$ Benzenemethanamine, N-methyl-N-[[3-meth- ylene-2-[methyl(phenylmethyl)amino]- 1-cyclohexen-1-yl]methyl]-	n.s.g.	238(4.27),289(3.36)	94-2494-76
$C_{24}H_{30}N_2O_{10}S_2$ 2,4(1H,3H)-Pyrimidinedione, 1-[3,5-di- O-acetyl-2-S-ethyl-6-O-[(4-methylphen- yl)sulfonyl]-2-thio-β-D-mannofurano- syl]-5-methyl-	EtOH	266(4.00),273s(3.95)	136-0041-76C
6-deoxy-6-iodo deriv.	EtOH	266(4.00)	136-0041-76C
$C_{24}H_{30}N_2Se$ Methanaminium, N,N'-[selenobis[3-(4- methylphenyl)-2-propen-3-yl-1-yli- dene]]bis[N-methyl-, diperchlorate	DMF	321(4.29)	118-0521-76

Compound	Solvent	λ_{max}(log ϵ)	Ref.
$C_{24}H_{30}N_4O_4S$ Benzenesulfonic acid, 4-methyl-, [1-(4-methyl-3,6-dioxo-2,5-di-1-pyrrolidin-yl-1,4-cyclohexadien-1-yl)ethylidene]-hydrazide	EtOH	229(4.31),273(4.30), 352(4.04),392(3.98), 535(3.00)	94-1731-76
$C_{24}H_{30}N_4O_6S$ Benzenesulfonic acid, 4-methyl-, [1-(4-methyl-2,5-di-4-morpholinyl-3,5-di-oxo-1,4-cyclohexadien-1-yl)ethyli-dene]hydrazide	EtOH	230(4.42),423(3.76)	94-1731-76
$C_{24}H_{30}N_4S_2$ Benzo[g]quinoxaline, 2,2'-dithiobis-[5,5a,6,7,8,9,9a,10-octahydro-, [5aS-[2(5aR*,9aR*),5aα,9aβ]]-	MeCN	234(3.89),300(3.90), 314s(3.86)	142-0299-76S
$C_{24}H_{30}N_8O_4S_2$ 1,2,4,5-Tetrazine-3,6-dicarboximidamide, N,N,N'',N''-tetraethyl-N'',N'''-bis(phen-ylsulfonyl)-	EtOH	242(4.57),525(2.68)	32-0001-76
$C_{24}H_{30}N_8O_5$ Glutamic acid, N-[4-[[2,4-diamino-6-pteridinyl]methyl]methylamino]benz-oyl]-, diethyl ester	EtOH	220(4.35),262(4.41), 296(4.43),375)3.89)	4-0727-76
$C_{24}H_{30}O$ Benzo[a]biphenylen-6-ol, 5,10b-bis(1,1-dimethylethyl)-5,6,6a,10b-tetrahydro-, (5-exo,6-endo)	EtOH	266(3.915),272(3.838)	24-1301-76
Indeno[1,2-a]inden-9-ol, 4b,10-bis(1,1-dimethylethyl)-4b,9,9a,10-tetrahydro-	EtOH	267(3.181),274(3.118)	24-1287-76
$C_{24}H_{30}O_2$ 2,4,6,8,10,12,14-Pentadecaheptaenoic acid, 15-(2,6,6-trimethyl-1-cyclo-hexen-1-yl)-	EtOH	415(4.93)	36-0580-76
$C_{24}H_{30}O_4$ Mogoltacin	n.s.g.	218(4.25),244(3.80), 290(4.00),326(4.24)	105-0084-76
Tadzhiferin	n.s.g.	243(3.65),251(3.51), 295(3.92),325(4.18)	105-0533-76
$C_{24}H_{30}O_6$ 1,2-Benzenedicarboxylic acid, 3-[2-(5-acetyloctahydro-7a-methyl-1-oxo-1H-inden-4-yl)ethyl]-, methyl ester, [3aS-(3aα,4α,5β,7aβ)]-	MeOH	232(3.91),280(3.26), 288(3.20)	23-3508-76
$C_{24}H_{30}O_{13}$ Sweroside, tetraacetate	MeOH	241(3.93)	95-0683-76
$C_{24}H_{31}ClO_5$ Androsta-1,4-dien-3-one, 6β-acetoxy-7α-chloro-17β-(1-oxopropoxy)-	EtOH	244(4.25)	56-0097-76
$C_{24}H_{31}ClO_6$ Pregn-4-en-3-one, 21-acetoxy-17,20-[carbonylbis(oxy)]-20α-chloro-	n.s.g.	240(4.22)	44-0078-76

Compound	Solvent	λ_{max}(log ϵ)	Ref.
Pregn-4-en-3-one, 21-acetoxy-17,20- [carbonylbis(oxy)]-20β-chloro-	n.s.g.	240(4.22)	44-0078-76
$C_{24}H_{31}NO_2$ 2-Aza-19-norpregna-1,3,5(10)-trien- 20-yn-17α-ol, 3-(cyclopentyloxy)-	MeOH	279(3.49)	44-2864-76
$C_{24}H_{31}NO_{12}$ Bakankosin, tetraacetate	MeOH	236(4.20)	94-1406-76
$C_{24}H_{31}N_3O_7S$ L-Phenylalanine, N-[(1,1-dimethyleth- oxy)carbonyl]-, 3'-ester with 4-thio- thymidine	n.s.g.	203(4.30),245(3.40), 333(4.23)	128-0351-76
$C_{24}H_{31}N_5O_5$ D-Mannitol, 1-deoxy-3,4:5,6-bis-O-(1- methylethylidene)-1-[6-(phenylmeth- yl)amino]-7H-purin-7-yl]-	CHCl$_3$	296(4.28)	136-0049-76G
$C_{24}H_{32}Cl_2O_5$ Androst-4-en-3-one, 6β-acetoxy-4,7α-di- chloro-17β-(1-oxopropoxy)-	EtOH	251(4.12)	56-1919-76
Androst-4-en-3-one, 7α-acetoxy-4,6β-di- chloro-17β-(1-oxopropoxy)-	EtOH	258(4.20)	56-1919-76
$C_{24}H_{32}N_2O_2$ Pentanoic acid, 4-cyano-2-(1-cyano-1- methylethyl)-2,4-dimethyl-4-cyclo- hexylphenyl ester	dioxan	263(2.66),270(2.59)	126-3481-76
Pentanoic acid, 4-cyano-2-(2-cyano-2- methylpropyl)-4-methyl-4-cyclohexyl- phenyl ester	dioxan	264(2.89),270(2.88)	126-3481-76
$C_{24}H_{32}N_2O_3$ Acetamide, N-(4-methoxyphenyl)-2-[[4- (nonyloxy)phenyl]imino]-	EtOH	237(3.08)	65-0702-76
$C_{24}H_{32}N_4O_2$ 4,14-Diazatricyclo[15.3.1.1^{7,11}]docosa- 1(21),7,9,11(22),17,19-hexaene-3,13- dione, 10,20-bis(dimethylamino)-	MeOH	251(3.98)	142-1095-76
$C_{24}H_{32}N_7$ 1H-Benzimidazolium, 2-[3-(1,3-diethyl- 1,3-dihydro-5-methyl-2H-benzimidazol- 2-ylidene)-1-triazenyl]-1,3-diethyl- 5-methyl-, tetrafluoroborate	MeOH	436(3.26)	33-0148-76
$C_{24}H_{32}N_7O_2$ 1H-Benzimidazolium, 2-[3-(1,3-diethyl- 1,3-dihydro-5-methoxy-2H-benzimidazol- 2-ylidene)-1-triazenyl]-1,3-diethyl- 5-methoxy-, tetrafluoroborate	MeOH	449(3.36)	33-0148-76
$C_{24}H_{32}N_8O_4S_2$ 1,2,4,5-Tetrazine-3,6-dicarboximidamide, N,N,N'',N''-tetraethyl-1,2-dihydro- N',N'''-bis(phenylsulfonyl)-	MeCN	252(4.50)	32-0001-76

Compound	Solvent	$\lambda_{max}(\log \epsilon)$	Ref.
$C_{24}H_{32}O_3S_2$ Spiro[1,3-dithiolan-2,2'(4'aH)-phenan-threne]-4'a-carboxylic acid, 1',3'-4',9'-tetrahydro-6'-methoxy-1',1'-dimethyl-7'-(1-methylethyl)-, methyl ester, (±)-	MeOH	238(3.27),280(3.09), 285(3.09)	18-1985-76
$C_{24}H_{32}O_4S$ Pregna-1,4-diene-3,20-dione, 11-hydroxy-16,17-[oxy(1-methylethylidene)thio]-, (11β,16α)-	MeOH	242(4.18)	24-0185-76
$C_{24}H_{32}O_5$ 5α-Androst-8-ene-7,11,17-trione, 3β-acetoxy-4,4,14α-trimethyl-	dioxan	266(3.94)	5-1206-76
Pregna-5,8-diene-7,11,20-trione, 3β,6-dihydroxy-4,4,14α-trimethyl-	dioxan	232(4.14),280(3.83), 355(3.51)	5-1206-76
$C_{24}H_{32}O_6$ Pregn-4-en-3-one, 20β-acetoxy-17,20α-[carbonylbis(oxy)]-	n.s.g.	239(4.23)	44-0078-76
$C_{24}H_{32}O_8$ Asatone	EtOH	275(3.83)	102-1567-76
Dibenzo[b,n][1,4,7,10,13,16,19,22]octa-oxacyclotetracosin, 6,7,9,10,12,13-20,21,23,24,26,27-dodecahydro-	MeOH	276(3.64)	64-0330-76B
sodium perchlorate complex	MeOH	275(3.67)	64-0330-76B
sodium picrate complex	MeOH	271.5(3.83)	64-0330-76B
Lyoniresinol, dimethyl ether, (-)-	EtOH	274(3.26)	94-2102-76
Rastronol D	EtOH	233(3.95)	33-0772-76
$C_{24}H_{33}ClO_5$ Androst-4-en-3-one, 6β-acetoxy-7α-chloro-17β-(1-oxopropoxy)-	EtOH	233(4.16)	56-1919-76
$C_{24}H_{33}FO_4$ Pregn-4-ene-3,20-dione, 21-fluoro-16α,17-[(1-methylethylidene)bis(oxy)]-	EtOH	240(4.23)	33-1027-76
$C_{24}H_{33}FO_4S$ Pregn-4-ene-3,20-dione, 9-fluoro-11-hydroxy-16,17-[(oxy)(1-methylethyli-dene)(thio)]-, (11β,16α)-	MeOH	238(4.19)	24-0185-76
$C_{24}H_{33}NO_5$ 17α-Pregn-4-ene-21-carboxylic acid, 17-hydroxy-7α-[(methoxycarbonyl)-amino]-3-oxo-, γ-lactone	MeOH	241(4.15)	87-0975-76
5α-Pregn-8-ene-6,7,11,20-tetrone, 3β-hydroxy-4,4,14α-trimethyl-, 6-oxime	dioxan	274(3.85)	5-1214-76
$C_{24}H_{34}INO_5$ Isochroman, 3-[2-[2-(dimethylamino)eth-yl]-4,5-dimethoxyphenyl]-6,7-dimeth-oxy-, methiodide	pH 1	228(4.31),283(3.69)	33-0949-76
$C_{24}H_{34}N_2O_3$ 2-Propanone, 1-[(3β,17β)-3-acetoxy-androst-5-en-17-yl]-3-diazo-	EtOH	204(3.75),247(3.97), 265s(3.90)	33-2125-76

Compound	Solvent	$\lambda_{max}(\log \epsilon)$	Ref.
$C_{24}H_{34}O_3$			
Chola-4,22-dien-24-oic acid, 3-oxo-	EtOH	220(2.20),240(4.19)	13-0431-76A
$C_{24}H_{34}O_3S$			
Pregn-4-ene-3,20-dione, 16,17-[oxy(1-methylethylidene)thio]-, (16α)-	MeOH	240(4.23)	24-0185-76
(16β)-	MeOH	240(4.25)	24-0185-76
$C_{24}H_{34}O_4$			
Pregn-4-ene-6α-acetic acid, 3,20-dioxo-, methyl ester	EtOH	239(4.18)	12-2077-76
5α-Pregn-8-ene-7,11,20-trione, 3β-hydroxy-4,4,14α-trimethyl-	dioxan	268(3.93)	5-1206-76
$C_{24}H_{34}O_5$			
19-Norpregn-5(10)-en-6-one, 3β,20β-diacetoxy-	MeOH	247(4.02)	44-0313-76
4a(2H)-Phenanthrenecarboxylic acid, 1,3,4,9,10,10a-hexahydro-5,6-dimethoxy-1,1-dimethyl-7-(1-methylethyl)-9-oxo-, ethyl ester, trans	EtOH	224(4.40),273(4.13)	44-1005-76
$C_{24}H_{34}O_5S$			
Pregn-4-ene-3,20-dione, 16,17-[oxy(1-methylethylidene)sulfonyl]-, (16α)-	MeOH	239(4.22)	24-0185-76
$C_{24}H_{34}O_6$			
4-Cyclohexene-1,3-dione, 4,4'-(2-methylpropylidene)bis[5-hydroxy-2,2,6,6-tetramethyl-	EtOH	270(4.27)	12-2525-76
	EtOH-acid	262(--)	12-2525-76
$C_{24}H_{34}O_7$			
Ajugarin I	MeOH	212(4.00)	77-0949-76
Kaur-16-ene-1,6,7,15-tetrol, 7,20-epoxy-, 6,15-diacetate (rastronol A)	EtOH	234(3.95)	33-0772-76
$C_{24}H_{34}O_8$			
Asatone A, hydroxy-	MeOH	218(3.65),235s(3.42)	18-1940-76
Asatone B, hydroxy-	MeOH	213(3.40),277(3.64)	18-1940-76
Rastronol B	EtOH	229(3.87)	33-0772-76
$C_{24}H_{35}NO_2$			
2H-Pyrrol-2-one, 1,5-dihydro-4-[[(3β,17β)-3-hydroxyandrost-5-en-17β-yl]methyl]-	EtOH	210(4.14)	33-2138-76
$C_{24}H_{35}NS$			
1(2H)-Isoquinolinethione, 2-[[2,6-bis-(1,1-dimethylethyl)cyclohexylidene]-methyl]-3,4-dihydro-	hexane	260(4.1),275s(--), 337(3.7),448(2.2)	24-0906-76
$C_{24}H_{36}N_2O_2$			
1'H-Androstano[3,2-c]pyrazol-17-ol, 1',5'-dimethyl-, acetate, (5α,17β)-	EtOH	234(3.83)	22-0255-76
$C_{24}H_{36}N_2S_4$			
Cyclohexanamine, N,N'-[dithiobis(carbonothioyl-2-cyclopentyl-1-ylidene)]-bis-	CHCl$_3$	240(4.11),310(4.14), 410(4.91)	39-1706-76C

Compound	Solvent	$\lambda_{max}(\log \epsilon)$	Ref.
$C_{24}H_{36}O_2$			
Pregn-4-ene-3,20-dione, 16-ethyl-17-methyl-	EtOH	240(4.23)	39-1558-76C
2-Propen-1-one, 1-[(3β,16α,17β)-3-hydroxy-16,17-dimethylandrost-5-en-17-yl]-	EtOH	218(3.89)	39-1558-76C
$C_{24}H_{36}O_4$			
Carnosic acid, dimethyl ether, ethyl ester, (+)-	EtOH	276(2.79)	44-1005-76
2H-Pyran-2-one, 5-[3-[2-(1,4-dioxaspiro[4.5]dec-8-yl)-1-ethylbutyl]cyclopentyl]-	EtOH	298(3.72)	5-1222-76
$C_{24}H_{36}O_5$			
Carnosic acid, 7β-hydroxy-, dimethyl ether, ethyl ester, (+)-	EtOH	274(2.83)	44-1005-76
24-Norchol-4-en-23-oic acid, 7α,12α-dihydroxy-3-oxo-, methyl ester	EtOH	244(4.17)	13-0655-76B
$C_{24}H_{36}O_7$			
5α-Decal-7-one, 10-(ethoxycarbonyl)-4,4-dimethyl-8-(hydroxymethylene)-9-[1-(methoxycarbonyl)-2-oxo-4-methylpentyl]-	EtOH base	285(3.84) 285(4.26),304(4.26)	44-1005-76 44-1005-76
$C_{24}H_{36}O_8$			
Asatone, tetrahydro-	MeOH	229(3.59),278(3.71)	18-1940-76
Isoasatone, tetrahydro-	MeOH	222(3.32)	18-1940-76
$C_{24}H_{38}$			
Tricyclo[12.10.0.0^{3,16}]tetracosa-1,3(16),14-triene	isooctane	240(3.81),280(2.51), 289s(2.41)	44-4081-76
$C_{24}H_{38}N_2O_3$			
Carbamic acid, [3-(hexyloxy)phenyl]-, 3-(1-piperidinyl)cyclohexyl ester	H_2O	236(4.07),282(3.36)	106-0024-76
Carbamic acid, [4-(hexyloxy)phenyl]-, 3-(1-piperidinyl)cyclohexyl ester	H_2O	239(4.18),282(3.15)	106-0024-76
$C_{24}H_{38}N_4O_4$			
4,4'-Bipyrimidine, 2,2',6,6'-tetrakis-(1,1-dimethylethoxy)-	hexane	248(3.47),291s(4.11), 299(4.16),311s(3.96)	19-0017-76
$C_{24}H_{38}O_8$			
Asatone, hexahydro-	MeOH	221(3.45)	18-1940-76
$C_{24}H_{40}N_2S_4$			
1-Hexanamine, N,N'-[dithiobis(carbonothioyl-2-cyclopentyl-1-ylidene)]bis-	benzene	308(4.18),410(4.91)	39-1706-76C
$C_{24}H_{40}O_8$			
Asatone, tetrahydrodihydroxy-	MeOH	217(3.38)	18-1940-76
$C_{24}H_{41}ClO_2S$			
2-Thiophenecarboxylic acid, 4-(chloromethyl)-5-heptadecyl-, methyl ester	heptane	255(3.92),277(4.00)	34-0233-76
$C_{24}H_{42}N_2$			
Cyclooctanamine, N,N'-(2,2,4,4-tetramethyl-1,3-cyclobutanediylidene)bis-	C_6H_{12} EtOH	260(2.11),274(2.13) 248(--),264(--)	59-1415-76 59-1415-76

Compound	Solvent	$\lambda_{max}(\log \epsilon)$	Ref.
$C_{24}H_{42}OS$			
1-Eicosanone, 1-(2-thienyl)-	heptane	258(3.97),275(3.87)	34-0233-76
$C_{24}H_{42}O_2S$			
2-Thiophenecarboxylic acid, 5-hepta-decyl-4-methyl-, methyl ester	heptane	254(3.93),284(3.99)	34-0233-76
2-Thiophenecarboxylic acid, 5-nona-decyl-	heptane	254(4.31),274(4.00)	34-0233-76
$C_{24}H_{43}NO$			
2,5-Cyclohexadien-1-one, 2,4,6-tris-(1,1-dimethylethyl)-4-ethyl-, O-(1,1-dimethylethyl)oxime	hexane	258(3.77)	18-1142-76
$C_{24}H_{44}S$			
Thiophene, 2-eicosyl-	heptane	234(3.87)	34-0233-76
Thiophene, 3-methyl-2-nonadecyl-	heptane	236(3.81)	34-0233-76
$C_{24}H_{48}Si_3$			
Silane, (1-methyl-1,2-pentadien-4-yne-1,3,5-triyl)tris[(1,1-dimethylethyl)-	MeOH	232s(4.05),240(4.11), 256(4.11)	35-8426-76

Compound	Solvent	$\lambda_{max}(\log \epsilon)$	Ref.
$C_{25}H_{13}Cl_3N_2O$ Anthra[1,2-b]phenazin-6(9H)-one, 1,4,7-trichloro-9-methyl-	CH_2Cl_2	268(4.69),292(4.52), 328(4.33),445(4.31), 699(4.31)	33-0688-76
$C_{25}H_{13}Cl_5N_2O_2$ 1,4-Anthracenedione, 2,5,8-trichloro- 3-[(6,7-dichloro-1,3-dimethyl-2(1H)- quinoxalinylidene)methyl]-	CH_2Cl_2	326(4.29),402(4.07), 605(4.22)	33-0688-76
$C_{25}H_{13}Mn_3O_{10}$ Manganese, nonacarbonyl[μ_3-[η^5:η^5:η^5- [(hydroxymethylidyne)tri-2,4-cyclo- pentadien-1-ylidene]]]tri-	CH_2Cl_2 acid	330(3.48) 380(4.45),720(4.21)	70-2241-76 70-2241-76
$C_{25}H_{13}NOS_2$ Bis[1]benzothieno[3,2-b:2',3'-e]pyri- dine, 6-(benzofuranyl)-	$C_2H_4Cl_2$	224(3.90),263(3.81), 292(3.72),333(3.73), 369(3.83)	103-0977-76
$C_{25}H_{14}N_2S_2$ Bis[1]benzothieno[3,2-b:2',3'-e]pyri- dine, 6-(1H-indolyl)-	$C_2H_4Cl_2$	245(3.86),286(3.77), 342(3.51)	103-0977-76
$C_{25}H_{15}ClN_2O$ Anthra[1,2-b]phenazin-6(9H)-one, 7-chloro-9-methyl-	CH_2Cl_2	306(4.17),402(4.05), 613(4.20)	33-0688-76
$C_{25}H_{15}NO$ 1-Naphthalenecarbonitrile, 4-[2-(3-di- benzofuranyl)ethenyl]-	DMF	368(4.59)	33-0819-76
$C_{25}H_{15}NO_3$ 2H-Pyrano[2,3-b]quinolin-2-one, 3-benzoyl-5-phenyl-	dioxan	<u>235(4.3),310s(4.1),</u> <u>375(4.1),430(3.7)</u>	56-0625-76
$C_{25}H_{16}BrNO_2$ 9,10-Anthracenedione, 1-amino-4-bromo- 2-[(2-naphthalenyl)methyl]-	n.s.g.	480(3.82)	146-0001-76
$C_{25}H_{16}O_2$ Benz[a]anthracene-7,12-dione, 2-methyl- 5-phenyl-	$CHCl_3$	245s(4.44),250(4.47), 293(4.56),337s(3.64), 371(3.51),424(3.65)	18-3361-76
Indeno[2,1-b]pyran-9-carboxaldehyde, 2,4-diphenyl-	dioxan	243(4.30),268(4.45), 307(4.25),392(4.21), 520(3.71)	78-2225-76
$C_{25}H_{16}O_3$ Benz[a]anthracene-7,12-dione, 2-methoxy- 5-phenyl-	$CHCl_3$	248(4.51),293(4.56), 430(3.73)	18-3361-76
$C_{25}H_{17}Cl_2N_5O$ 1,3,4-Oxadiazole, 2-[3-chloro-4-[2-[2- (4-chlorophenyl)-5-methyl-2H-1,2,3- triazol-4-yl]ethenyl]phenyl]-5-phenyl-	DMF	355(4.72)	33-2469-76
$C_{25}H_{17}N$ 1-Naphthalenecarbonitrile, 4-(2-[1,1'- biphenyl]-4-ylethenyl)-	DMF	365(4.56)	33-0819-76

Compound	Solvent	$\lambda_{max}(\log \epsilon)$	Ref.
$C_{25}H_{17}NO_2$			
9,10-Anthracenedione, 1-amino-2-(2-naph-thalenylmethyl)-	n.s.g.	475(3.25)	146-0001-76
Benzeneacetonitrile, α-[1-(benzoyloxy)-3-phenyl-2-propynylidene]-4-methyl-	EtOH	215(4.39),239(4.64), 335(4.52)	34-0115-76
$C_{25}H_{17}NO_2S_2$			
Bis[1]benzothieno[3,2-b:2',3'-e]pyri-dine, 6-(3,4-dimethoxyphenyl)-	$C_2H_4Cl_2$	245(5.83),268(5.45), 289(5.07),340(4.83)	103-0977-76
$C_{25}H_{17}NO_3$			
9,10-Anthracenedione, 1-amino-2-[(4-hy-droxy-1-naphthalenyl)methyl]-	n.s.g.	475(3.65)	146-0001-76
2H-Pyrano[2,3-b]quinolin-2-one, 3-benz-oyl-3,4-dihydro-5-phenyl-	dioxan	<u>240(4.7),385s(3.2), 430(3.3)</u>	56-0625-76
$C_{25}H_{17}N_3$			
Pyrido[2,3-b]pyrazine, 2,3,7-triphenyl-	EtOH	250(4.54),387(4.46)	22-0251-76
$C_{25}H_{17}N_3O$			
2-Naphthalenol, 1-[2-(2-quinolinyl)phen-ylazo]-	n.s.g.	486(4.33)	7-0379-76
2-Naphthalenol, 1-[3-(2-quinolinyl)phen-ylazo]-	n.s.g.	477(4.37)	7-0379-76
2-Naphthalenol, 1-[4-(2-quinolinyl)phen-ylazo]-	n.s.g.	494(4.55)	7-0379-76
$C_{25}H_{17}N_3O_2$			
5,8-Quinazolinedione, 4-methyl-2-(1-naphthalenyl)-6-(phenylamino)-	EtOH	222(4.84),285(4.40), 319s(--),396(4.04), 514s(--)	5-1809-76
$C_{25}H_{17}N_3O_4$			
9H-Benzo[e]quino[2,3-i]perimidine-9,15(14H)-dione, 8-(2-methoxyethoxy)-	DMF	270(4.17),366(3.66), 462(3.48),550(3.57), 590(3.73)	80-0257-76
$C_{25}H_{18}ClN_3$			
Pyrazino[2,3-b]quinoline, 8-chloro-3,4-dihydro-2-methyl-3-methylene-4,10-diphenyl-	DMF	284(4.67),305s(4.18), 341s(3.81),417(4.06), 438(4.02)	48-0039-76
$C_{25}H_{18}ClN_3O$			
2H-[1,4]Diazepino[2,3-b]quinolin-2-one, 9-chloro-1,5-dihydro-4-methyl-5,11-di-phenyl-	DMF	276s(4.21),288(4.24), 334s(4.00),360s(4.22), 372(4.27),395s(3.98)	48-0039-76
$C_{25}H_{18}ClN_5O$			
1,2,4-Oxadiazole, 5-[4-[2-[2-(4-chloro-phenyl)-5-methyl-2H-1,2,3-triazol-4-yl]ethenyl]phenyl]-3-phenyl-	DMF	351(4.71)	33-2469-76
1,3,4-Oxadiazole, 2-[3-chloro-4-[2-(5-methyl-2-phenyl-2H-1,2,3-triazol-4-yl)ethenyl]phenyl]-5-phenyl-	DMF	353(4.69)	33-2469-76
1,3,4-Oxadiazole, 2-(4-chlorophenyl)-5-[4-[2-(5-methyl-2-phenyl-2H-1,2,3-triazol-4-yl)ethenyl]phenyl]-	DMF	350(4.76)	33-2469-76
$C_{25}H_{18}N_2$			
Benzenamine, 4-(1-phenylbenzo[f]quino-3-yl)-	EtOH	295(4.53),333(4.42), 373(4.26)	103-0106-76

Compound	Solvent	$\lambda_{max}(\log \epsilon)$	Ref.
Benzenamine, 4-(3-phenylbenzo[f]quinolin-1-yl)-	EtOH	280(4.54),360(3.76)	103-0106-76
$C_{25}H_{18}N_2O_2$ 9,10-Anthracenedione, 1,4-diamino-2-(2-naphthalenylmethyl)-	n.s.g.	550(4.30),585(3.56)	146-0001-76
$C_{25}H_{18}N_2O_3$ 9,10-Anthracenedione, 1,4-diamino-2-[(4-hydroxy-1-naphthalenyl)methyl]-	n.s.g.	550(4.28),590(3.72)	146-0001-76
$C_{25}H_{18}N_4O$ 2H-1,2,3-Triazole, 2-phenyl-4-[2-[4-(5-phenyl-2-oxazolyl)phenyl]ethenyl]-	DMF	363(4.79)	33-2469-76
$C_{25}H_{18}N_6OS$ 1H-Pyrrole-3-carbonitrile, 2-amino-5-[(6-methoxy-2-benzothiazolyl)azo]-1,4-diphenyl-	MeCN	299(4.13),351(3.74), 518(4.58)	48-0663-76
$C_{25}H_{18}O$ Indeno[2,1-b]pyran, 4-methyl-2,9-diphenyl-	dioxan	230(4.31),264(4.40), 298(4.29),337(4.36), 371(4.15),537(3.47)	78-2219-76
Indeno[2,1-b]pyran, 9-methyl-2,4-diphenyl-	dioxan	229(4.31),272(4.37), 363(4.19),540(3.46)	78-2225-76
1H-Naphtho[2,1-b]pyran, 2,3-diphenyl-	EtOH	217(4.66),240(4.60), 270s(--)	22-1967-76
$C_{25}H_{18}O_2$ Ethanone, 2-(2,6-diphenyl-4H-pyran-4-ylidene)-1-phenyl-	MeCN	224(4.34),249(4.29), 309(4.25),412(4.46)	142-0739-76
$C_{25}H_{18}O_5Te$ 6H-Cyclohepta[c]tellurophene-5,7-dicarboxylic acid, 6-oxo-1,3-diphenyl-, dimethyl ester	CHCl$_3$	305(4.29),360(4.39)	24-3886-76
$C_{25}H_{18}O_9$ 8H-[ ]Benzopyrano[2,3-a][3]benzoxepin-8-one, 2,4,9,11-tetrahydroxy-6-(4-hydroxy-3-methoxyphenyl)-5-methyl-(dehydrosilychristin)	MeOH	268(4.22),295(4.23), 365(3.97)	88-2241-76
	MeOH-AlCl$_3$	274(4.32),298(4.23), 352(3.90),422(4.07)	88-2241-76
	MeOH-AlCl$_3$-HCl	275(4.31),298(4.24), 355(3.93),420(4.04)	88-2241-76
$C_{25}H_{19}BrN_2O_3$ 1H-Pyrazole-4-carboxylic acid, 3-(4-bromobenzoyl)-1,5-diphenyl-, ethyl ester	EtOH	263(4.48)	4-0989-76
$C_{25}H_{19}ClN_2O_2S$ 3-Pyridinecarboxamide, 6-chloro-1,2-dihydro-1-(3-methylphenyl)-N-phenyl-4-(phenylthio)-	MeOH	335(4.4)	120-0103-76
$C_{25}H_{19}ClN_4O$ Benzoxazole, 2-[4-[2-[2-(4-chlorophenyl)-5-methyl-2H-1,2,3-triazol-4-yl]-ethenyl]phenyl]-5-methyl-	DMF	357(4.80)	33-2469-76
Benzoxazole, 2-[4-[2-[2-(4-chlorophen-	DMF	360(4.81)	33-2469-76

Compound	Solvent	$\lambda_{max}(\log \epsilon)$	Ref.
yl)-5-methyl-2H-1,2,3-triazol-4-yl]-ethenyl]phenyl]-6-methyl- (cont.)			33-2469-76
Benzoxazole, 2-[4-[2-[2-(4-chlorophen-yl)-2H-1,2,3-triazol-4-yl]ethenyl]-phenyl]-5-ethyl-	DMF	355(4.85)	33-2469-76
$C_{25}H_{19}Cl_2N_3O_2$			
1H-Isoindole-1,3(2H)-dione, 2-[[7-chlo-ro-5-(2-chlorophenyl)-2,3-dihydro-1-methyl-1H-1,4-benzodiazepin-2-yl]-methyl]-	EtOH	231s(--),240s(--), 270s(--),372(3.28)	111-0501-76
$C_{25}H_{19}NOS$			
9-Thia-7-azatricyclo[4.2.1.0^{2,5}]non-3-en-8-one, 1,6,7-triphenyl-, (1α,2β,5β,6α)-	MeOH	245s(3.82)	39-2562-76C
$C_{25}H_{19}NO_2$			
9,10-Ethanoanthracene-11,12-dicarboxim-ide, 9,10-dihydro-N-methyl-9-phenyl-	ether	250(3.12),257(3.10), 261(3.09),270(2.92)	22-0493-76
4,11-Etheno-1H-naphtho[2,3-f]isoindole-1,3(2H)-dione, 3a,4,11,11a-tetra-hydro-2-methyl-5-phenyl-	ether	237(4.58),260(3.65), 270(3.74),280(3.81), 290(3.71)	22-0493-76
2-Pyridinol, 3,4,6-triphenyl-, acetate	EtOH	253(4.37),295(4.06)	4-0195-76
$C_{25}H_{19}NO_2S$			
9-Thia-7-azatricyclo[4.2.1.0^{2,5}]non-3-en-8-one, 1,6,7-triphenyl-, 9-oxide, (1α,2β,5β,6α)-	MeOH	226(4.36)	39-2562-76C
$C_{25}H_{19}NO_4$			
4,11[1',2']-Benzeno-1H-[1,6]benzodioxo-cino[3,4-c]pyrrole-1,3(2H)-dione, 3a,4,11,11a-tetrahydro-2-methyl-4-phenyl-, (3aα,4β,11β,11aα)-	ether	264(3.87),270(3.89), 278(3.79)	22-0493-76
7,11-Etheno-7H-bisoxireno[1,2:3,4]naph-tho[2,3-f]isoindole-8,10(7aH,9H)-dione, 1a,5b,10a,11-tetrahydro-9-methyl-1a-phenyl-	ether	253(3.69),259(3.77), 265(3.79),272(3.83), 278(3.83)	22-0493-76
$C_{25}H_{19}N_3$			
Benzenamine, 4,4'-benzo[f]quinoline-1,3-diylbis-	EtOH	297(4.60),329(4.56), 367(4.33)	103-0106-76
Pyrazino[2,3-b]quinoline, 3,4-dihydro-2-methyl-3-methylene-4,10-diphenyl-	DMF	278(4.56),310(4.19), 318(4.18),340s(3.95), 354s(3.84),390s(3.94), 413(4.11),427(4.05)	48-0039-76
$C_{25}H_{19}N_3O$			
1H-1,2-Diazepino[2,3-b]quinolin-2-one, 1,5-dihydro-4-methyl-5,11-diphenyl-	DMF	288(4.20),365(4.29), 390s(4.09)	48-0039-76
3,2':3',3"-Ter-[1H-indol]-2"(3"H)-one, 1"-methyl-	MeOH	219(4.81),258(4.09), 283(4.08),292(4.01)	103-0178-76
$C_{25}H_{19}N_3O_4$			
Spiro[2H-1-benzopyran-2,5'(4'H)-pyrido-[2,3,4,5-lmn]phenanthridine], 8-meth-oxy-4',10'-dimethyl-6-nitro-	EtOH	232(4.61),253s(4.49), 318s(4.01),360(4.10), 376(4.10),555(3.03)	103-0273-76
	$C_2H_4Cl_2$	256(4.58),300s(4.14), 305(4.11),318s(4.03),	103-0273-76

Compound	Solvent	$\lambda_{max}(\log \epsilon)$	Ref.
(cont.)		340s(4.02),360(4.16), 376(4.17)	103-0273-76
$C_{25}H_{19}N_3O_5$			
1H-Pyrazole-4-carboxylic acid, 3-benzoyl-1-(4-nitrophenyl)-5-phenyl-, ethyl ester	EtOH	258(4.160)	4-0989-76
1H-Pyrazole-4-carboxylic acid, 3-(4-nitrobenzoyl)-1,5-diphenyl-, ethyl ester	EtOH	260(4.463)	4-0989-76
$C_{25}H_{19}N_5O$			
1,2,4-Oxadiazole, 3-(4-methylphenyl)-5-[4-[2-(2-phenyl-2H-1,2,3-triazol-4-yl)ethenyl]phenyl]-	DMF	340(4.71)	33-2469-76
1,2,4-Oxadiazole, 5-[4-[2-(5-methyl-2-phenyl-2H-1,2,3-triazol-4-yl)ethenyl]phenyl]-3-phenyl-	DMF	352(4.69)	33-2469-76
1,3,4-Oxadiazole, 2-(3-methylphenyl)-5-[4-[2-(2-phenyl-2H-1,2,3-triazol-4-yl)ethenyl]phenyl]-	DMF	344(4.80)	33-2469-76
1,3,4-Oxadiazole, 2-[4-[2-(5-methyl-2-phenyl-2H-1,2,3-triazol-4-yl)ethenyl]phenyl]-5-phenyl-	DMF	348(4.75)	33-2469-76
5-Pyrazolol, 3-methyl-1-phenyl-4-[2-(2-quinolinyl)phenylazo]-	n.s.g.	402(4.28)	7-0379-76
5-Pyrazolol, 3-methyl-1-phenyl-4-[3-(2-quinolinyl)phenylazo]-	n.s.g.	391(4.42)	7-0379-76
5-Pyrazolol, 3-methyl-1-phenyl-4-[4-(2-quinolinyl)phenylazo]-	n.s.g.	413(4.56)	7-0379-76
$C_{25}H_{19}N_5O_2$			
1,2,4-Oxadiazole, 5-[4-[2-[2-(4-methoxyphenyl)-2H-1,2,3-triazol-4-yl]ethenyl]phenyl]-3-phenyl-	DMF	347(4.69)	33-2469-76
1,3,4-Oxadiazole, 2-(3-methoxyphenyl)-5-[4-[2-(2-phenyl-2H-1,2,3-triazol-4-yl)ethenyl]phenyl]-	DMF	345(4.80)	33-2469-76
1,3,4-Oxadiazole, 2-(4-methoxyphenyl)-5-[4-[2-(2-phenyl-2H-1,2,3-triazol-4-yl)ethenyl]phenyl]-	DMF	348(4.76)	33-2469-76
$C_{25}H_{19}O_2P$			
6-Phosphanthridinol, 5,6-dihydro-5,6-diphenyl-, 5-oxide	EtOH	212(4.62),266(3.85), 273(3.83)	39-2050-76C
$C_{25}H_{20}$			
Benzene, 1,1',1'',1'''-methanetetrayltetrakis-	C_6H_{12}	244(4.48)	24-3099-76
$C_{25}H_{20}BrN_3O_2$			
1H-Pyrazole-4-carboxamide, 3-(4-bromobenzoyl)-5-methyl-N-(4-methylphenyl)-1-phenyl-	EtOH	265(4.59)	4-0989-76
$C_{25}H_{20}ClN_3O_2$			
1H-Pyrazole-4-carboxamide, 3-benzoyl-1-(4-chlorophenyl)-5-methyl-N-(4-methylphenyl)-	EtOH	260(4.67)	4-0989-76

Compound	Solvent	$\lambda_{max}(\log \epsilon)$	Ref.
$C_{25}H_{20}ClN_3O_9S$			
1-Azuleneethanol, 2-chloro-, tosylate, trinitrobenzene complex	CH_2Cl_2	282(4.78),291(4.79), 333(3.64),347(3.64), 561(2.55),600(2.51), 660(2.09)	44-1811-76
$C_{25}H_{20}ClS_2$			
Thiopyrylium, 6-(4-chlorophenyl)-2-(ethylthio)-3,4-diphenyl-, tetrafluoroborate	MeOH	253(4.29),304(4.06), 436(4.04)	44-0818-76
$C_{25}H_{20}N$			
Pyridinium, 1-(7,12-dimethylbenz[a]anthracen-5-yl)-, perchlorate	MeOH	234(4.36),256(4.48), 279(4.50),290(4.68), 300(4.69),366(3.83), 406(3.49)	44-2679-76
Pyridinium, 1-(7-ethylbenz[a]anthracen-12-yl)-, perchlorate	MeOH	233(4.48),262(4.62), 272(4.71),282(3.86), 292(4.74),338(3.72), 348(3.86),360(3.88), 374(3.73),398(3.56)	44-2679-76
Pyridinium, 1-[(7-methylbenz[a]anthracen-12-yl)methyl]-, picrate	MeOH	237(4.51),266(4.56), 275(4.61),286(4.77), 297(4.76),352(4.31), 363(4.32),381(4.22), 400(4.08)	44-2679-76
Pyridinium, 1-[(12-methylbenz[a]anthracen-7-yl)methyl]-, picrate	MeOH	260(4.62),275(4.63), 284(4.79),295(4.80), 348(4.33),363(4.35)	44-2679-76
$C_{25}H_{20}N_2O_2$			
Ethanone, 1-[3-benzoyl-1-(4-methylphenyl)-5-phenyl-1H-pyrazol-4-yl]-	EtOH	258(4.548)	4-0989-76
Ethanone, 1-[3-(4-methylbenzoyl)-1,5-diphenyl-1H-pyrazol-4-yl]-	EtOH	262(4.41)	4-0989-76
3H-Fluorene-3,9(2H)-dione, 2-[1-ethyl-3-methyl-2(1H)-quinoxalinylidene]-1-methyl-	CH_2Cl_2	374(4.54),438(3.98), 664(4.57),725(4.65)	33-0692-76
Spiro[2H-1-benzopyran-2,5'(4'H)-pyrido-[2,3,4,5-lmn]phenanthridine], 6-methoxy-4',10'-dimethyl-	EtOH	217(4.63),231(4.57), 253(4.45),285s(4.06), 320(3.80),370(3.7)	103-0273-76
Spiro[2H-1-benzopyran-2,5'(4'H)-pyrido-[2,3,4,5-lmn]phenanthridine], 8-methoxy-4',10'-dimethyl-	EtOH	220(4.76),230s(4.72), 253(4.52),275s(4.29), 315s(3.88),378(3.78)	103-0273-76
$C_{25}H_{20}N_2O_3$			
Ethanone, 1-[3-benzoyl-1-(4-methoxyphenyl)-5-phenyl-1H-pyrazol-4-yl]-	EtOH	258(4.541)	4-0989-76
$C_{25}H_{20}N_2O_3S$			
3-Pyridinecarboxamide, 1,2-dihydro-6-hydroxy-1-(2-methylphenyl)-N-phenyl-4-(phenylthio)-	MeOH	339(4.51)	120-0103-76
3-Pyridinecarboxamide, 1,2-dihydro-6-hydroxy-1-(3-methylphenyl)-N-phenyl-4-(phenylthio)-	MeOH	339(4.50)	120-0103-76
3-Pyridinecarboxamide, 1,2-dihydro-6-hydroxy-1-(4-methylphenyl)-N-phenyl-4-(phenylthio)-	MeOH	337(4.44)	120-0103-76

Compound	Solvent	$\lambda_{max}(\log \epsilon)$	Ref.
$C_{25}H_{20}N_2O_4S$			
3-Pyridinecarboxamide, 1,2-dihydro-6-hydroxy-1-(3-methoxyphenyl)-N-phenyl-4-(phenylthio)-	MeOH	339.5(4.43)	120-0103-76
3-Pyridinecarboxamide, 1,2-dihydro-6-hydroxy-1-(4-methoxyphenyl)-N-phenyl-4-(phenylthio)-	MeOH	338(4.50)	120-0103-76
$C_{25}H_{20}N_4$			
1H-Pyrrole-3-carbonitrile, 4-methyl-5-phenyl-1-(phenylamino)-2-[(phenyl-methylene)amino]-	MeOH	229(4.52),270(4.37), 392(4.21)	48-0663-76
$C_{25}H_{20}N_4O$			
Benzoxazole, 5,6-dimethyl-2-[4-[2-(2-phenyl-2H-1,2,3-triazol-4-yl)ethen-yl]phenyl]-	DMF	355(4.78)	33-2469-76
Benzoxazole, 5,6-dimethyl-2-[2-[4-(2-phenyl-2H-1,2,3-triazol-4-yl)phenyl]-ethenyl]-	DMF	353(4.75),364(4.76)	33-2469-76
Benzoxazole, 5,6-dimethyl-2-[2-[4-(4-phenyl-2H-1,2,3-triazol-2-yl)phenyl]-ethenyl]-	DMF	367(4.78)	33-2469-76
Benzoxazole, 5,7-dimethyl-2-[2-[4-(2-phenyl-2H-1,2,3-triazol-4-yl)phenyl]-ethenyl]-	DMF	350(4.76),362(4.77)	33-2469-76
Benzoxazole, 5,7-dimethyl-2-[2-[4-(4-phenyl-2H-1,2,3-triazol-2-yl)phenyl]-ethenyl]-	DMF	363(4.78)	33-2469-76
Benzoxazole, 5-methyl-2-[4-[2-(5-methyl-2-phenyl-2H-1,2,3-triazol-4-yl)ethen-yl]phenyl]-	DMF	359(4.79)	33-2469-76
Benzoxazole, 6-methyl-2-[4-[2-(5-methyl-2-phenyl-2H-1,2,3-triazol-4-yl)ethen-yl]phenyl]-	DMF	360(4.79)	33-2469-76
$C_{25}H_{20}N_4O_2$			
Benzoxazole, 5-methyl-2-[4-[2-(5-methyl-2-phenyl-2H-1,2,3-triazol-4-yl)ethen-yl]phenyl]-	DMF	364(4.78)	33-2469-76
Benzoxazole, 2-[4-[2-[2-(4-methoxyphen-yl)-2H-1,2,3-triazol-4-yl]ethenyl]-phenyl]-5-methyl-	DMF	358(4.81)	33-2469-76
Benzoxazole, 2-[4-[2-[2-(4-methoxyphen-yl)-2H-1,2,3-triazol-4-yl]ethenyl]-phenyl]-6-methyl-	DMF	358(4.81)	33-2469-76
$C_{25}H_{20}N_4O_3$			
Benzoxazole, 5-methoxy-2-[4-[2-[2-(4-methoxyphenyl)-2H-1,2,3-triazol-4-yl]ethenyl]phenyl]-	DMF	361(4.81)	33-2469-76
$C_{25}H_{20}O$			
Indeno[2,1-b]pyran, 2,3,4-trimethyl-9-(2-naphthalenyl)-	dioxan	253(4.47),280(4.39), 323(4.47),513(3.52)	78-2225-76
$C_{25}H_{20}O_3$			
Cyclobuta[b]naphthalene-3,8-dione, 1,2,2a,8a-tetrahydro-8a-methoxy-1,1-diphenyl-	CCl$_4$	303(3.30),312(2.28), 352(2.23),365(2.18)	18-2596-76
	MeCN	303(3.28),312(3.23), 349(2.18),361(2.15)	18-2596-76

Compound	Solvent	$\lambda_{max}(\log \epsilon)$	Ref.
$C_{25}H_{20}O_9$ 8H-[1]Benzopyrano[2,3-a][3]benzoxepin-8-one, 7a,13a-dihydro-2,4,9,11-tetrahydroxy-6-(4-hydroxy-3-methoxyphenyl)-5-methyl-, trans	EtOH	215(4.61),297(4.53), 310s(--)	88-2241-76
$C_{25}H_{20}O_{10}$ 2,3-Dehydrosilybin	MeOH	285(4.37),325(4.34)	64-0876-76B
$C_{25}H_{21}BrN_2O_3$ Spiro[2H-1-benzopyran-2,2'-[2H]indole], 8-bromo-1',3'-dihydro-1',3',3'-trimethyl-6-nitro-5'-phenyl- merocyanine form	EtOH toluene EtOH dioxan	278(4.26) 386(4.14),598(4.43) 381(4.25),531(4.36) 385(4.19),580(4.44)	103-0417-76 103-0417-76 103-0417-76 103-0417-76
$C_{25}H_{21}ClINS$ Pyridinium, 2-(4-chlorophenyl)-1-methyl-4-(methylthio)-3,5-diphenyl-, iodide	MeOH	218s(4.57),272(4.16), 325(4.09)	44-0818-76
$C_{25}H_{21}NO_2$ 8-Quinolinol, 2-phenyl-3-propyl-, benzoate	EtOH	233(4.67),250(4.49), 280s(3.88),300s(3.82)	95-0968-76
$(C_{25}H_{21}NO_2)_n$ Poly[4-(vinylbenzyloxy)phenyl 1-isoquinolinyl carbinol]	THF	264s(3.87),273(3.92), 282s(3.84),308(3.64), 321(3.70)	47-1661-76
$C_{25}H_{21}NS_2Sn$ Stannane, triphenyl[[(phenylamino)thioxomethyl]thio]-	CHCl$_3$	252(4.30),293(2.34)	101-0055-76D
$C_{25}H_{21}N_2O_3S$ Sulfonium, [4-[(4-methoxyphenyl)amino]-3-nitrophenyl]diphenyl-, tetrafluoroborate	EtOH	302(4.28),400(3.69)	30-0235-76
$C_{25}H_{21}N_3O_2$ 1H-Pyrazole, 4,5-dihydro-3-[4-(4-nitrophenyl)-1,3-butadienyl]-1,5-diphenyl- 1H-Pyrazole-4-carboxamide, 3-benzoyl-5-methyl-1-(4-methylphenyl)-N-phenyl-	n.s.g. EtOH	468(5.51) 256(4.34)	124-0383-76 4-0989-76
$C_{25}H_{21}N_3O_2S$ 5H-1,3,4-Thiadiazolo[3,2-a]pyrimidin-5-one, 6,7-dihydro-7-(4-methoxyphenyl)-2-methyl-6,6-diphenyl-	EtOH	253(3.96)	94-2532-76
$C_{25}H_{21}N_3O_3$ 1-Isoquinolinecarbonitrile, 2-benzoyl-1,2-dihydro-6,7-dimethoxy-1-(2-pyridinylmethyl)- Methanone, [3-(1-methylethyl)-1-(4-methoxyphenyl)-5-phenyl-1H-pyrazol-4-yl]-phenyl-	MeOH EtOH	225(4.67),248(4.60), 315(4.33) 258(4.62)	83-0072-76 4-0989-76
$C_{25}H_{21}N_3O_9S$ 1-Azuleneethanol, tosylate, trinitrobenzeneadduct	CH$_2$Cl$_2$	278(4.73),283(4.70), 288(4.63),333(3.56),	44-1811-76

Compound	Solvent	$\lambda_{max}(\log \epsilon)$	Ref.
1-Azuleneethanol, tosylate, trinitro-benzene adduct (cont.)		342(3.70),357(3.42), 590(2.52),637(2.45), 701(2.03)	44-1811-76
$C_{25}H_{21}N_5O_2$ 1H-Pyrrole-3-carbonitrile, 2-amino-1-(4-methoxyphenyl)-5-[(4-methoxyphenyl)azo]-4-phenyl-	MeCN	285(4.21),431(4.39), 454s(4.36)	48-0663-76
$C_{25}H_{21}N_7O_4$ 1H-Pyrido[3,2-c][1,2,4]triazolo[1,2-a]-pyridazine-1,3(2H)-dione, 6-(3,5-di-oxo-4-phenyl-1,2,4-triazolidin-1-yl)-9-ethyl-5,6-dihydro-2-phenyl-	12M HCl	252(4.28),267s(4.16), 313(3.51)	103-0590-76
$C_{25}H_{22}BrNO_4$ Isoquinoline, 1-[(6-bromo-1,3-benzodiox-ol-5-yl)methyl]-3,4-dihydro-7-methoxy-6-(phenylmethoxy)-, hydrochloride	EtOH EtOH-NaOH	244(4.31),302(4.06), 360(3.99) 281(4.01),301(4.00)	73-1219-76 73-1219-76
$C_{25}H_{22}ClNO_4$ 1,3-Dioxolo[4,5-g]isoquinoline, 5-[[2-chloro-3-(phenylmethoxy)phenyl]meth-yl]-7,8-dihydro-9-methoxy-, hydro-chloride	MeOH	250(4.07),285(3.60), 350(3.97)	12-2003-76
$C_{25}H_{22}INO_4$ 1,3-Dioxolo[4,5-g]isoquinoline, 7,8-di-hydro-5-[[2-iodo-3-(phenylmethyl)phen-yl]methyl]-9-methoxy-	MeOH MeOH-acid	287(4.21),317(3.88) 350(4.21)	12-2003-76 12-2003-76
$C_{25}H_{22}NO$ Pyrylium, 4-[4-(dimethylamino)phenyl]-2,6-diphenyl-, perchlorate	MeCN MeCN CH_2Cl_2	233(4.09),263(4.18), 287(4.12),375(4.13), 537(4.80) 263(4.21),291(4.16), 377(4.34),536(4.81) 263(4.20),295(4.16), 382(4.42),550(4.99)	4-0073-76 4-1089-76 4-1089-76
$C_{25}H_{22}NS$ Thiopyrylium, 4-[4-(dimethylamino)phen-yl]-2,6-diphenyl-, perchlorate	MeCN CH_2Cl_2	243(4.21),259(4.28), 285s(4.14),380(4.20), 577(4.77) 260(3.73),387(4.15), 592(4.76)	4-1089-76 4-1089-76
$C_{25}H_{22}N_2$ 1H-Pyrazole, 4,5-dihydro-1,5-diphenyl-3-(4-phenyl-1,3-butadienyl)-	n.s.g.	403(4.63)	124-0383-76
$C_{25}H_{22}N_2O$ 1H-Pyrrole, 3-[6-(2-furanyl)-1,3,5-hexa-trienyl]-4,5-dihydro-1,5-diphenyl-	n.s.g.	427(4.67)	124-0383-76
$C_{25}H_{22}N_2O_3$ 4-Quinolinecarbonitrile, 3-(benzoyloxy)-1,2,3,4-tetrahydro-2-methoxy-3-methyl-2-phenyl-	EtOH	235(4.62),284s(4.00), 295(4.05)	95-0968-76

Compound	Solvent	$\lambda_{max}(\log \epsilon)$	Ref.
Spiro[2H-1-benzopyran-2,2'-[2H]indole], 1',3'-dihydro-1',3',3'-trimethyl-6-nitro-5'-phenyl-	EtOH	247(4.48)	103-0417-76
merocyanine form	toluene	384(4.42),592(4.67)	103-0417-76
	EtOH	388(4.06),534(4.43)	103-0417-76
	dioxan	377(4.62),557(3.85)	103-0417-76
Spiro[2H-1-benzopyran-2,2'-[2H]indole], 1',3'-dihydro-1',3',3'-trimethyl-6-nitro-7-phenyl-	octane	237(4.46),277(4.16)	103-0174-76
$C_{25}H_{22}N_2O_4$			
Benzoic acid, 4-methoxy-, 4-cyano-1,2,3,4-tetrahydro-2-hydroxy-3-methyl-2-phenyl-3-quinolinyl ester	EtOH	249(4.39),270s(4.06), 295(3.49)	95-0968-76
$C_{25}H_{22}N_2O_6$			
1,3-Dioxolo[4,5-g]isoquinoline, 7,8-di-hydro-9-methoxy-5-[[2-nitro-3-(phen-ylmethoxy)phenyl]methyl]-	MeOH	216s(4.62),287(3.95), 322s(3.74),466(2.87), 495s(2.83)	12-2003-76
	MeOH-acid	217s(4.54),250s(4.07), 348(3.99)	12-2003-76
$C_{25}H_{22}N_2S$			
1H-Pyrazole, 4,5-dihydro-1,5-diphenyl-3-[6-(2-thienyl)-1,3,5-hexatrienyl]-	n.s.g.	437(4.83)	124-0383-76
$C_{25}H_{22}N_3S_2$			
Benzothiazolium, 2-[(4-amino-3,5-diphen-yl-2(3H)-thiazolylidene)methyl]-3-ethyl-, perchlorate	n.s.g.	459(4.62)	103-0648-76
$C_{25}H_{22}O$			
2,4-Cyclopentadien-1-ol, 1,2-dimethyl-Indeno[2,1-b]pyran, 9-(3,4-dihydro-2-naphthalenyl)-2,3,4-trimethyl-	C_6H_{12} dioxan	246(4.342),331(3.55) 254(4.25),322(4.18), 517(3.48)	5-0793-76 78-2225-76
Indeno[2,1-b]pyran, 3-ethyl-9-(1H-inden-2-yl)-2,4-dimethyl-	dioxan	315(4.48),345(4.47), 550(3.53)	78-2225-76
$C_{25}H_{22}O_3$			
5H-1-Benzopyran-5-one, 2,6,7,8-tetrahy-dro-7,7-dimethyl-2-(2-oxo-2-phenyl-ethylidene)-4-phenyl-	MeOH	366(4.34),405(4.35)	24-1549-76
5H-1-Benzopyran-5-one, 4,6,7,8-tetrahy-dro-7,7-dimethyl-4-(2-oxo-2-phenyl-ethylidene)-2-phenyl-	MeOH	274(4.39),384(4.33)	24-1549-76
$C_{25}H_{22}O_4$			
Benzeneacetic acid, α-[3-(4-methoxyphen-yl)-3-oxo-1-phenylpropylidene]-, methyl ester, (E)-	EtOH	221s(4.30),273(4.33)	4-0195-76
$C_{25}H_{22}O_6$			
Cyclomorusin	EtOH	223(4.45),225(4.38), 283(4.43),383(4.19)	94-2898-76
	MeOH-NaOMe	270(4.42),409(4.41)	94-2898-76
	MeOH-AlCl$_3$	229(4.51),265(4.35), 285(4.41),379(4.24), 429(3.83)	94-2898-76
3H,7H-Pyrano[2',3':7,8][1]benzopyrano-[3,2-d][1]benzoxepin-7-one, 8,9-di-	EtOH	219(4.45),235(4.45), 278(4.47),334(4.16)	94-2898-76

Compound	Solvent	$\lambda_{max}(\log \epsilon)$	Ref.
hydro-6,12-dihydroxy-3,3-dimethyl-9-(1-methylethenyl)- (cont.)			94-2898-76
$C_{25}H_{22}O_{10}$ 1,2,4,5,10-Anthracenepentol, 7-methyl-, pentaacetate	$CHCl_3$	256s(4.94),264(5.25), 360(3.78),379(3.91), 400(3.81)	12-1535-76
4H-1-Benzopyran-4-one, 5,7-diacetoxy-6-[bis(acetyloxy)methyl]-8-methyl-2-phenyl-	EtOH	255(4.27),297(4.36)	2-0009-76
4H-1-Benzopyran-4-one, 5,7-diacetoxy-8-[bis(acetyloxy)methyl]-6-methyl-2-phenyl-	EtOH	257(4.44),294(4.33)	2-0009-76
Silychristin	MeOH	288(4.30),322s(--)	64-0876-76B
Silydianin	MeOH	288(4.29),325s(--)	64-0876-76B
$C_{25}H_{23}BrN_4$ 1-Piperazinamine, 4-(10-bromo-5H-dibenzo[a,d]cyclohepten-5-yl)-N-(2-pyridinylmethylene)-	MeOH	305(4.46)	87-1161-76
1-Piperazinamine, 4-(10-bromo-5H-dibenzo[a,d]cyclohepten-5-yl)-N-(3-pyridinylmethylene)-	MeOH	297(4.40)	87-1161-76
1-Piper zinamine, 4-(10-bromo-5H-dibenzo[a,d]cyclohepten-5-yl)-N-(4-pyridinylmethylene)-	MeOH	310(4.46)	87-1161-76
$C_{25}H_{23}F_3N_2O_3$ 4H-1-Benzopyran-4-one, 6-hexyl-7-hydroxy-3-(1-phenyl-1H-pyrazol-4-yl)-2-(trifluoromethyl)-	EtOH	260(4.48),320(4.15)	103-0914-76
$C_{25}H_{23}NO_4S$ Sulfonium, diphenyl-, 3-methoxy-1-(methoxycarbonyl)-3-oxo-2-[(phenylmethyl)-imino]propylide	EtOH	241(4.19),280(3.99), 355(3.79)	78-0431-76
$C_{25}H_{23}N_3O$ 3-Indolizineacetaldehyde, 1-acetyl-6-methyl-α,7-diphenyl-, hydrazone	EtOH	225(4.32),252(4.34), 310(4.24),380(4.16)	103-0424-76
$C_{25}H_{23}N_3O_3S$ Spiro[2H-1-benzopyran-2,2'(3'H)-benzothiazole], 8-methoxy-6-[(4-methoxyphenyl)azo]-3,3'-dimethyl-	EtOH	300s(4.00),310(4.04), 350s(4.02),372(4.18), 383(4.42)	28-0691-76A
$C_{25}H_{23}N_3O_4S$ Spiro[2H-1-benzopyran-2,2'(3'H)-benzothiazole], 3,8-dimethoxy-6-[(4-methoxyphenyl)azo]-3'-methyl-	EtOH	295s(4.10),308(4.10), 348s(4.10),378(4.30), 389(4.69)	28-0691-76A
$C_{25}H_{23}N_3O_6$ [1,4]Oxazino[3,4-b]quinazoline-3,6-(1H,4H)-dione, 4-[[3-(acetyloxy)-2,3-dihydro-2-oxo-1H-indol-3-yl]-methyl]-1-(1-methylethyl)-	MeOH	226(4.64),260(4.03), 268(4.04),280(3.92), 307(3.65),318(3.53)	88-2861-76
$C_{25}H_{24}$ Methanotris[2.2.2]paraxylylene, pseudogeminal	C_6H_{12}	258s(2.81),261s(2.84), 266(2.93),274(3.03), 279(2.92),282(3.03)	18-0529-76

Compound	Solvent	$\lambda_{max}(\log \epsilon)$	Ref.
$C_{25}H_{24}BrNO_4$ Isoquinoline, 1-[(6-bromo-1,3-benzodiox- ol-5-yl)methyl]-1,2,3,4-tetrahydro- 7-methoxy-6-(phenylmethoxy)-	EtOH	230s(4.02),290(3.87)	73-1219-76
$C_{25}H_{24}ClNO_5$ Benzeneacetamide, 2-chloro-N-[2-(4-meth- oxy-1,3-benzodioxol-5-yl)ethyl]- 3-(phenylmethoxy)-	MeOH	276(3.79),283(3.78)	12-2003-76
$C_{25}H_{24}INO_4$ 1,3-Dioxolo[4,5-g]isoquinoline, 5,6,7,8- tetrahydro-5-[[2-iodo-3-(phenylmeth- oxy)phenyl]methyl]-9-methoxy-	MeOH	278(3.69),286(3.71)	12-2003-76
$C_{25}H_{24}INO_5$ Benzeneacetamide, 2-iodo-N-[2-(4-meth- oxy-1,3-benzodioxol-5-yl)ethyl]- 3-(phenylmethoxy)-	MeOH	218(3.59),286(3.57)	12-2003-76
$C_{25}H_{24}N_2O_2S$ Benzenesulfonamide, 4-methyl-N-[3-[[4- (1-methylethyl)phenyl]methylene]- 3H-indol-2-yl]-	CHCl$_3$	270(4.34),355(4.20), 395s(4.0)	103-0292-76
$C_{25}H_{24}N_2O_6$ 1,3-Dioxolo[4,5-g]isoquinoline, 5,6,7,8- tetrahydro-9-methoxy-5-[[2-nitro- 3-(phenylmethoxy)phenyl]methyl]-, monohydrochloride, (+)-	MeOH	274(3.76),350(3.64)	12-2003-76
$C_{25}H_{24}N_2O_7$ Benzeneacetamide, N-[2-(4-methoxy- 1,3-benzodioxol-5-yl)ethyl]- 2-nitro-3-(phenylmethoxy)-	MeOH	278(3.43)	12-2003-76
$C_{25}H_{24}N_4O_2$ Benzo[g]pteridine-2,4(3H,4aH)-dione, 4a-(1,3-cyclopentadien-1-yl)-5,10- dihydro-7,8,10-trimethyl-3-(phenyl- methyl)-	50% MeOH- pH 0.5 50% MeOH- pH 7, 10	230s(--),269(4.36), 296(4.17),390(3.75) 275(4.15),362(3.78)	5-2037-76 5-2037-76
$C_{25}H_{24}O$ 3,7-(Ethano[1,4]benzenoethano)-5H-di- benzo[a,d]cyclohepten-5-ol, 10,11- dihydro- stereoisomer	C$_6$H$_{12}$ C$_6$H$_{12}$	265(2.95),272s(2.97), 274(3.00),279s(2.90), 281(2.96) 244s(3.69),265s(3.09), 274(3.12),284(3.05)	18-0529-76 18-0529-76
4,7-Methano-2H-inden-2-one, 4,5,6,7- tetrahydro-4,8,8-trimethyl-1,3-di- phenyl-, (4R)-	hexane	255(4.58),462(3.67)	78-0309-76
$C_{25}H_{24}O_2$ Benzenemethanol, α-(3-ethoxy-3,3-diphen- yl-1-propynyl)-α-methyl-	EtOH	247s(2.79),253(2.82), 258(2.82),264(2.76)	56-0203-76
Benzenemethanol, α-(3-ethoxy-3-phenyl- 1-butynyl)-α-phenyl-	EtOH	248s(2.80),253(2.81), 259(2.83),264(2.76)	56-0203-76
$C_{25}H_{24}O_6$ Morusin	EtOH	206(4.49),220s(4.43),	94-2898-76

Compound	Solvent	$\lambda_{max}(\log \epsilon)$	Ref.
Morusin (cont.)		270(4.60),300s(4.00), 320s(3.90),350(3.81)	94-2898-76
$C_{25}H_{24}O_7$ 3H,7H-Pyrano[2',3':7,8][1]benzopyrano- [3,2-d][1]benzoxepin-7-one, 8,9-di- hydro-6,12-dihydroxy-9-(1-hydroxy- 1-methylethyl)-3,3-dimethyl-	MeOH	218(4.49),234(4.49), 278(4.51),334(4.24)	94-2898-76
$C_{25}H_{24}S_3$ 1H-Indene, 1-methyl-5-(methylthio)- 1,3-bis[4-(methylthio)phenyl]-	MeOH	277(4.57)	44-0884-76
$C_{25}H_{25}Cl_3N_2O_2Si$ Urea, N'-(diphenylmethylene)-N-phenyl- N-[2,2,2-trichloro-1-[(trimethylsil- yl)oxy]ethyl]-	EtOH	255.8(4.01)	39-1523-76C
$C_{25}H_{25}NO_2$ 6H-Dibenzo[a,h]quinolizine, 5,8,9,13b- tetrahydro-2-methoxy-3-(phenylmeth- oxy)-	EtOH	220(4.29),287(3.60)	118-0719-76
1H-Indole-3-methanol, 5-methoxy-α,2-di- phenyl-1-propyl-	EtOH	227(4.37),295(4.07)	42-0606-76
4H-Indol-4-one, 1,5,6,7-tetrahydro- 1,6,6-trimethyl-3-(2-oxo-2-phenyl- ethyl)-2-phenyl-	EtOH	244(4.32),275(3.95)	94-1160-76
$C_{25}H_{25}NO_3$ Benzoic acid, 2-[benzoyl[2-(1,1-dimeth- ylethyl)phenyl]amino]-, methyl ester	MeOH	210(4.54),270(3.85)	39-0465-76C
Benzoic acid, 2-[benzoyl[4-(1,1-dimeth- ylethyl)phenyl]amino]-, methyl ester	MeOH	220(4.33),250s(4.09), 280s(3.81)	39-0465-76C
Benzoic acid, 2-[[[2-(1,1-dimethyleth- yl)phenyl]imino]phenylmethoxy]-, methyl ester	MeOH	208(4.64),230(4.44), 270(3.68)	39-0465-76C
Benzoic acid, 2-[[[4-(1,1-dimethyleth- yl)phenyl]imino]phenylmethoxy]-, methyl ester	MeOH	208(4.63),230(4.42), 270(3.68)	39-0465-76C
Benzoic acid, 5-(1,1-dimethylethyl)- 2-[phenyl(phenylimino)methoxy]-, methyl ester	MeOH	214(4.48),230s(4.39), 270(3.65)	39-0465-76C
$C_{25}H_{25}NO_5$ Benzeneacetamide, N-[2-(4-methoxy- 1,3-benzodioxol-5-yl)ethyl]- 3-(phenylmethoxy)-	MeOH	273(3.47),279(3.46)	12-2003-76
$C_{25}H_{25}NO_6$ 1,3-Benzodioxole-5-acetamide, N-[2-hy- droxy-2-[3-methoxy-4-(phenylmethoxy)- phenyl]ethyl]-	EtOH	231(4.16),282(3.82)	142-1263-76
$C_{25}H_{25}N_3O_2S$ Benzenesulfonamide, N-[3-[[4-(dimethyl- amino)phenyl]methylene]-1,3-dihydro- 1-methyl-2H-indol-2-ylidene]-4-methyl-	CHCl$_3$	271(3.20),332(2.41), 471(3.18)	103-0292-76
$C_{25}H_{25}N_3O_7$ 2(1H)-Pyrimidinone, 4-amino-1-[3,5-bis-	MeOH-acid	283(4.16)	87-0667-76

Compound	Solvent	$\lambda_{max}(\log \epsilon)$	Ref.
O-(phenylacetyl)-β-D-arabino-furanosyl]-, hydrochloride (cont.)			87-0667-76
$C_{25}H_{26}N_2O_3$ 4H-1-Benzopyran-4-one, 6-hexyl-7-meth-oxy-3-(1-phenyl-1H-pyrazol-4-yl)-	EtOH	270(4.12)	103-0914-76
$C_{25}H_{26}N_2O_4$ Acetamide, N-acetyl-N-[2-methyl-3-(phen-ylmethoxy)-5-[(phenylmethoxy)methyl]-4-pyridinyl]-	EtOH	280(3.70)	87-0999-76
Pyrido[4,3-b]benzo[g]quinolizin-13-one, 2-benzoyl-9,10-dimethoxy-1,2,3,4,6,7-12,12a-octahydro-	n.s.g.	287(3.90),330(3.21)	2-0784-76
$C_{25}H_{26}O$ 2,5-Cyclohexadien-1-one, 4-(diphenyl-methylene)-2,6-bis(1-methylethyl)-	isooctane	369(4.43)	73-1676-76
$C_{25}H_{26}O_5$ Uvaretin, dimethyl ether	EtOH	291(4.40)	44-1852-76
$C_{25}H_{27}Cl_3O_3Si$ Silane, tris(4-chloro-3,5-dimethylphen-oxy)methyl-	dioxan	278(3.45),281(3.45)	56-0363-76
$C_{25}H_{27}NO_4$ Benzamide, 2-(2-hydroxyethyl)-N-[2-[4-methoxy-3-(phenylmethoxy)phenyl]-ethyl]-	EtOH	220(4.23),278(3.68)	118-0719-76
$C_{25}H_{27}NO_5$ Pergularinine, O-acetyl-	MeOH	258(4.5),286(4.2), 313(3.7),340(3.9)	102-1561-76
$C_{25}H_{27}N_5O_5$ L-Alanine, 3-(1-naphthalenyl)-N-[N-(5-oxo-L-prolyl)-L-histidyl]-, methyl ester	HOAc	273(3.76),283(3.85), 293(3.69)	94-3149-76
L-Alanine, 3-(2-naphthalenyl)-N-[N-(5-oxo-L-prolyl)-L-histidyl]-, methyl ester	HOAc	269(3.62),276(3.64), 286(3.48)	94-3149-76
$C_{25}H_{27}OPS$ 7-Phosphabicyclo[2.2.1]hept-2-ene, 5-(3,4-dihydro-6-methoxy-1-naphthal-enyl)-2,3-dimethyl-7-phenyl-, 7-sulfide, (endo,anti)-(±)-	EtOH	268(4.16)	78-2427-76
15-Phospha-1,3,5(10),8,16-estrapentaene, 3-methoxy-17-methyl-15-phenyl-, 15-sulfide, (±)-	EtOH	275(4.23)	78-2427-76
$C_{25}H_{28}$ 1H-Cyclopenta[a]chrysene, 3-ethyl-2,3,8-9,10,11-hexahydro-8,8-dimethyl-	C_6H_{12}	257(4.63),265(4.72), 285(4.10),296(4.00), 308(4.10)	78-0257-76
$C_{25}H_{28}N_2O_{11}S_2$ 2,4(1H,3H)-Pyrimidinedione, 1-[2,6-bis-O-[(4-methylphenyl)sulfonyl]-β-D-glucopyranosyl]-5-methyl-	MeOH	221(4.21),265(4.03)	48-0079-76

Compound	Solvent	$\lambda_{max}(\log \epsilon)$	Ref.
$C_{25}H_{28}O_4$			
4H-1-Benzopyran-4-one, 6-(3,7-dimethyl-2,6-octadienyl)-2,3-dihydro-5,7-di-hydroxy-2-phenyl-	MeOH	206(4.50),211s(4.47), 229(4.23),294(4.22), 352(3.48)	83-0152-76
	MeOH-acid	206(4.49),212(4.52), 229(4.27),294(4.22), 332(3.46)	83-0152-76
	MeOH-base	250(3.91),327(4.42)	83-0152-76
Cycloglabrol	EtOH	279(4.19),308s(3.88)	94-0752-76
Glabrol	EtOH	281(4.35),313(4.06)	94-0752-76
D(17a)-Homo-18,26,27-trinorcholesta-2,13,15,17,22-pentaene-1,24-dione, 6,7-epoxy-5-hydroxy-, (5α,6α,7α,22E)-	EtOH	219(4.42)	39-0304-76C
Isocycloglabrol	EtOH	315s(4.00),368(4.46)	94-0752-76
$C_{25}H_{28}O_7$			
Tricoccin S_6	MeOH	211(4.21)	88-3295-76
$C_{25}H_{28}O_{12}$			
Centapicrin	MeOH	237(4.32),303(3.55)	95-0683-76
$C_{25}H_{28}O_{16}$			
Tricoccin S_4	MeOH	212(4.22)	88-3295-76
$C_{25}H_{29}N_2O_5$			
Dibenzo[a,g]quinolizinium, 9-[[2-(acet-yloxy)ethyl]amino]-5,6-dihydro-2,3,10-trimethoxy-13-methyl-, bromide	EtOH	232(4.52),283(4.49), 346(4.22),479(3.85)	88-1595-76
$C_{25}H_{29}N_3O_5S$			
4H-Indazol-4-one, 3-(3,4-dimethoxyphen-yl)-1,5,6,7-tetrahydro-1,6,6-trimeth-yl-, O-[(4-methylphenyl)sulfonyl]-oxime, (E)-	MeOH	220(4.50),264s(3.97), 272s(3.91),288s(3.81)	24-1898-76
$C_{25}H_{29}N_5O_4$			
1H,5H-Imidazo[1',2':1,2]pyrido[2,3-b]-indole-2,5(3H)-dione, 7a-(1,1-dimeth-ylpropyl)-7a,12-dihydro-3-(1H-imida-zol-4-ylmethylene)-6,12-dimethoxy-1-methyl-, [7aS-(3E,7aR*,12aS*)]-	MeOH	224(4.21),348(4.27)	78-2625-76
$C_{25}H_{30}N_2O_6$			
L-Phenylalanine, N-[N-[3-(4-hydroxy-3-methoxyphenyl)-1-oxo-2-propenyl]-glycyl]-, 1,1-dimethylethyl ester, (E)-	MeOH	236(4.21),293(4.24), 322(4.33)	20-0657-76
	MeOH-NaOH	245(4.06),300s(3.72), 307(3.81),372(4.45)	20-0657-76
$C_{25}H_{30}N_3$			
Crystal Violet	EtOH	589(5.05)	44-0221-76
	8% DMSO	603(5.04)	44-0221-76
$C_{25}H_{30}N_4O_2S$			
Benzo[g]pteridine-2,4(3H,4aH)-dione, 4a-[[(1,1-dimethylethyl)thio]methyl]-5,10-dihydro-7,8,10-trimethyl-3-(phenylmethyl)-	50% MeOH-pH -0.5	300(3.84),392(3.66)	5-2037-76
	50% MeOH-pH 7, 10	300s(--),367(3.91)	5-2037-76
$C_{25}H_{30}N_4O_6P_2$			
Phosphonic acid, (3,3a,4,4a,5,7a,8,8a-	MeOH	222(3.8),260s(--),	23-0044-76

Compound	Solvent	λ_{max} (log ϵ)	Ref.
octahydro-3,5-diphenyl-4,8-methano-benzo[1,2-c:5,4-c']dipyrazole-3,5-diyl)bis-, tetramethyl ester (cont.)		266s(--),274s(--),337(2.5)	23-0044-76
C2 anti-anti	MeOH	223(3.1),257s(--),263s(--),269s(--),274s(--),340(1.9)	23-0044-76
anti-syn	MeOH	215(4.2),336(2.8)	23-0044-76
$C_{25}H_{30}N_5O_8PS$ Adenosine, N-(1-oxobutyl)-8-[(phenyl-methyl)thio]-, cyclic 3',5'-(hydro-gen phosphate) 2'-butanoate	pH 7	299(4.27)	87-0899-76
$C_{25}H_{30}O_5$ 14α-Carda-5,16,20(22)-trienolide, 3β-acetoxy-7-oxo-	MeOH	225(4.11),270(4.06)	24-3304-76
D(17a)-Homo-18,26,27-trinorcholesta-2,13,15,17-tetraene-1,24-dione, 6,7-epoxy-5,22-dihydroxy-, (5α,6α,7α-22R)-	EtOH	216(4.18),276(3.15),295s(3.04),302(3.04)	39-0304-76C
$C_{25}H_{30}O_6$ Andibenin	n.s.g.	249s(3.59),312(2.91)	77-0270-76
$C_{25}H_{30}O_6P_2$ Phosphonic acid, (3,7-diphenyltetracy-clo[3.3.1.0^{2,4}.0^{6,8}]nonane-3,7-diyl)-bis-, tetramethyl ester, (anti,anti)	MeOH	224(4.1),258s(--),264s(--),271s(--),280s(--)	23-0044-76
(anti,syn)	MeOH	212(3.8),223(3.9),249(2.0),254(2.1),260(2.2),267(2.0)	23-0044-76
$C_{25}H_{30}O_7$ β-D-Glucopyranoside, [4-cyclohexylidene-(4-hydroxyphenyl)methyl]phenyl-	MeOH	243(4.26)	24-2259-76
$C_{25}H_{30}O_{11}$ Leucanthinin, acetate	MeOH	211(3.86)	44-3956-76
$C_{25}H_{30}O_{12}$ Melampodinin	MeOH	213(4.23)	44-3956-76
$C_{25}H_{30}O_{13}$ Grandifloroside	MeOH	219(4.2),230s(4.1),300s(4.0),330(4.1)	102-1305-76
	MeOH-NaOMe	264s(3.8),311(3.6),380(4.2)	102-1305-76
$C_{25}H_{31}NO_3$ Androsta-5,16-dieno[16,17-c]pyridine-5'-carboxaldehyde, 3β-acetoxy-	C_6H_{12}	240(3.92),275(3.45)	88-4065-76
$C_{25}H_{31}N_3O_9$ 1(2H)-Pyrimidinebutanoic acid, α-(benz-oylamino)-3,6-dihydro-3-[2,3-O-(1-methylethylidene)-β-D-ribofurano-syl]-2,6-dioxo-, (S)-	EtOH	222(4.21),260(4.00)	24-0082-76
$C_{25}H_{32}N_4O_{12}$ Cyclopentaneacetic acid, 4-carboxy-2,3-bis(2,2-dimethyl-1-oxopropoxy)-α-	EtOH	362(4.41)	23-2925-76

Compound	Solvent	$\lambda_{max}(\log \epsilon)$	Ref.
[(2,4-dinitrophenyl)hydrazono]-, 1-methyl ester, (1α,2α,3β,4α)- (cont.)			23-2925-76
$C_{25}H_{32}N_8O_5$ Hexanedioic acid, 2-[[4-[[[(2,4-diamino-6-pteridinyl)methyl]methylamino]benzoyl]amino]-, diethyl ester, (±)-	EtOH	220(4.35),262(4.40), 299(4.44),374(3.90)	4-0727-76
$C_{25}H_{32}O_2$ Indeno[1,2-a]inden-9a(4bH)-ol, 4b,9-bis-(1,1-dimethylethyl)-9,10-dihydro-10-methoxy-, (4bα,9α,9aα,10α)-	EtOH	267(3.187),274(3.101)	24-1301-76
$C_{25}H_{32}O_5$ 1,2-Benzenedicarboxylic acid, 3-[2-octahydro-7a-methyl-5-(1-methylethenyl)-1-oxo-1H-inden-4-yl]ethyl]-, dimethyl ester, [3aS-(3aα,4α,5β,7aα)]-	MeOH	232(3.91),280(3.26), 288(3.20)	23-3508-76
14α-Carda-5,20(22)-dienolide, 3β-acetoxy-7-oxo-	MeOH	221(4.33)	24-3304-76
3-Cyclohexene-1-acetaldehyde, 2-(2,6-diacetoxy-4-pentylphenyl)-4-methyl-α-methylene-, (1R-trans)-	EtOH	220(4.54),260(2.69), 270s(2.48)	39-0008-76C
2-Cyclohexen-1-one, 4-(2,6-diacetoxy-4-pentylphenyl)-2-methyl-5-(1-methylethenyl)-, (4R-trans)-	EtOH	237(4.09)	39-0008-76C
$C_{25}H_{32}O_6$ Pregna-3,5-diene-11,20-dione, 3,17-diacetoxy-	n.s.g.	234(4.29)	44-0078-76
$C_{25}H_{32}O_8$ Pregn-4-ene-3,11,20-trione, 21-acetoxy-17-[(methoxycarbonyl)oxy]-	n.s.g.	238(4.19)	44-0078-76
$C_{25}H_{32}O_9$ Glaucarubolone 15-tiglate	EtOH	220(4.15),238s(4.01)	87-1130-76
2-Naphthalenecarboxylic acid, 1,2,3,4-tetrahydro-6,7,8-trimethoxy-3-(methoxymethyl)-1-(3,4,5-trimethoxyphenyl)-, [1R-(1α,2β,3α)]-	EtOH	273.5(3.32)	94-2102-76
$C_{25}H_{33}BrN_4O_7S$ Inosine, 8-bromo-, 2'-[2,4,6-tris(1-methylethyl)benzenesulfonate]	pH 1	239(4.21),257s(4.10), 280s(3.87)	94-0672-76
	pH 13	237(4.16),263(4.08), 277s(4.04)	94-0672-76
	MeOH	238(4.20),257s(4.08), 280s(3.86)	94-0672-76
Inosine, 8-bromo-, 3'-[2,4,6-tris(1-methylethyl)benzenesulfonate]	pH 1	237(4.29),255s(4.20), 275s(4.00),280s(3.88)	94-0672-76
	pH 13	233s(4.16),263(4.11), 276s(4.06)	94-0672-76
	MeOH	237(4.23),255s(4.12), 277(3.90),280s(3.82)	94-0672-76
$C_{25}H_{33}NO_6$ Pregn-4-en-18-al, 21-acetoxy-3,20-dioxo-, 18-(O-acetyloxime)	MeOH	240(4.23)	44-3532-76

Compound	Solvent	$\lambda_{max}(\log \epsilon)$	Ref.
$C_{25}H_{33}N_3O$ 12-Azabicyclo[9.2.1]tetradeca-1(13),11-(14)-diene, 2-ethyl-13-[[3-methoxy-5(1H)-pyrrol-2-yl)-2H-pyrrol-2-yli-dene]methyl]- (metacycloprodigiosin) hydrochloride	MeOH-KOH MeOH	285(3.78),325(3.69), 467(4.49) 273(3.56),297(3.84), 362(3.70),500s(4.50), 530(4.88)	78-1855-76 78-1855-76
$C_{25}H_{33}N_3O_5$ Aspidospermidine-3-carboxamide, 4-acet-oxy-6,7-didehydro-3-hydroxy-16-meth-oxy-N,1-dimethyl-, (2β,3β,4β,5α,12β-19α)-	MeOH or EtOH	211(4.49),250(3.85), 304(3.63)	33-2858-76
$C_{25}H_{34}N_2O_6$ Pentanedioic acid, 2-[1-[[2-(1H-indol-3-yl)ethyl]amino]ethylidene]-3-(2-methoxy-2-oxoethyl)-, 1-(1,1-dimeth-ylethyl) 5-methyl ester	MeOH	225(4.42),294(4.28)	24-3825-76
$C_{25}H_{34}O_4$ Cyclopenta[c][1]benzopyran-3(2H)-one, 9-acetoxy-1,4-dihydro-4,4-dimethyl-7-(1,2-dimethylheptyl)-	EtOH	330(4.27)	4-1101-76
$C_{25}H_{34}O_5$ 16-Anhydrogitoxigenin, 3-acetate	EtOH	270(4.27)	94-1950-76
$C_{25}H_{34}O_6$ Pregna-1,4-dien-3-one, 20,21-diacetoxy-17-hydroxy-, (20R)-	EtOH	245(4.18)	94-0828-76
$C_{25}H_{34}O_7$ 24-Norchol-20(22)-ene-19,23-dioic acid, 3-acetoxy-14,22-dihydroxy-, γ-lactone, (3β,5α,14β)- Pregn-4-ene-3,20-dione, 21-acetoxy-17-[(methoxycarbonyl)oxy]-	MeOH n.s.g.	220(4.36) 241(4.20)	94-2886-76 44-0078-76
$C_{25}H_{34}O_8$ Lyoniresinol, trimethyl ether, (-)-	EtOH	274(3.28)	94-2102-76
$C_{25}H_{35}FO_5$ 6H-Dibenzo[b,d]pyran-1,10-diol, 9β-flu-oro-6a,7,8,9,10,10a-hexahydro-6,6,9α-trimethyl-3-pentyl-, diacetate, [6aR-(6aα,9α,10β,10aβ)]-	EtOH	284(3.30)	33-1963-76
$C_{25}H_{35}NO$ Acridan, 9-(11-methoxyundecyl)-	MeOH	216(4.65),297(4.59)	39-0465-76C
$C_{25}H_{35}NO_4S$ Acetamide, N-[(16β)-16-(acetylthio)-3,20-dioxopregn-4-en-17-yl]-	EtOH	240(4.18)	70-1101-76
$C_{25}H_{35}NO_5$ 17α-Pregn-4-ene-21-carboxylic acid, 7α-[(ethoxycarbonyl)amino]-17-hydroxy-3-oxo-, γ-lactone	MeOH	241(4.12)	87-0975-76

Compound	Solvent	$\lambda_{max}(\log \epsilon)$	Ref.
$C_{25}H_{35}N_3O$ 1H-Pyrrole, 2-[[3-methoxy-5-(1H-pyrrol-2-yl)-2H-pyrrol-2-ylidene]methyl]-5-undecyl-	MeOH-HCl MeOH-KOH	500s(4.70),530(5.00) 462(4.60),530s(3.85)	78-1851-76 78-1851-76
$C_{25}H_{36}INO_5$ Benzeneethanaminium, 2-(3,4-dihydro-6,7-dimethoxy-1-methyl-1H-2-benzo-pyran-3-yl)-4,5-dimethoxy-N,N,N-trimethyl-, iodide	pH 1	227(4.46),282(3.86)	33-0949-76
$C_{25}H_{36}O_3$ 7H-Dibenzo[b,d]pyran-7-one, 6,8,9,10-tetrahydro-1-hydroxy-3-(1,2-dimeth-ylheptyl)-6,6,9-trimethyl-	EtOH	335(4.20)	4-1101-76
$C_{25}H_{36}O_3S$ Androst-5-en-3-one, 17-(5,6-dihydro-1,4-oxathiin-2-yl)-, cyclic 1,2-ethanediyl acetal, (17β)-	MeOH	230(3.93)	44-2292-76
$C_{25}H_{36}O_5$ Card-17(20)-enolide, 3-acetoxy-14-hy-droxy-	EtOH	212(3.78)	94-1950-76
isomer	EtOH	210.5(3.68)	94-1950-76
5β-Chol-8-en-24-oic acid, 11α-hydroxy-3,7-dioxo-, methyl ester	EtOH	252(3.95)	44-3840-76
5H-Inden-5-one, 4-[2,2-dimethoxy-2-(3-methoxyphenyl)ethyl]-1-(1,1-dimethyl-ethoxy)-1,2,3,6,7,7a-hexahydro-7a-methyl-, (1S-cis)-	MeOH	215(4.01),253(3.98), 280(3.45)	24-2948-76
Pregn-4-en-3-one, 18,20β-diacetoxy-	MeOH	240(4.23)	44-3532-76
$C_{25}H_{36}O_6$ 14α-Card-20(22)-enolide, 6α-acetoxy-3β,5α-dihydroxy-	MeOH	218(3.84)	24-3304-76
14α-Card-20(22)-enolide, 6β-acetoxy-3β,5α-dihydroxy-	MeOH	216(4.18)	24-3304-76
$C_{25}H_{36}O_6S_3$ 2-Propenoic acid, 3-[2-[3-[(tetrahydro-2H-pyran-2-yl)oxy]butyl]-1,3-dithian-2-yl]-, 2-[(4-methylphenyl)sulfonyl]-ethyl ester	EtOH	222(4.29)	39-1718-76C
$C_{25}H_{36}O_7$ 14α-Card-20(22)-enolide, 6α-acetoxy-3β,5α,11α-trihydroxy-	MeOH	216(4.17)	24-3304-76
14α-Card-20(22)-enolide, 6β-acetoxy-3β,5α,11α-trihydroxy-	MeOH	216(4.16)	24-3304-76
14α-Card-20(22)-enolide, 6α-acetoxy-3β,5α,17α-trihydroxy-	MeOH	212(4.17)	24-3304-76
14α-Card-20(22)-enolide, 6β-acetoxy-3β,5α,17α-trihydroxy-	MeOH	216(4.11)	24-3304-76
$C_{25}H_{36}O_{10}$ Glaucarubinone, dihydro-	EtOH	239(1.84)	87-1130-76
$C_{25}H_{37}NO_2$ 1,4-Ethano-4H-[1]benzopyrano[3,4-d]pyri-	EtOH	280(3.99)	87-0445-76

Compound	Solvent	$\lambda_{max}(\log \epsilon)$	Ref.
din-10-ol, 1,2,3,5-tetrahydro-5,5-di-methyl-8-(1,2-dimethylheptyl)- (cont.)			87-0445-76
$C_{25}H_{39}N_3O_6$ 2,2'-Anhydro-1-(3',5'-dioctanoyl-β-D-arabinofuranosyl)cytosine, hydro-chloride	MeOH-acid	232(4.00),262(4.04)	87-0663-76
$C_{25}H_{40}CoN_5O_4$ Cobalt, [(10-methyl-2,3,17,18-nonadec-anetetrone tetraoximato(3-)](pyri-dine)-, (OC-6-24)-	n.s.g.	461(3.08)	89-0427-76
$C_{25}H_{41}ClO_2S$ Carbonothioic acid, O-(4-chlorophenyl) O-octadecyl ester	EtOH	234(3.52)	39-2112-76C
$C_{25}H_{41}N_3O_5$ 2,2'-Anhydro-1-(3'-palmitoyl-β-D-ara-binofuranosyl)cytosine, hydrochloride	MeOH-acid	233(4.00),264(4.04)	87-0654-76
$C_{25}H_{41}N_3O_7$ Cytosine, 1-(3',5'-dioctanoyl-β-D-ara-binofuranosyl)-, hydrochloride	MeOH-acid	211(4.00),283(4.16)	87-0667-76
$C_{25}H_{42}N_6O_{11}$ 1,1',1''-[(6-Methyl-1,4-benzoquinone-2,3,5-triyl)trimethyl]trispiperidin-ium trinitrate	H_2O	257(4.02),328(3.11)	12-1163-76
$C_{25}H_{42}O_5$ 4-Cyclohexene-1,3-dione, 2,5-dihydroxy-2,6,6-tris(3-methylbutyl)-4-(2-meth-yl-1-oxopropyl)-	MeOH-HCl MeOH-NaOH	240(4.00),280(4.14) 255(4.18),270s(4.09)	20-0293-76 20-0293-76
$C_{25}H_{43}N_3O_6$ 2(1H)-Pyrimidinone, 4-amino-1-[3-O-(1-oxohexadecyl)-β-D-arabinofuranosyl]- 2(1H)-Pyrimidinone, 4-amino-1-[5-O-(1-oxohexadecyl)-β-D-arabinofuranosyl]-	MeOH-acid MeOH-acid	212(4.00),284(4.17) 212(3.96),285(4.10)	87-0667-76 87-0667-76
$C_{25}H_{44}O_2S$ 2-Thiophenecarboxylic acid, 4-methyl-5-nonadecyl- 2-Thiophenecarboxylic acid, 5-nonadec-yl-, methyl ester	heptane heptane	253(3.94),285(3.99) 254(3.97),275(4.04)	34-0233-76 34-0233-76
$C_{25}H_{45}NO$ 2,5-Cyclohexadien-1-one, 2,4,6-tris-(1,1-dimethylethyl)-4-(1-methyleth-yl)-, O-(1,1-dimethylethyl)oxime	hexane	259(3.88)	18-1142-76
$C_{25}H_{46}Cl_3N_{13}O_9$ Tuberactinomycin O	pH 1 H_2O pH 13	268(4.42) 268(4.41) 286(4.24)	88-2343-76 88-2343-76 88-2343-76

Compound	Solvent	$\lambda_{max}(\log \epsilon)$	Ref.
$C_{26}H_{14}$ 1,14-Ethenonaphtho[1',8':5,6,7]cyclo-hepta[1,2,3,4-def]phenanthrene (hexa[7]circulene)	C_6H_{12}	220(4.76),236s(4.75), 243(4.74),268(4.72), 277s(4.65),304(4.59), 321(4.48)	12-0173-76
$C_{26}H_{14}N_2$ 1-Acenaphthylenecarbonitrile, head-head dimer	ether	225(4.99),284(4.09), 296(4.32),309(4.22), 318(3.68)	22-0217-76
head-tail dimer	ether	219(5.02),281(4.11), 290(4.11),302(3.96), 319(3.18)	22-0217-76
$C_{26}H_{15}NO_3$ 13H-Dibenzo[a,h]carbazole-7,12-dione, 5-hydroxy-13-phenyl-	EtOH	240(4.00),276(4.11), 338(3.51),470(3.45)	115-0185-76
$C_{26}H_{16}$ Tribenzo[a,e,i]cyclotetradecene, 17,18,19,20-tetradehydro-	THF	237(4.39),256(4.33), 291(4.54),330(4.36), 356(4.05)	18-2840-76
$C_{26}H_{16}Cl_2N_6$ 2H-Naphtho[1,2-d]triazole, 2-[3-chloro-4-[2-[2-(4-chlorophenyl)-2H-1,2,3-triazol-4-yl]ethenyl]phenyl]-	DMF	323(4.43),369(4.79)	33-2469-76
$C_{26}H_{16}O_4$ 1-Acenaphthylenecarboxylic acid, head-tail dimer	ether	218(5.02),273(3.90), 284(4.10),292(4.11), 305(3.96),310(3.88), 319(3.31),323(3.07)	22-0217-76
$C_{26}H_{17}ClN_6$ 2H-Naphtho[1,2-d]triazole, 2-[3-chloro-4-[2-(2-phenyl-2H-1,2,3-triazol-4-yl)-ethenyl]phenyl]-	DMF	311(4.38)	33-2469-76
$C_{26}H_{17}I$ Phenanthrene, 3-iodo-4,5-diphenyl-	MeOH	319s(4.19),361(2.88), 379(2.81)	39-1115-76B
$C_{26}H_{17}NO_2S$ [1]Benzothieno[3,2-b]pyridine, 2-benzo-furanyl-4-(4-methoxyphenyl)-	$C_2H_4Cl_2$	245(5.25),324(5.48), 367(5.06)	103-0977-76
$C_{26}H_{17}O_3S$ [1]Benzothieno[3,2-b]pyrylium, 2-benzo-furanyl-4-(4-methoxyphenyl)-, per-chlorate	$C_2H_4Cl_2$	250(3.99),300(3.70), 330(4.06),485(4.79)	103-0977-76
$C_{26}H_{18}$ Benzene, 1,2-bis[2-(2-ethynylphenyl)eth-enyl]-, (E,E)-	THF	259(4.42),315(4.31)	18-2840-76
2,7-(Etheno[1,3]benzeno[1,3]benzenoeth-eno)naphthalene	C_6H_{12}	224(4.72),250s(4.32), 273s(4.08)	12-0173-76
Naphthalene, 1-phenyl-7-(4-phenyl-1-buten-3-ynyl)-, (E)-	MeOH	220s(4.43),238(4.43), 259s(4.43),282(4.60), 287s(4.60),292(4.61), 332(4.61),348s(4.49)	39-1104-76B

Compound	Solvent	$\lambda_{max}(\log \epsilon)$	Ref.
Phenanthrene, 1,2-diphenyl-	MeOH	298s(4.24),330(2.72), 343(2.68),350(2.65)	39-1115-76B
Phenanthrene, 1,3-diphenyl-	EtOH	265(4.80),309(4.38)	2-0038-76
Phenanthrene, 3,4-diphenyl-	MeOH	299(4.17),336(2.83), 352(2.79)	39-1115-76B
Phenanthrene, 4,5-diphenyl-	MeOH	310s(4.21),355(2.82), 372(2.72)	39-1115-76B
$C_{26}H_{18}Br_2$			
Azulene, 1,3-bis[2-(3-bromophenyl)-ethenyl]-, cis-cis	hexane	289(4.30),334(4.38), 406(3.98),650(2.39), 688s(2.29)	117-0169-76
cis-trans	hexane	297(4.29),338(4.43), 400(4.15),616(2.34), 654(2.40)	117-0169-76
	DMF	617s(2.46),668(2.49)	117-0169-76
$C_{26}H_{18}Cl_4O_4$			
1H-Cyclopenta[b][1,4]benzodioxin, 1-[bis(4-methoxyphenyl)methylene]-5,6,7,8-tetrachloro-3a,9a-dihydro-, cis	C_6H_{12}	217(4.85),258(4.24), 309(4.41)	78-0147-76
$C_{26}H_{18}F_2N_2O_3$			
Benzoic acid, 5-fluoro-2-(phenylamino)-, anhydride	CH_2Cl_2	284(4.29),405(4.22)	5-0946-76
$C_{26}H_{18}I_2$			
Azulene, 1,3-bis[2-(3-iodophenyl)ethenyl]-, cis-cis	hexane	294(4.18),340(4.31), 403(3.96),652(2.84)	117-0169-76
cis-trans	hexane	224(4.60),299(4.33), 351(4.55),620s(2.39), 659(2.45),738s(2.27)	117-0169-76
trans-trans	DMF	672(2.65)	117-0169-76
$C_{26}H_{18}N_4S$			
Propanedinitrile, 2,2'-[thiobis[3-(4-methylphenyl)-2-propen-3-yl-1-ylid-ene]]bis-	HOAc	426(4.61)	48-0705-76
$C_{26}H_{18}OS$			
Indeno[2,1-b]pyran, 9-(1H-inden-2-yl)-4-methyl-2-(2-thienyl)-	dioxan	327(4.47),339(4.45), 376(4.45),587(3.47)	78-2225-76
$C_{26}H_{18}O_2$			
[1]Benzopyrano[4,3-b]naphtho[1,2-e]pyr-an, 6,7-dihydro-6-phenyl-	EtOH	227(4.85),297(3.74), 314(3.70)	22-1967-76
9,9'-Bi-9H-xanthene	EtOH	207(4.74),241s(--), 287(3.76)	83-0735-76
Methanone, [1,1'-biphenyl]-4,4'-diylbis-[phenyl-	EtOH	255s(4.2),300(4.5)	47-0319-76
3,3'-Spirobi[3H-naphtho[2,1-b]pyran], 2-methyl-	C_6H_{12}	242(4.87),246(5.06), 284(4.04),293(4.14), 310(4.10),333(3.90), 341(3.84),347(4.00)	22-2039-76
	EtOH	240(4.85),246(5.00), 286(4.00),297(4.08), 310(4.13),331(3.87), 341(3.84),347(3.95)	22-2039-76

Compound	Solvent	$\lambda_{max}(\log \epsilon)$	Ref.
$C_{26}H_{18}O_2S$ Naphtho[2,3-c]thiophene-5,8-dione, 6,7-dimethyl-1,3-diphenyl-	CH_2Cl_2	445(3.97),490s(3.80)	28-0607-76A
$C_{26}H_{18}O_3$ [9,9'-Bi-9H-xanthen]-9-ol	EtOH	208(4.69),240s(--), 290(3.76)	83-0735-76
Methanone, (oxydi-4,1-phenylene)bis- [phenyl-	EtOH	<u>257(4.3),290(4.4)</u>	47-0319-76
$C_{26}H_{18}O_4$ [9,9'-Bi-9H-xanthene]-9,9'-diol	EtOH	213(4.68),239s(--), 288(3.78)	83-0735-76
$C_{26}H_{19}ClN_4O$ 2H-1,2,3-Triazole, 2-(4-chlorophenyl)- 4-methyl-5-[2-[4-(5-phenyl-2-oxazol- yl)phenyl]ethenyl]-	DMF	366(4.79)	33-2469-76
$C_{26}H_{19}I$ Benzo[c]phenanthrene, 1-(3,5-dimethyl- phenyl)-2-iodo-	MeOH	314s(4.17),327s(3.99), 339s(3.75),369(2.57)	39-1115-76B
$C_{26}H_{19}NO_2$ Benzeneacetonitrile, α-[1-(benzoyloxy)- 3-phenyl-2-propynylidene]-4-ethyl-	EtOH	209(4.22),237(4.43), 337(4.59)	34-0115-76
$C_{26}H_{20}$ Azulene, 1,3-bis(2-phenylethenyl)-, cis- cis	C_6H_{12}	243s(4.39),300(4.42), 332(4.44),653(2.47)	117-0169-76
cis-trans	C_6H_{12}	243(4.50),248(4.48), 305(4.37),342(4.50), 673(2.44)	117-0169-76
trans-trans	C_6H_{12}	253(4.09),310s(4.24), 352(4.45),692(2.32)	117-0169-76
Benzene, 1,1',1'',1'''-(1,2-ethenediyli- dene)tetrakis-, dilithium salt	THF	385(4.48),495(4.42)	35-1461-76
disodium salt	THF	485(4.57)	35-1461-76
Benzo[c]phenanthrene, 1-(3,5-dimethyl- phenyl)-	MeOH	307s(4.07),321(3.94), 333s(3.81)	39-1115-76B
Phenanthrene, 3-[4-(3,5-dimethylphenyl)- 1-buten-3-ynyl]-, cis	MeOH	246(4.65),252(4.68), 273(4.38),284(4.33), 331s(4.33),340(4.37), 349s(4.30),360s(4.19)	39-1104-76B
trans	MeOH	244(4.49),252(4.56), 274(4.44),285(4.44), 342(4.72),360s(4.55), 368s(4.33)	39-1104-76B
$C_{26}H_{20}Br_2N_6O_2$ 2,5-Cyclohexadiene-1,4-dione, 2,5-di- bromo-3,6-bis[(2,3-dimethyl-6-quin- oxalinyl)amino]-	EtOH	214(4.14),255(4.34), 297(4.66)	56-0073-76
$C_{26}H_{20}N$ Pyridinium, 1-(1,2-dihydro-3-methylbenz- [j]aceanthrylen-1-yl)-, perchlorate	MeOH	222(4.57),233(4.47), 273(4.61),283(4.82), 295(4.86),327(3.66), 341(3.83),358(3.91), 375(3.76),393(3.15)	44-2679-76

Compound	Solvent	$\lambda_{max}(\log \epsilon)$	Ref.
$C_{26}H_{20}NO_2PS_2$ Phosphonium, [(dithiocarboxy)(4-nitro-phenyl)methyl]triphenyl-, hydroxide, inner salt	CHCl$_3$	273(4.26),362(3.86), 430(3.62)	39-1404-76B
$C_{26}H_{20}N_2$ 1,10-Phenanthroline, 2,9-dimethyl-4,7-diphenyl-, cuprous complex	MeOH	460(4.15)	118-0001-76
$C_{26}H_{20}N_2O$ 1-Isoquinolinecarbonitrile, 2-benzoyl-1-(ethenylbenzyl)-1,2-di-hydro-, homopolymer	THF	298(4.0),317(4.0)	116-0010-76
1-Isoquinolinecarbonitrile, 2-benzoyl-1-[(4-ethenylphenyl)methyl]-1,2-di-hydro-, homopolymer	THF	267s(3.93),287s(3.93), 300(3.96),317(3.96)	116-0010-76
$C_{26}H_{20}N_2O_2S$ 1H-Thieno[3,4-d]imidazole-6-carboxylic acid, 2,4-diphenyl-1-(phenylmethyl)-, methyl ester	MeOH	247(4.33),324(4.10), 366(4.36)	49-0299-76
$C_{26}H_{20}N_2O_2S_2$ Disulfide, bis[2-(phenylamino)benzoyl]-	CHCl$_3$	272s(4.35),283(4.43), 400(4.33)	48-0161-76
$C_{26}H_{20}N_2O_4$ Benzo[1,2-d:4,5-d']bisoxazole, 2,6-bis-[2-(4-methoxyphenyl)ethenyl]-	CHCl$_3$	380(4.80),394(4.84), 412s(4.60)	103-1331-76
Benzo[1,2-d:5,4-d']bisoxazole, 2,6-bis-[2-(4-methoxyphenyl)ethenyl]-	CHCl$_3$	343(4.77)	103-1331-76
$C_{26}H_{20}N_4O$ 5H-Indeno[5,6-d]oxazole, 6,7-dihydro-2-[4-[2-(2-phenyl-2H-1,2,3-triazol-4-yl)ethenyl]phenyl]-	DMF	359(4.84)	33-2469-76
2-Propanone, 1,3-diphenyl-1-(1-phenyl-1H-pyrazolo[3,4-d]pyrimidin-4-yl)-	EtOH	233s(4.10),264(4.20), 359(4.47),371(4.53)	95-1352-76
2H-1,2,3-Triazole, 4-methyl-2-phenyl-5-[2-[4-(5-phenyl-2-oxazolyl)phenyl]-ethenyl]-	DMF	365(4.79)	33-2469-76
$C_{26}H_{20}N_4O_2$ Acetamide, N-[2-(acetylamino)phenyl]-N-benzo[a]phenazin-5-yl-	dioxan	226(4.72),282(4.58), 363s(3.82),382(4.02), 392(4.00),402(4.09)	49-0879-76
	50% dioxan	225(4.67),284(4.58), 368s(3.81),386(4.02), 392(4.02),405(4.09)	49-0879-76
$C_{26}H_{20}N_4S$ 1,3,4-Thiadiazole, 2,3-dihydro-2,3,5-triphenyl-2-(phenylazo)-	CHCl$_3$	347(4.20)	39-0038-76C
$C_{26}H_{20}N_4S_2$ 1,2,4,5-Dithiadiazine, 3,4-dihydro-3,4,6-triphenyl-3-(phenylazo)-	CHCl$_3$	286(4.30)	39-0038-76C
$C_{26}H_{20}N_6O_5S$ Benzenesulfonamide, N-(4,6-dimethyl-2-pyrimidinyl)-4-[2-(4-nitrophenyl)-	EtOH-NaOH	210(4.46),250(4.23), 288s(4.10)	106-0574-76

Compound	Solvent	$\lambda_{max}(\log \epsilon)$	Ref.
4-oxo-3(4H)-quinazolinyl]- (cont.)			106-0574-76
$C_{26}H_{20}O$			
Indeno[2,1-b]pyran, 4-methyl-2-(4-methylphenyl)-9-phenyl-	dioxan	265(4.43),296(4.30), 340(4.35),368(4.20), 538(3.58)	78-2219-76
Methanone, [4-(diphenylmethyl)phenyl]-phenyl-	CHCl$_3$	340(2.13)	35-1526-76
$C_{26}H_{20}O_2$			
3H-Naphtho[2,1-b]pyran, 3-methoxy-2,3-diphenyl-	EtOH	210(4.43),240(4.66), 320(4.04),352(4.04)	22-1975-76
$C_{26}H_{21}As$			
1H-Benz[f]arsindole, 2,9-dimethyl-1,4-diphenyl-	CDCl$_3$	268(4.63),274(4.63), 310(4.51),322(4.28), 345(3.68),362(3.54)	78-2131-76
$C_{26}H_{21}ClN_4O$			
Benzoxazole, 2-[4-[2-[2-(4-chlorophenyl)-5-methyl-2H-1,2,3-triazol-4-yl]-ethenyl]phenyl]-5,6-dimethyl-	DMF	361(4.81)	33-2469-76
$C_{26}H_{21}ClSi$			
Silane, chloro(1,2-diphenylethenyl)di-phenyl-, (E)-	n.s.g.	232(4.54),272(4.53)	65-1044-76
$C_{26}H_{21}CuN_3O_6$			
Copper, [2-[(4-nitrophenyl)azo]-1-phenyl-1,3-butanedionato-O^1,O^3](1-phenyl-1,3-butanedionato-O,O')-	n.s.g.	254(4.22),294(4.12), 395(4.15),543(3.30)	65-2619-76
$C_{26}H_{21}N_3O_3$			
Rutaecarpine, 2-benzoyloxy-3-methoxy-	MeOH	254(4.48),336(4.48), 351(4.50),369(4.41)	102-1095-76
$C_{26}H_{21}N_3O_9S_2$			
Imidazo[1,2-a]pyrazin-3(7H)-one, 8-(phenylmethyl)-6-[4-(sulfooxy)-phenyl]-2-[[4-(sulfooxy)phenyl]-methyl]-	MeOH	258(4.38),350(3.82), 426(3.89)	88-2971-76
	MeOH-HCl	266(4.41),335(3.93)	88-2971-76
	MeOH-NaOH	266(4.47),348(3.99), 400(3.85)	88-2971-76
$C_{26}H_{21}N_5O$			
1,3,4-Oxadiazole, 2-(3-methylphenyl)-5-[4-[2-(5-methyl-2-phenyl-2H-1,2,3-triazol-4-yl)ethenyl]phenyl]-	DMF	348(4.76)	33-2469-76
$C_{26}H_{21}N_5O_2$			
Benzoic acid, 4-[(5-amino-4-cyano-1,3-diphenyl-1H-pyrrol-2-yl)azo]-, ethyl ester	MeCN	273(4.27),455(4.54)	48-0663-76
1,3,4-Oxadiazole, 2-(3-methoxyphenyl)-5-[4-[2-(5-methyl-2-phenyl-2H-1,2,3-triazol-4-yl)ethenyl]phenyl]-	DMF	348(4.76)	33-2469-76
1,3,4-Oxadiazole, 2-(4-methoxyphenyl)-5-[4-[2-(5-methyl-2-phenyl-2H-1,2,3-triazol-4-yl)ethenyl]phenyl]-	DMF	352(4.77)	33-2469-76

Compound	Solvent	$\lambda_{max}(\log \epsilon)$	Ref.
$C_{26}H_{21}P$ Acridophosphine, 5,5-dihydro-5-phenyl-5-(phenylmethyl)-	THF	277(4.15),392(3.80), 495(3.90),582(3.47)	88-0481-76
$C_{26}H_{21}PS_2$ Phosphonium, (2-mercapto-1-phenyl-2-thioxoethyl)triphenyl-, hydroxide, inner salt	$CHCl_3$	287(4.23),363(3.93)	39-0692-76C
$C_{26}H_{22}$ Dibenz[a,h]anthracene, 1,6,8,13-tetramethyl-	EtOH	243(4.41),285(4.70), 297(4.97),306(5.04), 325(4.36),338(4.38), 346(4.24),354(4.13), 361(4.12)	2-0943-76
$C_{26}H_{22}Br_4N_6O_2$ Bromanil, complex with 6-amino-2,3-dimethylquinoxaline (1:2)	$CHCl_3$	244(5.09),257(4.68), 314(4.23),373(3.97), 445(3.48)	56-0073-76
$C_{26}H_{22}N_2O_2S$ Spiro[2H-anthra[1,2-d]imidazole-2,1'-cyclohexane]-6,11-dione, 1,3-dihydro-4-(phenylthio)-	EtOH	555(4.08)	104-0172-76
$C_{26}H_{22}N_2O_3$ 1H-Pyrazole-4-carboxylic acid, 3-benzoyl-1-(4-methylphenyl)-5-phenyl-, ethyl ester	EtOH	253(4.20)	4-0989-76
$C_{26}H_{22}N_2O_3S$ 3-Pyridinecarboxamide, 1,2-dihydro-6-methoxy-1-(2-methylphenyl)-N-phenyl-4-(phenylthio)-2-oxo-	MeOH	330(4.5)	120-0103-76
$C_{26}H_{22}N_2O_4S$ Spiro[2H-anthra[1,2-d]imidazole-2,1'-cyclohexane]-6,11-dione, 1,3-dihydro-4-(phenylsulfonyl)-	EtOH	523(4.15)	104-0172-76
$C_{26}H_{22}N_4$ Benzenecarbohydrazonic acid, N-phenyl-, phenyl(phenylmethylene)hydrazide	$CHCl_3$	300(4.18),336(4.62)	39-0038-76C
$C_{26}H_{22}N_4O$ Benzoxazole, 5,6-dimethyl-2-[4-[2-(5-methyl-2-phenyl-2H-1,2,3-triazol-4-yl)ethenyl]phenyl]-	DMF	362(4.80)	33-2469-76
Benzoxazole, 5,7-dimethyl-2-[4-[2-(5-methyl-2-phenyl-2H-1,2,3-triazol-4-yl)ethenyl]phenyl]-	DMF	358(4.78)	33-2469-76
Benzoxazole, 5-(1-methylethyl)-2-[4-(2-phenyl-2H-1,2,3-triazol-4-yl)ethenyl]-phenyl]-	DMF	353(4.82)	33-2469-76
Benzoxazole, 2-[4-(2-phenyl-2H-1,2,3-triazol-4-yl)ethenyl]phenyl]-5-propyl-	DMF	353(4.82)	33-2649-76
$C_{26}H_{22}N_4O_2$ Benzoxazole, 2-[4-[2-[2-(4-methoxyphen-	DMF	360(4.83)	33-2469-76

Compound	Solvent	$\lambda_{max}(\log \epsilon)$	Ref.
yl)-2H-1,2,3-triazol-4-yl]ethenyl]- phenyl]-5,6-dimethyl- (cont.)			33-2469-76
$C_{26}H_{22}N_4O_5$ 4,14-Epoxy-5H,8H-1,3-dioxolo[5,6][1,3]- oxazocino[3,2-a]pteridin-8-one, 3a,4- 14,14a-tetrahydro-2,2-dimethyl-10,11- diphenyl-, [3aR-(3aα,4β,14β,14aα)]-	MeOH	227(4.45),252s(4.24), 271(4.19),361(4.15)	24-3159-76
$C_{26}H_{22}N_4S$ Benzenecarbohydrazonothioic acid, N- phenyl-, anhydrosulfide	EtOH	242(4.48),284(4.25), 335(4.18)	39-0038-76C
$C_{26}H_{22}N_4S_2$ Disulfide, bis[phenyl(phenylhydrazono)- methyl]-	EtOH	304(4.28),324(4.34), 400(4.12)	39-0038-76C
$C_{26}H_{22}O$ 7H-Cyclohept[b]indeno[1,2-e]pyran, 8,9,10,11-tetrahydro-5-(1H-inden- 2-yl)-	dioxan	226(4.15),252(4.23), 262(4.17),316(4.52), 349(4.54),581(3.51)	78-2225-76
Indeno[2,1-b]pyran, 3-ethyl-2,4-dimeth- yl-9-(2-naphthalenyl)-	dioxan	253(4.45),270(4.31), 280(4.47),312(4.44), 344(4.44),513(3.52)	78-2225-76
$C_{26}H_{22}O_4STe$ Tellurolo[3,4-d]thiepin-5,7-dicarbox- ylic acid, 1,3-diphenyl-, diethyl ester	$CHCl_3$	290(4.06),345(4.08)	24-3886-76
$C_{26}H_{22}O_8$ 5a,11a-[2]Butenonaphthacene-5,6,11,12- tetrone, 1,13-diacetoxy-1,4,4a,12a- tetrahydro-	MeOH	232(4.51),260(4.02), 305(3.43)	44-2296-76
$C_{26}H_{22}O_{10}$ Benzo[1,2-b:4,5-b']bisbenzofuran-6,12- diol, 2,3,8,9-tetramethoxy-, diacetate	EtOH	258(4.35),328(4.47), 341(4.59),360(4.74)	12-0179-76
$C_{26}H_{22}P$ Acridophosphinium, 5,10-dihydro-5-phen- yl-5-(phenylmethyl)-, bromide	EtOH	255(3.71),261(3.70), 269(3.71),276(3.74), 284(3.66)	88-0481-76
$C_{26}H_{23}As$ 1H-Arsole, 3-(diphenylmethylene)-2-eth- ylidene-2,3-dihydro-5-methyl-1-phenyl-	EtOH	240s(4.29),315(4.10), 324(4.15),335s(4.12)	78-2131-76
1H-Benz[f]arsindole, 9,9a-dihydro- 2,9-dimethyl-1,4-diphenyl-	EtOH	235s(4.34),322s(4.37), 332(4.43),343s(4.34)	78-2131-76
$C_{26}H_{23}BrN_2O_4$ Spiro[2H-1-benzopyran-2,2'-[2H]indole], 8-bromo-1',3'-dihydro-5'-(4-methoxy- phenyl)-1',3',3'-trimethyl-6-nitro- merocyanine form	EtOH	278(4.47)	103-0417-76
	toluene	387(4.19),597(4.42)	103-0417-76
	EtOH	387(4.24),530(4.33)	103-0417-76
	dioxan	388(4.31),582(4.57)	103-0417-76

Compound	Solvent	$\lambda_{max}(\log \epsilon)$	Ref.
$C_{26}H_{23}ClNS$ Pyridinium, 2-(4-chlorophenyl)-4-(ethyl- thio)-1-methyl-3,5-diphenyl-, tetrafluoroborate	MeOH	237s(4.25),276(4.08), 326(3.96)	44-0818-76
$C_{26}H_{23}ClN_4O_5$ 2,4(1H,3H)-Pteridinedione, 1-[5-chloro- 5-deoxy-2,3-O-(1-methylethylidene)- β-D-ribofuranosyl]-6,7-diphenyl-	pH 4.0	221s(4.39),272(4.16), 356(4.11)	24-3159-76
	pH 10.0	222s(4.40),264(4.29), 358(4.18)	24-3159-76
	MeOH	221(4.43),272(4.20), 357(4.13)	24-3159-76
$C_{26}H_{23}IN_4O_5$ 2,4(1H,3H)-Pteridinedione, 1-[5-deoxy- 5-iodo-2,3-O-(1-methylethylidene)- β-D-ribofuranosyl]-6,7-diphenyl-	pH 6.0	222s(4.36),272(4.14), 358(4.08)	24-3159-76
	pH 11.0	222s(4.34),265(4.29), 358(4.18)	24-3159-76
	MeOH	222(4.40),273(4.20), 358(4.10)	24-3159-76
$C_{26}H_{23}NS_2Sn$ Stannane, triphenyl[[[(phenylmethyl)- amino]thioxomethyl]thio]-	CHCl	247(3.13),267(2.47)	101-0055-76D
$C_{26}H_{23}N_3$ 1H-Perimidin-6-amine, 2,3-dihydro-4,5- dimethyl-2-phenyl-N-(phenylmethylene)-	EtOH	238(4.63),350(4.06), 400(3.59)	12-2247-76
$C_{26}H_{23}N_5$ 5H-Dibenzo[a,d]cycloheptene-10-carbo- nitrile, 5-[4-[(2-pyridinylmethyl- ene)amino]-1-piperazinyl]-	MeOH	306(4.55)	87-1161-76
5H-Dibenzo[a,d]cycloheptene-10-carbo- nitrile, 5-[4-[(3-pyridinylmethyl- ene)amino]-1-piperazinyl]-	dioxan	302(4.47)	87-1161-76
5H-Dibenzo[a,d]cycloheptene-10-carbo- nitrile, 5-[4-[(4-pyridinylmethyl- ene)amino]-1-piperazinyl]-	MeOH	311(4.51)	87-1161-76
$C_{26}H_{24}N_2$ 2,2'-Bipyridine, 6,6'-dimethyl-4,4'-bis- (4-methylphenyl)-, Cu(I) complex	EtOH	485(4.16)	118-0001-76
1H-Indole, 3,3'-(1,3,5,7-octatetraene- 1,8-diyl)bis[1-methyl-	CH_2Cl_2	230(4.34),234s(4.34), 246s(4.36),280s(4.03), 318s(3.89),390s(4.74), 410(4.89),430(4.83)	5-1060-76
oxidized form	MeCN	258(4.08),268(4.08), 317(3.74),327(3.74), 391(3.92),406(3.92), 575(4.96)	5-1060-76
$C_{26}H_{24}N_2O_2$ 2,2'-Bipyridine, 6,6'-dimethyl-4,4'-bis- (4-methoxyphenyl)-, Cu(I) complex	EtOH	485(4.15)	118-0001-76
$C_{26}H_{24}N_2O_3$ 4-Quinolinecarbonitrile, 3-(benzoyloxy)- 1,2,3,4-tetrahydro-2-hydroxy-2-phenyl- 3-propyl-	EtOH	236(4.36),275s(3.37), 285s(3.40),297(3.49)	95-0968-76

Compound	Solvent	λ_{max}(log ϵ)	Ref.
C$_{26}$H$_{24}$N$_2$O$_4$			
Spiro[2H-1-benzopyran-2,2'-[2H]indole], 1',3'-dihydro-5'-(4-methoxyphenyl)-1',3',3'-trimethyl-6-nitro-merocyanine form	EtOH	276(4.63)	103-0417-76
	toluene	382(4.37),597(4.72)	103-0417-76
	EtOH	376(4.23),537(4.46)	103-0417-76
	dioxan	388(4.40),583(4.84)	103-0417-76
Spiro[2H-1-benzopyran-2,2'-[2H]indole], 1',3'-dihydro-8-methoxy-1',3',3'-trimethyl-6-nitro-5'-phenyl-merocyanine form	EtOH	285(4.40)	103-0417-76
	toluene	401(4.43),609(4.70)	103-0417-76
	EtOH	398(4.30),557(4.42)	103-0417-76
	dioxan	397(4.57),600(4.77)	103-0417-76
C$_{26}$H$_{24}$N$_4$OS			
5H-1,3,4-Thiadiazolo[3,2-a]pyrimidin-5-one, 7-[4-(dimethylamino)phenyl]-6,7-dihydro-2-methyl-6,6-diphenyl-	EtOH	261(4.28)	94-2532-76
C$_{26}$H$_{24}$N$_4$O$_3$			
Benzo[g]pteridine-5(3H)-carboxaldehyde, 4a-(1,3-cyclopentadienyl)-2,4,4a,10-tetrahydro-7,8,10-trimethyl-2,4-dioxo-3-(phenylmethyl)-	50% MeOH-pH -0.5	267(4.18),393(3.61)	5-2037-76
	50% MeOH-pH 7	255s(--),320(4.12)	5-2037-76
C$_{26}$H$_{24}$N$_4$O$_5$S			
4(1H)-Pteridinone, 2,3-dihydro-1-[2,3-O-(1-methylethylidene)-β-D-ribofuranosyl]-6,7-diphenyl-2-thioxo-	pH 5.0	220s(4.22),301(4.25), 372(4.18)	24-3159-76
	pH 10.0	221(4.28),286(4.23), 377(4.20)	24-3159-76
C$_{26}$H$_{24}$N$_4$O$_6$			
2,4(1H,3H)-Pteridinedione, 1-[2,3-O-(1-methylethylidene)-β-D-ribofuranosyl]-6,7-diphenyl-	pH 6.0	220s(4.44),272(4.19), 358(4.15)	24-3159-76
	pH 11.0	220s(4.37),264(4.29), 358(4.17)	24-3159-76
	MeOH	223(4.54),274(4.14), 358(4.14)	24-3159-76
C$_{26}$H$_{24}$O			
Indeno[2,1-b]pyran, 3-ethyl-9-(3,4-dihydro-2-naphthalenyl)-2,4-dimethyl-	dioxan	255(4.23),322(4.26), 517(3.49)	78-2225-76
C$_{26}$H$_{24}$O$_6$			
1,3-Benzenediol, 4-(3,4-dihydro-5-methoxy-2-phenyl-2H-furo[2,3-h]-1-benzopyran-8-yl)-5-methoxy-6-methyl-, (S)-	EtOH	257(3.90),287s(4.07), 300(4.10),308s(4.09), 330s(3.86)	39-1570-76C
C$_{26}$H$_{24}$O$_{12}$			
Griseusin A, diacetate	MeOH	247(4.14),347(3.42)	78-2207-76
C$_{26}$H$_{25}$AsOSi			
3H-Arsol-3-one, 1,2-dihydro-1,5-diphenyl-2-(phenylmethylene)-4-(trimethylsilyl)-	C$_6$H$_{12}$	320(3.82)	78-2131-76
C$_{26}$H$_{25}$BrN$_4$			
1-Piperazinamine, 4-(10-bromo-5H-dibenzo[a,d]cyclohepten-5-yl)-N-[(6-methyl-2-pyridinyl)methylene]-	MeOH	307(4.43)	87-1161-76

Compound	Solvent	$\lambda_{max}(\log \epsilon)$	Ref.
$C_{26}H_{25}Cl_2N_3O_4S_2$			
Benzenesulfonamide, 4-chloro-N-[7-[[(4-chlorophenyl)sulfonyl]amino]-2,3,4,5-6,7-hexahydro-1-methyl-2,7-methano-1H-1-benzazonin-12-ylidene]-	MeCN	232(4.67),350(3.68)	39-0481-76C
Benzenesulfonamide, 4-chloro-N-[2-[[(4-chlorophenyl)sulfonyl]amino]-1'-methylspiro[cyclohexane-1,3'-[3H]indol]-2'(1'H)-ylidene]-	EtOH	224(4.56),283(4.25)	39-0481-76C
$C_{26}H_{25}F_3N_2O_3$			
4H-1-Benzopyran-4-one, 6-hexyl-7-methoxy-3-(1-phenyl-1H-pyrazol-4-yl)-2-(trifluoromethyl)-	EtOH	260(4.20),315(3.88)	103-0914-76
$C_{26}H_{25}NO_2$			
5(1H)-Quinolinone, 4,6,7,8-tetrahydro-1,7,7-trimethyl-4-(2-oxo-2-phenylethylidene)-2-phenyl-	MeOH	410s(4.10),434(4.28)	24-1549-76
$C_{26}H_{25}N_3O$			
3-Indolizineacetaldehyde, 3-acetyl-6-methyl-α-phenyl-7-(phenylmethyl)-, hydrazone	EtOH	242(4.35),302(4.32), 340(4.32),380(4.26)	103-0424-76
$C_{26}H_{25}N_3O_2$			
2H-Pyrrol-2-one, 4-ethyl-1,5-dihydro-5-phenyl-3-[[phenyl[(phenylmethylene)-amino]methoxy]amino]-	EtOH	209(4.54),245(4.50), 338(3.36)	40-1100-76
Spiro[2H-1-benzopyran-2,2'-[2H]indole], 1',3'-dihydro-6-[(4-methoxyphenyl)-azo]-1',3',3'-trimethyl-	EtOH	305(4.08),360(4.02), 382(4.20)	28-0691-76A
Spiro[2H-1-benzopyran-2,2'-[2H]indole], 1',3'-dihydro-8-methoxy-1',3',3'-tri-methyl-6-(phenylazo)-	EtOH	272(4.30),296(4.0), 370(4.10)	28-0691-76A
$C_{26}H_{25}N_5O_5$			
4(1H)-Pteridinone, 2-amino-1-[2,3-O-(1-methylethylidene)-β-D-ribofuranosyl]-6,7-diphenyl-	pH 0.0	226(4.43),277(4.15), 358(4.16)	24-3159-76
	pH 7.0	225s(4.40),264(4.31), 356(4.21)	24-3159-76
	pH 14.0	223(4.44),275(4.29), 372(4.18)	24-3159-76
	MeOH	223s(4.47),267(4.33), 358(4.21)	24-3159-76
$C_{26}H_{26}BrNO_4$			
Isoquinoline, 1-[(6-bromo-1,3-benzodioxol-5-yl)methyl]-1,2,3,4-tetrahydro-7-methoxy-2-methyl-6-(phenylmethoxy)-	EtOH	235s(4.14),290(3.87)	73-1219-76
$C_{26}H_{26}Cl_2O_2$			
4aH-Xanthen-4a-ol, 9-(4-chlorophenyl)-5-[(4-chlorophenyl)methylene]-1,2,3,4,5,6,7,8,9,9a-decahydro-	EtOH	210(4.21),299(4.42)	114-0051-76C
$C_{26}H_{26}Co$			
Cobaltocenium, 1,1'-bis(1-phenylethyl)-	MeOH	274(4.35),320(3.32), 403(2.59)	101-0189-76I

Compound	Solvent	$\lambda_{max}(\log \epsilon)$	Ref.
$C_{26}H_{26}N_2$			
1H-Indole, 3,3'-(1,3,5-hexatriene-1,6-diyl)bis[1,2-dimethyl-	CH_2Cl_2	246(4.46),285s(4.05), 380s(4.70),395(4.79), 416(4.67)	5-1060-76
oxidized form (two electrons lost)	DMF	252(4.22),347s(4.04), 358(4.08),374s(3.99), 518(4.77)	5-1060-76
Indolizine, 1,1'-(1,3,5-hexatriene-1,6-diyl)bis[2,3-dimethyl- oxidized form	CH_2Cl_2	280s(4.26),340(3.91), 444(4.64)	5-0317-76
	MeCN	287(3.99),365(3.88), 526(4.89)	5-0317-76
Indolizine, 3,3'-(1,3,5-hexatriene-1,6-diyl)bis[1,2-dimethyl- semiquinone	CH_2Cl_2	322(4.22),335s(4.14), 472(4.75)	5-0317-76
	1:1 CH_2Cl_2- MeCN	807(--),1018(--), 1258(--)	5-0317-76
oxidized form (dication diperchlorate)	MeCN	272(3.96),328(4.05), 340(4.03),532(4.75)	5-0317-76
$C_{26}H_{26}N_2O_3$			
Ethanone, 1-[4-[1-[(4-acetylphenyl)amino]-3-ethoxy-1,3-dihydro-2H-isoindol-2-yl]phenyl]-	EtOH	231(4.02),314(4.50)	83-0356-76
$C_{26}H_{26}N_2O_4$			
4H-1-Benzopyran-4-one, 7-acetoxy-6-hexyl-3-(1-phenyl-1H-pyrazol-4-yl)-	EtOH	270(4.31)	103-0914-76
$C_{26}H_{26}N_4O_2$			
[2,2'-Bipyridine]-5,5'-dicarboxamide, 1,1',2,2'-tetrahydro-1,1'-bis(phenylmethyl)-	MeOH	277(3.44),355(3.89)	18-3269-76
[2,4'-Bipyridine]-3',5-dicarboxamide, 1,1',2,4'-tetrahydro-1,1'-bis(phenylmethyl)-	MeOH	273(4.17),360(3.96)	18-3269-76
$C_{26}H_{26}N_4O_3$			
Benzo[g]pteridine-5(3H)-carboxaldehyde, 4a-(1-cyclopentenyl)-2,4,4a,10-tetrahydro-7,8,10-trimethyl-2,4-dioxo-3-(phenylmethyl)-	50% MeOH- pH -0.5	267(4.18),393(3.61)	5-2037-76
	50% MeOH- pH 7	255s(--),321(4.12)	5-2037-76
$C_{26}H_{26}N_5$			
Pyrido[2,3-b]pyrazinium, 1,2,3,5-tetrahydro-5-methyl-2,3-bis(2-methyl-1H-indol-3-yl)-	n.s.g.	279(4.30),289(4.20), 351(4.20)	103-0948-76
$C_{26}H_{26}O_2$			
Benzenemethanol, α,α'-(1,3-phenylenedi-2,1-ethenediyl)bis[4-methyl-	n.s.g.	254(4.53)	124-0842-76
2(1H)-Naphthalenone, 1-(1-methylethyl)-1-[[1-(1-methylethyl)-2-naphthalenyl]oxy]-	EtOH	234(4.92),283(3.96), 294(4.01),316(4.00)	39-2570-76C
$C_{26}H_{26}O_4$			
1,3-Benzenedimethanol, α,α'-bis[2-(4-methoxyphenyl)ethenyl]-	n.s.g.	265(4.65),292(3.94), 304(3.68)	124-0842-76
Benzenemethanol, α,α'-(1,3-phenylenedi-2,1-ethenediyl)bis[4-methoxy-	n.s.g.	254(4.81)	124-0842-76

Compound	Solvent	$\lambda_{max}(\log \epsilon)$	Ref.
$C_{26}H_{27}NO$ 4,7-Methano-1H-inden-1-one, 3a,4,5,6,7-7a-hexahydro-2,3-diphenyl-7a-(1-pyrrolidinyl)-, $(3a\alpha,4\beta,7\beta,7a\alpha)$-	MeOH	226(4.24),298(4.10)	87-0414-76
$C_{26}H_{27}NO_2$ 1H-Indole-3-methanol, 1-butyl-5-methoxy-α,2-diphenyl-	EtOH	230(4.54),295(4.30)	42-0606-76
1H-Indole-3-methanol, 5-methoxy-1-(2-methylphenyl)-α,2-diphenyl-	EtOH	226(4.46),295(4.16)	42-0606-76
$C_{26}H_{27}NO_5$ 1,4-Benzoxazepine, 5-(3,4-dimethoxyphenyl)-2,3-dihydro-8-methoxy-3-methyl-7-(phenylmethoxy)-	n.s.g.	215(4.49),255(4.19), 315(4.13)	20-0787-76
$C_{26}H_{27}NO_6$ Tylophorinidine, diacetyl-	MeOH	254(4.6),271(4.4), 287(4.2),310(3.9), 336(3.3),350(3.1)	102-1561-76
$C_{26}H_{27}NO_{11}$ 2-Naphthacenecarboxylic acid, 4-[(3-amino-2,3,6-trideoxy-α-L-lyxo-hexopyranosyl)oxy]-1,2,3,4,6,11-hexahydro-2,5,12-trihydroxy-7-methoxy-6,11-dioxo-, hydrochloride, (2S-cis)-	MeOH	234(4.54),252(4.42), 288(3.94),478(4.07), 495(4.08),530(3.83)	87-0395-76
$C_{26}H_{27}P$ Phosphine, (phenylethynyl)bis(2,4,6-trimethylphenyl)-	MeOH	215(3.24),258(3.24), 270(3.18),290(2.91)	139-0169-76A
$C_{26}H_{28}BrNO$ Ethanamine, 2-[4-(2-bromo-1,2-diphenylethenyl)phenoxy]-N,N-diethyl-, cis	MeOH	232(4.40),258s(4.09), 291(3.99)	36-1545-76
trans	MeOH	240(4.37),302(3.98)	36-1545-76
$C_{26}H_{28}ClNO$ Ethanamine, 2-[4-(2-chloro-1,2-diphenylethenyl)phenoxy]-N,N-diethyl-, cis	MeOH	231(4.31),246s(4.20), 292(4.10)	36-1545-76
trans	MeOH	242(4.34),300(4.06)	36-1545-76
$C_{26}H_{28}ClN_3O_2$ 1H-Pyrrolo[3,2-b]quinoline-3-carboxylic acid, 7-chloro-1-[3-(dimethylamino)-propyl]-2-methyl-9-phenyl-	EtOH	230s(4.45),256(4.81), 322s(3.81),336(4.08), 357(3.97),370s(3.91)	44-1743-76
$C_{26}H_{28}N_4O_5$ Pyrazinecarboxamide, N-methyl-3-[[2,3-O-(1-methylethylidene)-β-D-ribofuranosyl]amino]-5,6-diphenyl-	MeOH	230(4.29),282(4.33), 375(4.07)	24-3194-76
$C_{26}H_{28}O_2$ 4aH-Xanthen-4a-ol, 1,2,3,4,5,6,7,8,9,9a-decahydro-9-phenyl-5-(phenylmethylene)-	EtOH	209(4.32),293(4.40)	114-0051-76C
$C_{26}H_{28}O_6$ 1,3-Benzenediol, 4-[2-(3,4-dihydro-7-hydroxy-5-methoxy-2-phenyl-2H-1-benzopyran-8-yl)ethyl]-5-methoxy-6-methyl-, (S)-	EtOH EtOH-base	280(3.55) 290(3.90)	39-1570-76C 39-1570-76C

Compound	Solvent	$\lambda_{max}(\log \epsilon)$	Ref.
$C_{26}H_{28}O_8$ 1,4:5,8-Dimethanocyclobuta[1",2":3,4;3"-4":3',4']dicyclobuta[1,2:1',2']dibenz-ene-4b,4c,8b,8c-tetracarboxylic acid, 1,4,4a,4d,5,8,8a,8d-octahydro-, tetra-methyl ester, (1α,4α,4aβ,4bα,4cβ-4dα,5α,8α,8aα,8bβ,8cα,8dβ)-	EtOH	219(4.36),308(3.30)	89-0165-76
$C_{26}H_{28}O_{10}$ Wikstromol, triacetate	EtOH	226(4.04),274(3.72), 280(3.69)	102-1789B-76
$C_{26}H_{29}NO$ Pyrrolidine, 1-(4-cyclohexylidene-1-oxo-2,3-diphenyl-2-butenyl)-	MeOH	205(4.47),234(4.31), 295(4.16)	87-0414-76
$C_{26}H_{29}NO_2S$ Piperidine, 1-(4,5,6,7-tetrahydro-2,3-diphenyl-4,7-methanobenzo[b]thien-7a(3aH)-yl)-, S,S-dioxide, (3aα,4β,7β,7aα)-	MeOH	220(4.29),256(3.98)	87-0414-76
$C_{26}H_{29}NO_6$ Benzamide, 3,4-dimethoxy-N-[2-[3-meth-oxy-4-(phenylmethoxy)phenoxy]-1-methylethyl]-	n.s.g.	215(4.56),250(4.10), 285(3.96)	20-0787-76
$C_{26}H_{29}N_2$ Acridinium, 9-[4-(dipropylamino)phenyl]-10-methyl-, (triiodide)	CH_2Cl_2	260(5.05),292(4.70), 360(4.53),630(4.18)	104-1531-76
$C_{26}H_{30}N_4O_2$ Benzo[g]pteridine-2,4(3H,4aH)-dione, 4a-(2,3-dimethyl-2-butenyl)-5,10-dihydro-7,8,10-trimethyl-3-(phen-ylmethyl)-	MeOH	305s(--),346(3.72)	5-2037-76
$C_{26}H_{30}N_4O_4$ 1H-Pyrrole-3-carboxylic acid, 2-amino-1,5-bis(benzylcarbamoyl)-4-methyl-, tert-butyl ester	MeOH	219(4.53),295(4.16)	24-2983-76
$C_{26}H_{30}N_8O_4S_2$ Pyrrolidine, 1,1'-[(1,2-dihydro-1,2,4,5-tetrazine-3,6-diyl)bis[[(4-methylphen-yl)sulfonyl]carbonimidoyl]]bis-	MeCN	245(4.62),520(2.68)	32-0001-76
$C_{26}H_{30}O_7$ Tricoccin S_3	MeOH	214(4.30)	88-3295-76
$C_{26}H_{30}O_8$ 1,3-Benzenediol, 5,5'-(1,6-hexanediyl)-bis-, tetraacetate	MeOH	262(3.49)	12-1989-76
[1,1'-Biphenyl]-3,3'-diol, 2,2',4,4'-tetramethoxy-6,6'-di-2-propenyl-, diacetate	MeOH	227(4.25),277(3.33)	18-1940-76
$C_{26}H_{31}NO_5$ Vertaline, (±)-	MeOH	223s(4.26),241s(4.00), 280s(3.64),293(3.81)	94-1045-76

Compound	Solvent	$\lambda_{max}(\log \epsilon)$	Ref.
$C_{26}H_{31}N_3$			
1H-Indol-3-amine, 2-(2,2-dimethylprop-yl)-N-[2-(2,2-dimethylpropyl)-3H-indol-3-ylidene]-	MeOH	225(4.83),260(4.35), 340(3.88),530(4.18)	23-1020-76
	MeOH-HCl	223(4.80),283(4.34), 295(4.28),590(4.52)	23-1020-76
$C_{26}H_{32}N_8O_6S_2$			
Morpholine, 4,4'-[(1,2-dihydro-1,2,4,5-tetrazine-3,6-diyl)bis[[(4-methyl-phenyl)sulfonyl]carbonimidoyl]]bis-	MeCN	223s(--),255(4.54)	32-0001-76
$C_{26}H_{32}O_6$			
Tadzhikovin	n.s.g.	243(3.65),251(3.53), 325(4.19)	105-0533-76
$C_{26}H_{32}O_7$			
Androsta-5,8-diene-7,11,17-trione, 3β,6-diacetoxy-4,4,14α-trimethyl-	dioxan	270(4.03),325(3.17)	5-1206-76
β-D-Glucopyranoside, 4-[cyclohexylid-ene(4-hydroxyphenyl)methyl]phenyl 4-O-methyl	MeOH	243(4.26)	24-2259-76
$C_{26}H_{32}O_9$			
Kaur-16-en-20-oic acid, 1,11,14-triacet-oxy-18-hydroxy-16-oxo-, δ-lactone, (4α,7α,11β,14R)-	EtOH	236(4.03)	33-0772-76
Kaur-16-en-15-one, 11,14,18-triacetoxy-7,20:18,20-diepoxy-, (4α,7α,11β,14R-20S)-	EtOH	233(4.05)	33-0772-76
$C_{26}H_{32}O_{10}$			
2,3-Naphthalenedicarboxylic acid, 1,2,3,4-tetrahydro-6,7,8-trimethoxy-1-(3,4,5-trimethoxyphenyl)-, dimethyl ester	n.s.g.	278(3.237)	2-0127B-76
$C_{26}H_{32}O_{13}$			
Grandifloroside, methyl-	MeOH	219(4.2),234(4.1), 300s(3.9),327(4.1)	102-1305-76
	MeOH-NaOMe	216(4.2),232s(4.0), 300s(3.6),311(3.6), 381(4.2)	102-1305-76
$C_{26}H_{32}O_{14}$			
Tarennoside, pentaacetate	MeOH	248(4.15)	94-1216-76
$C_{26}H_{33}NO_5$			
24-Norchola-5,8,22-triene-2,3,7,11-tet-rone, 6-hydroxy-4,4,14-trimethyl-, 2-oxime	dioxan	232(4.30),273(3.88), 350(3.51)	5-1214-76
$C_{26}H_{34}N_4O_4S$			
Benzenesulfonic acid, 4-methyl-, [1-(4-methyl-3,6-dioxo-2,5-di-1-piperidinyl-1,4-cyclohexadien-1-yl)ethylidene]hy-drazide	EtOH	233(4.51),435(3.93)	94-1731-76
$C_{26}H_{34}N_4O_8$			
Cytidine, 4-N-(N-tert-butoxycarbonyl-L-phenylalanyl)-2',3'-O-isopropyli-dene-	n.s.g.	246(4.13),298(3.78)	128-0351-76

Compound	Solvent	$\lambda_{max}(\log \epsilon)$	Ref.
$C_{26}H_{34}N_4O_{12}$ Cyclopentaneacetic acid, 2,3-bis(2,2-di-methyl-1-oxopropoxy)-α-[(2,4-dinitro-phenyl)hydrazono]-4-(methoxycarbon-yl)-, methyl ester, $(1\alpha,2\alpha,3\beta,4\alpha)$-	EtOH	360(4.18)	23-2925-76
$C_{26}H_{34}N_8O_4S_2$ 1,2,4,5-Tetrazine-3,6-dicarboximidamide, N,N,N'',N''-tetraethyl-N',N'''-bis(4-methylphenyl)sulfonyl]-	EtOH	242(4.62),525(2.70)	32-0001-76
$C_{26}H_{34}N_8O_5$ Heptanedioic acid, 2-[[4-[[(2,4-diamino-6-pteridinyl)methyl]methylamino]benz-oyl]amino]-, diethyl ester	EtOH	220(4.31),262(4.41), 299(4.44),375(3.91)	4-0727-76
$C_{26}H_{34}O_2$ Benzenemethanol, α-(1,1-dimethylethyl)-α-(3-ethoxy-4,4-dimethyl-3-phenyl-1-pentynyl)-	n.s.g.	232s(2.08),252(2.65), 257(2.77),263(2.69)	56-1301-76
$C_{26}H_{34}O_5S$ Pregna-1,4-diene-3,20-dione, 11-acetoxy-16,17-[oxy(1-methylethylidene)thio]-, $(11\beta,16\alpha)$-	MeOH	240(4.16)	24-0185-76
$C_{26}H_{34}O_6$ Phorbol, 12,13,20-triacetate	MeOH	194(4.08),232(3.72), 334(1.85)	108-0046-76
$C_{26}H_{34}O_8$ [1,1'-Biphenyl]-3,3'-diol, 2,2',4,4'-tetramethoxy-6,6'-dipropyl-, diacetate	MeOH	232(4.14),279(3.39)	18-1940-76
$C_{26}H_{34}O_9$ Kaur-16-en-15-one, 7,11,14-triacetoxy-18,20-epoxy-20-hydroxy-, $(4\alpha,7\alpha,11\beta-$14R,20R)-	EtOH	235(3.99)	33-0772-76
$C_{26}H_{34}O_{15}$ Secoxyloganin, tetraacetate, methyl ester	MeOH	232(4.22)	39-0160-76C
$C_{26}H_{35}F_7O_2$ 2-Hepten-4-one, 2-[[(5α,17β)-androstan-17-yl]oxy]-5,5,6,6,7,7,7-heptafluoro-	dioxan	280(4.21)	20-0333-76
$C_{26}H_{35}NO$ 2,5-Cyclohexadien-1-one, 2,6-bis(1,1-di-methylethyl)-4-(phenyl-1-piperidinyl-methylene)-	hexane	430(4.51)	73-2614-76
$C_{26}H_{35}NO_7$ 4-Isoxazolecarboxylic acid, 5-[6-(acet-oxymethyl)tetrahydro-2,2-dimethyl-4H-cyclopenta-1,3-dioxol-4-yl]-3-(2,4,6-trimethylphenyl)-4,5-dihydro-, ethyl ester	EtOH	217(4.08)	23-0861-76

Compound	Solvent	$\lambda_{max}(\log \epsilon)$	Ref.
$C_{26}H_{36}$			
5-Decene-3,7-diyne, 5,6-bis(3,3-dimeth-yl-1-butynyl)-2,2,9,9-tetramethyl-	EtOH	238(3.93),252(4.04), 265(3.93),299s(4.18), 309(4.37),318(4.47), 331(4.53)	78-1293-76
5-Dodecene-3,7,9-triyne, 5-(3,3-dimeth-yl-1-butynyl)-6-(1,1-dimethylethyl)-2,2,11,11-tetramethyl-	EtOH	229(5.05),241(4.20), 257s(3.78),309s(4.27), 310(4.41),328(4.46), 340(4.44)	78-1293-76
$C_{26}H_{36}N_2O_8$			
[2,2'-Bipyridine]-3,3',5,5'-tetracarb-oxylic acid, 1,1',2,2'-tetrahydro-2,2',6,6'-tetramethyl-, tetraethyl ester	pyridine	387(3.86)	18-3269-76
[4,4'-Bipyridine]-3,3',5,5'-tetracarb-oxylic acid, 1,1',4,4'-tetrahydro-2,2',6,6'-tetramethyl-, tetraethyl ester	MeOH	240(4.31),275(4.00), 360(3.93)	18-3269-76
$C_{26}H_{36}O_3$			
24,25,26,27-Tetranor-5α-lanosta-8,22-di-ene-3,7,11-trione	dioxan	268(3.91)	5-1214-76
$C_{26}H_{36}O_3S$			
1,6,8-Nonatrien-3-ol, 3,7-dimethyl-5-(phenylsulfonyl)-9-(2,6,6-trimethyl-1-cyclohexen-1-yl)-, (6E,8E)-	EtOH	272(4.21)	22-0513-76
2,4,8-Nonatrien-1-ol, 3,7-dimethyl-7-(phenylsulfonyl)-9-(2,6,6-trimethyl-1-cyclohexen-1-yl)-, all-E	hexane	237(4.536)	33-0397-76
2,6,8-Nonatrien-1-ol, 3,7-dimethyl-4-(phenylsulfonyl)-9-(2,6,6-trimethyl-1-cyclohexen-1-yl)-	isoPrOH	215(4.26),273(4.23)	44-3287-76
$C_{26}H_{36}O_4$			
7H-Dibenzo[b,d]pyran-7-one, 1-acetoxy-6,8,9,10-tetrahydro-6,6-dimethyl-3-(1,2-dimethylheptyl)-7-oxo-	EtOH	300(4.19),350(3.97)	4-1101-76
21,24-Dinorchola-5,16,20(22)-trien-23-oic acid, 3-acetoxy-, ethyl ester	n.s.g.	275(4.14)	70-0150-76
24-Norchola-5,8,22-triene-7,11-dione, 3β,6-dihydroxy-4,4,14-trimethyl-	EtOH	234(4.04),282(3.78), 355(3.48)	5-1206-76
$C_{26}H_{36}O_5$			
5α-Pregn-8-ene-7,11,20-trione, 3β-acetoxy-4,4,14α-trimethyl-	dioxan	268(3.92)	5-1206-76
$C_{26}H_{36}O_9$			
Rastronol C	EtOH	228(3.98)	33-0772-76
$C_{26}H_{37}NO_2$			
2H-[1]Benzopyrano[4,3-c]pyridin-10-ol, 1,3,4,5-tetrahydro-8-(1,2-dimethyl-heptyl)-5,5-dimethyl-2-(2-propynyl)-	EtOH	280(4.15)	87-0445-76
$C_{26}H_{37}NO_3$			
2H-Pyrrol-2-one, 4-[(3β-acetoxyandrost-5-en-17β-yl)methyl]-1,5-dihydro-	EtOH	206(4.21)	33-2138-76

Compound	Solvent	$\lambda_{max}(\log \epsilon)$	Ref.
$C_{26}H_{37}NO_4$ 24-Norchola-8,22-diene-6,7,11-trione, 3-hydroxy-4,4,14-trimethyl-, 6-oxime, (3β,5α)-	dioxan	280(3.83)	5-1214-76
$C_{26}H_{38}N_2$ Diazene, bis(1,1,3-trimethyl-3-phenyl- butyl)-	toluene	374(1.38)	73-1557-76
$C_{26}H_{38}O_3$ Chola-4,22-dien-24-oic acid, 3-oxo-, ethyl ester	EtOH	218(4.16),238(4.13)	13-0431-76A
24-Norchola-8,22-diene-7,11-dione, 3-hy- droxy-4,4,14-trimethyl-, (3β,5α)-	dioxan	265(3.91)	5-1206-76
4,8-Undecadienal, 11-(3,4-dihydro-6-hy- droxy-2,5,7,8-tetramethyl-2H-1-benzo- pyran-2-yl)-4,8-dimethyl-, [R-(E,E)]-	EtOH	225s(4.05),293(3.51)	33-0290-76
$C_{26}H_{38}O_4$ 24-Nor-5α-chola-8,16-dien-23-al, 3β-hy- droxy-4β-(hydroxymethyl)-15-oxo-	EtOH	237(3.90)	33-1997-76
24-Norchola-5,8-diene-7,11-dione, 3β,6- dihydroxy-4,4,14-trimethyl-	EtOH	234(4.01),281(3.73), 355(3.49)	5-1206-76
4,8-Undecadienoic acid, 11-(3,4-dihydro- 6-hydroxy-2,5,7,8-tetramethyl-2H-1- benzopyran-2-yl)-4,8-dimethyl-, [R-(E,E)]-	EtOH	235s(4.03),291(3.49)	33-0290-76
$C_{26}H_{38}O_5$ 24-Nor-5α-chola-8,16-dien-23-oic acid, 3β-hydroxy-4β-(hydroxymethyl)- 4,14α-dimethyl-15-oxo-	EtOH	237(3.92)	33-1997-76
5α-Pregn-8-ene-7,11,20-trione, 3β-hy- droxy-4,4,14α-trimethyl-, cyclic 20-(1,2-ethanediyl acetal)	dioxan	269(3.93)	5-1206-76
$C_{26}H_{38}O_{10}$ Leuconolide A_3, 5,18-O,O-diacetyl- 9-dehydro-18-dihydro-	EtOH	280(4.24)	94-1749-76
Leuconolide A_3, 9,18-O,O-diacetyl-, 5,18-hemiacetal	EtOH	232(4.24)	94-1749-76
$C_{26}H_{38}Si$ Silane, tributyl(1,2-diphenylethenyl)-, (E)-	n.s.g.	235(4.25),266(4.28)	65-1044-76
$C_{26}H_{40}$ Cyclohexene, 1,1'-(3,4-dimethyl-1,3,5- hexatriene-1,6-diyl)bis[2,6,6-tri- methyl-	EtOH	312(4.29)	138-1127-76
$C_{26}H_{40}CuN_2O_{10}S_2$ Copper, bis[diethyl 5-[2-(methylthio)- ethyl]-4-oxo-1,3-pyrrolidinedicarb- oxylato]-	EtOH	275(4.45)	4-0113-76
$C_{26}H_{40}N_4O_4$ 2H-Pyrrole, 5,5',5'',5'''-(1,2-ethanediyl- idene)tetrakis[3,4-dihydro-2,2-dimeth- yl-, 1,1',1'',1'''-tetraoxide	EtOH	252(4.13),383(4.05)	39-1942-76C

Compound	Solvent	$\lambda_{max}(\log \epsilon)$	Ref.
$C_{26}H_{40}O_3$ 24-Norchol-8-ene-7,11-dione, 3-hydroxy- 4,4,14-trimethyl-, (3β,5α)-	dioxan	268(3.92)	5-1206-76
$C_{26}H_{40}O_4$ 2(3H)-Furanone, 3,5-bis(1,1-dimethyl- ethyl)-3-[3,5-bis(1,1-dimethylethyl)- 2-hydroxyphenoxy]-	MeOH	205(4.37),275(3.28)	35-1510-76
$C_{26}H_{40}O_7$ Deacetylkidjoladinin	EtOH	215(3.93)	94-2185-76
$C_{26}H_{42}N_4O_4$ 2H-Pyrrole, 5,5',5'',5'''-(1,2-ethanediyl- idene)tetrakis[3,4-dihydro-2,2-dimeth- yl-, 1,1',1'',1'''-tetraoxide	EtOH	245(4.01)	39-1942-76C
$C_{26}H_{42}O_2$ 1,5-Hexadiene-3,4-diol, 3,4-dimethyl- 1,6-bis(2,6,6-trimethyl-1-cyclo- hexen-1-yl)-	C_6H_{12}	233(4.28)	39-0561-76C
$C_{26}H_{42}O_5$ 2-Pentenoic acid, 4-[3-[1-ethyl-2-(4,4- ethylenedioxycyclohexyl)butyl]cyclo- pentyl]-4,5-epoxy-, ethyl ester, threo	EtOH	216(4.06)	5-1222-76
$C_{26}H_{42}O_6$ Deacetyltomentosin	EtOH	218(4.12)	94-1552-76
$C_{26}H_{43}NO$ 4-Azacholest-5-en-3-one	EtOH	235(4.1)	49-0511-76
$C_{26}H_{44}N_2O$ 4-Azacholest-5-en-3-one, 4-amino-	EtOH	243(4.10)	49-0511-76
$C_{26}H_{45}ClO_2S$ 2-Thiophenecarboxylic acid, 4-(chloro- methyl)-5-nonadecyl-, methyl ester	heptane	256(3.93),278(3.98)	34-0233-76
$C_{26}H_{46}N_2O$ 4-Aza-5α-cholestan-3-one, 4-amino-	EtOH	none above 220 nm	49-0511-76
$C_{26}H_{46}OS$ 1-Octanone, 1-(4-methyl-5-tridecyl- 2-thienyl)-	heptane	264(3.91),298(4.01)	34-0233-76
$C_{26}H_{46}O_2S$ 2-Thiophenecarboxylic acid, 4-methyl- 5-nonadecyl-, methyl ester	heptane	254(3.91),285(4.00)	34-0233-76

Compound	Solvent	$\lambda_{max}(\log \epsilon)$	Ref.
$C_{27}H_{16}Br_2O_2$ Phenanthro[9,10-d]-1,3-dioxole, 6,9-dibromo-2,2-diphenyl-	CHCl$_3$	266(4.72),284(4.18), 300(3.63),315s(--), 330(4.16),344(4.23), 372s(--),390(3.29), 411(3.29)	104-1298-76
$C_{27}H_{17}BrO$ Indeno[2,1-b]pyran, 2-(4-bromophenyl)-9-(1H-inden-2-yl)-	dioxan	242(4.26),277(4.35), 319(4.52),359(4.55), 598(3.51)	78-2225-76
$C_{27}H_{17}ClN_4O$ Naphtho[1,2-d]oxazole, 2-[3-chloro-4-[2-(2-phenyl-2H-1,2,3-triazol-4-yl)ethenyl]phenyl]-	DMF	323(4.36),375(4.75)	33-2469-76
$C_{27}H_{17}ClO$ Indeno[2,1-b]pyran, 2-(4-chlorophenyl)-9-(1H-inden-2-yl)-	dioxan	227(4.30),279(4.37), 361(4.52),574(3.51)	78-2225-76
$C_{27}H_{17}NO$ 1-Naphthalenecarbonitrile, 4-[2-(2-phenyl-6-benzofuranyl)ethenyl]-	DMF	387(4.63)	33-0819-76
$C_{27}H_{17}NO_3$ 13H-Dibenzo[a,h]carbazole-7,12-dione, 5-hydroxy-13-(phenylmethyl)-	EtOH	240(3.66),280(3.73), 334(3.26),485(3.20)	115-0185-76
$C_{27}H_{18}Cl_2N_6$ 2H-Naphtho[1,2-d]triazole, 2-[3-chloro-4-[2-[2-(4-chlorophenyl)-5-methyl-2H-1,2,3-triazol-4-yl]ethenyl]phenyl]-	DMF	374(4.80)	33-2469-76
$C_{27}H_{18}N_2O_2$ 9,10-Anthracenedione, 1-amino-2-[(diphenylmethylene)amino]-	EtOH	209(4.69),262(4.60), 508(3.91)	104-0172-76
1H-Anthra[1,2-d]imidazole-6,11-dione, 2,3-dihydro-2,2-diphenyl-	EtOH	546(4.08)	104-0172-76
$C_{27}H_{18}N_4O$ Naphtho[1,2-d]oxazole, 2-[4-[2-(2-phenyl-2H-1,2,3-triazol-4-yl)ethenyl]phenyl]-	DMF	321(4.43),369(4.78)	33-2469-76
$C_{27}H_{18}O$ 9(10H)-Anthracenone, 10-(diphenylmethylene)-	isooctane	353(3.95)	73-1676-76
Indeno[2,1-b]pyran, 9-(1H-inden-2-yl)-2-phenyl-	dioxan	269(4.42),323(4.60), 358(4.64),585(3.54)	78-2225-76
$C_{27}H_{18}OS$ Spiro[2H-1-benzothiopyran-2,9'-[9H]xanthene], 3-phenyl-	dioxan	240(4.56),310(4.06), 344(3.95)	88-3563-76
$C_{27}H_{19}ClN_6$ 2H-Naphtho[1,2-d]triazole, 2-[3-chloro-4-[2-(5-methyl-2-phenyl-2H-1,2,3-triazol-4-yl)ethenyl]phenyl]-	DMF	373(4.79)	33-2469-76

Compound	Solvent	$\lambda_{max}(\log \epsilon)$	Ref.
$C_{27}H_{19}ClN_6O$ 2H-Naphtho[1,2-d]triazole, 2-[3-chloro- 4-[2-[2-(4-methoxyphenyl)-2H-1,2,3- triazol-4-yl]ethenyl]phenyl]-	DMF	289(4.34),373(4.79)	33-2469-76
$C_{27}H_{19}N$ 1-Naphthalenecarbonitrile, 4-[2-[4-(2- phenylethenyl)phenyl]ethenyl]-	DMF	382(4.69)	33-0819-76
$C_{27}H_{19}NOS$ 1,4-Epithioisoquinolin-3(2H)-one, 1,4-dihydro-1,2,4-triphenyl-	MeCN	242s(4.103)	39-0672-76C
$C_{27}H_{19}NO_3S$ [1]Benzothieno[3,2-b]pyridine, 2-benzo- furanyl-4-(3,4-dimethoxyphenyl)-	$C_2H_4Cl_2$	247(5.28),326(5.88), 370(4.89)	103-0977-76
$C_{27}H_{19}N_3OS$ 3-Thiophenecarbonitrile, 4-amino-5-benz- oyl-2-(1-methyl-2-phenyl-1H-indol- 3-yl)-	EtOH	294(4.23),400(4.24)	142-0009-76
$C_{27}H_{19}O_4S$ [1]Benzothieno[3,2-b]pyrylium, 2-benzo- furanyl-4-(3,4-dimethoxyphenyl)-, perchlorate	$C_2H_4Cl_2$	255(3.88),278(3.71), 294(3.64),325(4.06), 390(5.79)	103-0977-76
$C_{27}H_{20}$ Phenanthrene, 9-methyl-1,2-diphenyl-	MeOH	301s(4.20),336(2.74), 348(2.67),356(2.66)	39-1115-76B
$C_{27}H_{20}N_2O$ 2H-Cyclopenta[gh]perimidin-2-one, 1,3- dihydro-1,3-bis(phenylmethyl)-	n.s.g.	385(4.08)	103-0455-76
4(3H)-Quinazolinone, 2-[2-(1-naphthal- enyl)ethenyl]-3-(phenylmethyl)-	EtOH	320(4.12)	115-0341-76
$C_{27}H_{20}N_6$ 1,2-Benzenediamine, N-(1-methyl-3-phen- ylpyrimido[5,4-a]phenazin-5-yl)-	dioxan	206(4.57),267s(--), 277s(--),305(4.63), 342(4.22),463(4.07)	5-1809-76
$C_{27}H_{20}N_6O_4S$ Benzenesulfonamide, N-(2,3-dihydro- 1,3-dioxo-1H-isoindol-4-yl)- 4-(3,5-diphenyl-1-formazano)-	DMF	301(4.39),466(4.15)	104-1340-76
$C_{27}H_{20}O$ Indeno[2,1-b]pyran, 9-(1-acenaphthylen- yl)-2,3,4-trimethyl-	dioxan	230(4.80),298(4.42), 318(4.37),517(3.84)	78-2225-76
$C_{27}H_{20}OS$ Indeno[2,1-b]pyran, 9-(3,4-dihydro-2- naphthalenyl)-4-methyl-2-(2-thienyl)-	dioxan	281(4.38),340(4.42), 360(4.38),573(3.52)	78-2225-76
$C_{27}H_{20}O_2Se$ Selenonium, diphenyl-, 1-benzoyl-2-oxo- 2-phenylethylide	$CHCl_3$	256(4.55),325s(3.16)	104-0033-76

Compound	Solvent	$\lambda_{max}(\log \epsilon)$	Ref.
$C_{27}H_{20}O_5$ Dibenzo[c,h]pyrano[3,2-a]xanthen-9(6H)-one, 10-hydroxy-11-methoxy-6,6-dimethyl-	EtOH	215(4.86),241(4.58), 267(4.49),280(4.53), 303(4.56),339s(4.04), 427(3.78)	78-0543-76
	EtOH-AlCl$_3$	218(4.85),238(4.77), 246(4.78),270s(4.51), 287s(4.43),327(4.76), 357s(4.31)	78-0543-76
$C_{27}H_{21}ClN_4O$ 5H-Indeno[5,6-d]oxazole, 2-[4-[2-[2-(4-chlorophenyl)-5-methyl-2H-1,2,3-triazol-4-yl]ethenyl]phenyl]-6,7-dihydro-	DMF	364(4.81)	33-2469-76
$C_{27}H_{21}Cl_2N_5O_6$ Adenine, 9-(2-O-acetyl-5-O-benzoyl-3-deoxy-3-C-(dichloromethylene)-1-β-D-pentofuranosyl)-N-benzoyl-	EtOH	210(4.28),275(4.28), 278(4.19)	111-0489-76
$C_{27}H_{21}IO_7$ D-ribo-Hex-1-enitol, 2,5-anhydro-6-deoxy-6-iodo-, tribenzoate	MeOH	230(4.63),260s(3.28)	44-1836-76
$C_{27}H_{21}N$ 2H-Isoindole, 2-(diphenylmethyl)-1-phenyl-	EtOH	252(3.2),320s(--), 354(3.2)	24-0906-76
$C_{27}H_{21}N_3O_3$ 1,3,5-Triazine-2,4,6(1H,3H,5H)-trione, 1,3,5-tris(2-phenylethenyl)-	THF	271(4.27)	95-0962-76
$C_{27}H_{21}N_3O_{10}$ Coumermic acid	EtOH-4% MCS	243s(4.57),305s(4.48), 343(4.71)	78-2057-76
$C_{27}H_{21}N_7O_4S$ Benzenesulfonamide, 4-(3,5-diphenyl-1-formazano)-N-(1,2,3,4-tetrahydro-1,4-dioxo-5-phthalazinyl)-	DMF DMF-NaOH	304(4.49),488(4.16) 355(4.16),550(4.68)	104-1340-76 104-1340-76
$C_{27}H_{22}BrNO_4$ 2-Azabicyclo[3.2.0]heptane-5-carboxylic acid, 2-(4-bromophenyl)-3,4-dioxo-1,7-diphenyl-, ethyl ester, (1α,5α,7α)-	dioxan	223(4.28),260s(3.68), 310(3.79)	142-1229-76
$C_{27}H_{22}FI_2N_5O_5$ Adenosine, N,N-dibenzoyl-5'-deoxy-4'-C-fluoro-5',5'-diiodo-2',3'-O-(1-methylethylidene)-	dioxan	249(4.41),275(4.28)	35-3346-76
Benzamide, N-benzoyl-N-[9-[5-deoxy-4-C-fluoro-5,5-diiodo-2,3-O-(1-methylethylidene)-α-L-lyxofuranosyl]-9H-purin-9-yl]-	dioxan	249(4.43),274(4.31)	35-3346-76
$C_{27}H_{22}IN_5O_5$ Adenosine, N,N-dibenzoyl-4',5'-didehydro-5'-deoxy-5'-iodo-2',3'-O-(1-methylethylidene)-	MeOH	250(4.37),272(4.25)	35-3346-76

Compound	Solvent	$\lambda_{max}(\log \epsilon)$	Ref.
$C_{27}H_{22}N_2$ 4H-Pyrazino[3,2,1-jk]carbazole, 5,6-di-hydro-2-phenyl-4-(phenylmethyl)-	EtOH	240(4.44),295(4.58), 350(3.90),418(3.34)	103-0662-76
$C_{27}H_{22}N_2O_3S$ 1H-Thieno[3,4-d]imidazole-4-carboxylic acid, 2-[(4-methoxyphenyl)phenylmeth-yl]-6-phenyl-, methyl ester	MeOH	232(4.42),323(4.13), 309(4.07),357(4.39)	49-0299-76
$C_{27}H_{22}N_2O_4S$ 1H-Thieno[3,4-d]imidazole-4-carboxylic acid, 2-[hydroxy(4-methoxyphenyl)-phenylmethyl]-6-phenyl-	MeOH	232(4.46),307(4.03), 321(4.08),357(4.41)	49-0299-76
$C_{27}H_{22}N_4O$ 5H-Indeno[5,6-d]oxazole, 6,7-dihydro-2-[4-[2-(5-methyl-2-phenyl-2H-1,2,3-triazol-4-yl)ethenyl]phenyl]-	DMF	365(4.81)	33-2469-76
$C_{27}H_{22}N_4O_7$ 5H-Furo[2',3':4,5]oxazolo[3,2-a]pterid-in-5-one, 8-acetoxy-9-(acetoxymethyl)-7a,8,9,10a-tetrahydro-2,3-diphenyl-,[7aS-(7aα,8α,9β,10aα)]-	MeOH	226(4.45),268(4.17), 358(4.16)	24-3217-76
$C_{27}H_{22}O$ 7H-Cyclohept[b]indeno[1,2-e]pyran, 8,9,10,11-tetrahydro-5-(2-naph-thalenyl)-	dioxan	228(4.61),262(4.42), 281(4.38),310(4.42), 330(4.42),539(4.38)	78-2225-76
$C_{27}H_{22}O_9$ D-Allonic acid, 2,5-anhydro-, tribenz-oate	EtOH	230(4.36),275(3.37)	138-1119-76
D-Altronic acid, 2,5-anhydro-,tribenzoate	EtOH	230(4.36),275(3.37)	138-1119-76
$C_{27}H_{22}O_{11}$ Benzo[1,2-b:4,5-b']bisbenzofuran-2,6,9-triol, 3,8,12-trimethoxy-	$CHCl_3$	246(4.33),307(4.26), 319(4.48),348(4.26)	12-0179-76
$C_{27}H_{23}BrN_4$ Benzonitrile, 4-[[[4-(10-bromo-5H-diben-zo[a,d]cyclohepten-3-yl)-1-piperazin-yl]imino]methyl]-	dioxan	322(4.45)	87-1161-76
$C_{27}H_{23}FIN_5O_5$ Adenosine, N,N-dibenzoyl-5'-deoxy-4'-C-fluoro-5'-iodo-2',3'-O-(1-methyleth-ylidene)-	dioxan	249(4.34),275(4.24)	35-3346-76
Benzamide, N-benzoyl-N-[9-[5-deoxy-4-C-fluoro-5-iodo-2,3-O-(1-methylethylid-ene)-α-L-lyxofuranosyl]-9H-purin-6-yl]-	dioxan	249(4.37),275(4.26)	35-3346-76
$C_{27}H_{23}NS$ Benzenamine, N,N-dimethyl-4-(9-phenyl-9H-thioxanthen-9-yl)-	EtOH	222(4.66),271(4.61)	88-4609-76
$C_{27}H_{23}N_3$ Benzenamine, 4-[1-(4-aminophenyl)benzo-[f]quinolin-3-yl]-N,N-dimethyl-	EtOH	298(4.57),345(4.63), 371(4.51)	103-0106-76

Compound	Solvent	$\lambda_{max}(\log \epsilon)$	Ref.
$C_{27}H_{23}N_5O_5$ Adenosine, N,N-dibenzoyl-4',5'-didehy- dro-5'-deoxy-2',3'-O-(1-methylethyl- idene)-	MeOH	250(4.32),272s(4.22)	35-3346-76
$C_{27}H_{23}PS_2$ Phosphonium, triphenyl-, 2-(methylthio)- 1-phenyl-2-thioxoethylide	EtOH	272(4.42),325(3.89)	39-0692-76C
$C_{27}H_{24}$ Acenaphthylene, 1,2-dihydro-1-phenyl- 2-(2,4,6-trimethylphenyl)-	hexane	222(4.79),293(3.78)	35-1860-76
$C_{27}H_{24}ClNO_5$ 1,3-Dioxolo[4,5-g]isoquinoline, 6-acet- yl-5-[[2-chloro-3-(phenylmethoxy)phen- yl]methylene]-5,6,7,8-tetrahydro- 9-methoxy-, (Z)-	MeOH	224s(4.61),315(4.25)	12-2003-76
$C_{27}H_{24}FN_5O_5$ Adenosine, N,N-dibenzoyl-5'-deoxy-4'-C- fluoro-2',3'-O-(1-methylethylidene)-	MeOH	250(4.36),273(4.24)	35-3346-76
$C_{27}H_{24}N_2$ 1H-Pyrazole, 4,5-dihydro-1,5-diphenyl- 3-(6-phenyl-1,3,5-hexatrienyl)-	n.s.g.	425(4.79)	124-0383-76
$C_{27}H_{24}N_2O_5S$ 3-Pyridinecarboxamide, 6-(ethoxycarbon- yloxy)-1,2-dihydro-1-(3-methylphenyl)- 2-oxo-N-phenyl-4-(phenylthio)-	MeOH	333(4.5)	120-0103-76
$C_{27}H_{24}N_4O$ Benzoxazole, 5-(1,1-dimethylethyl)-2-[4- [2-(2-phenyl-2H-1,2,3-triazol-4-yl)- ethenyl]phenyl]-	DMF	353(4.83)	33-2469-76
Benzoxazole, 5-(1-methylethyl)-2-[4- [2-(5-methyl-2-phenyl-2H-1,2,3-tri- azol-4-yl)ethenyl]phenyl]-	DMF	357(4.79)	33-2469-76
Benzoxazole, 2-[4-[2-(4-methyl-2-phenyl- 2H-1,2,3-triazol-4-yl)ethenyl]phenyl]- 5-propyl-	DMF	358(4.80)	33-2469-76
$C_{27}H_{24}N_4O_3$ Benzo[g]pteridine-2,4(1H,3H)-dione, 5-benzoyl-5,10-dihydro-7,8,10- trimethyl-3-(phenylmethyl)-	50% MeOH- pH -0.5	322(4.01)	5-2037-76
	50% MeOH- pH 7, 10	265s(--),312(3.89)	5-2037-76
$C_{27}H_{24}OS$ Ethanol, 2-[[4-(triphenylmethyl)phenyl]- thio]-	n.s.g.	263(4.14)	35-1065-76
4,8-Etheno-5H-indeno[5,6-b]thiophen- 5-one, 3a,4,4a,7a,8,8a-hexahydro- 4a,6-dimethyl-7,7a-diphenyl-	EtOH	216(4.16),242(3.89)	35-6405-76
2-Thiabicyclo[2.2.1]hept-5-en-7-one, 3-(2,4,6-cycloheptatrien-1-yl)- 1,4-dimethyl-5,6-diphenyl-	EtOH	250(3.03),266(3.02)	35-6405-76

Compound	Solvent	$\lambda_{max}(\log \epsilon)$	Ref.
$C_{27}H_{24}O_3$ Benzene, 1,3,5-tris(phenylmethoxy)-	EtOH	270(4.09)	35-5380-76
$C_{27}H_{24}O_6$ 4H-1-Benzopyran-4-one, 7-acetoxy-8-[(2-hydroxyphenyl)methyl]-5-methoxy-2-methyl-3-(phenylmethyl)-	EtOH	237(3.34),253(3.27), 262(4.30),295(3.77), 312(3.75)	44-1852-76
$C_{27}H_{24}O_{14}$ 3-Furanacetic acid, 4-(5,10-dihydro-2,3,7,8-tetramethoxy-5,10-dioxo-[1]-benzopyrano[5,4,3-cde][1]benzopyran-1-yl)tetrahydro-5-(methoxycarbonyl)-2-oxo-, methyl ester, (3α,4α,5α)-(±)-	dioxan	<u>255(4.7),362(4.0),</u> <u>375(4.0)</u>	5-2169-76
$C_{27}H_{25}Cl_2N_3O_4S_2$ Benzenesulfonamide, N,N'-(4,5,9,10-tetrahydro-7H-pyrrolo[3,2,1-de]phenanthridine-7,7a(8H)-diyl)bis[4-chloro-	EtOH	234(4.61),343(3.68)	39-2254-76C
$C_{27}H_{25}NO_5$ 1,3-Dioxolo[4,5-g]isoquinoline, 6-acetyl-5,6,7,8-tetrahydro-9-methoxy-5-[[3-(phenylmethoxy)phenyl]methylene]-, (Z)-	MeOH	250s(4.29),305(4.13)	12-2003-76
Elmerrillicine, N-acetyl-O-benzyl-	MeOH	232(4.53),282(4.18), 298(4.07)	12-2003-76
$C_{27}H_{25}N_3O_3S$ 4H-Indazol-4-one, 1,5,6,7-tetrahydro-1-(phenylmethyl)-3-phenyl-, O-[(4-methylphenyl)sulfonyl]oxime, (E)-	MeOH	228(4.58),262(4.04)	24-1898-76
$C_{27}H_{25}N_5$ 5H-Dibenzo[a,d]cycloheptene-10-carbonitrile, 5-[4-[[(6-methyl-2-pyridinyl)-methylene]amino]-1-piperazinyl-	MeOH	308(4.54)	87-1161-76
$C_{27}H_{26}ClN_2OS_2$ Benzothiazolium, 2-[2-[4-chloro-5-[2-(3-ethyl-2(3H)-benzothiazolylidene)ethylidene]-5,6-dihydro-2H-pyran-3-yl]-ethenyl]-3-ethyl-, iodide	EtOH	784(5.42)	103-0153-76
$C_{27}H_{26}ClN_2S_3$ Benzothiazolium, 2-[2-[4-chloro-5-[2-(3-ethyl-2(3H)-benzothiazolylidene)ethylidene]-5,6-dihydro-2H-thiopyran-3-yl]-ethenyl]-3-ethyl-, bromide	EtOH	780(5.39)	103-0153-76
$C_{27}H_{26}INO_5$ 1,3-Dioxolo[4,5-g]isoquinoline, 6-acetyl-5,6,7,8-tetrahydro-5-[[2-iodo-3-(phenylmethoxy)phenyl]methyl]-9-methoxy-	MeOH	212(4.88),280(3.70), 287(3.72)	12-2003-76
$C_{27}H_{26}N_2O_5$ Spiro[2H-1-benzopyran-2,2'-[2H]indole], 1',3'-dihydro-8-methoxy-5'-(4-methoxyphenyl)-1',3',3'-trimethyl-6-nitro-	EtOH	283(4.45)	103-0417-76

Compound	Solvent	$\lambda_{max}(\log \epsilon)$	Ref.
merocyanine form (cont.)	toluene	403(4.46),612(4.53)	103-0417-76
	EtOH	408(4.28),563(4.48)	103-0417-76
	dioxan	401(4.37),601(4.63)	103-0417-76
$C_{27}H_{26}N_2O_7$ 1,3-Dioxolo[4,5-g]isoquinoline, 6-acet-yl-5,6,7,8-tetrahydro-9-methoxy-5-[[2-nitro-3-(phenylmethyl)phenyl]-methyl]-, (±)-	MeOH	280(3.58)	12-2003-76
$C_{27}H_{26}N_4O_3$ Benzo[g]pteridine-2,4(3H,4aH)-dione, 5,10-dihydro-7,8,10-trimethyl-4a-(phenoxymethyl)-3-(phenylmethyl)-	50% MeOH-pH -0.5	270(4.14),300s(--), 390(3.71)	5-2037-76
	50% MeOH-pH 7, 10	300s(--),365(3.80)	5-2037-76
Spiro[2H-1-benzopyran-2,2'-[2H]indole], 1',3'-dihydro-3',3'-dimethyl-1'-(1-methylethyl)-6-[(4-nitrophenyl)azo]-	EtOH	295(4.05),400(4.20)	28-0691-76A
$C_{27}H_{26}N_4O_6$ 2,4(1H,3H)-Pteridinedione, 1-methyl-3-[2,3-O-(1-methylethylidene)-β-D-ribo-furanosyl]-6,7-diphenyl-	MeOH	224(4.43),279(4.24), 367(4.18)	24-3194-76
2,4(1H,3H)-Pteridinedione, 3-methyl-1-[2,3-O-(1-methylethylidene)-β-D-ribo-furanosyl]-6,7-diphenyl-	MeOH	226(4.42),270(4.21), 359(4.13)	24-3194-76
4(1H)-Pteridinone, 2-methoxy-1-[2,3-O-(1-methylethylidene)-β-D-ribofurano-syl]-6,7-diphenyl-	pH 7.0	225(4.42),269(4.18), 357(4.16)	24-3159-76
	MeOH	225(4.43),269(4.19), 357(4.14)	24-3159-76
$C_{27}H_{26}N_4O_8S$ 2,4(1H,3H)-Pteridinedione, 1-[2,3-O-(1-methylethylidene)-5-O-(methylsulfon-yl)-β-D-ribofuranosyl]-6,7-diphenyl-	pH 6.0	220(4.51),272(4.28), 356(4.23)	24-3159-76
	pH 10.0	220s(4.47),264(4.38), 356(4.28)	24-3159-76
	MeOH	222(4.53),272(4.30), 357(4.23)	24-3159-76
$C_{27}H_{26}O_6$ D-Glucitol, 1,3:2,4:5,6-tris-O-(phenyl-methylene)-, (1R,2S,5S)-	MeCN	266(2.79)	12-1859-76
$C_{27}H_{26}O_{10}$ 2-Propenoic acid, 3-(4-acetoxyphenyl)-2-acetoxy-1,3-propanediyl ester	n.s.g.	219(4.74),225(4.83), 285(4.99)	102-0811-76
$C_{27}H_{27}$ Pentacyclo[18.3.2.2^{9,12}.0^{4,15}.0^{6,17}]hep-tacosa-4,6(17),9,11,15,20,22,24,26-nonaen-1-ylium, (deloc-1,20,21,22-23,24,25)-, perchlorate	CH_2Cl_2	<u>232(4.3),260s(3.6), 315(3.5),425(3.5)</u>	88-3899-76
$C_{27}H_{27}Cl_2N_3O_4S_2$ Benzenesulfonamide, N,N'-(4,5,8,9,10,11-hexahydro-7H-pyrrolo[3,2,1-de]phenan-thridine-7a,11a-diyl)bis[4-chloro-	EtOH	235(4.49),261s(3.99), 318(3.54)	39-2254-76C
$C_{27}H_{27}F_3N_2O_3$ 4H-1-Benzopyran-4-one, 7-ethoxy-6-hexyl-	EtOH	260(4.20),315(3.90)	103-0914-76

Compound	Solvent	$\lambda_{max}(\log \epsilon)$	Ref.
3-(1-phenyl-1H-pyrazol-4-yl)-2-(tri-fluoromethyl)- (cont.)			103-0914-76
$C_{27}H_{27}NO_2$			
5(1H)-Quinolinone, 1-ethyl-4,6,7,8-tet-rahydro-7,7-dimethyl-4-(2-oxo-2-phen-ylethylidene)-	MeOH	419s(--),429(4.39)	24-1549-76
$C_{27}H_{27}N_2OS_2$			
Benzothiazolium, 3-ethyl-2-[2-[5-[2-(3-ethyl-2(3H)-benzothiazolylidene)eth-ylidene]-5,6-dihydro-2H-pyran-3-yl]-ethenyl]-, iodide	EtOH	764(5.43)	103-0153-76
$C_{27}H_{27}N_3O_2$			
2H-Pyrrol-2-one, 1,5-dihydro-4-(1-meth-ylethyl)-5-phenyl-3-[[phenyl[(phenyl-methylene)amino]methoxy]amino]-	EtOH	209(4.46),252(4.36), 262(4.27)	40-1100-76
2H-Pyrrol-2-one, 1,5-dihydro-5-phenyl-3-[[phenyl[(phenylmethylene)amino]-methoxy]amino]-4-propyl-	EtOH	209(4.35),243(4.35), 340(3.38)	40-1100-76
$C_{27}H_{27}N_3O_3$			
Spiro[2H-1-benzopyran-2,2'-[2H]indole], 1',3'-dihydro-8-methoxy-6-[(4-meth-oxyphenyl)azo]-1',3',3'-trimethyl-	EtOH	300(3.99),310(4.32), 380(4.25),392(4.20)	28-0691-76A
1,3,5-Triazine-2,4,6(1H,3H,5H)-trione, 1,3,5-tris(2-phenylethyl)-	THF	260(2.97)	95-0962-76
$C_{27}H_{27}N_5$			
Pyrido[2,3-b]pyrazine, 5-ethyl-1,2,3,5-tetrahydro-2,3-bis(2-methyl-1H-indol-3-yl)-, conjugate monoacid	n.s.g.	280(4.21),289(4.11), 355(4.11)	103-0948-76
$C_{27}H_{27}N_5O_5$			
4(1H)-Pteridinone, 2-(methylamino)-1-[2,3-O-(1-methylethylidene)-β-D-ribofuranosyl]-6,7-diphenyl-	pH 0.0	226(4.43),277(4.15), 358(4.16)	24-3159-76
	pH 7.0	225s(4.40),264(4.31), 356(4.21)	24-3159-76
	pH 14.0	227(4.42),281(4.21), 384(4.09)	24-3159-76
	MeOH	223s(4.35),271(4.21), 364(4.08)	24-3159-76
$C_{27}H_{28}N_2O_2$			
Methanone, [4-[(dimethylamino)methyl]-5-hydroxy-2-phenyl-1-propyl-1H-indol-3-yl]phenyl-	dioxan	231(4.39),288(4.16)	42-0017-76
$C_{27}H_{28}N_2O_4$			
Asperglaucide	CH_2Cl_2	238(3.9)	39-0578-76C
Benzenepropanamide, N-[1-[(acetyloxy)-methyl]-2-phenylethyl]-α-(benzoyl-amino)-, [S-(R*,R*)]-	MeOH	214(4.18),227s(4.07)	31-1106-76
$C_{27}H_{28}N_2O_5$			
4H-1-Benzopyran-2-carboxylic acid, 6-hexyl-7-hydroxy-4-oxo-3-(1-phenyl-1H-pyrazol-4-yl)-, ethyl ester	EtOH	275(4.36),330(3.99)	103-0914-76
1,3-Dioxolo[4,5-g]isoquinoline, 6-acet-	MeOH	267(3.71)	12-2003-75

Compound	Solvent	$\lambda_{max}(\log \epsilon)$	Ref.
yl-5-[[2-amino-3-(phenylmethoxy)phenyl]methyl]-5,6,7,8-tetrahydro-9-methoxy- (cont.)	MeOH-acid	255s(3.60),295(3.62)	12-2003-76
$C_{27}H_{28}O$ 2,5-Cyclohexadien-1-one, 2,6-bis(1,1-dimethylethyl)-4-(9H-fluoren-9-ylidene)-	isooctane	243(4.51),272(4.40), 280(4.40),424(4.59)	44-0214-76
$C_{27}H_{28}O_2$ Benzene, 1,1',1"-(1,4-diethoxy-1-methyl-2-butyn-1-yl-4-ylidene)tris-	EtOH	247s(2.83),253(2.87), 259(2.87),264(2.77)	56-0203-76
$C_{27}H_{28}O_4$ Propanoic acid, 2,2-dimethyl-, 1,2-bis-(4-methoxyphenyl)-2-phenylethenyl ester, cis	EtOH	246(4.35),303(4.21)	35-7726-76
trans	EtOH	239(4.31),300(4.24)	35-7726-76
$C_{27}H_{29}BrOS$ 2,5-Cyclohexadien-1-one, 4-[[(4-bromophenyl)thio]phenylmethylene]-2,6-bis-(1,1-dimethylethyl)-	hexane	387(4.44)	73-2614-76
$C_{27}H_{29}ClO$ 1(2H)-Acenaphthylenone, 5-chloro-2-[2,4,6-tris(1-methylethyl)phenyl]-	n.s.g.	201(4.80),218(4.73), 256(4.28),323(3.79), 342(3.70)	44-3599-76
$C_{27}H_{29}ClOS$ 2,5-Cyclohexadien-1-one, 4-[[(4-chlorophenyl)thio]phenylmethylene]-2,6-bis-(1,1-dimethylethyl)-	hexane	384(4.44)	73-2614-76
$C_{27}H_{29}FOS$ 2,5-Cyclohexadien-1-one, 4-[[(4-fluorophenyl)thio]phenylmethylene]-2,6-bis-(1,1-dimethylethyl)-	hexane	383(4.44)	73-2614-76
$C_{27}H_{29}NO$ Acridine, 2,7-bis(1,1-dimethylethyl)-9-phenoxy-	MeOH	209(4.57),260(5.18), 345(3.83),353(4.21)	39-0465-76C
Tricyclo[5.2.1.0^{2,6}]dec-4-en-3-one, 6-methyl-4,5-diphenyl-2-pyrrolidino-, 2,6-cis	MeOH	219s(4.18),294(4.13)	87-0414-76
$C_{27}H_{29}NO_{10}$ Dannorubicin, hydrochloride	pH 6.2	477(4.06),495(4.05)	69-2062-76
DNA complex	pH 6.2	506(3.83)	69-2062-76
$C_{27}H_{29}NO_{11}$ Adriamycin, hydrochloride	pH 6.2	476(4.06),495(4.06)	69-2062-76
DNA complex	pH 6.2	505(3.84)	69-2062-76
2-Naphthacenecarboxylic acid, 4-[(3-amino-2,3,6-trideoxy- -L-lyxo-hexopyranosyl)oxy]-1,2,3,4,6,11-hexahydro-2,5,12-trihydroxy-7-methoxy-6,11-dioxo-, methyl ester, hydrochloride, (2S-cis)-	MeOH	234(4.58),252(4.40), 288(3.93),478(4.08), 495(4.08),530(3.80)	87-0395-76

Compound	Solvent	$\lambda_{max}(\log \epsilon)$	Ref.
$C_{27}H_{29}N_3O_7S$ DL-Phenylalanine, N-[(phenylmethoxy)-carbonyl]-, 3'-ester with 4-thio-thymidine	n.s.g.	204(4.36),242(3.48), 333(4.25)	128-0351-76
$C_{27}H_{29}N_5O_5$ L-Alanine, 3-[1,1'-biphenyl]-4-yl-N-[N-(5-oxo-L-prolyl)-L-histidyl]-, methyl ester	HOAc	254(4.20)	94-3149-76
$C_{27}H_{30}N_2O_2$ 16,24-Cyclo-21-norchola-5,16,20(22),23-tetraene-20,23-dicarbonitrile, 3-acet-oxy-	MeCN	232(4.11),238s(4.10), 244s(4.04)	78-2997-76
$C_{27}H_{30}N_6O_3$ Adenosine, 5'-[bis(phenylmethyl)amino]-5'-deoxy-2',3'-O-(1-methylethylidene)-	EtOH	259(4.20)	87-0684-76
$C_{27}H_{30}O$ 1(2H)-Acenaphthylenone, 2-[2,4,6-tris-(1-methylethyl)phenyl]-	hexane	202(4.84),215(4.80), 256(4.31),315(3.81), 336(3.75)	35-1860-76
2,5-Cyclohexadien-1-one, 2,6-bis(1,1-di-methylethyl)-4-(diphenylmethylene)-	isooctane	368(4.41)	73-1676-76
$C_{27}H_{30}OS$ 2,5-Cyclohexadien-1-one, 2,6-bis(1,1-di-methylethyl)-4-[phenyl(phenylthio)-methylene]-	hexane	387(4.43)	73-2614-76
$C_{27}H_{30}O_2$ 1(2H)-Acenaphthylenone, 2-hydroxy-2-[2,4,6-tris(1-methylethyl)phenyl]-	hexane	218(4.92),255(4.20), 270(3.83),312(3.88), 337(3.77)	35-1860-76
$C_{27}H_{30}O_3S$ 2,5-Cyclohexadien-1-one, 2,6-bis(1,1-di-methylethyl)-4-[phenyl(phenylsulfon-yl)methylene]-	hexane	322(4.41)	73-2614-76
$C_{27}H_{30}O_9$ Glaucarubolone, 15-benzoate	EtOH	233(4.31),275(3.00), 282(2.83)	87-1130-76
$C_{27}H_{30}O_{15}$ Multiflorin B	MeOH	267(4.20),300s(3.96), 345(4.00)	95-1217-76
	MeOH-NaOMe	275(4.95),310s(3.89), 325s(3.90),386(4.33)	95-1217-76
	MeOH-AlCl$_3$	275(4.20),305s(3.91), 350(4.02),401(4.00)	95-1217-76
	MeOH-AlCl$_3$-HCl	276(4.18),303s(3.92), 344(4.01),398(3.92)	95-1217-76
$C_{27}H_{31}NO_4$ Ilicicolin H	EtOH	248(4.37),349(3.72)	88-3827-76
$C_{27}H_{31}N_2O_6$ Dibenzo[a,g]quinolizinium, 9-[acetyl-[2-(acetyloxy)ethyl]amino]-5,6-di-	EtOH	232(4.46),263(4.46), 339(4.42),420(3.72)	88-1595-76

Compound	Solvent	$\lambda_{max}(\log \epsilon)$	Ref.
hydro-2,3,10-trimethoxy-13-methyl-, bromide (cont.)			88-1595-76
$C_{27}H_{31}N_3O_2$ 16,24-Cyclo-21-norchola-5,16,20(22),23-tetraene-20,23-dicarbonitrile, 3-acetoxy-22-amino-, (3β)-	MeCN	229s(4.37),237(4.43), 252(3.91),265(3.75), 355(3.92)	78-2997-76
$C_{27}H_{32}N_4O_{15}$ 5H-1,2,3-Triazolo[4,5-b]pyridin-5-one, 2,4-dihydro-2,4-bis(2,3,5-tri-O-acetyl-β-D-ribofuranosyl)-	MeOH	234(3.84),301(4.17), 312(4.13)	23-1029-76
$C_{27}H_{32}O_2$ 1,2-Acenaphthylenediol, 1,2-dihydro-1-[2,4,6-tris(1-methylethyl)phenyl]-, trans	hexane	205(4.85),227(5.01), 287(3.99)	35-1860-76
$C_{27}H_{32}O_6$ Tricoccin S_7	MeOH	252(4.41)	88-3295-76
$C_{27}H_{32}O_9$ Glaucarubolone, dihydro-, 15-benzoate	EtOH	231(4.13),274(3.02), 281(2.95)	87-1130-76
$C_{27}H_{33}NO_3$ [1]Benzopyrano[3,4-c]pyrrol-4(1H)-one, 7-(1,2-dimethylheptyl)-2,3-dihydro-9-hydroxy-2-(phenylmethyl)-	EtOH	316(4.09)	87-0445-76
$C_{27}H_{34}O_2$ 1,2,3,5,7,11,13-Cyclotetradecaheptaen-9-yne-1-carboxylic acid, 4,8,11-tris-(1,1-dimethylethyl)-	THF	272s(3.74),283s(3.92), 309s(4.51),335(4.13), 476(4.28),537(3.03), 557(3.03),605(3.86)	88-1881-76
$C_{27}H_{34}O_7$ Tricoccin S_8	MeOH	256(4.42)	88-3295-76
$C_{27}H_{34}O_{15}$ Geniposide, pentaacetate	MeOH	237(4.10)	94-2644-76
$C_{27}H_{34}O_{17}$ Secogalioside, pentaacetate	MeOH	236(4.04)	1-0743-76
$C_{27}H_{35}N$ Benzeneacetonitrile, 4-heptyl-α-[(4-pentylphenyl)methylene]-, cis	n.s.g.	309(4.40)	65-2044-76
trans	n.s.g.	328(4.49)	65-2044-76
$C_{27}H_{35}NO_3$ 10(9H)-Acridineacetic acid, 9-oxo-, dodecyl ester	isoPrOH	214(4.36),253(4.70), 299(3.41),358s(3.64), 376(3.94),394(4.02)	44-3406-76
$C_{27}H_{36}O$ [14]Annulene, 1,8-bisdehydro-7,10,14-tris(1,1-dimethylethyl)-3-methoxy-	THF	244(3.89),249(3.97), 255(4.03),261(3.94), 304s(4.46),330(5.17), 438s(4.04),460(4.27),	88-2153-76

Compound	Solvent	λ_{max}(log ϵ)	Ref.
[14]Annulene, 1,8-bisdehydro-7,10,14-tris(1,1-dimethylethyl)-3-methoxy-(cont.)		528(2.62),549(2.68), 594(3.38)	88-2153-76
$C_{27}H_{36}O_3$ Violoxin	EtOH	230(0.56),262s(0.38), 300(0.08)	36-1549-76
$C_{27}H_{36}O_6$ Benzoic acid, 4-hydroxy-, 1,2,3,3a,4,5-8,8a-octahydro-3-hydroxy-6,8a-dimethyl-3-(1-methylethyl)-1-[(2-methoxy-1-oxo-2-butenyl)oxy]-4-azulenyl ester (akichenin)	n.s.g.	260(4.16)	105-0093-76
Cannabidiol, 6α-hydroxy-, triacetate	EtOH	265(2.56),273s(2.45)	39-0008-76C
Cannabidiol, 7-hydroxy-, triacetate	EtOH	265(2.83)	39-0008-76C
Cannabidiol, 10-hydroxy-, triacetate	EtOH	268(2.65),272s(2.54), 281(2.26)	39-0008-76C
$C_{27}H_{36}O_6S$ Pregna-3,5-dien-20-one, 3,16β-diacetoxy-17-(acetylthio)-	MeOH	234(4.33),241s(4.26)	24-0185-76
17α-	MeOH	235(4.31),242s(4.23)	24-0185-76
$C_{27}H_{36}O_{11}$ Quassimarin	EtOH	239(4.03)	44-3481-76
$C_{27}H_{36}O_{12}$ Nudiposide	EtOH	280(3.57)	94-2102-76
	EtOH-NaOH	256(--),297(--)	94-2102-76
$C_{27}H_{37}ClO_3$ Cyclohexanone, 2-[(4-chlorophenyl)hydroxymethyl]-3-methyl-6-[1-methyl-1-[6-methyl-3-(1-methylethylidene)-2-oxocyclohexyl]ethyl]-, [2S-[2α(S*),3α,6α(1R*,6S*)]]-	isooctane	223(4.13),253(3.93), 317(2.26)	22-0985-76
	EtOH	224(4.02),262(3.81), 315(2.23)	22-0985-76
$C_{27}H_{37}FO_4$ Spirost-9(11)-ene-3,12-dione, 11-fluoro-, (5α,25R)-	MeOH	247(4.17)	39-0101-76C
$C_{27}H_{37}NO_4$ Pregna-5,16,20-triene-16,20-dicarboxaldehyde, 3-acetoxy-21-(dimethylamino)-, (3β,20Z)-	isooctane	260(4.31),280(4.38)	88-4065-76
$C_{27}H_{38}O_3S$ Retinol, 11,12-dihydro-12-[(4-methylphenyl)sulfonyl]-	isoPrOH	228(4.41),262(4.23)	44-3287-76
$C_{27}H_{38}O_4$ Spirost-4-en-3-one, 9,11-epoxy-, (9β,11β,25R)-	EtOH	244.5(4.16)	95-0764-76
$C_{27}H_{38}O_4S$ Retinol, 11,12-dihydro-12-[(4-methoxyphenyl)sulfonyl]-	isoPrOH	242(4.39),260(4.22)	44-3287-76

Compound	Solvent	$\lambda_{max}(\log \epsilon)$	Ref.
$C_{27}H_{38}O_5$			
5β-Chola-8,11-dien-24-oic acid, 3,3-[1,2-ethanediylbis(oxy)]-	EtOH	300(3.88)	44-3840-76
Spirost-4-ene-3,11-dione, 4-hydroxy-, (25R)-	EtOH	277.5(4.09)	95-0863-76
$C_{27}H_{38}O_5S_2$			
Pregn-5-en-20-one, 3-acetoxy-16,17-bis-(acetylthio)-, (3β,16β)-	EtOH	235(3.75)	70-1947-76
$C_{27}H_{38}O_6$			
5β-Chol-9(11)-en-24-oic acid, 7α-acetoxy-3,12-dioxo-, methyl ester	EtOH	236(4.03)	44-3840-76
$C_{27}H_{38}O_7$			
14α-Card-20(22)-enolide, 3β,6α-diacetoxy-5α-hydroxy-	MeOH	216(4.20)	24-3304-76
14α-Card-20(22)-enolide, 3β,6β-diacetoxy-5α-hydroxy-	MeOH	217(4.17)	24-3304-76
Rastronol A acetonide	EtOH	234.5(3.96)	33-0772-76
$C_{27}H_{39}ClO_2$			
Cholesta-1,4-diene-3,6-dione, 7α-chloro-	EtOH	250(4.20)	56-0097-76
$C_{27}H_{39}NO_3$			
2H-[1]Benzopyrano[4,3-c]pyridin-10-ol, 2-(cyclopropylcarbonyl)-8-(1,2-dimethylheptyl)-1,3,4,5-tetrahydro-5,5-dimethyl-	EtOH	280(3.85)	87-0445-76
$C_{27}H_{39}NO_5S$			
Acetamide, N-[(3β,16β)-3-acetoxy-16-(acetylthio)-20-oxopregn-5-en-17-yl]-	EtOH	236(3.55)	70-1101-76
$C_{27}H_{40}NO_2$			
2H-[1]Benzopyrano[4,3-c]pyridinium, 8-(1,2-dimethylheptyl)-1,3,4,5-tetrahydro-10-hydroxy-2,5,5-trimethyl-2-(2-propynyl)-, iodide	EtOH	380(3.97)	87-0445-76
$C_{27}H_{40}O_2$			
Cholesta-1,4-dien-3-one, 6α,7α-epoxy-	EtOH	247(4.13)	56-0097-76
Cholesta-1,4-dien-3-one, 6β,7β-epoxy-	EtOH	250(4.18)	56-0097-76
Cholesta-4,6-dien-3-one, 8,9-epoxy-	hexane	313(4.17),328(4.26), 345(4.19)	65-1598-76
$C_{27}H_{40}O_4$			
5β-Spirost-2-en-11-one, 9α-hydroxy-	EtOH	311(1.54)	95-0764-76
Spirost-4-en-3-one, 4-hydroxy-, (25R)-	EtOH	278(4.11)	95-0863-76
4,8-Undecadienoic acid, 11-(3,4-dihydro-6-hydroxy-2,5,7,8-tetramethyl-2H-1-benzopyran-2-yl)-4,8-dimethyl-, methyl ester, [R-(E,E)]-	EtOH	215s(4.02),291(3.51)	33-0290-76
$C_{27}H_{40}O_5$			
5β-Chol-7-en-24-oic acid, 3α-acetoxy-6-oxo-, methyl ester	EtOH	245(4.17)	12-1815-76
Spirostane-2,11-dione, 9-hydroxy-, (5β,25R)-	EtOH	310(1.57)	95-0764-76

Compound	Solvent	$\lambda_{max}(\log \epsilon)$	Ref.
$C_{27}H_{40}O_6$			
5β-Chol-7-en-24-oic acid, 3α-acetoxy-14α-hydroxy-6-oxo-, methyl ester	EtOH	242(4.11)	12-1815-76
5β-Chol-8-en-24-oic acid, 3,3-[1,2-ethanediylbis(oxy)]-11α-hydroxy-7-oxo-, methyl ester	EtOH	250(3.93)	44-3840-76
Spirostan-11-one, 2,9-epidioxy-2-hydroxy-, (2α,5β,25R)-	EtOH	271(1.80)	95-0764-76
$C_{27}H_{40}O_8$			
5α-Androstan-6-one, 2β,3β,17β-triacetoxy-4-hydroxy-4,4a-dimethyl-	EtOH	310(1.52)	22-1865-76
$C_{27}H_{41}ClO_2$			
Cholesta-1,4-dien-3-one, 6α-chloro-7β-hydroxy-	EtOH	246(4.18)	56-0097-76
Cholesta-1,4-dien-3-one, 7α-chloro-6β-hydroxy-	EtOH	247(4.20)	56-0097-76
$C_{27}H_{41}NO_2$			
2H-[1]Benzopyrano[4,3-c]pyridin-10-ol, 2-(cyclopropylmethyl)-1,3,4,5-tetrahydro-8-(1,2-dimethylheptyl)-5,5-dimethyl-	EtOH	280(4.05)	87-0445-76
$C_{27}H_{41}NO_3$			
2(1H)-Pyridinone, 5-cyclohexyl-3-[(decahydro-4,7-dimethyl-1-propyl-2-naphthalenyl)carbonyl]-4-hydroxy-	EtOH	235(3.95),280(3.70), 340(3.90)	88-3827-76
Severtzidinedione	n.s.g.	250(2.60),305(2.07)	105-0320-76
$C_{27}H_{41}NO_4$			
2(1H)-Pyridinone, 3-[(decahydro-4,7-dimethyl-1-propyl-2-naphthalenyl)carbonyl]-4-hydroxy-5-(4-hydroxycyclohexyl)-	EtOH	235(3.95),280(3.70), 340(3.90)	88-3827-76
$C_{27}H_{42}$			
Cholesta-2,4,6-triene	isoPrOH	296(4.86),306(4.92), 320(4.72)	94-0375-76
$C_{27}H_{42}N_4O$			
3-Aza-A-homocholest-4a-eno[3,4-d]tetrazol-6-one	MeOH	260(4.16)	39-1210-76C
7a-Aza-B-homocholest-5-eno[7a,7-d]tetrazol-4-one	EtOH	272(4.42)	12-0447-76
$C_{27}H_{42}N_8$			
3,6-Diaza-AB-bishomocholest-4a-eno[3,4-d][6,7-d]bistetrazole	EtOH	246(4.29)	39-1210-76C
$C_{27}H_{42}O$			
Cholesta-5,7,25-trien-3β-ol	EtOH	272(--),282(4.04), 294(--)	39-0731-76C
3α,5α-Cyclocholest-7-en-6-one	EtOH	249(4.09)	44-1064-76
$C_{27}H_{42}O_2$			
Cholesta-1,7-dien-3-one, 2-hydroxy-, (3α)-	EtOH	268.5(3.93)	95-0863-76
Cholest-4-ene-3,6-dione	EtOH	255(4.04)	2-0802-76

Compound	Solvent	$\lambda_{max}(\log \epsilon)$	Ref.
5α-Cholest-8(14)-ene-3,15-dione	EtOH	259(4.22)	88-3775-76
deuterated	EtOH	259(4.22)	88-3775-76
Cholest-5-en-7-one, 3,4-epoxy-	EtOH	241(3.46)	2-0936-76
3α,5-Cyclo-5α-cholest-7-en-6-one, 7-hydroxy-	EtOH	235(3.54)	73-2788-76
$C_{27}H_{42}O_4$			
24-Nor-5β-cholest-7-ene-3β,25-diol, 6-oxo-, 3-formate	EtOH	246(4.05)	12-1815-76
$C_{27}H_{42}O_5$			
26,27-Dinorergost-7-en-6-one, 3-(formyloxy)-14,24-dihydroxy-, (3β,5β)-	EtOH	243(4.10)	12-1815-76
$C_{27}H_{42}O_5S$			
Cyclopentanemethanol, 3-[2-(1,4-dioxaspiro[4.5]dec-8-yl)-1-ethylbutyl]-, 4-methylbenzenesulfonate	EtOH	226(4.12)	5-1222-76
$C_{27}H_{42}O_7$			
1-Naphthaleneacetic acid, 8a-(ethoxycarbonyl)decahydro-2-(hydroxymethylene)-5,5-dimethyl-α-(3-methyl-1-oxobutyl)-3-oxo-, 1,1-dimethylethyl ester (isomeric mixture)	EtOH base	288(3.88) 306(4.20)	44-1005-76 44-1005-76
isomer B	EtOH	285(3.85)	44-1005-76
$C_{27}H_{43}ClN_4$			
7a-Aza-B-homocholest-5-eno[7a,7-d]tetrazole, 3β-chloro-	EtOH	240(4.13)	12-0447-76
$C_{27}H_{43}ClO$			
Cholest-1-en-3-one, 2-chloro-	EtOH	250(3.98)	56-2087-76
$C_{27}H_{43}NO_2$			
3-Aza-A-homocholest-4a-ene-4,6-dione	MeOH	234(4.20)	39-1210-76C
$C_{27}H_{43}NO_3$			
Imperialine	MeOH	290(1.46)	88-3161-76
Veratraman-11-one, 13,17-didehydro-5,6,12,13-tetrahydro-3,23-dihydroxy-, (3β,5α,23β)-	MeOH	205(4.00)	39-1312-76C
$C_{27}H_{43}N_3O_5$			
2,2'-Anhydro-1-(3'-elaidoyl-β-D-arabinofuranosyl)cytosine, hydrochloride	MeOH-acid	233(4.01),264(4.04)	87-0654-76
2,2'-Anhydro-1-(3'-oleoyl-β-D-arabinofuranosyl)cytosine, hydrochloride	MeOH-acid	233(4.01),264(4.05)	87-0654-76
$C_{27}H_{43}N_3O_6$			
2,2'-Anhydro-1-(3',5'-dinonanoyl-β-D-arabinofuranosyl)cytosine, hydrochloride	MeOH-acid	233(4.00),263(4.05)	87-0663-76
$C_{27}H_{44}$			
Cholesta-2,4-diene	hexane	266(3.78),275(3.76)	23-3508-76
Cholesta-3,5-diene	hexane	230(4.39),235(4.42), 244(4.24)	94-0487-76
	hexane	230(4.38),236(4.42), 244(4.24)	94-0859-76

Compound	Solvent	$\lambda_{max}(\log \epsilon)$	Ref.
Cholesta-3,5-diene (cont.)	EtOH	228(4.26),235(4.28), 243(4.09)	39-1442-76C
	n.s.g.	228(--),235(4.26), 243(--)	78-0797-76
Cholesta-4,6-diene	n.s.g.	234(--),240(4.31)	78-0797-76
$C_{27}H_{44}N_4$			
7a-Aza-B-homocholest-5-eno[7a,7-d]tetra- zole	EtOH	243(4.10)	12-0447-76
$C_{27}H_{44}O$			
Cholesta-5,7-dien-3α-ol	EtOH	252s(3.63),262s(3.86), 269(3.98),280(4.02), 291(3.79)	44-1067-76
5α-Cholest-7-en-6-one	MeOH	246(4.203)	13-0609-76A
5β-Cholest-7-en-6-one	EtOH	246(4.16)	12-1815-76
$C_{27}H_{44}O_2$			
Cholest-1-en-3-one, 2-hydroxy-	EtOH	270.5(3.93)	95-0764-76
Cholest-7-en-6-one, 5α-hydroxy-	MeOH	249(4.109)	13-0609-76A
Cholest-7-en-6-one, 14-hydroxy-, (5α,14α)-	MeOH	242(4.134)	13-0609-76A
(5β,14α)-	EtOH	243(4.09)	12-1815-76
Cholest-8(14)-en-15-one, 3-hydroxy-, (3β,5α)-	EtOH	263(4.06)	130-0001-76
$C_{27}H_{44}O_3$			
Cholesta-5,7-diene-3β,24ξ,25-triol	EtOH	272(--),282(4.07), 294(--)	39-0727-76C
$C_{27}H_{44}O_4$			
Cholest-7-en-6-one, 2β,3β,5α-trihydroxy-	EtOH	250(4.05)	13-0609-76A
$C_{27}H_{44}O_5$			
1,2-Secocholestane-1,2-dioic acid, 3-oxo-	EtOH	313(1.78)	56-0815-76
$C_{27}H_{44}O_7$			
Ecdysterone	EtOH	250(4.05)	105-0494-76
Inokosterone	EtOH	250(4.07)	105-0494-76
$C_{27}H_{45}NO_2$			
25-Isosolafloridine	EtOH	239(2.41)	88-3653-76
$C_{27}H_{45}NO_3$			
Edpetisinine derivative V	n.s.g.	305(1.63)	105-0615-76
$C_{27}H_{45}N_3O_5$			
2,2'-Anhydro-1-(3'-stearoyl-β-D-arabino- furanosyl)-, hydrochloride	MeOH-acid	232(4.00),264(4.03)	87-0654-76
$C_{27}H_{45}N_3O_6$			
Cytosine, 1-(3'-elaidoyl-β-D-arabino- furanosyl)-	MeOH-acid	212(3.98),284(4.16)	87-0667-76
Cytosine, 1-(3'-oleoyl-β-D-arabino- furanosyl)-	MeOH-acid	212(4.00),284(4.16)	87-0667-76
$C_{27}H_{45}N_3O_7$			
2(1H)-Pyrimidinone, 4-amino-1-[3,5-bis- O-(1-oxononyl)-β-D-arabino-furano- syl]-, hydrochloride	MeOH-acid	211(4.02),283(4.19)	87-0667-76

Compound	Solvent	$\lambda_{max}(\log \epsilon)$	Ref.
$C_{27}H_{46}N_2O$ Solacallinidine	EtOH	239(2.39)	88-3653-76
$C_{27}H_{47}N_3O_3$ A-Nor-3,5-secocholestan-3-oic acid, 5-[(aminocarbonyl)hydrazono]-	EtOH	229(4.12)	49-0511-76
$C_{27}H_{47}N_3O_6$ Cytosine, 1-(3'-stearoyl-β-D-arabino- furanosyl)-	MeOH-acid	212(4.01),283(4.15)	87-0667-76
$C_{27}H_{48}O_{10}$ Erythronolide A, 12-deoxy-3-O-(2,6-di- deoxy-α-L-arabino-hexopyranosyl)-	MeOH	280(1.69)	78-2375-76
Erythronolide A, 12-deoxy-3-O-(2,6-di- deoxy-α-L-ribohexopyranosyl)-	MeOH	287(1.56)	78-2375-76

Compound	Solvent	λ_{max}(log ϵ)	Ref.
$C_{28}H_{12}O_8$ Phenanthro[1,10,9,8-opqra]perylene-7,14-dione, 1,3,4,6,8,13-hexahydroxy-	DMSO-1% pyridine	340(4.68),381(4.22), 440(4.22),468(4.32), 502(4.17),537(4.52), 580(4.80)	12-1523-76
$C_{28}H_{14}BrN_3O_2$ 9H-Benzo[e]quino[2,3-i]perimidine-9,15-(14H)-dione, 11-bromo-6-phenyl-	DMF	278(4.190),352(3.939), 462(3.571),594(3.942)	80-0257-76
$C_{28}H_{14}N_4O_4$ 9H-Benzo[e]quino[2,3-i]perimidine-9,15-(14H)-dione, 12-nitro-6-phenyl-	DMF	270(3.929),604(3.341)	80-0257-76
$C_{28}H_{15}N_3O_2$ 9H-Benzo[e]quino[2,3-i]perimidine-9,15-(14H)-dione, 6-phenyl-	DMF	270(3.950),354(3.655), 542(3.392)	80-0257-76
$C_{28}H_{16}$ Benzo[a]naphtho[1,2-e]aceanthrylene	benzene	294(4.74),319(4.63), 385(3.91),407(4.06)	24-1991-76
Benzo[b]naphtho[1,2,3,4-def]chrysene	benzene	298(4.73),310(4.94), 323(4.53),340(4.65), 378s(--),397(3.95), 420(4.20),447(4.25)	24-1991-76
$C_{28}H_{16}Cl_2N_2$ Benzonitrile, 2,2'-(2,6-naphthalenediyldi-2,1-ethenediyl)bis[4-chloro-	DMF	364(4.79)	33-0819-76
Benzonitrile, 2,2'-(2,6-naphthalenediyldi-2,1-ethenediyl)bis[6-chloro-	DMF	378(4.75)	33-0819-76
Benzonitrile, 4,4'-(2,6-naphthalenediyldi-2,1-ethenediyl)bis[3-chloro-	DMF	386(3.79)	33-0819-76
$C_{28}H_{16}Cl_4O_2$ 11H-Benzo[b]indeno[1,2-e][1,4]dioxin, 6,7,8,9-tetrachloro-11-(diphenylmethylene)-4b,10a-dihydro-, cis	C_6H_{12}	217(5.17),234(4.74), 304(4.57)	78-0147-76
1,4-Ethano-1H-fluorene-10,11-dione, 1,2,3,4-tetrachloro-9-(diphenylmethylene)-4,4a,9,9a-tetrahydro-, (1α,4α,4aβ,9aβ)-	CH_2Cl_2	280(4.20)	78-0147-76
$C_{28}H_{16}O_2S$ Anthra[2,3-c]thiophene-5,10-dione, 1,3-diphenyl-	CH_2Cl_2	277(4.63),319(4.65), 460(4.13)	28-0607-76A
$C_{28}H_{17}ClO$ Indeno[2,1-b]pyran, 2-(4-chlorophenyl)-9-(2-naphthalenyl)-	dioxan	227(4.67),271(4.58), 349(4.44),565(3.48)	78-2225-76
$C_{28}H_{18}$ Acephenanthrylene, 4,5-diphenyl-	MeOH	206(4.64),232(4.58), 250(4.67),266s(4.60), 302(4.18),313(4.27), 339(4.05),355(4.10), 377(4.05),412(3.48)	54-0165-76
Benzo[g]chrysene, 9-phenyl-	MeOH	229(4.45),256s(4.51), 272s(4.65),289(4.75), 293(4.79),315s(4.11), 326(4.07),342s(3.89)	54-0165-76

Compound	Solvent	$\lambda_{max}(\log \epsilon)$	Ref.
Benzo[c]phenanthrene, 2-(4-phenyl-1-buten-3-ynyl)-, trans	MeOH	218(4.47),235(4.44), 249(4.43),274(4.57), 311(4.65),319s(4.61), 347(4.59)	39-1104-76B
$C_{28}H_{18}Br_2$ Benzene, 1,1'-(1,4-diphenyl-1,2,3-buta-triene-1,4-diyl)bis[4-bromo-	benzene C_6H_{12}	314s(3.59),433(4.63) 275s(4.59),282(4.62), 315s(3.58),427(4.64)	19-0089-76 19-0089-76
$C_{28}H_{18}Cl_2$ Benzene, 1,1'-(1,4-diphenyl-1,2,3-buta-triene-1,4-diyl)bis[4-chloro-	benzene C_6H_{12}	313s(3.59),428(4.64) 279(4.61),314(3.59), 423(4.66)	19-0089-76 19-0089-76
$C_{28}H_{18}Cl_2N_6$ 2H-Benzotriazole, 2-[3-chloro-4-[2-[2-(4-chlorophenyl)-5-phenyl-2H-1,2,3-triazol-4-yl]ethenyl]phenyl]-	DMF	364(4.71)	33-2469-76
$C_{28}H_{18}I_2$ Benzene, 1,1'-(1,4-diphenyl-1,2,3-buta-triene-1,4-diyl)bis[4-iodo-	benzene C_6H_{12}	436.2(4.63) 286(4.57),432(4.65)	19-0089-76 19-0089-76
$C_{28}H_{18}O_2S_2$ [2,2'-Bibenzo[b]thiophene]-3,3'(2H,2'H)-dione, 2,2'-diphenyl-	EtOH	207(4.75),253(4.70), 330(4.20)	111-0533-76
$C_{28}H_{18}O_4$ [2,2'-Bibenzofuran]-3,3'(2H,2'H)-dione, 2,2'-diphenyl-	EtOH	234(4.40),241(4.43)	111-0533-76
$C_{28}H_{18}O_8$ [1,1'-Bianthracene]-4,4',9,9',10,10'-(1H,1'H)-hexone, 2,2',3,3'-tetra-hydro-5,5'-dihydroxy-Bisdeoxynorrugulosin	dioxan dioxan	286(4.63),433(4.18) 244(4.07),275(3.83), 295(3.58),307(3.53), 387(3.98)	78-0333-76 78-0333-76
$C_{28}H_{18}S_2$ Dispiro[9H-fluorene-9,2'(3'H)-[1,4]di-thiin-3',9"-[9H]fluorene]	MeOH	224s(4.72),272(4.36), 298s(3.81),306s(3.70)	24-3775-76
$C_{28}H_{19}BrO$ Indeno[2,1-b]pyran, 2-(4-bromophenyl)-9-(3,4-dihydro-2-naphthalenyl)-	dioxan	273(4.31),337(4.53), 358(4.55),577(3.50)	78-2225-76
$C_{28}H_{19}ClN_6$ 2H-Benzotriazole, 2-[3-chloro-4-[2-(2,5-diphenyl-2H-1,2,3-triazol-4-yl)ethen-yl]phenyl]-	DMF	363(4.71)	33-2469-76
$C_{28}H_{19}ClO$ Indeno[2,1-b]pyran, 2-(4-chlorophenyl)-9-(3,4-dihydro-2-naphthalenyl)-	dioxan	230(4.35),271(4.34), 338(4.55),358(4.57), 585(3.50)	78-2225-76
$C_{28}H_{19}NO_2$ 1(3H)-Isobenzofuranone, 3-[10-(phenyl-amino)-9(10H)-anthracenylidene]-	dioxan	352(4.29)	104-0168-76

Compound	Solvent	λ_{max}(log ϵ)	Ref.
8-Quinolinol, 2,3-diphenyl-, benzoate	EtOH	233(4.73),262(4.67), 320s(3.86)	95-0968-76
$C_{28}H_{19}N_3$ 1H-Indol-3-amine, 2-phenyl-N-(2-phenyl-3H-indol-3-ylidene)-	MeOH	241(4.60),260(4.60), 330(4.26),570(4.08)	23-1020-76
	MeOH-HCl	230(4.63),291(4.56), 360(4.18),645(4.43)	23-1020-76
$C_{28}H_{19}N_3O_3S$ 4H-Pyrimido[2,1-b]benzothiazol-4-one, 2,3-dihydro-2-(4-nitrophenyl)-3,3-diphenyl-	EtOH	240(4.40),277s(4.10)	94-2532-76
$C_{28}H_{19}N_3O_4$ 4H-Pyrimido[2,1-b]benzoxazol-4-one, 2,3-dihydro-2-(4-nitrophenyl)-3,3-diphenyl-	EtOH	275(4.19)	94-2532-76
1H-Pyrrolo[3',4':4,5]pyrido[1,2-b]inda-zole-11-carboxylic acid, 2,3-dihydro-1,3-dioxo-2,4-diphenyl-, ethyl ester	MeOH	207(4.26),257(3.99), 293(4.35),398(3.87)	44-0129-76
$C_{28}H_{19}N_5O$ 1,3,4-Oxadiazole, 2-(1-naphthalenyl)-5-[4-[2-(2-phenyl-2H-1,2,3-triazol-4-yl)ethenyl]phenyl]-	DMF	351(4.79)	33-2469-76
1,3,4-Oxadiazole, 2-(2-naphthalenyl)-5-[4-[2-(2-phenyl-2H-1,2,3-triazol-4-yl)ethenyl]phenyl]-	DMF	350(4.77)	33-2469-76
$C_{28}H_{20}$ Benzene, 1,1',1'',1'''-(1,2,3-butatriene-1,4-diylidene)tetrakis-	benzene	317(3.46),422(4.59)	19-0089-76
1H-Indene, 1-(diphenylmethylene)-3-phen-yl-	EtOH	222(4.37),248(4.42), 298(4.10),349(4.21)	56-1301-76
Naphthalene, 1,2,3-triphenyl-	C_6H_{12}	248(4.70)	5-0022-76
$C_{28}H_{20}BrNO_3S$ Methanone, [1-(4-bromophenyl)-1-hydroxy-4-methyl-1H-pyrrolo[2,1-c][1,4]thiaz-ine-7,8-diyl]bis[phenyl-	MeOH	235(4.48),285(4.23)	44-0187-76
$C_{28}H_{20}Cl_8O_4$ 13H-6,13a-Ethenopentaleno[1,2-b:5,4-b']-bis[1,5]benzodioxepin, 1,2,3,4,8,9,10-11-octachloro-5a,5b,6,6a,13b,15-hexa-hydro-13,13,15,15-tetramethyl-, (5aα,5bα,6β,6aα,13aβ,13bα)-	MeCN	218(5.19),295(3.43)	78-0147-76
(5aα,5bβ,6α,6aβ,13aα,13bβ)-	MeCN	217(5.26),298(3.48)	78-0147-76
$C_{28}H_{20}N_2O$ 2,5-Imino-2H-2-benzazepin-1(5H)-one, 3,4,10-triphenyl-	EtOH	207(4.57),250(4.30), 280(4.11),297(4.08), 350(3.95)	39-2281-76C
Spiro[3H-naphtho[2,1-b]pyran-3,5'(4'H)-pyrido[2,3,4,5-lmn]phenanthridine], 4',10'-dimethyl-	dioxan	242(4.80),255s(4.64), 310s(4.11),362(4.00)	103-0273-76
	HOAc	293(4.13),342(4.02), 390(3.97),470(3.98)	103-0273-76

Compound	Solvent	$\lambda_{max}(\log \epsilon)$	Ref.
$C_{28}H_{20}N_2O_3$ 2H-1,4-Benzoxazine, 2,2'-oxybis[3-phenyl-	EtOH	235(4.36),286(4.50), 320(4.34)	78-2033-76
$C_{28}H_{20}N_2O_3S$ 3-Pyridinecarboxamide, 1,2-dihydro-6-hydroxy-1-(1-naphthalenyl)-N-phenyl-4-(phenylthio)-	MeOH	337(4.45)	120-0103-76
$C_{28}H_{20}N_4$ 1,2-Benzenediamine, N-(2,2-diphenylcyclobuta[b]quinoxalin-1(2H)-ylidene)-	EtOH	261(4.61),353(4.24), 410(3.78)	5-0284-76
	CHCl$_3$	263(4.67),354(4.30), 404(3.77)	5-0284-76
$C_{28}H_{20}N_4O$ Naphth[1,2-d]oxazole, 2-[4-[2-(5-methyl-2-phenyl-2H-1,2,3-triazol-4-yl)ethenyl]phenyl]-	DMF	322(4.37),374(4.75)	33-2469-76
$C_{28}H_{20}N_4O_2$ Methanone, (2,5-dihydro-2,5-diphenyl-1,2,4,5-tetrazine-3,6-diyl)bis[phenyl-	EtOH	265(4.542),290s(4.412), 390(2.749)	18-0321-76
Naphth[1,2-d]oxazole, 2-[4-[2-[2-(4-methoxyphenyl)-2H-1,2,3-triazol-4-yl]ethenyl]phenyl]-	DMF	322(4.35),371(4.82)	33-2469-76
$C_{28}H_{20}N_8Ni$ Nickel, [4,15-dihydro-1-methyl-3-phenyl-3H-tribenzo[c,f,j]pyrazolo[3,4-m]-[1,2,5,8,9,12]hexaazacyclotetradecinato(2-)-N^4,N^{10},N^{15},N^{21}]-, (SP-4-2)-	isooctane	230(4.21),260(4.18), 305(4.15),395(3.87), 470(4.08),710(3.34)	103-0562-76
$C_{28}H_{20}O$ Indeno[2,1-b]pyran, 9-(1H-inden-2-yl)-2-(4-methylphenyl)-	dioxan	274(4.36),322(4.55), 360(4.56),590(3.56)	78-2225-76
Indeno[2,1-b]pyran, 9-(1H-inden-2-yl)-4-methyl-2-phenyl-	dioxan	265(4.32),322(4.37), 356(4.26),577(3.48)	78-2219-76
$C_{28}H_{20}O_8$ 3,4,9,10-Perylenetetrol, tetraacetate	n.s.g.	251(4.48),259(4.48), 295(3.47),403(4.15), 425(4.46),452(4.54)	39-2149-76C
$C_{28}H_{20}S$ Thiophene, tetraphenyl-	EtOH	230(4.49),258(4.37), 306(4.24)	18-2158-76
$C_{28}H_{21}BrN_8Ni$ Nickel, [N-[2-[(2-aminophenyl)azo]phenyl]-4-[(2-bromophenyl)azo]-3-methyl-1-phenyl-1H-pyrazol-5-aminato(2-)]-, (SP-4-2)-	isooctane	254(4.23),370(4.05), 435(3.83),555(3.24), 665(3.16)	103-0562-76
$C_{28}H_{21}ClN_2O_8S$ 1,2,4-Thiadiazol-3(2H)-one, 5-chloro-2-(2,3,5-tri-O-benzoyl-β-D-ribofuranosyl)-	EtOH	235(3.59),280(3.10)	4-0169-76

Compound	Solvent	λ_{max}(log ϵ)	Ref.
$C_{28}H_{21}NO_3S$ Methanone, (1-hydroxy-4-methyl-1-phenyl- 1H-pyrrolo[2,1-c][1,4]thiazine-7,8-di- yl)bis[phenyl-	MeOH	245(4.40),287(4.10)	44-0187-76
$C_{28}H_{21}NO_7$ Spiro[1,3-dioxolane-2,12'(7'H)-[6H]naph- thaceno[1,12-bc]furan]-10-carboxamide, 6'a,12'a-dihydro-9',11'-dihydroxy-6'- oxo-1'-phenyl-, cis	EtOH	225(4.48),256s(4.36), 265(4.37),275(4.29), 295(4.11),307(4.10), 348(4.11)	39-0503-76C
$C_{28}H_{21}N_3O_4S$ 4,11-Epithiopyrrolo[3',4':4,5]pyrido- [1,2-b]indazole-11-carboxylic acid, 2,3,3a,4-tetrahydro-1-hydroxy-3-oxo- 2,4-diphenyl-, ethyl ester	MeOH	207(3.91),230s(3.45), 287(3.50),298s(3.38), 347(3.31)	44-0129-76
$C_{28}H_{22}$ Cyclopenta[def]fluorene, 4,8-dihydro- 4,8-bis(2-methylphenyl)-	EtOH	271(4.07),287(4.17), 295(4.05),305(3.54)	35-8268-76
Phenanthrene, 4-(3,5-dimethylphenyl)- 3-phenyl-	MeOH	290s(4.26),298(4.21), 336(2.44),352(2.55)	39-1115-76B
Tetrabenz[c,fg,j,mn]octalene, 4,9,13,18- tetrahydro-	C_6H_{12}	240s(3.98),265(3.26), 273(3.14)	35-8268-76
$C_{28}H_{22}N_2$ 1H-Imidazole, 2,4,5-triphenyl-1-(phenyl- methyl)-	EtOH	279(4.3)	24-0906-76
$C_{28}H_{22}N_4$ 5,5'-Bi-3H-indazole, 3,3'-dimethyl- 3,3'-diphenyl-	isoPrOH	224s(4.44),313(4.37)	4-0033-76
$C_{28}H_{22}N_4O_3$ Acetamide, N-acetyl-N-[2-(acetylbenzo- [a]phenazin-5-ylamino)phenyl]-	dioxan	225(4.71),284(4.59), 366s(3.83),384(4.04), 392(4.03),406(4.10)	49-0879-76
	50% dioxan	223(4.63),285(4.59), 368s(3.82),386(4.04), 392(4.04),407(4.11)	49-0879-76
$C_{28}H_{22}O$ Furan, 2,5-dihydro-2,3,4,5-tetraphenyl-	C_6H_{12}	240(4.19),253(4.21), 270s(3.94),346(2.18)	78-1549-76
Indeno[2,1-b]pyran, 9-(1-acenaphthylen- yl)-3-ethyl-2,4-dimethyl-	dioxan	230(4.75),288(4.39), 318(4.33),518(3.80)	78-2225-76
$C_{28}H_{22}O_7$ Isokhellactone, dibenzoate	EtOH	220(4.10),246(3.35), 257(3.50),300s(4.20), 324(4.30)	102-1293-76
$C_{28}H_{22}S_4$ 1,3-Butadiene, 1,4-bis[5-[4-(2-thienyl)- 1,3-butadienyl]-2-thienyl]-	DMF	460(4.54),482(4.62), 510s(4.43)	126-1857-76
$C_{28}H_{23}BrN_8$ 1H-Pyrazol-5-amine, N-[2-[(2-aminophen- yl)azo]phenyl]-4-[(2-bromophenyl)azo]- 3-methyl-1-phenyl-	isooctane	228(4.52),256(4.43), 328(4.41),400(4.38)	103-0562-76

Compound	Solvent	$\lambda_{max}(\log \epsilon)$	Ref.
$C_{28}H_{23}NO_2$ Pyridinium, 2,5-dimethyl-4-phenyl-, 1-benzoyl-2-oxo-2-phenylethylide	EtOH	284(4.33),320(4.10)	103-0424-76
$C_{28}H_{23}N_3O_7$ 1,2,4-Triazine-3,5(2H,4H)-dione, 2- [2,3-0-carbonyl-5-0-(triphenyl- methyl)-β-D-ribofuranosyl]-	MeOH	261(3.73)	73-2110-76
$C_{28}H_{23}N_4O_4P$ Phosphonium, triphenyl-, 3-[(2,4-di- nitrophenyl)azo]-2-butenylide	DMF	287(--),387(3.76), 567(4.75)	65-2135-76
$C_{28}H_{24}$ 1H-Indene, 2,3-dihydro-1,2-diphenyl- 3-(phenylmethyl)-, cis	MeOH	261(3.14),267(3.16), 275(3.03)	5-0022-76
trans	MeOH	260(3.10),264(3.09), 266(3.12),274(2.95)	5-0022-76
Naphthalene, 1,2,3,4-tetrahydro-1,2,3- triphenyl-, m. 114°	MeOH	255(2.89),261(2.98), 267(2.93),275(2.89)	5-0022-76
isomer m. 147°	MeOH	261(3.05),263(3.06), 270(2.99),276(2.79)	5-0022-76
Phenanthrene, 6-[4-(3,5-dimethylphenyl)- 1-buten-3-ynyl]-2,4-dimethyl-, cis	MeOH	246(4.67),254(4.76), 277(4.43),288(4.39), 332s(4.36),345(4.42), 363s(4.25)	39-1104-76B
trans	MeOH	248(4.54),254(4.60), 279(4.50),290(4.50), 334(4.61),347(4.70), 365s(4.55)	39-1104-76B
$C_{28}H_{24}NO_2PS_2$ Phosphonium, triphenyl-, 2-(ethylthio)- 1-(4-nitrophenyl)-2-thioxoethylide	$CHCl_3$	280(4.39),356(3.90), 425(3.63)	39-1404-76B
$C_{28}H_{24}N_2$ Benzenemethanamine, N,N'-(1,2-diphenyl- 1,2-ethanediylidene)bis-	EtOH	251(4.05)	78-0431-76
2,2'-Bipyridine, 6,6'-dimethyl-4,4'-bis- (2-phenylethenyl)-, Cu(I) complex	EtOH	505(4.18)	118-0001-76
$C_{28}H_{24}N_2O$ 4H-Pyrazino[3,2,1-jk]carbazole, 5,6-di- hydro-8-methoxy-2-phenyl-4-(phenyl- methyl)-	EtOH	244(4.36),306(4.60), 352(3.83),414(3.13), 436(3.09)	103-0662-76
$C_{28}H_{24}N_2O_2S$ 1H-Thieno[3,4-d]imidazole-4-carboxylic acid, 2-[(3,4-dimethylphenyl)phenyl- methyl]-6-phenyl-, methyl ester	MeOH	239(4.29),322(4.04), 357(4.35)	49-0299-76
$C_{28}H_{24}N_2O_2S_2$ Disulfide, bis[2-[(phenylmethyl)amino]- benzoyl]	$CHCl_3$	261s(4.07),275s(4.21), 393(4.29)	48-0161-76
$C_{28}H_{24}N_3O_2P$ Phosphonium, triphenyl-, 3-[(4-nitro- phenyl)azo]-2-butenylide	DMF	331(3.84),409(3.94), 603(4.61)	65-2135-76

Compound	Solvent	$\lambda_{max}(\log \epsilon)$	Ref.
$C_{28}H_{24}N_4O_4P$ Phosphonium, [3-[(2,4-dinitrophenyl)- hydrazono]-1-butenyl]triphenyl-	EtOH	257(4.045),375(4.323)	65-2135-76
$C_{28}H_{24}O$ Indeno[2,1-b]pyran, 9-(2-methylpropyl)- 2,4-diphenyl-	dioxan	230(4.25),264(4.35), 363(3.76),550(3.11)	78-2225-76
$C_{28}H_{24}O_2$ Azulene, 1,3-bis[2-(4-methoxyphenyl)eth- enyl]-, cis-cis	C_6H_{12}	248(4.40),305(4.52), 339(4.60),668(2.58)	117-0169-76
cis-trans	C_6H_{12}	255(4.30),307(4.53), 349(4.57),680(2.53)	117-0169-76
trans-trans	C_6H_{12}	255(4.29),308(4.56), 354(4.53),684(2.47)	117-0169-76
$C_{28}H_{24}O_5$ 1,4-Naphthalenedione, 2-[(5-hydroxy- 4,5',6-trimethyl-3-methylene-3'-oxo- spiro[benzofuran-2(3H),1'-[4]cyclopen- ten-2'-yl]methyl]-3-methyl-, trans	EtOH	234(4.46),268(4.38), 340(3.93)	39-2067-76C
$C_{28}H_{24}P$ 5,5'-Spirobi[5H-dibenzophospholium], 3,3',7,7'-tetramethyl-	MeOH	247(5.06),283(4.24), 364(3.65)	139-0151-76A
$C_{28}H_{25}NO_4S$ Benzo[f]quinolinium, 3-(4-hydroxyphen- yl)-1,4-dimethyl-, p-toluenesulfonate	EtOH	217(4.56),238(4.51), 281(4.42),312(4.07), 386(4.23)	65-0390-76
$C_{28}H_{25}N_2P$ Phosphonium, triphenyl-, 3-(phenylazo)- 2-butenylide	DMF	286(3.91),380(4.46), 489(3.37)	65-2135-76
$C_{28}H_{25}N_3O_2P$ Phosphonium, [3-[(4-nitrophenyl)hydraz- ono]-1-butenyl]triphenyl-, chloride	EtOH	254(4.11),410(4.61)	65-2135-76
$C_{28}H_{25}N_3O_5$ Cytidine, N-(benz[a]anthracen-7-ylmeth- yl)-	MeOH	260(4.59),270(4.72), 280(4.95),291(5.03), 319(3.71),335(3.93), 351(4.09),369(3.90), 388(2.95)	44-1597-76
Cytidine, 3-(benz[a]anthracen-7-ylmeth- yl)-, hydrobromide	MeOH	260(4.57),270(4.66), 281(4.87),292(4.95), 320(3.76),336(3.92), 353(3.99),370(3.87), 388(3.40)	44-1597-76
$C_{28}H_{25}N_5O$ 1,3,4-Oxadiazole, 2-[4-(1,1-dimethyleth- yl)phenyl]-5-[4-[2-(2-phenyl-2H-1,2,3- triazol-4-yl)ethenyl]phenyl]-	DMF	345(4.80)	33-2469-76
$C_{28}H_{26}IPS_2$ Phosphonium, [2,2-bis(methylthio)-1- phenylethenyl]triphenyl-, iodide	EtOH	222(4.61),272(4.37), 327s(3.80)	39-0692-76C

Compound	Solvent	$\lambda_{max}(\log \epsilon)$	Ref.
$C_{28}H_{26}N_2$			
1H-Indole, 3,3'-(1,3,5,7,9-decapentaene-1,10-diyl)bis[1-methyl-	MeCN	210(4.45),244s(4.22), 280s(3.93),423(4.72), 447(4.67)	5-1060-76
oxidized form	MeCN	258(3.98),430(3.90), 613(4.99)	5-1060-76
$C_{28}H_{26}N_2O_2$			
Indolizine, 3-acetyl-1-[p-(dimethylamino)cinnamoyl]-6-methyl-7-phenyl-	EtOH	226(3.52),310(3.22), 360(3.44),434(4.82)	103-0424-76
$C_{28}H_{26}N_2P$			
Phosphonium, triphenyl[3-(phenylhydrazono)-1-butenyl]-, chloride	EtOH	257(4.25),383(4.56)	65-2135-76
$C_{28}H_{26}N_2S_2$			
1,2-Ethenediamine, 1,2-bis(methylthio)-N,N,N',N'-tetraphenyl-	C_6H_{12}	265(4.1),304(4.2), 348(3.8)	24-1309-76
$C_{28}H_{26}N_4O$			
Benzoxazole, 5-(1,1-dimethylethyl)-2-[4-[2-(5-methyl-2-phenyl-2H-1,2,3-triazol-4-yl)ethenyl]phenyl]-	DMF	358(4.79)	33-2469-76
$C_{28}H_{26}N_4O_2S_2$			
Disulfide, bis[[(4-methoxyphenyl)hydrazono]phenylmethyl]	EtOH	302(4.48),407(4.18)	39-0038-76C
$C_{28}H_{26}N_4O_6S$			
4(1H)-Pteridinone, 1-(5-0-acetyl-2,3-0-(1-methylethylidene)-β-D-ribofuranosyl)-2,3-dihydro-6,7-diphenyl-2-thioxo-	pH 4.0	218s(4.22),300(4.40), 372(4.25)	24-3159-76
	pH 10.0	221(4.27),286(4.23), 378(4.20)	24-3159-76
$C_{28}H_{26}N_4O_7$			
2,4(1H,3H)-Pteridinedione, 1-[5-0-acetyl-2,3-0-(1-methylethylidene)-β-D-ribofuranosyl]-6,7-diphenyl-	pH 6.0	220(4.40),272(4.17), 357(4.13)	24-3159-76
	pH 11.0	220(4.34),263(4.28), 357(4.14)	24-3159-76
	MeOH	223(4.44),272(4.23), 357(4.16)	24-3159-76
$C_{28}H_{26}O$			
Benzenebutanol, α,δ,δ-triphenyl-	MeOH	260(2.99)	35-6711-76
$C_{28}H_{26}O_3$			
Spiro[7H-1-benzopyran-7,1'-cyclohexan]-5(6H)-one, 2,8-dihydro-2-(2-oxo-2-phenylethylidene)-4-phenyl-	MeOH	367(4.34),404(4.35)	24-1549-76
Spiro[7H-1-benzopyran-7,1'-cyclohexan]-5(6H)-one, 4,8-dihydro-4-(2-oxo-2-phenylethylidene)-2-phenyl-	MeOH	275(4.39),386(4.33)	24-1549-76
$C_{28}H_{26}O_6$			
5,9-Methano-1H-benzocycloheptene-1,4,6-trione, 10-[(1,4-dihydro-3-methyl-1,4-dioxo-2-naphthalenyl)methyl]-4a,5,9-9a-tetrahydro-9-hydroxy-2,5,7,9a-tetramethyl-, (4aα,5α,9β,9aα,10S*)-	EtOH	246(4.34),267(4.20), 273(4.19),334(3.51)	39-2067-76C
1,4-Naphthalenedione, 2-methyl-3-[(1,2,4a,9b-tetrahydro-4a,8-di-	EtOH	240(4.40),250s(4.38), 270(4.22),295s(3.82),	39-2067-76C

Compound	Solvent	$\lambda_{max}(\log \epsilon)$	Ref.
hydroxy-4,7,9,9b-tetramethyl-2-oxo-1-dibenzofuranyl)methyl]-, (1α,4aα,9bα)- (cont.)		330(3.59)	39-2067-76C
$C_{28}H_{26}O_7$			
2H,8H-Benzo[1,2-b:3,4-b']dipyran-2-one, 9-[(9,10-dihydro-9,9-dimethyl-2-oxo-2H,8H-benzo[1,2-b:3,4-b']dipyran-8-yl)oxy]-9,10-dihydro-8,8-dimethyl-isomer B	EtOH	214(4.50),246(3.85), 256(3.81),327(4.46)	39-1857-76C
	EtOH	212(4.53),246(3.87), 256(3.83),327(4.37)	39-1857-76C
$C_{28}H_{26}O_8$			
[1,1'-Binaphthalene]-5,5',8,8'-tetrone, 2,2'-diethoxy-4,4'-dihydroxy-6,6',7,7'-tetramethyl-	CHCl$_3$	279(4.38),292s(4.31), 455(3.90)	12-2247-76
$C_{28}H_{26}O_{10}$			
Licoflavanol, tetraacetate	EtOH	256(4.38),303(4.31)	94-1242-76
$C_{28}H_{26}O_{11}$			
Tenuiorin, 2-O-acetyl-	CHCl$_3$	268(3.92),308(3.60)	12-1147-76
$C_{28}H_{26}PS_2$			
Phosphonium, [2,2-bis(methylthio)-1-phenylethenyl]triphenyl-, iodide	EtOH	222(4.61),272(4.37), 327s(3.80)	39-0692-76C
$C_{28}H_{27}O_3P$			
7-Oxa-8-phosphabicyclo[4.2.0]octa-2,4-diene, 6-methyl-8-(1-methylethoxy)-1,3,5-triphenyl-, 8-oxide	EtOH	258s(4.26),268(4.29)	24-3099-76
$C_{28}H_{28}ClN_2O_2$			
Benzothiazolium, 3-ethyl-2-[2-[3-[(3-ethyl-2(3H)-benzothiazolylidene)-ethylidene]-1-(2-chloro-1-cyclo-hexen-1-yl)]ethenyl]-, iodide	EtOH	794(5.43)	103-0153-76
$C_{28}H_{28}NO_4$			
Dibenzo[a,g]quinolizinium, 5,6-dihydro-2,3,10-trimethoxy-8-methyl-11-(phen-ylmethoxy)-, iodide	MeOH	266(4.42),287(4.72), 308s(4.56),339(4.34), 383(3.94)	39-1218-76C
Dibenzo[a,g]quinolizinium, 5,6-dihydro-2,3,11-trimethoxy-8-methyl-10-(phen-ylmethoxy)-, iodide	MeOH	266(4.48),287(4.8), 308s(4.54),339(4.43), 383(4.0)	39-1218-76C
$C_{28}H_{28}N_2$			
Indolizine, 1,1'-(1,3,5,7-octatetraene-1,8-diyl)bis[2,3-dimethyl-	CH$_2$Cl$_2$	290s(4.24),465(4.79)	5-0317-76
oxidized form (dication)	MeCN	290(3.92),395(3.98), 554(4.97)	5-0317-76
Indolizine, 3,3'-(1,3,5,7-octatetraene-1,8-diyl)bis[1,2-dimethyl-	CH$_2$Cl$_2$	330(4.24),354s(4.12), 488(4.88),510s(4.85)	5-0317-76
semiquinone	1:1 MeCN-CH$_2$Cl$_2$	890(--),1150(--), 1420(--)	5-0317-76
oxidized form, diperchlorate	MeCN	284(3.98),353(4.09), 366(4.05),552(4.94)	5-0317-76
$C_{28}H_{28}N_2O_7$			
6H-Pyrido[4,3-b]carbazole, 5,11-dimeth-yl-6-(2,3,4-tri-O-acetyl-β-D-xylo-	n.s.g.	264(4.34),327(4.17)	28-1001-76A

Compound	Solvent	$\lambda_{max}(\log \epsilon)$	Ref.
pyranosyl)- (cont.)			28-1001-76A
$C_{28}H_{28}N_2O_7S_2$ Uridine, 2'-deoxy-5-(1,3-dithiolan-2-yl)-, 3',5'-bis(4-methylbenzoate)	$CHCl_3$	244(4.57)	87-0909-76
$C_{28}H_{28}N_4OS_3$ 1-Thia-4,6,8,10-tetraazaspiro[4.5]decan-2-one, 3,3-dimethyl-6,8,10-tris(phenylmethyl)-7,9-dithioxo-	EtOH	281(4.12)	33-2768-76
$C_{28}H_{28}N_4O_2$ 1H-Pyrido[3,4-b]indole, 2-acetyl-1-[(2-acetyl-2,3,4,9-tetrahydro-1H-pyrido-[3,4-b]indol-1-ylidene)methyl]-2,3,4,9-tetrahydro-1-methyl-	MeOH-acid	245(4.45),252s(4.41), 310(4.25),345s(4.13)	12-2023-76
$C_{28}H_{28}N_4O_4$ Spiro[2H-1-benzopyran-2,2'-[2H]indole], 1',3'-dihydro-8-methoxy-3',3'-dimethyl-1'-(1-methylethyl)-6-[(4-nitrophenyl)azo]-	EtOH	295(4.28),415(4.20)	28-0691-76A
$C_{28}H_{28}O_3P$ Phosphonium, triphenyl[(3,4,5-trimethoxyphenyl)methyl]-, bromide	EtOH	262(3.85),269(3.84), 276(3.76)	102-1057-76
$C_{28}H_{29}ClN_3S_2$ Benzothiazolium, 2-[2-[4-chloro-5-[2-(3-ethyl-2(3H)-benzothiazolylidene)ethylidene]-1,2,5,6-tetrahydro-1-methyl-3-pyridinyl]ethenyl]-3-ethyl-, bromide	EtOH	790(5.20)	103-0153-76
$C_{28}H_{29}NO_5$ Isoquinoline, 2-acetyl-1,2,3,4-tetrahydro-6,7-dimethoxy-1-[[(3-methoxy-4-(phenylmethoxy)phenyl]methylene]-, (Z)-	MeOH	332(4.55)	39-1218-76C
Isoquinoline, 2-acetyl-1,2,3,4-tetrahydro-6,7-dimethoxy-1-[[(4-methoxy-3-(phenylmethoxy)phenyl]methylene]-, (Z)-	MeOH	332(4.55)	39-1218-76C
$C_{28}H_{29}N_2S_2$ Benzothiazolium, 3-ethyl-2-[2-[3-[(3-ethyl-2(3H)-benzothiazolylidene)-ethylidene]-1-cyclohexen-1-yl]-ethenyl]-, iodide	EtOH	766(5.48)	103-0153-76
$C_{28}H_{29}N_3O_2$ Spiro[2H-1-benzopyran-2,2'-[2H]indole], 1',3'-dihydro-6-[(4-methoxyphenyl)-azo]-3',3'-dimethyl-1'-(1-methylethyl)-	EtOH	307(4.04),359(4.10), 384(4.23)	28-0691-76A
$C_{28}H_{29}N_5O_5$ 4(1H)-Pteridinone, 2-(dimethylamino)-1-[2,3-O-(1-methylethylidene)-β-D-ribofuranosyl]-6,7-diphenyl-	pH -2.0 pH 6.0	252(4.71),365(4.17) 223(4.49),267(4.61), 362(4.22)	24-3159-76 24-3159-76

Compound	Solvent	$\lambda_{max}(\log \epsilon)$	Ref.
(cont.)	MeOH	223(4.55),271(4.42), 360(4.24)	24-3159-76
$C_{28}H_{30}Br_2O_3$ 2,4,6,8,10,12,14,16-Heptadecaoctaenoic acid, 17-bromo-17-(4-bromo-3-methoxy-phenyl)-, 2-methylpropyl ester	acetone CHCl$_3$	423s(--),445(4.17), 473(4.09) 430s(--),453(5.12), 482(5.04)	88-4023-76 88-4023-76
$C_{28}H_{30}N_2O_2$ Methanone, [1-butyl-4-[(dimethylamino)-methyl]-5-hydroxy-2-phenyl-1H-indol-3-yl]phenyl-	dioxan	229(4.42),289(4.18)	42-0017-76
Methanone, [4-[(dimethylamino)methyl]-5-hydroxy-1-(2-methylpropyl)-2-phenyl-1H-indol-3-yl]phenyl-	dioxan	230(4.44),290(4.19)	42-0017-76
$C_{28}H_{30}N_2O_4$ 2-Butene-1,4-dione, 2,3-bis(3,4-dihydro-2,2-dimethyl-2H-pyrrol-5-yl)-1,4-di-phenyl-, N,N'-dioxide	EtOH	250(4.18),300(3.85)	39-1944-76C
$C_{28}H_{30}N_2O_9$ Thymidine, α,α-dimethoxy-, 3',5'-bis(4-methylbenzoate)	MeOH	243(4.58)	87-0909-76
$C_{28}H_{30}O_5$ Propanoic acid, 2,2-dimethyl-, tris(4-methoxyphenyl)ethenyl ester	EtOH	249(4.38),305(4.29)	35-7726-76
$C_{28}H_{31}NO$ Piperidine, 1-(3-bicyclo[2.2.2]oct-2-en-2-yl-1-oxo-2,3-diphenyl-2-propenyl)-	MeOH	231(4.27),290(4.09)	87-0414-76
Piperidine, 1-(3-bicyclo[3.2.1]oct-2-en-3-yl-1-oxo-2,3-diphenyl-2-propenyl)-	MeOH	228(4.28),283(4.03)	87-0414-76
$C_{28}H_{31}NO_6$ 4H-Pyran-4-one, 2-methoxy-3,5-dimethyl-6-(tetrahydro-4-(2,4,6-trimethyl-7-(4-nitrophenyl)-2,4,6-heptatrienyl-idene]-2-furanyl]-, (Z,E,E,E)-(+)-(spectinabilin)	EtOH	218(4.28),252(4.25), 268(4.26),367(4.19)	78-0217-76
$C_{28}H_{31}N_3O_4S_2$ Benzenesulfonamide, N-[2,3,4,5,6,7-hexa-hydro-1-methyl-7-[[(4-methylphenyl)-sulfonyl]amino]-2,7-methano-1H-1-benzazonin-12-ylidene]-4-methyl-	EtOH	215(4.57),228s(4.48), 252s(4.16),306(3.43)	39-0481-76C
$C_{28}H_{32}N_4O_2$ Benzo[g]pteridine-2,4(3H,4aH)-dione, 4a-(2,5-dimethyl-2,4-hexadienyl)-5,10-di-hydro-7,8,10-trimethyl-3-(phenylmeth-yl)-	50% MeOH- pH 7, 10	300s(--),370(3.74)	5-2037-76
$C_{28}H_{32}N_4O_6$ Tryptophan, N-acetyl-, methyl ester, dimer	MeOH	284(4.08),292(4.05)	138-0805-76

Compound	Solvent	$\lambda_{max}(\log \epsilon)$	Ref.
$C_{28}H_{32}OS$ 2,5-Cyclohexadien-1-one, 2,6-bis(1,1-di-methylethyl)-4-[[(4-methylphenyl)thio]-phenylmethylene]-	hexane	390(4.43)	73-2614-76
$C_{28}H_{32}O_2$ 4aH-Xanthen-4a-ol, 1,2,3,4,5,6,7,8,9,9a-decahydro-9-(4-methylphenyl)-5-[(4-methylphenyl)methylene]-	EtOH	209(4.28),294(4.42)	114-0051-76C
$C_{28}H_{32}O_2S$ 2,5-Cyclohexadien-1-one, 2,6-bis(1,1-di-methylphenyl)-4-phenyl-4-[(4-methoxy-phenyl)thio]-	hexane	387(4.43)	73-2614-76
$C_{28}H_{32}O_4$ 4aH-Xanthen-4a-ol, 1,2,3,4,5,6,7,8,9,9a-decahydro-9-(4-methoxyphenyl)-5-[(4-methoxyphenyl)methyl]-	EtOH	211(4.16),294(4.47)	114-0051-76C
$C_{28}H_{32}O_6$ 2,6-Dehydronicandrenone D(17a)-Homo-18-norergosta-2,13,15,17-tetraen-26-oic acid, 6α,7α:24,25-diepoxy-5,22-dihydroxy-1-oxo-, δ-lactone	MeOH EtOH	218(4.21),300(2.20) 216(4.22)	106-0647-76 39-0669-76C
$C_{28}H_{32}O_8$ Heptacyclo[10.4.2.2^{4,9}.0^{1,15}.0^{4,6}.0^{7,9}-0^{12,14}]eicosa-17,19-diene-17,18,19,20-tetracarboxylic acid, tetramethyl ester	EtOH	244(3.97)	89-0165-76
$C_{28}H_{32}O_{11}$ Physalin B, dihydroxy-	MeOH	223(3.85)	100-0405-76
$C_{28}H_{32}O_{12}S$ Breynogenin, triacetyl-	MeOH MeOH-NaOH	236(4.31) 300(4.45),354(3.90)	94-0114-76 94-0114-76
$C_{28}H_{33}N_2$ Acridinium, 9-[4-(dibutylamino)phenyl]-10-methyl-, (triiodide)	CH_2Cl_2	260(5.13),292(4.83), 360(4.65),633(4.14)	104-1531-76
$C_{28}H_{33}N_3O_8S_2$ Acetamide, N,N'-[5-[bis[2-[[(4-methyl-phenyl)sulfonyl]oxy]ethyl]amino]-1,3-phenylene]bis-	MeOH	214(3.46),245(4.59), 316(3.46)	87-0426-76
$C_{28}H_{34}N_2O_2$ Pyrrolo[1,2-a:4,5-b']diindol-13-one, 10-acetyl-5a,12-dibutyl-5a,10,10a,11-tetrahydro-10a-methyl-	MeOH	240(4.56),257(4.30), 290(3.54),320(3.15), 405(3.38)	23-1015-76
$C_{28}H_{34}N_2O_5S$ Compound X (unknown structure)	n.s.g.	199s(3.58),274(3.81), 297s(3.68)	33-0532-76
$C_{28}H_{34}N_2O_6$ Cyclobuta[1,2-c:4,3-c']diquinoline-5,8-	EtOH	207(4.62),233(4.40)	94-1400-76

Compound	Solvent	$\lambda_{max}(\log \epsilon)$	Ref.
dicarboxylic acid, 6,7-diethoxy-6,6a,6b,7,12b,12c-hexahydro-, diethyl ester (cont.)			94-1400-76
Propanedioic acid, (3-acetyl-1,4,6,7,12-12b-hexahydro-4-oxoindolo[2,3-a]quinolizin-2-yl)-, bis(1,1-dimethylethyl) ester	MeOH	225(4.65),275(4.00), 282(3.99),292(3.87)	24-3825-76
$C_{28}H_{34}N_4O_9P_2$			
Benzoic acid, 4-[3,5-bis(dimethoxyphosphinyl)-3,3a,4,4a,5,7a,8,8a-octahydro-5-(4-methoxyphenyl)-4,8-methanobenzo-[1,2-c:5,4-c']dipyrazol-3-yl]-, methyl ester, γ-anti-anti	MeOH	234(4.3),339(2.5)	23-0044-76
isomer	MeOH	234(4.2),339(2.8)	23-0044-76
$C_{28}H_{34}O_5$			
D(17a)-Homo-18-norergosta-13,15,17,24-tetraen-26-oic acid, 5,6α-epoxy-7α,22-dihydroxy-1-oxo-, δ-lactone	EtOH	216(4.21)	39-0669-76C
D(17a)-Homo-18-norergosta-13,15,17,24-tetraen-26-oic acid, 6α,7α-epoxy-5,22-dihydroxy-1-oxo-, δ-lactone	EtOH	213(4.28),261(3.18), 266(3.06),275(2.86)	39-0669-76C
$C_{28}H_{34}O_6$			
D(17a)-Homo-18-norergosta-2,13,15,17-tetraen-1-one, 6,7:22,26:24,25-tri-epoxy-5,26-dihydroxy-	EtOH	217(4.12),219s(4.11), 259(2.85),267(2.85), 276(2.83),293(2.46)	39-0304-76C
D(17a)-Homo-18-norergosta-13,15,17-trien-26-oic acid, 6α,7α:24,25-diepoxy-5,22-dihydroxy-1-oxo-, δ-lactone	EtOH	214(4.00),220s(3.94), 261(2.98)	39-0669-76C
Nicandrenone	MeOH	218(4.23),254(2.77), 260(2.74),267(2.74), 276(2.68),328(1.62)	106-0647-76
$C_{28}H_{34}O_7$			
Withaphysalin C, lactone	EtOH	227(4.24)	39-1244-76C
$C_{28}H_{34}O_9P_2$			
Benzoic acid, 4-[3,7-bis(dimethoxyphosphinyl)-7-(4-methoxyphenyl)tetracyclo-[3.3.1.0^{2,4}.0^{6,8}]non-3-yl]-, methyl ester	MeOH	211(3.9),238(4.4), 252(3.5)	23-0044-76
$C_{28}H_{34}O_{10}$			
Gomisin D	EtOH	216(4.57),256s(3.94), 294(3.68)	88-1359-76
$C_{28}H_{35}BF_2N_2O_6$			
Propanedioic acid, [3-(hydroxyethyl)-1,2,3,4,6,7,12,12b-octahydro-4-oxoindolo[2,3-a]quinolizin-2-yl]-, bis(1,1-dimethylethyl) ester, BF$_3$ product	MeOH	222(4.59),283(4.51), 290(4.47)	24-3825-76
$C_{28}H_{35}BrO_5$			
D(17a)-Homo-18-norergosta-13,15,17,24-tetraen-26-oic acid, 7β-bromo-5,6α-22-trihydroxy-1-oxo-, δ-lactone	EtOH	213(4.30)	39-0669-76C

Compound	Solvent	$\lambda_{max}(\log \epsilon)$	Ref.
$C_{28}H_{35}N_3O_5$ 4a(2H)-Phenanthrenecarboxylic acid, 1,3,4,9,10,10a-hexahydro-6-hydroxy-1,1-dimethyl-7-(1-methylethyl)-5-[(4-nitrophenyl)azo]-, ethyl ester, [4aα,5(E),10aβ]-(±)-	EtOH base	380(4.30),480(3.91) 380(4.20),515(3.92)	44-1005-76 44-1005-76
$C_{28}H_{35}N_5O_6$ Oxaline, N,N'-diacetylhexahydro-	MeOH	238(4.24),288(4.00)	78-2625-76
$C_{28}H_{36}N_2O_4$ 3H-2-Benzazepine, 1,1'-(1,4-butanediyl)-bis[4,5-dihydro-7,8-dimethoxy-	MeOH	262(4.04),295(4.00)	87-0202-76
$C_{28}H_{36}N_2O_5$ Propanedioic acid, (3-ethylidene-1,2,3,4,6,7,12,12b-octahydro-4-oxo-indolo[2,3-a]quinolizin-2-yl)-, bis(1,1-dimethylethyl) ester	MeOH	224(4.62),272(4.00), 278(3.98),290(3.87)	24-3825-76
$C_{28}H_{36}N_7O_4$ 1H-Benzimidazolium, 5-(ethoxycarbonyl)-2-[3-[5-(ethoxycarbonyl)-1,3-dihydro-1,3-diethyl-2H-benzimidazol-2-ylid-ene]-1-triazenyl]-1,3-diethyl-, tetrafluoroborate	MeOH	435(3.52)	33-0148-76
$C_{28}H_{36}N_8O_6S_2$ 1,2,4,5-Tetrazine-3,6-dicarboximidamide, 1,2-diacetyl-N,N,N'',N''-tetraethyl-1,2-dihydro-N',N'''-bis(phenylsulfonyl)-	EtOH	240(4.92),265(4.91)	32-0001-76
$C_{28}H_{36}O$ Ethanone, 1-[5,8,12-tris(1,1-dimethyl-ethyl)-1,3,5,6,7,9,11-cyclotetradeca-heptaen-13-yn-1-yl]-	THF	238(3.83),273s(3.91), 285(4.03),342(5.04), 481(4.32),542(3.15), 561(3.12),612(3.86)	88-1881-76
$C_{28}H_{36}O_2$ Pregn-4-ene-3,20-dione, 17-methyl-16-phenyl-	EtOH	240(4.23)	39-1558-76C
$C_{28}H_{36}O_3$ Trispiro[cyclopropane-1,1'-cyclopent-ane-17''(2''H)-[3H]cyclopenta[a]phen-anthrene-17''(2''H),2'''(5''')-furan]-2',5'''-dione, 1'',3''',4''',8'',9'',10''-11'',12'',14'',15'',16''-dodecahydro-10'',13''-dimethyl-	MeOH	232(4.33),240(4.36), 248(4.16)	88-2677-76
epimer	MeOH	232(4.41),240(4.44), 248(4.23)	88-2677-76
$C_{28}H_{36}O_6$ D(17a)-Homo-18-nor-5α-ergosta-13,15,17-trien-1-one, 6α,7α:22,26:24,25-triep-oxy-5,26-dihydroxy-	EtOH	215s(4.08),220s(4.02)	39-0669-76C
Nicandrenone, 2,3-dihydro-	MeOH	221(3.83),268(2.73), 276(2.69),292(1.74)	106-0647-76

Compound	Solvent	$\lambda_{max}(\log \epsilon)$	Ref.
$C_{28}H_{36}O_7$			
Pregna-5,8-diene-7,11,20-trione, 3β,6-diacetoxy-4,4,14α-trimethyl-	dioxan	273(4.04)	5-1206-76
Withaphysalin C	EtOH	225(4.23)	39-1244-76C
$C_{28}H_{36}O_{10}$			
Kaur-16-20-al, 1,7,11,14-tetraacetoxy-15-oxo-	EtOH	234.5(4.03)	33-0772-76
Rastronol D, half acetal acetate	EtOH	232(3.99)	33-0772-76
Rastronol G, 7,14,20-tri-0-acetyl-	EtOH	234(4.03)	33-0772-76
$C_{28}H_{36}O_{17}$			
1H,3H-Pyrano[3,4-c]pyran-1-one, 5-(1,2-diacetoxyethyl)-4,4a,5,6-tetrahydro-6-[(2,3,4,6-tetra-0-acetyl-β-D-glucopyranosyl)oxy]-	MeOH	243(3.92)	1-0743-76
epimer	MeOH	244(3.87)	1-0743-76
$C_{28}H_{37}NO_7$			
5α-Pregn-8-ene-6,7,11,20-tetrone, 3β-acetoxy-4,4,14α-trimethyl-, 6-(0-acetyloxime)	dioxan	235s(3.60),312(3.84)	5-1214-76
$C_{28}H_{38}N_6O_5$			
Adenosine, 5'-deoxy-5'-[[4-(1,1-dimethylethoxy)-4-oxobutyl](phenylmethyl)-amino]-2',3'-0-(1-methylethylidene)-	EtOH	259(4.24)	87-0684-76
$C_{28}H_{38}O_4$			
4,5-Secoestr-9-en-5-one, 17β-(1,1-dimethylethoxy)-3,3-o-phenylenedioxy-	MeOH	241(4.14),249(4.14), 280(3.61)	33-0999-76
$C_{28}H_{38}O_4S$			
Retinol, 9,10-dihydro-9-(phenylsulfonyl)-, acetate	EtOH	235(4.575)	33-0397-76
Retinol, 11,12-dihydro-11-(phenylsulfonyl)-, acetate	EtOH	272(4.15)	22-0513-76
	EtOH	215(4.38),245(4.14), 267(4.19),274(4.20), 281s(4.18)	33-0387-76
$C_{28}H_{38}O_5$			
5H-Inden-5-one, 1-(1,1-dimethylethoxy)-1,2,3,6,7,7a-hexahydro-7a-methyl-4-[6-(2-methyl-1,3-benzodioxol-2-yl-3-oxohexyl]-, (1S-cis)-	MeOH	242(4.06),250(4.08), 278(3.64),290(3.51)	33-0999-76
$C_{28}H_{38}O_6$			
2,5-Cyclohexadiene-1,4-dione, 2,2'-(1,14-tetradecanediyl)bis[6-methoxy-	MeOH	265(4.13),355(2.87)	12-1979-76
$C_{28}H_{38}O_7$			
Withanolide E	EtOH	226.5(4.13)	94-1403-76
Withaphysalin C, 2,3-dihydro-, $(2a_1)$	EtOH	227(3.98)	39-1244-76C
$(2a_2)$	EtOH	227(3.99)	39-1244-76C
$C_{28}H_{38}O_8$			
2,5-Cyclohexadiene-1,4-dione, 2,2'-(1,14-tetradecanediyl)bis[3-hydroxy-6-methyl-	MeOH	288(4.57),410(2.91)	12-1979-76
Withanolide E, 4β-hydroxy-	EtOH	219(4.17)	94-1403-76
	EtOH	215(4.25)	102-0340-76

Compound	Solvent	$\lambda_{max}(\log \epsilon)$	Ref.
$C_{28}H_{38}O_9$ Rastronol A, 7,14-di-O-acetyl-	EtOH	234(3.99)	33-0772-76
$C_{28}H_{39}NO_6$ 4H-Pyran-4-one, 2-methoxy-3,5-dimethyl-6-[tetrahydro-4-[2,4,6-trimethyl-7-(4-nitrophenyl)heptyl]-2-furanyl]-(octahydrospectinabilin)	EtOH	247s(4.39),268(4.50)	78-0217-76
$C_{28}H_{40}$ 1,4-Cyclohexadiene, 3,3'-(2-butene-1,4-diylidene)bis[1,2,4,5,6,6-hexamethyl-, (E)-	CH_2Cl_2	380s(--),399(4.83), 421(4.82)	5-1103-76
(Z)-	CH_2Cl_2	380s(--),398(4.78), 416(4.73)	5-1103-76
$C_{28}H_{40}N_2O_4$ 1H-2-Benzazepine, 1,1'-(1,4-butanediyl)-bis[2,3,4,5-tetrahydro-7,8-dimethoxy-, dihydrochloride	MeOH	237(4.20),283(3.75)	87-0202-76
$C_{28}H_{40}O_3$ Cyclohexanone, 2-[hydroxy(4-methylphenyl)methyl]-3-methyl-6-[1-methyl-1-[6-methyl-3-(1-methylethylidene)-2-oxocyclohexyl]ethyl]-, [2S-[2α(S*),3α-6α(1R*,6S*)]]-	isooctane	220(4.06),257(3.96), 300(2.54)	22-0985-76
	EtOH	218(3.98),259(3.75)	22-0985-76
$C_{28}H_{40}O_4$ Chola-4,22-dien-24-oic acid, 3-(carboxymethylene)-, 24-ethyl ester	EtOH	213(4.17),278(4.29)	13-0431-76A
Cyclohexanone, 2-[hydroxy(4-methoxyphenyl)methyl]-3-methyl-6-[1-methyl-1-[6-methyl-3-(1-methylethylidene)-2-oxocyclohexyl]ethyl]-, [2S-[2α(S*),3α-6α(1R*,6S*)]]-	isooctane	228(4.07),257(3.89), 315(2.22)	22-0985-76
	EtOH	229(4.15),263(3.97), 312(2.19)	22-0985-76
24-Norchola-8,22-diene-7,11-dione, 3-acetoxy-4,4,14-trimethyl-, (3β,5α)-	EtOH	270(3.85)	5-1206-76
$C_{28}H_{40}O_6$ Cochlioquinone B	MeOH	260(4.01),303s(3.30), 398(2.95)	32-0147-76
$C_{28}H_{40}O_7$ 5α,17α-Ergost-2-en-1-one, 6α,7α:17,24-22,26-triepoxy-5,25,26-trihydroxy-	EtOH	224(3.79)	39-0296-76C
$C_{28}H_{40}O_8$ Withanolide E, 2,3-dihydro-4β-hydroxy-	EtOH	228.0(3.87)	94-1403-76
$C_{28}H_{41}NO_3S$ Retinol, 12-[[4-(dimethylamino)phenyl]-sulfonyl]-11,12-dihydro-	isoPrOH	284(4.55)	44-3287-76
$C_{28}H_{42}CoN_4S_4$ Cobalt, bis[3-[(cyclohexylimino)methyl]-5-[(dimethylamino)methyl]-2(3H)-thiophenethionato-N^3,S^2]-	EtOH	263(4.59),350(3.81), 395(3.85),540(2.92)	70-1498-76

Compound	Solvent	λ_{max} (log ϵ)	Ref.
$C_{28}H_{42}N_4NiS_4$			
Nickel, bis[3-[(cyclohexylimino)methyl]-5-[(dimethylamino)methyl]-2(3H)-thiophenethionato-N^3,S^2]-	EtOH	267(4.63),555(3.55)	70-1498-76
$C_{28}H_{42}O_3$			
9,11-Dehydro-11-methyltigogenone	CHCl$_3$	277(1.43)	95-0863-76
$C_{28}H_{42}O_4$			
24-Norchol-8-ene-7,11-dione, 3-acetoxy-4,4,14-trimethyl-, (3β,5α)-	EtOH	271(3.88)	5-1206-76
4,8-Undecadienoic acid, 11-(3,4-dihydro-6-hydroxy-2,5,7,8-tetramethyl-2H-1-benzopyran-2-yl)-4,8-dimethyl-, ethyl ester, [R-(E,E)]-	EtOH	225s(3.99),292(3.51)	33-0290-76
$C_{28}H_{42}O_6$			
Cochlioquinone A, O^{20}-deacetyl-11-deoxy-	MeOH	270(4.00),380(3.02)	32-0147-76
$C_{28}H_{42}O_7$			
Dehydrotomentosin	EtOH	214(4.08)	94-0443-76
$C_{28}H_{42}O_8$			
Kidjoladinin	EtOH	214(4.14)	94-2185-76
$C_{28}H_{44}O$			
Epiergosterol	EtOH	261(3.90),270(4.06), 281(4.09),292(3.85)	39-0829-76C
5α-Ergosta-1,7-dien-3-one	EtOH	228(3.92)	39-0826-76C
Ergosta-4,7-dien-3-one	EtOH	238(4.05)	39-0826-76C
Toxisterol$_2$-D	C$_6$H$_{12}$	243(4.34),252(4.42), 262(4.28)	77-0659-76
Toxisterol$_2$-E	C$_6$H$_{12}$	243(4.34),252(4.42), 262(4.26)	77-0659-76
$C_{28}H_{44}O_2$			
Ergosta-5,7-dien-3-ol, 22,23-epoxy-	EtOH	271(4.08),282(4.10), 292(3.85)	39-0826-76C
Isoanhydrospergulatriol	EtOH	244(4.24),252(4.30), 261(4.15)	88-2327-76
$C_{28}H_{44}O_3$			
5β-Cholest-7-en-6-one, 3β-(formyloxy)-	EtOH	245(4.13)	12-1815-76
19-Norcholest-5(10)-en-6-one, 3β-acetoxy-	MeOH	246(4.06)	44-0313-76
Spirostan-3-one, 11-methyl-, (5α,11β-25R)-	CHCl$_3$	289(1.43)	95-0863-76
$C_{28}H_{45}N_5O$			
1H-Naphtho[2',1':4,5]indeno[1,2-f]pyrido[1,2-c]tetrazolo[1,5-a][1,3]diazepin-2-ol, 2,3,4,4a,4b,5,6,6a,6b,7,7a-8,9,10,11,11a,17,17a,17b,18,19,19a-docosahydro-4a,6a,7,10-tetramethyl-	EtOH	240(3.53)	49-0501-76
$C_{28}H_{46}$			
Cholesta-3,5-diene, 3-methyl-	n.s.g.	232(--),239(4.29), 248(--)	78-0797-76
Cholesta-3,5-diene, 4-methyl-	n.s.g.	232(--),238(4.30), 245(--)	78-0797-76

Compound	Solvent	$\lambda_{max}(\log \epsilon)$	Ref.
Cholesta-3,5-diene, 6-methyl-	n.s.g.	236(--),242(4.30), 251(--)	78-0797-76
Cholesta-3,5-diene, 7-methyl-	n.s.g.	237(--),243(4.31), 253(--)	78-0797-76
$C_{28}H_{46}N_2O_2$			
Acetamide, N-(3-oxo-4-azacholest-5-en-4-yl)-	EtOH	230(4.08)	49-0511-76
Solacasine	EtOH	end absorption	31-0415-76
$C_{28}H_{46}O$			
Ethanone, 1-A-norcholest-3(5)-en-3-yl-	EtOH	257(4.18)	44-1943-76
$C_{28}H_{46}O_2$			
Cholest-4-en-6-one, 3α-methoxy-	n.s.g.	254(3.73)	78-0797-76
Ergosta-5,7-diene-3β,22-diol, (22S)-	EtOH	272(3.92),282(3.93), 293(3.72)	39-0826-76C
(23RS)-	EtOH	271(3.87),281(3.90)	39-0826-76C

Compound	Solvent	$\lambda_{max}(\log \epsilon)$	Ref.
$C_{29}H_{16}N_4O_3$ 9H-Benzo[e]quino[2,3-i]perimidine- 8-carboxamide, 14,15-dihydro- 9,15-dioxo-6-phenyl-	DMF	270(3.981),622(3.543)	80-0257-76
$C_{29}H_{16}N_4O_5$ 9H-Benzo[e]quino[2,3-i]perimidine- 9,15(14H)-dione, 8-methoxy- 12-nitro-6-phenyl-	DMF	270(4.083),622(3.457)	80-0257-76
$C_{29}H_{17}NO_5$ 9,10-Anthracenedione, 1-amino-2-[(9,10- dihydro-4-hydroxy-9,10-dioxo-1-anth- racenyl)methyl]-	n.s.g.	450(3.61)	146-0001-76
$C_{29}H_{17}N_3O_3$ 9H-Benzo[e]quino[2,3-i]perimidine-9,15- (14H)-dione, 8-methoxy-6-phenyl-	DMF	270(4.10),361(4.06), 380(3.61),548(3.71), 596(3.51)	80-0257-76
$C_{29}H_{18}$ 2H-Cyclopenta[jk]fluorene, 2-(9H-fluo- ren-9-ylmethylene)-	EtOH-NaOEt- DMSO	486(4.18),572s(4.68), 597(4.92)	24-2956-76
$C_{29}H_{18}N_2O_5$ 9,10-Anthracenedione, 1,4-diamino-2- [(9,10-dihydro-4-hydroxy-9,10- dioxo-1-anthracenyl)methyl]-	n.s.g.	412(3.86),585(3.25)	146-0001-76
$C_{29}H_{18}N_4O_3$ 2H-Anthra[1,9,8-cdef][2,7]naphthyridine- 1,6,11(10H)-trione, 5,7-bis(phenyl- amino)-	DMF	299(4.35),449(4.75), 560(4.25)	103-1352-76
$C_{29}H_{19}ClN_2O_2$ Methanone, 1-[(4-chlorophenyl)-5-phenyl- 1H-pyrazole-3,4-diyl]bis[phenyl-	EtOH	257(4.66)	4-0989-76
$C_{29}H_{19}ClN_4O$ Benzoxazole, 2-[4-[2-[2-(4-chlorophen- yl)-5-phenyl-2H-1,2,3-triazol-4-yl]- ethenyl]phenyl]-	DMF	363(4.76)	33-2469-76
Benzoxazole, 2-[4-[2-[2-(4-chlorophen- yl)-2H-1,2,3-triazol-4-yl]ethenyl]- phenyl]-6-phenyl-	DMF	364(4.87)	33-2469-76
$C_{29}H_{19}ClN_8O_9$ Benzamide, 4,4'-dinitro-N,N'-[[(4-nitro- benzoylamino)[(4-chlorophenyl)imino]- methyl]carbonimidoyl]bis-	$MeOCH_2CH_2OH$	274(4.76)	40-0488-76
$C_{29}H_{19}NO$ Benz[c]oxireno[a]acridine, 1a,11b-di- hydro-1a,11b-diphenyl-	EtOH	262(4.57),272(4.62), 326(3.98),346(3.96)	4-0619-76
$C_{29}H_{19}N_3O_4$ Methanone, 1-(4-nitrophenyl)-5-phenyl- 1H-pyrazole-3,4-diyl]bis[phenyl-	EtOH	262(4.51)	4-0989-76

Compound	Solvent	$\lambda_{max}(\log \epsilon)$	Ref.
$C_{29}H_{20}BrN_3O_2$ 1H-Pyrazole-4-carboxamide, 3-(4-bromo-benzoyl)-N,1,5-triphenyl-	EtOH	260(4.61)	4-0989-76
$C_{29}H_{20}ClNO_2$ Spiro[cyclopropane-1,9'-[9H]fluorene]-2-carboxamide, 3-benzoyl-N-(4-chloro-phenyl)-	EtOH	255(4.56)	115-0001-76
Spiro[cyclopropane-1,9'-[9H]fluorene]-2-carboxamide, 3-(4-chlorobenzoyl)-N-phenyl-	EtOH	257(4.52)	115-0001-76
$C_{29}H_{20}ClN_3O_2$ 1H-Pyrazole-4-carboxamide, 3-benzoyl-1-(4-chlorophenyl)-N,5-diphenyl-	EtOH	255(4.65)	4-0989-76
$C_{29}H_{20}NO$ Pyrylium, 4-(9H-carbazol-9-yl)-2,6-di-phenyl-, perchlorate	MeCN	272(4.45),380(4.45), 445(4.33)	4-0073-76
$C_{29}H_{20}NOS$ Pyrylium, 4-(10H-phenothiazin-10-yl)-2,6-diphenyl-	MeCN	262(4.43),338(4.54)	4-0073-76
$C_{29}H_{20}N_4$ Quinazoline, 1,4-dihydro-2-phenyl-4-[(2-phenyl-4-quinazolinyl)methylene]-	CDCl₃	<u>255(4.5),415(4.2),</u> <u>445(4.5),470(4.5)</u>	5-2007-76
$C_{29}H_{20}N_4O$ Benzoxazole, 2-[4-[2-(2,5-diphenyl-2H-1,2,3-triazol-4-yl)ethenyl]phenyl]-	DMF	362(4.74)	33-2469-76
Benzoxazole, 5-phenyl-2-[4-[2-(2-phenyl-2H-1,2,3-triazol-4-yl)ethenyl]phenyl]-	DMF	355(4.80)	33-2469-76
Benzoxazole, 6-phenyl-2-[4-[2-(2-phenyl-2H-1,2,3-triazol-4-yl)ethenyl]phenyl]-	DMF	361(4.87)	33-2469-76
Benzoxazole, 5-phenyl-2-[2-[4-[2-phenyl-2H-1,2,3-triazol-4-yl)phenyl]ethenyl]-	DMF	362(4.79)	33-2469-76
Benzoxazole, 5-phenyl-2-[2-[4-(4-phenyl-2H-1,2,3-triazol-2-yl)phenyl]ethenyl]-	DMF	365(4.81)	33-2469-76
Benzoxazole, 6-phenyl-2-[2-[4-(2-phenyl-2H-1,2,3-triazol-4-yl)phenyl]ethenyl]-	DMF	361(4.80)	33-2469-76
Benzoxazole, 6-phenyl-2-[2-[4-(4-phenyl-2H-1,2,3-triazol-2-yl)phenyl]ethenyl]-	DMF	363(4.81)	33-2469-76
$C_{29}H_{20}N_4O_3Se$ 5H-1,3,4-Selenadiazolo[3,2-a]pyrimidin-5-one, 6,7-dihydro-7-(4-nitrophenyl)-2,6,6-triphenyl-	EtOH	262(4.43)	94-2532-76
$C_{29}H_{20}O$ 3-Butyn-1-one, 2-(diphenylmethylene)-1,4-diphenyl-	C_6H_{12}	251(4.49),318(4.28)	39-1257-76C
Indeno[2,1-b]naphtho[2,1-e]pyran, 5,6-dihydro-12-(1H-inden-2-yl)-	dioxan	243(4.29),277(4.46), 329(4.52),351(4.51), 600(3.57)	78-2225-76
Indeno[2,1-b]pyran, 4-methyl-2-(2-naph-thalenyl)-9-phenyl-	dioxan	222(4.62),258(4.59), 293(4.45),348(4.34), 369(4.37),543(3.54)	78-2219-76
Indeno[2,1-b]pyran, 4-methyl-9-(2-naph-thalenyl)-2-phenyl-	dioxan	268(4.53),278(4.46), 315(4.44),338(4.46), 545(3.51)	78-2225-76

Compound	Solvent	λ_{max}(log ϵ)	Ref.
$C_{29}H_{21}BrO_2$			
2(5H)-Furanone, 4-bromo-3-(diphenylmeth-yl)-5,5-diphenyl-	$CHCl_3$	245(3.98)	18-2645-76
$C_{29}H_{21}NO$			
2(1H)-Pyridinone, 1,4,5,6-tetraphenyl-	MeOH	243(4.28),340(4.05)	44-0818-76
$C_{29}H_{21}NOS_2$			
7,8-Dithia-2-azatricyclo[3.2.1.0^{2,4}]oct-an-6-one, 1,3,4,5-tetraphenyl-	MeOH	244(4.40)	44-1724-76
$C_{29}H_{21}NO_2$			
Benz[c]acridine-5,6-diol, 5,6-dihydro-5,6-diphenyl-, trans	EtOH	260(4.41),314(3.68), 330(3.73),345(3.81)	4-0619-76
1H-Pyrrol-3-ol, 1,2,5-triphenyl-, benzoate	EtOH	228(4.42),290(4.19)	39-0794-76C
Spiro[cyclopropane-1,9'-fluorene]-2-car-boxamide, 3-benzoyl-N-phenyl-	EtOH	251(4.56)	115-0001-76
$C_{29}H_{21}N_3O_2$			
1H-Pyrazole-4-carboxamide, 3-benzoyl-N,1,5-triphenyl-	EtOH	252(4.45)	4-0989-76
$C_{29}H_{21}N_5O$			
1,3,4-Oxadiazole, 2-[4-[2-(5-methyl-2-phenyl-2H-1,2,3-triazol-4-yl)ethenyl]-phenyl]-5-(1-naphthalenyl)-	DMF	357(4.77)	33-2469-76
1,3,4-Oxadiazole, 2-[4-[2-(5-methyl-2-phenyl-2H-1,2,3-triazol-4-yl)ethenyl]-phenyl]-5-(2-naphthalenyl)-	DMF	281(4.26),353(4.79)	33-2469-76
$C_{29}H_{22}ClN_3O_2$			
1H-Pyrazole-4-carboxamide, 3-benzoyl-N-(4-chlorophenyl)-4,5-dihydro-5,5-diphenyl-	EtOH	253(4.39),334(4.17)	115-0001-76
$C_{29}H_{22}ClN_5O_2$			
Benzamide, N,N'-[[(benzoylamino)[(4-chlorophenyl)imino]methyl]carbon-imidoyl]bis-	MeOCH$_2$CH$_2$OH	241s(4.54),262(4.60)	40-0488-76
$C_{29}H_{22}NO$			
Pyrylium, 4-(diphenylamino)-2,6-diphen-yl-	MeCN	335(4.63)	4-0073-76
$C_{29}H_{22}N_2O$			
Spiro[3H-naphtho[2,1-b]pyran-3,5'(4'H)-pyrido[2,3,4,5-lmn]phenanthridine], 4'-ethyl-10'-methyl-	EtOH	241(4.79),255s(4.59), 285s(4.22),314(4.02), 362(3.92),380s(3.75)	103-0273-76
	HOAc	295(4.22),344(4.08), 393(4.03),475(4.05)	103-0273-76
$C_{29}H_{22}N_2O_2S$			
4H-Pyrimido[2,1-b]benzothiazol-4-one, 2,3-dihydro-2-(4-methoxyphenyl)-3,3-diphenyl-	EtOH	278(3.84),285(3.84), 310s(3.29)	94-2532-76
$C_{29}H_{22}N_2O_3$			
4H-Pyrido[2,1-b]benzoxazol-4-one, 2,3-dihydro-2-(4-methoxyphenyl)-3,3-di-phenyl-	EtOH	275.5(3.94)	94-2532-76

Compound	Solvent	$\lambda_{max}(\log \epsilon)$	Ref.
4-Quinolinecarbonitrile, 3-(benzoyloxy)-1,2,3,4-tetrahydro-2-hydroxy-2,3-diphenyl-	EtOH	235(4.37),272s(3.36), 285s(3.39),298(3.47)	95-0968-76
$C_{29}H_{22}N_6O_5$			
Benzamide, N,N'-[[(benzoylamino)[(3-nitrophenyl)imino]methyl]carbonimidoyl]-bis-	MeOCH$_2$CH$_2$OH	245(4.54),262(4.61)	40-0488-76
$C_{29}H_{22}O$			
7H-Cyclohept[b]indeno[1,2-e]pyran, 5-(1-acenaphthylenyl)-8,9,10,11-tetrahydro-	dioxan	230(4.75),298(4.43), 529(3.73)	78-2225-76
Indeno[2,1-b]pyran, 9-(3,4-dihydro-2-naphthalenyl)-2-(4-methylphenyl)-	dioxan	264(4.30),335(4.46), 359(4.48),565(3.51)	78-2225-76
Indeno[2,1-b]pyran, 9-(3,4-dihydro-2-naphthalenyl)-4-methyl-2-phenyl-	dioxan	265(4.32),330(4.38), 349(4.30),555(3.48)	78-2225-76
Indeno[2,1-b]pyran, 9-(1H-inden-2-yl)-2,4-dimethyl-3-phenyl-	dioxan	250(4.11),318(4.33), 350(4.35)	78-2225-76
Indeno[2,1-b]pyran, 9-(1H-inden-2-yl)-4-methyl-2-(4-methylphenyl)-	dioxan	271(4.34),323(4.41), 361(4.33),579(3.41)	78-2225-76
$C_{29}H_{22}O_2$			
Benzenepropanoic acid, α-(diphenylethenylidene)-β-phenyl-	CHCl$_3$	243(4.28),253(4.28)	18-2645-76
Phenanthro[9,10-d]-1,3-dioxole, 6,9-dimethyl-2,2-diphenyl-	hexane	256(4.57),263(4.66), 278(4.17),305s(--), 318(4.04),333(4.12), 361(3.05),379(3.28), 400(3.32)	104-1298-76
$C_{29}H_{22}O_6$			
4H-1-Benzopyran-4-one, 5,6-dihydroxy-7-(phenylmethoxy)-2-[4-(phenylmethoxy)phenyl]-	EtOH EtOH-AlCl$_3$	276(4.1),333(3.9) 290(--),344(--)	2-0592-76 2-0592-76
$C_{29}H_{22}O_8$			
Benzoic acid, 3-(1,3-dioxolan-2-yl)-2,6-dihydroxy-4-[(5-oxo-2-phenyl-5H-naphtho[1,8-bc]furan-4-yl)methyl]-, methyl ester	EtOH	222(4.39),264(3.91), 327(4.38),404(4.46)	39-0503-76C
Spiro[1,3-dioxolane-2,12'(7'H)-[6H]naphthaceno[1,12-bc]furan]-10'-carboxylic acid, 6'a,12'a-dihydro-9',11'-dihydroxy-6'-oxo-1'-phenyl-, methyl ester, cis	EtOH	228(4.49),267(4.42), 295(4.08),307(4.04), 344(4.13)	39-0503-76C
$C_{29}H_{22}S_2$			
Azulene, 4,6,8-trimethyl-1,3-bis[(phenylethynyl)thio]-	hexane	630(2.96)	104-1966-76
$C_{29}H_{23}NO_2$			
Cyclopropanecarboxamide, 2-benzoyl-N,3,3-triphenyl-	EtOH	248(4.61)	115-0001-76
$C_{29}H_{23}NO_2S$			
Ethanone, 1-[5-benzoyl-4-methyl-2-(1-methyl-2-phenyl-1H-indol-3-yl)-3-thienyl]-	EtOH	290(4.35),380(3.97)	142-0009-76

Compound	Solvent	$\lambda_{max}(\log \epsilon)$	Ref.
$C_{29}H_{23}NO_4S_2$ 4,7-Epithio-2,1-benzisothiazole-5,6-di-carboxylic acid, 4,5,6,7-tetrahydro-3,4,7-triphenyl-, dimethyl ester stereoisomer	CH_2Cl_2 CH_2Cl_2	262.5(4.06) 274(4.18)	88-2163-76 88-2163-76
$C_{29}H_{23}N_3O_2$ 5-Pyrazolecarboxamide, 3-benzoyl-4,5-di-hydro-N,5,5-triphenyl-	EtOH	251(4.48),335(4.23)	115-0001-76
$C_{29}H_{23}N_5O_3$ Benzamide, N,N'-[[(benzoylamino)(phenyl-imino)methyl]carbonimidoyl]bis-	$MeOCH_2CH_2OH$	239(4.52),274(4.56)	40-0488-76
$C_{29}H_{24}N_2O$ 2-Butanone, 4-[[4-(1-phenylbenzo[f]quin-olin-3-yl)phenyl]amino]-	EtOH	298(4.53),341(4.51), 376(4.44)	103-0106-76
2-Butanone, 4-[[4-(3-phenylbenzo[f]quin-olin-1-yl)phenyl]amino]-	EtOH	282(4.65),343(4.02), 363(4.02)	103-0106-76
$C_{29}H_{24}N_2O_4$ Benzenepropanoic acid, β-(2-benzoxazol-ylamino)-4-methoxy-α,α-diphenyl-	EtOH	274(3.66)	94-2532-76
$C_{29}H_{24}N_2O_9S$ 2H-1,2,4-Oxadiazin-3(4H)-one, 5,6-di-hydro-5-thioxo-2-(2,3,5-tri-O-benz-oyl-β-D-ribofuranosyl)-	EtOH	229(4.61),276(4.34)	44-3128-76
$C_{29}H_{24}N_2O_{10}$ 2H-1,2,4-Oxadiazine-3,5(4H,6H)-dione, 2-(2,3,5-tri-O-benzoyl-β-D-ribo-furanosyl)-	EtOH	230(4.65)	44-3128-76
$C_{29}H_{24}O_6S_2$ 1,4-Epithio-5,8-etheno-1H-benzo[3,4]cy-clobuta[1,2-c]thiopyran-6,7-dicarbox-ylic acid, 3,4,4a,4b,5,8,8a,8b-octa-hydro-3-oxo-1,4-diphenyl-, dimethyl ester, 11-oxide	MeOH	217s(4.46)	39-2562-76C
$C_{29}H_{25}NO_2$ 1H-Indole-3-methanol, 5-methoxy-α,2-di-phenyl-1-(phenylmethyl)-	EtOH	224(4.46),298(4.16)	42-0606-76
Pyridinium, 2,5-dimethyl-4-(phenylmeth-yl)-, 1-benzoyl-2-oxo-2-phenylethyl-ide	EtOH	280(3.92),318(4.02)	103-0424-76
$C_{29}H_{26}N_2O$ Benzeneacetamide, N-(phenylmethyl)-α-[phenyl[(phenylmethyl)amino]-methylene]-	EtOH	246(4.15),306(3.72)	78-0431-76
$C_{29}H_{26}N_2O_2S$ Benzenepropanol, γ-(2-benzothiazolyl-amino)-4-methoxy-β,β-diphenyl-	EtOH	269(4.22),274(4.23), 298s(3.66)	94-2532-76
$C_{29}H_{26}N_2O_3$ Benzenepropanol, γ-(2-benzoxazolylami-no)-4-methoxy-β,β-diphenyl-	EtOH	229(4.32),246(4.28), 282(4.06)	94-2532-76

Compound	Solvent	$\lambda_{max}(\log \epsilon)$	Ref.
$C_{29}H_{26}N_4O$ Benzoxazole, 5-cyclohexyl-2-[4-[2-(2-phenyl-2H-1,2,3-triazol-4-yl)ethenyl]phenyl]-	DMF	353(4.79)	33-2469-76
$C_{29}H_{26}N_4O_9$ 2,4(1H,3H)-Pteridinedione, 1-(2,3,5-tri-O-acetyl-α-D-arabinofuranosyl)-6,7-diphenyl-	MeOH	223(4.39),274(4.17), 358(4.10)	24-3217-76
2,4(1H,3H)-Pteridinedione, 1-(2,3,5-tri-O-acetyl-β-D-arabinofuranosyl)-6,7-diphenyl-	MeOH	222(4.45),272(4.22), 357(4.16)	24-3217-76
2,4(1H,3H)-Pteridinedione, 3-(2,3,5-tri-O-acetyl-α-D-arabinofuranosyl)-6,7-diphenyl-	MeOH	224(4.45),287(4.27), 367(4.11)	24-3217-76
2,4(1H,3H)-Pteridinedione, 1-(2,3,5-tri-O-acetyl-β-D-ribofuranosyl)-6,7-diphenyl-	pH 6.0	222s(4.41),272(4.16), 357(4.13)	24-3184-76
	pH 10.0	222s(4.35),263(4.28), 357(4.18)	24-3184-76
	MeOH	222(4.42),272(4.19), 357(4.13)	24-3184-76
$C_{29}H_{26}O$ Benzene, 1-[2,3-diphenyl-2-(phenylmethoxy)cyclopropyl]-4-methyl-, (1α,2β,3β)-	C_6H_{12}	232(4.24)	12-2445-76
Benzene, 1-[3,3-diphenyl-3-(phenylmethoxy)-1-propenyl]-4-methyl-	C_6H_{12}	223(4.04),260(4.34)	12-2445-76
$C_{29}H_{26}O_7$ 6H-12,17a-Propenoindeno[2,1-b]naphtho-[2,1-e]pyran-5,6,8,11,16(17H)-pent-one, 7b,11a,12,12a-tetrahydro-12-hydroxy-9,10,11a,12a,14,15-hexamethyl-, (7aα,7bβ,12β,12aα)-(±)-	EtOH	246(4.57),252(4.58), 259(4.46),274s(4.26), 335(3.27),422(3.17)	39-0600-76C
	EtOH-NaOH	477(3.32)(changing)	39-0600-76C
$C_{29}H_{26}O_9$ [3,8'-Bi-4H-1-benzopyran]-4,4'-dione, 2,3-dihydro-5,5',7,7'-tetramethoxy-2-(4-methoxyphenyl)-	MeOH	225(4.73),258(4.45), 285(4.44)	39-1458-76C
$C_{29}H_{27}NO_2$ 1H-Benz[g]indole-3-methanol, 5-methoxy-α,2-diphenyl-1-propyl-	EtOH	222(4.42),280(4.61)	42-0606-76
$C_{29}H_{27}N_5O$ 1,3,4-Oxadiazole, 2-[4-(1,1-dimethylethyl)phenyl]-5-[4-[2-(5-methyl-2-phenyl-2H-1,2,3-triazol-4-yl)ethenyl]phenyl]-	DMF	352(4.75)	33-2469-76
$C_{29}H_{27}N_5O_2$ Benzo[g]pteridine-2,4(3H,4aH)-dione, 5,10-dihydro-4a-(1H-indol-3-ylmethyl)-7,8,10-trimethyl-3-(phenylmethyl)-	50% MeOH- pH -0.5	338s(--),357(3.86), 369s(--)	5-2037-76
	50% MeOH pH 7, 10	290s(--),361(3.60)	5-2037-76
$C_{29}H_{28}Cl_2N_2O_2$ 4H-Pyrrolo[2,1-c][1,4]benzoxazine, 1,1'-(1-methylethylidene)bis[8-chloro-4,4-dimethyl-	MeOH	270(4.32),310(4.36)	4-0311-76

Compound	Solvent	$\lambda_{max}(\log \epsilon)$	Ref.
$C_{29}H_{28}N_2O_2$ Pyridinium, 4,4'-[1,3-bis[(2-hydroxy-phenyl)methylene]-1,3-propanediyl]-bis[1-methyl-, diiodide	pH 3.24	<u>220(4.7),260(4.3), 325(4.3)</u>	39-1575-76B
$C_{29}H_{28}N_2O_4S$ Thymidine, 4-thio-5'-O-(triphenylmeth-yl)-	n.s.g.	208(4.48),231s(3.96), 335(4.22)	128-0351-76
$C_{29}H_{28}N_4O$ Benzoxazole, 5-(1,1-dimethylethyl)-7-methyl-2-[4-[2-(5-methyl-2-phenyl-2H-1,2,3-triazol-4-yl)ethenyl]phenyl]-	DMF	358(4.79)	33-2469-76
$C_{29}H_{28}O_{12}$ 9,10-Anthracenedione, 3-methyl-1-[(2,3,4,6-tetra-O-acetyl-β-D-glucopyranosyl)oxy]-	$CHCl_3$	259(4.56),276s(4.24), 343(3.66)	12-2231-76
$C_{29}H_{29}NO$ Benzenepropanol, γ-[4-(dimethylamino)-phenyl]-α,γ-diphenyl-	MeOH	262(4.18),300s(--)	35-6711-76
$C_{29}H_{29}N_2O$ Methanaminium, N-[4-[[4-(dimethylamino)-phenyl](4-phenoxyphenyl)methylene]-2,5-cyclohexadien-1-ylidene]-N-methyl-	EtOH 8% DMSO	616(5.01) 458(4.41),628(4.97)	44-0221-76 44-0221-76
$C_{29}H_{29}N_3O_3S$ 4H-Indazol-4-one, 1,5,6,7-tetrahydro-6,6-dimethyl-3-phenyl-1-(phenylmeth-yl)-, O-[(4-methylphenyl)sulfonyl]-oxime, (E)-	MeOH	228(4.57),260(4.04)	24-1898-76
$C_{29}H_{30}N_2O_2$ 4H-Pyrrolo[2,1-c][1,4]benzoxazine, 1,1'-(1-methylethylidene)bis[4,4-dimethyl-	MeOH	268(4.36),298(4.35)	4-0311-76
$C_{29}H_{30}N_4O_7$ Isotryptoquivaline	MeOH	228(4.60),233(4.58), 255(4.32),280(4.15), 307(3.62),320(3.43)	88-2861-76
$C_{29}H_{31}N_3O_3$ Spiro[2H-1-benzopyran-2,2'-[2H]indole], 1',3'-dihydro-8-methoxy-6-[(4-meth-oxyphenyl)azo]-3',3'-dimethyl-1'-(1-methylethyl)-	EtOH	300(3.99),310(4.32), 380(4.25),392(4.20)	28-0691-76A
$C_{29}H_{31}N_3O_4S_2$ Benzenesulfonamide, N,N'-(4,5,9,10-tet-rahydro-7H-pyrrolo[3,2,1-de]phenan-thridine-7,7a(8H)-diyl)bis[4-methyl-	EtOH	231(4.53),263s(4.15), 345(3.66)	39-2254-76C
$C_{29}H_{31}N_5OS_2$ Propanethioamide, N,N,2-trimethyl-2-[[tetrahydro-4-oxo-3-phenyl-1,5-bis(phenylmethyl)-6-thioxo-1,3,5-triazin-2(1H)-ylidene]amino]-	EtOH	280(4.46)	33-2768-76

Compound	Solvent	λ_{max}(log ϵ)	Ref.
$C_{29}H_{32}N_2O_{13}S_2$ 2,4(1H,3H)-Pyrimidinedione, 1-[3,4-di-O- acetyl-2,6-bis-O-[(4-methylphenyl)sul- fonyl]-β-D-glucopyranosyl]-5-methyl-	MeOH	221(4.22),265(4.04)	48-0079-76
$C_{29}H_{32}O_6$ 2H-1-Benzopyran, 3,4-dihydro-5,7-dimeth- oxy-2-phenyl-γ-[2-(2,4,6-trimethoxy- 3-methylphenyl)ethenyl]-, [S-(E)]-	EtOH	306(4.19),329s(4.03)	39-1570-76C
$C_{29}H_{32}O_{16}$ Multiflorin A	MeOH	267(4.54),342(4.16)	95-1217-76
$C_{29}H_{33}AsOSi_2$ 3H-Arsol-3-one, 1,2-dihydro-1,5-diphen- yl-2-[phenyl(trimethylsilyl)methyl- ene]-4-(trimethylsilyl)-	C_6H_{12}	320(3.87)	78-2131-76
$C_{29}H_{33}NO_3$ Benzoic acid, 2-[benzoyl[4-(1,1-dimeth- ylethyl)phenyl]amino]-5-(1,1-dimeth- ylethyl)-, methyl ester	MeOH	206(4.57),275(3.91)	39-0465-76C
Benzoic acid, 5-(1,1-dimethylethyl)-2- [[[2-(1,1-dimethylethyl)phenyl]imino]- phenylmethoxy]-, methyl ester	MeOH	208(4.73),230(4.49), 290(3.73)	39-0465-76C
Benzoic acid, 5-(1,1-dimethylethyl)-2- [[[4-(1,1-dimethylethyl)phenyl]imino]- phenylmethoxy]-, methyl ester	MeOH	207(4.77),230(4.50), 290(3.84)	39-0465-76C
1H-Isoindol-1-one, 7-[7-(2,5-dihydro- 5-oxo-2-furanyl)-4-methyl-1-hepten- yl]-2,3-dihydro-4,5-dimethyl-3-(phen- ylmethyl)-	EtOH	207(4.51),272(4.26), 282(4.04)	33-0914-76
$C_{29}H_{34}O_5$ D:C-Friedo-C,24,25-trinoroleana-1(10),3- 5,7,9(11)-pentaen-29-oic acid, 3-hy- droxy-11-methyl-2,22-dioxo-, methyl ester, (20α)-	MeOH	325s(3.11),448(4.17)	2-0131-76
$C_{29}H_{36}O_3S$ 1,4-Oxathiane, 2-[[(17β)-3-(phenylmeth- oxy)estra-1,3,5(10)-trien-17-yl]oxy]-	MeOH	238(4.26)	24-0185-76
$C_{29}H_{37}N_3O_5$ 4a(2H)-Phenanthrenecarboxylic acid, 1,3,4,9,10,10a-hexahydro-6-methoxy- 1,1-dimethyl-7-(1-methylethyl)-5- [(4-nitrophenyl)azo]-, ethyl ester	EtOH	279(4.16),347(4.05), 485(3.00)	44-1005-76
$C_{29}H_{38}N_2O_2$ 1'H-Androstano[3,2-c]pyrazol-17-ol, 5'- methyl-1'-phenyl-, acetate, (5α,17β)-	EtOH	260(4.29)	22-0255-76
$C_{29}H_{38}N_4O_4$ Orantin, O-methyl-	EtOH and EtOH-NaOH	229(4.26),247s(3.02), 253(2.95),260s(3.06), 276s(3.47),280(3.51), 287s(3.37)	133-0371-76

Compound	Solvent	$\lambda_{max}(\log \epsilon)$	Ref.
$C_{29}H_{38}N_4O_9S_2$			
Inosine, 7,8-dihydro-8-oxo-6-thio-, 2',5'-diacetate 3'-[2,4,6-tris-(1-methylethyl)benzenesulfonate]	pH 1	236(4.35),287s(3.91), 343(4.24)	94-0672-76
	pH 13	243s(4.29),288s(3.88), 323(4.37)	94-0672-76
	MeOH	236(4.36),287s(3.91), 315(4.17),340(4.09)	94-0672-76
$C_{29}H_{38}O_2$			
Pregn-4-ene-3,20-dione, 17-methyl-16-(phenylmethyl)-, (16α)-	EtOH	242(4.25)	39-1558-76C
$C_{29}H_{38}O_5$			
Spirosta-1,4,6-trien-3-one, 2-acetoxy-	EtOH	218(4.16),265(4.02), 304(4.08)	95-0863-76
$C_{29}H_{38}O_{11}$			
Baccharin	EtOH	259(4.27)	35-7092-76
Butanoic acid, 2-hydroxy-2,3-dimethyl-4-[(5,6,7,8-tetrahydro-5,6-dihydroxy-1,2,3-trimethoxy-6,7-dimethylbenzo-[3,4]cycloocta[1,2-f][1,3]benzodi-oxol-13-yl)oxy]-, methyl ester, [5α,6β,7β,13(2S*,3S*)]-(+)-	EtOH	217(4.68),255s(3.97), 279s(3.53)	88-1359-76
$C_{29}H_{38}O_{12}$			
Glaucarubinone, 1,12-diacetate	EtOH	239(4.03)	87-1130-76
$C_{29}H_{39}NO_6$			
2,10-Undecadienoic acid, 4-hydroxy-8-methyl-11-[octahydro-3a,5-dihydroxy-7-methyl-6-methylene-3-oxo-1-(phen-ylmethyl)-1H-isoindol-4-yl]-	EtOH	208(4.46),245(2.64), 251(2.55),258(2.58), 264(2.38),268(2.24)	33-0914-76
$C_{29}H_{40}N_2O_2$			
1'H-Androstano[3,2-c]pyrazol-17-ol, 2,5'-dihydro-1'-methyl-5'-phenyl-, acetate	EtOH	250(3.82)	22-0255-76
isomer	EtOH	246(3.68)	22-0255-76
isomer	EtOH	249(3.96)	22-0255-76
$C_{29}H_{40}N_8O_{10}$			
L-Argininamide, N-[(1,1-dimethylethoxy)-carbonyl]glycyl-N-[1-(5-O-benzoyl-β-D-arabinofuranosyl)-1,2-dihydro-2-oxo-4-pyrimidinyl]-, hydrochloride	H_2O	216(4.24),236(4.23), 282(3.71),300(3.74)	87-1013-76
$C_{29}H_{40}O_4$			
Fusida-17(20),24-dien-16β,21-olactone, 3,11-dioxo-	n.s.g.	232(4.10)	39-0710-76C
4,5-Secoandrost-9(11)-en-5-one, 17β-(1,1-dimethylethoxy)-3,3-o-phenyl-enedioxy-	MeOH	233(3.50),286(3.62)	33-0999-76
$C_{29}H_{40}O_5$			
Cyclopenta[f][1]benzopyran, 7-(1,1-di-methylethoxy)-1,2,3,5,6,6a,7,8-octa-hydro-2-methoxy-6a-methyl-3-[3-(2-phenyl-1,3-benzodioxol-2-yl)propyl]-	MeOH	242(4.19),248(4.19), 280(3.63)	33-0999-76
Spirosta-1,6-dien-3-one, 2-acetoxy-, (5α,25R)-	EtOH	238.5(3.98)	95-0863-76

Compound	Solvent	$\lambda_{max}(\log \epsilon)$	Ref.
Spirosta-1,7-dien-3-one, 2-acetoxy-, (5α,25R)-	EtOH	236(3.99)	95-0863-76
Spirosta-1,9(11)-dien-3-one, 2-acetoxy-, (5α,25R)-	EtOH	234.5(4.01)	95-0863-76
Spirosta-1,11-dien-3-one, 2-acetoxy-, (5α,25R)-	EtOH	238.5(4.01)	95-0863-76
Spirosta-2,7-dien-4-one, 3-acetoxy-, (5α,25R)-	EtOH	236.5(3.83)	95-0863-76
Spirosta-3,9(11)-dien-2-one, 3-acetoxy-, (5α,25R)-	EtOH	238.5(3.90)	95-0863-76
Spirosta-3,11-dien-2-one, 3-acetoxy-, (5α,25R)-	EtOH	240.5(3.91)	95-0863-76
Spirosta-4,9(11)-dien-3-one, 4-acetoxy-, (5α,25R)-	EtOH	246(4.20)	95-0863-76
$C_{29}H_{40}O_6$			
Spirost-1-ene-3,11-dione, 2-acetoxy-, (5α,25R)-	EtOH	235(4.01)	95-0764-76
Spirost-3-ene-2,11-dione, 3-acetoxy-	EtOH	235(3.93)	95-0764-76
Spirost-3-ene-2,11-dione, 3-acetoxy-, (5α,25R)-	EtOH	237.5(3.89)	95-0764-76
Spirost-4-ene-3,11-dione, 9α-acetoxy-, (25R)-	EtOH	238(4.21)	95-0764-76
$C_{29}H_{41}NO_2$			
Pyridinium, 1-benzoyl-2-oxoheptadecyl-ide	EtOH	284(4.13),368(3.07)	78-2647-76
	dioxan	279(4.01),420(3.20)	78-2647-76
	MeCN	282(4.09),416(3.26)	78-2647-76
$C_{29}H_{41}N_3O_2$			
1H-Pyrrole-2-carboxylic acid, 3-ethyl-5-[[3-ethyl-5-[(4-ethyl-3,5-dimethyl-2H-pyrrol-2-ylidene)methyl]-4-methyl-1H-pyrrol-2-yl]methyl]-4-methyl-, 1,1-dimethylethyl ester, hydrobromide	CHCl$_3$	490(4.94)	78-1793-76
$C_{29}H_{42}N_2O_3$			
Chol-5-en-24-one, 3β-acetoxy-24-(1H-imidazol-2-yl)-	THF	278(4.13)	33-2753-76
Chol-5-en-24-one, 3β-acetoxy-24-(1H-imidazol-4-yl)-	THF	259(4.14)	33-2753-76
$C_{29}H_{42}N_2O_5$			
3-Piperidinol, 1-acetyl-2-[1-(1'-formyl-3',4,5,5',6',7',8',8'a-octahydro-6'-hydroxy-7,8'-dimethylspiro[2,1-benz-isoxazole-3(3aH),2'(1'H)-naphthalen)-6-yl]ethyl]-5-methyl-	EtOH	259(4.30)	39-1297-76C
$C_{29}H_{42}O_2$			
2,5-Cyclohexadien-1-one, 4-[[3,5-bis-(1,1-dimethylethyl)-4-hydroxyphenyl]methylene]-2,6-bis(1,1-dimethylethyl)-radical	MeOH	399(4.55)	64-0965-76B
	MeOH	409(--),435(5.18), 770(--),870(--)	64-0965-76B
anion	MeOH	577(5.19)	64-0965-76B
$C_{29}H_{42}O_4$			
Fusida-17(20),24-dien-16β,21-olactone, 11α-hydroxy-3-oxo-	n.s.g.	235(4.08)	39-0710-76C

Compound	Solvent	$\lambda_{max}(\log \epsilon)$	Ref.
$C_{29}H_{42}O_5$			
Spirost-1-en-3-one, 2-acetoxy-	EtOH	239.5(3.94)	95-0764-76
5β-Spirost-2-en-1-one, 2-acetoxy-	EtOH	233(3.81)	95-0764-76
Spirost-3-en-2-one, 3-acetoxy-, (5α,25R)-	EtOH	240(3.90)	95-0764-76
5β-	EtOH	242(3.68)	95-0764-76
Spirost-4-en-3-one, 4-acetoxy-	EtOH	246.5(4.18)	95-0863-76
Spirost-8-en-11-one, 2-acetoxy-	EtOH	254.5(3.98)	95-0764-76
4,8-Undecadienoic acid, 11-(6-acetoxy-3,4-dihydro-2,5,7,8-tetramethyl-2H-1-benzopyran-2-yl)-4,8-dimethyl-, methyl ester, [R-(E,E)]-	EtOH	227s(4.00),278(3.25), 285(3.30)	33-0290-76
$C_{29}H_{42}O_5S$			
5β-Chol-8-en-24-oic acid, 11 -acetoxy-3,3-[1,2-ethanediylbis(thio)]-7-oxo-, methyl ester	EtOH	248(4)	44-3840-76
$C_{29}H_{42}O_6$			
25D-Spirost-4-en-3-one, 9α-acetoxy-11β-hydroxy-	EtOH	241.5(4.21)	95-0764-76
$C_{29}H_{42}O_7$			
Spirostan-11-one, 2-acetoxy-2,9-epidioxy-, (2α,5β,25R)-	EtOH	310(1.58)	95-0764-76
$C_{29}H_{42}O_9$			
Corotoxigenin, α-L-rhamnopyranoside	MeOH	218(4.20)	94-2886-76
Corotoxigenin, L-rhamnoside	MeOH	218(4.20)	94-0108-76
$C_{29}H_{42}O_{10}$			
Convallatoxin	EtOH	217(4.22),303(1.45)	105-0048-76
$C_{29}H_{43}ClO_3$			
Cholesta-1,4-dien-3-one, 6β-acetoxy-7α-chloro-	EtOH	247(4.23)	56-0097-76
$C_{29}H_{43}NO$			
2,5-Cyclohexadien-1-one, 4-[(1-hexadecyl-4(1H)-pyridinylidene)ethylidene]-(also other solvents)	MeOH EtOH pyridine	486(--) 400(4.3),513(4.5) 605(--)	3-0450-76 3-0450-76 3-0450-76
$C_{29}H_{44}N_2O_5$			
3-Piperidinol, 1-acetyl-2-[1-[3',4,5,5'-6',7',8,8'a-octahydro-6'-hydroxy-1'-(hydroxymethyl)-7,8'-dimethyl-spiro[2,1-benzisoxazole-3(3aH),2'(1'H)-naphthalenyl]-6-yl]ethyl]-5-methyl-	EtOH	205(4.10),259(4.12)	39-1297-76C
$C_{29}H_{44}O_2$			
α-Tocotrienol, (2R,3'E,7'E)-	EtOH	225s(4.01),291(3.51)	33-0290-76
$C_{29}H_{44}O_5$			
27-Nor-5α-lanosta-8,16-diene-15,24-dione, 3β,23,28-trihydroxy-, (23S)-	EtOH	239(4.06)	33-1997-76
27-Nor-5α-lanost-8-ene-15,24-dione, 17,23-epoxy-3β,28-dihydroxy-, (23S)- (eucosterol)	EtOH	197(4.03),280(2.10), 293(2.20)	33-1997-76
$C_{29}H_{44}O_6$			
27-Norlanost-8-ene-15,24-dione, 17,23-	EtOH	197(4.04),288(2.16)	33-1997-76

Compound	Solvent	$\lambda_{max}(\log \epsilon)$	Ref.
epoxy-3,16,28-trihydroxy-, (3β,4β,16β-23S)- (16β-hydroxyeucosterol) (cont.)			33-1997-76
$C_{29}H_{44}O_8$ Evomonoside	EtOH	217(4.22)	105-0048-76
$C_{29}H_{44}O_9$ Coroglaucigenin, L-rhamnoside	MeOH	220(4.02)	94-0108-76
Malloside	MeOH	220(4.25)	94-0108-76
Periplogenin, α-L-rhamnoside	EtOH	217(4.17)	105-0048-76
$C_{29}H_{44}O_{10}$ Panoside	MeOH	220(4.17)	94-0108-76
$C_{29}H_{45}NO$ 2,5-Cyclohexadien-1-one, 2,4,6-tris(1,1-dimethylethyl)-4-(phenylmethyl)-, O-(1,1-dimethylethyl)oxime	hexane	259(3.98)	18-1142-76
$C_{29}H_{46}N_2O_4$ 5'H-Cholest-5-eno[6,7-d][1,2,3]oxadiazol-3-ol, acetate. 2'-oxide	EtOH	225(3.80),292(3.83)	94-1795-76
$C_{29}H_{46}N_6O_7$ Adenosine, 5'-[bis[4-(1,1-dimethylethoxy)-4-oxobutyl]amino]-5'-deoxy-2',3'-O-(1-methylethylidene)-	EtOH	259(4.13)	87-0684-76
$C_{29}H_{46}O_3$ Cholest-1-en-3-one, 2-acetoxy-	EtOH	240(3.89)	95-0764-76
Cholest-7-en-6-one, 5 -acetoxy-	MeOH	248(4.08)	13-0609-76A
$C_{29}H_{47}N_3O_6$ 2,2'-Anhydro-1-(3',3'-didecanoyl-β-D-arabinofuranosyl)cytosine, hydrochloride	MeOH-acid	234(4.03),264(4.09)	87-0663-76
$C_{29}H_{48}$ A-Norcholesta-1,3-diene, 3,6α,6β-trimethyl-	EtOH	245(3.58)	44-1943-76
$C_{29}H_{48}O_2$ Stigmast-4-en-3-one, 6β-hydroxy-	EtOH	246(4.11)	102-0427B-76
	EtOH	246(4.11)	102-1313-76
$C_{29}H_{48}O_5$ 1,2-Secocholestane-1,2-dioic acid, 3-oxo-, dimethyl ester	EtOH	274(1.79),281(1.78), 312(1.76)	56-0815-76
$C_{29}H_{49}N_3O_5$ 2,2'-Anhydro-1-(3'-arachidoyl-β-D-arabinofuranosyl)cytosine, hydrochloride	MeOH-acid	232(3.99),265(4.03)	87-0654-76
$C_{29}H_{49}N_3O_7$ 2(1H)-Pyrimidinone, 4-amino-1-[3,5-bis-O-(1-oxodecyl)-β-D-arabinofuranosyl]-	MeOH-acid	212(4.00),284(4.15)	87-0667-76
$C_{29}H_{50}N_2O$ 4-Azacholest-5-en-3-one, 4-[(1-methylethyl)amino]-	EtOH	244(3.99)	49-0511-76

Compound	Solvent	$\lambda_{max}(\log \epsilon)$	Ref.
$C_{29}H_{51}N_3O_6$ 2(1H)-Pyrimidinone, 4-amino-1-[3-O-(1- oxoeicosyl)-β-D-arabinofuranosyl]-	MeOH-acid	213(3.99),284(4.16)	87-0667-76
$C_{29}H_{54}S$ Thiophene, 3-methyl-5-(2-methyldecyl)- 2-tridecyl-	heptane	241(3.90)	34-0233-76

Compound	Solvent	$\lambda_{max}(\log \epsilon)$	Ref.
$C_{30}H_{12}O_{12}$ Hypericindicarboxylic acid	DMSO-pyridine	348(4.36),382(3.93), 486(4.10),518(4.01), 555(4.23),600(4.48)	12-1509-76
$C_{30}H_{14}F_6O_2$ Bianthrone, 2,2'-bis(trifluoromethyl)-	CH_2Cl_2	234(4.79),260s(4.42), 285s(4.18),304s(4.09), 398(4.19)	35-0615-76
$C_{30}H_{14}O_8$ Cyclopseudohypericin	DMSO-1% pyridine	344(4.54),386(3.97), 440(4.01),468(4.10), 506(3.89),542(4.35), 585(4.69)	12-1523-76
$C_{30}H_{14}O_{10}$ Phenanthro[1,10,9,8-opqra]perylene-3- carboxylic acid, 7,14-dihydro-1,6,8- 10,11,13-hexahydroxy-4-methyl-7,14- dioxo- (hypericincarboxylic acid)	DMSO-pyridine	347(4.43),390(4.09), 485(4.09),518(3.90), 555(4.30),600(4.59)	12-1509-76
$C_{30}H_{16}O_9$ Phenanthro[1,10,9,8-opqra]perylene- 7,14-dione, 1,3,4,6,8,13-hexahydroxy- 10-(hydroxymethyl)-11-methyl- (pseudohypericin)	DMSO-1% pyridine	344(4.54),384(4.14), 448(4.11),480(4.14), 516(4.00),554(4.39), 598(4.64)	12-1523-76
$C_{30}H_{16}O_{10}$ Phenanthro[1,10,9,8-opqra]perylene- 7,14-dione, 1,3,4,6,8,13-hexahydroxy- 10,11-bis(hydroxymethyl)-	DMSO-1% pyridine	344(4.45),380(4.09), 446(3.95),480(4.03), 512(3.95),551(4.33), 595(4.61)	12-1523-76
$C_{30}H_{17}BrO_2$ Benz[c]indeno[2,1-a]fluorene-13,14-di- one, 8b-bromo-4b,8b-dihydro-4b-phenyl-	$CHCl_3$	310(4.19)	77-0848-76
$C_{30}H_{18}Cl_2N_2$ Benzonitrile, 2,2'-([1,1'-biphenyl]- 4,4'-diyldi-2,1-ethenediyl)bis[4- chloro-	DMF	367(4.86)	33-0819-76
Benzonitrile, 2,2'-([1,1'-biphenyl]- 4,4'-diyldi-2,1-ethenediyl)bis[5- chloro-	DMF	370(4.89)	33-0819-76
Benzonitrile, 2,2'-([1,1'-biphenyl]- 4,4'-diyldi-2,1-ethenediyl)bis[6- chloro-	DMF	369(4.84)	33-0819-76
Benzonitrile, 4,4'-([1,1'-biphenyl]- 4,4'-diyldi-2,1-ethenediyl)bis[3- chloro-	DMF	374(4.91)	33-0819-76
$C_{30}H_{18}O_2$ Benz[a]indeno[2,1-c]fluorene-5,10-dione, 5a,14c-dihydro-5a-phenyl-	$CHCl_3$	279(4.69),454(3.32)	77-0848-76
Benz[c]indeno[2,1-a]fluorene-13,14-di- one, 4b,4c-dihydro-4b-phenyl-, trans	$CHCl_3$	297(4.55),373(4.14), 604(3.45)	77-0848-76
Benz[c]indeno[2,1-a]fluorene-13,14-di- one, 4b,13b-dihydro-4b-phenyl-	$CHCl_3$	292(4.05),332s(3.67), 430(3.23)	77-0848-76

$C_{30}H_{18}O_{10}-C_{30}H_{20}O_2$

Compound	Solvent	$\lambda_{max}(\log \epsilon)$	Ref.
$C_{30}H_{18}O_{10}$			
[9,9'-Bianthracene]-1,1',4,4'-tetrone, 2,2',5,5',10,10'-hexahydroxy-7,7'-dimethyl-	CHCl$_3$	255(4.60),318(4.13), 520(4.12),555(4.13), 598(4.10)	12-1535-76
$C_{30}H_{18}O_{12}$			
[9,9'-Bianthracene]-2,2'-dicarboxylic acid, 9,9',10,10'-tetrahydro-4,4',5-5',7,7'-hexahydroxy-10,10'-dioxo-	MeOH-1% HCOOH	270(4.36),375(4.34)	12-1509-76
$C_{30}H_{19}BrO_2$			
Naphtho[1,2-c]furan-1(3H)-one, 4-bromo-3,3,5-triphenyl-	CHCl$_3$	320(4.00),332(3.94)	18-2503-76
$C_{30}H_{19}I$			
Triphenylene, 2-iodo-1,12-diphenyl-	MeOH	272(4.74),290s(4.60), 310s(4.32)	39-1115-76B
$C_{30}H_{20}$			
Benzene, 1,1'-[1,2-bis(phenylethynyl)-1,2-ethenediyl]bis-	n.s.g.	265(4.48),268(4.53), 280(4.58),288(4.49), 296(4.58)	104-0223-76
Benzo[c]phenanthrene, 1,2-diphenyl-	MeOH	325s(4.06),340s(3.80), 365(2.80),383(2.41)	39-1115-76B
Phenanthrene, 4-(1-naphthalenyl)-3-phenyl-	MeOH	291(4.29),299(4.32), 334(2.99),353(2.85)	39-1115-76B
Phenanthrene, 1-phenyl-9-(4-phenyl-1-buten-3-ynyl)-	MeOH	222(4.64),250s(4.50), 260(4.57),286(4.58), 341(4.43)	39-1104-76B
Triphenylene, 1,12-diphenyl-	MeOH	267(4.76),287(4.60)	39-1115-76B
$C_{30}H_{20}ClNOS_2$			
8-Thia-6-azabicyclo[3.2.1]oct-2-en-7-one, 5-(4-chlorophenyl)-2,3,6-triphenyl-4-thioxo-	MeOH	237s(4.41),294(4.43), 427(4.05)	44-0818-76
1,3-Thiazocin-1-ium, 2-(4-chlorophenyl)-3,4-dihydro-8-mercapto-4-oxo-3,6,7-triphenyl-	MeOH	244(4.39),328(3.95), 460(3.68)	44-0818-76
$C_{30}H_{20}N_2$			
Benzonitrile, 2,2'-([1,1'-biphenyl]-4,4'-diyldi-2,1-ethenediyl)bis-	DMF	363(4.86)	33-0819-76
Benzonitrile, 4,4'-([1,1'-biphenyl]-4,4'-diyldi-2,1-ethenediyl)bis-	DMF	370(4.94)	33-0819-76
Spiro[3H-indazolium-3,1'-[1H]indene], 3-phenyl-1H-inden-1-ylide	CH$_2$Cl$_2$	265(4.28),284(4.20), 302s(4.11),310(4.10), 543(4.52)	24-2596-76
$C_{30}H_{20}N_2O$			
Cyclobutanone, 2-diazo-3,4-bis(diphenyl-methylene)-	CHCl$_3$	262(4.00),271(4.04), 338(3.83),420(3.71)	18-3173-76
2-Naphthalenol, 1-[(9,10-dihydro-9,10-[1',2']benzenoanthracen-2-yl)azo]-	CHCl$_3$	268(2.40),317(2.74), 423(2.89),485(2.97)	104-0185-76
$C_{30}H_{20}N_2O_4$			
[2,4'-Bioxazole]-5,5'(2H,4'H)-dione, 2,2',4,4'-tetraphenyl-	n.s.g.	249(4.37)	33-2149-76
$C_{30}H_{20}O_2$			
Bianthrone, 2,2'-dimethyl-	CH$_2$Cl$_2$	268(4.50),290s(4.29), 392(4.21)	35-0615-76

Compound	Solvent	$\lambda_{max}(\log \epsilon)$	Ref.
$C_{30}H_{20}O_{10}$			
[9,9'-Bianthracene]-2-carboxylic acid, 9,9',10,10'-tetrahydro-2',4,4',5,5'-7-hexahydroxy-7'-methyl-10,10'-dioxo-	MeOH-1% HCOOH	274(4.24),366(4.36)	12-1509-76
[8,8'-Bi-4H-1-benzopyran]-4,4'-dione, 2,3-dihydro-5,5',7,7'-tetrahydroxy-2,2'-bis(4-hydroxyphenyl)-	EtOH	275(4.45),298(4.48), 342s(4.32)	88-4509-76
$C_{30}H_{20}S_4$			
1,3-Dithiole, 2-(4,5-diphenyl-1,3-dithiol-2-ylidene)-4,5-diphenyl-	THF	302(4.47),407(3.69)	97-0360-76
cation	MeCN	461(--),698(--)	97-0360-76
dication	MeCN	251(4.40),528(4.03)	97-0360-76
$C_{30}H_{21}BrO_2$			
2(5H)-Furanone, 3-bromo-4-(2,2-diphenylethenyl)-5,5-diphenyl-	CHCl$_3$	238(4.25),328(4.13)	18-2503-76
2(5H)-Furanone, 4-(1-bromo-2,2-diphenylethenyl)-5,5-diphenyl-	EtOH	248(4.50),318(3.80), 332(3.75)	18-2503-76
$C_{30}H_{21}ClN_4O$			
Benzoxazole, 2-[4-[2-[2-(4-chlorophenyl)-5-methyl-2H-1,2,3-triazol-4-yl]-ethenyl]phenyl]-5-phenyl-	DMF	360(4.83)	33-2469-76
Benzoxazole, 2-[4-[2-[2-(4-chlorophenyl)-5-methyl-2H-1,2,3-triazol-4-yl]-ethenyl]phenyl]-6-phenyl-	DMF	367(4.86)	33-2469-76
Benzoxazole, 2-[4-[2-[2-(4-chlorophenyl)-5-phenyl-2H-1,2,3-triazol-4-yl]-ethenyl]phenyl]-5-methyl-	DMF	365(4.76)	33-2469-76
Benzoxazole, 2-[4-[2-[2-(4-chlorophenyl)-5-phenyl-2H-1,2,3-triazol-4-yl]-ethenyl]phenyl]-6-methyl-	DMF	365(4.77)	33-2469-76
Benzoxazole, 2-[4-[2-[2-(4-chlorophenyl)-2H-1,2,3-triazol-4-yl]ethenyl]-phenyl]-5-(phenylmethyl)-	DMF	355(4.86)	33-2469-76
$C_{30}H_{21}NOS_2$			
8-Thia-6-azabicyclo[3.2.1]oct-2-en-7-one, 2,3,5,6-tetraphenyl-4-thioxo-	MeOH	239(4.32),293(4.38), 425(4.01)	44-0818-76
1,3-Thiazocin-1-ium, 3,4-dihydro-8-mercapto-4-oxo-2,3,6,7-tetraphenyl-	MeOH	240(4.56),328(4.11), 463(3.88)	44-0818-76
$C_{30}H_{21}NO_2S$			
8-Thia-6-azabicyclo[3.2.1]oct-2-ene-4,7-dione, 2,3,5,6-tetraphenyl-	MeOH	246(4.45),300(4.26), 353(4.17)	44-0818-76
$C_{30}H_{21}N_2$			
Spiro[3H-indazole-3,1'-[1H]inden]ium, 2-(1-phenyl-1H-inden-3-yl)-, perchlorate	CH$_2$Cl$_2$	475(4.03)	24-2596-76
$C_{30}H_{21}N_5O$			
1,3,4-Oxadiazole, 2-[1,1'-biphenyl]-4-yl-5-[4-[2-(2-phenyl-2H-1,2,3-triazol-4-yl)ethenyl]phenyl]-	DMF	350(4.81)	33-2469-76
1,3,4-Oxadiazole, 2-[4-[2-(2,5-diphenyl-2H-1,2,3-triazol-4-yl)ethenyl]phenyl]-5-phenyl-	DMF	355(4.71)	33-2469-76

Compound	Solvent	$\lambda_{max}(\log \epsilon)$	Ref.
$C_{30}H_{22}$ 1,1'-Bi-1H-indene, 1,3'-diphenyl-	dioxan	231(4.64),260(4.13), 298(3.25)	24-2596-76
	DMSO-EtOH- NaOEt	420(4.24)	24-2596-76
$C_{30}H_{22}NO$ Pyrylium, 4-(9-methyl-9H-carbazol-3-yl)- 2,6-diphenyl-, perchlorate	MeCN	270(4.40),338(4.19), 390(4.32),498(4.59)	4-0073-76
$C_{30}H_{22}NOS$ Pyrylium, 4-(10-methyl-10H-phenothiazin- 3-yl)-2,6-diphenyl-, perchlorate	MeCN	273(4.42),392(4.53), 585(4.38)	4-0073-76
$C_{30}H_{22}N_2$ 1H-Indole, 3,3'-ethenediylidenebis[2- phenyl-	MeOH	242(4.85),306(4.48)	23-1020-76
$C_{30}H_{22}N_2O_2$ Methanone, (4-benzoyl-1,5-diphenyl-1H- pyrazol-3-yl)(4-methylphenyl)-	EtOH	259(4.56)	4-0989-76
Methanone, [1-(4-methylphenyl)-5-phenyl- 1H-pyrazole-3,4-diyl]bis[phenyl-	EtOH	256(4.62)	4-0989-76
$C_{30}H_{22}N_2O_4$ Benzamide, N-(1-benzoyl-2,4-dioxo-3,5- diphenyl-3-pyrrolidinyl)-, trans	n.s.g.	228(4.45),268s(3.91)	33-2149-76
$C_{30}H_{22}N_4NiO_2$ Nickel, bis(3-phenyl-2-quinolinamine- 1-oxidato-N^2,O^1)-, (T-4)-	$CHCl_3$	244s(4.30),250s(4.39), 256(4.45),262(4.43), 317(3.69),362s(3.49), 434(3.30)	12-0357-76
$C_{30}H_{22}N_4O$ Acetamide, N-[2-[(2,2-diphenylcyclobuta- [b]quinoxalin-1(2H)-ylidene)amino]- phenyl]-	$CHCl_3$	264(4.71),364(4.37), 405(3.79)	5-0284-76
Benzoxazole, 2-[4-[2-(2,5-diphenyl-2H- 1,2,3-triazol-4-yl)ethenyl]phenyl]- 5-methyl-	DMF	363(4.76)	33-2469-76
Benzoxazole, 2-[4-[2-(2,5-diphenyl-2H- 1,2,3-triazol-4-yl)ethenyl]phenyl]- 6-methyl-	DMF	365(4.76)	33-2469-76
Benzoxazole, 2-[4-[2-(5-methyl-2-phenyl- 2H-1,2,3-triazol-4-yl)ethenyl]phenyl]- 5-phenyl-	DMF	363(4.83)	33-2469-76
Benzoxazole, 2-[4-[2-(5-methyl-2-phenyl- 2H-1,2,3-triazol-4-yl)ethenyl]phenyl]- 6-phenyl-	DMF	367(4.84)	33-2469-76
Benzoxazole, 5-(phenylmethyl)-2-[4-[2- (2-phenyl-2H-1,2,3-triazol-4-yl)eth- enyl]phenyl]-	DMF	354(4.84)	33-2469-76
$C_{30}H_{22}O$ 5H-Benzo[6,7]cyclohept[1,2-b]indeno- [1,2-e]pyran, 6,7-dihydro-13-(1H- inden-2-yl)-	dioxan	266(4.26),322(4.51), 355(4.52),595(3.54)	78-2225-76

Compound	Solvent	$\lambda_{max}(\log \epsilon)$	Ref.
$C_{30}H_{22}O_2$			
Cyclopropanecarboxylic acid, bis(diphenylmethylene)-	CHCl$_3$	263(4.41),373(4.56)	18-3173-76
2(3H)-Furanone, 4-(diphenylethenylidene)dihydro-5,5-diphenyl-	CHCl$_3$	270(4.16)	18-2503-76
2(5H)-Furanone, 4-(2,2-diphenylethenyl)-5,5-diphenyl-	CHCl$_3$	327(4.28)	18-2503-76
$C_{30}H_{22}O_6$			
Dibenzo[c,h]-1,4-dioxino[2,3-a]pyrano-[2,3-j]xanthene-3,18(2H,15H)-dione, 2,2,15,15-tetramethyl- (guayin)	EtOH	226(4.81),250s(4.51), 282(4.57),294s(4.53), 323(4.22),414(3.84)	78-0543-76
$C_{30}H_{22}O_8$			
[9,9'-Bianthracene]-10,10'(9H,9'H)-dione, 2,2',4,4',5,5'-hexahydroxy-7,7'-dimethyl-	MeOH-1% HCOOH	279(4.25),364(4.40)	12-1509-76
[1,1'-Bianthracene]-4,4',9,9',10,10'-hexone, 2,2',3,3'-tetrahydro-5,5'-dihydroxy-7,7'-dimethyl-	dioxan	288(4.21),440(3.82)	78-0333-76
Bisdeoxyrugulosin	dioxan	253(3.84),282(3.64), 300(3.41),313(3.34), 390(3.73)	78-0333-76
$C_{30}H_{22}O_{10}$			
Benzo[j]fluoranthene-3,4,7,8,9-pentol, pentaacetate	n.s.g.	248(4.80),274(3.96), 284(4.15),303(4.32), 315(4.47),323(4.54), 328(4.52),336(4.24), 358(3.75),374(4.02), 385(3.89),393(4.14), 420(3.66),444(3.53)	39-2149-76C
[6,8'-Bi-4H-1-benzopyran]-4,4'-dione, 2,2',3,3'-tetrahydro-5,5',7,7'-tetrahydroxy-2,2'-bis(4-hydroxyphenyl)-, [S-(R*,R*)]-	MeOH	208(4.68),223(4.65), 294(4.49),336(3.80)	39-0098-76C
	MeOH-NaOAc	257(4.38),271(4.42), 300(4.36),320(4.36)	39-0098-76C
	MeOH-AlCl$_3$	224(4.76),256(4.25), 315(4.63),384(3.88)	39-0098-76C
$C_{30}H_{22}S$			
4H-Thiopyran, 4-(diphenylmethylene)-3,5-diphenyl-	CH$_2$Cl$_2$	334(4.12)	44-2279-76
$C_{30}H_{23}NO$			
4(1H)-Pyridinone, 1-methyl-2,3,5,6-tetraphenyl-	MeOH	236s(4.36),276(4.12)	44-0818-76
$C_{30}H_{23}NO_2$			
Spiro[cyclopropane-1,9'-[9H]fluorene]-2-carboxamide, 3-benzoyl-N-(4-methylphenyl)-	EtOH	252(4.60)	115-0001-76
$C_{30}H_{23}NS$			
4(1H)-Pyridinethione, 1-methyl-2,3,5,6-tetraphenyl-	MeOH	238(4.18),272s(3.80), 360(4.18)	44-0818-76
$C_{30}H_{23}N_3O_2$			
1H-Pyrazole-4-carboxamide, 3-benzoyl-1-(4-methylphenyl)-N,5-diphenyl-	EtOH	255(4.68)	4-0989-76

Compound	Solvent	$\lambda_{max}(\log \epsilon)$	Ref.
$C_{30}H_{23}N_3O_2Se$ 5H-1,3,4-Selenadiazolo[3,2-a]pyrimidin-5-one, 6,7-dihydro-7-(4-methoxyphenyl)-2,6,6-triphenyl-	EtOH	262(4.37),294s(3.93), 321(3.81)	94-2532-76
$C_{30}H_{23}N_3O_3$ 5H-1,3,4-Oxadiazolo[3,2-a]pyrimidin-5-one, 6,7-dihydro-7-(4-methoxyphenyl)-2,6,6-triphenyl-	EtOH	261(4.22),292(4.08), 304s(4.03)	94-2532-76
$C_{30}H_{24}$ Benzene, 1,1'-(1,4-diphenyl-1,2,3-butatriene-1,4-diyl)bis[4-methyl-	benzene C_6H_{12}	311(3.86),429(4.54) 273(4.52),278(4.56), 310(3.87),425(4.56)	19-0089-76 19-0089-76
9H-Fluorene, 2,2'-(1,2-dimethyl-1,2-ethenediyl)bis-, cis	isooctane	271(4.658)	44-1935-76
trans	isooctane	278(4.671)	44-1935-76
$C_{30}H_{24}Cl_2N_2O_2$ Acridine, 6-chloro-9-(6-chloro-2-methoxy-10-methyl-9(10H)-acridinylidene)-9,10-dihydro-2-methoxy-10-methyl-	CH_2Cl_2	241(4.71),302(4.24), 446(4.20)	35-0615-76
$C_{30}H_{24}F_2$ Benzene, 1,1',1'',1'''-(2,5-difluoro-1,5-hexadiene-1,6-diylidene)tetrakis-	MeOH	250(3.45)	44-0940-76
$C_{30}H_{24}NO$ Pyrylium, 4-[4-(methylphenylamino)phenyl]-2,6-diphenyl-	MeCN	265(4.42),290(4.42), 378(4.43),530(4.81)	4-0073-76
$C_{30}H_{24}N_2$ 3,3'-Bi-1H-indole, 1,1'-dimethyl-2,2'-diphenyl-	$CHCl_3$	<u>300(4.4)</u>	39-0309-76B
$C_{30}H_{24}N_2O$ 1-Propanone, 1-phenyl-3-[2-phenyl-4-(phenylamino)-3-quinolinyl]-	dioxan	<u>235(4.8),420(3.8)</u>	56-0625-76
$C_{30}H_{24}N_2O_2$ Benzamide, N-benzoyl-N-(3-ethyl-3-phenyl-3H-indol-2-yl)-	MeOH	320(4.16)	56-0857-76
$C_{30}H_{24}N_2O_9$ 2,4(1H,3H)-Pyrimidinedione, 1-(2,3,5-tri-O-benzoyl-β-D-arabinofuranosyl)-	MeOH	235(4.69),259(4.12)	44-3138-76
$C_{30}H_{24}N_4O_3$ Spiro[2H-1-benzopyran-2,2'-[2H]indole], 1',3'-dihydro-3',3'-dimethyl-6-[(4-nitrophenyl)azo]-1'-phenyl-	EtOH	296(4.01),401(4.21)	28-0691-76A
$C_{30}H_{24}O$ 2,4-Cyclopentadien-1-ol, 2-methyl-1,3,4,5-tetraphenyl-	C_6H_{12}	246(4.445),341(3.699)	5-0793-76
Cyclopropanemethanol, 2,3-bis(diphenylmethylene)-	$CHCl_3$	260(4.38),375(4.46)	18-3173-76
Indeno[2,1-b]pyran, 9-(3,4-dihydro-2-naphthalenyl)-4-methyl-2-(4-methylphenyl)-	dioxan	267(4.34),331(4.36), 358(4.32),570(3.48)	78-2225-76

Compound	Solvent	λ_{max}(log ϵ)	Ref.
Indeno[2,1-b]pyran, 9-(1H-inden-2-yl)-2,4-dimethyl-3-(phenylmethyl)-	dioxan	254(4.36),315(4.34), 340(4.32),547(3.53)	78-2225-76
$C_{30}H_{24}OS$ 1H-Thiopyran, 1-(4-methoxyphenyl)-2,4,6-triphenyl-	C_6H_{12}	232(4.50),305(4.20), 355(3.70),530(3.70)	4-0237-76
$C_{30}H_{24}O_2$ Benzene, 1,1'-(1,4-diphenyl-1,2,3-butatriene-1,4-diyl)bis[4-methoxy-	benzene C_6H_{12}	440(4.55) 269s(4.39),284(4.43), 435(4.56)	19-0089-76 19-0089-76
Benzenepropanoic acid, α-(diphenylethenylidene)-β-phenyl-, methyl ester	$CHCl_3$	252(4.27)	18-2645-76
$C_{30}H_{24}O_6$ 4H-1-Benzopyran-4-one, 5-hydroxy-6-methoxy-7-(phenylmethoxy)-2-[4-(phenylmethoxy)phenyl]-	EtOH EtOH-AlCl$_3$	276(4.3),339(4.0) 290(--),350(--)	2-0592-76 2-0592-76
$C_{30}H_{25}NO_2$ Cyclopropanecarboxamide, 2-benzoyl-N-(4-methylphenyl)-3,3-diphenyl-	EtOH	250(4.58)	115-0001-76
Cyclopropanecarboxamide, 3-(4-methylbenzoyl)-N,2,2-triphenyl-	EtOH	254(4.56)	115-0001-76
$C_{30}H_{25}NO_3S$ 3-Thiophenecarboxylic acid, 5-benzoyl-4-methyl-2-(1-methyl-2-phenyl-1H-indol-3-yl)-, ethyl ester	EtOH	290(4.32),380(3.96)	142-0009-76
$C_{30}H_{25}N_3OS$ 4H-Pyrimido[2,1-b]benzothiazol-4-one, 2-[4-(dimethylamino)phenyl]-2,3-dihydro-3,3-diphenyl-	EtOH	240(4.44),261(4.33), 293s(3.98)	94-2532-76
$C_{30}H_{25}N_3O_2$ 1H-Pyrazole-4-carboxamide, 3-benzoyl-4,5-dihydro-N-(4-methylphenyl)-5,5-diphenyl-	EtOH	252(4.35),336(4.17)	115-0001-76
4H-Pyrimido[2,1-b]benzoxazol-4-one, 2-[4-(dimethylamino)phenyl]-2,3-dihydro-3,3-diphenyl-	EtOH	263(4.33)	94-2532-76
$C_{30}H_{25}N_5O_3$ Benzamide, N,N'-[[(benzoylamino)[(4-methylphenyl)imino]methyl]carbonimidoyl]bis-	MeOCH$_2$CH$_2$OH	240s(4.43),274(4.52)	40-0488-76
$C_{30}H_{25}N_5O_4$ Benzamide, N,N'-[[(benzoylamino)[(4-methoxyphenyl)imino]methyl]carbonimidoyl]bis-	MeOCH$_2$CH$_2$OH	239(4.48),273s(4.46)	40-0488-76
$C_{30}H_{26}N_2O_{10}S$ 2aH-2-Thia-6a,10c-diazabenzo[a]cyclopent[cd]azulene-1,3,4,5,6-pentacarboxylic acid, 2a-phenyl-, 1-ethyl 3,4,5,6-tetramethyl ester	MeOH	225(4.48),265s(4.21)	44-0129-76

Compound	Solvent	$\lambda_{max}(\log \epsilon)$	Ref.
$C_{30}H_{26}O_2S_2$ 1-Propanone, 3,3'-dithiobis[1,3-diphen-yl-	EtOH	240(4.52),285(3.88), 310(3.77)	32-0139-76
$C_{30}H_{26}O_2S_3$ 1-Propanone, 3,3'-trithiobis[1,3-diphen-yl-	EtOH	247(4.43)	32-0139-76
$C_{30}H_{26}O_6$ 1,4-Naphthalenedione, 2-[(5-acetoxy-4,5',6-trimethyl-3-methylene-3'-oxo-spiro[benzofuran-2(3H),1'-[4]cyclopen-ten-2'-yl)methyl]-3-methyl-, trans	EtOH	233(4.57),247s(4.41), 262s(4.30),320(3.80), 330(3.76)	39-2067-76C
1H-Naphtho[2'',3'':2',3']cyclobuta[1',2'-1,2]pentaleno[6a,1-b]benzofuran-1,10,15-trione, 7-acetoxy-9,9a,16,16a-tetrahydro-3,6,8,9a-tetramethyl-, (3aS*,8bS*,9aα,15aS*,16'aβ)-	EtOH	230(3.20),295(4.52)	39-2067-76C
$C_{30}H_{26}O_7$ 1,4-Dibenzofurandione, 8-acetoxy-4a-[(1,4-dihydro-3-methyl-1,4-dioxo-2-naphthalenyl)methyl]-4a,9b-dihydro-2,7,9,9b-tetramethyl-, cis	EtOH	238(4.48),252(4.35), 270(4.43),325(3.51)	39-2067-76C
$C_{30}H_{26}O_{10}$ 4H-1-Benzopyran-5-carboxylic acid, 7-methyl-4-oxo-8-(1,4,5-triacetoxy-7-methyl-2-naphthalenyl)-, ethyl ester	n.s.g.	234(4.48),266s(3.76), 301(3.84),331s(3.12)	39-2155-76C
$C_{30}H_{26}S_4$ 10,10'-Bidibenzo[b,f]thiepin, 10,10',11-11'-tetrahydro-8,8'-bis(methylthio)-	MeOH	263.5(4.27)	73-0443-76
$C_{30}H_{27}N_3O_3$ Benzenecarboximidamide, N-[2-(2,2-di-methyl-1-oxopropyl)-3,6-dioxo-4-(phenylamino)-1,4-cyclohexadien-1-yl]-N-phenyl-	EtOH	231(4.40),263s(--), 400(4.27)	5-1809-76
$C_{30}H_{28}N_2O_4S$ 2,4-Pentadienoic acid, 5,5'-thiobis[2-cyano-5-(4-methylphenyl)-, diethyl ester	HOAc	416(4.56)	48-0705-76
$C_{30}H_{28}N_6O_3$ 1H-Pyrrole-2-carboxamide, 5-[bis[[(phen-ylmethyl)amino]carbonyl]amino]-4-cya-no-3-methyl-N-(phenylmethyl)-	MeOH	240(4.08)	24-2983-76
$C_{30}H_{28}O_4$ 3H-Benz[e]inden-3-one, 2,3a,4,5-tetrahy-dro-7-methoxy-3a-methyl-2-(3,3a,4,5-tetrahydro-7-methoxy-3a-methyl-3-oxo-2H-benz[e]inden-2-ylidene)-, [R*,R*-(E)]-	EtOH	510(3.81)	25-1030-76
$C_{30}H_{28}O_7$ 5,9-Methano-1H-benzocycloheptene-1,4,6-trione, 9-acetoxy-10-[(1,4-dihydro-	EtOH	246(4.50),267(4.37), 273(4.36),334(3.57)	39-2067-76C

Compound	Solvent	$\lambda_{max}(\log \epsilon)$	Ref.
3-methyl-1,4-dioxo-2-naphthalenyl)-methyl]-4a,5,9,9a-tetrahydro-2,5,7,9a-tetramethyl- (cont.)			39-2067-76C
$C_{30}H_{29}NO$			
Piperidine, 1-[3-(3,4-dihydro-2-naphthalenyl)-1-oxo-2,3-diphenyl-2-propenyl]-	MeOH	226s(4.38),277(4.20), 312(4.24)	87-0414-76
$C_{30}H_{29}NO_2$			
1H-Benz[g]indole-3-methanol, 1-butyl-5-methoxy-α,2-diphenyl-	EtOH	222(4.54),275(4.69)	42-0606-76
1H-Benz[g]indole-3-methanol, 5-methoxy-1-(2-methylpropyl)-α,2-diphenyl-	EtOH	220(4.47),280(4.59)	42-0606-76
$C_{30}H_{30}N_2O_9$			
6H-Pyrido[4,3-b]carbazol-9-ol, 5,11-dimethyl-6-(2,3,4-tri-O-acetyl-β-D-xylopyranosyl)-, acetate	n.s.g.	266(4.42),332(4.27)	28-1001-76A
$C_{30}H_{30}N_8O_2$			
Pyrimido[4,5-g]quinazoline-4,9-diamine, 2,7-dimorpholino-N,N'-diphenyl-	EtOH	266(4.58),312(4.36), 425(3.71)	24-2921-76
$C_{30}H_{30}O_8$			
4,7-Etheno-4H-cyclobuta[1,5]cyclopenta-[1,2,3-ef]heptalene-1,2,5,6-tetracar-boxylic acid, 2a,7-dihydro-4,9,11,12-tetramethyl-, tetramethyl ester	dioxan	280(4.16),292(4.17), 347(3.53),426(3.15)	89-0108-76
2(3H)-Furanone, dihydro-4-[[2-(4-hydr-oxy-3-methoxyphenyl)-7-methoxy-3-methyl-5-benzofuranyl]methyl]-3-[(4-hydroxy-3-methoxyphenyl)methyl]-	MeOH	309(4.40)	88-3961-76
$C_{30}H_{31}NO_2$			
Spiro[cyclohexane-1,7'(1'H)-quinolin]-5'(6'H)-one, 1'-ethyl-4',8'-dihydro-4'-(2-oxo-2-phenylethylidene)-2'-phenyl-	MeOH	415s(4.39),430(4.40)	24-1549-76
$C_{30}H_{31}NO_8$			
8H-Dibenzo[a,g]quinolizin-8-one, 13-(3,4-dimethoxyphenyl)-5,6-dihydro-2,3,10,11,12-pentamethoxy-	n.s.g.	231(4.60),252(4.53), 334(4.36),354s(4.23), 370s(4.04)	44-2201-76
$C_{30}H_{31}NO_{10}$			
Viridicatumtoxin	EtOH	237(4.34),285(4.51), 317(3.44),331(3.40), 347(3.28),424(3.90)	77-0728-76
$C_{30}H_{32}O_9$			
Lappaol A	MeOH	283(3.89)	88-3961-76
$C_{30}H_{33}NO_8$			
8H-Dibenzo[a,g]quinolizin-8-one, 13-(3,4-dimethoxyphenyl)-5,6,13,13a-tetrahydro-2,3,10,11,12-pentameth-oxy-, 13,13a-cis	n.s.g.	220(4.75),262s(4.08), 276(4.11),310(3.48)	44-2201-76
Isoquinoline, 1-[(3,4-dimethoxyphenyl)-methylene]-1,2,3,4-tetrahydro-6,7-di-methoxy-2-(3,4,5-trimethoxybenzoyl)-, (Z)-	n.s.g.	230(4.51),281(4.27), 322(4.29)	44-2201-76

Compound	Solvent	$\lambda_{max}(\log \epsilon)$	Ref.
$C_{30}H_{33}N_3O_4$ D-Ribitol, 1-C-(5-amino-1-methyl-1H-pyrazol-4-yl)-1,4-anhydro-2,3,5-tris-O-(phenylmethyl)-, (S)-	MeOH	233(3.78)	44-3000-76
$C_{30}H_{33}N_5S_3$ Propanethioamide, N,N,2-trimethyl-2-[[tetrahydro-1,3,5-tris(phenylmethyl)-4,6-dithioxo-1,3,5-triazin-2(1H)-ylidene]amino]-	EtOH	285(4.62)	33-2768-76
$C_{30}H_{34}Cl_3N_4O_{14}P$ Thymidine, P-(2,2,2-trichloroethyl)thymidylyl(5'→3')-, 5'-(phenoxyacetate)	MeOH	264(4.28)	35-3655-76
$C_{30}H_{34}O_{13}S$ Breynogenin, tetraacetyl-	MeOH MeOH-NaOH	236(4.17) 300(4.36),352(3.76)	94-0114-76 94-0114-76
$C_{30}H_{36}N_6O_4S$ L-Phenylalanine, 4-[[(4-amino-2-methyl-5-pyrimidinyl)methyl]formylamino]-3-[(2-amino-1-oxo-3-phenylpropyl)thio]-3-pentenyl ester, trihydrochloride, (S)-	EtOH	246(4.17)	94-0852-76
$C_{30}H_{36}O_6$ 6-Dehydrosantonin dimer 1H,11H-3a,6:13a,16-Dimethano-10H,20H-biscyclopenta[1,3]cyclopropa[1,2-c-1',2'-1][1,10]dioxacyclooctadecin-1,8,11,18(3bH,13bH)-tetrone, 4,5,6,7,14,15,16,17-octahydro-3b,7,13b,17-tetramethyl-	EtOH EtOH	243(4.44) 234(3.96)	44-1256-76 44-1256-76
$C_{30}H_{36}O_7$ D(17a)-Homo-18-norergosta-2,13,15,17-tetraen-1-one, 26-acetoxy-6,7:22,26-24,25-triepoxy-5-hydroxy-, (5α,6α,7α,22R,24S,25S,26S)-	EtOH	222(4.20)	39-0669-76C
$C_{30}H_{36}O_8$ Tricyclo[11.3.1.1⁵,⁹]octadeca-1(17),5,7-9(18),13,15-hexaene-3,3,11,11-tetracarboxylic acid, tetraethyl ester	C_6H_{12}	265(2.63),271s(2.56)	138-1405-76
$C_{30}H_{37}NO_7$ 24,25,26,27-Tetranorlanosta-5,8,22-triene-3,7,11-trione, 6-acetoxy-2-(acetoximino)-	dioxan	270(4.06)	5-1214-76
$C_{30}H_{37}N_2$ Acridinium, 9-[4-(dipentylamino)phenyl]-10-methyl-, triiodide	CH_2Cl_2	260(4.97),291(4.62), 358(4.43),628(3.97)	104-1531-76
$C_{30}H_{38}O_6$ 2,5-Cyclohexadiene-1,4-dione, 2,2'-(8-hexadecyne-1,16-diyl)bis[6-methoxy-	MeOH	268(4.38),357(3.20)	12-1979-76
$C_{30}H_{38}O_{13}$ Wallichoside, pentaacetate	EtOH	220(4.56),256(4.19),	102-0995-76

Compound	Solvent	$\lambda_{max}(\log \epsilon)$	Ref.
Wallichoside, pentaacetate (cont.)		298(3.34)	102-0995-76
$C_{30}H_{40}O_5$			
Cucurbitacin S	EtOH	268(3.95)	102-0559-76
	EtOH-base	315(--)	102-0559-76
$C_{30}H_{40}O_6$			
24-Norchola-5,8,22-triene-7,11-dione,	dioxan	272(4.06)	5-1206-76
3,6-diacetoxy-4,4,14-trimethyl-			
Pregn-5-ene-3,12,14,17,20-pentol,	EtOH	217(4.30),223(4.28),	94-2185-76
12-(3-phenyl-2-propenoate),		279(4.15)	
(3β,12β,14β,17α,20S)-			
$C_{30}H_{40}O_8$			
5α-Ergost-2-ene-1,12-dione, 6α,7α:22,26-	EtOH	224(3.77)	39-0296-76C
24,25-triepoxy-5,26-dihydroxy-, 26-			
acetate, [(20S,22R,24S,25S,26R)]-			
Pregnane-3,8,12,14,17,20-hexol, 5,6-ep-	MeOH	217(4.18),223(4.11),	94-3085-76
oxy-, 12-(3-phenyl-2-propenoate),		281(4.30)	
[3β,5α,6α,12β(E),14β,17α,20S]-			
$C_{30}H_{40}O_9$			
Gnidilatidin methanolysis product	MeOH	231(4.48)	44-3850-76
Withanolide E, 4β-hydroxy-, acetate	EtOH	218(4.24)	102-0340-76
$C_{30}H_{40}O_{12}$			
Deoxowallichoside, pentaacetate	EtOH	220(4.06),270(2.97)	102-0995-76
$C_{30}H_{41}NO_2$			
2H-1-Benzopyrano[4,3-c]pyridin-10-ol, 8-	EtOH	280(3.90)	87-0445-76
(1,2-dimethylheptyl)-5,5-dimethyl-2-			
(phenylmethyl)-1,3,4,5-tetrahydro-			
hydriodide	EtOH	280(4.09)	87-0445-76
$C_{30}H_{41}NO_6$			
1H-Isoindol-1-one, octahydro-6,7a-di-	EtOH	195(4.70),200(4.52),	33-0914-76
hydroxy-4-methyl-5-methylene-7-[4-		247(2.36),253(2.42),	
methyl-7-(tetrahydro-3-methoxy-5-		259(2.50),264(2.46),	
oxo-2-furanyl)-1-heptenyl]-3-(phen-		268(2.42),287(2.35)	
ylmethyl)-			
24-Norchola-8,22-diene-6,7,11-trione,	dioxan	275(3.71),305s(3.59)	5-1214-76
3-acetoxy-4,4,14-trimethyl-,			
6-(O-acetyloxime), (3β,5α)-			
2,10-Undecadienoic acid, 4-hydroxy-8-	EtOH	195(4.78),208(4.48),	33-0914-76
methyl-11-[octahydro-3a,5-dihydroxy-		251(2.64),258(2.55),	
7-methyl-6-methylene-3-oxo-1-(phen-		264(2.39),268(2.25)	
ylmethyl)-1H-isoindol-4-yl]-, methyl			
ester, [1S-[1α,3aα,4β(2E,4R*,8S*-			
10E),5α,7β,7aα]]-			
$C_{30}H_{42}$			
1,4-Cyclohexadiene, 3,3'-(2,4-hexadiene-	CH_2Cl_2	400s(--),423(4.95),	5-1103-76
1,6-diylidene)bis[1,2,4,5,6,6-hexa-		445(4.95)	
methyl-, (E,E)-			
$C_{30}H_{42}N_2O_3S$			
Neothiobinupharidine, 6-hydroxy-	EtOH-acid	298(3.43)	44-0291-76
Neothiobinupharidine, 6'-hydroxy-	EtOH-acid	275(2.67)	44-0291-76

Compound	Solvent	λ_{max}(log ϵ)	Ref.
$C_{30}H_{42}O_4$			
Lanosta-7,9(11)-dien-18-oic acid, 20-hydroxy-3,23-dioxo-, γ-lactone	MeOH	236(4.07),244(4.10), 252(3.96)	78-2313-76
Lanosta-9(11),24-dien-18-oic acid, 20-hydroxy-3,23-dioxo-, γ-lactone	MeOH	238(4.13)	78-2353-76
	MeOH-KOH	244(3.98),276(3.97)	78-2353-76
Lanosta-9(11),25-dien-18-oic acid, 20-hydroxy-3,23-dioxo-, γ-lactone	MeOH	237(4.00)	78-2353-76
	MeOH-KOH	243(3.83),277(3.75)	78-2353-76
$C_{30}H_{42}O_5$			
Spirosta-3,9(11)-dien-2-one, 3-acetoxy-11-methyl-, (5α,25R)-	EtOH	236.5(3.90)	95-0863-76
Spirost-1-en-3-one, 2-acetoxy-11-methylene-, (5α,25R)-	EtOH	238(3.99)	95-0863-76
Spirost-1-en-3-one, 2-acetoxy-12-methylene-, (5α,25R)-	EtOH	237(4.02)	95-0863-76
Spirost-3-en-2-one, 3-acetoxy-11-methylene-, (5α,25R)-	EtOH	239.5(3.87)	95-0863-76
Spirost-3-en-2-one, 3-acetoxy-12-methylene-, (5α,25R)-	EtOH	240(3.93)	95-0863-76
Spirost-4-en-3-one, 4-acetoxy-11-methylene-, (25R)-	EtOH	245.5(4.12)	95-0863-76
$C_{30}H_{42}O_6$			
2,5-Cyclohexadiene-1,4-dione, 2,2'-(1,16-hexadecanediyl)bis[6-methoxy-	dioxan	265(3.79),355(2.75)	12-1979-76
24-Norchola-5,8-diene-7,11-dione, 3β,6-diacetoxy-4,4,14-trimethyl-	dioxan	272(4.04)	5-1206-76
$C_{30}H_{42}O_8$			
Cochlioquinone A, 11-deoxy-11-oxo-	MeOH	260s(3.92),278(3.96), 380(3.03)	32-0147-76
2,5-Cyclohexadiene-1,4-dione, 2,2'-(1,16-hexadecanediyl)bis[3-hydroxy-6-methoxy-	dioxan	278(4.58),407(2.97)	12-1979-76
$C_{30}H_{42}O_9$			
Glycopenupogenin	MeOH	217(4.20),223(4.18), 280(4.20)	94-3085-76
$C_{30}H_{43}NO_5$			
Lanosta-5,8-diene-2,3,7,11-tetrone, 6-hydroxy-, 2-oxime	dioxan	230(4.31),276(3.87), 344(3.51)	5-1214-76
$C_{30}H_{44}N_2O_3$			
Acetamide, N-[4-(heptyloxy)phenyl]-2-[[4-(nonyloxy)phenyl]imino]-	EtOH	242(3.47)	65-0702-76
$C_{30}H_{44}N_2O_5$			
5α-Lanost-8-ene-2,3,6,7,11-pentone, 2,6-dioxime	EtOH	246(3.97),275s(--)	5-1214-76
$C_{30}H_{44}O_3$			
Cyclohexanone, 2-[hydroxy[4-(1-methylethyl)phenyl]methyl]-3-methyl-6-[1-methyl-1-[6-methyl-3-(1-methylethylidene)-2-oxocyclohexyl]ethyl]-	isooctane	222(4.12),256(3.99), 315(2.32)	22-0985-76
	EtOH	222(3.96),262(3.82), 320(2.23)	22-0985-76
$C_{30}H_{44}O_4$			
Chola-4,22-dienoic acid, 3-(carboxymethylene)-, ethyl ester	EtOH	227(4.27),280(3.92)	13-0431-76A

Compound	Solvent	$\lambda_{max}(\log \epsilon)$	Ref.
$C_{30}H_{44}O_5$			
Spirost-1-en-3-one, 2-acetoxy-11-methyl-, (5α,11β,25R)-	EtOH	237.5(3.97)	95-0863-76
Spirost-3-en-2-one, 3-acetoxy-11-methyl-, (11β,25R)-	EtOH	240(3.90)	95-0863-76
$C_{30}H_{44}O_6$			
5H-Inden-5-one, 4-[3,3-dimethoxy-6-(2-methyl-1,3-benzodioxol-2-yl)hexyl]-1-(1,1-dimethylethoxy)-1,2,3,6,7,7a-hexahydro-7a-methyl-	MeOH	241(4.09),250(4.10), 280(3.64),290(3.51)	33-0999-76
Spirost-1-en-3-one, 2-acetoxy-11-hydroxy-11-methyl-, (5α,11β,25R)-	EtOH	235(3.83)	95-0863-76
Spirost-3-en-2-one, 3-acetoxy-11-hydroxy-11-methyl-, (5α,11β,25R)-	EtOH	238.5(3.88)	95-0863-76
$C_{30}H_{44}O_7$			
Cochlioquinone A, 11-deoxy-	MeOH	270(4.02),380(3.02)	32-0147-76
$C_{30}H_{44}O_8$			
Cochlioquinone A	MeOH	270(3.86),388(3.02)	32-0147-76
$C_{30}H_{46}O_2$			
Epiergosterol, acetate	EtOH	281(4.19)	39-0829-76C
$C_{30}H_{46}O_3$			
A(1)-Nor-18α-olean-2-ene-3-carboxylic acid, 19β,28-epoxy-	C_6H_{12}	230(4.02)	73-0271-76
Urs-12-en-30-oic acid, 3-oxo-	EtOH	205(4.09),271(2.75)	102-1997-76
$C_{30}H_{46}O_4$			
Lanost-8-en-18-oic acid, 3,20,23-trihydroxy-3-oxo-, γ-lactone	MeOH	237(4.09),244(4.11), 253(3.97)	78-2313-76
$C_{30}H_{46}O_5$			
29-Nordammara-17(20),24-dien-21-oic acid, 11α,16α-dihydroxy-3-oxo-, methyl ester, (17Z)-	n.s.g.	235(3.83)	39-0710-76C
$C_{30}H_{46}O_8$			
Uzarigenin, β-D-digitaloside	MeOH	218(4.20)	102-1275-76
$C_{30}H_{47}NO_4$			
5α-Lanost-8-ene-6,7,11-trione, 3β-hydroxy-, 6-oxime	EtOH	281(3.81)	5-1214-76
$C_{30}H_{48}CoI_2N_4S_4$			
Cobalt(2+), bis[4-[(cyclohexylimino)-methyl]-4,5-dihydro-N,N,N-trimethyl-5-thioxo-2-thiophenemethanaminiumato-N^4,S^5]-, diiodide	EtOH	218(4.70),263(4.61), 350(3.96)	70-1498-76
$C_{30}H_{48}N_2O_3$			
4-Azacholest-5-en-3-one, 4-(N,N-diacetylamino)-	EtOH	225(4.06)	49-0511-76
$C_{30}H_{48}O_2$			
Ergosta-5,7-dien-3β-ol, acetate	EtOH	281(4.05)	39-0826-76C
D:A-Friedooleanane-2,3-dione	EtOH	276(3.99)	102-0427B-76
D:A-Friedoolean-4(23)-en-3-one, 21α-hydroxy-	EtOH	232(4.13)	102-0797-76

Compound	Solvent	$\lambda_{max}(\log \epsilon)$	Ref.
Lanosta-7,9(11),24-triene-3β,15α-diol	EtOH	237(4.13),244(4.16), 253(3.97)	28-1045-76A
Lup-1-en-3-one, 2-hydroxy-	EtOH	273(3.88)	102-1417-76
	EtOH-HCl	273(3.88)	102-1417-76
	EtOH-KOH	312(3.76)	102-1417-76
$C_{30}H_{48}O_3$			
Cholest-4-ene-6β-acetic acid, 3-oxo-, methyl ester	EtOH	239(4.20)	12-2077-76
6-epimer	EtOH	239(4.11)	12-2077-76
Olean-13(18)-en-11-one, 3β,29β-dihydr-oxy-	EtOH	200(3.8)	65-0912-76
$C_{30}H_{48}O_4$			
11-Oxabicyclo[5.3.1]undeca-4,6-dien-2-ol, 5-[[6-hydroxy-3,7-dimethyl-10-(1-methylethyl)-11-oxabicyclo[5.3.1]un-dec-1-en-3-yl)methyl]-1-methyl-8-(1-methylethyl)-	MeOH	205(4.14),247(3.82)	18-3137-76
11-Oxabicyclo[5.3.1]undec-6-en-2-ol, 5-[[6-hydroxy-7-methyl-10-(1-methyleth-yl)-11-oxabicyclo[5.3.1]undec-1-en-3-ylidene]methyl]-1,5-dimethyl-8-(1-methylethyl)-	MeOH	203(4.07),254(4.35)	18-3137-76
$C_{30}H_{48}O_5$			
29-Nordammara-17(20),24-dien-21-oic acid, 3α,11α,16β-trihydroxy-, methyl ester, (17Z)-	n.s.g.	232(3.96)	39-0710-76C
$C_{30}H_{50}N_4O_4$			
2H-Pyrrole, 5,5'-[1,2-bis(3,4-dihydro-3,3-dimethyl-2H-pyrrol-5-yl)-1,2-eth-anediyl]bis[3,4-dihydro-2,2,4,4-tet-ramethyl-, N,N',1,1'-tetraoxide, isomer A	EtOH	262(4.05)	39-1942-76C
isomer B	EtOH	259(4.05)	39-1942-76C
$C_{30}H_{50}O$			
D:A-Friedours-4(23)-en-3β-ol (cymbopog-onol)	n.s.g.	222(3.58)	102-1074-76
$C_{30}H_{50}OSi$			
Silane, [[(3β,5α)-3,5-cyclocholesta-6,8(14)-dien-6-yl]-	isooctane	263(4.38)	44-1064-76
$C_{30}H_{50}O_2$			
Dammar-24-en-3-one, 20-hydroxy-, (20R)-	EtOH	292(1.60)	102-0785-76
$C_{30}H_{52}N_8O_{14}$			
1,1',1'',1'''-[(1,4-Benzoquinone-2,3,5,6-tetrayl)tetramethyl]tetrakis iperid-inium tetranitrate	EtOH	254s(3.72),313(3.56), 384s(2.88)	12-1163-76
$C_{30}H_{52}O_2$			
Dammar-24-ene-3,20-diol, (3β,20R)-	EtOH	204(3.61)	102-0785-76
$C_{30}H_{52}O_2Si$			
5α-Cholest-7-en-6-one, 14α-[(trimethyl-silyl)oxy]-	MeOH	242(3.996)	13-0609-76A

Compound	Solvent	$\lambda_{max}(\log \epsilon)$	Ref.
$C_{30}H_{54}OS$			
1-Dodecanone, 1-(4-methyl-5-tridecyl-2-thienyl)-	heptane	264(3.91),298(4.02)	34-0233-76
$C_{30}H_{54}S_2$			
Ethanedithione, bis[2,6-bis(1,1-dimethylethyl)cyclohexyl]-, cis	hexane	246(3.9),266s(--), 362(3.7),499(1.3)	24-0906-76
	CH_2Cl_2	260s(--),369(3.8), 480s(--)	24-0906-76
$C_{30}H_{61}NO_2$			
Neotridecanohydroxamic acid	heptane	222(2.60)	34-0504-76
	H_2O	190(3.76)	34-0504-76
	EtOH	203(3.48)	34-0504-76

Compound	Solvent	$\lambda_{max}(\log \epsilon)$	Ref.
$C_{31}H_{18}ClN_3O_3$ Anthra[1,2-b]phenazin-6(9H)-one, 7-chloro-9-[(4-nitrophenyl)methyl]-	CH_2Cl_2	266(4.79),397(4.40), 667(4.26)	33-0688-76
$C_{31}H_{20}O$ Indeno[2,1-b]pyran, 9-(1-acenaphthylenyl)-4-methyl-2-phenyl-	dioxan	232(4.73),301(4.33), 328(4.30)	78-2225-76
$C_{31}H_{20}O_2$ Methanone, (2,4-diphenylindeno[2,1-b]pyran-9-yl)phenyl-	dioxan	243(4.32),277(4.42), 357(4.13),392(4.16), 520(3.49)	78-2225-76
$C_{31}H_{21}N_3O_4$ 9H-Benzo[e]quino[2,3-i]perimidine-9,15(14H)-dione, 8-(2-methoxyethoxy)-6-phenyl-	DMF	270(4.358),470(3.301)	80-0257-76
$C_{31}H_{22}N_4O$ 2H-1,2,3-Triazole, 4-[2-[4-(4,5-diphenyl-2-oxazolyl)phenyl]ethenyl]-2-phenyl-	DMF	362(4.75)	33-2469-76
2H-1,2,3-Triazole, 2,4-diphenyl-5-[2-[4-(5-phenyl-2-oxazolyl)phenyl]ethenyl]-	DMF	369(4.75)	33-2469-76
$C_{31}H_{22}N_4O_3$ 2H-Anthra[1,9,8-cdef][2,7]naphthyridine-2,6,11(10H)-trione, 5,7-bis[(4-methylphenyl)amino]-	DMF	304(4.40),455(3.80), 540(4.25),572(4.30)	103-1352-76
$C_{31}H_{22}O$ 5H-Benzo[6,7]cyclohept[1,2-b]indeno-[1,2-e]pyran, 6,7-dihydro-13-(2-naphthalenyl)-	dioxan	268(4.55),279(4.48), 322(4.44),344(4.41), 566(3.39)	78-2225-76
$C_{31}H_{23}ClN_4O_7$ 1H-1,2,3-Triazolo[4,5-c]pyridine, 4-chloro-1-(2,3,5-tri-O-benzoyl-β-D-ribofuranosyl)-	pH 1 pH 11 MeOH	245(4.55),275s(4.38) 241(4.60),274s(4.33) 230(4.65),266(4.00)	44-1449-76 44-1449-76 44-1449-76
2H-1,2,3-Triazolo[4,5-c]pyridine, 4-chloro-2-(2,3,5-tri-O-benzoyl-β-D-ribofuranosyl)-	pH 1 MeOH	246(4.53),283s(4.28) 228(4.69),265(4.00), 275(3.98),282(3.98)	44-1449-76 44-1449-76
3H-1,2,3-Triazolo[4,5-c]pyridine, 4-chloro-3-(2,3,5-tri-O-benzoyl-β-D-ribofuranosyl)-	pH 1 MeOH	248(4.45),285s(4.23), 304s(4.18) 230(4.65),277s(3.84), 283(3.88),294(3.84)	44-1449-76 44-1449-76
$C_{31}H_{23}ClO_3$ 2H-1-Benzopyran-2-one, 3-[3-(4'-chloro-[1,1'-biphenyl]-4-yl)-1,2,3,4-tetrahydro-1-naphthalenyl]-4-hydroxy-	MeOH	266(4.57),309(4.15)	39-1190-76C
$C_{31}H_{23}ClO_3S$ 2-Propen-1-one, 1-(4-chlorophenyl)-3-[[3-(4-methoxyphenyl)-3-oxo-1-phenyl-1-propenyl]thio]-3-phenyl-	EtOH	226(4.42),277(4.44), 347(4.37)	4-0691-76
$C_{31}H_{23}Cl_3N_2O$ Anthra[1,2-b]phenazin-6(9H)-one, 1,4,7-trichloro-9-cyclohexyl-11-methyl-	CH_2Cl_2	293(4.56),336(4.34), 352(4.28),483(4.33), 709(4.33)	33-0688-76

Compound	Solvent	$\lambda_{max}(\log \epsilon)$	Ref.
$C_{31}H_{23}N_5O$ 1,3,4-Oxadiazole, 2-[1,1'-biphenyl]- 4-yl-5-[4-[2-(5-methyl-2-phenyl-2H- 1,2,3-triazol-4-yl)ethenyl]phenyl]-	DMF	353(4.82)	33-2469-76
$C_{31}H_{24}N_2O$ 1-Aziridinamine, N-[(2,5-diphenyl-3-fur- anyl)methylene]-2,3-diphenyl-, trans	C_6H_{12}	314(4.52)	39-1257-76C
1H-Perimidine-4-methanol, 1-methyl- $\alpha,\alpha,2$-triphenyl-	n.s.g.	344(4.11),405(3.22)	103-0586-76
$C_{31}H_{24}N_4O$ Benzoxazole, 2-[4-[2-(2,5-diphenyl-2H- 1,2,3-triazol-4-yl)ethenyl]phenyl]- 5,6-dimethyl-	DMF	367(4.77)	33-2469-76
Benzoxazole, 2-[4-[2-(5-methyl-2-phenyl- 2H-1,2,3-triazol-4-yl)ethenyl]phenyl]- 5-(phenylmethyl)-	DMF	359(4.80)	33-2469-76
$C_{31}H_{24}N_4O_7S$ β-D-Ribofuranosylamine, N-[1,2,3]thiadi- azolo[5,4-b]pyridin-7-yl-, 2,3,5-tri- benzoate	pH 1	236(4.65),275s(4.25), 327(4.03)	44-1449-76
	pH 11	237(4.65),275s(4.25), 341(4.02)	44-1449-76
	MeOH	232(3.72),275(4.03), 338(3.92)	44-1449-76
4H-1,2,3-Triazolo[4,5-c]pyridine-4-thi- one, 1,5-dihydro-1-(2,3,5-tri-O-benz- oyl-β-D-ribofuranosyl)-	CHCl$_3$	276(3.68),284(3.62), 334(4.25)	44-1449-76
$C_{31}H_{24}O$ 5H-Benzo[6,7]cyclohept[1,2-b]indeno- [1,2-e]pyran, 13-(3,4-dihydro- 2-naphthalenyl)-6,7-dihydro-	dioxan	236(4.27),265(4.24), 337(4.51),357(4.54), 585(3.53)	78-2225-76
Indeno[2,1-b]pyran, 2,4-dimethyl-9-(2- naphthalenyl)-5-(phenylmethyl)-	dioxan	253(4.48),279(4.38), 290(4.46),303(4.47), 510(3.53)	78-2225-76
$C_{31}H_{24}O_2$ Cyclopropanecarboxylic acid, bis(diphen- ylmethylene)-, methyl ester	CHCl$_3$	262(4.45),373(4.56)	18-3173-76
$C_{31}H_{24}O_3$ 2H-1-Benzopyran-2-one, 3-[3-[1,1'-bi- phenyl]-4-yl-1,2,3,4-tetrahydro- 1-naphthalenyl)-4-hydroxy-	MeOH	257(4.43),309(4.09)	39-1190-76C
$C_{31}H_{25}Cl_3O_4$ 3-Benzofuranethanol, 5-chloro-α-(5-chlo- ro-2,3-dihydro-2-methyl-3-methylene- 2-benzofuranyl)methyl]-2-methyl-	MeOH	285(3.85),293(3.83), 325(3.81),338(3.83)	87-1214-76
$C_{31}H_{25}N_3O_9$ 1H-1,2,4-Triazole-3-carboxylic acid, 1-(1,3,4-tri-O-benzoyl-6-deoxy-β- D-erythro-5-hexen-2-ulofuranosyl)-, methyl ester	dioxan	230(4.57),268(3.38), 275(3.46),283(3.46)	44-1836-76
$C_{31}H_{25}OP$ Phosphine oxide, (1,1'-biphenyl-4-yl- phenylmethyl)diphenyl- (carbanion)	DMSO	460(4.50)	70-0762-76

Compound	Solvent	$\lambda_{max}(\log \epsilon)$	Ref.
$C_{31}H_{26}ClNO_2S$			
Methanone, [4-[[(2-chlorophenyl)thio]-methyl]-5-hydroxy-2-phenyl-1-propyl-1H-indol-3-yl]phenyl-	dioxan	230(4.44),282(4.19)	42-0017-76
Methanone, [4-[[(3-chlorophenyl)thio]-methyl]-5-hydroxy-2-phenyl-1-propyl-1H-indol-?-yl]phenyl-	dioxan	288(4.48),290(4.28)	42-0017-76
Methanone, [4-[[(4-chlorophenyl)thio]-methyl]-5-hydroxy-2-phenyl-1-propyl-1H-indol-3-yl]phenyl-	dioxan	230(4.49),290(4.31)	42-0017-76
$C_{31}H_{26}ClN_3O_9$			
1H-1,2,4-Triazole-3-carboxylic acid, 1-(1,3,4-tri-O-benzoyl-6-chloro-6-deoxy-β-D-psicofuranosyl)-, methyl ester	dioxan	229(4.64),275(3.52), 282(3.43)	44-1836-76
$C_{31}H_{26}Cl_2N_2O_2$			
7,13-Epoxy-6H-quino[4,3-c][5,1]benzoxaz-ocine, 2,11-dichloro-5,6a,7,8,13,14a-hexahydro-5,8-dimethyl-13,14a-diphen-yl-	isoPrOH	259(4.37),317(3.69)	4-0131-76
$C_{31}H_{26}IN_3O_8$			
2(1H)-Pyrimidinone, 4-amino-1-(1,3,4-tri-O-benzoyl-6-deoxy-6-iodo-β-D-psicofuranosyl)-	MeOH-acid	230(4.66),276(4.24), 281(4.24)	44-1836-76
$C_{31}H_{26}IN_3O_9$			
1H-1,2,4-Triazole-3-carboxylic acid, 1-(1,3,4-tri-O-benzoyl-6-deoxy-6-iodo-β-D-psicofuranosyl)-, methyl ester	dioxan	231(3.63),270(3.48), 275(3.51),283(3.42)	44-1836-76
1H-1,2,4-Triazole-5-carboxylic acid, 1-(1,3,4-tri-O-benzoyl-6-deoxy-6-iodo-β-D-psicofuranosyl)-, methyl ester	dioxan	231(4.63),275(3.51), 282(3.43)	44-1836-76
$C_{31}H_{26}NS$			
Pyridinium, 1-methyl-4-(methylthio)-2,3,5,6-tetraphenyl-, iodide	MeOH	248(4.02),323(3.93)	44-0818-76
$C_{31}H_{26}N_4OSe$			
5H-1,3,4-Selenadiazolo[3,2-a]pyrimidin-5-one, 7-[4-(dimethylamino)phenyl]-6,7-dihydro-2,6,6-triphenyl-	EtOH	261(4.55),297s(4.16)	94-2532-76
$C_{31}H_{26}N_4O_2$			
5H-1,3,4-Oxadiazolo[3,2-a]pyrimidin-5-one, 7-[4-(dimethylamino)phenyl]-6,7-dihydro-2,6,6-triphenyl-	EtOH	244(4.34),262(4.37)	94-2532-76
$C_{31}H_{26}N_4O_4$			
Spiro[2H-1-benzopyran-2,2'-[2H]indole], 1',3'-dihydro-8-methoxy-3',3'-dimeth-yl-6-[(4-nitrophenyl)azo]-1'-phenyl-	EtOH	297(4.30),413(4.17)	28-0691-76A
$C_{31}H_{26}O$			
7H-Cyclohept[b]indeno[1,2-e]pyran, 5-(1,2-diphenylethenyl)-8,9,10,11-tetrahydro-	dioxan	228(4.46),298(4.51), 345(4.28),528(3.37)	78-2225-76

Compound	Solvent	$\lambda_{max}(\log \epsilon)$	Ref.
Indeno[2,1-b]pyran, 9-(3,4-dihydro-2-naphthalenyl)-2,4-dimethyl-3-(phenylmethyl)-	dioxan	252(4.27),322(4.46), 352(4.47),530(3.56)	78-2225-76
$C_{31}H_{27}NO_2S$ Methanone, [5-hydroxy-2-phenyl-4-[(phenylthio)methyl]-1-propyl-1H-indol-3-yl]phenyl-	dioxan	230(4.42),287(4.20)	42-0017-76
$C_{31}H_{27}N_3O_2$ Spiro[2H-1-benzopyran-2,2'-[2H]indole], 1',3'-dihydro-8-methoxy-6-(phenylazo)-3',3'-dimethyl-1'-phenyl-	EtOH	270(4.28),297(4.1), 369(4.12)	28-0691-76A
Spiro[2H-1-benzopyran-2,2'-[2H]indole], 1',3'-dihydro-6-[(4-methoxyphenyl)-azo]-3',3'-dimethyl-1'-phenyl-	EtOH	306(4.06),360(4.12), 384(4.22)	28-0691-76A
$C_{31}H_{28}N_2O_2$ Methanone, [4-[(dimethylamino)methyl]-5-hydroxy-2-phenyl-1-(phenylmethyl)-1H-indol-3-yl]phenyl-	dioxan	229(4.42),290(4.15)	42-0017-76
$C_{31}H_{29}NO$ 4,9-Methanobenz[f]azulen-1(3aH)-one, 4,9,10,10a-tetrahydro-2,3-diphenyl-10a-pyrrolidino-	MeOH	220s(4.36),268(3.89), 274(3.94),298(4.04)	87-0414-76
$C_{31}H_{29}NO_2$ 4H-Indol-4-one, 1,5,6,7-tetrahydro-6,6-dimethyl-3-(2-oxo-2-phenylethyl)-2-phenyl-1-(phenylmethyl)-	EtOH	243(4.46),273(4.06)	94-1160-76
$C_{31}H_{29}N_3O$ 2-Butanone, 4-[[4-[3-[4-(dimethylamino)-phenyl]benzo[f]quinolin-1-yl]phenyl]-amino]-	EtOH	298(4.80),345(4.89) 371(4.77)	103-0106-76
$C_{31}H_{29}N_3O_9$ 2H-1,2,4-Oxadiazin-3(6H)-one, 5-(dimethylamino)-2-(2,3,5-tri-O-benzoyl-β-D-ribofuranosyl)-	EtOH	231(4.67),250(4.34)	44-3128-76
$C_{31}H_{30}N_2O_5S$ Thymidine, 4-thio-5'-O-(triphenylmethyl)-, 3'-acetate	n.s.g.	206(4.50),231s(3.95), 332(4.20)	128-0351-76
$C_{31}H_{30}N_4O_6$ Pyrazolo[1,5-a]-1,3,5-triazine-2,4-(1H,3H)-dione, 8-[2,3,5-tris-O-(phenylmethyl)-β-D-ribofuranosyl]-	MeOH	235s(3.73),252(3.72), 257(3.72)	44-3000-76
$C_{31}H_{30}N_7$ 3H-Imidazo[1,2-a]benzimidazolium, 3-[[4-(dimethylamino)phenyl](2,9-dimethyl-9H-imidazo[1,2-a]benzimidazol-3-yl)-methylene]-2,9-dimethyl-, bromide	MeOH	365(3.61),560(3.77)	103-0114-76
$C_{31}H_{30}O_6$ Propiolic acid, methyl 3-(2,3-O-isopropylidene-5-O-trityl-α-D-ribofurano-syl)-	EtOH	253(2.76),259(2.83), 262(2.75),266(2.68), 269(2.49)	44-0084-76

Compound	Solvent	$\lambda_{max}(\log \epsilon)$	Ref.
Propiolic acid, methyl 3-(2,3-O-isopropylidene-5-O-trityl-β-D-ribofuranosyl)-	EtOH	253(2.88),260(2.90), 263s(2.85),266s(2.79), 269s(2.64)	44-0084-76
$C_{31}H_{31}NO_9$ 8H-Dibenz[a,g]quinolizin-8-one, 11-acetoxy-13-(3,4-dimethoxyphenyl)-5,6-dihydro-2,3,10,12-tetramethoxy-	n.s.g.	230(4.64),246s(4.52), 335(4.34),360s(4.18), 378s(3.98)	44-2201-76
$C_{31}H_{31}N_3O_5$ 1H-Pyrazole-3-carboxamide, 4-[2,3-O-(1-methylethylidene)-5-O-(triphenylmethyl)-β-D-ribofuranosyl]-	pH 13	232(4.13),248s(3.95)	44-0084-76
$C_{31}H_{32}N_2O_6$ 1H-Pyrazole-3-carboxylic acid, 4-[2,3,5-tris-O-(phenylmethyl)-α-D-ribofuranosyl]-, methyl ester	MeOH	230s(3.85)	44-3000-76
1H-Pyrazole-3-carboxylic acid, 4-[2,3,5-tris-O-(phenylmethyl)-β-D-ribofuranosyl]-, methyl ester	MeOH	236s(3.73)	44-3000-76
$C_{31}H_{32}N_4O_{10}$ 2,4(1H,3H)-Pteridinedione, 5-acetyl-5,6,7,8-tetrahydro-6,7-diphenyl-1-(2,3,5-tri-O-acetyl-β-D-ribofuranosyl)-	pH 6.0 pH 11.0 MeOH	240s(3.83),291(4.20) 255s(3.86),291(4.14) 240s(3.85),291(4.18)	24-3184-76 24-3184-76 24-3184-76
$C_{31}H_{33}NO_9$ 8H-Dibenzo[a,g]quinolizin-8-one, 11-acetoxy-13-(3,4-dimethoxyphenyl)-5,6,13,13a-tetrahydro-2,3,10,12-tetramethoxy-, trans	n.s.g.	237s(4.40),280(4.04), 293(3.88),308(3.54)	44-2201-76
Isoquinoline, 2-(4-acetoxy-3,5-dimethoxybenzoyl)-1-[(3,4-dimethoxyphenyl)-methylene]-1,2,3,4-tetrahydro-6,7-dimethoxy-, (Z)-	n.s.g.	281(4.20),321(4.30)	44-2201-76
$C_{31}H_{33}N_9O_{18}$ Garamine, 1,3,3'-tris-N-(2,4-dinitrophenyl)-	acetone	356(4.63)	39-1088-76C
$C_{31}H_{34}N_2O_7$ D-Ribitol, 1-C-[3-(methoxycarbonyl)-1H-pyrazol-4-yl]-2,3,5-tris-O-(phenylmethyl)-, (R)-	MeOH	247s(3.66)	44-3000-76
$C_{31}H_{34}O_9$ Lappaol B	MeOH	282(3.87)	88-3961-76
$C_{31}H_{34}O_{15}$ Centapicrin, triacetate	MeOH	231(4.23),280s(3.31)	95-0683-76
$C_{31}H_{36}N_4O_2$ 3H-Pyrido[3,4-b]indole, 1,1'-(1,7-heptanediyl)bis[4,9-dihydro-6-methoxy-	MeOH	211(4.77),230s(4.43), 326(4.51),360s(4.19)	87-0202-76
$C_{31}H_{39}NO_4$ Benzamide, N-[[(3β)-3-acetoxy-20-oxopregna-4,16-dien-6-yl]methyl]-	EtOH	239(4.23)	70-1098-76

Compound	Solvent	$\lambda_{max}(\log \epsilon)$	Ref.
Benzamide, N-[[(3β)-3-acetoxy-20-oxo-pregna-5,16-dien-6-yl]methyl]-	EtOH	241(4.24)	70-1098-76
$C_{31}H_{39}NO_7$ D-glycero-D-gulo-Heptopyranose, 2,3:6,7-di-O-cyclohexylidene-4-O-(phenylmethyl)-1-C-(2-pyridinyl)-	MeOH	260(3.36)	30-0613-76
$C_{31}H_{40}N_4O_2$ 1H-Pyrido[3,4-b]indole, 1,1'-(1,7-heptanediyl)bis[2,3,4,9-tetrahydro-6-methoxy-, dihydrochloride	MeOH	221(4.73),275(4.25), 288s(4.15),307s(3.93)	87-0202-76
$C_{31}H_{41}Cl_2N_3O_4$ 1H-Pyrrole-2-carboxylic acid, 3-(2-chloroethyl)-5-[[3-(2-chloroethyl)-5-[[4-(3-methoxy-3-oxopropyl)-3,5-dimethyl-2H-pyrrol-2-ylidene]methyl]-4-methyl-1H-pyrrol-2-yl]methyl]-4-methyl-, 1,1-dimethylethyl ester, hydrobromide	CHCl$_3$	486(4.96)	78-1793-76
$C_{31}H_{41}NO_3$ Acetamide, N-[(3β)-3-acetoxy-20-phenyl-pregna-5,17(20)-dien-21-yl]-	EtOH	247(2.48),251(2.52), 258(2.57),264(2.54), 284(2.49)	33-1850-76
$C_{31}H_{42}FeO_4$ Iron, tricarbonyl[(5,6,7,8-η)-ergosta-5,7,22-trien-3-one]-	CHCl	252(4.03)	39-0829-76C
Iron, tricarbonyl[(5,6,10,19-η)-(3β,5Z,7E,22E)-9,10-secoergosta-5,7,10(19),22-tetraen-3-one]-	EtOH	235(4.08),309(3.87)	39-0829-76C
$C_{31}H_{42}N_2O_7$ Hydrazinecarboxylic acid, [[2-(3,4-di-hydro-5,7-dimethoxy-1H-2-benzopyran-3-yl)-4,5-dimethoxyphenyl]methylene]-, 5-methyl-2-(1-methylethyl)cyclohexyl ester	dioxan	233(4.09),290(4.05), 311s(3.89)	33-0949-76
$C_{31}H_{42}O_6S$ 2-Propanone, 1-[(3β,17β)-3-acetoxy-androst-5-en-17-yl]-3-[[(4-methyl-phenyl)sulfonyl]oxy]-	EtOH	205s(3.85),224(4.08)	33-2138-76
$C_{31}H_{43}NO_5$ Spirosta-2,4-dieno[2,3-d]isoxazole-3'-carboxylic acid, ethyl ester, (25R)-	EtOH	256(4.18)	65-1381-76
$C_{31}H_{43}N_3O_4$ 1H-Pyrrole-2-carboxylic acid, 3-ethyl-5-[[3-ethyl-5-[[4-(3-methoxy-3-oxopropyl)-3,5-dimethyl-2H-pyrrol-2-yl-idene]methyl]-4-methyl-1H-pyrrol-2-yl]methyl]-4-methyl-, 1,1-dimethyl-ethyl ester, hydrobromide	CHCl$_3$	490(4.96)	78-1793-76
$C_{31}H_{44}FeO_4$ Iron, tricarbonyl-calciferol complex, α-	EtOH	217(4.12),240(4.08), 310(3.83)	39-0829-76C

Compound	Solvent	$\lambda_{max}(\log \epsilon)$	Ref.
Iron, tricarbonyl-calciferol complex, β-	EtOH	235(4.16),309(3.90)	39-0829-76C
$C_{31}H_{44}N_2O_4$ 2'H-Spirosta-2,4-dieno[3,2-c]pyrazole-5'-carboxylic acid, ethyl ester, (25R)-	EtOH	266(4.30)	65-1381-76
$C_{31}H_{44}O_6$ 16,23-Anhydro-23-deoxocucurbitacin S, 23-methoxy-	EtOH EtOH-base	269(4.09) 315(--)	102-0559-76 102-0559-76
$C_{31}H_{44}O_7$ Spirost-3-en-2-one, 3,11-diacetoxy-, (5α,11α,25R)-	EtOH	239(3.89)	95-0863-76
$C_{31}H_{44}O_7S_5$ 2-Propenoic acid, 1-methyl-3-[2-[3-[2-[(4-methylphenyl)sulfonyl]ethoxy]-3-oxo-1-propenyl]-1,3-dithian-2-yl]-propyl ester, (E,E)-	EtOH	215(4.45)	39-1718-76C
$C_{31}H_{45}NO_5$ Pregn-5-en-21-one, 3,16-diacetoxy-21-(4,5-dihydro-4-methyl-3H-pyrrol-2-yl)-20-methyl-, [3β,16β,20S,21R]-	EtOH	270(2.23)	94-0661-76
16,28-Secosolanida-3,22(28)-dien-23-one, 3,16-diacetoxy-	EtOH	277(2.42)	94-0661-76
$C_{31}H_{46}O_4$ Furosta-5,22(23)-diene, 27-acetoxy-23-acetyl-, cis	EtOH	277(4.03)	2-0073-76
Lanosta-8,24-dien-26-oic acid, 3,21-di-oxo-, methyl ester, (13α,14β,17α,20ξ-24Z)-	EtOH	212(4.10)	32-0785-76
$C_{31}H_{46}O_7$ Esculentic acid, 2-oxo-, 30-methyl ester	EtOH	204(3.67),282(2.63)	102-1315-76
$C_{31}H_{47}N_3O_6$ 10-Undecenoic acid, [2,3,3a,9a-tetrahy-dro-6-imino-3-[(1-oxo-10-undecenyl)-oxy]-6H-furo[2',3':4,5]oxazolo[3,2-a]pyrimidin-2-yl]methyl ester, hydrochloride	MeOH-acid	234(4.00),263(4.05)	87-0663-76
$C_{31}H_{48}O_3$ A(1)-Nor-18α-olean-2-ene-3-carboxylic acid, 19β,28-epoxy-, methyl ester	C_6H_{12}	230(4.04)	73-0271-76
$C_{31}H_{48}O_5$ Cholesta-5,7-diene-3β,24ξ,25-triol, 3,24-diacetate	EtOH	272(--),282(4.06), 293(--)	39-0727-76C
$C_{31}H_{48}O_6$ Cholest-7-en-6-one, 2β,3β-diacetoxy-5α-hydroxy-	EtOH	250(4.05)	13-0609-76A
$C_{31}H_{49}NO_{12}$ Demycarosyl-9-dehydromaridomycin I	EtOH	240.5(4.12)	94-0463-76
Demycarosyl-9-dehydromaridomycin III	EtOH	240.5(4.16)	94-0463-76

Compound	Solvent	λ_{max}(log ϵ)	Ref.
$C_{31}H_{49}N_3O_7$ 2(1H)-Pyrimidinone, 4-amino-1-[3,5-bis-O-(1-oxo-10-undecenyl)-β-D-arabino-furanosyl]-, hydrochloride	MeOH-acid	212(3.96),282(4.14)	87-0667-76
$C_{31}H_{50}O_3$ Stigmast-5-en-7-one, 3β-acetoxy-α-Tocopherol, acetate, (2R,4'R,8'R)-	EtOH EtOH	240(4.15) 227s(4.02),278(4.25), 285(3.31)	39-0023-76C 33-0290-76
$C_{31}H_{50}O_5$ Cycloart-24-en-26-oic acid, 3β,11α,15α-trihydroxy-, methyl ester	MeOH	200(4.11),277(1.58)	78-1077-76
$C_{31}H_{51}N_3O_5$ 13-Docosenoic acid, 2,3,3a,9a-tetrahy-dro-2-(hydroxymethyl)-6-imino-6H-furo-[2',3':4,5]oxazolo[3,2-a]pyrimidin-3-yl ester, hydrochloride, [2R-[2α,3β(Z),3aβ,9aβ]]-	MeOH-acid	234(4.03),265(4.06)	7-0654-76
$C_{31}H_{51}N_3O_6$ Undecanoic acid, [2,3,3a,9a-tetrahydro-6-imino-3-[(1-oxoundecyl)oxy]-6H-furo-[2',3':4,5]oxazolo[3,2-a]pyrimidin-2-yl]methyl ester, hydrochloride, [2R-(2α,3β,3aβ,9aβ)]-	MeOH-acid	234(4.00),264(4.04)	87-0663-76
$C_{31}H_{52}N_2O$ 4-Azacholest-5-en-3-one, 4-(1-piperidin-yl)-	EtOH	236(4.05)	49-0511-76
$C_{31}H_{52}O_5$ 1,4-Benzenediol, 2,3-dimethoxy-5-methyl-6-(3,7,11,15-tetramethyl-2-hexadecen-yl)-, 4-acetate	EtOH	282(3.29)	65-1350-76
$C_{31}H_{53}N_3O_5$ Docosanoic acid, 2,3,3a,9a-tetrahydro-2-(hydroxymethyl)-6-imino-6H-furo-[2',3':4,5]oxazolo[3,2-a]pyrimidin-3-yl ester, hydrochloride	MeOH-acid	232(3.92),263(3.96)	87-0654-76
$C_{31}H_{53}N_3O_6$ 2(1H)-Pyrimidinone, 4-amino-1-[3-O-(1-oxo-13-docosenyl)-β-D-arabino-furanosyl]-	MeOH-acid	213(3.99),283(4.17)	87-0667-76
$C_{31}H_{53}N_3O_7$ 2(1H)-Pyrimidinone, 4-amino-1-[3,5-bis-O-(1-oxoundecyl)-β-D-arabino-furanosyl]-	MeOH-acid	212(3.98),283(4.14)	87-0667-76
$C_{31}H_{55}N_3O_6$ 2(1H)-Pyrimidinone, 4-amino-1-[3-O-(1-oxodocosyl)-β-D-arabinofuranosyl]-	MeOH-acid	213(4.01),284(4.16)	87-0667-76
2(1H)-Pyrimidinone, 4-amino-1-[5-O-(1-oxodocosyl)-β-D-arabinofuranosyl]-	MeOH-acid	213(4.01),284(4.16)	87-0667-76

Compound	Solvent	$\lambda_{max}(\log \epsilon)$	Ref.
$C_{32}H_{18}N_4O_4S$ 5-Benzothiazolecarboxamide, N-(9,10-di-oxo-1-anthracenyl)-2-[(1-hydroxy-2-naphthalenyl)azo]-	DMF	483(4.160)	7-0593-76
$C_{32}H_{20}$ [2.2](3,6)-Phenanthrenophanediene	n.s.g.	250(5.11)	1-0369-76
$C_{32}H_{20}Br_4N_2O_2$ Dispiro[2H-1-benzopyran-2,5'(10'H)-pyri-do[2,3,4,5-lmn]phenanthridine-10',2"-[2H-1]benzopyran, 6,6",8,8"-tetra-bromo-4',9'-dihydro-4',9'-dimethyl-	50% HOAc-H_2SO_4	267s(4.41),276(4.44), 240(4.20),390s(4.09), 404(4.24)	103-0273-76
$C_{32}H_{20}Cl_2N_6$ 2H-Naphtho[1,2-d]triazole, 2-[3-chloro-4-[2-[2-(4-chlorophenyl)-5-phenyl-2H-1,2,3-triazol-4-yl]ethenyl]phenyl]-	DMF	374(4.77)	33-2469-76
$C_{32}H_{20}N_4OS$ 2-Naphthalenol, 1-[2-[4-(2-benzothiazo-lyl)quinolin-2-yl]phenylazo]-	n.s.g.	486(4.23)	7-0379-76
2-Naphthalenol, 1-[2-[6-(2-benzothiazo-lyl)quinolin-2-yl]phenylazo]-	n.s.g.	489(4.11)	7-0379-76
2-Naphthalenol, 1-[3-[4-(2-benzothiazo-lyl)quinolin-2-yl]phenylazo]-	n.s.g.	477(4.19)	7-0379-76
2-Naphthalenol, 1-[3-[6-(2-benzothiazo-lyl)quinolin-2-yl]phenylazo]-	n.s.g.	476(4.18)	7-0379-76
2-Naphthalenol, 1-[4-[4-(2-benzothiazo-lyl)quinolin-2-yl]phenylazo]-	n.s.g.	497(4.42)	7-0379-76
2-Naphthalenol, 1-[4-[6-(2-benzothiazo-lyl)quinolin-2-yl]phenylazo]-	n.s.g.	496(4.29)	7-0379-76
$C_{32}H_{20}O_4$ 2(5H)-Furanone, 5-(5-oxo-3,4-diphenyl-2(5H)-furanylidene)-3,4-diphenyl-	CHCl$_3$-EtOH	415(4.63)	22-1491-76
	MeOH-KOH	284(4.17),407(4.39)	22-1491-76
Maleic anhydride adduct of benzo[b]naph-tho[1,2,3,4-def]chrysene, potassium salt of diacid	H_2O	272(4.52),293s(--), 315s(--),326(4.24), 343(4.25),371(3.28)	24-1991-76
$C_{32}H_{21}ClN_6$ 2H-Naphtho[1,2-d]triazole, 2-[3-chloro-4-[2-(2,5-diphenyl-2H-1,2,3-triazol-4-yl)ethenyl]phenyl]-	DMF	375(4.77)	33-2469-76
$C_{32}H_{22}$ [18]Annulene, 1,14-dimethyl-5,10-diphen-yl-6,8,15,17-tetrakisdehydro-	THF	244(4.22),256(4.29), 270s(4.18),278(4.25), 292s(4.14),299(4.16), 324s(4.16),346s(4.48), 354s(4.49),368s(4.58), 385s(4.80),401(5.33), 542s(4.24),582(4.93), 666s(2.57),715s(2.69), 742(3.25)	18-0297-76
[18]Annulene, 5,10-dimethyl-1,14-diphen-yl-6,8,15,17-tetrakisdehydro-	THF	244(4.22),250(4.20), 256(4.26),261s(4.19), 270s(4.19),278(4.25), 298(4.14),299(4.15), 325s(4.17),348s(4.47),	18-0297-76

Compound	Solvent	$\lambda_{max}(\log \epsilon)$	Ref.
(cont.)		354s(4.48),370s(4.58), 384s(4.79),400(4.34), 544s(4.25),584(4.95), 650s(2.87),670(2.65), 715s(2.75),742(3.29)	18-0297-76
6,15-Ethanonaphtho[2,3-c]pentaphene, 6,15-dihydro-, (6R)-	EtOH-0.08% dioxan	267(5.43),353(3.95), 371(4.05),391(3.96)	35-5408-76
$C_{32}H_{22}Br_2N_2O_2$ Dispiro[2H-1-benzopyran-2,5'(10'H)-pyrido[2,3,4,5-lmn]phenanthridine-10',2"-[2H-1]benzopyran], 6,6"-dibromo-4',9'-dihydro-4',9'-dimethyl-	dioxan	255(4.90),323(4.00), 365(3.99)	103-0273-76
$C_{32}H_{22}I_2$ [2.2.2.2]Paracyclophanetetraene, 4,20(21)-diiodo-	n.s.g.	302(4.71)	1-0369-76
$C_{32}H_{22}N_2O_2$ 2H-Pyrrol-2-one, 5-(1,5-dihydro-5-oxo-3,4-diphenyl-2H-pyrrol-2-ylidene)-1,5-dihydro-3,4-diphenyl-	CHCl$_3$-EtOH	425(4.67),435s(4.65)	22-1491-76
$C_{32}H_{22}N_6OS$ 1H-Pyrazol-5-ol, 4-[2-[4-(2-benzothiazolyl)-2-quinolinyl]phenylazo]-3-methyl-1-phenyl-	n.s.g.	404(4.19)	7-0379-76
1H-Pyrazol-5-ol, 4-[2-[6-(2-benzothiazolyl)-2-quinolinyl]phenylazo]-3-methyl-1-phenyl-	n.s.g.	404(4.11)	7-0379-76
1H-Pyrazol-5-ol, 4-[3-[4-(2-benzothiazolyl)-2-quinolinyl]phenylazo]-3-methyl-1-phenyl-	n.s.g.	393(4.39)	7-0379-76
1H-Pyrazol-5-ol, 4-[3-[6-(2-benzothiazolyl)-2-quinolinyl]phenylazo]-3-methyl-1-phenyl-	n.s.g.	391(4.44)	7-0379-76
1H-Pyrazol-5-ol, 4-[4-[4-(2-benzothiazolyl)-2-quinolinyl]phenylazo]-3-methyl-1-phenyl-	n.s.g.	416(4.60)	7-0379-76
1H-Pyrazol-5-ol, 4-[4-[6-(2-benzothiazolyl)-2-quinolinyl]phenylazo]-3-methyl-1-phenyl-	n.s.g.	418(4.69)	7-0379-76
$C_{32}H_{22}O$ Indeno[2,1-b]pyran, 9-(1H-inden-2-yl)-4-methyl-2-(2-naphthalenyl)-	dioxan	260(4.53),294(4.40), 323(4.52),358(4.46), 582(3.50)	78-2225-76
$C_{32}H_{22}O_6S$ 2-Propen-1-one, 3,3'-thiobis[1-(1,3-benzodioxol-5-yl)-3-phenyl-, (E,Z)-	EtOH	237(4.51),280(4.31), 350(4.38)	4-0691-76
$C_{32}H_{22}O_{10}$ Cupressuflavone, 7,7"-dimethoxy-	MeOH	220(4.68),268(4.49), 320(4.50)	78-0503-76
$C_{32}H_{23}ClN_4O$ 5H-Indeno[5,6-d]oxazole, 2-[4-[2-[2-(4-chlorophenyl)-5-phenyl-2H-1,2,3-triazol-4-yl)ethenyl]phenyl]-6,7-dihydro-	DMF	370(4.79)	33-2469-76

Compound	Solvent	$\lambda_{max}(\log \epsilon)$	Ref.
$C_{32}H_{24}$			
Benzo[c]phenanthrene, 1-(3,5-dimethyl-phenyl)-2-phenyl-	MeOH	323s(4.06),339s(3.98)	39-1115-76B
Bicyclo[4.2.0]octa-2,4,7-triene, 2,5,7-8-tetraphenyl- (equimolar mixture with 1,2,4,7-tetraphenylcycloocta-tetraene)	hexane	255(4.38),350(4.05)	35-3247-76
1,1'-(1,2-Ethenediylidene)bis(5,10-di-methylcyclotrideca-2,4,10,12-tetra-ene-6,8-diyne)	CHCl	284(4.48),566(4.92)	88-3631-76
$C_{32}H_{24}N_2$			
Benzonitrile, 4,4'-([1,1'-biphenyl]-4,4'-diyldi-2,1-ethenediyl)bis[3-methyl-	DMF	367(4.95)	33-0819-76
$C_{32}H_{24}N_4O_8$			
4(3H)-Pteridinone, 3-(2,3,5-tri-O-benz-oyl-β-D-ribofuranosyl)-	MeOH	230(4.73),272(3.98), 304(3.88)	24-3208-76
$C_{32}H_{24}N_{10}Ni$			
Nickel, [1,8,9,20-tetrahydro-3,6-dimeth-yl-1,8-diphenyldibenzo[c,m]dipyrazolo-[3,4-f:4',3'-j][1,2,5,8,9,12]hexaaza-cyclotetradecinato(2-)-N^4,N^9,N^{14},N^{20}]-	heptane	238s(5.12),242(5.13), 248(5.13),257s(5.03), 268s(4.76),272s(4.70), 280s(4.57),291(4.63), 300s(4.58),319s(4.49), 362s(4.19),386s(4.17), 468(4.53),670s(3.42), 722(3.66)	103-0802-76
$C_{32}H_{24}O_2$			
9H-Xanthene, 9-(triphenylmethoxy)-	EtOH	288(4.67)	22-1136-76
$C_{32}H_{24}O_6$			
Benzenepropanoic acid, β-hydroxy-β-[hy-droxy(5-oxo-3,4-diphenyl-2(5H)-furan-ylidene)methyl]- -phenyl-	ether MeOH-KOH	290(4.29) 281(4.18),410(4.41)	22-1491-76 22-1491-76
$C_{32}H_{24}O_{12}$			
Benzo[j]fluoranthene-2,3,4,7,8,9-hexol, hexaacetate	n.s.g.	248(4.76),274(3.96), 295(4.16),309(4.41), 322(4.60),341(4.04), 376(3.99),395(4.11), 420(3.45),450(3.15)	39-2149-76C
$C_{32}H_{25}ClNOS_2$			
1,3-Thiazocin-1-ium, 2-(4-chlorophenyl)-8-(ethylthio)-3,4-dihydro-4-oxo-3,6,7-triphenyl-, perchlorate	MeOH	244(4.14),305s(3.64), 428(3.63)	44-0818-76
$C_{32}H_{25}ClN_{10}Ni$			
Nickel, 5-[[2-[(2-aminophenyl)azo]phen-yl]imino]-1,5-dihydro-3-methyl-1-phen-yl-4H-pyrazol-4-one (5-chloro-3-meth-yl-1-phenyl-1H-pyrazol-4-yl)hydrazon-ato](2-)]	heptane	239s(4.60),253(4.65), 281s(4.54),311s(4.39), 375(4.23),443(4.27), 550s(3.70),630s(3.53), 676(3.66)	103-0802-76
$C_{32}H_{25}NO_3$			
Poly[4-(vinylbenzyloxy)phenyl 1-isoquin-olinylcarbinyl benzoate]	THF	264s(3.95),273(3.99), 283s(3.92),309(3.61), 321(3.67)	47-1661-76

Compound	Solvent	$\lambda_{max}(\log \epsilon)$	Ref.
Poly[4-(vinylbenzyloxy)phenyl 2-quino-linylcarbinyl benzoate hemihydrate]	THF	272s(--),282s(--), 296s(--),304(3.73), 308s(3.68),317(4.88)	116-0221-76
$C_{32}H_{26}IN_5O_7$ 9H-Purin-6-amine, 9-(1,3,4-tri-O-benz-oyl-6-deoxy-6-iodo-α-D-psicofurano-syl)-	MeOH-acid	232(4.65),258(4.23), 264(4.21)	44-1836-76
β-	MeOH-acid	231(4.63),258(4.26), 265(4.22)	44-1836-76
$C_{32}H_{26}N_4O$ Benzoxazole, 5-(1-methyl-1-phenylethyl)-2-[4-[2-(2-phenyl-2H-1,2,3-triazol-4-yl)ethenyl]phenyl]-	DMF	354(4.84)	33-2469-76
$C_{32}H_{26}O_2$ [9,9'-Bianthracene]-10,10'(9H,9'H)-di-one, 2,2',3,3'-tetramethyl-, (±)-	EtOH	219(4.60),277(4.42), 324s(4.00)	12-2257-76
meso	EtOH	219(4.60),282(4.42), 312s(4.10)	12-2257-76
Cyclopropanecarboxylic acid, bis(diphen-ylmethylene)-, ethyl ester	CHCl$_3$	264(4.47),374(4.58)	18-3173-76
2,4,10,12-Cyclotridecatetraene-6,8-diyn-1-ol, 1,1'-(1,2-ethynediyl)bis[5,10-dimethyl-	CHCl$_3$	263(4.44),329(3.97)	88-3631-76
$C_{32}H_{26}O_2S$ 2-Propen-1-one, 3,3'-thiobis[1-(4-meth-ylphenyl)-3-phenyl-, (E,Z)-	EtOH	273(4.38),347(4.25)	4-0691-76
$C_{32}H_{26}O_{11}$ Dinaphtho[1,2-b:2',3'-d]furan-1,4,7,11-12-pentol, pentaacetate	n.s.g.	269s(4.44),277(4.52), 300s(3.93),308s(3.83), 321s(3.74),338s(3.72), 362s(3.58),384(3.59)	39-2155-76C
$C_{32}H_{26}S_3$ 2H-Thiopyran, 2-(diphenylmethylene)-4,5-bis(methylthio)-3,6-diphenyl-	CH$_2$Cl$_2$	285(4.08),328(4.04), 403(3.84)	44-2279-76
$C_{32}H_{27}ClN_{10}$ 1H-Pyrazol-5-amine, N-[2-[(2-aminophen-yl)azo]phenyl]-4-[(5-chloro-3-methyl-1-phenyl-1H-pyrazol-4-yl)azo]]-3-methyl-1-phenyl-	heptane	217(4.50),252s(4.37), 329(4.40),382(4.34), 405s(4.27),465s(4.01)	103-0802-76
$C_{32}H_{27}GeN$ Methanamine, N-(diphenylmethylene)-1-(triphenylgermyl)-	C_6H_{12}	259(4.26),264s(4.25)	101-0295-76G
$C_{32}H_{27}NSi$ Methanamine, N-(diphenylmethylene)-1-(triphenylsilyl)-	C_6H_{12}	256s(4.27),258(4.27)	101-0295-76G
$C_{32}H_{27}NSn$ Methanamine, N-(diphenylmethylene)-1-(triphenylstannyl)-	C_6H_{12}	260(4.17),266(4.18)	101-0295-76G

Compound	Solvent	λ_{max} (log ϵ)	Ref.
$C_{32}H_{28}$			
Bicyclo[4.2.0]oct-7-ene, 2,5,6,7-tetraphenyl-	hexane	228s(4.36),288(4.13)	35-3247-76
$C_{32}H_{28}ClNO_2S$			
Methanone, [1-butyl-4-[[(2-chlorophenyl)thio]methyl]-5-hydroxy-2-phenyl-1H-indol-3-yl]phenyl-	dioxan	228(4.51),288(4.32)	42-0017-76
Methanone, [1-butyl-4-[[(3-chlorophenyl)thio]methyl]-5-hydroxy-2-phenyl-1H-indol-3-yl]phenyl-	dioxan	229(4.50),289(4.29)	42-0017-76
Methanone, [1-butyl-4-[[(4-chlorophenyl)thio]methyl]-5-hydroxy-2-phenyl-1H-indol-3-yl]phenyl-	dioxan	231(4.50),291(4.32)	42-0017-76
Methanone, [4-[[(2-chlorophenyl)thio]methyl]-5-hydroxy-1-(2-methylpropyl)-2-phenyl-1H-indol-3-yl]phenyl-	dioxan	229(4.45),290(4.28)	42-0017-76
Methanone, [4-[[(3-chlorophenyl)thio]methyl]-5-hydroxy-1-(2-methylpropyl)-2-phenyl-1H-indol-3-yl]phenyl-	dioxan	228(4.47),288(4.25)	42-0017-76
Methanone, [4-[[(4-chlorophenyl)thio]methyl]-5-hydroxy-1-(2-methylpropyl)-2-phenyl-1H-indol-3-yl]phenyl-	dioxan	230(4.47),288(4.31)	42-0017-76
$C_{32}H_{28}NO$			
Pyridinium, 4-ethoxy-1-methyl-2,3,5,6-tetraphenyl-, tetrafluoroborate	MeOH	243(4.41),290s(3.94)	44-0818-76
$C_{32}H_{28}NS$			
Pyridinium, 4-(ethylthio)-1-methyl-2,3,5,6-tetraphenyl-, tetrafluoroborate	MeOH	248(4.18),322(4.06)	44-0818-76
$C_{32}H_{28}N_2$			
3,3'-Bi-1H-indole, 1,1'-diethyl-2,2'-diphenyl-	$CHCl_3$	<u>300(4.4)</u>	39-0309-76B
$C_{32}H_{28}N_2O$			
Pyrrolo[1,2-a:4,5-b']diindol-12(5aH)-one, 10,10a,11,11a-tetrahydro-10a-methyl-5a,11a-bis(phenylmethyl)-, (5aα,10aα,11aβ)-	MeOH	238(4.45),305(3.67), 410(3.32)	23-1015-76
$C_{32}H_{28}N_2O_4$			
Hydrazine, bis(β-benzoyl-α-phenylpropanoyl)-	EtOH	240(4.30)	115-0761-76
$C_{32}H_{28}N_4$			
1,2-Benzenediamine, N-(5,6-dimethyl-2,2-diphenylcyclobuta[b]quinoxalin-1(2H)-ylidene)-4,5-dimethyl-	$CHCl_3$	262(4.68),357(4.32), 400(3.80)	5-0284-76
$C_{32}H_{28}N_4O_2$			
2H-Indazole, 2-acetyl-5-(2-acetyl-2,3-dihydro-3-methyl-3-phenyl-5H-indazol-5-ylidene)-3,5-dihydro-3-methyl-3-phenyl-	isoPrOH	263(3.75),334(3.75), 446(3.65),478(4.37), 515(5.44),568(3.99), 619(3.73)	4-0033-76
$C_{32}H_{28}O_2$			
Cyclohexane, 1,4-trans-diphenyl-2,3-cis-dibenzoyl-	MeOH	247(4.36)	35-3247-76

Compound	Solvent	$\lambda_{max}(\log \epsilon)$	Ref.
isomer (cont.)	MeOH	245(4.31)	35-3247-76
$C_{32}H_{28}O_3S$ 4H-Thiopyran, 2,4,6-tris(4-methoxyphenyl)-4-phenyl-	EtOH	257(4.62)	4-0237-76
$C_{32}H_{28}O_4S_2$ Dispiro[9H-thioxanthene-9,1'-cyclobutane-3',9"-[9H]thioxanthene], 2,2",3,3"-tetramethoxy-	MeOH	278(4.29),335(3.69)	73-1396-76
$C_{32}H_{28}O_{16}$ 1,3,5-Benzenetriol, 2,2'-[[5-acetoxy-1,3-phenylene]bis(oxy)]bis-, hexaacetate	MeCN	229s(4.53),268(3.47), 273(3.41)	102-1082-76
$C_{32}H_{29}NO_2S$ Methanone, [1-butyl-5-hydroxy-2-phenyl-4-[(phenylthio)methyl]-1H-indol-3-yl]phenyl-	dioxan	230(4.50),290(4.30)	42-0017-76
Methanone, [5-hydroxy-1-(2-methylpropyl)-2-phenyl-4-[(phenylthio)methyl]-1H-indol-3-yl]phenyl-	dioxan	231(4.54),290(4.37)	42-0017-76
$C_{32}H_{29}N_3O_3$ Spiro[2H-1-benzopyran-2,2'-[2H]indole], 1',3'-dihydro-8-methoxy-6-[(4-methoxyphenyl)azo]-3',3'-dimethyl-1'-phenyl-	EtOH	302(4.02),309(4.30), 381(4.24),394(4.22)	28-0691-76A
$C_{32}H_{29}N_3O_9$ 1H-Imidazole-4-carboxylic acid, 5-amino-2-(2,3,5-tri-O-benzoyl-α-D-arabinofuranosyl)-, ethyl ester	EtOH	234(4.50),280(4.10)	111-0067-76
β-	EtOH	233(4.48),279(4.05)	111-0067-76
$C_{32}H_{30}ClN_3O_5$ Imidazo[1,2-c]pyrimidin-5(1H)-one, 7-chloro-1-[2,3,5-tris-O-(phenylmethyl)-β-D-arabinofuranosyl]-	pH 1 pH 7 pH 11	252(3.71),303(4.15) 255s(4.18),314(4.22) 255s(4.10),314(4.21)	87-0814-76 87-0814-76 87-0814-76
$C_{32}H_{30}N_2O_4$ Asterriquinone	CHCl$_3$	298(4.51),508(3.74)	94-1853-76
$C_{32}H_{30}N_2O_5$ 1,3(2H,9bH)-Dibenzofurandione, 7,9-dihydroxy-8,9b-dimethyl-2-[1-[(phenylmethyl)amino]ethylidene]-6-[1-[(phenylmethyl)imino]ethyl]-, (+)-	MeCN	301(4.23)	23-2795-76
$C_{32}H_{30}N_4O_2$ Pyrido[2,3-g]quinoline, 2,7-dimorpholino-4,9-diphenyl-	EtOH	246(4.51),280(4.54), 353(4.21)	24-2921-76
$C_{32}H_{30}OSi_2$ 6H-7,12[1',2']-Benzeno-6,13-etheno-5,7-(silanoxysilano)-benzo[5,6]cycloocta[1,2-b]naphthalene, 12,13-dihydro-23,23,25,25-tetramethyl-	ether	225(4.67),252(4.56), 273(3.82),283(3.75), 297(3.71),307(3.54), 320(2.98),348(2.82)	88-2277-76

Compound	Solvent	$\lambda_{max}(\log \epsilon)$	Ref.
$C_{32}H_{30}O_4$			
Tectol, dimethyl ether	MeOH	230(4.72),267(4.71), 276(4.78),353(3.96)	83-0829-76
$C_{32}H_{31}N_3O_5$			
Imidazo[1,2-c]pyrimidin-5(1H)-one, 1-[2,3,5-tris-O-(phenylmethyl)- β-D-arabinofuranosyl]-	pH 1 pH 7 pH 11	245(3.66),293(3.96) 250(3.75),303(4.05) 250(3.66),303(4.03)	87-0814-76 87-0814-76 87-0814-76
$C_{32}H_{32}Cl_4O_4Si$			
Silicic acid, tetrakis(4-chloro-3,5-di-methylphenyl) ester	dioxan	276(3.38),281(3.38)	56-0363-76
$C_{32}H_{32}N_2O_2$			
Benzo[1,2-d:4,5-d']bisoxazole, 2,6-bis-[2-[4-(1,1-dimethylethyl)phenyl]eth-enyl]-	$CHCl_3$	368(4.53),386(4.52), 408s(4.36)	103-1331-76
Benzo[1,2-d:5,4-d']bisoxazole, 2,6-bis-[2-[4-(1,1-dimethylethyl)phenyl]eth-enyl]-	$CHCl_3$	338(4.78)	103-1331-76
$C_{32}H_{32}N_2O_4$			
Hydrazine, bis(1-hydroxy-1-phenyl-3-ben-zoylpropyl)-	EtOH	240(4.52)	115-0761-76
$C_{32}H_{32}N_2O_6$			
1H-Pyrazole-3-carboxylic acid, 4-[2,3-O-(1-methylethylidene)-5-O-(triphen-ylmethyl)-β-D-ribofuranosyl]-, methyl ester	pH 13	232(4.10),250s(3.89)	44-0084-76
$C_{32}H_{32}N_4O_3$			
21H,23H-Porphine-2-propanoic acid, 8-ethenyl-13-ethyl-18-formyl-3,7,12,17-tetramethyl-	pyridine	416(5.09),520(3.79), 562(4.15),583(4.04), 637(3.36)	78-0275-76
	4:1 CH_2Cl_2-TFA	415(5.23),567(3.89), 620(3.97)	78-0275-76
$C_{32}H_{32}N_4O_5$			
Imidazo[1,2-c]pyrimidin-5(1H)-one, 7-amino-1-[2,3,5-tris-O-(phenyl-methyl)-β-D-arabinofuranosyl]-	pH 1 pH 7 pH 11	265(3.89),302(4.24) 293(4.26) 293(4.27)	87-0814-76 87-0814-76 87-0814-76
$C_{32}H_{32}O_6$			
9,10-Ethanoanthracene-11,12-dicarbox-ylic acid, 1-formyl-9,10-dihydro-2-methoxy-9-[2-methyl-4-(1-methyleth-yl)phenyl]-, dimethyl ester, (9R,10R,11R,12R)-	EtOH	212(4.60),272(3.49), 280(3.53),290(3.53)	18-1163-76
(9R,10R,11S,12S)-	EtOH	212(4.60),261(3.79), 270s(3.68),281s(3.51), 318(3.56)	18-1163-76
Propiolic acid, 3-(2,3-O-isopropylidene-5-O-trityl-α-D-ribofuranosyl)-, ethyl ester	EtOH	253(2.83),259(2.88), 263(2.80),266(2.72), 269(2.57)	44-0084-76
β-	EtOH	254(2.92),259(2.96), 263(2.89),266(2.83), 269(2.68)	44-0084-76

Compound	Solvent	$\lambda_{max}(\log \epsilon)$	Ref.
$C_{32}H_{32}O_7$ 9H-7a,11:8,10-Dimethano-7H-cyclobuta-[5,6]pentaleno[1,2-b]naphtho[2,1-e]-pyran-9,13,14-trione, 5-acetoxy-8,8a,9a,10,11,11a,11b,11c-octahydro-11b-hydroxy-6,8a,9a,10,11,11a,11c-heptamethyl-	n.s.g.	294s(3.55),300(3.71), 314(3.82),329(3.76)	39-0595-76C
$C_{32}H_{33}N_5O_7$ Cytidine, N-[[[[(triphenylmethyl)amino]-acetyl]amino]acetyl]-	n.s.g.	246s(4.02),296(3.50)	128-0351-76
$C_{32}H_{34}N_2O_6$ 1H-Pyrazole-5-carboxylic acid, 1-methyl-4-[2,3,5-tris-O-(phenylmethyl)-β-D-ribofuranosyl]-, methyl ester	MeOH	231(3.97)	44-3000-76
$C_{32}H_{34}N_4O_3$ 21H,23H-Porphine-2-propanoic acid, 8,13-diethyl-18-formyl-3,7,12,17-tetramethyl-	pyridine	415(5.23),519(3.85), 562(4.27),583(4.11), 637(3.11)	78-0275-76
	4:1 CH_2Cl_2-TFA	415(5.34),555(4.04), 620(4.07)	78-0275-76
$C_{32}H_{34}O_{15}$ Haplanthin, pentaacetyl-	EtOH	279(4.22),305s(3.72)	2-0644-76
$C_{32}H_{35}As$ 1H-Arsole, 5-(1,1-dimethylethyl)-2-(2,2-dimethylpropylidene)-3-(diphenylmeth-ylene)-2,3-dihydro-1-phenyl-	EtOH	232(3.93),241(4.23), 318(3.95)	78-2131-76
$C_{32}H_{36}N_2O_5$ Chaetoglobosin C	MeOH	222(4.56),273(3.83), 281(3.83),291(3.76)	88-1351-76
Chaetoglobosin D	EtOH	221(4.64),273(3.96), 281(3.96),290(3.88)	88-1351-76
$C_{32}H_{36}N_2O_{16}$ L-Threonine, N-[(phenylmethoxy)carbon-yl]-O-(2,3,4,6-tetra-O-acetyl-β-D-glucopyranosyl)-, 2-nitrophenyl ester	EtOH NaOH	258(3.27) 286(3.17)	136-0001-76E 136-0001-76E
$C_{32}H_{36}N_4O_2$ 21H,23H-Porphine-2-carboxylic acid, 7,12,18-triethyl-3,8,13,17-tetra-methyl-, methyl ester	$CHCl_3$	403(5.35),503(4.05), 541(4.23),564(3.99), 620(3.38)	78-1793-76
	$CHCl_3$-5% TFA	406(5.60),548(4.16), 594(4.06)	78-1793-76
$C_{32}H_{38}N_2O_5$ Chaetoglobosin E	EtOH	221(4.75),275(3.85), 281(3.85),291(3.80)	88-1351-76
Chaetoglobosin F	EtOH	222(4.68),276(3.84), 283(3.83),292(3.78)	88-1351-76
$C_{32}H_{38}N_8O_8$ Butanedial, [2-(1,2,3,4,4a,7,8,8a-octa-hydro-1,2,4a,5-tetramethyl-1-naphtha-lenyl)ethylidene]-, bis(2,4-dinitro-phenylhydrazone)	dioxan	258(4.45),363(4.59)	94-0294-76

Compound	Solvent	$\lambda_{max}(\log \epsilon)$	Ref.
$C_{32}H_{40}N_2O_3$ 6-Azabicyclo[3.2.1]octan-7-one, 1-(3-methoxyphenyl)-6-[[1-(3-methoxyphenyl)-6,8-dimethyl-6-azabicyclo[3.2.1]-oct-7-ylidene]methyl]-8-methyl-, [1α,5α,6(R*,R*),8R*]-	MeOH	217(4.32),274(4.11), 281(4.08)	94-1514-76
$C_{32}H_{40}N_2O_8$ Pithomycolide	MeOH	250(2.80),256(2.83), 262(2.74),266(2.58)	23-1360-76
$C_{32}H_{41}NO_7$ D-glycero-D-gulo-2-Octulopyranose, 3,4:7,8-di-O-cyclohexylidene-1-deoxy-5-O-(phenylmethyl)-1-(2-pyridinyl)-	MeOH	260(3.31)	30-0613-76
$C_{32}H_{42}O_6$ 8-Hexadecyne-3,14-dione, 1,16-bis(3,5-dimethoxyphenyl)-	MeOH	275(3.55),285(3.55)	12-1979-76
$C_{32}H_{42}O_8$ Pregn-5-ene-3,8,12,14,17,20-hexol, 20-acetate 13-(3-phenyl-2-propenoate)	EtOH	217(4.29),223(4.22), 278(4.38)	94-2185-76
$C_{32}H_{44}$ 1,4-Cyclohexadiene, 3,3'-(2,4,6-octatriene-1,8-diylidene)bis[1,2,4,5,6,6-hexamethyl-, (E,E,E)-	CH_2Cl_2	422s(--),443(4.99), 472(4.96)	5-1103-76
Pentacyclo[20.10.0.0^{3,24}.0^6,^{17}.0^8,^{19}]-dotriaconta-1,3(24),6,8(19),17,22-hexaene	isooctane	236(4.17),251(4.12), 306s(2.68)	44-4081-76
$C_{32}H_{44}N_2O_7$ Hydrazinecarboxylic acid, [[2-(3,4-dihydro-6,7-dimethoxy-1-methyl-1H-2-benzopyran-3-yl)-4,5-dimethoxyphenyl]methylene]-, 5-methyl-2-(1-methylethyl)-cyclohexyl ester	dioxan	233(4.28),290(4.26), 313(4.10)	33-0949-76
$C_{32}H_{44}O_7$ Tomentodin	EtOH	217(4.29),223(4.22), 278(4.38)	94-0443-76
	EtOH	217(4.27),223(4.25), 278(4.35)	94-1552-76
$C_{32}H_{45}ClO_5$ Olean-12-en-29-oic acid, 3-acetoxy-19-chloro-18-hydroxy-11-oxo-, γ-lactone, (3β,19α,20β)-	EtOH	249(4.13)	65-2273-76
$C_{32}H_{45}NO$ 4-Azacholest-5-eno[4,5,6-ab]indol-3-one	EtOH	243(4.30),267(4.01), 297(3.65)	49-0511-76
$C_{32}H_{45}N_5O_8$ Adenosine, 3',4'-didehydro-3'-deoxy-N-(2,2-dimethyl-1-oxopropyl)-, 2'-[3-(2,2-dimethyl-1-oxopropoxy)-4,4-dimethyl-2-pentenoate] 5'-(2,2-dimethylpropanoate)	MeOH	212(4.62),271(4.30)	35-8213-76

Compound	Solvent	λ_{max}(log ϵ)	Ref.

C$_{32}$H$_{46}$IN$_5$O$_8$
   2-Pentenoic acid, 3-(2,2-dimethyl-1-oxo-   MeOH   213(4.46),272(4.27)   35-8204-76
   propoxy)-4,4-dimethyl-, ester with N-
   [9-[2-deoxy-5-O-(2,2-dimethyl-1-oxo-
   propyl)-2-iodo-β-D-xylofuranosyl]-9H-
   purin-6-yl]-2,2-dimethylpropanamide

   2-Pentenoic acid, 3-(2,2-dimethyl-1-oxo-   MeOH   212(4.51),271(4.28)   35-8204-76
   propoxy)-4,4-dimethyl-, ester with N-
   [9-[3-deoxy-5-O-(2,2-dimethyl-1-oxo-
   propyl)-3-iodo-β-D-xylofuranosyl]-9H-
   purin-6-yl]-2,2-dimethylpropanamide

C$_{32}$H$_{46}$N$_2$O$_4$
   2'H-Spirosta-2,4-dieno[3,2-c]pyrazole-   EtOH   276(4.32)   65-1381-76
   5'-carboxylic acid, 2'-methyl-,
   ethyl ester, (25R)-

C$_{32}$H$_{46}$O$_4$
   Benzene, 1,1'-(8-hexadecyne-1,16-diyl)-   MeOH   226s(4.23),274(3.52),   12-1979-76
   bis[3,5-dimethoxy-   281(3.52)

C$_{32}$H$_{46}$O$_5$S
   Androst-5-ene-17β-methanol, 3β-[(tetra-   EtOH   203(3.85),223(4.10)   33-2125-76
   hydro-2H-pyran-2-yl)oxy]-, 4-methyl-
   benzenesulfonate

C$_{32}$H$_{46}$O$_6$
   29-Nordammara-17(20),24-dien-21-oic   n.s.g.   230(3.82)   39-0710-76C
   acid, 16β-acetoxy-3,11-dioxo-,
   methyl ester, (17Z)-

C$_{32}$H$_{46}$O$_8$
   Cucurbitacin E, 23,24-dihydro-   EtOH   270(3.90)   97-0322-76

C$_{32}$H$_{46}$O$_{10}$
   Gnidiglaucin   MeOH   241(3.88)   44-3850-76

C$_{32}$H$_{47}$BrO$_2$
   Oleana-9(11),12,15-trien-3-ol, 16-bro-   EtOH   279(3.78)   2-0403-76
   mo-, acetate, (3β)-

C$_{32}$H$_{47}$N$_5$O$_8$
   Adenosine, 2'-deoxy-N-(2,2-dimethyl-1-   MeOH   212(4.55),257s(4.14),   35-8204-76
   oxopropyl)-, 3'-[3-(2,2-dimethyl-1-   271(4.29)
   oxopropoxy)-4,4-dimethyl-2-penten-
   oate] 5'-(2,2-dimethylpropanoate)

   Adenosine, 3'-deoxy-N-(2,2-dimethyl-1-   MeOH   212(4.52),257s(4.12),   35-8204-76
   oxopropyl)-, 2'-[3-(2,2-dimethyl-1-   271(4.27)
   oxopropoxy)-4,4-dimethyl-2-penten-
   oate] 5'-(2,2-dimethylpropanoate)

C$_{32}$H$_{48}$N$_2$O
   4-Azacholest-5-en-3-one, 4-(phenyl-   EtOH   233(4.28)   49-0511-76
   amino)-

C$_{32}$H$_{48}$O$_4$
   Benzene, 1,1'-(8-hexadecene-1,16-diyl)-   MeOH   274(3.52),290(3.52)   12-1979-76
   bis[3,5-dimethoxy-, (Z)-

   Isoanhydrospergulatriol, diacetate   EtOH   244(4.28),252(4.33),   88-2327-76
       261(4.18)

Compound	Solvent	$\lambda_{max}(\log \epsilon)$	Ref.
Seychellogenin, acetate	MeOH	236(4.09),243(4.10), 252(3.94)	78-2313-76
$C_{32}H_{48}O_5$ Olean-13(18)-en-29-oic acid, 3-acetoxy-11-oxo-	EtOH	200(4.11)	65-2273-76
$C_{32}H_{48}O_6$ 29-Nordammara-17(20),24-dien-21-oic acid, 16β-acetoxy-11α-hydroxy-3-oxo-, methyl ester, (17Z)-	n.s.g.	234(3.62)	39-0710-76C
$C_{32}H_{48}O_7$ Cochlioquinone A, 9,11,14,17-tetra-dehydro-14,17-dideoxo-11-deoxy-14,17-dimethoxy-	MeOH	224(4.37),291(4.18)	32-0147-76
$C_{32}H_{49}BrO_2$ Oleana-9(11),12-dien-3-ol, 15-bromo-, acetate, (3β)-	EtOH	276(3.78)	2-0403-76
$C_{32}H_{49}BrO_3$ Olean-12-en-11-one, 3β-acetoxy-15-bromo-	EtOH	249.5(4.04)	2-0403-76
$C_{32}H_{50}O_3$ 16-Oxotaraxeryl acetate	EtOH	245(4.02)	2-0403-76
$C_{32}H_{50}O_4$ Benzene, 1-(1,16-hexadecanediyl)bis[3,5-dimethoxy-	MeOH	274(3.54),280(3.54)	12-1979-76

Compound	Solvent	$\lambda_{max}(\log \epsilon)$	Ref.
$C_{33}H_{20}N_2S_2$ 4aH-1-Pyrindine-4a-carbonitrile, 2,7-dihydro-3,4,5,6-tetraphenyl- 2,7-dithioxo-	EtOH	263(4.56),317(4.37), 394(4.59)	77-1045-76
$C_{33}H_{20}N_{10}$ Hydrazyl, 1,1'-(5-azido-2,4,6-tricyano- 1,3-phenylene)bis[2,2-diphenyl-	CH_2Cl_2	314(4.36),328(4.39), 424s(3.86),482(4.17), 625(3.67)	5-1739-76
$C_{33}H_{22}NOS$ Pyrylium, 4-(12H-benzo[a]phenothiazin- 5-yl)-2,6-diphenyl-, perchlorate	MeCN	255s(4.38),275(4.48), 398(4.31),655(4.10)	4-0073-76
also other solvents	MeOH	690(4.1)	4-0073-76
	CHCl$_3$	770(4.1)	4-0073-76
$C_{33}H_{22}N_8$ Hydrazyl, 1,1'-(5-amino-2,4,6-tricyano- 1,3-phenylene)bis[2,2-diphenyl-	CH_2Cl_2	319(4.38),328(4.45), 444(4.10),468(4.09), 627(3.73)	5-1739-76
$C_{33}H_{22}O$ 5H-Benzo[6,7]cyclohept[1,2-b]indeno- [1,2-e]pyran, 13-(1-acenaphthyl- enyl)-6,7-dihydro-	dioxan	233(4.69),300(4.43), 322(4.39),525(3.73)	78-2225-76
Indeno[2,1-b]pyran, 9-(1H-inden-2-yl)- 2,4-diphenyl-	dioxan	275(4.52),318(4.54), 363(4.54),593(3.53)	78-2225-76
$C_{33}H_{22}OTe$ 6H-Cyclohepta[c]tellurophen-6-one, 1,3,5,7-tetraphenyl-	CHCl$_3$	305(4.56),385(4.22)	24-3886-76
$C_{33}H_{22}O_3$ 2,4,6-Cycloheptatrien-1-one, 4,5-dibenz- oyl-2,7-diphenyl-	CHCl$_3$	295(4.35),370(4.02)	24-3886-76
$C_{33}H_{23}$ Methylium, diphenyl(10-phenyl-9-anthra- cenyl)-	n.s.g.	810(3.66)	138-0643-76
$C_{33}H_{23}Cl_3N_6$ 2H-Benzotriazole, 2,2'-[[2-(3-chloro- phenyl)-1,3-propanediyl]bis(3-chlo- ro-4,1-phenylene)]bis-	DMF	313(4.76),319(4.75)	33-0802-76
$C_{33}H_{24}BrN_4O_4P$ Phosphonium, triphenyl-, 3-(4-bromo- phenyl)-3-[(2,4-dinitrophenyl)azo]- 2-propenylide	DMF	389(3.785),584(4.720)	65-2135-76
$C_{33}H_{24}O$ Indeno[2,1-b]pyran, 9-(1-acenaphthylen- yl)-2,4-dimethyl-3-(phenylmethyl)-	dioxan	230(4.62),299(4.35), 536(3.72)	78-2225-76
$C_{33}H_{25}BrN_3O_2P$ Phosphonium, triphenyl-, 3-(4-bromophen- yl)-3-[(4-nitrophenyl)azo]-2-propenyl- ide	DMF	409(4.293),606(4.563)	65-2135-76

Compound	Solvent	$\lambda_{max}(\log \epsilon)$	Ref.
$C_{33}H_{25}BrN_4O_4P$ Phosphonium, [3-(4-bromophenyl)-3-[[2,4-dinitrophenyl)hydrazono]-1-propenyl]-triphenyl-, chloride	EtOH	253(4.19),390(4.47)	65-2135-76
$C_{33}H_{25}NO_2$ Indolizine, 1,3-dicinnamoyl-6-methyl-7-phenyl-	EtOH	226(4.42),260(4.18), 306(4.54),394(4.62)	103-0424-76
$C_{33}H_{25}N_4O_4P$ Phosphonium, triphenyl-, 3-[(2,4-di-nitrophenyl)azo]-3-phenyl-2-prop-enylide	DMF	388(3.753),580(4.602)	65-2135-76
$C_{33}H_{26}$ 8aH-Cyclopent[a]azulene, 8a,9-dimethyl-1,2,3-triphenyl-	CH_2Cl_2	220(4.17),330(3.97), 420(3.25)	88-4049-76
$C_{33}H_{26}BrN_2P$ Phosphonium, triphenyl-, 3-(4-bromophen-yl)-3-(phenylazo)-2-propenylide	DMF	287(4.088),375(4.167), 487(--)	65-2135-76
$C_{33}H_{26}BrN_3O_2P$ Phosphonium, [3-(4-bromophenyl)-3-[(4-nitrophenyl)hydrazono]-1-propenyl]-triphenyl-, chloride	EtOH	254(4.04),407(4.74)	65-2135-76
$C_{33}H_{26}N_3O_2P$ Phosphonium, triphenyl-, 3-[(4-nitro-phenyl)azo]-3-phenyl-2-propenylide	DMF	330(3.767),413(3.98), 610(4.498)	65-2135-76
$C_{33}H_{26}N_4O_3$ 2H-Anthra[1,9,8-cdef][2,7]naphthyridine-1,6,11(10H)-trione, 2,10-dimethyl-5,7-bis[(4-methylphenyl)amino]-	DMF	300(4.46),530(3.80), 592(4.62)	103-1352-76
$C_{33}H_{26}N_4O_4P$ Phosphonium, [3-[(2,4-dinitrophenyl)-hydrazono]-3-phenyl-1-propenyl]-triphenyl-, chloride	EtOH	253(4.36),385(4.50)	65-2135-76
$C_{33}H_{27}BrN_2P$ Phosphonium, [3-(4-bromophenyl)-3-(phen-ylhydrazono)-1-propenyl]triphenyl-, chloride	EtOH	266(4.39),377(4.13)	65-2135-76
$C_{33}H_{27}N$ Benzeneethanamine, N-(diphenylmethyl-ene)-β,β-diphenyl-	C_6H_{12}	251(4.22)	101-0295-76G
$C_{33}H_{27}N_2P$ Phosphonium, triphenyl-, 3-phenyl-3-(phenylazo)-2-propenylide	DMF	286(4.02),361(4.03), 494(3.95)	65-2135-76
$C_{33}H_{27}N_3O_2P$ Phosphonium, [3-[(4-nitrophenyl)hydra-zono]-3-phenyl-1-propenyl]triphenyl-, chloride	EtOH	254(4.18),405(4.68)	65-2135-76
$C_{33}H_{28}IN_3O_9$ Acetamide, N-[1,2-dihydro-2-oxo-	MeOH	231(4.66),283(3.90),	44-1836-76

Compound	Solvent	$\lambda_{max}(\log \epsilon)$	Ref.
1-(1,3,4-tri-O-benzoyl-6-deoxy-6-iodo-β-D-psicofuranosyl)-4-pyrimidinyl]-(cont.)		298(3.89)	44-1836-76
$C_{33}H_{28}N_2P$ Phosphonium, triphenyl[3-phenyl-3-(phenylhydrazono)-1-propenyl]-, chloride	EtOH	259(4.377),388(4.450)	65-2135-76
$C_{33}H_{28}O_7$ Cyclopropanecarboxylic acid, bis(diphenylmethylene)-, 1-methylethyl ester	CHCl$_3$	260(4.47),370(4.56)	18-3173-76
$C_{33}H_{28}O_7$ 2H-Furo[2,3-h]-1-benzopyran, 8-(2-benzoyloxy-4-hydroxy-5-methyl-6-methoxyphenyl)-3,4-dihydro-5-methoxy-2-phenyl-, (2S)-	EtOH EtOH-base	286s(4.00),305(4.03) 320(4.03)	39-1570-76C 39-1570-76C
$C_{33}H_{29}N_3O_{11}$ 1H-1,2,4-Triazole-3-carboxylic acid, 1-(6-O-acetyl-1,3,4-tri-O-benzoyl-β-D-psicofuranosyl)-, methyl ester	dioxan	212s(4.34),230(3.66), 270s(3.52),275(3.49), 283(3.40)	44-1836-76
$C_{33}H_{30}O$ 2,4-Cyclopentadien-1-ol, 2-methyl-3,4,5-triphenyl-1-(2,4,6-trimethylphenyl)-	C_6H_{12}	245(4.492),330(3.903)	5-0793-76
$C_{33}H_{30}O_7$ 1,3-Benzenediol, 4-[2-(3,4-dihydro-7-hydroxy-5-methoxy-2-phenyl-2H-1-benzopyran-8-yl)ethenyl]-5-methoxy-6-methyl-, 3-benzoate, [S-(E)]-	EtOH	233(4.49),316(4.40), 326s(4.38)	39-1570-76C
$C_{33}H_{31}N_3O_2$ 2-Butanone, 4,4-[benzo[f]quinoline-1,3-diyl]bis(4,1-phenyleneimino)]bis-	EtOH	299(4.32),340(4.32), 371(4.22)	103-0106-76
$C_{33}H_{33}F_6N_3O_8S_2$ Pyrrolo[3,2,1-de]phenanthridinium, 4,5,7a,8,9,10,11,11a-octahydro-7a,11a-bis(p-tolylsulfonamido)-, bis(trifluoroacetate)	EtOH	200(4.58),215s(4.49), 230(4.40),312(3.66)	39-2254-76C
$C_{33}H_{33}O_2PS$ Sulfur, (4,4-dimethyl-2,6-dioxocyclohexylidene)methylphenyl(triphenylphosphine)- (inner salt)	MeOH DMSO	226(4.67),263(4.36), 270(4.35),277(4.33), 285s(4.26) 264s(4.34),271(4.40), 276(4.40),283s(4.35)	30-0405-76 30-0405-76
$C_{33}H_{34}N_2O_6$ 1H-Pyrazole-3-carboxylic acid, 4-[2,3-O-(1-methylethylidene)-5-O-(triphenylmethyl)-β-D-ribofuranosyl]-, ethyl ester	pH 13	233(4.07),251s(3.90)	44-0084-76
$C_{33}H_{34}N_4O_3$ 21H,23H-Porphine-2-propanoic acid, 8-ethenyl-13-ethyl-18-formyl-3,7,12,17-tetramethyl-, methyl ester	CH$_2$Cl$_2$	413(5.26),519(3.86), 561(4.30),581(4.12), 637(3.15)	78-0275-76

Compound	Solvent	$\lambda_{max}(\log \epsilon)$	Ref.
(cont.)	CH_2Cl_2- 1% TFA	416(5.42),565(4.04), 617(4.06)	78-0275-76
$C_{33}H_{34}N_4O_6$ Pyrazolo[1,5-a]-1,3,5-triazine-2,4- (1H,3H)-dione, 8-[2,3,5-tris-O- (phenylmethyl)-β-D-ribofuranosyl]- 1,3-dimethyl-	MeOH	239(3.75),252(3.76), 258(3.78)	44-3000-76
$C_{33}H_{34}O$ 1(2H)-Acenaphthylenone, 2-phenyl-2- [2,4,6-tris(1-methylethyl)phenyl]-	n.s.g.	213(4.83),253(4.21), 318(3.76),340(3.76)	44-3599-76
2-Cyclopropen-1-one, 2,3-bis[3,8-di- methyl-5-(1-methylethyl)- 1-azulenyl]-	benzene	293(4.45),338(4.48), 460(4.65),484(4.70), 560s(3.43),600s(3.43), 650s(3.18)	44-2379-76
	C_6H_{12}	242(4.56),290(4.40), 325s(4.38),336(4.48), 456(4.62),480(3.66), 564(3.36),610s(3.36), 660s(3.18)	44-2379-76
	EtOH	244(4.60),290(4.43), 326(4.38),336(4.45), 456s(4.66),479(4.72), 591s(3.49),650(3.20)	44-2379-76
	dioxan	291(4.40),336(4.43), 459(4.62),482(4.66), 556s(3.42),600s(4.41), 660s(3.18)	44-2379-76
	MeCN	242(4.59),289(4.42), 335(4.48),455(4.67), 478(4.72),520s(3.18)	44-2379-76
$C_{33}H_{35}N_3O_3$ Spiro[2H-1-benzoxocin-2,2'-[2H]indole], 4-[(1,3-dihydro-1,3,3-trimethyl-2H- indol-2-ylidene)methyl]-1',3,3',4- tetrahydro-1',3',3'-trimethyl-8-nitro-	hexane MeOH dioxan DMSO	210(4.46),315(4.49) 315(4.37),570(3.46) 316(4.32) 318(4.55),420(3.20)	103-1227-76 103-1227-76 103-1227-76 103-1227-76
$C_{33}H_{35}N_5O_6$ 21H,23H-Porphine-2,18-dipropanoic acid, 3,7,8,12,17-pentamethyl-13-nitro-, dimethyl ester	$CHCl_3$	373(4.85),423(4.98), 522(3.81),568(4.18)	12-2325-76
$C_{33}H_{36}$ Acenaphthylene, 1,2-dihydro-1-phenyl- 2-[2,4,6-tris(1-methylethyl)phenyl]-, cis	hexane	224(4.81),293(3.95)	35-1860-76
$C_{33}H_{36}N_2O$ Spiro[2H-1-benzoxocin-2,2'-[2H]indole], 4-[(1,3-dihydro-1,3,3-trimethyl-2H- indol-2-ylidene)methyl]-1',3,3',4- tetrahydro-1',3',3'-trimethyl-	hexane dioxan DMSO	211(4.59),315(4.38) 229(4.65),316(4.65) 320(4.60)	103-1227-76 103-1227-76 103-1227-76
$C_{33}H_{36}N_4O_3$ 21H,23H-Porphine-2-carboxylic acid, 12- ethenyl-7-ethyl-18-(3-hydroxypropyl)- 3,8,13,17-tetramethyl-, methyl ester	CH_2Cl_2	406(5.30),514(3.91), 559(4.25),577(4.00), 635(2.95)	78-0275-76
	$CH_2Cl_2-5\%$ TFA	412(5.46),559(4.07), 610(3.98)	78-0275-76

Compound	Solvent	$\lambda_{max}(\log \epsilon)$	Ref.
21H,23H-Porphine-2-propanoic acid, 8-ethenyl-13-ethyl-18-(hydroxymethyl)-3,7,12,17-tetramethyl-, methyl ester	CH_2Cl_2	402(5.29),505(4.08), 541(4.12),571(3.92), 626(3.38)	78-0275-76
	CH_2Cl_2- 1% TFA	407(5.51),554(4.23), 600(3.87)	78-0275-76
$C_{33}H_{36}O_2$ 1,2-Acenaphthylenediol, 1,2-dihydro-1-phenyl-2-[2,4,6-tris(1-methylethyl)-phenyl]-, trans	n.s.g.	209(4.81),228(4.86), 292(3.97)	44-3599-76
$C_{33}H_{39}NO_7S$ D-glycero-β-D-gulo-Heptopyranose, 1-C-2-benzothiazolyl-2,3:6,7-di-O-cyclohex-ylidene-4-O-(phenylmethyl)-	MeOH	255(3.52)	30-0613-76
$C_{33}H_{40}O_{20}$ Polygonatin	MeOH	268(4.14),304s(4.21) 330(4.16)	95-1180-76
	MeOH-NaOMe	280(--),331(--), 395(--)	95-1180-76
$C_{33}H_{42}N_4O_2$ 22H-Biline-2-carboxylic acid, 7,12,18-triethyl-10,23-dihydro-1,3,8,13,17,19-hexamethyl-, methyl ester, dihydro-bromide	$CHCl_3$	450(4.45),517(5.32)	78-1793-76
21H-Biline-1,19-dione, 2,3,8,12,17,18-hexaethyl-22,24-dihydro-7,13-dimeth-yl-	$CHCl_3$	370(4.61),630(4.08)	31-1107-76 +118-0335-76
with p-toluenesulfonic acid added	$CHCl_3$	293(4.15),366(4.66), 670(4.43)	118-0335-76
$C_{33}H_{34}N_4O_6$ Orantin, N,N'-diacetyl-O-methyl-	EtOH	230(4.32),275s(3.49), 281(3.55),289s(3.38)	133-0371-76
$C_{33}H_{42}O_{13}$ Baccharin, diacetate	EtOH	262(4.32)	35-7092-76
$C_{33}H_{45}NO_5$ Acetamide, N-[(3β)-3,20-diacetoxy-20-phenylpregn-5-en-21-yl]-	EtOH	246(2.28),251(2.33), 257(2.37),264(2.26)	33-1850-76
$C_{33}H_{46}N_2O_7$ 3-Piperidinol, 1-acetyl-2-[1-[6'-acet-oxy-1'-formyl-3',4,5,6',7',8',8'a-octahydro-7,8'a-dimethylspiro[2,1-benzisoxazole-3(3aH),2'(1'H)-naph-thalen]-6-yl]ethyl]-5-methyl-, acet-ate	EtOH	203(4.10),259(4.17)	39-1297-76C
isomer	EtOH	203(4.12),257(4.20)	39-1297-76C
$C_{33}H_{47}IN_2O_6$ 3-Piperidinol, 1-acetyl-2-[1-[6'-acet-oxy-3',4,5,5',6',7',8',8'a-octahydro-1'-(iodomethyl)-7,8'a-dimethylspiro-[2,1-benzisoxazole-3(3aH),2'(1'H)-naphthalen]-6-yl]ethyl]-5-methyl-, acetate	EtOH	259(4.26)	39-1297-76C

Compound	Solvent	$\lambda_{max}(\log \epsilon)$	Ref.
$C_{33}H_{47}N_3O_7$ 3-Piperidinol, 1-acetyl-2-[1-[6'-acet-oxy-3',4,5,5',6',7',8',8'a-octahydro-1'-[(hydroxyimino)methyl]-7,8'-dimeth-ylspiro[2,1-benzisoxazole-3(3aH),2'-(1'H)-naphthalen]-6-yl]ethyl]-5-meth-yl-, acetate	EtOH	205(4.18),257(4.12)	39-1297-76C
$C_{33}H_{48}N_2O_7$ Dihydroisophoto-TVTTN 3-Piperidinol, 1-acetyl-2-[1-[6'-acet-oxy-3',4,5,5',6',7',8',8'a-octahydro-1-(hydroxymethyl)-7,8'a-dimethylspiro-[2,1-benzisoxazole-3(3aH),2'(1'H)-naphthalen]-6-yl]ethyl]-5-methyl-, acetate	EtOH EtOH	208(4.15),257(4.15) 205(4.13),259(4.16)	39-1297-76C 39-1297-76C
$C_{33}H_{48}O_2$ A-Norcholestan-2-one, 3-benzoyl- A-Norcholestan-2-one, 5-benzoyl-	EtOH EtOH	208(3.95),248(4.06), 278(3.22),320(2.23) 208(4.00),244(4.02), 274(3.24),325(2.31)	22-1953-76 22-1953-76
$C_{33}H_{48}O_6$ Olean-12-en-29-oic acid, 3-acetoxy-18β,19β-epoxy-11-oxo-, methyl ester	EtOH	249(4.14)	65-2273-76
$C_{33}H_{48}O_{12}$ Nudiposide, hexamethyl ether	EtOH	274(3.30)	94-2102-76
$C_{33}H_{49}ClO_6$ Olean-12-en-29-oic acid, 3-acetoxy-19α-chloro-18β-hydroxy-11-oxo-, methyl ester	EtOH	249(4.12)	65-2273-76
$C_{33}H_{50}N_8O_{10}$ L-Argininamide, N-[(1,1-dimethylethoxy)-carbonyl]glycyl-N-[1,2-dihydro-2-oxo-1-[5-O-(tricyclo[3.3.1.1^{3,7}]dec-1-yl-carbonyl)-β-D-arabinofuranosyl]-4-pyrimidinyl]-, hydrochloride	H_2O	213s(4.24),247(4.10), 301(3.85)	87-1013-76
$C_{33}H_{50}O_2$ 2,5-Cyclohexadien-1-one, 4-[1-[3,5-bis-(1,1-dimethylethyl)-4-hydroxyphenyl]-2,2-dimethylpropylidene]-2,6-bis(1,1-dimethylethyl)- anion radical	MeOH MeOH MeOH	320(4.36) 560(4.30) 470(3.70),626(--)	64-0965-76B 64-0965-76B 64-0965-76B
$C_{33}H_{50}O_4$ 19-Nor-9,10-secocholesta-1,3,5(10)-tri-ene-1,10-dicarboxylic acid, 9-(1-meth-ylethenyl)-, dimethyl ester, (9β)-	hexane	232(3.91),281(3.26), 290(3.20)	23-3508-76
$C_{33}H_{50}O_9$ Cochlioquinone A, 14-acetoxy-14,17-di-dehydro-14,17-dideoxo-17-methoxy-	MeOH	215(4.19),285(3.33)	32-0147-76

Compound	Solvent	$\lambda_{max}(\log \epsilon)$	Ref.
$C_{33}H_{51}NO_5$ 16,28-Secosolanid-22-ene-3,16-diol, 28- acetyl-, diacetate, (3β,5α,16α)-	EtOH	236(3.87)	88-3653-76
$C_{33}H_{51}NO_6$ 5α-Veratraman-3β,11β,23β-triol, 28-acet- yl-13,17-didehydro-5,6,12,13-tetra- hydro-, 3,23-diacetate	MeOH	217(4.04)	39-1312-76C
$C_{33}H_{54}O_3$ Ergosta-5,7-dien-22-ol, 3-[(tetrahydro- 2H-pyran-2-yl)oxy]-, (3β,22S)-	EtOH	271(4.07),282(4.09), 293(3.86)	39-0826-76C
Ergosta-5,7-dien-23-ol, 3-[(tetrahydro- 2H-pyran-2-yl)oxy]-, (3β,22R)-	CHCl$_3$	275(4.15),285(4.18), 296(3.91)	39-0826-76C
$C_{33}H_{55}N_3O_6$ Dodecanoic acid, [2,3,3a,9a-tetrahydro- 6-imino-3-[(1-oxododecyl)oxy]-6H-furo- [2',3':4,5]oxazolo[3,2-a]pyrimidin- 2-yl]methyl ester, hydrochloride	MeOH-acid	234(4.00),262(4.05)	87-0663-76
$C_{33}H_{57}N_3O_7$ 2(1H)-Pyrimidinone, 4-amino-1-[3,5-bis- O-(1-oxododecyl)-β-D-arabinofurano- syl]-	MeOH-acid	211(4.01),283(4.16)	87-0667-76
$C_{33}H_{60}OS$ 1-Tetradecanone, 2-methyl-1-(4-methyl- 5-tridecyl-2-thienyl)-	heptane	257(3.89),298(4.01)	34-0233-76

Compound	Solvent	$\lambda_{max}(\log \epsilon)$	Ref.
$C_{34}H_{17}Br$ Dibenzo[a,rst]naphtho[8,1,2-cde]penta- phene, 12-bromo-	u.s.g.	484(4.37)	104-0180-76
$C_{34}H_{20}N_2S_2$ 5H-Naphtho[1,8-bc]thiophen-5-one, 2- phenyl-, (2-phenyl-5H-naphtho[1,8- b]thien-5-ylidene)hydrazone	CH_2Cl_2	530(4.64)	88-4713-76
$C_{34}H_{20}N_6$ Benzimidazo[1,2-c]benzimidazo[1',2':1,6]- pyrimido[5,4-g]quinazoline, 6,10-di- phenyl-	H_2SO_4	235(4.43),267(4.20), 340(3.18),427(4.51), 452(4.53)	116-0626-76
$C_{34}H_{21}N_5O_3S_2$ Methanone, [4-[(6-nitro-2,7-diphenyl- thiazolo[4,5-b]pyridin-5-yl)amino]- 2-phenyl-5-thiazolyl]phenyl-	$CHCl_3$	260s(4.45),304(4.55), 395(4.45)	48-0039-76
$C_{34}H_{22}Br_2O_2$ Cyclobuta[1",2":3,4;3",4":3',4']dicyclo- buta[1,2-a:1',2'-a']diindene-4b,9b- diol, 4c,9c-dibromo-4c,4d,9c,9d- tetrahydro-4d,9d-diphenyl-	EtOH	235(2.70),260(2.84), 269(2.97),283(2.84), 318(2.50),332(2.75)	39-0695-76C
Pentacyclo[6.2.0.0^{1,4}.0^{3,6}.0^{5,8}]decane- 2,7-dione, 3,6-dibromo-4,5,9,10-tetra- phenyl-	EtOH	222(4.32),286(4.22)	39-0695-76C
Tricyclo[6.2.0.0^{3,6}]deca-3,8-diene-2,7- dione, 1,6-dibromo-4,5,9,10-tetra- phenyl-	EtOH	305(2.93)	39-0695-76C
$C_{34}H_{22}N_2O$ Furo[3,4-g]phthalazine, 1,4,6,8-tetra- phenyl-	CH_2Cl_2	240(4.52),293(4.55), 357(4.21),552(3.95)	28-0555-76A
$C_{34}H_{22}O$ Indeno[2,1-b]pyran, 9-(2-naphthalenyl)- 2,4-diphenyl-	dioxan	273(4.52),319(4.32), 326(4.34),565(3.47)	78-2225-76
$C_{34}H_{22}O_2$ 2-Propen-1-one, 3,3'-(1,3-butadiyne-1,4- diyldi-2,1-phenylene)bis[1-phenyl-	THF	267(4.46),290(4.33)	18-2840-76
$C_{34}H_{22}O_2S_4$ 1,4-Butanedione, 1,4-diphenyl-2,3-bis- (4-phenyl-3H-1,2-dithiol-3-ylidene)-	CH_2Cl_2	467(4.42)	1-0478-76
1,4-Butanedione, 1,4-diphenyl-2,3-bis- (5-phenyl-3H-1,2-dithiol-3-ylidene)-	CH_2Cl_2	473(4.50)	1-0478-76
$C_{34}H_{22}O_3$ Methanone, (1,3-diphenyl-5,5-isobenzo- furandiyl)bis[phenyl-	CH_2Cl_2	265s(4.53),280(4.57), 330(4.12),385(4.05), 440(4.02)	28-0555-76A
Unnamed compd. IV, m. 246-7°d.	dioxan	210(4.577),244(4.235), 275(4.204),373(4.485)	24-0576-76
$C_{34}H_{22}S_6$ 3,3'-Bi-1,6,6a-trithiapentalene, 2,2',5,5'-tetraphenyl-	CH_2Cl_2	480(4.42)	1-0478-76

Compound	Solvent	$\lambda_{max}(\log \epsilon)$	Ref.
$C_{34}H_{23}IN_4O_{17}$ D-Psicofuranose, 6-deoxo-6-iodo-, tetrakis(4-nitrobenzoate)	dioxan	257(4.73)	44-1836-76
$C_{34}H_{23}OP$ Phosphine oxide, phenylbis[2-(phenyl- ethynyl)phenyl]-	EtOH	227s(4.63),274s(4.45), 290(4.53),308(4.46)	24-2405-76
$C_{34}H_{23}P$ Phosphine, phenylbis[2-(phenylethynyl)- phenyl]-	C_6H_{12}	223(4.66),282(4.56), 301s(4.52)	24-2405-76
$C_{34}H_{24}Br_4N_2O_2$ Dispiro[2H-1-benzopyran-2,5'(10'H)-pyri- do[2,3,4,5-lmn]phenanthridine-10',2"- [2H-1]benzopyran', 6,6",8,8"-tetra- bromo-4',9'-diethyl-4',9'-dihydro-	dioxan	230(4.85),260(4.78), 330(4.06),366(3.97)	103-0273-76
$C_{34}H_{24}Cl_2$ 1,1'-Biphenyl, 4,4"-[(2,5-dichloro-1,4- phenylene)di-1,2-ethenediyl]bis-	DMF	385(4.86)	33-0802-76
$C_{34}H_{24}N_2$ 1,4-Benzenediamine, N,N'-bis(diphenyl- ethenylidene)-	C_6H_{12}	284(4.72),374(3.56)	54-0154-76
$C_{34}H_{24}N_6O_2$ Benzamide, N,N'-[4,6-bis(1H-benzimida- zol-2-yl)-1,3-phenylene]bis-	H_2SO_4	265(4.67),305(4.50)	116-0626-76
$C_{34}H_{24}O$ Indeno[2,1-b]pyran, 9-(3,4-dihydro- 2-naphthalenyl)-2,4-diphenyl-	dioxan	263(4.55),309(4.44), 575(3.07)	78-2225-76
$C_{34}H_{25}N_2O_4PS_3$ Phosphonium, triphenyl-, 1-(4-nitrophen- yl)-2-[[2-(4-nitrophenyl)-1-thioxo- ethyl]thio]-2-thioxoethylide	$CHCl_3$	265(4.31),324(4.07), 495(4.01)	39-1404-76B
$C_{34}H_{26}$ Bicyclo[4.2.0]octa-3,7-diene, 2,5-bis- (diphenylmethylene)-	$CHCl_3$	276s(4.05),349(4.54)	89-0487-76
1,3,6-Cyclooctatriene, 5,8-bis(diphen- ylmethylene)-	$CHCl_3$	339(4.42)	89-0487-76
$C_{34}H_{26}O_{10}$ [9,9'-Bianthracene]-10,10'(9H,9'H)-di- one, 3,3'-diacetyl-4,4',5,5',7,7'- hexahydroxy-2,2'-dimethyl-	MeOH-HOAc	282s(4.21),364(4.36)	12-2225-76
$C_{34}H_{27}NOS$ Sulfonium, diphenyl-, 1-benzoyl-2-phen- yl-2-[(phenylmethyl)imino]ethylide	EtOH	257(4.30),284s(4.13), 408(3.71)	78-0431-76
$C_{34}H_{27}NO_2$ 2(1H)-Naphthalenone, 1-[[3-(4-methoxy- phenyl)-2,4-dimethylbenzo[f]quino- lin-1(4H)-ylidene]ethylidene]-	EtOH acetone $CHCl_3$	668(3.0) 696(3.25) 706(4.60)	103-1012-76 103-1012-76 103-1012-76
Unknown compd., m. 218-220°	EtOH	228(4.56),301(4.34)	24-0576-76

Compound	Solvent	λ_{max}(log ϵ)	Ref.
$C_{34}H_{27}N_4O_5P$ Phosphonium, triphenyl-, 3-[(2,4-di- nitrophenyl)azo]-3-(4-methoxyphen- yl)-2-propenylide	DMF	267(--),378(3.645), 586(4.682)	65-2135-76
$C_{34}H_{27}PS_3$ 2-Phenyl-1-(phenyldithioacetyl)-2-tri- phenylphosphonioethenethiolate	$CHCl_3$	257(4.16),275(4.10), 325(4.13),370(3.93), 473(3.81)	39-0692-76C
$C_{34}H_{28}$ Phenanthrene, 1-(3,5-dimethylphenyl)- 9-[4-(3,5-dimethylphenyl)-1-buten- 3-ynyl]-, (E)-	MeOH	227s(4.61),252s(4.48), 260(4.54),280(4.54), 289(4.56),342(4.45)	39-1104-76B
Triphenylene, 1,12-bis(3,5-dimethyl- phenyl)-	MeOH	269(4.74),292(4.54)	39-1115-76B
$C_{34}H_{28}Cl_8O_4$ Dispiro[cyclohexane-1,13'-[13H-6,13a]- ethenopentaleno[1,2-b:5,4-b']bis[1,5]- benzodioxepin-15'(5'aH),1"-cyclohex- ane], 1',2',3',4',8',9',10',11'-octa- chloro-5'b,6',6'a,13'b-tetrahydro- isomer	C_6H_{12} C_6H_{12}	219(5.27),300(3.57) 220(5.54),300(3.78)	78-0147-76 78-0147-76
$C_{34}H_{28}NO_2$ Benzo[f]quinolinium, 1-[2-(2-hydroxy-1- naphthalenyl)ethenyl]-3-(4-methoxy- phenyl)-2,4-dimethyl-, iodide	EtOH	455(3.84),678(3.38)	103-1012-76
$C_{34}H_{28}N_2O_4$ Dispiro[2H-1-benzopyran-2,5'(10'H)-pyri- do[2,3,4,5-lmn]phenanthridine-10',2"- [2H-1]benzopyran], 4',9'-dihydro- 6,6"-dimethoxy-4',9'-dimethyl-	$CHCl_3$ 50% HOAc- H_2SO_4	255(4.78),350(4.05), 268s(4.41),276(4.44), 342(4.24),386s(4.12), 403(4.26),450s(3.57)	103-0273-76 103-0273-76
Dispiro[2H-1-benzopyran-2,5'(10'H)-pyri- do[2,3,4,5-lmn]phenanthridine-10',2"- [2H-1]benzopyran], 4',9'-dihydro- 8,8"-dimethoxy-4',9'-dimethyl-	dioxan	227(4.84),252(4.71), 275s(4.38),330(3.88), 365(3.95)	103-0273-76
$C_{34}H_{28}N_3O_3P$ Phosphonium, triphenyl-, 3-(4-methoxy- phenyl)-3-[(4-nitrophenyl)azo]- 2-propenylide	DMF	415(4.353),611(4.553)	65-2135-76
$C_{34}H_{28}N_4O_5P$ Phosphonium, [3-[(2,4-dinitrophenyl)- hydrazono]-3-(4-methoxyphenyl)-1- propenyl]triphenyl-, chloride	EtOH	256(4.40),378(4.31)	65-2135-76
$C_{34}H_{28}O_2$ Methanone, phenyl[2-[2-(triphenylmeth- oxy)ethyl]phenyl]-	MeOH	250(4.14),280(3.51), 325(2.08)	35-0541-76
$C_{34}H_{28}O_7$ 7H-1-Benzopyran-7-one, 4-(7-acetoxy-3,4- dihydro-5-methoxy-2-phenyl-2H-1-benzo- pyran-8-yl]-5-methoxy-2-phenyl-	n.s.g.	279(4.29),288(4.29), 313(3.99),328s(3.93), 390(4.04),480(4.08)	39-1392-76C

Compound	Solvent	$\lambda_{max}(\log \epsilon)$	Ref.
$C_{34}H_{29}N_3O_3P$ Phosphonium, [3-(4-methoxyphenyl)- 3-[(4-nitrophenyl)hydrazono]-1- propenyl]triphenyl-, chloride	EtOH	253(4.34),408(4.66)	65-2135-76
$C_{34}H_{29}N_5O_8$ 4(3H)-Pteridinone, 2-(dimethylamino)-8- (2,3,5-tri-O-benzoylribofuranosyl)-	MeOH	230(4.73),275(4.29), 400(3.68)	24-3208-76
α-D-Ribofuranoside, 2-(dimethylamino)- 4-pteridinyl-, 2,3,5-tribenzoate	MeOH	231(4.67),276(4.15), 390(3.66)	24-3208-76
β-D-Ribofuranoside, 2-(dimethylamino)- 4-pteridinyl-, 2,3,5-tribenzoate	MeOH	231(4.68),277(4.15), 390(3.72)	24-3208-76
$C_{34}H_{29}N_5O_9$ 4,7(1H,8H)-Pteridinedione, 2-(dimethyl- amino)-8-(2,3,5-tri-O-benzoyl-β-D- ribofuranosyl)-	MeOH	227(4.73),277(3.91), 295(3.89),362(4.10)	24-3228-76
$C_{34}H_{29}N_9O_6$ 1(3H)-Isobenzofuranone, 6,7-dimethoxy- 3-[1,2,3,4-tetrahydro-2-methyl-6,7- bis[(1-phenyl-1H-tetrazol-5-yl)oxy]- 1-isoquinolinyl]-, [S-(R*,S*)]-	EtOH	220(4.72),235s(4.08), 278s(3.43),310(3.64)	44-1657-76
$C_{34}H_{30}$ 5H-14a,18b-[1',2']Benzeno-6,9-etheno- 3,12-methenobenzo[e]cyclododec[jk]- as-indacene, 1,2,4,10,11,13,14,18c- octahydro-	THF	<u>225s(4.3),265s(3.5),</u> <u>290s(3.3),315(3.2)</u>	18-3300-76
$C_{34}H_{30}N_2OP$ Phosphonium, [3-(4-methoxyphenyl)- 3-(phenylhydrazono)-1-propenyl]- triphenyl-, chloride	EtOH	244(4.60),379(4.19)	65-2135-76
$C_{30}H_{30}N_4O_{10}$ 2,4(1H,3H)-Pteridinedione, 5-acetyl- 5,6,7,8-tetrahydro-1-(2,3,5-tri-O- benzoyl-β-D-ribofuranosyl)-	pH 7.0 pH 12.0 MeOH	232(4.59),284(4.21) 230(4.63),284(4.17) 228(4.64),283(4.24)	24-3184-76 24-3184-76 24-3184-76
$C_{34}H_{30}N_6O_5$ 2H-1,4-Benzodiazepin-2-one, 7,7'-azoxy- bis[1,3-dihydro-1-(methoxymethyl)-5- phenyl-	isoPrOH	220(4.65),260(4.50), 352(4.31),388s(4.04)	87-1378-76
$C_{34}H_{30}O_8$ D-Glucitol, 1,3:2,4-bis-O-(phenylmeth- ylene)-, dibenzoate, (1R,2S)-	1:1 MeCN- DMSO	252(3.24)	12-1859-76
$C_{34}H_{32}S_6$ Dispiro[1,3-benzodithiole-2,1'(2'H)- naphtho[2,1-c][1,2]dithiin-2',2"(1"H)- naphthalene], 1"-(hexahydro-1,3-benzo- dithiol-2-ylidene)-3a,4,5,6,7,7a- hexahydro-	CH_2Cl_2	263(4.29),354s(3.67), 450(3.98)	88-3815-76
$C_{34}H_{35}N_5O_7$ 1H-Pyrrole-3,4-dicarboxylic acid, 1,2- bis[[(phenylmethyl)amino]carbonyl]- 5-[[[(phenylmethyl)amino]carbonyl]- amino]-, diethyl ester	MeOH	248(3.84)	24-2983-76

Compound	Solvent	$\lambda_{max}(\log \epsilon)$	Ref.
$C_{34}H_{36}N_2O_2$			
Phenol, 2,2'-[[1,2-bis(2,4,6-trimethyl-phenyl)-1,2-ethanediyl]bis(nitrilo-methylidyne)]bis-, (R*,S*)-	MeOH	267(4.42)	24-0001-76
Phenol, 2,2'-[1,2-bis[[(2,4,6-trimethyl-phenyl)methylene]amino]-1,2-ethane-diyl]bis-, (R*,S*)-	MeOH	322(3.42)	24-0001-76
$C_{34}H_{36}N_2O_6$			
Lindolhamine	EtOH	205(4.65),220s(4.39), 280(3.91)	142-1073-76
1H-Pyrazole-5-carboxylic acid, 1-methyl-4-[2,3-O-(1-methylethylidene)-5-O-(triphenylmethyl)-β-D-ribofuranosyl]-, ethyl ester	EtOH	248s(3.85)	44-0084-76
$C_{34}H_{36}N_2O_6S_2$			
4,9-Diazapyrenium, 4,9-diethyl-5,10-di-methyl-, bis(p-toluenesulfonate)	EtOH	225(4.76),267(4.40), 332(3.93),500(3.35)	103-0273-76
$C_{34}H_{36}N_4O_3$			
21H,23H-Porphine-2-propanoic acid, 18-acetyl-8-ethenyl-13-ethyl-3,7,12,17-tetramethyl-, methyl ester	CH_2Cl_2	409(5.25),513(3.87), 554(4.19),578(3.96), 633(2.70)	78-0275-76
	CH_2Cl_2-5% TFA	414(5.39),561(4.02), 611(3.95)	78-0275-76
$C_{34}H_{36}N_4O_4$			
21H,23H-Porphine-2,18-dipropanoic acid, 8-ethenyl-3,7,12,17-tetramethyl-, dimethyl ester	CH_2Cl_2	401(5.25),501(4.23), 536(4.11),573(3.93), 626(3.63)	39-2501-76C
$C_{34}H_{36}O_{16}$			
Astringin, heptaacetate	MeOH	208(4.32),230(4.26), 299(4.5),310(4.5)	102-2006-76
$C_{34}H_{37}ClN_4O_4$			
21H,23H-Porphine-2,18-dipropanoic acid, 8-(2-chloroethyl)-3,7,12,17-tetra-methyl-, dimethyl ester	CH_2Cl_2	399(5.22),496(4.12), 529(3.78),568(3.75), 623(3.53)	39-2501-76C
$C_{34}H_{38}N_8O_4S$			
L-Tryptophan, 3-[[2-amino-3-(1H-indol-3-yl)-1-oxopropyl]thio]-4-[[(4-amino-2-methyl-5-pyrimidinyl)methyl]formyl-amino]-3-pentenyl ester, trihydro-chloride, (S)-	EtOH	245(4.43)	94-0852-76
$C_{34}H_{38}O_{16}$			
Dihydroastringin, heptaacetate	MeOH	267(3.18),271s(3.16)	102-2006-76
$C_{34}H_{39}N_3O_4$			
1'H-Androstano[3,2-c]pyrazol-17-ol, 5'-(4-nitrophenyl)-1'-phenyl-, acetate, (5α,17β)-	EtOH	264(4.06),315s(3.79)	22-0255-76
$C_{34}H_{39}N_3O_6$			
Carbamic acid, [4-[2,3,5-tris-O-(phenyl-methyl)-β-D-ribofuranosyl]-1H-pyrazol-3-yl]-, 1,1-dimethylethyl ester	MeOH	228s(3.81),258(2.93), 264(2.85)	44-3000-76

Compound	Solvent	$\lambda_{max}(\log \epsilon)$	Ref.
$C_{34}H_{41}N_3O_4$ 1'H-Androstano[3,2-c]pyrazol-17-ol, 2',5'-dihydro-5'-(4-nitrophenyl)- 1'-phenyl-, acetate	EtOH	247(4.16),302(4.20), 426(4.08)	22-0255-76
$C_{34}H_{42}$ 1,2,3,4,5,7,9,15,17-Cyclooctadecanona- ene-11,13-diyne, 1,6,10,15-tetrakis- (1,1-dimethylethyl)-, (E,E,Z,Z,Z)-	THF	220(4.99),238(4.29), 250(4.21),261(3.76), 275(3.63),337(4.55), 374(5.36),408s(4.11), 525(4.84),621(2.18), 656(2.27),707(2.97)	18-0302-76
$C_{34}H_{42}N_2O_2$ 1'H-Androstano[3,2-c]pyrazol-17-ol, 2',5'-dihydro-1',5'-diphenyl-, acetate	EtOH	252(3.80),286(3.96)	22-0255-76
stereoisomer 3	EtOH	262(4.25)	22-0255-76
stereoisomer 4	EtOH	272(4.05)	22-0255-76
$C_{34}H_{42}O_7S_2$ D-glycero-D-gulo-Heptitol, 2,3:6,7-di- O-cyclohexylidene-4-O-(phenylmethyl)- 1,1-di-C-2-thienyl-	MeOH	242(3.04)	30-0613-76
$C_{34}H_{42}O_9$ D-glycero-D-gulo-Heptitol, 2,3:6,7-di- O-cyclohexylidene-1,1-di-C-2-furanyl- 4-O-(phenylmethyl)-	MeOH	245(2.91)	30-0613-76
$C_{34}H_{45}NO_2$ 2,5-Cyclohexadien-1-one, 4-[[3,5-bis- (1,1-dimethylethyl)-4-hydroxyphenyl)- 3-pyridinylmethylene]-2,6-bis(1,1- dimethylethyl)-	MeOH	412(4.40)	64-0965-76B
anion	MeOH	385(--),612(4.83)	64-0965-76B
radical	MeOH	391(--),468(4.74), 980(--)	64-0965-76B
$C_{34}H_{46}N_6O$ 1H-Pyrazole, 1,1'-(azoxydi-4,1-phenyl- ene)bis[3,5-bis(1,1-dimethylethyl)-	EtOH	243(3.90),340(4.18)	22-0195-76
$C_{34}H_{46}O_2$ 4,6,8,14,16,18-Docosahexaene-10,12-di- yne-3,20-dione, 9,14-bis(1,1-dimeth- ylethyl)-2,2,21,21-tetramethyl-	EtOH	247(4.36),288s(4.35), 325s(4.60),343(4.66), 385(4.61),425s(4.39)	18-2306-76
$C_{34}H_{46}O_8$ Tomentodin 3-acetate	EtOH	215(4.35),275(4.30)	94-0443-76
$C_{34}H_{48}O_3$ 5α,14β-Cholest-7-en-15-one, 3β-(benzoyl- oxy)-	n.s.g.	230(4.45)	88-4401-76
$C_{34}H_{50}O_6$ Lanosta-9(11),25-dien-18-oic acid, 3,23- diacetoxy-20-hydroxy-, γ-lactone	MeOH	236(4.11),244(4.14), 252(4.00)	78-2313-76

Compound	Solvent	λ_{max}(log ϵ)	Ref.
$C_{34}H_{51}NO_6$ 5α-Lanost-8-ene-6,7,11-trione, 3β-acetoxy-, 6-(0-acetyloxime)	EtOH	248(3.72),308(3.48)	5-1214-76
$C_{34}H_{52}OSe$ 5α-Cholestan-3β-ol, selenobenzoate (0-ester)	CHCl$_3$	254(3.98),338(3.89)	39-2112-76C
$C_{34}H_{52}O_5$ Oleana-11,13(18)-diene-3β,16α,28-triol, 3β,28-diacetate	n.s.g.	242(4.47),251(4.52), 261(4.36)	12-1549-76
$C_{34}H_{52}O_8$ Phorbol, 12-0-decanoyl-, 13-acetate	MeOH	193(4.09),231(3.72), 325(2.15)	102-1070-76
$C_{34}H_{54}O_5$ 19-Norlanost-5-en-7-one, 3,11-diacetoxy-9-methyl-	EtOH	245(4.13)	88-4761-76
$C_{34}H_{62}OS$ 1-Hexadecanone, 1-(4-methyl-5-tridecyl-2-thienyl)-	heptane	263(3.92),297(4.02)	34-0233-76

Compound	Solvent	$\lambda_{max}(\log \epsilon)$	Ref.
$C_{35}H_{20}Cl_4O_5$ Ethanedione, (3-benzoyl-5,6,7,8-tetra-chloro-2,3-dihydro-2,3-diphenyl-1,4-benzodioxin-2-yl)phenyl-	C_6H_{12}	217(4.891),256(4.92), 380(2.02)	5-0793-76
$C_{35}H_{23}Cl_3N_2O_2$ Benzoxazole, 2,2'-[[2-(3-chlorophenyl)-1,3-propanediyl]bis(3-chloro-4,1-phenylene)]bis-	DMF	298(4.73),307(4.78), 322(4.56)	33-0802-76
$C_{35}H_{24}N_4O$ Benzoxazole, 2-[4-[2-(2,5-diphenyl-2H-1,2,3-triazol-4-yl)ethenyl]phenyl]-5-phenyl-	DMF	366(4.79)	33-2469-76
Benzoxazole, 2-[4-[2-(2,5-diphenyl-2H-1,2,3-triazol-4-yl)ethenyl]phenyl]-6-phenyl-	DMF	370(4.81)	33-2469-76
$C_{35}H_{24}N_8$ Hydrazyl, 1,1'-[5-(1-aziridinyl)-2,4,6-tricyano-1,3-phenylene]bis[2,2-di-phenyl-	CH_2Cl_2	314(4.49),326(4.50), 449s(4.22),473(4.24), 626(3.93)	5-1739-76
$C_{35}H_{25}N$ Benzenamine, N-[bis(diphenylmethylene)-cyclopropylidene]-	$CHCl_3$	272(4.57),395(4.51)	18-3173-76
$C_{35}H_{25}NO$ Pyridine, 4-[(2,6-diphenyl-4H-pyran-4-ylidene)methyl]-2,6-diphenyl-	MeCN	380(4.55)	4-0577-76
$C_{35}H_{25}NS$ Pyridine, 4-[(2,6-diphenyl-4H-thiopyran-4-ylidene)methyl]-2,6-diphenyl-	MeCN	392(4.46)	4-0577-76
$C_{35}H_{26}ClNO_2S$ Methanone, [4-[[(2-chlorophenyl)thio]-methyl]-5-hydroxy-2-phenyl-1-(phen-ylmethyl)-1H-indol-3-yl]phenyl-	dioxan	225(4.44),285(4.21)	42-0017-76
Methanone, [4-[[(3-chlorophenyl)thio]-methyl]-5-hydroxy-2-phenyl-1-(phen-ylmethyl)-1H-indol-3-yl]phenyl-	dioxan	226(4.49),287(4.27)	42-0017-76
Methanone, [4-[[(4-chlorophenyl)thio]-methyl]-5-hydroxy-2-phenyl-1-(phen-ylmethyl)-1H-indol-3-yl]phenyl-	dioxan	226(4.52),290(4.36)	42-0017-76
$C_{35}H_{26}NO$ Pyrylium, 4-[4-(diphenylamino)phenyl]-2,6-diphenyl-, perchlorate	MeCN	272(4.25),390(4.25), 535(4.62)	4-0073-76
$C_{35}H_{27}NO$ 3-Butenamide, 2-(diphenylmethylene)-N,4,4-triphenyl-	$CHCl_3$	245(4.33),337(4.15)	18-3173-76
$C_{35}H_{27}NO_2S$ Methanone, [5-hydroxy-2-phenyl-1-(phen-ylmethyl)-4-[(phenylthio)methyl]-1H-indol-3-yl]phenyl-	dioxan	230(4.45),288(4.27)	42-0017-76
Sulfonium, diphenyl-, 1-benzoyl-3-oxo-3-phenyl-2-[(phenylmethyl)imino]-propylide	EtOH	248(4.35),284s(4.16), 422(3.69)	78-0431-76

Compound	Solvent	λ_{max}(log ϵ)	Ref.
$C_{35}H_{28}N_2O$ 2-Azetidinone, 4-[(diphenylmethylene)-amino]-1-methyl-3,3,4-triphenyl-	EtOH	250(4.93)	39-1528-76C
$C_{35}H_{28}O$ Anthracene, 10-(diphenylmethylene)-9-ethoxy-9,10-dihydro-9-phenyl-	n.s.g.	290(3.90)	138-0643-76
Indeno[2,1-b]pyran, 9-(1,2-diphenyleth-enyl)-2,4-dimethyl-3-(phenylmethyl)-	dioxan	236(4.42),296(4.53), 318(4.21),510(3.37)	78-2225-76
$C_{35}H_{30}O$ 1,4-Methanonaphthalen-9-one, 1,4,4a,5,6-7,8,8a-octahydro-1,2,3,4-tetraphenyl-, (1α,4α,4aα,8aα)-	CH_2Cl_2	220(4.79),260s(4.23)	78-2053-76
(1α,4α,4aβ,8aβ)-	CH_2Cl_2	220(4.68),270(4.24)	78-2053-76
$C_{35}H_{30}OSi$ Silanol, methylphenyl(1,2,3,4-tetraphen-yl-1,3-butadienyl)-	C_6H_{12}	232s(3.40),267(3.23), 287(3.23),312s(3.04)	78-1549-76
$C_{35}H_{30}O_7$ 7H-1-Benzopyran-7-one, 4-(7-acetoxy-3,4-dihydro-5-methoxy-2-phenyl-2H-1-benzo-pyran-8-yl]-5-methoxy-6-methyl-2-phenyl-	n.s.g.	277(--),286(4.34), 316(3.95),330(3.95), 385(4.04),490(4.16)	39-1392-76C
$C_{35}H_{34}N_2O_5$ N-Norisotrilobine	EtOH	280(3.49)	2-0062-76
$C_{35}H_{34}O_7$ Phenol, 2-[2-(3,4-dihydro-5,7-dimethoxy-2-phenyl-2H-1-benzopyran-8-yl)ethen-yl]-3,5-dimethoxy-4-methyl-, benzoate, [S-(E)]-	EtOH	234(4.51),318(4.42), 326s(4.41)	39-1570-76C
$C_{35}H_{36}N_2O_8$ 1H-Pyrazole-3,4-dicarboxylic acid, 5-[2,3-O-(1-methylethylidene)-5-O-(triphenylmethyl)-β-D-ribofur-anosyl]-, 3-ethyl 4-methyl ester	pH 13	230s(4.06),253(3.96)	44-0084-76
4-ethyl 3-methyl ester	pH 13	230s(4.07),252(3.99)	44-0084-76
1H-Pyrazole-3,5-dicarboxylic acid, 4-[2,3-O-(1-methylethylidene)-5-O-(triphenylmethyl)-β-D-ribofur-anosyl]-, 3-ethyl 5-methyl ester	pH 13	254(3.98)	44-0084-76
$C_{35}H_{36}N_4O_4$ 21H,23H-Porphine-2-propanoic acid, 8-ethenyl-13-ethyl-18-(3-methoxy-3-oxo-1-propenyl)-3,7,12,17-tetra-methyl-, (E)-	pyridine	417(5.10),515(3.95), 557(4.23),579(4.08), 646(3.30)	78-0275-76
	CH_2Cl_2-1% TFA	415(5.39),561(4.10), 612(3.98)	78-0275-76
$C_{35}H_{36}N_4O_5$ 21H,23H-Porphine-2,18-dipropanoic acid, 8-ethenyl-13-formyl-3,7,12,17-tetra-methyl-, dimethyl ester	CH_2Cl_2	417(5.19),518(4.00), 559(4.13),586(3.91), 645(3.33)	39-2501-76C
$C_{35}H_{37}ClN_4O_5$ 21H,23H-Porphine-2,18-dipropanoic acid,	CH_2Cl_2	412(5.22),515(3.91),	39-2501-76C

Compound	Solvent	$\lambda_{max}(\log \epsilon)$	Ref.
8-(2-chloroethyl)-13-formyl-3,7,12,17-tetramethyl-, dimethyl ester (cont.)		556(4.12),582(3.90), 643(3.20)	39-2501-76C
$C_{35}H_{39}N_5O_4$ Ergoline-8-carboxamide, 9,10-didehydro-N-[1-[[hexahydro-1,4-dioxo-3-(phenylmethyl)pyrrolo[1,2-a]pyrazin-2(1H)-yl]carbonyl]-2-methylpropyl]-6-methyl-, [8β-[1S(3S,8aR)]]-	MeOH	219(4.404),308(3.957)	73-3415-76
$C_{35}H_{40}N_4O_3$ Bonafousine	EtOH	228(4.88),286(4.09), 294s(4.54),300(3.00)	77-0510-76
	EtOH-base	228(4.88),286(4.54), 294(4.54),320(4.49)	77-0510-76
$C_{35}H_{40}N_4O_4$ 21H,23H-Porphine-2-propanoic acid, 8-ethenyl-13-ethyl-2,3-dihydro-18-(methoxycarbonyl)-3,7,12,17,20-pentamethyl-, methyl ester, (2S-trans)-	CH_2Cl_2	401(5.24),501(4.04), 538(4.04),570(3.81), 625(3.36)	78-0275-76
	CH_2Cl_2-5% TFA	407(5.57),551(4.18), 598(3.77)	78-0275-76
$C_{35}H_{41}N_3O_6$ Carbamic acid, [1-methyl-4-[2,3,5-tris-O-(phenylmethyl)-β-D-ribofuranosyl]-1H-pyrazol-5-yl]-, 1,1-dimethylethyl ester	MeOH	227s(3.79),257(2.93), 263(2.85)	44-3000-76
$C_{35}H_{41}N_5O_4$ Ergoline-8-carboxamide, N-[1-[[hexahydro-1,4-dioxo-3-(phenylmethyl)pyrrolo[1,2-a]pyrazin-2(1H)-yl]carbonyl]-2-methylpropyl]-6-methyl-, [8β-[1S(3S,8aR)]]-	MeOH	222(4.46),273(3.84), 279(3.86),289(3.76)	73-3415-76
$C_{35}H_{42}HgN_4O_2$ Mercury, (acetato-O)(2,7,12,17-tetraethyl-3,8,13,18,21-pentamethyl-21H,23H-porphinato-N^{21},N^{22},N^{23},N^{24})-	$CHCl_3$	432(5.00),510s(3.48), 540(3.90),586(4.02), 631(3.34)	78-0597-76
$C_{35}H_{44}N_4O_4$ 21H-Biline-8,12-dipropanoic acid, 2-ethyl-10,24-dihydro-1,3,7,13,17,19-hexamethyl-, dimethyl ester, dihydrobromide	$CHCl_3$	450(4.40),517(5.31)	78-1793-76
$C_{35}H_{45}O_2$ 2,5-Cyclohexadien-1-one, 4-[[3,5-bis-(1,1-dimethylethyl)-4-hydroxyphenyl]phenylmethylene]-2,6-bis(1,1-dimethylethyl)-, ion(1-)-radical	MeOH	397(--),470(4.74), 990(--)	64-0965-76B
$C_{35}H_{46}O_2$ 2,5-Cyclohexadien-1-one, 4-[[3,5-bis-(1,1-dimethylethyl)-4-hydroxyphenyl]phenylmethylene]-2,6-bis(1,1-dimethylethyl)- anion	MeOH	412(4.43)	64-0965-76B
	MeOH	385(--),601(4.87)	64-0965-76B

Compound	Solvent	$\lambda_{max}(\log \epsilon)$	Ref.

$C_{35}H_{47}N_3O_3$
  5α,8α-[1',2']-1',2',4'-Triazolidinochol-
  esta-6,25-diene-3',5'-dione, 3β-hy-
  droxy-4'-phenyl-
	EtOH	255(3.63)	39-0731-76C

$C_{35}H_{47}N_3O_8$
  1H-Pyrrole-3-propanoic acid, 2-[(1,1-di-
  methylethoxy)carbonyl]-5-[[3-(3-meth-
  oxy-3-oxopropyl)-5-[[4-(3-methoxy-3-
  oxopropyl)-3,5-dimethyl-2H-pyrrol-2-
  ylidene]methyl]-4-methyl-1H-pyrrol-
  2-yl]methyl]-4-methyl-, methyl ester,
  hydrobromide
	CHCl	492(4.93)	78-1793-76

$C_{35}H_{49}N_3O_5$
  4a,13b-Etheno-1H,5H-benzo[c]cyclopenta-
  [h][1,2,4]triazolo[1,2-a]cinnolin-
  1,3(2H)-dione, 11-(4,5-dihydroxy-1,5-
  dimethylhexyl)-6,7,8,8a,8b,9,10,10a-
  11,12,13,13a-dodecahydro-6-hydroxy-
  8a,10a-dimethyl-2-phenyl-
	EtOH	255(3.66)	39-0727-76C

$C_{35}H_{50}N_2O_8$
  3-Piperidinol, 1-acetyl-2-[1-[6'-acet-
  oxy-1'-(acetoxymethyl)-3',4,5,5',6'-
  7',8',8'a-octahydro-7,8'a-dimethyl-
  spiro[2,1-benzisoxazole-3(3aH),2'-
  (1'H)-naphthalen]-6-yl]ethyl]-
  5-methyl-, acetate
	EtOH	206(4.12),258(4.18)	39-1297-76C

$C_{35}H_{50}O_2$ Ergosta-5,7-dien-3β-ol, benzoate	CHCl$_3$	272(4.03),282(4.03)	39-0826-76C

$C_{35}H_{50}O_8$
  27-Nor-5α-lanosta-8,16-diene-15,24-di-
  one, 3β,23,28-triacetoxy-, (23S)-
	EtOH	229(3.69)	33-1997-76

$C_{35}H_{52}O_{14}$
  Corotoxigenin, β-D-glucopyranosyl(1→4)-
  α-L-rhamnopyranoside
	MeOH
	MeOH
		218(4.25)
		218(4.18)
			94-0108-76
			94-2886-76

$C_{35}H_{54}O_4$
  Oleana-11,13(18)-diene-3β,16α,28-triol,
  cyclic 16,28-(1-methylethylidene
  acetal) 3-acetate
	n.s.g.	246s(4.38),254(4.44),
		  264(4.30)
			12-1549-76

$C_{35}H_{54}O_{14}$
  Coroglaucigenin, β-D-glucopyranosyl-
  (1→4)-α-L-rhamnopyranoside
	MeOH	220(4.17)	94-0108-76

Mallogenin, β-D-glucopyranosyl- (1→4)-α-L-rhamnopyranoside	MeOH	220(4.19)	94-0108-76

$C_{35}H_{54}O_{15}$
  Panogenin, β-D-glucopyranosyl-
  (1→4)-α-L-rhamnopyranoside
	MeOH	220(4.18)	94-0108-76

Compound	Solvent	$\lambda_{max}(\log \epsilon)$	Ref.
$C_{36}H_{22}Cl_4O_2$			
Benzo[b]cyclopenta[e][1,4]dioxepin, 5,6,7,8-tetrachloro-3a,10-dihydro-	C_6H_{12}	214(4.947),246s(4.516), 359(3.732)	5-0793-76
1H-Cyclopenta[b][1,4]benzodioxin, 5,6,7,8-tetrachloro-3a,9a-dihydro-	C_6H_{12}	220(4.89),300(4.25)	5-0793-76
$C_{36}H_{22}Cl_4O_3$			
Spiro[1,4-benzodioxan-2(3H),2'-[2H]pyran], 5,6,7,8-tetrachloro-3',4',5',6'-tetraphenyl-	C_6H_{12}	216(4.9111),300s(4.128), 320s(4.069)	5-0793-76
$C_{36}H_{22}Cl_4O_4$			
2H-1,5-Benzodioxepin-3(4H)-one, 6,7,8,9-tetrachloro-2-(3-oxo-1,2,3-triphenyl-1-propenyl)-2-phenyl-, (Z)-	C_6H_{12}	216(4.863),300(3.792)	5-0793-76
$C_{36}H_{22}Cl_4O_5$			
Compound, m. 273°	C_6H_{12}	220(4.812),290(3.312), 297(3.279),329(2.49)	5-0793-76
$C_{36}H_{22}N_8$			
1,2,4-Triazolo[4,3-c][1,2,4]triazolo-[4',3':1,6]pyrimido[5,4-g]quinazoline, 3,5,9,11-tetraphenyl-	H_2SO_4	260(4.716),322(4.415)	116-0626-76
$C_{36}H_{22}O_2$			
Benz[c]indeno[2,1-a]fluorene-13,14-dione, 4b,8b-dihydro-4b,8b-diphenyl-, cis	$CHCl_3$	298(4.29)	77-0848-76
$C_{36}H_{24}$			
Naphthalene, 1,1'-(1,4-diphenyl-1,2,3-butatriene-1,4-diyl)bis-	benzene	315(4.10),324(4.10), 405(4.55)	19-0089-76
	C_6H_{12}	256(4.41),314(4.14), 320(4.14),403(4.54)	19-0089-76
$C_{36}H_{24}ClN_5O$			
1,3,4-Oxadiazole, 2-[1,1'-biphenyl]-4-yl-5-[4-[2-[2-(4-chlorophenyl)-5-phenyl-2H-1,2,3-triazol-4-yl]ethenyl]-phenyl]-	DMF	360(4.79)	33-2469-76
$C_{36}H_{24}Cl_4O_3$			
1H-Cyclopenta[b][1,4]benzodioxin-1-ol, 5,6,7,8-tetrachloro-3a,9a-dihydro-1-methyl-2,3,3a,9a-tetraphenyl-, (1α,3aα,9aα)-	C_6H_{12}	217(4.873),300s(3.835)	5-0793-76
(1α,3aβ,9aβ)-	C_6H_{12}	219(4.928),303(3.638)	5-0793-76
$C_{36}H_{24}N_2O$			
4-Pyridineacetonitrile, α-(2,6-diphenyl-4H-pyran-4-ylidene)-2,6-diphenyl-	MeCN	390(4.33)	4-0577-76
$C_{36}H_{24}N_2O_3$			
Spiro[cyclobutane-1,3'-[3H]indazole]-2,4',7'-trione, 3,4-bis(diphenyl-methylene)-3'a,7'a-dihydro-	$CHCl_3$	283(4.42),362(4.23), 450(4.00)	18-3173-76
$C_{36}H_{24}O_3$			
Spiro[bicyclo[4.1.0]hept-ene-7,1'-cyclo-butane]-2,2',5-trione, 3',4'-bis(di-	$CHCl_3$	246(4.40),265(4.35), 350(4.15),455(3.87)	18-3173-76

Compound	Solvent	$\lambda_{max}(\log \epsilon)$	Ref.
phenylmethylene)- (cont.)			18-3173-76
$C_{36}H_{25}As$ 1H-Benz[f]arsindole, 1,2,4,9-tetraphenyl-	EtOH	267(4.71),305(4.69), 314(4.70),330(4.71), 346(4.71),348s(4.71), 361s(4.64),381(4.76)	78-2131-76
$C_{36}H_{25}NO_2$ Cyclobutanone, 2,3-bis(diphenylmethylene)-4-(phenylimino)-, N-oxide	CHCl$_3$	241(4.24),280(4.21), 408(4.00),505(4.07), 585s(3.57),640(3.36)	18-3173-76
$C_{36}H_{25}N_3O_2$ 6-Oxa-5,7,8-triazaspiro[3.4]oct-7-en-1-one, 2,3-bis(diphenylmethylene)-5-phenyl-	CHCl$_3$	242(4.43),315(4.16)	18-3173-76
$C_{36}H_{25}N_5O$ 1,3,4-Oxadiazole, 2-[1,1'-biphenyl]-4-yl-5-[4-[2-(2,5-diphenyl-2H-1,2,3-triazol-4-yl)ethenyl]phenyl]-	DMF	359(4.79)	33-2469-76
$C_{36}H_{26}N_8O_2$ Benzamide, N,N'-[4,6-bis(5-phenyl-1H-1,2,4-triazol-3-yl)-1,3-phenylene]bis-	H$_2$SO$_4$	263(4.8633)	116-0626-76
$C_{36}H_{26}O_2S_4$ 1,4-Butanedione, 2,3-bis[4-(4-methylphenyl)-3H-1,2-dithiol-3-ylidene]-1,4-diphenyl-	CH$_2$Cl$_2$	473(4.46)	1-0478-76
$C_{36}H_{26}O_3$ Maleic anhydride adduct	MeOH	292(4.11)	35-3247-76
$C_{36}H_{26}O_4S_4$ 1,4-Butanedione, 2,3-bis[5-(4-methoxyphenyl)-3H-1,2-dithiol-3-ylidene]-1,4-diphenyl-	CH$_2$Cl$_2$	472(4.56)	1-0478-76
$C_{36}H_{27}As$ 1H-Arsole, 3-(diphenylmethylene)-2,3-dihydro-1,5-diphenyl-2-(phenylmethylene)-	EtOH	240(4.34),311(4.18), 355(4.25),365(4.26)	78-2131-76
1H-Benz[f]arsindole, 9,9a-dihydro-1,2,4,9-tetraphenyl-, trans	CHCl$_3$	254(4.67),297s(4.60), 366s(4.68),379(4.72), 397s(4.68)	78-2131-76
$C_{36}H_{27}NO$ Cyclopropanecarboxamide, 2,3-bis(diphenylmethylene)-N-phenyl-	CHCl$_3$	254(4.52),373(4.53)	18-3173-76
Pyridine, 2,6-diphenyl-4-[1-(2,6-diphenyl-4H-pyran-4-ylidene)ethyl]-	MeCN	355(4.29)	4-0577-76
$C_{36}H_{28}NS$ Pyridinium, 4-[(2,6-diphenyl-4H-thiopyran-4-ylidene)methyl]-1-methyl-2,6-diphenyl-, perchlorate	MeCN	485(4.57)	4-0577-76
$C_{36}H_{28}N_2O_2$ 3H-Pyrrolo[1,2-a]indol-3-one, 1-[(2,3-	CHCl$_3$	244(4.59),373(4.13)	83-0937-76

Compound	Solvent	$\lambda_{max}(\log \epsilon)$	Ref.
dihydro-3-oxo-9-phenyl-1H-pyrrolo-[1,2-a]indol-1-yl)methyl]-1,2-di-hydro-1-methyl-9-phenyl- (cont.)			83-0937-76
$C_{36}H_{28}N_4$ 1,2-Benzenediamine, N^2,N^2-diphenyl-4-(10-phenyl-5(10H)-phenazinyl)- radical cation	MeCN	255(4.94),295(4.32), 367(3.68)	22-1567-76
	MeCN	259(4.87),283(4.30), 375(3.68),436(3.84), 464(4.01),676(3.37), 744(3.36)	22-1567-76
$C_{36}H_{29}N$ Pyridine, 1,4-dihydro-1-methyl-2,3,4,5,6-pentaphenyl-	MeOH	275(4.13),343s(3.81)	44-0818-76
$C_{36}H_{30}N_2O_8$ Diquinolizino[6,5,4,3-cde:3',4',5',6'-ghi][4,7]phenanthrolinediium, 1,3,8-10-tetrakis(ethoxycarbonyl)-, dibro-mide ($11\lambda,10\epsilon$)	H_2O	204(4.57),230s(4.43), 235(4.44),291(4.58), 303s(4.45),313s(4.37), 326(4.14),444s(4.39), 471(4.64),510(3.74), 534s(?)	39-0341-76C
$C_{36}H_{30}O$ 1,4:5,8-Dimethanonaphthalen-9-one, 1,2,3,4,4a,5,8,8a-octahydro-5,6,7,8-tetraphenyl-, ($1\alpha,4\alpha,4a\beta,5\beta,8\beta,8a\beta$)-	CH_2Cl_2	220(4.53),260s(4.00)	78-2053-76
$C_{36}H_{30}O_4$ 2-Propen-1-one, 3-[2,4-bis(phenylmeth-oxy)phenyl]-1-[4-(phenylmethoxy)-phenyl]-	n.s.g.	338(4.36)	39-0186-76C
$C_{36}H_{32}N_2O_8$ [1,1'-Bipyrido[2,1,6-de]quinolizine]-4,4',6,6'-tetracarboxylic acid, tetraethyl ester	CH_2Cl_2	248(4.25),295(4.88), 327(4.0),403(4.51), 484(4.85)	39-0341-76C
$C_{36}H_{32}N_4O_5$ 21-Phorbinecarboxylic acid, 3,4-dide-hydro-9,14-diethenyl-3-(3-methoxy-3-oxo-1-propenyl)-4,8,13,18-tetra-methyl-20-oxo-, methyl ester	$CHCl_3$	440(5.22),535(4.05), 580(4.09),598(4.10), 656(3.15)	77-0120B-76
$C_{36}H_{34}N_2O_7$ Benzaldehyde, 4-hydroxy-3-[2-methoxy-5-[(1,2,3,4,9,10,11,12-octahydro-6-meth-oxy-2,11-dimethyl-12-oxo-[1,4]dioxino-[2,3-g:6,5-h']diisoquinolin-1-yl)meth-yl]phenyl]-	EtOH	212(4.10),282(3.44), 310s(2.94)	44-1293-76
	EtOH-base	230(4.00),292(3.27), 340(3.21)	44-1293-76
$C_{36}H_{34}N_4O_{10}$ 2,4(1H,3H)-Pteridinedione, 5-acetyl-5,6,7,8-tetrahydro-6,7-dimethyl-1-(2,3,5-tri-O-benzoyl-β-D-ribo-furanosyl)-	pH 7.0	232(4.29),287(4.17)	24-3184-76
	pH 12.0	228(4.36),286(4.10)	24-3184-76
	MeOH	229(4.34),285(4.20)	24-3184-76
$C_{36}H_{34}N_6O_2$ Methanone, [5,13-bis(dimethylamino)-8,8a,16,16a-tetrahydropyrazino-	$CHCl_3$	252(4.53),305(3.41), 330(3.63)	22-0291-76

Compound	Solvent	$\lambda_{max}(\log \epsilon)$	Ref.
[2,1-a:5,4-a']diphthalazine-8,16-diyl]-bis[phenyl-, S_1 (cont.)			22-0291-76
S_4	$CHCl_3$	251(4.55),300(3.60), 325(3.70)	22-0291-76
$C_{36}H_{34}O_{13}$ 4H-1-Benzopyran-4-one, 2-[3-(β-D-gluco-pyranosyloxy)-5-hydroxy-4-(phenyl-methoxy)phenyl]-5-hydroxy-3-meth-oxy-7-(phenylmethoxy)-	MeOH	264(4.00),268(4.00), 344(3.95)	2-0384-76
	MeOH-AlCl$_3$	275(--),353(--), 395(--)	2-0384-76
$C_{36}H_{35}O_7P$ 1H-Phosphindolium, 3-butyl-1,2-diphen-yl-, 2-methoxy-3,4,5-tris(methoxy-carbonyl)-2,4-cyclopentadien-1-ylide	EtOH	294(3.81),344s(3.78)	4-0065-76
$C_{36}H_{36}Cl_2Pd$ Cyclobutadiene, tetrakis(2,4-dimethyl-phenyl)-, PdCl$_2$ complex dimer	CHCl$_3$	256(4.18),331(4.27), 405(4.18)	101-0385-76N
Cyclobutadiene, tetrakis(2,5-dimethyl-phenyl)-, PdCl$_2$ complex dimer	CHCl$_3$	262(4.24),328(4.26), 397(4.03)	101-0385-76N
$C_{36}H_{36}N_2O_4$ Hydrazine, bis(β-p-toluoyl-α-p-tolyl-propanoyl)-	EtOH	252(4.33)	115-0761-76
$C_{36}H_{36}N_2O_5$ Dinklacorine	MeOH	222(4.70),236s(4.68), 294(3.97)	100-0213-76
methiodide	MeOH	222(4.82),245s(4.65), 293(3.91)	100-0213-76
Isotrilobine	EtOH	280(3.50)	2-0062-76
$C_{36}H_{36}N_2O_8$ Hydrazine, bis(β-p-methoxybenzoyl-α-p-methoxyphenylpropanoyl)-	EtOH	283(4.71)	115-0761-76
$C_{36}H_{37}ClOPd$ Cyclobutenol, tetrakis(2,4-dimethyl-phenyl)-, PdCl$_2$ complex dimer	CHCl$_3$	290(4.59),403(3.97)	101-0385-76N
Cyclobutenol, tetrakis(2,5-dimethyl-phenyl)-, PdCl$_2$ complex dimer	CHCl$_3$	290(4.57),385(3.96)	101-0385-76N
$C_{36}H_{38}N_2O_6$ Krukovine	EtOH	285(3.82)	44-0317-76
	NaOH	291(3.97)	44-0317-76
$C_{36}H_{38}N_4$ [9,9'-Bianthracene]-2,2',7,7'-tetrone, N,N,N',N',N'',N'',N''',N'''-octamethyl-	C_6H_{12}	216(4.90),262s(5.20), 269(5.28),315(5.49), 406(4.59),432s(4.20), 389s(4.45)	12-1613-76
	10% H$_2$SO$_4$	227(4.88),250s(5.53), 256(4.65),352(4.26), 369(4.43),388(4.42)	12-1613-76
$C_{36}H_{38}N_4O_4$ 21H,23H-Porphine-2-propanoic acid, 8-ethenyl-13-ethyl-18-(3-methoxy-3-oxo-1-propenyl)-3,7,12,17-tetramethyl-, methyl ester, (E)-	CH_2Cl_2	413(5.21),513(3.91), 555(4.29),576(4.10), 642(3.18)	78-0275-76

Compound	Solvent	$\lambda_{max}(\log \epsilon)$	Ref.
(cont.)	CH_2Cl_2-1% TFA	414(5.42),560(4.13), 610(4.00)	78-0275-76
$C_{36}H_{38}O_2$ Methanone, (1,10-decanediyldi-4,1-phenylene)bis[phenyl-	EtOH	<u>263(4.5),330(2.6)</u>	47-0319-76
$C_{36}H_{39}BrN_4O_4$ 21H,23H-Porphine-2,18-dipropanoic acid, 7-(2-bromoethyl)-12-ethenyl-3,8,13,17-tetramethyl-, dimethyl ester	$CHCl_3$	401(5.25),500(4.19), 535(4.09),569(3.91), 624(3.67)	78-2753-76
21H,23H-Porphine-2,18-dipropanoic acid, 8-(2-bromoethyl)-13-ethenyl-3,7,12,17-tetramethyl-, dimethyl ester	$CHCl_3$	401(5.25),500(4.18), 535(4.07),569(3.89), 624(3.63)	78-2753-76
$C_{36}H_{40}N_4O_4$ 21H,23H-Porphine-2-propanoic acid, 8,13-diethyl-18-(3-methoxy-3-oxo-1-propenyl)-3,7,12,17-tetramethyl-, methyl ester, (E)-	CH_2Cl_2	413(5.18),510(4.00), 549(4.20),575(4.00), 636(3.42)	78-0275-76
	CH_2Cl_2-1% TFA	411(5.42),556(4.18), 604(3.92)	78-0275-76
$C_{36}H_{40}N_4O_5$ Bacteriomethylphaeophorbide e (mixture of three homologs)	dioxan	442(5.03),540(4.02), 574(3.93),606(3.94), 661(4.40)	5-0566-76
21H,23H-Porphine-2,18-dipropanoic acid, 7-ethenyl-12-(2-hydroxyethyl)-3,8,13-17-tetramethyl-, dimethyl ester	$CHCl_3$	404(5.28),504(4.35), 538(4.26),574(4.02), 625(3.46)	78-2753-76
21H,23H-Porphine-2,18-dipropanoic acid, 8-ethenyl-13-ethyl-β^{18}-hydroxy-3,7,12,17-tetramethyl-, dimethyl ester	CH_2Cl_2	402(5.29),502(4.07), 540(4.08),569(3.82), 626(3.32)	78-0275-76
	CH_2Cl_2-5% TFA	408(5.54),553(4.20), 599(3.81)	78-0275-76
21H,23H-Porphine-2,18-dipropanoic acid, 8-ethenyl-13-(2-hydroxyethyl)-3,7,12-17-tetramethyl-, dimethyl ester	$CHCl_3$	404(5.28),504(4.36), 538(4.26),574(4.01), 625(3.48)	78-2753-76
$C_{36}H_{40}N_4O_6$ 21H,23H-Porphine-2,18-dipropanoic acid, 7-(1,2-dihydroxyethyl)-12-ethenyl-3,8,13,17-tetramethyl-, dimethyl ester	$CHCl_3$	405(5.25),504(4.12), 539(4.09),574(3.84), 627(3.59)	78-2753-76
21H,23H-Porphine-2,18-dipropanoic acid, 8-(1,2-dihydroxyethyl)-13-ethenyl-3,7,12,17-tetramethyl-, dimethyl ester	$CHCl_3$	405(5.26),504(4.15), 539(4.11),574(3.87), 627(3.62)	78-2753-76
21H,23H-Porphine-2,7,18-tripropanoic acid, 3,8,13,17-tetramethyl-, trimethyl ester	$CHCl_3$	398(5.23),496(4.08), 531(3.86),568(3.72), 622(3.50)	39-2492-76C
dication	$CHCl_3$-CF_3COOH	416(5.45),553(4.18), 597(3.80)	39-2492-76C
21H,23H-Porphine-2,8,12-tripropanoic acid, 3,7,13,17-tetramethyl-, trimethyl ester	$CHCl_3$	398(5.22),498(4.14), 531(3.77),568(3.66), 621(3.49)	39-2492-76C
dication	$CHCl_3$-CF_3COOH	417(5.41),554(4.20), 597(3.81)	39-2492-76C
$C_{36}H_{42}N_2$ Diazene, bis(1,1-dimethyl-3,3-diphenylbutyl)-	toluene	374(1.41)	73-1557-76

Compound	Solvent	$\lambda_{max}(\log \epsilon)$	Ref.
$C_{36}H_{42}O_{18}$			
Pentaacetylmethylgrandifloroside	MeOH	215(--),223s(--), 280(4.2),316s(3.8)	102-1305-76
	MeONa	227(--),300s(3.5), 311(3.6),375(4.3)	102-1305-76
$C_{36}H_{44}FN_5OOs$			
Osmium, fluoronitrosyl[2,3,7,8,12,13,17- 18-octaethyl-21H,23H-porphinato(2-)- $N^{21},N^{22},N^{23},N^{24}$]oxo-, (OC-6-23)	CH_2Cl_2	342(4.49),419(4.98), 534(4.16),569(4.35)	24-1465-76
$C_{36}H_{44}N_4$			
21H,23H-Porphine, 2-ethenyl-3,7,8,12,13- 17,18-heptaethyl-	$CHCl_3$	403(5.11),503(4.06), 539(4.03),570(3.80), 626(3.56),663(2.97)	39-0794-76C
$C_{36}H_{44}N_6O_2Os$			
Osmium, dinitrosyl[2,3,7,8,12,13,17,18- octaethyl-21H,23H-porphinato(2-)- $N^{21},N^{22},N^{23},N^{24}$]-, (OC-6-12)-	benzene	350(4.65),423(4.70), 539(4.07),576(4.22)	24-1465-76
$C_{36}H_{51}N_3O_3$			
Ergosterol, 22,23-dihydro-, adduct with 4-phenyl-1,2,4-triazoline-3,5-dione	EtOH	215(3.98)	39-0826-76C
$C_{36}H_{52}$			
Pentacyclo[22.12.0.0^{3,26}.0^{6,19}.0^{8,21}]- hexatriaconta-1,3(26),6,8(21),19,24- hexaene	isooctane	233(4.23),241(4.19), 295s(2.57),318s(2.20)	44-4081-76
$C_{36}H_{56}O_6$			
Dibenzo[b,k][1,4,7,10,13,16]hexaoxacy- clooctadecin, 1,3,13,15-tetrakis(1,1- dimethylethyl)-6,7,9,10,17,18,20,21- octahydro-	MeOH	276(3.57),280(3.56)	64-0330-76B
potassium picrate complex	MeOH	271(3.68),278s(3.66)	64-0330-76B
sodium picrate complex	MeOH	271(3.74),278s(3.72)	64-0330-76B
$C_{36}H_{56}O_{13}$			
Uzarigenin, β-D-glucosyl(1→4)-β-D-digi- taloside	MeOH	218(4.25)	102-1275-76
$C_{36}H_{58}O_5$			
Mollugogenol D, diacetate A	EtOH	204(3.76)	2-0059-76
$C_{36}H_{66}OS$			
1-Hexadecanone, 1-(4-methyl-5-pentadec- yl-2-thienyl)-	heptane	264(3.90),298(3.96)	34-0233-76
1-Octadecanone, 1-(4-methyl-3-tridecyl- 2-thienyl)-	heptane	265(3.92),298(4.02)	34-0233-76
$C_{36}H_{68}S$			
Thiophene, 5-hexadecyl-3-methyl-2-penta- decyl-	heptane	241(3.90)	34-0233-76
Thiophene, 3-methyl-5-octadecyl-2-tri- decyl-	heptane	241(3.90)	34-0233-76

Compound	Solvent	$\lambda_{max}(\log \epsilon)$	Ref.
$C_{37}H_{24}Cl_4O_3$ 1H-Cyclopenta[b][1,4]benzodioxin-1-one, 5,6,7,8-tetrachloro-3a,9a-dihydro-3,3a-bis(4-methylphenyl)-2,9a-diphenyl-	C_6H_{12}	310(4.20)	5-0793-76
$C_{37}H_{27}O_4P$ Phosphorin, 6-benzoyl-1,1-dihydro-1-hydroxy-1-oxo-2,4,6-triphenyl-, benzoate	dioxan	371(3.61)	24-3099-76
$C_{37}H_{28}N_2O_3S$ Benzenesulfonic acid, 4-methyl-, [2,3-bis(diphenylmethylene)-4-oxocyclobutylidene]hydrazide	$CHCl_3$	281(4.45),398(4.38), 440s(3.92),515(4.00)	18-3173-76
$C_{37}H_{30}N_8$ Hydrazyl, 1,1'-[2,4,6-tricyano-5-[(1,1-dimethylethyl)amino]-1,3-phenylene]-bis[2,2-diphenyl-	CH_2Cl_2	316(4.52),331(4.50), 452(3.98),466(3.98), 611(3.58)	5-1739-76
$C_{37}H_{31}N_2$ Pyridinium, 1-methyl-4-[(1-methyl-2,6-diphenyl-4(1H)-pyridinylidene)methyl]-2,6-diphenyl-, perchlorate	MeCN	515(4.96)	4-0577-76
$C_{37}H_{34}N_2O_2Si$ Urea, N'-(diphenylmethylene)-N-[2,2-diphenyl-1-[(trimethylsilyl)oxy]ethenyl]-N-phenyl-	EtOH	258.5(4.66)	39-1523-76C
$C_{37}H_{34}O_9$ 1,3-Benzenediol, 4-[2-(7-acetoxy-3,4-dihydro-5-methoxy-2-phenyl-2H-1-benzopyran-8-yl)ethenyl]-5-methoxy-6-methyl-, 1-acetate 3-benzoate, [S-(E)]-	EtOH EtOH-base	230s(4.50),289s(4.11), 316(4.18) 346(4.21)	39-1570-76C 39-1570-76C
$C_{37}H_{36}$ Benzene, 1,1',1'',1'''-(2,2,8,8-tetramethyl-3,4,5,6-nonatetraene-1,3,7,9-tetrayl)tetrakis-	C_6H_{12}	252(4.47),264(4.54), 290(4.35),355(3.16)	88-1395-76
$C_{37}H_{36}N_2O_7$ Benzaldehyde, 4-methoxy-3-[2-methoxy-5-[(1,2,3,4,9,10,11,12-octahydro-6-methoxy-2,11-dimethyl-12-oxo-[1,4]-dioxino[2,3-g:6,5-h']diisoquinolin-1-yl)methyl]phenyl]-	EtOH	212(4.05),282(3.31), 310s(2.90)	44-1293-76
$C_{37}H_{38}N_2O_5$ Dinklacorine, O-methyl- dimethiodide	MeOH MeOH	212(4.84),245s(4.63), 291(4.04) 215(4.80),228s(4.67), 245s(4.56),291(3.77)	100-0213-76 100-0213-76
$C_{37}H_{39}N_5O_4$ 21H,23H-Porphine-2,18-dipropanoic acid, 7-(2-cyanoethyl)-12-ethenyl-3,8,13,17-tetramethyl-, dimethyl ester 21H,23H-Porphine-2,18-dipropanoic acid, 8-(2-cyanoethyl)-13-ethenyl-3,7,12,17-	$CHCl_3$ $CHCl_3$	405(5.27),503(4.16), 536(4.01),574(3.80), 627(3.63) 405(5.25),503(4.14), 536(4.00),574(3.80),	78-2753-76 78-2753-76

Compound	Solvent	$\lambda_{max}(\log \epsilon)$	Ref.
tetramethyl-, dimethyl ester (cont.)		627(3.61)	78-2753-76
$C_{37}H_{40}N_2O_6$			
Berbamine	EtOH	284(3.79)	102-0471-76
Sciadenine	EtOH	277(3.48),283(3.47)	142-0471-76
$C_{37}H_{40}N_4O_7$			
21H,23H-Porphine-2,7,18-tripropanoic acid, 12-formyl-3,8,13,17-tetramethyl-, trimethyl ester dication	CHCl$_3$	412(5.23),516(4.00), 556(4.23),581(4.03), 645(3.31)	39-2492-76C
	CHCl$_3$- CF$_3$COOH	431(5.23),569(4.08), 617(4.03)	39-2492-76C
21H,23H-Porphine-2,8,12-tripropanoic acid, 17-formyl-3,7,13,18-tetramethyl-, dimethyl ester dication	CHCl$_3$	412(5.23),516(3.99), 558(4.23),582(4.03), 645(3.36)	39-2492-76C
	CHCl$_3$- CF$_3$COOH	431(5.23),571(4.08), 618(4.02)	39-2492-76C
$C_{37}H_{40}O_9$			
Daphnetoxin, 6,7-deepoxy-6,7-didehydro-5-deoxy-21-dephenyl-21-(phenylmethyl)-, 20-(3-benzeneacetate)	MeOH	238(3.9),290(3.5)	102-0333-76
$C_{37}H_{42}O_{19}$			
Grandifloroside, hexaacetyl-	MeOH	218(--),278(4.0), 331s(3.6)	102-1305-76
	MeOH-NaOMe	212(--),234s(--), 264s(3.7),311(3.6), 373(4.1)	102-1305-76
$C_{37}H_{43}N_4O_{20}P$			
Uridine, 2',3'-di-O-acetyl-, P-[2-[(1,1-dimethylethoxy)carbonyl]phenyl]uridylyl(5'→3')-, 3',5'-diacetate	EtOH	260(4.31)	94-2903-76
$C_{37}H_{44}BrNO_{12}$			
Rifamycin, 3-bromo-1,4-dideoxy-1,4-dioxo-	MeOH	274(4.44),392(--), 401(3.64)	35-7064-76
$C_{37}H_{44}ClNO_{12}$			
Rifamycin, 3-chloro-1,4-dideoxy-1,4-dioxo-	MeOH	273(4.34),300(--), 504(3.72)	35-7064-76
$C_{37}H_{44}INO_{12}$			
Rifamycin, 3-iodo-1,4-dideoxy-1,4-dioxo-	MeOH	274(4.49),351(3.87), 410(3.72)	35-7064-76
$C_{37}H_{44}O_{10}$			
Gnidilatidin	MeOH	232(4.56)	44-3850-76
$C_{37}H_{46}ClCoN_4$			
Cobalt, (chloromethyl)[2,3,7,8,12,13,17-18-octaethyl-21H,23H-porphinato(2-)-N^{21},N^{22},N^{23},N^{24}]-, (SP-5-31)-	CHCl$_3$	392(5.30),524(3.97), 554(4.36)	18-2529-76
$C_{37}H_{46}Cl_2N_4O_4$			
22H-Biline-2,18-dipropanoic acid, 7,12-bis(2-chloroethyl)-10,23-dihydro-1,3,8,13,17,19-hexamethyl-, dimethyl ester, dihydrobromide	CHCl$_3$	454(4.43),566(5.35)	78-1793-76

Compound	Solvent	$\lambda_{max}(\log \epsilon)$	Ref.
$C_{37}H_{46}N_2O_{12}$ Rifamycin, 3-amino-1,4-dideoxy-1,4-di-hydro-1,4-dioxo-	MeOH	264(4.32),311(3.96), 377(3.63)	35-7064-76
$C_{37}H_{47}CoN_4$ Cobalt, methyl[2,3,7,8,12,13,17,18-octaethyl-21H,23H-porphinato(2-)-$N^{21},N^{22},N^{23},N^{24}]-$, (SP-5-31)-	$CHCl_3$	393(5.34),519(4.02), 552(4.37)	18-2529-76
$C_{37}H_{47}N_3O_6$ 4a,13b-Etheno-1H,5H-benzo[c]cyclopenta-[h][1,2,4]triazolo[1,2-a]cinnoline-1,3(2H)-dione, 6-acetoxy-6,7,8,8a-8b,9,10,10a,11,12,13,13a-dodecahydro-11-(5-hydroxy-1,5-dimethyl-4-oxo-2-hexenyl)-8a,10a-dimethyl-2-phenyl-	EtOH	255(3.64)	39-0727-76C
$C_{37}H_{47}N_5O_2Os$ Osmium, methoxynitrosyl[2,3,7,8,12,13-17,18-octaethyl-21H,23H-porphinato-(2-)-$N^{21},N^{22},N^{23},N^{24}]-$, (OC-6-23)-	CH_2Cl_2-MeOH	342(4.56),419(4.96), 533(4.23),568(4.44)	24-1465-76
$C_{37}H_{48}N_2O_4$ 2'H-Spirosta-2,4-dieno[3,2-c]pyrazole-5'-carboxylic acid, 2'-phenyl-, ethyl ester, (25R)-	EtOH	268(4.28)	65-1381-76
$C_{37}H_{48}O_{10}$ Gnidilatin	MeOH	231(4.18)	44-3850-76
$C_{37}H_{49}N_3O_5$ 5α,8α-[1',2']-1',2',4'-Triazolidinochol-esta-6,25-dien-22-ol, 3β-acetoxy-3',5'-dioxo-4'-phenyl-	EtOH	255(3.62)	39-0731-76C
$C_{37}H_{50}O_{10}$ Gnidilatin, dihydro-	MeOH	231(4.23)	44-3850-76
$C_{37}H_{50}O_{12}$ 5α-Cholesta-9(11),24-dien-6α-yl tetra-O-acetyl-β-D-glucoside, 3β-acetoxy-23-oxo-	MeOH	235(3.94)	39-1357-76C
$C_{37}H_{50}O_{14}$ Card-20(22)-enolide, 3-acetoxy-14-hydr-oxy-19-oxo-5-[(2,3,4-tri-O-acetyl-6-deoxy-α-L-mannopyranosyl)oxy]-, (3β,5α)-	EtOH	217(4.30)	105-0687-76
$C_{37}H_{54}O_8$ Oleana-11,13(18)-dien-28-oic acid, 2,3,23-triacetoxy-, methyl ester, (2α,3β,4α)-	EtOH	244(4.41),252(4.47), 261(4.27)	94-0178-76
$C_{37}H_{63}N_3O_6$ Tetradecanoic acid, [2,3,3a,9a-tetrahy-dro-6-imino-3-[(1-oxotetradecyl)oxy]-6H-furo[2',3':4,5]oxazolo[3,2-a]pyri-midin-2-yl]methyl ester, hydrochlor-ide, [2R-(2α,3β,3aβ,9aβ)]-	MeOH-acid	236(4.00),262(4.06)	87-0663-76

Compound	Solvent	$\lambda_{max}(\log \epsilon)$	Ref.
$C_{37}H_{65}N_3O_7$ 2(1H)-Pyrimidinone, 4-amino-1-[3,5-bis- O-(1-oxotetradecyl)-β-D-arabino- furanosyl]-	MeOH-acid	213(4.02),284(4.18)	87-0667-76
$C_{37}H_{68}OS$ 1-Octadecanone, 2-methyl-1-(4-methyl- 5-tridecyl-2-thienyl)-	heptane	257(3.91),297(4.00)	34-0233-76
$C_{37}H_{70}S$ Thiophene, 3-methyl-5-(2-methyloctadec- yl)-2-tridecyl-	heptane	242(3.90)	34-0233-76

Compound	Solvent	$\lambda_{max}(\log \epsilon)$	Ref.
$C_{38}H_{22}$ 7b,7'b-Bi-7bH-indeno[1,2,3-jk]fluorene	n.s.g.	226(4.78),285(4.42)	44-2120-76
$C_{38}H_{24}N_2S$ [2]Benzothieno[5,6-g]phthalazine, 1,4,7,9-tetraphenyl-	CH_2Cl_2	274(4.66),322(4.94), 398(3.79),656(3.73)	28-0555-76A
$C_{38}H_{24}O_2S$ Methanone, (1,3-diphenylnaphtho[2,3-c]- thiophene-6,7-diyl)bis[phenyl-	CH_2Cl_2	260(4.67),325(4.72), 421(3.87),532(3.91)	28-0555-76A
$C_{38}H_{24}O_4$ Methanone, 2,3,6,7-naphthalenetetrayl- tetrakis[phenyl-	CH_2Cl_2	269(4.91),344(3.68)	28-0555-76A
$C_{38}H_{26}Cl_4O_2$ Benzo[b]cyclopenta[e][1,4]dioxepin, 5,6,7,8-tetrachloro-3a,10-dihydro- 1,3a-bis(4-methylphenyl)-2,3-diphenyl-	C_6H_{12}	214(4.898),248s(4.47), 360(3.638)	5-0793-76
Benzo[b]cyclopenta[e][1,4]dioxepin, 5,6,7,8-tetrachloro-3a,10-dihydro- 2,3-bis(4-methylphenyl)-1,3a-diphenyl-	C_6H_{12}	214(4.933),250(4.472), 367(3.771)	5-0793-76
1H-Cyclopenta[b][1,4]benzodioxin, 5,6,7,8-tetrachloro-3a,9a-dihydro- 1-methylene-2,9a-bis(4-methylphenyl)- 3,3a-diphenyl-	C_6H_{12}	220(4.83),302(4.19)	5-0793-76
1H-Cyclopenta[b][1,4]benzodioxin, 5,6,7,8-tetrachloro-3a,9a-dihydro- 1-methylene-3,3a-bis(4-methylphenyl)- 2,9a-diphenyl-	C_6H_{12}	220(4.883),305(4.288)	5-0793-76
$C_{38}H_{26}Cl_4O_3$ Spiro[1,4-benzodioxin-2(3H),2'-[2H]pyr- an], 5,6,7,8-tetrachloro-3',6'-bis(4- methylphenyl)-4',5'-diphenyl-	C_6H_{12}	216(4.975),300s(4.13), 322s(4.15)	5-0793-76
Spiro[1,4-benzodioxin-2(3H),2'-[2H]pyr- an], 5,6,7,8-tetrachloro-4',5'-bis(4- methylphenyl)-3',6'-diphenyl-	C_6H_{12}	216(4.938),302s(4.161), 323s(4.179)	5-0793-76
$C_{38}H_{26}Cl_4O_4$ 2H-1,5-Benzodioxepin-3(4H)-one, 6,7,8,9- tetrachloro-2-[1,2-bis(4-methylphen- yl)-3-oxo-3-phenyl-1-propenyl]- 2-phenyl-	C_6H_{12}	214(4.843),300s(3.949), 322(3.843)	5-0793-76
2H-1,5-Benzodioxepin-3(4H)-one, 6,7,8,9- tetrachloro-2-(4-methylphenyl)-2- [3-(4-methylphenyl)-3-oxo-1,2- diphenyl-1-propenyl]-	C_6H_{12}	216(4.813),290s(3.886), 300(3.761)	5-0793-76
$C_{38}H_{26}Cl_4O_5$ 2H-1,5-Benzodioxepin-3(4H)-one, 6,7,8,9- tetrachloro-2-(4-methylphenyl)-2- [3-(4-methylphenyl)-3-oxo-1,2- diphenyl-1-propenyl]-, oxidation product	C_6H_{12}	219(4.832),298s(3.512), 327(2.672)	5-0793-76
Unknown compound	C_6H_{12}	220(4.863),290s(3.389), 300(3.333),330(2.591)	5-0793-76
$C_{38}H_{26}O_3$ Methanone, (4,9-dihydro-1,3-diphenyl- naphtho[2,3-c]furan-6,7-diyl)bis[phen- yl-	CH_2Cl_2	333(4.42)	28-0555-76A

Compound	Solvent	$\lambda_{max}(\log \epsilon)$	Ref.
$C_{38}H_{27}$ Benzene, 1-(3,4-diphenyl-1,3-cyclopenta- dien-1-yl)-4-(2,3-diphenylcyclopropen- 1-yl)-, ion(1-)	MeCN	230(4.47),300(4.50), 310(4.51),458(4.49)	88-4655-76
$C_{38}H_{28}$ Benzene, 1-(3,4-diphenyl-1,3-cyclopenta- dien-1-yl)-4-(2,3-diphenylcyclopropen- 1-yl)-	MeCN	390(4.54)	88-4655-76
$C_{38}H_{28}Cl_4O_3$ 1H-Cyclopenta[b][1,4]benzodioxin-1-ol, 5,6,7,8-tetrachloro-3a,9a-dihydro- 1-methyl-2,9a-bis(4-methylphenyl)- 3,3a-diphenyl-	C_6H_{12}	218(4.961),304(3.82)	5-0793-76
isomer	C_6H_{12}	218(4.96),304s(3.82)	5-0793-76
1H-Cyclopenta[b][1,4]benzodioxin-1-ol, 5,6,7,8-tetrachloro-3a,9a-dihydro- 1-methyl-3,3a-bis(4-methylphenyl)- 2,9a-diphenyl-	C_6H_{12}	219(4.923),302(3.748)	5-0793-76
isomer	C_6H_{12}	217(4.934),300s(3.94)	5-0793-76
$C_{38}H_{28}IN_5O_7$ Adenosine, N,N-dibenzoyl-5'-deoxy- 5'-iodo-, 2',3'-dibenzoate	MeOH	231(4.64),274(4.36)	44-1836-76
$C_{38}H_{28}O_2$ 9,20:10,19-Dimethanocyclooncta[1,2-1:5,6- 1']diphenanthrene-21,22-dione, 9,10,19,20-tetrahydro-9,10,19,20- tetramethyl-, (9α,10β,19β,20α)-	n.s.g.	254(4.85),277(4.51), 296(4.23)	77-0053-76
$C_{38}H_{29}N_5O_8$ Adenosine, N,N-dibenzoyl-, 2',3'-dibenz- oate	MeOH	231(4.52),273(4.23)	44-1836-76
$C_{38}H_{30}CuN_4O_2$ Copper, bis(N-hydroxy-N,N'-diphenylbenz- enecarboximidamidato-N',O)-	$CHCl_3$	274(4.36),340(4.16)	12-0357-76
$C_{38}H_{30}N_4NiO_2$ Nickel, bis(N-hydroxy-N,N'-diphenylbenz- enecarboximidamidato-N',O)-, (T-4)-	$CHCl_3$	307s(4.20),380(3.81)	12-0357-76
$C_{38}H_{30}N_4O$ Acetamide, N-[2-(diphenylamino)-4-(10- phenyl-5(10H)-phenazinyl)phenyl]-	MeCN	255(4.98),292(4.40), 364(3.75)	22-1567-76
radical cation	MeCN	259(4.93),285(4.33), 374(3.75),439(3.91), 468(4.08),678(3.42)	22-1567-76
dication	MeCN	311(4.02),427(4.24), 575(3.75),683(3.67)	22-1567-76
$C_{38}H_{30}N_8$ Hydrazyl, 1,1'-[2,4,6-tricyano-5-piperi- dino-1,3-phenylene]bis[2,2-diphenyl-	CH_2Cl_2	317(4.54),432(4.00), 468(3.97),619(3.56)	5-1739-76
$C_{38}H_{31}N_3O_{10}$ 2(1H)-Pyrimidinone, 4-amino-1-(1,3,4,6- tetra-O-benzoyl-β-D-psicofuranosyl)-	MeOH-acid	230(3.76),281(4.18)	44-1836-76

Compound	Solvent	$\lambda_{max}(\log \epsilon)$	Ref.
$C_{38}H_{32}N_2O_8S_2$ 1,4-Epithio-5,8-iminoisoquinoline-6,7-dicarboxylic acid, 1,2,3,4,4a,5,8,8a-octahydro-9-[(4-methylphenyl)sulfonyl]-3-oxo-1,2,4-triphenyl-, dimethyl ester, 10-oxide, ($1\alpha,4\alpha,4a\alpha,5\alpha,8\alpha,8a\beta$)-	MeOH	225s(4.505)	18-3314-76
$C_{38}H_{34}$ [18]Annulene, 6,8,15,17-tetrakisdehydro-5,10-di-tert-butyl-1,14-diphenyl-	THF	231s(4.46),245s(4.20), 257(4.25),270s(4.18), 278(4.25),299(4.15), 325s(4.18),347s(4.48), 368s(4.57),386s(4.79), 402(5.34),544s(4.21), 587(4.94),670s(2.68), 740(3.34)	18-0305-76
$C_{38}H_{34}O_6$ Chalcone, 2',3,4'-tribenzyloxy-4,5'-dimethoxy-	MeOH	277(4.3),327(4.01)	2-1011-76
$C_{38}H_{34}O_7$ Chalcone, 2',3,4'-tribenzyloxy-4,5'-dimethoxy-, epoxide	MeOH	235(4.2),280(4.06), 340(4.0)	2-1011-76
$C_{38}H_{36}IN_5O_8$ Hexanamide, N-[9-(1,3,4-tri-O-benzoyl-6-deoxy-6-iodo-α-D-psicofuranosyl)-9H-purin-6-yl]-	MeOH	231(4.61),273(3.30), 280s(4.20)	44-1836-76
Hexanamide, N-[9-(1,3,4-tri-O-benzoyl-6-deoxy-6-iodo-β-D-psicofuranosyl)-9H-purin-6-yl]-	MeOH	231(4.63),273(4.36), 281(4.24)	44-1836-76
$C_{38}H_{36}N_2O_8$ Benzaldehyde, 5-acetoxy-3-[2-methoxy-5-[(1,2,3,4,9,10,11,12-octahydro-6-methoxy-2,11-dimethyl-12-oxo-[1,4]dioxino[2,3-g:6,5-h']diisoquinolin-1-yl)methyl]phenyl]-	EtOH	212(4.01),270(3.44), 320s(2.86)	44-1293-76
$C_{38}H_{38}N_2O_6$ Dinklacorine, O-acetyl-	MeOH	218(4.61),229s(3.60), 294(3.74)	100-0213-76
$C_{38}H_{38}N_2O_7$ Tiliamosine, N-acetate	EtOH EtOH-base	292.5(4.08) 305(4.01)	88-4241-76 88-4241-76
$C_{38}H_{40}N_2O_5$ Dinklacorine, O-ethyl-	MeOH	212(4.49),246s(4.29), 292(3.68)	100-0213-76
$C_{38}H_{40}N_2O_8$ Thalibrunimine	EtOH	241s(4.48),283(4.02), 300s(3.91)	88-0513-76
$C_{38}H_{42}N_4O_6$ 21H,23H-Porphine-2,18-dipropanoic acid, β^{18}-acetoxy-8-ethenyl-13-ethyl-3,7,12,17-tetramethyl-, dimethyl ester	CH_2Cl_2	402(5.26),503(4.02), 540(4.08),570(3.81), 627(3.20)	78-0275-76

Compound	Solvent	$\lambda_{max}(\log \epsilon)$	Ref.
(cont.)	CH_2Cl_2- CF_3COOH	408(5.55),553(4.16), 602(3.83)	78-0275-76
$C_{38}H_{42}N_4O_8$ 21H,23H-Porphine-2,7,18-tripropanoic acid, 5-acetoxy-3,8,13,17-tetra-methyl-, trimethyl ester	$CHCl_3$	404(5.24),502(4.08), 511(3.68),536(3.68), 623(2.93)	39-2492-76C
	$CHCl_3-$ CF_3COOH	424(5.33),561(4.06), 602(3.49)	39-2492-76C
21H,23H-Porphine-2,7,18-tripropanoic acid, 3-(2-methoxy-2-oxoethyl)-8,13,17-trimethyl-, trimethyl ester	$CHCl_3$	402(5.30),499(4.14), 535(3.96),567(3.83), 622(3.55)	12-1561-76
$C_{38}H_{44}N_2O_{12}$ Rifamycin, 3-cyano-1,4-dideoxy-1,4-di-hydro-1,4-dioxo-	MeOH	272(4.29),340(3.86), 534(3.58)	35-7064-76
$C_{38}H_{44}N_4O_7$ 21H,23H-Porphine-2,7,18-tripropanoic acid, 12-(2-hydroxyethyl)-3,8,13,17-tetramethyl-, trimethyl ester	$CHCl_3$	400(5.21),501(4.10), 532(3.93),569(3.76), 620(3.57)	78-2757-76
21H,23H-Porphine-2,8,12-tripropanoic acid, 17-(2-hydroxyethyl)-3,7,13,18-tetramethyl-, trimethyl ester	$CHCl_3$	400(5.21),500(4.11), 532(3.92),568(3.79), 620(3.45)	78-2757-76
$C_{38}H_{46}$ 1,2,3,4,5,7,9,11,17,19,21-Cyclodocosaun-decaene-13,15-diyne, 1,6,12,17-tetra-kis(1,1-dimethylethyl)-	THF	222(4.57),231(4.57), 264(4.20),299(4.11), 311(4.04),369s(4.61), 382(4.67),414(4.54), 552s(4.15),596(4.66), 732(2.11),768(2.11), 858(2.69)	18-2306-76
$C_{38}H_{47}CoN_4$ Cobalt, ethenyl[2,3,7,8,12,13,17,18-octaethyl-21H,23H-porphinato(2-)-$N^{21},N^{22},N^{23},N^{24}$]-, (SP-5-31)-	$CHCl_3$	392(5.40),520(4.01), 554(4.45)	18-2529-76
$C_{38}H_{48}FeO_5$ Iron, tricarbonyl[(5,6,7,8-η)(3β,22E)-ergosta-5,7,22-trien-3-yl benzoate]	C_6H_{12}	230(4.42)	39-0821-76C
$C_{38}H_{48}N_4$ 21H,23H-Porphine, 5-ethenyl-2,3,7,8,12,13,17,18-octaethyl-	CH_2Cl_2	407(5.06),507(4.05), 540(3.74),575(3.72), 627(3.30)	5-1537-76
$C_{38}H_{48}O_2$ 7,9,11,17,19,21-Cyclodocosahexaene-2,4,13,15-tetrayne-1,6-diol, 1,6,12,17-tetrakis(1,1-di-methylethyl)-	EtOH	279s(4.66),292(4.92), 302(5.01),324s(4.10), 360s(4.10),385(4.24), 409(4.21)	18-2306-76
$C_{38}H_{49}CoN_4$ Cobalt, ethyl[2,3,7,8,12,13,17,18-octa-ethyl-21H,23H-porphinato(2-)-$N^{21},N^{22},N^{23},N^{24}$]-, (SP-5-31)-	$CHCl_3$	392(5.30),520(4.00), 551(4.37)	18-2529-76

Compound	Solvent	λ_{max}(log ϵ)	Ref.
$C_{38}H_{49}CoN_4O$ Cobalt, (2-hydroxyethyl)[2,3,7,8,12,13-17,18-octaethyl-21H,23H-porphinato-(2-)-$N^{21},N^{22},N^{23},N^{24}$]-, (SP-5-31)-	$CHCl_3$	392(5.31),520(4.01), 552(4.43)	18-2529-76
$C_{38}H_{50}CoN_5$ Cobalt, (2-aminoethyl)[2,3,7,8,12,13,17-18-octaethyl-21H,23H-porphinato(2-)-$N^{21},N^{22},N^{23},N^{24}$]-, (SP-5-31)-	$CHCl_3$	388(5.28),524(4.01), 554(4.43)	18-2529-76
$C_{38}H_{50}N_4O_2Os$ Osmium, dimethoxy[2,3,7,8,12,13,17,18-octaethyl-21H,23H-porphinato(2-)-$N^{21},N^{22},N^{23},N^{24}$]-, (OC-6-12)-	CH_2Cl_2-MeOH	370(5.09),497(3.97), 530(3.90)	24-1465-76
$C_{38}H_{50}O_2$ 4,6,8,14,16,18-Docosahexaene-1,10,12,21-tetrayne-3,20-diol, 3,9,14,20-tetra-kis(1,1-dimethylethyl)-	EtOH	224(4.47),263(4.44), 280s(4.50),293(3.63), 305(4.69),327(4.59), 347(4.58),369(4.50), 398(4.35)	18-2306-76
$C_{38}H_{53}N_2$ Acridinium, 9-[4-(dinonylamino)phenyl]-10-methyl-, triiodide	CH_2Cl_2	260(5.10),292(4.80), 360(4.62),635(4.29)	104-1531-76
$C_{38}H_{70}OS$ 1-Hexadecanone, 1-(5-heptadecyl-4-methyl-2-thienyl)-	heptane	264(3.90),296(4.01)	34-0233-76
1-Octadecanone, 1-(4-methyl-5-pentadec-yl-2-thienyl)-	heptane	263(3.89),297(4.00)	34-0233-76
1-Tetradecanone, 1-(4-methyl-5-nonadec-yl-2-thienyl)-	heptane	264(3.91),298(4.00)	34-0233-76
$C_{38}H_{72}S$ Thiophene, 5-eicosyl-3-methyl-2-tridec-yl-	heptane	241(3.89)	34-0233-76
Thiophene, 2-heptadecyl-5-hexadecyl-3-methyl-	heptane	241(3.89)	34-0233-76
Thiophene, 3-methyl-5-octadecyl-2-penta-decyl-	heptane	241(3.89)	34-0233-76

Compound	Solvent	$\lambda_{max}(\log \epsilon)$	Ref.
$C_{39}H_{26}N_8$ 1,3,5-Benzenetricarbonitrile, 2,4- bis(2,2-diphenylhydrazino)-6- (phenylamino)-	CH_2Cl_2	308(4.47),327(4.43), 422s(4.01),450s(4.09), 470(4.14),624(3.73)	5-1739-76
$C_{39}H_{30}N_8$ Hydrazyl, 1,1'-[5-(7-azabicyclo[4.1.0]- hept-7-yl)-2,4,6-tricyano-1,3-phenyl- ene]bis[2,2-diphenyl-	CH_2Cl_2	316(4.46),326(4.46), 453s(4.16),472(4.24), 622(3.80)	5-1739-76
$C_{39}H_{44}N_2O_7$ Thalrugosaminine	MeOH	262(3.80),282(3.88)	100-0065-76
$C_{39}H_{44}N_2O_{19}$ Pentanedioic acid, 2,4-bis[2,3-O-(1- methylethylidene)-5-O-(4-nitroben- zoyl)-α-D-ribofuranosyl]-3-oxo-, diethyl ester	EtOH pH 13	258(4.41) 278(4.55)	44-0287-76 44-0287-76
$C_{39}H_{44}N_4O_6$ 21H,23H-Porphine-2-propanoic acid, β-acetoxy-18-(3-acetoxypropyl)- 12-ethenyl-7-ethyl-3,8,13,17- tetramethyl-, methyl ester	CH_2Cl_2 CH_2Cl_2-5% CF_3COOH	402(5.24),502(4.01), 539(4.07),570(3.82), 626(3.30) 408(5.53),551(4.14), 598(3.82)	78-0275-76 78-0275-76
$C_{39}H_{46}N_4O_4$ 21H,23H-Porphine-2-propanoic acid, 18- [3-(1,1-dimethylethoxy)-3-oxo-1-pro- penyl]-8,13-diethyl-3,7,12,17-tetra- methyl-, methyl ester, (E)-	CH_2Cl_2 CH_2Cl_2-5% CF_3COOH	412(5.18),509(4.03), 548(4.20),574(3.97), 633(3.48) 411(5.41),555(4.18), 602(3.95)	78-0275-76 78-0275-76
$C_{39}H_{47}N_3O_{13}$ Garamine, 2'-O-acetyl-1,3,3'-tris- N-benzyloxycarbonyl- Garamine, 4-O-acetyl-1,3,3'-tris- N-benzyloxycarbonyl- Garamine, 4'-O-acetyl-1,3,3'-tris- N-benzyloxycarbonyl-	MeOH MeOH MeOH	208(4.39) 205(4.42) 207(4.42)	39-1088-76C 39-1088-76C 39-1088-76C
$C_{39}H_{48}CoN_5$ Cobalt, (2-cyanoethyl)[2,3,7,8,12,13,17- octaethyl-21H,23H-porphinato(2-)- $N^{21},N^{22},N^{23},N^{24}$]-, (SP-5-31)-	$CHCl_3$	391(5.25),519(4.02), 552(4.45)	18-2529-76
$C_{39}H_{50}HgN_4O_2$ Mercury, (acetato-O)(2,3,7,8,12,13,17- 18-octaethyl-21-methyl-21H,23H-por- phinato-N^{23})-	$CHCl_3$	434(5.24),511s(3.66), 541(4.13),586(4.03), 628(3.60)	78-0597-76
$C_{39}H_{51}N_3O_{12}$ 1H-Pyrrole-3-propanoic acid, 2-[(1,1-di- methylethoxy)carbonyl-5-[[4-(2-meth- oxy-2-oxoethyl)-5-[[4-(2-methoxy-2- oxoethyl)-3-(3-methoxy-3-oxopropyl)- 5-methyl-2H-pyrrol-2-ylidene]methyl- 3-(3-methoxy-3-oxopropyl)-1H-pyrrol- 2-yl]methyl-4-methyl-, methyl ester, hydrobromide	CH_2Cl_2	329(4.09),497(4.77)	5-1637-76

Compound	Solvent	$\lambda_{max}(\log \epsilon)$	Ref.
$C_{39}H_{54}O_7$ Querataroic (30)caffeate	n.s.g.	217(4.28),250(4.19), 302(4.25),333(4.35)	102-0430-76
$C_{39}H_{62}N_3O$ Acridinium, 9-(dodecylamino)-10-[2-(do- decylamino)-2-oxoethyl]-, chloride	isoPrOH	222(4.34),260s(4.55), 271(4.69),290s(3.85), 322s(3.36),417(4.03), 437(4.03)	44-3406-76
$C_{39}H_{62}O_{13}$ Acryloyl-α-cyclodextrin	H_2O	197(3.97)	116-0701-76
$C_{39}H_{72}OS$ 1-Eicosanone, 2-methyl-1-(4-methyl-5- tridecyl-2-thienyl)-	heptane	251(3.90),298(4.01)	34-0233-76
1-Tetradecanone, 2-methyl-1-(4-methyl- 5-nonadecyl-2-thienyl)-	heptane	257(3.91),298(3.99)	34-0233-76
$C_{39}H_{74}S$ Thiophene, 3-methyl-5-(2-methyltetradec- yl)-2-nonadecyl-	heptane	241(3.89)	34-0233-76

Compound	Solvent	λ_{max}(log ϵ)	Ref.
$C_{40}H_{24}N_2$ 3,5-Cyclopentadiene-1,3-dicarbonitrile, 2-[4-(2,3-diphenyl-2-cyclopropen- 1-ylidene)-2,5-cyclohexadien-1- ylidene]-4,5-diphenyl-	MeCN	274(4.58),301(4.58), 315s(4.55),540(4.60)	88-4655-76
$C_{40}H_{25}N_2$ 1,3-Cyclopentadiene-1,3-dicarbonitrile, 2-[4-(2,3-diphenylcyclopropen-1-yl)- phenyl]-4,5-diphenyl-, ion(1-), N,N,N-trimethanaminium	MeCN	265(4.38),311(4.47), 406(4.48)	88-4655-76
Cyclopropenylium, [4-(2,5-dicyano-3,4- diphenyl-1,4-cyclopentadien-1-yl)- phenyl]diphenyl-, trifluoroa, etate	CF$_3$COOH	259(4.34),324(4.75)	88-4655-76
$C_{40}H_{26}O_2$ Benzo[1,2:4,5]dicycloheptene-3,9-dione, 2,4,8,10-tetraphenyl-	CHCl$_3$	270s(4.31),339(4.97)	22-0914-76
$C_{40}H_{28}F_4P_2$ 1,4-Diphosphorinium, 1,1,4,4-tetrakis- (4-fluorophenyl)-1,4-dihydro-2,5-di- phenyl-, dichloride	MeOH	193(3.83),275(3.18)	139-0169-76A
$C_{40}H_{38}O_6$ [4,8'-Bi-2H-1-benzopyran]-7'-ol, 3,3',4- 4'-tetrahydro-5,5'-dimethoxy-6-methyl- 2,2'-diphenyl-7-(phenylmethoxy)-	n.s.g.	210(4.91),235s(4.31), 280(3.66)	39-1392-76C
$C_{40}H_{40}N_2O_8$ Asterriquinone, reductive acetylation product	CHCl$_3$	310(4.40)	94-1853-76
$C_{40}H_{43}N_3O_7S$ Thymidine, 3'-[O-(N-benzyloxycarbonyl)- DL-phenylalanyl]-5'-O-(triphenylmeth- yl)-4-thio-	n.s.g.	207(4.59),230s(3.93), 258(3.54),334(4.21)	128-0351-76
$C_{40}H_{44}N_4O_8$ 21H,23H-Porphine-2,7,18-tripropanoic acid, 12-ethenyl-3-(2-methoxy-2-oxo- ethyl)-8,13,17-trimethyl-, trimethyl ester	CHCl$_3$	404(5.27),504(4.10), 541(4.11),571(3.86), 627(3.46)	12-1561-76
Porphine-2,6,7-tripropanoic acid, 4-(3- methoxy-3-oxo-1-propenyl)-1,3,5,8- tetramethyl-, trimethyl ester	CHCl$_3$	413(5.08),510(3.97), 548(4.14),577(3.92), 635(3.44)	39-2492-76C
	CHCl$_3$- CF$_3$COOH	425(5.17),563(4.09), 616(3.92)	39-2492-76C
Porphine-4,6,7-tripropanoic acid, 2-(3- methoxy-3-oxo-1-propenyl)-1,3,5,8- tetramethyl-, trimethyl ester	CHCl$_3$	413(5.08),510(3.97), 548(4.12),577(3.92), 635(3.44)	39-2492-76C
	CHCl$_3$- CF$_3$COOH	425(5.17),563(4.09), 610(3.92)	39-2492-76C
$C_{40}H_{44}N_4O_9$ Porphine-8,13,17-tripropanoic acid, 3-acetyl-12-(methoxycarbonylmethyl)- 2,7,18-trimethyl-, trimethyl ester	CHCl$_3$	411(5.33),512(4.01), 552(4.18),577(3.97), 635(3.06)	12-1561-76
Porphine-8,13,17-tripropanoic acid, 3-acetyl-18-(methoxycarbonylmethyl)- 2,7,12-trimethyl-, trimethyl ester	CHCl$_3$	411(5.25),510(4.07), 548(4.04),578(3.87), 634(3.33)	12-1561-76

Compound	Solvent	$\lambda_{max}(\log \epsilon)$	Ref.
$C_{40}H_{46}Cl_3N_3O_{14}$ Garamine, 1,3,3'-tris-N-benzyloxycarbo- nyl-4-(2,2,2-trichloroethoxycarbonyl)-	MeOH	207(4.42)	39-1088-76C
$C_{40}H_{46}N_4O_8$ Isocoproporphyrin, tetramethyl ester	CHCl$_3$	401(5.20),499(4.12), 534(3.94),568(3.79), 622(3.63)	12-1561-76
	CHCl$_3$	398(5.30),493(4.16), 526(3.98),559(3.83), 609(3.65)	78-1793-76
	CHCl$_3$-5% TFA	403(5.60),343(4.21), 582(3.88)	78-1793-76
Porphine-2,4,6,8-tetrapropanoic acid, 1,3,5,8-tetramethyl-, tetramethyl ester	CHCl$_3$	400(5.20),497(4.14), 531(3.96),567(3.77), 592(3.11),620(3.63)	78-1793-76
Porphine-2,6,7-tripropanoic acid, 4-(2- acetoxyethyl)-1,3,5,8-tetramethyl-, trimethyl ester	CHCl$_3$	400(5.23),501(4.12), 532(3.93),568(3.81), 620(3.42)	78-2757-76
Porphine-4,6,7-tripropanoic acid, 2-(2- acetoxyethyl)-1,3,5,8-tetramethyl-, trimethyl ester	CHCl$_3$	400(5.25),502(4.16), 532(3.99),568(3.82), 621(3.64)	78-2757-76
$C_{40}H_{46}N_4O_9$ 21H,23H-Porphine-8,13,17-tripropanoic acid, 3-(1-hydroxyethyl)-12-(2-meth- oxy-2-oxoethyl)-2,7,18-trimethyl-, trimethyl ester	CHCl$_3$	403(5.28),500(4.12), 537(4.01),568(3.83), 622(3.50)	12-1561-76
$C_{40}H_{48}$ Bicyclo[10.10.2]tetracosa-1(23),2,4,8- 10,12(24),13,15,19,21,23-undecaene- 6,17-diyne, 5,8,16,19-tetrakis(1,1- dimethylethyl)-	THF	216(4.09),244(4.18), 250(4.24),263(4.31), 275(4.51),323(4.10), 371(4.57),384(5.06), 401(5.75),468(3.64), 492s(3.81),518(4.28), 553(4.74),637(2.68), 676(4.01),742(3.33)	35-6410-76
$C_{40}H_{48}N_4O_2$ Tabernamine	EtOH	235(4.53),287(4.02), 295(4.00)	88-0649-76
$C_{40}H_{48}N_4O_4$ Propanedioic acid, [(2,3,7,8,12,13,17- 18-octaethyl-21H,23H-porphin-5-yl)- methylene]-	CH$_2$Cl$_2$	407(5.05),507(4.03), 540(3.76),579(3.79), 630(3.45)	5-1537-76
$C_{40}H_{49}N_3O_{12}$ Garamine, 1,3,3'-tris-N-benzyloxycarbo- nyl-4,5-O-isopropylidene-	MeOH	208(4.39)	39-1088-76C
$C_{40}H_{49}N_5O_6$ Jubanine A	MeOH	265(3.96),320(3.79)	102-0541-76
$C_{40}H_{50}$ 1,2,3,5,7,9,11,17,19,21,23-Cyclotetra- cosaundecaene-13,15-diyne, 1,4,12,17- tetrakis(1,1-dimethylethyl)-	THF at -78°	<u>365(5.1),630(3.4)</u>	88-2623-76

Compound	Solvent	λ_{max} (log ϵ)	Ref.
$C_{40}H_{54}N_2O_9S$ Piperidine, 1-acetyl-3-acetoxy-2-[1-[6'-acetoxy-3',4,5,5',6',7',8',8'a-octaethyl-7,8'a-dimethyl-1'-[[[(4-methylphenyl)sulfonyl]oxy]methyl]spiro[2,1-benzisoxazole-3(3aH),2'(1'H)-naphthalen]-6-yl]ethyl]-5-methyl-	EtOH	205(4.23),226(4.26), 257(4.17)	39-1297-76C
$C_{40}H_{54}N_4O_8S$ Benzenesulfonic acid, 4-methyl-, [[6-[1-[1-acetyl-3-acetoxy-5-methyl-2-piperidinyl]ethyl]-6'-acetoxy-3',4,5,5',6'-7',8',8'a-octahydro-7,8'a-dimethyl-spiro[2,1-benzisoxazole-3(3aH),2'(1'H)-naphthalen]-1'-yl]methylene]hydrazide	EtOH	206(4.36),233(4.28), 256(4.17)	39-1297-76C
$C_{40}H_{56}O_{23}S$ Breynin A	pH 1, 7 pH 13	258(<u>4.2</u>) 290(<u>4.3</u>),350(<u>3.9</u>), 395(<u>3.7</u>)	94-0169-76 94-0169-76
$C_{40}H_{56}O_{27}S_3$ β-D-Glucopyranose, O-2,3,4,6-tetra-O-acetyl-α-D-glucopyranosyl-(1→4)-O-2,3,6-tri-O-acetyl-α-D-glucopyranosyl-(1→4)-1-thio-, 2,3-diacetate 1-(O-ethyl carbonodithioate) 6-methanesulfonate	MeOH	273(4.01)	136-0282-76C

Compound	Solvent	$\lambda_{max}(\log \epsilon)$	Ref.
$C_{41}H_{24}N_8O_2$ Hydrazyl, 1,1'-[2,4,6-tricyano-5-(1,3-dihydro-1,3-dioxo-2H-isoindol-2-yl)-1,3-phenylene]bis[2,2-diphenyl-	CH_2Cl_2	334(4.43),430s(3.89), 490(4.17),631(3.73)	5-1739-76
$C_{41}H_{28}Cl_2N_6$ 2H-Naphtho[1,2-d]triazole, 2,2'-[(2-phenyl-1,3-propanediyl)bis(3-chloro-4,1-phenylene)]bis-	DMF	342(4.48),350(4.42), 359(4.48)	33-0802-76
$C_{41}H_{30}NO$ Pyridinium, 4-[(2,6-diphenyl-4H-pyran-4-ylidene)methyl]-1,2,6-triphenyl-, perchlorate	MeCN	505(4.86)	4-0577-76
$C_{41}H_{30}O_{26}$ β-D-Glucopyranose, cyclic 4,6-(4,4',5-5',6,6'-hexahydroxy[1,1'-biphenyl]-2,2'-dicarboxylate) 1,2,3-tris-(3,4,5-trihydroxybenzoate)	EtOH	281(4.58),300s(--), 342s(--)	5-0112-76
	dioxan	277(4.60),295s(--)	5-0112-76
$C_{41}H_{34}N_7$ 3H-Imidazo[1,2-a]benzimidazolium, 3-[[4-(dimethylamino)phenyl](9-methyl-2-phenyl-9H-imidazo[1,2-a]benzimidazol-3-yl)methylene]-9-methyl-2-phenyl-, chloride	MeOH	440(3.78),622(4.57)	103-0114-76
$C_{41}H_{35}N_5O_9$ 7(8H)-Pteridinone, 2-(dimethylamino)-4-(phenylmethoxy)-8-(2,3,5-tri-0-benzoyl-β-D-ribofuranosyl)-	MeOH	228(4.63),274(3.70), 295s(3.50),364(4.14)	24-3228-76
β-D-Ribofuranoside, 2-(dimethylamino)-4-(phenylmethoxy)-7-pteridinyl-, 2,3,5-tribenzoate	MeOH	231(4.73),278(4.13), 368(4.08)	24-3228-76
$C_{41}H_{38}O_{11}$ Glaucarubolone, tribenzoate	EtOH	233(4.70),275(3.56), 282(3.45)	87-1130-76
$C_{41}H_{42}N_4O_5$ Scandomelonine	EtOH	235s(4.11),264(4.08), 296(4.03),329(4.18)	44-3275-76
$C_{41}H_{48}N_4O_3$ Accedinisine	MeOH	231(4.66),286(4.20), 293(4.18)	24-3527-76
$C_{41}H_{48}N_4O_4$ Accedinine	MeOH	231(4.67),286(4.21), 293(4.19)	24-3527-76
$C_{41}H_{49}N_2O_8$ Thalistyline, chloride	MeOH	276(3.86),283(3.84)	88-3687-76
$C_{41}H_{49}N_3O_{14}$ Garamine, 2',4'-di-O-acetyl-1,3,3'-tris-N-benzyloxycarbonyl-	MeOH	207(4.42)	39-1088-76C

Compound	Solvent	$\lambda_{max}(\log \epsilon)$	Ref.
$C_{41}H_{49}O_2$ 2,5-Cyclohexadien-1-one, 4-[[1,1'-bi-phenyl]-2-yl-[3,5-bis(1,1-dimethyl-ethyl)-4-hydroxyphenyl]methylene]-2,6-bis(1,1-dimethylethyl)-, ion(1-) radical	MeOH	385(--),606(4.88)	64-0965-76B
	MeOH	459(3.67)	64-0965-76B
$C_{41}H_{50}N_5$ Pyridinium, 1-(2,3,7,8,12,13,17,18-hexa-ethyl-21H,23H-porphin-5-yl)-	$CHCl_3$	402(4.18),505(4.16), 537(4.06),566(3.93), 616(3.92)	88-4009-76
	$CHCl_3$-1% CF_3COOH	420(4.38),556(4.21), 599(3.99)	88-4009-76
$C_{41}H_{50}O_2$ 2,5-Cyclohexadien-1-one, 4-[[1,1'-bi-phenyl]-2-yl-[3,5-bis(1,1-dimethyl-ethyl)-4-hydroxyphenyl]methylene]-2,6-bis(1,1-dimethylethyl)-	MeOH	416(4.42)	64-0965-76B
$C_{41}H_{52}N_4O_3$ Accedinisine, tetrahydro-	MeOH	231(4.60),286(4.20), 293(4.19)	24-3527-76
$C_{41}H_{52}N_4O_8$ 21H-Biline-3,7,18-tripropanoic acid, 13-ethyl-10,23-dihydro-2-(2-methoxy-2-oxoethyl)-1,8,12,17,19-pentamethyl-, trimethyl ester, dihydrobromide	$CHCl_3$	453(4.42),519(5.39)	78-1793-76
$C_{41}H_{53}CoN_4O$ Cobalt, [2,3,7,8,12,13,17,18-octaethyl-21H,23H-porphinato(2-)-N^{21},N^{22},N^{23}-N^{24}](4-oxopentyl)-, (SP-5-31)-	$CHCl_3$	392(5.24),520(3.96), 552(4.43)	18-2529-76
$C_{41}H_{53}CoN_4O_2$ Cobalt, (3-ethoxy-3-oxopropyl)[2,3,7,8-12,13,17,18-octaethyl-21H,23H-porphi-nato(2-)-N^{21},N^{22},N^{23},N^{24}]-, (SP-5-31)-	$CHCl_3$	392(5.27),517(4.02), 551(4.41)	18-2529-76
$C_{41}H_{53}NO_{12}S$ Rifamycin, 1,4-dideoxy-3-[(1,1-dimethyl-ethyl)thio]-1,4-dihydro-1,4-dioxo-	MeOH	273(4.34),330(3.68)	35-7064-76
$C_{41}H_{53}N_3O_{14}$ 15H-Tripyrrine, 1-tert-butoxycarbonyl-2,7,12-tris(2-methoxycarbonylethyl)-3,8,13-tris(methoxycarbonylmethyl)-14-methyl-5,16-dihydro-, hydrobromide	CH_2Cl_2	365(3.93),497(4.98)	5-1637-76
$C_{41}H_{53}N_5O_2$ β-Alanine, N-[(2,3,7,8,12,13,17,18-octa-ethyl-21H,23H-porphin-5-yl)methylene]-, methyl ester	CH_2Cl_2	405(5.25),506(4.12), 541(3.90),576(3.72), 630(3.54)	5-1537-76
protonated	CH_2Cl_2	445(5.38),575(3.65), 665(4.02)	5-1537-76
$C_{41}H_{56}O_{26}S_2$ β-D-Glucopyranose, O-2,3,4,6-tetra-O-acetyl-α-D-glucopyranosyl-(1 4)-O-2,3,6-tri-O-acetyl-α-D-glucopyrano-	MeOH	274(4.06)	136-0282-76C

Compound	Solvent	$\lambda_{max}(\log \epsilon)$	Ref.
syl(1→4)-1-thio-, 2,3,6-triacetate 1-(O-ethyl carbonodithioate) (cont.)			136-0282-76C
$C_{41}H_{62}O_{19}$ Corotoxigenin, β-D-glucopyranosyl-(1→6)- β-D-glucopyranosyl-(1→4)-α-L-rhamno- pyranoside	MeOH	218(4.20)	94-2886-76
$C_{41}H_{63}NO_{17}$ Maridomycin III, 9-dehydro-N-formyl-	EtOH	240.5(4.12)	94-0463-76
$C_{41}H_{65}NO_{15}$ Leucomycin V, 4-demethoxy-4,17-epoxy-, 3-acetate 4^B-(3-methylbutanoate)	MeOH	232(4.31)	78-0991-76
$C_{41}H_{65}NO_{16}$ Maridomycin III, 9-dehydro-	EtOH	240.5(4.16)	94-0463-76
$C_{41}H_{71}N_3O_6$ Hexadecanoic acid, [2,3,3a,9a-tetrahy- dro-6-imino-3-[(1-oxohexadecyl)oxy]- 6H-furo[2',3':4,5]oxazolo[3,2-a]pyr- imidin-2-yl]methyl ester, hydro- chloride	MeOH-acid	235(4.03),264(4.09)	87-0663-76
$C_{41}H_{73}N_3O_7$ 2(1H)-Pyrimidinone, 4-amino-1-[3,5-bis- O-(1-oxohexadecyl)-β-D-arabino- furanosyl]-	MeOH	212(3.99),283(4.15)	87-0667-76

Compound	Solvent	$\lambda_{max}(\log \epsilon)$	Ref.
$C_{42}H_{26}$ 1,2,3,4,5,7,9,15,17-Cyclooctadecanona-ene-11,13-diyne, 1,6,10,15-tetra-phenyl-	THF	245(4.37),261(4.39), 302(4.43),351(4.42), 373(4.58),472(5.44), 635(5.03),772(3.61)	18-0297-76
$C_{42}H_{30}N_2O$ 4,5-Diazaspiro[2.3]hexan-6-one, 1,2-bis-(diphenylmethylene)-4,5-diphenyl-	$CHCl_3$	270(4.43),404(4.45)	18-3173-76
$C_{42}H_{32}N_2O_2$ Dispiro[3H-naphtho[2,1-b]pyran-3,5'-(10'H)-pyrido[2,3,4,5-lmn]phenan-thridine-10',3"-[3H]naphtho[2,1-b]-pyran], 4',9'-diethyl-4',9'-dihydro-	$C_2H_4Cl_2$	259(4.97),305(4.19), 316(4.20),350s(4.20), 362(4.21)	103-0273-76
	HOAc	270s(4.38),308(4.30), 338(4.27),352(4.28), 375(4.19),394(4.22), 520(4.30)	103-0273-76
$C_{42}H_{34}O_{14}$ [9,9'-Bianthracene]-10,10'(9H,9'H)-di-one, 2,2',4,4',5,5'-hexaacetoxy-7,7'-dimethyl-	$CHCl_3$	289(4.38)	12-1535-76
$C_{42}H_{38}$ Benzene, 1,1',1",1"'-[3,8-bis(1,1-di-methylethyl)-3,4,6,7-decatetraene-1,9-diyne-1,5,6,10-tetrayl]tetrakis-	$CHCl_3$	276(4.77),316s(4.46)	18-2515-76
Benzene, 1,1',1"-[3,6-bis(1,1-dimethyl-ethyl)-3-(phenylethynyl)-4,5-octadi-ene-1,7-diyne-1,4,8-triyl]tris-	$CHCl_3$	280(4.49),297(4.42)	18-2515-76
Benzene, 1,1'-[3,4-bis[1-(1,1-dimethyl-ethyl)-3-phenyl-2-propynylidene]-1-cyclobutene-1,2-diyl]bis-	$CHCl_3$	318(4.61)	18-2515-76
Tricyclo[6.2.0.0^{3,6}]deca-1,3(6),4,7,9-pentaene, 2,7-bis(1,1-dimethylethyl)-4,5,9,10-tetraphenyl-	$CHCl_3$	275(4.72),297(4.61)	18-2515-76
$C_{42}H_{38}Br_4$ Tricyclo[6.2.0.0^{3,6}]deca-1,3(6),7-tri-ene, 4,5,9,10-tetrabromo-2,7-bis(1,1-dimethylethyl)-4,5,9,10-tetraphenyl-	$CHCl_3$	305(3.76)	18-2515-76
$C_{42}H_{38}O_2$ 1,4-Butanedione, 2,4-bis[1-(1,1-dimeth-ylethyl)-3-phenyl-2-propynylidene]-1,4-diphenyl-, (E,Z)-	$CHCl_3$	270(4.61)	18-2515-76
3,4-Octadien-7-yne-1,2-dione, 3,6-bis-(1,1-dimethylethyl)-1,5,8-triphenyl-6-(phenylethynyl)-	$CHCl_3$	255(4.73)	18-2515-76
$C_{42}H_{38}O_2S_4$ 1,4-Butanedione, 2,3-bis[5-[4-(1,1-di-methylethyl)phenyl]-3H-1,2-dithiol-3-ylidene]-1,4-diphenyl-	CH_2Cl_2	472(4.53)	1-0478-76
$C_{42}H_{38}O_4$ Methanone, [3,6-bis(1,1-dimethylethyl)-1,2,4,5-benzenetetrayl]tetrakis[phen-yl-	$CHCl_3$	300(4.46),330s(4.30), 350s(3.97)	18-2515-76

Compound	Solvent	λ_{max}(log ϵ)	Ref.
$C_{42}H_{42}ClN_4O_{16}P$			
Thymidine, P-(2-chlorophenyl)-5'-O-(phen-oxyacetyl)thymidylyl[3'→3']-, 5'-(phenoxyacetate)	MeOH	265(4.33)	35-3655-76
$C_{42}H_{42}O_2$			
1H,3H-Benzo[1,2-c:4,5-c']difuran, 4,8-bis(1,1-dimethylethyl)-5,7-dihydro-1,3,5,7-tetraphenyl-	CHCl$_3$	297(4.35)	18-2515-76
3,4-Octadien-7-yne-1,2-diol, 3,6-bis-(1,1-dimethylethyl)-1,5,8-triphenyl-6-(phenylethynyl)-	CHCl$_3$	250(4.54),257(4.54)	18-2515-76
$C_{42}H_{44}Cl_3N_4O_{13}P$			
Thymidine, P-(2,2,2-trichloroethyl)thy-midylyl[3'→5']-3'-O-[(4-methoxyphen-yl)diphenylmethyl]-	MeOH	232s(4.26),265(4.31)	35-3655-76
$C_{42}H_{46}N_4O_6$			
Scandomeline	EtOH	232s(3.93),259(3.94), 302(3.92),328(4.01)	44-3275-76
(episcandomeline has same spectra)	EtOH-HCl	234s(3.85),270(4.08), 296(3.92),327(3.94)	44-3275-76
$C_{42}H_{46}O_7S_2$			
D-glycero-D-gulo-Heptitol, 1,1-di-C-benzo[b]thien-2-yl-2,3:6,7-di-O-cyclohexylidene-4-O-(phenylmethyl)-	MeOH	264(3.73)	30-0613-76
$C_{42}H_{48}N_4O_{10}$			
Porphine-2,4,6,7-tetrapropanoic acid, 1-(2-methoxy-2-oxoethyl)-3,5,8-tri-methyl-, tetramethyl ester	CHCl$_3$	402(5.27),500(4.16), 536(4.01),569(3.84), 622(3.64)	12-0393-76
Porphine-2,4,6,7-tetrapropanoic acid, 3-(2-methoxy-2-oxoethyl)-1,5,8-tri-methyl-, tetramethyl ester	CHCl$_3$	402(5.28),500(4.16), 536(4.01),568(3.85), 622(3.64)	12-0393-76
Porphine-2,4,6,7-tetrapropanoic acid, 5-(2-methoxy-2-oxoethyl)-1,3,8-tri-methyl-, tetramethyl ester	CHCl$_3$	402(5.27),500(4.14), 536(3.99),568(3.83), 622(3.61)	12-0393-76
	CHCl$_3$	400(5.28),500(4.14), 534(3.98),568(3.83), 595(3.08),622(3.57)	78-1793-76
Porphine-2,4,6,7-tetrapropanoic acid, 8-(2-methoxy-2-oxoethyl)-1,3,5-tri-methyl-, tetramethyl ester	CHCl$_3$	402(5.26),500(4.14), 536(3.99),568(3.83), 622(3.60)	12-0393-76
Porphine-2,6,7-tripropanoic acid, β-acetoxy-4-(2-acetoxyethyl)-1,3,5,8-tetramethyl-, trimethyl ester	CHCl$_3$	403(5.24),501(4.20), 532(3.76),552(3.81), 624(3.15)	78-2757-76
Porphine-4,6,7-tripropanoic acid, β-acetoxy-2-(2-acetoxyethyl)-1,3,5,8-tetramethyl-, trimethyl ester	CHCl$_3$	403(5.24),501(4.17), 532(3.72),552(3.74), 622(3.15)	78-2757-76
$C_{42}H_{50}N_4O_5$			
Voacamine, N-demethyl-	MeOH	225(4.76),286(4.30), 293(4.29)	24-3527-76
$C_{42}H_{51}N_3O_{13}$			
Garamine, 2'-O-acetyl-1,3,3'-tris-N-benzyloxycarbonyl-4,5-O-ispropyli-dene-	MeOH	208(4.40)	39-1088-76C

Compound	Solvent	$\lambda_{max}(\log \epsilon)$	Ref.
$C_{42}H_{52}N_4O_4$			
Propanedioic acid, [(2,3,7,8,12,13,17-18-octaethyl-21H,23H-porphin-5-yl)-methylene]-, dimethyl ester	CH_2Cl_2	408(5.16),509(4.06), 543(3.85),580(3.79), 632(3.48)	5-1537-76
monoethyl ester	CH_2Cl_2	404(5.05),509(4.03), 542(3.58),580(3.79), 634(3.45)	5-1537-76
$C_{42}H_{56}N_2O_2$			
Staphinine	EtOH	268(4.24)	88-1055-76
$C_{42}H_{58}N_2O$			
Staphidine	EtOH	268(4.24)	88-1055-76
$C_{42}H_{64}O_{17}$			
Leucomycin V, 3^A-de(methylamino)-3^A-oxo-, 2^A,3-diacetate 4^B-(3-methyl-butanoate)	EtOH	232(4.03)	94-1749-76
$C_{42}H_{67}F_2NO_{15}$			
Leucomycin V, 17,17-difluoro-, 3-acet-ate 4^B-(3-methylbutanoate)	MeOH	231(4.42)	78-0991-76
$C_{42}H_{67}NO_{16}$			
Maridomycin II, 9-dehydro-	EtOH	240(4.18)	94-0450-76
$C_{42}H_{68}BrNO_{15}$			
Leucomycin V, 17-bromo-, 3-acetate 4^B-(3-methylbutanoate)	MeOH	231(4.42)	78-0991-76
$C_{42}H_{68}ClNO_{15}$			
Leucomycin V, 17-chloro-, 3-acetate 4^B-(3-methylbutanoate)	MeOH	231(4.41)	78-0991-76
$C_{42}H_{68}FNO_{15}$			
Leucomycin V, 17-fluoro-, 3-acetate 4^B-(3-methylbutanoate)	MeOH	231(4.34)	78-0991-76

Compound	Solvent	$\lambda_{max}(\log \epsilon)$	Ref.
$C_{43}H_{28}Cl_4N_2O_3$ Methanone, phenyl[5,6,7,8-tetrachloro- 3-(6,7-dimethyl-3-phenyl-2-quinoxal- inyl)-2,3-dihydro-2,3-diphenyl-1,4- benzodioxin-2-yl]-, cis	C_6H_{12}	217(4.98),246(4.79), 335(4.01)	5-0793-76
$C_{43}H_{31}NO$ 5-Azaspiro[2.3]hexan-4-one, 1,2-bis(di- phenylmethylene)-5,6-diphenyl-	CHCl$_3$	260(4.56),383(4.47)	18-3173-76
$C_{43}H_{33}N_3O_3$ Pyrimido[1,2-a]benzimidazol-4(3H)-one, 10-(diphenylacetyl)-2,10-dihydro- 2-(4-methoxyphenyl)-3,3-diphenyl-	EtOH	251s(4.35),287(4.00), 298(4.00)	94-2532-76
$C_{43}H_{38}O_6$ Methanone, [4-methoxy-2,6-bis(phenyl- methoxy)phenyl][2-methyl-4,6-bis- (phenylmethoxy)phenyl]-	EtOH	284(4.24)	35-5380-76
Methanone, 2-methoxy-6-methyl-4-(phenyl- methoxy)phenyl][2,4,6-tris(phenylmeth- oxy)phenyl]-	EtOH	241(4.43),273(4.58)	35-5380-76
$C_{43}H_{45}N_3O_7S$ L-Phenylalanine, N-[(1,1-dimethyleth- oxy)carbonyl]-, ester with 4-thio- 5'-O-(triphenylmethyl)thymidine	n.s.g.	203(4.80),232s(4.06), 333(4.30)	128-0351-76
$C_{43}H_{47}N_5O_6$ Jubanine B	MeOH	270(3.54),318(3.36)	102-0541-76
$C_{43}H_{50}N_4O_4$ Accedinisine, acetyl derivative	MeOH	231(4.62),286(4.16), 293(4.13)	24-3527-76
$C_{43}H_{51}N_3O_{15}$ Garamine, 2',4,4'-tri-O-acetyl-1,3,3'- tris-N-benzyloxycarbonyl-	MeOH	207(4.42)	39-1088-76C
Garamine, 2',4,5-tri-O-acetyl-1,3,3'- tris-N-benzyloxycarbonyl-	MeOH	207(4.42)	39-1088-76C
Garamine, 2',4',5-tri-O-acetyl-1,3,3'- tris-N-benzyloxycarbonyl-	MeOH	207(4.42)	39-1088-76C
$C_{43}H_{54}N_4O_{10}$ 22H-Biline-2,7,12,18-tetrapropanoic acid, 10,23-dihydro-3-(2-methoxy-2- oxoethyl)-1,8,13,17,19-pentamethyl-, tetramethyl ester, dihydrobromide	CHCl$_3$	453(4.56),526(5.21)	78-1793-76
$C_{43}H_{54}N_4O_5$ Tabernaelegantine A	EtOH	224(4.70),285(4.14), 293(4.10)	39-1432-76C
Tabernaelegantine B	EtOH	227(4.71),287(4.09), 296(4.23)	39-1432-76C
Tabernaelegantine C	EtOH	224(4.76),285(4.18), 293(4.16)	39-1432-76C
Tabernaelegantine D	EtOH	226(4.73),287(4.13), 296(4.16)	39-1432-76C

Compound	Solvent	$\lambda_{max}(\log \epsilon)$	Ref.
$C_{43}H_{62}O_3$			
2,5-Cyclohexadien-1-one, 4-[bis[3,5-(1,1-dimethylethyl)-4-hydroxyphenyl]-methylene]-2,6-bis(1,1-dimethylethyl)-	MeOH	448(4.46)	64-0965-76B
anion	MeOH	569(4.98)	64-0965-76B
radical anion	MeOH	417(--),442(5.26), 810(--)	64-0965-76B
$C_{43}H_{67}NO_{17}$			
Maridomycin I, 9-dehydro-N-formyl-	EtOH	239.5(4.14)	94-0463-76
$C_{43}H_{69}NO_{16}$			
Maridomycin I, 9-dehydro-	EtOH	240.5(4.14)	94-0463-76
$C_{43}H_{72}N_4O_{15}S$			
Maridomycin II, 18-thiosemicarbazone	EtOH	270(4.38)	94-0450-76

Compound	Solvent	$\lambda_{max}(\log \epsilon)$	Ref.
$C_{44}H_{24}$ Bi-2,13-pentahelicenylene (propellicine)	C_6H_{12}	225s(--),233s(--), 241(5.11),284(4.85), 305s(--),315s(--), 340s(--)	1-0688-76
$C_{44}H_{24}Fe_2N_8$ Biferrocenylenediiron, 1:2 product with TCNQ	MeCN	234(4.33),270(4.09), 394(4.95),743(4.23), 762(4.16),822(4.35), 842(4.51),1550(3.28)	35-3181-76
$C_{44}H_{26}Cl_4N_2O_4$ 15H-6,14a-Etheno[1,5]benzodioxepino- [3',2':3,4]pyrazolo[1,2-b]phthala- zine-8,13-dione, 1,2,3,4-tetrachloro- 5a,6-dihydro-5a,6,17,18-tetraphenyl-, (5aα,6α,14aα)-	C_6H_{12}	217(4.95),290(4.00), 298s(4.00),333(3.84)	5-0793-76
$C_{44}H_{27}Cl_4N_3O_4$ 5H-4a,12-Etheno-1H-[1,5]benzodioxepino- [3',2':3,4]pyrazolo[1,2-a][1,2,4]tri- azole-1,3(2H)-dione, 7,8,9,10-tetra- chloro-11a,12-dihydro-2,11a,12,14,15- pentaphenyl-, (4aα,11aα,12α)-	C_6H_{12}	219(4.875),287(4.000)	5-0793-76
$C_{44}H_{27}Cl_4N_3O_5$ Spiro[1,4-benzodioxin-2(3H),7'(8'H)- [5,8]etheno-1H,5H-[1,2,4]triazolo- [1,2-c][1,3,4]oxadiazine]-1',3'(2'H)- dione, 5,6,7,8-tetrachloro-2',5',8'- 10',11'-pentaphenyl-	C_6H_{12}	216(5.046),297(3.643)	5-0793-76
$C_{44}H_{28}HgN_4$ Mercury, [5,10,15,20-tetraphenyl- 21H,23H-porphinato(2-)-N^{21},N^{22},N^{23}- N^{24}]-, (SP-4-1)-	$CHCl_3$	426(5.62),487s(3.80), 519s(3.87),550(3.98), 590s(3.65),647(3.42)	78-0597-76
$C_{44}H_{29}N_3O_3$ 6,10-Epoxy-6H-isoindolo[5,6-g]phthala- zine-7,9(6aH,8H)-dione, 9a,10-di- hydro-1,4,6,8,10-pentaphenyl-	CH_2Cl_2	298(4.09)	28-0555-76A
$C_{44}H_{29}N_5O_2$ Porphine, 5,10,15,20-tetraphenyl-, nitro derivative	$CHCl_3$	425(5.24),534(4.12), 555(3.53),600(3.49)	88-4863-76
$C_{44}H_{32}N_4O_2$ 21H-Bilin-1(19H)-one, 23,24-dihydro- 19-(hydroxyphenylmethylene)-5,10,15- triphenyl-	$CHCl_3$ $CHCl_3$-1% TFA	345(4.57),565(4.37), 587s(4.33) 375(4.44),431(4.23), 585(4.44)	88-4863-76 88-4863-76
$C_{44}H_{32}N_4O_9$ 2,4(1H,3H)-Pteridinedione, 1-(2,3,5-tri- O-benzoyl-α-D-arabinofuranosyl)- 2,4(1H,3H)-Pteridinedione, 1-(2,3,5-tri- O-benzoyl-β-D-arabinofuranosyl)-	MeOH MeOH	228(4.80),272(4.29), 357(4.12) 230(4.77),273(4.21), 360(4.14)	24-3217-76 24-3217-76

Compound	Solvent	$\lambda_{max}(\log \epsilon)$	Ref.
$C_{44}H_{33}N$ 3H-Azepine, 2,3,4,5,7-pentaphenyl- 6-(2-phenylethenyl)-	CH_2Cl_2	283(4.51),315(4.26), 375(4.16)	44-0543-76
$C_{44}H_{35}N_2O$ Pyrylium, 4-[2-[4-(dimethylamino)phenyl]- 1-(2,6-diphenyl-4-pyridinyl)ethenyl]- 2,6-diphenyl-, perchlorate	MeCN	635(4.86)	4-0577-76
$C_{44}H_{36}N_4O_2$ Pyrimido[1,2-a]benzimidazol-4(3H)-one, 2-[4-(dimethylamino)phenyl]-10-(di- phenylacetyl)-2,10-dihydro-3,3-di- phenyl-	EtOH	254(4.37),297(3.98)	94-2532-76
$C_{44}H_{38}N_4$ Quinazoline, 4,4'-(1,4-butanediyl)bis- [1,4-dihydro-2,4-diphenyl-	EtOH	234(4.64),315(4.05)	44-0497-76
$C_{44}H_{38}N_4O_6$ 2,4(1H,3H)-Pteridinedione, 6,7-diphenyl- 1-[2,3,5-tris-O-(phenylmethyl)-α-D- arabinofuranosyl]-	pH 6.0	225s(4.54),278(4.19), 364(4.12)	24-3217-76
	pH 11.0	270(4.27),363(4.14)	24-3217-76
β-	pH 6.0	225s(4.53),275(4.21), 360(4.10)	24-3217-76
	pH 11.0	265(4.29),356(4.14)	24-3217-76
2,4(1H,3H)-Pteridinedione, 6,7-diphenyl- 3-[2,3,5-tris-O-(phenylmethyl)-α-D- arabinofuranosyl]-	pH 5.0	230s(4.50),275(4.30), 368(4.22)	24-3217-76
	pH 11.0	240s(4.26),294(4.38), 392(4.07)	24-3217-76
$C_{44}H_{40}O_4P_2$ 1,4-Diphosphorinium, 1,4-dihydro- 1,1,4,4-tetrakis(2-methoxyphenyl)- 2,3-diphenyl-	MeOH	220(3.78),293(3.54)	139-0169-76A
1,4-Diphosphorinium, 1,4-dihydro- 1,1,4,4-tetrakis(4-methoxyphenyl)- 2,3-diphenyl-	MeOH	200(4.14),245(3.64)	139-0169-76A
$C_{44}H_{40}P_2$ 1,4-Diphosphorinium, 1,4-dihydro- 1,1,4,4-tetrakis(2-methylphenyl)- 2,5-diphenyl-	MeOH	200(5.06),278(4.27)	139-0169-76A
1,4-Diphosphorinium, 1,4-dihydro- 1,1,4,4-tetrakis(4-methylphenyl)- 2,5-diphenyl-	MeOH	200(5.10),227(4.74), 270(4.35)	139-0169-76A
$C_{44}H_{44}O_2$ Tricyclo[6.2.0.0^{3,6}]deca-1,3,6,8-tetra- ene, 2,7-bis(1,1-dimethylethyl)-5,10- dimethoxy-4,5,9,10-tetraphenyl-	$CHCl_3$	260(3.84),358(4.48)	138-1323-76
$C_{44}H_{46}N_{10}$ 1,2,4,5-Tetrazine-1,4-dimethanamine, tetrahydro-N,N',2,5-tetraphenyl- N,N'-bis[2-(phenylazo)ethyl]-	EtOH	244(4.48),249(4.62), 244[sic](4.67), 261(4.54)	12-1853-76
$C_{44}H_{51}CoN_4$ Cobalt, [2,3,7,8,12,13,17,18-octaethyl- 21H,23H-porphinato(2-)-N^{21},N^{22},N^{23}- N^{24}](2-phenylethenyl)-, [SP-5-31-(E)]-	$CHCl_3$	393(5.38),524(4.02), 556(4.44)	18-2529-76

Compound	Solvent	$\lambda_{max}(\log \epsilon)$	Ref.
$C_{44}H_{52}N_2O_7$ D-glycero-D-gulo-Heptitol, 2,3:6,7-di-O-cyclohexylidene-1,1-di-C-(1-methyl-1H-indol-2-yl)-4-O-(phenylmethyl)-	MeOH	285(3.88)	30-0613-76
$C_{44}H_{52}N_4O_6$ Catharanthine, 19'-vindolylde(methoxy-carbonyl)-	MeOH or EtOH	222(4.53),250(3.99), 285(3.59),292(3.89), 300(3.89)	33-2858-76
$C_{44}H_{54}N_4O_6$ Vincaleukoblastine, 3',4'-didehydro-18'-de(methoxycarbonyl)-4'-deoxy-	MeOH or EtOH	220(4.63),250(4.15), 283(3.94),291(3.94), 300(3.19)	33-2858-76
$C_{44}H_{54}O_{12}$ Gnidimacrin	MeOH	229(4.36)	35-5719-76
$C_{44}H_{55}N_5OOs$ Osmium, carbonyl(α,γ-dimethyl-α,γ-di-hydrooctaethylporphinato)(pyridine)-	benzene	439(4.79)	101-0109-76G
$C_{44}H_{55}N_5O_5$ Vincaleukoblastine, 3',4'-didehydro-3,18'-bis[de(methoxycarbonyl)]-4'-deoxy-3-[(methylamino)carbonyl]-, (18'α)-	MeOH or EtOH	256(4.15),287(3.96), 293(3.98),300(3.93)	33-2858-76
isomer	MeOH or EtOH	225(4.30),237(4.28), 252(3.86),263(3.71), 280(3.73),303(3.68)	33-2858-76
$C_{44}H_{56}N_4O_4$ Propanedioic acid, [(2,3,7,8,12,13,17-18-octaethyl-21H,23H-porphin-5-yl)-methylene]-, diethyl ester	CH_2Cl_2	408(5.22),508(4.13), 543(3.89),580(3.81), 632(3.53)	5-1537-76

Compound	Solvent	$\lambda_{max}(\log \epsilon)$	Ref.
$C_{45}H_{34}N_4O_9$ 2,4(1H,3H)-Pteridinedione, 1-methyl-6,7-diphenyl-3-(2,3,5-tri-O-benzoyl-β-D-ribofuranosyl)-	MeOH	223(4.79),273(4.18), 279s(4.16),367(4.04)	24-3194-76
2,4(1H,3H)-Pteridinedione, 3-methyl-6,7-diphenyl-3-(2,3,5-tri-O-benzoyl-β-D-ribofuranosyl)-	MeOH	228(4.80),271(4.30), 280s(4.26),358(4.16)	24-3194-76
$C_{45}H_{52}N_4O_6$ Accedinine, acetyl derivative	MeOH	286(4.07),293(4.06), 320(4.01)	24-3527-76
$C_{45}H_{53}N_3O_{16}$ Garamine, 2',4,4',5-tetra-O-acetyl-1,3,3'-tris-N-benzyloxycarbonyl-	MeOH	207(4.42)	39-1088-76C
$C_{45}H_{54}N_4O_{14}$ 21H-Biline-3,8,13,17-tetrapropanoic acid, 2,7,18-tris(2-methoxy-2-oxo-ethyl)-10,23-dihydro-12-methyl-, tetramethyl ester	CH_2Cl_2	370(4.24),455(4.68), 529(4.96)	5-1637-76
$C_{45}H_{60}O_{10}$ Olean-12-en-28-oic acid, 3-acetoxy-29-[[3-(3,4-diacetoxyphenyl)-1-oxo-2-propenyl]oxy]-, [3β,20β,29(E)]-	n.s.g.	222(3.96),280(4.09)	102-0430-76
$C_{45}H_{71}NO_{16}$ Leucomycin A_3, 2'-O-acetyl-3'-N-demethyl-3'-N-acetyl-	EtOH	232(4.33)	94-1749-76
$C_{45}H_{72}O_{36}$ β-Cyclodextrin, mono-2-propenoate	H_2O	198(4.08)	116-0701-76
$C_{45}H_{73}NO_{32}$ α-Cyclodextrin, mono[6-[(1-oxo-2-propen-yl)amino]hexanoate]	H_2O	200(4.11)	116-0701-76
$C_{45}H_{75}N_3O_6$ 9-Octadecenoic acid, [2,3,3a,9a-tetra-hydro-6-imino-3-[(1-oxo-9-octadecen-yl)oxy]-6H-furo[2',3':4,5]oxazolo-[3,2-a]pyrimidin-2-yl]methyl ester, hydrochloride	MeOH-acid	234(4.02),264(4.05)	87-0663-76
isomer	MeOH-acid	235(4.02),264(4.05)	87-0663-76
$C_{45}H_{77}N_3O_7$ 2(1H)-Pyrimidinone, 4-amino-1-[3,5-bis-O-(1-oxo-9-octadecenyl)-β-D-arabino-furanosyl]-, (E,E)-, hydrochloride	MeOH-acid	211(4.00),283(4.15)	87-0667-76
(Z,Z)-	MeOH-acid	210(4.04),283(4.17)	87-0667-76
$C_{45}H_{79}N_3O_6$ Octadecanoic acid, [2,3,3a,9a-tetrahy-dro-6-imino-3-[(1-oxooctadecyl)oxy]-6H-furo[2',3':4,5]oxazolo[3,2-a]pyrim-idin-2-yl]methyl ester, hydrochloride	MeOH-acid	235(4.03),264(4.09)	87-0663-76
$C_{45}H_{81}N_3O_7$ 2(1H)-Pyrimidinone, 4-amino-1-[2,3-bis-O-(1-oxooctadecyl)-β-D-arabino-furano-syl]-	MeOH-acid	212(4.05),283(4.19)	87-0667-76

Compound	Solvent	$\lambda_{max}(\log \epsilon)$	Ref.
$C_{46}H_{28}Br_2O_2S_4$ 1,4-Butanedione, 1,4-bis(4-bromophenyl)-2,3-bis(4,5-diphenyl-3H-1,2-dithiol-3-ylidene)-	CH_2Cl_2	483(4.51)	1-0478-76
$C_{46}H_{28}Cl_3N_7O_3$ 2-Quinazolineacetonitrile, 6-chloro-α,α-bis[(6-chloro-4-phenyl-2-quinazolinyl)methyl]-4-phenyl-, N^3,N'^3,N''^3-trioxide	EtOH	234(4.36),267(4.40), 310s(3.94),360s(3.71)	4-0433-76
$C_{46}H_{30}$ 1,2,3,4,5,7,9,11,17,19,21-Cyclodocosaundecaene-13,15-diyne, 1,6,12,17-tetraphenyl-	THF	290(4.41),306(4.47), 400(4.38),419(4.58), 442(4.74),469(5.56), 682(4.78),942(3.36)	18-2579-76
$C_{46}H_{30}Cl_4N_2O_4$ 15H-6,14a-Etheno[1,5]benzodioxepino-[3',2':3,4]pyrazolo[1,2-b]phthalazine-8,13-dione, 1,2,3,4-tetrachloro-5a,6-dihydro-6,17-bis(4-methylphenyl)-5a,18-diphenyl-, (5aα,6α,14aα)-	C_6H_{12}	218(5.083),290s(4.008), 300(4.021),335(3.858)	5-0793-76
$C_{46}H_{30}O_2S_4$ 1,4-Butanedione, 2,3-bis(4,5-diphenyl-3H-1,2-dithiol-3-ylidene)-1,4-diphenyl-	CH_2Cl_2	475(4.43)	1-0478-76
$C_{46}H_{31}Cl_4N_3O_4$ 5H-4a,12-Etheno-1H-[1,5]benzodioxepino-[3',2':3,4]pyrazolo[1,2-a][1,2,4]triazole-1,3(2H)-dione, 7,8,9,10-tetrachloro-11a,12-dihydro-12,15-bis(4-methylphenyl)-2,11a,14-triphenyl-, (4aα,11aα,12α)-	C_6H_{12}	217(4.92),290(4.00)	5-0793-76
$C_{46}H_{32}O_2$ 7,9,11,17,19,21-Cyclodocosahexaene-2,4,13,15-tetrayne-1,6-diol, 1,6,12,17-tetraphenyl-	THF	255(4.47),262(4.47), 280(4.24),348(5.03), 394(4.13),416(4.24), 441(4.23)	18-2579-76
$C_{46}H_{34}N_4NiO_2$ Nickel, bis(N-hydroxy-N'-1-naphthalenyl-N-phenylbenzenecarboximidamidato-N',O)-, (T-4)-	$CHCl_3$	305s(4.28),400s(3.60)	12-0357-76
$C_{46}H_{37}N_5O_8$ β-D-Ribofuranoside, 2-[bis(phenylmethyl)amino]-4-pteridinyl, 2,3,5-tribenzoate	MeOH	230(4.72),278(4.25), 390(3.76)	24-3208-76
$C_{46}H_{40}O_{25}$ 1,2,3-Benzenetriol, 5-[2,6-diacetoxy-4-(2,4,6-triacetoxyphenoxy)phenoxy]-4-(3,4,5-triacetoxyphenoxy)-, triacetate	MeCN	210s(--),232(4.61), 274(3.94)	102-1279-76
$C_{46}H_{44}O_2P_2Pt$ Platinum, bis(3-hydroxy-3-methyl-1-but-	$CHCl_3$	322(3.76)	101-0113-76M

Compound	Solvent	$\lambda_{max}(\log \epsilon)$	Ref.
ynyl)bis(triphenylphosphine)-, (SP-4-1)- (cont.)			101-0113-76M
$C_{46}H_{44}O_4$ Tricyclo[6.2.0.0^{3,6}]deca-2,5,7,10-tetra- ene-4,9-diol, 2,7-bis(1,1-dimethyleth- yl)-4,5,9,10-tetraphenyl-, diacetate	CHCl$_3$	268(4.37),355(4.20)	138-1323-76
$C_{46}H_{46}ClN_4O_{13}P$ Thymidine, P-(2-chlorophenyl)thymidylyl- [3'→5']-3'-O-[(4-methoxyphenyl)diphen- ylmethyl]-	MeOH	232s(4.27),265(4.32)	35-3655-76
$C_{46}H_{46}O_{20}$ [9,9'-Bianthracene]-10,10'(9H,9'H)-di- one, 3,3'-diacetyl-5,5'-bis(-D-glu- copyranosyloxy)-4,4',7,7'-tetrahy- droxy-2,2'-dimethyl-	MeOH-HOAc	252(4.49),277s(4.29), 350(4.29)	12-2225-76
$C_{46}H_{48}O_2$ Tricyclo[6.2.0.0^{3,6}]deca-1,3,6,8-tetra- ene, 2,7-bis(1,1-dimethylethyl)-5,10- diethoxy-4,5,9,10-tetraphenyl-	CHCl$_3$	264(3.92),359(4.52)	138-1323-76
$C_{46}H_{50}O_4$ Tricyclo[6.2.0.0^{3,6}]deca-1,3(6),7-tri- ene, 2,7-bis(1,1-dimethylethyl)- 4,5,9,10-tetramethoxy-4,5,9,10- tetraphenyl-	CHCl$_3$	273(3.57),284(3.40)	18-2515-76
$C_{46}H_{52}Cl_3N_3O_{17}$ Garamine, 2',4',5-tri-O-acetyl-1,3,3'- tris-N-benzyloxycarbonyl-4-O-(2,2,2- trichloroethoxycarbonyl)-	MeOH	208(4.40)	39-1088-76C
$C_{46}H_{52}N_4O_{14}$ Porphine-2,4,6,7-tetrapropanoic acid, 1,3,7-tris(2-methoxy-2-oxoethyl)-8- methyl-, tetramethyl ester	CHCl$_3$	405(5.31),501(4.17), 537(3.99),570(3.83), 624(3.58)	12-0393-76
Porphine-2,4,6,7-tetrapropanoic acid, 1,3,8-tris(2-methoxy-2-oxoethyl)-5- methyl-, tetramethyl ester	CHCl$_3$	405(5.33),501(4.19), 537(4.00),570(3.86), 624(3.60)	12-0393-76
Porphine-2,4,6,7-tetrapropanoic acid, 1,5,8-tris(2-methoxy-2-oxoethyl)-3- methyl-, tetramethyl ester	CHCl$_3$	404(5.32),501(4.18), 537(4.00),570(3.85), 624(3.59)	12-0393-76
Porphine-2,4,6,7-tetrapropanoic acid, 3,5,8-tris(2-methoxy-2-oxoethyl)- 1-methyl-, tetramethyl ester	CHCl$_3$	405(5.32),501(4.18), 537(4.00),570(3.87), 624(3.61)	12-0393-76
$C_{46}H_{54}N_4O_8$ Aspidospermidine-3-carboxylic acid, 4- acetoxy-6,7-didehydro-15-[(2α,5β,6α- 18β)-3,4-didehydro-18-(methoxycarbo- nyl)-7,16-cyclo-7,8-secoibogamin-8- yl]-3-hydroxy-16-methoxy-1-methyl-, methyl ester, (3β,4β,5α,12β,19α)-	EtOH	222(4.57),226s(4.56), 258(4.09),288(3.99), 296(3.99)	35-7017-76
	EtOH-acid	222(4.59),224s(4.57), 255(4.08),287(3.95), 295(3.95)	35-7017-76
$C_{46}H_{55}IN_4O_{14}$ 21H-Biline-3,8,13,17-tetrapropanoic acid, 10,23-dihydro-19-iodo-2,7,12- tris(2-methoxy-2-oxoethyl)-1,18-di-	CH$_2$Cl$_2$	370(4.36),460(4.42), 528(5.12)	5-1637-76

Compound	Solvent	$\lambda_{max}(\log \epsilon)$	Ref.
methyl-, tetramethyl ester, dihydro-bromide (cont.)			5-1637-76
21H-Biline-3,8,13,17-tetrapropanoic acid, 10,23-dihydro-19-iodo-2,7,18-tris(2-methoxy-2-oxoethyl)-1,12-di-methyl-, tetramethyl ester, dihydrobromide	CH_2Cl_2	460(4.54),533(4.98)	5-1637-76
$C_{46}H_{56}N_4O_8$			
Vincaleukoblastine, 3',4'-didehydro-4'-deoxy- isomer	MeOH or EtOH	212(4.74),261(4.24), 285(4.13),292(4.08)	33-2858-76
	MeOH or EtOH	215(4.5),265(4.0), 285(3.9),296(3.8)	33-2858-76
$C_{46}H_{56}N_4O_9$			
Vincaleukoblastine, 3',4'-didehydro-4'-deoxy-19'-hydroxy-	MeOH or EtOH	213(4.78),262(4.28), 283(4.17),293(4.11), 307(3.86)	33-2858-76
$C_{46}H_{56}N_4O_{14}$			
21H-Biline-3,8,13,17-tetrapropanoic acid, 10,23-dihydro-2,7,18-tris(2-oxo-2-methoxyethyl)-1,12-dimethyl-, tetramethyl ester, dihydrobromide	CH_2Cl_2	453(4.49),526(4.98)	5-1637-76
$C_{46}H_{57}N_5O_7$			
Vincaleukoblastine, 3',4'-didehydro-3-de(methoxycarbonyl)-4'-deoxy-3-[(methylamino)carbonyl]-	MeOH or EtOH	264(4.16),285(4.03), 292(4.20),302(3.88)	33-2858-76
$C_{46}H_{58}HfN_4O_4$			
Hafnium, [2,3,7,8,12,13,17,18-octaethyl-21H,23H-porphinato(2-)-N^{21},N^{22},N^{23}-N^{24}]bis(2,4-pentanedionato-O,O')-	CH_2Cl_2	400(5.47),532(4.04), 569(4.33)	24-1477-76
$C_{46}H_{58}N_4O_4Zr$			
Zirconium, [2,3,7,8,12,13,17,18-octaeth-yl-21H,23H-porphinato(2-)-N^{21},N^{22},N^{23}-N^{24}]bis(2,4-pentanedionato-O,O')-	CH_2Cl_2	401(5.46),534(4.09), 570(4.38)	24-1477-76
$C_{46}H_{58}N_4O_6$			
Tabernaelegantinine A	EtOH	224(4.74),285(4.18), 293(4.15)	39-1432-76C
Tabernaelegantinine B	EtOH	227(4.75),287(4.14), 295(4.18)	39-1432-76C
$C_{46}H_{58}N_4O_8$			
Vincaleukoblastine, 4'-deoxy-	MeOH or EtOH	225(4.55),261(4.15), 285(4.08),297(4.02), 315s(3.70)	33-2858-76
4'-epi	MeOH or EtOH	225(4.41),257(3.97), 285(3.84),295(3.81), 310s(3.54)	33-2858-76
$C_{46}H_{58}O_{27}$			
Coptiside I	MeOH	270(4.15),326(4.16)	94-0407-76
	MeOH-NaOAc	270(--),326(--)	94-0407-76
	MeOH-AlCl₃	279(--),342(--)	94-0407-76
$C_{46}H_{73}NO_{17}$			
Leucomycin A₃, 9-dehydro-18-dihydro-	EtOH	280(4.42)	94-1749-76

Compound	Solvent	$\lambda_{max}(\log \epsilon)$	Ref.
$C_{47}H_{41}BrO_4$ 2H-Cyclobuta[2,3]naphtho[1,8-bc]furan-3-carboxylic acid, 6,8b-bis(1,1-dimethylethyl)-7,8b-dihydro-2-oxo-4,5,7,8-tetraphenyl-, methyl ester	$CHCl_3$	363(4.37),382s(4.27)	77-0030-76
$C_{47}H_{41}IN_2O_2$ Benzo[f]quinolinium, 3-(4-methoxyphenyl)-1-[3-[3-(4-methoxyphenyl)-2,4-dimethylbenzo[f]quinolin-1(4H)-ylidene]-1-propenyl]-2,4-dimethyl-, iodide	EtOH	768(4.38)	103-1012-76
$C_{47}H_{56}N_4O_3Zr$ Zirconium, [2,3,7,8,12,13,17,18-octaethyl-21H,23H-porphinato(2-)-N^{21},N^{22},N^{23}-N^{24}](2,4-pentanedionato-O,O')phenoxy-	CH_2Cl_2	405(5.40),534(4.13), 571(4.48)	24-1477-76
$C_{47}H_{56}N_4O_{12}$ Porphine-2,4,6,7-tetrapropanoic acid, α-acetoxy-8-(ethoxycarbonylmethyl)-1,3,5-trimethyl-, 6,7-diethyl 2,4-dimethyl ester	$CHCl_3$	405(5.31),502(4.21), 536(3.82),572(3.81), 627(3.21)	12-0393-76
$C_{47}H_{56}N_4O_{16}$ 21H-Biline-3,8,13,17-tetrapropanoic acid, 10,23-dihydro-2,7,12,18-tetrakis(2-methoxy-2-oxoethyl)-, tetramethyl ester, dihydrobromide	CH_2Cl_2	370(4.23),456(4.67), 528(4.96)	5-1637-76
$C_{47}H_{58}N_2O_{12}$ Rifamycin, 1,4-dideoxy-1,4-dihydro-1,4-dioxo-3-[(4-phenylbutyl)amino]-	MeOH	262(4.40),315(4.07), 370(3.79),534(3.17)	35-7064-76
$C_{47}H_{58}N_4O_{14}$ 21H-Biline-3,8,13,17-tetrapropanoic acid, 10,23-dihydro-2,7,18-tris(2-methoxy-2-oxoethyl)-1,12,19-trimethyl-, tetramethyl ester, dihydrobromide	CH_2Cl_2	368(4.11),451(4.90), 527(5.09)	5-1637-76
$C_{47}H_{80}O_{13}$ α-D-Glucopyranoside, (3β,11β,20β)-1,29-dihydroxyolean-12-en-3-yl 3,4-di-O-methyl-2-O-(2,3,4-tri-O-methyl-β-D-glucopyranosyl)-	EtOH	205(3.94)	65-0912-76

Compound	Solvent	$\lambda_{max}(\log \epsilon)$	Ref.
$C_{48}H_{28}N_2O_2$ Naphthaceno[2,3-g]phthalazine-8,13-di- one, 1,4,7,14-tetraphenyl-	CH_2Cl_2	253(4.58),293(4.89), 340(4.59),450(4.09), 472(4.07),504(3.74)	28-0555-76A
$C_{48}H_{28}O_4$ 5,14-Pentacenedione, 9,10-dibenzoyl- 6,13-diphenyl-	CH_2Cl_2	255(4.71),315(4.85), 424(4.08),443(4.08)	28-0555-76A
$C_{48}H_{30}O_5$ 6,13-Epoxypentacene-5,14-dione, 9,10- dibenzoyl-5a,6,13,13a-tetrahydro- 6,13-diphenyl-	CH_2Cl_2	263(4.86),347(3.63)	28-0555-76A
5,12[3',4']-Furanonaphthacene-6,11-di- one, 2,3-dibenzoyl-5,5a,11,12-tetra- hydro-14,16-diphenyl-	CH_2Cl_2	245(4.76),332(4.56)	28-0555-76A
$C_{48}H_{33}OP$ 9H-Tribenzo[b,d,f]phosphepin, 1,2,3,4,9- pentaphenyl-, 9-oxide, 9(a)	$CHCl_3$	257(4.56),285s(4.12)	24-2405-76
9(e)	$CHCl_3$	260(4.54),290s(4.10)	24-2405-76
$C_{48}H_{33}P$ 9H-Tribenzo[b,d,f]phosphepin, 1,2,3,4,9- pentaphenyl-, 9(a)	$CHCl_3$	257(4.79),285s(4.22)	24-2405-76
9(e)	$CHCl_3$	257(4.70),282s(4.33)	24-2405-76
$C_{48}H_{34}Hg_2N_4O_4$ Mercury, bis(acetato)[5,10,15,20-tetra- phenyl-21H,23H-porphinato(2-)]di-	CH_2Cl_2- 0.5% THF	447(5.63),525s(3.49), 558(3.92),581(4.06), 603s(4.02),641(3.68)	78-0597-76
$C_{48}H_{38}N_4$ Propanedinitrile, 2,2'-[[2,5-bis(1,1- dimethylethyl)-7,8-diphenylbicyclo- [4.2.0]octa-1,3,5,7-tetraene-3,4- diyl]bis(phenylmethylidyne)]bis-, anti	$CHCl_3$	266(4.58),309(4.50), 410s(3.34)	77-0030-76
syn	$CHCl_3$	258(4.53),310s(3.95), 410(3.11)	77-0030-76
$C_{48}H_{40}I_2P_2$ Phosphonium, [1,3-azulenediylbis(meth- ylene)]bis[triphenyl-, diiodide	EtOH	270s(4.32),278s(4.45), 288(4.54),339(3.74), 348(3.74),366(3.85), 575(3.00),610s(2.64), 680s(2.19)	117-0169-76
$C_{48}H_{42}N_2O_4S$ Thymidine, 4-thio-3',5'-bis-O-(triphen- ylmethyl)-	n.s.g.	205(4.81),233s(4.21), 334(4.27)	128-0351-76
$C_{48}H_{44}O_4$ Cyclobuta[b]naphthalene-4,5-dicarbox- ylic acid, 3,8-bis(1,1-dimethyleth- yl)-1,2,6,7-tetraphenyl-, dimethyl ester	$CHCl_3$	262s(4.50),335(4.53), 400(3.75)	77-0030-76
$C_{48}H_{46}O_4$ Cyclobuta[b]naphthalene-4,5-dicarbox- ylic acid, 3,8-bis(1,1-dimethyleth-	THF	414(3.32),316(4.58)	77-0177-76

Compound	Solvent	λ_{max}(log ϵ)	Ref.
yl)-4,5-dihydro-1,2,6,7-tetraphenyl-, dimethyl ester (cont.)			77-0177-76
Tetracyclo[4.4.2.0^{1,4}.0^{6,9}]dodeca-2,4,7-9-tetraene-11,12-dicarboxylic acid, 5,10-bis(1,1-dimethylethyl)-2,3,7,8-tetraphenyl-, dimethyl ester, anti	CHCl$_3$	250(4.31),320(4.49)	77-0177-76
syn	CHCl$_3$	248(4.32),320(4.55)	77-0177-76
$C_{48}H_{48}O_8P_2$ 1,4-Diphosphorinium, 1,1,4,4-tetrakis-(2,6-dimethoxyphenyl)-1,4-dihydro-2,5-diphenyl-, dichloride	MeOH	201(4.12),252(5.30)	139-0169-76A
$C_{48}H_{50}N_8$ 29H,31H-Phthalocyanine, 2,8,12,18-tetra-kis(1,1-dimethylethyl)-, radical cation	n.s.g.	690(4.95),750(5.29)	104-0119-76
oxidation product	n.s.g.	560(4.49),890(4.30), 910(4.48)	104-0119-76
$C_{48}H_{51}$ Cyclopropenylium, 1,2,3-tris[3,8-dimeth-yl-5-(1-methylethyl)-1-azulenyl]-, perchlorate	MeCN	240s(4.51),290(4.41), 335(4.31),483(4.42)	44-2379-76
$C_{48}H_{51}N_7O_9S_2$ 5-Thia-1-azabicyclo[4.2.0]oct-2-ene-2-carboxylic acid, 7-[[5-[[(1,1-dimeth-ylethoxy)carbonyl]amino]-6-(diphenyl-methoxy)-1,6-dioxohexyl]amino]-7-meth-oxy-3-[[(1-methyl-1H-tetrazol-5-yl)-thio]methyl]-8-oxo-, diphenylmethyl ester, [6R-(6α,7α,7(R*))]-	THF	259(3.90)	94-2629-76
$C_{48}H_{52}N_4P_2$ 1,4-Diphosphorinium, 1,1,4,4-tetrakis-[4-(dimethylamino)phenyl]-1,4-di-hydro-2,5-diphenyl-	MeOH	214(3.91),310(4.02)	139-0169-76A
$C_{48}H_{54}N_4O_{16}$ 21H,23H-Porphine-2,7,12,18-tetrapropan-oic acid, 20-acetoxy-3,8,13-tris(2-methoxy-2-oxoethyl)-17-methyl-, tetramethyl ester	CHCl$_3$	408(5.37),504(4.24), 538(3.79),575(3.81), 630(3.17)	12-0393-76
$C_{48}H_{57}IN_4O_{16}$ 21H-Biline-3,8,13,17-tetrapropanoic acid, 10,23-dihydro-19-iodo-2,7,12,18-tetrakis(2-methoxy-2-oxoethyl)-1-meth-yl-, tetramethyl ester, dihydrobrom-ide	CH$_2$Cl$_2$	455(4.50),530(4.95)	5-1637-76
$C_{48}H_{58}N_4O_{16}$ 21H-Biline-3,8,13,17-tetrapropanoic acid, 10,23-dihydro-2,7,12,18-tetra-kis(2-methoxy-2-oxoethyl)-1-methyl-, tetramethyl ester, dihydrobromide	CH$_2$Cl$_2$	524(5.04)	5-1637-76
$C_{48}H_{66}N_6O_2$ 38H-29,32-Imino-27,24-metheno-2,25-(metheno[2,5]-endo-pyrrolometheno)-	CHCl$_3$	401(4.85),500(3.97), 535(3.86),568(3.62),	88-4477-76

Compound	Solvent	$\lambda_{max}(\log \epsilon)$	Ref.
2H-pyrrolo[2,3-o][1,7,20]triazacyclo-dotriacontine-6,21-dione, 30,31,40,41-tetraethyl-4,5,7,8,9,10,11,12,13,14-15,16,17,18,19,20,22,23-octadecahydro-3,35-dimethyl-		622(3.56)	88-4477-76
zinc complex	$CHCl_3$	401(5.32),538(4.01), 571(4.11)	88-4477-76
$C_{48}H_{68}N_2O_4$			
Androst-5-en-3-ol, 17,17'-[2,5-pyrazine-diylbis(methylene)]bis-, diacetate, (3β)-(3'β)-	EtOH	210(4.07),279(3.88), 308s(3.23)	33-2138-76
$C_{48}H_{68}N_6O_2$			
21H,23H-Porphine-2,12-dipropanamide, 7,8,17,18-tetraethyl-N,N'-dihexyl-3,13-dimethyl-	$CHCl_3$	401(4.84),500(3.97), 535(3.88),568(3.62), 622(3.55)	88-4477-76
zinc complex	$CHCl_3$	401(5.33),538(4.01), 571(4.13)	88-4477-76

Compound	Solvent	$\lambda_{max}(\log \epsilon)$	Ref.
$C_{49}H_{32}N_5Zn$ Zinc(1+), [1-(5,10,15,20-tetraphenyl-21H,23H-porphin-2-yl)pyridiniumato-(2-)-$N^{21},N^{22},N^{23},N^{24}$]-, (SP-4-2)-, perchlorate	CH_2Cl_2	427(5.43),520s(3.76), 555(4.22),594(4.11)	77-0236-76
$C_{49}H_{34}N_5$ Pyridinium, 1-(5,10,15,20-tetraphenyl-21H,23H-porphin-2-yl)-, perchlorate	CH_2Cl_2	420(5.43),523(4.24), 600(3.87),655(4.06)	77-0236-76
$C_{49}H_{35}NO$ 5-Azaspiro[2.3]hexan-4-one, 1,2-bis(diphenylmethylene)-5,6,6-triphenyl-	$CHCl_3$	263(4.48),383(4.38)	18-3173-76
$C_{49}H_{36}$ 1,3,5-Cycloheptatriene, 1,2,3,4,5,6,7-heptaphenyl-	$C_2H_4Cl_2$	268(4.44),3υ0s(--), 312s(4.17)	44-3374-76
$C_{49}H_{36}P$ 9H-Tribenzo[b,d,f]phosphepinium, 9-methyl-1,2,3,4,9-pentaphenyl-, iodide	MeOH	220(4.90),260s(4.42), 290s(4.09)	24-2405-76
$C_{49}H_{47}N_{15}O_{27}$ Sisomicin, 1,2',3,3'',6'-pentakis-N-(2,4-dinitrophenyl)-	acetone	351(4.88)	39-1088-76C
$C_{49}H_{58}N_4O_{18}$ 21H-Biline-3,8,13,17-tetrapropanoic acid, 22,23-dihydro-2,7,18-tris(2-methoxy-2-oxoethyl)-1,19-bis(methoxycarbonyl)-12-methyl-, tetramethyl ester	CH_2Cl_2	307(4.26),397(4.44), 710(3.80)	5-1637-76
$C_{49}H_{60}N_4O_{16}$ 21H-Biline-3,8,13,17-tetrapropanoic acid, 10,23-dihydro-2,7,12,18-tetrakis(2-methoxy-2-oxoethyl)-1,19-dimethyl-, tetramethyl ester, dihydrobromide	CH_2Cl_2	367(4.18),450(4.90), 527(5.11)	5-1637-76
$C_{49}H_{87}N_3O_6$ Eicosanoic acid, [2,3,3a,9a-tetrahydro-6-imino-3-[(1-oxoeicosyl)oxy]-6H-furo[2',3':4,5]oxazolo[3,2-a]pyrimidin-2-yl]methyl ester, hydrochloride, [2R-(2α,3β,3aα,9aβ)]-	MeOH-acid	235(4.03),263(4.07)	87-0663-76
$C_{49}H_{89}N_3O_7$ 2(1H)-Pyrimidinone, 4-amino-1-[3,5-bis-O-(1-oxoeicosyl)-β-D-arabinofuranosyl]-	MeOH-acid	213(3.98),283(4.15)	87-0667-76

Compound	Solvent	$\lambda_{max}(\log \epsilon)$	Ref.
$C_{50}H_{42}Br_2P_2S_2$ 1,3-Butadienylenebis(2,5-thienylmethyl-ene)di(triphenylphosphonium) dibrom-ide	DMF	363(4.24),382(4.31), 404(4.23)	126-1857-76
$C_{50}H_{46}Br_4N_2$ Tricyclo[6.2.0.0^{3,6}]deca-2,5,7,10-tetra-ene-4,9-diacetonitrile, 4,5,9,10-tet-rakis(4-bromophenyl)-2,7-bis(1,1-di-methylethyl)-$\alpha,\alpha,\alpha',\alpha'$-tetramethyl-	CHCl$_3$	287(4.06),372(4.56)	138-0657-76
$C_{50}H_{48}N_2O_3$ 1H-Indole, 3,3'-[oxybis(phenylmethyl-ene)]bis[5-methoxy-2-phenyl-1-propyl-	EtOH	218(4.80),300(4.32)	42-0606-76
$C_{50}H_{50}Cl_3N_4O_{15}P$ Thymidine, 5'-O-(phenoxyacetyl)-P-(2,2,2-trichloroethyl)thymidylyl-(3'→5')-3'-O-[(4-methoxyphenyl)-diphenylmethyl]-	MeOH	232s(4.28),264(4.33)	35-3655-76
$C_{50}H_{50}N_2$ Tricyclo[6.2.0.0^{3,6}]deca-2,5,7,10-tetra-ene-4,9-diacetonitrile, 2,7-bis(1,1-dimethylethyl)-$\alpha,\alpha,\alpha',\alpha'$-tetramethyl-4,5,9,10-tetraphenyl-	CHCl$_3$	282(4.01),387(4.64)	138-0657-76
$C_{50}H_{54}N_4O_4$ 21H,23H-Porphine-5,15-diol, 2,3,7,8,12-13,17,18-octaethyl-, dibenzoate	CHCl$_3$	409(5.27),506(4.22), 536(3.58),580(3.76), 631(2.76)	39-0794-76C
$C_{50}H_{78}F_2N_2O_{17}$ Leucomycin V, 17,17-difluoro-18-(1-pyr-rolidinyl)-, 2^A,3,9-triacetate 4^B-(3-methylbutanoate)	MeOH	231(4.46)	78-0991-76
$C_{51}H_{35}N_8P$ Hydrazyl, 1,1'-[2,4,6-tricyano-5-[(tri-phenylphosphoranylidene)amino]-1,3-phenylene]bis[2,2-diphenyl-	CH$_2$Cl$_2$	311(4.65),444(4.13), 460s(4.12),604(3.80)	5-1739-76
$C_{51}H_{57}IO_{13}$ Gnidimacrin, 20-(4-iodobenzoate)	MeOH	229(4.41),255(4.24)	35-5719-76
$C_{51}H_{83}NO_{37}$ β-Cyclodextrin, mono[6-[(1-oxo-2-prop-enyl)amino]hexanoate]-	H$_2$O	200(4.11)	116-0701-76
$C_{52}H_{34}O_2$ 5,10:13,18-Diepoxybenzo[3,4]cycloocta-[1,2-b:5,6-b']dinaphthalene, 5,10,13,18-tetrahydro-5,10,13,18-tetraphenyl-	dioxan	228(4.26),240(4.39), 258(3.96),265s(3.91), 272s(3.77),282s(3.51)	89-0117B-76
$C_{52}H_{48}N_8S_2$ Benzenecarbohydrazonic acid, N-phenyl-, [dithiobis(phenylmethylene)]bis(phen-ylhydrazide)	CHCl$_3$	300(4.60),344(4.90)	39-0038-76C

Compound	Solvent	$\lambda_{max}(\log \epsilon)$	Ref.
$C_{52}H_{48}IrP_4$ Iridium, [1,2-bis(diphenylphosphino)- ethyl]-, chloride	pentanol	442(3.58),522(2.81)	32-1013-76
$C_{52}H_{52}N_2O_3$ 1H-Indole, 3,3'-[oxybis(phenylmethyl- ene)]bis[1-butyl-5-methoxy-2-phenyl-	EtOH	225(4.65),300(4.37)	42-0606-76
1H-Indole, 3,3'-[oxybis(phenylmethyl- ene)]bis[5-methoxy-1-(2-methylprop- yl)-2-phenyl-	EtOH	226(4.74),300(4.46)	42-0606-76
$C_{52}H_{52}O_2P_2Pt$ Platinum, bis[(1-hydroxycyclohexyl)eth- ynyl]bis(triphenylphosphine)-, (SP-4-1)-	$CHCl_3$	322(3.83)	101-0113-76M
$C_{52}H_{58}N_4O_2$ 41H-29,32-Imino-24,27-nitrilo-23,33- ([2,5]-endo-pyrrolomethenopyrrol[2]- yl[5]ylidene)-27H-dibenzo[a,b][1,14]- dioxacyclononacosin, 6,7,8,9,10,11,12-	CH_2Cl_2	409(5.26),507(4.12), 541(3.64),576(3.76), 628(3.08)	77-0881-76
13,14,15,16,17-dodecahydro-25,26,30- 31,37,38,43,44-octamethyl-	CH_2Cl_2- CF_3COOH	430(--),568(--), 610(--)	77-0881-76
$C_{52}H_{73}Cl_2CoN_4O_{14}$ Cobyrinic acid, dichloride, heptamethyl ester	CH_2Cl_2	264s(4.23),327s(4.21), 348(4.25),419s(3.77), 480(3.94),514s(3.91)	5-1150-76
$C_{52}H_{81}N_7O_{16}$ Echincandin B	MeOH	194(5.00),226s(4.12), 276(3.16)	88-4147-76
$C_{53}H_{68}ClCoN_6O_{14}$ Cobyrinic acid, 10-chloro-8-hydroxy- dicyanide, γ-lactone, hexamethyl ester	CH_2Cl_2	294(3.93),318(3.87), 334(3.87),355s(4.09), 370(4.40),500s(3.54), 535s(3.70),574(3.86), 615(3.93)	5-1150-76
$C_{53}H_{74}O_{11}$ Gnidilatidin, 20-hexadecanoate	MeOH	232(4.61)	44-3850-76
$C_{53}H_{78}O_{11}$ Gnidilatin, 20-hexadecanoate	MeOH	231(4.26)	44-3850-76
$C_{53}H_{91}N_3O_6$ 13-Docosenoic acid, [2,3,3a,9a-tetrahy- dro-6-imino-3-[(1-oxo-1,3-docosenyl)- oxy]-6H-furo[2',3':4,5]oxazolo[3,2-a]- pyrimidin-2-yl]methyl ester, hydro- chloride	MeOH-acid	233(4.00),263(4.04)	87-0663-76
$C_{53}H_{93}N_3O_7$ 2(1H)-Pyrimidinone, 4-amino-1-[3,5-bis- O-(1-oxo-13-docosenyl)-β-D-arabino- furanosyl]-	MeOH-acid	211(4.00),283(4.15)	87-0667-76
$C_{53}H_{95}N_3O_6$ Docosanoic acid, [2,3,3a,9a-tetrahydro- 6-imino-3-[(1-oxodocosyl)oxy]-6H-	MeOH-acid	235(4.03),263(4.09)	87-0663-76

Compound	Solvent	λ_{max}(log ϵ)	Ref.
furo[2',3':4,5]oxazolo[3,2-a]pyrimidin-2-yl]methyl ester, hydrochloride (cont.)			87-0663-76
$C_{53}H_{97}N_3O_7$ 2(1H)-Pyrimidinone, 4-amino-1-[3,5-bis-O-(1-oxodocosyl)-β-D-arabinofuranosyl]-	MeOH-acid	213(3.98),283(4.15)	87-0 67-76
$C_{54}H_{48}O_2$ Tricyclo[6.2.0.0^{3,6}]deca-1,3,6,8-tetraene, 2,7-bis(1,1-dimethylethyl)-5,10-diphenoxy-4,5,9,10-tetraphenyl-	CHCl$_3$	360(4.58)	138-1323-76
$C_{54}H_{50}N_2$ Tricyclo[6.2.0.0^{3,6}]deca-2,5,7,10-tetraene-4,9-diamine, 2,7-bis(1,1-dimethylethyl)-N,N',4,5,9,10-hexaphenyl-	CHCl$_3$	260s(4.48),358(4.62)	138-1323-76
$C_{54}H_{53}N_5O_{23}$ D-Allaric acid, 2-O-(2,3,6-tri-O-acetyl-α-D-glucopyranosyl)-, 1,4-lactone, 6-methyl ester, 3,5-dibenzoate, 4' 5'-anhydride with N-benzoyl-adenosine, 2',3'-diacetate	MeOH	232(4.16),278(4.08)	73-0800-76
$C_{54}H_{60}N_4O_4$ Propanedioic acid, [(2,3,7,8,12,13,17-18-octaethyl-21H,23H-porphin-5-yl)-methylene]-, bis(phenylmethyl) ester	CH$_2$Cl$_2$	408(5.18),508(4.10), 542(3.88),579(3.81), 632(3.54)	5-1537-76
$C_{54}H_{72}ClCoN_6O_{14}$ Cobyrinic acid, 10-chloro-, dicyanide, heptamethyl ester	CH$_2$Cl$_2$	290(3.94),308(3.90), 321(3.94),356s(4.09), 373(4.40),430s(3.48), 502s(3.66),535s(3.73), 570(3.87),608(3.94)	5-1150-76
$C_{54}H_{86}$ 3,3'-Bicholesta-2,4-diene	CHCl$_3$	272(4.55),280(4.61), 292(4.48)	94-0487-76
3,3'-Bicholesta-3,5-diene	CHCl$_3$	298(4.70),312(4.80), 328(4.67)	94-0487-76
	CHCl$_3$	298(4.68),312(4.80), 327(4.66)	94-0859-76
Cholesta-2,4,6-triene, 3-cholest-5-en-3-yl-	isoPrOH	302(4.82),315(4.90), 330(4.70)	94-0375-76
$C_{56}H_{44}Br_4O_4$ Tricyclo[6.2.0.0^{3,6}]deca-2,5,7,10-tetraene-4,9-diol, 4,5,9,10-tetrakis(4-bromophenyl)-2,7-bis(1,1-dimethylethyl)-, dibenzoate	CHCl$_3$	275(4.13),360(4.56)	138-0657-76
$C_{56}H_{48}O_4$ Tricyclo[6.2.0.0^{3,6}]deca-2,5,7,10-tetraene-4,9-diol, 2,7-bis(1,1-dimethylethyl)-4,5,9,10-tetraphenyl-, dibenzoate	CHCl$_3$	265(4.08),360(4.59)	138-0657-76 +138-1323-76

Compound	Solvent	$\lambda_{max}(\log \epsilon)$	Ref.
$C_{56}H_{48}O_{30}$ 1,2,3-Benzenetriol, 5-[2,6-diacetoxy-4-(2,4,6-triacetoxyphenoxy)phenoxy]-4-[3,5-diacetoxy-4-(3,4,5-triacetoxyphenoxy)phenoxy]-, triacetate	MeCN	215s(--),232(4.71), 275(4.03)	102-1279-76
$C_{56}H_{54}N_2$ Tricyclo[6.2.0.0^{3,6}]deca-2,5,7,10-tetraene-4,9-diamine, 2,7-bis(1,1-dimethylethyl)-N,N'-bis(4-methylphenyl)-4,5,9,10-tetraphenyl-	CHCl$_3$	356(4.60)	138-1323-76
$C_{56}H_{56}BN_4P$ Phosphonium, bis[4,4'-bis(dimethylamino)-2,2'-biphenylylene]-, tetraphenylborate	MeOH	260(4.93),300(4.69), 334(4.74),350s(4.72), 490(3.59)	139-0151-76A
$C_{57}H_{48}O$ Ethenone, [2,5-bis(1,1-dimethylethyl)-7,8-diphenyl-4-(1,2,3-triphenyl-2-cyclopropen-1-yl)bicyclo[4.2.0]octa-1,3,5,7-tetraen-3-yl]phenyl-	CHCl$_3$	456(3.11)	77-1010-76
1,2,3b-Metheno-3bH-cyclobuta[a]cyclopenta[1,3]cyclopropa[1,2-d]benzen-3(3aH)-one, 4,7-bis(1,1-dimethylethyl)-1,2-dihydro-1,2,3a,5,6,8-hexaphenyl-	CHCl$_3$	244(4.39),306(4.29)	77-1010-76
$C_{57}H_{58}N_4O_6$ 21H,23H-Porphine-5,10,15-triol, 2,3,7,8,12,13,17,18-octaethyl-, tribenzoate	CHCl$_3$	415(5.22),511(4.16), 544(3.49),588(3.66), 648(2.79)	39-0794-76C
$C_{57}H_{66}O_{26}$ β-D-Glucopyranose, O-2,3,4,6-tetra-O-acetyl-α-D-glucopyranosyl-(1→4)-O-2,3,6-tri-O-acetyl-α-D-glucopyranosyl-(1→4)-6-O-(triphenylmethyl)-, triacetate	MeOH	259(2.83)	136-0282-76C
$C_{58}H_{48}N_2O_3$ 1H-Indole, 3,3'-[oxybis(phenylmethylene)]bis[5-methoxy-2-phenyl-1-(phenylmethyl)-	EtOH	224(4.69),300(4.35)	42-0606-76
$C_{58}H_{52}N_2O_3$ 1H-Benz[g]indole, 3,3'-[oxybis(phenylmethylene)]bis[5-methoxy-2-phenyl-1-propyl-	EtOH	222(4.74),280(4.88)	42-0606-76
$C_{58}H_{52}O_2$ 2,4-Cyclohexadien-1-one, 2-(1,1-dimethylethyl)-6-[5-(1,1-dimethylethyl)-6-oxo-3-(triphenylmethyl)-2,4-cyclohexadien-1-ylidene]-4-(triphenylmethyl)-	isooctane at 70°	284s(3.82),353(3.80), 547(3.89)(changing)	88-4883-76
Spiro[3,5-cyclohexadiene-1,8'-[7]oxabicyclo[4.2.0]octa[1,3,5]trien]-2-one, 3,5'-bis(1,1-dimethylethyl)-3',5-bis(triphenylmethyl)-	isooctane at 70°	274s(4.02),284s(3.95), 322s(3.40)	88-4883-76
$C_{58}H_{70}O_{16}$ Gl 5 octaacetate	EtOH	236(4.38)	35-3007-76

Compound	Solvent	$\lambda_{max}(\log \epsilon)$	Ref.
$C_{58}H_{80}O_4$ 2,5-Cyclohexadien-1-one, 4,4'-[1,2-bis-[3,5-bis(1,1-dimethylethyl)-4-hydroxyphenyl]-1,2-ethanediylidene]bis[2,6-bis(1,1-dimethylethyl)-, ion(2-) radical anion	MeOH	445(4.68),507(--), 613(--)	64-0965-76B
	MeOH	389(--),415(--), 443(4.93),860(--)	64-0965-76B
$C_{58}H_{82}O_4$ 2,5-Cyclohexadien-1-one, 4,4'-[1,2-bis-[3,5-bis(1,1-dimethylethyl)-4-hydroxyphenyl]-1,2-ethanediylidene]bis[2,6-bis(1,1-dimethylethyl)-	MeOH	392(4.49),490(--)	64-0965-76B
$C_{59}H_{94}O_{27}$ Holotoxin A	EtOH	none above 210 nm	94-0266-76
$C_{60}H_{49}N_5O_{16}$ 7(8H)-Pteridinone, 2-(dimethylamino)-8-(2,3,5-tri-O-benzoyl-β-D-ribofuranosyl)-4-[(2,3,5-O-benzoyl-β-D-ribofuranosyl)oxy]-	MeOH	229(4.96),272(3.99), 279(3.98),364(4.17)	24-3228-76
β-D-Ribofuranoside, 2-(dimethylamino)-4,7-pteridinediylbis-, 2,2',3,3',5,5'-hexabenzoate	MeOH	230(4.92),277(4.13), 367(3.94)	24-3228-76
$C_{60}H_{56}N_2O_3$ 1H-Benz[g]indole, 3,3'-[oxybis(phenylmethylene)]bis[1-butyl-5-methoxy-2-phenyl-	EtOH	224(4.64),280(4.82)	42-0606-76
1H-Benz[g]indole, 3,3'-[oxybis(phenylmethylene)]bis[5-methoxy-1-(2-methylpropyl)-2-phenyl-	EtOH	225(4.65),280(4.84)	42-0606-76
$C_{60}H_{71}N_5O_{18}$ D-Streptamine, O-3-deoxy-4-C-methyl-3-(methylamino)-β-L-arabino-pyranosyl-5-C-methoxy-β-L-threo-hexpyranosyl-(1→4)-2-deoxy-, pentakis(N-benzyloxycarbonyl) derivative	MeOH	208(4.44)	39-1088-76C
$C_{60}H_{84}O_{13}$ Gnidimacrin, 20-hexadecanoate	MeOH	229(4.40)	35-5719-76
$C_{60}H_{84}O_{28}$ Oleandrigenin, β-gentiobiosyl(1→4)-β-D-digitaloside nonaacetate	MeOH	218(4.30)	102-1275-76
$C_{60}H_{86}N_4$ 21H,23H-Porphine, 5,10,15,20-tetra-9-decenyl-	CHCl$_3$	420(5.20),521(4.11), 558(3.94),600(3.54), 660(3.60)	5-2058-76
$C_{60}H_{86}N_4O_8$ 21H,23H-Porphine-5,10,15,20-tetranonanoic acid, tetramethyl ester	CHCl$_3$	420(5.36),518(3.86), 554(3.70),598(3.48), 660(3.60)	5-2058-76
$C_{60}H_{86}N_4O_{12}S_4$ 1-Decene-1-sulfonic acid, 10,10',10''-10'''-(21H,23H-porphine-5,10,15,20-	H$_2$O	421(5.34),523(4.26), 558(4.08),600(3.78),	5-2058-76

Compound	Solvent	$\lambda_{max}(\log \epsilon)$	Ref.
tetrayl)tetrakis-, tetrasodium salt (cont.)		660(3.63)	5-2058-76
$C_{61}H_{66}N_4O_{18}$ 21H-Biline-3,8,13,17-tetrapropanoic acid, 22,23-dihydro-2,7,18-tris-(2-methoxy-2-oxoethyl)-12-methyl-1,19-bis[(phenylmethoxy)carbonyl]-, tetramethyl ester	CH_2Cl_2	308(4.24),398(4.48), 700(3.69)	5-1637-76
$C_{62}H_{84}ClFeN_8O_8$ Heminbis(ω-aminoenanthyl)-ω-aminoenanthic acid	$CHCl_3$-EtOH	403(4.15),500(4.13), 540(4.05),630(3.79)	104-1859-76
$C_{63}H_{83}NO_4$ 2,5-Cyclohexadien-1-one, 4,4'-[2,4-pyridinediylbis[[3,5-bis(1,1-dimethylethyl)-4-hydroxyphenyl]methylidyne]]bis-[2,6-bis(1,1-dimethylethyl)-, ion(2-) dianion diradical	MeOH	630(4.95)	64-0965-76B
	MeOH	470(4.95)	64-0965-76B
2,5-Cyclohexadien-1-one, 4,4'-[2,5-pyridinediylbis[[3,5-bis(1,1-dimethylethyl)-4-hydroxyphenyl]methylidyne]]bis-[2,6-bis(1,1-dimethylethyl)-, ion(2-) dianion diradical	MeOH	424(--),625(4.90)	64-0965-76
	MeOH	399(--),468(4.82)	64-0965-76B
2,5-Cyclohexadien-1-one, 4,4'-[2,6-pyridinediylbis[[3,5-bis(1,1-dimethylethyl)-4-hydroxyphenyl]methylidyne]]bis-[2,6-bis(1,1-dimethylethyl)-, ion(2-) dianion diradical	MeOH	612(5.05)	64-0965-76B
	MeOH	454(4.93)	64-0965-76B
2,5-Cyclohexadien-1-one, 4,4'-[3,5-pyridinediylbis[[3,5-bis(1,1-dimethylethyl)-4-hydroxyphenyl]methylidyne]]bis-[2,6-bis(1,1-dimethylethyl)-, ion(2-) dianion diradical	MeOH	399(--),621(5.06)	64-0965-76B
	MeOH	468(4.99)	64-0965-76B
$C_{63}H_{85}NO_4$ 2,5-Cyclohexadien-1-one, 4,4'-[2,4-pyridinediylbis[[3,5-bis(1,1-dimethylethyl)-4-hydroxyphenyl]methylidyne]]bis-[2,6-bis(1,1-dimethylethyl)-	MeOH	406(4.63)	64-0965-76B
2,5-Cyclohexadien-1-one, 4,4'-[2,5-pyridinediylbis[[3,5-bis(1,1-dimethylethyl)-4-hydroxyphenyl]methylidyne]]bis-[2,6-bis(1,1-dimethylethyl)-	MeOH	417(4.62)	64-0965-76B
2,5-Cyclohexadien-1-one, 4,4'-[2,6-pyridinediylbis[[3,5-bis(1,1-dimethylethyl)-4-hydroxyphenyl]methylidyne]]bis-[2,6-bis(1,1-dimethylethyl)-	MeOH	405(4.63)	64-0965-76B
2,5-Cyclohexadien-1-one, 4,4'-[3,5-pyridinediylbis[[3,5-bis(1,1-dimethylethyl)-4-hydroxyphenyl]methylidyne]]bis-[2,6-bis(1,1-dimethylethyl)-	MeOH	410(4.67)	64-0965-76B
$C_{64}H_{62}N_4O_8$ 21H,23H-Porphine-5,10,15,20-tetrol, 2,3,7,8,12,13,17,18-octaethyl-, tetrabenzoate	$CHCl_3$	422(5.27),518(4.17), 556(3.69),597(3.66), 655(3.11)	39-0794-76C

Compound	Solvent	λ_{max} (log ϵ)	Ref.
...-1-one, 4,4'-[1,6-naph-[3,5-bis(1,1-dimethyl-xyphenyl]methylidyne]]-1-dimethylethyl)-, ion	MeOH	401(--),609(5.00)	64-0965-76B
...dical	MeOH	450(4.96)	64-0965-76B
	MeOH	431(--),610(4.99)	64-0965-76B
...-1-one, 4,4'-[2,6-naph-dien-1-one, 4,4'-[2,6-naph-bis[[3,5-bis(1,1-dimethyl-droxyphenyl]methylidyne]]-s(1,1-dimethylethyl)-, ion	MeOH	393(--),456(4.91)	64-0965-76B
...iradical	MeOH	414(4.71)	64-0965-76B
...hexadien-1-one, 4,4'-[1,6-naph-ediylbis[[3,5-bis(1,1-dimethyl-]-4-hydroxyphenyl]methylidyne]]-,6-bis(1,1-dimethylethyl)-	MeOH	427(4.70)	64-0965-76B
...clohexadien-1-one, 4,4'-[2,6-naph-enediylbis[[3,5-bis(1,1-dimethyl-yl)-4-hydroxyphenyl]methylidyne]]-[2,6-bis(1,1-dimethylethyl)-	MeOH	230(4.97),273(4.34),361(4.14)	24-3217-76
...N₄O₁₆ — N_4O_{16} (1H,3H)-Pteridinedione, 6,7-diphenyl-1,3-bis(2,3,5-tri-O-benzoyl-α-D-ara-binofuranosyl)-	MeOH	228s(4.31),275(4.00),364(3.95)	24-3217-76
...H₆₄N₄O₁₀ — $H_{64}N_4O_{10}$ 2,4(1H,3H)-Pteridinedione, 6,7-diphenyl-1,3-bis[2,3,5-tris-O-(phenylmethyl)-α-D-arabinofuranosyl]-	MeOH	235(4.43)	35-3007-76
$C_{70}H_{86}O_{38}$ 2H-Pyran-4-acetic acid, 3-ethylidene-2-3,4-dihydro-5-(methoxycarbonyl)-2-[(2,3,5,6-tetra-O-acetyl-β-D-gluco-pyranosyl)oxy]-4-[2-[[2,3,4-dihydro-acetyl-6-O-[[3-ethylidene-3,4-tetra-5-(methoxycarbonyl)-2-[2,3,4,6-tetra-O-acetyl-β-D-glucopyranosyl]oxy]-2H-pyran-4-yl]acetyl]-β-D-glucopyrano-syl]oxy]ethyl]phenyl ester	EtOH		64-0965-76B
	MeOH	377(--),592(4.84)	64-0965-76
$C_{70}H_{88}O_4$ 2,5-Cyclohexadien-1-one, 4,4'-[1,1'-bi-phenyl]-2,2'-diylbis[[3,5-bis(1,1-di-methylethyl)-4-hydroxyphenyl]methyli-dyne]]bis[2,6-bis(1,1-dimethylethyl)-, ion(2-) dianion diradical	MeOH	476(4.85)	64-0965-76
	MeOH	389(--),601(5.11)	64-0
2,5-Cyclohexadien-1-one, 4,4'-[1,1'-bi-phenyl]-3,3'-diylbis[[3,5-bis(1,1-di-methylethyl)-4-hydroxyphenyl]methyli-dyne]]bis[2,6-bis(1,1-dimethylethyl)-, ion(2-) dianion diradical	MeOH	393(--),473(4.95)	64-0
	MeOH	399(--),610(5.08)	64-
2,5-Cyclohexadien-1-one, 4,4'-[1,1'-bi-phenyl]-3,5-diylbis[[3,5-bis(1,1-di-methylethyl)-4-hydroxyphenyl]methyli-methylethyl)-4-hydroxyphenyl]methyli-			

$C_{64}H_{88}ClFeN_8O_8$
Heminbis(ω-aminoenanthyl-ω-aminoenan-thate), dimethyl ester

$C_{64}H_{102}N_4$
21H,23H-Porphine, 5,10,15,20-tetraundec-yl-

$C_{65}H_{104}O_{32}$
Holotoxin B

$C_{65}H_{108}O_{23}$
Lanosta-9(11),17(20)-dien-18-oic acid, 3-[(O-6-deoxy-2,3,4-tri-O-methyl-α-D-glucopyranosyl-(1→4)-O-[O-2,3,4,6-tetra-O-methyl-β-D-glucopyranosyl-(1→3)-2,4,6-tri-O-methyl-β-D-glucopyranosyl-(1→2)]-3-O-methyl-β-D-xylo-pyranosyl)oxy]-25-hydroxy-16-oxo-, methyl ester, (3β)-   MeOH   hexane

$C_{66}H_{62}ClN_4O_{14}P$
Thymidine, P-(2-chlorophenyl)-methoxyphenyl)diphenyl)-3'-O-[(4-yl][5' 5']-3'-O-[(4-methoxyphenyl)-diphenylmethyl]-   MeOH

$C_{66}H_{66}HfN_4O_4$
Hafnium, bis(1,3-diphenyl-1,3-propane-dionato-O,O')[2,3,7,8,12,13,17,18-octaethyl-21H,23H-porphinato(2-)-N21,N22,N23,N24]-   232s(4.52),26... CH₂Cl₂

$C_{66}H_{66}N_4O_4Zr$
Zirconium, bis(1,3-diphenyl-1,3-propane-dionato-O,O')[2,3,7,8,12,13,17,18-octaethyl-21H,23H-porphinato(2-)-N21,N22,N23,N24]-   CH₂Cl₂   401(5.43),534(4.15), 571(4.42)

$C_{66}H_{72}Cl9N_8O_{27}P_3$
Thymidine, P-(2,2,2-trichloroethyl)thy-midylyl(3'→5')-P-(2,2,2-trichloroeth-yl)thymidylyl(3'→5')-P-(2,2,2-tri-chloroethyl)thymidylyl-3'-O-[(4-methoxyphenyl)diphenylmethyl]-   CH₂Cl₂   401(5.45),534(4.03), 572(4.41)

$C_{66}H_{76}ClFeN_8O_8$
Heminbis(ω-aminoenanthyl-L-phenylala-nine)   MeOH   264(4.58)

$C_{66}H_{78}$
1,1':2',1''-Terphenyl, 4,4''-bis(1,1-di-methylethyl)-3',4',5',6'-tetrakis-[4-(1,1-dimethylethyl)phenyl]-   CHCl₃-EtOH   403(4.95),505(3.96), 537(3.91),636(3.63)   104-1859-76

$C_{66}H_{80}ClFeN_8O_8$
Heminbis(ω-aminoenanthyl-L-phenylala-nine), dimethyl ester   hexane   254(4.765),277(4.520)   101-0385-76N

CHCl₃-EtOH   404(4.01),475(4.16), 594(3.99)   104-1859-76

35-3655-76

$C_{68}H_{86}O_4$
2,5-Cyclohexadien...
thalenediylbis...
ethyl)-4-hydro...
bis[2,6-bis(1...
(2-)
dianion dira...

2,5-Cyclohexa...
thalenediyl...
ethyl)-4-h...
bis[2,6-bi...
(2-)
dianion c...

$C_{68}H_{88}O_4$
2,5-Cyclo...
thalen...
ethyl)...
bis[2...

2,5-Cy...
thal...
eth...
bis...

$C_{70}H_5...$
2,...

$C_7...$

24-...

Compound	Solvent	$\lambda_{max}(\log \epsilon)$	Ref.
dyne]]bis[2,6-bis(1,1-dimethylethyl)-, ion(2-) (cont.)			64-0965-76B
dianion diradical	MeOH	475(4.90)	64-0965-76B
2,5-Cyclohexadien-1-one, 4,4'-[1,1'-bi-phenyl]-4,4'-diylbis[[3,5-bis(1,1-di-methylethyl)-4-hydroxyphenyl]methyli-dyne]]bis[2,6-bis(1,1-dimethylethyl)-, ion(2-)	MeOH	419(--),607(5.03)	64-0965-76B
dianion diradical	MeOH	455(4.80)	64-0965-76B
$C_{70}H_{90}O_4$			
2,5-Cyclohexadien-1-one, 4,4'-[1,1'-bi-phenyl]-2,2'-diylbis[[3,5-bis(1,1-di-methylethyl)-4-hydroxyphenyl]methyli-dyne]]bis[2,6-bis(1,1-dimethylethyl)-	MeOH	400(4.63)	64-0965-76B
2,5-Cyclohexadien-1-one, 4,4'-[1,1'-bi-phenyl]-3,3'-diylbis[[3,5-bis(1,1-di-methylethyl)-4-hydroxyphenyl]methyli-dyne]]bis[2,6-bis(1,1-dimethylethyl)-	MeOH	414(4.71)	64-0965-76B
2,5-Cyclohexadien-1-one, 4,4'-[1,1'-bi-phenyl]-3,5-diylbis[[3,5-bis(1,1-di-methylethyl)-4-hydroxyphenyl]methyli-dyne]]bis[2,6-bis(1,1-dimethylethyl)-	MeOH	412(4.68)	64-0965-76B
2,5-Cyclohexadien-1-one, 4,4'-[1,1'-bi-phenyl]-4,4'-diylbis[[3,5-bis(1,1-di-methylethyl)-4-hydroxyphenyl]methyli-dyne]]bis[2,6-bis(1,1-dimethylethyl)-	MeOH	425(4.73)	64-0965-76B
$C_{71}H_{90}O_4$			
2,5-Cyclohexadien-1-one, 4,4'-[methyl-enebis[3,1-phenylene[3,5-bis(1,1-di-methylethyl)-4-hydroxyphenyl]methyl-idyne]]bis[2,6-bis(1,1-dimethyleth-yl)-, ion(2-)	MeOH	388(--),599(5.13)	64-0965-76B
dianion diradical	MeOH	473(4.95)	64-0965-76B
$C_{71}H_{92}O_4$			
2,5-Cyclohexadien-1-one, 4,4'-[methyl-enebis[3,1-phenylene[3,5-bis(1,1-di-methylethyl)-4-hydroxyphenyl]methyl-idyne]]bis[2,6-bis(1,1-dimethyleth-yl)-	MeOH	414(4.71)	64-0965-76B
$C_{72}H_{122}N_8$			
21H,23H-Porphine-5,10,15,20-tetranonam-ine, N,N,N',N',N'',N'',N''',N'''-octa-ethyl-	$CHCl_3$	421(5.20),523(4.10), 559(3.90),600(3.58), 660(3.62)	5-2058-76
$C_{73}H_{80}N_{16}$			
1,2,4,5-Tetrazin-1(2H)-yl, 2,2',2'',2'''-(methanetetrayltetra-4,1-phenylene)-tetrakis[6-(1,1-dimethylethyl)-3,4-dihydro-4-phenyl-	DMF	326(4.68),340(4.67), 384(4.57),687(4.43)	24-2389-76
$C_{74}H_{58}N_2$			
Cyclobuta[4,5]benz[1,2]cyclooctene-5,8-dicarbonitrile, 3,10-bis(1,1-dimethyl-ethyl)-6,7-bis(diphenylmethylene)-6,7-dihydro-5,10-diphenyl- $(2\lambda,3\epsilon)$	$CHCl_3$	?(4.65),334(4.70), 400s(4.30)	77-0030-76

Compound	Solvent	$\lambda_{max}(\log \epsilon)$	Ref.
$C_{76}H_{108}ClFeN_{10}O_{10}$ Heminbis(ω-aminoenanthyl-ω-aminoenan-thyl-ω-aminoenanthic acid)	$CHCl_3$-EtOH	406(4.95),508(3.96), 540(3.88),585(3.75)	104-1859-76
$C_{78}H_{114}ClFeN_{10}O_{10}$ Heminbis(ω-aminoenanthyl-ω-aminoenan-thyl-ω-aminoenanthic acid), dimethyl ester	$CHCl_3$-EtOH	405(4.10),468(4.29), 596(4.10)	104-1859-76
$C_{88}H_{56}Fe_2N_9$ Iron, μ-nitridobis[5,10,15,20-tetraphen-yl-21H,23H-porphinato(2-)- $N^{21},N^{22},N^{23},N^{24}$]di-	xylene	380(5.26),409(5.30), 533(4.25)	35-1747-76
$C_{92}H_{66}$ 1,3,5-Cycloheptatriene, 3,3'-(1,4-phen-ylene)bis[1,2,4,5,6,7-hexaphenyl-	$C_2H_4Cl_2$	268(4.76),268s[sic], 278s(4.70)	44-3374-76
$C_{93}H_{126}O_6$ 2,5-Cyclohexadien-1-one, 4,4',4"-[1,3,5-benzenetriyltris[[3,5-bis(1,1-dimeth-ylethyl)-4-hydroxyphenyl]methyli-dyne]]tris[2,6-bis(1,1-dimethylethyl)-	MeOH	413(4.85)	64-0965-76B
trianion	MeOH	404(--),612(5.27)	64-0965-76B
trianion triradical	MeOH	451(--),473(5.14)	64-0965-76B
$C_{98}H_{70}$ 1,3,5-Cycloheptatriene, 3,3'-[1,1'-bi-phenyl]-4,4-diylbis[1,2,4,5,6,7-hexa-phenyl-	$C_2H_4Cl_2$	268(4.78),285s(--), 320s(4.57)	44-3374-76
$C_{98}H_{70}O$ 1,3,5-Cycloheptatriene, 3,3'-(oxydi-4,1-phenylene)bis[1,2,4,5,6,7-hexaphenyl-	$C_2H_4Cl_2$	268(4.79),298s(--), 319(4.50)	44-3374-76
$C_{99}H_{72}$ 1,3,5-Cycloheptatriene, 3,3'-(methylene-di-4,1-phenylene)bis[1,2,4,5,6,7-hexa-phenyl-	$C_2H_4Cl_2$	265(4.78),299s(--), 319(4.46)	44-3374-76
$C_{99}H_{130}O_6$ 2,5-Cyclohexadien-1-one, 4,4',4"-[[1,1'-biphenyl]-2,2',4-triyltris[[3,5-bis-(1,1-dimethylethyl)-4-hydroxyphenyl]-methylidyne]]tris[2,6-bis(1,1-dimeth-ylethyl)-	MeOH	393(4.82)	64-0965-76B
trianion	MeOH	384(--),588(5.06)	64-0965-76B
trianion triradical	MeOH	465(5.00)	64-0965-76B
2,5-Cyclohexadien-1-one, 4,4',4"-[[1,1'-biphenyl]-2,2',5-triyltris[[3,5-bis-(1,1-dimethylethyl)-4-hydroxyphenyl]-methylidyne]]tris[2,6-bis(1,1-dimeth-ylethyl)-	MeOH	404(4.79)	64-0965-76B
trianion	MeOH	423(--),618(5.06)	64-0965-76B
trianion triradical	MeOH	405(--),469(5.02)	64-0965-76B
$C_{129}H_{172}O_8$ 2,5-Cyclohexadien-1-one, 4,4',4",4"'-[methylenebis[5,3,1-benzenetriylbis-[3,5-bis(1,1-dimethylethyl)-4-hydroxy-phenyl]methylidyne]]tetrakis[2,6-bis-(1,1-dimethylethyl)-	MeOH	410(4.94)	64-0965-76B
	MeOH	397(--),602(5.36)(tetra-anion)	64-0965-76B
	MeOH	458(5.19)(tetraradical)	64-0965-76B

1-   -76, Acta Chem. Scand. B, 30 (1976)
0024   S. Baklien et al.
0064   G. Andersson
0133   E.B. Pedersen
0189   O. Buchardt et al.
0198   E. Pohjala
0251   R. Lounasmaa et al.
0281   P. Krogsgaard-Larsen and S.B.
        Christensen
0297   J. Dahmen et al.
0353   K. Torssell
0368   J. Better and S. Gatenbeck
0369   B. Thulin and O. Wennerström
0403   G. Andersson
0466   O. Ceder et al.
0468   O. Ceder and B. Beijer
0478   C.T. Pedersen et al.
0512   E.K. Pohjala
0567   H. Hjeds and P. Krogsgaard-Larsen
0613   T. Reffstrup and P.M. Boll
0619   B.P. Nilsen and K. Undheim
0635   E.T. Østensen and M.M. Mishrikey
0675   C.L. Pedersen
0688   B. Thulin and O. Wennerström
0695   U. Berg
0705   T. Popoff and O. Theander
0743   K. Bock et al.
0781   J. Lykkeberg and P. Krogsgaard-
        Larsen
0796   R. Göthe and C.A. Wachtmeister
0853   J. Bergman et al.
0863   J. Becher and E.G. Frandsen
0884   P. Krogsgaard-Larsen and H. Hjeds
0904   J. Becher and E.G. Frandsen

2-   -76, Indian J. Chem. Sect. B, 14 (1976)
0001   K. Nagarajan and R.K. Shah
0009   B.S. Joshi and D.H. Gawad
0014   M. Mehta and A.B. Kulkarni
0038   R.R. Rao et al.
0054   S.A. Metwally
0057   C.M. Gupta et al.
0059   M.K. Choudhury et al.
0062   V.J. Tripathi et al.
0073   P. Balakrishnan and S.C. Bhatta-
        charyya
0077   P. Pal et al.
0084   V.P. Barve et al.
0088   M.A. Dave and A.B. Kulkarni
0101   A.A. Youssef and A.G. Abdel-Rehiem
0127   N. Adityachaudhury and A.K. Das
0127B   M. Metha and A.B. Kulkarni
0131   G.C.S. Reddy et al.
0134   M.S. Premila and B.R. Pai
0150   M. Sivakumaran and K.W. Gopinath
0152   P.N. Sarma et al.
0157   P.L. Kamat et al.
0176   P.B. Talukdar et al.
0222   M. Krishnamurti and S.N. Seshagiri
0229   S.B. Malik et al.
0230   S.S. Pandit and A.B. Kulkarni
0233   A. Chatterjee et al.
0238   T.K. Das et al.
0243   R.V. Venkateswaran et al.
0259   R. Pal et al.
0263   N. de Kimpe et al.

0300   B. Talapatra et al.
0312   D.K. Banerjee et al.
0323   D.S. Chothia et al.
0329   N.S. Narasimhan et al.
0336   N.A. Kudar et al.
0339   J.M. Rao et al.
0346   A.K. Sen and S. Ray
0351   A.K. Sen and S. Ray
0357   M.B. Devani et al.
0382B   P.B. Talukdar et al.
0384   S.C. Chhabra et al.
0389   C.P. Dutta and N. Banerjee
0403   K. Chattopadhyay
0430   N.S. Narasimhan and S.L. Kelkar
0462   G. Subrahmanyam and R. Srinivasan
0472   P.N. Chaudhari et al.
0477   D.N. Dhar et al.
0504   S.A.M. Metwally
0516   M. Bandopadhyay et al.
0547   P.P. Paranjpe and G. Bagavant
0548   M.K. Gurjar et al.
0551   A. Sharma and G.B. Behera
0579   B.S. Holla and S.Y. Ambekar
0584   S.N. Seshagiri and M. Krishna-
        murti
0592   V.K. Ahluwalia et al.
0613   B. Talapatra et al.
0616   V. Seshadri et al.
0618   H. Singh et al.
0620   B.V. Swaminathan
0644   N. Viswanathan and R.R. Sidhaye
0651   S.C. Chhabra et al.
0657   D.N. Bhedi et al.
0718   R.S. Varma
0731   T.R. Kasturi and V.K. Sharma
0759   V.P. Arya and S.T. Shenoy
0773   V.P. Arya et al.
0784   V.P. Arya and S.J. Shenoy
0793B   P.A. Reddy et al.
0799   C.V. Ananthanarayanan et al.
0802   K.K. Purushothaman and S. Chandra-
        sekharan
0816   A. Patra et al.
0817   P. Sengupta et al.
0819   D. Nasipuri and S.K. Ghosh
0823   T.R. Kasturi et al.
0841   H. Suguna and B.R. Pai
0849   A.K. Sen et al.
0861   U.M. Wagh and R.N. Usgaonkar
0879   V.P. Arya et al.
0883   V.P. Arya et al.
0887   B.K. Paul et al.
0901   V.K. Belavadi and S.N. Kulkarni
0936   M.S. Ahmad et al.
0939   K.S. Sharma et al.
0943   K.S. Sharma et al.
0951   M. Krishnamurti and S.N. Seshagiri
0964   T.R. Kasturi and V.K. Sharma
0994   S.N. Bannore and J.L. Bose
1011   S.N. Seshagiri et al.

3-   -76, Anal. Chem., 48 (1976)
0195   R.D. Arah and B. McDuffie
0450   G.L. Gaines, Jr.
0714   S.A. Abbasi
1915   M.H. Jones and J.T. Woodcock

4-   -76, J. Heterocyclic Chem., 13 (1976)

0033	G.S. Zenchoff et al.
0045	A. Sami Shawali and A. Osman Abdel-hamid
0057	I. Lalezari et al.
0065	A.N. Hughes et al.
0073	J. Van Allan and G. Reynolds
0083	N.R. El-Rayyes and A.H.A. Ali
0097	A. Da Settimo et al.
0107	C.O. Okafor
0111	I. Antonini et al.
0113	H.C. Wormser and H.N. Abramson
0117	A. Shafiee et al.
0131	A. Walser et al.
0141	A. Shafiee and A. Rashidbaigi
0169	G.R. Revankar and R.K. Robins
0195	F.G. Baddar et al.
0211	G. Romussi and G. Ciarallo
0225	P. Schenone et al.
0237	H. Pirelahi et al.
0253	C. Eskenazi et al.
0257	F.G. Baddar et al.
0269	A. Zirnis et al.
0285	N.R. El-Rayyes and N.A. Al-Salman
0291	R.L. Lauer and G. Zenchoff
0311	I. Jirkovsky et al.
0321	J.M. Kokosa et al.
0383	J.F. Wolfe et al.
0393	J. Iriarte et al.
0405	D.M. McDaniel and S. Benisch
0433	R.I. Fryer et al.
0439	J.I. Degraw and V.H. Brown
0491	J. Bourdais et al.
0497	E. Campaigne and G.M. Shutske
0509	M.P. Serve et al.
0517	B.M. Goldschmidt et al.
0521	S. Gelin and D. Hartmann
0577	J.A. Van Allan and G.A. Reynolds
0589	R.S. Klein et al.
0597	I.W. Elliott, Jr. and Y. Takekoshi
0619	D. Avnir and J. Blum
0649	D. Daniil and H. Meier
0661	T. Kurihara et al.
0691	F.G. Baddar et al.
0727	A. Rosowsky et al.
0741	L.H. Klemm et al.
0745	G.H. Milne and L.B. Townsend
0765	H. Agui and T. Nakagome
0793	G. Garcia-Munoz et al.
0797	H.L. Yale and E.R. Spitzmiller
0861	J.M. Kokosa et al.
0907	A. Walser and G. Zenchoff
0929	J.D. Fissekis et al.
0961	H.C. Van Der Plas and M. Wozniak
0989	A.S. Shawali and A.O. Abdelhamid
1001	M.L. Druelinger et al.
1015	Y.F. Shealy and C.A. O'Dell
1041	Y.F. Shealy and C.A. O'Dell
1049	G. Santroch and L.G. Humber
1063	T.K. Liao et al.
1067	J.K. Suzuki et al.
1089	G.A. Reynolds and J.A. Van Allan
1101	R.K. Razdan et al.
1137	A.S. Shawali et al.
1145	D.O. Cheng et al.
1155	G. Adembri et al.
1187	D.H. Kim
1201	L. Mosti et al.
1241	E. Garcia-Abbad et al.
1253	K.L. Kirk
1265	D. Florentin et al.
1273	A.M. Kiwan and H.M. Marafie
1297	F.E. Herkes and T.A. Blazer
1305	S.Y. Tam et al.
1325	D.T. Hill et al.
1327	P.M. Hadjimihalakis
1333	J.E. Prickett and B.T. Gillis
1337	S.Y. Dike and J.R. Merchant
1343	M. Higashi and H. Yamaguchi
1353	Y.F. Shealy and C.A. O'Dell
1359	A.F. Lewis et al.
1363	L.B. Townsend et al.
1365	B.L. Cline et al.

5-   -76, Ann. Chem. Liebigs (1976)

0022	H. Gusten and D. Schulte-Frohlinde
0112	H. Fursetnwerth and H. Schildknecht
0140	K.D. Gundermann and K.-D. Röker
0153	H. Neunhoeffer and V. Bohnisch
0269	K. Heyns and J. Heukeshoven
0284	W. Ried and H. Knorr
0317	S. Hünig and F. Linhart
0383	G. Pfeiffer and H. Bauer
0566	H. Brockmann, Jr. et al.
0745	H. Vorbrüggen and K. Krolikiewicz
0793	W. Friedrichsen et al.
0946	G. Ege et al.
0979	G. Kresze et al.
1039	S. Hünig and H.-C. Steinmetzer
1060	S. Hünig and H.-C. Steinmetzer
1090	S. Hünig and H.-C. Steinmetzer
1103	S. Hünig and P. Schilling
1150	A. Gossauer et al.
1206	W. Kreiser and W. Ulrich
1214	W. Kreiser and W. Ulrich
1222	W. Kreiser and H.A.F. Heinemann
1276	J. Koziol and P. Hemmerich
1295	H. Schildknecht et al.
1395	M. Kröger et al.
1435	E. Adler and G. Andersson
1514	G.P. Schiemenz and U. Schmidt
1531	G. Cleve et al.
1537	J.H. Fuhrhop et al.
1637	J. Engel and A. Gossauer
1689	G. Jager et al.
1739	J. Rieser and K. Wallenfels
1799	J. Schnekenburger et al.
1809	W. Schafer and C. Falkner
1850	D. Hoppe and M. Kloft
1873	K.D. Gundermann et al.
1972	H. Biere and W. Seelen
1997	M. Kloft and D. Hoppe
2007	W. Ried and L. Kaiser
2037	W.-R. Knappe and P. Hemmerich
2058	J.H. Fuhrhop and M. Buccouche
2169	W. Mayer et al.
2185	D. Hoppe
2206	W. Draber et al.

7-   -76, Ann. chim.(Rome), 66 (1976)

0379	E. Barni and G. DiModica

0593   S. Pasquino and G. DiModica

11-   -76A, Chemica Scripta, 9 (1976)
0220   G. Lindgren

11-   -76B, Chemica Scripta, 10 (1976)
0126   R. Lantz and A.B. Hornfeldt

12-   -76, Australian J. Chem., 29 (1976)
0163   J.R. Davy and J.A. Reiss
0173   P.J. Jessup and J.A. Reiss
0179   L.H. Briggs et al.
0327   D.J. Freeman et al.
0339   S. Nimgirawath et al.
0357   L.H. Briggs et al.
0393   P.S. Clezy et al.
0447   M.S. Ahmad et al.
0459   R.G. Dickinson and N.W. Jacobsen
0673   M.S. Allen et al.
0883   H. Blatt et al.
1023   J.T. Baker and C.C. Duke
1059   S. Huneck and M.V. Sargent
1087   J.A. Rideout et al.
1147   A.J. Bryan and J.A. Elix
1153   D. St.C. Black et al.
1163   D.W. Cameron and R.H. Cracknell
1295   N.K. Hart et al.
1375   U.M. Pagnoni et al.
1393   P.K. Dommen
1435   D.A. Burgess et al.
1509   H.J. Banks et al.
1523   D.W. Cameron and W.D. Raverty
1535   D.W. Cameron et al.
1549   O.D. Hensens et al.
1561   P.S. Clezy and T.T. Hai
1583   M.Y. Shandala and N.H. Al-Jobour
1613   P.J. Kissane et al.
1685   W.A. Bubb and S. Sternhell
1699   S.H. Goh, K.C. Chan and H.L. Chong
1735   D. St.C. Black and V.C. Davis
1815   J.F. Kinnear et al.
1853   J.A. Lamberton and E.R. Nelson
1859   D.J. Brecknell et al.
1865   D.W. Cameron and M.D. Sidell
1979   J.A. Croft et al.
1989   L. Cleaver et al.
2003   L. Cleaver et al.
2023   K. Picker et al.
2077   D.J. Collins et al.
2087   P.K. Dommen
2207   J.M. Coxon and N.B. Lindley
2225   H.J. Banks et al.
2231   H.J. Banks et al.
2247   H.J. Banks et al.
2257   D.W. Cameron et al.
2263   M.V. Sargent et al.
2271   D. St.C. Black and A.J. Hartshorn
2325   P.S. Clezy and V. Diakiw
2445   J.J. Brophy
2459   B.S. Deol et al.
2499   D.W. Cameron and E.L. Samuel
2511   D.St.C. Black and A.B. Boscacci
2525   W.D. Crow et al.
2533   R. Kazlauskas et al.
2561   D. St.C. Black and A.B. Boscacci
2621   R.K. Norris and D. Randles

2631   D.J. Freeman and R.K. Norris
2683   K.S. Ng et al.
2747   D.J. Gale et al.
2753   S.J. Gumbley and L. Main

13-   -76A, Steroids, 27 (1976)
0225   M. Sharma and V. Georgian
0431   E.D. Bergmann and A. Solomonovici
0609   K.T. Alston et al.
0717   V.S. Salvi et al.
0759   J.F. Grunwell et al.
0845   V.R. Mattox and A.N. Nelson

13-   -76B, Steroids, 28 (1976)
0101   P.N. Rao and P.H. Moore, Jr.
0197   S.R. Ramadas et al.
0655   Y. Shalon and W.H. Elliott

18-   -76, Bull. Chem. Soc. Japan, 49 (1976)
0245   H. Yamaoka et al.
0292   J. Ojima et al.
0297   T. Katakami et al.
0302   S. Tomita and M. Nakagawa
0305   T. Nomoto et al.
0321   A.S. Shawali and A.O. Abdelhamid
0499   A. Murai et al.
0529   F. Imashiro et al.
0737   T. Horaguchi et al.
0741   K. Kimura et al.
0748   A. Haneda et al.
0762   S. Ito et al.
0817   Y. Kawasaki
0825   T. Tatsuoka and I. Murata
0831   J. Tsunetsugu et al.
1101   K. Yamane amd F. Fujimori
1138   O. Tsuge and A. Torii
1142   Y. Inagaki et al.
1163   M. Iwata and S. Emoto
1171   K. Yamada et al.
1339   K. Kubo et al.
1381   N. Nishimura et al.
1395   T. Hirohashi et al.
1401   A. Kuboyama and H. Arano
1447   K. Uehara et al.
1521   S. Sekiguchi et al.
1650   S. Kurokawa and A.G. Anderson, Jr.
1653   M. Tsukayama
1725   M. Takahashi et al.
1893   M. Seno et al.
1937   S. Yamamura and Y. Terada
1940   S. Yamamura et al.
1955   K. Kurosawa and K. Nogami
1980   H. Ochi et al.
1985   P.K. Oommen
2000   A. Chatterjee et al.
2027   R. Gopal et al.
2158   R. Abu-Eittah and R. Hilal
2224   K. Ichimura and S. Watanabe
2230   M. Sato et al.
2292   M. Ohashi et al.
2306   M. Iyoda et al.
2333   Y. Omote and T. Komatsu
2495   A. Yabe and K. Honda
2503   F. Toda and E. Todo
2515   F. Toda and Y. Takahara
2529   H. Ogoshi et al.

		22-	-76, <u>Bull. soc. chim. France</u> (1976)

2560   T. Sonoda et al.
2579   S. Akiyama et al.
2596   T. Otsuki
2645   F. Toda, Y. Todo and E. Todo
2679   M. Kondo
2770   Y. Takahashi et al.
2790   A. Takuwa
2805   K. Yamada et al.
2817   K. Uehara et al.
2837   S. Kozima et al.
2840   J. Ojima et al.
3132   K. Maruyama and T. Takahashi
3137   M. Niwa et al.
3148   M. Niwa et al.
3165   T. Fuchigami and K. Odo
3170   T. Fuchigami et al.
3173   K. Ueda et al.
3196   Y. Moriyama and T. Takahashi
3224   T. Tanabe et al.
3239   J. Ide and Y. Kishida
3243   J. Ide et al.
3269   K. Kano and T. Matsuo
3300   T. Toyoda et al.
3314   H. Matsukubo and H. Kato
3333   H. Matsukubo and H. Kato
3339   T. Nishiwaki
3357   S. Shimokawa et al.
3359   M. Niwa et al.
3361   K. Maruyama et al.
3363   M. Ochi et al.
3552   O. Makabe et al.
3607   T. Fuchigami and K. Odo
3646   S. Akutagawa
3673   R. Noyori et al.
3685   A. Kuboyama et al.
3709   J. Ojima et al.
3713   T. Otsuki

19-   -76, <u>Bull. Acad. Polon. Sci.</u>, <u>24</u> (1976)
0017   L. Strekowski
0029   L. Strekowski
0037   M. Draminski and E. Frass
0089   W. Jasiobedzki and A. Zimniak
0175   S. Mejer and Z. Marcinow
0517   G. Wenska et al.

20-   -76, <u>Bull. soc. chim. Belges</u>, <u>85</u> (1976)
0035   Y. Van Haverbeke et al.
0141   A. Cornelis et al.
0147   R. Fuks et al.
0161   Y. Berger et al.
0293   D. De Keukeleire et al.
0333   L. Dehennin and J. Levisalles
0421   A. Waefelaer et al.
0503   P. de Clerq et al.
0573   A. Maquestiau et al.
0579   A. Maquestiau et al.
0647   H. de Pooter et al.
0657   L. Van Rompaey et al.
0663   H. de Pooter et al.
0707   M. Herin and B. Tursch
0787   A. Waefelaer et al.
0819   J.O. Madsen and S.-O. Lawesson
0898   A. Waefelaer et al.

22-   -76, <u>Bull. soc. chim. France</u> (1976)
0088   A. Claude-Lafontaine et al.
0142   F. Auclair et al.
0177   G. Morel et al.
0184   P. Bouchet et al.
0195   P. Bouchet and C. Coquelet
0217   A. Castellan et al.
0251   N. Vinot and P. Maitte
0255   J.B. Cazaux et al.
0291   A. Guingant and J. Renault
0493   J. Rigaudy et al.
0513   M. Julia and D. Uguen
0519   M. Julia et al.
0761   M. Robbe and M.C. de Sevricourt
0776   M. Pfau and C. Ribiere
0839   P. Bouchet et al.
0849   M. de Botton
0889   B. Cheminat and B. Mege
0901   G. Descotes et al.
0914   N. Soyer
0930   I. Al Adel et al.
0957   C. Metge and C. Bertrand
0978   F. Ghozland et al.
0983   F. Ghozland et al.
0995   C. Eskenazi and P. Maitte
1131   G. Reverdy
1136   G. Reverdy
1178   J. Daunis and M. Follet
1240   J.-C. Jacquesy et al.
1473   F. Titeux et al.
1491   G. Rio and B. Serkiz
1541   G.H. Labib and A.R.S.A. Mourra
1567   D. Serve
1599   R. Gruber et al.
1829   M. Cariou
1865   R. Hanna and B. Muckensturm
1889   R. Jacquesy and H.L. Ung
1913   G. Lhommet and P. Maitte
1916   O. Chartier et al.
1953   J. Muzart and J.-P. Pete
1961   D. Cartier et al.
1967   J. Andrieux et al.
1975   J. Andrieux et al.
1983   J.L. Barascut et al.
1993   D. Serve
2021   M. Maumy and J. Rigaudy
2039   P. Appriou et al.

23-   -76, <u>Can. J. Chem.</u>, <u>54</u> (1976)
0037   A.M. Osman et al.
0044   H. Cohen and C. Benezra
0097   H.J. Liu et al.
0415   G.M. Strunz and P.I. Kazinoti
0423   A. Fischer and R. Roederer
0448   M. Ashraf
0455   I.W.J. Still et al.
0488   J. Zauhar and B.F. Ladouceur
0861   G. Just and B. Chalard-Faure
1015   V. Dave and E.W. Warnhoff
1020   V. Dave and E.W. Warnhoff
1029   B.M. Lynch and S.C. Sharma
1074   M. Farnier et al.
1083   M. Farnier et al.
1105   J. Favre-Bouvin et al.

1197	J. Armand and L. Boulares
1205	A.S. Shawali and B.E. El-Anadouli
1262	S. McLean et al.
1360	R. Rahman et al.
1449	P.M. Burke et al.
1512	R.B. Kelly et al.
1599	N.L. Holy
1703	W.A. Ayer and D.R. Taylor
1752	C. Sabate-Alduy and J. Bastide
1795	A. Fischer and D.R.A. Leonard
1827	G. Dana et al.
1958	B.F. Anderson et al.
2089	G. Just et al.
2127	A. Deflandre et al.
2310	C. Schmidt
2385	R.A. Perry et al.
2436	E. Buncel and W. Eggimann
2563	J.W. Lown and S.-K. Kim
2681	O.S. Tee and M. Endo
2795	J.P. Kutney and I.H. Sanchez
2804	J.P. Devlin
2862	G.M. Strunz and W.Y. Ren
2925	G. Just and R. Ouellet
3026	P.R. Olivato et al.
3038	C.C. Lee and P.J. Smith
3089	A. Gandini et al.
3260	A.S. Shawali and B.M. Altahou
3508	P. Yates and F.M. Walliser
3545	C. Kavanagh-Caron et al.
3620	J.J. Fuentes et al.
3713	J.P. Kutney et al.
3721	J.P. Kutney et al.

24- -76, Chem. Ber., 109 (1976)

0001	F. Vögtle and E. Goldschmitt
0041	H. Wolf et al.
0082	F. Seela and F. Cramer
0139	J.C. Jochims and A. Abu-Taha
0154	J.C. Jochims and A. Abu-Taha
0185	H. Hofmeister et al.
0200	H.G. Aurich and U. Grigo
0253	A. Fehlauer et al.
0261	W. Sucrow and K.P. Grosz
0268	W. Sucrow and M. Meyer
0433	R. Mengel and H. Wiedner
0518	G. Vitt et al.
0562	M. Hirth et al.
0576	H. Krapf et al.
0659	H. Böshagen and W. Geiger
0705	R. Oehl et al.
0740	H. Gotthardt et al.
0793	K. Weinges and K. Klessing
0819	F. Bohlmann and C. Zdero
0855	K. Krohn and A. Mondon
0901	F. Bohlmann and C. Zdero
0906	E. Schaumann
1038	W. Girke and E.D. Bergmann
1106	H. Mohrle and M. Lappenberg
1269	H. Wamhaff and J. Hartlapp
1287	O. Constantinescu-Simon et al.
1294	O. Constantinescu-Simon et al.
1301	O. Constantinescu-Simon et al.
1309	D. Seebach et al.
1332	G. Quinkert et al.
1389	K. Krohn et al.
1395	R. Mengel and H. Wiedner

1429	P. Eilbracht
1465	J.W. Buchler and P.D. Smith
1477	J.W. Buchler et al.
1506	W. Ried et al.
1549	B. Eistert et al.
1584	F. Bohlmann and M. Grenz
1625	K.H. Büchel and H. Erdmann
1643	W. Ried and R. Schweitzer
1650	H. Meier et al.
1724	L. Szabo et al.
1737	C. Szantay et al.
1787	H. Junek et al.
1803	E. Weber and F. Vogtle
1898	V. Bardakos and W. Sucrow
1928	R.W. Hoffmann and J. Backes
1964	F. Bohlmann and C. Zdero
1991	G.P. Blümer et al.
2021	F. Bohlmann and C. Zdero
2126	H. Plieninger et al.
2140	H. Plieninger and D. Schmalz
2154	W. Sucrow and K.P. Grosz
2159	A. Roedig and M. Foure
2259	K. Kieslich et al.
2291	F. Bohlmann and H. Czerson
2338	G. Zolyomi and E. Berenyi
2389	F.A. Neugebauer et al.
2405	W. Winter
2469	L. Bucsis and K. Friedrich
2596	P. Luger et al.
2615	F.-S. Tjoeng et al.
2657	H.J. Vidic et al.
2691	G. Lang et al.
2805	H. Wolf et al.
2823	H. Prinzbach et al.
2921	W. Ried et al.
2948	U. Eder et al.
2963	D. Trautmann et al.
2983	H. Wamhoff and B. Wehling
3025	G. Neef et al.
3099	M. Constenla and K. Dimroth
3159	K. Kobayashi and W. Pfleiderer
3175	K. Kobayashi and W. Pfleiderer
3184	K. Kobayashi and W. Pfleiderer
3194	K. Kobayashi and W. Pfleiderer
3208	K. Eistetter and W. Pfleiderer
3217	W. Hutzenlaub et al.
3228	T. Itoh and W. Pfleiderer
3282	J. Gudjons et al.
3304	F. Wagner and S. Lang
3318	W. Kreiser
3379	U. Schwarzmaier
3432	G. Diderrich and A. Haas
3462	B. Eistert and G, Holzer
3505	H. Prinzbach et al.
3527	H. Achenbach and E. Schaller
3606	H.-J. Vidic et al.
3615	M. Kröger et al.
3626	G. Kinast and L.F. Tietze
3640	G. Kinast and L.F. Tietze
3661	W. Stadlbauer and T. Kappe
3668	T. Kappe and W. Golser
3735	C. Reichardt and E. Wurthwein
3775	A. Schönberg et al.
3817	G. Rackur et al.
3825	B. Hachmeister et al.
3886	E. Luppold et al.

3929   F.-G. Fick and K. Hartke
3939   F.-G. Fick and K. Hartke

25-   -76, Chem. and Ind.(London) (1976)
0029   A. Rahman et al.
0410   A. Chatterjee et al.
0954   V.H. Belgaonkar and R.N. Usgaonkar
1030   A.B. Turner

27-   -76, Chimia, 30 (1976)
0453   U. Widmer et al.

28-   -76A, Compt. rend., 282 (1976)
0195   A.P. Chatrousse and F. Terrier
0319   G. Mabon and G. Le Guillanton
0485   G. Assef et al.
0555   L. Lepage and Y. Lepage
0607   M. Peyrot et al.
0683   M. Goudard and J. Chopin
0691   J.-C. Le Duc et al.
0757   J. Herscovici et al.
0761   R. Lorne and G. Linstrumelle
0951   D. Bondon et al.
1001   M. Bessodes et al.
1045   M. Hamonniere et al.

28-   -76B, Compt. rend., 283 (1976)
0021   S. David and G. de Sennyey
0143   G. Assef et al.
0227   T.H. Dinh et al.
0343   J. Barbe et al.
0759   J. Le Men et al.

30-   -76, Doklady Akad. Nauk S.S.S.R., 226-
       231 (1976)(English translation
       pagination)
0004   N.S. Bystrov et al.
0072   V.A. Mironov et al.
0139   A.A. Akhrem et al.
0235   O.A. Ptitsyna et al.
0366   Y.V. Belkin et al.
0405   B.A. Arbuzov et al.
0613   Y.A. Zhdanov et al.
0629   A.N. Kost et al.

31-   -76, Experientia, 32 (1976)
0140   P.M. Scott et al.
0283   W.R. Chan et al.
0415   L.A. Mitscher et al.
0828   Y.H. Kuo et al.
0966   A. Matsuo et al.
1106   B.C. Maiti and R.H. Thomson
1107   D.A. Lightner and C.S. Pak
1111   M.S. Bhatia and Pawanjit
1236   A. Chatterjee et al.
1490   C. Rivera et al.

32-   -76, Gazz. chim. ital., 106 (1976)
0001   S. Auricchio et al.
0019   E. Chiacchierini et al.
0057   P. Esposito et al.
0065   G. Cignarella et al.
0119   F. Marletti et al.
0139   G. Purello and A. LoVullo
0147   L. Canonica et al.
0205   A. LoVullo and G. Purrello

0617   L. Bonsignore et al.
0625   M. Adinolfi et al.
0651   M. Marta et al.
0657   S. Grattadauria et al.
0681   F.D. Monache et al.
0725   A. Bianco et al.
0733   A. Bianco et al.
0773   C. Galeffi et al.
0779   E. Fattorusso et al.
0785   T. Pozzo-Balbi et al.
0871   A. Albini et al.
1013   A. Sacco and G. Vasapallo
1083   G. Grandolini et al.

33-   -76, Helv. Chim. Acta, 59 (1976)
0001   W. Skorianetz and G. Ohloff
0021   B. Szechner and A.S. Dreiding
0032   M.P. Zink et al.
0075   G. Ohloff et al.
0082   B. Frei and H.R. Wolf
0100   C.B. Chapleo et al.
0133   P. Bloch et al.
0148   H. Balli and R. Maul
0155   H. Balli and R. Low
0179   H. Fritz et al.
0190   U. Horn et al.
0211   U. Horn et al.
0222   F. Matterer and C.D. Weis
0229   F. Matterer and C.D. Weis
0248   B. Schircks et al.
0290   J.W. Scott et al.
0387   P.S. Manchand et al.
0397   A. Fischli et al.
0532   Y. Morita et al.
0551   M.H. Elnagdi et al.
0567   G.L. Olson et al.
0626   K. Hermann and A.S. Dreiding
0664   M. Jackson-Mülly
0688   D. Schelz and M. Priester
0692   D. Schelz and M. Priester
0695   D.G. Leppard et al.
0727   B.R. von Wartburg et al.
0747   H. Dugger and A.S. Dreiding
0755   H. Gerlach et al.
0765   C.B. Chapleo and A.S. Dreiding
0772   K. Nomoto et al.
0802   V. Coviello and A.E. Siegrist
0819   V. Coviello and A.E. Siegrist
0855   M. Juillerat and J.P. Bargetzi
0907   A.P. Alder et al.
0914   D. Scherling et al.
0949   W. Wiegrebe et al.
0999   U. Eder et al.
1018   W. Stegmann et al.
1027   P. Wieland
1203   H.P. Härter et al.
1213   W. Hofheinz et al.
1226   M. Karpf and A.S. Dreiding
1253   H. Eichenberger et al.
1427   Y. Hirose et al.
1438   H. Loewenschuss et al.
1469   C.O. Bender and J. Wilson
1621   H. Stünzi and G. Anderegg
1698   A. Gossauer and K. Suhl
1745   P. Schiess and P. Fünfschilling
1756   P. Schiess and P. Fünfschilling

1763	G. Mukherjee-Müller	1052	M. Sindler-Kulyk and W.H. Laar-
1809	P. Bollinger and T. Zardin-Tarta-		hoven
	glia	1099	J.H. Boyer and J. Kooi
1850	A. Tzikas et al.	1204	B.M. Trost and L.S. Melvin, Jr.
1917	J.-M. Cassal et al.	1461	G. Levin et al.
1925	W. Tadros et al.	1492	P.D. Cook et al.
1963	T. Petrzilka et al.	1510	Y. Moro-oka and C.S. Foote
1988	T. Petrzilka et al.	1526	J.F. Garst and C.D. Smith
1997	R. Ziegler and C. Tamm	1545	H. Hart and M. Kuzuya
2012	W. Oppolzer et al.	1551	H. Hart and M. Kuzuya
2021	W. Keller-Schierlein and J. Widmer	1569	D.H. Miles et al.
2048	W. Heller et al.	1587	D.E. Bergstrom and J.L. Ruth
2074	K. Dietliker et al.	1747	D.A. Summerville and I.A. Cohen
2125	T.W. Güntert et al.	1860	A.R. Miller and D.Y. Curtin
2138	T.W. Güntert et al.	1875	T.J. Levek and E.F. Kiefer
2149	N. Gakis et al.	1988	B.M. Trost and W.B. Herdle
2201	L. Faber and W. Wiegrebe	2027	H.J. Rodriguez et al.
2244	W. Fröstl and P. Margaretha	2295	S.M. Kupchan et al.
2261	A.F. Thomas and R. Guntz-Dubini	2301	E.C. Taylor and P.A. Jacobi
2268	G. Grethe et al.	2328	J.W. Keller and C. Heidelberger
2273	M. Gizin and J. Wirz	2398	R.J. Boyd et al.
2294	G. Schmid et al.	2600	R.C. Hahn and R.P. Johnson
2298	K.P. Prasad et al.	2605	A. Padwa et al.
2393	A.M. Becker and R.W. Rickards	2636	M.L. Casey et al.
2421	A. Eberle and R. Schwyzer	2683	P.S. Engel et al.
2469	A.E. Siegrist et al.	2826	J. Dabrowski and K. Kamienska-Trela
2499	B. Hardegger and S. Schatzmiller	2928	M.A. Winnik et al.
2551	M. Gerecke et al.	2947	G. Desimoni et al.
2566	S. Chaloupka et al.	3001	E.M. Kosower and H. Kanety-Londner
2635	E. Haselbach and M. Rossi	3007	R.T. LaLonde et al.
2704	O. Schindler et al.	3023	M. Hashimoto et al.
2724	R. Martin et al.	3049	M. Braun and G. Buchi
2727	W. Heinzelmann and P. Gilgen	3181	C. LeVanda et al.
2738	S. Iwasaki et al.	3242	A.M. Halpern and A.L. Lyons, Jr.
2753	S. Iwasaki et al.	3247	E.H. White et al.
2768	U. Schmid et al.	3262	T.D. Doyle et al.
2786	J. Streith et al.	3346	I.D. Jenkins et al.
2841	Y. Nakamura et al.	3393	K.B. Wiberg and M.E. Jason
2858	J.P. Kutney et al.	3555	A. Padwa et al.
2902	P. Margaretha	3564	A.G. Schultz and M.B. DeTar
2906	P. Skrabal and M. Hohl-Blumer	3579	R.L. Hillard, III and L.P.C. Voll-
2972	V. Skaric et al.		hardt
		3627	E. Ghera, Y. Gaoni and S. Shoua
34- -76, J. Chem. Eng. Data, 21 (1976)		3636	G.R. Krow et al.
0115	M.Y. Shandala et al.	3641	D. Eggerding and R. West
0118	M.Y. Shandala and N.H. Al-Jobour	3645	E. Wenkert et al.
0120	M.Y. Shandala and N.H. Al-Jobour	3655	R.L. Letsinger and W.B. Lunsford
0233	J.G. Pomonis et al.	3661	J.B. Heather et al.
0504	G.M. Gasparini and E. Polidori	3669	S.R. Wilson et al.
		3678	T.H. Barrows et al.
35- -76, J. Am. Chem. Soc., 98 (1976)		3721	U.R. Ghatak, B. Sanyal and S. Ghosh
0104	H.G. Kuivila et al.	3732	R.B. Banks and H.M. Walborsky
0109	H. Bock et al.	3910	S.W. Staley et al.
0171	P. Beak et al.	3987	N.J. Leonard et al.
0299	J.M. Liesch et al.	4174	R.M. Pollack and R.H. Kayser
0541	S.-S. Tseng and E.F. Ullman	4236	C.A. Bunton and R.J. Rubin
0558	H.G. Gardner and J.K. Kochi	4301	P.W. Jeffs and J.D. Scharver
0610	T. Fukunaga	4515	P.K. Wagner and D.A. Ersfeld
0611	T. Fukunaga	4577	P.T. Lansbury and R.W. Britt
0613	J. Bolte et al.	4668	R. Weiss et al.
0615	I. Agranat and Y. Tapuhi	4867	R.E. Lehr et al.
0628	R.P. Thummel	4975	S. Danishefsky and P. Cain
0635	M. Nakagawa, H. Okajima and T. Hino	5038	D.F. Covey and C.H. Robinson
0830	F. Yoneda et al.	5380	C.M. Harris et al.
0956	M.K. Kaloustian et al.	5408	N. Harada, Y. Takuma and H. Uda
1048	A. Padwa et al.	5412	C.J. Sih et al.

5414	Y. Shimizu et al.
5581	A. Padwa and A. Au
5597	W.S. Johnson and L.A. Bunes
5671	R.A. Abramovitch et al.
5689	K.E. Taylor and J.B. Jones
5706	G. Levin et al.
5715	E. Block et al.
5719	S.M. Kupchan et al.
5988	D.M. Jerina et al.
5996	D.J. Dawson et al.
6218	A. Couture et al.
6233	A.R. Gutierrez and D.G. Whitten
6350	W.H. Rastetter
6382	D. Seyferth et al.
6401	D.A. Jaeger
6405	A.G. Anastassiou et al.
6410	S. Akiyama et al.
6643	J. Meinwald et al.
6711	C.-I. Lin et al.
6714	J.L. Moniot and M. Shamma
6752	P. Warner and S.-L. Lu
7017	N. Langlois et al.
7040	J.R. Scheffer et al.
7064	M.F. Dampier et al.
7069	U. Hornemann et al.
7092	S.M. Kupchan et al.
7095	L.C. Dunn et al.
7108	R. Gottlieb and J.L. Newmeyer
7327	J.A. Goldstein et al.
7381	M.J. Robins et al.
7408	J.R. Barrio et al.
7680	H.E. Zimmerman et al.
7726	Z. Rappoport et al.
7733	T.M. Harris et al.
7777	W.R. Abrams and R.J. Kallen
7818	S.J. Lee et al.
7823	N.L. Holy et al.
7835	J.A. Bertrand et al.
7870	C.S. Cheng et al.
8125	P.J. Wagner et al.
8204	M.J. Robins et al.
8213	M.J. Robins et al.
8218	N.J. Leonard and D.F. Wiemer
8237	J.M. Liesch et al.
8268	A. Dagan and M. Rabinovitz
8277	S. Masamune et al.
8426	W. Priester and R. West
8446	J.G. O'Connor and P.M. Keehn

36-   -76, J. Pharm. Sci., 65 (1976)

0068	D.V.C. Awang and A. Vincent
0132	M.A. Elsohly et al.
0134	S.K. Gupta and S.A. Marathe
0274	D.V. Naik et al.
0294	V. Preininger et al.
0525	D. Alkalay et al.
0538	J.R. Dimmock et al.
0580	K.H. Lee et al.
0778	C.D. Hufford et al.
0908	J.W. Sowell, Sr. and C.D. Blanton
0918	R. Pal et al.
1545	A. Richardson, Jr. et al.
1549	S. Ghosal et al.
1733	J.J. Kaminski et al.
1794	L.J. Stoel et al.

39-   -76B, J. Chem. Soc., Perkin Trans. II, (1976)

0076	P.J. Baldry et al.
0211	M.J. Cook et al.
0309	M. Colonna et al.
0323	P.B. Koster and M.J. Janssen
0456	G. Aloisi and G. Favaro
0669	C.W. Funke and H. Cerfontain
0869	G. Favaro
1001	P. Beltrame et al.
1030	G. Bianchi and D. Maggi
1104	A.H.A. Tinnemans and W.H. Laarhoven
1115	A.H.A. Tinnemans and W.H. Laarhoven
1176	R.J. Badger and G.B. Barlin
1245	J. Arriau et al.
1249	N. Levi and D.S. Malament
1349	J.W. Pilkington and A.J. Waring
1404	G. Bombieri et al.
1428	J. Frank and A.R. Katritzky
1506	R.M. Banks and H. Maskill
1575	H.W. Gibson and F.C. Bailey
1594	R.P. Bell and P.E. Sørensen
1731	M.D. Cohen and R. Cohen
1773	I.R. Bellobono et al.
1829	G. Guiheneuf et al.
1845	A. Bigotto et al.
1850	P. Bentley and J.F. McKellar

39-   -76C, J. Chem. Soc., Perkin Trans. I, (1976)

0001	W.K. Anderson et al.
0005	J.M. Sprake and K.D. Watson
0008	N. Lander et al.
0023	W.H. Hui and M.M. Li
0038	D.H.R. Barton et al.
0042	J. Griffiths and B. Roozpeikar
0045	R.M. Acheson and G. Paglietti
0048	J. Griffiths and M. Lockwood
0063	S.T. Lu et al.
0075	G. Cooper and W.J. Irwin
0090	F. Bergmann et al.
0098	F. Chen and Y. Lin
0101	D.H.R. Barton et al.
0117	R. Sunder et al.
0125	J. Clark et al.
0146	B.A. Dadson and A. Minta
0147	P. Djura et al.
0160	R.T. Brown et al.
0162	G.M. Brooke et al.
0186	M.A. Fitzgerald et al.
0204	P.S. Steyn and R. Vleggaar
0228	R.M. Christie and D.H. Reid
0239	F. Bergmann and M. Rahat
0249	M.J. Bullivant and G. Pattenden
0264	P. Barraclough and D.W. Young
0296	M.J. Begley et al.
0304	M.J. Begley et al.
0312	J. Elguero et al.
0315	A.S. Afridi et al.
0323	I. Degani and R. Fochi
0336	R. Faragher and T.L. Gilchrist
0339	G.A. Swan et al.
0341	D. Farquhar et al.
0380	J. Larkin et al.
0394	J. Machin et al.

0399	D. Johnston and D.M. Smith	1269	P.J. Abbott et al.
0404	E. Suzuki and S. Inoue	1297	H. Suginome et al.
0407	S.M. Ali et al.	1312	H. Suginome et al.
0410	G.A. MacAlpine et al.	1327	R. Nutiu and A.J. Boulton
0421	G. Tennant and R.J.S. Vevers	1357	S.H. Nicholson and A.B. Turner
0433	D.S.R. East et al.	1361	B.A.J. Clark et al.
0442	T.B.H. McMurry and R.R. Talekar	1377	R.C. Ellis et al.
0456	A.J. Bellamy and J. Hunter	1389	M.M.E. Budran and W.B. Whalley
0459	D.N. Jones et al.	1392	E.O.P. Agbakwuru and W.B. Whalley
0465	R.M. Acheson and C.W.C. Harvey	1414	W.L.F. Armarego and P.A. Reece
0481	A.S. Bailey and P.A. Wilkinson	1424	G.B. Barlin
0484	N. Lander and R. Mechoulam	1432	E. Bombardelli et al.
0499	D.H.R. Barton et al.	1442	E. Glotter et al.
0503	D.H.R. Barton et al.	1458	W.H. Ansari et al.
0507	H.E. Foster and J. Hurst	1466	B.G. James et al.
0532	S. Kurata et al.	1476	B.G. James and G. Pattenden
0550	S.J. Fuerniss et al.	1492	E.J. Gray and M.F.G. Stevens
0561	R.E. Sioda et al.	1496	E.J. Gray et al.
0570	V. Bertini et al.	1507	G. Zvilichovsky and J. Feingers
0578	R.E. Cox et al.	1514	F.G. Schreiber and R. Stevenson
0584	A.G.W. Baxter and R.J. Stoodley	1523	I. Matsuda et al.
0592	D.E. Ames and C.J.A. Byrne	1528	I. Matsuda et al.
0595	F.M. Dean et al.	1538	P.H.J. Ooms et al.
0600	F.M. Dean et al.	1558	J. Cairns et al.
0613	J. Banville and P. Brassard	1570	L. Merlini and G. Nasini
0628	P.J. Machin and P.G. Sammes	1577	I. Fleming and T. Mah
0634	P.G. Duggan and W.S. Murphy	1639	S.T. Gore et al.
0669	E. Glotter et al.	1643	R.V. Cooms and R.P. Danna
0672	H. Kato et al.	1669	U.R. Ghatak et al.
0692	G. Purrello and P. Fiandaca	1688	M. Yamamoto
0695	C.W. Shoppee and Y. Wang	1696	H.E. Hogberg et al.
0710	S.S. Welankiwar and W.S. Murphy	1706	T. Takeshima et al.
0713	F. Yoneda et al.	1716	T. Kusumi et al.
0722	A.S. Sarma et al.	1718	E.W. Colvin et al.
0727	S.C. Eyley and D.H. Williams	1762	F.J. McQuillin and M. Wood
0731	S.C. Eyley and D.H. Williams	1784	D. Lloyd and H. McNab
0735	C.T. Bedford et al.	1792	C.W. Spangler and T.W. Hartford
0745	T. Hino et al.	1796	N. Harada et al.
0783	T. Sato et al.	1805	F. Yoneda et al.
0794	R. Bonnett et al.	1819	W.G. Wright
0821	D.H.R. Barton et al.	1838	P.A. Chaloner and A.B. Holmes
0826	J. Brynjolffssen	1847	W.F. Keir and H.C.S. Wood
0829	D.H.R. Barton and H. Patin	1852	J. Banville and P. Brassard
0850	R.E. Corbett and T.L. Chee	1857	V.S. Kamat et al.
0863	H. Kato et al.	1889	D. Nasipuri and S.K. Ghosh
0876	J.R. Hanson et al.	1901	N. Suzuki et al.
0880	R.M. Christie and D.H. Reid	1908	R.M. Acheson et al.
0909	A.R. Katritzky and D. Moderhack	1911	R.M. Acheson et al.
0913	R.P. Gellatly et al.	1915	M. Cavazza et al.
0967	G.B.V. Subramanian et al.	1942	D. St.C. Black et al.
0975	A.C.W. Curran	1944	D. St.C. Black et al.
0989	T.L. Gilchrist and D.P.J. Pearson	1951	D. St.C. Black et al.
0997	P.M. Brown and R.H. Thomson	1955	G.W. Alderson et al.
1056	A.R. Battersby et al.	1975	U.R. Ghatak et al.
1064	D.N. Kirk and M.S. Rajagopalan	1991	D.G. Doughty et al.
1073	D.E. Ames and O. Ribeiro	2004	J.W. Cornforth
1088	M. Kugelman et al.	2038	G. Cooper and W.J. Irwin
1155	M. Sainsbury and R.F. Schinazi	2050	D.W. Allen and A.C. Oades
1165	A.G. Brown et al.	2067	F.M. Dean et al.
1190	R.S. Shadbolt et al.	2079	T.G. Back et al.
1210	H. Singh et al.	2089	J.H. Adams et al.
1218	T. Kametani et al.	2103	S.R. Landor et al.
1241	G. Jones and J.R. Phipps	2112	D.H.R. Barton et al.
1244	I. Kirson et al.	2137	S.F. Tan and T.H. Tjia
1257	T.L. Gilchrist and D.P.J. Pearson	2144	M. Node et al.
1260	F.M. Dean and B.K. Park	2149	R.L. Edwards and H.J. Lockett

2155	T.J. Lillie et al.		0861	P. Bhattacharyya et al.
2197	H. Pande and D.S. Bhakuni		0913	D, Sen and P.K. Dutta
2207	J.V. Greenhill		1053	F.G. Baddar et al.
2241	J.I. Okogun		1067	B.K. Deshmukh and R.B. Kharat
2248	P.K. Brooke et al.		1223	H. Singh and R.K. Mehta
2254	A.S. Bailey et al.			

2277   C. Pene et al.

44-   -76, J. Org. Chem., 41 (1976)

2281   N. Dennis et al.   0010   W.F. Berkowitz and S.C. Grenetz
2285   N. Dennis et al.   0013   J.P. Ferris and R.W. Trimmer
2289   N. Dennis et al.   0019   J.P. Ferris and R.W. Trimmer
2296   N. Dennis et al.   0044   C.F. Bernasconi and R.H. de Rossi
2307   N. Dennis et al.   0066   W.J. McGahren et al.
2329   N. Dennis et al.   0078   M.L. Lewbart
2334   J. Banerji et al.   0084   F.G. De Las Heras
2338   N. Dennis et al.   0102   R.J. Owellen and C.A. Hartke
2363   A.M. Maione et al.   0110   A.L. Johnson and P.B. Sweetser
2386   B. Lythgoe et al.   0125   G.F. Koser and S.-M. Yu
2403   R.M. Harrison et al.   0129   K.T. Potts and J.L. Marshall
2407   M. Sato and H. Kakisawa 0158   I. Antonini et al.
2444   F.M. Dean et al.   0180   A. Padwa and P.H.J. Carlsen
2457   R.S. Atkinson and R.H. Green   0187   K.T. Potts et al.
2462   J.M.A. Al-Rawi and J.A. Elvidge 0214   H.D. Becker and K. Gustafsson
2470   E. Beska and P. Rapos   0221   M.L. Herz et al.
2484   D.A. Archer and B.W. Singer   0242   W.L. Mock and J.H. McCausland
2492   P.W. Couch et al.   0252   J.F. Johnson et al.
2501   D.E. Games et al.   0287   S. De Bernardo and M. Weigele
2521   K. Wasti and M.M. Joullie   0291   R.T. LaLonde and C. Wong
2533   J.E. McCormick and R.S. McElhinney 0294   R.C. Moschel and N.J. Leonard
2540   A.G.W. Baxter and R.J. Stoodley   0306   L.M. Lerner
2546   T.J. Lillie et al.   0313   F.H. Batzold and C.H. Robinson
2550   T.S. Burton et al.   0317   J.M. Saa et al.
2556   J.I. Grayson et al.   0351   R. Buchan et al.
2562   H. Matsukobo and H. Kato   0385   T. Novinson et al.
2565   H. Matsukobo and H. Kato   0388   M. Weigele et al.
2570   J. Carnduff and D.G. Leppard   0390   M. Weigele et al.
2587   M. Ikeda et al.   0400   Y. Kanaoka and Y. Hatanaka
2594   K.K. Purashothaman et al.   0491   R.A. Abramovitch and W.D. Holcomb
                                0497   J.G. Smith et al.

40-   -76, Nippon Kagaku Kaishi (1976)   0531   R.B. Garland et al.
0133   K. Fukunaga et al.   0543   A. Padwa et al.
0138   H. Iida et al.   0563   L.J. Dolby and G. Hanson
0144   K. Takahashi et al.   0567   B.S. Hahn and S.Y. Wang
0166   I. Naito et al.   0568   B.N. Holmes and N.J. Leonard
0451   H. Imai et al.   0593   C.R. Smith, Jr. et al.
0462   Y. Takanaka et al.   0600   F.W. Lichtenthaler and T. Kuli-
0488   Y. Yuki and K. Inoue   kowski
0497   Y. Muramoto and H. Asakura   0678   S. Borresen and J.K. Crandall
0631   Y. Takikawa et al.   0699   J.C. Bradley and G. Buchi
1100   C. Shin et al.   0707   B. Aweryn et al.
1513   H. Oda et al.   0711   A.M. van Leusen et al.
                                0729   F. Malek-Yazdi and M. Yalpani
42-   -76, J. Indian Chem. Soc., 53 (1976)   0730   H.K. Spencer et al.
0017   G.S. Gadaginamath and S. Siddappa   0743   D. Yang et al.
0038   G. Bagavant et al.   0745   J. Wolinsky et al.
0050   O.P. Vig et al.   0813   K.T. Potts et al.
0274   N.N. Ghosh and M.M. Nandi   0818   K.T. Potts et al.
0286   A.K. Koli   0836   A.L. Johnson
0321   P. Bhattacharyya et al.   0838   D.S. Tee and G.V. Patil
0331   N.N. Ghosh and M.M. Nandi   0855   H.O. House et al.
0454   M.M. Patel et al.   0863   H.O. House and L.F. Lee
0490   M. Gindy et al.   0870   T.J. Novak et al.
0531   A.K. Barua et al.   0875   B. Weinstein and A.R. Craig
0602   B.S. Patwa and A.R. Parikh   0879   M.V. Lakshmikanthan and M.P. Cava
0606   G.S. Gadginanath and S. Siddappa   0882   M.V. Lakshmikanthan and M.P. Cava
0812   D.N. Chatterjee and S.R. Chakra-   0884   S.H. Pines
        borty   0940   H.M. Walborsky and P.C. Collins

0977	H. Yagi et al.	2485	H.K. Spencer and R.K. Hill
0986	R.E. Ireland et al.	2496	Y. Becker et al.
1005	W.L. Meyer et al.	2509	N. Finch et al.
1021	W. Herz and R.P. Sharma	2514	R.T. Lalonde et al.
1041	T.C. Thurber and L.B. Townsend	2536	G.R. Newkome and J.M. Robinson
1064	M. Anastasia et al.	2542	T. Kametani et al.
1067	D.J. Aberhart et al.	2552	M.P. Li et al.
1089	U.R. Ghatak and S. Chakrabarty	2571	L.H. Klemm et al.
1095	A.D. Broom et al.	2663	J.S. Dutcher et al.
1100	T. Sasaki et al.	2670	J.E. Ellis et al.
1135	K.G. Taylor et al.	2679	E. Cavalieri and R. Roth
1141	K.G. Taylor and M.S. Clark, Jr.	2700	C. Levanda et al.
1146	K.G. Taylor et al.	2706	R.G. Harvey et al.
1229	M. Nakazaki et al.	2720	R.Y. Ning et al.
1244	S. Barcza et al.	2724	R.Y. Ning et al.
1256	K. Yamakawa and K. Nishitani	2732	S. Sunder et al.
1293	M. Shamma and J.E. Foy	2736	R.W. Leiby and N.D. Heindel
1320	A.L. Johnson et al.	2784	P.S. Venkataramani et al.
1389	T. Kajimoto et al.	2793	C.K. Chu et al.
1449	J.A. May, Jr. and L.B. Townsend	2850	J.F.W. Keana and P.E. Eckler
1466	J.W. Hines, Jr. et al.	2864	R.J. Chorvat and R. Pappo
1484	M. Mizuno et al.	2874	J.E. Baldwin et al.
1487	K. Okuhara	2943	R.R. Sauers et al.
1539	D. Caine et al.	2950	J.K. Stille et al.
1570	A. Kakehi et al.	2976	D.C. Dittmer et al.
1597	R. Shapiro and S.-J. Shiuey	2981	H. Feuer and L.F. Spinicelli
1635	P.V. Alston and R.M. Ottenbrite	2985	D.F. Wiemer and N.J. Leònard
1657	S. Teitel and J.P. O'Brien	2995	K. Kondo et al.
1724	K.T. Potts et al.	3000	C.M. Gupta et al.
1733	J.R. Becker and J.A. Yahner	3010	G.R. Owen et al.
1743	J. Szmuszkovicz et al.	3018	J. Banville and P. Brassard
1768	P.Y. Johnson and D.J. Kerkman	3027	A.D. Broom and D.G. Bartholomew
1799	M.E. Childs and W.P. Weber	3050	R.J. Boatman and H.W. Whitlock
1811	R.N. McDonald et al.	3067	H.O. House et al.
1822	R.N. McDonald et al.	3076	H.O. House and K.A.J. Snoble
1836	E.J. Prishe et al.	3083	H.O. House and C.-Y. Chu
1852	J.R. Cole et al.	3096	K.K. Andersen et al.
1855	S.J. Torrance et al.	3124	T. Huynh-Dihh et al.
1858	P.F. Wiley and F.A. Mackellar	3128	P.T. Berkowitz et al.
1886	M.J. Robins and T. Kanai	3133	B.A. Otter et al.
1889	D.W. Wiley et al.	3138	T. Sasaki et al.
1935	M. Minabe et al.	3191	R.S. Macomber and E.R. Kennedy
1943	S.K. Pradhan et al.	3210	S.M. Kupchan and C.-K. Kim
1952	W.J. Scott et al.	3275	M. Daudon et al.
1957	J.H. Sedon and P. Zuman	3287	G.L. Olson et al.
2006	M.W. Galley and R.C. Hahn	3312	L.R. Caswell et al.
2031	W.M. Welch	3374	J.A. Harvey and M.A. Ogliaruso
2112	D.B. Miller et al.	3377	H. Hart and E. Shih
2120	G. Baum and H. Schechter	3381	F.P. Tsui et al.
2124	J.F.W. Keana et al.	3389	P. Beak et al.
2162	K. Ohkata	3406	R.Y. Ning et al.
2201	G.R. Lenz	3438	E. Leete
2258	R. Weiss et al.	3441	J.N. Anderson and R.A. Martin
2279	N. Ishibe and M. Tamura	3468	S. Danishefsky et al.
2292	B.R. Samant and F. Sweet	3480	D.J. Vanderah and F.J. Schmitz
2296	W.W. Lee et al.	3481	S.M. Kupchan and D.R. Streelman
2303	S.M. Hecht et al.	3524	L.A. Paquette et al.
2312	J.E. Baldwin et al.	3529	G.E. Keyser and N.J. Leonard
2373	K.L. Kirk	3532	G.R. Lenz
2379	I. Agranat and E. Aharon-Shalom	3540	R.T. Murray and N.H. Cromwell
2396	D.C. Wigfield and D.J. Phelps	3546	A. Kascheres et al.
2401	H.O. House et al.	3549	E.S. Hand and W.W. Paudler
2435	A.P. Gray et al.	3583	E.J. Moriconi and C.F. Hummel
2465	S.J. Wratten and D.J. Faulkner	3599	A.R. Miller
2468	R.D. Stipanovic and A.A. Bell	3627	P. Jacob, III et al.
2477	S. Oh and C. Monder	3632	D. Caine et al.

3634   B. Weinstein et al.
3725   M. Nakazaki et al.
3760   T.R. Kowar and E. Le Goff
3775   M.K. Eberle and L. Brzechffa
3784   C. Temple, Jr. et al.
3820   J.D. Scribner
3824   J.D. Scribner
3827   J. Herscovici et al.
3840   J.F.W. Keana and R.R. Schumaker
3847   J.W. Huffman
3850   S.M. Kupchan et al.
3904   D. Eggerding and R. West
3931   G.W. Griffin et al.
3940   E.L. Shapiro et al.
3956   N.H. Fischer et al.
3967   I. Carelli et al.
4026   S.G. Levine et al.
4038   S. Hirano et al.
4041   E. Le Goff
4070   Z.H. Israili and E.E. Smissman
4074   M. Fuertes et al.
4081   M. Nakazaki et al.
4108   J.F. Blount et al.

46-   -76, J. Phys. Chem., 80 (1976)
0112   G. Gorin et al.
0122   Farhataziz and L.M. Perkey
0287   A.T. Jeffries, III and C. Parkanyi
0559   S.K. Vidyarthi et al.
0717   D. Nielsen et al.
0722   E.A. Nothnagel and R.N. Zitter
1804   M.P. Pileni et al.
2614   A. Castellano et al.

47-   -76, J. Polymer Sci., Polymer Chem.
       Ed., 14 (1976)
0319   D.J. Andrews and W.J. Feast
0845   H. Inouye and T. Otsu
1661   H.W. Gibson and F.C. Bailey
1901   K. Sugita et al.
2725   D. Bailey et al.
2983   G.J. Smets and S. Matsumoto

48-   -76, J. prakt. Chem., 318 (1976)
0001   E. Uhlig and D. Linke
0039   H. Schäfer et al.
0079   G. Etzold et al.
0127   E. Fanghängel et al.
0161   J. Faust and R. Mayer
0221   B. Bartho et al.
0291   P. Nuhn et al.
0298   R. Spitzner et al.
0313   K. Gewald and H.J. Jansch
0343   K. Gewald and M. Hentschel
0347   W. Schäfer and K. Gewald
0359   J. Wrubel et al.
0471   E. Schmitz et al.
0507   M. Mühlstädt et al.
0515   I.D. Postescu and D. Suciu
0663   K. Gewald and M. Hentschel
0705   J. Liebscher and H. Hartmann
0731   J. Liebscher and H. Hartmann
0745   J. Brunn et al.
0779   K. Gewald et al.
0816   N.R. El-Rayyes and N.A. Al-Salman
0823   B.M. Neumann et al.

0881   A. Gelleri et al.
0946   M. Schulz et al.

49-   -76, Monatsh. Chem., 107 (1976)
0195   C. Krieger et al.
0299   H. Berner and H. Reinshagen
0307   H. Falk and J.M. Ribo
0501   R. Franzmair
0511   R. Franzmair
0663   H. Bartsch
0783   O.S. Wolfbeis
0831   H. Falk et al.
0879   R. Ott et al.
0907   H. Falk et al.
1413   K. Gewald and G. Heinhold

54-   -76, Rec. trav. chim., 95 (1976)
0108   J.H. van Boom et al.
0154   D.J. Sikkema et al.
0165   W.H. Laerhoven and T.J.H.M. Cuppen
0223   U.K. Pandit and H. Bieraugel
0285   J. Verbeek et al.

56-   -76, Roczniki Chem., 50 (1976)
0073   H. Poradowska and J. Bajgrowicz
0097   M. Kocor et al.
0203   W. Jasiobedzki and J. Wozniak
0347   A.M. El-Nagger et al.
0363   R. Piekos and M. Sankowski
0451   M. Wozniak et al.
0499   S. Goszczynski et al.
0625   W. Zankowska-Jasinska and A. Kolasa
0759   J. Gronowska and H. Dabkowska
0815   W.J. Rodewald and W.J. Szczepek
0857   M. Jawdosink and M. Makosza
0967   M. Kuczek and K. Nowak
0973   A. Erndt and M. Zubek
1067   W. Basinski and Z. Jerzmanowska
1257   Z. Jankowski and R. Stolarski
1267   A. Ledochowski et al.
1301   W. Jasiobedzki et al.
1523   W.E. Hahn and B. Rybczynski
1603   A. Ledochowski and S. Skonieczny
1805   K. Golankiewicz and J. Langer
1919   M. Kocor and A. Kurek
1931   J. St. Pyrek et al.
2041   K. Golankiewicz and H. Koroniak
2087   M. Kocor et al.
2095   W.M. Daniewski et al.
2193   K. Krowicki

59-   -76, Spectrochim. Acta, 32A (1976)
0269   Y. Beauchamp and G. Durocher
0351   S. Trovato et al.
0501   S. Fisichella et al.
0837   F.G. Baddar et al.
1015   G. Peyronel et al.
1311   S. Badilescu and A.T. Balaban
1415   J.J. Worman et al.
1593   H.S. Blair and G.A.F. Roberts

60-   -76, J. Chem. Soc., Faraday Trans. II
1574   J. Metcalfe and D. Phillips

61-   -76, Ber. Bunsen-Gesellschaft. Phys.
           Chem., 80 (1976)
0288   H. Strahle et al.
0301   H. Bisle et al.

62-   -76A, Z. phys. Chem.(Leipzig), 257
           (1976)
0249   H. Khalifa et al.
0497   D.K. Hazra et al.

64-   -76B, Z. Naturforsch., 31b (1976)
0115   N.H. Pirzada et al.
0122   N.H. Pirzada and L.A. Summers
0251   G. Entenmann
0330   L.J. Tusek et al.
0371   G. Zacharias and H. Junek
0514   O.S. Wolfbeis and E. Ziegler
0533   A. Rahman et al.
0807   W.H. Gundel
0876   H. Wagner et al.
0965   W. Gierke et al.
0981   G. Entenmann et al.
1009   H. Gusten and E.F. Ullman
1248   H. Strahle et al.
1317   N. Wiberg and G. Hubler

64-   -76C, Z. Naturforsch., 31c (1976)
0001   O. Sciacovelli et al.
0101   R. Harcus et al.
0263   F. Seela and F. Hansske
0328   N.J.F. Dodd et al.
0389   F. Seela
0403   S. Imre and A. Oztunc

65-   -76, Zhur. Obshchei Khim., 46 (1976)
           (English translation pagination)
0174   N.P. Lyalyakina et al.
0239   V.P. Kukhar' and T.N. Kasheva
0390   N.S. Koslov et al.
0435   A.N. Smirnov et al.
0664   L.G. Tish
0702   N.A. Akmanova et al.
0807   O.S. Anisimova et al.
0912   G.A. Tolstikov et al.
1044   A.I. Nogaideli et al.
1108   N.N. Chipanina et al.
1141   N. Singh and K. Krishnan
1170   V.A. Mironov et al.
1350   E.A. Obol'nikova et al.
1358   A.V. Belotsvetov et al.
1370   A.A. Akhrem et al.
1378   A.A. Akhrem et al.
1381   M.P. Irismetov et al.
1424   R.D. Gareev and A.N. Pudovik
1598   M.K. Shakhova et al.
1642   L.I. Mizrakh et al.
1670   R.D. Gareev and A.N. Pudovik
1850   V.V. Alenin and V.D. Domkin
2037   V.M. Potapov et al.
2044   Y. Daugvila et al.
2057   V.D. Orlov et al.
2135   I.V. Megera and M.I. Shevchuk
2273   R.K. Gayanov et al.
2512   G.N. Yashchenko et al.
2619   K.I. Pashkevich et al.
2630   M.S. Miftakhov et al.

69-   -76, Biochemistry, 15 (1976)
0114   A.L. Maycock et al.
0217   J.P. Miller et al.
0576   K.O. Martin and C. Monder
0898   H. Kasai et al.
1005   S.M. Hecht et al.
1408   Y. Sasaki et al.
1791   S. Ghisla et al.
2062   E.J. Gabbay et al.
3724   W.L. Dills, Jr. et al.
3783   F.B. Howard et al.
4139   V.P. Marshall et al.
4730   F.L. Bellino et al.

70-   -76, Izvest. Akad. Nauk S.S.S.R.,
           25 (1976)
0103   G.I. Nikishin et al.
0150   A.A. Akhrem et al.
0174   R.R. Shagidullin et al.
0384   A.M. Kritsyn et al.
0396   A.K. Khusid et al.
0428   A.K. Khusid et al.
0455   G.T. Katvalyan et al.
0465   Y.A. Levin and M.S. Skorobogatova
0547   U.M. Dzhemilev et al.
0577   S.A. Krasnaya et al.
0602   O.M. Nefedov et al.
0652   G.V. Kryshtal' et al.
0683   U.M. Dzhemilev et al.
0762   E.S. Petrov et al.
0837   D.P. Del'tsova and N.P. Gombaryan
0841   U.M. Dzhemilev et al.
0909   G.V. Kryshtal' et al.
0917   N.V. Lyukshova et al.
0986   T.N. Filatova and Y.I. Khurgin
1068   N.S. Mirzabekyants et al.
1098   A.V. Kamernitskii et al.
1101   A.V. Kamernitskii et al.
1115   A.G. Kozlovskii et al.
1162   A.Y. Lazaris and A.N. Egorochkin
1163   A.Y. Lazaris and A.N. Egorochkin
1306   A.A. Akhrem et al.
1353   S.I. Zav'yalov et al.
1388   S.V. Larionov and L.A. Kosareva
1498   Y.L. Gol'dfarb and M.A. Kalik
1558   Y.N. Porshnev et al.
1780   G.F. Bannikov et al.
1906   A. Shevelev et al.
1913   A.N. Mirskova et al.
1947   A.V. Kamernitskii et al.
1997   S.A. Amitina and L.B. Volodarskii
2007   G.F. Bannikov et al.
2141   V.M. Karpov et al.
2214   B.M. Lerman et al.
2241   A.G. Ginzburg et al.
2353   B.M. Lerman et al.
2595   A.A. Akhrem et al.

73-   -76, Coll. Czech. Chem. Comm., 41 (1976)
0262   R.G. Salih and M.Y. Shandala
0271   E. Klintova et al.
0311   I. Basnak and J. Farkas
0443   J.O. Jilek et al.
0590   J. Kavalek et al.
0614   D. Vegh et al.
0800   L. Kalvoda et al.

REFERENCES

0881   I. Cervena et al.
0906   V. Valenta and M. Protiva
1035   M. Protiva et al.
1182   M. Uher et al.
1208   P. Jacquignon and O. Perin-Roussel
1219   H. Suguna and B.R. Pai
1225   K. Hejno and F. Sorm
1363   A. Mistr et al.
1388   M. Dzurilla and P. Kristian
1396   K. Sindelar et al.
1551   M. Uher et al.
1557   M. Prochazka
1565   J. Liebscher and H. Hartmann
1676   B. Koutek et al.
1692   F. Povazanec et al.
1954   M. Krojidlo et al.
2020   Z. Vejdelek et al.
2047   J. Harmatha et al.
2059   J. Farkas et al.
2096   A. Holy
2110   P. Drasar et al.
2216   L. Drobnica et al.
2422   D. Vegh et al.
2571   R. Kada et al.
2577   A. Krutosikova and J. Kovac
2614   B. Koutek et al.
2771   R. Smrz et al.
2788   V. Cerny et al.
2952   N.T. Hghia et al.
3085   E. Jedlovska et al.
3146   H. Potesilova et al.
3157   S. Pavelka and E. Smekal
3335   D. Cech and A. Holy
3371   A. Jurasek et al.
3391   M. Hrdina and A. Jurasek
3415   A. Cerny et al.
3420   R. Smrz et al.
3509   J. Plesek et al.
3607   V. Valenta et al.
3635   I. Busnak and J. Farkas
3654   S. Pavelka and J. Kovar

77-   -76, J. Chem. Soc., Chem. Comm. (1976)
0021   M. Wakselman and E. Guibe-Jampel
0030   F. Toda and D. Nan
0034   F. Bellesia et al.
0053   B. Fuchs et al.
0083   R. Ramasseul et al.
0091   S.M. Kupchan et al.
0120B   P.S. Clezy and C.J.R. Fookes
0148   E.M. Engler et al.
0173   R.W. Alder et al.
0177   F. Toda and K. Tanaka
0198   J. Libman
0235   J.L. Ripoll
0236   A.G. Padilla et al.
0241   T.J. King et al.
0261   D.E. Goldberg et al.
0269   M.J. Robins and W.H. Muhs
0270   A.W. Dunn et al.
0313   A.G. Anastassiou and E. Reichmanis
0331   J.C.A. Windhorst
0338   M. Nagai et al.
0340   K. Wada and T. Ishida
0402   R.W. Doskotch et al.
0430   Y. Kobayashi et al.

0446   Y. Kitahara et al.
0505   I. Tse and V. Snieckus
0510   M. Damak et al.
0566   N.N. Girotra et al.
0659   A.G.M. Barrett et al.
0728   C. Kabuto et al.
0729   N.C. Yang et al.
0787   K. Kimura et al.
0818   S.J. Martinez and J.A. Joule
0823   J.A. Barltrop et al.
0848   F. Toda and Y. Todo
0862   K.J. Dignam and A.F. Hegarty
0881   J.E. Baldwin et al.
0899   A.S. Afridi et al.
0949   I. Kubo et al.
0964   T.J. van Bergen and R.M. Kellogg
0966   H.K. Spencer et al.
1010   F. Toda and K. Tanaka
1045   K. Matsumoto et al.

78-   -76, Tetrahedron, 32 (1976)
0043   M. Ikehara, Y. Ogiso and T. Morii
0051   J. Klein et al.
0057   G. Doleschall
0065   A.K. Sarkar et al.
0073   A.K. Sarkar and A. Chatterjee
0079   M. Harnik et al.
0089   P.J. Barnes et al.
0107   G. Cardillo et al.
0109   S. Huneck
0147   R. Allmann et al.
0173   O. Ceder and B. Beijer
0211   M. Shamma and V. St. Georgiev
0217   K. Kaninuma et al.
0239   A. Rassat and J. Ronzaud
0257   A.C. Greiner et al.
0269   C.W. Bird et al.
0275   G.F. Griffiths et al.
0309   A.W. Burgstahler et al.
0333   D.-M. Yang et al.
0341   R. Faure et al.
0431   Y. Tamura et al.
0441   P. Courtot and R. Rumin
0467   V. Bhujle et al.
0487   S. Brenner and B. Alterovich
0503   S. Ahmad and S. Razaq
0507   T. Sato and T. Hino
0543   G.D. Manners
0571   G.C. Barrett and R. Walker
0579   G.C. Barrett and R. Walker
0583   G.C. Barrett and R. Walker
0597   M.F. Hudson and K.M. Smith
0615   H. Fischer and L.A. Summers
0619   M. Schneider and H. Strohacker
0661   R. Mukherjee and R.M. Moriarty
0665   K.L. Stevens and L. Jurd
0765   T.C. Jain et al.
0797   J. Pusset and R. Beugelmans
0839   K. Lempert et al.
0847   T. Polanski and K. Prajer
0919   M.M. Kadooka et al.
0991   N.N. Girotra et al.
1019   M. Barczai-Beke et al.
1025   J. Dabrowski et al.
1051   O. Achmatowicz, Jr. et al.
1077   R.H. Takata and P.J. Scheuer

1081	C.R. Ellefson et al.
1085	M.A. Wuonola and R.B. Woodward
1097	J.J. Kaminski and N. Bodor
1171	Y.M. Sheikh et al.
1189	C.C. McCarney et al.
1221	J. Skarzewski and Z. Skrowaczewska
1267	A.H. Beckett et al.
1293	H. Hauptmann
1331	N. Ishibe and S. Yutaka
1353	A.A. Bell et al.
1375	R. Jacquesy and H.L. Ung
1391	P.D. Gokhale et al.
1407	F. Chioccara et al.
1411	R. Kada et al.
1549	M.P. Mahajan et al.
1699	J.-C. Jacquesy et al.
1713	T. Tanaka et al.
1735	G. Doleschall and K. Lempert
1767	M.P. Carmody et al.
1773	M.J. Cook et al.
1793	J.A.P. Baptista de Almeida
1829	I. Barrow and A.E. Pedler
1839	J. Klein et al.
1851	H.H. Wasserman et al.
1855	H.H. Wasserman et al.
1863	H.H. Wasserman et al.
1867	H.H. Wasserman et al.
1881	B.A. Burke et al.
1899	A. Chiaroni et al.
1973	J. Imai, Y. Kondo and T. Takemoto
2021	L. Kutschabsky et al.
2033	F. Chioccara et al.
2053	J.M. Coxon and M.A. Battiste
2057	A.E. Wick et al.
2099	H. Molines and C. Wakselman
2131	G. Markl and H. Hauptmann
2153	M.S. Baird and C.B. Reese
2207	N. Tsuji et al.
2219	W. Schroth and G.W. Fischer
2225	G.W. Fischer and W. Schroth
2239	H. Schmidt et al.
2265	M. Akhtar et al.
2303	H.I.X. Mager and W. Berends
2313	A. Kelecom et al.
2339	D. Lloyd et al.
2353	A. Kelecom et al.
2375	P. Collum et al.
2379	P. Singh and M. Weinreb
2381	S. Hirano, T. Hiyama and H. Nozaki
2401	D. Bondon et al.
2407	A. Chatterjee, D. Ganguly and R. Sen
2409	S.N. Mikhailov et al.
2427	Y. Kashman et al.
2533	G. Rousseau et al.
2539	F. Khuong-Huu et al.
2545	B. Mompon and R. Toubiana
2571	H.P. Figeys et al.
2625	D.W. Nagel et al.
2633	J.P. Majoral et al.
2647	G. Surpateanu et al.
2681	C.W. Spangler
2693	H. Ishii et al.
2725	F. Bourelle-Wargnier and R. Jeanne-Carlier
2731	A. Mitra and M. Duttagupta

2753	G.W. Kenner et al.
2757	A.H. Jackson et al.
2795	R.C. Bleackley et al.
2805	A.Y. Meyer et al.
2827	P. Pasman et al.
2839	L. Diatta et al.
2843	F.X. Woolard et al.
2923	D. De Keukeleire et al.
2931	M.M. El-Abadelah et al.
2939	J.P.C.M. Van Dongen et al.
2967	A.A.M. Roof et al.
2991	D. Dunaway-Mariano
2997	B. Green et al.
3003	M. Moriyama et al.
3031	C. Hubschwerlen et al.
3041	J.Y. Becker

80- -76, Revue Roumaine Chim., 21 (1976)

0113	D. Suciu and G. Gsavassy
0193	V. Macovei
0241	A.T. Balaban
0257	F. Urseanu et al.
0781	R.M. Issa et al.
1073	I. Simiti and A. Muresan
1127	L. Ciohadaru and A. Meghea
1227	F. Urseanu
1345	A. Boldea et al.
1497	E. Dragu et al.
1543	I. Dragota et al.

83- -76, Arch. Pharm., 309 (1976)

0018	K. Gorlitzer
0034	G. Seitz and H. Braun
0072	J. Knabe and G. Link
0081	H.J. Roth and W. Kok
0092	H.J. Roth and W. Kok
0131	E. Röder and U. Franke
0152	J. Reisch et al.
0185	U. Franke and E. Röder
0197	H. Möhrle and K. Sieker
0316	J. Reisch and S.K. Kapoor
0333	G. Willuhn and H.-D. Herrmann
0356	K. Görlitzer
0376	K. Görlitzer et al.
0425	F. Eiden and C. Herdeis
0467	J. Reisch and W.F. Ossenkop
0512	R. Haller and U. Busser
0592	J. Schnekenburger and D. Heber
0631	P. Trommer et al.
0638	G. Rücker and E. Dyck
0679	G. Seitz and T. Kampchen
0735	M. Takacs et al.
0769	H. Weber
0829	C.H. Brieskorn and R. Pöhlmann
0928	K.C. Liu et al.
0937	U. Franke et al.

87- -76, J. Med. Chem., 19 (1976)

0040	D.T. Witiak et al.
0131	E.A. Coats et al.
0165	E.E. Smissman et al.
0192	R.B. Moffett et al.
0194	E.C. Ressner et al.
0202	L.J. Fliedner, Jr. et al.
0291	R.H. Springer et al.
0395	G. Tong et al.

0414	M.H. Rosen et al.
0419	H. Uno et al.
0426	G.H. Denny et al.
0445	H.G. Pars et al.
0521	F.I. Carroll et al.
0549	R.K. Razdan et al.
0555	C.I. Hong et al.
0643	T.L. Chwang et al.
0654	E.K. Hamamura et al.
0663	E.K. Hamamura et al.
0667	E.K. Hamamura et al.
0684	C.-D. Chang and J.K. Coward
0692	E.R. Clark et al.
0729	J. Olivard et al.
0814	D.G. Bartholomew et al.
0839	W.D. Kingsbury et al.
0882	R.K.Y. Zee-Cheng and C.C. Cheng
0892	M.J. Kornet and A.P. Thio
0899	G. Cehovic et al.
0903	A. Kampf et al.
0909	A. Kampf et al.
0915	T.-S. Lin et al.
0975	R.M. Weier and L.M. Hofmann
0999	W. Korytnyk et al.
1002	B. Paul and W. Korytnyk
1013	W.J. Wechter et al.
1020	P.C. Srivastava et al.
1026	G.R. Revankar et al.
1072	R.A. Long et al.
1094	R.T. Borchardt et al.
1130	S.M. Kupchan et al.
1161	C.R. Ellefson and J.W. Cusic
1168	T.L. Nguyen et al.
1186	R.D. Elliott and J.A. Montgomery
1195	J.M. Grisar et al.
1214	D.T. Witiak et al.
1265	A. Rosowsky et al.
1279	A. Hampton et al.
1330	D.S. Fullerton et al.
1345	C.R. Ellefson and J.W. Cusic
1378	A. Walser et al.

88- -76, Tetrahedron Letters (1976)

0045	A. Dhainaut et al.
0067	H. Wagner et al.
0109	S. Kogiso et al.
0153	G. Belot and C. Degrand
0209	A.C. Brouwer and H.J.T. Bos
0211	A. Hamon et al.
0251	S. Kuroda et al.
0275	D. Michelot and G. Linstrumelle
0297	J.-L. Fourrey
0301	J.-L. Fourrey and J. Moron
0313	R. Davies
0375	Y. Maki and M. Sako
0423	E.M. Engler and V.V. Patel
0435	R. Beugelmans et al.
0481	C. Jongsma et al.
0513	J.M. Saa et al.
0521	R.F. Abdulla
0593	S.S. Hecht et al.
0649	D.G.I. Kingston et al.
0665	G. Markl and H. Kellerer
0731	W. Fenical et al.
0767	S. Shiotani and T. Kometani
0779	J. C. Damiano et al.

0839	M. Oda et al.
0851	Y. Maki and M. Sako
0863	A. Kumar et al.
0887	W.R. Dolbier, Jr. et al.
0945	M. Ballester et al.
1013	N. Ojima et al.
1043	R. Luhowy and P.M. Keehn
1055	S.W. Pelletier et al.
1063	G. Just and S. Kim
1085	F. Reis et al.
1089	J. Rigaudy et al.
1117	Y. Fujise et al.
1129	K. Senga et al.
1223	I. Willner and M. Rabinovitz
1233	J.T. Baker and C.C. Duke
1243	E. Doomes and R.M. Beard
1251	C.L. Semmelhack
1265	L. Eisenhuth and H. Hopf
1351	S. Sekita et al.
1359	Y. Ikeya et al.
1389	L. Hevesi
1395	J.C. Jochims and G. Karich
1415	M. Shamma and L.A. Smeltz
1509	H. Horita et al.
1539	A.B. Holmes et al.
1557	J.V. Silverton et al.
1565	J.F.M. Oth et al.
1595	S. Naruto et al.
1601	H. Nagasawa et al.
1631	V. Vecchietti et al.
1637	D.G. Miller et al.
1659	R. Kaiser and D. Lamparsky
1661	R. Kreher and K.J. Herd
1687	T. Azuma et al.
1719	C. Berg et al.
1737	A.R. Cashmore et al.
1741	L. Jurd
1753	D.R. White
1825	T. Kurihara and M. Mori
1873	R. Besselievre and H.-P. Husson
1881	T. Satake et al.
1903	T. Reffstrup and P.M. Boll
1931	B.R. Vogt et al.
1987	H. Iwamura et al.
2005	J. Liebscher and H. Hartmann
2045	N. Kato et al.
2113	J.F.W. Keana and R.H. Morse
2129	H. Prinzbach et al.
2153	T. Satake et al.
2163	H. Gotthardt and F. Reiter
2167	Y. Gaoni
2241	R. Hansel et al.
2277	G. Felix et al.
2321	K. Harada and S. Suzuki
2327	I. Kitagawa et al.
2339	M. Hirama and S. Ito
2343	T. Teshima et al.
2347	M. Ikeda et al.
2391	M. Minhaj et al.
2405	M.T. Wu et al.
2427	J. Perman et al.
2441	J. Ficini and J. d'Angelo
2459	I. Saito et al.
2489	P.S. Manchand and J.F. Blount
2501	W.H. Watson et al.
2539	T. Furuya et al.

2541	W. Takagi et al.
2623	S. Nakatsuji et al.
2663	L.T. Scott and G.J. DeCicco
2677	M.J. Green et al.
2715	H.N.C. Wong et al.
2815	D. Davalian and P.J. Garratt
2861	M. Yamazaki et al.
2899	D. Clerin et al.
2909	D.H. Bremner et al.
2959	N. Dennis et al.
2971	S. Imoue et al.
2997	P.K. Ghosal et al.
3005	S. Daluge and R. Vince
3049	I. Addae-Mensah et al.
3069	K.E. Malterud et al.
3161	S. Ito et al.
3295	A. Mondon et al.
3299	W. Eberbach and J.C. Carré
3303	W. Eberback and J.C. Carré
3447	C. Besselievre et al.
3513	H. Straub
3515	H. Ehrhardt and S. Hunig
3559	M. Leboeuf et al.
3563	H. Gotthardt and S. Nieberl
3567	J. Bruneton et al.
3591	G. Maier and H.P. Reisenauer
3619	Y. Asakawa et al.
3631	L. Lombardo and F. Sondheimer
3649	K.H. Pauly and H. Dürr
3653	G.J. Bird et al.
3665	M.L. Kaplan and E.A. Truesdale
3687	W.-N. Wu et al.
3715	Y. Kobayashi et al.
3717	T. Yoshida et al.
3739	P.H. Bentley et al.
3775	E.J. Parish and G.J. Schroepfer, Jr.
3807	H. Abe et al.
3815	R. Okazaki et al.
3827	M. Matsumoto and H. Minato
3833	E. Tsankova and I. Ognyanov
3841	L. Lombardo and F. Sondheimer
3899	H. Horita et al.
3903	T. Miyashi et al.
3945	Y. Langlois et al.
3961	A. Ichihara et al.
3979	A. Brandt et al.
3989	G. Adam and T. v. Sung
4009	G.H. Barnett et al.
4019	S.D. Young and W.T. Borden
4023	A.G. Andrewes et al.
4049	H. Durr and V. Fuchs
4065	G. Dauphin and D. Planat
4095	T. Shibata et al.
4119	S. Jolad et al.
4125	R.P. Gajewski
4143	G. Markl and J. B. Rampal
4147	C. Keller-Juslen
4177	P.J. Garratt and W. Koller
4183	Z. Goldschmidt and S. Mauda
4227	H.H. Sun et al.
4241	K.P. Guha et al.
4271	A. Chollet et al.
4311	V.W. Armstrong et al.
4367	H. Seto et al.
4371	K. Matsui
4401	E.J. Parish et al.

4409	G.L. Lange et al.
4427	C.L. Liotta and A. Abidaud
4455	D.B. Stierle et al.
4473	M. Uyegaki et al.
4477	H. Ogoshi et al.
4485	M. Ikehara et al.
4509	M.S. Raju et al.
4521	J.E.L. McDonald and J.S. Roberts
4529	A. Dagan and M. Rabinovitz
4559	D.T. Longone and J.A. Gladysz
4609	H. Pirelahi et al.
4621	J. Bellan et al.
4633	A. Sarkar et al.
4655	K. Takahashi et al.
4669	V.R. Haddon and H. Chen
4713	R. Neidlein and A.D. Kraemer
4753	I. Saito et al.
4761	Z. Paryzek et al.
4825	V.N. Knyazev et al.
4851	E. Proksch and A. de Meijere
4859	J. Streith and J.-L. Schuppiser
4863	B. Evans et al.
4883	H.-D. Becker and K. Gustafsson
4887	E. Ali et al.

89-   -76, Angew. Chem., 15 (1976)
0052	G. Pfeiffer et al.
0053B	K. Heger and W. Grimme
0104	K. Hafner et al.
0108	K. Hafner et al.
0117B	H.N.C. Wong and F. Sondheimer
0160	A. Nishinaga et al.
0165	K. Menke and H. Hopf
0228	M. Schafer-Ridder et al.
0229	E. Vogel et al.
0240	K. Yamamoto and I. Murata
0365	U.E. Meissner et al.
0372	J. Leitich
0427	H. Flohr et al.
0441	G. Kaupp and I. Zimmermann
0442	G. Kaupp
0443	M. Janda and P. Hemmerich
0487	Y. Kitahara et al.
0487B	Y. Kitahara et al.
0611	K. Nakasuji et al.
0690	G. Markl and J.B. Rampal
0696B	R. Appel and M. Halstenberg
0704	A.H. Schmidt et al.
0766	S. Kato et al.
0775	R. Neidlein and H. Seel
0779B	G. Kaupp and E. Jostkleigrewe

90-   -76, J. Inorg. Nucl. Chem., 38 (1976)
| 0125 | T. Jarolim and J. Podlahova |
| 1357 | M. Tavares and M.A. Gouveia |

93-   -76, J. Applied Chem. U.S.S.R., 49 (1976)
| 0163 | A.V. El'tsov et al. |

94-   -76, Chem. Pharm. Bull.(Japan), 24 (1976)
0026	T. Hirayama et al.
0052	K. Ito et al.
0092	Y. Konda et al.
0108	H. Okabe et al.

0114	F. Sakai et al.
0147	S. Kadoya et al.
0169	H. Koshiyama et al.
0173	T. Murakami et al.
0178	T. Honda et al.
0181	M. Kimura and T. Miura
0220	M. Kozawa et al.
0235	Y. Maki
0266	I. Kitagawa et al.
0275	I. Kitagawa et al.
0294	I. Kitagawa et al.
0360	Y. Moriyama and T. Takahashi
0369	M. Sekiya et al.
0375	Y. Kurasawa et al.
0394	K. Honma et al.
0407	H. Fujiwara et al.
0443	H. Seto et al.
0450	H. Muroi et al.
0463	M. Muroi et al.
0487	Y. Kurasawa et al.
0507	T. Hirayama et al.
0549	T. Murakami et al.
0552	K. Murase et al.
0565	M. Ikehara and T. Maruyama
0580	M. Ito et al.
0591	T. Koyama et al.
0596	T. Ueda, I. Ito and Y. Itaka
0607	K. Suzuki et al.
0613	C. Takahashi et al.
0661	G. Kusano et al.
0672	M. Ikehara and M. Muraoka
0716	S. Kobayashi et al.
0752	T. Saitoh et al.
0782	H. Otomasu et al.
0815	M. Arisawa and N. Morita
0828	H. Sakamoto and M. Kato
0852	H. Yasuo et al.
0859	Y. Kurasawa et al.
0912	K. Matsumura et al.
0948	K. Matsumura et al.
0970	Y. Furukawa et al.
0991	T. Kinoshita et al.
1013	K. Shudo and T. Okamoto
1040	H. Kohda et al.
1045	M. Hanaoka et al.
1068	K. Ueno et al.
1106	T. Kurihara et al.
1160	Y. Tamura et al.
1189	I. Ito, N. Oda and T. Kato
1216	Y. Takeda et al.
1242	T. Saitoh et al.
1299	K. Mizuyama et al.
1331	M. Sekiya et al.
1390	T. Itoh et al.
1400	M. Ikeda et al.
1403	K. Sakurai et al.
1406	H. Inouye et al.
1419	N. Tanaka et al.
1451	T. Hisano and M. Ichikawa
1459	T. Koyama et al.
1485	M. Degawa and Y. Hashimoto
1514	M. Takeda et al.
1532	T. Ohmoto et al.
1544	T. Kato et al.
1552	H. Seto et al.
1561	N. Yamaji et al.

1581	T. Jojima et al.
1588	T. Jojima et al.
1609	M. Arisawa et al.
1668	H. Taguchi
1671	Y. Tominaga et al.
1724	S. Hagishita and K. Kuriyama
1731	M. Akiba et al.
1749	A. Nakagawa et al.
1795	M. Onda et al.
1813	I. Takeuchi and Y. Hamada
1853	Y. Yamamoto et al.
1950	D. Satoh and T. Hashimoto
1961	T. Murakami et al.
2052	T. Naka and M. Honjo
2089	T. Miyoshi et al.
2102	M. Ogawa and Y. Ogihara
2185	H. Seto et al.
2191	S. Kobayashi et al.
2219	Y. Kobayashi et al.
2241	T. Murakami et al.
2250	Y. Maki and M. Sako
2274	K. Kaji and M. Kawase
2305	T. Sawayama et al.
2317	C. Takahashi et al.
2409	O. Iwamoto et al.
2421	M. Watanabe et al.
2461	T. Kato, N. Oda and I. Ito
2494	O. Matsuda and M. Sekiya
2532	M. Sakamoto et al.
2568	Y. Tamura et al.
2585	T. Koyama et al.
2629	B. Shimizu et al.
2637	Y. Kimura et al.
2644	Y. Takeda et al.
2751	K. Kawashima and Y. Kawano
2810	K. Yamakawa and K. Nishitani
2886	H. Okabe and T. Yamauchi
2898	T. Nomura et al.
2903	M. Sato and Y. Mizuno
2955	T. Nakanishi et al.
2969	S. Ohta and S. Kimoto
2977	S. Ohta and S. Kimoto
3001	S. Kishimoto et al.
3011	T. Yamazaki et al.
3085	H. Bando et al.
3135	Y. Maki and T. Hiramitsu
3149	Y. Yabe et al.

95-   -76, J. Pharm. Soc. Japan, 96 (1976)

0044	K. Ogawa et al.
0154	S. Kimoto et al.
0160	Y. Fujiwara et al.
0199	H. Sone et al.
0246	T. Endo et al.
0254	M. Komatsu et al.
0381	M. Takido et al.
0527	C. Tani et al.
0631	H. Irie et al.
0683	S. Sakina and K. Aota
0700	S. Goya et al.
0725	H. Sawanishi and Y. Kamiya
0764	S. Nakajima et al.
0863	S. Nakajima et al.
0919	Y. Terada et al.
0927	T. Yamaguchi et al.
0962	C. Tanaka et al.

0968   E. Hayashi and A. Miyashita
0984   M. Takahashi et al.
1031   T. Kametani et al.
1089   Y. Nisikawa et al.
1094   M. Yoshimura et al.
1161   H. Ishii et al.
1180   N. Morita et al.
1217   S. Takagi et al.
1322   Y. Nishikawa et al.
1334   T. Okano et al.
1352   T. Higashino et al.
1426   T. Sakano et al.
1453   S. Fukushima et al.
1458   H. Ishii et al.

96-   -76, The Analyst, 101 (1976)
0469   M.R. Smyth et al.
0986   J.S. Burmicz et al.

97-   -76, Z. Chemie, 16 (1976)
0018B   J. Liebscher and H. Hortmann
0049   M. Muhlstadt et al.
0102   H. Dehne and M. Susse
0192   G. Heublein and G. Schubert
0268   J. Liebscher and H. Hartmann
0270   H. Schubert et al.
0318   T. Kissel et al.
0318B   M. Susse and E. Angrick
0322   H. Ripperger
0355   H. Viola and R. Mayer
0360   G. Schukat et al.
0365   B. Arventiev et al.

98-   -76, J. Agr. Food Chem., 24 (1976)
0724   G.A. Yost and L.L. Miller
0876   P.D. Schickedantz et al.
1062   L. Ruzo et al.
1085   J.G. Buta and J.F. Worley

99-   -76, Theor. Exptl. Chem., 12 (1976)
0081   V.E. Didkovskii et al.

100-   -76, Lloydia, 39 (1976)
0060   D. Dwuma-Badu et al.
0065   W.N. Wu et al.
0134   P.T.O. Chang et al.
0178   G. Kavalali
0204   W.N. Wu et al.
0213   D. Dwuma-Badu et al.
0218   S.K. Gupta et al.
0253   J.L. Cabrera and H.R. Juliani
0350   M. Nieto et al.
0385   D. Dwuma-Badu et al.
0399   J.H. Adams et al.
0405   R.R. Alluri et al.
0409   R.H. Takata and P.J. Scheuer
0412   M. Jacobsen et al.
0459   M. Leboeuf and A. Cave

101-   -76D, J. Organometallic Chem., 107
        (1976)
0055   T.N. Srivastava and V. Kumar

101-   -76G, J. Organometallic Chem., 110
        (1976)
0109   J.W. Buchler et al.

0295   E. Popowski et al.

101-   -76H, J. Organometallic Chem., 111
        (1976)
0235   F.-W. Grevels et al.

101-   -76I, J. Organometallic Chem., 112
        (1976)
0189   N. El Murr

101-   -76M, J. Organometallic Chem., 116
        (1976)
0039   P.V. Roling et al.
0113   A. Furlani et al.
0299   L. Parkanyi et al.

101-   -76N, J. Organometallic Chem., 117
        (1976)
0303   G. Zweifel et al.
0385   S. Staicu et al.

101-   -76Q, J. Organometallic Chem., 120
        (1976)
0369   A.R. Siedle

102-   -76, Phytochemistry, 15 (1976)
0029   M. Taniguchi et al.
0187   G.G. Freeman and R.J. Whenham
0191   W. Vichnewski et al.
0209   A.B. Zheleva et al.
0220   A. Kato and J. Takahashi
0225   G.D. Manners and L. Jurd
0229   N. Adityachaudhury et al.
0234   J. Reisch et al.
0240   J. Reisch et al.
0313   F. Fish et al.
0317   S. Imre et al.
0321   M. Koch et al.
0327   S. Nakajima et al.
0333   F.J. Evans and R.J. Schmidt
0336   W.H. Hui and M.M. Li
0340   I. Kirson et al.
0344   A.G. Gonzalez et al.
0347   P.G. Waterman
0355   P.S. Steyn and R. Vlegaar
0356   S. Ray and D.B. Chakraborty
0427B   W.H. Hui and M.M. Li
0430   J.S. Agarwal and R.P. Rastogi
0465   H.A. Stafford and M.A. Brown
0471   M. Akasu et al.
0525   A. Romo de Vivar et al.
0537   L. Camarda et al.
0541   R. Tschesche et al.
0543   J.C. Simoes et al.
0545   V.B. Pandey et al.
0547   C.L. Chen et al.
0551   F.J.A. Matos et al.
0555   H. Suzuki et al.
0557   D.K. Kulshreshtha and R.P. Rastogi
0559   P.J. Hylands and A.M. Salama
0567   R. Braz Fⁿ. et al.
0570   M. Weinberg et al.
0574   D.S. Bhakuni et al.
0576B   T. Sevenet et al.
0577   A. Ikuta and H. Itokawa
0578   P.G. Waterman

0609	M.A. Sprecker et al.		1799	D.A. Okorie
0743	J. Vaquette et al.		1802	M. Shamma et al.
0767	A. Chevolot-Maguer et al.		1953	G. Vidari et al.
0773	O.R. Gottlieb et al.		1973	N.G. Bisset and A.A. Khalil
0785	P.M. Baker et al.		1986	J. Dahmen and K. Keander
0797	W.-H. Hui et al.		1987	M. Miyazawa and H. Kameoka
0799	S. Huneck		1997	C.M. Chen and M.-T. Chen
0811	Y. Asakawa and E. Wollenweber		2006	M. Aritomi and D.M.X. Donnelly
0817	J. Bruneton et al.		2011	R.K. Gupta and M. Krisnamurti
0822	D. Dwuma-Badu et al.		2016	U. Kumar et al.
0827	S.C. Sharma et al.		2018	J. Singh et al.

0838 S. Mathuram et al.
0844 M.A. De Alvarenga et al.

103-   -76, **Khim. Geterosikl. Soedin.**, 12
        (1976)

0991 A.G. Gonzalez et al.
0995 P. Sengupta et al.
0999 E. Stahl and D. Herting
1019 S.H. Harper et al.
1025 M.R. Parthasarathy et al.
1029 R. Braz Fº. et al.
1033 J.B. Fernandes et al.
1037 D.G.I. Kingston et al.
1049 B.E. Nielsen and E. Jensen
1057 Y. Asakawa et al.
1074 S.W. Hanson et al.
1075 F. Bohlmann and C. Zdero
1080 N. Finkelstein and D.E.A. Rivett
1082 K.-W. Glombitza et al.
1089 M. Komatsu et al.
1090 M.S.R. Nair
1093 R.F. Garcia M.
1095 B. Danieli et al.
1119 J.A. Hargreaves et al.
1157 R.D. Hartley and E.C. Jones
1169 C.D. Hufford
1185 Y. Sashida et al.
1186 M.E.L. de Almeida et al.
1187 R. Braz Fº. et al.
1267 A.B. De Oliveira et al.
1275 T. Yamauchi et al.
1279 K.-W. Glombitza et al.
1283 M.E. Oberholzer et al.
1289 O.R. Gottlieb et al.
1293 A.B. Zheleva et al.
1295 S. Takahashi et al.
1303 A. Chatterjee and D. Ganguly
1305 J.P. Chapelle
1309 F. Bohlmann et al.
1313 W.H. Hui and M.M. Li
1315 W.S. Woo and S.S. Kang
1318 F. Bohlmann and C. Zdero
1323 R. Hansel and A. Leuschke
1413 A.C. Bazan and J.M. Edwards
1417 N. Kumar and T.R. Seshadri
1523 D.I.C. Scopes et al.
1531 W. Vichnewski et al.
1553 M.C. Do Nascimento et al.
1559 Y. Hashimoto and K. Kawanishi
1561 N.B. Mulchandani and S.R. Venka-
      tachalam
1567 N. Hayashi et al.
1745 F. Abe and T. Yamauchi
1753 R. Baute et al.
1775 W. Vichnewski et al.
1785 D. Broadbent et al.
1789B S. Tandon and R.P. Rastogi
1795 R.K. Gupta and M. Krishnamurti

0044 F.A. Trofimov et al.
0052 P.I. Abramenko et al.
0056 F.S. Mikhailitsyn and A.F. Bekhli
0076 A.B. Tomchin et al.
0086 V.M. Potapov et al.
0099 N.S. Prostakov et al.
0106 N.S. Kozlov et al.
0114 V.A. Anisimova et al.
0125 V.A. Loskutov and E.P. Fokin
0142 Z.I. Zelikman et al.
0153 A.I. Tolmachev et al.
0174 E.V. Braude and M.A. Gal'bershtam
0178 G.I. Zhungietu and L.P. Sinyavskaya
0191 G.Y. Dubur and E.L. Kharina
0205 Y.P. Andreichikov et al.
0220 A.Y. Ozola et al.
0245 A.N. Grinev et al.
0261 N.O. Saldabol et al.
0268 V.L. Savel'ev et al.
0273 E.R. Zakhs et al.
0292 A.N. Kost et al.
0300 N.L. Lifshits et al.
0312 N.S. Prostakov et al.
0329 K.I. Kuchkova et al.
0342 B.I. Buzykin et al.
0361 A.Y. Lazaris et al.
0402 V.P. Chernykh et al.
0417 M.A. Gal'bershtam
0424 N.S. Prostakove et al.
0428 V.M. Potapov et al.
0450 A.F. Pozharskii et al.
0455 T.I. Vinokurova and A.F. Pozharskii
0461 S.A. Giller et al.
0465 V.S. Mokrushin et al.
0471 L.A. Ignatova et al.
0502 D.O. Lolya et al.
0533 O.P. Shvaika and V.I. Fomenko
0562 V.M. Dziomko and U.A. Tomsons
0586 L.P. Smirnova et al.
0590 P.B. Terent'ev et al.
0596 Y.A. Sedov et al.
0600 A.Y. Lazaris et al.
0619 Z.V. Mezheritskii et al.
0648 E.D. Sych and L.T. Gorb
0650 E.D. Sych et al.
0662 A.N. Grinev et al.
0665 G.I. Zhungietu et al.
0669 N.A. Kogan
0680 M.A. Gal'bershtam et al.
0683 E.R. Zakhs et al.
0691 N.N. Chipanina et al.
0707 V.A. Azimov and L.N. Yakhontov

0711   O.P. Shvaika et al.
0717   G.G. Skvortsova et al.
0736   B.M. Krasovitskii and V.M. Shersh-
        ukov
0749   N.M. Turkewicz and M.I. Ganitkewicz
0751   Y.V. Svetkin et al.
0766   A.N. Kost et al.
0802   V.M. Dziomko et al.
0808   P.B. Terent'ev et al.
0817   P.B. Zikmanis
0831   A.N. Borisevich and P.S. Pel'kis
0834   G.I. Migachev et al.
0851   S.D. Sokolov and G.B. Tikhomirova
0854   G.A. Yugai and M.A. Mostoslavskii
0857   M.A. Mostoslavskii et al.
0862   V.G. Belin et al.
0868   I.Y. Postovskii et al.
0880   N.M. Przheval'skii et al.
0902   V.M. Potapov et al.
0910   R. Nowicki and A. Fabrycy
0914   V.P. Khilya et al.
0924   V.I. Ofitserov et al.
0938   N.A. Plashkina et al.
0942   I.Y. Postovskii et al.
0947   N.A. Kogan
0948   V.N. Charushin et al.
0977   G.N. Dorofeenko et al.
0981   M.G. Voronkov et al.
0999   R.E. Valter and G.A. Karlivan
1008   M.I. Vlasova and N.A. Kogan
1012   N.S. Kozlov and O.D. Lhikareva
1022   A.I. Mikhalev and M.E. Konshin
1026   A.A. Akhrem et al.
1029   A.A. Akhrem et al.
1091   V.N. Novikov and S.V. Borodaev
1100   V.G. Kul'nevich et al.
1128   L.S. Efros et al.
1144   V.A. Khaldeeva and M.E. Konshin
1162   N.A. Kogan and M.I. Vlasova
1165   I.A. Nasyr and V.M. Cherkasov
1186   A.V. Yarosh et al.
1214   V.T. Khilya et al.
1227   O.M. Babeshko et al.
1246   V.S. Rozhkov et al.
1249   N.A. Kogan and M.I. Vlasova
1280   V.S. Kobrin and L.B. Volodarskii
1331   E.M. Vernigor et al.
1343   B.P. Bespalov et al.
1352   S.V. Reznichenko et al.
1355   N.V. Latypov et al.
1379   V.I. Shvedov et al.
1386   Z.V. Pushkareva et al.

104-   -76, Zhur. Organ. Khim., 12 (1976)
            (English translation)
0022   N.N. Podgornova et al.
0033   N.N. Magdesieva et al.
0049   V.I. Sladkov et al.
0085   Y.N. Kreitsberg et al.
0089   K.M. Ermolaev
0107   T.I. Smirnova et al.
0119   S.F. Vul'fson et al.
0149   L.P. Rasteikene et al.
0161   Y.L. Yagupol'skii et al.
0168   R.E. Valter et al.
0172   M.V. Gorelik et al.

0180   N.S. Dokunikhin et al.
0185   V.K. Shalaev et al.
0223   O.I. Yurchenko et al.
0230   T.V. Mandel'shtam and N.S. Osmol-
        ovskaya
0237   R.E. Valter and G.A. Karlivan
0307   G.A. Tolstikov et al.
0339   L.T. Lantseva et al.
0395   A.P. Voropaeva et al.
0405   E.Y. Belyaev et al.
0436   S.D. L'vova and V.I. Gunar
0440   I.N. Chernyuk et al.
0458   V.M. Potapov et al.
0467   L.M. Yagupol'skii et al.
0544   R.G. Dubenko et al.
0613   N.S. Dokunikhin and G.A. Mezentseva
0625   V.A. Bren' et al.
0640   T.G. Tkhor et al.
0643   S.Y. Mel'nik et al.
0668   V.A. Buevich et al.
0681   A.V. Upadysheva et al.
0685   V.V. Razin and E.M. Rud'
0688   Y.A. Sharanin and K.Y. Lopatins-
        kaya
0796   V.I. Sladkov et al.
0802   M.I. Komendantov et al.
0818   V.E. Platonov et al.
0840   E.A. Medyantseva et al.
0857   T.N. Gerasimova et al.
0865   M.V. Gorelik and M.S. Kharash
0901   I.A. Korbukh et al.
0910   V.D. Domkin and L.A. Kur'yanovich
0914   V.A. Buevich et al.
0929   M.O. Lozinskii and V.S. Dmitrukha
0957   S.I. Yakimovich and V.A. Khrustalev
0964   I.A. Vasilenko et al.
1002   V.A. Mironov et al.
1057   N.I. Nogina and V.A. Koptyug
1087   V.S. Rozhkov et al.
1097   A.V. El'tsov et al.
1105   L.A. Khanina et al.
1114   L.D. Sadchenko and V.I. Gudzenko
1124   A.I. D'yachkov et al.
1131   A.N. Mirskova et al.
1141   O.V. Budanova et al.
1143   F.F. Blanko et al.
1146   T.P. Kosulina et al.
1245   I.K. Korobitsyna and V.A. Nikolaev
1252   I.K. Korobitsyna and V.A. Nikolaev
1277   L.A. Ostashevskaya et al.
1289   Y.S. Shabarov et al.
1298   G.I. Borodkin et al.
1340   V.M. Ostrovskaya et al.
1399   V.M. Potapov et al.
1440   L.Y. Sakhare et al.
1451   E.Y. Belyaev et al.
1531   O.N. Chupakhin et al.
1535   L.P. Sysoeva et al.
1538   I.A. Korbukh et al.
1544   S.V. Stepanova et al.
1550   I.Y. Kvitko et al.
1565   L.M. Yagupol'skii et al.
1575   I.Y. Kvitko et al.
1649   B.I. Buzykin et al.
1735   G.A. Chmutova and N.N. Vtyurina
1796   V.I. Koshutin and L.L. Koshutina

1797	I.V. Samartseva et al.
1798	V.N. Drozd et al.
1802	B.A. Ivin et al.
1804	M.A. Mostoslavskii et al.
1859	A.A. Khachatur'yan et al.
1912	R.G. Dubenko et al.
1920	L.M. Yagupol'skii et al.
1966	Y.N. Porshnev et al.
1968	A.K. Vaitkevichyus et al.
1974	T.D. Zheved' and K.V. Altukhov
1985	V.A. Buevich and V.V. Rudchenko
1988	M.V. Gorelik et al.
2034	A.A. Potekhin et al.
2039	R.I. Bodina et al.
2062	V.N. Odinokov et al.
2133	V.A. Lysanov et al.
2170	B.S. Drach et al.
2192	V.M. Potapov et al.
2193	V.M. Potapov et al.
2200	A.V. Upadysheva et al.
2207	B.V. Ioffe et al.
2220	A.A. Solov'yanov et al.
2286	K.M. Ermolaev
2333	V.A. Gailite et al.
2355	V.N. Bochenkov
2358	I.A. Aleksandrova et al.
2361	A.M. Sergeev et al.
2374	A.N. Kost et al.
2381	G.N. Lipunova et al.
2411	A.A. Solov'yanov et al.
2433	V.P. Tverdokhlebov et al.
2499	M.V. Gorelik et al.
2503	M.V. Gorelik et al.
2523	D.A. Oparin et al.

105-   76, Khim. Prirodn. Soedin., 12 (1976)

0048	N.S. Pal'yants et al.
0084	T.K. Khasunov et al.
0092	M.R. Nurmakhamedova and G.K. Nikonov
0093	A.S. Kadyrov et al.
0099	S.Z. Ivanova et al.
0100	V.A. Raldugin et al.
0110	S.K. Maekh et al.
0111	A. Karinove et al.
0116	A. Abdusamatov et al.
0118	N.N. Margvelashvili et al.
0147	V.G. Leont'eva et al.
0150	A.I. Sokolova et al.
0157	V.A. Raldugin and V.A. Pentegova
0193	I.A. Bessonova et al.
0201	G.V. Chkhikvadze et al.
0228	L.S. Teslov and G.G. Zapesochnaya
0245	M.M. Tadzhibaev et al.
0247	M.M. Tadzhibaev et al.
0262	K.S. Rybalko et al.
0266	V.A. Raldugin et al.
0278	V.A. Raldugin et al.
0282	V.I. Akhmedzhanova et al.
0313	M.K. Yusupov et al.
0320	K. Samikov et al.
0335	E.F. Ametova et al.
0344	S.V. Serkerov
0346	K.S. Rybalko et al.
0432	L.D. Yakhontova et al.
0461	K.S. Mikhailov et al.
0464	M.S. Marchenkov et al.

0491	G.V. Sugitdinova et al.
0494	E.A. Krasnov et al.
0499	K.A. Ubadullaev et al.
0533	M.E. Perel'son et al.
0539	M.A. Khasanova and G.K. Nikonov
0551	Y.I. Eidler et al.
0615	A. Nabiev et al.
0623	V.N. Dobrynin et al.
0652	O.E. Krivoshchekova et al.
0666	L.P. Nikonova and G.K. Nikonov
0675	A.M. Reinbol'd and D.P. Popa
0687	N.S. Pal'yants et al.
0702	D.A. Abdullaeva et al.
0709	D.M. Razakova et al.
0716	I.A. Israilov et al.
0736	K.M. Kamilov and G.K. Nikonov
0747	K.A. Kadyrov et al.

106-   -76, Die Pharmazie, 31 (1976)

0024	M. Pesak et al.
0051	I. Bornschein et al.
0051B	G. Toth et al.
0153	G. Wagner and B. Dietzsch
0279	P. Richter et al.
0363	J. Kracmar et al.
0540	H. Mohrle and K. Sieker
0574	A.R.E.N. Ossman and N.E. Taaii
0598	G. Wagner et al.
0603	H. Mohrle and K. Sieker
0647	G. Adam et al.

107-   -76, Synthetic Comm., 6 (1976)

0011	A. Casares and L.A. Maldonado
0039	J.A. Noguez and L.A. Maldonado
0299	J. Ide, K. Sakai and Y. Yura
0399	J.W. Ashmore et al.
0457	A.V. Zeiger and M.M. Joullie
0515	F.L. Malanco and L.A. Maldonado

108-   -76, Israel J. Chem., 15 (1976)

0039	C. Braud and M. Vert
0046	G. Snatzke et al.

110-   -76, Russian J. Phys. Chem., 50 (1976)

0703	A.V. Belotsvetov
1170	A.V. Belotsvetov et al.
1173	A.V. Belotsvetov et al.
1711	N.A. Shcheglova and D.W. Shigorin

111-   -76, European J. Med. Chem., 11 (1976)

0067	G. Barnathan et al.
0083	M. Flammang and C.-G. Wermuth
0241	G. Malesani et al.
0279	D. Bailey et al.
0489	J.M.J. Tronchet and D. Schwarzen-bach
0501	H. Liepmann et al.
0527	B. Kokel et al.
0533	J. Badin et al.
0555	S. Renault et al.
0571	I. Jirkovsky et al.

112-   -76, Spectroscopy Letters, 9 (1976)

0487	R.J. Sturgeon and S.G. Schulman
0777	J.S. Dixon et al.
0865	N.K. Narain

114-   -76A, Acta Chim. Acad. Sci. Hung.,
            88 (1976)
0053   J. Balog and J. Csaszar
0293   A. Levai and R. Bognar
0309   E.R. David et al.
0413   G. Hornyak et al.

114-   -76B, Acta Chim. Acad. Sci. Hung.,
            89 (1976)
0381   R.M. Issa et al.

114-   -76C, Acta Chim. Acad. Sci. Hung.,
            90 (1976)
0051   G. Oszbach et al.
0285   P. Benko et al.
0301   A. Gelleri et al.
0381   V. Szabo and E. Antal
0395   E. Berenyi et al.
0405   P. Benko et al.

114-   -76D, Acta Chim. Acad. Sci. Hung.,
            91 (1976)
0081   E. Tomorkeny et al.
0327   P. Benko and L. Pallos
0341   A.M. Abbady et al.

115-   -76, Egyptian J. Chem., 19 (1976)
0001   W.I. Awad et al.
0185   R.M. Issa et al.
0203   F.M. Abdel-halim et al.
0341   M.Z.A. Badr and H.A.H. El-Sherif
0379   R.M. Issa et al.
0621   A.R. Nada
0761   W.I. Awad et al.

116-   -76, Macromolecules, 9 (1976)
0010   H.W. Gibson and F.C. Bailey
0221   H.W. Gibson and F.C. Bailey
0626   V.V. Korshak et al.
0701   A. Harada et al.

117-   -76, Org. Preps. Procedures, 8 (1976)
0025   S.C. Cottrell et al.
0041   V. Dave
0125   L.L. Ansell et al.
0133   L.L. Ansell et al.
0163   L.H. Klemm and D.R. Taylor
0169   J.O. Curric, Jr. et al.
0223   J.W. Ashmore and G.H. Helmkamp
0287   H. Grinberg et al.

118-   -76, Synthesis (1976)
0001   F. Kröhnke
0025   G.L. Olson et al.
0051   H. Wamhoff and B. Wehling
0105   N. Dennis et al.
0108   T.W. Spangler and T.W. Hartford
0249B   H. Bader and Y.H. Chiang
0256   H.-D. Scharf et al.
0261   M. Ferrey et al.
0270   G. Bartoli and G. Rosini
0273   J. Liebscher and H. Hartmann
0274   W. Löwe
0307   D.K. Banerjee et al.
0323   R. Weiss and C. Schlierf
0335   D.A. Lightner et al.

0339   M. Furakawa and T. Okawara
0403   J. Liebscher and H. Hartmann
0414   V. Bocchi et al.
0445   G.H.R. Seitz and G. Arndt
0449   G. Descotes and A. Faure
0451   F. Marcuzzi and G. Melloni
0461   A. Kumar and D. Devaprabhakara
0469   K.-D. Kampe
0489   M. Narita and C.U. Pittman, Jr.
0521   J. Liebscher and H. Hartmann
0541   M.T. Langin-Lanteri and J. Huet
0604   A. Nishinaga et al.
0673   C. Jutz et al.
0675   D.N. Nicolaides
0705   E. Campaigne and S.W. Schneller
0719   W. Meise and H.-L. Muller
0748   G. Bettinetti et al.
0759   I. Degani and R. Fochi
0764   G. Coudert et al.
0805   K. Matsumoto et al.

120-   -76, Pakistan J. Sci. Ind. Research,
            19 (1976)
0101   S. Ahmad and G. Hussain
0103   A. Butt et al.
0120   I.H. Qureshi et al.
0214   B. Robinson and M.U. Zubair

123-   -76, Moscow U. Chem. Bull., 31 (1976)
0095   V.V. Men'shikov
0598   Y.P. Shvachkin et al.
0714   V.M. Potapov et al.

124-   -76, Ukrain. Khim. Zhur., 42 (1976)
0204   M.M. Kul'chickij et al.
0383   S.V. Cukerman
0842   V.M. Nikitchenko et al.
0857   A.T. Pilipenko et al.
0961   E.D. Sych and L.T. Gorb
1108   A.Z. Gershuns and P. Pustovar
1159   F.S. Babichev et al.
1162   S.N. Kukota et al.
1174   R.E. Koval'chuk and A.J. Id'chenko
1298   V.N. Skopenko et al.

126-   -76, Die Makromol. Chem., 177 (1976)
0395   S. Abuhantash et al.
1357   R. Kerber et al.
1857   G. Kossmehl et al.
3089   C. Pinazzi and A. Fernandez
3481   H. Kammerer and J. Pachta

128-   -76, Croatica Chem. Acta, 48 (1976)
0041   M. Lacan et al.
0341   V. Skaric et al.
0351   V. Skaric et al.

130-   -76, Bioorganic Chem., 5 (1976)
0001   G.N. Phillips, Jr. et al.
0025   F. Seela and F. Cramer
0031   A. Hampton et al.
0037   T.H. Fife and B.M. Benjamin
0283   E.E. Van Tamelen et al.

131-   -76B, J. Mol. Structure. 31 (1976)
0275   M. Moutin et al.

133-   -76, Pharm. Acta Helv., 51 (1976)
  0065   A. Hulshoff and J.H. Perrin
  0304   P.C. Rowbotham et al.
  0314   J. Topart et al.
  0343   M. Otagiri et al.
  0371   H. Bosshardt et al.

135-   -76A, J. Appl. Spectroscopy S.S.S.R.,
         24 (1976)
  0077   V.M. Treushnikov et al.
  0717   L.V. Volod'ko

135-   -76B, J. Appl. Spectroscopy S.S.S.R.,
         25 (1976)
  1054   V.P. Volkova et al.

136-   -76A, Carbohydrate Research, 46 (1976)
  0009   J.M.J. Tronchet et al.
  0019   J.M.J. Tronchet et al.
  0043   C. Chavis et al.
  0119   J.M.J. Tronchet and J. Poncet
  0127   J.M.J. Tronchet and E. Mihaly

136-   -76B, Carbohydrate Research, 47 (1976)
  0081   D. Horton and A. Liav

136-   -76C, Carbohydrate Research, 48 (1976)
  0041   D. Horton and M. Sakata
  0282   K. Takeo and T. Kuge

136-   -76D, Carbohydrate Research, 49 (1976)
  0049   A.M. Seldes et al.
  0093   G.S. Hajivarnava et al.
  0185   H.S. El Khadem et al.
  0209   P. Angibeaud et al.
  0275   P. Finch and A.G. Nagpurkar

136-   -76E, Carbohydrate Research, 50 (1976)
  0001   J. Martinez et al.

136-   -76G, Carbohydrate Research, 52 (1976)
  0049   Y. Ishido et al.
  0079   A. Rosenthal and K. Dooley

138-   -76, Chemistry Letters (1976)
  0019   H. Takaku et al.
  0203   H. Ishihara and Y. Hirabayashi
  0215   M. Oda et al.
  0219   Y. Kitahara et al.
  0363   M. Moriyama et al.
  0389   T. Sonoda et al.
  0413   A. Kakehi et al.
  0441   T. Katada et al.
  0519   I. Takeuchi et al.
  0543   A. Tajira et al.
  0643   M. Nojima et al.
  0657   F. Toda et al.
  0701   S. Aburaki and M. Kinoshita
  0793   O. Seshimoto et al.
  0805   Y. Omori et al.
  0975   A. Murai et al.
  0991   K. Yamakawa et al.
  1011   M. Oda et al.
  1119   K. Arakawa et al.
  1127   A. Ishida and T. Mukaiyama
  1215   S. Inouye et al.

  1297   S. Nishida and F. Kataoka
  1323   Y. Takehira et al.
  1405   T. Shiumyozu et al.

139-   -76A, Phosphorus and Related Group V
         Elements, 6 (1976)
  0151   D. Hellwinkel and H.-J. Wilfinger
  0169   J.C. Williams, Jr. et al.
  0207   R.J.W. Cremlyn et al.

139-   -76B, P and S and Related Elements,
         1 (1976)
  0087   S.S. Tratch et al.
  0289   A.G. Pinkus and J.K. Jones

139-   -76C, P and S and Related Elements,
         2 (1976)
  0105   M. Srinivasan and J.B. Rampal

140-   -76, J. Anal. Chem. S.S.S.R., 31 (1976)
  0837   D.I. Belkin et al.
  1691   V.P. Antonovich et al.

142-   -76, Heterocycles, 4 (1976)
  0009   Y. Tominaga et al.
  0131   S. Sakai
  0221   N. Abe
  0235   S.M. Kupchan et al.
  0471   K. Takahashi et al.
  0595   S. Ito and M. Kodama
  0705   K. Mizuyama et al.
  0739   Y. Suzuki et al.
  0753   J.M. Saa et al.
  0939   M. Fujito et al.
  0985   S. Sakai and N. Shinma
  1065   J.D. McChesney and R. Buchman
  1073   S.T. Lu and I.-S. Chen
  1089   P. Venturella et al.
  1095   N. Numao and O. Yonemitsu
  1229   T. Sano and Y. Tsuda
  1257   K. Kigasawa et al.
  1263   V. Simanek et al.
  1275   M. Nakagawa et al.
  1347   H. Agui and T. Nakagome
  1355   Y. Tsuda et al.
  1371   Y. Okuno and O. Yonemitsu
  1481   T. Momose et al.
  1493   Y. Tominaga et al.
  1637   T. Kametani et al.
  1655   M. Ruccia et al.
  1675   A. Kubo et al.
  1875   M. Watanabe et al.

142-   -76S, Heterocycles, 5 (1976)
  0025   Z. Bernstein and D. Ginsburg
  0035   H. Morita and S. Oae
  0041   M. Shamma et al.
  0053   I. Saito et al.
  0109   J. Palmer et al.
  0147   H. Hamberger et al.
  0167   T. Kato et al.
  0245   N. Cerletti
  0299   G. Snatzke and G. Hajos
  0427   Y. Nakamura et al.
  0445   E.J. Corey and R.D. Balanson
  0471   J. Streith and G. Wolff

144-   -76, <u>Mol. Photochem.</u>, <u>7</u> (1976)
    0325   K. Tsutsumi et al.
    0399   J. Brokken-Zijp
    0475   W.H. Waddell et al.

145-   -76, <u>Arzneimittel Forsch.</u>, <u>26</u> (1976)
    0478   M. Szekerke and M. Horvath
    0537   G. di Marchi et al.
    1907   C. Lorincz et al.

146-   -76, <u>J. Appl. Chem. Biotech.</u>, <u>26</u> (1976)
    0001   S.A.M. Metwally
    0131   A.T. Peters
    0667   A.S. Hammam

148-   -76, <u>Pharm. Res. Comm.</u>, <u>8</u> (1976)
    0435   P. Versace et al.

149-   -76A, <u>Photochem. Photobiol.</u>, <u>23</u> (1976)
    0037   G.W. Byers et al.

149-   -76B, <u>Photochem. Photobiol,</u>, <u>24</u> (1976)
    0217   J.H. Hong et al.
    0383   N. Duran et al.